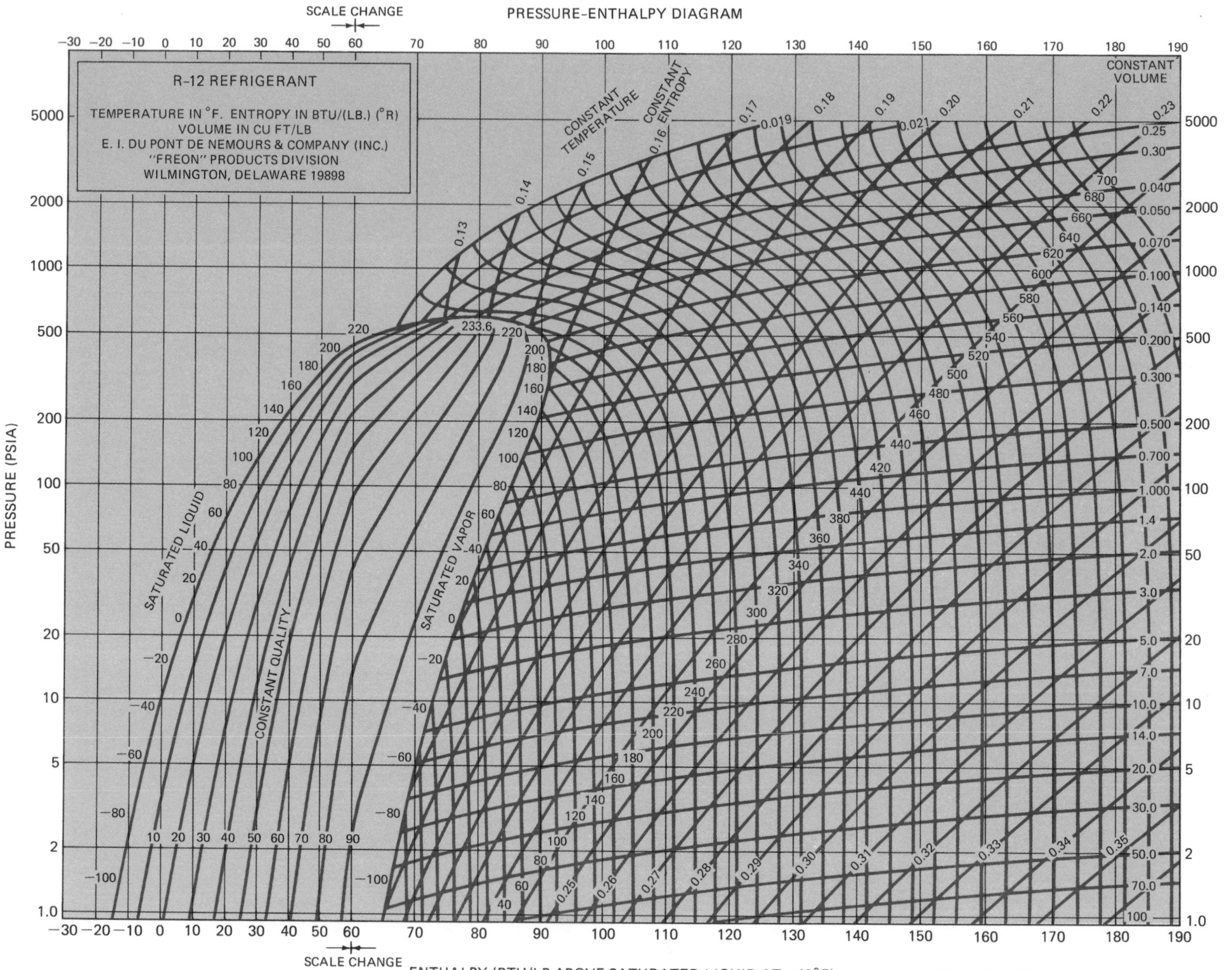

PRESSURE-ENTHALPY DIAGRAM
SCALE CHANGE
R-12 REFRIGERANT
TEMPERATURE IN °F. ENTROPY IN BTU/(LB.) (°R)
VOLUME IN CU FT/LB
E. I. DU PONT DE NEMOURS & COMPANY (INC.)
"FREON" PRODUCTS DIVISION
WILMINGTON, DELAWARE 19898
CONSTANT TEMPERATURE
CONSTANT ENTROPY
CONSTANT VOLUME
SATURATED LIQUID
SATURATED VAPOR
CONSTANT QUALITY
PRESSURE (PSIA)
ENTHALPY (BTU/LB ABOVE SATURATED LIQUID AT −40°F)

Refrigeration and Air Conditioning

Third Edition

AIR-CONDITIONING AND REFRIGERATION INSTITUTE
1501 Wilson Boulevard, 6th Floor
Arlington, VA 22209-2403

Refrigeration and Air Conditioning

Third Edition

PRENTICE HALL
Upper Saddle River, New Jersey Columbus, Ohio

Library of Congress Cataloging-in-Publication Data

Refrigeration and air conditioning/Air-Conditioning and
Refrigeration Institute.—3rd ed.
p. cm.

Includes Index.
ISBN 0-13-323775-3
1. Refrigeration and refrigerating machinery. 2. Air conditioning. I. Air-Conditioning and Refrigeration Institute.
TP492.R377 1997
621.5'6—dc21 97-344
CIP

Cover photos: Bristol Compressors, Copeland Corporation, Scotsman Ice Systems, Vilter Manufacturing Corp.

Editor: Ed Francis

Developmental Editor: Carol Hinklin Robison

Production Editor: Stephen C. Robb

Design Coordinator: Julia Zonneveld Van Hook

Text Designer: Elm Street Publishing Services, Inc.

Cover Designer: Brian Deep

Production Manager: Pamela D. Bennett

Illustrations: Clarinda Company

Marketing Manager: Danny Hoyt

Production Supervision: Elm Street Publishing Services, Inc.

This book was set in Times Roman and Helvetica Neue by Clarinda Company and was printed and bound by Von Hoffmann Press, Inc. The cover was printed by Von Hoffmann Press, Inc.

Photo Credit: Title page, section opener, and chapter opener photos by Terry Leininger, Moraine Park Technical College.

Carrier Corporation makes training widely available. To obtain a catalog, telephone 800-962-9212.

Printed in the United States of America

10 9 8 7 6 5 4 3 2 1

ISBN: 0-13-323775-3

Prentice-Hall International (UK) Limited, *London*
Prentice-Hall of Australia Pty. Limited, *Sydney*
Prentice-Hall of Canada, Inc., *Toronto*
Prentice-Hall Hispanoamericana, S. A., *Mexico*
Prentice-Hall of India Private Limited, *New Delhi*
Prentice-Hall of Japan, Inc., *Tokyo*
Simon & Schuster Asia Pte. Ltd., *Singapore*
Editora Prentice-Hall do Brasil, Ltda., *Rio de Janeiro*

Contents

SECTION 10

Troubleshooting 935

SECTION 11

Preventive Maintenance 997

Preface

Refrigeration and Air Conditioning was sponsored by the Manpower Development Committee of the Air-Conditioning and Refrigeration Institute (ARI), the national industry trade association whose members manufacture more than 90 percent of all air-conditioning and refrigeration equipment produced in the United States.

This textbook is the culmination of the ARI's longstanding involvement in the promotion and development of training programs designed to prepare beginning service technicians to install, service, and maintain air-conditioning and refrigeration equipment. More than 100,000 beginning service technicians have already learned their first skills with the help of this book.

The first edition was based on course outlines developed in cooperation with the U.S. Office of Education's Vocational Education Branch and, subsequently, was approved by that agency. The second edition reflected the results of a survey of countless instructors and their students who have used *Refrigeration and Air Conditioning* at more than three hundred institutions nationwide and, indeed, around the world. In order to make the second edition an even better learning tool, several basic changes were made in organization, content matter, and pedagogy to reflect the recommendations and suggestions of the students and educators who were surveyed.

The Third Edition

In keeping with the ARI's commitment of continually improving *Refrigeration and Air Conditioning*, this third edition, sponsored by ARI's Education and Training Committee, is completely revised, updated, and pedagogically enhanced. Based on input from the members and staff of ARI, industry experts, and many academic advisors, *Refrigeration and Air Conditioning, Third Edition,* reflects the state of the art as well as solid, fundamental concepts and practical applications.

Among key features of the third edition are:

- The text and lab manual mirror the new published ARI national curriculum guidelines.
- Special chapters focus on troubleshooting.
- The text draws extensively upon field-tested materials from industry sources, relating concepts to real-world situations.
- Refinements have greatly improved the coverage of electrical fundamentals and the new electronic control systems.
- Recent government regulations and industry environmental guidelines are incorporated.
- Featured are new pedagogical aids: Performance Objectives, Practicing Professional Services, Preventive Maintenance, and Knowledge Extension.

Chapters are logically organized in sections that include Refrigeration Principles, Refrigeration Service Techniques, Basic Electricity/Electronics, Domestic Refrigeration, Commercial Refrigeration, Air-Conditioning Systems, Heating Systems, Central Heat Pumps, Installation and Start-Up, Troubleshooting, and Preventive Maintenance. A glossary complements the chapter material.

This third edition is supported with a complete teaching and learning package. In addition to the improved textbook, available to instructors are the *Instructor's Manual,* Lab Manual, Transparency Masters, Prentice Hall Custom Test (Windows), ARI Curriculum Guide, and Study Guide for ARI/GAMA Competency Examinations in HVACR.

As noted in the previous editions, readers are advised that refrigeration, air conditioning, and the environmental control industry, in general, all have dramatically expanded in the variety of available equipment and applications to a level beyond the scope of a single text. This text is, therefore, limited to general principles and applications. ARI recommends that the use of specific product information, as well as installation and service information from individual manufacturers, become part of the learning process. In addition, federal and state regulations and guidelines regarding refrigerants and refrigerant recovery are under review and are subject to rapid change. This text is, therefore, as up to date as possible, but it obviously may not reflect regulatory or technological changes made after printing.

Acknowledgments

Prentice Hall and the Air-Conditioning and Refrigeration Institute express sincere appreciation to those who helped to put together this new edition:

Bob Cooper, who did most of the writing;
Bill Chaisson, for research and technical editing;
Dave Sullivan, for organizing, developing pedagogy, and editing;
Mary Anne Cooper, for helping with the illustration program;
Tom Wagner, for developing many of the sidebars;
Greg Jourdan, for producing the supplements and putting the finishing touches on the text;
Robert Chatenever, for technical consulting and review of edited manuscript and proofs;
Leslie Sandler, the staff of ARI, and ARI member companies for technical feedback and cooperation throughout the development process.

The following reviewers are gratefully acknowledged for their help in shaping the final product:

Michael D. Brock, Florida Community College at Jacksonville
Ron Stout, Northeast State Technical Community College
William Whitman, Triton College
Gregory E. Jourdan, Wenatchee Valley College
David R. Keefer, Tarrant County Junior College
Richard Jazwin, Universal Technical Institute
Albert Burman, Refrigeration School, Inc.
Tim Nicholson, OCM B.O.C.E.S. Career Training Center
Patrick L. Murphy, Union County Vocational Technical Schools
Bob Walters, Pima Community College
Tracey Hamilton, N.Y. Technical Institute
David Eishen, Cedar Valley College
Richard Shaw, Ferris State University
Irvin Easterling, Tarrant County Junior College
Greg Parakes, Murfreeseboro Area Vocational Technical College
Peter Honczar, Philadelphia Wireless Technical Institute
Carl Mulbeck, Lincoln Technical Institute
Don Tarasi, Gateway Technical Institute
Dan Hornsby, Denver Institute of Technology
Dan McLaughlin, Triangle Tech. Inc.
Bill Wilson, Belleville Area College
Richard Echard, Orleans Technical Institute
George Koshy, Sunnyvale, CA
Floyd Brownfield, Ivy Tech College
Elwin Hunt, San Joaquin Valley College
Benny Barnes, Live Oaks Vocational Tech College
Mike Schuster, Knowles, WI
Herb Haushahn, College of DuPage
Jack Campbell, Linn Benton Community College
Robert Nolan, Daytona Beach Community College
Joseph Hoag, I.T.T. Tech
Vern Rettig, Triton College
Irving Panzer, Houston Community College
Dennis Wash, Oakhurst, CA
John Niro, Palm Beach Technical Education Center
John Vucci, Landover, MD
Scott Posey, Floyd College
Tom Foshec, Ayers State Technical College
Andy Rhinefort, Tarrant County College
Dale Cook, North Seattle Community College
Norm Christopherson, San Jose City College
Jim Ward, Wichita Tech College

Finally, special thanks go to the following companies and organizations that very generously provided material for inclusion in this book:

ACCA (Air Conditioning Contractors of America)
Aeroquip Corporation
AES-NTRON, Inc.
Airserco
Allied-Signal, Inc.

Alnor Instrument Company
Amana Refrigeration, Inc.
American Gas Association
American Metal Products
Amprobe Instrument, Inc.
Anamet Industrial, Inc.
ASHRAE (American Society of Heating, Refrigeration, and Air Conditioning Engineers)
AVO International
Bachrach, Inc.
Baltimore Air Coil Co.
Bard Manufacturing Co., Inc.
The Black and Decker Corporation
Bristol Compressors
Calmac Manufacturing Corp.
Carlyle Compressor Co.
Carrier Air-Conditioning Company
Cooper Instrument Corp.
Copeland Corporation
Cutler Hammer, Eaton
Dole Refrigerating Company
Dunham-Bush
Duro Metal Products, Inc.
Dwyer Instruments, Inc.
Ebco Manufacturing Co.
Elkay Manufacturing Company
Energy Engineering
E-Tech, Inc.
Evapco, Inc.
Federal Pacific Electric
Fields Controls Co.
Fluke Corp.
Fostoria Industries, Inc.
Freon Products Division of E. I. du Pont de Nemours & Company
Frick Co.
Frigidaire
Fulton Boiler Works
GE Appliance
General Analysis Corp.
Glowcore A.C., Inc.
Halstead & Mitchell, Division of Climate Control
Heatcraft Refrigeration Products
Henry Valve Company
Honeywell, Inc.
Hussman Corp.
Imperial Eastman Company
Industrial Instruments and Supplies
Inter-City Products Corp.
ITT Bell and Gossett
J/B Industries, Inc.
Klein Tools
Koldwave Division of Mestek, Inc.
Kramer Trenton Company
Lennox Industries, Inc.
Manitowac Co., Inc.
Marley Cooling Tower Company
Marsh Instrument Co.
Master Bilt Products
Maytag Appliances
McQuay International
MSA Instrument Division
Nicholson Tools
Nor-Lake, Inc.
NORDYNE, Inc.
NRP (National Refrigeration Products)
Perfect Sense, Inc.
Quatro Consolidated Industries
Ranco North America
Refrigeration Research, Inc.
Research Products Corp.
Ridge Tool Company
Ritchie Engineering Company, Inc.
Roberts Gordon, Inc.
Robertshaw Controls
Robinair Division of SPX Corporation
Robur Corp.
Rotorex Company, Inc.
R. W. Beckett Co.
Scotsman Ice Systems
Selkirk Metalbestos
Shortridge Instrument
Skuttle Manufacturing Co.
SMACNA (Sheet Metal and Air Conditioning Contractors National Association)
Sporlan Valve Company
SRI International (Stanford Research Institute)
Standard Refrigeration Co.
Taco, Inc.
Tecumseh Products Co., Inc.
Thermal Engineering
Thermo King Corp.
TIF Instruments, Inc.
Toxalert, Inc.
The Trane Company
Tranter, Inc.
Trol-A-Temp
Turbo Refrigerating
Turbotec Products, Inc.
Tyler Refrigeration Corp.
U.S.A. Coil & Air, Inc.
Vilter Manufacturing Corp.
Virginia KMP Corp.
White-Rodgers Div. of Emerson Electric Co.
Worker's Compensation Board of British Columbia
York International Corp.

SECTION 1

Refrigeration Principles

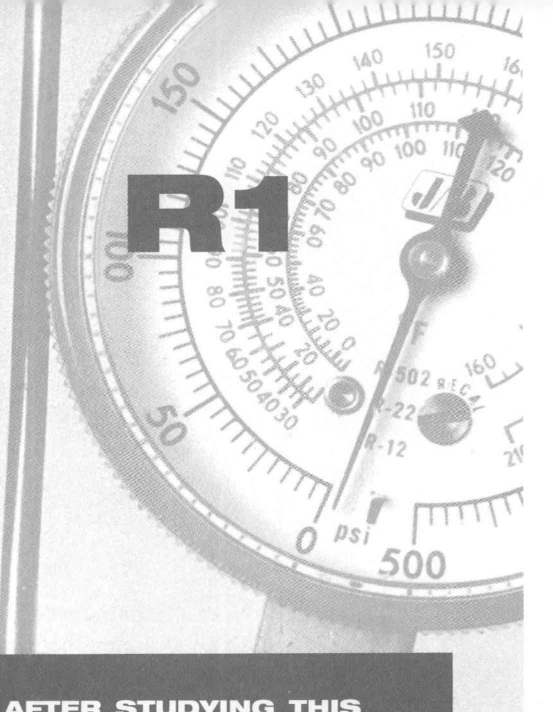

R1

Introduction to Refrigeration

AFTER STUDYING THIS CHAPTER, THE STUDENT WILL BE ABLE TO:

- Recognize/identify important events in the history and development of the heating, ventilation, air conditioning, and refrigeration (HVAC/R) industry.
- Explain the varied uses of refrigeration.
- Comprehend personnel requirements for the near future.
- Begin to form career goals and objectives in a chosen segment of the industry.
- Name the types of jobs that are available in this industry.
- List the types of employers who hire HVAC/R technicians.
- List the services available through ARI.
- List the benefits of passing the ARI/GAMA competency exam.
- Understand and use the Units of Measurement for refrigeration including the Inch-Pound (IP) English Systems or Standard International (SI).
- List and discuss trade associations.

R1-1 MARKETS FOR REFRIGERATION

Not only is food preserved in homes of today by mechanical refrigeration, but commercial preservation of food is one of the most important applications of refrigeration.

More than three-fourths of the food used is produced, packaged, shipped, stored, and preserved by refrigeration. Millions of tons of food are stored in refrigerated warehouses, private lockers, and packing and processing plants.

Without the many types of refrigeration in stores, warehouses, aircraft, railway cars (Fig. R1-1), trucks, and ships, the storage and transportation of all types of perishables from all over the world would be impossible.

Refrigeration has improved the economy of many areas by providing a means of preserving their products en route to remote consumers. It has contributed significantly to the development of agricultural and livestock-producing regions through greater demand for the products.

In addition to food, many other industries have benefitted from refrigeration. Prior to 1914, most of the tires for automobiles, trucks, and airplanes were made of natural rubber produced in plantations in southern Asia. When the shipment of latex from foreign plantations was curtailed during World War II, industry and the federal government established a cooperative synthetic rubber program. Scientists discovered manufacturing processes that could make artificial rubber more durable and wear-resistant through the use of low temperatures. Thus, refrigeration became vital to another industry.

A rapid increase in new products occurred after World War II. The petrochemical (plastics), textile, and data processing industries became heavy users of process refriger-

Figure R1-1 The application of refrigeration to railroad cars permitted the transportation of perishable foods.

ation and air conditioning. Without refrigeration many of these products and their applications would not have developed as they are known today.

Energy production also relies on refrigeration processes. Liquid natural gas obtained from foreign sources must be chilled to –270°F to change the gas into a liquid, which is then loaded onto a refrigerated tanker for shipment to a receiving port. The liquid must be kept refrigerated at –270°F until it is ready for vaporization back into a gas. Solar heating, refrigeration, and air conditioning are also expected to provide many opportunities for new products and for qualified application, installation and service personnel.

In 1980, homes, plants, and commercial buildings added new refrigeration and air conditioning equipment valued at $8 billion. In 1993, shipments of heating and cooling units reached a value of $18 billion. The dollar total was much larger, since the cost of many accessory products, such as duct, grilles, insulation, and controls were not included. Refrigeration and air conditioning together constitute one of the United States' major industries.

World markets are also experiencing rapid growth. Canada, Japan, Germany, the United Kingdom, France, Mexico and Venezuela are among the major users of refrigeration and air conditioning. This market will continue to expand as foreign countries improve their standard of living and engage in industrial development.

R1-2 JOB OUTLOOK

As a result of the expanding markets for refrigeration and air conditioning and changing technology, a dramatic need exists for qualified personnel both at home and abroad. The most needed jobs to be filled through the 1990s are for mechanically skilled technicians. For every $1 million installed value of equipment, the following employees are needed:

1 Graduate engineer
2 Technicians/master mechanics
11 Refrigeration and air conditioning mechanics
2 Refrigeration and air conditioning helpers
7 Sheet-metal mechanics
1 Sheet-metal helper
2 Salespeople

In actual practice it is common for a technician/mechanic to be qualified or required to work on refrigeration or air conditioning. Thus jobs are available in a broad category. The job opportunities not only result from a continuously expanding market, but from a turnover of many installers and service mechanics who will be retiring or moving up to supervisory positions.

R1-3 OCCUPATIONAL OPPORTUNITIES

Fields in which employment opportunities exist include high-temperature cooling (process air conditioning), medium-temperature cooling (product storage above freezing), low-temperature cooling (quick freezing and storage below freezing) and extremely low temperature cooling (medical, chemical, industrial, and scientific applications).

The largest employer classifications are wholesaler and contracting trades. The refrigeration wholesaler must have trained personnel capable of servicing a dealer/contractor organization on the proper application, installation, and maintenance of the product lines they carry.

The dealer/contractor installs and services the product to the end user. Depending on the extent of the contractor's business, a technician may be involved in installation, start-up, or customer service. In small organizations, they may do all three.

In larger metropolitan areas some refrigeration service companies are involved in nothing but service. They specialize in maintenance contracts with the user to provide scheduled inspections as well as emergency repairs.

Large manufacturing and industrial plants or institutions that use refrigeration and air conditioning products often have in-house maintenance departments that do

KNOWLEDGE EXTENSION

Significant events in the History of Refrigeration are listed below.

1824 Michael Faraday identifies the principles of vaporative cooling.

1834 Jacob Perkins, an American engineer, invented machinery for making artificial ice. His was the forerunner of modern compression systems.

1855 The first absorption type of refrigerating equipment, produced by a German engineer.

1865 Daniel Livingston Holden supervises installation of an aqua-ammonia absorption refrigeration system. A year later he obtained rights to "Chymogene", a petroleum spirit, as a vapor-compression system refrigerant and began the manufacture of ice-making plants.

1890 The mechanical ice-making industry develops over severe shortages of natural ice. Also, a new cold storage industry uses ice making machinery to provide central storage of refrigerated goods.

1894 Hugh J. Barron founds the American Society of Heating and Ventilating Engineers (ASHVE).

1904 Refrigeration engineers form the American Society of Refrigeration Engineers (ASRE)

1908 The First International Congress of Refrigeration is held in Paris, France.

1911 Willis Carrier publishes his psychrometric chart.

1912 J. M. Larsen produces a manually operated domestic refrigerator.

1913 The first international exposition devoted exclusively to refrigeration is held in Chicago.

1916 Margaret Ingels is the first woman in the world to earn a degree in mechanical engineering.

1918 Kelvinator produces the first automatic refrigerator for the American market.

1921 A package chiller using a centrifugal compressor with dichloroethylene refrigerant is developed by Willis Carrier.

1923 Fast freezing is introduced as a way of preserving food, the beginning of the modern frozen-foods industry.

1925 Clarence Birdseye patents the first of many processes.

1928 General Electric introduces the first "hermetic" automatic refrigeration unit, named the Monitor Top.
Chlorofluorocarbon (CFC) refrigerants are synthesized for Frigidaire.
Mechanical refrigeration systems for summer cooling are connected to heating plants.

1935 Frederick McKinley Jones produces an automatic refrigeration system for long-haul trucks.

1954 ASHVE and ASHRE merge to become ASHRAE.

1955 Mass marketing of frozen dinners begins.

1969 Neil Armstrong and Buzz Aldrin walk on moon in space suits with life-support and cooling systems.

1990 The Clean Air Act and its amendments schedule the phasing out of CFC refrigerants; world-wide concurrent.

most of the routine service work. Large food chain companies employ skilled technicians to assist in or direct the installation and service of refrigeration equipment. The modern supermarket cannot afford down-time on either its food preservation equipment or its air conditioning.

Equipment manufacturers also hire a number of trained technicians for various jobs. Laboratory technicians may be involved in actually building prototypes of new products or they may conduct tests on products for performance ratings, life-cycle testing, compliance with ARI certification, UL safety standards, or sound evaluation.

Factory service personnel prepare service instructions, training material, and renewal parts information. Some manufacturers employ large numbers of qualified service technicians for field assignment, which involves installation, start-up, and customer service. Other areas of employment include electric utilities, schools and certain industries.

Knowledge, training, and experience are the key ingredients of a successful technician. This text is designed to help prepare you to take advantage of these broad opportunities.

R1-4 TRADE ASSOCIATIONS

With the rapid growth and variety of interests, trade associations naturally evolved to represent specific groups. The list includes manufacturers, wholesalers, contractors, sheet metal dealers, and service organizations. Each is important and makes a valuable contribution to the field. Space does not permit a detailed examination of all of these organizations, or all of their activities, but throughout the book many of these associations will be acknowledged as specific subjects are covered.

R1-4.1 ASHRAE

The American Society of Heating, Refrigeration and Air Conditioning Engineers (ASHRAE) is an organization started in 1904 as the American Society of Refrigeration Engineers (ASRE) with 70 members. Today its membership is composed of thousands of professional engineers and technicians from all phases of the industry. ASHRAE also creates equipment standards for the industry. Its most important contribution probably has been a series of books that have become the reference books of the industry. These include the *Guide and Data Books for Equipment, Fundamentals, Applications and Systems.*

In cosponsorship with ARI, ASHRAE holds an annual international Air Conditioning-Heating-Refrigeration Exposition which may draw 30,000 to 50,000 people in the field. Product exhibits, technical displays, seminars, and business seminars highlight the event as shown in Fig. R1-2.

R1-4.2 ASME

The American Society of Mechanical Engineers (ASME) is an organization composed of engineers in a wide variety of industries. Among other functions, ASME writes standards related to safety aspects of pressure vessels.

R1-4.3 ACCA and RSES

The Air Conditioning Contractors of America (ACCA) is a service contractor's association concerned with the education of technicians, service managers, and business improvement techniques. In cooperation with Ferris State University, ACCA provides technician EPA certification.

The Refrigeration Service Engineers Society (RSES) is a service association dedicated to education in the HVAC/R industry. RSES also has a technician EPA certification program. RSES chapters conduct classroom training in technical areas. RSES is also a source for educational printed material and books.

R1-5 AIR CONDITIONING AND REFRIGERATION INSTITUTE (ARI)

The Air Conditioning and Refrigeration Institute (ARI) is a national trade association representing manufacturers of over 90% of United States-produced central air-conditioning and commercial refrigeration equipment. ARI was formed in 1954 through a merger of two related trade associations. Since that time several other trade associations have also merged into ARI. ARI traces its history back to 1903 when it started as the Ice Machine Builders Association of the United States. Today ARI has over 180 member companies.

Many services are provided by ARI to assist HVAC/R technicians. Some of these services which would supplement this text are listed below:

1. ARI/GAMA competency examination. These tests are made available to students of educational institutions, to test their knowledge of fundamental and basic skills necessary for entry level HVAC/R technician positions. The information in this text covers the topics in the ARI curriculum guide and would assist the student in taking this examination. A directory of those who pass the examination is published nationally to assist prospective employers in identifying job candidates.

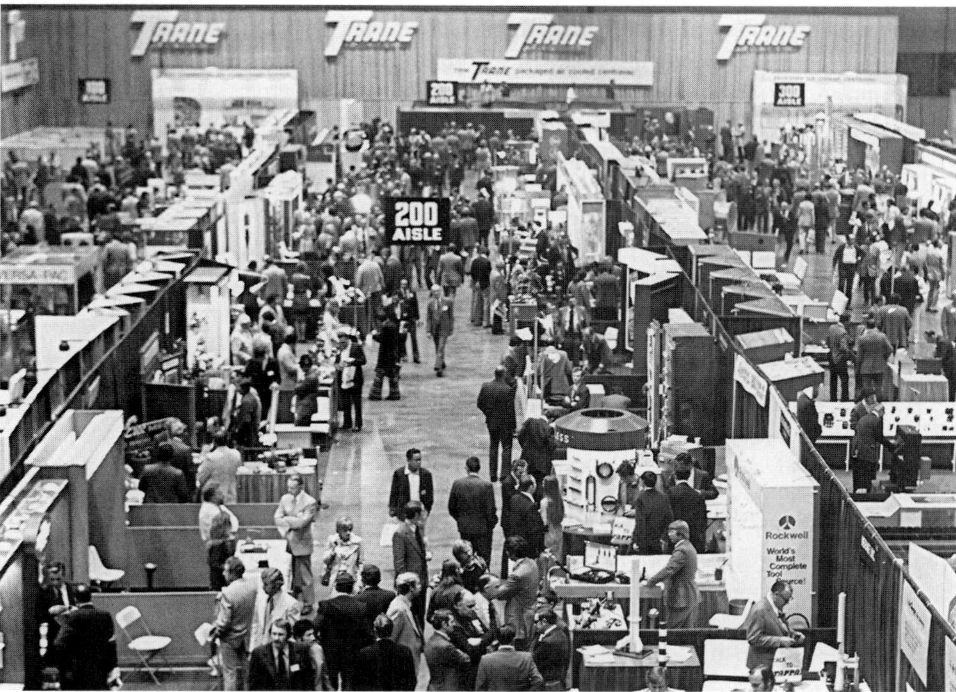

Figure R1-2
International air-conditioning, heating, and refrigeration exposition.

PRACTICING PROFESSIONAL SERVICES

Ask a local contractor or maintenance supervisor at your school if you can interview them for a short period of the day to learn more about the career opportunities within the refrigeration industry. Be frank, forward, and polite, expressing the sincerity of your interests and aspirations within the profession. Most managers will find the time in their busy schedule to help you with your career questions. Type a formal report of your interview for your teacher and submit a copy to the manager involved with a thank-you note. Upon nearing graduation, contact that employer and ask if they will be hiring soon or if they can direct you towards potential employment opportunities. Remember to always be persistent but courteous with all professionals, and they will do the same for you.

ON THE JOB

Contact your school counselor or advisor and find out about summer internships available within your profession. Working for an employer during the summer gaining information and hands-on experience helps your chances of finding a good job upon graduation.

Sometimes co-operative work experience jobs are also available to students who are concurrently enrolled in school. These give the student the opportunity to go to school and still work with the tools as a helper or apprentice when not in the classroom. Again, this will give you insight into your career choice and provide valuable work experience.

2. Equipment donations to schools participating in the above program. ARI contacts industry sources having no-cost or low-cost equipment available to supply a school's laboratory needs.
3. Technicians' certification program. In accordance with EPA's enforcement of the Clean Air Act, the sale of refrigerants is made only to those technicians who have been certified. ARI is among those approved by EPA to administer the test for certification. In addition, ARI provides study material to prepare for the test.
4. Reclaimer certification program. EPA also requires certification of any processor of recovered refrigerant for resale. ARI is among those assigned by EPA to carry out a certification program for companies which seek to reclaim refrigerants. The technicians when handling reclaimed refrigerant should become familiar with the *Directory of Certified Reclaimed Refrigerants,* published every March and September by ARI.
5. Certification program for equipment used to recover and recycle refrigerant. ARI is one of the companies approved by EPA to certify equipment used to recover and recycle refrigerants. The technicians should become familiar with the *Directory of Certified Refrigerant Recovery/Recycling Equipment,* published every March and September by ARI.
6. HVAC/R equipment certification program. ARI maintains a certification service which tests a wide variety of equipment and products to verify the performance described by the manufacturer. Certified directories for various products are published semi-annually and annually.

ARI has a full program of educational activities geared toward helping the nation's vocational and technical schools improve and expand their education and training programs. Under the direction of ARI's education director and its Education and Training Committee, ARI serves as the source from the manufacturers to school instructors, department heads, and guidance counselors. In addition to this textbook and its companion pieces, ARI produces the *Bibliography of Training Aids,* a career brochure and a promotional videotape for schools to use to recruit students into HVAC/R programs. Many schools around the country have adopted the ARI competency exams as final exams for their programs. ARI's most recent efforts involve participation in efforts to develop National HVAC/R competency standards.

Having students pass the ARI/GAMA competency exams and training toward national competency standards

KNOWLEDGE EXTENSION

The refrigeration student can become better acquainted with industrial personnel within the HVAC/R trade by becoming an active member of professional trade associations such as ACCA, ASHRAE, and RSES. Attending conferences and seminars sponsored by these trade organizations will increase your chances to meet individuals and learn more about professional employment opportunities.

will improve the quality of installations and service. New HVAC/R technicians will be better prepared, resulting in three basic advantages:

1. Limited training required for contractors
2. Limited rework or repeat calls due to error.
3. Limited warranty/replacement for manufacturers.

The costs of repeat service calls, which is borne by contractors, may be reduced substantially through the use of properly trained technicians. Every new technician receives training and serves as an apprentice for a period of time. That is essentially a period where contractors pay two people to do one job. A properly trained technician will generally require less training time and function sooner than a poorly trained technician.

R1-6 UNITS OF MEASUREMENT

In the mid-1970s most of the industrial nations of the world made the decision to adopt the International System of Measurements (SI), also referred to as the metric system. Prior to that, many of these countries were using either an earlier version of the metric system or the English system of measurements (Inch-Pound, abbreviated IP). With a great increase in world trade it was considered advantageous to have a uniform system of measurement. The SI system offered a simplification of terms, having only one unit for each quantity (for example, pascal for pressure).

The Metric Conversion Act of December 1975 in the United States allowed voluntary conversion. Since that time, the adoption of SI units has been slow but persistent. In 1992 the Omnibus Trade and Competitive Act was passed by Congress requiring that all federal agencies adopt the metric (SI) system for procurement, grants, and other federal business activities by the end of 1992. ASHRAE is promoting and assisting the HVAC/R industry's implementation of the use of SI units by the year 2000.

In order for this text to be most useful, both the Inch-Pound (IP) and the Standard International (SI) system values, wherever practical, will be shown.

R1-6.1 SI Units

In the SI system each physical quantity has only one unit. The seven base units are:

Base Unit	Symbol	Represents
metre	m	length
kilogram	kg	mass
second	s	time
ampere (amp)	A	electric current
kelvin	K	thermodynamic temperature
mole	mol	amount of substance
candela	cd	luminous intensity

Many derived units are combinations of base units. The base and derived units may be modified by prefixes such as those listed below.

Prefix	Symbol	Represents
giga	G	1,000,000,000
mega	M	1,000,000
kilo	k	1,000
milli	m	0.001
micro	μ	0.000,001
nano	n	0.000,000,001

Physical Quantities

Area. The unit of area is the square meter (m^2). Large areas are expressed in square kilometers (km^2) or hectares (ha). The hectare is used only for land or sea areas and is equivalent to 10,000 m^2.

Energy. The unit of energy, work, and quantity of heat is the joule (J). The unit of power and heat flow rate is the watt (W). 1 watt (W) = 1 joule per second (J/s).

Force. The unit of force is the newton (N). The newton is also used in derived units which include force.

Examples: pressure = N/m^2 = Pa (pascal)
work = N • m = (J) joule
power = N • m/s = W (watt)

Length. The unit of length is the meter.

Mass. The unit of mass is the kilogram (kg). This is the only base unit that contains a prefix. However, names of multiples of unit mass are formed by attaching prefixes to the term gram. The megagram, Mg, (1000 kg) is the metric ton. The term weight should not be used when mass is intended.

Weight and Mass

Mass is the amount or quantity of matter in a physical body. It is that property of matter which gives it inertia, which is a body's resistance to being moved, speeded up or slowed down—that is, its resistance to acceleration.

Weight is a measure of the amount of force with which a body is pulled toward the earth by gravity. If an object is said to weigh 10 kg, it simply means that the earth pulls on it to that extent.

Mass remains the same, while weight changes at different altitudes and at different places on the surface of the earth. As an object is moved farther from the earth, its weight decreases due to the reduced gravitational attraction, but the mass remains unchanged.

Pressure. The unit of pressure, force per unit of area, is newton per square metre. This unit is called the pascal (Pa). The SI system has no equivalent symbol for pressure per square inch gauge (psig) or pressure per square inch absolute (psia); however, if conversion to SI is desirable, it may be written Pa (absolute) or Pa (gauge).

Volume. The unit of volume is the cubic meter (m^3). Smaller units are the liter, L (m^3/1000); milliliter, mL (0.001 L); and microliter, μL (0.000,001 L). Liter per second (L/s) replaces gallons per minute (gpm) and cubic feet per minute (cfm) in HVAC usage. One liter of water has a mass of one kilogram at its maximum density of 4°C.

Temperature. The unit of absolute temperature is the Kelvin (K). The Celsius (C) scale is a specific range of the Kelvin scale. Temperature intervals are the same on both scales. Zero degrees Celsius (0°C) equals 273 K and the boiling point of 100°C is equal to 373 K. These temperatures are based on sea-level pressure.

Time. The unit of time is the second (s). Minute and hour are not used, except for revolutions per minute (rpm), which may be used, but revolutions per second is preferred. Day, week, or month can be used for longer periods but do not have any SI symbols.

REVIEW

- Important dates: Refer to the sidebar in this chapter for important dates in the development of HVAC/R.
- Potential markets and uses for refrigeration include: Preservation of food, food production, shipping, storage, packing & processing plants, aircraft, railway cars, trucks, and ships; industrial uses such as production of rubber and plastic, and air conditioning.
- Types of jobs available in the HVAC/R Industry:

Graduate Engineer	Service Manager
Technician/Master Mechanic	Refrigeration Technician—Domestic
HVAC Technician	Refrigeration Technician—Commercial
HVAC Helper/Apprentice	Maintenance Mechanic
Sheet Metal Mechanic	
Sheet Metal Helper/Apprentice	
Sales person	

- Occupational opportunities in the refrigeration industry include high temperature cooling (process air conditioning), medium temperature cooling (product storage above freezing), low temperature cooling (quick freezing and storage below freezing) and extremely low temperature cooling (medical, chemical, industrial and scientific applications).
- Other areas of employment include electric and gas utilities, schools, hospitals, and various related industries.
- Services available through ARI: ARI/GAMA competency exams; equipment donations to schools; EPA technician certification; reclaimer certification; certification for recovery & recycling equipment; HVAC/R equipment certification; standards; industry lobby.
- Benefits of passing the ARI/GAMA competency exam:
  Improves the quality of installations & service
  Limited rework or repeat calls due to errors
  Limited warranty replacements
  Improved job opportunities
- Trade Associations:
  ARI—Air Conditioning & Refrigeration Institute
  AHAM—Association of Home Appliance Manufacturers
  GAMA—Gas Appliance Manufacturers Association
  ASHRAE—American Society of Heating, Refrigeration, and Air Conditioning Engineers
  RSES—Refrigeration Service Engineers Society
  ACCA—Air Conditioning Contractors of America
  ASME—American Society of Mechanical Engineers
- The units of measurement for the refrigeration industry include the Inch-pound (IP) English System or Standard International (SI). The American Society of Heating, Refrigeration and Air Conditioning Engineers (ASHRAE) is promoting and assisting the HVAC/R industry's implementation of the use of SI units by the year 2000.

Problems and Questions

1. The modern way of life depends on refrigeration/air conditioning for: the _____ of food, _____ comfort, and industrial _____.
2. The air conditioning and refrigeration field affords many _____ for those interested in: _____, _____, _____, and _____.
3. _____ solving ability is of great importance to those who design, install, or service HVAC/R systems.
4. Electrical and _____ ability is another area of competence for success in this field.
5. The air conditioning and refrigeration industry encompasses many occupations such as _____ and _____.
6. List the services available through ARI.
7. What could passing the ARI/GAMA exam mean for you? What benefit could passing the exam provide for your future employer?
8. What employers of HVAC/R technicians are there in your area? (Possibly visit a local shop or job service,

or use a city directory, telephone book, or local newspapers.)

9. Write a job description, including work conditions and benefits, for a related occupation that interests you. (Suggested reference books are *Dictionary of Occupational Titles* and *Occupational Outlook Handbook.* Also a visit to a local shop might be helpful.)
10. ACCA stands for _____.
11. Who makes up the ARI membership?
12. RSES stands for _____.
13. The Refrigeration and Air Conditioning industry represents one of the major domestic and world industries. True or false?
14. For every $1,000,000 in installed value of equipment the following employees are needed: 1 graduate engineer, 2 master level technicians, 11 HVAC/R technicians, 2 apprentices, 7 sheet metal mechanics, 1 sheet metal apprentice, and 2 sales persons. True or false?

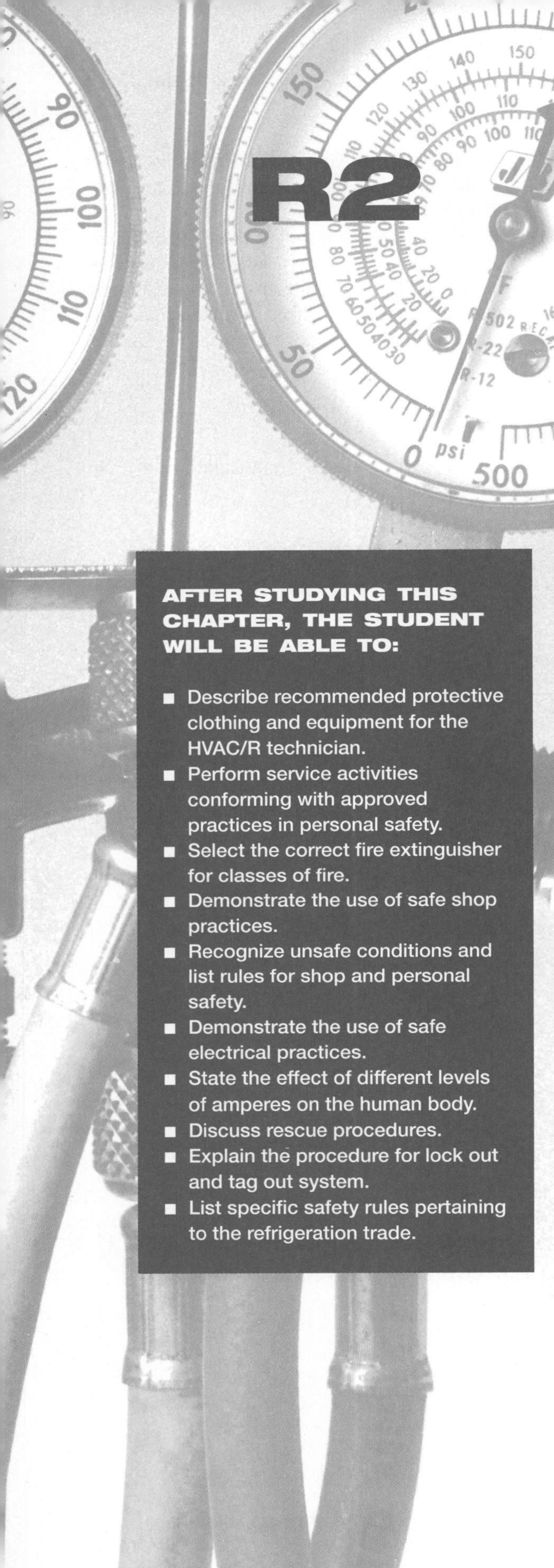

R2 Safety

AFTER STUDYING THIS CHAPTER, THE STUDENT WILL BE ABLE TO:

- Describe recommended protective clothing and equipment for the HVAC/R technician.
- Perform service activities conforming with approved practices in personal safety.
- Select the correct fire extinguisher for classes of fire.
- Demonstrate the use of safe shop practices.
- Recognize unsafe conditions and list rules for shop and personal safety.
- Demonstrate the use of safe electrical practices.
- State the effect of different levels of amperes on the human body.
- Discuss rescue procedures.
- Explain the procedure for lock out and tag out system.
- List specific safety rules pertaining to the refrigeration trade.
- Demonstrate the use of safe refrigerant practices.
- Understand first aid classifications of accidents and the proper steps to follow.
- Discuss the nature of environmental concern associated with ozone depletion and the effects of different refrigerants.
- Discuss the reasons why certain service techniques are no longer allowed.
- Discuss the reasons for recovery and reclamation of refrigerants.
- Describe creation and destruction of ozone in the stratosphere and troposphere.
- Identify refrigerant remnants in the stratosphere.

KNOWLEDGE EXTENSION

Landmarks in the development of refrigerants are listed below.

1834 The first practical refrigerating machine (Jacob Perkins') used ether in a vapor compression cycle.
1849 Ferdinand Carre used ammonia and water in his refrigeration machine.
1850 The first absorption machine (Edmond Carre's) used water and sulfuric acid.
1866 Vapor compression systems used a mixture of petrol ether and naphtha (patented as chemogene). The same year carbon dioxide was introduced as a refrigerant.
1873 Ammonia was first used in vapor compression systems.
1875 Sulfur dioxide and methyl ether were introduced as refrigerants.
1878 Methyl chloride emerged as a refrigerant, soon followed by dichloroethene (dilene) used by Willis Carrier in his first centrifugal compressor.
All of the above were either flammable, toxic, or both. Accidents were common.
1926 Carrier switched to methylene chloride for his centrifugal compressors.
1930 R-12, nontoxic and nonflammable, was synthesized. It was the first of the chlorofluorocarbon (CFC) refrigerants. R-12, R-11, R-114, and R-113 were produced from 1931 to 1934.
1936 The first hydrochlorofluorocarbon (HCFC) refrigerant, R-22 was produced.

By 1963 the above five refrigerants constituted 98 percent of the total production of a thriving organic fluorine industry.

R2-1 PERSONAL SAFETY

In all trades, safety is a major concern. Accidents are caused by carelessness, as well as lack of awareness of proper safety procedures. This chapter deals with some of the safety tips and procedures the refrigeration technician should follow—whether on the job site or on related locations where hazards could exist.

The following are some of the important areas in practicing personal safety that, if followed, will help in preventing accidents:

1. *Protective clothing and equipment,* appropriate to activity being performed.
2. Safety procedures for *handling harmful substances.*
3. *Safe work practices,* including proper care and use of tools.
4. Necessary precautions for *preventing electrical shocks.*
5. *Avoiding refrigerant contacts* with any part of your body. Always *keep the pressure of a confined gas within safe limits.*

R2-2 PROTECTIVE CLOTHING AND EQUIPMENT

The following clothing and equipment should be used by refrigeration technicians:

Head Protection

An approved hard hat (Fig. R2-1) or cap should be worn whenever there is a danger of things dropping on the head or where the head may be bumped. On a construction site, proper head gear for protection is a *must.*

Confine long hair and loose clothing before operating rotating equipment.

Ear Protection

Hearing protection devices (Fig. R2-2) must be worn whenever there is exposure to high noise levels of any duration. These devices are of two types: (1) ear plugs which are inserted in the ear, and (2) ear muffs which cover the ear. Either one must be properly selected on the basis of how much protection is required.

Eye and Face Protection

Approved eye or face protectors (Fig. R2-3) must be worn whenever there is a danger of objects striking the eyes or face. Eye and face protectors have various shapes and sizes, some of them very specialized. If prescription eye glasses are worn, they must have side shields.

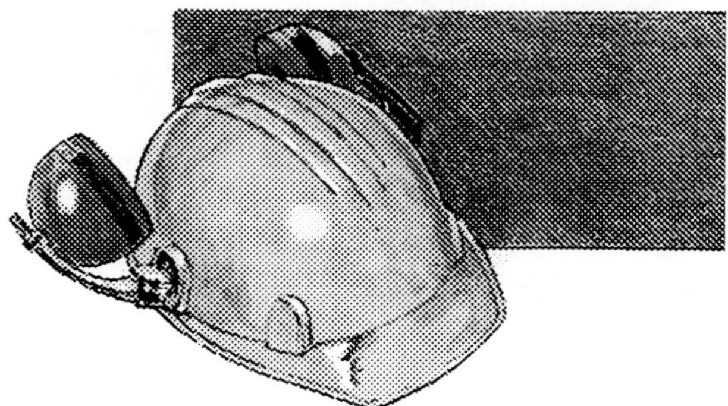

Figure R2-1 Hard hat for head protection (showing ear protectors). (Use granted by Worker's Compensation Board of British Columbia.)

Figure R2-2 Types of hearing protection equipment. (Use granted by Worker's Compensation Board of British Columbia.)

Special eye protectors must be worn when arc-welding, gas welding and burning to cut out harmful radiation. These come in various shades which filter out the harmful emissions. Take time to identify the right one for the job. For example, *never wear oxyacetylene welding goggles when arc welding.*

Respiratory Protection

There are two main types of respirators as shown in Fig. R2-4: (1) those that purify the air by filtering out harmful dusts, mists, metals, fumes, gases and vapors; and (2) those which supply clean breathing air from a compressed air source. The second type should always be worn when working in a confined space where concentrations of harmful substances are very high or where the concentration is unknown. Remember that most refrigerants are odorless, tasteless, and invisible, and can cause asphyxiation in a very short time.

Respirators must fit tightly against the skin so that there is no leakage from the outside into the face. Workers who are required to use respirators at any time must be instructed in their use, care, maintenance, and limitations.

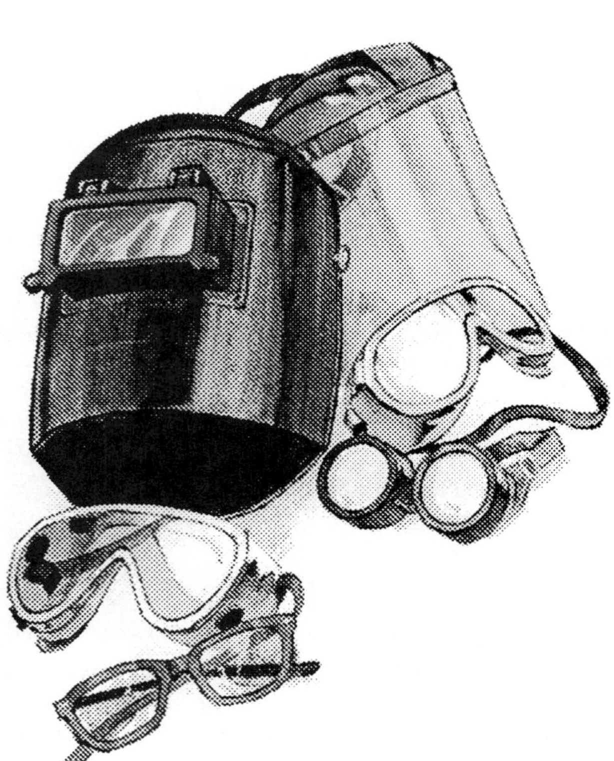

Figure R2-3 Eye and face protection equipment. (Use granted by Worker's Compensation Board of British Columbia.)

Figure R2-4 Various types of respirators. (Use granted by Worker's Compensation Board of British Columbia.)

PRACTICING PROFESSIONAL SERVICES

Lockout/Tagout is a safety procedure which has been implemented nationwide to establish an easy-to-understand guideline for locking or tagging the control of equipment and sources of energy to prevent unanticipated actuation while work repairs and maintenance are being performed.

When there is no place to physically attach a lock, a tag is substituted and attached to the machine's control switch, energy master switch, or main power panel. It is there to warn other workers that someone is working on a piece of equipment and could be injured if the equipment is actuated.

Machines or equipment must be stopped and isolated from all sources of energy and the controls must be locked before employees begin to perform service, replacement, or maintenance because unexpected start-up of the equipment could cause injury.

Hands and Feet Protection

Hands. There are many different kinds of gloves (Fig. R2-5). Some are made for special usages, such as gloves of steel mesh or Kevlar to protect against cuts and puncture wounds. Different glove materials are needed to protect against a variety of different chemicals. Choose the right kind from a dependable supplier who can supply this information. Discard the damaged ones.

Feet. When choosing foot protection (Fig. R2-5) use the following guidelines:

Figure R2-5 Hand and foot protection equipment.
(Use granted by Worker's Compensation Board of British Columbia.)

1. All footwear must be well constructed to support the foot and to provide secure footing.
2. Where there is danger of injury to the toes, top of the foot, or from electrical shock, the proper shoe or boot must have Construction Safety Approval (CSA) indicated.
3. Where there is danger of injury to the ankle, footwear must cover the ankle and have a built-in protective element/support.
4. If there is danger of harmful liquids dropping on the foot, the top of the shoe must be completely covered with an impervious material or treated to keep the dripped substance from contacting the skin.

Fall Protection

Two methods of preventing injury from falling are: (1) fall prevention equipment and (2) fall-arresting equipment. Either of these methods are required when working at heights over 10 feet above grade when no other means have been provided for preventing falls. Fig. R2-6 illustrates safety belts and harnesses.

In *fall prevention,* a worker is prevented from getting into a situation where he can fall. For example, a safety belt attached to a securely anchored lanyard will limit the distance a worker can move.

In *fall arresting* the worker must wear a safety harness attached to a securely anchored lanyard which will limit the fall to a safe distance above impact. The harness helps prevent the worker from suffering internal damage. Belts should not be used to arrest a fall. Where a fall-arresting system is not practicable, a safety net should be suspended below the work activity. The worker should be secured separately from the tools and equipment.

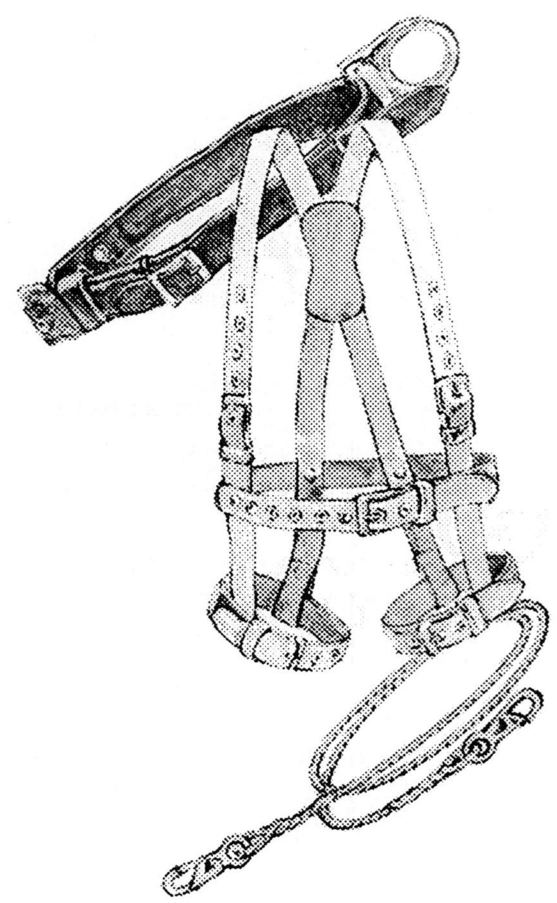

Figure R2-6 Safety belts and harnesses. (Use granted by Worker's Compensation Board of British Columbia.)

R2-3 HARMFUL SUBSTANCES

Workers in the mechanical trades can be exposed to a variety of harmful substances, such as dust, asbestos, carbon monoxide, refrigerants, resins, adhesives, and solvents.

All *dust* can be harmful. Where dust cannot be controlled by engineering methods, an approved respirator designed to filter out specific dust must be worn.

When *asbestos*-containing material (insulation) is being cut or shaped, the particles must be removed by a ventilation system which discharges them through a high efficiency particulate air (HEPA) filter. All waste materials that contain asbestos must be placed in impervious bags for transfer to an approved disposal site. These fibers, when inhaled, are considered carcinogenic.

Mobile equipment operating in an enclosed area can produce dangerous levels of *carbon monoxide* (CO). Oil- or gas-fired space heaters without suitable vents can also produce carbon monoxide. Areas must be well ventilated while being heated with these devices.

Some *refrigerants* are more dangerous than others. All refrigerants are dangerous if they are allowed to replace the oxygen in the air. Even the so-called "safe refrigerants" can produce a poisonous phosgene gas when heated to high temperatures. Refrigerants sprayed on any part of the body can quickly freeze tissue. The safe handling of refrigerants will be discussed in detail in a later part of this chapter.

Resins, adhesives, and *solvents* can be dangerous if not properly handled. Ensure that the workspace is continuously ventilated with large amounts of fresh air.

Never use *carbon tetrachloride* for any purpose, because it is extremely toxic, either inhaled or on the skin. Even slight encounters with it can cause chronic problems. Consult a physician if exposed to it.

R2-4 SAFE WORK PRACTICES

The refrigeration and air conditioning technician works in many areas: in the shop, in various types of buildings, in equipment rooms, on rooftops, and on the ground outside buildings. Each location requires different activities where safe performance is essential.

In addition, the worker deals with many potentially dangerous conditions, such as handling pressurized liquids and gases, moving equipment and machines, working with electricity and chemicals, and exposure to heat and cold. It is important, therefore, that the technician practice good safety procedures wherever the work is being done or whatever part of work in which he or she is engaged.

Hand Tools

1. Keep all hand tools sharp, clean, and in safe working order.
2. Defective tools should be repaired or replaced.
3. Use correct, proper fitting wrenches for nuts, bolts, and objects to be turned or held.
4. Do not work in the dark; use plenty of light.
5. Do not leave tools on the floor.

Power Tools

1. Only use power tools that are properly grounded.
2. Stand on dry non-conductive surfaces when using electrical tools.
3. Use only properly sized electrical cords in good condition.
4. Turn on the power only after checking to see that there is no obstruction to proper operation.
5. Disconnect the power from an electrical tool (or motor) before performing the maintenance task of oiling or cleaning.
6. Disconnect the power supply when equipment is not in use.

R2-4.1 Shop Safety

1. Keep the shop or laboratory floor clear of scraps, litter and spilled liquid.

ON THE JOB

How safe are refrigerants? All have concerns associated with toxicity, flammability, and physical hazards. Now, more so than in the past, these can be used safely if one recognizes there is not now, or likely to be discovered in the future, the "ideal refrigerant", free from the above concerns.

The new, alternative refrigerants can be used with comparable or higher safety than those they replace when applied following recommended selection, handling, installation, and operating practices in equipment conforming to recognized safety standards.

Inherent risks to refrigerants are:

1. *Quantity* or *concentration levels. Acute toxicity* refers to the impacts of single exposures, often at high concentrations. *Chronic toxicity* refers to the effects of repeated or sustained exposures over a long time. The Program for Alternative Fluorocarbon Toxicity (PAFT) Testing is a cooperative effort to reduce both risks. Sponsored by the major CFC producers from nine countries, this program has initiated intensive testing for R-123, R-134a, R-142b, R-124, R-125, R-225ca, R-225cb, and R-32 refrigerants, and is ongoing. These studies investigate acute, subchronic and chronic toxicity as well as genetic effects on man and other living organisms in the environment. Another program simply investigates effects on tumors, benign or malignant.
2. *Accidents, failures* or *incidents of short duration.*
   a. Direct exposure to refrigerants at low temperatures can cause frostbite. Use appropriate eye protection and gloves.
   b. Suffocation from leaks and spills is also a risk. Install leak sensors and use proper installation, handling, and storage procedures.
   c. Flammability increases with temperature and pressure, especially in the presence of combustible lubricants. Accidents have occurred even with completely nonflammable refrigerants. Never use R-22 mixed with air for leak testing; use dry nitrogen instead.
   d. Other preventive measures: Properly identify container and storage vessels. Install one or more pressure-relief devices on all refrigerating systems in case of fire or other abnormal conditions. Limit machinery room access to authorized personnel. Provide at least one self-contained breathing apparatus for emergency use.

2. Store oily shop towels or oily waste in metal containers in an open, airy place.
3. Clean the chips from a machine with a brush; do not use a towel, bare hands, or compressed air.
4. Keep safety glasses and gloves in a prominent location adjacent to machinery used for grinding, buffing or hammering and where material with sharp edges is handled.
5. Establish cleaning periods. Make sure everyone is clear when using compressed air to clean.

Steps in Maintaining an Orderly Shop

1. Arrange machinery and equipment to permit safe, efficient work practices and ease in cleaning.
2. Materials, supplies, tools, and accessories should be safely stored in cabinets, on racks, or in other readily available locations.
3. Working areas and work benches should be clear. Floors should be clean. Keep aisles, traffic areas, and exits free of materials and obstructions.
4. Combustible materials should be properly disposed of or stored in approved containers.
5. Drinking fountain and wash facilities should be clean and in good working order at all times.

Fire Extinguishers

The danger of fire is always present. Oil, grease or paint soaked rags can ignite spontaneously. Keep them in metal containers.

Sparks, open flames and hot metal can ignite many materials. Always have a fire extinguisher close at hand when welding or burning.

Extreme caution should be taken with highly flammable and volatile solvents. Due to its low flash point (the temperature at which vapors will ignite), gasoline should never be used as a cleaning solvent.

PREVENTIVE MAINTENANCE

Proper maintenance and care is required to extend the life and reliability of all tools and test equipment. This includes cleaning tools of grease and oils regularly upon job completion. Gauges should be hung up or securely fastened and recalibrated monthly. Test instruments should always be turned off when work is complete and reinstalled in carrying cases. Low batteries should be replaced immediately and discarded in appropriate containers. Vacuum pumps should be drained of oil and refilled with fresh oil when work is complete. Oxygen-acetylene welding units should have caps on containers and be secured when moving. All welding regulators should be back seated and bled free of any residual gas pressures.

Fire extinguishers should be readily accessible, properly maintained, regularly inspected, and promptly refilled after use. They are classified according to their capacity for handling specific types of fires (Fig. R2-7).

Class A Extinguishers. These are used for fires involving ordinary combustible materials such as wood, paper, and textiles, where a quenching, cooling action is required.

Class B Extinguishers. These are for flammable liquid and gas fires involving oil, gas, paint, and grease, where oxygen exclusion or flame interruption is essential.

Class C Extinguishers. These are for fires involving electrical wiring and equipment where non-conductivity of the extinguishing agent is critical. This type of extinguisher should be present whenever functional testing and system energizing take place.

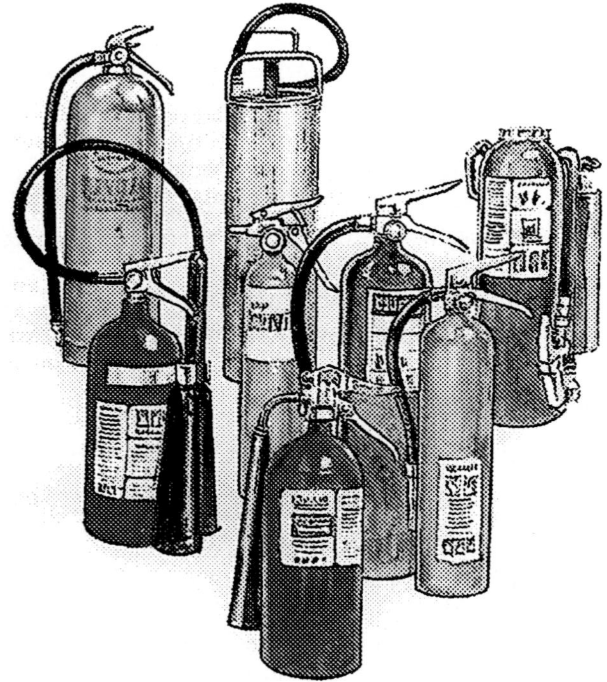

Figure R2-7 Types of fire extinguishers. (Use granted by Worker's Compensation Board of British Columbia.)

Material Handling

Use mechanical lifting devices whenever possible. Use a hoist line when lifting tools or equipment to a roof. If you are required to lift a heavy object, get help. In order not to strain your back, the following procedures should be observed when lifting heavy objects (Fig. R2-8):

1. Bend your knees and pick up the object, keeping your back straight up.
2. Gradually lift the weight using your leg muscles, continuing to keep your back straight up.

R2-4.2 Access Equipment

Access equipment refers to ladders and scaffolds that are used to reach locations not accessible by other means. The following precautions should be practiced in the use of *ladders* (Fig. R2-9):

1. Only use CSA- or ANSI-approved ladders. Maintain ladders in good condition. Inspect ladders before each use and discard ladders needing frequent repairs or showing signs of deterioration.

Figure R2-8 Properly lifting heavy loads. (Use granted by Worker's Compensation Board of British Columbia.)

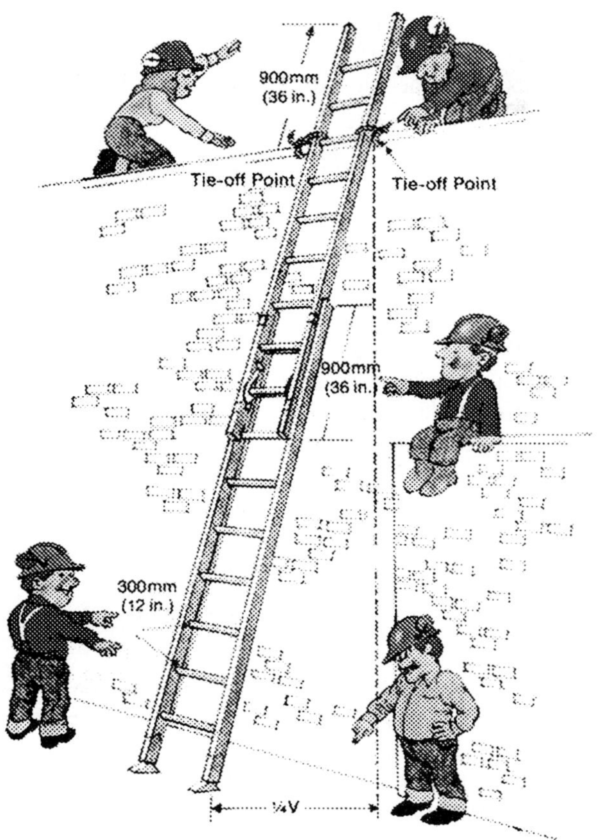

Figure R2-9 Proper use of a ladder. (Use granted by Worker's Compensation Board of British Columbia.)

2. All portable ladders must have non-slip feet.
3. Place ladders on a firm footing, no further out from the wall than ¼ the length of the ladder.
4. Ladders must be tied, blocked, or otherwise secured to prevent them from slipping sideways.
5. Never overload a ladder. Follow the maximum carrying capacity of the ladder, including the person and equipment. The American National Standards Institute (ANSI) sets the standard for ladders.
6. Only one person should be on a ladder, unless the ladder is designed to carry more people. Follow maximum load rating.
7. Never use a broken ladder, or a ladder on top of scaffolding.
8. Always face the ladder and use both hands when climbing or descending a ladder.
9. Use fiber glass or wood ladders when doing any work around electrical lines.
10. Ladders should be long enough so you can perform the work comfortably, without leaning or having to go beyond the two rungs below the top rung safety barrier.
11. Step ladders should only be used in their fully open positions.

The following recommendations apply to *scaffolds:*

1. Scaffolds must be supported by solid footings.
2. A scaffold having a height exceeding 3 times its base dimension must be secured to the structure.
3. Wheels of rolling scaffolds must be locked when used by workers and/or their materials.
4. No worker is to remain on the scaffold while it is being moved.
5. Access to the work platform must be a fixed vertical ladder or other approved means.

R2-4.3 Welding and Cutting

Welding and cutting is a specialized skill and requires special training. Many refrigeration and air conditioning technicians acquire this training due to the need to perform some of these operations as part of their work. It must be recognized, strictly from a safety standpoint, that this work should not be attempted without adequate knowledge and instruction.

R2-4.4 Air-Acetylene Torches

Air-acetylene torches are used by HVAC/R technicians to produce sufficient heat for silver (hard) soldering. These torches use a mixture of air and acetylene as a fuel. The following safety rules should be practiced when using this equipment:

1. Always use a regulator on the acetylene tank.
2. Always secure the cylinder against something solid to prevent it from being accidentally knocked over.
3. Wear the proper colored safety glasses.
4. Open the valve on the cylinder only one-quarter of a turn.
5. Light the air-acetylene torch with a striker.

R2-4.5 Commercial Job Site Safety

1. Always follow contractors safety guidelines. Most safety guidelines parallel Occupational Safety and Health Act (OSHA) or state safety rules.
2. Hard hat, steel-toe shoes, and safety glasses should be used when required for your own safety.
3. Every person who goes to work in the morning has a right to return home safe after his or her work is done.

R2-4.6 First Aid

Refrigeration and air-conditioning workers are advised to enroll in an approved first aid course. Prompt and correct treatment of injuries not only reduce pain but could also save lives. A classification of accidents which occur to

HVAC/R personnel, related to the hazards described, includes the following:

1. Injuries due to mechanical causes.
2. Injuries due to electrical shocks.
3. Injuries due to high pressure.
4. Injuries due to burns and scalds. Injuries due to explosions.
5. Injuries due to breathing toxic gases.

Steps to Be Followed in Case of an Accident

1. All accidents, injuries, and illnesses should be reported to whomever is in charge, no matter how minor injuries may seem.
2. First aid should be administered if needed, only by those qualified to do so. Posted emergency procedures should be followed, as applicable. For example: If liquid refrigerant is sprayed on the skin or in the eyes, flush the area with cold water and get treatment.
3. The victim may be sent or taken to receive medical services.
4. An accident report form should be filled out by the person in charge.
5. The area should be cleaned up to remove any contaminants causing the injury, before permitting the area to be used again.
6. An investigation of the accident should be conducted to determine the cause of the accident and to determine ways to prevent similar incidents.

R2-5 SAFETY WHEN WORKING WITH ELECTRICITY

All possible precautions must be practiced to prevent electrical shock—current passing through the body. Very few realize the damage that can be done even by a small amount of current.

The following information applies to low-voltage circuits where current is measured in milliamps. One amp is equal to 1000 milliamps.

The tabulation and the illustration in Fig. R2-10 indicate the effect on the body when various amounts of current pass through the body—all 100 milliamps or less.

Electric Current	Body Sensation
Less than .5 milliamps	no sensation
1 to 2 milliamps	muscular contraction
5 to 25 milliamps	painful shock, inability to let go
Over 25 milliamps	violent muscular contractions
50 to 500 milliamps	heat convulsions, death
Over 100 milliamps	paralysis of breathing, burns

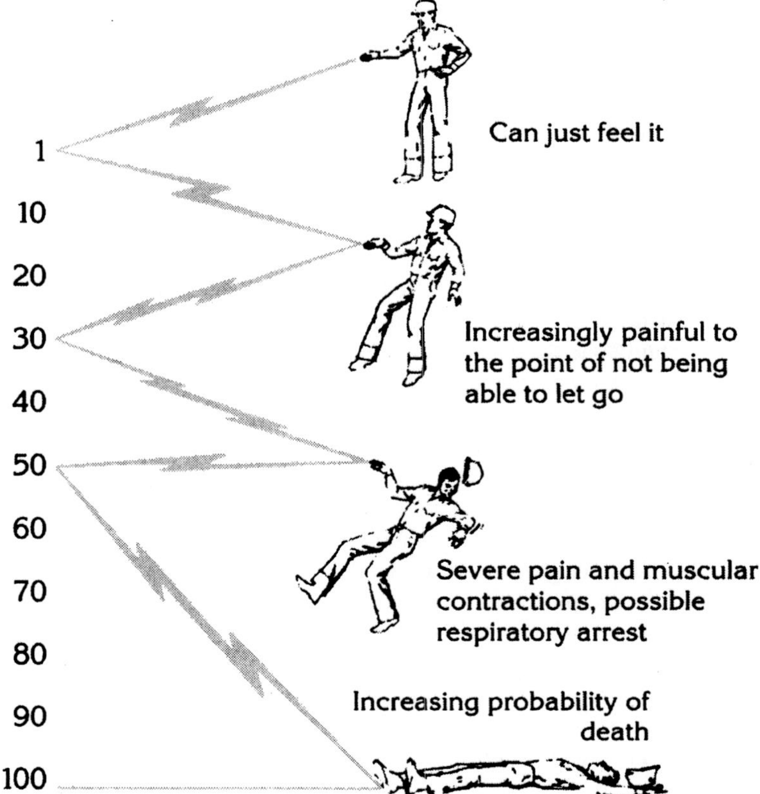

Figure R2-10 Amperage ratings of electric current creating various shock effects. (Use granted by Worker's Compensation Board of British Columbia.)

Electrical Safety Rules:

1. Check all circuits for voltage before doing any service work. Tag and lock all electrical disconnects when working on live circuits.
2. Stand on dry non-conductive surfaces when working on live circuits.
3. Work on live circuits only when absolutely necessary.
4. Never bypass an electrical protective device.
5. Properly fuse all electrical lines.
6. Properly insulate all electrical wiring.
7. Shut off and tag valves before working on refrigerant, steam and water lines.
8. Follow all codes when making modifications or repairing any system.
9. Never chip ice or frost from refrigeration line, coils, or sight glass.
10. Read Material Safety Data Sheets (MSDS).
11. Follow warning and caution signs.
12. Following refrigerant recovery, ensure the system is open to the atmosphere before brazing.

R2-6 REFRIGERATION SAFETY

The hazards associated with refrigeration service are principally associated with the proper use of refrigerants and their storage in closed containers and systems. A large improvement was made when the industry started using the so-called "safe refrigerants" (Class I or fluorocarbons) which were non-toxic and nonflammable. Dangers now relate to the use of pressurized gas or liquid and the fact that these chemicals when released accidentally can replace oxygen in a confined space without sensory detection.

The following rules can help decrease the hazards of refrigeration service.

R2-6.1 Use of Refrigerants

1. Good ventilation is essential where work is being done with refrigerants, and whenever welding, brazing, or using a cutting torch. Don't use torches in high concentrations of refrigerant.
2. Wear safety goggles and gloves when working with refrigerants. Liquid refrigerant can cause "frostbite" when in contact with eyes and skin.
3. Wrap cloth around hose fittings before removing them from a pressurized system or cylinder. Inspect all fittings before attaching hoses or working on them.
4. Ventilate a room containing refrigerants, pressurized gas, or liquid before entering to work. Breathing a mixture of air and refrigerant can cause unconsciousness because the refrigerant contains no oxygen.
5. Install oxygen monitors and alarm systems. These are required in machinery rooms with refrigeration equipment. Never enter an area where refrigerant is above exposure limits without proper breathing apparatus.
6. Wear gloves when servicing a system where a compressor has burned out. Refrigerant oil contained in the system can be very acidic. It should never be allowed to touch the skin.

R2-6.2 Handling Refrigerant Cylinders

1. Do not fill a cylinder with liquid refrigerant to more than 80% of its volume. Heat can expand the refrigerant and create a rupture pressure. Space must be available inside the cylinder for proper expansion to take place. In recovering refrigerants, this is particularly important. Special cylinders have been designed for recovery that have an automatic volume-limiting device.
2. In using a cylinder or transporting it, the cylinder must be secured with a chain or a rope in an upright position. Do not drop a cylinder.
3. Mixing refrigerants is dangerous. Cylinders are color coded to help identify each refrigerant. Each system has an identifying label. Do not mix refrigerants. Maintain the identification system.
4. Never apply a torch to a system containing refrigerant. If heat is needed to vaporize refrigerant, use hot water at a temperature not to exceed 125°F.
5. Do not refill disposable refrigerant cylinders.
6. Replace the cylinder cap when not using a cylinder. The cap protects the valve. Do not lift or carry a cylinder by the valve.

R2-6.3 System Safety

1. Never use oxygen or acetylene to pressurize a system. Use dry nitrogen from a tank properly fitted with a pressure regulator.
2. When isolating a section of piping or component of a system, exercise caution to prevent damage and potential hazard from liquid expansion.
3. Always charge refrigerant *vapor* into the low side of the system. *Liquid* refrigerant entering the compressor could damage the compressor or cause it to burst.
4. Never service a refrigeration system where an open flame is present. The flame must be enclosed and vented outdoors. If a fluorocarbon refrigerant comes in contact with intense heat, it can produce poisonous phosgene gas.
5. Always prevent moisture (water) from entering the refrigeration system. It can cause considerable damage. All parts must be kept dry. Containers of

oil must be tightly sealed to prevent contamination from the absorption of moisture.

R2-7 ENVIRONMENTAL SAFETY PRINCIPLES

One of the important factors in the development of this business has been the availability of "safe" refrigerants. By "safe" is meant refrigerants that are non-toxic, non-explosive, non-corrosive, non-flammable, and stable. Fortunately, these refrigerants have been plentiful, with suitable performance characteristics obtainable at a reasonable price.

Around 1970, scientists proposed that the ozone layer in the upper atmosphere of the earth was being depleted, affecting our protection against the sun's ultraviolet rays. By 1974 it was discovered that certain refrigerants containing chlorine, when released to the atmosphere, were a factor in this destructive process.

In 1987 an international group of scientists and government officials representing the major industrial nations met in Montreal and initiated world-wide controls on the production and use of CFCs, including such commonly used refrigerants as R-12 and R-22. Phaseout of CFC production was scheduled to occur by the year 2000. In 1992 this was accelerated to December 31, 1995. A transition schedule was set up to permit the shift to a hydrochlorofluorocarbon (HCFC-22), with limitations, and to hydrofluorocarbons (HFCs).

In the United States, similar regulations were legislated through the Clean Air Act amendments. The Environmental Protection Agency (EPA) provides implementing regulations and directives. The transition period of the 1990s is creating an entirely new set of restrictions.

These restrictions are causing a major change in the design and service of refrigeration equipment. This text is being revised during the period of transition. The information supplied is current with present industry practices; however, it is of the utmost importance that students keep closely in touch with developments in the use of new replacement refrigerants and the approved methods of handling the ones being phased out, as well as equipment adaptations.

R2-7.1 The Ozone Layer and the Atmosphere

From a human standpoint, the atmosphere is important because it is where the weather occurs. Atmospheric conditions are very complex, but one of the most striking facts is the variation of temperature at higher altitudes. After falling to –75°F at 26 kilometers (16 miles), it rises, just below 60 kilometers (36 miles), to very nearly the boiling point of water. At about 80 kilometers (48 miles) it falls to –30°F; at 100 kilometers (60 miles) it is zero and is warmer above. The temperature inversion is probably associated with the ozone layer.

Ozone (O_3), the triple-oxygen molecule, occupies an atmospheric layer from 10 to 30 miles up (17 to 50 kilometers). Its importance lies in the fact that it absorbs ultraviolet light almost completely.

Ozone abundances are maintained by a delicate balance between production, transport, and removal as shown in Fig. R2-11. Ozone is produced by sunlight in the tropics and transported to polar regions. In between, a certain amount is reduced in the whirlpools of atmo-spheric circulation which occur in winter and spring.

Also important (and much studied), however, is the chemical removal of ozone by chlorine-catalyzed cycles which occur in the polar regions during winter and spring. This activity originates in the polar stratospheric clouds (PSCs), solid nitric acid and ice clouds formed during the extremely cold winters over the poles. Although these clouds are familiar to scientists, their chemical behavior has been revealed only since the late 1980s.

Chemical reactions on the surface of cloud particles change chlorine from inactive to active status, which catalyzes ozone removal. An indicator of this occurrence is the formation of enhanced concentrations of chlorine

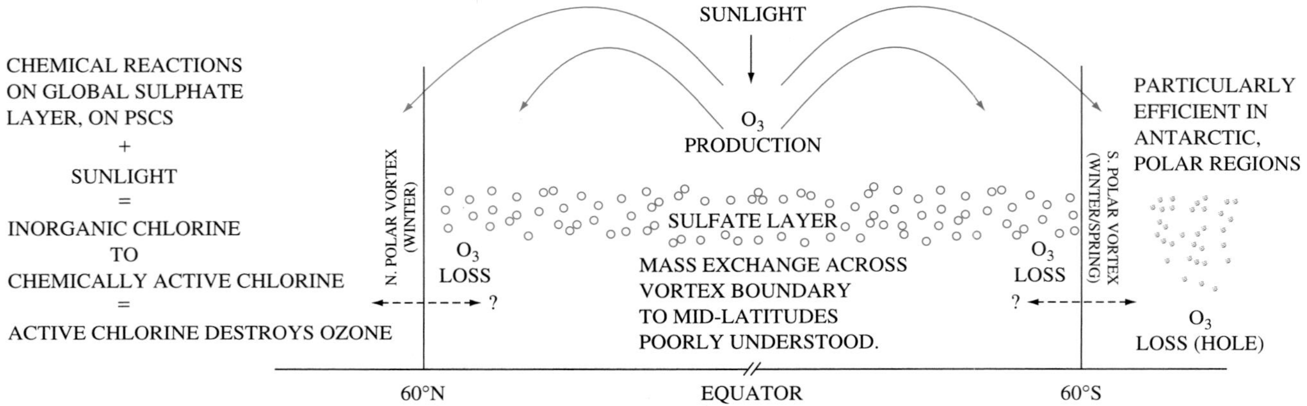

Figure R2-11 Probing the stratospheric ozone.

monoxide (ClO), which persists one to two weeks after the catalytic action of the clouds (PSCs). This, in turn, is associated with low, cloud-forming temperatures. The action starts in December as the atmosphere gets cold enough for PSCs to form, and continues through January, processing most of the vortex air and removing ozone. By February the warming of the vortex air causes disappearance of the PSCs. No hole in the ozone has appeared at the Arctic pole, because warming of the vortex air occurs before the arrival of sufficient sunlight for catalytic ozone removal.

Studies and data gathered in other regions of the earth are controversial because of the complexity of atmospheric events over time. Current studies indicate that depletion of ozone is occurring and increasing, and is not necessarily confined to the polar regions of the world. Chlorine compounds, such as CFCs used as refrigerants, are culprits. Although very stable in the atmosphere, when swept up into the stratosphere they are broken down by ultraviolet light into molecular fragments which destroy ozone.

R2-7.2 Global Warming and the HVAC/R Industry

Some alternative refrigerants or systems have shown a tendency toward more energy-intensive cooling systems (inefficiencies) which could result in higher quantities of carbon dioxide (CO_2) discharged into the environment from the combustion of fossil-fuel energy sources.

Higher quantities of carbon dioxide increase the group of gases called greenhouse gases. These gases allow solar radiation to pass through the earth's atmosphere but limit the amount of energy escaping, thus maintaining a range of temperatures compatible with living things. Excessive gases could result in excessive global warming.

Compared to the issue of ozone layer degradation, there is less world-wide consensus regarding the causes and impact of global warming. Both carbon dioxide and CFC refrigerants, however, have been identified as greenhouse gases. The removal of CFCs has the potential of reducing greenhouse gases and global warming.

Removing ozone-depleting CFCs tends to remove the greenhouse effect, thus lowering the potential for global warming. However, the replacement refrigerants may produce lower efficiency cooling systems, thus requiring the consumption of more electricity. The increased CO_2 that results from the additional electrical production tends to offset the benefit of removing the CFCs. Refrigerants that are ultimately considered acceptable should have both low ozone depletion potential (ODPs) and low global warming potential (GWPs). Fig. R2-12 shows these characteristics of various refrigerants.

Chemical	Chemical Formula	ODP	GWP[1]
Chlorofluorocarbons			
CFC-11	CCl_3F	1.0	1.0
CFC-12	CCl_3F_2	1.0	2.8
CFC-114	$CClF_2$-$CClF_2$	1.0	3.7
Hydrochlorofluorocarbons			
HCFC-22	$CHClF_2$	0.05	0.34
HCFC-123	$CHCl_2$-CF_3	0.02	0.02
HCFC-124	$CHClF$-CF_3	0.02	0.09
Hydrofluorocarbons			
HFC-125	CHF_2-CF_3	0.0	0.6
HFC-134a	CHF_2-CF_3	0.0	0.3
Refrigerant Mixtures			
R-502	51.1% CFC-115 48.8% HCFC-22	0.3	4.1
Ternary Blend:	36% HCFC-22 24% HFCl5a 40% HCFC 124	0.03	0.16
Other Refrigerants			
Water	H_2O	0.0	0.0
Ammonia	NH_3	0.0	0.0
Combustion Product			
Carbon dioxide	CO_2	0.0	1.0

[1]Calculated; compared to CFC-12 = 2.8

Figure R2-12 Ozone depleting and global warming potentials.

R2-8 FEDERAL REGULATION OF REFRIGERANTS

During the last twenty years, scientists have documented an "ozone hole" in the earth's atmosphere, and theories have been advanced that some of the chlorine-containing refrigerants were, at least in part, responsible for the atmospheric changes being observed.

Ozone is a molecule in the earth's atmosphere consisting of three oxygen atoms linked together (O_3). Its properties are different from oxygen molecules, which consist of two oxygen atoms linked together (O_2). Ozone is found in the air close to the earth (atmospheric ozone) and also in the outer reaches of our atmosphere eight to thirty miles above the earth (stratospheric ozone). Atmospheric ozone is harmful to breathe. It is the byproduct of ultraviolet rays from the sun and air pollution. However, stratospheric ozone is a naturally-occurring layer that has the beneficial property of absorbing and dissipating ultraviolet rays, preventing them from reaching the earth. Scientists are concerned with the depletion of this stratospheric ozone layer. The resulting increase in ultraviolet radiation that reaches the earth has been theorized to cause crop loss, eye and skin disorders, damage to marine life, damage to forests, and an increase in the level of atmospheric ozone. In addition, the loss of stratospheric ozone has been linked to global warming.

There is still some disagreement among scientists about whether or not the depletion of stratospheric ozone is actually caused by refrigerants. Some scientists attribute the loss of ozone to volcano eruptions or other natural causes. Other scientists observe that the increase of chlorine concentration measured in the stratosphere over the last twenty years matches the rise in the concentration of fluorine which has different natural sources than chlorine, but is used in the same refrigerants that contain chlorine. Those same scientists also note that the increase in chlorine concentrations corresponds to the increased use of the refrigerants that are suspected of being the cause.

R2-8.1 Effect of Various Refrigerants on the Stratospheric Ozone Layer

Not all refrigerants are equally harmful to the stratospheric ozone layer. The different refrigerants can be grouped according to the types of atoms that are found in the molecules of each.

CFC refrigerants contain atoms of **C**hlorine, **F**luorine, and **C**arbon. CFC refrigerants do the most damage to the ozone layer. They cause ozone molecules to break apart, and then reform as oxygen molecules.

HCFC refrigerants contain **H**ydrogen, **C**hlorine, **F**luorine, and **C**arbon. While these refrigerants also attack the ozone layer, their effect is only about $\frac{1}{20}$ as harmful as an equal quantity of CFC refrigerant.

HFC refrigerants have been invented recently as a long-term alternative to CFC and HCFC refrigerants. Their molecules contain **H**ydrogen, **F**luorine, and **C**arbon. Because they contain no chlorine, they do not contribute to ozone depletion.

Examples of CFC refrigerants are R-11, R-12, and R-502. R-22 is an HCFC. These four refrigerants were by far the most used in the fields of air conditioning and refrigeration until they were blamed for ozone depletion. Now, they are all being phased out of use.

R2-8.2 The Montreal Protocol

We can be proud of the role that our industry played in responding to the various theories that blamed the refrigerants we were using for ozone depletion. There was an international agreement signed in 1987, The Montreal Protocol. It took effect in 1989. It provided for a voluntary reduction and gradual phaseout of many ozone-depleting substances, including CFC and HCFC refrigerants. This agreement generally reduced production of CFC refrigerants to zero as of January 1, 1996. R-11, R-12, and R-502 are no longer being produced. However, there are millions of pounds already installed, and reclaimed refrigerant from discarded equipment will make these refrigerants available for the foreseeable future (probably at very high cost).

The phase-out of HCFC refrigerants under the Montreal Protocol was much slower. HCFC production is scheduled to be virtually eliminated after 2020, although discussion continues on the exact date.

R2-8.3 Federal Regulation of Refrigerants in the United States

The voluntary production limits and phaseout of harmful refrigerants of the Montreal Protocol was deemed to be insufficient by the Environmental Protection Agency (EPA). The EPA wrote additional requirements in the Federal Clean Air Act (Section 608). Among the provisions are the following points.

1. All persons who maintain, service, repair, or dispose of appliances must be certified by the EPA.
2. Requirements are set forth on the handling of refrigerants, including reclaiming, recovering, recycling, and the equipment that is used in these service techniques.
3. Intentional venting of CFC and HCFC refrigerants is prohibited.
4. Record-keeping is required to account for all refrigerant purchased, showing where each pound was used.
5. Repair of equipment that exceeds a certain size and leakage rate was made mandatory.

6. Violation of the certification requirement or refrigerant handling rules can result in fines of $25,000 per day, per violation. A bounty of $10,000 is offered to those who provide information concerning violations to EPA.
7. Onerous tax burdens were imposed (excise taxes and floor taxes), increasing the prices of refrigerants dramatically to discourage their use.

R2-8.4 Alternative Refrigerants

At this time, the industry response to replacing refrigerants is far from settled. The primary focus has been on finding replacements for the CFC refrigerants (R-11, 12, 502) because they are no longer being produced (other than that which is reclaimed). There have been many different responses.

R-134a. R-134a is an attractive alternative to R-12 because it is an HFC refrigerant that does not deplete the ozone. Its pressure-temperature characteristics are similar to R-12. R-12 has been the industry standard in refrigeration and automotive air-conditioning applications for many years. R-134a has been selected by automobile manufacturers as the refrigerant being used in newer automobile air conditioners. Some manufacturers of refrigeration equipment have also chosen R-134a for applications that previously would have used R-12. R-134a has also been used to simply replace the R-12 refrigerant in some applications, without actually changing the equipment. There is not universal agreement on what changes must be made to systems where R-134a is used to replace R-12. As a minimum, the refrigerant mineral oil in the R-12 system must be removed, and replaced with an ester-based (polyol ester) oil. Change-out of the metering device or other components might also be required. There is no "drop-in" replacement for any of the CFC refrigerants. Technicians working with R-134a are required to use a special set of hoses, gauges, vacuum pump, and recovery/recycling machine to be used with R-134a only. Older electronic leak detectors may also be unusable with R-134a.

R-123. R-123 is being used as a replacement for R-11 in low-pressure water chillers. R-123 is an HCFC and will also need to be replaced in time. However, some manufacturers of low-pressure chillers have chosen to design them with R-123 because of very low chiller operating costs. It is felt by those manufacturers that R-123 is a good solution for the next twenty to thirty years.

R-22. As with R-123, R-22 is an HCFC refrigerant that is being used in applications that formerly used CFCs. In small refrigeration applications that formerly relied on R-12, R-22 systems are now being used for applications where evaporator temperatures of −10°F or higher are acceptable.

Blends of HCFCs. HCFC blends are intended to be used in retrofit applications where the original refrigerant (R-12 and R-502) is no longer available or cost effective. These refrigerants use an alkylbenzene oil. A refrigerant blend is one that is a mixture of two (or more) refrigerants that have different pressure-temperature relationships. This results in some unique characteristics. For example, refrigerant blends have a temperature "glide." Consider the refrigerant blend consisting of two refrigerants, A and B, shown in Fig. R2-13. The compressor discharges both vapors at the same pressure, P1. In the condenser, heat is removed from both refrigerants. However, when the temperature of the mixture drops to T1, only refrigerant A begins to condense. As heat continues to be removed, A continues to condense until we are left with refrigerant A as a liquid, while refrigerant B is still a vapor. Then the temperature of the mixture begins to drop again, until refrigerant B reaches its saturation temperature, and it also condenses. The difference between the two temperatures, T1 and T2, over which the condensation takes place is referred to as temperature glide. The same effect happens in the evaporator.

At any temperature, the two (or more) refrigerants that comprise the blend will each exert their own different pressures on the tubing walls. If a leak occurs in the tubing, the higher pressure refrigerant will leak out at a faster rate than the other refrigerant component. When the tech-

ON THE JOB

Material Safety Data Sheets (MSDS) provide detailed information about various chemicals. MSDSs are divided into eight sections:

1. Product Identification
2. Hazardous Ingredients
3. Physical Data
4. Fire and Explosion Information
5. Health and Hazard Data
6. Reactivity Data
7. Spill or Leak Procedures
8. Safe Handling and Use

A master file of MSDSs for hazardous chemicals to which employees may be exposed is kept in the MSDS folder available for review as required.

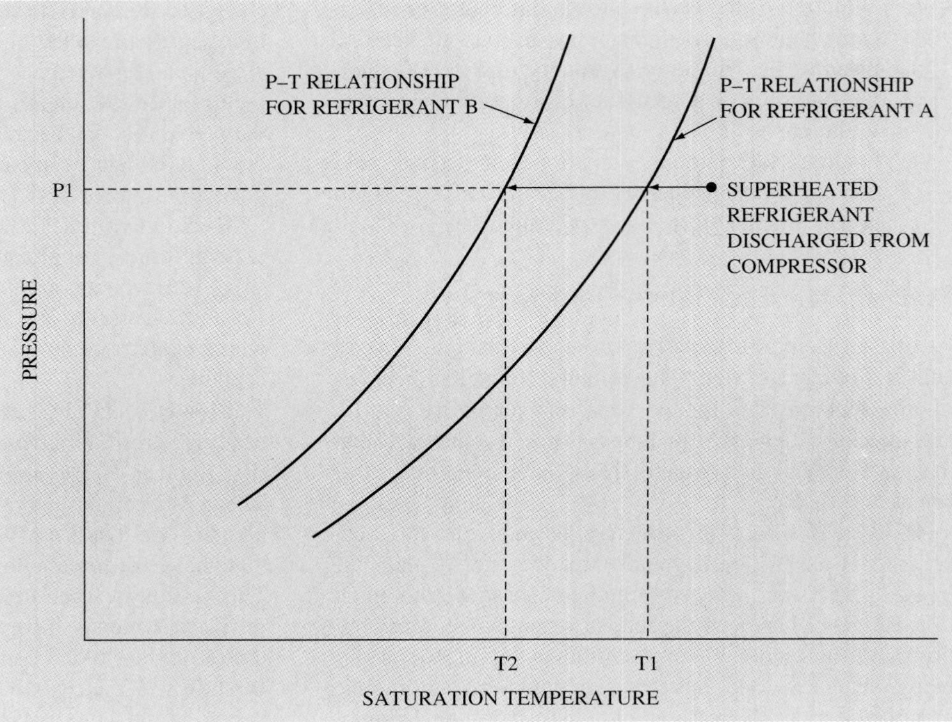

Figure R2-13 The refrigerant blend contains both refrigerant A and refrigerant B. In the condenser, condensation occurs at two different temperatures.

nician arrives and diagnoses a low charge condition, there will be no way to know how much of each refrigerant component leaked out. This change in composition is called fractionation. The only choice will be to remove all the refrigerant and charge new liquid refrigerant into the system. Some of the new blends (along with their earlier designations) now being marketed are: R-401A (MP39), an HCFC replacement for medium temperature replacement of R-12; R-404A (FX70/HP62), an HFC blend for low and medium temperature replacement of R-502; R-408A (FX10), an HCFC blend for low and medium temperature replacement of R-502; and R-409A (FX56), an HCFC blend for replacement for low and medium temperature replacement of R-12.

Other Replacements. There has been a proliferation of other refrigerants proposed to replace R-12 and R-502. None have achieved the popularity yet of the replacements described above. This alphabet soup of refrigerants will present future problems for technicians that are not yet being adequately addressed. Each refrigerant will be identified by its own color and by identifying tags on the equipment. But years from now when the tags are gone, it may be impossible for a technician to know what refrigerant to use in a system that has leaked out all of the refrigerant.

R2-8.5 Refrigerant Oils

When R-12, R-22, and R-502 were the refrigerants being used in almost all applications, the refrigerant oils that were in use were all mineral-based. However, most of the new refrigerants are not compatible with mineral oil. Two general classes of oils are now in use, alkylbenzene oil and polyol ester oil. Alkylbenzene oil and polyol ester oil are synthetic lubricants. They are backward compatible with mineral oil, which means that a compressor containing alkylbenzene oil or polyol ester oil can be installed in an R-12, R-22, or R-502 system that contains mineral oil.

R2-8.6 Recovery, Recycling, Reclaiming

The Clean Air Act makes it illegal to intentionally discharge refrigerants. So what is the technician to do to relieve the refrigerant pressure in a system that must be brazed to repair a leak? The terms recover, recycle, and reclaim sound similar. However, they have specific definitions, and the differences between them must be understood. Recovery is the process where the technician removes refrigerant from a system and stores it in a container. Recycling involves passing the refrigerant taken from a system through filter-driers to remove moisture and contaminants before reusing the refrigerant in the system where it originated. Reclaiming is a process done at a factory and not by a field technician. It is a reprocessing of the refrigerant so that it meets the same specifications as brand-new refrigerant. Generally, field technicians will take recovered refrigerant to a wholesaler who collects it and sends it on to a reclaiming facility. There will probably be a charge imposed by the wholesaler for handling the recovered refrigerant.

Equipment for recovering and recycling refrigerant has become a new industry in recent years. It must be certified by an EPA approved equipment testing organization to meet EPA standards. Two types of recovery units are shown in Figures R2-14 and R2-15.

Most recovery units consist of a compressor and condenser, and an assorted range of accessories. The compressor draws refrigerant out of a system, the condenser changes it to a liquid, and the outlet from the condenser is routed to a recovery tank (see Fig. R2-16). The tank may need to be placed in ice to reduce its pressure and speed the recovery process.

The accessories that may be supplied with a recovery unit include gauges, various valve arrangements, a heat exchanger to allow the recovery system to draw liquid refrigerant without causing damage to the recovery compressor, safety switches that shut down the recovery compressor on high pressure (if the recovery tank is over-filled), filters, driers, oil level indicators and drains. Figures R2-17 and R2-18 show different methods for recovering refrigerant from the disabled unit. Note that the special recovery tank has both a liquid connection and a vapor connection.

The levels of vacuum that must be produced by recovery equipment is legislated by EPA (and tested on the certification exam). The required levels are shown in Fig. R2-19. There are exceptions for small appliances and motor vehicle air conditioners.

R2-8.7 Technician Certification

The certification required by section 608 of the Clean Air Act is required before any technician is allowed to purchase or handle refrigerant. It has four separate categories (Fig. R2-20). The test is administered by a variety of wholesalers, contractors, schools, and others. Regardless of who administers the test, the questions come from the same test bank of questions maintained by the EPA. In re-

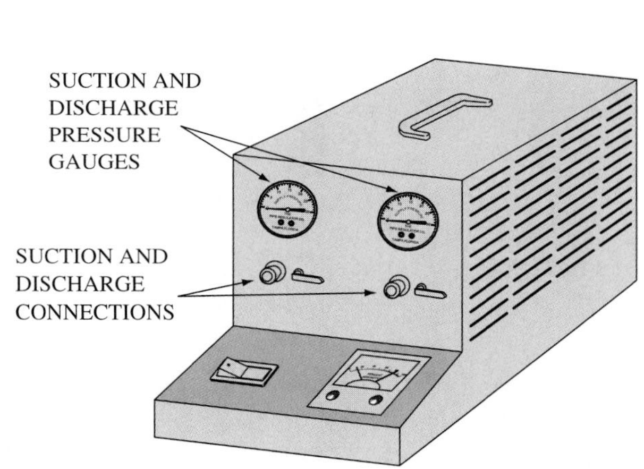

Figure R2-14 The portable refrigerant recovery/recycling unit weighs 25 to 50 lb.

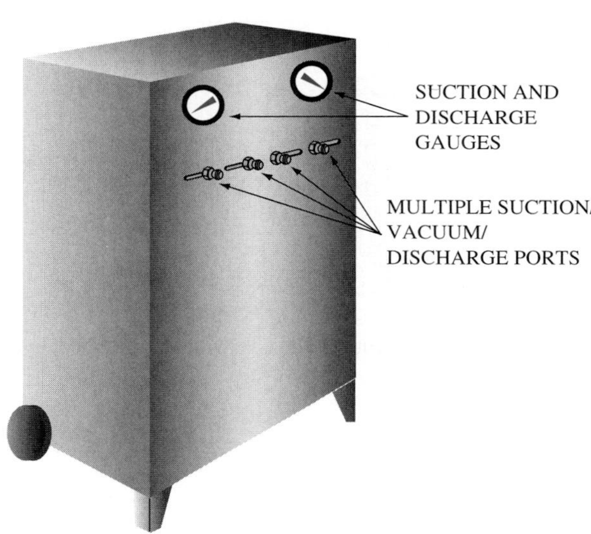

Figure R2-15 High capacity recycling unit for use in the contractor shop or equipment room.

Figure R2-16 Vapor recovery.

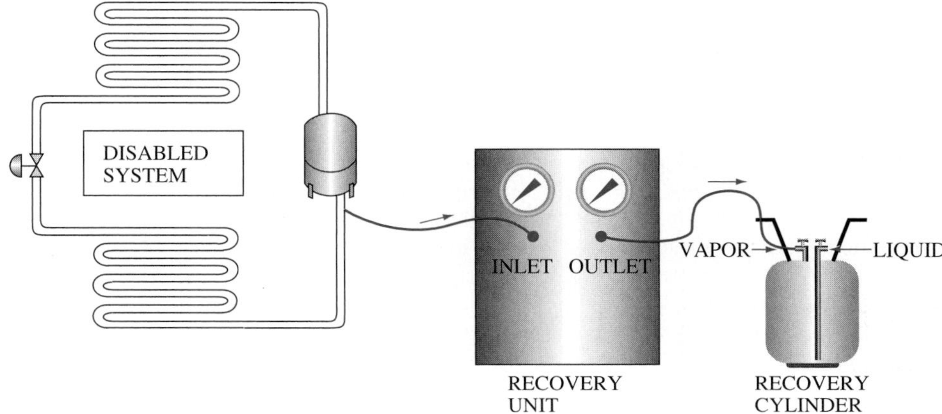

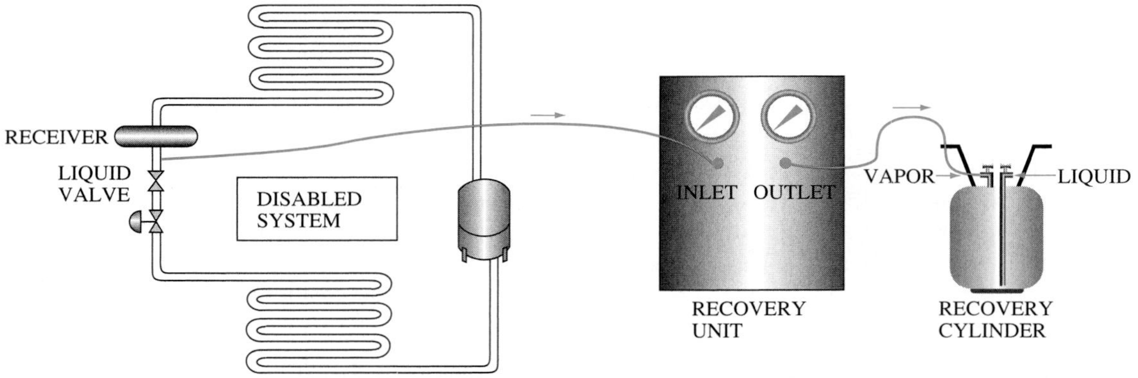

RECEIVER
LIQUID VALVE
DISABLED SYSTEM
INLET OUTLET
VAPOR
LIQUID
RECOVERY UNIT
RECOVERY CYLINDER

Figure R2-17 Liquid recovery. The recovery unit has a heat exchanger that allows recovery of liquid from the disabled unit, but boils it into vapor before it reaches the compressor in the recovery unit.

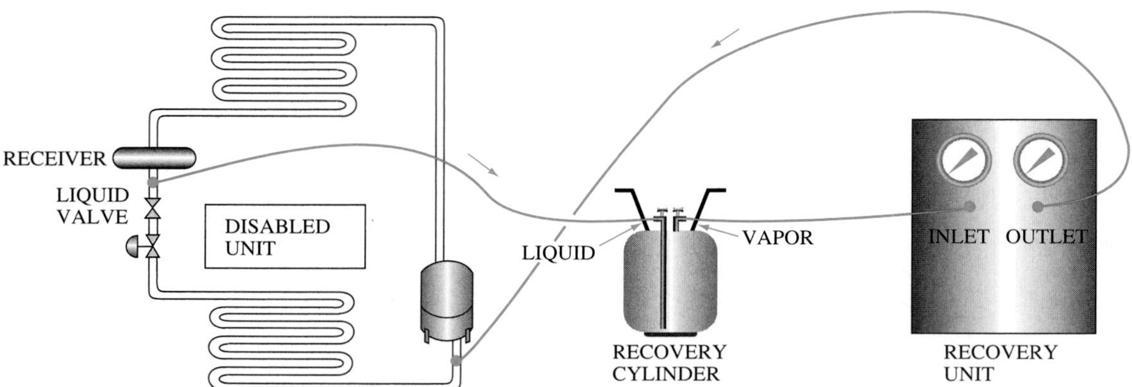

Figure R2-18 Liquid recovery with a recovery unit that does not have liquid-handling capability.

Type of System	Using Recovery or Recycling Equipment Manufactured or Imported Before Nov. 15, 1993	Using Recovery or Recycling Equipment Manufactured or Imported On or After Nov. 15, 1993
HCFC-22 appliances, or isolated component of such appliance, normally containing less than 200 lb of refrigerant.	0	0
HCFC-22 appliances, or isolated component of such appliance, normally containing 200 lb or more of refrigerant.	4	10
Other high pressure appliances, or isolated component of such appliance, normally containing less than 200 lb of refrigerant.	4	10
Other high pressure appliances, or isolated component of such appliance, normally containing 200 lb or more of refrigerant.	4	15
Very high pressure appliance	0	0
Low-pressure appliance	25	29 mm Hg absolute

Figure R2-19 Required evacuation levels.

Type I	Persons who service or repair small appliances (sealed system containing less than 5 lb of refrigerant)
Type II	Persons who service or repair or dispose of high-pressure or very-high-pressure appliances (except small appliances and vehicle air conditioning)
Type III	Persons who service or repair or dispose of low-pressure appliances
Universal	All of the above

Figure R2-20 Categories of technician certification.

cent years, the test questions have become public domain. The test administrators usually offer a short pretest training class that is specifically targeted towards the topics on the certification exam.

The exam consists of four sections (A, I, II, III) of twenty-five multiple-choice questions each. Section A is referred to as the "core" section, and must be passed to receive any type of certification. Section A contains questions about ozone depletion, the Clean Air Act, dates of only historical significance (when certain rules became effective), recovery/recycling/reclaiming generally, dehydration, and refrigerant cylinder handling and safety. Sections I, II, and III contain questions pertaining to those specific subject areas. A technician who passes any of these sections will receive a certification in that particular area (assuming that section A has also been passed). A technician who passes all four sections will receive a "Universal" certification.

REVIEW

- Recommended clothing and equipment:
  Safety glasses
  Gloves for refrigerant handling
  Hardhats for construction sites
  Work shoes with rubber soles and safety toes
  Long sleeve work shirt when brazing or soldering
  Ear (hearing) protection in machinery rooms
  Face shields where flying particles may be present
  Safety lines and belts to prevent falling
  Respiratory protection in confined spaces or areas of large refrigerant leaks
- Fire extinguisher for classes of fire:
  Class A—ordinary combustion material (paper, wood, etc.)
  Class B—flammable liquids
  Class C—electrical
- Rules for shop and personal safety:
  Keep shop and lab areas clean.
  Store oily shop towels in approved metal containers.
  Clean metal chips from machines with a brush.
  Wear protective clothing and use proper equipment.
  Keep work areas organized and aisles, traffic areas free of material and obstacles.
  Dispose of hazardous waste properly.
  Keep fire extinguishers present.
  Use only CSA- or ANSI-approved ladders.
  Follow recommended safety procedures.
- Material Safety Data Sheets:
  Provide safety information on all substances.
  Labs and shops should display and update this material for students and workers.
- Students should review electrical safety rules:
  Check for voltages present before servicing.
  Use the lockout-tagout procedure to disable electrical circuits when servicing.
  Stand on a dry surface.
  Use a high-voltage insulating pad when required.
  Work on live circuits only when necessary.
  Never bypass protective devices.
  Properly fuse or circuit breaker protect.
  Use one-handed "Hop-Skip" techniques when probing for voltages.
- Effects of current on the human body range from no sensation at 0.5 milliamps (ma) to heat convulsions and death at 50 to 500 milliamps. Refer to Section R2-5 for details.
- Rescue procedures:
  Turn off the power
  Use a nonconductive object to push the victim away
  Call for emergency assistance
  Perform CPR and/or rescue breathing until help arrives
- Refrigerant safety:
  Provide good ventilation.
  Wear safety goggles and gloves.
  Wrap cloth around hose fittings before removing them from pressurized system or cylinder.
  Ventilate room containing refrigerants before entering and/or wear self contained breathing apparatus.
  Install refrigerant alarms and monitors.
  Wear gloves when working with compressor burnouts.
  Do not fill a cylinder to more than 80% of its volume.
  Transport refrigerant in the upright secured position.
  Do not pressurize refrigeration system with air or oxygen; use only dry nitrogen or mixture of nitrogen and R-22.
  Do not mix refrigerants.
  Never apply a torch to a pressurized vessel.
  Do not refill a disposable cylinder.
  Replace cylinder caps.
  When isolating sections of refrigeration system, exercise caution to prevent liquid expansion.
  Use only approved refrigeration service techniques.
  Never service where an open flame is present.
  Prevent moisture from entering the sealed system.

- Environmental safety principles:
  Become properly trained and certified.
  Isolate and/or use recovery techniques before opening the sealed system.
  Never purge refrigerant charge to the atmosphere.
  Use approved equipment.
  Use approved service techniques.

Problems and Questions

1. What safety clothing and equipment is recommended for the air conditioning and refrigeration technician?
2. How do you clear your gauge hoses of air?
3. When is it acceptable to release a minimum amount of refrigerant during service or normal operation?
4. What type of fire extinguisher is required for fuel oil fires?
5. What type of fire extinguisher should be used for electrical fires?
6. What is the purpose of the "Lockout-tagout" system? Describe the procedure.
7. Is it OK to use two hands when probing an electrical circuit for voltages with a multimeter? Why or why not?
8. When should you use electrical rubber gloves?
9. How many amps of electrical current could kill a person?
10. Describe a safe procedure for using a multimeter to check for voltages in an energized electrical circuit.
11. If one of your fellow workers is being shocked by electricity, how can you safely rescue him? Discuss the rescue procedure.
12. What safety rules are required when using refrigerants?
13. When charging a refrigeration/air conditioning system, what safety equipment should you use? Why?
14. When working on refrigeration systems, when is a self-contained breathing apparatus required?
15. Why are the world governments concerned over the problem with ozone depletion in the stratosphere?
16. Which refrigerants have the greatest ODP?
17. Why are these service techniques no longer permitted?
    a. Purging
    b. Flushing with refrigerant
    c. Blowing the charge
18. What environmental concern is associated with the "Greenhouse Effect", known as the Global Warming Potential (GWP)?
19. List reasons why a service technician needs to recover refrigerant. What technique may be used as an alternate method?
20. What is meant by the term "reclamation" of refrigerants?
21. When is recycling of refrigerants allowed?
22. How is natural ozone formed in the stratosphere? How is it destroyed?
23. What remnants of refrigerants are found in the stratosphere that confirm that CFCs are destroying ozone?
24. What produces ozone in the lower atmosphere? Is this good? Why or why not?

Fundamentals R3

R3-1 DEFINITIONS

Refrigeration is the process of removing heat energy from a place it is not wanted and disposing of it in a place where it is wanted or not objectionable. **Mechanical refrigeration** makes use of mechanical components to produce work energy and transfer heat from an area of lower temperature to an area of higher temperature, such as from the interior of the refrigerator cabinet to the surrounding area of the kitchen in the case of the household refrigerator. **Absorption refrigeration** is the use of heat energy to produce the conditions necessary to transfer heat energy from one place to another. The heat energy is converted to work energy and the desired results are obtained, in the same manner as in a mechanical refrigeration system.

A **refrigerant** is a fluid that picks up heat by evaporating at a low temperature and pressure and gives up heat by condensing at a higher temperature and pressure. Two common types of refrigerants are the Freon or halogenated hydrocarbon refrigerants—R-12 (Refrigerant 12) and R-22 (Refrigerant 22). As the heat is removed from the space, the area appears to become cooler. Therefore, **cold** can be defined as the relative absence of heat energy. Theoretically, there is always some heat energy present down to absolute zero (–460°F), where no heat energy is said to be present.

Temperature is the measurement of the speed or intensity of the molecules of a substance. Temperature is measured by a thermometer using either the **Fahrenheit** or **Celsius** scales. Each scale has a boiling and a freezing point established. **One ton of refrigeration** is the refrigeration produced by melting one ton of ice at a temperature of 32°F in 24 hours. It is the refrigerating effect expressed as: 288,000 Btu/24 hours, 12,000 Btu/hour, or 200 Btu/minute. One **Btu,** or *British thermal unit,* is the quantity of heat required to raise the temperature of one pound of water one degree Fahrenheit.

An **air conditioning system** is a refrigeration system used to cool, dehumidify, filter, and/or heat the air of a space. The refrigeration system removes the heat from the air in the space, reducing the temperature of the space. Water vapor contained in the air is collected on the cool surface of the evaporator and drained out, thus dehumid-

AFTER STUDYING THIS CHAPTER, THE STUDENT WILL BE ABLE TO:

- Define terms and abbreviations used in HVAC/R.
- Explain the refrigeration process in terms of heat transfer.
- Define refrigeration cycle terminology.
- Identify system components.
- Explain the importance of latent heat as compared to sensible heat.
- State the purpose of each of the basic cycle components.
- Identify the state, pressure, temperature, and condition of the fluid at key points in the cycle.
- Identify absorption cycle components.
- State the purpose of each of the absorption cycle components.
- Identify the areas of heat transfer, the condition, temperature, state, and pressure of the fluid at key points in the refrigeration and absorption cycle.

ifying the air in the space also. Mechanical or electronic filters can be used to remove dust or other contaminants from the conditioned space.

Everywhere in this universe exist objects that consist of matter or substance. **Matter** is anything that has weight and takes up space. All matter is made up of molecules. **Heat** is not matter or substance, but rather a form of energy which may be produced by other forms of energy such as: light, electricity, chemical means, mechanical means or atomic energy. Heat does not have weight nor does it occupy a space. A quart of water weighs no more or no less whether is it hot or cold.

Since heat is energy, it is subject to the **law of conservation of energy.** This means that the heat energy cannot be created or destroyed. It is the by-product of other forms of energy. The adding to or removing of heat energy from a substance determines the speed or intensity of the molecules. The intensity of speed of molecules determines the physical state of matter—solid, liquid, or gas.

Cold is a relative term to describe the energy level or temperature of an object or area as compared to a known energy level or temperature. Although some definitions describe cold as the "absence of heat," this implies that there is no heat present, in fact, there is nothing known in the world today from which heat is totally absent. No process yet devised has been capable of achieving "absolute zero," the removal of all heat energy (molecular movement) from a space or an object. Theoretically, this zero point would be 459.69° below zero (–459.69°) on the Fahrenheit thermometer scale or 273.16° below zero (–273.16°) on the Celsius thermometer scale.

Change of state is the change from a solid to a liquid and liquid to a gas, or vice versa. A large amount of heat must be added or removed to cause a change of state. Heat energy is measured in Btu.

The amount of Btu of heat energy necessary to change one pound of any material one degree Fahrenheit is called its **specific heat.** The specific heat of water is, therefore, one.

Heat flow is the movement of heat from a warmer to a cooler body. Heat moves by: conduction, convection, or radiation. **Conduction** is the transfer of heat from particle to particle of a substance without movement of the particles themselves. **Convection** is the transfer of heat by a flowing medium. It takes place only in liquids and gases, since solids do not flow. **Radiation** is a wave form of heat movement similar to light, except it cannot be seen. Like light, it requires no medium to travel.

In referring to heat, we need to talk about heat and temperature in terms of specific heat, latent heat, super heat, and sensible heat.

Latent heat is heat energy absorbed in the process of changing the forms of a substance (melting, vaporization, or fusion) without a change in temperature or pressure. There are three types of latent heat: latent heat of vaporization, latent heat of fusion, and latent heat of condensation.

Latent heat of vaporization is the amount of heat energy required to change a substance from a liquid to a vapor without changing its temperature or pressure. *Latent heat of fusion* is the amount of heat energy required to change a substance from a liquid to a solid without a change in temperature or pressure at standard atmospheric pressure.

Latent heat of condensation is related to latent heat of vaporization. It is the heat energy that must be removed from a substance to change state from a vapor to a liquid without changing its temperature or pressure. **Superheat** is the temperature of a vapor above the boiling temperature of its liquid at that pressure. Heat that is added to a gas after the liquid has vaporized is measured in degrees.

Sensible heat is heat added to or removed from a substance that causes a change in the temperature of the substance. Saturation temperature is another name for boiling point. The *boiling point* of a liquid is affected by pressure—*higher* pressure raises the boiling point; *lower* pressure lowers the boiling point.

Pressure is an impact on a unit area; force or thrust exerted on a surface, measured in pounds per square inch (psi). Pressure is referred to in one of three ways: atmospheric, gauge, or absolute. **Atmospheric pressure** is

PREVENTIVE MAINTENANCE

While surveying equipment rooms during routine maintenance, learn to look for potential problems such as boxes in front of cooler, over-stacked display cases, oil puddles around piping, and dirt collecting in one area. Inspect sight glasses, oil levels, time clocks, pressure switches, and primary operating controls. It is the obvious and routine maintenance procedures which will challenge the novice technician to remedy the problem. Learn to look for the simple solutions before tackling the complex problems. Patience will help to develop good service practices, and time will create skilled maintenance troubleshooting competence.

stated as 14.7 psi at sea level. **Absolute pressure** is pressure measured on the absolute scale. The zero point is at zero atmospheric pressure. The total of the gauge and atmospheric pressures is called *absolute pressure.* **Gauge pressure** is measured on the gauge scale. The zero point is 14.7 psi, or atmospheric pressure at sea level. Gauge pressure is the measured pressure above or below atmospheric. Pressure below atmospheric is measured in inches of mercury.

A liquid at its boiling point is referred to as *saturated vapor.* A liquid above its boiling point is referred to as *superheated vapor.* A liquid below its boiling point is referred to as *subcooled liquid.*

R3-2 REFRIGERATION PROCESS

The transfer of heat in the refrigeration system is performed by a refrigerant operating in a closed system. The refrigeration process has its application in both refrigerated systems and air-conditioning systems. Refrigerated systems are chiefly concerned with cooling products, whereas air-conditioning systems cool (or heat) people. Air-conditioning systems use refrigeration to provide comfort cooling and dehumidification of air.

One of the very useful properties of the refrigerant is the pressure-temperature relationship of the saturated vapor. A refrigerant vapor is said to be saturated whenever both liquid and vapor are present in the same container, in stable equilibrium.

Under these conditions a fixed relationship exists between the temperature of the refrigerant in the container and its pressure. A typical temperature-pressure table is shown in Fig. R3-1.

R3-2.1 Components of the Refrigeration Cycle

In the simple **vapor-compression refrigeration system** shown in Fig. R3-2, there are four essential parts:

1. The compressor
2. The condenser
3. The metering device
4. The evaporator

The **compressor** is a mechanical device for pumping refrigerant vapor from a low pressure area (the evaporator) to a high pressure area (the condenser). Since pressure, temperature, and volume of a gas are related, a change in pressure from low to high causes an increase in temperature and a decrease in volume or a compression of

KNOWLEDGE EXTENSION

To gather more knowledge about the basic components of the refrigeration cycle, begin to look for the major refrigeration components in every aspect of your life. Go into your kitchen and observe the refrigerator. Notice the type of evaporator. Is it finned, forced air, or automatic defrost? Feel for the heat rejection from the condenser. Is it coming from underneath or around the back? This will indicate if it is using a forced air cooled condenser or a static cooled.

Next time you go to your local grocery store, observe the coolers in the medium temperature and frozen foods section. Look for case thermometers, drain lines, ice buildup, and pipe chases. Listen for any noises associated with the equipment. Try to locate where the mechanical room and all of the compressors are located. Are there any units which are self-contained? (Unit with its own compressor, condenser, and evaporator.)

Observing and listening wherever you are will give indications of the refrigeration and air conditioning systems. This includes convenience stores, offices, and department stores. Look for ductwork, piping, thermostats, coolers, or any other equipment associated with the mechanical refrigeration and air conditioning systems in the building. The methods modern building owners use to conceal their mechanical systems and equipment will sometimes amaze the astute observer.

ON THE JOB

When you go to a job site for the first time, learn to locate the four major components of the refrigeration cycle—the compressor, evaporator, condenser, and metering device. Think about the basics you learned in school regarding heat transfer principles. Look, listen, and feel for obvious problems such as dirty coils, noisy bearings, and proper air flows. Remember that heat always moves from hot to cold and that hot air will naturally rise and cold air will fall.

Temperature °F.	Refrigerant—Code				
	12-F	22-V	500-D	502-R	717-A
−60	*19.0*	*12.0*	*17.0*	*7.2*	*18.6*
−55	*17.3*	*9.2*	*15.0*	*3.8*	*16.6*
−50	*15.4*	*6.2*	*12.8*	*0.2*	*14.3*
−45	*13.3*	*2.7*	*10.4*	**1.9**	*11.7*
−40	*11.0*	**0.5**	*7.6*	**4.1**	*8.7*
−35	*8.4*	**2.6**	*4.6*	**6.5**	*5.4*
−30	*5.5*	**4.9**	*1.2*	**9.2**	*1.6*
−25	*2.3*	**7.4**	**1.2**	**12.1**	**1.3**
−20	**0.6**	**10.1**	**3.2**	**15.3**	**3.6**
−18	**1.3**	**11.3**	**4.1**	**16.7**	**4.6**
−16	**2.0**	**12.5**	**5.0**	**18.1**	**5.6**
−14	**2.8**	**13.8**	**5.9**	**19.5**	**6.7**
−12	**3.6**	**15.1**	**6.8**	**21.0**	**7.9**
−10	**4.5**	**16.5**	**7.8**	**22.6**	**9.0**
− 8	**5.4**	**17.9**	**8.8**	**24.2**	**10.3**
− 6	**6.3**	**19.3**	**9.9**	**25.8**	**11.6**
− 4	**7.2**	**20.8**	**11.0**	**27.5**	**12.9**
− 2	**8.2**	**22.4**	**12.1**	**29.3**	**14.3**
0	**9.2**	**24.0**	**13.3**	**31.1**	**15.7**
1	**9.7**	**24.8**	**13.9**	**32.0**	**16.5**
2	**10.2**	**25.6**	**14.5**	**32.9**	**17.2**
3	**10.7**	**26.4**	**15.1**	**33.9**	**18.0**
4	**11.2**	**27.3**	**15.7**	**34.9**	**18.8**
5	**11.8**	**28.2**	**16.4**	**35.8**	**19.6**
6	**12.3**	**29.1**	**17.0**	**36.8**	**20.4**
7	**12.9**	**30.0**	**17.7**	**37.9**	**21.2**
8	**13.5**	**30.9**	**18.4**	**38.9**	**22.1**
9	**14.0**	**31.8**	**19.0**	**39.9**	**22.9**
10	**14.6**	**32.8**	**19.7**	**41.0**	**23.8**
11	**15.2**	**33.7**	**20.4**	**42.1**	**24.7**

Temperature °F.	Refrigerant—Code				
	12-F	22-V	500-D	502-R	717-A
12	**15.8**	**34.7**	**21.2**	**43.2**	**25.6**
13	**16.4**	**35.7**	**21.9**	**44.3**	**26.5**
14	**17.1**	**36.7**	**22.6**	**45.4**	**27.5**
15	**17.7**	**37.7**	**23.4**	**46.5**	**28.4**
16	**18.4**	**38.7**	**24.1**	**47.7**	**29.4**
17	**19.0**	**39.8**	**24.9**	**48.8**	**30.4**
18	**19.7**	**40.8**	**25.7**	**50.0**	**31.4**
19	**20.4**	**41.9**	**26.5**	**51.2**	**32.5**
20	**21.0**	**43.0**	**27.3**	**52.4**	**33.5**
21	**21.7**	**44.1**	**28.1**	**53.7**	**34.6**
22	**22.4**	**45.3**	**28.9**	**54.9**	**35.7**
23	**23.2**	**46.4**	**29.8**	**56.2**	**36.8**
24	**23.9**	**47.6**	**30.6**	**57.5**	**37.9**
25	**24.6**	**48.8**	**31.5**	**58.8**	**39.0**
26	**25.4**	**49.9**	**32.4**	**60.1**	**40.2**
27	**26.1**	**51.2**	**33.2**	**61.5**	**41.4**
28	**26.9**	**52.4**	**34.2**	**62.8**	**42.6**
29	**27.7**	**53.6**	**35.1**	**64.2**	**43.8**
30	**28.4**	**54.9**	**36.0**	**65.6**	**45.0**
31	**29.2**	**56.2**	**36.9**	**67.0**	**46.3**
32	**30.1**	**57.5**	**37.9**	**68.4**	**47.6**
33	**30.9**	**58.8**	**38.9**	**69.9**	**48.9**
34	**31.7**	**60.1**	**39.9**	**71.3**	**50.2**
35	**32.6**	**61.5**	**40.9**	**72.8**	**51.6**
36	**33.4**	**62.8**	**41.9**	**74.3**	**52.9**
37	**34.3**	**64.2**	**42.9**	**75.8**	**54.3**
38	**35.2**	**65.6**	**43.9**	**77.4**	**55.7**
39	**36.1**	**67.1**	**45.0**	**79.0**	**57.2**
40	**37.0**	**68.5**	**46.1**	**80.5**	**58.6**
41	**37.9**	**70.0**	**47.1**	**82.1**	**60.1**

Temperature °F.	Refrigerant—Code				
	12-F	22-V	500-D	502-R	717-A
42	**38.8**	**71.4**	**48.2**	**83.8**	**61.6**
43	**39.8**	**73.0**	**49.4**	**85.4**	**63.1**
44	**40.7**	**74.5**	**50.5**	**87.0**	**64.7**
45	**41.7**	**76.0**	**51.6**	**88.7**	**66.3**
46	**42.6**	**77.6**	**52.8**	**90.4**	**67.9**
47	**43.6**	**79.2**	**54.0**	**92.1**	**69.5**
48	**44.6**	**80.8**	**55.1**	**93.9**	**71.1**
49	**45.7**	**82.4**	**56.3**	**95.6**	**72.8**
50	**46.7**	**84.0**	**57.6**	**97.4**	**74.5**
55	**52.0**	**92.6**	**63.9**	**106.6**	**83.4**
60	**57.7**	**101.6**	**70.6**	**116.4**	**92.9**
65	**63.8**	**111.2**	**77.8**	**126.7**	**103.1**
70	**70.2**	**121.4**	**85.4**	**137.6**	**114.1**
75	**77.0**	**132.2**	**93.5**	**149.1**	**125.8**
80	**84.2**	**143.6**	**102.0**	**161.2**	**138.3**
85	**91.8**	**155.7**	**111.0**	**174.0**	**151.7**
90	**99.8**	**168.4**	**120.6**	**187.4**	**165.9**
95	**108.2**	**181.8**	**130.6**	**201.4**	**181.1**
100	**117.2**	**195.9**	**141.2**	**216.2**	**197.2**
105	**126.6**	**210.8**	**152.4**	**231.7**	**214.2**
110	**136.4**	**226.4**	**164.1**	**247.9**	**232.3**
115	**146.8**	**242.7**	**176.5**	**264.9**	**251.5**
120	**157.6**	**259.9**	**189.4**	**282.7**	**271.7**
125	**169.1**	**277.9**	**203.0**	**301.4**	**293.1**
130	**181.0**	**296.8**	**217.2**	**320.8**	—
135	**193.5**	**316.6**	**232.1**	**341.2**	—
140	**206.6**	**337.2**	**247.7**	**362.6**	—
145	**220.3**	**358.9**	**264.0**	**385.0**	—
150	**234.6**	**381.5**	**281.1**	**408.4**	—
155	**249.5**	**405.1**	**298.9**	**432.9**	—

Figure R3-1 Pressure-temperature table. *Italic Figures* represent vacuum-inches of mercury. **Bold Figures** represent pressure-pounds per square inch. (Courtesy of Sporlan Valve Co.)

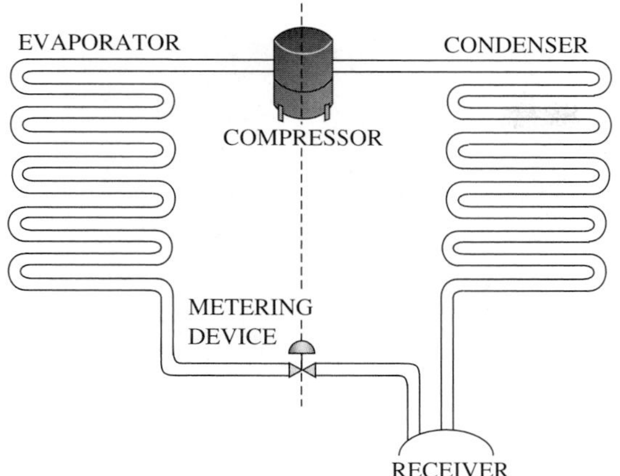

Figure R3-2 Simple vapor-compression refrigeration cycle.

the vapor. The main types of compressors are: reciprocating (piston), rotary, centrifugal, screw, and scroll. See Fig. R3-3.

Compressors get their names from their mechanical parts. In the reciprocating compressor, a piston travels back and forth in a cylinder. The rotary has a vane that rotates within a cylinder. The centrifugal compressor has a very high speed centrifugal impeller with multiple blades. The impeller rotates within a housing. The screw compressor uses a rotating screw within a tapered housing. The scroll compressor has a stationary and an orbiting scroll that moves within the stationary scroll.

The **condenser** is a device for removing heat from the refrigeration system. In the condenser, the high temperature, high pressure vapor transfers heat through the condenser tubes to the surrounding medium (usually air or water). When the temperature of the vapor reaches the saturation temperature, the additional latent heat removed causes condensation of the refrigerant, producing liquid refrigerant.

There are three types of condensers: air-cooled, water-cooled, and evaporative. (Fig. R3-4) The air-cooled condenser uses air as the condensing medium, the water-cooled condenser uses water as the condensing medium, and the evaporative condenser uses both air and water.

Air-cooled condensers consist of two types: forced-air condensers and natural-draft (static) condensers. Air-cooled condensers can be further classified by their construction: (1) fin and tube, and (2) plate. There are four types of water-cooled condensers: (1) double pipe, (2) open vertical shell and tube, (3) horizontal shell and tube and (4) shell and coil.

A **metering device** controls the flow of refrigerant to the *evaporator* (described later). It separates the high-pressure and the low-pressure parts of the system. High-pressure, medium-temperature liquid in the metering device enters the low-pressure, low-temperature evaporator. The pressure is low because the compressor is continuously pumping vapor from the evaporator. The metering device controls the flow of refrigerant into the evaporator.

Two actions occur in the metering device: (1) the refrigerant liquid is cooled to the evaporator temperature by actual evaporation of some of the liquid refrigerant, and (2) the pressure of the refrigerant is reduced to a pressure corresponding to the evaporator temperature at the saturated condition.

PRACTICING PROFESSIONAL SERVICES

Now that you know the basics of refrigeration, it is time to apply your knowledge to a real world problem.

It is July 4th at one o'clock in the afternoon and you are on call for your company. One of your company's customers has called, and their air conditioner is not cooling the house. It is 92°F outside. You climb into your van and leave for the call. Arriving at the customer's house, you notice that the outside unit appears to be running. A check of inside conditions reveals that the house is 85°F and humid. Air is blowing out of the registers, but it does not feel cool. A check of the filter shows a dirty filter, but you have seen worse. The blower could use cleaning also. The evaporator coil is clean, however, and the filter appears to have done its job. You change the filter, clean the blower, and head outside where you find a condenser clogged with grass clippings and fuzzy residue from a nearby cottonwood tree. You clean the condenser, recheck the unit, and now it appears to be working well. Your customer asks, "What was the problem?" What do you say?

What caused the problem: the dirty filter, the dirty blower, or the dirty condenser? Explain your answer in terms of heat transfer and the refrigeration cycle.

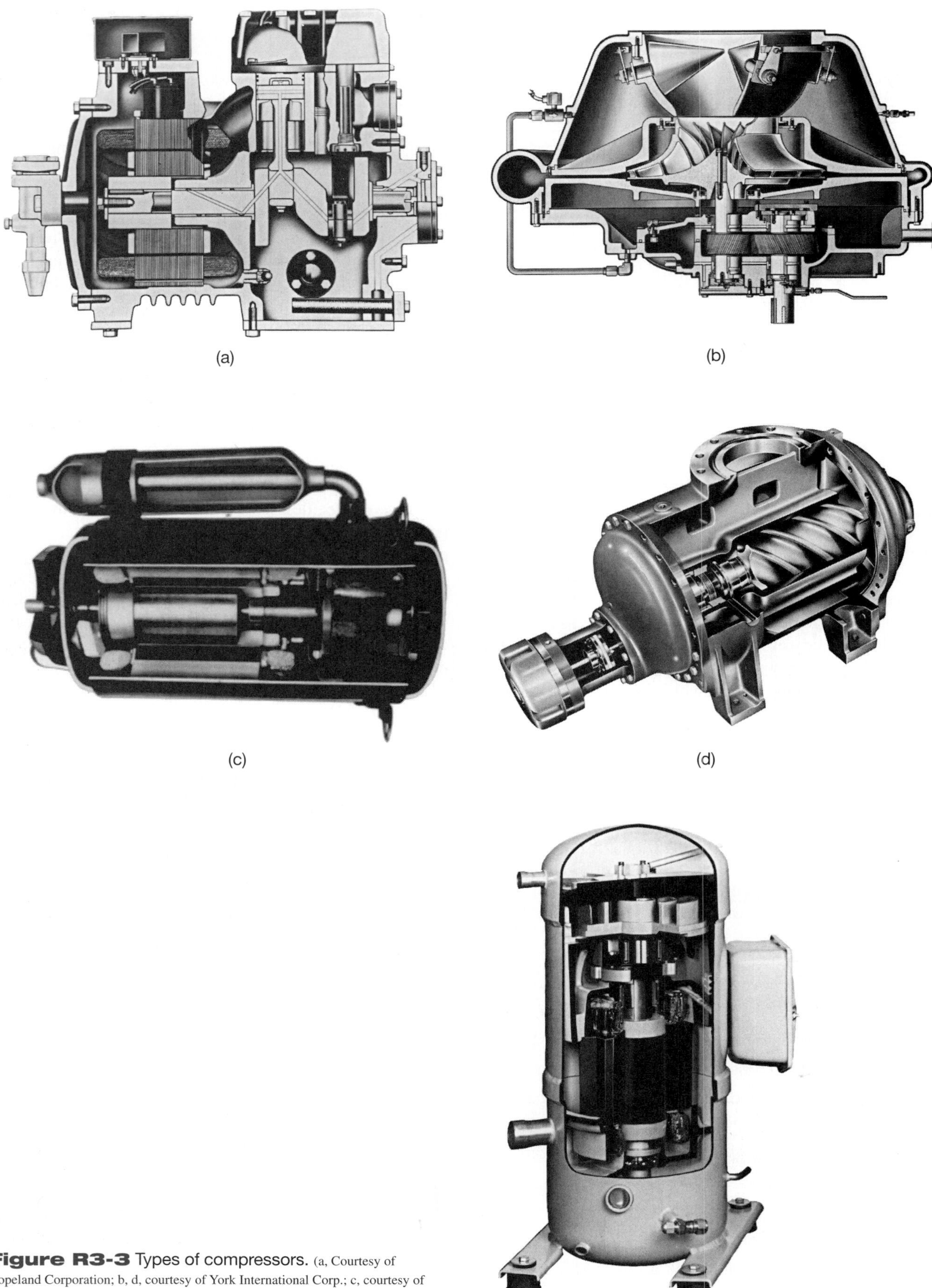

(a) (b) (c) (d) (e)

Figure R3-3 Types of compressors. (a, Courtesy of Copeland Corporation; b, d, courtesy of York International Corp.; c, courtesy of Rotorex Company, Inc.; e, The Trane Company)

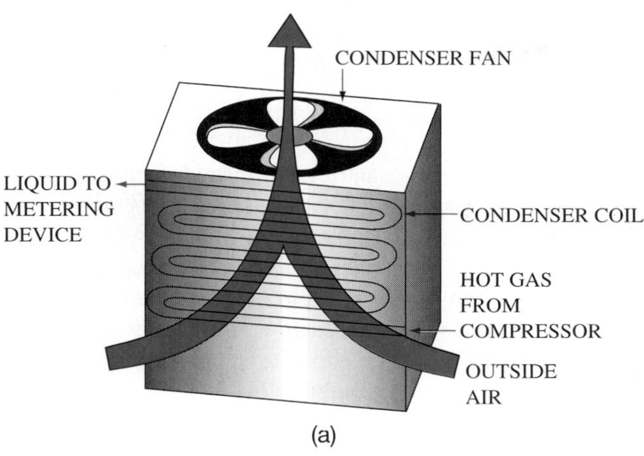

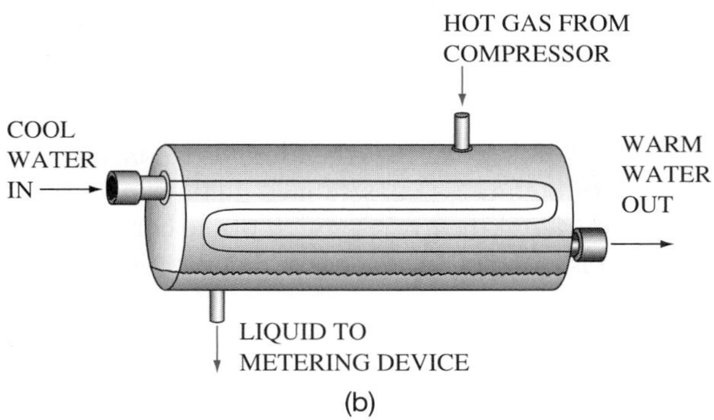

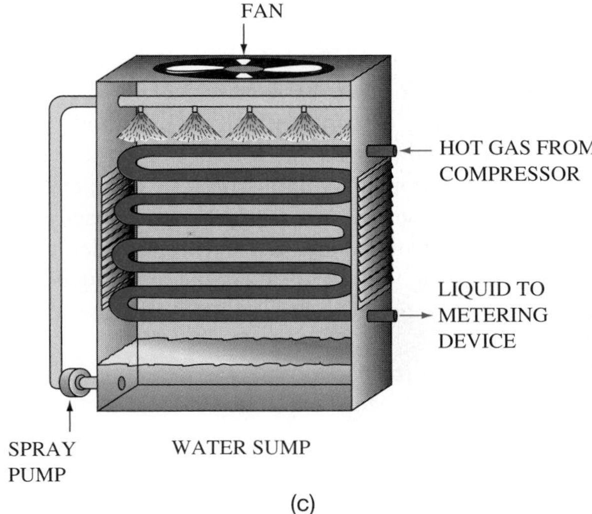

Figure R3-4 Types of condensers.

There are six major types of metering devices: (1) hand-operated expansion valve, (2) low-side float, (3) high-side float, (4) automatic expansion valve, (5) thermostatic expansion valve and (6) capillary tube. Metering devices are selected by application. There are other types of metering devices used for special applications.

The **evaporator** is a device for absorbing heat into the refrigeration system. In the evaporator, the saturated refrigerant absorbs heat from its surroundings and boils into a low pressure vapor. Liquid refrigerant boiling and vaporizing in a cold evaporator can be compared to water boiling on a stove. The refrigerant is heated by the product load, whereas the water is heated by the stove burner. The action and results are comparable. The difference is that the stove is hot and the refrigerator is cold; however, in both cases heat energy is being transferred.

Some superheating of the vapor takes place before the suction gas reaches the compressor. This is desirable since unevaporated liquid refrigerant could damage the compressor.

Although there are many variations and modifications of evaporators, there are three basic types: bare pipe, finned tube, and plate. (Fig. R3-5) Like condensers, evaporators can be forced air or natural draft (static). Evaporators are designed on the basis of their intended usage. Application determines which type is the best suited.

Action in a Typical Refrigeration Cycle

There are a number of different kinds of refrigeration cycles. The most common type is the vapor-compression cycle, shown in Fig. R3-6.

To review the action that occurs in a typical system, first note the dotted vertical line in the center of the diagram. It separates the high side (on the right) from the low side (on the left). The pressure difference between these two areas is maintained by the operation of the compressor and the restriction produced by the metering device.

The performance of the cycle in various stages of its operation can be observed in four ways:

1. The changes in state of the refrigerant
2. The changes in pressure of the refrigerant
3. The changes in temperature of the refrigerant
4. The change in heat content of the refrigerant.

The following text has been simplified by assuming no pressure loss through the evaporator, condenser, and interconnecting piping. In actual practice there will be some losses; however, this does not detract from the principles explained below.

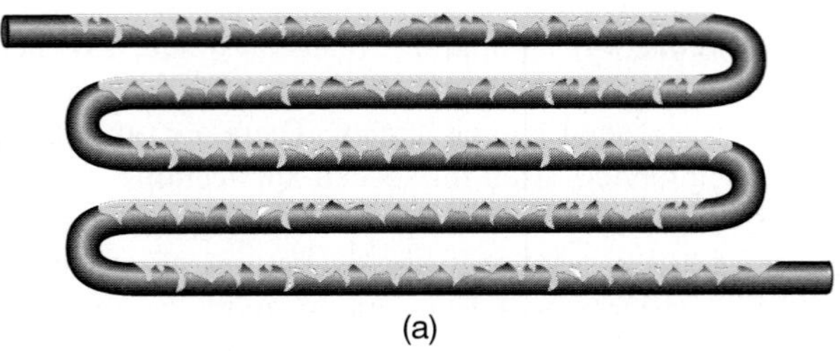

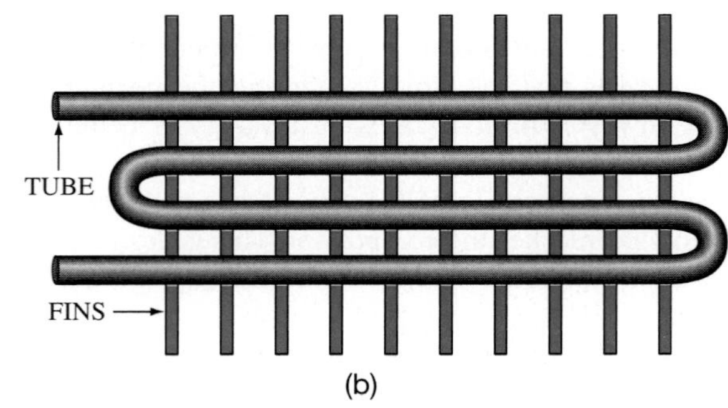

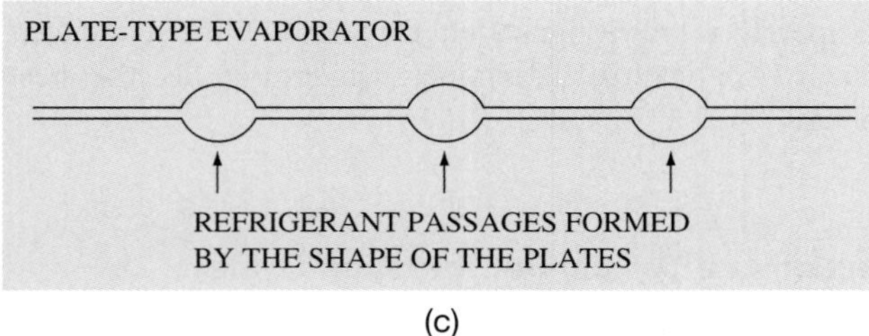

Figure R3-5 Types of evaporators.

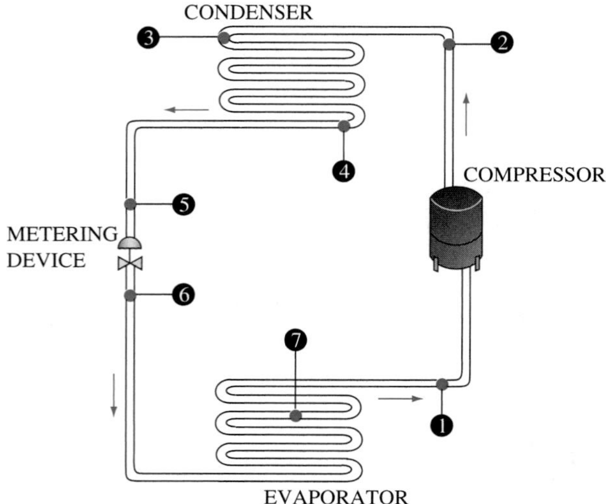

Figure R3-6 Vapor compression refrigeration cycle (color-coded, numbered positions).

R3-3 CHANGE OF STATE, PRESSURE, TEMPERATURE, AND HEAT CONTENT

Referring to Fig. R3-6 and the numbered positions in the diagram, the changes in state, pressure, temperature, and heat content of the refrigerant are as follows.

(Position 1.) The refrigerant has picked up some superheat in the final evaporator circuit. Superheating is the process of continuing to heat the refrigerant after sufficient latent heat has been added to vaporize all the liquid. Superheating assures no liquid slugs will reach the compressor and cause damage to the valves and pistons. The refrigerant enters the compressor as a low-temperature, low-pressure, superheated vapor.

(Position 2.) The refrigerant leaves the compressor as a high-pressure, high-temperature, superheated vapor. Heat of compression is also absorbed into the refrigerant.

(Position 3.) As the refrigerant enters the condenser, in the first portion of the condenser heat is removed and the temperature of the refrigerant falls to the saturation temperature. As additional latent heat is removed, the vapor condenses. At this point the condition of the refrigerant is a mixture of high-pressure, saturated liquid and vapor.

(Position 4.) At the lower portion of the condenser, the refrigerant is completely condensed and is a high-pressure liquid.

(Position 5.) This is the same refrigerant condition as Position 4. The refrigerant is all liquid; however, some subcooling has taken place in the final pass through the condenser. As additional heat is removed from the refrigerant, it becomes subcooled. Subcooling is the process of continuing to remove heat from the refrigerant after all the latent heat has been extracted and the vapor has been changed to a liquid. Subcooling reduces the temperature of the liquid below its boiling point at a particular pressure. Adequate subcooling will prevent the refrigerant from starting to boil as it experiences small pressure drops through the piping or components. Such boiling causes *flash gas* (the flashing of the refrigerant to a gas due to the sudden pressure drop and volume increase at the entrance to the evaporator) and can reduce the system capacity. It is desirable to subcool the liquid refrigerant either in the condenser or in the liquid line before the metering device. Subcooling the liquid refrigerant reduces flash gas, and increases mass flow.

(Position 6.) In passing through the metering device to the low-pressure zone, some refrigerant is evaporated, cooling the remaining liquid. The refrigerant is a mixture at this point.

(Position 7.) Heat from the air or the product being cooled in the evaporator is absorbed by the liquid refrigerant and causes the refrigerant to boil or vaporize. As the compressor draws the vaporized gas from the evaporator, the metering device admits more refrigerant, to continue the process. The refrigerant at this point is a mixture, as in position 6.

For all practical purposes these are the two pressures in the system: the low-side pressure and the high-side pressure. From the metering device, evaporator, and suction line up to the compressor inlet represent the low side of the system. The compressor, discharge line, condenser, liquid line, up to the metering device are considered the high side of the system. The compressor is considered to be in the high side and the metering device in the low side of the system. The compressor and the metering device work in partnership to maintain this pressure difference. The metering device controls the flow into the evaporator and the expansion of the refrigerant causes a pressure drop. The compressor pumps the refrigerant out of the evaporator and maintains the pressure.

Heat is added to the refrigerant in absorbing the product load in the evaporator. This constitutes primarily the net refrigeration effect, plus a small gain that occurs in the piping up to where the refrigerant enters the compressor. The compressor adds heat to the refrigerant in a sizable quantity. This is equivalent to the work done in compressing the refrigerant. In a suction gas cooled semi-hermetic or hermetic motor-compressor unit, the motor heat is also transferred to the refrigerant.

The heat added in the evaporator and by the compressor is removed in the condenser. Some relatively small additional losses occur in the receiver and the liquid line piping up to the metering device.

R3-4 THE MECHANICAL REFRIGERATION CYCLE

A mechanical refrigeration system must have several basic components which can circulate the refrigerant and transfer heat. These components are the evaporator (cooling coil), compressor, condenser, metering device (liquid refrigerant control), and interconnecting tubing (suction, discharge, and liquid lines). Auxiliary devices such as accumulators, filter-driers, receivers, etc., may be installed on certain systems but are not considered a basic cycle component.

R3-4.1 The Evaporator

The evaporator or cooling coil is fabricated from metals such as copper or aluminum or both. These metals are selected because of their good conductivity. The evaporator is fabricated to the desired size and shape. The tubing is interconnected by aluminum fins which serve to both direct the air flow through the coil and increase transfer of heat by conduction. (Fig. R3-7)

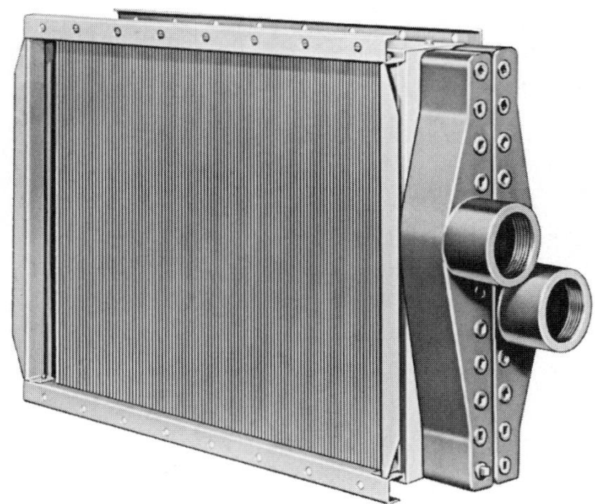

Figure R3-7 Fin and tube construction. (The Trane Company)

The evaporator coil is a storage container for the boiling (saturated) liquid refrigerant. It is located in the area to be cooled or air from this space is drawn by a blower fan through this coil and redistributed back into the conditioned space. (Fig. R3-8)

A low pressure is maintained inside the evaporator coil, thus lowering the boiling point of the refrigerant. Upon entering the evaporator the refrigerant liquid begins to flash to saturation and boil. The refrigerant, in order to vaporize, must absorb heat. Heat is transferred from the fins and tubes of the evaporator to the refrigerant circulating inside. This provides a cold surface for the absorption of heat from the surrounding air. As air is pulled into the evaporator coil by the blower fan (see Fig. R3-8), heat is removed from the air and the moisture in the air collects on the cold surface of the evaporator coil. The collected water vapor drains into a collection pan and is routed to a drain by tubing (see Fig. R3-9).

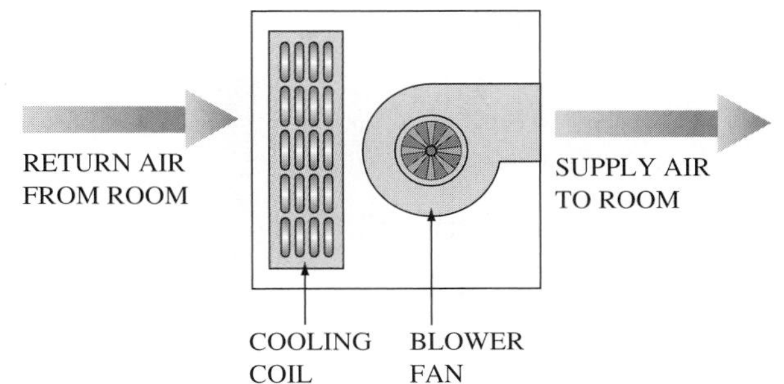

Figure R3-8 Evaporator coil and fan.

Figure R3-9
Condensate drain system.
(The Trane Company)

R3-4.2 The Compressor

The compressor (Fig. R3-10) pumps the heat-laden vapor from the evaporator. This causes the low pressure that is maintained in the evaporator by the compressor, and the restricted flow into the evaporator by the liquid refrigerant control (metering device) on the inlet to the evaporator. The resulting temperature and pressure of the saturated refrigerant is lowered. The refrigerant boils and vaporizes, absorbing latent heat at this low temperature and pressure.

The heat-laden vapor is then compressed by the compressor, increasing the pressure and the temperature of the vapor. The temperature of the refrigerant vapor leaving the compressor (discharge line) and entering the condenser must be higher than the temperature of the condensing medium.

The compressor, which is driven by an electric motor lubricated with a special refrigeration oil and mounted in a welded steel shell, is referred to as a welded hermetic compressor. The electrical components associated with the compressor, as shown in Fig. R3-11, are the start relay (current operating) and the bimetal overload. Replacement compressors like the one pictured may also have a starting capacitor. The replacement compressor may also come with an installed suction service valve.

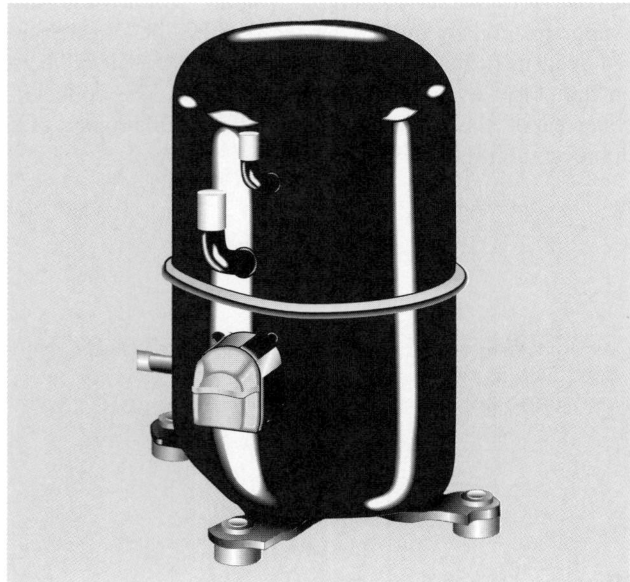

Figure R3-10 Compressor.

R3-4.3 The Condenser

The condenser, as shown in Fig. R3-12, is similar to the evaporator in construction. It is a series of tubes through which the hot, high-pressure vapor passes. Air is forced through the condensing coil by the blower fan, and heat is given up by the refrigerant to the surrounding air, causing the vapor to condense.

Once the vapor's sensible temperature is reduced to its condensing temperature, latent heat will continue to be removed as the vapor is condensed into a liquid refrigerant. The condenser must be sized in accordance with the size of the compressor and evaporator. Heat that is absorbed in the cooling coil, superheat, plus miscellaneous heat caused by electricity or friction, a by-product of the process of compression, must be removed by the condenser. The suction line, returning to the compressor, is insulated to prevent picking up any excessive superheat through it.

The liquid refrigerant leaving the condenser passes through the *filter-drier,* as shown in Fig. R3-13, before it enters the metering device (capillary tube). The **filter-drier** is a mechanical device with a screen to block foreign particles and a desiccant, or drying agent, to remove moisture from the refrigerant. It is extremely important not to use any antifreeze compounds with the refrigerant in the system, as these will clog the drier.

R3-4.4 The Metering Device

The most common type of metering device used on refrigeration systems is the restrictor type of flow control,

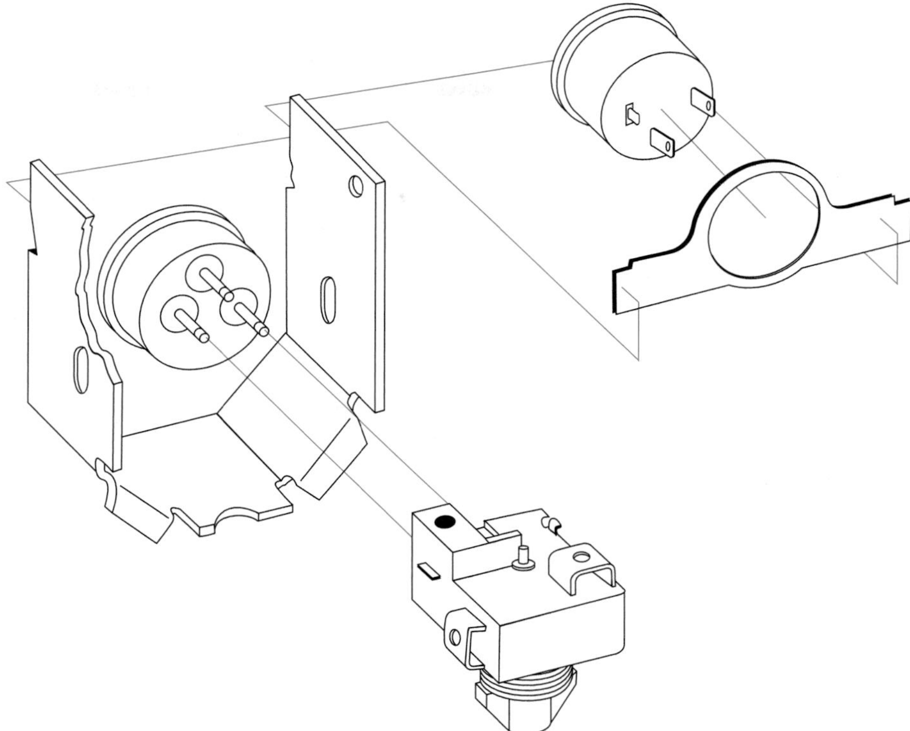

Figure R3-11 Start relay, overload, compressor pins.

commonly known as the *capillary tube*. The capillary tube (see Fig. R3-14) is located after the filter-drier and connects the condenser and the evaporator. The capillary tube meters the flow of liquid refrigerant from the liquid line and provides a pressure drop from the high-pressure side of the system to the low-pressure side of the system. The capillary tube's length and inside diameter determine the correct amount of liquid flow into the evaporator and the correct pressure drop.

Figure R3-12 Condenser. (The Trane Company)

The *accumulator* is a storage cylinder located between the evaporator and the suction line to the compressor. The purpose of the accumulator is to store liquid refrigerant. This prevents spillover into the suction line, thus obtaining maximum efficiency from the evaporator coil by permitting it to be fully flooded with refrigerant.

R3-4.5 Interconnection Tubing

The components of the refrigeration system are connected by three refrigerant lines: the *liquid line,* connecting the condenser and evaporator, the *discharge line,* from the compressor to the condenser and the *suction line,* from the evaporator back to the compressor. The liquid line carries the high pressure, warm liquid from the condenser to the capillary tube (metering device). The discharge line

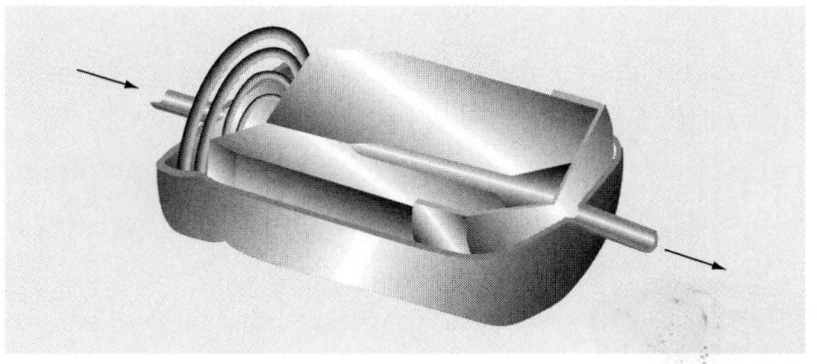

Figure R3-13 Filter-drier.

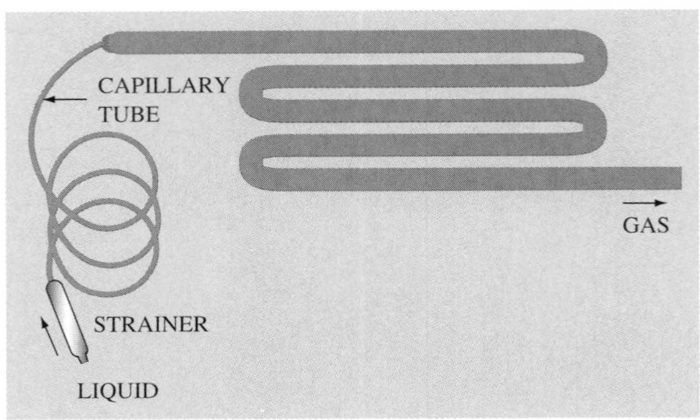

Figure R3-14 Capillary tube.

is normally smaller than the suction line. The discharge line carries the high-pressure, hot vapor from the compressor to the condenser. The suction line carries the low-pressure, cool vapor from the evaporator coil to the compressor inlet.

The operation of the compressor removes the vapor from the evaporator, reducing the pressure above the liquid refrigerant. Lowering the pressure lowers the boiling point (saturation temperature) of the liquid. Heat from the area to be refrigerated or cooled flows by convection to the evaporator, then by conduction through the fans and tubing to the liquid refrigerant.

This latent heat causes the liquid to vaporize without changing the temperature or pressure of the boiling liquid. The heat-laden vapor from the evaporator is compressed by the compressor. This increases the temperature and pressure of the refrigerant vapor above the condensing medium.

This highly superheated vapor enters the condenser through the discharge line. Condenser heat is removed from the high temperature, high-pressure vapor by conduction to the tubes and fans, then by convection to the air being forced through the condenser by the blower fan. Thus, heat is carried away from the refrigeration system at the condenser.

As heat is removed from the vapor, the sensible temperature of the conditioned air is reduced. The cool surface of the evaporator also collects water vapor from the conditioned air and drains it.

As latent heat is removed from the vapor at the condenser, the vapor condenses to a liquid. The liquid flows to the outlet from the condenser, through the liquid line to the metering device (capillary tube). The liquid is subcooled as it travels this path.

The compressor maintains a pressure differential between the evaporator coil and the condenser, with high pressure in the condenser and low pressure in the evaporator coil. This pressure differential causes the liquid refrigerant to flow through the filter-drier, to the capillary tube restriction to the evaporator coil. The filter-drier removes moisture and contaminants from the refrigerant. The metering device (capillary tube) meters the correct amount of liquid to the evaporator and provides the desired pressure drop.

The cycle of refrigerant flow from the evaporator coil, where heat is absorbed through the refrigeration system and returns to the evaporator after the heat is released at the condenser, ready to absorb more heat, is known as the mechanical refrigeration cycle.

R3-5 THE ABSORPTION REFRIGERATION CYCLE

(Adapted from Carrier Training Module, *Basic Refrigeration Cycles.*)

The absorption cycle is slightly different than the mechanical vapor compression cycle. There is no compressor; it is replaced by an *absorber, concentrator,* and *pump.* It does, however, have a high-pressure and low-pressure side. It can be located in a single shell or may be separated into several shells, similar to the centrifugal chiller, as shown in Fig. R3-15. Another difference is in the fluids used as refrigerants. In an absorption system, water is used as the refrigerant. Lithium bromide, a type of salt, is mixed with the water as an absorbing solution.

Depending on the amount of lithium bromide in the solution, the absorbent is said to be "concentrated," "inter-

PREVENTIVE MAINTENANCE

Educate your customer. Explain when and how to change the filter. Explain why this is important maintenance for their unit. Tell them to be careful when cutting grass, to point the lawn mower away from the condenser or use a grass catcher when possible. Explain how the customer can avoid this type of service call in the future by keeping the condenser clean with a garden hose. Be sure to point out safety precautions; turn off the power to the unit outside, cover electrical components and boxes with plastic. Spray the coil from the inside to the outside. Explain how blocked air flow can affect the performance of the air conditioner.

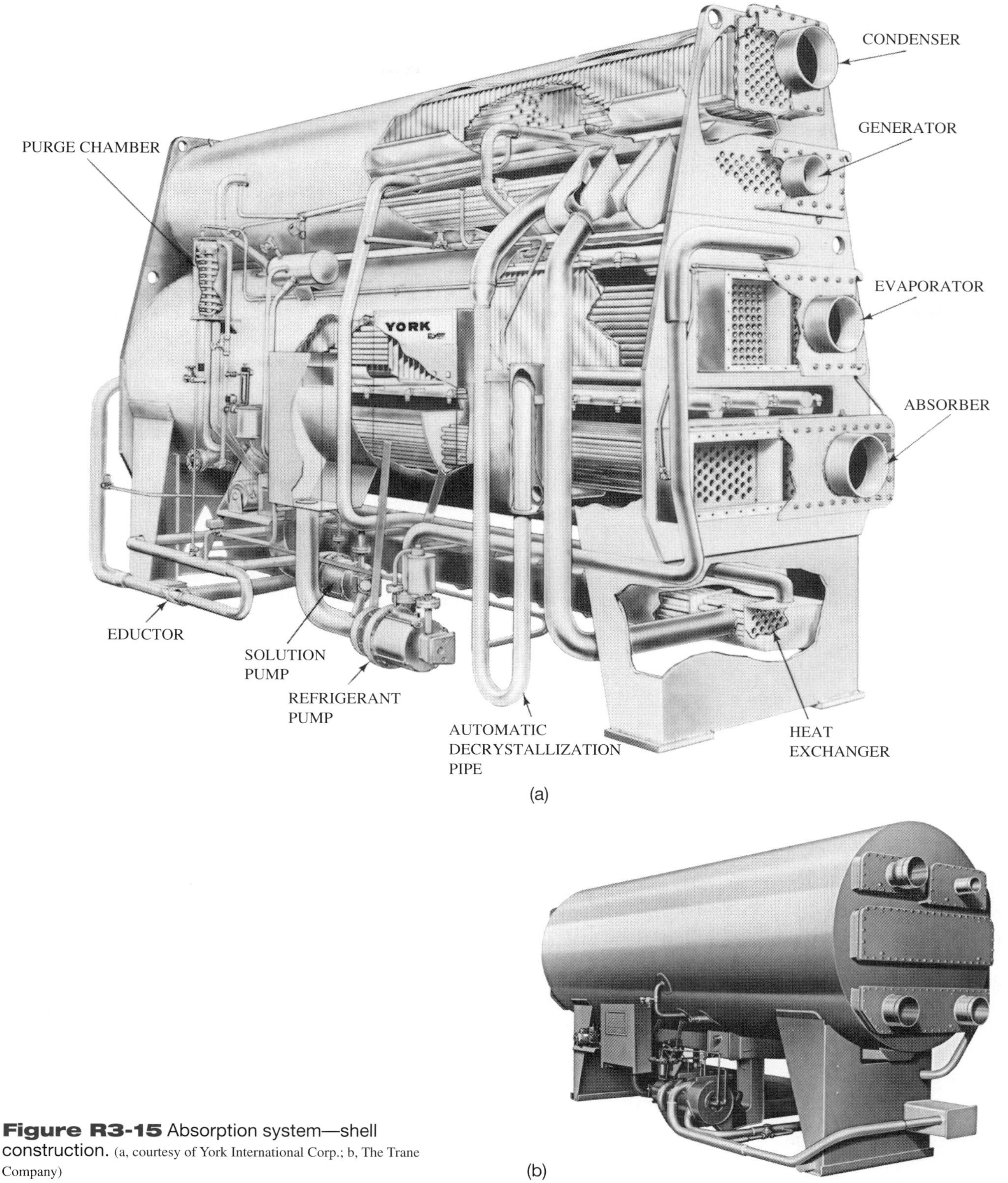

CONDENSER
GENERATOR
PURGE CHAMBER
EVAPORATOR
ABSORBER
EDUCTOR
SOLUTION PUMP
REFRIGERANT PUMP
AUTOMATIC DECRYSTALLIZATION PIPE
HEAT EXCHANGER

(a)

(b)

Figure R3-15 Absorption system—shell construction. (a, courtesy of York International Corp.; b, The Trane Company)

mediate," or "diluted." An absorption cycle is more complex than the vapor-compression cycle. It contains several sections and cycles. Sections include an absorber, concentrator, condenser, evaporator, heat exchanger, several pumps, and spray nozzles. It also requires a heat source to produce the pressure difference, usually steam from a boiler. Like a centrifugal chiller, it uses a secondary refrigerant (it cools system water) which is used for refrigerating purposes.

The first component in the cycle is the evaporator, which is filled with water used as a refrigerant. (Fig. R3-16) Inside of the evaporator is a coil through which

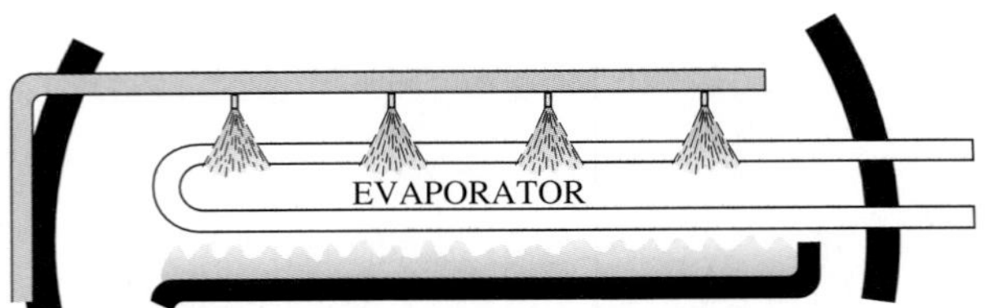

Figure R3-16 Evaporator section. (The Trane Company)

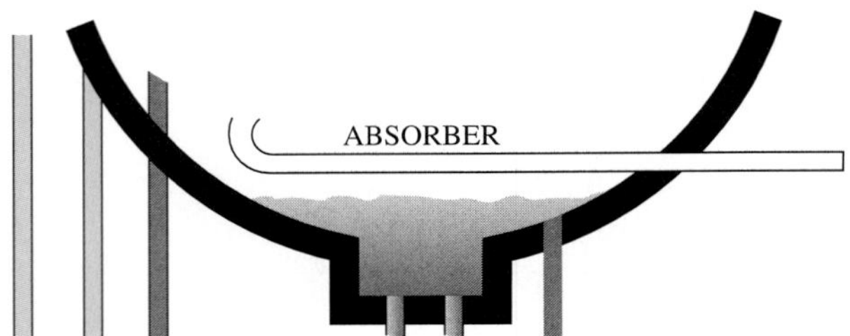

Figure R3-17 Absorption section. (The Trane Company)

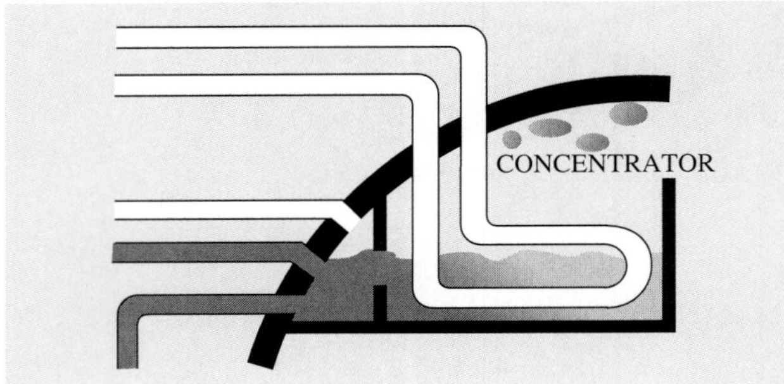

Figure R3-18 Concentrator. (The Trane Company)

chilled water (system water) runs. This water is pumped to remote locations and used as the secondary refrigerant (it provides the refrigeration). The evaporator section is under high vacuum which allows the refrigerant water to boil at low temperature. As this refrigerant water boils, it absorbs heat from the chilled water circulating through the coil. An evaporator pump circulates the refrigerant water to spray nozzles above the coil. Water spraying across the coil increases heat transfer.

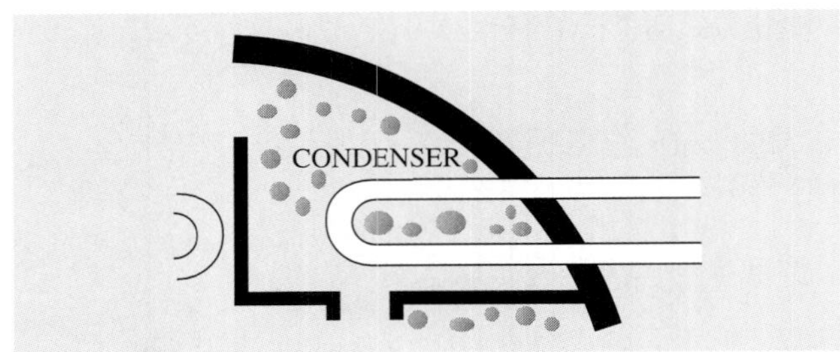

Figure R3-19 Condenser. (The Trane Company)

Connected to the evaporator by a pipe is the absorber containing the lithium bromide solution, which attracts the water vapor from the evaporator. (Fig. R3-17) In order to separate the refrigerant water from the lithium bromide solution, a solution pump moves the fluid to a concentrator.

The concentrator contains a heat source (a heating coil, or steam heat exchanger) which causes the water to evaporate from the salt. (Fig. R3-18)

The refrigerant water vapor is piped to a condenser, where it condenses back into a liquid as heat is removed from the fluid. (Fig. R3-19) The concentrated lithium bromide solution flows back to the absorber and the refrigerant water flows from the condenser to the evaporator, to start the cycle over again.

R3-6 SYSTEM COMPONENTS

Let's look at a more sophisticated absorption system (Fig. R3-20). The cylinder is separated into four separate compartments. The evaporator, which is under a high vacuum, the absorber, which has a pressure of 0.15 psia, the concentrator, which has a pressure of 1.5 psia, and the condenser. Between the high and low side, we have the cycle dealing with the absorber and the concentrator.

In the bottom of the shell lays the intermediate absorbing solution. This solution is pumped to the concentrator by the concentrator pump. Before it reaches the concentrator, the solution passes through a heat exchanger, where the temperature is raised by heated solution returning from the concentrator. This increases the efficiency of the system.

The solution flows into the concentrator, where it is heated by a coil containing steam from a boiler system.

ON THE JOB

Suppose you have a friend who is planning to remodel their kitchen. He plans to build cabinets over the refrigerator, and build countertops with cabinets on both sides of the unit. His refrigerator has the condenser on the back of its cabinet. What advice would you give him?

Explain your answer in terms of heat transfer and the refrigeration cycle.

PRACTICING PROFESSIONAL SERVICES

A customer calls and complains that the refrigerator runs *all* of the time. Upon inspecting it, you notice that the condenser is located under the cabinet. You also see that someone has removed the cardboard on the back of the refrigerator cabinet, and you notice a considerable amount of dust and lint underneath the refrigerator. It looks as if someone did try to clean the condenser but could not get to the hard-to-reach areas. You ask the customer about the cardboard back. They admit removing it and trying to clean the condenser. They tell you, "it seemed to work better without the cardboard." How do you explain to them what is happening in terms of heat transfer and the refrigeration cycle?

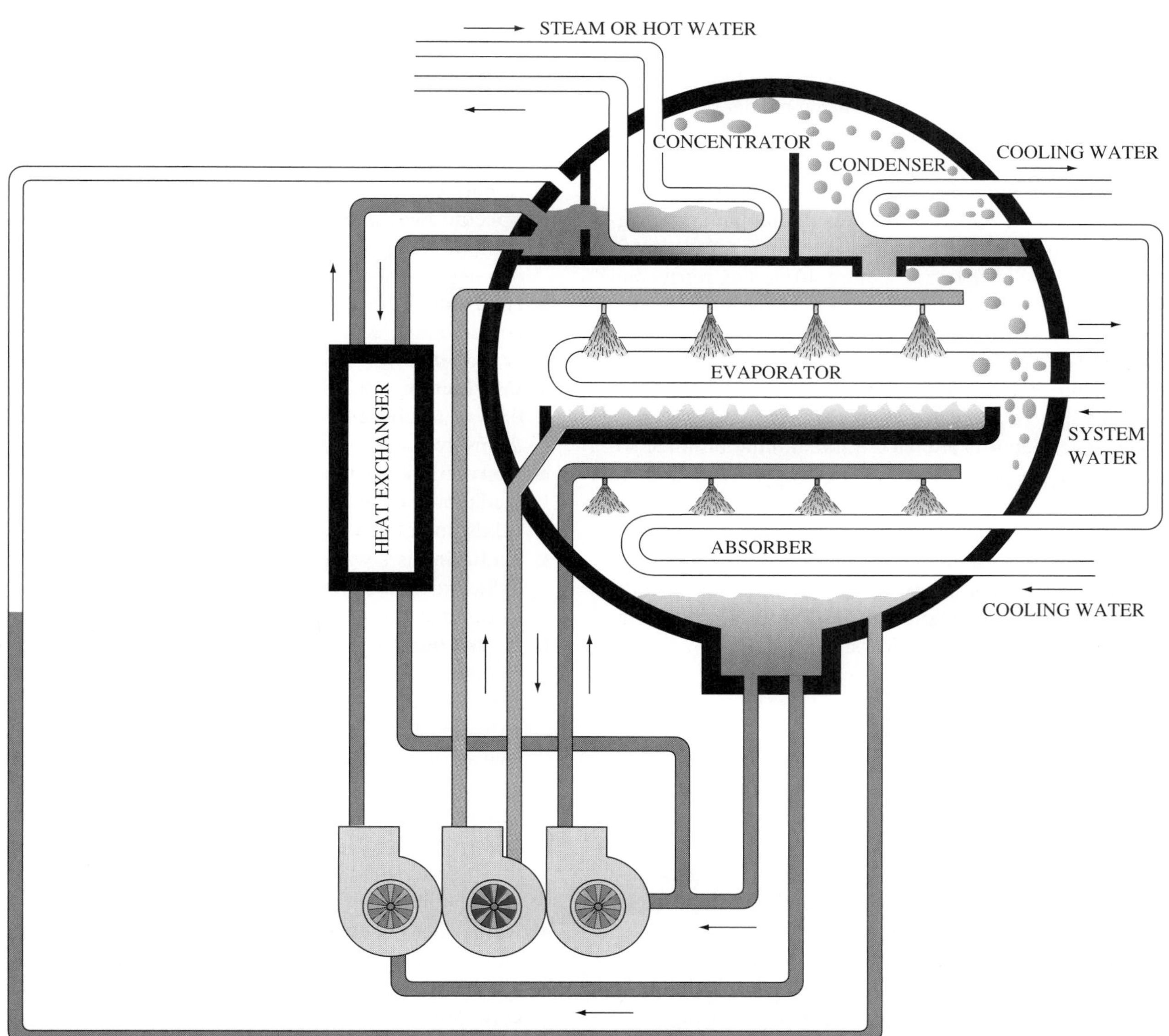

Figure R3-20 Absorption system. (The Trane Company)

The heat separates the salt (lithium bromide) from the refrigerant water by the process of distillation.

The water vapor then flows into the condenser section, thus leaving behind a concentrated absorbing solution. The heated concentrated absorbing solution is drawn or pumped back through the heat exchanger, to exchange heat with the intermediate solution. After leaving the heat exchanger, it meets a line of diluted solution from the absorber, mixing into an intermediate solution.

This solution is pumped up to a spray nozzle, where it is sprayed past the absorber coil, containing condensing water from the cooling tower. This also increases system efficiency.

Some of the intermediate cooled solution then flows back past the heat exchanger. It is pumped back to the concentrator. The other part is mixed with the concentrated solution to be sent through the spray nozzles of the absorber section.

Water that has evaporated from the concentrator section is drawn into the condenser section, where a cooling coil condenses it. This coil is supplied by the cooling tower.

The water sits on the bottom of the condenser section, where a hole allows it to flow into the low side of the system—the evaporator.

As the refrigerant water enters the evaporator, some water flashes to a vapor. This cools the remaining refrigerant to evaporator temperature. The liquid refrigerant is pumped by the evaporator pump, from the refrigerant sump, to the spray nozzles located over the tube bundle. The tube bundle is where system water (chilled water) is flowing. The spray wets the tube bundle for maximum heat transfer. Heat transfers to the refrigerant water, causing it to vaporize and carry away the heat.

The water vapor is attracted to the lithium bromide solution in the absorber section. The heated water adds to the diluted lithium bromide solution. The solution is drawn by a pump, where it mixes with the concentrated solution returning from the concentrator. This intermediate solution is then sprayed over the absorber coils, cooled, and is ready to be pumped back to the concentrator to start the whole cycle over again.

REVIEW

- **Refrigeration** is the process of removing heat energy from a place it is not wanted and disposing of it in a place where it is not objectionable.
- **Refrigerant** is a fluid that picks up heat by evaporating at a low temperature and pressure and gives up heat by condensing at a higher temperature and pressure.
- **Temperature** is the measurement of the *speed* or *intensity* of the molecules of a substance.
- Temperature is measured by a thermometer using either the **Fahrenheit** or **Celsius** scale. Each scale has a *boiling* and a *freezing point* established.
- **One ton of refrigeration** is the refrigeration produced by melting one ton of ice at a temperature of 32 degrees Fahrenheit in 24 hours. It is the refrigerating effect expressed as: 288,000 Btu/24 hours, 12,000 Btu/hour, 200 Btu/minute.
- An **air conditioning system** is a refrigeration system used to cool, dehumidify, filter and/or heat the air of a space.
- **Matter** is anything that has mass and weight and takes up *space*. All matter is made up of molecules.
- **Heat** is a form of energy which may be produced by other forms of energy such as: light, electricity, chemical means, mechanical means, or atomic energy.
- **Conservation of energy** means that heat energy cannot be created or destroyed. It is the *by-product* of other forms of energy. The *adding* or *removing* of heat energy to a substance determines the *speed* or intensity of the molecules. The speed or intensity of molecules determines the physical state of matter—*solid, liquid* or *gas*.
- **Change of state** is the change from a solid to a liquid and liquid to a gas, or vice versa. A large amount of heat must be added or removed to cause a change of state.
- **Btu** is the amount of heat required to raise one pound of pure water, one degree Fahrenheit.
- **Specific heat** is the quantity of heat (Btu) required to raise the temperature of one pound of a substance by one degree Fahrenheit.
- **Heat-flow** is the movement of heat from a warmer to a cooler body. Heat moves by: conduction, convection, or radiation.
- **Conduction** is the transfer of heat from particle to particle of a substance without movement of the particles themselves.
- **Convection** is the transfer of heat by a flowing medium. It takes place only in liquids and gases since solids do not flow.
- **Radiation** is a wave form of heat movement similar to light, except it cannot be seen. Like light, it requires no medium to travel.
- **Latent heat** is heat energy absorbed in the process of changing the forms of a substance (melting, vaporization, or fusion) without a change in temperature or pressure. There are three types of latent heat: latent heat of vaporization, latent heat of fusion, and latent heat of condensation.
- **Latent heat of vaporization** is the amount of heat energy required to change a substance from a liquid to a vapor without changing its temperature or pressure.
- **Latent heat of fusion** is the amount of heat energy required to change a substance from a liquid to a solid without a change in temperature or pressure at standard atmospheric pressure.
- **Latent heat of condensation** is related to latent heat of vaporization. It is the heat energy that must be removed

from a substance to change state from a vapor to a liquid without changing its temperature or pressure.

- **Superheat** is the temperature of a vapor above the boiling temperature of its liquid at that pressure. Heat added to a gas after the liquid has vaporized is measured by the number of degrees of superheat that results.
- **Sensible heat** is heat added to or removed from a substance that causes a change in the temperature of the substance.
- *Saturation temperature* is another name for *boiling point.* The **boiling point** of a liquid is affected by *pressure—higher* pressure *raises* the boiling point; *lower* pressure *lowers* the boiling point.
- **Pressure** is an impact on a unit area; force or thrust exerted on a surface, measured in pounds per square inch (psi). Pressure is referred to in one of three ways: *atmospheric, gauge,* or *absolute.*
- **Atmospheric pressure** is stated as 14.7 psi at sea level.
- **Absolute pressure** is pressure measured on the absolute scale. The zero point is at zero atmospheric pressure.
- The *total* of the *gauge* and *atmospheric* pressures is called *absolute.* **Gauge pressure** is measured on the gauge scale. The zero point is 14.7 psi or atmospheric pressure at sea level. Gauge pressure is the *measured pressure* above or below atmospheric. Pressure *below atmospheric* is measured in *inches of mercury.*
- A *fluid at its boiling point* is referred to as *saturated vapor/liquid.*
- A *fluid above its boiling point* is referred to as *superheated vapor.*
- A *fluid below its boiling point* is referred to as *subcooled liquid.*

Problems and Questions

1. Define the following terms:

Refrigeration	Change of state
Cold	Refrigerant
Heat	Latent heat of vaporization
Temperature	
Btu	Latent heat of fusion
Latent heat	Radiation
One ton of refrigeration	Conduction
	Heat-flow
Pressure	Convection
Superheat	Subcooling
Saturation temperature	Saturated
Gauge pressure	Specific heat
Absolute pressure	Temperature
Sensible heat	

2. State the purpose of each of the basic cycle components: Compressor, Condenser, Metering device, and Evaporator.
3. Draw the mechanical vapor-compression refrigeration cycle. Identify system components and the state, pressure, temperature, and condition of the refrigerant at various key points in the refrigeration cycle.
4. State the purpose of each of the absorption cycle components: Generator, Condenser, Heat exchanger, Absorber, and Evaporator.
5. Draw the absorption refrigeration cycle. Identify system components, the areas of heat transfer, and the condition, temperature, and state of the fluid used as a refrigerant at various key points in the absorption refrigeration cycle.
6. Atmospheric pressure is 14.7 at sea level. True or false?
7. The metering device controls the flow of liquid from the evaporator to the condenser. True or false?
8. The compressor is a mechanical device for pumping refrigerant liquid from a low pressure area to a high pressure area. True or false?
9. The finned tube evaporator is a device for absorbing heat into the refrigeration system. True or false?
10. The receiver is a storage cylinder located on the low side of the refrigeration system to store excess liquid. True or false?
11. Latent heat is heat
    a. that changes temperature without changing state.
    b. that changes state without changing temperature.
    c. is normally measured with an electronic thermometer.
    d. None of the above.
12. One ton of refrigeration is
    a. 12,000 Btu/hour.
    b. 200 Btu/minute.
    c. 288,000 Btu/24 hours.
    d. All of the above.
13. Subcooled liquid is a
    a. liquid below its boiling point, at a given pressure.
    b. liquid above its boiling point, at a given pressure.
    c. liquid subcooled at the outlet of evaporator to improve overall efficiency.
    d. condition at which the refrigerant reaches its saturation state.
14. Refrigerant is a fluid that picks up
    a. heat by evaporating at a higher pressure and temperature.
    b. heat by evaporating at a lower pressure and temperature.
    c. cold air by evaporating at a lower pressure and temperature.
    d. None of the above.
15. Refrigeration is the process of removing
    a. heat energy from a place it is not wanted and disposing of it.
    b. cold energy from a place where it is not wanted.
    c. liquid refrigerant from the evaporator and pumping it into the accumulator.
    d. None of the above.

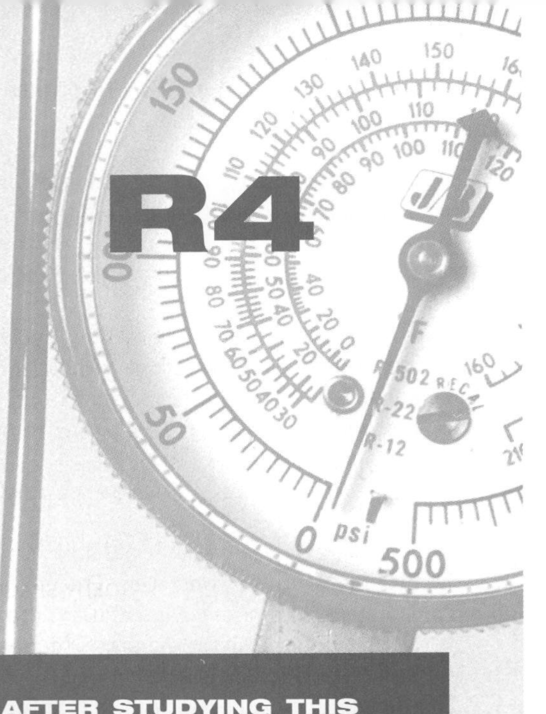

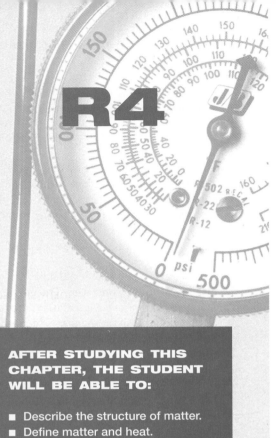

R4 Principles of Heat Transfer

AFTER STUDYING THIS CHAPTER, THE STUDENT WILL BE ABLE TO:

- Describe the structure of matter.
- Define matter and heat.
- Distinguish between the three states of matter.
- Explain the arrangement and movement of molecules.
- Identify the states in which matter exists.
- Define temperature.
- Identify different types of thermometers.
- Compare Fahrenheit, Centigrade (Celsius) and their absolute temperature scales.
- Identify the reference points of temperature.
- Convert from one temperature scale to another.
- Explain the difference between heat and temperature.
- Define and describe the differences between latent, sensible and specific heat.
- Describe the system for measuring heat quantity.
- Describe the three methods of heat transfer—convection, conduction, and radiation.
- Define an insulator.
- Define a conductor.
- Identify the units of measurement for insulation and conduction of heat.
- Explain the five principle changes of states.

R4-1 STATES OF MATTER

The physical state of a substance can be controlled by temperature and pressure. For example, many outdoor ice rinks are designed for when the weather goes below freezing temperatures. The weight of a skater on the surface of the ice causes the ice under the hard surface of the skates to melt, making the ice "slippery". In nature, heavy blocks of ice, such as glaciers, pressing on the hard surface of the earth can cause the bottom of the ice to melt. This process lubricates glaciers resting on slopes, allowing them to move.

Matter exists in three states: *solid, liquid* and *vapor* or gas, as illustrated in Fig. R4-1.

A common example is water, which exists in all three states. Water is in a solid (ice) stage below 32°F (0°C), a liquid state from temperatures of 32°F (0°C) to 212°F (100°C) and a vapor or gas at 212°F (100°C) and above.

The various states of matter have unique characteristics (Fig. R4-1).

A **solid** is a substance that has a definite shape, which it will hold under a certain degree of stress or pressure, depending on the material and the type of disturbance. It must be supported or it will fall to the next level of support. This condition requires the design of an adequate foundation when dealing with solids.

Solids of sufficient density will retain their size and weight. Solids of light density will lose molecular quantity under certain conditions and lose weight and quantity. Carbon dioxide (CO_2) in solid state (known as dry ice) will pass from the solid state into the gaseous state under certain conditions.

A **liquid** is a substance that can take the shape of any enclosure when it is allowed to move freely, but the volume remains constant. It is considered incompressible. When a liquid fills an enclosure it exerts both a horizontal and a vertical pressure on the enclosure. For example,

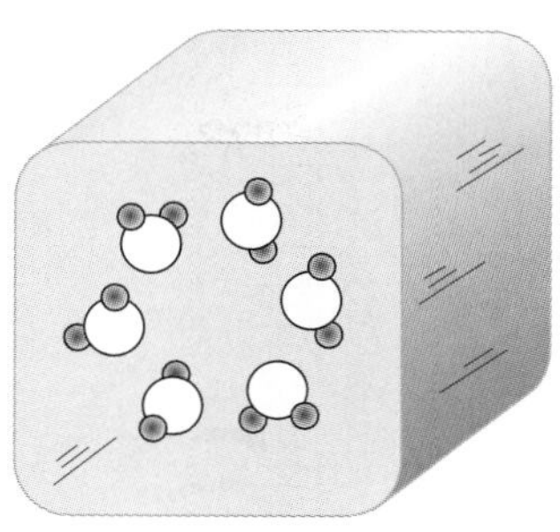
IN SOLID FORM, WATER MOLECULES REMAIN RIGIDLY IN PLACE AND FORM IN HOLLOW RINGS, GIVING ICE ITS LOW DENSITY.

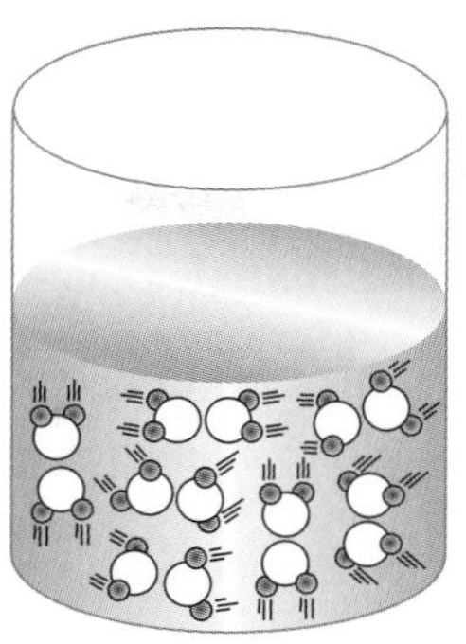
WATER MOLECULES ARE CLOSE TOGETHER YET FREELY SLIP OVER ONE ANOTHER, GIVING LIQUID ITS FLOW.

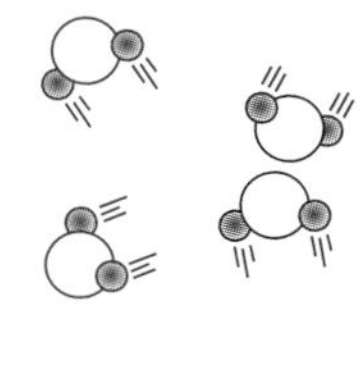
CHAOS! MOLECULES IN GAS (STEAM) ARE WIDELY SPACED, DART RAPIDLY, AND COLLIDE WITH ONE ANOTHER.

Figure R4-1 Three states of matter. Water is a unique substance which exists naturally in all three physical states.

valves located in the basement piping of a hot-water heating system for a multi-story building would need to be capable of withstanding the pressure exerted by the column of water above them.

If a cubic foot of water in a container measuring 1 foot on each side is transferred to a container of different dimensions, the quantity and weight of the water will remain the same, although the dimension will change (see Fig. R4-1). In time, liquids of light density, such as water, will gradually lose quantity and weight by molecular loss to the gaseous state.

A **vapor** or **gas** is a substance that has no fixed shape or volume, and therefore must be contained in an enclosure or it will escape into the atmosphere. A good example is the enclosure of a refrigerant in vapor form. Certain refrigerants that are destructive to the environment are being replaced and must not be purged into the atmosphere.

If a 1-cubic-foot cylinder containing gaseous state water, called "steam," or some other vapor, is connected to a 2-cubic-foot cylinder of which, theoretically, a perfect vacuum has been drawn, the vapor will expand to occupy the volume of the larger cylinder as well as the original. Other changes will occur in the vapor which will be discussed later.

R4-1.1 Density, Specific Volume and Specific Gravity

Density is a measurement that can be used to compare the mass of various substances. Density is substance's mass per unit of volume. In the I-P system a convenient unit is pounds per cubic foot. In SI terms the unit is kilograms per cubic meter.

Specific volume (SV) is the volume occupied by a unit mass of a substance, under standard conditions. In the I-P system it is expressed in cubic feet per pound. The standard conditions are 68°F and 29.92 inches of mercury pressure. In the SI system it is expressed in cubic meters per kilogram. The standard conditions are 20°C and 101.3 kPa pressure. For example, one pound of air at standard atmospheric conditions occupies 13.45 cubic feet.

Specific gravity is the ratio of the mass of a given volume of a liquid or solid to the mass of an equal volume of water. Where gases are used, air or hydrogen is the comparing substance. Specific gravity is an absolute term; use density for measurement.

Although these specific differences exist in the three states of matter, quite frequently, under changing conditions of pressure and temperature, the same substance may exist in any one of the three states. For example, water could exist as a solid (ice), a liquid (water) or a gas (steam). Solids always have a definite shape, whereas liquids and gases have no definite shape of their own and so will conform to the shape of their container.

R4-2 THERMODYNAMIC PRINCIPLES

Thermodynamics is the branch of science dealing with mechanical action of heat. Two main laws of thermodynamics are of interest in HVAC/R applications.

R4-2.1 First Law of Thermodynamics

The first law of thermodynamics states that "energy can neither be created nor destroyed, it can only be converted from one form to another." (With the development of nuclear energy, this is no longer correct; however, for the purposes of this text, the first law of thermodynamics applies totally.)

Energy itself is defined as the ability to do work, and heat is one form of energy. It is also the final form, as ultimately all forms of energy end up as heat. Other common forms of energy are: mechanical, electrical, and chemical, which may be converted easily from one form to another. The steam-driven turbine generator of a power plant is a device that converts heat energy into electrical energy. Chemical energy may be converted into electrical energy by the use of a battery. Electrical energy is converted into mechanical energy through the use of an electric magnetic coil to produce a push-pull motion or the use of an electric motor to create rotary motions. Electrical energy may be changed directly to heat energy by means of heating resistance wires such as in an electric toaster, grill, furnace, and anticipator.

R4-2.2 Second Law of Thermodynamics

The second law of thermodynamics states that "to cause heat energy to travel, a temperature difference must be established and maintained." Heat energy travels downward on the intensity scale. Heat from a higher-temperature (intensity) material will travel to a lower-temperature (intensity) material, and this process will continue as long as the temperature difference exists. The rate of travel varies directly with the temperature difference. The higher the temperature difference (commonly called the delta temperature or ΔT), the greater the rate of heat travel. Commonly, the lower the ΔT, the lower the rate of heat travel. Heat and heat transfer are expressed in different forms and terms that are important to the refrigeration industry.

R4-3 TEMPERATURE MEASUREMENT AND CONVERSION

We are all acquainted with common measurements such as those referring to length, width, volume, etc. We must also acquaint ourselves with the methods of measuring heat energy.

The quantity of heat energy in a substance starts with the size of the substance as well as the intensity or level of heat energy in the substance. The level of heat energy is measurable on a comparison basis by means of a thermometer. The thermometer was developed using the principle of the expansion and contraction of a liquid, such as mercury, in a tube of small interior diameter which includes a reservoir for the liquid. Upon being subjected to a temperature (heat intensity) change, the liquid will rise up the line upon increase of temperature, or fall upon decrease of temperature.

The two most common standards of temperature (intensity) measurement are the Fahrenheit and Celsius (formerly centigrade) scales. Fig. R4-2 shows a direct comparison of the scales of a Fahrenheit and a Celsius thermometer.

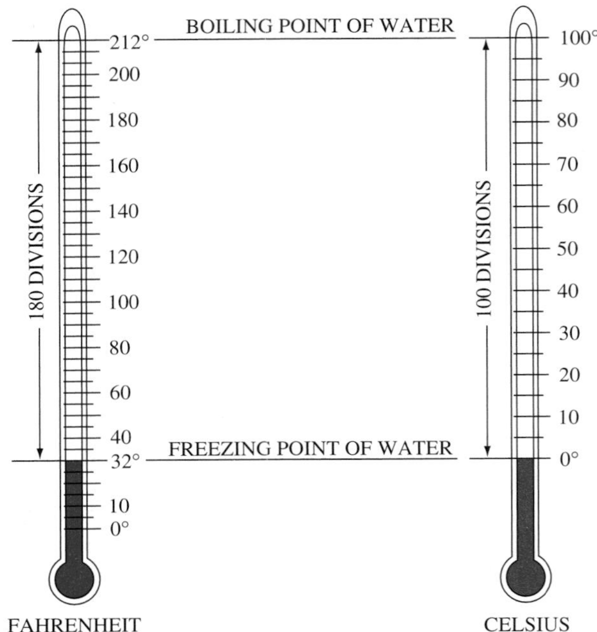

Figure R4-2 Comparison of the Fahrenheit and Celsius temperature scales.

R4-3.1 Measurement of Temperature

Two definitions of temperature are:

1. "Temperature is a relative term that can be defined as something that will produce a sensation of hot or cold."
2. "Temperature is a thermal state of two adjacent substances that determines their ability to exchange heat." A conclusion of this definition is that substances in contact that do not exchange heat are at the same temperature.

The first definition defines temperature in terms of a human sensation. This is an important aspect of temperature since a great deal of attention is given to keeping people comfortable. Detailed information on this subject is included in the air-conditioning section.

The second definition defines temperature in terms of its heat-transfer quality. The whole system of refrigeration is involved in moving heat from places where it is not wanted to places where it can be tolerated.

Temperatures can be measured using a thermometer as shown in Fig. R4-3.

Glass-stem thermometers have been commonly used by refrigeration service people. The temperature is measured by the expansion or contraction of a liquid (mercury or alcohol) with changes in temperature. In a hollow glass tube the fluid moves up and down a scale that is calibrated to register the ambient temperature. Pocket sizes are available as well as longer, more accurate models for laboratory work. Breakage is a problem since glass thermometers are fragile.

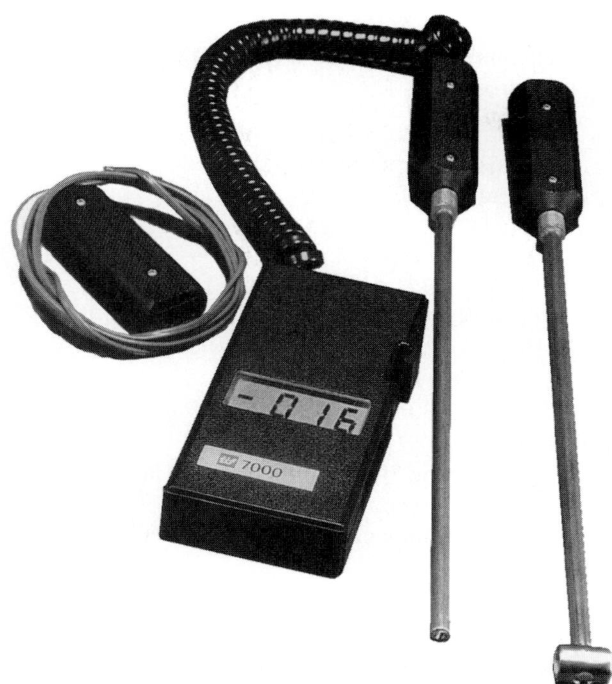

Figure R4-3 Digital thermometer with retractable probe, selector switch for Centigrade or Fahrenheit. Range: –40° to 1999°F (–40°C–1100°C). (Courtesy of TIF Instruments, Inc.)

Dial thermometers indicate temperature by a pointer moving over a circular scale. They are more rugged and can be used for a wide variety of applications.

Digital thermometers with compact sensing elements are popular due to their speed in sensing a change in temperature and the ease in reading the alphanumeric display instead of a dial or a scale. They are useful in measuring surface temperatures, such as suction superheat in adjusting an expansion valve.

R4-3.2 Temperature Conversion

Most frequently a conversion from one temperature scale to the other is made by the use of a conversion table, but if one is not available, the conversion can be done easily by using formulas based on a definite reference point—absolute zero. This is the point where, it is believed, all molecular action ceases. On the Fahrenheit temperature scale this is about 460° below zero, –460°F, while on the Celsius scale it is about 273° below zero, or –273°C.

Certain basic laws are based on the use of absolute temperatures. If a Fahrenheit reading is given, the addition of 460° to this reading will convert it to degrees Rankine or °R; whereas if the reading is from the Celsius scale, the addition of 273° will convert it to degrees Kelvin, °K. These conversions are shown in Fig. R4-4.

The Fahrenheit scale is associated with the Inch-Pound (IP) system of measurement, which is still predominant in the United States. The Celsius scale, formerly referred to as the centigrade scale, is part of the Standard International (SI) system of measurement, which is considered desirable for universal usage.

The Fahrenheit scale is based on using 32° as the melting point of ice and 212° as the boiling point of water, at standard atmospheric pressure. The range between these two temperatures is 180°. Scales can go higher or lower depending on the use of the instrument. For example, outdoor thermometers usually have scales from –60°F to 120°F. Indoor thermometers usually range from 50°F to 90°F.

On the Celsius scale the melting point of ice is 0° and the boiling point of water is 100°. The range between these two temperatures is 100°. On an outdoor thermometer the scale is usually from – 50°C to + 50°C. On an indoor thermometer the scale is usually from 10°C to 30°C.

These reference points (melting point of ice and boiling point of water) are based on one atmosphere of pressure, which is standard at sea level. This reference pressure in SI units is 101.325 kPa, which is exactly 1013.25 millibars. In I-P units the value is approximately 14.696 psi, or 29.921 inches of mercury at 32°F.

Referring to Fig. R4-4, it is interesting to note that –40°F is equivalent to –40°C. This is the only place where the two scales coincide.

Two other scales are used for scientific work, the Rankine (I-P) scale and the Kelvin scale (SI). Both of these scales start at a theoretical value called absolute zero. This is the lowest hypothetically possible temperature. There is no heat in a substance at this point.

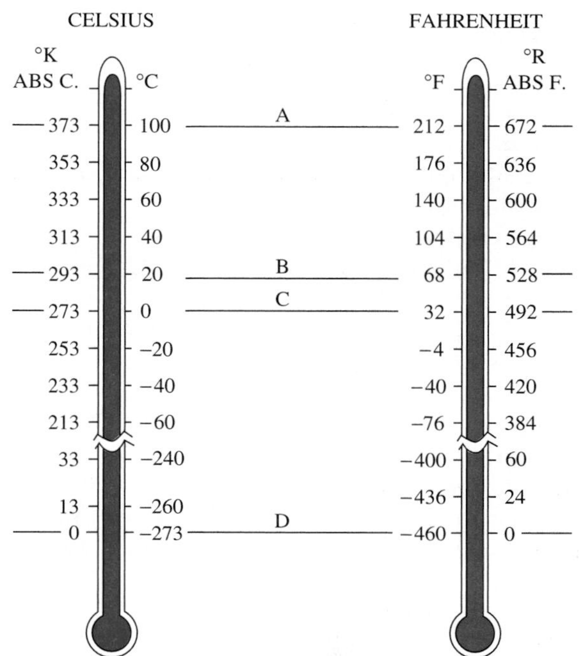

Figure R4-4 Fahrenheit, Celsius, Rankine, and Kelvin thermometer scales: (a) boiling temperature of water; (b) standard conditions temperature; (c) freezing temperature of water; (d) absolute zero.

To convert Fahrenheit temperatures to Celsius temperatures or the reverse, the following three methods can be used:

1. Use the conversion tables in Appendix 1
2. Calculate the conversion using the following formula:

$$°F = \frac{9}{5}(°C) + 32$$

or

$$°C = \frac{5}{9}(°F - 32)$$

3. Estimate the converted value, where precise accuracy is not required, by doubling the Celsius temperature and adding 30° to obtain the Fahrenheit temperature.

EXAMPLE

Suppose you observe a weather report that indicates a temperature in Chicago of 20°C. You are more familiar with Fahrenheit temperatures, so you would like to know the equivalent.

Solution

To convert 20°C to ?°F:

1. Using the tables, refer to the table in Appendix 1:
   Note that the table is made up of three columns.
   - The center column gives "Temperature to be converted"
   - The first column gives "Degrees F"
   - The third column gives "Degrees C"

   First, find 20° in the center column.
   Then project horizontally to the first column.
   Note 68.0°F.
   Therefore, 20°C = 68°F
2. Using the formula: °F = 9⁄5(°C) + 32°F =
   9⁄5(20°) + 32° = 36° + 32°F = 68°
   This is the same answer found using the tables.
3. Using an estimate: It would be done as follows:
   Double the Celsius degrees and add 30°.
   Doubling 20° is 40°. Adding 30° is 70°.
   Although 2 degrees off, it is reasonable that the Chicago temperature is within the comfort range. This value may be close enough for some uses.

Let's convert in the other direction. Suppose you are a Canadian looking at a U.S. weather report. You are familiar with Celsius temperatures. They report the weather in Seattle as 60°F. Would you feel comfortable without a coat? To convert 60°F to degrees C:

1. Using the table: Find 60° in the center column.
   Project over to the third column.
   Read 15.6°C.
   Therefore, 60°F = 15.6°C.
2. Using the formula: °C = 5⁄9(°F – 32) °C =
   5⁄9(60° – 32°) = 5⁄9(28°) °C = 15.6°
   This is the same answer found by using the tables.
3. Using the estimate:
   Subtract 30° from the degrees F then take half of it.
   Subtracting 30° from 60° is 30°. Half of 30° is 15°. °C = approximately 15°.
   This is a little cool. Better wear a coat.

EXAMPLE

What is the freezing point of water in degrees Rankine (°R)?

Solution

Since the freezing point of water is 32°F, adding 460° makes the freezing point of water 492° Rankine (32° + 460° = 492°).

R4-4 MATTER AND HEAT BEHAVIOR

All matter is composed of small particles known as molecules, and the molecular structure of matter (as studied in chemistry) can be further broken down into atoms.

In Chapter E-19, you will learn how the atom is further divided into electrons, protons, and neutrons. In that area you will study electron theory, which concerns electric current flow—the movement of electrons through a conductor.

For the present we concern ourselves only with the molecule, the smallest particle into which any matter or substance can be broken down and still retain its identity. For example, a molecule of water (H_2O) is made up of two atoms of hydrogen and one atom of oxygen. If this molecule of water were broken down more, or divided further, into subatomic particles, it would no longer be water.

Molecules vary in shape, size, and weight. In physics we learn that molecules have a tendency to cling together. The character of the substance of matter itself is dependent on the shape, size, and weight of the individual mol-

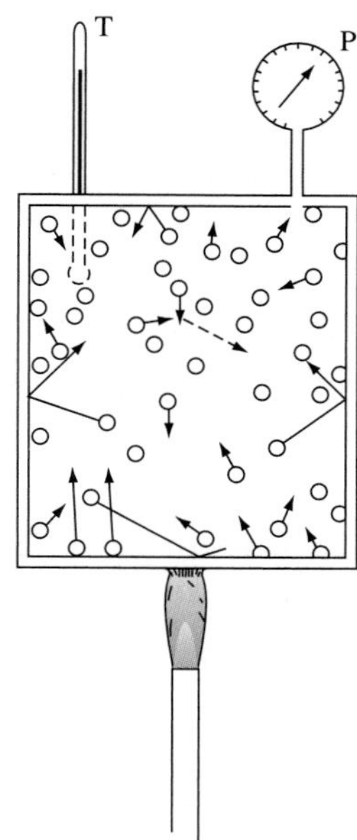

Figure R4-5 When heat is applied, the velocity of the molecules will increase.

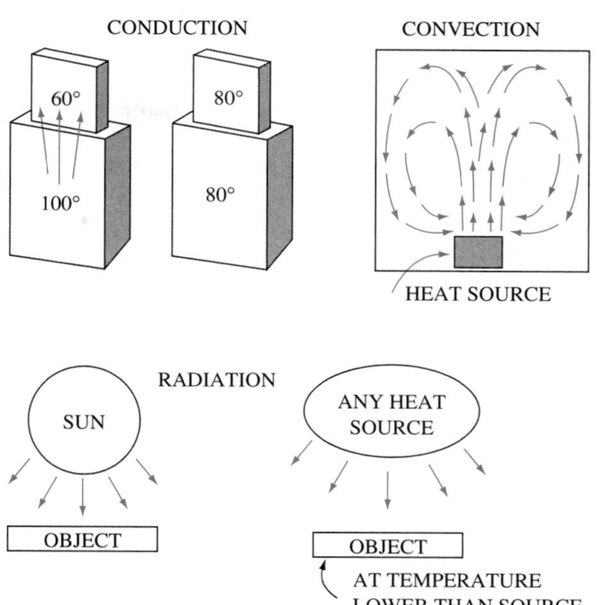

Figure R4-6 Three methods for transferring heat.

ecules of which it is made and also the space or distance between them, for they are, to a large degree, capable of moving about.

When heat energy is applied to a substance (Fig. R4-5), it increases the internal energy of the molecules, which increase their motion or velocity of movement. With this increase in movement of the molecules, there is also a rise or increase in the temperature of the substance.

When heat is removed from a substance, it follows that the velocity of the molecular movement will decrease and also that there will be a decrease or lowering of the internal temperature of this substance.

Heat is a form of energy that causes an increase in temperature of a body or parts of a body or its environment when it is added, and a decrease in temperature when it is removed, provided there is no change in state. When heat is used to change the state of a substance no temperature change takes place.

The transfer of heat takes place between two bodies of different temperature. The movement is usually considered to be from hot to cold.

R4-4.1 Transferring Heat

As we saw in Chapter R3, there are three principal ways that heat is transferred (Fig. R4-6): convection, conduction and radiation. Most refrigeration systems utilize all three methods. Further discussion of heat transfer is given later in this chapter.

In the I-P system the unit of heat is the British thermal unit (Btu). This is the amount of heat required to raise 1 pound of water 1 degree F at sea level, between the temperature of 59° and 60°F.

In the SI system, the unit used for heat is the Joule, which is the same as the unit for energy. Since the Joule is small it is common practice to use kiloJoules (kJ), equivalent to 1000 Joules. The amount of heat required to raise 1 kg of water 1°C is 4.187 kJ. A comparison of the two units is shown in Fig. R4-7.

Two useful conversion factors are:

$$1 \text{ J/g} = 0.4299 \text{ Btu/lb}$$
$$1 \text{ Btu/lb} = 2.326 \text{ J/g}$$

It is important in measuring heat to note the rate of change, since the amount of heat required to perform an operation is directly affected by the speed of the process. In I-P terms we use Btu per hour or abbreviated "Btuh" to include time as a factor. In the SI system the element of time is included in the term of Watts, a Watt being equivalent to Joules per second (J/s). Either Btuh or Watts, can, therefore, be used to rate the capacity of a refrigeration, heating or cooling unit. The conversion factors are:

$$1 \text{ Watt} = 3.412 \text{ Btuh}$$
$$1 \text{ Btuh} = 0.293 \text{ Watts}$$

If a building has a heat loss of 100,000 Btuh, in SI units this would be equivalent to 100,000/3.412 which is equal to 29,300 W.

The units Btuh and Watt are also used in the measurement of conductivity, expressed as Btuh/ft/°F or W/m/°C.

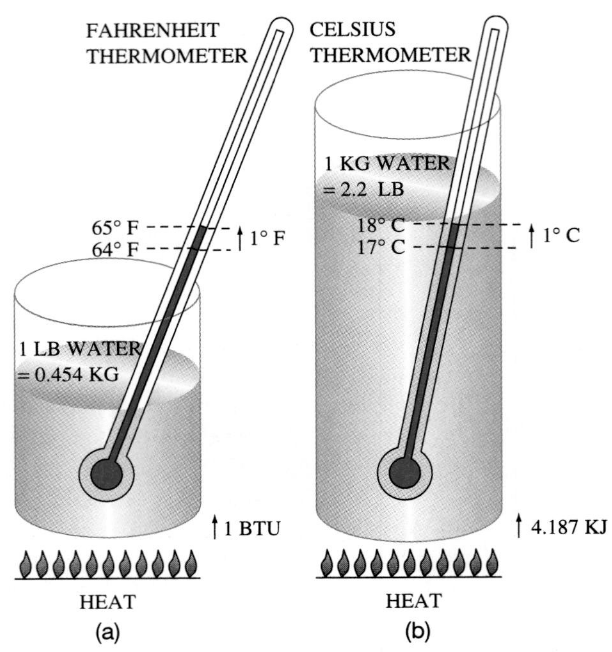

Figure R4-7 Comparison of Btu and kiloJoules (kJ). (a) Raising water 64°F to 65°F requires 1 Btu; (b) 1 kg of water raised from 17°C to 18°C requires 4.187 kJ.

R4-4.2 Sensible and Latent Heat

There are two distinctly different types of heat: *sensible heat,* where the change can be sensed by a thermometer and *latent heat,* which is not sensed by a thermometer but is required to change the state of a substance.

For an example of latent heat refer to Fig. R4-8.

The heat added to ice at 32°F (0°C) to change it into water at the same temperature is 144 Btu/lb (335 kJ/kg). This added heat changes the state but not the temperature.

The formula used to determine the sensible heat added or removed from a substance is:

$$Q = Wt \times SH \times TD$$

where

Q = Quantity of heat
Wt = Weight of the material
SH = Specific heat
TD = Temperature difference

Specific heat in the I-P system is the amount of heat required to raise 1 pound of a substance 1°F. In the SI system, specific heat is the amount of heat required to raise 1 kg of a substance 1°C. Commonly used specific heat values are shown in Fig. R4-9.

The formula used to determine the latent heat added or removed from a substance is:

$$Q = Wt \times LH$$

where

LH is the change in latent heat per unit of weight.

For example, if 10 lbs (4.5 kg) of water at 212°F (100°C) is changed to steam at 212°F (100°C), the latent heat that is added is 970 Btu per pound of water (I-P) or 2260 kJ/kg (SI).

The total heat added is:

$$Q = Wt \times LH = 10 \text{ lb} \times 970 \text{ Btu} = 9700 \text{ Btu}$$

or

$$Q = \text{Mass} \times LH = 4.5 \text{ kg} \times 2260 \text{ kJ/kg} = 10{,}170 \text{ kJ}$$

R4-4.3 Temperature-Heat Diagrams

Temperature-heat diagrams for water are shown in Fig. R4-8 for both the I-P and SI units.

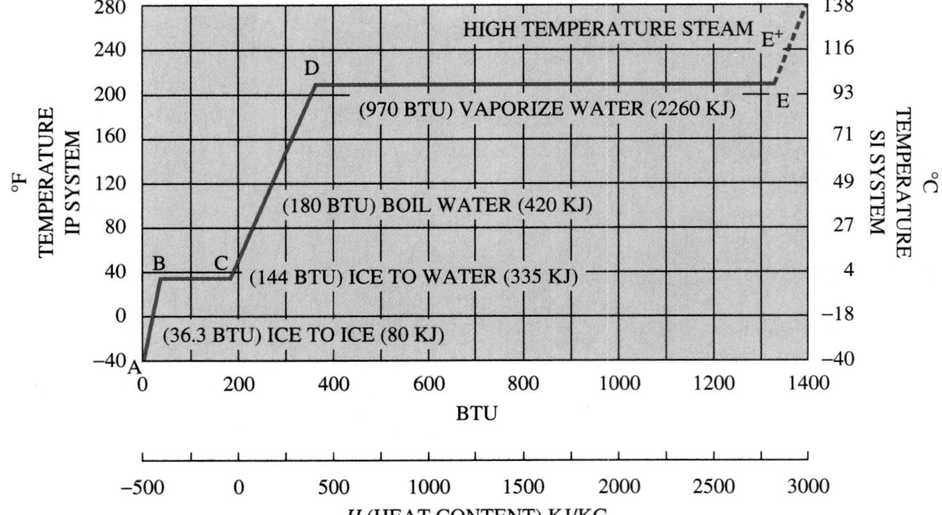

Figure R4-8 Temperature/Heat diagram for one pound (kg) water at atmospheric pressure, −40 to vaporization, in both IP and SI systems.

Water	1.00
Ice	0.50
Air (dry)	0.24
Steam	0.48
Aluminum	0.22
Brass	0.09
Lead	0.03
Iron	0.10
Mercury	0.03
Copper	0.09
Alcohol	0.60
Kerosene	0.50
Olive oil	0.47
Glass	0.20
Pine	0.67
Marble	0.21

Figure R4-9 Specific heats of common substances (Btu/lb/°F).

Referring first to the I-P diagram:

A—B represents the sensible heat added to ice at –40°F to raise it to the melting temperature of 32°F.

Btu = 0.504 (specific heat) × 72°F (*TD*) = 36.3

B—C represents the latent heat required to melt the ice.

Btu = 144

C—D represents the sensible heat added to the melted water at 32° to raise it to steam temperature at 212°F.

Btu = 1 (spec. heat of water) × 180° (*TD*) = 180 Btu

D—E represents the latent heat required to change 212°F water to steam at 212°F.

Btu = 970

E^+ represents superheated steam.

Referring to the SI system diagram:

A—B represents the heat added to ice at –40°C to raise it to 0°C.

kJ = 2 kJ/kg (specific heat of ice) × 40 (*TD*) = 80

B—C is the latent heat added to melt ice at 0° C.

kJ = 335

C—D represents the sensible heat required to raise water at 0°C to the steaming temperature of 100°C.

kJ = 4.2 kJ/kg (specific heat) × 100 (*TD*) = 420 kJ

D—E represents the latent heat required to change water to steam at 100°C.

kJ = 2260

E^+ represents superheated steam.

R4-5 HEAT QUANTITY AND MEASUREMENT

Heat quantity is different from heat intensity, because it takes into consideration not only temperature of the fluid or substance being measured, but also its weight. The unit of heat quantity is the British thermal unit (Btu). As we have seen, a Btu is the amount of heat required to change the temperature of 1 pound of water 1°F at sea level.

Two Btu will cause a change in temperature of 2°F of 1 pound of water; or it will cause a change in temperature of 1°F of 2 pounds of water; therefore, when considering a change in temperature of water, the following equation may be utilized:

$$\text{Btu} = W \times \Delta T \qquad \text{(R4-1)}$$

where

Change in heat (in Btu)
= weight (in pounds) × temperature difference

EXAMPLE

Calculate the amount of heat necessary to increase 10 lb of water from 50°F to 100°F.

Solution

Since the temperature difference (ΔT) = 50°, then heat = Btu = $W \times \Delta T = 10 \times 50 = 500$ Btu.

In the example above, heat is added to the quantity of water, but the same equation is also used if heat is to be removed.

EXAMPLE

Calculate the amount of heat removed if 20 lb of water is cooled from 80°F to 40°F.

Solution

Btu = $W \times \Delta T = 20 \times 40 = 800$ Btu

R4-5.1 Specific Heat

The specific heat of a substance is the quantity of heat in Btu required to change the temperature of 1 pound of the substance 1 degree Fahrenheit. Earlier we were presented with the information that 1 Btu was the amount of heat necessary to change the temperature of 1 pound of water

1 degree Fahrenheit, or to change the temperature of the same weight of water by the same unit of measurement on a thermometer.

The specific heat of water is therefore 1.0; and water is the basis for the specific heat table in Fig. R4-9. You will see that different substances vary in their capacity to absorb or to give up heat. The specific heat values of most substances will vary with a change in temperature; some vary only a slight amount, whereas others can change considerably.

Suppose that two containers are placed on a heating element or burner side by side, one containing water and the other an equal amount, by weight, of olive oil. You would soon find that the temperature of the olive oil increases at a more rapid rate than that of the water.

If the rate of temperature increase of the olive oil were approximately twice that of the water, it could be said that olive oil required only half as much heat as water to increase its temperature 1°F. Based on the value of 1.0 for the specific heat of water, the specific heat of olive oil must be approximately 0.5, or half that of water. (The table of specific heats of substances shows that olive oil has a value of 0.47.)

Equation R4-1 can now be stated as

$$\text{Btu} = W \times c \times \Delta T \qquad \text{(R4-2)}$$

where

c = specific heat of a substance

EXAMPLE

Calculate the amount of heat required to raise the temperature of 1 lb of olive oil from 70°F to 385°F.

Solution

Since $\Delta T = 315°$ and c of olive oil = 0.47, then

$$\begin{aligned} \text{heat} = \text{Btu} &= W \times c \times \Delta T \\ &= 1 \times 0.47 \times 315 = 148 \text{ Btu} \end{aligned}$$

The specific heat of a substance will also change with a change in the state of the substance. Water is a very good example of this variation in specific heat. We have learned that, as a liquid, its specific heat is 1.0, but as a solid (ice) its specific heat approximates 0.5, and this same value is applied to steam (the gaseous state).

Within the refrigeration circuit we will be interested, primarily, with substances in liquid or gaseous form, and their ability to absorb or give up heat. Also, in the distribution of air for the purpose of cooling or heating a given area, we will be interested in the possible changes in the values for specific heat—more about this later.

Air, when heated and free or allowed to expand at a constant pressure, will have a specific heat of 0.24. Refrigerant-12 (R-12) vapor at approximately 70°F and at constant pressure, has a specific heat value of 0.148, whereas the specific heat of R-12 liquid is 0.24 at 86°F. When dealing with metals and the temperature of a mixture, a combination of weights, specific heats, and temperature differences must be considered in the overall calculations of heat transfer. In order to calculate the total heat transfer of a combination of substances, it is necessary to add the individual rates as follows:

$$\begin{aligned} \text{Btu} = &(W_1 \times c_1 \times \Delta T_1) \\ &+ (W_2 \times c_2 \times \Delta T_2) \\ &+ (W_3 \times c_3 \times \Delta T_3) \ldots \text{etc.} \end{aligned} \qquad \text{(R4-3)}$$

EXAMPLE

How much heat must be added to a 10-lb copper vessel, holding 30 lb of water at 70°F to reach 185°F if the specific heat of copper is 0.095°F?

Solution

Equation (R4-3) may be used as follows:

$$\text{Btu} = (W_1 \times c_1 \times \Delta T_1) + (W_2 \times c_2 \times \Delta T_2)$$

where

W_1, c_1 and ΔT_1 pertain to the copper vessel, and W_2, c_2, and ΔT_2 pertain to the water. Therefore,

$$\begin{aligned} \text{Btu} &= (10 \times 0.095 \times 115) + (30 \times 1.0 \times 115) \\ &= 109.2 + 3450 \\ &= 3559.2 \text{ Btu} \end{aligned}$$

R4-5.2 Change of State

We are now ready to look at the five principal changes of state:

Solidification: a change from a liquid to a solid
Liquefaction: a change from a solid to a liquid
Vaporization: a change from a liquid to a vapor
Condensation: a change from a vapor to a liquid
Sublimation: a change from a solid to a vapor without passing through the liquid state.

When a solid substance is heated, the molecular motion is chiefly in the form of rapid motion back and forth, the molecules never moving far from their normal or original position. But at some given temperature for that particular substance, further addition of heat will not necessarily increase the molecular motion within the substance; instead, the additional heat will cause some solids to liquefy (change into a liquid). Thus the additional heat causes a change of state in the material. The temperature at which this change of state in substance takes place is called its melting point. Let us assume that a container of water at 70°F, in which a thermometer has been placed, is left in a freezer for hours (Fig. R4-10). When it is taken from the freezer, it has become a block of ice—solidification has taken place.

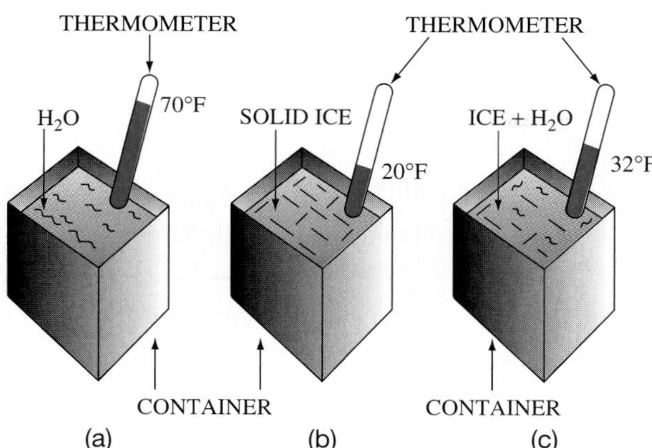

Figure R4-10 H_2O changes temperature and changes state as it absorbs sensible and latent heat.

KNOWLEDGE EXTENSION

Observing dry ice is a good way to learn about sublimation—a change of state from a solid to a vapor. Order some dry ice for your next camping trip for your food cooler. (*Caution:* Don't seal your cooler too tight or it might rupture when the dry ice changes state.) Observe the ice during the initial cool-down period. Wait twelve hours and look for water as the dry ice evaporates. You should not see any water from the dry ice evaporation.

Let us further assume that the thermometer in the block indicates a temperature of 20°F. If it is allowed to stand at room temperature, heat from the room air will be absorbed by the ice until the thermometer indicates a temperature of 32°F, when some of the ice will begin to change into water.

With heat continuing to transfer from the room air to the ice, more ice will change back into water; but the thermometer will continue to indicate a temperature of 32°F until all the ice has melted. Liquefaction has now taken place. (This change of state without a change in the temperature will be discussed later.)

As mentioned, when all the ice is melted the thermometer will indicate a temperature of 32°F, but the temperature of the water will continue to rise until it reaches or equals room temperature.

If sufficient heat is added to the container of water through outside means (Fig. R4-11), such as a burner or a torch, the temperature of the water will increase until it reaches 212°F. At this temperature, and under "standard" atmospheric pressure, another change of state will take place—vaporization. Some of the water will turn into steam, and with the addition of more heat, all of the water will vaporize into steam, yet the temperature of the water will not increase above 212°F. This change of state—without a change in temperature—will also be discussed later.

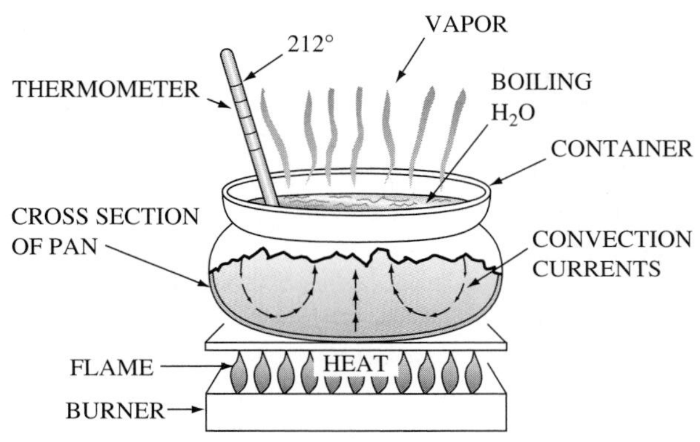

Figure R4-11 Convection currents caused by temperature differential.

If the steam vapor could be contained within a closed vessel, and if the source of heat were removed, the steam would give up heat to the surrounding air, and it would condense back into a liquid form—water. What has now taken place is condensation—the reverse process from vaporization.

Oxygen is a gas above −297°F, a liquid between that temperature and −324°F, and a solid below that point. Iron is a solid until it is heated to 2800°F, and vaporizes at a temperature approximating 4950°F.

Thus far we have learned, with some examples, how a solid can change into a liquid, and how a liquid can change into a vapor. But it is possible for a substance to undergo a physical change through which solid will change directly into a gaseous state without first melting into a liquid. This is known as *sublimation.*

All of us probably have seen this physical change take place without fully recognizing the process. Damp or wet clothes, hanging outside in the freezing temperature, will speedily become dry through sublimation; just as dry ice (solid carbon dioxide, or CO_2) sublimes into a vapor under normal temperature and pressure.

R4-5.3 Sensible Heat

Heat that can be felt or measured is called *sensible heat.* It is the heat that causes a change in temperature of a substance, but not a change in state. Substances, whether in a solid, liquid, or gaseous state, contain sensible heat to some degree, as long as their temperatures are above absolute zero. Equations used for solutions of heat quantity, and those used in conjunction with specific heats, might be classified as being sensible heat equations, since none of them involve any change in state.

As mentioned earlier, a substance may exist as a solid, liquid, or as a gas or vapor. The substance as a solid will

ON THE JOB

Tons of refrigeration and Btu/hr is the standard nomenclature for indicating the amount of heat transfer ability for refrigeration equipment. When you order replacement compressors, evaporators, or condensers from your local refrigeration supply house be prepared to give the sales staff information in Btu's, tonnage, operating pressures, temperatures, horsepower, manufacturer, refrigerant type, and model number. Most larger equipment is rated in tonnage and not btu/hr. It is not uncommon for the tonnage to be eliminated on equipment identification tags. The manufacturer's model number will usually provide a crossreference for determining tonnage available at specified pressures and temperatures.

contain some sensible heat, as it will in the other states of matter. The total amount of heat needed to bring it from a solid state to a vapor state is dependent on (1) its initial temperature as a solid, (2) the temperature at which it changes from a solid to a liquid, (3) the temperature at which it changes from a liquid to a vapor, and (4) its final temperature as a vapor. Also included is the heat that is required to effect the two changes in state.

R4-5.4 Latent Heat

Under a change of state, most substances will have a melting point at which they will change from a solid to a liquid without any increase in temperature. At this point, if the substance is in a liquid state and heat is removed from it, the substance will solidify without a change in its temperature. The heat involved in either of these processes (changing from a solid to a liquid, or from a liquid to a solid), without a change in temperature, is known as the *latent heat of fusion.*

Fig. R4-12 shows the relationship between temperature in Fahrenheit degrees and both sensible and latent heat in Btu.

As pointed out earlier, the specific heat of water is 1.0 and that of ice is 0.5, which is the reason for the difference in the slopes of the lines denoting the solid (ice) and the liquid (water). To increase the temperature of a pound of ice from 0°F to 32°F requires only 16 Btu of heat; the

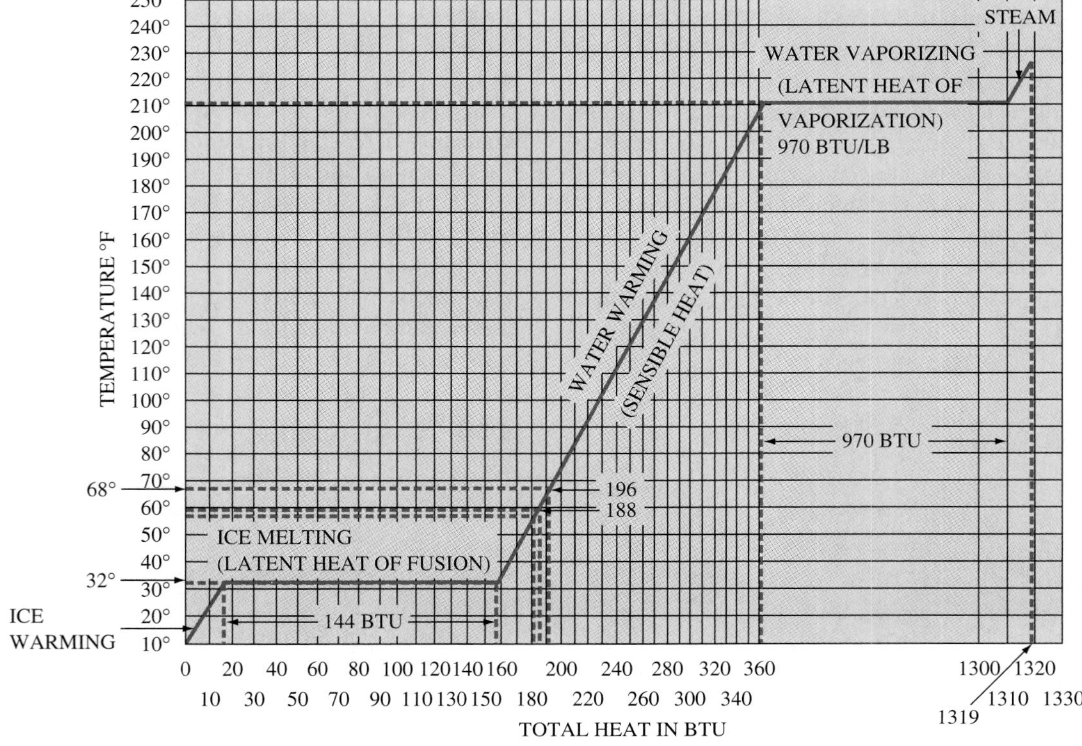

Figure R4-12 Chart demonstrating sensible and latent heat relationships in melting ice, changing ice to water and water to steam.

other line shows that it takes only 8 Btu to increase the temperature of a pound of water 8°F (60°F to 68°F), from 188 to 196 Btu/lb.

Fig. R4-12 also shows that a total of 52 Btu of sensible heat is involved in the 196 Btu necessary for converting a pound of 0°F ice to 68°F water. This leaves a difference of 144 Btu, which is the latent heat of fusion of water or ice, depending on whether heat is being removed or added.

The derivation of the word *latent* is from the Latin word for hidden. This is hidden heat, which does not register on a thermometer, nor can it be felt. Needless to say, there is no increase or decrease in the molecular motion within the substance, for it would show up in a change in temperature on a thermometer.

$$\begin{aligned}\text{Btu} = &(W_1 \times c_1 \times \Delta T_1)\\ &+ (W_1 \times \text{latent heat})\\ &+ (W_2 \times c_2 \times \Delta T_2)\end{aligned} \qquad \text{(R4-4)}$$

EXAMPLE

Calculate the amount of heat needed to change 10 lb of ice at 20°F to water at 50°F.

Solution

Utilizing Eq. (R4-4), we have

$$\begin{aligned}\Delta T_1 &= (32° - 20°) = 12°\\ \Delta T_2 &= (50° - 32°) = 18°\\ \text{Btu} &= (10 \times 0.5 \times 12) + (10 \times 144)\\ &\quad + (10 \times 1.0 \times 18)\\ &= 60 + 1440 + 180\\ &= 1680\end{aligned}$$

Another type of latent heat that must be taken into consideration when total heat calculations are necessary is called the *latent heat of vaporization.* This is the heat that 1 pound of a liquid absorbs while being changed into the vapor state. It can also be classified as the latent heat of condensation, for when sensible heat is removed from the vapor to the extent that it reaches the condensing point, the vapor condenses back into the liquid form.

The latent heat of vaporization of water as 1 pound is boiled or evaporated into steam at sea level is 970 Btu. That amount also is the heat that 1 pound of steam must release or give up when it condenses into water. Fig. R4-12 also shows the relationship between temperature and both sensible heat and latent heat of vaporization.

The absorption of the amount of heat necessary for the change of state from a liquid to a vapor by evaporation, and the release of that amount of heat necessary for the change of state from a vapor back to a liquid by condensation are the main principles of the refrigeration process, or cycle. Refrigeration is the transfer of heat by the change in state of the refrigerant.

Fig. R4-12 shows the total heat in Btu necessary to convert 1 pound of ice at 0°F to superheated steam at 230°F under atmospheric pressure. This total amounts to 1319 Btu. Only 205 Btu is sensible heat; the remainder is made up of 144 Btu of latent heat of fusion and also 970 Btu of latent heat of vaporization.

R4-5.5 Refrigeration Effect Measurement

A common term that has been used in refrigeration work to define and measure capacity or refrigeration effect is called a ton or ton of refrigeration. It is the amount of heat absorbed in melting a ton of ice (2000 lb) over a 24-hour period.

The ton of refrigeration is equal to 12,000 Btu per hour. This may be calculated by multiplying the weight of ice (2000 lb) by the latent heat of fusion (melting) of ice (144 Btu/lb). Thus

$$2000 \text{ lb} \times 144 \text{ Btu/lb} = 288{,}000 \text{ Btu}$$

in 24 hours or 12,000 Btu per hour (288,000/24). Therefore, one ton of refrigeration equals 12,000 Btu/hr.

A 10-ton refrigerating system will have a capacity of 10 × 12,000 Btu/hr = 120,000 Btu/hr.

R4-6 HEAT TRANSFER

The second law of thermodynamics, discussed earlier, states that heat transfers (flows) in one direction only, from a high temperature (intensity) to a lower temperature (intensity). This transfer will take place using one or more of the following basic methods of transfer:

1. Conduction
2. Convection
3. Radiation

R4-6.1 Conduction

Conduction is described as the transfer of heat between the closely packed molecules of a substance, or between substances that are in good contact with one another. When the transfer of heat occurs in a single substance, such as a metal rod with one end in a fire or flame, movement of heat continues until there is a temperature balance throughout the length of the rod.

If the rod is immersed in water, the rapidly moving molecules on the surface of the rod will transmit some heat to the molecules of water, and still another transfer of heat by conduction takes place. As the outer surface of the rod cools off, there is still some heat within the rod, and this will continue to transfer to the outer surfaces of the rod and then to the water, until a temperature balance is reached.

PRACTICING PROFESSIONAL SERVICES

Heat is transferred by either conduction, convection, or radiation. Very often it is transferred by a combination of all three methods. When you are on a job site, learn to look and feel for heat transfer from the equipment. Use pocket thermometers to give approximate indications of coil supply and return air and/or temperature differences. There should be substantial temperature differences between the two temperatures if the coils are working properly. Consult with the manufacturer's technical reference materials to determine exact coil design information. If the coil is not providing the temperature difference as specified, then it is an early indication of potential heat transfer problems. Look for obstructed air flow, dirty coils, broken fans, low refrigerant levels, or possibly noncondensibles in the refrigerant. Remedy the problem, and always check the air temperature upon job completion.

The speed with which heat will transfer by means of conduction will vary with different substances or materials if the substances or materials are of the same dimensions. The rate of heat transfer will vary according to the ability of the material or substances to conduct heat. Solids, on the whole, are much better conductors than liquids; in turn, liquids conduct heat better than gases or vapors.

Most metals, such as silver, copper, steel, and iron, conduct heat fairly rapidly, whereas other solids, such as glass, wood, or other building materials, transfer heat at a much slower rate and therefore are used as insulators.

Copper is an excellent conductor of heat, as is aluminum. These substances are ordinarily used in the evaporators, condensers, and refrigerant pipes connecting the various components of a refrigerant system, although iron is occasionally used with some refrigerants.

The rate at which heat may be conducted through various materials is dependent on such factors as (1) the thickness of the material, (2) its cross-sectional area, (3) the temperature difference between the two sides of the material, (4) the heat conductivity (*k* factor) of the material, and (5) the time duration of the heat flow. Fig. R4-13 is a table of heat conductivities (*k* factors) of some common materials.

Note: The *k* factors are given in Btu/hr/ft^2/°F/in. of thickness of the material. These factors may be utilized correctly through the use of this equation:

$$\text{Btu} = \frac{A \times k \times \Delta T}{X} \qquad \text{(R4-5)}$$

where

A = cross-sectional area, ft^2
k = heat conductivity, Btu/hr/ft^2/°F/in.
ΔT = temperature difference between the two sides, °F
X = thickness of material, in.

Metals with a high conductivity are used within the refrigeration system itself because it is desirable that rapid heat transfer occur in both evaporator and condenser. The evaporator is where heat is removed from the conditioned space or substance or from air that has been in direct contact with the substance; the condenser dissipates this heat to another medium or space.

In the case of the evaporator, the product or air is at a higher temperature than the refrigerant within the tubing and there is a transfer of heat downhill. In the condenser, however, the refrigerant vapor is at a higher temperature than the cooling medium traveling through or around the condenser, and here again there is a downhill transfer of heat.

Plain tubing, whether copper, aluminum, or another metal, will transfer heat according to its conductivity or *k* factor, but this heat transfer can be increased through the addition of fins on the tubing. They will increase the area of heat transfer surface, thereby increasing the overall efficiency of the system. If the addition of fins doubles the surface area, it can be shown by the use of Eq. (R4-5) that the overall heat transfer should itself be doubled compared to that of plain tubing.

Material	Conductivity *k*
Plywood	0.80
Glass fiber—organic bonded	0.25
Expanded polystyrene insulation	0.25
Expanded polyurethane insulation	0.16
Cement mortar	5.0
Stucco	5.0
Brick (common)	5.0
Hard woods (maple, oak)	1.10
Soft woods (fir, pine)	0.80
Gypsum plaster (sand aggregate)	5.6

Figure R4-13 Conductivities for common building and insulating materials. *k* values expressed in Btu/hr/ft^2/°F/inch thickness of material.

R4-6.2 Convection

Another means of heat transfer is by motion of the heated material itself and is limited to liquid or gas. When a material is heated, *convection* currents are set up within it, and the warmer portions of it rise, since heat brings about the decrease of a fluid's density and an increase in its specific volume.

Air within a refrigerator and water being heated in a pan are prime examples of the results of convection currents, as shown in Fig. R4-14. The air in contact with the cooling coil of a refrigerator becomes cool and therefore more dense, and begins to fall to the bottom of the refrigerator. In doing so, it absorbs heat from the food and the walls of the refrigerator, which, through conduction, has picked up heat from the room.

After heat has been absorbed by the air it expands, becoming lighter, and rises until it again reaches the cooling coil where heat is removed from it. The convection cycle repeats as long as there is a temperature difference between the air and the coil. In commercial-type units, baffles may be constructed within the box in order that the convection currents will be directed to take the desired patterns of airflow around the coil.

Water heated in a pan will be affected by the convection currents set up within it through the application of heat. The water nearest the heat source, in absorbing heat, becomes warmer and expands. As it becomes lighter, it rises and is replaced by the other water, which is cooler and more dense. This process will continue until all of the water is at the same temperature.

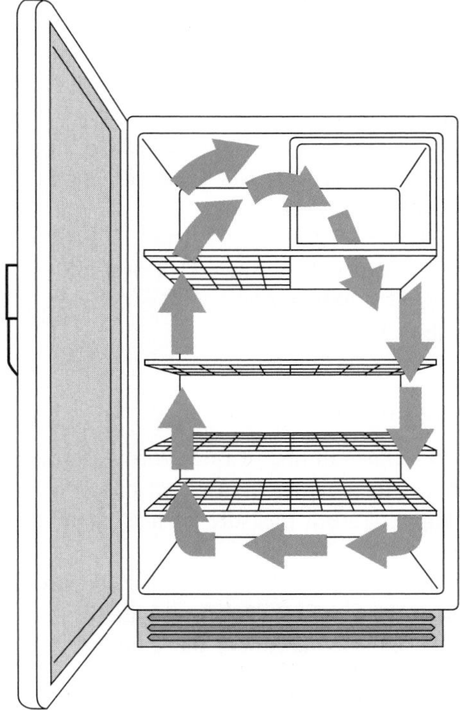

Figure R4-14 Convection currents caused by temperature differential.

Convection currents as explained and shown here are natural (passive), and as in the case of the refrigerator, a natural (passive) flow is a slow flow. In some cases, convection must be increased through the use of fans or blowers and, in the case of liquids, pumps are used for forced (active) circulation to transfer heat from one place to another.

R4-6.3 Radiation

A third means of heat transfer is through *radiation* by waves similar to light or sound waves. The sun's rays heat the earth by means of radiant heat waves, which travel in a straight path without heating the intervening matter or air. The heat from a light bulb or hot stove is radiant in nature and felt by those near them, although the air between the source and the object, which the rays pass through, is not heated.

If you have been relaxing in the shade of a building or a tree on a hot sunny day and move into direct sunlight, the direct impact of the heat waves will hit like a sledgehammer, even though the air temperature in the shade is approximately the same as in the sunlight.

At low temperatures there is only a small amount of radiation, and only minor temperature differences are noticed; therefore, radiation has very little effect in the actual process of refrigeration itself. But results of radiation from direct solar rays can cause an increased refrigeration load in a building in the path of these rays.

Radiant heat is readily absorbed by dark or dull materials or substances, whereas light-colored surfaces or materials will reflect radiant heat waves, just as they do light rays. Wearing-apparel designers and manufacturers make use of this proven fact by supplying light-colored materials for summer clothes.

This principle is also carried over into the summer air-conditioning field, where light-colored roofs and walls allow less of the solar heat to penetrate into the conditioned space, thus reducing the size of the overall cooling equipment required. Radiant heat also readily penetrates clear glass in windows, but will be absorbed by translucent or opaque glass.

When radiant heat or energy (since all heat is energy) is absorbed by a material or substance, it is converted into sensible heat—that which can be felt or measured. Every body or substance absorbs radiant energy to some extent, depending on the temperature difference between the specific body or substance and other substances. Every substance will radiate energy as long as its temperature is above absolute zero and another substance within its proximity is at a lower temperature.

If an automobile has been left out in the hot sun with the windows closed for a long period of time, the temperature inside the car will be much greater than the ambient air temperature surrounding it. This demonstrates that radiant energy absorbed by the materials of which the car is constructed is converted to measurable sensible heat.

R4-7 INSULATION

Certain substances are excellent conductors of heat, while others are poor conductors, and may be classified as insulators. Any material that deters or helps to prevent the transfer of heat by any means is called and may be used as *insulation.* Of course, no material will stop the flow of heat completely. If there were such a substance, it would be very easy to cool a given space down to a desired temperature and keep it there.

Such substances as cork, glass fibers, mineral wool, and polyurethane foams are good examples of insulating materials, but numerous other substances are used in insulating refrigerated spaces or buildings. The compressible materials, such as fibrous substances, offer better insulation if installed loosely packed or in blanket or batt form than if they are compressed or tightly packed.

The thermal conductivity of materials, the temperature to be maintained in the refrigerated space, the ambient temperature surrounding the enclosed space, permissible wall thicknesses of insulating materials, and the cost of the various types of insulation are all points to consider in selecting the proper material for a given project. Most service personnel are not involved in the selection or the installation of insulating material in a refrigeration application, but they may come in contact with different types of insulation, and under various conditions.

Insulation should be fire- and moisture-resistant, and also vermin-proof. Large refrigeration boxes or walk-in types of coolers are usually insulated with a rigid type of insulation such as corkboard, fiberglass, foam blocks, and the like, whereas smaller boxes or receptacles might be filled or insulated with a foam type that flows like a liquid and expands to fill up the available cavity with foam. Low-temperature boxes require an insulation that is also vapor resistant, such as unicellular foam, if the walls of the refrigerated enclosure are not made of metal on the outside. Otherwise, water vapor could penetrate the insulation and condense there, reducing the insulating efficiency.

REVIEW

- **Matter** is anything that has mass and weight and occupies a space.
- **Heat** is a form of energy, the result of molecular action.
- **Cold** is the absence of heat.
- Three States of Matter: solid, liquid, and gas (vapor).
- Temperature is a measure of the intensity of sensible heat.
- Reference Points of Temperature:
  Boiling point of water: 212 degrees
  Freezing point of water: 32 degrees
  Critical temperature: temperature at which vapor and liquid have the same properties.
- **Latent heat** is heat energy absorbed in the process of changing state, without a change in temperature or pressure.
- **Specific heat** is the ratio of heat required to raise the temperature of a substance one degree to that required to raise the temperature of an equal mass of water one degree.
- A **conductor** is a substance that easily conducts heat.
- A **Btu** is a British thermal unit, the amount of heat required to raise the temperature of one pound of water one degree Fahrenheit.
- **Calorie** is a unit of heat energy, specific heat in the SI system; the amount of heat required to increase the temperature of one gram of water, one degree Celsius.
- A **Joule** is the SI system unit of measurement for heat.
- **Conduction** is the transfer of heat between molecules within a substance, or between substances that are touching in good contact with one another.
- **Convection** is the transfer of heat using the flow of another medium to transfer the heat.
- **Insulation** is any material that deters or helps to prevent the transfer of heat by any means.
- **Solidification** is a change from a liquid to a solid.
- **Liquefaction** is a change from a solid to a liquid.
- **Vaporization** is a change from a liquid to a vapor.
- **Condensation** is a change from a vapor to a liquid.
- **Sublimation** is a change from a solid to a vapor without passing through the liquid state.
- Conversion Formulas:
  $°C = \frac{5}{9}(°F - 32)$; $°C = (°F - 32)/1.8$
  $°F = \frac{9}{5}(°C) + 32$; $°F = 1.8(°C) + 32$
  $Btu = (W) \times (c) \times \Delta T$
  Btu = kJ/1.055; kJ = 1.055 × Btu
  Calories = Btu × 252; Btu = Calories/252
  1 Joule/g = .4299 Btu/lb
  1 Btu/lb = 2.326 J/gram

Problems and Questions

1. Matter exists in one of three forms. Name them.
2. What is the smallest particle of matter?
3. Is temperature a measure of heat quantity or intensity?
4. What is meant by the term "absolute zero"?
5. Define the first law of thermodynamics.
6. Define the second law of thermodynamics.
7. Define a British thermal unit (Btu).
8. Convert 68°F to the Celsius scale.
9. How much heat is required to raise the temperature of 100 lb of water from 70°F to 120°F?

10. If 750 Btu is supplied to 15 lb of water at 72°F, what will be the resulting temperature?
11. What is the specific heat value of water?
12. Latent heat can be measured on a thermometer. True or false?
13. Heat energy travels by one or more of three ways. Name them.
14. The heat that changes the temperature of a substance is called _____.
15. The heat that changes the state of a substance is called _____.
16. A substance that can change from a solid directly into a vapor is called a _____.
17. The heat that changes a liquid to a vapor is called the heat of _____.
18. Define a "ton" of refrigerating effect.
19. How many Btu are there in a standard ton of refrigerating effect?
20. Is insulation a good or a poor conductor of heat?
21. The method of heat transfer in which heat travels through a substance or is transferred between objects by direct contact is
    a. radiation
    b. conduction
    c. convection
    d. None of the above.
22. Sublimation
    a. is a change from a vapor to a solid without passing through the liquid state.
    b. is a change of state without a change of temperature, typically found at ice temperatures.
    c. is a change from a solid to a vapor without passing through the liquid state.
    d. None of the above.
23. A British thermal unit is defined as
    a. the amount of heat required to change the temperature of 1 pound of water 1 degree Fahrenheit at sea level.
    b. the amount of heat required to change the temperature of a product 1 degree Fahrenheit at sea level.
    c. the amount of heat necessary to change 1 pound of ice at 32 degrees to 1 pound of water at 32 degrees Fahrenheit.
    d. None of the above.
24. The formula to convert 20 degrees C to Fahrenheit is:
    a. $\frac{9}{5}$(C degrees) + 32
    b. $\frac{5}{9}$(F − 32)
    c. $\frac{9}{5}$(C + 32)
    d. None of the above.
25. The first law of thermodynamics states that
    a. to cause heat energy to travel, a temperature difference must be established and maintained.
    b. energy can neither be created nor destroyed; it can only be converted from one form to another.
    c. energy will always move from warmer bodies of energy to cooler bodies of energy.
    d. None of the above.

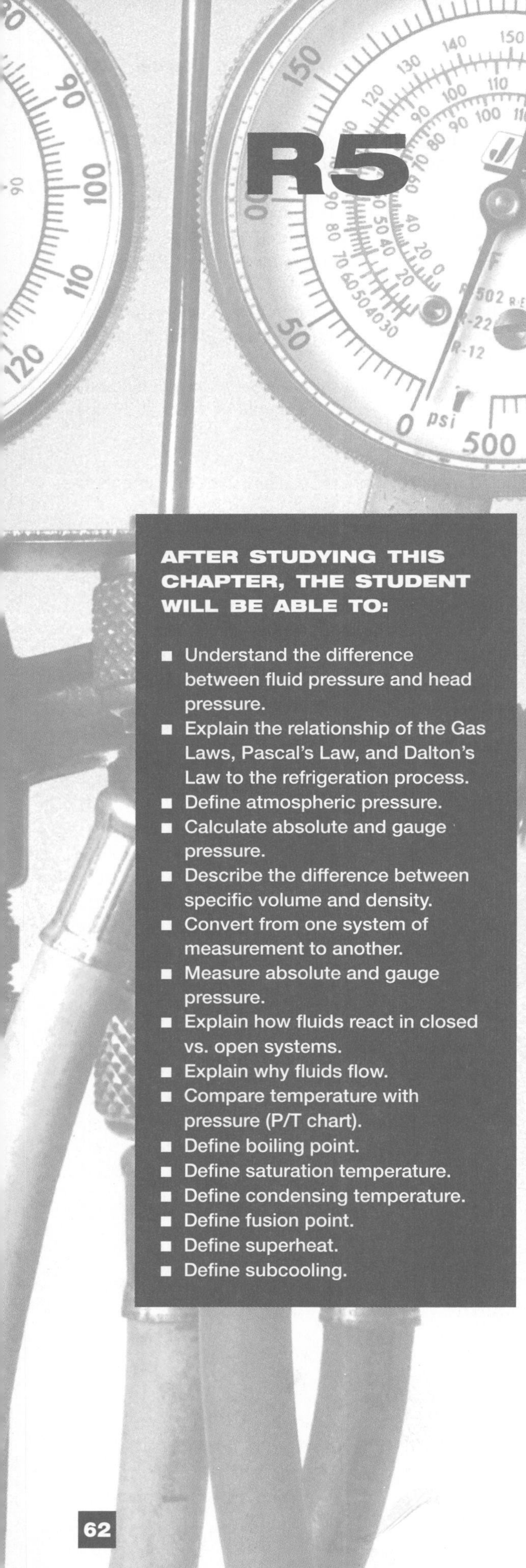

R5 Fluids and Pressure

AFTER STUDYING THIS CHAPTER, THE STUDENT WILL BE ABLE TO:

- Understand the difference between fluid pressure and head pressure.
- Explain the relationship of the Gas Laws, Pascal's Law, and Dalton's Law to the refrigeration process.
- Define atmospheric pressure.
- Calculate absolute and gauge pressure.
- Describe the difference between specific volume and density.
- Convert from one system of measurement to another.
- Measure absolute and gauge pressure.
- Explain how fluids react in closed vs. open systems.
- Explain why fluids flow.
- Compare temperature with pressure (P/T chart).
- Define boiling point.
- Define saturation temperature.
- Define condensing temperature.
- Define fusion point.
- Define superheat.
- Define subcooling.

R5-1 GENERAL

The dictionary describes a fluid as "any substance that can flow, liquid or gas." A refrigerant may therefore, be classified as a fluid, since within the refrigeration cycle, it exists both as a liquid and as a vapor or gas. Although, as previously mentioned, ice—a solid—is also used in heat removal, its use in refrigeration has been overshadowed somewhat by the discovery of the versatility of the chemicals and chemical combinations used as refrigerants today.

R5-2 FLUID PRESSURE

The weight of a block of wood or any other solid material acts as a force downward on whatever is supporting it. The force of this solid object is the overall weight of the object, and the total weight is distributed over the area on which it lies. The weight of a given volume of water, however, acts not only as a force downward on the bottom of the container holding it, but also as a force laterally on the sides of the container. If a hole is made in the side of the container below the water level shown in Fig. R5-1, the water above the hole will be forced out because of its force acting downward and sideways.

Fluid pressure is the force per unit area that is exerted by a gas or a liquid. It is usually expressed in terms of psi (pounds per square inch). It varies directly with the density and the depth of the liquid, and at the same depth below the surface, the pressure is equal in all directions. Notice the difference between the terms used: force and pressure. **Force** means the total weight of the substance; **pressure** means the unit force or pressure per square inch.

If the tank in Fig. R5-1 measures 1 foot in all dimensions, and it is filled with water, we have a cubic foot of water, the weight of which approximates 62.4 lb. Therefore, we would have a total force of 62.4 lb being exerted on the bottom of the tank, the area of which

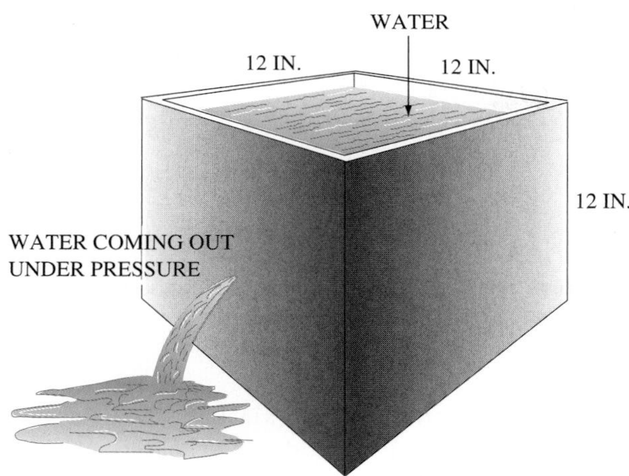

Figure R5-1 Water pressure in a container exerts pressure in all directions.

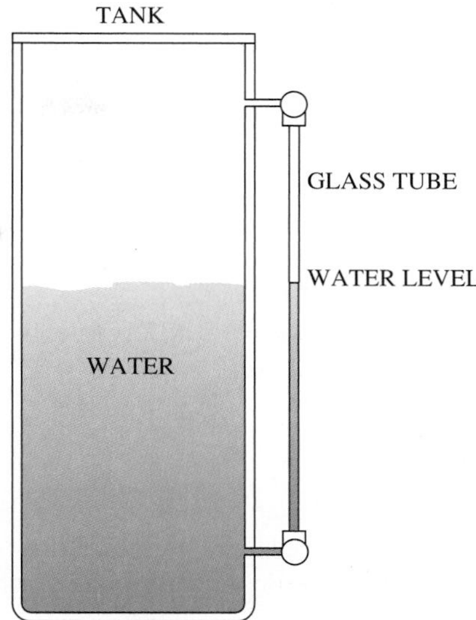

Figure R5-2 Fluid pressure same in tube as in tank.

equals 144 square inches (12 in. × 12 in. = 144 in.^{2}). Using the equation

$$\text{pressure} = \frac{\text{force}}{\text{area}} \tag{R5-1}$$

or

$$p = \frac{F}{A}$$

the unit pressure of the water will be

$$\text{unit pressure} = \frac{\text{force}}{\text{area}} = \frac{62.4 \text{ lb}}{144 \text{ in.}^2} = 0.433 \text{ psi}$$

This pressure of 0.433 psi is exerted downward and also sideways. If a tank is constructed as in Fig. R5-2, the pressure of the water would cause the tube to be filled with water to the same level as in the tank. Fluid pressure is the same on each square inch of the walls of the tank at the same depth, and it will act at right angles to the surface of the tank.

EXAMPLE

A tank is 4 ft square and is filled with water to its depth of 3 ft. Find (a) the volume of water, (b) the weight of the water, (c) the force on the bottom of the tank, and (d) the pressure on the bottom of the tank.

Solution

(a) Volume = 4 ft × 4 ft × 3 ft = 48 ft^3
(b) Weight = 48 ft^3 × 62.4 lb/ft^3 = 2995 lb
(c) Force = weight = 2995 lb
(d) Pressure = weight ÷ area = 2995 ÷ 2304 = 1.3 psi
(area = 4 ft × 4 ft × 144 in.^{2}/ft^2 = 2304 in.^{2})

R5-3 HEAD

Pressure and depth have a close relationship when a fluid is involved. In hydraulics, a branch of physics that has to do with properties of liquids, the depth of a body of water is called the **head** of water. *Water pressure* varies directly with its depth. As an example, if the tank in Fig. R5-1 was 2 ft high and filled with water, it would contain a volume of 2 ft^3 of water and would weigh 2 × 62.4, or 124.8 lb. Now the force of the water on the bottom of the tank would still be distributed over 144 in.^{2}, and the unit pressure would be 0.866 psi (124.8 ÷ 144). This is twice the amount of pressure that was exerted when the head of water was only 1 ft. Therefore, in an open-top container, the pressure of the water will equal 0.433 psi for each foot of head.

If there is a decrease or increase in the head of a body of water, there will be a corresponding decrease or increase in the pressure involved, as well as in the weight of the water, provided that the other dimensions stay the same. If the head of water was only 6 in. (½ ft), the pressure would equal 0.433 × ½, or 0.217 psi. If there were 10 ft of water in an open tank, the pressure would be 0.433 × 10 or 4.33 psi.

This relationship can be expressed in the equation

$$p = 0.433 \times h \tag{R5-2}$$

where

p = pressure, psi
h = head, feet of water

The tank in Fig. R5-1 has an area of 1 ft^2 with 1 ft of head; therefore, the pressure on the bottom of the tank is 0.433 psi. If there is a fish pond covering an area of 50 ft^2, and the depth of water in it is 1 ft, pressure on the bottom of the pond will still be just 0.433 psi, even though there is a larger volume of water. This points out the relationship between pressure and depth and demonstrates that there is not necessarily a relationship between pressure and volume.

With this relationship between pressure and depth established, we can transpose the equation so that the depth of water in a tank can be found if we know the pressure reading at the bottom of the tank.

If $p = 0.433 \times h$, then $h = p/0.433$ is true, by transposition.

EXAMPLE

If a pressure gauge located at the bottom of a 50-ft-high water tower showed a reading of 13 psi, (a) what is the depth of the water in the tank, and (b) what would the gauge indicate if the tower were filled with water?

Solution

(a) $h = \dfrac{p}{0.433} = \dfrac{13}{0.433} = 30$ ft of water

(b) $p = 0.433h = 0.433 \times 50 = 21.65$ psi

R5-4 PASCAL'S LAW

In the middle of the seventeenth century, a French mathematician and scientist named Blaise Pascal was experimenting with water and air pressure. His scientific experiments led to the formulation of what is known as Pascal's law, namely, that pressures applied to a confined liquid are transmitted equally throughout the liquid, irrespective of the area over which the pressure is applied. The application of this principle enabled Pascal to invent the hydraulic press, which is capable of a large multiplication of force. In Fig. R5-3, the unit pressure is equal in all vessels regardless of their shape, for pressure is independent of the shape of the container.

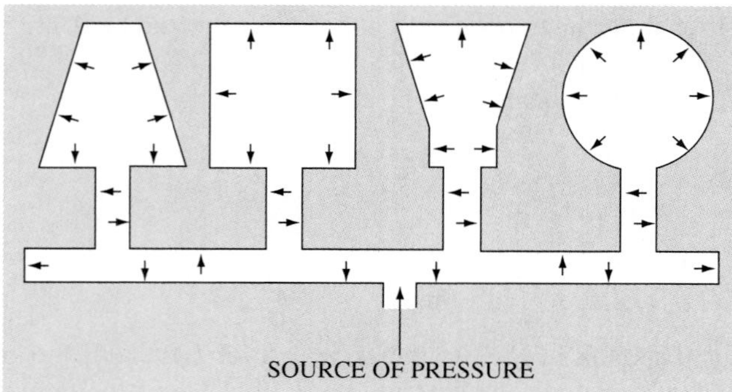

Figure R5-3 Pascal's principle of pressure equalization in various-shaped containers.

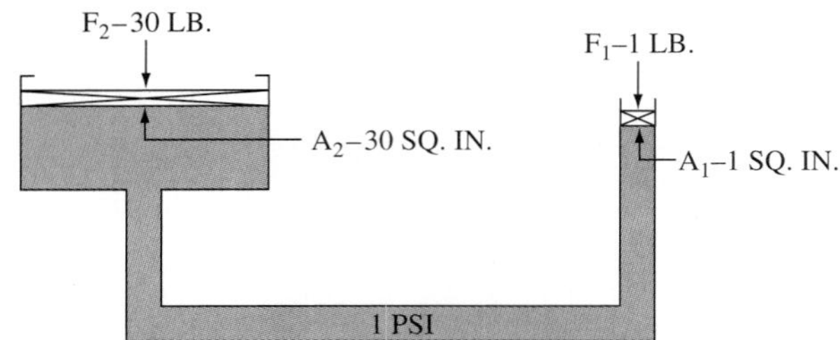

Figure R5-4 Pascal's principle of pressure transmission by hydraulic action.

Fig. R5-4, illustrating this principle, shows a vessel containing a fluid such as oil; the vessel has a small and a large cylinder connected by a pipe or tubing, with tight-fitting pistons in each cylinder. If the cross-sectional area of the small piston is 1 in.^{2} and the area of the large one is 30 in.^{2}, a force of 1 lb when applied to the smaller piston will support a weight of 30 lb on the larger piston, because a pressure of 1 psi throughout the fluid will be exerted.

Since we found earlier that pressure equals force divided by area, a force of 1 lb applied to an area of 1 in.^{2} will create a pressure of 1 psi. By transposition, force equals pressure multiplied by the area; therefore, a weight of 30 lb will be supported when a pressure of 1 psi is applied over an area of 30 in.^{2}.

R5-5 DENSITY

From a scientific or physics viewpoint, **density** is the weight per unit volume of a substance, and it may be expressed in any convenient combination of units of weight and volume used, such as *pounds per cubic inch* or *pounds per cubic foot.* An equation can be formulated which expresses this relationship:

$$D = \frac{W}{V} \qquad \text{(R5-3)}$$

where

D = density
W = weight
V = volume

As mentioned previously density of water is approximately 62.4 lb/ft^3, and it can be expressed as 0.0361 lb/in.^{3} (1 ft^3 contains 1728 in.^{3}, and $62.4 \div 1728 = 0.0361$). The density of some other common substances are listed in Fig. R5-5.

The **specific gravity** of any substance is the ratio of the weight of a given volume of the substance to the weight of the same volume of another given substance (where solids or liquids are concerned, water is used as a basis for specific gravity calculations that may have to

KNOWLEDGE EXTENSION

To extend your knowledge about applied physics regarding fluids and pressures for refrigeration applications, go to your local library and locate any introductory physics textbook. This should provide further insight, interesting readings, and additional examples to help clarify the concepts introduced throughout Chapter R5.

be made; air or hydrogen is used as a standard for gases).

density (solid or liquid)
= specific gravity × density of water (lb/ft^3)

The specific gravity of water is considered as 1.0 and values for other substances are listed in Fig. R5-5.

Pressure within a fluid is directly proportional to the density of the fluid. Consider the tank in Fig. R5-1, which, if filled with water weighing 62.4 lb, has a force on the bottom of the tank of 62.4 lb and a unit pressure of 0.433 psi. If this tank were filled instead with gasoline, which has a specific gravity of 0.66, the force on the bottom would be only 66% as great as when water is in the tank, and the pressure of the gasoline would be only 66% as great. Therefore, the relationship can be expressed as

pressure = head × density

$$p = h \times D \qquad \text{(R5-4)}$$

Substance	Density (lb/ft^3)	Specific Gravity
Water (pure)	62.4	1
Aluminum	168	2.7
Ammonia (liquid, 60°F)	38.5	0.62
Brass	530	8.5
Brick (common)	112	1.8
Copper	560	8.98
Cork (average board)	15	0.24
Gasoline	41.2	0.66
Glass (average)	175	2.8
Iron (cast)	448	7.2
Lead	705	11.3
Mercury	848	13.6
Oil (fuel) (average)	48.6	0.78
Steel (average)	486	7.8
Woods		
Oak	50	0.8
Pine	34.2	0.55

Figure R5-5 Density and specific gravity of some common substances.

where

p = pressure, lb/ft^2
h = head or depth below the surface, ft
D = density, lb/ft^3

or *where*

p = pressure, psi
h = head or depth below the surface, in.
D = density, $lb/in.^3$

Care must be taken whenever this equation is used to make sure that the proper units of weight and measurement are placed in the equation.

R5-6 SPECIFIC VOLUME

The specific volume of a substance is usually expressed as the number of cubic feet occupied by 1 lb of the substance. In the case of liquids, it will vary with temperature and pressure. The volume of a liquid will be affected by a change in its temperature, but since it is practically impossible to compress liquids, the volume is not affected by a change in pressure.

The volume of a gas or vapor is definitely affected by any change in either its temperature or the pressure to which it is subjected. In refrigeration, the volume of the vapor under the varying conditions involved is most important in the selection of the proper refrigerant lines.

The appropriate specific volumes and pressures for refrigerants will be covered later, but as an example of the effect temperature has on a refrigerant vapor, refer to Fig. R5-6 for the properties of Refrigerant-12 (R-12).

Note that at +5°F the specific *volume of vapor* is 1.46 ft^3/lb, whereas at 86°F it is only 0.38 ft^3/lb. Correspondingly, there is an increase in pressure (psia) from 26.48 psia to 108.04 psia. Note also the change in density from 90.14 lb/ft^3 to 80.67 lb/ft^3.

Under the pressure columns, two sets of data are shown: psia for absolute pressure and psig for gauge pressure. The following discussion explains the difference between them.

R5-7 ATMOSPHERIC PRESSURE

The earth is surrounded by a blanket of air called the atmosphere, which extends 50 or more miles upward from the surface of the earth. Air has weight and also exerts a pressure known as *atmospheric pressure.* It has been shown that a column of air with a cross-sectional area of one square inch and extending from the earth's surface at sea level to the limits of the atmosphere, would weigh approximately 14.7 lb. As was pointed out earlier in the

Temp °F	Pressure		Volume (Vapor)	Density (Liquid)	Heat Content (Btu/lb)	
	psia	psig	(ft/lb)	(lb/ft^3)	Liquid	Vapor
–150	0.154	*29.61*	178.65	104.36	–22.70	60.8
–125	0.516	*28.67*	57.28	102.29	–17.59	83.5
–100	1.428	*27.01*	22.16	100.15	–12.47	66.2
–75	3.388	*23.02*	9.92	97.93	–7.31	69.0
–50	7.117	*15.43*	4.97	95.62	–2.10	71.8
–25	13.556	2.32	2.73	93.20	3.17	74.56
–15	17.141	2.45	2.19	92.20	5.30	75.65
–10	19.189	4.49	1.97	91.70	6.37	76.2
–5	21.422	6.73	1.78	91.18	7.44	76.73
0	23.849	9.15	1.61	90.66	8.52	77.27
5	**26.483**	**11.79**	**1.46**	**90.14**	**9.60**	**77.80**
10	29.335	14.64	1.32	89.61	10.68	78.335
25	39.310	24.61	1.00	87.98	13.96	79.9
50	61.394	46.70	0.66	85.14	19.50	82.43
75	91.682	76.99	0.44	82.09	25.20	84.82
86	**108.04**	**93.34**	**0.38**	**80.67**	**27.77**	**85.82**
100	131.86	117.16	0.31	78.79	31.10	87.63
125	183.76	169.06	0.22	75.15	37.28	88.97
150	249.31	234.61	0.16	71.04	43.85	90.53
175	330.64	315.94	0.11	66.20	51.03	91.48
200	430.09	415.39	0.08	60.03	59.20	91.28

Pressures in italic are inches of mercury below 1 atm.

Figure R5-6 Properties of R-12 refrigerant. Note pressures corresponding to standard evaporating temperature of 5°F and condensing temperature of 86°F.
(*Source:* The DuPont Company)

chapter, force also means the weight of a substance, and pressure means unit force per square inch; therefore, standard atmospheric pressure is considered to be 14.7 psi at sea level.

This pressure is not constant; it will vary with altitude or elevation above sea level, and there will be variations due to changes in temperature as well as water vapor content of the air. This atmospheric pressure can be demonstrated by construction of a simple barometer, as shown in Fig. R5-7, using a glass tube about 36 in. long and closed at one end, an open dish or bowl, and a supply of mercury. Fill the tube with mercury and invert it in the bowl of mercury, holding a finger at the open end of the tube so that the mercury will not spill out while the tube is being inverted. Upon removal of the finger, the level of the mercury in the tube will drop somewhat, leaving a vacuum at the closed end of the tube. The atmospheric pressure bearing down on the open dish or bowl of mercury will force the mercury in the tube to stand up to a height determined by the pressure being exerted on the open surface of the mercury, which will be approximately 30 in. at sea level.

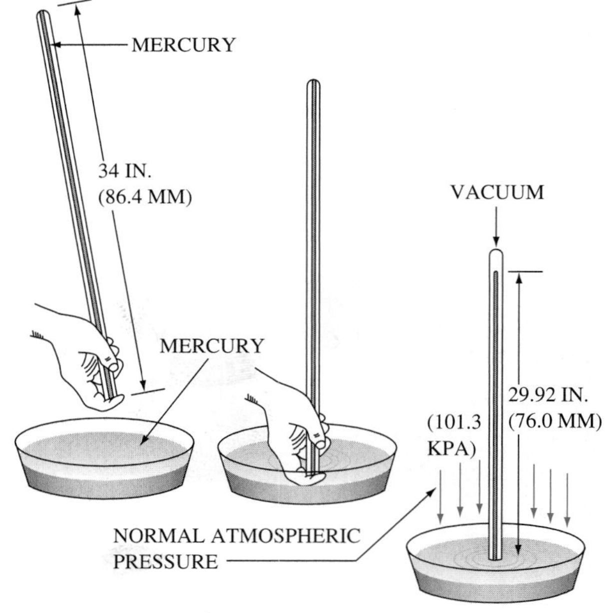

Figure R5-7 Column of mercury supported by normal atmospheric pressure.

As previously shown in Fig. R5-5, the specific gravity of mercury is 13.6; that is, mercury weighs and exerts pressure 13.6 times that of an equal volume of water. Since a 1-in. high column of water exerts pressure of 0.0361 psi, a similar column of mercury will exert pressure that is $0.0361 \times 13.6 = 0.491$ psi. A 30-in. column of mercury will exert pressure of $0.491 \times 30 = 14.7$ psi, or an amount equal to atmospheric pressure. Conversely, atmo-spheric pressure at sea level (14.7 psi) bearing down on the open dish of mercury divided by the unit pressure that is exerted by a 1-in. column of mercury (0.491) should cause the column of mercury in the tube to stand approximately at a height of 30 in. (actually, 29.92 in.).

$$\frac{14.7}{0.491} = \text{30-in. column of mercury (Hg)}$$

Of course, water could possibly be used in this barometer instead of mercury; but the tube would have to be about 34 ft. high, and this is not practical.

$$\frac{14.7}{0.036} = \text{407 in. of water (33.9 ft)}$$

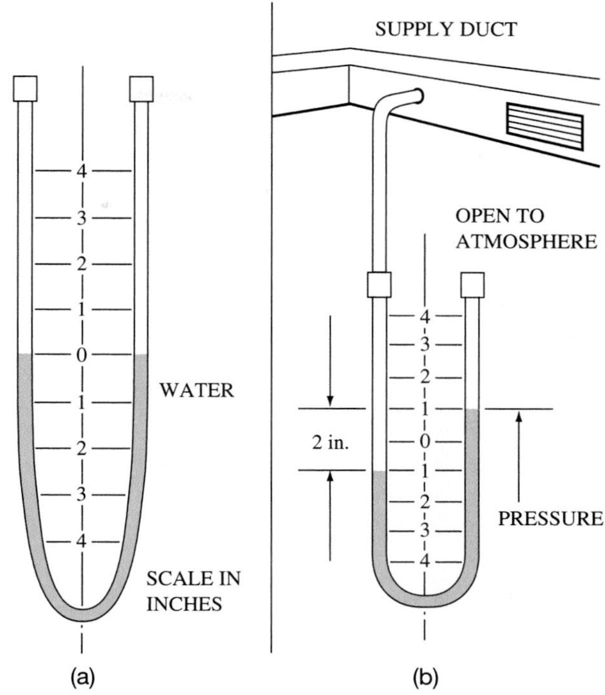

Figure R5-8 Water filled manometer used to measure air pressure.

R5-8 MEASUREMENT OF PRESSURE

A manometer is one type of device used in the refrigeration and air-conditioning field for the measurement of pressure. This type of pressure gauge uses a liquid, usually mercury, water, or gauge oil, as an indicator of the amount of pressure involved. The water manometer or water gauge is customarily used when measuring air pressures, because of the lightness of the fluid being measured.

A simple open-arm manometer is shown in Fig. R5-8. The U-shaped glass tube is partially filled with water, as shown in Fig. R5-8a, and is open at both ends. The *water is at the same level* in both arms of the manometer, because both arms are open to the atmosphere, and no external pressure is being exerted on them.

Figure R5-8b shows the manometer in use with one arm connected to a source of positive air pressure that is being measured. The water is at different levels in the arms, and the difference denotes the amount of pressure being applied.

A space that is void, or lacking any pressure, is described as having a *perfect vacuum.* If the space has pressure less than atmospheric pressure, it is defined as being a *partial vacuum.* It is customary to express this partial vacuum in inches of mercury, and not as negative pres-

ON THE JOB

Chapter R-5 explains the importance of comparing temperature with pressure when using the refrigerant Pressure/Temperature (P/T) chart. Checking the superheat setting on the Thermostatic Expansion Valve (TXV) is a common on-the-job application for this principle. Using an electronic thermometer, the technician takes a temperature reading close to the thermobulb at the outlet of the evaporator. The suction pressure is taken and then converted to temperature using the refrigerant P/T chart. A standard superheat is approximately 8–12 degrees, depending upon the equipment used and application. If there is a long run of suction pipe from the evaporator to the compressor, a pressure drop is probably present, and the calculation should be adjusted accordingly and the superheat rechecked.

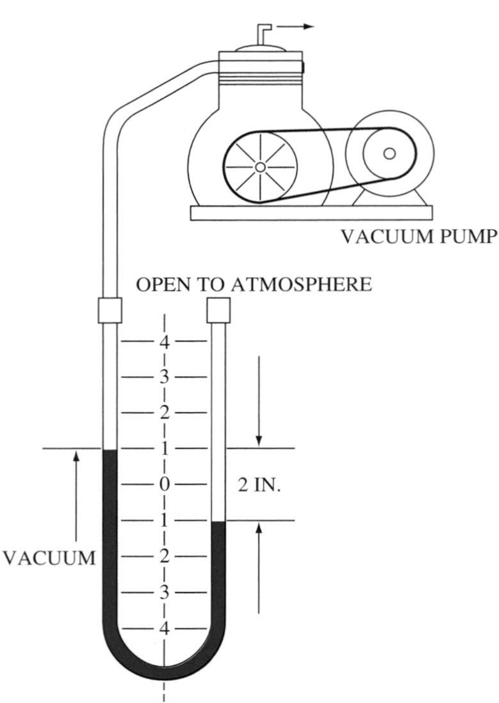

Figure R5-9 Mercury fill manometer measuring "vacuum" pressure.

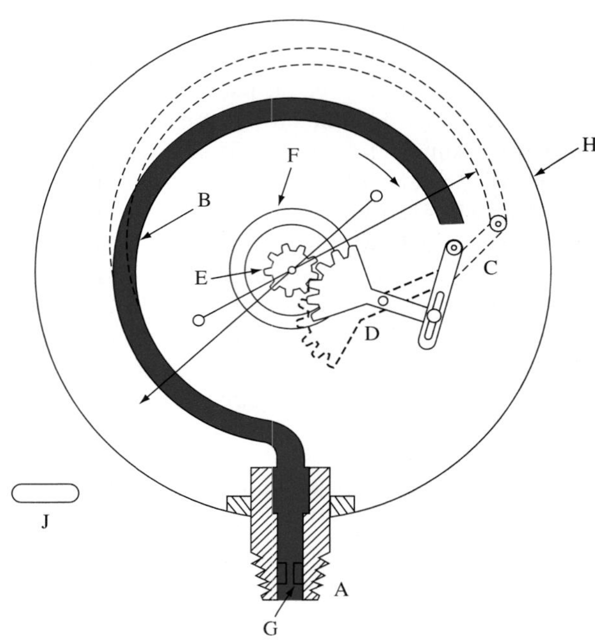

Figure R5-10 Internal construction of a pressure gauge: (a) adapter fitting, usually an ⅛-in. pipe thread; (b) Bourdon tube; (c) link; (d) gear sector; (e) Pinter shaft gear; (f) calibrating spring; (g) restricter; (h) case; (j) cross section of the Bourdon tube. The dashed lines indicate how the pressure in the Bourdon tube causes it to straighten and operate the gauge.

sure. In some instances it is also referred to as a given amount of absolute pressure, expressed in psia, and this will be covered later in this chapter.

If a partial vacuum has been drawn on the left arm of the manometer by means of a vacuum pump, as shown in Fig. R5-9, the mercury in the right arm will be lower, and the difference in levels will designate the partial vacuum in inches of mercury.

Pressure gauges most commonly used in the field by service technicians, to determine pressure within the refrigeration system, are of the Bourdon-tube type. As is shown in Fig. R5-10, an internal view, the essential element of this type of gauge is the Bourdon tube. This oval metal tube is curved along its length and forms an almost complete circle. One end of the tube is closed, and the other end is connected to the equipment or component being tested.

As shown in Fig. R5-11, the gauges are preset at 0 lb, which represents the atmospheric pressure of 14.7 psi. Any additional pressure applied when the gauge is connected to a piece of equipment will tend to straighten out the Bourdon tube, thereby moving the needle or pointer and its mechanical linkage, thus indicating the amount of pressure being applied.

Pressures below atmospheric are customarily expressed in inches of mercury. There is an indication of the range between 0 gauge and 30 in. of mercury (Hg) on the compound gauge.

Figure R5-11 Refrigeration gauges with temperature scales. *(Courtesy of Marsh Instrument Company)*

R5-9 ABSOLUTE PRESSURE

Fig. R5-12 shows a definite relationship among absolute, atmospheric, and gauge pressures. For many problems, atmospheric pressure does not need to be considered, so the

	Gauge Pressure	Absolute Pressure
Above Atmospheric Pressure	40 psig	54.7 psia
	30 psig	44.7 psia
	20 psig	34.7 psia
	10 psig	24.7 psia
	0 psig	14.7 psia
Below Atmospheric Pressure	10" Hg	9.7 psia
	20" Hg	4.7 psia
	30" Hg	0.0 psia

Figure R5-12 Relationship between absolute and gauge pressure.

customary pressure gauge is calibrated and graduated to read zero under normal atmospheric conditions. Yet when gases are contained within an enclosure away from the atmosphere, such as in a refrigeration unit, it is necessary to take atmospheric pressure into consideration, and mathematical calculations must be in terms of the absolute pressures involved.

R5-10 PRESSURE OF GAS

The volume of a gas is affected by a change in either the pressure or temperature, or both. There are laws that govern the mathematical calculations in computing these variables.

Boyle's law states that the volume of a gas varies inversely as its pressure if the temperature of the gas remains constant. This means that the product of the pressure times the volume remains constant, or that if the pressure of a gas doubles, the new volume will be one-half of the original volume. Or it may be considered that, if the volume is doubled, the absolute pressure will be reduced to one-half of what it was originally.

This concept may be expressed as

$$p_1V_1 = p_2V_2 \tag{R5-5}$$

where

p_1 = original pressure
V_1 = original volume
p_2 = new pressure
V_2 = new volume

It must be remembered that p_1 and p_2 have to be expressed in the *absolute pressure* terms for Eq. (R5-5) to be used correctly.

EXAMPLE

If the gauge pressure on 2 ft^3 of gas is increased from 20 psig to 50 psig while the temperature of the vapor remains constant, what will be the new volume?

Solution

$p_1 = 20 + 14.7$
$= 34.7$ psia

$p_2 = 50 + 14.7$
$= 64.7$ psia

$V_1 = 2\ \text{ft}^3$

$V_2 = ?$

Since

$$p_1V_1 = p_2V_2$$

then

$$V_2 = \frac{p_1V_1}{p_2}$$

Therefore,

$$V_2 = \frac{34.7 \times 2}{64.7} = 1.072\ \text{ft}^3$$

PRACTICING PROFESSIONAL SERVICES

You can apply Boyle's law to your professional practice when working with compressors. The law states that if the pressure of a gas doubles, the new volume will be one-half of the original volume. Conversely, when applied to refrigeration compressors, the law indicates that the lower the suction pressure the more volume the compressor will have to pump. This means the compressor will work harder during low suction periods and have less capacity. Thus, to increase the capacity of the refrigeration compressor, raise the suction pressure as high as possible and observe the results. The net overall capacity can be verified with the compressor manufacturer's technical data sheets.

The same basic equation may be used to determine the given new pressure.

EXAMPLE

If additional pressure is applied to a volume of 2 ft^3 of gas at 20 psig so that the volume is lessened to 1.072 ft^3 and the temperature of the gas remains constant, what is the new pressure in psig?

Solution

$p_1 = 34.7$ psia (20 + 14.7)

$V_1 = 2$ ft^3

$V_2 = 1.072$ ft^3

$$p_2 = \frac{p_1 V_1}{V_2}$$

$$p_2 = \frac{34.7 \times 2}{1.072} = 64.7 \text{ psia}$$

$$p_2 = 64.7 - 14.7 = 50 \text{ psig}$$

R5-11 EXPANSION OF GAS

Most gases will expand in volume at practically the same rate with an increase in temperature, provided that the pressure does not change. If the gas is confined so that its volume will remain the same, the pressure in the container will increase at about the same rate as an increase in temperature.

Theoretically, if the pressure remains constant, a gas vapor will expand or contract at the rate of 1/492 for each degree of temperature change. The result of this theory would be a zero volume at a temperature of –460°F, or at 0° absolute.

Charles' law states that the volume of a gas is in direct proportion to its absolute temperature, provided that the pressure is kept constant; and the absolute pressure of a gas is in direct proportion to its absolute temperature, provided that the volume is kept constant. That is,

$$\frac{V_1}{V_2} = \frac{T_1}{T_2} \qquad \text{(R5-6)}$$

and

$$\frac{p_1}{p_2} = \frac{T_1}{T_2} \qquad \begin{array}{l} T = \text{absolute temperature} \\ p = \text{absolute pressure} \end{array} \qquad \text{(R5-7)}$$

To clear the fractions, these may also be expressed as

$$V_1T_2 = V_2T_1 \quad \text{and} \quad p_1T_2 = p_2T_1$$

EXAMPLE

If the temperature of 2 ft^3 of gas is increased from 40°F to 120°F, what would be the new volume if there was no change in pressure?

Solution

$$V_2 = \frac{V_1T_2}{T_1} = \frac{2 \times (120 + 460)}{40 + 460} = \frac{1160}{500} = 2.32 \text{ ft}^3$$

EXAMPLE

If a container holds 2 ft^3 of gas at 20 psig, what will be the new pressure in psig if the temperature is increased from 40°F to 120°F?

Solution

$$p_2 = \frac{p_1T_2}{T_1} = \frac{(20 + 14.7) \times (120 + 460)}{40 + 460} = 40.25 \text{ psig}$$

$$40.25 - 14.7 = 25.55 \text{ psig}$$

In numerous cases dealing with refrigerant vapor, none of the three possible variables remains constant, and a combination of these laws must be utilized, namely the general law of perfect gas:

$$\frac{p_1V_1}{T_1} = \frac{p_2/V_2}{T_2} \quad \text{or} \quad p_1V_1T_2 = p_2V_2T_1 \qquad \text{(R5-8)}$$

in which the units of p and T are always used in the absolute.

EXAMPLE

If a volume of 4 ft^3 of gas at a temperature of 70°F and at atmospheric pressure is compressed to one-half its original volume and increased in temperature to 120°F, what will be its new pressure?

Solution

Transposing Eq. (R3-8), we have

$$p_2 = \frac{p_1V_1T_2}{V_2T_1}$$

Therefore,

$$p_2 = \frac{14.7 \times 4(120 + 460)}{2 \times (70 + 460)} = \frac{34{,}104}{1{,}060} = 32.17 \text{ psia}$$

$$32.17 - 14.7 = 17.47 \text{ psig}$$

R5-12 BOILING POINT

The most important point to understand when dealing with the action in a refrigeration system is the "boiling point" of the liquid (refrigerant) in the system. Lowering the boiling point causes the refrigerant to absorb heat and vaporize or "boil." Conversely by raising the "boiling point" the vapor gives up the heat and condenses. Basically, the refrigeration system operates by "control of the boiling point."

In Chapter R3, boiling point was defined as the temperature at which a liquid turns from a liquid to a vapor or condenses from a vapor to a liquid depending on the *absorption* or rejection of heat energy.

The chart used was based on using water at standard atmospheric pressure of 29.92 in. of mercury and 70°F. At these conditions water will boil at 212°F or 100°C with the addition of heat energy, or condense at this same temperature with the removal of heat energy.

When referring to "boiling point," the pressure that the liquid is subjected to must also be considered. When referring to the boiling point of water as 212°F or 100°C, the assumption is made that the water is subjected to standard barometric pressure of 29.92 in. of mercury.

In reality, the boiling point of a liquid will change in the same direction as the pressure to which the liquid is subjected. This is a very important basic law of physics that must be remembered. In a later chapter this law will be applied to various refrigerants, but for discussion in this chapter water will be used as the refrigerant. Fig. R5-13 gives examples of the boiling point of water at sea level (212°F) and at 14,100 ft above sea level (167°F). Obviously, this is because the boiling point of water drops as the atmospheric pressure drops. Again, the boiling point of a liquid will vary in the same direction as the pressure to which the liquid is subjected.

If the boiling point of water is determined at various pressures, both above (pressure) and below atmospheric pressure (vacuum), and the temperatures are plotted on a graph, the results would be as pictured in Fig. R5-14. Here we can see that the boiling point of water can be raised to 276°F with a pressure of 30 psig or lowered to 40°F at a pressure of 29.67233 in. Hg. Therefore, to obtain a desired boiling point of water it is only necessary to maintain an equivalent pressure based on the pressure-temperature curve of the liquid.

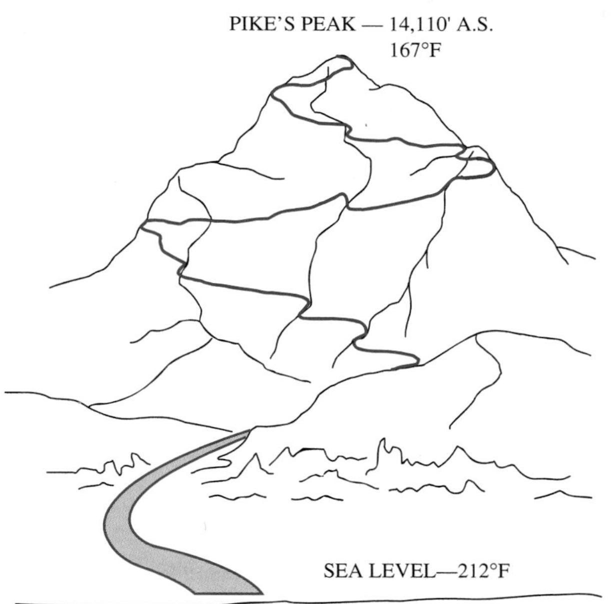

Figure R5-13 Boiling point of water at different elevations.

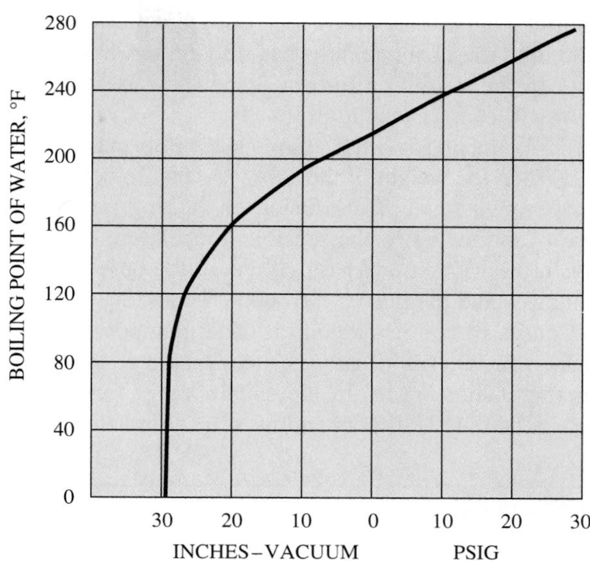

Figure R5-14 Pressure-temperature curve for water.

R5-13 CONDENSING TEMPERATURE

With the liquid in a vapor state, to remain as a vapor, the sensible temperature must be higher than the condensing temperature. If the heat energy is removed from the vapor to the point where the sensible temperature is attempting to fall below the condensing temperature of the vapor, the vapor will liquify or condense.

The boiling point and the condensing temperature for liquid are the same. Only a difference in the action taking place is implied. Boiling point implies liquid to vapor; condensing temperature implies vapor to liquid. In each case, the temperature at which the change takes place varies as the pressure to which the liquid is subjected. Lowering the pressure lowers the boiling point or condensing temperature. Raising the pressure raises the boiling point or condensing temperature.

R5-14 FUSION POINT

The fusion point (temperature at which solidification of liquid or melting of a solid takes place) is also affected by the pressure to which the solid is subjected. If the fusion point of a solid is raised by increasing the pressure on the solid until the fusion point is equal to the sensible temperature of the solid, any further attempt to raise the fusion point will cause the solid to liquefy.

This principle is what allows glaciers to move downhill. When the weight of the snow and ice in the glacier is high enough to raise the fusion or melting point of the base of the ice up to the sensible temperature of the ice, any further rise in pressure will cause the base of the ice to liquify, and the glacier moves on a layer of water.

Control of the fusion point is of little importance in refrigeration, as it is usually an undesirable result of temperature maintenance. In air conditioning, formation of ice is a definite detriment and must be eliminated.

R5-15 SATURATION TEMPERATURE

In Section R5-12 the statement was made that the boiling point and condensing temperature of a liquid at a given pressure are the same. This means that the liquid has reached the point where it contains all the heat energy it can without changing to a vapor. This condition is described by referring to it as a "saturated liquid." This means that if any more heat energy is added, the liquid will boil. Commonly, if the vapor is cooled to a point where the vapor is so dense that any further reduction in heat energy causes it to condense to a liquid, the condition is referred to as a "saturated vapor."

In sections R4-4 and R4-5, where sensible heat and latent heat were discussed, sensible heat was said to change temperature and latent heat was said to change state. Therefore, liquid at the boiling point is saturated with sensible heat and any heat added would be latent heat to vaporize the liquid. Vapor at the condensing temperature has been reduced to that temperature by removing the sensible heat until the density of the vapor is to a point where any further removal of heat will cause condensation of the vapor and removal of the latent heat of vaporization. At this point the vapor is said to be a "saturated vapor."

R5-16 SUPERHEAT

Superheat is "the heat added to a vapor after it becomes a vapor," a simple rise in temperature of the vapor above the boiling point. If, for example, water were to boil at 212°F and before leaving the passages in the boiler it were to take on more heat and the steam temperature rises to 220°F, the steam would be superheated 8°F, the difference between the boiling point of the liquid and the actual physical (sensible) temperature of the vapor. This is called superheated steam. To remove the heat and condense the steam to a liquid, the first action required is to desuperheat the steam to the saturation point (condensing temperature) and then remove the latent heat of vaporization to produce the liquid.

Superheat is very important in refrigeration systems to produce the maximum system capacity together with heat equipment life, and will be referred to throughout the various chapters.

R5-17 SUBCOOLING

When a liquid is at a sensible temperature below its boiling point, it is said to be "subcooled." For example, water at standard atmospheric conditions with a sensible temperature of 70°F will be subcooled 142°F (212°F – 70°F = 142°F subcooled). The subcooling of the liquid in the refrigeration or air-conditioning system is important for maximum capacity and efficiency. This will be discussed further in later chapters.

REVIEW

- **Fluid pressure** is the force per unit area that is exerted by a gas or a liquid.
- **Head** is pressure exerted by a fluid due to its depth.
- **Pascal's law**—Pressures applied to a confined liquid are transmitted equally throughout the liquid, irrespective of the area over which the pressure is applied.
- **Density** is the weight per unit volume of a substance, expressed in units of weight per volume.
- **Atmospheric pressure**—14.7 pounds per square inch at sea level. Pressure exerted on the earth by a column of air rising to the top of the atmosphere.
- **Manometer** is a device used to measure low pressures.
- **Pressure gauges**—Used in the field by service technicians to measure system pressures, calibrated to 0 gauge pressure at sea level. Gauges operate off of the Bourdon tube principle.
- **Absolute pressure** equals gauge pressure + 14.7 psi.
- **Boyle's law**—The volume of a gas varies inversely as its pressure if the temperature is held constant.
- **Charles' law**—The volume of a gas varies directly with its absolute temperature, provided the pressure is held constant, and the absolute pressure of a gas varies directly with its absolute temperature provided the volume is held constant.
- **Boiling point**—The temperature at which a fluid will begin to boil, at a given pressure. When pressure changes, boiling point also changes. Raising pressure increases boiling point.
- **Condensing temperature** is the temperature at which a fluid will condense into a liquid. Boiling point and condensing temperature are the same value for a given pressure.
- **Saturation temperature** — A temperature at which a fluid has exchanged all of the heat it can without

changing state. Refrigerant is at saturation in the evaporator and the condenser. Vapor and liquid are in contact with each other.

- **Fusion** is the temperature at which solidification of liquid or melting of a solid takes place. This process is also affected by the pressure to which the solid is subjected.
- **Subcooling** is when a liquid is a sensible temperature below its boiling point.
- **Superheat** is heat added to a vapor after it has changed state to a vapor.
- To tie various terms together, review Fig. R5-15 while reading the following points. Starting with ice at 0°F and adding heat energy, the following changes take place:

1. Sensible heat is added to the subcooled ice raising its sensible temperature to the saturated temperature of 32°F or 0°C.
2. Upon reaching the saturation temperature, additional heat is latent heat that changes the saturated ice to saturated liquid.
3. With all the solid (ice) now liquid (water), further heat energy addition is sensible heat that raises the temperature of the liquid (water) to the boiling point, where it is now a saturated liquid.
4. Additional heat (latent heat) changes the liquid from a saturated liquid to a saturated vapor.
5. After the liquid is completely vaporized, additional heat will raise the temperature of the vapor (sensible heat) and superheat the vapor.

Starting with superheated steam and removing heat energy will produce the following changes:

1. Removing sensible heat from the vapor will lower the temperature and desuperheat the vapor to the saturated vapor point, referred to as the condensing temperature.
2. Further removed latent heat will reduce the saturated vapor to a saturated liquid still at the same temperature. Condensing temperature (saturated vapor containing latent heat) and boiling point (saturated liquid that does not contain the latent heat) are the same temperature.
3. Further removal of sensible heat from the liquid will lower the temperature of the liquid (subcool the liquid below the boiling point) until the temperature of the liquid reaches the fusion point.
4. Although never discussed, this liquid at the fusion point could also be classified as a saturated liquid based on the liquid-to-solid reaction.

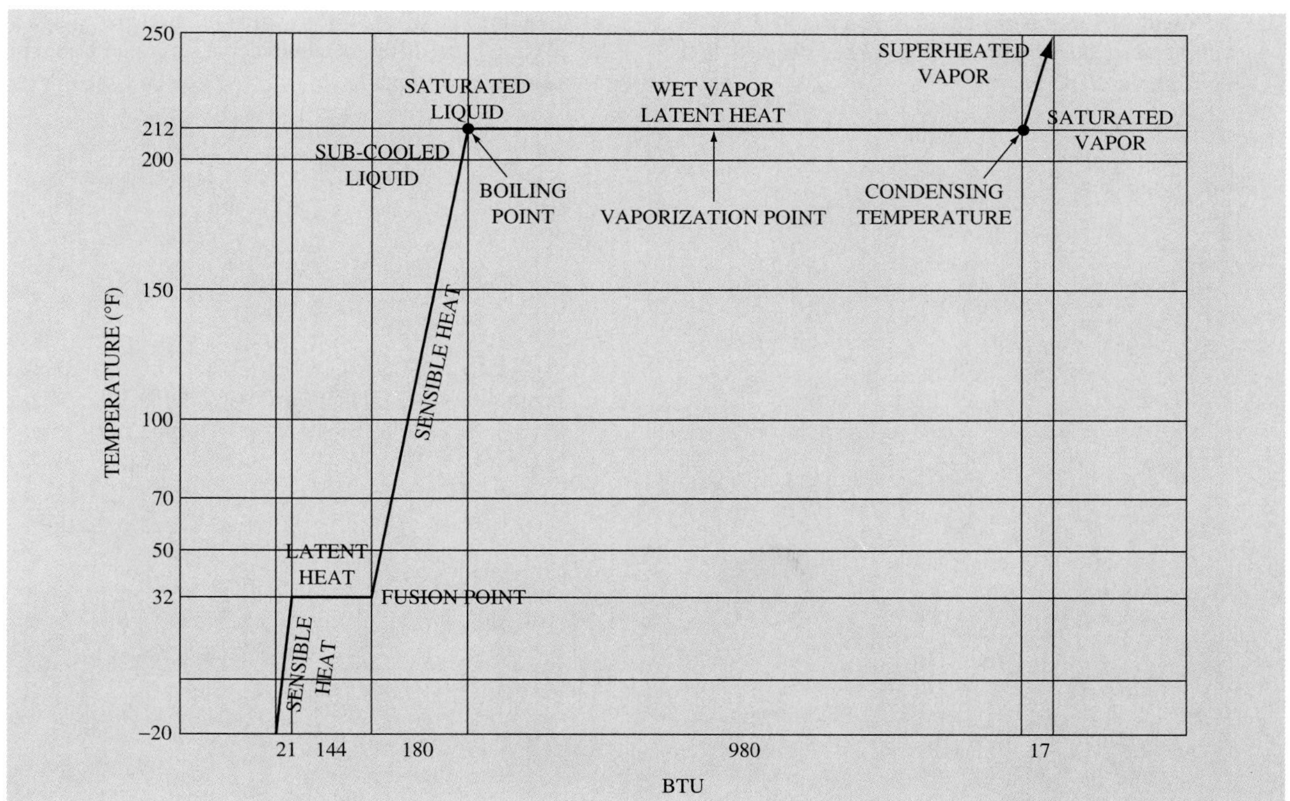

Figure R5-15 Chart demonstrating all terms concerning changes in physical conditions and temperature of solid, liquid, and vapor of water.

5. Removal of latent heat from the liquid changes the liquid to a solid (latent heat of fusion).
6. Further removal of sensible heat will reduce the temperature or subcool the solid (ice) below the melting point.

All these terms must be understood before proceeding into discussions of refrigeration systems.

Problems and Questions

1. Define "fluid pressure."
2. What term is usually used to express fluid pressure?
3. Define the term "force" when discussing fluid pressure.
4. Find the total force and also the unit pressure exerted on the bottom of a tank filled with water if the tank measures 3 ft × 3 ft × 1 ft high.
5. Define the term "head."
6. If a tank with a flat base is filled with water to a level of 8 ft, what pressure is exerted on the bottom of the tank?
7. What pressure is exerted halfway between the bottom of the tank and the surface of the water?
8. Find the unit pressure exerted on the roof of a building on which is located a cooling tower. The tower weighs 1580 lb when filled with water and operating. The size of the base of the tower is 3 ft × 4 ft.
9. In a hydraulic press, what force must be exerted on the small piston having an area of 2 $in.^2$ if a 600 lb weight must be supported on a larger piston which has an area of 16 $in.^2$?
10. What would be the pressure of the liquid in the press in Problem 9?
11. Define "density."
12. The density of water is ______.
13. Define "specific gravity."
14. Define "specific volume."
15. What factor affects the specific volume of a liquid?
16. What factors affect the specific volume of a vapor?
17. Define "Boyle's law."
18. If the volume of 10 ft^3 of gas at a pressure of 25 psi is to be compressed to 2 ft^3 with no change in temperature, what will be the new pressure in psi?
19. If the volume of 2 ft^3 of gas at a pressure of 183.8 psi is to be compressed to 1 ft^3 with no change in temperature, what will be the new pressure in psi?
20. Define "Charles' law."
21. With the pressure remaining constant, find the new volume of 4 ft^3 of gas when its temperature is increased from 60°F to 250°F.
22. Give the formula for the general law of perfect gas.
23. If a volume of 10 ft^3 of gas at a temperature of 60°F and a pressure of 57.7 psig were compressed to 5 ft^3 and the temperature raised to 127 psig, what would be its new pressure?
24. What would be the new volume if 20 ft^3 of gas at 70°F and 70.2 psig were raised to 115°F and 146.8 psig?
25. Explain the differences between superheat and subcooling.
26. What is the difference between a saturated liquid and a saturated vapor?

Tools and Equipment

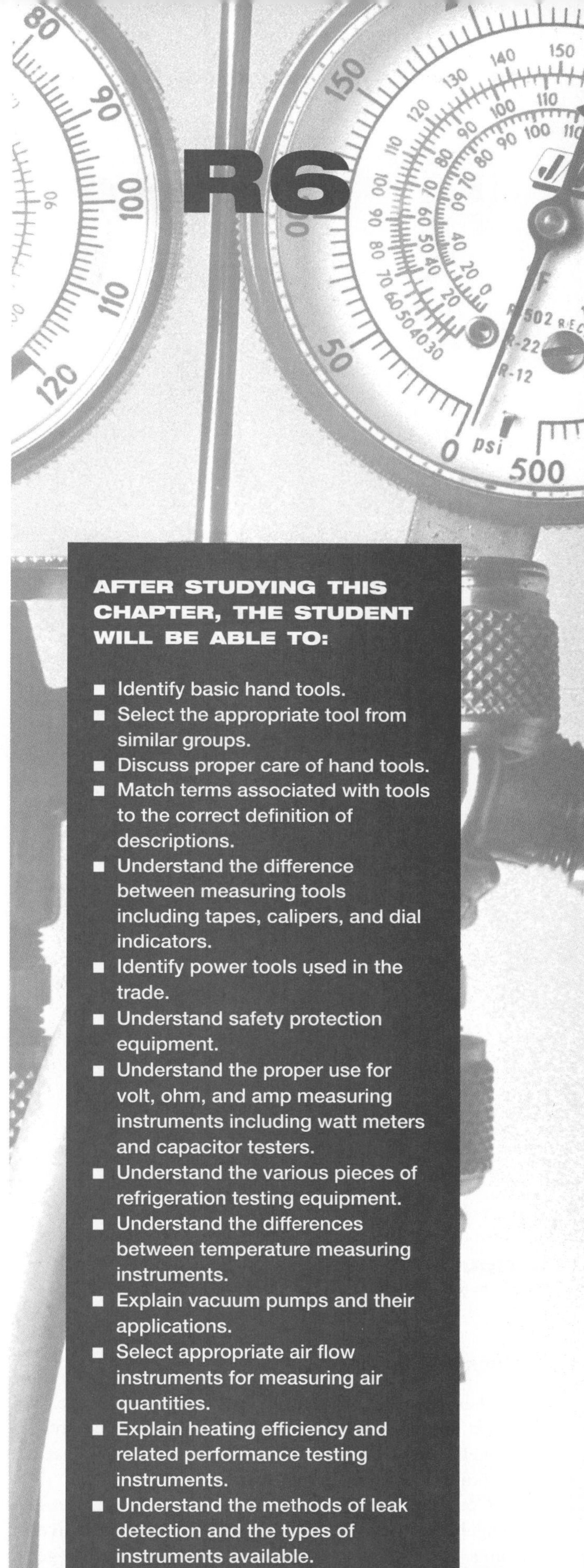

R6

AFTER STUDYING THIS CHAPTER, THE STUDENT WILL BE ABLE TO:

- Identify basic hand tools.
- Select the appropriate tool from similar groups.
- Discuss proper care of hand tools.
- Match terms associated with tools to the correct definition of descriptions.
- Understand the difference between measuring tools including tapes, calipers, and dial indicators.
- Identify power tools used in the trade.
- Understand safety protection equipment.
- Understand the proper use for volt, ohm, and amp measuring instruments including watt meters and capacitor testers.
- Understand the various pieces of refrigeration testing equipment.
- Understand the differences between temperature measuring instruments.
- Explain vacuum pumps and their applications.
- Select appropriate air flow instruments for measuring air quantities.
- Explain heating efficiency and related performance testing instruments.
- Understand the methods of leak detection and the types of instruments available.

R6-1 HAND TOOLS AND ACCESSORIES

The careful selection, care, and knowledge of the use of tools are important considerations for the technician. Poor workmanship or injury can frequently be traced to a lack, or improper use, of hand tools.

The common hand tools needed by a refrigeration technician are described in this chapter.

R6-1.1 Wrenches

Wrenches are probably the most frequently used tool.

The steel ratchet wrench shown in Fig. R6-1 has a square drive opening and is well suited for opening and closing valves with square stem ends. Examples are refrigeration service valves, compressor gas tanks, etc. The ratchet permits rapid changing of direction so the operator can switch movement to open or close a valve, etc. The wrench handle openings vary from 3/16 in. to 3/8 in. Some also have a 1/2-in. hex socket cast into one end.

Rachet drivers are also available with a swivel head. Drive extensions are available with a wobble end which permits offset in hard to reach situations.

Service Valve Wrenches

Service valves on compressors are usually constructed with a 1/4-in. square drive and require a special wrench as shown in Fig. R6-2. One end of the wrench has a 1/4-in. square ratchet to fit the valve stem. The other end has fixed square openings usually 3/16 in., 1/4 in., and 5/16 in.

For quickly opening a valve, the ratchet end is used. To only crack open the valve, as required for some service operations, the fixed-opening end is used. Using the fixed opening, the operator has better control of the opening amount.

Some wrenches also have a lever for reversing the rotation of the wrench as required for either opening or closing the valve.

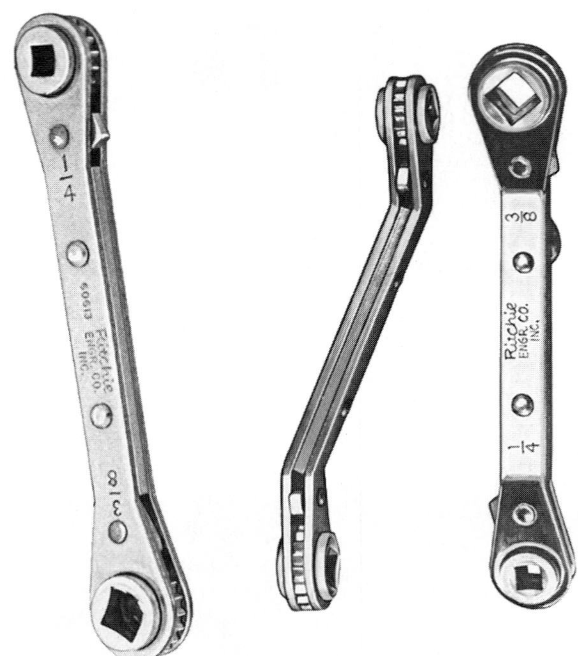

Figure R6-1 Ratchet wrenches. (Courtesy of Yellow Jacket Division, Ritchie Engineering Company)

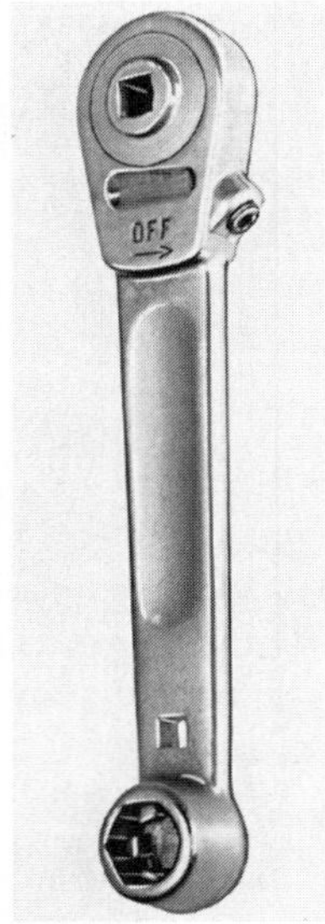

Figure R6-2 Refrigeration service valve wrench. (Courtesy of Robinair Division, SPX Corporation)

Socket Wrenches

Socket wrenches (Fig. R6-3) are used to slip over bolt heads and nuts. They are made of steel, and the array of sockets vary from square to 6-point hexagonal shape to 12-point double hexagonal shape.

Common sizes run from 5⁄32 in. to 2 in. Metric sizes are available in the same range. Swivel sockets and universal joints are particularly useful for reaching hard-to-get-to nuts and bolts. The more points a socket has the easier it is to use in a restricted area.

The handle of a socket wrench may vary from a straight, fixed tee drive to a ratchet operation to a special torque-wrench handle that includes a gauge to measure the force being applied.

Torque Wrenches

Depending on the type of wrench used, there is always a danger of tightening a bolt beyond the breaking point. Breaking off a bolt could result in major damage to the equipment, which could be difficult to repair. It is often, therefore, recommended by equipment manufacturers that a torque wrench be used. A typical torque wrench is shown in Fig. R6-4. It uses standard square drive sockets.

The dial on the wrench indicates the amount of pressure being applied to turn the bolt. Torque is a twisting action. Torque is measured in pound-feet, being the product of the length of the handle (in feet) times the force applied (in pounds).

Box Wrenches

Box wrenches, as shown in Fig. R6-5, are useful in certain close-quarter situations. The ends are usually in the form of a 12-point double hexagonal shape. Ends may be of the same or different sizes. The handle may be straight or offset. Common sizes range from ¼ in. to 1½ in.

Flare-Nut Wrenches

The flare-nut wrench (Fig. R6-6) is a special variation of the box wrench in that the heads are slotted to allow the wrench to slip over the tubing and then onto a flare nut.

Open-End Wrenches

Open-end wrenches (Fig. R6-7) are needed where it is impossible to fit a socket or box wrench on a nut, bolt, or fitting from the top. The open-end wrench permits access to the object from the side.

Combination Box and Open-End Wrenches

Some service mechanics prefer the combination box and open-end wrench (Fig. R6-8) to gain the advantage of each type. The ends are made to fit the same size nut or bolt so the technician can use the best selection.

Common sizes are ¼ in. to 2½ in., (I-P). In SI units, the sizes are 8 mm to 50 mm.

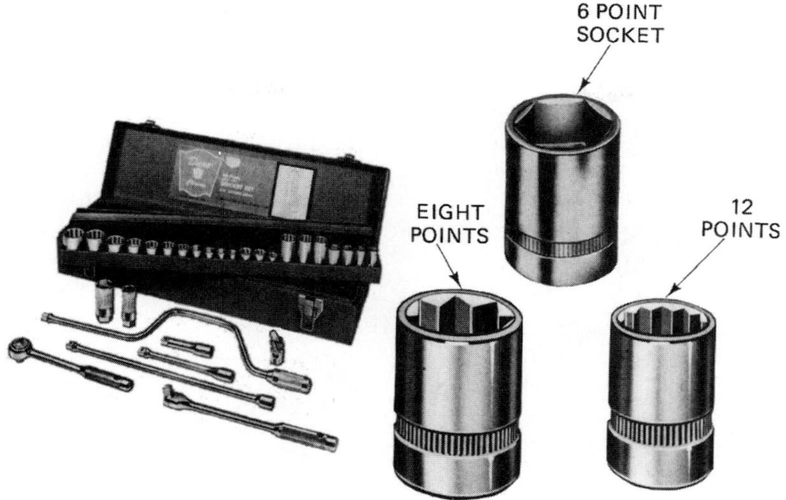

Figure R6-3 Socket wrench. (Source: Duro Metal Products, Inc.)

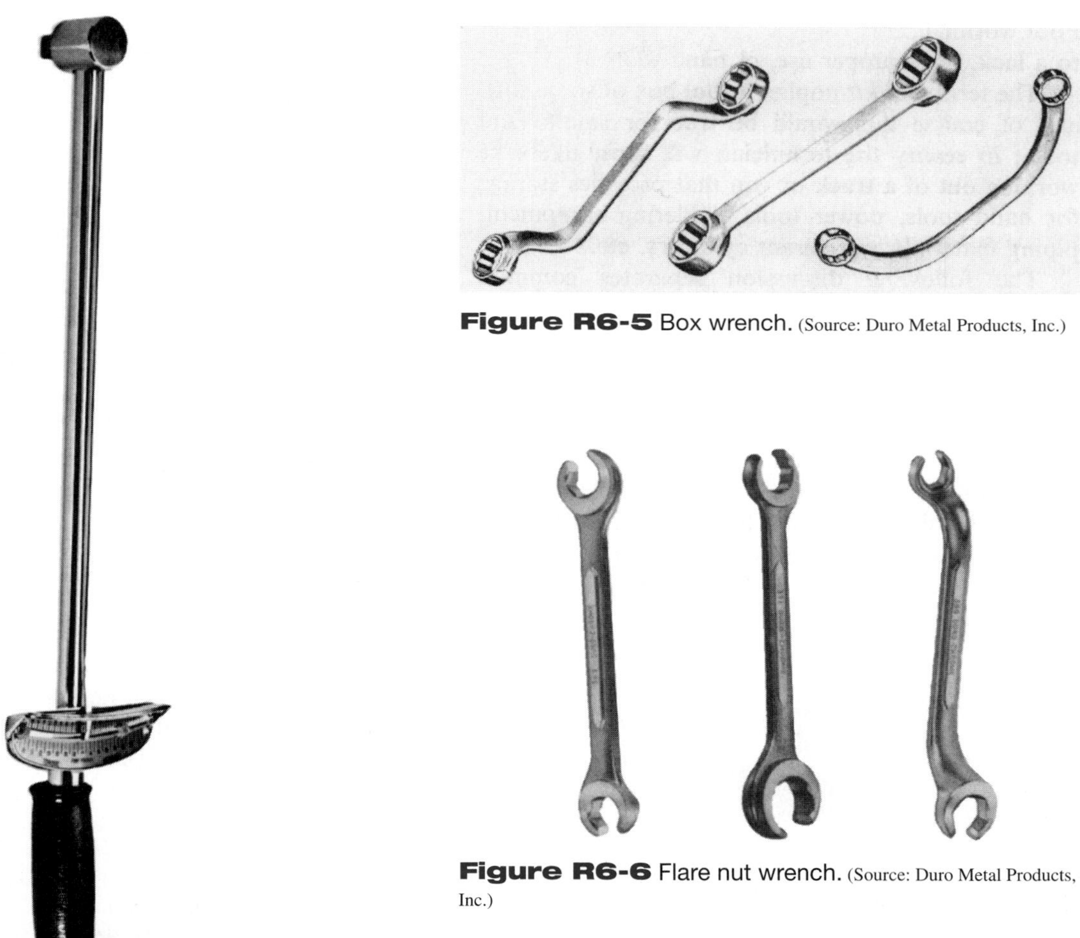

Figure R6-4 Torque wrench. (Courtesy of Klein Tools)

Figure R6-5 Box wrench. (Source: Duro Metal Products, Inc.)

Figure R6-6 Flare nut wrench. (Source: Duro Metal Products, Inc.)

Figure R6-7 Open end wrench. (Source: Duro Metal Products, Inc.)

Figure R6-8 Combination box/open-end wrench. (Source: Duro Metal Products, Inc.)

Adjustable Wrenches

The familiar adjustable wrench as illustrated in Fig. R6-9 is useful where a regular open-end wrench could be used, but the adjustable screw permits fitting the flat to any size object within the maximum and minimum opening. Always use this type of wrench in a manner such that the force is in a down or clockwise direction when tightening a bolt. This keeps the force against the head. If the wrench were used in an opposite manner, it might suddenly loosen and injure the operator.

Typical sizes are 4 in. to 12 in.

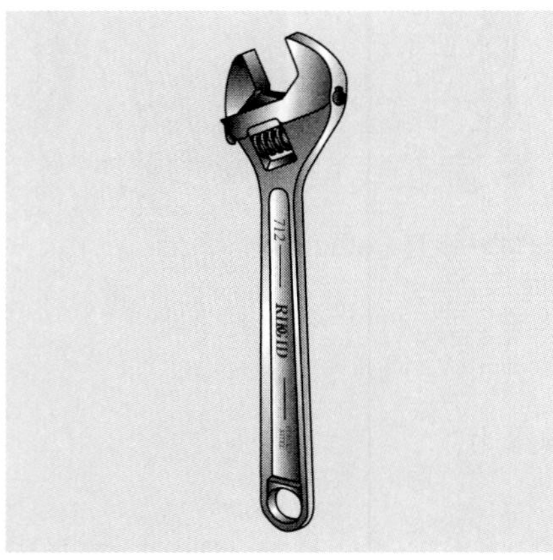

Figure R6-9 Adjustable wrench.

Pipe Wrenches

The pipe wrench (Fig. R6-10) is a common tool used in refrigeration installation and service work to assemble or disassemble threaded pipe.

Figure R6-10 Pipe wrenches. (Courtesy of Ridge Tool Company)

At least two sizes are recommended—an 8-in. size, which can handle up to 3-in.-diameter pipe and a 14-in. size for up to 8-in.-diameter pipe. Some have replaceable jaw inserts to extend the life of the tool.

Another form of adjustable pipe wrench is called the chain wrench. This wrench can make work easier in a confined area or on round, square, or irregular shapes.

Allen Wrenches

Allen wrenches (Fig. R6-11) are necessary for removing or adjusting fan pulleys, fan blade hubs, and other components that are held in place or adjusted by Allen set screws. The wrench goes inside the set screw and may be used at either end.

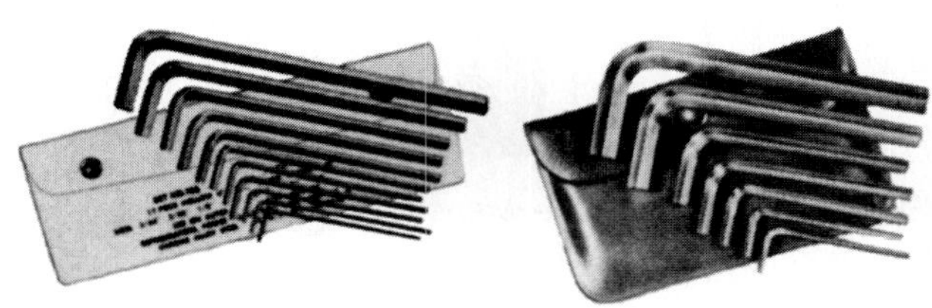

Figure R6-11 Allen wrenches. (Source: Duro Metal Products, Inc.)

Nut Drivers

Nut drivers (Fig. R6-12) are actually a type of small socket wrench. The nut driver consists of a plastic handle that has different drive sockets to fit over the screw or nut head. The nut driver is useful in tightening or removing hex head sheet-metal screws or machine screws that hold equipment panels in place or fasten control-box covers.

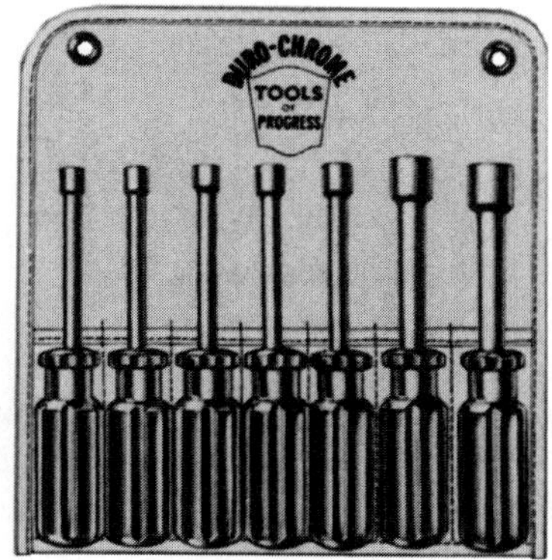

Figure R6-12 Nut drivers. (Source: Duro Metal Products, Inc.)

R6-1.2 Other Hand Tools

Pliers

Pliers are also one of the most frequently used tools and are available in many types (Fig. R6-13). The familiar slip-joint pliers (Fig. R6-13a) are handy for general use. The arc-joint (curved-joint) pliers (Fig. R6-13b) can be used as pliers or in a fashion similar to a pipe wrench. Two or three sizes are recommended for the technician's tool kit. Locking or vise-grip pliers (Fig. R6-13c) are often needed to clamp objects during soldering, for example, thus freeing the hand of the operator.

Several types of pliers are useful for electric and electronic work. In electrical work several different styles of pliers are needed. Lineman's pliers are useful for general electrical work and for cutting wire. Small diagonal-cutting pliers (Fig. R6-13d) and long-nose pliers (Fig. R6-13e) are handy for control and electronic work.

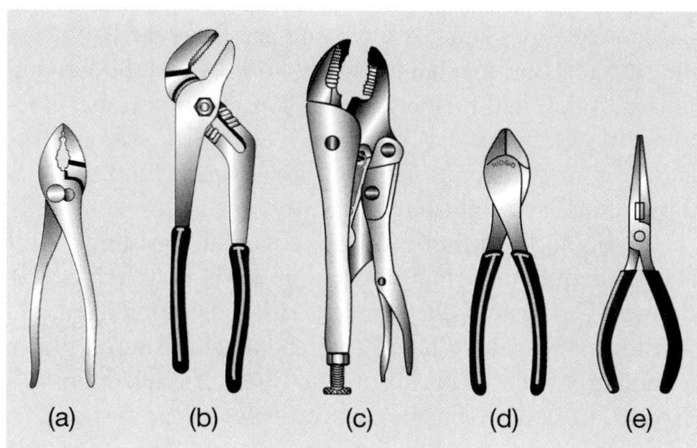

Figure R6-13 Pliers.

Screwdrivers

The well-equipped tool kit must have a variety of screwdrivers. The most common is the flat-blade design (Fig. R6-14a), and a complete set is recommended. Flat blades fit the common single-slot screw and it is important to have a tight fit; otherwise, it is possible to strip the screw slot.

Phillips-tip screwdrivers (Fig. R6-14b) are common in the electrical phases of refrigeration work. Again, a complete set is recommended.

The technician will customize his needs with specialized items like screwdrivers with magnetized blades or blades with a screw-grasping clip. Screwdrivers, in addition to their primary purpose, can be used (with discretion) for light-duty prying, wedging, or scraping, but should never be pounded with a hammer.

Other special drivers such as the torx driver may be needed to fit fasteners that the standard flat blade or Phillips blades will not fit.

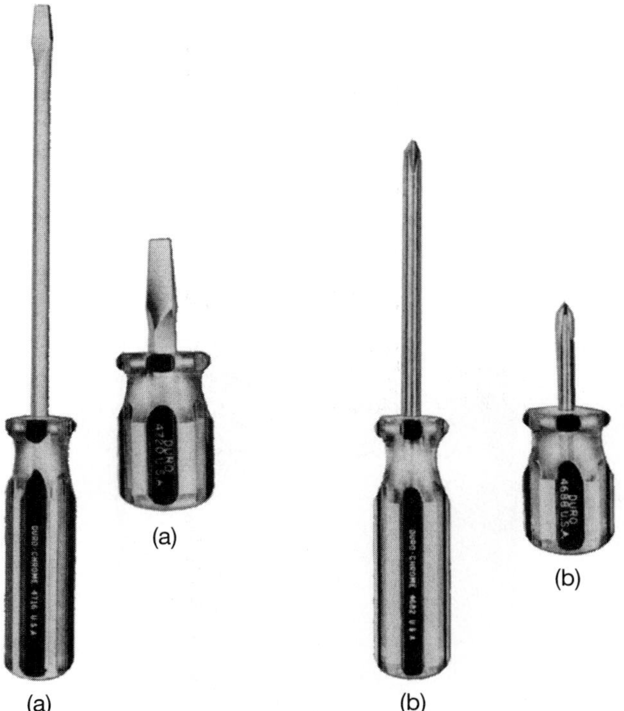

Figure R6-14 Screwdrivers. (Source: Duro Metal Products, Inc.)

Impact Driver

For freeing frozen or rusted screws, an impact driver is recommended. Different types and sizes fit into a hex socket, and the unit can be set to loosen or tighten fasteners. Constructed of heavy metal, a sharp tap with a hammer is usually all that is needed to break loose rusted fasteners that have resisted all other attempts.

Brushes

The use of a wire brush (Fig. R6-15) is recommended to clean the inside of tubing and fittings. These brushes come in sizes from ¼ to 2⅛ in. to fit the outside diameter (OD) of the soldered fitting.

A solder flux brush is also recommended in applying paste. A paintbrush is useful to brush dirt or dust out of a control box, for example, or it may be used to apply cleaning solvent to an object.

Figure R6-15 Brushes.

Files

Files come in several shapes: flat or rectangular, round, half-round, triangular, square, etc. In refrigeration work the common flat file and the half-round are used in preparing tubing for soldering—squaring the end or removing burrs.

Fig. R6-16 illustrates single-cut and double-cut direction files. A single-cut file is used for finishing a surface such as in preparing copper pipe for soldering. A double-cut file is more coarse and would be used where deeper and faster metal removal is needed. A rasp is an extremely coarse double-cut file intended for very rough work.

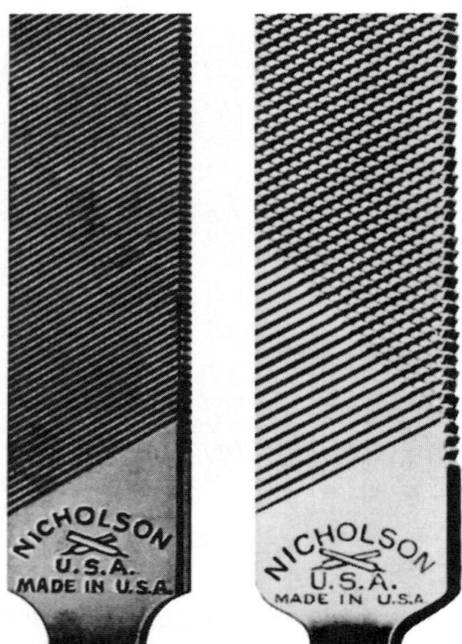

Figure R6-16 Files. (Source: Nicholson)

Vises

A machinist's vise (Fig. R6-17a) can be very useful mounted in a service van. The pipe vise on a tripod stand (Fig. R6-17b) is a requirement both in the field and in the shop to hold tubing or pipe while cutting or threading operations are taking place. *Caution:* When holding copper or other soft metal, be sure that the jaws are soft so as not to mar the surface, and that the clamping pressure does not squeeze tubing out of round.

Tapes and Hand Rules

Start with the familiar 10- or 12-ft (2- or 3-meter) flexible steel tape. It is invaluable for measuring tube diameters, short tube lengths, filter sizes, duct sizes, etc. A 50- or 100-ft (15- or 30-m) steel or fiberglass tape is essential for measuring long piping runs. Fiberglass tapes are recommended when working near electrical equipment as they are nonconductive. A nonrusting steel machinists ruler for more precise measurements and a six-foot folding wood ruler are also useful.

Calipers

Sometimes a service technician must check dimensions of parts very accurately (in thousandths of an inch). A simple caliper (Fig. R6-18) can be very useful for determining (together with a ruler or scale) the outside diameter of tubing or a fan shaft which would be difficult to measure otherwise.

Dial-indicator calipers are relatively inexpensive but are simple to use and can read to 0.001 in. In SI the graduations are 0.05 mm. These are used to align the compressor shaft with the motor shaft when the motor is supplied separately from the compressor.

The typical measuring range is 0–6 in. (0–150 mm in SI). Some typical uses would be accurate measures of pipe or tubing wall thickness and sheet metal thickness.

Drills

Drills (Fig. R6-19) are a must and are frequently used by the refrigeration mechanic in the work of installation and repair. In the field it is assumed a hand-held portable electric drill will be used. A ¼-in. or ⅜-in. chuck size is sufficient for small operations, such as drilling wood, plastic, thin metals, and light-duty masonry work.

Heavy-duty drilling in brick, concrete, and thick steel is often required during installation work. For such operations a ½-in. heavy-duty model drill is recommended.

Either drill should have a variable-speed control, plus a reversing switch to back out stuck bits. Bit selection will depend, of course, on the nature of the material, but as a minimum, a set of high-speed alloy steel bits for metal is a must. Such bits can also be used for drilling wood and

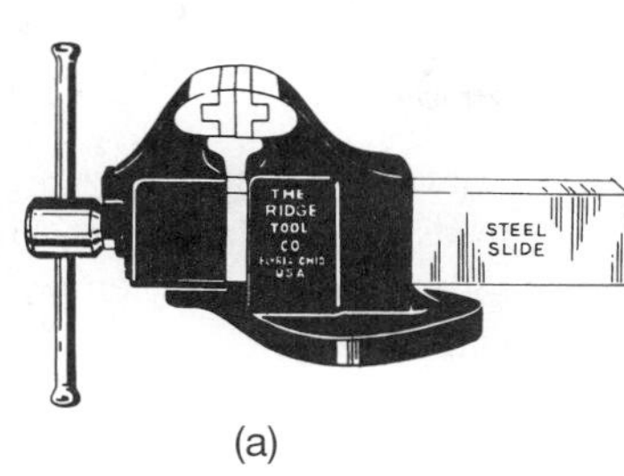

(a)

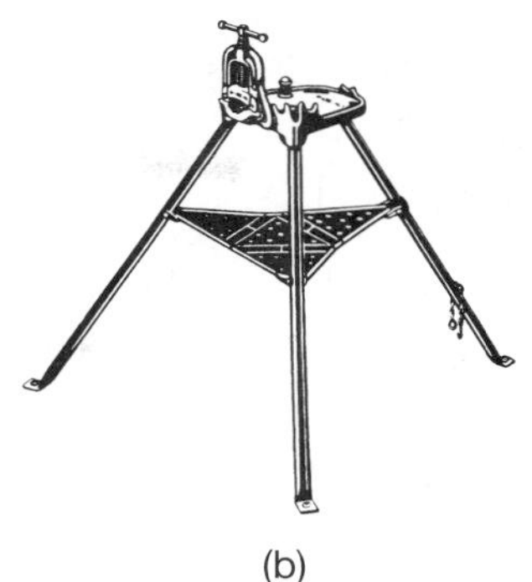

(b)

Figure R6-17 Vises. (Courtesy of Ridge Tool Company)

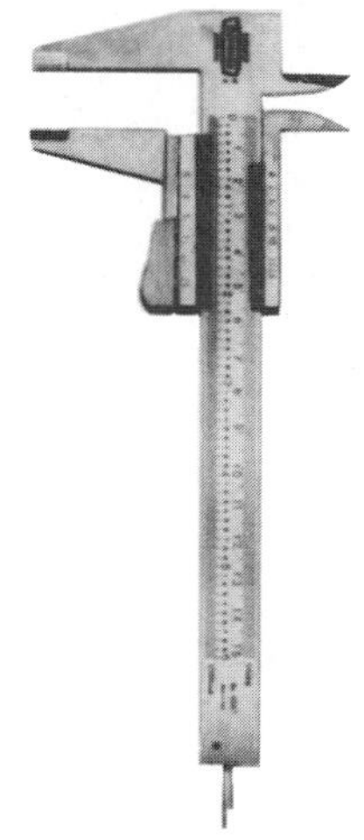

Figure R6-18 Caliper.

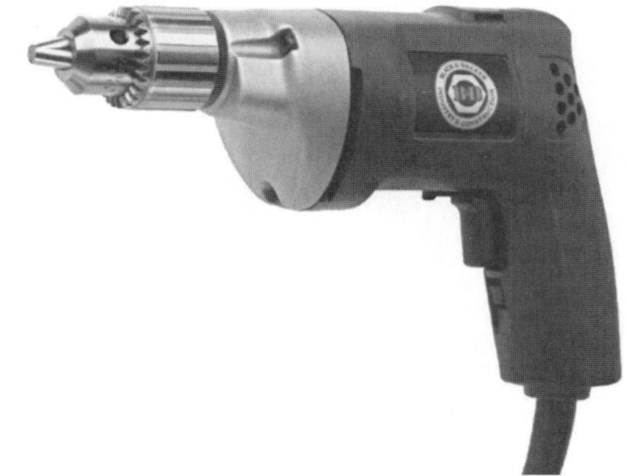

Figure R6-19 Electric drill. (Courtesy of The Black and Decker Corporation)

plastic. However, masonry bits and wood-boring bits may be added for more intensive job requirements.

Rechargeable, battery-operated drilling equipment can be useful time savers where some light drilling or driving of screws is required and no power outlet is handy. These units typically have variable-speed controls and a reversing switch.

For more than occasional drilling of concrete, masonry, etc. for anchors and shields, nothing beats a rugged rotary hammer drill with the proper carbide masonry bits. The savings in time can pay for the tool in a short time.

Accessory Items

In addition to hand tools, the refrigeration technician will need an array of accessory items. Most of these are expendable and will need periodic replacement. Following is a partial listing:

- Abrasive sand cloth for preparing copper tubing and fittings
- Steel wool and cleaning pads
- Cleaning solvent
- Clean rags or paper towels
- Electrical tape
- Duct tape
- Pipe "dope" for sealing threads
- Teflon pipe tape for sealing threads
- Assorted sheet metal and machine screws, nuts, and washers
- Gasket stock, assorted
- Non-detergent lubricating oil for motor and fan bearings
- Fuses: control and power to suit service needs
- Grease
- Stepladder
- Extension ladder
- Small mirror

Safety Equipment

Minimum safety equipment should include:

Hard hats
Safety glasses
Safety shoes
Gloves
Fire extinguisher
First-aid kit
List of phone numbers (hospital, doctor, fire department, police, etc.)

R6-2 ELECTRICAL AND ELECTRONIC INSTRUMENTS

Testing electrical circuits is an important skill that each technician needs to develop. Although practice and experience are significant, a high degree of success is obtainable by following a proven procedure, such as the following:

1. Know the unit electrically. This means understanding the proper function of each control and the sequence of the control operation.
2. Be able to read schematic wiring diagrams and have them available.
3. Be able to use the proper electrical test instruments. Know the instrument. Read instructions carefully before using.

In the following sections, various electrical and electronic instruments will be described along with information on how they are best used. For electrical troubleshooting, the instruments most commonly used are:

1. A voltmeter to measure electrical potential
2. An ammeter to measure electrical current
3. An ohmmeter to measure electrical resistance
4. A wattmeter to measure electrical power
5. A capacitor-checker to measure electrical capacitance

R6-2.1 General Principles

Before specific types of meters and their applications are described, a number of general principles should be observed in the use of meters, as follows:

1. Always use the highest scale on the meter first, then work down to the appropriate scale. This prevents damaging a meter by applying excessive power. An auto-ranging voltmeter will do this automatically.
2. Always check the function of a meter before using it. If it is a battery operated meter, the batteries may be run down. The meter could be damaged during transportation. Other things could happen to affect the readings.
3. In using a clamp-on ammeter, be sure the jaws are around only one wire. If it is around two "hot" wires, the reading could be meaningless. Start at the high range and work down as described in (1) above.
4. Always have an extra set of meter fuses on hand for replacement. Sometimes the proper fuses are difficult to obtain when you need them. The same applies for batteries.
5. Never use an ohmmeter in a circuit that is powered. An ohmmeter has its own power supply and can be destroyed by connecting to a live power source.
6. Some tests require the use of an adapter such as for measuring temperature with a thermocouple, thermistor, or remote temperature device (RTD). Be sure an adapter and accessory sensors are included with the testing tools.

R6-2.2 Multimeters

Many common meters combine a number of functions. This is particularly true for meters measuring volts, amperes, and ohms. These are called multimeters. They are an essential tool for the technician.

Fig. R6-20 illustrates one type of digital multimeter. The features of this instrument are as follows:

- Rotary switch
- Measures volts (AC and DC), ohms, and amps
- Overload protection provided in each range

Fig. R6-21 illustrates another model of a digital multimeter similar to the model shown in Fig. R6-20, with the addition of the following features:

- Auto-ranging—meter will select the range with the best resolution
- Memory-comparison function—stores readings and calculates the difference between two readings
- Locks readings onto the display for easier viewing
- Meter signals with an audible tone when continuity is established.

Figure R6-20 Digital multimeter selections positioned by rotary switch. (Courtesy of TIF Instruments, Inc.)

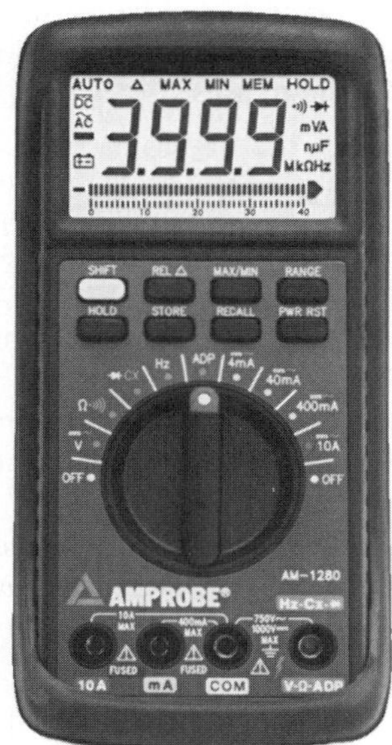

Figure R6-21 Digital multimeter with auto ranging. (Courtesy of Amprobe Instrument®)

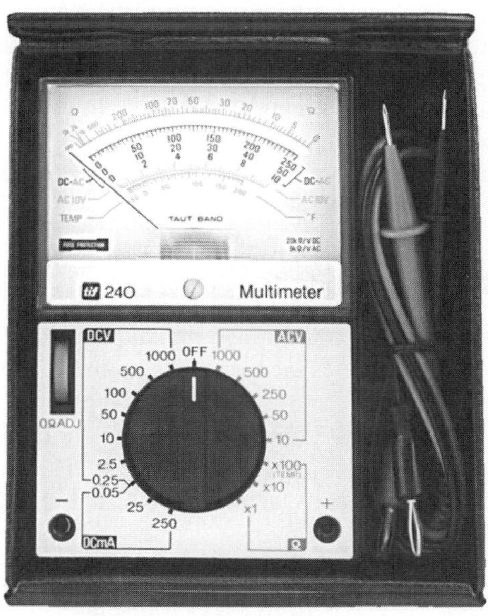

Figure R6-22 Analog multimeter with temperature scale and parallax mirror. (Courtesy of TIF Instruments, Inc.)

Fig. R6-22 illustrates an analog type multimeter with a temperature scale and parallax mirror.

R6-2.3 Clamp-on Ammeters

The clamp-on ammeter is designed for measuring current flow through a single wire, as shown in Fig. R6-23. It ranks with the multimeter as an essential tool for the technician.

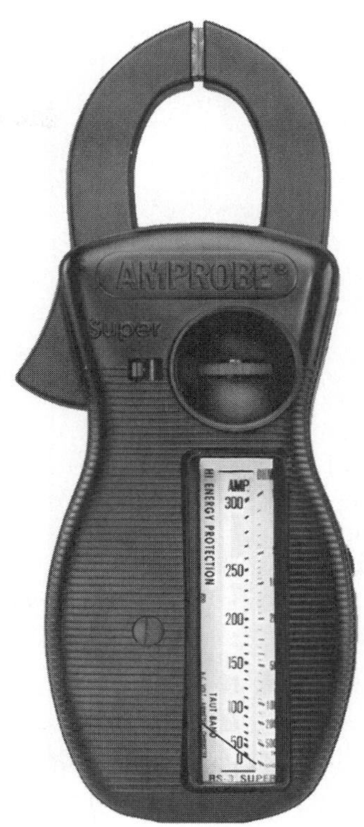

Figure R6-23 Analog clamp-on ammeter with provision for reading volts and ohms. (Courtesy of Amprobe Instrument®)

The clamp jaws can be opened and placed around the wire to be measured. They close automatically by spring pressure when the trigger is released. Fig. R6-23 shows an analog-type clamp-on ammeter. The range scale is also positioned by a thumb wheel revolving the scales viewed through the window. Some ammeters also have volt and ohm scales. To use the voltmeter, leads are inserted in the bottom of the instrument. One of the ohmmeter leads includes the required battery and is plugged into the side of the meter. The shape of the instrument makes it convenient to hold and operate with one hand. A pointer lock is provided when the meter is used in hard-to-see locations. A number of different range scales are available, as shown in Fig. R6-23.

Fig. R6-24 shows a pair of digital type clamp-on ammeters. The principal difference between the two models shown is the size of the jaws and the AC current range.

Fig. R6-25 shows a pocket sized clamp-on ammeter.

R6-2.4 Selecting the Proper Instrument

The three instruments commonly used for electrical troubleshooting are the voltmeter, ammeter and ohmmeter. The question arises, "How does the technician determine which one to use?"

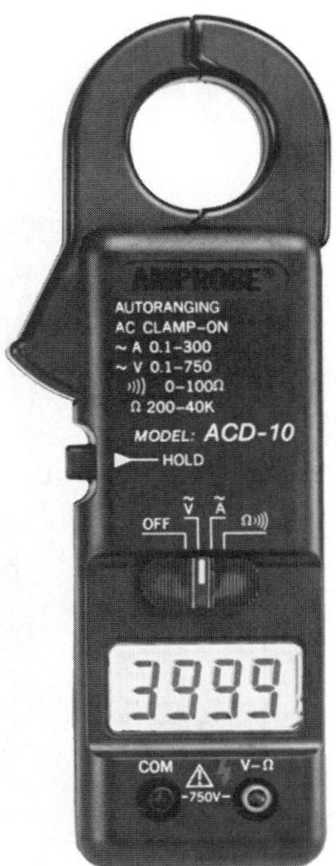

Figure R6-24 Digital clamp-on ammeter with auto-ranging. (Courtesy of Amprobe Instrument®)

The following guidelines are helpful:

1. If any part of the unit operates, the voltmeter and ammeter are normally used.
2. If no portion of the unit operates, check to make sure power is present at the unit terminals. If present, a short circuit could be the problem and after disconnecting power the ohmmeter is probably the best instrument to use.

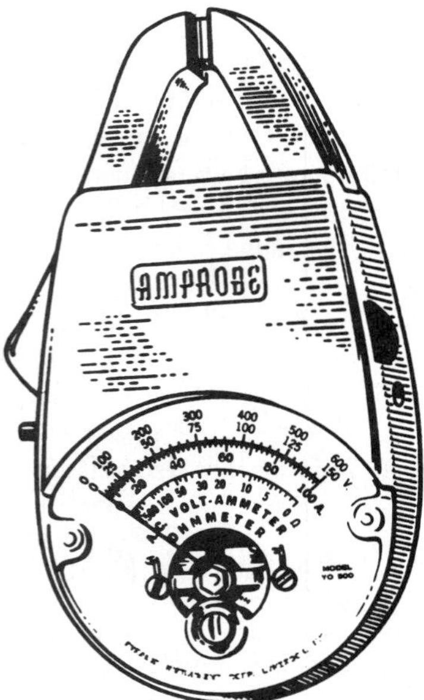

Figure R6-25 Junior analog clamp-on ammeter. (Courtesy of Amprobe Instrument®)

R6-2.5 Wattmeters

Fig. R6-26 shows a digital multimeter with a clamp-on ammeter attachment (also called current transducer). This makes possible the measuring of watts. A digital multimeter is also available with a recorder for monitoring the power being consumed over a period of time.

Fig. R6-27 shows a clamp-on digital wattmeter, battery operated.

This instrument reads true power, including an allowance for the power factor. The meter makes the necessary calculations for power, in accordance with the power formula:

$$\text{Watts} = \text{volts} \times \text{amps} \times \text{power factor}.$$

The meter automatically zeros when not in use and automatically selects the proper range scale when measuring watts. It can be used to measure single, split-phase and three-phase power sources.

Fig. R6-28 shows a clamp-on kW-kWH meter. It can be used to measure three-phase unbalanced loads. It measures the true power, taking into consideration the power factor. It independently displays amps, volts, and kilowatts.

A power factor meter is illustrated in Fig. R6-29. For calculating watts, given the volts and amps, it is necessary to know the power factor.

R6-2.6 Capacitor Tester

Fig. R6-30 shows a multi-range digital capacitance meter designed to measure capacitor values up to 20,000 microfarads.

It will also indicate open, leaking and shorted capacitors. It can be used to check continuity in fuses, wire resisters, transformer coils, alarm circuits, and relays.

R6-3 REFRIGERATION—SERVICING AND TESTING EQUIPMENT

R6-3.1 Temperature Measuring Instruments

Digital Thermometer

Fig. R6-31 shows a digital thermometer (pyrometer) with a retractable sensing probe. The unit will read temperatures in either Fahrenheit or Celsius degrees. It is powered by a replaceable 9V battery and can be purchased in kit form with three different types of sensor probes and a carrying case. It includes an over range and low-battery indicator.

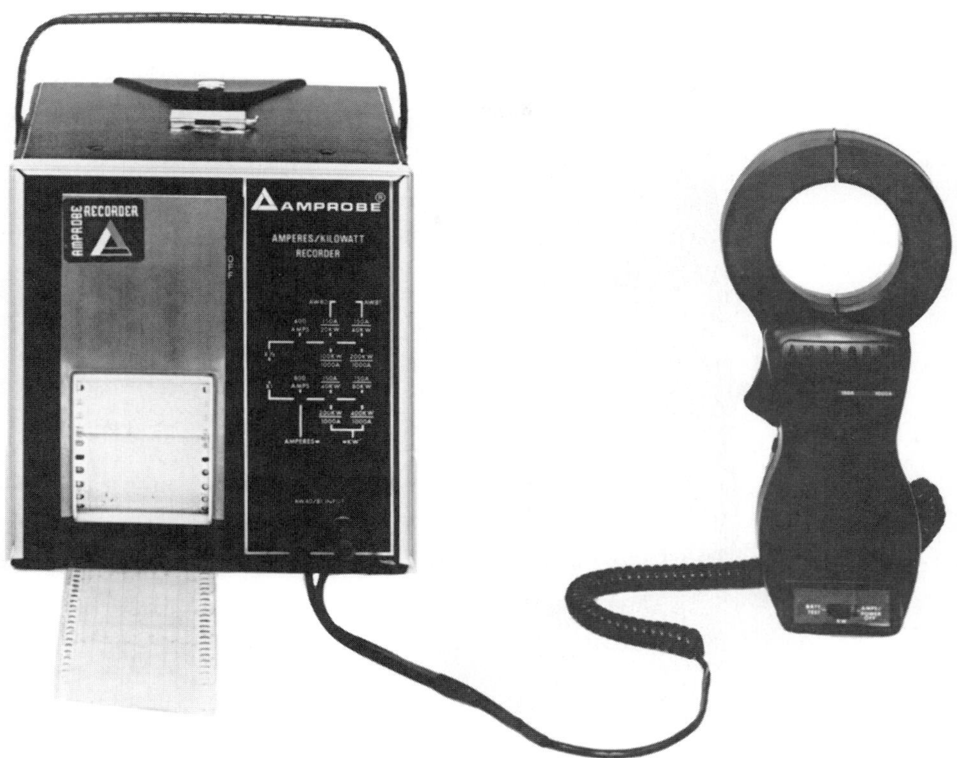

Figure R6-26 Digital multimeter with a clamp-on arrangement for reading watts. (Courtesy of Amprobe Instrument®)

Figure R6-27 Clamp-on digital wattmeter used for single phase, split phase or 3-phase. (Courtesy of TIF Instruments, Inc.)

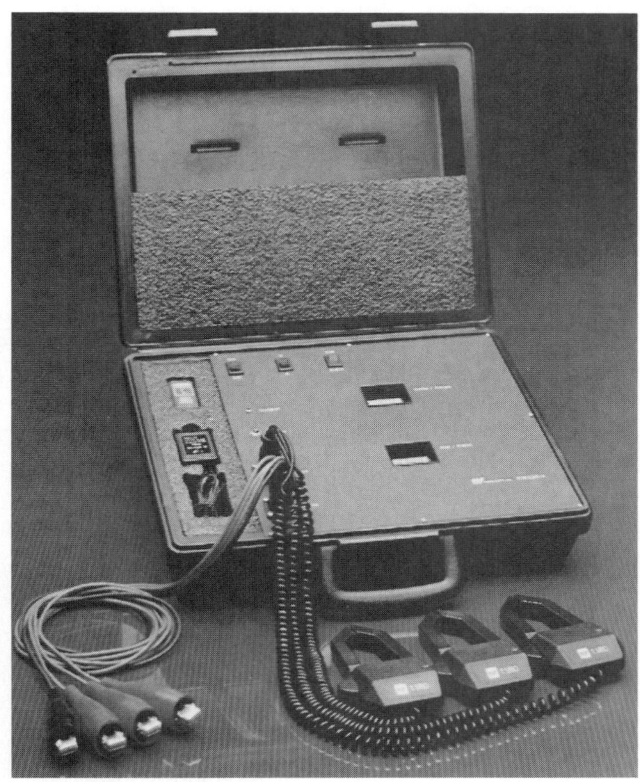

Figure R6-28 Digital clamp-on kW-kWH meter measures true power, not kVA. (Courtesy of TIF Instruments, Inc.)

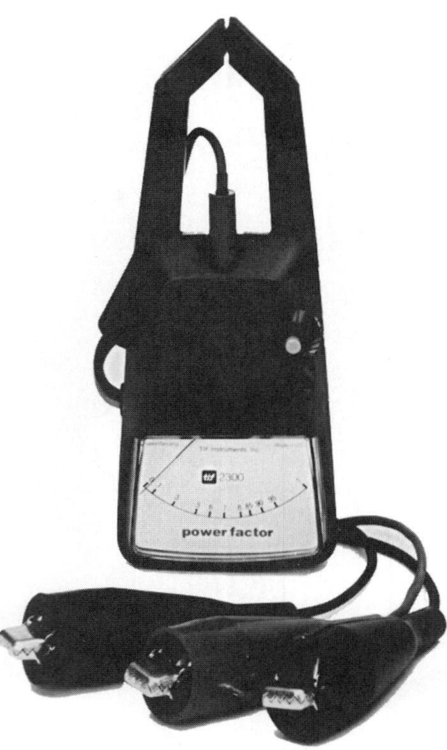

Figure R6-29 Direct reading power-factor meter suitable for single-phase, split-phase and 3-phase power supplies. (Courtesy of TIF Instruments, Inc.)

Figure R6-31 Digital pyrometer with retractable sensor. (Courtesy of Amprobe Instrument®)

R6-3.2 Leak Detectors

Pump-style Electronic Leak Detector

Fig. R6-32 shows a pump-style electronic halogen leak detector. The electronic leak detector draws air over a platinum diode.

This unit is capable of detecting leaks as low as 0.4 ounce per year. It can be used for HCFC, CFC, and HCFC gases. The pump located in the handle draws air directly to the sensing tip. No calibration is required. It is battery operated. It has both a visual and an audible signal which increase in frequency as the leak source is approached.

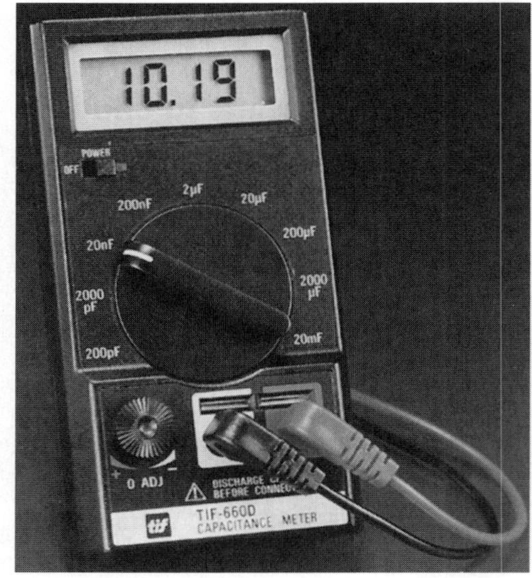

Figure R6-30 Digital capacitor tester for measuring ratings. Also checks leaking capacitors. (Courtesy of TIF Instruments, Inc.)

Figure R6-32 Pump style electronic leak detector. (Courtesy of TIF Instruments, Inc.)

Figure R6-33 Automatic halogen leak detector. Uses the Corona technology. (Courtesy of Robinair Division, SPX Corporation)

Pump-style Leak Detector, Using Corona Discharge Technology

This technology creates a high-voltage corona between the inner tip and surrounding shell. When refrigerant interrupts the electronic field the alarm is triggered. The instrument, as shown in Fig. R6-33, will sense HFC/CFC/HCFC refrigerants by operating a selection switch on the face of the instrument.

Ultrasonic-type Leak Detector

An ultrasonic type leak detector is shown in Fig. R6-34. This instrument will detect any gas leaking through an orifice.

The features and specifications for this meter are as follows:

- Detects pressure or vacuum leaks
- Unaffected by windy, roof-top conditions
- Unaffected by background noise
- Detects ultrasonic noise from arcing electrical switchgear
- Can be used for finding leaks in duct work

Tips for Using a Leak Detector

- In leak-testing a packaged unit, remember that the refrigerant is heavier than air. If there is a leak near the top of the unit, the refrigerant will tend to puddle near the bottom. Therefore, start testing at the top and gradually work your way down.
- If there is a strong wind, such as on the roof of a building, shield the area you are checking to secure a more accurate reading.
- When working with refrigerants, always be sure there is adequate ventilation. Be concerned about the possibility of oxygen depletion.

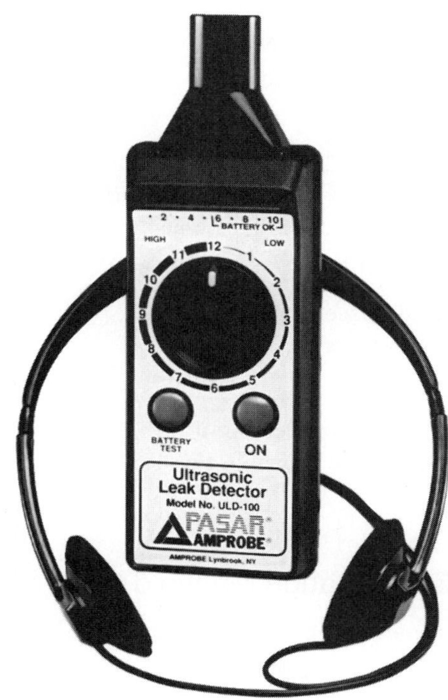

Figure R6-34 Ultrasonic leak detector. (Courtesy of Amprobe Instrument®)

PRACTICING PROFESSIONAL SERVICES

It is extremely important to observe safety procedures when you work with electrical and mechanical refrigeration equipment. Do not work on electrical equipment with live power applied while standing in puddles of water. If you must work on equipment in wet, moist, or humid conditions, turn off the power first. Use insulated tools and rubber gloves to take off covers whenever chances are high for electrical shock.

When using electrical test measuring instruments, be sure to only touch the rubber test probes with your hands. If possible, keep one hand behind your back, ground one end with an alligator clip, and use the other hand to test the electrical circuits. In case of an accidental electrical shock, this will help to prevent electricity from going through your heart.

Inspect power tools for worn electrical cords or poor grounding. When an extension cord is required, be sure to ground the equipment. Avoid using small gauge extension cords. They create too much voltage drop and generate heat, thus overloading circuit protectors. Don't use accessories that convert three-prong plugs into two-prong adaptors. They do not provide the necessary grounding required for electrical safety.

PREVENTIVE MAINTENANCE

Test equipment and electronic instruments such as gas analyzers, power meters, vacuum pumps, and micron gauges need maintenance and calibration on an annual basis, sometime twice a year if they have been used rigorously. This usually requires shipping the equipment and instruments off to the manufacturer's factory or an authorized dealer. Slow seasons, such as spring and fall, are good times to do this. Equipment that is not properly maintained is prone to damage and can have a short life span. Develop a schedule with your supervisor or peers to rotate all test equipment and instruments regularly.

R6-3.3 Vacuum Pump

Fig. R6-35 shows a typical high-performance vacuum pump, capable of fast pulldown and thorough evacuation. The specifications for this pump are given as follows:

Free air displacement:	6 cfm
Number of stages:	Two
Factory micron rating:	20 microns

Typical hose connections using a vacuum valve for positive shutoff are shown in Fig. R6-36.

Figure R6-35 Two-stage vacuum pump, 6 cfm capacity. (Courtesy of Robinair Division, SPX Corporation)

R6-3.4 Gauge Manifold Sets

Three types of gauge manifold sets are illustrated in Fig. R6-37.

Fig. R6-37a shows the digital calibratable type. The combination low-side gauge reads from 30 in. Hg to 99.9 psi. The high-side gauge reads from 0 to 500 psi. The 4-way block is equipped with three refrigerant hoses and one evacuation hose.

Fig. R6-37b shows a different gauge manifold set. These liquid-filled analog gauges are filled with glycer-

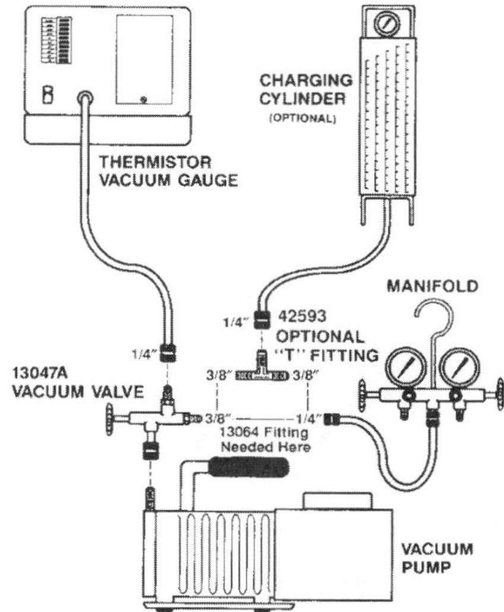

Figure R6-36 Vacuum valve assembly. Provides positive shutoff of the vacuum pump. (Courtesy of Robinair Division, SPX Corporation)

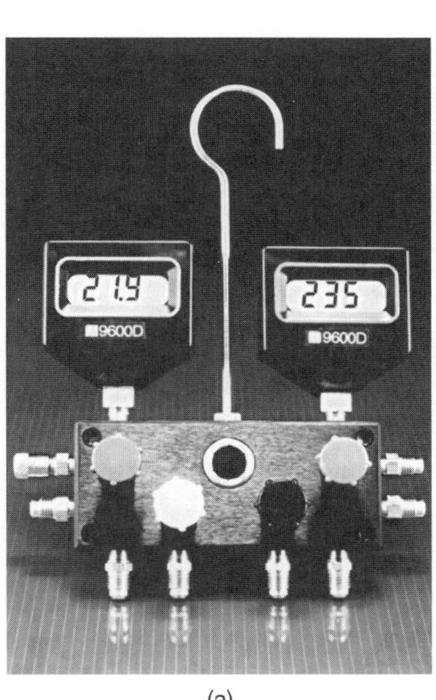

(a)

(b)

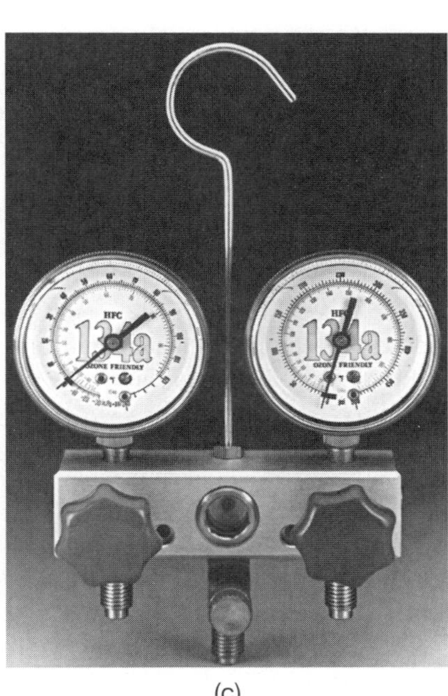

(c)

Figure R6-37 Three types of gauge manifolds. (Courtesy of TIF Instruments, Inc.)

ine to dampen pulsations, lengthen service life, and improve accuracy. The low-side gauge reads pressures from 30 inches of Hg to 120 psi. The high pressure gauge reads from 0 to 500 psi. The 4-way block is equipped with three refrigerant hoses and one evacuation hose.

Fig. R6-37c shows a gauge manifold set with dry gauges mounted on a 2-way refrigerant block.

All digital, glycerine-filled and standard dry gauges are also available with Metric Bar, kPa and kg/cm^2 scales.

The gauge manifold is one of the handiest tools in the technician's kit, used for checking operating pressures, adding or removing refrigerant, adding oil, and performing other necessary operations.

The manifold has five connections. The two top connections hold the compound and the high pressure gauges. The compound gauge, placed on the left side, reads pressures on the low pressure side of the system from 30 in. Hg vacuum to usually 250 psig pressure. The high-pressure gauge reads pressures on the high side of the system from 0 to 500 psig. Older gauges may have lower maximum pressures.

The bottom of the manifold has three connections. To these openings are attached high-vacuum hoses, usually capable of being leak-tight down to 50 microns or less.

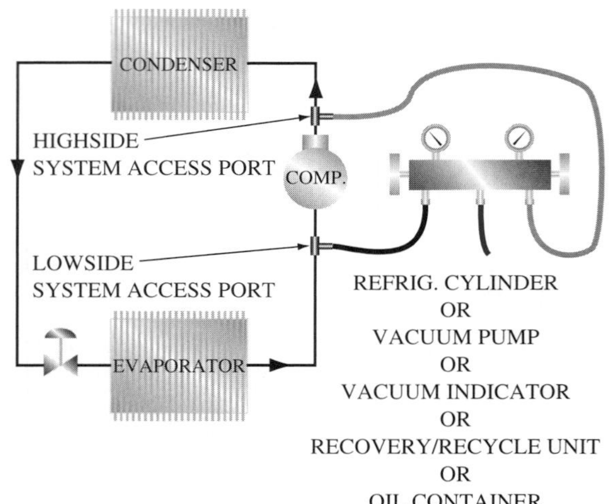

Figure R6-38 Gauge manifold attached to a typical system.

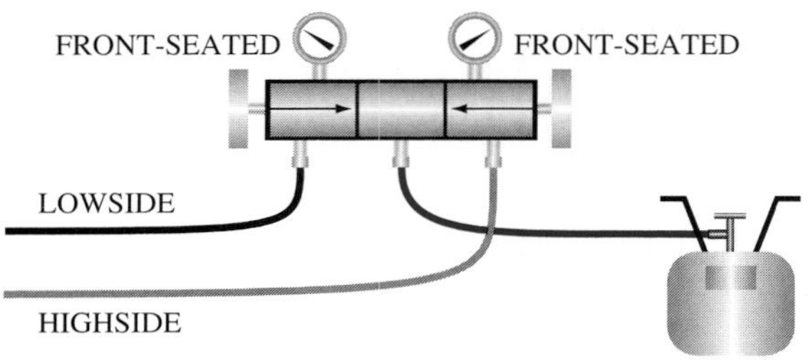

Figure R6-40 Gauge manifold, both valves front-seated.

The left hose is connected to the low side of the refrigeration system being serviced. The right hose is connected to the high side of the system. The center hose has a number of uses, all associated with servicing the system.

Refrigerant hoses are available with anti-blow-back valves to prevent venting of refrigerant.

A gauge manifold attached to a typical system is shown in Fig. R6-38.

The exteriors of gauge manifolds are often color-coded. The compound gauge and low-side hose are blue. The high-side gauge and high-side hose are red. The center utility hose is usually white or yellow. The center hose is useful for connecting to the charging cylinder, vacuum pump, vacuum indicator, recovery/recycling unit or container. The container may be a cylinder for collecting the refrigerant during the recovery process or an oil container used for charging the system with lub ricating oil.

Two positions of the service valves for adjusting the gauge manifold are shown in Fig. R6-39.

The valve on the left in Fig. R6-39 is front-seated. In this position the compound gauge reads the pressure in the low side of the system. The opening to the utility port is shut off.

The valve on the right side is back-seated. In this position the high side of the system is open to the utility port for whatever use is desired.

There is also a third possible position of the valve (not shown) where the connection to the utility port is only partially open—called the cracked position. In this position the flow through the utility port can be controlled by minor adjustments of the valve.

When both service valves are front-seated (Fig. R6-40), the following service operations can be performed:

1. Connecting or disconnecting the gauge manifold from the system.
2. After the gauge manifold is in place, measuring both the low-side and the high-side system pressures without disturbing the system (no flow).

Fig. R6-41 shows the connections at the bottom of the gauge manifold when set for a charging operation.

The left valve in Fig. R6-41 is back-seated; the right valve is front-seated. The hose from the utility connection

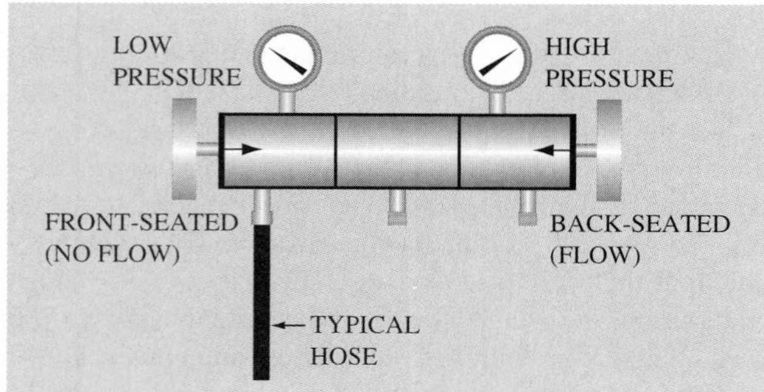

Figure R6-39 Manifold with one valve open and one valve closed.

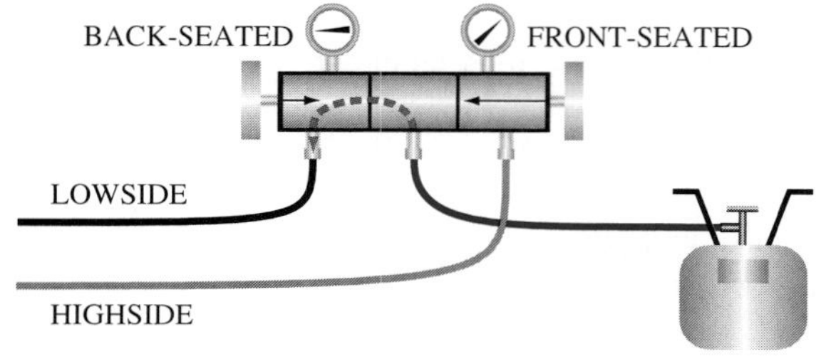

Figure R6-41 Gauge manifold for charging refrigerant into the low side.

goes to a refrigerant cylinder. When the valve on the cylinder is opened, the refrigerant can flow from the cylinder into the low side of the system. This is a typical arrangement for charging refrigerant vapor with or without the compressor running.

The low pressure gauge reads the pressure at the outlet of the cylinder. The high-pressure gauge reads the pressure on the high-side of the system.

Fig. R6-42 shows the service valve settings for removing the refrigerant from the high side of the system.

The left valve in Fig. R6-42 is front-seated and the right valve is back-seated. These are the settings used when a vacuum pump or a recovery unit is evacuating the high side of a system.

The student/technician should be cautioned not to operate a hermetic compressor while evacuating the system because the loss of refrigerant would reduce the normal cooling of the seated-in motor.

With the compressor off and the refrigerant cylinder connected to the utility opening the refrigerant would migrate to the cylinder, provided it was at a lower temperature. When both service valves are back-seated and either a vacuum pump or a recovery unit is attached to the utility connection (Fig. R6-43), both gauges register the pressure in the utility hose.

This setting allows the evacuation system to remove refrigerant from both the low side and the high side at the same time. Based on there being a restriction at the metering device, such as a closed expansion valve, this arrangement is effective in performing complete evacuation.

When both of the service valves are back-seated with the utility connection capped (Fig. R6-44), the difference in pressure between the two sides causes a flow through the valve until the pressures equalize. This can occur accidentally. If it does, both gauges will read the same.

Fig. R6-45 shows the position of the valves for liquid charging into the receiver on the high side of the system, with the compressor shut off. The left gauge is front-seated and the right valve is back-seated.

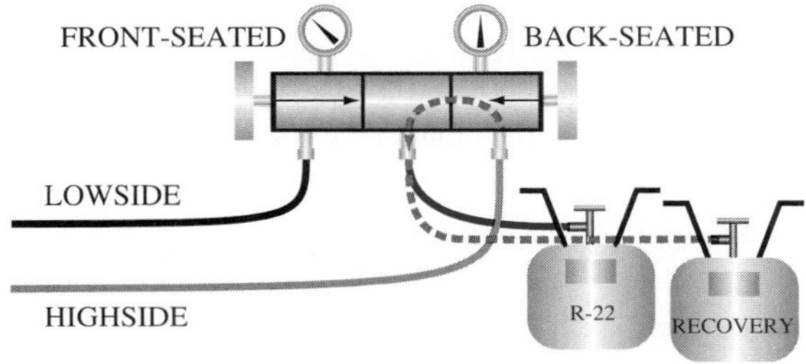

Figure R6-42 Manifold valve settings for removing refrigerant from the high side.

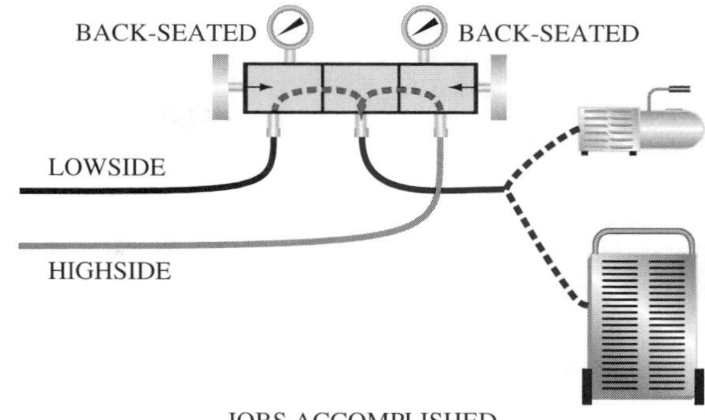

Figure R6-43 Gauge manifold, both valves back-seated.

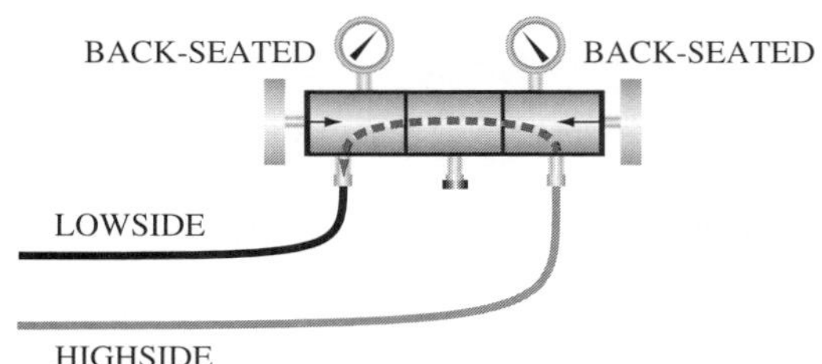

Figure R6-44 Gauge manifold, bypass position.

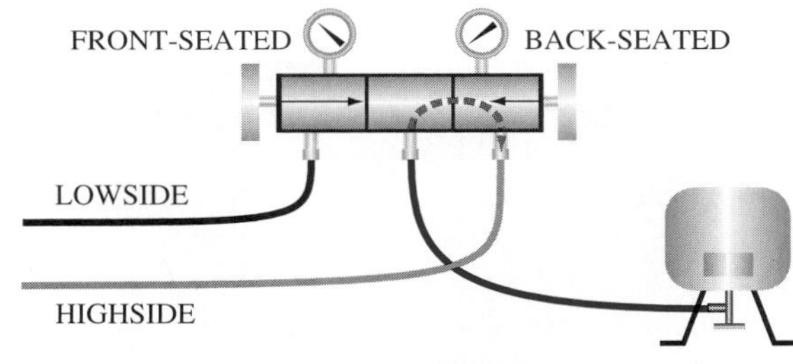

Figure R6-45 Manifold valve settings for liquid charging into the receiver.

It is also feasible to charge liquid to the high side of the system, entering on the inlet side of the expansion valve, with the compressor running. For this arrangement, the valve positions are the same as shown in Fig. R6-45.

A modified gauge manifold that has been designed especially for evacuating and dehydrating a system (Fig. R6-46) has larger hose connections and uses larger hose. It has multiple openings on the utility connection to accommodate various devices using this connection. These innovations serve to speed up the evacuation process.

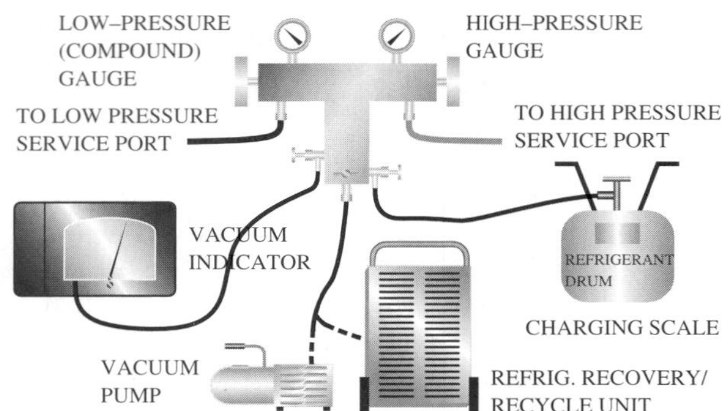

Figure R6-46 Manifold gauges hooked up to a vacuum pump or a refrigerant recovery unit.

A multiple port utility connection can be fabricated in the field that would facilitate use of the standard gauge manifold. Some manufacturers offer these utility connections as an accessory for the gauge manifold.

R6-3.5 Charging Cylinder

A charging cylinder (Fig. R6-47) provides an excellent way to measure a refrigerant charge. The calibrated shrouds make it simple to compensate for volume fluctuations due to temperature variations. Cylinders with heaters make charging faster and more complete. All cylinders are protected by a pressure-relief valve. Gauges are available in psi, kgcm^2 and kPa. Sizes range from 2½ to 10 pounds of refrigerant.

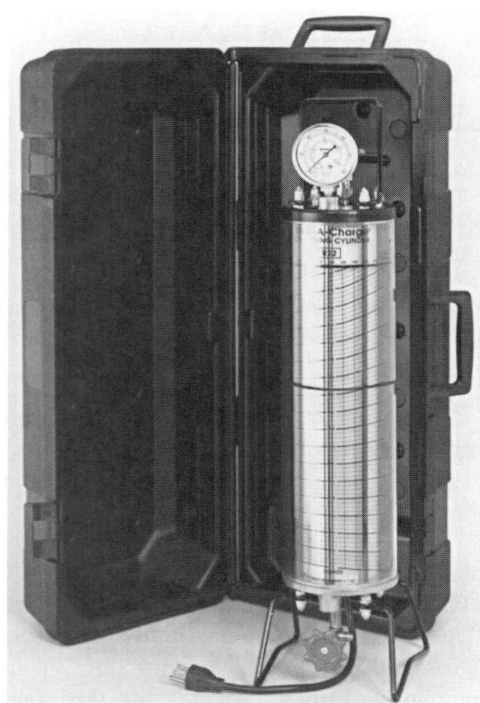

Figure R6-47 Dial-a-Charge cylinder. Gauges are psi or kgcm^2/kPa. (Courtesy of Robinair Division, SPX Corporation)

Figure R6-48 Electronic charging scale. (Courtesy of Robinair Division, SPX Corporation)

R6-3.6 Charging Scale

An electronic charging scale, which will measure the charge by weight, is shown in Fig. R6-48. It is designed for refrigerant tanks up to 110 pounds (50 kg). A solenoid valve stops the charge when the programmed weight has been dispensed.

R6-3.7 Micron Vacuum Gauge

A digital micron gauge (Fig. R6-49), plus a vacuum indicator, instantly recognizes an increase or decrease in vacuum, moisture presence, and leaks.

A 4-digit LCD provides exact readings of vacuum level. It has dual power supply, $9V_{DC}$ or AC. It has auto ranging and a field-replaceable sensor.

R6-3.8 Electronic Sight Glass

An electronic sight glass is shown in Fig. R6-50. This ultrasonic instrument has both visual and audible bubble detection. The LED display illuminates actual bubbles passing between the sensor clamps. Transducer clamps fit tubing from ⅛ to 1¼ in. in diameter.

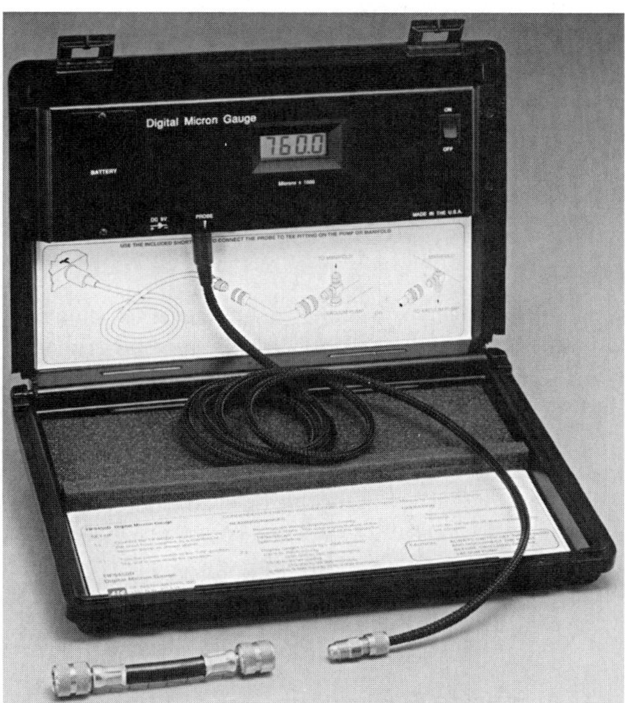

Figure R6-49 Digital micron gauge with 9V battery or AC. (Courtesy of TIF Instruments, Inc.)

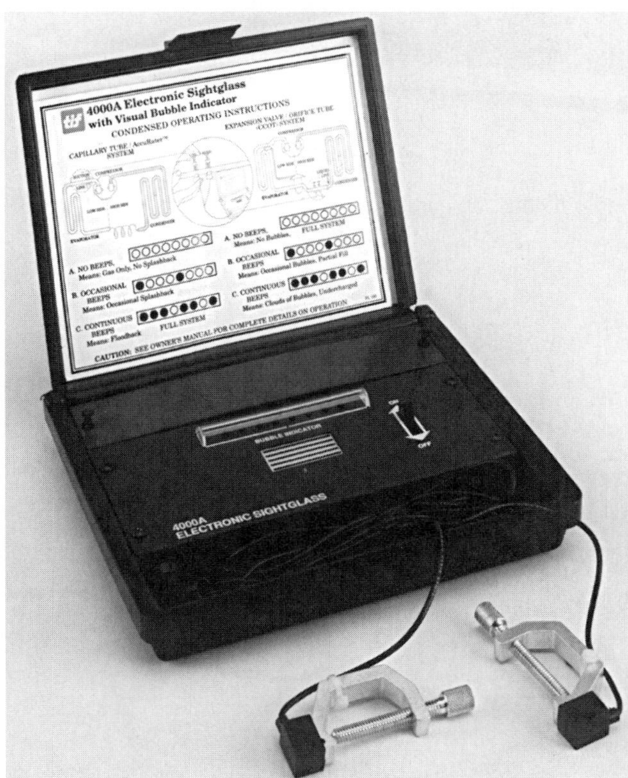

Figure R6-50 Electronic sight glass with both visual and audible bubble detection. (Courtesy of TIF Instruments, Inc.)

R6-4 AIR FLOW—MEASURING AND TESTING EQUIPMENT

Diaphragm Type Differential Pressure Gauge

These gauges (Fig. R6-51) are suitable for reading low pressure (inches of water) applications such as measuring the pressure drop across an air filter. Filters are typically changed when the pressure drop reaches certain limits.

Pitot tube and manometer (Fig. R6-52) are used for measuring air velocity (and airflow) in duct work. The pitot tube is installed in the duct and connected to a manometer which reads both total pressure and static pressure. From these two readings the velocity can be calculated.

Electronic flow hood is shown in Fig. R6-53. It is used to read the air volume supplied by a grille or diffuser. This device balances air flow in large commercial air-conditioning systems.

Figure R6-51 Diaphragm-type differential pressure gauge. (Courtesy of Dwyer Instruments, Inc.)

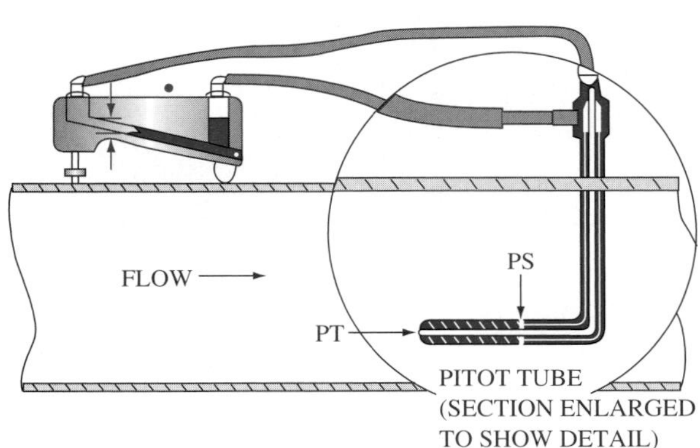

Figure R6-52 Pitot tube and manometer.

Figure R6-53 Electronic flow hood. (Courtesy of Shortridge Instrument)

Volume Air Balancer

An air balancer (Fig. R6-54) makes direct measurements of cfm in residential air-conditioning systems.

It can be used on grilles, registers, and diffusers. It averages the outlet conditions. It also reads air velocity in fpm. It requires no power or special maintenance.

Anemometer

A hand-held, battery-powered anemometer is shown in Fig. R6-55. It has a low-battery indicator and ranges of 0–600 fpm, 500–6000 fpm.

Airflow Indicator

A pocket-sized airflow indicator (Fig. R6-56) has a range of 0–1000 fpm, readability within 50 fpm, instant readings from any position, with no leveling or balancing required. A carrying case is available.

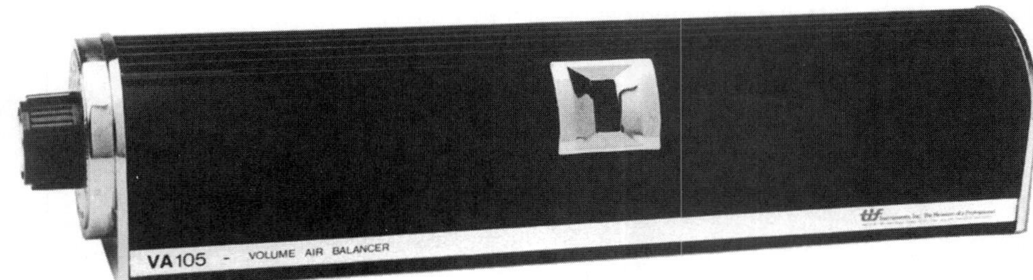

Figure R6-54 Volume air balancer. No power required. (Courtesy of TIF Instruments, Inc.)

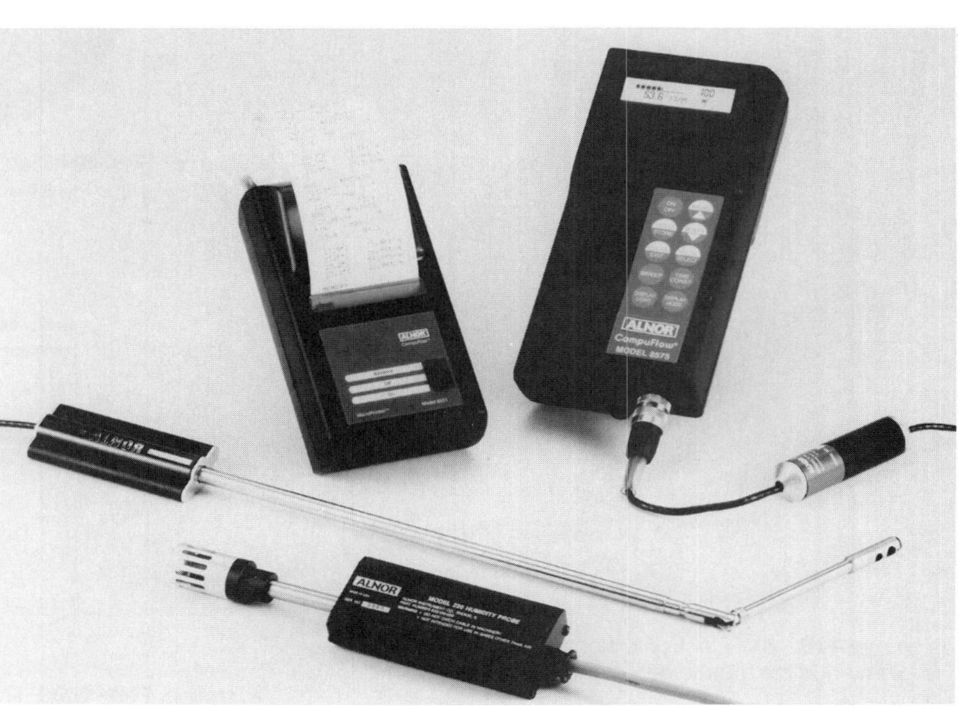

Figure R6-55 Hand held thermal anemometer, battery powered. (Courtesy of Alnor Instrument Company)

Figure R6-56 Pocket sized airflow indicator. Range, 0–1000 fpm. (Courtesy of Bacharach, Inc.)

Air Meter Kit

A direct-reading air meter kit is shown in Fig. R6-57, complete with meter, probes, and air-velocity calculator.

Psychrometers

Two types of psychrometers (Fig. R6-58) measure wet and dry bulb temperatures, required to determine relative humidity. Wet and dry bulb readings are also used in plotting air-conditioning values on the psychrometric chart.

The sling psychrometer is a hand-operated instrument. The air movement across the bulbs is created by rotating the thermometer around the handle. In a psychrometer, a fan is used to move the air across the thermometer bulbs.

R6-5 HEATING—SERVICING AND TESTING EQUIPMENT

The following is a list of the types of instruments which are most useful to the technician in testing heating system efficiencies:

1. Draft gauge (Fig. R6-60)
2. Smoke tester (Fig. R6-61)

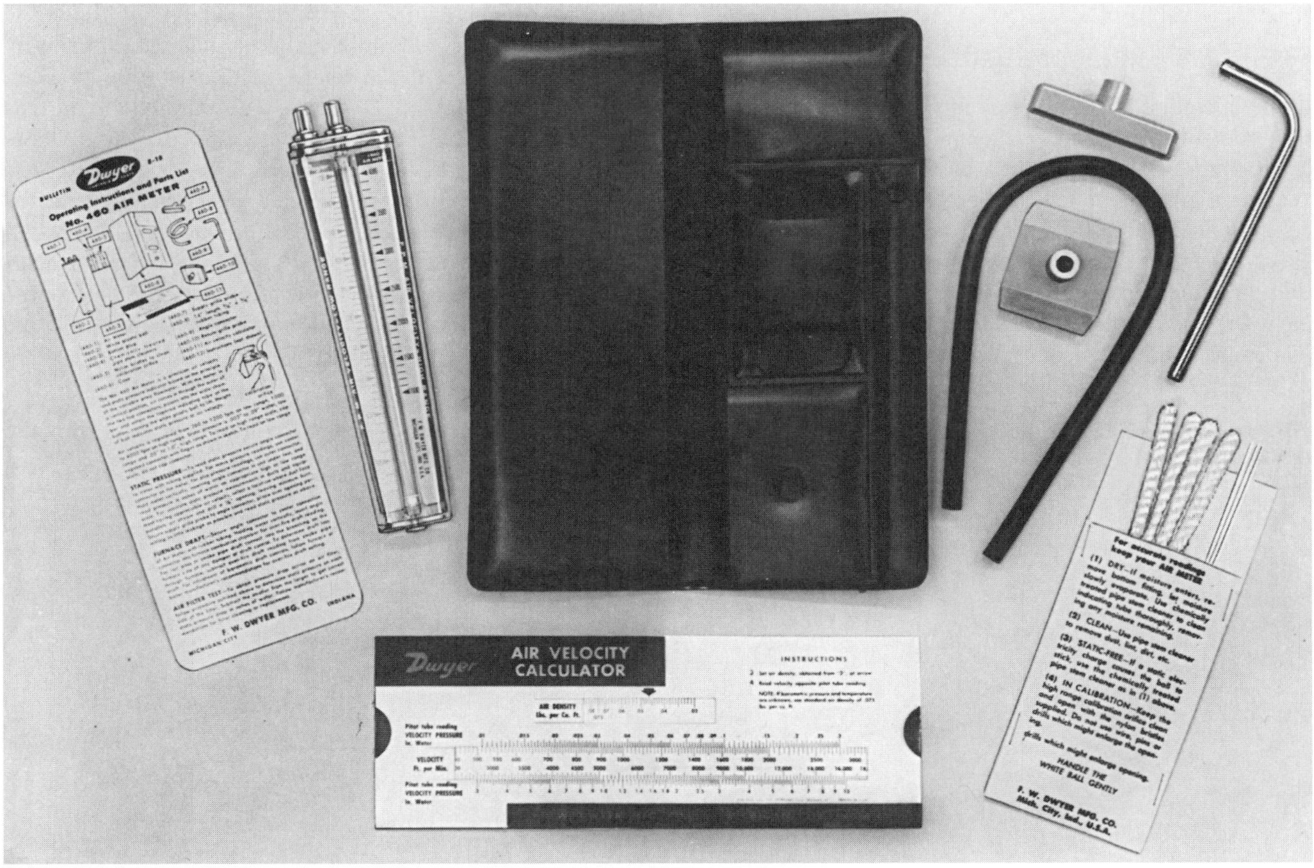

Figure R6-57 Air meter kit for reading air velocity and static pressure. (Courtesy of Dwyer Instruments, Inc.)

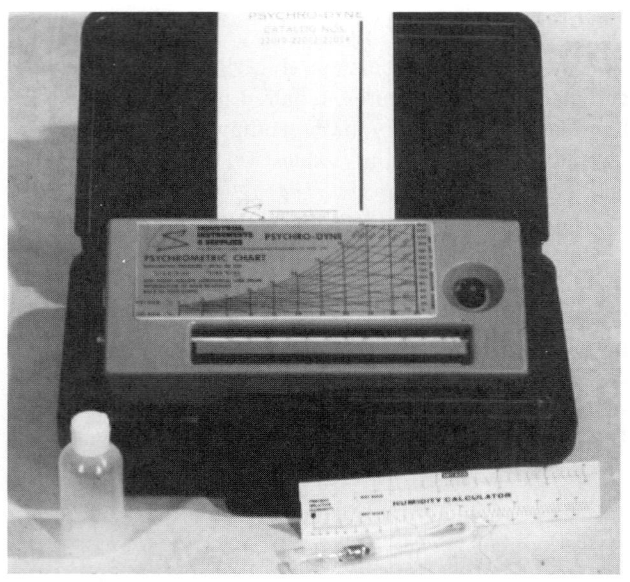

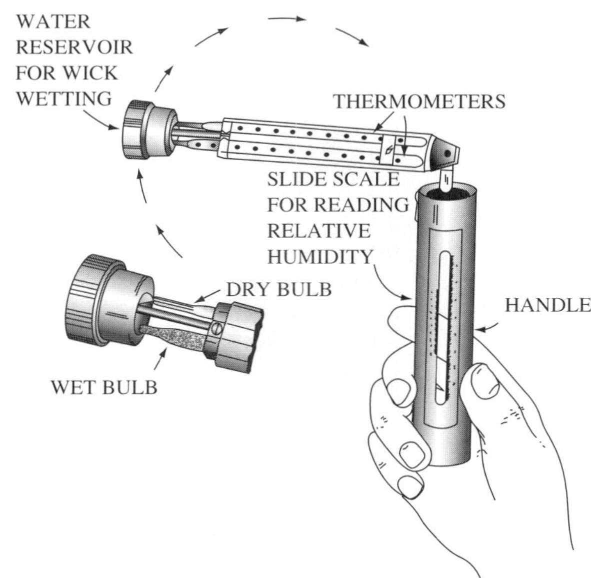

Figure R6-58 Two types of psychrometers. The model on the left has a power-drive fan. (Left photo, copyright 1966, Industrial Instruments and Supplies, Inc., Southhampton, PA)

3. Flue gas analyzer (Fig. R6-62)
4. Stack thermometer (Fig. R6-63)
5. Complete combustion kit (Fig. R6-64)

R6-5.1 Measurement of CO_2

In the operation of heating systems, it is essential that the CO_2 content of flue gases be maintained as high as practical to improve the efficiency of the heating unit while keeping a low smoke level. In practice, as far as oil burners are concerned, 10–12% CO_2 is desirable. For natural gas furnaces, the CO_2 range should be 8.25–9.50%. The CO_2 measurement, plus the stack temperature measurement, are used by the combustion slide rule to indicate the efficiency of the furnace.

Draft Gauge

Fig. R6-59 shows a draft gauge, dry type, with a range of +0.05 to −0.25 inches of WC (water column). This instrument is supplied with a 5-inch draft tube to be inserted in the flue pipe, with a 9-foot rubber tubing extension.

There are two places where the draft is usually taken. One is in the flue pipe at the furnace exit and the other is in the door of the furnace (over the fire). By taking these two measurements, draft problems can be analyzed. The draft tube is inserted in the flue pipe opening and the meter registers the amount of draft, in inches of water column. The probe is then inserted in the door of the furnace to read the draft at this location. If there is less draft at the furnace door, it is an indication that there is a leak in the heat exchanger that needs to be corrected.

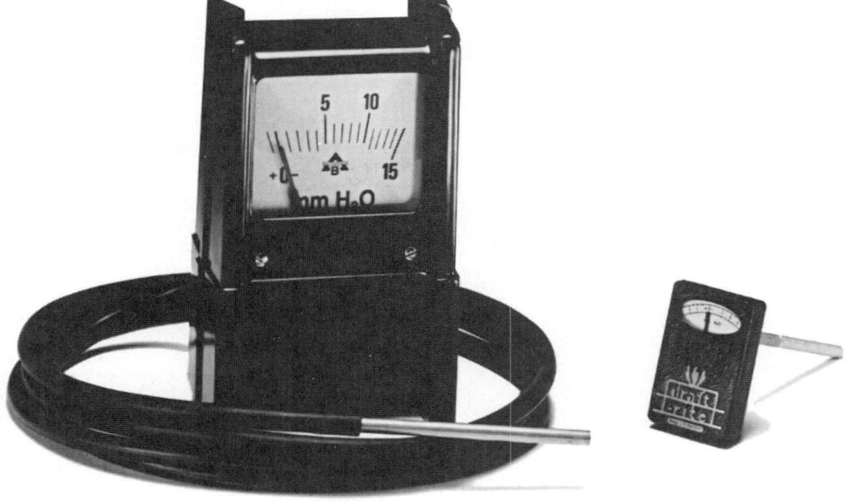

Figure R6-59 Analog draft gauge with a range from +0.05 to −0.25 in. WC. (Courtesy of Bacharach, Inc.)

Fig. R6-60 shows the location of the test hole in the flue pipe for measuring the draft. On an oil burner installation, such as shown in Fig. R6-60, the measurement is made on the furnace side of the draft regulator. The second test hole is for measuring the flue temperature.

Smoke Tester

The tube from the tester shown in Fig. R6-61 is inserted in one of the test holes during operation of the furnace and a sample of the flue gas is passed through a filter. Ten pump strokes of the tester are required to get a proper sample. The color of this filter, as compared to the standard smoke scale, identifies the smoke density in the flue gases (for oil burners this is usually scale number 1 or 2 on a scale 0–9). Any reading above 2 would require adjusting the burner.

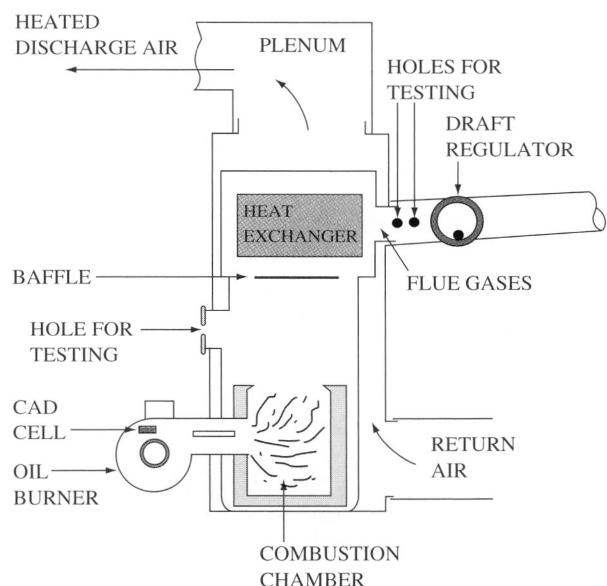

Figure R6-60 Heating plant equipment showing the location of instrument testing holes. (Courtesy of Bacharach, Inc.)

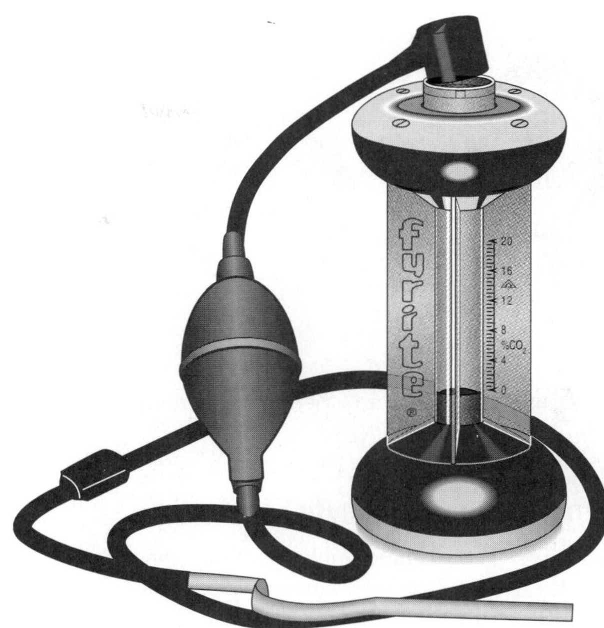

Figure R6-62 Flue gas analyzer, indicates percentage of CO_2 in flue gases. (Courtesy of Bacharach, Inc.)

Flue Gas Analyzer

This instrument (Fig. R6-62) measures the CO_2 (carbon dioxide) in the flue gases. The probe is inserted in the flue pipe during the operation of the furnace and the rubber bulb at the end of the tubing is squeezed 18 times to obtain a representative sample. A similar tester is available to measure O_2 (oxygen) in the flue gases. These gases can also be measured by electronic sensors and instruments.

Stack Temperature Thermometer

This is a bimetal type of thermometer with a range from –40°F to 1,000°F (Fig. R6-63). It is possible to recalibrate this thermometer if necessary.

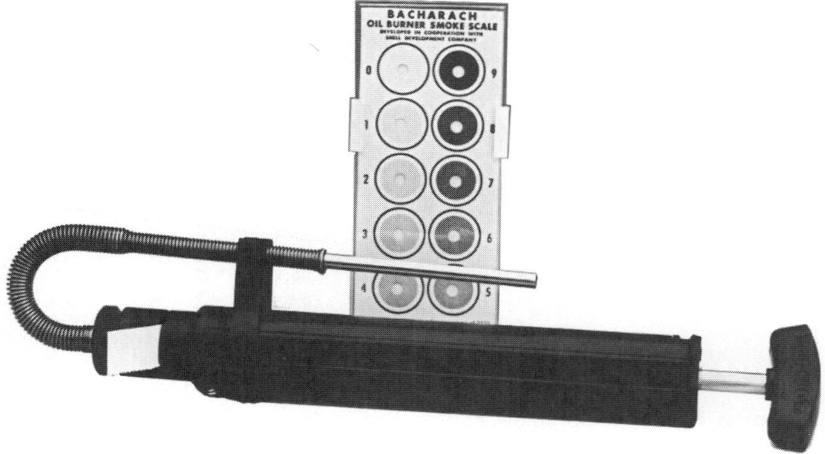

Figure R6-61 Smoke tester with filter strips for evaluating smoke density. (Courtesy of Bacharach, Inc.)

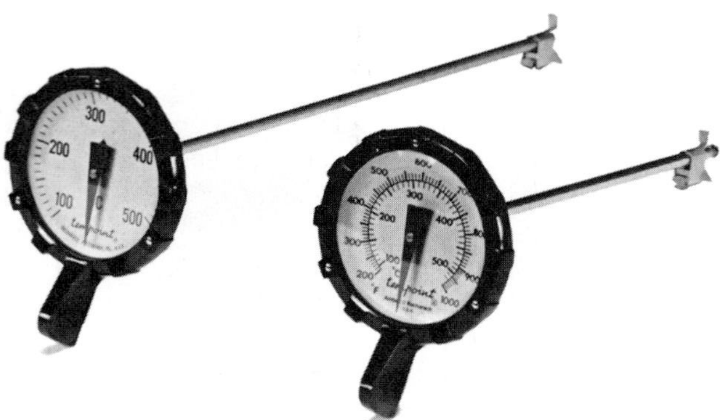

Figure R6-63 Stack thermometer, range –40°F to 1000°F. (Courtesy of Bacharach, Inc.)

Complete Combustion Kit

Fig. R6-64 shows the complete combustion kit. This kit includes a CO_2 indicator, smoke tester, efficiency-calculating slide rule, dial thermometer, and draft gauge. This particular kit shown is for oil burners, and a kit for gas burners is also available.

R6-5.2 Gas Identifiers and Monitors

There are a number of uses for gas identifiers and monitors:

1. To notify the owner of a refrigerant leak.
2. To protect anyone entering a room where refrigerants are stored or used. Refrigerants can displace oxygen, and therefore become dangerous. Most refrigerants do not have an odor and are not normally detected.
3. To identify the CO_2 content in the air, and in some cases, to automatically control the addition of outside air. There is an increasing amount of concern about indoor air quality and particularly the need for ventilation to prevent the buildup of CO_2 in enclosed spaces.

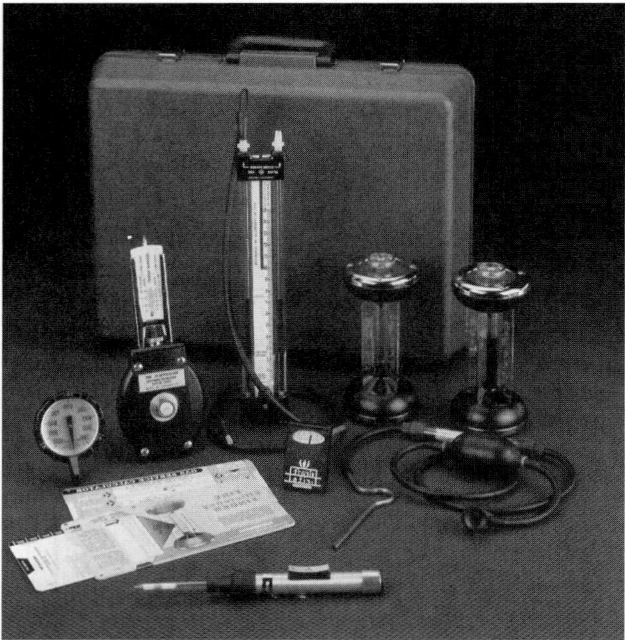

Figure R6-64 Complete combustion kit for testing heating equipment. (Courtesy of Bacharach, Inc.)

The refrigerant identifier shown in Fig. R6-65 is a microprocessor-based instrument which extracts a sample of the refrigerant and analyzes it to determine the type of refrigerant. In some cases, systems have been charged with refrigerants without properly tagging the installation with the refrigerant number. The service technician MUST be certain of the refrigerant that has been installed in order to perform any services necessary.

Refrigerant Leak Detection System

Fig. R6-66 shows a refrigerant leak detection system for monitoring specific refrigerants. The system shown iden-

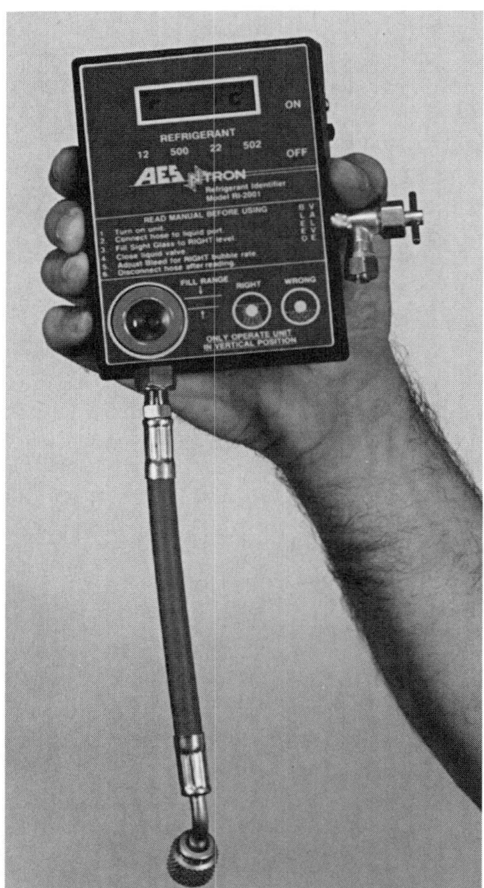

Figure R6-65 Refrigerant identifier for R-12, R-22, R-500, and R-502. (Courtesy of AES-NTRON, Inc.)

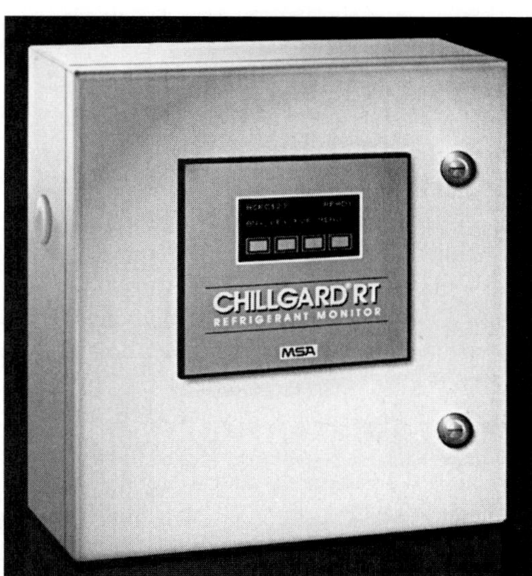

Figure R6-66 Refrigerant leak detection system and monitor for refrigerants R-11, R-12, R-22, ammonia, HCFC-123, and HCFC-134a. (Courtesy of MSA Instrument Division)

tifies R-11, R-12, R-22, Ammonia, HCFC-123 and HCFC-134a. This instrument complies with ASHRAE Standard 15-1992 which requires the use of an instrument of this type where air-conditioning and refrigeration systems are installed. The unit has a visible alarm and an LCD readout showing the actual gas concentration. It relays each alarm level. Using a multipoint sequencer it can be used to monitor six different locations. The power requirement is $120V_{AC}$.

Oxygen-depletion Monitor

The oxygen-depletion monitor shown in Fig. R6-67 has a remote oxygen sensor and a controller which are normally mounted outside the mechanical equipment room door.

It is designed to continuously monitor the oxygen level in mechanical equipment rooms where class A, group I refrigerants are used. Oxygen is normally 20% of the atmosphere, and when levels drop to 19.5% an amber warning light goes on, starting a mechanical exhaust. If the oxygen level continues downward to 18.5%, a red warning light goes on and an alarm will sound. Relays provide an option for sending the warning to remote locations. A "purge" switch permits the fan to run continuously or only when signalled by the monitor. The alarm turns off automatically when normal levels of oxygen return, or may be turned off manually—at which time there is automatic reset.

The first stage, or warning stage, activates an amber LED and a set of relay contacts to operate a fan or other mechanical equipment. The second stage, or alarm stage, activates a red LED, an audible alarm, and auxiliary set of contacts for optional remote alarm indication. After the problem is solved and the alarm is off, the system automatically resets.

Other Instruments

An insulation-resistance tester (Fig. R6-68) is useful in testing open and semi-hermetic motor insulation leakage to ground. This instrument uses high voltage (500—1000 volts) to test for this leakage.

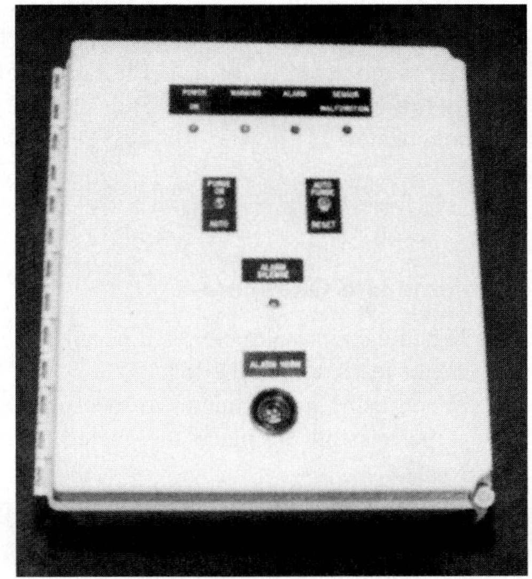

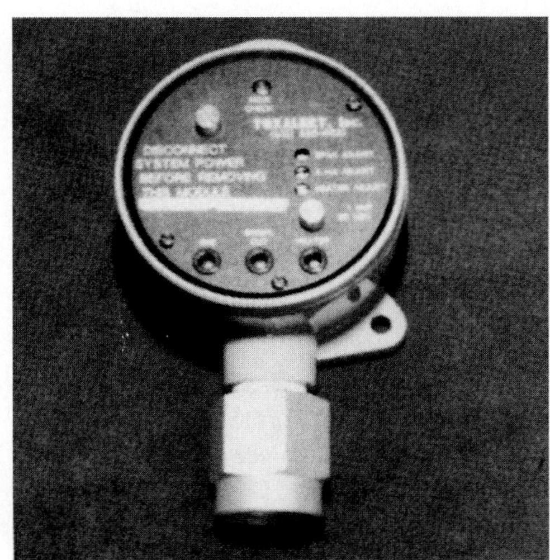

Figure R6-67 Oxygen-depletion monitor as required by ASHRAE Standard 15-1992. (Courtesy of Toxalert, Inc.)

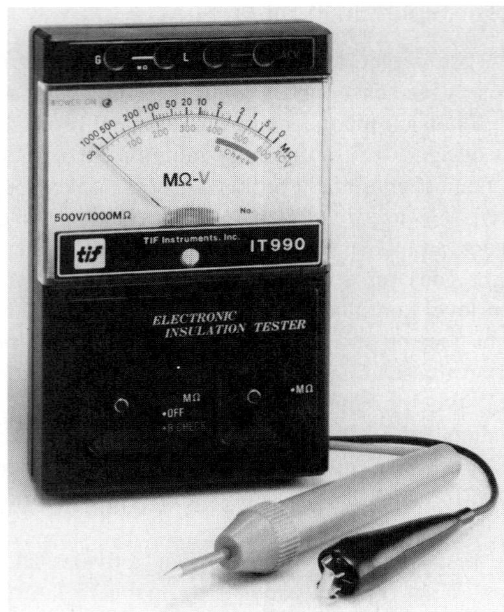

Figure R6-68 Insulation-resistance tester. (Courtesy of TIF Instruments, Inc.)

REVIEW

Air-conditioning, heating, and refrigeration technicians should be familiar with the operation and care of each of the tools and test equipment covered in this chapter. You will need to know how to operate and care for the tools and equipment to be successful on the job.

- Basic Hand Tools:
  - Ratchet wrench
  - Torque wrench
  - Open-end wrench
  - Pipe wrench
  - Screwdrivers
  - Vises
  - Caliper
  - Sheet metal vise grips
  - Wire cutter
  - Service valve wrench
  - Box wrench
  - Combination wrench
  - Allen wrenches
  - Hammers
  - Wire brush
  - Tapes and hand rulers
  - Vise grips
  - Inspection mirror
  - Socket wrenches
  - Flare nut wrench
  - Adjustable wrench
  - Nut driver
  - Pliers
  - Files
  - Impact driver
  - Tin snips
  - Wire stripper
- Power Tools:
  - Electric drills
  - Soldering gun
  - Light cords
  - Rotary hammer drill
  - Jig saw
  - Cordless drill
  - Power cords
- Fasteners:
  - Nails
  - Tapping screws
  - Wall anchors
  - Pipe straps
  - Cotter pins
  - Staples and rivets
  - Machine screws
  - Toggle bolts
  - Pipe hooks
  - Nylon straps (wire ties)
  - Wood screws
  - Set screws
  - Threaded rod
  - Plumbers strap
  - Grille clips
- Tubing and Pipe Tools:
  - Tubing cutter
  - Lever bender
  - Flaring tools
  - Hacksaw
  - Chain vise
  - PVC cutter
  - Tubing vise
  - Torches
  - Swaging tools
  - Pipe cutter
  - Yoke vise
  - Spring bender
  - Tubing brush
  - Reamers
  - Pipe threader
  - Pipe wrench
- Electrical Test Instruments and Meters:
  - Multimeter
  - Continuity tester
  - Wattmeter
  - Clamp-around ammeter
  - Capacitance tester
  - Voltage tester
  - Megohmmeter
- Refrigeration Servicing Tools and Equipment:
  - Hard hat
  - Safety glasses
  - Gloves
  - Safety shoes
  - Gauges
  - Deep vacuum gauge
  - Electronic thermometer
  - Vacuum pump
  - Hermetic compressor tester
  - Leak detectors
  - Manometer
  - Charging scale
  - Charging cylinder
- Air Flow Measuring and Monitoring Equipment:
  - Pitot tube
  - Anemometers
  - Differential pressure gauge
  - Flow hood
  - Oxygen indicator
  - Psychrometer
  - Refrigerant detector
  - Inclined manometer
  - Air pressure gauge
  - Velocimeter
  - Carbon dioxide indicator
  - Carbon monoxide indicator
  - Volume air balancer
- Heating Servicing Equipment:
  - Draft gauge
  - Stack thermometer
  - Water manometer
  - Millivoltmeter
  - Smoke tester
  - Combustion kit
  - Orifice drill index
  - Flue gas analyzer
  - Belt tension gauge
  - Cad cell tester

Problems and Questions

1. When using a voltmeter, is it permissible to place the meter leads across the load?
2. When using a multimeter to measure DC current, is it permissible to place the meter leads across the load?
3. DC milliamps are measured on what type of meter?
   a. Voltmeter
   b. Clamp-on ammeter
   c. Ohmmeter
   d. Multimeter

4. How many wires should the jaws of a clamp-on ammeter enclose?
   a. 1 c. 3
   b. 2 d. 4
5. What powers an ohmmeter?
   a. The circuit under test
   b. Electrolytic capacitor
   c. No power needed
   d. Meter batteries
6. What meter will quickly disclose an overloaded motor?
   a. Clamp-on ammeter
   b. Insulation resistance tester
   c. Capacitance tester
   d. Low-range ohmmeter
7. In what type of meter is the power factor included?
   a. Voltmeter c. Ammeter
   b. Wattmeter d. Ohmmeter
8. If an ohmmeter reads infinity (∞), what does it mean?
   a. Open circuit
   b. Continuity
   c. Ohmmeter is defective
   d. No power
9. What is the adjustment knob used for on an ohmmeter?
   a. To test continuity
   b. To turn the instrument on
   c. To change the range
   d. To calibrate the instrument
10. In leak-testing a unit, where should the technician start?
    a. At the top of the unit
    b. At the bottom of the unit
    c. At the power source
    d. Near the greatest load
11. Flare nut wrenches are a special variation of the box wrench. True or false?
12. The chain wrench is another form of an adjustable pipe wrench. True or false?
13. Newer electronic volt/ohm meters don't require turning off the power when measuring ohms. True or false?
14. Rotary hammers with carbide bits work well for concrete drilling. True or false?
15. Wattmeters make necessary calculations in accordance with the power formula which is: Watts = volts × amps × power factor. True or false?
16. Electronic pump style refrigerant leak detectors are able to detect leaks up to _____ ounces per year.
17. The ultrasonic type leak detector will detect _____ gas leaking through an orifice.
18. The electronic sight glass is actually an _____ instrument.
19. The oil burner gas analyzer measures carbon _____ in the flue gases.
20. A refrigeration leak detection system is used to comply with ASHRAE Standard 15-1992 and measures refrigerant where?

R7 Refrigeration System Components

AFTER STUDYING THIS CHAPTER, THE STUDENT WILL BE ABLE TO:

- Identify basic refrigeration cycle components.
- Identify all connecting lines.
- Explain the function of each of the components of the vapor-compression refrigeration cycle.
- Indicate the condition, state, pressure, and temperature of refrigerant at various key points in the cycle.
- Identify the types of metering devices.
- Describe the operating characteristics of metering devices.
- Identify types of evaporators.
- Explain the benefits of superheating and subcooling.
- Identify applications for various types of evaporators.
- Identify the types of compressor casings and compressors.
- Explain the major differences for each compressor type.
- State the function of the compressor in the vapor-compression refrigeration cycle.
- Identify major working parts of various types of compressors.
- Define and identify the types of condensers.
- Explain the relationship between the refrigerant and the condensing medium.
- Explain the reason for system accessories and identify them.

GENERAL

The vapor-compression refrigeration cycle as discussed in Chapter R3 is the most common method of heat energy transfer. There are four major components in the compression cycle: pressure-reducing device, evaporator, compressor, and condenser.

R7-1 REFRIGERANT FLOW CONTROLS

A fundamental and indispensable component of any refrigeration system is the *flow control,* or pressure-reducing device. Its main purposes are:

1. To maintain the proper pressure and boiling point in the evaporator to handle the desired heat load
2. To permit the flow of refrigerant into the evaporator at the rate needed to remove the heat of the load

The pressure-reducing device is one of the dividing points in the system.

The principal means of refrigerant flow control, in the early stages of refrigeration, was a basic hand valve. Knowing their work and their equipment, early operators of units in ice plants and similar operations having constant loads knew how far to open the hand valve for the work to be performed. However, in modern applications that have frequently varying loads, this is impractical because the hand-valve setting would have to be changed as the load changed.

The eight main types of pressure-reducing devices now used in various phases of refrigeration are:

1. Hand-operated expansion valve
2. Automatic expansion valve

PRACTICING PROFESSIONAL SERVICES

The first step upon arriving at any unfamiliar job site is to identify the major components, piping, and associated accessories, including controls, valves, receivers, accumulators, etc. Learn to look and listen for unusual noises, equipment cycling times, oil leaks, sight glasses, and any indications of abnormal conditions. Become acquainted with the building manager responsible for the equipment. Ask if they know of any peculiar noises or symptoms that occurred prior to your arrival.

3. Electronic expansion valve
4. Thermostatic expansion valve
5. Capillary tube
6. Fixed orifice
7. Low-side float
8. High-side float

All are used to reduce the liquid refrigerant pressure and in some cases, to control the volume of flow.

The hand valve is obviously not suited for automatic operation, since any variation in requirements needs manual adjustment, so the automatic expansion valve came into being.

The *automatic expansion* or *constant-pressure valve,* shown in Fig. R7-1, maintains a constant pressure in the cooling coil while the compressor is in operation. In this diaphragm type of constant-pressure expansion valve, the pressure in the evaporator affects the movement of the diaphragm, to which the needle assembly is attached.

A condition of stability in refrigerant flow and evaporation is necessary for the correct operation of the constant-pressure expansion valve. Like the hand valve, its use is limited to conditions of more or less constant loads on the evaporator, a situation that applies to the automatic expansion valve as well. In either valve there is a screw that applies pressure to the spring above the bellows or diaphragm. When the screw is adjusted clockwise, it causes more pressure on the bellows or diaphragm, forcing the valve to open more, admitting additional refrigerant to the evaporator, and resulting in higher operating pressure. If a lower operating pressure in the cooling coil is desired, the screw is turned counterclockwise, releasing pressure on the spring and therefore on the bellows or diaphragm. This allows the valve to close and curtails the flow of refrigerant.

Following any adjustment, ample time should be allowed for a controlling device to settle down before any further change is made in its setting. For a given load on the evaporator coil being fed refrigerant, there is only one correct setting of the automatic expansion valve: when the coil is completely frosted. If the pressure is lowered, there will be a curtailment in the refrigerant flow, and the heat absorption capability of the coil will be lessened. If the pressure is raised, the flow of refrigerant will increase, with the possibility of liquid refrigerant flooding into the suction line, from which the refrigerant might reach the compressor and damage it.

Because all refrigeration loads do not remain constant, and someone cannot always be present at every installation to make compensating adjustments, another type of valve, the *thermostatic expansion valve,* was developed. Like the automatic expansion valve, the thermostatic expansion valve may be either the bellows type or the diaphragm type shown in Fig. R7-2. Both are equipped with a capillary tube and feeler bulb assembly, which transmits to the valve the pressure relationship of the temperature of the suction vapor at the outlet of the evaporator coil, where the feeler bulb is attached.

The basic purpose of the thermostatic expansion valve is to maintain an ample supply of refrigerant in the evaporator, without allowing liquid refrigerant to pass into the suction line and the compressor. When the metering device is a thermostatic expansion valve, its operation will depend on superheated vapor leaving the evaporator, since a portion of the evaporator is used for the superheating of the vapor about 5 to 10°F above the temperature corresponding to the evaporative pressure.

*VALVE IS USED WITH EITHER INTERNAL OR EXTERNAL EQUALIZER, BUT NOT WITH BOTH.

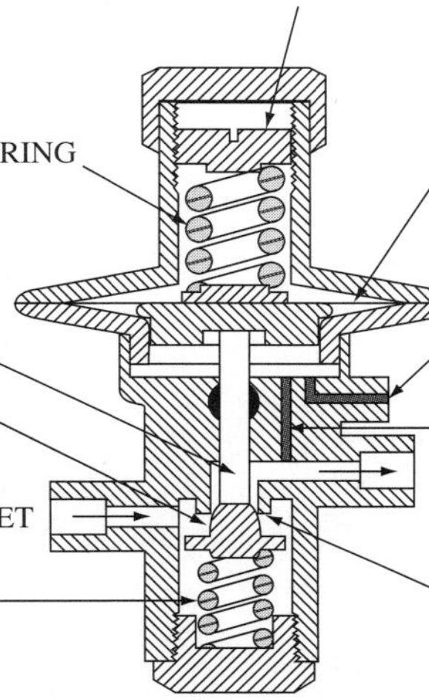

Figure R7-1 Constant-pressure expansion valve.

Knowledge Extension

To learn more about the various pieces of equipment mentioned in this chapter, visit your local library or contact your HVAC/R supply house and request any catalogs or technical data sheets of your interest. Build your own reference library from the material you gather and organize into a portable, 3-ring binder, reference and troubleshooting catalog. This will enable you to be better prepared when you are on a job site looking for possible solutions to a problem area.

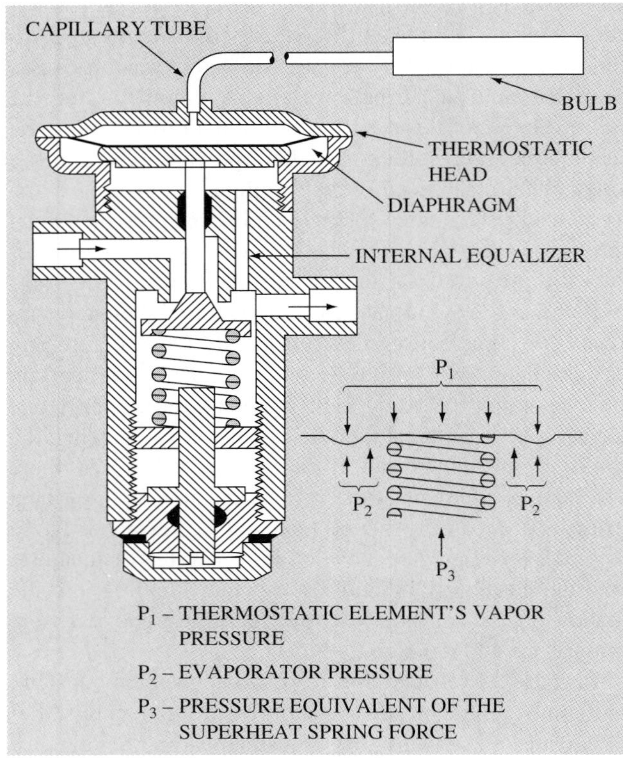

Figure R7-2 Diaphragm-type thermostatic-expansion valve.

The capillary tube, which is based on the principle described above, is the simplest form of refrigerant control or metering device and generally the least expensive. There are no moving parts to wear out or require replacing, since it is a small-diameter tube of the right length for the refrigeration load it is designed to handle. This pressure-reducing device, like any other, is located between the condenser and evaporator, at the end of the liquid line or instead of a liquid line. One type of capillary refrigerant control is shown in Fig. R7-3. The advantages of this control have just been discussed; however, its disadvantages are that it is subject to clogging, requires an exact refrigerant charge, and is not as sensitive to load changes as other metering devices. Its internal cross-sectional area is so small that it takes only a minute dirt particle to plug the tube, or a small amount of moisture to freeze in it. A drier and filter or strainer should be installed at the inlet to the capillary tube to prevent this clogging possibility.

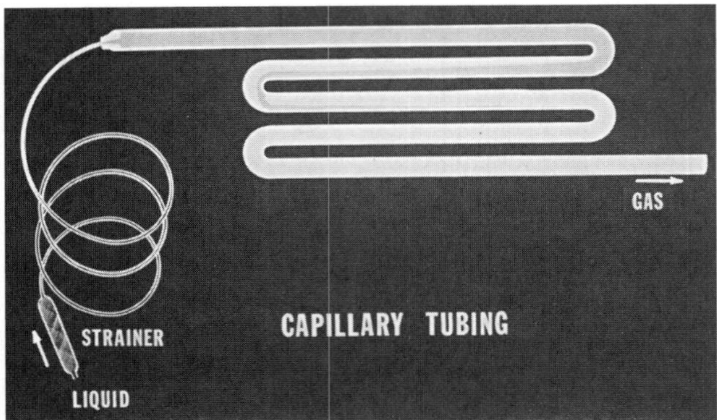

Figure R7-3 Capillary tubing.

Another type of refrigerant control device is the *float arrangement,* which also meters the flow of refrigerant into the evaporator. The float itself is made of metal that will not cause a reaction with the refrigerant used in the system. It is constructed in the shape of a ball or an enclosed pan, which will rise or fall within the float chamber with the level of the refrigerant. It is connected through an arm and linkage to a needle valve, which opens and closes against a seat, allowing and curtailing the flow of refrigerant into the chamber.

A *high-side float,* as its name indicates, is located in the high-pressure side of the system. It may be of a vertical or horizontal design and construction, and it may be located near either the condenser or the evaporator. A typical high-side float is shown in Fig. R7-4. Its design is such that, as the float chamber fills with refrigerant, the buoyancy of the float lifts it and raises the valve pin away from the seat. This permits the refrigerant to flow or be metered to the low-pressure side of the system and into the evaporator.

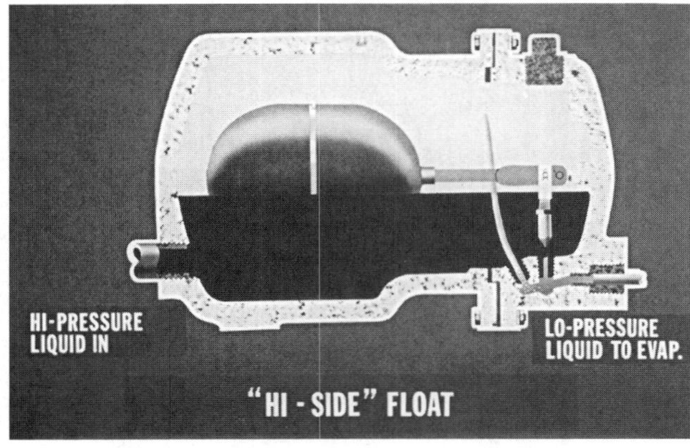

Figure R7-4 High-side float.

Since the float, through the pivot arrangement, is set to open at a given level, only a small amount of liquid stays in the high-side float chamber; most of the refrigerant in the system is in the evaporator. Therefore, the system refrigerant charge is critical, to the extent that only enough refrigerant to maintain the proper level in the flooded evaporator is desirable, with no liquid flooding over into the suction line and compressor. If there is an overcharge, flooding will result, whereas a shortage of refrigerant will cause the evaporator to be starved and the system will be inefficient.

A *low-side float* metering device is one in which the float is located in the evaporator, or in a chamber adjacent to the cooling coil that is flooded, maintaining a definite liquid level within the evaporator. It is constructed somewhat like the high-side float, with the exception that as the float rises it closes off the flow of the refrigerant. Its action is shown in Fig. R7-5: the high-pressure liquid is at the inlet to the float chamber, and the float assembly itself is in the low-pressure liquid, maintaining a definite level within the float chamber. As the load on the evaporator increases, liquid is evaporated and the liquid level in the evaporator and the float chamber drops. As the float lowers, the needle is pulled away from the seat, allowing additional refrigerant to enter until the desired level is reached.

If the load on the evaporator decreases, less evaporation will take place, and the liquid level will be maintained, the float causing the needle to close against the seat. In this manner the low-side float assembly can maintain the correct flow of refrigerant as it is needed, by a fluctuating load condition. Usually, the same float arrangement cannot be used if a different refrigerant is desired in the system because of its operating characteristics, for the refrigerants will have different specific gravities as other characteristics change. A float with the correct buoyancy must be used.

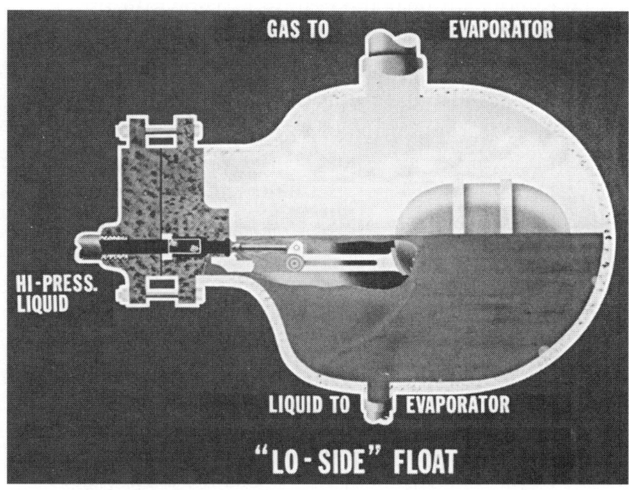

Figure R7-5 Low-side float.

R7-2 EVAPORATORS

The evaporator or cooling coil is the part of the refrigeration system where heat is removed from the product: air, water, or whatever is to be cooled. As the refrigerant enters the passages of the evaporator it absorbs heat from the product being cooled, and as it absorbs heat from the load, it begins to boil and vaporizes. In this process, the evaporator accomplishes the overall purpose of the system—refrigeration.

Manufacturers develop and produce evaporators in several different designs and shapes to fill the needs of prospective users. The blower coil or forced-convection type of evaporator (Fig. R7-6) is the most common design; it is used in both refrigeration and air-conditioning installations.

Figure R7-6 Blower coil. (Source: Kramer Trenton)

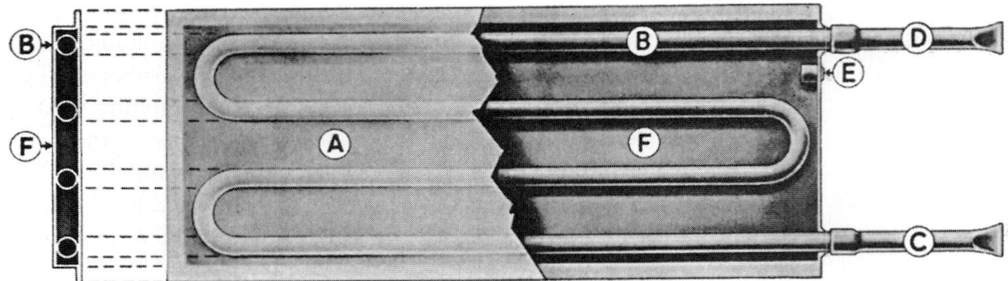

Figure R7-7 The Dole vacuum plate: (a) Outside jacket of plate. Heavy, electrically welded steel. Smooth surface. (b) Continuous steel tubing through which refrigerant passes. (c) Inlet from compressor. (d) Outlet to compressor. Copper connections for all refrigerants except ammonia where steel connections are used. (e) Fitting where vacuum is drawn and then permanently sealed. (f) Vacuum space in dry plate. Space in holdover plate contains eutectic solution under vacuum. No maintenance required due to sturdy, simple construction. No moving parts; nothing to wear or get out of order; no service necessary. (Courtesy of Dole Refrigerating Company)

Specific applications may require the use of flat plate surfaces for contact freezing. Continuous tubing is formed or placed between the two metal plates, which are welded together at the edges, and a vacuum is drawn on the space between the plates. These plates also may be assembled in groups arranged as shelving, utilizing refrigerant in a series flow pattern, as shown in Fig. R7-7.

Other shapes of plate-type evaporators are shown in Fig. R7-8. They are widely used in small refrigerators, freezers, and soda fountains, where mass production is economical, and the plates can easily be formed into a variety of shapes.

Plate-type evaporators also are assembled in groups or banks for installation in low-temperature storage rooms and are mounted near the ceiling as shown in Fig. R7-9. This type may be connected for either series or parallel refrigerant flow, depending on usage requirements. Plate-type coils are also used in refrigerated trucks and railway cars for the transportation of refrigerated food and frozen-food products, as shown in Fig. R7-10. Frequently, the space between the plates is filled with a solution that retains its refrigeration if the unit is not in operation for short periods.

The bare-tube type of coil may be used for the cooling of either air or a liquid, with the smaller evaporators being constructed of copper tubing. Steel pipe is used for evaporators in systems using ammonia as the refrigerant and in the larger evaporators containing other refrigerants.

Figure R7-8 Plate evaporators in soda fountain. (Courtesy of Dole Refrigerating Company)

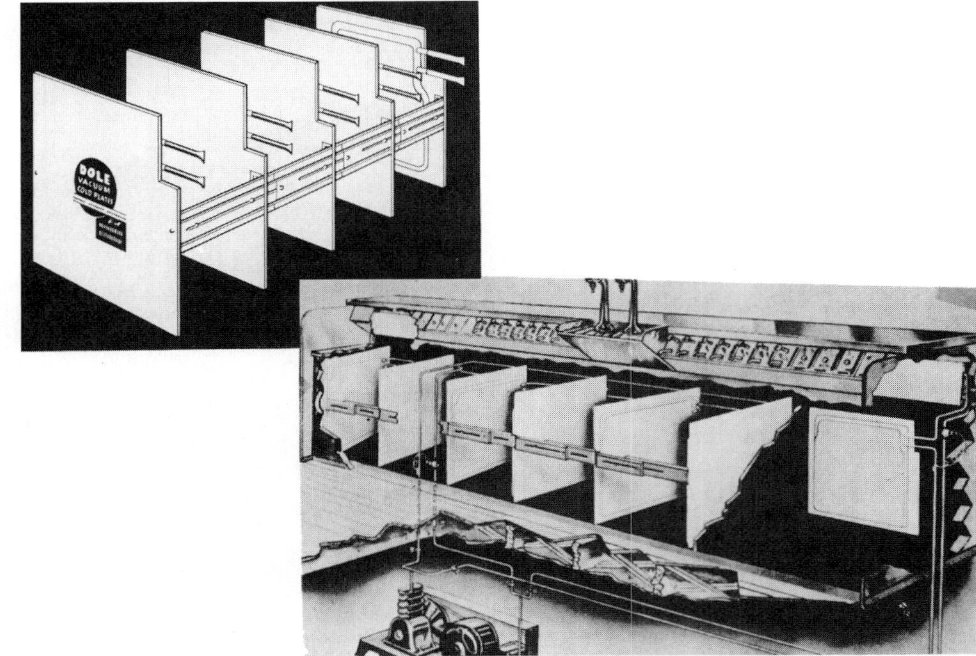

Figure R7-9 Plate evaporators for storage rooms. (Courtesy of Dole Refrigerating Company)

An air film adheres to the outside surface of a coil, acting as an insulator and slowing down the heat-transfer process, which is dependent primarily on surface area and temperature differential. One of the methods used to overcome or compensate for the conduction loss due to the air film is to increase the surface area. This may be accomplished through the addition of fins to the evaporator pipe or tubing, as shown in Fig. R7-11. The addition of fins does not eliminate air film, for it furnishes more area to which air film will cling or adhere; however, it does afford more surface area for heat transfer, without increasing the size of the coil to any great extent.

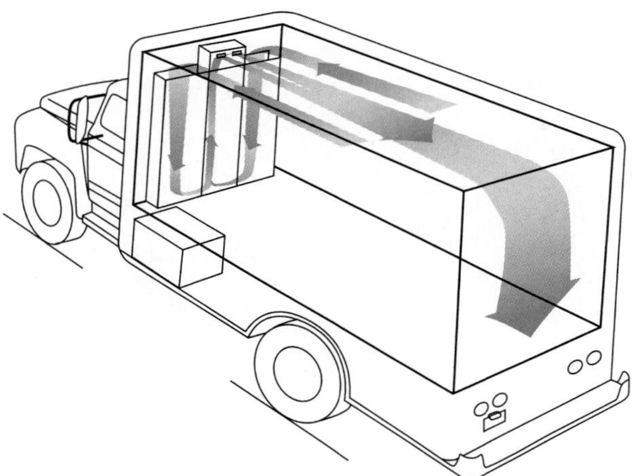

Figure R7-10 Refrigerated truck.

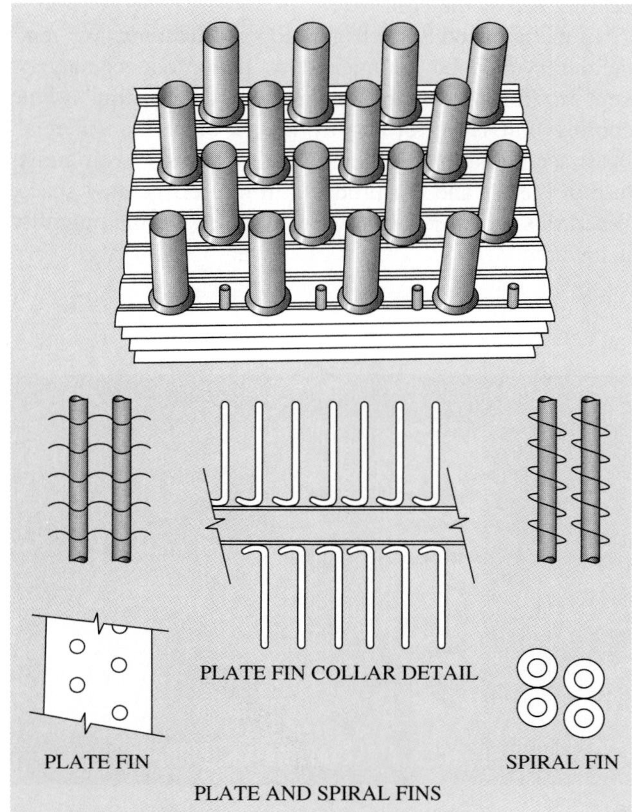

Figure R7-11 Finned tube evaporator.

Another method of overcoming the heat-transfer loss caused by air film is through the addition of a fan or blower, which will cause rapid movement of air across the evaporator. Such a type of forced convection coil is shown in Fig. R7-12. Depending on the design and usage of the coil, the fan may be located for the movement of air across the coil either by means of an induced or drawing action of the air or by a forced circulation or blowing action of the air across the evaporator coil.

The use of a fan improves the airflow and transfer of heat from the air to the refrigerant within the coil, since a greater proportion of the air will come in contact with the coil's surface area. Many coil manufacturers have designed their heat-transfer units with staggered rows of tubing, thus permitting, with the use of a blower, a large volume of air to come in contact with either the tubing surface or the fins connected to it. Forced or induced air motion across the coil usually will result in a greater portion of the air giving up heat to the refrigerant within the coil over a specific period.

In the early days of mechanical refrigeration, cooling coils were constantly maintained at a temperature below freezing. Since these evaporators did not reach a temperature above 32°F, the frost accumulating on them did not have an opportunity to melt off while the equipment was in operation. The units had to be shut off and manually defrosted, since frost accumulating on the evaporator curtailed the amount of heat it could remove from the air passing across the coil.

In many of today's refrigeration applications, low temperatures must be maintained so that products may be kept in frozen storage condition. But defrosting of the cooling unit is performed by means other than manual. Frost accumulation on the cooling unit comes from moisture in the air and the products in the refrigerated space. When this moisture is removed from the air, the humidity is lowered.

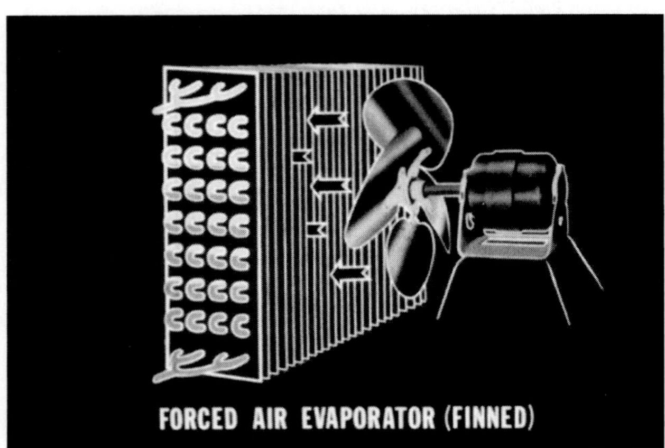

Figure R7-12 Forced-air evaporator (finned).

Conditions may be such that an extremely low temperature or a low moisture content of the air surrounding the cooling unit is not desirable. If the temperature in the refrigerated space needs to be maintained at approximately 35°F, a coil or evaporator in which the refrigerant is at a temperature below this desired temperature must be used. As the air comes in contact with the cooling coil at a temperature below 32°F, some frost will form on the surface of the evaporator. When the desired temperature is achieved, the control mechanism will stop operation of the refrigeration unit. With the surrounding air temperature at 35°F, this warm air will melt the frost on the cooling unit and thereby defrost the cooling unit. This will occur naturally, particularly if it is a forced air coil and the warmer temperature air is forced across the evaporator surface.

The OFF-cycle period of the refrigeration unit should be long enough to assure complete defrosting of the cooling coil. If not, only partial defrosting may occur, which results in moisture collecting on the lower section of the unit. If this occurs, an icing condition on the coil may result. If this condition is allowed to continue, ice may cover the entire surface of the coil and may develop into a complete blockage of the coil.

For air-conditioning applications, the evaporator operates at a temperature above 32°F. These nonfrosting evaporators condense moisture out of the air, but it does not freeze. It is collected in a drain pan, and removed through a condensate drain (Fig. R7-13).

The evaporators described thus far have been of the *dry-expansion type,* as compared with the flooded type. The direct or dry-expansion type coil is designed for complete evaporation of the refrigerant in the coil itself, with only a vapor leaving the coil outlet. This vapor is

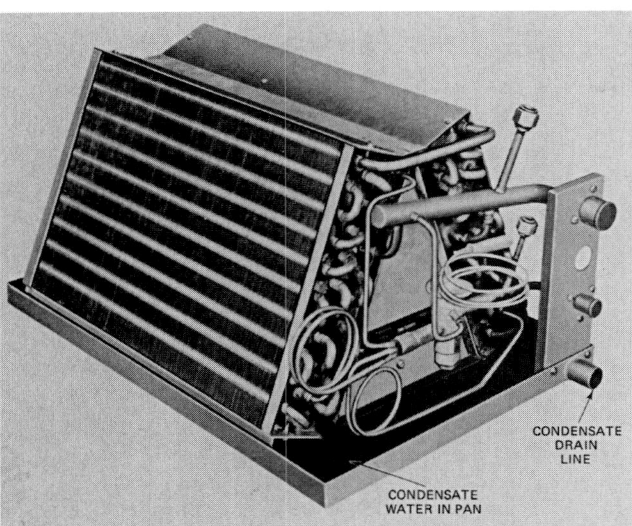

Figure R7-13 Condensate drain. (Courtesy of York International Corp.)

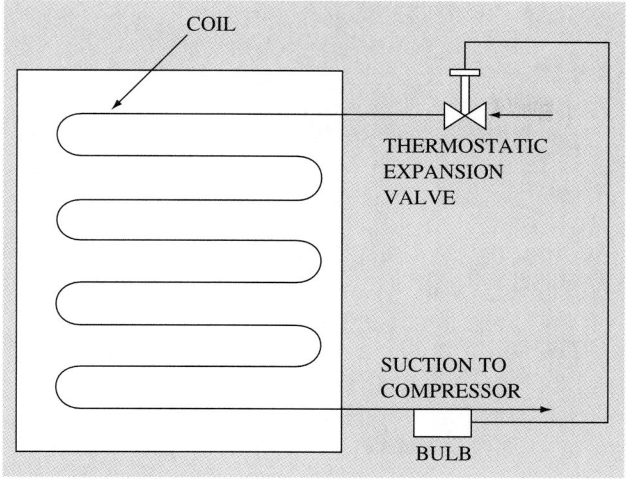

Figure R7-14 Dry expansion coil with thermostatic expansion valve.

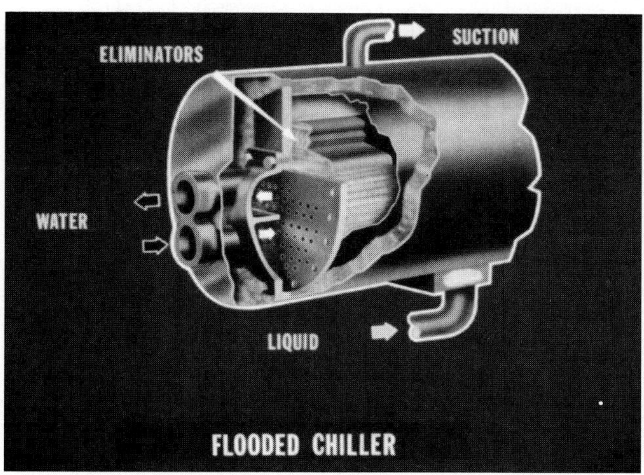

Figure R7-15 Flooded chiller.

usually superheated in the last part of the cooling coil. (*Superheating* means raising the temperature of the refrigerant vapor above that temperature required to change it from a liquid to a vapor.) It will reach the compressor in a superheated condition, picking up additional heat as it passes through the suction line. Fig. R7-14 is a schematic showing a direct-expansion coil with a thermostatic expansion valve. The coil contains a mixture of liquid and gaseous refrigerant at all times when the unit is in operation. A constant superheat is maintained by the modulating of the valve, which is caused by the sensitivity of the thermal bulb to temperature changes at its location.

The characteristics of the dry or direct-expansion coil can be maintained by the automatic expansion valve, which maintains a constant pressure within the evaporator. This type of valve is usually used when a steady load is anticipated. Refrigerant controls are discussed more fully later in this chapter.

The *flooded* type of evaporator is filled with liquid refrigerant. It is designed so that the refrigerant (liquid) level is maintained by a float arrangement located in an accumulator situated outside the evaporator coil itself. A typical design is shown in Fig. R7-15. Part of the liquid refrigerant evaporates in the coil, and this vapor goes to the accumulator. From there the vapor is drawn from the top into the suction line and then to the compressor, while any liquid left in the accumulator is available for recirculation in the evaporator coil. When the equipment is properly calibrated, the remaining liquid is minimal.

As the refrigerant in the flooded coil evaporates as a result of the heat it has absorbed, the liquid level lowers in the coil. As the float lowers with the liquid level, it permits more refrigerant to flow into the accumulator so that a fairly constant liquid level is maintained. A flooded coil has excellent heat-transmission efficiency because its interior surfaces are liquid-wetted instead of being vapor-wetted.

Liquid-cooling coils vary in their design depending on their application and usage, just as do air-cooling coils. Since there is a greater heat transfer between liquids and metals than between air and metals, a submerged coil has the capability of removing several times as many Btu as an air-cooled coil under similar conditions. Submerged coils are used in a water-bath type of cooler, in which the "cold-holding" capacity is put to good use when cans filled with warm milk or other liquids are placed in the cooler.

Shell-and-tube and *shell-and-coil* are other types of arrangements for the cooling of one or more liquids, even in the cooling of brine solutions. A shell-and-coil water cooler is shown in Fig. R7-16. It is a direct-expansion type of system with the refrigerant circulated within the coil as the water is circulated within the shell at a temperature not much below 40°F to prevent freezing.

Tube-in-tube, sometimes classified as *double-pipe evaporator,* is a liquid-cooling coil that provides high heat-transfer rates between the refrigerant and the liquid being cooled. The path of refrigerant flow may be through either of the tubes, although usually the brine or liquid to be cooled is circulated through the inner tubing, and the refrigerant removing the heat is between the two tubes. This type of heating-exchange coil is also used in condenser design, described later in this chapter.

A *Baudelot cooler,* shown in Fig. R7-17, has several applications. It may be used for cooling water or other liquids for various industrial uses, and it is frequently used as a milk cooler. The evaporator tubing is arranged vertically, and the liquid to be cooled is circulated over the cooling coils by gravity flow from the trough type of arrangement located above the coils. The liquid gathers in

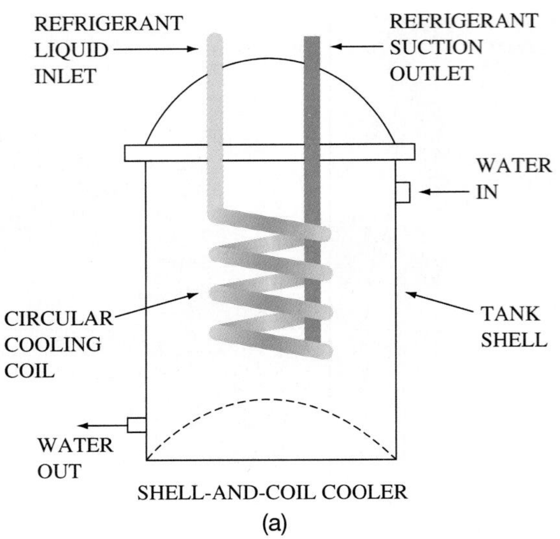

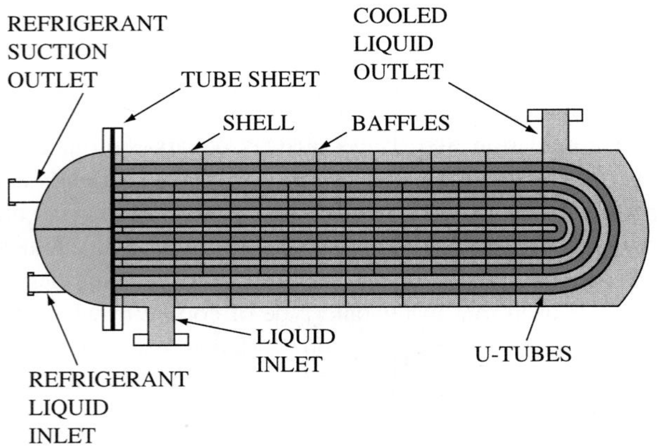

Figure R7-16 Shell-and-coil water cooler and direct (U-tube type) expansion liquid cooler.

a collector tray at the bottom of the coil, from which it may be recirculated over the Baudelot cooler or pumped to its destination in the industrial process.

R7-3 COMPRESSORS

After it has absorbed heat and vaporized in the cooling coil, the refrigerant passes through the suction line to the next major component in the refrigeration circuit, the compressor. This unit, which has two main functions within the cycle, is frequently classified as the *heart* of the system, for it circulates the refrigerant through the system. The functions it performs are:

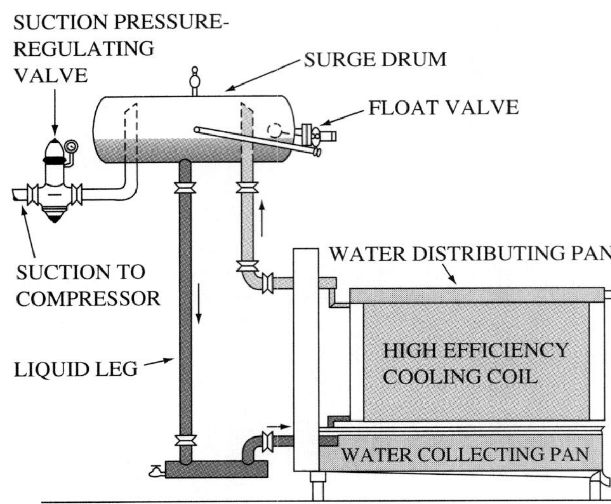

Figure R7-17 Baudelot cooler.

1. Receiving or removing the refrigerant vapor from the evaporator, so that desired pressure and temperature can be maintained
2. Increasing the pressure of the refrigerant vapor through the process of compression, and simultaneously increasing the temperature of the vapor so that it will give up its heat to the condenser cooling medium

Compressors are usually classified into five major types: reciprocating, rotary, centrifugal, screw, and scroll. (Screw and scroll compressors are discussed in Chapter R-11.) The *reciprocating compressor* is used in the majority of domestic, small commercial and industrial condensing unit applications. This type of compressor can be further classified according to its construction, according to whether it is open and accessible for service in the field, or fully hermetic, not able to be serviced in the field.

Reciprocating compressors vary in size from that required for one cylinder and its operating piston to one large enough for 16 cylinders and pistons. The body of the compressor may be constructed of one or two pieces of cast iron, cast steel, or, in some cases, aluminum. The arrangement of the cylinders may be horizontal, radial, or vertical, and they may be in a straight line or arranged to form a V or a W.

Fig. R7-18 shows an external view of a common type of reciprocating compressor used in commercial applications. As compressors differ in design and construction, so do the individual components within the compressors. But their main goal remains the same—the compression of the refrigerant vapor to high temperature and high pressure, so that its heat content can be reduced and it will condense into a liquid to be used over again in the cycle.

Figure R7-18 Typical reciprocating compressor.
(Courtesy of York International Corp.)

Pistons within the compressors may have the suction valve located in the top of the piston; this is classified as a *valve-in-head* type, or the piston may have a solid head, with the suction and discharge valves located in a valve plate or cylinder head. A typical valve plate, showing the suction and discharge internal valves of a two-cylinder reciprocating compressor is shown in Fig. R7-19.

Fig. R7-20 presents sketches of a compressor piston and the internal suction and discharge valves in different stages of the compression cycle.

Fig. R7-21 shows an assembly consisting of the piston, wrist pin, connecting rod, and crankshaft. All components of the reciprocating piston arrangement are finely machined, balanced carefully to eliminate vibration, and fitted with close tolerances to assure that the compressor will have a high efficiency in pumping the refrigerant vapor. A different type of crankshaft, one of an eccentric design, is shown in Fig. R7-22. The connecting rod is assembled on an off-center eccentric fastened with balance weights. If the crankshaft is not almost completely machined, it should be dynamically balanced.

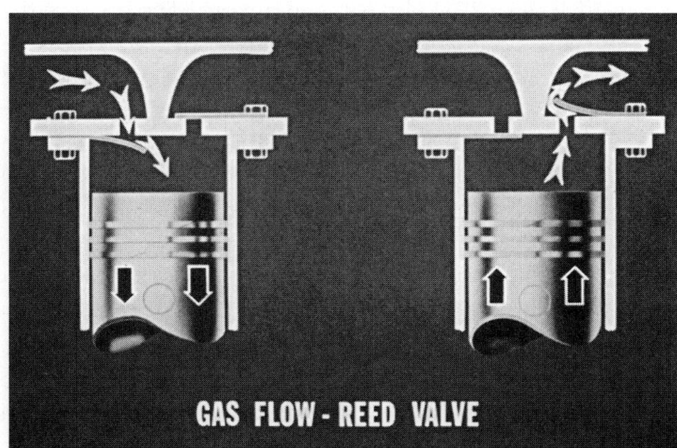

Figure R7-19 Gas-flow-reed valves.

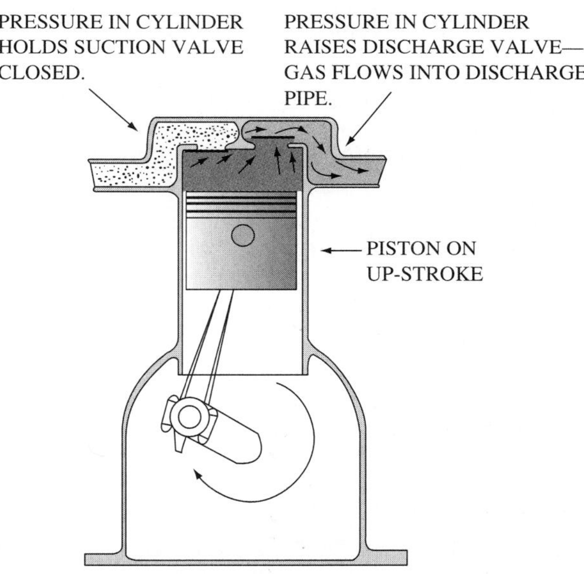

Figure R7-20 How differential pressures work the valves of the reciprocating compressor.

The internal valves of a compressor receive quite a bit of wear and tear in normal operation, since they must open and close hundreds of times each minute the com-

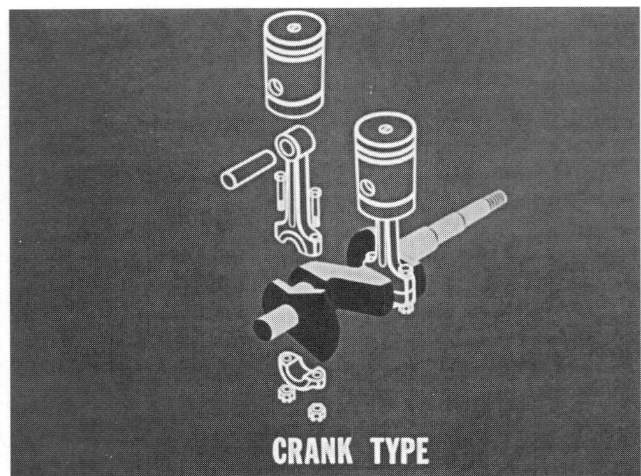

Figure R7-21 Crank-type assembly.

pressor is running. Small commercial units usually have a high-grade steel disk or reed-type of valve, both of which are quieter operating, more efficient, simpler in construction, and longer lasting than the nonflexing ringplate type of valve. Figure R7-23 shows some of the various designs of internal compressor valves. The proper operation of the valves is very important to the overall efficiency of the compressor.

If the suction valves do not seat properly and allow refrigerant vapor to escape from the cylinder, the piston cannot pump out all of the compressed vapor into the hot-gas line. If the suction valve leaks, the compressed vapor, or part of it, will go into the suction line and heat up the low-pressure, low-temperature vapor there. If the discharge valve leaks, some of the high-pressure, high-temperature vapor in the hot-gas line will leak back into the cylinder on the down stroke of the piston, limiting the volume of suction vapor entering the cylinder.

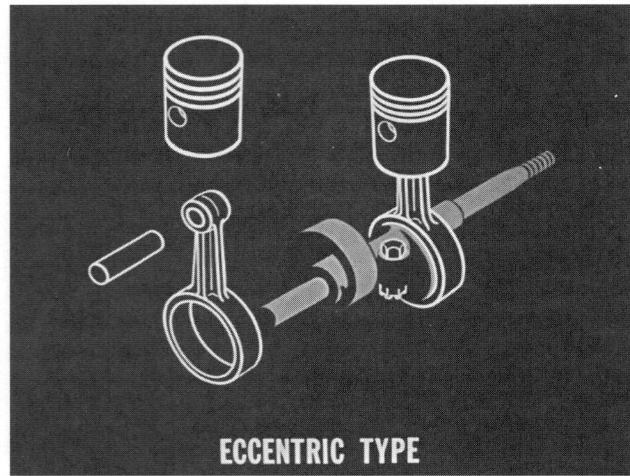

Figure R7-22 Eccentric-type crankshaft.

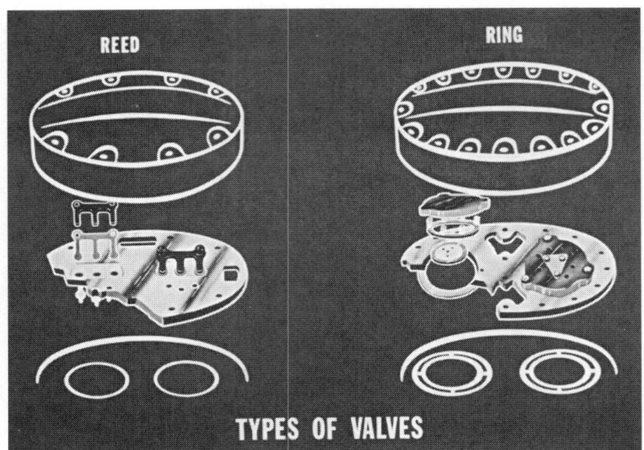

Figure R7-23 Types of valves.

In an open-type compressor, one end of the crankshaft extends through the crankcase housing for connection directly to an outside drive motor, or it may have a pulley attached for belt drive by an external motor. Some provision must be made to prevent the leakage of gas and oil around the crankshaft where it extends through the compressor shell; this is accomplished through the addition of a shaft seal.

One type of shaft seal is pictured in Fig. R7-24. The type of crankshaft shown has a seal shoulder built into it, against which a neoprene washer and self-lubricating seal ring are held stationary by the seal cover plate. A gas- and oil-tight seal is maintained between the seal ring and the crankshaft shoulder seal by the neoprene washer, which fits tightly on the shaft. Seals on reciprocating compressors are on the low-pressure or suction side. It is desirable that as nearly perfect a seal as possible be maintained, since if conditions require that the low-pressure side of the system operate in a vacuum, a leak at the seal or elsewhere in the low side would draw air and moisture into the system.

In most reciprocating compressors gaskets are utilized between mating parts to assure leakproof conditions, be-

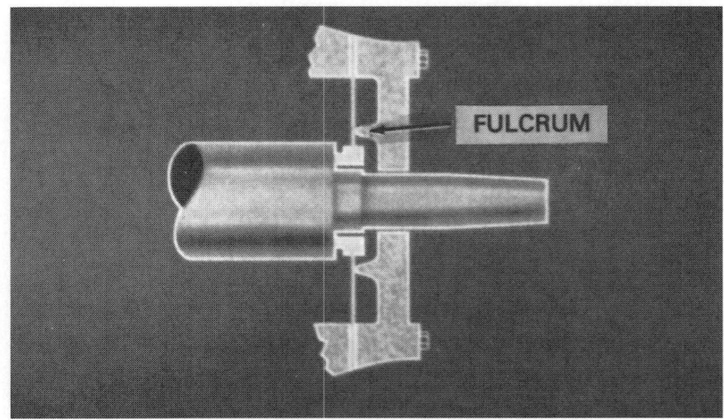

Figure R7-24 Crankshaft seal (diaphragm type).

cause most surfaces are not so finely machined as to provide metal-to-metal leakproof joints. Primarily, gaskets are used between the cylinder head and the valve plate, between the valve plate and the compressor housing, between the compressor body and the bottom plate (if there is one), and also between the exterior service valves and their mounting bases.

When the mating parts are tightly secured, they impress their form and outline on the gasket material, which is usually soft and resilient enough to take the impression and thus seal off any gas or oil from possibly leaking to the atmosphere and prevent taking in air. The gasket material must be such that there will be no chemical reaction when it comes in contact with the oil and refrigerant in the system. When gaskets need replacing after some component has been removed and possibly replaced, replacement gaskets should be of the same material originally used by the manufacturer and of the same thickness as those removed, whether they are of aluminum, cork, rubber, asbestos, or composition. A variation in thickness will affect the efficiency and operation of the compressor. Too thick a gasket between the compressor housing and the valve plate will increase the clearance space above the piston and cause a loss in the volumetric efficiency. Too thin a gasket may permit the piston to hit against the valve plate, damaging the compressor.

The purpose of the hermetic compressor is the same as that of the open compressor, to pump and compress the vapor, but it differs in construction in that the motor is sealed in the same housing as the compressor. A typical fully hermetic compressor cutaway is shown in Fig. R7-25. Note the vertical crankshaft, with the connecting rod and piston in a horizontal position. The fully hermetic unit has an advantage in that there is no projecting crankshaft; therefore no seal is necessary, and there is no possibility of leakage of refrigerant from the compressor or of air being drawn in when the system is operating in a vacuum. A compressor of this design cannot be serviced in the field; internal repairs must be made in a regional repair station or at the factory where it was manufactured.

Some hermetic compressors are constructed with internally mounted springs to absorb vibration caused by the pulsation of the refrigerant vapor being pumped by the pistons. Some hermetic compressors also have springs or hard-rubber vibration mounts located on the outside to absorb shock and vibration.

The bottom portion of the hermetic compressor acts as an oil sump, like the crankcase of an open-type compressor. As the oil circulates and lubricates the internal moving parts, it picks up some of the compressor heat caused by friction of the moving parts. The oil transfers some of this heat to the external shell of the compressor.

Most hermetic compressors are constructed so that the suction vapor is drawn across the motor windings before it is taken into the cylinder or cylinders. This, of course, helps to remove some of the heat from the motor windings and also helps to evaporate any liquid refrigerant that may have entered the compressor.

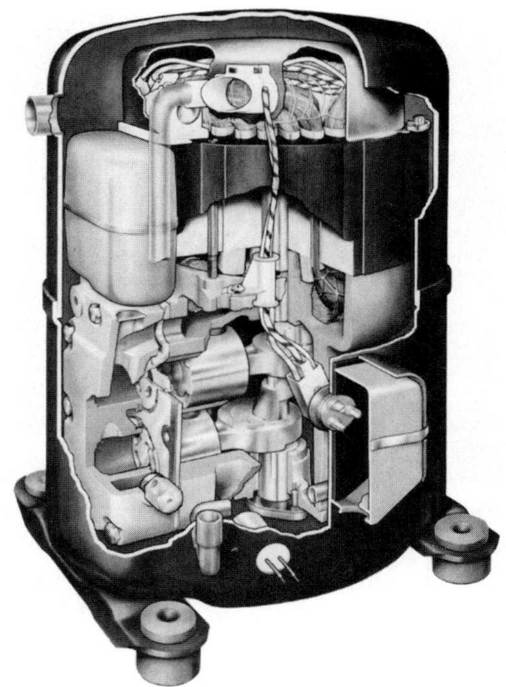

Figure R7-25 Typical reciprocating compressor. (Courtesy of Tecumseh Products Company)

Suction and discharge mufflers are built into some of the smaller hermetic compressors, to absorb or lessen the sound caused by the pulsing vapor as it is pumped through the compressor. A suction muffler can be seen in the cutaway view shown in Fig. R7-25.

Fig. R7-26 shows a typical external-type discharge muffler arrangement used on some compressors. Since the crankshaft and motor shaft are the same unit, high-speed operation causes considerably more noise than the slower operating open compressors, thus requiring mufflers.

Another type of compressor is shown in Fig. R7-27. It combines the motor in the same shell as the compressor,

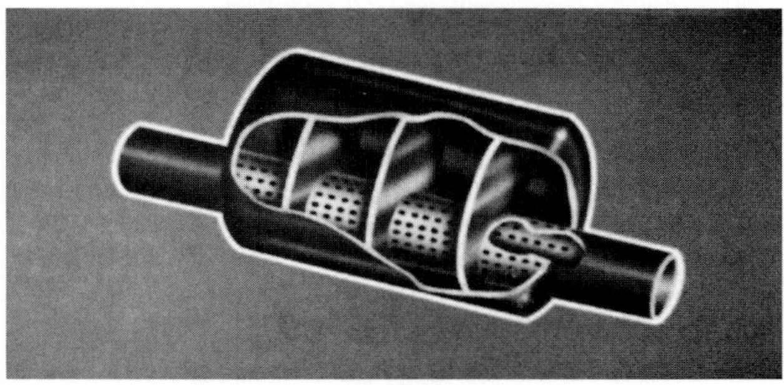

Figure R7-26 External discharge muffler.

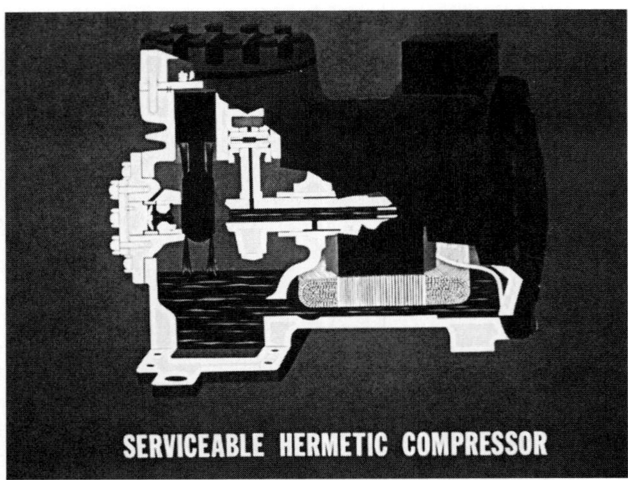

Figure R7-27 Serviceable hermetic compressor.

but, unlike the fully hermetic unit, this type provides access to the compressor for repair in the field. This unit is called by several names such as *semihermetic, accessible,* and *serviceable hermetic.*

Rotary compressors are so classified because they operate through application of a rotary, or circular, motion, instead of the reciprocating operation previously described. A rotary compressor is a positive displacement unit, and usually can be used to pump a deeper vacuum than a reciprocating compressor.

There are two primary types of rotary compressors used in the refrigeration field: the *rolling-piston* type, with a stationary blade, and the *rotating-blade* or *vane* type. Both are similar in capacity, variety of applications, physical size, and stability, but they differ in manner of operation.

The rolling-piston type, as shown in Fig. R7-28, has the roller mounted on an eccentric shaft. The blade is located in a keeper in the housing of the compressor. As the rolling piston rotates, vapor is drawn into the space ahead of the spring-loaded blade, as shown, and is compressed by the roller into a continually smaller space until it is forced out the discharge port, and the compression cycle begins again. The roller does not make a metal-to-metal contact with the cylinder, because a film of oil, in normal operation, provides a clearance between the two surfaces.

Fig. R7-29 shows the other primary type of rotary compressor. This unit consists of a cylinder and an eccentric roller or rotor having several blades, held in place by either springs or centrifugal force. As the roller turns in the cylinder, suction vapor is trapped in the crescent-shaped space between two of the blades. As the roller continues to turn, the suction gas is compressed in volume, and its pressure and temperature are increased until it is discharged from the cylinder.

As mentioned earlier, a film of oil prevents leakage of vapor from the cylinder, or between the spaces separated by the blades, while the unit is in operation. To prevent hot gas from leaking back into the cylinder from the discharge port when it is uncovered by the roller, a check valve is usually placed in the discharge line. During the OFF cycle or period of shutdown, warm vapor is prevented from leaking back into the evaporator by the check valve.

Rotary compressors are well-balanced units, and those enclosed in a hermetic shell are usually spring-supported or spring-mounted. Usually, they are very quiet in operation. Since hermetic rotary compressors are direct-driven, they operate at the motor speed—usually 3450 rpm—and, although the sound level is in direct ratio to speed and horsepower, they operate comparatively quietly by any standards. Discharge mufflers are widely used to prevent the pulsations of discharged gas from causing vibrations to be carried over to the discharge line and condenser. The principles of design and operation given here pertain to household and small commercial units, although, in general, the same principles also apply to larger rotary compressors, some of which have primary usage in the low-temperature field where they are used as low-stage or booster compressors.

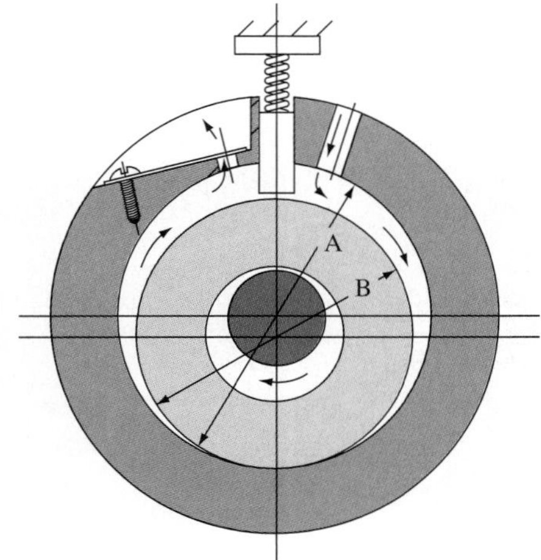

Figure R7-28 Rolling-piston rotary compressor.

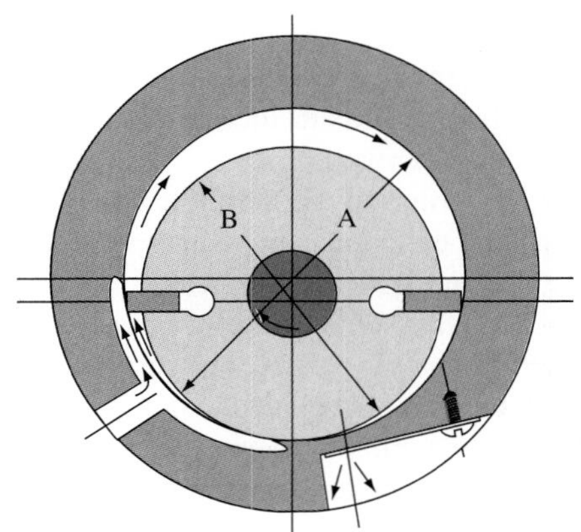

Figure R7-29 Rotating-vane rotary compressor.

PREVENTIVE MAINTENANCE

Preventive maintenance for refrigeration systems is equally important to troubleshooting and repair. Good maintenance practices will extend equipment life, detect early potential problems, and eliminate costly down times during heavy refrigeration-load periods.

A complete centrifugal refrigeration unit compresses the refrigeration vapor, as its name implies, through centrifugal action or force. This action is performed mainly by the impeller or rotor. Vapor is drawn in at the intake near the shaft of the rotor and discharged from the exhaust openings at the outer edge of the rotor. With the rotation of the impeller, suction vapor is drawn rapidly into the impeller chambers, where it is forced to the outside of the housing sections through centrifugal action. To maintain the centrifugal force, the impeller is operated at a high rate of speed by an outside driving force, such as an electric motor, gasoline engine, or steam turbine. The pressure differential between inlet and outlet vapor is small. Therefore, it is not a positive-displacement unit like the types described previously, and it is not capable of building up pressure against a closed valve in the system.

A centrifugal compressor may have one or more impellers. Compressors with several stages are constructed so that the discharge of one impeller or stage enters the suction inlet of the next. If the speed of the driving motor does not provide the desired operating speed of the compressor, speed-increasing gears or fluid couplings may be used to obtain optimum operating conditions. Since no pistons or internal suction and discharge valves are present, little wear and tear occurs on the unit. The main bearings in the housing, supporting the drive shaft, are the components most subject to wear.

R7-4 CONDENSERS

The next major component in the refrigeration system, following the compression stage, is the condenser. Basically, the condenser is another heat-exchange unit in which the heat picked up by the refrigerant in the evaporator—as well as that added to the vapor in the compression phase—is dissipated to some condensing medium. High-pressure, high-temperature vapor leaving the compressor is superheated, and this superheat customarily is removed in the hot-gas discharge line and in the first portion of the condenser. As the temperature of the refrigerant is lowered to its saturation point, the vapor condenses into a liquid for reuse in the cycle.

Condensers may be air-cooled, water-cooled, or cooled by evaporation. Domestic refrigerators usually have an air-cooled condenser, which depends on the gravity flow of air circulated over it. Other air-cooled units use fans to blow or draw large volumes of air across the condenser coils.

Fig. R7-30 depicts a typical small commercial condensing unit using an air-cooled condenser. It is dependent on an ample supply of relatively "cool" air, for, in order to have a heat transfer from the refrigerant in the condenser to the coolant, the air must be at a lower temperature than the refrigerant. Even when the surrounding temperature is above 100°F, the air is still cooler than the refrigerant in the condenser, which must give up some heat to return to its liquid state.

Air-cooled condensers are constructed somewhat like other types of heat exchangers, with coils of copper or aluminum tubing equipped with fins. Evaporators usually have filters in front to reduce clogging by dust, lint, and other matter, but condensers are not so equipped, and so must be cleaned frequently to prevent reduction of their capacities.

Remote air-cooled condensers usually have wider fin spacing to prevent clogging as quickly as in those directly mounted on the condensing unit. Also, they can be located away from the compressor, which is a distinct advantage. Occasionally, a complete condensing unit is placed somewhere inside the building in which it is to be used, where the heat dissipated from the condenser and motor can cause an increase in temperature within the storage or mechanical equipment room. As a result, the unit might have

Figure R7-30 Small air-cooled condensing unit. (Courtesy, Copeland Corporation)

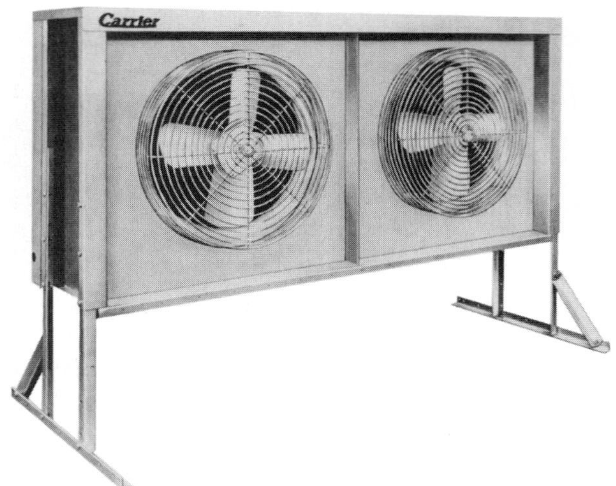

Figure R7-31 Remote air-cooled condenser.

a higher operating discharge temperature and pressure, which would decrease its efficiency.

Fig. R7-31 shows a remote air-cooled condenser, which may be located outdoors—beside a building or on a flat roof. In such an open, outdoor location, an adequate supply of air as a coolant is readily available at the ambient outdoor temperature, thus avoiding undesirable temperatures in the building. The air movement across the coil is created by either a belt-driven centrifugal fan or a direct-drive propeller-type fan. The slow-speed, wide-blade propeller fan moves the required volume of air without creating unreasonable noise.

This type of condenser may be assembled in any combination of units that may be required for the necessary heat removal. The air may either be drawn through or blown through the coils. In another design, a single condenser may have more than one circuit in its coil arrangement, so that it may be used with several separate evaporators and compressors.

In most installations of remote air-cooled condensers in this country, the difference between the ambient air temperature and the condensing temperature of the refrigerant is approximately 30°F. Therefore, if the outdoor temperature is 95°F, the refrigerant will condense at approximately 125°F.

Some difficulty may arise with remote air-cooled condensers when they are operated in low ambient temperatures, unless proper precautions are taken to maintain head pressures that are normal for the unit. Earlier in this chapter it was stated that a too-high condensing temperature and pressure lessens the overall operating efficiency of the unit. Conversely, a too-low condensing temperature and pressure will affect the efficient operation of the system by causing a reduction in pressure difference across the metering device, thus resulting in a loss of refrigerant flow into the cooling coil. Later in this chapter the effect of pressure drop through a metering device on the overall capacity and efficiency of a system will be discussed.

Some remote air-cooled condensers equipped with multiple fans have controls for the cycling of one or more of the fans during periods of low ambient temperatures. The flow of air across other types of condensers may be controlled by adjustable louvers. Still other manufacturers install controls to allow partial flooding of the condenser with liquid which, in turn, will lessen the condensing capacity. This is another means of keeping the head pressure within allowable limits.

Water-cooled condensers permit lower condensing temperatures and pressures, and also afford better control of the head pressure of the operating units. They may be classified as follows:

1. Shell-and-tube
2. Shell-and-coil
3. Tube-in-a-tube

As pointed out in an earlier chapter, water usually is an efficient medium for transferring heat, since the specific heat of water is 1 Btu per pound per °F change in temperature. If 25 lb of water increases 20°F, and if this heat is removed in 1 minute, 500 Btu is removed from the source of heat each minute. Also, if this rate of heat transfer continues for 1 hour (60 minutes), this means that the water is absorbing 30,000 Btu/hr.

If the example above involved a water-cooled condenser, it would mean that 3 gallons of water were being circulated per minute. Also, if the heat of compression amounted to 6000 Btu/hr, it would mean that the evaporator load was 24,000 Btu/hr—a 2-ton refrigeration load.

Water coming from a well or other underground source will be quite a bit cooler than the ambient outdoor air. If cooling-tower water is used, its temperature can be lowered in the cooling tower, after it has picked up heat in the condenser, to within 5 to 8°F of the outdoor wet-bulb temperature. The use of a cooling tower and circulating pump permits the reuse of the water, except for a slight loss due to evaporation, and keeps the consumption and cost of water to a minimum.

The shell-and-tube type of water-cooled condenser consists of a cylindrical steel shell containing several copper tubes running parallel with the shell. Water is pumped through the tubes by means of the inlet and outlet connections on the end plates. The hot refrigerant vapor enters the shell at the top of the condenser, as shown in Fig. R7-32, and the liquid refrigerant flows as needed from the outlet at the bottom of this combination condenser-receiver.

The end plates are bolted to the shell of the condenser for easy removal to permit the rodding or cleaning of the water tubes of minerals that may be deposited on the inside of the tubes, causing a restricted water flow, a reduction in the rate of heat transfer, or both. A control of the water flow, that is, the number of times it travels the

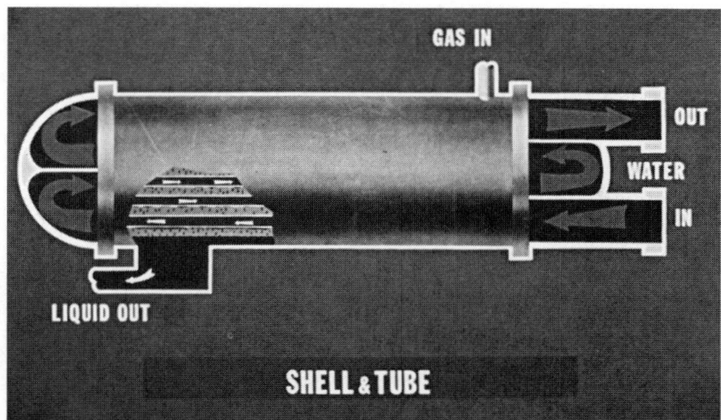

Figure R7-32 Shell-and-tube condenser.

length of the condenser or the number of passes it makes, is built into the end plates of the condenser. If the water enters one end plate, passes through all the tubes once, and leaves the condenser at the other end plate, it is called a *one-pass condenser.* If the water inlet and outlet are both in the same end plate, it is a *two-pass* or some other even-numbered type of pass condenser.

If, instead of a number of tubes within the condenser shell, there are one or more continuous or assembled coils through which water flows to remove heat from the condensing vapor, it is classified as a *shell-and-coil* type of condenser. Fig. R7-33 shows a condenser of this type. It is a compact unit and usually serves as a combination condenser-receiver within the circuit. Usually, this type of condenser is used only on small-capacity units and when there is an assurance of reasonably clean water, for the only means of cleaning it is by flushing with a chemical cleaner.

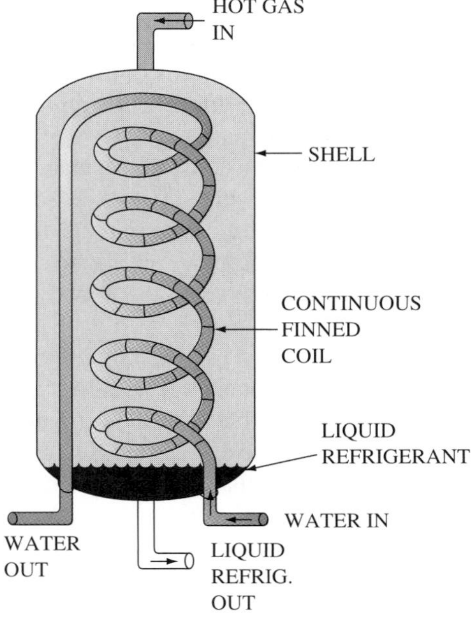

Figure R7-33 Shell-and-coil condenser.

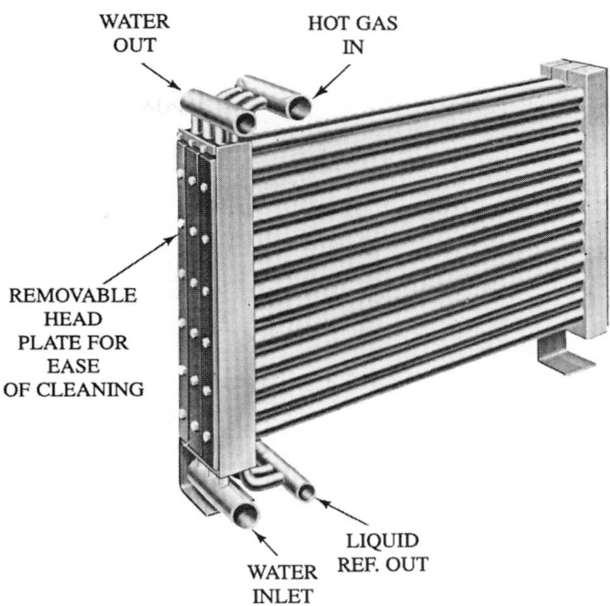

Figure R7-34 Water-cooled tube-in-tube condenser. (Courtesy of Heatcraft Refrigeration Products)

The *tube-in-a-tube,* or *double-tube* as it is also known, may be classified as a combination air-and-water-cooled type of condenser. As pictured in Fig. R7-34, it has the refrigerant flowing through the outer tubing where it is exposed to the cooling effect of air flowing naturally over the outside of the outer tubes while water is being circulated through the inner tubes. Generally, water enters the bottom tubes of water-cooled condensers, and leaves at the top. In this manner peak efficiency is obtained, for the coolest water is capable of removing some heat from the refrigerant in a liquid state, thereby subcooling it. Then the warmer water still is able to absorb heat from the vapor, assisting in the condensation process.

When the ambient temperature is such that a satisfactory condensing temperature cannot be obtained with an air-cooled condenser, and when the water supply is inadequate for heavy usage, an *evaporative condenser* may be used to advantage. A diagram of this type of condenser is shown in Fig. R7-35 which shows the combined use of air and water for the purpose of heat removal from the refrigerant vapor within the condenser coil.

There is actually a double heat transfer in this unit: The heat of the vapor and the coil containing it is transferred to the water, wetting the outer surface of the coil, and then is transferred to the air as the water evaporates. The air can either be forced or drawn through the spray water.

When air is blown through the unit, the fan and motor are in the dry entering airstream. When the system has a draw-through fan, it is essential that eliminators be installed before the air mover. Otherwise, there would quickly be an accumulation of scale on all of the air-mover components. Even with blow-through units, there is a possibility of some of the spray water blowing out of the evaporative condenser, and a set of eliminator plates should be installed to prevent this.

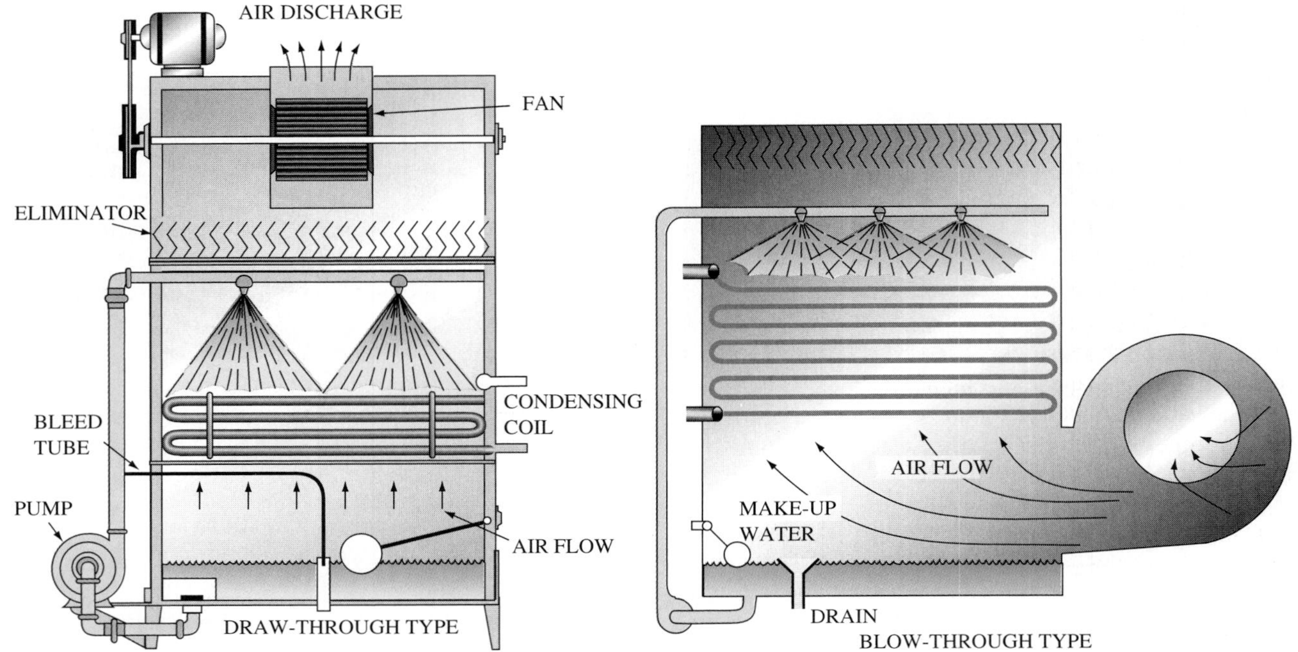

Figure R7-35 Functional view of evaporative condensers.

R7-5 RECEIVERS

As mentioned earlier, some water-cooled shell-and-tube condensers also act as receivers, with liquid refrigerant occupying the space in the bottom of the condenser where there are no water tubes. If there is too much liquid in this type of condenser-receiver, some of the water tubes may be covered by the liquid level. This reduces the area of heat transfer surface in the condenser.

In systems other than those having condenser-receivers and those operating with a critical refrigerant charge, a receiver is needed, which is actually a storage container for the refrigerant not in circulation within the system. Receivers that are part of small, self-contained commercial units usually are large enough to hold the complete operating charge of refrigerant in the systems. This applies to a number of larger systems as well. Yet, in some cases, the receiver may not be large enough to hold the entire refrigerant charge if a pump-down becomes necessary for repair or replacement of a component. An auxiliary receiver would be necessary to provide pump-down capacity. If this is not provided, the surplus refrigerant would have to be pumped into an empty refrigerant drum or wasted to the atmosphere.

Precautionary measures are usually taken by manufacturers of receivers against the possibility of too much pressure or too high a temperature build-up in the receiver or in a combination condenser-receiver. These safety measures usually include the installation of pressure-relief valves, customarily spring-loaded, that will open if excessive pressure should build up within the receiver. A fusible-plug type of relief valve may be installed, which is designed to melt at a preselected temperature and thus release the refrigerant if, for any reason whatsoever, that temperature is reached within the receiver.

R7-6 REFRIGERANT LINES

Previous portions of this chapter have been devoted to the proper selection, balancing, and operation of the major components of a refrigeration system. But regardless of how well these have been selected and balanced, system operation is dependent on the means of moving the refrigerant, both liquid and vapor, from one component to another in the refrigeration circuit.

Just as a highway must be constructed, maintained, and kept open between communities to provide adequate access for the vehicles that must use it, so must the piping in a refrigeration system be properly sized and installed so that there will be no restrictions to the flow of the refrigerant.

Refrigerant oil, necessary for the proper lubrication of the moving parts in the compressor and the metering device, must be readily miscible with the refrigerant in the liquid state. The oil will travel with the liquid as long as the liquid line is sized so that the refrigerant will travel through its entire length at a proper velocity.

If the system is self-contained, proper sizing and installation of the refrigerant lines is the manufacturer's responsibility. But a system built up in the field, using products of several manufacturers, becomes the problem and responsibility of the person designing the complete system, as well as those who install and connect the components. A liquid line that is sized too small or with too many restrictions (fittings and bends) could easily cause too great a pressure drop in the line, which could result in a loss in the capacity of the pressure-reducing device when compared to the capacity required in the cooling coil.

Lines through which refrigerant vapor flows are the most critical, and these are the suction and hot-gas or discharge lines. The velocity of the vapor should be at least 750 ft/min in horizontal lines and above 1500 ft/min in vertical lines, so that the refrigerant oil will be entrained with the vapor and returned to the compressor. If the lines are too large, the desired velocity cannot be maintained and the oil may not return, and the compressor may become short of its proper oil charge.

If the evaporator is located above the compressor, the oil usually will return to the compressor through gravity flow, provided that no risers or possible traps are built into the line.

If the evaporator is situated below the compressor, or when the condenser is located a distance above the compressor, the vapor must have the proper velocity in order to carry the oil droplets with it. Oil traps may have to be built into the piping, or it may be necessary to use dual pipes, if the capacity control of the compressor varies because of changing load conditions. These measures ensure that even in the case of minimum load conditions of 10 to 25% of full capacity, the refrigerant vapor will have adequate velocity to carry the oil along with it.

REVIEW

Basic refrigeration cycle components include:

- Compressor—The heart of the refrigeration system. It pumps refrigerant, works in partnership with the metering device to control flow, and provides the pressure difference and "lift" to reach the condensing pressure.
- Condenser—A heat-exchange unit, designed to remove the heat from the system by the condensing action of the refrigerant, and heat is dissipated to some condensing medium.
- Metering device—A flow-control and pressure-reducing device. It maintains the proper pressure and boiling point in the evaporator to handle the desired heat load. It meters the flow of liquid refrigerant into the evaporator at a rate needed to remove the heat of the product load.
- Evaporator—A heat exchange unit, where heat is absorbed into the system, by the action of the boiling refrigerant. The refrigerant then carries this heat through the system to the condenser, where it is rejected.
- Suction line—The interconnecting line between the evaporator and the compressor.
- Liquid line—The interconnecting line between the condenser and the metering device.
- Discharge line—The interconnecting line between the compressor and the condenser.

Metering devices include:

- Hand operated needle valve—Hand operated variable orifice to provide constant flow.
- Capillary tube—Constant-flow device made of long, small interior-diameter (ID) tubing.
- Automatic expansion valve—Constant pressure device.
- Thermostatic expansion valve—Modulates to provide constant superheat in evaporator.
- High-side float—Located in the liquid side of the system, controls flow from the high side to the evaporator, while maintaining constant liquid level.
- Low-side float—Located in the evaporator, controlling the amount of liquid in the evaporator to maintain constant liquid level.
- Electric/electronic expansion valve—Interfaced with microprocessor control to provide constant superheat in evaporator.

Types of evaporators include:

- Dry (Direct) expansion—Designed for complete evaporation of the refrigerant, leaving only a superheated vapor exiting the coil outlet. Typically constructed of bare pipe, plate type, or fin-and-tube.
- Flooded—Filled with a liquid and controlled by a float mechanism, used for liquid cooling or large industrial applications. Construction is shell-and-tube, shell-and-coil, or fin-and-tube.

Types of compressors include:

- Reciprocating (piston)—A piston compressor used in domestic, commercial, air-conditioning, and industrial applications.
- Rotary—Operates through application of a rotary or circular motion. Uses a rotating or fixed vane having positive displacement, with the ability to pump a deeper vacuum than a piston compressor. Used in domestic, small commercial, and industrial applications to increase efficiency.
- Rotary-screw—A machine with interlocking male and female spiral lobes rotating together within a chamber. Designed for large commercial and industrial refrigeration and air conditioning applications.
- Scroll—Operates on the principle of a stationary and an orbiting scroll. Introduced at first in the air-conditioning market, it is now available in commercial refrigeration applications. Provides high efficiency, durability, and reduced noise.
- Centrifugal—Compresses the refrigerant, as its name implies, by centrifugal force. Utilizes one or more impellers to force the refrigerant to accelerate and is discharged at the exhaust opening at the outer edge of the rotor. These compressors have the ability to move large amounts of refrigerant, therefore are used in large commercial and industrial applications.

Compressor casing designs:

- Hermetic—A welded, air-tight construction, fully sealed off from the outside.
- Open—Driven by an external motor, which is coupled to the compressor. The shaft of the compressor has a rotary seal to prevent refrigerant leakage.
- Semi-hermetic—Sealed compressor and motor assembly, but may be disassembled for repairs or inspection.

Types of condensers include:

- Air cooled—Can be either forced air, fin-and-tube, static-tube or plate construction.
- Water cooled—Can be either shell-and-tube, shell-and-coil, or tube-within-a-tube.
- Evaporative—Uses water and air to cool the condensing refrigerant with either draw-through or blow-through.

Refrigeration system accessories:

- Liquid receiver—Used to store refrigerant and ensure the flow of a solid stream of liquid enter the metering device.
- Check valve—Prevents reverse flow of refrigerant.
- Discharge muffler—Reduces discharge noises from the compressor.

Problems and Questions

1. Name the four major components in the compression-type refrigeration cycle.
2. What are the most frequently used methods for compensating for conduction loss due to an air film around the evaporator-coil tubing?
3. Under what circumstances should nonfrosting evaporator coils be installed?
4. What are some types of cooling coils used for cooling liquids?
5. What are the five major classifications (by method of compression) of compressors?
6. What is the difference between a hermetic and an open compressor?
7. What are the two primary types of rotary compressors used in the refrigeration field?
8. How many valves are used in the compression operation of a centrifugal compressor?
9. What is the basic purpose of the condenser?
10. What are the primary types of condensers?
11. What are the principal types of pressure-reducing devices used to control the flow of refrigerant to the evaporator?
12. Which type of pressure-reducing device is the simplest? Why?

Metering Devices and Accessories

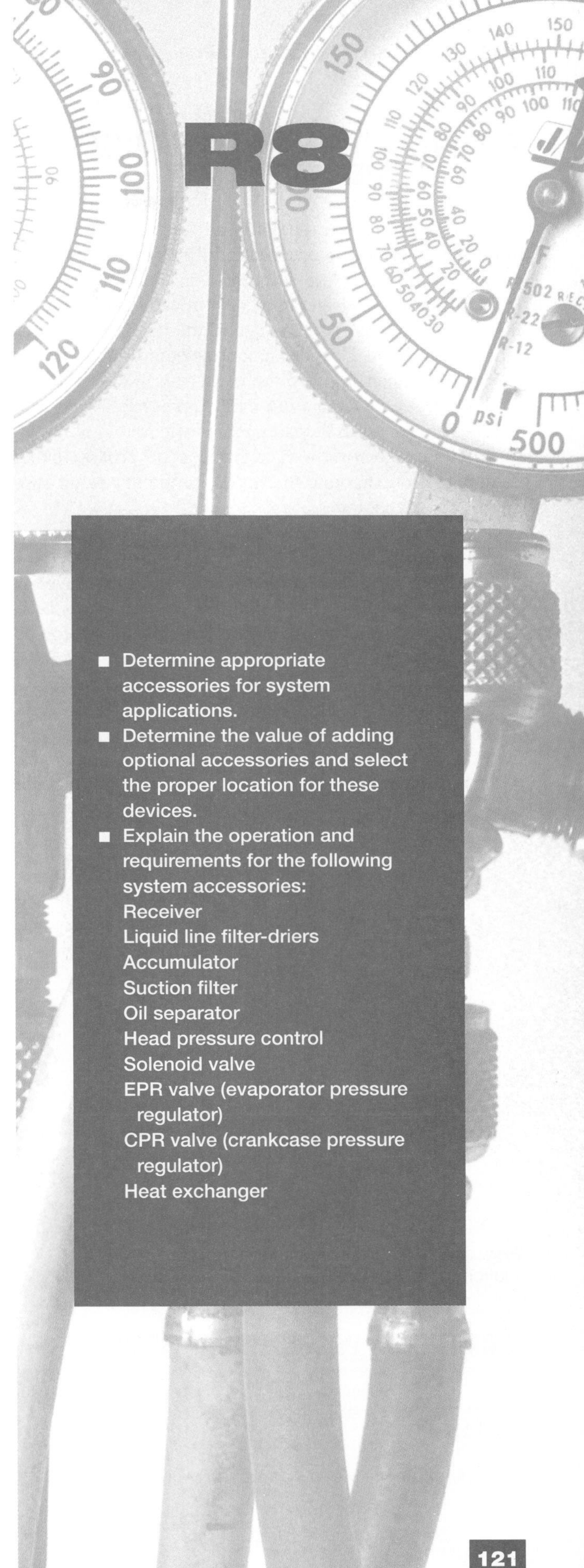

AFTER STUDYING THIS CHAPTER, THE STUDENT WILL BE ABLE TO:

- Identify and describe the various types of metering devices and accessories and how they function in the refrigeration system.
- Evaluate system performance under varying load conditions using different types of flow-control (metering) devices.
- Adjust and size metering devices.
- Explain the benefits of superheating and subcooling.
- Describe the operating characteristics of various metering devices.
- Install metering devices.
- Describe procedures to measure, test, and adjust metering devices for proper operation.
- Measure and adjust superheating and/or subcooling.
- Describe possible malfunctions in the operation of metering devices and suggest a correction.
- Select the proper location in a cycle for system accessories.
- Determine appropriate accessories for system applications.
- Determine the value of adding optional accessories and select the proper location for these devices.
- Explain the operation and requirements for the following system accessories:
  Receiver
  Liquid line filter-driers
  Accumulator
  Suction filter
  Oil separator
  Head pressure control
  Solenoid valve
  EPR valve (evaporator pressure regulator)
  CPR valve (crankcase pressure regulator)
  Heat exchanger

R8-1 HOW A METERING DEVICE OPERATES

A *metering device* is a type of restrictor placed in the liquid line between the condenser and the evaporator, to produce a difference in pressure between the high side and the low side of a refrigeration system and to regulate the flow of refrigerant. The amount of the restriction is provided to maintain a condensing temperature high enough above the condensing medium (water or air) to condense the high-pressure vapor from the compressor. The restriction is also provided to maintain an evaporating temperature low enough to absorb heat from the product being cooled and to evaporate the liquid refrigerant being supplied to the evaporator.

The location of the metering device in the system is shown in Fig. R8-1.

Fig. R8-2 shows a pressure-heat diagram for a typical refrigeration process. The action of the metering device is shown between the points 4 and 1. At 4, the subcooled high-pressure liquid enters the metering device. The line between 4 and 1 is exactly vertical since no heat is absorbed from the outside during the process. In order to lower the saturation temperature between the inlet and outlet of the device, some refrigerant evaporates. Although the refrigerant enters the valve as a liquid, it leaves as a mixture of liquid and vapor. It is normal for about 20% of the liquid to be evaporated when the refrigerant enters the evaporator, at point 1 on the diagram.

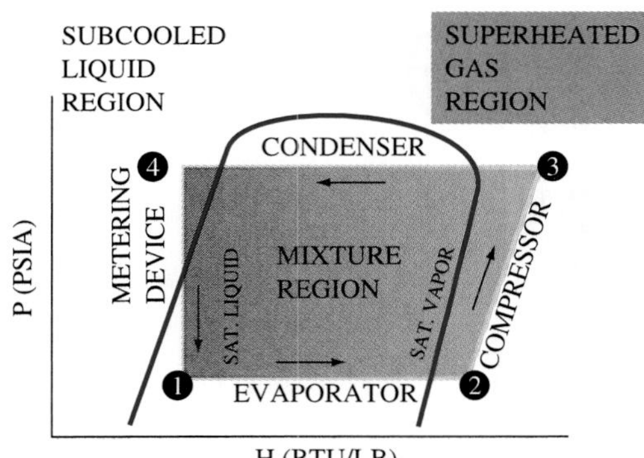

Figure R8-2 View of the pressure-enthalpy (heat) diagram showing the action that takes place in the metering device.

The evaporated liquid is called "flash gas" as shown in Fig. R8-3. The heat required to lower the evaporating temperature of the refrigerant, in going from the high side of the system to the low side, is absorbed from the liquid refrigerant creating "flash gas."

The table in Fig. R8-4 gives a summary of the condition of the refrigerant before and after passing through the

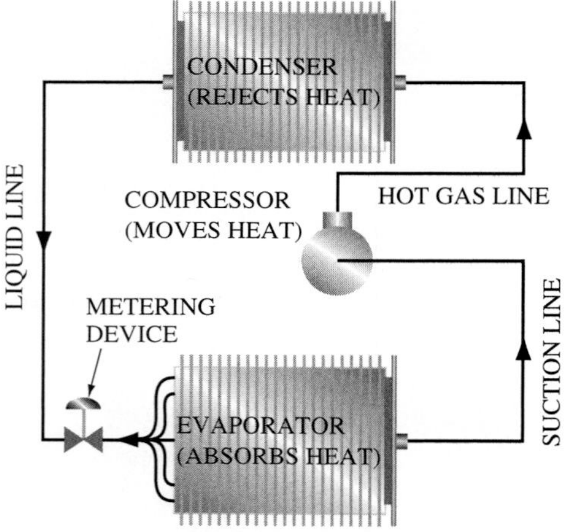

Figure R8-1 Location of the metering device in relation to the other components of the system.

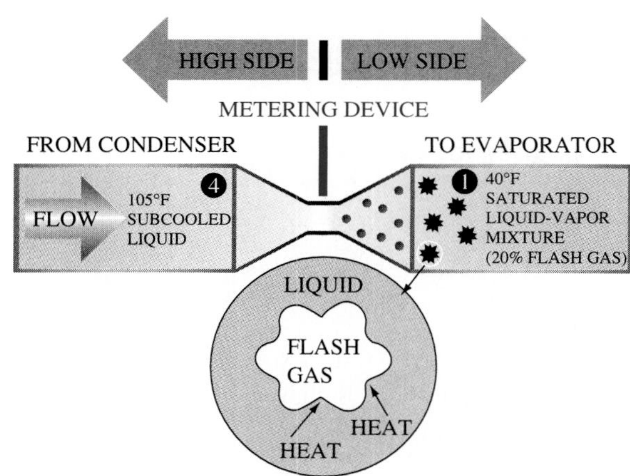

Figure R8-3 How flash gases are formed when the liquid refrigerant flows through the metering device.

PRACTICING PROFESSIONAL SERVICES

Whenever any metering device or valve has been adjusted or calibrated, it is standard operating procedure to clean and re-install all caps, covers, sensing thermobulbs, and equipment lids properly as originally intended. This includes securely fastening the device with the proper tools and leak checking upon completion to verify no refrigerant leaks exist.

Figure R8-4 Inlet and outlet conditions of the refrigerant passing through a typical metering device.

	AT INLET TO METERING DEVICE	AT OUTLET TO METERING DEVICE
SATURATION TEMPERATURE (°F)	120	40
ACTUAL REFRIGERANT TEMP. (°F)	105	40
SUPERHEAT (°F)	0	10
SUBCOOLING (°F)	15	0
PRESSURE (PSIA)	277.7	83.7
PRESSURE (PSIG)	263.0	69.0
ENTHALPY (BTU/LB)	42.0	42.0

metering device. Note the numbers in this summary are based on using R-22 refrigerant in the system. The tabulation is based on the performance of an air-conditioning unit with an air-cooled condenser operating at an outside temperature of 95°F.

R8-2 TYPES OF METERING DEVICES

There are eight different types of metering devices. They are divided into two groups—those that are fixed and those that are adjustable to provide regulation matching the load.

R8-2.1 Fixed Metering Devices

There are two types of fixed metering devices: the capillary tube and the fixed orifice.

The *capillary tube* is probably the simplest of all metering devices. The capillary tube is commonly used as a metering device on domestic refrigerators and other small appliances. It is constructed of a single tube with an inside diameter in the size range of .026 in. to .090 in. It is normally located near the entrance to the evaporator and sometimes coiled to conserve space, as shown in Fig. R8-5. The size of the tube and the length are carefully selected to match the pumping capacity of the compressor at full load.

Due to the critical selection of the capillary tube, most systems that use this type of metering device are packaged units, such as domestic refrigerators and freezers, where the system is installed at the factory and critically charged with refrigerant. It operates best where the load is more nearly constant and is usually used on small systems not over 3 tons in capacity. It has been applied successfully to room coolers, close-coupled split systems and small heat pumps. Due to the small size of the tube, it can be easily plugged. Many systems use a built-in liquid line filter at the entrance of the tube.

One advantage of the capillary tube is that on shutdown the high-side and the low-side system pressures are equalized. Not having to start against a pressure differential permits the use of a low-starting-torque compressor motor. This can also be a disadvantage since on shutdown the evaporator can be filled with liquid and possibly damage the compressor on subsequent startup. To counteract this problem, many systems use a suction line accumulator at the entrance to the compressor, as shown in Fig. R8-6. This captures the liquid refrigerant and returns it to the system as vapor.

The *fixed orifice,* shown in Fig. R8-7, is the second type of fixed metering device. This orifice is built into a rugged assembly and has the advantage for heat pump application of including a built-in check valve. During the reverse cycle on the heat pump, this metering device will function with the flow in either direction. It is like the capillary tube in a number of ways:

a. It must be carefully selected to match the load.
b. The system must be critically charged.
c. It permits refrigerant migration into the evaporator during the OFF-cycle and requires all of the same

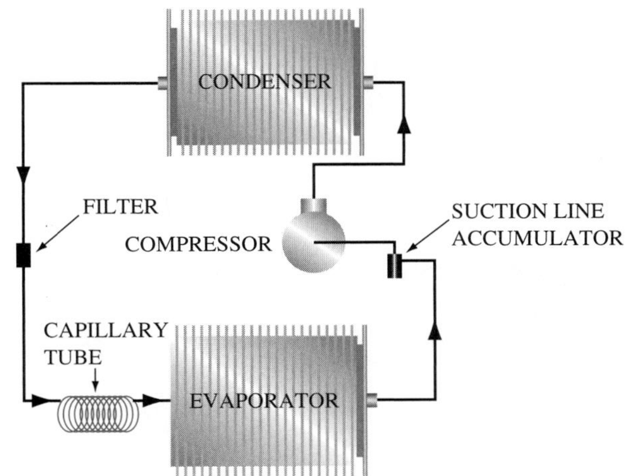

Figure R8-5 Typical location of a capillary-tube-type metering device.

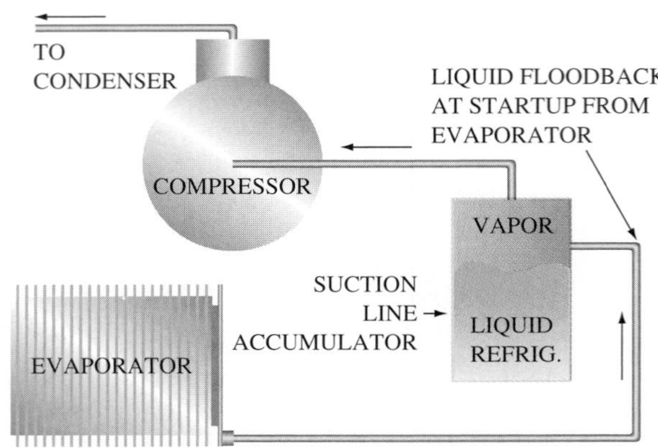

Figure R8-6 Location and function of the suction-line accumulator.

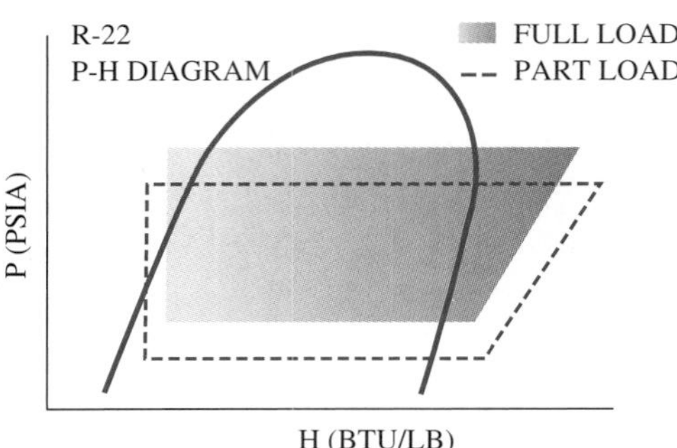

Figure R8-8 Function of the fixed orifice in a pressure-enthalpy diagram. Note that there is no adjustment of this type of metering device.

protective accessories as the capillary tube. It is factory selected for the application. For split systems it is shipped with the outdoor unit to match the unit's capacity.

Fig. R8-8 show the performance of the fixed orifice in the pressure-enthalpy diagram. *Enthalpy* is the scientific term for heat. The orifice works best at full-load conditions. For most systems under normal operating conditions it is satisfactory. Problems can occur if the unit is oversized and operates for long periods of time at low load. As a precaution, the unit should not be oversized.

Fig. R8-9 shows a diagrammatic view of the orifice installed. The fixed metering orifice automatically adjusts to normal changes in load. The orifice regulates the flow of refrigerant in a manner described as "floating with the load." When the compressor moves less heat, the discharge pressure drops. This reduces the pressure across the orifice, lowering the flow of refrigerant. When the load is increased, the opposite condition takes place.

R8-2.2 Adjustable Metering Devices

There are six types of adjustable metering devices:

1. Hand-operated expansion valve
2. Low-side float (LSF)
3. High-side float (HSF)
4. Automatic expansion valve (AEV)
5. Thermostatic expansion valve (TXV)
6. Electric and electronic expansion valve

All of the above mechanically adjust to the changes in load.

The *hand-operated expansion valve* is illustrated in Fig. R8-10. The rate of flow through the valve is determined by:

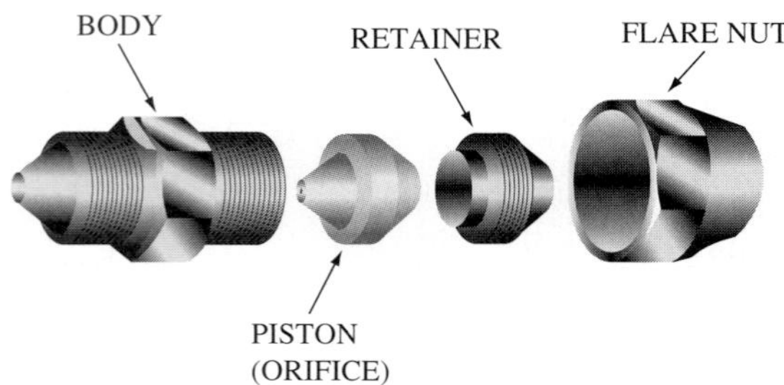

Figure R8-7 Construction of a fixed-orifice-type metering device.

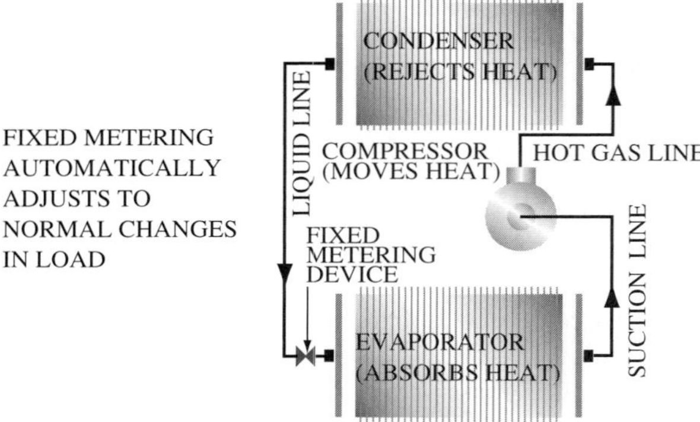

Figure R8-9 The fixed orifice floats with the load and automatically adjusts to changes in the load.

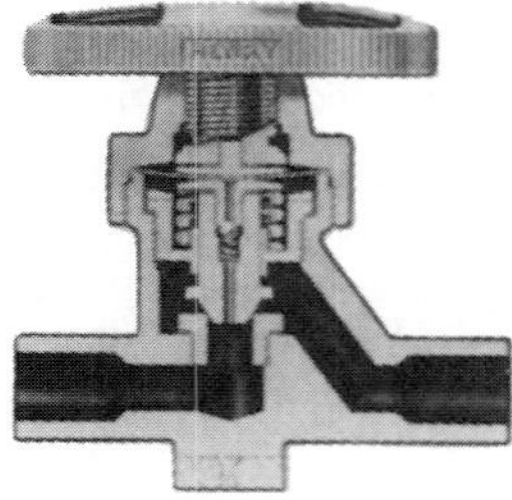

Figure R8-10 A typical hand expansion valve showing the internal construction. (Courtesy of Henry Valve Company)

KNOWLEDGE EXTENSION

Manufacturers of metering devices and associated accessories will often provide technical manuals and data free of charge. To gain more specific knowledge regarding this equipment, contact your local refrigeration supply house and request the name and phone number of your territorial metering device factory sales representative. Contact them and ask for any educational materials available. Frequently, they will visit with potential product users and distribute free literature, charts, graphs, and materials to interested individuals and groups. Additionally, they will often give technical presentations and talks to student groups and field personnel.

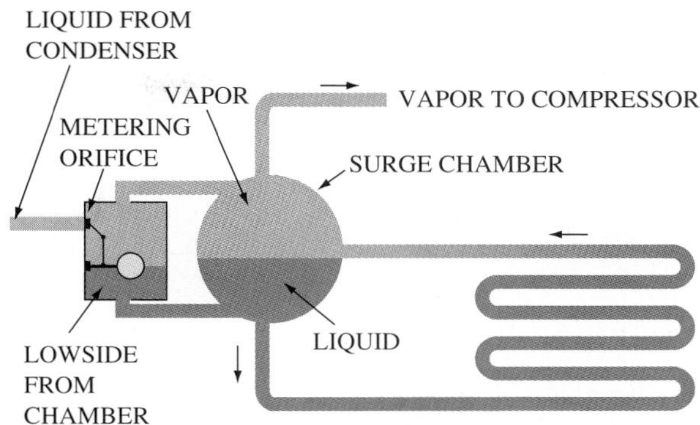

Figure R8-12 Alternate location for the low-side float in a separate float chamber.

a. The size of the valve port opening or orifice.
b. The pressure difference across the orifice.
c. How far the valve is opened.

An increase in any of these conditions will increase the flow. A decrease will reduce the flow.

The obvious disadvantage of the hand expansion valve is that it has no automatic arrangement to control the size of the orifice to match the load. These valves were used in the past on applications where the load was fairly constant and an operator was present to manually make adjustments when necessary.

The hand expansion valve was often used on ammonia systems that had a nearly constant load, where an operator was available to make adjustments when needed. It is seldom used today, except for laboratory tests to explore optimum flow rates.

The *low-side-float* (LSF) type metering device is illustrated in Fig. R8-11. Using this arrangement, the evaporator is flooded with refrigerant liquid and the level in the evaporator is maintained by a float. If the load is increased, the valve opens wider, passing more refrigerant to the coil. If the load is reduced, less refrigerant boils away and the valve is moved toward the closed position.

The low-side float can be installed in the evaporator or in a separate float chamber as shown in Fig. R8-12.

One advantage of this arrangement is that the heat transfer rate from a flooded coil is higher than with a mixture of liquid and vapor. One disadvantage is the possibility of light loads of oil collecting in the evaporator and not returning to the compressor. These valves have been used widely on systems using ammonia refrigerant.

The low-side float has been rather commonly used on ammonia systems that use flooded evaporators.

The *high-side float* (HSF), illustrated in Fig. R8-13, modulates refrigerant flow to the evaporator based on liquid level. Unlike the low-side float, this float assembly is located on the high-pressure side of the metering device's orifice.

Fig. R8-14 shows the action that takes place in the system using the high-side float. As the load increases, more refrigerant is condensed in the condenser and flows into the high-side float assembly. As the liquid level rises in the chamber (1), the valve opens, permitting greater flow

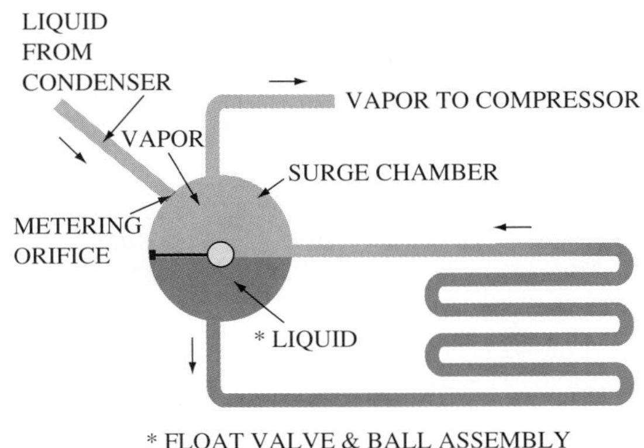

Figure R8-11 Operation of the low-side float valve.

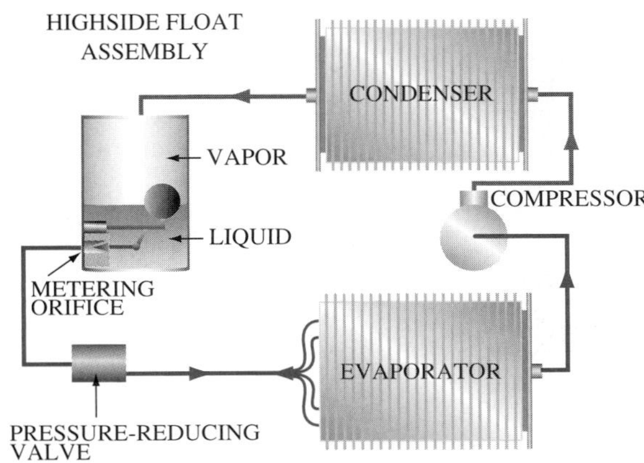

Figure R8-13 Construction and location of a high-side float valve.

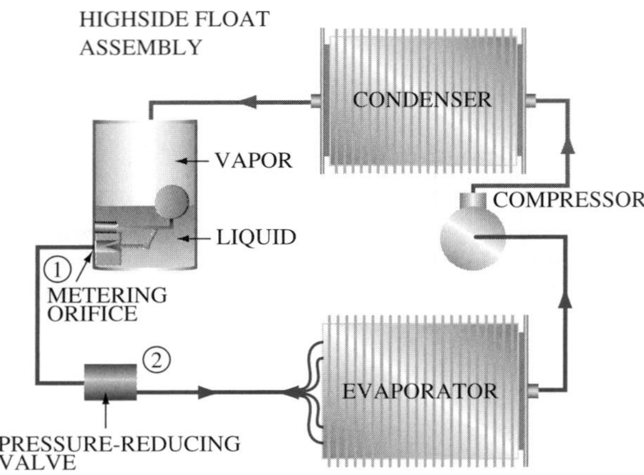

Figure R8-14 High-side float assembly with its pressure-reducing valve.

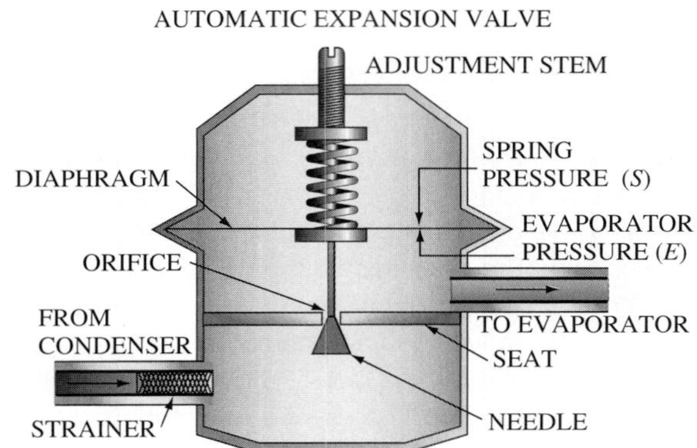

Figure R8-16 Construction of an automatic expansion valve.

into the evaporator. The pressure-reducing valve, or weight valve, (2) is used in a long liquid line to prevent evaporation before the liquid reaches the evaporator. If the load is decreased, the float is lowered, reducing the flow of refrigerant.

Fig. R8-15 shows a variation of the high-side float, feeding refrigerant into a flooded evaporator assembly. The high-side float system is critically charged, meaning that it works best with a certain refrigerant charge and any deviation from this reduces the efficiency.

The most common application of the high-side float is in the flooded cooler of a centrifugal chiller.

The *automatic expansion valve* (AEV) is illustrated in Fig. R8-16. The valve consists of a mechanical arrangement for metering the liquid refrigerant into the evaporator to maintain a constant evaporator pressure. Note in the illustration that the valve has an adjustment at the top for setting the evaporating pressure to produce a desired evaporating temperature. The evaporator pressure exerts a force against the bottom of the diaphragm. An adjustable spring exerts a pressure on the top of the diaphragm. As the evaporator pressure increases, it overcomes the spring pressure and moves the diaphragm up, thus closing the valve. As the evaporator pressure decreases, the spring pressure overcomes the evaporator pressure and pushes the valve open.

Figs. R8-17 and R8-18 show how the valve operates to maintain a constant evaporator pressure. The needle valve is attached to the plunger. The top of the plunger is attached to the diaphragm and to the spring. The diaphragm moves to position the plunger in the orifice. The difference in pressure between the spring and the evaporator control the orifice opening.

Referring to Fig. R8-17, if the load is light, less liquid will boil in the evaporator. This causes the evaporator pressure to fall and opens the valve. This excess refrigerant in the evaporator can overflow into the compressor and cause a serious hazard.

Referring to Fig. R8-18, if the load increases, more liquid refrigerant will boil into vapor, and the needle valve will move toward the closed position. This causes the refrigerant vapor going into the compressor to have an increased amount of superheat. This may cause the compres-

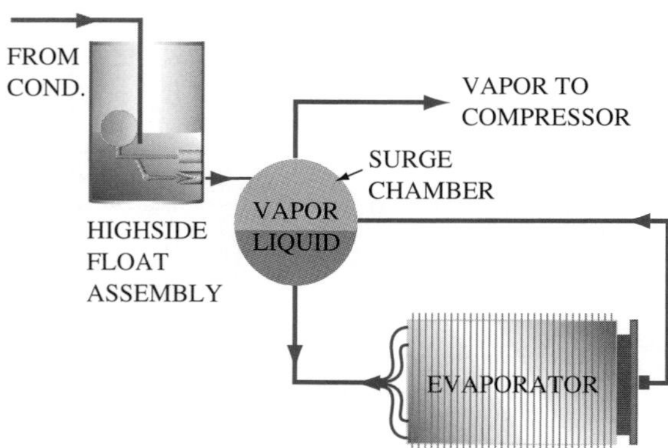

Figure R8-15 Alternate location for the high-side float valve.

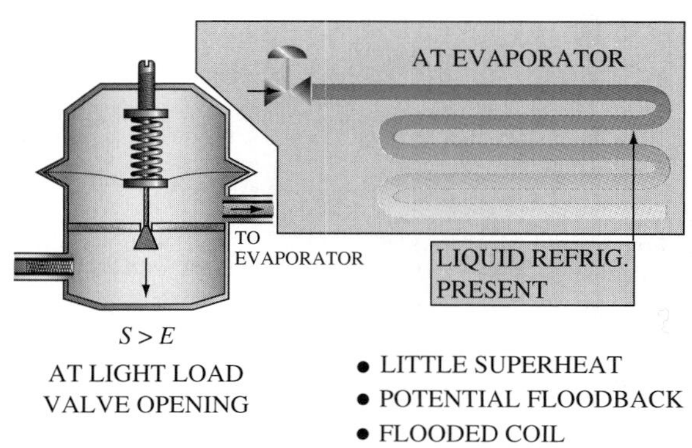

Figure R8-17 Operation of the automatic expansion valve under light load condition.

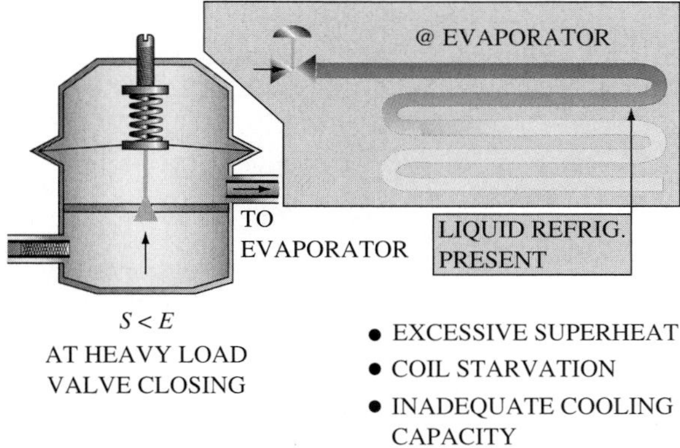

Figure R8-18 Operation of the automatic expansion valve under heavy load condition.

sor to overheat, resulting in high discharge temperature and pressure, oil breakdown, carbonizing of the valves, and poor efficiency.

The problem of the valve is that it is not load-oriented and therefore has increasingly limited applications. It is usually used on small units where the load is relatively constant.

Applications of this valve include some types of domestic refrigerators and freezers, and small retail freezer cabinets.

The *thermostatic expansion valve* (TXV) is illustrated in Fig. R8-19. It supplies the evaporator with enough refrigerant for any or all load conditions. Although it is considered a control, it is not classified as a temperature, suction-pressure, humidity, or operating control.

The TXV is constructed in many respects the same as the AEV, except that it has a remote bulb that senses the superheat of the refrigerant leaving the evaporator. The position of the orifice is controlled by three pressures: (1) the evaporator pressure; (2) the spring pressure acting on the bottom of the diaphragm; and (3) the bulb pressure opposing these two pressures and acting on the top of the diaphragm.

In Fig. R8-20 the system is in equilibrium. The pressures on the bottom of the diaphragm exactly balance the bulb pressure on the top of the diaphragm. The flow of refrigerant through the valve matches the load and produces a satisfactory superheat in the refrigerant leaving the evaporator coil, usually 10°F.

This illustration is based on using R-22 refrigerant. The evaporating temperature is 40°F and the pressure is 69.0 psig. The temperature of the bulb is 50°F, allowing 10°F superheat. The bulb exerts a pressure of 84.0 psig on the top of the diaphragm. The spring, which acts on the bottom of the diaphragm, is set for a pressure of 15.0 psig, equivalent to 10°F of superheat. Thus, the pressure exerted on the bottom of the diaphragm is

$$69 \text{ psig} + 15 \text{ psig} = 84 \text{ psig}$$

This is equal to the pressure exerted on the top and the valve is in equilibrium.

If the load increases, the superheat drops, the valve supplies more refrigerant and gradually returns to an equilibrium condition. If the load decreases, the superheat rises and the valve moves in the direction of the closed position to match the new load, then gradually returns to an equilibrium condition. The advantage of the TXV is its ability to automatically adjust its orifice to match the load.

The TXV is commonly used for most all types of air conditioning and refrigeration systems, except for small appliances.

For larger applications (Fig. R8-21), externally equalized TXV valves measure the suction pressure using ¼-in. OD tubing at the evaporator outlet for more accurate control. This avoids a pressure drop across the liquid distributor.

The *electronic expansion valve* is activated by an electronically controlled stepper motor, as shown in Fig. R8-22. The motor shaft moves in and out in tiny steps. When the sleeve attached to the shaft moves upward, it exposes more metering slots. This increases the refrigerant flow to the evaporator. When it moves downward, it

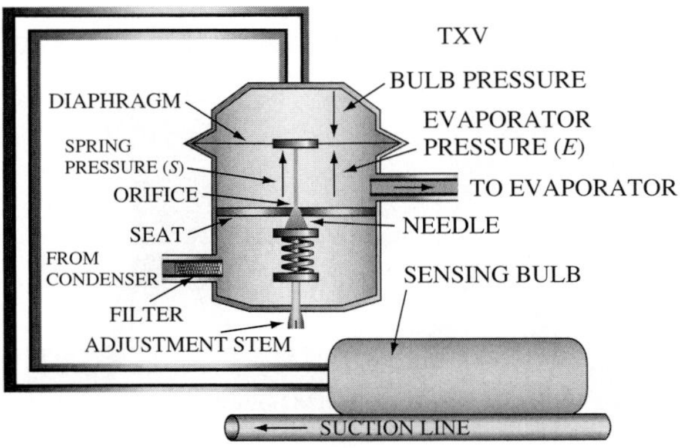

Figure R8-19 Construction of thermostatic expansion valve.

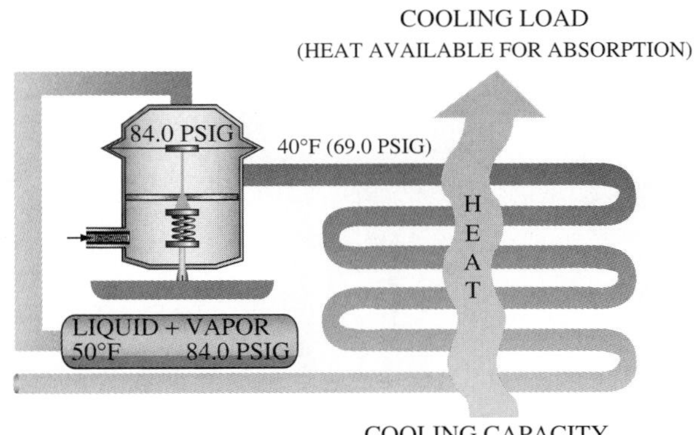

Figure R8-20 Operation of a thermostatic expansion valve in equilibrium.

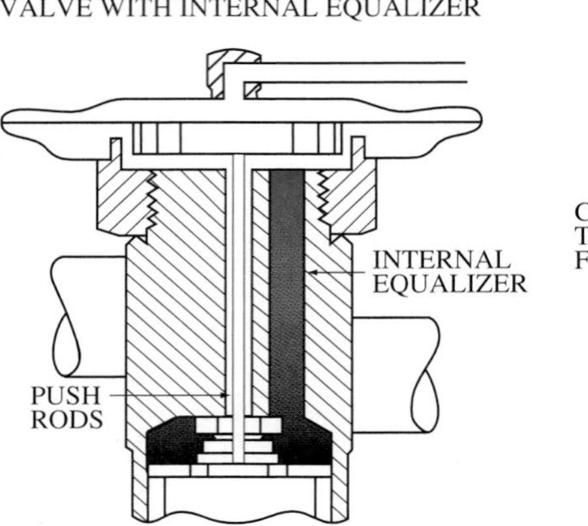

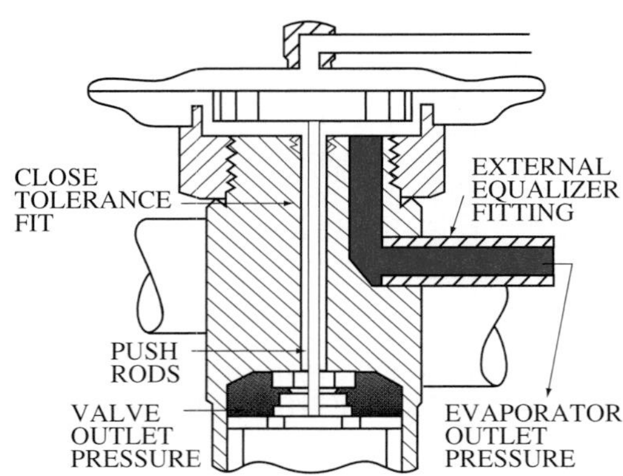

Figure R8-21 Difference in construction of internal and external equalized valves.

covers the slots, thus reducing flow and cooling capacity. Like the TXV, this type of valve is designed to maintain a constant superheat. The stepper motor gets its signal from an electronic control panel that is attached to an electronic sensor that measures refrigerant superheat. The stepper motor valve can have a long stroke and corresponding low pressure drop when compared to a conventional TXV. Since it is controlled independently of pressure, this valve is able to provide safe startup, shutdown, and operation, and high energy efficiency through its full range of operating conditions.

Another electronic control alternative to the TXV is the pulsating solenoid valve, shown in Fig. R8-23. Unlike the TXV, which closes or opens by pressure, the solenoid valve opens and closes electrically. Pulse-modulated valves control flow by increasing or decreasing open time during each cycle. When coupled with the proper electronic control system, a solenoid valve can pulse rapidly, opening and closing quickly in response to the cooling load. The illustration on the left shows the valve closed, and the solenoid valve is de-energized; the view on the right shows the energized condition. This valve also serves the function of a liquid line solenoid valve, blocking flow to the evaporator during the OFF-cycle and at other times when refrigerant flow is not required.

One manufacturer uses stepper-motor-type TXV valves on its air-cooled packaged chillers. The large sizes, using more than four compressors, use a two-circuit chiller with two stepper-motor-type TXV valves.

As electronic control becomes more popular, more electronically controlled valves will be put into use.

R8-2.3 Externally Equalized Valves

When the pressure drop in the evaporator (including the distributor) is sizable, in order to obtain an accurate superheat setting, it is necessary to use an *externally equalized thermal expansion valve,* as shown in Fig. R8-24. In the valve body itself, the evaporator inlet from the bottom of the valve is sealed off and a piping connection is arranged from above the diaphragm to the end of the evaporator. This connection not only improves the superheat setting, but also makes possible full use of the evaporator, improving the efficiency.

The equalizer connection normally penetrates the suction line 6 or 8 in. downstream from the sensing bulb, unless the manufacturer's instructions advise differently.

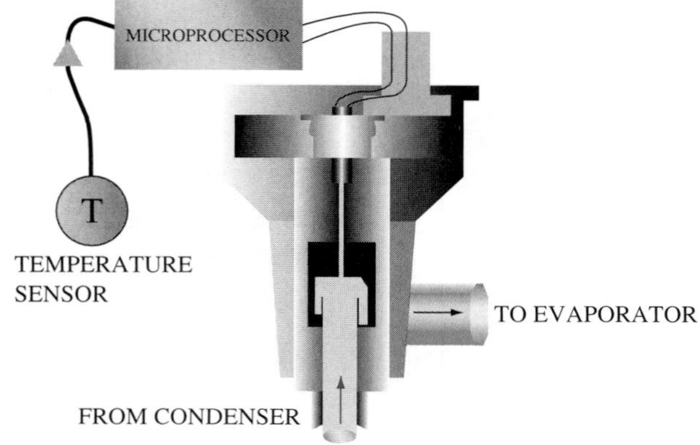

Figure R8-22 Construction of an electronic expansion valve.

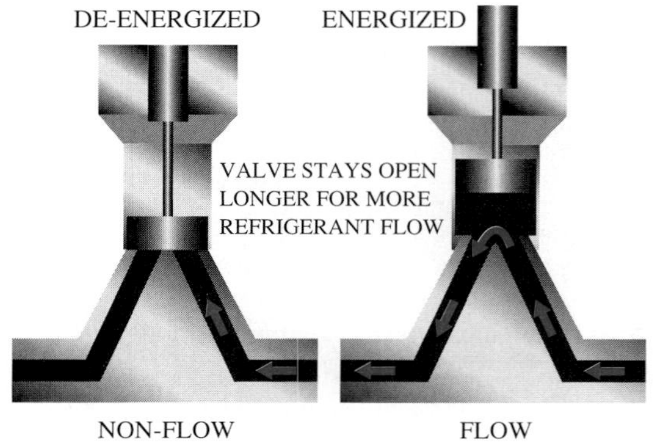

Figure R8-23 Pulsating solenoid valve, both energized and de-energized.

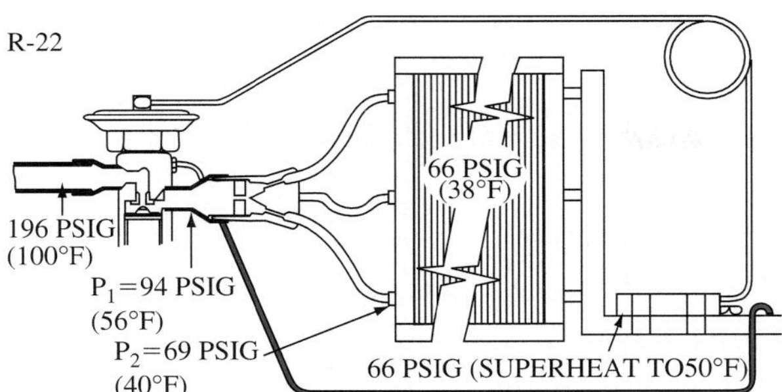

Figure R8-24 Effect of an external equalized thermostatic expansion valve on the operation of the system.

R8-2.4 Maximum-operating-pressure (MOP) Valves

The conventional thermostatic expansion valve uses a remote sensing bulb with some liquid refrigerant in it. This can cause operating problems during the OFF-cycle as well as on startup. This type of charge can cause excess pressure buildup in the evaporator during the OFF-cycle and high suction-discharge temperatures which will overload the motor on compressor startup. This can be prevented by using a *maximum-operating-pressure* (MOP) *valve,* shown in Fig. R8-25, which has a gas charged sensing bulb. Most all air-conditioning systems using thermostatic expansion valves limit the amount of pressure that can develop in the evaporator.

Valves that have some liquid in the bulb at all times are called liquid-charged bulbs. The liquid charged bulbs have some liquid below the pre-selected maximum operating pressure (and temperature). Above that pressure and temperature, the refrigerant in the bulb and in the space above the diaphragm is all vapor. Increase in temperature of a vapor produces only a slight increase in pressure. This solves the problem of liquid residual at compressor shutdown and startup.

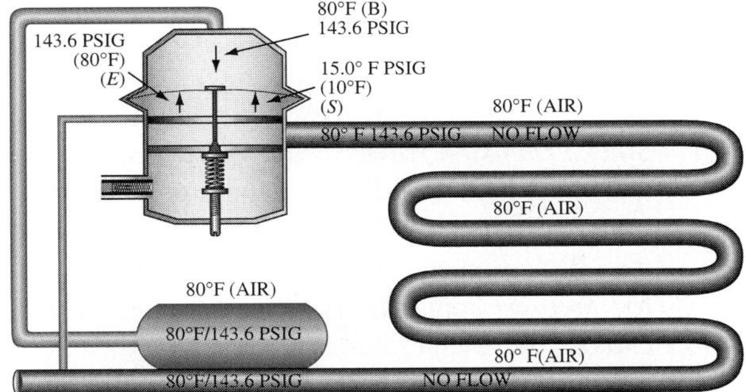

Figure R8-25 Use of a maximum operating pressure (MOP) thermostatic expansion valve.

R8-3 MEASURING SUPERHEAT

It is important to accurately measure the superheat to determine if the expansion valve is operating properly or needs adjustment. On close coupled installations, the operating suction pressure can be measured at the compressor service valve, as shown in Fig. R8-26. This value can be converted to saturated evaporator temperature by referring to a pressure-temperature chart. The temperature of the vapor at the expansion-valve bulb can be read using an electronic thermometer. The difference between these two temperatures is the superheat.

If limited adjustment needs to be made, the valve stem or adjuster can be turned in small increments to change spring tension, which in turn will change the superheat setting.

On many valves the adjustment is clockwise (CW) to increase superheat and counterclockwise (CCW) to decrease superheat. Valve instructions should be checked to be sure of correct adjustments.

R8-4 LIQUID DISTRIBUTORS

Liquid distributors are placed between the metering device and the evaporators with multiple circuit coils to equally distribute the refrigerant to each circuit, as shown in Fig. R8-27 and Fig. R8-28. They are usually supplied by the coil manufacturer and are used where one metering device serves from two to 40 evaporator circuits, using connecting tubes ranging from 5⁄32 in. OD to 3⁄8 in. OD.

Some of these distributors have a thick flat washer inside, called a "nozzle," as shown in Fig. R8-29. The purpose of this nozzle is to thoroughly mix the liquid and vapor refrigerant before it enters the coil tubes. Some distributors have a special internal construction to accomplish the same thing.

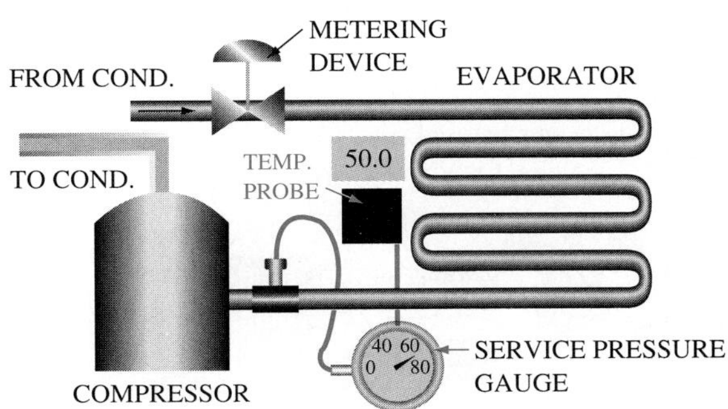

TO MEASURE SUPERHEAT:
1. FIND SUCTION PRESSURE
2. FIND MATCHING SATURATION TEMPERATURE
3. READ TEMPERATURE LEAVING EVAPORATOR
4. SUPERHEAT = TEMP. LEAVING − SATURATION TEMP.

Figure R8-26 Method of measuring superheat on a system using thermostatic expansion valve.

PREVENTIVE MAINTENANCE

Measuring for superheat is a simple preventive maintenance technique to predict potential problems. To measure superheat, use the following procedures:

1. Install your pressure gauges, read suction pressure, and convert to saturation temperature using a P/T chart. (Allow for pressure drop if suction pipe is located over 20 feet away.)
2. Measure the suction pipe temperature with an accurate thermometer.
3. Compare the saturated refrigerant temperature to the pipe temperature.
4. If low superheat exists, (less than 10°F) it could indicate a potential problem and possibly liquid slugging at the compressor. (See common metering device problems in chapter review.)
5. If high superheat exists, (greater than 20°F) it could also indicate a potential problem and future compressor overheating. (See common metering device problems in chapter review.)

R8-5 LOW LOAD LIMITS

Both distributor nozzles and thermostatic expansion valves have low load limitations. If the entire system is purchased as a package usually there is no problem. However on field-assembled units, difficulties can occur.

The load limit for most nozzles is usually 50% to 200% of its maximum capacity. The load limit for most thermostatic expansion valves is 35% to 50% of its maximum capacity.

When the flow through the nozzle drops below 50%, the gas and liquid refrigerant are not equally mixed. The superheat in the top circuits of the evaporator can be excessive due to liquid "starving." The liquid floods back to the compressor from the lower circuits, as shown in Fig. R8-30. This can easily be checked by feeling or measuring the temperature difference between the top and the bottom of the coil.

This problem can be solved by replacing the nozzle with the next smaller size or arranging the control system to cut off one section of the evaporator when the low limit is exceeded.

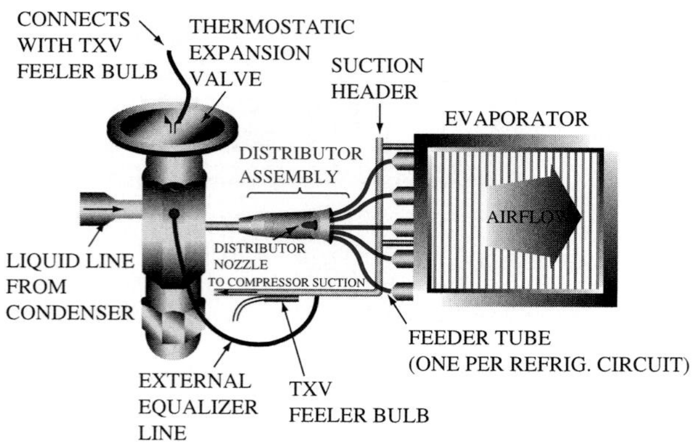

When the flow through a thermostatic expansion valve drops below 30%, the valve will not hold constant settings, as shown in Fig. R8-31. This condition is called *hunting*. Under these conditions superheat values fluctuate.

To correct this condition, a smaller expansion valve can be substituted or the control system can be modified to cut off a section of the evaporator when the capacity reaches the low limit.

On small-capacity systems, it is sometimes advisable to use a double-ported expansion valve. This valve can accommodate a two-circuit evaporator without the need for a separate distributor nozzle on the coil. An illustration of this valve is shown in Fig. R8-32.

Figure R8-27 Construction and use of a liquid distributor assembly.

R8-6 COMMON METERING DEVICE PROBLEMS

Problems can be incurred with the use of metering devices. If they occur, it is important to find a solution as quickly as possible to prevent damage to the compressor. Most of these problems can be easily corrected if they are properly identified.

Below is a list of common metering device problems and the symptom that can be observed for each:

1. *Stuck expansion valve* If the valve is stuck open, the valve will flood the evaporator resulting in liquid refrigerant entering the compressor. There may be ice on the suction line at the compressor for freezer applications. If the valve is stuck closed, the evaporator will be starved, and the suction line will be too warm.
2. *Oversized expansion valve* The valve will overreact to changes in load, causing hunting. The

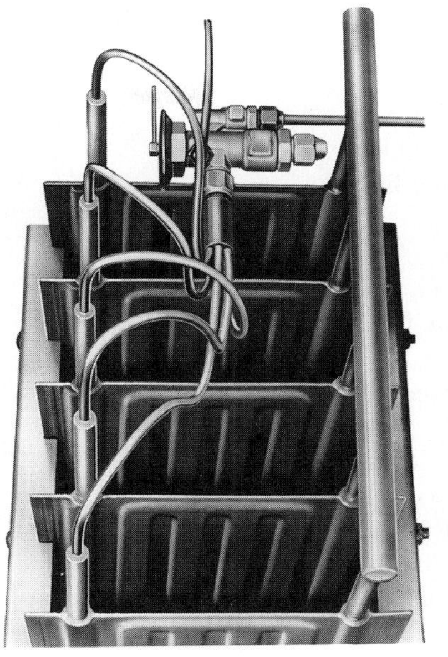

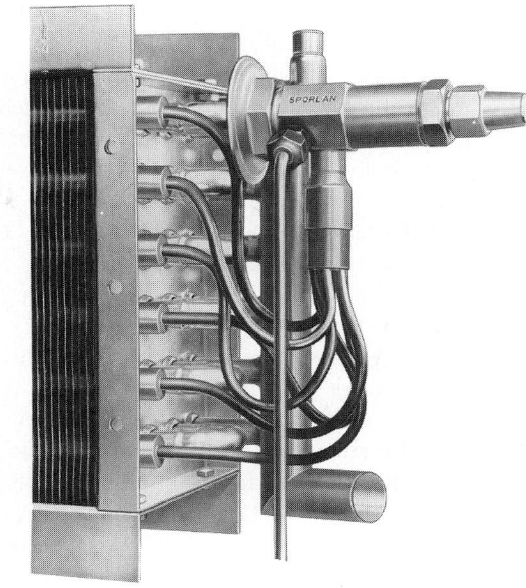

Figure R8-28 Application of the distributor nozzle to both fin coils and plate evaporators. (Courtesy of Sporlan Valve Co.)

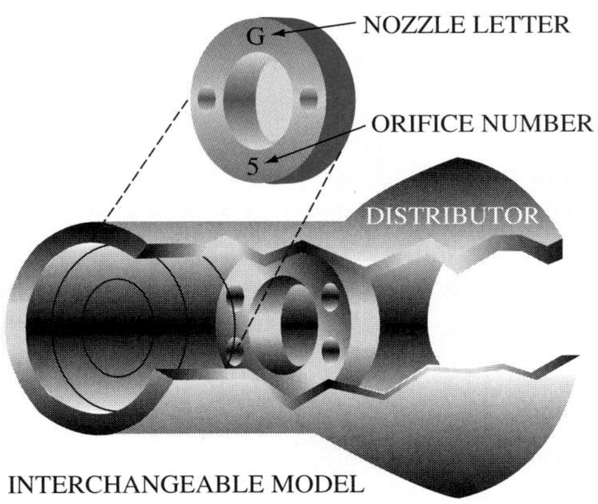

Figure R8-29 Detailed view of the nozzle used with a liquid distributor.

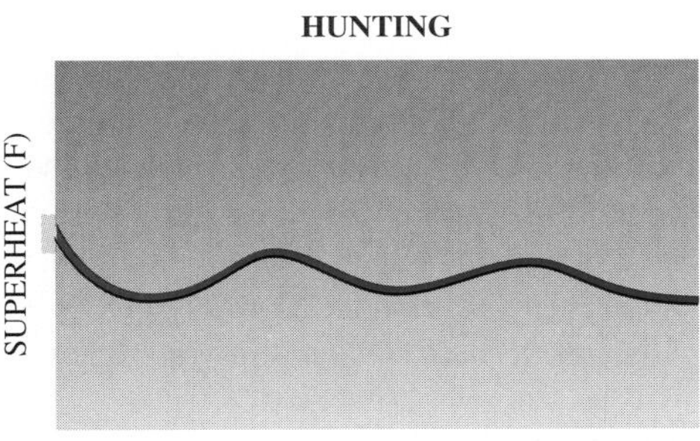

Figure R8-31 How the superheat fluctuates when the thermostatic expansion valve is *hunting.*

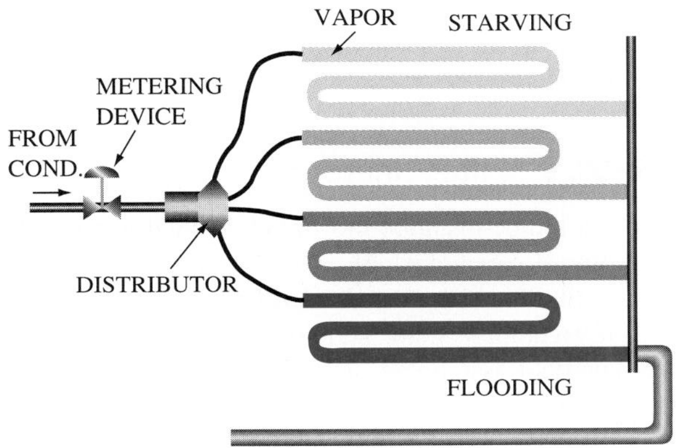

Figure R8-30 The effect of flooding the evaporator when a larger-than-necessary nozzle is used.

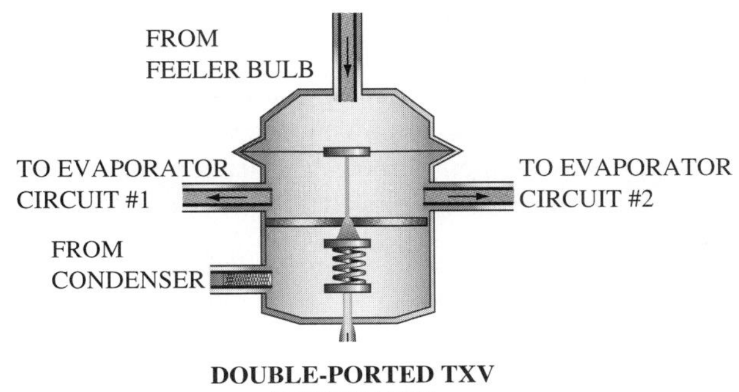

Figure R8-32 A double ported expansion valve used for a two-circuit evaporator.

suction pressure will continually rise and fall, never becoming stable.

3. *Improper sensing bulb location* On horizontal suction lines, the bulb should be located between 8 o'clock and 4 o'clock. On vertical suction lines, the bulb should be located so that the sensing line emerges from the top of the bulb.
4. *Valve superheat setting too low or too high* The evaporator will be flooded or starved. If a valve adjustment does not solve the problem, look for a different problem.
5. *Wrong type of valve* TXVs are color-coded on the power head to correspond to the type of refrigerant that is in the system. The standard color codes are orange for R-11, white for R-12, green for R-22, purple for R-113, light blue for R-134a, dark blue for R-114, yellow for R-500, orchid for R-502, and silver for R-717. Also, if the evaporator has a refrigerant distributor, the valve must be externally equalized.
6. *Plugged equalizer line* The end of the evaporator circuit being supplied by a plugged equalizer line may be warmer than the other circuits due to the restricted refrigerant flow.
7. *Plugged distributor* This will have the same symptoms as an expansion valve that is stuck closed.
8. *Improper sized distributor nozzle* Some manufacturers use different size nozzles in the same expansion valve body. If the nozzle used is too small, the evaporator will be starved. If the nozzle is too large, the expansion valve will hunt, or may flood the evaporator.
9. *Loss of thermal element charge* The evaporator will be starved.
10. *Thermal bulb loosely fastened to suction line* The evaporator will be flooded.

R8-7 REFRIGERATION CYCLE ACCESSORIES, NON-ELECTRIC

An accessory is a device added to a basic system to enhance its operation. The following accessories will be described in terms of their purpose, application, and location in the system operation—as well as their benefits and disadvantages. These are shown in Fig. R8-33:

1. Filter-drier
2. Sight glass/moisture indicator
3. Suction-line accumulator
4. Crankcase heater
5. Muffler
6. Oil separator

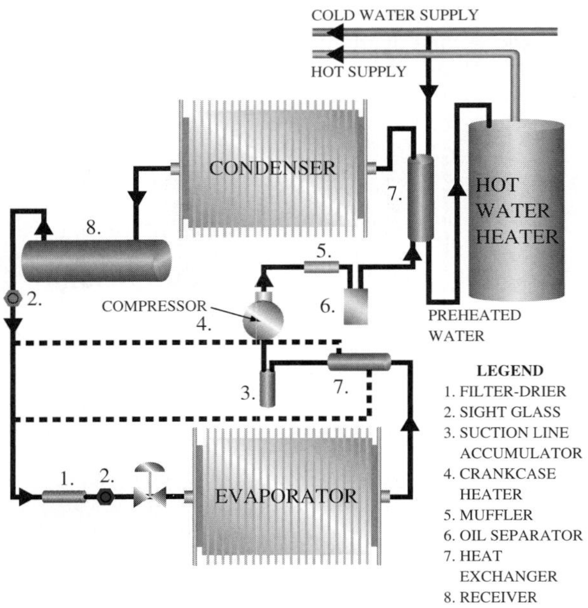

Figure R8-33 The location of eight different refrigerant cycle accessories.

7. Heat exchanger
8. Receiver

On factory-assembled equipment, any of these accessories would be installed at the factory. On field assembled systems they must be installed on the job.

The following list of electric and non-electric devices, sometimes considered accessories, have been included in the section on controls:

1. Solenoid valve (electric)
2. Check valve (non-electric)
3. Condenser water valve (non-electric)
4. Back pressure regulator (non-electric)
5. Oil safety switch (electric)
6. Filter-driers are illustrated in Fig. R8-34 and Fig. R8-35. It is a dual-purpose accessory, functioning as a strainer to collect foreign matter such as dirt, pipe and brazing scale, rust and metal chips. As a drier it removes moisture from the system and stores it where it can do no harm.

R8-7.1 Filter-driers

Filters should be installed in front of all metering devices. Most compressors are supplied with a filter located on the suction side of a compressor, with provision for cleaning.

All refrigerant systems should have a permanent filter-drier. For field-assembled units the filter-drier is selected and installed on the job.

Fig. R8-36 shows a non-refillable filter-drier. These units are often installed on packaged equipment and brazed into the liquid line between the condenser outlet

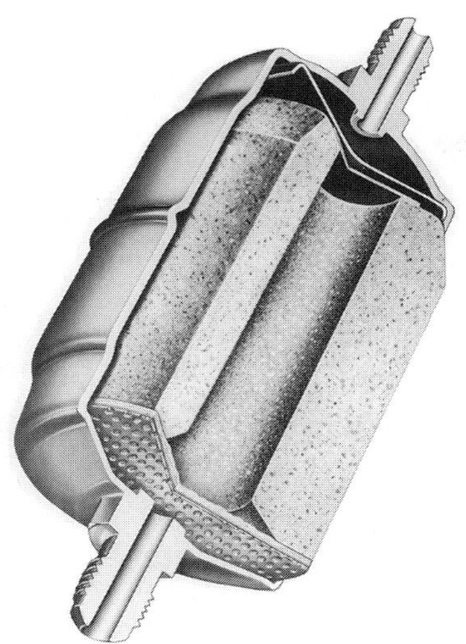

Figure R8-34 Construction of a typical filter-drier. (Courtesy of Sporlan Valve Co.)

and the metering device. There is an advantage in locating the filter-drier upstream from the moisture-indicating sight glass, as shown. In case there is any blockage in the filter-drier it will show up as bubbles in the sight glass, indicating a pressure drop.

In larger systems it is advisable to use a filter-drier with a removable core, as shown in Fig. R8-37. This illustration shows a piping arrangement that permits changing the core without interrupting the operation of the system. In case of a burnout, as many filter-drier cores as necessary can be used to clean up the system. If you notice that the outlet line of a filter-drier is colder than the inlet, it is an indication that the filter is partially plugged and must be changed. The pressure drop caused by a plugged filter-drier causes flashing of the refrigerant similar to a metering device.

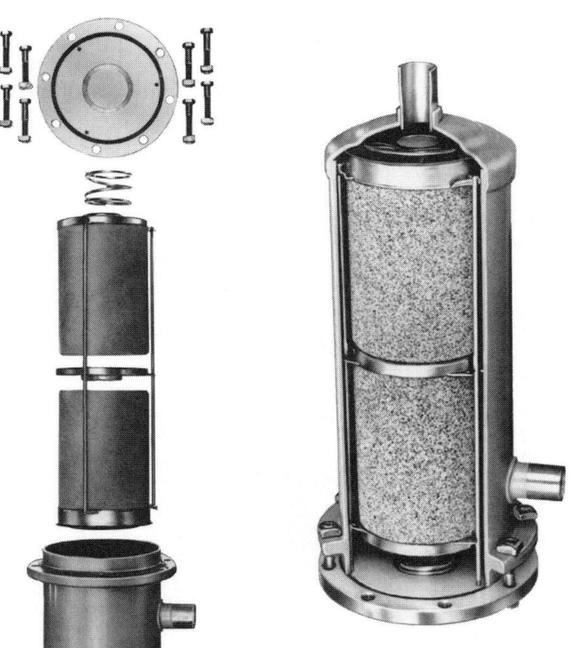

Figure R8-35 Construction of a replaceable core drier. (Courtesy of Sporlan Valve Co.)

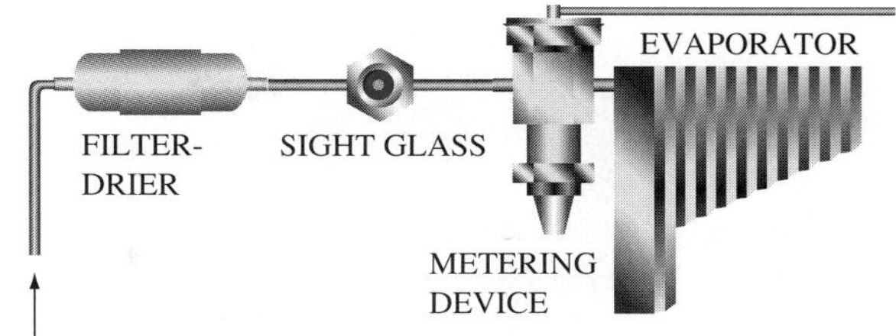

Figure R8-36 Location of the filter-drier in relation to other accessories.

It is extremely important to keep the system clean and free of moisture. The filter protects the compressor from damage caused by solid particles. The drier removes moisture that can cause internal ice formation, which causes metering devices to freeze. Removal of moisture prevents internal metal corrosion as well as compressor motor insulation breakdown. Moisture in the system can cause copper from the piping to be deposited on moving compressor parts.

R8-7.2 Sight Glass/Moisture Indicator

A *sight glass* is a type of window placed in the liquid line to determine the condition of the refrigerant. The sight glass shown in Fig. R8-38 is a moisture-indicating sight glass. It makes possible determining two conditions of the refrigerant:

a. whether the liquid line contains only liquid or a mixture of liquid and vapor refrigerant;
b. whether the refrigerant is dry or if it contains moisture.

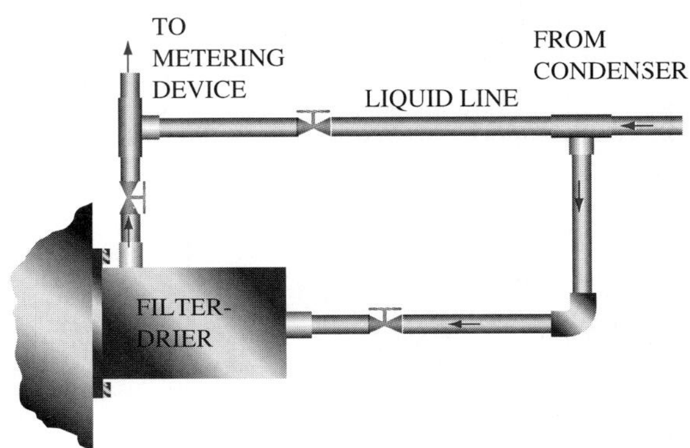

Figure R8-37 Location and construction of a replaceable core type filter-drier.

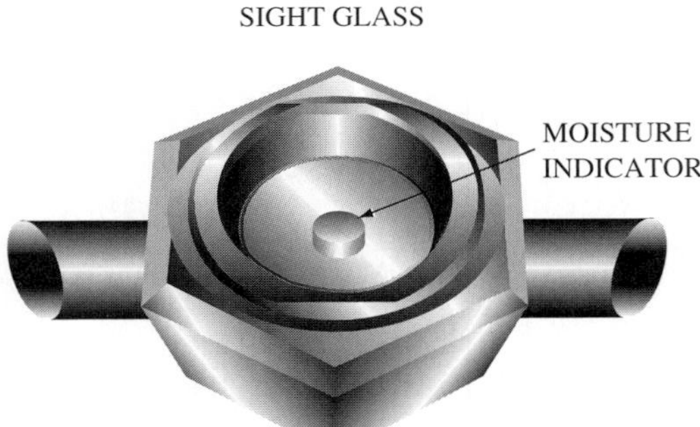

Figure R8-38 Construction of a moisture-indicating sight glass.

The sight glass is placed in the liquid line between the condenser and the evaporator, before the liquid line enters the metering device, as shown in Fig. R8-39. It is necessary on a properly operating system, at this point, to have a solid stream of dry (moisture-free) refrigerant. If this is not the case, certain service operations need to be performed to correct the condition.

The presence of bubbles in the sight glass indicates a mixture of liquid and vapor. In most circumstances, by adding more refrigerant to the system, the bubbles can be made to disappear.

The presence of moisture in the refrigerant is indicated by the color of the moisture indicator. When the refrigerant changes from wet to dry, the color of the indicator changes. The use of a drier (desiccant) in the liquid line will remove the moisture from wet refrigerant.

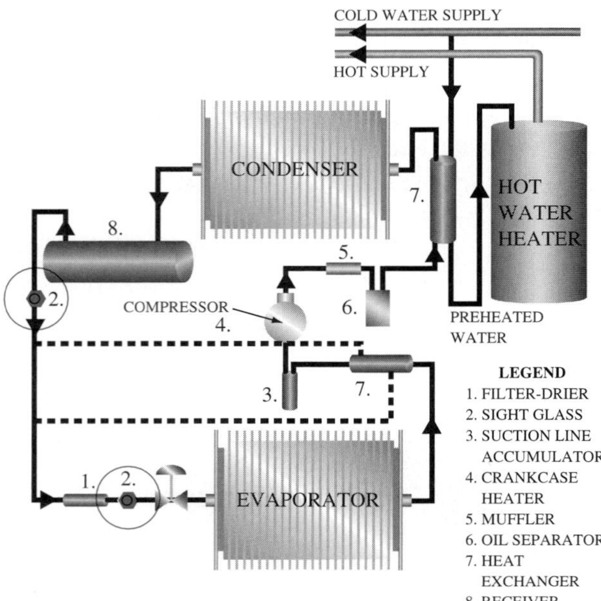

Figure R8-39 The sight glass is located in the liquid line.

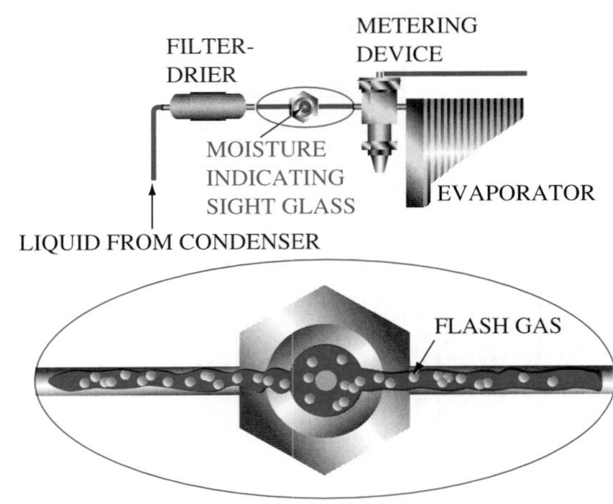

Figure R8-40 Sight glass with bubbles in the refrigerant, indicating flash gas.

On some systems two sight glasses are used—one where the refrigerant leaves the condenser and one in the liquid line before the refrigerant enters the metering device, as shown in Fig. R8-40. If bubbles do not appear in the first glass but do appear in the second, it is an indication that there is a restriction in the line, causing a pressure drop and producing flash gas. The condition is corrected by providing additional subcooling.

One thing more about using sight glasses: since most refrigerants are clear and colorless, it is easy to mistake an empty sight glass for a full one. If the system is producing any cooling at all, the sight glass is not empty.

R8-7.3 Suction-Line Accumulator

A *suction-line accumulator* (Fig. R8-41) is a container placed in the suction line ahead of the compressor to catch any liquid refrigerant that has not evaporated before it reaches the compressor. It is a very simple device, but very important in certain types of installations. Some of the systems that routinely require suction-line traps are systems which

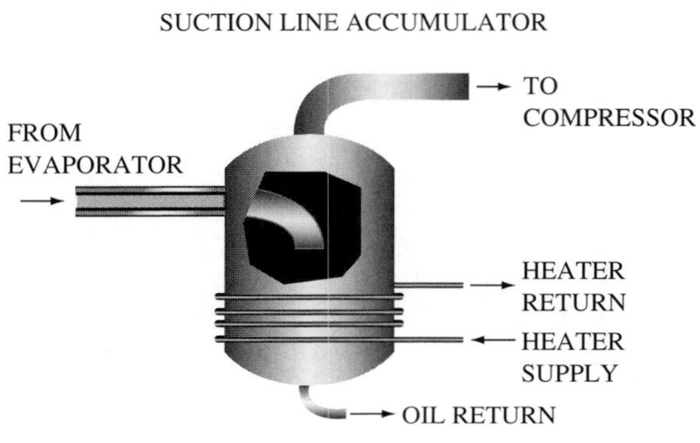

Figure R8-41 Construction of a suction-line accumulator.

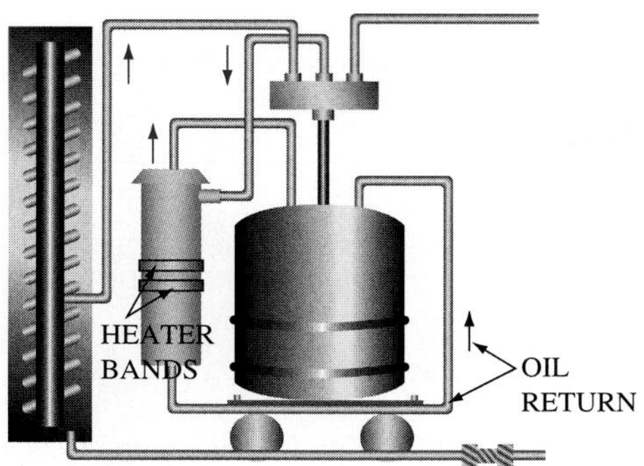

Figure R8-42 Location of a suction-line accumulator.

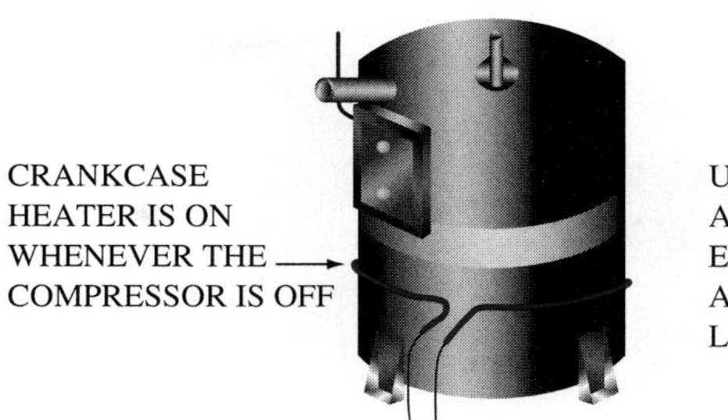

Figure R8-44 Location of a crankcase heater at the base of a compressor shell.

a. have wide or rapid load changes;
b. utilize capacity-control devices;
c. are heat-pump systems;
d. incorporate an evaporator defrost system, including a heat-pump-type defrost system;
e. require compressor replacement due to liquid slugging.

Accumulators are installed in the piping as near the compressor inlet as possible, as shown in Fig. R8-42. Due to the increased volume in this area and the reduction of the refrigerant velocity, the liquid refrigerant and oil drop to the bottom of the container rather than enter the suction opening to the compressor. The diagram shows the pipe at the bottom that permits the oil to return to the compressor crankcase. The liquid refrigerant in the accumulator gradually evaporates, entering the compressor as vapor. The accumulator shown in the illustration has a heater which speeds up the evaporation process.

Fig. R8-43 shows another type of accumulator with a different type of construction. The suction gas enters at the upper right. Liquid slugs are directed toward the right-hand wall and run to the bottom. The metering orifice at the bottom of the return pipe slowly sends oil back to the compressor through the suction line, plus a limited amount of liquid refrigerant.

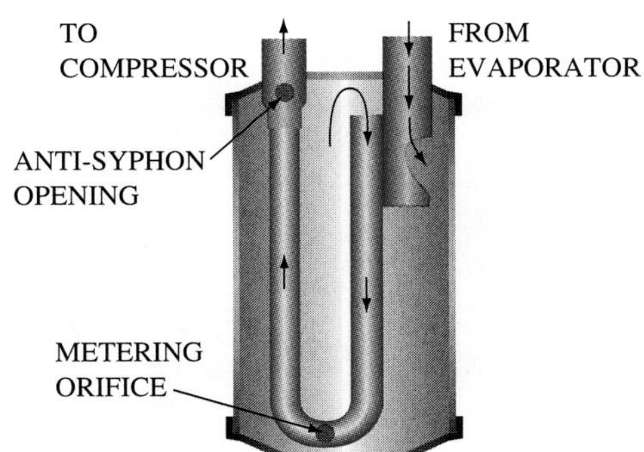

Figure R8-43 Suction-line accumulator with anti-syphon opening and oil-return orifice.

R8-7.4 Crankcase Heater

The *crankcase heater* (Fig. R8-44) is an electrical device that is placed either in the compressor crankcase or around the lower part of a hermetically sealed compressor shell. It provides heat to evaporate any liquid refrigerant that reaches the crankcase during the OFF-cycle. During shutdown, the compressor often becomes the coolest part of the system, attracting the liquid refrigerant by the process called migration.

The crankcase heater protects the compressor in starting by vaporizing any liquid refrigerant that may enter the compressor during the OFF-cycle. On many compressors the crankcase heater is on only during the OFF-cycle. On others it is on all of the time.

Fig. R8-45 shows the various types of crankcase heaters. The heater at the left is fastened to the bottom of the crankcase. The heater in the center is inserted in a tube located in the crankcase. The one on the right is wrapped around the shell.

Liquid refrigerant in the crankcase can be very damaging to the compressor. If refrigerant collects in the crankcase it mixes with the oil, as shown in Fig. R8-46.

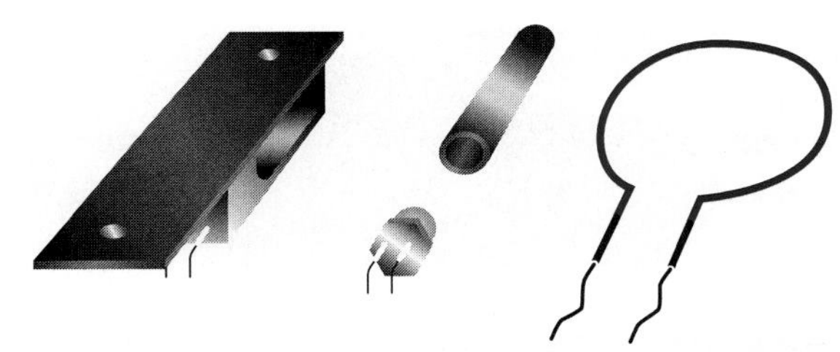

Figure R8-45 Various types of crankcase heaters.

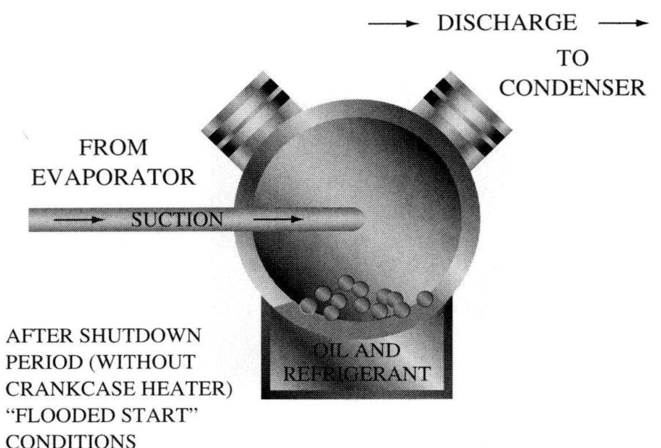

Figure R8-46 Liquid refrigerant migrating to the crankcase will dilute the oil and cause serious compressor damage on start up.

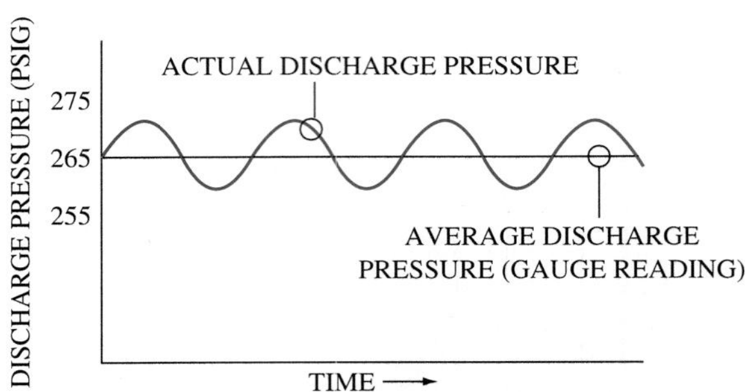

Figure R8-48 Fluctuation in discharge pressure from a compressor which can cause vibration and noise.

This creates a condition known as flooded start. When the compressor cycles on, the crankcase pressure quickly falls to the level of the suction pressure. This causes the liquid refrigerant mixed with oil to boil violently, creating a foam which results in compressor slugging and the oil leaving the compressor crankcase. The liquid refrigerant in the oil dilutes the oil, causing improper lubrication.

R8-7.5 Hot-Gas Muffler

A *hot-gas muffler* (Fig. R8-47) is a mechanical device (with no moving parts) placed in the hot-gas line to dampen out pulses created by the compressor that may cause objectionable noise or vibration. Each pulse of pressure forces the hot gas through holes in the pipe. These pulses enter chambers where they are dampened out. The internal construction of the muffler varies, depending on the frequency of the pulses and their source. They are also designed to prevent oil trapping when properly installed.

Fig. R8-48 illustrates the pressure fluctuation of the gas leaving a reciprocating compressor. This fluctuation is not indicated on a test pressure gauge because it occurs faster than the gauge can read. The straight line or average is the reading shown on the gauge. Nevertheless, the fluctuation does occur and can cause noise and vibration problems. Most reciprocating compressors pulsate enough to justify the use of a muffler.

The best location for a compressor discharge muffler is shown in Fig. R8-49. It should be installed in a horizontal or downflow portion of a hot-gas line, immediately after the compressor outlet in such a way that it does not trap oil.

R8-7.6 Oil Separator

An *oil separator* (Fig. R8-50) is a mechanical device for collecting oil being carried away from the compressor in the discharging refrigerant, and returning it to the compressor. Oil separators are most commonly used on very low temperature systems, and on systems that use a refrigerant that is nonmiscible (non-mixable) with oil.

Relative to miscibility, ammonia is the best example of a non-miscible refrigerant. R-11, R-22, R-114, and R-500 are all miscible. R-22 and R-502 have limited miscibility. The advantage in oil-trapping a non-miscible refrigerant is to contain the entrained oil as near to the compressor as possible rather than dropping it someplace else in the system, where it will be less convenient to return.

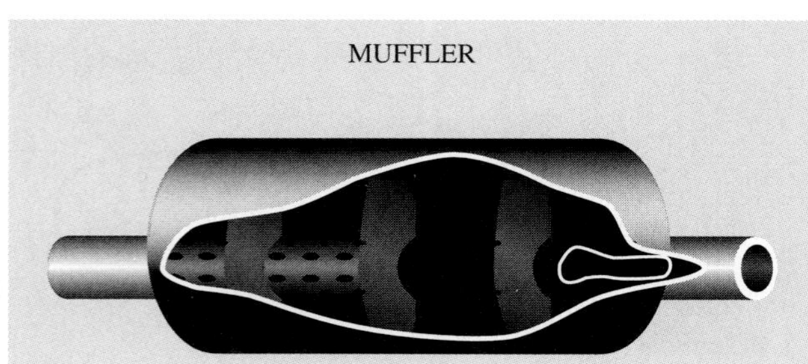

Figure R8-47 Construction of a hot-gas muffler.

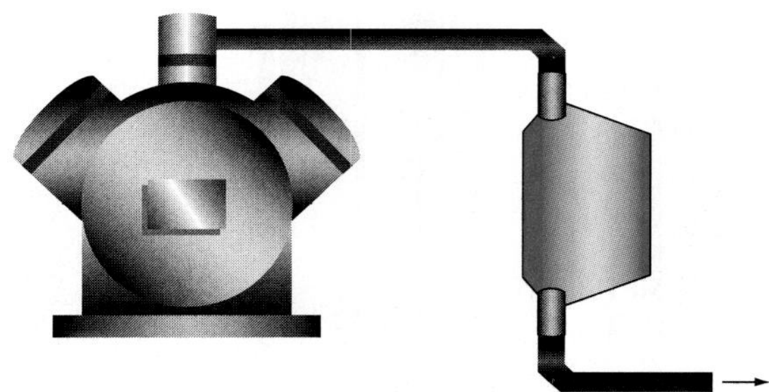

Figure R8-49 Location of the muffler.

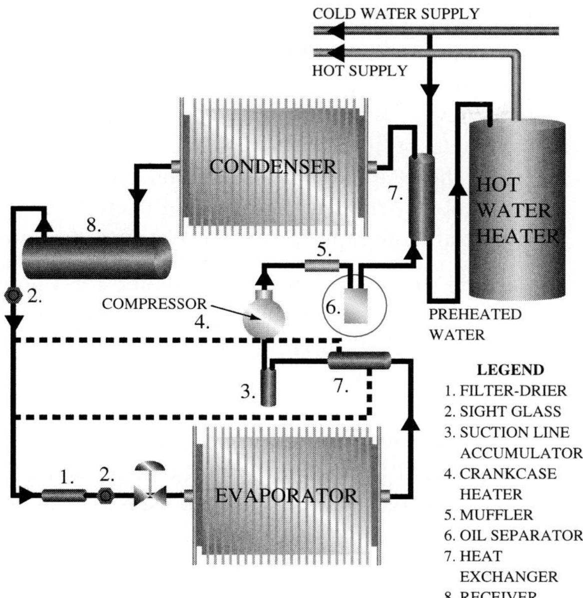

Figure R8-50 Location of the oil separator.

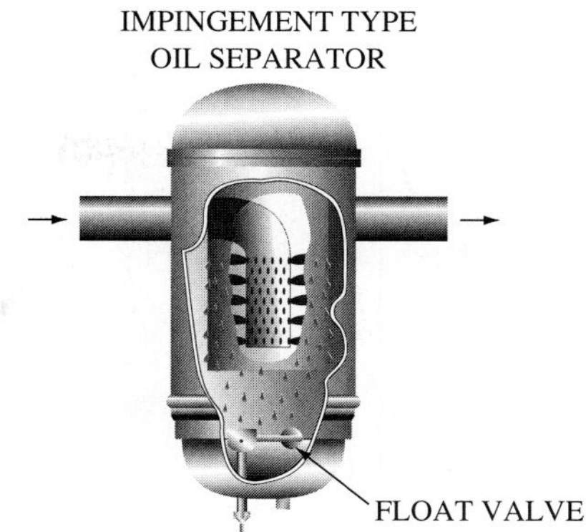

Figure R8-52 Impingement-type oil separator.

The oil separator can best be located in the discharge line near the compressor, but following a muffler, if one is used, as shown in Fig. R8-51.

There are two principal types of oil separators: the impingement type and the chiller type.

An impingement-type oil separator is shown in Fig. R8-52. The oil-laden discharge gas is sent through a series of screens or baffles. Since the area inside the separator is larger than the discharge line, the hot-gas flow slows down. This allows the oil to impinge on and stick to the screens. The oil drains off the screens into the sump at the bottom of the container. A float valve in the sump maintains a liquid seal and automatically returns the collected oil to the compressor. Some oil separators return oil to the compressor crankcase, while others slowly enter it into the compressor suction.

The chiller-type oil separator can be water-cooled as shown in Fig. R8-53, or refrigerant-cooled. It is constructed like a water-cooled condenser. The water is circulated through the tubes and the oil-laden refrigerant is circulated through the shell. The oil precipitates on the water-cooled tubes and drops down into the sump. From there it is automatically returned to the compressor through a float valve. The water flow through the separator must be carefully controlled so that the refrigerant vapor is not condensed. If it is, refrigerant will be sent back to the compressor inlet, causing slugging, oil foaming, and oil dilution.

The advantages of an oil separator are:

1. It minimizes oil return problems.
2. It improves heat transfer efficiency.
3. It keeps oil at the compressor.

The main disadvantages of an oil separator is that it may allow liquid refrigerant to get into the compressor crankcase. This is most likely to occur during the OFF-cycle when the separator and the discharge line are cool.

It is important that the float assembly and the orifice, shown in Fig. R8-54, be kept clean. Otherwise the sludge will accumulate and cause the float to stick open, allowing

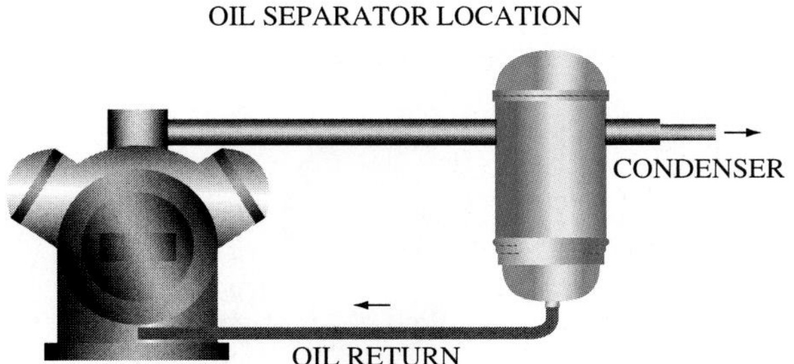

Figure R8-51 An enlarged view of the oil separator in the piping.

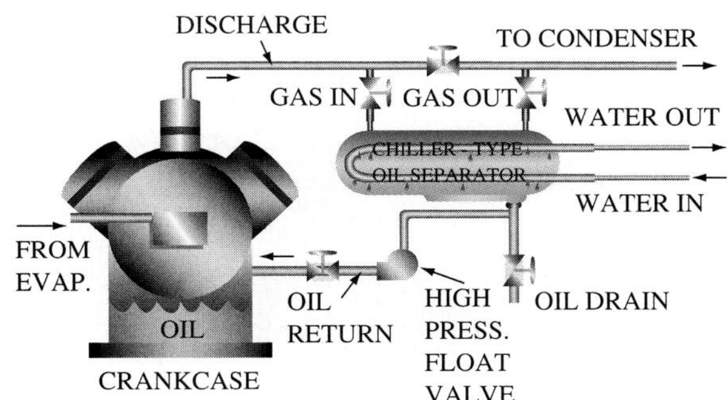

Figure R8-53 Chiller-type oil separator.

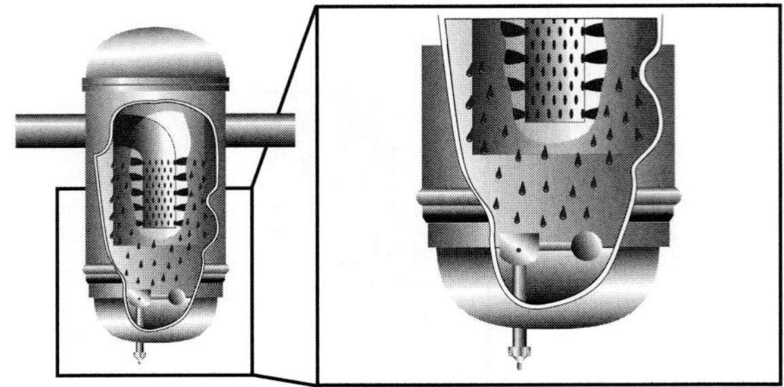

Figure R8-54 Float assembly and orifice of an oil separator.

hot gas to pass directly into the compressor inlet. If the float valve sticks closed, oil will not be returned to the compressor. Either problem can be damaging to the compressor.

R8-7.7 Heat Exchangers

A *heat exchanger* is a device for transferring heat between two separated fluids flowing through it, which have a temperature difference.

Two different types of heat exchangers are used in mechanical refrigeration systems: the suction-liquid refrigeration heat exchanger and the refrigerant-water preheater.

The suction-liquid refrigerant heat exchanger, shown in Fig. R8-55, transfers heat from the liquid refrigerant leaving the condenser to the suction gas leaving the evaporator. There are three reasons for doing this:

a. To increase the efficiency of the refrigeration cycle, particularly on low-temperature applications;
b. To subcool the liquid refrigerant coming from the condenser to prevent flash gas at the entrance to the metering device;
c. To prevent liquid slugging by evaporating small amounts of liquid refrigerant expected to return from the evaporator under certain conditions. For example, systems with rapid load fluctuations drop the load more rapidly than the system can respond.

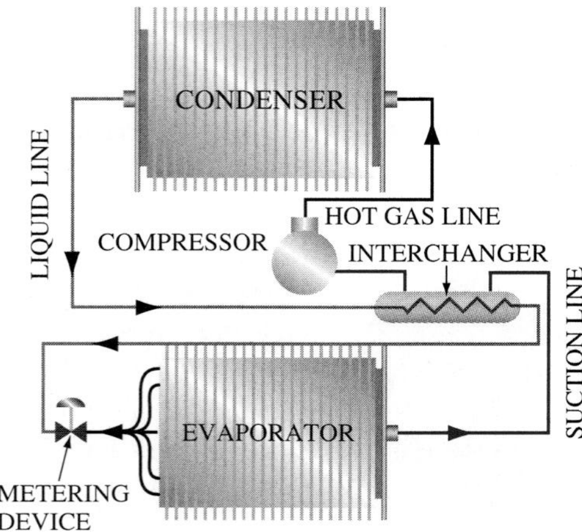

Figure R8-55 Liquid to suction-line heat exchanger.

Fig. R8-55 shows the location and piping for a liquid-suction heat exchanger. Note that the flow through the two sections of the exchanger are in opposite directions. This counterflow arrangement increases the efficiency of the unit.

Note that the rise in suction temperature is greater than the drop in liquid temperature. For example, consider a reach-in refrigerated display case that uses R-502 and is maintained at about 28°F. A 24°F rise in suction vapor temperature will be matched by a reduction in liquid temperature of 12°F.

If the purpose of the unit is to exchange heat, it should be located as close to the condenser as possible. If its purpose is to clean up the excess liquid in the suction line, it should be located as close to the evaporator as possible. Three types of liquid-suction heat exchangers are shown:

1. Suction line and liquid line bonded together (Fig. R8-56). The longer the run, the greater the heat exchange.
2. Tube-in-tube (Fig. R8-57). This is a more compact design. It can be purchased assembled or added in the field.
3. Shell and finned coil type (Fig. R8-58). Due to the increased surface provided by the finned tube, it is more compact than the other two types for a similar capacity.

The disadvantages of a liquid-suction heat exchanger are:

1. It increases the suction temperature which increases power consumption.

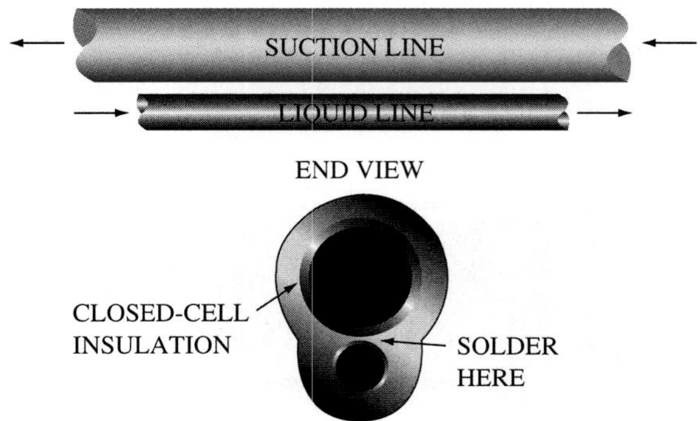

Figure R8-56 Heat exchanger made by bonding together the suction line and the liquid line.

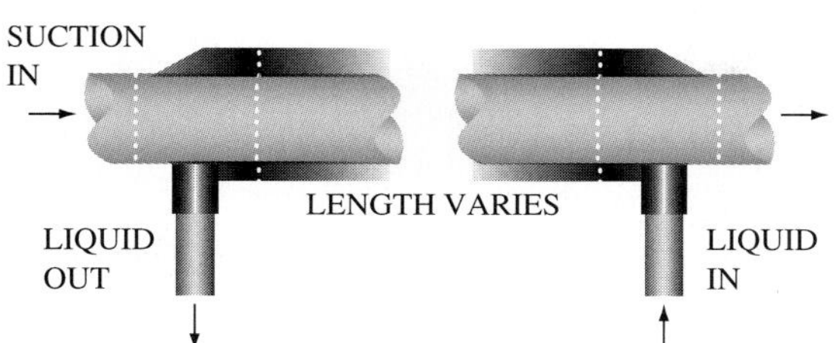

Figure R8-57 Construction of a tube-in-tube heat exchanger.

2. The upper suction-temperature limits for safe compressor operation must be observed.
3. It increases the piping complexity.
4. It must be properly drained to avoid oil-return problems.

The refrigerant-water preheater is shown in Fig. R8-59. This is a type of heat-recovery system. With this device, water is heated as the discharge gas leaving the compressor is de-superheated. A tube-in-tube type heat exchanger is sometimes used for this purpose. With this arrangement, the counterflow principle is applied with the hot gas running on the outside and the water flowing on the inside tube.

A typical installation for a residence is shown in Fig. R8-60. During the cooling cycle, water is heated for the domestic supply. When the water in the storage tank reaches 140°F, any excess heat is rejected by the air-cooled condenser. When required, the standard water heater supplements the heat-recovery unit to bring the water up to the required temperature. Note that plumbing codes may require a vented double-wall heat exchanger to avoid any possibility of a leak contaminating potable water supply with refrigerant or oil, as shown in Fig. R8-61.

R8-7.8 Receivers

A *receiver* is a storage area for refrigerant, placed in the system to hold refrigerant that is not immediately required

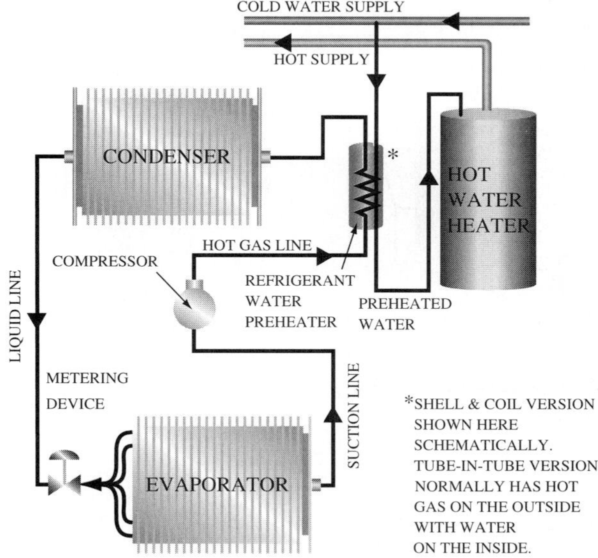

Figure R8-59 Piping arrangement for using a refrigerant-water preheater.

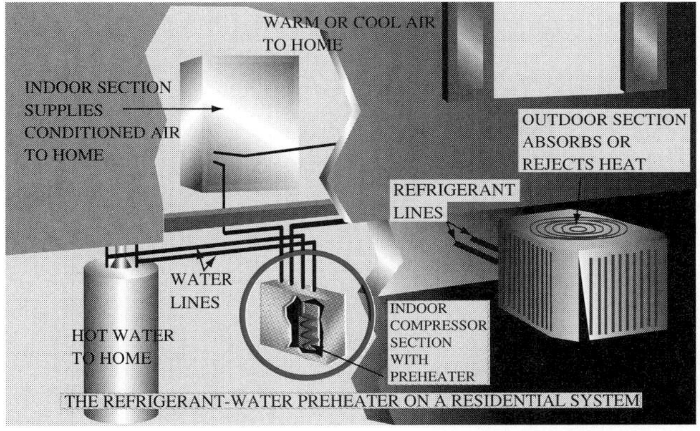

Figure R8-60 Residential application of a water preheater.

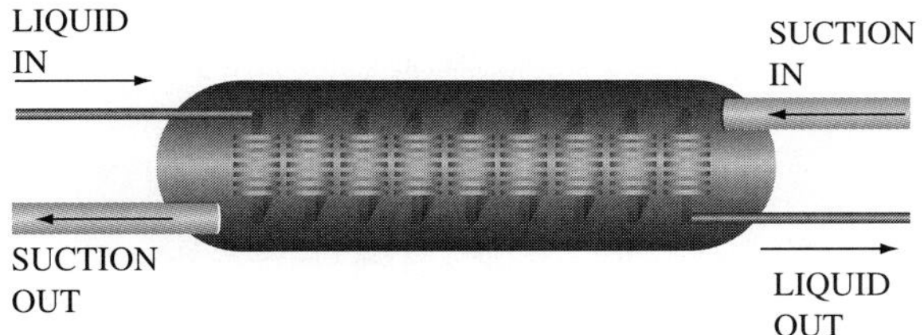

Figure R8-58 Construction of shell and finned-coil heat exchanger.

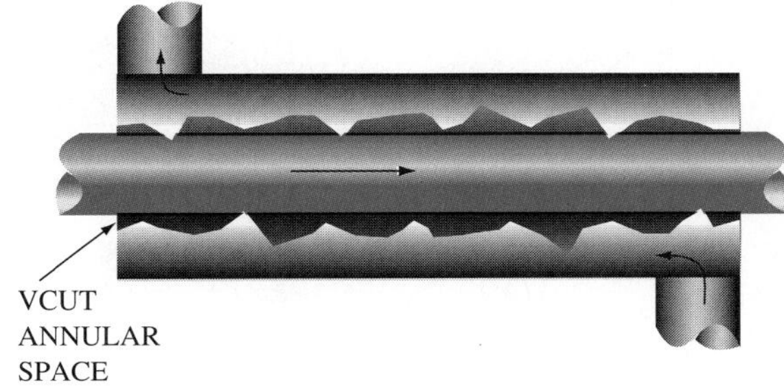

Figure R8-61 Vented double-wall heat exchanger.

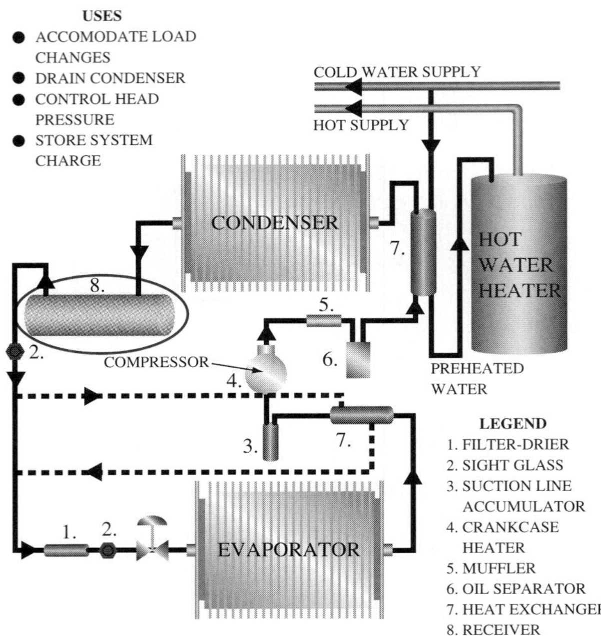

Figure R8-62 Location of a refrigerant liquid receiver.

in the cycle. The additional storage space is needed for the following purposes:

1. Accommodate changes during operation.
2. Freely drain the condenser of refrigerant.
3. Use a refrigerant floodback method of head pressure control.
4. Provide a place to store the system charge during system service procedures or prolonged shutdown periods.

The position of the receiver in the system is shown in Fig. R8-62. It is normally located near or below the condenser. All systems do not require receivers, and they should not be used unless required.

Systems using hand expansion valves, automatic expansion valves, thermostatic expansion valves, and low-side floats need some means of refrigerant storage during operation. The amount of refrigerant used varies inversely with the load. The surplus refrigerant required by larger loads must be stored somewhere. Some systems store the excess refrigerant in the condenser. When this is done no receiver is necessary. On other systems, however, the full condenser surface is needed for condensing and the addition of a receiver is required (Fig. R8-63).

There are two types of receivers: the flow-through receiver and the surge-type receiver.

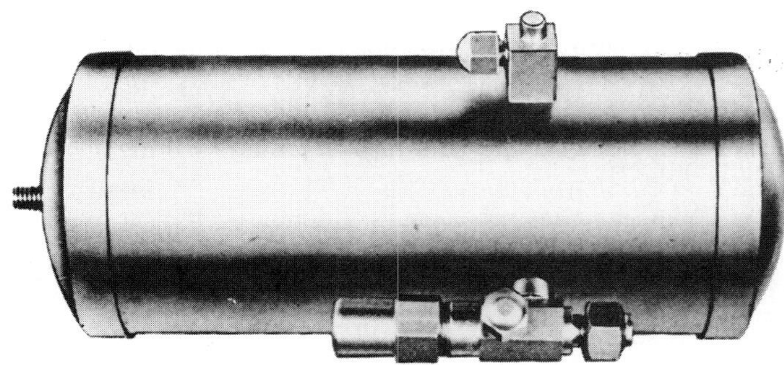

Figure R8-63 Receivers shown are used to store refrigerant. (Courtesy of Refrigeration Research, Inc.)

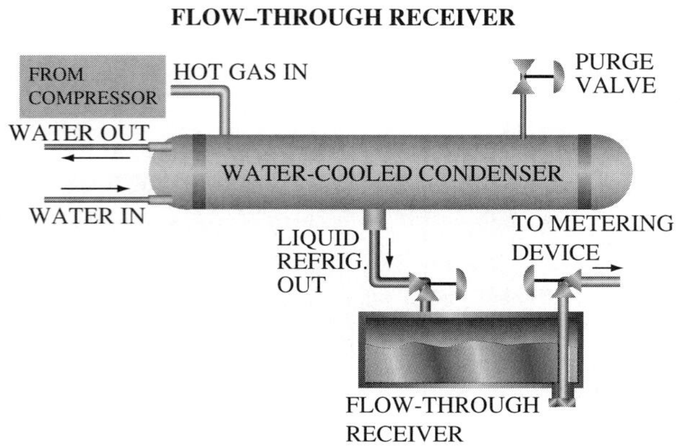

FLOW–THROUGH RECEIVER

FROM COMPRESSOR
HOT GAS IN
PURGE VALVE
WATER OUT
WATER-COOLED CONDENSER
WATER IN
LIQUID REFRIG. OUT
TO METERING DEVICE
FLOW-THROUGH RECEIVER

Figure R8-64 Piping arrangement using a flow-through receiver.

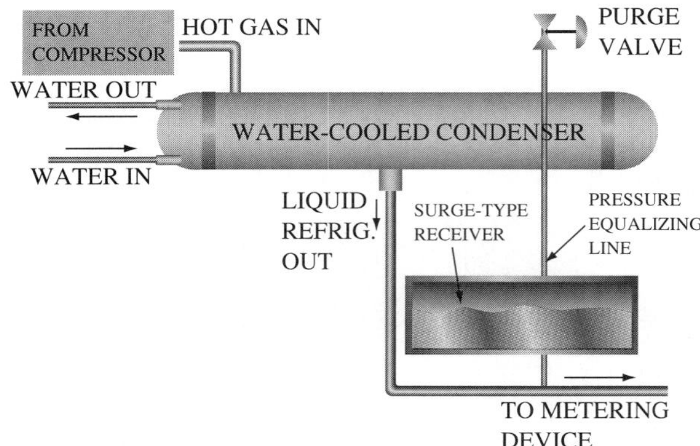

Figure R8-65 Piping arrangement for a surge-type receiver.

The piping for the flow-through receiver is shown in Fig. R8-64. Using this arrangement the liquid flows through the receiver before entering the metering device. The liquid will be saturated since both liquid and vapor are present. Note that the liquid is drawn off the bottom of the tank.

The piping for the surge-type receiver is shown in Fig. R8-65. The surge-type receiver releases only the amount of refrigerant needed. This allows subcooled refrigerant from the condenser to pass directly to the metering device, without losing its superheat.

Condensers, liquid receivers, and other vessels which may be valved off to contain liquid refrigerant must be protected by a refrigerant pressure-relief valve. A vessel filled by warming liquid can be subject to tremendous hydrostatic pressures, possibly rupturing the vessel at great danger to life and limb.

REVIEW

- Metering device—a restrictor device placed in the liquid line to produce a difference in pressure between the high side and low side, expanding the refrigerant and causing a drop in pressure and temperature in the evaporator. It regulates the flow of refrigerant into the evaporator.
- Flash gas—occurs at the outlet of the metering device, which lowers the evaporating temperature of the refrigerant. Decreasing "flash gas" increases "mass flow" into the evaporator. It is desirable to minimize "flash gas" to less than 25%, which can be accomplished by subcooling the liquid refrigerant prior to entering the metering device.
- Types of metering devices include:
  Hand-operated expansion valve
  Capillary tube, fixed orifice
  Automatic expansion valve
  Thermostatic expansion valve
  Electric/electronic expansion valve
  High-side float
  Low-side float
- Superheat (at outlet of evaporator) = evaporator outlet temperature − evaporator saturation temperature. It can be adjusted at the thermostatic expansion valve.
- Subcooling = condenser saturation temperature − liquid line temperature.
- Equalizer line—when the pressure drop in an evaporator is sizeable, a thermostatic expansion valve with an external equalizer line should be installed. This line should be installed 6–8 in. downstream of the sensing bulb.
- MOP valves (Maximum operating pressure)—have a gas-charged sensing bulb; prevent overloading the compressor during a warm startup.
- Liquid distributors—placed between the metering device and the evaporator; distribute refrigerant to multiple circuit evaporators.
- Common metering device problems include:
  Restricted or stuck expansion valve
  Oversized or improperly sized nozzle
  Improper sensing-bulb location
  Valve set too high or too low
  Wrong type of valve installed
  Plugged equalizer line or plugged distributor
  Loss of charge at thermobulb
  Thermobulb loose or improperly secured
- Accessories—any devices or articles that add to convenience or effectiveness of the system
  Liquid-line filter-drier—a device installed in the liquid line to remove moisture and/or acid.
  Suction-line filter—a device installed in the suction line within close proximity of the compressor to protect the compressor from solid particles, acid, and moisture.
  Sight glass/moisture indicator—located in the liquid line as an aid to determine the condition of the refrigerant within the system. The moisture indicator is designed to warn the service technician when moisture in dangerous quantities is present within the system.
  Suction-line accumulator—prevents the flooding of liquid to the compressor from the evaporator.
  Crankcase heater—increases compressor crankcase temperature during shutdown to minimize oil dilution during startup.
  Muffler—dampens or removes the hot-gas pulsations set up by a reciprocating compressor.
  Oil separator—reduces the flow of oil through the system, and returns the oil to the crankcase of the compressor.
  Heat exchanger—a device to transfer heat from the liquid refrigerant to the suction gas.
  Receiver—a vessel used to store liquid refrigerant within the refrigeration system.
  Solenoid valve—used to control the flow of liquid or gas.
  Check valve—a device designed to allow the flow of liquid or gas in one direction only.

Problems and Questions

1. Name the eight major metering (pressure reducing) devices used in refrigeration systems.
2. What is the disadvantage of an automatic expansion valve?
3. A thermostatic expansion valve regulates the refrigerant boiling point according to the load and the coil maintaining a fairly constant _____.
4. What three pressures are used to operate a thermostatic expansion valve?

5. Of the three pressures that operate a thermostatic expansion valve, which two pressures oppose the third?
6. The sensing bulb of the TXV should always be placed on the suction line between the 4 o'clock and 8 o'clock. True or False?
7. Using capillary tubes as a metering device has several advantages. The two major ones are _____ and _____.
8. Using a capillary tube as a metering device has two disadvantages. What are they?
9. The thermostatic expansion valve regulates refrigerant flow to the evaporator based on:
   a. heat of compression
   b. suction line superheat
   c. external equalizer
   d. evaporator pressure
10. The capillary tube is a type of:
   a. external equalizer
   b. metering device
   c. TXV sensor
   d. device that senses superheat
11. The TXV sensing bulb is mounted:
   a. on the liquid line between the condenser and the TXV
   b. at the inlet to the condenser
   c. on the suction line
   d. on the compressor
12. An external equalizer is used when:
   a. the pressure drop across the evaporator is excessive
   b. the liquid line pressure becomes too great
   c. the receiver is too full of refrigerant
   d. the pressure drop at the automatic
13. When the load on the evaporator increases, the TXV:
   a. increases the refrigerant flow
   b. decreases the refrigerant flow
   c. maintains the flow of the refrigerant
   d. decreases the pressure across the evaporator
14. The liquid-charged bulb is:
   a. a sensing bulb on an automatic expansion valve
   b. a bulb attached to one end of a capillary tube
   c. a sensing bulb at the inlet of the expansion valve
   d. a TXV sensing bulb charged with a fluid similar to that of the system refrigerant
15. Superheat is:
   a. the sensible heat absorbed by the refrigerant after it has boiled to a vapor
   b. the heat of compression at the compressor
   c. heat used to boil the liquid in the evaporator
   d. latent heat given off by the condenser
16. The sensing bulb and transmission tube are a part of the:
   a. automatic expansion valve
   b. capillary tube
   c. king valve
   d. TXV
17. A small amount of superheat in the suction line is desirable with a TXV to:
   a. keep the proper amount of refrigerant in the receiver
   b. insure the proper pressure at the condenser
   c. keep from starving the evaporator
   d. ensure that no liquid refrigerant leaves the evaporator
18. The thin metal disc connected to the needle in the TXV is called the:
   a. spring
   b. diaphragm
   c. seat
   d. sensing bulb
19. A refrigeration system accessory is an article or device that:
   a. adds to the convenience of a system
   b. comes in many form
   c. is not essential to the refrigeration system
   d. adds to the effectiveness of a system
   e. all of the above
20. A moisture indicator reveals the presence of moisture in the system by changing _____ .
21. What is the purpose of the crankcase heater?
22. The principal uses of a heat exchanger are _____ .
23. An oil separator can cure all oil return problems. True or False?
24. Why are mufflers installed in a refrigeration system?
25. What is the primary purpose of a suction-line accumulator?
26. Name two types of filter-driers.
27. What is the purpose of a check valve?
28. Which type of metering device maintains a constant evaporator pressure?
   a. Thermostatic expansion valve
   b. Capillary tube
   c. Automatic expansion valve
   d. Electronic expansion valve
29. Which type of metering device includes a check valve?
   a. Fixed orifice
   b. Capillary tube
   c. Thermostatic expansion valve
   d. Automatic expansion valve
30. What type of metering device uses a stepper motor?
   a. Thermostatic expansion valve
   b. Automatic expansion valve
   c. Electric and electronic expansion valve
   d. Low side float
31. What type of metering device can use an external equalizer line?
   a. TXV
   b. AEV
   c. LSF
   d. HSF

32. When is the crankcase heater needed?
    a. During the running cycle.
    b. During the off-cycle.
33. What type of device can be used as a moisture indicator?
    a. Receiver
    b. Suction line accumulator
    c. Filter-drier
    d. Sight glass
34. What type of accessory temporarily stores refrigerant?
    a. Filter-drier
    b. Suction line accumulator
    c. Heat exchanger
    d. Crankcase heater

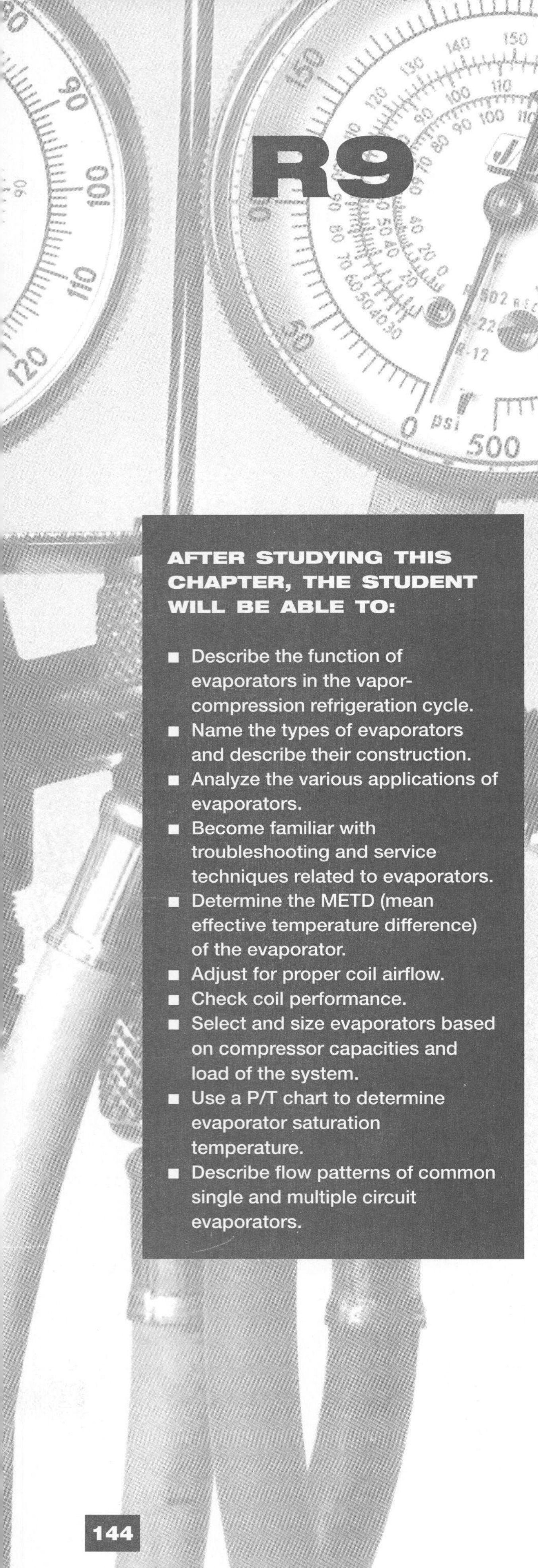

R9 Evaporators

AFTER STUDYING THIS CHAPTER, THE STUDENT WILL BE ABLE TO:

- Describe the function of evaporators in the vapor-compression refrigeration cycle.
- Name the types of evaporators and describe their construction.
- Analyze the various applications of evaporators.
- Become familiar with troubleshooting and service techniques related to evaporators.
- Determine the METD (mean effective temperature difference) of the evaporator.
- Adjust for proper coil airflow.
- Check coil performance.
- Select and size evaporators based on compressor capacities and load of the system.
- Use a P/T chart to determine evaporator saturation temperature.
- Describe flow patterns of common single and multiple circuit evaporators.

The *evaporator* is a heat exchanger with refrigerant contained within tubes, passages, or a vessel. The fluid (air, water, brine, etc.) or product to be cooled is separated from the refrigerant by the heat exchanger walls or shell.

The evaporator is that portion of the vapor-compression refrigeration cycle where heat flows into the system (Fig. R9-1).

The heat flow is caused because the temperature of the refrigerant is lower than the temperature of the air or water that is being cooled. The refrigerant temperature in the evaporator is maintained at the saturation temperature that corresponds to the evaporator pressure.

When the compressor starts, the pressure in the evaporator drops, causing the refrigerant liquid in the evaporator to evaporate (flash) and cool (Fig. R9-2).

R9-1 TYPES OF EVAPORATORS

There are three general types of evaporators:

1. Bare pipe
2. Extended surface
3. Plate

In the vapor-compression refrigeration cycle, all of the refrigerants have to be confined within the system by components and interconnecting tubing strong enough to withstand the pressures, temperatures, and vibration imposed by the system. This ordinarily means metal construction. Compressors are generally housed in steel or cast-iron enclosures. Interconnecting piping and tubing are usually copper (or aluminum) in small sizes and copper or steel in larger sizes. Heat-exchanger tubes are usually copper (or aluminum in small sizes) in halocarbon systems, although steel may be used, especially in larger sizes. A halocarbon system is one with a refrigerant where the refrigerant molecule contains carbon and either chlorine or fluorine. Ammonia systems must be all iron, steel or aluminum (no copper or brass).

R9-1.1 Bare-Pipe Evaporators

Bare-pipe applications (Fig. R9-3) are seldom seen in common air-conditioning applications. They may be used

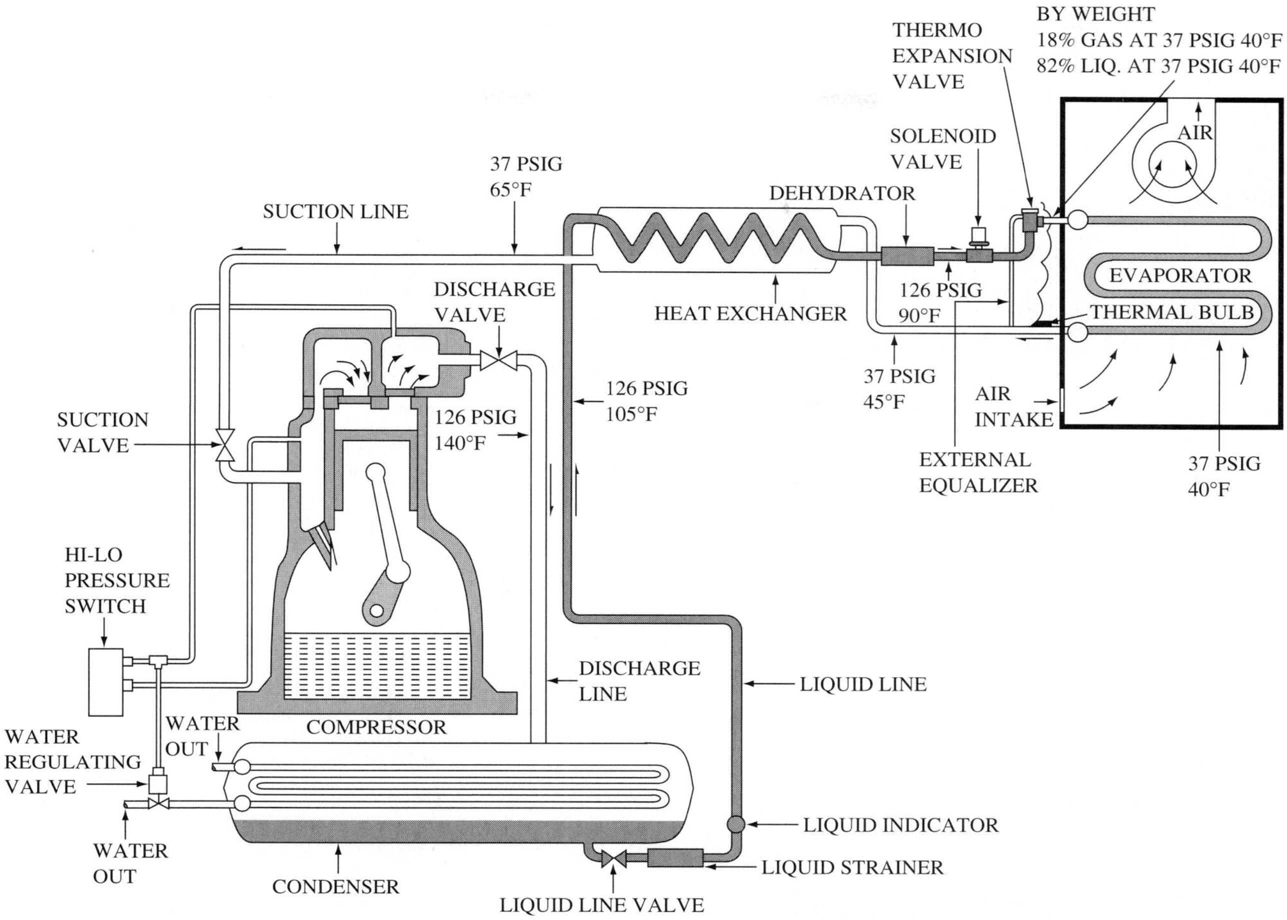

Figure R9-1 Location of the evaporator in relation to other major components.

PREVENTIVE MAINTENANCE

When at a job site, learn to do a visual inspection of evaporators to determine if they are clean, completely defrosted, and that all fans are running. If it is not possible to view the evaporator due to product loads, duct work, or any other reasons, look at and feel the suction line. If it is abnormally cold or ice-covered, it might indicate possible airflow problems with the evaporator.

Change air filters at the evaporator on a regular schedule. This will improve the efficiency and prevent liquid slugging back to the compressor.

ON THE JOB

Remember to put all covers, screws, and hardware back on the evaporators as was originally intended before leaving the job site. Check for proper airflow and air patterns from diffusers, grilles, and evaporator coils. A few minutes of casual inspections can help secure customer relations and eliminate future problems.

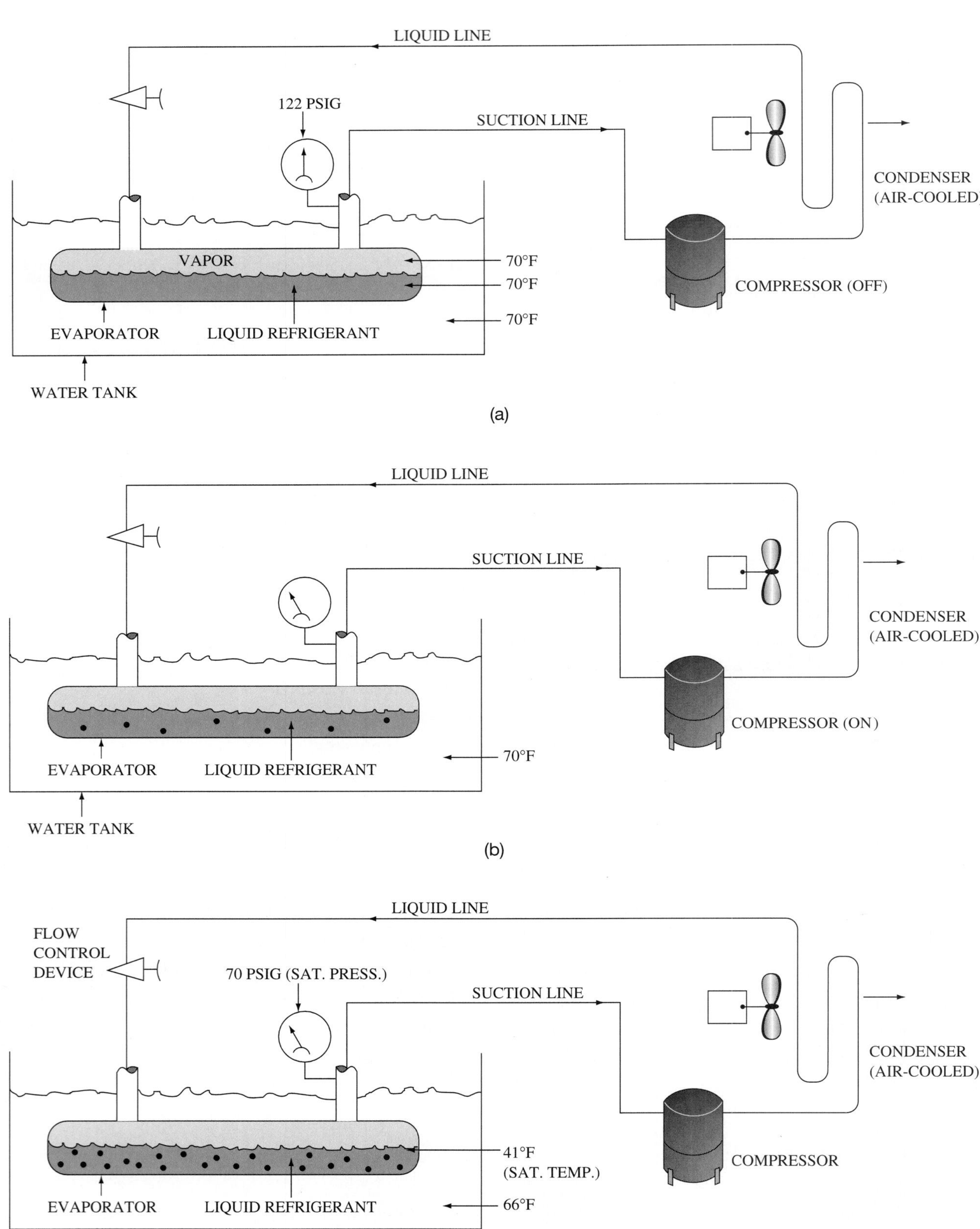

Figure R9-2 Diagrams of the refrigeration process (R-22 refrigerant): (a) system off; (b) system startup; (c) system running.

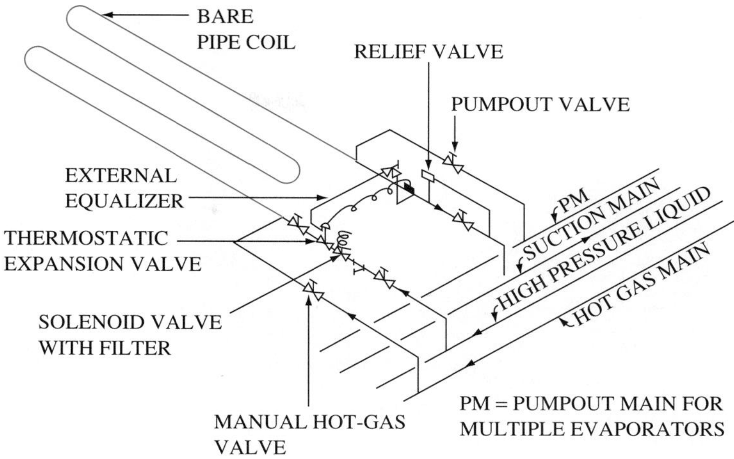

Figure R9-3 Bare-pipe evaporator coil with gravity air circulation. (Copyright by the American Society of Heating, Refrigeration, and Air-Conditioning Engineers, Inc. Used by permission.)

for some larger refrigeration work and immersed tank coolers. They are also known as prime-surface evaporators. (See more under Applications in this chapter.)

R9-1.2 Extended Surface Evaporators

Extended surfaces (Figs. R9-4 and R9-5) help equalize the heat transfer capabilities of refrigerant-evaporator tubes when cooling air. The heat transfer capacity on the refrigerant side is high due to turbulent boiling liquid refrigerant. The air side is much more limited. Air is not as good a conductor of heat. By adding tight-fitting aluminum fins to the copper tubes, the air-side surface and corresponding heat transfer can be greatly increased, while the size and cost of the evaporator may be reduced, compared to a prime surface unit. To assure good heat transfer, the fins must be tightly bonded to the tubes. With plate-fin construction, the tubes are mechanically or hydraulically expanded into the tube collar. Multi-row construction is available and the fins can be shaped or dimpled to provide additional turbulence on the air side for increased heat transfer.

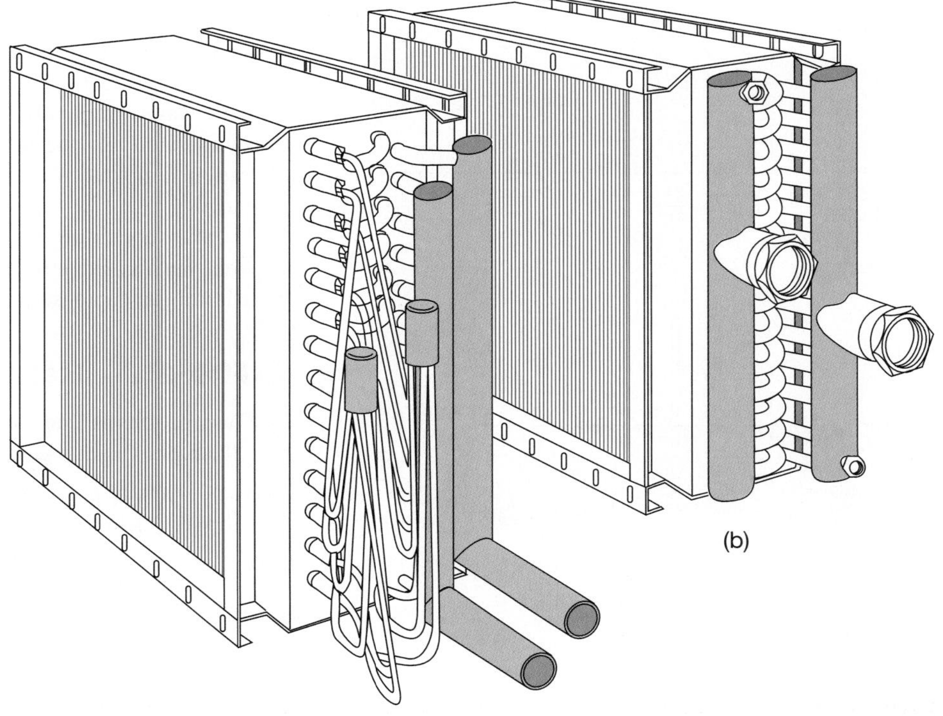

Figure R9-4 Plate-fin cooling coils for (a) direct expansion; (b) chilled water. (The Trane Company)

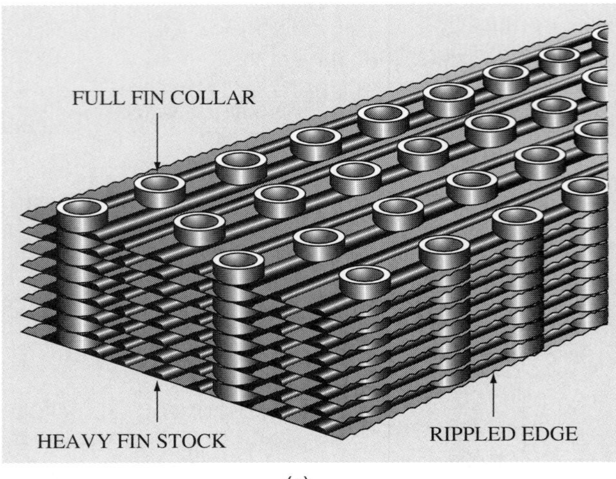

(a)

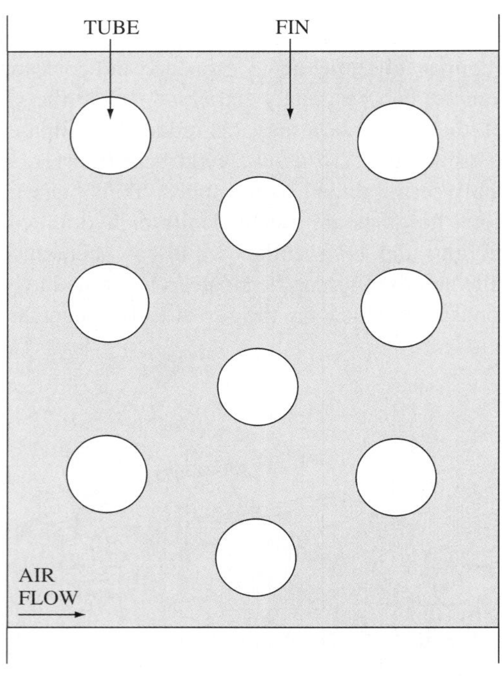

(b)

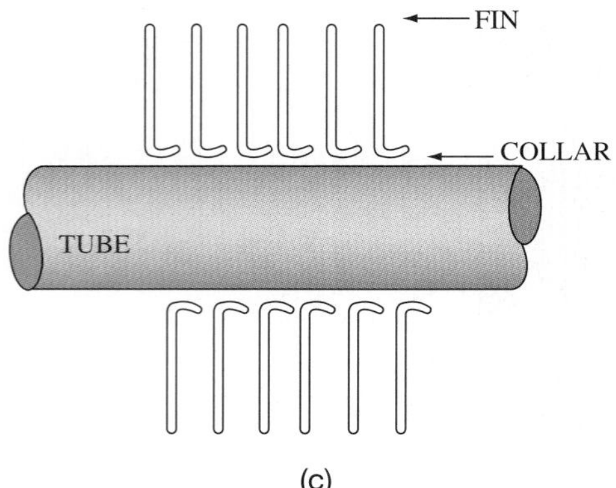

(c)

Figure R9-5 Plate-fin coil details: (a) four-row coil construction; (b) staggered tubes; (c) cross section of finned-tube assembly.

KNOWLEDGE EXTENSION

Visit your local library and reference the ASHRAE manuals. Research possible evaporator configurations and piping designs. Contact ARI for performance specifications on manufacturers' equipment. The ARI directory will give design Btu criteria and performance standards.

R9-1.3 Plate Evaporators

Plates (Fig. R9-6) are a special form of extended heat transfer surface used in refrigeration and freezer applications. The flat plate surfaces may be fabricated in a variety of shapes and the refrigerant passages may be integral or attached.

R9-2 PRESSURE-HEAT DIAGRAM

The pressure-heat diagram (Fig. R9-7) affords a good view of the cooling process in the evaporator. Initially a high-pressure liquid should be subcooled 8–10°F or more if possible.

When subcooled liquid from point *A* flows through the expansion device, its flow is controlled and its pressure drops to evaporator pressure. Approximately 20% of the liquid boils off to gas, cooling the remaining liquid-gas mixture. Its total heat (enthalpy) at point *B* is unchanged from *A*. No external heat energy has been exchanged. From points *B* to *C*, the remainder of the liquid boils off, absorbing the heat flowing in from the evaporator load (air, water, or product). At point *C*, all of the liquid has evaporated and the refrigerant is 100% vapor at the saturation temperature corresponding to the evaporator pressure.

The subcooling increases cycle efficiency and can prevent flash gas due to pressure loss from components, pipe friction, or increase in height.

R9-2.1 Flash Gas Problems

Two problems can occur, both due to the formation of "flash gas." Flash gas can be formed as a result of:

1. A restricted filter-drier, as shown in Fig. R9-8.
2. A rise in the liquid line, lowering the pressure, causing some of the refrigerant to boil, as shown in Fig. R9-9.

The problem with Fig. R9-8 can be solved by changing the restricted drier. If the condensing unit had sufficient condenser surface to provide additional liquid subcooling, the likelihood of flash gas problems would greatly diminish.

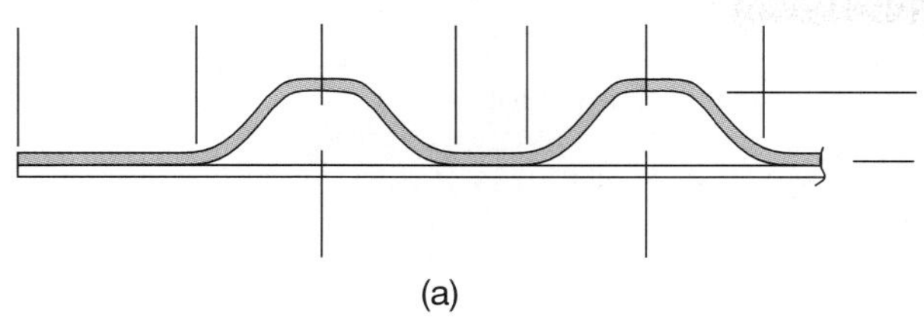

(a)

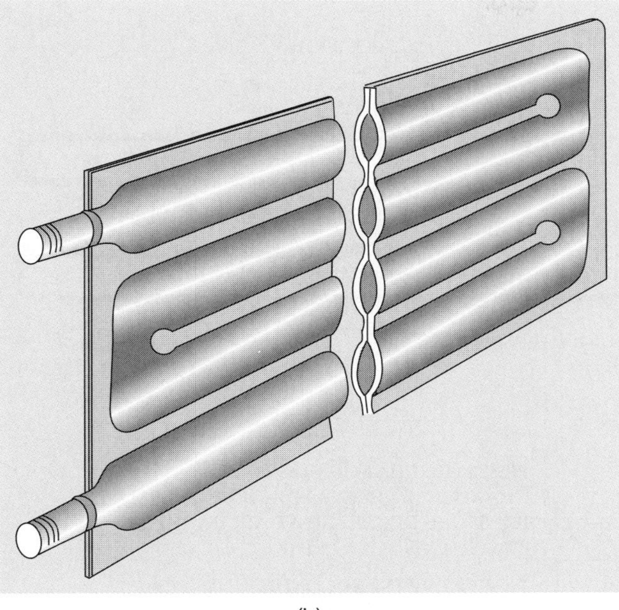

(b)

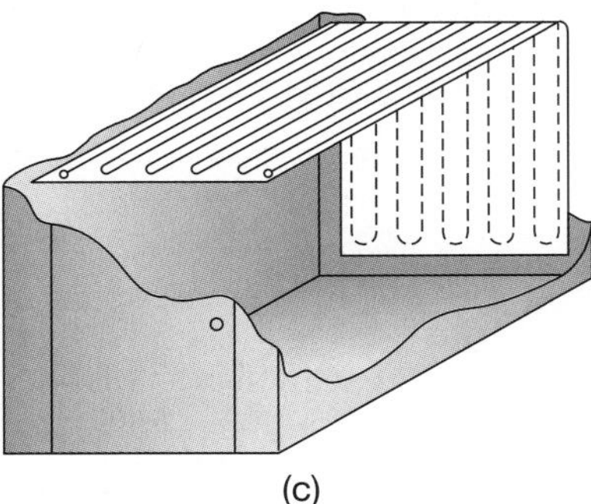

(c)

Figure R9-6 Plate-type evaporators: (a) single embossed design; (b) serpentine circuiting; (c) cabinet installation. (Courtesy of Tranter, inc.)

Fig. R9-9 points to the need for caution with systems where the evaporator is significantly above the condenser. The R-22 liquid-line pressure will drop 1 psig for every 2-ft increase in altitude. A 30-ft rise equals a 15 psig drop in pressure, which is equivalent to a 5°F drop in saturation temperature. This is before any friction losses are considered.

A possible solution would be the addition of a suction-to-liquid heat exchanger placed near the condensing unit or providing suction-to-liquid heat exchange by running the lines tightly together (soldered every 2 feet and insulated) to achieve the needed liquid subcooling.

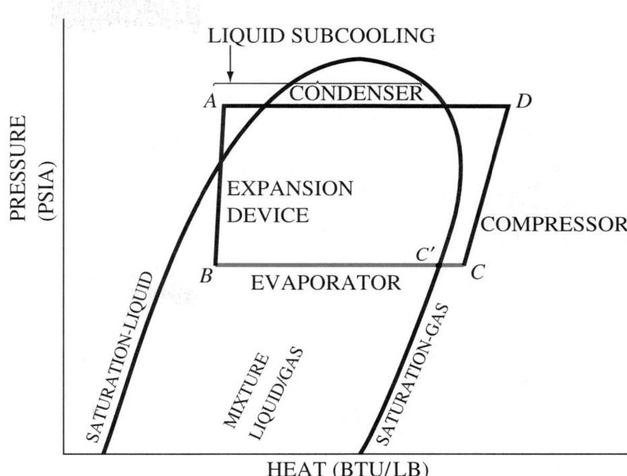

Figure R9-7 Pressure-heat diagram showing the refrigerating effect in the evaporator.

The subcooling increases cycle efficiency and can prevent flash gas due to pressure loss from components, pipe friction or from increase in height.

R9-2.2 Direct Expansion (DX) Evaporators

Most smaller refrigeration systems are designed to have the expansion device control the refrigerant flow so the evaporator will heat the vapor beyond saturated conditions and ensure no liquid droplets will enter and possibly damage the compressor (liquid slugs). It is assumed here for the sake of simplicity there is no pressure drop through the evaporator. (In reality there are pressure drops which would slightly shift the evaporating and condensing processes from the constant pressure lines shown.)

This additional heating of the gas (at constant pressure) is called *superheating*. It simply means heated above saturation temperature. The last pass of the heat exchanger tubes usually provides the superheating (Fig. R9-10), and at somewhat less efficiency because the refrigerant being heated is a gas (not a boiling liquid), and the temperature difference is reduced. A common range of superheat is 8–10°F which carries the process to point *C*. At this point the cold, low-pressure, super-heated refrigerant vapor is ready to flow to the compressor and complete the closed cycle.

R9-2.3 Flooded Evaporators

If an evaporator does not have to superheat refrigerant vapor, it can produce more cooling capacity. On small systems the difference is negligible and it is important to protect the compressor. On very large systems, an increase in evaporator performance can be important. A flooded evaporator absorbs heat from points *B* to *C*. It can circulate more pounds of refrigerant (more cooling capacity) per square foot of heat transfer surface. No surface needs to be used to superheat the suction vapor.

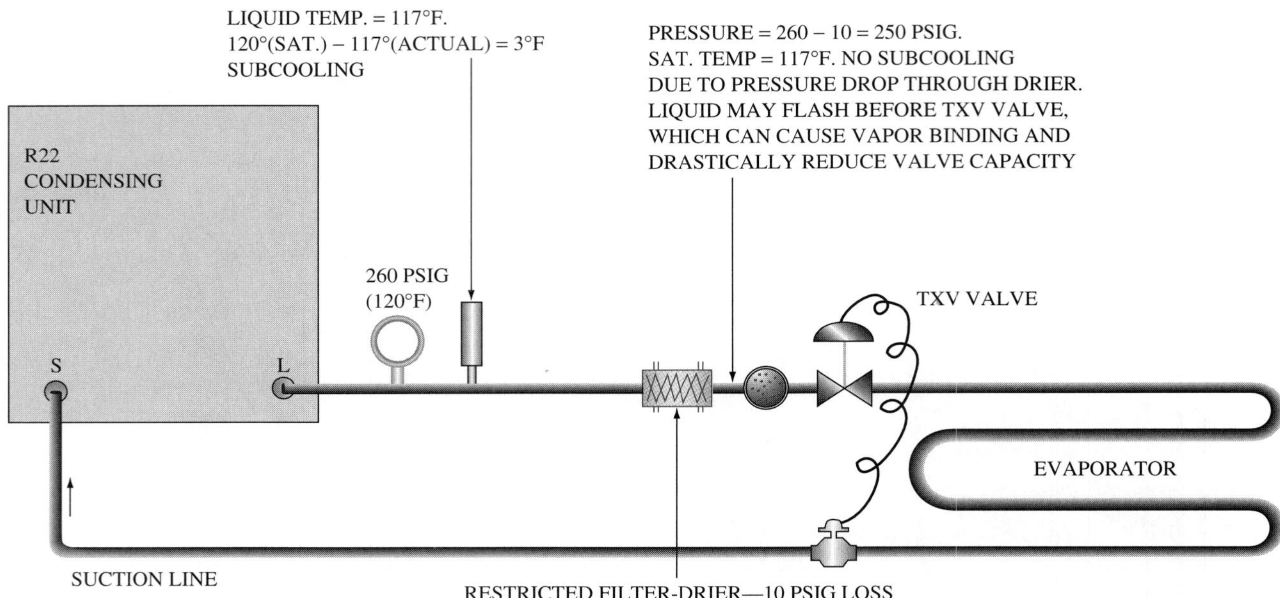

Figure R9-8 Flash gas caused by restricted filter-drier.

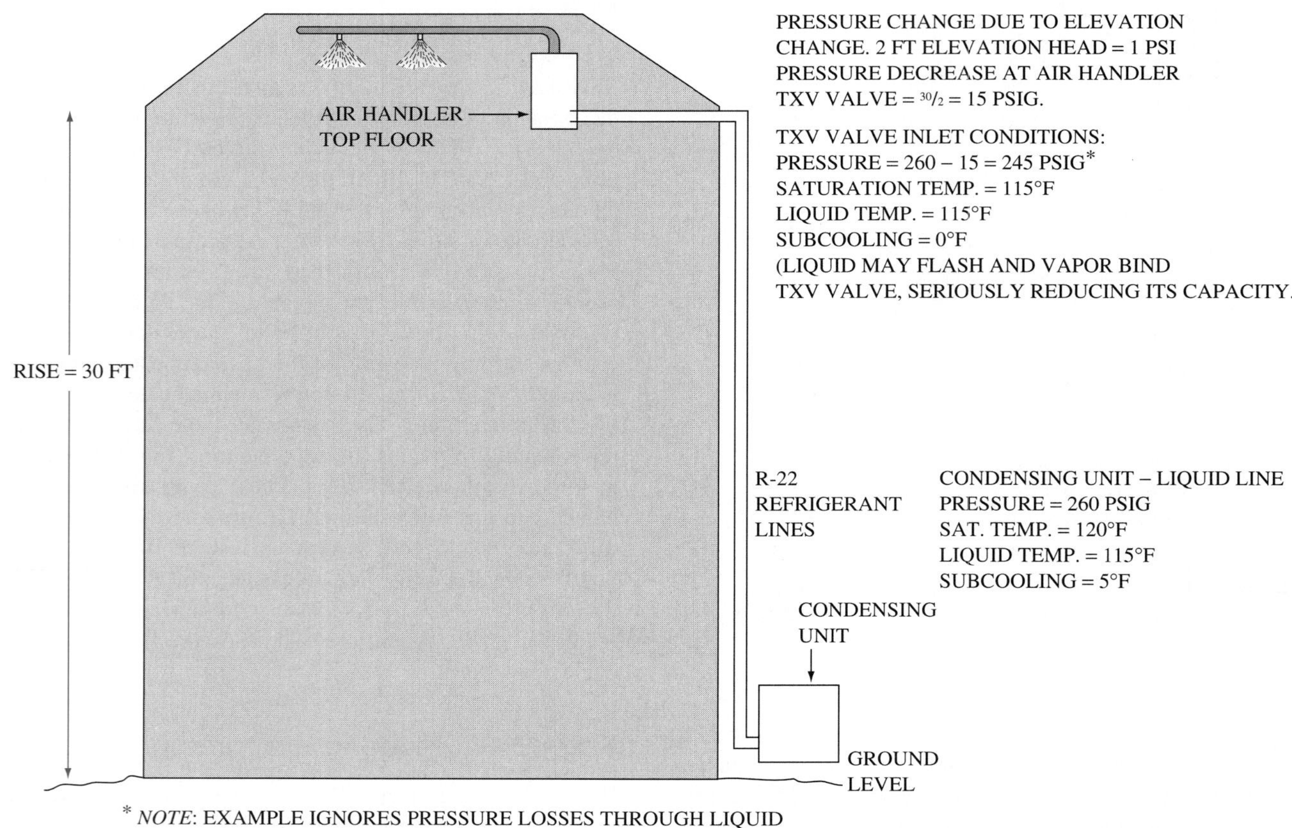

Figure R9-9 Flash gas caused by rise of liquid line.

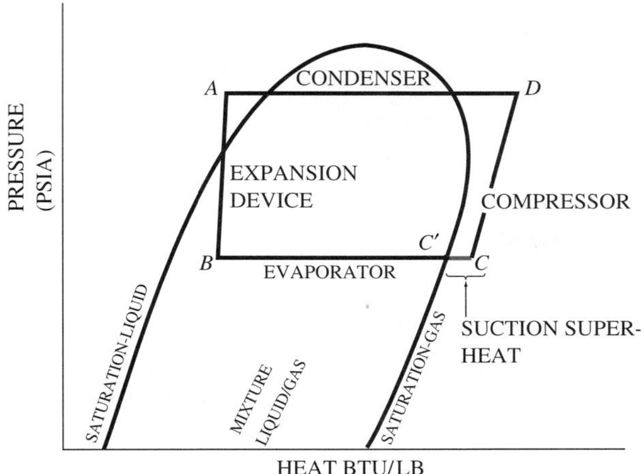

Figure R9-10 Pressure-heat diagram showing the suction superheat.

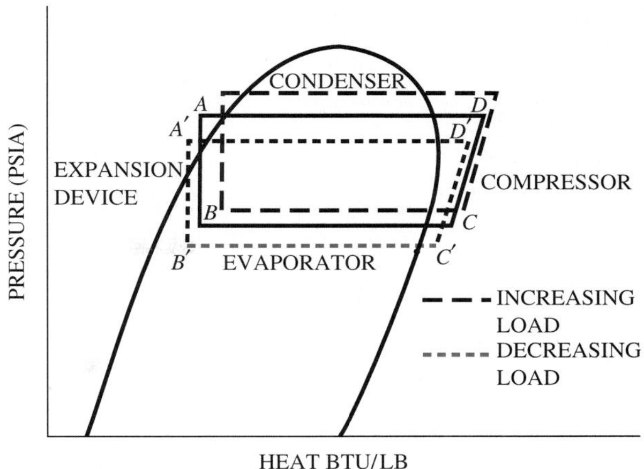

Figure R9-12 Effect of increasing and decreasing the refrigerant load.

Large water chillers use flooded evaporators (Fig. R9-11). It is very important to ensure that the saturated refrigerant flowing to the compressor does not contain quantities of liquid which could cause mechanical damage.

R9-2.4 Load Variation

What happens when the load changes on an evaporator with a thermal expansion valve control? As the load diminishes, the evaporator will lower the temperature of whatever it is cooling while handling the same refrigerant flow rate. Due to the diminished temperature difference, this will reduce the heat flow into the evaporator and the superheat of the outlet vapor will drop. The thermal expansion valve (TXV) will sense this reduction and throttle the refrigerant flow until a new balance point is reached and the load can again superheat the suction vapor 8 or 10°F to satisfy the valve setting. The evaporator will be running at a lower suction pressure and temperature and capacity. If the compressor has capacity control, it will reduce capacity when the suction pressure drops to the unloader setting. The new operating points will be *A′*, *B′*, *C′*, and *D′* (Fig. R9-12).

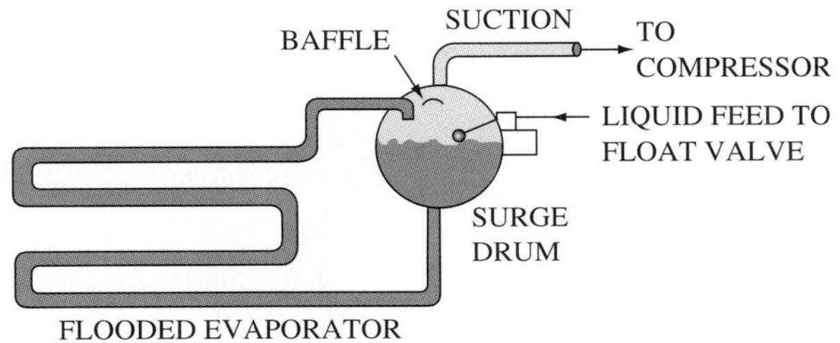

Figure R9-11 Flooded evaporator with float-valve metering device.

When a partially loaded evaporator has its load increased, the temperature difference between the load and refrigerant will increase, the suction superheat will increase, and the TXV will open to increase flow and reestablish proper superheat. As this happens, the refrigerant flow rate picks up and the suction pressure increases. The operating points on the pressure-heat chart will shift back toward *A—B—C—D* in Fig. R9-12.

R9-2.5 Undersized Evaporator

What is the effect of reducing the size of the evaporator on a system? As shown in Fig. R9-13, a smaller evaporator with less heat-transfer surface will not handle the same heat load at the same temperature difference. With a given condensing unit, a new balance point will be reached with a lower suction pressure and temperature. The load will be reduced and the discharge pressure and temperature will be lower also. It is important to match the capacities of system components, to achieve the desired results.

R9-3 EVAPORATOR CONSTRUCTION—AIR COOLING

The most common halocarbon evaporator for cooling air is the plate-fin coil. This applies to small refrigerated reach-in and walk-in boxes, as well as to large comfort air-conditioning units. The most common tube material is copper. Aluminum tubes also have been used. Aluminum and steel are suitable for ammonia refrigerant, but copper is not. Aluminum tubes are difficult to patch in the field if they develop a leak.

The most common fin material is aluminum. For special applications such as sprayed-coil dehumidifiers,

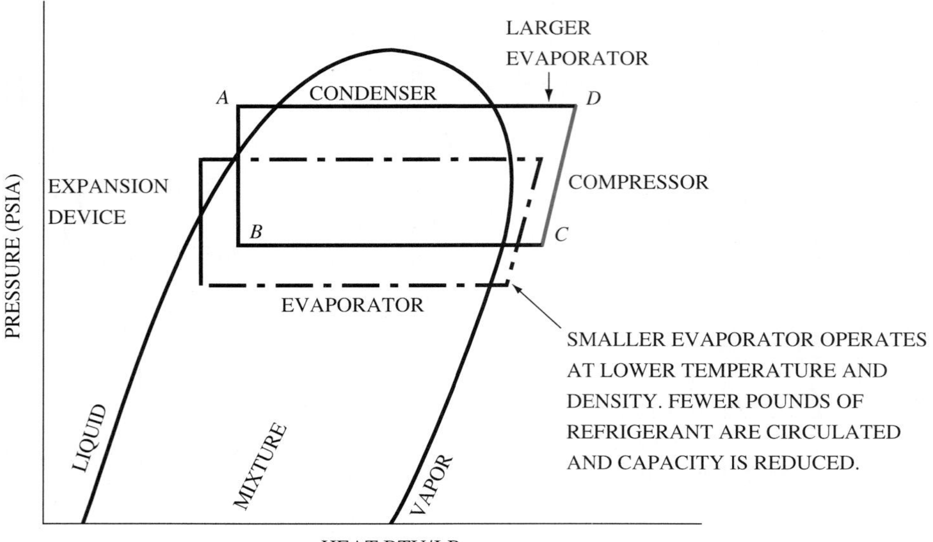

Figure R9-13 Effect of reducing the evaporator size.

where water sprays on the coil fins are used to maintain close temperature and humidity conditions, copper fins would be recommended. They are costly and not ordinarily used.

It is possible to order coils with special protective coatings for marine applications or industrial use where ordinary fins are attacked and corroded. The coatings may reduce capacity somewhat, but lengthen equipment life.

For air-conditioning applications, fin spacing ranges from 8 to 14 fins per in. (see Fig. R9-14). The closer fin spacing produces increased performance per square foot of coil, but the additional fins produce a higher air-side pressure drop and more energy is required to operate the fan. This type of trade-off in design is not unusual. Tube rows in the direction of air flow typically range from two to six with three and four most common in direct-expansion coils.

Face velocities typically range from 300 to 500 fpm. Most designers limit maximum cooling coil velocity to 500–550 fpm to reduce the possibility of condensate

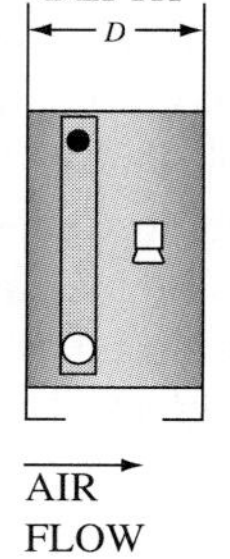

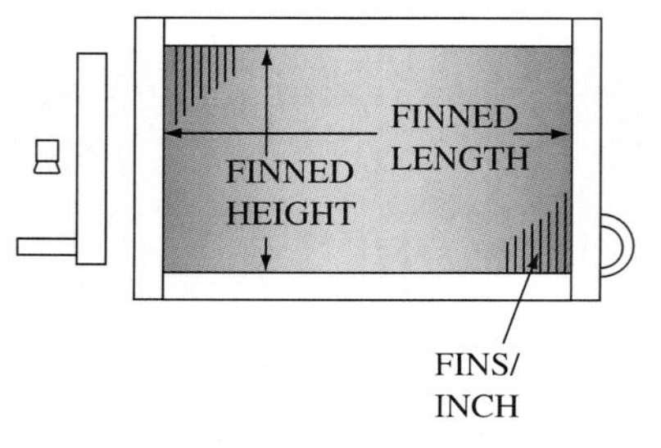

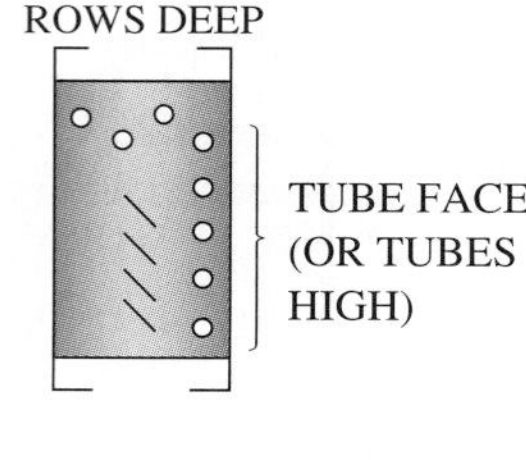

Figure R9-14 Finned-tube coil used for air cooling. (Courtesy of U.S.A. Coil & Air, Inc.)

carry-over from the coil fins during humid weather. If the condensate carries past the drain pan, it can leak from the air handler cabinet and may cause expensive water damage to the building or its contents.

To ensure maximum coil performance, adjacent rows of tubes are staggered to cause good air mixing and turbulence. The number of tubes in a row is designated as the tube face. The coil face dimensions open to airflow are designated as tube length (or finned length), and the height as finned height.

Counterflow

Heat exchangers generally should be installed and piped *counterflow*—that is, with the two fluid streams flowing in opposite directions (see Fig. R9-15). This produces the greatest mean effective temperature difference (METD), therefore, the greatest heat transfer capacity. In the case of DX coil, the final refrigerant circuits where the gas must be superheated are in contact with the warmest entering air.

R9-4 EVAPORATOR CONSTRUCTION—WATER COOLING

Water chillers (evaporators) are made in a variety of sizes and designs. Small units can be tube-in-tube coaxial design (Fig. R9-16). A common construction is copper water tube and steel refrigerant tube.

DX water chillers of the shell-and-tube design range from 5 tons to 350 tons (and up). The shell is usually steel, although brass pipe has been used in some smaller diameters. The refrigerant tubes are copper (some with an integral rolled fin to increase the heat transfer surface). Some designs utilize metal inserts or turbulators in the tubes to enhance heat transfer on the refrigerant side. The shell holds the water and baffles provide for even water flow and loading of all circuits. Refrigerant circuits normally number one or two. (See Fig. R9-17.)

Shell-in-tube DX chillers with two refrigerant circuit construction are shown in Figs. R9-17 and R9-18.

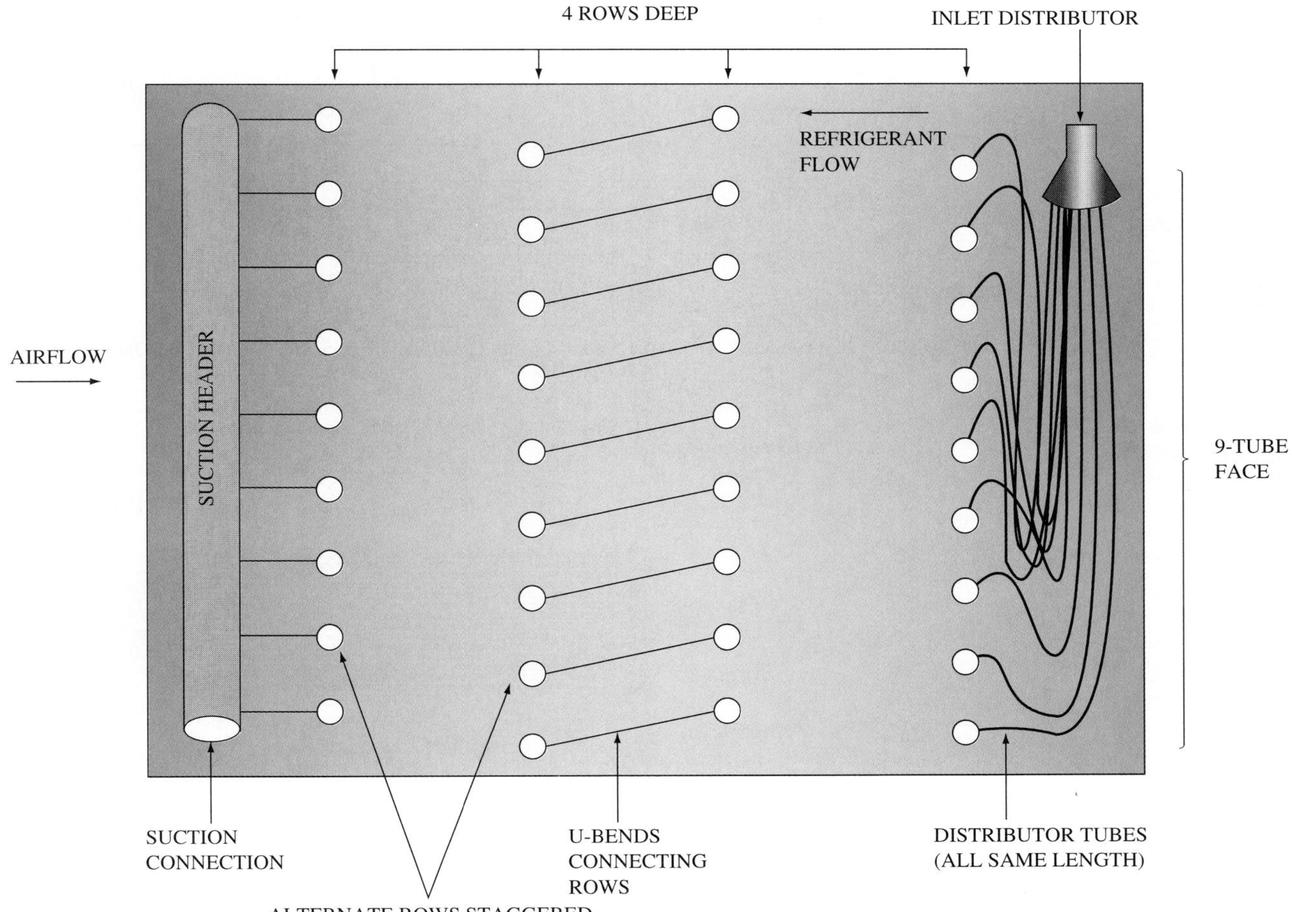

Figure R9-15 Direct-expansion evaporator coil showing counterflow airflow.

Figure R9-16 Coaxial-tube heat exchanger. (Courtesy of Turbotec Products, Inc. Windsor CT)

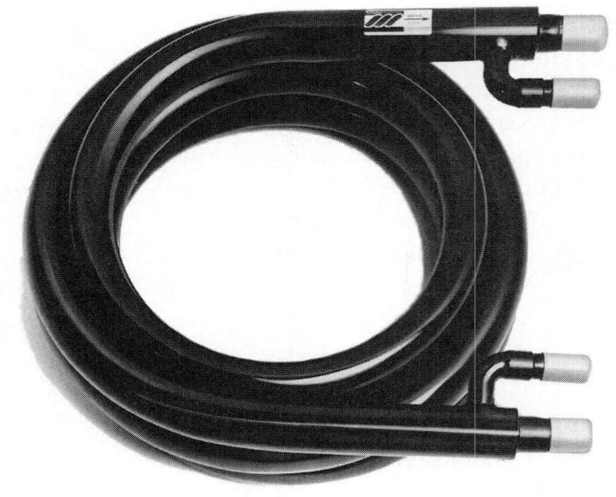

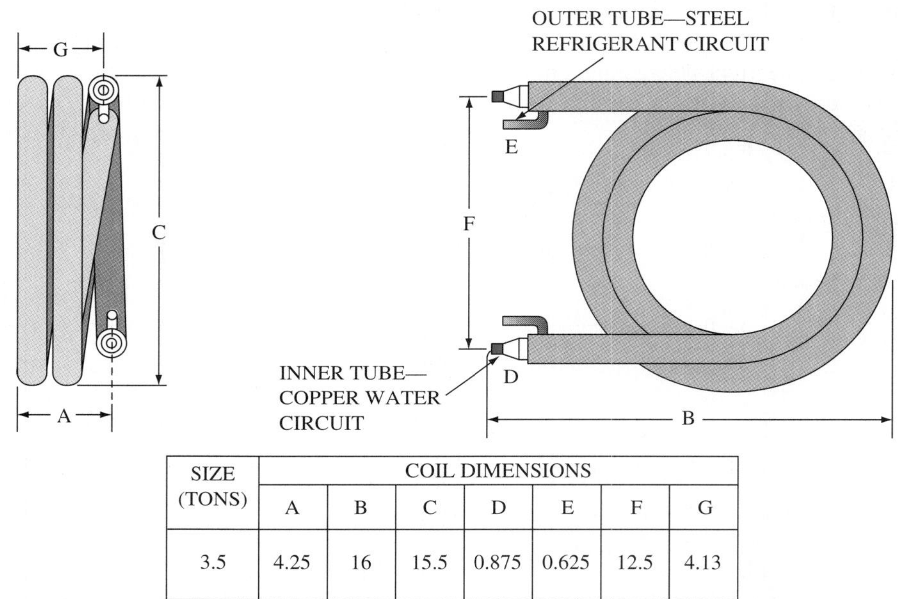

SIZE (TONS)	COIL DIMENSIONS						
	A	B	C	D	E	F	G
3.5	4.25	16	15.5	0.875	0.625	12.5	4.13

NOTE: ALL DIMENSIONS ARE GIVEN IN INCHES. DIMENSION D IS AN OD. DIMENSION E IS AN OD.

Figure R9-17 Direct-expansion chiller. Refrigerant in tubes, water in the shell. (Copyright by the American Society of Heating, Refrigerating, and Air-Conditioning Engineers, Inc. Used by permission.)

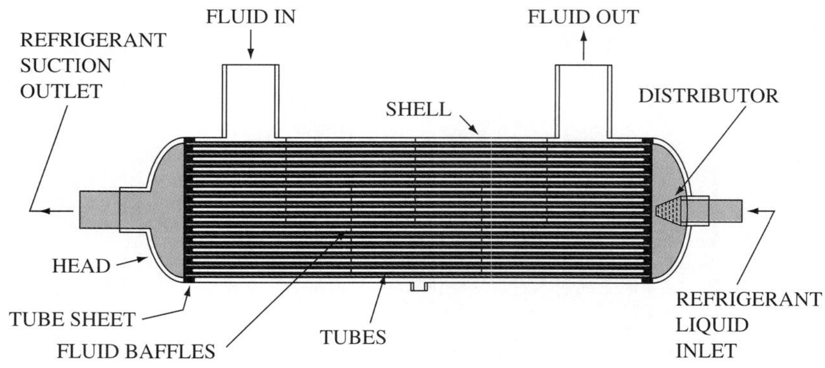

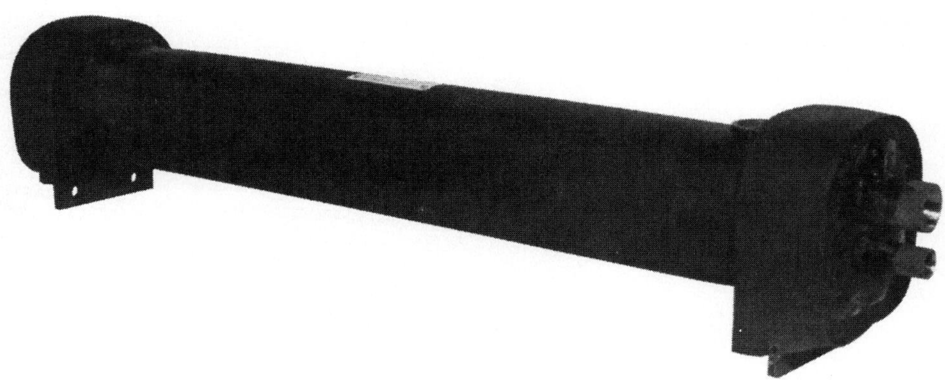

Figure R9-18 Direct-expansion chiller with two refrigeration circuits. (Courtesy of Standard Refrigeration Company)

R9-5 APPLICATIONS

R9-5.1 Bare-Pipe Coils

Steel pipe coils were installed overhead in older ammonia freezer plants. Defrost was manual—the ice was knocked off the pipes. Construction was rugged; it had to be to withstand the abuse.

Fig. R9-19 shows a forced-air evaporator using a bare-pipe coil. This type of unit is used for commercial refrigerated storage rooms. A circulating brine solution is sprayed on the coils to remove frost and increase their cooling capacity.

R9-5.2 Plate-Type Evaporators

A domestic chest (Fig. R9-20) or upright freezer is a good example of a plate-type evaporator. The walls, floor, shelves, and dividers may be plate coils. Heat transfer is effective. Defrost is usually manual. Frost buildup depends on ambient humidity and the number of door openings.

Plate evaporators also come in large economy sizes. Thermal storage systems are designed to make ice late at night and in the early morning when electrical power plants have surplus capacity and metered rates are low. These are known as off-peak plants. Ice is made in ¼-in. thick sheets, released on defrost, and stored in large ice tanks located below the ice maker as shown in Fig. R9-21. During the day ice water is circulated to air conditioning cooling coils, slowly melting the stored ice. These so-called "ice shucker" systems have been encouraged by electric utilities looking to shift some of their load off-peak and avoid building new generating plants.

The ice-making refrigeration may also run during the peak hours, generating chilled water to assist in cooling the building (see Fig. R9-22). This type of system is known as a "load leveler" instead of a "load shifter". There are other types of cooling storage systems using ice, chilled water, and phase change materials.

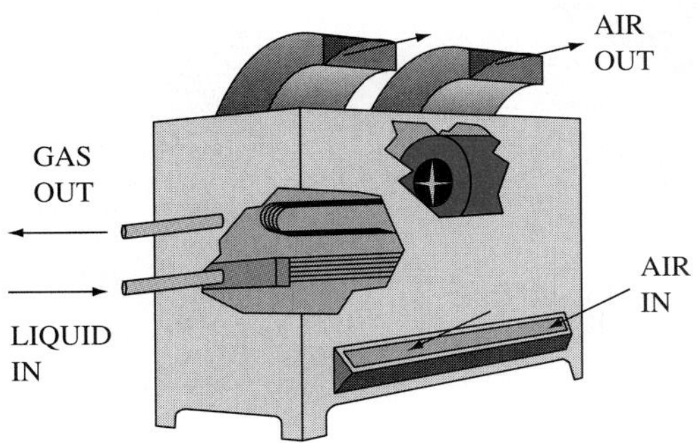

Figure R9-19 Forced-air evaporator with bare tube coils.

R9-5.3 Extended Surface Evaporators—Plate Fin

Direct expansion gravity coils, located in the upper part of the enclosure, chill the air in contact with the coil fins. The cold air is denser and sinks to the bottom of the case, replaced by lighter, more buoyant air. The convection currents provide effective gentle circulation.

The gravity coils are used for applications such as florists' display cases as well as closed-service meat and fish display cases (Fig. R9-23).

R9-5.4 Extended-Surface Evaporators—Refrigerated Forced Air

Spiral-fin, spine-fin, and plate-fin forced-air evaporators pack a lot of performance in a relatively compact space. The plate-fin designation is not to be confused with plate-type evaporators. Plate-fin coils have multiple fins for forced air application.

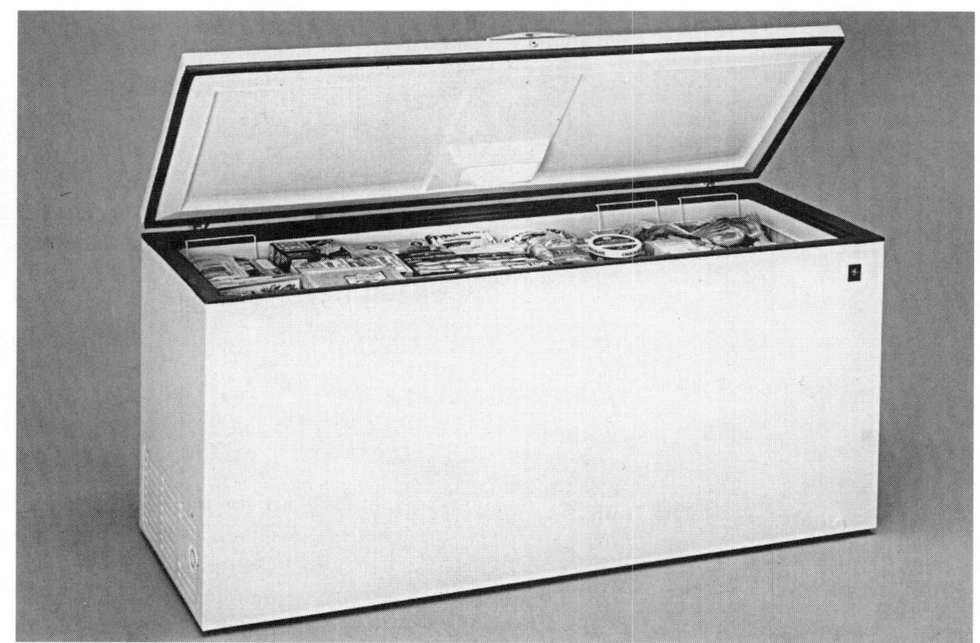

Figure R9-20 Chest-type freezer with plate evaporator coils. (Courtesy of GE Appliances)

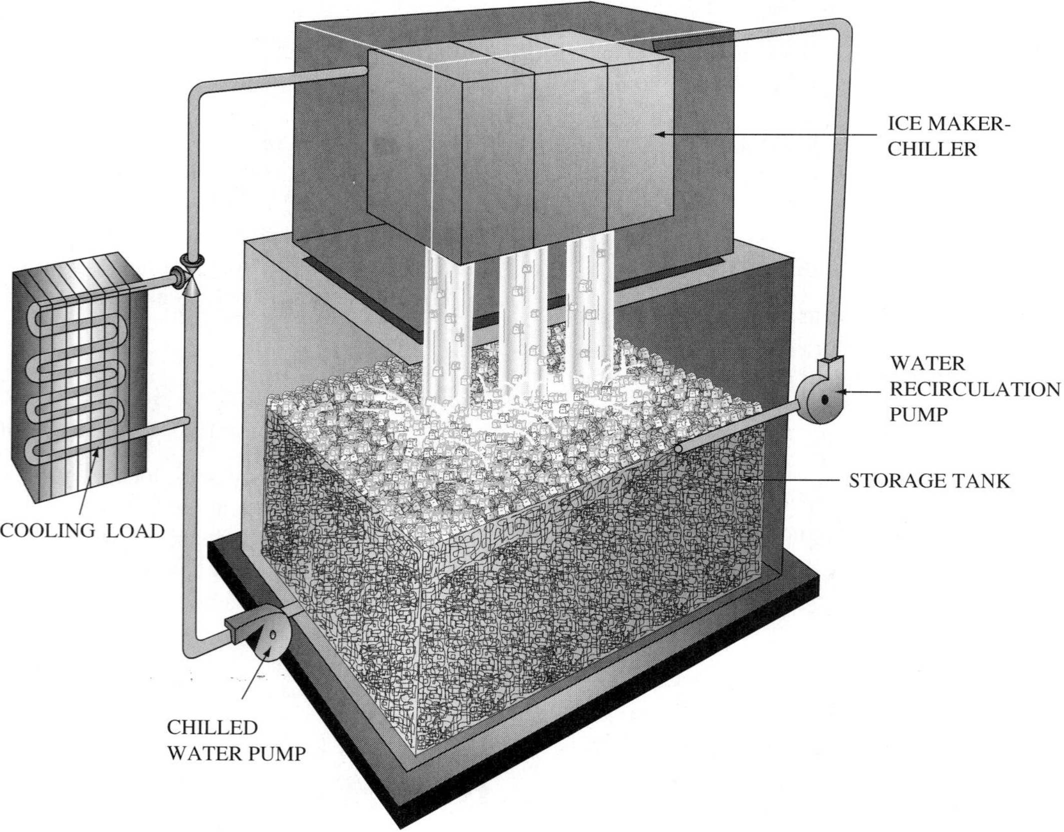

Figure R9-21 Ice that was made during off-peak (electrical) periods is used to provide cooling during peak periods. (Courtesy of Turbo Refrigerating)

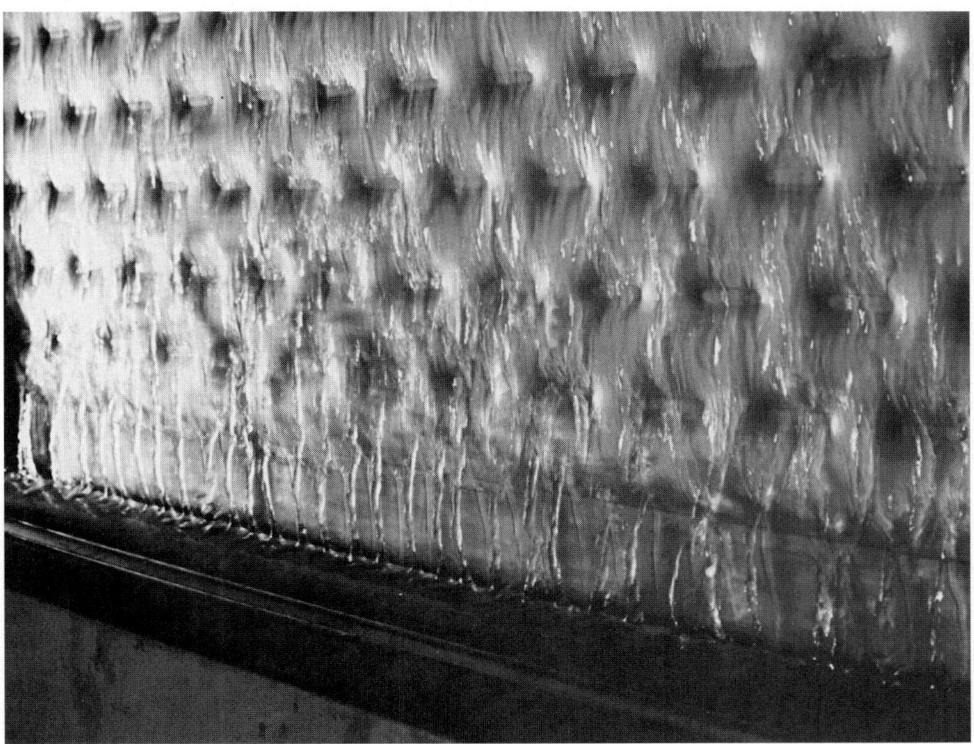

Figure R9-22 Stainless steel plate evaporator. (Courtesy of Turbo Refrigerating)

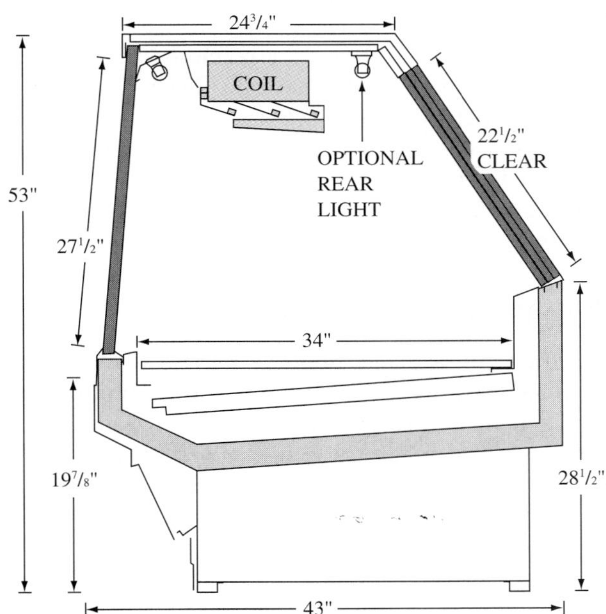

Figure R9-23 Display case with gravity DX cooling coil. (Courtesy of Tyler Refrigeration Corp.)

The following illustrations and drawings (pp. 157–160) will provide an idea of some of the variety of direct-expansion refrigeration and air-conditioning applications for these evaporator coils:

Description	Figure Number
Refrigerated display case	R9-24
Frozen food display case	R9-25
Unit cooler for a reach-in refrigerator	R9-26
Unit cooler for a walk-in refrigerator	R9-27
"A" coil for a residential air conditioner	R9-28
Horizontal fan coil unit for a split system	R9-29

This is just a sample of the great array of applications and types of extended-surface forced-air evaporator coils. (See Chapters R-8 and R-12 for details on refrigerant metering and control devices.)

R9-6 CAPACITY CONTROL

Evaporator coils can be arranged in multiple circuits and the refrigerant flow controlled by solenoid valves in larger systems. Row and face control are two general choices. Face control is usually preferred for better humidity control (dehumidification), as shown in Fig. R9-30 (p. 160).

When unloading face control coils, the lower coil section should always be the first on and last off. This ensures that condensate will not drain over an inactive section of coil and re-evaporate.

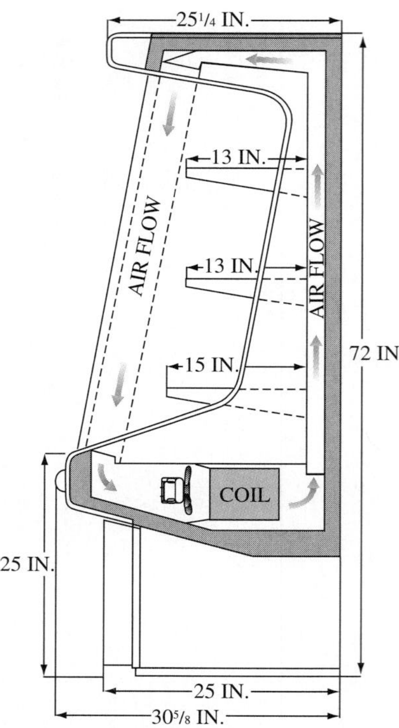

Figure R9-24 Refrigerated display case with forced-air circulation. (Drawing Courtesy of Hussmann Corp.)

R9-7 CONDENSATE

DX coils operating below the dewpoint of the entering airstream will cool the water vapor (humidity) in the air and condense it as droplets on the coil. The droplets drain by gravity into the drain pan and flow through a condensate drain line to a drain or sink disposal (Fig. R9-31, p. 161). The drain line should be provided with a water seal trap and never be connected directly to a sanitary drain.

Condensate on refrigerator-freezers can be drained to a tray under the appliance and evaporated by warm air from the air-cooled condenser (Fig. R9-32, p. 161).

Condensate on package-terminal air conditioners (PTAC) and window air conditioners is drained to a sump, picked up by a slinger on the condenser fan, and hurled onto the condenser coil to evaporate (Fig. R9-33, p. 162).

When condensate cannot be drained by gravity, small pumps with a reservoir and float-operated switch can be used to pump the condensate to a nearby drain (Fig. R9-34, p. 162).

R9-8 TYPES OF DEFROST

While air-conditioning units remove humidity from the air as water, refrigeration and freezer units run colder and will freeze the condensed water as frost or ice on the coil. To allow for some frost buildup, the fins on a refrigeration coil are spaced much wider than an air conditioning coil, ranging from 4 to 8 fins per in., depending on the application and other considerations.

Although the wide fin spacing allows air flow with some frost on the fins, the frost will eventually build up sufficiently to block the airflow. Also, the frost acts as an insulator and inhibits heat transfer. The coil will require periodic defrosting to ensure sufficient airflow and cooling. The frost buildup depends on air conditions and evaporating temperature.

R9-8.1 Air Defrost

Coolers that operate at 28–30°F suction temperature will build up frost on the plate fins. They can be defrosted simply by shutting off the refrigeration and letting uncooled air circulate over the coil (Fig. R9-35, p. 162). Air at 35–39°F will melt the frost. Periodic defrosting can be initiated by a time switch. Air defrosting may go slower than other means.

R9-8.2 Electric Defrost

Refrigeration units running below 28°F with box temperatures below freezing (32°F) will not air-defrost. Electric defrost uses electric resistance heaters to melt ice off the coil, drain pan and drain line. The electric defrost cycle is simple and widely used (Fig. R9-36 and Fig. R9-37, p. 163).

Defrost is usually initiated by a time switch although pressure or temperature sensors also have been used. The liquid refrigeration solenoid valve closes and the compressor pumps down. The fan shuts down. The coil and drain pans defrost heaters activate. The cycle is usually temperature terminated with a backup time termination provision to prevent the box from over-heating should the temperature control fail for any reason. On re-starting the cooling cycle, there may be a fan on delay period.

R9-8.3 Hot-Gas Defrost

Using hot gas available from the refrigeration cycle can be an attractive means of defrost. To successfully implement it in practice requires some care. With a single evaporator, a means is needed to re-evaporate hot-gas refrigerant condensed in the evaporator during the defrost cycle. One such system uses a stored heat reservoir to re-evaporate the refrigerant.

In the refrigeration cycle (Fig. R9-38a, p. 164), the compressor discharges refrigerant vapor, which passes through water-storage-tank heating coils. Some heat is extracted from the hot gas and stored in the water (*A*). The refrigerant passes through a normally open (N.O.) solenoid valve (*G*) to the condenser (*B*). Liquid refrigerant drains from the condenser through a check valve into the receiver (*C*). From there it passes through a filter drier, liquid-suction heat exchanger (*D*), liquid solenoid valve (*K*), thermal expansion valve (*E*), and into the evaporator (*F*). After va-

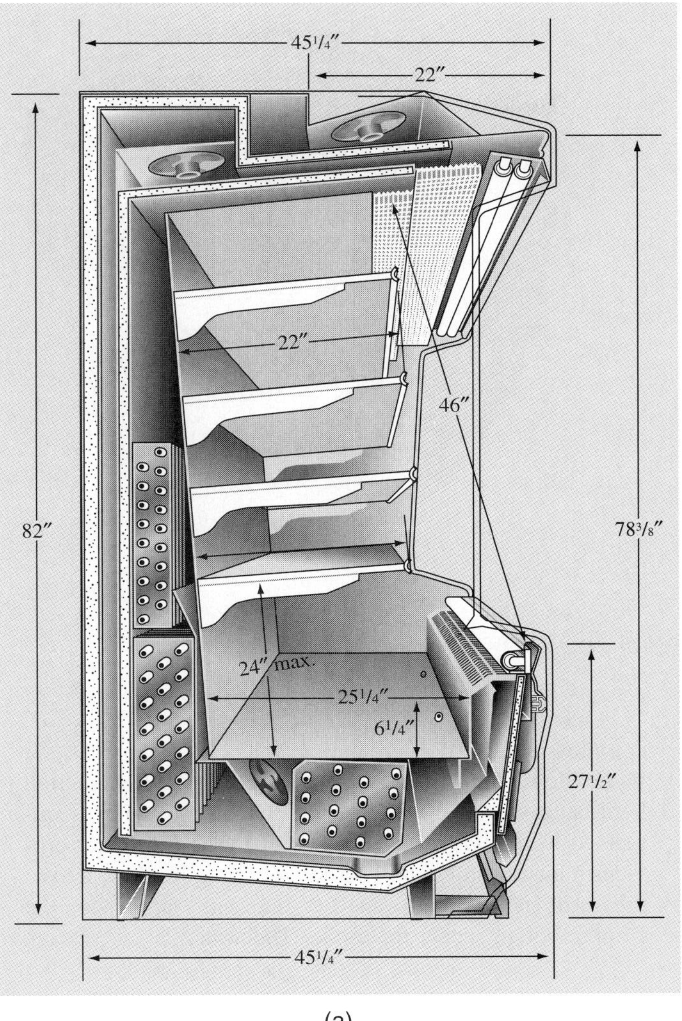

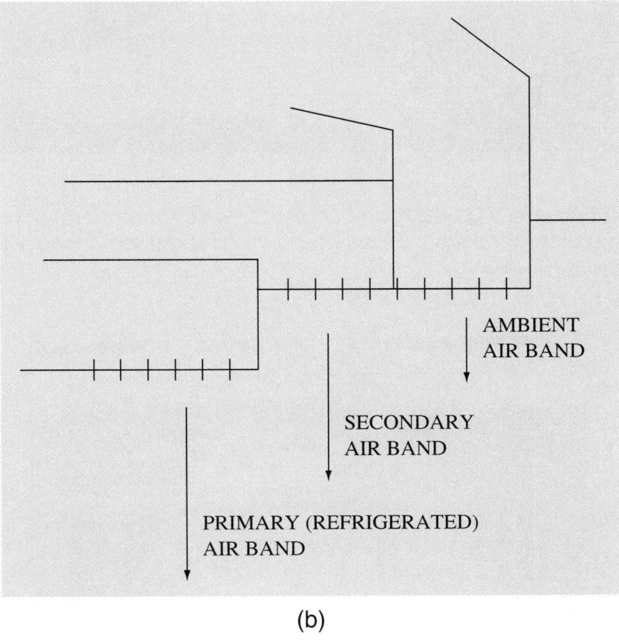

(a) (b)

Figure R9-25 (a) Frozen food display case; (b) the curtain of air blanketing the case opening from top to bottom consists of three bands. (Courtesy of Tyler Refrigeration Corp.)

Figure R9-26 Unit cooler for reach-in refrigerator. (Courtesy of Heatcraft Refrigeration Products)

porizing and picking up the evaporator heat load, the suction gas passes through the liquid-suction heat exchanger (*D*) and through a suction solenoid valve (*L*) back to the compressor.

When the defrost cycle is initiated (Fig. R9-38b, p. 164) the following events occur:

1. Discharge solenoid (*G*) is energized closed.
2. Evaporator fans (*H*) are shut down.
3. Evaporator hot gas solenoid (*J*) is energized open.
4. Liquid solenoid (*K*) is de-energized closed.
5. Suction solenoid (*L*) is closed.

Discharge gas is blocked from the condenser and flows through a spring-loaded check valve (*M*) (15 psi), through the receiver (*C*) and liquid line to the evaporator via the hot gas solenoid (*J*). The hot gas quickly defrosts the drain pan and evaporator coil and condenses to a high-pressure liquid which flows through the suction line toward the

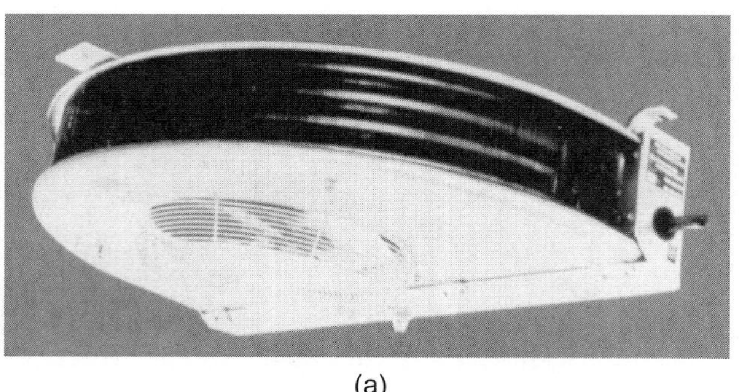

(a)

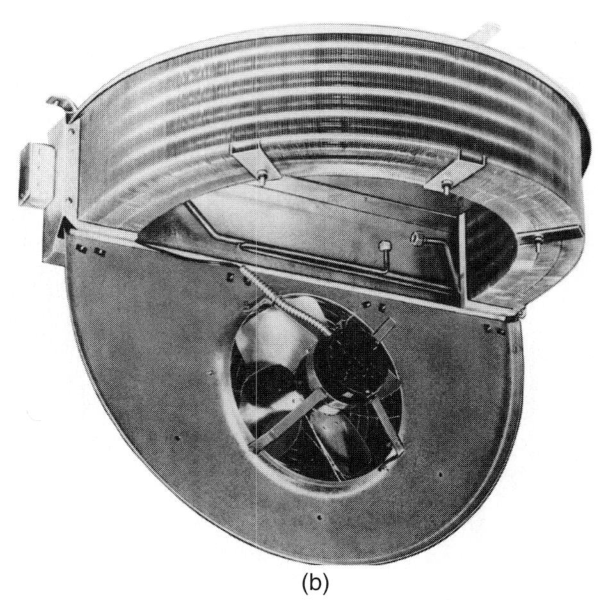

(b)

Figure R9-27 Unit cooler for walk-in refrigerator: (a) blow-thru-type; (b) swing-down arrangement for service access. (Courtesy of Heatcraft Refrigeration Products)

Figure R9-28 "A"-type evaporator for residential air-conditioning unit. (Courtesy of U.S.A. Coil & Air, Inc.)

Figure R9-29 Horizontal-style fan-coil unit for split-system installation. (Courtesy of U.S.A. Coil & Air, Inc.)

compressor. The closed suction solenoid (*L*) blocks the liquid flow, causing it to be metered through a suction hold-back valve into the re-evaporator coil which is immersed in the water storage tank (*A*). The hot water provides a load to safely boil off the returning liquid to a low-pressure gas which can be returned safely to the compressor to repeat the cycle. The water in the storage

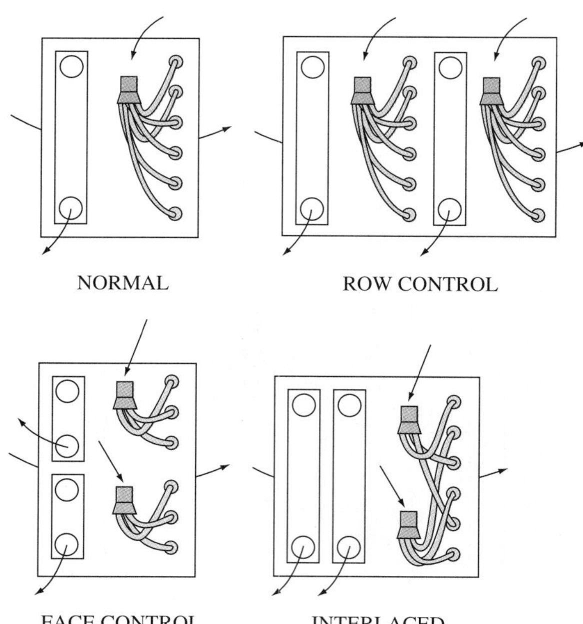

Figure R9-30 Direct expansion coil, row and face control.

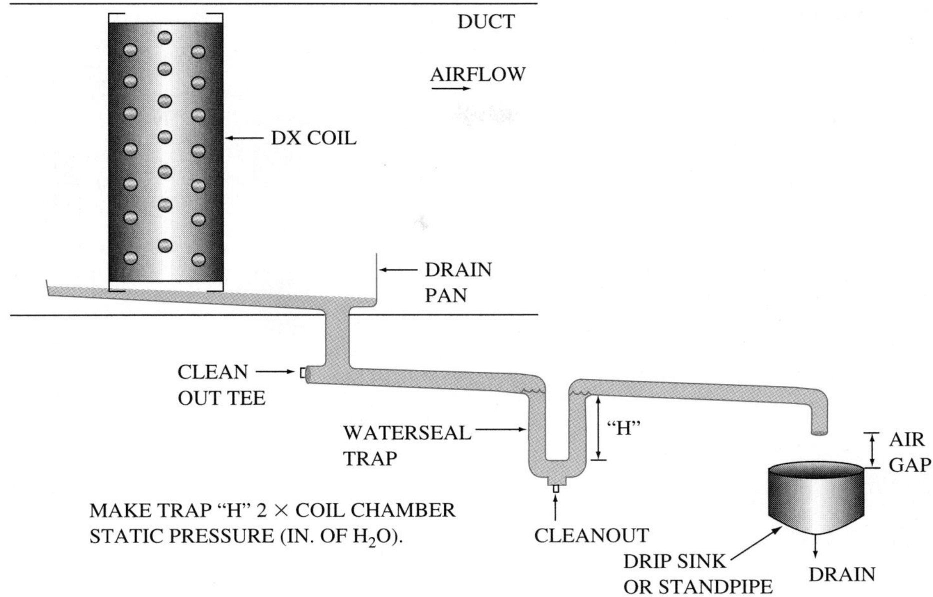

Figure R9-31 Condensate drain pan and piping.

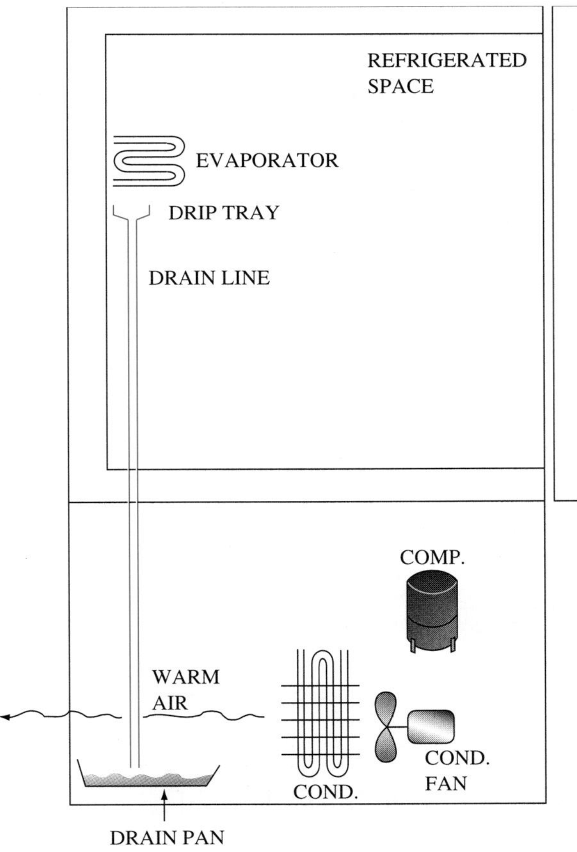

Figure R9-32 Condensate disposal for a refrigerator.

tank will cool and then freeze during the re-evaporation process.

The defrost cycle is terminated when the evaporator pressure rises to a set pressure, causing the discharge solenoid to open to the condenser and the evaporator hot-gas solenoid to close (see wiring diagram, Fig. R9-39, p. 165). The liquid and suction solenoids stay closed and the fans off until the evaporator pressure drops to about 20 psig sensed by the psi-defrost-pressure switch. The cycle is more complicated than electric defrost, but has the advantage of relatively fast defrost times and a saving of electric power.

Other hot-gas defrost arrangements are possible. With two evaporators on a single condensing unit it is possible to defrost one evaporator while continuing to utilize the other as an evaporator. Solenoid valves provide the means of re-routing the refrigerant flow. Still another system uses a reversing valve to swap the evaporator and condenser in a fashion similar to a heat pump. The main requirement of any defrost system is that it be reliable, effective, and as simple as possible for ease of service.

Drain lines from condensate drain pans in freezers must be free and clear to handle the drainage during the defrost cycle. They can be continuously heated by an electric heater cable (see Fig. R9-40, p. 166). They should have a steep pitch (4 in. per 12 in.) for good drainage and be insulated. Line sizes are typically a minimum of $\frac{7}{8}$ in. OD tubing with cleanout tees provided for maintenance.

Heater-cable capacity should be sized in accordance with supplier recommendations. A typical capacity is about 6 watts per lineal foot. It is also possible to run a hot-gas line strapped to the drain line as a heat source.

R9-8.4 Other Means of Defrost

In addition to air, electric, and hot-gas defrost, ice can be melted from evaporator coils by water or water-glycol sprays. These are not as common as the former.

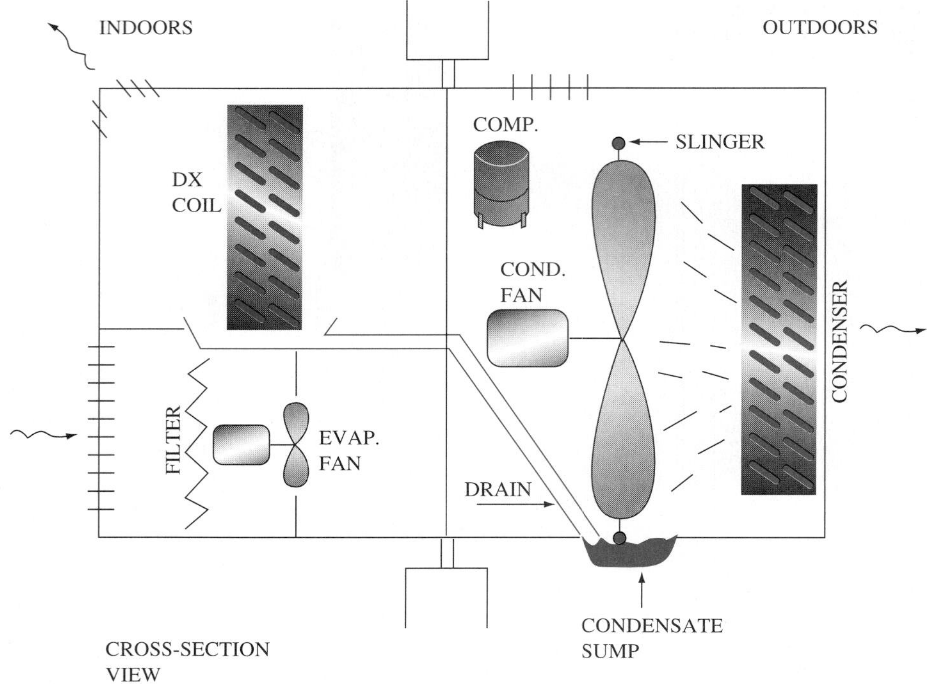

Figure R9-33 Condensate disposal for window air conditioner.

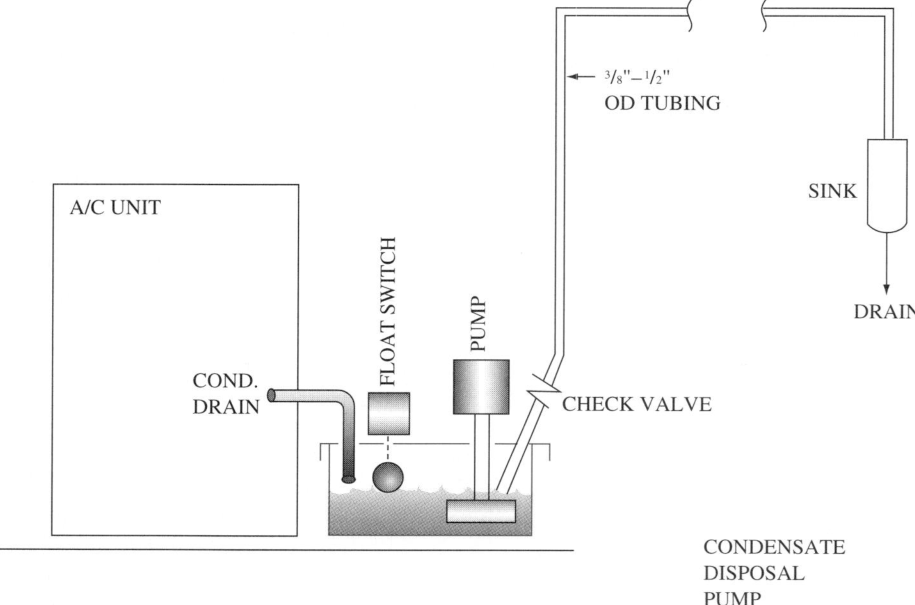

Figure R9-34 Condensate disposal using pump.

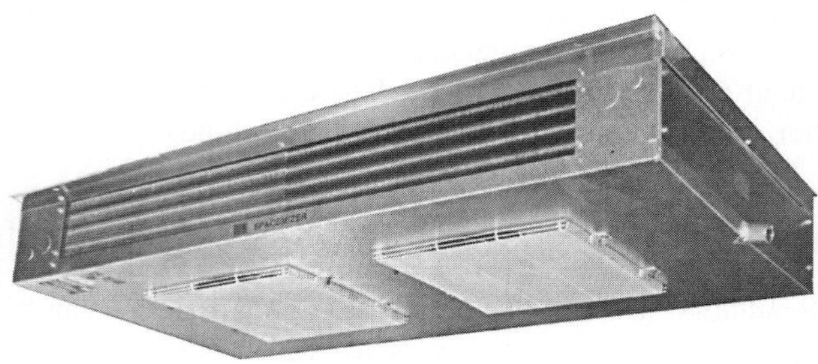

Figure R9-35 Air defrost evaporator unit. (Courtesy of Heatcraft Refrigeration Products)

R9-9 LIQUID COOLERS

Refrigerated liquid coolers can range from small drinking fountains to huge water chillers capable of cooling large buildings.

R9-9.1 Drinking-Water Fountains

The evaporator consists of a small storage tank cooled by an external coil of tubing wrapped around the tank, as

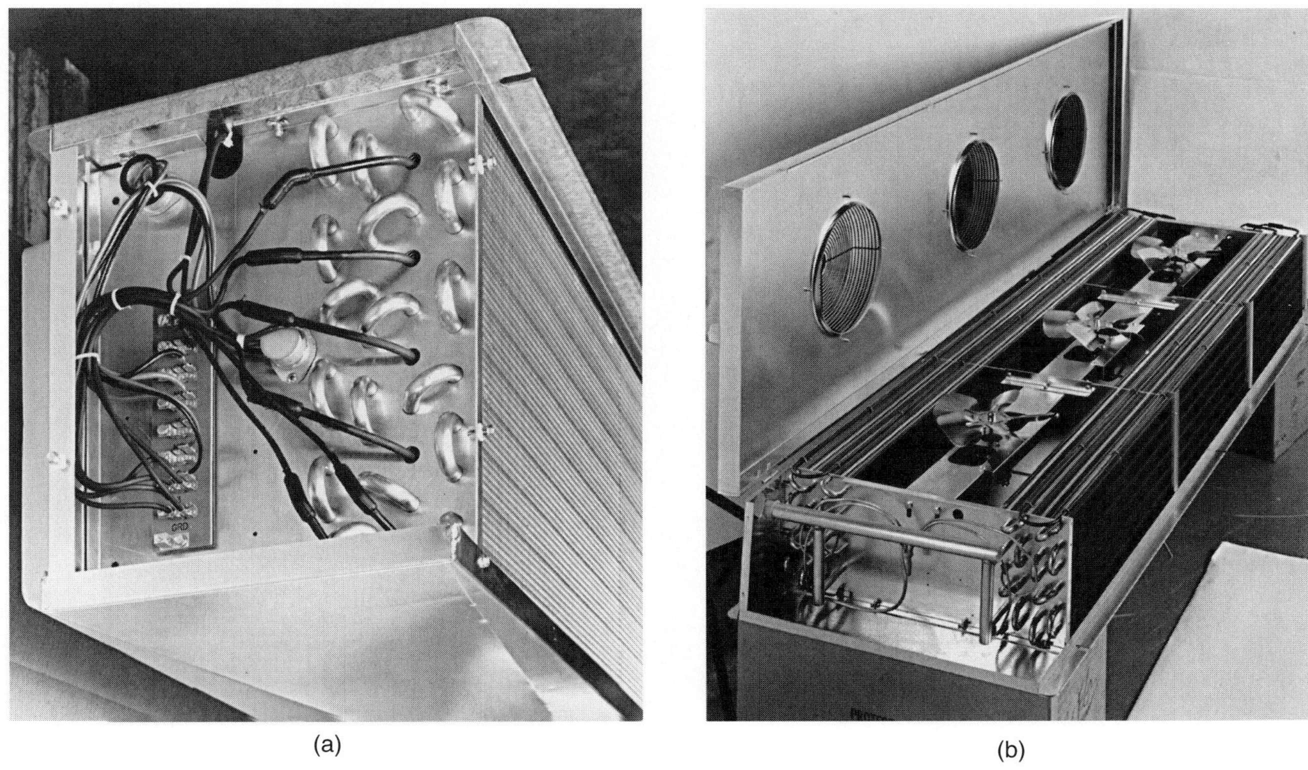

Figure R9-36 Electric defrost evaporator unit: (a) end cover removed; (b) bottom cover removed. (Courtesy of Tyler Refrigeration Corp.)

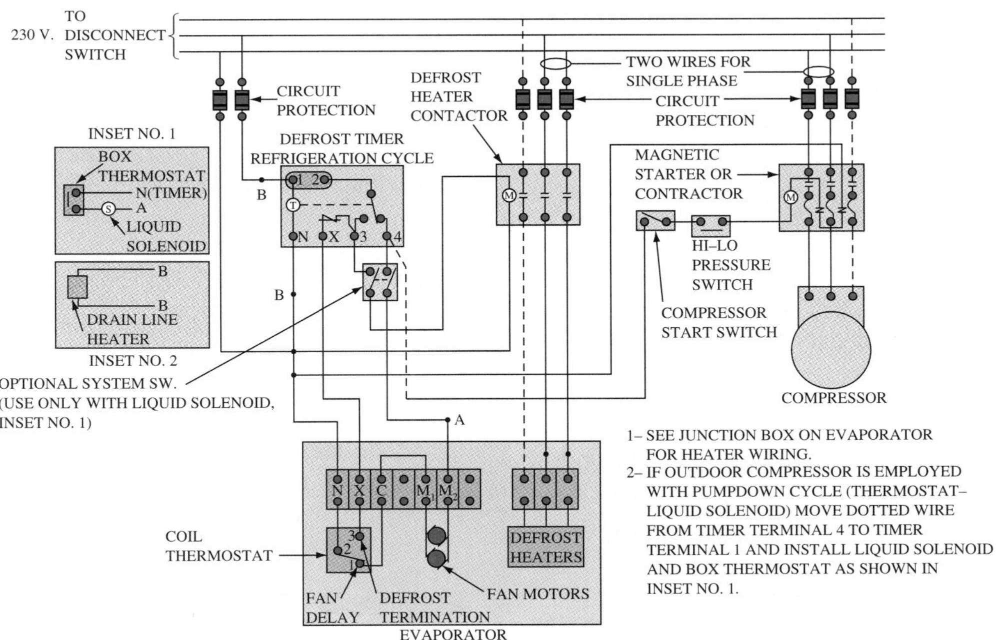

Figure R9-37 Typical wiring for electric defrost unit. Timer initiates, thermostat terminates defrost. (Source: Kramer Trenton)

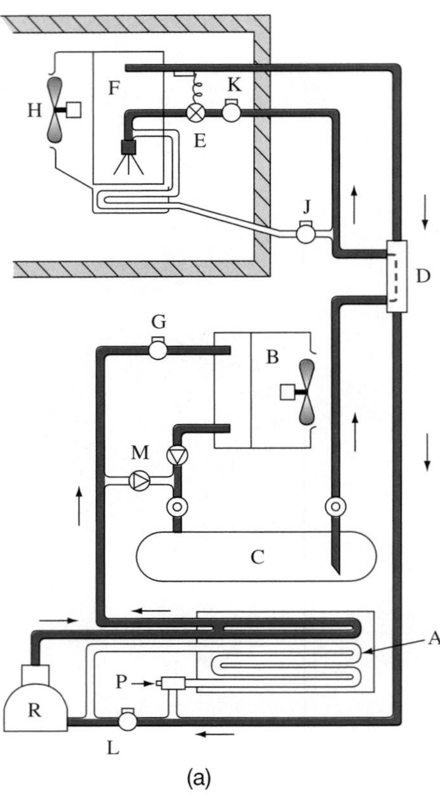

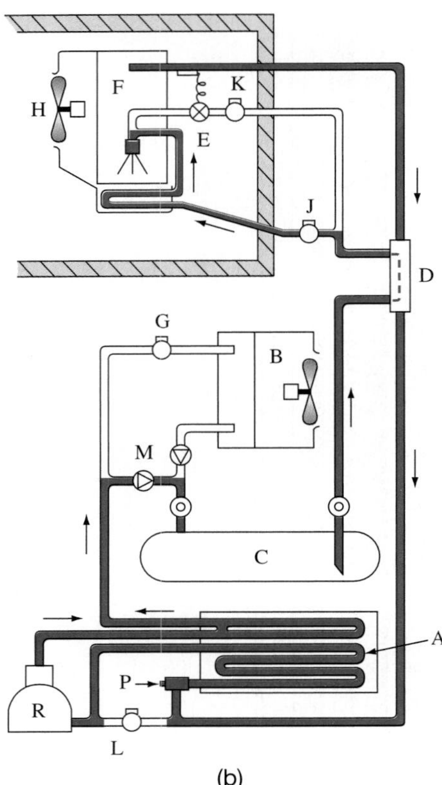

Figure R9-38 Thermal-bank hot-gas defrost system: (a) normal refrigeration cycle; (b) defrost cycle. (Source: Kramer Trenton)

shown in Fig. R9-41, p. 167. Construction can be considered double wall for safety. To enhance capacity, incoming water may be precooled by cold water draining from the fountain.

R9-9.2 Water Chillers

The construction of shell-and-tube water chillers has been covered earlier in this chapter. The most common application is to provide a secondary coolant, namely chilled water.

Chilled water is used in air conditioning for cooling multiple remote air handling units where it would be impractical to run long multiple refrigerant lines. The chilled water can be pumped wherever required through insulated piping (Fig. R9-42, p. 168). The air handlers have water coils instead of direct expansion refrigerant coils. Water coils are usually several rows deeper than an equivalent DX coil to get equivalent cooling performance. They are normally the same size in face area.

R9-9.3 Flywheel Effect

Control of cooling capacity is important. Water flow to small and large cooling coils can be controlled by modulating valves controlled by room thermostats. The chillers can usually be unloaded down to 25% capacity or less. Beyond that, the water circulating in the system provides a reserve load to be cooled and provides a flywheel effect, permitting satisfactory operation during light load conditions without excessive on/off cycling.

R9-9.4 Brine Coolers

Chillers are also used to cool brines (salt-water solutions) and other secondary coolants such as ethylene glycol or propylene glycol-water solutions below 32°F (Fig. R9-43, p. 168). The latter are commonly referred to as antifreeze solutions. These are found in skating rink applications and can also be used to produce ice for thermal storage.

R9-9.5 Chillers—Flooded

Reference has been made to direct expansion (DX) chillers. These are shell-and-tube heat exchangers (evaporators) with refrigerant in the tubes and water filling the shell. For larger capacity systems (several hundred tons and up) it is common to use flooded chillers instead of DX chillers (Fig. R9-44, p. 169). These reverse the fluids, with the refrigerant filling the shell and water (or brine) passing

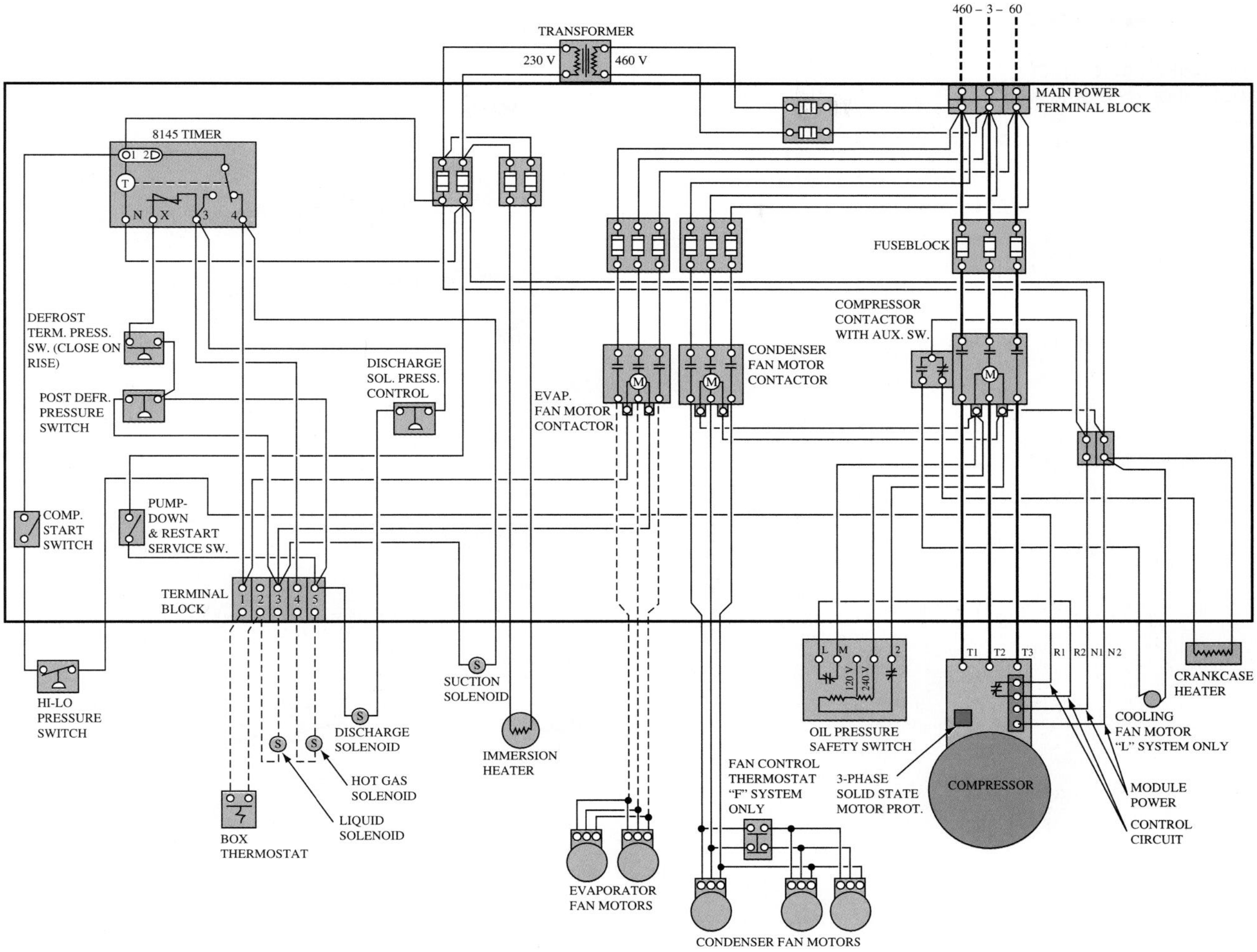

(a)

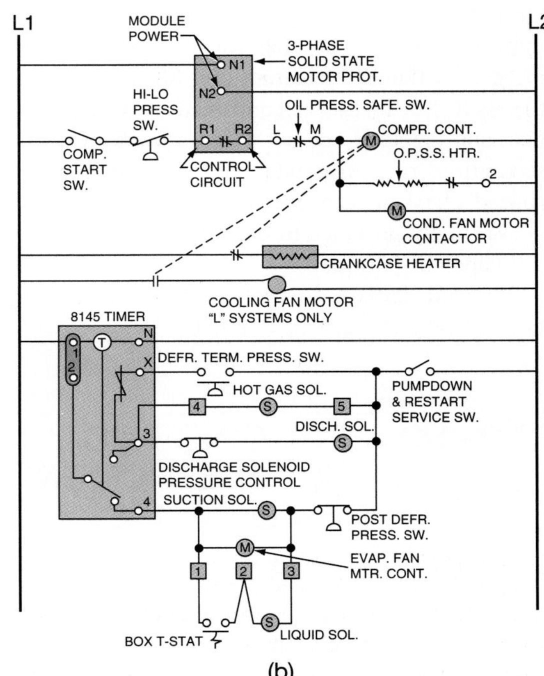

(b)

Figure R9-39 Thermal bank wiring diagram: (a) panel; (b) schematic. (Source: Kramer Trenton)

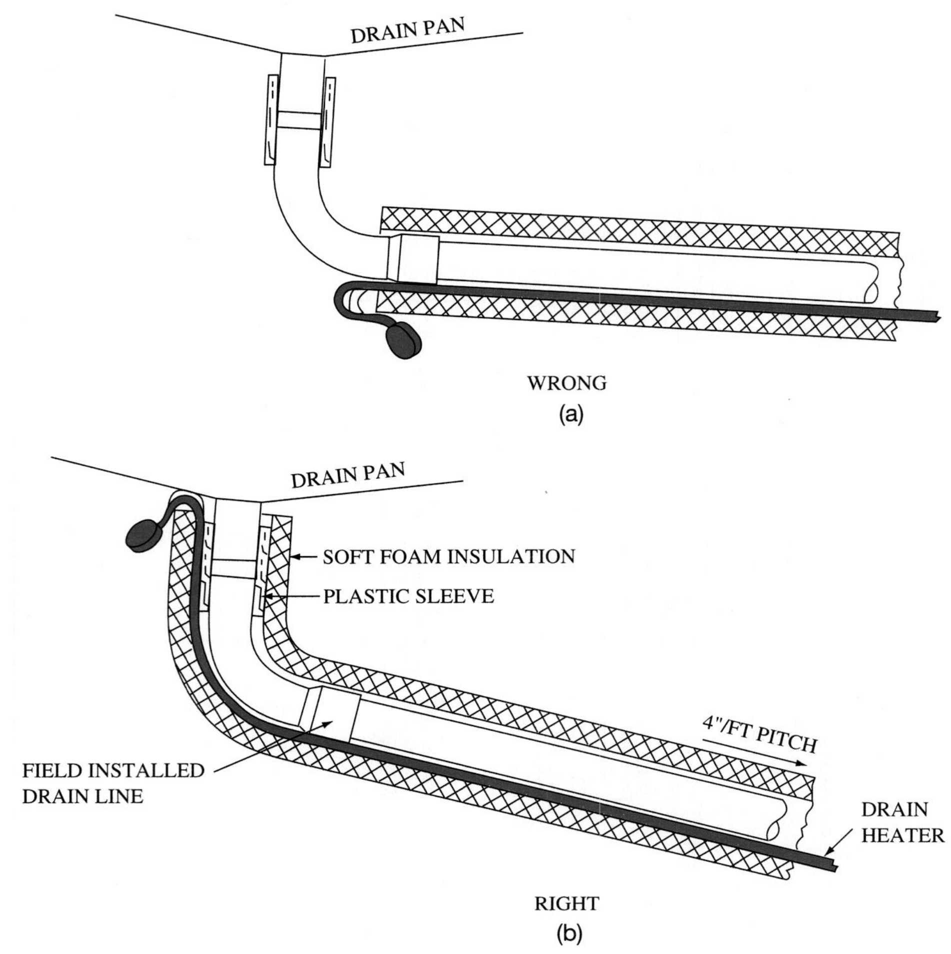

Figure R9-40 Drain line heaters: (a) wrong application; (b) correct application. (Source: Kramer Trenton)

through the tubes. They use enhanced surface tubing for improved refrigerant heat transfer. The tubes are submerged in liquid refrigerant and fill the lower portion of the shell. Refrigerant is controlled by a float valve or orifice, and the shell can act as a surge chamber, permitting separation of refrigerant vapor from liquid. Closely spaced chevron-type eliminator baffles or mesh screens are normally provided to ensure satisfactory separation of dry vapor from liquid.

Flooded chillers do not superheat suction vapor, and provide very efficient heat transfer.

R9-10 SERVICING AND TROUBLESHOOTING EVAPORATORS

Evaporator problems can result in unsatisfactory operation and possible compressor failures, so the technician will want to be alert to these problems. Listed below are some of the problems and their possible causes:

Air Cooling Evaporators

Problems	Possible Causes
Insufficient cooling (Insufficient air flow)	Plugged air filters or dirty coil Loose or broken fan belt Fan blades plugged with dirt Fan running backwards Coil iced (from any of above) Slipped or closed damper
Insufficient cooling (Airflow normal)	Refrigerant charge low Filter-drier plugged Plugged capillary tube Compressor tripping off on safety Damaged compressor valves Thermal expansion valve, loss of charge TXV valve gas binding due to flash gas in liquid line Frozen water in TXV valve, wet system

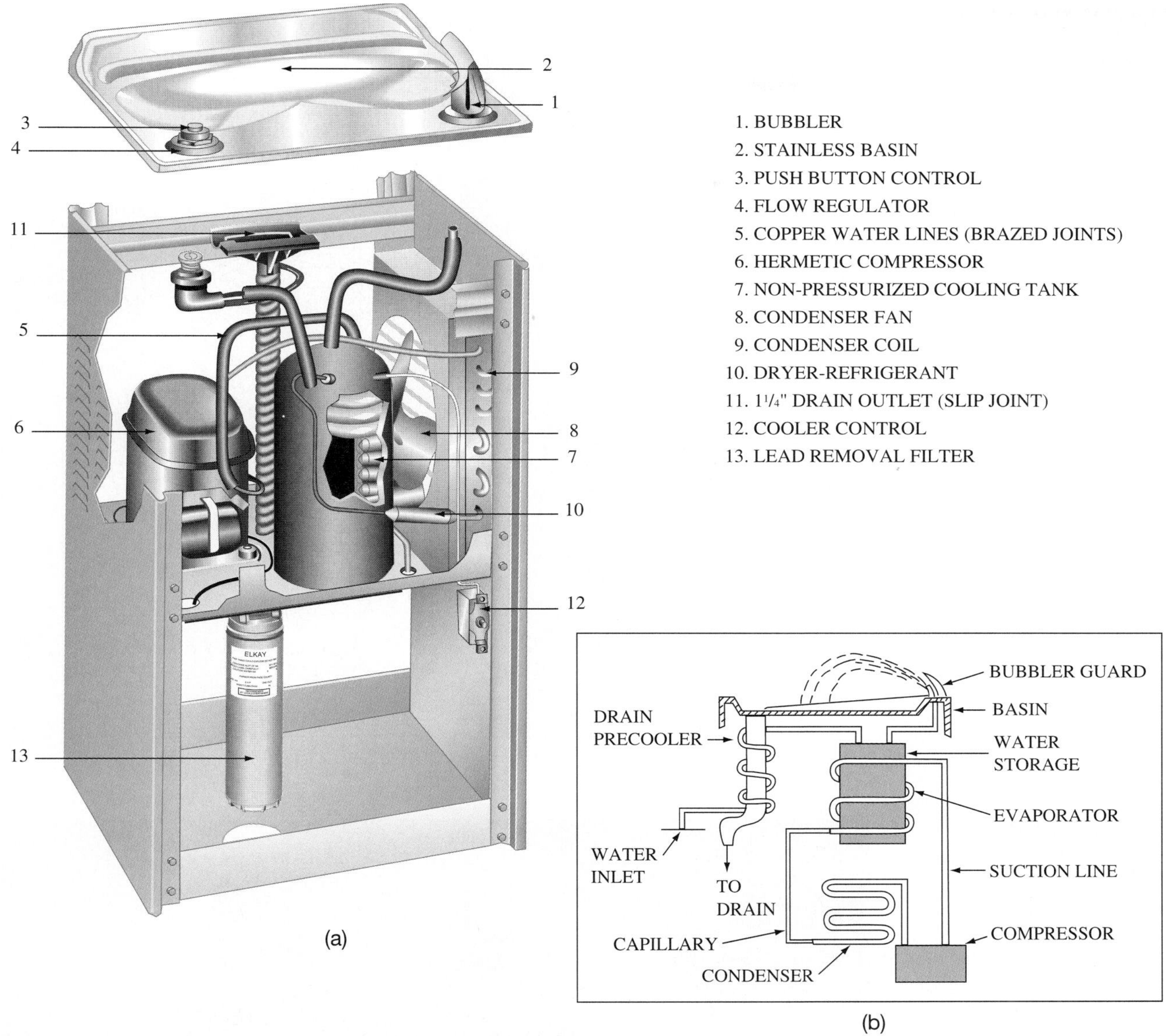

Figure R9-41 Self-contained drinking water cooler: (a) internal construction; (b) refrigeration piping. (a, Courtesy of Elkay Manufacturing Company; b, Copyright by the American Society of Heating, Refrigerating, and Air-Conditioning Engineers, Inc. Used by permission.)

Coil flooding liquid to compressor	TXV valve jammed open by foreign material (solder, slag, etc.) TXV valve bulb loose or making poor contact with suction line
Odors	Check for mold and mildew Clean coil, pan and drain lines

All of the above items assume a properly sized and installed system that has performed satisfactorily in the past.

If a high-efficiency (SEER) air-conditioning condensing unit has been installed and the old evaporator coil left in service, the coil may be inadequate, as the newer units are designed to run at higher evaporating temperatures and need more evaporator coil surface to achieve a satisfactory balance. Manufacturers recommend changing the DX coil when a new high-efficiency unit is installed, to ensure a proper equipment match.

Another hazard is having someone improperly charge the wrong refrigerant into a system. The technician should be aware of recent repairs made to the system.

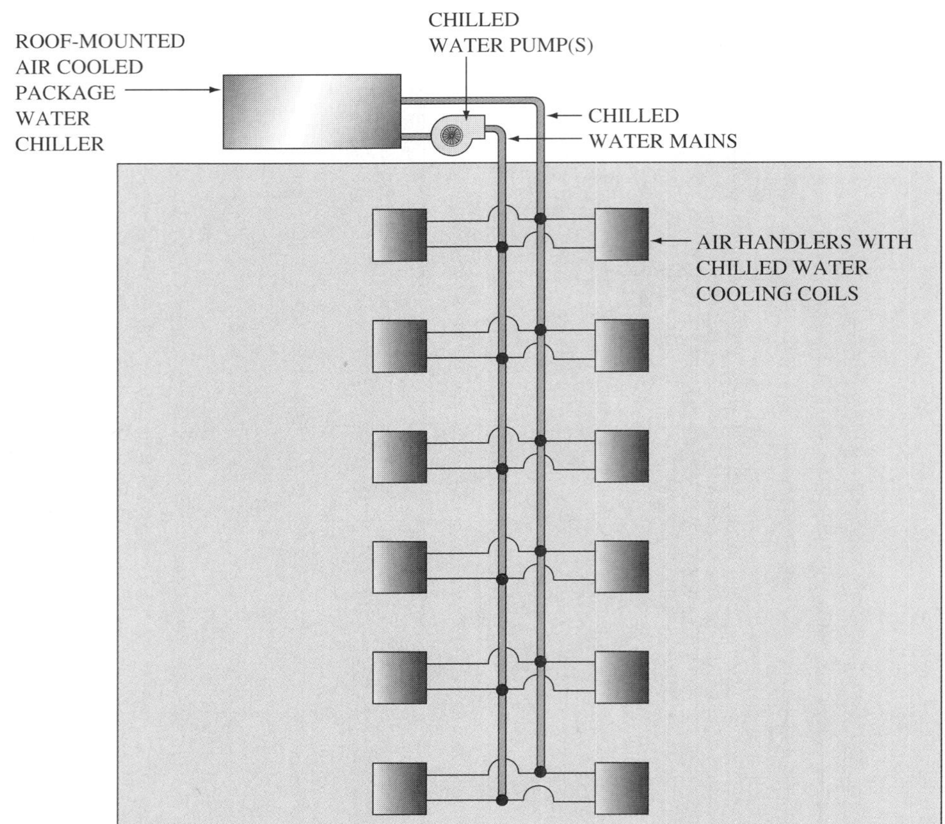

Figure R9-42 Roof-mounted, air-cooled packaged water chiller.

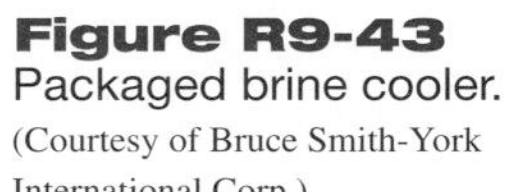

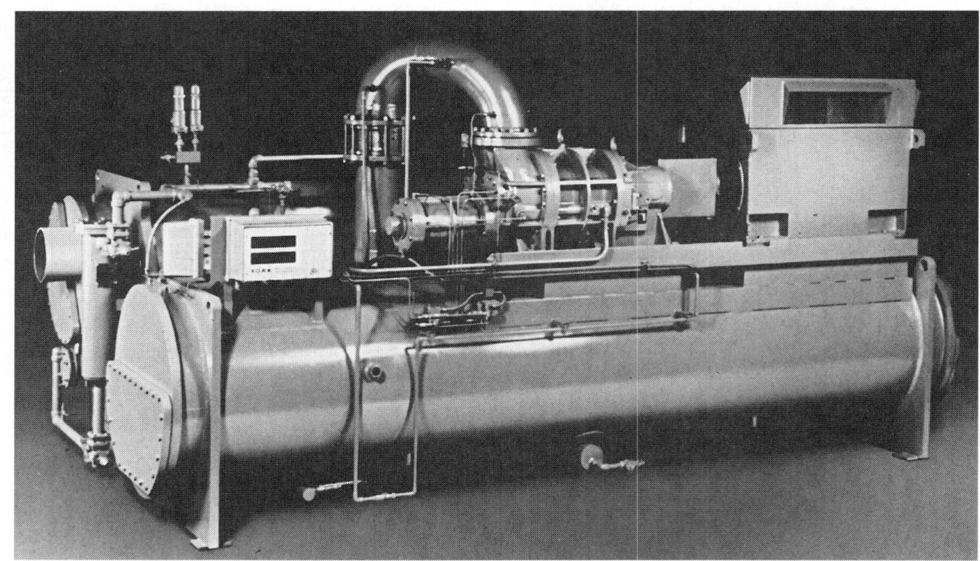

Figure R9-43 Packaged brine cooler. (Courtesy of Bruce Smith-York International Corp.)

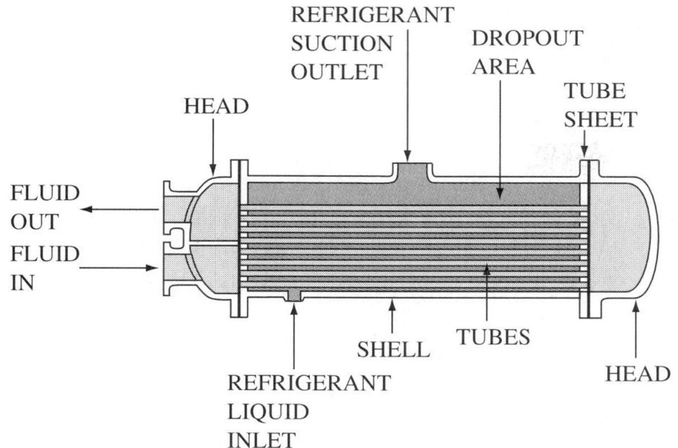

Figure R9-44 Flooded chiller; water in tubes, refrigerant in the shell. (Copyright by the American Society of Heating, Refrigerating, and Air-Conditioning Engineers, Inc. Used by permission.)

Problems	Possible Causes
Insufficient cooling (Low water flow)	Plugged pump strainers Plugged chiller, system piping, and/or coils due to scale/muck/rust System low on water or air bound Valve closed or throttled improperly
Insufficient cooling (Water flow normal)	Refrigerant charge low Filter-drier plugged Compressor tripping off on safety Damaged compressor valves Thermal expansion valve, loss of charge TXV valve gas binding due to flash gas in liquid line

When servicing chillers care should be taken to avoid possible damaging freezeups. Make sure:

1. Flow interlocks are properly wired so the chiller cannot operate until water flow is established.
2. Chilled water thermostat is correctly set. Usual minimum exiting water temperature is about 44°F.
3. Chiller freezestats, low-suction pressure cutouts and low refrigerant temperature cutouts are set as recommended by manufacturer to prevent freezeups.
4. Do not bypass any safety or operating controls.
5. If the chiller is exposed to freezing temperatures:
   a. Make sure electrical heaters and controls are functioning. If possible, protect secondary coolant by adding an antifreeze to lower the freeze point.
   b. At shutdown, drain chiller and piping. Circulate with antifreeze to ensure no unprotected pockets of water remain.

REVIEW

- The evaporator is a heat exchanger with refrigerant contained within tubes, passages, or a vessel.
- Heat flows from the product into the refrigerant within the evaporator, and the heat begins to evaporate the refrigerant.
- Types of evaporators include either the direct expansion or the flooded design. The direct expansion can be either 1.) bare pipe, 2.) extended surface, or 3.) plate. The flooded design is normally used with larger water-chiller vessels and industrial ammonia applications.
- Flash gas occurs at the outlet of the metering device, reducing the efficiency of the evaporator. It can be caused by a restricted filter-drier or a rise in the liquid line which lowers the pressure.
- Liquid subcooling increases cycle efficiency and can help avoid performance problems while helping to reduce the level of flash gas.
- Superheating is the additional heating of the refrigerant gas (constant pressure) above saturation temperature.
- As product load diminishes, the evaporator will be running at lower suction pressure, temperature, and capacity.
- Coil-fin spacing directly affects evaporator performance. The closer the fins, the greater the performance per square foot of coil.
- Counterflow is the term used with heat exchangers to indicate that two fluid streams are flowing in opposite directions, which produces the greatest heat transfer capacity.
- Evaporators remove humidity from the air as water. Refrigeration and freezer units which operate with suction temperatures below freezing will accumulate the moisture as frost. The frost acts as an insulator to the heat transfer process and must be defrosted periodically.
- Purpose of defrost is to remove frost from the evaporator to improve heat transfer, and to return oil to the compressor. Heat applied during defrost drives refrigerant and oil out of the evaporator and returns the oil down the suction line with refrigerant vapors to the compressor. The oil will drop out in the suction port of the compressor and return to the crankcase. If the oil were allowed to remain in the evaporator it would take up space that should be used for a boiling refrigerant and refrigeration efficiency would decrease.
- Defrost methods can be either air, electric, hot-gas, or water/glycol mixtures.
- Defrost is usually initiated by a time clock mechanism, although pressure or temperature sensors are also used. Defrost is terminated through either coil temperature, suction pressure, or time.

- Drain lines from condensate drain pans in freezers are continuously heated by electric heater cable, typically with a capacity of 6 watts per lineal foot.
- Water chiller construction is made of a shell-and-tube design, typically flooded on large industrial applications. The refrigerant is in the shell and the water or brine is in the tubes.
- Smaller commercial shell-and-tube water chillers that use direct expansion have the refrigerant in the tubes and water in the shell.
- Whenever high-efficiency condensing units are installed, manufacturers recommend changing the indoor DX coil to a high-efficiency model.
- Refrigeration effect is the amount of heat each pound of refrigerant picks up from the product being cooled and the air as it travels through the evaporator.
- Capacity of a direct-expansion coil is dependent upon:
  The temperature of the refrigerant being circulated.
  The temperature of the air (dry bulb and wet bulb) being circulated through the coil.
  The volume of the air being circulated.
- Capacity (Btu/h) of an evaporator depends upon:
  Surface area.
  "U" Factor
  METD (Mean Effective Temperature Difference)
- Factors to consider when selecting an evaporator include:
  Btu/h load requirements.
  Evaporator design temperature.
  METD (Mean Effective Temperature Difference)
  Refrigerant
  Air flow patterns
  Physical size and mounting
- Pressure-temperature relationships—saturation temperature can be determined by comparing the evaporator design temperature with the suction pressure converted to saturated temperature using a (P/T) chart. Condensing temperatures can also be determined using this process.

Problems and Questions

1. The simplest and most frequently used evaporator is called a _____ evaporator.
2. In a parallel evaporator, the refrigerant flow is divided into _____ or more _____ as it flows through the evaporator.
3. On multiple circuit evaporators with TXV, a _____ nozzle may be used to direct the flow of refrigerant to the parallel paths.
4. Why are fins or splines added to the tubing of evaporators?
5. How is heat transferred from the air entering the evaporator to the refrigerant? Why?
6. What are some of the methods manufacturers use to speed up the transfer of heat at the evaporator?
7. Why do evaporators require defrosting?
8. What is meant by METD?
9. What will happen to an evaporator if its fan fails?
10. What is the saturated evaporator temperature of a system using R-134a operating at –5°F? How cold will the air be at design temperature?
11. List the types of evaporator defrost.
12. What is the purpose of defrost cycles?
13. Explain how to correctly size and select an evaporator.
14. What is the best way to clean an evaporator?
15. Explain how to check performance of an evaporator.
16. What selection criteria must be considered when selecting an evaporator?
17. What is the saturated temperature of R-12 at zero degrees? How does this compare to R-401a?
18. Explain the significance of temperature difference in the diagnosis of evaporators and refrigeration systems.
19. What could happen if the seals around evaporator covers are not properly installed or removed?

Please select the letter that best represents the correct answer:

20. What causes heat to flow into an evaporator?
    a. Difference in pressures
    b. Boiling refrigerants
    c. Difference in temperature
    d. Saturated refrigerant vapor
21. Which of the following is NOT a type of evaporator?
    a. Plate
    b. Bimetal
    c. Extended surface
    d. Bare pipe
22. What is the most common material used to fabricate air-coil tubes/fins?
    a. Aluminum/copper
    b. Steel/steel
    c. Copper/copper
    d. Copper/aluminum
23. Why do direct-expansion evaporators superheat the suction gas?
    a. To protect the compressor
    b. To protect the evaporator and meter device
    c. To increase efficiency
    d. To reduce compressor discharge superheat
24. What is commonly used to control refrigerant flow to a flooded evaporator?
    a. Capillary tube
    b. Constant pressure valve
    c. Float valve
    d. Thermal expansion valve
25. Counterflow piping for heat exchangers provides increased capacity because it:
    a. Increases face velocity
    b. Increases oil return
    c. The METD (mean effective temperature difference) is minimized.
    d. The METD is increased.

26. Off peak refrigeration plants:
    a. Reduce electrical demand during off peak hours
    b. Run at minimum capacity during night and early morning
    c. Are limited to ammonia refrigerant applications
    d. Store cooling capacity for daytime use
27. Condensate drains should NOT be:
    a. Connected directly to a sanitary drain or vent
    b. Trapped
    c. Provided with an air gap
    d. Piped with cleanout tees in place of elbows
28. During electric defrost of a freezer box evaporator:
    a. The fans run continuously
    b. The fans are shut off
    c. The compressor runs continuously
    d. The time switch is bypassed
29. Hot-gas defrost is:
    a. Slower than air defrost
    b. Higher on energy cost than electric heaters because the compressor is running
    c. Generally fast
    d. Considered to be the most suitable system for 36°F spaces

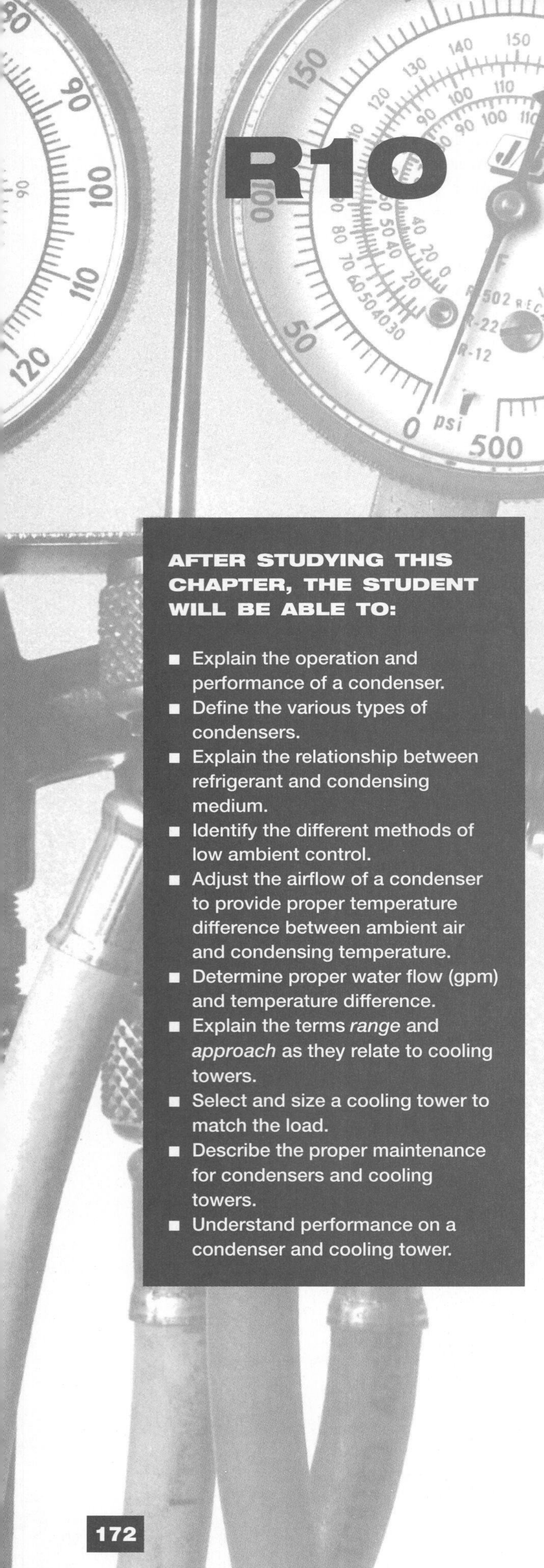

R10 Condensers

AFTER STUDYING THIS CHAPTER, THE STUDENT WILL BE ABLE TO:

- Explain the operation and performance of a condenser.
- Define the various types of condensers.
- Explain the relationship between refrigerant and condensing medium.
- Identify the different methods of low ambient control.
- Adjust the airflow of a condenser to provide proper temperature difference between ambient air and condensing temperature.
- Determine proper water flow (gpm) and temperature difference.
- Explain the terms *range* and *approach* as they relate to cooling towers.
- Select and size a cooling tower to match the load.
- Describe the proper maintenance for condensers and cooling towers.
- Understand performance on a condenser and cooling tower.

R10-1 OPERATION OF CONDENSERS

The *condenser* is located on the discharge side of the compressor, as shown in Fig. R10-1. The hot refrigerant vapor enters the condenser from the compressor and leaves the condenser as subcooled liquid refrigerant.

The function of the condenser is to transfer heat that has been absorbed by the system to air or water. In an air-cooled condenser the outside air passing over the condenser surface dissipates the heat to the atmosphere. Using a water-cooled condenser, the water is pumped to a cooling tower where the heat is transferred to the atmosphere by means of evaporation.

The pressure-heat diagram shown in Fig. R10-2 illustrates the action performed by the condenser. The hot discharge gas from the compressor enters the condenser at point 3. First the superheat is removed; then the vapor is condensed; then the liquid is subcooled until it reaches point 4.

The table shown in Fig. R10-3 summarizes the action that takes place in the condenser. This information is based on using a typical air-cooled condenser with R-22 refrigerant and 95°F air passing over the outside surface of the condenser. The condensing temperature is 120°F. The vapor enters the condenser at 165°F and the liquid leaves at 105°F. About 14% of the heat is removed as superheat, 81% of the heat is removed by condensation, and 5% of the heat is removed by subcooling. This entire action takes place in the condenser.

Even though the subcooling accounts for only a small part of the total heat rejection, it is important for two reasons:

1. It ensures that a solid stream of liquid will enter the metering device.
2. It adds to the cooling capacity of the system at a rate of about 0.5% of the total cooling capacity per degree of subcooling. For example: with 10°F of subcooling, 5% (.5% × 10°F) additional capacity is added to the system.

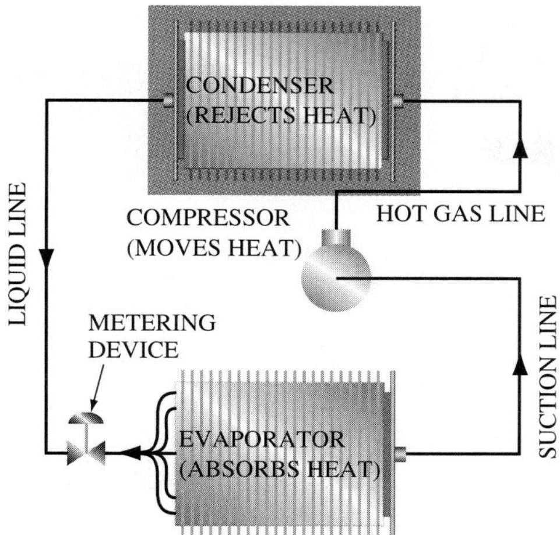

Figure R10-1 Refrigeration cycle showing the position of the condenser.

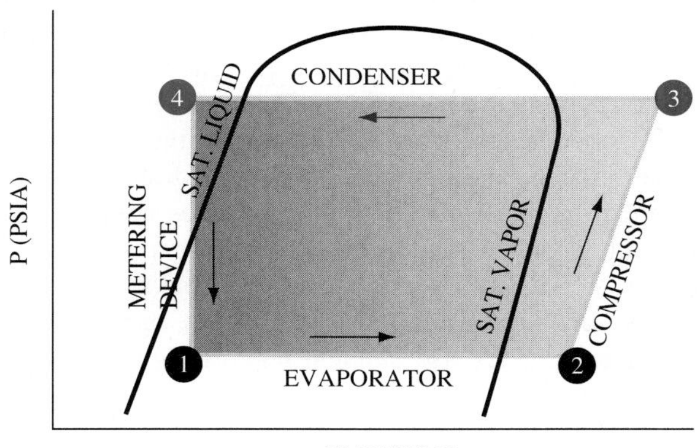

Figure R10-2 Pressure-enthalpy (heat) diagram showing the heat removed by the condenser.

	AT CONDENSER INLET	AT CONDENSER OUTLET
SATURATION TEMPERATURE (F)	120	*120
ACTUAL REFRIGERANT GAS TEMPERATURE (F)	165	105
SUPERHEAT (F)	45	0
SUB-COOLING (F)	0	15
PRESSURE (PSIA)	277.7	*277.7
PRESSURE (PSIG)	263.0	*263.0
ENTHALPY (BTU/LB)	125	42

*ASSUMES NEGLIGIBLE PRESSURE LOSS THROUGH HOT GAS LINE AND CONDENSER.

Figure R10-3 Typical effect of an air-cooled condenser on a system using R-22, outside temperature 95°F.

R10-2 TYPES OF CONDENSERS

Three types of condensers are used in HVAC/R systems (Fig. R10-4):

1. air-cooled
2. water-cooled
3. evaporative, which is a combination of the other two.

Actually, a condenser is a very useful application of a heat exchanger. The heat picked up from various sources within the system is expelled by means of the condenser. Heat exchangers are made of metal to permit fast and efficient heat transfer. The hot refrigerant vapor is in contact with one side of the heat-exchanger surface and the transfer medium such as air or water is on the other side.

The application of an air-cooled condenser is usually the simplest arrangement, particularly if the condenser is located outside the system or unit. Although the water-cooled condenser is more efficient, it is more costly to install. Evaporative condensers are mainly used on industrial applications. The examples that follow demonstrate some of the common uses of condensers.

R10-3 AIR-COOLED CONDENSERS

Air-cooled condensers eject heat to the outdoors. At normal peak load conditions, the temperature of the refrigerant in the condenser is 25° to 30°F higher than ambient temperature. This means that on a 95°F day the condensing temperature is between 120° to 125°F, as shown in Fig. R10-5.

With a water-cooled condenser, the temperature difference between the medium and the refrigerant is lower. For example, it is common for the water cooled unit to use

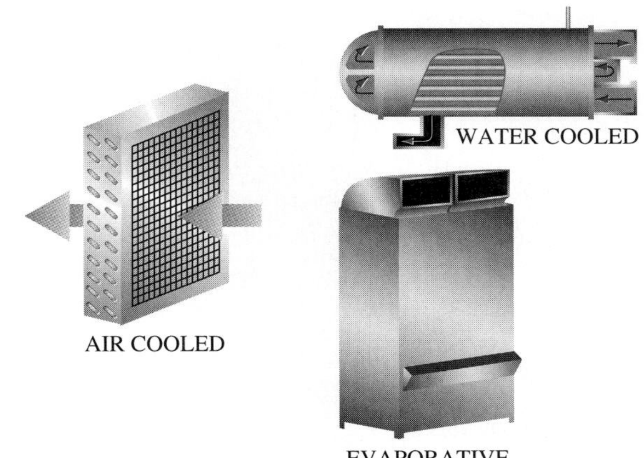

Figure R10-4 Three types of condensers.

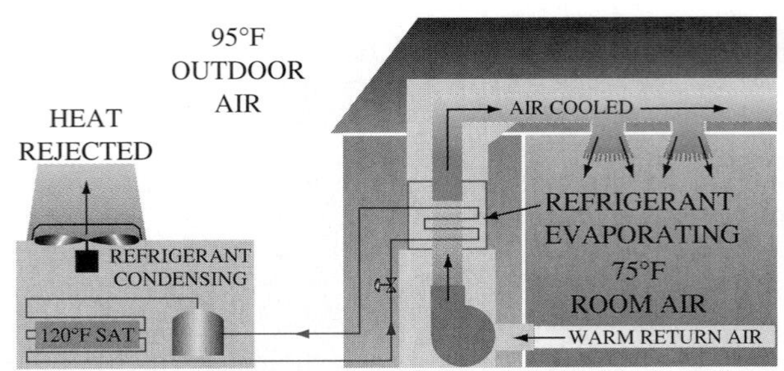

Figure R10-5 Condenser used on a residential split air-conditioning system.

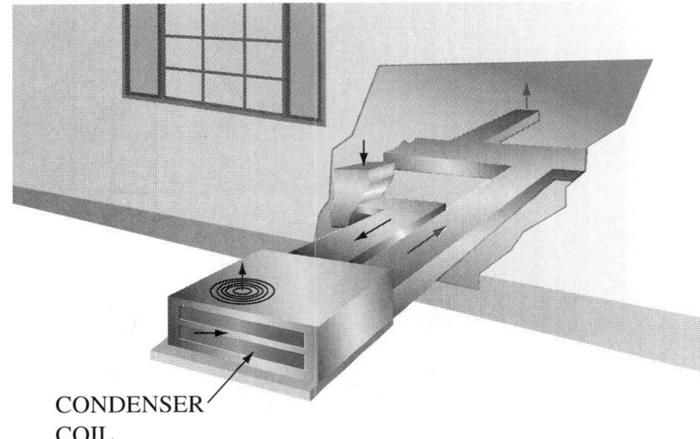

Figure R10-7 Condenser on a residential packaged air-conditioning system.

85°F water and operate at 105°F condensing temperature. This makes the water-cooled unit more energy efficient, but its cost is higher.

The air-cooled units are simple and easy to install and maintain. They are therefore commonly used for most residential air-conditioning systems up to 5 tons and commercial systems up to about 50 tons in capacity.

One common use of the air-cooled condenser is on domestic refrigerators and upright freezers, as shown in Fig. R10-6. The condensers are located in the lower part of the cabinet and a small fan is used to pull room air over the surface of the condenser coil and exhaust it back out into the room. Some refrigerators use a natural-draft-type condenser coil on the back of the cabinet.

Serious compressor problems can result if the condenser gets dirty. The dirt will reduce the heat transfer rate and the compressor head pressure can rise to damaging levels. The condenser can usually easily be cleaned with a brush and vacuum cleaner.

Another application of the air-cooled condenser is in the residential packaged air-conditioning unit, shown in Fig. R10-7. These are available for central air-conditioning in sizes from 1½ to 5 tons of cooling capacity. These units can also include provisions for a system of warm air heating or for cooling—only where the source of heating is hot water.

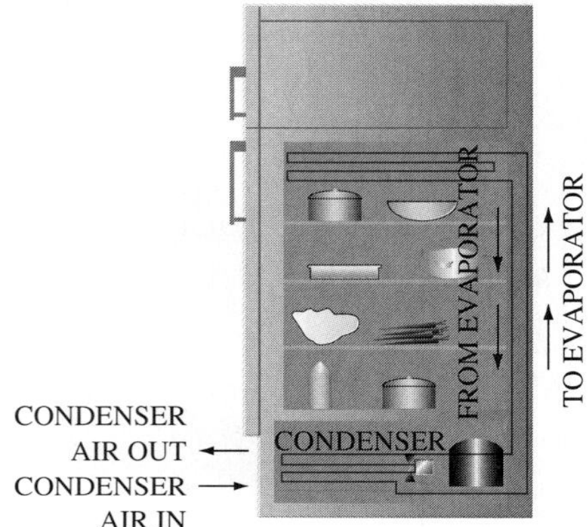

Figure R10-6 Condenser on a domestic refrigerator.

Packaged room coolers and through-the-wall air conditioners both use air-cooled condensers. A typical unit of this type is shown in Fig. R10-8. These units are manufactured in sizes ranging from ½ to 3½ tons of cooling capacity. They are commonly used in houses, apartments, town houses, condominiums, offices, schools and motels.

The residential condensing unit is a popular use of air cooled condensers. This unit includes the compressor, condenser, fan, and controls which are located outdoors, as shown in Fig. R10-9. Refrigerant lines connect to an evaporator coil located on the top of a forced-air furnace indoors or to a fan coil unit. Due to the fact that these air-conditioning systems consist of two parts, they are called *split systems.* They use ambient air to remove heat from the condenser.

Fig. R10-10 shows a commercial field-assembled air-conditioning unit with a remote air-cooled condensing unit. The portion of this system inside the building is primarily an air-handling unit, including the evaporator, the metering device, and a blower-filter unit. The portion on the outside includes the compressor, condenser, and condenser fan. It can be labeled a split system because it is similar to the residential split systems, except larger.

Fig. R10-11 shows an air-cooled condenser with a centrifugal fan. Condensers of this type can be located within the building and use duct work to exhaust the heat-laden air to the outside. This can provide a considerable advantage in keeping the refrigerant line connections to the other parts of the system as short as possible.

All of the systems shown in Figs. R10-5 through R10-11 use some type of air-cooled condenser.

R10-3.1 Air-cooled vs. Water-cooled Condensers

Air-cooled condensers are used on most residential air-conditioning systems and on commercial systems up to 20 tons in capacity. They are even used on some commercial

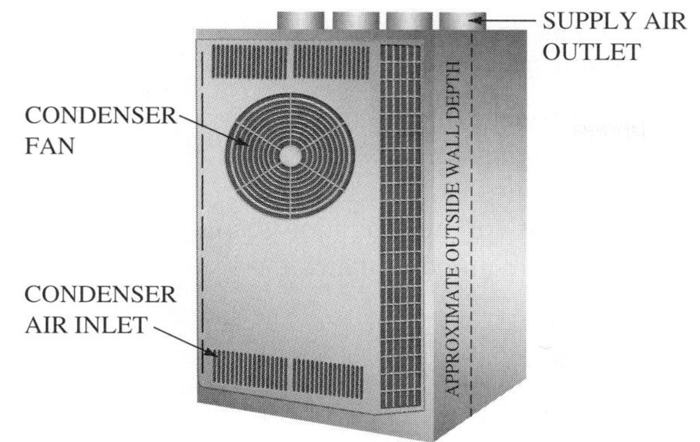

Figure R10-8 Condenser on a through-the-wall air conditioner.

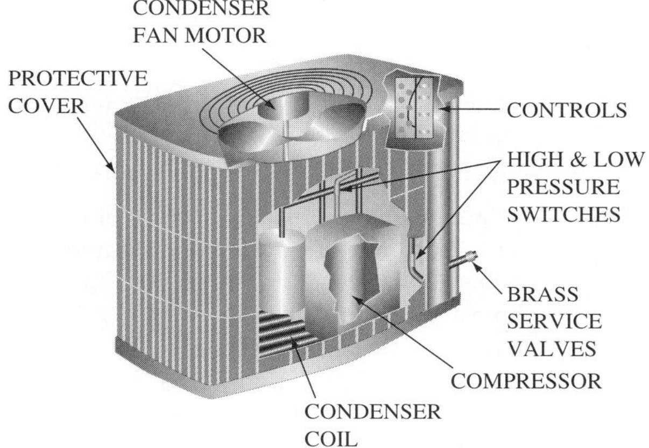

Figure R10-9 Condenser on a residential outside condensing unit.

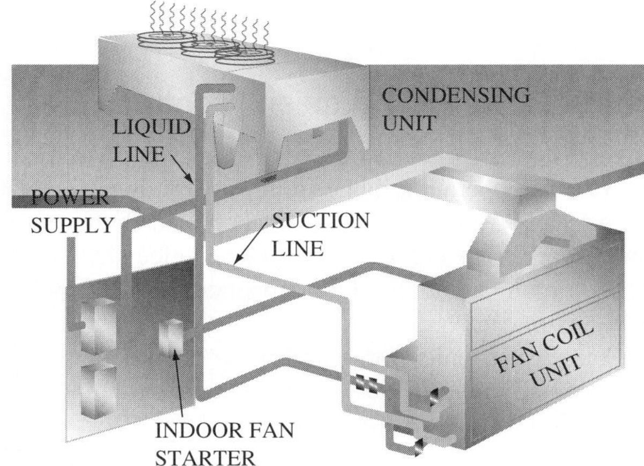

Figure R10-10 Commercial condenser on a field-assembled air-conditioning system.

systems up to 100 tons. The reason for their popularity is their simplified application and low maintenance requirement. It is necessary to run the compressor discharge line to their location and the refrigerant liquid line back to the evaporator. This presents no problem if the distance is relatively short.

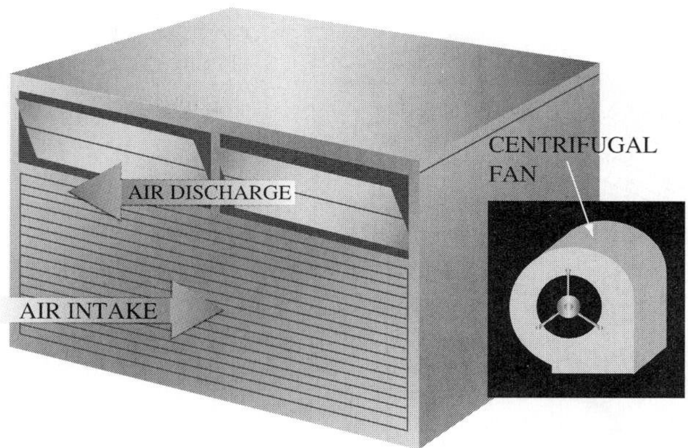

Figure R10-11 Air-cooled condenser using a centrifugal fan.

When a water-cooled condenser is used, a cooling tower needs to be added to conserve water. This requires a constant supply of makeup water as well as water treatment to prevent corrosion and scaling, as well as the formation of algae. A tower must include a pump which also requires service and maintenance. Periodically the tower needs to be thoroughly cleaned, and even the condenser tubes may collect deposits that must be removed. There is also the danger of water freezing in the piping during cold weather unless it is drained or receives special treatment.

Air-cooled condensers operate at higher condensing temperatures on design days than water cooled condensers; however, the advantage in using air-cooled condensers is that they are cost-effective, considering all factors.

The air-cooled condenser can be installed as a separate component, as shown in Fig. R10-12. With this arrangement the compressor is placed inside the building. The alternate is to purchase a packaged condensing unit with the compressor and condenser both in one package as shown in Fig. R10-13. This is called an air-cooled condensing unit.

The advantages of air-cooled condensing units as compared to separate air-cooled condensers are as follows:

1. Compressor noise and vibration is outside of the building.
2. The complete condensing unit can be factory built and tested.
3. Less valuable building space may be used.

R10-3.2 Selection of Air-cooled Equipment

To select an air-cooled condenser, the following information is needed:

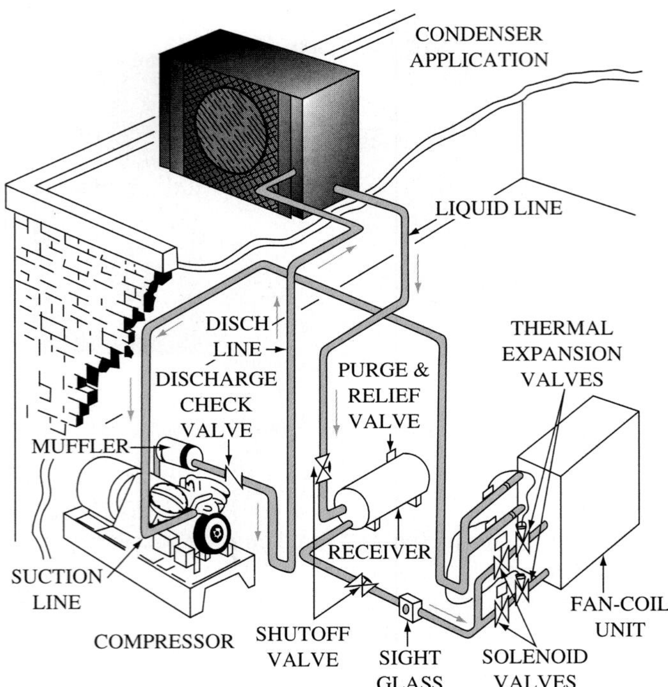

Figure R10-12 Remote-mounted air-cooled condenser with all other refrigeration cycle components inside the building.

1. The temperature difference between design ambient air temperature and condensing temperature. This should range between 20° and 25°F.
2. The total heat rejection based on published ratings. This allows for the total cooling load plus the heat of compression. The total heat rejection for an air cooled condenser of refrigeration is 14,500 to 18,000/Btu/ton or higher, depending upon whether the system is used for low-temperature refrigeration, commercial refrigeration or air conditioning.
3. The refrigerant being used.

The following example illustrates the method of selecting an air cooled condenser.

EXAMPLE

Given the following requirements: a nominal 20 ton semi-hermetic condenser system, using R-22, would have a total heat rejection of 293,000 Btu or 290 MBH, when operating at 40°F evaporating temperature and 120°F condensing temperature. Select a condenser.

Solution

Fig. R10-14 shows a typical rating table for air cooled condensing units. Using the table, the nearest standard unit meeting these requirements would be a Model BRH031 with a capacity of 298 MBH at 20°F TD. This is a direct-drive, vertical-airflow unit.

To select an air cooled condensing unit, the following information is needed:

1. The refrigerant being used.
2. The design ambient temperature.
3. The design suction temperature.
4. The tons of refrigeration capacity (at 12,000 Btu/ton) required to match the load.

EXAMPLE

An air-cooled condensing unit, using R-22, is required to supply 10 tons of refrigeration capacity at a suction temperature of 40°F. The design ambient temperature is 95°F. Find the model number of the unit and the kW of power required, using the capacity shown in Fig. R10-15, p. 179.

Solution

Referring to the table, a Model RCU-010T is suitable, with an input of 13.4 kW.

This unit is supplied with the following accessories:

1. Head pressure control valve, capable of operation to −20°F, outside ambient temperature.
2. Receiver with heater to maintain 90°F during compressor-off periods.
3. Compressor crankcase heater.
4. Unloader solenoid valve to reduce compressor capacity 50%, controlled by suction pressure. Optional accessories include hot gas bypass solenoid, and hot gas regulator for additional step of capacity reduction. A drawing showing the arrangement of these parts is given in Fig. R10-16, p 180.

R10-3.3 Low-ambient Control

Condensing units for commercial refrigeration have an approximate operating range from +35°F down to −20°F when matched with appropriate evaporators. Outdoor conditions vary from 115°F down to zero degrees and below. Refrigeration piping for low-temperature work also requires extra care.

Condensing units for comfort air-conditioning duty are nominally rated at 95°F outdoor ambient temperature to function with evaporators operating at 40°F with incoming air at 67° WB. However, there is an increasing need for comfort air conditioning to operate when outside ambients fall below 75°F. High internal heat loads from people, lights and electronic equipment may demand cooling when outside conditions go down to 35°F and below.

Where possible, an economizer providing "free" cooling should be used if outside air at the proper temperature is available. If low-ambient conditions are to be encoun-

PRACTICING PROFESSIONAL SERVICES

Always clean condenser coils with the condenser fans and compressor electrical circuits turned off. Remember to practice lock-out/tag-out procedures for safety purposes. When working with cooling towers and evaporative condensers, turn off the compressor and pump circuits. Again, lock-out/tag-out. If the condenser surface appears abnormally dirty with scale, corrosion, algae or other water problems, consult a water-treatment expert for proper chemical analysis and treatment.

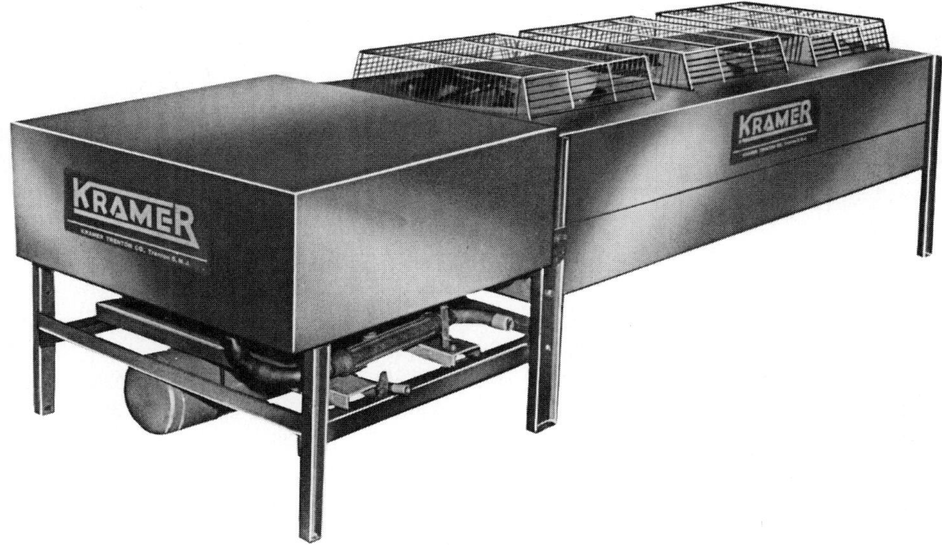

Figure R10-13 Typical commercial type air-cooled condensing unit. (Source: Kramer Trenton)

tered, proper controls need to be provided to enable operation.

In the case of both commercial refrigeration and air conditioning, low-ambient operation is an important need. In earlier discussions we learned that in order to have proper refrigerant vaporization in a direct-expansion (DX) evaporator coil, it was necessary to maintain a reasonable pressure differential across the expansion device. In a normal air-cooled condenser operating between 80 and 115°F ambient, condensing pressures are sufficiently high, but in winter they can drop 100 psi or more; thus the pressure across the expansion device may be insufficient to maintain control of liquid flow. Evaporator operation becomes erratic. The thermal expansion valve will alternately open and close, first causing flooding back of liquid refrigerant and then starving the coil when the valve closes. The capillary tube (if used) is worse, because it is a fixed metering device, and as pressure difference falls, the flow of refrigerant is severely reduced. Below outdoor air temperatures of 65°F, a capillary-tube system is in real trouble.

The solution to low-ambient operation is to maintain a minimum head pressure in order to assure proper feeding of refrigerant to the evaporator. There are several different ways of doing this.

Where multiple condenser fans are employed on a single coil, the control system of the condensing unit can be equipped with devices to switch or cycle "off" the fans in stages. These controls are usually air stats that sense outdoor ambient temperature or pressure controls that sense actual head pressure. As fans are turned off the airflow across the coil is reduced. Condensing temperatures therefore rise. All fans can be turned off, and the coil acts as a static condenser with only atmospheric air movement. Its capacity at this point is usually ample to continue operating below freezing conditions. Where only one condenser fan is used, air capacity can be reduced by employing a two-speed motor or solid-state speed controls, which have infinite speed control.

Another technique of restricting airflow across the condenser coil is the use of dampers on the fan discharge. These are used on nonoverloading centrifugal type fans, not propeller fans. Dampers modulate from a head pressure controller to some minimum position, at which time the fan motor is shut off.

One common and important characteristic of the airflow restriction methods is that the full charge of refrigerant and entrained oil is in motion at all times, ensuring positive motor cooling and oil lubrication to the compressor.

BRH Model	Fan Config.	R-404A, R-502 and R-507 Total Heat of Rejection, MBH					*R-22 Total Heat of Rejection, MBH					Maximum No. of Circ. Avail.
		1°TD	10°TD	15°TD	20°TD	30°TD	1°TD	10°TD	15°TD	20°TD	30°TD	
Single Row of Fans												
023	1 × 2	11.07	111	166	221	332	11.30	113	170	226	339	14
027	1 × 2	13.00	130	195	260	390	13.27	133	199	265	398	14
031	1 × 2	14.60	146	219	292	438	14.90	149	224	298	447	14
035	1 × 2	17.14	172	257	343	515	17.50	175	263	350	525	14
041	1 × 3	19.53	195	293	391	586	19.93	199	299	399	598	21
045	1 × 3	21.89	219	328	438	657	22.33	223	335	447	670	21
049	1 × 3	24.24	242	364	485	727	24.73	247	371	495	742	28
053	1 × 4	26.04	260	391	521	781	26.57	266	399	531	797	21
061	1 × 4	29.20	292	438	584	876	29.80	298	447	596	894	21
065	1 × 4	32.34	323	485	647	970	33.00	330	495	660	990	28
071	1 × 4	34.30	343	515	686	1029	35.00	350	525	700	1050	28
075	1 × 5	36.95	369	554	739	1108	37.70	377	566	754	1131	21
079	1 × 5	39.36	394	590	787	1181	40.17	402	603	803	1205	28
089	1 × 5	43.48	435	652	870	1304	44.37	444	666	887	1331	28
097	1 × 6	47.24	472	709	945	1417	48.20	482	723	964	1446	28
107	1 × 6	52.14	521	782	1043	1564	53.20	532	798	1064	1596	28
Double Row of Fans												
046	2 × 2	22.15	221	332	443	664	22.60	226	339	452	678	2 @ 14
054	2 × 2	26.04	260	391	521	781	26.57	266	399	531	797	2 @ 14
060	2 × 2	29.20	292	438	584	876	29.80	298	447	596	894	2 @ 14
066	2 × 2	32.34	323	485	647	970	33.00	330	495	660	990	2 @ 14
070	2 × 2	34.30	343	515	686	1029	35.00	350	525	700	1050	2 @ 14
080	2 × 3	39.04	390	586	781	1171	39.83	398	598	797	1159	2 @ 21
086	2 × 3	41.81	418	627	836	1254	42.67	427	640	853	1280	2 @ 21
090	2 × 3	43.77	438	657	875	1313	44.67	447	670	893	1340	2 @ 21
098	2 × 3	48.48	485	727	970	1454	49.47	495	742	989	1484	2 @ 28
106	2 × 4	52.07	521	781	1041	1562	53.13	531	797	1063	1594	2 @ 21
120	2 × 4	58.38	584	876	1168	1751	59.57	596	894	1191	1787	2 @ 21
132	2 × 4	64.65	646	970	1293	1939	65.97	660	990	1319	1979	2 @ 28
140	2 × 4	68.60	686	1029	1372	2058	70.00	700	1050	1400	2100	2 @ 28
152	2 × 5	73.86	739	1108	1477	2216	75.37	754	1131	1507	2261	2 @ 21
162	2 × 5	78.73	787	1181	1575	2362	80.33	803	1205	1607	2410	2 @ 28
168	2 × 5	82.39	824	1236	1648	2472	84.07	841	1261	1681	2522	2 @ 28
178	2 × 5	86.93	869	1304	1739	2608	88.70	887	1331	1774	2661	2 @ 28
194	2 × 6	94.47	945	1417	1889	2834	96.40	964	1446	1928	2892	2 @ 28
202	2 × 6	98.85	988	1483	1977	2965	100.87	1009	1513	2017	3026	2 @ 28
212	2 × 6	104.30	1043	1565	2086	3129	106.43	1064	1597	2129	3193	2 @ 28

*For R-134A capacity, multiply R-22 capacity by 0.95; for 50 HZ capacity multiply by 0.92.

Figure R10-14 Typical air-cooled condenser rating table. (Courtesy of Heatcraft Refrigeration Products)

Another technique of artificially raising the head pressure is to back the liquid refrigerant up into the condenser tubes. The normal free-draining condenser has very little liquid in the coil; it is mostly all vapor. But if a control valve is placed in the liquid outlet of the condenser (Fig. R10-17) and is actuated by inlet pressure, some of the discharge gas is allowed to bypass the condenser and enter the liquid drain. This restricts drainage of the liquid refrigerant from the condenser, flooding it exactly enough to maintain the head and receiver pressure. This method is not as common as restricting the air flow, because it is associated with systems that use an external receiver. Critically charged DX systems for air conditioning normally do not employ receivers.

Capacity Data* (60 HZ.)**
Condensing Units—R22

Model	Suction Temp °F	Ambient Temperature °F												EER @ ARI Base Rating Cond
		90°F		95°F		100°F		105°F		110°F		115°F		
		TONS	K.W.	TONS	K.W.	TONS	K.W.	TONS	K.W.	TONS	K.W.	TONS	K.W.	
RCU-008S	30	5.8	7.4	5.6	7.5	5.3	7.6	5.1	7.7	4.7	7.8	4.6	7.9	11.2
	35	6.6	7.8	6.5	8.0	6.1	8.1	5.7	8.2	5.5	8.3	5.3	8.5	
	40	7.4	8.2	7.0	8.4	6.8	8.6	6.6	8.7	6.3	8.9	5.9	9.2	
	45	8.0	8.6	7.9	8.9	7.6	9.1	7.4	9.3	7.0	9.6	6.7	9.8	
RCU-008SS†	30	6.4	7.0	6.2	7.1	5.8	7.2	5.6	7.4	5.5	7.6	5.2	7.7	12.4
	35	7.0	7.4	6.8	7.6	6.6	7.8	6.4	7.9	6.2	8.0	5.8	8.2	
	40	7.7	7.7	7.5	7.9	7.0	8.1	6.9	8.3	6.7	8.5	6.5	8.7	
	45	8.4	8.1	8.0	8.4	7.8	8.6	7.6	8.8	7.4	8.9	7.0	9.1	
RCU-010SS†	30	7.7	8.4	7.5	8.6	7.3	8.7	6.9	8.9	6.7	9.1	6.4	9.3	12.2
	35	8.5	8.9	8.3	9.1	8.0	9.2	7.8	9.4	7.5	9.6	7.3	9.8	
	40	9.2	9.3	9.0	9.5	8.8	9.8	8.6	10.0	8.3	10.2	8.0	10.4	
	45	10.0	9.8	9.8	10.1	9.6	10.3	9.4	10.5	9.0	10.8	8.8	11.0	
RCU-010T	30	8.5	11.5	8.2	11.7	7.9	11.9	7.5	12.0	7.2	12.1	7.0	12.2	9.7
	35	9.4	12.4	9.1	12.6	8.7	12.8	8.4	13.0	8.0	13.1	7.8	13.2	
	40	10.3	13.2	10.0	13.4	9.6	13.6	9.3	13.8	8.9	13.9	8.6	14.1	
	45	11.3	14.0	10.9	14.2	10.5	14.6	10.2	14.7	9.9	14.8	9.5	15.1	
RCU-015SS†	30	10.9	12.6	10.6	12.8	10.2	13.0	9.7	13.4	9.5	13.6	9.4	13.8	11.8
	35	12.1	13.2	11.8	13.7	11.4	13.9	11.1	14.2	10.8	14.6	10.5	14.8	
	40	13.3	14.1	12.8	14.2	12.4	14.6	12.0	14.9	11.9	15.4	11.4	15.6	
	45	14.5	14.8	14.5	15.2	13.8	15.4	13.3	16.0	12.9	16.2	12.5	16.4	
RCU-015T	30	11.2	14.8	10.8	15.1	10.2	15.4	9.7	15.5	9.2	15.7	8.8	15.9	10.3
	35	12.7	15.7	12.1	16.1	11.6	16.4	11.1	16.7	10.7	17.0	10.1	17.2	
	40	14.1	16.8	13.5	17.2	13.0	17.7	12.4	17.9	11.9	18.2	11.3	18.6	
	45	15.5	18.0	14.9	18.4	14.3	18.8	13.8	19.2	13.3	19.6	12.8	19.9	
RCU-020T	30	14.5	16.8	14.1	17.1	13.5	17.5	13.0	17.9	12.4	18.3	11.9	18.7	11.6
	35	16.1	17.8	15.6	18.1	15.0	18.5	14.5	19.0	14.0	19.2	13.4	19.6	
	40	17.4	18.5	16.9	19.0	16.3	19.5	15.8	20.0	15.4	20.3	14.6	20.4	
	45	19.0	19.7	18.5	21.0	17.8	20.6	17.3	21.0	16.8	21.4	16.4	21.9	
RCU-020SS†	30	15.7	19.1	15.2	19.5	14.9	19.9	14.3	20.2	14.0	20.5	13.4	21.0	11.1
	35	17.4	20.0	16.8	20.5	16.4	20.9	15.8	21.4	15.5	21.9	14.9	22.4	
	40	19.0	21.3	18.5	21.7	17.9	22.4	17.4	22.8	17.1	23.2	16.3	23.7	
	45	20.7	22.4	20.1	22.8	19.6	23.5	19.0	24.1	18.5	24.5	17.8	25.0	

Notes: [ ] ARI Base rating conditions 90° ambient, 45° suction temperature.

†All models with the suffix 'SS' denote single D/B-Metic accessible Hermetic compressors.

*For capacity ratings at 85°F ambient temperature multiply the ratings of 90°F ambient by 1.03 × Tons and .97 × KW.

**For 50 hertz capacity ratings, derate above table by .85 multiplier.

Figure R10-15 Typical condensing unit rating table. (Courtesy of Dunham-Bush, Inc.)

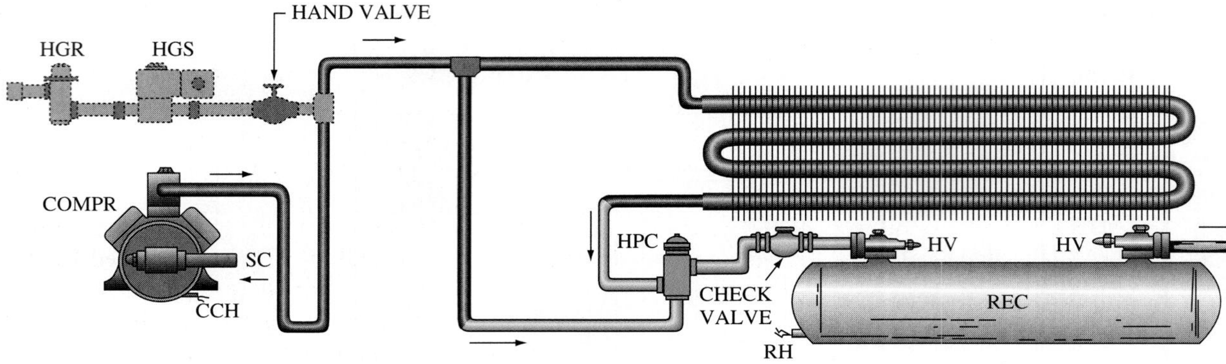

Figure R10-16 Condensing unit with –20°F head pressure control using semi-hermetic compressor with one-step capacity control. (Courtesy of Dunham-Bush, Inc.)

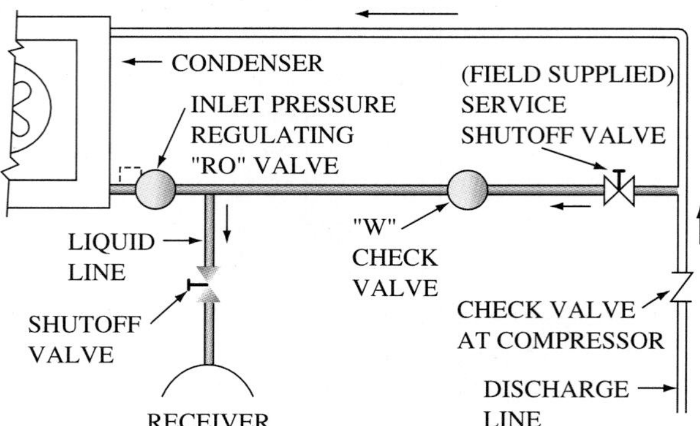

Figure R10-17 Head pressure control system for low-ambient operation using liquid flood-back arrangement.

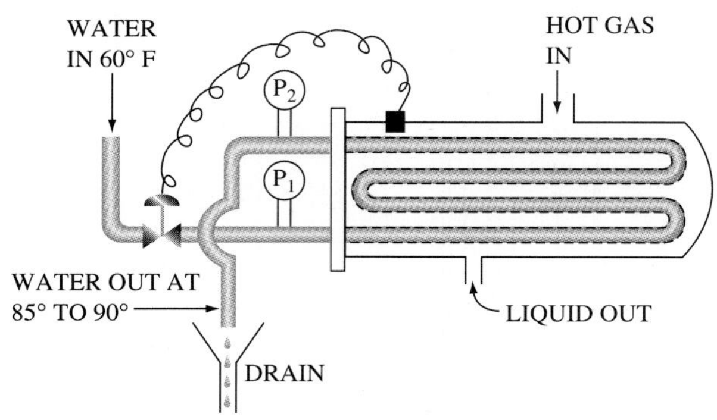

Figure R10-18 Series flow condenser for city water.

R10-4 WATER-COOLED CONDENSERS

In earlier applications of water-cooled condensers to refrigeration and air conditioning, it was common practice to tap the city water supply and then waste the discharge water to a drain connection as shown in Fig. R10-18. An adjustable automatic water regulating valve was placed in the line and the flow of incoming water was controlled by the condenser operating head pressure through a pressure tap.

The temperature of the incoming water would naturally affect the condenser performance and flow rate of any heat load. Depending on the geographic location, water temperatures in city water mains rarely rise above 60°F in summer and frequently drop to much lower temperatures in winter.

Condensers piped for city water flow were always arranged for series flow, circuited for several water passes to achieve maximum heat rejection to the water which was then wasted. Condensers drawing on city water could use only 1 or 1½ gallons per minute (gpm) per ton of refrigeration. The multipass circuit created high water-

PREVENTIVE MAINTENANCE

Perform routine subcooling and approach temperature measurements as part of the preventive maintenance program. This could indicate non-condensible gases, restricted airflow, and inefficient condenser performance. When necessary to clean condenser surface, and chemicals are being used, wear personal hygiene protection including goggles and rubber gloves.

pressure drops ($P_1 - P_2$) of 20 psi or more; however, most city pressures were able to supply the minimum pressure requirement (usually 25 psig).

In time, the cost and scarcity of city water (unless drawn from a lake or wells and returned) became prohibitive and was even outlawed for refrigeration and air conditioning use by many local city codes. Ordinances restricted the use of water to the point where such installations were forced to use all air-cooled equipment, water-saving devices such as the evaporative condenser described above, or the so-called water tower.

R10-4.1 Water-cooled Condensers and Water Towers

The condenser, when used with a recirculated flow such as in a water tower system (Fig. R10-19), is usually designed for parallel tubes with fewer water passes to accommodate a greater water quantity (3 gpm/ton) and lower pressure drop ($P_1 - P_2$) at 8 to 10 psi (18 to 23 ft of head) pressure drop. The nominal cooling tower application will involve a water temperature rise of 10°F through the condenser, with a condensing temperature approximately 10°F above the water outlet temperature. Another consideration in using the water-cooled condenser for open recirculated flow is the fouling factor, which affects heat transfer and water pressure drop. Fouling is essentially the result of a buildup on the inside of the water tubes, which comes from mineral solids (scale), biological contaminants (algae, slime, etc.) and entrained dirt and dust from the atmosphere. The progressive buildup on the condenser tubes creates an insulating effect that retards heat flow from the refrigerant to the water. As the internal diameter of the pipe is reduced, so is the water flow, unless more pressure is applied. Reduced water flow cannot absorb as much heat and thus condensing temperatures rise—as do operating costs.

In selecting condensers the application engineers will usually allow for the results of fouling so that the condenser will have sufficient excess tube surface to maintain satisfactory performance during normal operation, with a reasonable period of service between cleanings. For conditions of extreme fouling and poor maintenance, higher fouling factors are used. Proper maintenance depends on the type of condenser; mechanical or chemical cleaning—or both—may be needed or employed to remove scale deposits.

Water-cooled condensers are an efficient type of heat exchanger used to transfer the heat absorbed by the refrigeration cycle to water. In most installations using water-cooled condensers, water is supplied to the condenser from a cooling tower and returned to the tower to dissipate the heat picked up from the refrigerant to the air, as shown in Fig. R10-20. There are four steps in the process.

1. Heat is transferred from the refrigerant to the water in the condenser.
2. The water is pumped from the indoor condenser to the (usually) outdoor cooling tower.
3. At the tower, the heat is rejected to the outdoor air, cooling the water.
4. The cooled water is returned to the condenser to pick up more heat, making a continuous process.

Water-cooled condensers permit about 15°F lower compressor discharge temperature than air-cooled units. This allows the compressor to operate at a lower discharge pressure, increasing its capacity and lowering the power requirements. These advantages, however, must be considered along with an increase in installation and maintenance costs.

Water-cooled systems in North America are most popular in systems over 100 tons of cooling capacity. In other parts of the world greater use is made of them. Sizes manufactured range from 5 to 3000 tons.

A typical performance for a water cooled condenser system, operating under normal conditions, is shown in Fig. R10-21. The refrigerant condensing temperature is about 105°F. The water entering the condenser is about 20°F lower. The water temperature rises about 10°F in

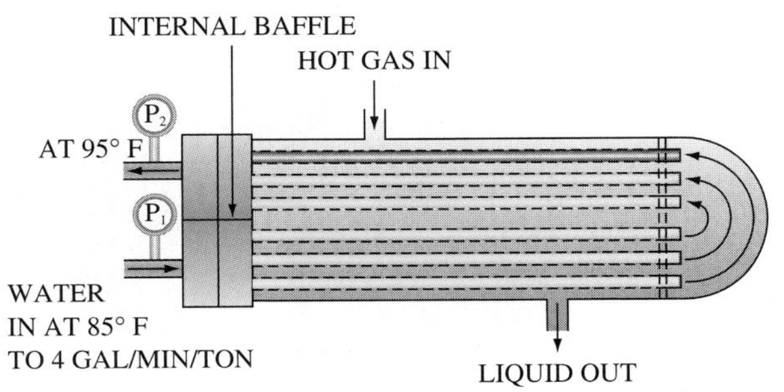

INTERNAL BAFFLE
HOT GAS IN
P_2
AT 95° F
P_1
WATER
IN AT 85° F
TO 4 GAL/MIN/TON
LIQUID OUT

Figure R10-19 Parallel flow condenser for cooling-tower operation.

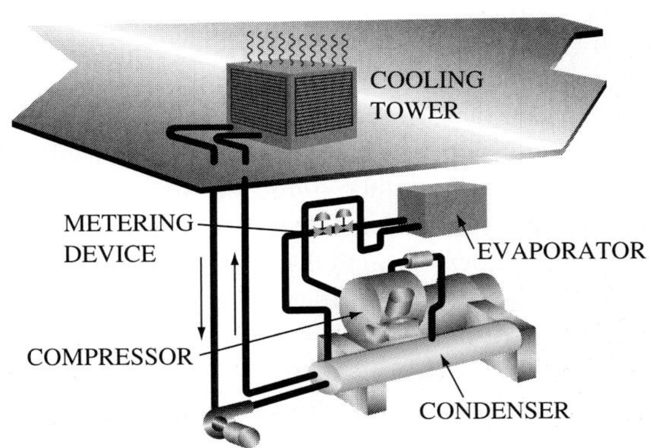

Figure R10-20 Refrigeration cycle showing location of water-cooled condenser.

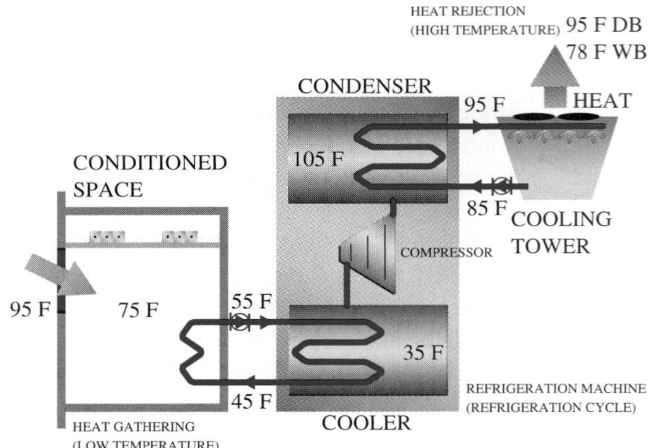

Figure R10-21 Comfort air-conditioning system showing normal temperatures for operation with water-cooled condenser and cooling tower.

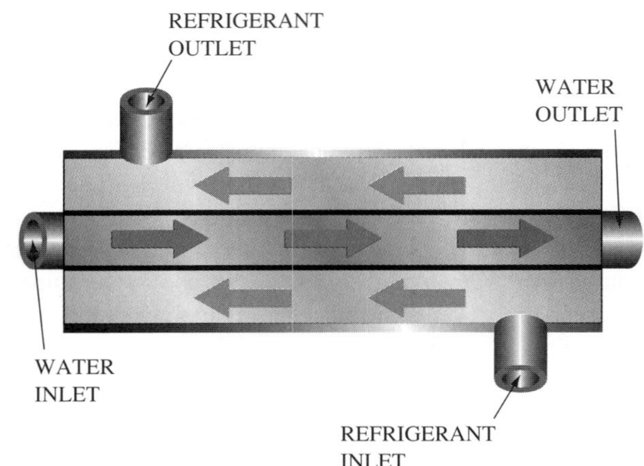

Figure R10-22 Typical refrigerant and water flow for tube-in-tube condenser.

passing through the condenser, leaving the condenser at about 95°F, which is only 10°F below the saturated refrigerant temperature.

The cooling tower is supplied water at 95°F. The warm water drops over the cooling tower's PVC wetted decking, while outside air moves past it in the opposite direction, causing evaporation. Each pound of water that is evaporated removes 1000 Btu from the remaining water, cooling it. A 10°F drop in temperature occurs by the time the water reaches the bottom of the tower. This 85°F water is returned to the condenser to absorb more heat. Water that is lost by evaporation, windage, and blowdown is replaced by make up water which is regulated by a float valve in the sump of the tower.

R10-4.2 Types of Water-cooled Condensers

There are 3 basic types of water-cooled condensers:

1. Tube-in-tube (coaxial)
2. Shell-and-coil
3. Shell-and-tube

Each type is constructed to accomplish the same thing, but in a different way.

The tube-in-tube or double type is illustrated in Fig. R10-22. The usual arrangement is for a smaller tube to be placed inside a larger tube, which is sealed at the end. When multiple tubes are used, they connect to the headers. The water flows through the inner tube and the refrigerant flows through the angular space between the tubes. An advantage of this design is that it can be wrapped into a shape to fit the space available. For example, the compact packaged condensing unit shown in Fig. R10-23 uses the tube-in-tube condenser.

The shell-and-coil design uses a welded steel shell with a coil of continuous finned tubing inside, as shown in Fig. R10-24. Water flows in the tubes and refrigerant flows in the shell. The illustration shows a vertical shell, although the shell can run horizontally to fit the available space better.

Shell-and-coil condensers are sometimes combined with the compressor to form a condensing unit package as shown in Fig. R10-25. These units are usually limited to 20 tons or less. Vertical packaged air-conditioning units from 20 to 60 tons use shell-and-coil condensers due to their compact dimensions, as shown in Fig. R10-26.

The shell and tube type condenser is illustrated in Fig. R10-27. This construction is used in the largest condensers. Capacities range from 10 to 1,000 tons and up. As in the previous type, water flows through the tubes (the tube side of the condenser) and refrigerant flows on the outside of the tubes (shell side of the condenser). These condensers have long, straight finned tubes, connected to a steel plate (tube sheet) at each end. At each end, water

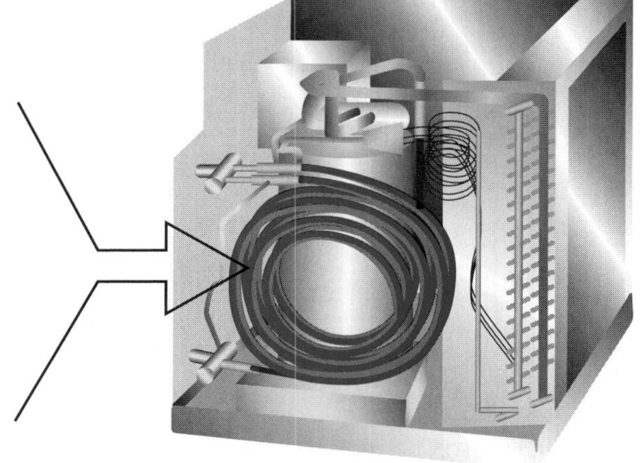

Figure R10-23 Typical use of a tube-in-tube condenser designed for in-the-room installation.

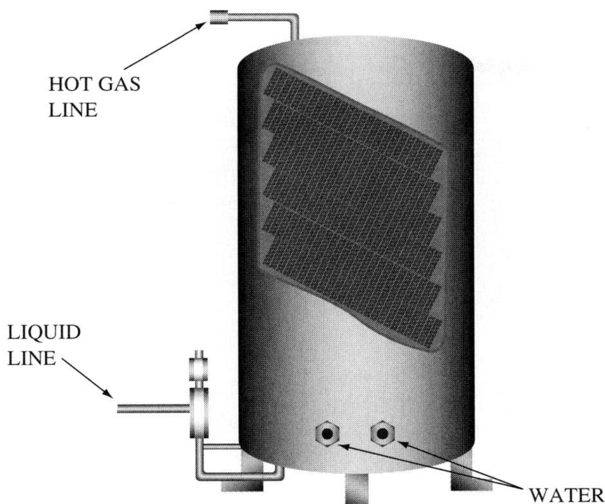

Figure R10-24 Typical design of a shell-and-coil condenser.

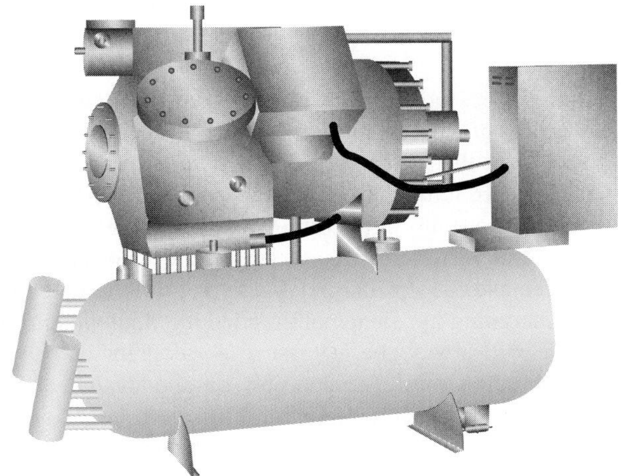

Figure R10-25 Typical water-cooled condensing unit using shell-and-coil condenser.

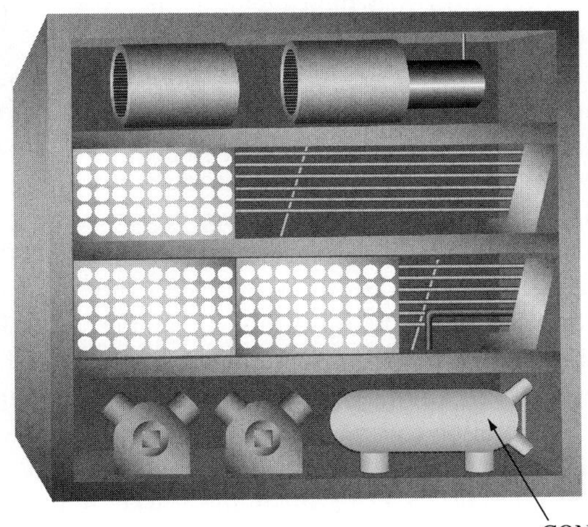

Figure R10-26 Typical packaged air conditioner using shell-and-coil condenser.

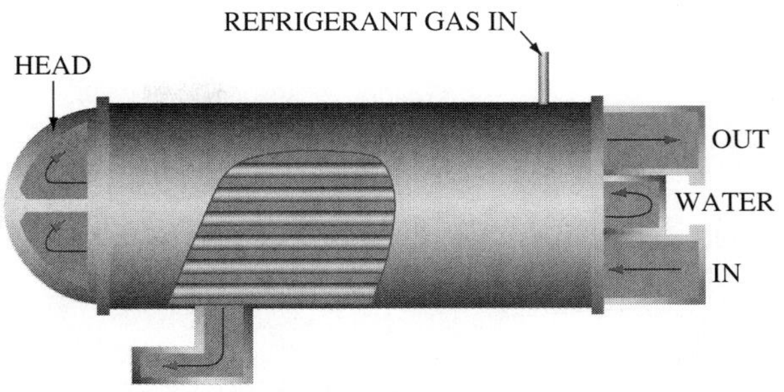

Figure R10-27 Typical design of a shell-and-tube condenser.

manifolds, called heads, are bolted to the shell. These heads direct the water to make from one to eight passes, depending on the size and design of the condenser. The heads can be removed to permit cleaning the individual tubes. Rubber gaskets provide a water tight seal. The other two types of water-cooled condensers, tube-in-tube and shell-and-coil, must be chemically cleaned.

Fig. R10-28 shows a typical water connection in the head. The lower shell view shows the hot-gas inlet and purge valve on the top, the liquid outlet valve on the bottom and the pressure-relief valve on the side.

Many of these shell-and-tube condensers are assembled with compressors to form a water-cooled condensing unit, as shown in Fig. R10-29. They range in size from 5 to 150 tons.

A common use of the shell-and-tube condenser is as a component of a water chiller, as shown in Fig. R10-30. This is a complete packaged refrigerated system including all piping and controls.

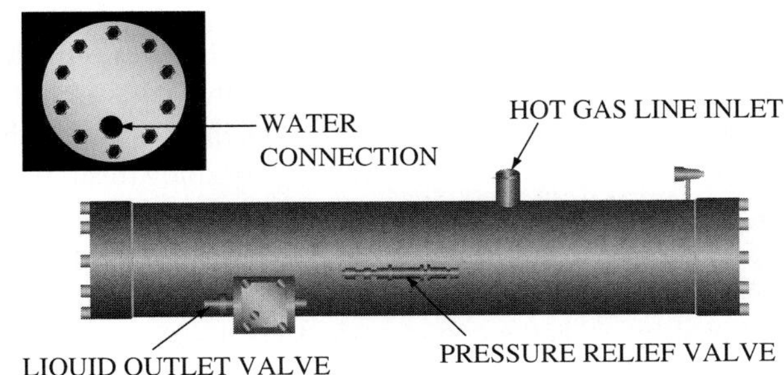

Figure R10-28 End and side views showing the connections on a shell-and-tube condenser.

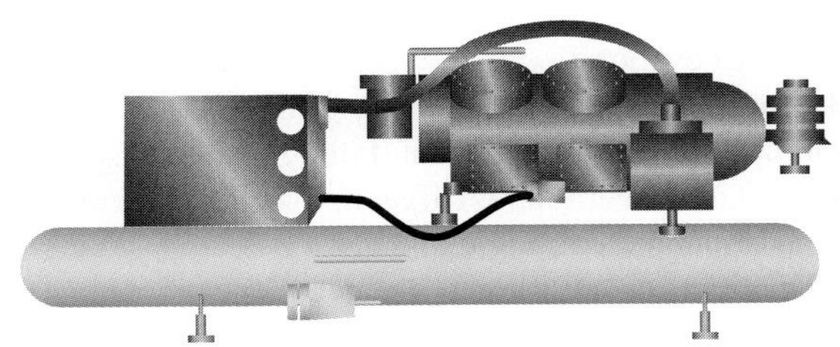

Figure R10-29 Typical water-cooled condensing unit with shell-and-tube condenser.

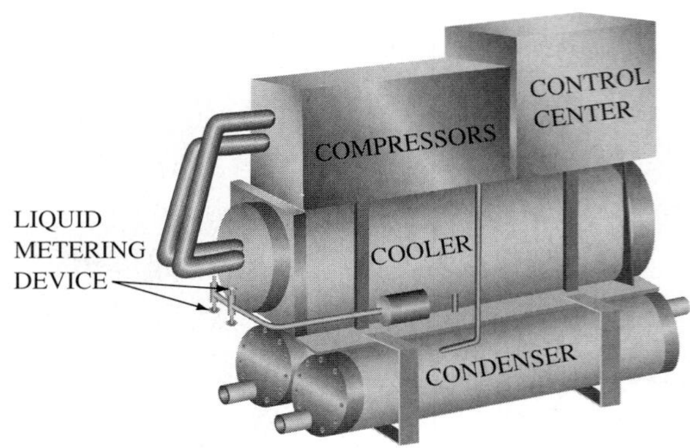

Figure R10-30 Typical packaged liquid chiller unit with a shell-and-tube condenser.

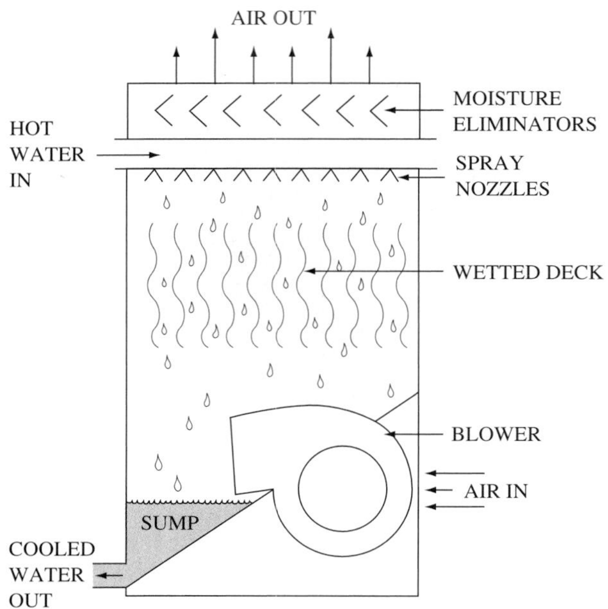

Figure R10-31 Schematic view of cooling-tower construction.

R10-4.3 Cooling Towers

The function of the water tower is to pick up the heat rejected by the condensers and discharge it into the atmosphere by the process of evaporation.

Fig. R10-31 is a schematic drawing showing the principle of operation used by a cooling tower. The water from the heat source is distributed over the wet deck surface by spray nozzles. Air is simultaneously blown upward over the wet deck surface, causing a small portion of the water to evaporate. This evaporation removes the heat from the remaining water. The cooled water is collected in the tower sump and returned to the heat source.

R10-4.4 Effect of Wet-bulb Temperature on Tower Operation

The evaporation of water from any surface requires the removal of a certain amount of heat from the water in order to bring about this change of state. This heat is called the *latent heat of vaporization.* When absorbing heat from water in this manner, air is capable of cooling water below the ambient dry-bulb temperature. It takes approximately 1000 Btu to evaporate 1 lb of water. This removal of latent heat by air is the cooling effect that makes it possible to cool the water in a cooling tower. *Relative humidity* is the ratio of the quantity of water vapor actually present in a cubic foot of air to the greatest amount of vapor that air could hold if it were saturated. When the relative humidity is 100%, the air cannot hold any more water and, therefore, evaporation does not take place. But when the relative humidity of the air is less than 100%, water will evaporate from the drops of falling water in a cooling tower. The lower the relative humidity, the greater the evaporation and cooling.

The drier the air, the more water will evaporate and the greater will be the difference between the dry bulb and wet bulb temperatures. It follows, then, that cooling-tower operation does not depend on the dry-bulb temperature. The ability of a cooling tower to cool water is a measure of how close the tower can bring the water temperature to the wet-bulb temperature of the surrounding air. The lower the wet-bulb temperature the lower the tower can cool the water. It is important to remember that no cooling tower can ever cool water below the wet-bulb temperature of the incoming air. In actual practice the final water temperature will always be at least a few degrees above the wet-bulb temperature, depending on the design conditions. The wet-bulb temperature selected in designing cooling towers for refrigeration and air-conditioning service is usually close to the average maximum wet bulb for the summer months at the given location.

There are two main types of cooling towers: mechanical draft and atmospheric draft (Fig. R10-32). A mechanical draft tower utilizes a motor-driven fan to move air through the tower, the fan being an integral part of the tower. They are available in many configurations and arrangements, including forced draft, induced draft, vertical discharge, and horizontal discharge.

Typically, the water enters the tower at the top or upper distribution basin. It then flows through holes in the distribution basin and into the tower filling, which retards the fall of the water and increases its surface exposure. The concept of splash-type filling is shown in Fig. R10-33. Meanwhile, the fan is pulling air through the filling. This air passes over and intimately contacts the water, and the resulting evaporation transfers heat from the warm water into the air. Finally, the falling water is cooled and collects in the lower (cold-water) basin of the tower. It is then pumped back to the water cooled condenser to pick up more heat.

(a)

(b)

Figure R10-32 Two types of cooling towers. (a) Mechanical draft and (b) Atmospheric draft. (Courtesy of Marley Cooling Tower Company)

The atmospheric draft tower depends on the spray nozzles to break up the water and effect air movement. This tower has no filling or fan, and its size, weight and location requirements (compared to mechanical draft towers) reduce its use considerably. They are seldom encountered; however, in normal refrigeration service work there may be occasions when the technician is called upon for service or maintenance on such units, and thus it is important to be familiar with the operation of these natural draft towers.

The following terms and definitions apply to all cooling towers (see Fig. R10-34).

Cooling range: the number of degrees (Fahrenheit or Celsius) through which the water is cooled in the tower. It is the temperature difference between the hot water entering the tower and the cold water leaving the tower.

Approach: the difference in degrees (Fahrenheit or Celsius) between the temperature of the cold water leaving the cooling tower and the wet-bulb temperature of the air entering the tower.

Heat load: the amount of heat "thrown away" by the cooling tower in Btu per hour (or per minute). It is equal to the pounds of water circulated multiplied by the cooling range.

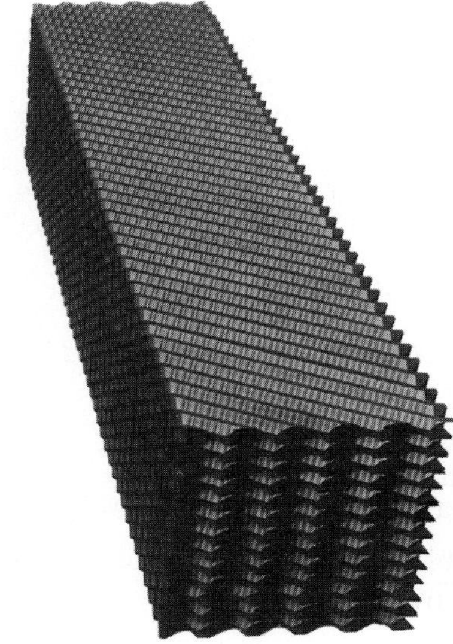

Figure R10-33 PVC fill material for cooling tower. (Courtesy of Evapco, Inc.)

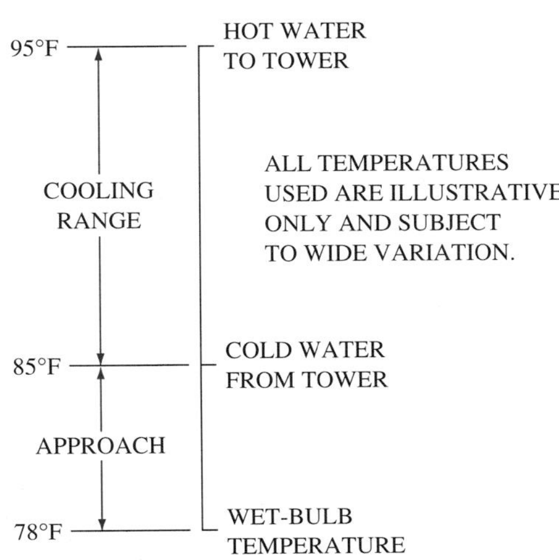

Figure R10-34 Range and approach as applied to cooling tower selection.

EXAMPLE

Given a tower circulating 18 gpm with a 10°F cooling range, what would be the capacity?

Solution

Use the following formula:

$$Q = \text{gpm} \times 500 \times \text{TD}$$

where:

Q = heat rejection in Btu/hr
gpm = flow in gallons per minute
500 = factor derived from, specific heat of water (1) × 60 min/hr × 8.33 lb/gal
TD = temperature difference in °F

Therefore,

$$Q = 18\ \text{gpm} \times 500 \times 10°\text{F}$$
$$= 90{,}000\ \text{Btu/hr}$$

Or, based on 1 ton tower capacity, equivalent to 15,000 Btu/hr,

$$Q = 6.0\ \text{cooling tower tons}$$

The above formula is useful for calculating heat rejection of hot water, condenser water, or chiller water systems.

R10-4.5 Piping Hookup for Cooling Tower

Fig. R10-35 represents a typical mechanical draft tower piping arrangement to the condenser of a packaged refrigeration or air-conditioning unit.

Cooling-tower pump head refers to the pressure required to lift the returning warm water from the cold-water basin operating level to the top of the tower and force it through the distribution system. This data is found in the manufacturer's specifications and is usually expressed in feet of head (1 lb of pressure = 2.31 ft of head). **Drift** is the small amount of water lost in the form of fine droplets carried away by the circulating air. It is independent of, and in addition to, evaporation loss. **Bleed-off** (often called blow-down) is the continuous or intermittent wasting of a small fraction of circulating water to prevent the buildup and concentration of scale-forming minerals and other non-volatile impurities in the water. **Makeup** is the water required to replace water lost by evaporation, drift, and bleed-off.

The first step in the design of the piping system is to determine the water flow to be circulated, based on the heat load given above. Normally, towers run between 3.0 and 4.0 gpm/ton.

Water supply lines should be as short as conditions permit. Standard weight steel pipe (galvanized), type L copper tubing, and CPVC plastic pipe are among the satisfactory materials, subject to job conditions and local codes.

Piping should be sized so that water velocity does not exceed 8 ft/sec. Fig. R10-36 lists approximate friction losses in standard steel pipe and type L copper tubing. Plastic pipe will have the same general friction loss as copper. The data are based on clear water, reasonable corrosion and scaling, and velocity flow at or below the 5 ft/sec range.

EXAMPLE

Based on Fig. R10-36, 100 feet of 1¼-in. standard steel pipe would have a pressure loss of 4.31 ft at 12 gpm. Determine the pressure losses for 50- and 200-ft lengths.

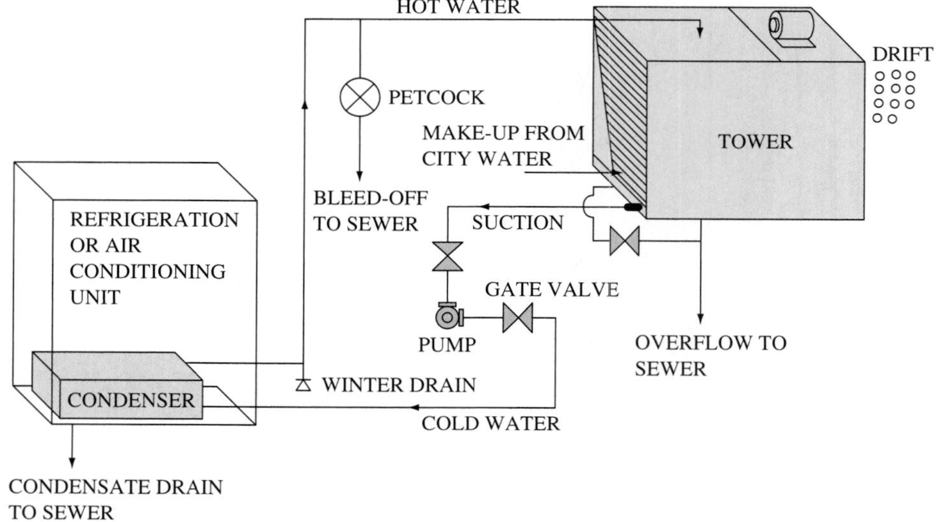

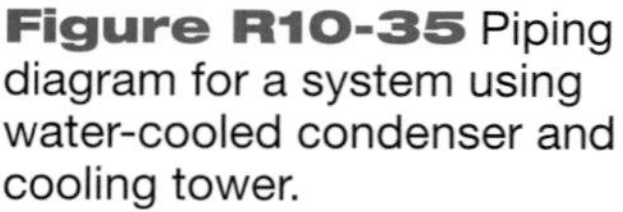

Figure R10-35 Piping diagram for a system using water-cooled condenser and cooling tower.

Water Flow (gal/min)	Type of Pipe or Tubing	¾ in.		1 in.		1¼ in.		1½ in.		2 in.	
		Velocity (ft/sec)	Head Loss (ft/100ft)	Velocity (ft/sec)	Head Loss (ft/100ft)	Velocity (ft/sec)	Head Loss (ft/100ft)	Velocity (ft/sec)	Head Loss (ft/100ft)	Velocity (ft/sec)	Head Loss (ft/100ft)
6	Std. steel	3.61	14.7	2.23	4.54						
	Copper type L	3.98	11.5	2.34	3.13						
9	Std. steel	5.42	31.1	3.34	9.72	1.93	2.75				
	Copper type L	5.96	24.2	3.50	6.63	2.30	2.38				
12	Std. steel			4.46	16.4	2.57	4.31	1.89	2.04		
	Copper type L			4.67	11.3	3.06	4.04	2.16	1.73		
15	Std. steel			5.57	24.9	3.22	6.35	2.36	3.22		
	Copper type L			5.84	17.1	3.83	6.12	2.70	2.62		
22	Std. steel					4.72	13.2	3.47	6.25	2.10	1.85
	Copper type L					5.21	12.5	3.96	5.57	2.28	1.40
30	Std. steel							4.73	11.1	2.87	3.29
	Copper type L							5.41	9.44	3.11	2.45
45	Std. steel									4.30	6.96
	Copper type L									4.66	5.20

Note: Data on friction losses based on information published in *Cameron Hydraulic Data* by Ingersoll Rand Company. Data based on clear water and reasonable corrosion and scaling.

Figure R10-36 Friction losses for steel pipe and copper tubing.

Solution

50 ft would have a pressure loss of

$$4.31 \times 50/100 = 2.15 \text{ ft}$$

and 200 ft would have a pressure loss of

$$4.31 \times 200/100 = 8.62 \text{ ft}$$

In using the table (Fig. R10-36), the smallest pipe size should be selected that will provide proper flow and velocity, to keep installation costs to a minimum. Friction loss is expressed in feet of head per 100 ft of straight pipe length.

The entire piping circuit should be analyzed to establish any need for proper valves for operation and maintenance of the system. A means of adjusting water flow is desirable; shutoff valves should be placed so that each piece of equipment can be isolated for maintenance.

Valves and fittings (elbows, tees, etc.) create added friction loss and pumping head. Fig. R10-37 lists the approximate friction loss expressed in equivalent feet of pipe.

The following is an example of calculating pipe sizing.

EXAMPLE

Determine the total pump head required for a 5-ton installation requiring 75 ft of steel pipe, 10 standard elbows, 4 gate valves, and a net tower static lift of 60 in. Water circulation will be 15 gpm and the pressure drop across the condenser is 13 psi (data obtained from the manufacturer).

Solution

1¼-in. pipe, since the velocity of 15 gpm is less than 5 ft/sec.

Quantity of 1¼-in. Pipe	Equivalent Pipe Length
75 ft of 1¼-in. standard steel pipe =	75.0 ft
(10) 1¼-in. standard elbows × 3.5 =	35.0 ft*
(4) 1¼-in. open gate valves × 0.74 =	2.96 ft*
Total	112.96 ft

From Fig. R10-36, we find that for 100 ft of 1¼-in. pipe, the loss is 6.35 ft; and for 112.96 ft:

$$\text{loss} = \frac{112.96 \times 6.35}{100} = 7.17 \text{ ft}$$

Pressure loss due to piping and fittings =	7.17 ft
Pressure loss due to condenser** = 13 × 2.31	= 30.00 ft
Pressure loss due to static lift—cooling-tower pump head	= 5.00 ft
Total Head	42.17 ft

Pipe size is adequate since velocity is less than 5 ft/sec.

*See Fig. R10-37.

**To convert pounds per square inch pressure to feet of head, multiply by 2.31.

In the above example, a tower pump may be selected based on the water flow (e.g., 15 gpm) and the total head (42.17 ft, or 43 ft). The pump manufacturer's catalog will rate the pump capacity in gpm versus feet of head and the horsepower size needed to do the job.

There are many types of pumps from which to choose. For most air-conditioning applications an iron-body, bronze-fitted, end-suction, centrifugal pump with mechanical seals will do the job. Close-coupled, 3500-rpm pumps are economical and do not have to be aligned. If continuity of service is important, install a standby pump. By locating the condenser pump outdoors below the tower, a leaking seal will be less of a problem. For some applications, 1750-rpm base-mounted pumps are specified. Motor replacements are easier and the motors run quietly.

The installation of the pumps includes the following:

1. The pump should be located between the tower and the refrigeration or air-conditioning unit so that the water is "pulled" from the tower and "pushed" through the condenser. See Fig. R10-35 for a typical piping diagram. It is good practice to place a flow-control valve (a gate valve is satisfactory) in the pump discharge line.
2. The pump should be installed so that the pump suction level is lower than the water level in the cold-water basin of the tower. This assures pump priming.
3. If the pump is located indoors, consideration should be given to noise and water leakage should the seal fail.
4. If an open, drip-proof motor is used outdoors, a rain cover will provide additional protection. Make sure ventilation is adequate.
5. The pump should be accessible for maintenance and installed to permit complete drainage for winter shutdown.

R10-4.6 Cooling Tower Wiring

The most desirable wiring and control arrangement varies, depending on the size of the equipment being installed. In every case, the objective is to provide the specified results with optimum operating economy and protection to the equipment involved.

For small refrigeration and air-conditioning equipment, the ideal arrangement is based on a sequence beginning with the cooling-tower pump. The starter controlling the fan and pump would then activate the compressor motor starter through an interlock. This method, illustrated in

Pipe Size (in.)	Gate Valve Full Open	45° Elbow	Long Sweep Elbow or Run of Std. Tee	Std. Elbow or Run of Tee Reduced One-Half	Std. Tee through Side Outlet	Close Return Bend	Swing Check Valve Full Open	Angle Valve Full Open	Globe Valve Full Open
¾	0.44	0.97	1.4	2.1	4.2	5.1	5.3	11.5	23.1
1	0.56	1.23	1.8	2.6	5.3	6.5	6.8	14.7	29.4
1¼	0.74	1.6	2.3	3.5	7.0	8.5	8.9	19.3	38.6
1½	0.86	1.9	2.7	4.1	8.1	9.9	10.4	22.6	45.2
2	1.10	2.4	3.5	5.2	10.4	12.8	13.4	29.0	58.0

Note: Data on fittings and valves based on information published by Crane Company.

Figure R10-37 Friction losses of valves and fittings used for cooling tower piping.

Fig. R10-38, assures sufficient condenser water flow so that compressor short-cycling is eliminated in the event of pump motor failure. In other words, the compressor cannot run unless the tower is operating. The tower fan is wired to allow cycling by a tower thermostat to maintain condenser water temperature.

There are other, more economical methods based on using the compressor starter to activate the pump and fan, but water temperature or flow-sensing devices should be incorporated as protection for the compressor. Where multiple refrigeration units are used on a common tower, the first unit that is turned on activates the cooling tower.

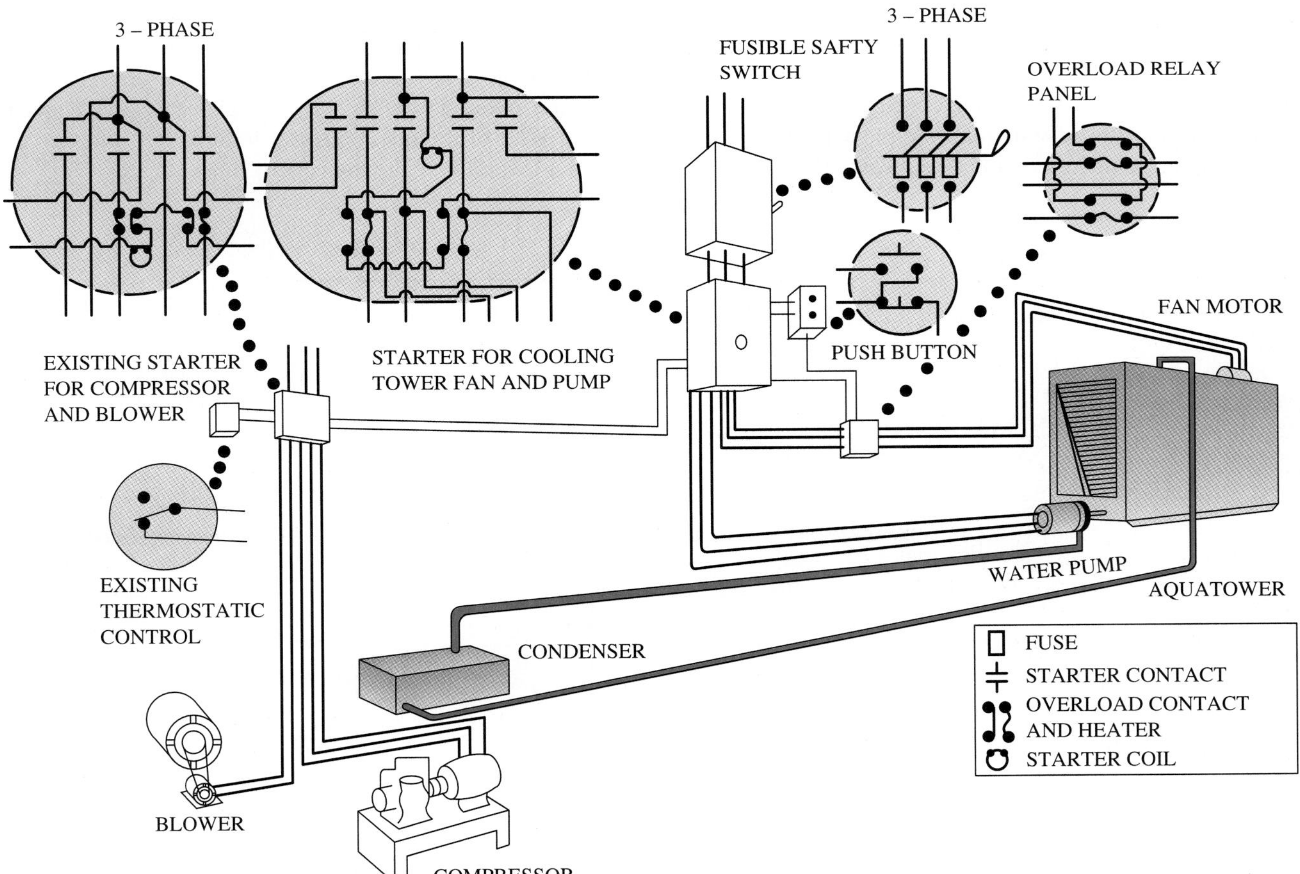

NOTE: ACTUAL WIRING DIAGRAM FOR EQUIPMENT PURCHASED SHOULD BE OBTAINED FROM STARTER OR RELAY MANUFACTURER.

Figure R10-38 Typical wiring diagram for water-cooled condensing unit and cooling tower.

R10-4.7 Protective Measures for Cooling Towers in Cold Weather

Winter operation or low-ambient operation of a cooling tower is subject to special treatment for temperatures near freezing. Water that exits the tower too cold may cause thermal shock to the condenser and result in a very low condensing temperature. One method of preventing the temperature of the water leaving the tower from falling too low is to turn off or reduce the speed of the tower fan. This will reduce the tower's thermal capability, causing the water temperature to rise. If this is not sufficient, a bypass valve can be placed in the tower piping to permit dumping warm water directly into the tower base, thus bypassing the spray or water-distribution system. A combination of both fan and total bypass may be needed to assure leaving water temperatures during cold weather, but reducing the water flow will contribute to tower freeze-up. Prolonged below-freezing conditions usually require total shutdown and draining of the tower sump and all exposed piping.

A piping and control arrangement to provide winter operation is shown in Fig. R10-39. This arrangement provides an inside sump. During normal operation, when the ambient temperature is above freezing, the water flows from the inside sump, through the condenser into the tower and back to the sump. When the thermostat senses near-freezing water temperature, the three-way valve is repositioned to direct the flow of water from the condenser directly to the inside sump, bypassing the tower. When there is no flow to the tower, the tower fan is cycled off.

R10-4.8 Condensing Unit Performance

The foregoing discussion on water-cooled condensers was presented as if most refrigeration and/or air conditioning equipment had separate condensers. While separate components are sold for built-up systems, these are relatively few. Most condensers are assembled at the factory as part of a water-cooled condensing unit (Fig. R10-40) or as part of a complete water-chilling unit. The performance rating of the equipment therefore becomes a system rating similar to the capacity rating table (Fig. R10-41), wherein the temperature of the water leaving the condenser and the saturated suction temperature to the compressor determine the unit tonnage, kilowatt power consumption, resulting condensing temperature, and condenser flow rate.

Additionally, the manufacturer's data will list the water pressure drop across the condenser (in psi or ft of head) for various flow rates and numbers of passes; the refrigerant charge in lb (maximum and minimum) when the condenser is used as a receiver or with an external receiver; and the pump-down capacity, which is usually equal to about 80% of the net condenser volume but does not exceed a level above the top row of tubes.

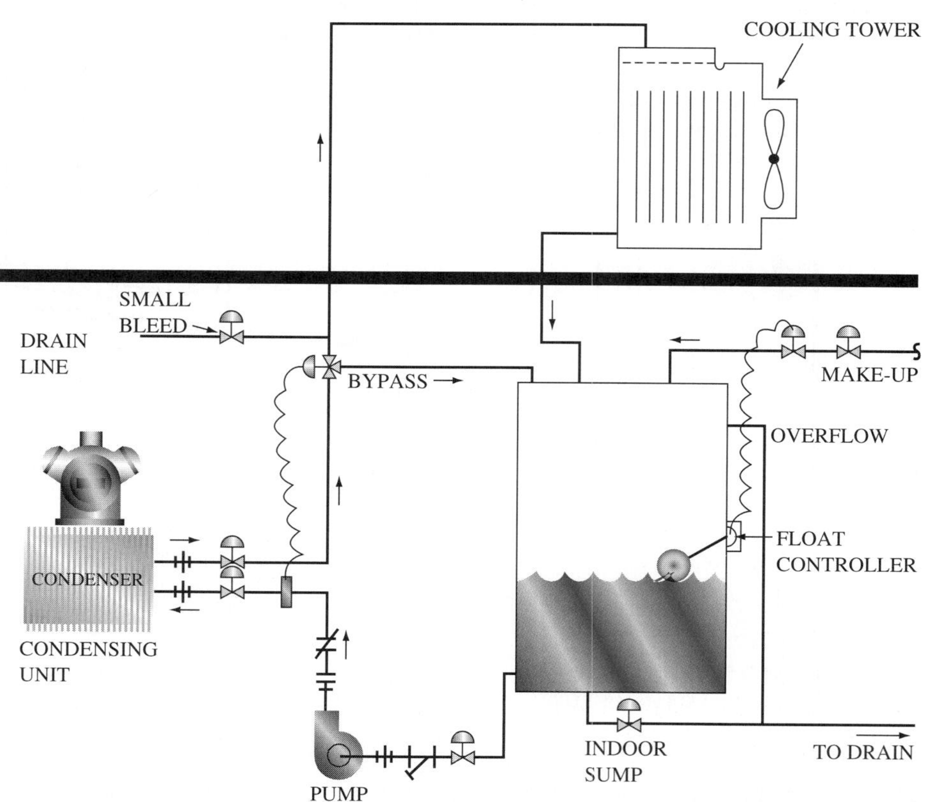

Figure R10-39 Winter operation using a cooling tower.

Figure R10-40 Typical water-cooled condensing unit used for commercial installations. (Courtesy of York Internatinal Corp.)

Pump-down is used to store or contain all the system refrigerant in the condenser, so as to be able to perform service or maintenance operations on other components and not lose the refrigerant. Shutoff valves on the condenser permit this operation.

Water-cooled condensers are usually equipped with a purge connection to vent noncondensable gases, and permit connection of a water-regulating valve or pressure stat. Separate provision is made for a pressure-relief valve to meet certain national and/or local code requirements. Setting of the relief valve varies somewhat with the refrigerant being used. The valve setting should be less than the maximum working pressure of the condenser.

R10-4.9 Maintenance for Water Cooled Equipment

Peak performance of a water-cooled condenser and cooling tower system depends heavily on regular maintenance. Growths of slime or algae, which reduce heat transfer and clog the system, should be prevented.

The technician should become familiar with the companies, chemicals, and cleaning techniques recommended locally. The success of any water treatment lies in starting it early and using it regularly. Once scale deposits have formed it can be costly to remove them.

In addition to preventing scale, water treatment should protect the system components against corrosion. This is critical on systems using steel pipe. The water tower continuously aerates the water, adding oxygen. Open-circuit cooling tower loops are much more subject to corrosion than closed-circuit chilled or hot water systems.

In addition to chemical treatment, regular draining, cleaning and flushing of the tower basin is recommended. Also, a bleed line valve should be adjusted to cause a small amount of overflow that trickles down the drain; this is called bleed-off, or sometimes blow-down. It is the continuous or intermittent removal of a small amount of water (1% or less) from the system. Dissolved concentrates are continually diluted and flushed away.

ON THE JOB

Always inform your customer when you will be shutting down his condensers and refrigeration equipment to do service and maintenance. Remember to turn all equipment back on, re-install covers, and remove lock-out/tag-out protection devices. Before leaving the job site, explain to your customer that all systems are back on and operating normally.

Condenser Leaving Water Temp. (F)	Saturated Suction Temp. (F)	Condensing Unit Model Number											
		JS43L-12W413						JS53M-12W523					
		TONS CAP.	KW	Cond. Temp. (F)	Heat Rej. MBH	Cond. GPM	Cond. Δ P	TONS CAP.	KW	Cond. Temp. (F)	Heat Rej. MBH	Cond. GPM	Cond. Δ P
80	20	27.0	25.0	86.8	409	81.8	9.5	31.6	30.9	86.4	485	97.0	8.6
	30	33.5	26.6	88.3	493	98.6	13.3	39.5	32.8	87.8	586	117.2	12.2
	40	41.0	28.0	90.2	588	117.6	18.3	48.1	34.5	89.6	695	139.0	16.6
	50	—	—	—	—	—	—	58.3	36.2	91.8	824	164.8	22.7
85	20	26.0	26.0	91.6	401	80.2	9.1	30.5	32.1	91.2	476	95.2	8.3
	30	32.4	27.8	93.1	484	96.8	12.9	38.1	34.4	92.7	574	114.8	11.7
	40	39.8	29.4	94.9	578	115.6	17.8	46.7	36.3	94.4	684	136.8	16.1
	50	48.1	31.1	97.1	683	136.6	24.3	56.7	38.2	96.4	810	162.0	22.0
90	20	25.1	26.8	96.4	393	78.6	8.8	29.5	33.2	96.0	467	93.4	8.0
	30	31.5	29.0	97.9	477	95.4	12.6	37.0	35.9	97.5	567	113.4	11.4
	40	38.5	30.9	99.6	568	113.6	17.2	45.3	38.1	99.1	674	134.8	15.7
	50	46.8	32.8	101.7	673	134.6	23.6	55.0	40.2	101.1	797	159.4	21.4
95	20	24.2	27.6	101.1	384	76.8	8.5	28.5	34.2	100.9	459	91.8	7.8
	30	30.4	30.1	102.5	468	93.6	12.2	35.7	37.2	102.2	555	111.0	11.0
	40	37.4	32.3	104.3	559	111.8	16.7	44.0	39.9	103.8	664	132.8	15.3
	50	45.5	34.5	106.4	664	132.8	23.0	53.5	42.3	105.8	786	157.2	20.8
100	20	23.3	28.3	106.0	377	75.4	8.2	27.4	35.1	105.7	449	89.8	7.5
	30	29.4	31.2	107.3	460	92.0	11.8	34.5	37.6	107.0	542	108.4	10.5
	40	36.0	33.7	109.0	547	109.4	16.1	42.5	41.6	108.6	652	130.4	14.8
	50	44.0	36.1	111.0	651	130.2	22.1	51.9	44.5	110.4	775	155.0	20.3
105	20	22.3	28.9	110.8	367	73.4	7.8	26.3	35.8	110.5	438	87.6	7.1
	30	28.3	32.2	112.2	450	90.0	11.3	33.2	39.8	111.8	534	106.8	10.3
	40	35.0	35.0	113.7	540	108.0	15.7	41.2	43.3	113.3	642	128.4	14.4
	50	42.7	37.8	115.7	641	128.2	21.5	50.5	46.7	115.1	765	153.0	19.8

Figure R10-41 Typical rating data for commercial condensing units. (Courtesy of York International Corp.)

Δ P = Pressure Drop (Ft. H_2O)

Water treatment and maintenance can be expensive and time consuming. Once commonly used compounds, chromates are no longer used. Chemicals used for treatment are being increasingly regulated. A treatment should be selected that complies with local regulations.

R10-5 EVAPORATIVE CONDENSERS

In an evaporative condenser (Fig. R10-42), the hot-gas piping from the compressor passes through the condenser like it does on an air-cooled condenser, except the tubes are bare, and no fins are needed. In addition, the evaporative condenser has water nozzles that spray water over the tubes. As water flows over the hot tubes, it evaporates and picks up heat from the refrigerant, causing it to condense. At the same time, the blower in the evaporative condenser exhausts the humid heat-laden air to the outside. The water absorbs approximately 1000 Btu per lb of moisture evaporated.

The evaporative condenser has a sump in the bottom with a float valve. As the water is evaporated, make-up water is added to the sump to replace the water that has evaporated. Water from the sump is pumped to the spray nozzles at the top of the condenser and recirculated to provide continuous flow.

When the water evaporates, it leaves a mineral residue in the sump that can accumulate. To prevent this a continuous bleed-off is provided. This residue also will adhere to the tubes and needs to be periodically cleaned off. If it accumulates on the tubes it will reduce the transfer rate. Water treatment is also recommended.

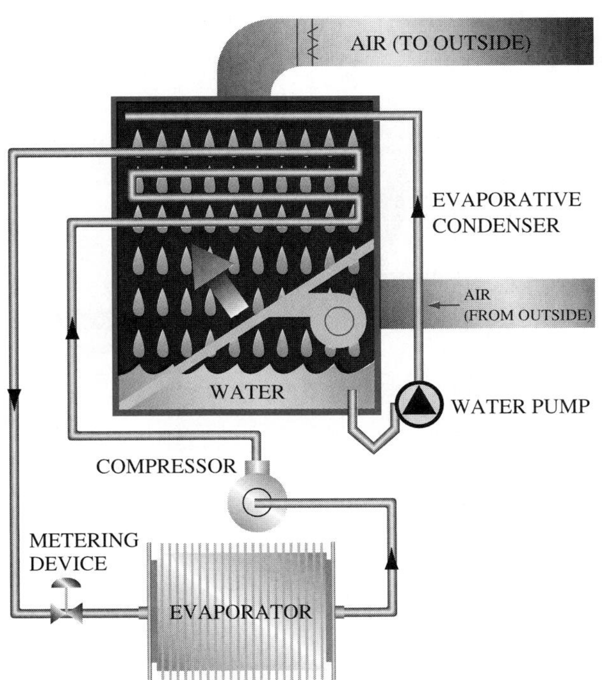

Figure R10-42 Refrigeration system showing position of the evaporative condenser.

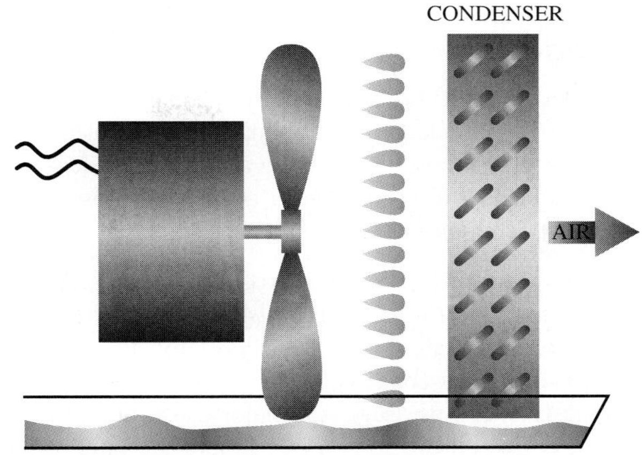

Figure R10-43 Typical construction of the evaporative condenser arrangement used on a room cooler.

Evaporative condensers can be located inside or outside the building. If they are located inside, outside air must be supplied to them. The heat-laden air they exhaust needs to be ducted to the outside of the building.

They are more efficient than either the air-cooled or water-cooled condenser. But of the three types, they are the least popular. Internal corrosion causes them to have the highest maintenance requirements. Their use is more frequent on industrial jobs. They range in capacity from about 10 to 1000 tons.

One desirable feature of the evaporative condenser is that under standard conditions it can operate at a low 105°F condensing temperature, the same as the water-cooled units.

The capacity of cooling towers and evaporative condensers is governed by the lower wet-bulb temperatures, whereas air-cooled condensers are related to the higher ambient air dry-bulb temperatures.

An interesting application of the evaporative condenser principle is included in the design of a room cooler, Fig. R10-43. The condensed water vapor from the evaporator is directed to a small sump in the bottom of the condenser fan housing. A slinger ring on the fan splashes this condensate onto the surface of the condenser coil. When this moisture evaporates from the surface of the hot condenser coil, it removes heat, similar to the action that takes place using an evaporative condenser. When the condensation is light or nonexistent, the condenser acts strictly as an air-cooled condenser.

R10-6 CONDENSER CAPACITY

Since the condenser is one of the major components of the refrigeration system, any problem in providing proper condenser capacity can affect the operation of the entire system. A number of factors can cause the condensing temperature to be too high, resulting in reduced capacity:

1. Design error: undersizing any type.
2. Airflow blockage or recirculation: air-cooled condenser, cooling tower and evaporative condenser.
3. Water flow blockage: water cooled condenser.
4. Dirty condenser coil: any type.

If the system has previously worked properly, the selection is not the problem. High discharge temperature is usually the problem. It can be caused by air or water blockage, or a dirty condenser.

Excessive head pressure is a major cause of compressor failure. Even if the compressor failure does not result, high head pressure can produce high bearing loads and poor efficiency. Excessive temperature can increase the chemical reaction between contaminants and cause damage to essential parts.

R10-6.1 Controlling Condenser Capacity

All air-cooled condensing units that must operate under low ambient conditions require special control equipment for proper operation. The units must be equipped with head pressure control.

Fig. R10-44 shows how the head pressure drops during low ambient conditions on an air-cooled system. Likewise a lower wet-bulb temperature reduces the head pressure on systems using cooling towers or evaporative condensers.

There is a benefit from lower head pressure in producing greater compressor efficiency; however, there is a limit to how low the pressure can drop without affecting the proper operation of the metering device. To prevent this condition, condenser capacity control systems are used to maintain a predetermined minimum condensing pressure.

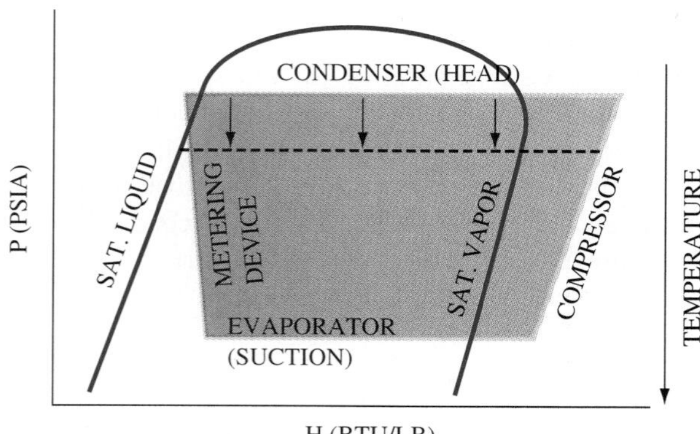

Figure R10-44 Drop in head pressure in an air-cooled system caused by decrease in ambient temperature shown on *P-H* diagram.

There are a number of methods for doing this on air cooled condensers:

1. Cycle the fans on and off.
2. On a single-fan condenser, use a variable speed fan motor.
3. On a multiple-fan condenser, cycle all but the one fan and modulate the last fan to produce smooth control. This arrangement can control the head pressure for ambient temperatures as low as −20°F.
4. Control floodback. This is a control arrangement that will cause liquid refrigerant to fill some of the condenser tubes. This effectively reduces the size of the condenser, since only the unflooded tubes will condense refrigerant.

On water-cooled condenser systems, a temperature-controlled water valve that bypasses the tower and mixes cooling-tower water with condenser water can be used, as shown in Fig. R10-45. With this arrangement the desired condenser water temperature can be maintained. The temperature sensing element for the bypass valve is placed at the water entrance to the condenser.

Another simple, inexpensive method is to cycle the cooling tower fan to maintain leaving water temperature in the 75 to 85°F range.

There are a number of arrangements for controlling evaporative condensers:

1. Shut off the water sprays. This reduces the evaporative condenser to an air-cooled condenser. The air-cooled capacity is about 50 or 60% of the wetted capacity. This, however, is a huge step in ca-

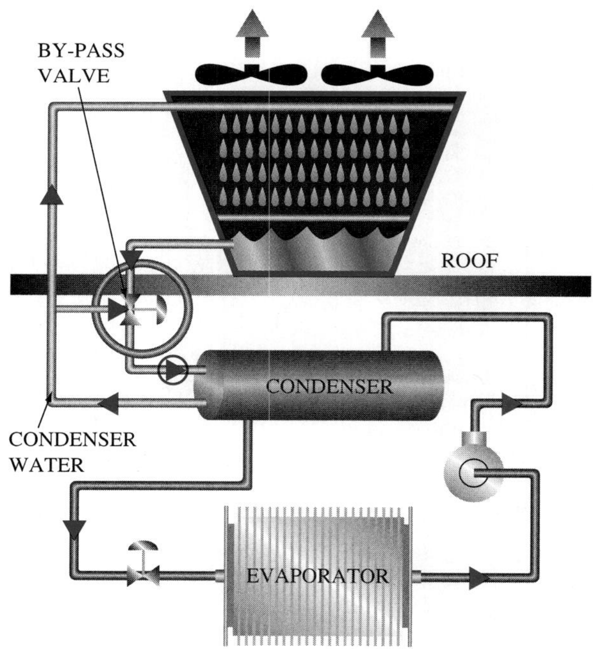

Figure R10-45 Typical piping arrangement for tower water to provide head pressure control.

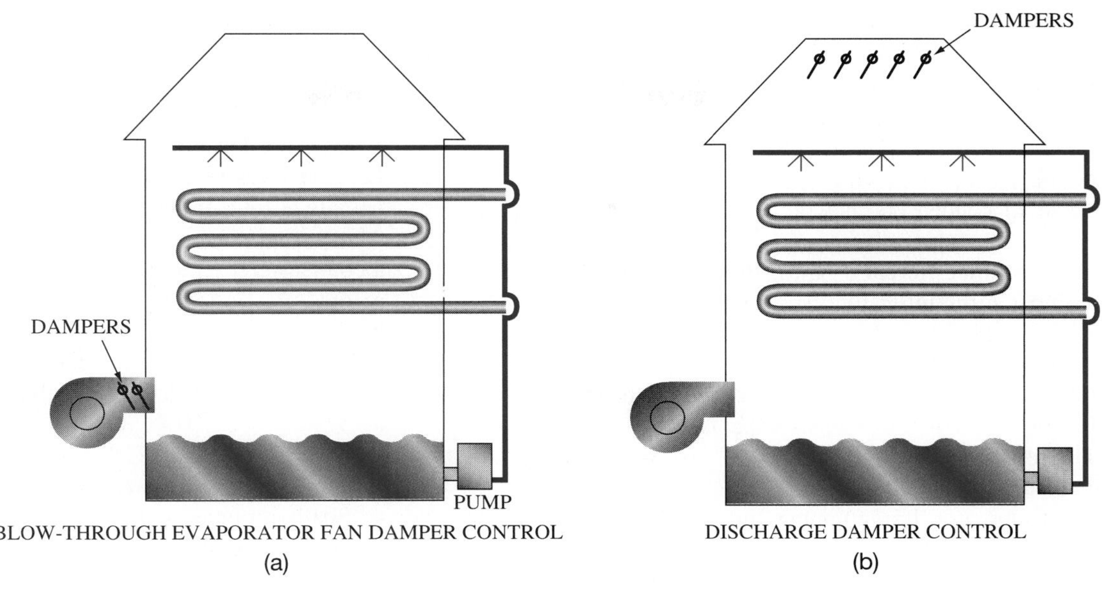

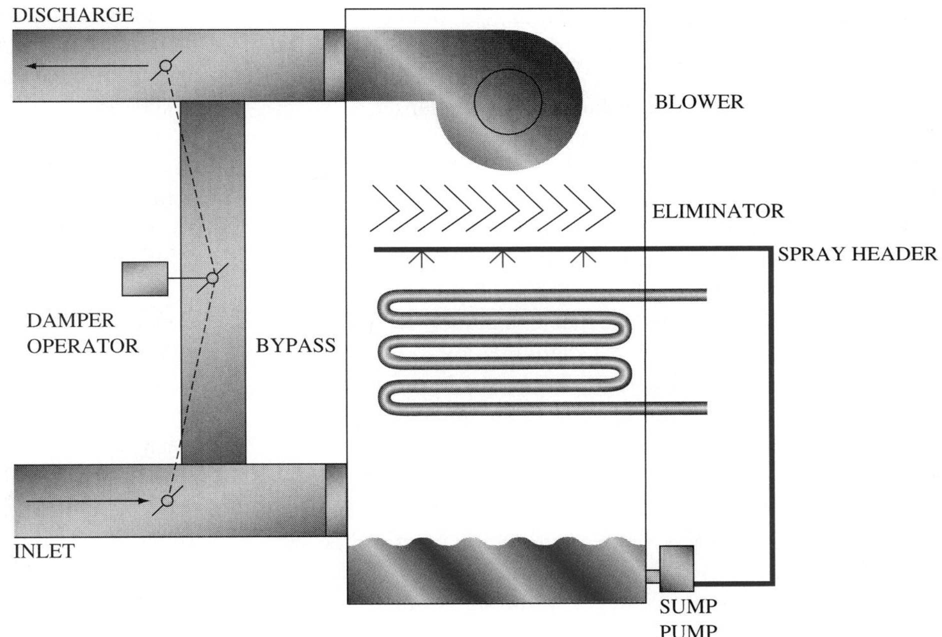

Figure R10-46 Three ways to control the airflow on the evaporative condenser.

pacity reduction and evaporative condensers have a hard time carrying the load dry in the 35 to 40°F range. An added disadvantage is that it may scale the coil.

2. The fan can be cycled. This can shorten the belt life. The speed may be modulated.
3. Dampers can be installed in the blower discharge or unit outlet and modulated to produce the desired flow (see Fig. R10-46).
4. Warm humid air from the condenser discharge can be mixed with outdoor air by the use of bypass dampers and duct work. This increases the relative humidity of the air entering the unit, reducing evaporation and, therefore, reducing the head pressure. This is a relatively easy means of control.

REVIEW

- The function of the condenser is to transfer heat that has been absorbed by the system to air or water.
- Total heat of rejection—about 14% of heat removed by the condenser is superheat, 81% of the heat is removed

by condensation, and 5% of the heat is removed by subcooling.

- Subcooling—ensures a solid stream of liquid will enter the metering device. It adds to the cooling capacity of the system at a rate of .5% of the total cooling capacity per degree of subcooling.
- Types of condensers include: 1.) air-cooled, 2.) water-cooled, 3.) evaporative.
- Temperature difference—the difference between design ambient air temperature and condenser temperature; should range 20–35°.
- Total heat of rejection: is based on ARI published ratings. Use published tables to determine selection.
- Factors that affect condenser capacity:
  The refrigerant being used.
  The design ambient temperature.
  The evaporator design temperature.
  The capacity required to match the load.
- Low ambient control—required on air-cooled condensers to maintain head pressure as outdoor temperatures go low. Methods of maintaining minimum head pressures during low-ambient outside air conditions include:
  Multiple- or variable-speed fans on the condenser.
  Cycling off the condenser fans.
  Flooding the condenser.
  Installing dampers to restrict condenser airflow.
- Water-cooled condenser—maintains head pressure by controlling the flow rate of water entering the condenser, with a water valve operated by head pressure.
- Types of water-cooled condensers include: 1.) shell-and-tube, 2.) tube-in-tube, 3.) shell-and-coil.
- Cooling tower—function is to pick up the heat rejected by the condenser and discharge it to the atmosphere by the process of evaporation of water.
- Approach—a term used to describe the difference in temperature between the water leaving the cooling tower and the wet-bulb temperature of air entering the tower.
- Range—the number of degrees through which the water is cooled in the cooling tower.
- Determining cooling tower capacity (use the following formula):

$$Q = \text{gpm} \times 500 \times \text{TD}$$

Q = Heat of Rejection in Btu/hr
gpm = Flow in gallons per minute
500 = Constant Factor (Specific heat of water (1) × 60 min/hr × 8.3 lb/gal)
TD = Temperature difference in degrees Fahrenheit

Problems and Questions

1. What effect will loss of the condensing fan have upon system operation?
2. What is the purpose of heat reclaim?
3. What is the condensing temperature of an air conditioner (R-22) with an ambient temperature of 85°F assuming a TD of 25°F?
4. What is the TD for water-cooled condensers?
5. List the ways to control head pressure in units that operate at low ambient temperatures.
6. About what percentage of the total heat removed by a condenser is removed by condensation?
   a. 60% c. 80%
   b. 70% d. 90%
7. About what percent of the capacity of the system is supplied by subcooling in the condenser?
   a. 57% c. 10%
   b. 7.5% d. 12.5%
8. Which type of condenser uses both air and water?
   a. Shell-and-tube
   b. Shell-and-coil
   c. Tube-in-tube
   d. Evaporative
9. Under normal conditions, using an air-cooled condenser, if the ambient temperature is 95°F what should be the condensing temperature?
   a. 110°F c. 130°F
   b. 120°F d. 140°F
10. On what type of condenser does the wet-bulb temperature affect the capacity?
    a. Air-cooled
    b. Water-cooled
    c. Evaporative
11. Condensers using city water normally use how many gpm?
    a. 1 c. 3
    b. 2 d. 4
12. What would you expect the normal head pressure to be on a water-cooled condenser with inlet water temperature at 85°F and leaving water at 95°F?
    a. 95°F c. 115°F
    b. 105°F d. 125°F
13. What is the range of sizes of water-cooled condenser systems in tons of refrigeration?
    a. 5 to 100 tons c. 5 to 1000 tons
    b. 5 to 500 tons d. 5 to 3000 tons
14. Which one of the following is NOT one of the types of water-cooled condensers?
    a. Shell-and-tube
    b. Cascade
    c. Shell-and-coil
    d. Tube-in-tube
15. What condensing temperature would be normal for an evaporative condenser operating on a 78° WB day?
    a. 95° c. 115°
    b. 105° d. 125°

Compressors R11

R11-1 FUNCTIONS OF COMPRESSORS

Annual production quantities of compressors are staggering, numbering many millions of units. Applications range from small refrigerator-freezers, water coolers, package terminal air conditioners, commercial refrigeration systems, split air-conditioning systems, rooftop air-conditioning systems, large water chillers for major building air-conditioning systems, as well as industrial process-cooling and refrigeration/freezer systems for food processing and storage. There are many more smaller systems than giant ones, but their refrigeration cycles all function in a similar manner. Most common types of refrigeration systems utilize the vapor-compression cycle.

The function of the compressor is to take the refrigerant vapor from the evaporator at a low pressure and temperature and raise it to a higher pressure and temperature, discharging it into the condenser. Thus, the compressor functions in two ways:

1. It lowers the temperature of the refrigerant in the evaporator so that it can absorb heat and
2. It raises the temperature of the refrigerant in the condenser so that it can discharge the heat.

In Chapters R-1 and R-2 the basic refrigeration cycle and pressure/heat (*P-H*) diagram were discussed. The compressor is the heart of this system. The compressor pumps refrigerant vapor from the evaporator, lowering the pressure in the evaporator. As the pressure is lowered, so is the corresponding saturation temperature of the refrigerant. As the refrigerant picks up heat in the evaporator, it vaporizes and enters the compressor.

As a comparison, the heat energy to boil water in a kettle on a stove comes from the gas or electric burner. In the same way, the heat to boil liquid refrigerant in an evaporator comes from the air (or fluid) surrounding the evaporator. It may be cold, but it still contains heat energy. As heat flows to the cold refrigerant liquid, more vapor boils away, carrying heat energy with it and keeping the evaporator cold.

AFTER STUDYING THIS CHAPTER, THE STUDENT WILL BE ABLE TO:

- Describe the function of a compressor in a vapor-compression refrigeration cycle.
- Describe how the capacity of compressors are changed to match the variations in system load.
- Understand lubrication requirements for compressors using new HFC and existing HCFC/CFC refrigerants.
- Identify the five principal types of compressors and describe their operation and application features.
- Select the compressor based on cooling load.
- Explain the methods to unload cylinders on reciprocating compressors for capacity control.
- Calculate and explain compression ratio.
- Identify major working parts of the various types of compressors.
- Discuss how the design temperature of the evaporator affects compressor size and capacity.
- Troubleshoot (test) a compressor.
- Replace (install) a compressor.

The compressor increases the pressure and temperature of the refrigerant to a higher level where the heat can be transferred to air or water flowing through the condenser, thus condensing the high-pressure refrigerant vapor to a high-pressure liquid and readying the cycle to repeat again. Heat is absorbed in the evaporator at a low temperature and rejected in the condenser to a cooling medium which is at a noticeably higher temperature. Thus, heat has been made to flow "uphill" (from a lower to a higher temperature).

In Fig. R11-1, starting at Point *A,* refrigerant flowing from *A* to *B* through the expansion valve or capillary tube undergoes a substantial pressure drop. The refrigerant changes from a liquid to a mixture of liquid and flash gas while maintaining the same heat content (enthalpy) in Btu/lb. Because no heat is gained or lost by the fluid in the process, it is known as an adiabatic process.

From *B* to *C* the refrigerant mixture is boiled off in the evaporator and superheated enough to assure no significant amounts of liquid can reach and damage the compressor. The refrigerant effect can be measured by the difference in heat from *B* to *C.*

From *C* to *D* the compressor boosts the refrigerant pressure, increasing the temperature and superheat. The theoretical amount of energy required for the compressor is the difference in enthalpy from *C* to *D.*

The condensing process takes place from *D* to *A.* First, the refrigerant hot gas is cooled to the saturation point, then condensed until it is 100% liquid and finally, the liquid is subcooled below the saturation temperature. Subcooling is desirable because it increases cycle efficiency and reduces the likelihood of gas bubbles forming in the liquid line (due to pressure losses from pipe friction or increase in elevation) which would gas-bind the expansion valve or capillary tube. The heat rejected by the cycle to the condenser can be measured by the heat (enthalpy) change from *D* to *A.*

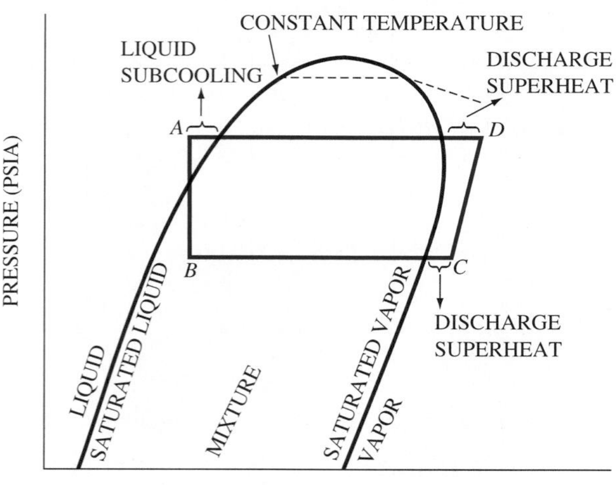

Figure R11-1 Diagram of vapor-compression cycle on pressure-heat plot (simplified to assume no pressure losses).

The suction pressure, or the pressure of the gas entering the compressor, is represented by *B—C* in Fig. R11-1. The discharge pressure, or pressure of the gas leaving the compressor, is represented by *D—A.* Both are measured in absolute units. *Compression ratio* is the ratio of discharge over suction pressure. Some types of compressors are capable of producing high compression ratios.

Operating compressors near or at their high limits can result in loss of efficiency and excessive discharge superheat. The latter can cause over-heating and lubrication failures. If an application requires a high "lift," it might be better to handle it in stages or with a cascade system. All common refrigeration and air-conditioning applications are single stage. The equipment manufacturers rate the capacity of their products for the permissible temperature ranges and type of duty for which they are suited.

R11-2 CAPACITY VERSUS LOAD

Compressors are selected to have sufficient *capacity* to meet the maximum cooling *load.* Fig. R11-2 shows pressure and temperature readings of an actual refrigeration cycle, represented on the *P-H* plot in Fig. R11-1.

In the case illustrated, the refrigerant is HCFC-22 and the application is a residential air-conditioning system.

A sample calculation of the compression ratio using the data from Fig. R11-2, is shown in Fig. R11-3.

Because most loads vary, the compressor will be oversized much of the time. So what happens when the compressor produces more cooling than needed? It "rides the curve."

A plot of compressor cooling capacity in relation to evaporating and discharge temperatures is shown in Fig.

	Compressor Suction	Compressor Discharge
Saturation Temperature (F)	40	120
Actual Gas Temperature (F)	50	165
Superheat (F)	10	45
Pressure (PSIA)	83.7	277.7
Pressure (PSIG)	69.0	263.0
Enthalpy (Btu/lb)	110	125

Figure R11-2 Typical comfort air-conditioning compressor suction/discharge readings.

	Compressor Suction	Compressor Discharge
Saturation Temperature (F)	40	120
Actual Gas Temperature (F)	50	165
Superheat (F)	10	45
Pressure (PSIA)	**83.7**	**277.7**
Pressure (PSIG)	69.0	263.0
Enthalpy (Btu/lb)	110	125
Compression Ratio		
277.7 PSIA	Absolute Discharge Pressure	
83.7 PSIA	Absolute Suction Pressure	
277.7 PSIA ÷ 83.7 PSIA = 3.32		

Figure R11-3 Compression ratio, expressed as a ratio of the absolute discharge pressure to absolute suction pressure.

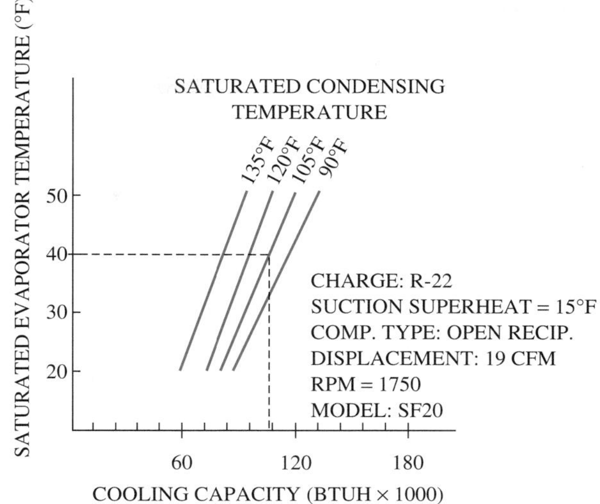

Figure R11-4 Typical capacity curves for compressors.

R11-4. When compressors have less work to do, they run at a lower suction pressure and their capacity decreases. When the load increases, the suction pressure rises and the capacity of the compressor increases.

The typical compressor encountered by a service technician is a positive-displacement device which runs at a constant speed. This means it displaces a constant flow volume, say 10 cubic feet of refrigerant gas per minute (cfm). This does NOT mean it is a constant-capacity device. Capacity depends on mass flow rate—pounds of refrigerant per minute (lb/min).

What happens when a compressor handles a fluctuating load? Assume the 10 cfm compressor had been maintaining a 40°F evaporator temperature until the load is diminished and the evaporator temperature drifts down to 30°F. As the load drops, the thermal expansion (TXV) valve senses a drop in superheat and throttles down to a new position to restore the constant superheat. The flow rate stabilizes at a new lower rate, the compressor continues to pump 10 cfm and the suction temperature stabilizes at a new lower temperature (see Fig. R11-5).

The change in mass rate of flow (lb/min) with a continuing constant volumetric flow (cfm) results in fewer pounds of refrigerant per cubic feet of displacement. In other words, the specific volume changes. Specific volume and density are reciprocals:

Specific volume in cf/lb = 1/density in lb/cf

At 40°F evaporating temperature and 10 cfm the refrigerant had a specific volume of approximately 0.7 cf/lb. The evaporator was handling approximately 14.3 lb/min of refrigerant. When the load dropped, the TXV valve throttled down and the temperature leveled off at 30°F. The gas had a specific volume of 0.8 cf/lb, and the mass flow rate was 12.5 lb/minute, a capacity reduction of 12.5%.

Fig. R11-6 shows the relationship of evaporator temperature to the specific volume for R-22 refrigerant. As the evaporator temperature drops, the specific volume increases, the density of the refrigerant decreases and fewer pounds of refrigerant are pumped.

For small changes in load the cycle tends to self-regulate its capacity as described above; however, other

PRACTICING PROFESSIONAL SERVICES

Remember to re-install all caps on suction and discharge service valves at the compressor. Check for refrigerant leaks at the valve, and if leaks exist, retighten the packing nut on the valve; then re-install cap.

Whenever servicing compressor equipment, always check compressor amps during run periods and compare readings to factory FLA (Full Load Amps) or RLA (Run Load Amps). Verify machine is fully loaded when checking amps, or reading may appear low due to partial load conditions.

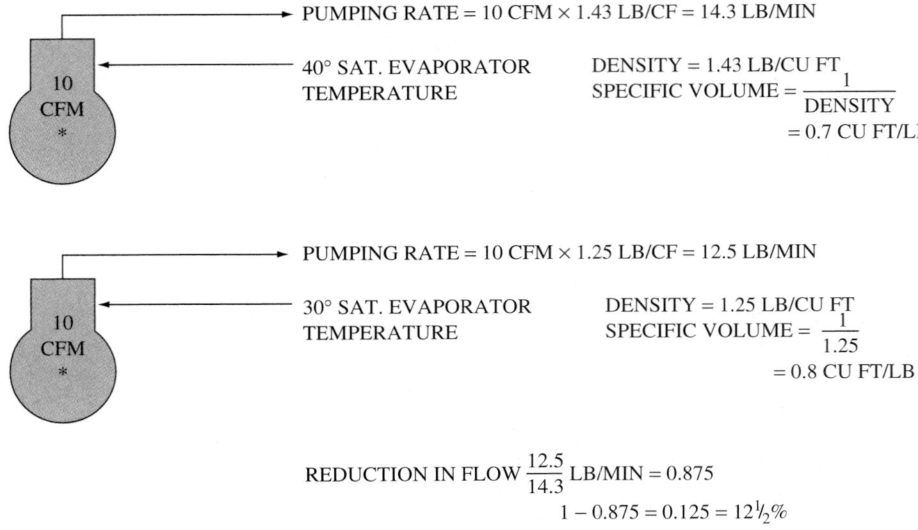

Figure R11-5 Mass flow rate (lb/min) rather than volume pumped (cfm), determines the compressor's capacity.

arrangements need to be considered for large fluctuations in loads.

R11-2.1 Capacity Control Factors

A number of factors control compressor capacity:

1. The choice of the refrigerant
2. Changes in the compressor displacement

The choice of refrigerant for a given size compressor can produce a wide variation in capacity. This is a design choice and once made, is not likely to be changed. It would not be practical on a given system to deal with a fluctuating load by changing the refrigerant.

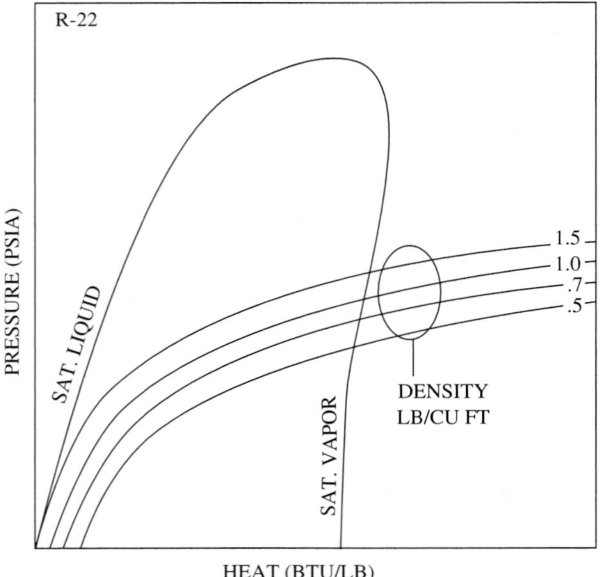

Figure R11-6 R-22 *P-H* diagram, showing refrigerant density.

Changing refrigerants in the field to convert a system may change power requirements dramatically. So if a system was to have the refrigerant changed, in addition to making sure the component materials are compatible, pressure ratings adequate, control devices adequate and properly set, and oil compatible, the motor capacity would have to be checked. A larger motor might be possible for an open machine; it is unlikely that a semi-hermetic motor could be increased in size and certainly it would not be possible with a welded hermetic.

Displacement offers a variety of ways of changing refrigeration capacity to match system load:

1. Cycling on/off
2. Cycle multiple machines on/off
3. Cylinder unloading
4. Hot-gas bypass
5. Speed control
   - Two speed
   - Variable-frequency drives (VFD)
6. Other

On/off cycling is simple and basic: it gives the choice of 0% and 100% capacity. Often it is all that is needed. It is not desirable to excessively short-cycle any equipment, and in some applications the temperature variations of on/off control are too wide to be acceptable. If so, better capacity control is needed.

Multiple machines with either common or independent refrigerant circuits are sometimes used for a variety of reasons, including closer temperature control, better humidity control, etc. In a twin compressor (Fig. R11-7), installation would give the choice of 0%, 50%, and 100% capacity steps.

Figure R11-7 Capacity control by using two compressors on a common refrigerant circuit. (Courtesy of Bristol Compressors)

Cylinder unloading is available on reciprocating compressors, 15–20 ton capacity and larger. Compressor unloaders can be actuated by gas, oil pressure (Fig. R11-8), or electrically. They can unload one or more cylinders by holding the suction valves open, bypassing the cylinder discharge to the suction manifold, or blocking off the suction inlet. Depending on the compressor, multiple stages of control can be provided and power savings achieved. Overheating can be a problem.

Hot-gas bypass is sometimes used on small machines where cylinder unloading is unavailable or on larger machines when additional capacity reduction is required below that possible with unloaders. Care must be taken not to overheat the compressor. It can be a practical solution in dealing with light loads. This process is not energy efficient.

Some compressors are available with two speed motors. This technique *(two-speed control)* is not widely used at present.

Variable-frequency drives (VFD) enable continuous variable speed (capacity) control on many machines. These have been applied to large centrifugal compressors as well as to smaller comfort cooling applications. Variable speed control works very well and is likely to be increasingly used as cost of VFD's (also called inverters) continues to drop. It can be very energy efficient.

Slide valves are commonly used on screw compressors to regulate capacity. The valves work well and save energy. They function based on diverting gas to the suction manifold before it is compressed.

Inlet vane capacity controls are commonly used on centrifugal chillers. These vanes can control down to 10–25% capacity. They function by giving the entering gas a pre-rotation swirl.

R11-3 COMPRESSOR LUBRICATION

All refrigeration compressors require lubrication of moving surfaces. Delivery of lubricant can be by simple splash systems, crankshaft or separate oil pump (Fig. R11-9). As some oil can be expected to carry over with the discharge gas, the oil should be miscible with the refrigerant and be compatible with the system designed to carry the oil through the condenser and evaporator and return it to the compressor crankcase.

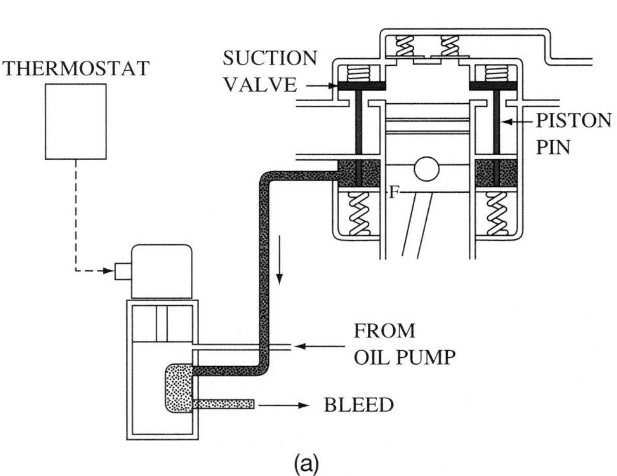

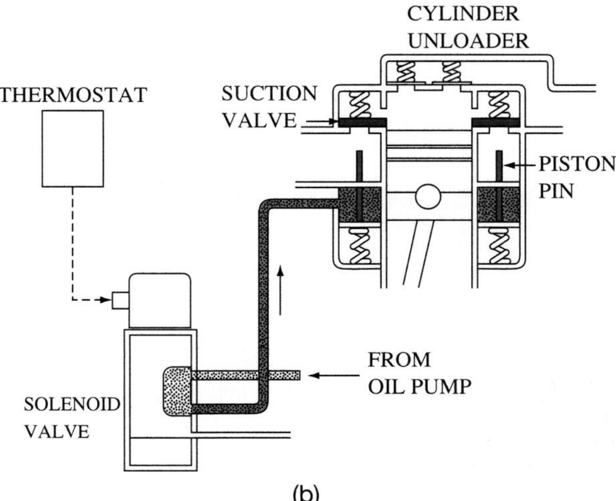

Figure R11-8 Capacity control using oil pressure to unload a cylinder: (a) cylinder unloaded; (b) cylinder loaded. (The Trane Company)

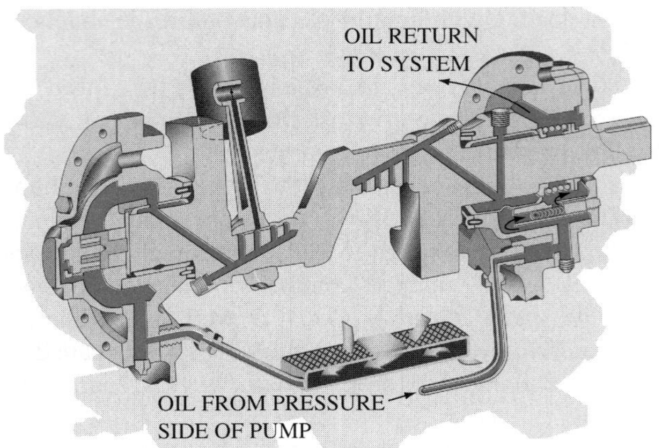

Figure R11-9 Lubrication of the compressor bearings using drilled oil passages. (Courtesy of Frick Company)

Some machines are designed to have high oil circulation rates and are always provided with discharge oil separators. Oils used in hermetic and semi-hermetic systems have to be compatible with the electrical components and be good insulators. Mineral oils should be wax free. All oils should have a low moisture content to ensure long corrosion-free system life, and freedom from electrical insulation deterioration and freeze-ups.

New HFC chlorine-free refrigerants will require use of special polyol ester synthetic oils. Older CFC machines used mineral-oil-based lubricants. Current recommendations for HCFC refrigerant installation and conversions include mineral oil, alkyl-benzene refrigeration oil, or a combination of the two. The choices have become more complicated. Synthetic refrigeration oil is more expensive but can be used with all halocarbon refrigerants. It must be used with HFC refrigerants as specified.

Pressure lubricated compressors usually are equipped with a differential pressure sensor to measure actual oil pressure and a time delay safety lock-out circuit to shut the compressor down should lubrication fail (see Fig. R11-10).

Refer to manufacturers' data for minimum pressure settings, normal operating pressures and high relief valve settings. Oil and refrigerant liquid are miscible so it is possible to have refrigerant migrate during long OFF-cycles to a compressor in a cold location, condense, and raise the oil level above the oil sight glass (if so equipped).

On startup the crankcase pressure drops, the liquid refrigerant flashes to gas and the mixture turns to foam which the oil pump cannot handle well and which is a poor bearing lubricant. Where such problems are encountered, crankcase heaters can be used as well as timed pumpout cycles and piping changes.

Although it is possible to drive a refrigeration compressor by a stationary engine, almost all applications the

PREVENTIVE MAINTENANCE

Whenever working on compressors, look at crankcase oil sight-glass reservoirs for proper oil level. If it is below ⅓ full, add oil; if greater than ⅔ full, consider draining the excess. If the oil is foaming excessively, then evaluate if any problems exist.

Once a year, when possible on larger compressors, perform an acid analysis of the oil. If heavy acids exist drain the oil and install fresh oil. Fill to a ½ sightglass or manufacturer's recommendations.

ON THE JOB

Never take ohm measurements or continuity checks on compressors until you confirm that all power is off, including the crank case heater.

Lockout and tagout the electrical supply.

Cut out or unbolt the compressor. Never use a torch to remove the compressor. Oil could ignite and start a fire. The pressure in the system should be 0–2 psig before opening the system.

Never run a compressor that is under a vacuum; this can cause arcing internally to the terminals and damage the compressor.

Never run a compressor until the discharge service valves are open to the system.

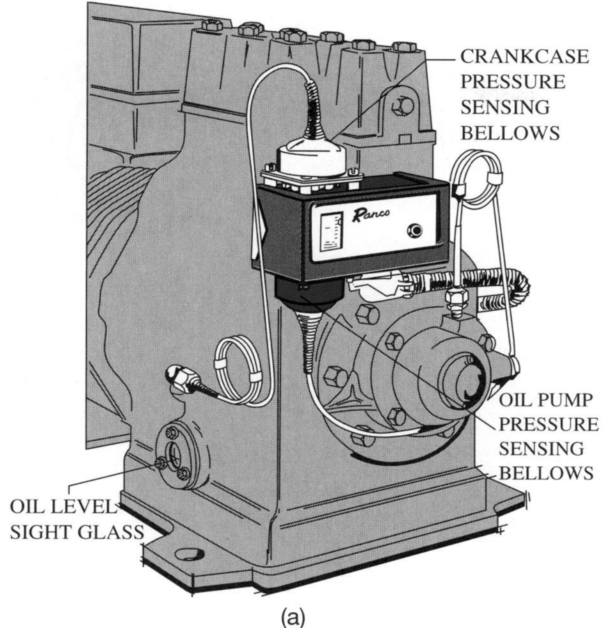

(a)

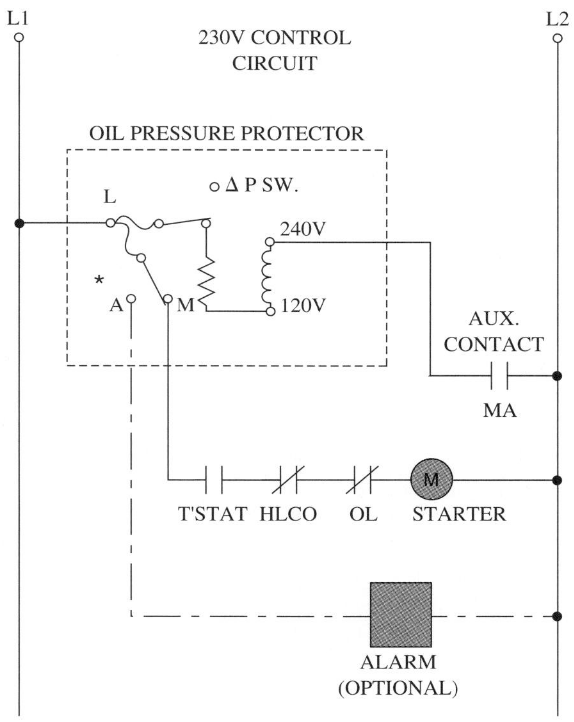

(b)

Figure R11-10 Oil pressure protector for pressure-lubricated compressors: (a) time delay ranges 45, 60, 90 and 120 seconds; (b) simplified wiring schematic. (Courtesy of Ranco North America)

service technician will encounter will be electric drive. An open compressor with separate motor will require ventilation to keep the motor from over-heating, if enclosed in a confined space. If the compressor is a welded hermetic or a semi-hermetic, the motor cooling is normally provided by the refrigerant suction-gas stream, and ventilation usually is not a concern. Engine-driven compressors are a special case, need ample ventilation, and should be installed according to manufacturers' recommendations.

Knowledge Extension

Contact your local refrigeration supply house and request technical service manuals for any compressors you expect to be maintaining or repairing. Always reference the service manual for troubleshooting and standard service procedures. If there is no technical information available, contact ARI for the compressor manufacturer's phone number and address. Request service literature from the manufacturer before experimenting on unfamiliar equipment.

Several compressor manufacturers offer three- to five-day factory training seminars in various regions throughout the year. Consider attending these every three years to stay current with new product releases and service information.

R11-4 COMPRESSOR TYPES

Having dealt with the general information on compressors, it is time to look at the various types available for refrigeration, and to examine the characteristics of each type—how they work and how they are serviced.

Fig. R11-11 illustrates the five different types of compressors used for the halocarbon refrigerants. Their uses depend on the application and the size (tonnage) of the project. The general types are as follows:

1. Reciprocating compressors
   a. Open, direct-drive
   b. Open, belt-drive
   c. Open, for transporting vehicles
   d. Semi-hermetic
   e. Hermetically sealed
2. Rotary
3. Scroll
4. Screw
5. Centrifugal

Fig. R11-12 indicates the various applications for these types of compressors.

R11-4.1 Reciprocating Compressors

Reciprocating compressors have been used in refrigeration service for a long time, dating from times when ammonia (NH_3) and carbon dioxide (CO_2) were the prevalent

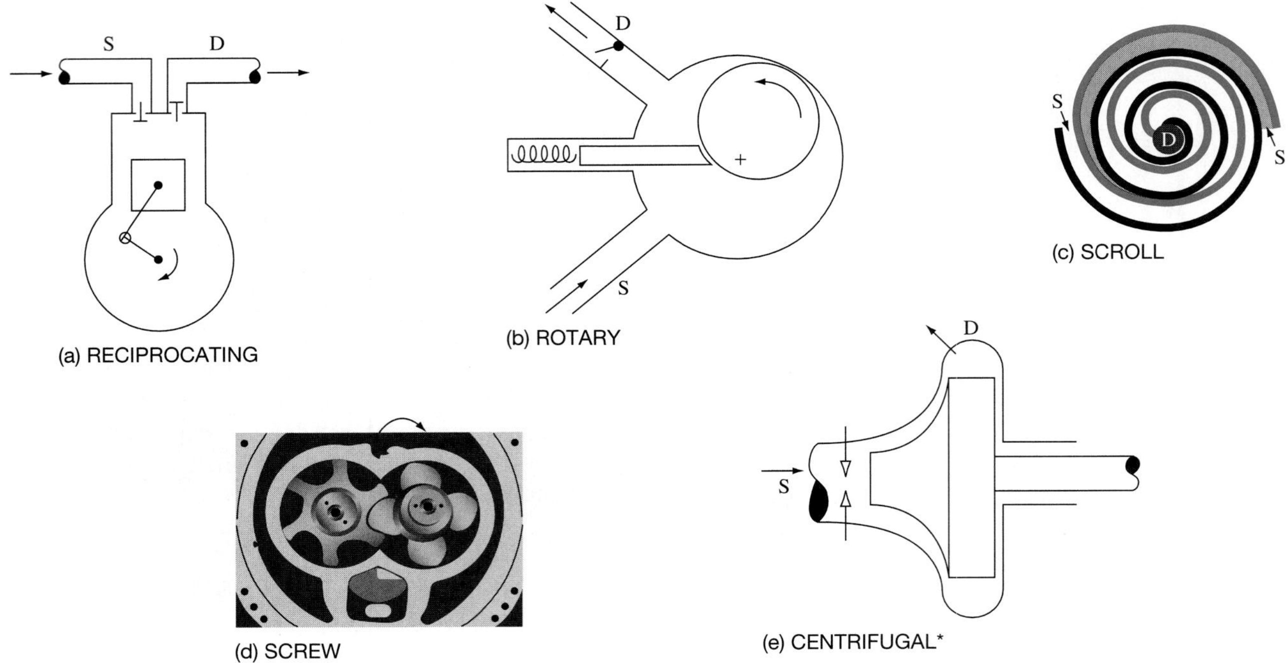

* ALL ARE POSITIVE DISPLACEMENT EXCEPT CENTRIFUGAL.

NOTE: OPEN TYPE RECIPROCATING, SCREW, AND ROTARY VANE COMPRESSORS ARE ALSO USED WITH AMMONIA REFRIGERANT.

Figure R11-11 Common types of halocarbon refrigeration compressors.

refrigerants. Like the automobile engine, these compressors are relatively complicated, have many moving parts, and yet are economical, perform well, and are reliable. The oldest compressor design is the so-called open type. It is driven by a separate motor usually mounted on a common steel base.

Open compressors have performed reliably over the decades, but have been largely superseded by newer de-

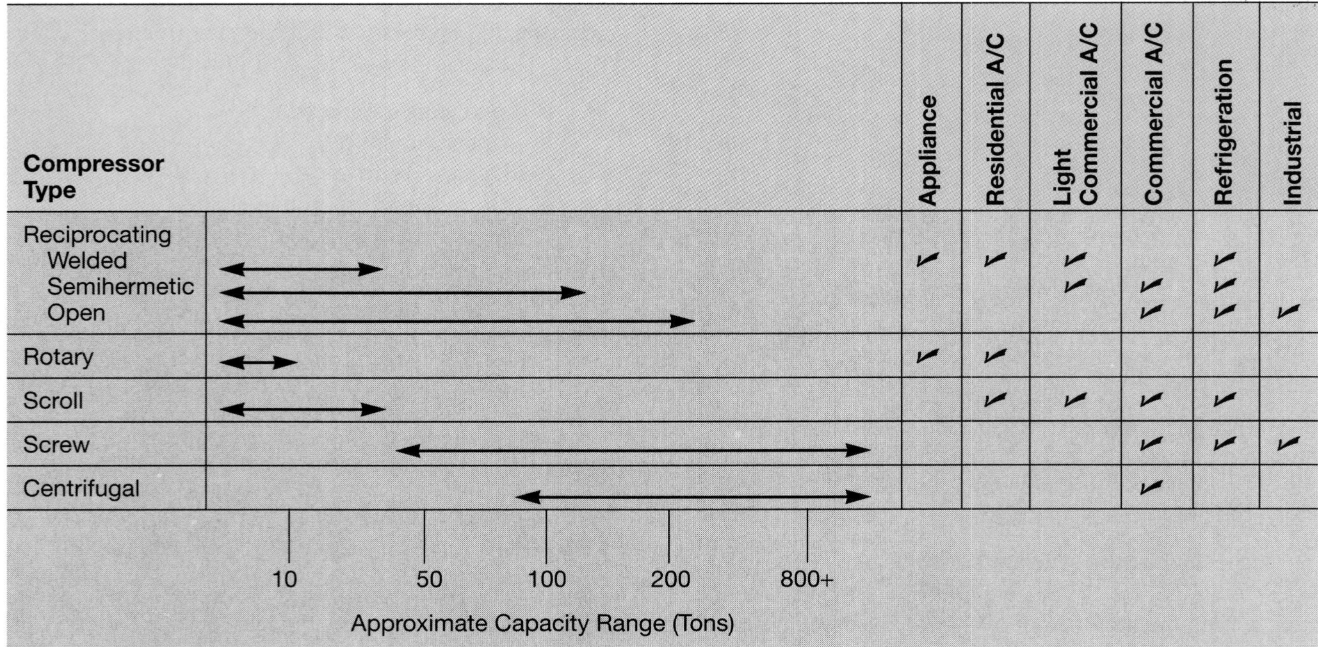

Compressor Type	Appliance	Residential A/C	Light Commercial A/C	Commercial A/C	Refrigeration	Industrial
Reciprocating						
Welded	✓	✓	✓		✓	
Semihermetic			✓	✓	✓	
Open				✓	✓	✓
Rotary	✓	✓				
Scroll		✓	✓	✓	✓	
Screw				✓	✓	✓
Centrifugal				✓		

Figure R11-12 Compressor applications by types.

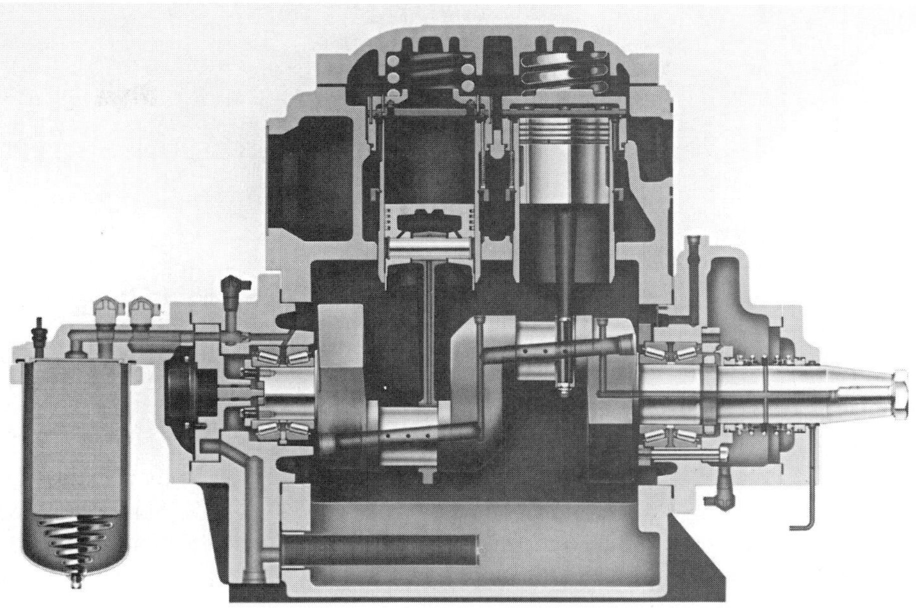

Figure R11-13 Cutaway view of open type industrial duty compressor. (Courtesy of Vilter Manufacturing Corp.)

signs. Open machines (unlike hermetics) can be used with ammonia (NH_3) refrigerant as long as all components and accessories are made of iron or steel (or other suitable materials). Sizes range from 5 to 150 tons and larger. They are often used for refrigeration and industrial applications where rugged, serviceable machinery is required.

Fig. R11-13 shows a cutaway view of an industrial-type open compressor. Fig. R11-14 and Fig. R11-15 show details of essential components.

All open machines use a shaft seal to prevent refrigerant from leaking around the crankshaft. This is also a maintenance consideration.

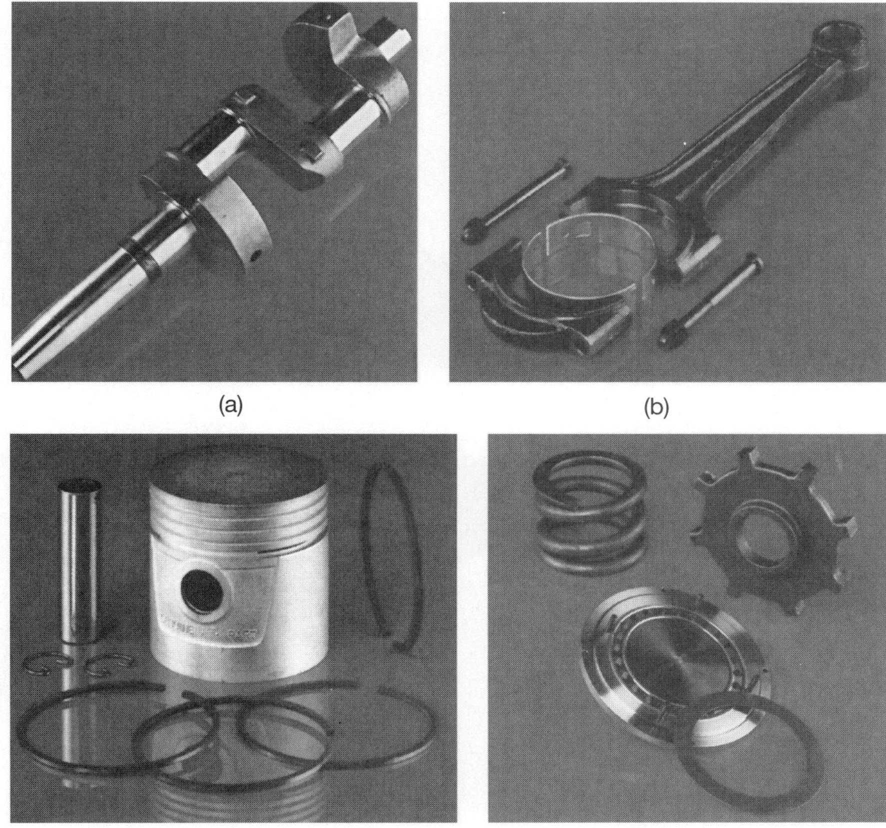

Figure R11-14 Industrial duty compressor components: (a) Ductile iron crankshaft, (b) Die forged steel connecting rod, (c) Tin plated aluminum piston, (d) Spring-loaded ring valve assembly. (Courtesy of Vilter Manufacturing Corp.)

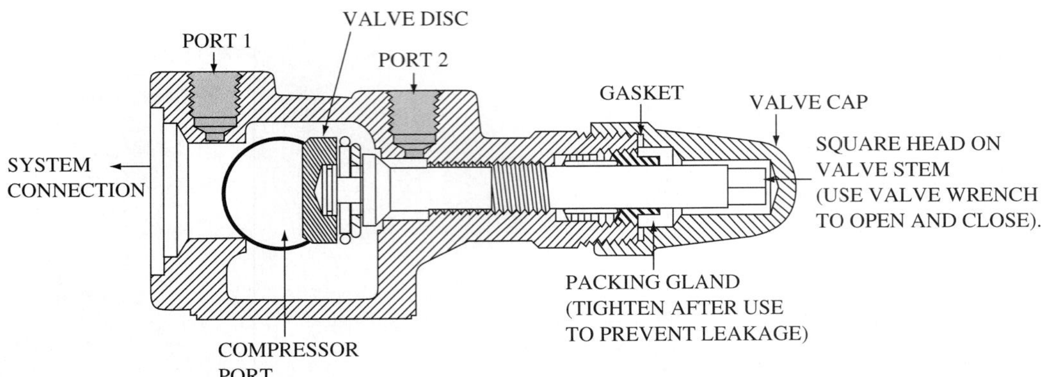

Figure R11-15 Cross section of compressor service valve. (Courtesy of Frick Company)

If *direct drive* (Fig. R11-16), the motor shaft is coupled to the compressor shaft and driven at motor speed.

Belt-driven machines (Fig. R11-17) offer the flexibility of selecting a compressor speed to match the load. The belts require additional space and a protective guard. They require increased maintenance and increased power due to operational losses.

The open-transport types (Fig. R11-18) were developed for the transportation industry to be engine driven.

Hermetic compressors were developed in the 1920s and 30s when refrigerants came into use that were compatible with electric motor components, especially wire insulation. The benefits for use on a domestic refrigerator were immense and obvious:

- Direct connected, reliable, no belts or couplings to wear;
- Sealed in welded steel shell, no shaft seal to leak;
- Very compact, permitting increased refrigerated storage space.
- Lower sound level, ideal for domestic appliances

Larger sealed machines were developed for air-conditioning applications by the late 1930s and by the mid 1950s the open machine had been largely supplanted in most appliance and packaged HVAC/R applications.

Hermetically sealed electric motor-compressor units, (Fig. R11-19), are made in a variety of sizes, from tiny fractional horsepower units meant for small appliances, to larger units up to about 20 tons for air-conditioning use. They are sometimes called welded hermetics, full hermetics or sealed hermetics. For the sake of simplicity, they shall simply be referred to hereafter as hermetics.

Figure R11-16 Open-type direct-drive compressor. (Courtesy of York International Corp.)

Figure R11-17 Open-type belt-driven compressor. (Courtesy of York International Corp.)

The welded steel shell prevents any field service access so there are no service procedures to replace damaged internal components such as motors, bearings, valves, etc. If damaged or defective, the entire hermetic compressor is replaced. They are usually internally spring isolated, (Fig. R11-20), to reduce the inherent vibration caused by the reciprocating action of the pistons.

Fig. R11-21 shows a high-efficiency hermetic compressor with the suction valves located in the crown of the piston. Fig. R11-22 shows a larger sized vertical hermetic compressor with a centrifugal oil pump.

Figure R11-18 Open compressor for transportation applications. (Courtesy of Thermo King)

Figure R11-19 Small welded hermetic compressor using HFC-134a refrigerant. (Courtesy, Copeland Corporation)

In addition to the (welded) hermetic, the *semi-hermetic* motor-compressor unit has evolved over the past 50–60 years which is field serviceable by virtue of its bolted construction. It may have a bolted cast-iron construction or bolted, flanged, drawn-steel shell. It is field- or factory-

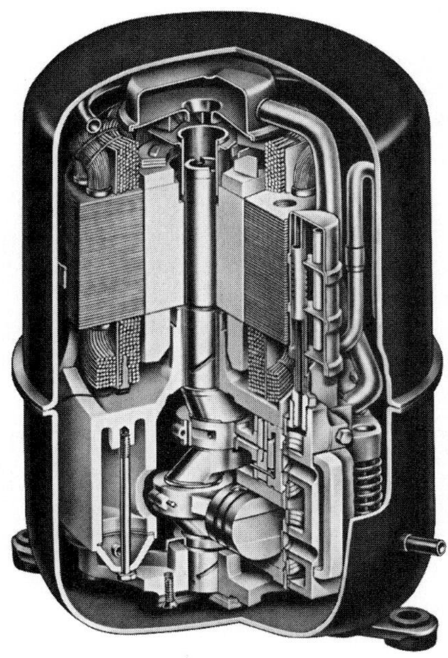

Figure R11-20 Cutaway view of hermetic compressor showing internal isolation springs. (Courtesy of Tecumseh Products Company)

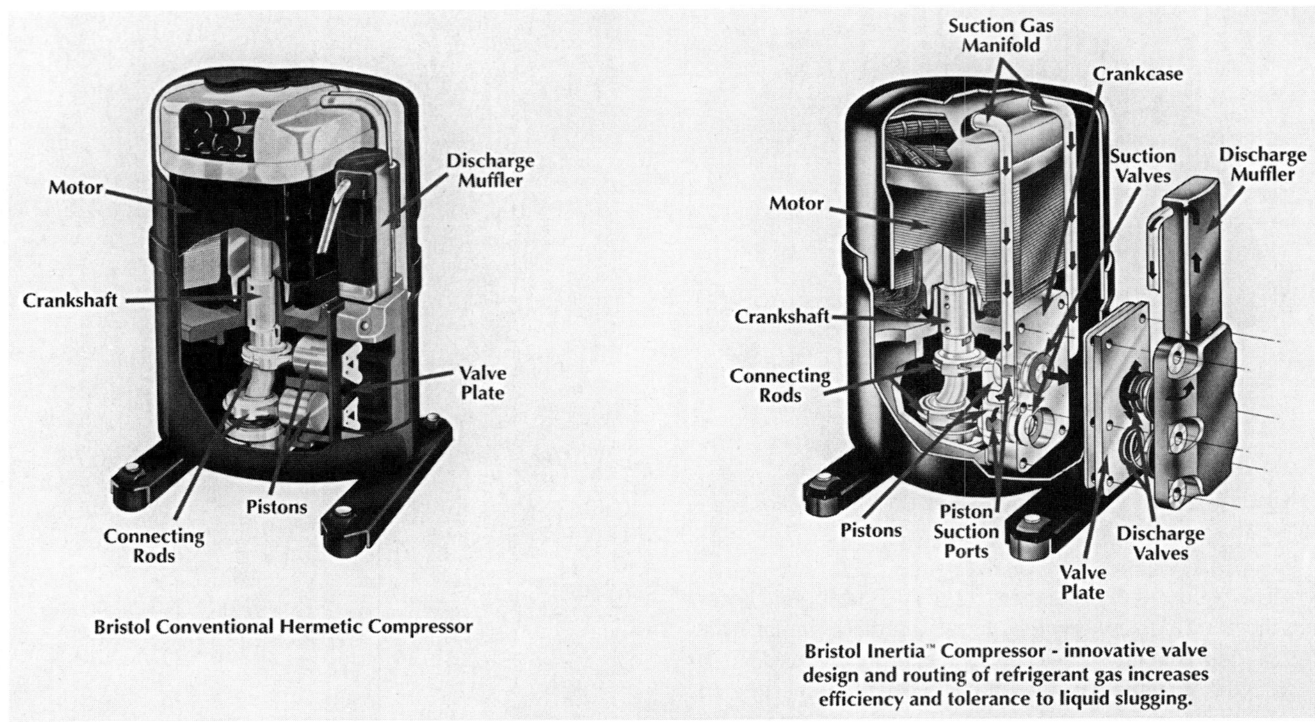

Figure R11-21 High-efficiency hermetic compressor with suction valves in crown of piston. (Courtesy of Bristol Compressors)

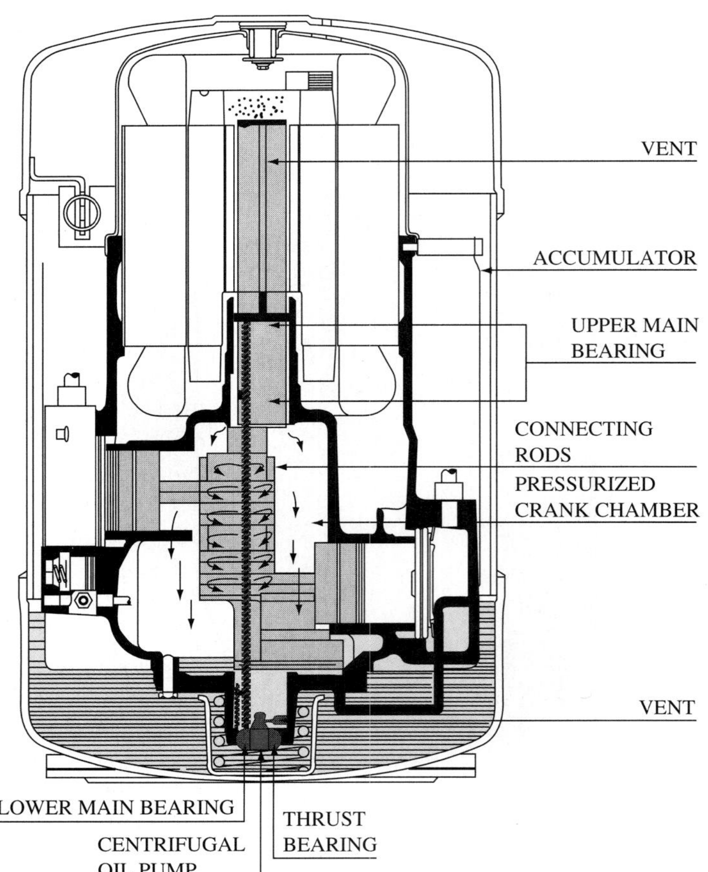

Figure R11-22 Section of hermetic compressor showing centrifugal oil pump. (Courtesy of Cutler-Hammer, Eaton)

Figure R11-23 Smaller sized semi-hermetic compressor using HFC-404a refrigerant. (Courtesy, Copeland Corporation)

repairable, and parts such as valve reeds, gaskets, bearing inserts, or motor stators may be replaced on various units. These machines are known by a variety of names: semi-hermetics, serviceable hermetics, accessible hermetics, and bolted hermetics. For the sake of consistency, they shall be referred to hereafter as semi-hermetics. The units are available in sizes that range from approximately 7.5 to 125 tons cooling capacity. Fig. R11-23 shows a semi-hermetic compressor used for medium and low temperature refrigeration.

Fig. R11-24 shows the pressure lubrication system for a refrigeration duty semi-hermetic compressor. Fig. R11-25

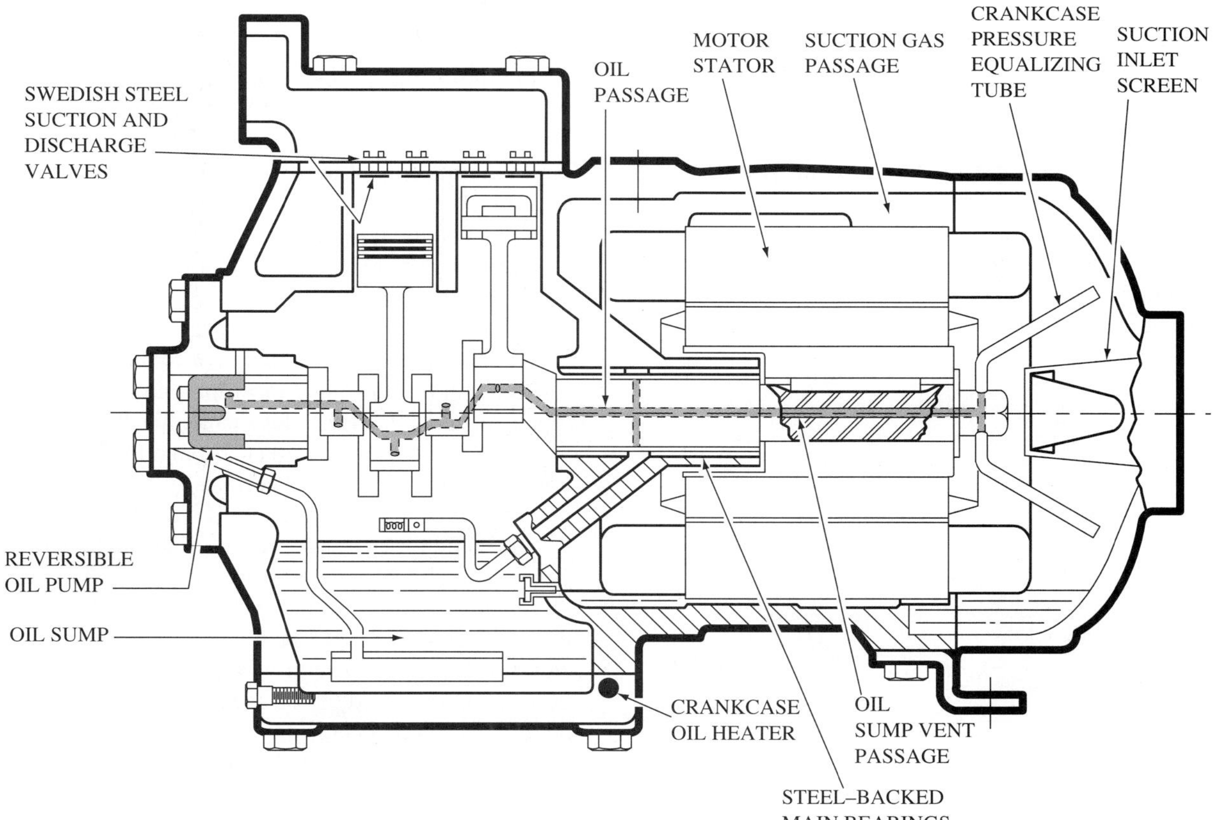

Figure R11-24 Refrigeration duty semi-hermetic compressor showing pressure lubrication. (Courtesy of Carlyle Compressor Company)

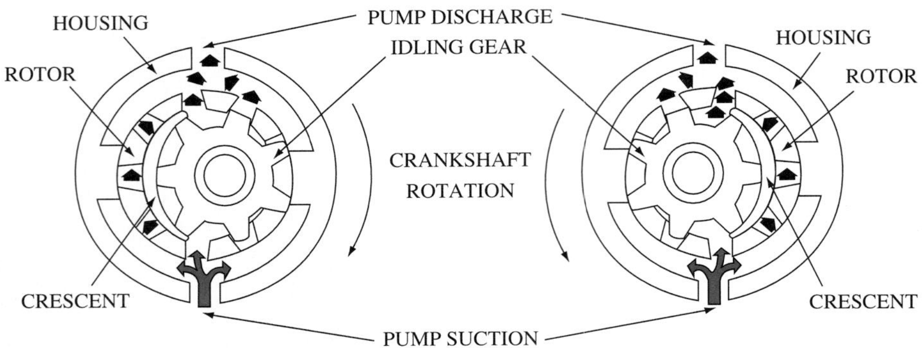

Figure R11-25 Reversible oil pump for larger semi-hermetic and open-type compressors. (The Trane Company)

shows a reversible oil pump that is used in both the larger semi-hermetic and open compressors. Fig. R11-26 shows the isolation arrangement for the electric power terminals on a semi-hermetic compressor.

Operation of Reciprocating Compressors

In a reciprocating compressor, (Fig. R11-27), the piston is driven up and down in the cylinder by the connecting rod and crankshaft. The valves are self-opening and closing due to pressure difference. They sometimes are spring loaded, but the principle is the same. Note that this is quite different from an automobile engine where the valves are mechanically opened and closed by a cam shaft synchronized to the crankshaft.

Fig. R11-28 shows the types of running gears used in the reciprocating compressors: (a) the crank type with split connecting rods; (b) the eccentric type with a one-piece strap; (c) the double Scotch yoke with a slide block.

Fig. R11-29 shows the various types of reciprocating cylinder valves: (a) the flexing-reed valve; (b) the floating-reed valve; (c) the ring valve; (d) the reduced-clearance poppet valve used for the Discus compressor.

As the piston strokes down in the cylinder (Fig. R11-30), the pressure will drop to a point where it is

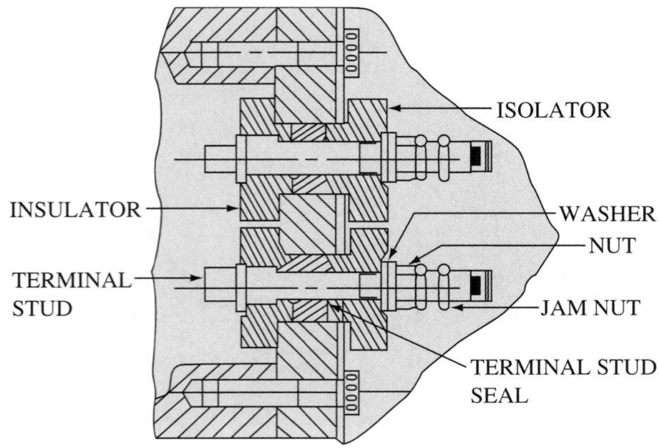

Figure R11-26 Electric motor terminals on semi-hermetic compressor. (The Trane Company)

below the pressure in the suction manifold. At that point the pressure difference will force the suction valve(s) to open, and low-pressure, cold (but superheated) suction gas from the evaporator will flow into the cylinder.

At the bottom of the stroke the piston reverses its travel (Fig. R11-31). The pressure difference that held the suction valve open disappears and the valve(s) starts to close. As the piston rises, compressing the gas and building up pressure, the suction valve(s) closes tightly. The discharge valve will stay closed until the cylinder pressure exceeds discharge manifold pressure.

At this point, (Fig. R11-32, p. 213) the discharge valve(s) opens and the compressed gas flows out into the discharge manifold, its pressure and superheat increased.

Clearance Volume

At the top of the stroke, piston travel will reverse again. All the gas that can has passed out into the discharge manifold. The so-called clearance volume space between the top of the piston and the cylinder head is filled with compressed gas that is not going anywhere. This gas will have to expand on the downstroke before the pressure can drop to a point to admit fresh suction gas. This reexpansion gas represents wasted work and lost capacity. The compressor designer strives to minimize the clearance volume while at the same time leaving adequate room for thermal expansion and for unwanted incompressible slugs of oil and liquid refrigerant. These slugs can enter the cylinder, and possess the ability to do great damage to the valves, piston, and cylinder head. Theoretically, 0% clearance volume would be good; practically it would be disastrous.

The action that takes place in the cylinder can be graphically shown on a pressure-volume (*P-V*) diagram.

The horizontal axis in Fig. R11-33 (p. 213) represents the cylinder volume and the cylinder pressure is displayed on the vertical axis. The previously discussed cylinder action can be reviewed on the *P-V* diagrams to gain a better picture of what is occurring.

Starting again, point *A* (Fig. R11-34, p. 213) represents top dead center of the piston stroke and shows the clearance volume and discharge pressure. As the piston travels on the downstroke the trapped compressed clearance vol-

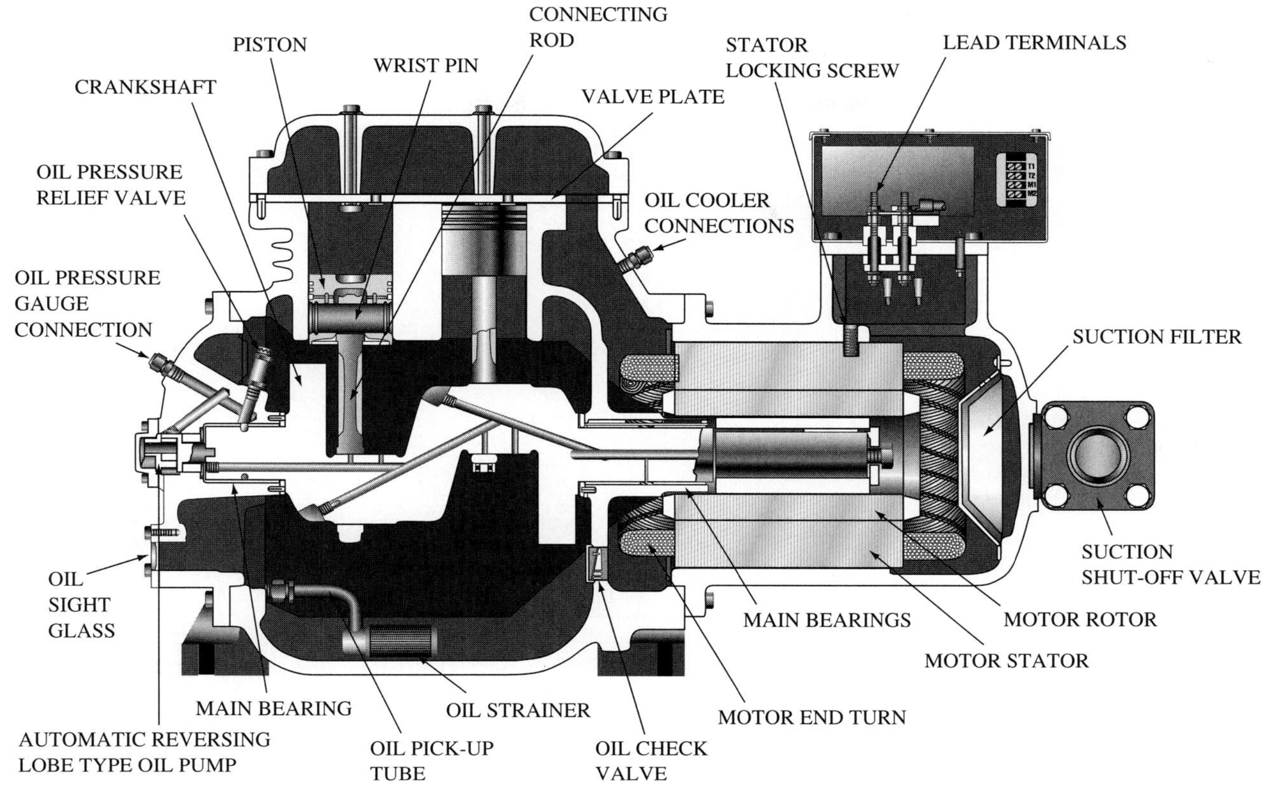

Figure R11-27 Cutaway view, semi-hermatic reciprocating motor compressor unit. (Courtsey of Dunham-Bush, Inc.)

ume gas has to re-expand until it reaches suction pressure at point *B*. This represents perhaps 20% of the intake stroke and points to the inefficiency involved. The other types of machines (scrolls, rotaries, etc.) do not have trapped re-expansion gas to deal with and, thus, have enhanced efficiencies.

As the downstroke continues from *B* to *C* the cylinder pressure is below the suction manifold pressure and the pressure difference opens the suction valve(s) and fills the cylinder with new refrigerant vapor, (Fig. R11-35). At the bottom of the stroke (*C*) the suction valve(s) closes.

The compression stroke from bottom dead center (*C*) to point *D* compresses the gas to a point about equal to discharge manifold pressure (Fig. R11-36, p. 214). To this point, both suction and discharge valves have remained

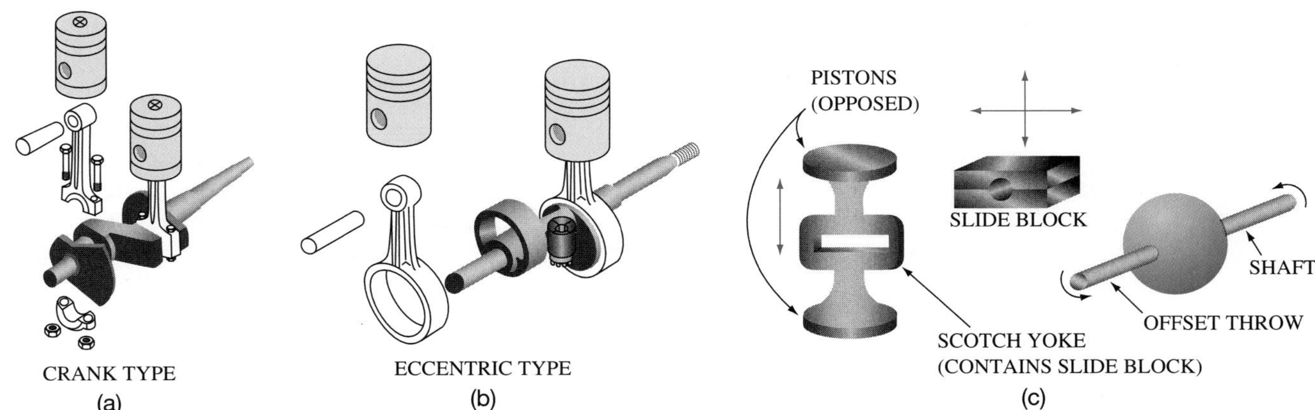

Figure R11-28 Types of reciprocating compressor running gear: (a) crank type, (b) eccentric type (c) double Scotch yoke. (Courtesy, Copeland Corporation)

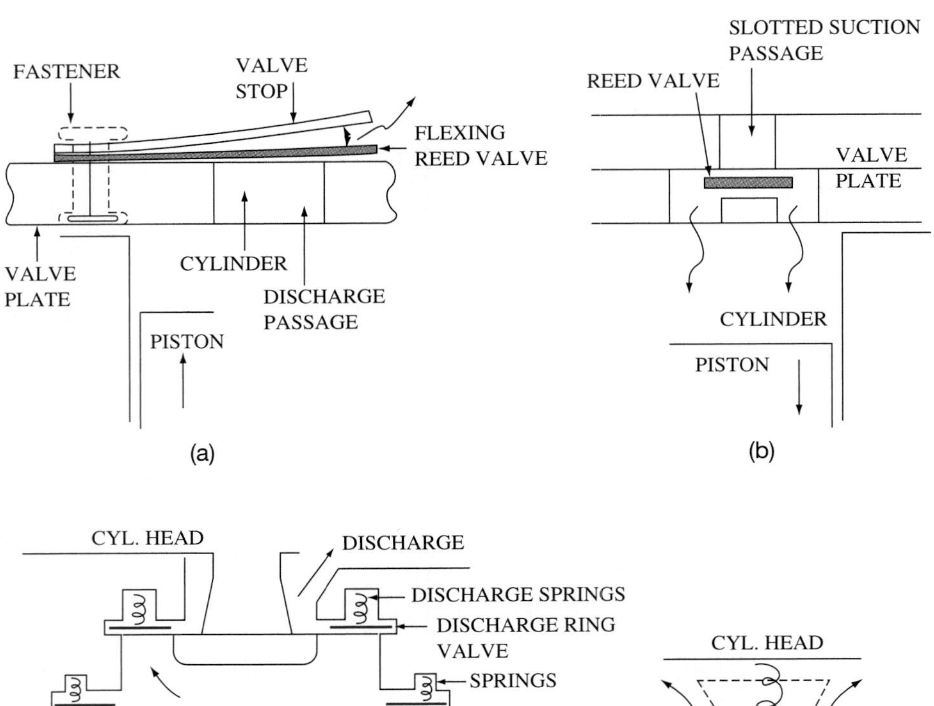

Figure R11-29 Typical reciprocating compressor cylinder valves: (a) Flexing reed valve, (b) Floating reed valve, (c) Ring valves, (d) Reduced clearance poppet valve, Discus compressor.

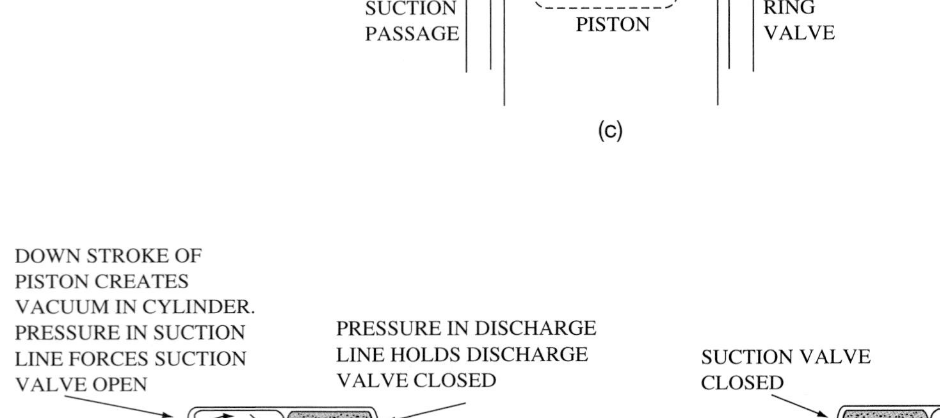

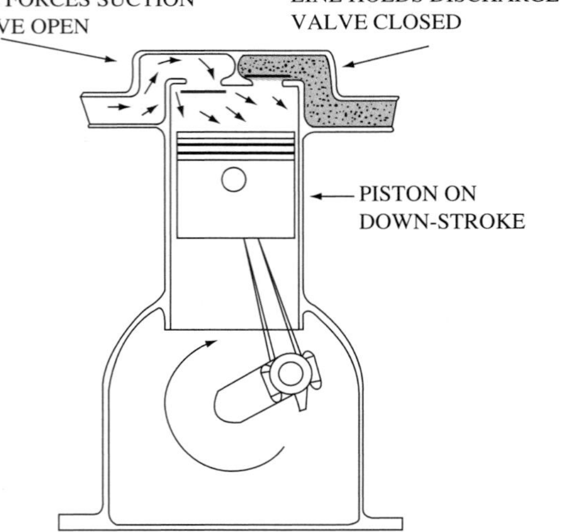

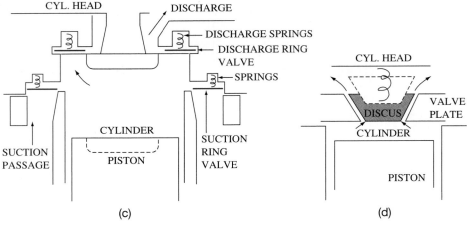

Figure R11-30 Down-stroke. Permits suction gas to open valve and to flow into lower pressure area.

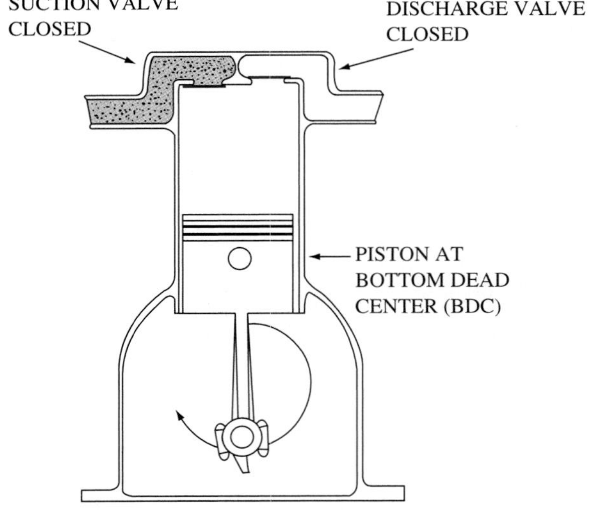

Figure R11-31 Bottom of stroke. Both suction and discharge valves close.

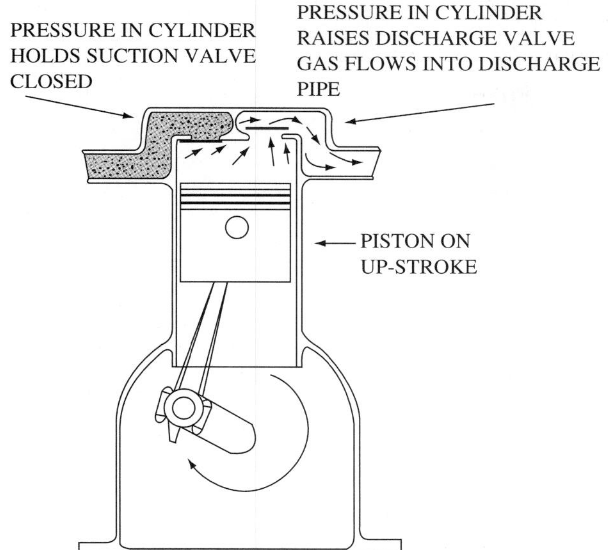

Figure R11-32 Up-stroke. Discharge valve will open when cylinder pressure exceeds discharge manifold pressure.

closed. The piston has traveled from 100% of cylinder volume to a point about 20% of cylinder volume. On air-conditioning R-22 applications, the suction pressure would be about 69 psi (40°F saturation) and the discharge pressure about 263 psig (120°F saturation).

As the piston completes its compression stroke from point *D* to *A* (top dead center), the cylinder pressure exceeds the discharge manifold pressure and forces the gas

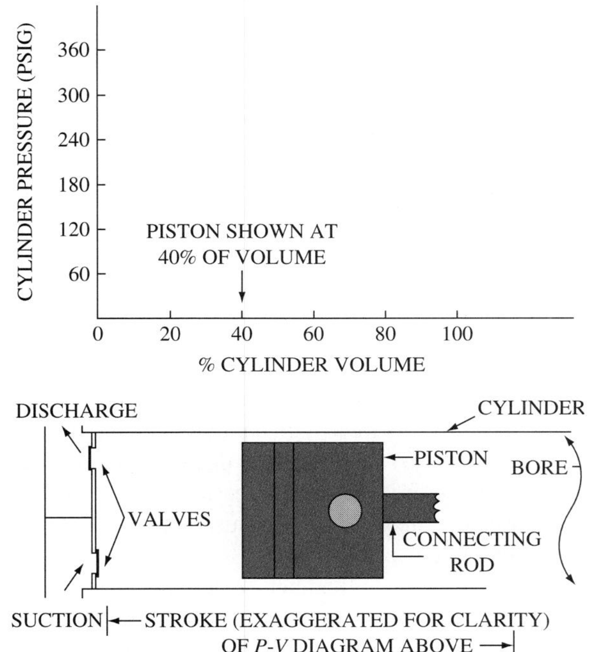

Figure R11-33 The two elements of a pressure-volume plot, used in the study of reciprocating-compressor performance.

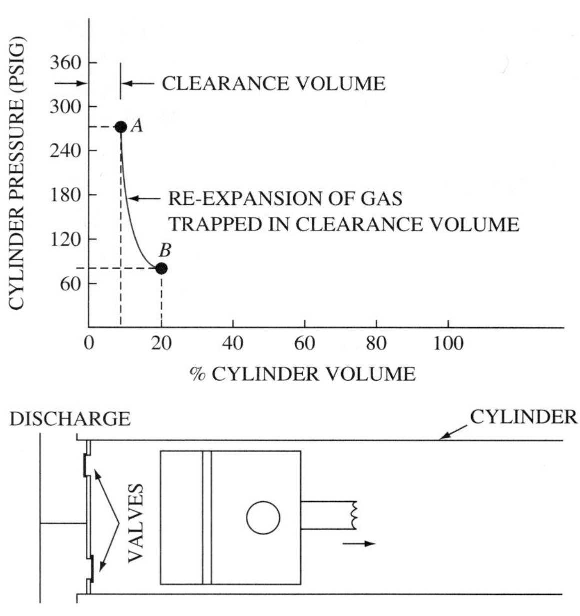

Figure R11-34 Beginning of the intake stroke. *A*: top dead center; *B*: suction valve ready to open.

charge out with the exception of the gas occupying the clearance volume space (approximately 4% volume) (Fig. R11-37). This trapped gas will remain and re-expand with the next intake stroke.

The previous illustrations assume an ideal world with no pressure losses across the valves, minimum turbulence, etc. A more realistic view is found in Fig. R11-38, which shows pressure losses. The area enclosed by points *A, B,*

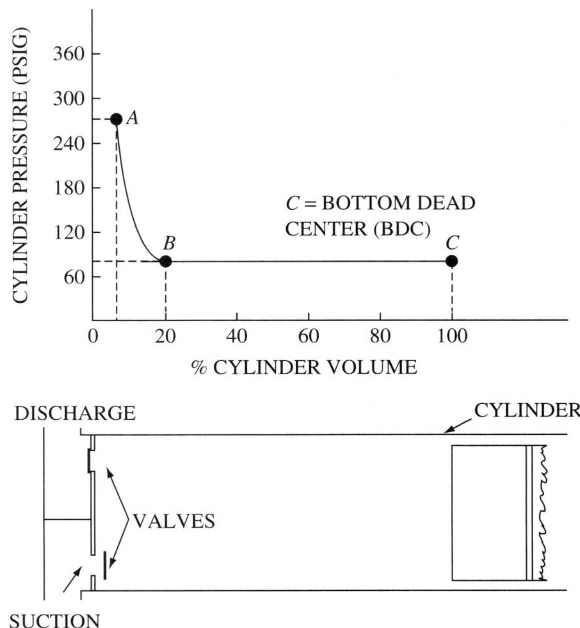

Figure R11-35 Useful portion of intake stroke, *B* to *C*.

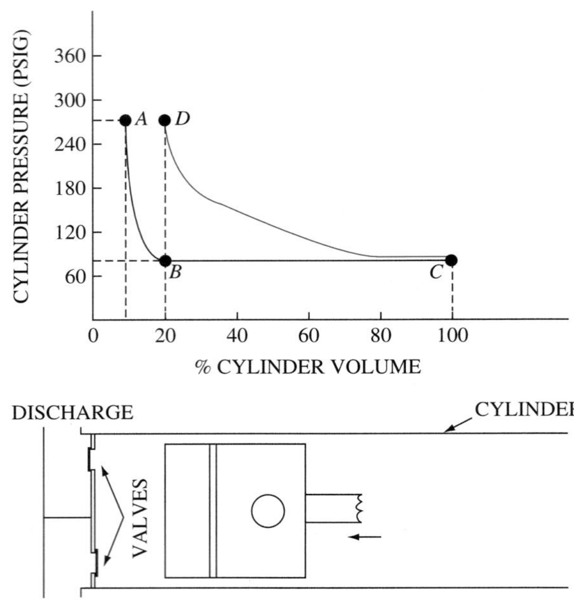

Figure R11-36 Compression stroke, *C* to *D,* with cylinder valves closed.

C and *D* on the pressure-volume (*P-V*) plot represents the work done by the compressor on the gas passing through the cylinder.

The wasted effort represented by re-expansion of clearance volume gas from points *A* to *B* can be increased if small amounts of liquid (a wet suction) are compressed, flash and re-expand. This will reduce the enclosed area, *A-B-C-D,* and reduce output. Increasing discharge pressure and/or reducing suction pressure likewise reduce the output of the cylinder. Other things being equal, a machine

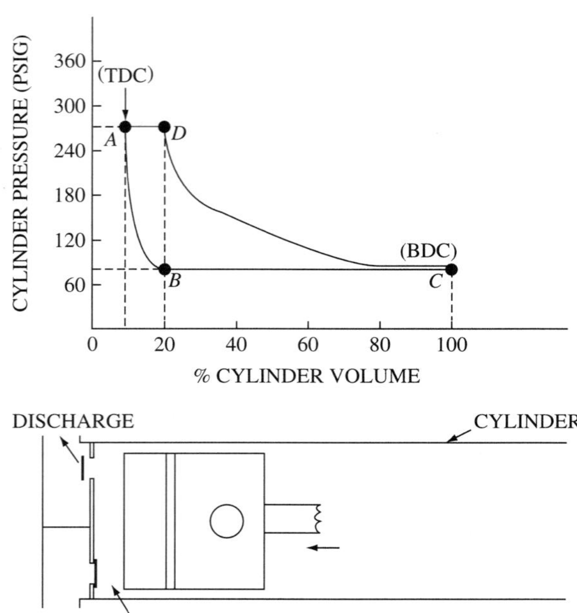

Figure R11-37 Discharge portion of compression stroke, *D to A.*

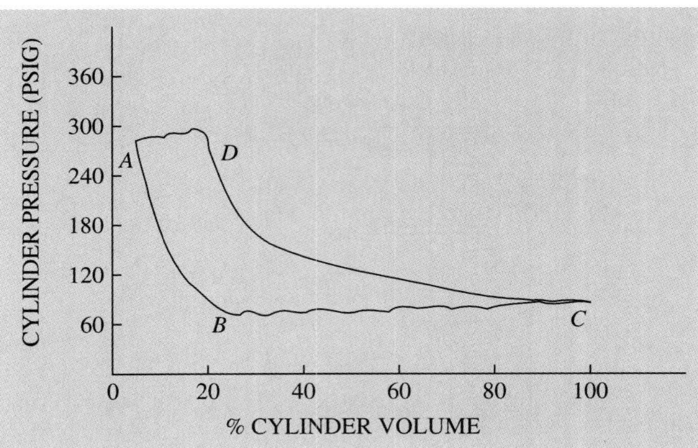

Figure R11-38 Complete *P-V* diagram for a one-cycle compressor cylinder.

will produce the greatest output by running at the highest suction temperature and the lowest discharge temperature practical: the most work at the lowest energy rate.

R11-4.2 Rotary Compressors

Rotary compressors are widely used in small welded hermetic sizes to power small refrigerated appliances, window air conditioners, packaged terminal air conditioners, and heat pumps in sizes up to about 5 tons. They are simple in design, efficient, and run smoothly and quietly. They are described as rolling-piston or single-valve rotaries (Fig. R11-39). They are not internally spring isolated.

In addition to the rolling piston type unit is a rotary vane compressor which includes a multivane type used for higher capacity design.

Operation of Rotary Compressors

The *rolling piston* (Fig. R11-40) is a rotating off-center cam or lobe which sweeps a path inside a round cylinder. The drive shaft is centered in the cylinder. A close-fitting spring-loaded sliding vane follows the piston and separates the cylinder openings for suction (inlet) gas and discharge (outlet) gas.

As the rolling piston rotates, gas will flow in and fill the suction cavity due to pressure difference, much the same as in a reciprocating compressor. The piston continues to rotate, closes off and passes the suction port and compresses the gas until its pressure is high enough to flow out through the discharge (reed) valve into the discharge manifold. No suction valve with its inherent pressure loss is needed, so volumetric efficiency is high. Because of the rotating rather than reciprocating piston motion, vibration levels are reduced and are easier to dampen (externally).

Compact hermetic versions are built in sizes from small fractional horsepower up to about 5 Hp. Close production tolerances are required and lubricating oil helps provide a seal between the active surfaces. Application is common in window air conditioners, package terminal units and appliances.

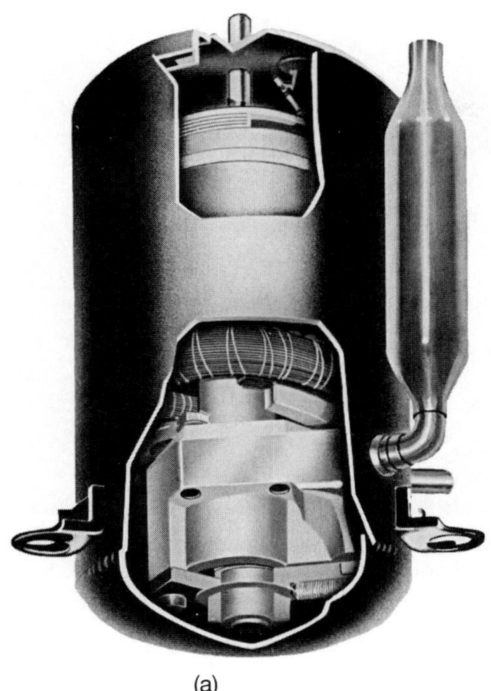

(a)

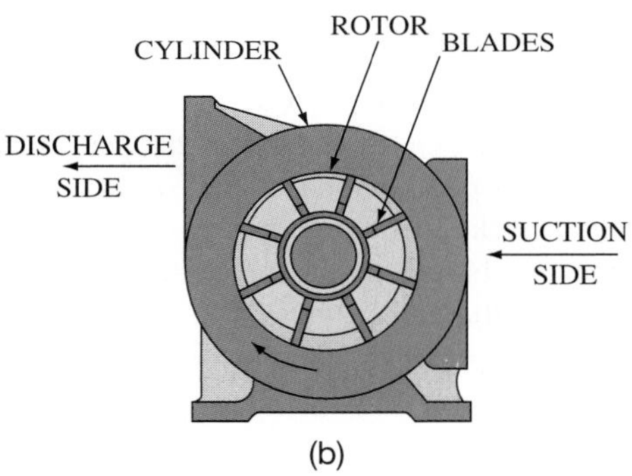

(b)

Figure R11-39 Rolling piston rotary compressor. (Courtesy of Rotorex Company, Inc.)

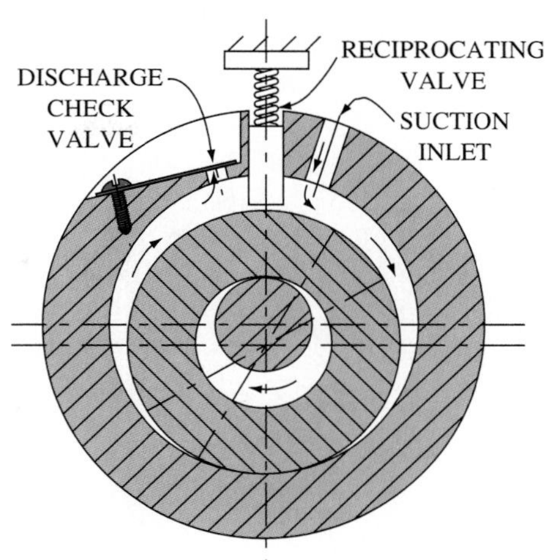

Figure R11-40 Diagram of a rolling-piston rotary compressor. (Copyright by the American Society of Heating, Refrigerating, and Air-Conditioning Engineers, Inc. Used by permission.)

Rotary-vane compressors are sometimes called fixed-vane rotary compressors (Fig. R11-41) and are used in larger specialized applications.

A large off-center rotating shaft carries multiple vanes in slots. The vanes have a reciprocating action as they move in and out, following the cylinder walls. The moving cavities (cells) formed between adjacent vanes fill with gas as they sweep by the suction port, then compress the gas as the moving space between the vanes, cylinder wall, and shaft is reduced. Finally, the compressed gas is released through a discharge port and the process repeats. An oil-flooded design provides sealing, and construction is of the open type. Typically, these machines are used as booster compressors for low-temperature ammonia systems as well as for single-stage compressors for industrial and process applications.

R11-4.3 Scroll Compressors

Scroll compressors (Fig. R11-42) are simple hermetic rotary machines which compress gas between two close-

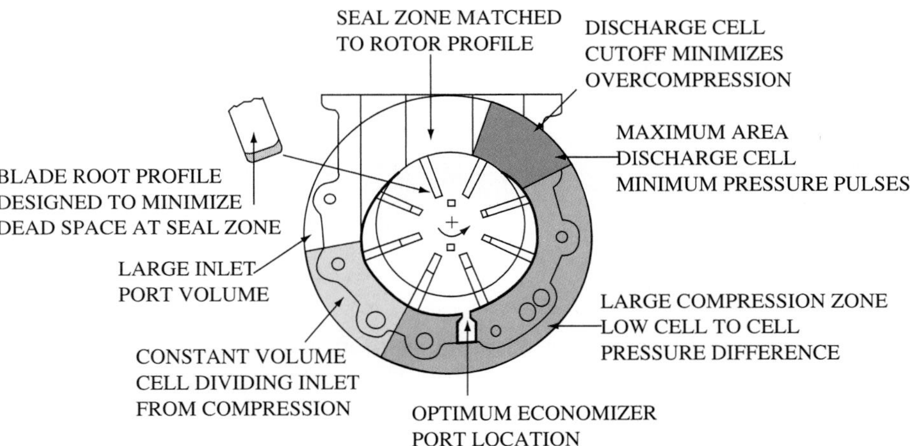

Figure R11-41 Cross section of a rotary-vane compressor. (Copyright by the American Society of Heating, Refrigerating, and Air-Conditioning Engineers, Inc. Used by permission.)

Figure R11-42 Partial cutaway of a scroll compressor. (Courtesy, Copeland Corporation)

fitted spiral scroll members. One is fixed and the other moves (but does not rotate) in an orbital path. Progressively reducing cavities compress the refrigerant gas with little or no vibration.

They are positive displacement machines with high volumetric efficiencies, currently available in sizes from about 1 to 12 tons air-conditioning capacity. Because they are approximately 10% more efficient than a comparable reciprocating compressor, they have experienced extraordinary growth in U.S. residential air-conditioning applications where manufacturers must meet increasingly higher federal energy efficiency (SEER) regulations.

They are built in a welded hermetic design.

Figure R11-43 Scroll plate: one of the two matched units that comprise the scroll compressor. (The Trane Company)

Operation of Scroll Compressors

For a scroll machine to work well, it must be finished to very close tolerances so the contact between the flanks and tips of the scroll members is very snug. Suction gas is captured in pockets at the periphery of the scroll members and then as the orbiting scroll motion moves the gas pockets towards the center, their size is progressively reduced. This compresses the gas which is finally discharged at the center through an opening in the fixed scroll.

Fig. R11-43 shows one of the matched units that comprise the scroll compressor. Fig. R11-44 shows further de-

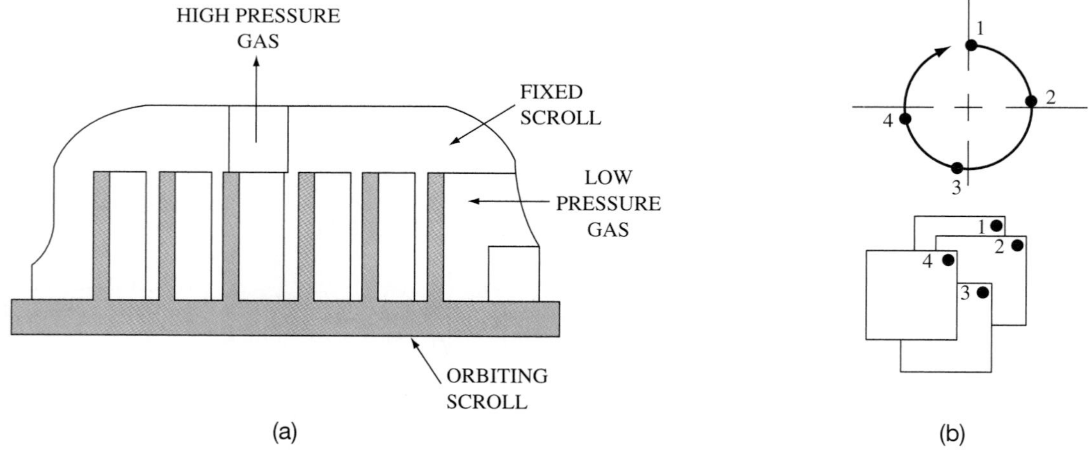

Figure R11-44 Details of the scroll compressor: (a) cross section showing inlet and discharge locations; and (b) the orbital path of a scroll compressor. (Copyright by the American Society of Heating, Refrigerating, and Air-Conditioning Engineers, Inc. Used by permission.)

tail of the design: (a) a simplified cross section through a scroll member shows the suction inlet and discharge locations; (b) the movable scroll member follows an orbital path, but does not rotate.

Scrolls are classed as compliant or non-compliant designs depending on the method used to accomplish the critical task of gas sealing. This can be accomplished in non-compliant designs with oil flooding and flexible tip seals.

Compliant designs can provide radial compliance by use of an Oldham coupling. This allows the scroll flanks to maintain close contact and yet the orbiting scroll can separate and unload to pass a slug of liquid or foreign material. Axial compliance is a feature whereby an adjustable force is applied to maintain sealing between the scroll tips during compressor operation and released on shutdown. A compliant design that will enable a compressor to handle liquid accumulations from whatever circumstances should enhance the reliability of system operation, especially under adverse conditions.

Fig. R11-45 shows a cross section of a scroll compressor. The Oldham coupling is used to drive the orbiting scroll providing radial compliance, permitting the compressor to handle liquid slugs.

Fig. R11-46 shows the scroll compression sequence, in diagrammatic form. One pair of cavities is shown in the compression sequence. Fig. R11-47 gives a comparison of the scroll and the reciprocating compressor capacities for a residential air conditioning application.

Fig. R11-48 shows tandem scroll compressors for a commercial air-conditioning application. This arrangement provides 0–50–100% capacity steps at a high efficiency level.

Scroll compressors have low vibration and sound levels. The compressor is not spring-isolated inside the welded shell. Isolation is external. It is sometimes possible to pick up low beat level frequencies from the compressor motor and scroll pump. Enclosure design and

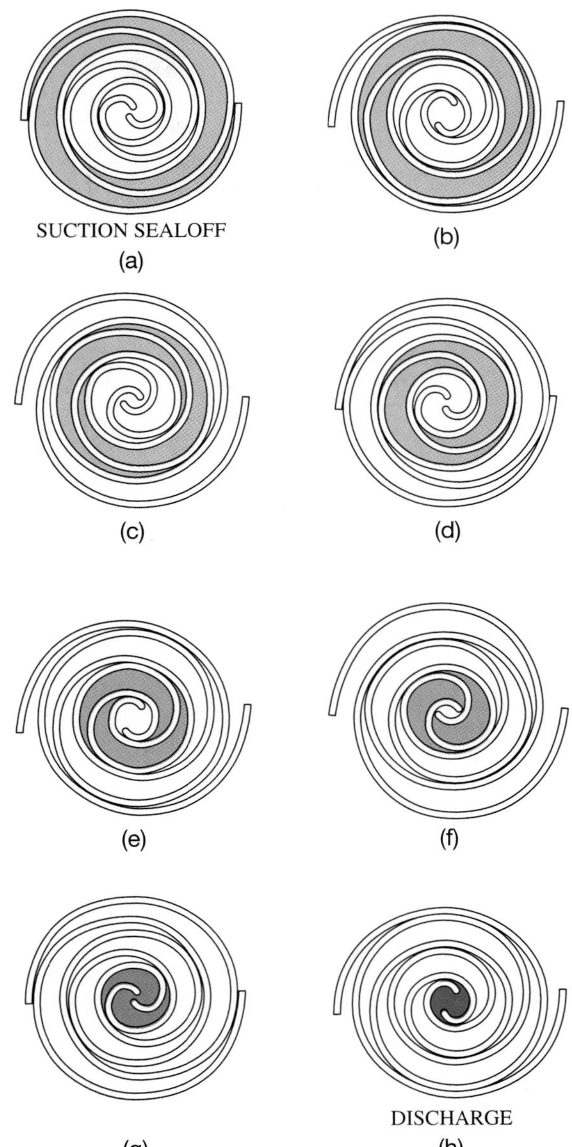

Figure R11-46 Eight views of the scroll compression sequence. (Copyright by the American Society of Heating, Refrigerating, and Air-Conditioning Engineers, Inc. Used by permission.)

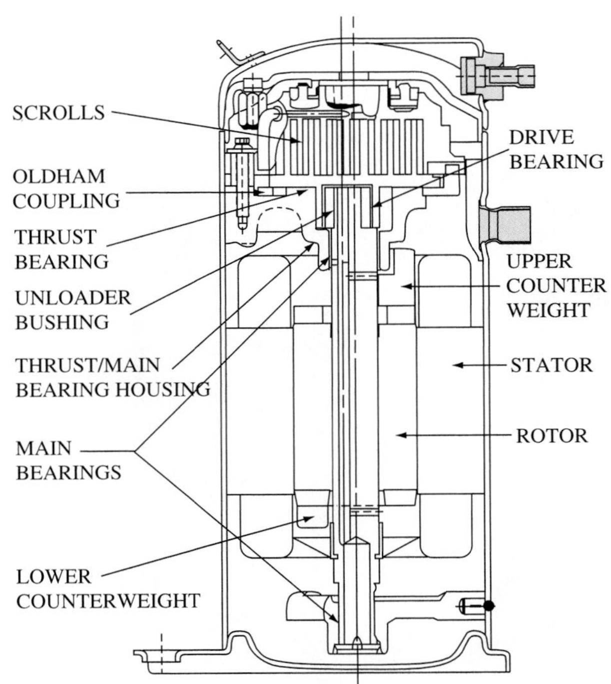

Figure R11-45 Cross section view of a scroll compressor using the Oldham coupling to drive the orbiting scroll. (Copyright by the American Society of Heating, Refrigerating, and Air-Conditioning Engineers, Inc. Used by permission.)

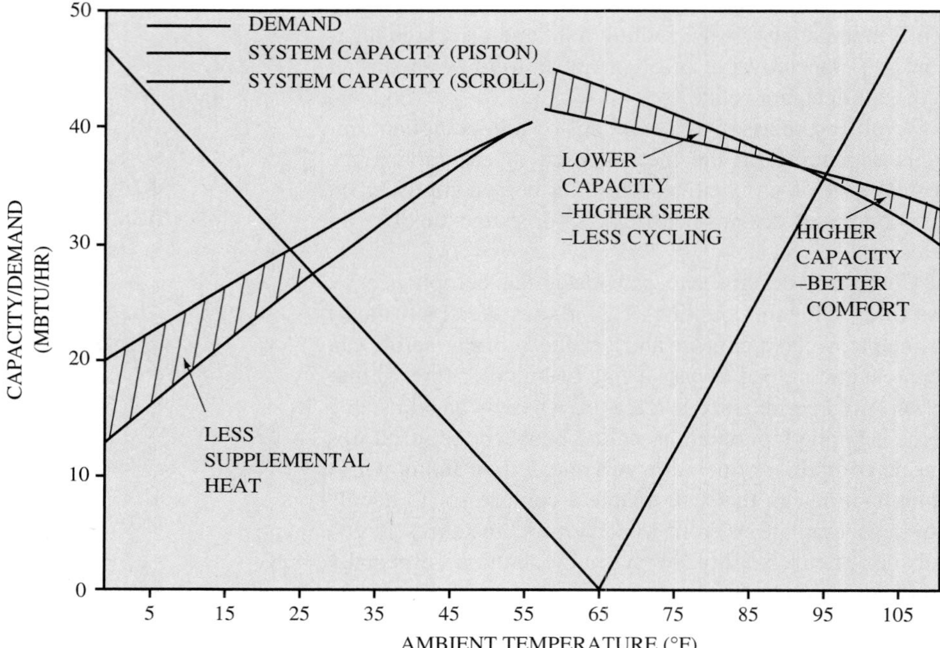

Figure R11-47 A comparison of scroll and reciprocating compressor capacities. (Courtesy, Copeland Corporation)

soundproofing can deal with this noise which is not found in other compressors.

Some scroll compressor observations:

1. Do not run the machine with the suction valve closed. Pulling a high vacuum could damage the machine.
2. Check for proper rotation of three-phase machines with gauge manifolds. If normal suction and discharge pressures do not quickly develop, shut off the machine and reverse any two-compressor power leads. Then re-check.
3. A discharge line check valve prevents continued backflow through a scroll when shutting down. A momentary sound will be heard for one or two seconds as the internal pressures equalize backward through the scroll.
4. It is possible for a scroll to seal. Always be sure to vent both the high and the low sides before working on the lines, especially if brazing or soldering.

Figure R11-48 Tandem scroll compressor used for commercial air conditioning application. (Courtesy, Copeland Corporation)

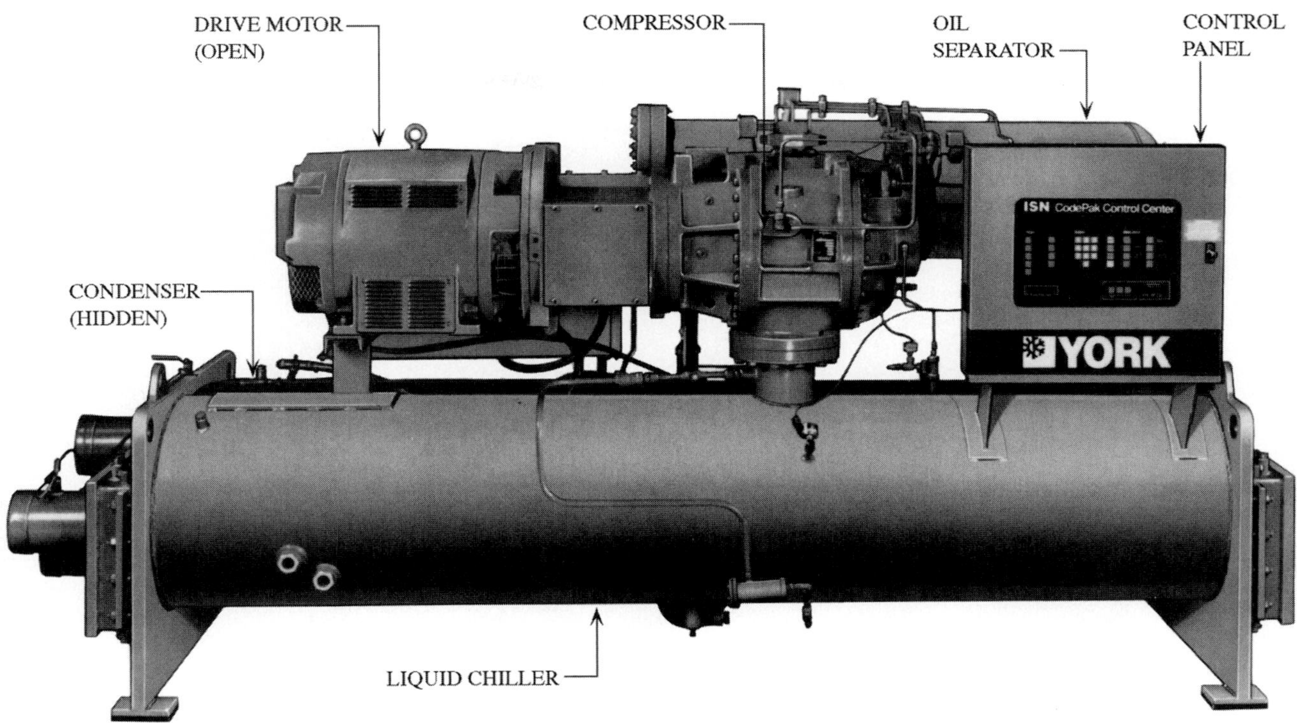

Figure R11-49 Open twin-screw rotary compressor, liquid chiller package. (Courtesy of Bruce Smith-York International Corp.)

R11-4.4 Screw Compressors

Helical rotary *screw compressors* are positive-displacement machines made for larger applications in both twin-rotor and single-rotor types. They are relatively simple in concept but, like the scroll, demand advanced manufacturing capability for producing complex shapes to very close tolerances. They are smoothly running and can deliver high compression ratios. Their volumetric efficiency is high, and they can handle a wide variety of refrigerants and gases. Consequently, they have many applications in refrigeration and air conditioning.

Twin screw type compressors are available in open (Fig. R11-49), semi-hermetic or hermetic (Fig. R11-50) construction. Hermetic designs are available from 35 to 175 Hp and are often used in multiples for larger capacity equipment.

In addition to the twin screw, a single screw design (Fig. R11-51), using gate rotor(s) is available. It shares the same attributes as the twin screw and is available in open construction for larger refrigeration applications up to about 1800 cfm displacement (twins go up to about 3500 cfm).

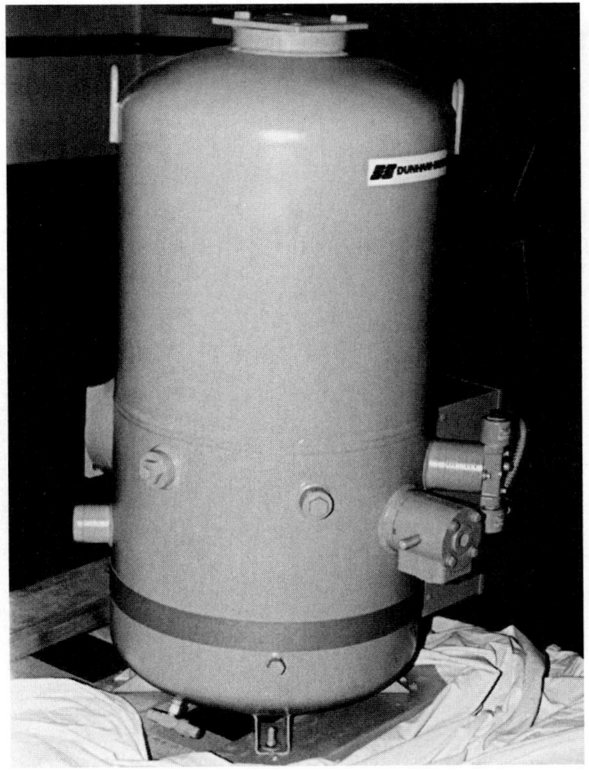

Figure R11-50 Welded hermetic dual rotary-screw compressor. (Courtesy of Dunham-Bush, Inc.)

Operation of Screw Compressors

The *twin-screw* models, (Fig. R11-52), range in capacity up to 3500 cubic feet per minute (cfm) and are used for

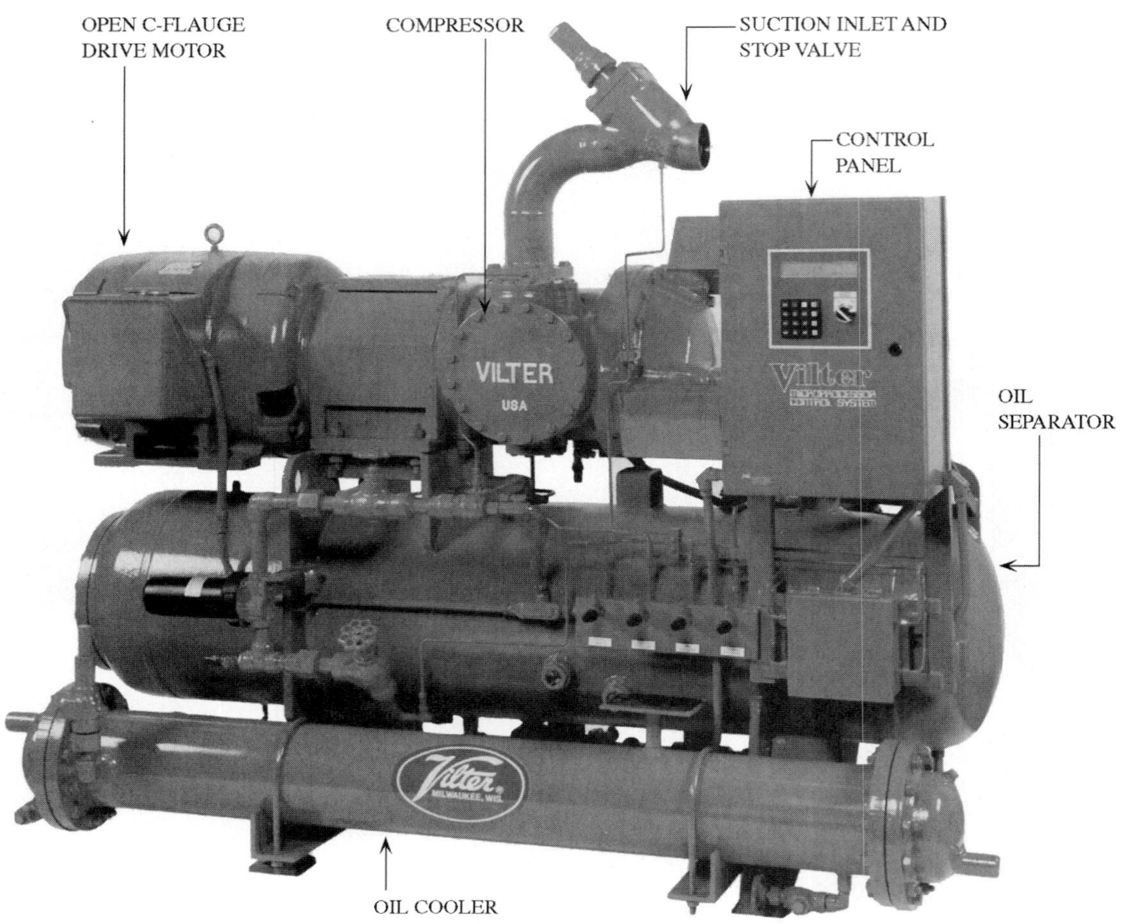

Figure R11-51 Single-screw open compressor. (Courtesy of Vilter Manufacturing Corp.)

Figure R11-52 Open-drive twin-screw compressor. Note the size of the discharge oil separator. (Courtesy of Frick Company)

larger refrigeration and air conditioning (usually water chilling) applications. They often consist of a driven male rotor and mating female rotor carried on bearings and rotating within a stationary housing containing gas inlet and discharge ports. Either rotor can be the driven component.

Sealing and lubrication are usually accomplished by oil flooding, although it is possible to use synchronized timing gears to drive both rotors. Typical combinations of male rotor lobes and female rotor flutes are 4:6, 5:6, and 5:7. The rotors do not run at the same speeds.

The refrigerant gas is drawn through the inlet port into voids created as the rotors turn and unmesh, as shown in Fig. R11-53. The entire length of the rotor space will fill with gas. As the rotors continue to rotate, they re-engage one another on the suction end. This progressively compresses the trapped gas as they mesh, moving it toward the discharge end of the machine. The compressed gas is released through the discharge port as the rotors turn and uncover the port. The cycle repeats as inter-lobe spaces are created, filled with gas, diminished (compressing the gas), and vented to the discharge. The individual discharges blend into a smooth flow of gas with little pulsation compared to a comparable reciprocating compressor.

Rotor bearings may be sleeve or anti-friction type, depending on machine design and size. The oil separator is an important component in oil-flooded machines and may be integrated into the design.

While capacity control is possible by speed adjustment, slot, or lift valves, the most commonly used device is the slide valve, as shown in Fig. R11-54. It is a simple arrangement which bypasses part of the gas back to the suction and regulates the discharge port. It permits satisfactory unloading over a wide capacity range and the mechanism is uncomplicated and reliable. Another view of the slide valve is shown in Fig. R11-55.

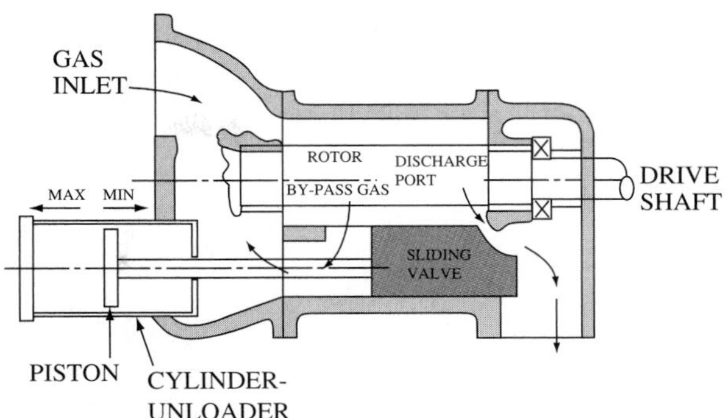

Figure R11-54 Capacity control slide valve. (Copyright by the American Society of Heating, Refrigerating, and Air-Conditioning Engineers, Inc. Used by permission.)

Because of the simplicity of the screw machine, one manufacturer is building smaller sizes (35-175 Hp) in fully welded steel enclosures, as shown in Fig. R11-56. Larger sizes (to 1000 tons+) are available as semi-hermetic machines for halocarbon refrigerants. Open models are built in a wide range of capacities and are used for ammonia and industrial applications, as well as for air conditioning.

Single-screw compressors use a single helical main rotor and either one or two gate rotors as shown in Fig. R11-57. The principle of operation is shown in Fig. R11-58.

The compression volume is formed by the rotor grooves, by the cylinder wall, and by the meshing gate-rotor tooth. The rotor groove fills with gas, turns, and compresses the gas as the volume is reduced by the meshing action with the gate rotor. Oil flooding is used for sealing as well as lubrication and cooling.

Main rotors are typically cast iron. One manufacturer builds the gate rotors from a reinforced-fiberglass engineered plastic material. Sizes range from 150 to 1800 cfm.

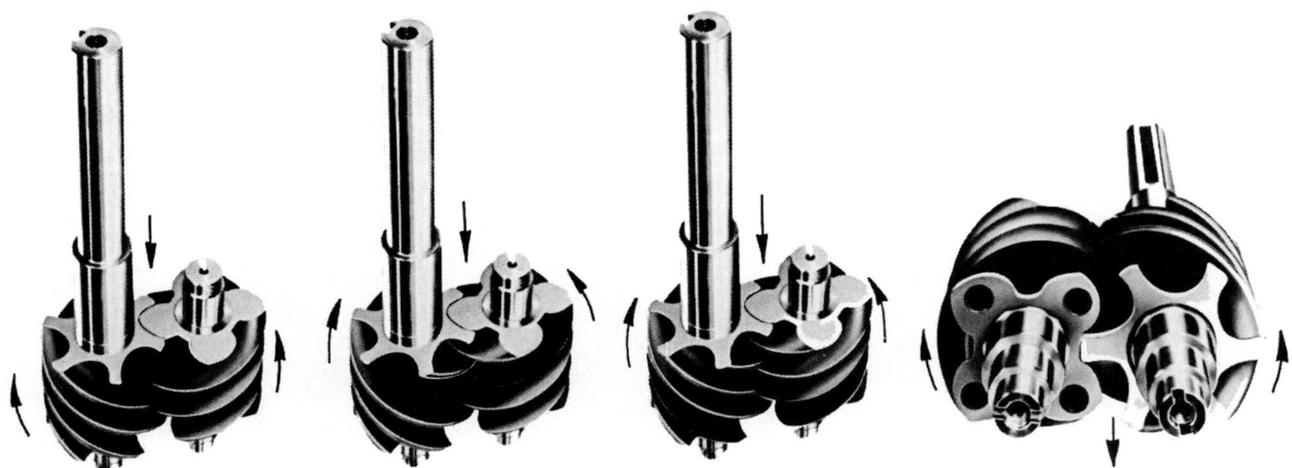

Figure R11-53 Gas flow through the dual-screw rotors. (Courtesy of Dunham-Bush, Inc.)

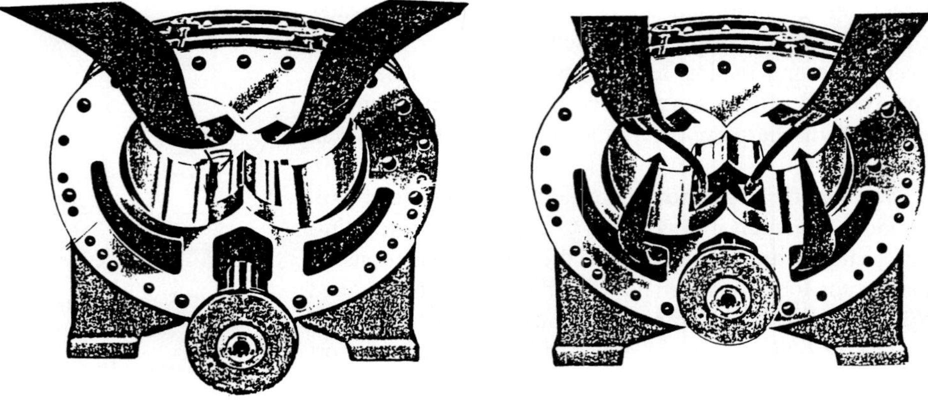

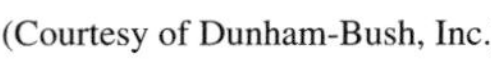

Figure R11-55 A view of the capacity-control slide valve for a screw compressor showing the actuator's travel. (Courtesy of Dunham-Bush, Inc.)

Capacity is controlled by slide valves in the compressor casing.

R11-4.5 Centrifugal Compressors

Centrifugal compressors are widely used in single-stage and multi-stage open and semi-hermetic (Fig. R11-59) configurations to chill water for building air conditioning in capacities from 100 to 8000 tons and higher. Until recently, they generally used R-11, a low pressure CFC refrigerant which is being phased out.

These compressors handled large volumes of vapor at low pressures (vacuum to 10 psig). A HCFC interim replacement refrigerant, R-123, is available, and a portion

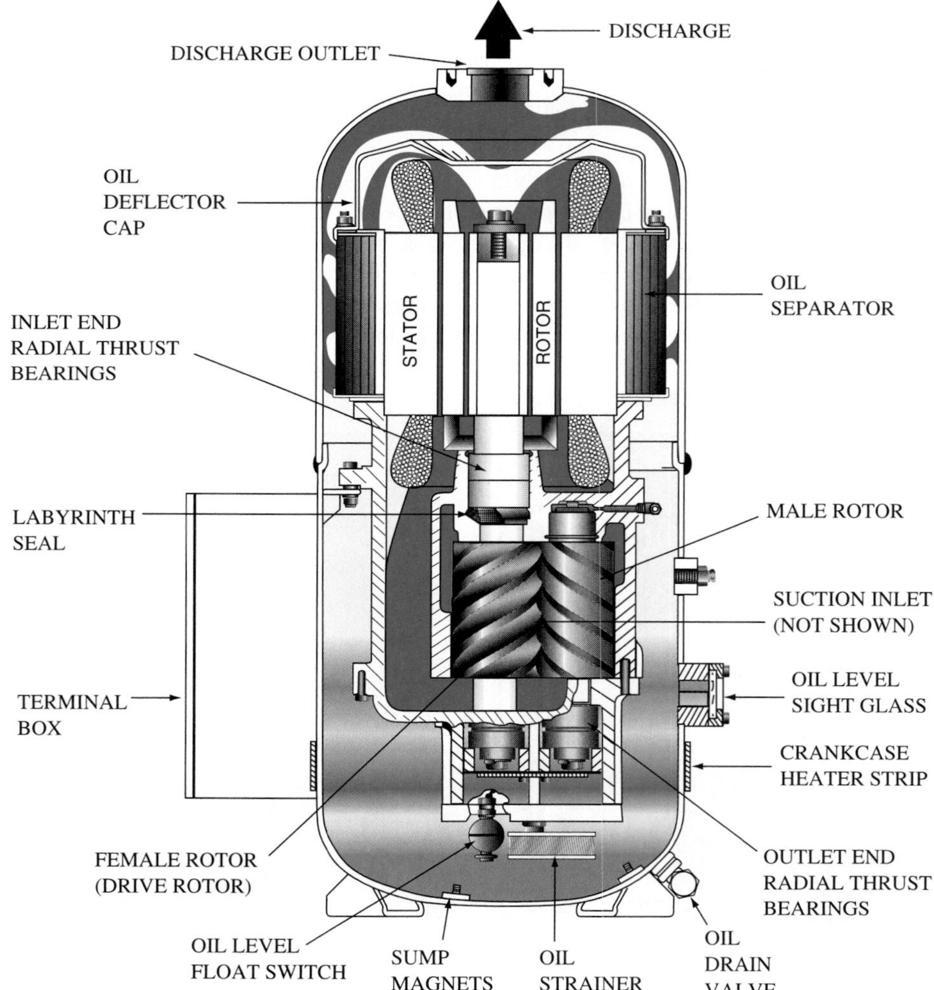

Figure R11-56 Sectional view of a welded hermetic twin-screw compressor. (Courtesy of Dunham-Bush, Inc.)

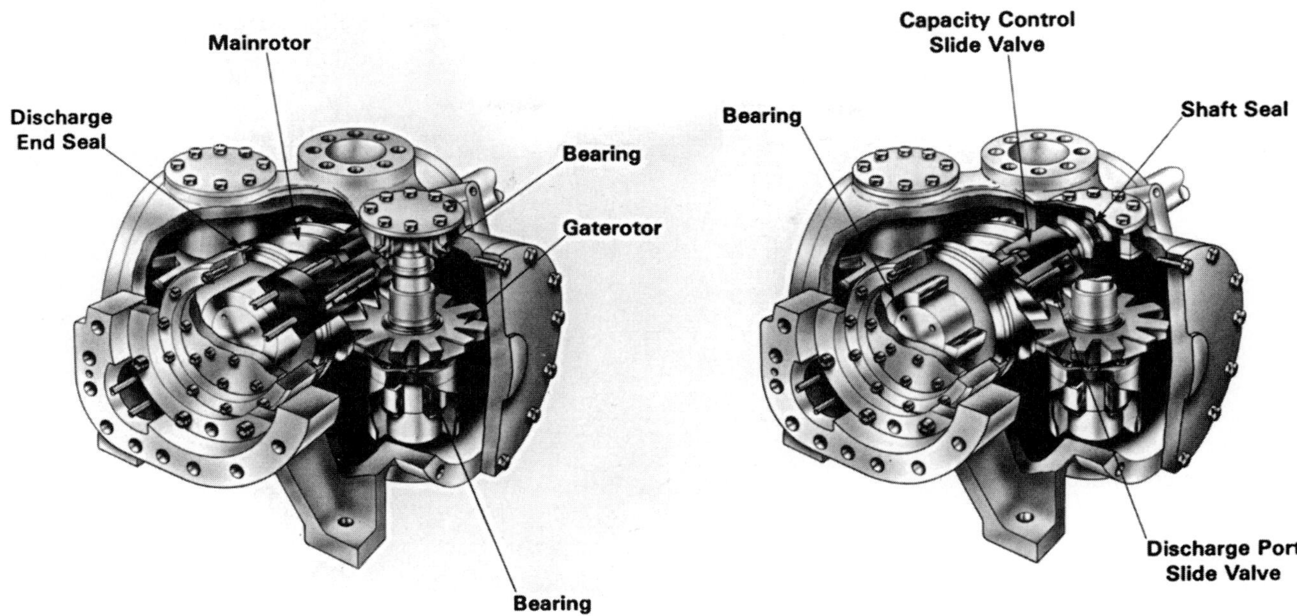

Figure R11-57 Cutaway views of a single-rotary open-drive screw compressor. (Courtesy of Vilter Manufacturing Corp.)

GAS ENTERS THE COMPRESSOR THROUGH THE SUCTION CONNECTION AND FLOWS INTO THE OPEN ENDS OF THE AVAILABLE GROOVES. ROTATION OF THE MAIN ROTOR CAUSES THE TEETH OF THE GATE ROTOR TO ENTER THE OPEN ENDS OF THE GROOVES IN SEQUENCE AND TRAP THE SUCTION VAPOR IN CHAMBERS FORMED BY THE THREE SIDES OF THE GROOVE, THE CYLINDRICAL CASING AND THE GATE ROTOR TOOTH ITSELF.

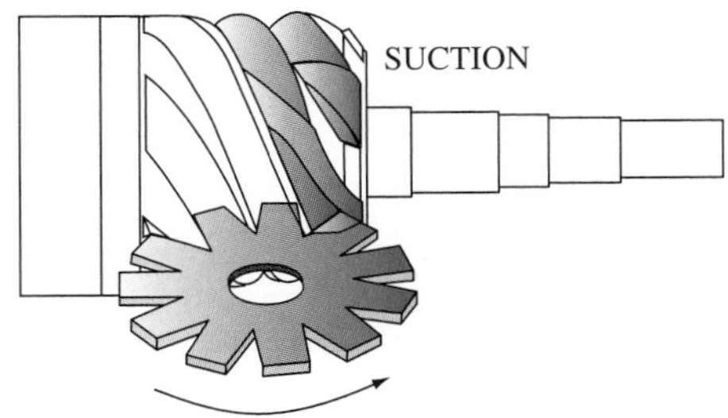

AS ROTATION CONTINUES, THE GROOVE VOLUME DIMINISHES AND THE VAPOR TRAPPED IS PROGRESSIVELY COMPRESSED UNTIL THE LEADING EDGE OF THE GROOVE PASSES THE EDGE OF THE DISCHARGE PORT IN THE CASING.

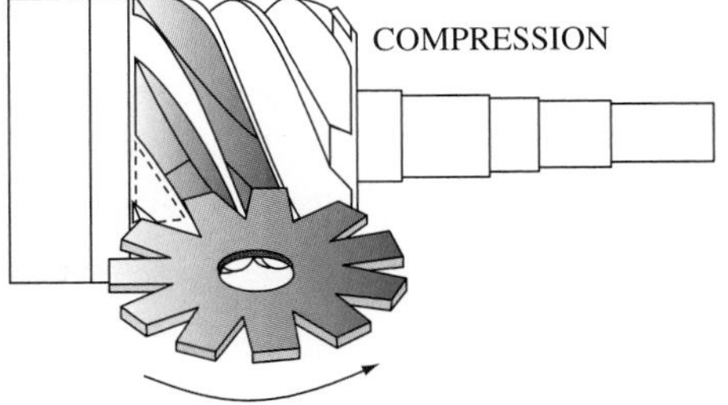

AT THIS POINT COMPRESSION CEASES AND THE VAPOR IS DISCHARGED INTO THE DISCHARGE CHAMBER UNTIL THE GROOVE VOLUME HAS BEEN REDUCED TO ZERO.

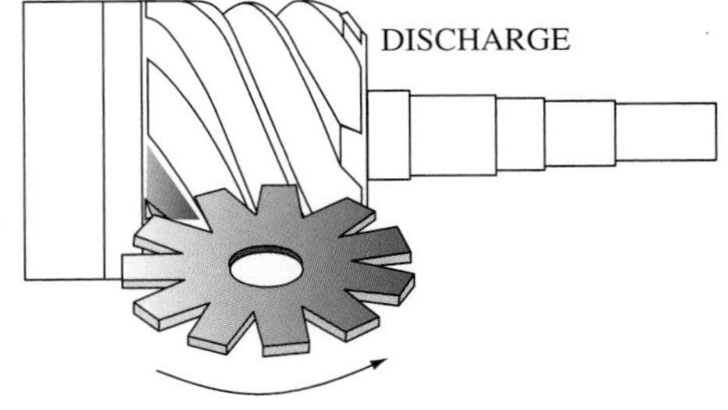

Figure R11-58 Three views showing the operating principle of a single rotary screw compressor. (Courtesy of Vilter Manufacturing Corp.)

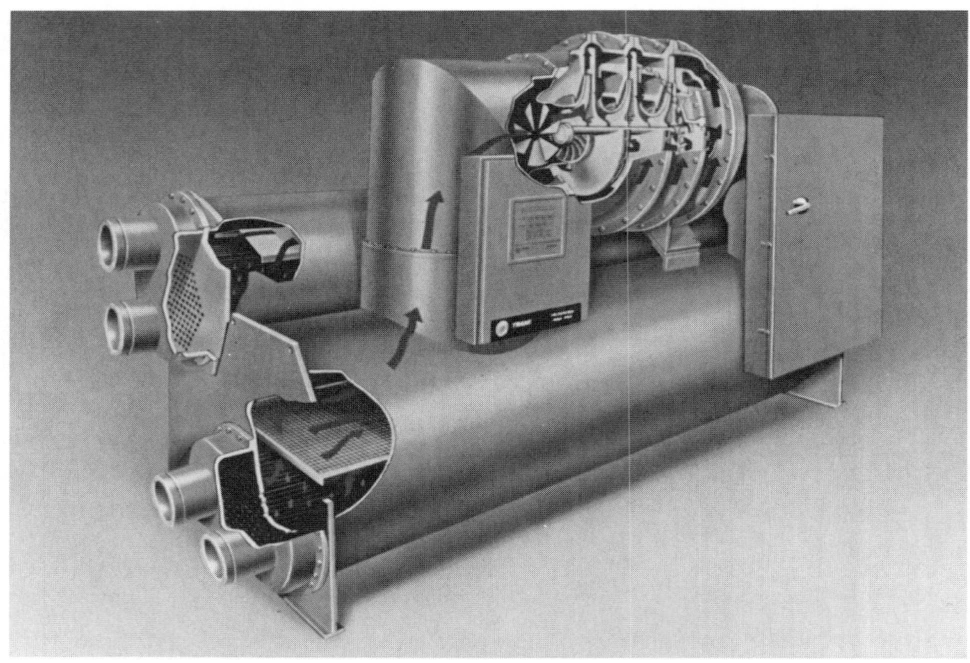

Figure R11-59 Centrifugal water chiller, multi-stage, low-pressure. (The Trane Company)

of the estimated 80,000 U.S. installations will be retrofitted. At present no HFC refrigerant is available for these machines, leaving a very doubtful future.

High-pressure centrifugal compressors are available in both open and semi-hermetic configurations. Some were especially designed for CFC R-12. They are now being supplied to run on HFC-134a refrigerant. Other machines are designed to use HCFC R-22. Presently their capacities are limited to under 1000 tons. Large chiller units will utilize multiple compressors. These machines will compete with the screw machines for large capacity chilled water cooling plants.

Figure R11-60 Semi-hermetic centrifugal compressor. (Reprinted with permission of McQuay International)

Figure R11-61 Lightweight high-speed centrifugal impeller. (Courtesy of Bruce Smith-York International Corp.)

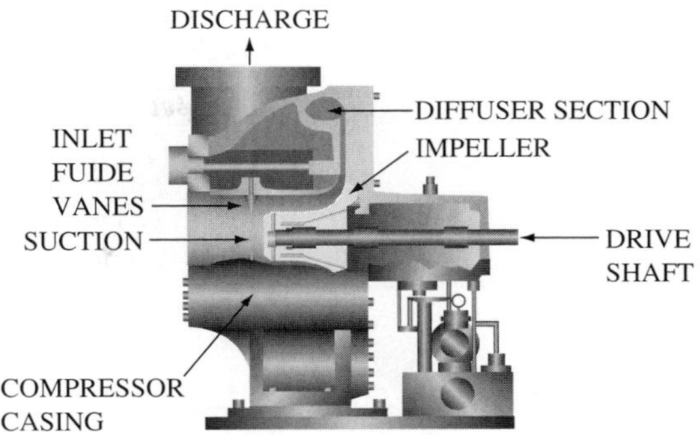

Figure R11-63 Centrifugal compressor, showing inlet vanes, used for capacity reduction. (Courtesy of Carrier Corp.)

Operation of Centrifugal Compressors

Centrifugal compressors, (Fig. R11-60), were once considered low-pressure machines until a semi-hermetic model was introduced for R-12 applications some years ago. It used a relatively small, lightweight, precision diecast impeller (Fig. R11-61). A two-pole 3600-rpm motor through a gear-type speed increaser (Fig. R11-62), drove the impeller at speeds of 8000 to 10,000 rpm and higher. To reduce sound levels, the impeller vanes were asymmetrically arranged but fully balanced.

Overall size is reduced and, because the machine always operates in a positive pressure, there is no need for an air-purge system. Purge systems and air leakage always have been high maintenance problems on the older low-pressure machines. Other manufacturers now build high-pressure centrifugals. They are available in both open and semi-hermetic configurations and are being designed for use with HCFC-22 as well as the newer HFC refrigerants.

Capacity control is commonly by inlet-vane controls (Fig. R11-63), but is also available with variable motor speed control.

Fig. R11-64 shows a flow diagram for a centrifugal chiller using a high-pressure refrigerant. Note the separate oil pump and the use of a float-valve metering device.

Figure R11-62 Gear-type speed increaser. (Reprinted with permission of McQuay International)

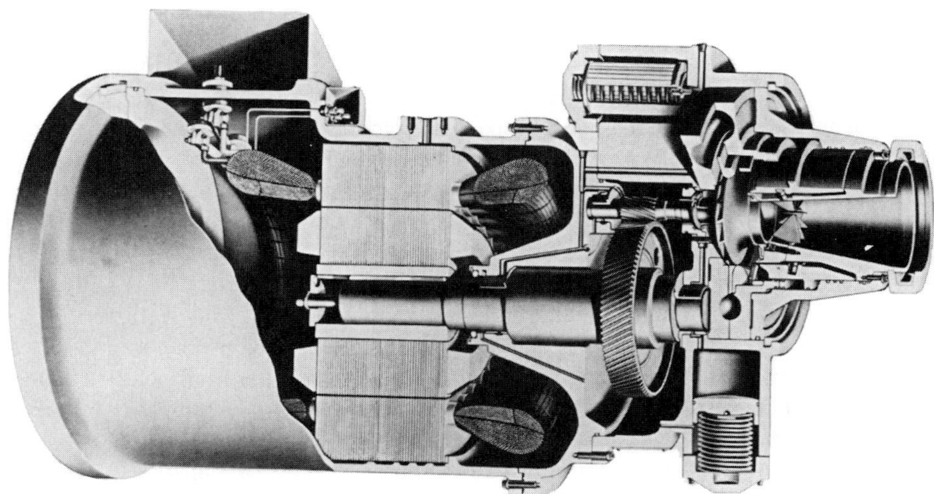

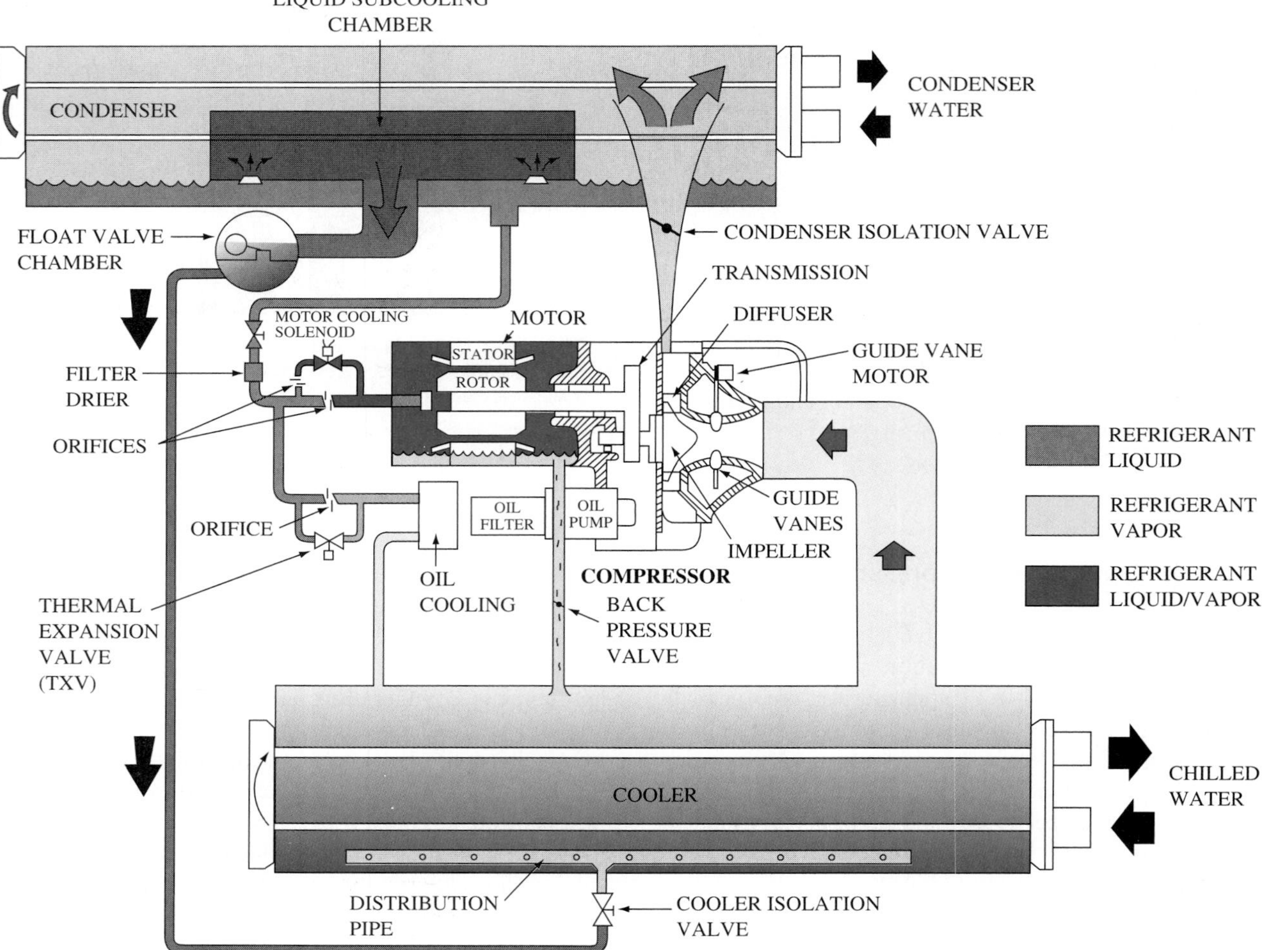

Figure R11-64 Centrifugal water chiller using high-pressure refrigerant. Refrigerant flow shown.

REVIEW

The vapor-compression refrigeration cycle is the most common in use. At its heart is the mechanical compressor, most often electrically driven. The smaller and intermediate sizes are reciprocating, rotary, and scroll compressors—all positive-displacement types. Hermetic construction is favored, followed by semi-hermetic in the larger sizes (above 5-10 Hp). Open machines are most often found in ammonia and industrial applications.

In the larger sizes (above 100 tons), screw and centrifugal machines do the job. The screw is a positive-displacement machine, capable of a very broad range of applications, including those requiring high compression ratios. It is highly regarded for its reliability and low maintenance requirements.

Another competitor in the high tonnage range is the high-speed centrifugal compressor designed for high-pressure HFC refrigerants. These machines have a much more limited pressure lift capability, but will find application as water-cooled package water chillers for large comfort cooling installations. The centrifugal, too, is well regarded for simplicity, reliability, and low maintenance requirements.

Review the pressure-enthalpy and pressure-volume plots at the beginning of the chapter to fix the basics in mind. They apply to these systems, large or small!

The following terms and concepts were presented in this chapter:

- Refrigeration compressors—refrigerant vapor pumps, commonly referred to as the heart of the refrigeration system.
- Functions of compressor—takes the refrigerant vapor from the evaporator, where it absorbed the heat at a low pressure and temperature, and raises it to a higher pressure and temperature, pumps it to the condenser, allowing the refrigerant to expel its heat.

- Adiabatic process—the refrigerant changes from a liquid to a mixture of liquid and gas, while maintaining the same heat content (enthalpy) in Btu/lb. No heat is gained or lost in the process.
- Compressors commonly used for refrigeration and air conditioning are single stage.
- Positive displacement compressor designs and applications include:
  - reciprocating—residential, commercial, industrial air conditioning and refrigeration.
  - rotary—residential and industrial refrigeration
  - helical screw—industrial refrigeration
  - scroll—residential and light commercial air conditioning
- Kinetic compressor designs include:
  - centrifugal compressor—large commercial/industrial air conditioning
- Compressor types include: 1.) open, 2.) hermetic, 3.) semi-hermetic
- Lubrication methods include: 1.) splash or 2.) force-feed pressure system (uses mechanical pump).
- Oil separator—located between the compressor and the condenser in the discharge line. It separates oil from the high-temperature, high-pressure refrigerant vapor, and returns the oil to the compressor crankcase.
- New HFC/HCFC refrigerants require use of special synthetic oils.
- Capacity control factors—depend upon the refrigerant and changes in compressor displacement.
- Capacity control methods include:
  - Cycling on/off—single or multiple machines
  - Cylinder unloading (on reciprocating compressors)
  - Hot-gas bypass
  - Speed control
  - Slide valve (screw compressors)
  - Modulating inlet vane (centrifugal compressors)

Problems and Questions

1. A centrifugal compressor is a positive-displacement design. True or False?
2. What are the two principle methods of compressor lubrication?
3. What is the greatest enemy of a compressor motor?
4. Name the six essential qualities of compressor lubricants.
5. Name the types of compressors.
6. What is the purpose of an oil separator?
7. Most refrigeration and air conditioning compressors have what type of motor?
8. Why is the scroll compressor so efficient?
9. In what application are rotary compressors likely to be found?
10. Where would two-stage compressors be used?
11. What does "EER" stand for?
12. What is the operating voltage tolerance of dual-voltage compressor motors?
13. How do compressors fail? (Give in terms of electrical and mechanical failures.)
14. What safety precautions must be followed on a compressor burnout?
15. Explain "Performance Factor."
16. What is one of the advantages of liquid subcooling?
    a. Reduces possibility of flash gas at expansion device.
    b. Boosts suction superheat.
    c. Reduces suction superheat.
    d. Reduces OFF-cycle migration.
17. Running at increased head pressure (condensing temperature):
    a. Increases system capacity.
    b. Stabilizes oil temperature.
    c. Increases refrigerant mass rate of flow.
    d. Reduces compressor capacity.
18. What is the most common type of capacity modulation for fractional-horsepower hermetic motor compressor units?
    a. Multi-speed induction motors.
    b. Floating head pressure control.
    c. On/off cycling.
    d. Gas-actuated cylinder unloaders.
19. What type of lubricant will a compressor be charged with if it is designed for use with HFC refrigerants?
    a. Wax-free refrigerant grade mineral oil.
    b. Polyol ester lubricant.
    c. Alkyl-benzene (AB) refrigerant oil.
    d. 50-50% mixture of AB and mineral oil.
20. Reciprocating compressors are:
    a. Immune to possible damage from liquid slugs.
    b. Limited in capacity to approximately 15 tons.
    c. Should not be used for low-temperature applications.
    d. Used for all types of high/medium/low temperature applications.
21. In a welded hermetic compressor
    a. field service repairs are possible.
    b. field service repairs are not possible.
    c. the motor windings are sealed from oil and gas exposure.
    d. the motor and crankshaft are connected by an in-line coupling.
22. Rotary compressors
    a. are most often used for capacities over 150 tons.
    b. run with very little vibration.
    c. have rolling-vane actuators.
    d. need extra isolation to reduce inherent vibration.

23. The scroll compressor, when compared to a similar sized reciprocating compressor:
    a. Is just as efficient, but has more vibration and noise.
    b. Is only 90% as efficient and is used only in low cost equipment.
    c. Is 10% more efficient and growing in use.
    d. Is 50% more efficient and growing in use.
24. Screw compressors
    a. are powered by one or two orbiting rotors.
    b. find their most significant use with low pressure R-11 refrigerant.
    c. find wide use in small appliances, window air conditioners, etc.
    d. can be designed to handle a wide range of temperature and refrigerants in high-capacity sizes.
25. High-speed centrifugal compressors designed for use with high-pressure HFC refrigerants
    a. do not need an auxiliary purge unit.
    b. are provided with cylinder unloaders in sizes above 15 tons.
    c. must have a multi-stage oil separator due to flooded lubrication.
    d. are classed as positive displacement chillers.

Controls

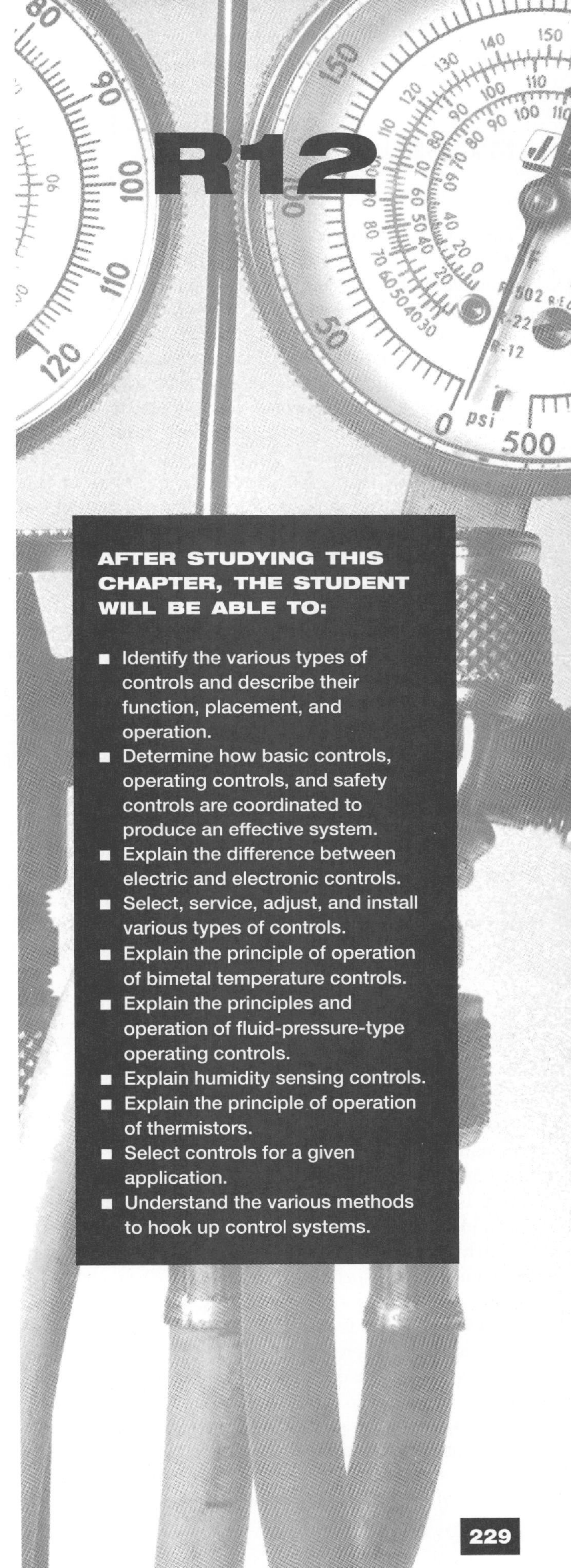

R12
AFTER STUDYING THIS CHAPTER, THE STUDENT WILL BE ABLE TO:
■ Identify the various types of controls and describe their function, placement, and operation.
■ Determine how basic controls, operating controls, and safety controls are coordinated to produce an effective system.
■ Explain the difference between electric and electronic controls.
■ Select, service, adjust, and install various types of controls.
■ Explain the principle of operation of bimetal temperature controls.
■ Explain the principles and operation of fluid-pressure-type operating controls.
■ Explain humidity sensing controls.
■ Explain the principle of operation of thermistors.
■ Select controls for a given application.
■ Understand the various methods to hook up control systems.

R12-1 DEFINITION OF CONTROLS

All of the controls described in this chapter can be termed **system controls.** This refers to their function, which is to start, stop, regulate, and protect the refrigeration cycle and its components. **Basic controls** are those which start and stop the equipment. **Operating controls** regulate processes, while those which protect the cycle are called **safety controls.**

The major components of the cycle can be modified and regulated to form a wide variety of applications. For example, if the application is a drinking-water cooler, a basic control (thermostat) starts and stops the refrigeration unit to maintain a desirable drinking water temperature. A safety control (motor overload) could be used to protect the compressor motor from damage from excess-current conditions.

Electric and electronic controls are operated by electricity and are connected by wires. **Electronic controls** differ from **electrical controls** in that they use semiconductor materials and solid-state controls. **Pneumatic controls** are non-electric and are operated by air pressure.

There are three types of materials used in electrical control systems:

1. Conductors, such as copper, aluminum and iron.
2. Semiconductors, such as metal oxides and metal compounds.
3. Nonconductors, such as glass, plastic, rubber and wood.

Conductors permit the free flow of electrons and offer the minimum amount of resistance to the flow of electric current.

Semiconductors restrict the flow of electrons and permit the limited flow of electricity under certain conditions. They perform unique electrical functions, which are useful in solid-state electrical systems. A separate section in this text describes their applications.

Nonconductors do not permit the flow of electrical energy and are used as insulators.

R12-2 SELECTING THE CONTROL SYSTEM

The type of controls selected for any system depends on many factors: the age of the equipment, the size of the equipment, the location of the components, the relative cost, and even the designer's preference. Older units in the small- and medium-sized class almost all use electrical controls. Larger HVAC/R systems use pneumatic controls when the cost of the air compressor can be justified. There is a preference for electronic controls due to their low cost, accuracy, and the extra performance features they are capable of executing.

Regardless of the type of controls, the function performed can be the same. A different type of control can perform the same function in a different way.

One of the big advantages of the pneumatic control systems is their ability to modulate a damper or a valve to produce an infinite number of positions between fully open and fully closed. This makes it possible to provide close control of temperature and humidity where required.

The availability of microprocessor chips has greatly increased the popularity of solid-state controls. A tiny chip contains a whole system of controls which replace complete circuits that formerly used electro-mechanical devices. Further, these chips are energy efficient and consume far less energy than the replaced devices.

A technician who is planning to service a wide variety of refrigeration and air-conditioning equipment must be knowledgeable on electrical and electronic control systems. Both types of control are commonly used in small and medium sized jobs, which constitute the largest part of the market. In this text the emphasis will be on these systems.

R12-3 TYPES OF BASIC CONTROLS

There are four types of basic controls. The first is operated by temperature and is called a **thermostat.** The second is operated by pressure and is called a **pressurestat.** The third is operated by moisture or humidity and is called a **humidistat.** The fourth is operated by time and is called a **switch** or time clock. Each of these controls can be used to regulate cycle operation. For example, when a home refrigerator becomes too warm, that is, the temperature becomes too high for food storage, the thermostat senses this and starts the compressor.

Pressurestats are often used to control temperature conditions in a display case by controlling evaporator pressure. When the evaporator pressure and corresponding temperature become high, the pressurestat operates and starts the compressor. In storage rooms where humidity is all-important, the humidistat is designed to start the refrigeration cycle when the humidity rises to a predetermined level. Conversely, all the above controls will stop cycle operation when the conditions are satisfied.

Some systems are operated by a time clock, particularly where energy conservation is concerned.

R12-3.1 Thermostats

Thermostats respond to temperature. They can do this because of the warping effect of a *bimetallic strip* or because of *fluid pressure.*

Bimetallic Elements

The thermostat illustrated in Fig. R12-1 is a bimetallic thermostat. The bimetallic element is composed of two different metals bonded together. As the temperature surrounding the element changes, the metals will expand or contract. As they are dissimilar metals, having different coefficients of expansion, one will expand or contract faster than the other. In the illustration, the open contacts are on the left. No current is flowing. If the temperature surrounding the bimetallic element is raised, metal A and B will both start to expand. However, metal A is chosen because it will expand faster than metal B. This will cause the bimetallic strip to bend and close the contacts as shown on the right. As the temperature drops, A would contract faster than B, thereby straightening the element and opening the contacts.

This thermostat is widely used, especially in residential heating and cooling units and in refrigeration systems that work above freezing conditions, where the stat is actually located in the controlled space. It is simple and inexpensive to construct, yet dependable and easily serviced.

One of the common variations of the bimetallic thermostat is a mercury bulb thermostat (Fig. R12-2).

Although the bimetallic principle is used, the contacts are enclosed in an airtight glass bulb containing a small amount of mercury as shown. The action of the bimetallic element tilts the bulb. By tilting the bulb to the left, mercury in the bulb will roll to the left and complete the electrical circuit. By tilting the bulb to the right, the mercury will flow to the end of the bulb and break the electrical

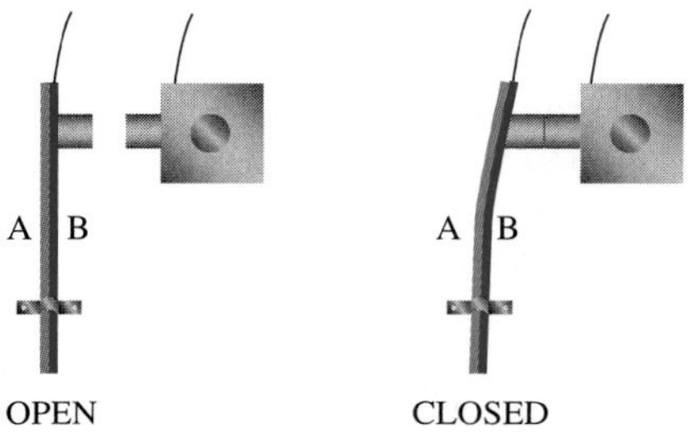

Figure R12-1 Bimetal thermostat showing opening and closing switch in response to temperature.

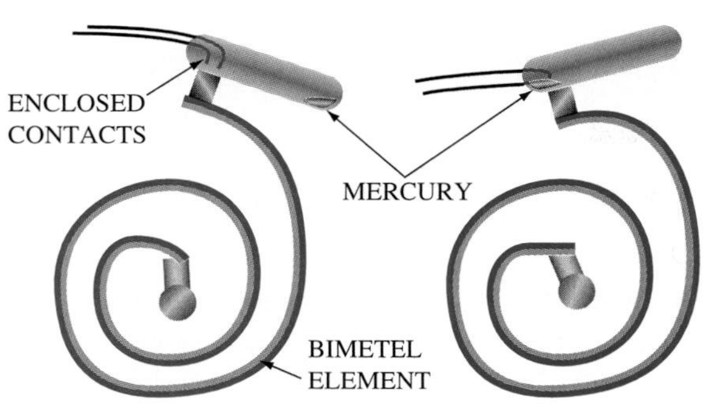

Figure R12-2 Mercury-bulb thermostat, a mercury switch protected by glass enclosure.

circuit. Extra flexible insulated leads connect the mercury bulb to the circuit without interfering with the bulb movement. This thermostat is used extensively in space heating and cooling. Although a little more expensive than the bimetallic open-contact thermostat, its operation is more dependable because dirt cannot collect on the contacts.

Fluid Pressure

The second method of thermostat control is the fluid-pressure type shown in Fig. R12-3.

With a liquid and gas in the bulb, the pressure in the bellows will increase or decrease as the temperature at the bulb varies. The thermostat illustrated is a heating thermostat. As the pressure in the bulb increases with the rise in bulb temperature, the bellows expands and, through a mechanical linkage, opens the electrical contacts.

As the pressure in the bellows decreases with a lowering of bulb temperature, the bellows contracts and closes the electrical contacts. This thermostat is sometimes referred to as a remote-bulb thermostat or temperature-control thermostat. The controlling bulb can be placed in a location other than that used for the operating switch mechanism. For example, with a cooling thermostat in a refrigerated room, the bulb can be placed inside the room and the capillary tube run through the wall to the operating switch mechanism located outside the room. Thermostat adjustments can be made without entering the refrigerated room, and the switch mechanism is not exposed to the extreme conditions inside the room.

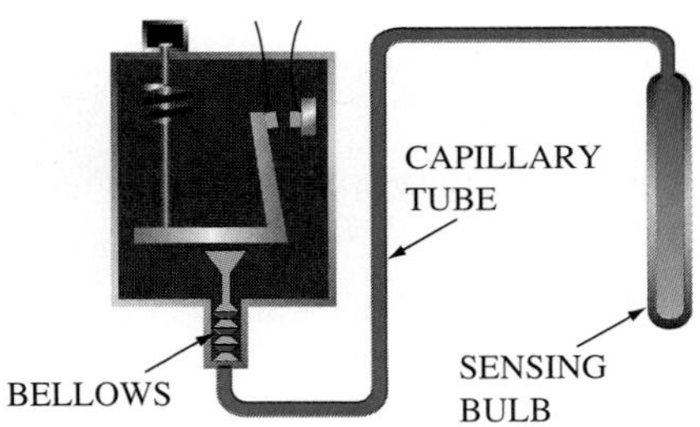

Figure R12-3 Remote-bulb thermostat using a bellows to operate the switch.

Another version of the pressure-actuated thermostat is the *diaphragm* type, shown in Fig. R12-4.

In this control the diaphragm is completely filled with liquid. The liquid expands and contracts with a change in temperature. The movement of the diaphragm is very slight, but the pressure that can be exerted is tremendous. Because the coefficient of expansion of liquid is small, relatively large-volume bulbs are used on the control. This results in sufficient diaphragm movement and also ensures positive control from the bulb.

When installing a remote bulb temperature controller, make **absolutely sure** the maximum bulb temperature will not be exceeded during any part of the operating or off cycle.

If the fluid leaks from the thermal system of a high temperature limit control, the control will not shut off and an unsafe condition could arise. How can you avoid this possibility? Use a subatmospheric fill control. Normally the thermal system operates in a partial vacuum. Any leakage will let air in, raise the pressure, and cause the control to shut off, thus “failing safe.”

Set Point Adjustment

Fig. R12-5 shows a dial arrangement for selecting the *set point* on a single-bulb mercury-tube thermostat. The dial is rotated until the indicator is placed at the setting. Note that this thermostat has other settings: a cool-off-heat system switch and a fan-on-auto switch. These switches are part of the sub-base that is used in mounting the thermostat.

Fig. R12-6 shows a heating-cooling thermostat, with two levers at the top for selecting the set points for both heating and cooling. The system can also be set to either supply automatic or manual changeover. Another switch is available on the base for selecting either continuous or intermittent fan operation. This control is available with two to four mercury bulbs and can provide up to two-stage heating and cooling controls.

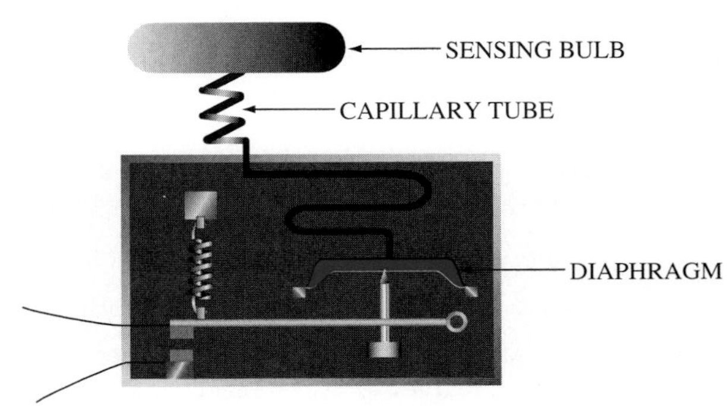

Figure R12-4 Diaphragm-type thermostat, using a liquid-filled sensing element and diaphragm to operate the switch.

Figure R12-5 Set-point adjustment for a thermostat with individual settings.

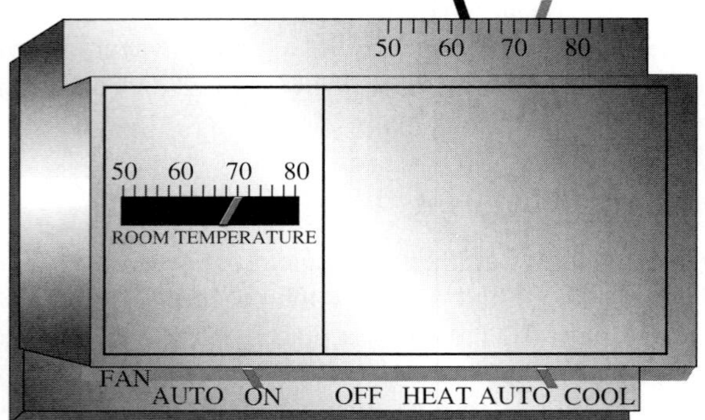

Figure R12-6 A combination heating/cooling thermostat with individual settings.

Electronic Thermostats

Fig. R12-7 shows a common sensing element for an electronic thermostat called a *thermistor*. The thermistor has a resistance element that is affected by temperature. Using an electronic circuit, the thermostat can be set so that when the temperature set point is reached, a switch is opened, turning off the power to the fuel operating device such as a gas valve.

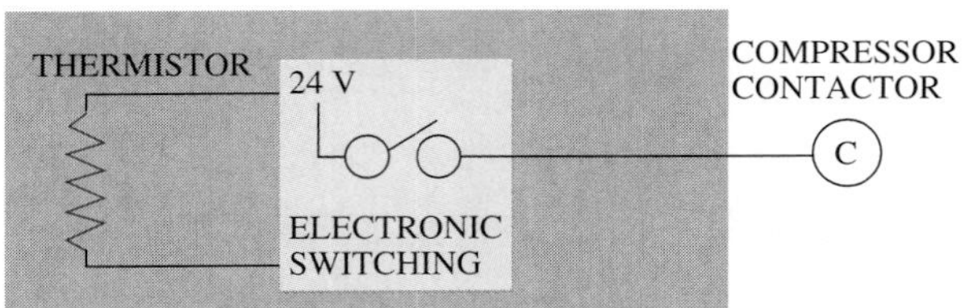

Figure R12-7 Electronic thermostat, using a thermistor as a sensing element.

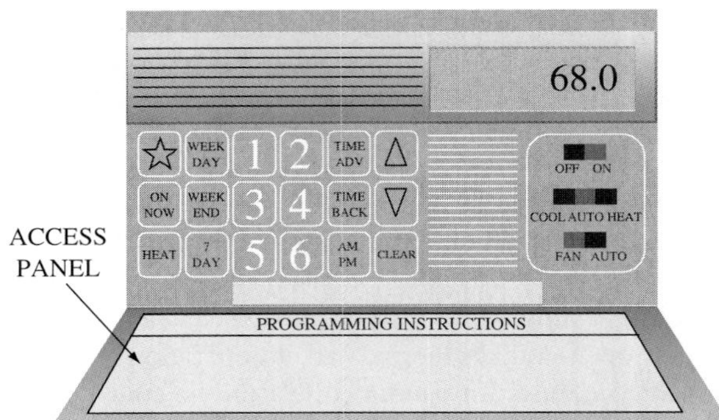

Figure R12-8 Programmable electronic thermostat, accurate within 1°F.

Fig. R12-8 shows an electronic programmable thermostat where temperature settings can be made for various periods throughout the week. These settings will take effect automatically. The great advantage of the electronic thermostat is its accuracy. It will control the temperature within one degree. The bimetal thermostat controls to an accuracy of about two degrees.

Fig. R12-9 shows an electronic thermostat that includes the additional feature of heating night setback. It can be programmed to reduce the set point at a certain time, for example 11:00 p.m., and automatically resume the day time setting in the morning, for example, 6:00 a.m. Most programmable thermostats have an over-ride arrangement so that the occupant can manually change the setting at any time.

Setback and *setup* are two common terms in thermostats. A heating thermostat with night (or unoccupied)

PRACTICING PROFESSIONAL SERVICES

Securely fasten all covers and caps on controls after calibration, repairs, or replacement. Start up and cycle equipment upon completing all work to verify proper and normal operation. Upon your customer's request, display and explain all broken and worn equipment prior to replacement. When job is completed, discard and clean up all defective parts and packaging materials from the mechanical room or work areas.

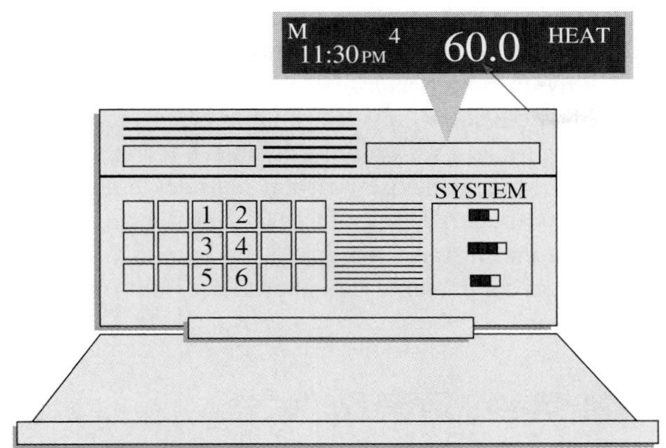

Figure R12-9 Programmable electronic thermostat, with night setback.

setback will automatically reduce the control setpoint during preset periods when lower than normal temperatures are acceptable.

Setup is the same feature applied to the cooling thermostat wherein the setpoint is raised during scheduled periods. The intent of setup and setback is to reduce cooling and heating energy usage.

R12-3.2 Pressurestats

Pressure controls can also be divided into two categories: the bellows type and the Bourdon-tube type. By far the most common is the bellows type, illustrated in Fig. R12-10.

The bellows is connected directly into the refrigerant system through a (capillary) tube. As the pressure within the system changes, so does the pressure within the bellows, causing the bellows to move in and out in conjunction with pressure variation. As shown, the electrical connections are broken as the pressure rises.

These controls are available with normally open (NO) and normally closed (NC) and single-pole double-throw (SPDT) switching. Depending on the control-pressure range, this type of control can be used as a low- or high-pressure control. That is, it can be connected into the high- or low-side of the system. Because of its simplicity, dependability, and adaptability, this control is found on almost every air-conditioning or refrigeration system.

The low-pressure control, shown in Fig. R12-11, is very similar in design to the high-limit pressurestat shown in Fig. R12-10. The difference is that this control is used for low-limit operation and the contacts break on a drop in temperature below set point.

Fig. R12-12 shows a Bourdon-tube pressure control. Note that a mercury bulb switch is shown here. The Bourdon-tube control is ideally suited for mercury bulb operation and is frequently found in applications requiring enclosed contacts. As pressure inside the tubing increases, the tube will tend to straighten out. This in turn moves the

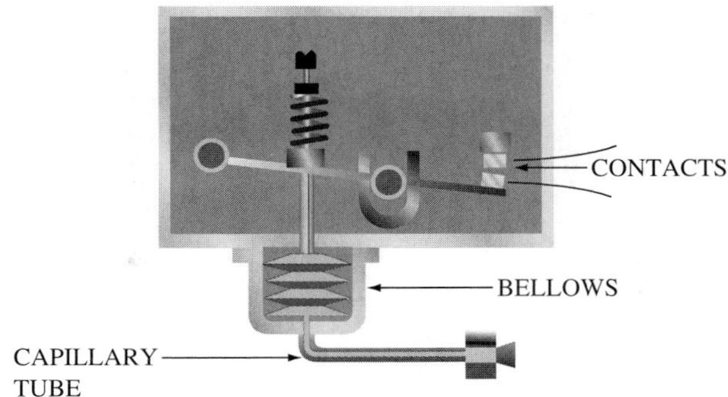

Figure R12-10 Bellows type pressurestat which connects directly with the refrigeration cycle.

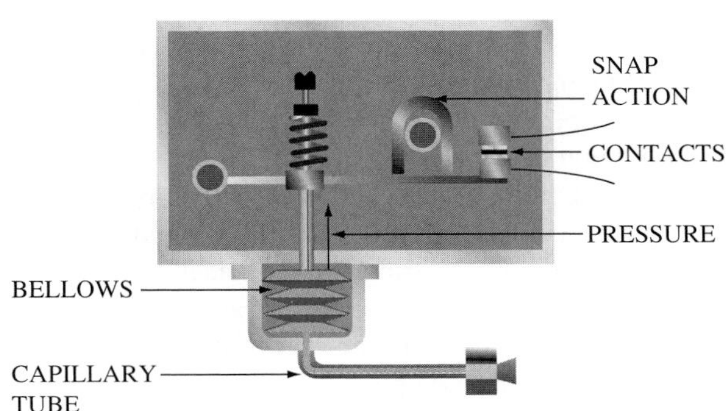

Figure R12-11 Bellows-type pressurestat acting as a low-pressure switch.

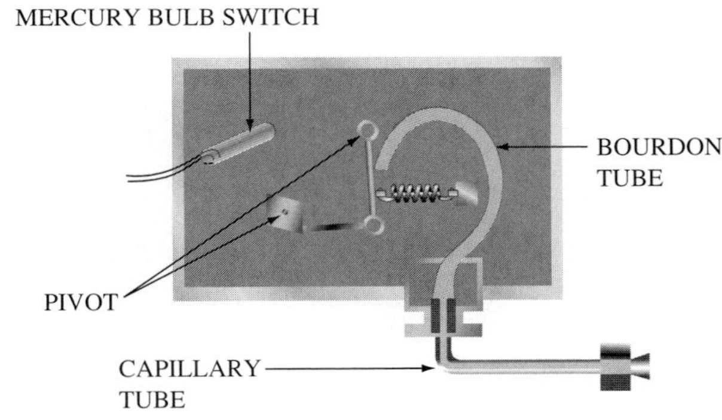

Figure R12-12 Bourdon tube pressurestat with a mercury bulb switch.

linkage attached to the mercury bulb, causing it to move over center, thus moving the mercury from one end of the bulb to the other and making electrical contacts.

The Bourdon-tube design can be supplied either as a low-pressure or high-pressure range switch. A pressure rise of either type extends the tube. In Fig. R12-12, a reduction of pressure in the tube causes the mercury-tube switch to close. In Fig. R12-13 the increase in pressure in the tube causes the mercury-bulb switch to open. Mercury-tube switches can be either single-pole single-throw (2-wire) or single-pole double-throw (3-wire) types.

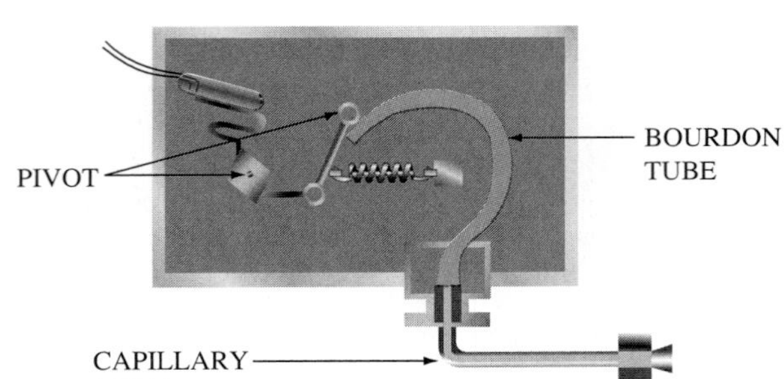

Figure R12-13 Bourdon tube pressurestat in pressurized position.

R12-3.3 Humidistats

The third type of basic control is the humidity control or humidistat (Fig. R12-14).

Hydroscopic elements are used on these controls, the most common being human hair. As the air becomes more moist and the humidity rises, the hair expands and allows the electrical contacts to close (or open). As the hair dries it contracts, thus transferring the electrical contacts. This type of control is susceptible to dirt and dust in the air. Although accurate, it must be carefully maintained and the calibration rechecked periodically.

Another form of humidistat, using a nylon element, is shown in Fig. R12-15.

The nylon is bonded to a light metal in the shape of a coil spring. The expanding and contracting of the nylon creates the same effect as that found in the spiral bimetallic strip used in thermostats. Another type uses a thin, treated, nylon ribbon as a sensor.

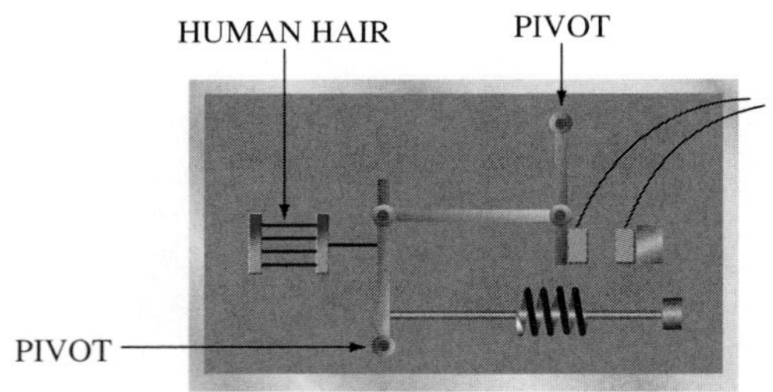

Figure R12-14 Human-hair humidistat.

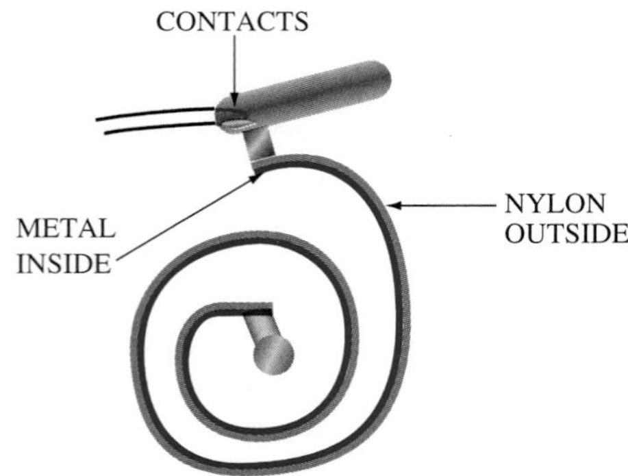

Figure R12-15 Nylon humidistat with a mercury bulb switch.

Fig. R12-16 shows an electronic circuit board with hydroscopic properties. Lithium salt is used for this purpose. Another arrangement uses carbon particles embedded in hydroscopic material. In both cases, the sensing element acts like a thermistor. Changes in the humidity affect the resistance of the material and alter the current in the electronic circuit.

R12-3.4 Time Switches

With the advent of microprocessors, time switches can be built into the electronic circuits and furnished at an affordable price.

Time clocks make possible automatic night setback functions in a programmable thermostat as well as such options as warm-up cycles that precede early morning occupancy times. They permit programming a weekly scheduling of temperature settings (Fig. R12-17).

Fig. R12-18 shows a commercial programmable time switch. It can turn the air conditioning off or on at preset times. Controllers of this type usually include batteries to

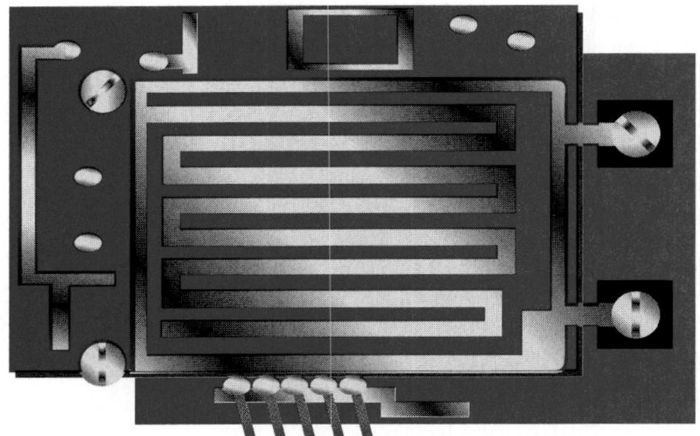

Figure R12-16 Electronic circuit board using electronic sensing elements.

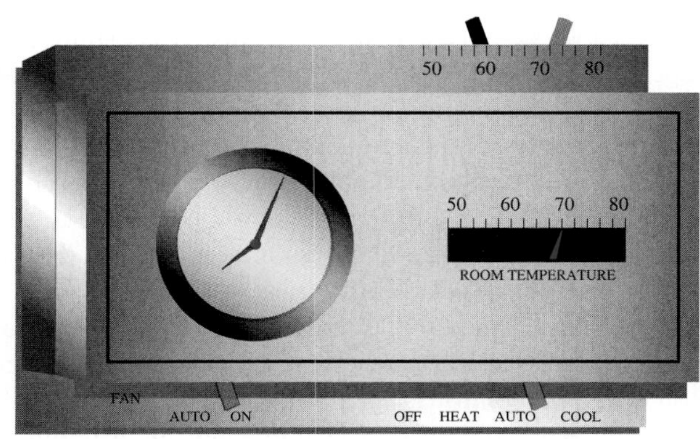

Figure R12-17 Night-setback thermostat used to cycle on time settings.

Figure R12-18 Seven-day electric clock used to preset on/off operations.

operate the clock during power outages. A manual override permits the owner to change the schedule for special events.

R12-3.5 Range Adjustment

There are almost as many possibilities for correct settings on controls as there are applications. Therefore, all controls have some way of being adjusted to compensate for the different conditions under which they may be required to operate. These may be field-adjustable or fixed at the factory. The first such adjustment is range. **Range** is the difference between the minimum and the maximum operating points within which the control will function accurately.

For example, a high-pressure control may be used to stop the compressor when head pressure becomes too high. This control may have an adjustment that will allow a maximum *cutout,* or stopping point, of 300 psig and a minimum cutout of 100 psig. It may be set at any cutout point within this range. A control should never be set outside its range, as it will always be inaccurate and frequently will not even function.

Fig. R12-19 shows one of the simpler methods of range control. As the pressure in the bellows increases, the lever is pushed to the left, thus opening the contacts. By varying the spring pressure, the bellows pressure can be increased or decreased as necessary to open the contacts—thus raising or lowering the system pressure at which the control will operate. Many controls use this principle and usually have some external adjustment for field use.

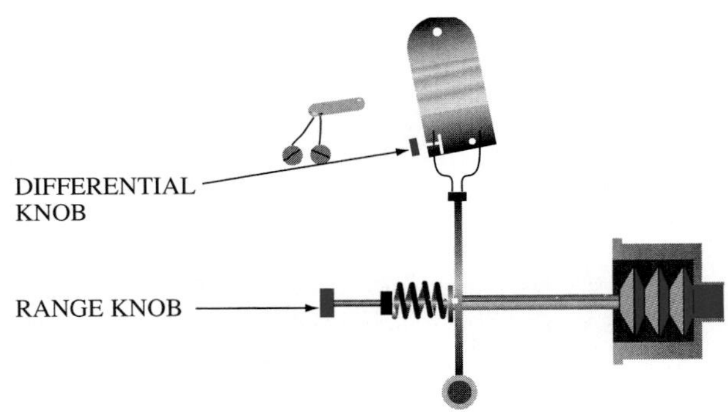

Figure R12-19 Control adjustment.

R12-3.6 Differential Adjustment

If a control cuts out or breaks the circuit, it is just as important that it cuts in, or remakes the circuit. The cut-out and cut-in points cannot be the same or the control would chatter. The difference between these points is the differential. **Differential** can be defined as the difference between the cut-out and cut-in points of the control. For example, if the high-pressure control discussed previously cuts out or breaks the electrical circuit at 250 psig and cuts in or remakes it at 200 psig, the pressurestat differential is 50 psig.

Referring to Fig. R12-19, the control also has a differential adjustment. By opening or closing the effective distance between the prongs of the operating fork, the pressure at which the control will cut in can be varied. As the pressure on the bellows increases, the operating fork moves to the right, and the left prong of the fork will tilt the bulb and remake the electrical circuit. When the adjustable stop on the left prong of the fork is moved away from the prong on the right, the pressure in the bellows must decrease further before the bulb will be tilted. The electrical distance between the prong and stop, the difference between the cut-out and the cut-in points, can then be varied.

Range and differential can be adjusted in many other ways, depending on the application, size and manufacturer's preference.

R12-3.7 Detent

Another feature of electrical controls that should be mentioned briefly is called *detent* or snap action. For technical reasons, all electrical contacts should be opened and closed quickly and cleanly. Detent is built into most controls to accomplish this purpose. Figs. R12-20 through R12-23 show four common examples.

The magnet in Fig. R12-20 should require little explanation. The pull of the magnet merely accelerates the closing and opening rate of the contacts. The closer the magnet is to the bimetallic strip in the closed position, the more positive the snap action.

The bimetallic disk, as shown in Fig. R12-21, is normally in a convex position. As the temperature rises, the two metals expand at different rates until the disk snaps to a concave position, sharply breaking the electrical contacts.

The mercury bulb shown in Fig. R12-22 is another method of obtaining detent action. As the element moves

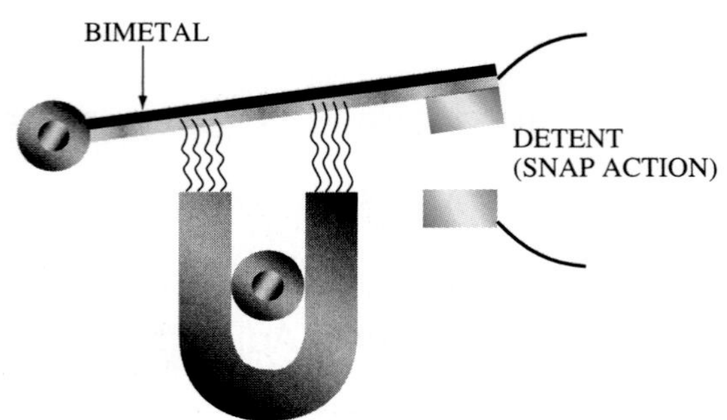

Figure R12-20 Detent (snap-action) magnetic element used to open/close a switch quickly.

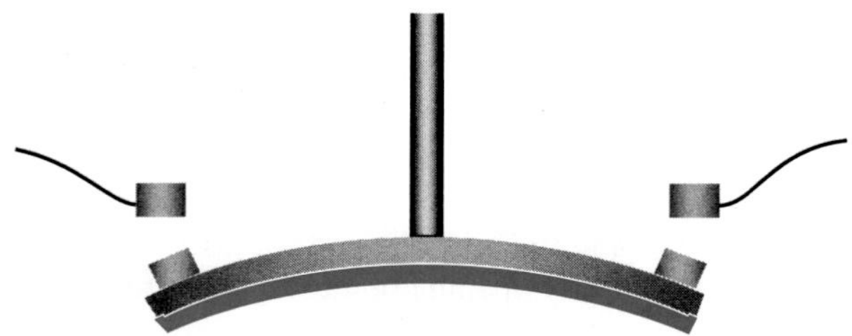

Figure R12-21 Bimetal disc which creates snap-action operation of a switch.

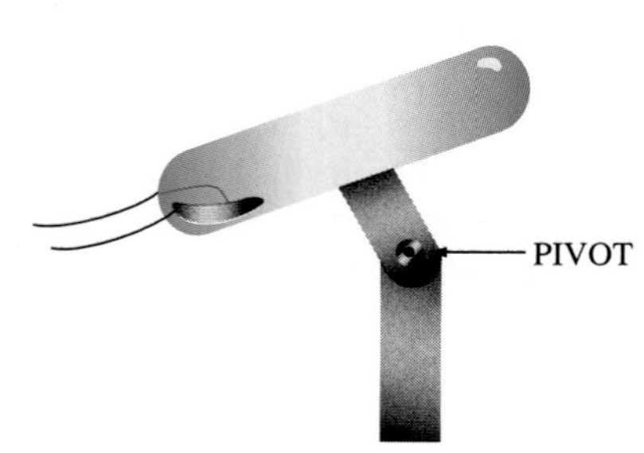

Figure R12-22 Mercury-bulb type switch.

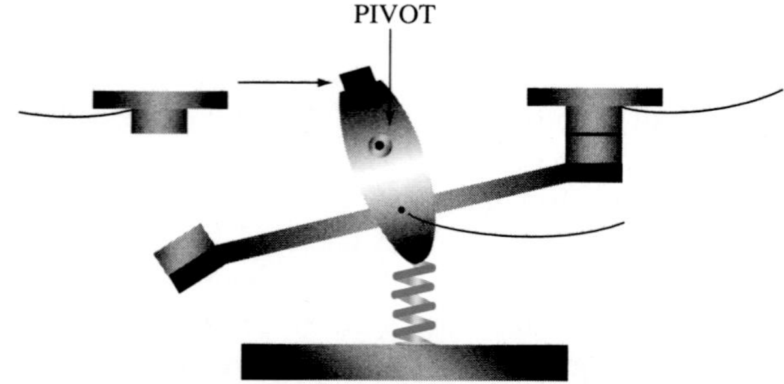

Figure R12-23 Compressed-spring type of detent action.

the bulb over the top center position, the heavy mercury runs from one end of the bulb to the other. This shifting of weight causes the bulb to move quickly from one side of the center to the other, thereby rapidly making or breaking contacts.

Shown in Fig. R12-23 is a fourth method of obtaining detent action. Here the action is induced by using a compressed spring, commonly found in a household toggle switch. Force is applied to the operating arm in the direction of the arrow. This rotates the toggle plate, which starts to compress the spring. As the switch approaches the center position the spring is at maximum compression. The moment the switch passes dead center position the compressed spring will force rapid completion of the switching action.

When the basic control sends a message to start or stop a unit, it usually acts through a relay, starter, or contactor (Fig. R12-24).

These devices consist of two parts: a coil that is energized when the primary control closes and one or more switches that change position when this occurs. The main difference between the three devices is mainly size. However, the starter is a contactor with the addition of overload protection.

Fig. R12-25 shows a more detailed view of a contactor. The starter has one or more sets of contacts located on the

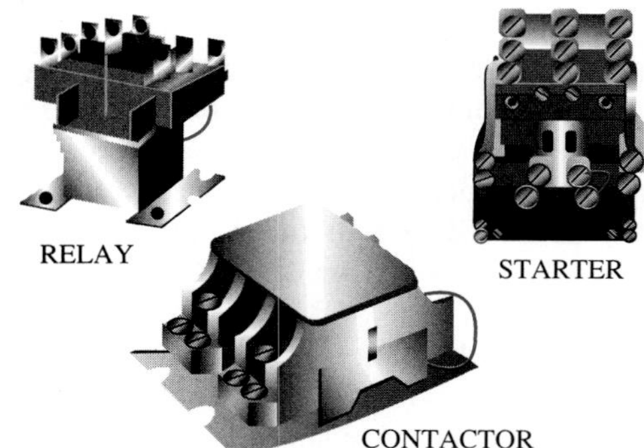

Figure R12-24 Relays, contactors, and starters used to start and stop the cycle.

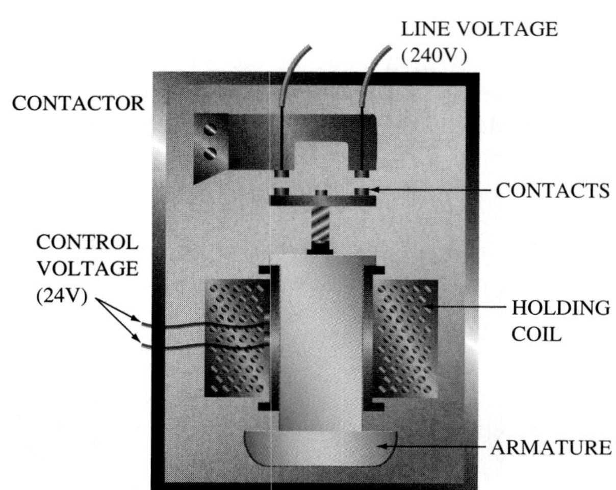

Figure R12-25 Contactor showing armature action to close switch contacts.

	Low Voltage	Line Voltage
Residential and light commercial	24 volt	115 volt 230 volt
Medium–large commercial and industrial	115 volt	240 volt 480 volt 575 volt

Figure R12-26 Tabulation of voltages used for power circuit and control circuit.

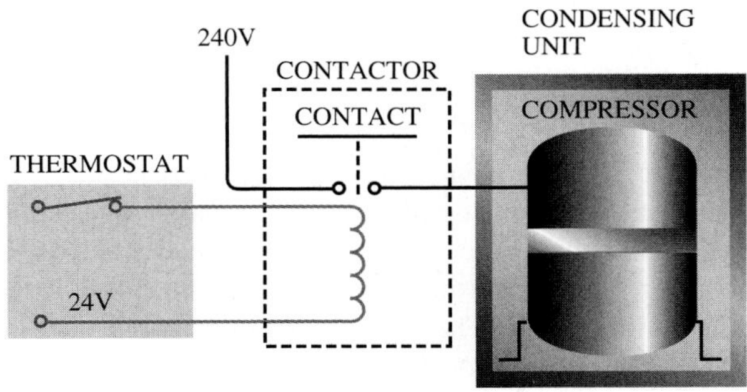

Figure R12-27 Thermostat in a 24V circuit energizes the contactor holding coil closing a 240V line switch.

armature. The armature moves up and down in the holding coil. When the coil is energized the armature moves and closes the contacts that start the motor. The motor contacts are considered normally open (NO) since they are open when no current is applied to the coil. Starters do sometimes have normally closed (NO) contacts that are closed when the coil is de-energized.

The control voltage can be different for various sized jobs as shown in Fig. R12-26.

The control voltage is used to operate the holding coil on a starter as well as for many other control functions. The lower control voltage is used for safety and to permit the use of more sensitive controls on these circuits as compared to the use of line voltage controls.

Fig. R12-27 shows a typical control circuit used to start a condensing unit through a contactor. Note that a thermostat operating in a low-voltage circuit closes the contacts in a high-voltage circuit to operate the condensing unit. When the thermostat is satisfied, the contactor coil is de-energized and the condensing unit stops.

R12-4 OPERATING CONTROLS

In this section, we will discuss electrical and mechanical controls which are used to operate the system, after it is started. We will refer to them as *operating controls.*

Metering Devices

Details on the various types of metering devices are covered in Chapter R-8.

Relays, Contactors, and Starters

When the basic controls such as a thermostat or pressurestat call for starting the system, the signal may go to a relay, contactor, or starter to place into operation a compressor or motor. These controls and their uses are described in detail under the subject of Electrical Components in Chapter E-20.

Condenser-Water Valve

When a water-cooled condenser is used, particularly where it uses city water, the valve regulates the quantity of water that flows through the condenser. It accomplishes two things. It regulates and conserves the flow to maintain a set condensing pressure and temperature. It also acts to conserve city water by shutting down the condenser flow when the refrigeration cycle shuts off.

Referring to the diagram in Fig. R12-28, the condensing pressure is transmitted to the valve through the bellows. The bellows operates the valve against a spring which acts to close the valve. By adjusting the spring tension on the top of the valve, the flow is regulated to maintain the condenser pressure setting.

Refrigerant Solenoid Valve

Solenoid valves, as shown in Fig. R12-29, are used for many purposes. Valves are available to operate on either line voltage or control circuit voltage. Some type of switch is operated to send power to the valve; for example, a thermostat. When the solenoid coil is energized the plunger is moved to control the flow of some fluid, such as refrigerant. In the diagram shown (Fig. R12-29), the valve is being used to start and stop the flow of refrigerant to the evaporator. The solenoid valve is placed in the liquid line ahead of the expansion valve. This type of control arrangement is often used where multiple evaporators

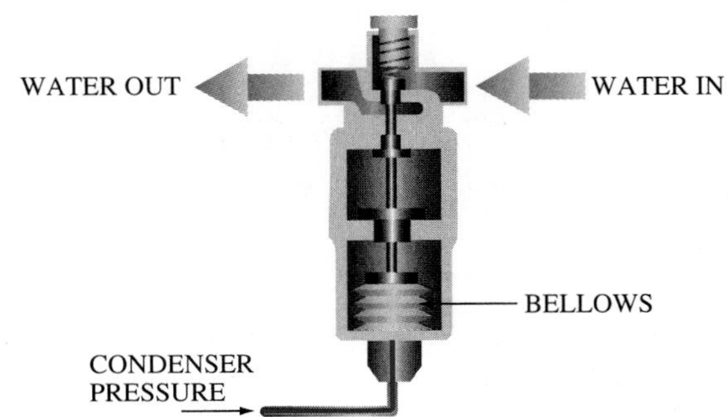

Figure R12-28 Condenser-water valve regulates head pressure at a constant setting.

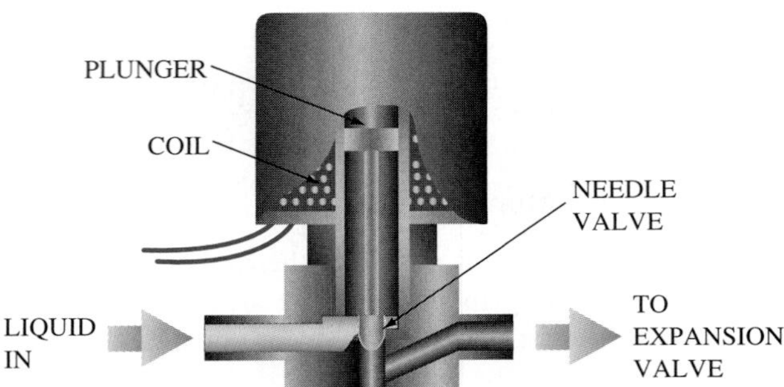

Figure R12-29 Refrigerant solenoid valve used to start and stop fluid flow.

operate from the same compressor and individual rooms require different temperatures.

Four-Way (Reversing) Valve

The four-way reversing valve is an essential part of the heat pump. The valve is used to switch the unit from cooling to heating (or the reverse). It is a type of solenoid valve, since the two positions are controlled by a plunger which is moved by energizing or de-energizing a solenoid coil. Fig. R12-30 shows the valve directing the flow of refrigerant through the unit to produce cooling. The refrigerant discharge from the compressor goes through the valve to the outside coil (condenser), through the metering device to the inside coil (evaporator), through the valve again, and back to the suction side of the compressor. This is basically a normal cycle for any cooling system, with the exception of the passage of the refrigerant vapors through the four-way valve.

In Fig. R12-31 the valve is positioned for heating. In this position the hot discharge gas from the compressor is directed to the inside coil to provide heating. Since the metering device will operate with the refrigerant flow in either direction, after the vapor has condensed in the inside coil it goes through the metering device and to the

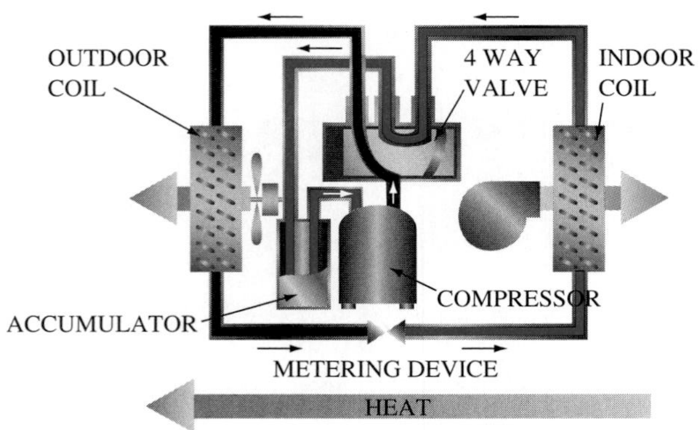

Figure R12-30 Four-way reversing valve used on a heat pump, cooling cycle.

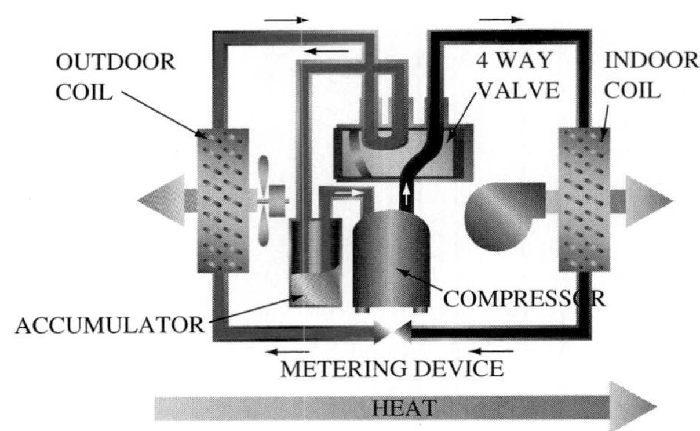

Figure R12-31 Four-way reversing valve used on a heat pump, heating cycle.

outside coil. Here it vaporizes as it absorbs heat and returns to the suction side of the compressor as a gas.

Note the changed usage of the four connections on the top of the valve. Starting from the left, the first connection is blocked off. The second and third connections (a loop) direct the suction gas back to the compressor. The fourth connection directs the discharge gas to the inside coil. Note that in both positions the inlet connection of the valve, shown on the bottom, is connected directly to the compressor. So basically what the valve does is to direct the compressor discharge gas either to the outside coil in the cooling cycle or to the inside coil in the heating cycle.

Back-Pressure Valves

This valve has other names which may be more descriptive, such as suction-pressure regulator or evaporator-pressure regulator. Regardless of which name is used, its function is to provide a constant suction pressure for a single coil or a group of coils, and thus maintain a constant evaporation temperature. The valve shown in Fig. R12-32 is a self-contained pressure-activated valve.

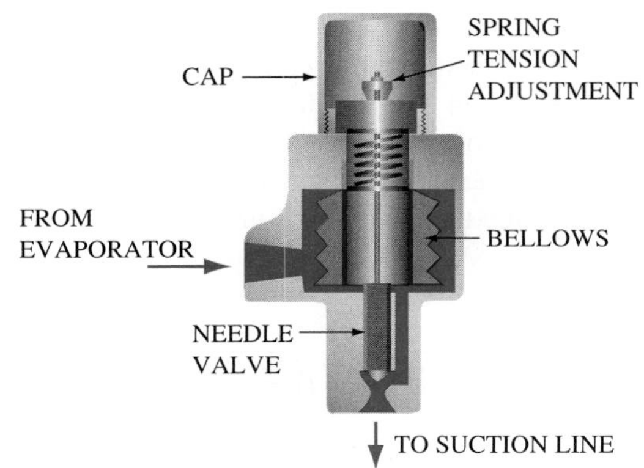

Figure R12-32 Back-pressure valve, used to maintain constant temperature in the evaporator.

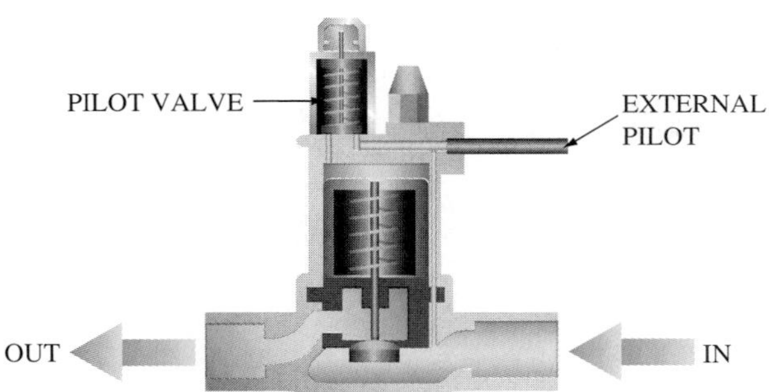

Figure R12-33 A pilot-operated evaporator pressure regulator, used on larger systems.

The spring tension of the valve stem can be adjusted at the top for whatever suction pressure is desired (within the limits of the valve). It is usually placed in the suction line leaving the evaporator, as close to the coil(s) as possible. One variation of the valve is shown in Fig. R12-33.

This version of the valve has a pilot-operated remote sensor that can be located at a more favorable control point. On larger jobs it is often desirable to control from the evaporator itself rather than from the suction line. Fig. R12-34 shows another variation of the evaporator-pressure-regulating valve (EPR), called the two-temperature valve.

It is used on applications where two evaporators with different evaporating temperatures are operated from the same compressor. The two-temperature valve is placed in the suction line from the highest temperature evaporator. The two-temperature valve is a snap-action valve, in that it is either fully open or fully closed. It therefore cycles the evaporator as though it were directly connected to its own condensing unit and controlled by a pressurestat or thermostat.

Check Valve

The check valve is a device that permits a flow of a fluid through piping in only one direction. Fig. R12-35 shows the use of a swing type check valve.

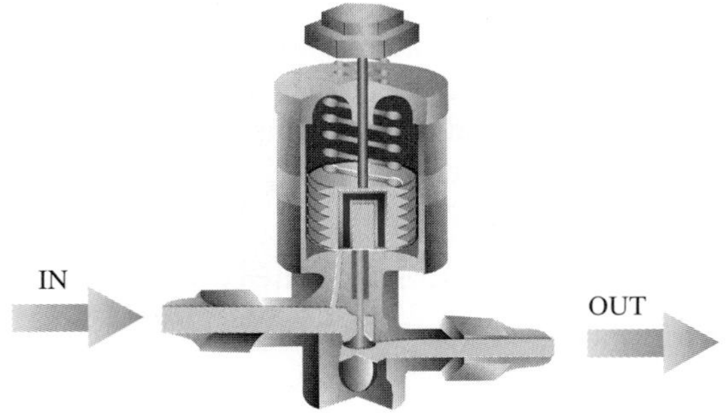

Figure R12-34 A two-temperature snap-action valve, used when two evaporators operate at different temperatures connected to the same compressor.

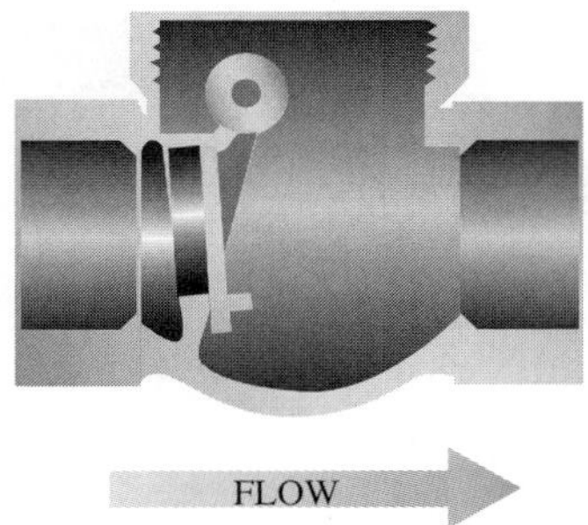

Figure R12-35 Check valve showing the direction of flow and the closed position.

When the flow moves in the direction of the arrow, the flapper opens and the flow is free to move. If the flow is reversed, the flapper slams shut and the flow is blocked. Fig. R12-36 shows a ball-type check valve.

It operates on the same principle as other types of check valves. When the flow is stopped, the ball falls into its seat. If the flow is in the direction of the arrow, the ball moves away from its seat and permits the flow. Reversing the flow causes the ball to block the opening.

Check valves have many uses. They are required on heat pump systems and two-temperature systems. They are used in compressor discharge lines to prevent backflow of refrigerant and oil to the compressor during shutdown. This prevents damage that could occur to the compressor at the time of start-up.

Timed Devices

Programmable thermostats incorporating time functions such as those shown in Fig. R12-37 permit the system owner to set at least two temperatures for the day's system operation.

This resetting function would be considered an operating type of control. Also, defrost timers are used on some commercial freezer room systems to prevent excessive ice buildup on the evaporator. Fig. R12-38 shows the arrangement used on heat pumps to initiate the defrost cycle.

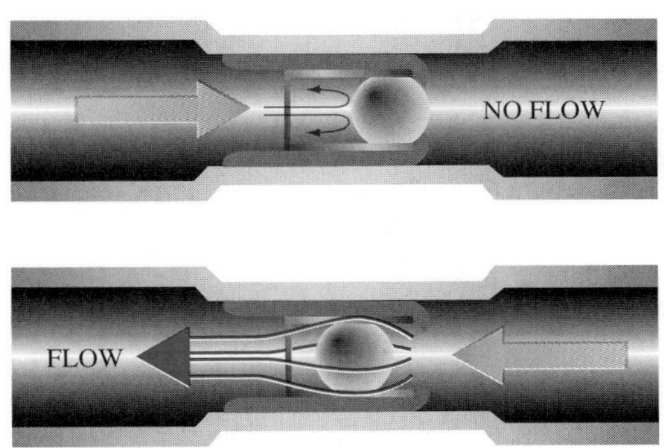

Figure R12-36 Ball-type check valve showing the two positions.

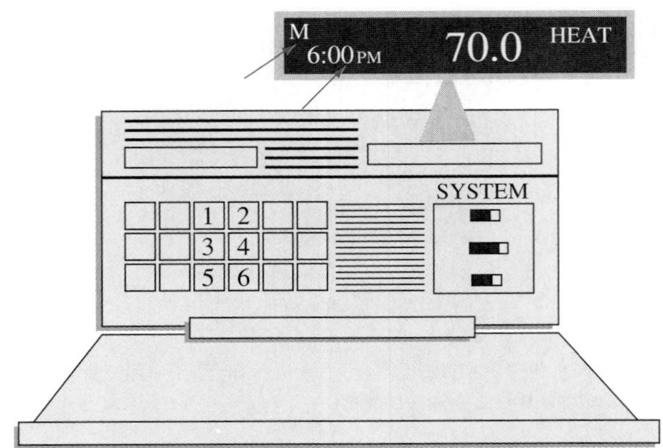

Figure R12-37 Clock thermostat functioning as a basic control.

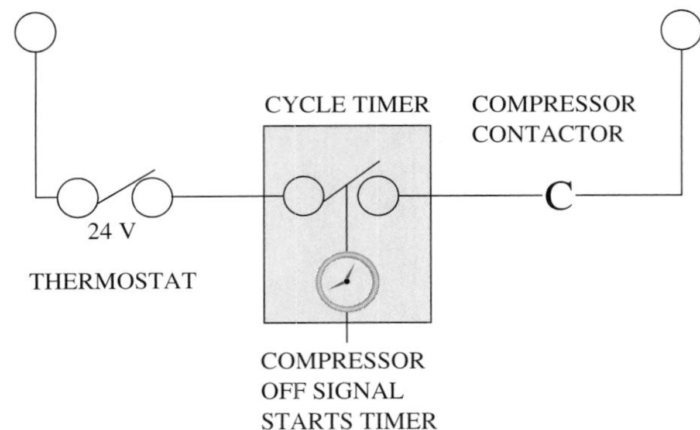

Figure R12-39 Compressor-cycle timer to protect against compressor short cycling.

When the unit is absorbing heat from the outside during the heating cycle, if the temperature outside is below freezing, ice can accumulate on the coil. Two conditions are necessary for the defrost cycle to take place: (a) The timer must be calling for defrost and (b) the temperature sensor on the condenser coil must be below freezing. When the defrost cycle is initiated, the unit is returned to the cooling cycle, causing the hot gas to flow to the outside coil. Fig. R12-39 shows a circuit using a timer to prevent the compressor from short cycling.

Short cycling means rapid turning on and off, which can occur on an air-conditioning system in mild weather. It can occur also during servicing the unit. In any event, such action can be damaging to the compressor. To prevent this, a short-cycling control is used.

The timer shown in Fig. R12-39 starts when the compressor shuts down. The timer opens a switch in the circuit that prevents the compressor from restarting until a certain time period has elapsed, usually five minutes. On some of the solid-state controls of this type, the timer is constructed to compensate for power outages.

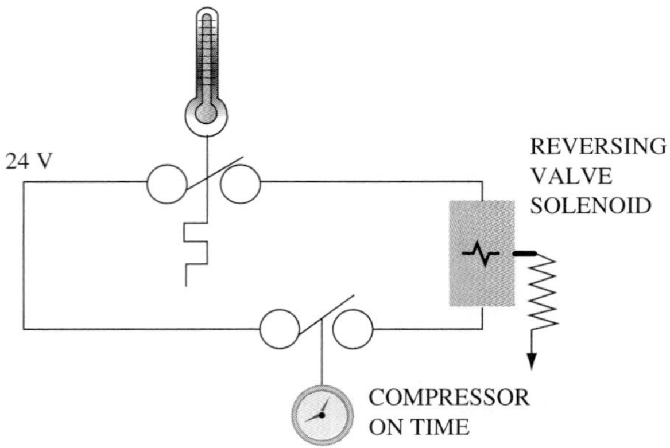

Figure R12-38 Heat-pump defrost arrangement using a reversing-valve solenoid and compressor time clock.

R12-5 SAFETY CONTROLS

Safety controls protect the cycle and its components from damage. They monitor the operation, but do not change the operation unless something goes wrong. Some safety controls, however, prevent the unit from starting under adverse conditions.

If the safety controls are brought into action due to some problem, they must be reset after the necessary repair has been made and the unit is to resume normal operation. These resets can either be automatic or manual, depending on the nature of the problem and the particular design of the control. It is advisable to consult the manufacturer's information relative to the type of reset.

R12-5.1 Reset

Temperature-actuated safety controls can be designed with a wide differential setting so that automatic reset would not result in damaging short cycling. An example of this is a compressor-motor-winding thermostat.

Pressure-actuated safety controls might recycle faster than desired. A manual-reset high-pressure switch could be used. Another option would be using a lockout relay so that the unit would not reset until the control circuit is interrupted by: (a) resetting the thermostat or (b) switching the unit's power off, then on. This permits the operator to try resetting the unit manually without gaining access to the equipment. The advantage: This procedure may avoid occasional nuisance service calls.

R12-5.2 Types of Safety Controls

Electrical Overloads

In all electrical circuits, some kind of protection against excessive current must be provided. The common household fuse is a good example of this type of protection. In motor circuits, however, a reset type of protector is frequently used. The overload shown in Fig. R12-40 is one

PREVENTIVE MAINTENANCE

Good preventive maintenance for controls includes visually inspecting all controls for contact pitting, loose wiring or connectors, and copper tubing with signs of premature fatigue. Electrical contact cleaner can help remove dust, dirt, and grease. Loose wiring should be securely fastened and worn connectors replaced. Control tubing which has been vibrating or rubbing against equipment should be repaired to prevent unexpected failures.

method by which this is accomplished. It is used in conjunction with the contactor previously discussed. When a contactor is used, there is both a control circuit, in which the primary control is inserted, and a load circuit which is opened and closed by the contactor. When excessive current is drawn in the load circuit, this device will break the control circuit.

The control circuit to the contactor coil passes through the controls at the lower right of Fig. R12-40. The load circuit is passing through the bimetallic element shown in the center. If the current through the bimetallic element becomes too high, the element bends to the left. This action forces the contact arm to the left and breaks the contacts. This breaks the control circuit and allows the contactor to open, thereby interrupting the power to the load. The bimetallic element will now cool and return to its original position, but because of the slot in the bottom arm, the contacts will not be remade. The control circuit will remain broken until the reset button is pushed. This moves the arm to the right and closes the control-circuit contacts. Because this control must be reset by hand, it is usually referred to as a manual-reset overload. Some overload relays may be field adjusted for manual- or auto-reset function.

Current/Temperature Devices

A second type of electrical overload protector is the bimetallic disk (Fig. R12-41).

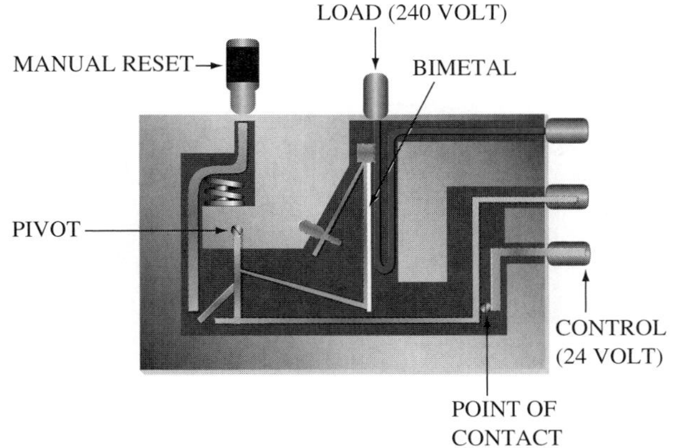

Figure R12-40 Electrical overload used in a circuit breaker.

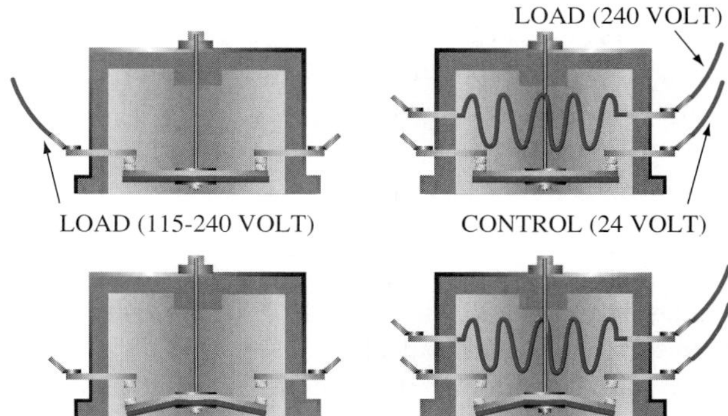

Figure R12-41 Bimetal-disc electrical overload, automatically resets.

When the temperature is raised, the disk will warp and break the electrical circuit. On the left is shown the simplest device in which heat from an external source activates the disk. On the right, external heat is supplemented by resistance heat from an electrical load. The device on the right will react much faster on electrical overload than the device on the left because the supplemental heat is within the device itself. The bimetallic-disk overload protector is widely used in the protection of electrical motors.

The current relay (Fig. R12-42) is another type of electrical overload that can be manually or automatically reset. The greater advantage of this type of relay is that it is only slightly affected by ambient temperatures, thereby avoiding nuisance trip-outs.

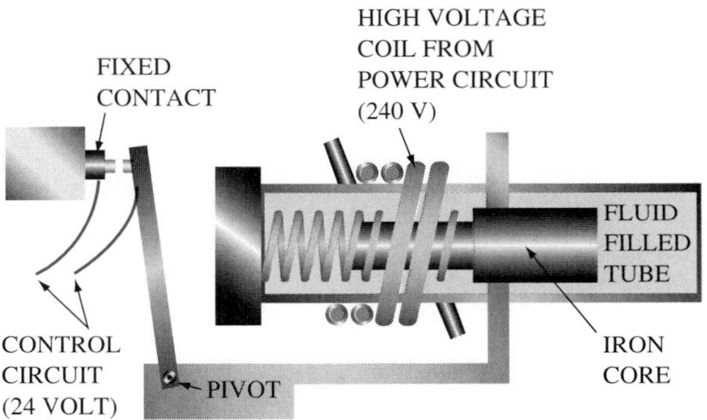

Figure R12-42 Current relay used as an electrical overload, available with automatic or manual reset.

The current relay is made up of a sealed tube completely filled with a fluid and holding a movable iron core. When an overload occurs, the movable core is drawn into the magnetic field, but the fluid slows its travel. This provides a necessary time delay to allow for momentary high current during motor start-up (locked rotor amps) without tripping the overload. When the core approaches the pole piece, the magnetic force increases and the armature is actuated, thus breaking the control circuit.

On short circuits, or extreme overloads, the movable core is not a factor because the strength of the magnetic field of the coil is sufficient to move the armature without waiting for the core to move. The time-delay characteristics are built into the relay and are a function of the core design and fluid selection.

Thermostats

The only difference between the thermostatic operating control previously discussed and the safety thermostat involves the application in which they are used; the controls themselves may be identical.

A good illustration of a thermostat used as a safety control can be found in most water-chilling refrigeration systems (Fig. R12-43).

In water chilling systems it is important that the water not be allowed to freeze because it could then do physical damage to the equipment. In such systems a thermostat may be used with the temperature-sensing element immersed in the water at the coldest point. The thermostat is set so that it will break the control circuit at some temperature above freezing. This will stop the compressor and prevent a further lowering of water temperature and possible freezing. The temperature control used for safety purposes is physically the same as that used for operational purposes.

Pressurestats

The high-pressure cutout is probably the best example of a pressure control used for safety purposes (Fig. R12-44).

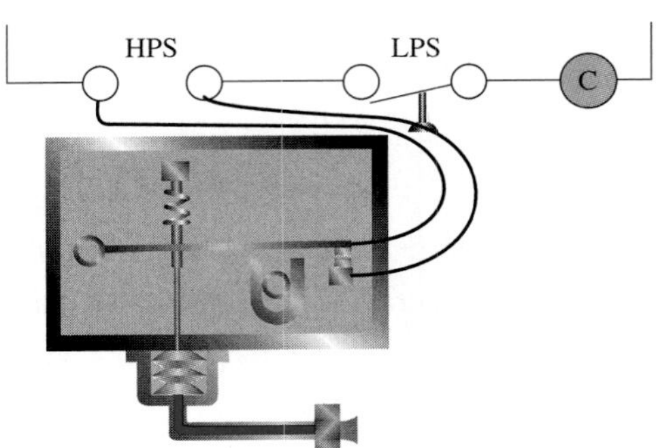

Figure R12-44 High-pressure cutout protects compressor against high-discharge pressure.

It can be set to stop the compressor before excessive pressures are reached. Such conditions might occur because of a water-supply failure in water-cooled condensers or because of a fan motor stoppage on air-cooled condensers.

The operation of the low-pressure limit control (Fig. R12-45) is the same as discussed under the subject of basic controls.

Mechanically, there is no difference between the low-pressure control used as an operating control and the one used as a safety control except for settings. The low-pressure control is used to stop the compressor at a predetermined minimum operating pressure. As a safety device, the low-pressure control can protect against loss of charge, high compression ratios, evaporator freeze-ups, and entrance of air into the system through low-side leaks.

Pressure-relief Valve

Fig. R12-46 shows the pressure-relief valve that can be placed in the hot-gas line, condenser, or liquid receiver as a high-pressure safety control. These controls are required by the installation codes on most water-cooled systems. This valve automatically resets when normal pressures are reestablished.

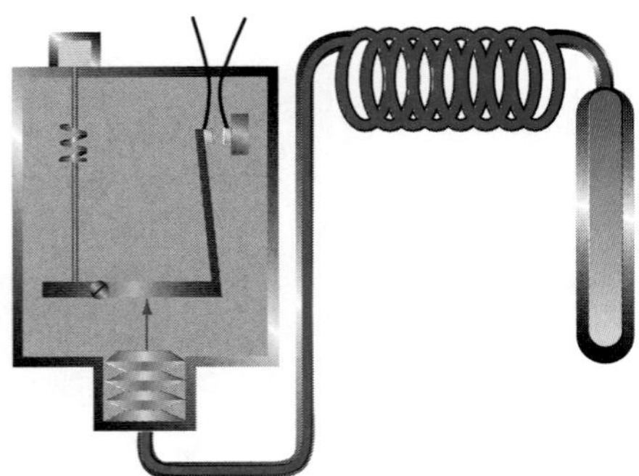

Figure R12-43 Remote-bulb thermostat used for temperature protection to prevent air, oil, water, or motor temperatures from being too low.

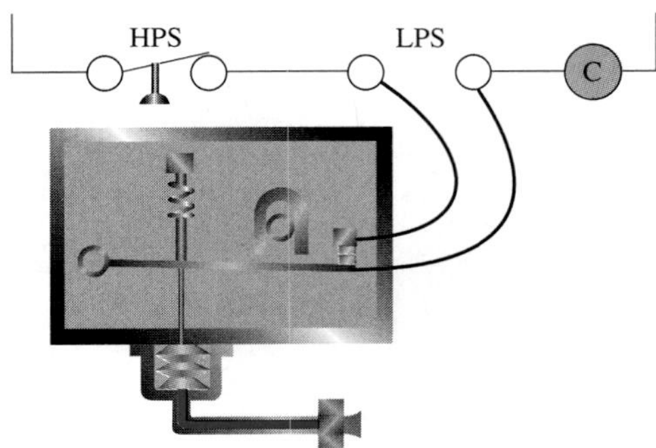

Figure R12-45 Low-pressure cutout opens when minimum safe suction pressure is reached.

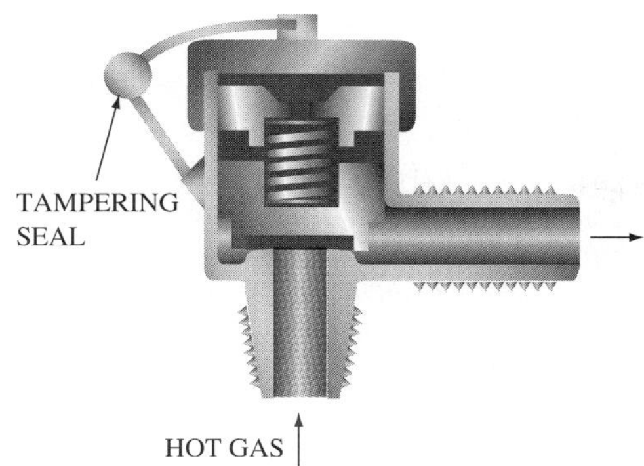

Figure R12-46 Pressure-relief valve prevents physical damage from excessive pressure.

Fusible Plug and Rupture Disc

Fig. R12-47 shows two other safety fittings, the fusible plug and the rupture disc. The fusible plug melts open on excessive temperature and the rupture disc opens on excessive pressure.

It is advantageous to use the pressure-relief valve in place of these fittings wherever possible since the relief valve automatically resets and does not permit losing the full refrigerant charge.

Oil-pressure Safety Switch

The oil safety switch is activated by pressure; however, it is operated by a pressure *differential* rather than by straight system pressure. It is designed to protect against the loss of oil pressure. Inasmuch as the lubrication system is contained within the crankcase of the compressor, a pressure reading at the oil-pump discharge will be the sum of the actual oil pressure plus the suction pressure. The oil safety switch measures the pressure difference between the oil-pump discharge pressure and compressor crankcase pressure and shuts down the compressor if the oil pump does not maintain an oil pressure as prescribed by the compressor manufacturer. For example, if a manufacturer indicated that a 15-lb net oil pressure was required, a compressor with a 40-lb back pressure must have at least a 55-lb pump-discharge pressure. At any lower discharge pressure from the pump, this switch would stop the compressor. An examination of the operating force inset on the right of Fig. R12-48 will show that since suction pressure is present on both sides, its effect is canceled. The control function depends on the new oil pressure overcoming the predetermined spring pressure. The spring pressure must equal the minimum oil pressure allowed by the manufacturer.

When a compressor starts, there is no oil pressure. Full oil pressure is not obtained until the compressor is up to speed. Therefore, a time-delay device is built into this type of control to allow the compressor enough time to start. One of the methods used to create this time delay is a small resistance heater. When the compressor is started, because the pressure switch is normally closed, the resistance heater is energized. If the oil pressure does not build up to the cutout setting of the control, the resistance heater warps the bimetallic element, breaks the control circuit and stops the compressor. The switch in Fig. R12-48 is automatically reset; however, these switches are normally reset manually.

If the oil pressure rises to the cutout setting of the oil safety switch within the required time after the compressor starts, the control switch is opened and de-energizes the resistance heater, and the compressor continues to operate normally.

If oil pressure should drop below the cut in setting during the running cycle, the resistance heater is energized and, unless oil pressure returns to cut out pressure within the time delay period, the compressor will be shut down. The compressor can never be run longer than the predetermined time on subnormal oil pressure. The time-delay setting on most of these controls is adjustable from approximately 60 to 120 seconds. Sometimes the timing and safety shutdown function will be accomplished by an electronic control module using a signal from an oil-pressure differential switch.

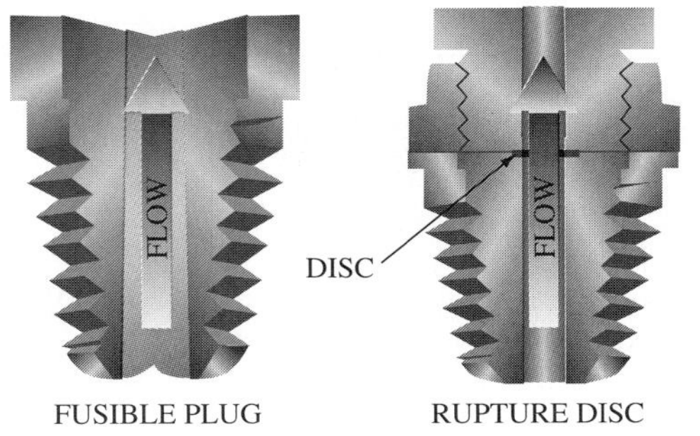

FLOW
DISC
FLOW
FUSIBLE PLUG
RUPTURE DISC

Figure R12-47 Fusible plug and rupture disc. These safety fittings protect the system against excessive temperature or pressure, respectively.

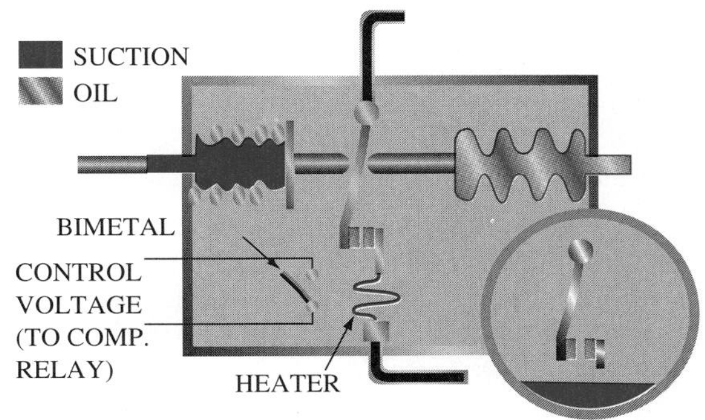

Figure R12-48 Oil-pressure safety switch protects the compressor against damage from loss of oil pressure.

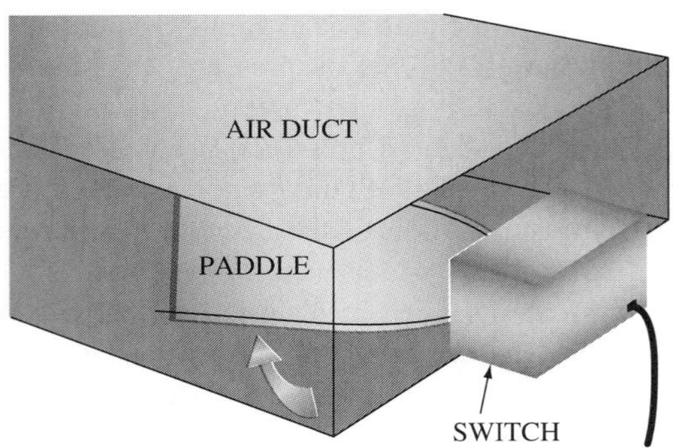

Figure R12-49 Sail switch senses flow, protects the system when air or water flow is inadequate.

KNOWLEDGE EXTENSION

The controls industry is one of the fastest growing and changing professions within the air-conditioning and refrigeration business. To learn more about specific controls, equipment, or a career in the controls industry, contact a controls manufacturer representative and ask for specific literature regarding their business. Most control company representatives are very willing to help their customers become better acquainted with their product and services.

Flow Switch

The sail switch, shown in Fig. R12-49, is a protective device to prevent the operation of a unit when there is inadequate fluid flow.

In an air system, the sail switch is placed in the duct to sense the flow of air. Unless there is an adequate supply of air over the coil, the unit is either not started or shut down. Switches of this type can also be placed in a water-line feed of a water-cooled condenser. If there is an inadequate supply of water, or no water, the unit is prevented from running.

R12-6 OPERATING CONTROL SYSTEMS

There are as many control methods for operating refrigeration systems as there are engineers to design them, but there are two simple methods that are standard and will be discussed briefly.

R12-6.1 Thermostats

The first control method, illustrated in Fig. R12-50, is the simple thermostat start-stop method. As the thermostat calls for cooling and closes the contacts, the control circuit is completed through the contactor coil. This control circuit includes a Klixon, a pressure control, and a motor-starter thermal-overload relay as safety/limit controls. When the contactor coil is energized, it closes the contacts and completes the power circuit to the motor. The thermostat is the operating control and the motor cycles as the thermostat dictates. Any of the safety controls will break the control circuit and stop the compressor in the same manner as the thermostat.

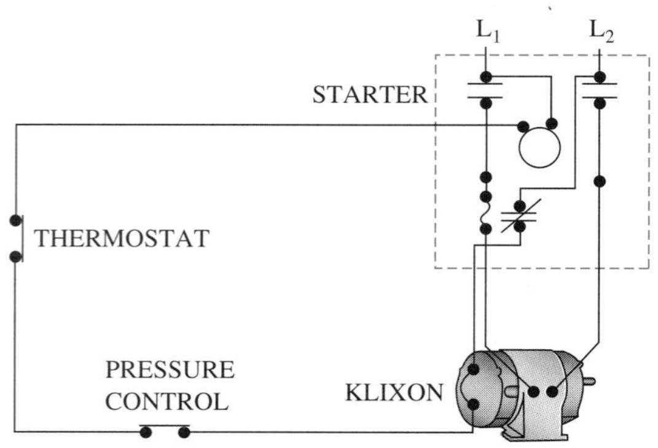

Figure R12-50 Simple control circuit, using thermostat connected to the starter control circuit to start and stop the system.

R12-6.2 Pump-down Cycle

A slightly more complex but still widely used control method is shown in Fig. R12-51. This is known as pump-down control. In this method the thermostat operates a solenoid valve. When the thermostat calls for cooling, it completes a circuit through the solenoid valve, which is in the liquid line ahead of the metering device. By opening the valve, high-pressure liquid flows through the metering device to the evaporator.

A pressure control is connected to the low side of the system. As the pressure rises in the evaporator coil, the

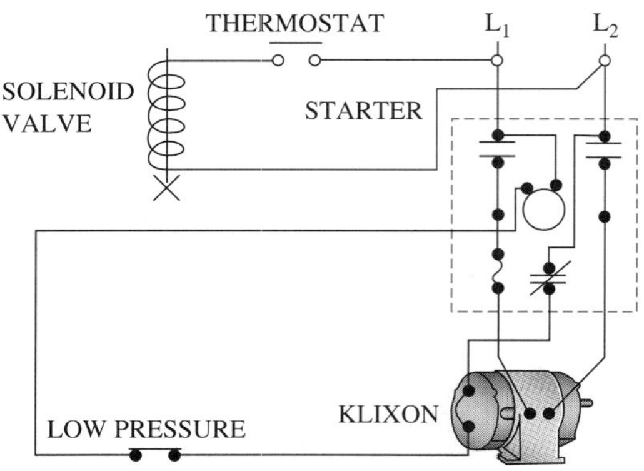

Figure R12-51 Pump-down control used to stop the system on low pressure after refrigerant in the evaporator is pumped out. Protects the compressor on start-up against liquid slugging.

pressure control closes and completes the circuit through the safety device to the contactor coil. The compressor starts and the system operates normally. When the thermostat is satisfied, the solenoid coil is de-energized and the flow of liquid is cut off to the evaporator. Because the compressor is still running, the evaporator pressure is reduced to the cutout point of the low-pressure control. The control contacts open and the compressor stops.

This system is designed to keep liquid refrigerant from filling the evaporator during shutdown. Regardless of the position of the thermostat, the low-pressure control will operate the compressor upon sensing a rise in pressure in the evaporator.

REVIEW

- Types of Controls
  Basic Controls—Start and stop system components.
  Operating Controls—Regulate processes.
  Safety Controls—Protect the cycle.
- Basic Controls
  Thermostat—operated by temperature
  Pressurestat—operated by pressure
  Humidistat—operated by moisture (humidity)
- Operating Controls
  Relays
  Contactors
  Starters
- Safety Controls
  Disc overload
  Current overload
  High-pressure cutout
  Low-pressure cutout
  Oil safety switch
- Bimetal Thermostat—An element composed of two different metals bonded together. As temperature surrounding the element changes, the metals expand or contract. One will expand or contract faster and the element will warp, opening or closing a switch contact.
- Fluid Pressure Operating Control—A fluid-filled bulb in which the fluid expands when heated and exerts a pressure through a capillary tube to a diaphragm or bellows to open or close a circuit.
- Thermistor—A temperature-sensing device used in electronic controls; it has a resistance affected by temperature. This resistance represents a temperature value to the electronic controller which opens or closes a circuit.
- Types of Pressurestats
  Bellows
  Bourdon tube
- Types of Humidistats
  Human or horse hair
  Nylon element
  Electronic
- Operating controls can include metering devices, relays, contactors, condenser water valve, refrigerant valves, check valves, and timed devices.
- Safety controls can include electric overloads, current temperature devices, thermostats, pressurestats, fusible plug, rupture disc, oil safety, and flow switches.

Problems and Questions

1. Which of the following is a type of control system?
   a. Semiconductor
   b. Electronic
   c. Pressurestat
   d. Bourdon tube
2. Which of the following do you consider basic controls?
   a. Thermostat
   b. Evaporator-pressure regulator
   c. Electrical overload
   d. Solenoid valve
3. Which of the following is used in the construction of a pressurestat?
   a. Bimetal
   b. Mercury tube
   c. Bourdon tube
   d. Spring return
4. Which of the following features is included in a programmable thermostat?
   a. Low-pressure control
   b. Humidity control
   c. 30-day schedule
   d. Night set-back
5. Which of the following materials can be used in a humidity measuring instrument?
   a. Bourdon tube
   b. Nylon element
   c. Copper filings
   d. Spring pressure
6. What is meant by *detent* action?
   a. Spring tensions
   b. Snap action
   c. Bellows operation
   d. Mercury bulb action
7. What is meant by *differential?*
   a. Rate at which contacts are opened.
   b. The adjustable limit setting.
   c. Size of the bimetal element.
   d. Difference between cut in and cut out positions.
8. When a thermostat operates a contactor, do the contacts open or close on a call for cooling?
   a. Open
   b. Close

9. Which of the following is NOT considered an operating control?
   a. Humidistat
   b. Check valve
   c. Low-side float
   d. Condenser-water valve
10. Which of the following is NOT considered a safety control?
    a. Fusible plug
    b. Automatic expansion valve
    c. Flow switch
    d. Rupture disc
11. What is the definition of range when used with thermostats?
12. Explain the purpose of the back-pressure valve.
13. What is a thermistor and where is it used?
14. Pneumatic controls are operated by _____ _____.
15. The oil-pressure switch measures the pressure difference between what two locations?
16. A pump-down cycle includes what important controls?

Refrigerants

R13

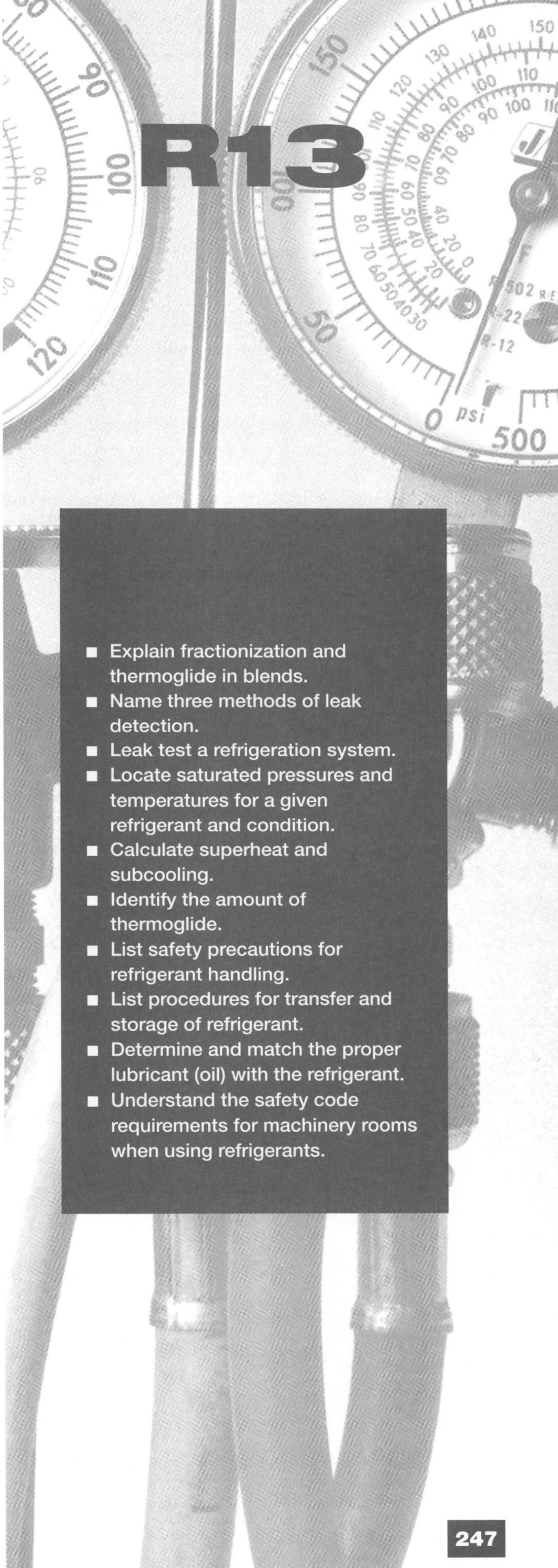

AFTER STUDYING THIS CHAPTER, THE STUDENT WILL BE ABLE TO:

- Select refrigerants that are not destructive to the ozone layer.
- Explain the physical and chemical properties of refrigerants.
- Match terms associated with refrigerants to the correct definition.
- Match the common types of refrigerants to their chemical name and ASHRAE number.
- Identify pure refrigerant compounds, azeotropic, and zeotropic mixtures.
- Match types of refrigerants to their applications.
- Explain classifications of refrigerants.
- Identify the cylinder color code to the refrigerant .
- List seven desirable characteristics of a refrigerant.
- Explain fractionization and thermoglide in blends.
- Name three methods of leak detection.
- Leak test a refrigeration system.
- Locate saturated pressures and temperatures for a given refrigerant and condition.
- Calculate superheat and subcooling.
- Identify the amount of thermoglide.
- List safety precautions for refrigerant handling.
- List procedures for transfer and storage of refrigerant.
- Determine and match the proper lubricant (oil) with the refrigerant.
- Understand the safety code requirements for machinery rooms when using refrigerants.

R13-1 DEFINITION OF A REFRIGERANT

A refrigerant is a medium (fluid) for heat transfer, used in a refrigeration system to pick up heat by evaporating at a low temperature and pressure, and to give up heat upon condensing at a higher temperature and pressure.

Fig. R13-1 shows a typical room cooler using the refrigerant R-22. Note that the evaporator of this unit is in the room being cooled, where the refrigerant picks up heat. The condenser is extended outside the building where the refrigerant gives up heat. The use of the compressor and the connecting piping is to form a closed system so that the refrigerant can be reused again and again.

The refrigerant characteristics for this application must be such that the evaporator boiling temperature is below room temperature and the condensing temperature is above the outside air temperature. These conditions are necessary so that proper heat transfer can take place to provide room cooling.

Many different substances can be used for refrigerants. Under certain conditions even water could be used as a refrigerant. Referring to Fig. R13-2, at standard conditions water boils at 212°F (100°C). In a vacuum of 29.75 in. of mercury, water boils at 40°F (4.4°C). Water would be an undesirable refrigerant for most applications because of the expense in producing and maintaining such a low vacuum.

R13-2 NUMBER DESIGNATION

Refrigerants are identified by number, preceded by the letter "R" (Refrigerant). This number designation has been established by the American Society of Heating, Refrigeration and Air Conditioning Engineers (ASHRAE) and is used throughout the industry.

Certain designations have been provided for the refrigerants in abbreviated form, to indicate the chemical composition, as well as to relate to the Ozone Depletion Factor (ODF) for the refrigerant. (ODF is a rating with the range of 1-0). For example, refrigerant R-12 can be designated CFC-12, R-22 can be designated HCFC-22 and R-134a can be designated HFC-134a.

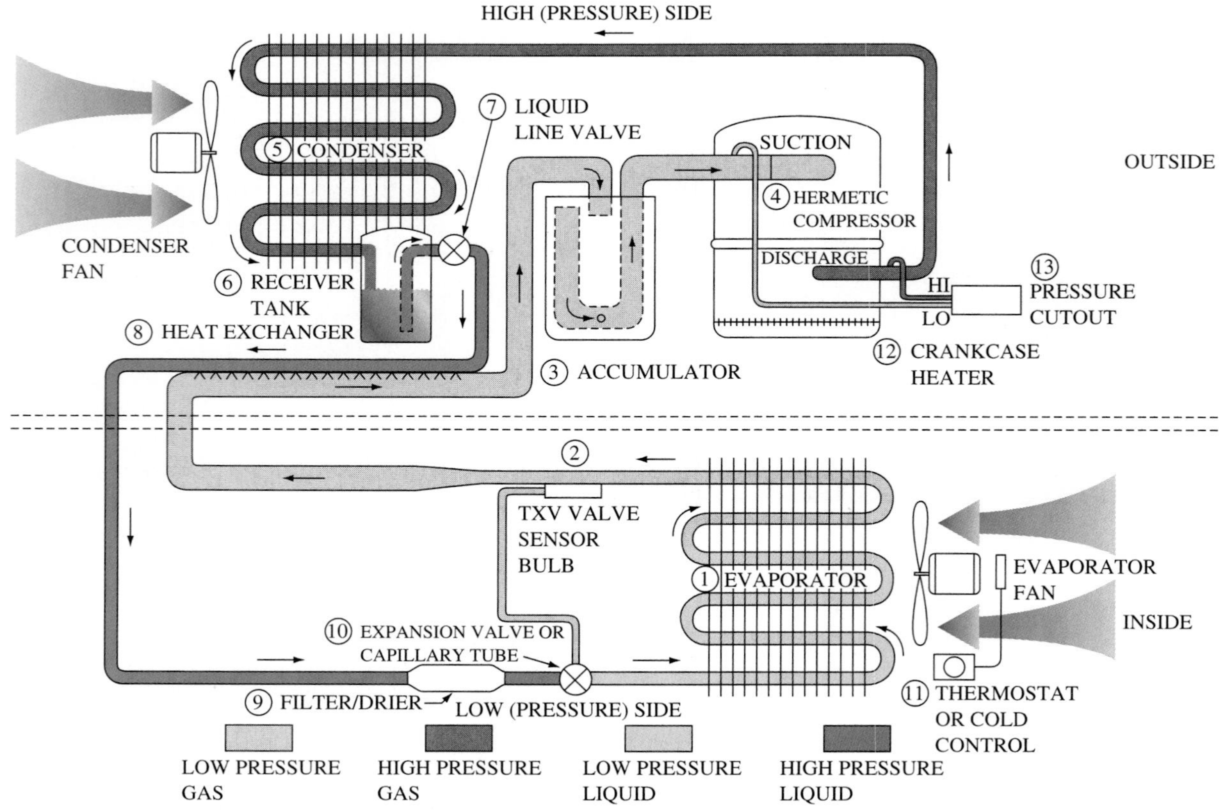

Figure R13-1 Typical refrigeration system showing the use of refrigerant to transfer heat from the inside to the outside. (Courtesy of Tecumseh Products Company)

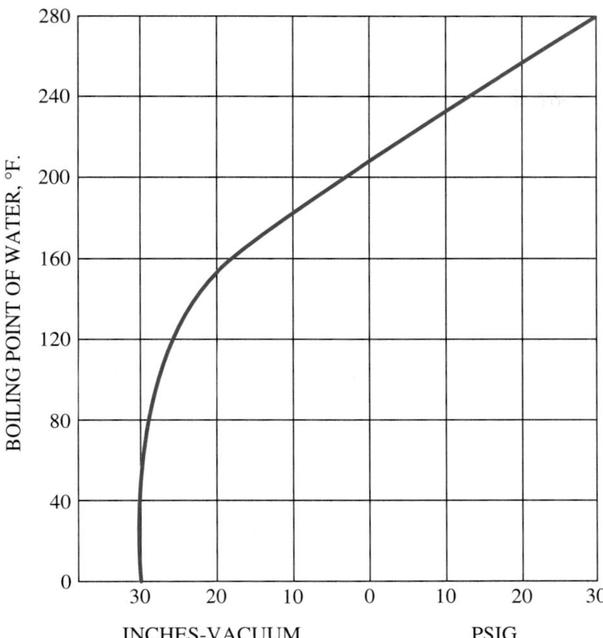

Figure R13-2 Pressure-temperature curve for water.

R13-3 CHEMICAL COMPOSITION OF REFRIGERANTS

In Fig. R13-3, the common refrigerants are listed, along with their chemical name and ODF. The prefix "CFC" refers to the family of refrigerants containing chlorine, fluorine, and carbon. Compounds that also contain hydrogen precede the abbreviation with the letter "H" to signify an increased deterioration potential before reaching the stratosphere—for example, the prefix "HCFC". The "FC" family does not contain chlorine and can also be preceded with an "H," as in "HFC."

Note that the CFC refrigerants have a high ODF, the HCFC refrigerants have a low ODF and the HFC refrigerants have a zero ODF.

Certain refrigerants which are mixtures, use the R number and also have a designation that shows the constituents. For example, R-502 is composed of HCFC-22/115. The ODF given is a 0.28 rating.

Azeotropic and Zeotropic Blends

A number of refrigerants, such as R-502, are made up of blends or chemically prepared mixtures of refrigerants.

The azeotropic blends consist of multiple components of different volumes that, when used in refrigeration cycles, do not change volumetric composition or saturation temperature as they evaporate or condense at constant pressure. These refrigerants have numbers in the 500 series, such as R-502.

The zeotropic blends consist of multiple components of different volumes that, when used in refrigeration cycles, change volumetric composition and saturation temperatures as they evaporate or condense at constant pressure.

R13-4 REFRIGERANT REQUIREMENTS

There are many properties that are characteristic of a "good" refrigerant. An entire group of features needs to be considered. In selecting a refrigerant there are usually some compromises that need to be considered due to conflicts which arise between desirable characteristics. For example, ammonia refrigerant has a high performance factor in terms of net refrigerating effect (Btu/lb) but should not be selected for use in systems with a proximity to people due to its potential toxicity and flammability.

On p. 250 is a list of some of the most important qualities to consider when selecting refrigerants.

ON THE JOB

PROBLEM: A service technician went to do a routine check on an air conditioner. The technician found the suction pressure was 38 and the discharge pressure 175. Checking the temperature drop across the coil, he found it to be 18°F. The first reaction was to find out why the pressures were so low. Before making adjustments he checked the name plate and found the refrigerant to be R-500.

SOLUTION: To make the mistake of adding a small amount of refrigerant 22 to adjust a charge could contaminate the entire charge. The new laws would require the complete charge to be recovered and disposed of, and the appropriate charge of R-500 added. Be sure to evacuate to proper levels and allow for appropriate time to ensure degassing of existing refrigerants. The recommended vacuum is 500 microns for at least 45 minutes; large systems may require more time.

Characteristics of Some Common Refrigerants					
"R" Number	Chemical Type	ODP	Similiar to	Components	Container Color
R-11	CFC	1.0		11	Orange
R-12	CFC	1.0		12	White
R-22	HCFC	0.050		22	Green
R-123	HCFC	0.02		123	Light gray
R-134a	HFC	0.0	R-12	Single	Sky blue
R-401a	HCFC	0.05	R-12	22,152a,124	Coral
R-402b	HCFC	0.02	R-502	125,290,22	Mustard
R-404a	HFC	0.0	R-502	125,143a,134a	Orange
R-500	CFC	0.74		12,152a	Yellow
R-502	HCFC	0.28		22,115	Orchid
R-507	HFC	0.0	R-502	125,143a	

Figure R13-3 "R" numbers are assigned by the American Society of Heating, Refrigeration and Air Conditioning Engineers (ASHRAE) after certification procedures. Ozone-depleting refrigerants are being replaced by many new ones. Cylinder colors as well as accurate and prominent labeling are required by the U.S. Dept. of Transportation for safety in shipment.

Performance

The refrigerant needs to operate within the temperature and pressure range required to perform the task assigned to it. For example, consider the requirements of a reach-in refrigerator. The box must be capable of maintaining temperatures in the range of 40° to 45°F. The refrigerant evaporating temperature needs to be in the range of 25° to 30°F (−3.9° to −1.1°C), which is 15°F (8.3°C) below box temperature. Using an air-cooled condenser, the condensing temperature typically will be in the range of 120° to 130°F (48.9° to 54.5°C.), which is 30°F (16.7°C) higher than ambient air. At these temperatures, the pressures must be in the range obtainable using a standard compressor.

Let us assume a compressor is available that can operate at pressures not to exceed 250 psig (1823.8 kPa). If R-12 is the refrigerant originally selected, the evaporating pressure would be 24.6 to 28.4 psig (270.9 to 297.6 kPa). The condensing pressure would be 157.6 to 181.0 psig (1188.6 to 1381.6 kPa). This is satisfactory.

If the new refrigerant R-134a is selected for replacement, the evaporating pressure would be 36.7 to 40.7 psig (261.4 to 281.3 kPa), with a condensing pressure of 185.8 to 199.2 psig (1281.6 to 1473.3 kPa). This is satisfactory.

Safety

The refrigerant shall be safe to use. It should be non-poisonous, non-explosive, non-flammable and non-toxic.

The ASHRAE Safety Code For Mechanical Refrigeration classifies the refrigerants in respect to their toxicity and flammability. The refrigerants are divided into groups: A for low toxicity and B for elevated toxicity. The number following the letter indicates the flammability of the refrigerant. The lower the number, the lower the flammability. A refrigerant designated as A1 has the lowest toxicity and the lowest flammability. Fig. R13-4 shows this information in diagrammatic form.

Environmental Impact

The refrigerant shall be free of any chemicals, which when released to the atmosphere, will damage the ozone layer. (Ozone Depletion Factor (ODF) = 0).

Efficiency

The refrigerant shall have a high latent heat (liquid to vapor phase) and a low vapor volume per pound. Under

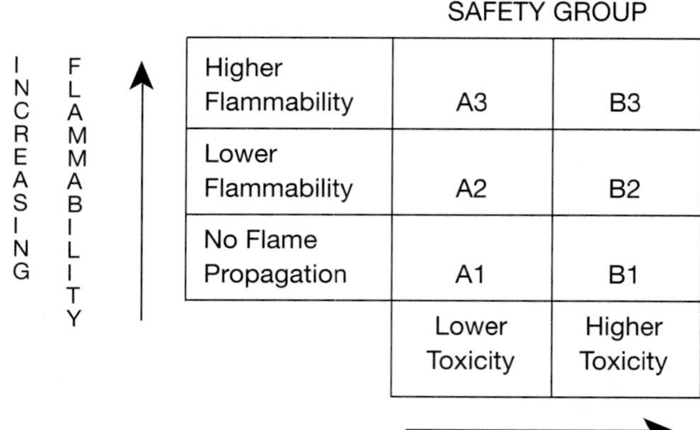

Figure R13-4 Safety classification for refrigerants. (Copyright by the American Society of Heating, Refrigerating, and Air-Conditioning Engineers, Inc. Used by permission.)

these conditions, less refrigerant needs to be pumped, resulting in a smaller compressor and reduced piping sizes.

Stability

Both the liquid and the vapor shall be stable so that the refrigerant will not decompose at normal operating temperatures and pressures.

Compatibility

The refrigerant must not react with or deteriorate materials it comes in contact with during the operation. These include metallic compressor parts, gaskets, "O" rings, seals, motor insulation and windings, piping, and condenser and evaporator heat transfer surfaces.

Lubrication

The refrigerant shall be compatible with the lubricating oil required for the refrigeration compressor. It is also desirable to have liquid refrigerant and oil miscible (soluble or capable of mixing) with one another. This will assure that the small percentage of the oil that is carried away from the compressor by the discharge gas will be carried through the condenser and evaporator and returned to the compressor.

Leaks

Refrigerant leaks should be easy to detect and locate.

Price

The refrigerant should be reasonably priced in the quantities required for normal systems. Rating is in dollars per pound.

Lifetime of the Refrigerant

The lifetime of the refrigerant should be equal to or greater than the lifetime of the equipment it serves.

R13-5 CHANGING REFRIGERANTS

In view of the restrictions placed on the use of many of the common refrigerants, the refrigerant manufacturers have been spending millions of dollars to find suitable alternates that meet the new requirements. Two of these new refrigerants now being used by some equipment manufacturers are R-123 (HCFC-123) and R-134a (HFC-134a). Fig. R13-5 gives a comparison of some of the characteristics of these alternate refrigerants with the common refrigerants they are designed to replace. R-123 is a replacement for R-11, and R-134a is a replacement for R-12.

Note that the table also shows R-22 (HCFC-22), ammonia (R-717) and the absorption refrigerant water/lithium bromide. R-22 is an HCFC type refrigerant, with an ODF = 0.05, and according to the present schedule, production will continue until the year 2030.

Ammonia is an organic refrigerant with an ODF = 0, which has limited use due to its toxic and flammable characteristics.

The absorption system refrigerant has an ODF = 0, and is limited to applications of absorption refrigeration. In view of the restrictions on the use of the common refrigerants R-11 and R-12, it is believed that the market for ammonia and absorption systems will be increased.

R-123 has a satisfactory efficiency, a low global warming potential (GWP), a better ODF than R-22, and its cost is in the middle range. It has a B1 safety rating and, there-

Attribute	R-11	R-123	R-12	R-134a	R-22	Ammonia (R-717)	Absorption Water/LiBr
Environmental							
efficiency (ideal COP)	7.78	7.63	6.91	6.77	7.06	7.28	1.4 to gas
(ideal kW/ton 40/100)	0.45	0.46	0.51	0.52	0.50	0.48	gas fired
greenhouse gas (GWP)	1,500	29	4,500	420	510	≈0	≈0
(GWP 100 yr)	3,500	85	7,300	1,200	1,500	≈0	≈0
ozone depletion (ODP)	1.00	0.02	1.00	0.00	0.05	0.00	0.00
Safety							
flammability (LFL %)	none	none	none	none	none	14.8	none
toxicity (TLV or ≈)	1,000	5–10	1,000	1,000	1,000	25	1,000
safety group (Std 34)	A1	B1	A1	A1	A1	B2	A1
Cost (OEM)							
current (relative $/lb)	2.00	3.75	2.40	9.50	0.90	0.40	—
estim 1995 (1991 $/lb)	3.75	2.75	4.25	4.75	0.90	0.40	—

Figure R13-5 Attributes of alternate refrigerants. (Copyright by the American Society of Heating, Refrigerating, and Air-Conditioning Engineers, Inc. Used by permission.)

fore, manufacturers that use it recommend the use of leak sensors and increased provision for ventilation.

R-134a has a favorable efficiency, an improved GWP compared to R-12, and a safety rating of A1, but is costly at the present time. A sizable cost reduction is anticipated when larger quantities are produced.

The tabulation below indicates how the common refrigerants are rated for ODF and suggests substitutions.

Refrig. #	Use	How It Meets the Requirements
CFC-11	For centrifugal water chiller systems.	Does not meet the environmental requirement; ODP = 1.0. For systems replacement, see R-123 below.
CFC-12	For domestic refrigerators and auto air conditioning.	Does not meet the environmental requirement: ODP ≅ 1.0. For replacement, see R-401a, R-401b, R-134a below.
R-401a (MP-39)	Is a medium replacement for R-12. For domestic freezers, refrigerators, and air conditioners, commercial refrigeration.	R-401a and R-401b are HCFC blends designed for retrofit of R-12 systems with ODPs of 0.22 and 0.24 respectively.
R-401b (MP-66)	Is a replacement for R-12 systems operating at an evaporating temperature below −10°F.	See R-401a.
HCFC-22	Comfort air conditioning and some use as refrigerant as a low-temperature replacement.	ODP = 0.05. Being used as an interim refrigerant, but will be phased out in 2010 to 2030.
HCFC-123	Used for centrifugal chillers, as an R-11 replacement.	ODP = 0.02. Can be used as an interim refrigerant to replace R-11; however, it has a toxicity factor of B1 and some manufacturers warn against using it without further training.
HFC-134a	For domestic refrigerators and freezers, commercial, auto air conditioning. Long term replacement for R-12.	A long-term replacement for R-12; ODP = 0. All other properties are favorable. Compressors using R-134a require polyol ester-based oils.
HCFC-502	For low temperature systems.	Azeotropic mixture consisting of 48.8% R-22 and 51.2% CFC-115; ODP = 0.28. Because CFCs are phased out as of December 31, 1995, this refrigerant is also being phased A typical out. replacement refrigerant is R-404a, which has an ODP of 0.
R-717	For commercial low temperature.	Ammonia is an inorganic compound with excellent properties for processing and refrigeration and a high latent storage heat content; ODP = 0. Available at a low price, but toxic and flammable so ASHRAE rating is B2.

R13-6 GUIDELINES FOR SELECTING REFRIGERANTS

For a new system, select equipment that uses a refrigerant that meets as many of the characteristics of good refrigerants as possible.

For example, a new reach-in refrigerator is being purchased. Previously this type of system used R-12 refrigerant, which is being phased out. Which one of the new refrigerants should it use?

From a performance standpoint the best selection would be HFC-134a. This refrigerant has an ODF = 0 and meets all safety requirements.

For an existing system, if a system is not leaking and is operating satisfactorily there is no technical reason to replace CFC refrigerants. In fact, it may void the U.L. listing of the unit.

For some types of service, such as supermarkets, the owner (or the manufacturer) is concerned about the availability of replacement refrigerant to maintain essential refrigeration. Under these conditions it may be advisable to replace the existing "CFC" refrigerant with a suitable substitute.

Whenever a replacement refrigerant is selected, either an interim HCFC refrigerant or a more permanent HFC

PREVENTIVE MAINTENANCE

Refrigerant leak detection is now an important component of any preventive maintenance program. Whenever doing any routine or scheduled preventive maintenance, begin to look at all sight glasses in liquid lines and compressors. Observe spots of dirt or oil around compressors, piping, valves, or coils. Dirt will accumulate where oil and refrigerant has been leaking. Early detection of small leaks will help prevent larger leaks in the future.

ON THE JOB

The following scenario could happen to you. What would you do? You go out on a service call to a grocery store. Your customer says that his frozen vegetable walk-in cooler is down and it hasn't worked properly since the last company repaired it. You go into the machinery room and locate the compressor that serves the broken unit. You observe that the compressor is short cycling on and off on its pressure switch. You inspect the liquid line sight glass and notice bubbles, indicating a leak. You get your electronic leak detector and locate the leak at a flare nut fitting on the suction line and repair it. You then decide to add refrigerant to the system but cannot locate any identification indicating what type of refrigerant is in the system. What procedure should you follow to determine the actual refrigerant which has been installed in the system?

refrigerant can be considered. The best selection is usually the HFC replacement, if it is available; however, the expected lifetime of the equipment being serviced needs to be examined. If the lifetime of the equipment is within the availability time of the interim refrigerant, the HCFC may be the best choice.

The lubrication requirements also need to be considered because they affect the cost of the changeover. If an HCFC refrigerant is selected, an Alkyl Benzene (AB) lubricant can be used.

If an HFC refrigerant is used, the lubricant must be a Polyol ester (POE) type. They absorb moisture from the air so must be handled and packaged with much more care than conventional oils. At least 95% of the original lubricant must be removed before the new lubricant can be left in the system. This usually requires three flushings. Each flushing requires operating the compressor in the system for at least 24 hours.

As an example, assume that the refrigerant in a commercial freezer using R-502 is being replaced with a new refrigerant.

If HCFC-402a is the new refrigerant, an AB lubricant can be used with one oil change. The ODF = 0.02, which is satisfactory for an interim refrigerant.

If HFC-404a is the new refrigerant, it would be a permanent replacement. A POE lubricant can be used with three oil changes. The ODF = 0.

R13-7 PROPERTIES OF COMMON REFRIGERANTS

The following information provides descriptions and tables for the properties of common refrigerants.

R-11

This is a refrigerant that was developed for centrifugal chillers. The centrifugal compressor handles a large volume of refrigerant vapor and has a low pressure rise. R-11 at saturated conditions of 5°F (–15°C) has an evaporating pressure of 2.931 psia (81.3 kPa). At 86°F (30.0°C) it has a condensing pressure of 18.3 psia (126.0 kPa). The volume of vapor at 5°F (–15°C) is 12.2 ft^3/lb (0.763 m^3/kg). The latent heat of vaporization is 84 Btu/lb (195 J/g).

Unfortunately this is one of the refrigerants that needs to be replaced since it is destructive to the ozone layer when released. Note in the figures above that the centrifugal chiller operates at an evaporator pressure below atmospheric pressure. Therefore, there is a tendency for air to leak in the system, which then requires purging.

An interesting characteristic of R-11 is that at atmospheric pressure the refrigerant is a liquid. It has been used for flushing the moisture out of systems which shortens the evacuation time. It also was used to flush out acids and products of decomposition after motor burnouts in semi-hermetic systems. This is no longer legal under EPA regulations.

Diagrams and tables for R-11, found in Appendix A, are as follows:

I-P Pressure-heat Diagram, Appendix A, Fig. A-1.

I-P Properties of Saturated Liquid and Vapor, Appendix A, Fig. A-2

R-12

This refrigerant has out sold all other refrigerants for a number of years, mainly because it is used in such a variety of equipment, including domestic refrigerators. The performance characteristics have been favorable to the products being cooled. At an evaporating temperature of 5°F (−15°C) the saturated liquid boils at a pressure of 11.9 psig (183.0 kPa) and at a condensing temperature of 86°F (30°C), the saturated vapor pressure is 93.3 psig (745 kPa). The latent heat of evaporation at 5°F is 68.2 Btu/lb (159 J/g). The toxicity and flammability rating is A1.

R-12 is soluble in oil down to −90°F, which is above normal operating temperatures. This permits the oil to be returned to the compressor, based on providing proper piping.

R-12 will absorb some moisture, which can make it corrosive to the hermetic-compressor-motor windings and other metal parts. Special precautions need to be made to keep the system dry.

Unfortunately this refrigerant is one of those that destroys the ozone layer when released, and therefore must be replaced. The best replacement now available appears to be R-134a.

The diagrams and tables available for R-12 are as follows:

I-P Pressure-heat Diagram, Appendix A, Fig. A-3

I-P Properties of Saturated Liquid and Vapor, Appendix A, Fig. A-4

R-22

This refrigerant has been popular for use in domestic freezers and packaged air-conditioning units. Fortunately, it has only a small effect on the ozone layer, with an OPF of 0.05. It has a low phase-out priority compared to R-11 or R-12.

At an evaporating temperature of 5°F (−15°C), the saturated vapor pressure is 28.19 psig (296 kPa) and at a condensing temperature of 86°F (30°C) the saturated vapor pressure is 158.17 psig (1190 kPa). The latent heat of vaporization at 5°F (−15°C) is 93.2 Btu/lb (217 J/g). It has a boiling temperature of −41°F (−41°C) at atmospheric pressure.

R-22 will absorb more water than R-12, and therefore a larger drier is needed. Oil will mix with the R-22 at most operating temperatures; however, at temperatures somewhat below −40°F the oil separates out.

The diagrams and tables available for R-22 are as follows:

I-P Pressure-heat Diagram, Appendix A, Fig. A-5

I-P Properties of Saturated Liquid and Vapor, Appendix A, Fig. A-6

R-123

This refrigerant was designed to replace R-11. The pressure-temperature curves show close performance characteristics between these two refrigerants.

At an evaporating temperature of 5°F (−15°C) the saturated vapor pressure is 2.03 psia (14.0 kPa) and at a condensing temperature of 86°F (30°C) saturated vapor pressure is 15.9 psia (109.6 kPa). The latent heat of vaporization at 5°F (−15°C) is 82.9 Btu/lb (192.8 J/g). It has a boiling temperature of 82.1°F (27.8°C) at atmospheric pressure.

R-123 has a safety group classification of B1, making it objectionable from the standpoint of toxicity. As a result some service companies refuse to use it. As yet no new universally acceptable refrigerant has been developed to replace R-11.

The ODF for R-123 is 0.02, which is low. Based on the need, therefore, it can serve as an interim replacement for R-11.

The diagrams and tables available for R-123 are as follows:

I-P Pressure-heat Diagram, Appendix A, Fig. A-7

I-P Properties of Saturated Liquid and Vapor, Appendix A, Fig. A-8

R-134a

This is a chlorine-free fluorinated refrigerant designed to replace R-12. It has an ODF of 0.0, and therefore meets the necessary environmental requirements. It is suitable for use in domestic refrigerators, automotive air conditioning, and medium- and high-temperature commercial applications. It is a suitable replacement for R-12 wherever the evaporation temperature is −10°F or higher.

At an evaporating temperature of 5°F (−15°C) the saturated vapor pressure is 9.1 psig (161.9 kPa) and at a condensing temperature of 86°F (30°C) saturated vapor pressure is 98.1 psig (680.7 kPa). The latent heat of vaporization at 5°F (−15°C) is 90.1 Btu/lb (209.6 J/g). It has a boiling temperature of −15.7°F (−26.5°C) at atmospheric pressure.

The diagrams and tables available for R-134a are as follows:

I-P Pressure-heat Diagram, Appendix A, Fig. A-9

I-P Properties of Saturated Liquid and Vapor, Appendix A, Fig. A-10

R-502

This refrigerant is an azeotropic mixture (blend of two or more component refrigerants whose equilibrium vapor-phase and liquid-phase compositions are the same at a given pressure). It is made up of 48.8% R-22 and 51.2% R-115. It is primarily used for low temperature refrigeration where the evaporator temperature is between 0 and −60°F (−18 to −51°C.). It is a safe refrigerant, non-toxic and non-flammable. It has an ODF of 0.28 and, therefore, has a limited period of use.

At an evaporating temperature of 5°F (−15°C) the saturated vapor pressure is 35.8 psig (348 kPa) and at a condensing temperature of 86°F (30°C) saturated vapor pressure is 177 psig (1320 kPa). The latent heat of vaporization at 5°F (−15°C) is 67.3 Btu/lb (157.0 J/g). It has a boiling temperature of −50°F (−46°C) at atmospheric pressure.

The diagrams and tables available for R-502 are as follows:

I-P Pressure-heat Diagram, Appendix A, Fig. A-11

I-P Properties of Saturated Liquid and Vapor, Appendix A, Fig. A-12

R-717

Ammonia is one of the oldest refrigerants in use today. It has a much higher latent heat of evaporation than the other common refrigerants. At 5° the value is 565 Btu/lb (1310 J/g). This means that smaller piping can be used. It is corrosive to copper, but not to iron, steel, or aluminum. It therefore requires an all-iron or steel or aluminum system including the compressor, condenser, evaporator, controls, and piping. It is not destructive to the ozone layer. Its greatest defect is its toxicity and flammability. It has a safety rating of B2.

Ammonia does not absorb oil. It requires the use of an oil separator in the discharge line. Oil that is carried over to the evaporator must be drained off. It is extremely soluble in water.

Due to ammonia's safety hazards it is not found in appliances or normal comfort cooling applications. Typically it is used in large commercial or industrial applications where its operating efficiencies (lower horsepower per ton) are important and where plant engineers are available to operate the system. Examples would be dairies, ice cream plants, and large cold-storage facilities.

Due to its toxicity, an operator must take special precautions to limit the quantity inhaled. Leaks are detected by use of litmus paper which changes color in the presence of ammonia, or a sulfur candle which creates smoke in contact with ammonia. Technicians must also take care to avoid ammonia contact with the skin.

At an evaporating temperature of 5°F (−15°C) the saturated vapor pressure is 19.6 psig (236 kPa) and at a condensing temperature of 86°F (30°C) the saturated vapor pressure is 155 psig (1170 kPa). The latent heat of vaporization at 5°F (−15°C) is 565 Btu/lb (1310 J/g). It has a boiling temperature of −28°F (−33°C) at atmospheric pressure.

R13-8 PRESSURE-TEMPERATURE RELATIONSHIPS

Fig. R13-6 shows a series of pressure-temperature curves for some commonly used refrigerants. This chart shows the refrigerants in a saturated condition. One very good use of these curves is to observe the refrigerants that are closely matched to determine the suitability of substituting one for the other.

For example, note that R-502 and R-22 are closely matched. Since R-22 has only about 7% of the ozone-depleting capability of R-502, it can be used as an interim replacement.

R13-9 CONTAMINANTS IN THE REFRIGERANT

As the system operates, certain impurities tend to collect in the refrigerant, including water/vapor, acid, high-boiling-point residues, particulates, and non-condensibles such as air. When a refrigerant is recycled, many of these impurities are removed. In order to resell the refrigerant it must be reclaimed by a certified source and brought back to the original level of purity indicated by ARI Standard 700-88 (or update) which gives maximum allowable contamination levels, shown in Fig. R13-7a and b.

R13-10 REFRIGERANT CYLINDERS

Fig. R13-8 shows cylinders for various refrigerants.A disposable cylinder should never be refilled. A hand valve may be clamped to the top of the cylinder for shutoff.

Refillable (service) cylinders should only be used if adequate equipment is available for determining the amount of refrigerant transferred. No cylinder should be filled with liquid to more than 80% of its volume. A fill-limit device is often furnished with these cylinders, particularly when they are used for refrigerant recovery. This device automatically cuts out the flow of refrigerant into the cylinder when the fill level is reached. These cylinders may also be supplied with two-stem valves to permit drawing either liquid or vapor from the cylinder.

Storage cylinders are used where large quantities of refrigerant are purchased for transfer to smaller cylinders. This should only be done when facilities are available for following the proper procedures.

VAPOR PRESSURE OF HFCS, HCFCS AND CFCS

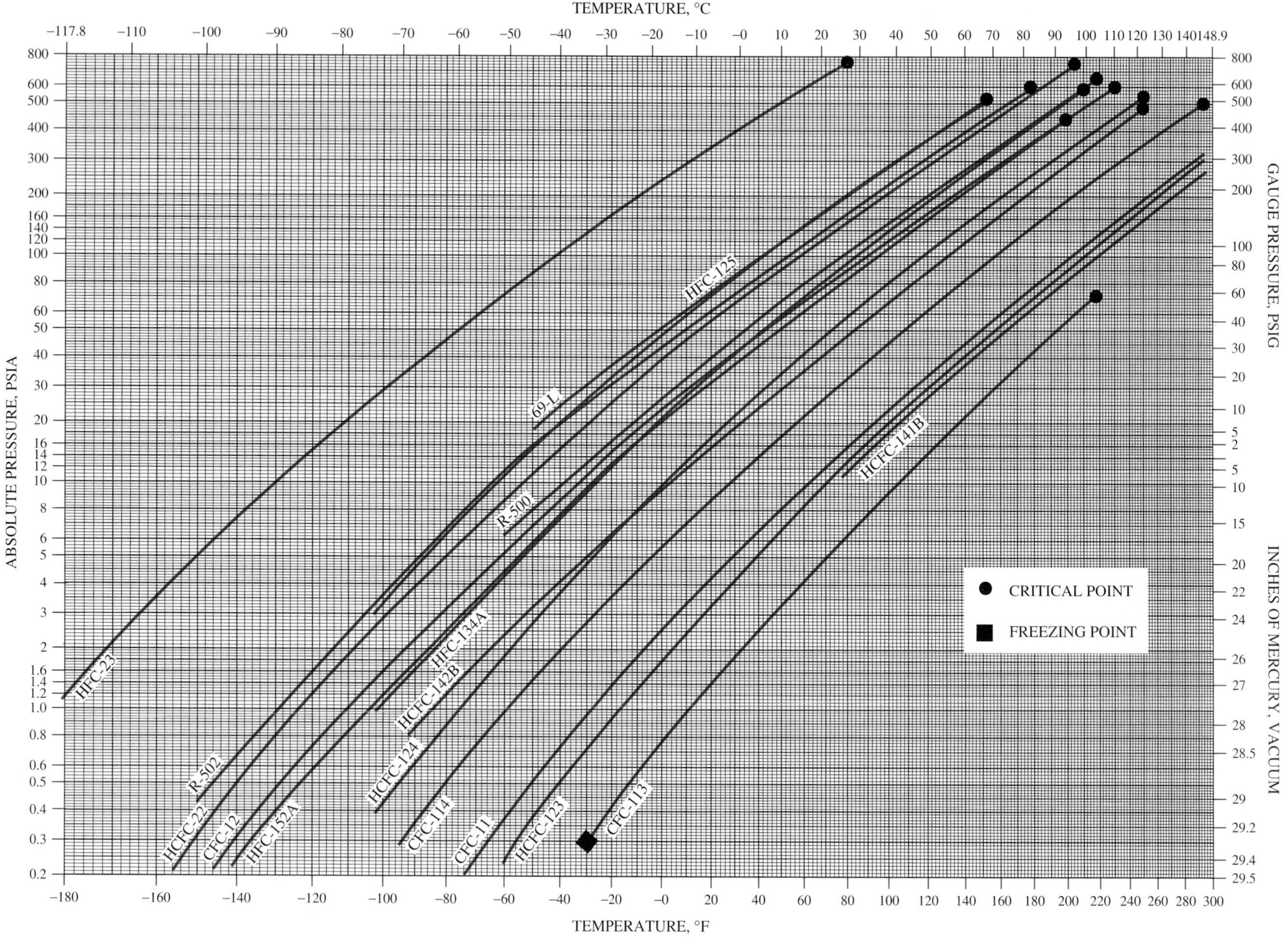

Figure R13-6 Pressure-temperature curves.

Characteristics of Refrigerants and Maximum Contaminant Levels

	Reporting Units	Reference (Subclause)	R-11	R-12	R-13	R-22	R-23	R-32	R-113	R-114	R-123	R-124	R-125	R-134a	R-143a
Characteristics*: Boiling Point*	°F @ 1.00 atm °C @ 1.00 atm	—	74.9 23.8	−21.6 −29.8	−114.6 −81.4	−41.4 −40.8	−115.7 −82.1	−61.1 −51.7	117.6 47.6	38.8 3.8	82.6 27.9	12.2 −11.0	−55.3 −48.5	−15.1 −26.2	−52.6 −47.0
Boiling Point Range*	K	—	0.3	0.3	0.5	0.3	0.5	0.3	0.3	0.3	0.3	0.3	0.3	0.3	0.3
Typical Isomer Content	by weight	—							0–1% R-113a	0–30% R-114a	0–8% R-123a	0–5% R-124a	N/A	0–5000 ppm R-134	0–100 ppm R-143
Vapor Phase Contaminants: Air and other non-condensibles	% by volume @ 25°C	5.9	N/A**	1.5	1.5	1.5	1.5	1.5	N/A**	1.5	N/A**	1.5	1.5	1.5	1.5
Liquid Phase Contaminants: Water	ppm by weight	5.4	20	10	10	10	10	10	20	10	20	10	10	10	10
All other impurities including refrigerants	% by weight	5.10	0.50	0.50	0.50	0.50	0.50	0.50	0.50	0.50	0.50	0.50	0.50	0.50	0.50
High boiling residue	% by volume	5.7	0.01	0.01	0.05	0.01	0.01	0.01	0.03	0.01	0.01	0.01	0.01	0.01	0.01
Particulates/solids	Visually clean to pass	5.8	pass	pass	pass	pass	pass	pass	pass	pass	pass	pass	pass	pass	pass
Acidity	ppm by weight	5.6	1.0	1.0	1.0	1.0	1.0	1.0	1.0	1.0	1.0	1.0	1.0	1.0	1.0
Chlorides***	No visible turbidity	5.5	pass	pass	pass	pass	pass	pass	pass	pass	pass	pass	pass	pass	pass

*Boiling points and boiling point ranges, although not required, are provided for informational purposes.

**Since R-11, R-113 and R-123 have normal boiling points at or above room temperature, non-condensible determinations are not required for these refrigerants.

***Recognized Chloride level for pass/fail is 3ppm.

Figure R13-7a ARI Refrigerant Standards #700-88. Maximum contamination levels.

Characteristics of Refrigerants and Maximum Contaminant Levels

	Reporting Units	Reference (Subclause)	R-401A	R-401B	R-402A	R-402B	R-500	R-502	R-503
Characteristics*: Refrigerant Components			R-22/152a/124	R-22/152a/124	R-125/290/22	R-125/290/22	R-12/152a	R-22/115	R-23/13
Nominal Comp, weight%			53/13/34	61/11/28	60/2/38	38/2/60	73.8/26.2	48.8/51.2	40.1/59.9
Allowable Comp, weight%			51–55/11.5–13.5/ 33–35	59–63/9.5–11.5/ 27–29	58–62/1–3/ 36–40	36–40/1–3/ 58–62	72.8–74.8/ 25.2–27.2	44.8–52.8/ 47.2–55.2	39–41/ 59–61
Boiling Point*	°F @ 1.00 atm °C @ 1.00 atm		–27.6 to –16.0 –33.4 to –26.6	–30.4 to –18.5 –34.7 to –28.6	–56.5 to –52.9 –49.1 to –47.2	–53.3 to –49.0 –47.4 to –45.0	–33.5	–45.4	–88.7
Boiling Point Range*	K	—	—	—	—	—	0.5	0.5	0.5
Vapor Phase Contaminants: Air and other non-condensibles	% by volume @ 25°C	5.9	1.5	1.5	1.5	1.5	1.5	1.5	1.5
Liquid Phase Contaminants: Water	ppm by weight	5.4	10	10	10	10	10	10	10
All other impurities including refrigerants	% by weight	5.10	0.50	0.50	0.50	0.50	0.50	0.50	0.50
High boiling residue	% by volume	5.7	0.01	0.01	0.01	0.01	0.05	0.01	0.01
Particulates/solids	Visually clean to pass	5.8	pass	pass	pass	pass	pass	pass	pass
Acidity	ppm by weight	5.6	1.0	1.0	1.0	1.0	1.0	1.0	1.0
Chlorides**	No visible turbidity	5.5	pass	pass	pass	pass	pass	pass	pass

*Boiling points and boiling point ranges, although not required, are provided for informational purposes.

**Recognized Chloride level for pass/fail is 3ppm.

Figure R13-7b ARI Refrigerant Standards #700-88. Maximum contamination levels. (continued)

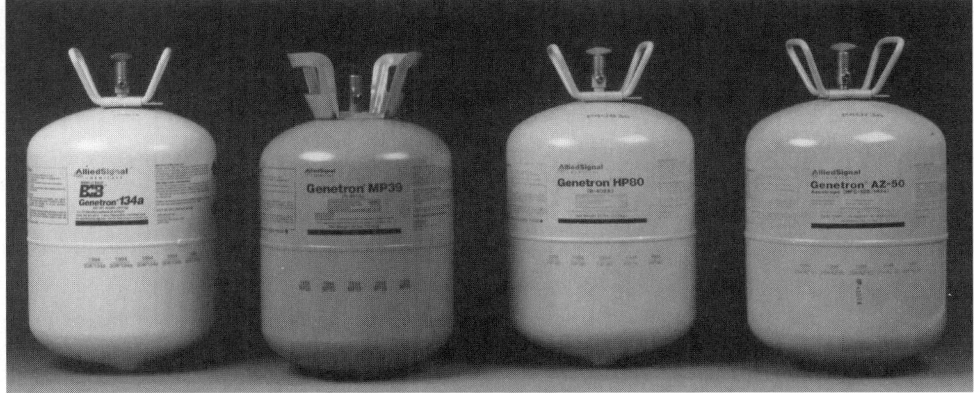

Figure R13-8 Refrigerant cylinders. (Courtesy of Allied Signal Genetron® Refrigerants Inc.)

Cylinders used for refrigerant must have a Department of Transportation (DOT) stamp of approval. Cylinders are made of steel or aluminum. DOT requires that cylinders that have contained a corrosive refrigerant must be tested every five years. Cylinders containing non-corrosive refrigerants must be tested every five years. Any cylinders over 4½ in. in diameter or over 12 in. long must have some type of pressure-relief device.

Refrigerant cylinders are color coded in accordance with the schedule in Fig. R13-9.

Refrigerant storage cylinders have a valve that uses two valve stems. One is used when liquid is drawn from the cylinder and the other for vapor. On the disposable cylinder the valve clamps to the top of the cylinder.

Product	Color
R-11	Orange
R-13	Light Blue
AZ-20	Rose
R-22	Light Green
MP-39	Coral
AZ-50	Teal
MP-66	Mustard
HP-80	Sand
HP-81	Green
R-114	Navy Blue
R-123	Light Gray
R-124	DOT Green
R-125	Medium Brown
R-134a	Light Blue
R-141b	Beige
R-142b	Gray
R-404A	Orange
R-407C	Chocolate Brown
R-500	Yellow
R-502	Light Purple
R-503	Blue-Green

Figure R13-9 Color coding for cylinders. (Courtesy of Allied Signal Genetron© Refrigerants Inc.)

R13-11 SAFETY

Due to the potential dangers involved, it is extremely important that anyone handling refrigerants observe proper safety procedures.

R13-11.1 Pressure

Most of the time the service person is dealing with confined liquids or vapors that have the capability of building up high pressures which have the potential of causing serious injury if carelessly handled. Precautions must be taken.

In storing refrigerant in approved cylinders, the liquid level must not exceed 80% of the volume of the container. This allows some space for expansion if heat is added.

Whenever a service person is called upon to examine a piece of refrigerating equipment, the pressures in the system should immediately be checked with suitable gauges. Depending on the R number of the refrigerant, an evaluation can be made of the condition of the system. In case there is any doubt of the type of refrigerant being used, test instruments are available to determine it. In no instance is it permissible to mix refrigerants.

R13-11.2 Leaks

Refrigerants are heavier than air and, with the exception of ammonia and sulfur, have little or no odor, thus giving no warning of their presence. The air we breathe consists of about 21% oxygen and 78% nitrogen plus moisture vapor and traces of other gases such as carbon dioxide. If refrigeration vapors, from an accidental discharge or rupture, displace air in a confined space, pit or low area, there can be great danger of asphyxiation in a short time to unsuspecting persons occupying or entering the space, causing loss of consciousness and possibly death. The risk is real and the technician must be aware of the hazard. The *Safety Code for Mechanical Refrigeration* requires oxygen depletion monitors for potentially hazardous conditions.

R13-11.3 Charging

Whenever the refrigerant system is being serviced, especially when charging or evacuating, the service person should wear goggles and gloves. Contact with cold refrigerant liquid or spray can cause freezing or frostbite. Cold refrigerant in an eye can cause loss of sight. Proper precautions need to be taken.

R13-11.4 Odor

Many of the common refrigerants have no odor and cannot be detected by this means. In case of doubt, assume that refrigerant is present in the atmosphere and protect yourself.

It is strongly recommended that a technician avoid working on systems or equipment using flammable or toxic refrigerants (other than those with an A1 safety rating) unless the technician has had special training and has proper equipment to deal safely with these materials.

Certain refrigerants have a pungent odor, such as ammonia (R-717) and sulfur dioxide (R-764). They can be dangerous if inhaled in sufficient quantities. Their odor even in small quantities provides a warning. If a service person must work in areas where these refrigerants are released, the person must wear a proper respirator.

Warning instruments are available to place in equipment rooms and rooms where refrigerant is stored or used. An alarm is sounded when a refrigerant is released. These instruments are also used in refrigerated areas of large systems to advise operators at an early stage when leaks occur.

R13-11.5 Oil

It is important the oil be stored in airtight containers, since oil will absorb moisture if exposed to the air. Moisture in the oil can be harmful to the equipment in which it is used. Unless moisture is removed it can cause hermetic compressor motor burnouts.

When a hermetic compressor motor burns out, the oil can become acidic. If it comes in contact with the skin, it can produce a severe burn. Precautions need to be taken.

R13-11.6 Safety Code Requirements

The *Safety Code of Mechanical Refrigeration* gives certain safety requirements for the machinery room that must be followed.

1. For group A1 refrigerants, machinery rooms shall be equipped with an oxygen sensor to warn of oxygen levels below 19.5% volume since there is insufficient odor warning.
2. For other refrigerants, a refrigerant vapor detector shall be located in an area where refrigerant from a leak is likely to concentrate, and an alarm shall be employed.
3. The minimum mechanical ventilation required to exhaust a potential accumulation of refrigerant due to leaks or a rupture of the system shall be capable of moving air from the machinery room in the following quantity: (as specified in ANSI/ASHRAE 15-1992, *Safety Code for Mechanical Refrigeration,* Section 11.13.74).

$$Q = 100 \times G0.5\ (Q = 70 \times G0.5)$$

where

Q = the airflow in cubic feet per minute (liters per second)
G = the mass of refrigerant in pounds (kilograms) in the largest system, any part of which is located in the machinery room.

A sufficient part of the mechanical ventilation shall be:

(a) operated, when occupied, at least 0.5 cfm per square foot (2.54 L/s per square meter) of machinery room area or 20 cfm per person (9.44 L/s) and

PRACTICING PROFESSIONAL SERVICES

It is now customary practice to tag and identify all refrigeration systems with the system refrigerant and lubricant (if it is not already tagged or identified) to avoid the accidental mixture of the wrong refrigerants and oils during repair and service.

During a changeout of CFC refrigerants with the new HFC refrigerants, all old mineral oils must be thoroughly removed. This normally means flushing and changing the refrigerant three times during the transition to verify all residual mineral oils are removed.

Safety during the handling of any refrigerants or oils should include the wearing of protective gloves and eyewear by the technician. It is also imperative that all retrofitted machine rooms with new refrigerants are in compliance with the new mechanical safety code.

(b) operable, if necessary for operator comfort, at a volume required to maintain a maximum temperature rise of 18°F (10°C) based on all of the heat-producing machinery in the room.

Note: The minimum ventilation rates prescribed may not prevent temporary accumulations of flammable refrigerants above the lower flammability limit (LFL) in the case of catastrophic leaks or ruptures. The designer may consider the provisions of *National Fire Protection Association (NFPA)* code for such cases.

4. Familiarity with the Safety Code is a must.

REVIEW

- Refrigerant—medium (fluid) for heat transfer, used in a refrigeration system to pick up heat by evaporating at a low temperature and pressure, and give up heat by condensing at a higher temperature and pressure.
- Azeotropic Mixtures—consist of multiple components of different volumes that, when used in the refrigeration cycle, do not change volumetric composition or saturation temperatures as they condense or evaporate at a constant pressure.
- Zeotropic Mixtures—consist of multiple components of different volumes that, when used in the refrigeration cycle, change volumetric composition or saturation temperatures as they condense or evaporate at a constant pressure.
- Fractionation—when a mixture separates back into its original components.
- Thermoglide—the difference between the saturated temperatures of component refrigerants in a zeotropic mixture.
- Leak Detection—soap bubbles, Halide torch, Halogen leak detector.
- Superheat—heat absorbed by the refrigerant after it has evaporated.
- Subcooling—heat removed from the refrigerant after it has condensed into a liquid.
- Common Refrigerants

R-11 CFC-11	Compound
R-12 CFC-12	Compound
R-22 HCFC-22	Compound
R-123 HCFC-123	Compound
R-717 Ammonia	Compound
R-134a HFC-134a	Compound
R-500	Azeotropic mixture
R-502	Azeotropic mixture
R-503	Azeotropic mixture
R-507 AZ-50	Azeotropic mixture
R-401A MP-39	Zeotropic mixture
R-401B MP-66	Zeotropic mixture
R-402A HP-80	Zeotropic mixture
R-402B HP-81	Zeotropic mixture
R-404A HP-62 FX-70	Zeotropic mixture
R-403A 69S	Zeotropic mixture
R-407A 60	Zeotropic mixture
R-407B 61	Zeotropic mixture
R-408A FX-10	Zeotropic mixture
R-406A GHG	Zeotropic mixture
R-405A Green cool 2015	Zeotropic mixture
R-407C 9000 66	Zeotropic mixture
R-410A Az-20	Zeotropic mixture
R-410B 9100	Zeotropic mixture

- Important Qualities of Refrigerants to Consider
  Performance
  Safety
    Toxicity
    Flammability
  Environmental impact
  Efficiency
  Stability
  Compatability
  Lubrication
  Leak tendency
  Price
  Lifetime of the refrigerant
- Lubrication Requirements—When using a new refrigerant, follow the manufacturers recommendation. Typically, when using an HFC refrigerant, use a Polyol Ester-type lubricant. When using an HCFC mixture, such as 402a, an alkylene blend lubricant is required. All older CFC refrigerants can use conventional mineral oils, until a new replacement refrigerant is installed.
- ODP—Ozone depletion potential
- Pressure-temperature curves or pressure-temperature charts can be used to determine saturated temperature at various pressures. Refrigerants are ONLY saturated in the evaporator or condenser.
- Contaminants such as water and air MUST NOT be allowed to enter the sealed refrigeration system as they will eventually destroy it.
- A disposable cylinder MUST NOT be refilled.
- DOT-approved cylinders must be hydrostatically tested every five years.
- Safety code for mechanical refrigeration requires that cylinders be color coded. Study this code!
- Liquid level in approved cylinders MUST NOT exceed 80% of the volume of the cylinder.
- DO NOT mix refrigerants.
- Use adequate ventilation and DO NOT ENTER areas of heavy refrigerant leaks without a self-contained breathing apparatus.
- Safety code requires monitors to sense refrigerant leaks in equipment spaces.
- When handling refrigerants, wear gloves and safety glasses.
- Familiarity with safety code is a MUST!

Problems and Questions

1. Using the pressure-heat diagram for R-134a (Fig. A-9 in Appendix A), plot the refrigeration cycle for the following set of conditions.
   Condensing temperature 130°F
   Evaporating temperature 20°F
   Superheat in evaporator 10°F
   Subcooling in the condenser 10°F
   Note that when the compressor performance is plotted, it follows the constant-entropy (heat) line.
2. From the diagram, determine the following:
   a. Latent heat of evaporation _____ Btu/lb
   b. Net refrigerating effect _____ Btu/lb
   c. Heat added by the compressor _____ Btu/lb
   d. Superheat removed by the condenser _____ Btu/lb
   e. Total heat removed by the condenser _____ Btu/lb
3. A refrigerant is a medium used for what purpose?
   a. To pump through the system.
   b. For heat transfer.
   c. To reject hot and cold air.
   d. To lubricate the compressor.
4. The number designation for refrigerants has been developed by the:
   a. Refrigeration Service Engineers.
   b. Air Conditioning and Refrigeration Institute.
   c. International Refrigeration Institute.
   d. ASHRAE.
5. Which of the following refrigerant compounds does not contain chlorine?
   a. HFC
   b. CFC
   c. HCFC
   d. NH_3
6. Which refrigerant is an approved replacement for R-12?
   a. 123
   b. 502
   c. 134a
   d. 22
7. If an existing CFC refrigerant system is not leaking and is operating satisfactorily, what should be done?
   a. No technical reason to replace it.
   b. Replace it with a HFC refrigerant.
   c. Recover/recycle the refrigerant and install a new ozone friendly refrigerant.
   d. Replace the oil for future use with a new refrigerant.
8. R-123 is used as the replacement for R-11 on centrifugal compressors. True or False?
9. Polyol Ester oils are used as the lubricant for HFC refrigerants. True or False?
10. No cylinder should be filled with liquid to more than 90% of its volume. True or False?
11. Refrigerants are lighter than air and have little odor. True or False?
12. When a hermetic compressor motor burns out, the compressor oil can become acidic. True or False?
13. List 10 important qualities when selecting a replacement refrigerant.
14. What are the safety requirements that all machinery rooms must comply to?
15. How often must DOT-approved refrigerant cylinders be re-tested?

Recovery, Recycling, Reclaiming and Replacing Refrigerants

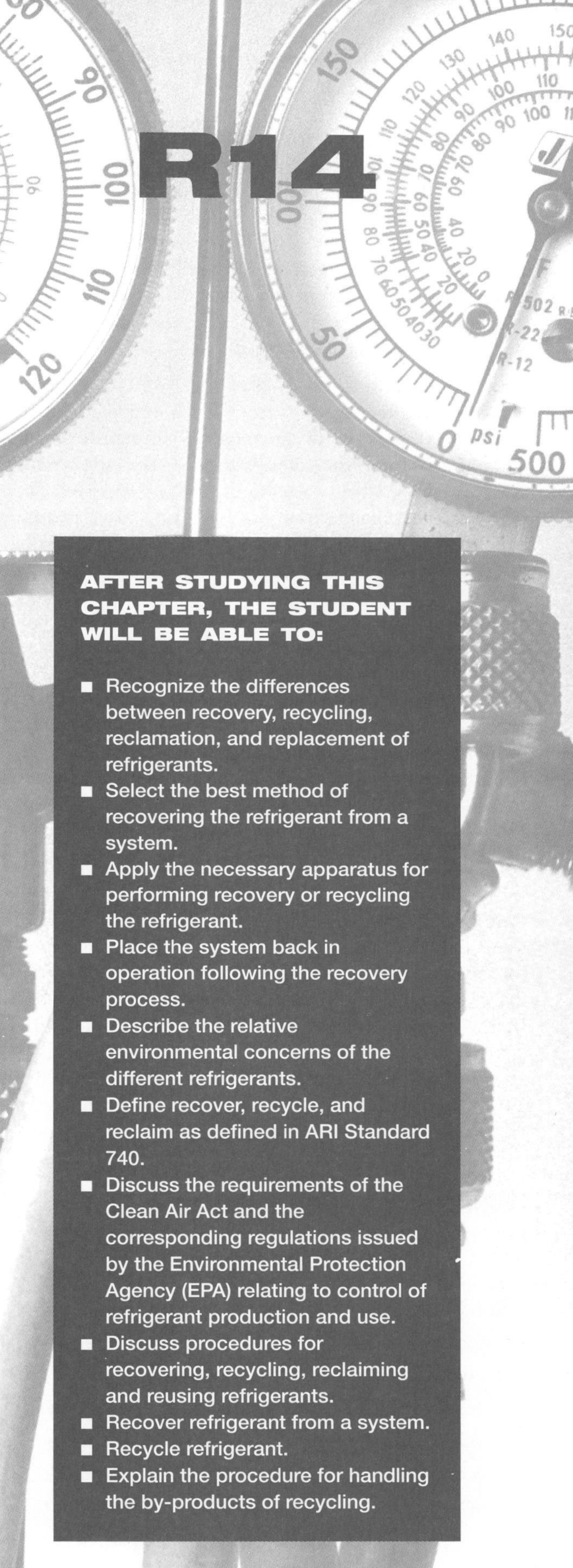

AFTER STUDYING THIS CHAPTER, THE STUDENT WILL BE ABLE TO:

- Recognize the differences between recovery, recycling, reclamation, and replacement of refrigerants.
- Select the best method of recovering the refrigerant from a system.
- Apply the necessary apparatus for performing recovery or recycling the refrigerant.
- Place the system back in operation following the recovery process.
- Describe the relative environmental concerns of the different refrigerants.
- Define recover, recycle, and reclaim as defined in ARI Standard 740.
- Discuss the requirements of the Clean Air Act and the corresponding regulations issued by the Environmental Protection Agency (EPA) relating to control of refrigerant production and use.
- Discuss procedures for recovering, recycling, reclaiming and reusing refrigerants.
- Recover refrigerant from a system.
- Recycle refrigerant.
- Explain the procedure for handling the by-products of recycling.

R14-1 WHY SHOULD REFRIGERANT BE RECOVERED?

1. It is economical.

It is a common misconception that the recovery of refrigerant from a system is a new development, resulting from the restrictions placed on the use of these refrigerants. Not so! Technicians have been saving refrigerants from systems for reuse for many years. Rising costs of replacing refrigerants as well as the availability factor have created an increasing awareness of the economy of recovery.

2. Refrigerant recovery is required by law.

The Clean Air Act of 1990 limits the use of chlorine-based refrigerants, and taxes their production and use. The Act, administered by the Environmental Protection Agency (EPA), provides for the phasing out of certain common refrigerants, some by the end of 1995. EPA regulations cover five main sections:

(a) No venting of refrigerants into the atmosphere;
(b) Certification required for those purchasing and handling refrigerants;
(c) Certification of equipment used for recovery, reclaiming, and recycling refrigerants;
(d) Requirement to repair substantial leaks; and
(e) Implementation of safe disposal procedures.

3. It is environmentally sound.

The harmful effects of atmospheric ozone depletion by chlorine-based refrigerants have been accepted internationally to the extent of agreeing to ban them from production at the earliest possible time. This is discussed in Chapter R1.

Like many changes that are forced upon us, there are benefits that we might not have introduced ourselves had not new emphases taken place. For example, along with the recovery process, the recycling action now requires cleaning up the refrigerant before its reuse which probably adds years of life to many systems. Contamination can get into the system even when using the best installation procedures. Flux or carbon gets left inside during the brazing process, for example. Even using all precautions, some air or moisture may remain in the system. When a good recycling process is performed these contaminants are removed.

R14-2 DEFINITION OF TERMS

Certain definitions have been set up by EPA to describe terms that are used in their regulations. It is advantageous for anyone involved with refrigerants to be familiar with these definitions to better understand the regulations. Below are some of the EPA definitions that particularly apply to this text.

Appliance: Any device that contains and uses a Class I (CFC) or a Class II (HCFC) substance as a refrigerant and which is used for household or commercial purposes, including any air conditioner, chiller or freezer. EPA interprets the definition to include all air conditioning and refrigeration equipment except those designed and used exclusively for military purposes.

Small Appliance: Any of the following products that are fully manufactured, charged, and hermetically sealed in a factory with 5 lbs or less of refrigerant: refrigerators and freezers designed for home use, room air conditioners (including window air conditioners and packaged terminal air conditioners), packaged terminal heat pumps, dehumidifiers, under-the-counter ice makers, vending machines, and drinking water coolers.

Technician: Any person who performs maintenance, service, or repair that could reasonably be expected to release Class I (CFC) or Class II (HCFC) substances into the atmosphere including, but not limited to, installers, contractor employees, in-house service personnel, and (in some cases) owners. Technician also refers to any person disposing of any appliances except small appliances.

Recover: To remove refrigerant in any condition from an appliance and store it in an external container without necessarily testing or processing it in any way.

Recycle: To extract refrigerant from an appliance and clean the refrigerant for reuse without meeting all the requirements for reclamation. In general, recycled refrigerant has been cleaned by using an oil separation process and by single or multiple passes through devices (such as replaceable core filter-driers) which reduce moisture, acidity, and particulate matter.

After recycling the refrigerant, the technician can:

1. Charge the refrigerant back into the original system, or
2. Charge the refrigerant into a comparable system used by the same owner.

Reclaim: To reprocess refrigerant to the purity specified in ARI Standard 700-1988, *Specifications for Fluorocarbon Refrigerants,* and to verify the purity using the analytical methodology prescribed in the Standard.

PRACTICING PROFESSIONAL SERVICES

It is standard practice to test the oil from the recovery machine whenever recovering refrigerant from a refrigeration system with a burned-out compressor. If any traces of acids are found, the refrigerant should be recycled thoroughly and retested again. If the recycling equipment does not remove all of the measurable amounts of acid, then the refrigerant should not be reinstalled into the system, but it should be sent off to an approved reclaim facility. Acid-core filters should be installed and virgin refrigerant should be charged back into the system and retested after the new compressor has run for one hour. Again, the refrigerant should be tested for acid and the above cycle repeated until the refrigerant is free of contaminants. The acid filter should be measured for pressure drop. If more than 2 psi pressure drop appear across the filter it should also be replaced until the system is rid of any potential acids.

ON THE JOB

When working with chemicals used in the HVAC/R field, read the material safety data sheet (MSDS). Be familiar with the storage, handling, shipping, and waste procedures for each one. When working on a commercial job, the general contractor will require you to supply them with copies of the MSDS for all chemicals you bring on the site.

This process is usually carried out by an independent EPA-approved reclaiming center. It can then be re-used or sold.

Relative to the sale of the refrigerant, EPA regulations state that after November 15, 1994, CFC and HCFC refrigerants can only be sold:

1. To certified technicians
2. To wholesalers for resale to certified technicians
3. To a manufacturer for an appliance
4. Contained in an appliance.

Replacement involves the removal of a CFC refrigerant having an Oxygen Depletion Factor (ODF) of one, and replacing it with an interim refrigerant having a low ODF, or a long-term refrigerant having an ODF of zero.

R14-3 RECOVERY REQUIREMENTS

EPA has set up different recovery requirements for various classifications of equipment. HVAC/R equipment is divided into the following groups:

1. Very-high-pressure equipment: Equipment that uses a refrigerant having a boiling point below –50°C (–58°F) at atmospheric pressure. This includes refrigerants CFC-13, HFC-23 and CFC-503.
2. High-pressure equipment: Equipment that uses a refrigerant with a boiling point between –50°C (–58°F) and 10°C (50°F) at atmospheric pressure. This includes CFC refrigerants 12, 114, 500, 502, and HCFC-22.
3. Low pressure equipment: Equipment that has a boiling point above 10°C (50°F) at atmospheric pressure which includes refrigerants CFC-11, CFC-113 and HCFC-123.

EPA also divides the requirements for equipment by size and usage. For small appliances the recovery equipment would be required to reduce the system pressure to 4 in. of mercury. The recovery requirements for high pressure appliances and low pressure appliances are as follows:

Required Levels of Evacuation for Appliances for Recovery and Recycling Equipment

	Inches of Mercury Vacuum* Using Equipment Manufactured	
	Before 11/15/93	**After 11/15/93**
High-pressure Appliances		
HCFC-22 charge (under 200 lbs)	0	0
HCFC-22 charge (over 200 lbs)	4	10
Other high-pressure appliances CFC-12, -500, -502, -114 charge (under 200 lbs)	4	10
Other, as above, charge 200 lbs (and over)	4	15
Very-high-pressure appliances (CFC-13, -503)	0	0
Low-pressure Appliances		
CFC-11, HCFC-123	25	25 mm Hg** Absolute

*Relative to standard atmospheric pressure of 29.9 in. Hg (Negative pressure, or vacuum)
**25 mm Hg Absolute = 25,000 microns
= 29 in. Hg Gauge (vacuum)
(Negative pressure or vacuum)

R14-4 RECOVERY OF REFRIGERANTS

There are three basic methods of recovery:

1. Charge migration
2. Use of the system's compressor
3. Use of a refrigerant recovery/recycling unit.

R14-4.1 Charge Migration

Fig. R14-1 illustrates the use of charge migration. The process can be performed without the use of either the refrigeration compressor or an external vacuum pump. Usually the storage cylinder is evacuated before the process takes place.

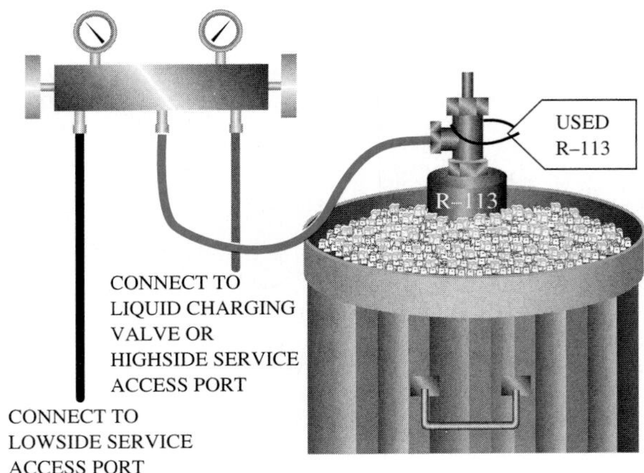

Figure R14-1 Recovery by charge migration showing connection of gauge manifold and storage cylinder.

The movement of refrigerant takes place due to the difference in pressure between the system and the cylinder. To speed up the process it is desirable to place the cylinder in an ice-water bath (tank), and also to supply some heat to the system to increase the pressure difference. The refrigerant will flow as long as a pressure difference exists. Some useful instructions for performing this process are shown in Fig. R14-2.

A major disadvantage in using the migration method is that part of the refrigerant is left in the system. This process works best for the low-pressure refrigerants such as R-113, R-11 and R-114. Medium- and high-pressure refrigerants have progressively lower boiling points. The higher the normal operating pressure—that is, the lower the boiling point—the more refrigerant will be left behind. All of the refrigerant cannot be removed. Some other process needs to be used to extract the rest of the refrigerant. The normal boiling points of common refrigerants at atmospheric pressure are shown in Fig. R14-3.

REFRIGERANT	BOILING POINT (°F)
R – 113	118
R – 123	85
R – 11	75
R – 114	39
R – 134A	–15
R – 12	–22
R – 500	–28
R – 22	–41
R – 502	–50

Figure R14-3 Boiling points of common refrigerants at atmospheric pressure.

In some systems the refrigeration compressor can be used to recover a portion of the refrigerant. Referring to Fig. R14-4, the hook-up is much the same as for migration (Fig. R14-1).

After hook up the compressor discharge valve is front seated so that the refrigerant is discharged into the cylinder rather than into the condenser.

For high-pressure refrigerants such as R-12, R-22, and R-502 and their replacements, the recovery machine should be capable of not less than 15 in. mercury (Hg) vacuum. For low-pressure refrigerants such as R-11 and R-113 and their replacements, the recovery machine must

- HAVE ENOUGH CYLINDERS ON HAND TO HOLD THE CHARGE.
- WEIGH THE CYLINDERS, BUCKET AND ICE BEFORE THE PROCESS IS STARTED.
- PUT A "USED REFRIGERANT" TAG ON THE CYLINDER.
- PACK ICE AROUND THE CYLINDER.
- CONNECT THE GAUGE MANIFOLD AS SHOWN, CONNECTING THE CYLINDER TO THE UTILITY OPENING ON THE GAUGE MANIFOLD.
- PURGE THE HOSE AND OPEN THE HIGH PRESSURE VALVE ON THE GAUGE MANIFOLD.
- STOP THE PROCESS WHEN THE FIRST CYLINDER IS 80% FULL. REPEAT WITH ADDITIONAL CYLINDERS IF ADDITIONAL STORAGE SPACE IS NECESSARY.

Figure R14-2 Instructions for performing recovery by charge migration.

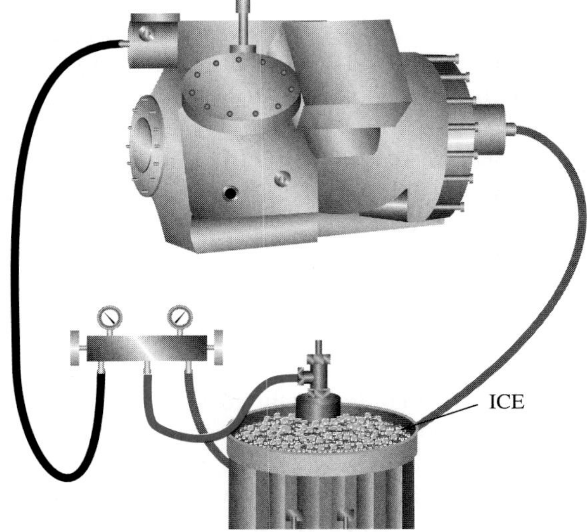

Figure R14-4 Recovery using a system compressor connection of components.

PREVENTIVE MAINTENANCE

Recovery/recycle equipment is prone to rigorous use, neglect, and abuse. It is recommended that the oil in the compressor of the recovery equipment be changed regularly, according to manufacturer's recommendations. Filters and strainers associated with the equipment should also be changed regularly. If the recovery unit has been used to recover refrigerant from any system with acids or other contaminants, then the recovery units should be drained and filters changed upon completion of the job. Leaving residual acidic oils in the sump of the recovery compressor could prematurely destroy its compressor.

be capable of 25 mm Hg absolute (29 in. Hg gauge), close to an absolute vacuum.

This arrangement is not satisfactory for most systems. Those using hermetically sealed compressors rely on the refrigerant vapor to cool the motor windings. As the amount of vapor is reduced, proper motor cooling is not feasible.

The other problem with this arrangement is that it does not remove all of the refrigerant and some other means needs to be provided to complete the job.

R14-5 REFRIGERANT RECOVERY/RECYCLING UNITS

The use of a recovery/recycling unit is the most practical means of removing the refrigerant (Fig. R14-5).

All that is required in addition to the unit is a gauge manifold, which includes both a high-side gauge and a low-side (vacuum) gauge. Recovery units can remove most of the refrigerant. The balance needs to be removed with a vacuum pump. Fig. R14-6 shows the recovery/recycling unit connected to a system.

The gauge manifold is connected to the system in the normal way, connecting the low-pressure port to the low side of the system and the high-pressure port to the high-pressure side of the system. The recovery unit is attached to the utility connection on the gauge manifold.

The recovery unit has a special internal or external refrigerant-storage cylinder. Most of these cylinders hold 50 lb of refrigerant, which is sufficient capacity to hold the charge for a system of up to 25 tons of cooling capacity. Additional cylinders can be added as required. For centrifugal units that have large refrigerant charges, properly sized recovery units are available.

R14-5.1 Types of Recovery Units

Basically there are three types of recovery units:

1. Vapor-recovery units
2. Liquid-recovery units
3. Units that will recover both vapor and liquid.

Vapor-Recovery Units

Fig. R14-7 shows the hook-up for a vapor recovery machine. A typical unit uses a compressor in the recovery unit below the pressure in the system containing the refrigerant. The discharge vapor from the compressor flows to a heat exchanger where the refrigerant condenses and moves to the recovery cylinder for storage. The heat exchanger-condenser can be air cooled or water cooled.

The recovery/recycling machine has a replaceable filter-drier for cleaning the refrigerant. It also usually has two circuits for separating oil, one for the equipment being

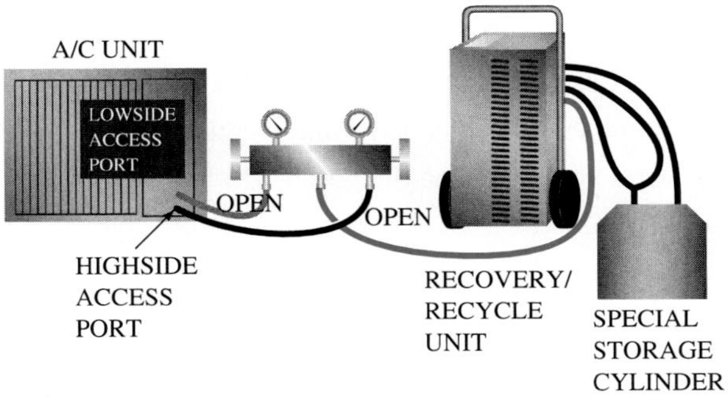

Figure R14-5 Recovery/recycling unit, connection of components.

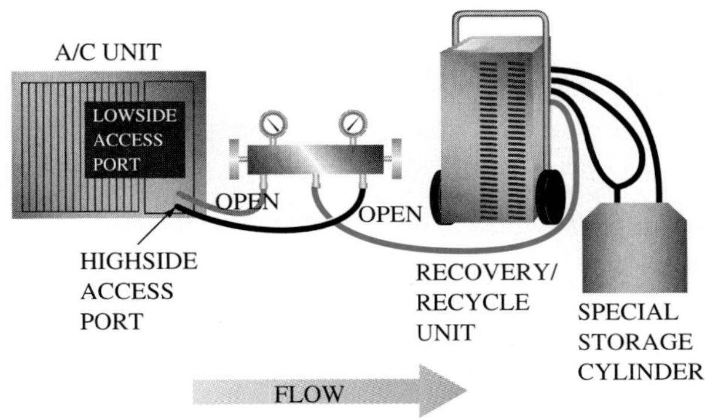

Figure R14-6 Flow of refrigerant in the recovery operation.

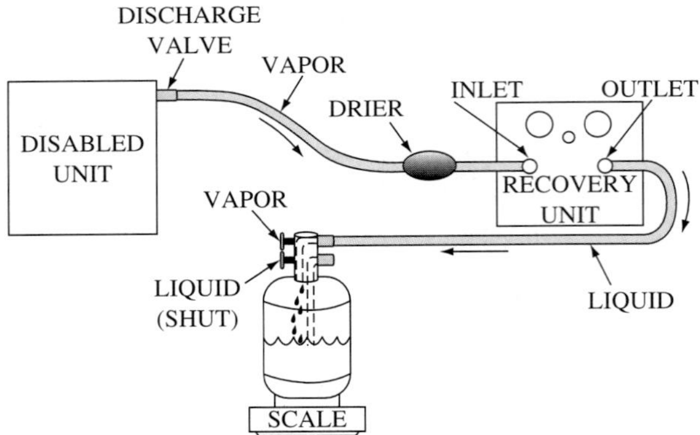

Figure R14-7 Piping arrangement for vapor recovery.

serviced and one for the recovery-unit compressor. The oil from the recovery-unit compressor is returned to the compressor. Oil separated from the recovered refrigerant is usually drained and not re-used. Vapor-recovery time can be reduced by immersing the recovery cylinder in ice water.

The units are equipped with a low-pressure cutout with a typical factory setting of 29 in. Hg (vacuum), which must not be changed. If a higher cutoff pressure is desired it can be controlled manually. The units also have a high-pressure cutout, which is factory set, to turn off the machine in case there is any obstruction.

The general rule is to let the machine cycle off once, then wait 10 or 15 minutes and start it up again. When the machine has cycled off two times the system is considered empty of refrigerant.

The EPA recommends that the technician have a separate recovery machine for each type of refrigerant: one for CFCs, one for HCFCs, and one for HFCs.

Liquid Recovery

There are several benefits from recovering liquid rather than vapor refrigerant. The biggest advantage is the greatly increased speed of recovery. A second benefit is that it puts less strain on the recovery equipment.

The disadvantage is that all the refrigerant cannot be recovered by this means. For the final recovery process it is necessary to use the vapor method. Also, if direct transfer is used, contamination in the system can be carried with the liquid to the recovery cylinder.

There are two primary techniques for liquid recovery:

1. The liquid-pump method (see Fig. R14-8).
2. The differential-pressure method (see Fig. R14-9). This method is used for both vapor and liquid recovery.

With the liquid-pump method, the pump inlet is connected to the lowest liquid location. In larger systems, this is usually the receiver. Liquid pumps are generally applied to large systems, as shown in the diagram. Care must be exercised to remove the liquid from all parts of the system. There is always a possibility that some refrigerant will be trapped in a low area. The use of an equalizer line as shown in the diagram is helpful. It may be necessary to recover the liquid from each component separately.

Both Vapor and Liquid Recovery

Fig. R14-9 shows the hook up using the differential method of liquid recovery (also known as the push/pull method) for both vapor and liquid recovery. With this method, the recovery machine pulls a vacuum in the recovery cylinder and, at the same time, increases the pressure in the disabled unit. This quickly forces liquid out of the disabled unit into the recovery cylinder.

Using this arrangement, the intake of the recovery machine (suction) is connected to the vapor fitting on the recovery cylinder. The outlet of the recovery machine (discharge) is connected to the vapor connection on the disabled unit. The liquid fitting of the recovery cylinder is connected to the liquid fitting on the disabled unit.

ON THE JOB

Never use a torch on a refrigerant container to warm it; instead put the drum in warm water, about 110°F.

Open the valve slowly when adding refrigerant to a system.

Store cylinders in an upright position and secure them to a wall or solid object.

Never reuse a one-time disposable cylinder. Don't drop or abuse the cylinder. Never use empty cylinders for a purpose other than what was intended.

Always verify that the refrigerant inside a container corresponds with what is marked on the outside.

When reclaiming refrigerant, never fill the reclaiming cylinder beyond the maximum fill weight marked on the tank.

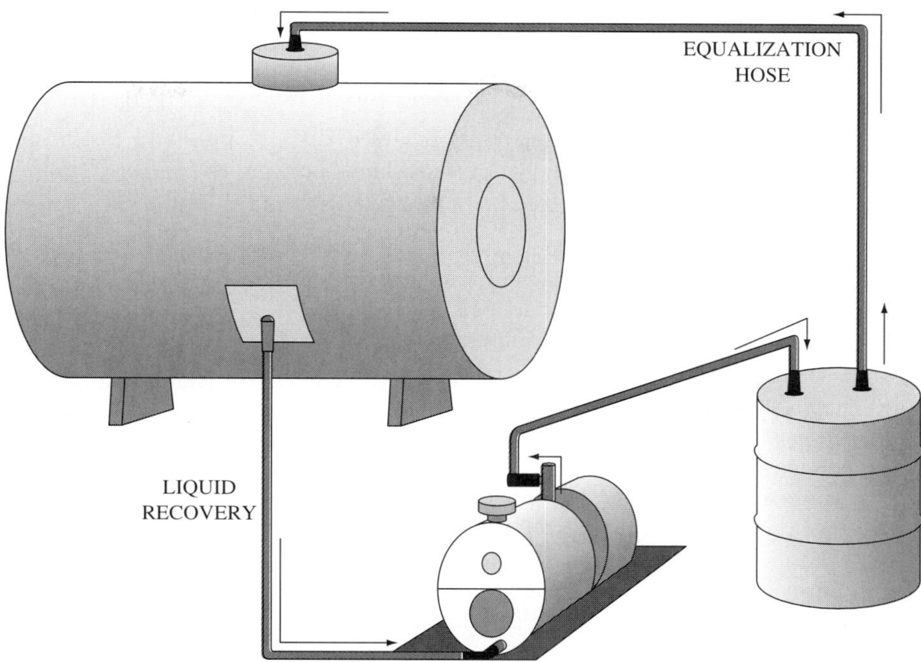

Figure R14-8 Piping arrangement for liquid recovery, using liquid pump.

R14-5.2 Removing Non-condensibles

A check should be made to see that there are no non-condensibles (air or moisture) in the recovery cylinder that might be transferred into a system when the refrigerant is reused. The procedure is as follows:

1. Measure the pressure in the recovery cylinder.
2. Measure the temperature of the cylinder.
3. Refer to the pressure-temperature chart for the refrigerant in the cylinder. Since the refrigerant in the cylinder is in a saturated condition, the pressure at the temperature measured in step 2 above should be nearly equivalent to the pressure measured in step 1 above.
4. If the pressure is higher than it should be, there are non-condensibles in the recovery cylinder that need to be removed.
5. A small quantity of vapor can be released from the cylinder. The pressure should drop to the allowed deviation. If it does not, the refrigerant will need to be recycled or reclaimed.

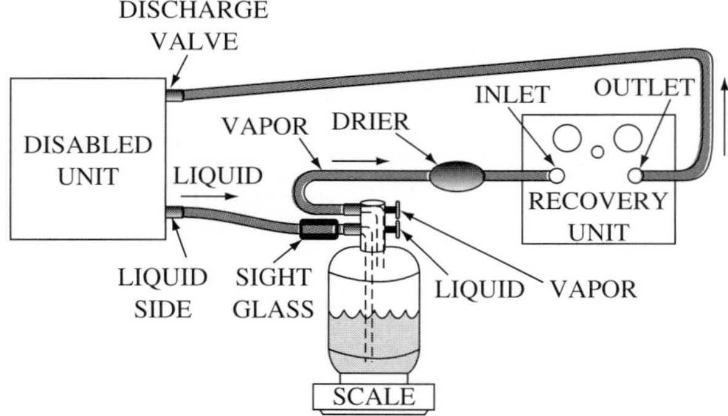

Figure R14-9 Piping arrangement for push-pull liquid recovery.

R14-6 RECLAIMING REFRIGERANTS

There are three reasons for reclaiming refrigerants:

1. The refrigerant may be contaminated, as in the case of an equipment burn out.
2. The quality and type of refrigerant is unknown and cannot be used.
3. The refrigerant is to be resold.

Central processing stations have been set up and are certified by EPA. If the refrigerant can be reclaimed, the stations take full responsibility for returning the refrigerant to its original condition, ready for resale. The refrigerant then can be used in any system. In case the unit receiving the refrigerant is still in warranty, the manufacturer should be contacted to be certain that using reclaimed refrigerant does not void the warranty.

If for any reason the refrigerant is too contaminated to be reclaimed, it will be rejected and destroyed by the processing station—usually by incineration—at the owner's expense. In general, reclaiming can be an expensive process.

R14-7 REPLACEMENT OF REFRIGERANTS

When a different refrigerant is chosen to replace the existing refrigerant for retrofitting, a number of steps should be taken ahead of the installation (see also Chapter R2):

1. Select a replacement refrigerant which can reasonably be expected to be available for the anticipated life of the equipment being serviced.
2. Match the replacement refrigerant with the one it is to replace. Replacement refrigerants have specific applications. Quite often the equipment or compressor manufacturer will be able to provide information on acceptable conversions.
3. Make a record of the performance of the equipment, using the existing refrigerant to compare with the performance using the new refrigerant.
4. Check the manufacturer's specification sheet for any characteristics of the new refrigerant which may be incompatible with that of the existing refrigerant. The types of oil used, for example, may be different.

R14-8 TYPES OF RECOVERY AND RECYCLING EQUIPMENT

The following are some features of available recovery and recycling equipment.

R14-8.1 Multi-refrigerant Recovery and Recycling System

Due to its size and weight this equipment is best used in the shop or machinery location. It is an excellent example of the operation of a recovery unit, as shown in Fig. R14-10.

Fig. R14-11 shows the piping arrangement when using this equipment.

R14-8.2 Recovery Sequences

During recovery, refrigerant is pulled from the system, through the intake filter and a manual selector valve. Vapor passes through, while liquid is metered through a capillary tube. The line includes a sight glass, a low-pressure cutout, and a solenoid valve which is energized during the recovery process. The refrigerant then flows through an evaporator, where the liquid will boil and vaporize using the heat from the system condenser.

The vapor then passes through an oil separator to remove unwanted oil from the refrigerant. Acid that may have developed within the system is concentrated in the oil. By separating the oil from the refrigerant, over 95% of the acid is also removed. Even with a compressor burnout, the remaining acid can be handled by the filter-drier in the recycling sequence.

KNOWLEDGE EXTENSION

Hundreds of manufacturers have entered into the refrigerant recovery/recycle equipment market. Prior to purchasing any equipment, obtain a technical sheet from the company, verify it is UL approved, and has a warranty. It is also important to verify ease of service, weight, and capacity, and then compare the product to other comparable pieces of equipment. Contact your local supply house to learn more about the various types of refrigerant recycle/recovery equipment and products available or attend the ARI/ASHRAE Conference and Exhibits.

To store refrigerant, it must be condensed to a liquid. After passing through the compressor, refrigerant proceeds to the return-oil separator, through the condenser, and into the storage tank.

The discharge-oil separator is necessary to trap and return oil to the recovery compressor, since this is a closed circuit. It is important to follow the manufacturer's instructions regarding checking and changing oil. The high

Figure R14-10 Multi-refrigerant recovery/recycling unit. (Courtesy of Robinair Division, SPX Corporation)

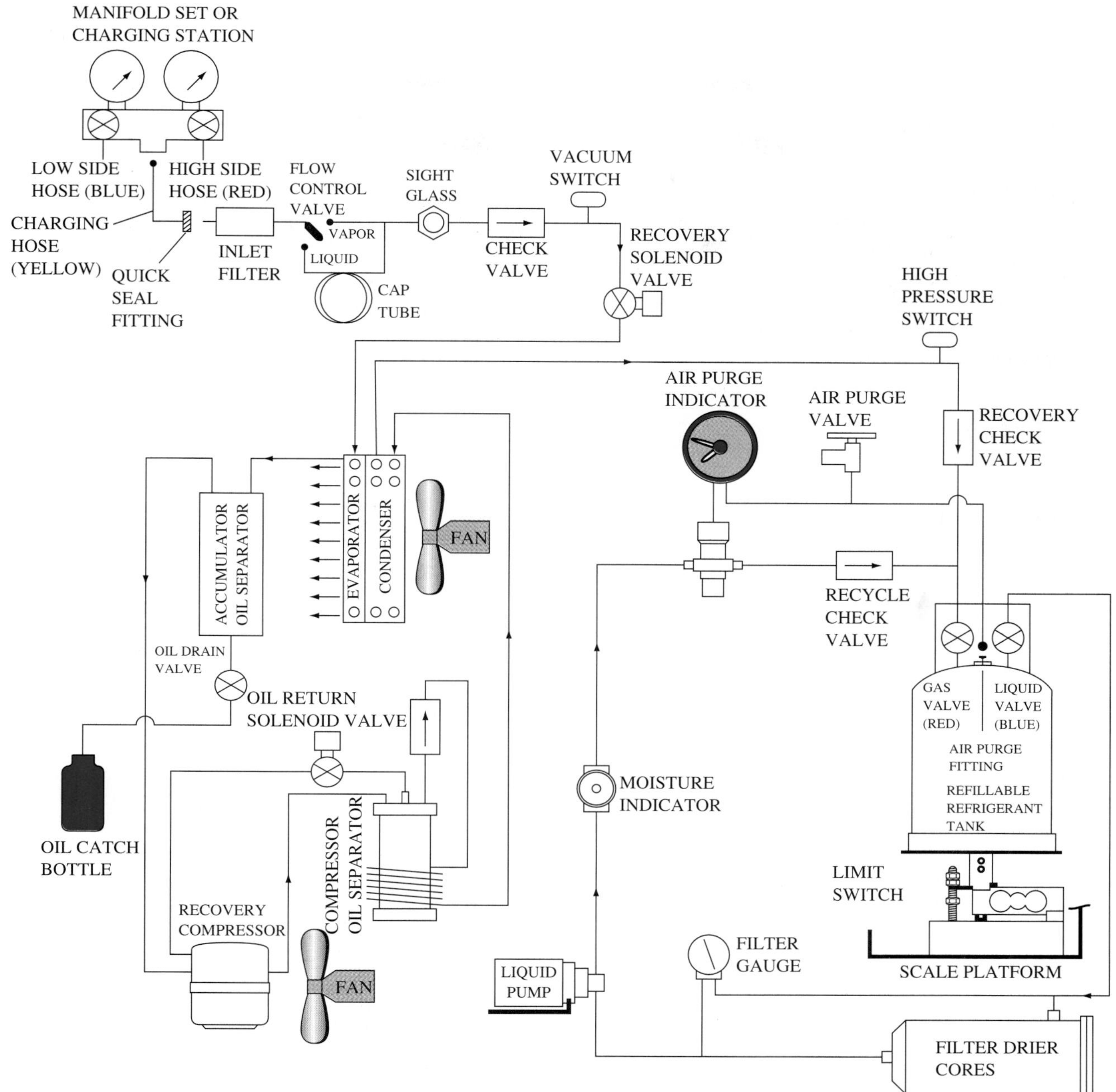

Figure R14-11 Piping for recovery/recycling unit.

side of the circuit includes a high-pressure cutout and a check valve. A scale weighs the deposit and a limit switch can be set to prevent over filling.

R14-8.3 Recycling Sequence

This process removes moisture, air, and the remaining acid from the refrigerant. The liquid pump, magnetically coupled to prevent leakage and reduce heat generation, circulates the refrigerant through a filter-drier unit. This recirculation process is repeated until the refrigerant is clean, dry, and ready for reuse. The air purge indicator detects pressure differences created by the presence of air in the tank. The air is purged manually.

R14-8.4 Extractor for Removing Refrigerant from Systems

This extractor, as shown in Fig. R14-12, features a heavy-duty ½-Hp compressor with an oil drain instead of the oil return found on most models. The compressor-oil drain allows the technician to recharge compressor oil in the

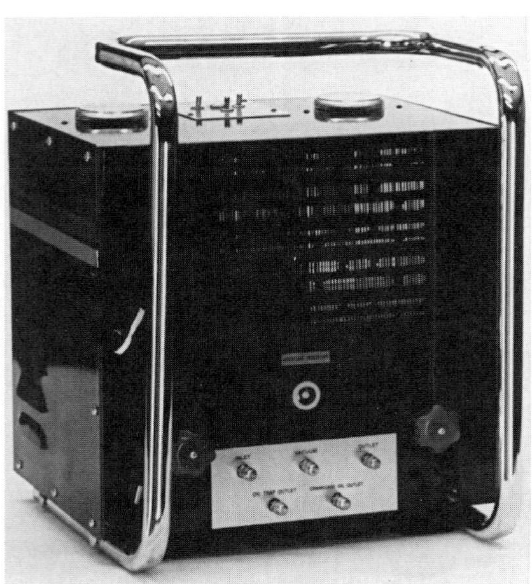

Figure R14-12 Extractor unit. (Courtesy of Thermal Engineering Company)

event of a burnout, removing all possible contaminants from the unit.

Fig. R14-13 shows the extractor's piping diagram. Note the use of a solenoid valve (SV) in a bypass pipe between the inlet and outlet. This valve is opened for a warm-up cycle before refrigerant extraction begins.

R14-8.5 Recycler

This unit, as shown in Fig. R14-14, uses a single-pass method of refrigerant recycling. It has an automatic crankcase-pressure regulator to eliminate the need for adjusting for variations in incoming refrigerant flow. It uses a sight glass to monitor the condition of the refrigerant prior to entering the storage cylinder.

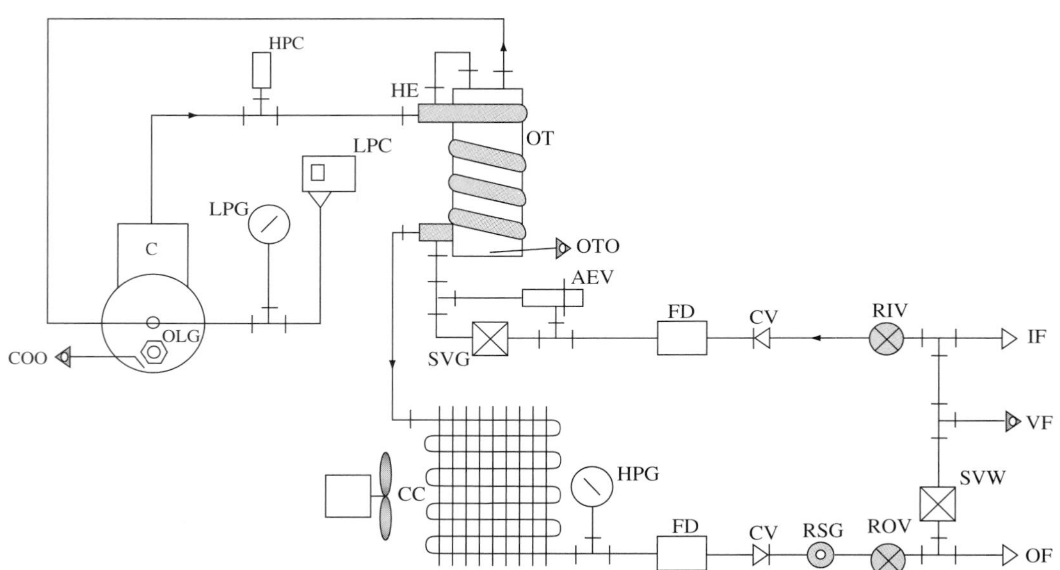

AEV = AUTOMATIC EXPANSION VALVE
C = COMPRESSOR
CC = CONDENSER COIL
COO = CRANKCASE OIL OUTLET
CV = CHECK VALVE
FD = FILTER DRIER
HE = HEAT EXCHANGER
HPC = HIGH PRESSURE CONTROL
HPG = HIGH PRESSURE GAUGE
IF = INLET FITTING
LPC = LOW PRESSURE CONTROL
LPG = LOW PRESSURE GAUGE
OF = OUTLET FITTING
OLG = OIL LEVEL GLASS
OT = OIL TRAP
OTO = OIL TRAP OUTLET
RIV = REFRIGERANT INLET VALVE
ROV = REFRIGERANT OUTLET VALVE
RSG = REFRIGERANT SIGHT GLASS
SVG = SOLENOID VALVE (GAS)
SVW = SOLENOID VALVE (WARMUP)
VF = VACUUM FITTING

Figure R14-13 Piping for the extractor unit.

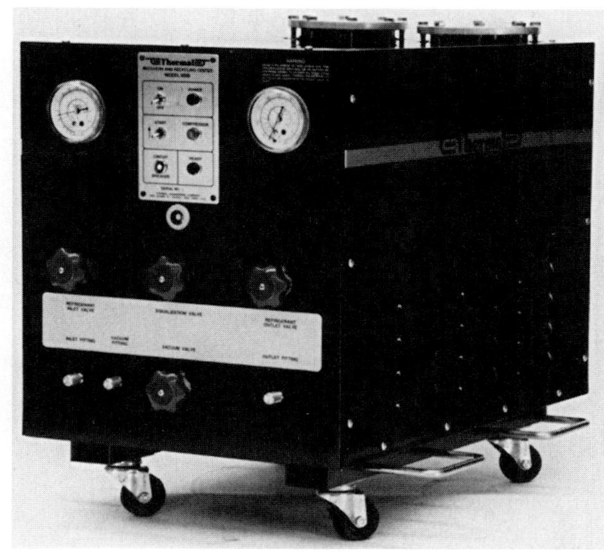

Figure R14-14 Typical recycling unit. (Courtesy of Thermal Engineering Company)

R14-8.6 Recycling Unit, Toolbox Version

This unit, as shown schematically in Fig. R14-15, can be used with any two-port DOT storage cylinder. A DOT cylinder is one that is approved by the U.S. Department of Transportation for refilling with refrigerant. Use of an unapproved cylinder could result in a large fine plus a lengthy imprisonment.

The standard unit has the following cleaning specifications: 70% removal of particles, 30% acid removal, 30% moisture removal (at 120 ppm), in a single pass. When used with a pre-filter burnout kit it will remove 85% of the acids, moisture, and particles.

The controls include automatic shutoff at 4 in. Hg vacuum, high-pressure cutoff at 350 psi, anti-liquid-slugging check valve, and compressor-overload lockout. The unit can empty a 200-lb receiver in 1 hour (liquid mode).

The following accessories are available:

DOT-approved refillable storage vessels
Burnout pre-filter kit
Wheel kit for unit mobility
On-board storage cylinder, with moisture-indicating sight glass, graduated for recharge.

R14-8.7 Recovery Unit with Two-port Cylinder (DOT-approved)

This unit is shown in Fig. R14-16. The connections between the disabled unit, the cylinder, and the recovery unit are shown in Fig. R14-17 for both liquid and vapor recovery.

The recovery unit pumps the refrigerant vapor from the top of the cylinder, and at the same time pressurizes the air-conditioning unit. By maintaining a pressure differential between the units, the liquid refrigerant is transferred to the cylinder, as shown in Diagram 1 of Fig. R14-17.

A scale is shown under the cylinder for weighing the liquid in the cylinder to prevent filling the cylinder more than 80% full. After the liquid has been removed, the remaining vapor is extracted during the vapor recovery, as shown in Diagram 2 of Fig. R14-17.

In the vapor-recovery phase, when the suction pressure gauge reads 0 psig, the vapor recovery is complete. This model has a built-in 80% safety shut off switch.

Figure R14-15 Recovery unit with piping diagram.

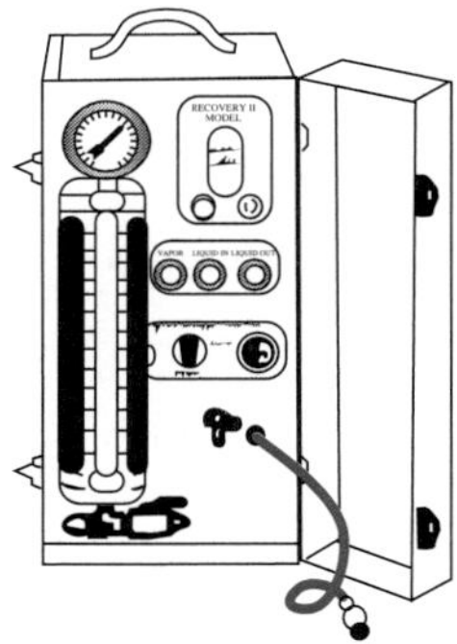

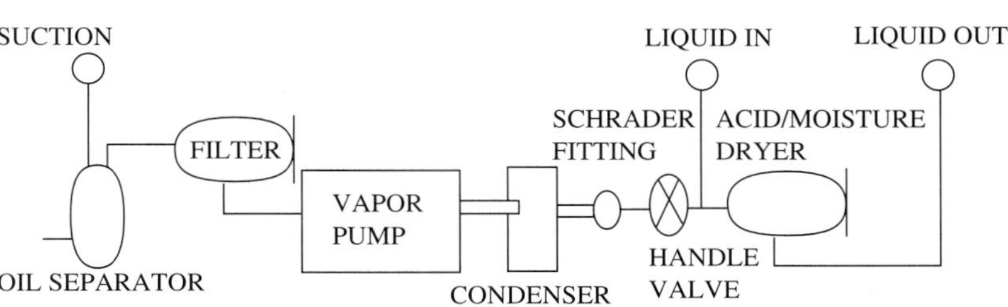

Figure R14-16 Typical recovery unit. (Courtesy of National Refrigeration Products, Bensalem, PA. Model LV1CUL)

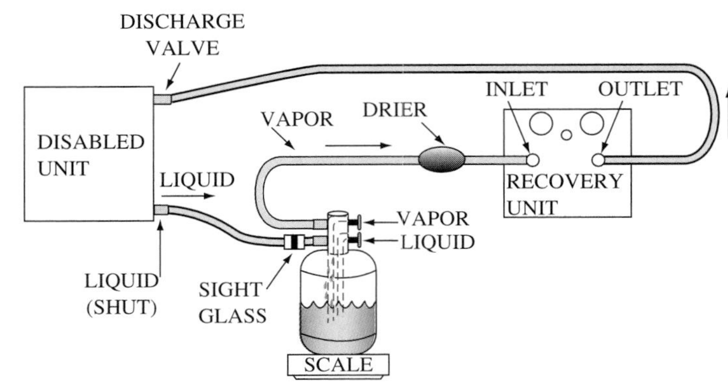

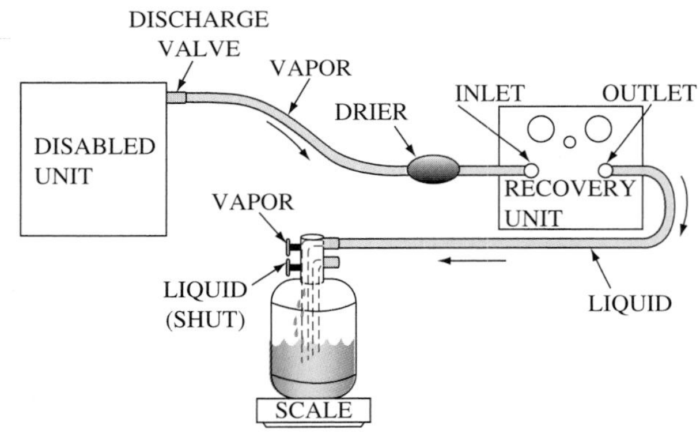

Figure R14-17 Two types of recovery unit piping.

ON THE JOB

When you are working on rooftop air-conditioning/heating equipment and it is necessary to recover and recycle the refrigerant, but the recycle equipment is too heavy to carry onto the roof, then recover the refrigerant into a portable DOT-approved tank. Carry the tank off the roof and proceed to thoroughly recycle the refrigerant on the ground until it is clean. When recycling is complete, recharge the unit with the refrigerant.

R14-8.8 Lightweight Refrigerant-recovery Device

A convenient but limited solution for refrigerant recovery, designed for the HVAC technician, is a lightweight (30 lb) unit. It is used primarily for R-12 and R-22 where the compressor is operatable. This unit essentially functions first as a parallel evaporator and then as a high-side receiver to accomplish the recovery.

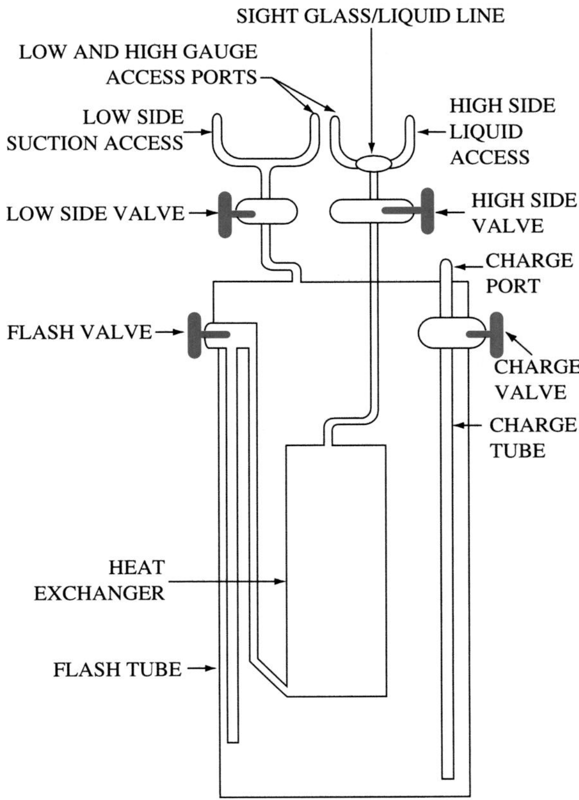

Figure R14-18 Internal diagram of a recovery device.

R14-9 PHASES OF REFRIGERANT RECOVERY

Refrigerant recovery is accomplished in three phases: *evaporation, equalization,* and *evacuation.*

Phase 1—evaporation—subcools and stabilizes the liquid refrigerant by means of its internal heat exchanger. Phase 1 ends when the charge remaining in the system is not able to condense further; only vapor remains (see Fig. R14-18).

Phase 2—equalization—condenses vapor by equalization of the temperature with the subcooled liquid from phase 1. This causes the oil vapor to be absorbed into the liquid refrigerant. This process maximizes the contact between the gas and the liquid. The newly condensed liquid is collected at this point.

Phase 3—evacuation—receives the remaining refrigerant vapor and again maximizes oil-vapor absorption. This pressurizes the cylinder to aid in transferring the refrigerant either back to the system or to an external cylinder. Fig. R14-19 shows the recovery device connected to a refrigeration system.

This unit is for use in the majority of service operations, such as leak repair and components replacement, where the system still has a functioning compressor. Repair of a system without a functioning compressor requires a recovery device as described in the previous examples.

A number of companies are building recovery and recycling devices. Their products must indicate they have been approved by the EPA.

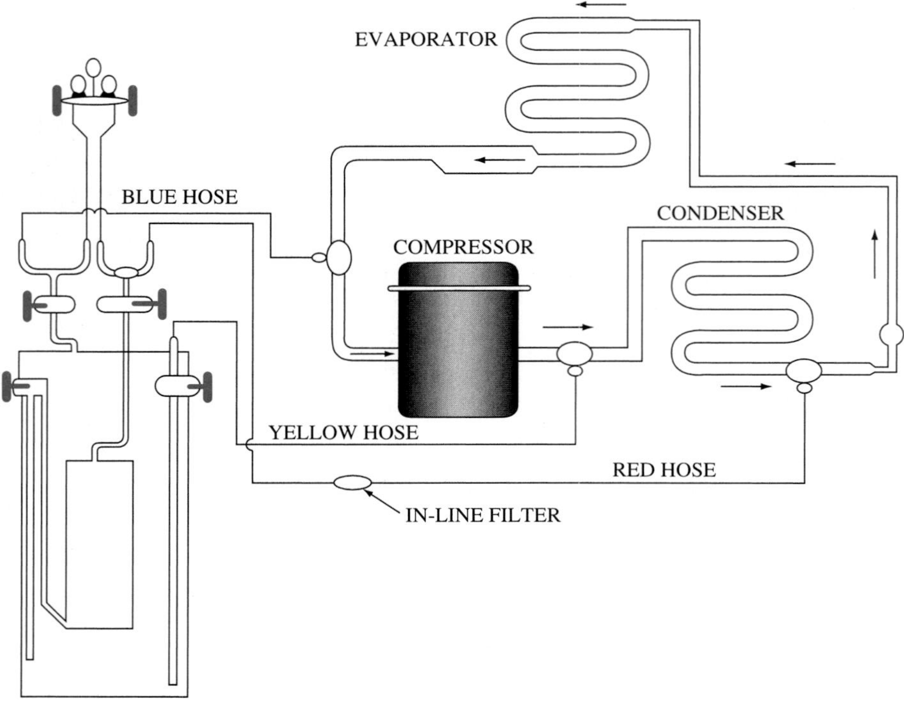

Figure R14-19 Connection diagram of a recovery device.

REVIEW

- Why should refrigerant be recovered?
  It is economical.
  It is required by law.
  It is environmentally sound.
- Recovery—Removal of refrigerant in any condition from an appliance and storage in an external container without necessarily being tested or processed in any way.
- Recycle—Extraction of refrigerant from an appliance and cleaning for reuse by filtering, distillation or other process without meeting the requirements for reclamation.
- Reclaim—Reprocessing of refrigerant to the purity specified in ARI Standard 700-1988, and verification of the purity with laboratory tests and documentation.
- Appliance—Any device that contains Class I or Class II substances as a refrigerant and is used for household or commercial purposes.
- Small Appliance—An appliance that is manufactured fully charged, and hermetically sealed in a factory with five pounds or less of refrigerant.
- Technician—Any person who performs maintenance, service, or repair that could reasonably be expected to release Class I or Class II substances into the atmosphere. Technician also refers to any person disposing of any appliances except small appliances.
- After November 15, 1994, CFC and HCFC refrigerants can only be sold to certified technicians, wholesalers, and manufacturers.
- Very-high-pressure equipment: Equipment that uses a refrigerant having a boiling point below 50°C at atmospheric pressure. This includes: CFC-13, HFC-23, and CFC-503. To work on this equipment requires a Type II certification.
- High-pressure equipment: Equipment that uses a refrigerant with a boiling point between 50°C and 10°C at atmospheric pressure. This includes: CFC-12, -114, -500, -502, and HCFC-22. Type I and II certifications are required depending on the charge and type of appliance.
- Low-pressure equipment: Equipment that has a refrigerant with a boiling point above 10°C at atmospheric pressure. This includes: CFC-11, CFC-113, and HCFC-123. This type of work requires a Type III certification.
- A technician that is certified to work on all equipment (Type I, II, and III) has a Type IV universal certification.
- Methods of Recovery
  Passive
    Charge migration
  Active
    Use of systems compressor
    Use of refrigerant-recovery unit
- Types of Recovery Units
  Vapor recovery
  Liquid recovery
  Vapor/Liquid recovery

Problems and Questions

1. What is the recommended percentage of the total volume of a cylinder that can safely be filled with refrigerant?
   a. 60%
   b. 70%
   c. 80%
   d. 90%
2. If R-22 were to escape from a cylinder while you were servicing a system near the ocean, at what temperature would you expect it to boil?
   a. −41°F
   b. −50°F
   c. −22°F
   d. −15°F
3. What would be the correct position of the valves on a gauge manifold, to charge refrigerant into the low side of a compressor?
   a. Both valves back seated
   b. Both valves front seated
   c. Low pressure valve back seated
      High pressure valve front seated
   d. Low pressure valve front seated
      High pressure valve back seated
4. Is it permissible to release refrigerant to the atmosphere in order to purge the gauge manifold hoses?
   a. Yes
   b. No
5. Where is the best place to find the amount of the refrigerant charge for a unit?
   a. On the door of the equipment room.
   b. On the nameplate or label attached to the unit.
   c. On the tag attached to the compressor
   d. From the previous service technician
6. When the migration process is used to recover the refrigerant, should the compressor be running?
   a. Yes
   b. No
7. The Clean Air Act of 1990 limits the use of chlorine-based refrigerants. True or False?
8. High-pressure equipment uses a refrigerant with a boiling point between −50°F and 10°F. True or False?
9. An R-11 centrifugal is an example of low-pressure equipment. True or False?
10. The biggest advantage of liquid recovery is the greatly increased speed of recovery. True or False?
11. When testing for non-condensibles, measure the ambient temperature, the tank temperature, and the tank pressure. True or False?
12. What is the difference between recycle and recover?
13. Why do some recovery/recycle pieces of equipment have a discharge oil separator?
14. How does a Type III and a universal technician certification differ?

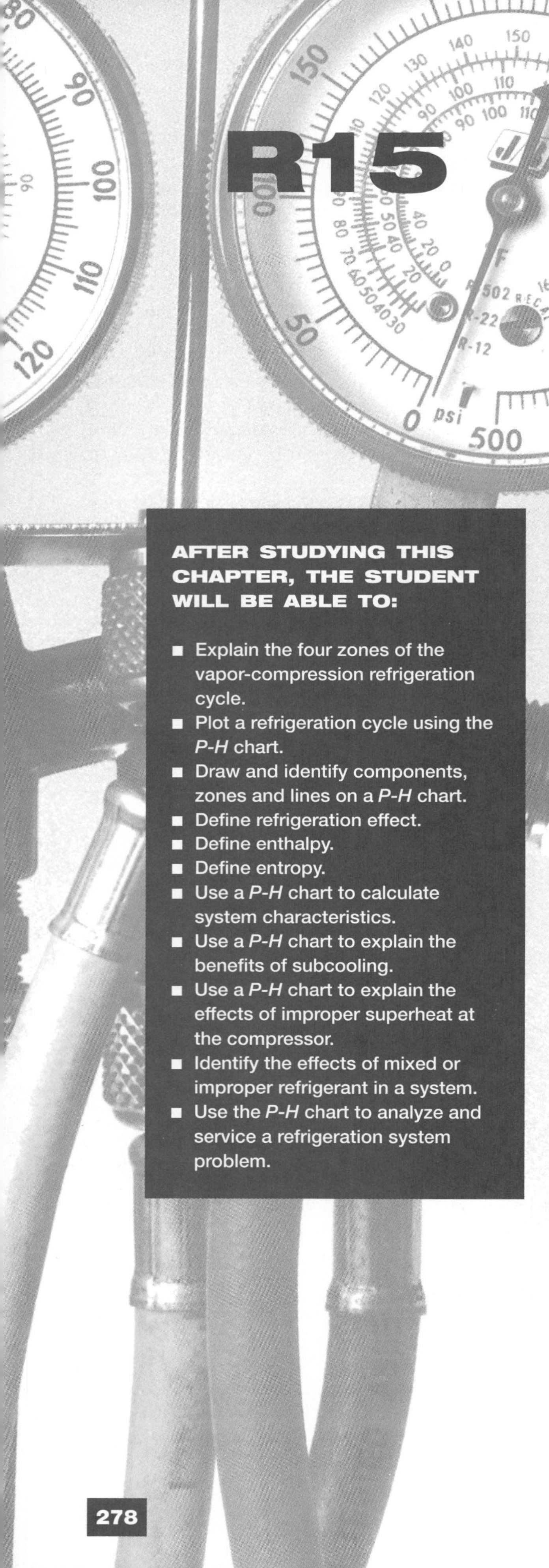

R15 Pressure-Enthalpy Diagrams

AFTER STUDYING THIS CHAPTER, THE STUDENT WILL BE ABLE TO:

- Explain the four zones of the vapor-compression refrigeration cycle.
- Plot a refrigeration cycle using the *P-H* chart.
- Draw and identify components, zones and lines on a *P-H* chart.
- Define refrigeration effect.
- Define enthalpy.
- Define entropy.
- Use a *P-H* chart to calculate system characteristics.
- Use a *P-H* chart to explain the benefits of subcooling.
- Use a *P-H* chart to explain the effects of improper superheat at the compressor.
- Identify the effects of mixed or improper refrigerant in a system.
- Use the *P-H* chart to analyze and service a refrigeration system problem.

R15-1 REFRIGERATION EFFECT

If a specific job is to be done in a refrigeration system or cycle, each pound of refrigerant circulating in the system must do its share of the work. It must absorb an amount of heat in the evaporator or cooling coil, and it must dissipate this heat—plus some that is added in the compressor—out through the condenser, whether air-cooled, water-cooled, or evaporatively cooled. The work done by each pound of the refrigerant as it goes through the evaporator is reflected by the amount of heat it picks up from the refrigeration load, chiefly when the refrigerant undergoes a change of state from a liquid to a vapor.

As mentioned previously, for a liquid to be able to change to a vapor, heat must be added to or absorbed by it. This is what happens—or should happen—in the cooling coil. The refrigerant enters the metering device as a liquid and passes through the device into the evaporator, where it absorbs heat as it evaporates into a vapor. As a vapor, it makes its way through the suction tube or pipe to the compressor. Here it is compressed from a low-temperature, low-pressure vapor to a high-temperature, high-pressure vapor; then it passes through the high-pressure or discharge pipe to the condenser, where it undergoes another change of state—from a vapor to a liquid—in which state it flows out into the liquid pipe and again makes its way to the metering device for another trip through the evaporator. Shown in Fig. R15-1 is a schematic of a simple refrigeration cycle, describing this process.

When the refrigerant, as a liquid, leaves the condenser it may go to a receiver until it is needed in the evaporator; or it may go directly into the liquid line to the metering device and then into the evaporator coil. The liquid entering the metering device just ahead of the evaporator coil will have a certain heat content (*enthalpy*), which is dependent on its temperature when it enters the coil, as

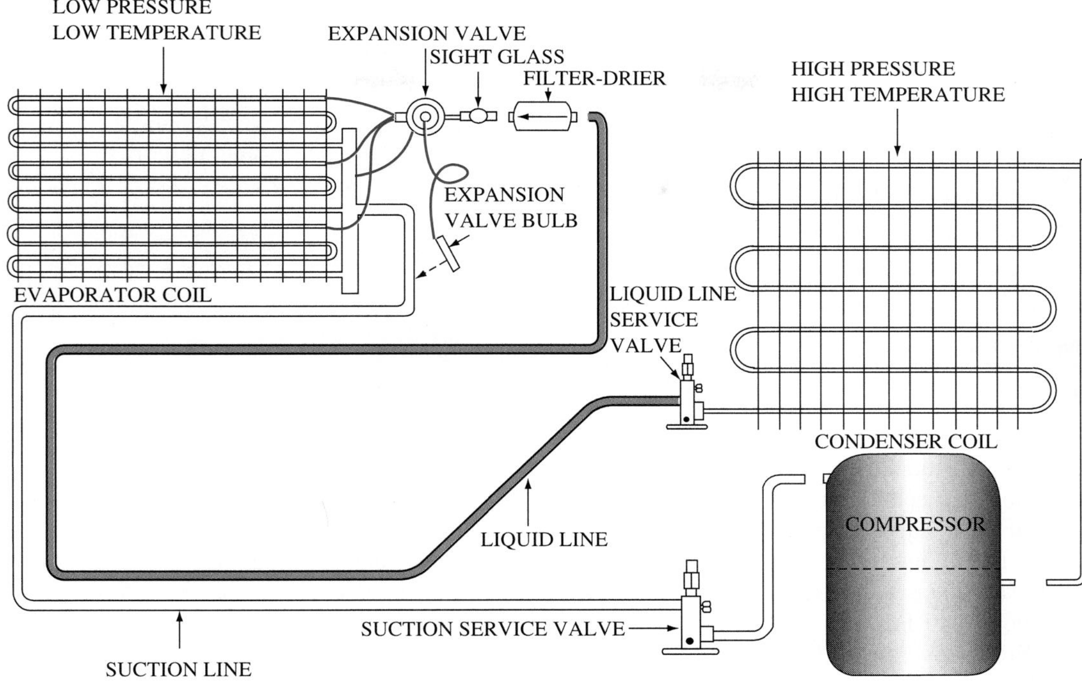

Figure R15-1 Schematic diagram of simple refrigeration cycle.

shown in the refrigerant tables in the Appendix. The vapor leaving the evaporator will also have a given heat content (enthalpy) according to its temperature, as shown in the refrigerant tables.

The difference between these two amounts of heat content is the amount of work being done by each pound of refrigerant as it passes through the evaporator and picks up heat. The amount of heat absorbed by each pound of refrigerant is known as the *refrigerating effect* of the system, or of the refrigerant within the system.

This refrigerating effect is rated in Btu per pound of refrigerant (Btu/lb); if the total heat load is known (given in Btu/hr), we can find the total number of pounds of refrigerant that must be circulated each hour of operation of the system. This figure can be broken down further to the amount that must be circulated each minute, by dividing the amount circulated per hour by 60.

EXAMPLE

If the total heat to be removed from the load is 60,000 Btu/hr and the refrigerating effect in the evaporator amounts to 50 Btu/lb, then

$$\frac{60{,}000 \text{ Btu/hr}}{50 \text{ Btu/lb}} = 1200 \text{ lb/hr or } 20 \text{ lb/min}$$

Since 12,000 Btu/hour equals the rate of 1 ton of refrigeration, the 60,000 Btu/hr in the example above amounts to 5 tons of refrigeration, and the 20 lb of refrigerant that must be circulated each minute is the equivalent of 4 lb/min/ton of refrigeration. (One ton of refrigeration for 24 hours equals 288,000 Btu.)

In this example, where 20 lb of refrigerant having a refrigerating effect of 50 Btu/lb is required to take care of the specified load of 60,000 Btu/hr, the results can also be obtained in another manner. As mentioned previously, it takes 12,000 Btu to equal 1 ton of refrigeration, which is equal to 200 Btu/min/ton.

Therefore, 200 Btu/min, when divided by the refrigerating effect of 50 Btu/lb, amounts to 4 lb/min. This computation can be shown by the equation

$$W = \frac{200}{\text{NRE}} \tag{R15-1}$$

where

W = weight of refrigerant circulated per minute, lb/min
200 = 200 Btu/min—the equivalent of 1 ton
NRE = net refrigerating effect, Btu/lb of refrigerant

Because of the small orifice in the metering device, a fact that will be discussed more thoroughly in a later chapter, when the compressed refrigerant passes from the smaller opening in the metering device to the larger tubing in the evaporator, a change in pressure occurs together with a change in temperature. This change in temperature occurs because of the vaporization of a small portion of the refrigerant (about 13%) and, in the process of this vaporization, the heat that is involved is taken from the remainder of the refrigerant.

ON THE JOB

When working at the job site, it is imperative to remember the basic principles of the pressure-enthalpy diagram. This includes maintaining a cool subcooled liquid refrigerant to the metering device to increase evaporator efficiency. It also means keeping condenser surfaces clean to provide as low of head pressure as possible to minimize the amount of refrigerant circulated throughout the system.

It is possible on persistent refrigeration problems to analyze the system by measuring temperatures and pressures and then recording them onto a pressure-enthalpy diagram for the respective refrigerant. A thorough analysis of the system will probably indicate abnormal conditions and provide insight into a necessary remedy often over looked under the stress of the normal daily routine.

From the table of saturated R-12 in the Appendix, it can be seen that the heat content of 100°F liquid is 31.10 Btu/lb and that of 40°F liquid is 17.27 Btu/lb; this indicates that 13.83 Btu/lb has to be removed from each pound of refrigerant entering the evaporator. The latent heat of vaporization of 40°F R-12 (from the Appendix tables) is 64.17 Btu/lb, and the difference between this amount and that which is given up by each pound of refrigerant when its liquid temperature is lowered from 100°F to 40°F (13.83 Btu/lb) is 50.34 Btu/lb. This is another method of calculating the refrigerating effect—or work being done—by each pound of refrigerant under the conditions given.

The capacity of the compressor must be such that it will remove from the evaporator that amount of refrigerant which has vaporized in the evaporator and in the metering device in order to get the necessary work done. The previous examples have shown that 3.07 ft^3/min must be removed from the evaporator to perform 1 ton of refrigeration under the conditions specified; therefore, when the total load amounts to 5 tons, then 5 times 3.07 ft^3/min, or 15.35 ft^3/min, must be removed from the evaporator. The compressor must be able to remove and send on to the condenser the same weight of refrigerant vapor, so that it can be condensed back into a liquid and so continue in the refrigeration circuit or cycle to perform additional work.

If the compressor, because of design or speed, is unable to move this weight, some of the vapor will remain in the evaporator. This, in turn, will cause an increase in pressure inside the evaporator, accompanied by an increase in temperature and a decrease in the work being done by the refrigerant, and design conditions within the refrigerated space cannot be maintained.

A compressor that is too large will withdraw the refrigerant from the evaporator too rapidly, causing a lowering of the temperature inside the evaporator, so that design conditions will not be maintained in this situation either.

In order for design conditions to be maintained within a refrigeration circuit, there must be a balance between the requirements of the evaporator coil and the capacity of the compressor. This capacity is dependent on its displacement and on its volumetric efficiency. The measured displacement of a compressor depends on the number of cylinders, their bore and stroke, and the speed at which the compressor is turning. Volumetric efficiency depends on the absolute suction and discharge pressures under which the compressor is operating. A thorough and elaborate presentation of these facts concerning displacement, as well as the variables pertaining to volumetric efficiency, will be offered in a later chapter, together with equations and other data.

R15-2 CYCLE DIAGRAMS

Fig. R15-1 shows a schematic flow diagram of a basic cycle in refrigeration, denoting changes in phases or processes. First the refrigerant passes from the liquid stage into the vapor stage as it absorbs heat in the evaporator coil. The compression stage, where the refrigerant vapor is increased in temperature and pressure, comes next; then the refrigerant gives off its heat in the condenser to the ambient cooling medium, and the refrigerant vapor condenses back to its liquid state where it is ready for use again in the cycle.

Fig. R15-2 is a reproduction of a Mollier diagram (commonly known as a *P-î chart*) of R-12, which shows the pressure, heat, and temperature characteristics of this refrigerant. Pressure-enthalpy diagrams may be utilized for the plotting of the cycle shown in Fig. R15-1, but a basic or skeleton chart as shown in Fig. R15-3 might be used as a preliminary illustration of the various phases of the refrigerant circuit. There are three basic areas on the chart denoting changes in state between the saturated liquid line and saturated vapor line in the center of the chart. The area to the left of the saturated liquid line is the subcooled area, where the refrigerant liquid has been cooled below the temperature corresponding to its pressure; whereas the area to the right of the saturated vapor line is the area of superheat, where the refrigerant vapor has been heated beyond the vaporization temperature corresponding to its pressure.

The construction of the diagram, or rather a knowledge and understanding of it, may bring about a clearer inter-

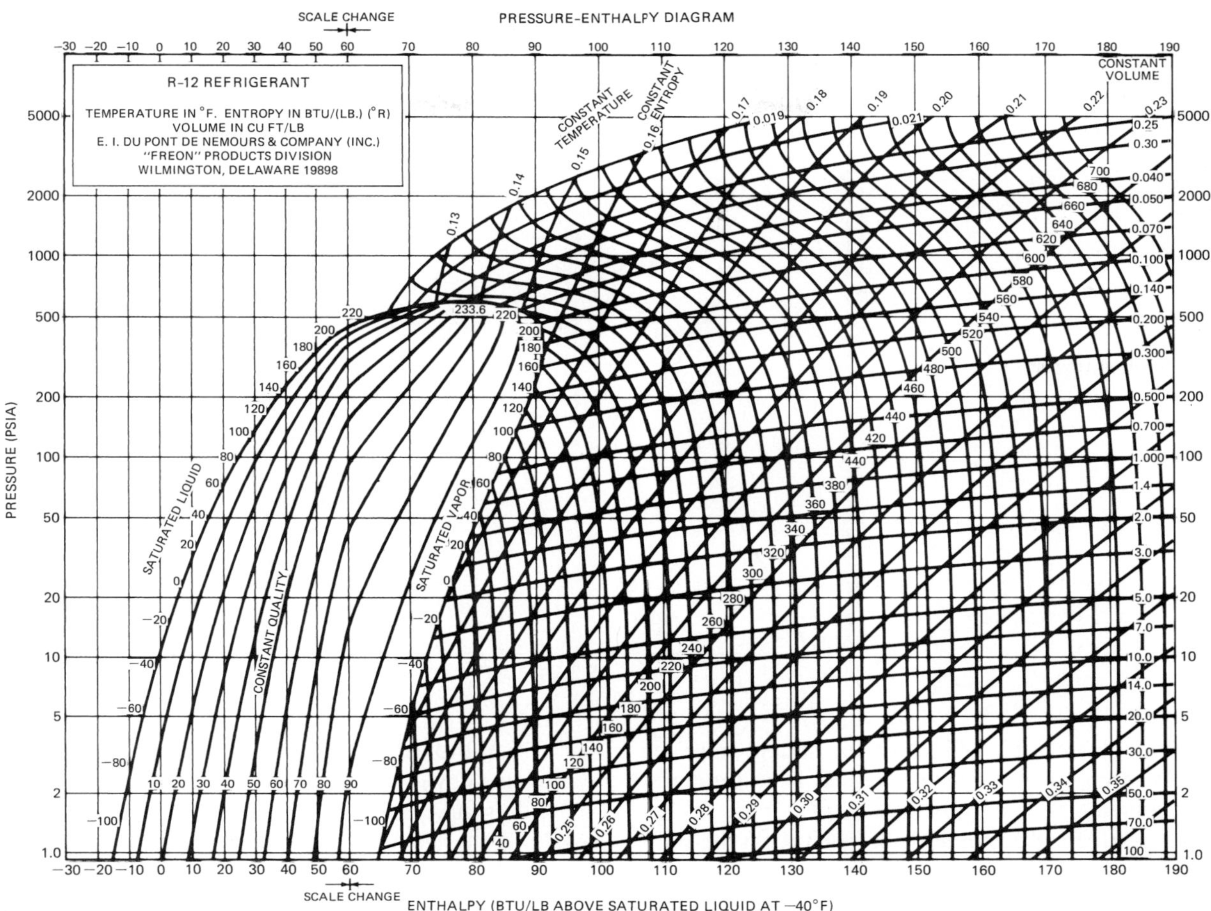

Figure R15-2 Pressure-enthalpy diagram for R-12. Note that 0 of enthalpy scale is taken at −40°F. (Courtesy of DuPont Chemicals)

pretation of what happens to the refrigerant at the various stages within the refrigeration cycle. If the state and any two properties of a refrigerant are known and this point can be located on the chart, the other properties can easily be determined from the chart.

If the point is situated anywhere between the saturated liquid and vapor lines, the refrigerant will be in the form of a mixture of liquid and vapor. If the location is closer to the saturated liquid line, the mixture will be more liquid than vapor, and a point located in the center of the area at a particular pressure would indicate a 50% liquid-50% vapor situation.

Referring to Fig. R15-3, the change in state from a vapor to a liquid—the condensing process—occurs as the path of the cycle develops from right to left; whereas the change in state from a liquid to a vapor—the evaporating process—travels from left to right. Absolute pressure is indicated on the vertical axis at the left, and the horizontal axis indicates heat content, or enthalpy, in Btu/lb.

The distance between the two saturated lines at a given pressure, as indicated on the heat content line, amounts to the *latent heat of vaporization* of the refrigerant at the given absolute pressure. The distance between the two lines of saturation is not the same at all pressures, for they do not follow parallel curves. Therefore, there are variations in the latent heat of vaporization of the refrigerant, depending on the absolute pressure. There are also variations in pressure-enthalpy charts of different refrigerants and the variations depend on the various properties of the individual refrigerants.

R15-3 REFRIGERATION PROCESSES

Based on the examples presented earlier in the chapter, it will be assumed that there will be no changes in the temperature of the condensed refrigeration liquid after it leaves the condenser and travels through the liquid pipe on its way to the expansion or metering device, or in the temperature of the refrigerant vapor after it leaves the evaporator and passes through the suction pipe to the compressor.

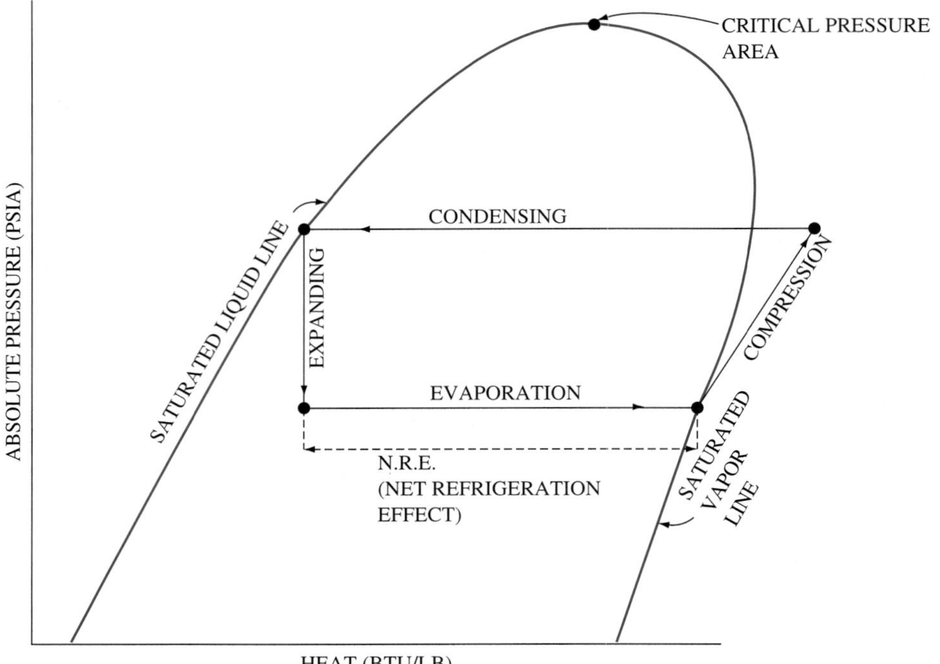

Figure R15-3 Pressure-enthalpy changes through a refrigeration cycle.

Fig. R15-4 shows the phases of the simple saturated cycle with appropriate labeling of pressures, temperatures, and heat content or enthalpy. A starting point must be chosen in the refrigerant cycle; let it be point *A* on the saturated-liquid line where all of the refrigerant vapor at 100°F has condensed into liquid at 100°F and is at the inlet to the metering device. What occurs between points *A* and *B* is the expansion process as the refrigerant passes through the metering device; and the refrigerant temperature is lowered from the condensation temperature of 100°F to the evaporating temperature of 40°F.

When the vertical line *A–B* (the expansion process) is extended downward to the bottom axis, a reading of 31.10 Btu/lb is indicated, which is the heat content of 100°F liquid. To the left of point *B* at the saturated-liquid line is point *Z,* which is also at the 40°F temperature line. Taking a vertical path downward from point *Z* to the heat content line, a reading of 17.27 Btu/lb is indicated, which is the heat content of 40°F liquid; this area between points *Z* and *B* is covered later in the chapter.

The horizontal line between points *B* and *C* indicates the vaporization process in the evaporator, where the 40°F

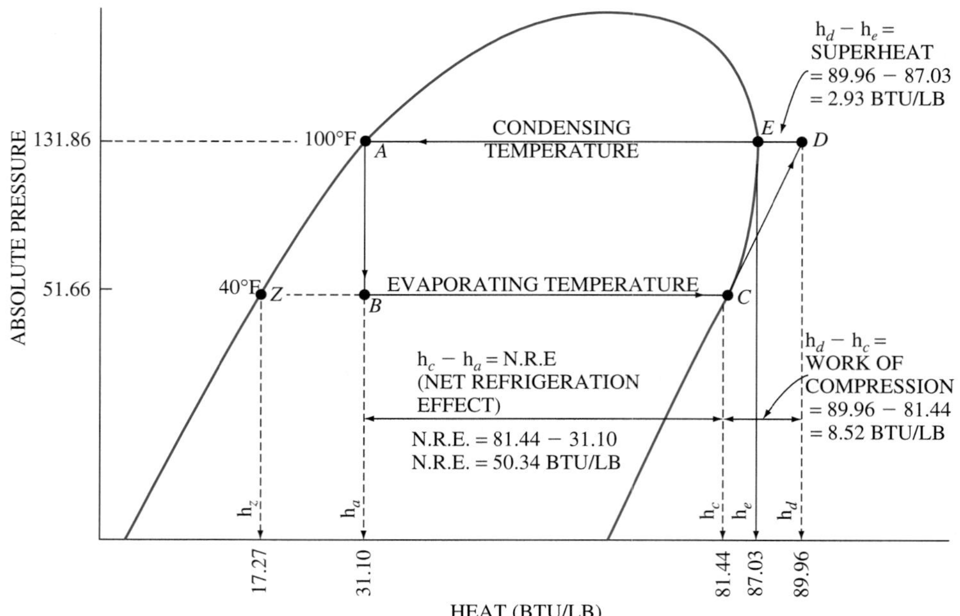

Figure R15-4 Pressure, heat, and temperature values for a refrigeration cycle operating with a 40°F evaporator.

liquid absorbs enough heat to completely vaporize the refrigerant. Point *C* is at the saturated vapor line, indicating that the refrigerant has completely vaporized and is ready for the compression process. A line drawn vertically downward to where it joins the enthalpy line indicates that the heat content, shown at h_c is 81.44 Btu/lb, and the difference between h_a and h_c is 50.34 Btu/lb, which is the refrigerating effect, as shown in an earlier example.

The difference between points h_a and h_c on the enthalpy line amounts to 64.17 Btu/lb, which is the latent heat of vaporization of 1 lb of R-12 at 40°F. This amount would also exhibit the refrigerating effect, but some of the refrigerant at 100°F must evaporate or vaporize in order that the remaining portion of each pound of R-12 can be lowered in temperature from 100°F to 40°F.

The various properties of refrigerants, mentioned earlier, should probably be elaborated on before we proceed with the discussion of the compression process. All refrigerants exhibit certain properties when in a gaseous state; some of them are: *volume, temperature, pressure, enthalpy* or *heat content,* and *entropy.* The last property—entropy—is really the most difficult to describe or define. It is the ratio of the heat content of the gas to its absolute temperature in degrees Rankine, and it relates to internal energy of the gas.

The Mollier chart plots the line of constant entropy, which stays the same providing the gas is compressed and no outside heat is added or taken away. When the entropy is constant, the compression process is called *adiabatic,* which means that the gas changes its condition without the absorption or rejection of heat either from or to an external body or source. It is common practice, in the study of cycles of refrigeration, to plot the compression line either along or parallel to a line of constant entropy.

In Fig. R15-4, line *C–D* denotes the compression process, in which the pressure and temperature of the vapor are increased from that in the evaporator to that in the condenser, with the assumption that there has been no pickup of heat in the suction line between the evaporator and the compressor. For a condensing temperature of 100°F, a pressure gauge would read approximately 117 psig; but the *P-H* chart is rated in absolute pressure and the atmospheric pressure of 14.7 must be added to the psig, making it actually 131.86 psia.

Point *D* on the absolute-pressure line is equivalent to the 100°F condensing temperature; it is not on the saturated vapor line, it is to the right in the superheat area, at a junction of the 131.86 psia line, the line of constant entropy of 40°F, and the temperature line of approximately 116°F. A line drawn vertically downward from point *D* intersects the heat content line at 89.96 Btu/lb, which is h_d; the difference between h_c and h_d is 8.52 Btu/lb—the heat of compression that has been added to the vapor. This amount of heat is the heat energy equivalent of the work done during the refrigeration compression cycle. This is the theoretical discharge temperature, assuming that saturated vapor enters the cycle; in actual operation, the discharge temperature may be 20 to 35° higher than that predicted theoretically. This can be checked in an operating system by strapping a thermometer or a thermocouple to the outlet of the discharge service valve on the compressor.

During the compression process the heat that is absorbed by the vapor is a result of friction caused by the action of the pistons in the cylinders and by the vapor itself passing through the small openings of the internal suction and discharge valves. Of course, the vapor is also heated by the action of its molecules being pushed or compressed closer together, commonly called *heat of compression.* Some of this overall additional heat is lost through the walls of the compressor. A lot depends, therefore, on the design of the compressor, the conditions under which it must operate and the balance between the heat gain and heat loss to keep the refrigerant at a constant entropy.

Line *D–E* denotes the amount of superheat that must be removed from the vapor before it can commence the condensation process. A line drawn vertically downward from point *E* to point h_e on the heat content line indicates the distance h_d–h_e, or heat amounting to 2.93 Btu/lb, since the heat content of 100°F vapor is 87.03 Btu/lb. This superheat is usually removed in the hot-gas discharge line or in the upper portion of the condenser. During this process the temperature of the vapor is lowered to the condensing temperature.

Line *E–A* represents the condensation process that takes place in the condenser. At point *E* the refrigerant is a saturated vapor at the condensing temperature of 100°F and an absolute pressure of 131.86 psia; the same temperature and pressure prevail at point *A,* but the refrigerant is now in a liquid state. At any other point on line *E–A* the refrigerant is in the phase of a liquid-vapor combination; the closer the point is to *A,* the greater the amount of the refrigerant that has condensed into its liquid stage. At point *A,* each pound of refrigerant is ready to go through the refrigerant cycle again as it is needed for heat removal from the load in the evaporator.

R15-4 COEFFICIENT OF PERFORMANCE

Two factors mentioned earlier in this chapter are of the greatest importance in deciding which refrigerant should be used for a given project of heat removal. Ordinarily, this decision is reached during the design aspect of the refrigeration and air-conditioning system, but we will explain it briefly now, and elaborate later.

The two factors that determine the *coefficient of performance* (COP) of a refrigerant are *refrigerating effect* and *heat of compression.* The equation may be written as

$$\text{COP} = \frac{\text{refrigerating effect}}{\text{heat of compression}} \qquad \text{(R15-2)}$$

Substituting values from the *P-H* diagram of the simple saturated cycle previously presented, the equation would be

$$\text{COP} = \frac{h_c - h_a}{h_d - h_c} = \frac{50.34}{8.52} = 5.91$$

The COP is therefore a rate or a measure of the efficiency of a refrigeration cycle in the utilization of expended energy during the compression process in ratio to the energy that is absorbed in the evaporation process. As can be seen from Eq. (R15-2), the less energy expended in the compression process, the larger will be the COP of the refrigeration system. Therefore, the refrigerant having the highest COP would probably be selected—provided other qualities and factors are equal.

R15-5 EFFECTS ON CAPACITY

The pressure-enthalpy diagrams in Figs. R15-4 and R15-5 show a comparison of two simple saturated cycles having different evaporating temperatures, to bring out various differences in other aspects of the cycle. In order that an approximate mathematical calculation comparison may be made, the cycles shown in Figs. R15-4 and R15-5 will have the same condensing temperature, but the evaporating temperature will be lowered 20°F. Data can either be obtained or verified from the table for R-12 in the Appendix; but we will take the values of *A, B, C, D,* and *E* from Fig. R15-4 as the cycle to be compared to that in Fig. R15-5 (with a 20°F evaporator). The refrigerating effect, heat of compression, and the heat dissipated at the condenser in each of the refrigeration cycles will be compared. The comparison will be based on data about the heat content or enthalpy line, rated in Btu/lb.

For the 20°F evaporating temperature cycle shown in Fig. R15-5:

$$\text{net refrigerating effect } (h_{c'} - h_a) = 48.28 \text{ Btu/lb}$$
$$\text{heat of compression } (h_{d'} - h_{c'}) = 10.58 \text{ Btu/lb}$$

In comparing the data above with those of the cycle with the 40°F evaporating temperature (Fig. R15-4), we find that there is a decrease in the NRE of 4% and an increase in the heat of compression of 28%. There will be some increase in superheat, which should be removed either in the discharge pipe or the upper portion of the condenser. This is the result of a lowering in the suction temperature, the condensing temperature remaining the same.

By utilizing Eq. (R15-1), it will be found that the weight of refrigerant to be circulated per ton of cooling, in

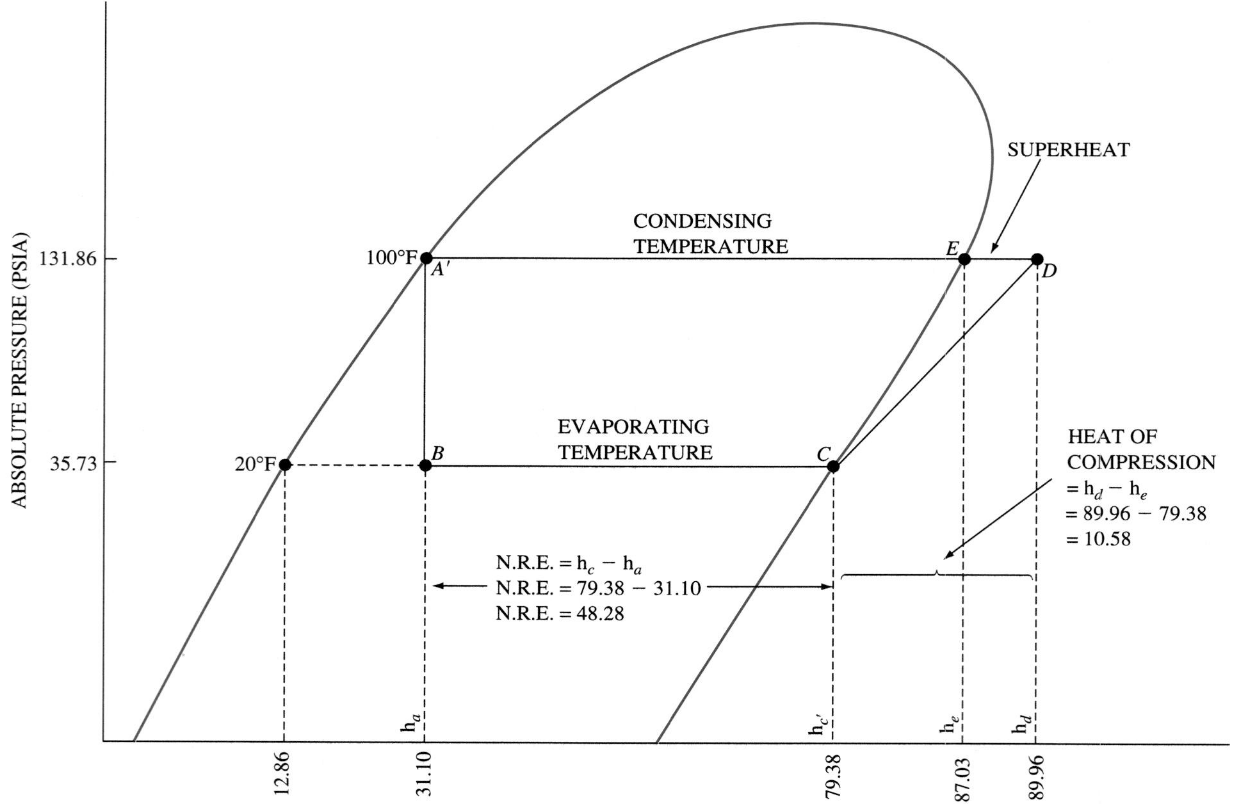

Figure R15-5 Pressure, heat, and temperature values for a refrigeration cycle operating with a 20°F evaporator.

Knowledge Extension

To learn more about the applications of the pressure-enthalpy diagram do the following procedures: Record temperatures and pressures on three different pieces of lab or shop equipment and then plot them out on the appropriate refrigerant pressure-enthalpy diagram. Evaluate possible methods to increase performance and efficiencies. Lower the head pressure and sub-cool the refrigerant and then measure temperatures and pressures again. Re-evaluate if the change helped the overall performance of the system. Record the differences before and after. Your results and findings may surprise you.

a cycle with a 20°F evaporating temperature and a 100°F condensing temperature, is 4.14 lb/min/ton:

$$W = \frac{200 \text{ (Btu/min)}}{\text{NRE (Btu/lb)}}$$

$$= \frac{200 \text{ Btu/min}}{48.28 \text{ Btu/lb}}$$

$$= 4.14 \text{ lb/min}$$

This of course would also necessitate either a larger compressor, or the same size of compressor operating at a higher rpm.

Fig. R15-6 shows the original cycle with a 40°F evaporating temperature, but the condensing temperature has been increased to 120°F.

Again taking the specific data from the heat content or enthalpy line, we now find for the 120°F condensing temperature cycle that $h_a = 36.01$, $h_c = 81.43$, $h_d = 91.33$, and $h_e = 88.61$.

$$\text{net refrigerating effect } (h_c - h_{a'}) = 45.42 \text{ Btu/lb}$$

$$\text{heat of compression } (h_{d'} - h_c) = 9.90 \text{ Btu/lb}$$

$$\text{condenser superheat } (h_{d'} - h_{e'}) = 2.72 \text{ Btu/lb}$$

In comparison with the cycle having the 100°F condensing temperature, it can be calculated that by allowing the temperature of the condensing process to increase 20°F, there is a decrease in the NRE of 9.8%, an increase in heat of compression of 16.2%, and a decrease of superheat to be removed either in the discharge line or in the upper portion of the condenser of 7.1%.

Through the use of Eq. (R15-1) it is found that with a 40°F evaporating temperature and a 120°F condensing temperature the weight of refrigerant to be circulated will

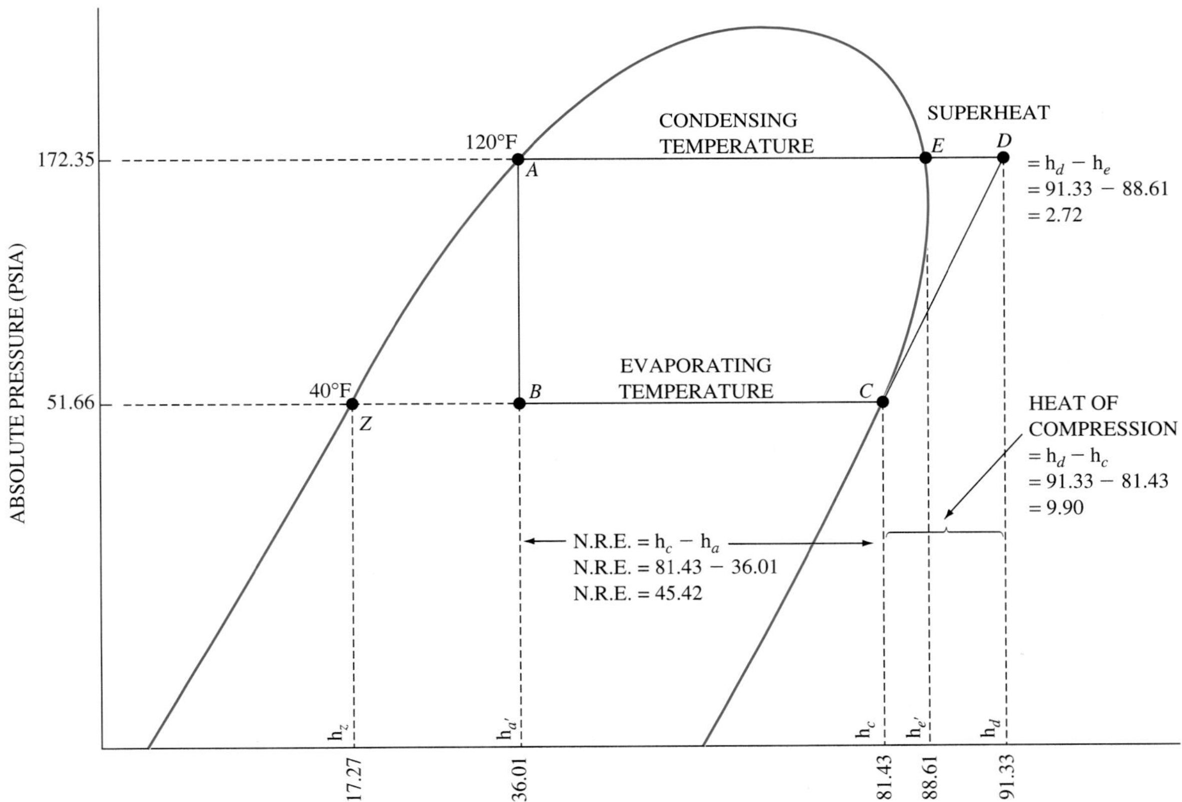

Figure R15-6 The C.O.P. is reduced when the system has a condensing temperature of 120°F.

be 4.4 lb/min/ton. This indicates that approximately 11% more refrigerant must be circulated to do the same amount of work as when the condensing temperature was 100°F.

Both of these examples show that for the best efficiency of a system, the suction temperature should be as high as feasible, and the condensing temperature should be as low as feasible. Of course, there are limitations as to the extremes under which systems may operate satisfactorily, and other means of increasing efficiency must then be considered. Economics of equipment (cost + operating performance) ultimately determine the feasibility range.

Referring to Fig. R15-7, after the condensing process has been completed and all of the refrigerant vapor at 120°F is in the liquid state, if the liquid can be subcooled to point A' on the 100°F line (a difference of 20°F), the NRE ($h_c - h_{a'}$) will be increased 4.91 Btu/lb. This increase in the amount of heat absorbed in the evaporator without an increase in the heat of compression will increase the COP of the cycle, since there is no increase in the energy input to the compressor.

This subcooling may take place while the liquid is temporarily in storage in the condenser or receiver, or some of the liquid's heat may be dissipated to the ambient temperature as it passes through the liquid pipe on its way to the metering device. Subcooling may also take place in a commercial-type water-cooled system through the use of a liquid subcooler, which, in a low-temperature application, may well pay for itself through the resulting increase in capacity and efficiency of the overall refrigeration system.

Another method of subcooling the liquid is by means of a heat exchanger between the liquid and suction lines, whereby heat from the liquid may be transferred to the cooler suction vapor traveling from the evaporator to the compressor. This type is shown in Fig. R15-8, a refrigeration cycle flowchart, using a liquid-suction heat exchanger. True, heat cannot be removed from the liquid and then added to the suction vapor without some detrimental effects to the overall refrigeration cycle; for example, the vapor would become superheated, which would in turn cause an increase in the specific volume of each pound of refrigerant vapor and consequently a decrease in its density. Thus, any advantage of subcooling in a saturated cycle would be negated; but, in an actual cycle, the conditions of a simple saturated cycle do not exist.

In any normally operating cycle, the suction vapor does not arrive at the compressor in a saturated condition. Superheat is added to the vapor after the evaporating process has been completed, in the evaporator and/or in the suction line, as well as in the compressor. If this superheat is added only in the evaporator, it is doing some useful cooling; for it too is removing heat from the load or

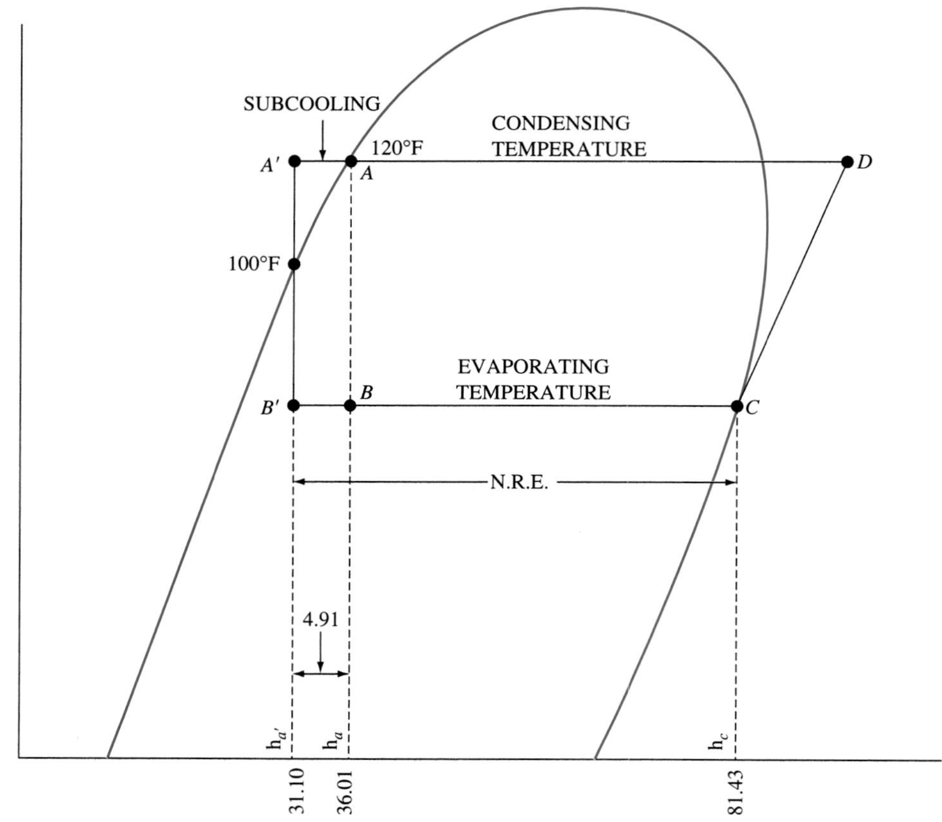

Figure R15-7
Subcooling in the condenser improves the refrigerating effect and the C.O.P.

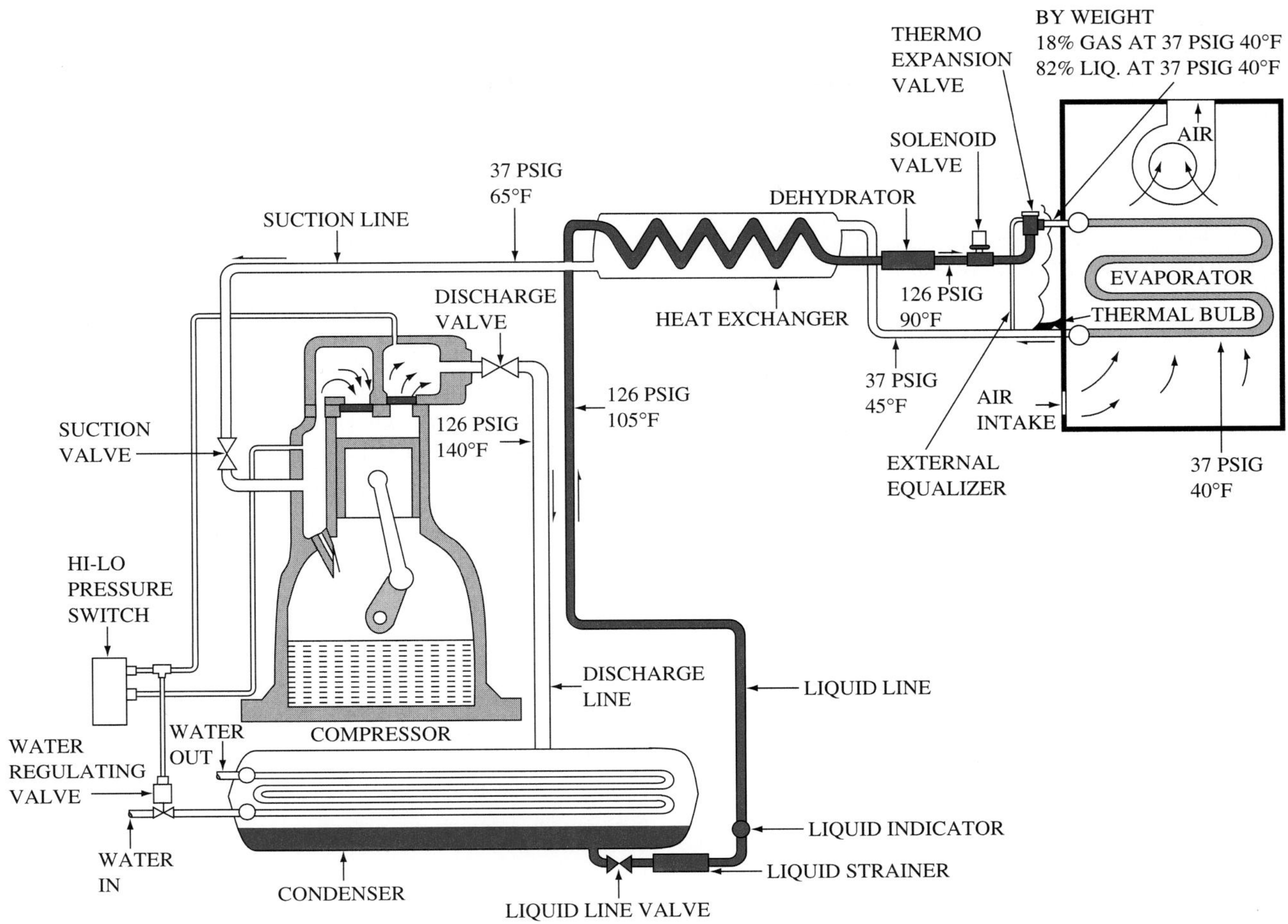

Figure R15-8 Flow diagram of R-12 refrigeration system.

product, in addition to the heat that was removed during the evaporating process. But if the vapor is superheated in the suction line located outside of the conditioned space, no useful cooling is accomplished; yet this is what takes place in the majority—if not all—of refrigeration systems.

Now, were some of this superheating in the suction pipe curtailed through the use of a liquid-suction heat exchanger, this heat added to the vapor would be beneficial, for it would be coming from the process of subcooling the liquid. As an example, suppose that the suction temperature in the evaporator is at 40°F; the superheated vapor coming out of the evaporator may be about 50°F, and the temperature of the vapor reaching the compressor may be 75°F or above, depending on the ambient temperature around the suction. This means that the temperature of the vapor has been increased 25°F, without doing any useful cooling or work, because this heat has been absorbed from the ambient air outside of the space to be cooled.

If some or most of this 25°F increase in the vapor temperature were the result of heat absorbed from the refrigerant liquid, it would be performing useful cooling, since the subcooling of the liquid will result in a refrigerating effect higher than it would be if the refrigerant reached the metering device without any subcooling. It is possible to reach an approximate balance between the amount of heat in Btu/lb removed by subcooling the liquid and the amount of heat added to the refrigerant vapor in the suction pipe without the heat exchanger.

REVIEW

- Four zones of the vapor compression refrigeration cycle:
  Compression
  Condensing
  Expanding
  Evaporation
- Using the R-22 pressure-enthalpy diagram in Appendix A (Fig. A-5), plot the following refrigeration cycle: Locate a point equal to suction pressure (P_1) and suction temperature (T_1). Locate a point equal to discharge pressure (P_2) and discharge temperature (T_2). Draw a line to connect these two points; this represents the compression stroke of the compressor. Locate on the discharge pressure line a point that represents the tem-

perature of the liquid line exiting the condenser (T_3); the distance from the saturated liquid line to T_3 represents the amount of "subcooling." Locate a temperature (T_4) entering the metering device; T_3 and T_4 should be close in temperature. Draw a line from T_3 to the suction pressure line; this represents the metering device. Locate a temperature (T_5) on the suction line 6 to 12 in. from the compressor; the distance from T_5 to the saturated vapor line represents "superheat."

- Refrigeration Effect—The amount of heat absorbed by each pound of refrigerant circulated in the system.
- Enthalpy—Total heat content of a substance calculated from an accepted temperature base (Btu/lb).
- Entropy—Mathematical factor used in engineering calculations. It is the ratio of the heat content of the gas to its absolute temperature in degrees Rankine, relates to internal energy of the gas.
- COP (Coefficient of Performance) = Refrigeration effect/heat of compression
- Purpose of subcooling—To increase refrigerant mass flow into the evaporator by minimizing "flash gas."
- Superheat—The calculated saturated temperature for a given pressure subtracted from the actual temperature measured at that point in the system.

Problems and Questions

1. What is meant by "net refrigerating effect"?
2. What two factors are involved in finding the NRE?
3. What is the equation used to find the weight of refrigerant to be circulated for a given load?
4. Define "superheat."
5. What is the difference between saturated and superheated vapor?
6. Upon what factors does the capacity of a compressor depend?
7. What is a *P–H* chart of a system?
8. What are the divisions shown in a *P–H* chart or diagram?
9. List the refrigerant properties that can be determined from a *P–H* chart or diagram.
10. What is the benefit of subcooling the refrigerant?
11. One ton of refrigeration equals 288,000 Btu/24 hours. True or False?
12. At the outlet of the metering device, a change of pressure occurs with a change in temperature. True or False?
13. A compressor that is too large will cause higher than normal temperatures in the evaporator. True or False?
14. The compression process is called adiabatic. True or False?
15. Higher condensing temperatures force more refrigerant to be circulated, providing a more efficient refrigeration system. True or False?

Piping and Tools

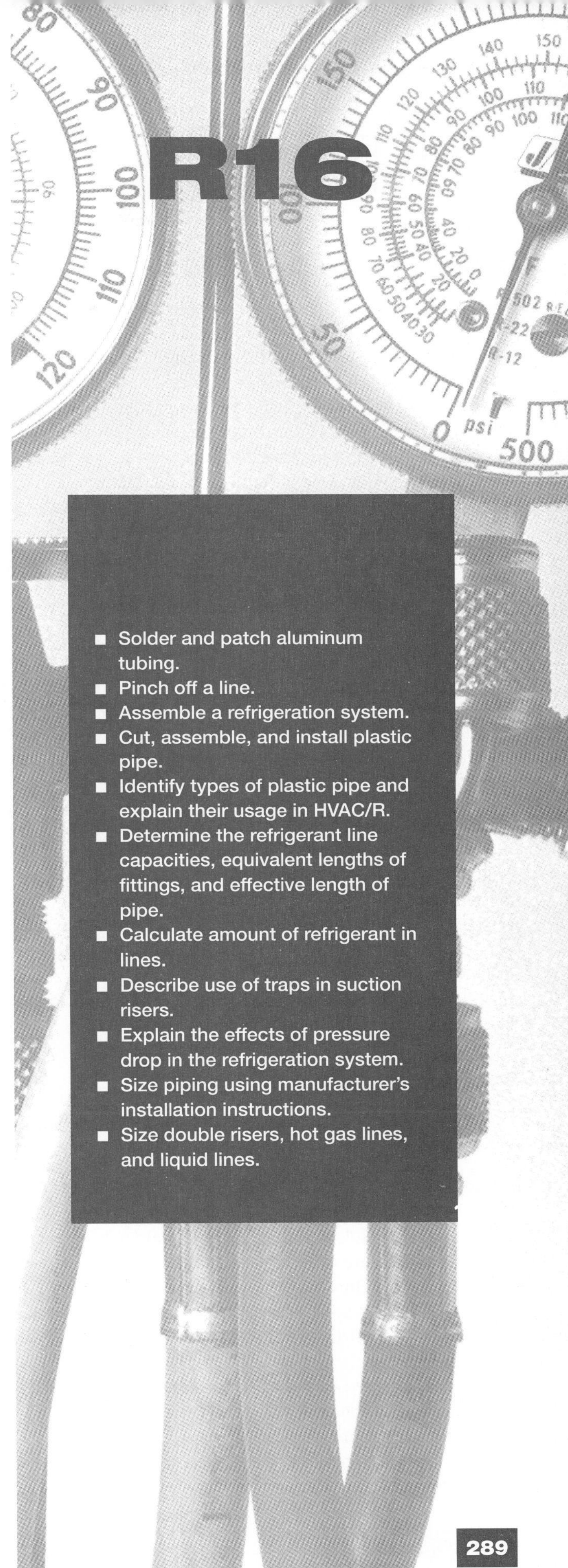

AFTER STUDYING THIS CHAPTER, THE STUDENT WILL BE ABLE TO:

- Identify types of pipe, tubing, and fittings used in refrigeration work.
- Describe methods of insulating pipe and tubing.
- Identify soldering and brazing alloys used in HVAC/R and explain their applications.
- Cut copper tubing with a tubing cutter, hacksaw, and sawing vise.
- Make single- and double-flare connections.
- Swage copper tubing.
- Soft-solder and silver braze copper tubing.
- Bend copper tubing.
- Construct copper tubing connections.
- Set up and use the oxygen-acetylene, acetylene, and mapp gas torches.
- Describe the application and installation of vibration eliminators.
- Solder and patch aluminum tubing.
- Pinch off a line.
- Assemble a refrigeration system.
- Cut, assemble, and install plastic pipe.
- Identify types of plastic pipe and explain their usage in HVAC/R.
- Determine the refrigerant line capacities, equivalent lengths of fittings, and effective length of pipe.
- Calculate amount of refrigerant in lines.
- Describe use of traps in suction risers.
- Explain the effects of pressure drop in the refrigeration system.
- Size piping using manufacturer's installation instructions.
- Size double risers, hot gas lines, and liquid lines.

R16-1 PIPING FOR REFRIGERATION SYSTEMS

R16-1.1 Packaged Units

Many refrigeration units, such as water coolers and ice makers, are manufactured in self-contained packages. They have been piped and charged with refrigerant at the factory. Unless some accident occurs or major service problem arises, the piping remains sealed. Packaged equipment has been a great asset to the reliability of refrigeration units. Piping, charging, and testing can be performed under controlled conditions at the factory. During shipment, the movement of refrigeration components is usually restricted by shipping braces and/or hold-down bolts that need to be loosened before the equipment is put in operation.

One of the important new requirements pertaining to self-contained units is that the manufacturer must provide a service fitting to facilitate refrigerant recovery. Small appliances must have a process tube. These access arrangements are used to remove the refrigerant from the system in case it is necessary to dispose of the unit.

R16-1.2 Field-assembled Units

Many types of refrigeration systems do not lend themselves to factory assembly. For example, a walk-in refrigerator is fabricated or assembled on the job. The refrigeration components are placed where space is available. The refrigeration installer properly locates the evaporator. The condensing unit, if it is air-cooled, must be located in an area where a proper air supply is available. Then it is the installer's responsibility to make the piping connections between the components of the system.

R16-1.3 Basic Principles

Certain basic functions of refrigeration piping must be observed when these lines are constructed:

1. The piping must assure a proper supply of refrigerant to the evaporator.
2. The sizes selected must be practical without excessive pressure drop while maintaining sufficient velocity to return oil with refrigerant vapor.
3. The piping must prevent excessive amounts of lubricating oil from being trapped in any part of the system.
4. The compressor must be protected at all times from loss of lubricating oil. An oil separator in the discharge line may be required.
5. Liquid refrigerant and oil slugs must be prevented from entering the compressor during operating or idle time. A liquid-line filter-drier unit is recommended on field-assembled systems.
6. It is important to maintain a clean, dry system. A suction accumulator may be required.

R16-2 REFRIGERANT PIPING MATERIALS

Most tubing used in refrigeration and air-conditioning piping is made of copper; however, aluminum is used by some manufacturers for fabrication of the evaporator and condenser internal-coil circuits. It has not become popular for field fabrication of the connecting refrigerant lines, principally because it cannot be worked as easily as copper and is more difficult to solder.

Steel piping is used in some larger factory-assembled units as well as in the assembly of the very large refrigeration systems where pipe sizes of 6-in. diameter and above are needed. Threaded-steel-pipe connections are seldom used in modern refrigeration work since they may leak. These systems are welded, and couplings and flanges are bolted to the equipment.

The term tubing generally applies to thin-wall materials, typically copper, which are joined together by means other than threads cut into the tube wall. Piping, on the other hand, is the term applied to thick-pipe-wall material (e.g., iron and steel) into which threads can be cut and which is joined by fittings that screw onto the pipe. Piping can also be welded. Another distinction between tubing and piping is the method of size measurement (see Fig. R16-1).

In the refrigeration trades, tubing sizes are expressed in terms of the outside diameter (OD), while pipe sizes are expressed by the plumbing trade as nominal inside diameters (ID). Thus in Fig. R16-1 the ½-in.-OD (type L) copper tube will have an inside diameter of 0.43 in. The ½-in.-ID nominal steel pipe will have an inside diameter of 0.50 in. and an outside diameter of 0.75 in.

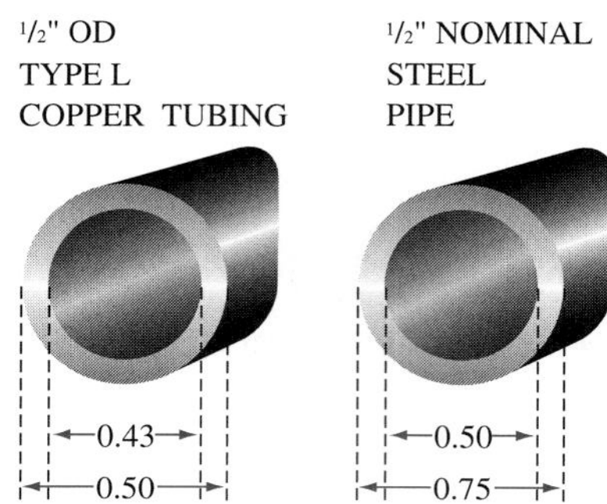

Figure R16-1 Method of sizing tubing and pipe.

R16-2.1 Confusion in Copper Sizes

Copper tubing is used by the refrigeration industry and is specified by outside diameter, whereas copper or iron piping used by plumbers is identified by nominal inside diameter. This has resulted in a certain amount of confusion and is the source of some installation errors. Further confusion results when SI units are used. Sizes in SI units refer to nominal (ID) pipe sizes. There are no SI units for OD tube sizes. Therefore, it is necessary to identify the inside diameter of the tubing, then use tables for the conversion to sizes in SI units. When the needed size tubing in a non-shelf item, the manufacturers will provide the tubing in the size required, in SI measurements.

The following chart is an example of common pipe and tubing sizes, showing differences in their actual measurement.

Refrigeration Size: OD	Plumber's Size: Nominal ID
½ in.	⅜ in.
⅝ in.	½ in.
⅞ in.	¾ in.
1⅛ in.	1 in.

Because of the special treatment of aluminum tubing and welded steel piping, the technique of fabrication will not be covered in this discussion.

R16-2.2 Copper Tubing

The tubing used in all domestic refrigeration systems is specially annealed copper. Copper tubing when formed has a tendency to harden, and this hardening action could cause cracks in the tubing ends when they are flared or formed. The copper may be softened by heating to a red surface color and allowing it to cool. This process is called annealing and is done at the factory.

Copper tubing manufactured for refrigeration and air-conditioning work is designated as ACR tubing, meaning that it is intended for use in air-conditioning and refrigeration work and has been specially manufactured and processed for this purpose. ACR tubing is purged by the manufacturer with nitrogen gas to seal the metal against air, moisture, and dirt, and also to minimize the harmful oxides that are normally formed during brazing. The ends are plugged in the process, and these plugs should be replaced after cutting a length of tubing.

R16-2.3 Copper Tubing Classification

Copper tubing has three classifications: K, L, and M, based on the wall thickness:

K—heavy wall; ACR approved
L—medium wall; ACR approved
M—thin wall; not used in refrigeration systems

Type-M thin-wall tubing is not used on pressurized refrigerant lines, for it does not have the wall thickness to meet the safety codes. It is, however, used on water lines, condensate drains, and other associated system requirements.

Type-K heavy-wall tubing is meant for special use where abnormal conditions of corrosion might be expected.

Type-L is most frequently used for normal refrigeration applications. Fig. R16-2 provides a table of specifications for both types-K and -L tubing. Both K and L copper tubing are available in soft- or hard-drawn types.

R16-2.4 Soft-drawn Copper Tubing

Soft-drawn copper tubing, as the name implies, is annealed to make the tubing more flexible and easier to bend and form. It is commercially available in sizes from ⅛ to 1⅝ in. OD and is usually sold in coils of 25-, 50-, and 100-ft lengths. It is ACR tubing, dehydrated and sealed at the factory. Soft copper tubing may be soldered or used with flared or other mechanical-type fittings. Since it is easily bent or shaped, it must be held by clamps or other hardware to support its weight. The more frequent application is for line sizes from ¼ to ¾ in. OD. Forming becomes rather difficult for sizes larger than ¾ in. OD.

Type	Diameter Outside (in.)	Diameter Inside (in.)	Wall Thickness (in.)	Weight per foot (lb)
K	½	0.402	0.049	0.2691
	⅝	0.527	0.049	0.3437
	¾	0.652	0.049	0.4183
	⅞	0.745	0.065	0.6411
	1⅛	0.995	0.065	0.8390
	1⅜	1.245	0.065	1.037
	1⅝	1.481	0.072	1.362
	2⅛	1.959	0.083	2.064
	2⅝	2.435	0.095	2.927
	3⅛	2.907	0.109	4.003
	3⅝	3.385	0.120	5.122
L	½	0.430	0.035	0.1982
	⅝	0.545	0.040	0.2849
	¾	0.666	0.042	0.3621
	⅞	0.785	0.045	0.4518
	1⅛	1.025	0.050	0.6545
	1⅜	1.265	0.055	0.8840
	1⅝	1.505	0.060	1.143
	2⅛	1.985	0.070	1.752
	2⅝	2.465	0.080	2.479
	3⅛	2.945	0.090	3.326
	3⅝	3.425	0.100	4.292

Figure R16-2 Specifications of common copper tubing sizes.

US Inch (ID)	ISO B36 mm	Diameter, mm Inside	Outside	Volume L/m
⅛	6	6.8	10.3	0.0363
¼	8	9.2	13.7	0.0665
⅜	10	12.5	17.1	0.123
½	15	15.8	21.3	0.196
¾	20	20.9	26.7	0.343
1	25	26.6	33.4	0.556
1¼	32	35.1	42.2	0.968
1½	40	40.9	48.3	1.31
2	50	52.5	60.3	2.16
2½	65	62.7	73.0	3.09
3	80	77.9	88.9	4.77
3½	90	90.1	101.6	6.38
4	100	102.3	114.3	8.22
5	125	128.2	141.3	12.9
6	150	154.1	168.3	18.6
8	200	202.7	219.1	32.3
10	250	253.2	273.1	50.4
12	300	304.8	373.9	73.0

Figure R16-3 Copper tubing in SI units (Copyright by the American Society of Heating, Refrigerating, and Air-Conditioning Engineers, Inc. Used by permission.).

R16-2.5 Hard-drawn Copper Tubing

Hard-drawn copper tubing is also used extensively in commercial refrigeration and air-conditioning systems. Unlike soft-drawn, it is hard and rigid and comes in straight lengths. It is intended for use with formed fittings to make the necessary bends or changes in direction. Because of its rigid construction it is more self-supporting and needs fewer supports. Sizes range from ¼ in. OD to over 6 in. OD. Hard-drawn tubing comes in standard 20-ft lengths that are dehydrated, charged with nitrogen, and plugged at each end to maintain a clean, moisture-free internal condition. The use of hard-drawn tubing is most frequently associated with large line sizes of ⅞ in. OD and above or where neat appearance is desired. Hard-drawn tubing is not suitable for flare connections.

R16-2.6 Tube Sizes for SI Systems

The nominal tube sizes for SI systems are shown in Fig. R16-3. This table indicates the inch equivalent and the metric size in millimeters (mm). Both the inside and outside diameters are given in millimeters. This table also shows the volume of the tube in liters per meter (L/m).

R16-3 USING REFRIGERANT TUBING

R16-3.1 Cutting Copper Tubing

There are two methods of cutting copper tubing. The first uses the hand-held tube cutters shown in Fig. R16-4.

Figure R16-4 Typical tubing cutters (Courtesy of Yellow Jacket Division, Ritchie Engineering Company).

These cutters are suitable for cutting soft- or hard-drawn tubing. Hand-held cutters may be obtained in different models to cut from ⅛ in. OD to as much as 4⅛ in. OD.

The hand-held cutter is positioned on the tubing at the proper cut point. Tightening the knob forces the cutting wheel against the tube. Then by rotating the cutter around the tube and continually tightening the knob, the cut is made. A built-in reamer blade is used to remove burrs from inside the tube after cutting.

A second method of cutting larger size hard-drawn tubing is the use of a hack saw and a sawing fixture to help square the end and make more accurate cuts (see Fig. R16-5). This is less desirable because of the problems that can be caused by unwanted filings.

The saw blade should have at least 32 teeth per inch to ensure a smooth cut. Try to keep saw filings from entering the tubing to be used. Filing the end of the tube provides a smooth surface. Wipe the inside of the tube with a clean cloth to remove filings. For contractors who do considerable refrigeration piping, there are portable power-operated machines that use abrasive wheels to cut the tubing,

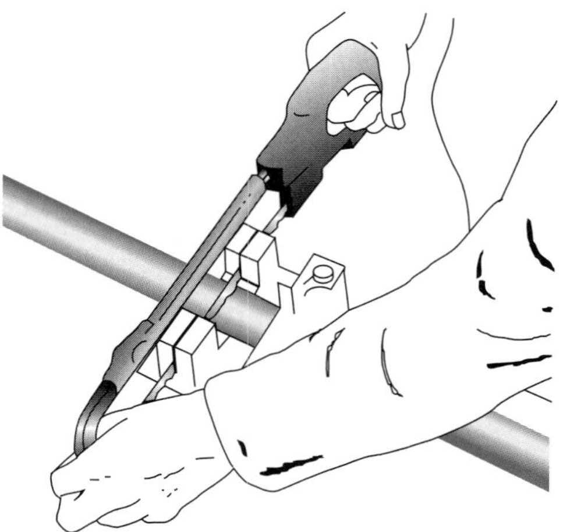

Figure R16-5 Cutting tube with hacksaw.

ream the inside of the tube, and clean the outside tube ends and insides of fittings ready for fluxing and soldering (brazing). Such machines are not required for the average installation.

R16-3.2 Tube Bending

Where smaller sizes of soft-drawn tubing are used, it is generally more convenient and economical to simply bend the tubing to fit the application requirements without using formed fittings. This can be done by hand without special tools, but it takes practice to not make too sharp or too tight bends and consequently flatten the tube.

As a rule of thumb, the minimum bending radius in which a smaller tube may be curved is about five times the tube diameter, as illustrated in Fig. R16-6. Larger tubing may require a radius of up to 10 times the diameter.

To make a hand bend, start with a larger radius and gradually work the tubing into the proper shape while decreasing the radius. Never try to make the first bend as small as the final radius desired.

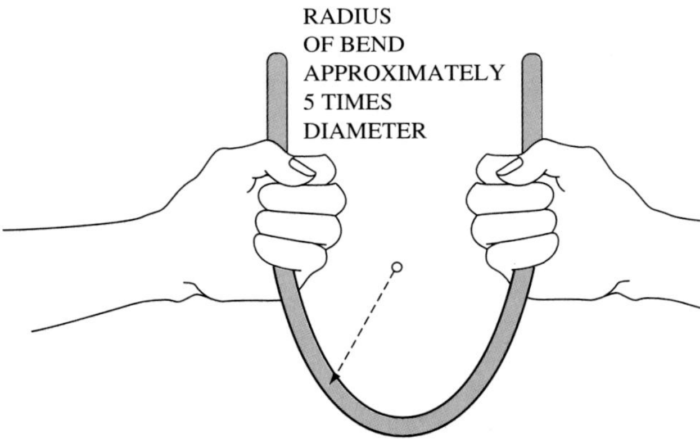

Figure R16-6 Recommended technique in bending tubing by hand.

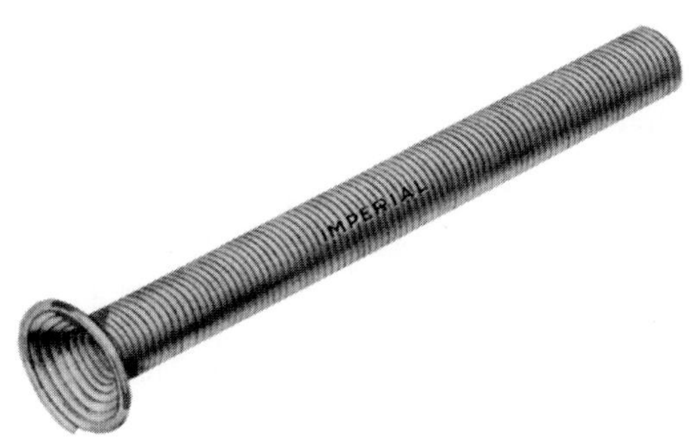

Figure R16-7 Spring-type tube bender (Courtesy of "IMPERIAL EASTMAN." "IMPERIAL" and the "I" within a diamond-shaped outline appearing on the products are trademarks owned by Imperial Eastman Acquisition Corp.).

Figure R16-8 Lever-type tube bender (Courtesy of "IMPERIAL EASTMAN." "IMPERIAL" and the "I" within a diamond-shaped outline appearing on the products are trademarks owned by Imperial Eastman Acquisition Corp.).

Tube-bending springs as illustrated in Fig. R16-7 are available to insert inside or on the outside of the tube so as to prevent the tube from collapsing. These springs are relatively inexpensive and come in sizes to fit most common tubing requirements.

The most accurate and reliable method of tube bending is a lever-type tube-bending tool kit as illustrated in Fig. R16-8. Various sizes of forming wheels and blocks are furnished up to about ⅞ in. OD. Bends can be made at any angle up to 180°.

R16-4 METHOD OF JOINING TUBING

As mentioned earlier, the walls of copper tubing are too thin for threading, so other means must be used to connect the tubing. These may be divided into two broad categories:

1. Mechanical Couplings: Flared and compression fittings that are semi-permanent in that they can be mechanically taken apart.

2. Soldering/Brazing: Soldering and brazing form leak-tight joints.

R16-4.1 Mechanical Couplings

Flared Connections

Since about 1890 the flared connection has been one of the most widely used techniques to join soft-drawn copper tubing. A properly made flare is most important if leak-proof joints are to be achieved—and this requires the right tools and practice.

Although convenient, flare fittings have historically been a source of refrigerant leaks. The causes are varied: loose flare nuts, poorly made flares, failure at the flare due to excessive vibration, etc. With the emphasis on refrigerant containment, the installer and service technician will have to pay close attention to these potential trouble points.

Flares are made with special tools, which expand the end of the copper tubing into a cone shape as shown in Fig. R16-9.

This cone is formed at a 45° angle that mates against the face of a flare fitting. The flare nut when tightened will press the soft copper against the machined fitting seat, thus forming a tight seal. The flare illustrated in Fig. R16-9 is called a single-thickness flare. Others are called double-thickness flares. Both forming techniques will be shown.

Single-Thickness Flares

Fig. R16-10 represents a typical flaring tool consisting of a flaring block or base and a slip-on yoke that holds the screw-driven flaring cone.

Note that the base-block dies slide to allow the tubing to be inserted into the particular size hole. The end wing nut is then tightened to hold the material in place during flaring.

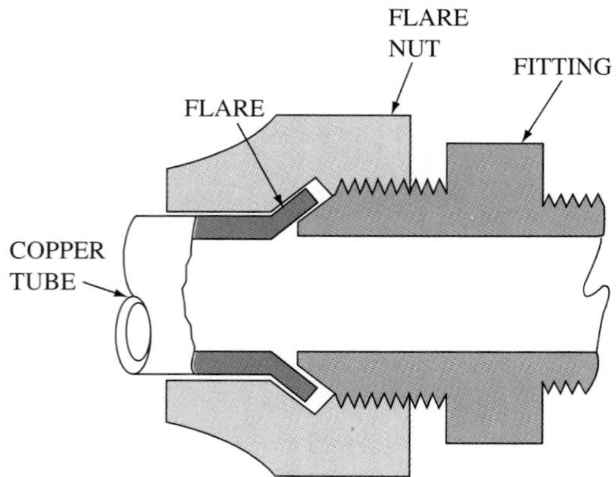

Figure R16-9 Cross section of 45° flared fitting.

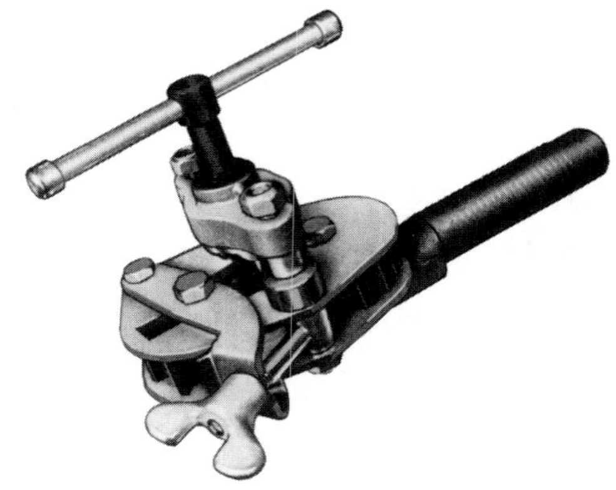

Figure R16-10 Typical flaring tool (Courtesy of Yellow Jacket Division, Ritchie Engineering Company).

Fig. R16-11 represents a typical flaring operation. The chamber in the block is at a 45° angle. The tubing should extend slightly above the flare block— aproximately one-third the height of the flare. This is necessary to provide enough material to fill the opening after the flaring cone is driven into the tubing and also to ensure the proper surface area against the fitting face. If it is too small, the flare nut may not hold the tubing; if too large, the flare nut may not fit over it. This operation takes practice.

Before making the flare it is important to prepare the tubing. Use a good grade of tubing (not all types of tubing are recommended for flaring). Follow proper cutting and deburring practices. Cut the tube squarely (use file if necessary) and remove internal and external burrs, using a deburring tool on the cutter.

To form the flare, first put the flare nut on the tubing. Insert the tubing into the proper size hole (adjusting the height above the block) and clamp with wing nuts. Put a drop of oil on the forming cone. Tighten the forming cone into the tubing using an initial one-half turn. Then back off one-quarter turn. Retighten three-quarters turn and then back off one-quarter turn. Continue this back and forth procedure until a flare is formed; with practice this motion becomes routine. Continuous turning is not recommended, since it does not give the operator a feel of the progress being made and may tend to harden the metal.

Remove the flare from the block and carefully examine the completed flare to see that the sides have no splits or other imperfections.

Maintain tools in a clean and well-lubricated condition at all times for ease of operation and extended tool life. Never over-torque the feed mechanism when making a flare—flare washout will result.

Note: 45° flares are the standard of the refrigeration and air-conditioning industry. In other industries, such as automotive, steel or brass tubing is used. These metals do not form as easily as copper, so 37½° angle flares are used. Therefore, flare fittings and tools are not interchangeable.

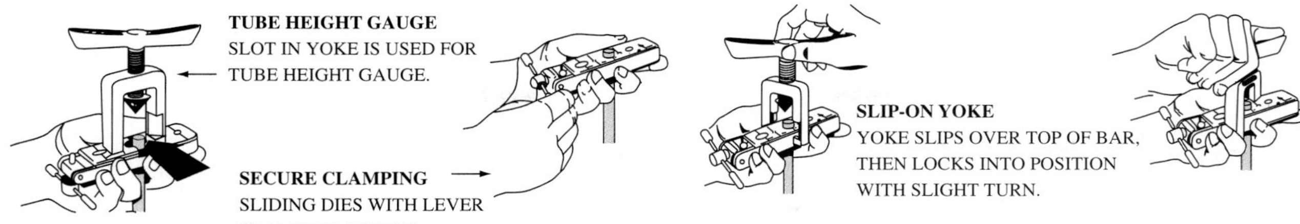

Figure R16-11 Typical flaring operation.

Double Thickness Flares

A double flare consists of a twin-wall sealing surface. Double-flare fabrication is accomplished through use of a conventional flaring tool provided with double-flare adapters.

As illustrated in Fig. R16-12, the tubing is clamped in the flaring bar and allowed to protrude a measured distance above the top of the bar. The adapter is positioned in the tube, the yoke is engaged with the bar, and the feed screw is advanced until a positive resistance is encountered. This completes the preform operation, which folds the tube inward. The adapter is removed and the forming cone is advanced until moderate resistance is encountered. Then, final forming is done as described for a single flare. This operation completes the two steps required in forming a double flare.

Why double flares? They are used mainly on larger-size tubing where a single flare may weaken upon excessive expansion. Double flares offer greater resistance to fracture on installations that are subjected to much vibration. Also, they can be assembled and disassembled more frequently without flare washout.

Flared Fittings

To accommodate the many types of flared fittings needed in refrigeration and air-conditioning systems, a variety of elbows, tees, unions, etc. are available to select from, as illustrated in Fig. R16-13. The fittings are usually drop-forged brass and are accurately machined to form the 45° flare face.

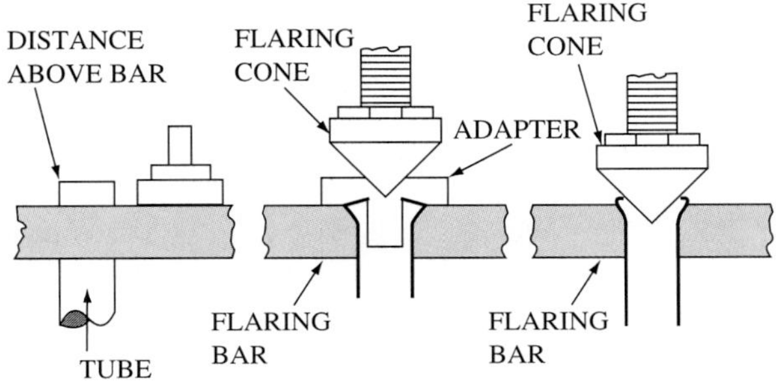

Figure R16-12 Method of making double flare.

Threads for attaching the flare nuts are SAE (Society of Automotive Engineers) National Fine Thread. Some fittings such as the flared half-union are made to join to national pipe threads. All fittings are based on the size of tubing to be used. The flare nuts are hexagon-shaped for

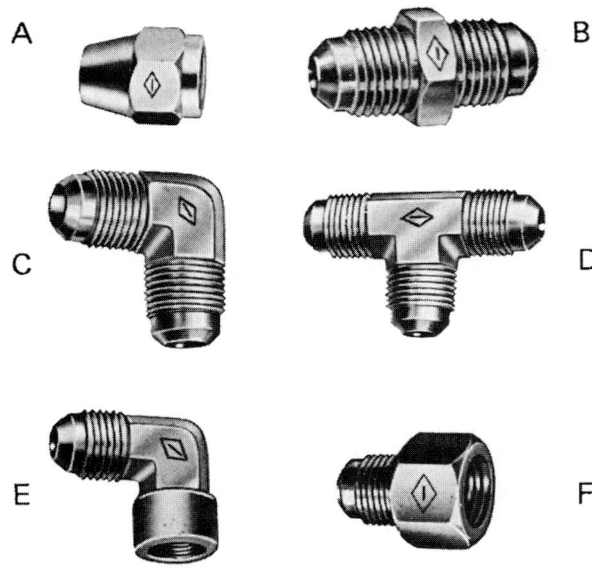

A = STANDARD FLARE NUT

B = FLARE COUPLING

C = FLARE 90° ELBOW

D = FLARE TEE

E = FLARE TO FEMALE PIPE ELBOW

F = FLARE TO FEMALE PIPE ADAPTER

Figure R16-13 Common flared fittings (Courtesy of "IMPERIAL EASTMAN." "IMPERIAL" and the "I" within a diamond-shaped out line appearing on the products are trademarks owned by Imperial Eastman Acquisition Corp.).

easy wrench tightening. The fitting body also usually has a flat surface to take an open-end wrench attachment.

Compression Fittings

In recent years there has been a trend toward using compression-type fittings for joining refrigerant tubing. This method has gained popularity in the residential air-conditioning field because it reduces the field labor requirement in making flared connections or soldered piping. Fig. R16-14 represents a typical mechanical coupling concept as manufactured by the Aeroquip Corporation.

A gas-tight seal is obtained by simply connecting the tubing to the coupler with a coupling nut, crimp collar, and O-ring to assure a leakproof assembly. Once assembled, the joint may be disconnected and reassembled without lessening its sealing effectiveness. This type of connection is generally found in automotive applications.

As long as the tubing is straight and round in order to be inserted into the tube entrance, a proper connection can be made with minimum mechanical skills. This particular coupling comes in sizes from ¼ in. to 1⅛ in. OD. Adapters are available to join these couplings to other fittings such as soldered pipe or threaded fittings.

Diaphragm Fittings

The development of flexible, quick-connect, pre-charged refrigerant lines has played a major role in the growth and reliability of home air conditioning. The liquid and suction lines are made of bendable copper or convoluted steel tubing. Suction lines are factory covered with a foam rubber insulation material. Each end of the tubing is fitted with a coupling half that mates with a coupling half on the equipment (Fig. R16-15).

Both coupling halves have diaphragms that provide a seal, which prevents refrigerant loss before connection. The male half (on the equipment) contains a cutter blade, the metal refrigerant sealing diaphragm, and an intermediate synthetic rubber seal to prevent loss of refrigerant while the coupling is being connected. The female half (on the tubing) contains a metal diaphragm, which is a leak-proof metal closure.

Tightening the union nut draws the coupling halves together (Fig. R16-16), piercing and folding both metal diaphragms back and opening the fluid passage.

When fully coupled (Fig. R16-17) a metal seal forms a permanent leak-proof joint between the two coupling halves. Note the service port for checking refrigerant pressure. The port is equipped with a valve (not shown), similar to a tire valve, that opens when depressed by gauge lines.

When installing quick-connect pre-charged refrigerant lines, follow the manufacturer's installation directions on the radius of the bend for the size of tubing, lubrication of the coupling, and proper torque values when couplings are joined together.

If excess tubing is present, form a loop or coil in a flat, horizontal manner. Do not form vertical loops, which will create an oil trap.

Pre-charged lines are manufactured in various sizes and lengths from 10 to 50 ft. It is necessary to plan the installation in order to have sufficient but not excessive tubing, which adds cost and pressure drop to the system.

The coupling described above is called a "one-shot connection," as it is designed to be coupled and remain coupled. It can be undone, however, to revise piping, change equipment location, etc. Uncoupling and recoupling in demonstration use revealed the ability to recouple this type of connection several times when proper tools are used.

Figure R16-14 Precharged refrigeration lines, evaporator, and condensing unit (Courtesy of Aeroquip Corporation).

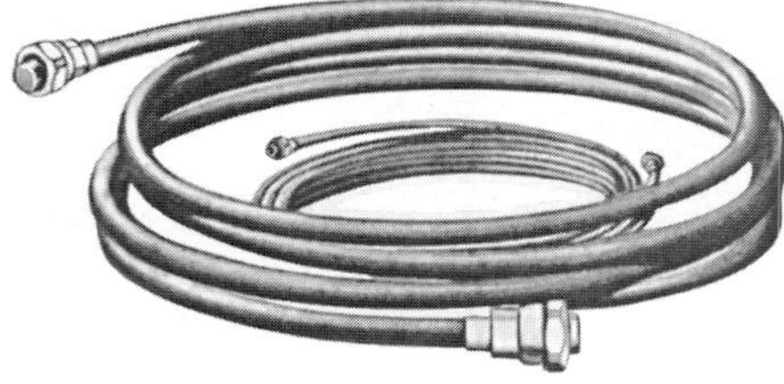

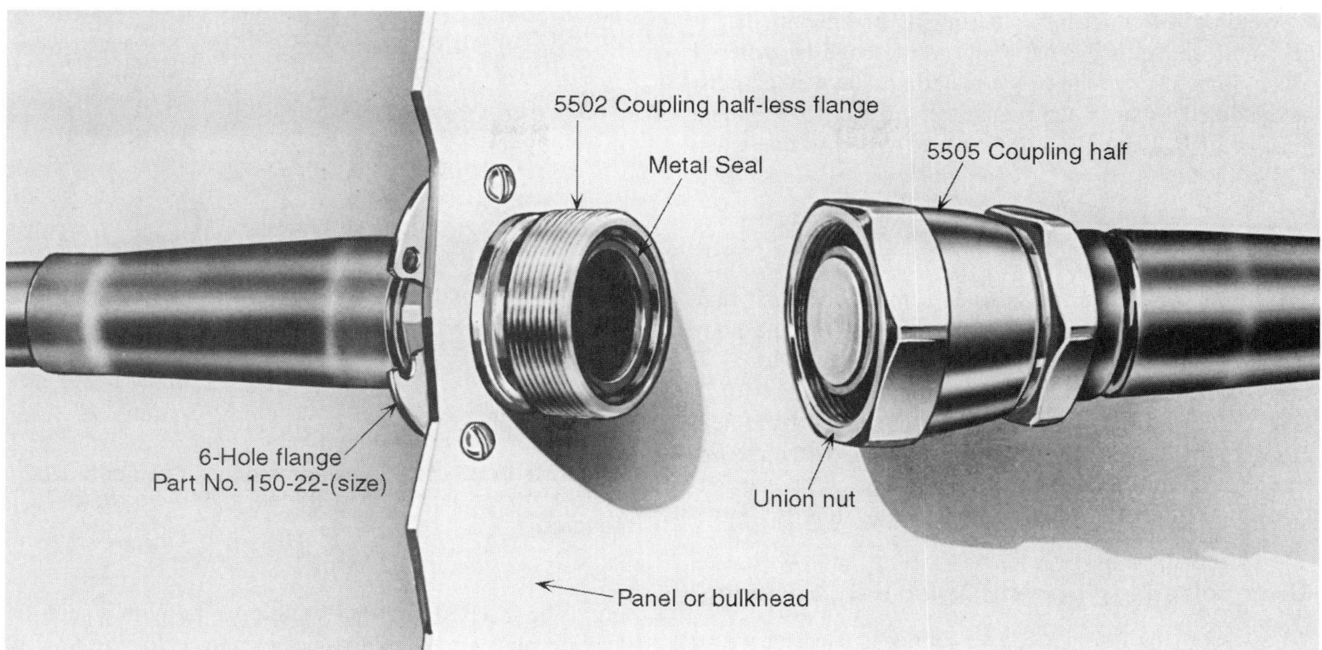

Figure R16-15 Line to unit coupling (Courtesy of Aeroquip Corporation).

The only factor that must be considered is the loss of refrigerant from the system. This coupling, as well as compression-type fittings, are subject to a high leak-failure rate when the wrong tools are used. Only open-end type wrenches that fit snugly on the wrench pads of the nuts should be used. Adjustable-type open-end wrenches are permissible if the wrench is in good condition with little worm wear or jaw spring. Do not use pliers, electrician's pliers, pipe wrenchs, etc., on these fittings. Any tool that produces a squeezing pressure causes the nut to collapse into an egg shape. When this happens, the fitting is destroyed, cannot be made to hold, and must be replaced.

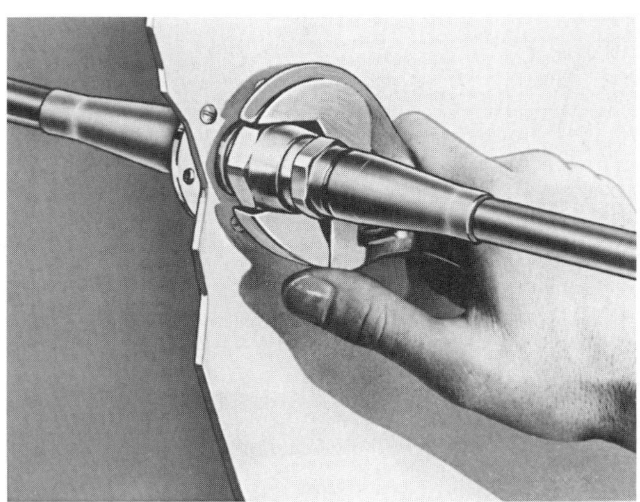

Figure R16-16 Tightening fitting (Courtesy of Aeroquip Corporation).

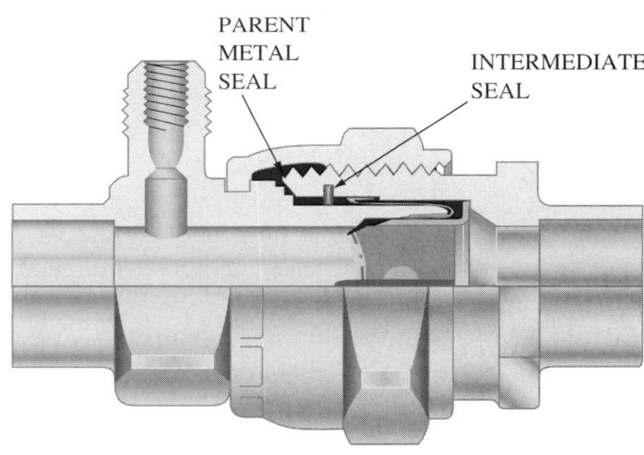

Figure R16-17 Completed diaphragm fitting.

R16-4.2 Soldering, Brazing and Welding

Soldering or brazing consists of joining two pieces of metal together with a third metal (or solder), which melts at a lower temperature than the pieces to be joined. When melted, the solder or brazing alloy flows between the two pieces of metal. The molten metal adheres to the surfaces of the two metals and forms a good bond between them. The solder usually has less strength than the metals it joins, so for greatest strength of the soldered joint, the layer of solder must be very thin.

The essential difference between soldering and brazing is the temperature at which the molten solder or brazing alloy flows. When the melting point of the alloy is above 800°F (425°C) the process is called brazing.

Welding differs from soldering or brazing in that it does not use another metal to act as the bonding material of the joint. The two pieces of material to be welded must be "puddled," that is, their edges or surfaces to be joined must be melted and allowed to mix together, so that when they cool they are a part of one another.

Soft Soldering

A form of soft solder, one that is somewhat harder than tin-lead solders, is composed of 95% tin and 5% antimony and is commonly referred to as ninety-five-five (95-5). It starts to melt at 450°F (232°C) and is fully liquid at 465°F (241°C). It is easily worked with a small hand-held propane or acetylene torch. Ninety-five-five is the recommended soft solder to use on refrigeration work, particularly on small-diameter soft-drawn copper tubing.

Silver Soldering (Formerly Called Hard Soldering)

When larger hard-drawn copper tubing is used, or where local building codes require the use of silver solder, the silver soldered joint makes a much stronger bond. Over the past twenty years or so, hard solders (silver solders or silver brazing alloys, as they are now called) have been widely accepted by the refrigeration and air-conditioning industry for joining metal. The selection of these alloys over others was prompted by the need for high-strength, corrosion-resistant, vibration-proof, and leak-tight joints. Silver brazing satisfied all these requirements, and also offered the additional advantage of joining similar or dissimilar metals with greater ease of application and at lower brazing temperatures.

Silver solder in general flows at a temperature of approximately 1100 to 1200°F (593 to 649°C). Copper melts at 1981°F (1083°C), so this alloy flows at some 800°F (427°C) below the melting point of copper, making it a very safe alloy to use on copper tubing and fittings. When properly made, the silver solder joint will have a tensile strength above the material it joins.

Some of the lower-temperature silver solders may be applied with air-acetylene or air-propane torches, but the oxygen-acetylene torch illustrated later is most commonly used. Techniques of proper soldering and brazing, and pointers on brazing equipment are also discussed later.

Solder Fittings

For use with both soft- and hard-soldered systems, a number of sweat fittings are available, as illustrated in Fig. R16-18. Some sweat fittings are for connecting tubing to tubing, and others may have a connection on one end for National pipe threads.

These fittings are brass rod, brass forging, or wrought copper and are accurately manufactured to permit the tubing to be inserted into the fitting opening with a snug fit, leaving only a very thin solder clearance for the flow of solder. If solder clearance is too large due to the fitting or the tube end being out of round, the joint will be weakened. Cast sweat fittings are not recommended for refrigeration service.

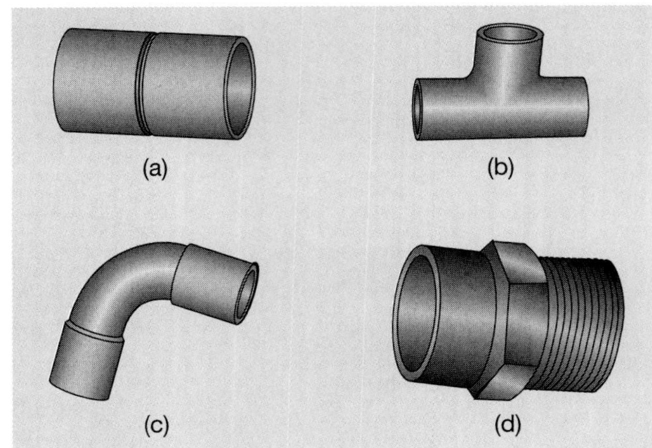

Figure R16-18 Some common soldered or brazed copper fittings: (a) coupling; (b) tee; (c) 90° elbow; (d) adapter.

Swaging Copper Tubing

Sometimes in the assembly of copper tubing with the same diameter, some fabricators feel that it is more reliable to join the two pieces of tubing by making a swaged connection with only one solder joint, as illustrated in Fig. R16-19a. This is done with a swaging tool (Fig. R16-19b), which is somewhat similar to the flaring block.

The block holds the tube, and the correct size punch is forced into the end of the tubing until the bell shape is

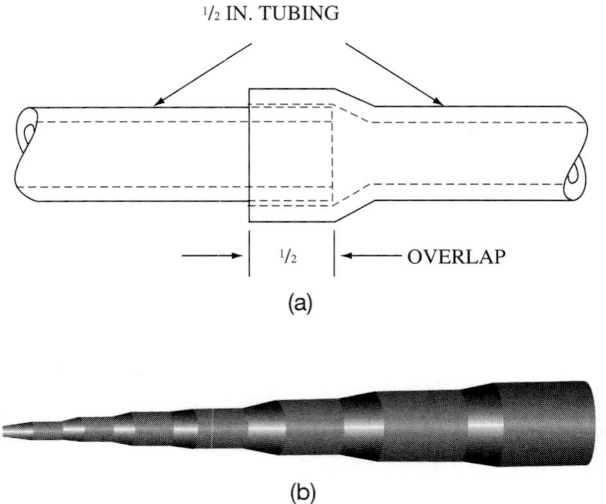

Figure R16-19 (a) Swaged connection. (b) Swaging tool.

PREVENTIVE MAINTENANCE

Whenever soldering or brazing on any pipe, apply a slight pressure (1–5 psig) of nitrogen to the system during the welding process. This will eliminate oxidation and the potential of residual debris collecting at filters, metering devices, etc. Be sure to leave a port open at the access valve to allow for the nitrogen pressure to escape; otherwise you will never fully seal off your pipefitting.

It is also good preventive maintenance to deburr your piping, clean it, and sand it prior to beginning welding.

produced. Some tools use the screw mechanism for force. Others require a blow by a hammer to force the punch into the tube.

Swaging, when properly done and soldered, reduces the number of soldered joints and thereby reduces leak hazards; however, it does take more time than using preformed fittings, and thus the option becomes one of individual choice.

Procedures for Soldering and Brazing

In the refrigeration industry, certain fundamental procedures of silver-alloy brazing have been arrived at through long experience. In many respects, these fundamentals are similar to those recommended for soft soldering, and a technician proficient at soft-soldering generally will make a good silver-brazing operator.

There are six simple steps to follow in producing strong, leak-tight joints that are basic to both soft soldering and silver brazing:

1. Good fit and proper clearance
2. Clean metal
3. Proper fluxing
4. Assembling and supporting
5. Heating and flowing the alloy
6. Final cleaning

Four steps are necessary in preparation for soldering. Each step has several substeps, as discussed below:

1. Cutting and fitting the pipe or tubing (Fig. R16-20).
    a. Cut the tube to the proper length. Make sure that the ends are cut square; a tube cutter is by far the best tool. If a hacksaw is used, tilt the tube downward so that the cuttings fall out.
    b. Remove burrs with a reamer or half-round file, as shown in Fig. R16-21.
    c. Try the end of the tube in the fitting to be sure that it has the proper close fit. Clearance should be uniform all around. Where necessary, on soft-drawn copper, a sizing tool may be used to round out the tubing.
2. Cleaning and fluxing (Figs. R16-22 and R16-23).
    a. Manufacturers may provide recommendations on cleaning fittings and tubing to remove oxide

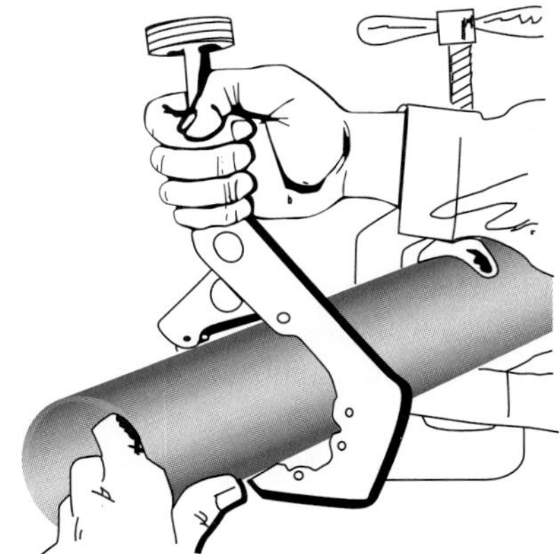

Figure R16-20 Cutting tube.

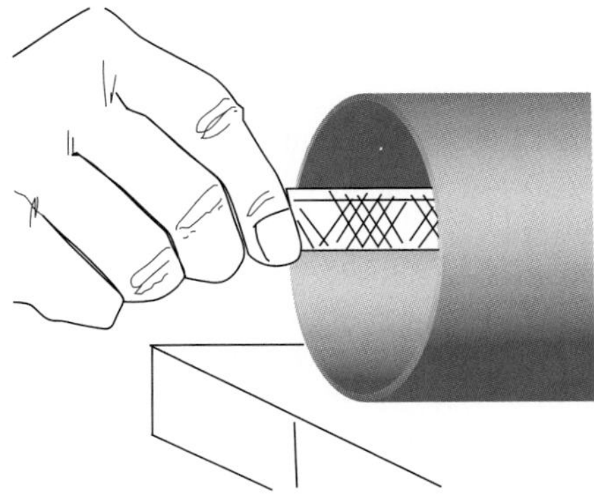

Figure R16-21 Deburring.

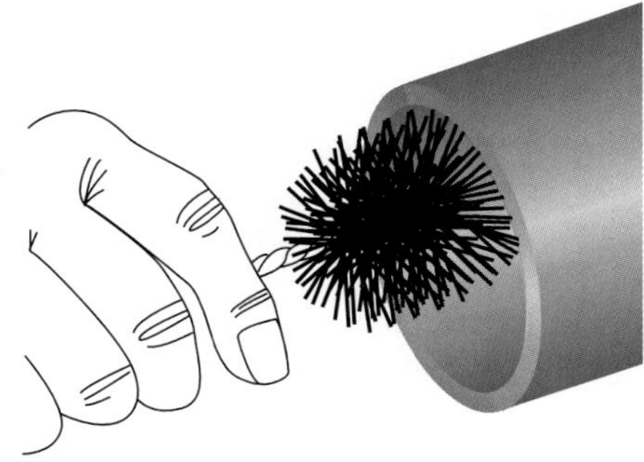

Figure R16-22 Wire brushing.

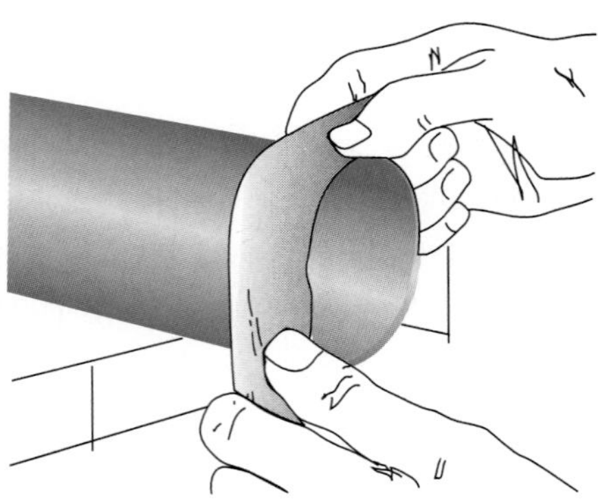

Figure R16-23 Sanding.

film and dirt, using abrasive cloth. Use only the flux(es) recommended as being compatible with the solder or brazing material used.

b. Clean socket and end of fitting and end of tubing with a clean wire brush and fine sand cloth made for cleaning tubing. Do not use steel wool (shreds may adhere to the fitting and cause a void). Do not use emery cloth, since it often contains oils and undesirable abrasives that can cut too deeply. Surfaces to be joined must be free of oil, grease, rust, and oxides.
c. Do not handle the surfaces after cleaning.
d. Cleaning should be done just before soldering so that oxidation is reduced to a minimum.

3. Proper fluxing. Flux does not clean the metal. It keeps the metal clean once it has been mechanically cleaned as mentioned above. Flux is available in both paste and liquid form, but in general the paste form is preferred by most refrigeration installers.
   a. The first rule is to select the proper flux, depending on whether the job is soft soldering or silver brazing. For brazing, use a good quality low-temperature silver brazing flux. This is very important and requires special attention in refrigeration work; consult a local welding supply jobber.
   b. Always stir the flux before using (see Fig. R16-24), since when flux stands the chemicals tend to settle to the bottom, especially in hot weather. Use a brush—never apply soldering paste with your fingers; perspiration and oils may prevent solder from sticking.
   c. In most ordinary work, both the end of the tube and the inside of the fitting are fluxed (Fig. R16-25), but in refrigeration work the end of the tube is inserted part way into the fitting and the paste flux is brushed all around the outside of the joint.
   d. The tube is then inserted to full depth in the socket (Fig. R16-26).

Figure R16-24 Stirring flux.

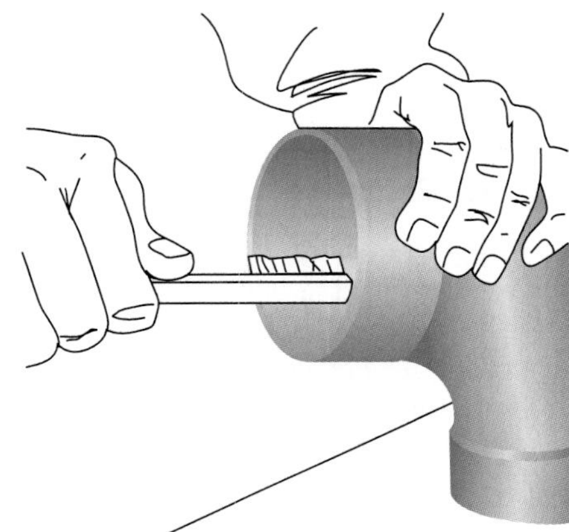

Figure R16-25 Applying flux.

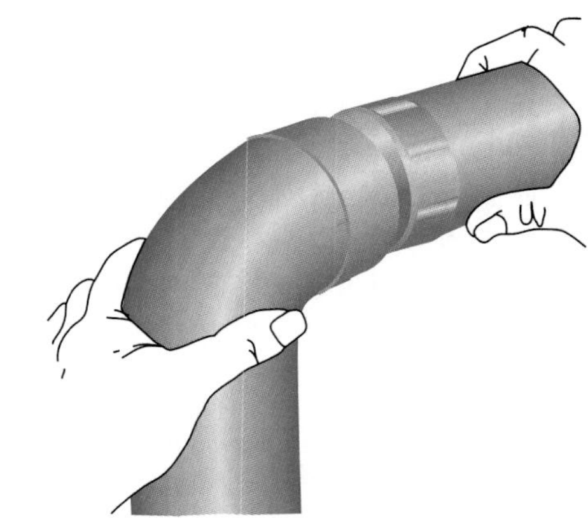

Figure R16-26 Joining.

   e. Where possible, revolve the fitting or tubing to spread the flux uniformly.
   f. Caution: Too much flux can be harmful to the internal components and operation of the refrigeration system.

g. Note: With certain silver brazing alloys, copper-to-copper brazing can be done without fluxing. The technician will develop a preference as practice and skill grow. On copper tube-to-brass fittings, flux is always required.

4. Supporting the assembly.
   a. Before soldering or brazing, the assembly should be carefully aligned and adequately supported.
   b. Arrange supports so that expansion and contraction will not be restricted.
   c. See that no strain is placed on the joint during brazing and cooling.
   d. A plumber's pipe strap makes an excellent temporary support until permanent arrangements are made.

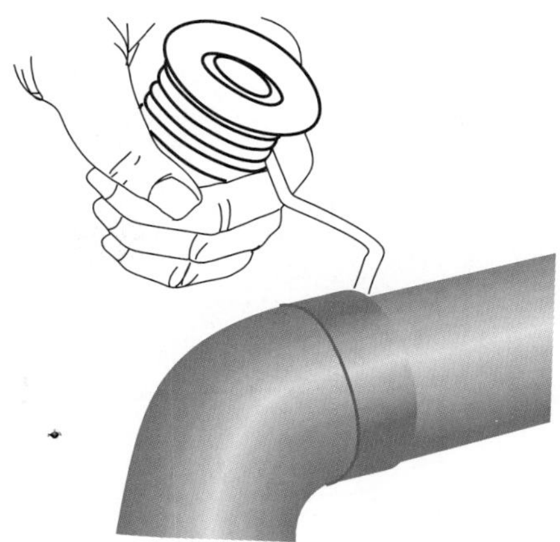

Figure R16-28 Applying solder.

Trouble Spots in Brazing

Sometimes an alloy fails to flow properly and has a tendency to form a ball. This is due to oxidation of the metal surfaces, or insufficient heating of the parts being joined. If the parts tend to oxidize when heat is applied, more flux should be added, because the coating is too thin.

If the alloy flows over the side of the part and does not enter the joint, it indicates that too much heat has been applied to one member and the other member is under heated. If this occurs, it is necessary to stop the operation, disassemble the joint and flux the parts.

Heating and Flowing Soft Solder

To soft-solder, apply the flame to the shoulder of the fitting so that the heat will flow in the direction of the tube (Fig. R16-27).

Be careful not to allow the flame to enter directly into the opening where the solder will be drawn. With an ell or tee, heat the heaviest part of the fitting first and then move toward the opening where the solder will enter. In this way, heat is distributed uniformly to both the fitting and tube. Occasionally, remove the flame momentarily and touch the joint with solder to see whether the metal is hot enough to melt solder.

When applying the soft solder (Fig. R16-28), never push the solder into the joint. When the joint is hot enough, merely touch the solder, and capillary attraction will draw it into the clearance space between the surfaces to be joined.

When a ring of solder appears all around the circumference, you have made a joint that is leak-proof and tremendously strong. Wipe the joint clean with a clean cloth while the solder is still molten (Fig. R16-29). This is called "smoothing" the solder.

When working with large fittings (2 in. or over) it is helpful to use two torches or, better yet, a Y-shaped torch tip (Fig. R16-30), which gives a double or quadruple flame. This ensures a sufficient and even distribution of heat.

Also, when working with larger fittings 2 in. and over, each fitting should be tapped (with a small mallet) at two or three points around its circumference while the solder is being fed (Fig. R16-31). This settles the joint and also

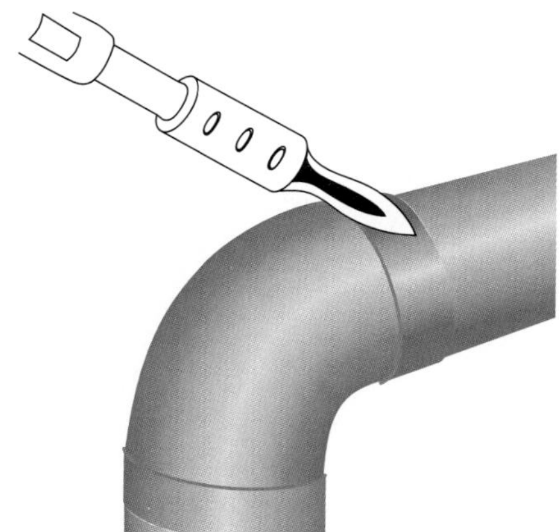

Figure R16-27 Applying flame.

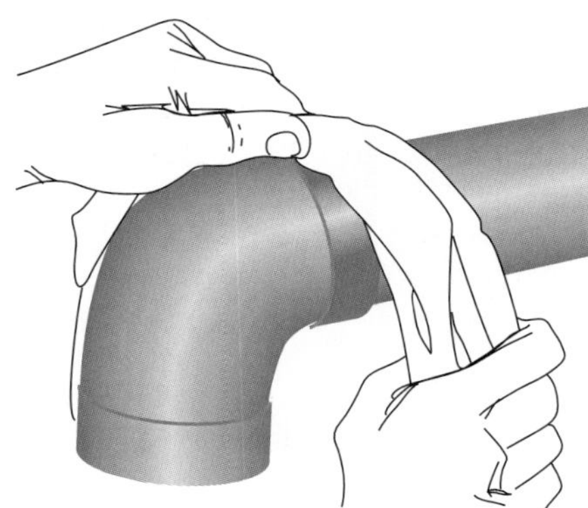

Figure R16-29 Smoothing solder.

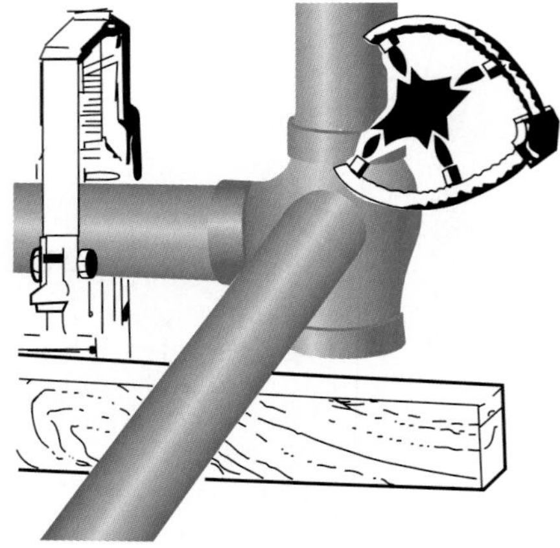

Figure R16-30 Large-size work.

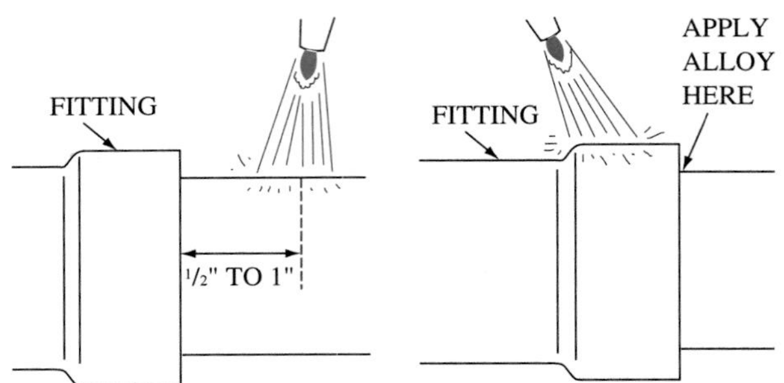

Figure R16-32 Brazing horizontal joints.

releases any trapped gases that might hinder the flow of the solder.

Heating and Flowing Silver Brazing Alloy

For rapid, efficient silver brazing, a soft bulbous oxy-acetylene flame provides the best type of heat. Air-acetylene or air-propane torches have been used successfully for silver brazing fittings and tubing up to about 1 in. in size. Adjust the oxyacetylene for a slightly reducing (less oxygen) flame. Start heating the tube about ½ in. to 1 in. away from the end of the fitting (Fig. R16-32). Heat evenly all around to get uniform expansion of the tube and to carry the heat uniformly to the end inside the fitting.

When the flux on the tube adjacent to the joint has melted to a clear liquid, transfer heat to the fitting.

Sweep the flame steadily back and forth from fitting to tube, keeping it pointed toward the tube. Avoid letting the flame impinge on the face of the fitting, as this can easily cause overheating.

When the flux is a clear liquid on both fittings and tube, pull the flame back a little and apply alloy firmly against the tube and the fitting. With proper heating the alloy will flow freely into the joints.

The technique in making vertical joints (Fig. R16-33) is essentially the same: start with the preliminary heating of the tube, and then move to the fitting. (Note: This is slightly different from the soft-solder technique.) When the tube and fitting reach a black heat the flux will become viscous and milky in appearance. Continued heating will bring the material up to brazing temperature, and the flux will become clear. At that point apply the silver brazing alloy and sweat it.

Brazing larger diameter pipe requires using the above technique but selecting only a 2-in. (50 mm) segment at a time and overlapping the braze from segment to segment. This operation requires a degree of practice once the basic points are learned.

Cleaning After Brazing

Do not quench. For cast fittings especially, allow to air-cool until the brazing alloy has set. Then apply a wet brush or swab to the joint to crack and wash off the flux. All flux must be removed before inspection and pressure testing. Use a wire brush if necessary.

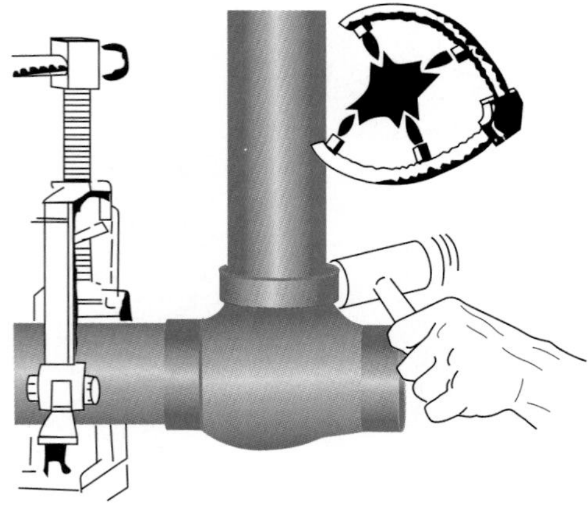

Figure R16-31 Tapping.

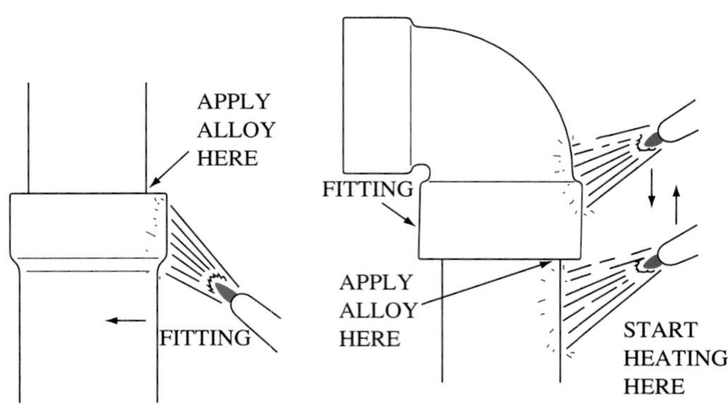

Figure R16-33 Brazing vertical joints.

ON THE JOB

Here are some oxyacetylene welding safety tips:

- Store oxygen cylinders away from combustible materials, especially oil and grease.
- Inspect hoses before each use.
- Use the proper hose connection made especially for oxygen and acetylene.
- Store cylinders in an upright position and strap securely in place.
- Wear flame retardant clothing when welding or using a cutting torch.
- If welding in a passage way, put up signs or barricade the area to keep people away.
- Leak testing with nitrogen in a confined space or area can displace enough oxygen to cause suffocation. This has been happening during testing. Symptoms are dizziness, and shortness of breath. Never enter a confined area without proper ventilation, or proper personal protection breathing apparatus.

Safety Hints for Soldering and Brazing

Never attempt to solder or braze while the system is under pressure or within a vacuum. Many fluxes contain chloride or fluoride and should be handled carefully to prevent excessive skin contact or inhalation of fumes.

Many silver solders contain cadmium in varying degrees. Cadmium fumes are very poisonous, especially with over heating, and can be fatal. Levels permitted in the workspace atmosphere are 100 times lower than welding fume levels. Fluxes without cadmium are available and should be selected.

The work space must be well ventilated. In an unventilated area, wear an approved respirator. Use safety goggles designed for brazing. Wear suitable clothing, gloves, and footwear for brazing.

Brazing Equipment

As mentioned earlier, it is possible to achieve melting temperatures for some low-temperature silver solders by using air-acetylene or air-propane torches, but where extensive installation or repairing of refrigeration equipment is involved, the use of oxyacetylene brazing equipment has proven to be the most satisfactory. The introduction of pure oxygen along with acetylene produces a very hot flame. Fig. R16-34 illustrates the components involved in a typical oxyacetylene rig.

The efficient use of oxyacetylene equipment depends on a constant metered flow of oxygen and acetylene in correct proportions. Therefore, the operator must become thoroughly familiar with the particular equipment being used.

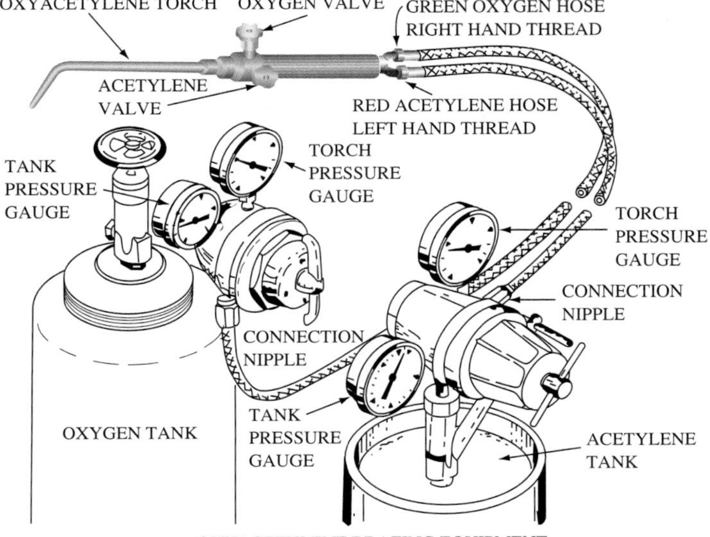

Figure R16-34 Oxyacetylene brazing equipment.

PRACTICING PROFESSIONAL SERVICES

Tendencies for inexperienced pipefitters after soldering or brazing a pipefitting joint are to rapidly cool down the material with a wet rag or cloth. This will tend to make your pipe brittle and easier to crack. If time allows, always cool down the pipe slowly.

If your area to be brazed is close to a valve or other critical component, then apply a wet rag or other heat sink to the device. There are paste-type heat-sink products on the market to protect components from excess heat without damage.

Remember, avoid trying to cool off your pipe too fast. Be patient; it will pay off.

Both the oxygen and acetylene tanks will have pressure regulators and two sets of gauges: one to register tank pressure and one to indicate pressure being furnished to the torch. The required torch pressure will vary with the particular torch and torch tip used.

While operating the oxyacetylene equipment never point the unlit torch toward any open flame or source of sparks. The acetylene is highly flammable—and the oxygen supports combustion very actively.

Lighting the torch calls for a torch lighter or sparker. Do not use matches. Open the torch acetylene valve approximately one-quarter turn. The torch oxygen valve is then "cracked" open as the spark or torch lighter is used to ignite the flame. Once ignition is achieved, the acetylene valve is adjusted to obtain the desired flame size. The oxygen valve is turned slowly to establish the type of flame, as illustrated in Fig. R16-35.

The correct flame is called a neutral flame. It has a luminous blue cone with a touch of reddish purple at the tip. A carbonizing flame will be evidenced by a greenish flame caused by too much acetylene. It produces soot or carbon that will restrict the flow of solder. An oxidizing flame results from too much oxygen; it will cause pitting of the metal and tends to harden the joint, making it more susceptible to breakage from vibration.

Soldering Is an Art

Beginners should test a few joints that they have soldered by reheating them and taking a joint apart to examine it. They can then determine whether some places in the joint were not soldered. If this is true, the cause of the faulty soldering should be analyzed to determine how to correct it.

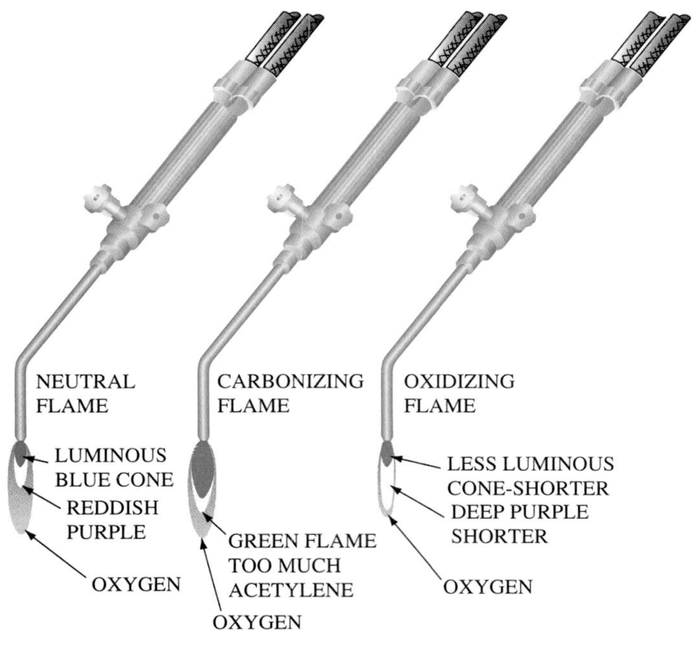

Figure R16-35 Oxyacetylene flames.

There is probably no work that service technicians are called on to do that requires more skill than soldering. They may know all the principles of soldering, but will find that considerable experience is required before they become adept at making consistently tight, strong, and neat soldered joints.

R16-5 REFRIGERANT PIPING REQUIREMENTS

The reliability of a field-assembled refrigeration or air-conditioning system is greatly influenced by proper design and installation of the various parts of the piping system. Proper refrigeration piping is essential to the successful operation of the system. Improper layout or sizing can change the effectiveness of the various components, and thus alter the system's capacity and performance. It can also result in equipment damage and failures.

The piping layout is usually made by an application engineer, but the refrigeration technician who installs and services the system must also become involved with this layout because of the possibility of difficulties and system faults. Also, the application engineer's layout may be diagrammatic only, with little regard for distances involved, either horizontal or vertical. Therefore, the technician is frequently in the position of needing to interpret the engineer's intent and then apply sound technical modifications to complete the installation properly.

R16-5.1 Function of Refrigeration Piping

The piping that connects the four major components of the system (Fig. R16-36) has two major functions: (1) It provides a passageway for the circulation of refrigerant, either in liquid or vapor form depending on the portion of the system involved; and (2) it provides a passageway through which lubricating oil which was carried over from the compressor is returned to the compressor. Each section of the piping should fulfill these two requirements with a minimum pressure drop of the refrigerant.

The second function, the return of oil to the compressor, is usually considered of secondary importance, but experience has proven that it is of equal importance to the function of carrying refrigerant. Reciprocating compressors, as well as rotary and centrifugal types, use various means to deliver oil to those bearings and surfaces requiring lubrication. These range from simple splash systems to pumped force-feed systems. Regardless of the method, some oil is swept out of the compressor with the discharged refrigerant, resulting in the unavoidable loss of oil within the compressor.

For example, in the reciprocating-type compressor, some of the oil in the crankcase gets on the cylinder walls during the down or intake stroke of the piston and is blown out with the compressed gaseous refrigerant through the discharge valve parts on the up or compression stroke (Fig. R16-37).

Some compressors pump much less oil than others, depending on the design and manufacturing methods. There is no way, however, to design a compressor so that none of the oil escapes into the refrigerant piping. This oil serves no other useful purpose in the system except to lubricate the compressor.

Presence of oil in the heat exchangers (evaporator and condenser) can reduce the capacity of the heat-exchange surfaces as much as 20%, as the excess oil forms an insulating coating on the interior surfaces of the refrigerant tubes. This, in turn, can cause thermal expansion valves to feed excess amounts of liquid refrigerant. Therefore, the presence of oil in the piping must be taken into consideration when installing piping. A piping system that is not correctly constructed to allow oil return can cause the following problems:

1. Seized compressor bearings due to insufficient oil returning to the compressor for lubrication.

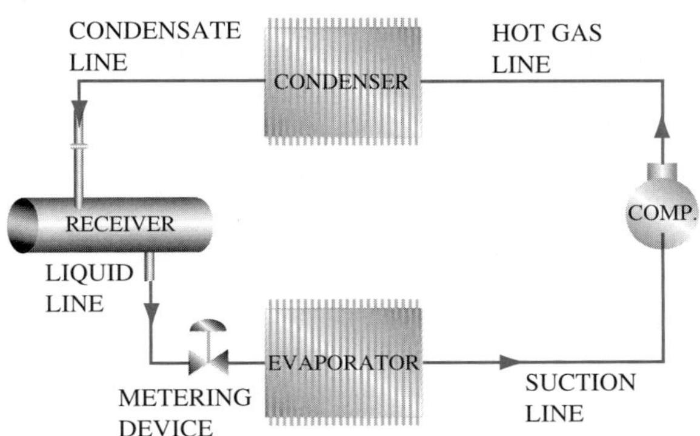

Figure R16-36 Refrigerant piping, showing connections to four major components.

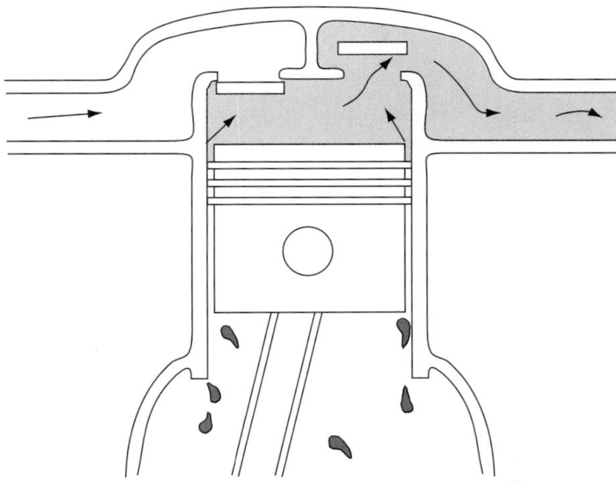

Figure R16-37 Compressor oil pumping.

2. Broken compressor valves, valve plates, pistons, and/or connecting rods due to liquid refrigerant and/or large quantities (slugs) of oil entering the compressor. It is important to understand that the compressor is designed to pump vapor and will not pump liquid.
3. Loss of capacity caused from the oil occupying portions of the evaporator, thus reducing the amount of effective surface and the overall system capacity.

R16-5.2 Basic Piping Precautions

A refrigeration technician must learn this subject well enough to avoid costly mistakes. There are five basic rules to keep in mind when piping a system:

1. Keep it clean. Cleanliness is a key factor in the actual installation. Dirt, metal filings, sludge, and moisture will cause breakdown in the system and must be avoided. Neat, clean work will avoid many service difficulties.
2. Proper sizing. Each section of the piping system must be properly sized to ensure proper oil return as well as maximize system capacity and efficiency. In those installations where one of these functions is to be reduced, proper oil return takes precedence in being fully retained.
3. Use as few fittings as possible (see Fig. R16-38). Fewer fittings mean less chance for leaks, and more importantly, less needless pressure drop.
4. Take special precautions in making every connection. Use the right material and follow the method recommended by the equipment manufacturer.
5. Pitch horizontal lines in the direction of refrigerant flow (see Fig. R16-39). To aid in forcing oil to travel through lines that contain vapor (suction line, hot gas line), horizontal lines should be pitched in the direction of refrigerant flow. This pitch, which helps the oil flow in the right direction, should be a minimum of ½ in. or more for each 10 ft of run. Pitch also helps to prevent backflow of the oil during shutdown.

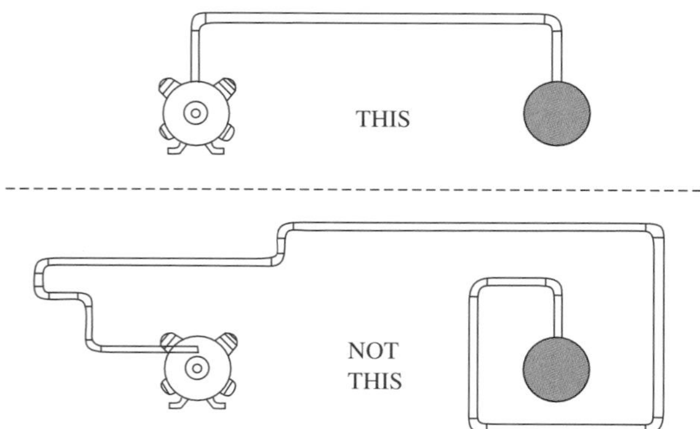

Figure R16-38 Proper and improper piping arrangement.

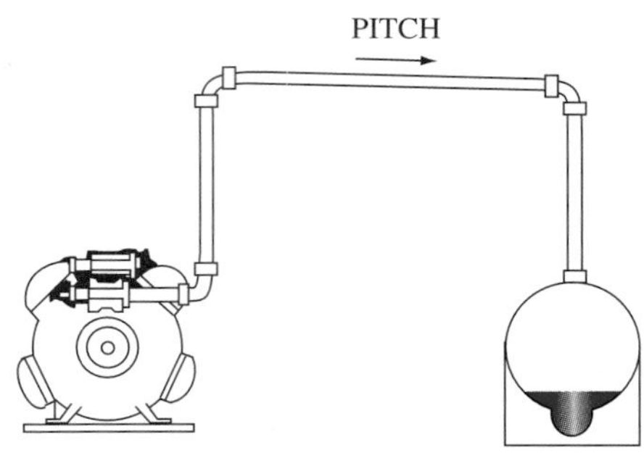

Figure R16-39 Proper pitch.

In piping systems where sufficient return gas velocity can be assured at all times, it is satisfactory to run the horizontal suction lines "dead" level. This may be desirable where head room is at a premium or where a sloping run will interfere with other piping.

R16-5.3 Suction Line

Oil circulates throughout the system and must be returned to the compressor to prevent damage to the compressor, as stated above. The most critical line in performing this function is the suction line.

For example, observe the behavior of two common refrigerants used in refrigeration and air conditioning, R-12 and R-22. In their liquid form these refrigerants will mix with oil and carry it along the piping with ease. Therefore, few, if any, oil problems exist in the liquid line. However, in their gaseous state the refrigerants are poor carriers of oil.

Oil in the suction line is at a lower temperature than the rest of the system, and therefore has higher viscosity, which slows down the flow over pipe surfaces. Also, the refrigerant is in vapor form and has only a mechanical effect on the oil. The vapor does not absorb the oil.

The suction line, therefore, must be carefully designed to ensure a uniform return of dry refrigerant gas as well as sufficient oil to the compressor.

The minimum load gas velocities within the suction line must be maintained at 500 ft/min (2.5 m/s) in horizontal runs and 1000 ft/min (5 m/s) through vertical rises with upward gas flow. The maximum recommended temperature drop is 2°F for R-12, R-22, and R-502 refrigerants. The temperature drop can be converted into equivalent pressure drop if desirable; however, the equivalent pressure drop is different for various refrigerants.

The reason the suction pressure and temperature drop are so important is because they relate to the ratio between the discharge and suction pressures. Any increase in this ratio reduces the capacity of the compressor to pump refrigerant vapor and increases the power required.

R16-5.4 Thermostatic Expansion Valve Systems

This system uses a thermostatic expansion valve (TXV) as a metering device to regulate the flow of refrigerant into the evaporator.

When the compressor is above the evaporator, liquid flood-back to the compressor is not a problem. In all thermostatic expansion valve systems, especially in those where the compressor is at the same level or below the evaporator, a trap and riser to at least the top of the coil must be placed in the suction line (see Fig. R16-40).

This trap is to prevent liquid flowing from the direct expansion (DX) coil into the compressor during shutdown. The trap and riser combination also promote free drainage of liquid refrigerant away from the thermostatic expansion valve bulb, thus permitting the bulb to sense suction gas superheat instead of evaporating liquid temperature.

Although it is important to prevent liquid refrigerant from draining from the evaporator to the compressor during shutdown, it is just as important to avoid unnecessary traps in the suction line near the compressor. Such traps would collect oil, which on start-up might be carried to the compressor in the form of slugs, thereby causing serious damage.

R16-5.5 Piping for Multiple Evaporators

Fig. R16-41 illustrates the use of multiple evaporators installed below the compressor. Notice that the piping is arranged so that refrigerant cannot flow from the upper evaporator into the lower evaporator.

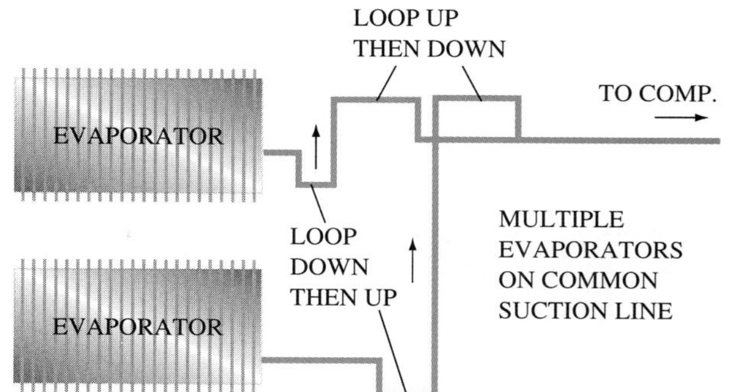

Figure R16-41 Off-cycle protection.

Where the vertical rise is over 20 ft (6 m) on either suction or discharge lines, as illustrated in Fig. R16-42, it is recommended that line traps be installed approximately every 20 ft (6 m) so that the storage and lifting of oil can be done in smaller stages.

R16-5.6 Preventing Liquid Slugging

To learn how to prevent liquid slugging, it is necessary to first understand what happens in a simple system when the compressor stops operating after its cooling requirements are satisfied. The evaporator is still filled with refrigerant—part liquid and part gas. There will also be some oil present. The liquid refrigerant and oil may drain by gravity to points where, when the compressor starts running again, the liquid will be drawn into the compressor and cause liquid slugging. The piping design must prevent liquid refrigerant or oil from draining to the compressor during shutdown. When the compressor is above the evaporator, this is not a problem.

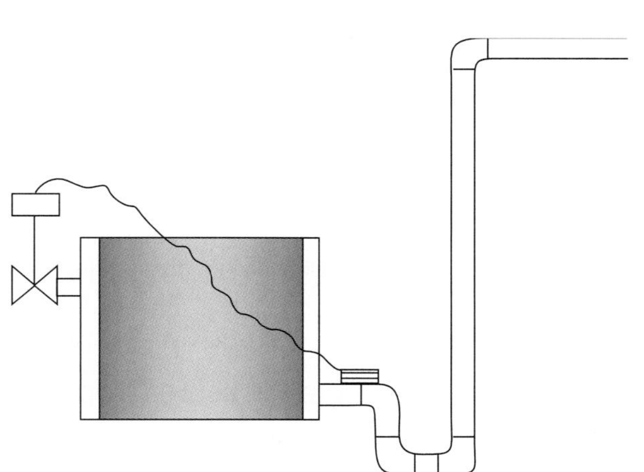

Figure R16-40 Suction line trap.

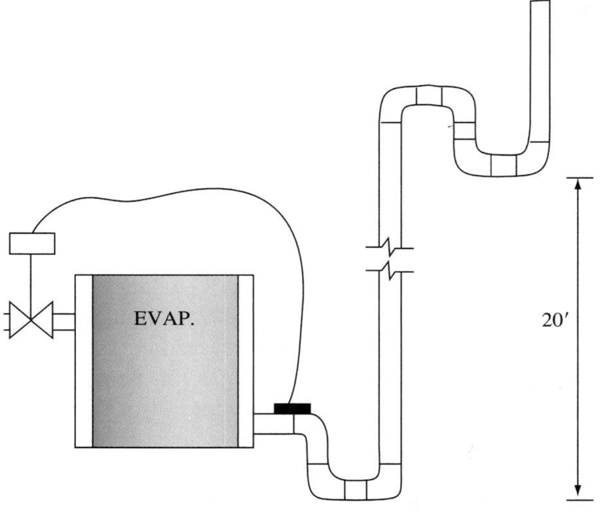

Figure R16-42 Multiple suction-line traps.

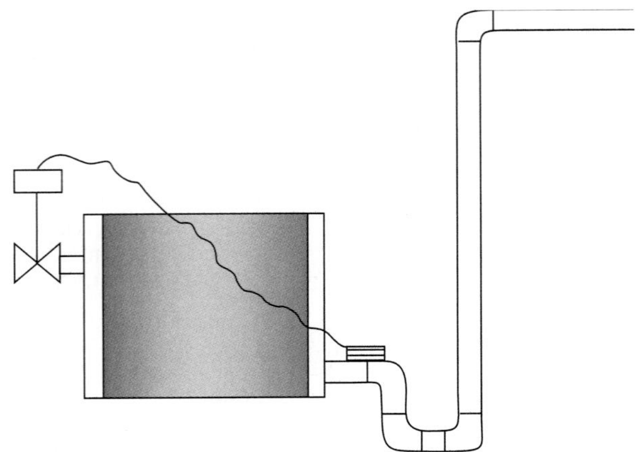

Figure R16-43 Proper TXV valve bulb location.

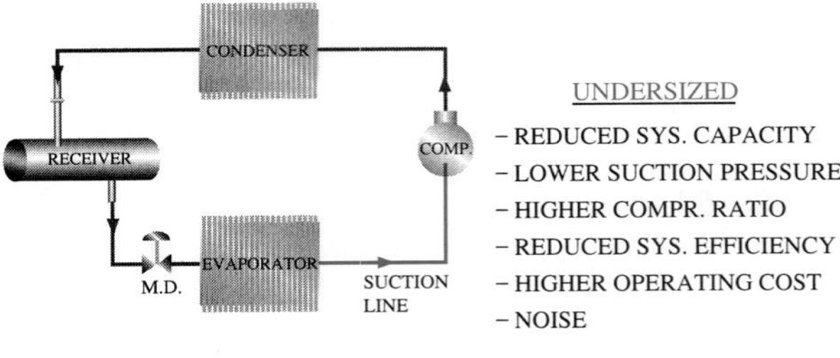

OVERSIZED

- OIL TRAPPING IN SYS.
- OIL STARVATION AT COMP.
- EXCESS. POWER CONSUMPTION
- SYSTEM CONTAMINATION
- HIGHER INSTALLED COST
- COMP. WEAR & OVERHEATING
- COMP. FAILURE
- SHORTENED SYSTEM LIFE

Figure R16-44 Suction-line sizing, giving effects of over sized and under sized piping.

When the compressor is located in an area where the ambient temperature is cooler than the condenser and/or evaporator, OFF-cycle refrigeration migration can be a problem. Traps in the discharge piping may be required as well as suction line accumulators and crankcase heaters. It is best to avoid these conditions where possible rather than to try to design around them.

The greater the system refrigerant charge, the more likely compressor failures may be experienced. Common sense dictates close-coupled systems with clean, simple piping layouts and minimal refrigerant charges.

If the compressor is on the same level or below the evaporator, as in Fig. R16-43, a riser to at least the top of the evaporator must be placed in the suction line. This inverted loop is to prevent liquid draining from the evaporator into the compressor during shutdown. The sump at the bottom of the riser promotes free drainage of liquid refrigerant away from the thermostatic expansion valve bulb, thus permitting the bulb to sense suction gas superheat instead of evaporating liquid refrigerant.

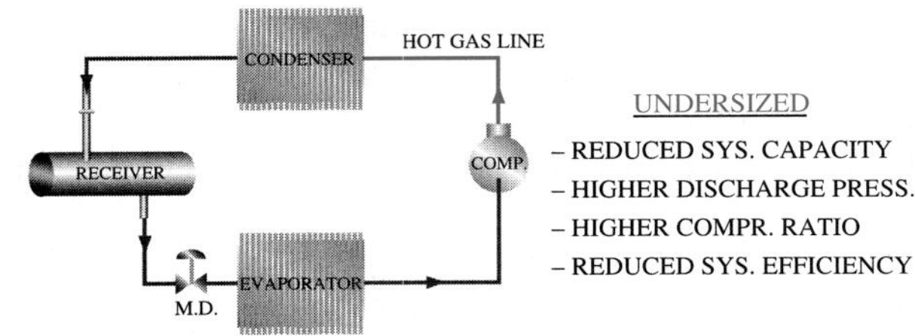

OVERSIZED

- OIL TRAPPING IN SYS.
- OIL STARVATION AT COMP.
- EXCESS. POWER CONSUMPTION
- SYSTEM CONTAMINATION
- HIGHER INSTALLED COST
- COMP. WEAR & OVERHEATING
- COMP. FAILURE
- SHORTENED SYSTEM LIFE

Figure R16-45 Hot-gas line sizing, giving effects of over sized and under sized piping.

R16-6 PIPE SIZING

The technician must be able to spot and modify obvious field piping errors caused by improper design or installation. If piping sizes are incorrect, either undersized or oversized, they can cause poor system performance or even damage to the equipment. Some of the common piping problems that are encountered for suction line sizing are shown in Fig. R16-44; in Fig. R16-45, for hot-gas (discharge) line sizing; in Fig. R16-46 for liquid line sizing; and in Fig. R16-47, for condensate line sizing. The suction line is most critical, the hot-gas line next, then the liquid line, and last, the condensate line.

The objective in sizing condensate lines is to provide ample size, allowing free drainage of the liquid refrigerant to the receiver while the gas refrigerant flows over the liquid in the opposite direction, thus serving as an equalizer line as well as a liquid drain.

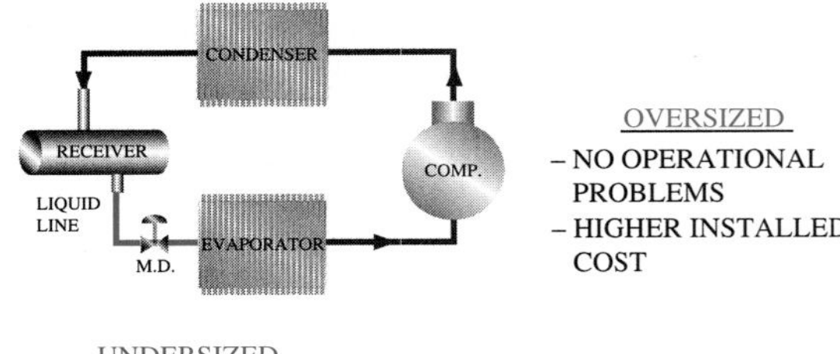

UNDERSIZED

- FLASH GAS AT MET. DEV.
- REDUCED EVAP. CAP.
- EVAP. STARVATION
- NOISY MET. DEV.
- UNSTABLE TXV CONTROL
- POSSIBLE LIQUID FLOODBACK
- REDUCED SYSTEM EFFICIENCY

Figure R16-46 Liquid line sizing, giving effects of over sized and under sized piping.

R16-6.1 Allowable Pressure Drop

In sizing refrigerant lines, it is desirable to keep the pressure drop within allowable limits. If the lines are too large the cost can be excessive; if the lines are too small the capacity of the equipment is reduced. A good com-

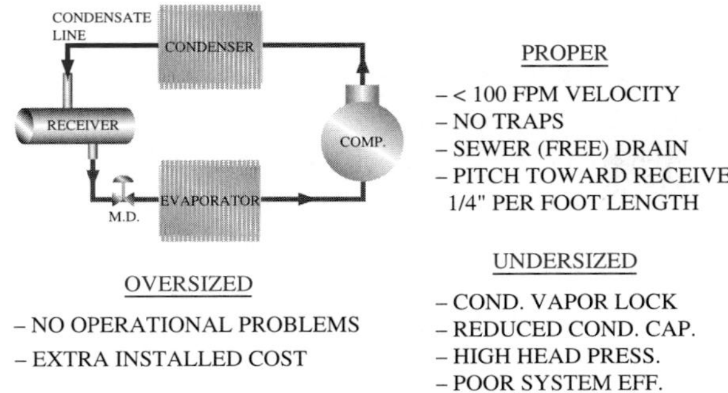

Figure R16-47 Condensate line sizing, giving effects of over sized and under sized piping.

promise is to use a pressure drop roughly equal to 2°F drop in saturation temperature. The equivalent change in pressure drop amounts to almost twice as much on the high side of the system as on the low side, as shown in Fig. R16-48.

Where the system capacity is variable because of capacity control or some other arrangement, a short riser will usually be sized smaller than the remainder of the suction line (Fig. R16-49) for a velocity of not less than 1000 ft/min (5 m/s) to ensure oil return up the riser. Although this smaller pipe has a higher friction, its short length adds a relatively small amount to the overall suction line friction loss.

In general, the pressure drop for the total suction line should be a maximum of 2°F (1°C), or 2 psi (14 kPa), to avoid loss of system capacity. The compressor cannot pump or draw gas nearly as effectively as pushing or compressing the gas.

R16-6.2 Riser Sizes

Riser sizes for both suction and hot gas lines are critical to permit carrying the oil upward with the force of the flowing gas, as shown in Fig. R16-50.

If the compressor is equipped with capacity control, the vertical runs must be sized properly to return the oil at minimum tonnage (Fig. R16-51). Vertical drops (flow down) will return the oil by gravity and are not critical as to size or velocity.

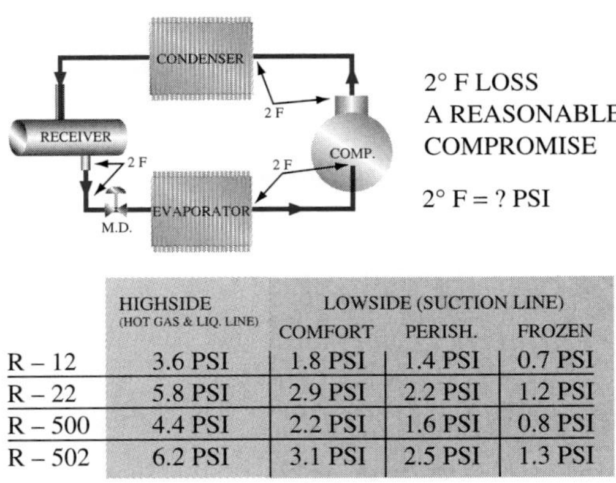

	HIGHSIDE (HOT GAS & LIQ. LINE)	LOWSIDE (SUCTION LINE) COMFORT	PERISH.	FROZEN
R – 12	3.6 PSI	1.8 PSI	1.4 PSI	0.7 PSI
R – 22	5.8 PSI	2.9 PSI	2.2 PSI	1.2 PSI
R – 500	4.4 PSI	2.2 PSI	1.6 PSI	0.8 PSI
R – 502	6.2 PSI	3.1 PSI	2.5 PSI	1.3 PSI

Figure R16-48 Pressure drop in piping based on suction using 2°F temperature drop.

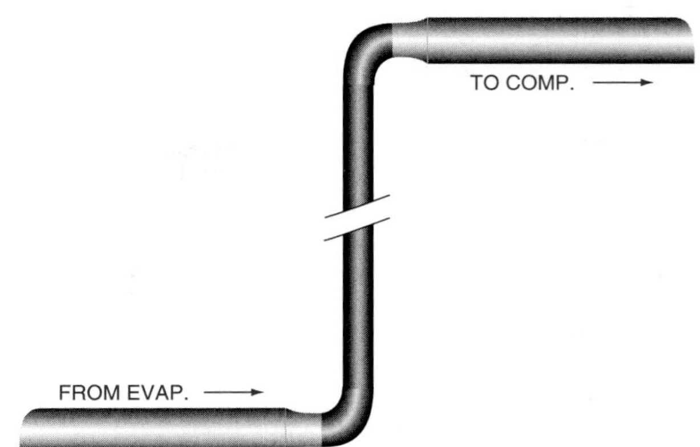

Figure R16-49 Reduced size suction-gas pipe riser.

An alternative to sizing the riser for minimum capacity is to provide a double riser in the suction line, as illustrated in Fig. R16-52. If a double riser is used, additional oil is required in the system to fill the trap during periods when the overall capacity of the system is reduced, and only one riser is used. When full capacity is resumed this extra oil can return to the compressor and over load the oil capacity of the crankcase, causing oil slugging. It is,

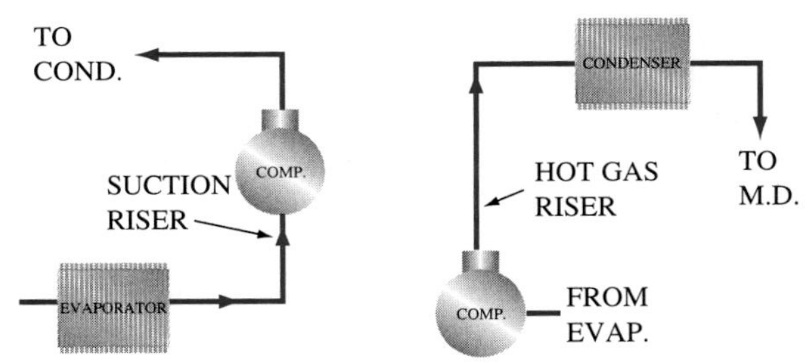

Figure R16-50 Piping risers for suction and hot-gas lines. Sizing for both is critical.

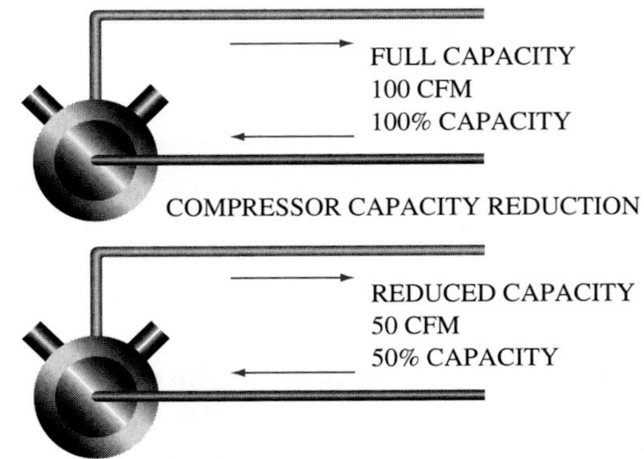

Figure R16-51 50% capacity reduction reduces the refrigerant flow 50%.

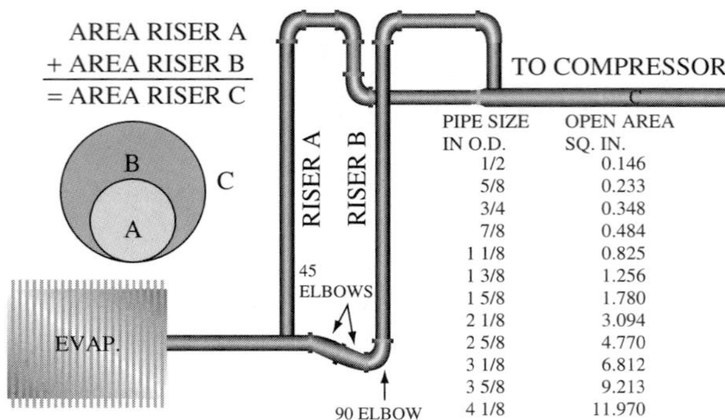

PIPE SIZE IN O.D.	OPEN AREA SQ. IN.
1/2	0.146
5/8	0.233
3/4	0.348
7/8	0.484
1 1/8	0.825
1 3/8	1.256
1 5/8	1.780
2 1/8	3.094
2 5/8	4.770
3 1/8	6.812
3 5/8	9.213
4 1/8	11.970

Figure R16-52 Double suction riser.

therefore, desirable to avoid the use of double risers wherever possible.

Caution: Small high-speed compressors have relatively small crankcase capacities. If double risers with oil seals are to be utilized, it may be necessary to add an auxiliary oil receiver. Be sure the system has sufficient oil to permit proper compression lubrication at full and minimum capacities.

The operation of the double-suction riser is similar to the hot-gas double riser. When maximum cooling is required, the system will run at full capacity, and both risers will carry refrigerant and oil. On part load, as the amount of refrigerant being evaporated decreases, the gas velocity will also decrease to a point where it will not carry oil upward through the vertical risers.

The oil trap, which is located at the bottom of the large riser, will fill with oil. All the refrigerant vapor will then pass up through the smaller riser, the pressure carrying oil with it. As the system load increases and more refrigerant is passed through the evaporator, this increased pressure will break the oil seal in the trap and carry oil upward through both risers.

Referring to Fig. R16-52, riser A is used to carry the suction gas at minimum load. Riser A plus riser B are used to carry the suction gas at maximum load. The area of riser A plus the area of riser B is made equal to the area of the main suction line, (labeled "C").

R16-6.3 Capillary Tube Systems

Sizing of the suction line depends on the amount of suction vapor to be handled. Therefore, suction lines in capillary tube systems are sized the same way as in thermostatic expansion valve (TXV) systems.

The difference in suction-line design between the two types of systems is in the use of traps. Traps are not recommended in capillary-tube installations, as they only add resistance to the flow of refrigerant. The trap is composed of three 90° bends, and each bend is equal to 5 ft of straight pipe. Adding a trap is the same as adding 15 ft of straight pipe.

Knowledge Extension

In larger metropolitan areas the plumber and pipefitter unions have dominated the installation of refrigeration piping. They will put their people through four to five years of apprentice-level pipefitting training. Wages are moderate at first, but typically very good upon completion of apprenticeship. Contact the local union shop steward to learn more about the opportunities within these organizations.

In smaller communities where unions are not prevalent, mechanical contracting companies will have their own installation and service crews. Visit with these contractors and express an interest in their company and they will probably give you more avenues of information and resources (possibly even offer you a job upon graduation).

Practically all direct expansion (DX) valve coils using capillary tubes are bottom feed, which reduces the possibility of gravity drain during the OFF-cycle of the system. Also, the refrigerant charge is limited to what the coil will hold during the OFF-cycle.

When properly charged with refrigerant, the capillary-tube system has less chance of liquid slugging. In addition, oil in the coil is lifted out into the suction header of the coil on each OFF-cycle because the coil fills with liquid refrigerant when the pressures balance. This prevents oil accumulation in the coil.

Traps are also not required in vertical risers because of the high gas velocities on startup. Standing idle, the DX coil, full of liquid refrigerant, will acquire considerable thermal energy at the balance pressure. Immediately on start-up of the compressor, when the suction pressure drops, a high quantity of vapor is produced in the DX coil, resulting in suction-line velocities in excess of 6000 ft/min (30 m/s). This velocity is sufficient to lift oil in vertical riser suction lines up to 65 ft. It is not recommended that installations be made of capillary-tube systems where a rise in the suction line in excess of 65 ft is required. In these installations, TXV valves and suction line traps should be used. It is important to follow manufacturers' recommendations where limitations of length of run, riser height, or charge are to be provided.

R16-7 INSULATION CONSIDERATIONS

R16-7.1 Suction Line Insulation

Insulation on the suction line is an absolute requirement. This eliminates the following:

1. Sweating of the suction line. Water condensing on the suction line can drip on occupants and can cause damage to ceilings, floors, furnishings, and electronics.
2. High-suction temperature gain. Hermetic and semi-hermetic motor compressor assemblies are suction-gas cooled. Therefore, the lower the returning suction-gas temperature, the better the heat removal from the motor and the lower the motor operating temperature and compressor discharge temperature; also, a lower oil temperature results, enabling better lubrication and bearing heat removal. This promotes longer compressor life.

R16-7.2 Hot Gas Line

In sizing and arranging hot gas lines, select tubing with a diameter small enough to provide the velocity to carry the hot vaporized oil to the condenser. On the other hand, the diameter must be large enough to prevent excessive pressure drop. In suction lines, the maximum allowable temperature drop is 2°F. In hot-gas lines 2°F is recommended, but a greater loss can be accommodated if conditions demand it.

If a higher pressure drop is used, the velocity of gas flow through the line can be excessive, causing noise, vibration, and serious reduction in system capacity. There would also be an increase in operating cost due to the higher compressor discharge pressure required.

The installation of an oil separator is an alternate to moving oil up the discharge riser (Fig. R16-53). It is installed downstream from the hot gas muffler. The oil from the separator is usually returned to the crankcase of the compressor. With this arrangement the hot-gas riser can be sized for low pressure drop.

Another arrangement would be to put the oil separator close to the compressor on the upstream side. Unless specifically required, it is also satisfactory to eliminate the muffler where there is a separator. A separator can serve as a muffler, although it is not as efficient as a well designed muffler.

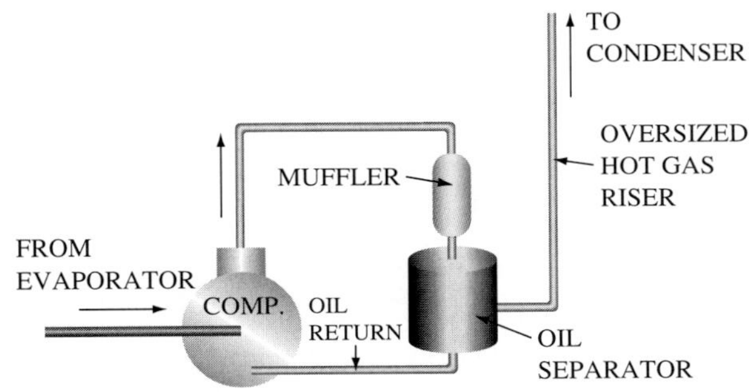

Figure R16-53 Discharge piping with oil separator.

R16-7.3 Hot Gas Line Insulation

In package units and condensing units with short hot-gas lines between compressor and condenser, no insulation should be used on the hot-gas line.

On remote condensers, insulating the hot-gas line is advisable. If the unit is expected to operate in low outside temperatures, it is possible to reach the condensing temperature of the discharge refrigerant before the refrigerant reaches the condenser. This can cause liquid slugs to fall backward down the hot-gas line into the superheated vapor from the compressor. Violent expansion of the slug vaporizing can cause "steam hammer," resulting in noise and vibration, even to the point of line breakage. Insulating the hot-gas line would prevent this action.

Where hot-gas lines are run indoors in machinery rooms, they should be insulated or otherwise protected to prevent accidental burns to operating personnel.

R16-7.4 Liquid Line

Liquid lines do not present a problem from an oil standpoint because liquid refrigerant and oil mix easily and the oil is carried through the liquid line with ease. Liquid lines, however, are critical for pressure loss both from a pressure drop due to pipe size and, in the case of the vertical upflow line, vertical lift of the refrigerant.

To limit the amount of refrigerant in the system, most manufacturers use ⅜-in. and even ¼-in. liquid lines in split systems using remote condensing units. This will vary with tonnage and line length. Do not attempt to change the liquid line size from the original specifications, as this will seriously affect performance and equipment life.

Some systems also depend on the pressure drop in the liquid line to add to the pressure drop when using a capillary tube for the pressure-reduction device. In these cases, both the size and the length of the liquid line are important for peak performance. Do not attempt to change either one.

R16-7.5 Preventing Flash Gas

Liquid lines can present a problem if the design causes a radical temperature change or pressure drop, since refrigerant leaving the condenser remains a liquid only as long as its boiling point (condensing temperature) is higher than its actual temperature. If a drop in pressure (drop in boiling point) or a rise in the liquid temperature were to reverse (boiling point fall below sensible temperature or sensible temperature rise above boiling point), vaporization of some of the liquid would occur before it passes through the pressure-reducing device. The vapor formed is called pre-expansion flash gas.

An example of pre-expansion flash gas production is laying a liquid line on the roof of a building, where it is on tar that has been heated by the sun. Refrigerant lines should be raised at least 18 in. above such a roof to minimize the effects of the sun. Flash gas within the liquid line is undesirable, since it displaces liquid at the port of the expansion valve, greatly reducing its capacity. Known as gas binding, it also affects the capacity of capillary tubes, as the ability of the tubes to carry vapor is considerably less than the ability to carry liquid. Liquid leaving the condenser is subcooled although generally at a higher temperature than that of the surrounding air. Flash gas is therefore not likely to occur.

Static loss refers to the pressure difference that exists between the bottom and the top of a liquid-filled pipe due to the weight of liquid in the line. Static loss is a frequent cause of the creation of pre-expansion flash gas. Friction and static losses do occur in properly sized lines.

For example, a standing column of liquid R-22 at 100°F and 210.6 psig, due to its weight, exerts a pressure of approximately 0.50 psi for every foot of height of the column (R-12 exerts 0.55 psi per foot of height). The pressure at the bottom of a 10-ft column of R-22 is 5 psi greater than the pressure at the top of the column. Conversely, for every 10 ft that R-22 is lifted in a vertical riser, the pressure at the top of the column is reduced by 5 psi.

Avoid, where possible, installations placing the condensing unit low and the evaporator high. An example is a condensing unit on grade and the air handler in the attic of a multi-story dwelling. Flash gas is likely to be a problem.

If you cannot avoid this situation, provide liquid subcooling by stacking the liquid and suction lines together and insulating both lines inside a common wrap to promote heat exchange from the liquid to the suction line.

Fig. R16-54 shows both a liquid lift and a liquid drop. The liquid lift is the condition where flash gas can exist at the top of the column, due to the drop in pressure.

Assume that the liquid R-22 leaves the condenser at standard conditions of 105°F condensing temperature with 3°F of subcooling. The pressure leaving the condenser would be 210.8 psig. Therefore, considering the 3°F of subcooling, the liquid leaves the condenser at 102°F. If the liquid-line riser has a 30-ft lift, the pressure at the top of the riser would be 15 psi less due to the weight of the refrigerant. Therefore, the final pressure would be:

$$210.8 \text{ psig} - 15 \text{ psig} = 195.8 \text{ psig}$$

$$195.8 \text{ psig} - 6 \text{ psig line loss} = 189.8 \text{ psig}$$

This will produce a boiling point of 97.86°F. Since the liquid left the condenser at 102°F, enough liquid will vaporize to cool the remaining liquid to the 97.9°F. To avoid the formation of flash gas in this riser, liquid subcooling leaving the condenser of 8°F or more is required instead of 3°F to compensate for lift and pressure drop in the line. This can be accomplished by the use of a liquid-to-suction-line heat exchanger.

Again, assume that the same conditions are applied to a typical air-conditioning unit which works at peak efficiency with 19°F of subcooling. Instead of leaving the condenser at 102°F the liquid would leave at 86°F. Again, assuming 6 psig line loss for a properly sized line, the 30 ft of vertical lift would not produce flash gas. The vertical lift could increase to 105 ft before flash gas would occur.

The 19°F of subcooling assumes that a water-cooled condenser, rather than an air-cooled condenser is used. It would not be practical on a hot day to obtain 86°F subcooled liquid using an air-cooled condenser.

This demonstrates the value of a proper amount of subcooling. It not only provides latitude for the designer when laying out the liquid line, but the proper amount of subcooling provides the maximum system capacity at the lowest operating cost.

R16-7.6 Locating Flash Gas

Referring to Fig. R16-55, a method of spotting flash gas is to compare (1) the temperature of the liquid line where a solid steam of liquid exists to (2) the temperature of the liquid line at a point where flash gas is suspected. A typical location for the second thermometer would be at

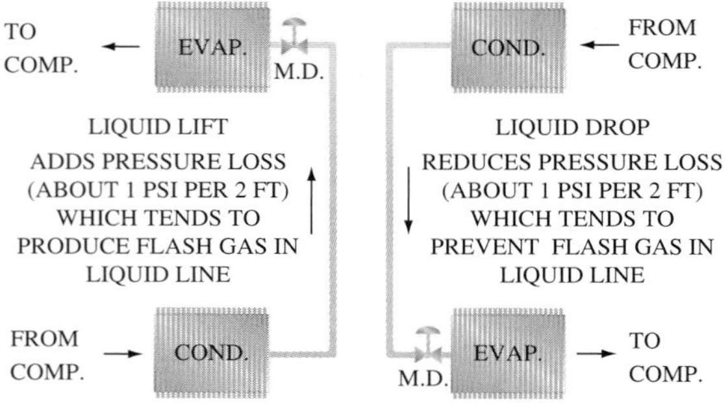

Figure R16-54 Liquid-line static head with lift or drop conditions.

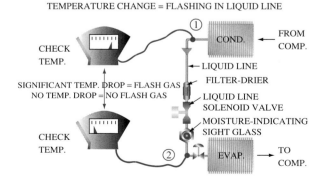

TEMPERATURE CHANGE = FLASHING IN LIQUID LINE
CHECK TEMP.
① COND. FROM COMP.
LIQUID LINE
FILTER-DRIER
SIGNIFICANT TEMP. DROP = FLASH GAS
NO TEMP. DROP = NO FLASH GAS
LIQUID LINE SOLENOID VALVE
MOISTURE-INDICATING SIGHT GLASS
CHECK TEMP.
② EVAP. TO COMP.

Figure R16-55 Liquid line piping showing test for flash gas.

the entrance to a TXV valve. If there is a noticeable temperature drop at point 2, flash gas could be present. An exception would be where the presence of gas binding and low liquid flow would cause the liquid line to be warm at the TXV valve and mask the presence of the flash gas.

Some installers place a sight glass at the entrance to the TXV valve. The sight glass will show bubbles if there is flash gas in the line. This is a highly recommended procedure.

R16-7.7 Liquid-line Insulation

Normally, no insulation is used on the liquid line because the greater the heat loss from the line, the lower the temperature of the liquid entering the pressure-reducing device, and the less system flash gas produced. However, when the liquid line runs through a hot space such as an attic, insulation of the liquid line may be required to prevent boiling from occurring prior to the expansion valve.

Insulation on the liquid line of a computer-room air-conditioning system is desirable. Liquid can get quite cold in winter and lines running through humidified space to the air conditioning unit will sweat and drip if not covered.

R16-7.8 Condenser Drain Line

The line between a condenser and a liquid receiver, called the condenser drain line, must be carefully sized. Although it is almost impossible to oversize the line (0 psig pressure drop is desirable), undersizing or installing too long a line is to be avoided. An undersized or excessively long line can restrict the flow of refrigerant to the TXV valve to the extent that some refrigerant will be held back in the condenser. The residue reduces the effective condenser surface and reduces condenser capacity. This causes the head pressure to rise, decreasing the overall system capacity. At the same time, power requirements and operating costs increase.

For proper operation, the length of the line between the condenser and receiver must be kept as short as possible, preferably not more than 18 in. The top of the receiver should be level with or (preferably) below the bottom of the condenser. This eliminates the need for the liquid to be forced uphill to enter the receiver. Except in very special applications, the condenser drain line is not insulated.

R16-7.9 Insulation Summary

The insulation needs of the piping are summarized in Fig. R16-56.

Generally the liquid line is not insulated. There are exceptions, however. If a liquid-to-suction line heat exchanger is used to avoid losing the subcooling, it is desirable to insulate it.

INSULATION NEEDS

	LIQUID LINE	SUCTION LINE	HOT GAS LINE
FREQUENCY	SOMETIMES	ALWAYS	SELDOM
THERMAL	✓	✓	✓
VAPOR BARRIER		✓	

Figure R16-56 Summary of piping insulation requirements.

It is always good practice to insulate the suction line since it is usually lower in temperature than the surrounding air and can condense moisture. Insulation should be applied and thoroughly sealed with a good moisture barrier to prevent condensation on the outside of the pipe or in the insulation.

The hot gas line is seldom insulated since any heat in the hot gas that is rejected benefits the system, reducing the amount of heat the condenser must reject.

The condenser drain line is usually not insulated since any loss of heat is desirable.

R16-8 PIPING SUPPORTS

All piping must be properly supported (Fig. R16-57). The supports must allow for the expansion and contraction of the pipe. The recommended allowance is ¾ in. movement per 100 ft of pipe. Hangers should be placed not more than 10 ft apart. They should always be placed near a bend in the piping, preferably on the longest straight connection to the bend. Where the pipe is insulated the hanger must have enough surface not to crush the insulation.

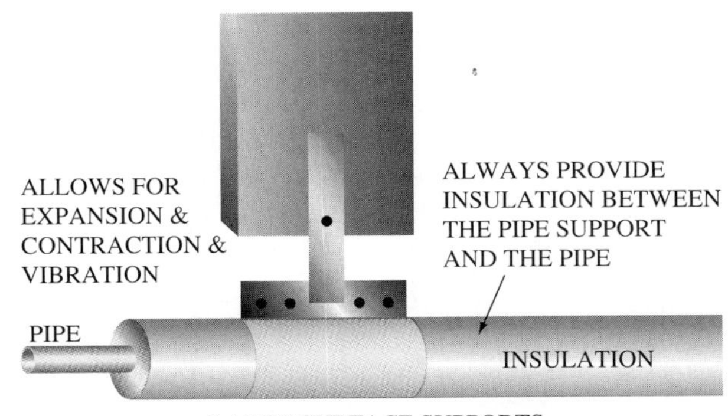
ALLOWS FOR EXPANSION & CONTRACTION & VIBRATION
ALWAYS PROVIDE INSULATION BETWEEN THE PIPE SUPPORT AND THE PIPE
PIPE
INSULATION
LARGE SURFACE SUPPORTS INSULATION ADEQUATELY TO AVOID PUNCTURING IT

Figure R16-57 Proper support for piping.

Sheet metal saddles or dense insulation block inserts (on large piping) will serve the purpose.

R16-9 VIBRATION DAMPENERS

It is usually desirable to isolate vibrating equipment to reduce noise and to prevent damage to the piping or other equipment. With soft copper tubing, loops or a coil of tubing can be connected to the moving part as shown in Fig. R16-57. For hard copper tubing a similar dampening effect can be produced by running the suction and discharge lines 15 times the pipe diameter in each of two or three directions before securing the pipe hanger, as shown in Fig. R16-58. This will provide some "give" to the piping without undue strain.

For larger equipment, rubber-in-shear or spring vibration eliminator mounts and flexible piping connections can be provided for the compressor to supply the necessary isolation. Concrete inertia blocks are sometimes incorporated into the base. Piping and electrical lines must be securely anchored beyond the isolators to be effective.

R16-9.1 The Hot Gas Muffler

The pulsations from a compressor, usually a reciprocating type, can cause serious vibrations and noise in the hot gas line. This can be most noticeable where a remote air-cooled condenser requires a long vertical hot-gas line. Usually the larger the compressor, the more noticeable the pulsations, although this is dependent on speed and number of cylinders.

The best way to solve this problem is to install a hot gas muffler in the compressor discharge line, as shown in Fig. R16-59.

It should be placed in a vertical position, so that it does not trap oil, as close to the compressor as possible, and securely mounted to the compressor. This usually destroys the resonance that the compressor has built up in the hot gas line. If this does not help, it may be necessary to enlarge the discharge line and relocate it, to destroy the resonance pattern.

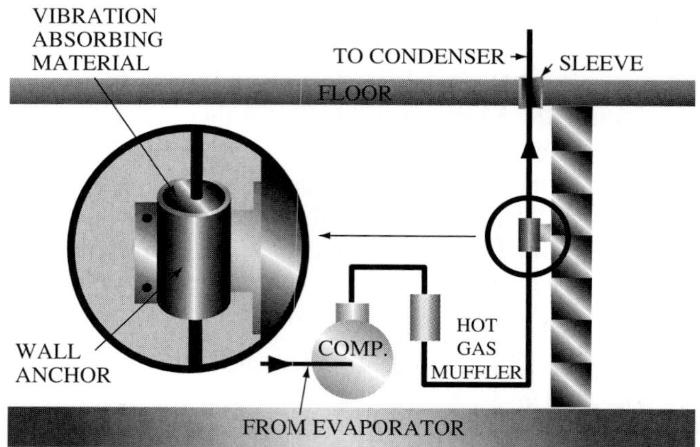

Figure R16-59 Using a hot-gas muffler to dampen discharge gas pulsations.

R16-10 OFF-CYCLE PROTECTION

Using a vertical discharge riser directly from the compressor could permit oil and liquid refrigerant to drain back down (or migrate) into the compressor head during the OFF-cycle. This can break valves on a reciprocating compressor on start-up. To prevent this, a discharge loop can be placed in the hot-gas line and a check valve (optional), as shown in Fig. R16-60. Modest quantities of oil and liquid refrigerant can accumulate in this trap when the compressor shuts down and be dissipated on start-up without damage to the compressor.

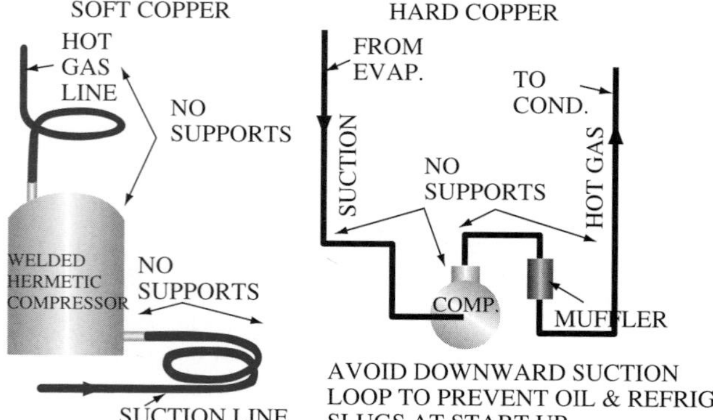

Figure R16-58 Piping loops to isolate vibrations and noise.

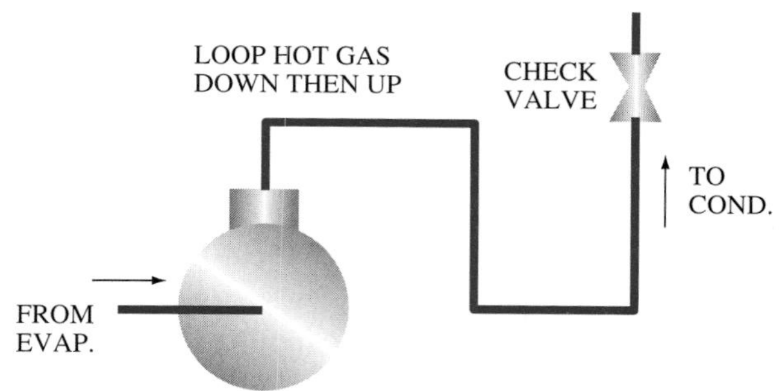

Figure R16-60 Discharge line OFF-cycle protection.

REVIEW

- Basic functions of refrigeration piping:
  The piping must assure a proper supply of refrigerant to the evaporator.
  Sizes selected must be practical without excessive pressure drop, while maintaining velocity to return oil with refrigerant vapor.
  The piping must not trap oil.
  The compressor must be protected from loss of oil.
  Liquid refrigerant and oil slugs must be prevented from entering the compressor.
  The system must be dry and clean.
- Tubing materials:
  Copper
  Aluminum
  Steel
- Tubing sizes are expressed in terms of outside diameters (OD).
- Pipe and tube sizes are expressed by the plumbing trades as nominal inside diameters (ID).
- Sizes in SI units refer to nominal ID pipe sizes. There are no SI units for OD tubing sizes.
- ACR tubing is pressurized by the manufacturer with dry nitrogen to seal the metal against air, moisture, dirt, and also to minimize the harmful oxides normally formed during brazing.
- Copper tubing has three classifications:
  K—Heavy wall; ACR approved; meant for special use where abnormal conditions of corrosion might be expected.
  L—Medium wall; ACR approved; most frequently used for normal refrigeration applications.
  M—Thin wall; Not used on refrigeration systems; will not meet safety codes.
- Type K and L copper tubing are available in soft- and hard-drawn types.
- Methods of cutting copper tubing:
  Hand-held tubing cutter
  Hack saw and sawing fixture
- Methods of bending tubing:
  Spring bender
  Lever bender
  Hand bend
- Minimum bending radius:
  Small—5 times the diameter
  Large—10 times the diameter
- Methods of joining copper:
  Mechanical couplings
    Flares
    Compression fittings
  Soldering/Brazing
- Double flares—used mainly on larger size tubing and on installations subject to much vibration. They can be assembled and disassembled more frequently without flare washout.
- The essential difference between soldering and brazing is the temperature at which the molten solder or brazing alloy flows. When the melting temperature is above 800°F, the process is called *brazing.*
- Six steps to produce strong, leaktight joints in soft soldering and silver brazing:
  Good fit and proper clearance
  Clean metal
  Proper fluxing
  Assembling and supporting
  Heating and flowing the alloy
  Final cleaning
- Oil trapping can cause:
  Seized compressor bearing
  Broken compressor valves
  Loss of capacity
- Piping precautions:
  Keep it clean
  Size properly
  Use as few fittings as possible
  Keep runs as short and straight as possible
  Follow recommended procedures
  Pitch the lines in the direction of refrigerant flow.
  Size to maintain velocity
  Avoid excessive pressure drop
  Prevent oil traps
  Use risers where necessary

Problems and Questions

1. When type L copper tube is purchased, the dimension indicating the size is:
   a. Outside diameter
   b. Inside diameter
   c. Wall thickness
   d. Weight per pound
2. One-half inch nominal sized pipe is equivalent to what SI dimension?
   a. 8 mm
   b. 10 mm
   c. 15 mm
   d. 20 mm
3. When bending copper tubing, the minimum radius of the bend should be:
   a. 3 times the diameter
   b. 5 times the diameter
   c. 7 times the diameter
   d. 10 times the diameter
4. The melting temperature of soft solder is:
   a. 360°F
   b. 415°F
   c. 1100°F
   d. 1200°F

5. The two metals that make up soft solder are:
   a. Tin and zinc
   b. Tin and copper
   c. Tin and aluminum
   d. Tin and lead
6. When swaging copper tubing, the overlap of the joint should be:
   a. 2 times the diameter
   b. ½ of the diameter
   c. 3 times the diameter
   d. Same as the diameter
7. What chemical gas is sometimes flowed through copper piping to prevent corrosion inside the pipe during brazing?
   a. Carbon dioxide
   b. Sulfur dioxide
   c. Oxygen
   d. Nitrogen
8. What chemical is sometimes present in silver solder that is poisonous, requiring the technician to have good ventilation?
   a. Lead
   b. Cadmium
   c. Tin
   d. Nickel
9. What is the range of sizes for open-end wrenches?
   a. ¼ to 1⅝ inch
   b. ⅛ to ¾ inch
   c. 1 to 5 inches
   d. ½ to 2 inch
10. The valve stem size on a compressor service valve is usually:
    a. ½-inch square
    b. ¼-inch square
    c. ⅜-inch hexagonal
    d. ½-inch hexagonal
11. Indicate which of the following is a function of the refrigerant piping:
    a. Passage-way for oil to be returned to the compressor.
    b. A support for the compressor.
    c. A means of charging the system.
    d. A vent space for air in the system.
12. A poorly constructed piping system can cause:
    a. System contamination.
    b. Higher installed cost
    c. Burned out compressor bearings.
    d. Too much oil return to compressor.
13. Which of the following is NOT a piping precaution?
    a. Cleanliness
    b. Use of few fittings as possible
    c. Pitch of horizontal lines
    d. Use of only hard-drawn copper tubing
14. The most critical line for returning oil to the compressor is:
    a. Hot-gas line
    b. Liquid line
    c. Suction line
    d. Condensate line
15. What is the minimum gas velocity to be maintained in a horizontal suction line?
    a. 300 ft/min
    b. 500 ft/min
    c. 700 ft/min
    d. 1,000 ft/min
16. A short, vertical riser in the suction line should be sized smaller than the horizontal suction line. True or false?
17. A double hot-gas riser should be used when the system includes:
    a. Capacity control
    b. A TXV valve
    c. A receiver
    d. Pressure control
18. When the vertical rise of the suction line is over 20 feet, what piping device should be used to assure oil return?
    a. A pressure-reducing valve
    b. A line trap
    c. A muffler
    d. A properly sized filter-drier

SECTION 2

Refrigeration Service Techniques

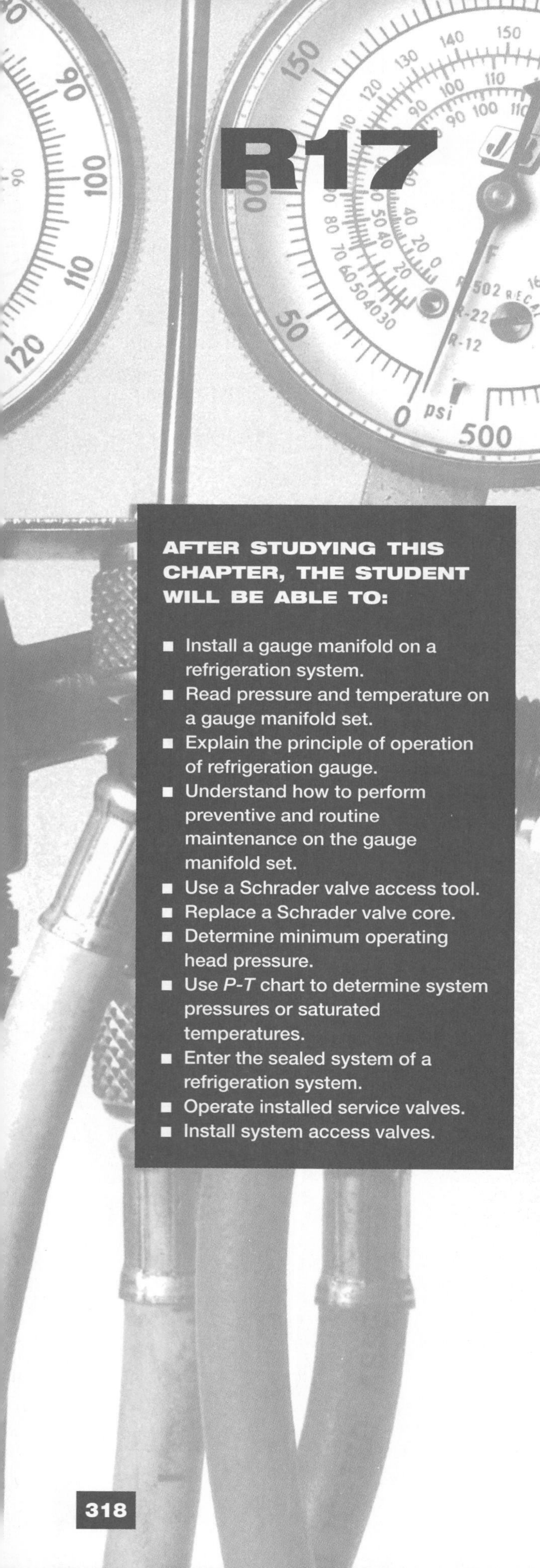

R17

Gaining Access to the Sealed System

AFTER STUDYING THIS CHAPTER, THE STUDENT WILL BE ABLE TO:

- Install a gauge manifold on a refrigeration system.
- Read pressure and temperature on a gauge manifold set.
- Explain the principle of operation of refrigeration gauge.
- Understand how to perform preventive and routine maintenance on the gauge manifold set.
- Use a Schrader valve access tool.
- Replace a Schrader valve core.
- Determine minimum operating head pressure.
- Use *P-T* chart to determine system pressures or saturated temperatures.
- Enter the sealed system of a refrigeration system.
- Operate installed service valves.
- Install system access valves.

R17-1 INSTALLATION OF GAUGE-MANIFOLD TEST SET

The gauge-manifold test set is a valuable service tool to the refrigeration servicer. It allows the service technician a quick method of installing pressure gauges for checking system conditions, charging, and adding oil to the compressor.

Just as a doctor uses the measurement of blood pressure to diagnose many illnesses, the qualified technician may use the gauge manifold to diagnose trouble in a refrigeration system. It allows the operator to watch both gauges simultaneously during purging or charging operations, and saves time on almost any work that must be done on the system. A cut-away view of a gauge manifold is shown in Fig. R17-1.

The gauge-manifold test set contains two shut off valves, three external connections, and two pressure gauges. The gauges and the flexible hoses that connect to the manifold to connect it to the system are color-coded; blue is the low side of the system, red is the high side. The left-hand gauge is called a compound or suction-pressure gauge. The right hand gauge is called the high- or discharge-pressure gauge. When both valves are closed (front seated) the center or utility port is isolated. The parts above and below each valve are interconnected so the gauges will read at all times when connected to the system. Opening the high-side valve opens the high-side port to the center port. Opening the low-side valve opens the low-side port to the center port. To determine system

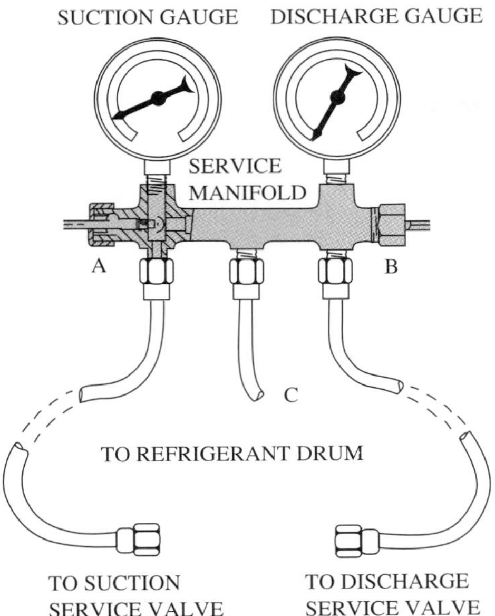

Figure R17-1 Cut-away view of a gauge manifold.

operation, add charge, purge, equalize, or evacuate, service gauges should have the best degree accuracy that is available.

This is a test instrument and should be treated as such:

1. Never drop or abuse the gauge manifold.
2. Keep ports or charging lines capped when not in use.
3. Never use any fluid other than clean oil and refrigerant.

Fig. R17-2 shows the pressure gauges on the manifold.

The high-pressure gauge (Fig. R17-3) has a single continuous scale, usually calibrated (marked off) to read 0 to 500 psi. The scale may be marked in either 2-lb or 5-lb increments and is usually connected to the high side of the refrigeration system. The black scale is the pressure scale and the red scales indicate the temperature of the various refrigerants at respective pressures. For example, if the gauge pointer indicated 200 psi pressure for R-12, the temperature of the refrigerant would be approximately 139°F.

The compound gauge (Fig. R17-4) measures both pressure and vacuum. It is usually calibrated from 0 to 30 in. of mercury and from 0 to 250 psi. Like the high-pressure gauge, the compound gauge also has scales calibrated to read temperatures of various refrigerants such as R-12, R-22, and R-502. With these scales it is not necessary to refer to pressure-temperature tables or curves to calculate pressure-temperature relationships.

Gauges use a Bourdon tube as the operating element (Fig. R17-5). The Bourdon tube is a flattened metal tube (usually a copper alloy) sealed at one end, curved and soldered to the gauge fitting at the other end.

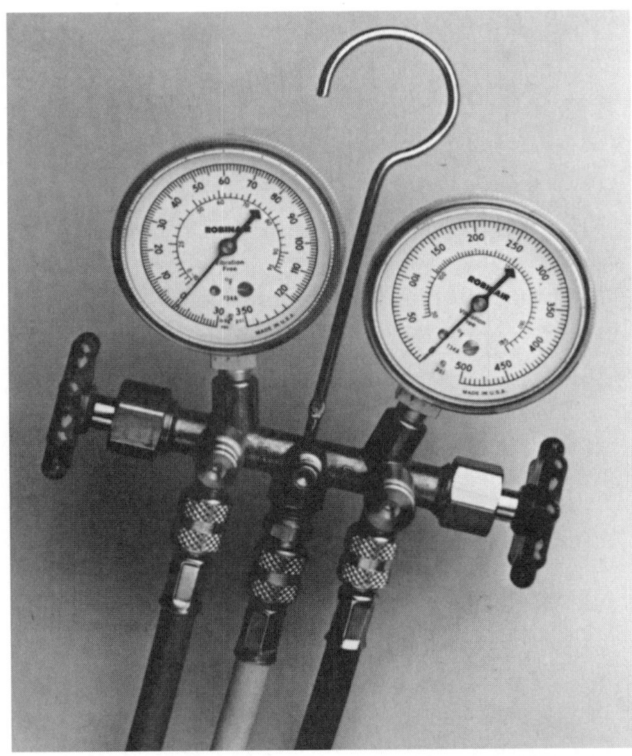

Figure R17-2 Testing manifold with gauges. (Courtesy of Robinair Division, SPX Corporation)

A pressure rise in a Bourdon tube tends to make it straighten. This movement will pull on the link, which will turn the gear sector counterclockwise. The pointer shaft will move clockwise to move the needle. On a decrease in pressure, the Bourdon tube moves towards its original (clockwise) position, and the pointer moves counterclockwise to indicate a decrease in pressure.

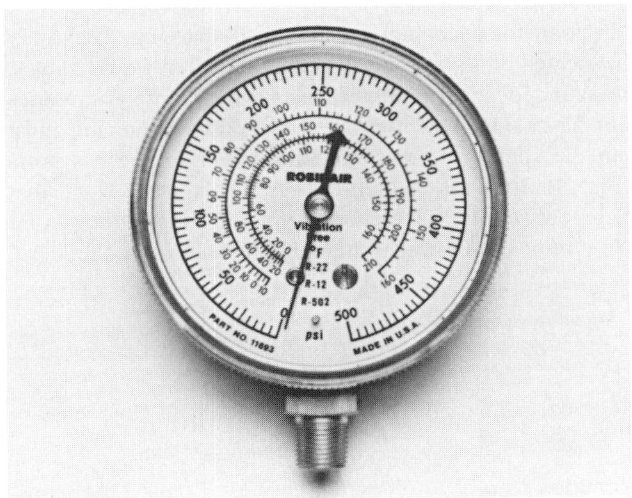

Figure R17-3 High-pressure gauge for service manifold. (Courtesy of Robinair Division, SPX Corporation)

PRACTICING PROFESSIONAL SERVICES

When installing manifold gauges on large reciprocating compressors, if the discharge gauge pulses rapidly, crack the gauge off the back seat slightly. This will reduce the pulsations and extend the life of your gauges.

If there is ever any oil in your gauges after removing them, be sure to drain the oil into an approved container to avoid re-introducing the wrong oil into a new system with different lubricants.

Whenever you are working with Schrader valves, use a quick connect line fitting on the discharge line. This will eliminate a large loss of refrigerant on a critically charged unit or avoid having refrigerant burn your hands or face.

Figure R17-4 Compound gauge for low-pressure side of service manifold. (Courtesy of Robinair Division, SPX Corporation)

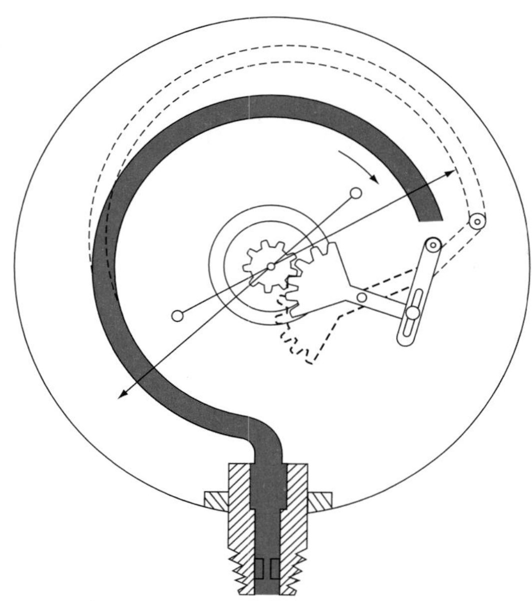

Figure R17-5 Bourdon tube.

Many systems have a high- and low-pressure service tap point for checking pressures and charging. They may be located on the compressor shut off valve, liquid valves, or as independent points. If the refrigeration system does not have "installed" service valves, the technician must tap the suction or discharge lines. Service access ports (Fig. R17-6) are sometimes used for this. Note that Schrader service valve ports require an adaptor fitting or a core remover between the service valve and the hose. These ports should always be leak-proof and capped or plugged when not in use.

The low side of the gauge manifold test set is attached to the suction side of the system and air is purged out of the lines with refrigerant. You can also evacuate the lines up to the service valve.

The high side of the gauge-manifold test set is attached to the discharge side of the system, and air is purged out of the lines again as before.

PREVENTIVE MAINTENANCE

Gauge manifolds should be routinely calibrated for accuracy. A simple test is to install your gauges on a new refrigerant cylinder. Convert the pressure to temperature using your PT chart and then measure the tank temperature. The two temperatures should match. If not, remove the cover from your gauges and calibrate the dial accordingly.

Figure R17-6 ¼″ refrigerant access valve. (J/B Industries Inc., Aurora, IL, USA)

Pressure readings are taken with the system operating. (The compressor is running.) Normal pressure readings of a medium-temperature R-12 refrigeration unit may be 0–10 psi pressure on the low side (compound gauge), and 125–175 psi on the high side (Fig. R17-7).

The temperature of the refrigerant may be calculated by observing the red scale for that particular refrigerant, in this case R-12, and the corresponding pressure. The pointer will also point toward its corresponding temperature. Let's look at the compound gauge. This gauge indicates 7 psig, the evaporating temperature of R-12 scale or red scale as –4°F. The high-pressure gauge indicates 136 psig; the corresponding temperature for R-12 at this pressure is 110°F.

Now to remove the gauge-manifold test set from the system: If the refrigeration system has two-way service valves installed or the technician has used a core removal tool, or has installed a valve to the flexible hoses connected to the suction and discharge service valves, shut off the service ports (suction and discharge), and purge the refrigerant remaining in the gauge lines, out through the center port of the gauge manifold. Remove the hoses and valves on the system and return to the storage position.

R17-2 DETERMINING MINIMUM OPERATING HEAD PRESSURE

Every liquid has its own boiling (saturation) point at atmospheric pressure. The saturation temperature of any liquid may be changed by increasing or decreasing the pressure above the liquid.

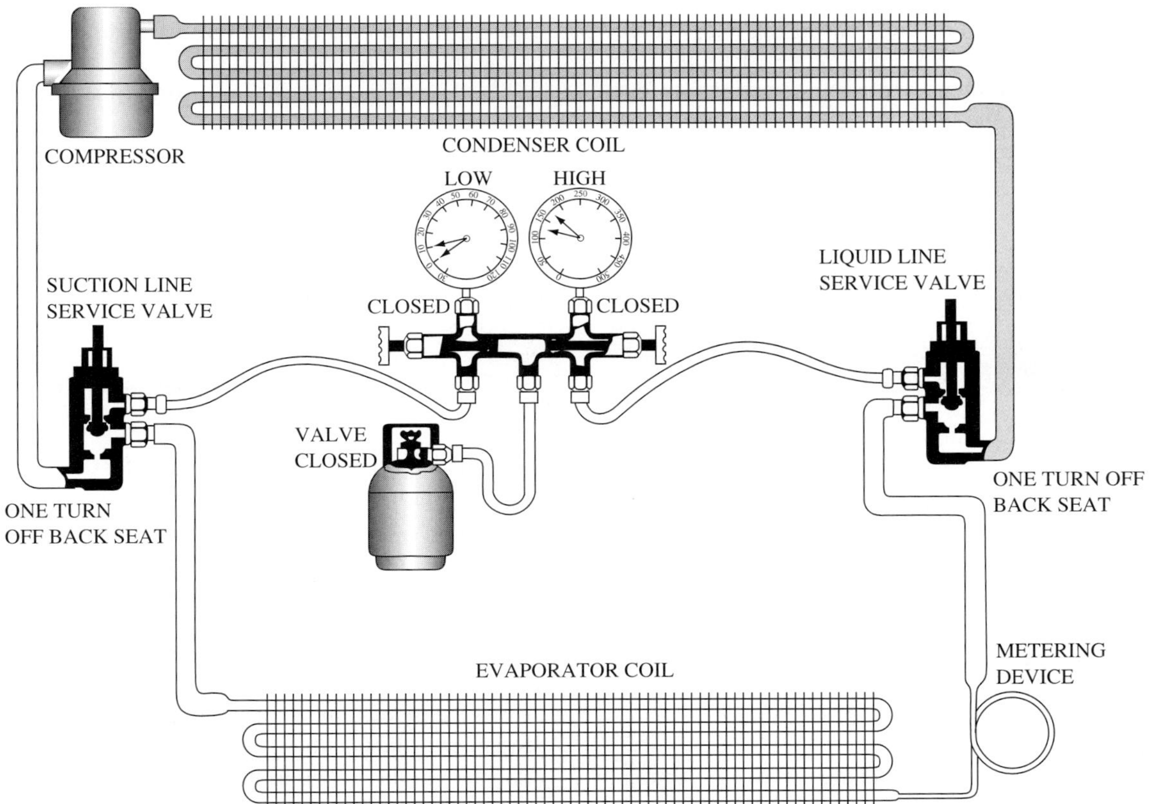

Figure R17-7 Typical gauge readings for a medium temperature R-12 unit.

A pressure-temperature (*P-T*) relationship chart is available free of charge through most refrigeration wholesale supply outlets (Fig. R17-8). This chart provides the technician with the temperature of various refrigerant liquids at any given temperature.

The *P-T* chart can be used by the technician, in conjunction with test manifold gauges to diagnose a refrigeration system malfunction.

The first step is to determine the type of refrigerant used in the refrigeration system you are diagnosing. To accomplish this, look at the UL (Underwriters Laboratory) tag on the unit. If a compound gauge installed on the suction service port of an R-12 unit shows a pressure of 37 psi, the temperature of the cooling coil (evaporator) is 40°F, the saturation temperature that corresponds to 37 psi for R-12.

The same pressure-temperature *(P-T)* relationship also applies to the high-pressure side of the refrigeration system. With a high-pressure gauge reading of 147 psi, the temperature of the refrigerant is 115°F. A rule of thumb for checking the high side for proper operation is to take a temperature reading of the ambient air temperature. Ambient air temperature is the temperature of the air that flows across the condensing unit. Use an electronic temperature gauge or a "superheat" thermometer for an accurate reading.

Let's suppose that the ambient temperature is 80°F. In order to maintain a proper condensing temperature at the condenser, there must be approximately a 20–35°F difference in ambient temperature and condensing temperature. This means that the ambient temperature (°F) plus 20°F will give you the minimum condensing temperature, for a low- or medium-temperature unit. The constant 35° is used for high-temperature units.

EXAMPLE

For 80°F ambient temperature and a 20°F temperature difference, the condensing temperature is 100°F. Looking at the *P-T* chart, this means that the high-side gauge should read at least 117.2 psi pressure.

In analyzing the low-side (compound-gauge) pressure, suppose the unit is designed for an evaporator temperature of 35°F at 70°F ambient and 45°F at 90°F ambient. Using R-12, the pressure of boiling refrigerant in the evaporator at 32°F would be 32.6 psi and at 45°F it would be 41.7 psi.

R17-3 ENTERING THE SEALED SYSTEM

Refrigeration problems can be categorized as electrical failures or mechanical failures or a combination of both. Approximately 80% of the problems are electrical in nature, and of the remaining 20%, only a small portion require extensive sealed-system service. One of the first things to make quite clear is to never enter the sealed system unless it is absolutely necessary.

If one or more of the electrical load devices is not functioning, the problem is electrical. If all components are functioning but the unit does not refrigerate, the problem is mechanical.

Mechanical problems include: poor-capacity compressor (does not pump properly, bad valves), low-side leak, low-side restriction, high-side leak, high-side restriction, dirty condenser, and dirty evaporator.

R17-3.1 Field-Installed Service Valves

One of the easiest devices available for sealed-system access is the saddle, piercing valve, or tap-a-line. Still other terms for this are used by some manufacturers. These piercing valves are clamped to the tubing, sealed by a bushing gasket, and pierce the tube with a tapered needle. Most contain some sort of shut-off control. The technician should keep in mind that these valves should be used to gain temporary access to a hermetically sealed system for checking system-operating pressures or for pressurizing for leak testing. The piercing valve (Fig. R17-9) allows quick access to system pressures to immediately start diagnosing the refrigeration problem.

Access piercing valves should be removed once the source of the sealed system malfunction has been located. Saddle valves, with Schrader valve cores that are brazed on the tubing, are often used (Fig. R17-10).

The Schrader valve core is a spring-loaded device, for position seating (see Fig. R17-11).

The valve is like those used on automobile tires. The stem must be depressed to force the valve's seat open against spring pressure. If a valve core leaks, it can be replaced by using a core-removal tool to unscrew it. Some tools allow this to be done while the system is under pressure (see Fig. R17-12).

R17-3.2 Using the Process Tube for Service

Another accepted method of access to the refrigeration sealed system is the process tube adaptor. The process tube is located on the hermetically sealed compressor. The equipment manufacturer uses this tube to evacuate and charge the system. This generally gains access to the low side of the system. A process tube adapter kit is shown in Fig. R17-13 (p. 325), and a process tube adaptor in Fig. R17-14 (p. 325).

Cut the process tube at the brazed pinched end. Use the proper size adaptor (these are marked according to outside diameters of the tubing to fit directly on the process tube). Tighten the handles of the tool, which will tighten the gasket ring to the tubing, and provide a compression seal.

Temperature °F.	Refrigerant—Code					Temperature °F.	Refrigerant—Code					Temperature °F.	Refrigerant—Code				
	12-F	22-V	500-D	502-R	717-A		12-F	22-V	500-D	502-R	717-A		12-F	22-V	500-D	502-R	717-A
−60	*19.0*	*12.0*	*17.0*	*7.2*	*18.6*	12	15.8	34.7	21.2	43.2	25.6	42	38.8	71.4	48.2	83.8	61.6
−55	*17.3*	*9.2*	*15.0*	*3.8*	*16.6*	13	16.4	35.7	21.9	44.3	26.5	43	39.8	73.0	49.4	85.4	63.1
−50	*15.4*	*6.2*	*12.8*	*0.2*	*14.3*	14	17.1	36.7	22.6	45.4	27.5	44	40.7	74.5	50.5	87.0	64.7
−45	*13.3*	*2.7*	*10.4*	1.9	*11.7*	15	17.7	37.7	23.4	46.5	28.4	45	41.7	76.0	51.6	88.7	66.3
−40	*11.0*	0.5	*7.6*	4.1	*8.7*	16	18.4	38.7	24.1	47.7	29.4	46	42.6	77.6	52.8	90.4	67.9
−35	*8.4*	2.6	*4.6*	6.5	*5.4*	17	19.0	39.8	24.9	48.8	30.4	47	43.6	79.2	54.0	92.1	69.5
−30	*5.5*	4.9	*1.2*	9.2	*1.6*	18	19.7	40.8	25.7	50.0	31.4	48	44.6	80.8	55.1	93.9	71.1
−25	*2.3*	7.4	1.2	12.1	1.3	19	20.4	41.9	26.5	51.2	32.5	49	45.7	82.4	56.3	95.6	72.8
−20	0.6	10.1	3.2	15.3	3.6	20	21.0	43.0	27.3	52.4	33.5	50	46.7	84.0	57.6	97.4	74.5
−18	1.3	11.3	4.1	16.7	4.6	21	21.7	44.1	28.1	53.7	34.6	55	52.0	92.6	63.9	106.6	83.4
−16	2.0	12.5	5.0	18.1	5.6	22	22.4	45.3	28.9	54.9	35.7	60	57.7	101.6	70.6	116.4	92.9
−14	2.8	13.8	5.9	19.5	6.7	23	23.2	46.4	29.8	56.2	36.8	65	63.8	111.2	77.8	126.7	103.1
−12	3.6	15.1	6.8	21.0	7.9	24	23.9	47.6	30.6	57.5	37.9	70	70.2	121.4	85.4	137.6	114.1
−10	4.5	16.5	7.8	22.6	9.0	25	24.6	48.8	31.5	58.8	39.0	75	77.0	132.2	93.5	149.1	125.8
− 8	5.4	17.9	8.8	24.2	10.3	26	25.4	49.9	32.4	60.1	40.2	80	84.2	143.6	102.0	161.2	138.3
− 6	6.3	19.3	9.9	25.8	11.6	27	26.1	51.2	33.2	61.5	41.4	85	91.8	155.7	111.0	174.0	151.7
− 4	7.2	20.8	11.0	27.5	12.9	28	26.9	52.4	34.2	62.8	42.6	90	99.8	168.4	120.6	187.4	165.9
− 2	8.2	22.4	12.1	29.3	14.3	29	27.7	53.6	35.1	64.2	43.8	95	108.2	181.8	130.6	201.4	181.1
0	9.2	24.0	13.3	31.1	15.7	30	28.4	54.9	36.0	65.6	45.0	100	117.2	195.9	141.2	216.2	197.2
1	9.7	24.8	13.9	32.0	16.5	31	29.2	56.2	36.9	67.0	46.3	105	126.6	210.8	152.4	231.7	214.2
2	10.2	25.6	14.5	32.9	17.2	32	30.1	57.5	37.9	68.4	47.6	110	136.4	226.4	164.1	247.9	232.3
3	10.7	26.4	15.1	33.9	18.0	33	30.9	58.8	38.9	69.9	48.9	115	146.8	242.7	176.5	264.9	251.5
4	11.2	27.3	15.7	34.9	18.8	34	31.7	60.1	39.9	71.3	50.2	120	157.6	259.9	189.4	282.7	271.7
5	11.8	28.2	16.4	35.8	19.6	35	32.6	61.5	40.9	72.8	51.6	125	169.1	277.9	203.0	301.4	293.1
6	12.3	29.1	17.0	36.8	20.4	36	33.4	62.8	41.9	74.3	52.9	130	181.0	296.8	217.2	320.8	—
7	12.9	30.0	17.7	37.9	21.2	37	34.3	64.2	42.9	75.8	54.3	135	193.5	316.6	232.1	341.2	—
8	13.5	30.9	18.4	38.9	22.1	38	35.2	65.6	43.9	77.4	55.7	140	206.6	337.2	247.7	362.6	—
9	14.0	31.8	19.0	39.9	22.9	39	36.1	67.1	45.0	79.0	57.2	145	220.3	358.9	264.0	385.0	—
10	14.6	32.8	19.7	41.0	23.8	40	37.0	68.5	46.1	80.5	58.6	150	234.6	381.5	281.1	408.4	—
11	15.2	33.7	20.4	42.1	24.7	41	37.9	70.0	47.1	82.1	60.1	155	249.5	405.1	298.9	432.9	—

Figure R17-8 Pressure-temperature table. *Italic Figures* represent vacuum-inches of mercury. **Bold Figures** represent pressure-pounds per square inch. (Courtesy of Sporlan Valve Co.)

Figure R17-9 Piercing valve. (Courtesy of Robinair Division, SPX Corporation)

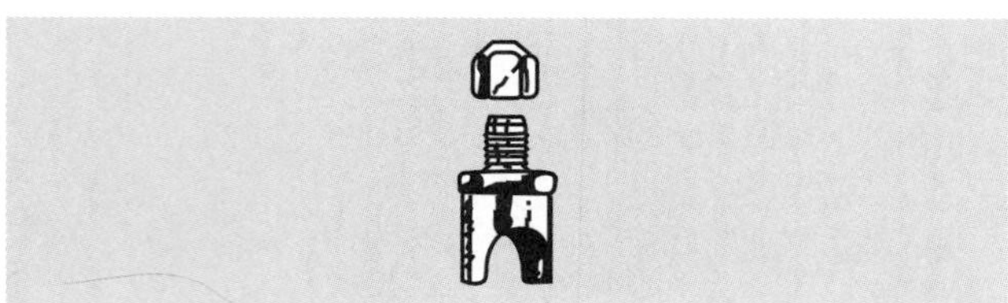

Figure R17-10 Saddle valve.

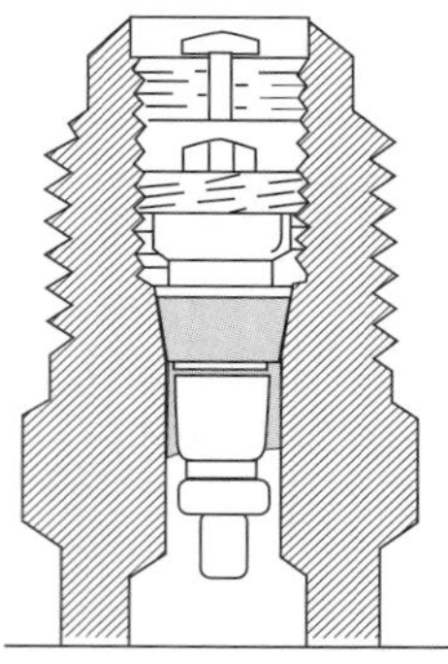

Figure R17-11 Schrader valve core.

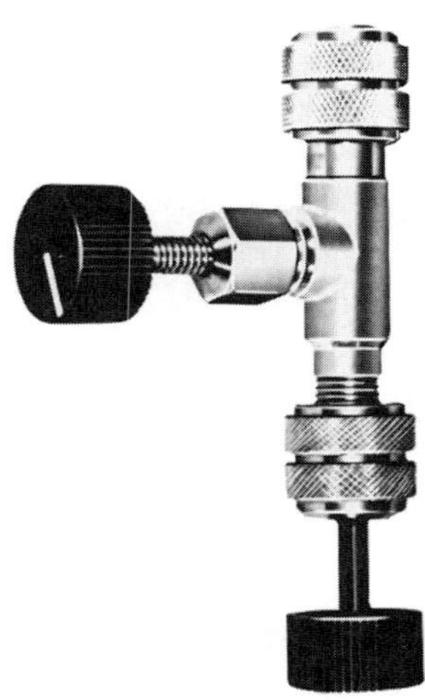

Figure R17-12 Valve core remover/installer. (Courtesy of Robinair Division, SPX Corporation)

ON THE JOB

Remember to always leak test your manifold gauges when leak testing any refrigeration system. They are more prone to leak than the system due to frequent use. If leaks occur, they will probably be found around the o-rings at the end of the hoses. Carry extra o-rings in your service vehicle and repair as needed. Over tightening of these o-rings is often the cause of the problem. Snug but not tight is a better policy to prevent gauges leaking. If the valve or stem on the gauge is found to be leaking, then order a repair kit. It is often much cheaper to repair your own gauges than buy a new pair. Proper care and maintenance will extend the life of your gauge over many years.

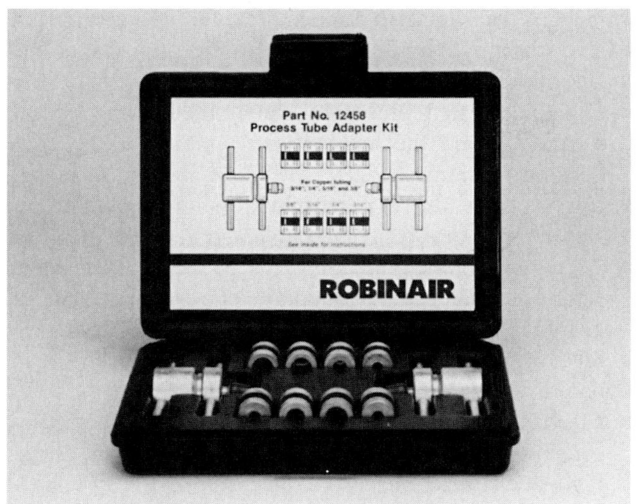

Figure R17-13 Process tube adapter kit. (Courtesy of Robinair Division, SPX Corporation)

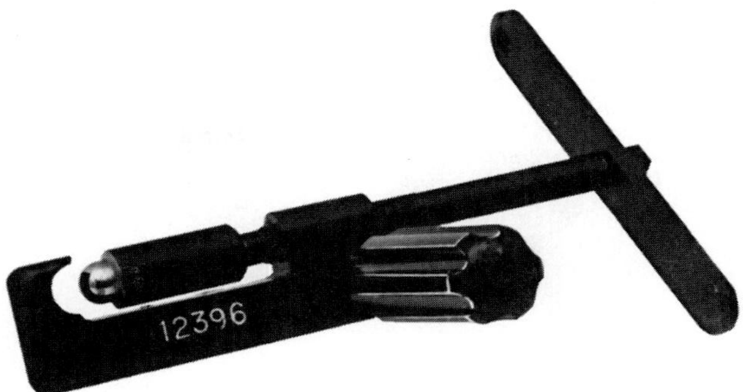

Figure R17-15 Pinch-off tool. (Courtesy of Robinair Division, SPX Corporation)

The charging hose is attached to the process tube adaptor, and the system can be repaired, evacuated, and charged. When the system has been completely serviced, a pinch-off tool (see Fig. R17-15) can be used to seal the hermetic system from the adaptor. The tool is left in place, while the end of the process tube is brazed shut again.

Be sure to use the temporary access valves to determine operating pressure prior to using the process tube adaptor, as these are important to accurate diagnosis. Recover the refrigerant before cutting off the pinched end of the process tube to attach the adaptor; otherwise, system refrigerant would be lost, also making it impossible to determine operating pressures at that line. When attaching gauge hoses, be sure that they are purged with refrigerant to remove air.

R17-3.3 Factory-Installed Service Valves

With the implementation of the Clean Air Act, all manufacturers, with the exception of Type I equipment manufacturers, are required to install factory-installed service valves. Many commercial refrigeration and air-conditioning systems have had factory-installed service valves for years.

Factory-installed service valves may be either a manually operated stem shutoff valve (see Fig. R17-16), or a Schrader type valve (see Fig. R17-17). Installed valves are usually located at the compressor as suction and discharge valves, and at the outlet of the receiver (the "King" valve). These service valves are equipped with a gauge service port. Operating refrigerant pressures may be observed on the service gauge manifold when hoses are connected to these ports.

The valve will be in the backseated (the stem turned all the way out, counterclockwise) position when you attach your gauges. This closes the gauge port, and the valve is open to the line connection. The valve is frontseated (the stem is turned all the way in, clockwise) to isolate the compressor from the system. The gauge port is open to the compressor but the line connection is closed. In order to read system pressures, the technician first checks to ensure that the service valve is backseated, then turns the stem in one or two turns (the service position, cracked), in order to slightly open the connection to

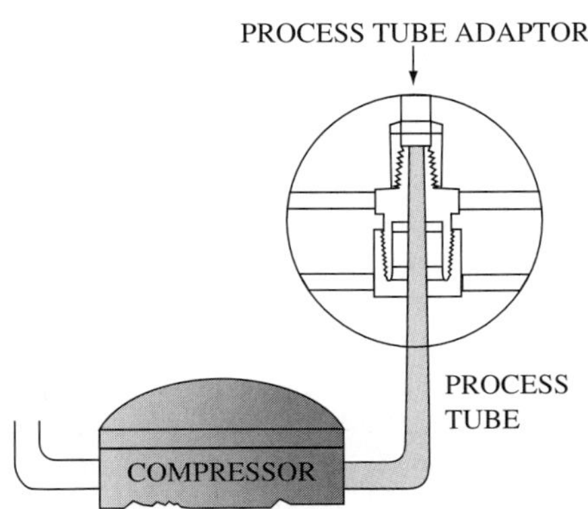

Figure R17-14 Process tube adapter installed.

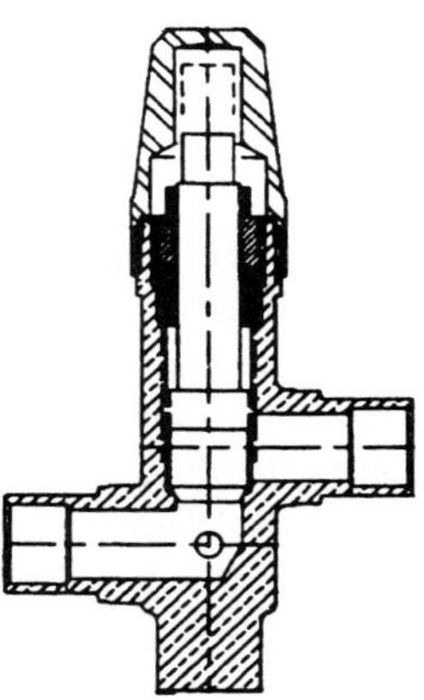

Figure R17-16 Manual shut-off valve. (Courtesy of Henry Valve Company)

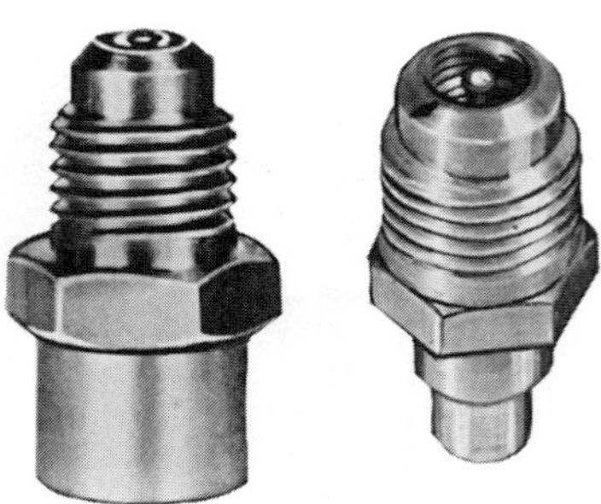

Figure R17-17 Schrader valve. (Courtesy of Robinair Division, SPX Corporation)

both line and gauge port. Service valves are returned to the backseated position after service and the gauges are removed. All caps are reinstalled on the system. Note that the compressor is always open to either the line connection or the gauge port, or both, if the valve is in the "cracked" position.

SAFETY NOTE: Be sure that internal pressure in the compressor is relieved by recovery and vacuum procedures before attempting to remove an isolated compressor from the system.

Schrader valves provide a convenient method of checking system pressures or servicing the system, where it is not economical or convenient to use the compressor service valves with gauge ports. Some manufacturers provide Allen wrench stop valves used in conjunction with the Schrader valve. The Schrader valve is similar to the air valves used on bicycle and automobile tires; however, they are not the same. The rubber used in tire valves is not compatible with refrigerants and would dissolve. The Schrader valve used in refrigeration and air-conditioning systems must also have a cap for the fitting to ensure a leak-proof operation.

This type of service valve enables the technician to quickly check system operation without disrupting the unit's operation. Technicians should use "quick-connect" low-loss hose adapters to attach to the Schrader valves, greatly reducing the refrigerant loss, when connecting or disconnecting the service hoses.

REVIEW

- A gauge manifold is a valuable service tool which allows the service technician a quick method of installing pressure gauges, checking the system conditions, charging, purging, evacuating, and charging oil to the compressor. A gauge manifold is a tool for diagnosis of trouble in a refrigeration sealed system.
- Never drop or abuse the gauge manifold.
- Keep the ports or charging lines capped when not in use.
- Never use any fluid other than clean refrigerant or oil.
- Gauges measure both temperature and pressure.
- The refrigeration gauge works on the principle of the Bourdon tube.
- High-side system pressure is determined by the type of refrigerant and the ambient temperature of the condensing medium.
- Low-side pressure is determined by the evaporator design temperature.
- A pressure-temperature chart can be used to determine saturated condensing temperature at a given pressure or evaporator pressure given the design temperature.
- 80% of refrigeration system problems are electrical in nature.
- Mechanical problems include: poor-capacity compressor, low-side leak, low-side restriction, high-side leak, high-side restriction, dirty condenser, and dirty evaporator or filters.
- The saddle valve and the process-tube method provide permanent field-installed access.
- Factory-installed valves are required on all new equipment, with the exception of Type I.

Problems and Questions

1. How often should the service technician change the seals in the gauge manifold?
2. How often are hose seals changed?
3. For reading pressures is it necessary to open any valves on the gauges?
4. What are the three positions of the gauge-manifold valves?
5. If you are servicing a sealed system, when are both valves open on the manifold?
6. If you are servicing a sealed system, when is the low-side valve open?
7. If you are servicing a sealed system, when is the high-side valve open?
8. What is the purpose of the Schrader valve?
9. If an R-12 system, with an air-cooled condenser has an ambient temperature of 80°F, what is the minimum operating head pressure (given a temperature difference of 20°F)?
10. What is the suction pressure of a refrigeration system with an evaporator design temperature of −10°F and R-502 as the refrigerant?
11. Why do some manufacturers only approve saddle valves or the process-tube access for warranty work?
12. What is the principle of operation for refrigeration gauges?
13. Where are temperature readings found on the gauges?
14. What is the color code for a service gauge manifold?
15. Whenever a sealed refrigeration system fails, there is an 80% chance it is electrical in nature and refriger-

ation gauges will not need to be installed. True or false?

16. Compound gauges measure both high-side and low-side parameters. True or false?
17. EPA requires that all refrigeration equipment that is Type I, have factory-installed access valves. True or false?
18. The *P-T* chart is a useful tool to determine the minimum operating pressure. True or false?
19. The refrigeration gauge works on the principle of the pitot tube. True or false?

R18 Mechanical Service Operations

AFTER STUDYING THIS CHAPTER, THE STUDENT WILL BE ABLE TO:

- Understand the EPA regulations and directives.
- Practice the necessary procedures for refrigeration containment.
- Explain the effects of the EPA regulations on new and existing refrigeration systems.
- Understand liquid and vapor refrigerant-recovery methods.
- Properly use measuring and testing equipment.
- Perform the proper techniques for evacuating a system.
- Explain the proper procedures for using gauge manifolds for determining pressure measurements.
- Understand the various methods of refrigerant leak detection.
- Charge the system with the proper amount of refrigerant and compressor oil.
- Explain how to charge refrigerant into the system as a vapor or as a liquid.
- Know how to use charging charts and calculators.
- Understand the importance of subcooling while charging refrigerant.
- Know how to add oil to the refrigeration system.

R18-1 PRACTICING REFRIGERATION CONTAINMENT

In 1990 the Congress of the United States passed a series of amendments to the Clean Air Act that greatly affected the refrigeration and air-conditioning industry. The act establishes a set of standards and requirements for the use and disposal of certain common refrigerants containing chlorine. These refrigerants, when allowed to escape into the atmosphere, are believed to be damaging the ozone layer. This stratospheric layer protects life forms from the harmful ultraviolet radiation from the sun.

EPA directives cover five main regulations:

1. Refrigerant emissions to the atmosphere must be reduced. (The no-venting rules)
2. A certification program is established for technicians. Only certified technicians can purchase restricted refrigerants or service equipment that contains these refrigerants.
3. Refrigerant reclaiming equipment must be certified.
4. Substantial leaks must be repaired.
5. Safe disposal regulations must be followed.

R18-1.1 Refrigerant Leaks

Certain refrigerant leaks are difficult to prevent. EPA, therefore, will permit leaks under the following conditions:

1. "De minimis" releases. These are releases that occur even when the approved procedures are followed, including the use of certified recovery and recycling machines.

2. Refrigerant released in the normal operation of equipment. Regulations require that certain leaks must be repaired, however.
3. R-22 mixed with nitrogen for leak testing. Nitrogen cannot be added to a charged system for leak testing. The refrigerant in the system must first be recovered. Pure CFCs or HCFCs are not to be released for leak testing.
4. The release of refrigerant from purging hoses or hose connections during charging or service is permitted. Low-loss hose fittings are required on all recovery and recycling equipment manufactured after November 15, 1993.

R18-1.2 Effect of EPA Regulations

In evaluating the effects of the new regulations, two conditions need to be considered:

1. New systems being evacuated and charged for the first time; and
2. Existing systems that have been operated for some time and require service.

Each needs to be treated differently to comply with the regulations and is described below.

New Systems

The system has been constructed recently. Assume all of the components in the system are assembled and that the system is ready for leak testing, evacuation, and charging. The technician shall proceed, in support of the EPA regulations, as follows:

1. The gauge manifold is installed with hose connections to both the high and the low sides of the system in the regular manner.
2. The system is then pressurized using nitrogen and a tracer of R-22 for leak testing. The pressures in the system should be comparable to the operating pressures unless local codes specify other testing pressures. Leak testing can be done with a halide torch or an electronic leak detector. If leaks are found, they must be repaired and the system retested.
3. The system is then evacuated, venting the test charge to the atmosphere.
4. It is advisable at this point to run a *high-vacuum* test. This means pulling down the system to at least 29 in. Hg and closing off the valves on the gauge manifold, letting the system stand overnight in a deep vacuum. If the pressure in the system rises during the test period, there is a leak. If a leak is indicated, be certain that it is not in the gauge manifold hose connections.
5. Based on the system being leak-tight, it is ready for charging. For most built-up systems, refrigerant is added to the system until the liquid-line sight glass shows a solid stream.

Existing Systems

This is a system that has been in operation but requires repair. It may have a refrigerant leak or a defective system component that needs replacing. The refrigeration circuit needs to be opened up for service.

Under these conditions, EPA has very strict rules. If there is still refrigerant in the system, the refrigerant

PRACTICING PROFESSIONAL SERVICES

Since venting refrigerant intentionally to the atmosphere is now illegal according to the EPA federal regulations, refrigerant recovery is mandatory. Allow extra time during your service procedures to pull refrigerant from a system if necessary. Utilize the system as much as possible to save time. If the system has a liquid receiver and a king valve, then pump down as much refrigerant as possible and store it in the system.

To avoid cross-contamination of refrigerant, identify all refrigerant storage containers and refrigeration systems with a permanent marking system. Be sure to evacuate empty cylinders prior to use. If you are unsure of the type of refrigerant in a recovery cylinder, then use the following procedures.

1. Test the temperature of the tank.
2. Measure the pressure of the tank.
3. Compare the temperature and pressure on a *P-T* chart.
4. If there is a match, then you determined the correct refrigerant within the cylinder.
5. If there is not a match using this method, you may have a contaminated cylinder and it will have to be sent to a reclamation site.

charge must be recovered before the system is opened. The kind of recovery operation is dependent on the type of system. The categories of systems are as follows:

1. Small appliances, such as home refrigerators and freezers, having a refrigerant charge of 5 pounds or less.
2. High-pressure appliances, such as residential air conditioners and heat pumps and most commercial air-conditioning equipment.
3. Low-pressure appliances, such as centrifugal chillers.
4. Very-high-pressure equipment, with a boiling point below –50°C (–58°F) at atmospheric pressure.

R18-1.3 Evacuation Requirements

Each one of the appliance groups has its own requirement for level of evacuation, as follows:

Small Appliances

There are two methods of evacuation. The recovery requirements differ for each method.

The *active* method uses self-contained recovery equipment. This method has its own compressor to pump out the system. It is required to reduce the system pressure to 4 in. Hg.

The *passive* method uses system-dependent recovery equipment. This method uses only the small-appliance compressor to recover the refrigerant. With this system, 90% of the refrigerant must be recovered if the compressor is running and 80% for a non-running unit. Recovery that uses only a chilled recovery tank would qualify for this method.

High-Pressure Appliances

The recovery requirements depend on the refrigerant used and the amount of the charge, as shown in the following table:

Required Levels of Evacuation for High-pressure Appliances

	Vacuum Levels Expressed in Inches of Mercury	
	Manufacturing Date	
Recovery Equipment Refrigerant and Charge	Before 11/15/93	After 11/15/93
HCFC-22 Appliance, < 200 lb Charge	0	0
HCFC-22 Appliance, > 200 lb Charge	4	10
Other High-pressure Appliance < 200 lb Charge	4	10
Other High-pressure Appliance > 200 lb Charge	4	15
Very-high-pressure Equipment	0	0

Referring to the table above, a zero (0) vacuum is atmospheric pressure. A perfect vacuum is 30 inches Mercury.

Low-Pressure Appliances

To recover the refrigerant from centrifugal chillers and other low-pressure units, a combination of liquid- and vapor-recovery methods is used. The liquid-recovery procedure will recover about 70% of the charge (Fig. R18-1).

The balance of the refrigerant is removed by the vapor method (Fig. R18-2).

The compressor, in the vapor-recovery machine, creates a vacuum which draws out the vapor from the system. Some of these machines will also remove part of the liquid, which is separated out by an accumulator or another arrangement.

R18-1.4 The Recovery Cylinder

The *recovery cylinder* is made especially for recovering refrigerants (see Figs. R18-2 and R18-3).

Figure R18-1 Liquid recovery using a liquid pump.

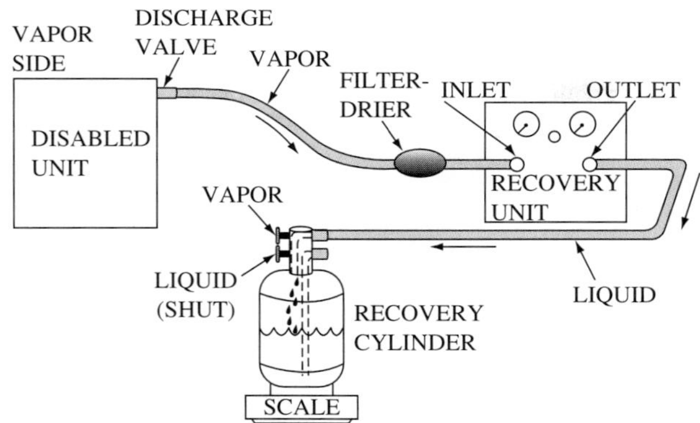

Figure R18-2 Vapor recovery.

1. It must have DOT (Department of Transportation) approval.
2. It should be color coded, gray with a yellow top, for all recovered refrigerants.

The recovery cylinder is usually supplied with the recovery machine. A fill-limit device is often part of the recovery equipment.

R18-2 LIQUID REFRIGERANT RECOVERY

For liquid recovery either a liquid pump (Fig. R18-1) can be used or the differential-pressure method (Fig. R18-3). This latter method is also known as the "push-pull" technique and is the most common method.

Referring to Fig. R18-3, the liquid is forced out of the unit using the recovery machine to pull a vacuum on the recovery container. At the same time, the pressure is increased in the unit. This causes rapid movement of the liquid.

The advantage of the liquid method of transfer is that it is much faster than the vapor-recovery method. In order for the liquid transfer method to be feasible, suitable access fittings must be provided to access liquid as well as vapor in the equipment being serviced. The final evacuation must be done by the vapor transfer method.

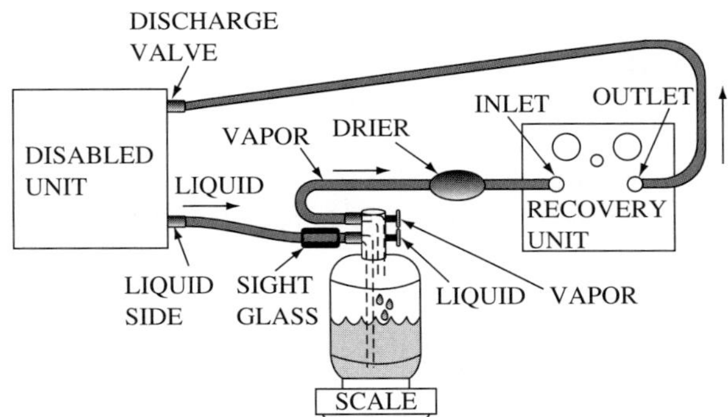

Figure R18-3 Liquid recovery using differential-pressure method.

R18-3 VAPOR REFRIGERANT RECOVERY

A typical recovery procedure with a vapor-type machine is as follows:

1. First, it is important to be certain which refrigerant is being recovered and that the machine being used is certified for this service.
2. The system must be turned off during this process and have all control valves open.
3. The recovery cylinder must be evacuated prior to the recovery operation and be standing in an upright position.
4. Connections are made to the system as shown in Fig. R18-2. Whatever type of fill limit device is used must be in service.
5. The valves accessing the gauges must be open.
6. The vapor valve on the recovery cylinder is open.
7. The recovery machine is turned on.

The vapor-recovery machine is equipped with a low-pressure cutout. The setting on this control is part of the certification. These controls are usually set below 29 in. Hg for low-pressure refrigerants and at 10 in. Hg for high-pressure refrigerants. The low-pressure control setting is made at the factory and cannot be altered in the field. If the control setting is too low for a particular operation, the technician can shut off the machine manually and save operating time.

The machine also has a high-pressure control. The setting is non-adjustable and is based on the refrigerants for which the equipment is used. The high-pressure control will shut off the machine in case there is an obstruction in the high-pressure lines, loss of cooling, an over filled cylinder or air in the system. The setting of this control is part of the certification.

The recovery machine is usually equipped with filters and separators which are used primarily to protect the recovery equipment. The machine usually has two circuits for oil separation: one to remove system oil from incoming vapor, and the other to recover compressor oil coming from the recovery machine. The oil from the system is drained and properly disposed of, and the oil from the recovery unit flows from the separator back to the recovery compressor.

When the machine shuts off on the low-pressure control, the evacuation is not necessarily completed. If the machine remains idle for a short period of time the pressure may creep up. The recovery machine should be run again. Usually, when the machine cuts off twice, the recovery is considered completed.

R18-4 MEASURING AND TESTING EQUIPMENT

In refrigeration (and air-conditioning) work, many different kinds of instruments and test equipment are used. Previous chapters have detailed basic hand tools of the toolkit and certain test equipment for general use. This chapter reviews measuring and testing equipment needed in the evacuation and charging operations.

R18-4.1 Temperature Measurements

When analyzing a refrigeration system, accurate temperature readings are important. The most common temperature measuring device is the pocket glass thermometer, illustrated in Fig. R18-4.

Note that it fits into a protective metal case. The thermometer head has a ring for attaching a string to suspend it if needed. The temperature ranges of glass thermometers vary, but –30 to +120°F is a common scale for refrigeration systems, and the thermometer is calibrated in 2° marks. Some have a mercury fill, but others use a red fill that is easier to read.

To check the calibration of a pocket glass thermometer, insert it into a glass of ice water for several minutes. It should read 32°F, plus or minus 1°F. Should the fill separate, place the thermometer in a freezer, and the resulting contraction will probably rejoin the separated column of fluid. Another way to connect the separated fill is to carefully heat the stem, not the bulb. In most cases the liquid will *coalesce,* or grow together, as it expands.

Another form of pocket thermometer is the dial type, shown in Fig. R18-5. It, too, has a carrying case with pocket clip. The dial thermometer is more convenient or more practical for measuring air temperatures in a duct. The stem is inserted into the duct, but the dial remains visible. Again, several ranges are available depending on the accuracy needed and the nature of the application. A common refrigeration range is –40 to +160°F. A different type of dial thermometer is the superheat thermometer, illustrated in Fig. R18-6.

The highly accurate expansion-bulb thermometer is used to measure suction-line temperature(s) in order to calculate, check, and adjust superheat. A common range is –40 to 65°F. The sensing bulb is strapped or clamped to the refrigerant line and covered with insulating material (such as a foam rubber sheet) to prevent air circulation over the bulb while a reading is taken. The superheat bulb thermometer can be used to measure air or water temperatures as well.

Figure R18-5 Dial thermometer. (Courtesy of Robinair Division, SPX Corporation)

The thermometers just described have been the basic temperature-measuring tools for many years; however, they do have certain limitations. For example, the operator is required to be present in the immediate area during the time of reading. An example of this limitation would be trying to measure the inside temperature of your home refrigerator without opening the door.

The thermometer illustrated in Fig. R18-7 is also a hand-held device, but it has a probe length of 30 in. for "remote" temperature measurement.

Different probes are available for measuring surface temperatures. In refrigeration use, the surface probe determines superheat settings of expansion valves, motor temperature, condensing temperature, and water temperature. This is a useful instrument for many applications but it is limited since only one reading at a time can be taken.

As a result of the rapid development of low-cost electronic devices, the availability and use of electronic thermometers is now very common. The electronic thermometer as illustrated in Fig. R18-8 consists of a tester with provision to attach one or several (three to six) sensing leads.

The sensing-lead tip is a thermistor element. Its resistance changes with temperature changes, varying the electrical current in the test circuit. The changes in electrical current are then converted to temperature readings. The

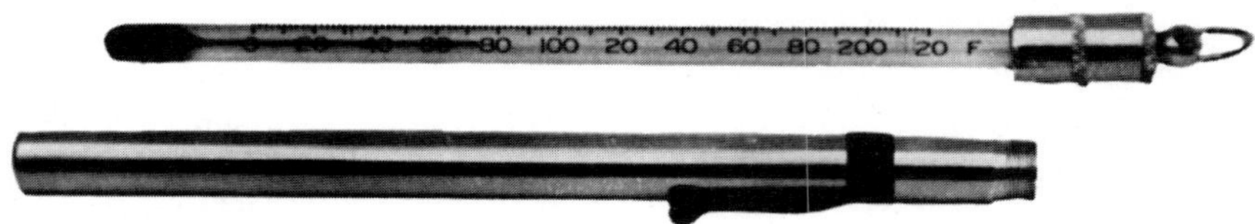

Figure R18-4 Glass thermometer. (Courtesy of Robinair Division, SPX Corporation)

Figure R18-6 Dial thermometer with sensing tube. (Courtesy of Marsh Instrument Company)

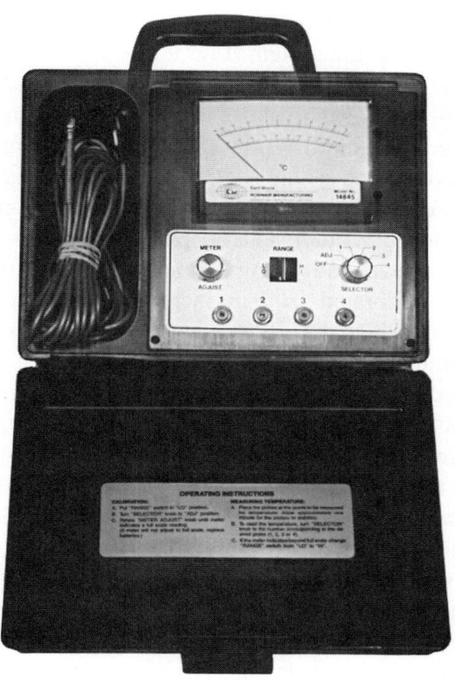

Figure R18-8 Electronic thermometer. (Courtesy of Robinair Division, SPX Corporation)

sensing leads vary in length depending on the make of the unit, but extensions can be used for remote testing.

Once the sensing probes are positioned in the areas to be tested, the technician may record temperatures without actually going into each test area, such as the refrigerator, walk-in cooler, freezer, or air duct. The sensing probe can also be used to check superheat.

Sometimes it may be necessary to record temperatures over long periods of time such as a day or even a week, in order to examine the changes in system conditions. Recording thermometers and compact, portable recording thermometers, (Fig. R18-9), which consist of a hand-wound chart-driving unit, take the guess work out of setting the system operating conditions or diagnosing and locating trouble areas—as well as providing permanent records of the results.

The final selection of temperature-measuring instruments will depend on the scope of work with which a technician is involved. In servicing HVAC/R equipment, the technician will need a variety of thermometers. It is important to remember that these are sensitive devices and require consistent care and calibration to provide accuracy and reliability.

R18-4.2 Pressure Measurements

Temperature measurements are usually taken outside the operating system, but it is also necessary for the service technician to know what is going on inside the system. This must basically be learned from pressure measurements.

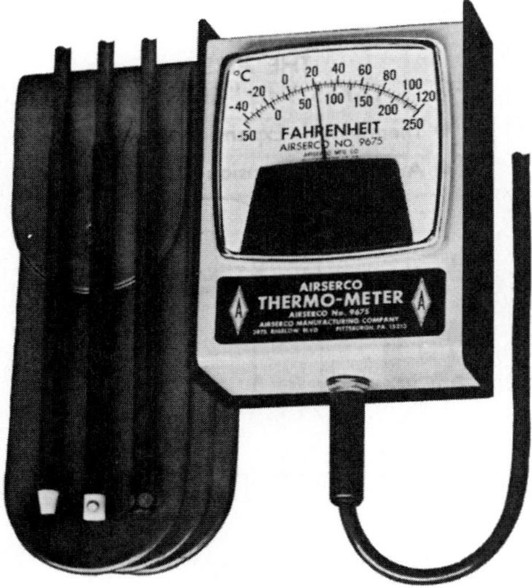

Figure R18-7 Remote thermometer. (Source: Airserco)

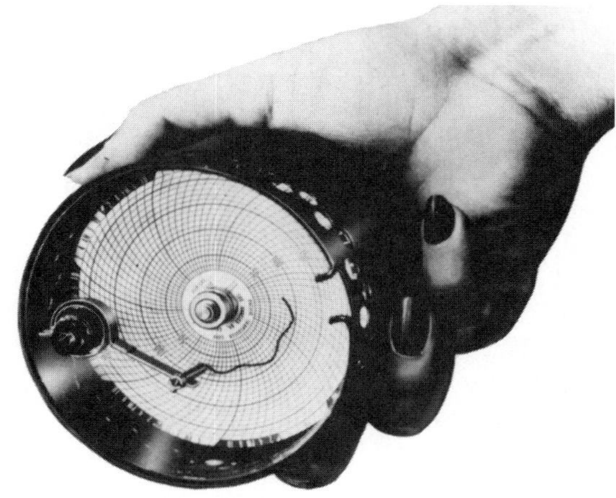

Figure R18-9 Recording thermometer. (Source: Airserco)

Two pressure gauges are necessary to determine the performance of the system (Fig. R18-10).

On the right (Fig. R18-10) is the high-pressure gauge, which measures high-side or condensing pressures. It is normally graduated from 0 to 500 psi in 5-lb increments.

The compound gauge (left, in Fig. R18-10) is used on the low side (suction pressures) and is normally graduated from a 30-in. vacuum to 120 psi; thus, it can measure pressure above and below atmospheric pressure. This gauge is calibrated in 1-lb increments. Other pressure ranges are available for both gauges, but these two are the most common.

Note that on the dial faces there is an *intra-scale* which gives the corresponding saturated refrigerant temperatures at a particular pressure. In Fig. R18-10 the dials are marked for R-12 and R-22 refrigerants. Gauges for other refrigerants and for SI measurements are also available.

A service device that includes both the high-pressure and compound gauges is called a *gauge manifold.* It enables the serviceman to check system operating pressures, add or remove refrigerant, add oil, purge noncondensibles, bypass the compressor, analyze system conditions, and perform many other operations without replacing gauges or trying to operate service connections in inaccessible places.

The testing manifold as illustrated in Fig. R18-11 consists of a service manifold containing service valves.

On the left the compound gauge (suction) is mounted and on the right is the high-pressure gauge (discharge). On the bottom of the manifold are hoses which lead to the equipment suction service valve (left), refrigerant drum (middle), and the equipment discharge, or liquid-line valve (right).

Many equipment manufacturers color code the low-side gauge casing and hose blue and the high-side gauge and hose red. The center or refrigerant hose is white. This system is very helpful to avoid crossing hoses and damaging gauges. A hook is provided on which to hang the assembly and free the operator from holding it.

By opening and closing the refrigerant valves on gauge manifold A and B (Fig. R18-12), different refrigerant flow patterns can be obtained.

The valving is so arranged that when the valves are closed (front seated), the center port on the manifold is closed to the gauges (Fig. R18-12a). When the valves are in the closed position, gauge ports 1 and 2 are still open to

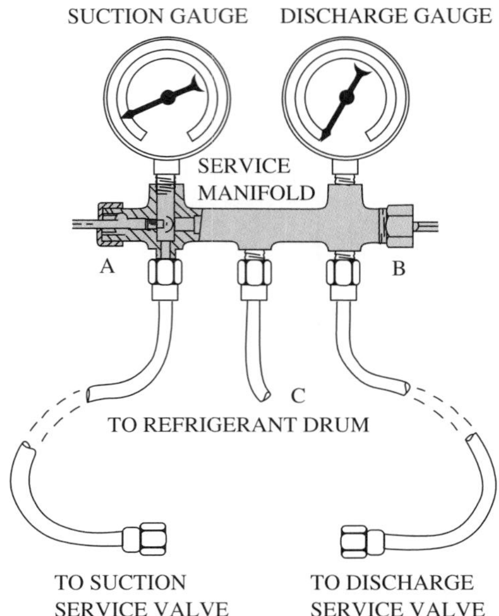

Figure R18-11 Testing manifold.

Figure R18-10 Pressure gauges. (Courtesy of Robinair Division, SPX Corporation)

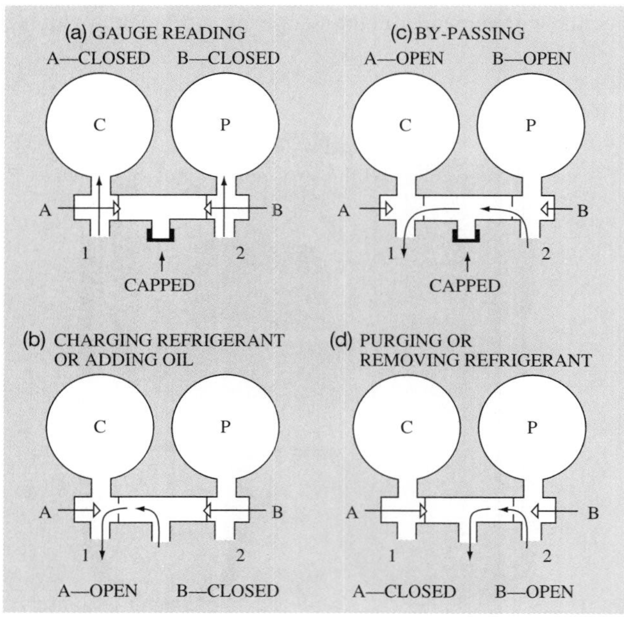

Figure R18-12 Manifold valve operation.

the gauges, permitting the gauges to register system pressures.

With the low-side valve (1) open and the high-side valve (2) closed (Fig. R18-12b), the refrigerant is allowed to pass through the low side of the manifold and the center port connection. This arrangement might be used when refrigerant or oil is added to the system.

Fig. R18-12c illustrates the procedure for bypassing refrigerant from the high side to the low side. Both valves are open and the center port is capped. Refrigerant will always flow from the high-pressure area to a lower-pressure area.

Fig. R18-12d shows the valving arrangement for purging or removing refrigerant. The low-side valve is closed. The center port is connected to an empty refrigerant drum. The high-side valve is opened, permitting a flow of high pressure out of the center port.

The method of connecting the gauge manifold to a refrigerant system depends on the state of the system—that is, whether the system is operating or just being installed. For example, let's assume that the system is operating and equipped with back seating in-line service valves (Fig. R18-13).

The first step is to purge the gauge manifold of contaminants before connecting it to the system. Purging and connecting is done with the following procedure:

1. Remove the valve stem caps from the equipment service valves and check to be sure that both service valves are back seated.
2. Remove the gauge-port caps from both service valves.
3. Connect the center hose from the gauge manifold to a refrigerant cylinder, using the same type of refrigerant that is in the system, and open both valves on the gauge manifold.
4. Open the valve on the refrigerant cylinder for about two seconds, and then close it. This will purge any contaminants from the gauge manifold and hoses.
5. Next, connect the gauge manifold hoses to the gauge ports—the low-pressure compound gauge to the suction service valve and the high-pressure gauge to the liquid-line service valve, as illustrated in Fig. R18-14.
6. Front seat or close both valves on the gauge manifold. Crack (turn clockwise) both service valves one turn off the back seat. The system is now allowed to register on each gauge. With the gauge manifold and hoses purged and connected to the system, we are free to perform whatever service function is necessary within the refrigeration cycle.

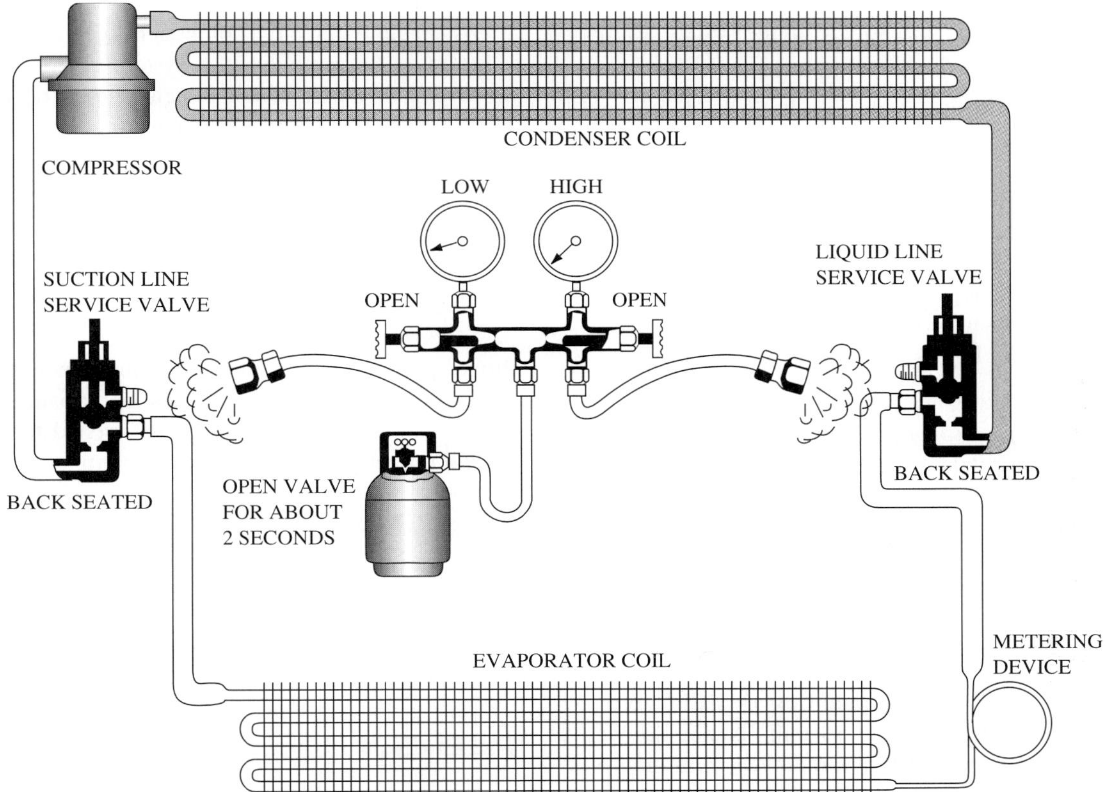

Figure R18-13 Purging gauge manifold.

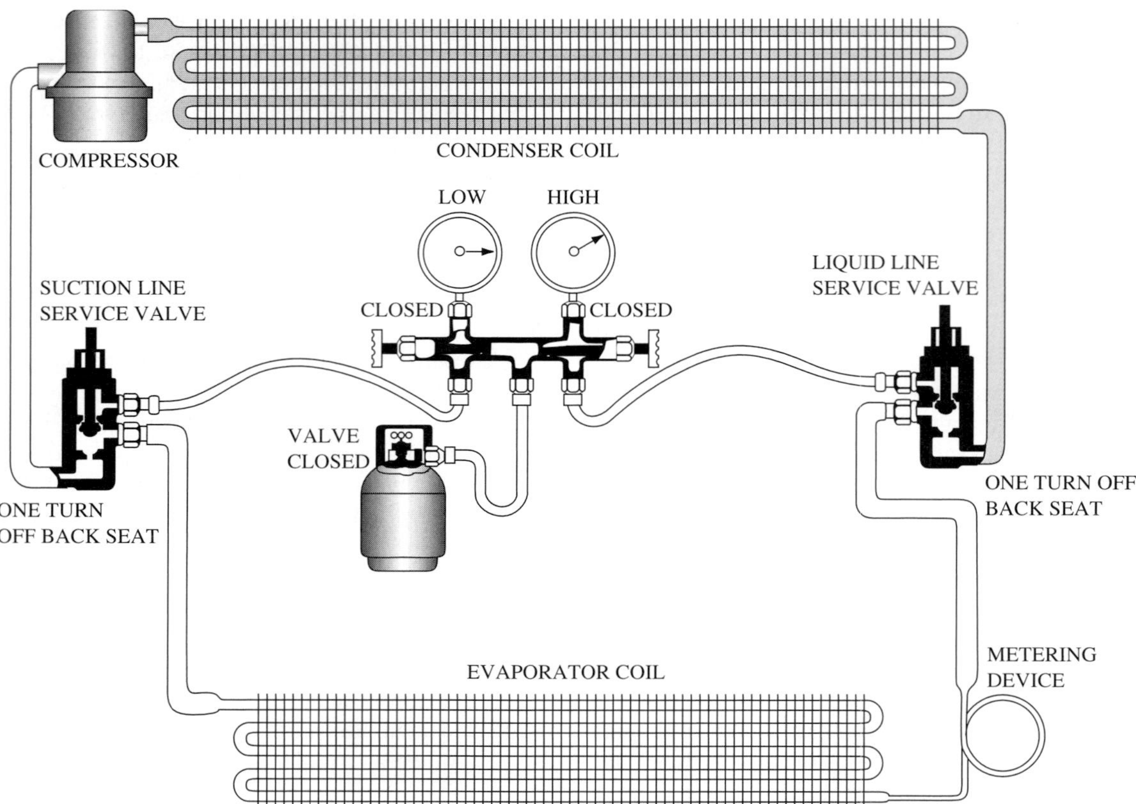

Figure R18-14 Connecting manifold.

To remove the gauge manifold from the system, follow this procedure:

1. Back seat (counterclockwise) both the liquid and suction service valves on the compressor.
2. Remove hoses from gauge ports and seal ends of hoses with ¼-in. flare plugs to prevent hoses from being contaminated. (Some manifold assemblies have built-in hose seal fittings).
3. Replace all gauge-port and valve-stem caps. Make sure that all caps contain the gaskets provided with them and are tight.

The manifold and gauges are necessary tools to perform many system operations. Once the system has been completed and cleaned of most of the air by purging, it must be tested for leaks. Also, whenever a component has been repaired or replaced, it is imperative that the entire system be checked for leaks. Leak testing is discussed further later in this chapter.

R18-4.3 Purging

Purging the entire system with pure CFC or HCFC refrigerants, following servicing the system, is not allowed under EPA regulations. If they are removed from the system, they must be recovered. If the system has been opened for any reason, and air and moisture permitted to enter, the system should be evacuated before recharging.

Whenever a defective component such as an expansion valve is to be removed, the system should be pumped down and that part of the system should be isolated by means of the service valves. Then when the new component is installed the lines should be purged from both sides. Pumping down means to store the refrigerant in the receiver or condenser.

R18-5 EVACUATION

Proper *evacuation* of a unit will remove noncondensibles (mainly air, water, and inert gases) and assure a tight, dry system before charging. The tools needed to evacuate a system properly are a good vacuum pump and vacuum indicator.

Fig. R18-15 shows a tool used to remove and replace the valve core from Schrader or Dill valves. By removing the core, a vacuum can be pulled much quicker. The valve core is a restriction to gas flow and should be removed. This can be done with this tool without opening the system to the atmosphere.

R18-5.1 Vacuum Pump

A vacuum pump, illustrated in Fig. R18-16, is somewhat like an air compressor in reverse. Most pumps are driven by a direct- or belt-driven electric motor, but gasoline-

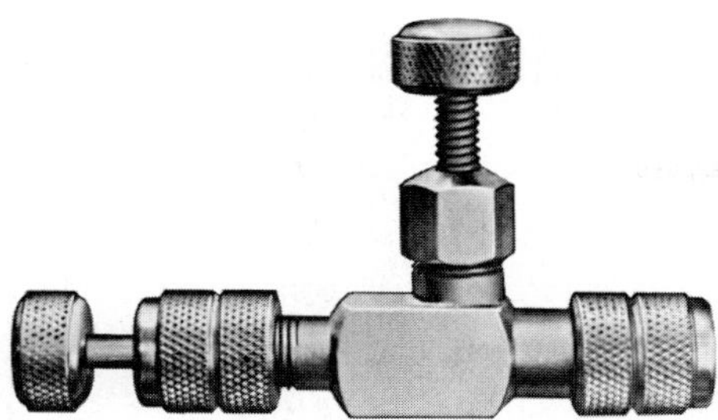

Figure R18-15 Valve core remover. (Courtesy of Yellow Jacket Division, Ritchie Engineering Co.)

engine-driven pumps are also available. The pump may be single or two stage depending on the design. Most pumps for normal field service are portable. They have carrying handles or are mounted on wheel dollies.

Vacuum pump sizes are rated according to the free air displacement in cubic feet per minute (cfm), or liters per minute (lpm) in the SI system. Specifications may also include a statement as to the degree of vacuum the pump can achieve, expressed in terms of microns.

R18-5.2 Microns

When the vacuum pressure approaches 29.5 to 30 in. on the compound gauge, the gauge is working within the last half-inch of pressure, and the readout beyond 29.5 in. is not reliable for the single deep vacuum method. The industry has therefore adopted another measurement, called the *micron*. The micron is a unit of linear measure equal to 1/25,400 of an inch and is based on measurement above total absolute pressure, as opposed to gauge pressure, which can be affected by atmospheric pressure changes. Fig. R18-17 shows a comparison of measurements starting at standard atmospheric conditions and extending to a deep vacuum.

Figure R18-16 Vacuum pump. (Courtesy of Robinair Division, SPX Corporation)

Boiling Point of Water		Unit of Absolute Pressure		
°F	°C	psia	Microns of Mercury	Units of Vacuum (in. Hg)
212	100	14.7	—	0
79	26	0.5	25,400	29.0
72	22	0.4	20,080	29.8
32	0	0.09	4,579	29.99
−25	−31	0.005	250	29.99
−40	−40	0.002	97	29.996
−60	−51	0.0005	25	29.999

Figure R18-17 Comparison of three different pressure measuring systems.

Fig. R18-17 not only demonstrates the comparison in units of measure but dramatically shows the changes in the boiling point of water as the evacuation approaches the perfect vacuum. This is the main purpose of evacuation—to reduce the pressure or vacuum enough to boil or vaporize the water and then pump it out of the system. It will be noted the compound gauge could not possibly be read to such minute changes in inches of mercury.

R18-5.3 High Vacuum Indicators

To measure these high vacuums the industry developed electronic instruments, such as the one shown in Fig. R18-18.

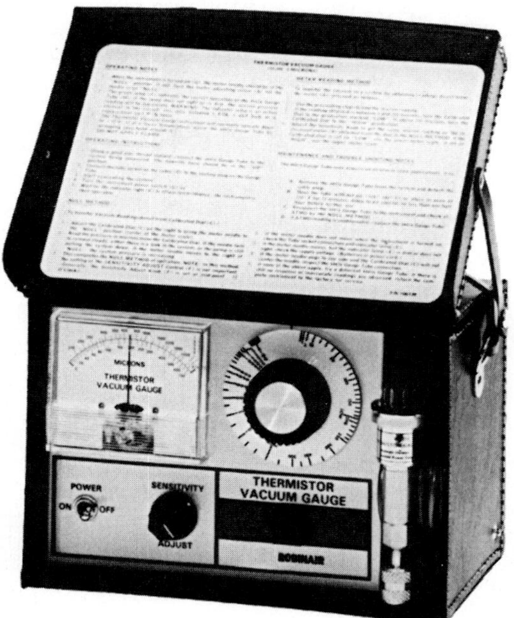

Figure R18-18 High vacuum gauge. (Courtesy of Robinair Division, SPX Corporation)

PREVENTIVE MAINTENANCE

Modern evacuation and charging equipment needs routine preventive maintenance similar to any piece of equipment. This includes changing the oils in the vacuum pumps and refrigerant-recovery equipment, replacing batteries in portable scales and leak detection instruments, and calibrating the refrigeration gauge manifold set.

The recovery machine normally has refrigerant filters and oil separators that may need replacing or cleaning also. Pressure cut out switches should be checked against a known working switch to determine accuracy. If the technician is unsure of the proper maintenance for the recovery machine, then consult the factory or wholesale house where you purchased the equipment.

Electronic thermometers need to be calibrated in an ice bath at 32°F to check for accuracy. If they cannot be adjusted within an acceptable tolerance, they should be sent to the factory for calibration.

In general, these are heat-sensing devices in which the sensing element, which is mechanically connected to the system being evacuated, generates heat. The rate at which heat is carried off changes as the surrounding gases and vapors are removed. Thus, the output of the sensing element (either thermocouple or thermistor) changes as the heat dissipation rate changes, and this change in output is indicated on a meter calibrated in microns of mercury.

The degree of accuracy of these instruments is approximately 10 microns, thereby approaching a perfect vacuum as shown in Fig. R18-17.

R18-5.4 Deep Method of Evacuation

The single deep-vacuum method is the most positive method of assuring a system free of air and water. It takes longer but the results are far more positive. A vacuum pump should be selected that is capable of pulling at least 500 microns and have a reliable electronic vacuum indicator. The procedure is illustrated in Fig. R18-19 and described below.

1. Install the gauge manifold as described earlier.

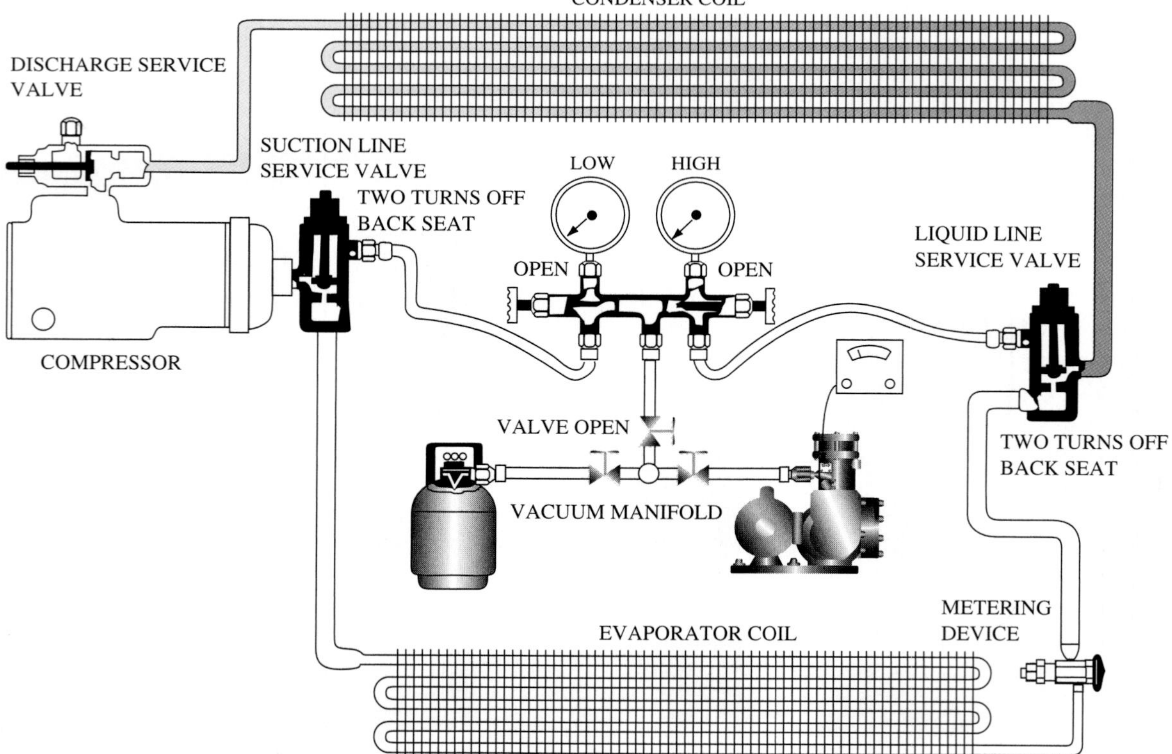

Figure R18-19 Deep vacuum method of evacuating the system.

2. Connect the center hose to the vacuum manifold assembly. This is simply a three-valve operation for attaching the vacuum pump and vacuum indicator and a cylinder of refrigerant, each with a shutoff valve.
3. Open the valves to the pump and indicator. Close the refrigerant valve. Follow the pump manufacturer's instructions for pump suction line size, oil, indicator location, and calibration.
4. Open wide both valves on the gauge manifold and midseat both equipment service valves.
5. Start the vacuum pump and evacuate the system until a vacuum of at least 500 microns is achieved.
6. Close the pump valve and isolate the system. Stop the pump for 5 minutes and observe the vacuum indicator to see if the system has actually reached 500 microns and is holding. If the system fails to hold, check all connections for tight fit and repeat evacuation until the system does hold.
7. Close the valve to the indicator.
8. Open the valve to the refrigerant cylinder and raise the pressure to at least 10 psig or charge the system to the proper level.
9. Disconnect the pump and indicator.

R18-6 REFRIGERANT CHARGING

Whether the system is a new one or an existing one that has been repaired, the final step in putting the system in operation is to charge it with refrigerant. In any event, the process is the same.

R18-6.1 How Much Refrigerant?

Some systems are more critical than others about the amount of the refrigerant charge. Systems that have a receiver—usually the larger systems—are not as critical since extra refrigerant can be stored in the receiver, and the expansion valve feeds the refrigerant into the evaporator as required to match the load.

In the smaller systems that do not have a receiver, any excess refrigerant will be stored in some part of the system where it reduces the effectiveness of that part and reduces the capacity of the system. If the system is short of refrigerant, the metering device is not supplied with a solid stream of refrigerant on full load and the evaporator will be starved for refrigerant. This will also reduce the capacity of the system. So, in the smaller systems it is critical that the proper charge be determined.

One way to determine the proper charge is to read it on the nameplate as shown in Fig. R18-20. This is the charge specified by the manufacturer.

On some systems, the length of the refrigerant lines is determined by the conditions of the installation. For example, on a split system the amount of charge required is affected by the length of the refrigerant lines. Most manufacturers give specific information for determining the charge on this basis. For example, the manufacturer may state, "charge adequate for matched system including up to 25 ft. of lines." Be sure to refer to the manufacturers' information.

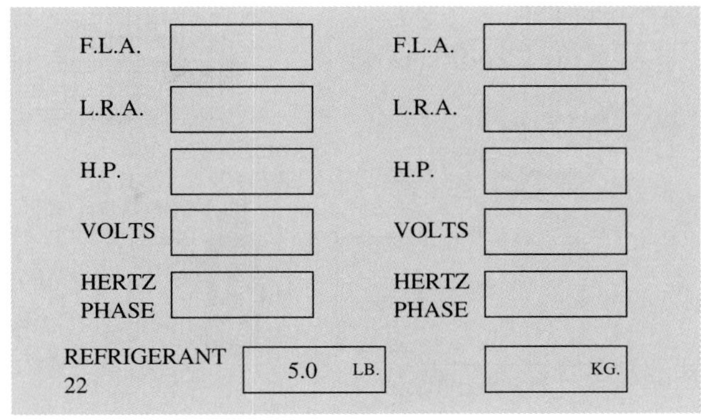

Figure R18-20 Unit nameplate.

R18-6.2 A Liquid or Vapor Charge

The unit can be charged with either liquid or vapor refrigerant. Whichever method is used, it is important to charge the system with the right amount of refrigerant and to protect the compressor from any damage that might be caused by liquid slugging the compressor.

Since the refrigerant cylinder is not filled over 80%, in an upright position the vapor is at the top and the liquid at the bottom. To charge with vapor, the refrigerant cylinder must be in the upright position. To charge with liquid, the cylinder is turned upside down, as shown in Fig. R18-21.

It is usually considered good practice to charge with vapor (Fig. R18-22) rather than liquid to prevent any danger of slugging the compressor with liquid. However, when using refrigerant blends, sometimes the tank will be constructed so that when the cylinder is right-side-up, it will discharge liquid. Refrigerant blends should not be charged as a vapor because the refrigerant will fractionate.

R18-6.3 Vapor Charging

Prior to charging, the system must be leak tested and evacuated. When the charging is started, the system is under vacuum so that when the refrigerant enters the system it will be drawn into the unit due to the difference in pressures.

To charge with vapor refrigerant, the system is connected to the gauge manifold in the usual way, as shown in Fig. R18-23. Both valves on the gauge manifold are backseated. When the refrigerant is released from the cylinder, the vapor flows into both sides of the system.

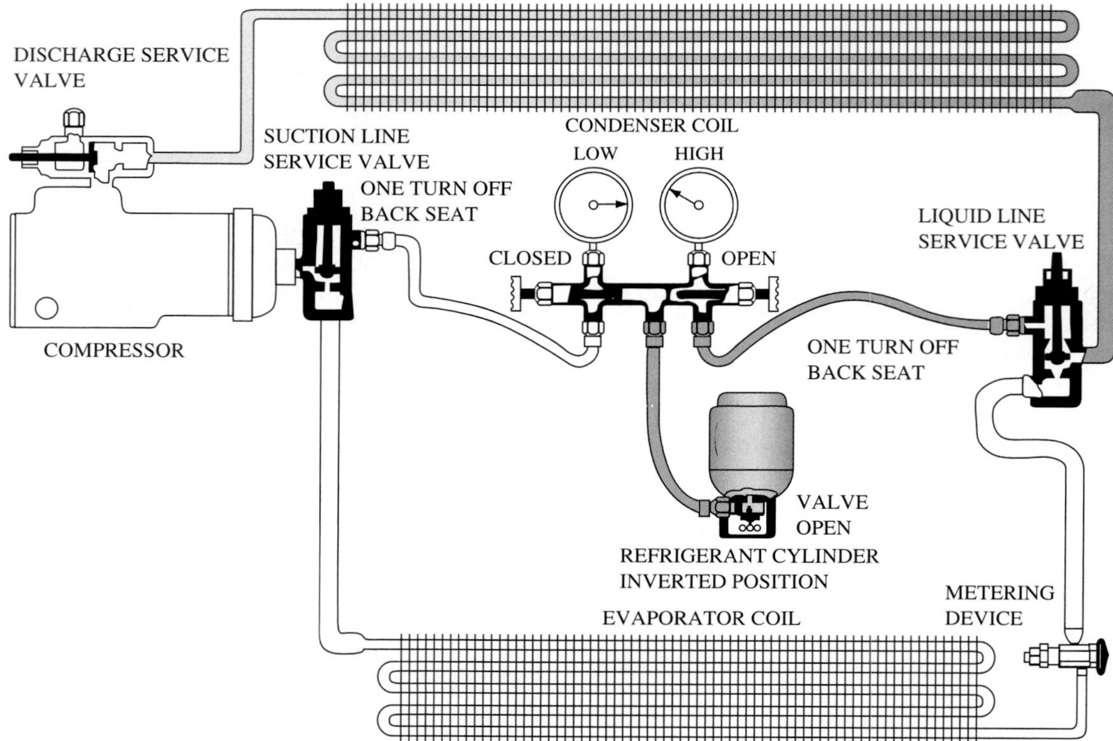

Figure R18-21 Liquid charging.

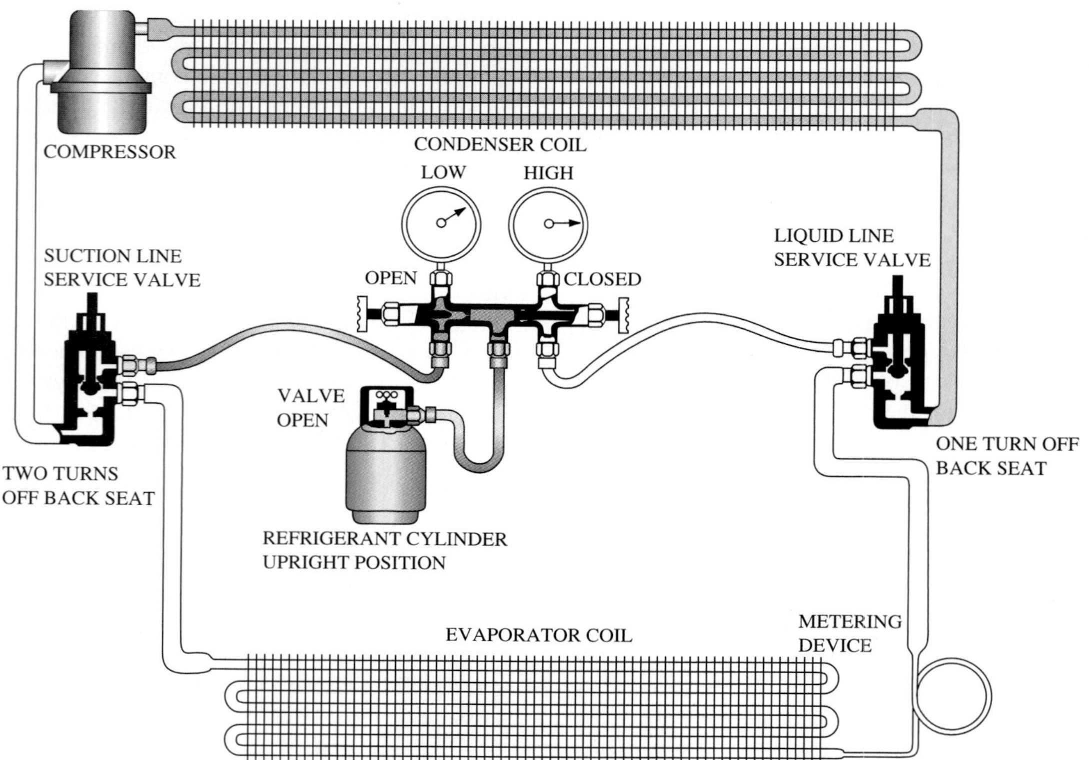

Figure R18-22 Vapor charging

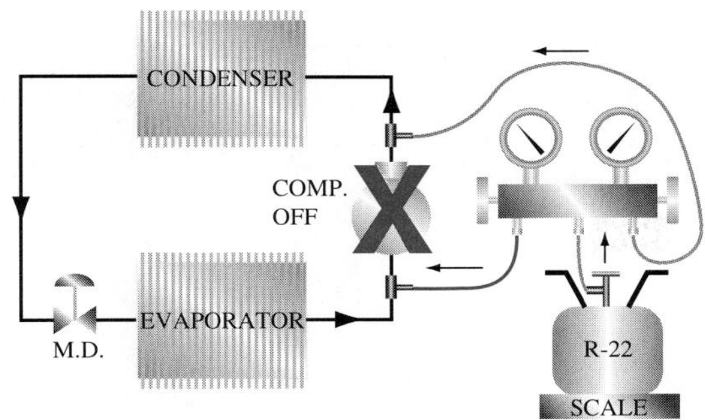

Figure R18-23 Both valves back seated.

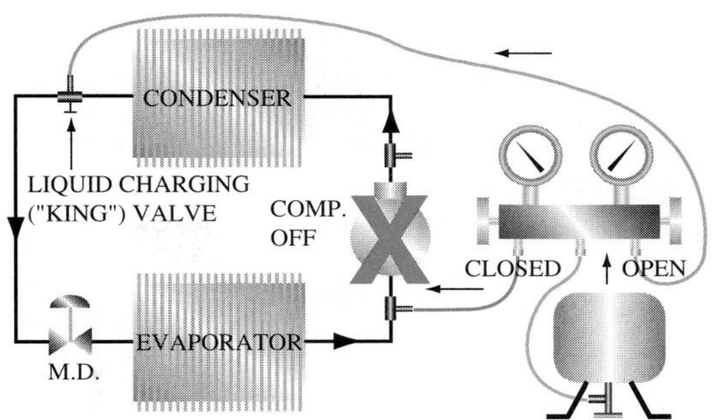

Figure R18-25 Liquid charging.

The refrigerant will stop flowing when the pressures equalize and no more refrigerant will enter the system.

The next step is to close (front seat) the high-side valve as shown in Fig. R18-24 and to continue the balancing of the vapor charging through the low side of the system.

At this point check the scales or charging cylinder to see how much of the charge has entered the system and how much more is needed. Normally about 50 to 75% of the required charge is completed.

The balance of the vapor refrigerant will be charged with the compressor running. Even with the compressor running the process slows down. One way to speed up the process is to add heat to the cylinder. This is done by placing it in a water bath at a temperature not to exceed 125°F.

The refrigerant measuring device should be watched closely. When the full charge has been added, turn off the compressor and disconnect the gauge manifold.

R18-6.4 Liquid Charging

Liquid charging is always much faster than charging with vapor. Liquid is charged on the high side of the system. On larger systems, 20 tons or more, a "king valve" located between the condenser and the metering device offers a convenient means of charging the system on the high side, with the compressor running.

On smaller systems the charging arrangement is shown in Fig. R18-25. The charging is done with the compressor off. The disadvantage of this method is that the full charge is seldom completed and the vapor method must be used to complete the process.

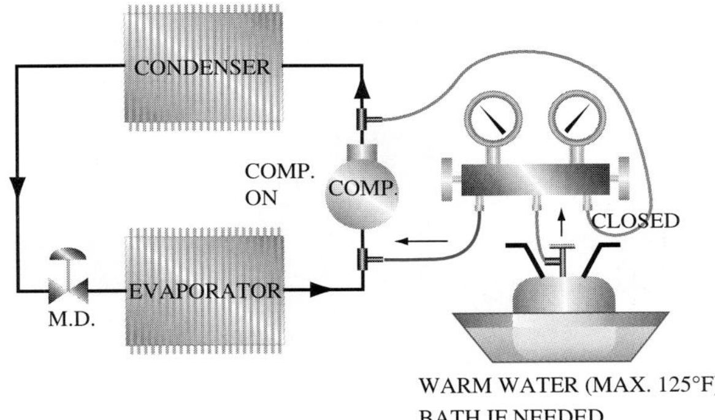

Figure R18-24 High side valve closed.

R18-6.5 Correct Charging

Many problems occur if the system is over charged or under charged, as shown in Fig. R18-26. The correct charge that is required and the charge that is installed should be carefully compared.

An undercharge can cause flash gas to form ahead of the metering device. This causes a low evaporator temperature and excess superheating of the suction gas. It can also cause poor cooling of a hermetically sealed gas-cooled motor. It can cause the discharge gas to be over heated. Sludge and harmful chemicals can form and the valves may carbonize.

An over charge can cause the metering device to over feed the evaporator. This causes liquid to flood back to the compressor. Liquid slugging will cause compressor damage. High-side pressure will increase, causing loss of capacity.

R18-6.6 Charging with a Charging Cylinder

The use of a charging cylinder, shown in Fig. R18-27, offers an accurate method of charging a small system or adding a make-up charge to a larger system. The scale on

CHARGE ACCURATELY

UNDER-CHARGE	PROPER	OVER-CHARGE
LOW LOWSIDE PRESSURE	LONG LIFE	HIGH HIGHSIDE PRESSURE
HIGH SUPERHEAT	SAFE OPERATION	HIGH DISCHARGE TEMP.
OVERHEATED COMP. MOTOR	DESIGN CAPACITY	FLOODBACK
OVERHEATED COMPRESSOR	PEAK EFFICIENCY	LOW SYS. CAPACITY
LOW SYS. CAPACITY		POOR EFFICIENCY
POOR EFFICIENCY		SLUDGE/CARBONIZATION
SLUDGE/CARBONIZATION		

Figure R18-26 Charging accurately.

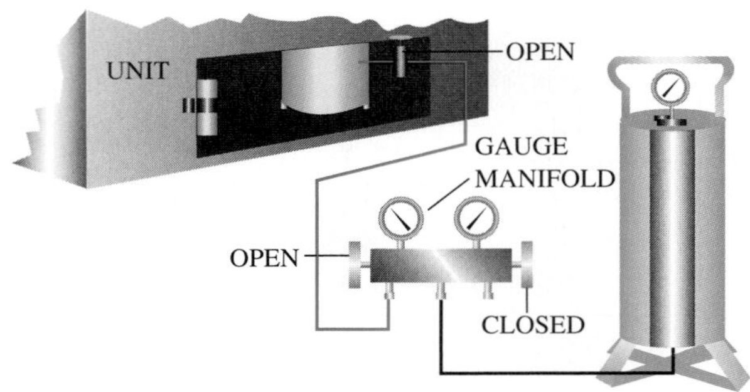

Figure R18-27 A charging cylinder.

the side of the container is graduated in pounds and ounces. The amount of the charge can be accurately measured.

R18-6.7 Charging by Weight

For a system that is too large for a charging cylinder, the best method is to charge by weight. Scales like the one shown in Fig. R18-28 can be used to support the refrigerant cylinder and to weigh it and the refrigerant in it before and after the charge has been made. The difference between the two readings will give the weight of the refrigerant used.

For example, if the initial weight of the cylinder and refrigerant (before charging) in Fig. R18-29 is 190 pounds and the end weight (after charging) is 150 pounds, 40 pounds of refrigerant have been charged. The position of the hoses should not be changed during charging as this can cause inaccurate weighing.

A battery-operated electronic scale is preferred. These scales vary in size according to the amount of refrigerant being used. For small charges, scales are available that read down to fractions of an ounce. For medium-sized jobs scales are available to the nearest ounce. For jobs of 50 tons or more, scales reading to the nearest pound are satisfactory.

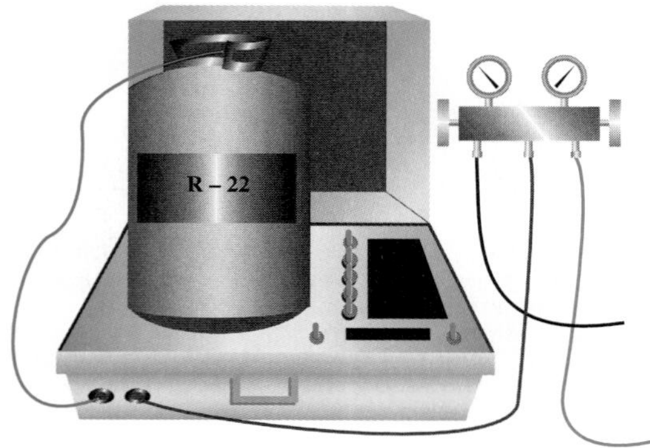

Figure R18-28 Weighing a charge.

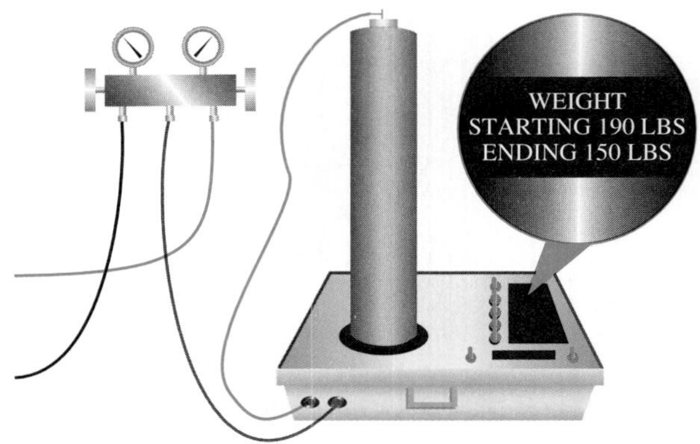

Figure R18-29 Calculating the amount of charge added to the system.

R18-6.8 Charging by Sight Glass

If a sight glass is installed in the liquid line, it can be used to determine when the charging is completed (Fig. R18-30). When the system is only partly charged, bubbles of refrigerant gas can be observed in the sight glass.

Charging is continued until the bubbles disappear and saturated liquid refrigerant is in the sight glass, as shown in Fig. R18-31. It is easy to mistake a full glass for an empty one. The glass looks the same when it is full or empty, since most refrigerants are colorless.

In following procedures for charging, it is important to be familiar with instructions provided by the manufacturer of the equipment. For example, even though there is a solid stream of liquid refrigerant going into the metering device, there may be a need for more refrigerant if specified in the manufacturer's instructions. Some manufacturers desire that an extra charge be added to provide addi-

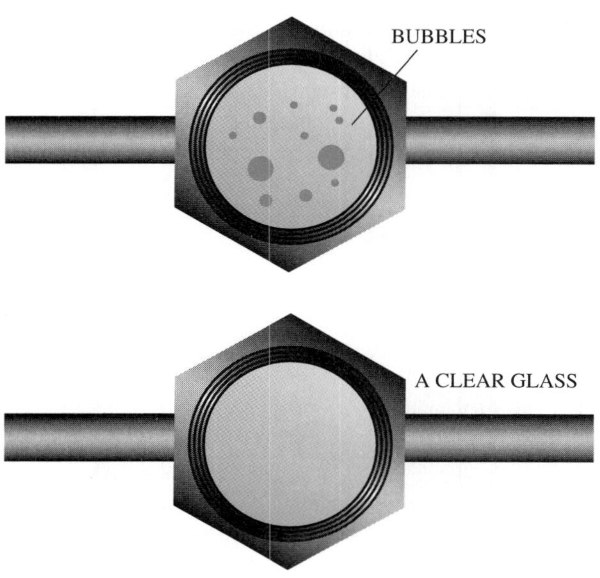

Figure R18-30 Sight glass indication of charge.

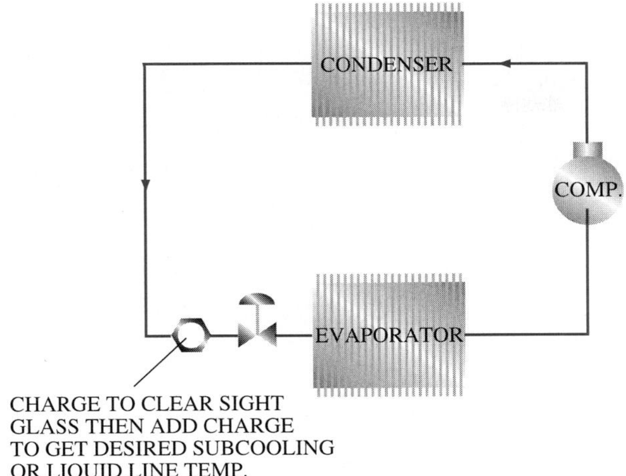

Figure R18-31 Charging by sight glass.

tional subcooling of the refrigerant leaving the condenser. Other manufacturers may require a certain liquid-line temperature, the amount being dependent on the operating conditions.

R18-6.9 Charging Charts and Calculators

Some manufacturers provide charging charts, charging calculators, or maximum-performance charts for checking the charge while the system is in operation. It is important the technician use whichever method the manufacturer provides to adjust the charge in factory-charged units (Fig. R18-32).

Some manufacturers' procedures require reading the outdoor dry-bulb temperature, the indoor wet-bulb temperature and the system pressures. With this information, tables are available for making any necessary adjustments.

R18-6.10 Charging for Proper Superheat

Manufacturers may recommend checking the vapor superheat after the system has been put in operation, to determine the proper refrigerant charge (Fig. R18-33).

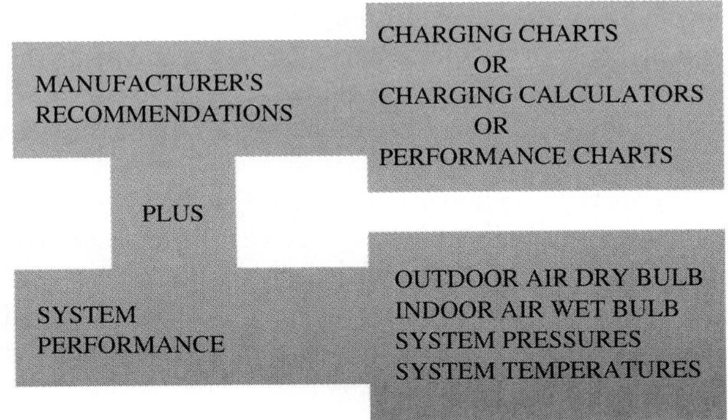

Figure R18-32 Adjusting charge.

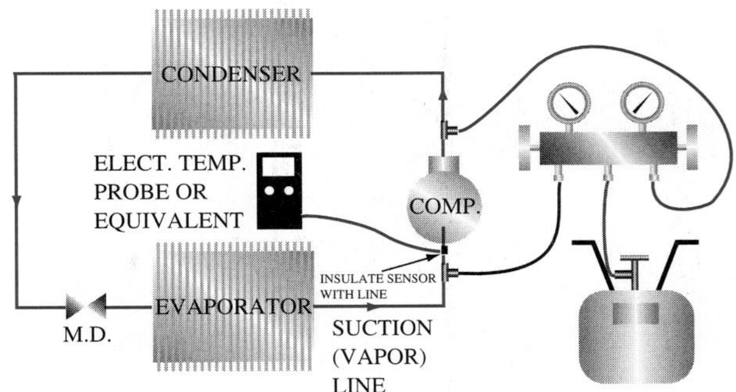

Figure R18-33 Checking the charge with superheat.

This applies only to fixed-metering-device systems, such as a capillary-tube system. The full-load superheat is usually 10°F to 15°F above the saturated refrigerant temperature in the evaporator.

The superheat on fixed-metering-device systems varies greatly under partial load conditions so it is important to follow the manufacturer's instructions, using the charts provided for the system being tested.

The superheat method is a very accurate means of checking the refrigerant charge (Fig. R18-34). About a 1% change in refrigerant charge will change the superheat 3°F or more.

In making these tests it is important to use a fast-reading resistance thermometer or an electronic temperature probe. Place the sensing element in good contact with a metal vapor line or suction line and insulate the element, along with the line.

Run the system for about 10 minutes to allow the temperatures and pressures to stabilize. Also, record the indoor and outdoor temperatures, since these are required in using the manufacturer's charts. Then read the pressure and temperature of the vapor line if it is a heat pump or suction line, or if it is a cooling-only line.

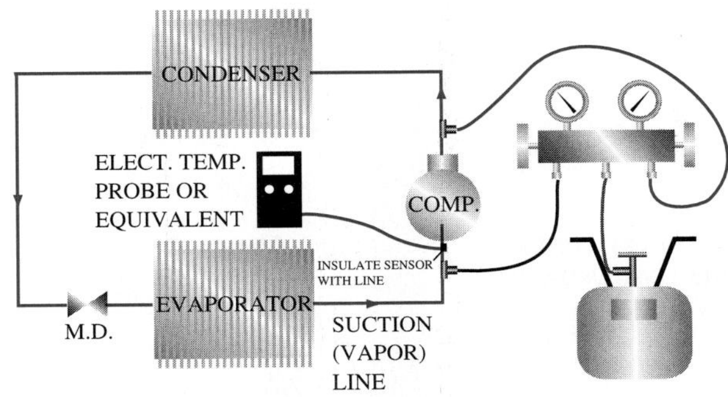

• 1% CHANGE IN CHARGE = 3 F° CHANGE IN SUPERHEAT

• SENSING ELEMENT MUST MAKE GOOD CONTACT

Figure R18-34 Relationship between charge and superheat.

ON THE JOB

Never use oxygen to leak test, purge, or pressure test piping or machinery.

Always use a gauge to meter gases when pressure testing systems.

Never exceed maximum field leak-test pressure of a system. Verify maximum allowable test pressures in the installation literature.

The full pressure from a nitrogen tank can cause a refrigerant tank to rupture. Always put the trace amount of refrigerant in first and valve off, before adding the nitrogen when leak testing a refrigeration system.

Use the pressure-temperature chart for the refrigerant being used and determine the saturation temperature that matches the pressure in the vapor line, or low side. Then subtract the saturation temperature from the actual vapor line temperature to obtain the superheat. Compare it with the manufacturer's recommendation. Make any adjustments necessary. Adding more charge gives less superheat; less charge gives more superheat.

For an example, refer to Fig. R18-35. Note that the superheat is too high by more than about 5°F. To correct this, add vapor refrigerant through the low-side port with the compressor running. If the superheat is too low by more than 5°F, remove vapor refrigerant by the proper recovery procedures.

After any charge adjustment, repeat the test procedure to be certain that the charge is within the proper range.

R18-6.11 Charging for Proper Subcooling

If the system uses a thermostatic expansion valve, the device will regulate the refrigerant flow over a wide range of load and charge conditions. The superheat method is therefore not satisfactory for testing the charge.

Some manufacturers recommend using subcooling to check the charge. Normally under full-load conditions, an air conditioner will have 5° to 15°F of subcooling in the condenser before the refrigerant reaches the metering device. The amount of subcooling can be determined by measurements similar to those used in measuring superheat, as shown in Fig. R18-36.

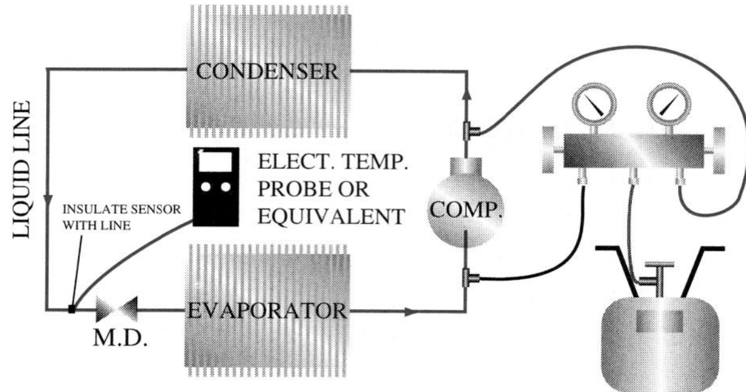

Figure R18-36 Measuring subcooling.

Compare the actual amount of subcooling with the manufacturer's recommendations. If the charge needs adjustment, more charge will increase the subcooling and less charge will decrease the subcooling.

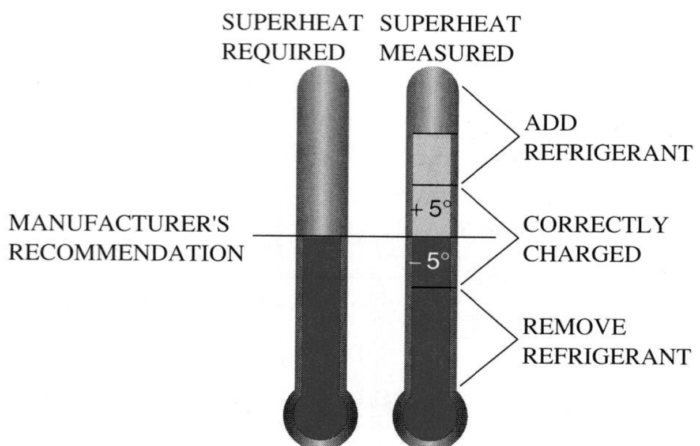

Figure R18-35 Adjusting superheat.

R18-7 LEAK TESTING

After a system has been repaired and before the final charge has been installed, the system needs to be leak tested. This is done by charging the system with an inert gas such as nitrogen or carbon dioxide mixed with 5% or 10% R-22 (Fig. R18-37). Oxygen should not be used, since it can cause an explosion.

A halide leak detector or an electronic leak detector can be used to check for leaks. A bubble test can also be made. The bubble test is not affected by the type of refrigerant used.

The inert gas cylinder must be equipped with a suitable regulator. Install a pressure-relief valve in the pressure-feed line with a setting not to exceed the manufacturer's safe limit. Put the refrigerant gas in first. Turn the valve off and remove the refrigerant cylinder. Connect the inert gas cylinder and charge the system to the desired pressure.

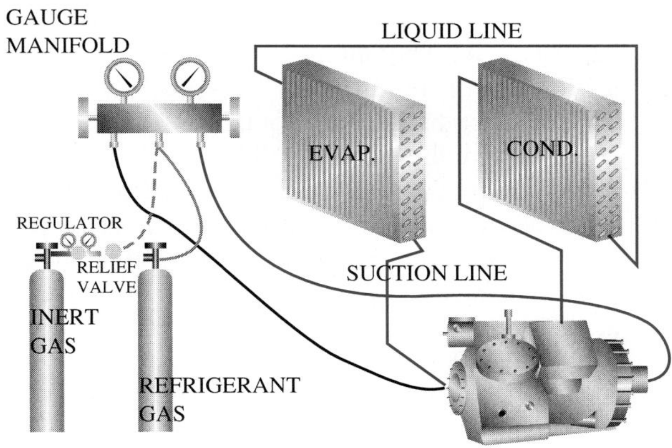

Figure R18-37 Testing for leaks.

R18-8 OIL CHARGING

Sometimes oil needs to be added to the system. Fig. R18-38 shows two methods.

On the left, the compressor is stopped, the crankcase is opened and oil is poured into the compressor. This is done prior to evacuating and final charging, since air and moisture can get into the system during this operation.

On the right, the compressor suction service valve is closed and the compressor is evacuated to a slight vacuum. The oil hose is fitted to the compressor, making an airtight connection and the needed oil is drawn into the compressor. The oil line must have some means of starting and stopping the flow, and the inlet must be below oil level.

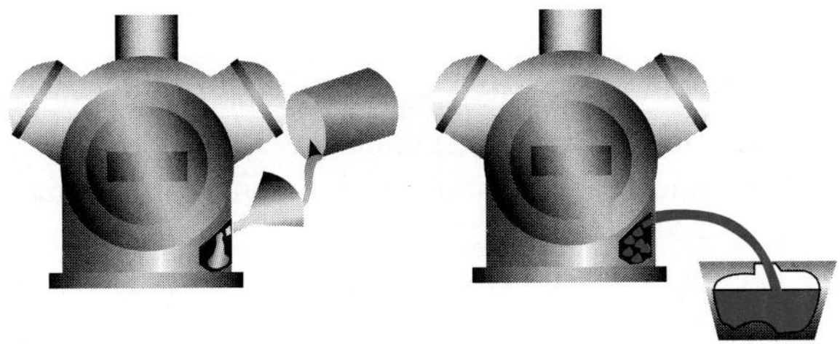

Figure R18-38 Adding oil.

Figure R18-39 Charging to the oil sight glass.

On compressors that have a crankcase sight glass (Fig. R18-39), the correct oil level can be accurately observed and the quantity adjusted to the level indicated in the sight glass. It is important, however, to observe the oil level after the system has been in operation, since some oil remains in the system and does not get back to the compressor.

REVIEW

- EPA directives:
  No venting
  Technician certification
  Refrigerant reclaiming equipment certification
  Recovery/recyling equipment certification
  Substantial leaks in systems containing 50 lbs charge must be repaired.
  Safe disposal
- Allowable release of refrigerants:
  "De Minimus" releases.
  Leaks that occur using an approved procedure or during normal operation.
  R-22 mixed with dry nitrogen, used for leak detection.
  Purging refrigerant hoses.
  Purging air from refrigerant cylinders.
- Required vacuum levels for recovery must be maintained based on amount of charge, type of refrigerant, and type of system.
- DOT-approved cylinders:
  Hydrostatically tested every five years
  80% max fill by volume
  Gray with yellow top
  Have a fill limit device installed
- Deep vacuum methods include:
  Evacuation of 500 microns
  Triple evacuation
- Types of thermometers:
  Dial
  Liquid fill
  Thermistor
  Recording-type thermometers
- Micron gauge—used to measure deep vacuums
- Refrigerant charging:
  Vapor charge
  Liquid charge
- Determining amount of charge:
  Weight
  Superheat & subcooling—charging charts and calculators
  Sight glass
- Oil charging is accomplished by either:
  Pouring or pumping it in
  Viewing the sight glass on compressor

Problems and Questions

1. To protect the ozone layer, EPA issued a series of directives. Which of the following was not included?
   a. No vent rules
   b. Technician's certification
   c. Equipment certification
   d. Indoor air quality
2. Which of the following refrigeration leaks are permitted by EPA?
   a. Leaks that occur using the approved procedure
   b. R-22 mixed with nitrogen, used for leak detection
   c. Purging refrigerant hoses
   d. Testing for leaks with pure CFC-type refrigerant
3. What is the purpose of the zero-vacuum test?
   a. To test the capacity of the vacuum pump
   b. For use in evacuating a system
   c. To test the system for leak-tightness
   d. To measure the refrigerant charge
4. Which of the following is considered a high-pressure appliance?
   a. A centrifugal chiller
   b. A domestic refrigerator
   c. A 25-ton packaged water chiller with a reciprocating compressor
   d. A home freezer
5. What is the necessary evacuation level for very high pressure equipment?
   a. 0 in. Hg
   b. 4 in. Hg
   c. 10 in. Hg
   d. 15 in. Hg
6. The DOT mark on a recovery cylinder means:
   a. Duty Of Technician
   b. Department Of Transportation, approved
   c. Date Of Transfer
   d. Don't Open Till ready to use
7. The vapor-recovery machine is equipped with:
   a. High- and low-pressure controls
   b. A pressure-regulating valve
   c. An automatic refrigerant type indicator
   d. A freeze-protection control
8. What is the purpose of filters and separators on a vapor-type recovery unit?
   a. To clean the refrigerant
   b. To protect the recovery equipment
   c. To prepare the refrigerant for resale
   d. To prepare the refrigerant for processing
9. What is the common range for an accurate expansion-bulb thermometer used for measuring suction line temperatures?
   a. –40 to 65°F
   b. –20 to 100°F
   c. 0 to 212°F
   d. –40 to 212°F
10. The pressure range on a compound gauge of a gauge manifold usually is:
    a. 0 to 250 psi
    b. 10 in. Hg to 150 psi
    c. 20 in. Hg to 200 psi
    d. 30 in. Hg to 120 psi
11. Only certified technicians can purchase refrigerants. True or false?
12. The advantage of the liquid method of transferring refrigerant is that it is much faster than the vapor-recovery method. True or false?
13. The most common temperature-measuring device is the electronic remote-bulb type. True or false?
14. Thermistor temperature-measuring instruments change resistance with temperature. True or false?
15. The superheat on fixed metering devices varies greatly on part load conditions. True or false?
16. The compound gauge is calibrated in _____ pound increments.
17. Purging minimum amounts of gases from the gauge hoses and manifold is: (legal, illegal).
18. A deep vacuum is when the vacuum pump pulls down to at least _____ microns.
19. The micron is a unit of linear measure equal to 1/25,400 of a(n) _____.
20. The main purpose of evacuation is to reduce the pressure enough to _____ off the water and then pump it out of the system.

SECTION 3

Basic Electricity/ Electronics

E19 Basic Electricity

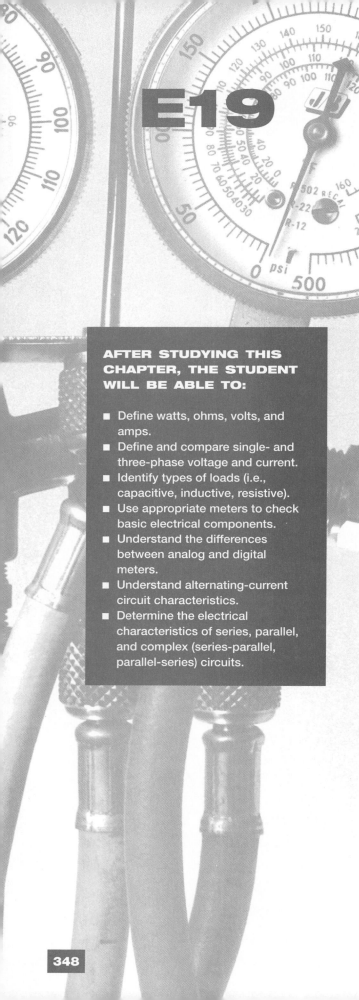

AFTER STUDYING THIS CHAPTER, THE STUDENT WILL BE ABLE TO:

- Define watts, ohms, volts, and amps.
- Define and compare single- and three-phase voltage and current.
- Identify types of loads (i.e., capacitive, inductive, resistive).
- Use appropriate meters to check basic electrical components.
- Understand the differences between analog and digital meters.
- Understand alternating-current circuit characteristics.
- Determine the electrical characteristics of series, parallel, and complex (series-parallel, parallel-series) circuits.

E19-1 TYPES OF ELECTRICITY

There are various ways to classify electricity. The following types relate to applications of this text:

1. Static electricity
2. Current electricity

E19-1.1 Static Electricity

Static electricity is associated with the concept of electric charge, either negative or positive, possessed by bodies. If a body contains equal amounts of both, it will exert no force on other bodies and is considered uncharged. If an uncharged body acquires or loses charges, it is now charged and exerts a force on other, similarly charged, bodies. If both bodies have acquired the same kind of charge (both negative or both positive), they will repel each other. If the two charged bodies each have opposite charges, they will attract each other (Fig. E19-1).

The amount of charge that a body has is measured by the *coulomb* and may be the amount either gained or lost by an uncharged body.

Probably the most common type of static electricity is lightning, occurring during a storm when cloud particles acquire large amounts of electric charge and discharge them to dissimilarly charged objects or bodies. The mag-

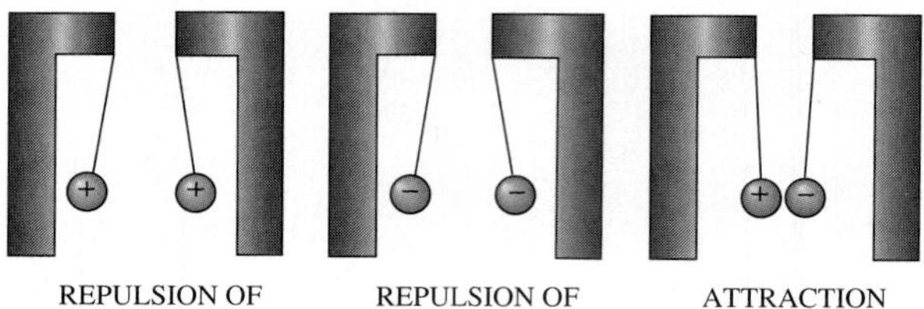

Figure E19-1 Static electricity negative and positive charges: similar charges repel and opposite charges attract.

nitude of the charge can cause serious damage because it is a natural, uncontrolled phenomenon. A smaller, more controlled charge is acquired in dry weather by a person by scuffing along a thick rug, collecting negative charges from the rug by friction. The negatively charged person exerts a force when touching a metal object, causing a spark to jump as the static electricity is discharged.

This type of electricity has only a few uses; however, one rather common electrical HVAC/R device that uses static electricity is the *electrostatic air cleaner.* Air is passed through a "charging" section containing fine ionizer wires (charged with a very high DC voltage) alternating with grounded electrodes. This "high" field charges dust particles in the air. They are then passed through a collector section consisting of alternating charged and grounded plates which attract and hold the charged dust particles. Sometimes an oily adhesive coating is used to bind the dust to the collector plates until the unit is cleaned with a water spray wash cycle. The process is very effective in removing fine dust particles and pollen from the air. Maintenance is judged to be medium to high.

E19-1.2 Current Electricity

Current refers to the flow of electrons through a conductor such as a wire, or through a given space, or past a given point. The flow direction is from negatively to positively charged terminals and occurs because of the potential difference (in the charge) between terminals. The potential difference creates a force, called an *electromotive force* (EMF), as shown in Fig. E19-2. This force is measured in volts (V). The letter symbol for voltage is *E,* referring to EMF, in circuits.

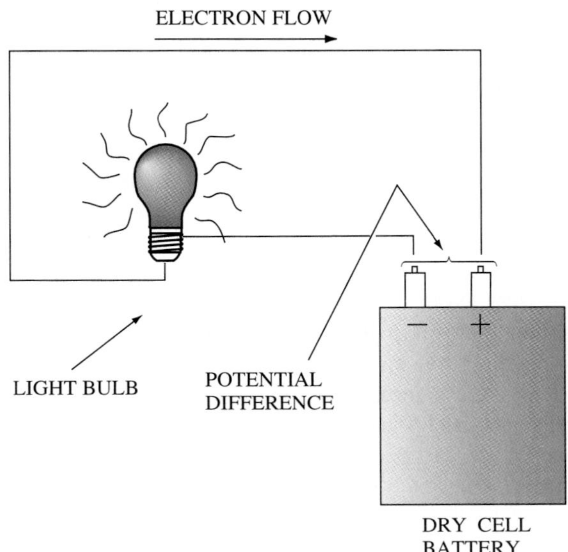

Figure E19-2 A simple electric circuit using a battery power source and a light bulb for the load.

The unit for electron quantity, the coulomb, is rarely used; however, one definition for a volt is one coulomb of electrons passing a fixed point in the conductor per second.

The *current,* or rate of flow of the charges, is measured by amperes. The rate depends on the amount of voltage applied (the difference of potential between the two ends of the conductor), the size of the conductor, and the material of which the conductor is made. The symbol for the current in amperes (A) is *I.* An electrical system illustrating current flow is shown in Fig. E19-3.

The conductor may permit more or less current to flow because of resistance associated with its physical state. It is identified by applying a known value of voltage and measuring the resulting current. The ratio of voltage to current is the *resistance* and is stated in ohms (Ω). For conductors, this depends on its dimensions and the material of which the conductor is made:

$$R = \rho l/A$$

in which

R = resistance in ohms
ρ = (rho), resistivity of the material in ohm-meters (a constant for a given material at a given temperature)
l = length in meters
A = cross sectional area of conductor

The resistivities of conducting materials commonly used are:

Aluminum	2.62
Copper	1.72
Iron	9.71

From the above it can be seen why copper is the most frequently used for wires of electrical circuits. For example, a piece of copper wire 0.5 mm (.02 inches) in diameter and one meter (39.37 inches) long has a resistance of only about 0.09 ohm.

In addition to the resistance of conductors, circuit elements called resistors are also used for various applications (Fig. E19-4).

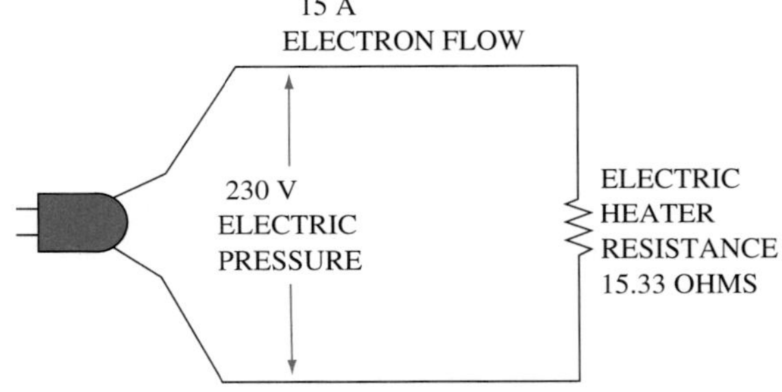

Figure E19-3 Simple electric circuit using an AC power source and an electric heater for the load.

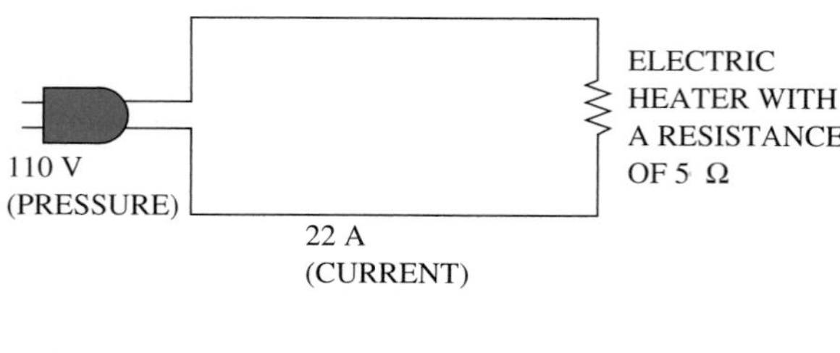

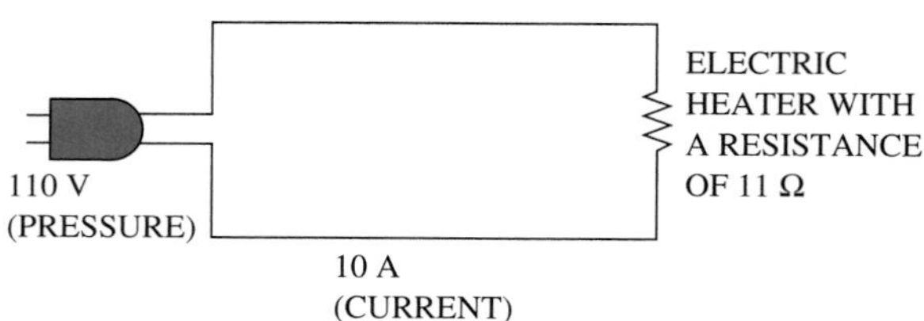

Figure E19-4 The effect of different loads on the current.

Putting it all together, current in a circuit is found to be directly proportional to the applied voltage and inversely proportional to the resistance of the circuit. This is known as *Ohm's law* and will be discussed in more detail later in the chapter.

The flow of electrons through a circuit dissipates energy in the form of heat in passing through the resistances that make up the circuit. Calculated amounts of this energy is called power, *P,* measured in watts (W). A watt of electricity is one ampere (A) of current, *I,* flowing with the force of one volt (V) of voltage, *E.* The power in watts in an electric circuit can be expressed by the equations for single-phase AC power:

$$P = EI\ (\text{PF})$$
$$P = I^2R$$
$$P = E^2/R$$

where

PF (Power Factor) = phase angle between E and I

and

PF = True power/apparent power as explained later in the chapter.

These equations vary for DC power, single-phase AC power, and three-phase single-phase power as will be explained later on in the chapter.

The amount of power required by electrical devices determines the design of the circuit. The electrical motor is the largest current-consuming device in most heating, cooling, and refrigeration systems. Other units which measure the output of such devices are the horsepower, equal to 746 watts (power output), and the British thermal unit (Btu) where one watt = 3.41 Btu/hour (heat output). A more abstract term commonly used by air conditioning manufacturers is the *energy efficiency rate* (EER) which identifies the amount of heat per watt of power consumed by equipment. It is becoming common to specify power ratings of many devices such as refrigeration systems, boilers, etc., in standard kilowatt (kW) terms.

Current electricity is produced as two types:

1. Direct current and
2. Alternating current

Direct Current

Direct current (DC) is the type supplied by batteries. It is a continuous flow of electrical particles in one direction through a conductor. The source has a positive and a negative terminal. It is used mostly for transportation refrigeration equipment, electronic air cleaners, in computers, and for electronic circuits.

For example, a thermocouple, used on a gas furnace as a safety device, generates its own DC current for keeping the gas valve open. This device consists of a pair of wires made of different metals joined together at one end. When the joined end and the other end of the wires are at different temperatures (one end being heated by the gas flame), direct current is produced between the wires and flows through the circuit. When the heat from the pilot light goes off, electricity is no longer generated and the gas valve closes.

Transmission of large amounts of power involves unavoidable line losses due to the resistance of the wires. By utilizing high voltages, the power can be transmitted by using relatively low line current, thus minimizing the line losses (which may be calculated as I^2R). Power is produced from AC generators. It is expensive converting AC voltages to DC and DC voltages to AC; therefore, alternating current (AC) is used for almost all ordinary power applications.

The most common method of obtaining AC voltage is by using a generator to create a magnetic field which moves in and out of a circuit to induce voltage in the circuit. The amount of voltage produced is proportional to the rate at which the magnetic flux linking the circuit changes. The AC voltage in turn produces AC current.

Alternating Current

Fig. E19-5 shows a typical cycle of alternating current. The AC waveform has a *sinusoidal* shape like that of a sine wave. Note that half the time the current is positive and half the time it is negative. Alternating current flows in one direction and then the other at regular intervals through the conductor. The frequency of AC current is the number of cycles per second. Frequency is measured in *Hertz* (Hz): one Hz is equivalent to one cycle per second. In North America 60 Hz power is standard. In many other parts of the world the standard is 50 Hz.

Since most devices are designed to run on AC current, this unique source of power has universal usage. Alternating current will be discussed in detail later in this chapter.

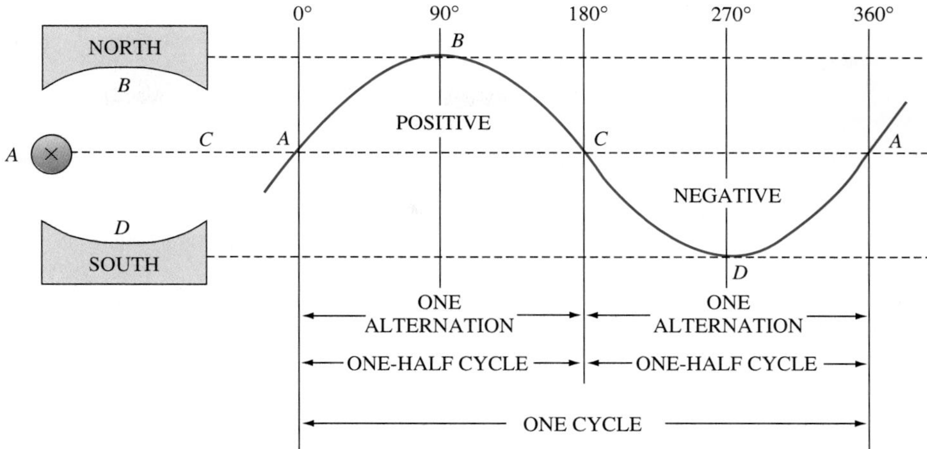

Figure E19-5 One cycle of alternating current.

Ohm's law, for DC circuits or AC single-phase circuits with purely resistive loads is as follows:

The relationship between electrical potential, *E*, measured in volts (V); current flow, *I*, measured in amperes (A); and resistance, *R*, measured in ohms (Ω), is expressed in Ohm's law. In simple terms it states that the greater the voltage, the greater the current; and the greater the resistance, the lesser the current flow. Ohm's law is expressed mathematically as "current is equal to electrical potential divided by resistance." Stated in symbols:

$$I = \frac{E}{R}$$

where

I = Current
E = Electrical potential
R = Resistance

The equation can be stated a number of ways. Use is made of whichever one applies. Other versions of Ohm's law are as follows:

$$E = IR \qquad R = E/I$$

This formula is very helpful in analyzing a circuit, since when any two of the terms are known or can be measured, the third value can be calculated using one of the above equations.

It should be noted that Ohm's law was first applied to circuits using direct current. It does not apply, without some modification, to AC circuits since coils of wire produce different effects with alternating current. The flow of current in AC circuits can be influenced by such factors as *inductance* and *inductive reactance* (the "resistance" offered by inductance to the flow of alternating current), and *capacitance* and *capacitive reactance,* which will be discussed later in this chapter. If modifications are made to resistance, it will be shown how the general principles contained in Ohm's law do apply to alternating current.

The following examples illustrate the relationships of voltage, current and resistance in a simple electrical circuit. In the examples a DC circuit or a single-phase AC circuit with purely resistive loads is used.

EXAMPLE

A simple electrical circuit (Fig. E19-6) has a power source of 120 V and a resistance load of 10 ohms. How much is the current flow in amperes?

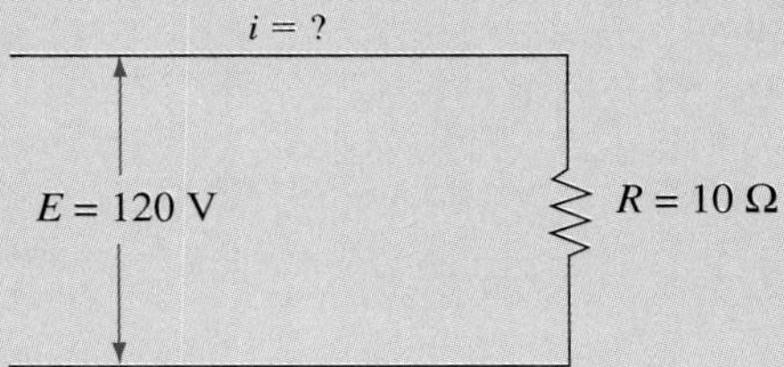

Figure E19-6 Calculating the current using Ohm's law, with voltage and resistance known.

Solution

$$I = E/R$$
$$I = 120 \text{ V} / 10 \text{ ohms}$$
$$I = 12 \text{ A}$$

EXAMPLE

Referring to Fig. E19-7, what is the resistance of an electric heater, using a power source of 120 V and drawing a current of 10 A?

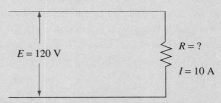

Figure E19-7 Calculating the resistance, using Ohm's law, with voltage and current known.

ON THE JOB

Applying Ohm's law principles on the job occurs on an everyday basis. While working on equipment, learn to look for loads in parallel utilizing the same amount of voltage. Notice that temperature controls are placed in series with the loads for a purpose—to control all action through the circuit.

Several power circuits can be wired for 230 V or 115 V. Whenever possible, wire up circuits for 230 V. This keeps the size of the wire down, which keeps initial costs down, while still using the same amount of power.

Solution

$$R = E/I$$
$$R = 120 \text{ V} / 10 \text{ A}$$
$$R = 12 \text{ ohms}$$

EXAMPLE

Assuming a circuit as shown in Fig. E19-8, with a resistance of 23 ohms and a current flow of 10 A, what is the voltage of the power supply?

Solution

$$E = I\,R$$
$$E = (10 \text{ A})\,(23 \text{ ohms})$$
$$E = 230 \text{ V}$$

Note in the above examples that by using Ohm's law, if two factors are known, the third can be calculated.

Calculating electric power, for DC circuits or single-phase AC systems with purely resistive loads can be done using the following formula:

$$P = E\,I$$

where E and I have the same meaning as used in calculations involving Ohm's law and P = power in watts.

In general, for single-phase AC systems, $P = EI$ (PF), where PF = phase angle between E and I. For three-phase AC systems, $P = \sqrt{3}EI$ (PF).

Variations of the Power Formula

The power formula can be stated in three different ways with variations for DC, single-phase AC and three-phase AC, shown in the table on p. 353.

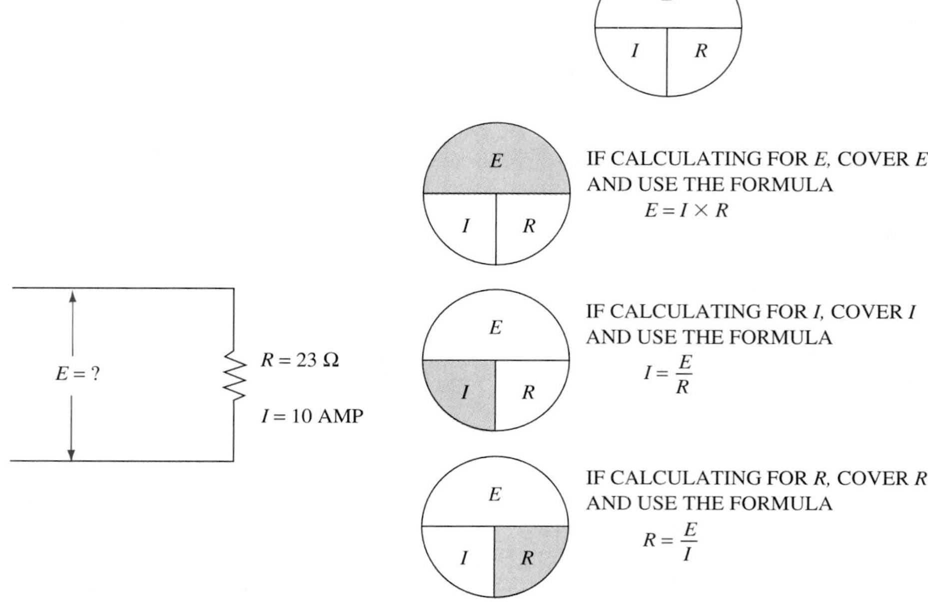
E
I
R
E = ?
R = 23 Ω
I = 10 AMP
IF CALCULATING FOR E, COVER E AND USE THE FORMULA
E = I × R
IF CALCULATING FOR I, COVER I AND USE THE FORMULA
I = E/R
IF CALCULATING FOR R, COVER R AND USE THE FORMULA
R = E/I

Figure E19-8 Various ways to use Ohm's law.

	DC	Single-Phase AC	Three-Phase AC
Power Formula 1	$P = EI$	$P = EI$(PF)	$P = \sqrt{3}EI$(PF)*
Power Formula 2	$P = E^2/R$	$P = E^2/R$	$P = \sqrt{3}E^2/R$
Power Formula 3	$P = I^2R$	$P = I^2R$	$P = \sqrt{3}I^2R$

*$\sqrt{3} = 1.732$

Some examples to illustrate the use of the power formulae are:

EXAMPLE

(Single-phase AC, assuming a PF = 1.0)
What is the power consumption in a circuit operating with 230 V and 5 A?

Solution

$$P = E\,I$$
$$P = (230\text{ V})\ (5\text{ A})$$
$$P = 1150\text{ W}$$

EXAMPLE

(Single-phase AC, assuming a PF = 1.0)
What is the current draw of an 8000-W electric heater operating on a 230-V power supply?

Solution

$$I = P/E$$
$$I = 8000\text{ W} / 230\text{ V}$$
$$I = 34.8\text{ A}$$

EXAMPLE

(DC or single-phase AC)
What is the power consumed by an electric resistance of 20 ohms, using 6 A?

Solution

$$P = I^2R$$
$$P = (6\text{ A})\ (6\text{ A})\ (20\text{ ohms})$$
$$P = 720\text{ W}$$

EXAMPLE

(Three phase AC)
What is the power consumed by a three-phase circuit with a voltage of 230 V, a current of 10 A, and a power factor of 0.80?

Solution

$$P = \sqrt{3}\ EI(\text{PF})$$
$$P = (1.732)\ (230)\ (10)\ (.80)$$
$$P = 3186\text{ W}$$

E19-2 ELECTRICAL POWER SYSTEMS

The basic element that supplies power for all HVAC/R systems is the electrical power system. The electrical circuit is a means of delivering electrical power to the HVAC/R equipment. An electrical power system has three essential requirements and one optional requirement:

1. A source of power (could be a transformer)
2. An electrical load device
3. An electric circuit: a path for the current to flow
4. (Optional) A switch to control the flow

E19-2.1 Power Sources

There are two types of power sources:

a. Direct current (DC)
b. Alternating current (AC)

The most common source of direct current is the battery. Batteries are used to supply power to many different types of electrical testing instruments, making them portable and convenient to use. DC power is used for the controls on automotive air-conditioning systems. DC power obtained from AC power supplies is also used on certain solid-state modules for defrost and overcurrent protection.

Alternating current is the most common source of power for most HVAC/R systems. AC power is generated by all the power companies. In residences, 120 V AC is used to power most small appliances. Larger appliances such as electric stoves and residential air-conditioning units use 240 V. The power company supplies residential users with 240 V over the incoming lines. A portion of it is tapped to supply the 120-V requirement. Commercial and industrial customers normally use higher voltages in single-phase and three-phase systems.

Transformers increase or decrease incoming voltages to meet the requirements of the load. For control circuits, it is common to use a transformer to obtain 24 V from line voltages of 120 or 220 V. More detailed explanations of how transformers work and are used will be found in the next chapter.

E19-2.2 Loads

The second condition for an electrical power system is that it must have a load. A *load* is any electrical device that requires power to operate. The most common loads for HVAC/R systems are electric motors. Motors drive the compressors, fans, and pumps. Motors also drive damper motors and zone valves. Many other electrical components require power such as resistance heaters and solenoid valves, to name a few.

E19-2.3 Electrical Circuit: Path for the Current

The third condition for a power system is that there must be a path for the current to travel (Fig. E19-9). Every electric circuit has at least two wires, often indicated as line terminals, L_1 and L_2. In order for there to be a complete circuit, the path of the electrical service (wire connections) flows through one wire of the electric circuit, passes through the load, and returns through the other wire of the electric circuit. In AC systems, the direction of power flow reverses 60 times per second.

E19-2.4 Control Device or Switches

The fourth (optional) condition for the power system is the *control device* or *switch.* The switch is a device to turn the load off and on. It may be manual or automatic, as in the case of a thermostat that turns a unit on and off in response to the surrounding temperature. The switch permits the circuit to be open (Fig. E19-10) or closed (Fig. E19-11). No current flows in the open circuit. When the circuit is closed the load receives power.

E19-2.5 Types of Circuits

There are several types of path arrangements for circuits, as follows:

1. The series circuit, which allows only one path for the current to flow;
2. The parallel circuit, which has more than one path;
3. The series-parallel circuit, which is a combination of series and parallel circuits.

Series Circuits

In a series circuit, there is only one path for the current to follow. The power must pass through each electrical device in succession in that circuit to go from one side of the power supply to the other. An example of a series circuit is shown in Fig. E19-12, where four resistance heaters are placed end-to-end in a single circuit.

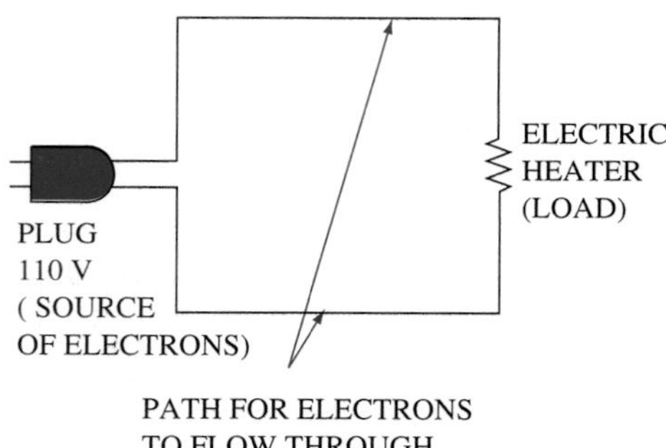

Figure E19-9 Principal components of a simple circuit.

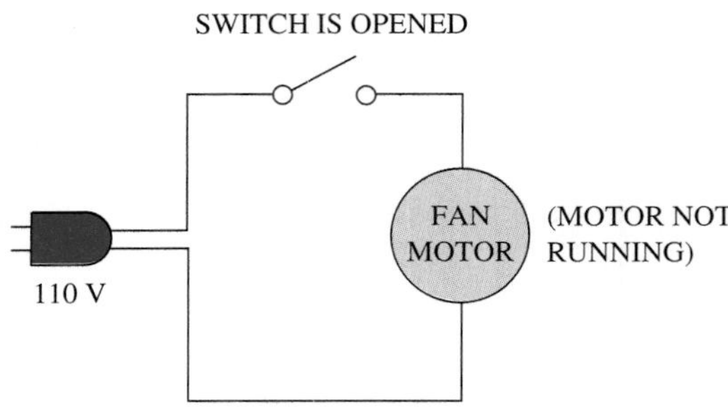

Figure E19-10 An open circuit: no current flows.

Series circuits are common on HVAC/R systems. Usually there is one load controlled by a series of switches, as shown in Fig. E19-13. In this diagram the 208-V power-supply terminals are indicated with the symbols L_1 and L_2. The one load is a compressor motor. The switches placed in series with the compressor motor are used to control its operation. These switches, shown in this diagram, are all safety switches and therefore are all normally closed (NC).

In a series circuit, all switches must be closed in order for current to flow through the circuit. Types of switches include the following:

1. *High-pressure switch,* which senses compressor discharge pressure, opens on a rise in pressure. It is set to cut out at protective high-limit pressure, but remain closed at normal operating pressures. It is also called a high-pressure cutout.
2. *Low-pressure switch,* which senses compressor suction pressure, opens on a drop in pressure. It is set to cut out at a protective low-limit pressure, but remains closed at normal operating pressures. It is also known as a low-pressure cutout.
3. *Compressor internal thermostat,* which senses compressor motor-winding temperature, opens on a rise in temperature. It is set to cut out at a protective high temperature, but remain closed under normal operating conditions.
4. An *operating control* (not shown), also placed in series with the compressor motor, starts and stops the compressor in response to temperature, pressure, humidity or a time clock.

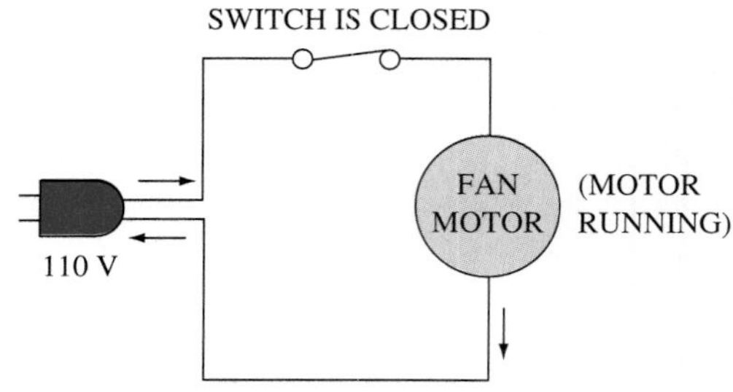

Figure E19-11 A closed circuit, with current flowing through the load.

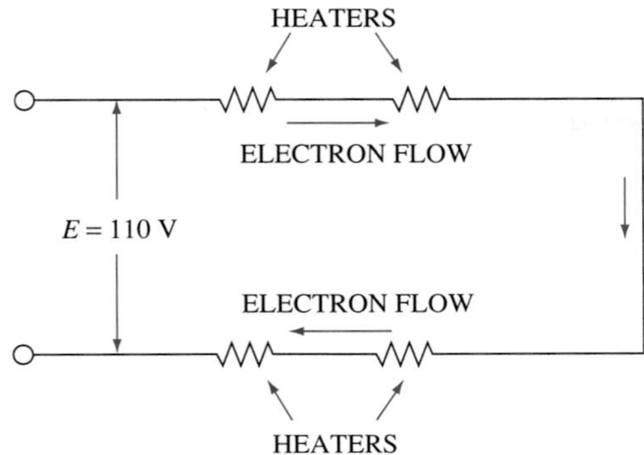

Figure E19-12 A series circuit with four resistances.

Calculations for a Series Circuit

For the following calculations a DC circuit or a single-phase AC circuit with purely resistive loads is used.

The current flowing through a series circuit is the same for each load in the circuit. For example, in the circuit shown in Fig. E19-14, which has four loads, the current is equal through each load, and expressed in symbols:

$$I_1 = I_2 = I_3 = I_4$$

The total resistance of a series circuit is the sum of all the individual resistances that are placed in series between L_1 and L_2. To state this in symbol form:

$$R_T = R_1 + R_2 + R_3 + R_4$$

The voltage drop across each one of these resistances can be calculated using Ohm's law, and the total of the individual voltage drops should add up to the circuit voltage. Thus,

$$E_T = E_1 + E_2 + E_3 + E_4$$

EXAMPLE

Using Fig. E19-14, calculate the individual voltage drops to see if their total equals the circuit voltage.

Solution

Step 1. Determine the total resistance of the circuit.

$$R_T = R_1 + R_2 + R_3 + R_4$$
$$R_T = 4 \text{ ohms} + 10 \text{ ohms} + 12 \text{ ohms} + 14 \text{ ohms}$$
$$R_T = 40 \text{ ohms}$$

Step 2. Determine the current flow in the circuit, using Ohm's law.

$$I = \frac{E}{R}$$
$$I = \frac{120 \text{ V}}{40 \text{ ohms}}$$
$$I = 3 \text{ A}$$

Step 3. Determine each voltage drop, using Ohm's law.

$$E_1 = I_1 \times R_1 \qquad E_2 = I_2 \times R_2$$
$$E_1 = 3 \times 4 \qquad E_2 = 3 \times 10$$
$$E_1 = 12 \text{ V} \qquad E_2 = 30 \text{ V}$$
$$E_3 = I_3 \times R_3 \qquad E_4 = I_4 \times R_4$$
$$E_3 = 3 \times 12 \qquad E_4 = 3 \times 14$$
$$E_3 = 36 \text{ V} \qquad E_4 = 42 \text{ V}$$

Step 4. Check the accuracy of these calculations. Their total should equal the circuit voltage of 120 volts. The procedure is as follows:

$$E_T = E_1 + E_2 + E_3 + E_4$$
$$E_T = 12 + 30 + 36 + 42$$
$$E_T = 120 \text{ V}$$

Parallel Circuits

Parallel circuits are used for most HVAC/R equipment wiring. Each load has its own separate path for the current to flow. Most equipment loads are connected directly to the voltage source. The control circuits normally use a lower voltage; 24 V is common. The control system also has load devices, such as relays, which require separate circuits. Control voltage is usually provided by a step-down transformer fed from the power source.

Fig. E19-15 illustrates a parallel circuit with four loads. No controls are shown in this diagram, which would be necessary if the circuit were operational. In parallel cir-

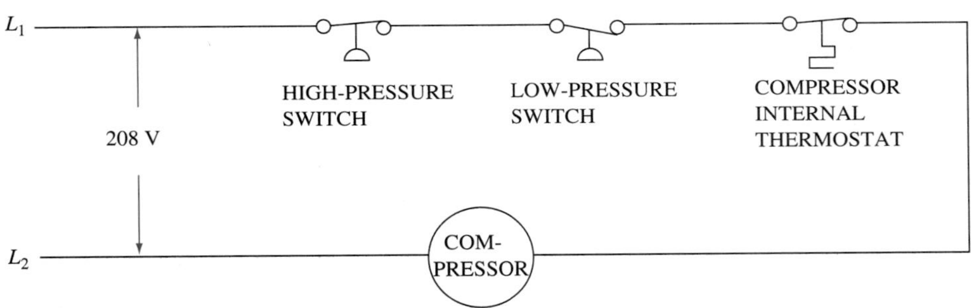

Figure E19-13 Three safety switches in series with a single load.

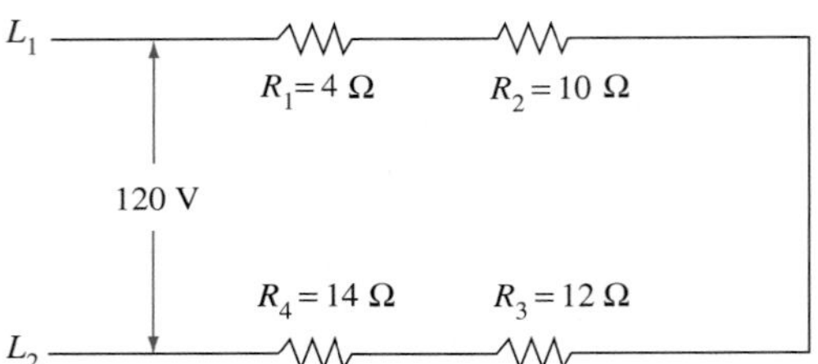

Figure E19-14 Calculating the current flow through a series circuit.

cuits, each circuit has its own, independent connection to the power source. If the switch is opened in one circuit, the other circuits will continue to operate.

Fig. E19-16 shows a number of parallel load circuits, with switches or controls in series with the loads, which could be used for an air-cooled condensing unit. There are three parallel circuits, C_1, C_2, and C_3. Going from top to bottom, they could be described as follows:

Circuit 1. The condenser fan motor #2 has a separate thermostat that turns it on and off.

Circuit 2. The compressor contactor coil (C) is energized when the primary thermostat calls for cooling, provided the two safety switches, LPS and HPS, are closed.

Circuit 3. The compressor contactor has two normally open switches which are in series with the two loads. When the contactor coil in circuit 2 is energized, the two "C" switches in circuit 3 close, supplying power to the compressor motor and the condenser fan #1 at the same time. In effect the two loads, condenser fan motor #1 and the compressor motor, are themselves in parallel and both receive line voltage when the contactor switches close.

In actual practice, a unit may have many parallel circuits for individual loads, all operated in accordance with the design specifications of the control system.

Calculations Using a Parallel Circuit

The following calculations use a DC circuit or a single-phase AC circuit with purely resistive loads. The current draw for a parallel circuit is determined for each of its parts. The current consumed by the entire parallel system is the sum of the individual circuits. The calculation is made using Ohm's law. To obtain the current flowing in the circuit, both the voltage and the resistance of the load(s) must be known. Thus, the total current is calculated as follows:

$$I_T = I_1 + I_2 + I_3 + I_4 + \ldots$$

The resistance of a parallel circuit gets smaller as more resistances are added. If there are only two resistances, the total resistance can be calculated by the following formula:

$$R_T = (R_1 \times R_2) / (R_1 + R_2)$$

If there are more than two resistances, use the following formula and solve for R_T.

$$1/R_T = 1/R_1 + 1/R_2 + 1/R_3 + 1/R_4 + \ldots$$

The voltage drop in a parallel circuit is the line voltage supplied to the loads, or simply stated,

Figure E19-15 Parallel circuit with four different loads.

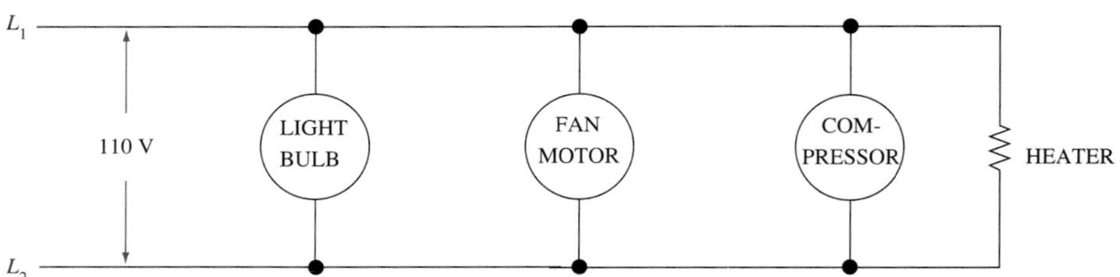

Figure E19-16 Condensing-unit wiring diagram showing three separate circuits.

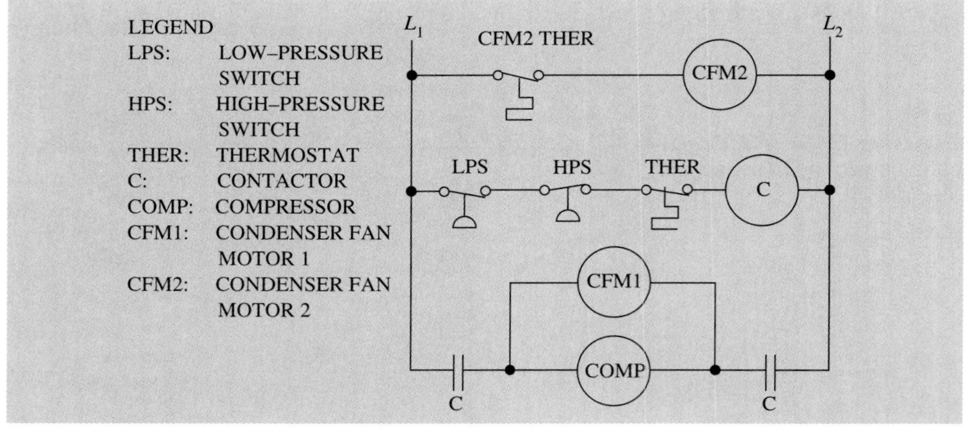

$$E_T = E_1 = E_2 = E_3 = E_4 = \ldots$$

Ohm's law can be used to calculate voltage, amperage, or resistance, if the other two values are known.

EXAMPLE

From the information given in Fig. E19-17, calculate the current draw for each of the individual circuits and the total current draw.

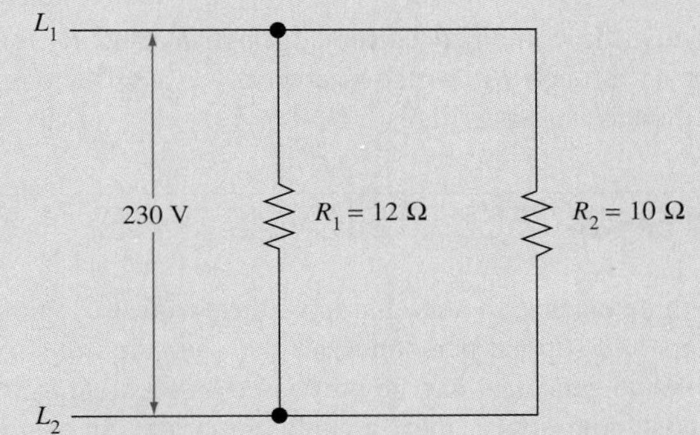

Figure E19-17 Calculating the current through parallel circuits.

Solution

$$I = E/R$$
$$I_1 = 120\text{ V} / 12\text{ ohms}$$
$$I_1 = 10\text{ A}$$
$$I_2 = 120\text{ V} / 10\text{ ohms}$$
$$I_2 = 12\text{ A}$$
$$I_T = I_1 + I_2$$
$$I_T = 12\text{ A} + 10\text{ A}$$
$$I_T = 22\text{ A}$$

EXAMPLE

Find the total resistance of the complete circuit shown in Fig. E19-17.

Solution

$$R_T = (R_1 \times R_2) / (R_1 + R_2)$$
$$R_T = (12\text{ ohms} \times 10\text{ ohms}) / (12\text{ ohms} + 10\text{ ohms})$$
$$R_T = 120 / 22$$
$$R_T = 5.4\text{ ohms}$$

EXAMPLE

What is the resistance with parallel resistances of 3 ohms, 4 ohms, and 5 ohms?

Solution

$$1/R_T = 1/R_1 + 1/R_2 + 1/R_3$$
$$1/R_T = 1/3\text{ ohms} + 1/4\text{ ohms} + 1/5\text{ ohms}$$

Converting to a common denominator:

$$1/R_T = 20/60 + 15/60 + 12/60$$
$$1/R_T = 47/60$$
$$R_T = 60/47$$
$$R_T = 1.28\text{ ohms}$$

Series-Parallel Circuits

A series-parallel circuit, as the name implies, combines both a series and a parallel arrangement of electrical loads. A typical diagram is shown in Fig. E19-18a.

Calculations

As in previous calculations, a DC circuit or single-phase AC circuit with purely resistive loads is used.

Both the voltage of the circuit and the values of all resistances are known (Fig. E19-18a). With two of the factors provided, the third (current flow) can be determined using Ohm's law from the following steps:

Step 1. Calculate the resistance through the parallel circuit consisting of R_1 and R_2 (Fig. E19-18b).

$$R_{1,2} = (R_1 \times R_2) / (R_1 + R_2)$$
$$R_{1,2} = (100 \times 100) / (100 + 100)$$
$$R_{1,2} = 10{,}000 / 200$$
$$R_{1,2} = 50\text{ ohms}$$

Therefore, 50 ohms can be substituted for the parallel resistances. The main circuit has now become a strictly series circuit (see Fig. E19-18c).

Step 2. Calculate the current flow though the revised main circuit (Fig. E19-18c).

$$I_T = E / (R_{1,2} + R_3)$$
$$I_T = 230\text{ V} / (50\text{ ohms} + 50\text{ ohms})$$
$$I_T = 230/100$$
$$I_T = 2.3\text{ A}$$

Step 3. Calculate the current flow through R_1 and R_2. Since R_3 is in series with $R_{1,2}$, in Fig. E19-18c, we calculate the voltage drop across R_3 and $R_{1,2}$ (Fig. E19-18b).

$$E_3 = I \times R_3$$
$$E_3 = 2.3\text{ amperes} \times 50\text{ ohms}$$
$$E_3 = 115\text{ V}$$
$$E_{1,2} = E - E_3$$
$$E_{1,2} = 230 - 115$$
$$E_{1,2} = 115\text{ V}$$

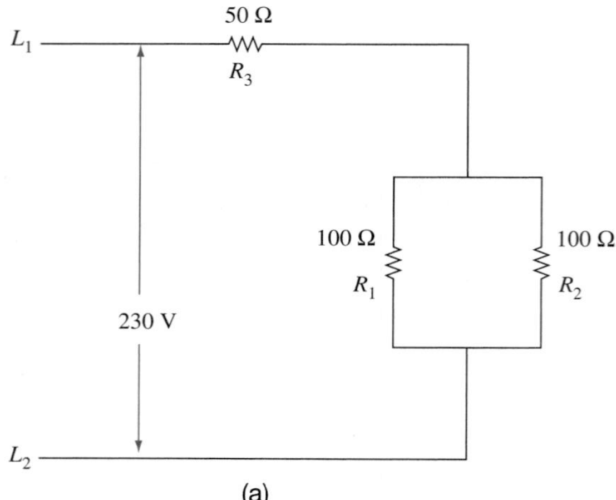

Figure E19-18a Parallel-series circuit showing values of three resistances.

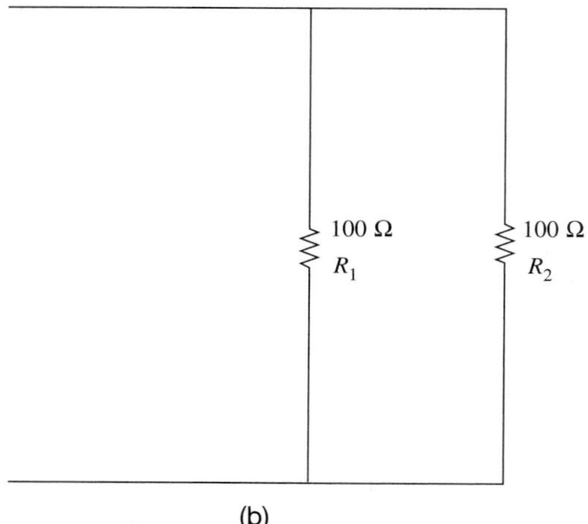

Figure E19-18b Determine the combined resistance of the two parallel resistances.

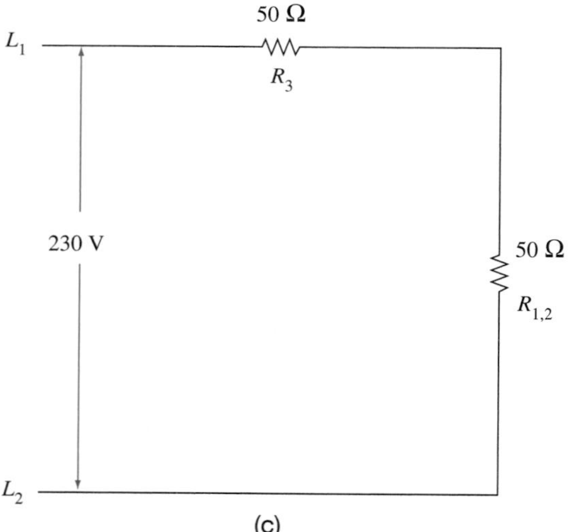

Figure E19-18c Parallel-series circuit combining two parallel resistances to form a simple series circuit.

Step 4. Calculate the current through R_1 and R_2.

$$I_1 = E_{1,2} / R_2$$
$$I_1 = 115 \text{ V} / 100 \text{ ohms}$$
$$I_1 = 1.15 \text{ A}$$
$$I_2 = I_1$$
$$I_2 = 1.15 \text{ A}$$
$$I_{1,2} = I_1 + I_2$$
$$I_{1,2} = 1.15 \text{ A} + 1.15 \text{ A}$$
$$I_{1,2} = 2.3 \text{ A}$$

Thus, the combined current through R_1 and R_2 is the same as through R_3, which is correct for a series circuit, and the answer is verified.

E19-3 ELECTRIC METERS

The three electric meters that have the greatest use for installers and service personnel are the *voltmeter, ammeter,* and *ohmmeter.* They can be purchased as separate meters, or most commonly, they are all combined in a single meter called a multimeter.

There are two basic types of meters, the *analog* and the *digital.* These terms are familiar since they also apply to watches. The digital meter is solid-state and gives a direct numerical readout of the measured value. The analog meter has a needle that points to the measured value. Each is shown in Fig. E19-19.

E19-3.1 Analog Meters

All analog meters operate on the same principle. When the current flows through a conductor, it produces a magnetic field around the conductor. If a magnetic needle is placed close to the current, the needle will attempt to line up with the field, as shown in Fig. E19-20.

In order to conserve space, the current-carrying wire is coiled and a scale is provided to indicate the position of the needle. The mechanism is so constructed that the greater the current flow, the greater the deflection of the needle on the scale.

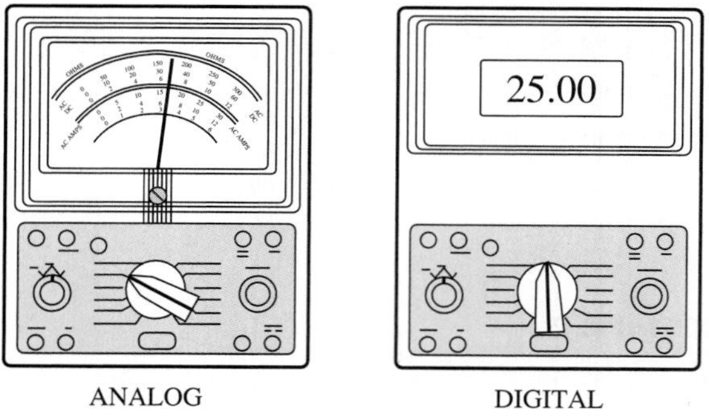

Figure E19-19 Two types of electrical meters.

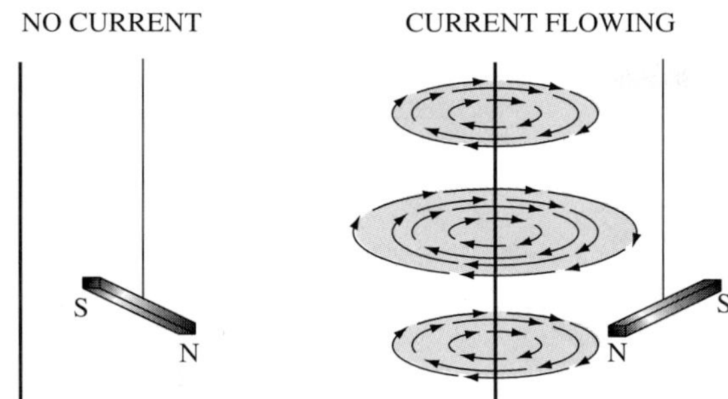

Figure E19-20 The effect of the magnetic field on the position of the meter needle.

Three important characteristics of analog meters need to be considered by those who use them:

1. The most accurate reading is at or near the midpoint of the scale. This is because the spring that opposes the deflection of the meter does not exert constant pressure across the scale. The meter may be inaccurate at either end of the scale. So whenever the operator has a choice of scales, the one selected should place the pointer in the most favorable (central) position.
2. Analog meters periodically need to re-calibrated. Most meters include some type of adjustment and instructions for calibration.
3. The small coil of wire that forms part of the meter movement is sensitive to excessive current. The meter may be made completely inoperable if subjected to excessive current. In using a multiple-scale meter, always use the higher scale first and move down to the scale required.

Analog meter accuracy is normally specified as a percent of full scale reading, so readings should be taken in the upper two-thirds of the scale for testing accuracy.

More expensive (and more accurate) analog meters are often furnished with a mirror scale to enable more accurate readings by avoiding *parallax,* an apparent difference in readings taken from different perspectives. The pointer and scale are aligned so the mirror image of the pointer disappears behind the pointer. This gives the most accurate reading.

Quality analog meters will no doubt become more expensive as digital meters become more popular. Many service technicians will want to have both. Just like analog watches, analog meters have advantages such as ease of reading changes and variations in the measurements.

E19-3.2 Digital Meters

Digital meters offer a number of advantages, although they are usually more expensive. Rugged versions are available and recommended. Because of the expense of digital meters, it is important that the technician be thoroughly familiar with the use of both types. Some of the advantages of digital meters are:

1. They are direct reading. There is no need to interpret the scale.
2. Digital meters can be obtained that will give accurate readings to three decimal places
3. They have no moving parts and are less likely to fail or get out of calibration than analog meters.
4. They often have automatic scaling features.

E19-3.3 Ammeters

The clamp-on ammeter is one of the most useful of the electric meters for HVAC/R technicians. It is used to measure current flow through a single wire by enclosing the wire within the jaws of the instrument, as shown in Fig. E19-21.

This instrument functions like a transformer. The primary coil is the test wire enclosed by the instrument. The secondary is a coil of wire within the instrument that is connected to the current-indicating mechanism. The current in the primary wire induces a flow of current in the secondary winding, measuring the current flow. The greater the current flow through the test wire, the greater the induced current and the greater the deflection of the needle reading on the scale. It is not necessary to disconnect or make contact with any wires to obtain a reading. This is very convenient.

Care of the Instrument

More accurate tests and longer life for the clamp-on ammeter are obtained by proper care of the instrument. Some of the ways the instrument should be treated are:

1. Keep the jaws clean and aligned.
2. When taking a reading always start on the highest possible scale and then work down to the most appropriate scale.

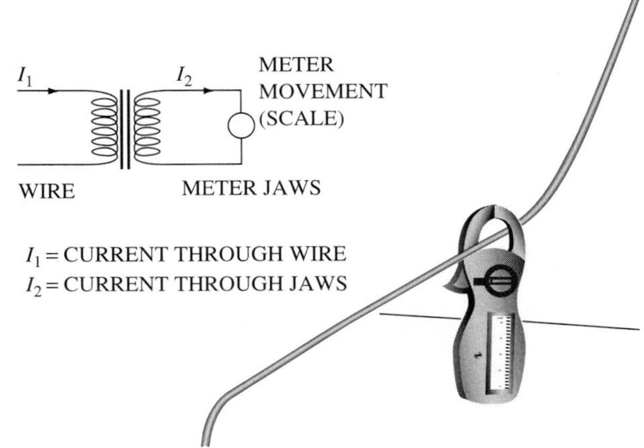

Figure E19-21 Construction of a clamp-on ammeter.

3. Do not cycle a motor off and on while taking readings unless the meter is first set to the highest scale.
4. Never put the clamp around two wires at the same time. If the current is flowing in opposite directions, the meter will read the difference between the two. If the current is moving in the same direction in both wires, the meter will add the two.

Reading Small Amounts of Current

The clamp-on ammeter is useful in reading small amounts of current. The procedure is to loop the wire around the jaws, as shown in Fig. E19-22. Passing the wire twice through the jaws doubles the strength of the magnetic field. It is therefore necessary to divide the meter reading by two to determine the actual current. For two passes, the actual current is ½ the reading; for three passes the actual current is ⅓ the reading, etc.

One application of the clamp-on ammeter is to adjust the anticipator setting of a thermostat. The *anticipator* supplies false heat to the thermostat, causing it to shut off the heat to the space being heated before the temperature in the room reaches the set point. This prevents the residual heat in the furnace, supplied by the fan after the burners are off, from overheating the room.

For the anticipator adjustment, 10 passes of wire are wrapped around the jaws of the instrument. To obtain the actual current flow through the anticipator, the scale reading is divided by 10. The diagram for the anticipator circuit is shown in Fig. E19-23. The location of the ammeter with its coil of wire is shown in Fig. E19-24. The wire that passes through the jaws of the ammeter is connected to the "W" terminal of the thermostat.

The In-Line Ammeter

Occasionally it is desirable to use an in-line ammeter. The proper location for its connections in a circuit are shown in Fig. E19-25. Note that it is connected in series with the circuit being tested. In DC circuits, verify the correct polarity of the ammeter used before energizing the circuit.

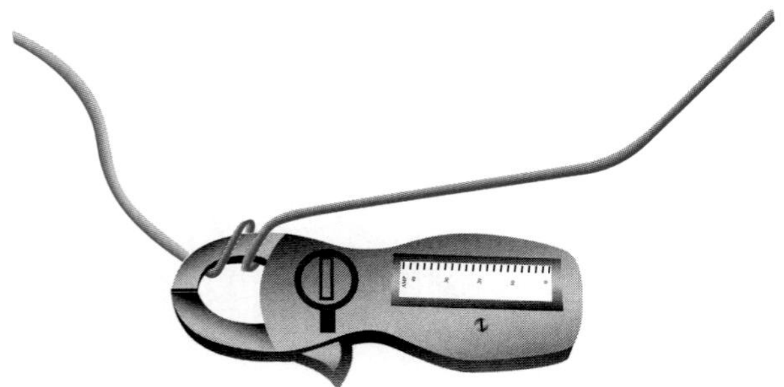

Figure E19-22 Method of measuring low current using clamp-on ammeter.

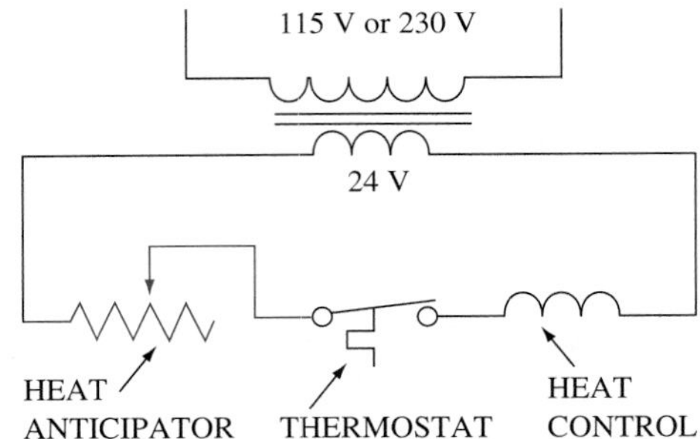

Figure E19-23 Position of the heat anticipator in the heating circuit.

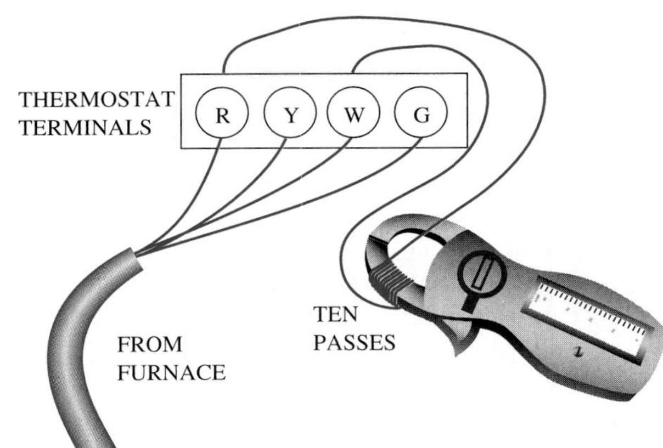

Figure E19-24 Procedure for testing the current flow through the heat anticipator.

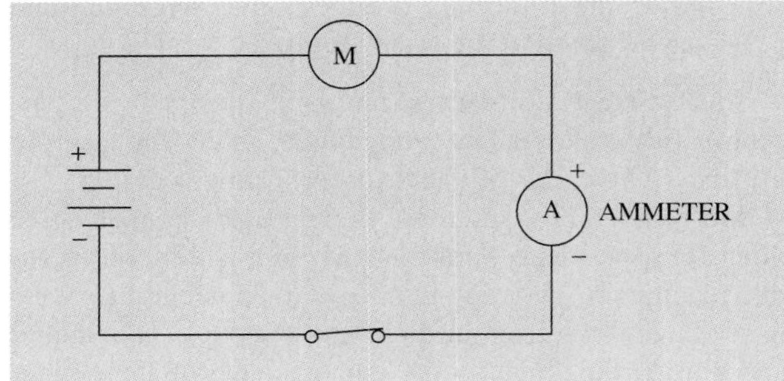

Figure E19-25 Testing position of the in-line ammeter.

Never connect an ammeter across a load. It will be destroyed by line current as there is no load to limit it!

E19-3.4 Voltmeters

Two leads from the voltmeter are connected to the circuit being tested, as shown in Fig. E19-26. Voltmeters are connected in parallel with the load to read the voltage drop.

Fig. E19-27 shows a voltmeter that has multiple scales. Each scale has a different resistance in the meter placed in

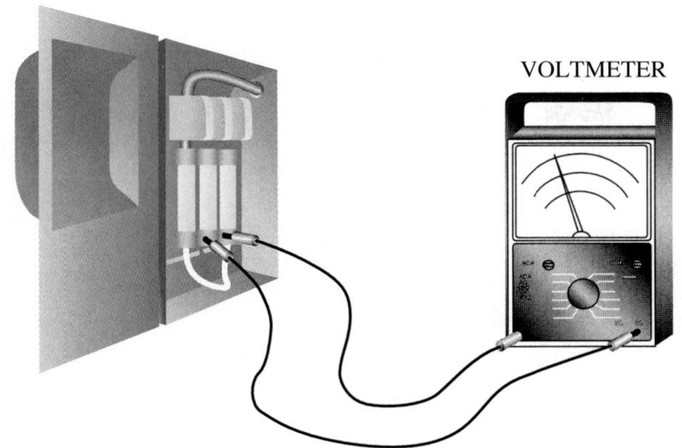

Figure E19-26 Method of testing the voltage of the incoming power service.

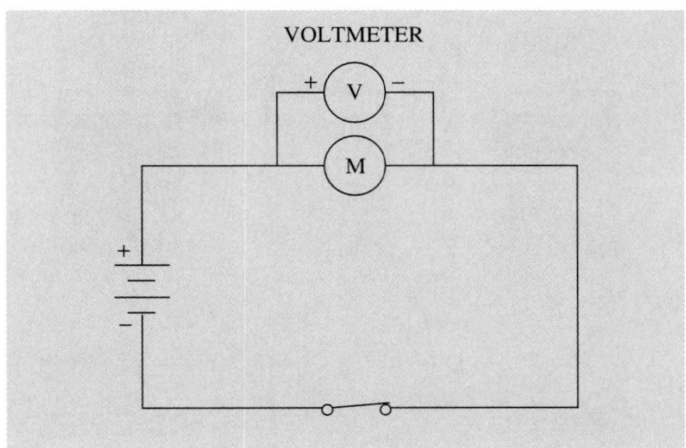

Figure E19-28 Position of the voltmeter for measuring the voltage drop.

series with the circuit being tested. A knob in the center face of the meter adjusts the meter to the scale being used. When using a multi-range meter, always start to measure voltage using the highest range on the meter. When the approximate voltage is read, the meter range can then be reduced to the proper range for greater reading accuracy. Also, using the meter with a higher voltage than the range of the meter could cause burnout or otherwise damage the meter. Some digital voltmeters provide an auto-scaling function.

In DC circuits, verify the correct polarity of the probes that are used before connecting the meter to the DC circuit.

In testing a motor to see that it has proper voltage, one lead goes on each side of the load, as shown in Fig. E19-28. The illustration is a DC circuit in which the polarity of the leads must be observed.

The voltmeter can also be used to determine if a hidden switch is open or closed. This is very helpful in troubleshooting. If there is power in the circuit, and the leads of the voltmeter are placed on each side of the switch, a voltage reading indicates the switch is open and a zero reading indicates the switch is closed (if no other open switches are in the circuit).

E19-3.5 Ohmmeters

The ohmmeter is different from the other two popular meters in that it uses a battery as a power supply. The battery furnishes the current needed for resistance measurements. The wiring for a typical meter is shown in Fig. E19-29.

It is a direct application of Ohm's law. The higher the resistance, the lower the current flow and the less the meter deflection. The resting place for the needle is on the left of the scale. For a high resistance the deflection is small. For a small resistance the deflection is large.

One thing that is extremely important is the power to the circuit being tested must be turned off. Further, if there are any capacitors in the circuit, they must be discharged before the meter is used. There can be only one source of power to the meter and that must be from the battery within the meter itself.

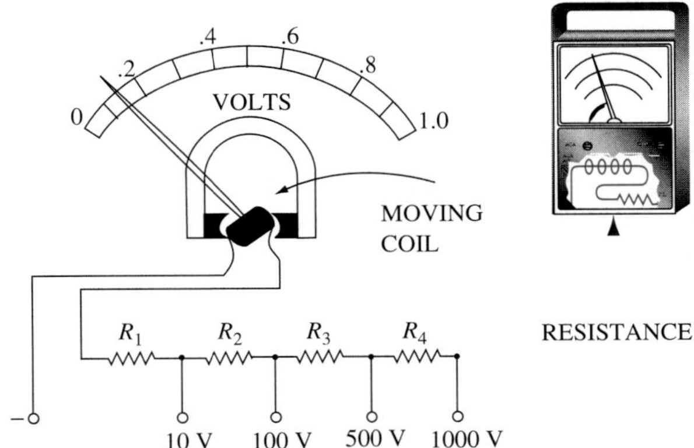

Figure E19-27 Electrical construction of a multimeter, reading volts.

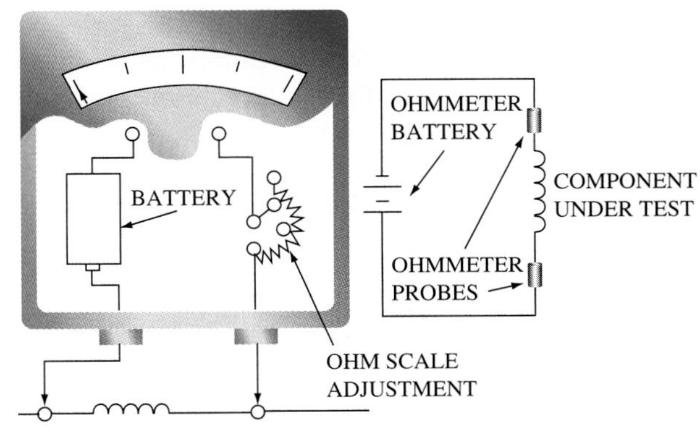

Figure E19-29 Construction of an ohmmeter, showing use of battery power.

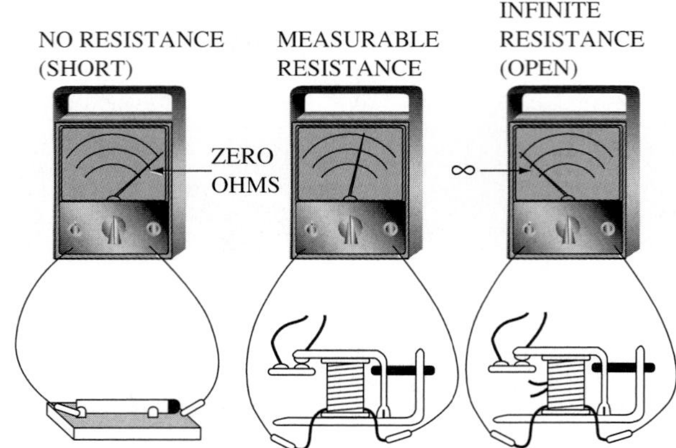

Figure E19-30 Use of an ohmmeter for continuity checking.

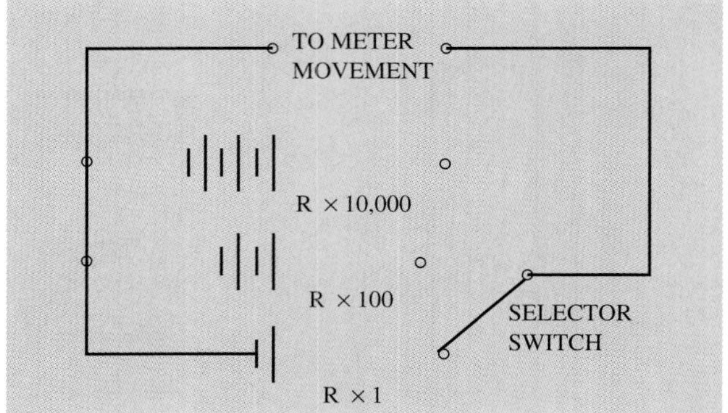

Figure E19-31 Use of a number of batteries in an ohmmeter to increase range.

Using the meter to check for open circuits is called *continuity testing*. Fig. E19-30 shows three diagrams representing the three possible responses that the meter can give.

1. In the diagram at the left, the meter is measuring the resistance through a closed switch and it registers zero. This indicates maximum current flow or zero ohms (a dead short).
2. In the diagram in the center, the meter is measuring the resistance of a coil of wire, which has a measurable resistance, which is read on the meter.
3. In the diagram on the right, the meter is measuring the resistance of an open circuit, which is read on the meter as infinity. Infinity means that the resistance is so large that it cannot be measured. It means that at this point there is a lack of continuity or no current flow.

An ohmmeter must be able to read resistances from a few ohms to tens of millions of ohms (megaohms). In order to do this, more than one battery is used, as shown in Fig. E19-31. The higher amount of power is required for higher resistances.

One very handy feature on an analog ohmmeter is the zero ohm adjustment, Fig. E19-32. This knob makes possible a quick and easy method of calibrating the instrument each time it is used. To test the adjustment, the two leads are touched together and if the reading is not zero, it is adjusted to zero before the instrument is used.

Unlike the voltmeter, one scale is used to read all resistance ranges. To determine the resistance value, multiply the meter reading by the number shown next to the selector switch setting. For example, in Fig. E19-33, the reading 5 is multiplied by the selector setting (×1), giving a resistance of 5 ohms.

Some ohmmeters have a selector position of R × 100,000 which is used in measuring very high resistances such as motor windings to ground.

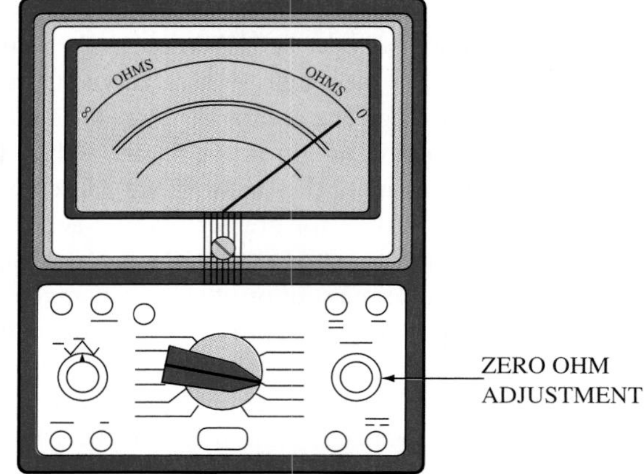

Figure E19-32 Zero ohm adjustment knob.

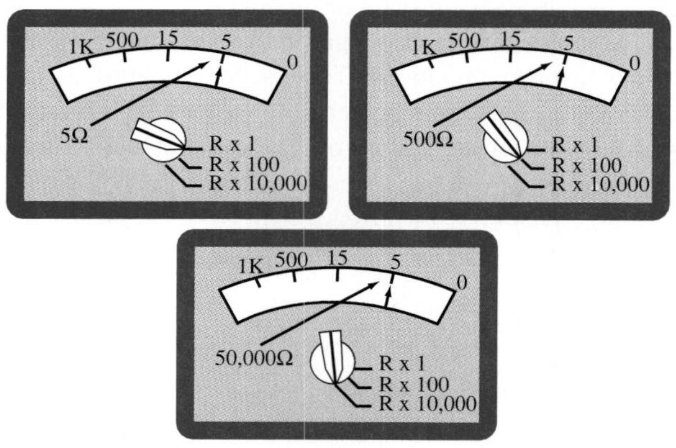

Figure E19-33 Reading the ohms on various range scales.

Care must be taken to prevent errors in reading resistance when two or more circuits are connected in parallel, as shown in Fig. E19-34. The meter in the illustration is actually reading the combined resistance of two parallel resistances, C1 and IFR.

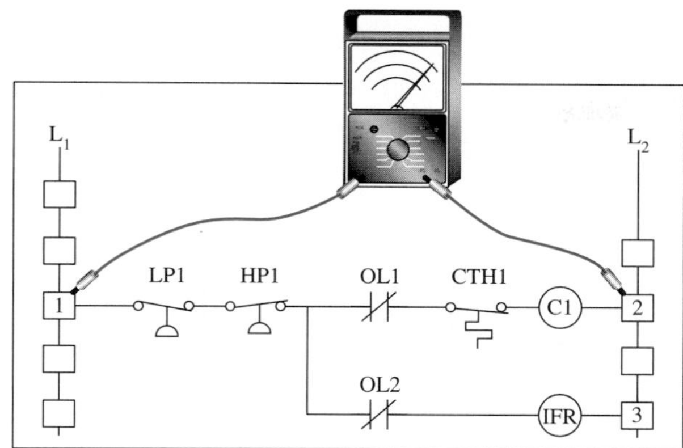

Figure E19-34 Possible error in reading ohms when two circuits are connected in parallel.

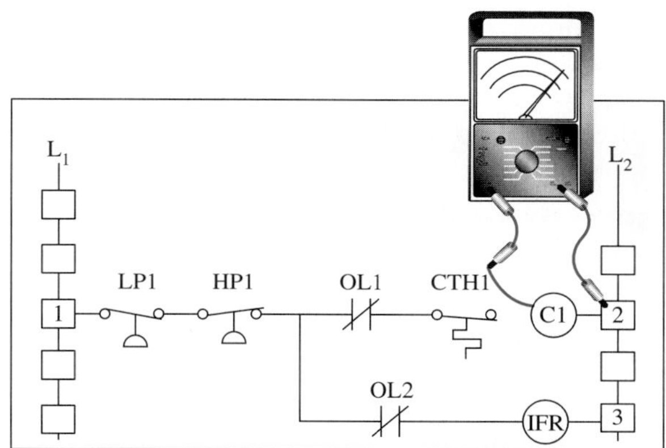

Figure E19-35 Proper method of reading resistances in parallel circuits.

In order to read only one resistance, one side of the component being tested is disconnected, as shown in Fig. E19-35.

One caution that needs to be followed: do not use an ohmmeter to test a solid-state circuit unless the manufacturer specifically requires it. The internal battery voltage of the ohmmeter can damage the solid-state circuit (Fig. E19-36).

E19-4 ALTERNATING CURRENT

Alternating current is a potential difference with a sinusoidal waveform that alternates polarity continuously at a fixed rate or frequency. The number of times per second the polarity is reversed is the *frequency* expressed in Hertz (Hz), or cycles per second.

E19-4.1 Magnetic Induction

Magnetic induction, as illustrated in Fig. E19-37, is used to generate voltage for commercial and residential use.

Figure E19-36 Possible circuit damage testing solid-state circuits.

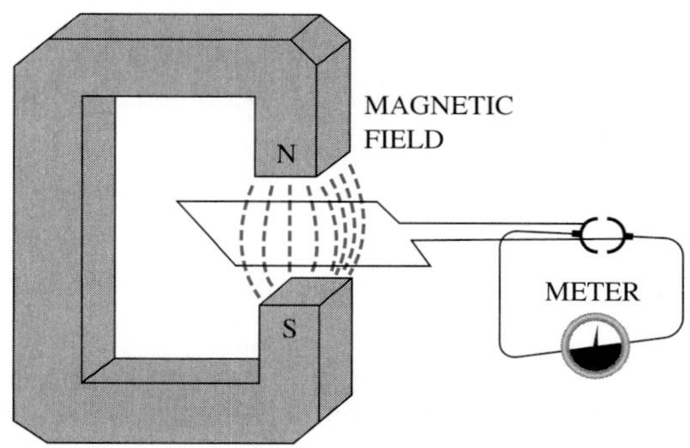

Figure E19-37 Magnetic induction demonstrated.

Induction occurs when a conductor is placed in a magnetic field. As the conductor is moved through the field, a potential difference is created in the conductor. The size of this potential difference is dependent on the strength of the field and the speed of the conductor through the field.

Alternating current is produced by revolving a coil, as shown, producing an output voltage, for one revolution of the coil, as illustrated in Fig. E19-38. This is a diagram of

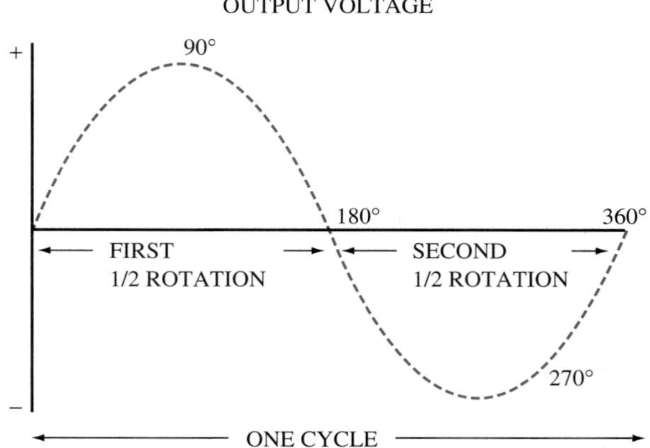

Figure E19-38 One-cycle output voltage from a generator.

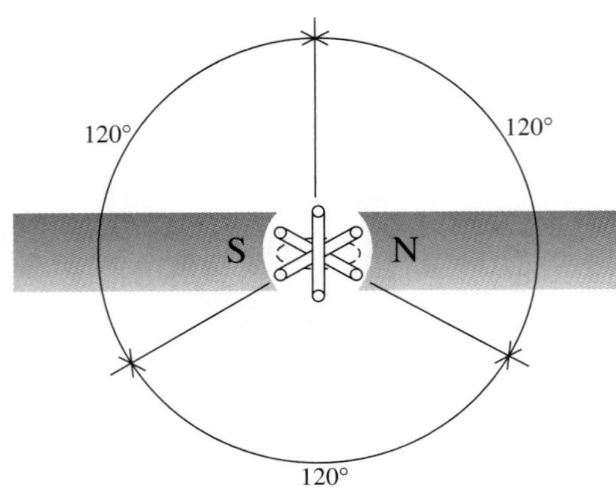

Figure E19-39 Three-phase AC power generation.

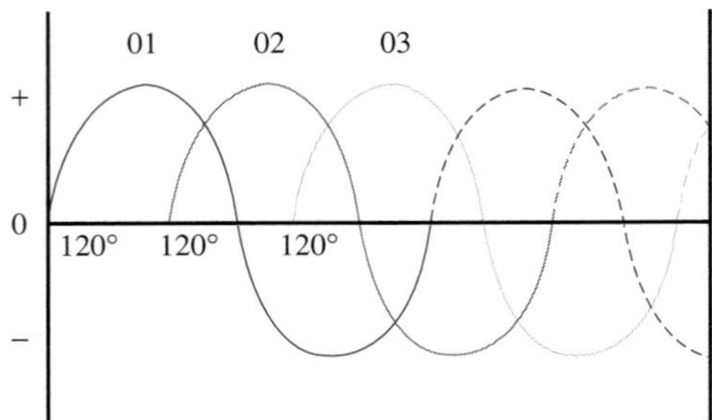

Figure E19-40 Waveforms for three-phase voltage.

a single cycle and is repeated 60 times per second (60 Hz), on a continuous basis. This is called single-phase voltage.

Single-phase power is commonly used for residential and small commercial systems. For larger motors, generally 10 Hp and above, three-phase power is normally used. This requires the addition of one more conductor to the generator, as shown in Fig. E19-39. The three conductors are positioned 120° from each other.

A diagram of the three-phase wave forms is shown in Fig. E19-40. They have the same shape, but they are 120° out of phase with each other.

E19-4.2 AC Circuit Characteristics

To review, when current passes through a conductor, a magnetic field is produced. The coiling of the conductor concentrates the lines of force, as shown in Fig. E19-41. The polarity at the ends of the coil will reverse as the current flow through the coil changes direction.

If an iron bar is placed inside the coil (Fig. E19-42), current flow through the coil causes lines of force to magnetize the core with polarity opposite to that of the coil. Because unlike poles attract and like poles repel, the iron bar or core attempts to center itself in the coil.

This principle is applied to the construction of relays and contactors. In the resting position, gravity keeps the contacts separated. Referring to Fig. E19-43, when current

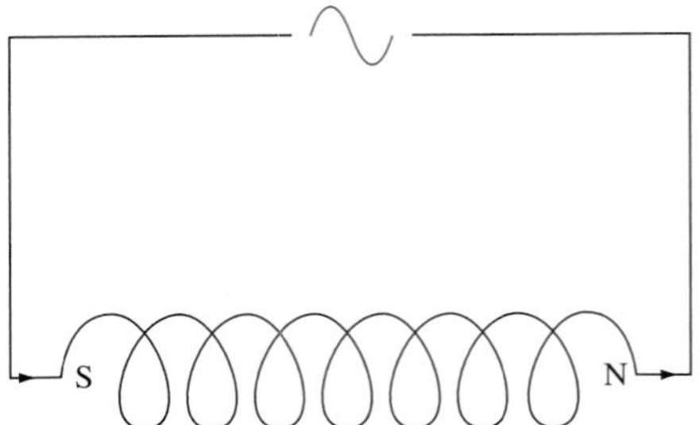

Figure E19-41 Coiling a conductor to increase the magnetic field.

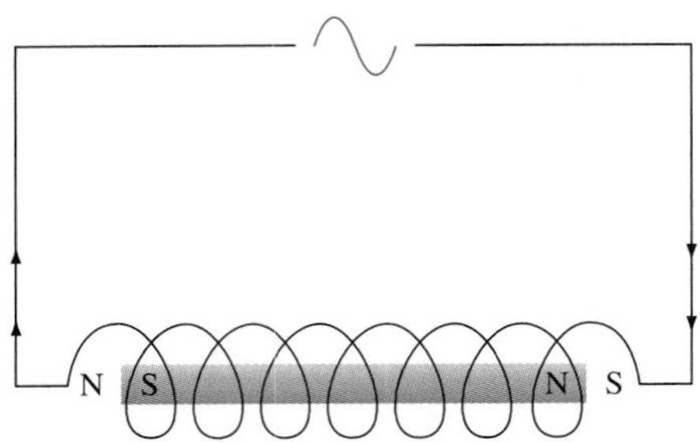

Figure E19-42 Adding an iron core to concentrate the magnetic field.

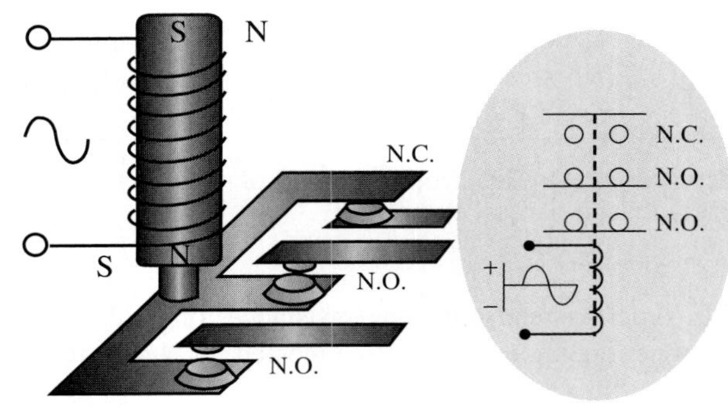

Figure E19-43 Using the solenoid principle to energize relays and contactors.

flows through the coil, the core is magnetized and tries to center itself in the coil. This raises the core and changes the position of the relay contacts, as shown in the diagram. A solenoid valve is another application of this principle.

This induction principle is also applied to the construction of AC motors, Fig. E19-44. The basic motor has two coils wrapped around stationary cores called *stator poles.* As AC current flows through the coils, the stator poles are magnetized. The stator field induces an opposing field in the rotor and the principle of attraction and repulsion causes the motor to run.

Another application of the induction principle is used in the construction of a transformer, Fig. E19-45. Transformers are used to step down the line voltage to 24 V for

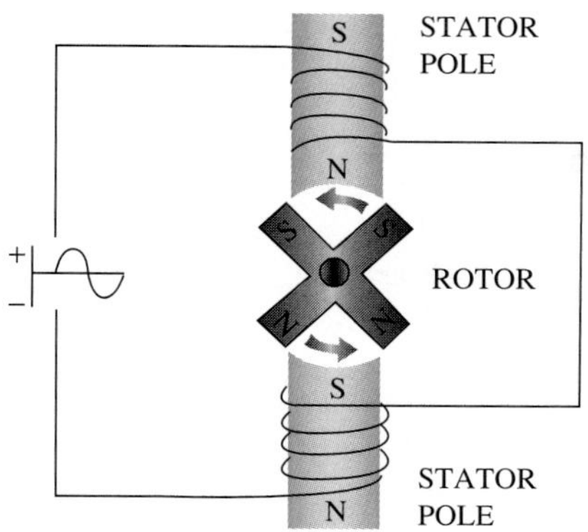

Figure E19-44 Application of magnetic induction to construction of electric motors.

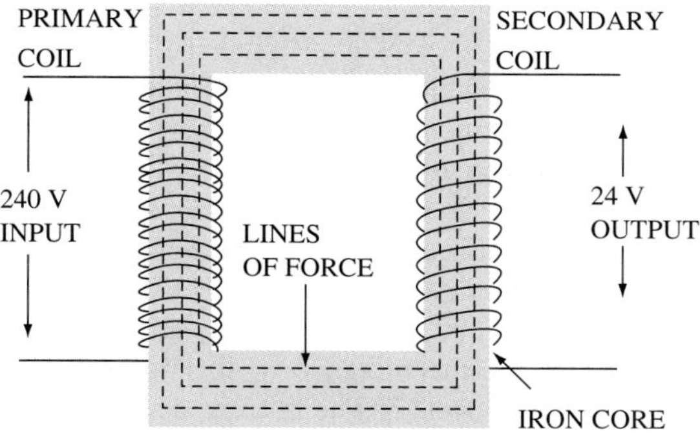

Figure E19-45 Application of magnetic induction to construction of transformers.

controls used in many HVAC/R units. Transformers contain a single iron core that is wrapped with two separate coiled conductors known as primary and secondary windings. When AC voltage is applied to the primary winding, the resulting lines of force are carried through the core. These lines create a current flowing through the secondary winding, inducing a voltage in that winding.

There are two types of transformers: step-up and step-down, as shown in Fig. E19-46. The amount of voltage induced in the secondary winding depends on the ratio of the number of turns in the primary winding to the number of turns in the secondary winding.

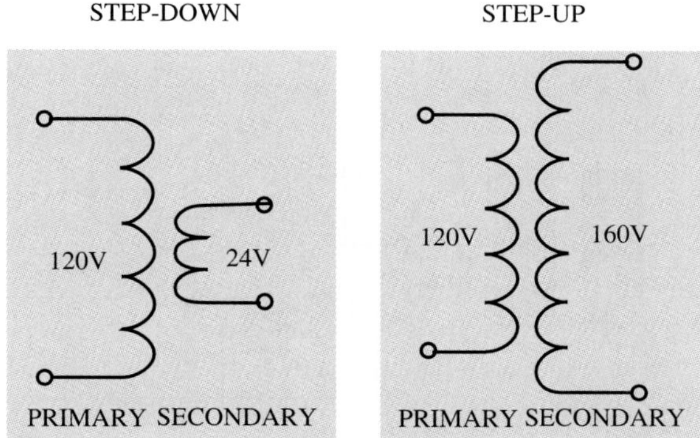

Figure E19-46 Construction of step-down and step-up transformers.

E19-4.3 Phase Shift

In working with AC power, certain characteristics affect the power calculation, $E \times I$, which are known as *phase shift* factors. The phase shift factors relate to:

1. Resistive circuits
2. Inductive circuits
3. Capacitive circuits

The *resistive circuit,* (Fig. E19-47), contains at least one resistive load, such as an electric heater or lamp. The current rises and falls with the voltage and the two are considered to be synchronized or "in phase". The maximum voltage occurs at the same time (or same phase angle) as the maximum amperage.

Motors, relays, transformers, and some other AC loads are constructed using coils of wire. These coils produce magnetism and are called inductive loads, as shown in Fig. E19-48. The voltage and current in circuits containing inductive loads are phase shifted, or out of phase, sometimes as much as 90°.

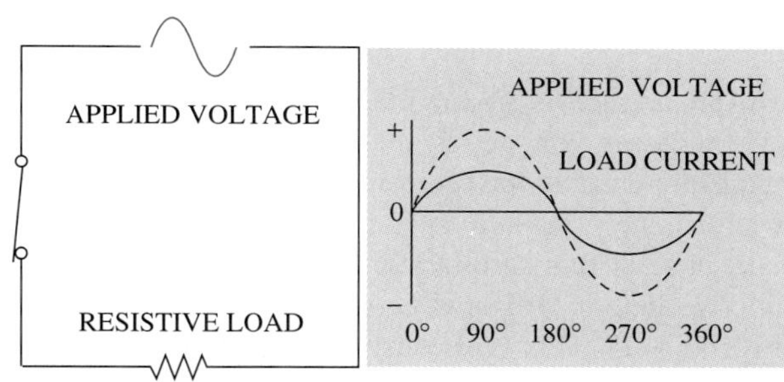

RESISTIVE LOAD = VOLTAGE/CURRENT IN PHASE
POWER = VOLTS × AMPS

Figure E19-47 Resistive circuit containing at least one resistance.

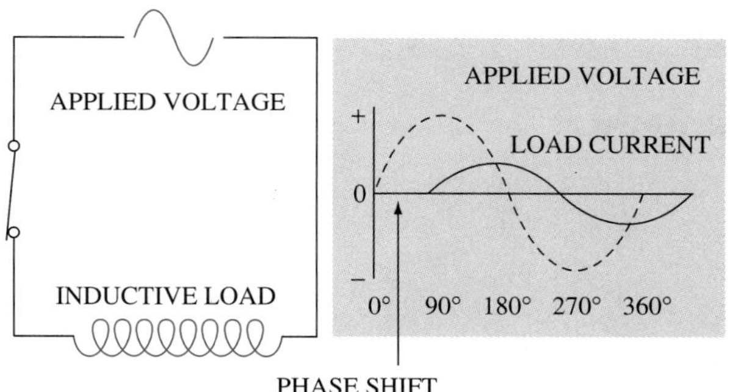

Figure E19-48 An inductive circuit formed by coiled conductor.

Knowledge Extension

Visit your local power generating utility and ask for a tour of the system to learn more about electrical generation. Ask if they have any blueprints or technical data of the plant. Look for turbines, step-up transformers, switching gear, and any other related equipment that can help clarify the concepts discussed in this chapter. Many facilities will have scheduled tours available to the public or specialized interest groups. Go to your local museum and ask if they have any old motors or generation equipment on display. Somewhere in your local community you will find a wealth of supplemental information to help expand your knowledge on electricity.

Inquire at the personnel or human relations office at the electric utilities administration building to determine if there are employment opportunities for students of refrigeration, heating, and air conditioning. Due to the heat generated by electricity at transmission stations, utilities will hire and train in-house personnel to perform maintenance and service rather than contracting the work out. There are also often opportunities as energy estimators performing air conditioning and heating load analysis on residential and commercial buildings to help conserve power.

In the *inductive circuit* the current lags, or is out of phase with the voltage, as shown in the diagram. In this circuit, the current waveform peaks 90° after the voltage waveform.

Because of this current lag, the measured power in an inductive circuit will always be less than the calculated power ($E \times I$). This is because measured power is an instantaneous reading, and at any particular time, one or the other or both voltage and current readings are not at their peak.

The term power factor (PF) is used to indicate this difference.

Power factor = (True power) / (apparent power)
or
Power factor = (Wattmeter reading) / ($E \times I$)

EXAMPLE

If a circuit with a single-phase AC electric motor (inductive) load has the following meter readings, how much is its power factor: 115 V, 5 A, 517.5 W?

Solution

$$PF = 517.5\ W / (115\ V \times 5\ A)$$

$$PF = 517.5 / 575$$

$$PF = 0.90$$

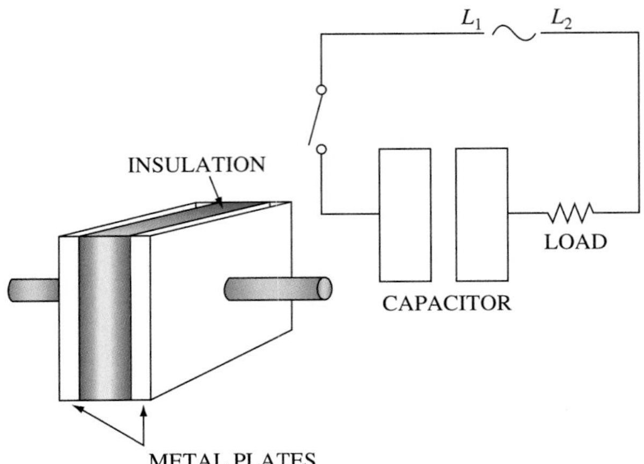

Figure E19-49 A capacitor creates a useful phase shift.

A *capacitor* is an electrical device that is used to change the phase relationship between the current and the voltage. This effect can be used to increase the starting power (torque) of an electric motor. A capacitor consists of a layer of insulation (a dielectric) placed between two plates of highly conductive metal. A capacitor is usually connected in series with the load, as shown in Fig. E19-49.

Obviously, no current can flow through the capacitor because of the dielectric. The current does flow through the series circuit, as shown in Fig. E19-50. When the switch is closed, the supply voltage is applied across the capacitor. At that instance the electrons flow rapidly from the source to the right side of the capacitor and from the left side of the capacitor to the source, causing a current to flow through the load. The capacitor quickly reaches peak current. It is described as charging during this period.

Following the initial rapid flow of electrons, the rate of current flow reduces, as shown in the downward movement of the current waveform in Fig. E19-51. As electrons leave one plate and accumulate on the other, a potential difference (voltage) begins to develop across the capacitor. This difference is created by current flow and there-

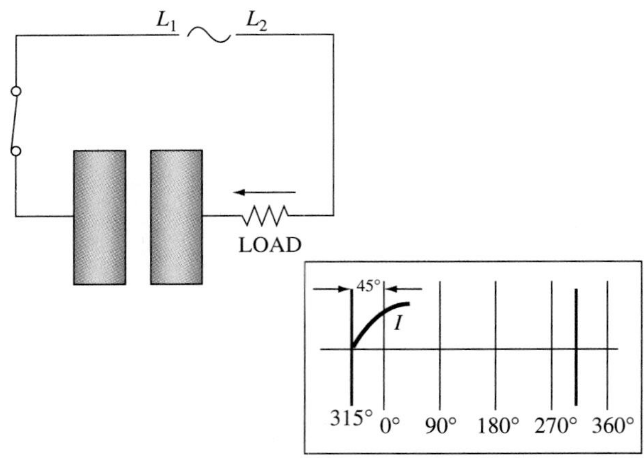

Figure E19-50 Capacitor charging.

ON THE JOB

Not ON doesn't necessarily mean OFF

Be careful working on and testing electrical equipment with the power ON. Motors, solenoids, and relays that are not energized may still be connected to a live (ungrounded) line and be HOT. Examples: Single Phase Motor; Three Phase Motor; Relay or Contactor; Solenoid.

Electric shock can be lethal! Take care and treat all electrical wires and terminals as potentially LIVE. Current leads the voltage as shown in Fig. E19-53.

In actual practice, the utility tries to maintain plus or minus 5% at the secondary of the transformer. There is an additional 5% allowable voltage drop in the building wiring, according to ANSI C84.1-1982.

fore lags the current. In this circuit, which contains a resistive load, current leads the voltage by 45° (Fig. E19-52).

As the supply voltage waveform crosses the baseline, the polarity across the capacitor changes and electrons leave the right-hand plate. The capacitor discharges and the current flows through the load in the opposite direction, as shown in Fig. E19-53. The capacitor continues in this manner, as long as the source voltage is applied.

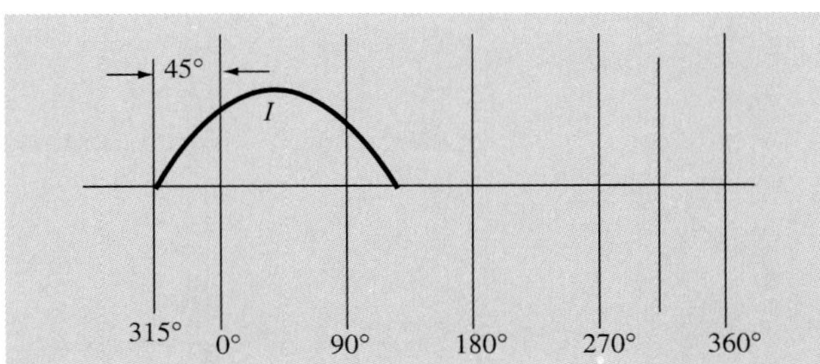

Figure E19-51 Current flow in a capacitor circuit during the first half-cycle.

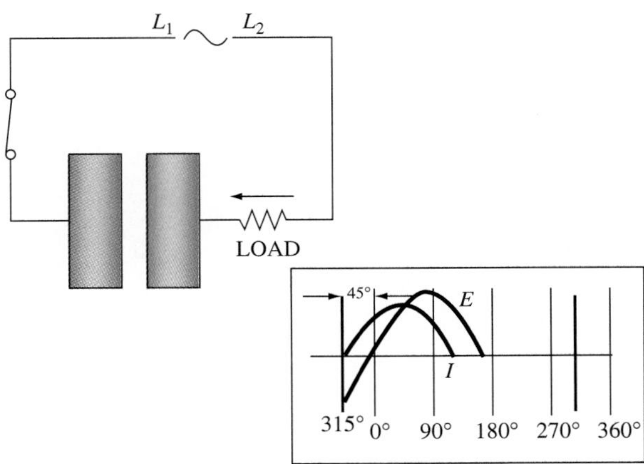

Figure E19-52 Voltage produced in a capacitor circuit during the first half-cycle.

E19-4.4 Impedance

Impedance is the opposition to the current flow in an AC circuit. Impedance is to an AC circuit what resistance is to a DC circuit; however, multiple impedances in a circuit cannot be added like resistances are added in direct current because the currents through inductive, capacitive, and resistance loads in an AC circuit are out of phase with each other.

The following is an example of calculating impedance in an inductive circuit. The formula that is used is:

$$Z = E/I$$

where

Z = impedance

EXAMPLE

What is the impedance of a single-phase inductive circuit having a voltage of 240 V and a current flow of 10 A?

Solution

$$Z = \frac{240 \text{ V}}{10 \text{ A}} = 24 \text{ ohms}$$

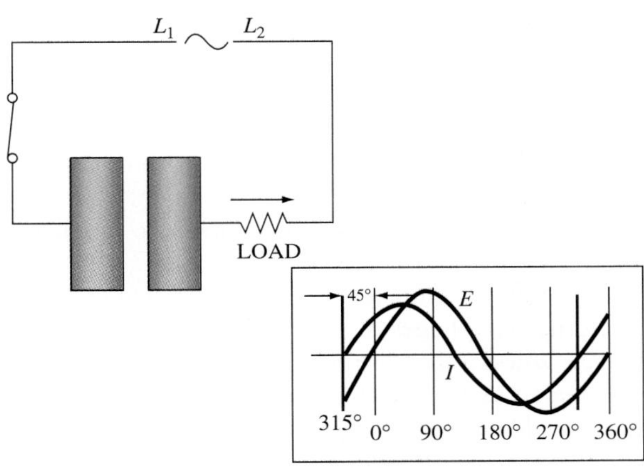

Figure E19-53 Capacitor discharging.

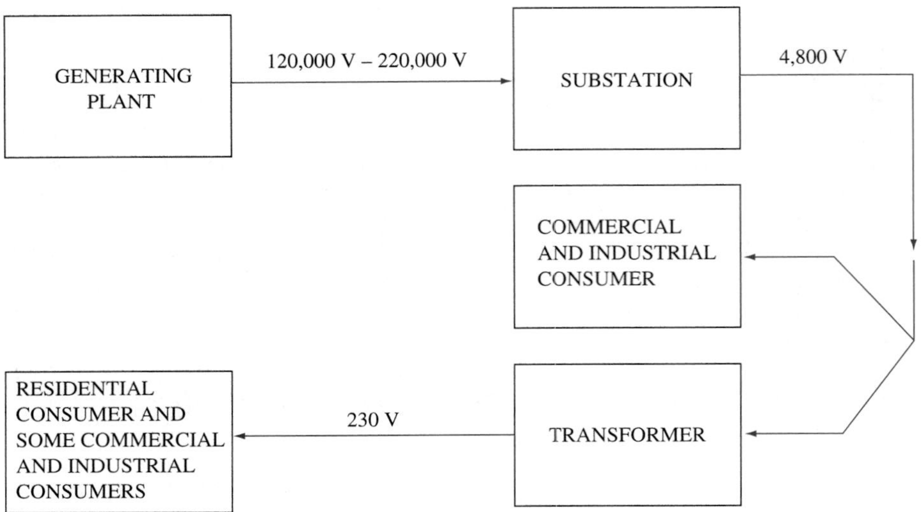

Figure E19-54 Voltage reduction in the power distribution system.

E19-5 POWER DISTRIBUTION

Almost all of the electrical power used by consumers today is alternating current. It is supplied to substations at voltages as high as 120,000 V or more. There it is reduced to voltages between 4800 V and 34,000 V for distribution to areas of commercial or residential users. Most power companies are now using 23,000 V or 34,999 V. Local transformers lower the voltage to user requirements, as shown in Fig. E19-54.

Four common low-voltage systems are available to consumers:

1. 230 V, single-phase, 60-Hz systems
2. 230 V, three-phase, 60-Hz systems
3. 208 V, three-phase, 60-Hz systems
4. 480 V, three-phase, 60-Hz systems

E19-5.1 230-V, Single-phase, 60-Hz Systems

Single-phase current is used for almost all residences. Any electric appliance that operates on 120 V is single-phase equipment.

The most common service supplied to residential and small commercial users is the 230 V, single-phase, 60-Hz system. The system uses three wires, two hot wires and one grounded neutral. A schematic diagram of this 230 V system is shown in Fig. E19-55.

The electric utility uses a transformer to produce this service, as shown in Fig. E19-56.

All HVAC/R equipment is manufactured to operate satisfactorily on voltages of plus or minus 10% of the rated voltage unless otherwise specified. For example, if the equipment has a voltage rating of 230 volts, the equipment should be able to operate at any voltage between 207 and 253 V. HVAC/R equipment has a tendency to operate more satisfactorily on maximum voltage than on minimum voltage.

The utility attempts to maintain a voltage at the load within this plus or minus 10% range. At peak load times, the line voltage may drop to near the permitted minimum. In case there are any voltage problems, tests of the line voltage at the HVAC/R load should be measured at these times.

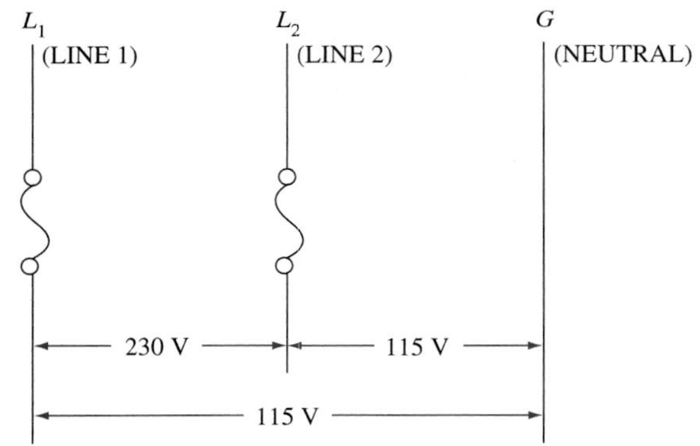

Figure E19-55 Line voltages for three-wire 230-V single-phase power supply.

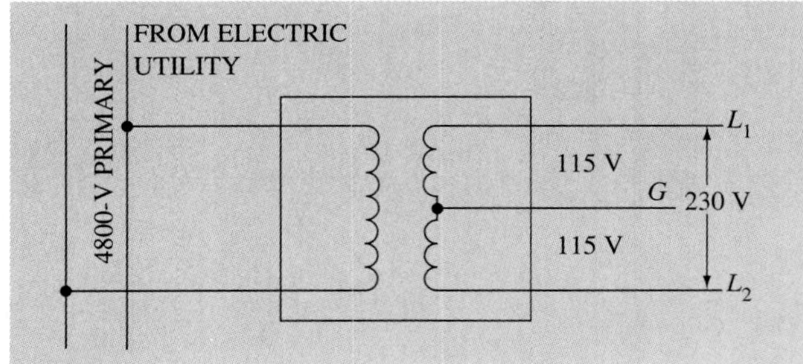

Figure E19-56 Step-down transformer for a 230-V three-wire single-phase power supply.

E19-5.2 230-V, Three-phase, 60-Hz Systems

Three-phase systems are commonly used for sizable commercial and industrial installations. These transformers have three hot legs of power and one neutral leg, as shown in Fig. E19-57. This type of power supply is obtained from a delta transformer secondary hookup, as shown in Fig. E19-58.

Three phase 230-V power is obtained connecting to the three hot legs. Single-phase 230-V power can be obtained by connecting to any two of the hot legs. Single phase 120-V power can be obtained by connecting to either of the adjacent hot legs and the midpoint neutral. Single-phase 208-V power can be obtained by connecting to the non-adjacent hot leg and the ground. This is called the wild leg.

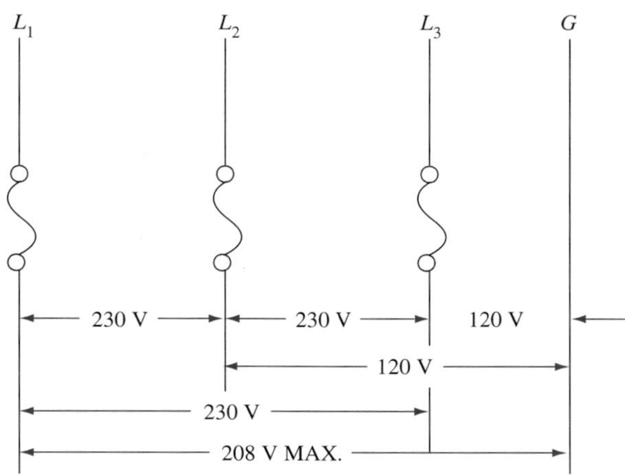

Figure E19-57 Line voltages for a four-wire 230-V three-phase power supply.

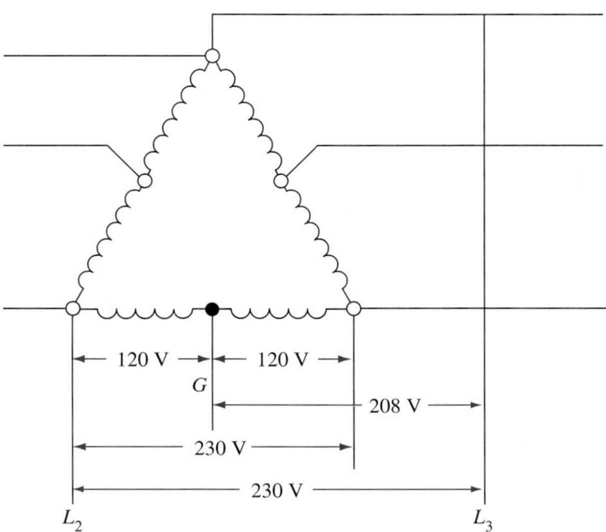

Figure E19-58 Line voltages for a four-wire 230-V three-phase system using a delta transformer secondary.

E19-5.3 208-V, Three-phase, 60-Hz Systems

These network systems are common in schools, hospitals and office buildings where 208-V three-phase motors and 120-V single-phase lighting and convenience circuits are required, as shown in Fig. E19-59.

From this type of system, 208-V three-phase, 208-V single-phase and 120-V single-phase services are available. It is recommended all motors used be rated for 208 V (not 230 V) for operation on 208-V systems.

The schematic for the transformer secondary hookup is shown in Fig. E19-60.

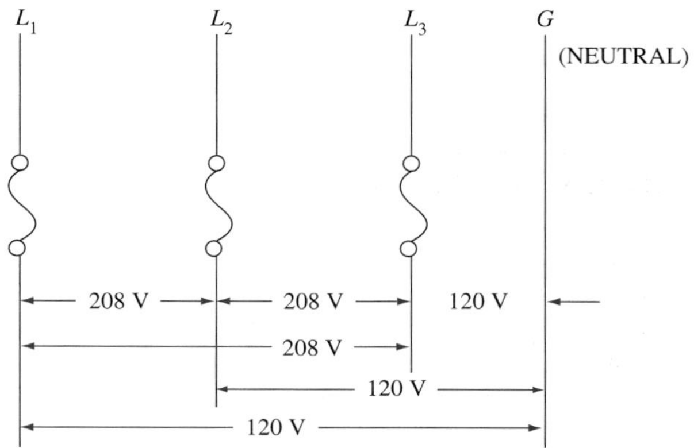

Figure E19-59 Line voltages for a four-wire, 208-V three-phase power supply.

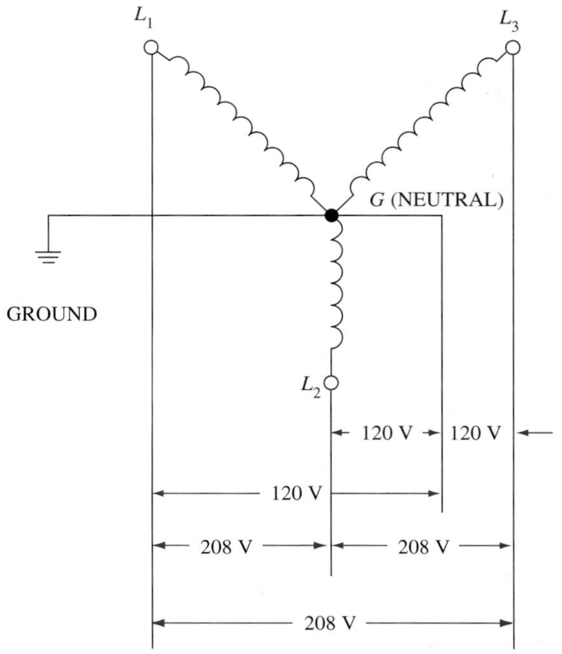

Figure E19-60 Line voltages for a four-wire, 208-V three-phase power supply using a wye transformer secondary.

E19-5.4 480-V Systems

The 208-V / 120-V three-phase four-wire wye-connected network system (Fig. E19-59 and Fig. E19-60) has been generally superseded on large buildings by the 480-V/277-V three-phase four-wire wye-connected network (Fig. E19-61 and Fig. E19-62). It was made possible by the development of 277-V fluorescent lighting. Standard 460-V three-phase motors can be used on 480-V systems. Convenience outlet circuits at 120 V are provided for by 480-V/208-V or 480-V/120-V stepdown transformers. The higher supply voltage permits larger loads to be serviced by smaller wires. In addition, voltage drops are reduced.

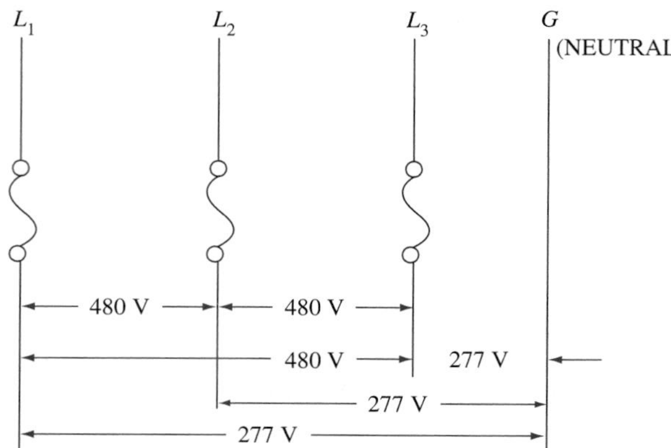

Figure E19-61 Line voltages for a four-wire, 277-V/480-V three-phase power supply.

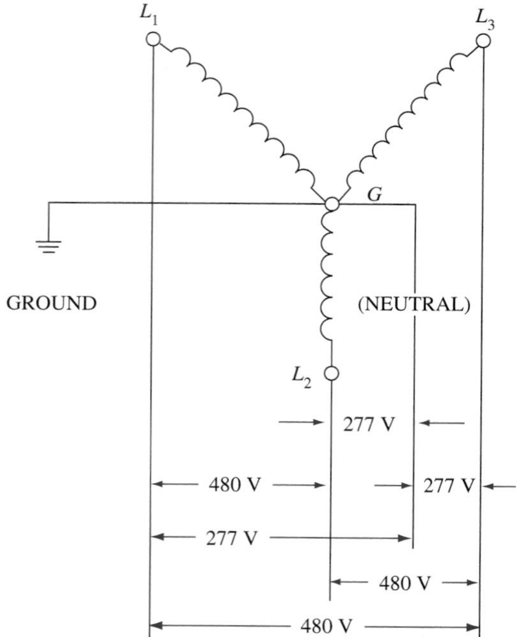

Figure E19-62 Line voltages for a four wire, 277-V/480-V three-phase power supply, using a wye transformer secondary.

For example, a 50-Hp, 460-V, three-phase, 60-Hz motor would have a full load of 57.5 A, while the same motor at 208 V would have a full load of 122 A.

E19-5.5 Advantages of Three-phase Power

Most larger HVAC/R equipment operates on three-phase power for three reasons. First, line current is reduced almost one-half for the same size motor load. This reduces the wire size and voltage drop. (Note: Line current is reduced by one-third, almost one-half). Second, there is less power loss in transformers in using it. Third, three-phase motors are smaller, less expensive, more reliable, and more efficient than single-phase. They do not require special capacitors to increase their starting torque.

REVIEW

- Static electricity—An electrical charge, either positive or negative, possessed by bodies. When a difference in charge occurs, current can flow between the bodies.
- Potential difference—creates a force, called electromotive force (EMF) measured in volts.
- 1 volt = one coulomb (10^{18} electrons) passing a fixed point in the circuit in one second.
- 1 volt = 1 ampere current × 1 ohm of resistance
- Conductor—A material that readily passes current flow.
- Insulator—A material that does not readily pass current flow.
- Conductivity—A measure of how well material conducts current. It is the inverse of resistance.
- Ohm's law: $E = IR$
- Watt's law: $P = IE$
- Watt = One ampere of current flowing with a force of one volt.
- PF (Power Factor) = Phase angle between E and I
- PF = True power/Apparent power
- 1 Horsepower = 746 watts
- 1 watt = 3.41 Btu/hr
- EER = Energy Efficiency Rate = number of Btu per watt.
- Kilowatt = 1000 watts
- Direct current (DC)—Current flowing in one direction through the load only.
- Alternating current (AC)—Varies in magnitude and polarity, current alternates in both directions through the load.
- Frequency—Number of cycles per second, measured in Hertz (Hz).
- Period—Time required to complete one cycle.
- Elements of a "simple circuit":
  A power source
  An electrical load
  A path for current to flow
  A switching device (control)

- Types of circuits:
  series circuit
  parallel circuit
  series-parallel circuit
  parallel-series circuit
  bridge circuit
- Kirchhoff's law:
  Series

$$E_T = E_1 + E_2 + E_3 \ldots$$
$$I_T = I_1 = I_2 = \ldots$$

  Parallel

$$E_T = E_1 = E_2 \ldots$$
$$I_T = I_1 + I_2 + I_3 \ldots$$

- Resistance:
  Series

$$R_T = R_1 + R_2 + R_3 \ldots$$

  Parallel

$$R_T = \frac{R_1 \times R_2}{R_1 + R_2}$$

Problems and Questions

1. Which of the following types of light is caused by static electricity?
   a. Candlelight
   b. Lightning
   c. Flashlight
   d. Fluorescent light
2. What type of power is supplied to a residence?
   a. Single-phase AC
   b. Three-phase AC
   c. Direct current
   d. Static electricity
3. What is the meaning of EMF?
   a. Electrical machine factor
   b. Efficiency microfarads
   c. Energy manufacturing facility
   d. Electromotive force
4. Electrical current is expressed in what terms?
   a. Amperes
   b. Volts
   c. Ohms
   d. Watts
5. An open circuit occurs when:
   a. The path permits the flow of current
   b. The path of the current is broken
   c. The loads are put in series
   d. Two loads are put in parallel
6. Ohm's law is as follows:
   a. $I = E/R$
   b. $R = EI$
   c. $R = I/E$
   d. $E = E/R$
7. In a purely resistance load, if $V = 120$ V and $R = 10$ ohms, what is I?
   a. 10 A
   b. 1.2 A
   c. 1200 A
   d. 12 A
8. If two 10-ohm resistances are in parallel, what is their combined resistance?
   a. 5 ohms
   b. 10 ohms
   c. 15 ohms
   d. 20 ohms
9. If two 10-ohm resistances are in series, what is their combined resistance?
   a. 5 ohms
   b. 10 ohms
   c. 15 ohms
   d. 20 ohms
10. A 100-ft extension cord, with a resistance of 5 ohms, is plugged into a 120-V single-phase AC power source. Assume the extension cord is purely resistive. What is the voltage at the other end of the cord supplying a 1-A load?
    a. 100 V
    b. 110 V
    c. 115 V
    d. 130 V
11. Voltage remains the same in a series circuit. True or false?
12. Amps remain the same in a parallel circuit. True or false?
13. To determine resistance total in a series circuit, just add up the individual resistors. True or false?
14. Three phase powered electric motors require no start capacitors. True or false?
15. Electrical power or watts is equal to amps times ohms. True or false?
16. What do electrical power companies do with transmission lines to avoid line losses due to the resistance of the wires?
17. The frequency of alternating current is measured in ______.
18. What is the equation for calculating AC power in three-phase systems?
19. There are conditions for power systems. What are they?
20. What is the primary difference between the analog and digital meter?

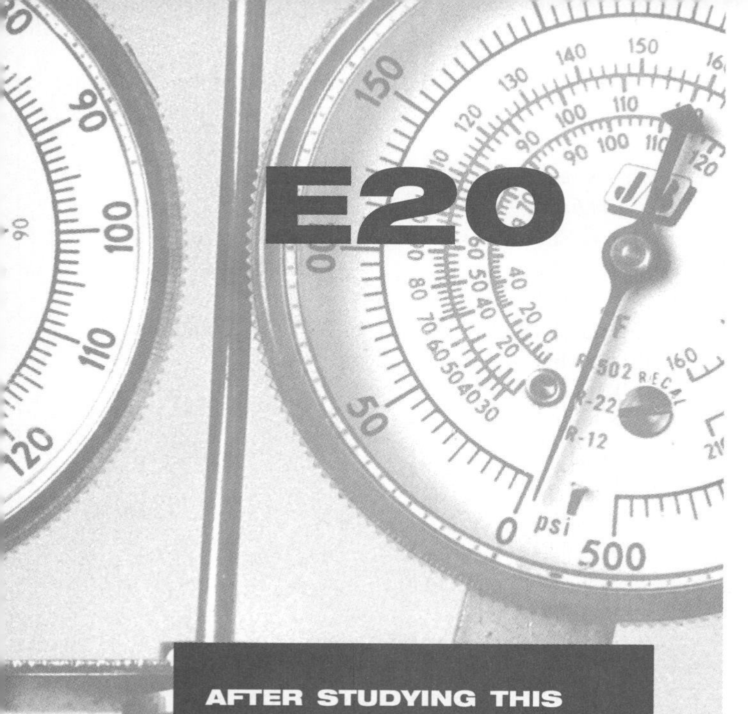

E20 Electrical Components and Wiring Diagrams

AFTER STUDYING THIS CHAPTER, THE STUDENT WILL BE ABLE TO:

- Understand the purpose and applications of electromagnetism.
- Locate the power sources for an electrical system and determine their characteristics.
- Identify the electrical loads and switches included in each circuit.
- Trace each circuit and determine the function of the electrical components it includes.
- Describe the various types of wiring diagrams.
- Read a schematic wiring diagram and determine the sequence of operation of the system.
- Make a schematic diagram from the information on a connection diagram or by tracing the circuits on the unit.
- Interpret detailed instructions for wiring diagrams.
- Draw electrical circuits that conform to standard industry logic and symbols using appropriate loads and controls.
- Understand how to wire actual electrical circuits from wiring diagrams.
- Use electrical meters to test circuits.
- Explain the difference between the label, installation, and connection diagrams.
- Identify and draw all electrical symbols used by the HVAC/R industry in diagrams.

E20-1 ELECTRICAL CIRCUITS

All HVAC/R electrical systems are made up of electrical circuits. An electrical circuit has three essential parts and one optional part as follows (Fig. E20-1):

1. A source of power
2. A load
3. A path for the current to follow
4. A control (optional)

As a result of these electrical components, electrical current is transformed into heat, light, sound, or mechanical motion. Although the control is optional, meaning that the circuit will operate without it, most systems have con-

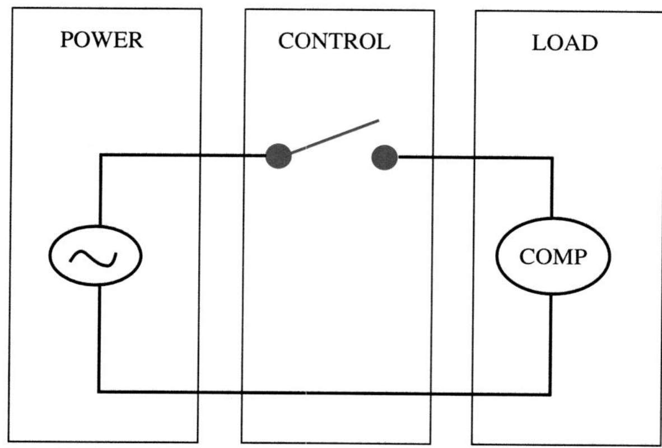

Figure E20-1 Electrical circuit showing a source of power, a control device, and a load that consumes the power.

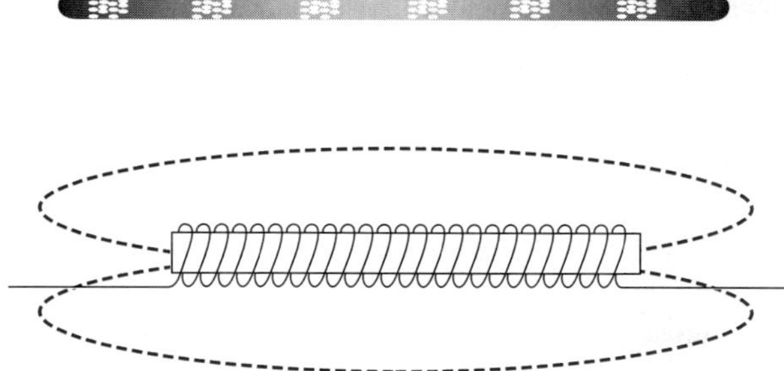

Figure E20-2 Electromagnet produced by current flowing through a conductor.

trols to regulate the supply of power to, or remove the power from, the load.

Since many electrical components operate using an electrical force termed *magnetism,* the technician should be familiar with some of its characteristics.

E20-2 MAGNETISM

Magnetism can be produced by electricity. Any time that current flows through a conductor it creates a magnetic field around it. To intensify the field, the conductor is coiled as shown in Fig. E20-2. When this occurs, magnetic poles form at each end which change polarity as the current alternates.

The magnetic force is further intensified by placing an iron rod in the center of the coil as shown in the figure. When the current flows through the coil the rod tends to center itself in the coil. This principle can be used to operate a switch, as shown in Fig. E20-3. This arrangement is useful in the construction of relays, contactors, solenoid valves, and motors.

Another characteristic of electromagnetism is its ability to induce a current flow in another conductor that passes through its magnetic field. This principle is known as *induction.* Induction is useful in the design of transformers, motors and generators.

E20-3 COMMON ELECTRICAL COMPONENTS AND THEIR SYMBOLS

In this section, some common electrical devices that are parts of standard units will be described to show how they are connected in the system to perform their proper function. As an example, a simplified packaged air-conditioning unit is shown in Fig. E20-4.

Where the power comes into the building, it enters a service entrance panel for distribution to the various electrical loads. Each electrical circuit that comes from this panel is electrically protected by either

1. A fuse, or
2. A circuit breaker

Two types of fuses and their symbols are shown in Fig. E20-5. A fuse is a special electrical conductor that is placed in series with a load and melts when excessive current flows through it, breaking the circuit. Fuses are available in various types and sizes so that they can be selected to match the requirements of specific loads. If they are too small they melt before they should. If they are too large they do not offer the proper protection. Their selection follows the rules set forth in the electrical code or in the specifications accompanying the load.

Where fuses are used to protect motors in the circuit, a special type of fuse is used called a *time-delay fuse* or *dual-element fuse.* This type of fuse has a built-in delayed action which will tolerate momentary heavy starting current on motor power-up, but functions the rest of the time to protect the motor against excessive running current.

All residences have some type of *electrical panel,* where the electrical service enters the building and is distributed to

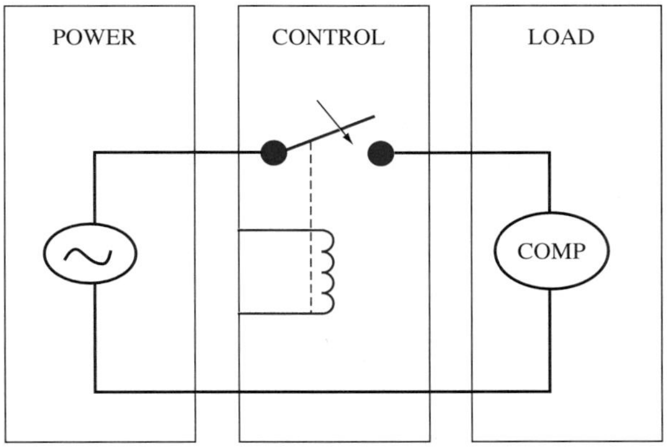

Figure E20-3 Magnetic switch used to send power to and remove power from the compressor.

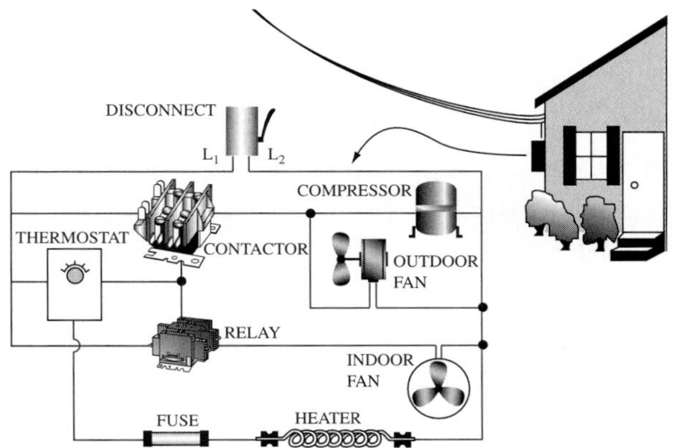

Figure E20-4 Electrical loads required for a packaged air conditioner.

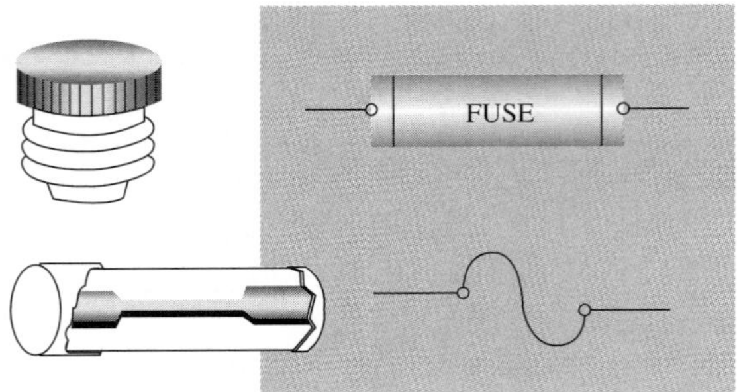

Figure E20-5 Protective fuses in the power circuit to prevent excessive current flow.

the various circuits in the building. Each circuit has some type of protective device to automatically disconnect the power in case the circuit is overloaded. This protection can either be a fuse as described above or a *circuit breaker,* as shown with its electrical symbol in Fig. E20-6.

The advantage of a circuit breaker over a fuse is that it can be manually reset at the electrical service panel after an overload, rather than replaced. Also, the circuit can be manually opened in case there is a need to perform service on the circuit. The circuit breaker can either have a thermal or a magnetic trip mechanism. Note the symbols used to represent a circuit breaker in an electrical wiring diagram.

In addition to the protective equipment in the entrance panel, the electrical codes usually require that each circuit be protected at a distribution or subpanel. In addition, a service-disconnect switch should be provided at the equipment being supplied current, as shown in Fig. E20-7. This disconnect is conveniently located to provide an easy way to disconnect the unit for service.

Inside the unit itself the power supply is connected to a terminal strip where the power is distributed to the various circuits within the unit. Some circuits such as fan motors and compressors use the full power source voltage. To produce the 24-V control voltage, a step-down transformer is required, as shown in Fig. E20-8.

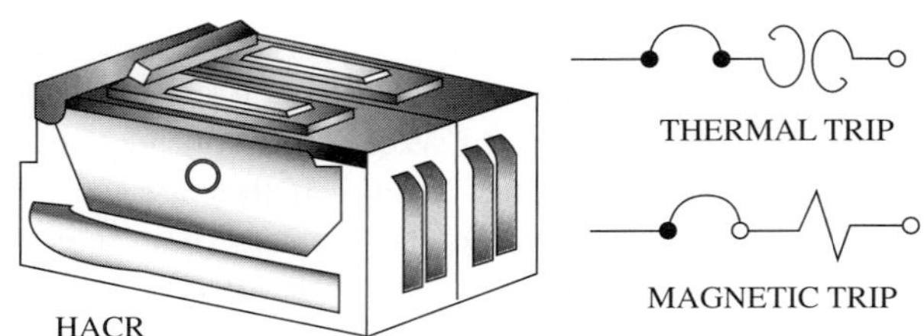

Figure E20-6 Circuit breaker which automatically opens the power circuit on excessive current flow, manually reset.

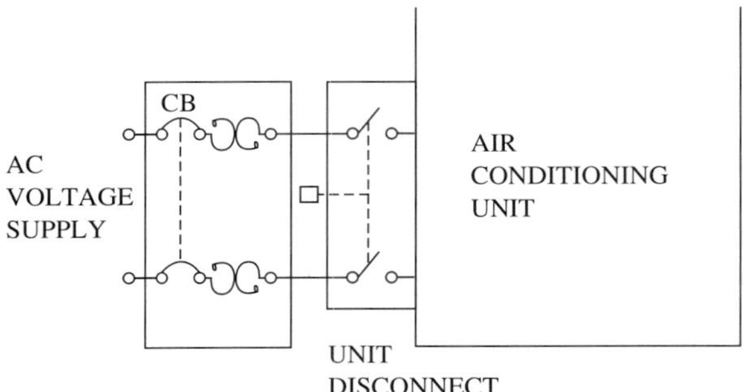

Figure E20-7 Power to individual air conditioner is usually supplied through a disconnect located on or near the unit.

PRACTICING PROFESSIONAL SERVICES

Prior to troubleshooting electrical circuits, the technician needs to learn to first visually check the obvious conditions. This includes inspecting the thermostat to verify if it is on and adjusted to actuate the circuit. Next, the technician can check circuit breakers and disconnects. If all appears normal, then open the control panel for the equipment and check the fuse(s).

When it becomes time to check the electrical circuit using the diagram, troubleshooting can be easier if the technician simplifies the circuit to three major areas: line voltage, low voltage, and thermostat control. This means to first locate on the schematic the thermostat circuit and analyze it. It is the primary control and without it nothing else will work. Using the voltmeter will instantly determine if the control circuit is sending a signal to the unit. Next, check the line-voltage side of the transformer and the secondary side of the transformer for control voltage. If all appears normal as expected, then begin to hop-scotch or jump through the circuit looking for open switches or failed components. Remember, an open switch will read potential voltage difference and a closed switch will not.

PREVENTIVE MAINTENANCE

Good preventive maintenance while working on electrical controls includes inspecting for loose wiring, worn connections, and dirty or pitted contacts. Wiring and connections should be repaired, but contacts which are pitted should be replaced. Occasionally, a dirty contact can be cleaned with an industrial-grade contact solvent, but this is often only a temporary fix to keep equipment running.

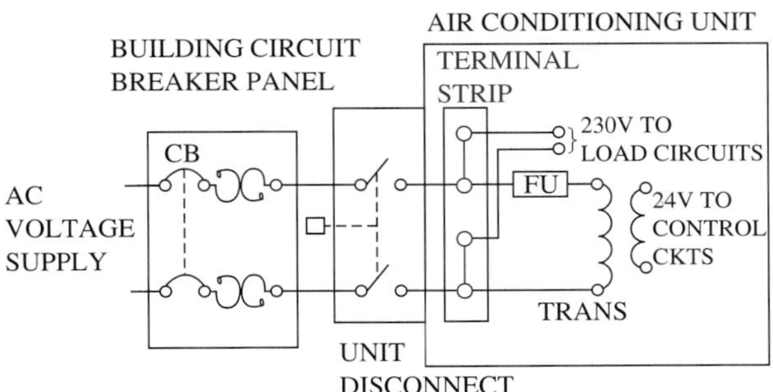

Figure E20-8 Inside the unit, power is connected to a terminal strip. From the strip, loads and transformers are connected.

E20-3.1 Transformers

HVAC/R equipment often requires more than one voltage. One or more transformers are often used to step down the line voltage to supply load or control requirements. Occasionally a step-up transformer may be used.

Transformers are constructed using the induction characteristic of AC power. When current flows through a coil, a magnetic field is produced. When a second coil is placed in the field of the current-carrying coil (primary), electric current can be transferred to the second coil (secondary), as shown in Fig. E20-9. The process is made more efficient by wrapping the coils around a common metal core. The voltage transferred is directly in proportion to the ratio of the number of turns on the primary coil to the number of turns on the secondary coil. More than one secondary coil can be used if additional voltages or circuits are required.

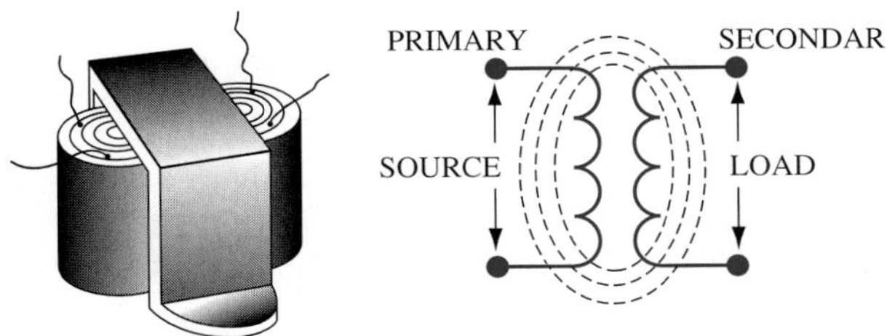

Figure E20-9 Typical control transformer showing primary and secondary windings.

E20-3.2 Electrical Load Devices

The whole purpose of the electrical system is to supply power to the electrical load devices in combination with a control sequence to produce a desired output. The proper operation of the load devices is of primary importance; therefore, the operation of some of the more common load devices will be examined to see how they fit into the wiring system.

The most useful type of diagram for observing the sequence of operation is the ladder diagram. In this diagram, the power is represented by vertical lines shown as L_1 and L_2 in Fig. E20-10. The electrical circuits are represented by horizontal lines, stretching between L_1 and L_2, each circuit containing a load device. The controls (switches) are added in series with the loads to turn the loads on or off as the proper operation of the system requires.

The first major load that will be discussed is the *compressor.*

Compressors and Electric Motors

A typical compressor is shown in Fig. E20-11 and some of the symbols that are used to represent them in the diagrams are shown in Fig. E20-12.

Referring to Fig. E20-12, the symbol at the top, consisting of a circle between two horizontal lines, represents

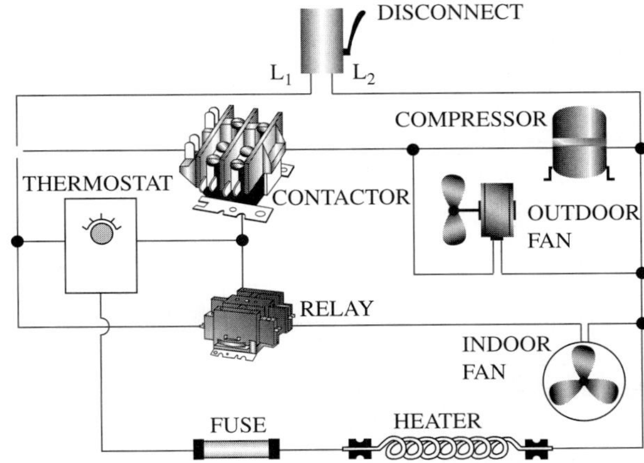

Figure E20-10 Elementary ladder diagram showing the various loads.

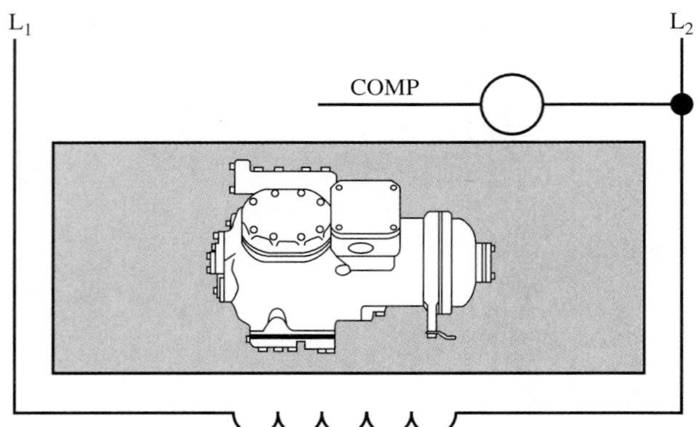

Figure E20-11 Compressor and symbol placed in the diagram.

a motor with two power-line connections. The caption COMP near the circle indicates that the motor is a compressor motor. Referring to the center diagram in Fig. E20-12, the symbol shows another type of compressor motor in more detail. This symbol indicates that the motor has both a start winding and a run winding which need to be properly wired into the system. The bottom diagram in Fig. E20-12 shows the three windings of a three-phase compressor motor, and the three power connections required.

Fig. E20-13 shows how the various motor symbols fit into the schematic wiring diagram. L_1 and L_2 represent the supply side of the power supply. Power is also being furnished to the primary connections of a 24-V control-circuit transformer.

The circuits for three types of motor loads have been connected to L_2. The connections to L_1 will be added later when the control devices are known. The legend for this diagram would show that OFM stands for an outside fan motor and IFM stands for an inside fan motor.

Note that the fuse with its symbol FU has been placed in series with the secondary of the control transformer to protect this circuit against overload.

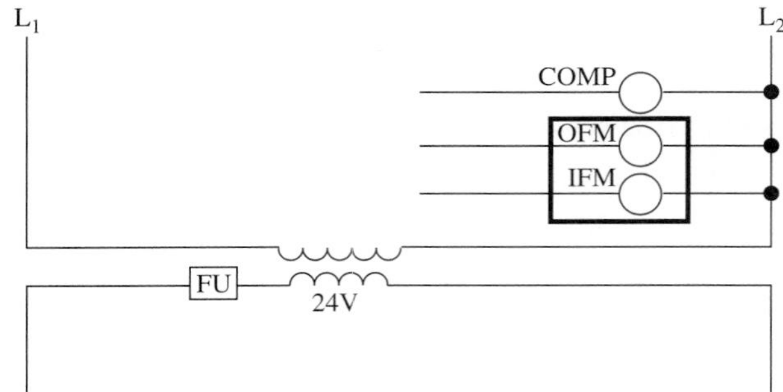

Figure E20-13 Schematic diagram partially completed showing position of line-voltage loads.

Heaters

Another type of load commonly used on air-conditioning units and heat pumps is an electric heater, shown in Fig. E20-14. As indicated in the illustration, the heater symbol is a zig-zag line with the letters HTR. The heater symbol is the same as the resistor symbol since these electric heaters are actually high-wattage resistance units. As current flows through them they give off heat. The illustration shows the configuration for three different types of resistance heaters.

Fig. E20-15 shows where the heater and its protective fuse are placed in the schematic diagram. Electric heaters may draw a high amount of current and should meet specific electrical requirements regarding fusing and overload protection in the heater circuit.

E20-3.3 Load Devices in the Control Circuit

Two commonly used load devices are placed in the control circuit. These are magnetic coils used in *relays* and *contactors* (Fig. E20-16). Basically, relays and contactors have the same function except the contactor is larger, more rugged, and carries more current. A starter is a contactor with motor overload protection added.

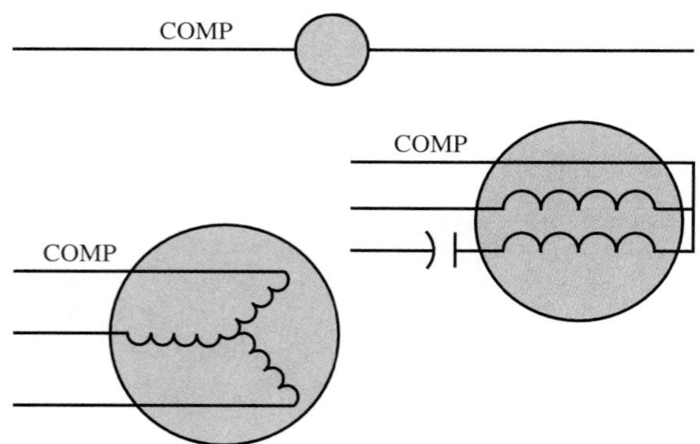

Figure E20-12 Other symbols for compressors showing more detail.

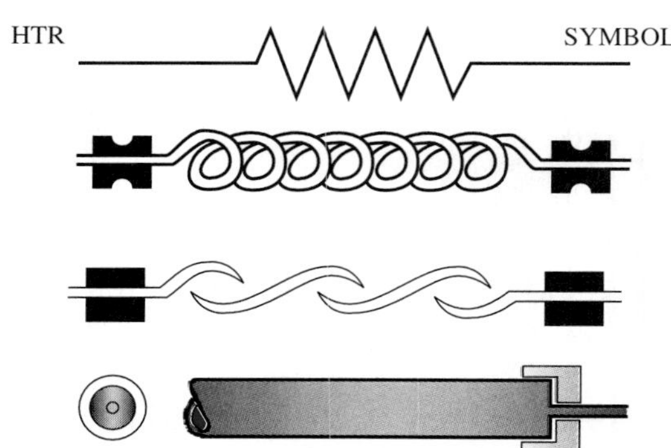

Figure E20-14 Various symbols for electric heaters.

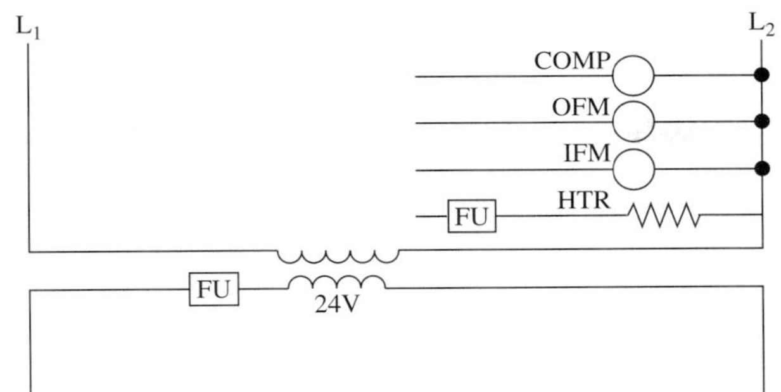

Figure E20-15 Schematic diagram showing position of the line-voltage electric heater and its fuse.

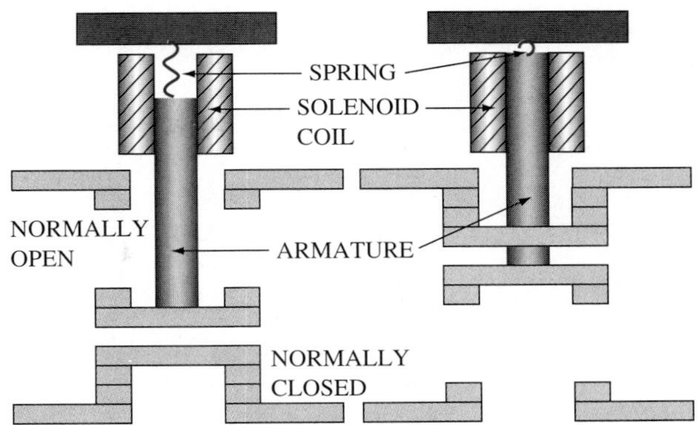

Figure E20-17 A magnetic relay or contactor in both Deenergized and Energized positions.

Actually these devices are combination units. They are composed of both a load device (coil) and one or more switches.

Fig. E20-17 shows how relays and contactors operate. When the coil is energized (supplied with current), a magnetic field is set up which attracts a metallic armature. As the armature moves to the center of the coil it closes or opens the contacts (switches) which are attached to it. As long as the current flows through the coil the relay or contactor is energized and the switches are in the energized position.

These relay type devices are designed with both normally open (NO) switches and normally closed (NC) switches. The normal position of the switches is always the position of the switches when the relay coil is de-energized. The NC switch has a diagonal line through it as shown in Fig. E20-18.

Note the letters that identify the relay, such as CR, appear both on the coil and on the switches. In this way switches that belong to each relay can be identified when they appear in different parts of a diagram. Very often the coil may operate at one voltage and the switches are located in another circuit operating at a different voltage.

A good example of the use of a relay in a schematic diagram is shown in Fig. E20-19. Examine the diagram to determine the function of the control relay (CR). The coil is located in the low-voltage part of the diagram and is controlled by the thermostat switches. One normally open switch, operated by the relay, is also located in the low-voltage part of the diagram and is in series with the compressor contactor coil (C).

The other two switches operated by the relay are located in the high-voltage part of the diagram. One of these switches is normally open and is in series with the outside fan motor (OFM). The other switch is normally closed and is in series with the crankcase heater (CH).

The control relay can also be diagrammed as shown in Fig. E20-20. This view shows a separate coil circuit and three separated switches, two NO and the third NC. There is no electrical connection between any of these parts of the relay, only a mechanical connection.

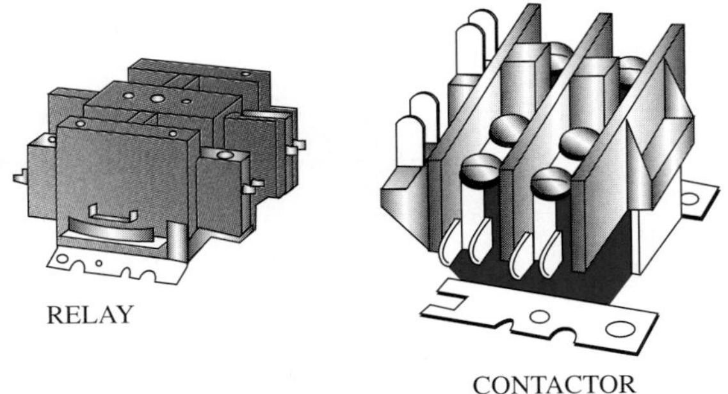

Figure E20-16 A relay and contactor are similarly constructed except the contactor handles higher currents.

E20-3.4 Heating Circuit Diagram—Electric Heat

To illustrate the application of electrical heat, refer to a simplified diagram of the cooling cycle shown in Fig. E20-21. The addition of electric heating requires the following:

1. Inserting a resistance heater (HTR) in the high-voltage area of the diagram with its fuse (FU) in series with the heater (Fig. E20-21). It is common practice to use 230-V power with resistance heaters, which means that both L_1 and L_2 will be "hot" wires and all the loads in the high-voltage area will be operating on the same voltage and will

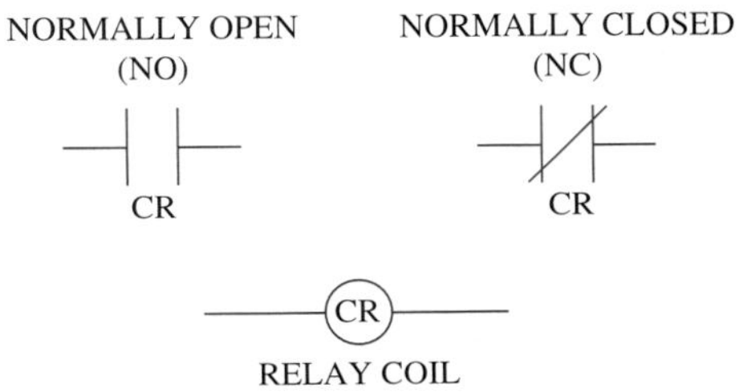

Figure E20-18 Symbols used for control relay with one NO switch and one NC switch.

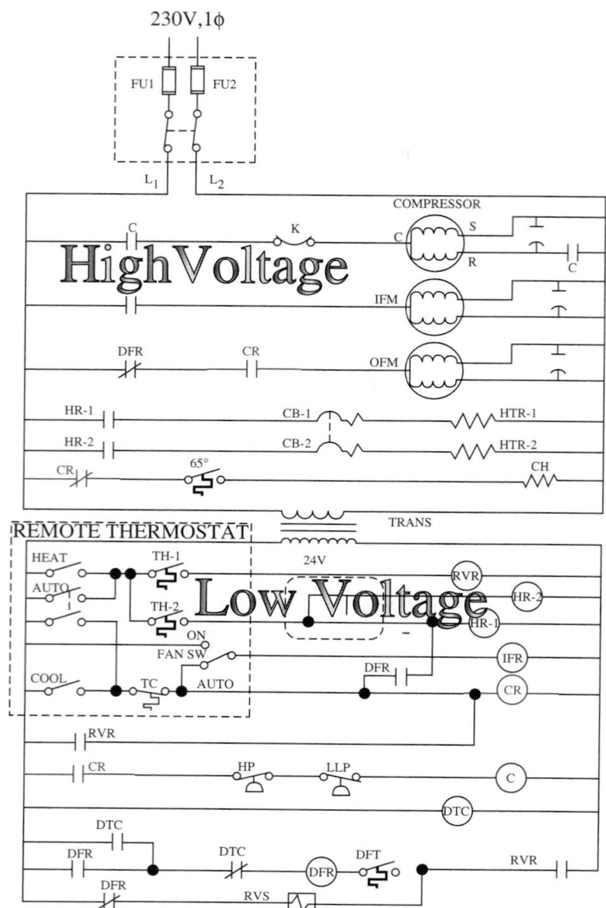

Figure E20-19 Schematic diagram showing the position of the control relay coil and three switches.

need to be selected on this basis. The transformer will have 230 V on the primary and 24 V on the secondary. Note that a fuse has been added in series with the secondary side of the transformer to protect it against overload.

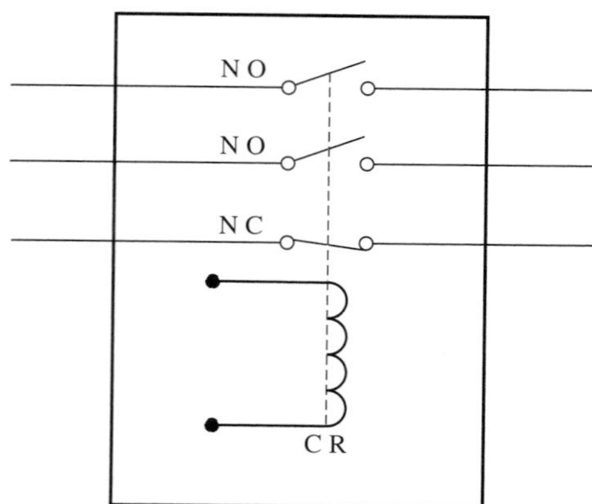

Figure E20-20 Schematic for the control relay.

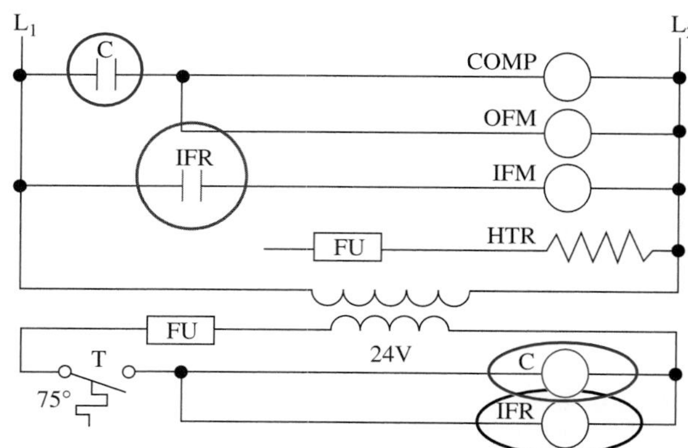

Figure E20-21 Contactor is added to control the compressor and the outdoor fan motor. A relay is added to control the inside fan motor.

2. A heating thermostat (T) has been added, along with a heater contactor (HC), Fig. E20-22. In a low-voltage circuit, the thermostat is placed in series with the heater contactor coil. In a high-voltage circuit, the heater contactor NO switch is placed in series with the resistance heater. When the thermostat calls for heating, the heater contactor coil is energized and the contacts are closed, energizing the heater.
3. An extra set of NC contacts (C) have been added to the compressor contactor (Fig. E20-23) to lock out the heater whenever the compressor is operating. The contacts open when the compressor motor is energized.
4. An interlock (Fig. E20-24) has been added between the heater circuit and the inside fan motor circuit to start the inside fan when the heater is operating. When the heater contactor switch is closed,

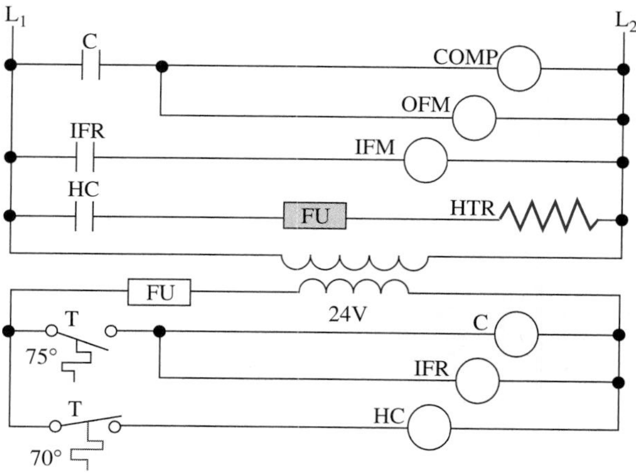

Figure E20-22 Electric heater and fuse has been added in the line-voltage circuit and thermostat to control a heating contactor coil.

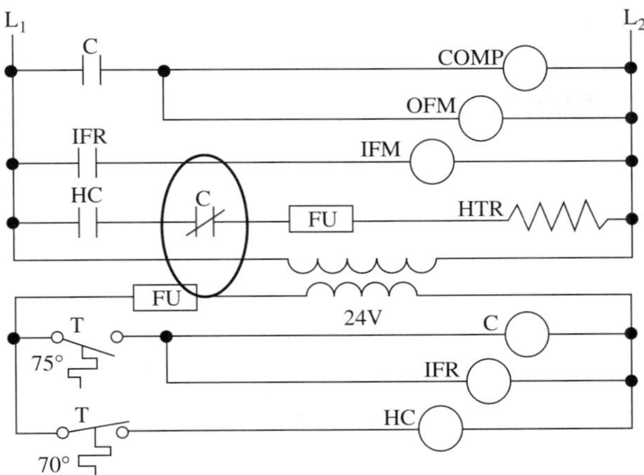

Figure E20-23 A lockout switch has been added in a heater circuit to prevent the heater from operating during the cooling cycle.

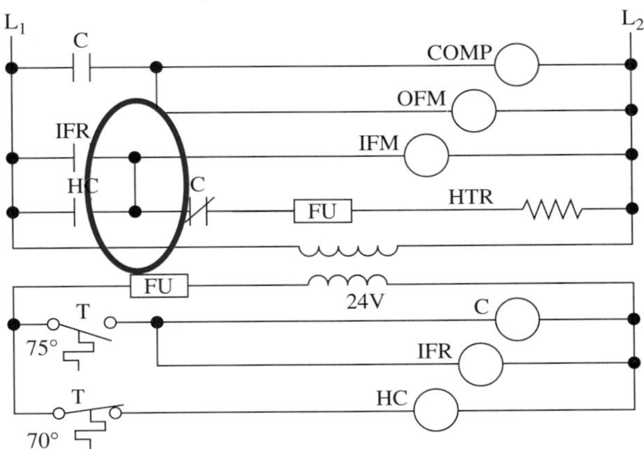

Figure E20-24 An interlock has been added to activate fan motor when thermostat calls for heating.

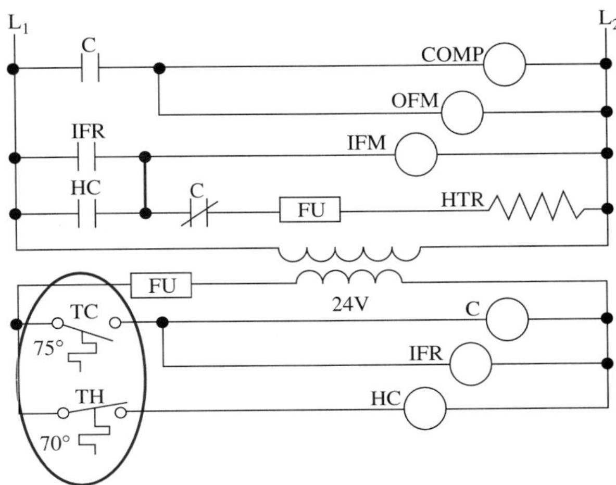

Figure E20-25 Cooling thermostat closes on a rise in temperature, and heating thermostat closes on a drop in temperature.

current from L_1 bypasses the IFR switch, energizing the inside fan motor.

5. In Fig. E20-25 the legends on the two thermostats have been changed to "TC" and "TH" to distinguish the cooling from the heating.

E20-4 WIRING DIAGRAMS

Wiring diagrams are a type of map, supplying complete information on how the electrical parts are connected to operate, control, and protect the unit. Certain standards have been set up relative to the use of electrical symbols for representing electrical components, but they are not always followed by every manufacturer. The technician, therefore, becomes more skillful with experience in interpreting individual manufacturer's offerings.

Basically there are three types of wiring diagrams:

1. External
2. Connection (or panel)
3. Schematic

The *external diagram* is supplied by many manufactures with the installation instructions to show the type of electrical service required and how the unit is connected to it. A diagram of this type would show the location(s) of terminals on the equipment, the type of external fusing needed and the type of power that must be furnished for the unit to operate properly. A typical external wiring diagram is shown in Fig. E20-26.

It is common practice for these diagrams to show both the external wiring supplied by others as well as the external wiring that is field installed by the installation crew or their electrician. The drawing may also specify wire sizes, the type of insulation required, and may instruct the installer to obtain certain information from local codes.

This is an extremely important part of the job. If power is not properly applied, the unit will not function properly.

If the unit is factory assembled, the *connection diagram* is the one used at the factory for wiring the unit (Fig. E20-27). The actual location of all electrical connections is shown on this drawing. If you were to examine the unit with a connection diagram in your hand, you would be able to locate all of the connections and be able to trace wires from one connection to the next. Regardless of what other wiring diagrams are available, this one would show as nearly as possible the way the wires are run on the actual unit. If you wanted to trace the wiring to locate a possible loose connection or a wire that possibly had been left out, the connection diagram would be the one to use.

To know how the unit is controlled or the sequence of operation of the various components, the diagram to use is the *schematic diagram.* If troubleshooting is needed, the schematic wiring diagram is practically a necessity. This diagram divides the system up into a series of individual

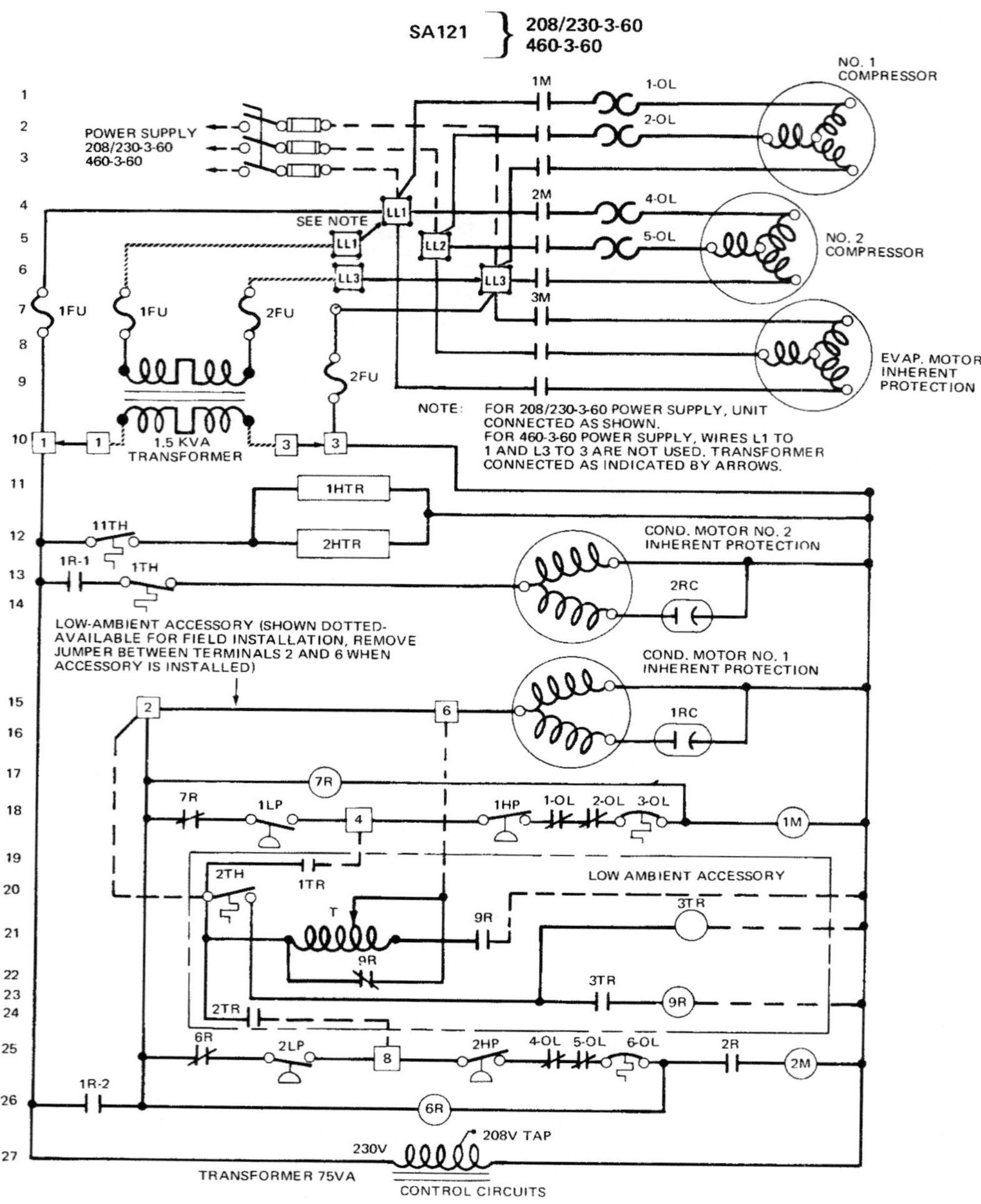

Figure E20-26 Typical external wiring diagram used for unit installation.

electrical circuits. Each circuit has a load (a component that requires power), a path for the power to travel, and one or more switches (optional).

The schematic may be drawn as a ladder diagram. These are clear and easy to follow. Each circuit appears as a ladder rung between two power sources (L_1 and L_2), as shown in Fig. E20-28. Due to the need for developing skill in using the schematic diagram, most of this chapter will be devoted to the use of ladder diagrams. Although a ladder drawing is usually made vertical, it can be made with the power lines horizontal (with the ladder lying on its side), if this is more convenient.

E20-4.1 Wiring Diagram Symbols

In order to keep the diagram compact and meaningful, most of the electrical components are represented by symbols. A standard list of symbols is shown in Fig. E20-29 (p. 383). These symbols are primarily used in making schematic wiring diagrams.

The wiring diagram symbols fall into the following categories:

1. *Loads.* These include any electrical device that uses power; including motors, transformers, resistance heaters, relays, lamps, and solenoid valves.

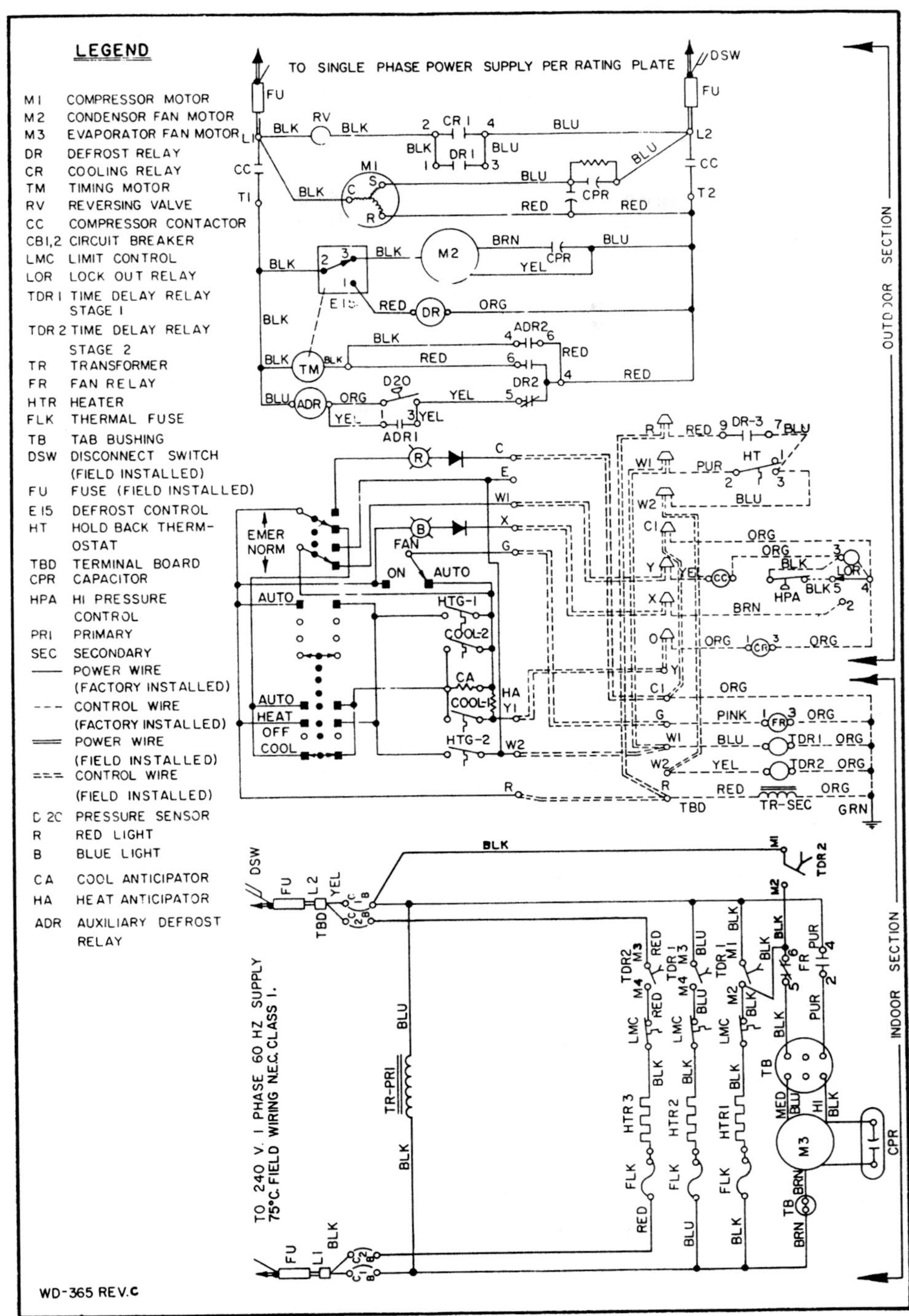

Figure E20-27 Typical connection diagram showing internal wiring of the unit.

Each electrical circuit must have some type of load.

2. *Switches.* A switch is any device placed in the electrical circuit which turns on or off the power supply to a load. There are many types of switches, as indicated in Fig. E20-29. Some of them are operated mechanically such as: thermostats, pressurestats, humidistats, flow switches, and float switches. Some of them are operated by electrical power such as relays, contactors, and starters. Some are protective in design, such as fuses and overloads.

All switches have a normally open (NO) or a normally closed (NC) position. These are usually the positions of the switch when the circuit is deenergized; however, there is a variation to this. A thermostat is usually shown in an open position. Other mechanical switches are usually shown in

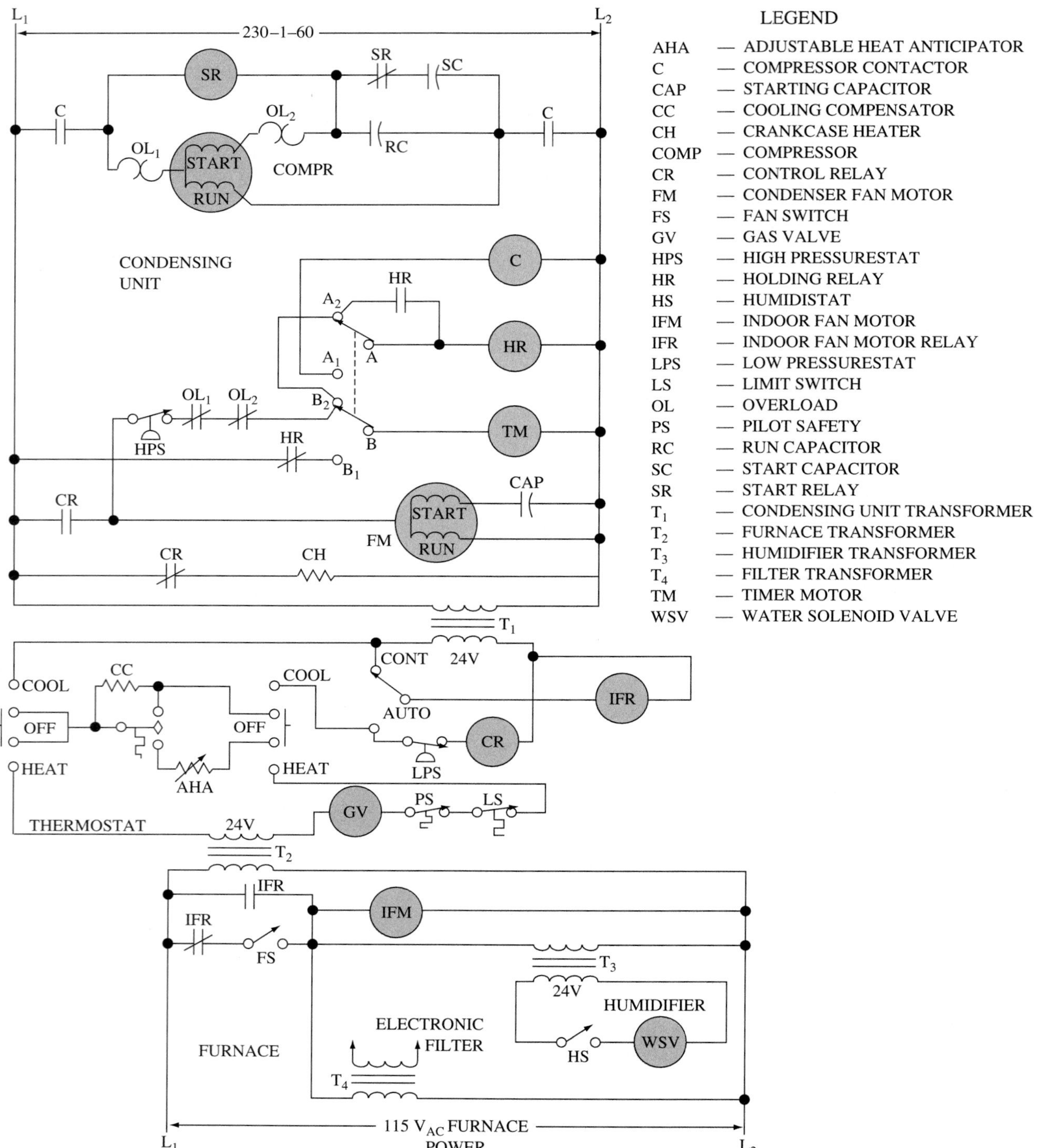

Figure E20-28 Schematic wiring diagram showing the various electrical circuits.

the position they would normally be if the unit was operating properly. For example, a high-pressure cut out switch is shown in a closed position.

3. *Combination load/switches.* All switches that require electrical power to operate are combination load/switches. For example, a relay has a coil that requires power to operate it and it has one or more NO or NC switches.
4. *Special electrical devices.* This group includes all other electrical devices that do not fit in the above categories, such as capacitors and thermocouples. Capacitors will be described in detail in the section

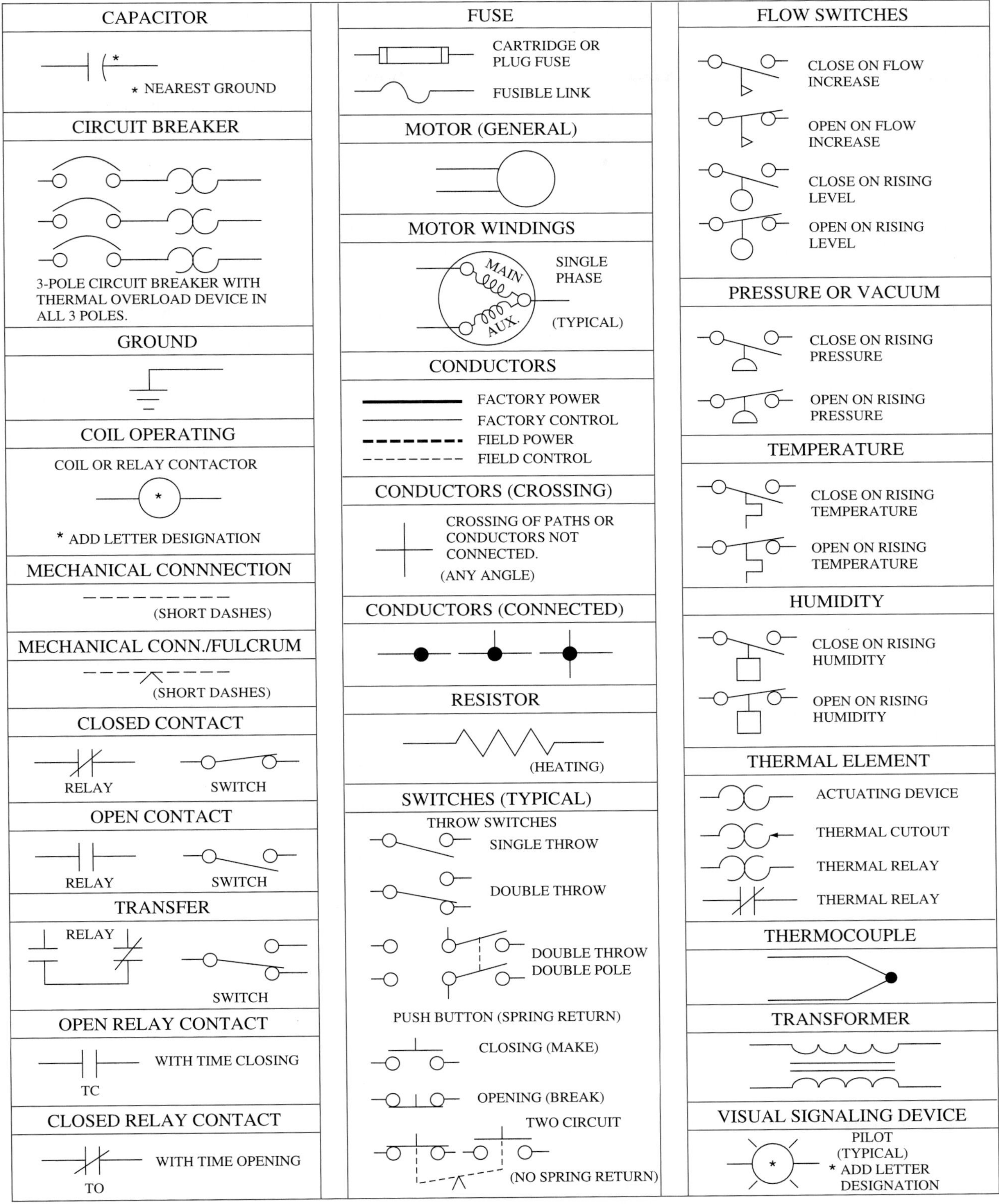

Figure E20-29 Recommended electric diagram symbols.

covering single-phase motors. A thermocouple is a unique device, constructed by joining the ends of two dissimilar pieces of metal. When the two junction points are at different temperatures, current flows in the circuit. This device will be fully described in the warm air heating section. It is used as a safety device to turn off the gas in case a standing gas pilot goes out.

5. *Wiring conventions.* The standard list of symbols shown in Fig. E20-29 gives a number of drawing conventions that are useful, particularly the ones used for crossing wires on a diagram.

Unfortunately all manufacturers do not adhere to the standard list of symbols, shown in Fig. E20-29. However, most manufacturers are consistent in their use of symbols. It is therefore recommended that the technician keep some type of notebook to record any new symbols encountered for future reference.

E20-5 CONSTRUCTION OF SCHEMATIC WIRING DIAGRAMS

A schematic wiring diagram consists of a group of lines and electrical symbols arranged in ladder form to represent individual circuits controlling or operating a unit. The electrical symbols represent loads or switches. The rungs of the ladder represent individual electrical circuits. The unit can be an electrical-mechanical device such as an air-conditioning system.

All schematic wiring diagrams are made up of one or more individual electrical circuits. Fig. E20-30 shows a single circuit made up of a battery (power source), a light bulb (load), a switch (control) and connecting wiring (path). The top diagram in the illustration is a pictorial diagram of the circuit and the diagram below it is the schematic.

Most circuits only have one load but as many switches as necessary to properly control the load. Since the power supply for the load shown in Fig. E20-30 is a battery, the two sides of the power supply are indicated as "+" and "–". Note in the schematic diagram that symbols are used to represent the various electrical devices in the circuit. Since the simple circuit in Fig. E20-30 shows the switch in closed position, the circuit is completed and the lamp is lighted.

E20-5.1 The Ladder Diagram

Fig. E20-31 shows a preliminary view of a ladder-type wiring diagram. Note that the power supply (symbols L_1 and L_2) are indicated by two vertical parallel lines. The three horizontal lines between them and connecting them are electrical circuits. Each circuit has a load: the first circuit, a condenser fan motor (CFM); the second circuit, a control relay (CR) coil; and the third circuit, a compressor (COMP) motor. Note that the third circuit, in addition to the compressor, has two switches in series with the load, an operating switch (on-off) and a high pressure (HP) safety switch.

A few other electrical devices need to be added to make this an operating system. Referring to Fig. E20-32, a circuit with a red light has been added at the top of the diagram to indicate when power is being supplied to the system. A manual switch has been added in series with the fan to turn it on. A thermostat switch for cooling has been added in series with the control relay (CR).

Note that the symbol for the thermostat shows that it is a cooling control since it closes on a rise in temperature.

It is also interesting to note that the control relay (CR) with its coil in circuit 3 has one set of NO contacts in circuit 4. These contacts (also indicated as CR) when closed will start the compressor, providing the on-off switch is closed.

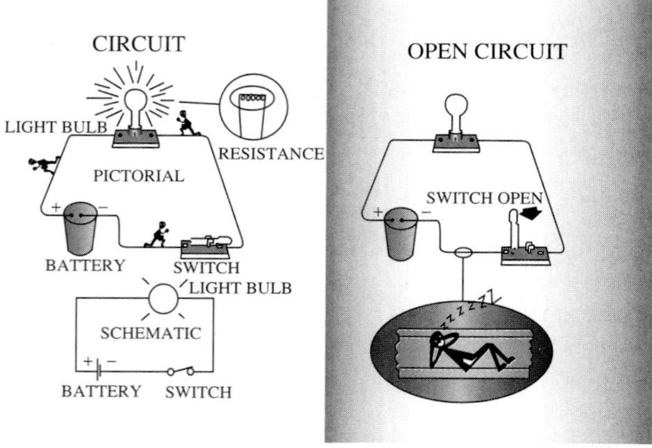

Figure E20-30 Simple electric circuit with battery power, a switch, and a lamp.

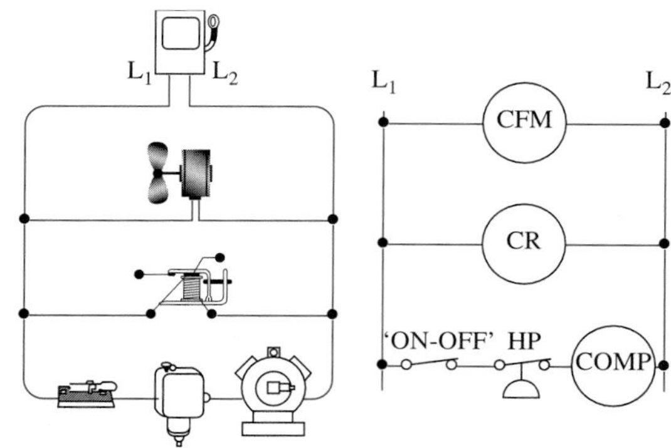

Figure E20-31 Ladder diagram showing a power supply and three loads.

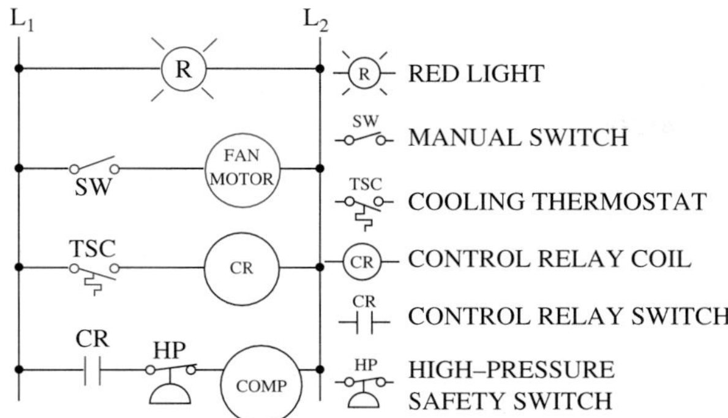

Figure E20-32 Ladder diagram showing the addition of necessary control switches.

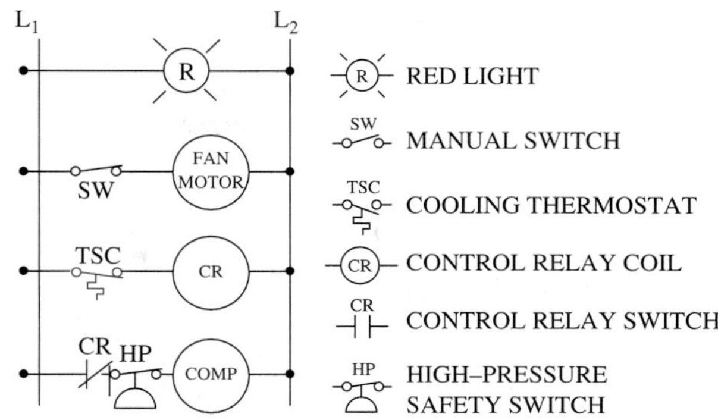

Figure E20-34 Schematic diagram showing thermostat calling for cooling.

E20-5.2 The Sequence of Operation

From the ladder diagram in Fig. E20-33, we can determine the sequence of operation of the system, or, the order in which the loads are energized.

Referring to Fig. E20-33, when power is supplied to L_1 and L_2 the red light comes on, verifying the power supply. Then the fan switch (SW) is closed manually, starting the fan motor.

In Fig. E20-34, the cooling thermostat (temperature setting for cooling, or TSC) closes, indicating a call for cooling, energizing the control relay (CR) coil. When this occurs, the NO relay switch (also indicated as CR) in circuit 4 closes, starting the compressor (COMP) motor. The high pressure (HP) safety switch, also in series with the compressor motor, remains closed since it only opens on a malfunction.

E20-5.3 Drawing a Schematic

We will now examine a more complex system and construct a schematic. Then, using the information from the schematic, we will determine the sequence of operations for the system.

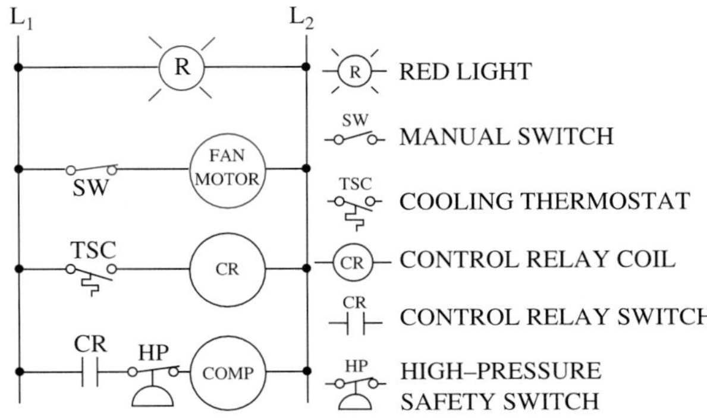

Figure E20-33 When the manual switch is closed, the fan motor runs.

Fig. E20-35 shows a connection diagram for an air-cooled condensing unit. A condensing unit is a packaged unit consisting primarily of a compressor, air-cooled condenser, and control system. The control system consists of a wiring panel enclosing (1) the compressor starter (in the upper center of the diagram), (2) the start relay (SR in the upper right of the diagram), (3) the run capacitor (RC in the lower right), (4) a thermostat (T) and switch (SW) combination (in the lower center), (5) a start capacitor (SC in the lower left), and (6) a junction box (in the upper left).

External to the wiring panel are the fan motor and power supply on the left of the diagram, and junction box, compressor motor, and high/low pressure control on the right of the diagram. The start capacitor and start relay are for alternate use if a start capacitor is required. The only place where there is field wiring is in bringing up the 230–1–60 power supply to the junction box (JB).

Fig. E20-35 also shows the schematic. The legends giving the meanings for the symbols and other notes are shown with the diagrams.

E20-5.4 Step-By-Step Procedure

1. The first step in preparing the schematic is to locate the loads and determine the number of circuits. There are five loads in Fig. E20-35, and therefore five circuits. The loads are:
   a. Green test lamp
   b. Compressor motor
   c. Start relay coil
   d. Fan motor
   e. Contactor for compressor
2. Locate the switches in each circuit.
   a. None
   b. Two contactor switches and compressor overload in series with the run winding, and start relay (SR) switch (NC) in series with the start winding.

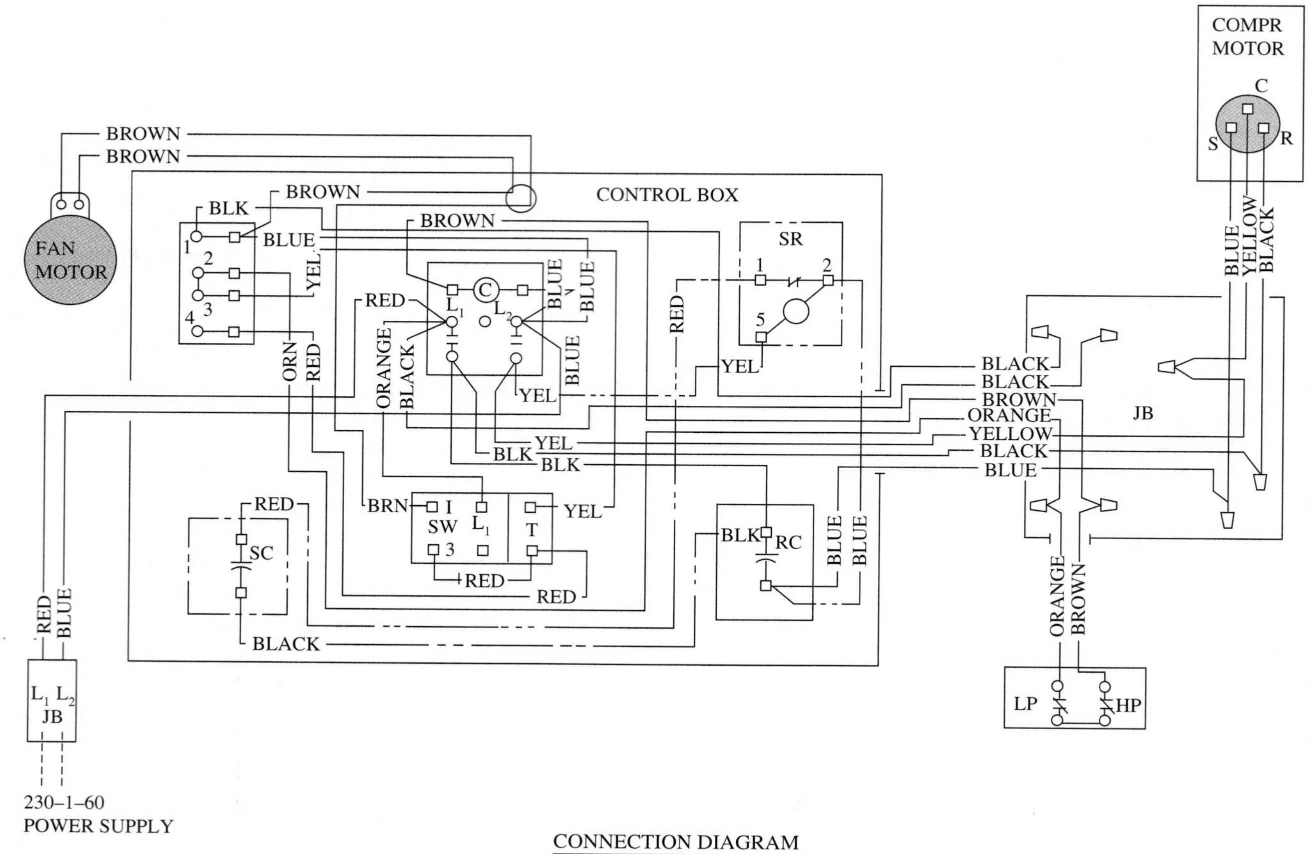

CONNECTION DIAGRAM

LEGEND

C	CONTACTOR	S	START
RC	RUN CAPACITOR	R	RUN
SC	START CAPACITOR	C	COMMON
SR	START RELAY	----	FIELD WIRING
T	THERMOSTAT	——	FACTORY WIRING
SW	SWITCH	-‑‑‑	ALTERNATE CSR WIRING
HP	HIGH-PRESSURE SWITCH		
LP	LOW-PRESSURE SWITCH		
JB	JUNCTION BOX		

NOTES

1. FAN MOTOR PROVIDED WITH INHERENT THERMAL PROTECTOR.
2. COMPR. MOTOR PROVIDED WITH INHERENT OVERLOAD PROTECTOR.
3. MAX. FUSE SIZE 30-AMP DUAL ELEMENT.

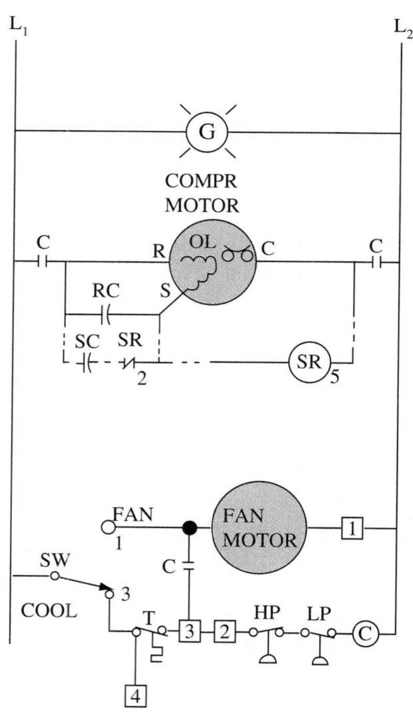

Figure E20-35 Diagrams for an air-conditioning unit showing both the connection and schematic.

c. Two contactor switches (same as in b above) in series with the start relay coil.
d. One manual switch, or the manual switch and the thermostat, in series with the fan motor.
e. The manual switch, the thermostat, and the high-pressure and low-pressure switches in series with the compressor contactor coil.

3. Draw each circuit, showing the position of each component. In the case of switches, show the normal position. The result of these three steps is shown in Fig. E20-35 in the schematic.
4. Describe the sequence of operation for the system. This will be accomplished using the information shown in the schematic to trace the circuit, as described below.

Methods of Tracing Circuits

Three conditions can occur in tracing circuits:

1. Start at one side of the power supply (L_1), go through the resistance (load), and return to the other side of the power supply (L_2). This is a complete circuit.
2. If the technician starts at L_1 and goes through the resistance and cannot reach L_2, this is an open circuit.
3. If the technician starts at L_1 and reaches L_2, without passing through a resistance (load), this is a short circuit.

E20-5.5 Rules for Making Vertical Schematic Diagrams

1. Assign letters in the legend for each symbol to represent the names of the components.
2. When using a 120-V AC power supply, show the hot line (L_1) on the left side and the neutral line (L_2) on the right side of the diagram.
3. When using a 120-V power supply, the switches must be placed on the hot side (L_1) of the load.
4. Relay coils and their switches should be marked with the same (matched) symbol letters.
5. Numbers can be used to show wiring connections to controls or terminals.
6. Always show switches in their normal, de-energized position.
7. Thermostats with switching sub-bases, primary controls for oil burners, and other packaged electrical devices, can be shown by terminals only in the main diagram. Sub-diagrams are used when necessary to show the internal control circuits.
8. It is common practice to start the diagram showing line voltage circuits first and low voltage (control) circuits second.

E20-5.6 Cooling Circuit Diagram

This is an exercise to expand your knowledge and ability to read schematic wiring diagrams. Although the diagram that we will be drawing is for an imaginary unit, it contains many of the control and protective devices found in actual residential and small commercial air conditioners.

Referring to Fig. E20-36, start by installing the primary load devices: the compressor motor (COMP), the outdoor fan motor (OFM) and the indoor fan motor (IFM). Since the compressor has two windings—a start winding and a run winding—both are included in the symbol shown. All three loads are first attached to L_2, allowing space on the left side of each circuit for the switches required.

In Fig. E20-37, the connections of the primary loads to L_1 are completed. A set of NO contacts from the compressor contactor (C) is placed in series with the compressor motor (COMP) and the outdoor fan motor (OFM). A set of NO contacts from the inside fan relay (IFR) is placed in series with the inside fan motor. Note that both the compressor motor and the outside fan motor are on line voltage and are placed in the top portion of the diagram.

Also in the diagram, the IFR coil and the compressor contactor (C) coil are connected to one side of the low-voltage transformer output.

When the compressor starter coil (C) is energized, the contactor switch (C) is closed, starting both the compressor and the outside fan motor (OFM). When the IFR coil is energized, the IFR switch is closed, starting the inside fan motor (IFM).

In Fig. E20-38, two selector switches have been added to the primary controls. One of these switches is in the cooling thermostat circuit. It gives the occupant the choice of manually activating the cooling or turning it off so that it will not operate. The other rotary switch gives the occupant the choice of running the fan continuously or only when there is a call for cooling. For example, by properly adjusting these two switches, the fan on the unit could be operated to provide circulation of air without using the cooling.

In the illustration, the off/cool switch is set to the off position and the fan off/auto switch is set to auto. The

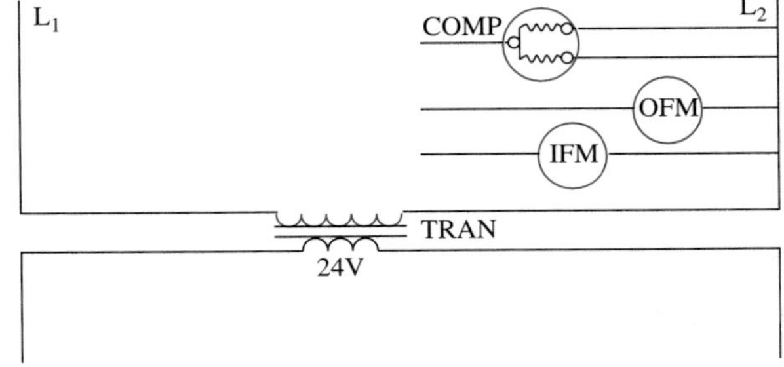

Figure E20-36 The construction of a schematic diagram showing the line-voltage loads.

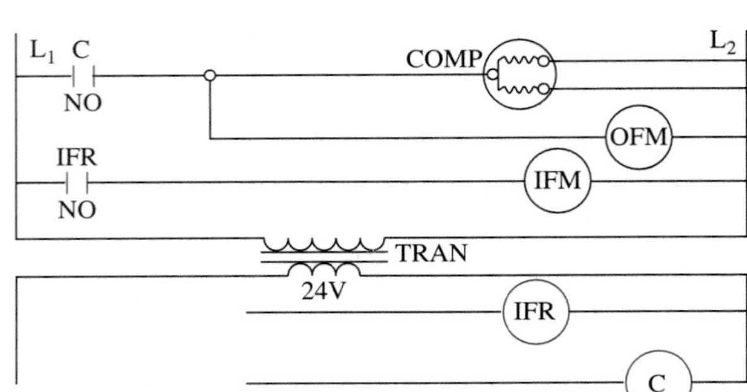

Figure E20-37 Schematic diagram showing the completed connections to the primary loads.

thermostat switch is open; therefore, the unit is not running. These switches are usually part of the thermostat or thermostat sub-base.

Note the wire connecting the output side of the thermostat with the fan auto terminal. This interlock assures that the fan and cooling will operate together when the thermostat switch is closed and the rotary switches are in the position shown.

In Fig. E20-39, a *start capacitor* (SC) and a *start relay* (SR) have been added to the compressor circuit to assist in starting the compressor motor. The start relay coil is placed parallel to the motor start windings and its NC switch is placed in series with the motor start windings. With this arrangement, the start windings are energized on start-up and are released when the motor reaches nearly full speed.

Fig. E20-40 shows the addition of *run capacitors* to the outside and the inside fan motors. These capacitors make the motors run more efficiently. Run capacitors are installed in series with the start windings.

The equipment will now operate in the cooling mode. When the thermostat calls for cooling, the compressor and the outside fan will start. Assuming the fan switch is in the auto position, when the thermostat calls for cooling the inside fan contacts will close and start the inside fan. If the fan switch had been in the on position, the fan would have

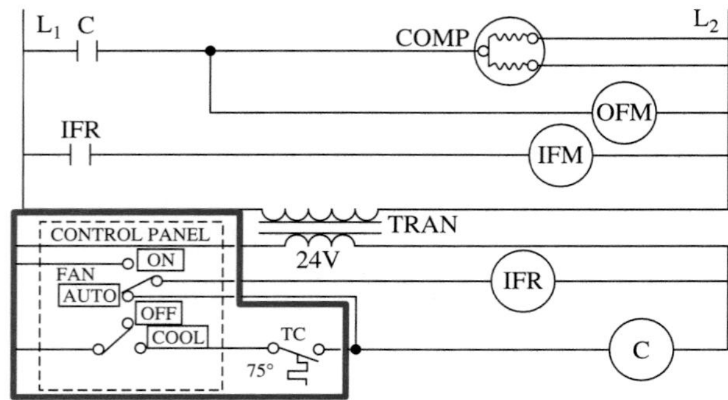

Figure E20-38 Schematic diagram showing the addition of two rotary switches and cooling thermostat in the control circuit.

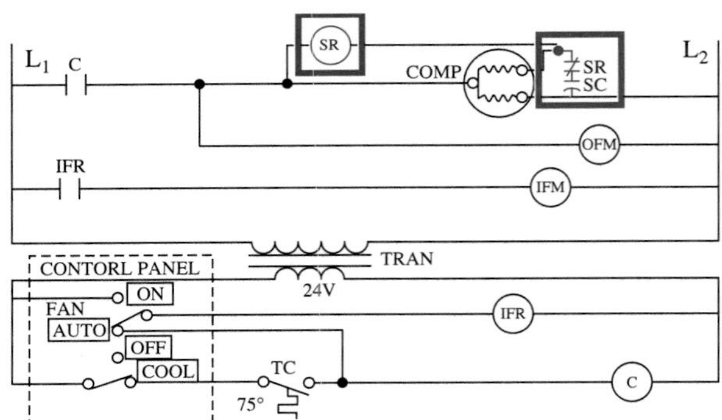

Figure E20-39 Schematic diagram showing the addition of a start relay and start capacitor used for starting the compressor.

been already running to provide continuous air circulation.

Certain protective devices need to be added. Fig. E20-41 shows the addition of a high-pressure switch (HPS) and a low-pressure switch (LPS) in series with the compressor contactor coil (C). These switches shut down the compressor when the refrigerant discharge pressure is too high or the suction pressure is too low. The low-pressure switch also protects the compressor in case there is a loss of refrigerant.

To provide further protection for the motors, overload switches are placed in series with each motor, as shown in Fig. E20-42. The ones shown in the figure are thermal types. They are made of bimetal, and warp when excess current flows through them, opening the switch. Usually these thermal switches are imbedded in the motor windings.

A magnetic-type overload can also be used. When excess current flows through the magnetic overload device, it acts like a relay, opening a switch in the motor circuit.

Fig. E20-43 shows an additional compressor contactor switch that has been added to the compressor circuit where it connects to L_2. This provides a second set of contacts for this circuit, breaking both sides of the line to the motor. This is safer for servicing personnel.

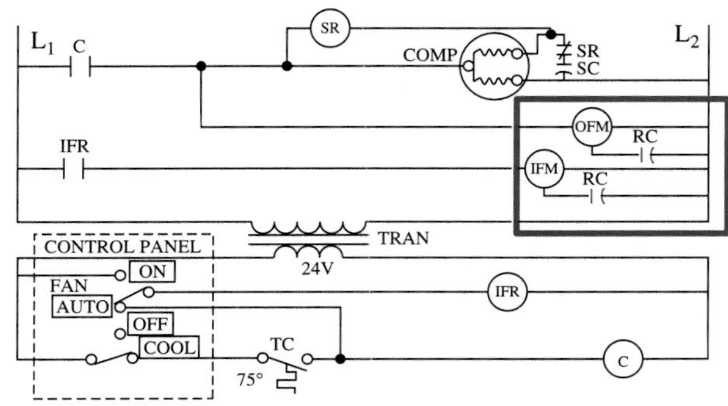

Figure E20-40 Schematic diagram showing the addition of run capacitors in the outside and inside fan motor circuits.

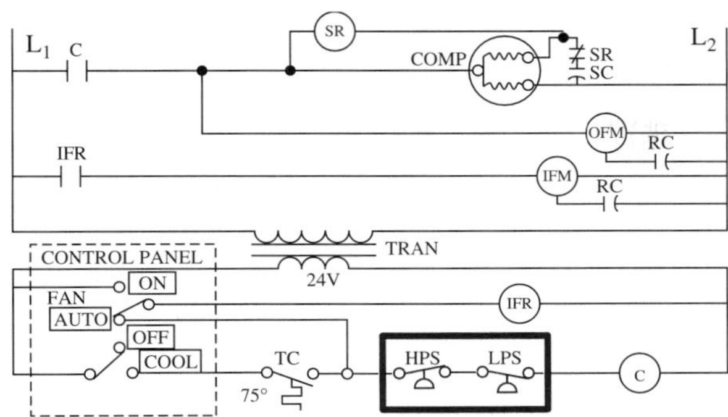

Figure E20-41 Schematic diagram showing the addition of high- and low-pressure safety switches in the compressor contactor coil circuit.

E20-5.7 Heating Circuit Diagram—Gas Heat

Many of the systems used for cooling also have some provision for heating. Where cooling is the primary use of the equipment, electric heating can be added, occupying only a small amount of additional space such as on a room cooler; however, where natural gas is available and heating is a major requirement, the best heating selection is usually gas. In the following exercise we will add gas heating to the cooling circuit diagrams illustrated earlier.

The first addition needed is a heating thermostat (TH), shown in Fig. E20-44. Also, it will be necessary to modify the rotary selection switch to include a heating position.

A series of controls need to be added to provide the system with heating capability. These include:

1. The *limit switch* (LS) shown in Fig. E20-45. The limit switch is heat sensitive and opens the gas valve circuit if the output temperature from the heat exchangers becomes excessive.
2. The *rollout switch* (RS) shown in Fig. E20-46 is another heat-sensitive device. It will open the circuit if the flames roll out of the burner area due to flue blockage or down draft. This can be dangerous and requires correction. After the correction is made the rollout switch can be manually reset.
3. The *ignition pack* (IGN) shown in Fig. E20-47 controls the flame ignition. As long as the flame is present, the ignitor signals the gas valve to remain open during the heating cycle. If the flame is extin-

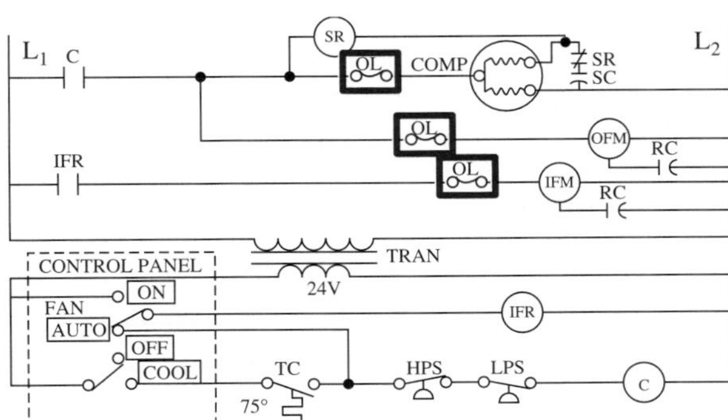

Figure E20-42 Schematic diagram showing the addition of thermal overloads in series with each motor.

Knowledge Extension

The largest power-consuming device in an air conditioner is normally the compressor, followed by the condenser and evaporator fans. The life of the power controllers for these motors depends on the number of switching cycles (short cycling = short life), voltage (low voltage = higher currents to interrupt) and the load (heavy loads = higher currents to interrupt). These components can be expected to deteriorate or fail and to need replacement.

The technician should become familiar with different standards under which contactors and starters may be rated. *NEMA* is the National Electrical Manufacturers Association (US). NEMA ratings are intended for equipment used on a wide variety of applications and are conservative.

Definite-purpose (DP) *ratings* are used for controls rated for specific applications. They tend to be lighter and less expensive than NEMA rated devices.

IEC is the International Electrotechnical Commission, based in Geneva, Switzerland. It is a multinational group composed of forty-four members each representing their country's electrical and electronic interests and composed of manufacturers, users, trade associations, government bodies, scholars, and engineers. They develop and publish standards and have also published a multilingual dictionary of electricity.

It is informative to observe sample contacts (shown below) from a manufacturer rated for the same motor load under the three standards.

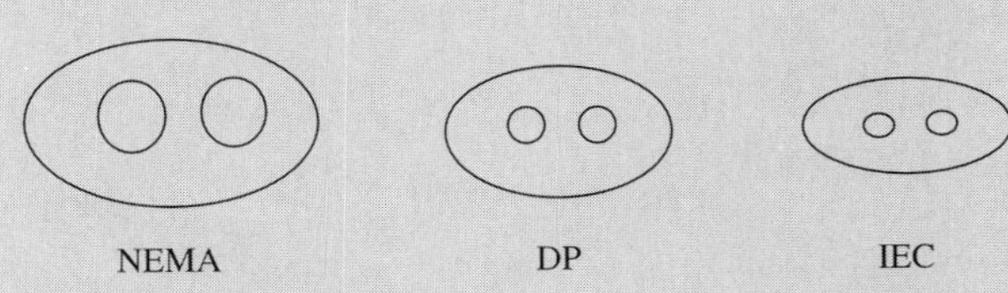

Generally speaking, assuming a satisfactory service life, replacing a contactor with an identical rated product should be satisfactory. Common sense should tell the technician that replacing a larger, heavier NEMA- or DP-rated device with a lighter IEC-rated device might result in a shorter service life.

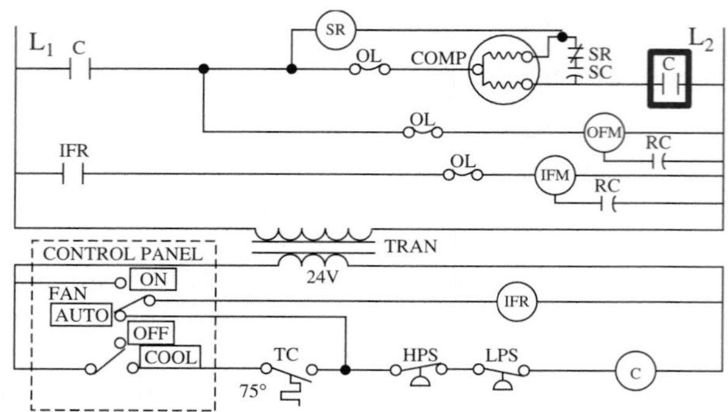

Figure E20-43 Schematic diagram showing an extra set of contacts on the compressor circuit.

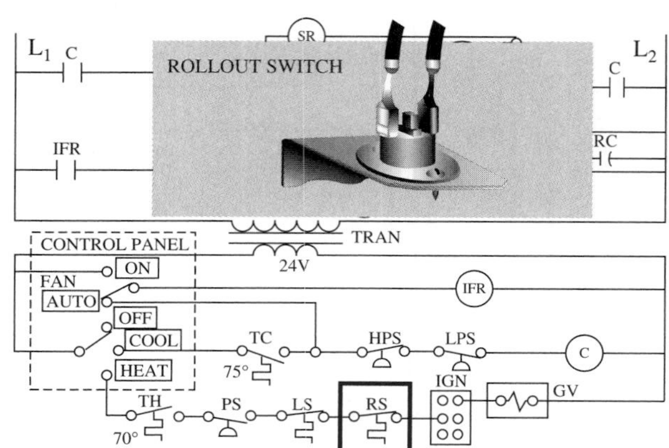

Figure E20-46 Addition of rollout switch in gas valve circuit.

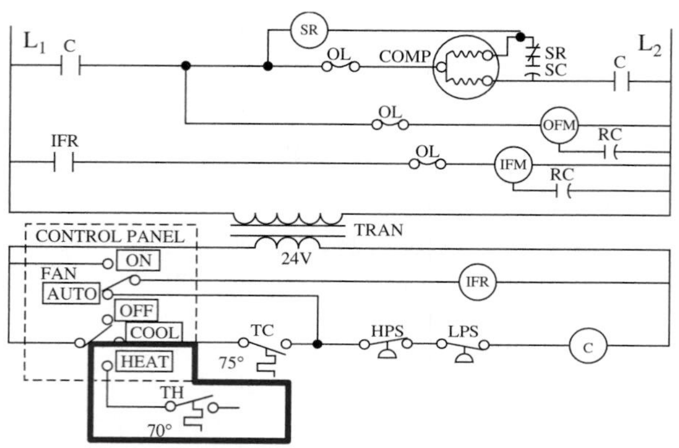

Figure E20-44 Addition of heating thermostat.

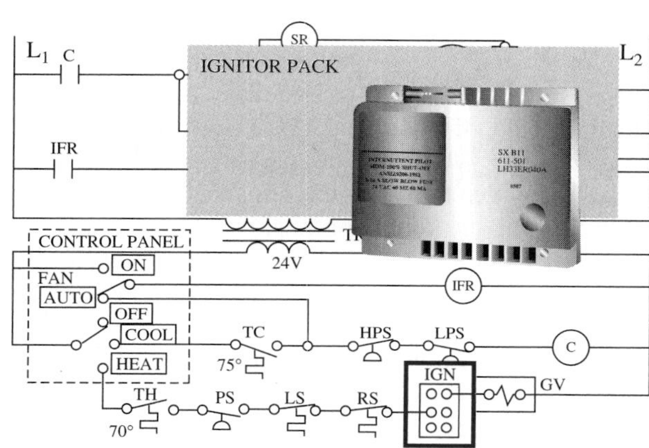

Figure E20-47 Addition of ignitor pack in the gas valve circuit.

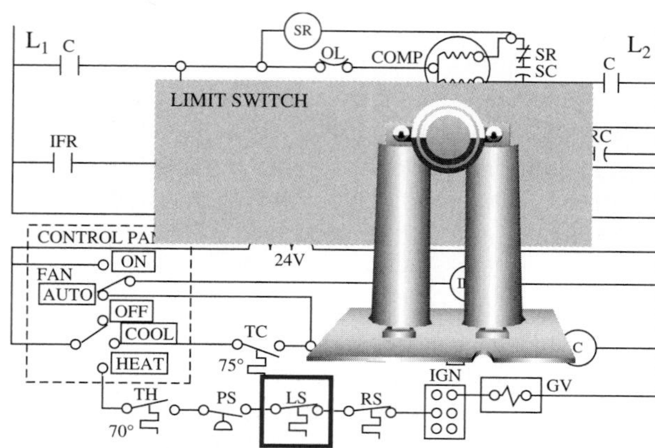

Figure E20-45 Addition of high-limit switch in gas valve circuit.

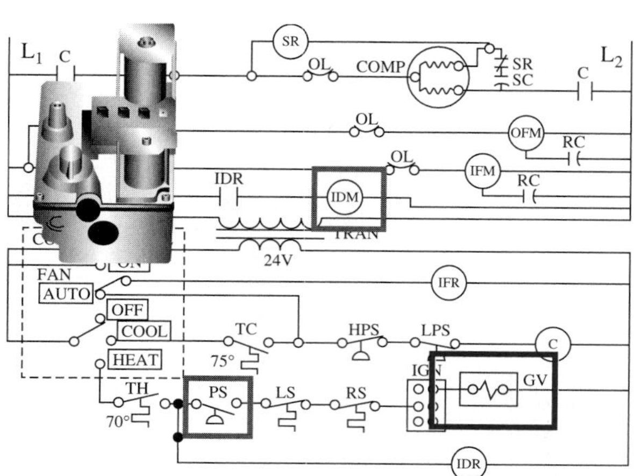

Figure E20-48 Addition of the gas valve, induced draft fan motor and induced draft for relay.

guished for any reason, the ignitor will signal the gas valve to shut off the gas supply.

4. The *gas valve* (GV) is highlighted in Fig. E20-48. The gas valve regulates and controls the supply of gas to the burners. The inducer draft fan motor (IDM) has been added. It drives a fan that draws air into the unit to support combustion. It is started through the inducer relay (IDR). When the inducer fan is operating properly, it will close the pressure switch (PS), completing the circuit to the ignitor and the gas valve.
5. The same fan used for cooling is also used for heating, as shown in Fig. E20-49. For heating, it is activated through the *time-delay relay* (TDR). The coil of this relay is energized when the thermostat closes, but the contacts do not close for about 45 seconds to allow the heat exchangers to warm up. Without the delay the fan would initially blow cold air into the conditioned space.

   The time delay also works at the end of the cycle. After the thermostat opens, the fan continues to run for a short time to take advantage of the residual heat in the heat exchangers.

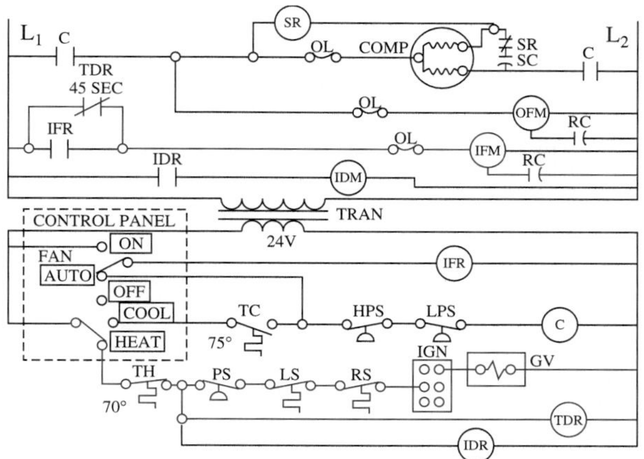

Figure E20-49 Addition of time-delay relay to operate the indoor fan motor.

E20-5.8 Equipment Wiring Diagrams

Most manufacturers supply comprehensive electrical circuit diagrams with the equipment when it is shipped. In many cases this is the only electrical information available to assist the technician in installing and servicing the unit. It is important that this material be retained on the job for immediate and future reference.

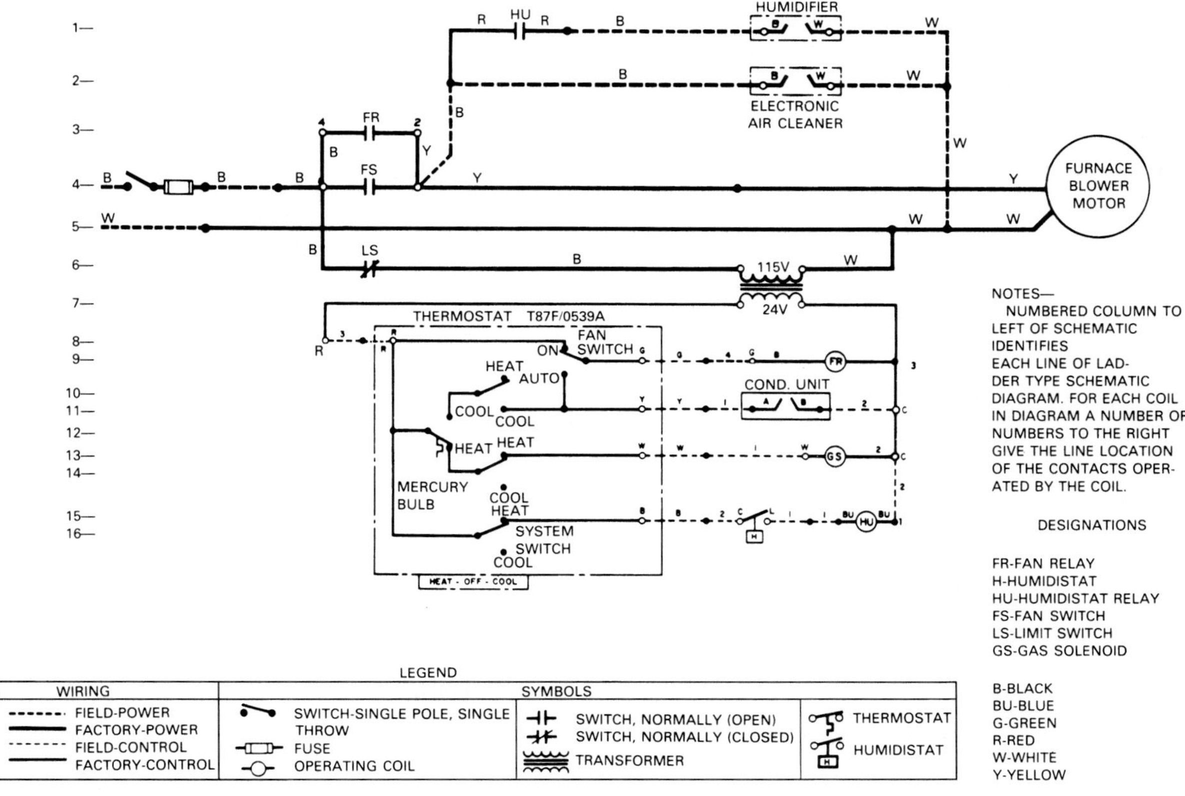

Figure E20-50 Label diagram comes with the equipment, pasted inside one of the panels.

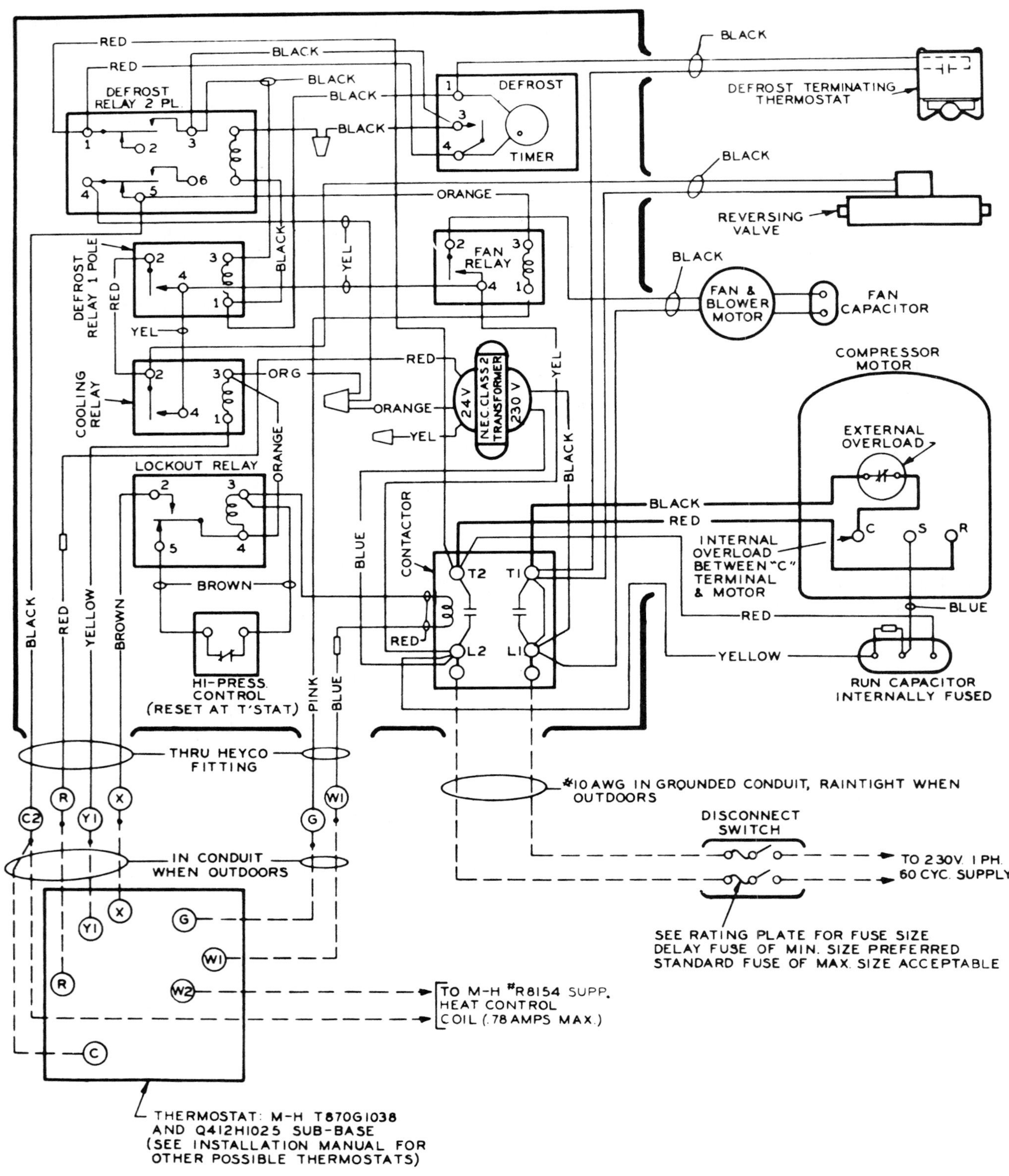

Figure E20-51 Component arrangement enlarged from label diagram.

Here are some of the types of diagrams that the technician is likely to find.

Label Diagrams

A typical label diagram is shown in Fig. E20-50 (p. 391). This is usually printed on peel-and-stick paper and attached to the unit in some convenient place, such as inside the control box cover or door of the equipment. It includes a component-arrangement diagram, a wiring diagram, a legend, and notes.

The part of the drawing indicated as the component arrangement diagram has been highlighted and enlarged in Fig. E20-51. This figure shows the actual location of the components inside the unit. In troubleshooting, a schematic drawing is helpful in determining the function(s) not performing. To test the circuits, however, it is absolutely necessary to locate actual parts and connections within the unit itself. The component-arrangement diagram is designed to help the technician in this respect.

The wiring diagram portion of the label diagram is also shown in Fig. E20-51. This drawing shows the actual internal wiring of the unit. It may show the color code for the wires used and the terminals where they are connected. It can be considered a schematic, however, since the components are not in their exact location. This liberty was taken in making the drawing to be able to adequately show all the necessary connections, some of which would be hidden from view on the actual unit.

The label wiring diagram is usually, but not always, organized showing the high-voltage circuits in the upper portion of the diagram, as shown in Fig. E20-26, and the control circuits (low voltage) in the lower portion of the diagram, as shown in Fig. E20-27.

Referring to high-voltage circuits (Fig. E20-26) the primary power is shown in the upper left and the load circuits in the upper right.

Referring to the low-voltage circuits (Fig. E20-27), they almost always operate from a step-down transformer. Note there are two elements of the control circuits. In the lower section are the contactor coils and switches that control them. In the high-voltage section are the contacts for the contactors.

The legend and notes on label diagrams are important. The legend provides the meanings of abbreviations. The notes provide useful information that cannot be supplied elsewhere.

Installation Wiring Diagrams

An installation wiring diagram (Fig. E20-52) shows how to connect power from the building supply to the unit. It also shows unit-to-unit wiring, if applicable, as well as

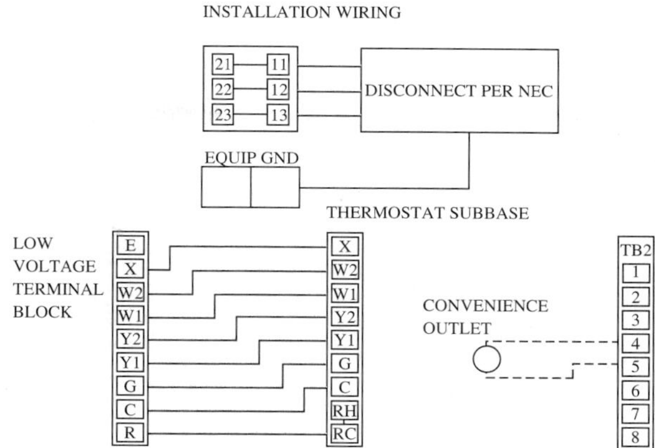

Figure E20-52 Installation wiring showing terminal connection from disconnect box to unit, and from low-voltage terminal block to thermostat sub-base.

wiring to a remote-control device such as a thermostat. This drawing does not show the internal wiring unless it is necessary for the installation.

Connection Wiring Diagrams

The wiring on connection diagrams is shown in a number of ways:

1. *Pictorial.* On these diagrams all of the components are shown in pictorial form in their actual location and the wiring is shown as it is actually used. A good illustration of this technique is shown in Fig. E20-53.
2. *Wiring Harness.* Using this method, groups of wires are placed in harnesses, with individual wires connected to each component. This technique is illustrated in Fig. E20-54.
3. *Terminal Numbers.* Using this method, each terminal is given a number. Few wires are shown. To find where wires are placed, the numbers are matched. This technique is illustrated in Fig. E20-55. An increasing number of manufacturers are using this method. The absence of many wires on the diagram makes it simpler and easier to read.

For tracing the control circuit to understand how the circuit is wired and operates, the ladder diagram is still one of the easiest diagrams to follow. If the technician is having difficulty understanding a schematic or label diagram, it can be worthwhile redrawing it as a ladder diagram to more clearly understand the circuit.

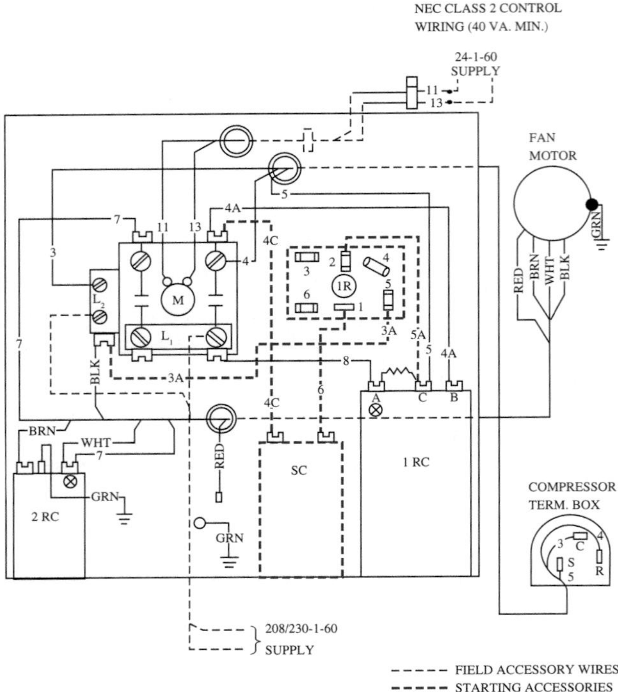

Figure E20-53 Connection or label wiring diagram.
(Courtesy of York International Corp.)

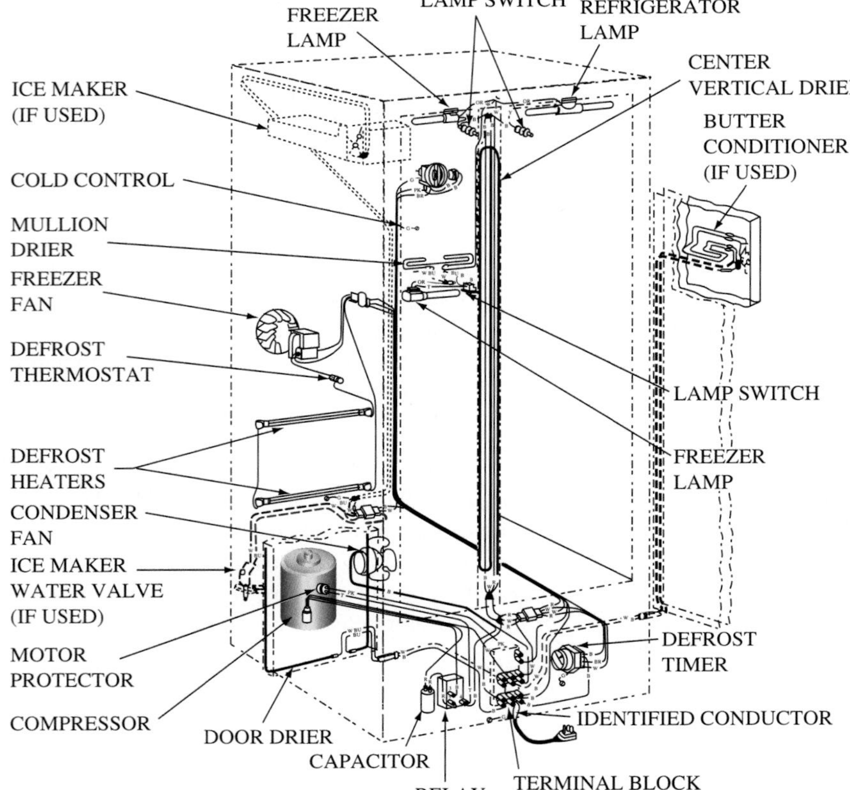

COLOR	SYM
BLACK	B
WHITE	W
RED	R
YELLOW	Y
GREEN	G
TAN	T
BROWN	BR
BLUE	BU
GRAY	GY
ORANGE	OR
PINK	PK

Figure E20-54 Wiring harness-type wiring diagram.

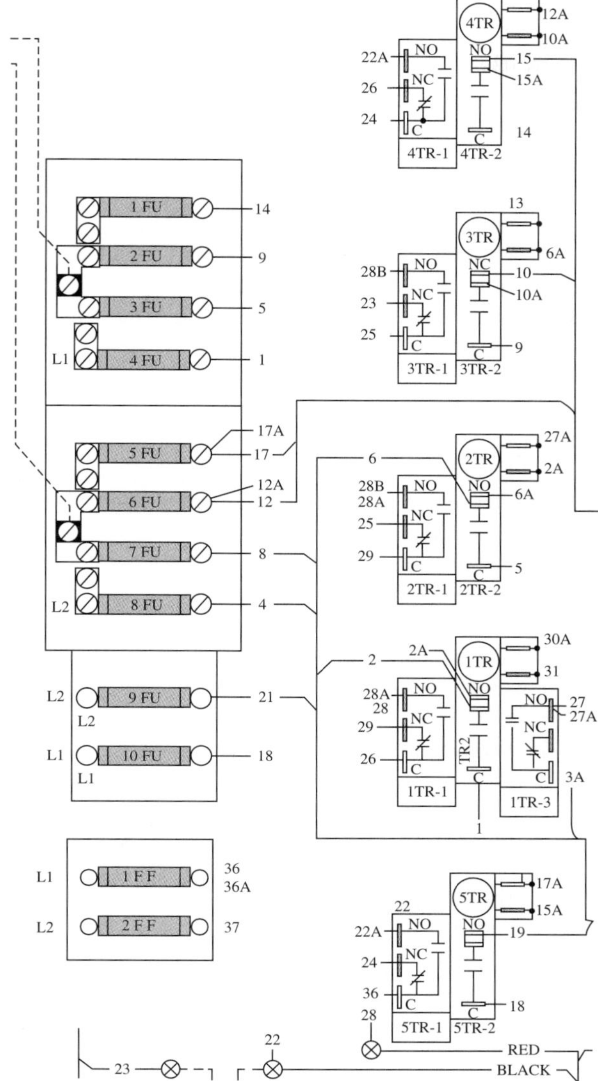

Figure E20-55 Terminal number-type wiring diagram. (Courtesy of York International Corp.)

REVIEW

- Magnetism—Any time that current flows through a conductor, it causes a magnetic field around it. When the conductor is coiled, the magnetic field intensifies. Magnetic poles form at each end which change polarity as the current alternates. An iron core further intensifies the magnetic field. Two magnetic fields of opposite polarity attract each other. This magnetic attraction is the principle of operation for motors, relays, contactors, and solenoid valves.
- Induction—When a current flow is induced in a conductor that passes through a magnetic field or when a magnetic field is passed through a conductor.
- Types of wiring diagrams:
  External
  Connection
    Pictorial
    Wiring harness
    Terminal numbers
  Schematic
    Wiring
    Ladder
- Ladder diagrams are used for troubleshooting.
- Wiring (connection) diagrams are used to locate components, wires, and terminals, determine location, hook up (wire) the equipment.
- Legend—Notes, codes, changes, or options; tells what symbols mean.
- Electrical circuits are protected by fuses and circuit breakers.

FUSE

CIRCUIT BREAKER

- Transformers are used to supply the low-control voltage.

120 V PRI. SEC. 24 V

- Load—Any electrical device that uses power and converts it to another form of energy such as light, heat, or magnetism (motion).

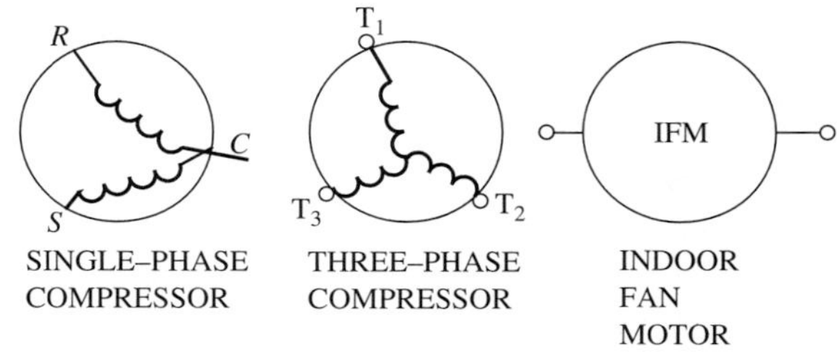

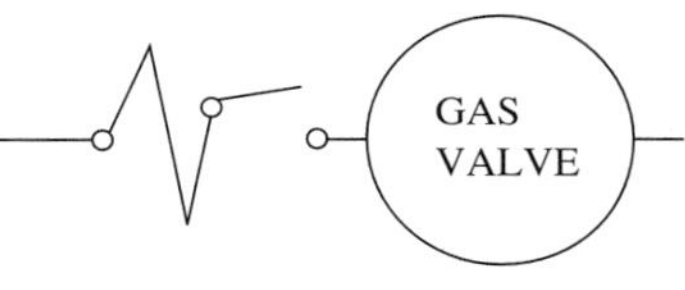

GAS
VALVE
SOLENOID

- Switch—A control device that breaks or makes an electrical circuit.

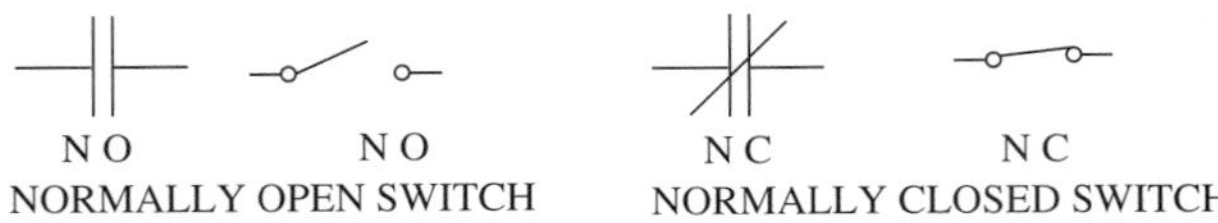

H P.

HIGH–PRESSURE SWITCH

L P

LOW-PRESSURE SWITCH

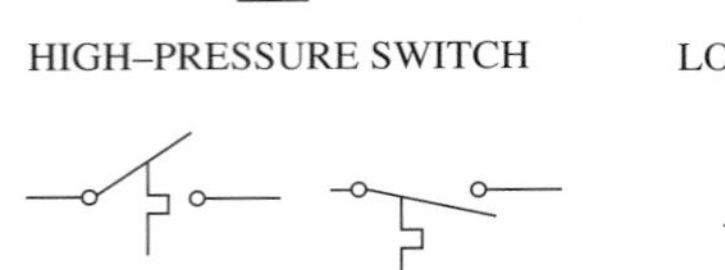

THERMOSTATS

PUSH-BUTTON SWITCH

- Combination load/switches—Relays, contactors, and starters.

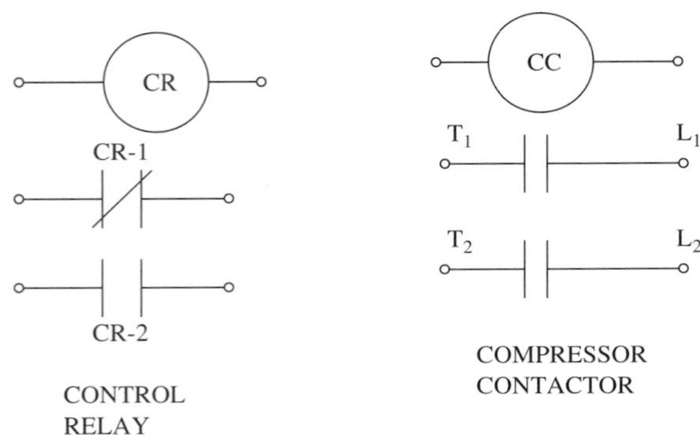

- Step-by-step procedure to read a schematic:
  Locate the loads.
  Locate the switches.
  Determine the circuits.
  Determine the sequence of operation.

Problems and Questions

1. Which of the following is a type of wiring diagram?
   a. Symbol
   b. Connection
   c. Legend
   d. Sequence
2. Which of the following is a load?
   a. Fuse
   b. Thermostat
   c. Limit control
   d. Compressor
3. Which of the following is a switch?
   a. Pressurestat
   b. Fan
   c. Heater
   d. Solenoid
4. What is another name for a schematic wiring diagram?
   a. Pictorial
   b. Installation
   c. Ladder
   d. Connection
5. Which of the following is a primary control?
   a. Gas valve
   b. Thermostat
   c. Disconnect
   d. Blower motor
6. Which of the following is a secondary operating control?
   a. Condenser water valve
   b. Overload
   c. Pressurestat
   d. Thermostat
7. Which of the following is a secondary safety control?
   a. Flow switch
   b. Check valve
   c. Four-way valve
   d. Relay
8. How would you classify a relay?
   a. Switch
   b. Load
   c. Combination switch/load
   d. Metering device
9. What is the first step in preparing a schematic diagram of an existing refrigeration system?
   a. Locate the switches.
   b. Measure the voltage.
   c. Locate the loads.
   d. Turn off the power.
10. If a relay coil has a legend of C, what legend should be used for the switches?
    a. R
    b. B
    c. C
    d. S
11. Any time current flows through a conductor it creates a magnetic field. True or false?
12. Induction is the ability to induce a current flow in another conductor that passes through its magnetic field. True or false?

13. A dual element fuse is designed to protect resistive loads from surges of electrical power. True or false?
14. The most common type of electrical diagram is the schematic (ladder) diagram. True or false?
15. The normal position of the switch, when shown on a schematic, is the position of the switches when they are powered up in the circuit. True or false?
16. Basically, there are three types of wiring diagrams. What are they?
17. There are three conditions that can occur when tracing circuits. Explain them.
18. What protection devices are added in series with compressor contactor coils on large refrigeration units?
19. What is the purpose of the legend?
20. Draw the symbols for a heating thermostat and cooling thermostat.

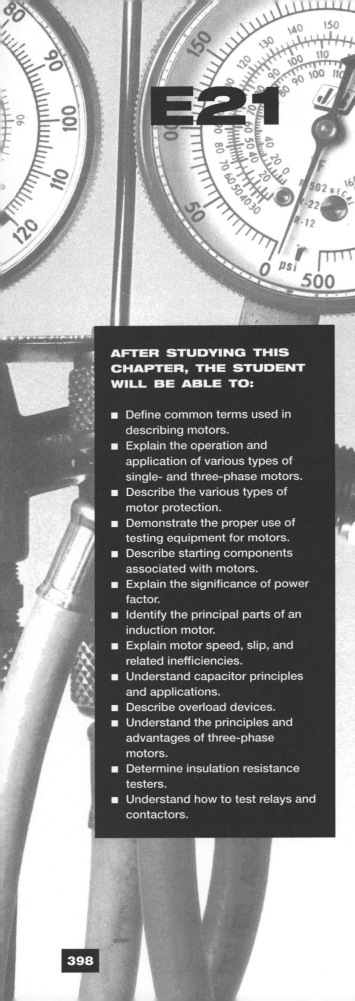

E21 Electric Motors

AFTER STUDYING THIS CHAPTER, THE STUDENT WILL BE ABLE TO:

- Define common terms used in describing motors.
- Explain the operation and application of various types of single- and three-phase motors.
- Describe the various types of motor protection.
- Demonstrate the proper use of testing equipment for motors.
- Describe starting components associated with motors.
- Explain the significance of power factor.
- Identify the principal parts of an induction motor.
- Explain motor speed, slip, and related inefficiencies.
- Understand capacitor principles and applications.
- Describe overload devices.
- Understand the principles and advantages of three-phase motors.
- Determine insulation resistance testers.
- Understand how to test relays and contactors.

E21-1 AC INDUCTION MOTORS

Electric motors are the most important load device in the various types of HVAC/R units. They convert large portions of electrical energy to useful work. It is important, therefore, for the technician to understand how they operate and how they can be protected.

Motors are designed for various types of service. Important qualities are torque, speed, and power usage.

Torque is the twisting (or turning) force that must be developed by a motor to turn its load. The power required to power a load is directly related to the torque required and the speed. A greater amount of torque is required to start a motor than to run it. The starting torque requirements for a fan are low. The starting torque requirements for a reciprocating compressor are high. Providing extra starting torque is expensive. Therefore, to keep the cost down the motor is selected for as small a torque as will adequately perform the work for which it is intended.

E21-2 INDUCTION MOTOR PRINCIPLES

The two principal parts of a motor are the stator and the rotor. The *stator* is the stationary part and the *rotor* is the rotating part or armature, as shown in Fig. E21-1. The sta-

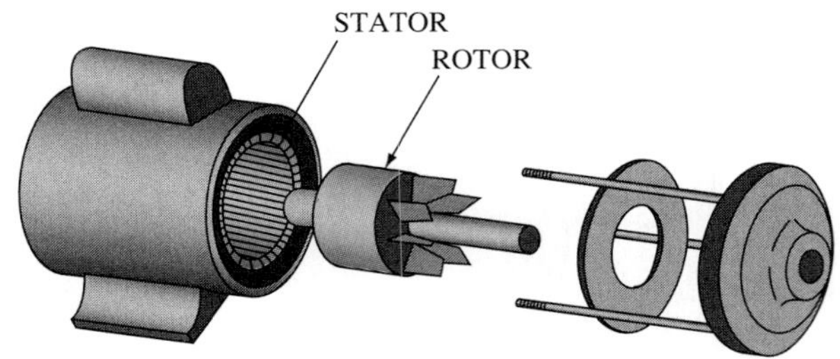

Figure E21-1 Essential motor parts, the rotor and stator.

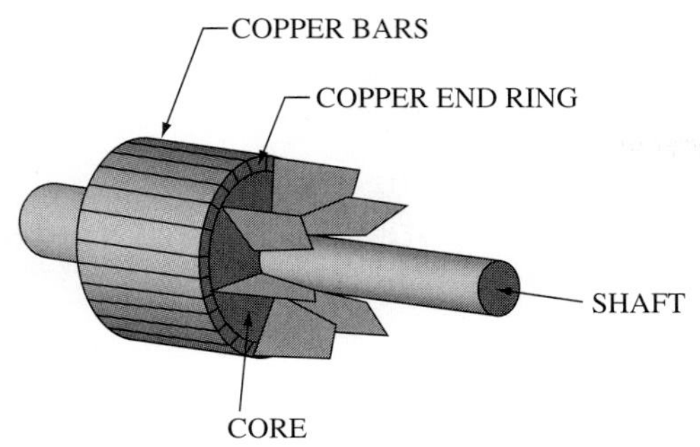

Figure E21-2 Construction of the rotor of an electric motor, showing copper bars.

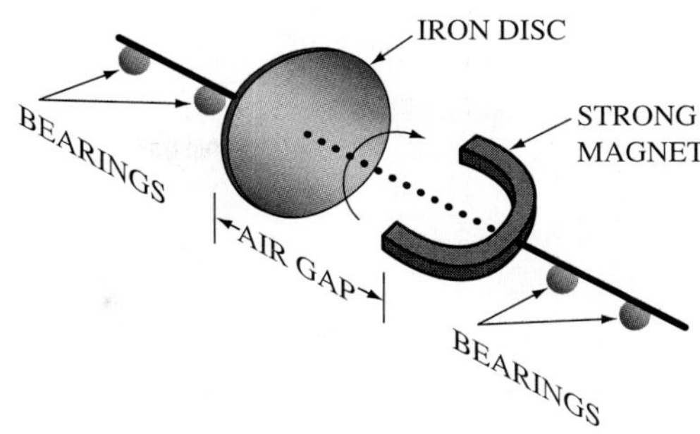

Figure E21-4 Magnetic induction causes rotation. Bar magnet replaces stator; iron disc replaces rotor.

tor is a coil of wire wrapped around a magnetic casing. When alternating current is applied to the stator coil, a rotating magnetic field is produced.

The rotor is a series of copper (or aluminum) bars mounted on a soft iron core. The core provides a path for the magnetic field of the rotor. The conductor bars are shorted together by an end ring (Fig. E21-2), permitting current to flow.

The current passing through the stator creates a powerful magnetic field. The repulsion and attraction of the rotor and the stator parts cause the motor to turn, as shown in Fig. E21-3.

No current is actually supplied to the rotor. The magnetic field of the rotor is produced by induction from the stator. The current induced in the rotor is in the opposite direction from the stator current. This opposing magnetic field of the rotor reacts with the stator field, producing rotation.

As shown in Fig. E21-4 when the magnet is rotated, the iron disc in the illustration will also rotate. This is due to the rotating magnetic field which has induced current in the disc. In Fig. E21-4 the rotating magnet represents the stator and the iron disc represents the rotor.

In Fig. E21-5, the rotor is shown as a permanent magnet pivoting on a shaft. The two stator poles have wire coiled around them. Since like poles repel and opposite poles attract, with the polarity shown, the permanent magnet rotor will move toward the stator poles.

As the alternating current in the stator coils changes direction, the polarity of the stator poles changes, as shown in Fig. E21-6. The magnet is repelled and with further changes in the alternating current, continuous rotation is produced.

A problem can occur if the rotor is stopped in the position shown in Fig. E21-7. In this position, regardless of polarity, no motion can occur. This position is sometimes described as "dead center".

To correct this condition, a *start winding* is added, as shown in Fig. E21-8. The original winding is called a *run winding.* The start winding is made using many more turns of smaller wire, causing a slower current buildup. The two windings—the run winding and start winding—are therefore out of phase with each other and enough torque is created to begin the turning. Motors of this type are called *split-phase motors.* Because of their low starting torque they are used only on fractional horsepower applications.

E21-2.1 Motor Speed

The speed of a motor is determined by the number of poles and the frequency (Hertz) of the current. The greater

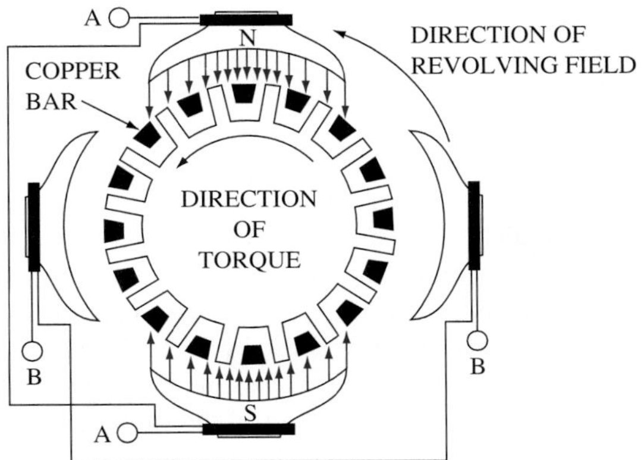

Figure E21-3 Detailed view of the rotor and stator construction showing the forces causing rotation.

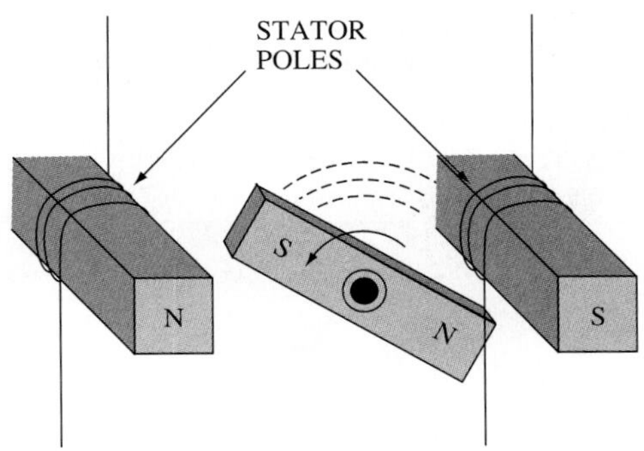

Figure E21-5 Rotor represented by permanent magnet pivoting on a shaft. Stator has two poles. Opposite poles attract.

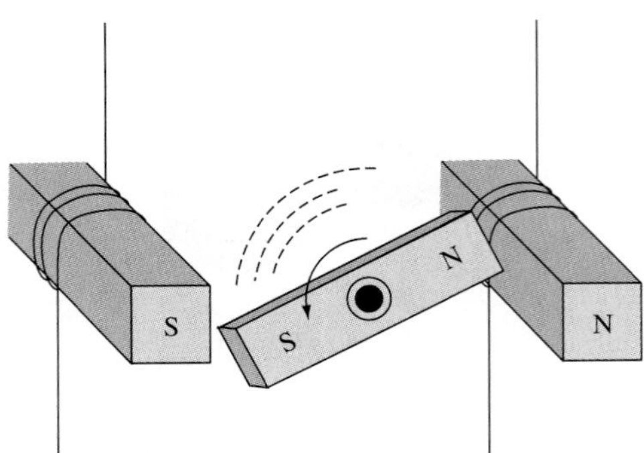

Figure E21-6 Alternating current causes stator poles to change polarity. Like poles repel.

the number of poles the slower the speed. The higher the frequency the faster the speed. Speed is measured in revolutions per minute (rpm). The maximum speed of a motor is known as *synchronous speed.*

In actual performance there is some slippage, or inefficiency, in the motor operation. For motors used to power HVAC/R equipment, the actual speed is usually 95 to 97% of synchronous speed (Fig. E21-9).

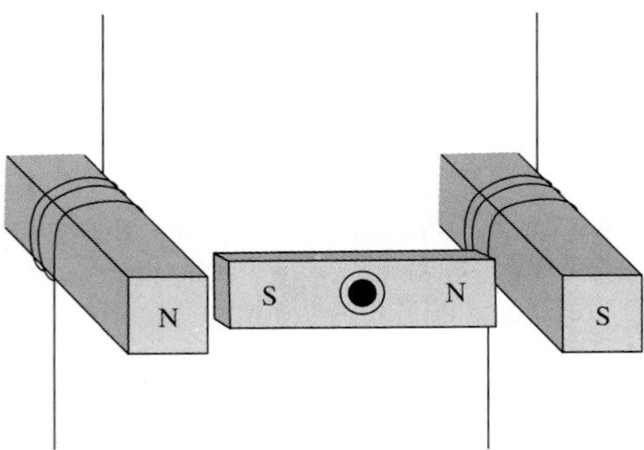

Figure E21-7 The condition where the rotor stops on "dead center" and no motion occurs.

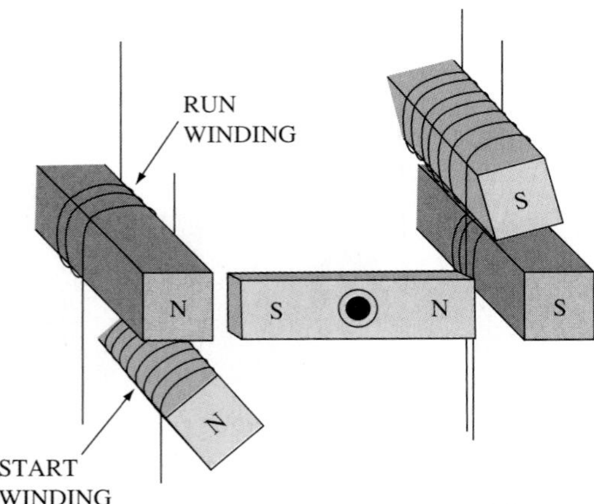

Figure E21-8 To prevent dead-center condition, extra set of start winding poles are added.

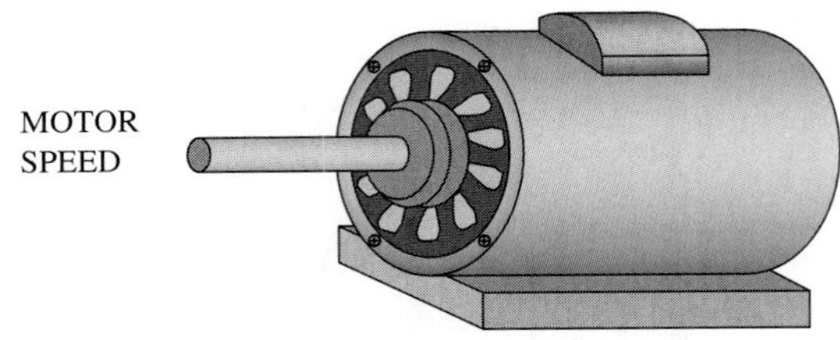

MOTOR
SPEED
• DETERMINED BY:
-NUMBER OF STATOR POLES
-FREQUENCY OF APPLIED VOLTAGE
• SYNCHRONOUS SPEED = MAXIMUM SPEED
• MOTORS ARE NOT 100 PERCENT EFFICIENT

Figure E21-9 Synchronous speed determined by number of poles.

The following is an example of calculating synchronous motor speed. The formula is:

$$\text{rpm} = \frac{\text{Hz} \times 60 \text{ sec/min}}{\frac{1}{2}p} \text{ or } \frac{\text{Hz} \times 120}{p}$$

where

rpm = revolutions per minute

Hz = frequency in cycles/sec

p = number of poles

EXAMPLE

What is the speed of a four-pole motor at 60 Hz?

Solution:

$$\text{rpm} = \frac{60 \times 120}{4}$$

$$\text{rpm} = \frac{7200}{4}$$

$$= 1800 \text{ revolutions per minute}$$

E21-3 CAPACITOR PRINCIPLES

The use of capacitors to increase motor torque was discussed previously. In order to provide strong starting torque for a split-phase motor, a start capacitor is placed in series with the start winding.

Referring to Fig. E21-10, when voltage is applied, the current through the start winding will lead the current through the run winding. This causes the magnetism in the start winding to occur earlier than in the run winding and provides an initial push to start the motor.

Fig. E21-11 shows a capacitor in series with the start winding. The illustration shows the rotor stopped between

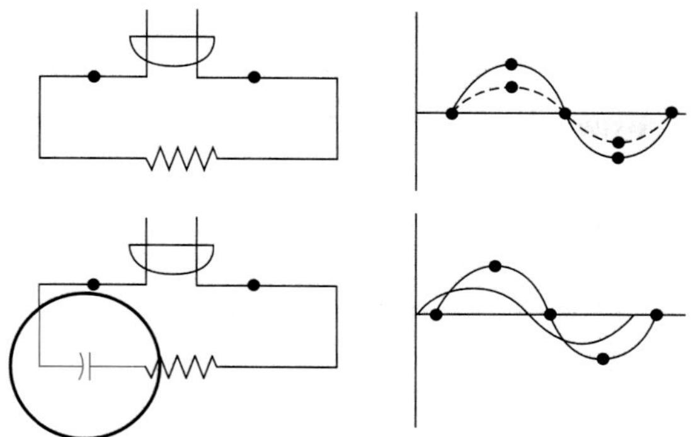

Figure E21-10 Capacitor in series with load. Load current leads voltage.

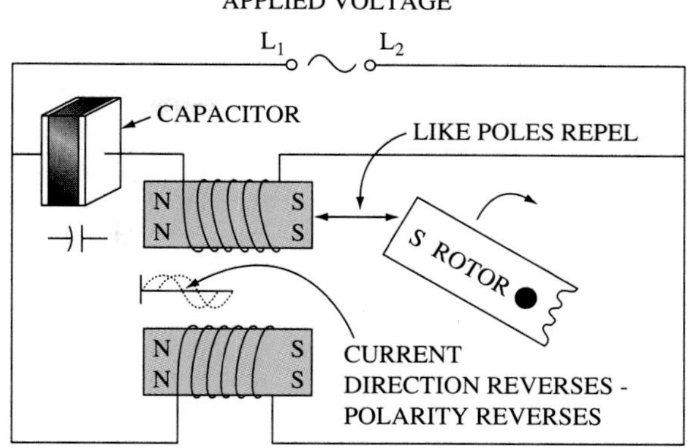

Figure E21-12 The effect of the capacitor in providing motor rotation.

the two poles. When voltage is applied, the magnetism in the start winding will occur earlier than in the run winding and will provide the push needed to start the motor.

When the current direction in the start winding reverses, as shown in Fig. E21-12, the polarity of start winding also changes. The rotor south magnetic pole has rotated to the point where it is repelled by the stator south pole, causing continuous rotation.

A capacitor is rated by its capacity and voltage limit. The unit of capacity is the *microfarad* (mfd and μF are commonly used abbreviations). A high mfd rating is obtained by either using large plates or a small amount of insulation. A low mfd is obtained by using smaller plates or more insulation. The voltage stamped on the outside of the capacitor is the maximum voltage that can be connected safely across the capacitor. If this voltage is exceeded the capacitor is likely to fail.

Capacitors are used to achieve the desired phase-angle shift and to obtain the required current through the series load. Both of these qualities are obtained by selecting the proper mfd rating.

Fig. E21-13 shows the wiring of a start capacitor and its relay, connected to a compressor motor. The start capacitor (SC) and the relay NC switch (SR) are in series with the start winding. The start relay coil (SR) is con-

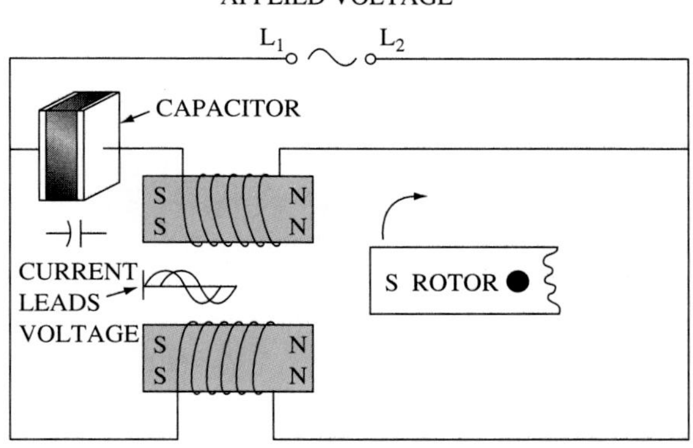

Figure E21-11 Use of a capacitor in series with start winding to change its phase relationship.

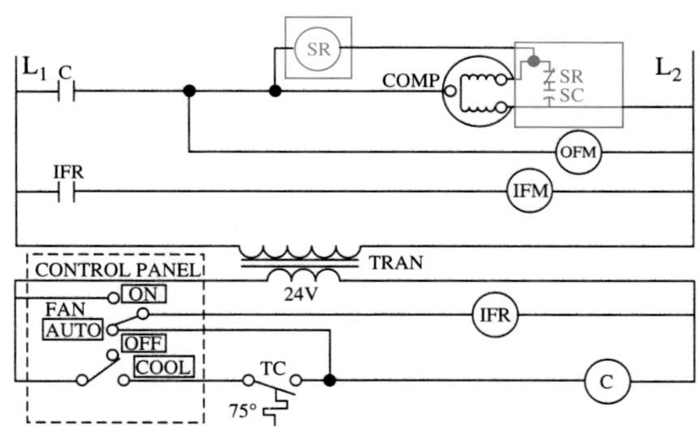

Figure E21-13 Diagram of a compressor motor using a start capacitor and a start relay to increase the starting torque.

PRACTICING PROFESSIONAL SERVICES

Whenever it is necessary to replace a failed electric motor in the field, be sure to write down all pertinent data associated with the motor including: supply voltage, phase, rpm, shaft size, model number, manufacturer, and service factor. This will help ensure that you get the right replacement the first time without costly callbacks to the job site.

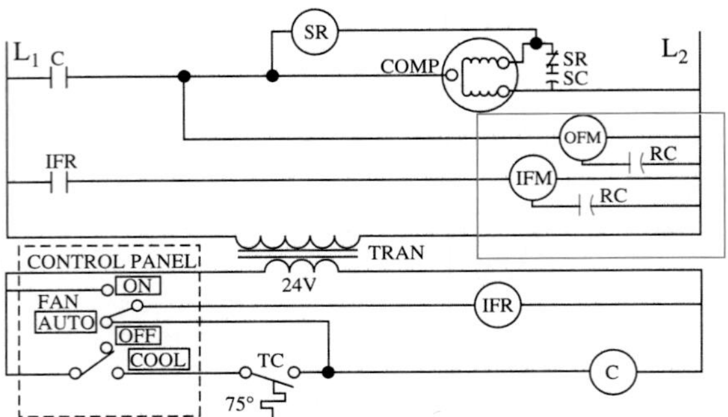

Figure E21-14 Run capacitors added to both outside and inside fan motor circuits.

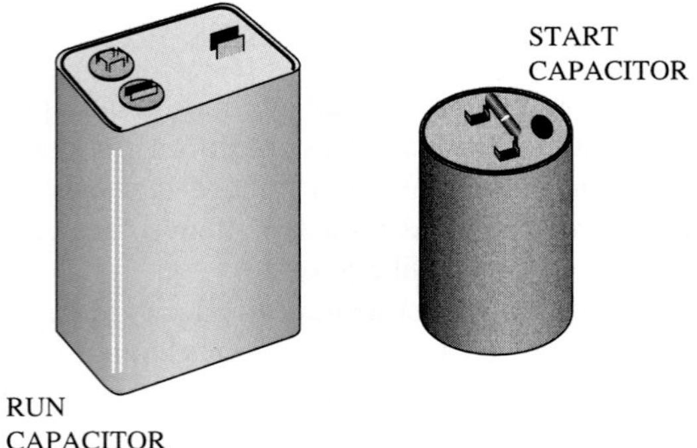

Figure E21-15 A comparison of the physical shape of start and run capacitors.

nected in parallel with the start winding. With this arrangement, the starting torque of the motor is increased. The relay removes the start capacitor from the circuit as soon as the motor is running.

Referring to Fig. E21-14, run capacitors (RC) have been connected to power (L_2) and start (S) terminals of the outside fan motor (OFM) and the inside fan motor (IFM). These run capacitors stay in the circuit all the time and increase the efficiency of the motor. If the run capacitor should fail, the current draw of the motor would be increased about 10% and the motor could overheat.

Fig. E21-15 shows the normal appearance of the run and start capacitors. The run capacitor, which stays in the circuit continuously, is made of large plates and a large amount of insulation (dielectric) to dissipate the heat. The run capacitor may also be round. The start capacitor does not stay in the circuit long, and therefore does not have a heat dissipation problem. It is typically made in rolled form "sandwiching" metal foil and insulating material.

E21-4 SINGLE-PHASE MOTORS

There are a number of different types of single-phase motors. They differ mainly from each other by the amount of starting and running torque. The following types are the most commonly encountered:

1. Permanent-split capacitor (PSC)
2. Capacitor start (CS)
3. Capacitor start/capacitor run (CSR)
4. Shaded pole

The *permanent-split capacitor (PSC) motor* has a run capacitor (RC) in series with the start winding (S) as shown in Fig. E21-16. This capacitor stays connected at all times. It starts the motor and then is left in the circuit to improve the efficiency of the motor after it is running.

The run winding has the number of turns of wire required to give the best motor performance at a given line voltage. The start winding has more turns of smaller wire, which gives it a higher resistance and lower current-carrying capacity than the run winding.

The two drawings shown in Fig. E21-16 are various ways the PSC motor can be represented on a schematic wiring diagram. Note the common (C) connection for the two windings is attached to power L_2, in the right figure. Actually, the run capacitor (RC) is connected between the start (S) winding and the run (R) winding as shown in the left figure. It therefore improves the performance of the motor both in starting and running.

The PSC motor has moderate starting torque and good running efficiency. It is used to power fans and small compressors. The low mfd rating of the run capacitor results in a small phase-angle shift, creating only a moderate starting torque. A diagram of the *capacitor start (CS) motor* is shown in Fig. E21-17. This motor has a high starting torque, but is not as efficient as the PSC motor. The reason for its lower efficiency is that the capacitor is switched out of the circuit immediately after starting. This motor has a high mfd start capacitor.

There are two ways that the start capacitor can be removed from the circuit:

1. Using a mechanical switch. This is a centrifugal switch, attached to the motor shaft. When the motor reaches ⅔ or ¾ of its rated speed, centrifugal force opens the switch, shown in Fig. E21-17.

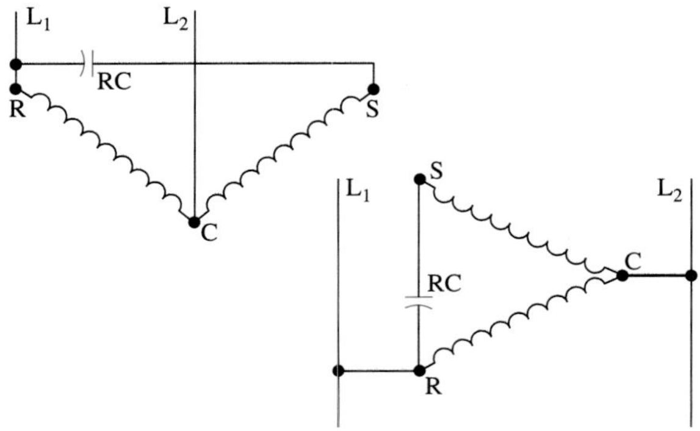

Figure E21-16 Permanent-split capacitor (PSC) motor has a run capacitor (RC) between R and S terminals.

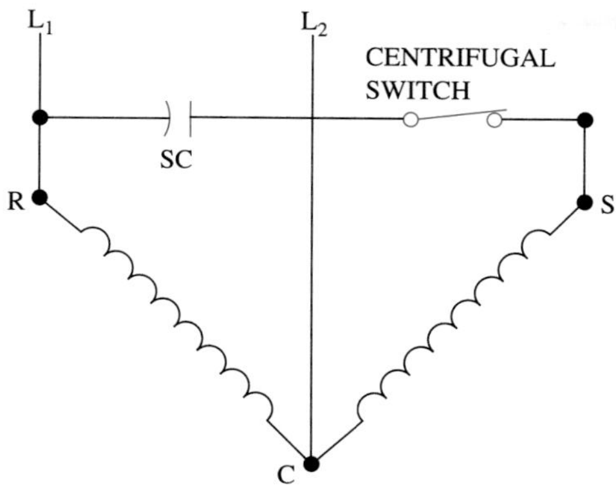

Figure E21-17 Schematic diagram of a capacitor-start motor (CS). Start capacitor is removed from circuit by centrifugal switch.

2. The electromagnetic method, as shown in Fig. E21-18. A start relay (SR) is placed across the start winding. Its contacts are placed in series with the high mfd capacitor. When the motor is started, the capacitor produces a high starting torque. As the motor speed increases, the induced voltage across the start-relay coil increases until it reaches a preset value. At this induced voltage value, the relay is energized and the NC switch is opened, removing the start capacitor from the circuit.

The *capacitor start/capacitor run (CSR) motor* has both a start capacitor (SC) and a run capacitor (RC), as shown in Fig. E21-19. It has excellent starting and running torque, but is not as efficient as the PSC motor. It is used to drive most compressors.

E21-4.1 Start Kits

These kits are available for the technician's use whenever it is necessary to improve the starting torque of a motor. When a motor keeps tripping out on overload, the addition of a start kit may solve the problem.

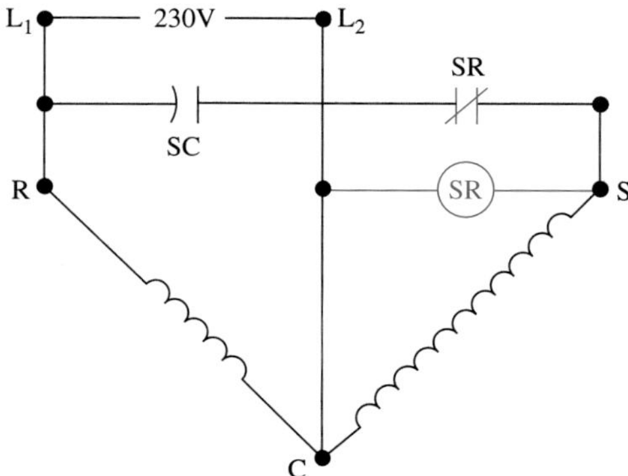

Figure E21-18 Alternate arrangement for removing start capacitor using start relay.

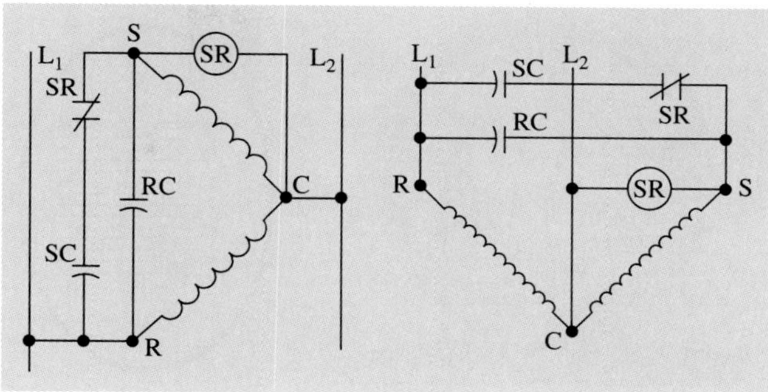

Figure E21-19 Capacitor start/capacitor run (CSR) motor diagram.

Where a high starting torque is required, a hard-start kit can be installed. Where low voltage or a voltage lag is experienced, a soft-starting kit is used. Both kits contain a start capacitor and a start relay. The difference is in the size of the capacitor.

A different soft-starting kit is available to be applied to PSC motors. This kit includes a *positive-temperature-coefficient* (PTC) *thermistor.* The PTC is a temperature-sensitive device whose electrical resistance will increase as its temperature increases. This PTC is placed across the run and start terminals, parallel to the run capacitor of a PSC motor, as shown in Fig. E21-20.

At room temperature, the PTC thermistor has a low resistance, about 25 or 50 ohms. When voltage is supplied, an initial surge of high current passes through the start winding, because the thermistor is effectively shorting out the capacitor. The surge causes an increased starting torque to start the motor. The temperature increase that results causes the PTC thermistor resistance to rise, removing the short from across the run capacitor. The motor then runs as a normal PSC motor.

The *shaded-pole motor,* as shown in Fig. E21-21, has a modified stator pole as shown in the inset. A groove separates a small portion of the stator pole from the rest of the pole. A band of metal is placed around the smaller section of the pole, which provides a phase shift needed to start the motor. Shaded-pole motors have a low starting torque and their speed control under varying load conditions is

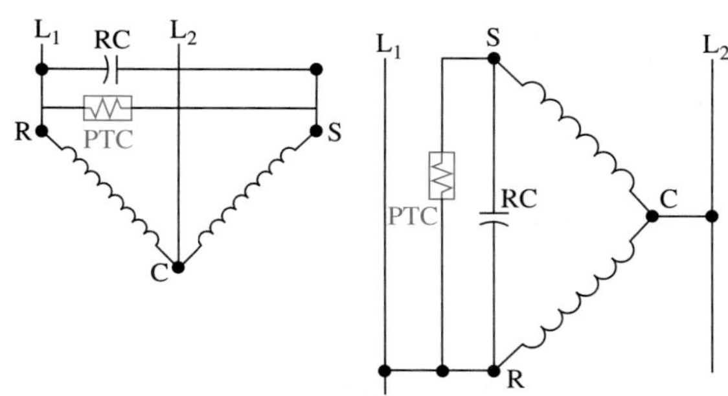

Figure E21-20 Alternate design for a permanent-split capacitor motor using a thermistor.

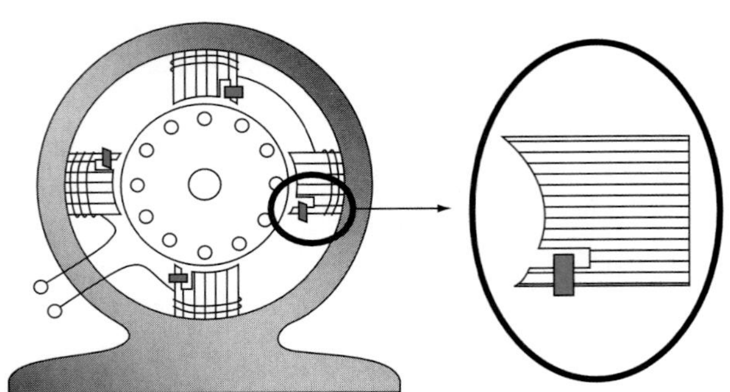

Figure E21-21 Construction of a shaded-pole motor.

Motor	Starting Torque	Running Efficiency	Cost	Use
Shaded Pole	Low	Good	Low	Small Fans & Pumps
PSC	Good	Excellent	Moderate	Fans & Some Compressors*
CS	High	Good	Moderate	Fans
CSR	High	Excellent	High	All Compressors

*When PSC used as compressor motor, positive refrigerant pressure equalizing system required.

Figure E21-22 Summary chart comparing characteristics of single-phase motors.

poor. They offer a low-cost motor for light-duty applications such as running blowers on small air-handling and heating units.

Fig. E21-22 gives a summary of the characteristics of the various types of single-phase motors.

E21-5 THREE-PHASE MOTORS

Three-phase motors have the following advantages over single-phase motors:

1. They are easily reversible. The direction of rotation can be changed by interchanging any two supply-voltage lines.
2. There is less running torque pulsation because at least one phase is always producing an induced rotational effect on the rotor.
3. They have higher starting torque because each winding is out of phase with the other windings and produces rotational torque.
4. They have higher efficiency because they have a built-in supply of rotational torque.

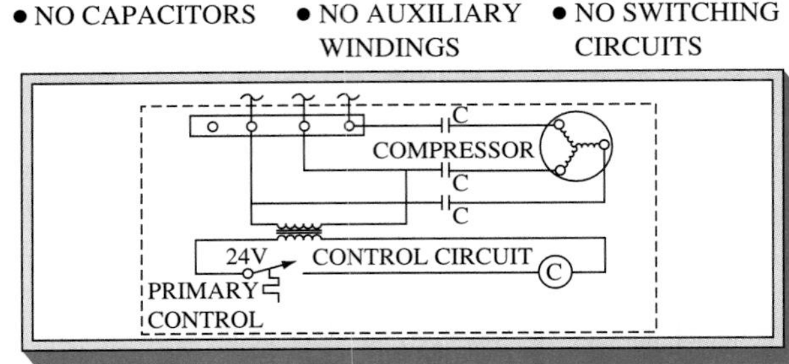

Figure E21-23 Self-starting three-phase motor. No special accessories are required.

The windings of a three-phase motor are 120 electrical degrees apart, which facilitates starting. In Fig. E21-23 a schematic drawing shows the use of a contactor in starting a three-phase motor. No capacitors, no auxiliary windings and no switching circuits are required. They create their own starting torque.

Three phase motors are wired in either a delta (Fig. E21-24) or a wye (sometimes called star) configuration (Fig. E21-25).

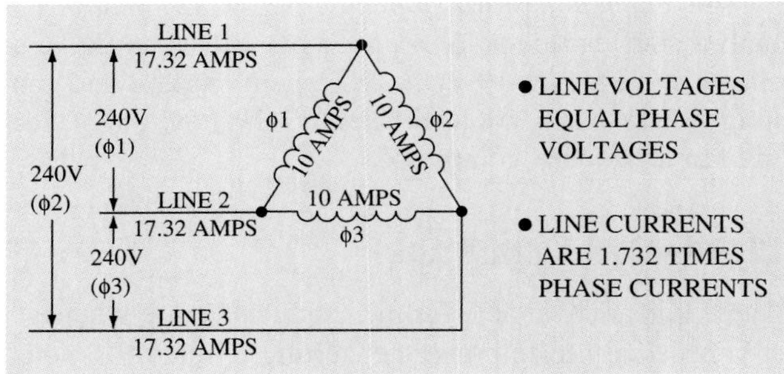

Figure E21-24 Delta-connected three-phase motor.

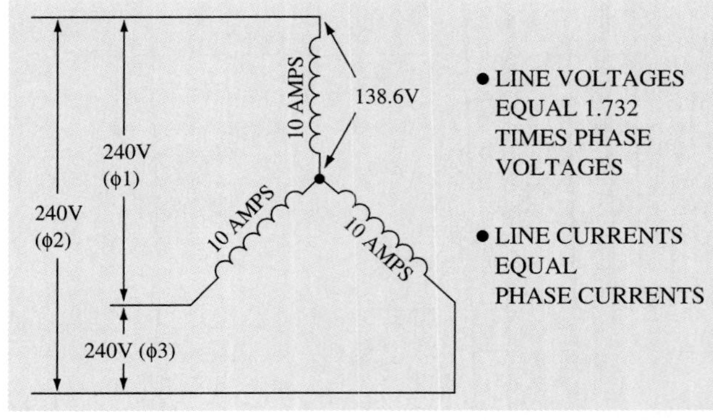

Figure E21-25 Wye-connected three-phase motor.

E21-6 MOTOR PROTECTION

There are a number of causes of motor failure. The most common problem is excessive heating. Among the causes of excessive heating are:

1. Defective start relay (single-phase)
2. Excessive load
3. Loss of refrigerant on a gas-cooled motor
4. Excessive time on locked-rotor current
5. Operation at too high or too low voltage
6. Single phasing (three-phase)

All motor protective devices are designed to cut off the power to the motor before damage has occurred. Repeated cycling on the protective device can damage a motor.

The locked-rotor current is the momentary starting current. This current can be three to five times the running current of the motor. If the motor is blocked from starting, this excessive current will be prolonged, causing heating and serious damage. A protective device is therefore required to disconnect power on either excessive temperature or excessive current. Fuses or circuit breakers are most often the devices that protect against excessive starting current.

All motors are designed to operate between certain voltage limits. If voltages outside these limits are applied, the life of the motor can be seriously reduced. The table below shows the limits of voltage that common motors are designed to handle:

Voltages Handled by Common Motors

Nameplate	Upper Limit	Lower Limit
208	228	188
230	253	207
460	506	414
575	633	518

Typically motors are designed to be applied within plus or minus 10% of nameplate voltage. Dual-rated motors such as 208-V/220-V may be rated plus 10%, minus 5%. Manufacturer's information should be referenced.

On three-phase motors, if the voltage varies more than 2% between phases, the life of the motor will be reduced. This is caused by unequal currents and heating.

Single phasing is another serious problem common to only three-phase motors. If the motor loses power in one of the three lines, the motor may continue to operate. This is called single phasing. This will cause sufficient unbalancing to increase the motor temperature and require motor protection.

E21-6.1 Types of Motor Protection

There are three types of protective devices that are used to protect the motor against excessive heat:

1. Temperature actuated
2. Current actuated
3. Combination of current and temperature

These devices can be either pilot-duty or line-duty arrangements, as shown in Fig. E21-26.

E21-6.2 Motor Overload Devices (Line Break)

For a single-phase motor, the overload device must interrupt one of the motor leads. For a three-phase motor, the device must interrupt two or three of the motor leads. On a wye-connected motor with a built-in overload device, this is often where the three windings are connected in common.

In a three-phase motor, a pilot-duty device senses current overload or excessive temperature within the motor, and opens the contactor circuit to remove power to the motor. A line-duty device senses current and/or temperature in the motor winding, and if an overload occurs, will disconnect power by directly opening the motor winding circuit.

The line-duty arrangement is commonly found on compressor motors used for domestic service. Most of these reset themselves automatically. However, since they are imbedded in the motor winding, as shown in Fig. E21-27, it takes some amount of time for the motor to cool

PREVENTIVE MAINTENANCE

Good preventive maintenance for motors includes lubrication (but not in excess), V-belt inspection for wear and proper tension, and cleaning of air flow passages which cool the motor. Inspections should also include checking the supply voltage and amperage at the motor and comparing it to the manufacturer's tag on the motor. Low voltage or high amperage will indicate that potential problems exist and repair is needed.

Proper motor maintenance can not only extend the life of the motor but it can also prevent untimely failures during heavy load periods.

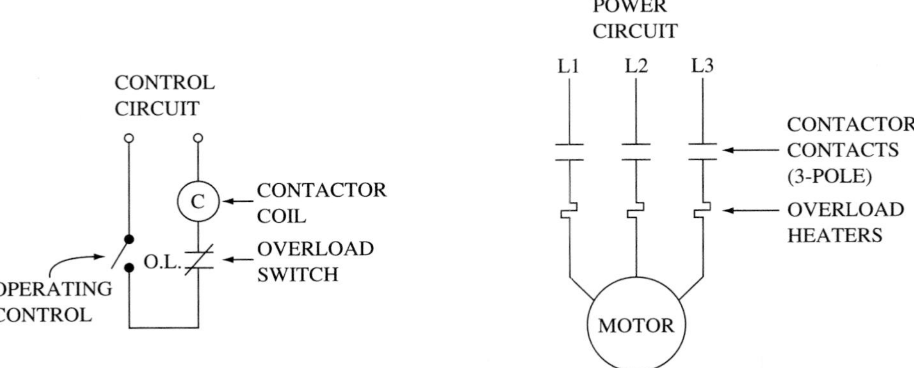

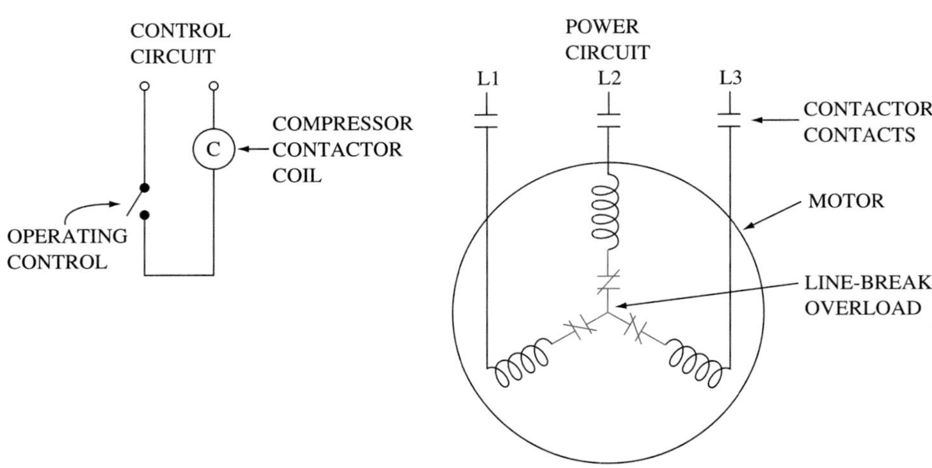

Figure E21-26 Two types of motor protection.

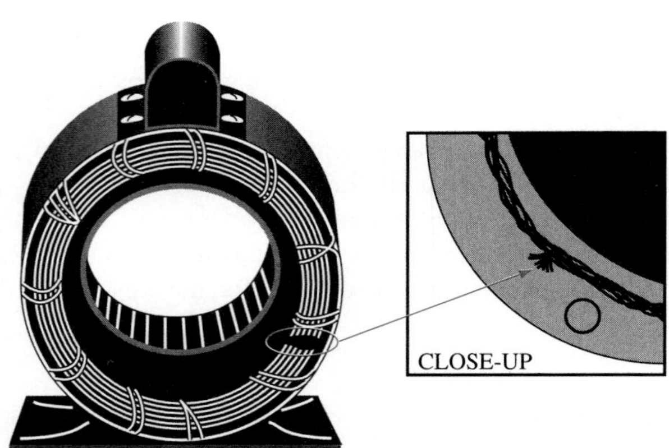

Figure E21-27 Line-duty motor protection.

down. During this waiting period it is easy to incorrectly diagnose the problem as compressor failure.

E21-6.3 Types of Reset

After a protective device has opened or tripped out, it needs to be reset. This can be done manually or automatically.

The advantage of manual reset is that a service technician has the opportunity to examine the cause of the problem before resetting the protective device.

The advantage of the automatic reset is that the unit can automatically go back into service in the case of a nuisance trip-out.

General automatic-reset devices are employed only where the time to reset is sufficient to insure the motor won't short cycle and be damaged by the protective device itself. Sometimes the protective device will reset automatically, but the control circuit will require re-setting by switching the unit off then on from the thermostat or control switch.

E21-6.4 Types of Overload Devices

The various types of overload devices that will be discussed are as follows:

1. Temperature overloads
2. Thermal-overload relay
3. Heater-element current overload
4. External supplemental overload
5. Magnetic overload (Heinemann)

6. Internal current-and-temperature overload
7. General Electric Thermotector®
8. Three-phase overload

Two types of temperature-overload devices are shown in Fig. E21-28.

On the left is the external shell thermostat for pilot duty and on the right is the internal motor thermostat for line duty. Both of these operate by the principle that heating a bimetal disc or strip will cause it to warp and open a power switch. The external device is mounted on the motor shell and resets automatically when the temperature returns to normal. The internal device is wound into the motor windings and automatically resets when the motor cools down.

A motor starter is a contactor with motor overload protection added. Starters made for three-phase motors will have protection on each leg (total of three overloads).

A thermal-overload relay is shown in Fig. E21-29. This relay is current sensitive with automatic reset and is connected in series with the motor winding. On an increase in heat due to excessive motor current, the upper bimetal strip will warp, opening the pilot-duty contacts wired in series with the contactor. Reset is automatic upon cooling. The lower bimetal strip prevents nuisance trip-outs caused by high ambient temperature.

A heater-element current overload device for pilot duty is shown in Fig. E21-30. This device uses a bimetal element (spiral or disc type) that responds to heat generated by a heater element. When the bimetal heats sufficiently, the contacts open, disconnecting the contactor. It automatically resets on cooling. The heater elements are interchangeable and calculated for a specific range of current. They should be matched to the motor current.

Another version of this same device is shown in Fig. E21-31. Instead of a bimetal sensing element, the "solder-pot" relay has a ratchet and spindle construction and requires manual reset. It uses interchangeable heaters similar to the bimetal overloads. Heaters are not interchangeable between makes of equipment.

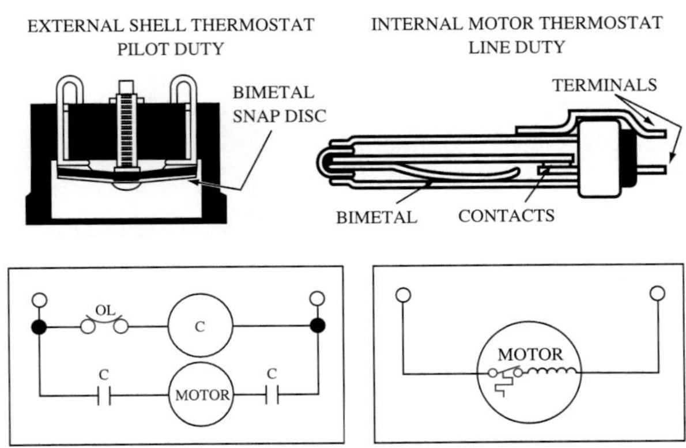

Figure E21-28 Two types of motor temperature overloads.

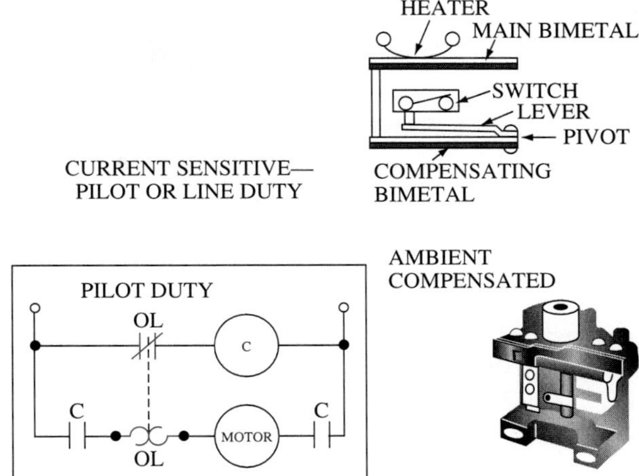

Figure E21-29 Thermal-overload relay used for pilot-duty motor protection.

External supplemental-overload devices are shown in Fig. E21-32. These devices provide either current protection or current-temperature protection. They use a bimetal disc and reset automatically after recovering from the overload condition. The current-temperature device is usually connected for line duty, but some versions are used for pilot duty.

The magnetic overload (Heinemann) is a current-sensitive device and is shown in Fig. E21-33. A core placed inside a coil will attempt to center itself in the magnetic field created when current flows through the coil. The sensing coil is wrapped around a sealed tube filled with silicone oil containing the core. Excessive current will pull the core toward the armature, causing the contacts which are attached to the armature to open. The silicone has a dampening effect which prevents operation of the device for temporary overload conditions such as occur on startup. Current rating is fixed and cannot be field-adjusted.

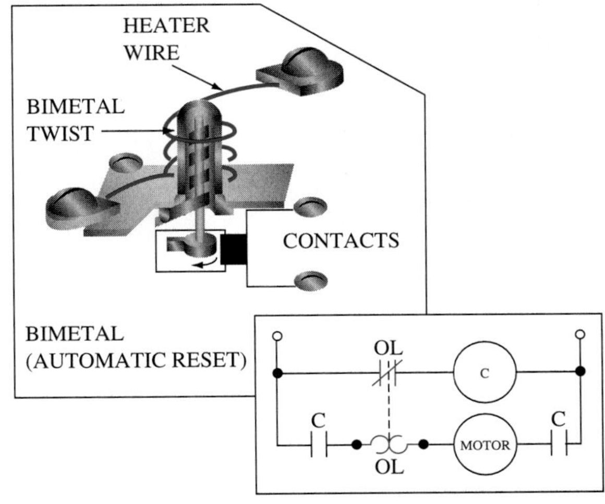

Figure E21-30 Heater-bimetal type current overload for pilot-duty motor protection.

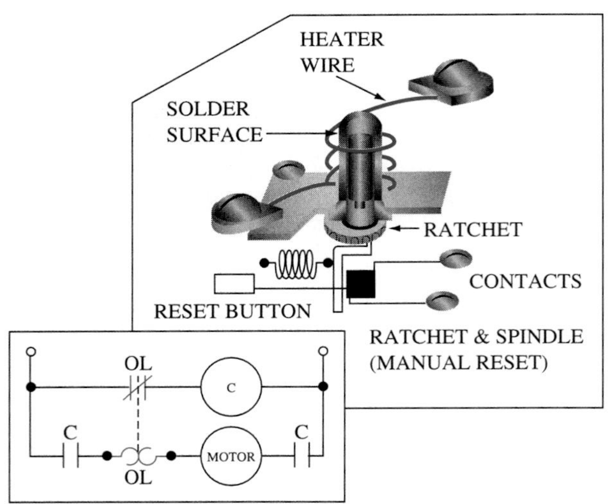

Figure E21-31 Heater/solder pot-type current overload for pilot-duty motor protection.

Another device which is sensitive to both internal current and temperature and is wound into the motor windings is shown in Fig. E21-34. As the overload occurs, the bimetal warps, opening the contacts. Reset is automatic.

The General Electric Thermotector® device is shown in Fig. E21-35. This is also an internally actuated temperature motor protector. On this device the case and the internal strip expand at different rates. It will trip early if the temperature rises rapidly. If the temperature rise is gradual it will trip at its normal setting (rate-compensated).

Three-phase overloads are shown in Fig. E21-36. These are current-temperature line-duty devices. They can be located internally or externally to the motor.

When an overload occurs the bimetal disc warps, breaking the electrical circuit. They automatically reset when the overload no longer exists.

Knowledge Extension

To learn more about electric motors and related accessories, visit your local commercial motor repair business. Ask the shop foreman for a tour of the shop to see how motors are rewound. You will find that small motors are not typically repaired, but discarded and replaced with new ones. The larger expensive motors, however, are rewound and rebuilt with new conductors and bearings. It may appear to be a tedious business, but the people who do the work are highly skilled and normally well paid for their services. Ask the foreman if you can have an old discarded motor for further studies. When you get home or back to your school, see if you can determine the motor type and number of poles, and locate the run and start windings. The motor repair business plays an important role in the refrigeration and air-conditioning industries.

E21-6.5 Motor Circuit Testing

Since motors are so important in HVAC/R systems they are carefully protected and can be thoroughly tested to see that they are operating properly. The tests that follow begin with single-phase motors. Since many single-phase motors require capacitors, a word of warning needs to be added about handling capacitors. Capacitors can be dangerous. Safety precautions are required.

Capacitors can hold a high-voltage charge after the power is turned off. Always discharge capacitors before touching them.

Fig. E21-37 shows the various types of instruments used for testing motors. The ohmmeter is used to check the resistance of the motor windings and the capacitor.

Figure E21-32 External supplemental overload providing either current protection or current-temperature protection.

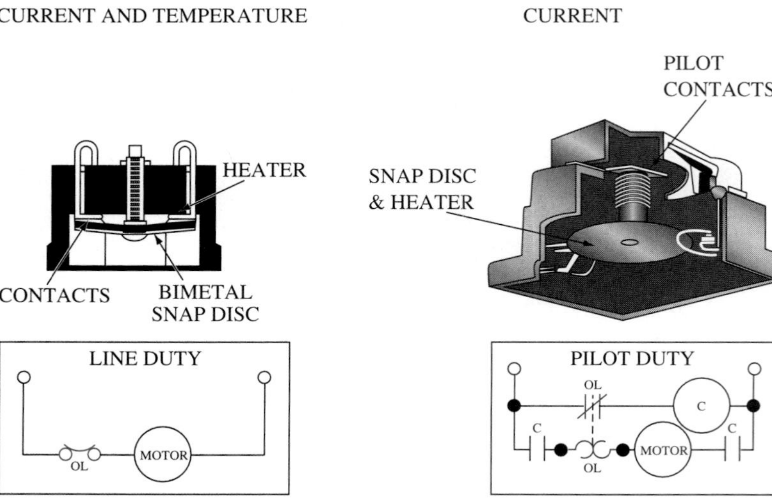

Figure E21-33 Heinemann magnetic overload, a current-sensitive device.

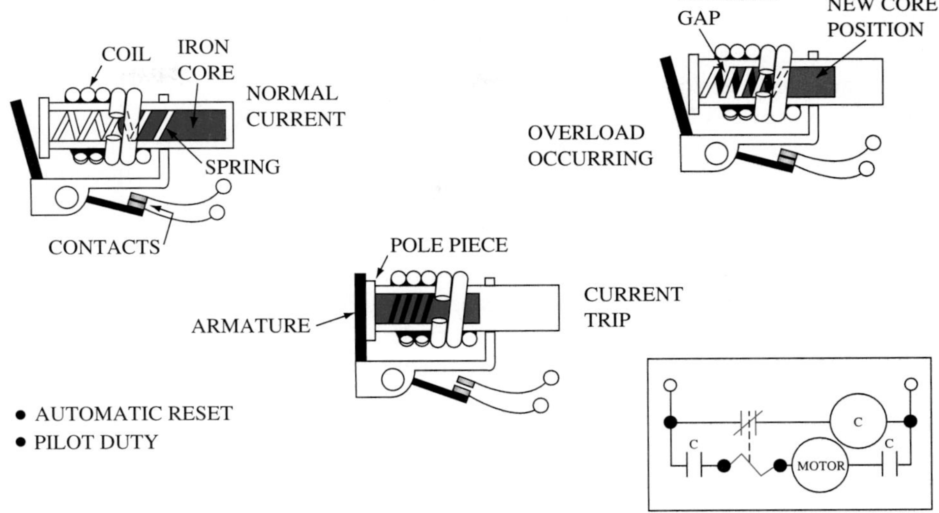

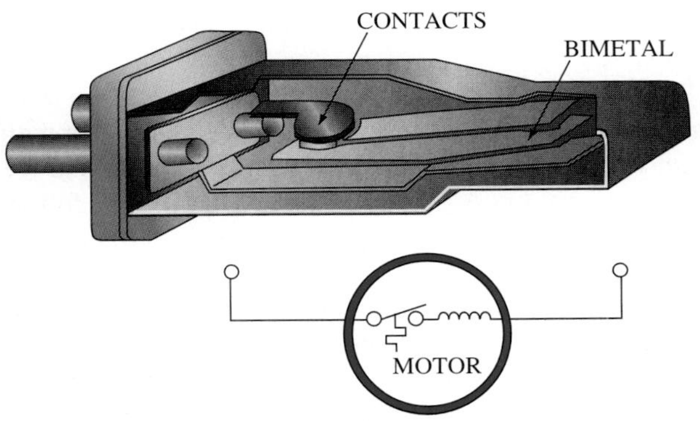

Figure E21-34 Internal line-duty overload device used for motor overloads with automatic reset.

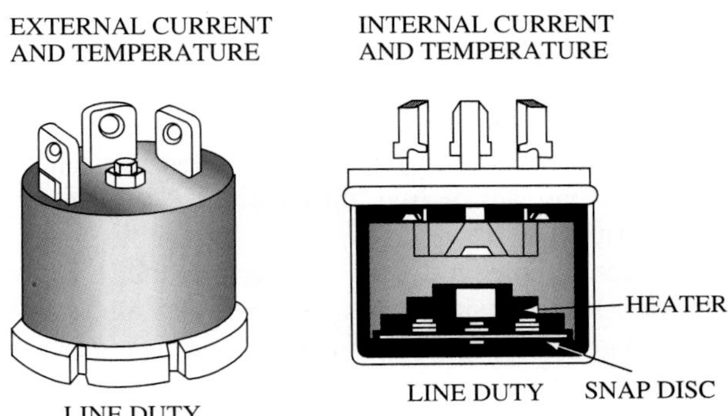

Figure E21-36 Three-phase motor overloads, both internal and external.

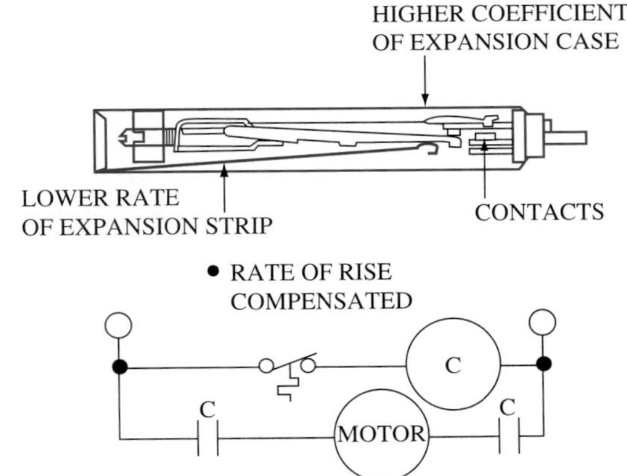

Figure E21-35 General Electric Thermotector® includes rate-of-rise temperature compensation. (Courtesy of GE Appliance)

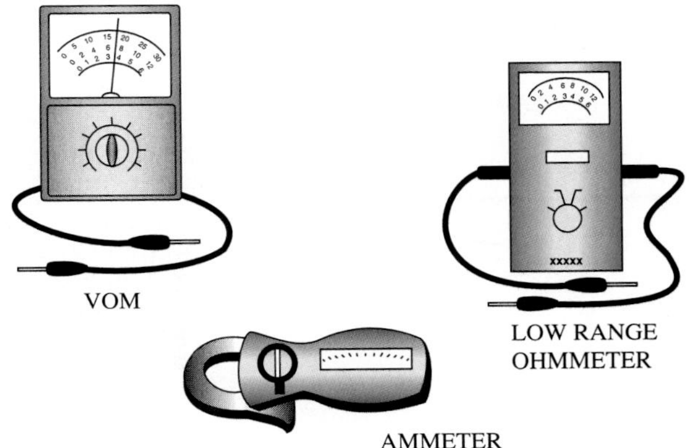

Figure E21-37 Various types of motor testing meters.

The ammeter measures the current in the circuit. The voltmeter measures the operating voltages. A capacitor analyzer measures the actual mfd rating of the capacitor.

The first thing to do in testing is to check the motor windings, as shown is Fig. E21-38. It may be necessary to remove a guard that encloses the terminals on a hermetic unit. The terminals should be marked C, S and R, the same as they are shown on the schematic drawing.

Caution! Occasionally the terminal has been damaged and on a pressurized system the terminal can blow out. To avoid possible injury, unless the charge has been removed, use terminal points some distance away from the compressor. Be sure that all accessories such as capacitors and relays are disconnected and that the refrigerant charge has been removed.

Use an ohmmeter to read the resistance in both the run and start winding. Be sure good contact is made with the proper terminals.

If the terminals are not marked they can be identified by a simple test. First, measure the resistance between each pair of terminals. For example, assume that these readings are 5½, 4, and 1½ ohms. By diagramming them as shown in Fig. E21-39, the terminals can be identified. In the example, the greatest reading is between 1 and 2. The common terminal is therefore 3, the one not being touched. Then, reading from C, the greatest resistance is to 2. Therefore, 2 is S. And, finally, since 2 is S and 3 is C, 1 must be R.

The motor is then tested for an open winding, a broken wire, or a shorted winding, as shown in Fig. E21-40. To do this, a good low-range ohmmeter ($R \times 1$) is required. A zero resistance means a shorted winding. Low resistance means the winding is good. An infinity reading means it is an open winding. As a rule the start winding has a resistance three to five times the resistance of the run winding.

Large motors (5–10 Hp and higher) have heavy copper windings to carry the motor current. They may therefore indicate a reading very close to a short when winding resistance is read—depending on the ohmmeter.

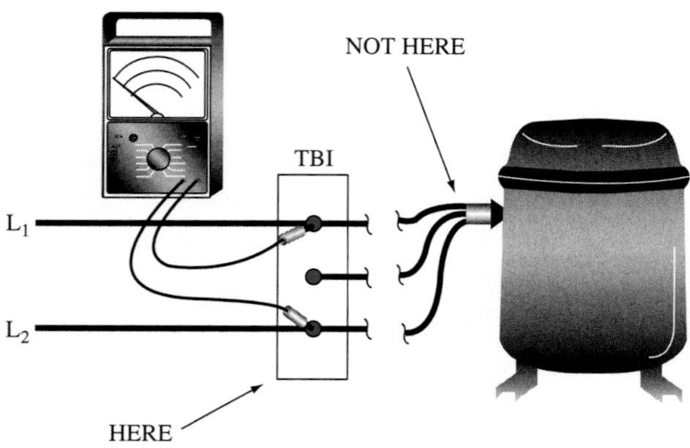

Figure E21-38 Proper testing location for motor windings.

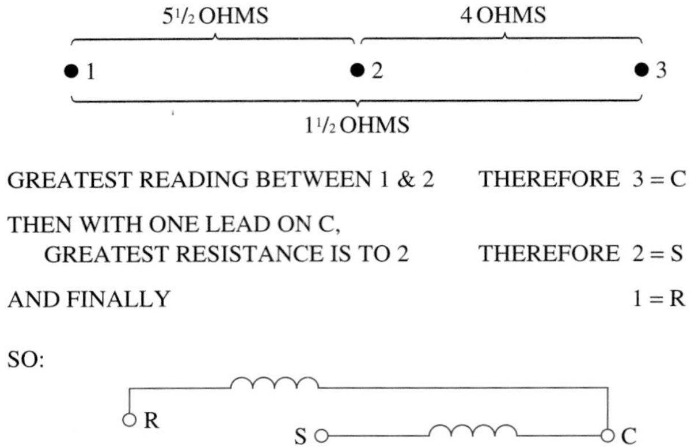

Figure E21-39 Method of determining run, start, and common motor terminals.

Fig. E21-41 shows the motor being tested for a grounded winding. The ohmmeter needs to be capable of measuring very high resistance ($R \times 100{,}000$). For an ungrounded winding the resistance is generally 1 to 3 megaohms. This applies to both single- and three-phase motors.

The temperature of the compressor is important in testing for a partially grounded winding. If the compressor will run, it should be run for about 5 minutes before testing.

E21-6.6 Insulation Resistance Testers

These are special testers which are invaluable for testing leakage resistance from motor windings to ground. They are often used to periodically test semi-hermetic motor insulation. The meters can test leakage at high voltage (500 V for 208–240-V motors and 1000 V for 480-V motor). They may be battery-operated or use a hand-cranked generator. They are often referred to as "meggers". They may detect insulation faults, while an ordinary multimeter using a few volts DC, will show a satisfactory reading.

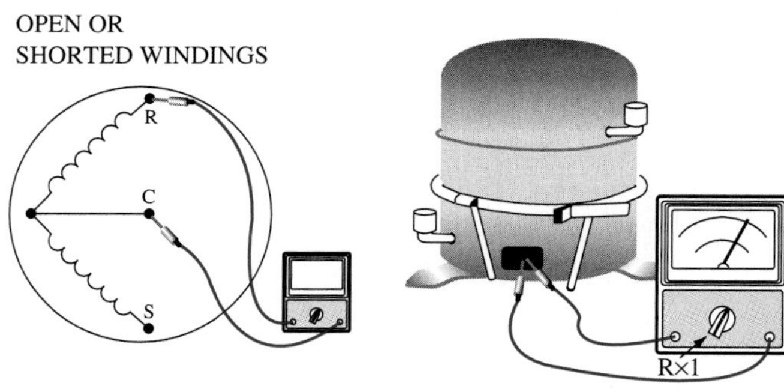

Figure E21-40 Method of testing for shorted windings and open windings.

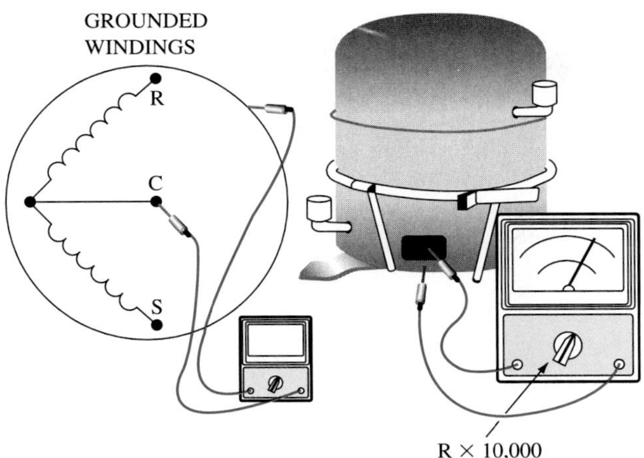

Figure E21-41 Method of testing for grounded windings.

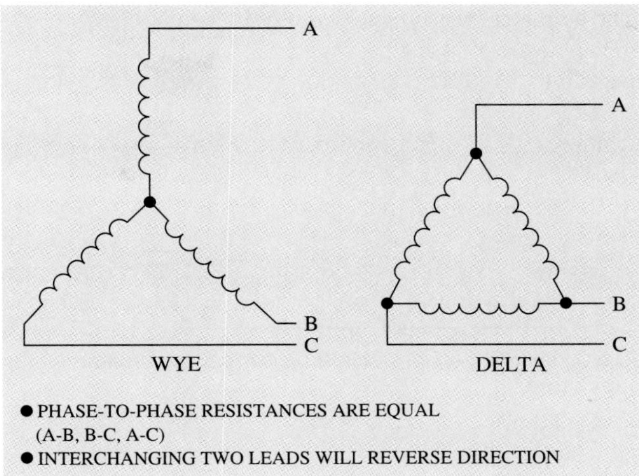

Figure E21-43 Three-phase motor windings.

Caution! When testing for a grounded winding, one test lead is placed on the terminal and the other on the outer shell of the compressor. Be careful to make good contact with the motor shell. A coat of paint or a layer of dirt can hide a grounded winding.

The diagram shown in Fig. E21-42 shows the use of an ammeter in testing a single-phase induction motor.

The conditions found in each one of the test locations help determine the cause of the problem. The quick reference table is given in Fig. E21-42.

Fig. E21-43 shows the terminals of the two types of three-phase motors. The windings of a three-phase motor all should have the same resistance. Remember when checking a three-phase motor to be sure to reconnect it in the same manner as originally connected. Interchanging any two connections can reverse the rotation of the motor.

E21-6.7 Testing Capacitors

The first operation in testing a capacitor is to discharge it. Do not discharge it by shorting out the terminals. This can damage the capacitor. To avoid electrical shock, the technician should never place fingers across the terminals before properly discharging the capacitor.

The proper way to discharge a capacitor is to put it in a protective case and connect a 20,000-ohm, 2-watt resistor across the terminals, as shown in Fig. E21-44. Most start capacitors have a bleed resistor; however, it is good practice to make sure the charge has been bled off.

Capacitors can be roughly checked by using an ohmmeter. The ohmmeter used in testing capacitors should have at least an $R \times 10{,}000$ scale. To test the capacitor, disconnect it from the wiring and place the ohmmeter leads on the terminals, as shown in Fig. E21-45.

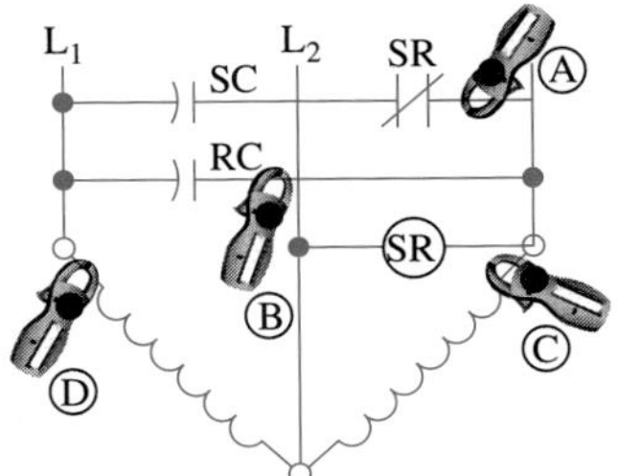

METER	ACCEPTABLE READING	BAD READING	PROBABLE CAUSE
A	HIGH CURRENT AS MOTOR STARTS, THEN DROP TO ZERO.	1. NO CURRENT WHEN MOTOR STARTS. 2. HIGH CURRENT UNTIL OVERLOADS TRIP.	1A. OPEN START CAPACITOR 1B. OPEN START RELAY CONTACTS 1C. OPEN START WINDING 2A. EITHER CAPACITOR SHORTED 2B. START WINDING SHORTED

Figure E21-42 Motor testing procedure, using a clamp-on ammeter.

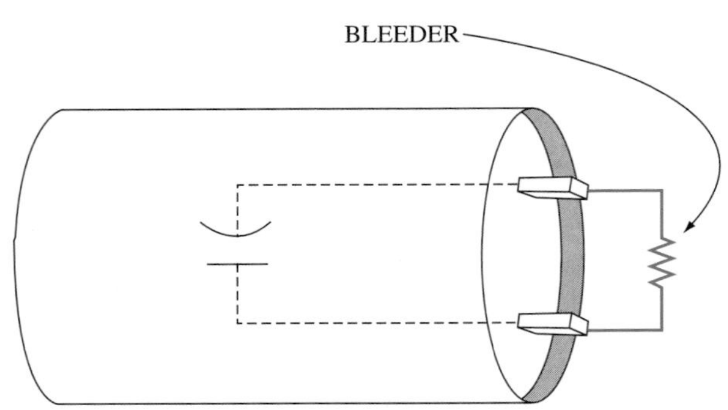

Figure E21-44 Using a bleed resistor on start capacitors.

If the capacitor is good, the needle will make a rapid swing toward zero and slowly return toward infinity. If the capacitor has an internal short, the needle will stay at zero, indicating that the instrument will not take the charge. What you are actually doing is attempting to charge the capacitor using the battery in the ohmmeter (be sure the battery in the ohmmeter is good). An open capacitor will read high with no dip and no recovery.

The use of a capacitor analyzer (Fig. E21-46) is highly recommended. This instrument will read the microfarad (mfd) rating and detect any breakdown in the insulation under load conditions. It will detect any capacitors that have failed to hold their ratings. It also is useful in measuring the rating of a capacitor that has an unreadable marking.

Almost all run capacitors have some sort of a mark, usually a red dot (Fig. E21-47), to indicate the terminal that should be connected to the run terminal.

With this arrangement, an internal short circuit to the capacitor case will blow the system fuses without passing the current through the motor start winding.

Where two run capacitors are used on the same equipment, multiple-section capacitors with a common terminal are used. One of these is illustrated in the top of Fig. E21-48.

These two capacitors have different ratings as required by the equipment they serve. Testing is similar to that for single run capacitors.

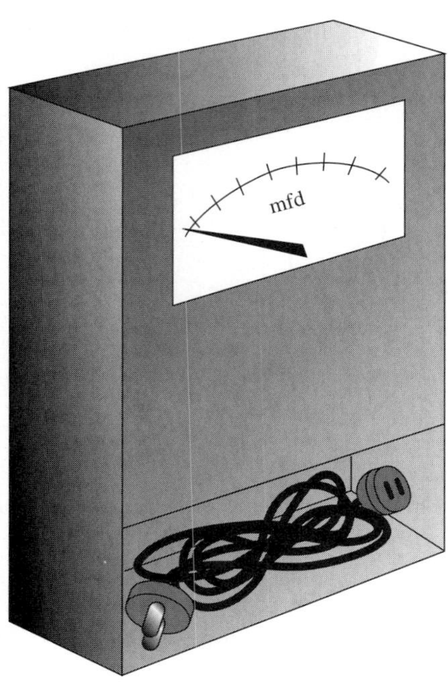

Figure E21-46 Capacitor analyzer.

In Fig. E21-48, the bottom device shows a start capacitor with a bleed resistor. The capacitor can be tested with this bleed resistor in place. The "pop-out" hole on the start capacitor allows insulation expansion if the capacitor is overheated. If the hole is ruptured, the capacitor must be replaced.

In replacing a capacitor, it is desirable to use an exact replacement—a capacitor with the same mfd rating and voltage-limit rating. When this is not possible, substitutions can be made as long as the rules for substitutions are carefully followed.

Do not interchange start and run capacitors. Start capacitors are high capacity (100–800 mfd) electrolytic units which are intended for momentary use in starting motors. They are normally encased in plastic.

Run capacitors have much lower capacitance ratings (2–40 mfd) but are made for continuous-duty use. They are normally sealed in a metal can.

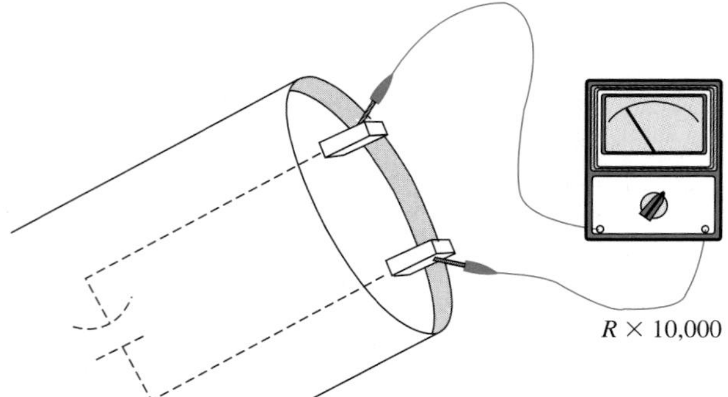

Figure E21-45 Using an ohmmeter to test capacitors.

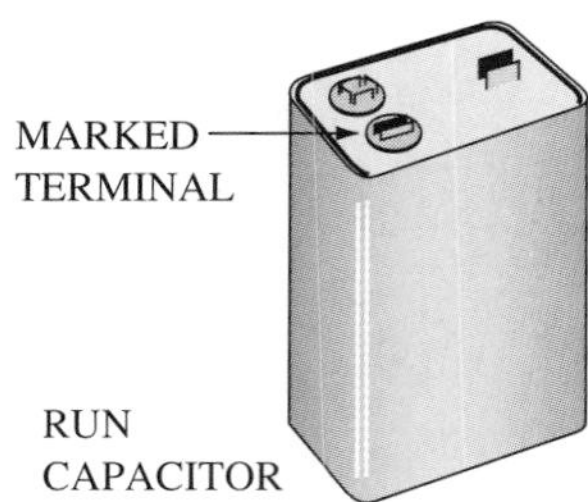

Figure E21-47 The marked terminal on the run capacitor should be connected to the run terminal of the compressor.

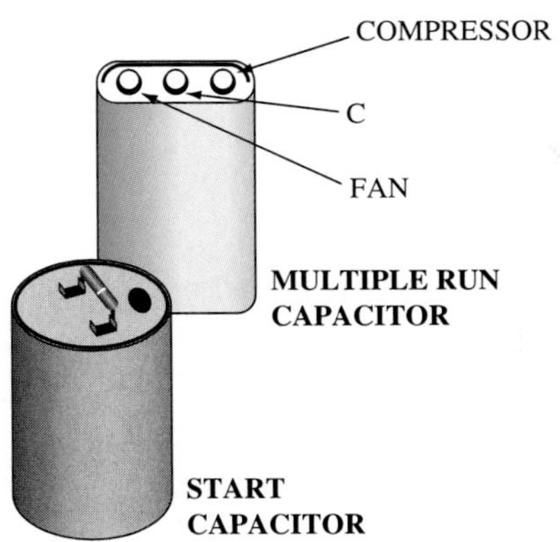

Figure E21-48 Multiple section or dual run capacitor and start capacitor.

E21-7 TESTING START RELAYS

In replacing start relays, it is important to do the following:

1. Always use the identical replacement. An improper substitution can damage the motor.
2. The replacement must be mounted in the same position as the original and connected the same way. The method of testing the relay coil and contacts is shown in Fig. E21-49.

Caution! Many of these start-relay coils have higher resistance than average control circuit relays. Be sure to test the coil on the $R{\times}100$ scale before deciding that the relay is defective. Note that the pull-in/drop-out voltage of the relay is unique. Do not attempt to replace it with an ordinary, similar voltage relay.

When testing start-relay contacts, the contacts should be closed and the ohmmeter will read zero resistance. Sometimes the contact sticks closed and under these conditions the start winding of the motor is often damaged. Sometimes the normally closed contacts will be badly burned and make a poor connection. If such is the case, replace the relay. Do not attempt to clean the contacts.

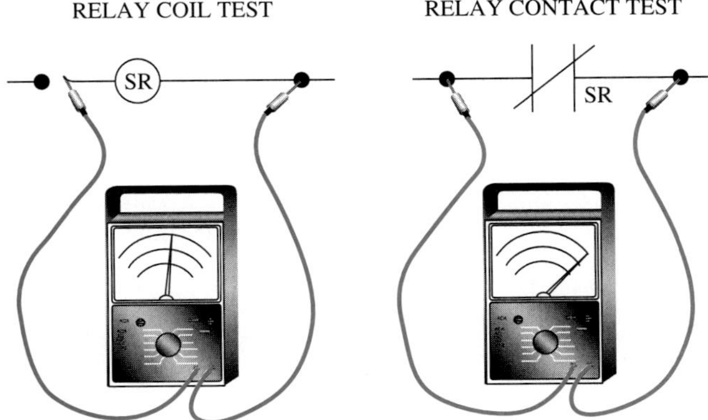

Figure E21-49 Using an ohmmeter to test a relay coil and its switch.

REVIEW

- Electric motors are the most important load device in HVAC/R units. They convert large amounts of electrical energy to useful work.
- Torque—The twisting (turning) force that must be developed by a motor to turn its load.
- Load—Directly related to the torque required and the speed.
- Two principal parts of a motor:
  Stator—stationary windings; coils of wire
  Rotor—rotating part; armature; series of copper or aluminum bars mounted on a soft iron core, shorted together by an end ring, permitting current to flow.
- Motor speed is determined by the number of poles and the frequency (Hz) of the current. Four-pole motors have a synchronous speed of 1800 rpm. Two-pole motors have a synchronous speed of 3600 rpm.

$$\text{rpm} = \frac{\text{Hz} \times 120\ \text{sec/min}}{\#\ \text{of Poles}}$$

- Types of single-phase motors:
  Shaded-pole
  Capacitor-start Induction Run (CS, CSIR)
  Permanent-split Capacitor (PSC)
  Capacitor Start/Capacitor Run (CSR)
- Types of three-phase motors:
  Wye
  Delta
  Dual-Voltage
  Multiple Motors
- Three-phase motor advantages:
  Easily reversible
  Less running torque pulsation
  Higher starting torque
  Higher efficiency
- Causes of motor failure (excessive heating):
  Defective start relay (single-phase)
  Excessive load
  Loss of refrigerant
  Excessive time on locked rotor current
  Operation at too high or too low a voltage
  Single phasing (three-phase)
- Locked rotor current = 3 to 5 times the running current of the motor.
- Voltage limits:

Nameplate	Upper Limit	Lower Limit
208	228	188
230	253	207
460	506	414
575	633	518

Dual rated motors are rated at + 10%, – 5%

- Types of motor protection:
  Temperature actuated
  Current actuated
  Combination of current and temperature
- Single-phase motor overloads must interrupt one of the motor leads.
- Three-phase motor overloads must interrupt two or three motor leads.
- Protective overloads may be pilot-duty or line-duty.
- Motor starter—A contactor with motor-overload protection.
- An ohmmeter may be used to field-test a motor and all of its associated components.

Problems and Questions

1. What is the synchronous speed of a four-pole motor?
   a. 900 rpm
   b. 1200 rpm
   c. 1800 rpm
   d. 3600 rpm
2. Which capacitor stays in the circuit after the motor is in full operation?
   a. Run capacitor
   b. Start capacitor
3. What does the voltage stamped on a capacitor mean?
   a. Minimum voltage that can be applied
   b. Maximum voltage that can be applied
4. A permanent-split capacitor motor (PSC) uses what type of capacitor?
   a. Run capacitor
   b. Start capacitor
   c. One run and one start capacitor
   d. Two run capacitors
5. On which type of starter kits is a thermistor used?
   a. LS motor start kit
   b. CSR motor start kit
   c. Hard-starting kit
   d. Soft-starting kit
6. What type of capacitor is used on a shaded-pole motor?
   a. Run
   b. Start
   c. One run and one start
   d. None
7. Which type of motor requires no capacitor and yet has a high starting torque?
   a. CS
   b. CSR
   c. PSC
   d. Three-phase
8. How much higher is locked rotor current than running current for most motors?
   a. 1½ to 2 times
   b. 2 to 2½ times
   c. 3 to 5 times
   d. 10 to 15 times
9. Normally, what is the upper voltage limit on a 208-V motor?
   a. 218 V c. 228 V
   b. 224 V d. 264 V
10. What is the condition of a three-phase motor that is single phasing?
    a. Motor rotation is reversed.
    b. Motor operates only on two legs of power.
    c. Single-phase motor is substituted.
    d. Used for starting only.
11. Torque is the force that must be developed by a motor to turn its load. True or false?
12. The stator on the electric motor is a series of aluminum bars mounted on a soft iron core. True or false?
13. Current passing through the stator creates a magnetic field. True or false?
14. The start winding is made up of many turns of smaller wire. True or false?
15. The greater the number of electrical poles inside the motor, the faster the speed of the motor. True or false?
16. What is the formula to calculate rpm?
17. Start capacitors are used to increase motor _____.
18. The PSC motor has a run capacitor in _____ with the start winding.
19. Three-phase motors can be easily reversed by doing what?
20. A motor starter is a contactor with what additional accessory?

SECTION 4

Domestic Refrigeration

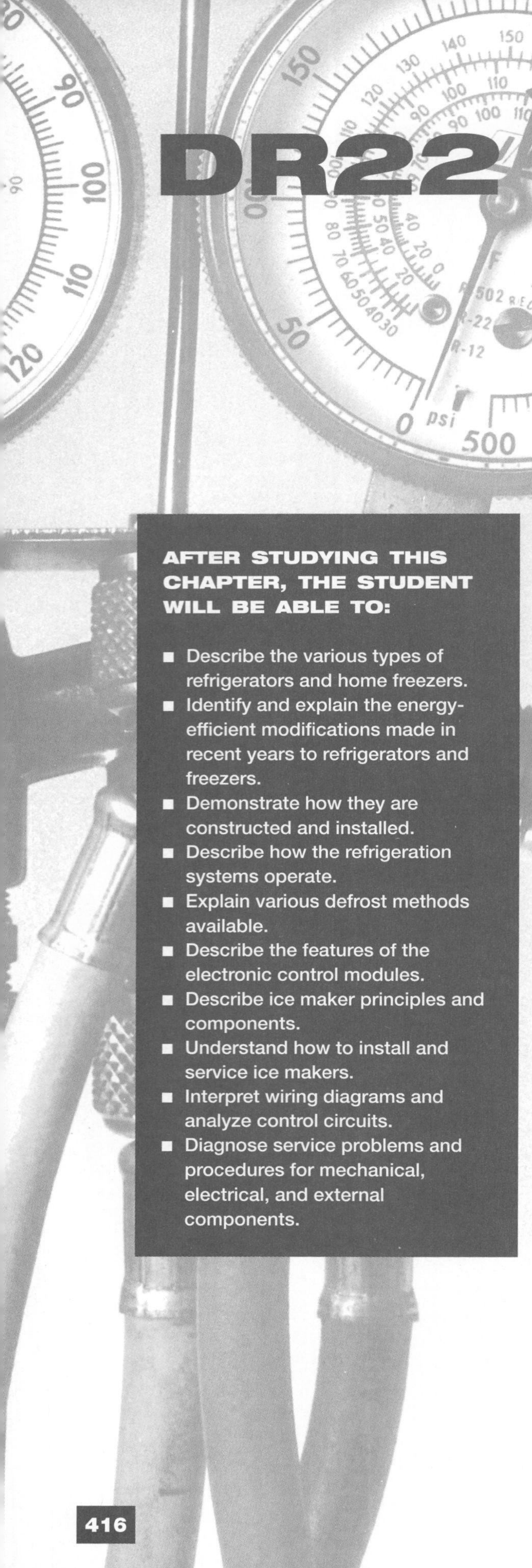

AFTER STUDYING THIS CHAPTER, THE STUDENT WILL BE ABLE TO:

- Describe the various types of refrigerators and home freezers.
- Identify and explain the energy-efficient modifications made in recent years to refrigerators and freezers.
- Demonstrate how they are constructed and installed.
- Describe how the refrigeration systems operate.
- Explain various defrost methods available.
- Describe the features of the electronic control modules.
- Describe ice maker principles and components.
- Understand how to install and service ice makers.
- Interpret wiring diagrams and analyze control circuits.
- Diagnose service problems and procedures for mechanical, electrical, and external components.

DR22 Refrigerators and Freezers

DR22-1 DOMESTIC REFRIGERATORS AND FREEZERS

DR22-1.1 Overview

The most common types of full-sized refrigerators and freezers for domestic uses are shown in Fig. DR22-1. These use the *vapor-compression* refrigeration cycle.

A number of smaller refrigerators have been developed using *absorption* and *thermoelectric cycles.* These fit the requirements of certain markets. For example, the absorption cycle is commonly used for recreation vehicles. The thermoelectric units are commonly used for small coolers in hotels.

One reason for the popularity of the vapor-compression cycle can be seen from the relative efficiencies of the various types of systems. The following is a comparison of the amount of heat removed by the three types of systems at 0°F freezer conditions in an ambient temperature of 90°F.

Thermoelectric—approximately 0.3 Btu/W-hr
Absorption—approximately 1.5 Btu/W-hr
Vapor-compression—approximately 4.5 Btu/W-hr

Domestic refrigerators and freezers are mainly used for the storage of food, with ice making an essential secondary function. In the refrigerator portion of the cabinet, it is usually considered good practice to maintain temperatures between 35 and 45°F. In the freezer compartment a temperature of near 0°F is maintained, except for short-term storage where the temperature is kept below 8°F. Most models will cool down in a reasonable amount of time with an ambient temperature as high as 110°F.

Types of domestic refrigerators and freezers shown in Fig. DR22-1, include the following types of refrigerators: single-door, side-by-side, and top-mount. Freezers include chest and upright types.

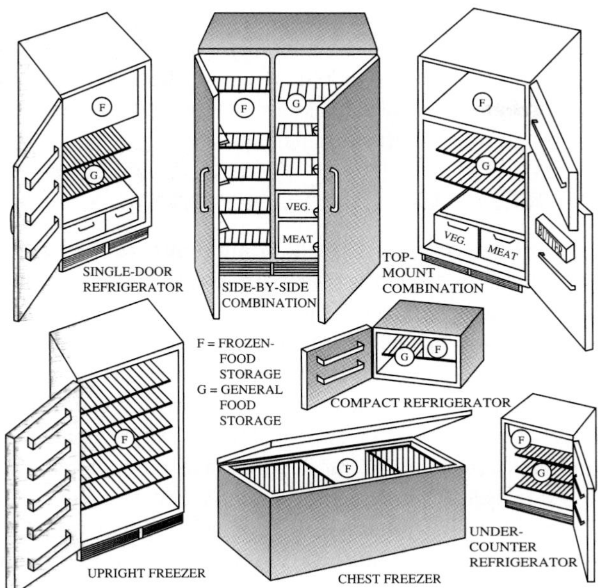

Figure DR22-1 Types of domestic refrigerators and freezers. (Copyright by the American Society of Heating, Refrigerating, and Air-Conditioning Engineers, Inc. Used by permission.)

DR22-1.2 Energy Conservation

In recent years manufacturers have given considerable attention to energy conservation and efficiency in the design of domestic refrigerators and freezers. This has been due to regulations published by the U.S. Department of Energy (DOE) and standards established by the Association of Home Appliances Manufacturers (AHAM). Testing procedures followed by manufacturers establish an estimated annual cost of energy for labeling electrical refrigerators, refrigerator-freezers and freezers under the Federal Trade Commission's (FTC's) Energyguide Program. For example, the energy now used by a 22 cu ft refrigerator has been reduced approximately 50%, compared to earlier models. A 22-cu ft refrigerator with automatic defrost can now be operated with as little power as it takes to operate a 75-W lamp.

Some of the things done to achieve these savings are:

1. A more efficient compressor is used. Older units are equipped with reciprocating compressors. Through the use of scroll-type compressors, refrigerators are not only more efficient, but quieter in operation.
2. Thicker insulation is placed in the door. The door formerly was 1½ in. thick and is now 2¼ in. thick.
3. A more efficient fan motor is used. The less efficient shaded-pole (SP) motor is replaced with a permanent-split capacitor (PSC) motor. This new motor uses a shaped air gap and internal capacitor and saves about 10% of the energy used previously.
4. A trap is placed in the drain tube from the evaporator. Formerly, moist hot air was blown up the tube to the evaporator coil by the condenser fan, requiring extra refrigeration.
5. A microprocessor-type defrost control is substituted for the time clock to initiate the defrost cycle. This new control defrosts the evaporator coil only when necessary.

DR22-1.3 Construction

The exterior cabinets of domestic refrigerators and freezers are made of *welded steel.* The outside surface is smooth and tight, to prevent any entrance of air or moisture. The inside shell is usually made of steel; however, some manufacturers use molded plastic.

The area between the two enclosures is filled with *urethane foam insulation,* expanded in place, to completely fill the space. The relative insulating value of the new synthetic insulation compared to the former mineral or glass fiber is shown as follows:

Thermal Conductivity

Insulation	Btu/in./hr/sq ft/°F
Mineral or glass fiber	0.22 to 0.28
Foamed-in-place urethane	0.13 to 0.16

The *breaker strip* between an inner and outer steel cabinet, where the doors fit, is made of plastic insulation, so that there is no temperature conduction between the two surfaces of the cabinet. Where the interior of the cabinet and door are molded plastic, no breaker strip is necessary.

Special-purpose compartments are provided in the cabinet construction. A warmer space for butter storage maintains the temperature of butter suitable for spreading. Some refrigerators have a meat storage compartment maintaining a temperature just above freezing.

A high humidity compartment is provided in most cabinets for the storage of leafy vegetables and fresh fruit. This is usually an enclosed drawer that protects the product from the drying effect of moving air.

Door gaskets are made of flexible vinyl with an air cushion which is compressed when the door is closed to provide a tight fit. Recent regulations require that the door can be opened from the inside with a pressure not to exceed 15 psi. A magnet is placed in the vinyl to hold the door closed in place of a latch.

Door hinges are made with an adjustable mechanism to provide even spacing on all sides. Door seals should be tight and can be tested by placing a thin piece of paper between the gasket and the cabinet in several places. With the door closed, it should require a small pull to remove the paper if the door is fitting properly.

Leveling screws are provided on the base for leveling the cabinet during installation. It is important to have a level unit so that the condensate collecting pan will properly hold the condensate prior to evaporation or disposal to the drain.

DR22-1.4 Refrigeration Load

The predictable portion of the load on refrigeration systems is shown in Fig. DR22-2.

A large part of the load comes from door openings which are variable and unpredictable quantities dependent on customer use. In a 90°F room, the surface temperature of a cabinet is usually 5 or 6°F below room temperature to prevent condensation on the outside of the cabinet. Cabinets should be located in the interior of the building away from the direct sun to minimize heat load.

The condenser must be cleaned periodically to maintain its normal efficiency. Defrosting of the evaporator to prevent ice accumulation is essential in maintaining proper temperatures within the refrigerated space.

DR22-1.5 Method of Cooling

In the *refrigerator*, air is circulated over the evaporator and distributed throughout the cabinet either by gravity or forced circulation. Shelves are usually made of heavy wire to hold the weight of the product and permit air movement within the box. The light switch is controlled by the opening and closing of the door(s).

The *upright freezer* uses a low-temperature evaporator with a circulating air fan and can be either manually or automatically defrosted.

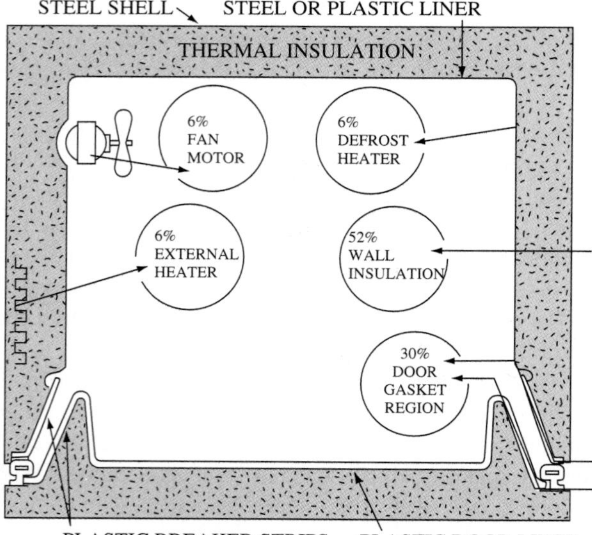

Figure DR22-2 Cabinet cross section showing distribution of heat load. (Copyright by the American Society of Heating, Refrigerating, and Air-Conditioning Engineers, Inc. Used by permission.)

The *chest freezer* has refrigerated side panels (plate-type evaporators), and is manually defrosted. Metal objects, such as knives, should never be used in removing the ice accumulation because of potential damage to the refrigerant tubes in the side panels.

DR22-1.6 Cycle Components

Evaporators are usually made of aluminum tubing, either with integral extruded fins or with extended surfaces mechanically attached to the tubing.

Condensers are usually made of steel tubing with extended surfaces of sheet steel or wire. Steel tubing is used on the high-pressure side of the system, which is normally dry. Copper is used for suction tubing, where condensation can occur. Due to its ductility and ease of brazing, copper is used for the capillary tube. Whenever aluminum tubing comes in contact with copper or iron, it must be protected against moisture to avoid electrolytic corrosion.

The *refrigerant* in the system is critically charged. This means that there is a definite quantity of refrigerant placed in the cycle to produce the best operating performance. Any charge more or less than the critical charge will reduce the capacity of the system. This charge is generally determined through factory testing.

Compressor capacities range from 300 Btu/hr to about 2000 Btu/hr when measured under the standard rating conditions by the manufacturer.

Standard Rating Conditions	
Evaporator	–10°F
Condenser	130°F
Ambient temperature	90°F
Suction gas superheated temperature	90°F
Liquid subcooled temperature	90°F

Various types of compressors are used, including reciprocating, rotary, and scroll. The rotary compressors are somewhat more compact and quieter than the reciprocating compressors. Scroll compressors offer an energy-saving feature. Starting torque is provided by the use of split-phase motors with a relay or permanent-split capacitor motors.

Temperature controls start and stop the condensing units. The location of the sensing bulb varies with different systems. For a simple gravity-cooled system, the sensing bulb of the thermostat is normally clamped to the evaporator. In a combination refrigerator-freezer with a split-air system, the sensing bulb is usually located to sense the temperature of the air leaving the evaporator. In a manual-damper-controlled system, the sensing bulb is placed in the cold air stream to the fresh food compartment. Where microprocessor control systems are used, thermistor sensing devices are often placed in both the refrigerator and freezer compartments. These systems provide a high degree of independent adjustment for the two main compartments.

ON THE JOB

Critically charged refrigerators and freezers require exceptional care when evacuating the system prior to adding refrigerant after repairs or any time the system has been accessed. This includes a good vacuum pump capable of pulling deep vacuums, and an electronic micron gauge to measure the depth of the vacuum. Industry standards and equipment manufacturers often indicate that 200–500 microns is the minimum evacuation level to maintain warranty agreements on replacement parts.

The refrigeration cycle for most domestic refrigerators and freezers have been completely *hermetically sealed.* This prevents outside contamination and produces a cycle that has long life and generally trouble-free operation. The cycle uses a hermetically sealed compressor, a capillary-tube metering device and welded refrigerant-line connections. The system is constructed, evacuated, charged with refrigerant and oil, and sealed at the factory. This type of construction gives the cycle an expected life of 15 to 20 years. Many run for over 30 years.

Due to the long expected life of the hermetically sealed units, very little provision has been made for servicing them in the field. The older hermetically sealed compressors have no access valves, and the only available provision for entering the cycle is through the sealed pinch-off tube. This is the access that the manufacturer used in originally evacuating and charging the system. On the newer compressors the manufacturer is required to provide access so that the refrigerant can be removed if the unit is scrapped. This access provision is useful if it is necessary to service the cycle.

Environmental Conservation

The new refrigerators use HFC-134a refrigerant or one having an ozone depletion factor (ODF) of zero. If a refrigerant leak should occur there should be no adverse effects to the ozone layer of the atmosphere. Older refrigerators use the CFC, R-12 refrigerant which has an ODF = 1 (in a refrigerant rating scale of 0–1) and is banned. Urethane foam insulation is no longer blown in place with CFC-11 refrigerant but has been replaced with HCFC-141b refrigerant or an equivalent environmentally safe one for this purpose. CFC-11 has an ODF = 1, whereas HCFC-141b has an ODF = 0.02.

DR22-2 DOMESTIC REFRIGERATORS

DR22-2.1 Mechanical Operation

A typical cycle diagram of a single-door domestic refrigerator with manual defrost is shown in Fig. DR22-3.

This is the simplest type of refrigerator. The evaporator is located in the upper inside of the cabinet. The compressor is below the refrigerator attached to the cabinet in an exposed area. The condenser is outside attached to the back of the cabinet. Heat is removed from the condenser by gravity air circulation.

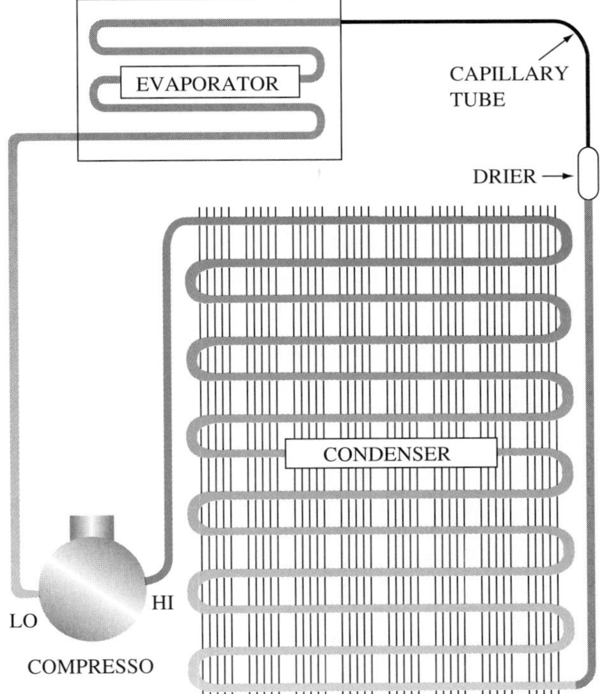

Figure DR22-3 Single-door refrigerator cycle diagram. (Courtesy of Frigidaire Company)

Referring to the diagram:

1. Low-pressure liquid refrigerant enters the *evaporator* and boils, absorbing heat. The accumulator separates out any liquid, permitting only low-pressure vapor refrigerant to enter the compressor.
2. The *compressor* accepts the low-pressure vapor and compresses the refrigerant to a high-pressure vapor and delivers it to the condenser.
3. In the *condenser*, the refrigerant is condensed into high-pressure liquid. The superheat is removed in the first rows and subcooling is done in the final rows of the condenser.

4. The liquid refrigerant flows through the liquid line, through the filter-drier and into the *capillary-tube metering device* connection.
5. The capillary tube carrying liquid refrigerant then enters the suction-to-liquid heat exchanger. The heat exchanger is constructed by brazing a portion of the capillary tube to the suction line. The heat exchanger has the effect of subcooling the liquid and superheating the suction gas, both of which improve the performance of the cycle.
6. The metering device reduces the pressure of the liquid to the evaporator pressure, and the cycle is repeated.

DR22-2.2 Manual-Defrost Systems

For the refrigerator-freezer combination units, some units have *manual defrost* and some have *automatic defrost.* Most of the automatic type use electric defrost; however, some use hot-gas defrost.

There are two methods of *manually* removing the ice accumulation on the evaporator:

1. Turn off the refrigerator at night and let the ice gradually melt and fall into a container provided.
2. Place a pan of hot water on the surface of the evaporator to quickly melt the ice and provide a container to catch the ice and water. Plastic tools can be used to assist in removing the ice accumulation. Metal tools should not be used since they may injure the refrigerant passages.

The electrical circuit for the manual-defrost single-door refrigerator is relatively simple, shown in Fig. DR22-4.

The connections to the compressor are indicated as "S" (start winding terminal), "R" (run winding terminal) and "C" (common terminal). The relay operates on the run winding current. When the compressor starts, the contacts on the relay are closed, allowing the compressor to start using the start winding. When the compressor reaches about ¾ speed, the relay switches the operation to the run winding. An overload is provided in the control circuit to protect the compressor.

The door switch turns on the interior light when the door is opened.

Fig. DR22-5 shows a refrigeration system for a two-door refrigerator with a top mounted freezer compartment, using manual defrost.

One different feature of this system is the use of an *oil cooler.* Note that the discharge gas from the compressor passes through a section of the condenser, then through a piping loop in the compressor to cool the oil before returning to the remainder of the condenser piping.

The lowest evaporator temperature is reached in the freezer compartment by cooling this section before passing the refrigerant into the refrigerator compartment. The final section of the evaporator is in the bottom of the refrigerator which is used for vegetable storage.

The wiring ladder diagram, (Fig. DR22-6) for the two-door refrigerator with a top-mounted freezer and manual defrost is similar to the single-door model; however, there are a few differences.

An ambient compensator for the thermostat is shown. This provides heat to the thermostat whenever the compressor is running. This causes the thermostat to be more sensitive to changes in box temperatures. A heater is installed in the door trim to prevent the door from freezing shut.

DR22-2.3 Automatic-Defrost Systems

Automatic-defrost systems are available for most box styles. The following two systems are representative.

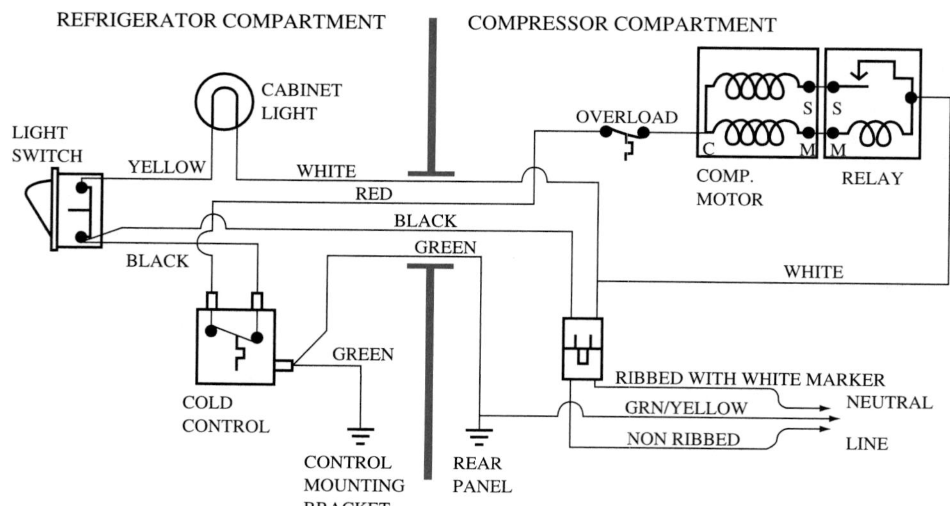

Figure DR22-4 Single-door refrigerator wiring diagram. (Courtesy of Frigidaire Company)

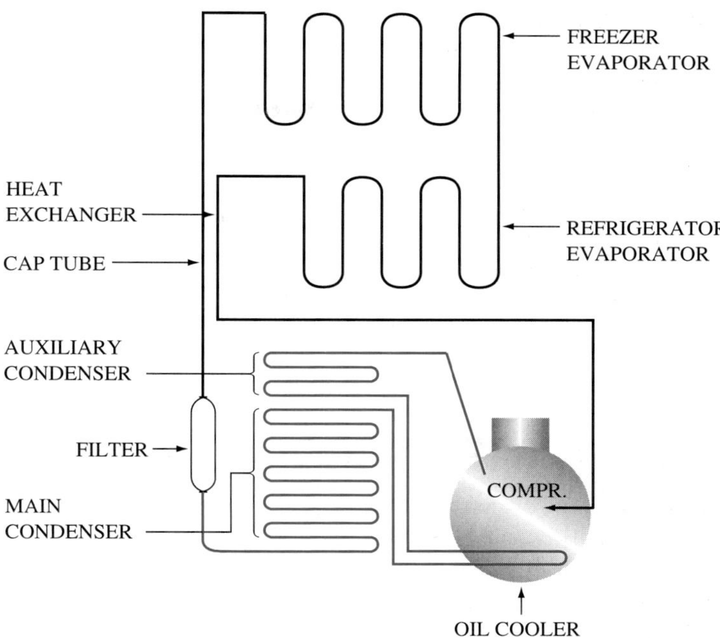

Figure DR22-5 Two-door top-mounted refrigerator cycle diagram.

For the two-door refrigerator with the top-mounted freezer, the diagrammatic illustration in Fig. DR22-7 shows the location of the components and the airflow within the cabinet.

The automatic defrost requires some differences in cabinet construction. Defrost water is collected in a condensate pan under the evaporator and is conducted through tubing provided in the cabinet to a heated drain pan in the compressor compartment. Heat is applied to evaporate the moisture and to allow it to pass into the room. The heat for evaporation is supplied from the hot-gas line of the compressor or by an electric resistance heater.

The *air system* uses an evaporator fan, which passes the air through the coil and circulates the cooled air in the freezer compartment. Ducts with adjustable dampers supply a controlled amount of cooled air into the refrigerator storage space, permitting return for re-cooling.

The evaporator is defrosted every 6 hours. The length of the defrost cycle is controlled by a termination thermostat attached to the evaporator, which shuts off the defrost heater at a box temperature of 50°F. The time clock initiates the defrost and limits the defrost time to about 30 minutes.

The thermostat in the freezer section maintains a temperature of about 0°F. The thermostat in the refrigerator regulates the dampers to maintain a temperature of 38 to 45°F.

The wiring for the system is shown in Fig. DR22-8. The operation of the compressor, the condenser, and evaporator fans are controlled by the freezer thermostat operating through the defrost timer. The ice maker, when used, has its own control. The door switch regulates the lighting. An energy-saving manual switch is provided to close off the *mullion* (door frame heater) and freezer flange heaters.

Another type of automatic defrost known as the "frost free" system is used in side-by-side refrigerators, as shown in the diagram, Fig. DR22-9 (p. 424).

The compressor, the condenser, and the evaporator are located in the back of the freezer compartment. A forced-air evaporator fan circulates air through the coil in the freezer section and down to the return. A duct with a con-

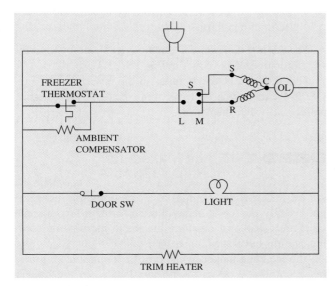

Figure DR22-6 Two-door refrigerator wiring diagram.

ENERGY SAVING 32-IN. TOP-MOUNT REFRIGERATOR AIRFLOW

Figure DR22-7 Air distribution in a two-door refrigerator with automatic defrost.
(Courtesy of Amana Refrigeration, Inc.)

trolled damper arrangement allows some of the cool air from the freezer section to flow into the refrigerator and down to the return near the bottom. The freezer compartment is maintained at 0°F or below, and the refrigerator between 28 and 45°F.

DR22-2.4 Electronic Control Modules

Some models have an electronic control console in the front above the ice dispenser, as shown in Fig. DR22-10.

The options at the control module are as follows:

1. Cubed or crushed ice selection.
2. Light switch for illumination of the control panel at night.
3. Door alarm which will operate whenever the freezer door is left open for 30 seconds or more.
4. Door-open monitor which will indicate when either door is left open.
5. Warm-temperature monitor which will indicate when the freezer temperature is above normal for an hour or more.
6. System monitor indicates status of the diagnostic system.
7. Diagnostic code which will flash if there is an abnormal condition.
8. System check/reset reviews diagnostic codes. Normal light will appear briefly if there are no abnormal conditions.

The wiring diagram for the frost-free side-by-side refrigerator-freezer is shown in Fig. DR22-11 (p. 425).

This diagram is very similar to the one shown in Fig. DR22-4 except this unit has more heaters, more lights, and a cold-water dispenser.

DR22-2.5 Ice Makers for Refrigerators

An ice maker is an essential part of a domestic refrigerator. Usually the ice is made in cube form, although some units are capable of crushing the ice to increase its cooling rate for beverages.

The ice-cube maker, whether it be a cube-making tray or an automatic device, is always located in the freezer portion of the refrigerator. The temperature in this area is usually held at 0 to 8°F. Time must be allowed for the freezing process for a single tray, usually 15 to 20 minutes

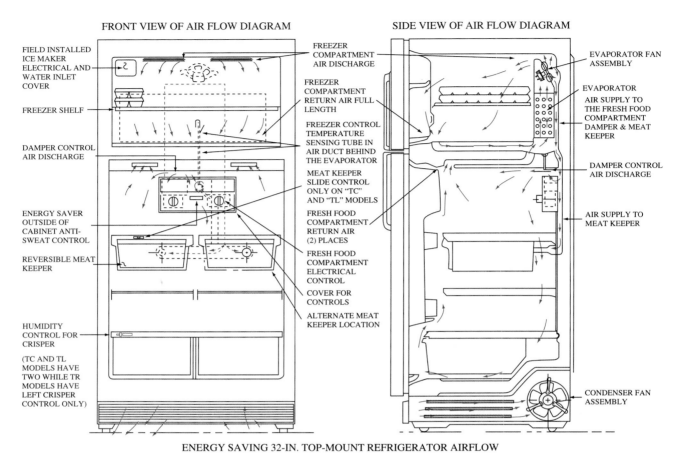

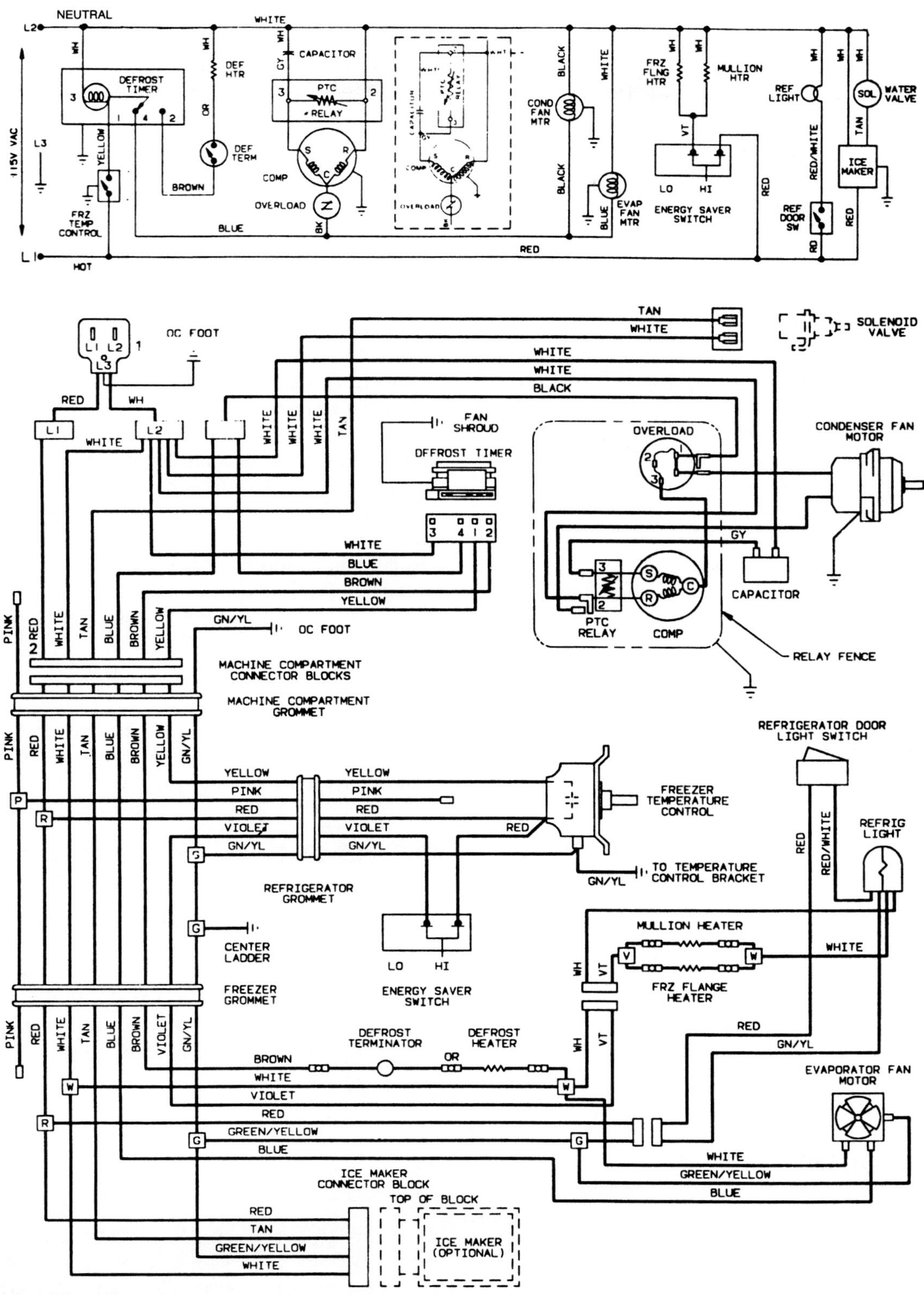

Figure DR22-8 Wiring diagram for a two-door refrigerator, with automatic defrost.
(Courtesy of Amana Refrigeration, Inc.)

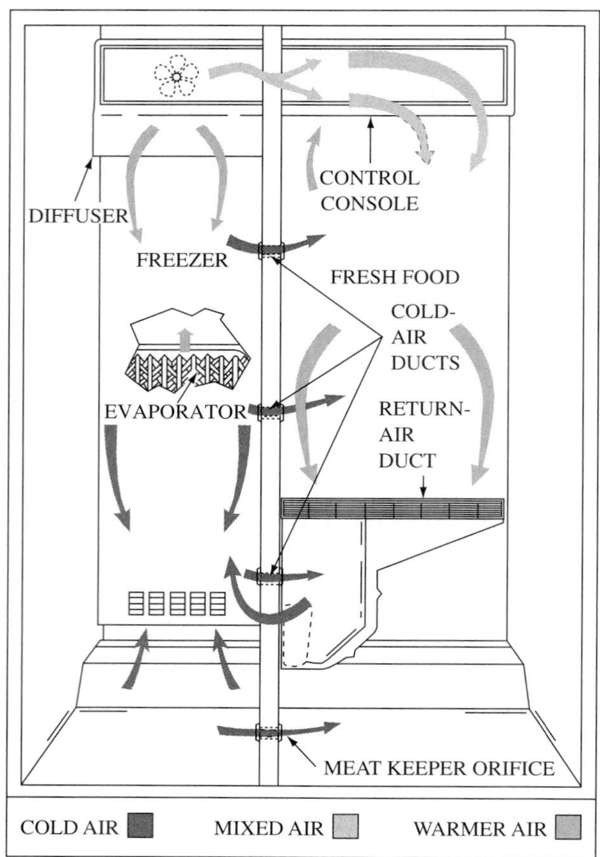

Figure DR22-9 Air distribution in a side-by-side refrigerator, frost-free. (Courtesy of GE Appliances)

from the time tap water is placed in the tray to the time the ice is ready to use.

Automatic Ice Makers

Automatic ice makers are of two general types, which vary mainly in their method of removing the ice from the freezing tray.

The *mechanical ice ejection* process is controlled by a timer and illustrated in Fig. DR22-12. A solenoid valve in the water line opens long enough to fill the tray with water. The timer controls the length of the freezing cycle. For harvesting the ice, the plastic tray is twisted and turned, dropping the ice in the bin. When the bin is full, an ice-level switch stops the process until there is a need for more ice.

An expanded view of this type of ice maker, using a solid-state control system, is shown in Fig. DR22-13 (p. 427). Referring to the diagram, "A" is the solid state circuit board, "B" is the cam gear that operates to release the ice during the harvest cycle, "C" is the plastic tray where the cubes are frozen, "D" is the temperature-sensing thermistor that stops the freezing when the ice temperature reaches 13°F and starts the harvest cycle, and "E" is the motor that twists and turns the tray to eject the ice.

In the *heat-operated ice-ejection* process, ice is formed in a metal tray with a bottom heater for loosening the ice for harvesting. A timer controls the amount of water to fill the tray and a temperature-sensing element is used to terminate the freezing cycle. For harvesting, the heater is activated and mechanical fingers push the ice from the trays into the bin, as shown in Fig. DR22-14 (p. 428). A bin-level control stops the process when the bin is full and restarts the process when there is a need for more ice.

Dispensers

Ice is collected in a bin located in the freezer compartment or allowed to flow to an external dispenser located on the front of the cabinet, as shown in Fig. DR22-15 (p. 428).

DR22-2.6 Servicing Refrigerators

The servicing of refrigerators has many similar procedures to that of freezers because they both contain small hermetically sealed refrigeration systems.

We will first discuss servicing domestic refrigerators, although much of the information also applies to freezers as well. In our discussion of freezer servicing, where similarities exist, the student may refer to the coverage under refrigerators.

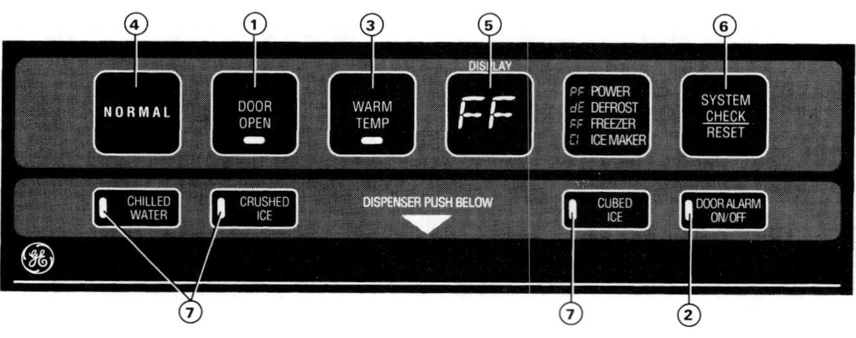

Figure DR22-10 Electronic control console. (Courtesy of GE Appliances)

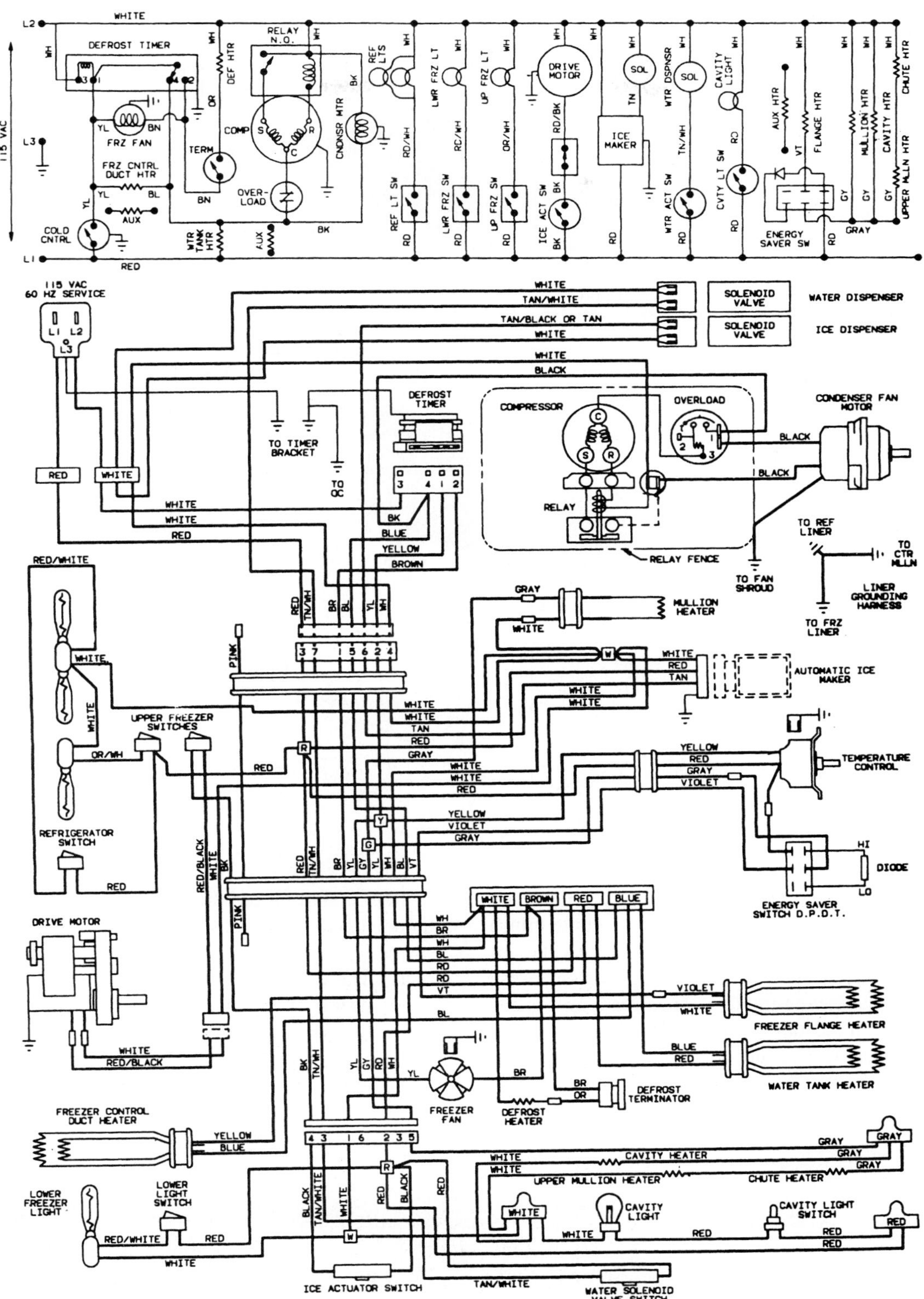

Figure DR22-11 Wiring diagram for a side-by-side refrigerator, frost free. (Courtesy of Amana Refrigeration, Inc.)

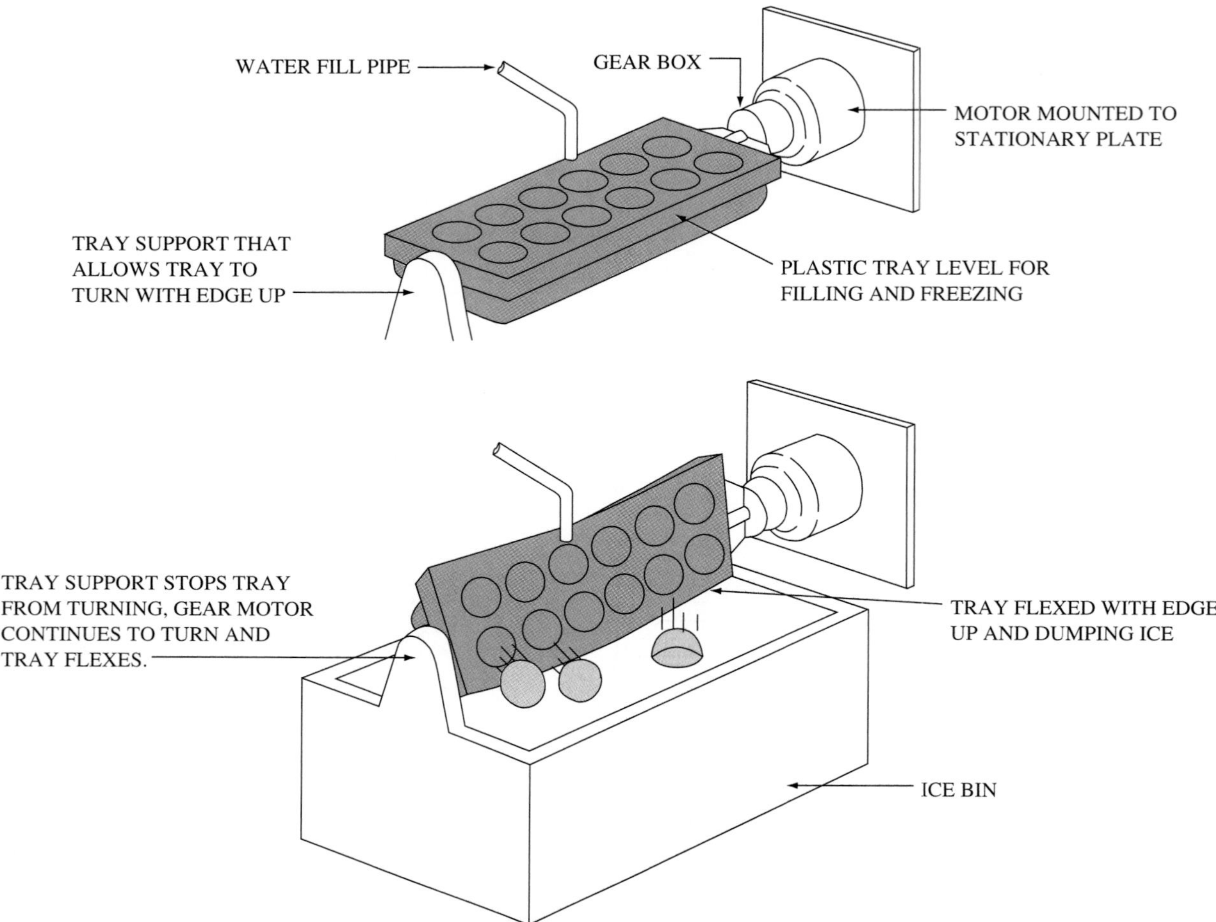

Figure DR22-12 Mechanical type ice maker.

Most of the common refrigerator service problems are external to the sealed system. They involve simple things like door alignment, torn door gaskets, and electrical component failures. They also may relate to customer use problems, such as leaving the door open for long periods of time on hot, humid days, causing condensation problems. In diagnosing refrigerator service problems, it is important to solve the external malfunctions first.

Before starting the diagnosis, a quick review of some fundamentals about refrigerators is as follows:

1. The main purpose of the refrigerator is to remove heat from the interior of the cabinet and deposit the heat outside the cabinet.
2. This is done by boiling the refrigerant in the evaporator. Liquid refrigerant is vaporized due to the

PREVENTIVE MAINTENANCE

Residential customers often neglect to do routine preventive maintenance on their household refrigerators and freezers. A method to increase business sales and service is to offer your customers annual scheduled preventive maintenance on their appliance. This can include cleaning the condenser of lint and dust, washing out the condensate drain, and verifying the defrost controls are working properly. Remember to always clean your work area prior to leaving the job site.

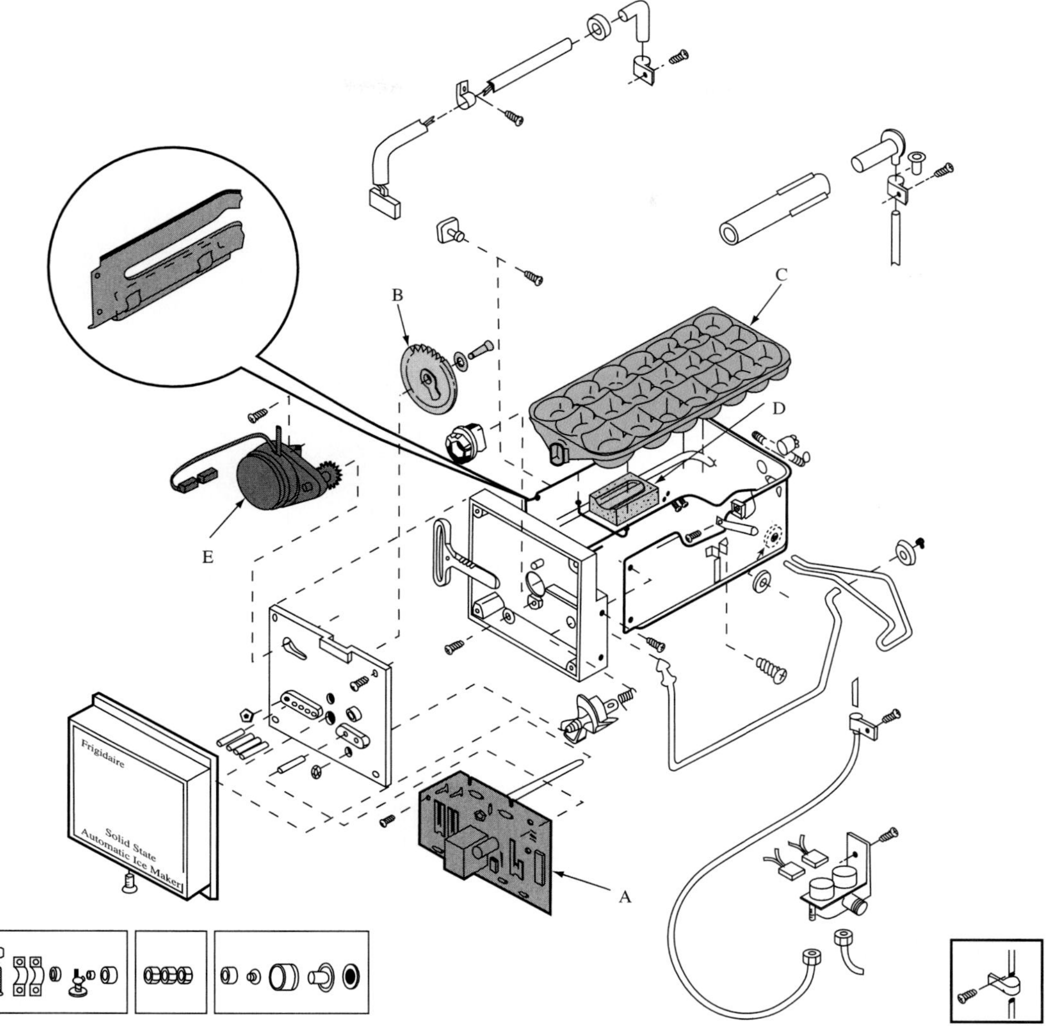

Figure DR22-13 Expanded view, automatic ice maker. (Courtesy of Frigidaire Company)

large reduction in pressure after passing through the small-diameter capillary tube into a larger-diameter evaporator tube.

3. The vaporizing of the refrigerant allows it to absorb a great amount of heat.
4. The compressor sucks the vapor from the evaporator, compressing it and raising its temperature.
5. This high-pressure, high-temperature gas is routed to the condenser where it exhausts its heat and returns back to liquid refrigerant to repeat the process.

Referring to the cycle diagram, Fig. DR22-3, a few accessories in the refrigeration cycle have special functions. These include the drier, heat exchanger, and accumulator.

The *drier* is added ahead of the capillary tube to remove moisture that could freeze and cause a restriction in the flow of refrigerant.

The *heat exchanger* increases the efficiency of the cycle. It is constructed by soldering the capillary tube to the suction line. The hot liquid refrigerant heats the suction gas to keep the suction line from sweating. The cool suction line reduces the temperature of the liquid so that it will pick up more heat in the evaporator.

The *accumulator* is located in the suction line ahead of the compressor. It holds liquid refrigerant and prevents it from entering and damaging the compressor.

DR22-2.7 Electrical Components

Electrical component failures should be eliminated prior to the analysis of the cycle operation. There are four basic electrical circuits for no-frost refrigerators:

1. The refrigeration cycle
2. The defrost mechanism

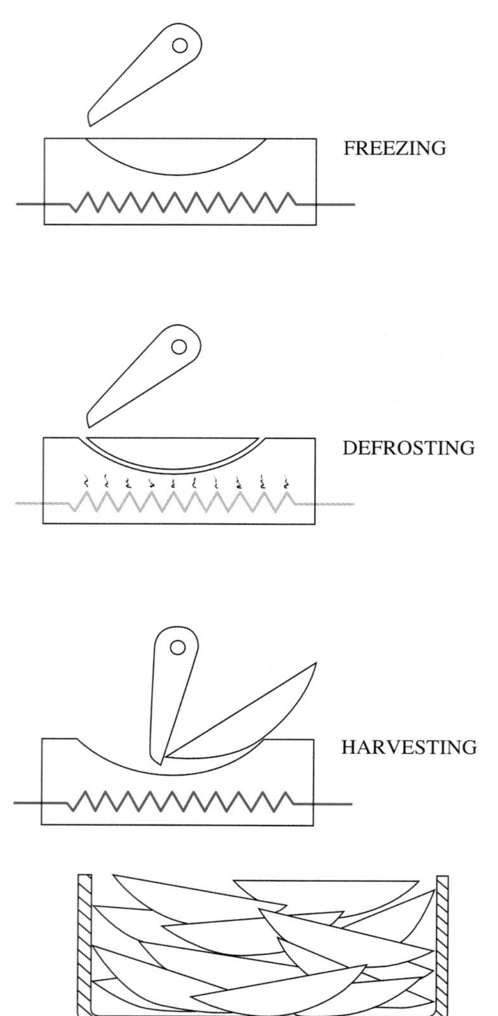

Figure DR22-14 Ice maker with heated metal tray.

Figure DR22-15 External ice dispenser. (Courtesy of Maytag)

3. The heater circuits that prevent sweating, located in the mullions and stiles
4. The icemaker equipment

Three diagrams of the internal mechanism of typical refrigerators are shown in Fig. DR22-16.

A *pictorial wiring diagram* (Fig. DR22-16a), shows the location of the wiring and wiring connections. Also shown is the wire color coding, indicating the color of the various wires and the color abbreviations that are indicated on the diagram.

A *schematic wiring diagram* (Fig. DR22-16b) makes it easy to trace the circuits and to understand the operation of the system.

A *refrigerant flow diagram* (Fig. DR22-16c) shows the principal components of the refrigeration system and the connecting piping.

Referring specifically to the schematic wiring diagram, Fig. DR22-16b, note that all electrical components are grounded to the cabinet. The green power cord is attached to the cabinet and runs to the electrical outlet ground connection. It is important to check the wiring to be certain a good ground is provided.

The various electrical devices in the system can be classified into two groups: *loads* and *switches.* The loads consume power and perform work, and include the compressor relay, compressor, condenser fan motor, evaporator fan motor, defrost timer motor, and defrost heater. The switches turn the loads on and off as required by the control system and include the defrost timer, thermostat, compressor overload, compressor relay, and defrost termination bimetal.

Compressor Circuit

The compressor circuit with the PTC relay (Fig. DR22-17) is shown in Fig. DR22-18. At the instant of start the resistance increases to the switch point. At this point the PTC relay no longer conducts electricity and allows the run capacitor to function in the circuit. Thus the motor runs as a permanent-split capacitor motor, which is very efficient.

The PTC relay is tested by the process of elimination. If all the other components of the circuit are satisfactory, and the compressor does not run when power is applied, the problem must be in the relay.

Normal Running

In the normal running condition, when the thermostat is calling for cooling, power passes through the timer switch

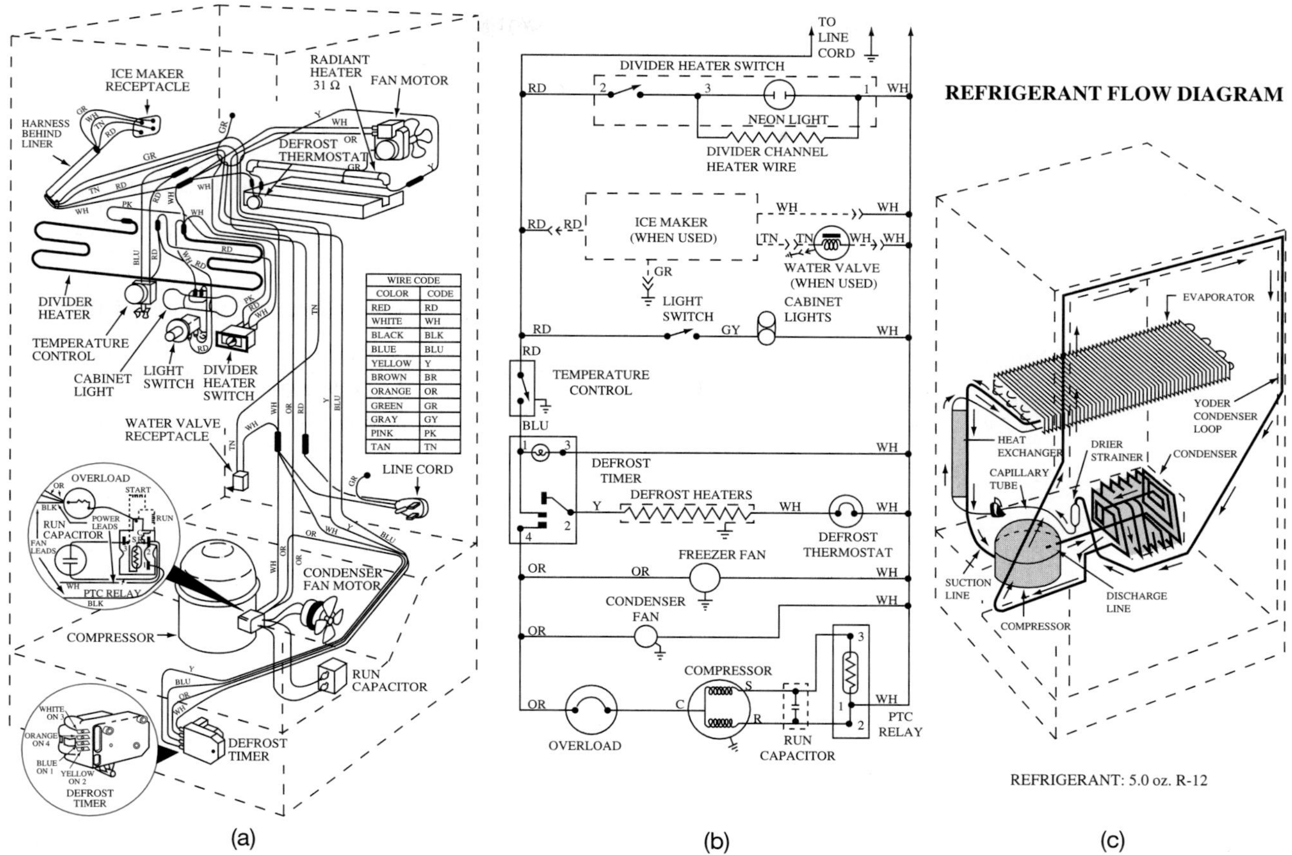

Figure DR22-16 Wiring diagrams and refrigerant flow schematic for a no-frost refrigerator. (Courtesy of Maytag)

operating the compressor, condenser fan, and freezer fan. The freezer fan circulates air throughout the fresh-food and freezer compartment. The defrost heaters do not operate because the defrost timer switch (1–2) is open (Fig. DR22-16b).

Defrosting

Again referring to the compressor circuit (Fig. DR22-16b), after 8 hours of normal operation the timer opens contacts 1–4 and closes contacts 1–2. This turns off the compressor, the condenser fan, and the freezer fan, and activates the defrost heaters. Once the temperature of the defrost termination thermostat reaches 47°F (the cut out point) the termination thermostat will open the circuit to the radiant heater; however, the compressor circuit remains open for the duration of the defrost interval. The timer returns its switch to the normal running position after a 23-minute defrost cycle.

If both fans are operating and the defrost circuit shows continuity, but the system is not defrosting, the problem is the defrost timer.

DR22-2.8 Diagnosing Problems in the Refrigerant Cycle

In diagnosing problems in the operation of the refrigeration cycle, it is important to have a clear understanding of how the unit should operate under normal conditions. From this information as a base, any deviations can be measured and compared. The types of failure generally fall into one of the types listed below:

1. Refrigerant leak or low charge.
2. Partial restriction in the refrigerant circuit.
3. Complete restriction in the refrigerant circuit.
4. An overcharge of refrigerant.

Before deciding that there is a refrigerant-cycle problem, be certain that there are no external problems. For example, there may be restrictions inside the cabinet preventing proper air flow, or the condenser coil may be dirty. Carefully check for external malfunctions before examining the refrigeration cycle.

The manufacturer will usually supply performance data for the equipment under normal operating conditions. If

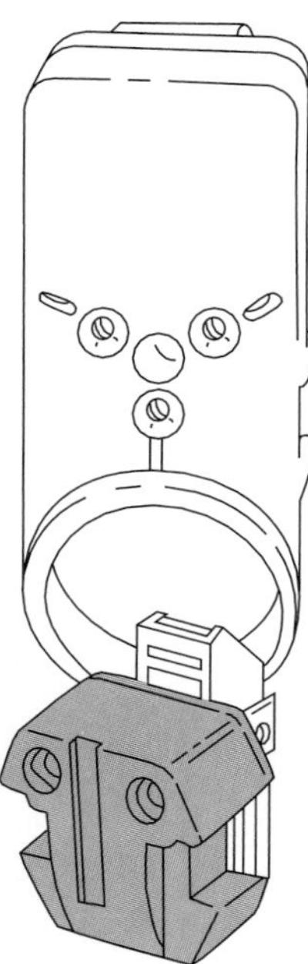

Figure DR22-17 PTC compressor relay. (Courtesy of Maytag)

not, the information can be recorded by examining and testing a properly operating unit. A discussion of the various failure types follows.

Refrigerant Leak or Low Charge

Under this condition there is less refrigerant in the system to absorb heat. Probably the most noticeable effect is the rise in the fresh-food storage temperature. The freezer storage temperature may not be affected, depending on the amount of refrigerant lost. The condenser inlet temperature drops considerably and the amperage slightly. The frost pattern is "partial."

Partial Restriction in the Refrigerant Circuit

This could be caused by a partially kinked or mashed tube, partially blocked tubing or a partially blocked drier screen. This condition means that the movement of the refrigerant in the system has been slowed down, causing the fresh-food storage temperature to rise to an unsatisfactory level. The freezer storage temperature is also higher than normal. The condenser temperatures drop and there is no difference between the inlet and outlet temperatures of the condenser. The amperage is low and the low side of the system is operating under a deep vacuum.

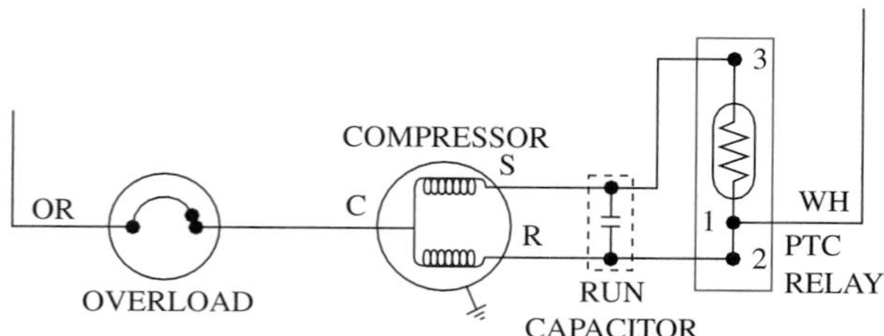

Figure DR22-18 Compressor starting circuit. (Courtesy of Maytag)

Look for a restriction, particularly at the filter-drier. The location of the restriction should be evident by a drop in temperature across the restriction.

Complete Restriction in the Refrigerant Circuit

The compressor runs, but there is no refrigerant flowing. The temperatures are unsatisfactory in both the freezer and the fresh-food storage. There is no frost pattern on the evaporator. The amperage drops and the compressor is operating in its lowest vacuum condition.

The location of the restriction should be evident. If the system has been opened and there is a chance that moisture is in the system, ice may have formed, blocking the metering device.

Overcharge of Refrigerant

An overcharge could be caused by improper recharging during service. It is recommended that if a charge is lost, the system be completely evacuated and accurately recharged with the kind and amount of refrigerant specified by the manufacturer.

A good indicator for detecting an overcharge is sweat or frost on the suction line. The high side pressure and amperage draw are excessive.

DR22-3 INSTALLING AND SERVICING ICE MAKERS

DR22-3.1 Installing

A number of manufacturers supply a packaged automatic ice maker that can be installed in the field (by the dealer). An illustration of one of these optional units is shown in Fig. DR22-19. Refrigerators that accept these units are provided with a removable panel and an opening for inserting the ice maker.

The location of one of these panels is shown in Fig. DR22-7 in the upper left hand corner of the freezing compartment. The electrical and water inlet connections are on the cover.

A wiring diagram showing how the ice maker and water solenoid valve are connected in the electrical circuit is shown in Fig. DR22-16b.

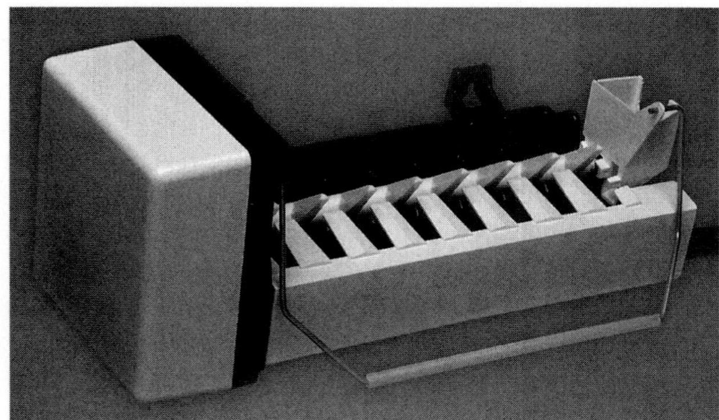

Figure DR22-19 Field-installed automatic ice maker. (Courtesy of Maytag Customer Service)

DR22-3.2 Servicing

For the *mechanical ejection unit,* such as shown in Figs. DR22-12, DR22-13 and DR22-19, these packaged systems can be removed if defective and exchanged for a rebuilt or new unit. Usually this is a lower cost solution for custom rather than field repair. If repair is selected, parts can be separated and examined, as shown in the expanded view. New parts can be ordered to replace defective ones. The unit can be re-assembled and put back into service.

For the *heat-operated ejector units,* since these have mainly electrical components, the following test procedure can be used:

1. Examine the cycle sequence for the blade operation. It should follow the sequence shown in Fig. DR22-20.
   a. Water valve is energized for 7.5 seconds
   b. Ejector blade is in stop position at about 1:30 o'clock.
   c. Ejector stalls on ice (½ to 5 minutes)

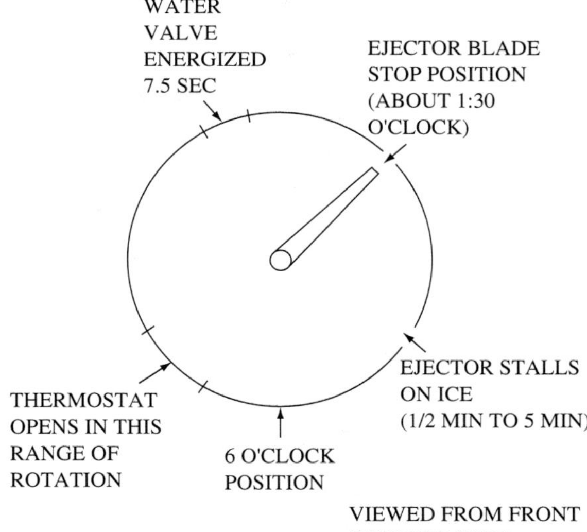

WHAT HAPPENS DURING BLADE ROTATION
(DO NOT TEST IF BLADES ARE PAST REST POSITION)

Figure DR22-20 Ice maker cycle sequence.

   d. Thermostat opens, after ice is ejected, at about 7 o'clock.
2. Using the wiring diagram shown in Fig. DR22-21, and the test points shown in Fig. DR22-22, use a jumper wire and perform the following tests (ice maker should be plugged into power, shut off arm should be down, and freezer cold):
   a. Test points L and N. Verify that 120 V is available to the ice maker module.
   b. Test points T and H. Verify the thermostat is open or closed.
   c. Short T and H with an insulated piece of wire to run the motor. If the motor does not run, re-

Figure DR22-21 Schematic wiring diagram of a heat release ice maker.

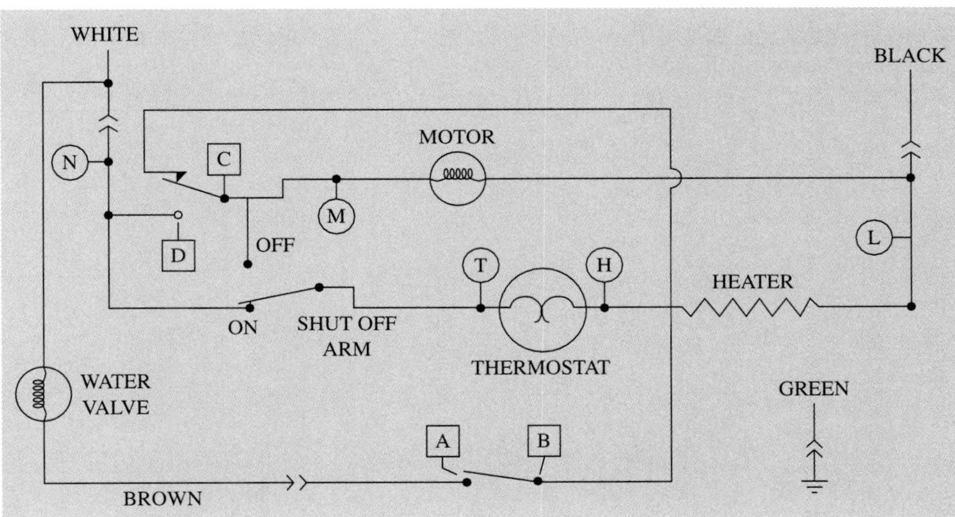

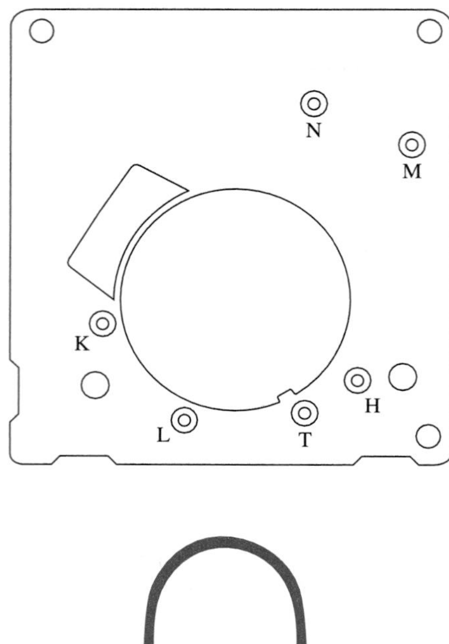
N
M
K
L
T
H

Figure DR22-22 Test points for an ice maker.

place the module assembly. If the motor runs, replace the bimetal thermostat. The thermostat can be replaced by removing two retaining clips.

d. To test the heater, leave the jumper in for ½ a revolution. If the heater is good it will heat up.
e. To test the water valve, remove the jumper and the water valve will be energized.
f. To adjust the water level, center the hole indicated in Fig. DR22-23, in the larger hole. The fill time can be adjusted for 7.5 seconds, by turning the adjustment screw.

DR22-4 DOMESTIC FREEZERS

DR22-4.1 Chest Freezers

Chest freezers are more efficient than upright types since most of the cold air stays in the cabinet when it is opened; however, the chest occupies more floor space than the upright model and is less convenient to use.

The refrigeration cycle for a chest freezer is shown in Fig. DR22-24.

The evaporator consists of piping built into the inner surfaces of the outside cabinet (#10, Fig. DR22-24). This produces a plate-type freezing arrangement, supplying even cooling for the freezing and storage areas.

The condenser also consists of continuous piping attached to the inside surface of the outside cabinet (#7). This prevents the exterior cabinet from sweating.

The compressor is located in an exposed compartment in the bottom corner of the cabinet (#1). The refrigerant circuit is completely hermetically sealed using a capillary-tube (#9) metering device. A control well (#15) is used for inserting the thermostat sensor tube.

The compressor uses an oil cooler. Discharge gas from the compressor is first cooled in a section of the air-cooled condenser (#3), then routed through a loop in the crankcase of the compressor (#4), then out (#5) to the remainder of the condenser tubing.

The space between the inner and outer cabinet is filled with urethane foam or fiberglass. A nonmetallic breaker strip is provided between the inner and outer cabinet to prevent conduction between the two surfaces. It is important that there be no openings in the outer cabinet that could allow moisture to enter and freeze in the insulation.

The wiring diagram for a chest freezer is shown in Fig. DR22-25.

Control devices provide the following operational and safety functions:

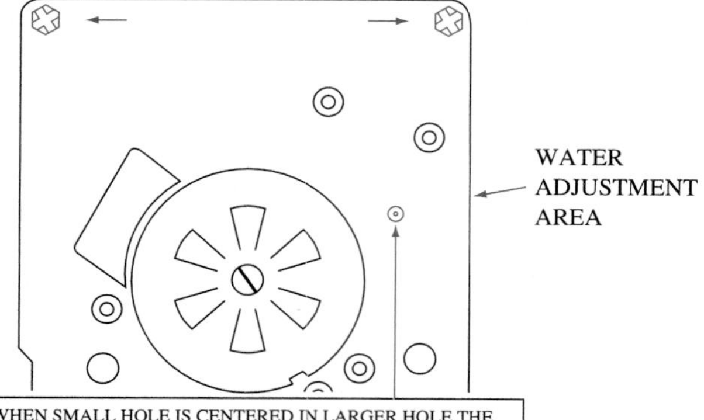
TURNING THE SCREW CLOCKWISE DECREASES WATER FILL
*1/2 TURN EQUALS 20CC (.67 OZ)
*1 TURN EQUALS 40CC (1.2 OZ)
WATER ADJUSTMENT AREA
WHEN SMALL HOLE IS CENTERED IN LARGER HOLE THE WATER FILL ADJUSTMENT IS FOR 7.5 SECONDS FILL TIME (NORMAL).

Figure DR22-23 Water level adjustment.

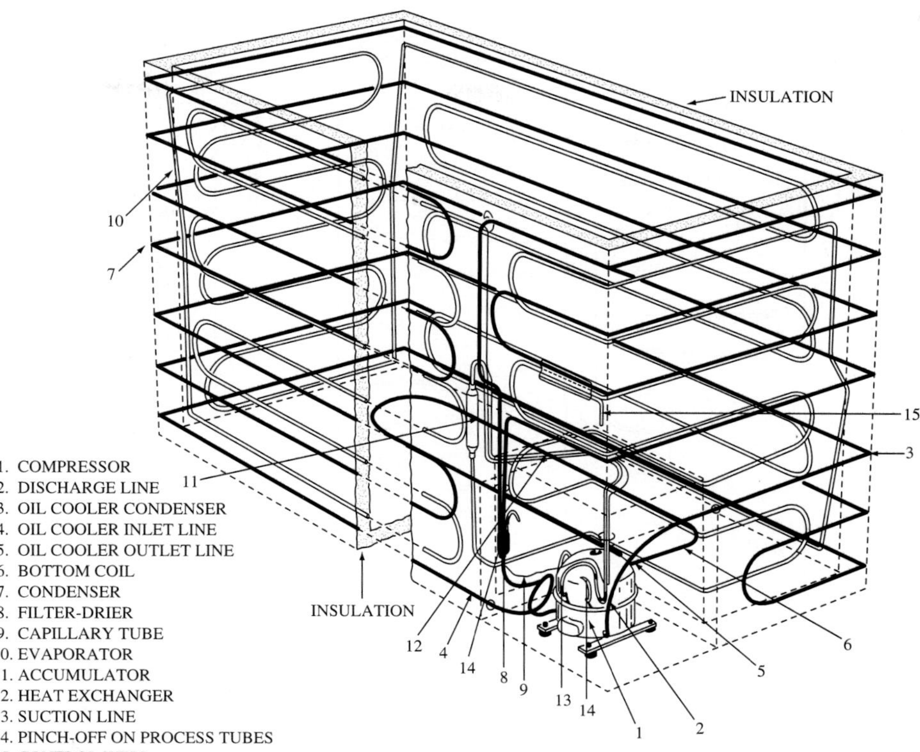

Figure DR22-24 Refrigeration cycle for a chest freezer. (Courtesy of Frigidaire Company)

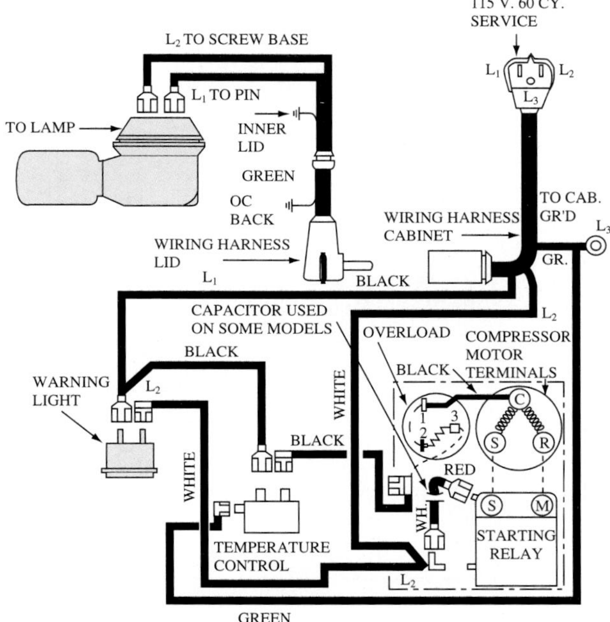

Figure DR22-25 Wiring for a chest freezer. (Courtesy of Amana Refrigeration, Inc.)

The thermostat automatically turns the compressor on and off.

The compressor uses a relay (to switch from the start winding to the run winding) and an overload protector.

The interior lamp operates through a door switch.

Some units are equipped with a warning light to indicate whether the power is on or off.

Another option is an alarm for signalling when the proper cabinet temperature is not being maintained. The box temperature should be between 0° and −10°F.

Chest freezers are manually defrosted. The procedure is to disconnect the power, remove the products from the cabinet, place pails of hot water in the storage compartment, and close the door. This should be done once or twice a year, depending on the frost buildup. Cabinets are provided with drain connections for removing the condensate.

DR22-4.2 Upright Freezers

A cycle diagram for an upright domestic freezer, with manual defrost, is shown in Fig. DR22-26.

The evaporator (located inside the cabinet), and the compressor (in an exposed compartment), are shown in the bottom of the cabinet. A fan draws air from the lower part of the cabinet, forces it through the evaporator, and discharges it at the top of the refrigerated space.

The condenser is attached to the inside surface of the outside cabinet. Heat is released from the exterior surface, which also prevents sweating. In locating the unit, space should be provided between the cabinet and surrounding walls to permit air movement.

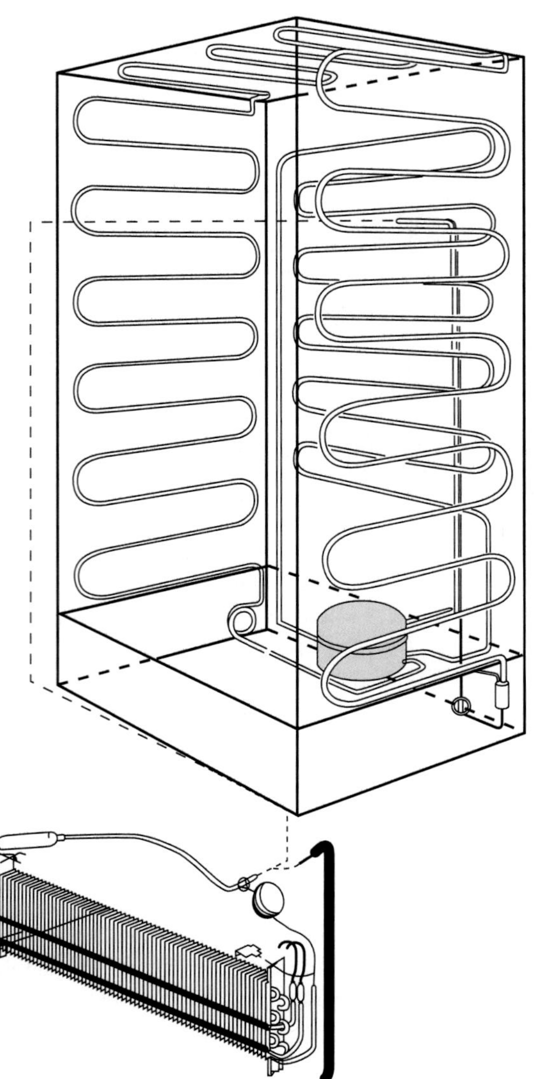

Figure DR22-26 Refrigeration cycle for an upright freezer with manual defrost. (Courtesy of Frigidaire Company)

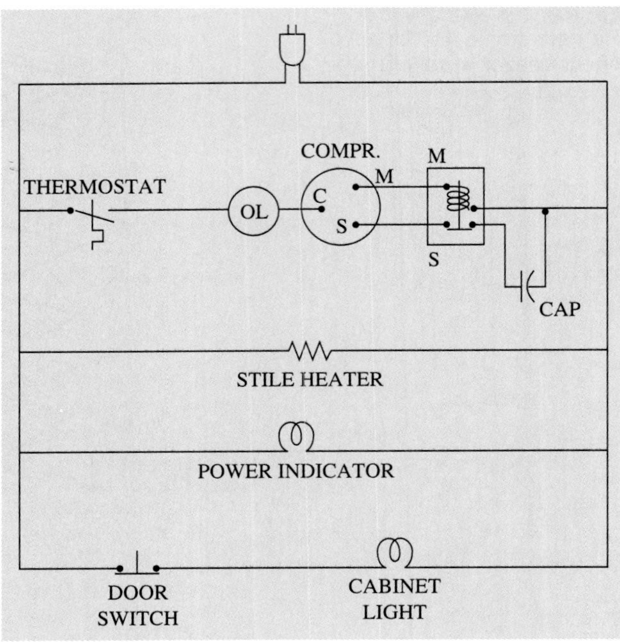

Figure DR22-27 Wiring diagram for an upright freezer, manual defrost.

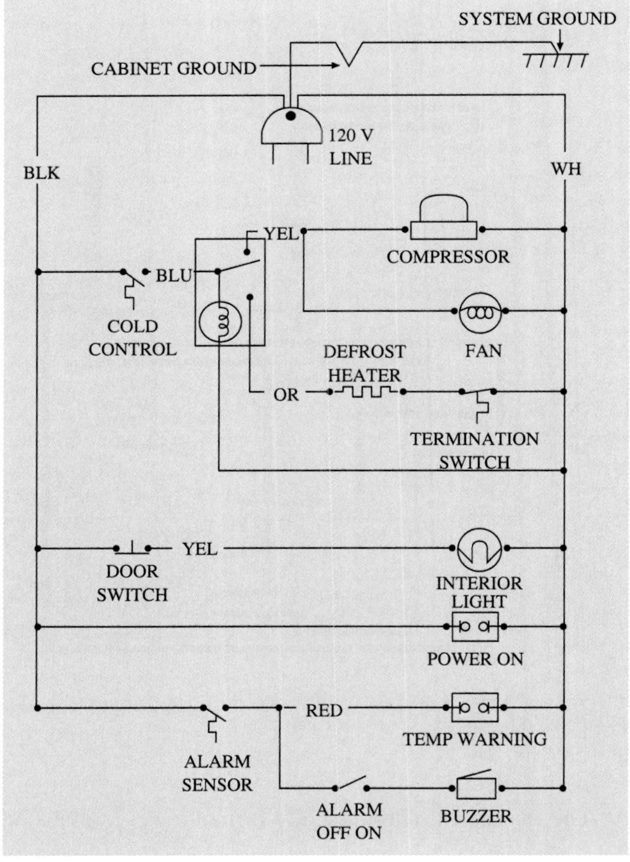

Figure DR22-28 Wiring diagram for an upright freezer, automatic defrost. (Courtesy of Frigidaire Company)

A ladder wiring diagram for the manual defrost vertical freezer is shown in Fig. DR22-27.

Figure DR22-27 shows a thermostat-operated compressor. A stile heater (door-opening heater) and a signal light are also energized when the thermostat is calling for cooling.

Upright freezers with automatic defrost use a finned tube condenser located in the compressor compartment. A condenser fan draws air in through a supply grille and out through a return grille located in the lower front of the cabinet. The evaporator is located behind a baffle with an evaporator fan to circulate cold air throughout the cabinet. Automatic defrost is provided using an electric-resistance heater and wired as shown in Fig. DR22-28.

DR22-4.3 Servicing Freezers

Many of the techniques used in servicing refrigerators also apply to freezers. The one principal difference is that the freezer has only a low-temperature storage compartment. Servicing will depend on the type of freezer, with the chest and the upright models being the two main types. Each of these models can have certain variations that must be carefully examined.

The following conditions may cause problems.

Location

The freezer must be located in a space that has normal temperatures. If the area is too cold or too hot it will not function properly. For example, some people place a chest freezer in an unheated garage or on a back porch where the sun shines on it a large portion of the day. Then too, the freezer must be located away from obstructions that would block the airflow over the condenser coil. Since all home freezers have air-cooled condensers, heat that is picked up inside the box must be transferred to the surrounding air.

Refer to the manufacturer's information for the clearances required around the cabinet. An upright freezer usually needs 3 in. on each side and back, and 4 in. at the top. A chest freezer usually needs 3 in. on each side and 1 in. in back.

Use

The home freezer is designed for the storage of frozen food. It will freeze a limited quantity of food, but basically it is built to accept food that has already been frozen and properly wrapped in a moisture-tight package. For example, there have been cases where a user has placed a side of beef in a chest freezer and has not been able to freeze it in a couple of days. This action is not conducive to good practice in preserving food. For best results, food should be *quickly* frozen to retain its flavor and consistency.

Power Supply

For the equipment to run properly, the freezer must be supplied with the power voltage as listed on the nameplate of the unit. Most domestic freezers require a power supply of 115 V. They are usually designed to run on voltages within 10% of the design rating, which means from 103.5 to 126.5 V. Obviously, there will be problems if the available voltage is only 95 V. These low voltages can occur particularly in rural areas where feeder lines are inadequate and usually at a time of high power requirements.

The other condition that can produce low voltage is the situation where the circuit being used for the freezer is overloaded by addition of other appliances. The technician should check the voltage at the freezer during a high-usage period when all other appliances on the circuit are in operation.

Defrosting

All chest and many upright freezers require manual defrosting. Any sizable frost or ice accumulation on the evaporator surface reduces the efficiency of the system. Usually the upright freezers require defrosting more often than the chest type, due to the fact that they lose cool air when the door is opened, which is replaced with moist room air.

Frozen food must be removed from the cabinet during defrosting and stored in an insulated bag or box. Usually the defrosting can be completed before the food reaches the melting stage and returned to the freezer without any loss.

Pans of hot water can be placed on the shelves of the upright freezer and in the bottom of the chest type to speed up the defrosting process.

In a number of instances, homeowners have used sharp metal tools to help remove the ice which has punctured the evaporator and caused a refrigerant leak. If a tool is used it should be of plastic material to prevent damage to the evaporator.

Odors

If the unit has been out of operation for some time and there is food spoilage, odors may have permeated the cabinet insulation and cannot be removed by cleaning the interior. Odor-absorbing materials such as charcoal, ground coffee, or baking soda may be placed in a dish with the unit operating and the doors closed. As a last resort, if the odor cannot be removed, the unit may need to be replaced.

DR22-4.4 Storage of Frozen Food During Service

If any major amounts of service need to be performed, it is usually best to move the unit to the shop where complete facilities are available. If the food is still in good condition, it may be necessary to move it into temporary storage. Here are some of the arrangements that have been used:

1. The dealer has a loan freezer that can be moved to the jobsite.
2. The food is moved to a storage locker or cold storage plant.
3. If the freezer is a chest type and external service is required, dry ice can be placed in the cabinet and work performed on the job. Dry ice should not

come in direct contact with the food. The dry ice can be placed in a layer at the top of the cabinet over a protective layer of paper. Twenty pounds of dry ice should be used for every 5 cu ft of box space to keep the food frozen for 24 hours.

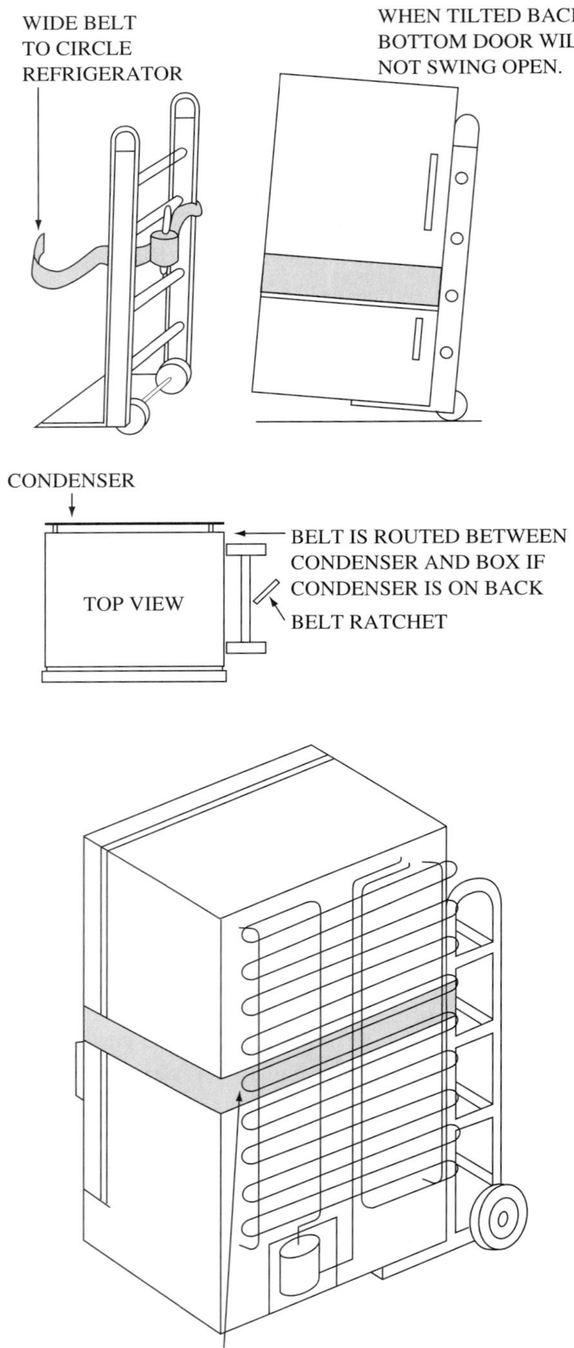

Figure DR22-29 Handtruck for moving refrigerators and upright freezers.

DR22-4.5 Moving the Freezer

A number of service items can be performed at the customer's location without removing the food, provided that the box can be moved to allow access to the working parts. Moving the cabinet must be done without damage to the freezer and without physical strain to the technician. Loaded cabinets can weigh as much as 1000 lb, and therefore often require two workmen using the proper trucks and/or dollies to perform the task.

In moving an upright freezer it is good practice to use a hand truck with a wide holding belt, such as shown in Fig. DR22-29. The base of the truck slips under the side of the cabinet, so that it does not damage the condenser. The belt is placed around the cabinet and under the gravity condenser. It secures the truck to the box and holds the doors shut. The truck is tilted back to transport the freezer.

The chest freezer requires a different technique due to its shape and construction. The base lip of the truck can be used as a pry bar to raise one end at a time (Fig.

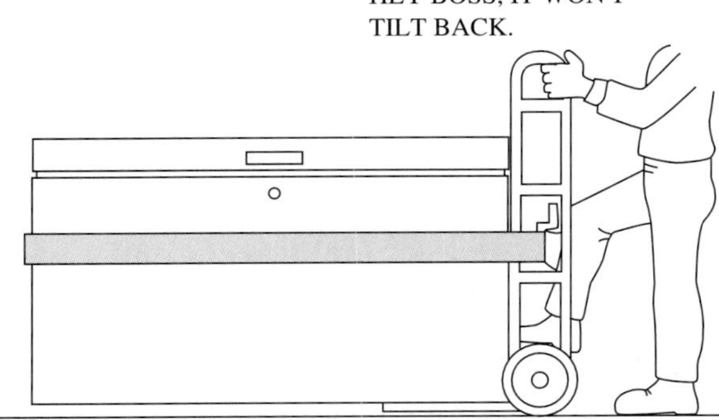

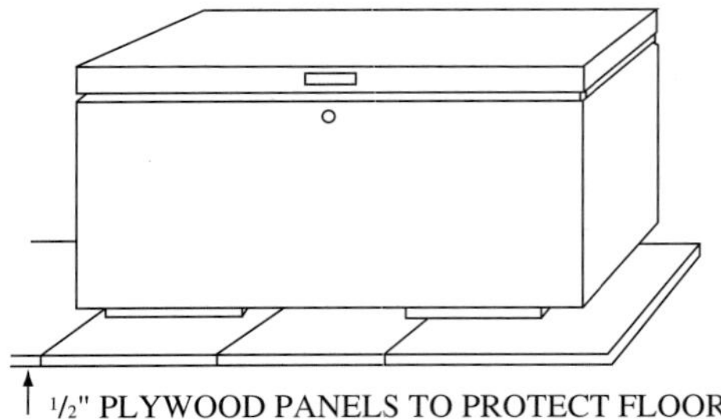

Figure DR22-30 Moving a chest freezer.

PRACTICING PROFESSIONAL SERVICES

Servicing refrigerators and freezers within customer's homes often requires moving the equipment away from the wall to access the compressor in the back. Caution must be exercised prior to sliding the refrigerator out onto a finished floor to avoid scratching or damaging the customer's floor. First inspect the base of the equipment; it may have been provided with wheels already. If wheels are not provided, then lay down cardboard or a similar type of material to protect against scratching the floor. When ready to move the equipment, slide it out slowly, one corner a few inches at a time, until it is moved enough to access the back service area. After the service work is complete, re-install the refrigerator by sliding it into the opening slowly.

If you are going to be servicing residential refrigerators and freezers on a regular basis, then invest in equipment specifically designed to move heavy appliances without damaging customer's floors. This equipment is costly, but litigation costs in court cases from damaging customer's floors can be avoided.

DR22-30). On a smooth floor, a small chest can often be moved by inserting a rug under the feet and pushing the cabinet, being careful not to damage the floor. For larger chests a four-wheel dolly is needed.

REVIEW

- Common types of types of domestic refrigerators:
  Single door
  Top-mounted freezer combination
  Bottom-mounted freezer combination
  Side-by-side combination
- Common types of domestic freezers:
  Chest
  Upright
- Classification by defrost cycle:
  Manual defrost
  Cycle defrost
  Frost-free (automatic defrost)
- Most full-sized refrigerators and freezers are vapor compression refrigeration cycle. Some smaller types use absorption and thermoelectric cycles. The absorption cycle is commonly used for recreation vehicles. The thermoelectric units are commonly used for small coolers in hotels and hospitals.
- Refrigerator temperatures:
  Fresh food compartment—35 to 45°F
  Freezer— −5 to +5°F
  Short term freezer—below 8°F
- To achieve energy conservation standards, refrigerators and freezers use:
  A more efficient compressor
  Better insulation
  A more efficient fan motor
  A trap in the drain tube from evaporator
  More efficient defrosting methods
- Typical cycle components and ratings:
  Evaporators— −10°F
  Condensers—105 to 130°F
  Ambient design temperature—90°F
  Liquid—subcooled to 90°F
  Accumulator—located at outlet of evaporator in suction line
  Metering device—capillary tube
  Heat exchanger—capillary tube soldered to suction line
  Refrigerant—R-12, R134a
  Compressors—300 Btu/hr to 2000 Btu/hr
- Common service problems:
  Door alignment/swing change
  Torn door gaskets
  Defrost problems
  Electrical component failures
- Four basic electrical circuits:
  Refrigeration cycle
  Defrost
  Heater/lights
  Ice makers
- Sealed system problems:
  Refrigerant leak or low charge
  Partial restriction
  Complete restriction
  Overcharge of refrigerant
- Compressor problems:
  Poor pumping capacity
  Stuck compressor
  Electrical problem with compressor
- Automatic ice maker types:
  Mechanical ice ejection
  Heat operated ice ejection

- Ice dispensers—Ice is collected in bin, then augered or allowed to flow to an external dispenser located in front of the refrigerator.
- Potential freezer problems and conditions:
  Location
  Use and application
  Power supply
  Odors/smells

Problems and Questions

1. What are typical temperatures in the fresh-food portion and freezer portion of a household refrigerator?
2. List the types of domestic refrigerators and freezers.
3. Who publishes the regulations for energy conservation and efficiency in the design of domestic refrigerators and freezers? Who establishes standards?
4. What makes today's refrigerators and freezers more energy efficient?
5. What is the purpose of the mullion and stile heaters?
6. What is the purpose of a high-humidity compartment?
7. At what pressure do safety regulations require that doors can be opened from the inside?
8. How are refrigerator and freezer cabinets leveled?
9. What periodic maintenance should the homeowner do?
10. What are the components used in the defrost cycle?
11. What does a "critical charge" mean?
12. What is the purpose of the accumulator?
13. Why do drain pans have heaters?
14. On top-mounted, frost-free refrigerators, how is cooling provided in the fresh-food compartment?
15. Some newer refrigerators and freezers have a continuous loop off the condenser that is attached to the inside surface of the outside cabinet. What is the purpose of this loop, called a "yoder" loop?
16. How often should chest freezers be defrosted?
17. Why is defrosting a freezer evaporator important?
18. What kind of problems would the loss of a condenser fan for a freezer produce?
19. What kind of problems would the loss of an evaporator fan cause?
20. Why are "breaker" strips made of plastic?
21. In the refrigerator portion of the cabinet, it is usually considered good practice to maintain temperatures between 40 and 50°F. True or false?
22. Newer, more efficient refrigerators use shaded-pole fan motors. True or false?
23. A trap is placed in the drain tube from the evaporator to prevent drain water from going back into the refrigerator. True or false?
24. A high-humidity compartment is provided in most cabinets for the storage of butter and related dairy products. True or false?
25. A large part of the refrigeration load comes from the door being opened. True or false?
26. The upright freezer uses:
    a. A low-temperature evaporator.
    b. A circulating evaporator fan.
    c. Manual or automatic defrosting.
    d. All of the above.
27. When a refrigerator or freezer has a critical charge of refrigerant, it:
    a. Means that there is a definite quantity of refrigerant placed in the cycle.
    b. Means that any charge more or less than the critical charge will reduce the capacity of the system.
    c. Is generally determined through factory testing.
    d. All of the above.
28. New refrigerators use ____ refrigerants.
    a. R-134a
    b. High-ODF
    c. R-12
    d. None of the above.
29. The suction-to-liquid heat exchanger:
    a. Subcools the liquid
    b. Superheats the suction gas
    c. Improves performance of the cycle.
    d. None of the above.
30. The freezer location:
    a. Must be in a space that has normal inside ambient temperatures.
    b. Can be located outside in the garage.
    c. Can be in the kitchen with custom cabinets built around it.
    d. None of the above

SECTION 5

Commercial Refrigeration

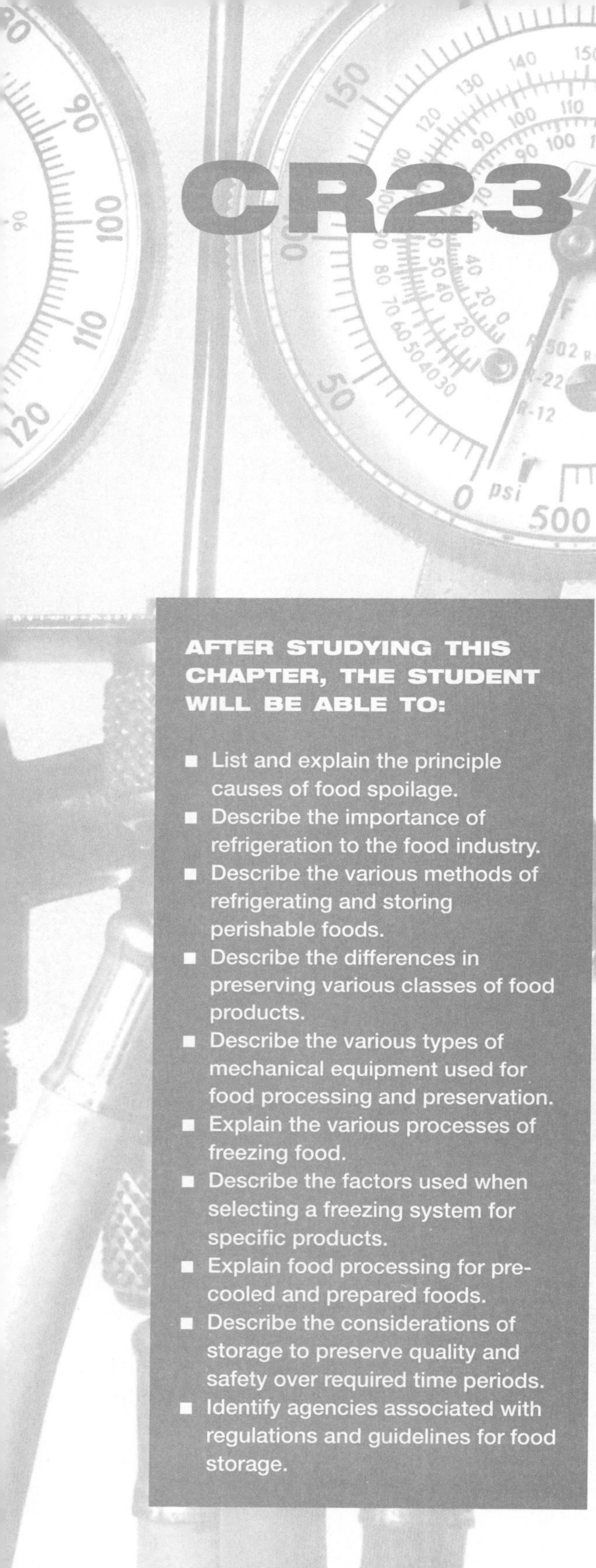

CR23 Food Preservation

AFTER STUDYING THIS CHAPTER, THE STUDENT WILL BE ABLE TO:

- List and explain the principle causes of food spoilage.
- Describe the importance of refrigeration to the food industry.
- Describe the various methods of refrigerating and storing perishable foods.
- Describe the differences in preserving various classes of food products.
- Describe the various types of mechanical equipment used for food processing and preservation.
- Explain the various processes of freezing food.
- Describe the factors used when selecting a freezing system for specific products.
- Explain food processing for pre-cooled and prepared foods.
- Describe the considerations of storage to preserve quality and safety over required time periods.
- Identify agencies associated with regulations and guidelines for food storage.

CR23-1 PRESERVATION OF PERISHABLE FOODS

The perishable food industry is one of the largest industries in the country. An industry of this size is extremely important. Proper refrigeration is an important factor in the success of this business.

Perishable foods can be classified into six groups:

1. Meats
2. Poultry
3. Seafood
4. Fruits
5. Vegetables
6. Dairy products

These perishable foods can be divided into three groups: animal, vegetable and dairy products. Each group requires separate treatment to preserve the products and to keep them palatable. The storage of animal products requires the prevention of deterioration of the non-living products. Fruits and vegetables, however, are as much alive while they are being transported as they were while growing and require an entirely different set of preservation conditions.

The principal causes of spoilage in foods are:

1. *Microbiological.* These include bacteria, molds, and fungi.
2. *Enzymes.* These are chemical in nature and do not deteriorate.
3. *Oxidation changes.* These are caused by atmospheric oxygen coming in contact with the food, producing discoloration and rancidity.
4. *Surface dehydration.* In freezing this is called "freezer burn".
5. *Wilting.* This applies to vegetables that lose their crispness.
6. *Suffocation.* Certain fresh vegetables must have air. When sealed in cellophane bags, the bags must have holes.

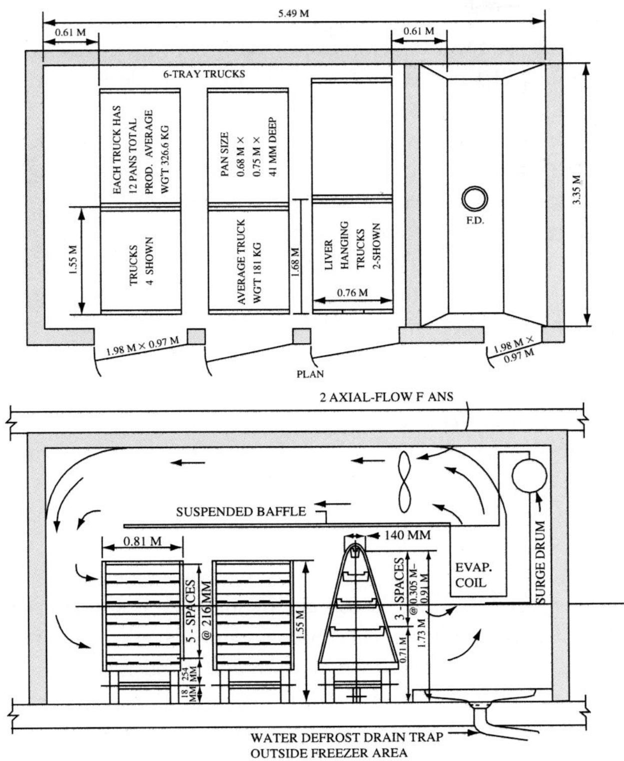

Figure CR23-1 Chilling room for variety meats. Trimmings are placed on the shelf trucks in pans. Air temperatures vary from −29 to −17.8°C. The desired meat temperature is −1.7 to −1.1°C. (Copyright by the American Society of Heating, Refrigerating, and Air-Conditioning Engineers, Inc. Used by permission.)

CR23-2 MEATS

These products deteriorate through the action of bacteria. The enzymes serve to tenderize the meat. Aging is the process of utilizing the good effects of the enzymes without the harmful effects of the bacteria. Sanitation is the most important factor in controlling bacteria. Air has many forms of bacteria present. One of the best ways of controlling this infection is through the use of germicidal or ultraviolet lamps. Oxidation is detrimental to meats, causing undesirable appearance and deterioration of the flavor. Dehydration can be controlled to a large extent by maintaining high humidity in the storage room. High humidity also protects against moisture loss which lowers the weight of the meat.

Good practice requires that pork be rapidly cooled after it is cut. This prevents destructive enzymatic action, which causes discoloration, rancidity, and poor flavor. Fig. CR23-1 shows a chill cabinet in a truck loaded with pork trimmings, which can be unloaded into the chiller room through side-opening doors. Fans circulate the chilled air over the meat at low velocity. Air temperature is in the 0°F range with a TD (temperature difference) of 10°F. A continuous trimming chiller is shown in Fig. CR23-2. Spray-type coolers are used in this type of chiller.

The layout for a sausage dry room is shown in Fig. CR23-3. The keeping qualities of a variety of dry sausages produced depend on curing ingredients, spices, and removal of moisture from the product by drying. The purpose of this room is to remove about 30% of the moisture, to a point where the sausage will keep for a long time virtually without refrigeration. This process is used as an alternative to the smoking process. The U.S. Department of Agriculture (USDA) requires that this room be maintained at temperatures above 45°F, and the length of time in the room depends on the diameter of the sausage after stuffing and method of preparation.

CR23-3 POULTRY

Problems associated with the preservation of poultry are similar to those of meat in many respects except that poultry spoils much faster. Poultry, however, can be precooled by the use of cold water without detrimental effects. This is a relatively simple and effective process and therefore quite generally used. Bacteria and enzyme action is useful only in preserving game birds, as such action has a tendency to enhance the "game flavor."

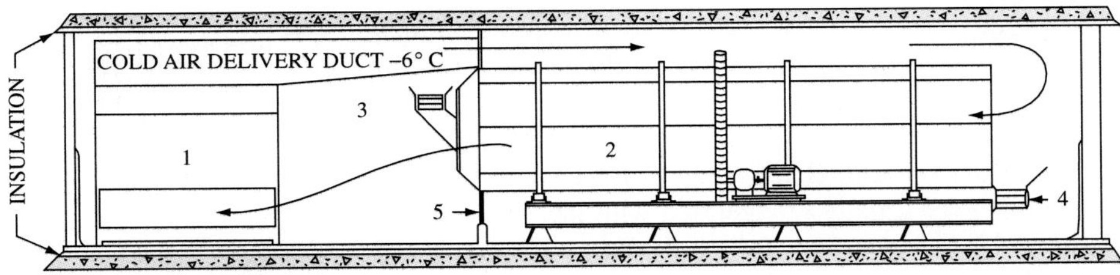

1 - SPRAY TYPE COOLING UNIT - 2 REQ.
2 - TRIMMINGS CHILLING CYLINDER
3 - TRIMMINGS FEED CONVEYOR
4 - TRIMMINGS REMOVAL CONVEYOR
5 - PARTITION BETWEEN COOLING UNITS AND CHILL CYLINDER

Figure CR23-2 Continuous trimming chiller. The air is chilled by two brine-spray unit coolers. A temperature difference of 5.56°C is maintained between room temperature and the refrigerant evaporating temperature. Coils are sprayed with antifreeze such as propylene glycol to remove the frost. (Copyright by the American Society of Heating, Refrigerating, and Air-Conditioning Engineers, Inc. Used by permission.)

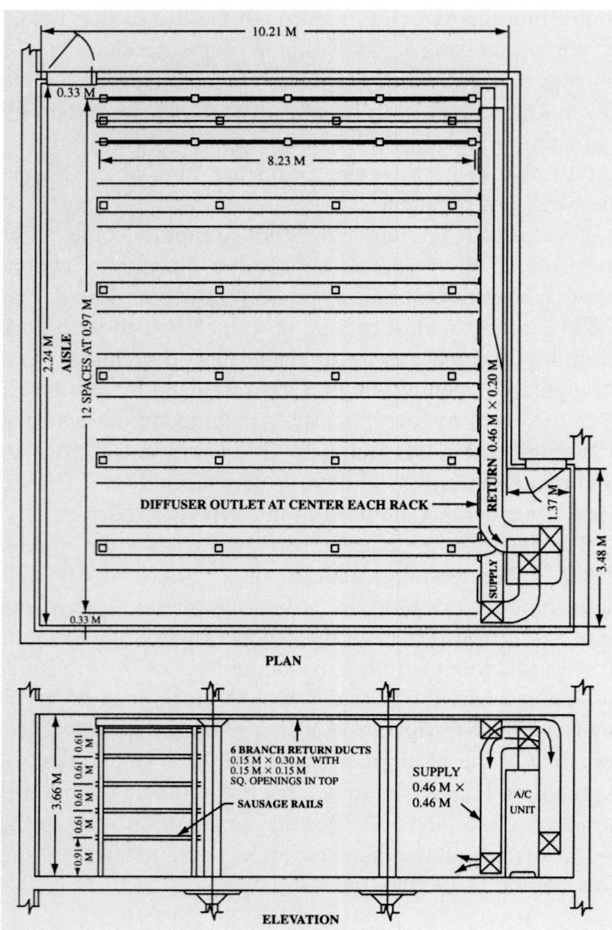

Figure CR23-3 Sausage dry room. Conditions in this room remove moisture from the sausage. The keeping quality of sausage is dependent on the use of curing ingredients and spices, and removal of moisture. Thirty percent of the moisture is removed. Typical conditions in the room are 7.2 to 12.8°C and 70 to 75% relative humidity. (Copyright by the American Society of Heating, Refrigerating, and Air-Conditioning Engineers, Inc. Used by permission.)

CR23-4 SEAFOOD

This product is the most perishable of all the animal foods, yet there is a vast difference in the keeping quality of different kinds of fish. For example, swordfish can be kept refrigerated for 24 days and be in a more palatable condition than mackerel refrigerated for 24 hours. Commercial fish are usually refrigerated with ice.

CR23-5 FRUITS AND VEGETABLES

The unique situation with fruits and vegetables is that they are still alive after they are picked. They grow, breathe, and ripen. Most fruits and vegetables are picked in an unripened condition. The best tasting products are vine-ripened. The purpose of refrigeration is to slow down the ripening process so that these products can reach consumers before spoiling.

Vegetables quickly lose their vitamin content when surface drying takes place. It is interesting to note that products shipped from California to Chicago that have been properly iced after harvest will be fresher than produce supplied from Illinois farms and shipped to a Chicago market without being iced.

Another way to improve the product when it reaches the user is to package the product. This cuts down on surface drying. Packages are usually made of cellophane or some similar plastic product. These containers must have holes so that the product can breathe (exchange oxygen and CO_2). Otherwise, the product will die, and a dead product will spoil rapidly.

A number of products require special treatment—bananas, for example. These are picked green and must be ripened for marketing. Banana ripening is initiated by the introduction of ethylene gas. For this to be effective, banana rooms must be airtight. Refrigeration is provided by using R-134a refrigerant, since an ammonia leak would be destructive to the fruit. Rooms such as those shown in Fig. CR23-4 are cooled using 45 to 65°F air. A design temperature difference of 15°F and a refrigerant temperature of 40°F are considered good practice.

An interesting application of refrigeration is the processing of fruit-juice concentrate (Fig. CR23-5). Hot-gas discharge from the compressor is used to supply heat for the evaporation of the juices. The water vapor is condensed by evaporating liquid refrigerant in a shell-in-tube condenser. Water vapor is used to superheat the suction gas. This type of apparatus provides a continuous process.

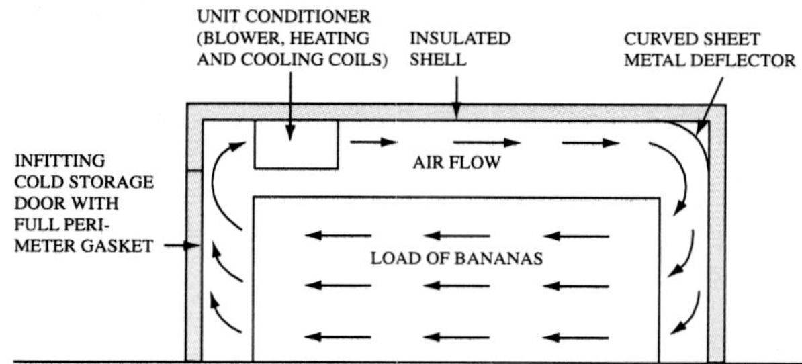

Figure CR23-4 Banana room. Ripening is initiated by exposure to ethylene gas. Rooms must be airtight. Exposure time is usually 24 hours. Floor drain must be trapped to prevent gas leakage. Air temperature range is 7.2 to 18.3°C. Recommended design temperature difference is 8.3°C with 4.4°C evaporating temperature. (Copyright by the American Society of Heating, Refrigerating, and Air-Conditioning Engineers, Inc. Used by permission.)

PRACTICING PROFESSIONAL SERVICES

The storage of fruits and vegetables in commercial and industrial facilities requires that the product be stored in high-humidity environments to protect against weight losses and product shrinkage. To maintain or increase the relative humidity, adjust the suction pressure through the evaporator-pressure regulator to as high as possible to keep the temperature difference low. The lower the temperature difference between the refrigerant and the return air, the higher the relative humidity, and the lower the product shrinkage.

Conversely, the higher the temperature difference between the refrigerant and the return air, the lower the relative humidity and the greater the product will shrink and lose weight. Product weight is a very critical factor during the retail sales process, due to the fact that most agricultural products are sold in stores by weight, not volume.

CR23-6 DAIRY PRODUCTS AND EGGS

Sanitation is extremely important in all stages of handling milk. The bacteria content of milk must be controlled. Mechanical refrigeration begins to cool it even during milking, from 90 to 50°F within the first hour and from 50 to 40°F within the next hour. As more milk is added, the blended liquid must not rise above 45°F. Limits are set for the number of bacteria (the bacteria count) for milk supplied by the producer.

Milk is stored in insulated or refrigerated silo-type tanks which maintain a 40°F temperature. After milk is pasteurized and homogenized it is again cooled in a heat exchanger (a plate or tubular unit) to 40°F or lower, and packaged.

CR23-6.1 Butter

Butter is manufactured from 30–40% cream obtained from the separation of warm, acidified milk. It is cooled to 46–55°F, then churned to remove excess water. The average composition of butter on the market has these ranges:

Fat:	80–81.2%	Salt:	1.0–2.5
Moisture:	16–18	Curd, etc.:	0.5–1.5

Butter keeps better if stored in bulk. If kept for several months, the temperature should not be above 0°F and preferably below –20°F. For short periods, 32–40°F is sat-

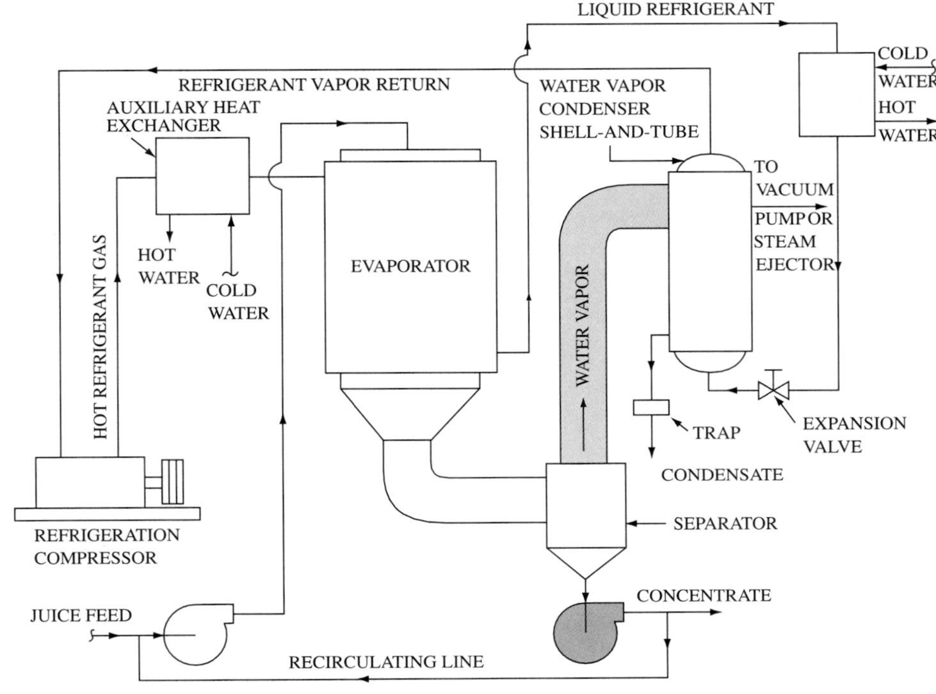

Figure CR23-5 Equipment for fruit juice concentrates. Hot refrigerant gas is used to supply heat for evaporation of juices. Water vapor is condensed by evaporating liquid refrigerant. Vapor and concentrate are usually separated by an arrangement of cyclones and baffles. Pressure in the system is maintained by steam ejector or vacuum pump. (Copyright by the American Society of Heating, Refrigerating, and Air-Conditioning Engineers, Inc. Used by permission.)

isfactory. If stored improperly, the quality of butter deteriorates from: (1) absorption of atmospheric odors, (2) loss of weight through evaporation, (3) surface oxidation, (4) growth of microorganisms and resulting activity of enzymes, and (5) low pH (high acid) of salted butter. Low temperatures, a clean environment, use of a good quality cream, avoidance of light, copper or iron, and adjustment of the pH to 6.8–7.0 eliminates most of these problems.

CR23-6.2 Cheeses

Cheeses are refrigerated to prevent too rapid mold growth. The surface must be kept moist or the cheese will become hard and brittle. Moisture, meanwhile, facilitates mold. While some mold enhances the flavor, too much mold creates waste because it must be removed before sale. The ideal storage temperature range for various types of cheese is in the range of 30–34°F for natural cheeses and 40–45°F for process cheeses. Maximum temperatures range from 45–60°F for the natural cheeses while the process cheeses may be kept on open shelves at 75°F.

CR23-6.3 Eggs

Eggs should be refrigerated at all stages of handling and storage. Shell eggs, which account for about 75% of use, are stored at the following temperatures and relative humidities:

°F Temperature	Relative Humidity (%)	Storage Period
50–60	75–80	2–3 weeks
45	75–80	2–4 weeks
29–31	85–92	5–6 months

Research has shown that microbial growth associated with *Salmonella* can be controlled by holding eggs at less than 40°F. This has led to major changes in storage and display areas not refrigerated or inefficiently refrigerated. All egg storage areas should maintain ambient temperatures at 45°F; however, with mechanized processing and packaging procedures which insulate the eggs within cartons, it may require up to one week of storage before the eggs reach the temperature of the storage room. If shipped earlier to sell a "fresh" product, eggs are only partially

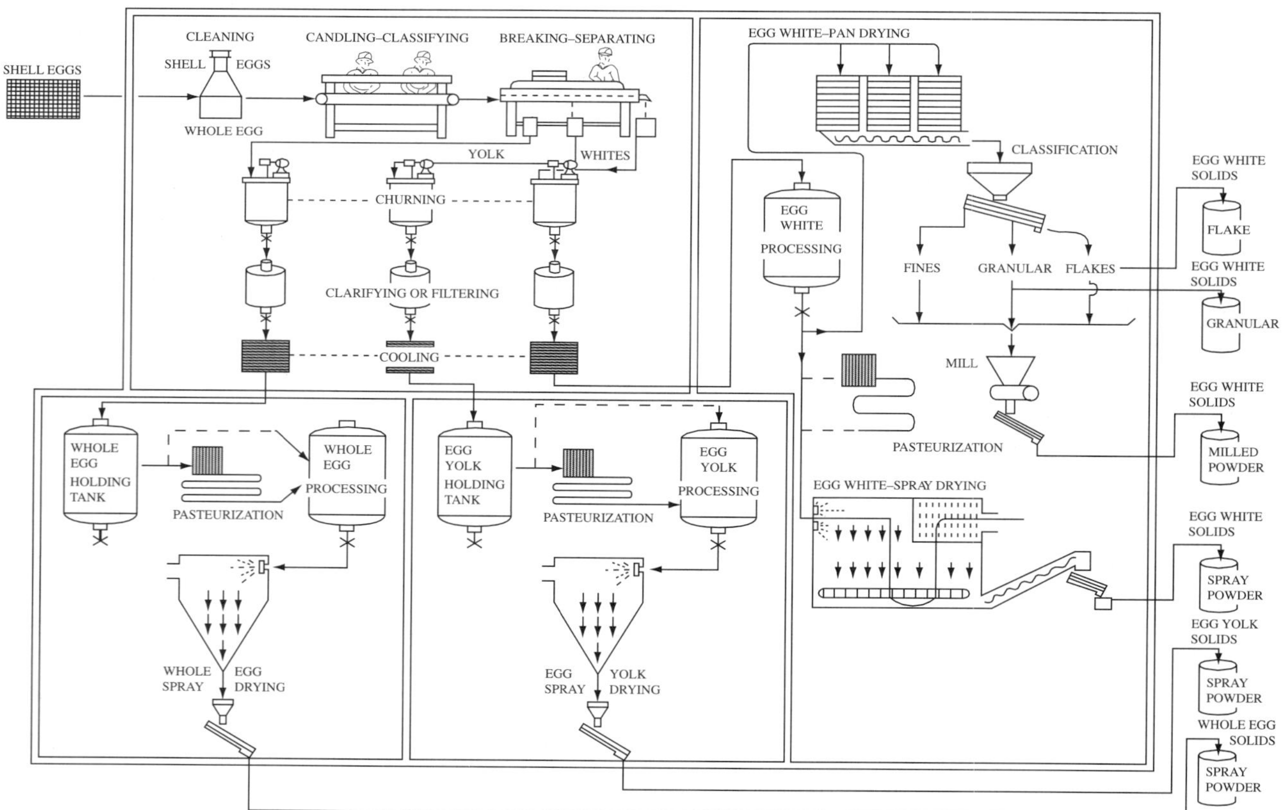

Figure CR23-6 Egg product processing. Various unit processes are involved in the manufacture of egg solids or dried egg products. Practically all egg products are required by law to be pasteurized. The USDA requires certain holding temperatures for egg products ranging from 4.2 to 18.3°C. (Copyright by the American Society of Heating, Refrigerating, and Air-Conditioning Engineers, Inc. Used by permission.)

cooled. Methods are being developed to improve cooling in the processing plant.

Shipment in refrigerated trucks is mandatory. Problems which arise are associated with the truck design, manner of loading, size of the shipment, and distance. A 1993 USDA survey found that over 80% of the trucks used to deliver eggs were unsuitable to maintain 45°F.

Fig. CR23-6 shows the use of mechanized equipment in egg product processing. The various products are shown to the right.

CR23-7 FROZEN FOODS

The freezing of food, although highly product-specific, is basically a time-temperature related process of three phases: (1) cooling to freezing point, (2) changing the water in the product to ice, and (3) lowering the freezing temperature to optimum frozen storage temperature. Product differences as well as quality relate to the specific values and to the rates of these stages.

The following factors are considered in selecting a freezing system for a specific product: (1) special handling requirements, (2) capacity, (3) freezing times, (4) quality consideration, (5) yield, (6) appearance, (7) first cost, (8) operating costs, (9) automation, (10), space availability, and (11) upstream/downstream processes.

CR23-7.1 Quick Freezing

This essential process produces small ice crystals which are less damaging to the product. Ideally, the ripe produce should be frozen immediately after harvest—the sooner the better. Small packages are better to freeze than large packages because the interior freezes more quickly.

Small packages may be frozen on or between refrigerated plates or in a "blast" freezer. Foods are frozen at temperatures between –5 and –20°F. Freezer burn should be avoided. This is the condition of surface oxidation that causes discoloration of the product. It is prevented by packaging in airtight containers or by waxing or glazing the product. Ice glazing is used to prevent surface drying of fish. Fruits are often glazed with a sugar syrup to prevent oxidation.

Vegetables must be *blanched* before freezing. This consists of placing the product in boiling water or steam to kill bacteria and to stop enzyme action. Air is removed from citrus juice before freezing.

Commercial freezing systems can be divided into four groups:

1. Air-blast freezers
2. Contact freezers
3. Immersion freezers
4. Cryogenic freezers

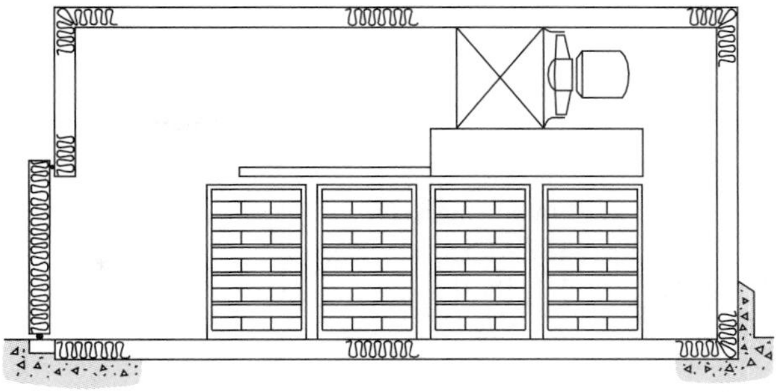

Figure CR23-7 Stationary freezing tunnel. Practically all products can be frozen in a stationary tunnel. Conditions are determined in relation to the product being frozen. (Copyright by the American Society of Heating, Refrigerating, and Air-Conditioning Engineers, Inc. Used by permission.)

Air-blast freezing can best be described as a convection system, where cold air at high velocities is circulated over the product. The air removes heat from the product and releases it to an air-refrigerant heat exchanger before being recirculated.

The air-blast freezer using a stationary tunnel (Fig. CR23-7), produces satisfactory results for practically all products, in or out of packages. Products are placed in trays which are held in racks, placed so that air bypass is minimized. Air blast freezers can be mechanized to provide a continuous process, as shown in Figs. CR23-8 and CR23-9. Two-stage belt freezers permit precooling at 15–25°F before transferring to the second belt for freezing at temperatures of –25 to –40°F.

Belt-type freezers use vertical airflow and greatly improve the contact between air and product. Two types are shown in Figs. CR23-10 and CR23-11.

The fluidization principle is illustrated in Fig. CR23-12. Solid particulate products such as peas, sliced and diced carrots, and shredded potatoes or cheese are floated upward and through the freezer by streams of air. The product is frozen in 3 to 11 minutes by refrigerant temperatures of –40°F. These freezers are packaged, factory-assembled units (Fig. CR23-13).

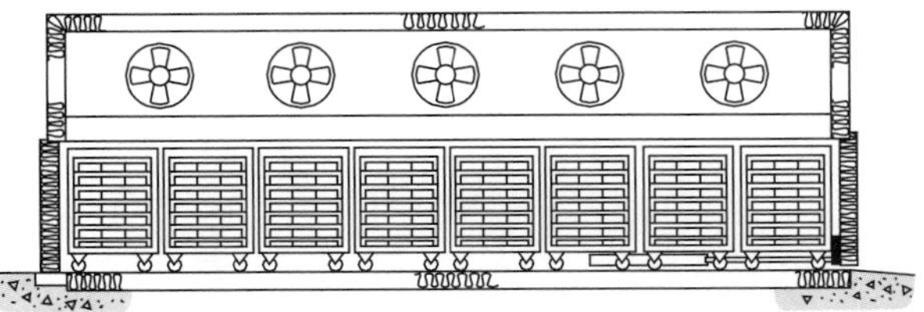

Figure CR23-8 Push-through tunnel. This type of freezer offers a convenient means of placing and removing freezing racks. It can be used for a variety of products. Conditions depend on the product being frozen. (Copyright by the American Society of Heating, Refrigerating, and Air-Conditioning Engineers, Inc. Used by permission.)

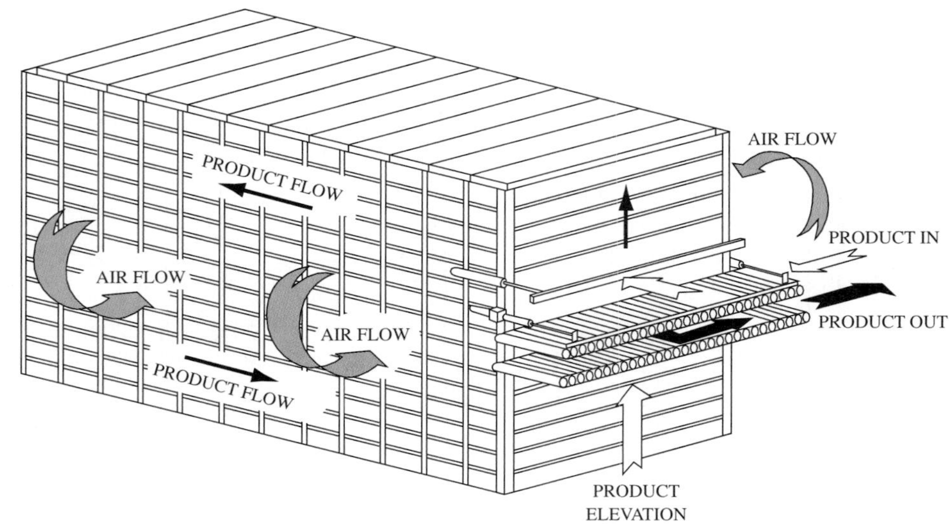

Figure CR23-9 Carrier freezer. This freezer arrangement is similar to two push-through tunnels on top of one another. In the top section carriers are pushed forward; in the lower section they are returned. In both ends there are elevating mechanisms. (Copyright by the American Society of Heating, Refrigerating, and Air-Conditioning Engineers, Inc. Used by permission.)

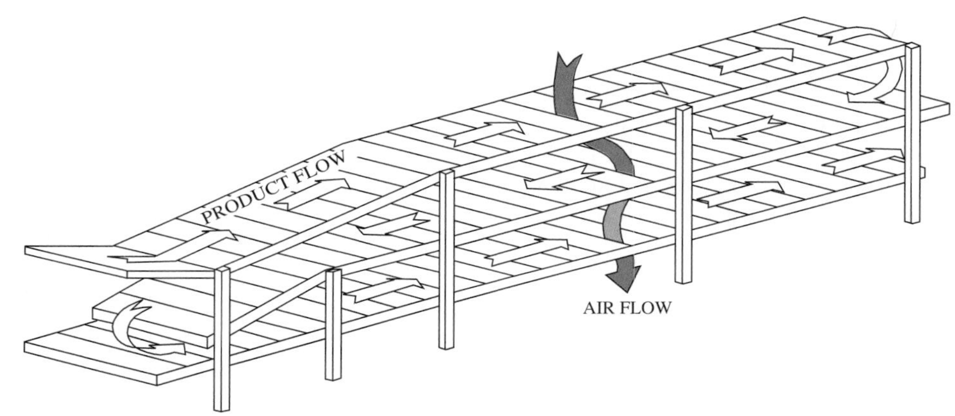

Figure CR23-10 Multiple-belt freezer. These freezers are suitable for individual quick freezing of fried fish sticks, fish portions, bakery items, and similar products. (Copyright by the American Society of Heating, Refrigerating, and Air-Conditioning Engineers, Inc. Used by permission.)

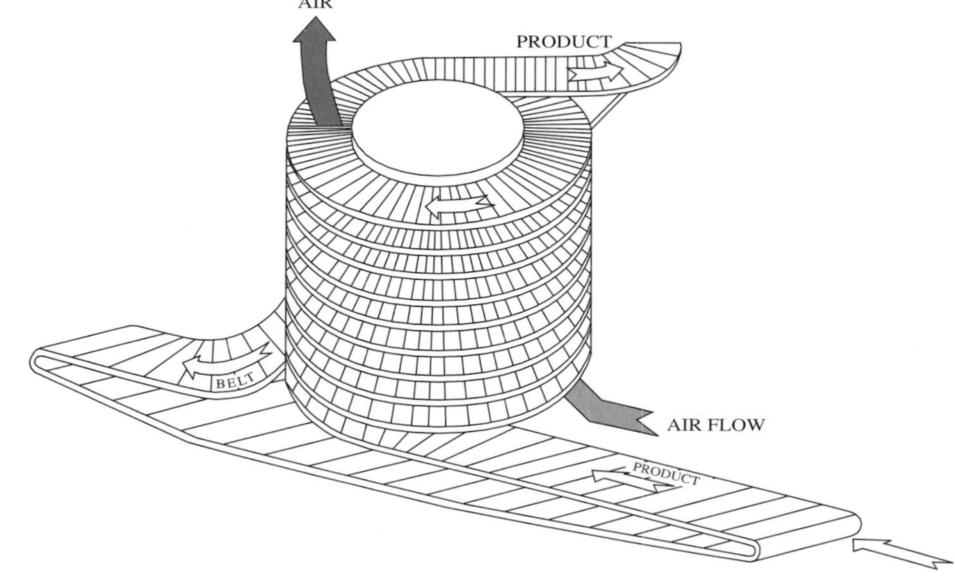

Figure CR23-11 Spiral belt freezer. This freezer is designed to save floor space. It is placed in a refrigerated room and the freezing time is regulated by the speed of the belt. It is suitable for a wide variety of unpackaged meat products. (Copyright by the American Society of Heating, Refrigerating, and Air-Conditioning Engineers, Inc. Used by permission.)

PREVENTIVE MAINTENANCE

Maintaining equipment for frozen foods includes performing periodic inspections of the evaporator and condenser coils to verify they are not plugged or restricting airflow due to dirt or ice buildup. Good maintenance for the industrial evaporative condenser or air-cooled condenser requires periodic cleaning using job-specific water treatment. Evaporators are not typically as prone to getting dirty as condensers, but ice accumulation is always a potential problem and will cause loss of efficiency and production levels. Often, the only adjustment for the evaporator defrost is to extend the defrost periods for longer times. It is also possible that the evaporator needs to defrost more times throughout the day.

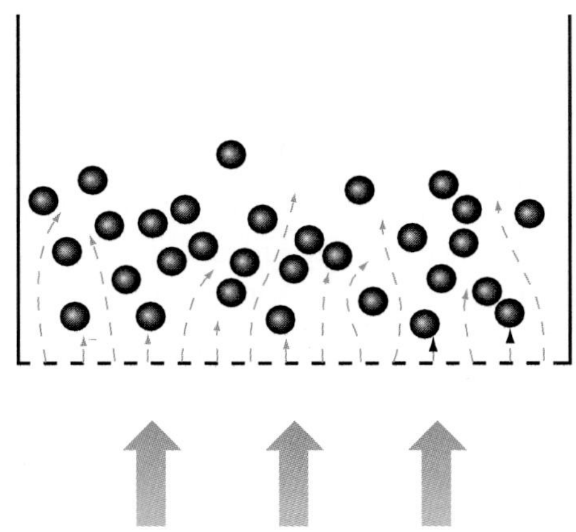

Figure CR23-12 Fluidization principle. Fluidization is the process of freezing small items such as peas by an upward flow of freezing air. At a certain air velocity the particles being frozen will float and maintain separation. The principle can be used in a continuous freezer as shown in Figure CR23-13. (Copyright by the American Society of Heating, Refrigerating, and Air-Conditioning Engineers, Inc. Used by permission.)

Contact freezers are conduction-type freezers. Products are placed on or between horizontal or vertical refrigerated plates which provides efficient heat transfer and short freezing time. Relatively thin packages of food or products, 2 to 3 in. thick, are used in this system. A typical freezer is shown in Fig. CR23-14.

Cryogenic freezers utilize both convection and/or conduction by exposing food to temperatures below −76°F in the presence of liquid nitrogen or liquid carbon dioxide refrigerants. Liquid nitrogen boils at −320°F and carbon dioxide boils at approximately −110°F. The freezers may be cabinets, straight-belt freezers, spiral conveyors, or liquid-immersion freezers. The boiling liquid comes in direct contact with the product. After use, the refrigerant is wasted to the atmosphere. A freezer of this type is shown in Fig. CR23-15. While operating costs are high, the small initial investment makes this economical for certain foods.

Some products, such as shrimp, are frozen by *immersion* in a boiling, highly purified refrigerant. The surface of a sticky or delicate product is "set" by this rapid freezing, reducing dehydration and improving the handling characteristics of the product. The product is then removed and the freezing process completed in a mechanical freezer. In these systems the refrigerant is re-

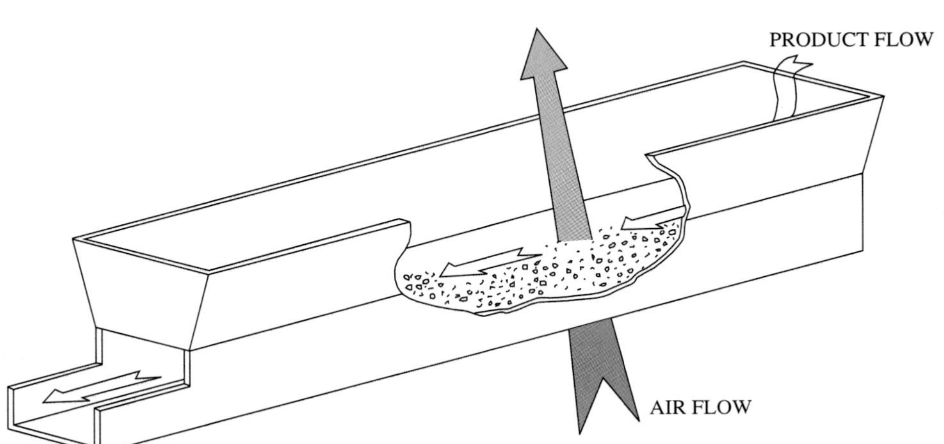

Figure CR23-13 Fluidized bed freezer. This is an automatic continuous-process freezer using the fluidization principle. It is suitable for small vegetable pieces, French-fried potatoes, cooked meatballs, and many similar products. (Copyright by the American Society of Heating, Refrigerating, and Air-Conditioning Engineers, Inc. Used by permission.)

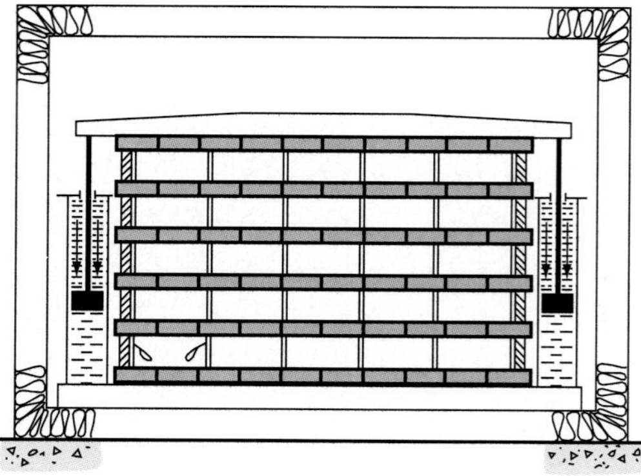

Figure CR23-14 Plate freezer. The product to be frozen is firmly pressed between two refrigerated plates. This process is limited to products with good transfer rates and thicknesses not to exceed 50 mm. (Copyright by the American Society of Heating, Refrigerating, and Air-Conditioning Engineers, Inc. Used by permission.)

covered by condensing on the surface of refrigerated coils.

CR23-7.2 Refreezing

When a vegetable product is in a frozen state, for all practical purposes it is considered "dead;" however, there may remain microbes and enzymes in an inactive state. When thawed, a large amount of water is present from the ruptured tissues to provide a favorable environment for the growth of microbes and deterioration due to the enzymes. All of these processes serve to lower the quality of the product and could continue after refreezing. The product should be heated sufficiently to kill these destructive agents before refreezing. Some canneries freeze products to prevent spoilage until they can schedule the final canning process.

CR23-8 FOOD PROCESSING

The production of precooked and prepared foods developed into an important industry the last half of the twentieth century. These foods, which include ready-to-use foods for main dishes and meals, vegetables, and potato production, require refrigeration and air-conditioning facilities.

Main dishes, which constitute the largest group of products in this area, include complete dinners, lunches and breakfasts, soups/chowders, low-calorie/diet specialties, and ethnic meals. They are characterized by having a large number of ingredients, several unit operations, an assembly-type packaging line, and final refrigerating or freezing of individual packages or cartons. Production falls into the following operations in the processing plant:

1. *Preparation, processing and unit operations.* This involves the initial preparation of all ingredients to be assembled, including refrigeration and/or freezing needs. These require specific attention to individual requirements for selecting processes, equipment, space, controls, and safeguards.
2. *Assembly, filling and packaging.* This includes all handling of components for putting into containers or packages, packaging, and placing the containers and packages in refrigerators or freezers. It is considered good practice to air condition filling and packaging lines to control bacteria and increase worker productivity.
3. *Cooling, freezing, and casing.* There is a constant effort to improve the economy and efficiency of production, often altering original design conditions. Space and equipment-capacity allowances should be 25–50% for increasing requirements. Maintenance should include checking temperatures and other specific conditions of the particular product involved. Defrosting on conveyor systems should be checked when there are hangups or stoppages. When the position or place of packages is changed in the storage processes, the quality of the product should be closely monitored.
4. *Finishing: storage and shipping.* Infiltration and product pull-down loads occur when there is negative air pressure due to exhausting more air than is supplied by ventilation. This causes a serious load on the refrigeration, making it difficult to maintain proper storage temperatures.

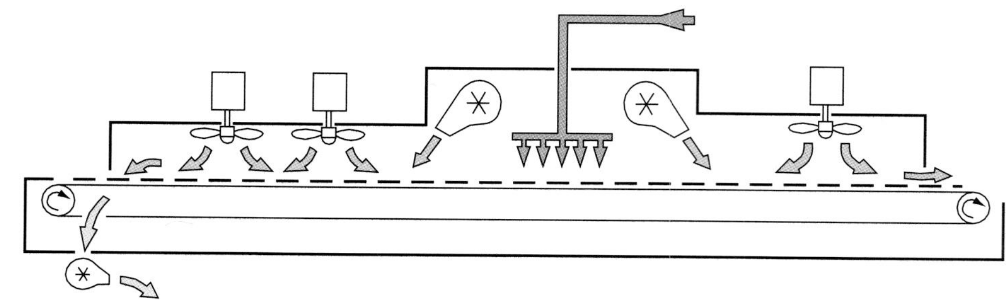

Figure CR23-15 Cryogenic freezer. This freezer uses a liquid nitrogen spray (–196°C). The product passes through the freezer on a moving belt. Vapor refrigerant is discharged at –30 to –100°C. This process provides fast freezing. Vaporized nitrogen is wasted. (Copyright by the American Society of Heating, Refrigerating, and Air-Conditioning Engineers, Inc. Used by permission.)

ON THE JOB

If the best results are to be obtained in the cold storage of agricultural products, it is highly important that the temperature in the room be held fairly constant. Large temperature fluctuations and variations tend to promote premature product deterioration. In addition, fluctuations in temperature often cause moisture on stored products, which is undesirable because it may favor the growth of surface mold and the development of decay.

To verify that temperatures are constant throughout the storage room, place recording thermometers in the four corners of the room, several feet from the wall. Also locate a thermometer in the center of the room within the product. Temperatures should be within 3–4° of each other and constant over a 24-hour period.

CR23-8.1 Refrigeration Loads

Records need to be kept of operating conditions to identify poor performance and to provide guidance for new systems. These records should show conditions for time of day, season, on/off-shift production, evaporator temperature, and equipment type/function.

CR23-8.2 Refrigeration Equipment

The refrigerant used for most refrigeration systems is ammonia. To avoid the hazard of a potential ammonia spill to workers in the plant, glycol chillers are used by some plants to circulate propylene glycol to evaporators located in the production areas. With high compression ratios required for freezing, two-stage compressors are used with evaporative condensers. Direct-expansion evaporators are seldom used.

Energy saving measures include floating-head pressure controls with over sized evaporative condensers coupled with two-speed fans, single-stage refrigeration for small areas and loads, variable-speed pumps for glycol chiller systems, ice builders to compensate for peak loads, door-infiltration protection devices, insulation, and computerized control systems.

CR23-8.3 Vegetables

All prepared vegetables are precooked and cooled before freezing as discussed in the previous section. Refrigeration is used for raw product cooling and storage, product cooling after blanching, freezing, process equipment located in freezer storage facilities, and freezer storage warehouses. Loads vary widely depending on the product.

In vegetable facilities that operate only for short periods at peak capacity—1500 to 2500 hours per year—spare equipment cannot be economically justified and good maintenance is important to avoid down-time losses.

CR23-8.4 Potatoes

Prepared potatoes products in various forms dominate the frozen, ready-to-use vegetable group and are processed year around. Products include French fries, hash browns, twice-baked potatoes, potato skins, and boiled potatoes. French fries are probably the most popular.

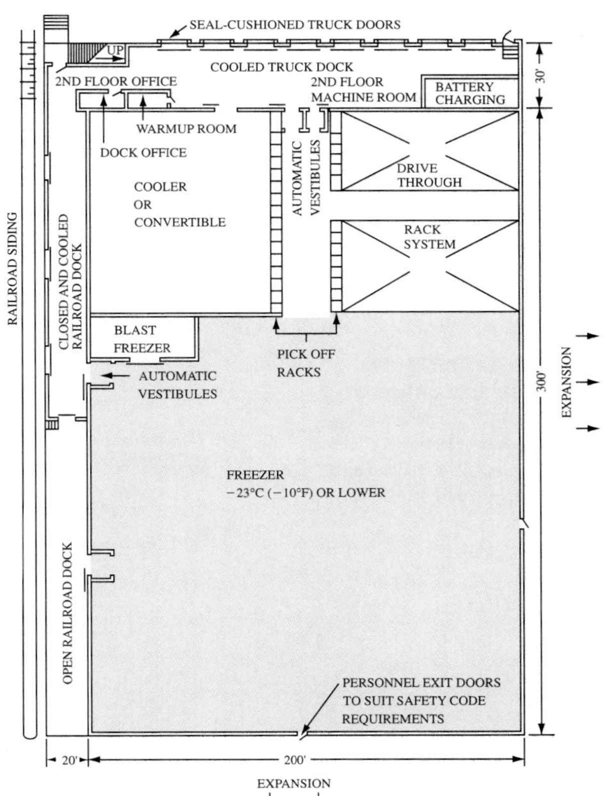

Figure CR23-16 One-story warehouse. In the layout, note refrigerated shipping platforms; conditioned vestibules; first-in, first-out rack arrangement; blast freezer; office; and shop. (Copyright by the American Society of Heating, Refrigerating, and Air-Conditioning Engineers, Inc. Used by permission.)

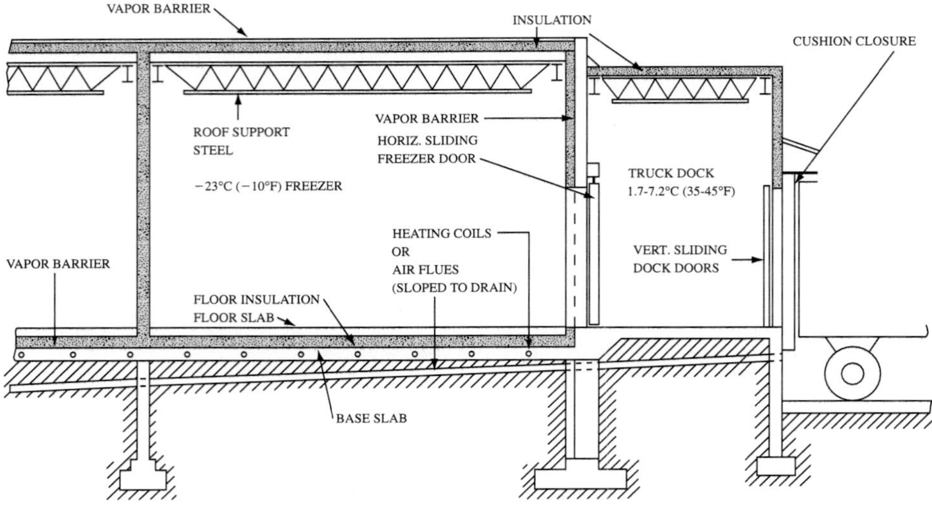

Figure CR23-17 Typical one-story construction with hung insulated ceiling and underfloor warming pipes. (Copyright by the American Society of Heating, Refrigerating, and Air-Conditioning Engineers, Inc. Used by permission.)

Raw potatoes for fries are steam peeled and trimmed, then cut into desired shapes. The slivers are graded out for use as puffs, tots, and wedges. The fries are blanched, then partially dried, and oil fried. They are frozen on a straight-belt freezer system with three separate conveyers for precooling and totally freezing the fries to 5–10°F. Sorting is done at 15°F and packaged in an air-conditioned area.

CR23-8.5 Storage

Freezing and thawing temperatures of animal, vegetable, dairy, and egg products vary widely. Temperatures must be maintained which preserve the quality and safety of products over time periods required. As a result of ongoing research, lower temperatures are being recommended.

Most modern refrigerated warehouses are one-story structures. A typical floor plan is shown in Fig. CR23-16 (p. 449). The building construction is shown in Fig. CR23-17. Note that the outer walls are independent of the rest of the building. It is usually convenient to use penthouse refrigeration equipment rooms (see Fig. CR23-18). Consideration must be given to:

a. Entering temperatures
b. Duration of storage
c. Required product temperature for maximum/minimum protection
d. Uniformity of temperatures
e. Air movement and ventilation
f. Humidity
g. Traffic in and out of storage space

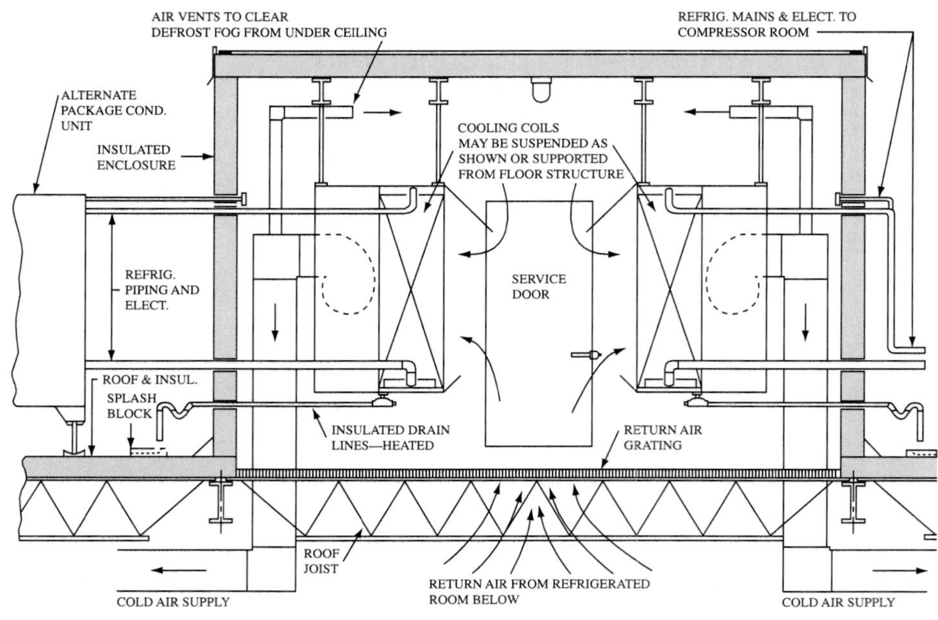

Figure CR23-18 Penthouse application of cooling units. (Copyright by the American Society of Heating, Refrigerating, and Air-Conditioning Engineers, Inc. Used by permission.)

h. Sanitation
i. Light

Freezer storage at vegetable processing plants must also consider potential additional loads: (1) extra reserve capacity needed for product pull-down during peak processing; (2) negative pressure which can increase infiltration by direct flow-through; and (3) the process machinery load (particularly pneumatic conveyors) associated with repack operations.

Regulations and guidelines for the refrigerated storage of foods have been established by the following agencies and should be familiar to the service person working in these areas:

1. The Association of Food and Drug Officials (AFDO).
2. Occupational Safety and Health Act (OSHA)
3. U.S. Department of Agriculture (USDA)
4. Environmental Protection Agency (EPA)

REVIEW

- Perishable foods:
  Meats
  Poultry
  Seafood
  Fruits
  Vegetables
  Dairy products
- Principal causes of food spoilage:
  Microbiological
    Bacteria
    Fungi
    Enzymes
  Oxidation changes
  Surface dehydration
  Wilting
  Suffocation
- Storage requirements:
  Meat deteriorates through the action of bacteria. Aging is the process of utilizing the good effects of enzymes without the harmful effects of bacteria. Sanitation is the most important factor. Dehydration can be controlled by maintaining a high humidity. Air should be circulated at a low velocity with temperature in the 0°F range with a TD of 10°F. U.S. Department of Agriculture (USDA), sets standards for storage and processing areas.
  Poultry is similar to meat but spoils much faster. Precooled by use of cold water.
  Seafood is the most perishable of the animal foods. Various types exhibit vast differences in keeping quality. Stored in ice at 32°F.
  Fruits and vegetables are still alive after they are picked. They grow, breathe, and ripen, and most are picked in an unripened condition. Refrigeration slows down the ripening process. Sprays are often required for proper humidity. Containers must allow products to breathe. Design temperatures are 45 to 65°F with a TD of 15°F and a refrigeration temperature of 40°F.
  Dairy products—Milk should be stored at 40 to 45°F; butter for short periods at 32 to 40°F, and long periods at −20 to 0°F; cheeses at 30 to 34°F.
  Storage of eggs varies as follows, with optimum storage at 45°F:

Temp.	RH%	Storage Period
50–60	75–80	2–3 weeks
45	75–80	2–4 weeks
29–31	85–92	5–6 months

- Commercial freezing systems:
  Air-blast freezers
  Contact freezers
  Immersion freezers
  Cryogenic freezers
  Liquid refrigerant freezers
- Considerations for storage:
  Entering temperature
  Duration of storage
  Required product temperature (max/min)
  Uniformity of temperature
  Air movement and ventilation
  Humidity
  Traffic in and out of storage area
  Sanitation
  Light
- Agencies for regulations and guidelines:
  Association of Food & Drug Officials (AFDO)
  Occupational Safety & Health Act (OSHA)
  U.S. Department of Agriculture (USDA)
  Environmental Protection Agency (EPA)

Problems and Questions

1. How is dehydration of food products in a refrigerator cabinet prevented?
2. Approximately what temperature should be maintained in frozen-food storage or display cabincts?
3. Generally speaking, there are two primary types of bacteria that affect foods in refrigerated storage. What are they?
4. Refrigerating fruits or vegetables immediately upon harvesting preserves them for longer periods than if there is a time lag between the harvesting and refrigeration. Why?
5. On what does the rate at which ice will cool water in a glass depend?
6. Why is it necessary to store different types of produce under different conditions of temperature and relative humidity?

7. The storage of different products in the same storage area is called _____.
8. List the six classes of perishable foods.
9. How are the six classes of perishable foods grouped?
10. What are the principal causes of spoilage in foods?
11. What is meant by "aging" the meat?
12. How can bacteria be controlled in food preservation?
13. What is the most perishable of all the "animal" foods, and must have the most controlled environment?
14. Fruits and vegetables are still alive after they are picked, and therefore grow, breathe, and ripen. They require special refrigeration processes. Describe these processes.
15. Temperatures for dairy products vary with each product. What is the storage temperature for milk, butter, eggs, and processed cheeses?
16. At what temperature must eggs be stored to control Salmonella?
17. How is the refrigeration process used in a food-processing plant?
18. Who sets regulations and guidelines which affect refrigeration storage of foods?
19. In the refrigeration storage of foods, what are some considerations?
20. In the storage of foods, what is the major concern of temperature control?
21. Certain fresh vegetables must have air to sustain shelf life. True or false?
22. Surface dehydration is sometimes called freezer burn. True or false?
23. Poultry is the most perishable of all the animal foods. True or false?
24. Banana ripening during refrigerated storage is initiated by the introduction of ethylene gas. True or false?
25. Vegetables must be blanched before freezing to kill bacteria and to stop enzyme action. True or false?
26. Freezing foods requires what process?
    a. Cooling to freezing point.
    b. Changing the water in the product to ice.
    c. Lowering the freezing temperature to optimum frozen storage temperature.
    d. All of the above.
27. Cryogenic freezers:
    a. Expose food to temperatures below −76°F.
    b. Use liquid nitrogen or liquid carbon dioxide refrigerants.
    c. Utilize both convection or conduction air currents.
    d. All of the above.
28. The refrigerant used for most large industrial equipment is:
    a. Ammonia
    b. R-22
    c. R-11
    d. None of the above.
29. Regulations and guidelines for the refrigerated storage of foods should follow:
    a. OSHA
    b. USDA
    c. EPA
    d. None of the above
30. Fruits and vegetables:
    a. Are still alive after being picked.
    b. Grow, breathe, and ripen after being picked.
    c. Are picked in an unripened condition.
    d. None of the above

Refrigeration Systems

CR24

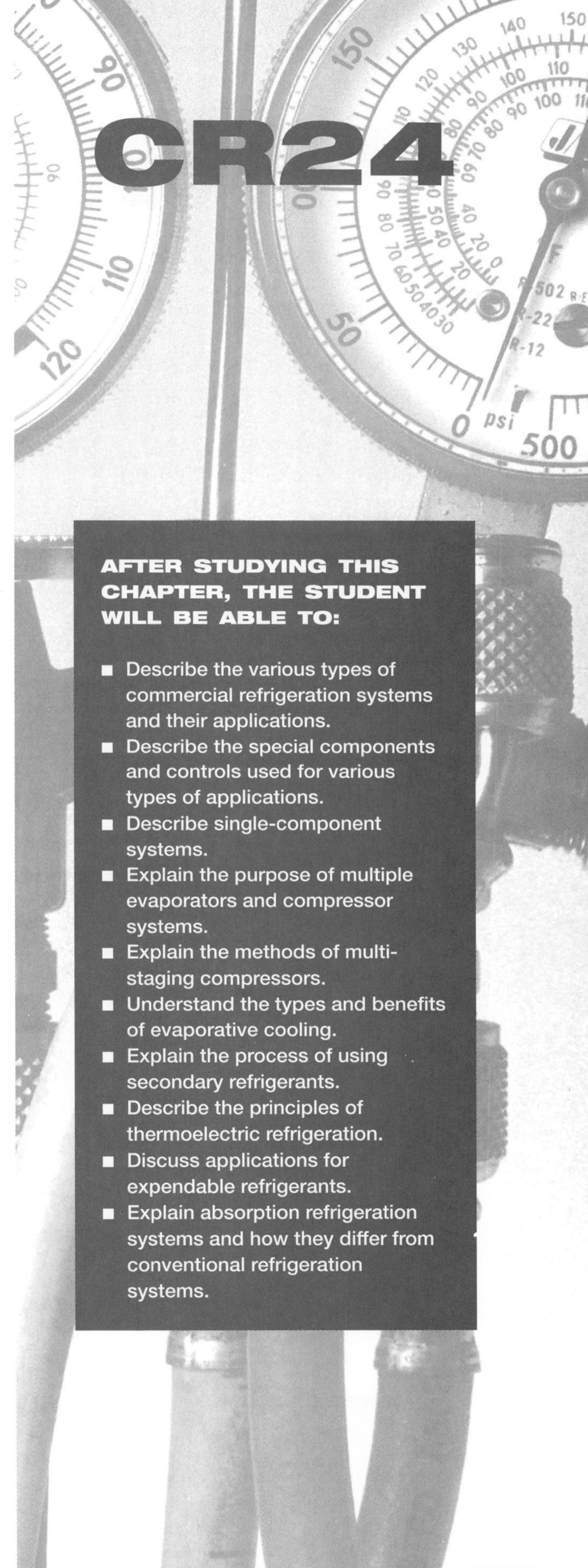

AFTER STUDYING THIS CHAPTER, THE STUDENT WILL BE ABLE TO:

- Describe the various types of commercial refrigeration systems and their applications.
- Describe the special components and controls used for various types of applications.
- Describe single-component systems.
- Explain the purpose of multiple evaporators and compressor systems.
- Explain the methods of multi-staging compressors.
- Understand the types and benefits of evaporative cooling.
- Explain the process of using secondary refrigerants.
- Describe the principles of thermoelectric refrigeration.
- Discuss applications for expendable refrigerants.
- Explain absorption refrigeration systems and how they differ from conventional refrigeration systems.

CR24-1 SYSTEMS

Refrigeration is the process of moving heat from a substance or area where it is not wanted to an area where it is not objectionable. The effect of the process is to produce cooling by lowering the temperature in the refrigerated space and transferring the heat to air or water located outside the refrigerated region.

There are many types of refrigeration systems. Each type of system is designed for specific applications. The current equipment being produced must meet the following criteria:

1. The quality of the product being cooled must remain satisfactory.
2. A minimum amount of energy must be used to perform the operation.
3. The process must be operated to comply with the laws relating to protecting the environment.

For example, a simple refrigeration system is selected to cool food in a reach-in refrigerator. It is designed to continuously produce box temperatures in the 40°F range. A rotary-type compressor is chosen for its high efficiency. The refrigerant selected is R-134a, which has an ozone depletion factor (ODF) of zero.

The "state of the art" is continually improving, and it is important to become familiar with equipment that incorporates the latest technology included in this text. This text, however, also describes earlier units that are still commonly used and may require service. Since there are many ways to produce refrigeration, it is important for the technician to understand each of the processes.

CR24-1.1 Types of Systems

The following are various types of systems used in commercial refrigeration, and are described in this section:

1. Single-component systems
2. Multiple-evaporator systems
3. Multiple-compressor systems

4. Multi-stage-compressor systems
    a. Compound
    b. Cascade
5. Evaporative-cooling systems
6. Secondary refrigerants
7. Thermoelectric refrigerators
8. Expendable refrigerants
9. Absorption-refrigeration systems

Single-component Systems

The essential elements of a *single-component system* (Fig. CR24-1) are the: compressor, condenser, metering device, and evaporator. Also shown are some of the important accessories common to commercial refrigeration systems: a high-low pressure control, condenser fan, receiver tank, suction-to-liquid-line heat exchanger, filter-drier, evaporator fan, thermostat, and suction-line accumulator. Each of these accessories help maintain an efficient system.

There are many variations of these basic elements which are useful in a wide variety of applications. For example, Fig. CR24-2 shows a water-cooled condenser used in place of the air-cooled condenser and fan-coil unit evaporator of an air-conditioning system. Note also the additional accessories shown: compressor and receiver service valves, a liquid-line solenoid, and a water-regulating valve for the condenser.

Multiple-evaporator Systems

This system is common in supermarkets. It makes possible using a single compressor to control a number of different case or fixture temperatures. Fig. CR24-3 shows a three-temperature system.

Evaporator pressure regulators (EPRs) are placed in the suction lines to the two lower temperature evaporators. These are adjusted to maintain the desired evaporator temperature.

A check valve is installed in the suction line from the lowest temperature coil. This prevents migration of the refrigerant from the higher-temperature coils to the low-temperature coil when the compressor is producing cooling for the higher-temperature coils.

When all of the coils require refrigeration, such as on start-up, the compressor will operate at the suction pressure required to cool the highest-temperature coil first, then the middle-temperature coil, and last, the low-temperature coil. The compressor must be sized to produce the entire cooling load at the evaporator pressure of the lowest-temperature coil.

The low-temperature coil will receive very little refrigeration until the higher-temperature coils are satisfied. For this reason the load on the low-temperature coil must account for 60% of the total load, otherwise it may not receive adequate refrigeration.

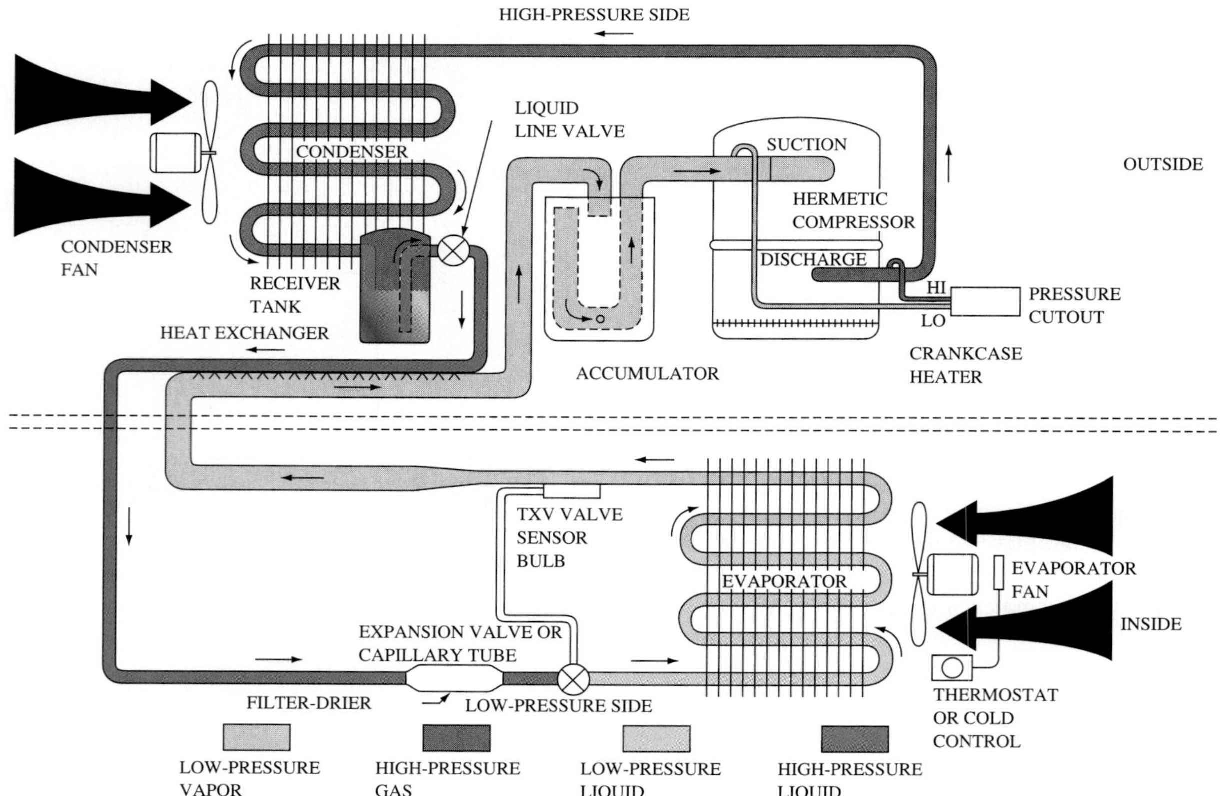

Figure CR24-1 A simple refrigeration system with a single compressor and air-cooled condenser. (Courtesy of Tecumseh Products Co.)

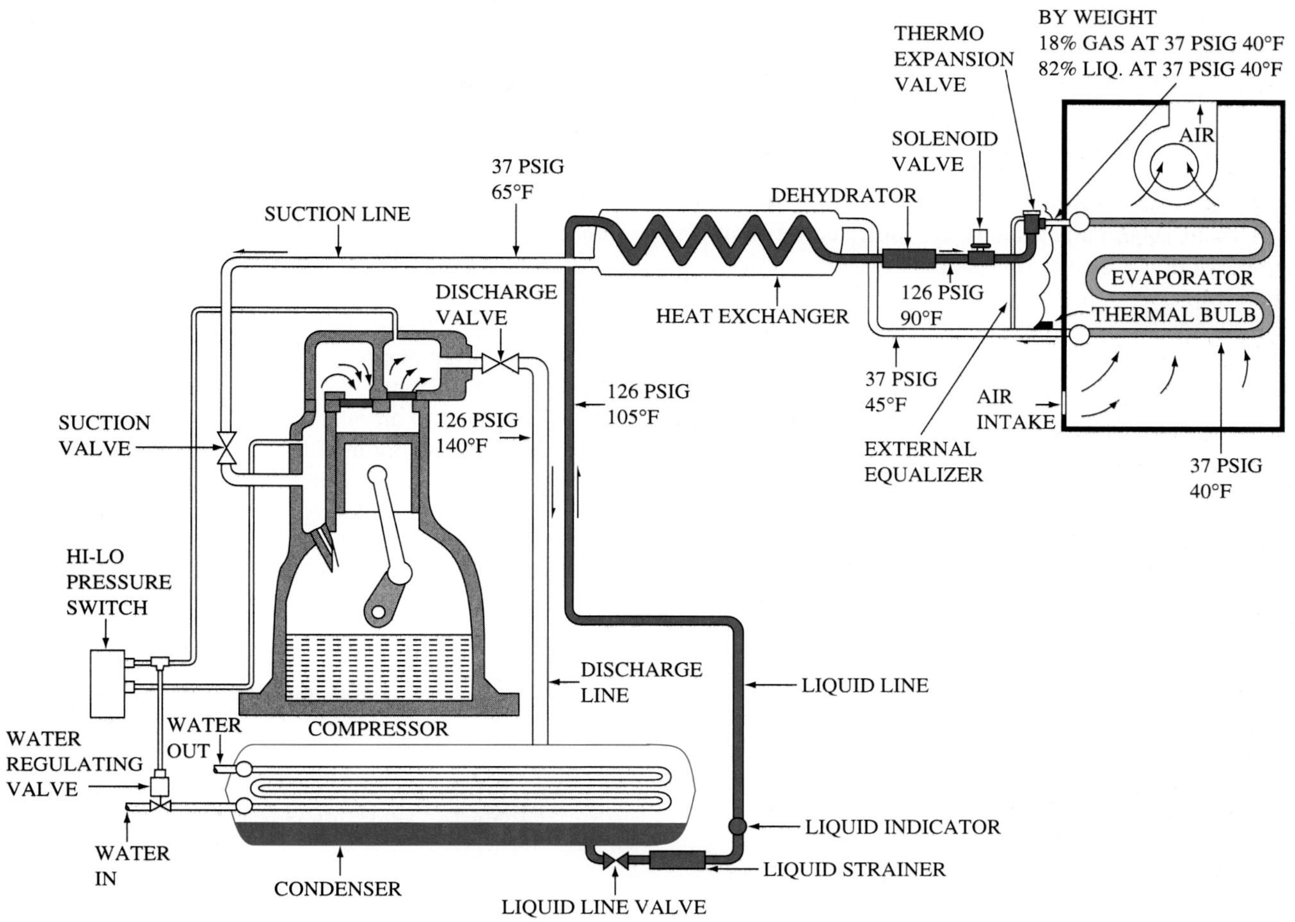

BY WEIGHT
18% GAS AT 37 PSIG 40°F
82% LIQ. AT 37 PSIG 40°F
THERMO EXPANSION VALVE
SOLENOID VALVE
AIR
37 PSIG 65°F
DEHYDRATOR
SUCTION LINE
EVAPORATOR
DISCHARGE VALVE
HEAT EXCHANGER
126 PSIG 90°F
THERMAL BULB
126 PSIG 105°F
37 PSIG 45°F
AIR INTAKE
SUCTION VALVE
126 PSIG 140°F
EXTERNAL EQUALIZER
37 PSIG 40°F
HI-LO PRESSURE SWITCH
DISCHARGE LINE
LIQUID LINE
WATER REGULATING VALVE
WATER OUT
COMPRESSOR
LIQUID INDICATOR
WATER IN
CONDENSER
LIQUID STRAINER
LIQUID LINE VALVE

Figure CR24-2 A simple refrigeration system with a single compressor and water-cooled condenser.

ELECTRIC POWER
EVAPORATOR
EVAPORATOR
EVAPORATOR
CHECK VALVE
0°F
35°F
45°F
EVAPORATOR-PRESSURE REGULATOR
TO COMPRESSOR MOTOR
LOW-PRESSURE MOTOR CONTROL
CONDENSER

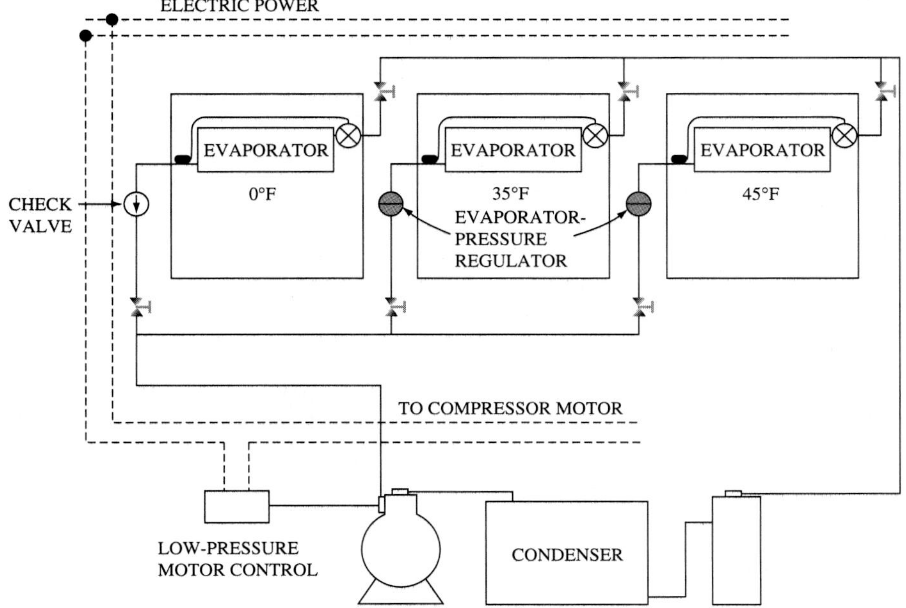

Figure CR24-3 Multiple evaporator system using evaporator-pressure regulators to maintain different conditions in each box.

On a small system it is general practice to install a surge tank in the suction line of the low-temperature evaporator to prevent short cycling.

Multiple-compressor Systems

Where the refrigeration load varies over a wide range, such as in supermarkets, it is desirable to use multiple compressors connected in parallel, as shown in Fig. CR24-4.

A suction pressure control is used to turn on and off individual compressors as required to match the load. These controls also have the ability to change the lead compressor to obtain approximately equal running time on each compressor. A single condenser and receiver is used for all units.

The refrigerant piping must be done properly, as shown in Figs. CR24-5 and CR24-6. Referring to Fig. CR24-5, the suction piping should be brought in above the level of the compressor. With multiple compressors, a common suction header should be used and the piping should be designed so that the oil return to several compressors is as nearly equal as possible.

If an oil level control is not used, the discharge piping as shown in Fig. CR24-6 should be used with the discharge piping running to a header near to floor level. With this arrangement, a discharge line trap is not required since the header serves this purpose.

It is also important to provide equalization of the oil level in all compressors using an arrangement such as shown in Fig. CR24-7. Oil is pumped by the compressor to a common discharge header and into an oil separator. Since the oil separator has a large holding capacity, oil is then transferred to an oil reservoir. Pressure in the reservoir is reduced by boiling the refrigerant contained in the oil and relieving the pressure above the oil through a vent line to the suction header. An oil-level control meters the oil to the compressors equal to their pumping rate and thereby maintains the oil level specified by the manufacturer.

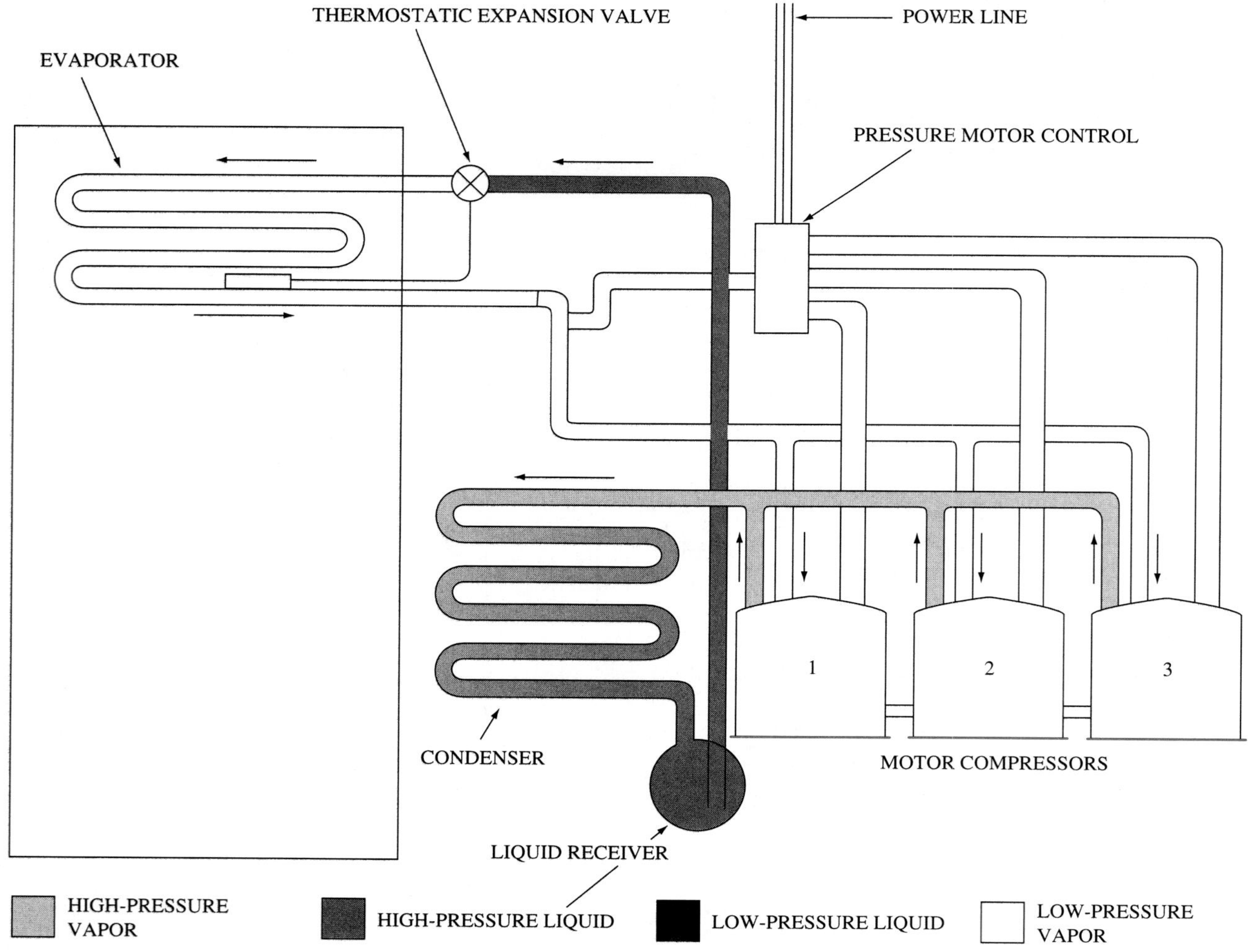

Figure CR24-4 Multiple compressor system using three compressors connected in parallel.

Knowledge Extension

Modern refrigeration systems commonly use multi-compressor rack systems. Learning about this type of equipment can be difficult at first. If they appear overwhelming upon the first visit to the mechanical room, then a systematic analysis should be done. Initially, you should become more familiar with this type of system by locating the blueprints or engineer's drawings. If drawings are unavailable, then begin to familiarize yourself by identifying one system at a time by tracing piping from the condenser to the metering device, evaporator, and back to the compressor. If you are on an emergency service call, then locate the single compressor or zone that is not refrigerating. Look and listen for obvious indications of problems. This includes compressors with low oil level, cool discharge lines, bubbles in sight glass, and compressors short cycling on and off. Remember, the majority of refrigeration problems are electrical in nature. This means the technician should check for electrical problems also, including circuit breakers, fuses, manual resets, and broken control circuits. If your knowledge of these types of systems is limited and you need more help, contact the equipment manufacturer for additional technical support.

Multistage-compressor Systems

There are two general methods of multistaging compressors, compound and cascade.

A *compound system* offers a method for producing low-temperature refrigeration. In order to achieve a low suction pressure, two compressors are connected in series as shown in Fig. CR24-8.

The suction line from the evaporator feeds the first compressor. Then the discharge from the first compressor enters the suction of the second compressor. The discharge from the second compressor goes to the condenser. Although both compressors handle about equal loads, the first compressor handles the greatest volume of gas, since the density of the gas in the first stage is less.

An important additional device in this system is the *intercooler* between the two stages that reduces the superheat in the discharge gas from the first compressor. This superheat, produced by the work done by the first compressor, must be removed to keep the gas temperature of the second compressor within limits. A single temperature control operates both compressor motors. On start-up, both motors are started at the same time.

It would be possible to use a single compressor to achieve the multi-stage effect; however, the disadvantage of a single multi-stage compressor is that there is a fixed ratio between the high-side and the low-side volumes which only applies to applications where this ratio produces a satisfactory evaporating temperature. Compound systems using two compressors gives the added flexibility of producing a wider range of low evaporating temperatures. Compound arrangements of compressors is seldom considered unless the desired evaporating temperature is below −20°F.

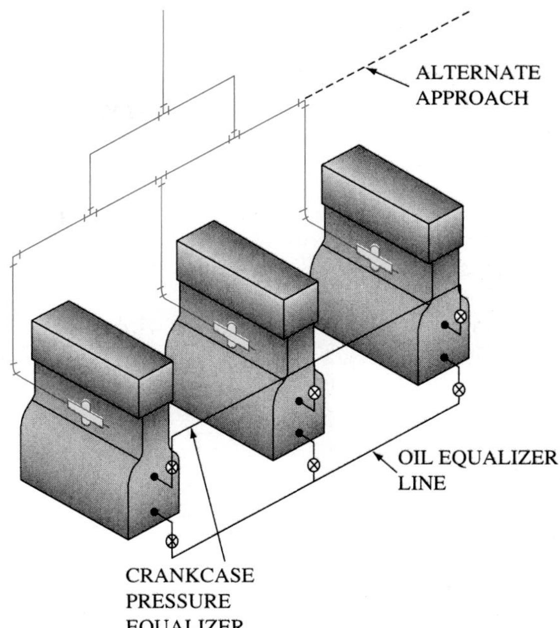

Figure CR24-5 Suction line piping for parallel compressors.

A single refrigerant is used throughout the entire compound system, in contrast to a cascade system which may use separate refrigerants for each compressor.

A *cascade system* is also used for low-temperature refrigeration, but has the capability of reaching much lower temperatures than the compound system. The cascade system shown in Fig. CR24-9 (p. 460) uses two compressors;

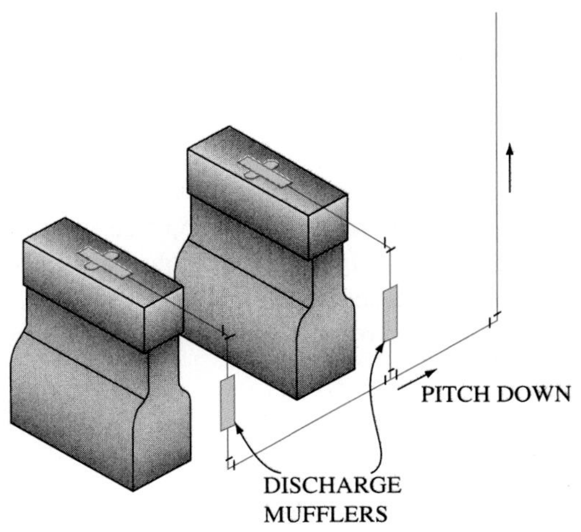

Figure CR24-6 Discharge line piping for parallel compressors.

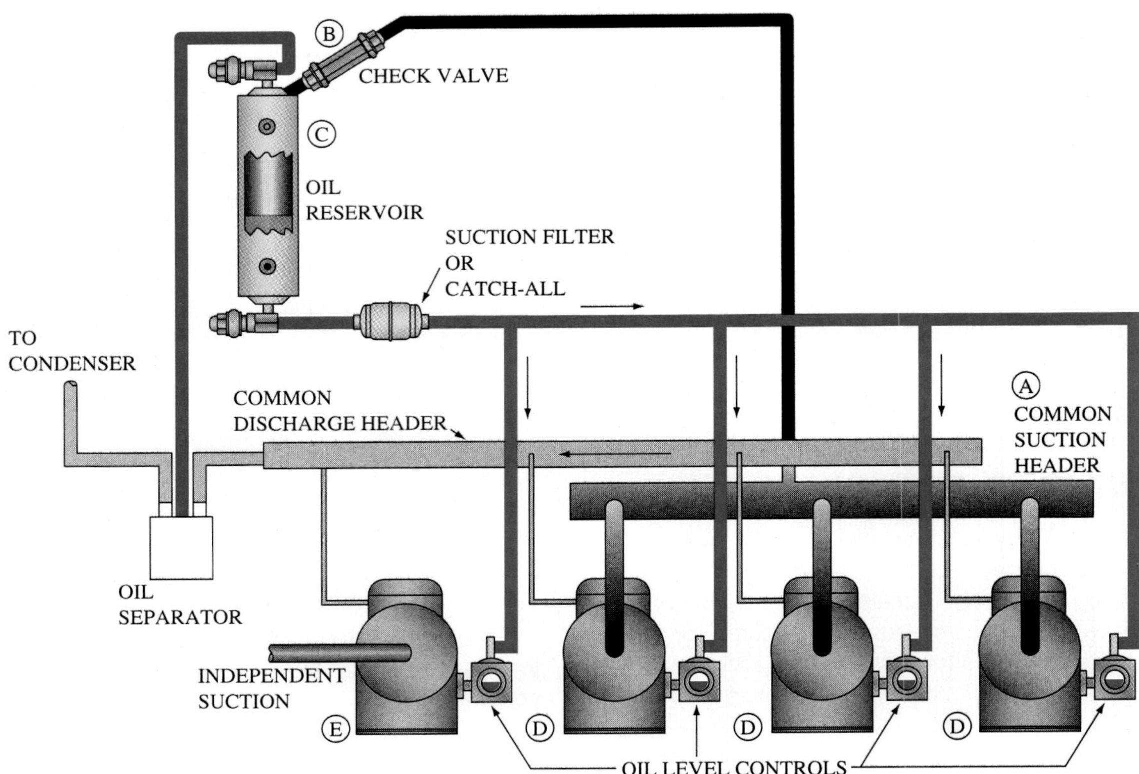

Figure CR24-7 Oil level control system for multiple compressors. (Courtesy of Sporlan Valve Company)

however, for lower temperatures, three or more stages (compressors) are used.

The system has an interesting configuration. Each stage has a separate refrigeration circuit. Each stage can use a different refrigerant if there is an advantage in doing so. The interconnection is through a common heat exchanger. In the system shown in Fig. CR24-9, the evaporator of the second-stage machine cools the condenser of the first-stage machine.

When the system is started up, the high-stage compressor is started first. After it lowers the evaporating temperature to the required level, the low-stage compressor is started. To facilitate shutdown, an expansion tank is added to the low-stage circuit, as shown in Fig. CR24-10 (p. 461).

This is done to prevent the pressures from becoming excessive when the refrigerant is subjected to room temperatures. A relief valve on the high-side circuit discharges into the expansion tank. Otherwise it would be necessary to transfer refrigerant to storage cylinders during shut down.

Evaporative-cooling Systems

Evaporative cooling is the process of absorbing heat by the evaporation of water. Approximately 970 Btu of heat is required to evaporate a pound of water. Two general types of evaporative cooling are:

a. An open system
b. A closed circuit system.

A diagram of an *open system* is shown in Fig. CR24-11 (p. 461).

A cooling tower such as shown in Fig. CR24-12 (p. 462) uses evaporative cooling to reduce the temperature of the hot water from the condenser so that it can be reused at a lower temperature.

Greenhouses (Fig. CR24-13, p. 462) use evaporative cooling to provide a safe temperature for plants. Roof sprays are used to reduce roof temperatures to lower the building heat load. All of these systems are effective as long as the water temperature provided for evaporation is higher than the wet-bulb temperature of the air.

A *closed-circuit* evaporative cooler is built like an evaporative condenser, as shown in Fig. CR24-14 (p. 462). Instead of refrigerant circulating through the coil, the coil is supplied with hot water from the condensers. Water is cooled by the evaporative action of the water on the outside of the tubes and is pumped back to the condensers for reuse. The evaporative cooler coil, the circulating pump, the water-cooled condensers, and an expansion tank are piped in series to form a closed loop.

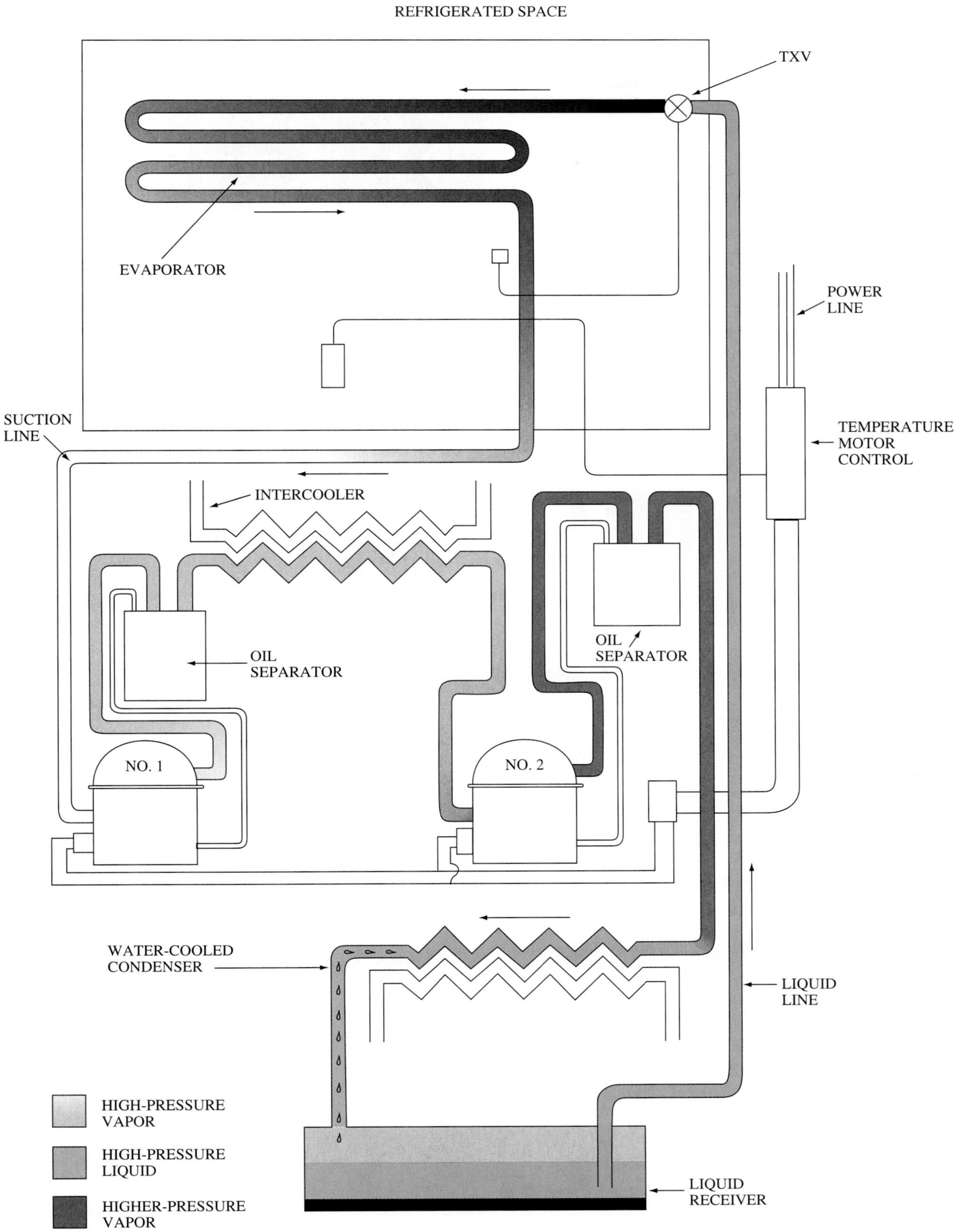

Figure CR24-8 Compound refrigeration system using two compressors.

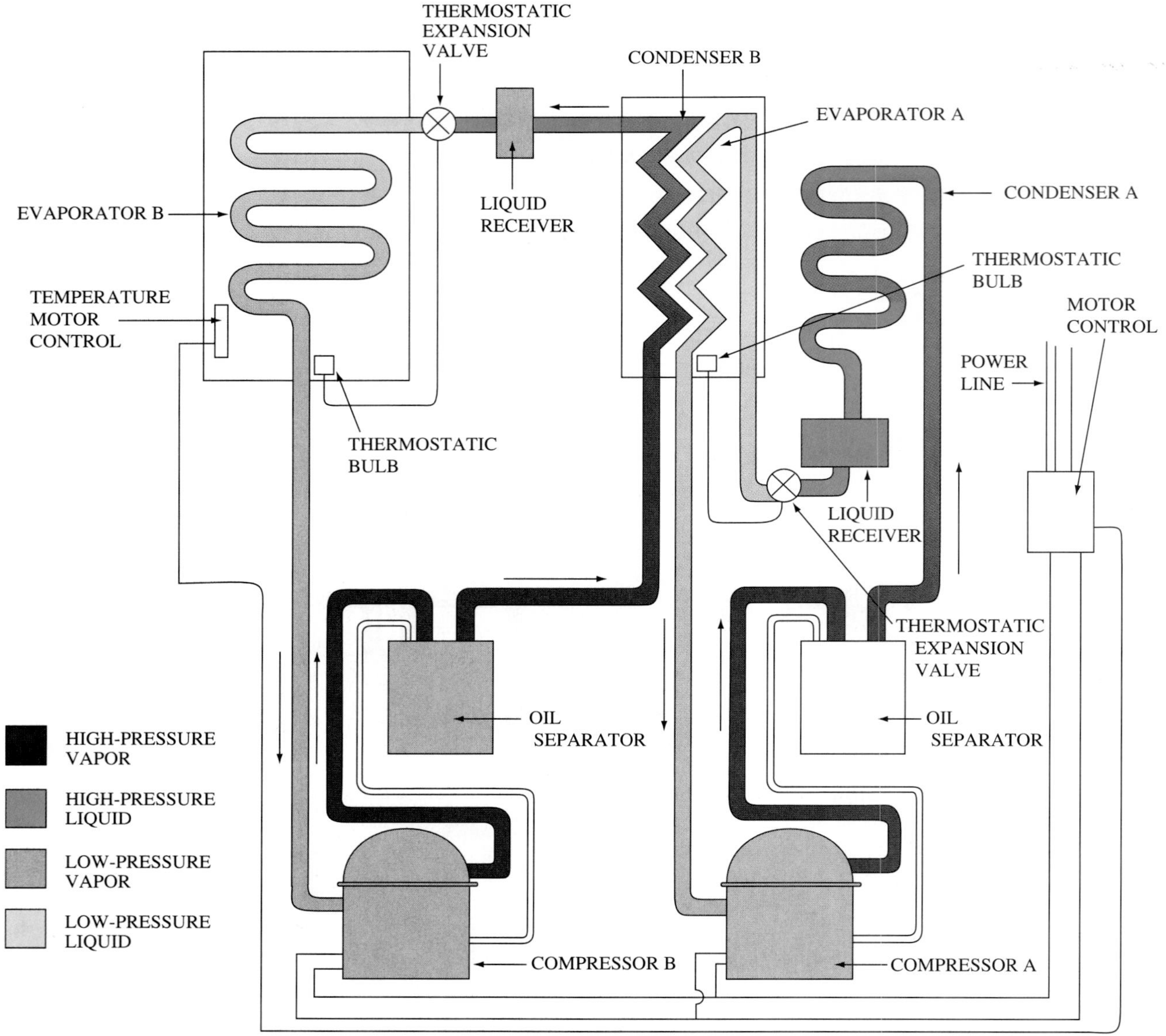

Figure CR24-9 Cascade refrigeration system using two compressors.

PRACTICING PROFESSIONAL SERVICES

When working with commercial refrigeration equipment, it is professional courtesy to always enter into your customer's office or work place with clean clothes and a neat appearance. If you are in a situation where you expect to get extremely dirty while working on the equipment and do not have a change of clothes, than wear a cover-all type of protective clothing.

When you are done with the job, clean up all waste materials in the mechanical room including empty cylinders, boxes, and related items that you generated. Be sure to explain to your customer the remedy and costs associated with your repair. Invoice your customer if necessary.

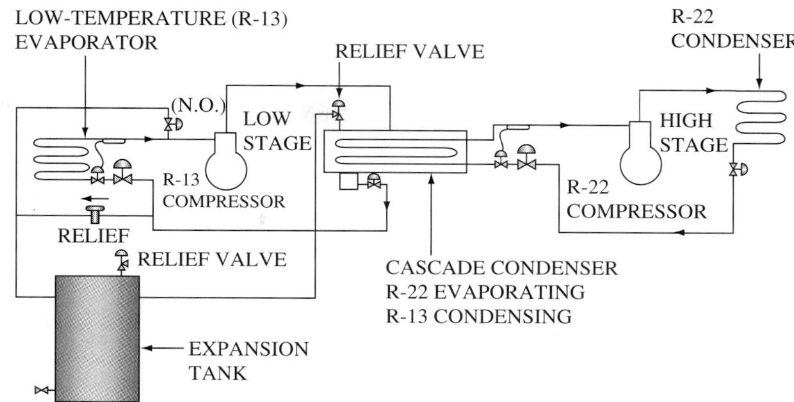

Figure CR24-10 Cascade refrigeration system showing expansion tank added to the low-stage circuit.

Secondary Refrigerants

A secondary refrigerant is a fluid cooled by direct refrigeration and used to transfer cooling to a distant area where cooling is needed and long direct-expansion lines are not practical or economical. A good example of the common use of a secondary refrigerant is in the application of *water chillers* to provide cooling for a large building.

A very effective use of water as the refrigerant for cooling is the *ice bank* application, as shown in Figure CR24-15.

These systems are used for comfort-cooling installations where cooling is required only during the day. The refrigeration system builds up ice banks during the night when electric rates are low. During the day, water is circulated through the ice banks and through coils in the air-conditioning system to provide cooling.

Where the secondary coolants are needed below the freezing temperature of water, a brine is used. *Brine* is the name given a solution of a substance in water that lowers the freezing point of the water. Common additives are calcium chloride, sodium chloride, ethylene and propylene glycols, methyl alcohol, and glycerin. The amount of additive used affects the freezing point of the solution as shown in Fig. CR24-16.

A good example of the use of a brine for secondary cooling is the *ice rink* application. The general plan for the piping is shown in Fig. CR24-17 (p. 464).

The brine used in modern rinks is a glycol solution. The layout includes supply and return headers with piping loops under the ice to carry the brine. The brine is pumped through the piping. It enters at a temperature of 12°F (−11°C) and returns to the chiller at 14°F (−10°C). Chillers are usually in the 100- to 125-Hp range.

Thermoelectric Refrigeration

The thermoelectric refrigerator uses the *Peltier effect,* a physical principle discovered by Jean Peltier in 1834. He found that if direct current was passed through the junction of two dissimilar metals, the junction point would be-

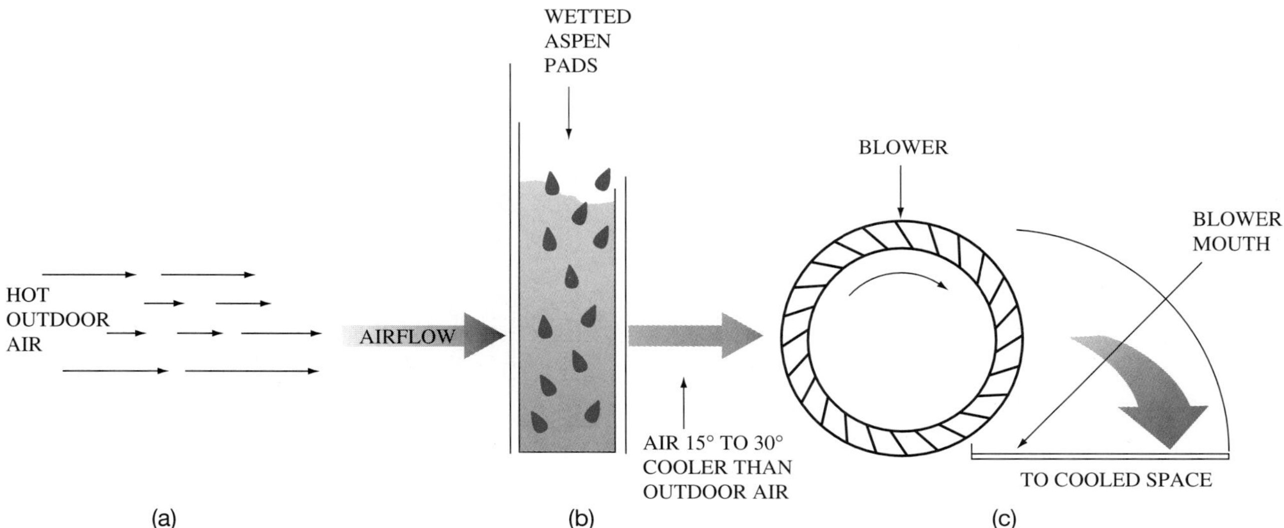

Figure CR24-11 Schematic drawing of an evaporative cooling system.

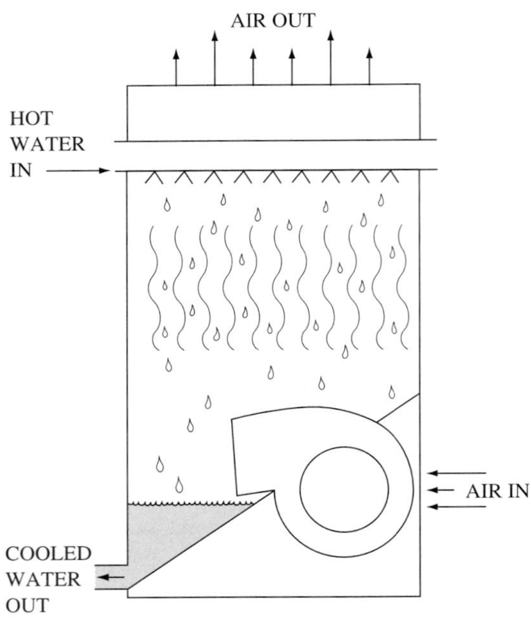

Figure CR24-12 Schematic drawing of a cooling tower.

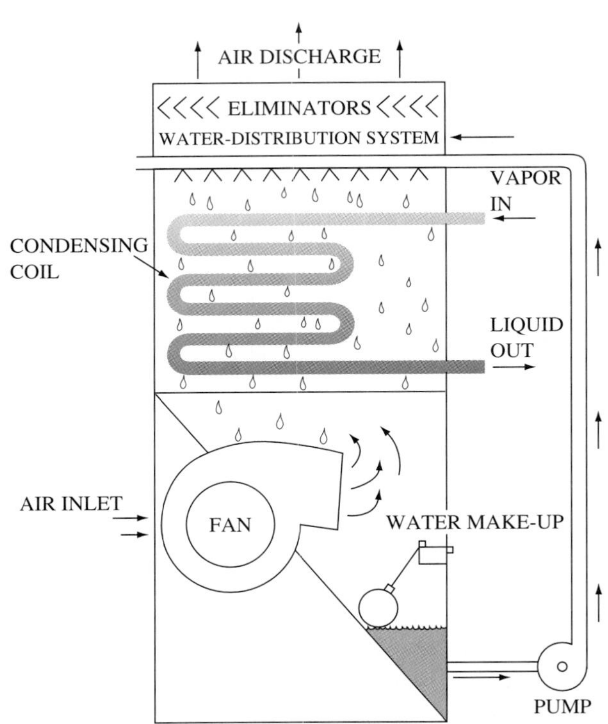

Figure CR24-14 Diagram of a closed-circuit evaporative cooler.

come hot or cold depending on the direction of current flow.

Fig. CR24-18 illustrates the construction of a thermoelectric refrigerator using semiconductors.

There are two types of semiconductors, the *N-type* which conducts electricity by the flow of negatively charged particles, and the *P-type* that conducts electricity by the flow of positively charged particles. When current flows from the N-type into the P-type, the junction absorbs heat. When current flows from the P-type to the N-type, the junction becomes warm and gives off heat. A single thermocouple produces a small amount of cooling so a number of N-P junctions put in series produce a greater cooling effect. The diagram in Fig. CR24-18 shows a series of thermoelectric couples which pick up heat from the inside of the box and transfer the heat to the outside of the box.

Thermoelectric refrigerators are available in small portable battery-operated units. Electrically operated units are often found in hotel guest rooms. Submarines and space ships are often equipped with thermoelectric refrigerators.

One advantage of these units is that they have no moving parts requiring service. A disadvantage is that they are not as efficient as other types of electric refrigerators.

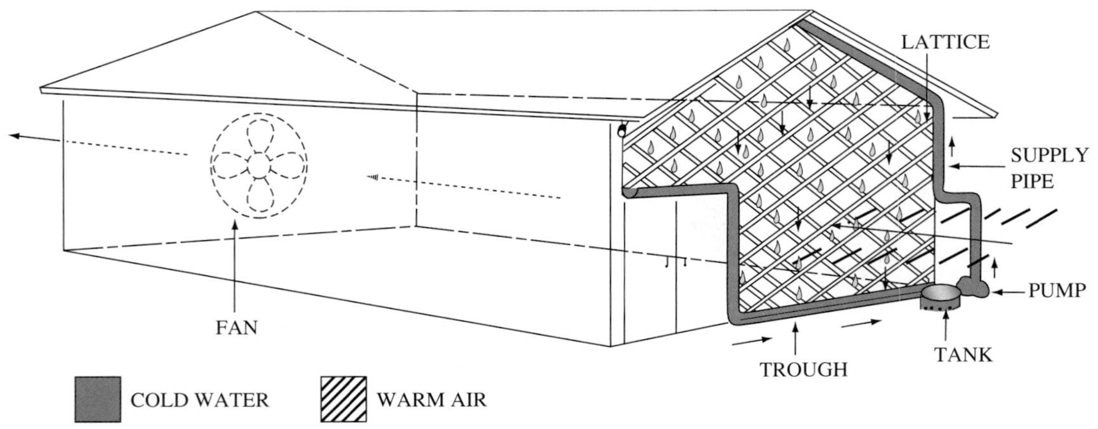

Figure CR24-13 Application of evaporative cooling to a greenhouse.

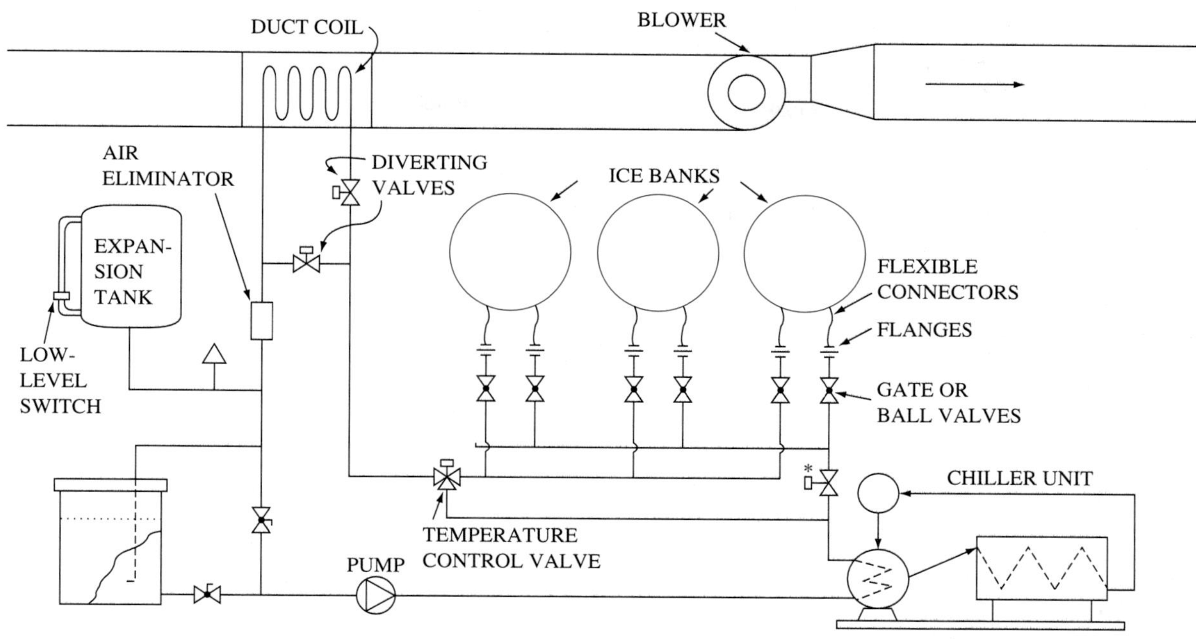

Figure CR24-15 Ice bank used for cooling storage. (Courtesy of Calmac Manufacturing Corp.)

Expendable Refrigerants

An expendable refrigerant is one that performs cooling by permitting the liquid refrigerant to boil at atmospheric pressures and is released to the atmosphere after the cooling operation is completed. Liquid nitrogen, with a boiling temperature of –320°F, and liquid carbon dioxide, with a boiling temperature of –108°F, are the most common refrigerants used. These refrigerants, when released to the atmosphere, are not considered destructive since they are both normally present.

The most common usages are for transportation vehicles, truck bodies, and railroad cars, and for shipping containers used to transport perishable items.

There are two basic ways to apply the refrigerants: (1) using cold plates (Fig. CR24-19), and (2) spray cooling (Fig. CR24-20).

Alcohol % by Wt	°F	Glycerine % by Wt	°F	Ethylene Glycol % by Vol	°F	Propylene Glycol % by Vol	°F
5	28.0	10	29.1	15	22.4	5	29.0
10	23.6	20	23.4	20	16.2	10	26.0
15	19.7	30	14.9	25	10.0	15	22.5
20	13.2	40	4.3	30	3.5	20	19.0
25	5.5	50	–9.4	35	–4.0	25	14.5
30	–2.5	60	–30.5	40	–12.5	30	9.0
35	–13.2	70	–38.0	45	–22.0	35	2.5
40	–21.0	80	–5.5	50	–32.5	40	–5.5
45	–27.5	90	+29.1			45	–15.0
50	–34.0	100	+62.6			50	–25.5
55	–40.5					55	–39.5
						59	–57.0
						Above 60% fails to crystallize at – 99.4°F	

Figure CR24-16 Freezing points of aqueous solution used for secondary cooling.
(Copyright by the American Society of Heating, Refrigerating, and Air-Conditioning Engineers, Inc. Used by permission.)

PREVENTIVE MAINTENANCE

Whenever working with secondary refrigerants and water chillers, be sure to calibrate and test the thermometers, operating temperature controls, and safety freeze circuits associated with controlling the water temperature. Good preventive maintenance requires checking these at the start-up of each cooling season. Thermometers or controls out of calibration can give false indications of water temperature, which could lead to chiller vessels freezing. If the vessel freezes it is normally ruined, which is costly to the customer and time-consuming to replace.

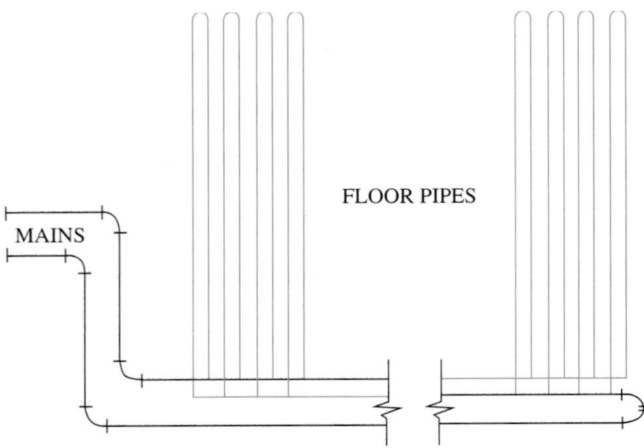

Figure CR24-17 Piping schematic for an ice rink.

A diagram showing a typical spray-type system for a truck body is shown in Fig. CR24-21. The storage cylinder is filled through the fill valve. The refrigerant is released at a rate to satisfy the thermostat which regulates a liquid control valve in the refrigerant line. Temperatures are maintained which range between –20 and 60°F depending on the product being cooled. Gauges record the tank pressure and the liquid level in the storage cylinder. A safety-relief valve and a vent valve are provided for the storage tank. A safety vent is installed in the truck body which will automatically open if the refrigerant pressure in the truck increases.

Safety precautions must be observed. Physical contact with the low-temperature refrigerants can cause immediate freezing of body tissue.

Figure CR24-18 Thermoelectric module used for cooling.

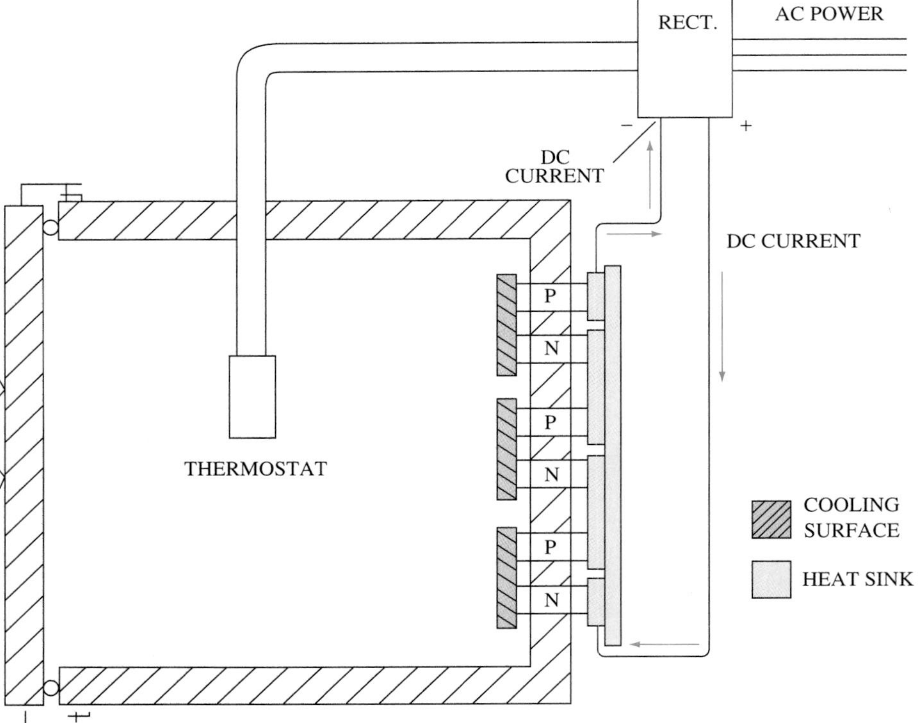

Figure CR24-19 Expendable refrigeration system using cold-plate evaporator.

Absorption Systems

Absorption systems have certain unique characteristics that give them a special use:

1. They use a heat source for power to operate the cycle rather than mechanical energy as required for compression refrigerating systems,
2. The absorption systems have very few moving parts. On the smaller systems the moving parts are in the controls. The larger systems may use fans or pumps.
3. Most systems are sealed at the factory and the refrigeration cycle is not normally serviced in the field.

Many of these systems are used in recreation vehicle (RV) units and for refrigerators where electricity is not available. The systems are also used for comfort air conditioning, although they are less popular than mechanical systems. The heat source can be liquid petroleum, kerosene, steam, solar heat, or an electrical heater.

Fig. CR24-22 shows a diagram of the operation of a typical absorption unit using ammonia as the refrigerant and an aqueous ammonia solution as the absorbent.

A pump is used to maintain the pressure difference between the high and low sides of the system. The same pump transfers the strong solution of water and ammonia from the absorber to the generator.

In the generator the solution of water and ammonia is heated, boiling off the ammonia vapor. The ammonia vapor then goes to the condenser (water-cooled or air-cooled) where the ammonia is condensed. The liquid ammonia then enters the evaporator where it boils and picks up heat from the product load, which in this case is water. The ammonia vapor that has boiled off then goes to the absorber where it enters into solution with the water. From there, the pump forces the strong solution into the generator and the cycle is repeated.

Larger units use water as a refrigerant and a lithium bromide solution for an absorber. A diagram of this absorption circuit is shown in Fig. CR24-23 (p. 468).

This system uses a solution pump, a refrigerant pump and a solution heat exchanger in addition to the standard

Figure CR24-20 Expendable refrigeration system using a spray header.

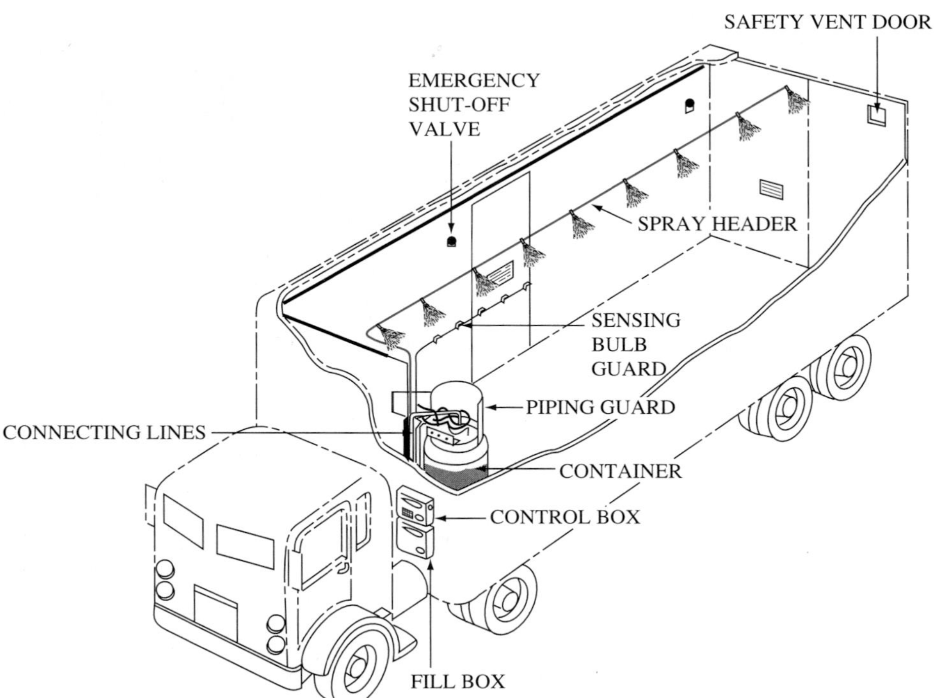

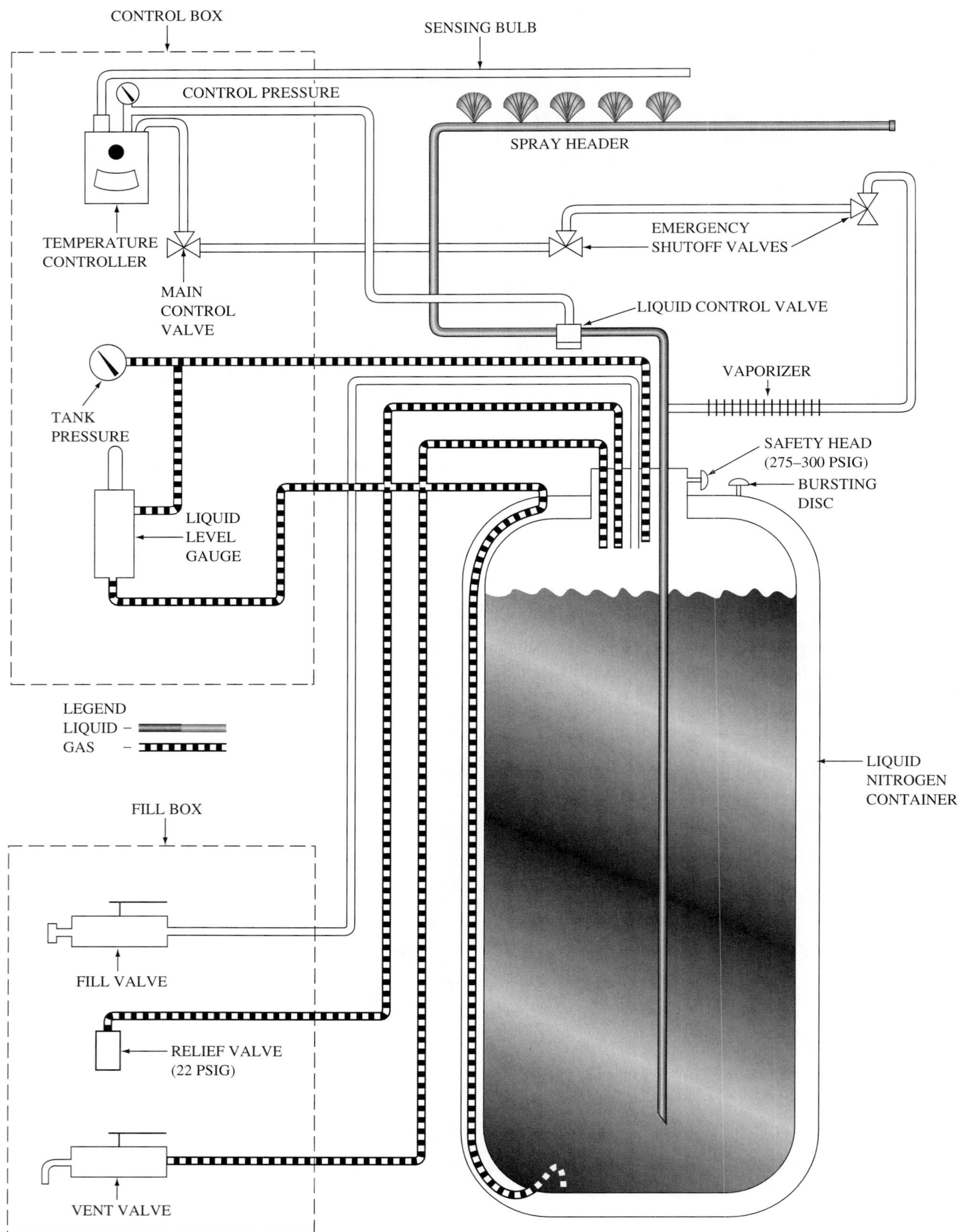

Figure CR24-21 Piping schematic for an expendable refrigeration system.

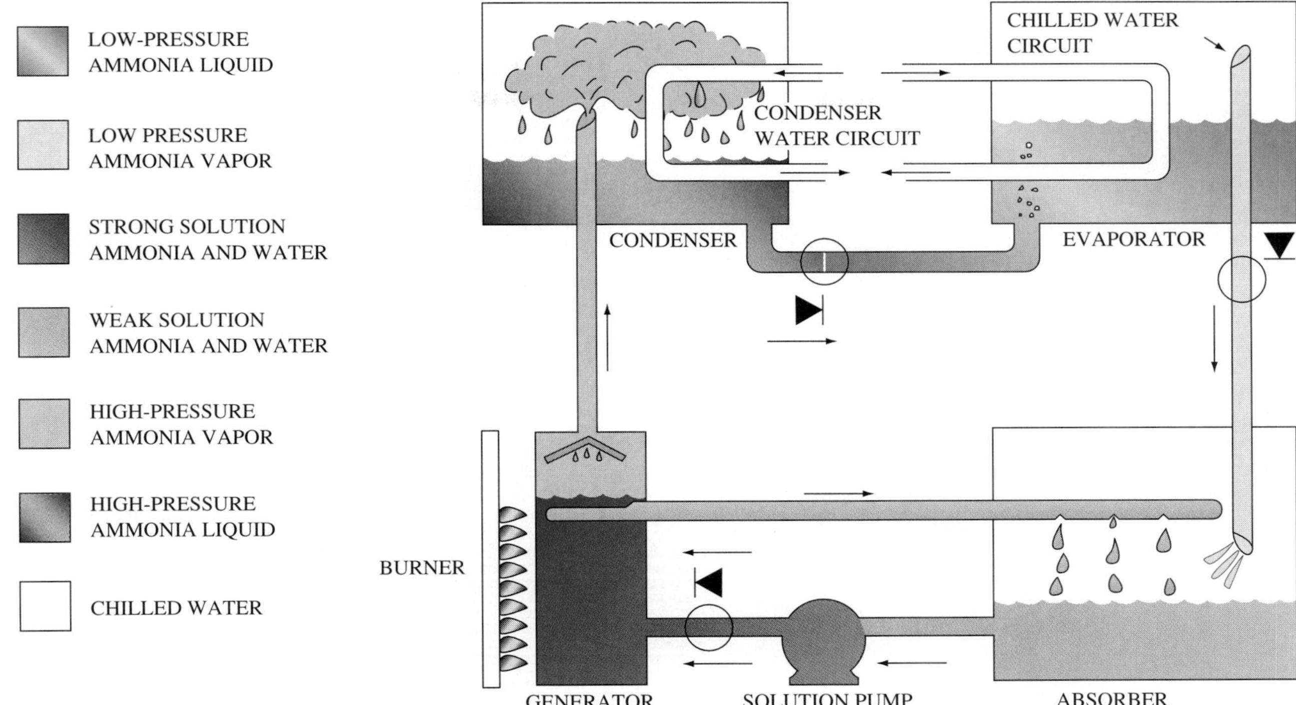

Figure CR24-22 Schematic diagram of an absorption system using ammonia as a refrigerant and water as an absorbent.

parts. The generator-condenser section is separated from the evaporator-absorber section, as shown.

CR24-2 SPECIAL COMPONENTS

The following is a brief review of the operation of some of the special components that form an important part of many types of refrigeration products. These are discussed in detail in other chapters of the text.

CR24-2.1 Defrost Systems

The most popular types of defrost systems are air, electric, and hot gas. Other defrost systems include glycol, water, and reversed cycle.

The glycol type requires special apparatus to heat the glycol solution, remove the excess water that is absorbed, and maintain the proper concentration of glycol.

The water-defrost system uses a relatively large quantity of water, relying on the water to quickly move through the unit, melting the ice and reaching the drain before it freezes.

Both the glycol- and water-defrost arrangements have gradually been replaced by the electric and hot-gas methods.

The reverse-cycle defrost operates like a heat pump, using a four-way valve to divert discharge gas to the evaporator for defrosting. The hot-gas method is preferred over the reverse-cycle since it offers a closer control of defrost temperatures.

A typical electric-defrost unit cooler is shown in Fig. CR24-24.

CR24-2.2 Condenser Pressure-temperature Control

For *air-cooled condensers* some of the options for controlling condenser pressure are: fan cycling, air-volume control, fan-speed control and refrigerant flooding. For *water-cooled condensers* a water-regulating valve (Fig. CR24-25) can be used.

On a cooling-tower system the water temperature is controlled by cycling the fan or water bypass. On evaporative condensers head pressure is controlled by air supply dampers.

CR24-2.3 Evaporator Pressure-temperature Control

This arrangement (Fig. CR24-26) usually requires an evaporator-pressure regulator (EPR) valve. It is used

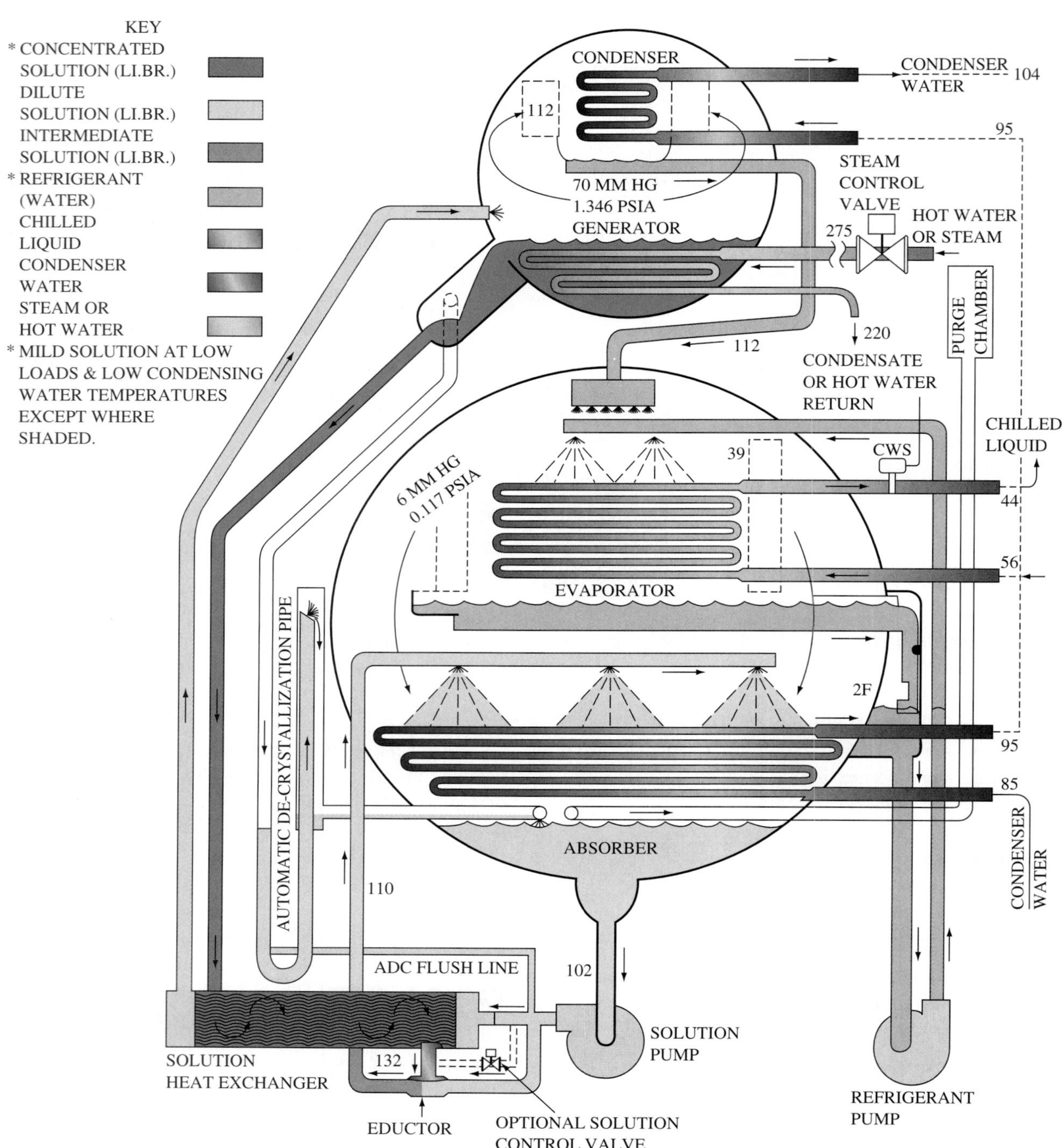

Figure CR24-23 Schematic diagram of an absorption system using water as the refrigerant and lithium bromide solution as the absorbent.

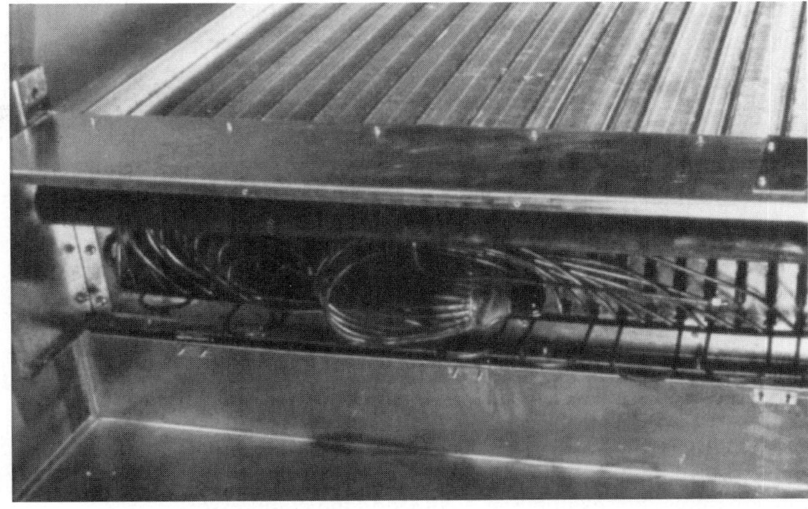

Figure CR24-24 Electric defrost-type unit cooler. (Courtesy of Heatcraft Refrigeration Products)

Figure CR24-25 Condenser water-regulating valve used for head-pressure control.

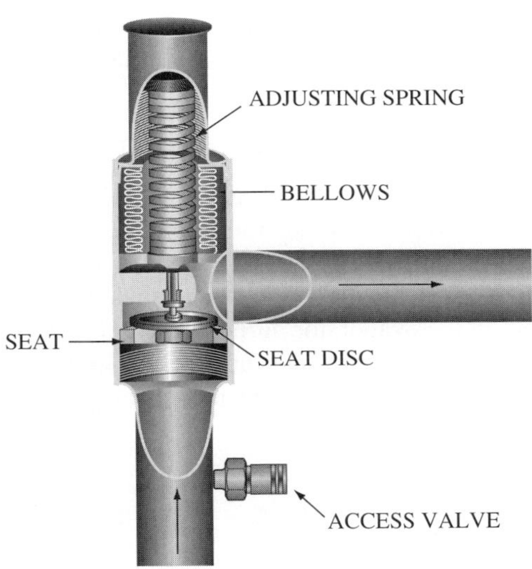

Figure CR24-26 Evaporator-pressure regulating (EPR) valve. (Courtesy of Sporlan Valve Company)

primarily on multiple evaporator systems where different temperatures are maintained in each unit. A two-temperature control can be applied where only two box temperatures are needed.

CR24-2.4 Low Ambient Controls

A diagram of a low-ambient-control system using the flooded-condenser principle is shown in Fig. CR24-27.

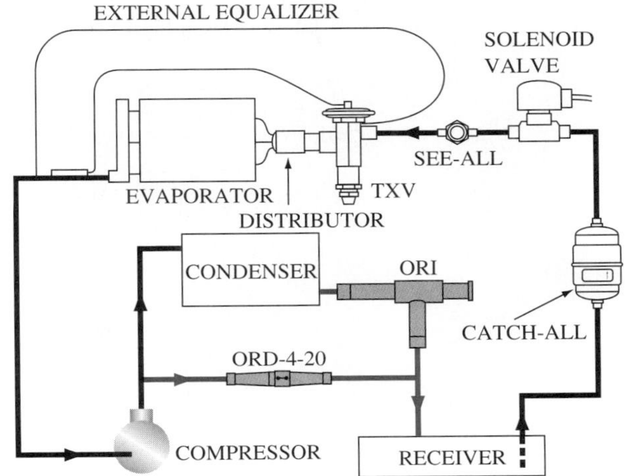

Figure CR24-27 Low ambient control for air-cooled condenser. (Courtesy of Sporlan Valve Company)

Where air-cooled condensers are subjected to low outside temperatures, the condensing pressure can drop below the setting of the low limit control and prevent starting. To eliminate this problem, some type of head-pressure control needs to be employed. Under extremely cold conditions the partially flooded evaporator needs to be used to maintain high enough condensing pressure to permit the compressor to start.

CR24-2.5 High-low Pressure Controls

This is a safety control (Fig. CR24-28) and not used as an operating control. It is usually considered to be standard equipment on most commercial compressor systems. It stops the compressor on excessive pressures or extremely low pressures. For example, it would stop the compressor on low pressure if there was a loss of refrigerant. It would stop the compressor on high pressure if the discharge of the compressor was blocked for any reason.

CR24-2.6 Suction Accumulator

This container (Fig. CR24-29) is placed in the suction line close to the compressor suction inlet to catch any liquid refrigerant which would cause slugging in the compressor. It is important particularly on start-up since liquid may accumulate in the evaporator during the OFF-cycle. Using a suction accumulator not only protects the compressor but also permits low superheat, thus increasing the capacity of the evaporator.

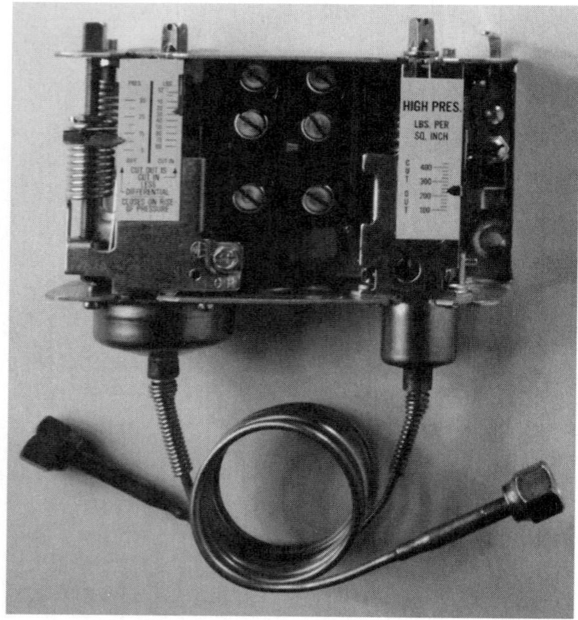

Figure CR24-28 High-low pressure limit control. (Photo Courtesy of Johnson Controls/PENN)

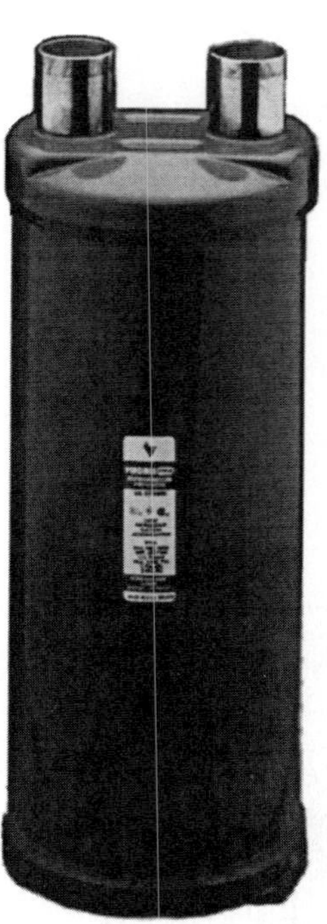

Figure CR24-29 Suction-pressure accumulator used to prevent liquid refrigerant from entering the suction of the compressor. (Courtesy of Virginia KMP Corporation)

CR24-2.7 Suction-to-liquid Heat Exchanger

These units (Fig. CR24-30) serve to subcool the liquid refrigerant and superheat the suction gas, as well as increase the capacity of the system. They are commonly used in commercial refrigeration evaporators.

CR24-2.8 Oil Separator

Oil separators (Fig. CR24-31) are placed on the discharge of the compressor to remove excess oil being carried with the discharge gas. The oil that is removed is automatically returned to the crankcase of the compressor. Certain compressors, such as the screw type particularly, require their use.

CR24-2.9 Receivers and Valves

Liquid receivers (Fig. CR24-32) provide a storage tank for the refrigerant in the system.

The receiver needs to have a sufficient volume to contain the refrigerant when the system needs to be opened up for service. Fig. CR24-33 shows the recommended receiver sizes for various systems. The receiver outlet pipe inside the receiver runs to the bottom of the tank, as

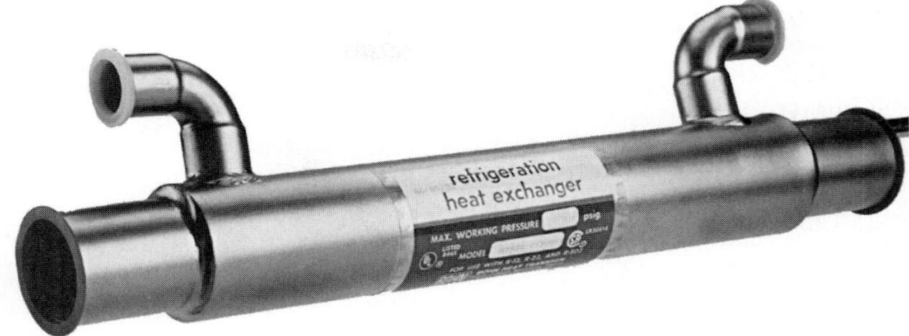

Figure CR24-30 Suction-to-liquid-line heat exchanger used to subcool the liquid and superheat the suction gas. (Courtesy of Heatcraft Refrigeration Products)

Figure CR24-31 Oil separators used to remove oil from the discharge gas. (Courtesy of Virginia KMP Corporation)

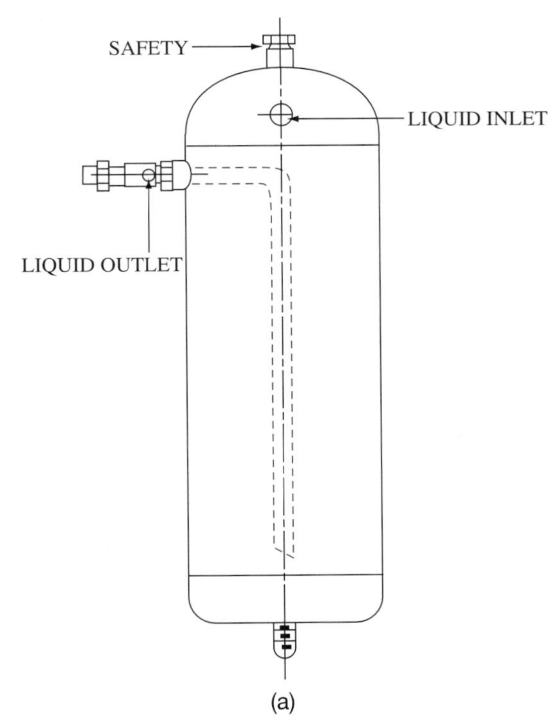

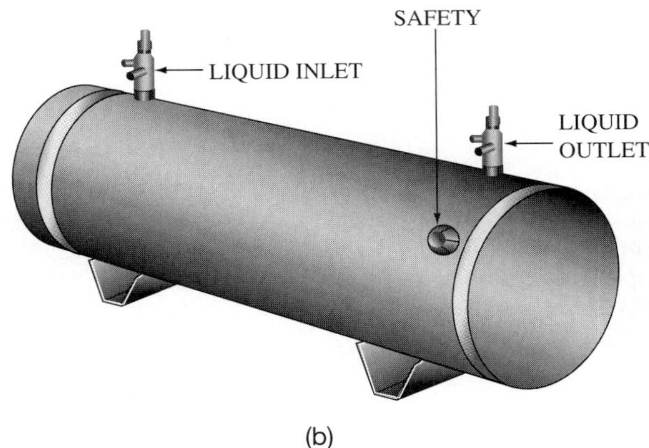

Figure CR24-32 Liquid receiver showing inlet and outlet valves and safety relief. (Courtesy of Standard Refrigeration Co.)

shown in Fig. CR24-34. This opening has a fine meshed screen to prevent contamination from entering.

Receivers usually are equipped with both an inlet and an outlet valve. The outlet valve is called a *king valve*. A system can be charged with liquid refrigerant through the gauge port on the king valve. When this is done the inlet valve is closed or throttled down to permit the system to draw refrigerant from the attached liquid cylinder.

Receivers are equipped with safety devices. The minimum requirement is a fusible plug, although receivers may also have a pressure-relief valve. If the receiver is located inside the building, a pipe to the outside is connected to the relief valve.

On systems with air-cooled condensers, using a flooded-condenser arrangement for low ambient control, the receiver may be located under the condenser. It has sufficient capacity to hold the added refrigerant charge.

Recommended Liquid Receiver Volumes					
		Weight (lb)			
		Refrigerant			
Hp	Volume (cu. in.)	R-12	R-22	R-500	R-502
½	150	6.8	6.2	5.9	6.3
¾	225	10.3	9.3	8.9	9.4
1	300	13.7	12.4	11.9	12.9
1½	450	20.5	18.6	17.9	19.3
2	600	27.4	24.8	23.8	25.8
3	750	35.0	32.0	31.8	33.0
5	900	41.0	37.0	35.5	38.5
7½	1500	70.0	64.0	61.6	66.0

Figure CR24-33 Recommended receiver sizes. (Courtesy of Standard Refrigeration Co.)

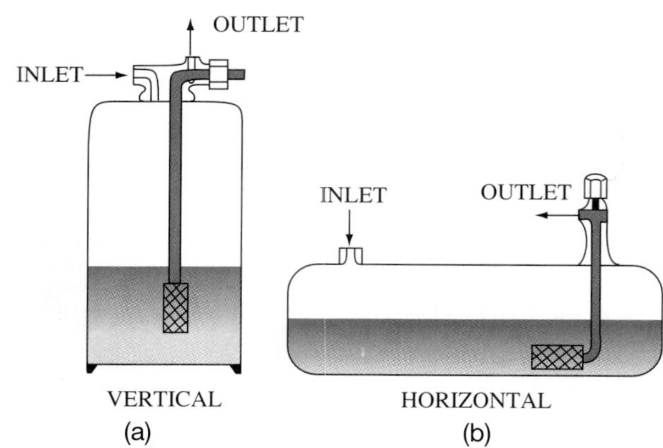

Figure CR24-34 Common types of receivers showing the outlet connection going to the bottom of the tank: (a) vertical liquid receiver; (b) horizontal liquid receiver. (Courtesy of Standard Refrigeration Co.)

CR24-2.10 Hot-gas Mufflers

Mufflers (Fig. CR24-35) are used in the discharge piping from compressors with remote air-cooled condensers located outside. To some extent they are valuable for reducing sound, but their greatest value is in dampening vibration from the compressor that is carried along the discharge line.

CR24-2.11 Heat-recovery Units

A typical heat-recovery system for a supermarket is shown in Fig. CR24-36. Since the refrigeration equipment is used all year, excess heat that would normally be dissipated by the condenser can be diverted to heat the market in the winter.

The diagram shows the hot gas from the compressor bypassed to the heat-recovery air-handling unit. A pressure regulator is used to maintain a sufficient discharge gas temperature for effective heating.

CR24-2.12 Crankcase Pressure Regulator

The installation of a crankcase regulator valve in the suction line to the compressor is shown in Fig. CR24-37. This valve offers an automatic means of preventing the over loading of the compressor on start-up. The function of this valve is similar to the practice of manually throttling the suction service valve until the machine can handle the load. Either of these arrangements will increase the pull-down time, but they are important for protecting the compressor.

Figure CR24-35 Hot-gas-line muffler used primarily with remote air-cooled condensers. (Courtesy of AC & R Components, Inc.)

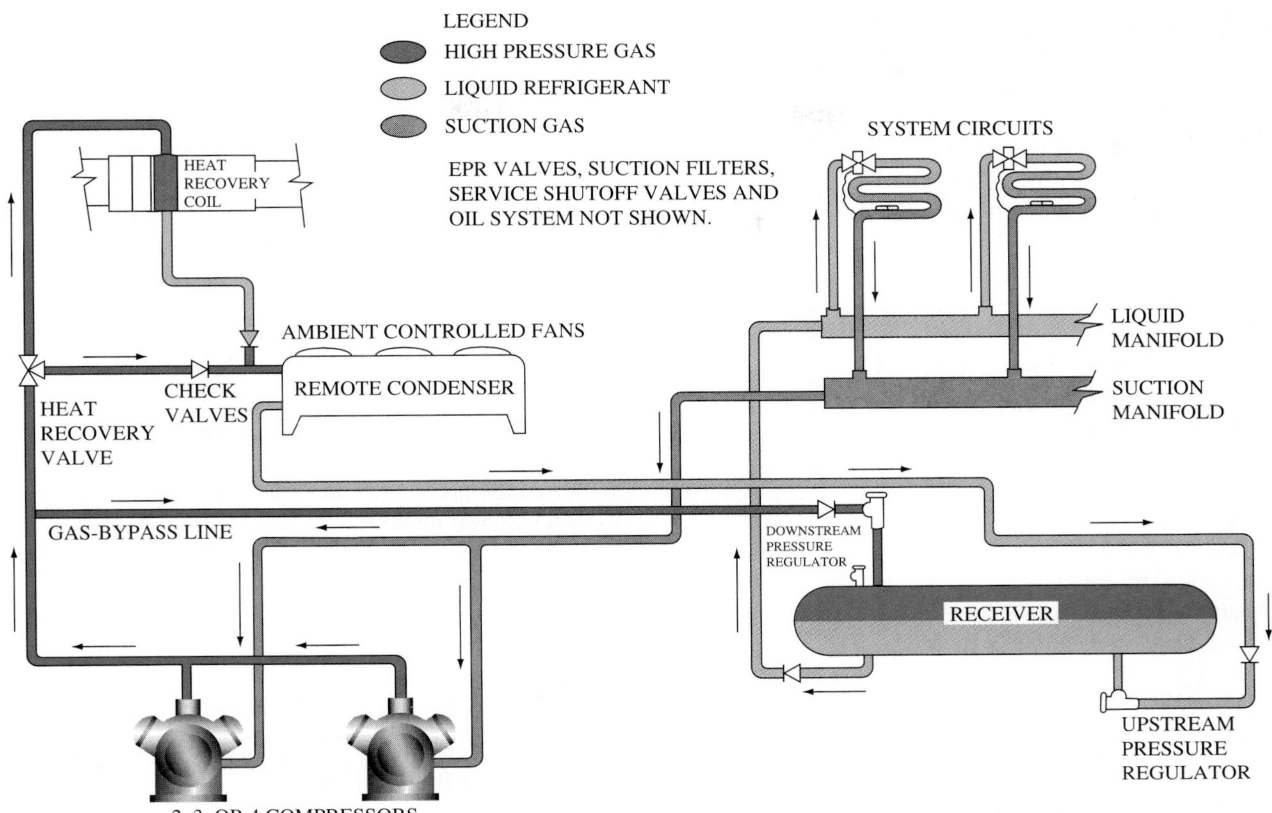

Figure CR24-36 Heat recovery system for a supermarket-type installation. (Copyright by the American Society of Heating, Refrigerating, and Air-Conditioning Engineers, Inc. Used by permission.)

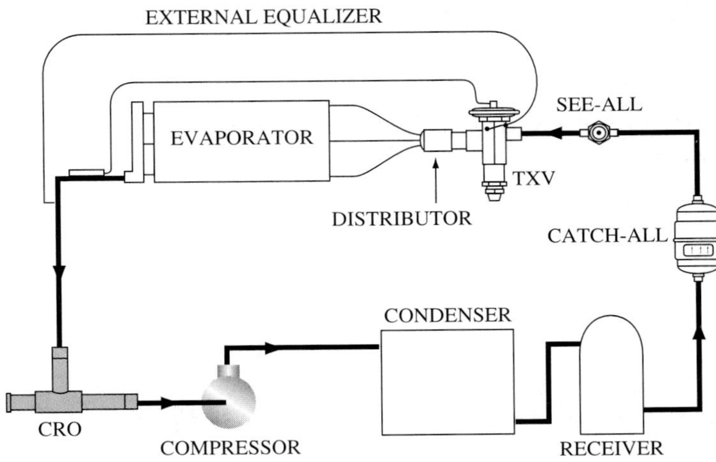

Figure CR24-37 Crankcase pressure regulator used to limit capacity on start-up. (Courtesy of Sporlan Valve Company)

REVIEW

- Equipment being produced must meet the following criteria:
  The quality of the product being cooled must be satisfactory.
  A minimum amount of energy must be used to perform the operation.
  The process must be operated to comply with environmental laws.
- Types of systems:
  Single component systems
  Multiple evaporator systems
  Multiple compressor systems
  Multi-stage systems
    Compound
    Cascade
  Evaporative cooling systems
  Secondary refrigerants
  Thermoelectric refrigerators
  Expendable refrigerants
  Absorption-refrigeration systems
- Single-component system—Consists of a single compressor, evaporator, condenser, and metering device.
- Multiple-evaporator system—Common in supermarkets, it uses EPR valves to control individual evaporator pressures and temperatures higher than the suction pressure of the system. The system requires a check valve in the suction line from the coldest evaporator to prevent migration of refrigerant. The lowest temperature evaporator must be 60% of the total refrigeration load.
- Multiple-compressor system—Two or more compressors connected in parallel. A suction-pressure control is used to cycle individual compressors for capacity control. The system has a single condenser and receiver. An oil-level control is used or the header is run near the floor.
- Multistage-compressor systems:
  Compound system—Two compressors connected in series. An intercooler is used between the stages to reduce superheat. The desired evaporating temperature is below –20°F.
  Cascade system—Two separate refrigeration systems, with two different refrigerants. The evaporator for the high stage serves as the condenser for the low stage.
- Evaporative cooling—There is no mechanical refrigeration involved. The cooling effect is accomplished by allowing water to evaporate.
- Secondary refrigerants—A refrigeration system is used to chill a liquid, and the chilled liquid is then pumped long distances for use as a cooling medium.
- Thermoelectric refrigerators—Use the Peltier effect to produce small amounts of refrigeration.
- Expendable refrigerants—Liquid nitrogen or liquid carbon dioxide is permitted to flash through a metering device into an evaporator. The vaporized refrigerant is allowed to escape to atmosphere.
- Absorption refrigeration—An ammonia/water or water/lithium bromide chemical system that uses heat as the primary energy source for producing cooling.
- Special components:
  Defrost—Air, electric, hot-gas, and glycol.
  Condenser pressure-temperature control methods:
    Fan cycling
    Air-volume control
    Fan-speed control
    Refrigerant condenser flooding
    Water-regulating valve
  EPR—Evaporator-pressure regulator
  Low ambient controls
  High-low pressure controls
  Suction accumulator
  Suction-to-liquid line heat exchanger
  Oil separator
  Liquid receivers
  Hot-gas mufflers
  Heat-recovery units
  CPR—Crankcase pressure regulators

Problems and Questions

1. Define "refrigeration."
2. What are three concerns in the manufacture of commercial refrigeration equipment?
3. List the types of commercial refrigeration systems.
4. What are the essential elements of a single-compressor (basic-cycle) system?
5. What are some common refrigeration accessories?
6. What is the purpose of system accessories?
7. What control (refrigerant) device is required on multiple evaporator systems?
8. What is the purpose of the check valve in the multiple evaporator system?
9. How much of the total load of the system on multiple-evaporator systems must the lowest temperature coil account for?
10. Multiple parallel compressors are installed in supermarket systems for what purposes?
11. What are the two types of multiple-staging systems? What are they used for?
12. Greenhouses use what type of system to maintain temperatures for plants?
13. What is the principle behind secondary refrigerant systems?
14. What is the physical principle behind thermoelectric refrigeration called?
15. What are some of the unique characteristics of absorption systems?
16. The multiple-evaporator system makes it possible to use a single compressor to control a number of different case or fixture temperatures. True or false?

17. When using multiple evaporators, a check valve is installed in the suction line of the lowest temperature coil. True or false?
18. Multiple compressors are connected in parallel when the refrigeration load varies over a wide range. True or false?
19. Compound systems utilize two compressors with the discharge line from the first compressor entering the suction of the second compressor. True or false?
20. The intercooler reduces the superheat between the cascade compressors. True or false?
21. Evaporative cooling is the process of absorbing heat by the:
    a. Evaporation of refrigerant.
    b. Evaporation of water.
    c. Evaporation of specially designed heat-absorbing chemicals.
    d. None of the above.
22. The cascade system utilizes:
    a. Two to three compressors.
    b. Different stages with different refrigerants.
    c. An evaporator which cools the condenser of the low-temperature unit.
    d. All of the above.
23. A secondary refrigerant is a fluid which is:
    a. Cooled by direct refrigeration.
    b. Used to transfer cooling to a distant area where cooling is needed.
    c. An application for centrifugal water chillers.
    d. All of the above.
24. Thermoelectric refrigeration uses:
    a. Thermal energy from the earth when using a ground-source heat pump.
    b. The principle that electrical energy will move from positive to negative with potential difference applied.
    c. The principle that if DC is passed through the junction of two dissimilar metals, the junction points will become hot or cold depending on the direction of current.
    d. None of the above.
25. Absorption systems:
    a. Use a heat source for power, such as natural gas, LP, or electricity.
    b. Have very few moving parts.
    c. Are sealed at the factory on smaller units and not serviced in the field.
    d. All of the above.

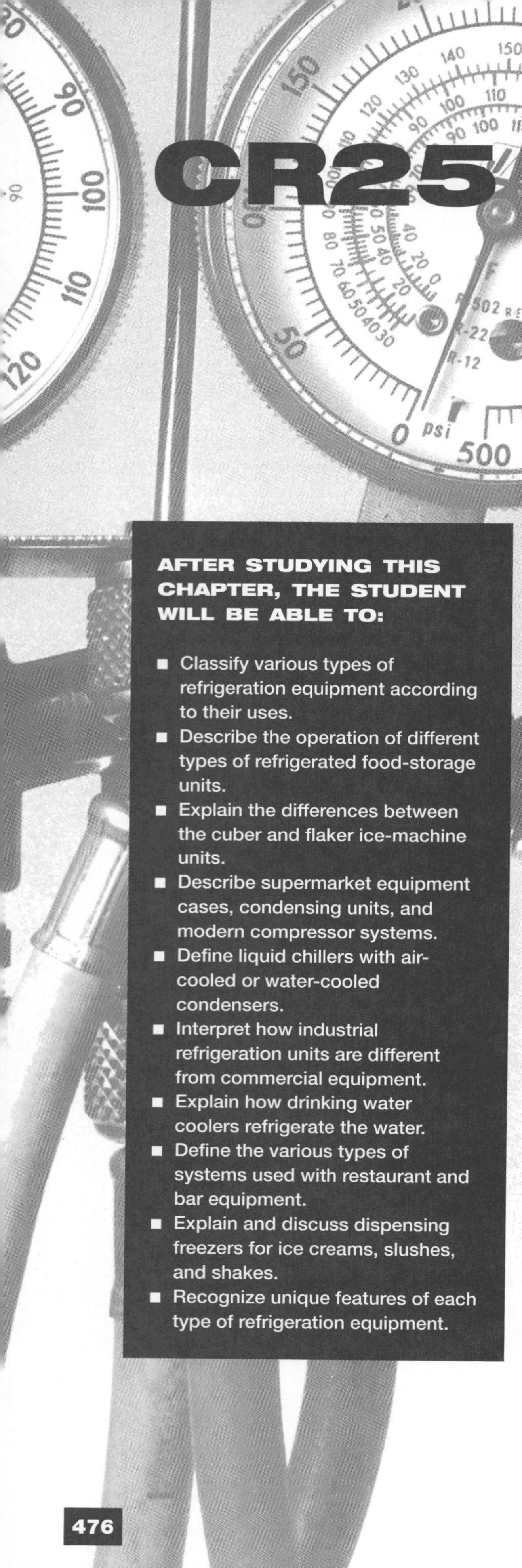

CR25 Refrigeration Equipment

AFTER STUDYING THIS CHAPTER, THE STUDENT WILL BE ABLE TO:

- Classify various types of refrigeration equipment according to their uses.
- Describe the operation of different types of refrigerated food-storage units.
- Explain the differences between the cuber and flaker ice-machine units.
- Describe supermarket equipment cases, condensing units, and modern compressor systems.
- Define liquid chillers with air-cooled or water-cooled condensers.
- Interpret how industrial refrigeration units are different from commercial equipment.
- Explain how drinking water coolers refrigerate the water.
- Define the various types of systems used with restaurant and bar equipment.
- Explain and discuss dispensing freezers for ice creams, slushes, and shakes.
- Recognize unique features of each type of refrigeration equipment.

CR25-1 CLASSIFICATION OF EQUIPMENT

There are many ways to classify refrigeration equipment. One of the simplest ways is to separate packaged equipment from field-assembled equipment since installation requirements are different. Another method is to separate all food-refrigeration equipment from industrial processing equipment. This, however, results in considerable overlapping. Equipment can also be separated by the temperature it maintains, such as medium temperature for fresh-food storage, low temperature for frozen foods and ultra-low temperature for industrial processing. In each category the same equipment can be used for various applications and the applications govern the modifications of the equipment to perform the assigned task.

The discussion that follows presents mainly factory-packaged units and the applications to which they apply. Some of these packages perform the entire task, such as a drinking-water cooler. Other packages are connected to components on the job, such as a cooling brine for an ice rink. The important element is that the technician view the equipment in terms of the entire system when performing his/her work, so that the system operates to accomplish its design requirements. The following equipment is presented in this context:

1. Refrigerated food-storage units
   - Reach-in cabinets
   - Walk-in coolers
2. Ice-making units
   - Cube ice machines
   - Flake ice machines
3. Supermarket equipment
   - Display cases
   - Rack-mounted condensing units
   - Parallel compressor units
   - Protocol system
4. Packaged liquid chillers
   - Air-cooled
   - Water-cooled

5. Industrial refrigeration units
6. Drinking water coolers
7. Restaurant/bar equipment
8. Dispensing freezers
   - Soft-serve ice cream
   - Slush machine
   - Shake machine
   - Hard-serve ice cream

CR25-2 REFRIGERATED FOOD-STORAGE UNITS

The use of refrigerated storage of both fresh- and frozen-food products constitutes a large part of the refrigeration industry. This was the first use of refrigeration and this segment of the business now offers many variations.

In commercial refrigeration these uses can be divided into the following product groups: (1) reach-in refrigerators and freezers, (2) walk-in refrigerators and freezers, and (3) storage warehouses.

CR25-2.1 Reach-in Refrigerators and Freezers

Three configurations of reach-in refrigerator-freezers are shown in Fig. CR25-1.

In this type of unit the refrigeration is contained in the top section. With the single-door arrangement the door can be installed to open either to the left or to the right. Care must be exercised to locate the unit where the floor offers adequate support, since some models can weigh as much as 3000 pounds.

Adjustable legs are furnished to permit leveling the unit. Leveling is important to permit proper draining of the evaporator condensate. A floor-drain connection is needed or a condensate vaporizer can be used. The unit needs to be wired with a plug-in connector located on the electrical box. The unit must be connected to a separately fused circuit on the electrical panel. A drain-line heater is used on freezer models.

The storage capacity of the three sizes shown are: 231, 478, and 737 cu ft, respectively. The compressors for the refrigerators are $\frac{1}{4}$, $\frac{1}{3}$ and $\frac{1}{2}$ Hp and for the freezers $\frac{1}{2}$, $\frac{3}{4}$ and 1 Hp, respectively. All units can run on 115-V/60-Hz/1-phase power. The 1-Hp size is also available for 208–230-V/60-Hz/1-phase power.

The refrigeration systems use an HFC (SUVA HP-81) refrigerant. The thermal sensing bulb for the refrigerator is located in the evaporator coil and, for the freezer, in the air. The refrigerator is controlled to maintain 38°F and the freezer 0°F. The refrigerator uses air defrost during the OFF-cycle. The freezer uses electrical defrost. A high-pressure cut off is standard on all models with a limit setting of 440 psi.

The wiring diagrams are as follows: refrigerator cooling cycle (Fig. CR25-2); freezer cooling cycle (Fig. CR25-3); freezer defrost cycle (Fig. CR25-4, p. 480).

The sequence of operation is as follows.

Refrigerator Cooling

With the main switch in the "on" position, the current flows through the high-pressure cut out to the relay. When the relay is energized the NO contacts close and with the

Figure CR25-1 Typical commercial reach-in refrigerator-freezers. (Courtesy of The Manitowac Company)

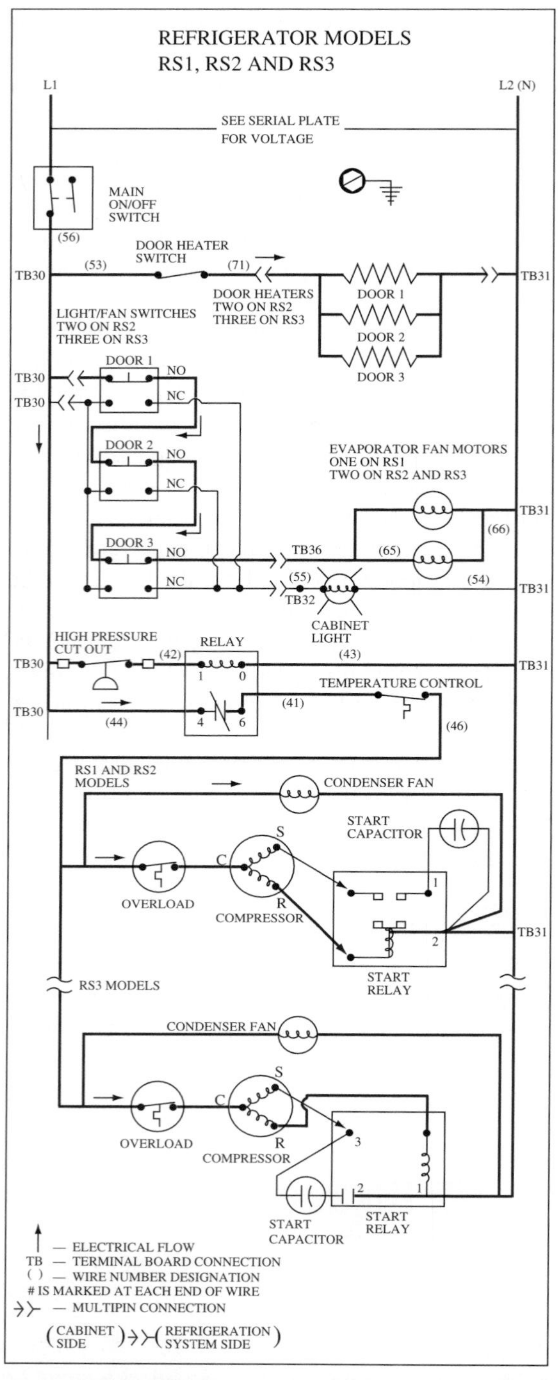

Figure CR25-2 Wiring diagram for reach-in refrigerator cooling cycle. (Courtesy of The Manitowac Company)

1. Main ON/OFF Switch .... On Position
2. Door Heater Switch (Optional On/Off) .... On
   A. Door Heater .... On
3. Light/Fan Switch(es) (Door Switches) .... Door Closed
   A. Normally Open (N.O.) Contact .... Closed
      1. Evaporator Fan Motor .... Energized
   B. Normally Closed (N.C.) Contact .... Open
      1. Interior Cabinet Light .... Off
4. High Pressure Cut-Out .... Closed
5. Relay .... Energized
   A. Normally Open (N.O.) Contacts .... Closed
6. Temperature Control .... Closed
   A. Compressor .... Energized
   B. Condenser Fan Motor .... Energized

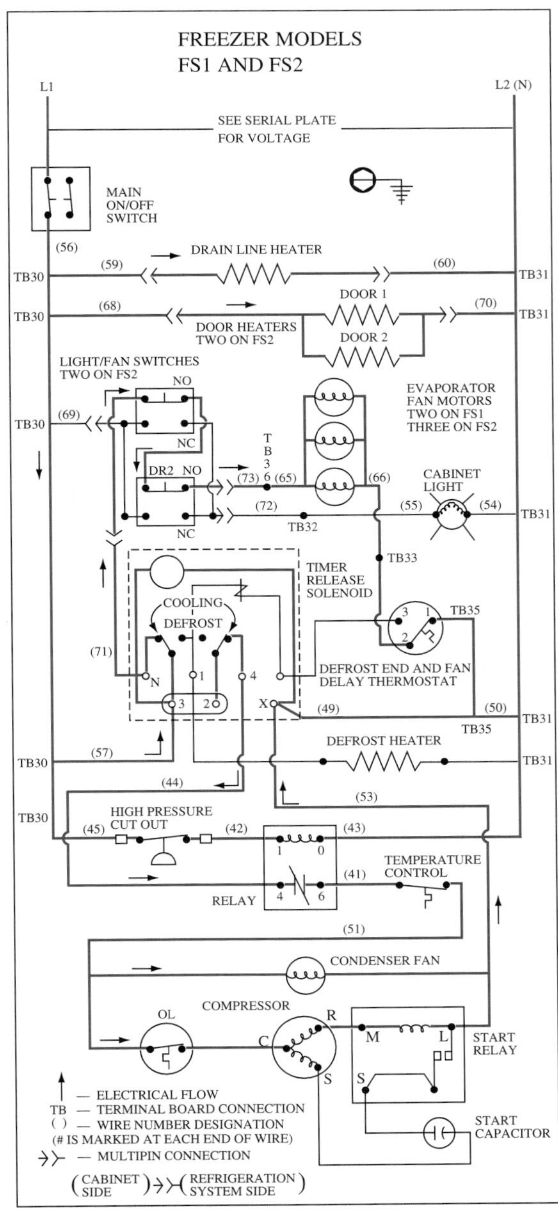

Figure CR25-3 Wiring diagram for reach-in freezer cooling cycle. (Courtesy of The Manitowac Company)

1. Main ON/OFF Switch	On Position
A. Drain Line Heater	On
B. Door Heater(s)	On
2. Automatic Defrost Time Clock	Cooling cycle
A. Contact "N"	Closed
1. Light/Fan Switch(es) (Door Switch(es))	Door Closed
a. Normally open (N.O.) contact	Closed
Evaporator fan motor(s)	Energized
b. Normally closed (N.C.) contact	Open
Interior cabinet light	Off
B. Contact #4	Closed
1. Temperature Control	Closed
a. Compressor	Energized
b. Condenser fan motor(s)	Energized
C. Contact #1	Open
1. Defrost Heater	Off
3. Defrost End and Fan Delay Thermostat	
A. Normally Closed (N.C.) Contacts	Closed
B. Normally Open (N.O.) Contacts	Open
4. High Pressure Cut-out	Closed
5. Relay	Energized
A. Normally Open (N.O.) Contact	Closed

thermostat calling for cooling, the compressor and condenser fan start. The evaporator fan stays on unless a door is opened. When a door opens the fan goes off and the interior light goes on.

When the thermostat is satisfied, the compressor and condenser fan are stopped. They re-start when the thermostat reaches 38°F.

Freezer Cooling

With the main switch in the "on" position, the current energizes the automatic defrost time clock. At the same time the current flows through the high-pressure cut out, energizing the relay. With contact "4" on the time clock closed and the thermostat calling for cooling, the refrigeration will start. With contact "N" closed on the time clock, current will flow through the NO contacts of the light/fan switch (with the door closed). When the evaporator-coil temperature reaches 30–35°F the defrost end and fan delay thermostat close, energizing the evaporator fan.

Freezer Defrosting

At preset times the time clock contacts will switch to defrost the evaporator. Contact "N" opens, stopping the evaporator fan. Contact "4" opens on the time clock, stop-

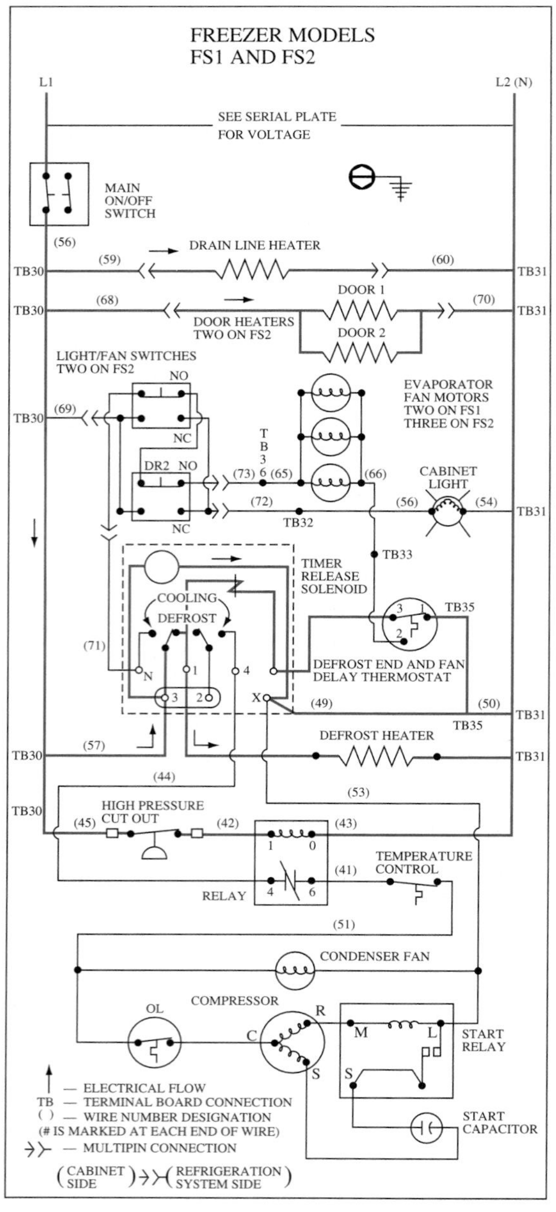

1. Main ON/OFF Switch	On Position
A. Drain Line Heater	On
B. Door Heater(s)	On
2. Automatic Defrost Time Clock	Defrost Cycle
A. Contact "N"	Open
1. Light/Fan Switch(es) (Door Switch(es))	Door Closed
a. Normally open (N.O.) contacts	Closed
Evaporator fan motor(s)	De-Energized
b. Normally closed (N.C.) contacts	Open
Interior cabinet light	Off
B. Contact #4	Open
1. Temperature Control	Closed
a. Compressor	De-Energized
b. Condenser fan motor(s)	De-Energized
C. Contact #1	Closed
1. Defrost heater	On
3. Defrost End and Fan Delay Thermostat (end of defrost)	
A. Normally Closed (N.C.) Contacts	Open
B. Normally Open (N.O.) Contacts	Closed
4. High Pressure Cut-out	Closed
5. Relay	Energized
A. Normally Open (N.O.) Contact	Closed

Figure CR25-4 Wiring diagram for reach-in freezer defrost cycle. (Courtesy of The Manitowac Company)

ping the refrigeration system. Contact "1" closes energizing the defrost heater. As the evaporator temperature rises, the defrost end and fan delay thermostat will open, terminating the defrost cycle. Contacts "N" and "4" will close. Contact "1" will open, de-energizing the defrost heater and starting the refrigeration system. The system will then be under the control of the thermostat.

CR25-2.2 Walk-in Refrigerators and Freezers

Walk-in coolers and freezers are made with modular insulated metal-clad panels. Using corner sections and standard wall and ceiling panels, various sized boxes can be made. Panels are assembled on the job using special eccentric cam fasteners. Insulated panels are used for the floors. The interior surface of wall and ceiling panels are usually clad with aluminum, floors with galvanized iron.

An assembled walk-in cooler is shown in Fig. CR25-5. The sectional panels are shown in Fig. CR25-6. The installation of a partition panel is shown in Fig. CR25-7. This makes it possible to include a cooler and a freezer in the same structure. By entering the cooler first and then the freezer, more economical operation is possible since a common wall is used.

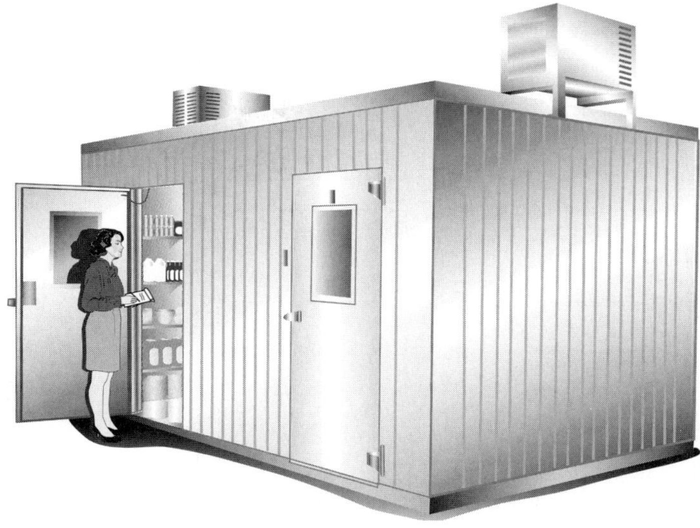

Figure CR25-5 Walk-in cooler-freezer.

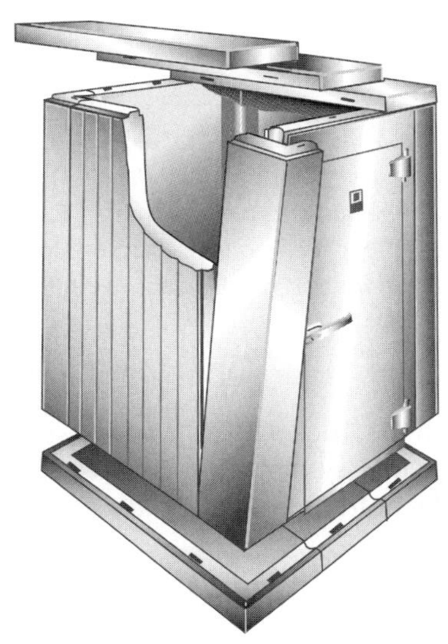

Figure CR25-6 Sectional panels for a walk-in cooler.

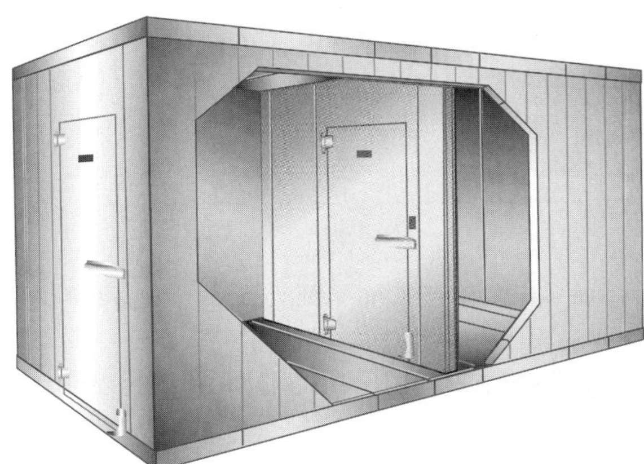

Figure CR25-7 Partition location in a walk-in cooler.

PRACTICING PROFESSIONAL SERVICES

Many new pieces of commercial refrigeration equipment have original equipment manufacturer (OEM) refrigerants sent pre-charged from the factory. Whenever you are working on new equipment and are unsure of the type of refrigerant in the system, look for model numbers with technical data indicating the type. If you are not 100% sure of the refrigerant type, then consult the equipment installer to verify refrigerant in the system.

If you go to a job site where several other technicians have worked on the equipment, and you are not sure of the refrigerant type, consider a couple of options: (1) request service data from the customer from the most recent work performed, (2) recover and reclaim the refrigerant, or (3) ship off a sample to your local contractors supply house. Never add refrigerant to a system without being sure of the type of refrigerant in the system.

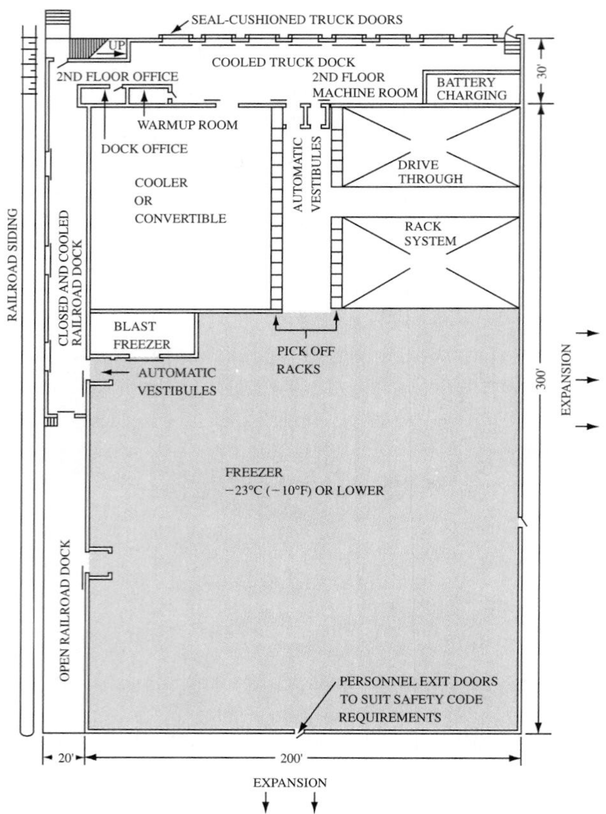

Figure CR25-8 One-story refrigerated warehouse.

CR25-2.3 Storage Warehouses

Most modern refrigerated warehouses are one story structures. A typical floor plan is shown in Fig. CR25-8. The building construction is shown in Fig. CR25-9. Note that the walls are constructed so that as few structural members as possible penetrate the insulated envelope. Insulated panels applied to the outside structural frame prevent conduction through the framing. It is usually convenient to locate the refrigeration equipment in a penthouse room, as shown in Fig. CR25-10.

> **Knowledge Extension**
>
> Learning about modern commercial refrigeration equipment is one of the greatest challenges that will face the new technician. Most new equipment comes with electronic controllers, special valves, and remote control interfaced into the office computer. To learn more about these types of systems, contact the specific manufacturer of the equipment or visit their office and request technical service data sheets and books. Often these companies will offer intense training seminars on their equipment at the factory. The costs to attend are high, but the educational benefits are excellent.

CR25-3 ICE-MAKING UNITS

There are two general types of ice-making machines: those that make clear *ice cubes* used to cool beverages and those that make *flake ice* used to cool products like fish.

CR25-3.1 Ice-cube Machines

Ice-cube machines (Fig. CR25-11) are usually self-contained refrigeration systems that produce and store clear ice cubes, which are preferred by bars, restaurants, and hotels.

The cubes are frozen using running water, which removes the air to create a clear cube. If there is air in the ice, the carbonation in a drink can be destroyed. Various machines produce different sizes and shapes of

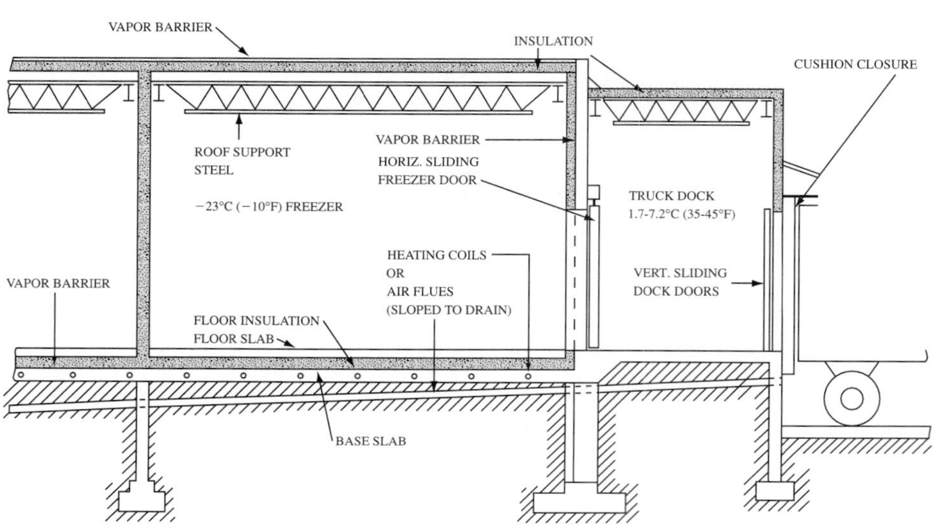

Figure CR25-9 Refrigerated warehouse construction.

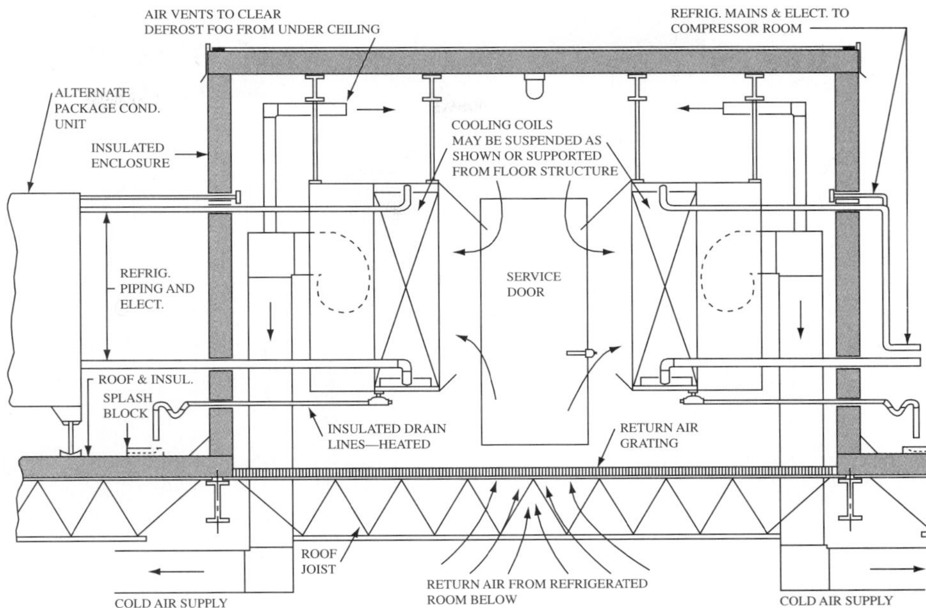

Figure CR25-10 Penthouse equipment room for a refrigerated warehouse.

the cubes, depending on the shape and construction of the evaporator.

Cubers are available having capacities to produce 450 to 1800 pounds of ice per day. Bins that collect the ice cubes fit under the cuber and can be selected in various sizes to fit the needs of the user.

An automatic self-cleaning system is available as an accessory which cleans the water passages of the machine on a selected schedule. Periodic cleaning is essential in maintaining a supply of clear ice (Fig. CR25-12).

To reduce the maintenance, the air to the condenser is filtered to remove lint, dust, and grease. To comply with environmental requirements these machines are converting to HCFC type refrigerants, such as DuPont SUVA HP81 refrigerant, which have a very low or zero ozone depletion factor (ODF).

The sequence of operations of the evaporator in producing ice cubes is as follows.

1. *Initial start-up.* Prior to the time that the refrigeration system is started, the water pump and the water-dump solenoid are energized to flush the water passages for 45 seconds. At the completion of this cycle, the compressor and condenser fan (on air-cooled models) are energized.

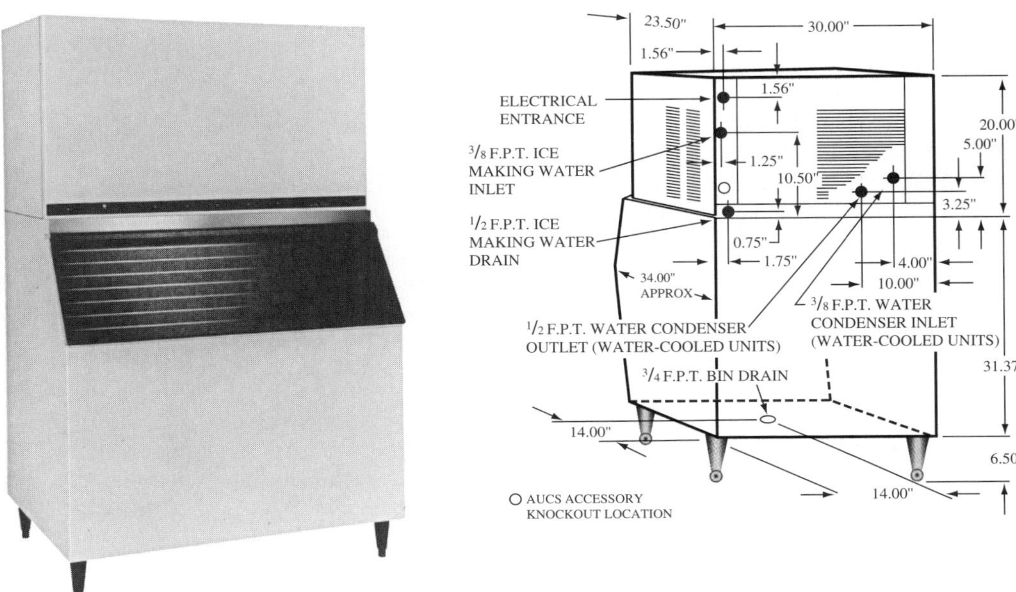

Figure CR25-11 Ice-cube machine with ice bin. (Courtesy of The Manitowac Company)

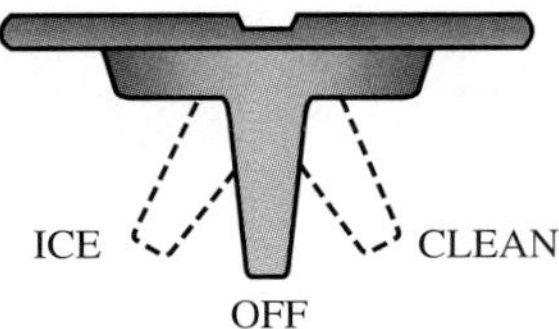

Figure CR25-12 Switch used to operate the water pump with the refrigeration system not running. (Courtesy of The Manitowac Company)

2. *Freeze cycle.* After a 30-second delay, the water pump is restarted and water is supplied to each freeze cell, where it freezes. When the required amount of ice is formed, the ice thickness probe is contacted, and after 7 seconds the harvest cycle begins.
3. *Harvest cycle.* The hot-gas valve opens, directing hot gas into the evaporator, loosening the cubes. The water dump solenoid is opened for 45 seconds, purging the excess water into the trough. The water pump and dump solenoid are then de-energized. The cubes slide into the bin. The sliding cubes move the water curtain out, activating the bin switch. This terminates the harvest cycle, returning the machine to the freeze cycle.
4. *Automatic shut off.* When the bin is full of ice, no more cubes can slide into the bin and the water curtain is held open. If this occurs for more than 7 seconds, the machine automatically shuts off. As ice is used from the bin, space develops for accepting more ice. The water curtain slips back into operating position and the machine restarts.

The machine can use a remote air-cooled condenser as shown in Fig. CR25-13.

The ice dispenser shown in Fig. CR25-14 is a variation of the cuber, which is commonly used in self-service restaurants.

Water filters (Fig. CR25-15) can be effectively used in the incoming water lines. These units offer a means of reducing scale formation, filtering sediment, and removing chlorine taste and odor.

Fig. CR25-16 shows the refrigeration system components and piping for a typical unit. Fig. CR25-17 shows a typical wiring diagram.

For other servicing information, consult the manufacturer's service manual.

CR25-3.2 Flake-ice Machines

Flake-ice machines have always been popular for contact cooling, such as cooling fish, where maintaining high humidity is required. These machines are available in small

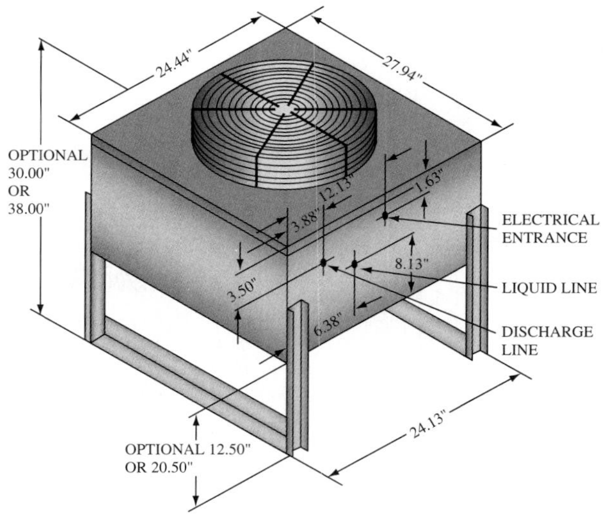

Figure CR25-13 Remote air-cooled condenser used with ice-cube machine. (Courtesy of The Manitowac Company)

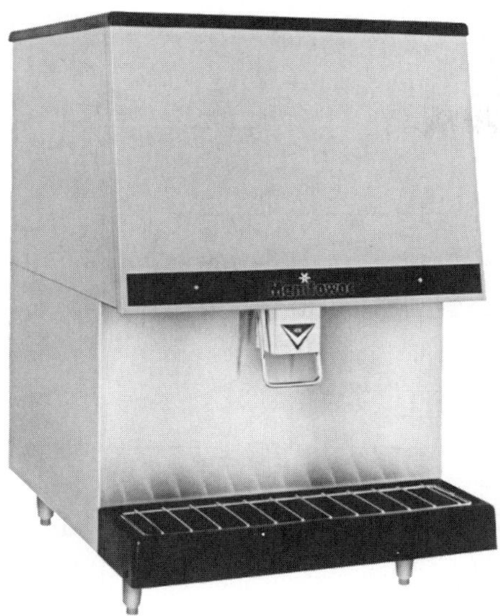

Figure CR25-14 Ice dispenser for self-serve beverage machine. (Courtesy of The Manitowac Company)

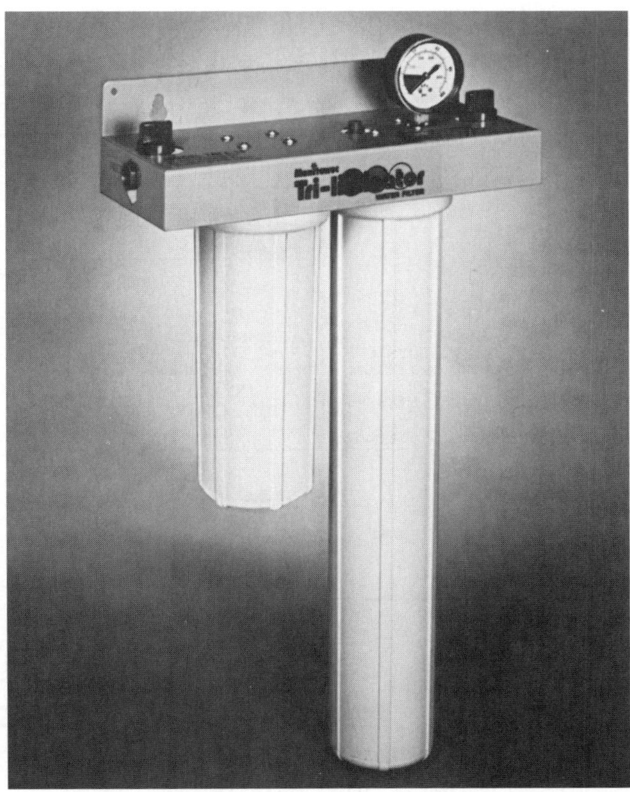

Figure CR25-15 Water filter for an ice-cube machine. (Courtesy of The Manitowac Company)

Figure CR25-16 Refrigeration cycle for an ice-cube machine. (Courtesy of The Manitowac Company)

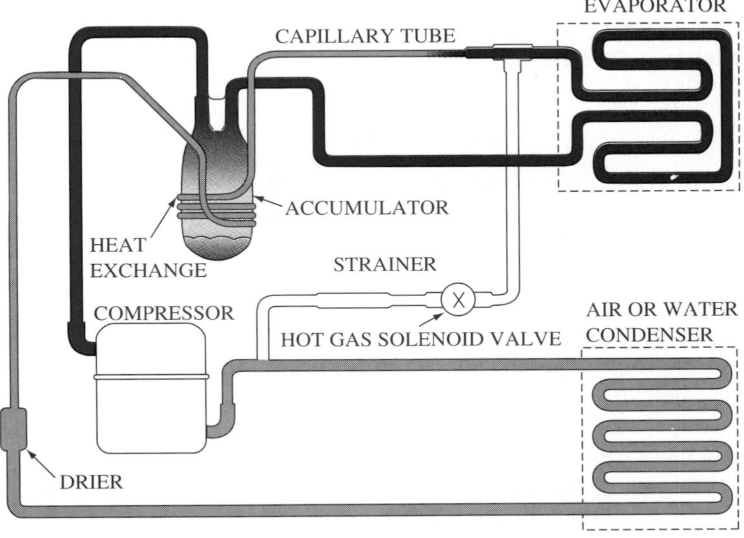

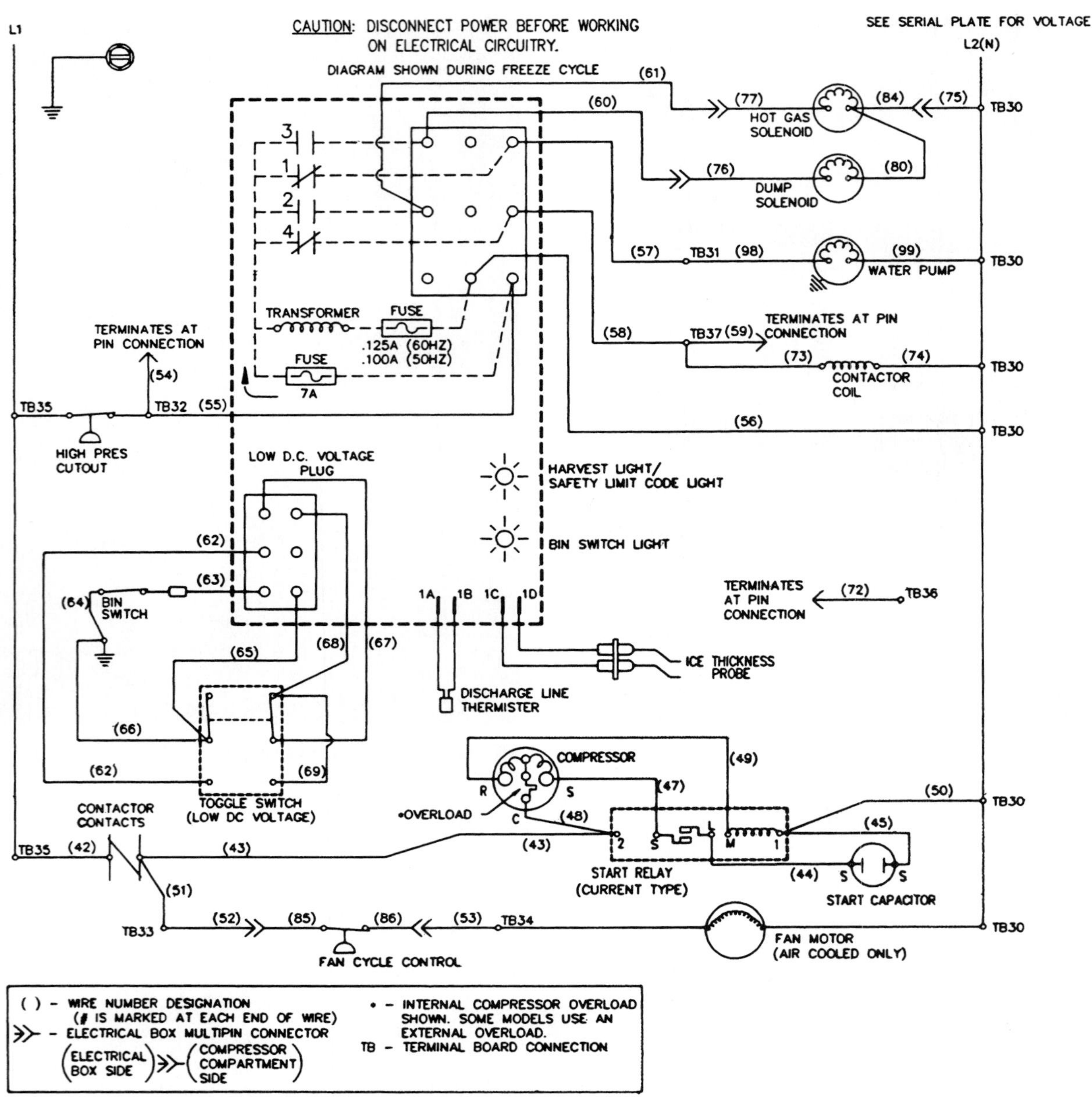

Figure CR25-17 Wiring diagram for an ice-cube machine. (Courtesy of The Manitowac Company)

sizes, under-the-counter and floor-mounted dispenser models, and in industrial sizes for making 1 to 10 tons of flake ice per day.

An innovation in the design of these machines is shown in Fig. CR25-18. A cut away of the evaporator is shown in Fig. CR25-19.

Referring to the illustrated evaporator (Fig. CR25-19), the *water distribution system* at the top evenly distributes water over the subzero evaporator surface. Below this cast aluminum tray is the squeegee and ice blade.

The *rubber squeegee* removes excess water from the ice. The excess water is recirculated to be used for further ice making. No water is wasted. The stainless-steel *ice blade* wedges the ice off the evaporator without touching the surface.

The *shaft* holds the major components and is the only moving part. It rotates at only 2 rpm. The shaft is supported by two bronze sleeve bearings.

Ice falls by gravity into the bin. The ice is 100% dry and subcooled. The flaker operates continuously and there is no need for hot gas defrost.

The refrigeration-cycle component piping is illustrated in Fig. CR25-20. A typical wiring diagram is shown in Fig. CR25-21 (p. 489).

Industrial flake-ice machines are available, as shown in Fig. CR25-22 (p. 489) using a remote condensing unit,

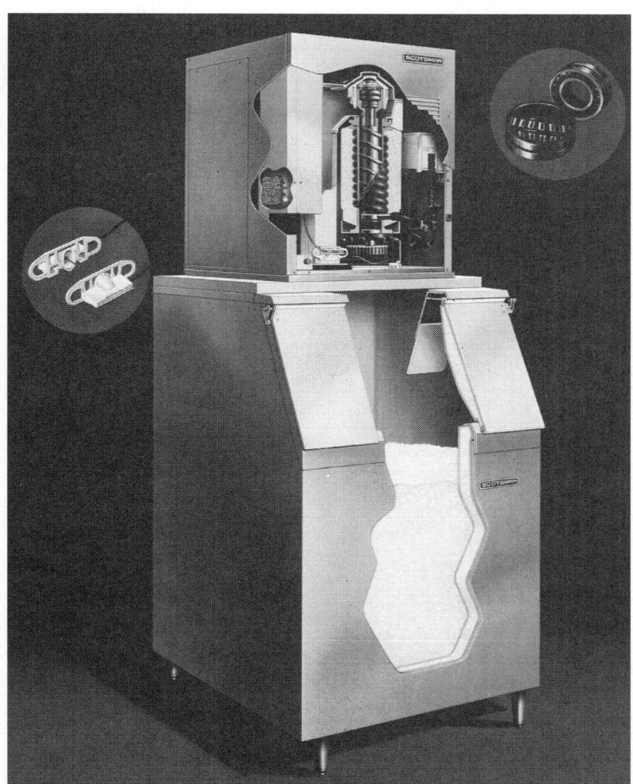

Figure CR25-18 Flake ice machine with bin. (Courtesy of Scotsman Ice Systems)

that will freeze either freshwater or seawater. A number of different refrigerants are available including R-22, R-404A, R-134A, and ammonia.

CR25-4 SUPERMARKET EQUIPMENT

Food-store systems vary depending on the size of the store, the food products being refrigerated, and the area in which the equipment is located. Generally speaking, there are two types of fixtures:

1. *Self-service equipment,* where the customers select their own products; and
2. *Service equipment,* where the employees select the products.

Within these categories are found fixtures made in open single-deck, closed single-deck, open multi-deck, and closed multi-deck styles.

For *produce,* the most generally used style is the open single-deck. A commonly used produce case is shown in Fig. CR25-23 (p. 490). It is sometimes referred to as a "vision-type" case because mirrors are used to enhance the appearance of the product.

The produce sections in supermarkets also use a walk-in cooler, not accessible to the public, maintained at about 38°F and 80% RH for storage, and a preparation room maintained at about 55°F.

Figure CR25-19 Cutaway view of the evaporator of a flake-ice machine. (Courtesy of Scotsman Ice Systems)

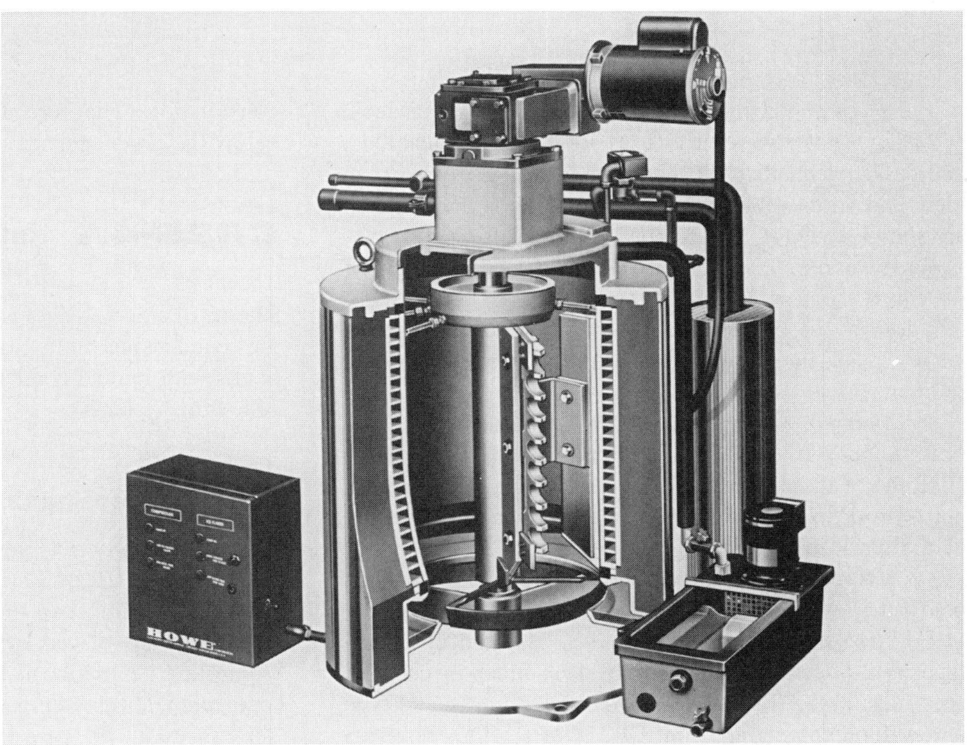

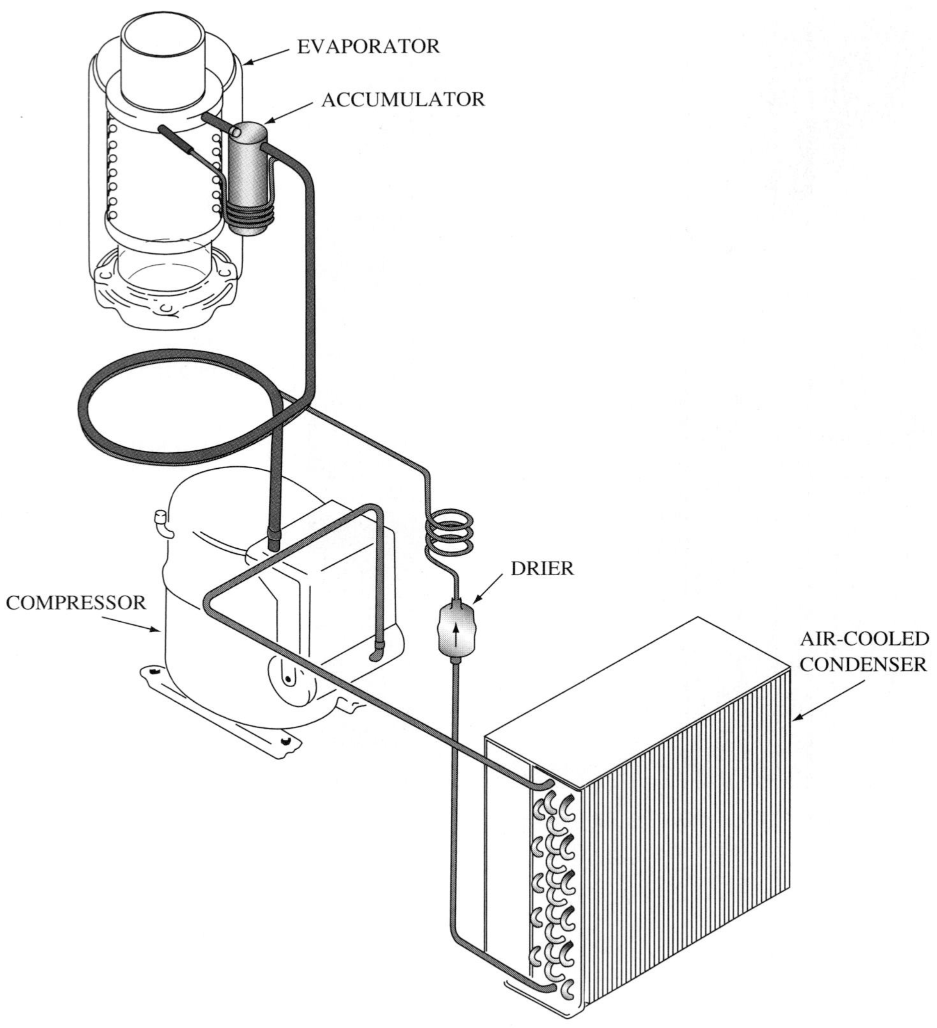

Figure CR25-20
Refrigeration cycle for a flake-ice machine. (Courtesy of Scotsman Ice Systems)

For *meat and dairy products,* single-deck, multi-deck, and island-type self-service cases are used. A typical meat and deli merchandiser is shown in Fig. CR25-24. A plan view and an elevation view of the cabinet's construction are shown in Fig. CR25-25 (p. 491).

An evaporator using gravity coil is located in the top of the cabinet to provide a gentle circulation of cool air without drying out the product. Meat markets also usually have a walk-in storage cooler and a meat preparation room.

For *dairy products* a multi-deck refrigerator as shown in Fig. CR25-26 (p. 492) is commonly used.

Some of these have a rear access for loading the cabinet from racks. An illustration of the typical construction of the dairy merchandiser is shown in Fig. CR25-27 (p. 493). A forced-air refrigerated evaporator coil is located below the lower deck.

For *frozen foods and ice cream,* single-deck island type, multi-deck open, and multi-deck reach-in cabinets are used. Fig. CR25-28 (p. 494) shows an island-type frozen-food cabinet and Fig. CR25-29 (p. 495) shows its construction. The forced-air evaporator coil is located below the product space.

CR25-4.1 Refrigeration Systems for Food Stores

The refrigeration systems for food stores range from single condensing units to multiple parallel compressor banks with remote condensers. These options include the following systems.

Single Condensing Unit per Fixture

This usually consists of a remote condensing unit as shown in Fig. CR25-30 (p. 496).

Where a number of these are used in a single installation, they are placed on racks in an equipment room. Controlled ventilation is provided to maintain a satisfactory ambient air temperature for the air-cooled condensers.

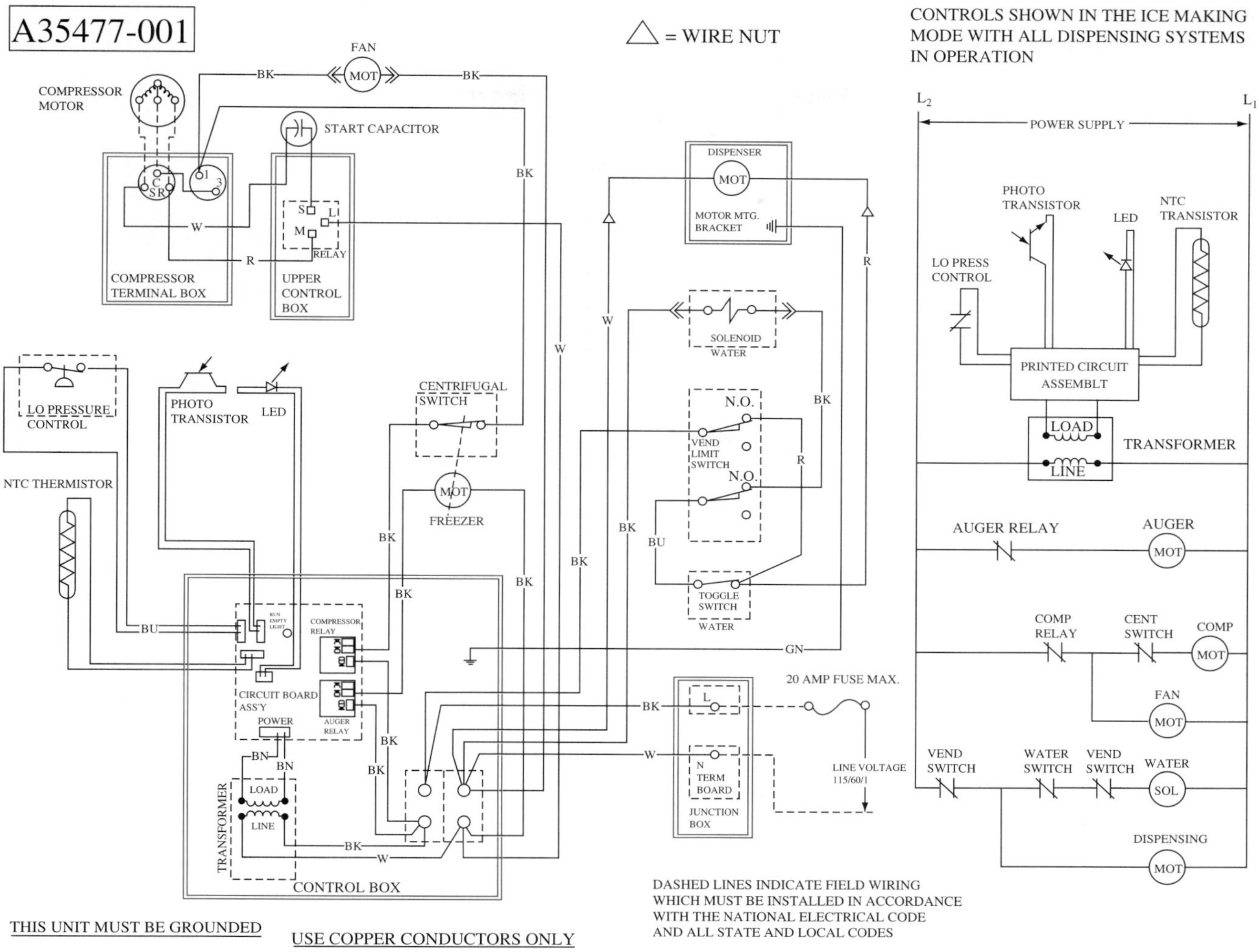

A35477-001

△ = WIRE NUT

CONTROLS SHOWN IN THE ICE MAKING MODE WITH ALL DISPENSING SYSTEMS IN OPERATION

FAN
COMPRESSOR MOTOR
START CAPACITOR
COMPRESSOR TERMINAL BOX
UPPER CONTROL BOX
RELAY
DISPENSER
MOTOR MTG. BRACKET
SOLENOID WATER
LO PRESSURE CONTROL
PHOTO TRANSISTOR
LED
CENTRIFUGAL SWITCH
VEND LIMIT SWITCH
N.O.
NTC THERMISTOR
FREEZER
TOGGLE SWITCH WATER
COMPRESSOR RELAY
CIRCUIT BOARD ASS'Y
POWER
AUGER RELAY
TRANSFORMER
LOAD
LINE
CONTROL BOX
20 AMP FUSE MAX.
TERM BOARD
JUNCTION BOX
LINE VOLTAGE 115/60/1

L_2 POWER SUPPLY L_1
PHOTO TRANSISTOR
LED
NTC TRANSISTOR
LO PRESS CONTROL
PRINTED CIRCUIT ASSEMBLT
LOAD
LINE
TRANSFORMER
AUGER RELAY
AUGER
COMP RELAY
CENT SWITCH
COMP
FAN
VEND SWITCH
WATER SWITCH
VEND SWITCH
WATER
SOL
DISPENSING

THIS UNIT MUST BE GROUNDED

USE COPPER CONDUCTORS ONLY

DASHED LINES INDICATE FIELD WIRING WHICH MUST BE INSTALLED IN ACCORDANCE WITH THE NATIONAL ELECTRICAL CODE AND ALL STATE AND LOCAL CODES

Figure CR25-21 Wiring diagram for a flake-ice machine. (Courtesy of Scotsman Ice Systems)

Figure CR25-22 Industrial flake-ice machines with remote condensing units. (Courtesy of Scotsman Ice Systems)

Figure CR25-23 Single deck produce display cabinet. (Courtesy of Hussmann Corporation, Division of Whitman Corporation)

Each individual refrigeration circuit has its own temperature and defrost control. This arrangement offers the greatest flexibility, but requires the greatest amount of compressors, piping, and electrical connections.

Single Condensing Unit for Multiple Cases (Multiplexing)

This arrangement is used where a number of cases are all maintained at the same conditions. Individual cases are controlled by thermostats operating liquid-line solenoid valves and evaporator-pressure regulators (EPRs). Installations are more complex; however, less energy is used and the overall cost is reduced.

Parallel Systems

These consist of a bank of parallel connected compressors and accessories mounted on a common base, piped and wired at the factory, ready for field connection to individual cases (Fig. CR25-31, p. 497).

The advantage of this type of arrangement is that individual compressors can be cycled on and off to provide the required capacity. In case of compressor failure, other compressors carry the load.

This equipment configuration is advantageous for latent-heat defrosting, heat recovery, and electrically operated total-control systems. Using this arrangement, defrosting can be selective, with some units operating while others defrost. The operating units can provide the necessary heat for the defrosting units.

Figure CR25-24 Meat and deli merchandiser cabinet. (Courtesy of Hussmann Corporation, Division of Whitman Corporation)

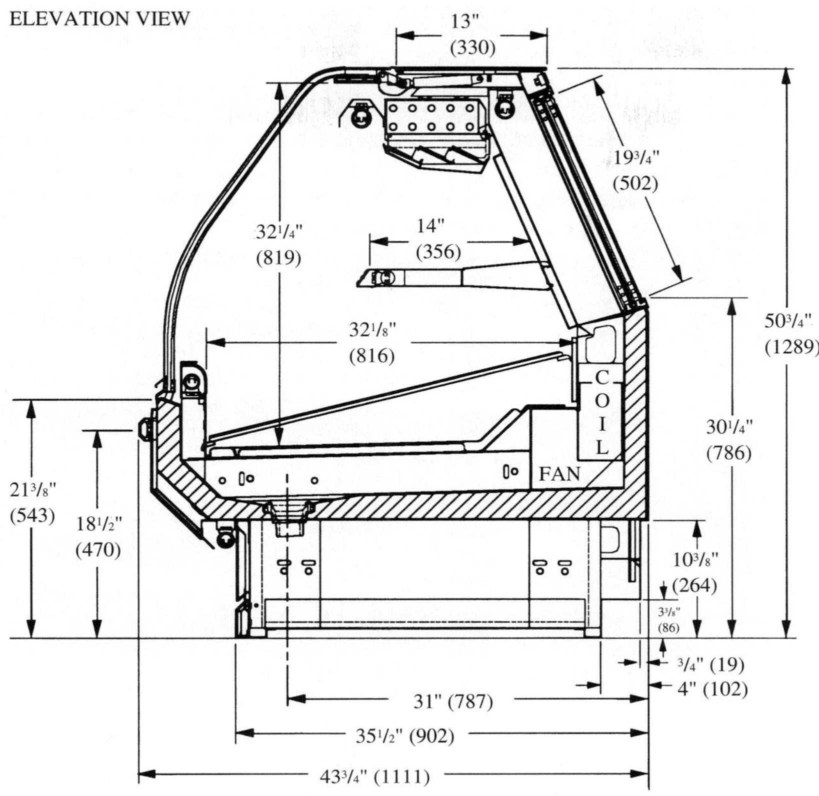

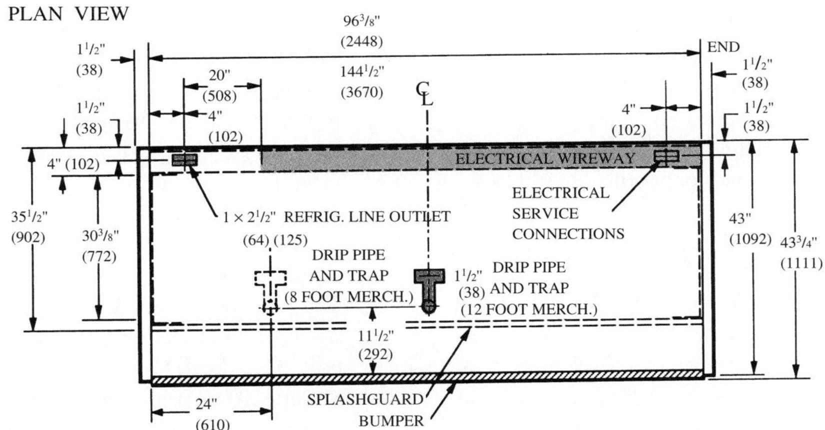

Figure CR25-25 Schematic diagram of meat and deli cabinet showing the gravity-airflow evaporator coil. (Courtesy of Hussmann Corporation, Division of Whitman Corporation)

Figure CR25-26 Multi-deck dairy merchandiser. (Courtesy of Hussmann Corporation, Division of Whitman Corporation)

For maximum efficiency a bank of compressors should be connected to evaporators operating at similar temperatures. From an operational standpoint, fixtures of different evaporating temperatures can be connected to the bank; however, this causes a substantial energy loss.

The disadvantage of this arrangement is danger of refrigerant leak in the system that would shut down a sizable number of cases. A good refrigerant-monitoring system must be used to alert responsible people should a refrigerant leak occur.

A schematic refrigerant piping diagram for a parallel system is shown in Fig. CR25-32 (p. 498).

A typical supermarket system produces refrigeration in the low- or medium-temperature ranges. A low-temperature rack maintains a suction temperature of –25°F with a satellite suction temperature of –33°F. A medium-temperature rack maintains a suction temperature of +16°F and a satellite suction temperature of +7°F. The system operates as follows, starting at the parallel compressors:

1. Compressed gas flows into the *Turba-shed,* which separates the vapor refrigerant from the liquid oil.
2. The three-way *heat-reclaim valve* directs the refrigerant to the condenser or reclaim coil.
3. The *flooding valve* maintains head pressure at low ambient conditions.
4. *Twin receivers* act as a vapor trap and supply high-quality liquid refrigerant to the main liquid solenoid.
5. The *main differential-pressure valve* (MS) functions during *Koolgas* defrost to reduce pressure to the liquid manifold. *Koolgas* is the hot gas used by this system for defrost and is usually in the range of 85°F. This is cooler than the hot gas coming from the compressor since it is removed from the top of the receiver. When the *Koolgas* performs its defrost function it is returned as a liquid to the receiver. By using lower defrost-gas temperatures, less stress is placed on the piping.
6. For the defrost cycle, the *branch line solenoid valve* (S) closes off refrigerant supply to the evaporator.
7. The *heat reclaim three-way valve* (HS) routes the discharge heat-laden vapor to the remote reclaim coil. The check valve prevents flooding of the reclaim coil when the heat reclaim cycle is off.
8. Receiver pressure is maintained by the *back-pressure regulating valve* (BP).
9. The receiver vapor flows directly to the *Koolgas manifold* and in the reverse direction through the evaporator to defrost it.
10. The *Koolgas vapor* condenses and flows into the reduced-pressure liquid line through a *bypass check valve* around the TXV, returning to the liquid-line manifold.
11. When the defrost is called for, the suction-line valve closes and a two-way *Koolgas valve* opens to permit the flow of *Koolgas* from the manifold to the evaporator.
12. The oil level in the compressor crankcases is maintained by the *oil level regulators.*

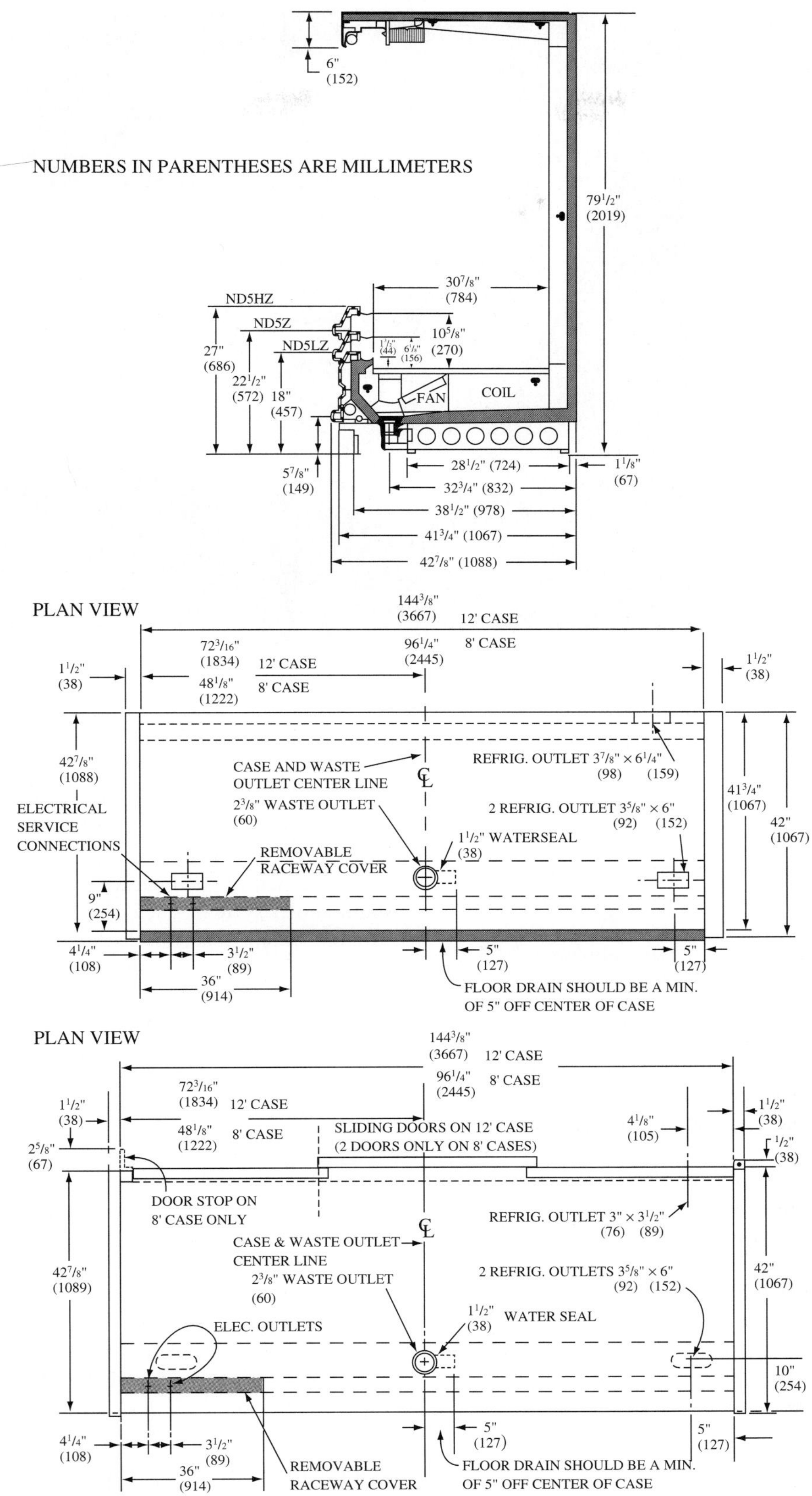

Figure CR25-27 Schematic diagram of the dairy-merchandiser cabinet showing the location of the forced-air evaporator. (Courtesy of Hussmann Corporation, Division of Whitman Corporation)

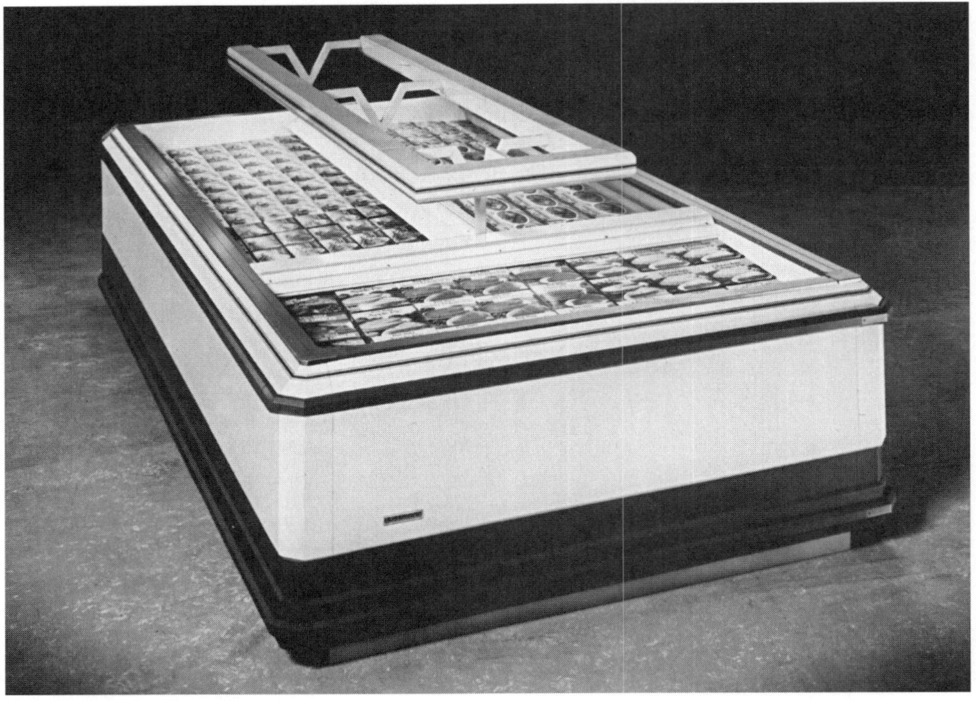

Figure CR25-28 Frozen-food/ice-cream island-type merchandiser. (Courtesy of Hussmann Corporation, Division of Whitman Corporation)

13. The *autosurge valve* directs the flow of liquid refrigerant either through the receiver or around the receiver to obtain the required amount of subcooling.

A *suction-pressure controller* is used to maintain a nearly constant suction pressure for the bank of compressors. One of these control units is shown in Fig. CR25-32 (p. 498).

For the frozen-food and ice-cream cabinets a bank of *two-stage compressors* can be used, as shown in Fig. CR25-33 (p. 500).

The advantage of the two-stage compressors is that they will operate at a lower suction pressure. This is advantageous where lower temperatures are desirable and there is a sufficient load to economically warrant equipment of this type.

These banks of compressors can be connected to a remote air-cooled condenser.

Piping from the central unit to the individual cases is usually carried in trenches underneath the floor. Piping must be properly insulated and isolated to prevent the possibility of electrolytic action.

The most common type of defrost is the *latent-heat method*. In larger systems, defrost cycles are staggered so that some units are always operating to produce heat.

CR25-4.2 The Protocol System

This is an advanced approach to supermarket refrigeration. The system uses multiple compressor units located near the cases being cooled. A typical store might have as many as 10 to 15 multiplexed compressors located throughout the store, as shown in Fig. CR25-34 (p. 500).

Each unit (Fig. CR25-35, p. 500) uses scroll compressors that provide quiet operation, and an HFC refrigerant, HP62 or R-507 (AZ-50), that has a zero ODF to meet environmental requirements. A compact-type plate condenser is part of each package, cooled from a closed-loop fluid-cooler system. Each unit includes an electronic controller which manages both compressor cycling and scheduled defrosts. Compressors are cycled to match the load requirements.

A central pedestal-mounted power-distribution panel furnishes electrical power to each unit by means of a four-wire drop cord that supplies power for the compressors, fans, lights, and anti-sweat heaters.

The advantages of this system are:

1. It reduces the refrigerant charge.
2. It reduces the refrigerant piping.
3. It reduces the possibility of refrigerant leaks.
4. It decreases or eliminates the need for EPR valves.
5. It eliminates the need for a central area for refrigeration equipment.
6. It lowers installation costs.
7. It provides load matching with multiplexed compressors.

CR25-5 PACKAGED LIQUID CHILLERS

Packaged water chillers are used to cool water or brine, as a secondary refrigerant, for both air-conditioning and re-

NFWE-90, NCWE-90, NFW,
NFWE-138, NCWE-138, NCW CROSS SECTION

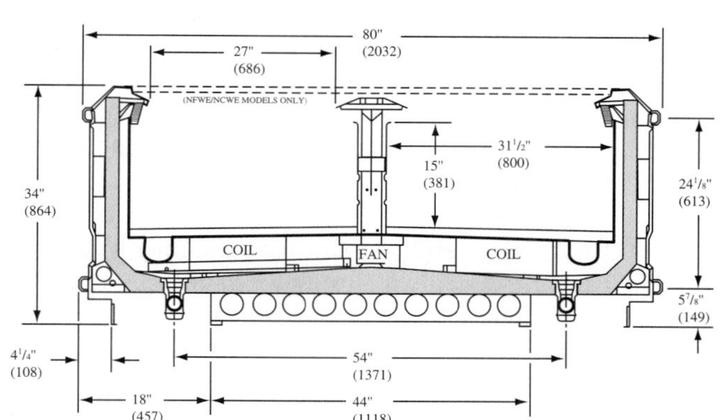

NCFS, NFFS CROSS SECTION

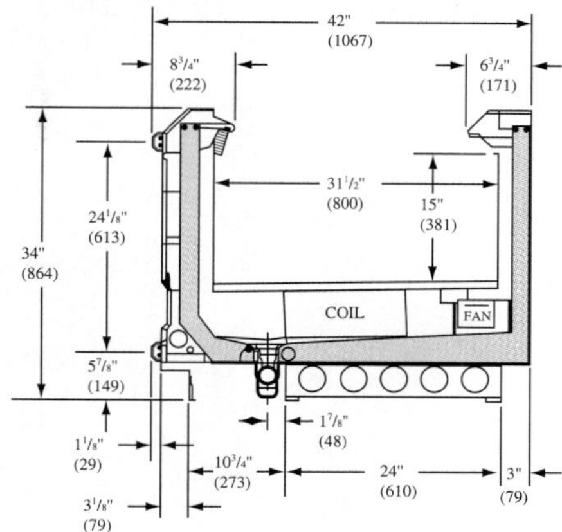

NFCWE-90, NFCWE-138, NFCW CROSS SECTION

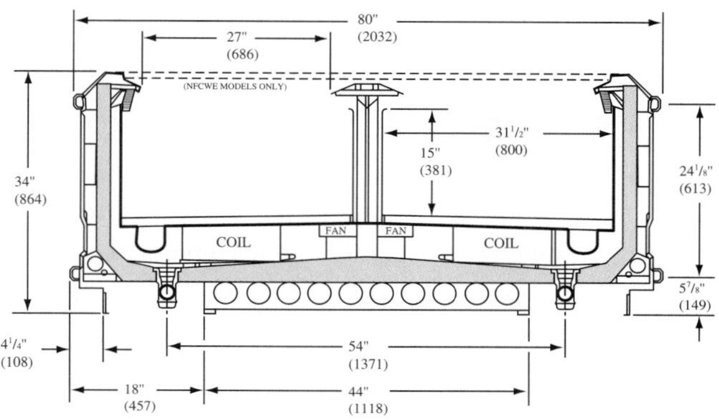

NFFS, NCFS FLOOR PLAN

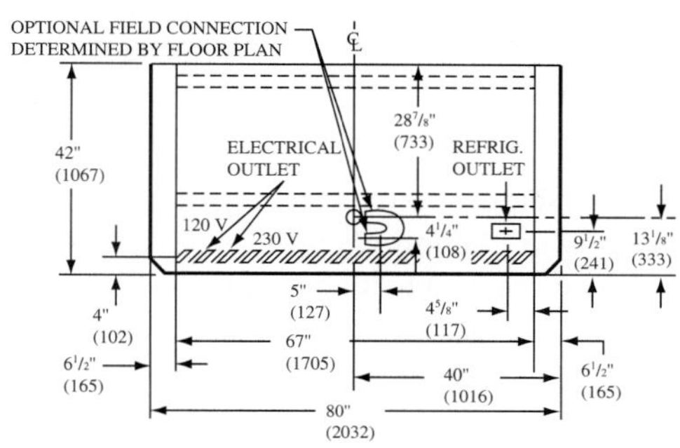

NFWE-90, NCWE-90, NFCWE-90,
NFWE-138, NCWE-138, NFCWE-138 FLOOR PLAN

NFW, NCW, NFCW FLOOR PLAN

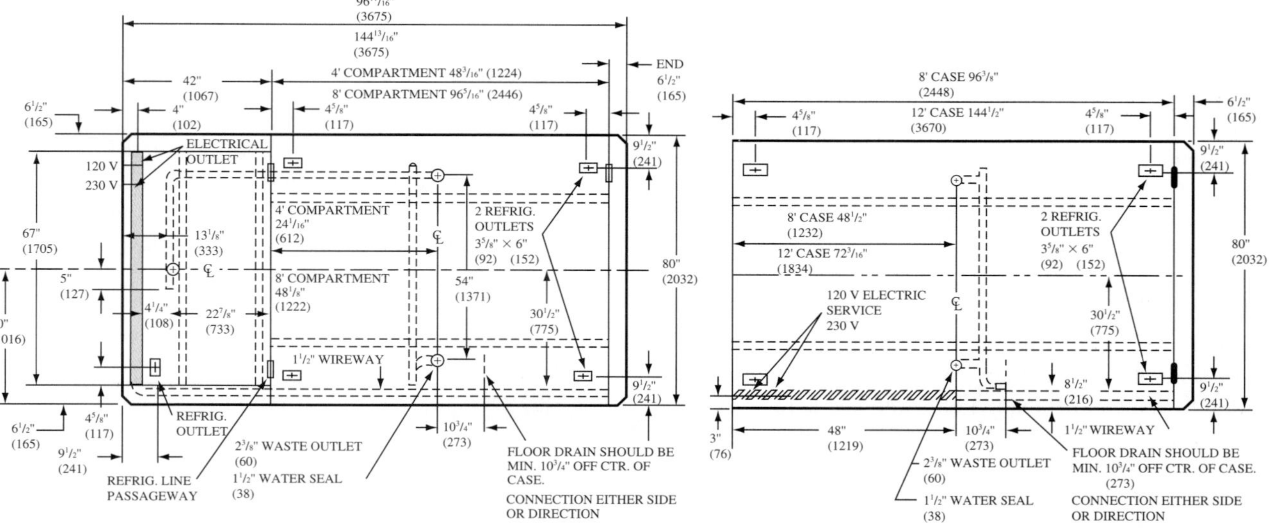

Figure CR25-29 Schematic diagram of the island-type merchandiser cabinet.
(Courtesy of Hussmann Corporation, Division of Whitman Corporation)

Figure CR25-30 Indoor air-cooled condensing unit for high, medium and low applications. (Courtesy of Hussmann Corporation, Division of Whitman Corporation)

frigeration applications. A good example of the refrigeration use is in cooling glycol brine for freezing ice rinks. The basic components include a motor-driven compressor, a liquid cooler (evaporator), a water-cooled or air-cooled condenser, a refrigerant metering device and a control system.

A schematic view of a basic refrigeration cycle is shown in Fig. CR25-36 (p. 501). Chilled water enters the cooler at 54°F and leaves at 44°F. Condenser water leaves a cooling tower at 85°F and returns to the tower at 95°F. Air-cooled condensers can be used in place of the water-cooled type. Liquid chillers are available in capacity sizes ranging from 25 to 3000 tons.

Chillers are available with reciprocating, scroll, screw, or centrifugal compressors. Chillers are also constructed using the absorption refrigeration cycle.

The type of control system varies somewhat depending on the type of compressor used. All chillers have a leaving liquid-temperature sensor that can be adjusted to maintain the desired liquid temperature. Most chillers have some arrangement for changing the capacity of the chiller to match the load. For example, a reciprocating compressor may have unloaders to reduce its capacity by 25, 50, and 75%.

Two popular types of modern chillers are: (1) an air-cooled chiller with multiple reciprocating compressors, and (2) a water-cooled chiller with a rotary screw-type compressor.

CR25-5.1 Air-cooled Chiller

An air-cooled chiller can range in size from 75 to 200 tons. It uses *multiple reciprocating compressors* and multiple condenser fans to achieve capacity reduction and maintain efficient operation at partial loads.

A diagram of an electronic expansion valve is shown in Fig. CR25-37 (p. 501). The valve uses a stepper motor controlled by a microprocessor to regulate the refrigerant flow. It is capable of operating at lower condensing temperatures than a normal TXV valve system, thus improving the efficiency.

A two-circuit shell-and-tube chiller is shown in (Fig. CR25-38, p. 502). At half the load, only half the chiller is used. Since units are usually running at reduced capacity, this feature also helps to improve partial-load efficiency.

The control system is self-diagnostic. In case there is a service problem the electronic circuit will flash a code number on the control panel to indicate the nature of the problem. During normal operation, the operator can read eight different system temperatures (see Fig. CR25-38) to indicate the performance of the unit.

CR25-5.2 Water-cooled Chiller

A typical water-cooled chiller using *multiple rotary screw-type* compressors is shown in Fig. CR25-39 (p. 502). The advantages of using the screw compressors are many.

1. The efficiency is higher than reciprocating units. Full-load efficiencies of approximately 0.80 kW/ton are offered.
2. They have about 1/10 the number of moving parts, and therefore require less maintenance.
3. They are quieter in operation.
4. They have infinite capacity modulation, using a slide-valve operation. Leaving water temperature can be controlled within 0.5°F, from full load to 30%.

The size range for the series illustrated is from 50 to 300 tons. The number of condensing units varies from one to four depending on the total capacity required. Multiple condensing units offer increased efficiency at partial load and simplify capacity reduction.

CR25-6 INDUSTRIAL REFRIGERATION UNITS

Certain manufacturers specialize in producing refrigeration for industry. Many of these units are highly specialized in design to supply specific needs in a production process. Often these units are individually tailored, assembled, tested at the factory, and shipped as a package.

Examples of industrial refrigeration units are shown in the next three figures. Fig. CR25-40 (p. 503) shows an in-

Table 1 Length, Base Weight, Load Points.

Number of Compressors			Rack Length (in.)	Base Rack Weight (lbs.)	Load Points Center to Center	
Copeland 2 & 3 Cyl.	Copeland 4 & 6 Cyl.	Carlyle			Length (in.)	Width (in.)
2–3	2	2	77	1,600	51	30
4	3	3	90.5	1,700	64.5	30
5	4	4	113	1,800	87	30
6	5	5	135.5	1,900	109.5	30
7	6	6	158	2,200	66 & 66	30
8	6	6	158	2,200	66 & 66	30

Figure CR25-31 Parallel compressors mounted on a single base complete with wiring, piping, and controls. (Courtesy of Hussmann Corporation, Division of Whitman Corporation)

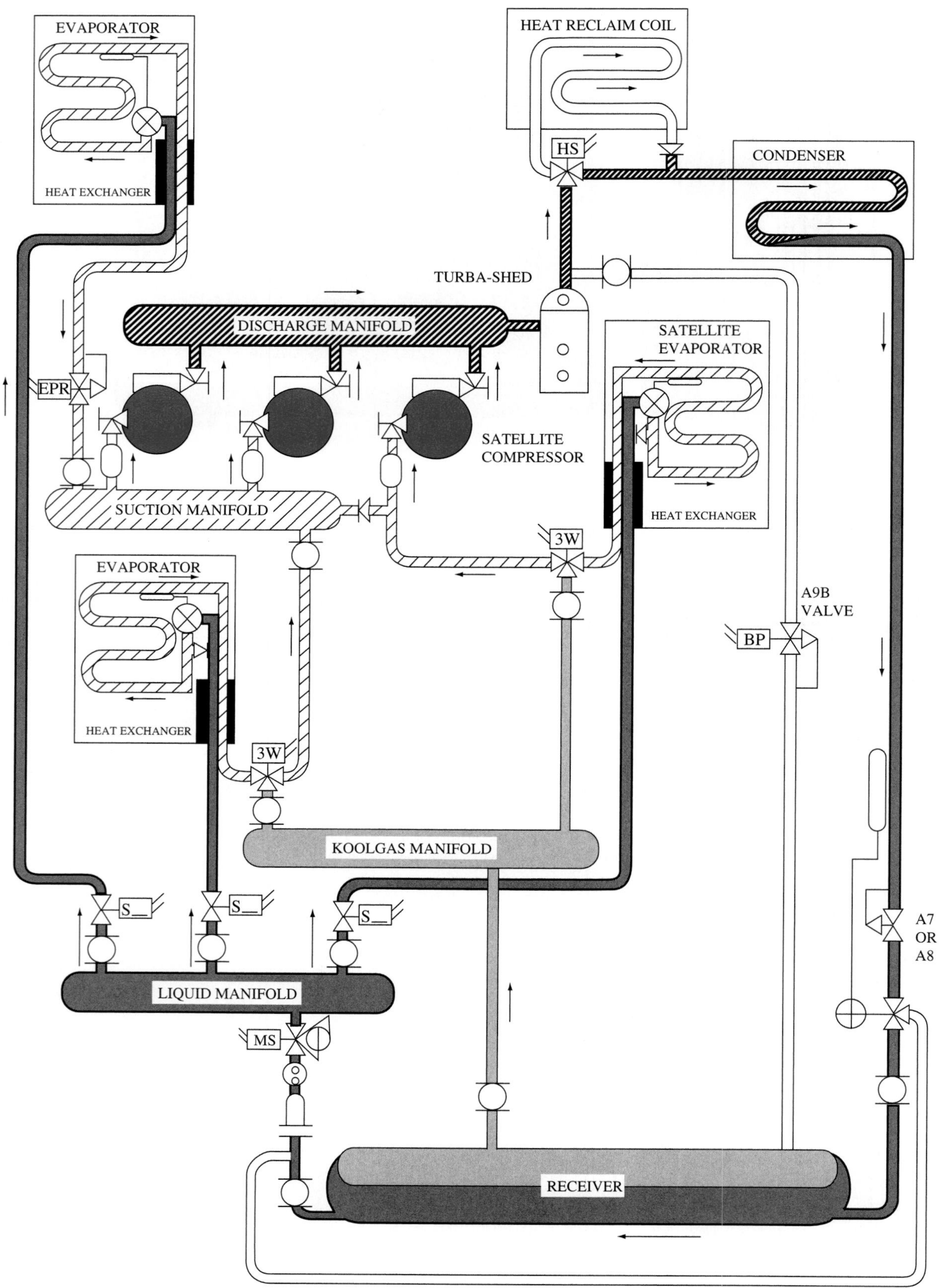

Figure CR25-32 A typical refrigeration system using parallel compressors for food-store application. (Courtesy of Hussmann Corporation, Division of Whitman Corporation)

BALL VALVE
CHECK VALVE
COMPRESSOR SERVICE VALVE
2-WAY VALVE
3-WAY VALVE
VALVE SOLENOIDS
SIGHT GLASS
SUCTION FILTER
LIQUID-LINE DRIER
THERMAL EXPANSION VALVE (TXV)
PRESSURE-REGULATING VALVE
A7 OR A8 FOR CONDENSER
A9 FOR RECEIVER
EPR FOR EVAPORATOR
MS
MAIN LIQUID-LINE PRESSURE DIFFERENTIAL VALVE
AUTOSURGE VALVE (OPTIONAL)
HEAT EXCHANGER (OPTIONAL)
HIGH-PRESSURE HOT VAPOR
HIGH-PRESSURE WARM VAPOR
HIGH-PRESSURE WARM LIQUID
LOW-PRESSURE COOL VAPOR
CUT-AWAY NOT SHOWN (EMPTY)

Figure CR25-32 Continued

Figure CR25-33 Two-stage refrigeration system using parallel compressors for low temperature application. (Courtesy of Hussmann Corporation, Division of Whitman Corporation)

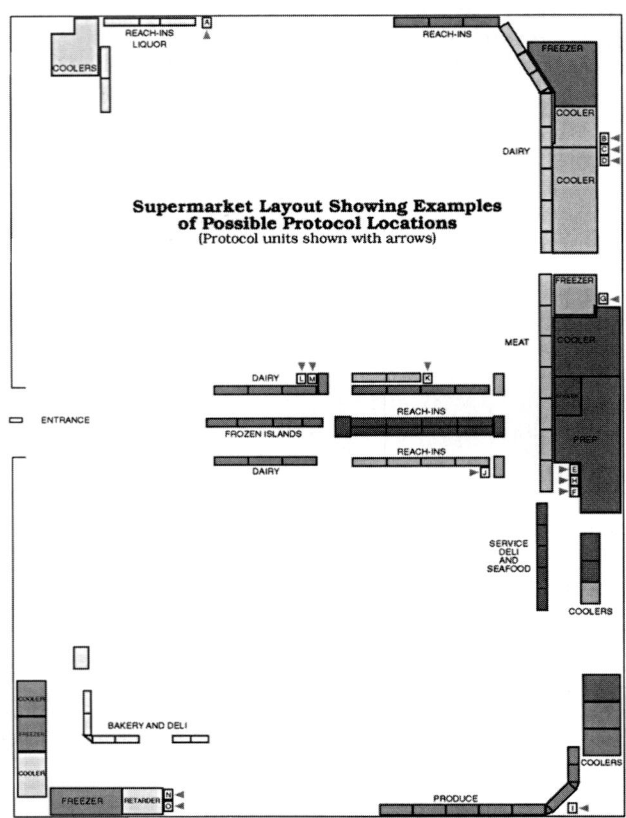

Figure CR25-34 Store layout using protocol refrigeration units. (Courtesy of Hussmann Corporation, Division of Whitman Corporation)

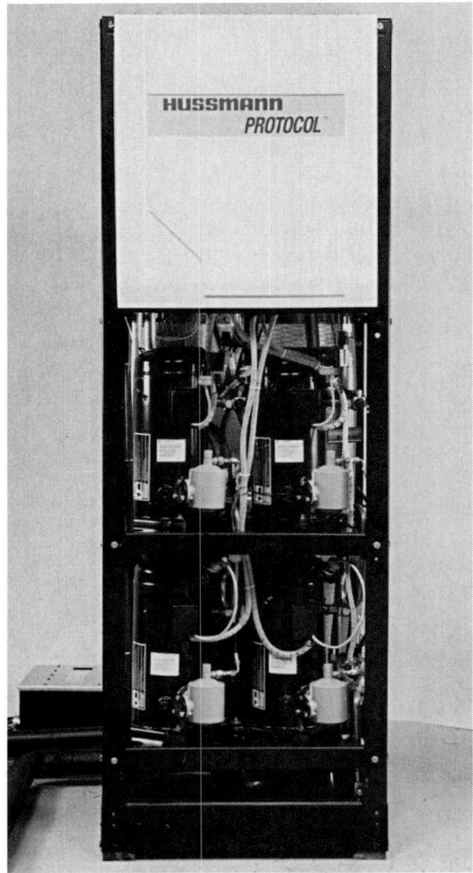

Figure CR25-35 A typical condensing unit for a protocol refrigeration system. (Courtesy of Hussmann Corporation, Division of Whitman Corporation)

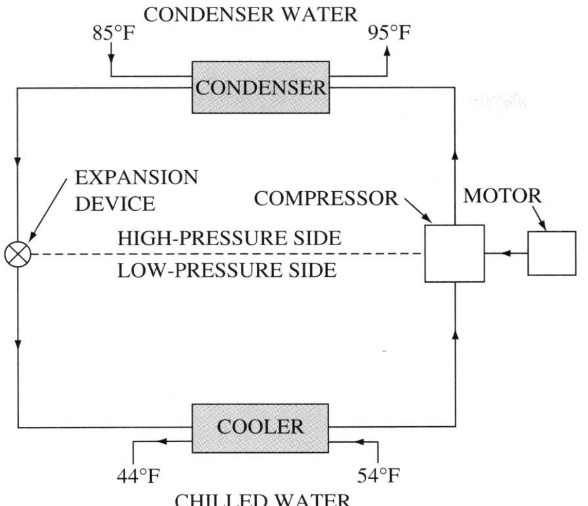

Figure CR25-36 Cycle diagram for a water chiller. (Copyright by the American Society of Heating, Refrigerating, and Air-Conditioning Engineers, Inc. Used by permission.)

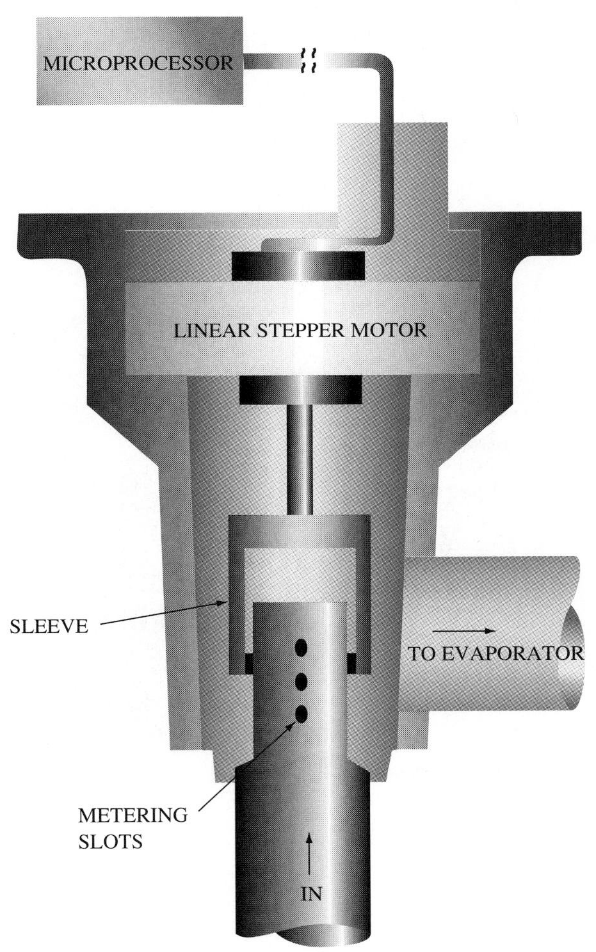

Figure CR25-37 Stepper-motor metering device.

dustrial quality compressor unit for process use. Fig. CR25-41 shows a custom-built 300-ton water chiller that is used in the manufacture of new refrigerant. Fig. CR25-42 (p. 504) shows a 70-ton ammonia system designed to contain only about 25% as much refrigerant ammonia as a conventional system.

CR25-7 DRINKING-WATER COOLERS

There are two styles of free-standing drinking-water coolers: the bubbler (Fig. CR25-43, p. 504) and the bottle (Fig. CR25-44, p. 504) types. The uses of these two models are almost equal in popularity. Both coolers operate with a similar refrigeration cycle, with the exception that the bottle cooler has no pre-cooler. The bottle cooler is more efficient, since all the cooled water is used.

The cut away view, Fig. CR25-45 (p. 505), shows the component parts that make up the bubbler styles. The bubbler mechanism includes the shutoff valve and the stream height adjustment. The precooler is a heat exchanger located in the drain line which effectively increases the capacity of the cooler about 40%. The cooling not only cools the incoming water but also can store cooling by creating an ice bank.

Many of these "coolers" also provide a source of hot water for making tea or coffee, using a resistance-heater accessory.

The capacities of water coolers are based on ARI standard conditions of 90°F room temperature, 90°F supply-water temperature, and 50°F drinking-water temperature.

The refrigerant systems are hermetically sealed using a capillary-tube metering device and are normally not serviced in the field. The compressor motors are fractional horsepower, split-phase, with a relay to cut out the start winding when the compressor is near full speed, as shown in Fig. CR25-46 (p. 505). The system is started and stopped by a water-temperature thermostat.

The refrigerants used have been R-12 and R-500; however, most manufacturers now use an HFC type refrigerant, such as R-134a. The amount of the refrigerant charge is usually less than one pound. To meet environmental regulations, EPA requires the removal of any CFC refrigerant if the unit is to be discarded or destroyed.

CR25-8 RESTAURANT/BAR EQUIPMENT

Normally restaurants use reach-in refrigerators and freezers, and depending on the size, walk-in coolers. Where the restaurant is also a bar, serving draft beer, special refrigeration equipment needs to be provided.

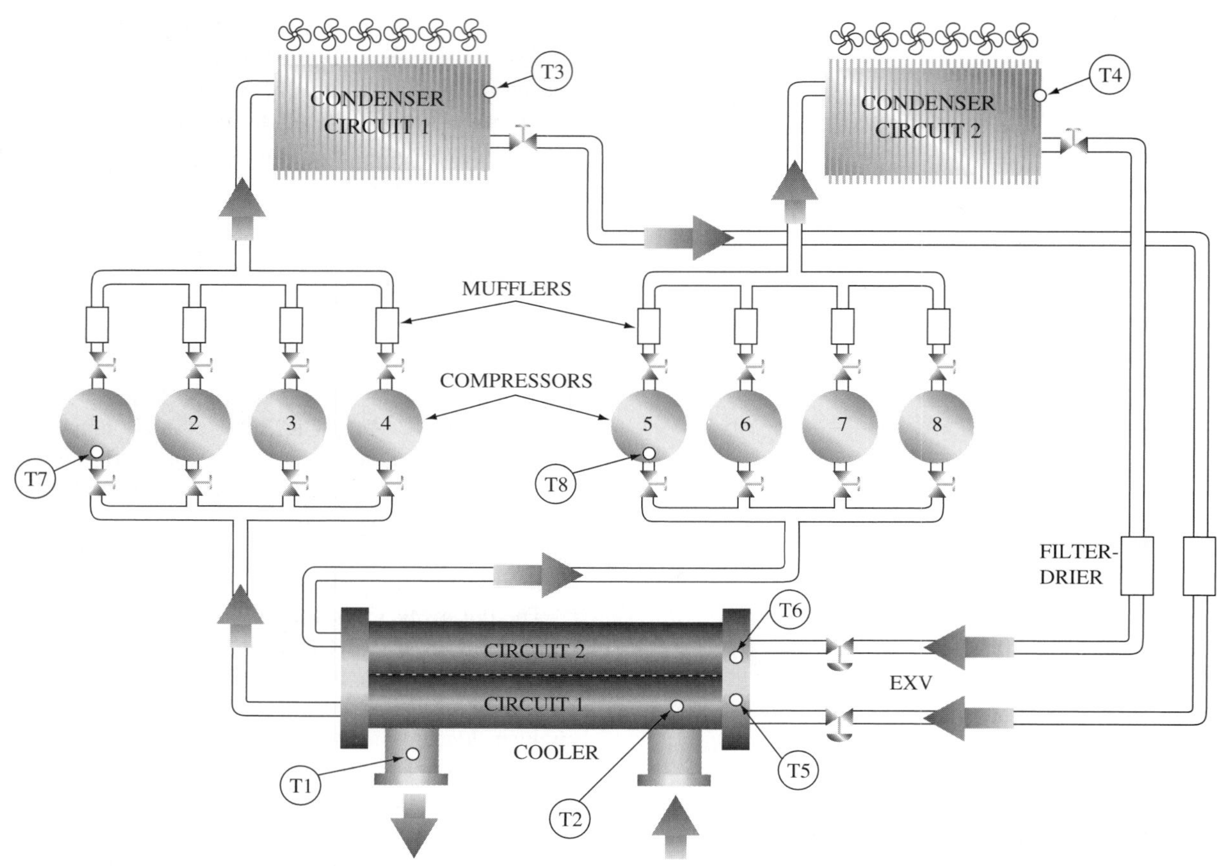

Figure CR25-38 Cycle diagram for air-cooled chiller showing two refrigeration circuits.

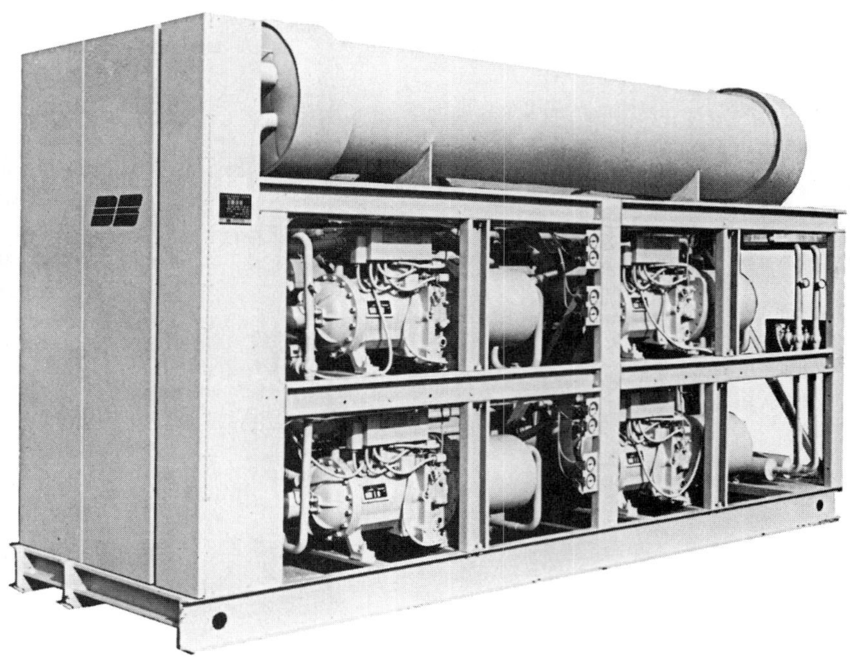

Figure CR25-39 Water-cooled chiller using multiple rotary screw compressors. (Courtesy of Dunham-Bush, Inc.)

Figure CR25-40
Industrial quality compressor unit. (Photo provided courtesy of Vilter Manufacturing Corporation, Cudahy, Wisconsin, U.S.A.)

Figure CR25-41
Custom-built water chiller. (Photo provided courtesy of Vilter Manufacturing Corporation, Cudahy, Wisconsin, U.S.A.)

Figure CR25-42 Ammonia refrigeration system. (Photo provided courtesy of Vilter Manufacturing Corporation, Cudahy, Wisconsin, U.S.A.)

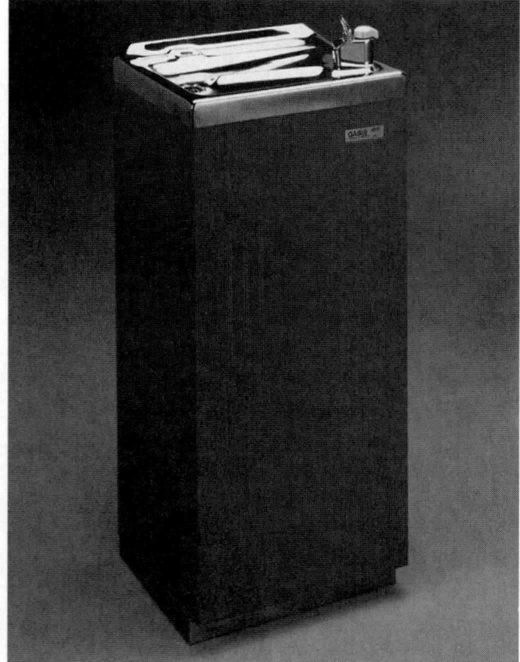

Figure CR25-43 Bubbler-type drinking water cooler. (Courtesy of Ebco Manufacturing Company)

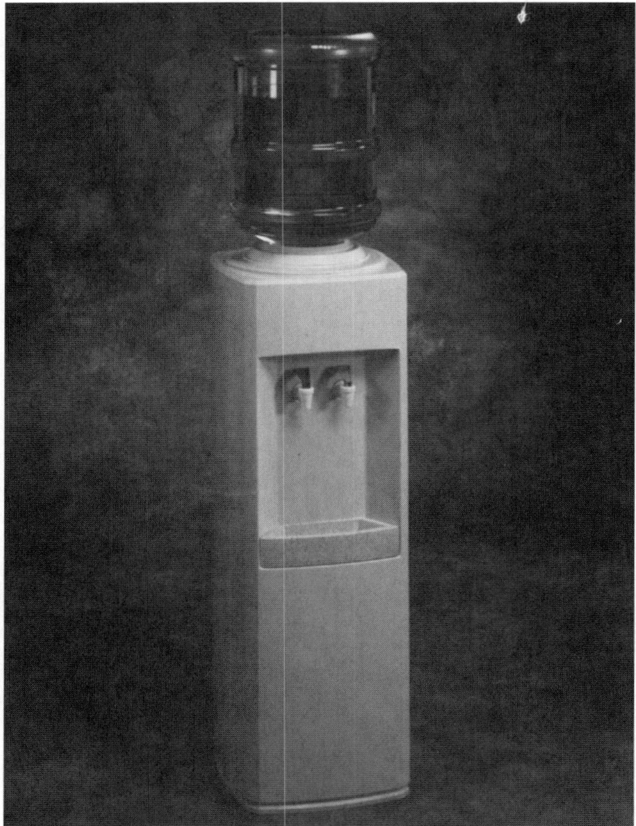

Figure CR25-44 Bottle-type drinking water cooler. (Courtesy of Ebco Manufacturing Company)

Figure CR25-45 Cutaway view showing internal parts of a bubbler-type drinking-water cooler. (Courtesy of Ebco Manufacturing Company)

Beer is a food product and must be properly handled to preserve its quality. Bottled beer is pasteurized and can be stored in dry beverage coolers which maintain a temperature of 38 to 40°F. Draft beer, which is preferred by most beer drinkers, is a perishable product and requires special refrigeration equipment for storing and dispensing it.

There are a number of kinds of draft beer dispensers. Regardless of the type, there are three fundamentals for satisfactory equipment: (1) *cleanliness,* (2) *temperature control,* and (3) *pressure control.*

Beer is a delicate food product which is easily contaminated. It is of the utmost importance that the service person follow strict cleanliness procedures in all phases of handling. All equipment, lines, and vessels coming in contact with the beer, right up to the time the beverage is set before the customer, must be absolutely clean.

It is important that the beer be maintained at the correct temperature. It should be drawn at 38°F, to assure that when the customer receives the glass, the temperature is between 40 and 42°F.

The pressure controls the speed of the draw at the faucet. Barrels are delivered at a pressure of 12 to 15 psi. Some carbon dioxide gas needs to be added to most direct-draw systems to maintain a pressure between 14 and 16 psi.

CR25-8.1 Dispensing Equipment

Bottle coolers are usually self-contained with the refrigeration equipment located in a lower compartment at one end. A back-bar cooler, (Fig. CR25-47) with front-door access is available for storing bottles or cans of beer and other items that need refrigeration.

A direct draw beer cooler is used for small bars that have space available under the counter. The refrigeration equipment is placed in a separate compartment at the end of the cabinet. Forced circulation of cold air keeps the beer cold right up to the faucet shanks.

There are two types of remote systems: (1) air-cooled beer lines, and (2) coolant-cooled beer lines.

The air-cooled remote system is shown in Fig. CR25-48. Two separate insulated ducts run side-by-side from the

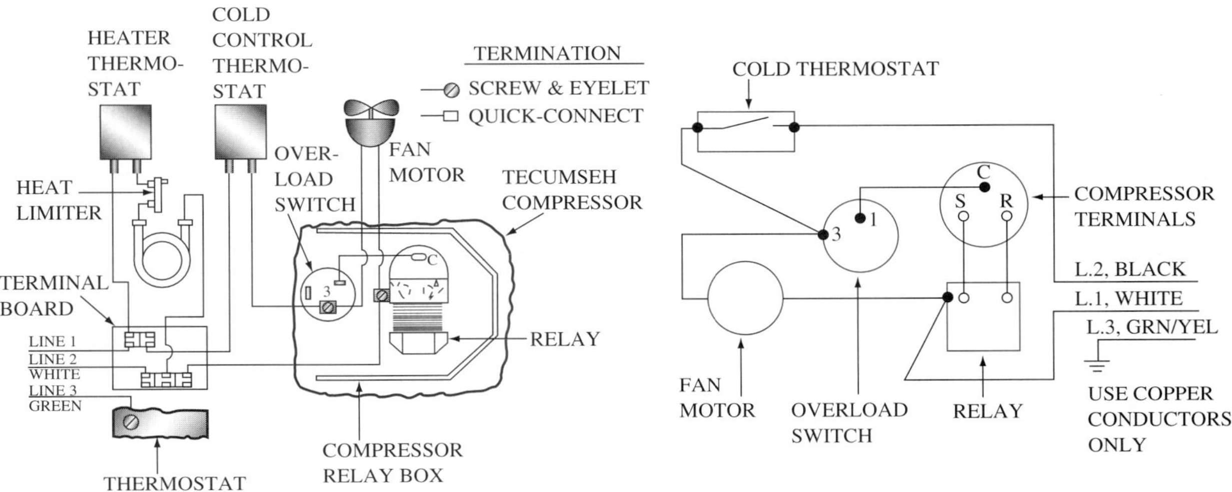

Figure CR25-46 Wiring diagram for a drinking-water cooler. (Courtesy of Ebco Manufacturing Company)

Figure CR25-47 Back bar for storing beverages and food products. (Courtesy of Nor-Lake, Inc.)

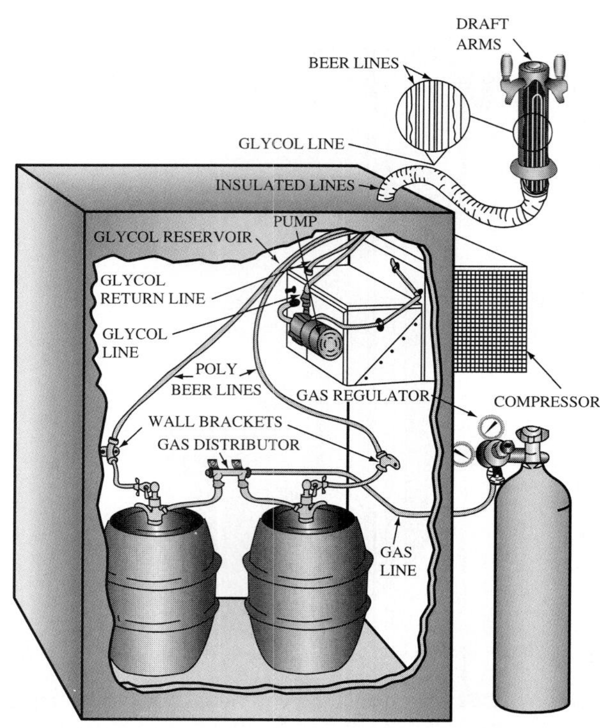

Figure CR25-49 Coolant-cooled draft beer system. (Courtesy of Master-Bilt Products, a Division of Standex International Corporation)

storage cooler to the dispensing cabinet. One carries the beer lines and the cold air to the tapping cabinet. The other carries the cold air back to the storage cooler. This type of system requires a sealed dispenser cabinet to prevent cold-air leakage.

The coolant-cooled remote system is shown in Fig. CR25-49. This system uses a single, well-insulated, hard rubber tube to carry the beer lines, supply, and return coolant lines. Beer lines to the dispenser are cooled by contact with the cold coolant lines.

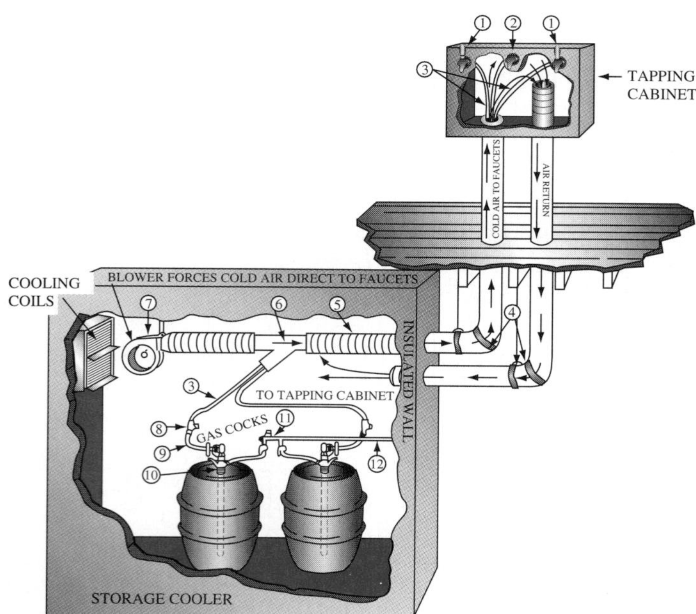

Figure CR25-48 Air cooled remote draft beer system. (Courtesy of Master-Bilt Products, a Division of Standex International Corporation)

CR25-9 DISPENSING FREEZERS

This group of products includes: *soft-serve ice-cream, slush,* and *shake machines.* All of these products are made by cooling prepared mixes. Soft-serve ice-cream and shake machines use mixes which include dairy products, and were developed first. Slush machines came later, using non-dairy products, and principally involved freezing carbonated beverages. Both the dairy and non-dairy-product freezing processes are essentially the same. They differ primarily in the constituents of the mix and serving temperature of the product.

The proper serving temperature is an essential condition of the product when it is to be consumed. The difference in serving temperatures is as follows:

Mix Description	Serving Temperature
Dairy	
Soft ice cream	20 to 22°F
Milkshake	26 to 28°F
Hard ice cream	5 to 9°F
Non-dairy	
Slush or FCB*	24 to 26°F

*Frozen Carbonated Beverages

If there is a problem with the appearance of the product, the freezing time and the product temperature should be checked. If these conditions are correct, the problem is in the mix.

CR25-9.1 Dairy Product Mixes

Ice-cream and *milkshake* basic mixes consist of the following ingredients:

1. Butterfat
2. Milk solids, non-fat
3. Sugar
4. Emulsifier-stabilizers (substances having a high water-holding capacity such as gelatin or vegetable gums).
5. Water

Because of the standardization of the mixes, problems which affect the quality of the product may be operational; however, the service person must not assume that the appearance of the finished product is due to mechanical problems of the refrigeration system unless product temperature is not being maintained. The following are some defects in ice cream:

1. *Sandy or grainy texture.* A quick test of the quality of the mix is to take a "lick" of the ice cream then rub your tongue against the roof of your mouth. If the ice cream is grainy (ice crystals), the mix is poor quality. It should be creamy and smooth.
2. *Butter churning,* which imparts a greasy taste to the ice cream, is a result of improper homogenization of the mix. Homogenizers are high-pressure pumps for breaking down the fat globules of the mix into tiny particles. For soft-serve mixes, because of repeated whipping, pressures of 3,000 to 3,560 psig need to be maintained, while 2,000 psig is satisfactory for hard ice creams. The supplier of the mix should be contacted and the operator should empty the freezing cylinder and wash all traces of butter clinging to the cylinder walls or dasher assembly.
3. *Wet, sloppy product.* Serum solids (milk solids non-fat) are important for improving firmness or "body" of the finished product. If the product does not "stand up" at the desired temperature or is wet and sloppy, or does not absorb air or overrun during churning, the quality of the serum solids in the mix can be questioned. Variations in this ingredient occur naturally and the mix manufacturer should have procedures in place for identifying and compensating for these variations. However, the service person must first determine that the product temperature is being satisfactorily maintained.
4. *Sugar content of the mix may vary.* The higher the content, the lower the temperature required to obtain the same firmness of the product. Small variations can often be compensated for by adjusting the temperature control either higher or lower. A large amount of sugar, requiring two or three degrees below normal range could result in a considerable loss of capacity for the soft-serve dispenser. The service person can use a thermometer to check product temperature and quality over a period of time to know whether the problem is in the mix or in the manner in which the freezer is operated.

CR25-9.2 Non-dairy Product Mixes

Slush mix, as it is now dispensed, consists of carbonated beverages which have variations in flavor, stabilizers, sugar, and water.

Overrun

To the mix ingredients, air must be added, called *overrun,* which is essential to the consistency of the product. Most machines have a fixed overrun, set at the factory for the particular mix being handled.

Function of the Operator

At the end of the day, when the machine is shut down, a number of moving parts of the freezer need to be removed by the operator and thoroughly cleaned. The operator must be trained not only to carefully remove them without damage, but also to properly re-assemble them and replace them for the next day's run.

The service technician needs to be able to check the proper assembly of these parts. Where the owner has a number of machines, it is possible to mix the parts. The same part must be used in the same machine each time the parts are reassembled. Particularly the dasher assemblies and scraper blades must be returned to the same machine from which they were removed.

The draw rates for the machines are limited by their individual capacities. If the designed draw rate is exceeded by the operator, the product will be too soft to stand up, and therefore unsatisfactory. This condition can be checked by the technician during a peak operating time.

Functions of the Refrigeration System

There are two unique parts to this machine: (1) the freezing cylinder evaporator, and (2) the control system.

The evaporator is usually a cylinder within a cylinder. The fully flooded refrigerant occupies the space between the two cylinders (Fig. CR25-50). Typical refrigeration systems are shown in Figs. CR25-51, CR25-52, and CR25-53.

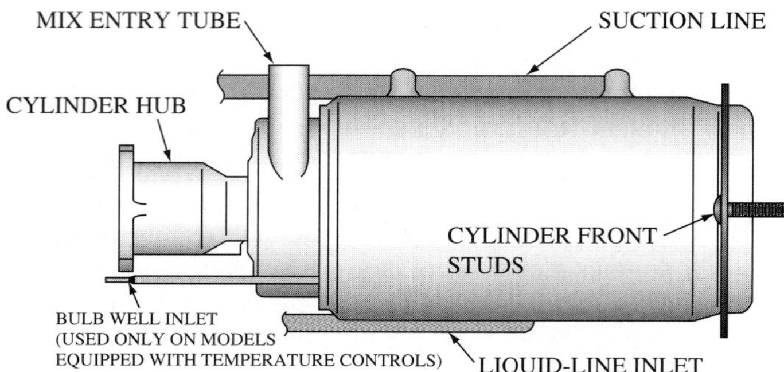

Figure CR25-50 Evaporator or freezing cylinder.

Refrigeration cylinders can be damaged by using water that is too hot during cleaning. Cleaning should be done with cold or cool water.

Measure the temperature of the product. If the product temperature is higher than normal for the mix being used, it is likely that a mechanical problem exists.

Function of the Control System

Temperature, pressure, and torque controls (consistency controls) are used. The temperature sensor is located as close to the product as possible. The torque control requires proper adjustment of the belt tension.

Many systems use an anticipator circuit which functions every time a serving of the product is made. The anticipator functions to start the machine, even though it has just turned off.

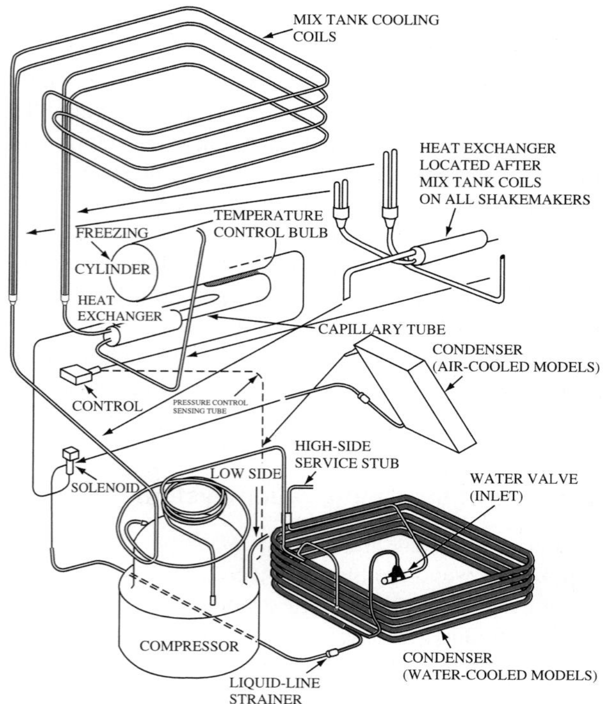

Figure CR25-51 Refrigeration system of ice-cream freezer using single capillary-tube flow control.

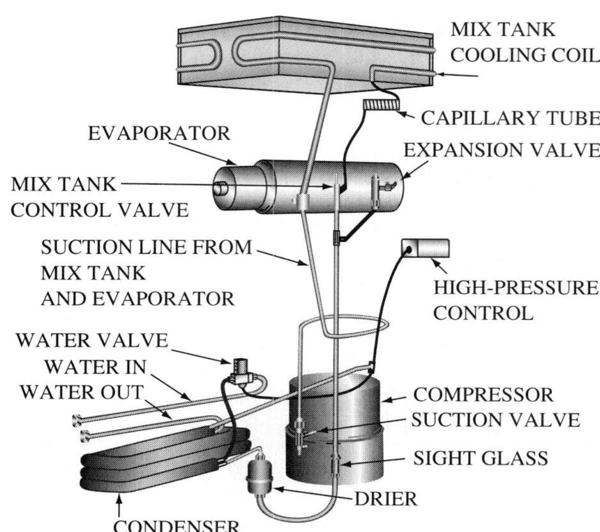

Figure CR25-52 Ice-cream freezer refrigeration system using capillary tube.

The basic function of the controls for all types of machines are as follows:

1. The refrigeration system must cut in as quickly as possible after the product has been drawn from the machine. This is necessary since fresh mix is automatically supplied to the freezer to replace product that has been drawn out. If the refrigeration does not come on after three or four servings, the cut-in point is too high.
2. The refrigeration must remain on during heavy draw conditions. If the compressor cycles off during heavy loads, the cut-out point is set too high.
3. The product temperature must be as even as possible between cut-in and cut-out points.

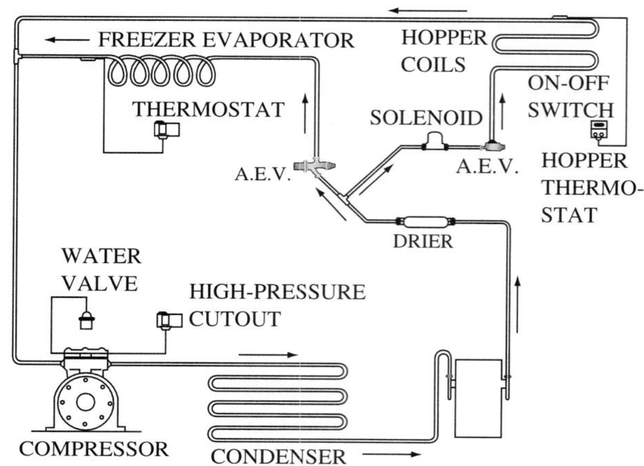

Figure CR25-53 Refrigeration circuit using two automatic expansion valves for refrigerant control.

4. ON-cycles should be as short as possible and OFF-cycles should be as long as possible, to avoid unnecessary whipping of the product.

Function of the Dasher Assembly

The dasher assembly has three functions: (1) to scrape the frozen product from the cylinder, (2) to whip the product, and (3) to eject the product.

The dasher blades must be kept in good condition and not be damaged during the cleaning operation. Most modern machines use dashers with plastic blades (Fig. CR25-54). The whipping action puts the necessary air into the product. The CO_2 or air pressure within the freezer also assists the ejection process.

Care must be taken to remove and replace the same dasher assembly in the machine it was taken out of during the cleaning operation. These units "wear in" and should not be interchanged.

Function of the Mix-Feed System

Mix-feed systems automatically meter liquid mix into the machine to replace that which has been drawn off (Fig. CR25-55). Most mix-feed systems require some adjustment at the time of original start-up and whenever a different mix is used.

Some of the basic points relating to mix-feed systems are as follows:

1. Mechanical-feed systems must be adjusted to meter an equivalent amount of liquid mix by weight into the cylinder, to replace the product that has been drawn off.
2. The gravity-feed systems do not require such precise settings. An orifice size is selected that permits the proper flow of mix.
3. Some models have different valve assemblies for soft ice cream and for shake makers. Milk shake machines require larger quantities of mix. These valve assemblies should not be interchanged.
4. The freezer cylinder must be preloaded with the correct amount of mix at the start of the day's operation. Consult the service manual for the machine being used for the initial loading and valve adjustment.

General Guide to Service

The three general categories of service are: (1) mix, (2) operational, and (3) mechanical. Always separate the problem into one of these three categories.

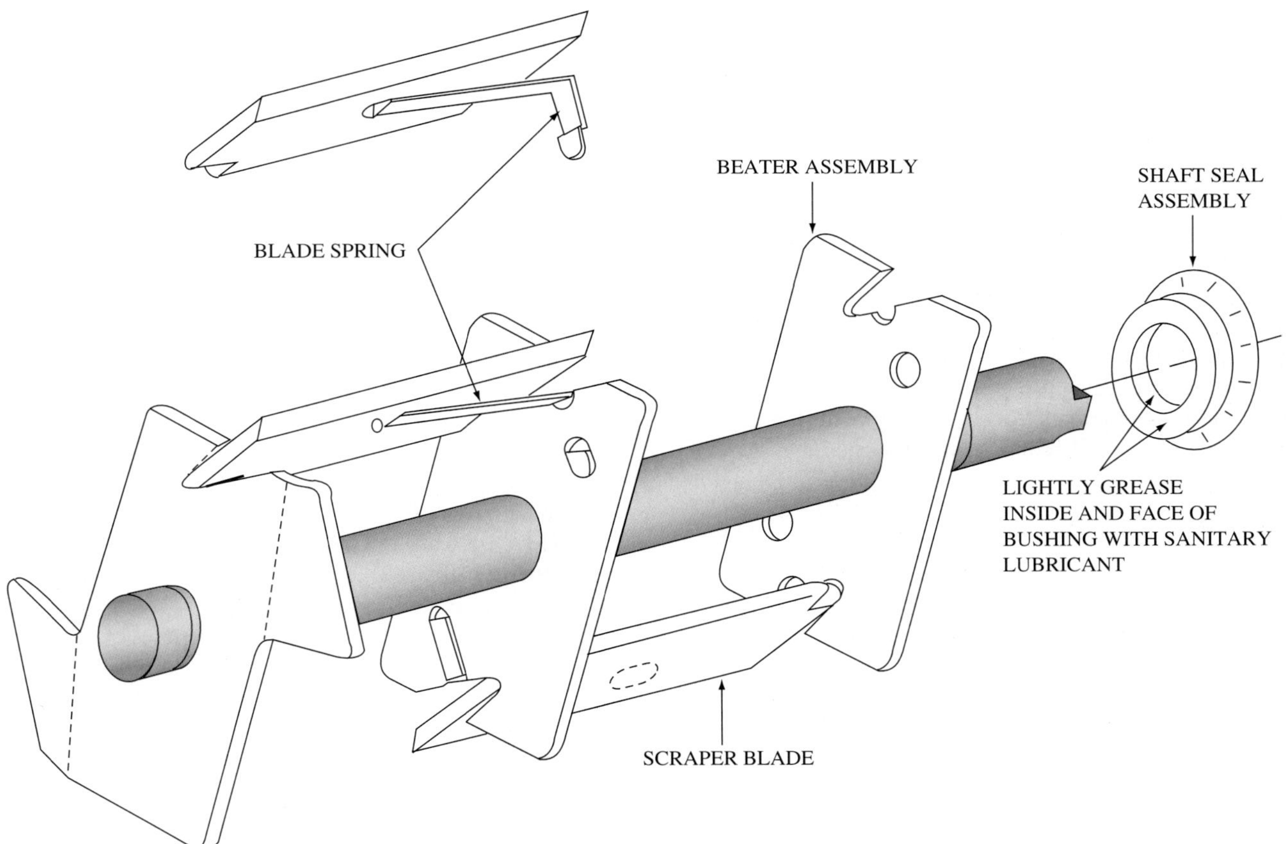

Figure CR25-54 Construction of the dasher which beats the mix.

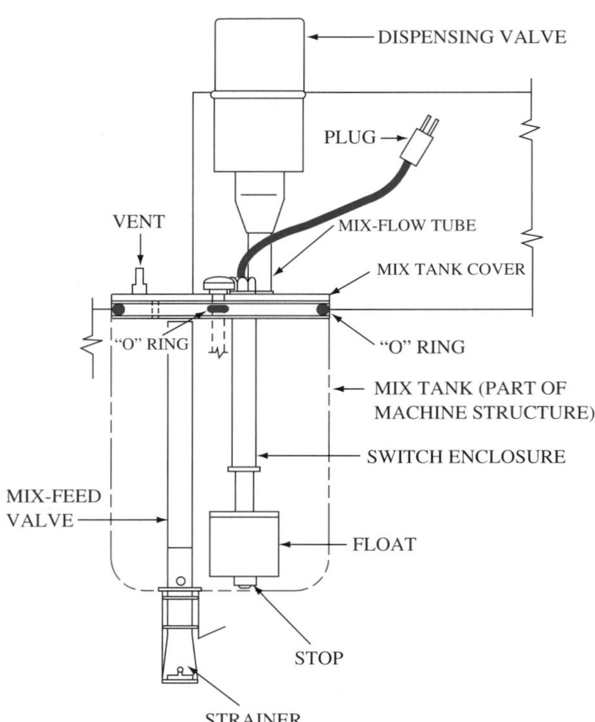

Figure CR25-55 Typical mix-feed system designed to operate the electric dispensing valves as well as automatically meter liquid mix into freezing cylinder.

Mix problems relate to the appearance and taste of the product. Solving these problems requires jointly working with the operator and mix supplier.

Operational problems require actually observing the use and cleaning of the equipment. It is important to explain to the operator the necessity of following accepted procedures.

Solving *mechanical* problems requires a thorough knowledge of the equipment being used. There are many different models—each having its own construction and service requirements. A careful study of the manufacturer's service manuals for the equipment being used makes the analysis of problems much easier.

REVIEW

- Classification of equipment:
  - Packaged equipment
  - Field-installed equipment
- Temperature application:
  - Ultra-low temperature—below −40°F
  - Low temperature—10 to −40°F
  - Medium temperature—32 to 10°F
  - High temperature—above 32°F
- Applications:
  - Refrigerated food storage units
    - Reach-in cabinets
    - Walk-in cabinets
    - Storage warehouse
  - Ice-making units
    - Cube
    - Flake
  - Supermarket equipment
    - Display cases
    - Rack-mounted condensing units
    - Parallel compressor units
    - Protocol system
  - Packaged liquid chiller
    - Air-cooled
    - Water-cooled
  - Industrial refrigeration units
    - Dairy equipment
    - Chemical processing
    - Ice-rink chillers
    - Fruit chilling
  - Drinking water coolers
    - Bubble style
    - Bottle style
  - Restaurant/bar equipment
    - Draft beer dispensers
    - Bottle coolers
    - Remote systems
    - Dispensing equipment
  - Dispensing freezers
    - Soft serve ice cream
    - Slush machine
    - Shake machine
    - Non-dairy product mixes

Problems and Questions

1. What are two classification schemes for commercial refrigeration equipment?
2. Equipment can be distinguished by usage or application. List at least four.
3. In commercial refrigeration, refrigerated food storage can be divided further into three product groups. What are these groups?
4. How are walk-in coolers and freezers assembled?
5. Where is the refrigeration equipment usually located for refrigerated warehouses?
6. What are the two types of ice machines?
7. What are the cycles of cube ice makers called?
8. What type of defrost cycle does the flake-ice maker have?
9. How would you classify the types of supermarket fixtures?
10. Supermarket equipment may be either _____ compressor systems or _____ compressor parallel systems.

11. What is the advantage of a multiple-compressor parallel system?
12. What types of defrost cycles are used for supermarket low-temperature systems?
13. What are typical low-temperature rack temperatures?
14. What are typical medium-temperature rack temperatures?
15. What is the advantage of two-stage compressors?
16. What are the advantages of the protocol system (multiplexed scroll compressors)?
17. What are the advantages of screw compressors?
18. Give some examples of industrial refrigeration applications.
19. What are the two styles of standing drinking-water coolers?
20. What are three requirements for beverage (liquid) coolers and dispensing freezers in use in restaurants and bars (regardless of the type)?
21. The important concept when working on commercial or industrial refrigeration equipment is to view the equipment in terms of the entire system. True or false?
22. Ice-making units make only cubes of ice for the beverage industries. True or false?
23. Commercial-size reach-in refrigerator units are designed to maintain 38°F. True or false?
24. High-pressure cut out switches are considered standard equipment on commercial equipment. True or false?
25. Whenever defrosting a freezer, the unit will stop defrosting only after the time clock is disengaged by the control circuit. True or false?
26. Ice-cube machines remove air from the cubes by:
    a. Using running water.
    b. Using special evaporators designed to eliminate the air bubbles.
    c. Using a special water filter.
    d. None of the above.
27. Multiplexing refrigeration equipment is:
    a. Using multiple compressors with multiple evaporators.
    b. Used on electronic controls with multiple refrigeration compressors.
    c. Using a single condensing unit for multiple cases.
    d. None of the above.
28. Parallel refrigeration systems:
    a. Consist of parallel compressors mounted on a common base.
    b. Consist of parallel compressors mounted with individual suction lines.
    c. Are piped to provide individual control.
    d. None of the above.
29. The term "Turba-shed" is used to describe a:
    a. Type of oil cooler.
    b. Type of oil separator.
    c. Type of liquid receiver.
    d. None of the above.
30. Packaged liquid chillers are used to:
    a. Cool water
    b. Cool brine
    c. Cool a secondary refrigerant.
    d. All of the above

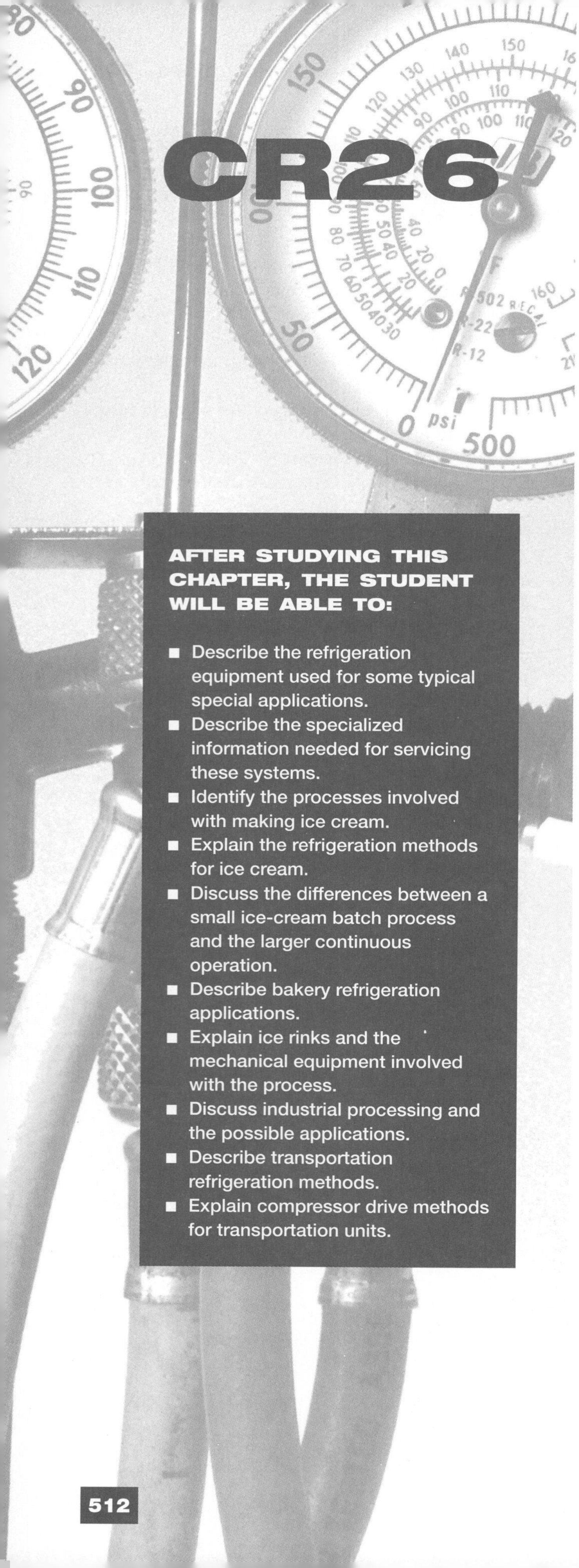

CR26 Special Refrigeration Applications

AFTER STUDYING THIS CHAPTER, THE STUDENT WILL BE ABLE TO:

- Describe the refrigeration equipment used for some typical special applications.
- Describe the specialized information needed for servicing these systems.
- Identify the processes involved with making ice cream.
- Explain the refrigeration methods for ice cream.
- Discuss the differences between a small ice-cream batch process and the larger continuous operation.
- Describe bakery refrigeration applications.
- Explain ice rinks and the mechanical equipment involved with the process.
- Discuss industrial processing and the possible applications.
- Describe transportation refrigeration methods.
- Explain compressor drive methods for transportation units.

CR26-1 ICE CREAM

Hard ice cream is defined legally by its butterfat content. The minimum, set by federal standards, is 8% for bulky flavored ice-cream mixes (chocolate) and 10% or above for other flavors (vanilla). For the specialty trade, the richer ice creams are 16–18% fat, but the average is 10–12%. Butterfat accounts for the rich flavor of the product. Refrigeration is a major cost in the making of ice cream, and its effectiveness depends on how well it transfers heat throughout the production processes. The making of ice cream begins with making the "mix", followed by freezing and storing the mix, and then using the mix to produce various kinds of ice cream. The service person should know the processes involved in order to recognize the source of problems which may arise, maintain efficient operations, avoid extra power use, and finish with a satisfactory product.

CR26-1.1 Creating the Mix

Pasteurizing

The basic ingredients for a typical ice cream are milkfat (12.5%), serum solids (non-fat solids of milk, 10.5%), sugar (15.0%), a stabilizer/emulsifier (0.3%), and (optionally) egg solids. The ingredients are placed in a vat, mixed, and heated (pasteurized) at 155°F for 0.5 hours to dissolve the solids and to destroy any pathogenic organisms. They are then precooled in a plate section using cool water. The final cooling to just above the freezing point is done in a second plate section using chilled water or glycol.

In the large plants, having automated and computerized processes, all liquid ingredients are used. These are blended, preheated, homogenized, then heated with plate

PRACTICING PROFESSIONAL SERVICES

Anytime you enter into a food processing plant or anticipate employment in the ice-cream or food industries, expect to practice professional sanitary health conditions. This includes wearing a hair net, shaving beards (most companies will not allow beards), wearing sanitary shoe covering, and washing down equipment regularly. Government regulations dictate that all food-handling personnel will practice standard sanitary procedures within the facility. If this does not appeal to the individual, then they should consider redirecting their career opportunities into a less stringent sanitary work environment.

equipment, a heat exchanger, or a direct-steam injector or infuser. This is followed by vacuum-chamber treatment for pre-cooling, which improves the flavor of the mix. It is then cooled through a regenerative plate section and additionally cooled indirectly to 40°F or less with chilled water.

Homogenizing

The heated mix is blended to reduce the size of the fat globules so the fat will not churn out during freezing. Should this happen, the ice cream has a greasy taste.

Cooling and Holding the Mix

The mix should be as cold as possible. In smaller plants the vats are equipped with both pre-cooling and final cooling sections. The mix is pre-cooled to about 10°F warmer than the entering water temperature, using city, well, or cooling-tower water. The final cooler using chilled water reduces the mix to about 40°F, and brine, or direct-expansion refrigerant, reduces the mix temperature to 30–33°F.

Larger plants usually use separate equipment for the final cooling. An ammonia-jacketed, scraped surface chiller is often used.

Where the mix can be held overnight, part of the final-mix cooling is done by means of a refrigerated surface built into the tanks. The rate of cooling is about 1°F per hour.

CR26-1.2 Freezing Ice Cream

Two kinds of freezers, for making either batch or continuous ice cream, use a cylinder with annular space or coils around the cylinder for direct refrigerant cooling. The space is either flooded, with an accumulator, or the cooling controlled by a thermostatic expansion valve. A dasher inside the cylinder has sharp metal blades which scrape the ice cream off the sides as it freezes. For efficient operation the blades should be kept sharpened.

Small operations use batch freezers which have an average maximum output of eight batches per hour depending on: the sharpness of the blades, the refrigeration supplied, and the overrun desired. At optimum operation the ice cream is drawn from the freezer at about 24°F, with refrigerant temperature around the freezer cylinder of about −15°F.

Large operations use one or more continuous freezers. The ice cream is discharged from several machines connected together to facilitate packaging. The mix is continuously pumped into the freezer cylinder. Air pressure for the overrun is drawn into the cylinder at 20 to over 100 psig, either with the mix or from a separate air compressor. A dasher with freezer blades agitates the mix and air as it moves through the cylinder. The ice cream is discharged at the other end of the cylinder at an average temperature of 22°F. The flooded system surrounding the cylinder, when operating with ammonia, is −25°F. Ice-cream temperatures as low as 16°F can be obtained with some mixes by regulating the evaporator temperature around the freezer cylinder using a suction pressure-regulating valve.

CR26-1.3 Hardening Ice Cream

The semi-solid ice cream leaving the cylinder is further refrigerated to a solid condition at 0°F for storage and distribution. Rapid freezing to obtain a smooth texture is achieved in hardening rooms kept at −20 to −30°F. Forced air is circulated from a unit cooler or a remote bank of coils. Ice cream in containers up to 5 gallons will harden in about 10 hours when spaced to allow air circulation. Systems using overhead coils or coil shelves with gravity circulation take twice this time.

Some larger plants use air-blast-type ice-cream hardening tunnels with operating temperatures ranging from −30 to −50°F to shorten hardening time to 4 hours.

Horizontal continuous-plate hardening systems automatically load and unload ice-cream packages, synchronized with the filler machines. They save space and power and eliminate package bulging, but can only be used for rectangular packages.

CR26-1.4 Refrigeration Equipment

Most of the larger commercial ice-cream plants use ammonia systems while smaller operations use single-stage halogen refrigerant compressors. These compressors are usually operated at conditions above the maximum compression ratio recommended by the manufacturer.

For maintaining −20°F freezer rooms, multi-stage compression is the most economical. One or more booster compressors are used at the same suction pressure, discharging into second-stage compressors. Where a hardening tunnel is used, at least two booster compressors should be used for the tunnel and for the freezer/storage room. Both units discharge into the second-stage compressor system. For plants with larger volumes, a three-stage system may be the most economical. A low-temperature booster is used for the tunnel and a second-stage booster discharges into the third-stage compressor system.

CR26-1.5 Maintenance of Operating Efficiencies

Conditions which lower the efficiency and effectiveness of the refrigeration system in ice-cream plants are related to the following:

1. *Heat transfer:* Air films, frost and ice, scale, noncondensible gases, abnormal temperature differentials, clogged sprays, slow liquid circulation, poor air circulation, and foreign particles.
2. *Ice-cream mix:* Viscous mix, low overrun percentage, high mix temperature, low ice-cream discharge temperature, and high sugar content in the mix.
3. *Dull scraper blades*
4. *Refrigeration equipment:* Low evaporator efficiency caused by rapid frost and ice development and an oil film.

Automatic defrost is recommended. Regular oil purges and an oil separator in the discharge line of the compressor will minimize oil film.

Neglected condensers lead to high head pressures and higher electrical requirements. Condenser surfaces should be kept clean of mineral deposits and other forms of scale or debris.

Noncondensible gases in compressor operation should be purged in order to avoid extra power use. Automatic purgers are available for this.

Door and conveyor openings are the cause of significant refrigeration losses. Insulation in storage rooms should be sufficient.

Cold rooms and ice banks should be serviced when the ambient air temperature is 26°F or lower.

CR26-2 BAKERY REFRIGERATION

Proper refrigeration is an essential part of bakery production. Refrigeration is used primarily in three areas:

1. Storage of the ingredients prior to use.
2. Controlling the temperature of the dough during the mixing process.
3. Storing and freezing the products.

CR26-2.1 Storing Ingredients

Bulk quantities require temperature protection during shipment or storage. Certain items, such as corn syrup, liquid sugar, lard, and vegetable oil, are held at 125°F to prevent crystallization or congealing. Fructose syrups are often used because they can be stored at a lower temperature, 84°F, to maintain fluidity and require less refrigeration during mixing. Smaller amounts and specialized sugars and shortenings in drums, bags, or cartons are stored at conditions for preventing mold and rancidity. Yeast should always be stored below 45°F. Cocoa, milk products, spices, and other raw materials subject to insect infestation should be stored at the same temperature as yeast.

Total-plant air conditioning is often used in new plant construction and includes filters to protect the equipment from airborne flour dust.

CR26-2.2 Mixing Process

During mixing, heat is produced in two ways: (a) heat of friction from the electrical energy of the mixer and (b) heat of hydration, produced when the material in the mixer absorbs water. The temperature rise in the dough is also dependent on the specific heat of each ingredient.

The temperature at which yeast is most active is quite critical. It is extremely active when mixed with water and fermentable sugars in a range of 80 to 100°F. The cells are killed at temperatures around 140°F and below 26°F, the freezing point of yeast. To maintain the proper mixing temperature, excess heat is removed by chilled water in contact with the dough.

For many years the principal refrigerant used in bakeries was ammonia. Other refrigerants are now in use for cooling the water to the desired temperature.

Fermentation

Two methods are used for processing the dough: the straight-dough process and the sponge-dough process.

In the straight dough process all the ingredients are mixed at once (yeast, flour, malt, and liquid) in an unjacketed tank at a temperature of 69 to 80°F and allowed to ferment for 1 hour. The second-hour fermentation takes place in a jacketed tank at about 84°F. A continuous process allows the dough to be drawn from one tank while the other is fermenting.

In the sponge-dough process, only part of the flour and water are mixed with the yeast for the first fermentation period. After fermentation takes place, the remaining flour and water are added. The total process takes 3½ to 5 hours.

CR26-2.3 Bread Cooling and Processing

Bread is cooled to 90 to 95°F for handling and slicing. This prevents condensation within the wrapper, which can cause mold to develop.

Refrigeration is required for the bread-wrapping machine. Evaporator temperature is usually around 10°F. The plate surface temperature must be held to about 16°F. Refrigeration is not required if bags are used.

Bread is frozen at temperatures between 16 and 20°F and stored at temperatures below 0°F to prevent moisture loss and crystallization of the starch. Freezing rooms maintain storage temperatures between 0 and −30°F. Two-stage refrigeration equipment is generally used.

CR26-3 ICE RINKS

The category of "ice rinks" includes all types of ice sheets, indoors or outdoors, created for recreation such as skating, hockey, and curling.

The rinks are usually made by laying down a series of pipe coils below the level of the design surface of the ice. A secondary refrigerant (brine), such as glycol, methanol, or calcium chloride solution, circulates through the coils from a central chilling system. The system usually uses R-22, although other refrigerants have been used.

The amount of refrigeration is usually dependent on the location of the rink (indoors or outdoors), the use of the rink, and the conditions of the building for indoor rinks (Fig. CR26-1).

Knowledge Extension

Ice rinks and ice cream processing plants constitute a large part of the refrigeration business but are often overlooked as employment opportunities. Most cities and communities, especially those in colder climates, have multiple ice-rink facilities inside and outside. Ice-cream plants are especially popular in warmer climates. To learn more about these facilities, contact the maintenance refrigeration operating engineer and schedule an appointment to walk through the equipment room to study the equipment. Look for thermometers indicating the product temperature, the brine heat exchanger, controls, valves, condensers, booster compressors, and suction and discharge pressures. Ask if there are any blueprints available to review and study. Express interest in training courses available or possibly internships to learn more about the business. Many of the larger industrial facilities will have full-time staff dedicated to the refrigeration process and will employ personnel in a training capacity.

CR26-3.1 Ice-rink Conditions

Indoor rinks are operated from 6 to 11 months a year, depending mainly on profitability. Rooms for heated rinks are usually maintained at 50 to 60°F with relative humidities ranging from 60 to 90%.

Four to Five Winter Months, above 37° Latitude	ft²/ton	m²/kW
Outdoors, unshaded	85 to 300	2.2 to 7.8
Outdoors, covered	125 to 200	3.3 to 5.2
Indoors, uncontrolled atmosphere	175 to 300	4.6 to 7.8
Indoors, controlled atmosphere	150 to 350	4.0 to 9.1
Curling rinks, indoors	200 to 400	5.2 to 10.4
Year-round (Indoors), Controlled Atmosphere	**ft²/ton**	**m²/kW**
Sports arena	100 to 150	2.6 to 4.0
Sports arena, accelerated ice making	50 to 100	1.3 to 2.6
Ice recreation center	130 to 175	3.4 to 4.6
Figure skating clubs and studios	135 to 185	3.5 to 4.8
Curling rinks	150 to 225	4.0 to 5.9
Ice shows	75 to 130	2.0 to 3.4

Figure CR26-1 Ice-rink refrigeration requirements for a variety of applications.

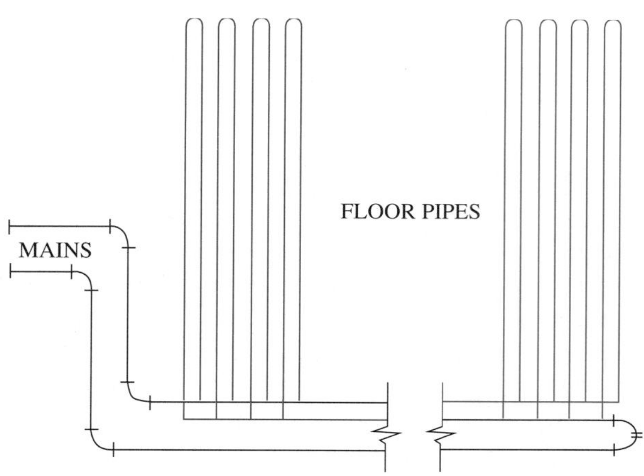

Figure CR26-2 Piping diagram for ice rink to provide balanced flow.

The temperature of the ice is controlled closely either by the brine temperature or the ice-surface temperature. The temperature of ice for hockey is 22°F, for figure skating 26°F, and for recreational skating 26 to 28°F. A 1-in. thickness of ice is usually maintained.

Average brine temperatures are 10°F lower than the required ice temperature. Usually two compressors are used for pull-down and one for normal operation. The temperature differential between supply and return brine is approximately 2°F.

CR26-3.2 Construction and Equipment

Brine systems generally use 1-in. steel or polyethylene pipe 4 in. on centers. Pipes are set dead level, with sand fill around and over the pipes. A balanced-flow distribution system is shown in Fig. CR26-2. An expansion tank must be installed in the piping.

The construction of the space below the rink is very important (Fig. CR26-3). There must be proper drainage to prevent "heaving" caused by water freezing below the rink. Many new rinks install heater cables below the floor of the rink or circulate warm brine to prevent heaving. A "header trench" is located at one end of the rink to house the piping header. A snow-melting pipe is usually located at the opposite end of the rink for melting the scraped ice, removed by scraping the surface of the rink.

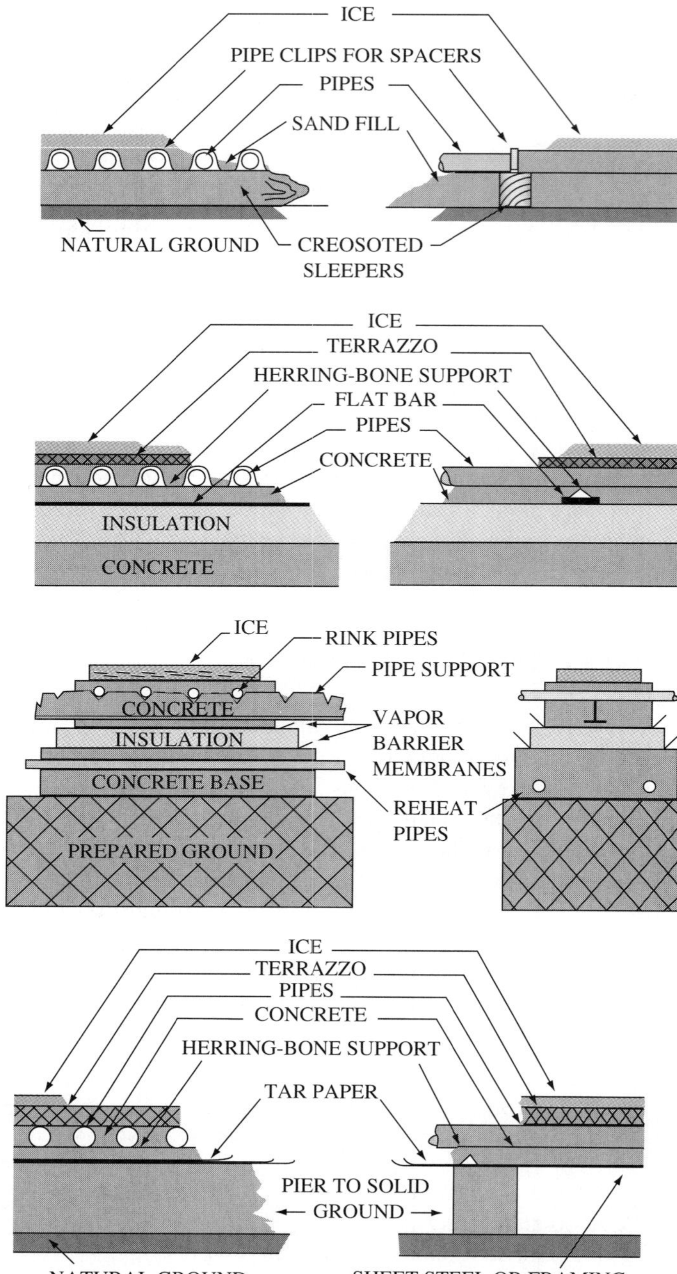

Figure CR26-3 Types of ice rink surface floors.

CR26-4 INDUSTRIAL PROCESSING

Some of the industries where refrigeration is most prevalent are: (1) *commercial ice plants* that produce large quantities of flake ice for refrigerating fresh fish and other perishable products, (2) *soil-stabilization systems* which prevent problems associated with "frost heave" encountered in cold climates like Alaska, (3) *chemical processes* that require refrigeration, usually for the removal of unwanted heat generated during manufacturing processes, and (4) *environmental test facilities* that use refrigeration to simulate environments where their products will be used. For example, one of the large automobile factories tests car operation in a –40°F room.

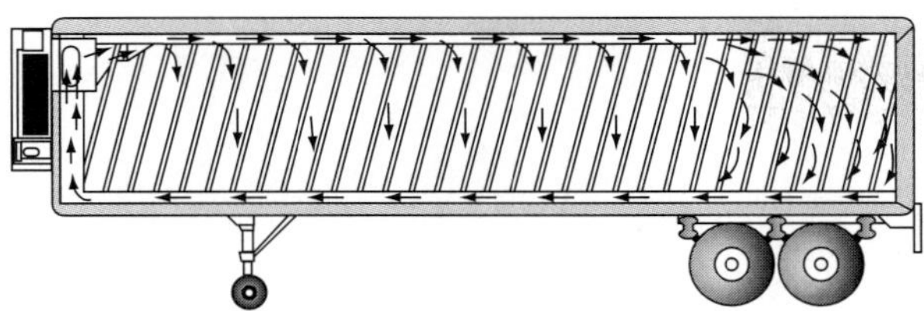
NOTE FALSE BULKHEAD WHICH FORMS RETURN AIR PLENUM TAKING AIR FROM FLOOR.

Figure CR26-4 Section of a trailer showing air circulation.

CR26-5 TRANSPORTATION REFRIGERATION

The transportation of refrigerated products is a large and important aspect of commercial refrigeration. Many truck bodies are well insulated, and reliable refrigeration units are available to maintain the required temperatures. There are a number of types of refrigerated trucks and trailers. There is some difference between those used for long and short hauls, as well as for the different types of products being transported.

CR26-5.1 Long Haul Systems

Long-distance hauls are usually handled by trailer trucks. These trailers can be detached from the engine or tractor portion of the assembly. The trailer is usually equipped with a stand-alone refrigeration system. An illustration of a typical insulated trailer is shown in Fig. CR26-4. Notice the bulkhead in the front for return air coming back to the evaporator through a raised floor. This illustration shows a "piggyback" or clip-on refrigeration unit.

Some trailers use refrigeration consisting of the evaporation of liquid nitrogen (N_2) or liquid carbon dioxide (CO_2). In these systems the evaporated refrigerant is wasted to the outdoors.

The semitrailer with stand-alone refrigeration can be placed on a railroad flat car and transported. Local delivery is accomplished by connecting a tractor and making the local delivery.

Four different temperature conditions are maintained in refrigerated transport, depending on the type of product being handled. These are:

Type of service	Temperature (°F)
1. Air conditioning for floral products, candy, etc.	55 to 70
2. Medium temperature for perishable foods	32 to 40
3. Fresh meats	28 to 32
4. Frozen foods	0 to –5

Some systems that are used both winter and summer have heating equipment as well as cooling equipment. Reverse-cycle refrigeration is popular for these applications. Reverse-cycle refrigeration can also be used for defrosting evaporators where conditions cause the accumulation of ice on the coils.

There are basically two different types of transport refrigeration units: one type for trucks (Fig. CR26-5) and the other type for trailers (Fig. CR26-6).

The most common type of trailer unit is mounted at the top front of the trailer (Fig. CR26-7). Note that the complete unit is factory assembled, ready for installation. All units are available with diesel-engine drive, a complete refrigeration system for cooling and heating, a control system, and accessories. Optional electric motors are available for stationary operation.

The net cooling capacity depends on the temperature maintained. Capacities range from 4,000 to 19,500 Btu/hr at 0°F and 100°F ambient. The fuel consumption varies from 0.13 to 5 gph, depending on the type of service.

Selection of the proper unit is based on the length of the truck body, the temperature maintained inside the body, the amount of insulation, and the number of door closings per day.

Figure CR26-5 Front-mount diesel-powered cooling/heating unit for large trucks. (Courtesy of Thermo King Corporation)

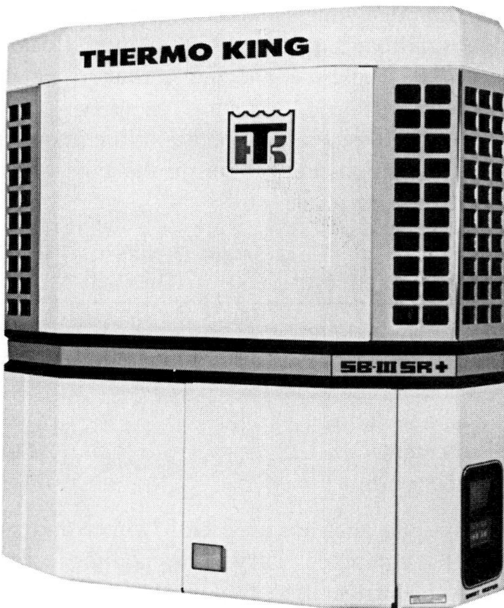

Figure CR26-6 Fuel-saver heating and cooling, high speed and low speed. Runs continuously. Automatic or manual defrost. (Courtesy of Thermo King Corporation)

CR26-5.2 Local Delivery Equipment

Where smaller trucks are used for local deliveries, refrigeration systems with *eutectic plates* are advantageously used. Eutectic plates are constructed with a provision for refrigeration storage capacity. Plates are made with an interior volume that holds a special liquid called a *eutectic solution.* This liquid stores the cold by changing the state of the solution. Just as ice freezes at 32°F, these eutectic solutions freeze at various temperatures depending on the desired conditions in the truck. Plates are available for operating temperatures of −58, −29, −14, −12, −10, −8, −6, +18, +23, and +26°F.

The plates are usually connected to the refrigeration compressor at night, when the truck is not being used. The refrigeration capacity of the plates is sufficient to carry the load during the truck's next-day use. Some trucks carry their own compressor so that they can be easily plugged into electric service wherever they may be. Illustrations of the eutectic plates are shown in Fig. CR26-8.

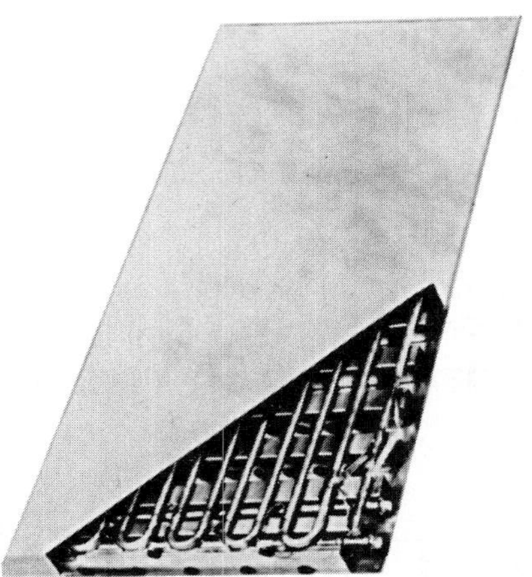

Figure CR26-8 Plate-type evaporator with eutectic solution for holdover cooling capacity. (Courtesy of Dole Refrigerating Company)

Another arrangement for ice-cream or frozen-food service consists of equipping the truck with auxiliary drive or an electric motor to operate the compressor while the truck is in service.

CR26-6 MECHANICAL REFRIGERATION

Many types of systems are available for furnishing the power to drive the compressor, including:

1. Independent gasoline (or propane or butane) engines
2. Power takeoffs from vehicle engines
3. Electric motors

Figure CR26-7 Front-mount cooling/heating unit. (Courtesy of Thermo King Corporation)

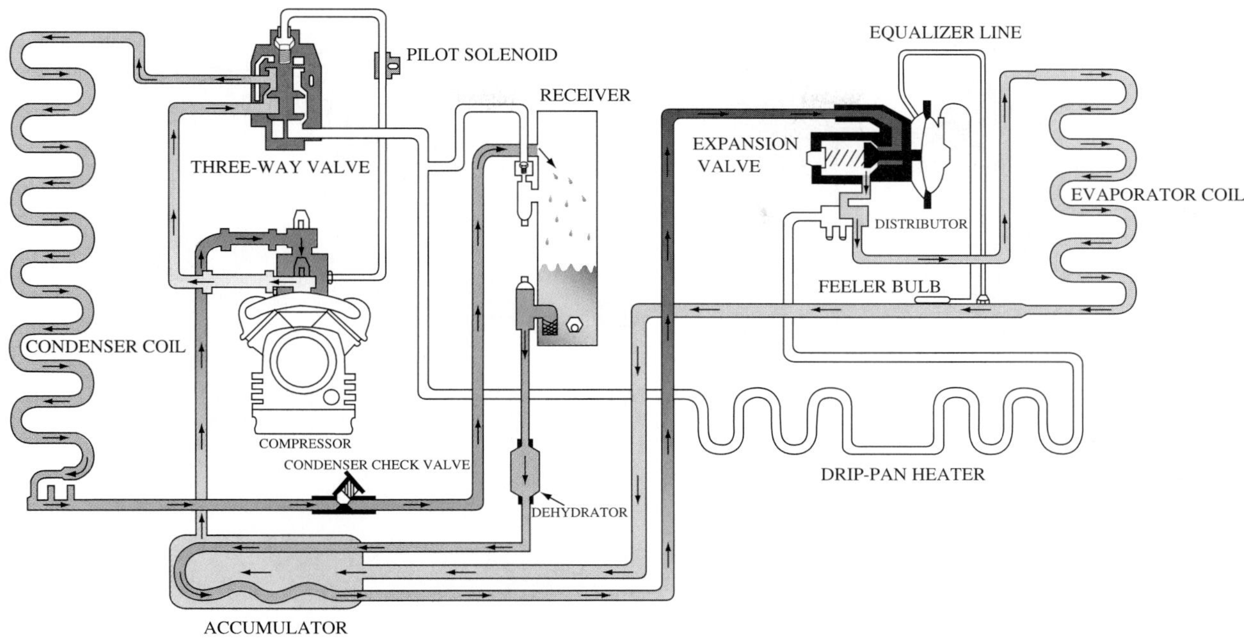

Figure CR26-9 Cooling cycle for a truck refrigeration unit.

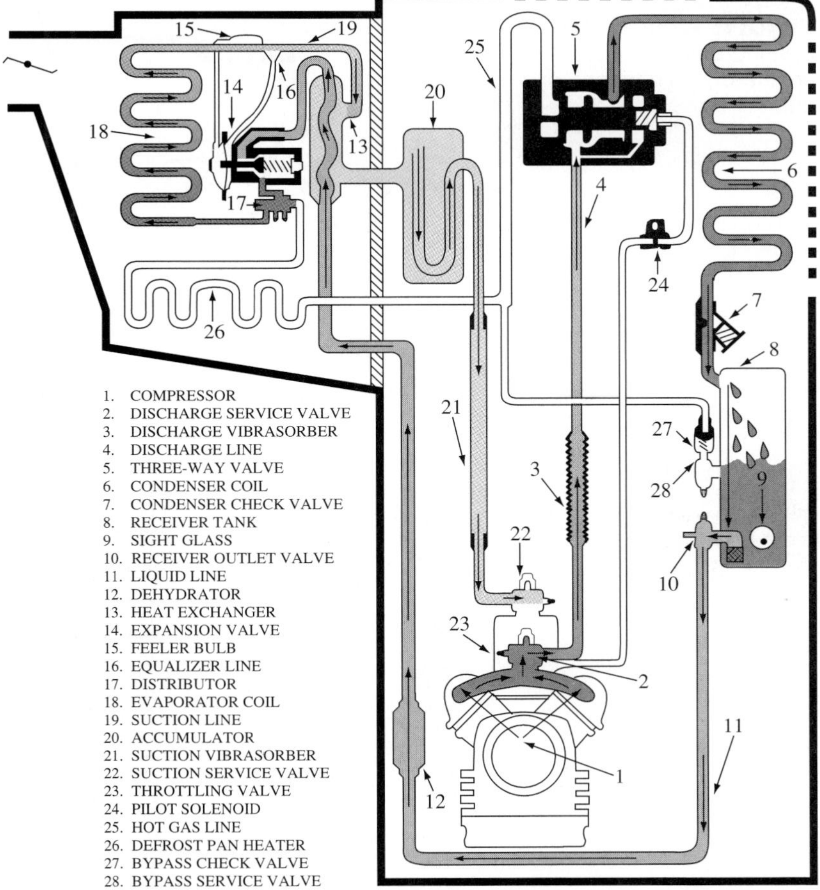

Figure CR26-10 Cooling cycle for an engine-driven refrigeration unit.

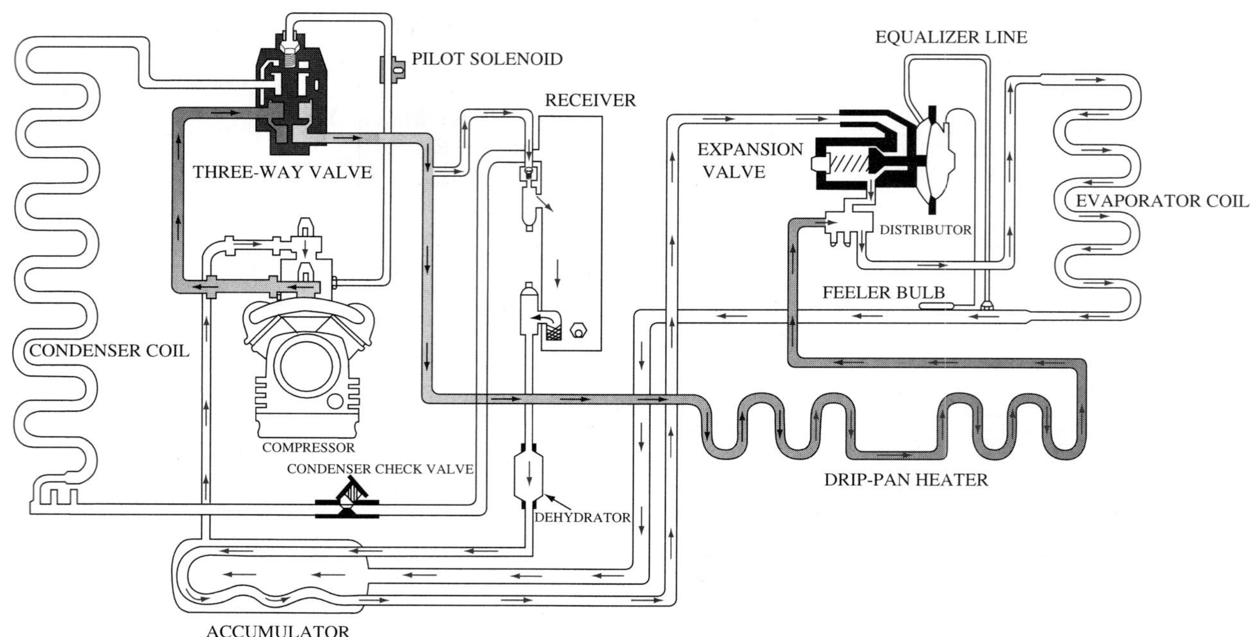

Figure CR26-11 Heating and defrost cycle. Heating is by reverse-cycle principle using hot-gas discharge from compressor.

Figure CR26-12 Heating/defrost cycle for an engine-driven refrigeration unit.

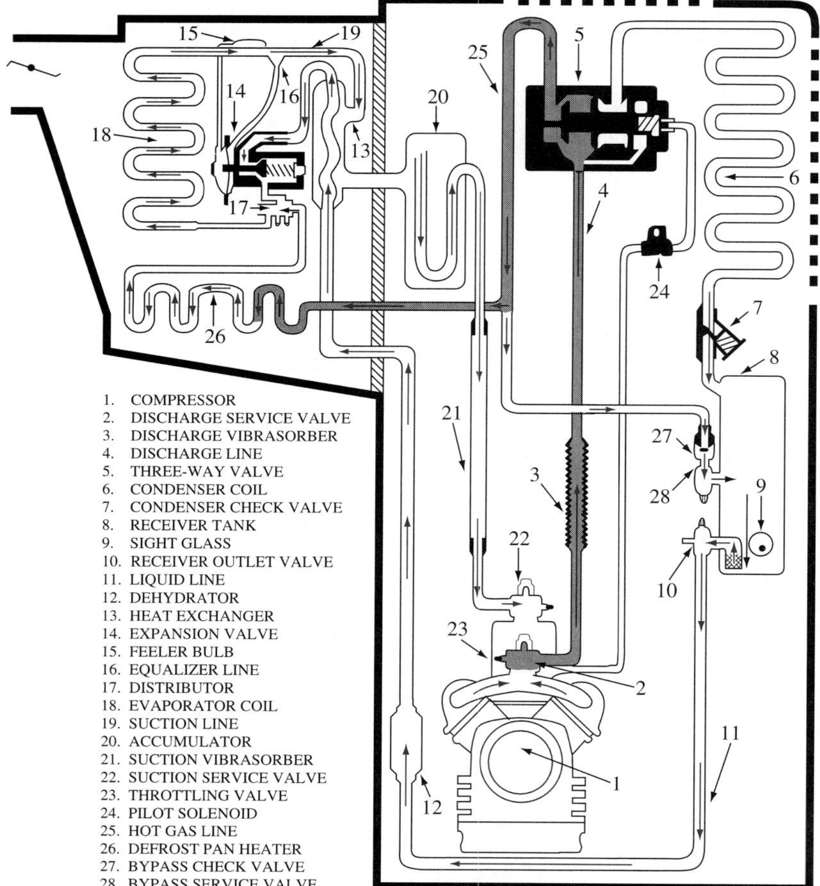

The compressor may be either on board the vehicle or at the garage for out-of-service use only. Evaporators are of many types: finned-and-pipe coils for forced or gravity circulation, standard plates, or eutectic plates.

Equipment must be designed to withstand severe motion, shock, and vibration. Short, rigid lines are subject to "cold working". They become hard and brittle through continual motion. Thus, flexible hose connections are often used. Air-cooled condensers need to be designed so that normal airflow due to movement of the truck does not prevent adequate air circulation. Units with independent engines require batteries for cranking, fuel tanks and pumps.

The typical refrigeration cycle is shown in Figs. CR26-9 and CR26-10 (p. 519), and the typical heating and defrost cycle is shown in Figs. CR26-11 and CR26-12.

The units are factory assembled with engine compressor and condenser in the outside enclosure and the evaporator projecting inside the trailer. The engine runs continuously and the operator has the option of high-speed or low-speed cooling and high-speed or low-speed heating. When required, the unit will automatically defrost with hot gas. All units have an hour meter to determine service periods for the engine.

REVIEW

Low-temperature applications:

- Ice cream:
  Minimum set by federal standards is 8% butterfat
  10% or above for flavors
  Richer ice creams use 16–18%
- The exact refrigeration requirements are complicated by the specific heat of the mix, which varies with its composition. A value of 0.80 Btu/lb is assumed for ice cream mixes.
- Two kinds of ice-cream freezers:
  Batch (small operations)—At optimum operation the ice cream is drawn from the freezer at about 24°F with a refrigerant temperature around the freezer cylinder of about −25°F.
  Continuous (large operations)—Mix is pumped into the freezer cylinder, air pressure for the overrun is drawn into the cylinder at 20 to 100 psi either with the mix or from a separate air compressor. A dasher with freezer blades agitates the mix and air as it moves through the cylinder. The ice cream is discharged at the other end at an average temperature of 22°F. Refrigerant temperature is about −25°F when operating with ammonia. Ice-cream temperatures as low as 16°F can be obtained with some mixes.
- Hardening ice cream:
  Refrigerated to solid condition at 0°F for storage and distribution.
  Rapid freezing to obtain a smooth texture is achieved in a hardening room kept at −20 to −30°F.
- Maintenance and efficiency of ice cream plants depends upon:
  Heat transfer
  Ice cream mix
  Scraper blades quality
  Refrigeration equipment
- Bakery refrigeration:
  Storage of ingredients prior to use
  Controlling the temperature of the dough during mixing
  Storing and freezing the products
- Ice rinks—Ice sheets indoors or outdoors, created for recreation such as skating, hockey, and curling.
- Industrial processing:
  Commercial ice plants
  Soil stabilization
  Chemical processes
  Environmental test facilities
- Transportation refrigeration:
  Long haul systems
  Local delivery equipment
- Compressor drive methods:
  Independent gasoline (or propane or butane) engines
  Power takeoffs from vehicle engines
  Electric motors

Problems and Questions

1. What distinguishes "hard ice cream" from "soft serve"?
2. What does the term "homogenized" mean in terms of ice-cream making?
3. What refrigerant is often used in dairy processing?
4. How does percentage of butterfat affect the freezing point?
5. Ice-cream freezers use what type of evaporators?
6. Rapid freezing rooms maintain temperatures of ______.
7. What type of refrigeration system works best for applications below −30°F?
8. Where is refrigeration used in the baking industry?
9. Ice rinks usually use what type of refrigeration system?
10. Before the use of refrigeration in railroad cars for fresh produce transportation, what method was used?
11. What are eutectic plates?
12. Some transport refrigeration systems and industrial process units use liquid nitrogen or liquid carbon dioxide to cool to extremely low temperatures. What are these systems called?
13. What is unique about trailer (truck) refrigeration systems?

14. What are the three ways of furnishing power to drive the compressor?
15. What is the major consideration for refrigerant lines on mobile equipment?
16. Richer ice cream products have a higher percentage of fat content. True or false?
17. Larger production ice cream plants utilize R-22 refrigerant. True or false?
18. Small ice cream operations use batch freezers which have an average maximum output of eight batches per hour. True or false?
19. Hardening ice cream requires rapid freezing at –20 to –30°F. True or false?
20. For maintaining –20°F room temperatures, multistage compressors are more economical. True or false?
21. Booster compressor systems utilize:
    a. Two or more compressors.
    b. A booster liquid pump to get the refrigerant to the top of the system into the receiver.
    c. A new ozone-friendly refrigerant.
    d. None of the above.
22. Efficiency during the ice cream production is affected by:
    a. The ice cream mix.
    b. Dull scraper blades.
    c. Low evaporator efficiency.
    d. All of the above.
23. Automatic purgers are available on low-pressure systems to:
    a. Purge the room of toxic gases in case of a leak.
    b. Remove non-condensibles.
    c. Purge the compressor of ammonia refrigerant after repairs.
    d. None of the above.
24. Total-plant air conditioning is often used in new large modern bakeries to:
    a. Protect and filter equipment from airborne flour dust.
    b. Provide a higher-quality product.
    c. Keep employees comfortable.
    d. None of the above.
25. Ice rinks:
    a. Are made by laying down a series of pipe coils below the level of the ice.
    b. Use a secondary refrigerant such as a brine.
    c. Operate from 22 to 28°F depending upon the activity.
    d. All of the above.

SECTION 6

Air-conditioning Systems

A27 Introduction to Air Conditioning

AFTER STUDYING THIS CHAPTER, THE STUDENT WILL BE ABLE TO:

- Describe a brief history of air conditioning.
- Discuss the early technical developments of air conditioning.
- Define "air conditioning" and relate it to human comfort conditions.
- Discuss the differences between air conditioning and heating.
- Explain comfort cooling and its impact on modern man.
- Describe trade associations and the role they play within the industry.
- Discuss the historical market and its related growth.
- Explain usage by types of equipment available.
- Describe opportunities in the international air-conditioning market.
- Describe the principal products that were developed in the order of their sequence.
- Name the leading markets for air-conditioning products.

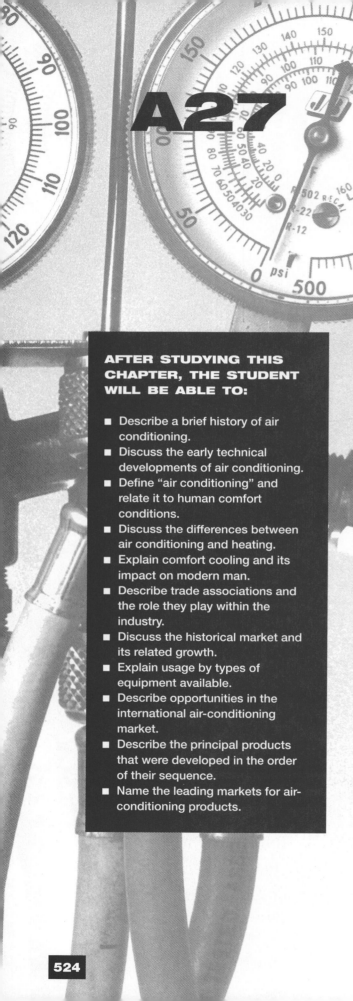

A27-1 HOW OLD IS AIR CONDITIONING?

Conditioning of air, or air conditioning, is as old as man himself. Primitive people wore skins to contain their body heat and lived in caves to moderate their harsh environment. The discovery and use of fire was perhaps the most important advance in this era (Fig. A27-1).

Egyptian history shows comfort air conditioning for a privileged few by the use of fans operated by slaves (Fig. A27-2). Thus, the art of evaporative cooling provided some relief from the desert heat. Romans engineered ventilation and panel heating into their famous baths. They also brought ice from the northern mountains to chill wine and possibly also to chill water for other conditioning of their environment.

In the middle ages, the remarkable Leonardo da Vinci built a water-driven fan to ventilate rooms of a

Figure A27-1 Ancient couple around fire.

Figure A27-2 Egyptian slave cooling master.

house. Other early innovations included a rocking chair with bellows to produce spot ventilation for the occupant, and a clock mechanism that activated a fan device above the bed.

A27-2 EARLY TECHNICAL DEVELOPMENTS

During the 19th century fans, boilers, and radiators had been invented and were in fairly common use. Early warm-air furnaces were coal-fired cast iron with gravity air distribution. Mechanical fans were added for forced circulation of air through ducts. Modern concepts of furnaces bear little resemblance to some of those "iron monsters." The major changes, however, were brought about by the gradual conversion from coal to oil and gas, and from manual to automatic firing.

One of the first American buildings to have a complete and scientifically planned heating and ventilating system was the U.S. Capitol. The heating system used 26 indirect heaters built into brick chambers in the sub-basement of each wing. These chambers were all connected by means of large subterranean ducts, with a fan chamber in each wing of the Capitol. Outside air was brought in through tunnels from an open-topped tower in a park west of the capitol.

In the 1880s, applications of steam heating were advanced by the introduction of remote, central steam plants. The New York Life Insurance building had four water-tube boilers, direct and indirect radiant heat, thermostatic controls, and air filters. Other applications used large fans to distribute and exhaust air.

Engineers involved in heating and ventilating began to share their experiences and formed the American Society of Heating and Ventilating Engineers in 1894 for the purpose of "honest cooperation" to increase incomes, enhance the dignity of businesses, and elevate the work of the heating and ventilating engineers.

Alfred R. Wolff designed many of the most important projects at the turn of the century, including a 300-ton cogeneration HVAC system for comfort cooling and humidity control of the New York Stock Exchange.

Willis H. Carrier (Fig. A27-3) was an outstanding engineer and leader (1876–1950) in the development of the air-conditioning industry. He contributed many basic, milestone innovations starting in 1902, when he designed a spray-type temperature- and humidity-control system. In 1911 he presented his historic paper on the properties of air, *Rational Psychrometric Formulae* and published his psychrometric chart, used today in plotting and calculating the performance of air-conditioning systems. In 1921, Carrier designed a packaged chiller using a centrifugal compressor with dichloroethylene refrigerant.

Carrier pioneered the use of induction systems for multi-room office buildings, hotels, apartments, and hospitals. During World War II, he supervised the design, installation and start-up of a system for the National Advisory Committee for Aeronautics (NACA) in Cleveland for cooling 10,000,000 ft^3 of wind-tunnel air down to −67°F (−19.45°C). Many industry professionals and historians consider him to be the "father of air conditioning."

In 1904 *Stuart Cramer,* a North Carolina textile engineer, coined the term "air conditioning" as a comprehensive designation for the heating, ventilating, and humidifying treatments of air.

Figure A27-3 Willis Carrier, a leader in the rapid development of the air conditioning industry (1876–1950). (Courtesy of Carrier Corporation)

A27-3 COMFORT COOLING COMES OF AGE

Comfort air conditioning had its first major use in *motion picture theaters* during the early 1920s. Famous New York City movie houses like the Rivoli, the Paramount, the Rosy and Loew's Theaters in Times Square (Fig. A27-4) were among the first. By the late twenties several hundred theaters throughout the country had been air conditioned. These were custom-designed, custom-manufactured, field-installed systems assembled on the job site.

The first *self-contained room air conditioner* was introduced in the late 1920s. It was the industry's first attempt to "package" products that could be mass-produced, factory tested, and operated prior to shipment.

The next milestone was the development of a *safe refrigerant.* In 1930 Thomas Midgley, of DuPont Company, developed the fluorocarbon Freon refrigerant. In 1931 Freon-12 (Fig. A27-5) was introduced as a commercial refrigerant. The characteristics of F-12 opened new vistas for compressor and system component designs. A whole family of Freon refrigerants soon followed where specific operating variations were needed. In 1955 other firms joined DuPont in manufacturing the fluorocarbon refrigerants. In 1956 a numbering system, the current R-12, R-22, etc., was adopted by the industry to provide a uniform label for the same refrigerant manufactured by different companies.

The first *hermetic compressor* was introduced for air conditioning in the mid-1930s. It was considerably larger than today's model having equivalent capacity. Its motor speed was 1750 rpm compared to the common 3600 rpm of today. The outer shell was bolted rather than completely welded (Fig. A27-6). The concept of passing cool suction gas to and over the motor windings started a trend in sizes up to 7½ tons that has now become almost universal among compressor manufacturers. Full-welded hermetics (Fig. A27-7) are currently offered in capacities up to 20 tons in a single shell.

Window air conditioners, store conditioners, and *applied machinery systems* (Fig. A27-8) *for large buildings* were mainly the products of post World War II.

Window units (Fig. A27-9) were used extensively to cool residences, smaller offices, small stores, and many other places where access to window or through-the-wall mounting could be accomplished.

The "main street" commercial comfort air-conditioning market for the drugstores, restaurants, dress shops, barber shops, grocery stores, etc., were handled by the self-

Figure A27-4 Early theater advertising "frosted air."

REMEMBER WHEN . . . summer meant straw hats and Saturday night talkies in a movie palace cooled with "frosted air." As air conditioning caught on, this 1929 marquee was followed by more flamboyant signs featuring penguins and igloos. Temperatures in today's theaters are less of a novelty than in the past, a Loew's executive stated, because people are accustomed to air conditioning in their homes and places of work. In addition, industry product certification programs of the air-conditioning industry in effect for 13 years have helped air-conditioning engineers in assuring cool comfort in homes and public places.

Figure A27-5 Refrigerant drum containing Freon-12. (Courtesy of DuPont Chemicals)

Figure A27-6 Bolted semi-hermetic compressor. (Courtesy, Copeland Corporation)

Figure A27-8 Water chilling system used to supply chilled water to air-conditioning units. (Courtesy of York International Corp.)

contained store conditioner (Fig. A27-10). These units were primarily water cooled and were usually located in the conditioned space. Although they had many deficiencies, these units created public appreciation and awareness for central comfort air conditioning. Some manufacturers offered variations that could be adapted to residential use, but the number of installations was relatively few.

Air-cooled operation instead of water for condensing purposes was a major improvement in 1953 and started industry sales skyrocketing. New technology in system design permitted pushing head pressures high enough so that machines could operate safely and reasonably efficiently up to 115°F outdoor ambient conditions. Early packaged units (Fig. A27-11) were mostly horizontal in configuration, for mounting in attics or on a slab at ground level or even on the roof. Installation consisted of mounting the unit, electrical hookup, and a simple duct system to distribute air into the space. The main disadvantage of these units was their lack of flexibility in adapting to all kinds of applications, particularly where heating and cooling were to be combined.

The *split system* (Fig. A27-12) was introduced to provide greater adaptability for residential applications. It consisted of two parts: an indoor cooling component and an outdoor condensing section. The two were connected by liquid and suction refrigerant lines. Although the units were an immediate success, installation costs and the need for highly trained installers were disadvantages.

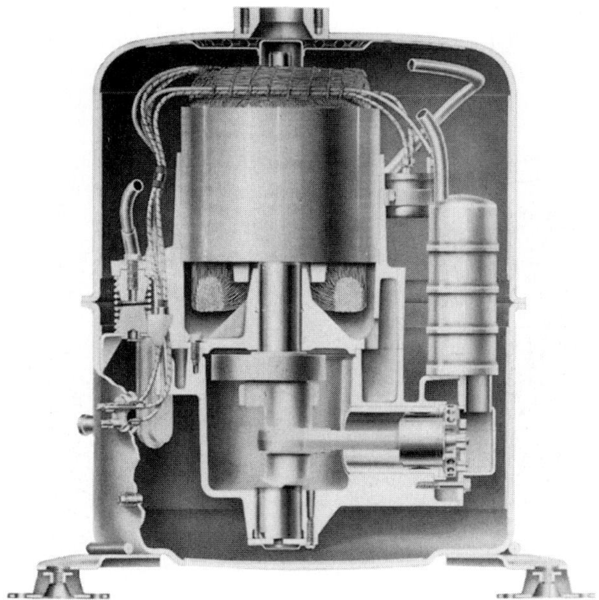

Figure A27-7 Cutaway view of a three-ton hermetic compressor.

Figure A27-9 One of the first window air conditioners.

Figure A27-10 Self-contained packaged air-conditioning unit (Unitaire, with water-cooled condenser). (Courtesy of Rheem Air Conditioning Division)

KNOWLEDGE EXTENSION

Individuals interested in the international market of the air-conditioning business should learn at least one foreign language and take courses in international trade and business. To learn more about gross sales and market shares, contact the ARI office and request a breakdown of international sales and growth opportunities. To gain employment in this market, contact some of the larger companies selling this equipment, and request employment opportunities available. Attending the ARI/ASHRAE annual conference will also enlighten the student of air conditioning of the impact the industry has had world wide.

Precharged lines (Fig. A27-13) and *quick connect couplings* were the next improvement. Precharged lines were cut and fabricated in various lengths and the suction line preinsulated with continuous covering. Both ends of the lines were equipped with mechanical couplings having a thin metal diaphragm to form a seal. A complete refrigeration system could be put together with a minimum of labor and minimum risk of contamination. The reliability of these systems also increased.

Heat pumps (Fig. A27-14) were introduced during the same period of development as the package and split-system cooling equipment. Cooling-only systems were converted to reverse-cycle operation, meaning that heat could be pumped into or out of the space as needed. They were often sold in extremely cold climates where operational performance was marginal. Popularity of heat pumps diminished further because qualified installation and service personnel were few or non-existent to handle application problems as they arose.

The *rooftop gas-heating and electric-cooling unit* (Fig. A27-15) was the final major innovation that occurred during the late 1950s and early 1960s and had a phenomenal growth rate. It started with 2-, 3-, 4-, and 5-ton systems being installed on rooftops of low-rise commercial structures of all descriptions, as well as in residential buildings. Today's capacities range upward of 100 tons of cooling. Both gas and electricity are available for heating.

Trade associations have also played a significant role in the rapid development of air conditioning. In 1935, two related manufacturers' associations merged to form the *Air-Conditioning and Refrigeration Institute* (ARI) to solve their common problems. The scope of its activities expanded to include not only technological assistance to

Figure A27-11 Air-cooled packaged conditioner installed in a residence.

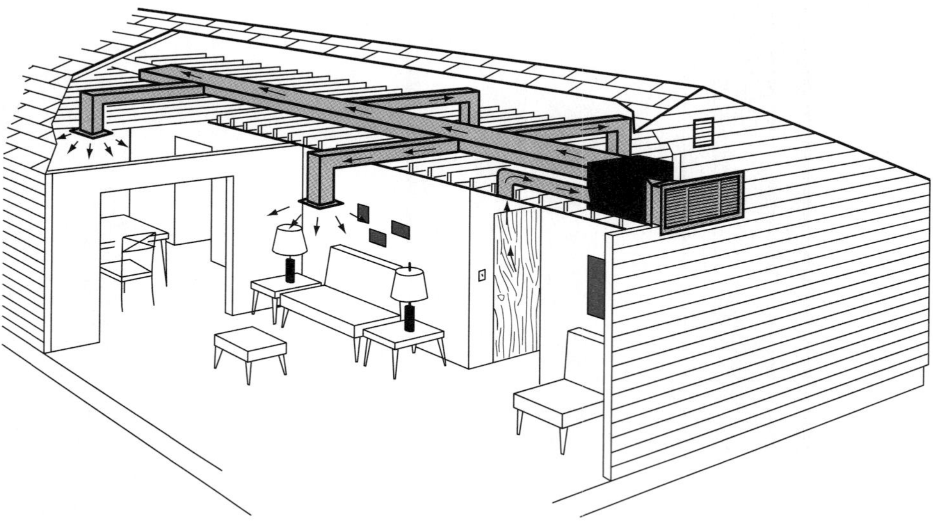

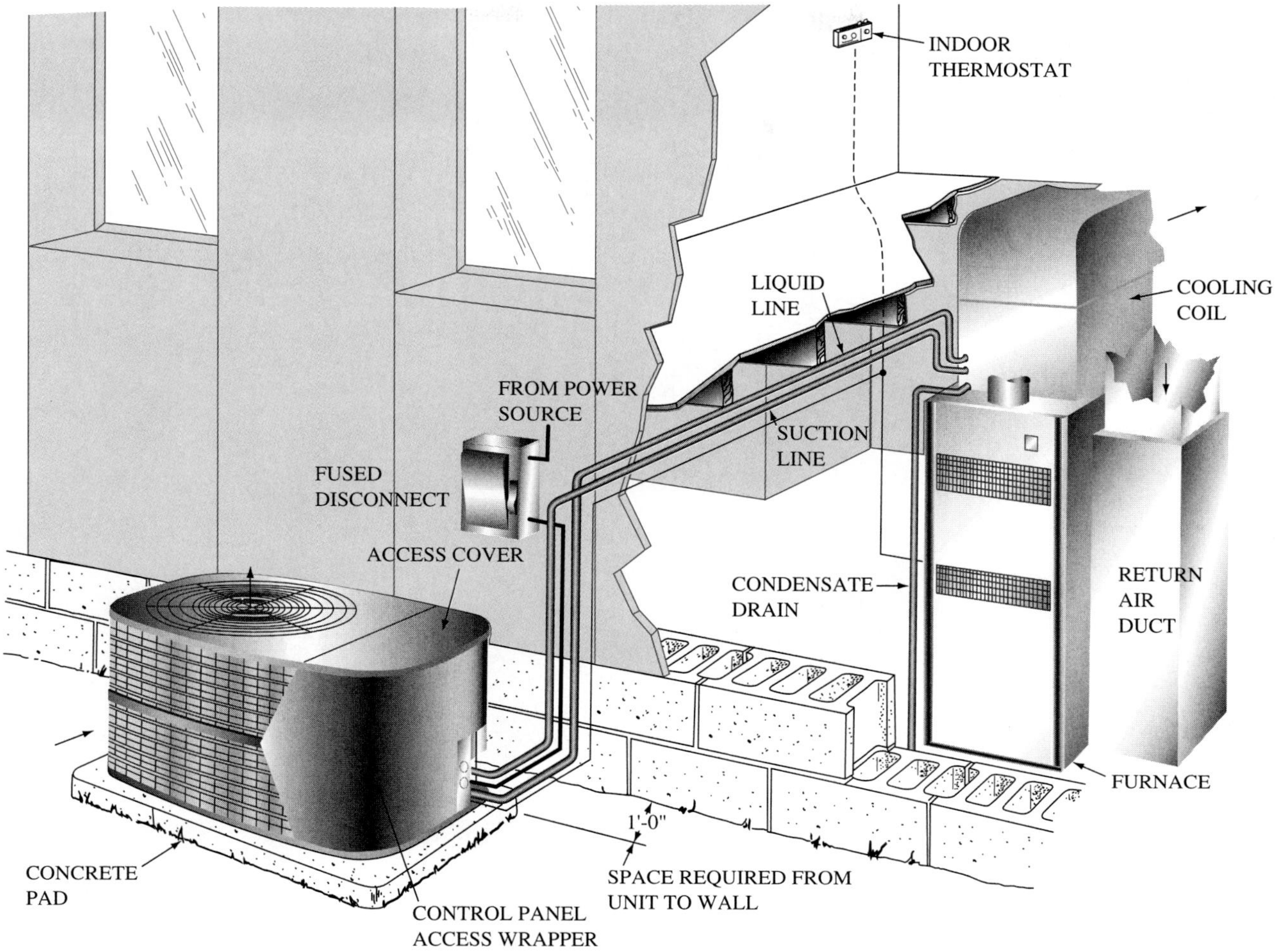

INDOOR
THERMOSTAT
LIQUID
LINE
COOLING
COIL
FROM POWER
SOURCE
SUCTION
LINE
FUSED
DISCONNECT
ACCESS COVER
CONDENSATE
DRAIN
RETURN
AIR
DUCT
FURNACE
1'-0"
CONCRETE
PAD
SPACE REQUIRED FROM
UNIT TO WALL
CONTROL PANEL
ACCESS WRAPPER

Figure A27-12 Split-system installation for heating and cooling.

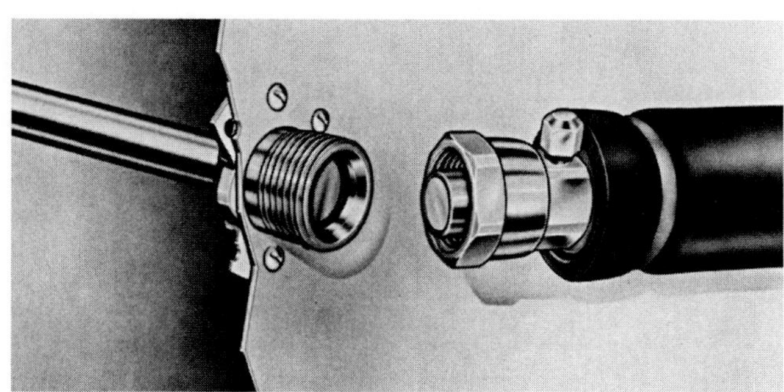

Figure A27-13 Pre-charged refrigerant line connection. (Courtesy of Aeroquip Corporation)

HEAT PUMP SYSTEM
ONE UNIT ENERGY
(ELECTRICITY) IN ▷
TWO UNITS ENERGY
(HEAT) OUT ▷▷

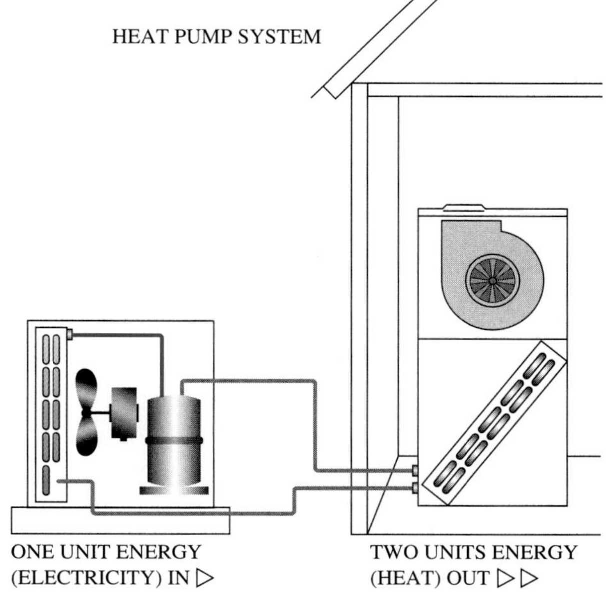

Figure A27-14 Split-system heat pump installation.

Figure A27-15 Rooftop heating and cooling unit being set in place. (Courtesy of York International Corp.)

its members, but also educational and research services in support of developing industry.

In the late 1950s the Field Investigation Committee and the Research Advisory Committee of the National Warm Air Heating and Air Conditioning Association, in cooperation with the University of Illinois, established the Engineering Experiment Station (Fig. A27-16). Many important studies were conducted on heat-flow characteristics, the effects of insulation, design, and the efficiency of air-distribution systems, etc. The name of the association has now been changed to the Air Conditioning Contractors of America (ACCA).

Figure A27-16 Engineering Experiment Station, University of Illinois. (Courtesy of Air-Conditioning Contractors of America)

PRACTICING PROFESSIONAL SERVICES

Professionals in the air-conditioning business are actively involved with specific industry associations to gather knowledge and resources for continued growth. It is a good practice to become active with a local chapter of a professional association such as the Air Conditioning Contractors of America (ACCA), the American Society of Heating, Refrigeration, and Air-Conditioning Engineers (ASHRAE), or the Refrigeration Service Engineers Society (RSES). Participating with these professional groups will help develop additional skills and opportunities through meetings, discussions, seminars, and related activities on educational topics and industry trends.

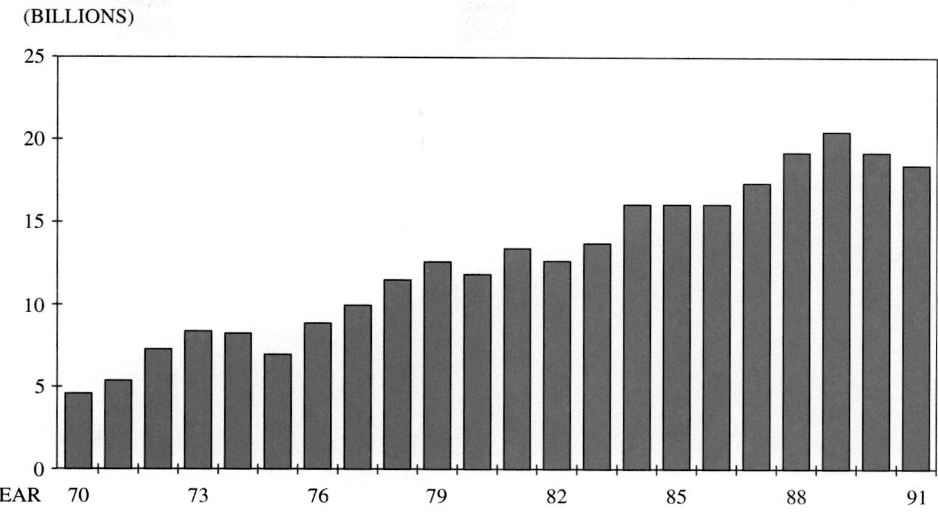

Figure A27-17 Value of shipments.

A27-4 HISTORICAL MARKET GROWTH

Those who contemplate a career in air conditioning are naturally interested in the industry's past track record of growth and most frequently ask, "Just how big is this industry? What kind of a growth rate has it experienced? What can be expected in the future?"

ARI keeps track of the industry shipments and breaks them down into several classifications. First we will take a look at value of shipments. This covers air-conditioning and warm-air heating equipment, along with commercial and industrial refrigeration equipment (Fig. A27-17). Note that the rise from 1970 to 1991 indicates a healthy growth.

Next we will look at shipments of unitary air conditioners and air-source heat pumps (Fig. A27-18). The growth here rises from 1,616,018 units in 1970 to 4,634,642 units in 1989, with a slight decrease in the next three years due to a drop in new construction and weather patterns. Many estimate this segment to be over 50% of total sales.

It is also interesting to note that the shipment-weighted seasonal-energy-efficiency ratios (SEER) of unitary air conditioners increased from 7.78 in 1981 to 10.53 in 1992 (Fig. A27-19). This is an indication of the improvement in performance that took place during those years. It also indicates that with the newer designs, fewer units were required to produce the same capacity.

Fig. A27-20 shows shipments of room air conditioners from 1970 to 1992. The gas furnace business has also been large for a good many years.

The charts in Fig. A27-21 show the rapid growth of air-conditioning systems in one-family homes. There is, of course, a regional difference due to climatic factors as shown.

The number of passenger cars with air conditioning increased from 4,825,999 units in 1970 to 7,563,000 in

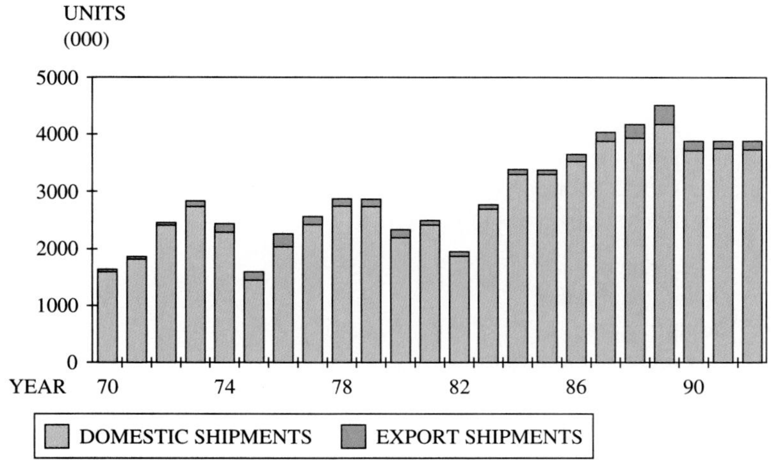

Figure A27-18 Shipments of unitary air conditioners and air-source heat pumps.

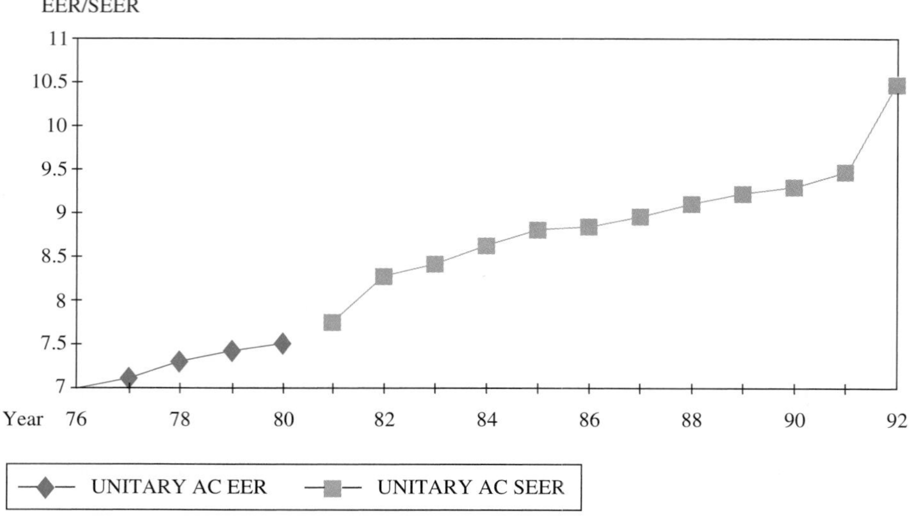

Figure A27-19 Shipment-weighted SEERS for unitary air conditioners.

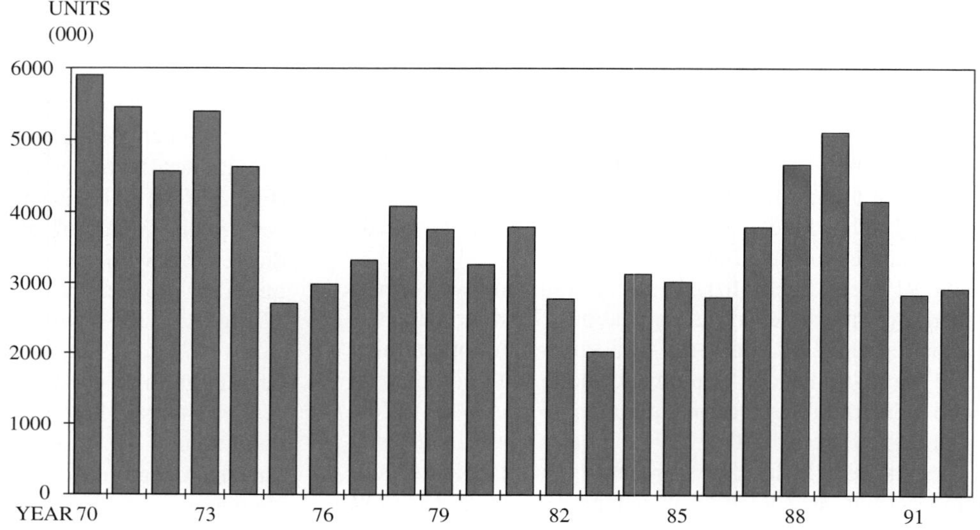

Figure A27-20 Shipment of room air conditioners.

1986. Light trucks with air conditioning increased from 723,000 in 1972 to 3,450,000 in 1992.

A27-5 USAGE BY TYPE

A federal government publication, *U.S. Industrial Outlook,* reported that the value of shipments of heating and cooling products for 1993 was $18,116,000. The proportion of this amount for various types of equipment was:

Heat transfer	25%
Unitary equipment	22%
Compressors	15%
Commercial refrigeration	12%
Warm-air furnaces	6%
Room coolers, dehumidifiers	5%
Parts	7%
Other	8%

ARI also reports the number and types of units shipped in 1993 as follows:

Central-station air handlers	57,000
Liquid chillers	18,434
Cooling towers	9,297
Fan-coil units	140,000
Auto air conditioning (Factory-installed)	6,446,832
Packaged terminal conditioners and heat pumps	220,000

Heating equipment unit shipments for 1993, reported by Gas Appliance Manufacturing Association (GAMA), indicated total shipment of all furnaces to be 3,081,862 units, which included:

Gas furnaces	2,548,563
Electric furnaces	348,496
Oil furnaces	148,803

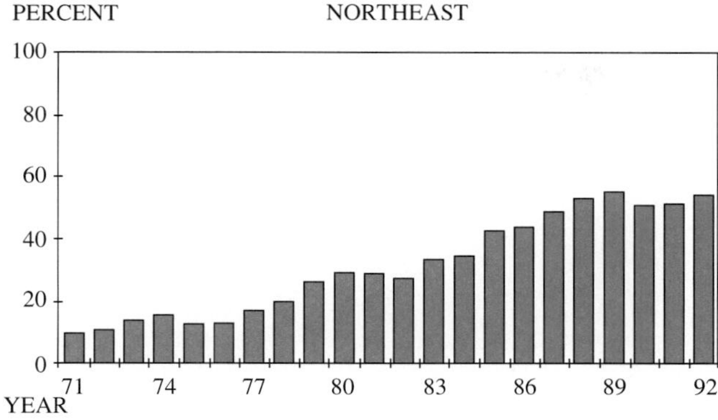

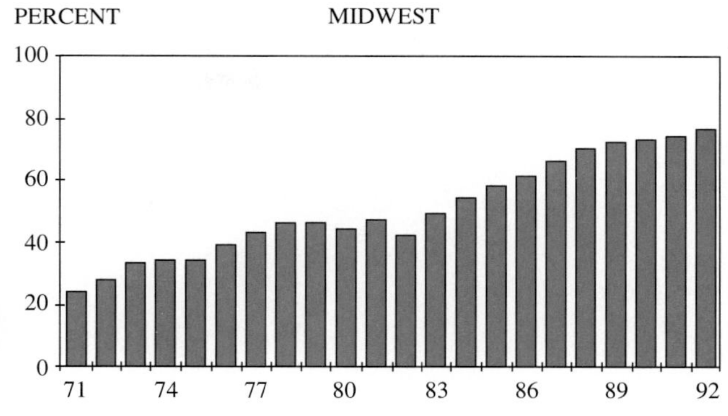

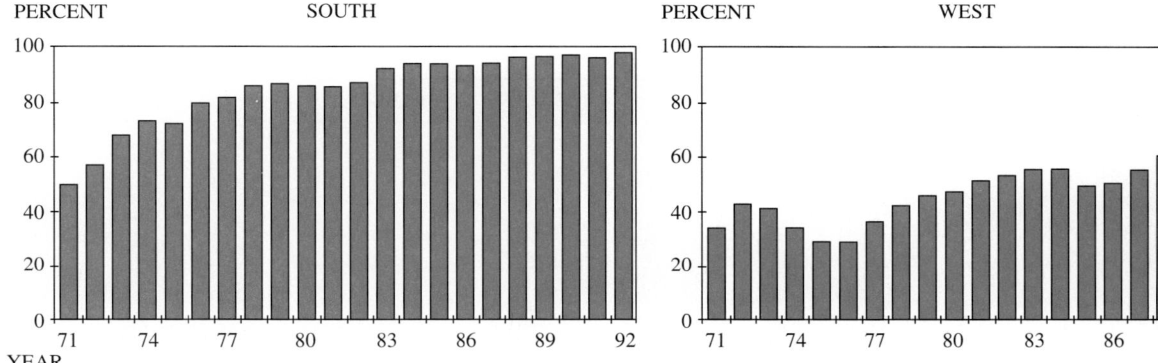

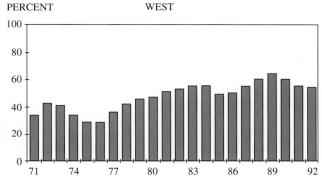

Figure A27-21 New one-family homes with central air conditioners.

For home heating alone, ARI reported an addition of 882,341 air-to-air and 92,280 water-source heat pumps to the above figures, which gives a picture of a robust home-heating industry.

There is also a sizable market for two heating and air-conditioning accessories, *air cleaners* and *humidifiers.* There were 155,600 electrostatic air cleaners and 600,000 power humidifiers shipped in 1993.

Another important product of the heating industry that offers a large market are *cast-iron and steel boilers.* In 1993, the number of boilers shipped was 347,334 units. The majority of these were used for commercial installations, but a portion of them (54,000 units) were placed in new single-family homes.

In a single year, using 1992 as a norm, the estimated installed value of air-conditioning shipments was $23 billion. And this is only a portion of the overall marketable goods and services in this growing industry.

For estimates of the manpower needs and occupational opportunities in this industry, please refer to chapter R1, "Introduction to Refrigeration."

A27-6 THE INTERNATIONAL MARKET

The U.S. air-conditioning and refrigeration equipment exports have more than doubled since 1988 to a level of over $4 billion annually. North America remains the largest export market followed by the Asian Pacific region, which continues to grow. The Latin American region, where market liberation is under way, will likely become a bigger export market through the remainder of the decade.

The major export markets for air conditioning and commercial/industrial refrigeration are shown in Fig. A27-22. The exports of selected air-conditioning equipment by product category is shown in Fig. A27-23, and for refrigeration in Fig. A27-24.

With expanding markets in Asia, Latin America, and Eastern Europe, the international trade outlook for the air-conditioning and refrigeration industry is extremely bright.

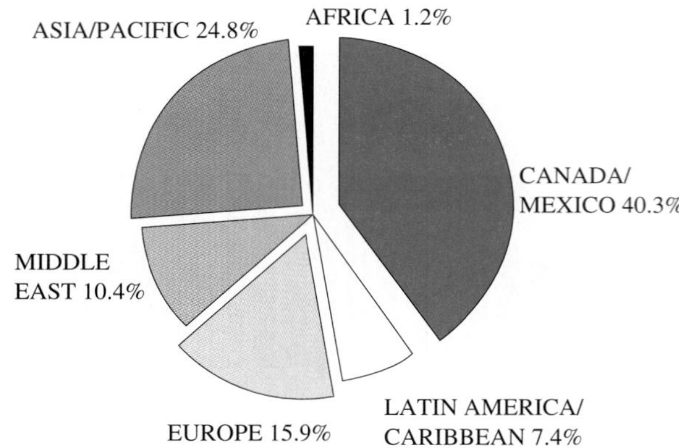

Figure A27-22 Major export markets, U.S. air conditioning and commercial/industrial refrigeration equipment—1992.

Product	1992	1991	1990	1989	1988	1987
Heat Pumps						
Self-contained:						
<60,000 Btu/hr	29.9	35.5	41.3	24.2	15.1	10.1
>60,000 Btu/hr	9.0	9.2	9.0	8.6	4.8	4.2
Split System	69.6	72.3	58.7	35.4	17.8	9.2
Year-round A/C;						
<60,000 Btu/hr	26.0	35.2	15.1	16.9	25.9	20.8
>60,000 Btu/hr	35.4	16.9	26.3	52.8	27.9	22.8
A/C Self-contained:						
<60,000 Btu/hr	16.4	10.5	9.8	22.7	28.9	19.1
>60,000 Btu/hr	26.6	19.9	29.3	35.9	37.5	32.4
Central Station Units	26.8	15.0	14.8	10.2	9.8	7.3
Room Fan Coils	7.7	8.2	7.5	9.5	11.4	6.0
Condensing Units	61.9	49.7	43.9	61.5	45.5	31.4
Evaporator Coils	29.8	25.3	23.9	26.4	19.0	11.2
Room A/C	92.4	80.2	109.8	140.0	77.7	40.1
Auto A/C	145.0	98.4	83.8	187.8	365.0	374.4
Other A/C and parts	799.0	741.0	1,000.6	752.2	305.5	220.3
Total:	1,613.1	1,473.4	1,473.8	1,384.1	1,088.1	878.4

Source: U.S. Bureau of the Census, Report EM 545.

Figure A27-23 U.S. exports of selected air-conditioning machinery by product category ($ millions).

Product	1992	1991	1990	1989	1988	1987
Chillers for Refrigeration Applications:						
Centrifugal:						
open	3.3	2.4	5.5	3.2	2.6	2.6
hermetic	58.3	41.6	21.4	9.9	3.8	3.7
Absorption	8.2	6.3	3.4	5.7	2.0	.7
Reciprocating	64.4	35.1	10.7	14.4	7.0	6.5
Refrigeration Condensing Units:						
<1 Hp	4.2	3.6	6.5	3.6	2.4	1.6
2–3 Hp	2.4	2.6	1.6	2.9	2.3	2.0
4–10 Hp	1.8	1.9	1.7	2.2	2.1	.7
11–30 Hp	1.7	2.2	1.6	1.3	1.3	.8
>30 Hp	2.2	2.4	1.8	2.0	1.8	1.3
Ice Makers:						
<500 lbs.	22.7	25.4	19.4	16.1	18.2	15.6
other	29.4	23.8	21.5	28.7	19.1	15.3
Water Coolers	26.3	33.0	31.8	31.9	27.2	19.0
Other Commercial Refrigeration Equipment	243.0	170.9	179.0	225.0	235.8	136.8
Refrigeration Equipment Parts	195.2	172.7	148.7	177.0	180.4	137.7
Total:	943.4	766.0	454.6	492.0	506.0	344.3

Source: U.S. Bureau of the Census, Report EM 545.

Figure A27-24 U.S. exports of selected refrigeration equipment by product category ($ millions).

REVIEW

- Air conditioning—History shows numerous accounts of civilizations attempting to condition their environment.
  Primitive people wore skins, lived in caves, and built fires.
  Egyptians used comfort air conditioning—evaporative cooling.
  Romans engineered ventilation and panel heating. They used ice to chill wine, and chilled water for other conditioning of their environment.
  Leonardo da Vinci built a water-driven fan to ventilate rooms of a house.
  A rocking chair with bellows produced spot ventilation.
  A clock mechanism activated a fan.
- Early technical developments:
  19th century—Fans, boilers, radiators, and warm-air furnaces (coal-fired with gravity air distribution).
  Mechanical fans were added—forced circulation of air.
  Gradual conversion occurred from coal to oil and gas, conversion from manual to automatic firing.
  U.S. Capitol was the first American building to have a scientifically planned heating system.
  1880s—Central steam-heating plants were installed in buildings like the New York Life Insurance building.
  1894—Engineers formed "American Society of Heating and Ventilating Engineers."
  Alfred R. Wolff—Designed a 300-ton cogeneration HVAC system for the New York Stock Exchange.
  Willis H. Carrier (1876–1950)—Developed a spray-type temperature and humidity control, psychrometrics, packaged chiller system, induction heating system. Considered the "father of air conditioning."
  Stuart Cramer—Coined the term, "air conditioning" as a comprehensive designation for heating, ventilation, cooling, and humidifying treatments of air.
- Comfort cooling comes of age:
  1920s—Comfort air conditioning is used in motion-picture theaters.
  Late 1920s—First self-contained room air conditioner developed.
  1930—Thomas Midgley of the DuPont Co. developed fluorocarbon freon refrigerants.

1930s—First hermetic compressor developed.
1931—R-12 introduced as a commercial refrigerant.
1953—Air-cooled condensing units used instead of water-cooled; split systems and heat pumps were introduced.
1955—Other manufacturers of fluorocarbon refrigerants emerged.
1956—A numbering system for refrigerants was adopted. Uniform labeling, window air conditioners, store air conditioners, and applied machinery rooms were the products of post-World War II.
Late 1950s—Rooftop gas-heating and electric-cooling units became popular.

- Air-conditioning industry continues to grow. There will be a demand for trained technicians continuing into the next century.
- Trade associations which have impacted the air-conditioning business include:
  Air-Conditioning and Refrigeration Institute (ARI)
  American Society of Heating, Refrigeration, and Air-Conditioning Engineers (ASHRAE)
  Air Conditioning Contractors of America (ACCA)
- Usage by type in 1993 (according to U.S. Industrial Outlook) was $18,116,000 in shipments. The types of equipment shipped include:
  Heat transfer
  Unitary equipment
  Compressors
  Commercial refrigeration
  Warm-air furnaces
  Room coolers
  Parts
  Central station air handlers
  Liquid chillers
  Cooling towers
  Fan coil units
  Auto air conditioning
  Packaged terminal and heat pumps
  Gas furnaces
  Electric furnaces
  Oil furnaces
- International market sells over $4 billion annually with large markets in:
  North America
  Asian Pacific region
  Latin America
  Eastern Europe

Problems and Questions

1. Name some ways in which early man tried to control the environment.
2. Who was the first person to patent a "refrigeration machine"?
3. Who is thought to be the first person to use the "refrigeration machine" to cool air?
4. The Du Pont Company developed the first "safe" refrigerant, called _____.
5. Hermetic compressors for air conditioning were first introduced in about the year _____.
6. ARI stands for _____.
7. ARI is made up of _____.
8. "ACCA" stands for _____.
9. ACCA is made up of _____.
10. The industry growth potential is limited to _____.
11. For every $1,000,000 in installed equipment, how many air-conditioning and heating technicians are needed?
12. The English were the first people to use comfort cooling with fans operated by slaves. True or false?
13. In the middle ages, Leonardo da Vinci built a steam-powered fan to ventilate rooms of a house. True or false?
14. Boilers and radiators were invented and in common use during the early part of the 20th century. True or false?
15. Willis Carrier developed the psychrometric chart. True or false?
16. Comfort cooling first became popular when used in motion picture theaters. True or false?
17. Who was the individual responsible for the development of Freon (R-12)?
    a. Willis Carrier
    b. Thomas Midgley
    c. Alfred Wolff
    d. None of the above.
18. The first hermetic compressor was introduced in the:
    a. 1920s.
    b. 1930s.
    c. 1940s.
    d. None of the above.
19. Window air conditioners and store air conditioning didn't become popular until:
    a. After World War I.
    b. After World War II.
    c. After the Korean War.
    d. None of the above.
20. The responsibilities of ARI are:
    a. Technological assistance.
    b. Educational service.
    c. Research service and support of developing industries.
    d. None of the above.
21. The largest export market for the air-conditioning industries currently is the:
    a. U.S. market.
    b. North America.
    c. Asia Pacific Regions.
    d. None of the above.

Fundamentals of Air Conditioning

A28

AFTER STUDYING THIS CHAPTER, THE STUDENT WILL BE ABLE TO:

- Select the air-conditioning conditions required to make people comfortable.
- Explain the principles of psychrometrics.
- Describe the terminology used with psychrometrics.
- Use the psychrometric chart to plot the performance of an air-conditioning system.
- Understand equipment necessary for air distribution and balancing.
- Determine the proper quantity and distribution of air for a system.
- Describe the types of fans and blowers used in air-conditioning applications.
- Select and apply appropriate fan laws to air-conditioning equipment.
- Determine the performance for a system blower in terms of air delivery, static pressure and brake horsepower.
- Evaluate the supply- and return-air duct system.
- Explain the various duct systems and common types used.
- Use duct design procedures implementing the equal-friction method.

A28-1 CONDITIONS FOR COMFORT

Comfort is the feeling of physical contentment with the environment. The air-conditioning process is used to produce human comfort. Air conditioning provides control of temperature, relative humidity, air motion, radiant heat, removal of airborne particles, and contaminating gases.

The study of human comfort relates to:

1. How the body functions with respect to heat and
2. How the area around a person affects the feeling of comfort.

The body has a remarkable ability to adjust to temperature change. When a person goes from a warm house into the cold outdoors, some compensation needs to be made to prevent excessive heat loss. Involuntary shivering can occur, providing body heat. When a person goes from an air-conditioned space into an outside temperature of 95°F, an adjustment in the circulatory and respiratory system takes place. The blood vessels dilate to bring the blood closer to the surface of the skin to provide better cooling. If this is not enough, sweating occurs, evaporating moisture, producing a body cooling.

The body behaves like a heating system. Fuel is consumed in the form of food, producing energy and heat. The body temperature is closely controlled at a temperature of 98.6°F, winter and summer. The proper functioning of the body is dependent on constantly maintaining this temperature. To be able to do this requires constantly losing heat to the surrounding area. Fig. A28-1 illustrates a comfortable condition. With room conditions of 73°F, the body loses heat at about the correct rate for a feeling of comfort.

The surrounding conditions must be cooler than the body temperature in order to dissipate heat. Fig. A28-2

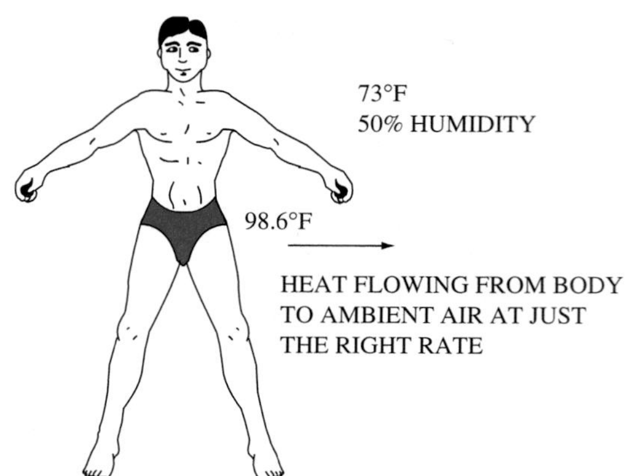

Figure A28-1 With ambient conditions of 73° and 50% RH, the body is able to give off the "right" amount of heat for comfort.

shows a condition where the surroundings are warmer than body temperature. This condition cannot be tolerated for any length of time without some adjustment being made. There are four ways the body can lose heat to the environment:

1. Conduction (Fig. A28-3),
2. Convection of cool air (Fig. A28-4),
3. Radiation (Fig. A28-5), and
4. Evaporation (Fig. A28-6).

Many tests have been run to determine comfort conditions in winter as well as summer, due partly to the use of warmer clothing. The surrounding temperature, relative humidity, and air movement are influencing factors. The results of test data have been incorporated in comfort charts, one for summer conditions (Fig. A28-7a) and one for winter conditions (Fig. A28-7b).

The enclosed areas on both of these charts indicate the range of conditions at which people are comfortable.

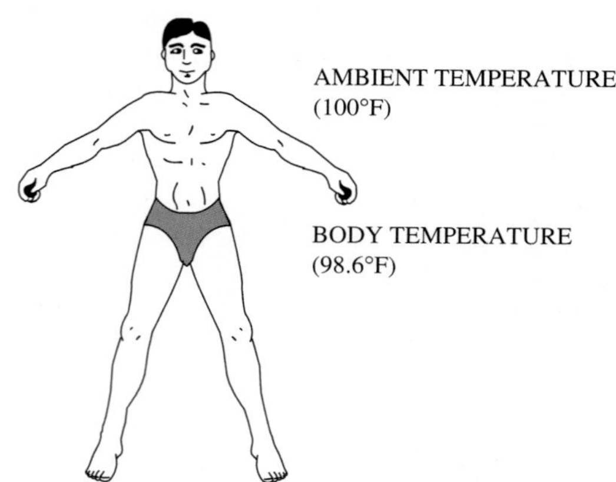

Figure A28-2 The body must dissipate heat or it will overheat in high ambient temperatures.

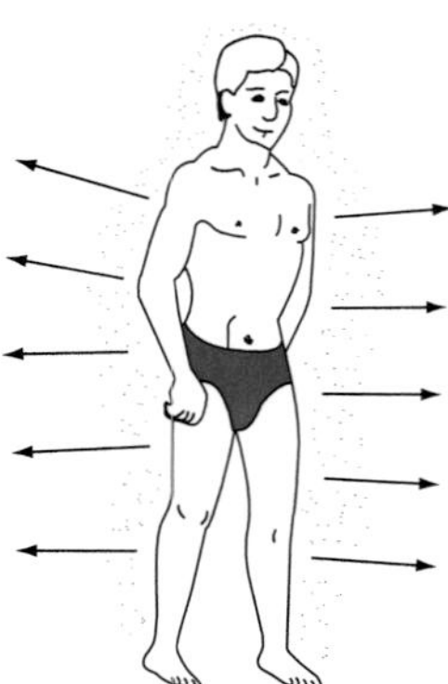

Figure A28-3 Heat is lost from the body by conduction to other objects.

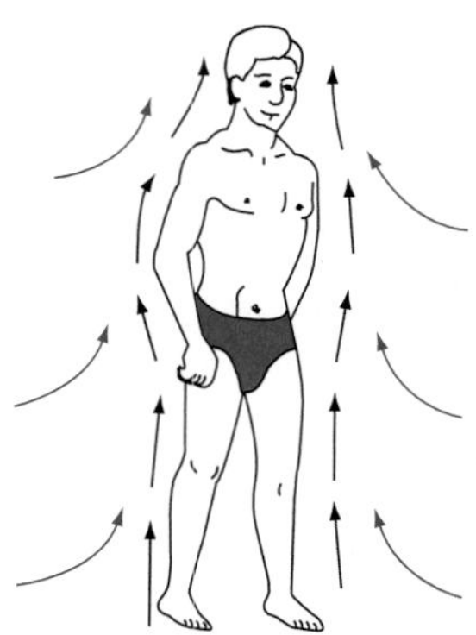

Figure A28-4 Heat is lost from the body by convection of the air around the body.

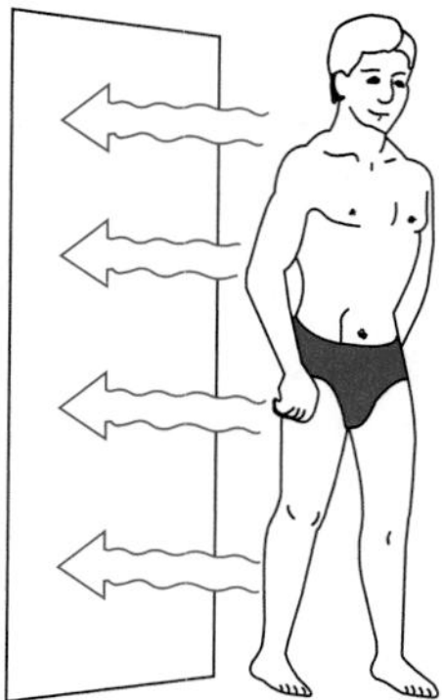

Figure A28-5 The body radiates heat to the surrounding surfaces/objects.

A28-2 PSYCHROMETRICS

The study of air and its properties is called *psychrometrics*. The psychrometric chart is used to show the relationship of these properties to each other.

Air has weight and occupies space. At standard conditions of 70°F and sea-level atmospheric pressure (29.92

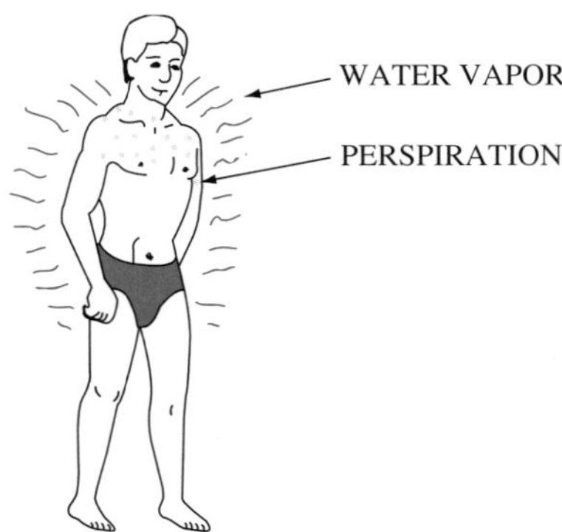

Figure A28-6 The body loses heat by evaporation of water from the skin.

in. Hg), dry air weighs 0.075 lb/cu ft (density). The reciprocal of this figure is useful. One pound of air at standard conditions occupies 13.33 cu ft (specific volume). These figures can be used effectively in air conditioning calculations.

EXAMPLE

An air handling unit is capable of delivering air at the rate of 1000 cu ft/min (cfm). If a heating coil can raise the temperature of the air 10°F, how much heat is added?

Solution:

$$Q_s = 1.08 \times \text{cfm} \times TD$$

where

Q_s = Sensible heat added in Btu/hr
= 1.08 conversion factor
cfm = 1000
TD = 10°F

Then

$$Q_s = 1.08 \times 1{,}000 \times 10 = 10{,}800 \text{ Btu/hr}$$

All air contains some moisture *(humidity)*. The amount is directly related to the temperature: the higher the temperature, the greater the amount of water will evaporate. The amount of moisture in the air can be measured in terms of relative humidity or percentage of saturation at a given temperature.

The process of vaporization takes two forms: evaporation, in which only part of the surface liquid turns into gas, and boiling, in which all parts of the liquid turn into gas. Boiling, which takes place at higher temperatures, can be considered an extreme case of evaporation.

The vapor formed above a liquid exerts pressure in direct proportion to the temperature. At standard conditions (70°F) the evaporating moisture from a pan of water would exert a vapor pressure of 0.74 in. Hg. Evaporation will only exist if the vapor pressure in the air above the liquid water has a lower vapor pressure. For example, if the air above the pan is 72°F and at 40% relative humidity, the moisture in the air would have a vapor pressure of 0.32 in. Hg and evaporation from the pan would take place.

When the temperature of the liquid is raised until its vapor pressure is higher than the prevailing atmospheric pressure, the vapor pressure is able to overcome the atmospheric pressure and the liquid changes into gas. This is known as the *boiling point* of the liquid. Water boils at 100°C (212°F), expanding more than 1500 times as it turns into saturated water vapor, or steam.

Partial pressures exerted by water vapor can be read from "Thermodynamic Properties of Water at Saturation", *ASHRAE Handbook on Fundamentals.*

In order to use the psychrometric chart, we need to review some of the fundamental properties of air.

A28-2.1 Moisture in Air

Air can hold only a relatively small amount of water vapor or moisture. Moisture in air is measured in grains (gr) per pound of dry air. One pound equals 7000 grains. For example, a sample of air at 75°F and 50% relative humidity, at atmospheric pressure, holds 64 grains (0.00914 lb) of moisture per pound of air.

The percent of relative humidity is a convenient means of representing the amount of water vapor in the air. It is the ratio of the actual density of water vapor to the saturated density, at the same dry-bulb temperature and barometric pressure.

A28-2.2 Dry-Bulb and Wet-Bulb Temperatures

The dry-bulb temperature of air is the temperature measured on an ordinary thermometer. The wet-bulb temperature is measured by placing a wick soaked with water around the thermometer bulb and moving it rapidly through the air. Water evaporates from the wick, cooling the bulb and lowering the temperature.

The relation of the wet-bulb temperature to the dry-bulb temperature is a measure of the relative humidity. Instruments used to measure these conditions are called "sling psychrometers" (Fig. A28-8, p. 542).

The relative humidity value can be obtained by plotting the readings on a psychrometric chart. Tables and slide rules are available for the same purpose.

A28-2.3 Dew-Point Temperature

The *dew-point temperature* of a vapor is the temperature at which the vapor reaches the point of 100% humidity

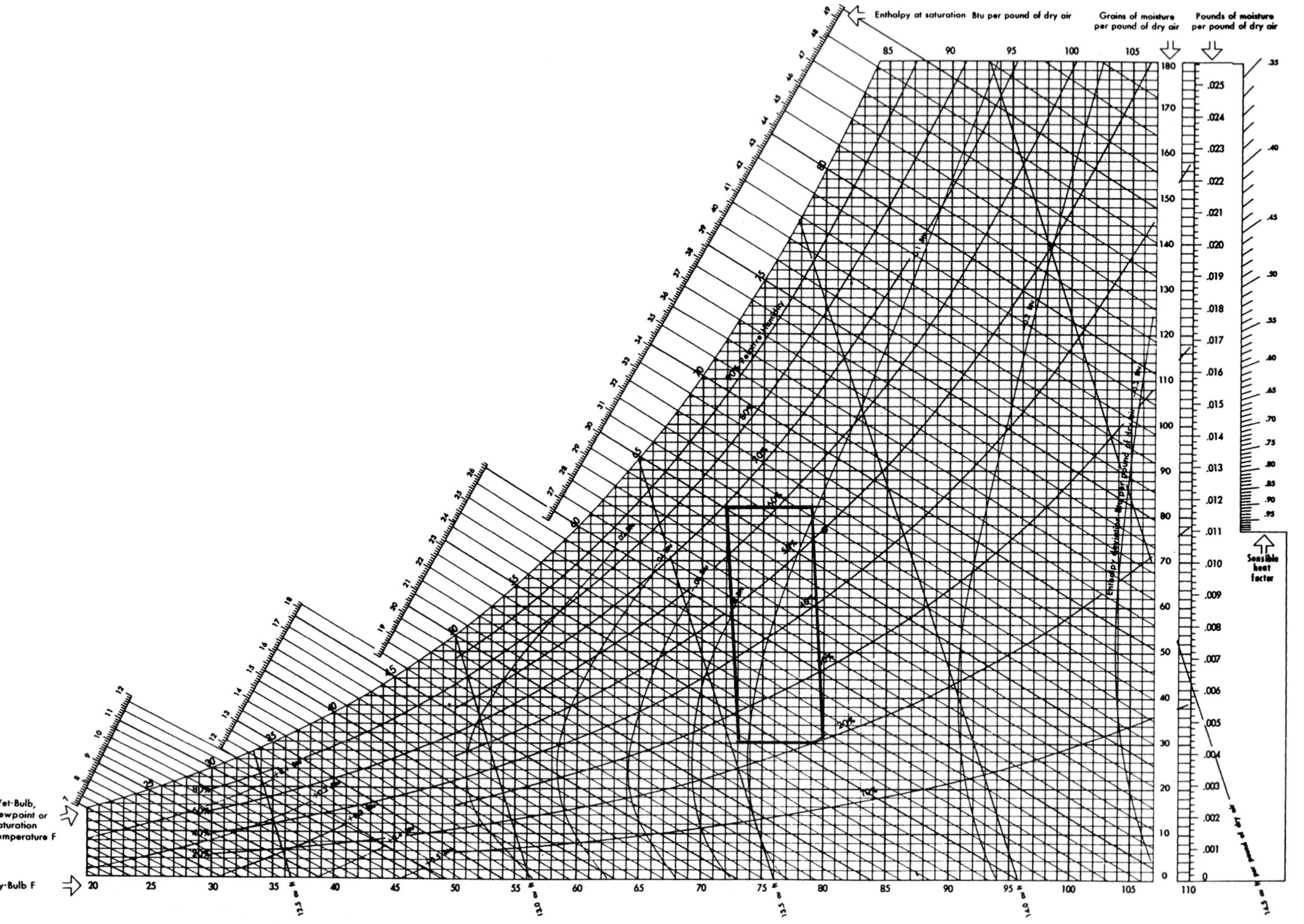

Figure A28-7a Comfort chart showing the range of conditions. Dry-bulb, wet-bulb, and relative-humidity ranges for comfortable summer conditions are shown in the rectangle. (Courtesy of Carrier Corporation)

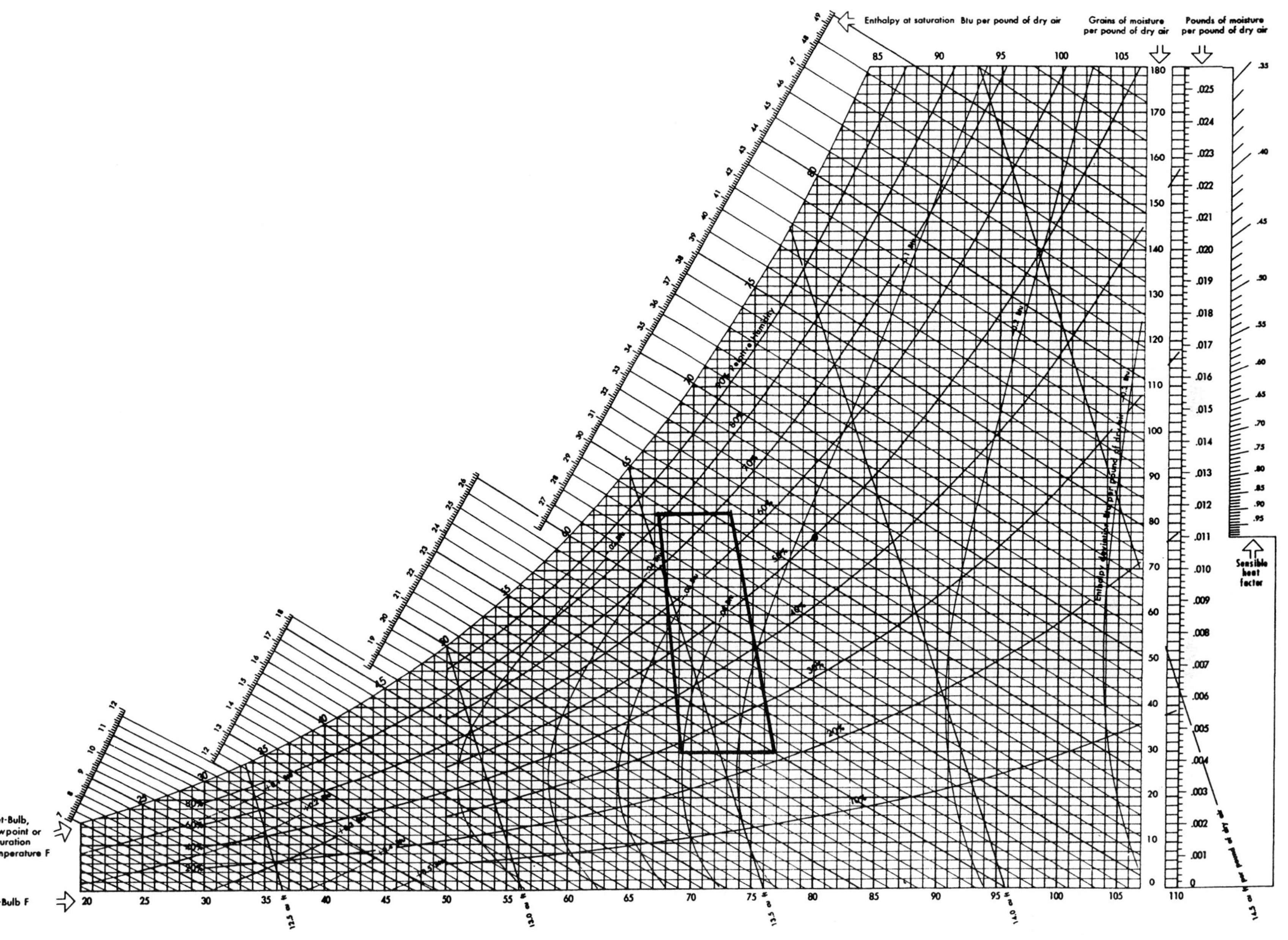

Figure A28-7b Comfort chart for winter, showing range of dry bulb, wet bulb and relative humidity for the most comfortable conditions. (Courtesy of Carrier Corporation)

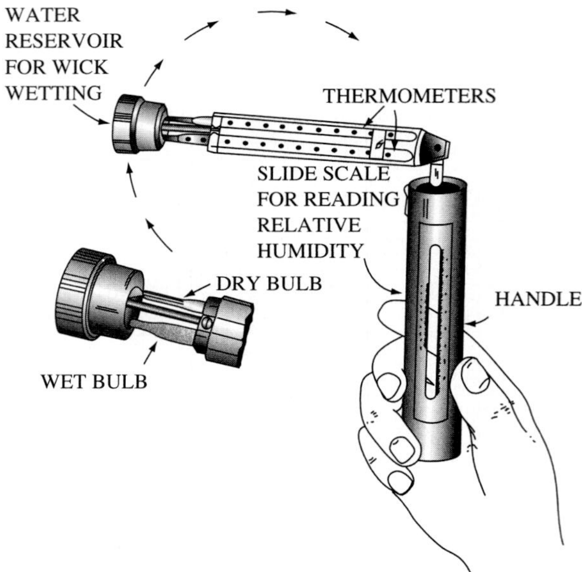

Figure A28-8 Sling psychrometer for measuring dry-bulb and wet-bulb temperatures to determine relative humidity.

(saturation). It is not correct to state that the dew point is the temperature at which condensation starts to occur. Condensation at dew point requires the removal of latent heat, and this can occur only if the vapor is cooled below dew-point temperature.

In an air-conditioning system, dehumidification takes place when the air passes over a coil whose temperature is below the dew-point temperature of the air. For an example, refer to the diagram of a refrigeration cycle of a room air conditioner, Fig. A28-9.

Assume that the evaporator temperature is 40°F. Air enters the coil at 75°F DBT (dry-bulb temperature), 50% RH (relative humidity) and at 55°F DPT (dew-point temperature). Since the coil temperature is below the dew point of the air, condensation will take place and excess water will drop into the drain pan.

A28-3 PSYCHROMETRIC CHART

The psychrometric chart is useful in plotting the performance of air-conditioning equipment. It provides a visual picture of the changes that take place in the properties of air passing through an air-conditioning unit. It looks complicated because it incorporates so much information; however, it is easy to use, as described below.

The simplest method of starting is to find the intersection of any two of the four principal air properties: dry-bulb temperature, wet-bulb temperature, dew-point temperature, or relative humidity. From this intersection point, the value of the other two properties (and more) can be determined.

For example, assume that a sample of room air has a dry bulb temperature of 80°F and wet-bulb temperature of 63.4°F. This represents a point on the psychrometric chart, Fig. A28-10.

The following readings can then be taken directly from the chart:

1.	Dry-bulb temperature	80°F
2.	Wet-bulb temperature	63.4°F
3.	Dew-point temperature	53.5°F
4.	Relative humidity	40%
	and	
5.	Total heat	28.8 Btu/lb
6.	Moisture	62 gr/lb
7.	Specific volume	13.7 ft^3/lb

Figs. A28-11 to A28-17 show the direction of the lines on the chart where each of the above readings is taken.

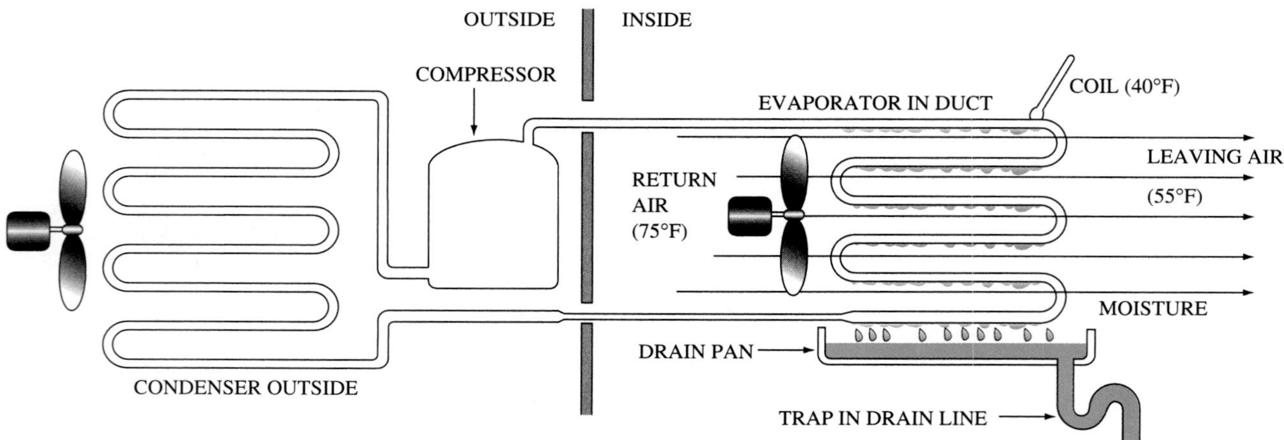

Figure A28-9 Diagram of a room cooler showing condensation of moisture on the evaporator, from the air.

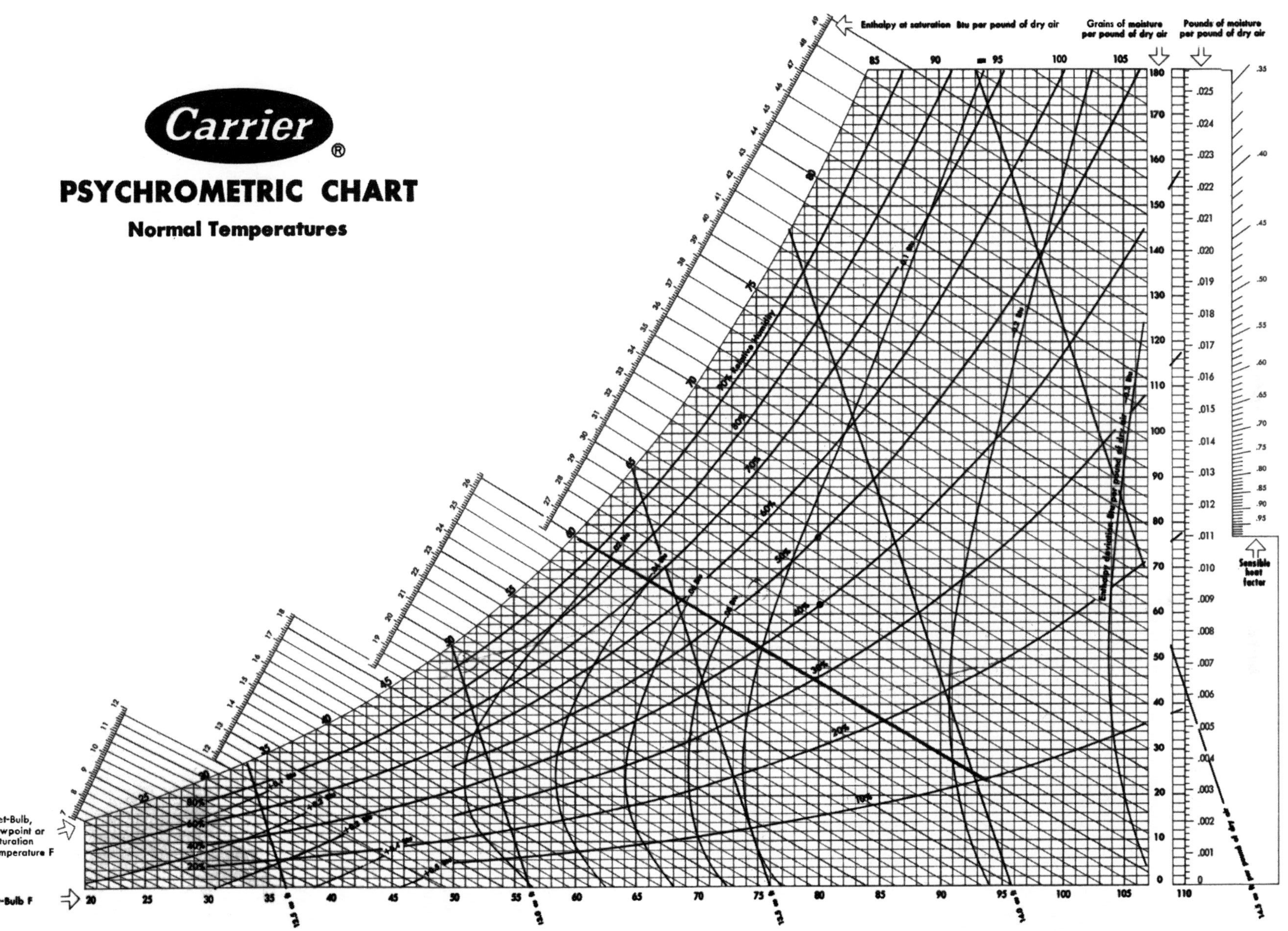

Figure A28-10 Psychrometric chart with 80°F dry bulb and 63.4°F wet bulb plotted, to identify other air properties. (Courtesy of Carrier Corporation)

Knowledge Extension

Watching the weather forecast via the newspapers and nightly television news can help the air-conditioning technician become more aware of the conditions which influence human comfort. As outside-air dew point changes, so does the relative humidity. If relative humidity increases, the load on the air-conditioning equipment increases if the dry bulb remains constant. Whenever the outside-air temperature reaches dew point, then moisture will begin to condense on lawns, windows, etc. Remember, the greater the latent load or relative humidity and moisture levels outside the house, the greater the load inside the house. A good forecaster will predict the weather accurately and a skilled technician will balance the air-conditioning equipment to respond to the ambient loads accordingly.

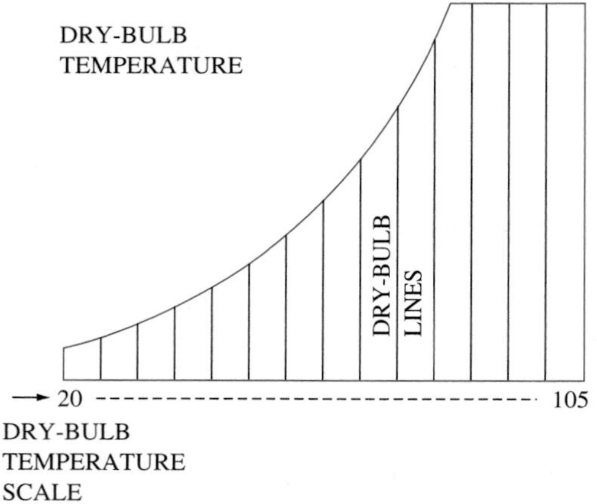

Figure A28-11 Dry-bulb temperature lines on psychrometric chart.

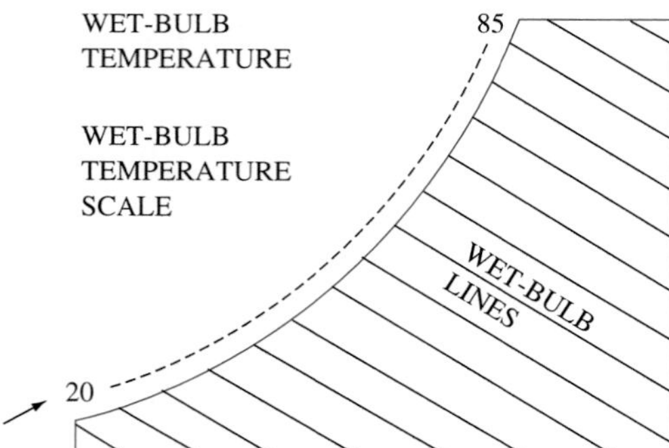

Figure A28-12 Wet-bulb temperature lines on psychrometric chart.

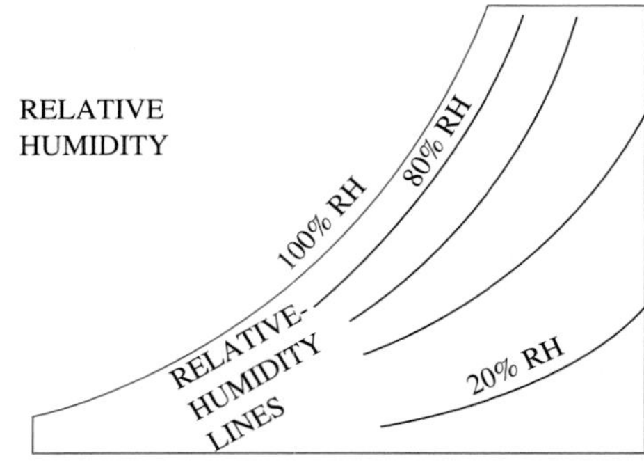

Figure A28-13 Relative humidity lines on the psychrometric chart.

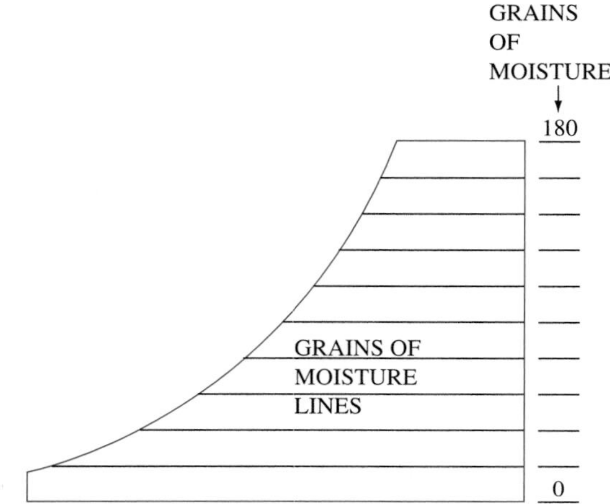

Figure A28-14 Grains of moisture lines on the psychrometric chart.

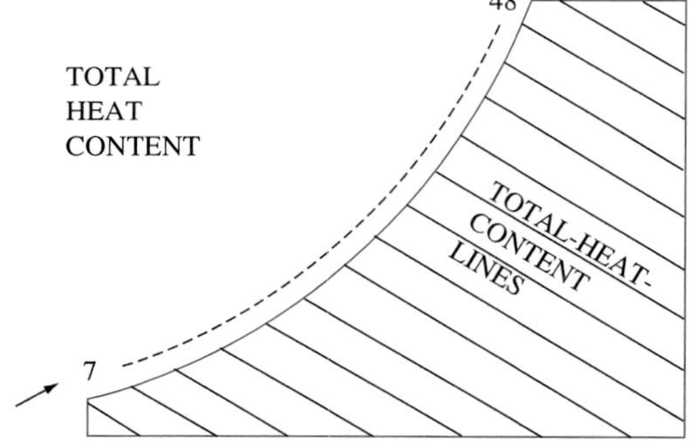

Figure A28-15 Total-heat lines on the psychrometric chart.

PRACTICING PROFESSIONAL SERVICES

When working on air-conditioning equipment used to cool and dehumidify in high relative-humidity conditions, learn to check the efficiency of the equipment by checking the entering-air dry-bulb and wet-bulb temperatures. Many air-conditioning equipment manufacturers will provide refrigerant charging charts using these temperatures as performance criteria. The service technician will therefore need to carry a wet-bulb and dry-bulb thermometer as a part of their tool package. Remember, a high latent load of relative humidity will require the air conditioner to work extra hard to dry the air. Checking this information will indicate if the equipment is performing properly as designed.

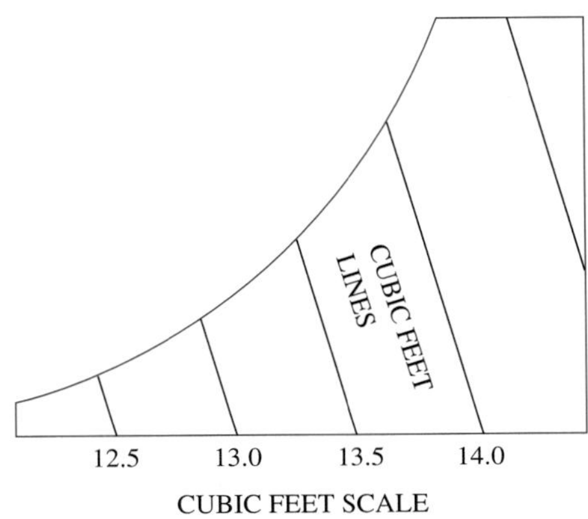

Figure A28-16 Specific-volume lines on the psychrometric chart.

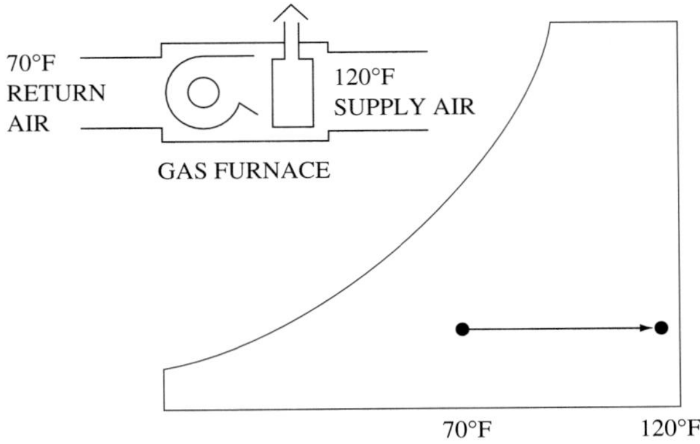

Figure A28-18 Plotting the addition of sensible heat by a furnace, on the psychrometric chart.

For example, the vertical lines indicate dry-bulb temperature values (Fig. A28-11) and the horizontal lines indicate moisture-content values.

Figures. A28-18 to A28-23 show plots on the chart representing various air-conditioning processes.

The changes that take place in the properties of the air during these processes are described as follows:

1. *Heating* (Fig. A28-18), adding sensible heat. The process is shown moving along a horizontal line to the right. The wet-bulb temperature and total heat content increase and the relative humidity decreases. The dew-point temperature and the moisture content do not change.
2. *Cooling* with a dry evaporator (Fig. A28-19), reducing sensible heat. The process is also shown moving along a horizontal line, but in the opposite direction. The wet-bulb temperature and total heat content decrease and the relative humidity in-

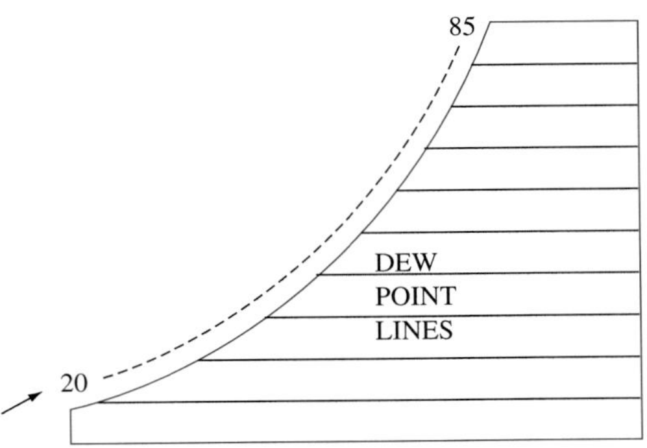

Figure A28-17 Dew-point temperature lines on the psychrometric chart.

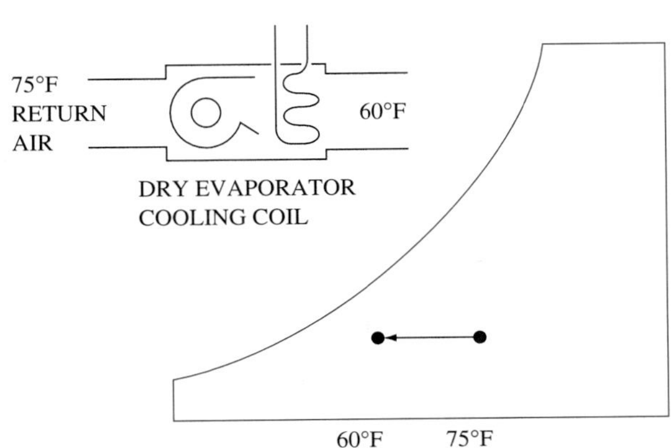

Figure A28-19 Plotting the temperature of air being cooled without dehumidification on the psychrometric chart.

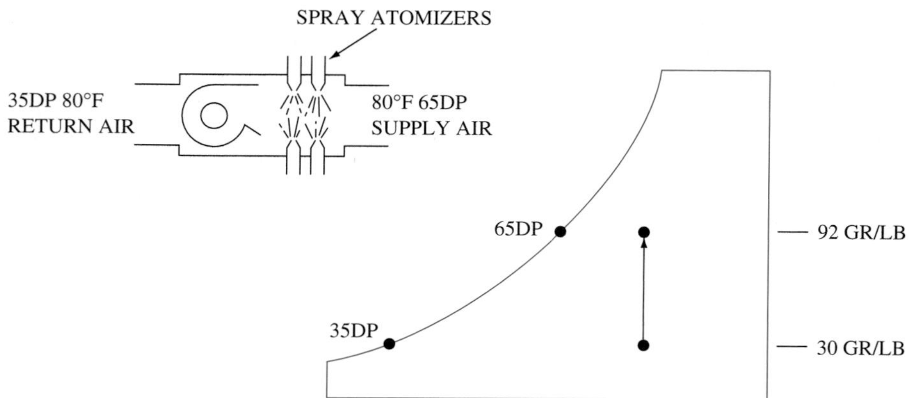

Figure A28-20 Plotting the addition of moisture to the air produced by a humidifier.

creases. Again, there is no change in dew point or moisture content.

3. *Humidifying* (Fig. A28-20), adding moisture. This action is represented by a vertical line moving upward with moisture increase. The only property that does not change is the dry-bulb temperature, which is held constant. Heat is added as evidenced by the increase in total heat, but the heat is used to evaporate the moisture. The wet-bulb temperature, dew point and relative humidity increase.
4. *Heating and humidifying* (Fig. A28-21), used for winter operation. The line representing the operation moves upward to the right. The air picks up both heat and moisture as it moves through the conditioner. The total heat and wet-bulb and dew-point temperatures all increase. The relative humidity could increase, remain the same, or decrease, depending on how much moisture is added. In actual operation the amount of heat and moisture added must represent the losses of the structure in order to maintain the entering air conditions (return air or room temperature).
5. *Cooling and dehumidification* (Fig. A28-22), used for summer operation. The line representing the operation moves downward to the left. The cooling coil both cools the air and condenses moisture as the air passes through it. The total heat and wet-bulb and dew-point temperatures all decrease. The relative humidity increases, since the air is nearly saturated when it leaves the coil. Again, the reduction of temperature and moisture in the process must be related to the load requirements.
6. *Cooling and humidifying* (Fig. A28-23). The process is called *evaporative cooling*. It uses evaporation of moisture to cool the air. The process follows the wet-bulb line in the chart. The dry-bulb temperature decreases as the moisture in the air and the relative humidity increase. Total heat is constant.

A28-4 CALCULATING PERFORMANCE

The psychrometric chart can be useful in roughly checking the performance of air-conditioning equipment, as shown in the following example.

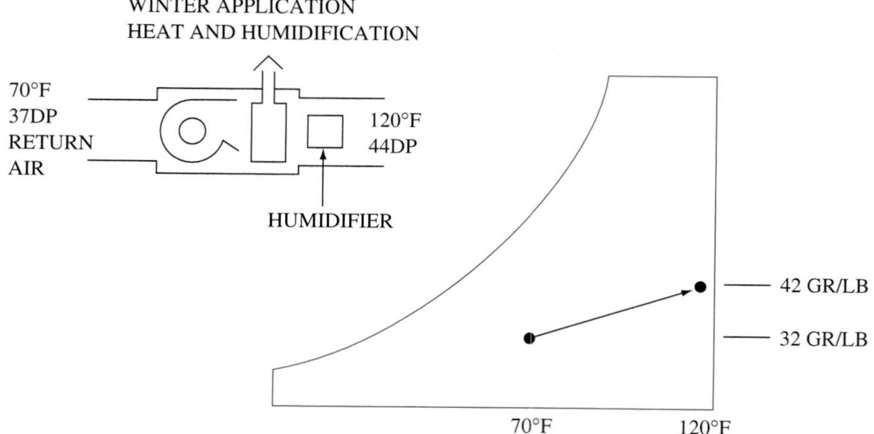

Figure A28-21 Heat and moisture added by a hot-air furnace with a humidifier.

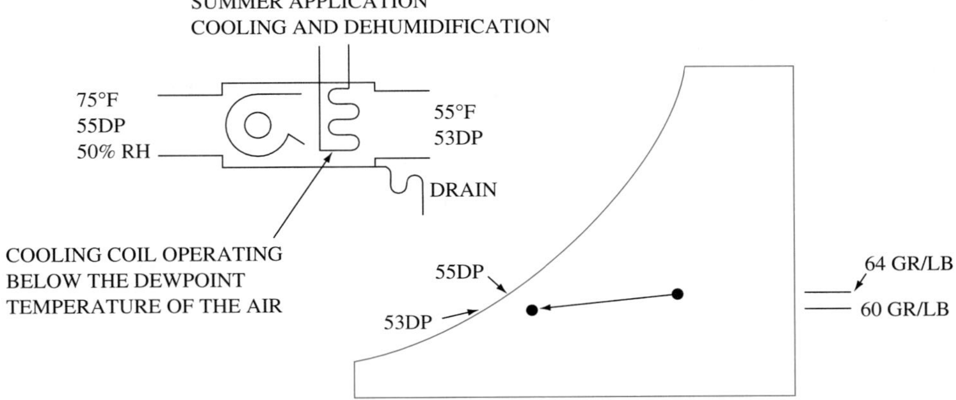

Figure A28-22 Cooling and dehumidification produced by an air conditioner, plotted on the psychrometric chart.

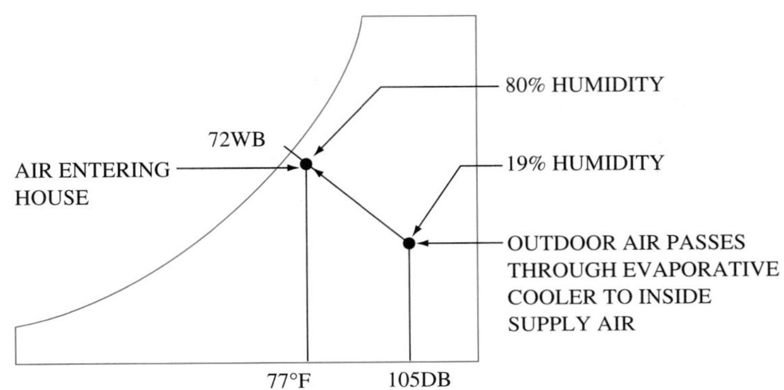

Figure A28-23 Cooling and humidification produced by an evaporative cooler plotted on the psychrometric chart.

EXAMPLE

Assume that a 3-ton (1 ton = 12,000 Btu/hr) air-conditioning system is installed in a residence. To check to see if the unit is producing its rated capacity, the following information is needed from the installation:

1. Dry-bulb and wet-bulb temperatures, inside air (*A*)
2. Dry-bulb and wet-bulb temperatures, outside air (*B*)
3. Air circulated by the blower, in cfm
4. Percent of ventilation air (or cfm)

For this example we will refer to Fig. A28-24 and assume that the following readings were taken at the job site:

1. Inside temperatures: 75°F DBT, 62.5°F WBT
2. Outside temperatures: 95°F DBT, 67°F WBT
3. Total air circulated: 1100 cfm
4. Ventilation air: 25% (or 247 cfm)

To plot these conditions on the chart (Fig. A28-24) and obtain further data, proceed as follows:

1. Locate the inside conditions. Mark this point *A*.
2. Locate the outside conditions. Mark this point *B*.
3. Connect points *A* and *B* with a straight line.
4. Determine the mixed-air temperature. The dry-bulb temperature of the mixture is found by taking 25% of the difference between inside and outside DBT (25% of 20° = 5°), adding this to the inside DBT (75 + 5 = 80°F), then locating this temperature on the line *AB*. Mark this point *C*. This is the condition of the air entering the unit.
5. For this example, we will assume that both cooling and dehumidification take place and that the condition of the air leaving the unit is measured to be 53°F DBT and 51°F WBT. Plot this on the chart and mark it point *D*.
6. Connect points *C* and *D* with a straight line. This represents the changes in the properties of air that take place when the air passes through the conditioning unit. The total amount of cooling and dehumidifying that takes place must conform to the load requirements.

A summary of information that can now be read from the chart is as follows:

Entering air conditions, point *C*
80°F DBT, 63.5°F WBT
Heat content of the air = 29.0 Btu/lb
Moisture content of the air = 62 gr/lb

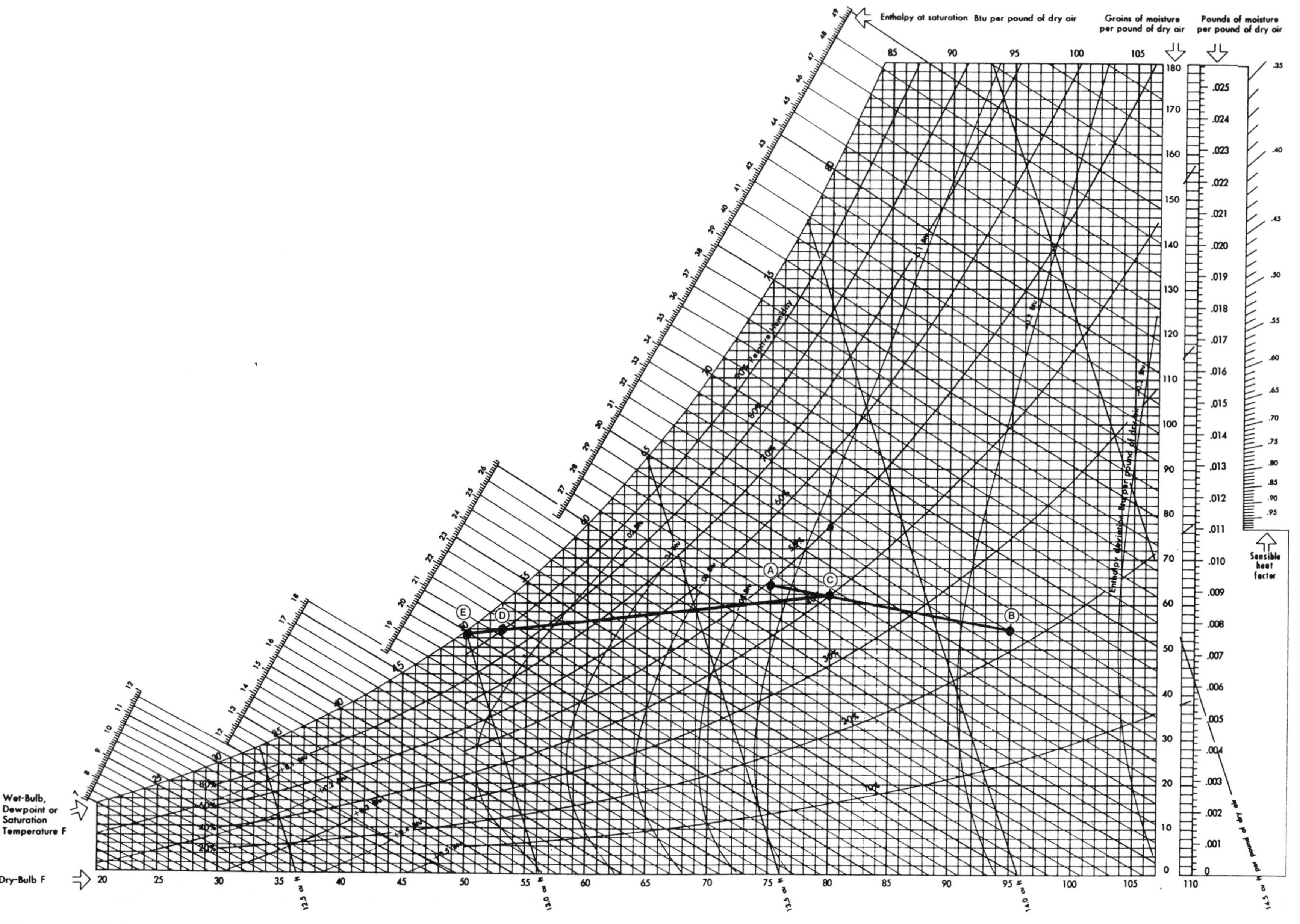

Figure A28-24 Plot of an air-conditioning system using 25% outside air with temperature reduction and dehumidification. (Courtesy of Carrier Corporation)

Leaving air conditions, point *D*
53°F DBT, 51°F WBT
Heat content of the air = 21.1 Btu/lb
Moisture content of the air = 54 gr/lb

The following formulas are used in making the performance calculations:

Sensible heat removed

$$Q_s = 1.08 \times \text{cfm} \times \text{DBT difference} \quad (1)$$

Latent heat removed

$$Q_l = 0.68 \times \text{cfm} \times \text{gr moisture difference} \quad (2)$$

Total heat removed

$$Q_t = Q_s + Q_l, \text{ or} \quad (3a)$$

$$Q_t = 4.5 \times \text{cfm} \times \text{total heat difference} \quad (3b)$$

If the chart is marked with a fine pen, the calculation of the total heat removed using formula (3a) should be nearly the same as using formula (3b). There will be some difference because there is a limit to how close the values can be read on the chart.

EXAMPLE

Below are the calculations based on using the data indicated above.

Sensible heat removed

$$Q_s = 1.08 \times \text{cfm} \times \text{DBT difference}$$
$$Q_s = 1.08 \times 1100 \times 27 = 32{,}076 \text{ Btu/hr}$$

Latent heat removed

$$Q_l = 0.68 \times \text{cfm} \times \text{gr difference}$$
$$Q_l = 0.68 \times 1100 \times 8 = 5{,}984 \text{ Btu/hr}$$

Total heat removed

$$Q_t = 4.5 \times 1100 \times 7.9 = 39{,}105 \text{ Btu/hr}$$

The slope of the line *CD* represents the *sensible-heat ratio* (*SHR* ratio) of the air in passing through the cooling coil (Fig. A28-24). In this example both the DBT and the moisture content are lowered. By definition:

Sensible-heat ratio =

$$\frac{\text{sensible heat removed in Btu/lb of air}}{\text{total heat removed in Btu/lb of air}}$$

In this example,

$$SHR \text{ ratio} = \frac{32{,}076}{39{,}105} = 0.82$$

The sensible-heat ratio performed by the operating equipment must conform to the load requirements. Otherwise design conditions will not be maintained.

The point *E* on the chart is called the *apparatus dew point.* This is the point on the 100% saturation line intersected by an extension of the *SHR* ratio line *CD*. The evaporator temperature of the cooling coil must be significantly below this point in order for moisture removal to take place.

In plotting the performance of a system, if the extension of the *SHR* ratio line does not intersect the saturation line, no moisture removal will take place.

A simplified method of calculation for finding the total is to operate the unit on the heating cycle. For example, if the gas input to the heating unit is 100,000 Btu/hr, with an efficiency of 80%, the output is 80,000 Btu/hr.

Using the formula for sensible heat, values can be inserted and the formula can be solved for cfm, as follows:

$$Q_s = 1.08 \times \text{cfm} \times \text{DBT difference}$$

$$\text{cfm} = \frac{Q_s}{1.08 \times \text{DBT difference}}$$

Assuming that the entering DBT = 70°F and the leaving DBT = 110°F

$$\text{cfm} = \frac{80{,}000}{1.08 \times 40}$$

$$\text{Total air} = 1852 \text{ cfm}$$

A28-5 AIR DISTRIBUTION AND BALANCING

Air-conditioning equipment is selected to control heating, cooling, ventilating, humidifying or dehumidifying, and cleaning the air in buildings. The effect is to produce comfortable conditions for the occupants. It also can serve to protect the contents of a building against damage due to improper temperature and humidity conditions.

A28-5.1 System Components

Forced-air systems are used to distribute heating and/or cooling in residential and small commercial buildings. An essential element in the process is the distribution of conditioned air.

For example, Fig. A28-25 shows the required heating and cooling for various areas of a small building.

These are the loads under design conditions. The distribution system must be able to adequately supply conditioned air to each area during peak periods, as well as during partial-load conditions. The distribution system is designed to handle peak requirements and the controls adjust the system to handle partial loads.

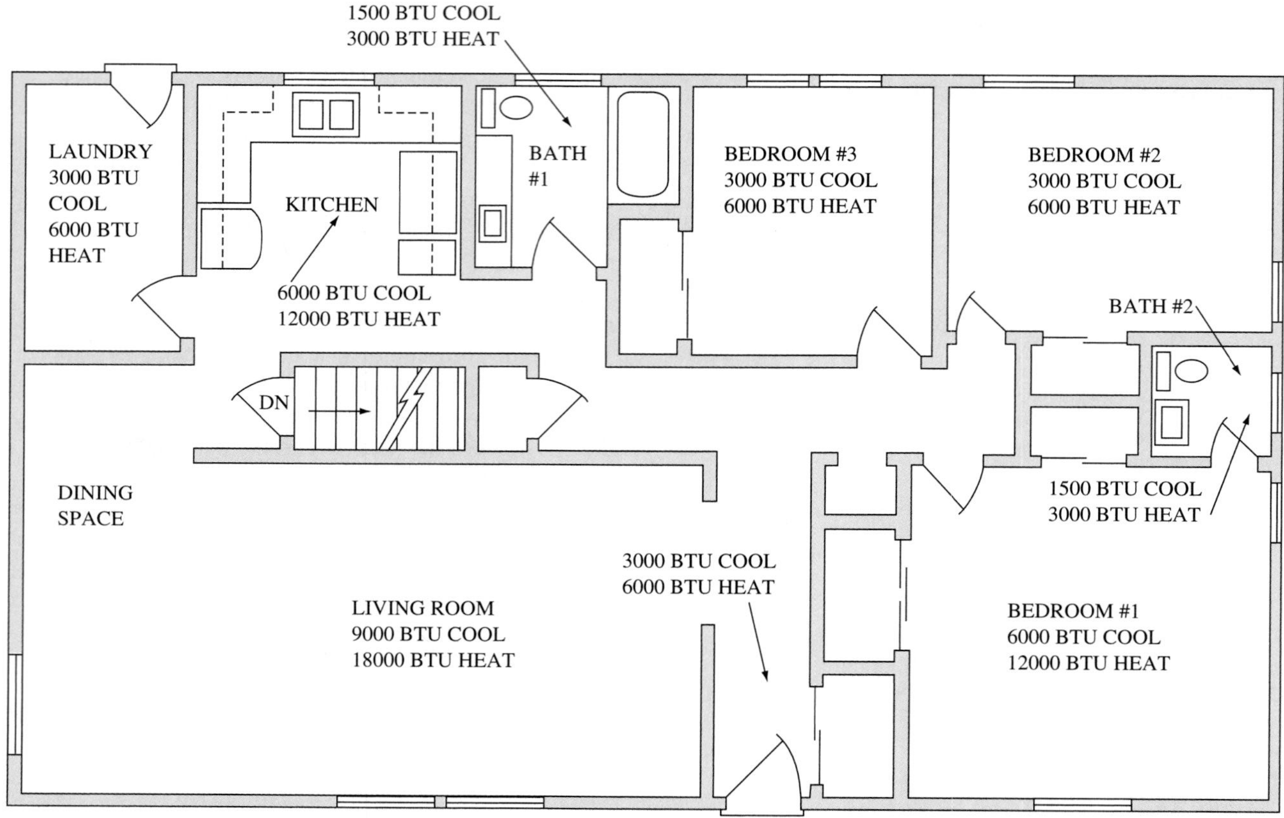

Figure A28-25 Floor plan showing heating and cooling load for each space.

Assuming that a forced-air system is used, the components consist of a blower, filters, humidifier, supply-air ducts, return-air ducts, ventilation-air ducts, registers, and grilles.

A28-5.2 Air Quantity

In order for the system to operate properly, the correct air quantities must be delivered to each area in proportion to the load. In the example shown, the total cooling load is 35,000 Btu/hr and a 3-ton air conditioning unit is selected. Since a typical conditioner delivers 400 cfm/ton, assume that a fan is selected that will supply a total of 1200 cfm. Based on proportioning the air supply to each area according to the cooling loads shown, the cfm requirements for each area are selected and shown in Fig. A28-26.

A28-5.3 The Blower

The blower must be capable of delivering the total air quantity against the external resistance of the conditioner, including the duct work and grilles. The air pressure required by a blower is relatively small. Instead of measuring this pressure in psi, it is measured in inches of water column (in. WC).

Atmospheric pressure can be indicated as 14.7 psi, or in terms of an equivalent column of water it will support, 34 ft. For small pressure differences such as a blower would create, pressures are measured in in. WC. For example, an air-conditioning unit for a small job, such as shown in Fig. A28-26, could be capable of delivering 1200 cfm against an external static pressure of 0.12 in. WC. These small pressures are measured using an inclined manometer, as shown in Fig. A28-27.

A28-5.4 Types of Pressures

In dealing with air distribution, and particularly duct systems, there are three types of pressure readings: *static, velocity,* and *total.* Static pressure is the force of the air against the sides of a vessel or duct, as shown in Fig. A28-28.

Total pressure is read by placing the manometer so that air flows directly into the sensing-tube opening, as shown in Fig. A28-29. The velocity pressure is the difference between the total and static pressures and can be read as shown in Fig. A28-30.

The velocity pressure is useful in measuring the velocity of air in a duct. Some manometers are calibrated to read duct velocity directly in feet per minute (fpm). A single instrument that can be inserted in the duct and connected to the manometer for reading velocity pressure, known as the *Pitot tube* is shown in Fig. A28-31.

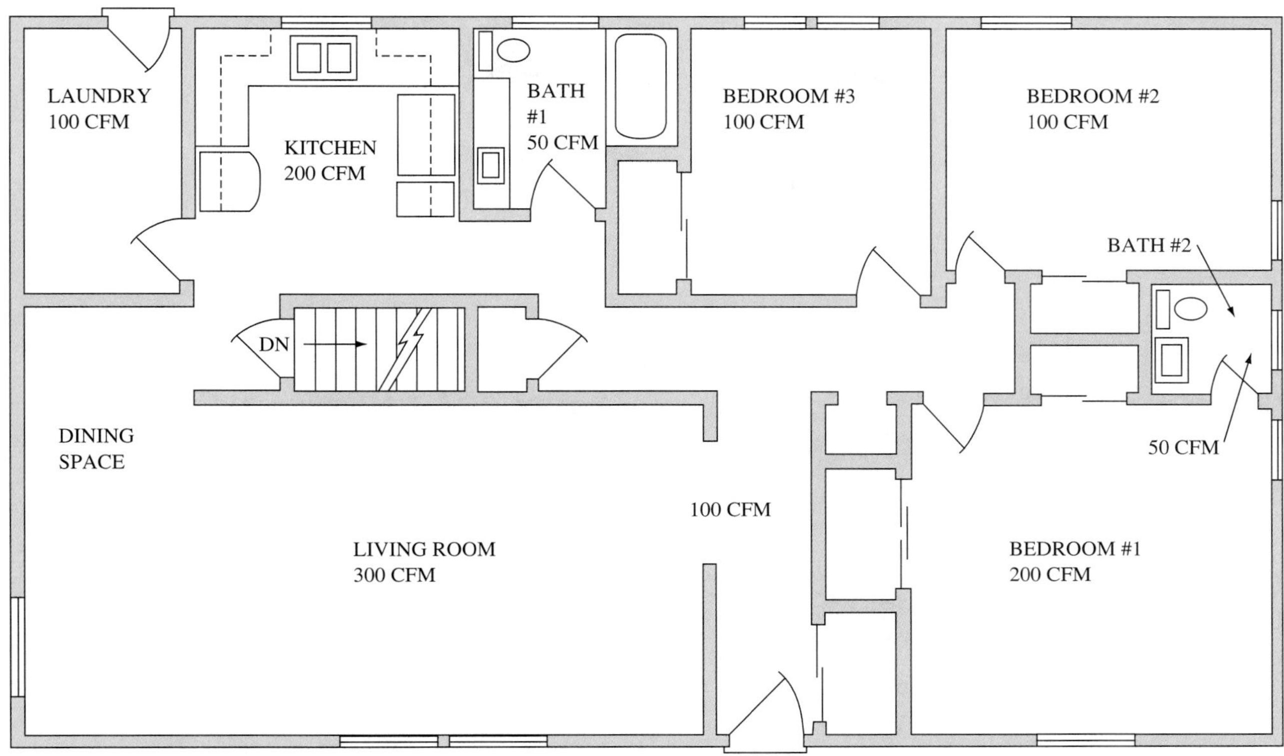

Figure A28-26 Floor plan showing air quantities (cfm) for each space.

A28-5.5 Types of Fans

The terms *fans* and *blowers* are often used interchangeably; however, the accepted term for this group of equipment is *fan.* There are two distinctly different types of fans: *centrifugal* and *axial,* based on the direction of air flow through the impeller.

Fig. A28-32 shows an exploded view of the centrifugal fan. These fans are sometimes called "squirrel cage" fans due to the shape of the impeller wheel. Two principal types are used in air-conditioning work: the forward-curved blade, which is most common, and the backward-curved blade. The advantage of the backward-curved blade is that it is non-overloading. The disadvantage is that it is noisier. These fans are widely used to distribute conditioned air through duct work.

Fig. A28-33 shows an exploded view of the axial fan. There are two principal types of axial fans: the vane and the propeller. The propeller type will handle large volumes of air for low-pressure applications. It has high usage for exhaust fans and condenser fans. Vane-type fans are highly efficient but noisy. The blade pitch can be adjusted to control the amount of air they handle.

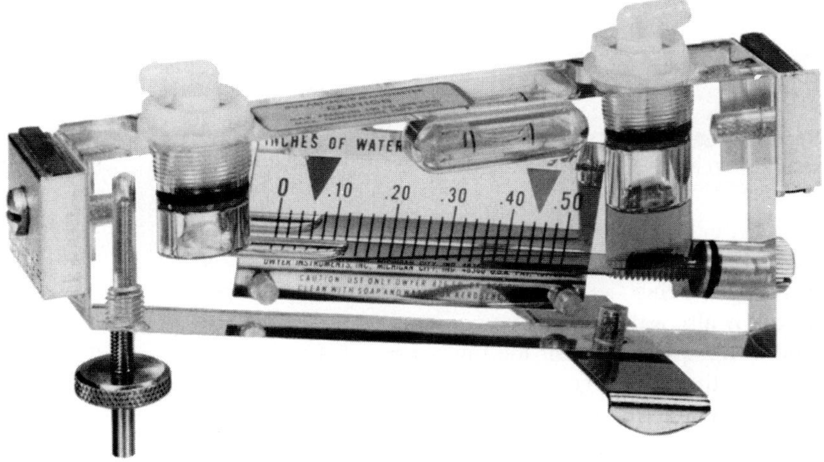

Figure A28-27 Inclined manometer for measuring air pressure. (Courtesy of Dwyer Instruments, Inc.)

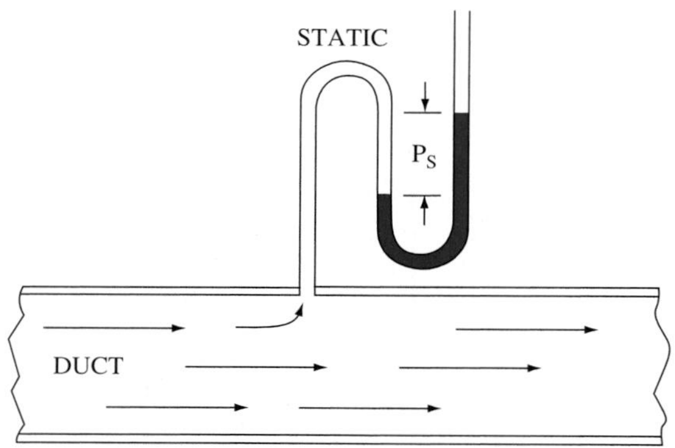

Figure A28-28 Manometer measuring static pressure, in inches of water column.

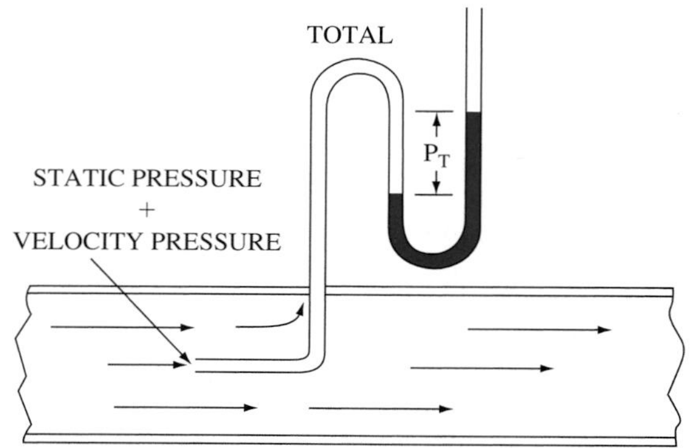

Figure A28-29 Manometer measuring total air pressure, in inches of water column.

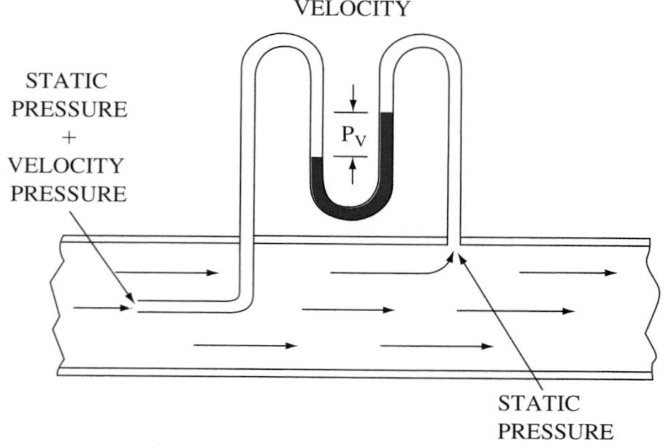

Figure A28-30 Manometer measuring velocity pressure in inches of water column.

It is interesting to note, that if the inlet to a fan is blocked, reducing the air quantity, the power required to drive it is also reduced. The power required is directly related to the quantity of air the fan pumps. If the outlet is blocked, the discharge pressure is increased, but the power required to operate the fan is reduced since it pumps fewer pounds of air. The greatest load on a fan is

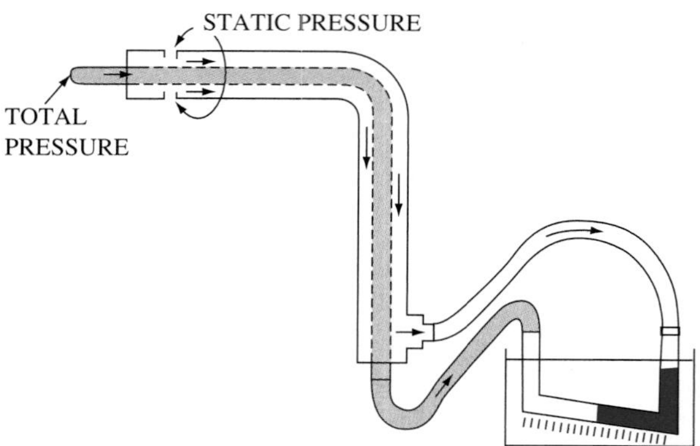

Figure A28-31 Pitot tube used in an air stream for measuring velocity pressure.

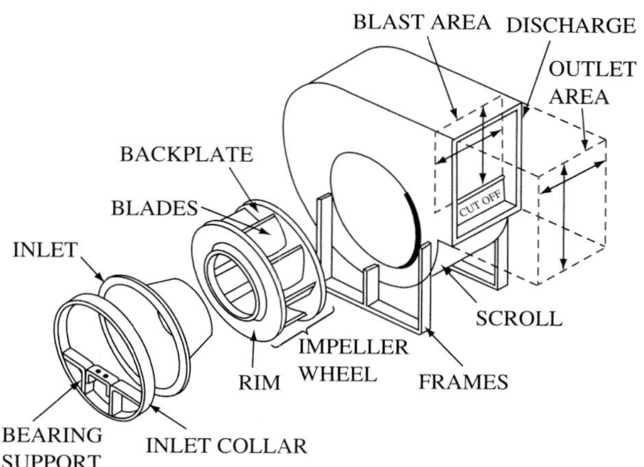

Figure A28-32 Expanded view of a centrifugal fan. (Copyright by the American Society of Heating, Refrigerating, and Air-Conditioning Engineers, Inc. Used by permission.)

created when the fan operates in the open without duct work restrictions, pumping the maximum quantity of air. There is always a danger under these conditions of overloading the fan motor.

Fans operate according to a set of formulas called *fan laws*. These laws show the relationship between changes in speed (rpm), air volume (cfm), static pressure (*SP*) and power in brake horsepower (Bhp). The fan laws are as follows:

1. $\dfrac{\text{cfm}_2}{\text{cfm}_1} = \dfrac{\text{rpm}_2}{\text{rpm}_1}$

2. $\dfrac{\text{SP}_2}{\text{SP}_1} = \left(\dfrac{\text{rpm}_2}{\text{rpm}_1}\right)^2$

3. $\dfrac{\text{Bhp}_2}{\text{Bhp}_1} = \left(\dfrac{\text{rpm}_2}{\text{rpm}_1}\right)^3$

These formulas are extremely useful in determining the effect of changing the speed of a fan. For example, if the speed is reduced 10%, the static pressure is reduced 19%

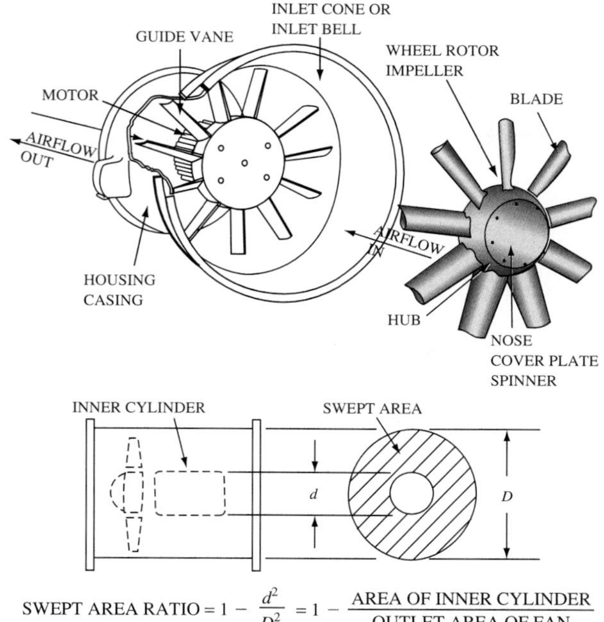

Figure A28-33 Cutaway view of an axial-type fan showing the component parts. (Copyright by the American Society of Heating, Refrigerating, and Air-Conditioning Engineers, Inc. Used by permission.)

and the brake horsepower is reduced 27%. A substantial amount of energy can be saved by slowing down a fan if it is delivering an excessive air volume.

A28-5.6 Types of Fan Drives

There are two types of drive arrangements for air conditioning fans: belt and pulley drives (Fig. A28-34) and direct drives (Fig. A28-35).

Motors are available with synchronous speeds of 1800 rpm (actual speed about 1750 rpm) and 3600 rpm (actual speed about 3450 rpm).

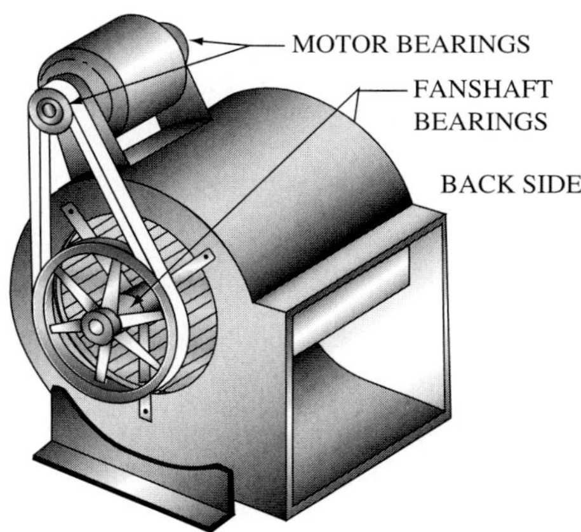

Figure A28-34 Belt-driven blower with two pulley wheels for motor and blower.

Figure A28-35 Direct-drive fan motor for a blower.

For *belt-driven motors* the speed of the fan is determined by the ratio of the motor pulley to the fan pulley. For example, if the motor pulley is 3 in. in diameter, the fan pulley 6 in. in diameter and the motor speed 1750 rpm, the speed of the fan is calculated as follows:

$$\text{Fan rpm} = \frac{\text{diameter of motor pulley} \times \text{motor rpm}}{\text{diameter of fan pulley}}$$

$$\text{Fan rpm} = \frac{3}{6} \times 1750$$

$$= 875 \text{ rpm}$$

Direct-drive motors are available with multiple-speed windings. The speed can be changed by switching wires in the motor terminal box. This can be done by the control system so that a different speed can be used for heating and cooling. Small fans use shaded-pole motors. Direct-drive fan motors are usually the PSC (permanent-split capacitor) type.

A28-6 DUCT SYSTEMS

All duct-work designs should start with an accurate load calculation. The amount of air supplied to each area depends upon the space temperature, the supply-air temperature, and the space heating and cooling loads.

The equipment used for forced-air heating and cooling systems in residential and small commercial systems consist of furnaces, air conditioners, and heat pumps. Furnaces can be up-flow, down-flow, horizontal-flow, or a packaged unit as shown in Fig. A28-36.

The most common air-conditioning system uses a split-system configuration with the evaporator located in the furnace plenum and a remote condensing unit. A packaged

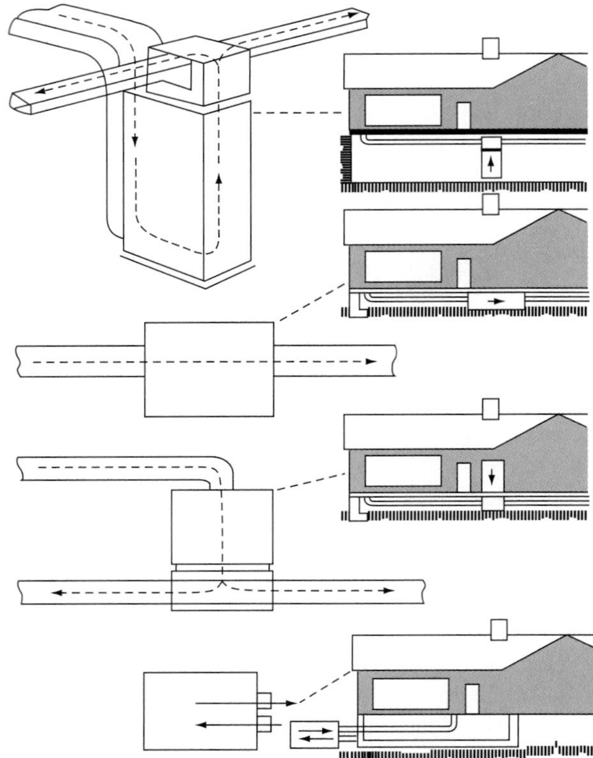

Figure A28-36 Various configurations of warm-air furnaces.

air conditioner contains all the components. The heat pump can be a split system or a complete package.

The accessories include humidifiers, air cleaners (filters), and economizer controls. There are several types of humidifiers, including self-contained steam, atomizing, evaporative, and heated-pan. Filters can be cleanable or throw-away types, with electronic air cleaners becoming increasingly popular. Economizers automatically use outside air for cooling or pre-cooling when outside conditions permit it. All of these accessories affect the system airflow and pressure requirements.

A28-6.1 System Design Procedure

The size and the performance of the system components (Fig. A28-37) are interrelated.

Following the calculation of heating and cooling loads, the following procedure should be followed:

1. Determine preliminary duct-work location.
2. Determine heating- and cooling-unit location.
3. Select accessory equipment. Provision may be necessary to add this equipment later.
4. Select control components.
5. Determine maximum airflow (heating and cooling) for each supply and return location.
6. Select heating and cooling equipment.
7. Finalize duct design and size system.
8. Select supply- and return-air grilles.

A28-6.2 Locating Outlets, Ducts and Equipment

The structure and layout of the building determine the available location for system components. In a residence, a full basement is an excellent location for the trunk ducts and conditioning equipment. Equipment can also be located in a closet space or utility room. All enclosures must meet local fire and safety codes.

For slab-type construction, ducts and even equipment can be placed in the attic (Figs. A28-38 and A28-39).

Ducts located in unconditioned areas must be properly insulated, and allowance for heat loss included in the load calculation. Ducts for perimeter (heating-only) systems can be located in the slab (Fig. A28-40).

Other types of duct layouts for small systems are shown in Fig. A28-41.

Supply outlets can be classified in four different groups, defined by their air-discharge patterns: (1) horizontal high, (2) vertical non-spreading, (3) vertical spreading, and (4) horizontal low. Fig. A28-42 (p. 557) lists the general characteristics of supply outlets. Note that there is no single outlet that is best for both heating and cooling.

Fig. A28-43 (p. 558) shows the preferred return locations for different supply-outlet positions. These return locations are based on a stagnant layer in the room, which is beyond the influence of the supply outlet. The stagnant layer develops near the ceiling during cooling and near the floor on heating. Cooling returns should be placed high and heating returns low. For year-round systems a compromise must be made by placing returns where the largest stagnant area will occur.

A28-6.3 Selecting Equipment

A furnace selection should either match or be slightly larger than the load. A 40% limit on oversizing is recommended by the Air Conditioning Contractors Association (ACCA). This limit minimizes venting problems and improves part load performance. *ASHRAE Handbook on Fundamentals* recommends that cooling units not be oversized more than 25% of the sensible load.

A28-6.4 Airflow Requirements

After selecting the equipment, the following determinations need to be made:

1. The air quantity (cfm) for each area.
2. The number and type of supply and return grilles and registers for each area.

The next step is to proceed with the duct layout.

The design of the system must be based on airflow and the static pressure limitation of the fan. The cfm for each

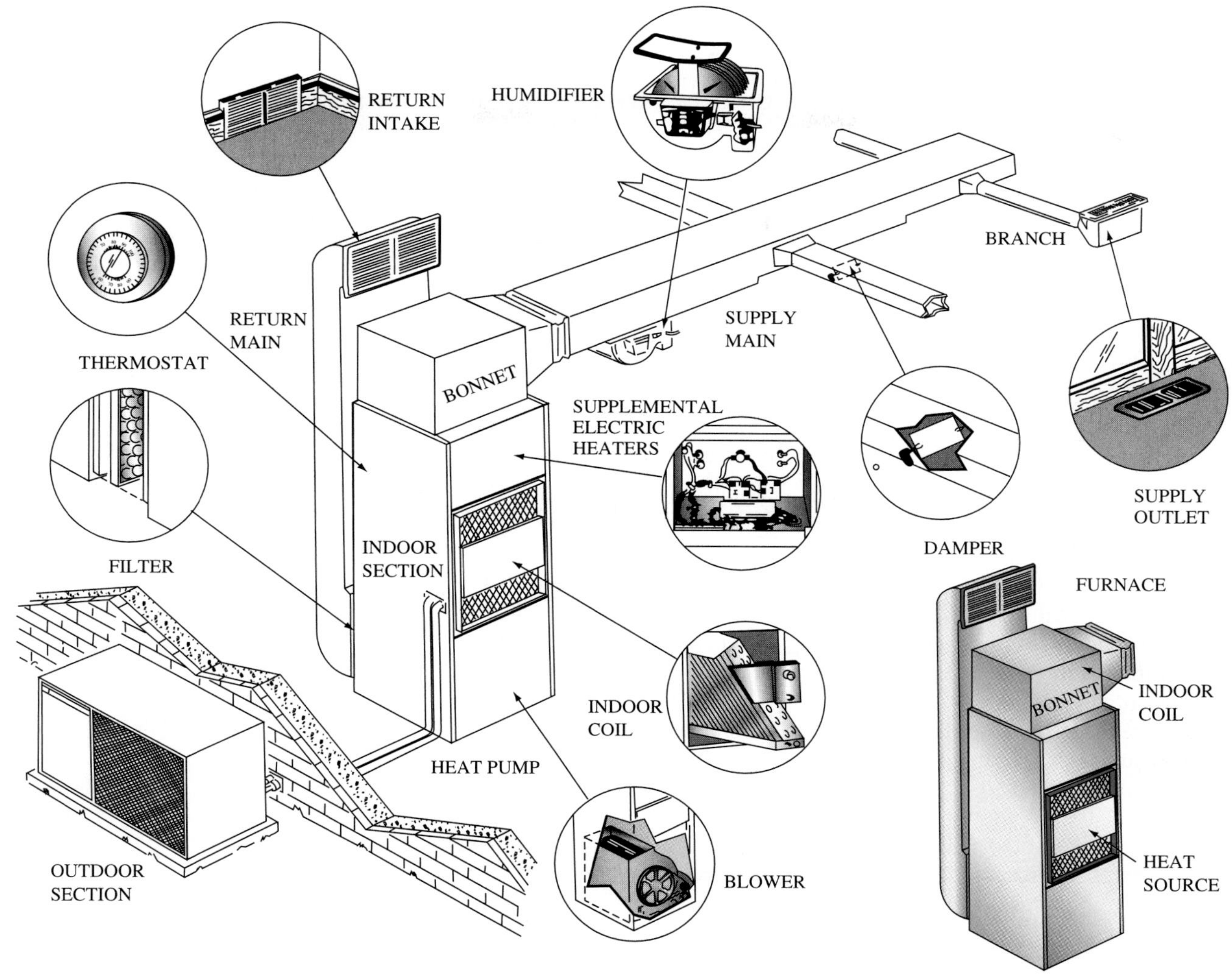

RETURN INTAKE
HUMIDIFIER
BRANCH
THERMOSTAT
RETURN MAIN
SUPPLY MAIN
BONNET
SUPPLEMENTAL ELECTRIC HEATERS
FILTER
INDOOR SECTION
DAMPER
SUPPLY OUTLET
FURNACE
BONNET
INDOOR COIL
INDOOR COIL
HEAT PUMP
HEAT SOURCE
OUTDOOR SECTION
BLOWER

Figure A28-37 Components which make up a complete heating and cooling system. (Copyright by the American Society of Heating, Refrigerating, and Air-Conditioning Engineers, Inc. Used by permission.)

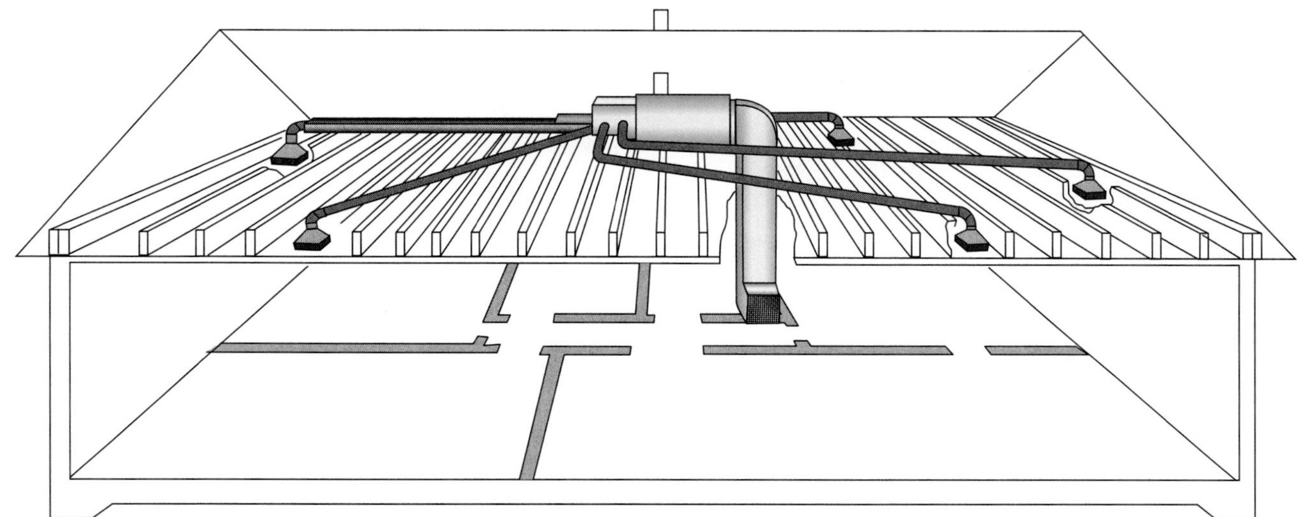

Figure A28-38 Heating and cooling system, with unit and duct work located in the attic for conditioning the floor below.

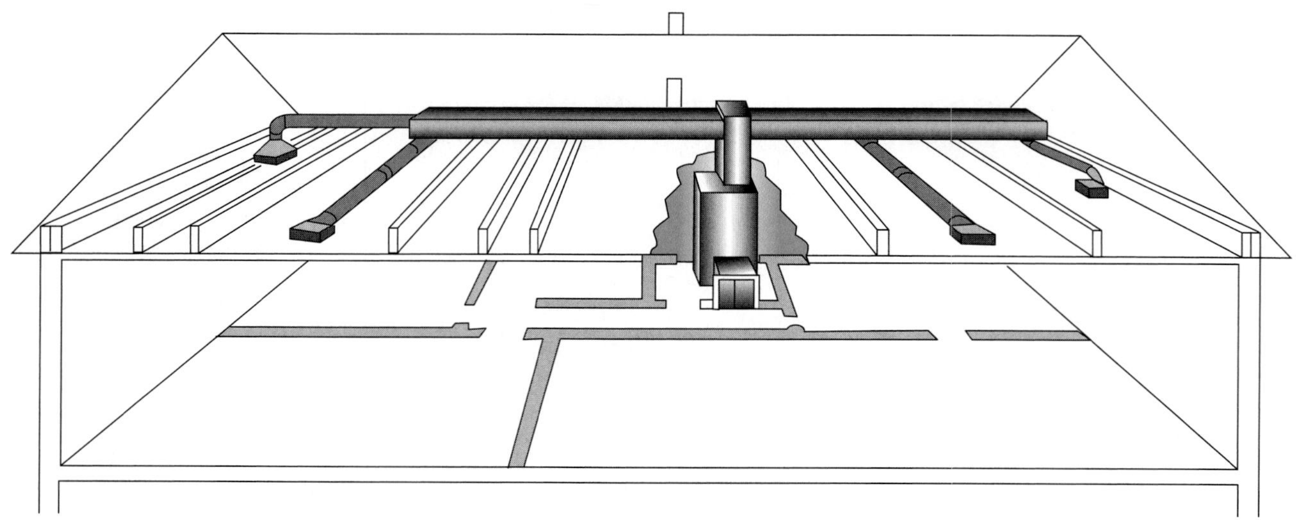

Figure A28-39 Heating and cooling system with duct work in the attic and unit located in the lower floor.

area is proportioned to the load of each area. It is recommended that the system be designed using the medium speed of the fan to allow excess capacity to handle accessories when they are added.

For systems designed for heating only, the air temperature rise must be within the manufacturer's recommendations (usually 40 to 80°F).

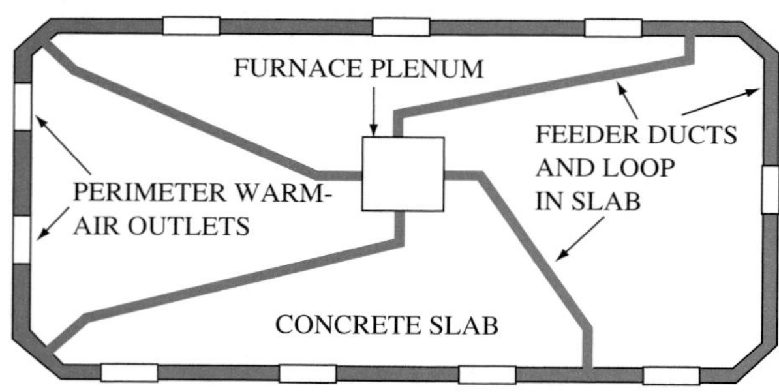

Figure A28-40 Perimeter loop system, with duct work in concrete slab floor.

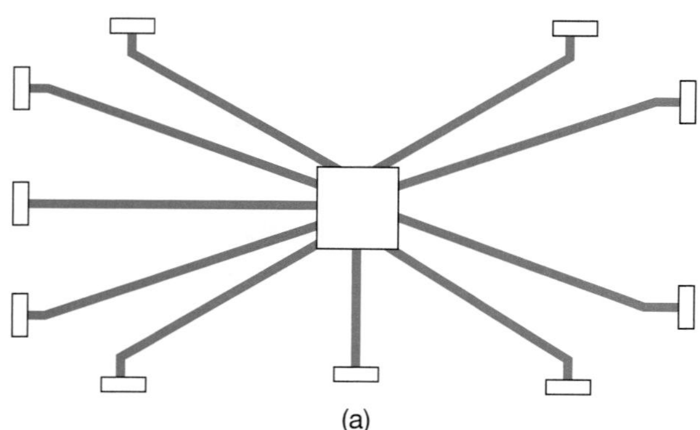

Figure A28-41a Central-plenum duct system located in basement for heating the space above.

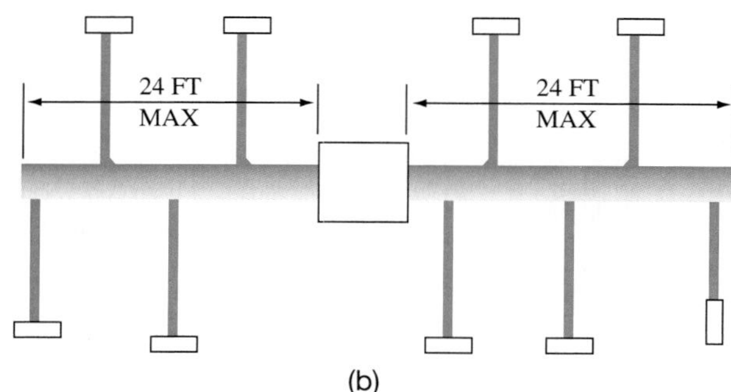

Figure A28-41b Extended-plenum or trunk-duct system, located in basement.

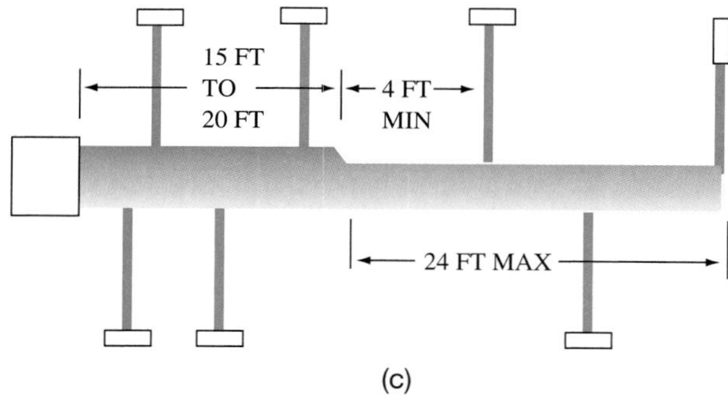

Figure A28-41c Extended-plenum-with-reducing-trunk duct system (located in basement).

For cooling-only or heating-and-cooling systems, the flow rate can be determined by the following equation:

$$Q = Q_s\,(1.08 \times TD)$$

where

Q = flow rate in cfm
Q_s = sensible-heat load in Btu/hr
TD = dry-bulb temperature difference in °F

Group	Outlet Type	Outlet Flow Pattern	Most Effective Application	Specific Use	Selection Criteria
1	Ceiling and high sidewall	Horizontal	Cooling	*Ceiling outlets* Full circle or widespread type	Select for throw equal to distance from outlet to nearest wall at design flow rate and pressure limitations.
				Narrow spread type	Select for throw equal to 0.75 to 1.2 times distance from outlet to nearest wall at design flow rate and pressure limitations.
				Two adjacent ceiling outlets	Select each so that throw is about 0.5 distance between them at design flow rate and pressure limits.
				High sidewall outlets	Select for throw equal to 0.75 to 1.2 times distance to nearest wall at design flow rate and pressure limits. If pressure drop is excessive, use several smaller outlets rather than one large one to reduce pressure drop.
2	Floor diffusers, baseboard, and low sidewall	Vertical, nonspreading	Cooling and heating	For cooling only	Select for 6 to 8 ft throw at design flow rate and pressure limitations.
3	Floor diffusers, baseboard, and low sidewall	Vertical, spreading	Heating and cooling	For heating only	Select for 4 to 6 ft throw at design flow rate and pressure limitations.
4	Baseboard and low sidewall	Horizontal	Heating only		Limit face velocity to 300 ft/min.

Figure A28-42 Tabulation of supply-air outlet characteristics. (Copyright by the American Society of Heating, Refrigerating, and Air-Conditioning Engineers, Inc. Used by permission.)

For preliminary design, an approximate *TD* is listed as follows:

Sensible Heat Ratio (*SHR*)	*TD* °F
0.75 to 0.79	21
0.80 to 0.85	19
0.85 to 0.90	17

$$SHR = \frac{\text{calculated sensible load}}{\text{calculated total load}}$$

For example, if $Q_s = 23{,}000$ Btu/hr

$$Q_l = 4900 \text{ Btu/hr}$$

where

$$Q_l = \text{latent heat load}$$

then

$$SHR = \frac{23{,}000}{(23{,}000 + 4900)} = 0.82$$

and

$$Q = \frac{23{,}000}{(1.08 \times 19)} = 1121 \text{ cfm}$$

The exact cfm value will be determined when the equipment is selected.

A28-6.5 Duct Design

The duct design procedure shown here is a modified *equal-friction method,* and should be used for systems requiring 2250 cfm or a 60,000 Btu/hr or less cooling load. This method has the advantage of simplifying the calculations and maintaining sufficient accuracy within this range.

The pressures available for the supply and return ducts are found by deducting coil, filter, grille, and accessories from the manufacturer's specified blower pressure. The

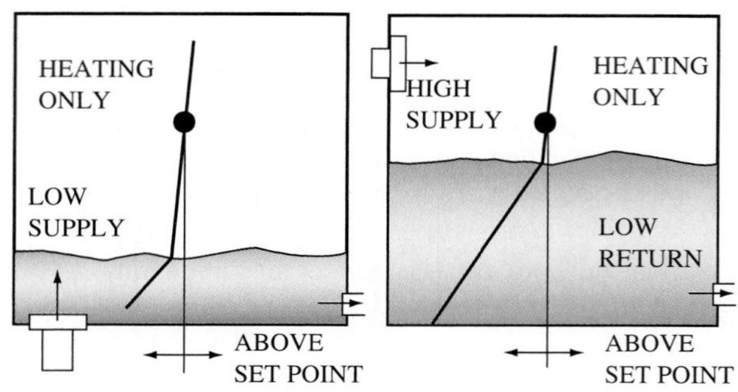

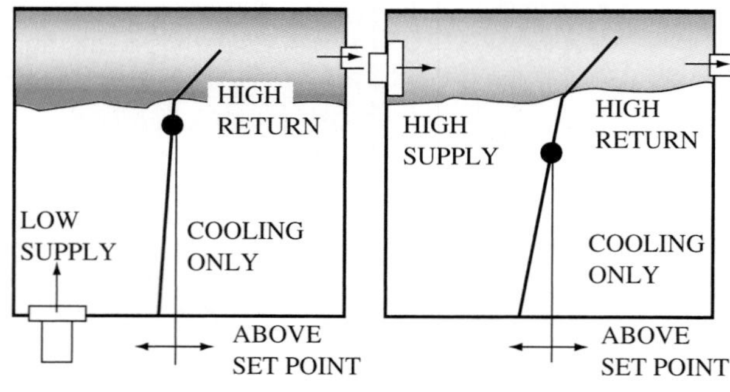

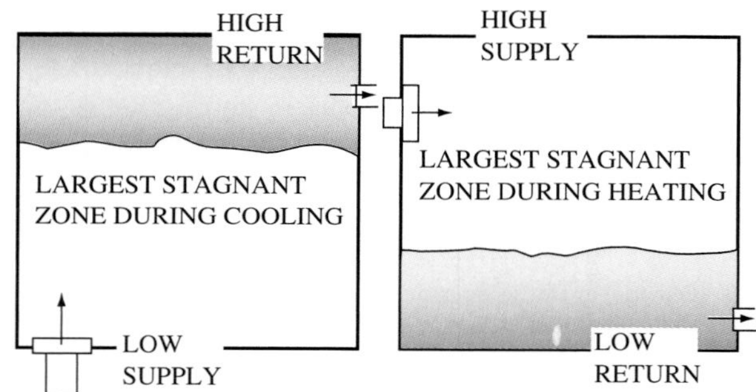

Figure A28-43 Preferred return-air locations in relation to supply-air outlets. (Copyright by the American Society of Heating, Refrigerating, and Air-Conditioning Engineers, Inc. Used by permission.)

remaining pressure is divided between the supply and return ducts as shown in Fig. A28-44.

General rules which apply to the design of duct work are:

1. Keep main ducts as straight as possible.
2. Streamline transitions.
3. Design elbows with an inside radius of at least $\frac{1}{3}$ the duct width, or use turning vanes.
4. Make ducts tight and sealed to limit air loss.
5. Insulate and/or line ducts, to conserve energy and limit noise.
6. Locate branch take-offs at least 4 ft downstream from a fan or transition joint.
7. Separate air-moving equipment from ducts, using flexible connectors to isolate noise.
8. Install a volume damper in each branch for balancing air flow.

System Characteristics	Supply, %	Return, %
A Single return at blower	90	10
B Single return at or near equipment	80	20
C Single return with appreciable return duct run	70	30
D Multiple return with moderate return duct system	60	40
E Multiple return with extensive return duct system	50	50

Figure A28-44 Recommended division of duct-pressure loss between supply- and return-air systems. (Copyright by the American Society of Heating, Refrigerating, and Air-Conditioning Engineers, Inc. Used by permission.)

The noise level for a small system is controlled by limiting the duct velocities as follows:

Main ducts	700 to 900 fpm
Branch ducts	600 fpm
Branch risers	500 fpm

For combination heating and cooling systems calculations produce a separate cfm for heating and cooling. Since the cooling loads require greater cfm, they are usually used to size the duct work. When two-speed fans are used, the air quantity can easily be reduced during the heating season.

A28-6.6 Duct Materials and Standards

For many years galvanized sheet steel was used exclusively for air-conditioning duct work; however, the material is expensive and costly to install, so other materials are sometimes used. These include fiberglass ductboard, aluminum, spiral-metal duct and flexible duct.

Aluminum duct is fabricated in the same manner as galvanized steel duct. Its higher cost limits its use, however.

Fiberglass duct has the advantage of being insulated. This reduces the duct losses and provides sound-absorbing qualities. It is particularly useful for duct running through cold areas, such as an unheated attic space. It is usually 1 in. thick, fabricated into round ducts and made in flat sheets for custom fabricating.

Spiral duct is made from long strips of narrow metal and fabricated with spiral seams. Machines are available for making ducts on the job, to fit required diameters and lengths.

	Comfort Heating or Cooling			Comfort Heating Only
	Galvanized Steel			
	Nominal Thickness (In Inches)	Equivalent Galvanized Sheet Gauge No.	Approximate Aluminum B & S Gauge	Minimum Weight Tin Plate Pounds Per Base Box
Round Ducts and Enclosed Rectangular Ducts				
14″ or less	0.016	30	26	135
Over 14″	0.019	28	24	—
Exposed Rectangular Ducts				
14″ or less	0.019	28	24	—
Over 14″	0.022	26	23	—

Figure A28-45 Gauges recommended for sheet-metal duct work.

Flexible duct, without the insulation, comes compressed in a box. When opened it expands lengthwise into ducts. It is easy to route around corners. Special care needs to be exercised during handling and installation not to flatten the duct and reduce the internal area.

Many localities adopt the metal duct standards developed by the Building Officials and Code Administrators International (BOCA). A guideline for selecting metal thickness is shown in Fig. A28-45.

Sheet metal ducts are usually held together by the use of drive straps and "S" hooks, as shown in Fig. A28-46. It is extremely important that air leakage from duct work be held to a minimum. For this purpose, joints can be sealed with tape.

A problem, shown in Fig. A28-47, illustrates the principles outlined above. The step-by-step procedure for making the calculations is as follows:

Duct Design Procedures Using the Equal-friction Method

1. Determine the heating and cooling load to be supplied by each outlet. Include duct losses or gains.
2. Make a simple diagram of the supply and return duct systems—at least one outlet in a room or area for each 8000 Btu/hr loss or 4000 Btu/hr sensible heat gain, whichever is greater.
3. Label all fittings and transitions to show equivalent lengths on the drawing (Figs. A28-48 through A28-56, at end of problem).
4. Show measured lengths of ductwork on the drawing.
5. Determine the total effective length of each branch supply. Beginning at the air handler,

$$\text{effective length} = \text{equivalent length} + \text{measured length.}$$

Branch R in the example, Fig. A28-47, is 92 ft.

6. Proportion the total airflow rate to each supply outlet for both heating and cooling:

Supply outlet flow rate =

$$\frac{\text{Outlet heat loss (gain)}}{\text{Total heat loss (gain)}} \times \text{Total system flow rate}$$

$$\text{Branch } R \text{ (Fig. A28-45)} = \frac{5400}{60{,}000} \times 900$$
$$= 81 \text{ cfm}$$

7. The flow rate required for each supply outlet is equal to the heating or cooling flow rate requirement (normally the larger of the two rates).
8. Label the supply outlet flow rate requirement on the drawing for each outlet.
9. Determine the total external static pressure available from the unit at the selected airflow rate.
10. Subtract the supply and return register pressure, external coil pressure, filter pressure, and box plenum (if used) from the available static pressure to determine the static pressure available for the duct design. Refer to Fig. A28-47 for the following example.

Equipment		0.50 in. water
Subtract:		
Cooling coil	0.24 in. water	
Supply outlet	0.03 in. water	
Return grille	0.03 in. water	
	0.30 in. water	

Total available static pressure = 0.20 in. water

11. Proportion the available static pressure between the supply and return systems. Refer to Fig. A28-47 for the following example.

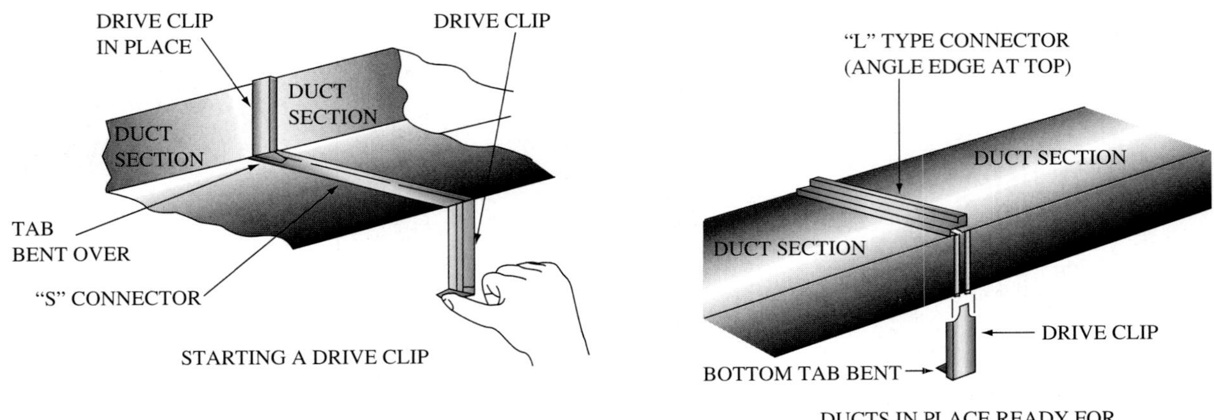

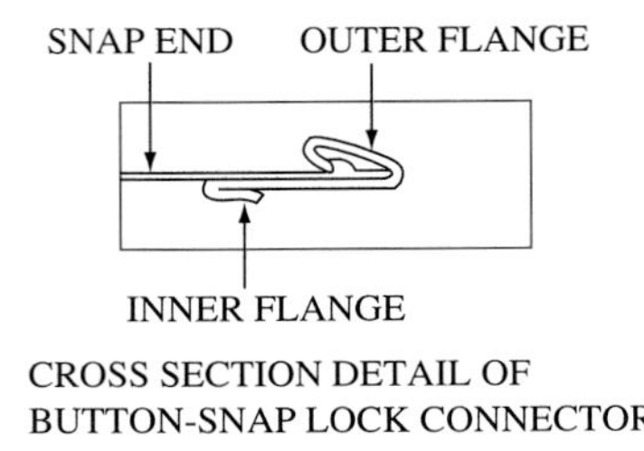

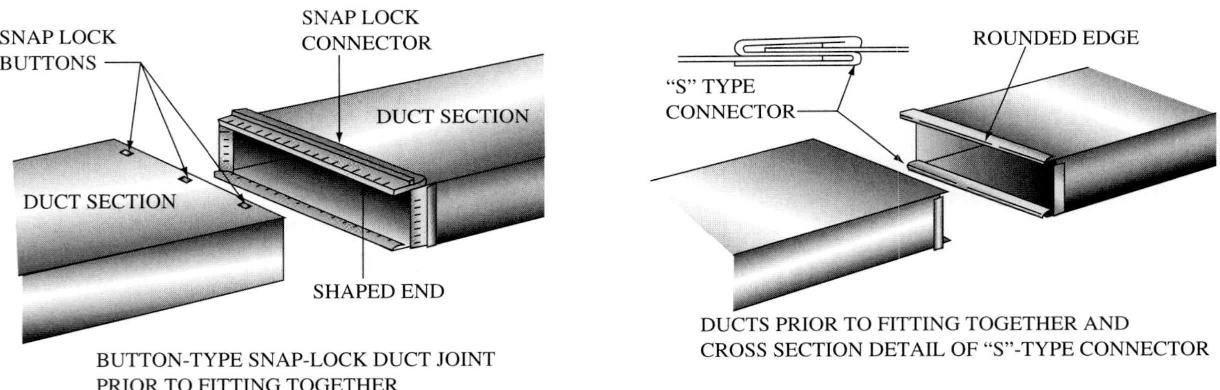

Figure A28-46 Sheet-metal duct work connections using drive straps and "S" hooks.

Supply (75%)	0.15 in. water
Return (25%)	0.05 in. water
Total =	0.20 in. water

Sizing the Branch Supply Air System

12. Use the supply static pressure available to calculate each branch design static pressure for 100 ft of equivalent length. For example:

Branch R design static pressure rate =

$$\frac{100\ \text{(Supply static pressure available)}}{\text{Effective length of each branch supply}}$$

$$100 \times \frac{0.15}{92} = 0.16\ \text{in. water/100 ft (Branch } R\text{)}$$

13. Enter the friction chart (Fig. A28-57, p. 567) at the branch design static pressure (0.163) opposite the flow rate for each supply, and read the round duct size (5 in.) and velocity (600 fpm).

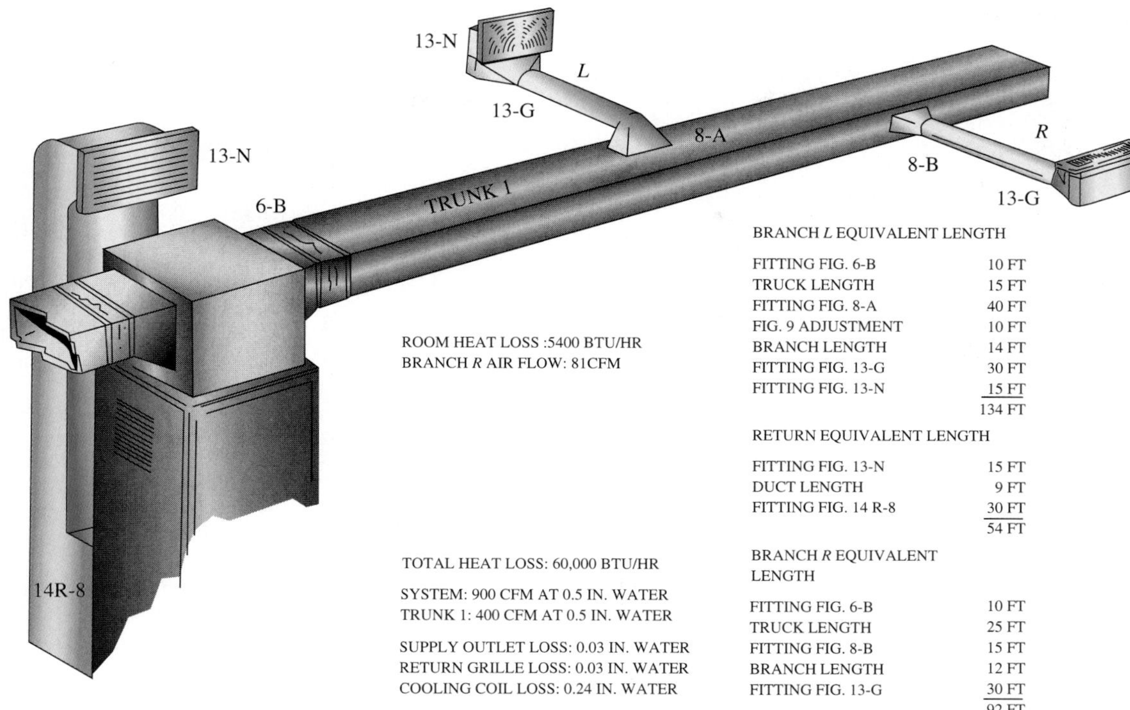

Figure A28-47 Equal-friction duct design for sample calculations. (Copyright by the American Society of Heating, Refrigerating, and Air-Conditioning Engineers, Inc. Used by permission.)

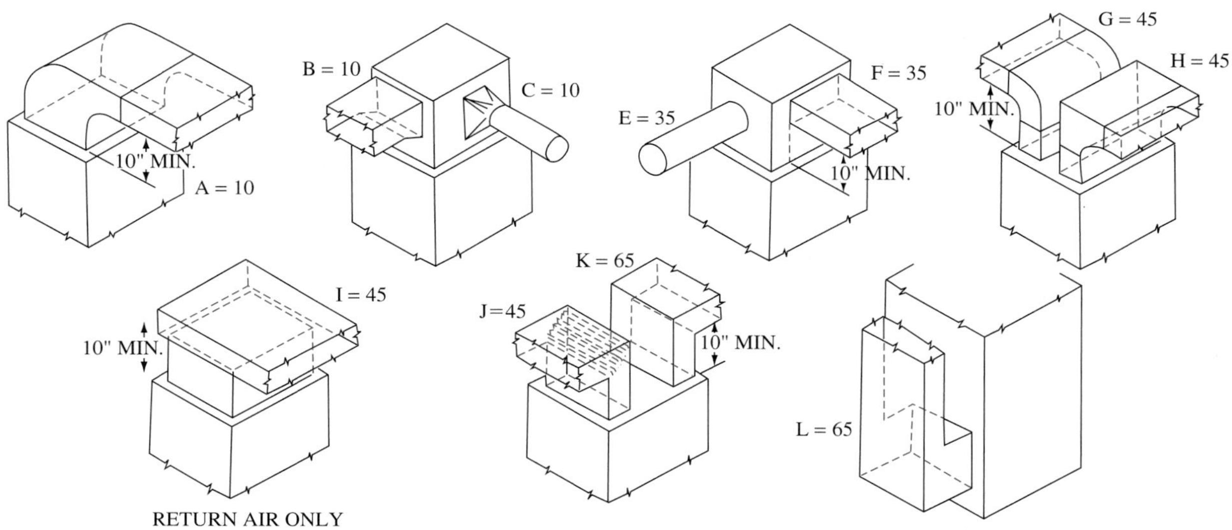

Figure A28-48 Equivalent lengths: supply- and return-air plenum fittings. (Copyright by the American Society of Heating, Refrigerating, and Air-Conditioning Engineers, Inc. Used by permission.)

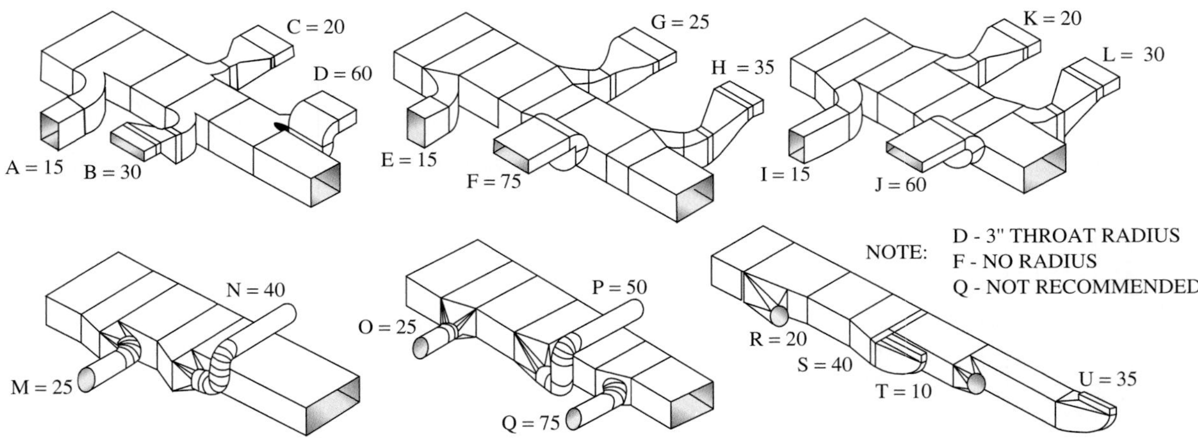

Figure A28-49 Equivalent lengths: reducing-trunk-duct fittings. (Copyright by the American Society of Heating, Refrigerating, and Air-Conditioning Engineers, Inc. Used by permission.)

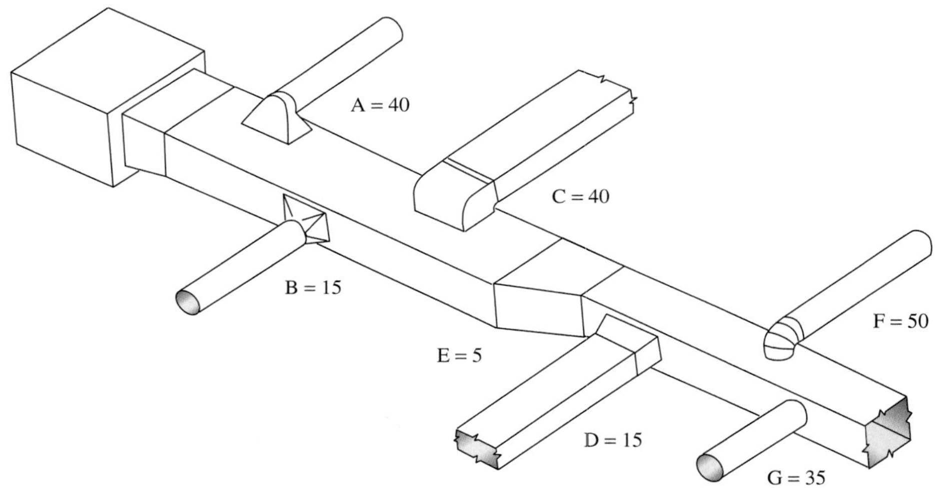

Figure A28-50 Equivalent lengths: extended-plenum fittings. (Copyright by the American Society of Heating, Refrigerating, and Air-Conditioning Engineers, Inc. Used by permission.)

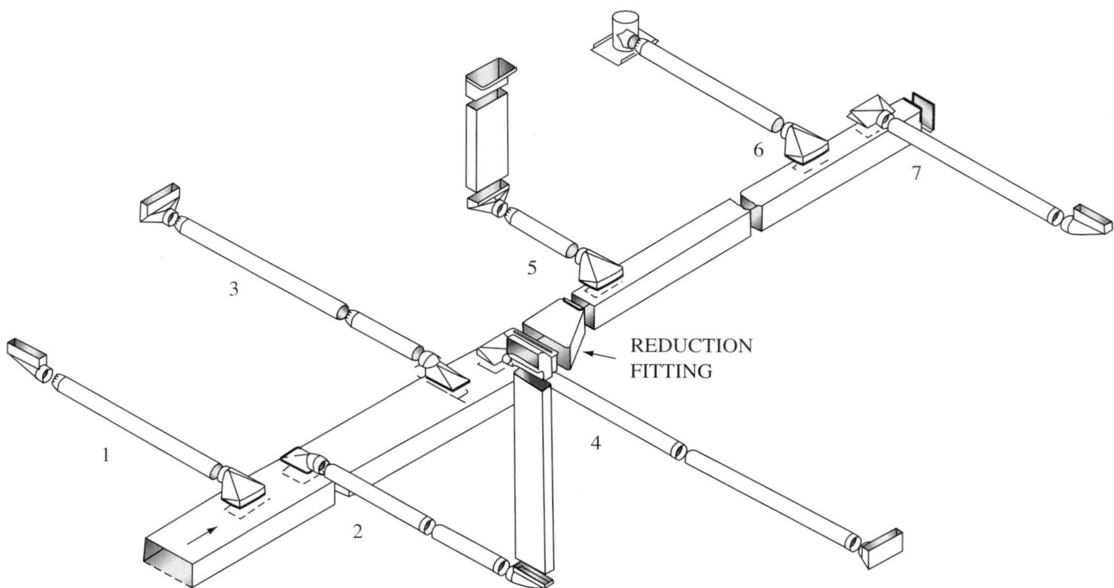

The pressure loss (expressed in equivalent length) of a fitting in a branch takeoff of an extended plenum depends on the location of the branch in the plenum. Takeoff branches nearest the furnace impose a higher loss, and those downstream, a lower loss. The equivalent lengths listed in the table are for the lowest possible loss—that is, the values are based on the assumption that the takeoff is the last (or only) branch in the plenum.

To correct these values, add velocity factor increments of 10 ft to each upstream branch, depending on the number of branches remaining downstream. For example, if two branches are downstream from a takeoff, add 2 × 10 or 20 ft of equivalent length to the takeoff loss listed.

In addition, consider a long extended plenum that has a reduction in width as two (or more) separate plenums—one before the reduction and one after. The loss for each top takeoff fitting in this example duct system is shown in the following table.

Branch Number	Takeoff Loss, ft	Downstream Branches	Velocity Factor, ft	Design Equiv. Length, ft
Before plenum reduction				
1	40	3	30	70
2	40	2	20	60
3	40	1	10	50
4	40	0	0	40
After plenum reduction				
5	40	2	20	60
6	40	1	10	50
7	40	0	0	40

Figure A28-51 Duct system example showing equivalent lengths. (Copyright by the American Society of Heating, Refrigerating, and Air-Conditioning Engineers, Inc. Used by permission.)

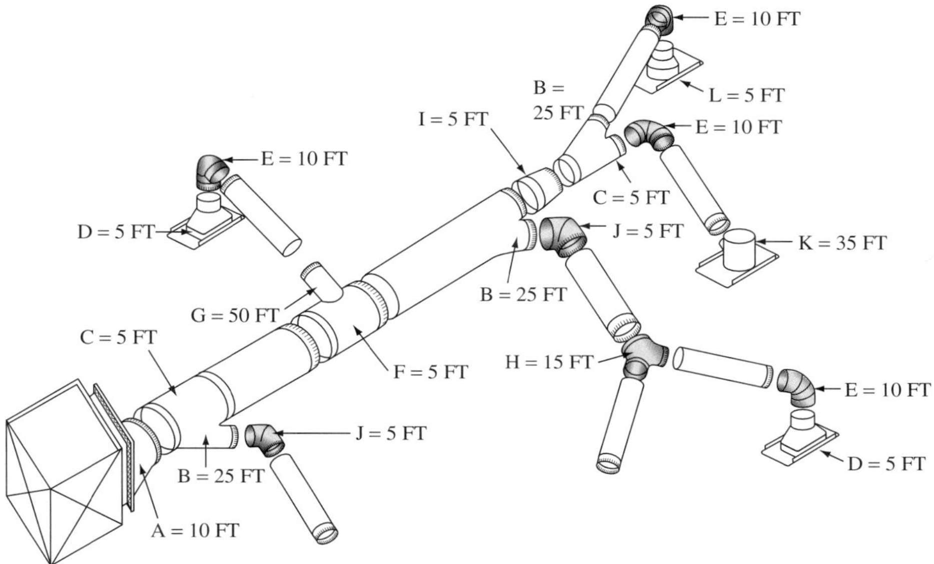

Figure A28-52 Equivalent length of round supply-system fittings. (Copyright by the American Society of Heating, Refrigerating, and Air-Conditioning Engineers, Inc. Used by permission.)

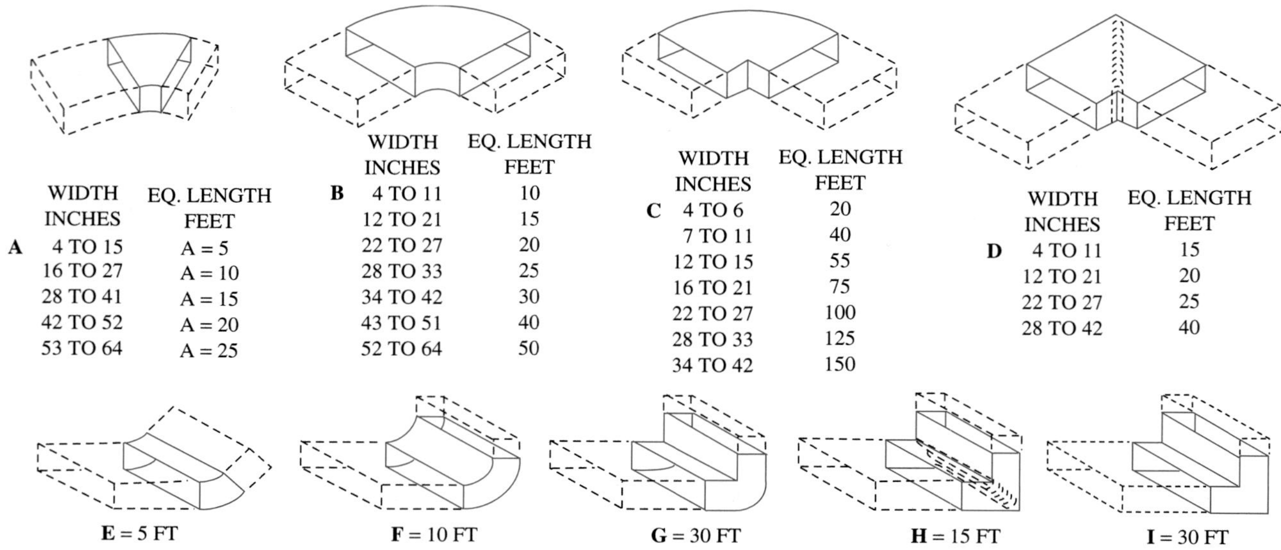

WIDTH INCHES
EQ. LENGTH FEET
A 4 TO 15 A = 5
16 TO 27 A = 10
28 TO 41 A = 15
42 TO 52 A = 20
53 TO 64 A = 25
B 4 TO 11 10
12 TO 21 15
22 TO 27 20
28 TO 33 25
34 TO 42 30
43 TO 51 40
52 TO 64 50
C 4 TO 6 20
7 TO 11 40
12 TO 15 55
16 TO 21 75
22 TO 27 100
28 TO 33 125
34 TO 42 150
D 4 TO 11 15
12 TO 21 20
22 TO 27 25
28 TO 42 40
E = 5 FT
F = 10 FT
G = 30 FT
H = 15 FT
I = 30 FT

Figure A28-53 Equivalent lengths of angles and elbows for trunk ducts. (Copyright by the American Society of Heating, Refrigerating, and Air-Conditioning Engineers, Inc. Used by permission.)

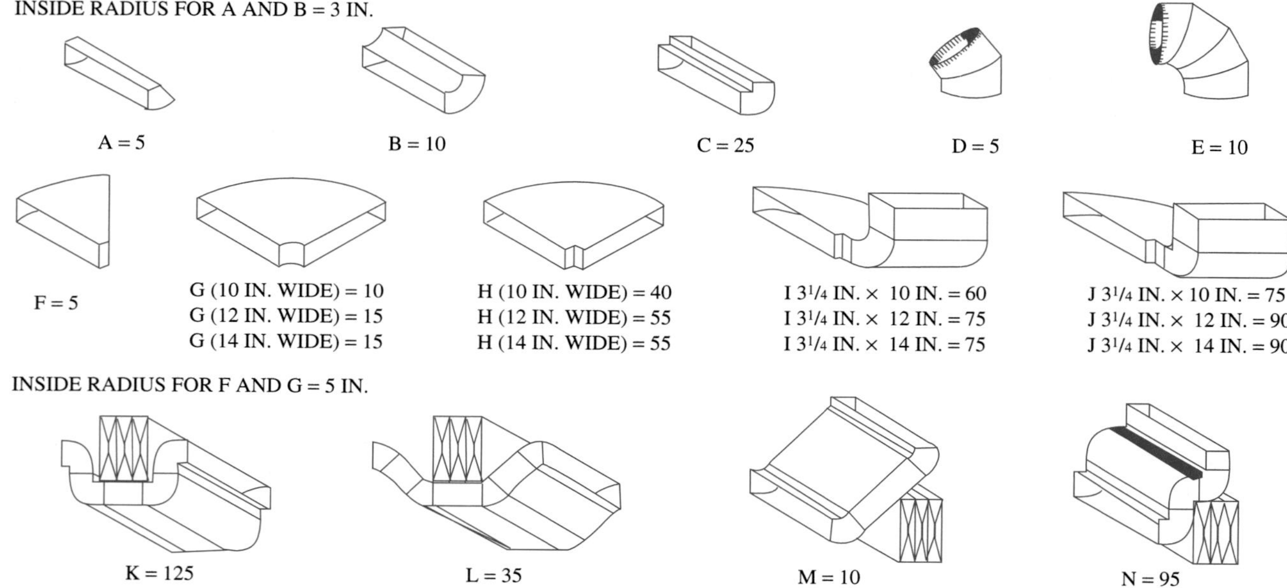

Figure A28-54 Equivalent lengths of angles and elbows for branch ducts. (Copyright by the American Society of Heating, Refrigerating, and Air-Conditioning Engineers, Inc. Used by permission.)

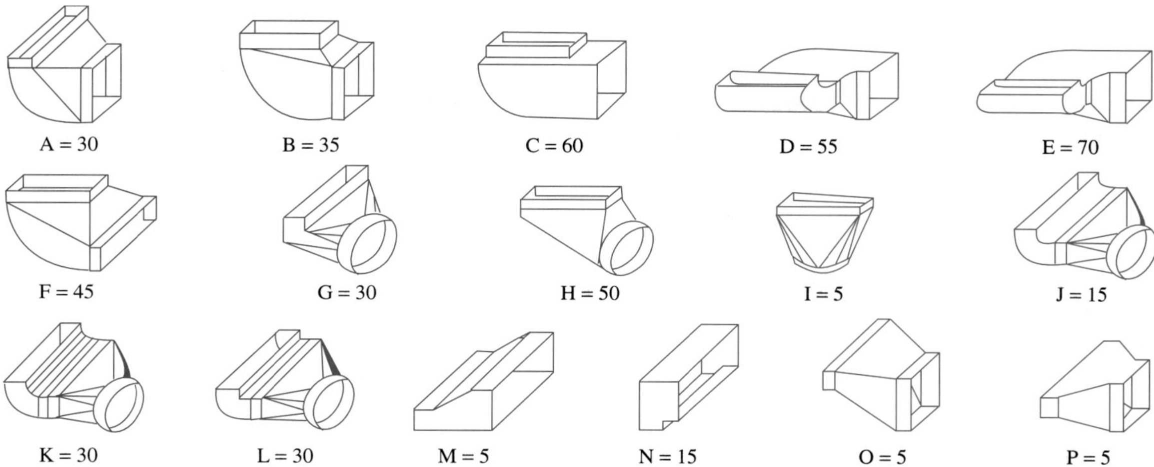

Figure A28-55 Equivalent lengths of boot fittings. (Copyright by the American Society of Heating, Refrigerating, and Air-Conditioning Engineers, Inc. Used by permission.)

14. If velocity exceeds maximum recommended values, increase the size and specify roundup damper.
15. Convert the round duct to rectangular, where needed (Fig. A28-58, p. 568).

Sizing the Supply Trunk System

16. Determine the branch supply with the longest total effective length, and from this, determine the static pressure to size the supply trunk duct system. For example:

$$\text{Supply trunk design static pressure rate} = \frac{100\text{ (Total supply static pressure available)}}{\text{Longest effective length of branch duct supplies}}$$

$$0.15 \times \frac{100}{134\text{ (Branch } L)} = 0.112 \text{ in. water/100 ft}$$

17. Total the heating airflow rate and the cooling airflow rate for each trunk duct section. Select the larger of the two flow rates for each section of duct between roundups or groups of roundups.
18. Design each supply trunk duct section by entering the friction chart at the supply trunk static pressure (0.112) and sizing each trunk section for the appropriate air volume handled by that section of duct. (Trunk 1—400 cfm; duct size = 9.5 in. at 800 fpm.)

Trunks should be checked for size after each roundup and reduced, as required, to maintain velocity above branch duct design velocity.

19. Convert round duct size to rectangular, where needed.

Sizing the Return-air System

20. Select the number of return-air openings to be used.
21. Determine the volume of air that will be returned by each of the return-air openings.
22. From step 11, select the return-air static pressure.
23. Determine the static pressure available per 100 ft effective length for each return run, and design the same as for the supply trunk system. For example:

$$\text{Return trunk design static pressure} = \frac{100\text{ (Total return static pressure available)}}{\text{Longest effective length of return duct runs}}$$

$$0.05 \times \frac{100}{54} = 0.093 \text{ in. water/100 ft}$$

The trunk design for the return air is the same as the trunk design for the supply system.

Example: Return duct in Fig. A28-47 (p. 561) is as follows:

Enter friction chart at 900 cfm and 0.093 in. of water. Duct size is 13.1 in. The velocity, however, is too high at 950 fpm, so use a larger duct.

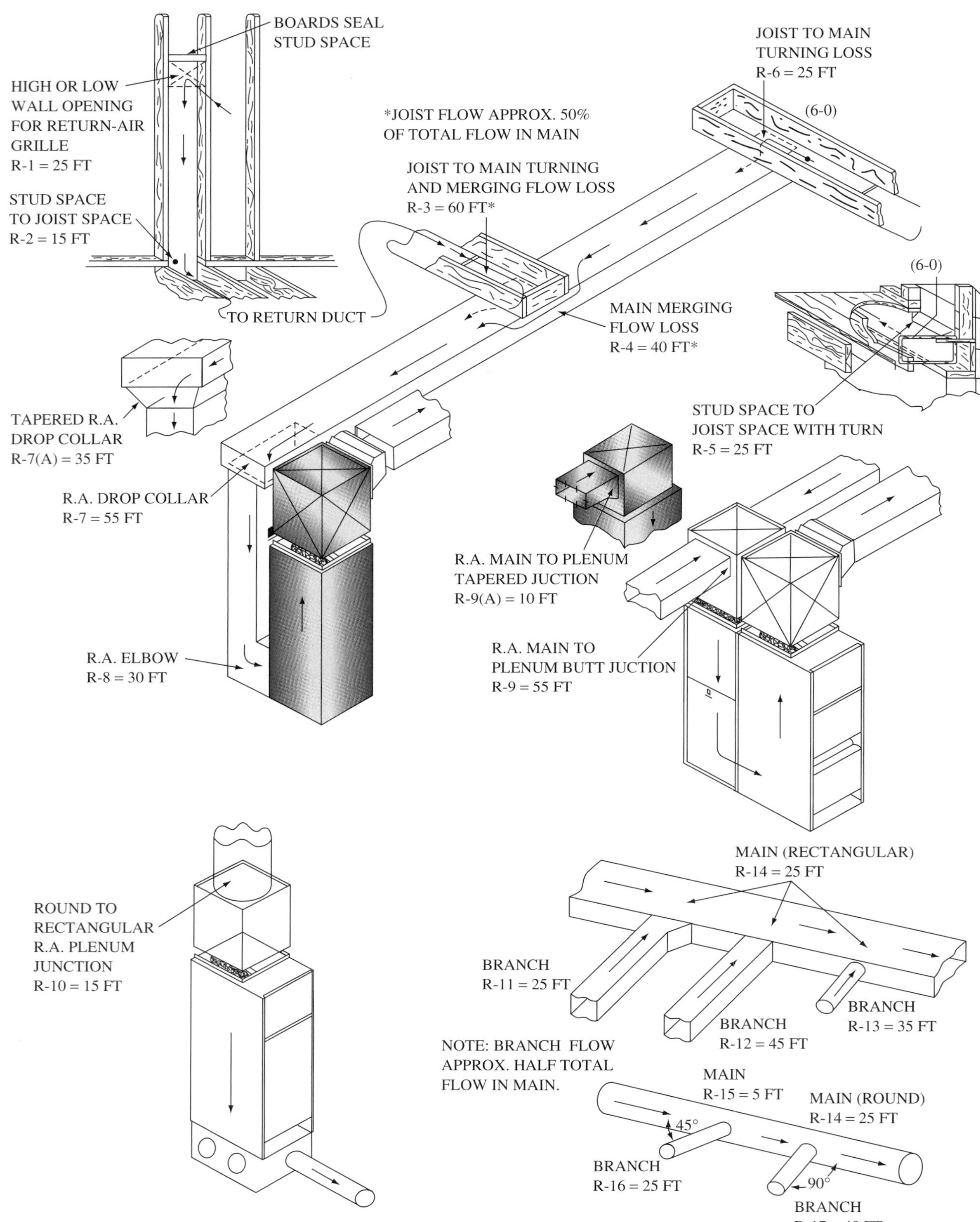

Figure A28-56 Equivalent lengths of special return fittings. (Copyright by the American Society of Heating, Refrigerating, and Air-Conditioning Engineers, Inc. Used by permission.)

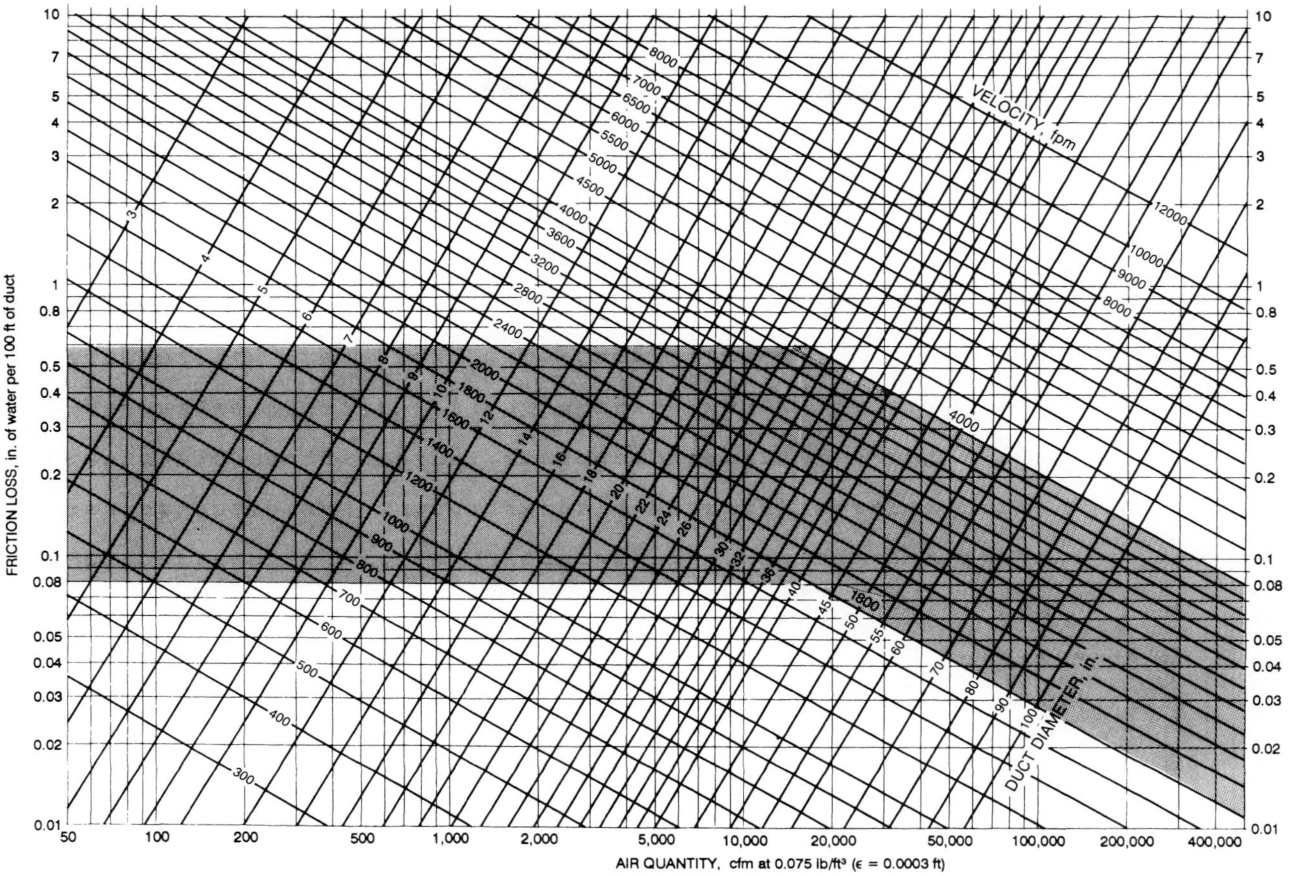

Figure A28-57 Friction chart for sizing ductwork. (Copyright by the American Society of Heating, Refrigerating, and Air-Conditioning Engineers, Inc. Used by permission.)

Lgth Adj[b]	Length of One Side of Rectangular Duct (a), in.																
	4.0	4.5	5.0	5.5	6.0	6.5	7.0	7.5	8.0	9.0	10.0	11.0	12.0	13.0	14.0	15.0	16.0
3.0	3.8	4.0	4.2	4.4	4.6	4.7	4.9	5.1	5.2	5.5	5.7	6.0	6.2	6.4	6.6	6.8	7.0
3.5	4.1	4.2	4.6	4.8	5.0	5.2	5.3	5.5	5.7	6.0	6.3	6.5	6.8	7.0	7.2	7.5	7.7
4.0	4.4	4.6	4.9	5.1	5.3	5.5	5.7	5.9	6.1	6.4	6.7	7.0	7.3	7.6	7.8	8.0	8.3
4.5	4.6	4.9	5.2	5.4	5.7	5.9	6.1	6.3	6.5	6.9	7.2	7.5	7.8	8.1	8.4	8.6	8.8
5.0	4.9	5.2	5.5	5.7	6.0	6.2	6.4	6.7	6.9	7.3	7.6	8.0	8.3	8.6	8.9	9.1	9.4
5.5	5.1	5.4	5.7	6.0	6.3	6.5	6.8	7.0	7.2	7.6	8.0	8.4	8.7	9.0	9.3	9.6	9.9

Lgth Adj[b]	Length of One Side of Rectangular Duct (a), in.																				Lgth Adj[b]
	6	7	8	9	10	11	12	13	14	15	16	17	18	19	20	22	24	26	28	30	
6	6.6																				6
7	7.1	7.7																			7
8	7.6	8.2	8.7																		8
9	8.0	8.7	9.3	9.8																	9
10	8.4	9.1	9.8	10.4	10.9																10
11	8.8	9.5	10.2	10.9	11.5	12.0															11
12	9.1	9.9	10.7	11.3	12.0	12.6	13.1														12
13	9.5	10.3	11.1	11.8	12.4	13.1	13.7	14.2													13
14	9.8	10.8	11.4	12.2	12.9	13.5	14.2	14.7	15.3												14
15	10.1	11.0	11.8	12.6	13.3	14.0	14.6	15.3	15.8	16.4											15
16	10.4	11.3	12.2	13.0	13.7	14.4	15.1	15.7	16.4	16.9	17.5										16
17	10.7	11.6	12.5	13.4	14.1	14.9	15.6	16.2	16.8	17.4	18.0	18.6									17
18	11.0	11.9	12.9	13.7	14.5	15.3	16.0	16.7	17.3	17.9	18.5	19.1	19.7								18
19	11.2	12.2	13.2	14.1	14.9	15.7	16.4	17.1	17.8	18.4	19.0	19.6	20.2	20.8							19
20	11.5	12.6	13.5	14.4	15.2	16.0	16.8	17.5	18.2	18.9	19.5	20.1	20.7	21.3	21.9						20
22	12.0	13.0	14.1	15.0	15.9	16.8	17.6	18.3	19.1	19.8	20.4	21.1	21.7	22.3	22.9	24.0					22
24	12.4	13.5	14.6	15.6	16.5	17.4	18.3	19.1	19.9	20.6	21.3	22.0	22.7	23.3	23.9	25.1	26.2				24
26	12.8	14.0	15.1	16.2	17.1	18.1	19.0	19.8	20.6	21.4	22.1	22.9	23.5	24.2	24.9	26.1	27.3	28.4			26
28	13.2	14.5	15.6	16.7	17.7	18.7	19.6	20.5	21.3	22.1	22.9	23.7	24.4	25.1	25.8	27.1	28.3	29.5	30.6		28
30	13.6	14.9	16.1	17.2	18.3	19.3	20.2	21.1	22.0	22.9	23.7	24.4	25.2	25.9	26.6	28.0	29.3	30.5	31.7	32.8	30
32	14.0	15.3	16.5	17.7	18.8	19.8	20.8	21.8	22.7	23.5	24.4	25.2	26.0	26.7	27.5	28.9	30.2	31.5	32.7	33.9	32
34	14.4	15.7	17.0	18.2	19.3	20.4	21.4	22.4	23.3	24.2	25.1	25.9	26.7	27.5	28.3	29.7	31.0	32.4	33.7	34.9	34
36	14.7	16.1	17.4	18.6	19.8	20.9	21.9	22.9	23.9	24.8	25.7	26.6	27.4	28.2	29.0	30.5	32.0	33.3	34.6	35.9	36
38	15.0	16.5	17.8	19.0	20.2	21.4	22.4	23.5	24.5	25.4	26.4	27.2	28.1	28.9	29.8	31.3	32.8	34.2	35.6	36.8	38
40	15.3	16.8	18.2	19.5	20.7	21.8	22.9	24.0	25.0	26.0	27.0	27.9	28.8	29.6	30.5	32.1	33.6	35.1	36.4	37.8	40
42	15.6	17.1	18.5	19.9	21.1	22.3	23.4	24.5	25.6	26.6	27.6	28.5	29.4	30.3	31.2	32.8	34.4	35.9	37.3	38.7	42
44	15.9	17.5	18.9	20.3	21.5	22.7	23.9	25.0	26.1	27.1	28.1	29.1	30.0	30.9	31.8	33.5	35.1	36.7	38.1	39.5	44
46	16.2	17.8	19.3	20.6	21.9	23.2	24.4	25.5	26.6	27.7	28.7	29.7	30.6	31.6	32.5	34.2	35.9	37.4	38.9	40.4	46
48	16.5	18.1	19.6	21.0	22.3	23.6	24.8	26.0	27.1	28.2	29.2	30.2	31.2	32.2	33.1	34.9	36.6	38.2	39.7	41.2	48
50	16.8	18.4	19.9	21.4	22.7	24.0	25.2	26.4	27.6	28.7	29.8	30.8	31.8	32.8	33.7	35.5	37.2	38.9	40.5	42.0	50
52	17.1	18.7	20.2	21.7	23.1	24.4	25.7	26.9	28.0	29.2	30.3	31.3	32.3	33.3	34.3	36.2	37.9	39.6	41.2	42.8	52
54	17.3	19.0	20.6	22.0	23.5	24.8	26.1	27.3	28.5	29.7	30.8	31.8	32.9	33.9	34.9	36.8	38.6	40.3	41.9	43.5	54
56	17.6	19.3	20.9	22.4	23.8	25.2	26.5	27.7	28.9	30.1	31.2	32.3	33.4	34.4	35.4	37.4	39.2	41.0	42.7	44.3	56
58	17.8	19.5	21.2	22.7	24.2	25.5	26.9	28.2	29.4	30.6	31.7	32.8	33.9	35.0	36.0	38.0	39.8	41.6	43.3	45.0	58
60	18.1	19.8	21.5	23.0	24.5	25.9	27.3	28.6	29.8	31.0	32.2	33.3	34.4	35.5	36.5	38.5	40.4	42.3	44.0	45.7	60
62		20.1	21.7	23.3	24.8	26.3	27.6	28.9	30.2	31.5	32.6	33.8	34.9	36.0	37.1	39.1	41.0	42.9	44.7	46.4	62
64		20.3	22.0	23.6	25.1	26.6	28.0	29.3	30.6	31.9	33.1	34.3	35.4	36.5	37.6	39.6	41.6	43.5	45.3	47.1	64
66		20.6	22.3	23.9	25.5	26.9	28.4	29.7	31.0	32.3	33.5	34.7	35.9	37.0	38.1	40.2	42.2	44.1	46.0	47.7	66
68		20.8	22.6	24.2	25.8	27.3	28.7	30.1	31.4	32.7	33.9	35.2	36.3	37.5	38.6	40.7	42.8	44.7	46.6	48.4	68
70		21.1	22.8	24.5	26.1	27.6	29.1	30.4	31.8	33.1	34.4	35.6	36.8	37.9	39.1	41.2	43.3	45.3	47.2	49.0	70
72			23.1	24.8	26.4	27.9	29.4	30.8	32.2	33.5	34.8	36.0	37.2	38.4	39.5	41.7	43.8	45.8	47.8	49.6	72
74			23.3	25.1	26.7	28.2	29.7	31.2	32.5	33.9	35.2	36.4	37.7	38.8	40.0	42.2	44.4	46.4	48.4	50.3	74
76			23.6	25.3	27.0	28.5	30.0	31.5	32.9	34.3	35.6	36.8	38.1	39.3	40.5	42.7	44.9	47.0	48.9	50.9	76
78			23.8	25.6	27.3	28.8	30.4	31.8	33.3	34.6	36.0	37.2	38.5	39.7	40.9	43.2	45.4	47.5	49.5	51.4	78
80			24.1	25.8	27.5	29.1	30.7	32.2	33.6	35.0	36.3	37.6	38.9	40.2	41.4	43.7	45.9	48.0	50.1	52.0	80
82				26.1	27.8	29.4	31.0	32.5	34.0	35.4	36.7	38.0	39.3	40.6	41.8	44.1	46.4	48.5	50.6	52.6	82
84				26.4	28.1	29.7	31.3	32.8	34.3	35.7	37.1	38.4	39.7	41.0	42.2	44.6	46.9	49.0	51.1	53.2	84
86				26.6	28.3	30.0	31.6	33.1	34.6	36.1	37.4	38.8	40.1	41.4	42.6	45.0	47.3	49.6	51.7	53.7	86
88				26.9	28.6	30.3	31.9	33.4	34.9	36.4	37.8	39.2	40.5	41.8	43.1	45.5	47.8	50.0	52.2	54.3	88
90				27.1	28.9	30.6	32.2	33.8	35.3	36.7	38.2	39.5	40.9	42.2	43.5	45.9	48.3	50.5	52.7	54.8	90
92					29.1	30.8	32.5	34.1	35.6	37.1	38.5	39.9	41.3	42.6	43.9	46.4	48.7	51.0	53.2	55.3	92
96					29.6	31.4	33.0	34.7	36.2	37.7	39.2	40.6	42.0	43.3	44.7	47.2	49.6	52.0	54.2	56.4	96

Figure A28-58 Circular equivalents of rectangular duct for equal friction and capacity.

Lgth Adj[b]	Length of One Side of Rectangular Duct (*a*), in.																				Lgth Adj[b]
	32	34	36	38	40	42	44	46	48	50	52	56	60	64	68	72	76	80	84	88	
32	35.0																				32
34	36.1	37.2																			34
36	37.1	38.2	39.4																		36
38	38.1	39.3	40.4	41.5																	38
40	39.0	40.3	41.5	42.6	43.7																40
42	40.0	41.3	42.5	43.7	44.8	45.9															42
44	40.9	42.2	43.5	44.7	45.8	47.0	48.1														44
46	41.8	43.1	44.4	45.7	46.9	48.0	49.2	50.3													46
48	42.6	44.0	45.3	46.6	47.9	49.1	50.2	51.4	52.5												48
50	43.6	44.9	46.2	47.5	48.8	50.0	51.2	52.4	53.6	54.7											50
52	44.3	45.7	47.1	48.4	49.7	51.0	52.2	53.4	54.6	55.7	56.8										52
54	45.1	46.5	48.0	49.3	50.7	52.0	53.2	54.4	55.6	56.8	57.9										54
56	45.8	47.3	48.8	50.2	51.6	52.9	54.2	55.4	56.6	57.8	59.0	61.2									56
58	46.6	48.1	49.6	51.0	52.4	53.8	55.1	56.4	57.6	58.8	60.0	62.3									58
60	47.3	48.9	50.4	51.9	53.3	54.7	60.0	57.3	58.6	59.8	61.0	63.4	65.6								60
62	48.0	49.6	51.2	52.7	54.1	55.5	56.9	58.2	59.5	60.8	62.0	64.4	66.7								62
64	48.7	50.4	51.9	53.5	54.9	56.4	57.8	59.1	60.4	61.7	63.0	65.4	67.7	70.0							64
66	49.4	51.1	52.7	54.2	55.7	57.2	58.6	60.0	61.3	62.6	63.9	66.4	68.8	71.0							66
68	50.1	51.8	53.4	55.0	56.5	58.0	59.4	60.8	62.2	63.6	64.9	67.4	69.8	72.1	74.3						68
70	50.8	52.5	54.1	55.7	57.3	58.8	60.3	61.7	63.1	64.4	65.8	68.3	70.8	73.2	75.4						70
72	51.4	53.2	54.8	56.5	58.0	59.6	61.1	62.5	63.9	65.3	66.7	69.3	71.8	74.2	76.5	78.7					72
74	52.1	53.8	55.5	57.2	58.8	60.3	61.9	63.3	64.8	66.2	67.5	70.2	72.7	75.2	77.5	79.8					74
76	52.7	54.5	56.2	57.9	59.5	61.1	62.6	64.1	65.6	67.0	68.4	71.1	73.7	76.2	78.6	80.9	83.1				76
78	53.3	55.1	56.9	58.6	60.2	61.8	63.4	64.9	66.4	67.9	69.3	72.0	74.6	77.1	79.6	81.9	84.2				78
80	53.9	55.8	57.5	59.3	60.9	62.6	64.1	65.7	67.2	68.7	70.1	72.9	75.4	78.1	80.6	82.9	85.2	87.5			80
82	54.6	56.4	58.2	59.9	61.6	63.3	64.9	66.5	68.0	69.5	70.9	73.7	76.4	79.0	81.5	84.0	86.3	88.5			82
84	55.1	57.0	58.8	60.6	62.3	64.0	65.6	67.2	68.7	70.3	71.7	74.6	77.3	80.0	82.5	85.0	87.3	89.6	91.8		84
86	55.7	57.6	59.4	61.2	63.0	64.7	66.3	67.9	69.5	71.0	72.5	75.4	78.2	80.9	83.5	85.9	88.3	90.7	92.9		86
88	56.3	58.2	60.1	61.9	63.6	65.4	67.0	68.7	70.2	71.8	73.3	76.3	79.1	81.8	84.4	86.9	89.3	91.7	94.0	96.2	88
90	56.8	58.8	60.7	62.5	64.3	66.0	67.7	69.4	71.0	72.6	74.1	77.1	79.9	82.7	85.3	87.9	90.3	92.7	95.0	97.3	90
92	57.4	59.3	61.3	63.1	64.9	66.7	68.4	70.1	71.7	73.3	74.9	77.9	80.8	83.5	86.2	88.8	91.3	93.7	96.1	98.4	92
94	57.9	59.9	61.9	63.7	65.6	67.3	69.1	70.8	72.4	74.0	75.6	78.7	81.6	84.4	87.1	89.7	92.3	94.7	97.1	99.4	94
96	58.4	60.5	62.4	64.3	66.2	68.0	69.7	71.5	73.1	74.8	76.3	79.4	82.4	85.3	88.0	90.7	93.2	95.7	98.1	100.5	96

[a]Table based on $D_e = 1.30(ab)^{0.625}/(a + b)^{0.25}$
[b]Length of adjacent side of rectangular duct (*b*), in.

Figure A28-58 Continued

REVIEW

- Air conditioning—Provides control of temperature, relative humidity, air motion, radiant heat, removal of airborne particles, and contaminating gases.
- Psychrometrics is the study of air and its properties.
- Changes in the properties of air due to conditioning:
  Heating—Adding sensible heat.
  Cooling with a dry evaporator—Reducing sensible heat.
  Humidifying—Adding moisture.
  Heating and humidifying—Used for winter operation.
  Cooling and dehumidification—Used for summer operation.
  Cooling and humidifying—Evaporative cooling.
- Performance of air-conditioning equipment can be checked using the psychrometric chart by knowing:
  Inside air—dry-bulb and wet-bulb temperatures
  Outside air—dry-bulb and wet-bulb temperatures
  Air circulated by the blower (cfm)
  Percent (or cfm), ventilation air.
- Forced-air system components:
  Blower
  Filters
  Humidifier
  Supply/return ducts
  Ventilation air
  Registers and grilles
- Types of duct pressure:
  Static pressure—The force of the air against the sides of a vessel or duct.
  Total pressure—Static pressure plus velocity pressure.
  Velocity pressure—The difference between the total pressure and static pressure. Useful in measuring velocity of air in a duct. Manometers are cali-

brated to read duct velocity direct in feet per minute (fpm).

- Types of fans:
  Centrifugal (squirrel cage)—Backward curved or forward curved
  Axial—vane and propeller type
- Types of fan drives:
  Belt/pulley—Speed of fan is determined by the ratio of fan and motor pulley.
  Direct—Multiple speed motors.
- System design:
  Calculate heating and cooling loads.
  Determine preliminary duct-work location.
  Determine heating- and cooling-unit location.
  Select accessory equipment.
  Select control components.
  Determine maximum airflow.
  Select heating and cooling equipment.
  Finalize duct design and size system.
  Select supply- and return-air grilles.
- Selecting equipment:
  A furnace selection should either match or be slightly larger than the load.
- Duct design:
  Keep main ducts as straight as possible
  Streamline transitions
  Design elbows with an inside radius of at least ⅓ of the duct width or use turning vanes
  Make ducts tight and sealed to limit air loss
  Insulate or line ducts to conserve energy and limit noise.
  Locate takeoffs at least 4 ft downstream from a fan or transition joint.
  Separate air moving equipment from ducts using flexible connections to isolate noise.
  Install a volume damper in each branch for balancing air flow.
- Maximum duct velocities:
  Main ducts — 700 to 900 fpm
  Branch ducts — 600 fpm
  Branch risers — 500 fpm
- Ducts are sized to greatest cfm, typically cooling.
- Duct materials:
  Sheet metal
  Fiberglass ductboard
  Aluminum
  Spiral metal duct
  Flexible duct.

Problems and Questions

1. What is the vapor pressure of a pan of water at 70°F?
   a. 0.74 in. Hg c. 0.54 in. Hg.
   b. 0.64 in. Hg. d. 0.44 in. Hg.
2. What is the specific heat of air at standard conditions?
   a. 0.34 Btu/lb c. 0.14 Btu/lb
   b. 0.24 Btu/lb d. 0.04 Btu/lb
3. What is the weight of air per cubic foot at standard conditions?
   a. 0.075 lb/ft^3 c. 0.055 lb/ft^3
   b. 0.065 lb/ft^3 d. 0.045 lb/ft^3
4. How many grains of moisture equal one pound?
   a. 9000 c. 7000
   b. 8000 d. 6000
5. Is the wet-bulb temperature of air equal to the dew-point temperature at 100% relative humidity?
   a. Yes b. No.
6. Is it possible to heat and dehumidify air at the same time?
   a. Yes b. No
7. If the sensible heat is 10,000 Btu/hr and the latent heat is 5,000 Btu/hr, what is the Sensible Heat Ratio?
   a. 0.37 c. 0.57
   b. 0.47 d. 0.67
8. In heating, if the heat added is 80,000 Btu/hr and the temperature rise is 40°F, what is the cfm?
   a. 1752 c. 1952
   b. 1852 d. 2052
9. According to the fan laws, if the speed of the fan is increased 10%, how much is the cfm increased?
   a. 10% c. 33%
   b. 20% d. 46%
10. ASHRAE standards set the limit for oversizing cooling equipment as:
    a. 50% c. 33%
    b. 40% d. 25%
11. The human body has a remarkable ability to adjust to temperature change. True or false?
12. The study of air and its properties is called air conditioning. True or false?
13. All air contains some moisture or humidity. True or false?
14. The relation of the wet-bulb temperature to the dry-bulb temperature is a measure of the relative humidity. True or false?
15. A cooling coil in the summer will cool and dehumidify the air. True or false?
16. Air conditioning provides control of:
    a. Temperature.
    b. Relative humidity and air motion.
    c. Radiant heat and removal of airborne particles and contaminating gases.
    d. All of the above.
17. Evaporative cooling is the process of:
    a. Evaporating refrigerant to cool the air.
    b. Evaporating water moisture to cool the air.
    c. Outside air for free cooling.
    d. None of the above.

18. The components of a residential air conditioner include:
    a. Blower and filters.
    b. Humidifier.
    c. Supply and return air ducts, and grills/registers.
    d. All of the above.
19. Total pressure within the duct is the sum of the:
    a. Velocity plus static pressure.
    b. Velocity plus fan pressure.
    c. Velocity minus static pressure.
    d. None of the above.
20. Furnace blowers are installed:
    a. Up-flow.
    b. Down-flow.
    c. Horizontal-flow.
    d. All of the above.

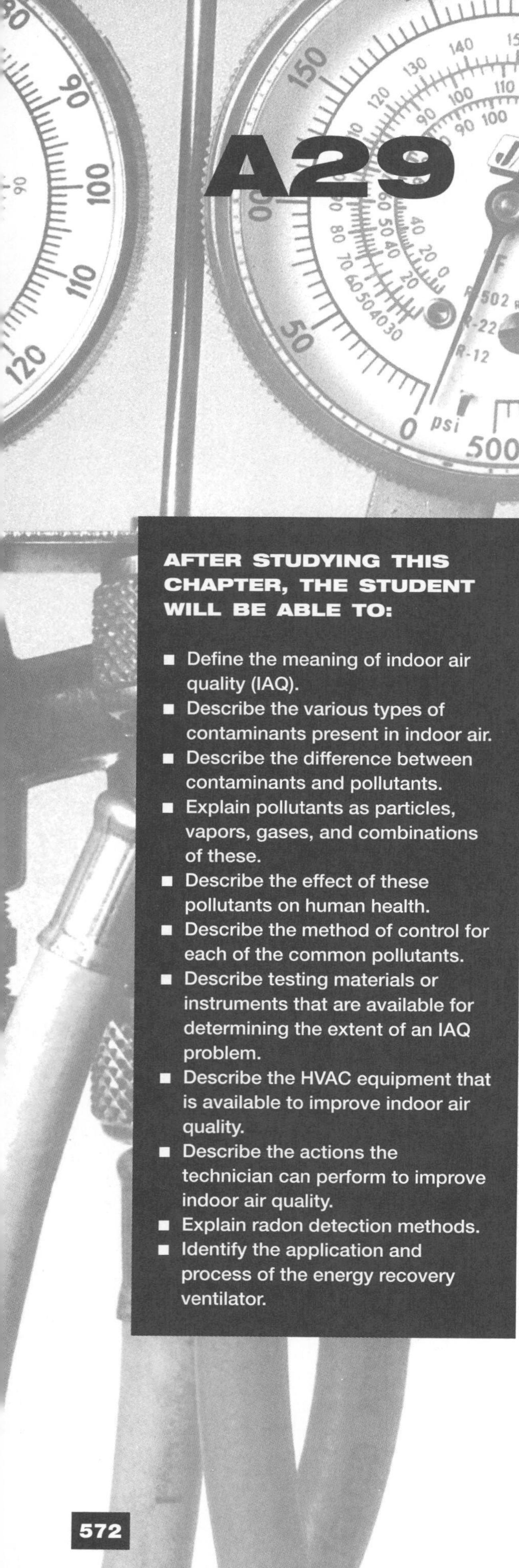

A29 Indoor Air Quality

AFTER STUDYING THIS CHAPTER, THE STUDENT WILL BE ABLE TO:

- Define the meaning of indoor air quality (IAQ).
- Describe the various types of contaminants present in indoor air.
- Describe the difference between contaminants and pollutants.
- Explain pollutants as particles, vapors, gases, and combinations of these.
- Describe the effect of these pollutants on human health.
- Describe the method of control for each of the common pollutants.
- Describe testing materials or instruments that are available for determining the extent of an IAQ problem.
- Describe the HVAC equipment that is available to improve indoor air quality.
- Describe the actions the technician can perform to improve indoor air quality.
- Explain radon detection methods.
- Identify the application and process of the energy recovery ventilator.

A29-1 OVERVIEW

Air is composed of 21% oxygen, 78% nitrogen and 1% carbon dioxide, argon, and trace amounts of other rare gases. In addition, it contains varying amounts of water vapor and small quantities of microscopic solid matter called *permanent atmospheric particulates.*

The availability of oxygen is essential to human health. When we breath, we inhale oxygen and exhale carbon dioxide. At atmospheric pressure, oxygen concentrations of less than 12% and carbon dioxide concentrations of greater than 5% are dangerous—even for short periods.

The permanent atmospheric particulates arise from natural processes such as wind erosion, sea-spray evaporation, volcanic eruption, and biological processes. They usually create far less contamination than man-made activities. Some of the man-made activities that cause air contamination are power plant operation, industrial processing, and various types of transportation and agricultural activities. These suspended contaminants can be roughly classified as solids, liquids, and gases.

The size of the suspended particles is important, since it relates to the method of removal, usually through filtration. Fig. A29-1 gives the size in microns of various common particulates.

A micron is a millionth of a meter, or one thousandth of a millimeter. To give some conception of the sizes shown, note that human hairs range in size from 25 to 300 microns. Particles 10 microns or more are visible to the naked eye. Smaller particles such as smoke or clouds are visible only in high concentrations. Health authorities are concerned with particles 2 microns or less since they can be retained in the lungs. Note the size range of a few common pollutants: pollen, 10 to 100 microns; bacteria, 0.4 to 5.0 microns; tobacco smoke 0.1 to 0.3 microns; and viruses 0.003 to 0.06 microns.

A distinction needs to be made between a contaminant and a pollutant. A *contaminant* is any unwanted material that gets into the air. A contaminant may or may not be destructive to human health. A *pollutant* is a substance

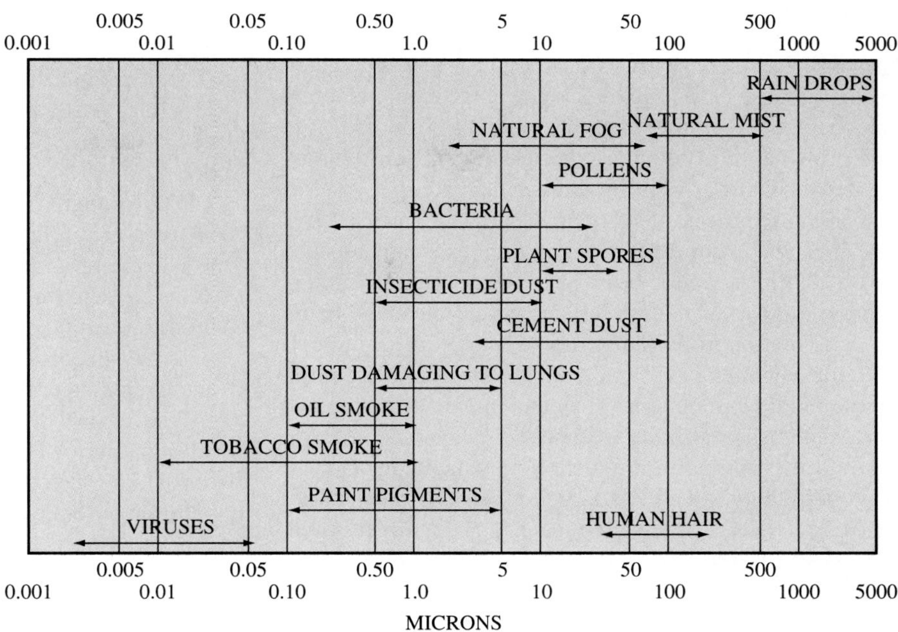

Figure A29-1 Sizes and characteristics of airborne solids and liquids.

which has entered the air and can cause health problems. The study of air quality is particularly concerned with pollutants. The technician should know their source, the amount the body will tolerate and how to control them.

Indoor air quality is the condition of air in an enclosed space, with respect to the presence of pollutants. It is desirable to maintain pollutant-free air or at least an acceptable pollutant level.

The two general methods of pollutant control are:

1. Elimination at the source
2. Reduction to an acceptable level.

A29-2 THE POLLUTANTS

The nature of pollutants and their effect on health can be separated into three groups that have related characteristics:

A. Particles
   1. Bioaerosols
      a) Allergens: pollen, fungi, mold spores, insect parts, and feces
      b) Pathogens: bacteria and viruses (almost always carried in or on other particle matter)
   2. Respirable particles: 10 microns or less in size
   3. Asbestos fibers
B. Vapors and gases
   1. Formaldehyde (HCHO)
   2. Radon: Naturally occurring soil gas produced from radioactive decay of radium-uranium, widely found.
   3. Volatile organic compounds (VOCs)
C. Combination particles/vapors
   1. Environmental tobacco smoke
   2. Combustion products

A29-2.1 Particles

Bioaerosols (Allergens and Pathogens)

Bioaerosols are airborne biological agents. *Allergens* relate to those microorganisms that cause allergies and include pollen and fungi (yeasts and molds). *Pathogens* are microorganisms such as bacteria and viruses that cause diseases.

Biological growth sources of these microorganisms include wet insulation or moisture-laden dirty duct work, ceiling tile, furniture, and stagnant water in air conditioners, humidifiers, cooling towers, or cooling coils. Pets also bring disease-bearing microorganisms into the house. Frequently the pollutants settle in the air-distribution system of the house since they are too small to be removed by the ordinary filter.

These biological agents can cause sneezing, watery eyes, coughing, shortness of breath, lethargy, fever, and digestive problems. Prolonged exposure to mold-spore allergens may wear down the immune system and increase an individual's susceptibility to infectious disease.

Methods of control involve examining potential sources such as porous wet surfaces and water-damaged material. Most remedies include good maintenance or preventive maintenance procedures, such as cleaning filters and wet areas located in the airstream, replacing water-damaged carpets and insulation, maintaining relative humidities between 40 and 60%, and cleaning and disinfecting drain pans.

Respirable Particles

These particles are usually considered to be 10 microns or smaller and are carried into the system through breathing. They can be either biological or non-biological. The non-biological particles often are carriers for bioaerosols and can deliver harmful substances to critical areas.

Common sources of respirable particles can be kerosene heaters, humidifiers, wood stoves, and tobacco smoke. Particles can be brought in the house on clothing, through the use of household sprays and cleaners, or by the deterioration of building materials and furnishings.

Particles and the substances they carry can cause nose, throat, and lung infections. They can impair breathing and increase the susceptibility to cancer.

Control is through introduction of outside air and use of good-quality air filters. *High-efficiency particulate-arresting* (HEPA) filters and *electrostatic air cleaners* (EACs) can remove particulates very effectively. Appropriate filter maintenance, by cleaning and replacing at regular intervals, is essential.

Asbestos

Asbestos is the name given to a group of naturally occurring silicates that occur in fiber bundles having unusual tensile strength and fire resistance. About 95% of the material is a substance called chrysolite. Once in place the material does not deteriorate rapidly. The danger occurs to the installer or to someone who damages the material when removing it. Its primary usages are for building materials, boiler insulation, sprayed-on fireproofing coatings, pipe insulation, and floor and ceiling tiles.

A number of diseases have been associated with the use of asbestos:

1. *Asbestosis,* or scarring of the lungs, leading to respiratory failure.
2. *Mesothelioma,* a damaging effect to the linings of the lungs and abdomen.
3. *Lung cancer,* a chronic, progressive, generally incurable disease.

There are no immediate symptoms of asbestos exposure. Most deaths have occurred as a result of high exposure in industrial locations. Studies of exposures in schools and homes have shown the exposure to be well below critical levels of 0.001 fiber/ml.

The primary health risk is to operations and maintenance people who, in the course of their duties, disturb the asbestos-containing materials. Therefore, the best method of control may be to not disturb it.

A29-2.2 Vapors and Gases

Formaldehyde

Formaldehyde is a colorless gas at room temperatures and has a pungent odor. It is used in the manufacturing of a number of building products and usually emerges as a pollutant during construction or remodeling.

It is used in the manufacturing of plywood, fiberboard, paneling, particleboard, urea-foam insulation, fiberglass, and wallboard. It also is a product of incomplete combustion.

The strength of the formaldehyde as well as the length of exposure influence the response to the material. Even low concentrations, besides the disagreeable odor, can cause eye and throat irritation and a biting sensation in the nose. A medium dose can cause a flow of tears in the eyes. A strong exposure can cause inflammation of the lungs and even danger of death.

The critical level is usually considered 0.1 parts per million (ppm). Homes older than 5 years without urea-formaldehyde insulation have levels in the 0.02 to 0.05 ppm range. Special equipment is required to monitor the concentration accurately; however, the odor can indicate most critical levels.

Control is usually accomplished through careful selection of building materials and through an "off-gasing" procedure. This is simply allowing time, after installation, for the undesirable vapors or gases to be emitted from the construction materials before the area is occupied. This process can be greatly speeded up by a "bake-out" procedure. A bake-out consists of three procedures: the indoor temperature is raised, maximum ventilation is provided, and adequate time is allowed for the process to do its work.

Materials that emit formaldehyde can also be coated to prevent leakage of gas. The National Aeronautics and Space Administration (NASA) found that ordinary house plants can significantly reduce formaldehyde levels.

Radon

Radon is a colorless odorless gas that is always present to some degree in the air supply. It is formed by the radioactive decay of radium and uranium. It produces radiation or radioactive elements which can be a major health concern. These elements attach to particles in the air and are inhaled during breathing. About 30% of the charged particles are retained in the lungs, providing a source of injury to the tissues.

The principal source of radon is from soil. The gas enters a building through cracks and crevasses in the foundation or through piping and other conduits in the foundation directly in contact with the soil. Tight construction and insulation "capture" the gases. This can be particularly significant if construction occurred on or near fill sites.

No immediate symptoms are noticed. Radon decay products increase the susceptibility to lung cancer. The critical level is very small, about 4 trillionths parts per liter.

The most effective means of controlling radon is to prevent it from entering the building. Cracks in the basement floor must be sealed. Foundations should provide some means of ventilating soil gases outside the structure. In residences, mechanical ventilation of the crawl space has proved satisfactory. In one location, increased ventilation at the first-floor level and exhaust out of the basement took care of the problem.

Radon Detectors

Kits for testing for the presence of radon are readily available, relatively inexpensive, and easy to use. Due to the dangers involved in breathing radon vapors into the lungs, it is recommended that every house have a radon test.

A number of manufacturers make these kits, which are often sold at local hardware stores or home centers. *Consumer Reports,* July 1995, lists seven kits which have been tested and found satisfactory.

There are two main types of detectors: short-term and long-term.

Short-term detectors track the radon level for up to seven days. These consist of a carbon-containing canister, envelope, or tray. The device is placed in the lowest occupied living area. It is then opened for a specified time to expose activated charcoal granules that trap the radon. The device is re-sealed and sent to a lab for analysis.

If the test results received back from the lab show results higher than the acceptable level, further testing using long-term detectors could be used to analyze the problem. Remedial action should be taken immediately, however, whenever significant radon levels are identified.

Long-term detectors smooth out quick fluctuations brought about by changes in temperature, barometric pressure and air currents. Instead of charcoal, these detectors cause the alpha particles emitted from the radioactive radon to leave imprints on a small piece of specially formulated plastic.

Volatile Organic Compounds (VOCs)

Some VOCs are always present in the indoor atmosphere. They constitute vapors which are emitted from many household products. Good examples are formaldehyde and products of tobacco smoke, such as benzene and phenols.

Some of the common sources are building materials, furniture materials, photocopying materials, disinfectants, gasoline, paints, and refrigerants. The ventilation system itself may bring in outside pollutants.

Some of the health effects include eye irritation, sore throat, nausea, drowsiness, fatigue and headaches. Studies have shown a relationship between certain VOCs and respiratory ailments, allergic reactions, heart disease, and cancer. The critical level seems to be in the area of 3 mg/m^3.

Where the source can be determined and removed or isolated from the building, this is the best method of control. Where this is not practical, increased ventilation is used to control the concentrations. Selective purchase of construction materials is always helpful. In new or remodelled construction, a bake-out procedure, previously described, can be effective.

VOC monitors are available with a cable for connecting the sensor to a datalogger or computer. The instruments will monitor VOCs and CO_2 at the same time. Additional temperature and humidity probes and specific gas monitors are available.

A29-2.3 Combination Pollutants

Environmental Tobacco Smoke

Environmental tobacco smoke is the emission from the burning end of a cigarette, cigar, or pipe. It is the second-hand smoke such as is exhaled by the smoker. It is sometimes termed “involuntary smoking”. It contains irritating gases and tar products. Since tobacco does not burn completely, pollutants are given off such as nitrogen oxide, sulfur dioxide, ammonia, formaldehyde, benzene, and arsenic, to name a few.

Second-hand smoking has been shown to increase the risk of cancer in adults. Studies have indicated that involuntary smoking increases respiratory illness in children. The Environmental Protection Agency (EPA) has shown that inhaling tobacco smoke can cause permanent damage to the blood cells.

Control of involuntary smoking can take many forms:

1. Eliminate smoking entirely.
2. Confine it to certain areas where separate ventilation is provided.

PRACTICING PROFESSIONAL SERVICES

The Environmental Protection Agency (EPA) and other related agencies are available for extended testing and evaluations of potential air-quality areas. They often provide a free or low-cost on-site air-quality testing and evaluation with follow-up recommendations and procedures to prevent future potential problems. Often, as part of their service, they will also offer literature and brochures to further clarify indoor air-quality issues and concerns.

3. Provide adequate ventilation to dilute the smoke to acceptable levels.
4. Filter the pollutants, using HEPA filters and electronic precipitators.

Since odor and irritation remain where the smoke is produced, ventilation and filtering do not completely solve the problem.

Combustion Products

Combustion products include respirable products, carbon monoxide, nitrogen oxides, and VOCs. Some products are exceedingly small and lodge in the lungs. Carbon monoxide is a colorless, odorless, highly poisonous gas. Nitrous oxides are all irritating gases that can have an impact on human health. Other VOCs include hydrocarbons which act in an additive fashion with other pollutants.

Combustion products are released as a result of incomplete combustion that can occur during the burning of wood, gas, and coal stoves, unvented kerosene heaters, fireplaces during downdraft conditions, and environmental tobacco smoking.

Carbon monoxide reduces the ability of the blood to carry oxygen to the body tissues. Common symptoms are dizziness, dull headache, nausea, ringing in the ears, and pounding of the heart. An extreme dose can cause death. Nitrous oxide can cause irritation to the eyes, nose and throat, respiratory infections, and some lung impairment. Combustion particles can affect lung function. The OSHA-recommended limit is 50 ppm in an 8-hour period or about 40 mg/m^3 in a 1-hour period. Acute exposure can be fatal.

Control includes maintaining and properly adjusting fuel-burning equipment. House plants can serve as a living air purifier for VOCs. When unusually high levels of combustion products are expected, additional ventilation can be used as a temporary measure. Do not use unvented heaters.

A29-3 THE ENERGY RECOVERY VENTILATOR

The energy-recovery ventilator (ERV) is an accessory for a heating and/or cooling system to supply and control the introduction of outside air and to exhaust stale air. It includes an air-to-air heat exchanger that is used to conserve energy (Fig. A29-2).

The heat exchanger usually consists of a desiccant-coated wheel that rotates between the two airstreams. In winter it recovers heat and moisture from the stale air and transfers them to the cold, dry, incoming air. A preheater adds extra heat when needed to the cold entering air. In summer, heat and moisture are removed from incoming

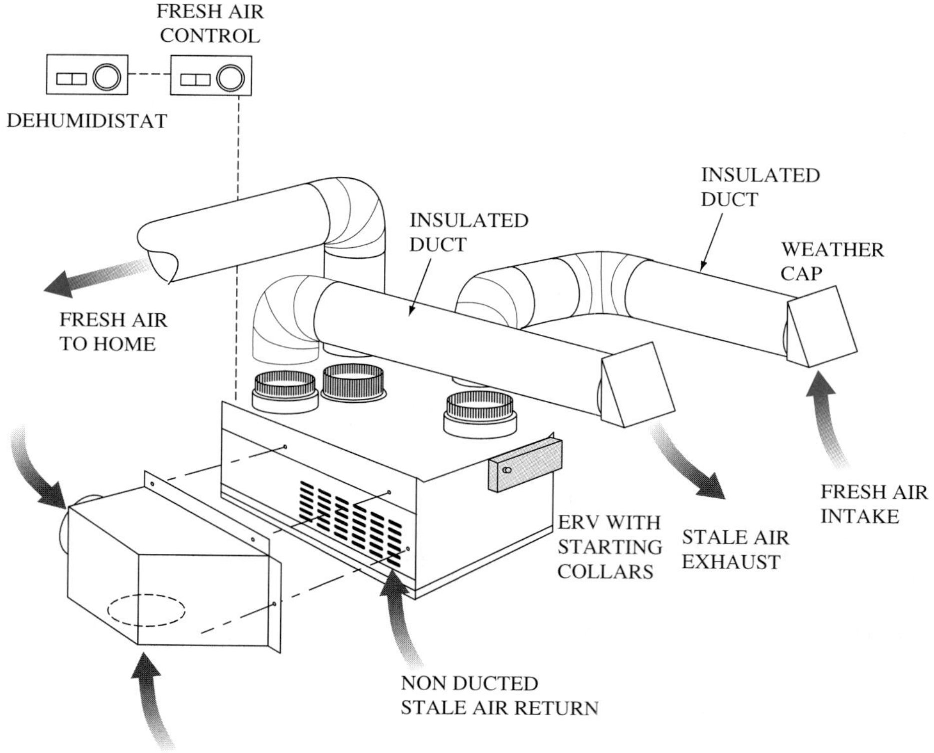

Figure A29-2 Energy-recovery ventilator.

air and transferred to the stale exhaust air. A standard air filter removes most airborne particles down to 5 microns in size, before they enter the ERV unit. These filters are readily replaceable as required.

A *humidistat* is used to control the relative humidity in the occupied space and prevent condensation on windows in winter. An outside air control permits the occupants to adjust the speed of the fan to regulate the ventilation rate.

A29-4 THE SERVICES OF THE HVAC TECHNICIAN

The value of technician training in providing IAQ service can be illustrated by citing a few examples.

EXAMPLE

A school is facing drastically reduced budgets for operation and maintenance. The maintenance staff was requested to do everything in their power to reduce the cost of energy. The first step the staff took was to seal some of the outside air intakes with plastic and duct tape to save heat. This led to a drastic reduction in ventilation throughout the system. There were many complaints throughout the building of health problems associated with lack of proper ventilation.

Coupled with the energy cutbacks, the school also reduced the cleaning budget. One of the first areas they eliminated was cleaning the drain pan underneath the refrigerator in the teacher's lounge. When the service people examined the pan they found an accumulation of about an inch of mold over the entire surface. Every time the refrigerator door was opened and closed some of the mold spores spread into the room. Several people became seriously ill as a result.

EXAMPLE

A real-estate sales person was showing a house in an area where severe flooding had occurred a few years earlier. As they entered the house they noticed a strong odor. It so happened that they were examining a house that had not been properly treated after the flood. As they descended into the basement the prospect's wife became suddenly very ill and had to be practically carried from the house.

A contractor was called in to examine the property and found the source of the smell was rotting material that produced biological pollution.

Technicians with a knowledge of pollution sources are able to prevent serious problems by recognizing dangerous conditions and informing the customer so that the necessary corrective measures can be taken.

A29-4.1 Treatment for Pans and Coils

Condensate pans and cooling coils are two locations likely to contribute to IAQ problems. Cooling coils are usually cleaned with brushes and a proper biocide. When this material is removed it must be taken away from the area before the operation of the system can be resumed.

In addition to cleaning, an anti-foulant needs to be used to reduce the buildup of scale, dirt, and biological matter. The condensate which forms on the coil surface can then wash the surface during normal operation. A product such as First Strike, which is available through Indoor Environmental Quality Alliance (IEQA), serves this purpose. The products used on the coils should be checked for excessive alkalinity or acidity. Either extreme can cause coil pitting.

The condensate pan is fertile ground for many microbial pollutants. The condensate in the pan must drain properly. It is important to check the slope and the tilt of

ON THE JOB

Public and commercial buildings are required to maintain a certain amount of outside air for ventilation purposes. Normally, this outside air is brought into the building through the economizer duct work. When on the job, to verify that the air-conditioning system is working properly, inspect the outside air dampers, protective screening and damper actuators. It is normally the economizer control circuits which modulate the dampers as needed to maintain a pre-determined minimum of fresh air into the building. To verify operation, adjust the set point of the mixed-air controller from fully open to fully closed. The outside air dampers should never close down completely during normal daytime running conditions. If they do close down, readjust the minimum outside air to at least 10% fresh air or whatever is prescribed in the local building code.

PREVENTIVE MAINTENANCE

Evaporator-coil drain pans, vent lines, condensate pumps, and humidifiers are a source of potential indoor air-quality problems if not checked when doing preventive maintenance on air-conditioning and heating equipment. Scale and algae growth can prohibit proper drainage and water flow if not cleaned regularly. Excess algae growth can also contribute to poor indoor air quality and your customer's health. Whenever working on this type of air-conditioning equipment, the technician should wear protective gloves and respirator protection devices while working in these areas to prevent potential contamination.

the pan with respect to the location of the drain. Be sure that the drain is fully functional. The pan drain must be properly trapped so that the fan will not pull air back through the drain and prevent the condensate from draining out.

The pan-cleaning agent must be designed to function for a long period of time to maintain the scale in suspension so it will properly drain out. A pan guard is recommended by IEQA and operates to keep the pan clean. Condensate-pan treatment products should also contain a wetting agent to reduce the possibility of contaminants sticking to the surface and building up. Look for a product with the lowest possible hazard rating on the material safety data sheet (MSDS).

If condensate is fed directly into the waste line, be sure the trap is kept full during long periods of disuse during the winter. This helps to reduce pollution that can migrate into the building from dry traps.

A29-4.2 Measuring Other Sources of Pollution

Organizations like EPA, IEQA and OSHA recognize carbon dioxide (CO_2) as a suitable measurement of IAQ. The CO_2 must be kept down to an acceptable level below 1000 ppm or 800 ppm as recommended by some groups. Basically, CO_2 is an indicator of outside air ventilation. Monitors are now readily available to measure CO_2.

As part of the inspection and service work on any HVAC/R system, the technician should determine that outside air intakes and the rest of the outside air-supply system are working properly.

Since the presence of carbon monoxide (CO) is very serious, it also can be measured by monitors readily available. CO should be kept at levels of a few parts per million or zero levels.

A29-5 PRODUCING THE BEST IAQ

As a special service to valued customers, the technician can provide IAQ inspection and preventive maintenance service. Measurements can be made where critical elements are involved, and corrections made where needed. These services can help the client to provide a cleaner, healthier, and more productive environment.

REVIEW

- Air—Composed of 21% oxygen, 78% nitrogen, and 1% carbon dioxide, argon, and other gases, varying amounts of water vapor, and small quantities of microscopic solid water.
- Oxygen concentrates of less than 12% and carbon dioxide concentrates of greater than 5% are dangerous even for short periods. Acceptable levels of CO_2 are below 1000 ppm or 800 ppm.
- Permanent atmosphere particulates result from natural processes such as wind erosion, sea-spray evaporation, volcanic eruption, and biological processes. They create far less contamination than man-made. Some of the man-made contamination comes from power plant operation, industrial processing, transportation, and agricultural activities.
- Contamination particulates are measured in "microns." A micron is a millionth of a meter. Objects 10 microns or more can be seen. Objects 2 microns or less are health hazards as they can be retained in the human body.
- A contaminant is any unwanted material that gets into the air.
- A pollutant is a substance that has entered the air and can cause health hazards.
- Indoor air quality is the condition of air in an enclosed space, with respect to the presence of pollutants.
- Methods of pollutant control:
  Elimination at the source
  Reduction to an acceptable level
- Pollutants:
  Particles
    Bioaerosols
      Allergens: Pollen, fungi, mold spores, insect parts, and feces
      Pathogen: Bacteria and viruses

Respirable particles: 10 microns or less in size
Asbestos fibers
Fiberglass
Vapor and gases
Formaldehyde (HCHO)
Radon
Volatile organic compounds
Combination particles/vapors
Environmental tobacco smoke
Combustion products

- Control of bioaerosols:
  Examine potential sources.
  Clean filters.
  Clean up wet areas in airstream.
  Replace water-damaged carpets.
  Maintain relative humidity between 40–60%.
  Clean and disinfect drain pans.
- Control of respirable particles:
  Introduction of outside air
  Use of good quality air filters
  High-efficiency particulate-arresting (HEPA) filters
  Electrostatic air cleaners (EAC)
  Appropriate filter maintenance, by cleaning and replacing at regular intervals.
- Control of formaldehyde:
  Careful selection of building materials
  "Off-gassing" procedure
  Bake-out
- Control of vapor and gas pollutants:
  Introduction of outside air
  Increased ventilation
  Sensors and alarms
  Eliminate smoking in building
  Filtering
  Maintain and properly adjust heating equipment
- Methods to provide indoor air quality service:
  Scheduled inspections
  Equipment preventive maintenance
  CO_2 measurement and testing

Problems and Questions

1. What is the meaning of indoor air quality?
2. What is a micron?
3. What is the size range of pollen?
4. What is the size range of bacteria?
5. What is the size range of tobacco smoke?
6. What is the size range of viruses?
7. Electronic air cleaners are capable of filtering airborne particles in the size range of 50 microns down to .03 microns. What airborne pollutants can be filtered out?
8. What is a pollutant?
9. Name some common pollutants.
10. List the two general methods of pollution control.
11. What pollutants can a mechanical filter remove?
12. Name a few allergens.
13. What are pathogens?
14. What causes radon gas? What can be done to lower the levels?
15. What are the combustion products that could cause health problems?
16. Air is composed of 78% oxygen. True or false?
17. When we breathe, we inhale oxygen, and exhale carbon monoxide. True or false?
18. Health authorities are concerned with particles of 2 microns or less since they can be retained in the lungs. True or false?
19. A contaminant is any unwanted material that gets into the air and can cause health problems. True or false?
20. Pollen, fungi, and mold spores are bioaerosol particles. True or false?
21. Individuals exposed to asbestos can experience:
    a. Scarring of the lungs.
    b. Mesothelioma, a damaging effect to the linings of the lungs and abdomen.
    c. Lung cancer.
    d. All of the above.
22. Radon is a:
    a. Colorless gas, formed by radioactive decay of radium.
    b. Colorless, pungent gas, formed by radioactive decay of radium/uranium.
    c. Colorless odorless gas, formed by radioactive decay of radium/uranium.
    d. None of the above.
23. Carbon monoxide reduces the ability of the:
    a. Blood to carry oxygen to the body tissues.
    b. White blood cells to actively regenerate as needed.
    c. Body to sleep as needed.
    d. None of the above.
24. The energy-recovery ventilator:
    a. Includes an air-to-air heat exchanger that is used to conserve energy.
    b. Controls and provides outside air and exhausts stale air.
    c. Filters the air.
    d. All of the above.
25. The EPA and OSHA recognize _____ _____ as a suitable measurement of indoor air quality.
    a. Carbon monoxide.
    b. Carbon dioxide.
    c. Nitrogen gas.
    d. None of the above.

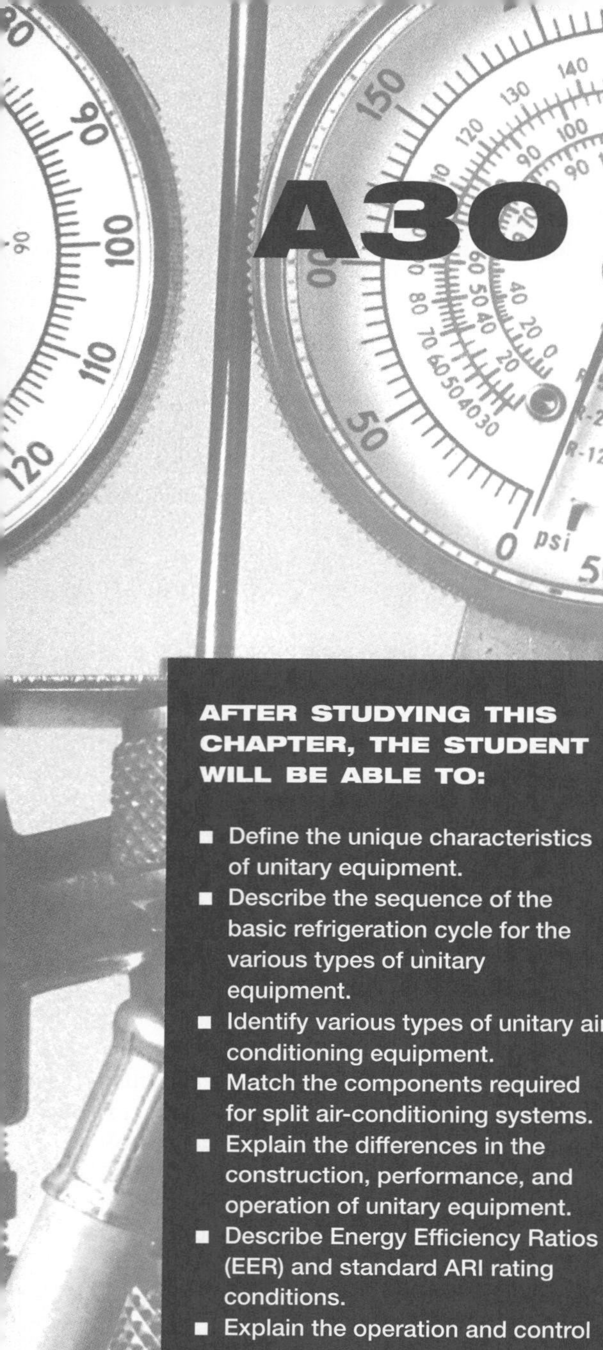

A30 Unitary Air-Conditioning Systems

AFTER STUDYING THIS CHAPTER, THE STUDENT WILL BE ABLE TO:

- Define the unique characteristics of unitary equipment.
- Describe the sequence of the basic refrigeration cycle for the various types of unitary equipment.
- Identify various types of unitary air conditioning equipment.
- Match the components required for split air-conditioning systems.
- Explain the differences in the construction, performance, and operation of unitary equipment.
- Describe Energy Efficiency Ratios (EER) and standard ARI rating conditions.
- Explain the operation and control sequence for various types of unitary equipment.
- Evaluate the application of a residential air-conditioning system.
- Describe the equipment options available for rooftop units.
- Select the proper refrigeration piping methods and installation criteria.
- Explain the different components available on commercial split systems.
- Describe the sequence of the basic refrigeration cycle and operation of the various types of air-conditioning systems.

A30-1 DEFINITIONS

Unitary air-conditioning equipment has been developed to provide factory-built and tested systems, complete as much as possible, with piping, controls, wiring, and refrigerant. These packages are usually simple to install, requiring only service connections and in some cases duct work for field applications.

Unitary air-conditioning equipment consists of one or more factory-made assemblies which normally include an evaporator or cooling coil, a compressor and condenser combination, and possibly include a heating unit. When the air conditioner is connected to a remote-condensing unit, such as in residential applications, the system is often referred to as a split system. A *packaged air conditioner,* where all components are included in one assembly, such as a room cooler, is referred to as a self-contained system.

The sizes of unitary equipment range from small fractional-tonnage room coolers to large packaged rooftop units in the 100-ton category.

A30-2 TYPES OF UNITARY EQUIPMENT

The various types of units described as unitary equipment include the following:

1. Room air conditioners
2. Console-type through-the-wall conditioners
3. Dehumidifier units
4. Single-package conditioners
5. Split-system conditioners
6. Rooftop conditioners
7. Desiccant cooling systems

Figure A30-1 Window air-conditioning unit for cooling.

Heat pumps are also classified as unitary equipment. However, due to their unique characteristics, they will be discussed in a separate chapter.

A30-3 ROOM AIR CONDITIONERS

Room air conditioners were primarily developed to provide a simplified means of adding air conditioning to an existing room. These units are considered semi-portable in that they can easily be moved from one room to another or from one building to another. They provide cooling, dehumidifying, filtering, and ventilation, and some units provide supplementary heating.

In numbers sold, room air conditioners outsell all other types of unitary equipment. They are relatively low in cost, easy to install, and can be used in almost any type of structure.

The disadvantage of room air conditioners is that they may either block part of the window area and prevent the window from being opened or require a special hole through the wall. Some people object to operating noise that they produce close to the occupants. They are best used to condition a single room; however, the spillover can supply some conditioning to adjacent areas. An external view of a room cooler is shown in Fig. A30-1.

A30-3.1 Construction and Installation

The general construction of the units is shown in Fig. A30-2. There are basically two parts to the unit. One section goes inside the room where the evaporator fan draws in return air through the filter and cooling coil, delivering conditioned air to the room. The other section extends outside the room where the condenser fan forces outside air through the condenser, exhausting the heat absorbed by the evaporator. One fan motor operates both fans, with the motor shaft extending through the separating partition. Condensate from the evaporator coil flows into the drain pan which extends below the condenser fan. The condenser-fan tip dips into condensate, splashing it onto the hot condenser, where it evaporates and is blown into the outside air.

The window-mounted units are supplied with a kit of parts for installation. Sill brackets, window-mounting strips, and sealing strips are set in place for installations in double-hung windows as shown in Fig. A30-3. Side curtains fold out to fill up the extra window space (Fig. A30-4). A sponge rubber seal is provided for the opening

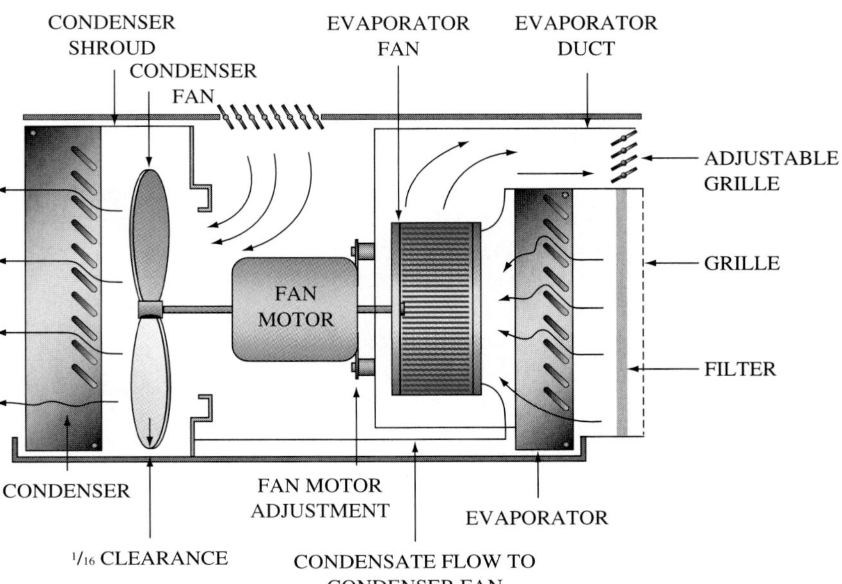

Figure A30-2 Air circuits and principal components of a window air conditioner. (Courtesy of Carrier Corp.)

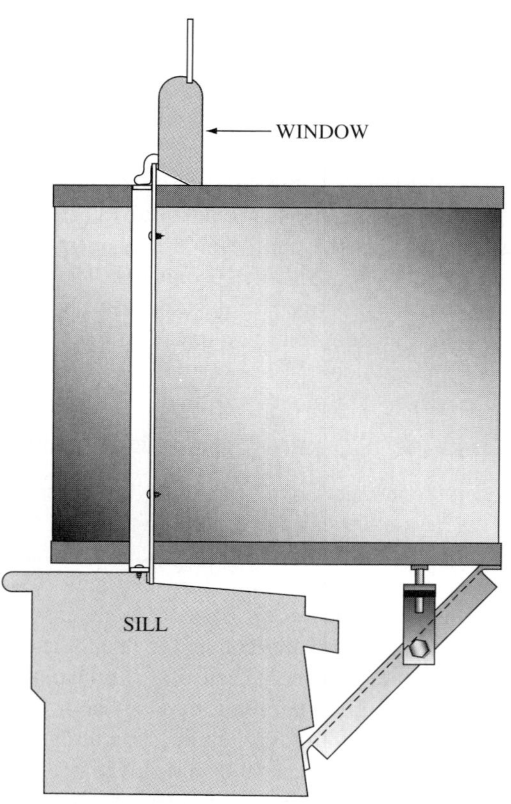

Figure A30-3 Installation diagram of a room cooler showing the supporting bracket, from the outside.

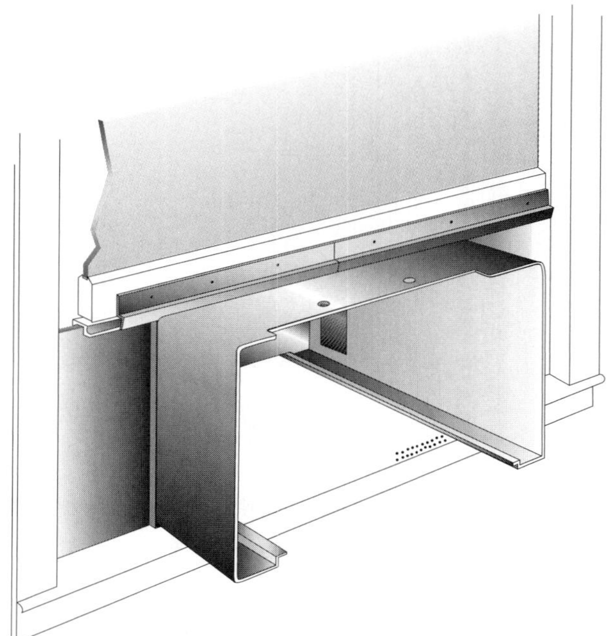

Figure A30-4 Installation diagram of a room cooler showing the window sealing strips and filler boards.

where the sash overlaps, and a sash bracket is installed to lock the lower sash in place (Fig. A30-5).

For through-the-wall type installations, a metal sleeve is provided to be placed in the masonry opening. The unit is designed with a slide-out chassis.

A30-3.2 Performance and Operation

A schematic diagram of the refrigeration cycle is shown in Fig. A30-6. The system uses a capillary-tube metering device. A typical wiring diagram for a cooling unit is shown in Fig. A30-7.

A selection switch offers the following modes of operation: FAN ONLY (for ventilation), LOW COOL (using the low evaporator fan speed), HIGH COOL (using the high speed of the evaporator fan), and OFF.

An electric-heater strip is sometimes provided to supply heat during mild weather. Some units use a heat pump for heating (heat pumps are described in a separate chapter).

The specifications for a typical series of room coolers are shown in Fig. A30-8 (p. 585). The size range is from 5000 Btu/hr to 20,000 Btu/hr. These ratings are based on inside air at 80°F DBT, 67°F WBT, and outside air at 95°F DBT, 75°F WBT. All of these units will operate on 115-V current, except the three large models, which operate on 230/208 V.

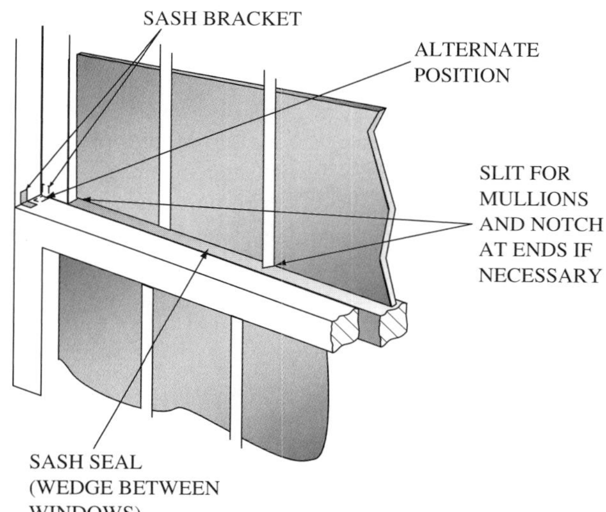

Figure A30-5 Installation diagram of a room cooler showing the sponge rubber seal between the upper edge of the lower sash and the upper sash of a double-hung window.

A30-4 CONSOLE THROUGH-THE-WALL CONDITIONERS

A console through-the-wall conditioner is a type of room cooler that is designed for permanent installation. It was developed to provide individual room conditioning for ho-

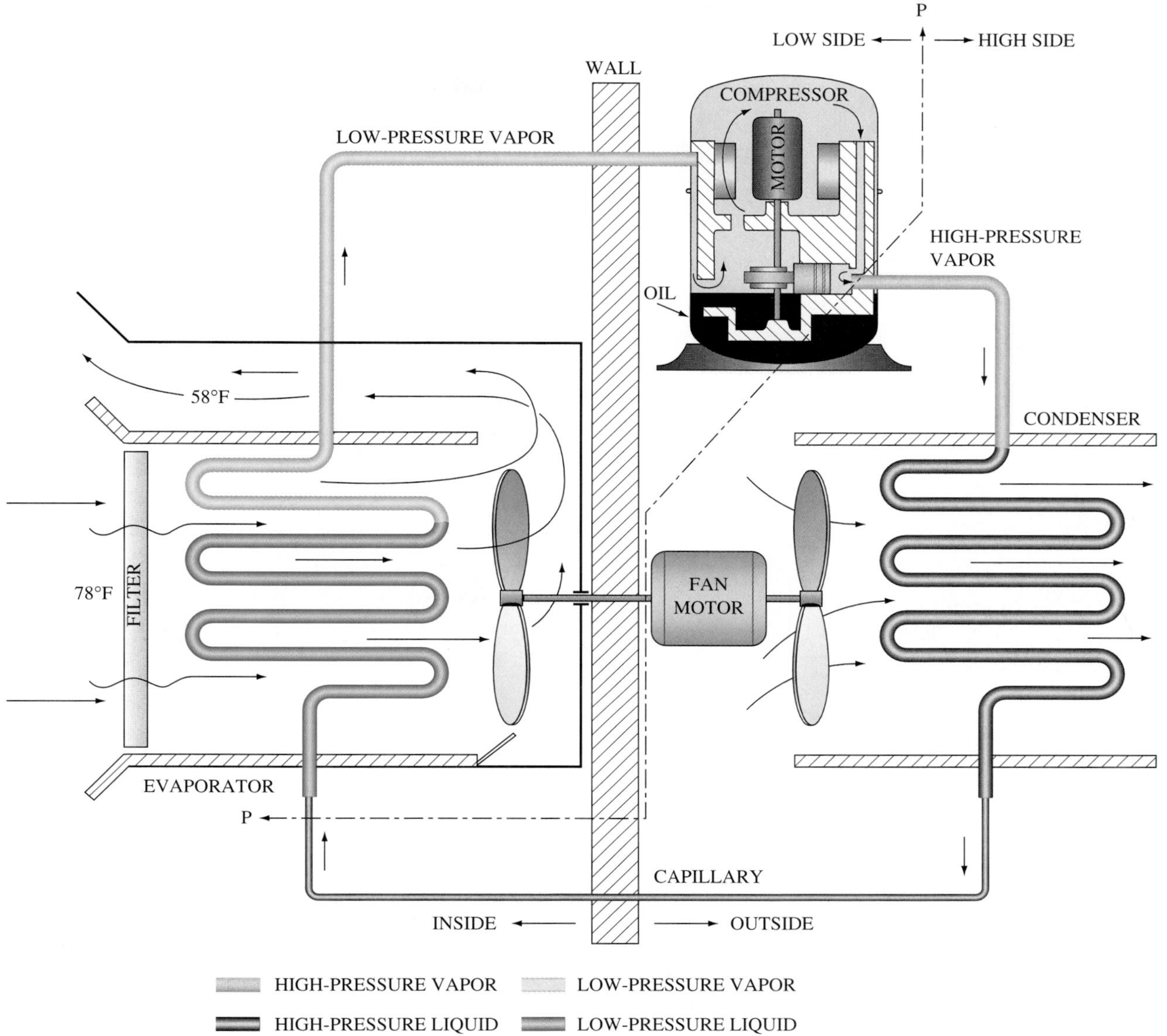

Figure A30-6 Refrigeration cycle of a room cooler.

tels, motels, and offices where it is impractical or uneconomical to install a central-plant system. An opening needs to be made in the outside wall adjacent to the unit for condenser air and ventilation.

These units are also known as packaged terminal air conditioners (PTAC). A pictorial view is shown in Fig. A30-9 (p. 586) and a schematic view of one installed, in Fig. A30-10 (p. 586).

Units are efficient, quiet and easy to install. The efficiency is usually stated in terms of *energy efficient ratio* (EER). The EER is equal to the cooling output in Btu/hr divided by the power input in watts under standard rating conditions. Standard rating conditions set up by ARI are based on 80°F DBT, 67°F WBT entering air and 95°F DBT outdoor ambient air. For example, a unit with an output of 6600 Btu/hr and 660 watts input, under standard conditions, would have an EER of 10 (6600/660), which is considered a good rating.

These units should comply with standards set up by the following associations:

- Canadian Standards Association (performance standards).
- Underwriters Laboratory (electrical and safety standards).
- Air-Conditioning and Refrigeration Institute (ARI): Standard #310 (for packaged terminal air conditioners and heat pumps); and Standard #380 (for refrigerant cooled liquid coolers, remote type).

PTAC units should also meet ASTME (American Society of Testing and Materials Engineers) wind and rain infiltration standards.

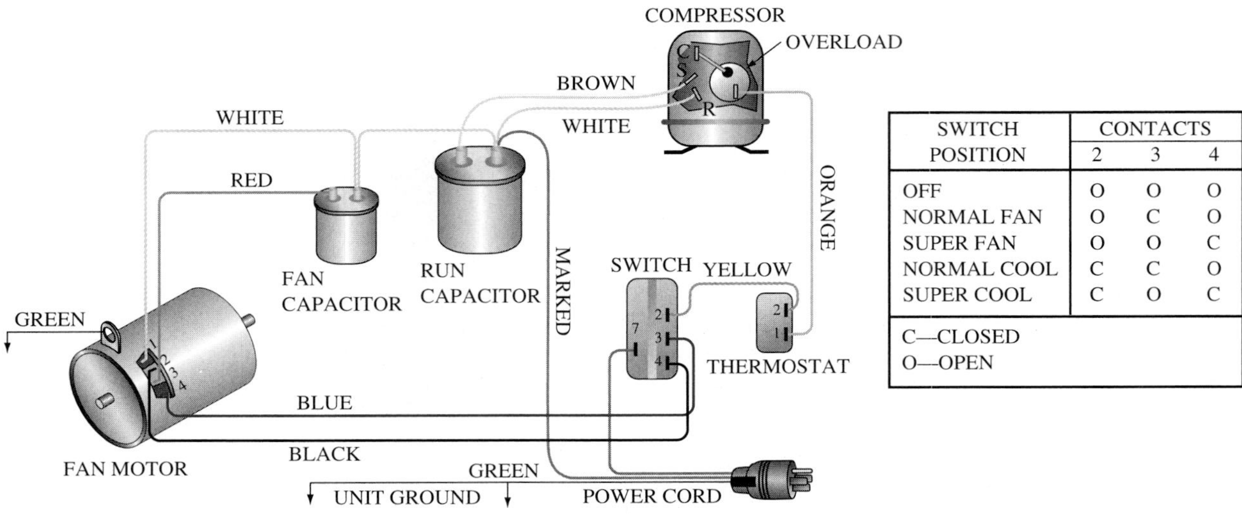

SWITCH POSITION	CONTACTS 2	3	4
OFF	O	O	O
NORMAL FAN	O	C	O
SUPER FAN	O	O	C
NORMAL COOL	C	C	O
SUPER COOL	C	O	C

C—CLOSED
O—OPEN

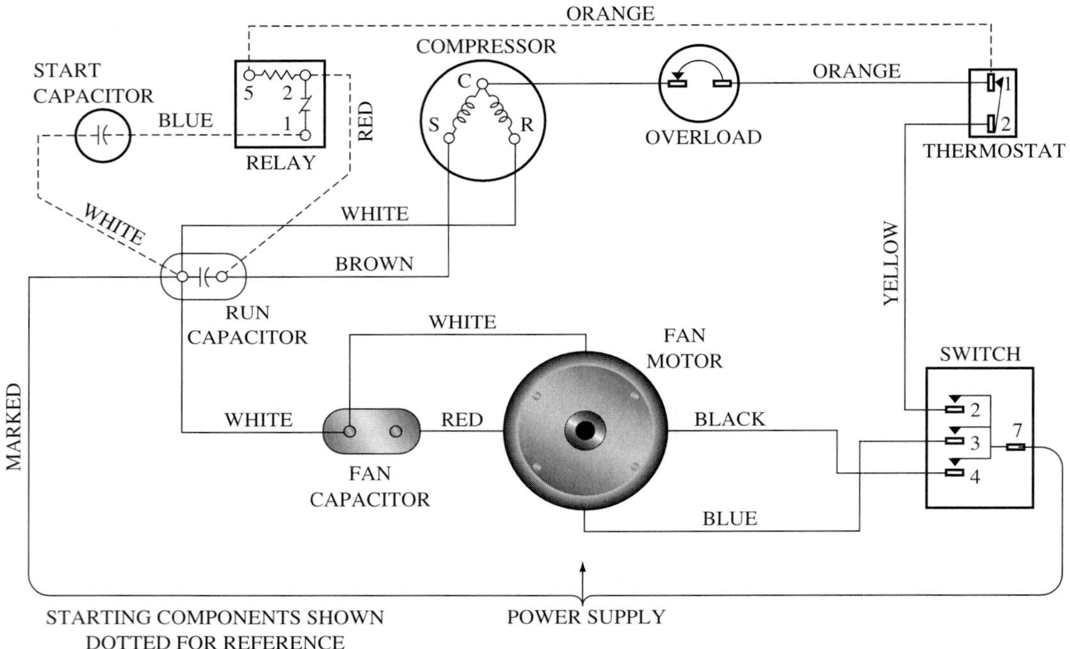

Figure A30-7 Wiring diagram for a room cooler. (Courtesy of Carrier Corp.)

The information in this section applies to PTAC console conditioners with electric heat. PTAC units are also manufactured using heat pumps, which are described in a separate chapter.

A30-4.1 Performance

The performance data for PTAC units are shown in Fig. A30-11 (p. 587). The sizes range from 6,800 to 14,900 Btu/hr at standard rating conditions. The quantity of outside air that the units can admit ranges from 40 to 55 cfm depending on the size of the unit.

Units are available for 115 and 230/208 V, single-phase, AC power. A sample rating table is shown in Fig. A30-12 (p. 588). The power input varies from 660 to 1715 watts for the various sizes shown. Referring to Fig. A30-13 (p. 589), electric-heat capacity from 1.5 to 5.0 kW can be installed in any sized unit. Power receptacle configurations (Fig. A30-14, p. 590) depend on the amperage drawn.

A30-4.2 Components and Installation

An expanded view of a typical unit is shown in Fig. A30-15 (p. 590). This illustration shows the location of

Room Air Conditioners

	Portable					Compact					High Capacity		
Models	**5P2MY**	**7P2MY**	**5P2MC**	**7P2MC**	**9P2MC**	**10C2MA**	**12C2MB**	**12C3V**	**10C2MT**	**12C2MT**	**14C2MA**	**18C3MA**	**21C3MS**
Capacity[†]													
Cooling (Btu)	5,000	6,500	5,300	6,600	8,600	10,000	11,800	11,800/11,700	10,000	11,800	13,700	17,900/17,400	20,500/20,300
Energy Efficiency Ratio													
EER	9.0	9.2	9.6	10.0	9.0	9.6	9.7	10.0	9.6	9.7	10.4	9.0	8.5
Dehumidification Pints Per Hour	1.1	2.0	1.4	2.0	3.0	2.0	3.3	3.3/3.3	2.0	3.3	3.5	5.1/5.1	6.8/6.8
Electrical													
Voltage	115	115	115	115	115	115	115	230/208	115	115	115	230/208	230/208
Amps	5.2	6.5	5.1	5.9	8.3	9.2	10.5	5.2/5.6	9.2	10.5	12.0	9.0/9.6	10.0/11.6
Watts	555	705	550	660	955	1040	1215	1180/1170	1040	1215	1315	1990/1940	2400/2390
Plug Type	PAR	PAR	PAR	PAR	PAR	PAR	PAR	TAN	PAR	PAR	PAR	TAN	TAN
Features													
Fan Speeds	2	2	3	3	3	3	3	3	3	3	3	3	3
RPM's	750/1000	750/1000	750/875/1000	750/875/1000	1000/1125/1250	850/795/1100	850/975/1100	850/1100	850/975/1100	850/975/1100	800/950/1115	800/950/1115	800/950/1115
Air Flow (CFM)[††]	180	180	180	180	190	340	340	340	340	340	420	450	450
Ventilation	—	—	Exhaust	Exhaust	Exhaust	Exhaust	Exhaust	Exhaust	Exhaust	Exhaust	Exhaust	Exhaust	Exhaust
Controls	Rotary	Rotary	Touch Cooling™	Touch Cooling™	Touch Cooling™	Touch Cooling™	Touch Cooling™	Rotary	Rotary	Rotary	Touch Cooling™	Touch Cooling™	Rotary
Easy Filter Access	Yes	Yes	Yes	Yes	Yes	Yes	Yes	Yes	Yes	Yes	Yes	Yes	Yes
Rotary Compressor	Yes	Yes	Yes	Yes	Yes	Yes	Yes	Yes	Yes	Yes	Yes	Yes	Yes
Copper Tubing	Yes	Yes	Yes	Yes	Yes	Yes	Yes	Yes	Yes	Yes	Yes	Yes	Yes
24 Hour Timer	—	—	—	—	—	—	—	—	Yes	Yes	—	—	—
Installation													
Window Mounting	Instamount	Instamount	Instamount	Instamount	Instamount	Instamount*	Instamount*	Optional**	Instamount*	Instamount*	Instamount***	Instamount***	Instamount***
Slide-Out Chassis	—	—	—	—	—	Yes	Yes	Yes	Yes	Yes	Yes	Yes	Yes
Fits Window Widths	22 1/2–40″	22 1/2–40″	22 1/2–40″	22 1/2–40″	22 1/2–40″	28–42″	28–42″	28–42″	28–42″	28–42″	30–44″	30–44″	30–44″
(Min.-Max. Inches)	57.2–101.6 cm	57.2–101.6 cm	57.2–101.6 cm	57.2–101.6 cm	57.2–101.6 cm	71.1–106.7 cm	71.1–106.7 cm	71.1–106.7 cm	71.1–106.7 cm	71.1–106.7 cm	76.2–111.8 cm	76.2–111.8 cm	76.2–111.8 cm
76.2–111.8 cm													
Thru-Wall (Max. Wall Thick.)	—	—	—	—	—	10 1/4″/26 cm	10 1/4″/26 cm	10 1/4″/26 cm	10 1/4″/26 cm	10 1/4″/26 cm	10 1/2″/26.7 cm	10 1/2″/26.7 cm	10 1/2″/26.7 cm
Dimensions													
Height	13 3/8″/34 cm	13 3/8″/34 cm	13 3/8″/34 cm	13 3/8″/34 cm	13 3/8″/34 cm	15 7/8″/40.3 cm	15 7/8″/40.3 cm	15 7/8″/40.3 cm	15 7/8″/40.3 cm	15 7/8″/40.3 cm	17 1/8″/43.5 cm	17 1/8″/43.5 cm	17 1/8″/43.5 cm
Width	19″/48.3 cm	19″/48.3 cm	19″/48.3 cm	19″/48.3 cm	19″/48.3 cm	24 1/2″/62.2 cm	24 1/2″/62.2 cm	24 1/2″/62.2 cm	24 1/2″/62.2 cm	24 1/2″/62.2 cm	26 1/2″/67.3 cm	26 1/2″/67.3 cm	26 1/2″/67.3 cm
Depth (with front on)	20 1/16″/51 cm	20 1/16″/51 cm	20 1/16″/51 cm	20 1/16″/51 cm	20 1/16″/51 cm	23 7/8″/60.6 cm	23 7/8″/60.6 cm	23 7/8″/60.6 cm	23 7/8″/60.6 cm	23 7/8″/60.6 cm	28 1/4″/71.8 cm	28 1/4″/71.8 cm	28 1/4″/71.8 cm
Shipping Weight	71 lbs./32.2 kg	71 lbs./32.2 kg	71 lbs./32.2 kg	71lbs./32.2 kg	84 lbs./38.1 kg	118 lbs./53.5 kg	118 lbs./53.5 kg	117 lbs./53.1 kg	118 lbs./53.5 kg	118 lbs./53.5 kg	165 lbs./74.8 kg	165 lbs./74.8 kg	165 lbs./74.8 kg

[†]Rating conditions are 80°F/27°C db, 67°F/19°C wb indoor air and 95°F/35°C db, 75°F/24°C wb outdoor air.
[††]Wet coil with fan on high speed.

*To mount in storm or mobile home window use adapter kit RISAK.
**With optional mounting kit (RSMK1).
Temperature control has approximately 4° differential on cooling.
Dual voltage units based on operation from 253 to 197 V.
***Fold-out side curtains require some assembly.

Figure A30-8 Performance specifications for a room cooler.

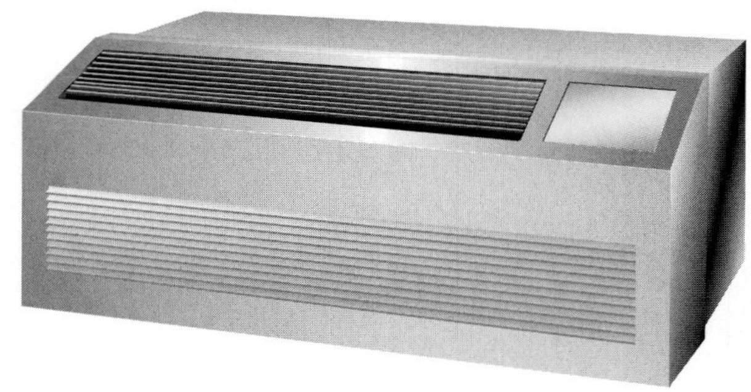

Figure A30-9 Exterior view of a packaged terminal air conditioner (PTAC). (The Trane Company)

the following essential parts: (a) wall sleeve, (b) outdoor fan, (c) dual-fan motors, (d) insulated outdoor and indoor sections, (e) hidden-fan cycle switch, (f) control panel, (g) outdoor air damper, (h) sloped diffuser panel, (i) electroplated paint coat, (j) condenser coil, (k) evaporator coil, (l) nichrome coiled heating elements, (m) insulated bulkhead, (n) rubber-isolated compressor, and (o) capillary-tube metering device. The air filter is located behind the front return-air panel and can be easily changed.

Five different types of installation arrangements are shown in Fig. A30-16. Interior legs are provided for installations using a subbase or the unit can be flush mounted on the floor. Sleeve installation information is shown in Fig. A30-17.

An optional duct package can be supplied for a PTAC unit, as shown in Fig. A30-18. This makes it possible to condition two rooms with one unit.

A30-4.3 Controls

A rotary switch is provided with the following choices:

OFF	Turns unit off.
FAN ONLY	Indoor fan operates.
COOL	Provides cooling with indoor fan.
HEAT	Provides heating with indoor fan.

A rocker switch provides the following choices:

HI	High fan speed
LOW	Low fan speed

The following additional controls are provided:

Adjustable temperature limiting device—Limits the range of the room thermostat.

Outside air damper—Control lever can be positioned to permit zero to fully-open supply of outside air.

Fan cycle switch—Allows continuous or intermittent fan operation.

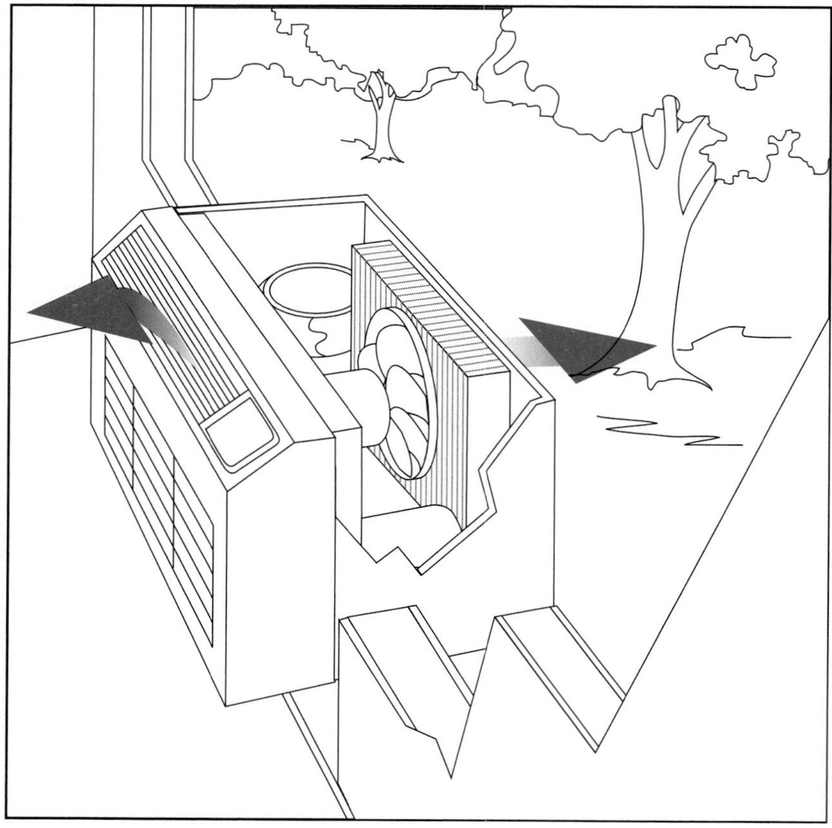

Figure A30-10 Installation diagram of a PTAC unit. (The Trane Company)

General Data—Air Conditioner with Electric Heat Models

Model Type	Air Conditioner											
Model No.	**PTEB 07**			**PTEB 09**			**PTEB 12**			**PTEB 15**		
Voltage[1]	230	208	265	230	208	265	230	208	265	230	208	265
Capacity[2] (Btuh)	6800	6800	6800	9200	9100	9200	11900	11600	11800	14900	14600	14900
Indoor Fan CFM												
High/Low (Wet coil)	240/210	215/185	240/210	240/210	215/185	240/210	280/260	260/240	280/260	280/260	260/240	280/260
High/Low (Dry Coil)	300/260	280/240	300/260	300/260	280/240	300/260	350/325	325/300	350/325	350/325	325/300	350/325
Fresh Air CFM (Dry Coil)[3]		40			40			55			55	
Approx. Ship Wt (Lbs)		129			133			148			157	
Refrig. Charge (oz)		20.0			30.0			29.0			36.0	
Oil Charge (oz)		8.8			7.4			10.8			13.9	

Minimum operating ambient temperature for cooling is 45 degrees.
[1]Minimum voltage on 230/280 V units is 197 V; maximum is 253 V.
[2]All capacities based on approved ARI rating point 80/67 entering air temperature, 95°F outdoor ambient.
[3]Cfm rating is with the unit in the "fan only" setting.

Figure A30-11 Performance data on a PTAC unit. (The Trane Company)

Cooling Performance—Air Conditioner with Electric Heat Models

Model Type	Air Conditioner											
Model No.	**PTEB 07**			**PTEB 09**			**PTEB 12**			**PTEB 15**		
Voltage[1]	230	208	265	230	208	265	230	208	265	230	208	265
Capacity[2] (Btuh)	6800	6800	6800	9200	9100	9200	11900	11600	11800	14900	14600	14900
Watts[3]	660	660	660	890	875	920	1190	1160	1180	1675	1640	1715
EER*	10.2	10.2	10.2	10.3	10.4	10.0	10.0	10.0	10.0	8.9	8.9	8.7
SHR	.77			.70			.67			.61		
Dehumidification (Pts/Hr)	1.4			2.4			3.8			4.7		

Minimum operating ambient temperature for cooling is 45°F.
*EER Based on DOE Sampling Program
[1]Minimum voltage on 230/280 V units is 197 V; maximum is 253 V.
[2]All capacities based on approved ARI rating point 80°F/67°F entering air temperature, 95°F/75°F outdoor ambient.
[3]Cooling watts includes fan motors.

Figure A30-12 Electrical data for a PTAC unit. (The Trane Company)

Electric Heat Capacity and Electrical Data (For PTEB and PTHB Models)

Electric Heater Size	1.5 kW (1-Stage)			2.5 kW (1-Stage)			3.0 kW (1-Stage)			3.5 kW (1-Stage)			4.0 kW (1-Stage)			5.0 kW (1-Stage)		
Voltage	230	208	265	230	208	265	230	208	265	230	208	265	230	208	265	230	208	265
Capacity (Btuh)	5200	4300	5200	8600	7100	8700	10400	8500	10400	12000	9900	12000	13800	11300	13800	17200	14100	17200
Watts (Max)	1540	1240	1540	2540	2085	2540	3040	2500	3040	3540	2900	3540	4040	3300	4040	5040	4130	5040
Amps (Max)	7.0	6.4	6.2	11.4	10.3	9.9	13.5	12.3	11.8	15.7	14.3	13.7	17.9	16.2	15.6	22.2	20.2	19.4
Minimum Ampacity	8.7	8.7	7.6	14.1	14.1	12.3	16.8	16.8	14.7	19.5	19.5	17.0	22.2	22.2	19.4	27.7	27.7	24.1
Maximum Electrical Protective Device[1]	15 Amp	15 Amp	15 Amp	15 Amp	15 Amp	15 Amp	20 Amp	20 Amp	15 Amp	20 Amp	20 Amp	20 Amp	25 Amp	25 Amp	20 Amp	30 Amp	30 Amp	25 Amp
NEMA Rating (Plug)[2]	6-15	6-15	7-20	6-15	6-15	7-20	6-20	6-20	7-20	6-20	6-20	7-20	6-25	6-25	7-20	6-30	6-30	7-30

Notes:

1. On permanently wired units, if compressor FLA is > 6 amps or if unit MCA is > 15 amps, then fuses must be used (per N.E.C.).
2. 230/208 V cord connected units are approved for UL installation into wall subbase receptacles. They can also be hard wired per Local Code Regulation. 265 V cord connected units are UL approved for installation into subbase receptacle only. The units can also be hard wired per Local Code Regulation.
3. Electrical requirements for units without electric heaters are governed by cooling F.L.A.
4. 3.5 kW heater @ 230 V is stock model for all PTEB/PTHB units.
5. 5.0 kW heater @ 230 V is stock model for all PTEB/PTHB 1501 units.
6. All other models are built on a sales order production run basis.

Figure A30-13 Electrical heating data for a PTAC unit. (The Trane Company)

Voltage	230/208			265	
Unit-supplied Plug					
Amps	15	20	30	20	30
NEMA Rating	6-15P	6-20P	6-30P	7-20P	7-30P
Receptacle					
Amps	20	20	30	20	30
NEMA Rating	6-20R	6-20R	6-30R	7-20R	7-30R

Notes:

1. All wiring, including receptacles, must be made in accordance with local electrical codes and regulations.
2. NEMA 6-15 and 6-20 Plugs fit in NEMA 6-20 receptacles.
3. Receptacles shown for 230/208 applications are factory supplied with subbase, per local codes for wall mounted receptacles use or the corresponding NEMA receptacles.
4. Receptacles shown for 265 volt applications are factory supplied with subbase. Codes do not allow 265 volt units to mate with wall mounted receptacles. They may however be hard wired per Local Codes.

Figure A30-14 Power receptacles for PTAC units. (The Trane Company)

Remote thermostat (optional)—Unit can be wired to use a remote wall-mounted thermostat rather than the one normally supplied in the unit.

Front-desk control interface (optional)—Units may be individually started and stopped with an energy-management panel from a central location.

Room-freeze protection (optional)—Overrides off signal when room thermostat goes below 40°F and turns off equipment automatically.

A30-5 DEHUMIDIFIER UNITS

Dehumidifier units are small, portable, self-contained refrigeration systems designed to extract moisture from the air. They are used in localities where high humidity can cause damage to stored materials. They are usually installed where cooling is not required. They consist of a motor-compressor unit, a condenser, an evaporator, an air-

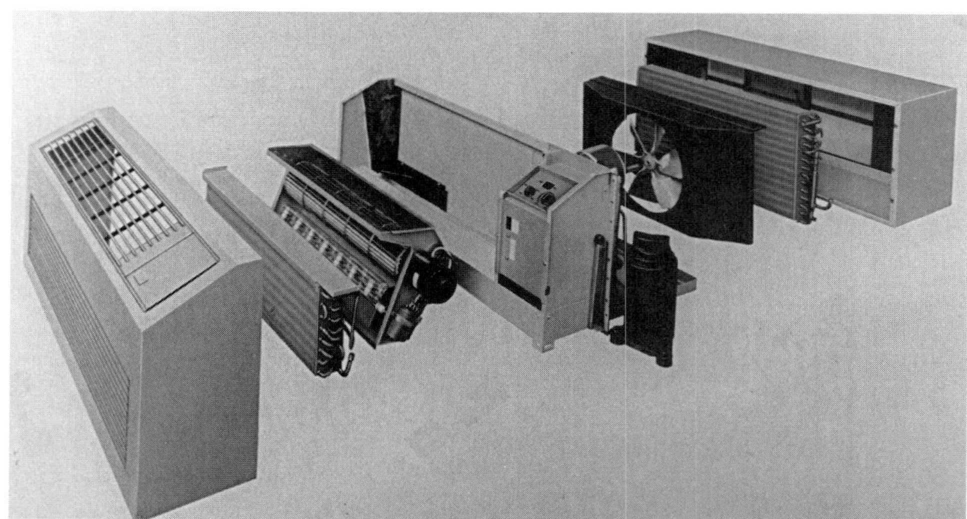

Figure A30-15 Expanded view of a PTAC unit. (The Trane Company)

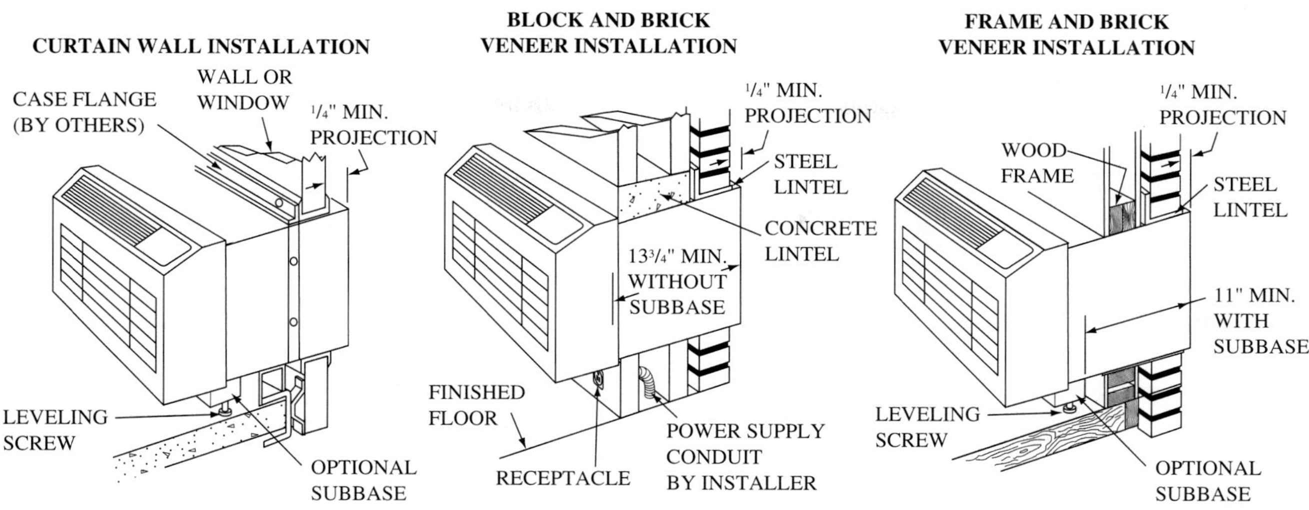

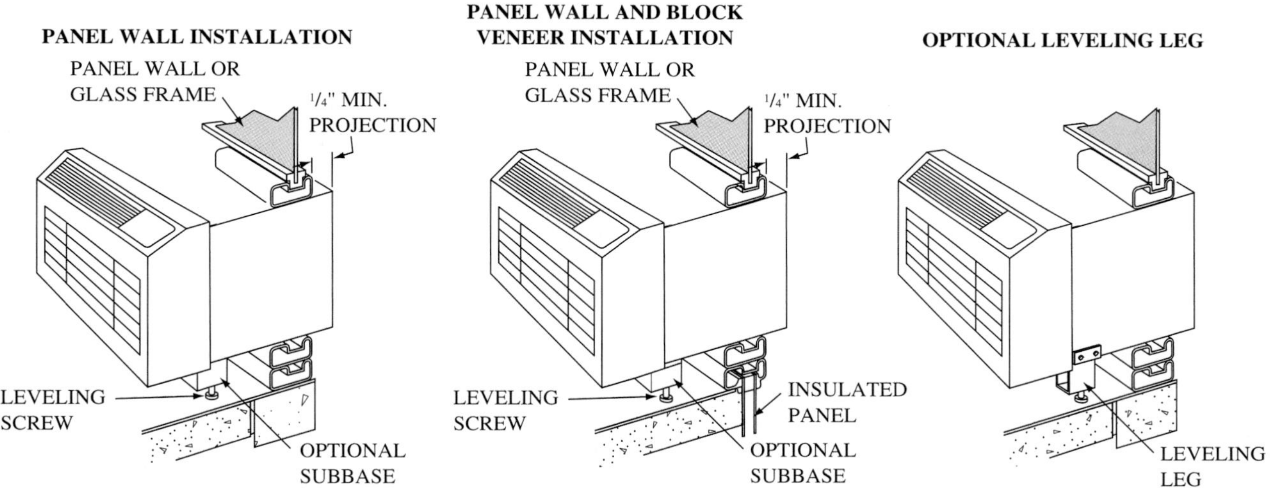

Figure A30-16 Installation arrangements for PTAC units. (The Trane Company)

ON THE JOB

PROBLEM: You have arrived on the job to find water all over the floor and some standing water in the blower compartment. There was no water coming from the drain line. The drain was hooked up with PVC and could not be cleaned without cutting the line. What do you do next?

SOLUTION: The technician cut the line and cleaned out the drain. He chose to put on a PVC union instead of a coupler. Some technicians choose to put in a tee so the line can be cleaned or blown out.

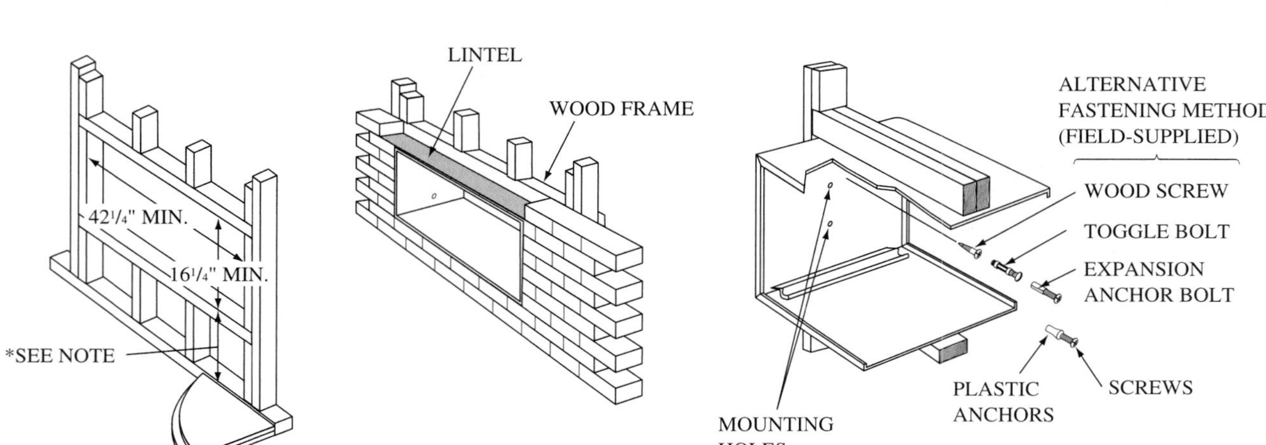

1. 3¼" MINIMUM WITH SUBBASE
2. ON APPLICATIONS NOT REQUIRING SUBBASE OR LEVELING LEGS, UNIT MAY BE FLUSH MOUNTED TO FLOOR.

Figure A30-17 Wall-sleeve installation arrangements, PTAC units. (The Trane Company)

circulating fan, and a means of collecting and/or disposing of the condensate, and a cabinet (Fig. A30-19).

The fan draws moist room air over the evaporator and cools it below dew-point temperature. The removed moisture drains into a collecting pan or drops into an open drain. The cooled air then passes over the condenser coil where it is heated and discharged into the room at a higher dry-bulb temperature and at a lower relative humidity. Continuous circulation gradually reduces the relative humidity in the room.

The compressor is a fractional-horsepower hermetically sealed unit, typically requiring from 200 to 700 W of power input. Most refrigeration systems use a capillary-type metering device, although some larger units use thermal expansion valves. The airflow rate is usually 125 to 250 cfm. The capacity ranges from 11 to 50 pints per day at standard test conditions of 80°F DBT and 60% RH.

The controls supplied with the unit vary depending on the manufacturer and the model. The following controls are desirable and come as either standard equipment or options:

1. On-off switch.
2. Humidity-sensing control to cycle the unit automatically.
3. Automatic-sensing switch to turn the unit off when the water receptacle is full and requires emptying.
4. Defrost controls which cycle the compressor off under frosting conditions.

Figure A30-18 Duct package for PTAC units. (The Trane Company)

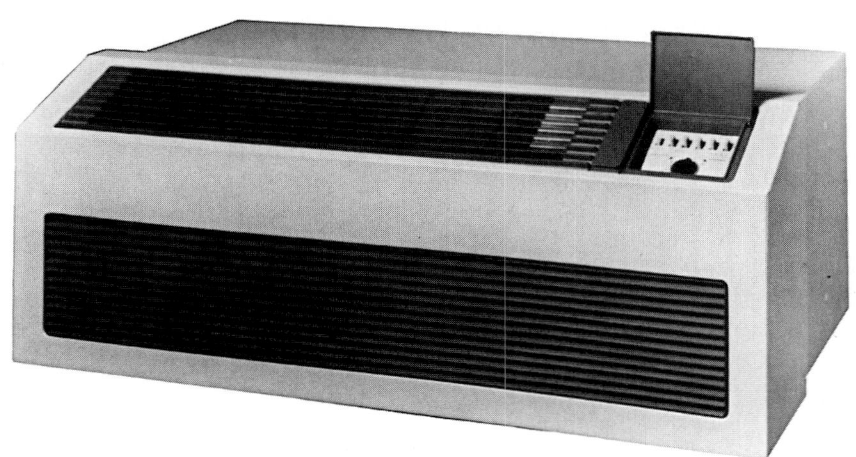

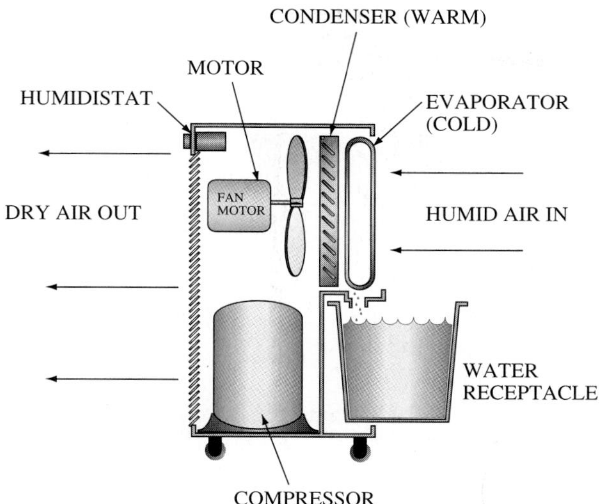

Figure A30-19 Diagrammatic view of a dehumidifier. (Copyright by the American Society of Heating, Refrigerating, and Air-Conditioning Engineers, Inc. Used by permission.)

A30-5.1 Servicing Dehumidifiers

A room dehumidifier is a portable device for removing moisture from an area where cooling is not required. Due to its portability and relatively light weight it can be taken into the shop when service is required. It consists of a hermetically sealed refrigeration unit that extracts moisture from room air and reheats it by passing the air through the condenser coil. The discharged air is therefore about the temperature of the room air, and no cooling is taking place.

The proper functioning of the following controls on dehumidifiers is important.

1. A manual switch turns the unit on and off.
2. A humidistat permits the unit to run only when the humidity level is above the setting of the control.
3. A frost control stops the compressor before the coil temperature reaches freezing temperature.

The following are types of service problems that can occur:

1. Water leakage. The customer reports that water is leaking from the unit onto the floor. This may be caused by not emptying the water receptacle often enough or leakage in the water run off system.
2. Continuous running. This may be due to dirty coils, which cause the unit to run inefficiently. If the unit is operating properly, it may be that the unit is too small to handle the load. Resetting the humidistat may be necessary. Dirty coils can also cause the compressor to cycle on overload.
3. Evaporator collects ice. This may point to a defective frost control or a frost control that is not making proper contact with the coil. It could also indicate a restriction in the refrigerant line, requiring hermetically sealed system service.
4. Unit will not operate. This may be due to a blown fuse, low voltage, a broken wire, or a defective on/off switch. The electrical system needs to be thoroughly checked.

A30-6 SINGLE-PACKAGE CONDITIONERS

A single-package conditioner, often called a Unitaire, is a complete self-contained factory-built unit, for permanent installation, to condition larger spaces than practical using room coolers.

In this category there are two variations of available equipment:

1. Horizontal conditioner: horizontal-packaged heating and cooling units, with integral air-cooled condensers, in the size range of 1½ to 5 tons
2. Vertical conditioner: vertical self-contained air conditioners, with water-cooled or remote air-cooled condensers, in the size range of 3 to 15 tons

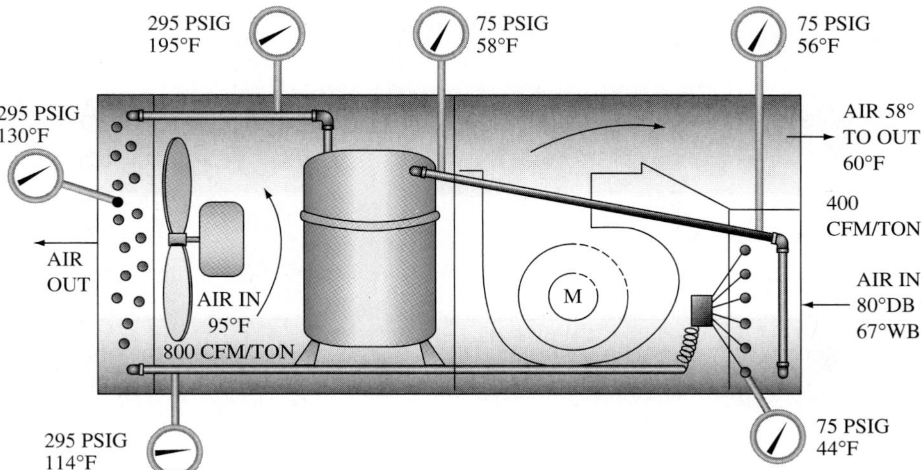

Figure A30-20 Component arrangement for a single-package conditioner.

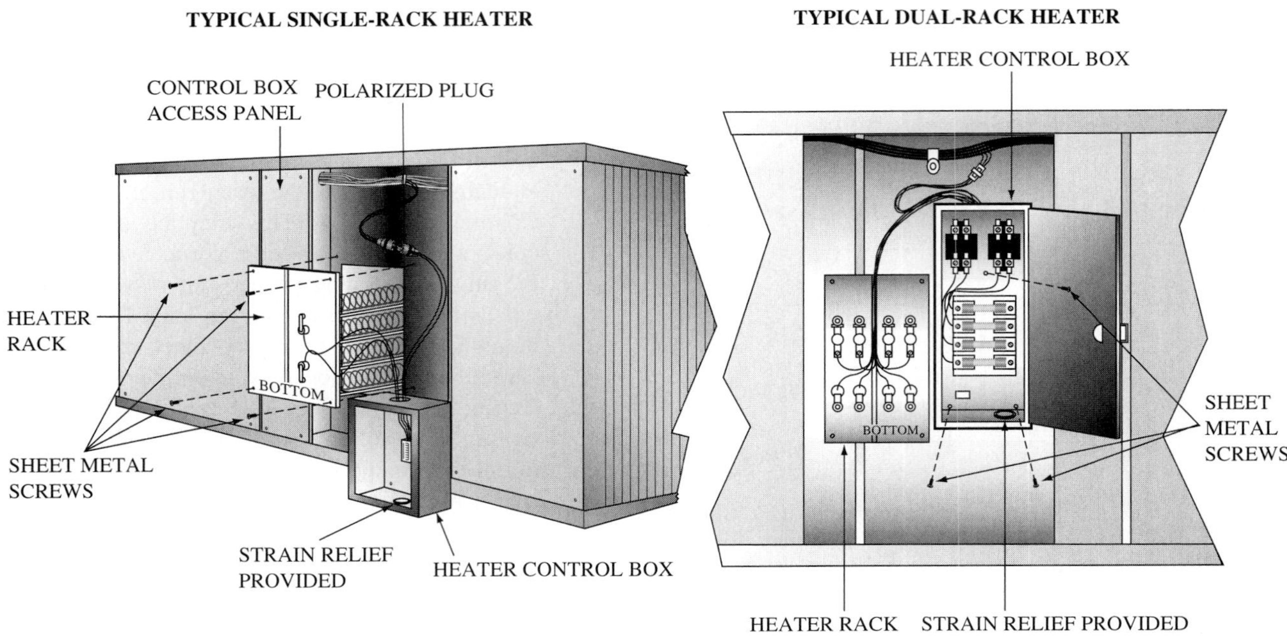

Figure A30-21 Electric heaters for a package air-conditioner. (The Trane Company)

A30-6.1 Horizontal Conditioner

The horizontal unit usually uses duct work for air distribution. In residences it can be used to supply cooling, humidification, and ventilation for a radiation-heated (hydronic) house. These systems are sometimes called split systems.

The horizontal unit is completely self-contained, including the air-cooled condenser. It therefore must be placed either entirely outside or at least with the condenser section outside. A schematic diagram of the arrangement of parts is shown in Fig. A30-20.

The unit is primarily a cooling unit although electric heaters can be installed in the unit as shown in Fig. A30-21.

The unit has supply- and return-air duct connections (Fig. A30-22) which can be entered from the side or bottom, depending on the application.

For residential use, the horizontal unit can be arranged for a ground-level installation, as shown in Fig. A30-23. The unit and ducts can also be installed in the attic space as shown in Fig. A30-24.

For commercial use the unit can be installed on the roof (Fig. A30-25, p. 597). This type of application is used for shopping malls, factories, and other commercial buildings.

The horizontal unit comes equipped with the following features:

PRACTICING PROFESSIONAL SERVICES

Since most smaller packaged unitary air-conditioning units do not come provided with a refrigeration access port, do not penetrate the system until it is deemed necessary to repair the unit. If the technician is called to the job site to repair the unit, first determine the problem and cause. If the problem area is electrical in nature, repair or replace the broken part. When it becomes mandatory to access the system, carefully install a leak-tight line tap or similar device to gain access to the refrigerant. Upon completion of the job, install caps, leak test, and replace all covers as originally installed.

If the technician is performing routine maintenance on the same packaged unit, only clean and visually inspect the unit. Never penetrate a sealed system unless it is required to repair the system.

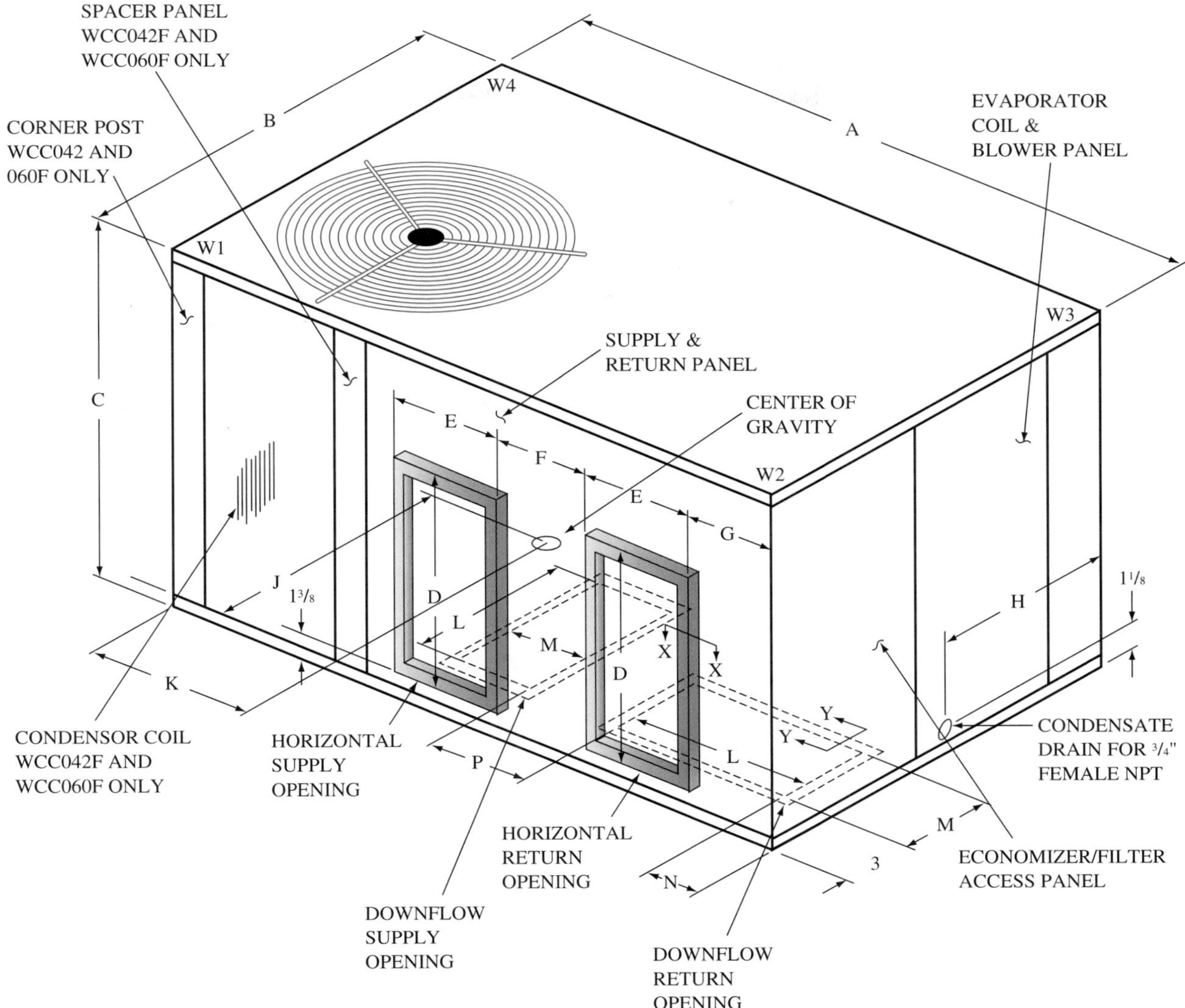

Figure A30-22 Duct connections for a single-package conditioner. (The Trane Company)

1. Water protection. A weather resistant cabinet along with a water-shedding base pan with elevated downflow openings and a perimeter channel prevent water from draining into the duct work.
2. Low ambient control (optional). Kits are available which control the condenser head pressure to permit the unit to start even though the ambient temperature goes as low as 0°F.
3. Economizer (optional). An economizer and dry-bulb temperature sensor can be supplied for downflow installations. This makes possible using outside air for cooling when temperatures permit it.
4. Enthalpy control kit (optional). This can be supplied in place of the dry-bulb sensor, or two enthalpy controls can be paired to provide differential enthalpy control.
5. Fresh air (25%) kit (optional). This kit can be mounted over the horizontal return-air openings for downflow requirements. It also can be used on horizontal applications by cutting a hole in the return-air duct or in the unit-filter access panel.
6. Fan-delay relay kit (optional). This control keeps the indoor blower on for about 90 seconds and improves the EER.
7. Anti-short-cycle timer. A time-off device insures a maximum of 5 minutes off between compressor cycles.

A30-6.2 Performance

Units range in capacity from 18,000 to 60,000 Btu/hr and range in air quantity from 600 to 2,000 cfm at standard

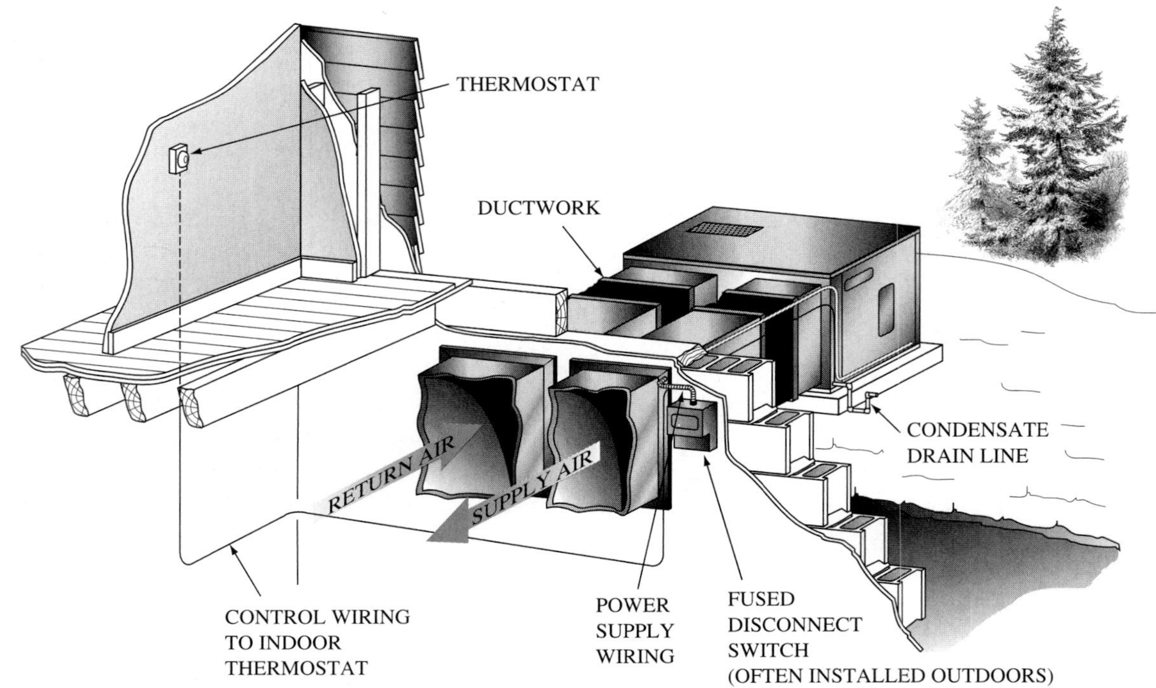

Figure A30-23 Ground level installation of a single-package conditioner.

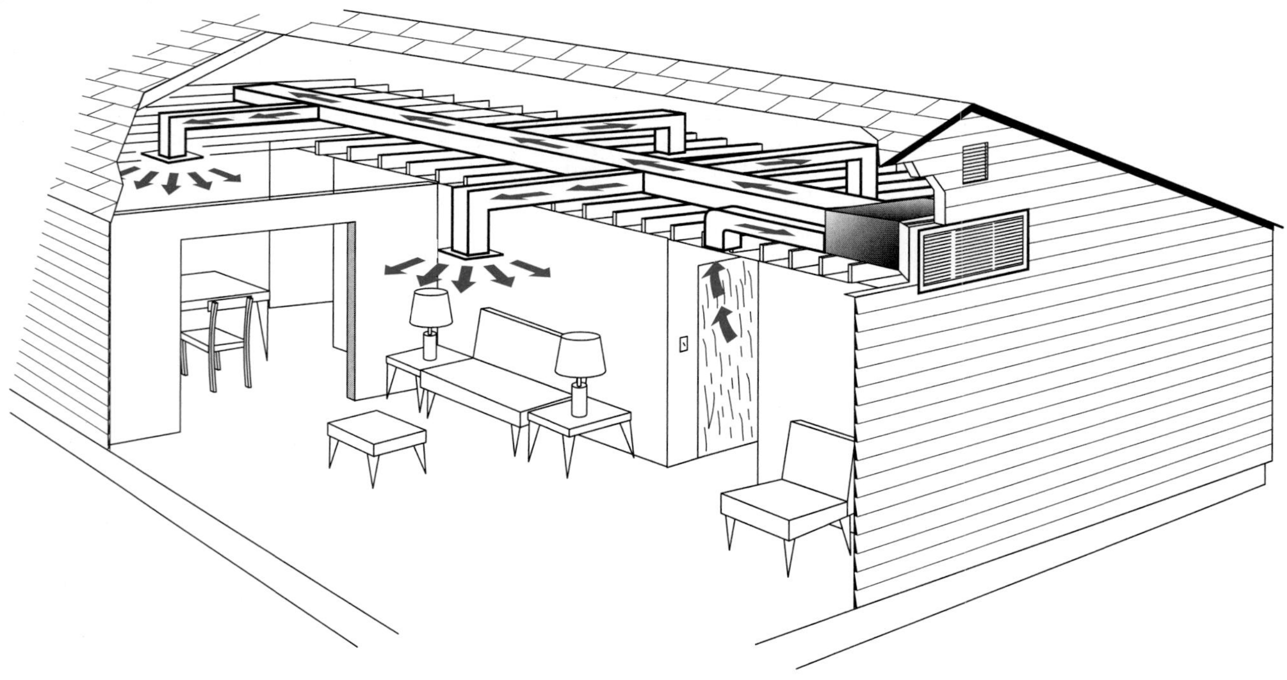

Figure A30-24 Attic installation of a single-package conditioner.

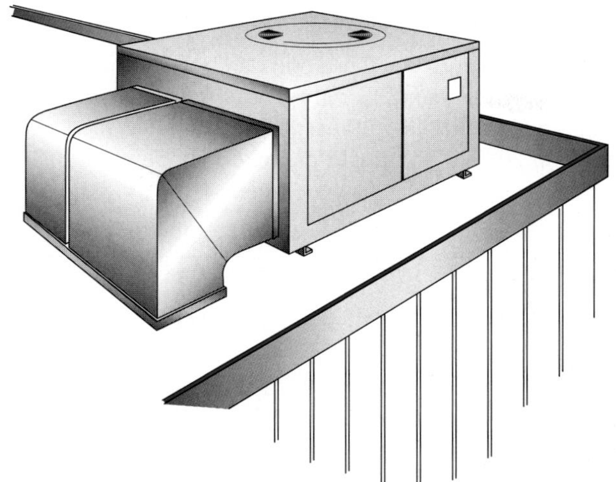

Figure A30-25 Roof installation of a single-package conditioner.

rating conditions. All units are available for 208–230-V/1-phase/60-Hz power. Efficiencies range from 8.6 to 10.0 EER, depending on the size. Optional electric heaters range in size from 3.74 to 29.80 kW.

A30-6.3 Vertical Conditioner

This is a commercial packaged unit for installation usually inside the space being conditioned. The condenser is water cooled. Water from a remotely located cooling tower needs to be piped in and out of the shell and coil type condenser(s), as shown in Fig. A30-26.

The unit can be supplied with a connection for a remote air-cooled condenser in place of the water-cooled arrangement.

In case the unit has two refrigeration circuits, an air-cooled condenser with a two-circuit coil needs to be supplied.

The unit is primarily designed for cooling; however, either hot water or steam coils can be installed in the unit. The extra resistance of these supplementary coils needs to be deducted from the rated external static pressure that the evaporator fan will handle.

Discharge air can either be free-throw or ducted horizontally or vertically. An accessory plenum and grille that fits on top of the unit is supplied for the free-throw arrangement. When duct work is used, a number of discharge configurations can be supplied, as shown in Fig. A30-27.

Application features that are incorporated into the unit are as follows:

1. The evaporator fan speed is adjustable, affecting the air-delivery cfm and the available duct-work static pressure.
2. Thermostats can be supplied as an integral part of the unit or provision can be made for remote mounting.
3. An anti-short-cycle timer is provided to protect the compressor from excess cycling.
4. If an air-cooled condenser is used, a low-ambient control can be provided to permit the unit to start with ambient temperatures as low as 0°F.

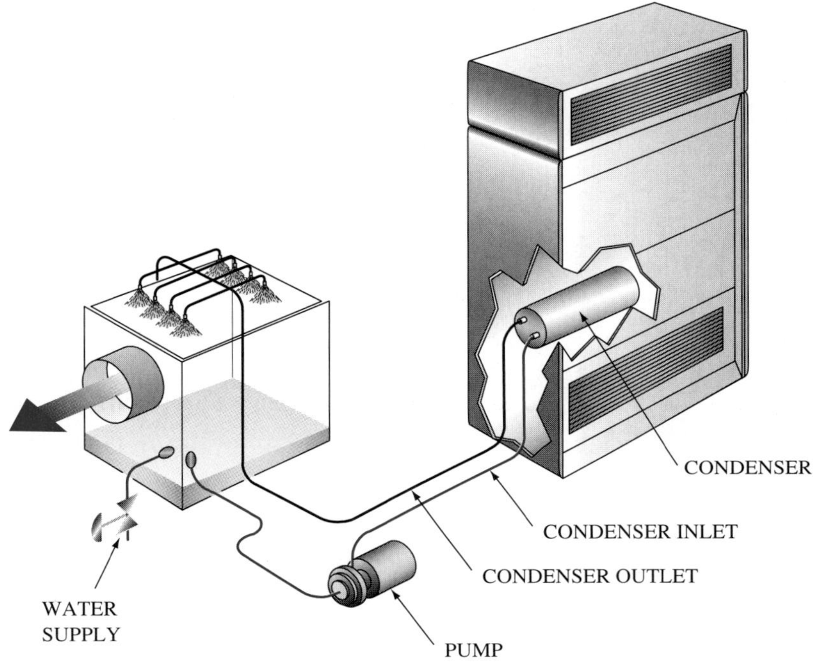

Figure A30-26 Vertical single-package conditioner with a water-cooled condenser and cooling tower.

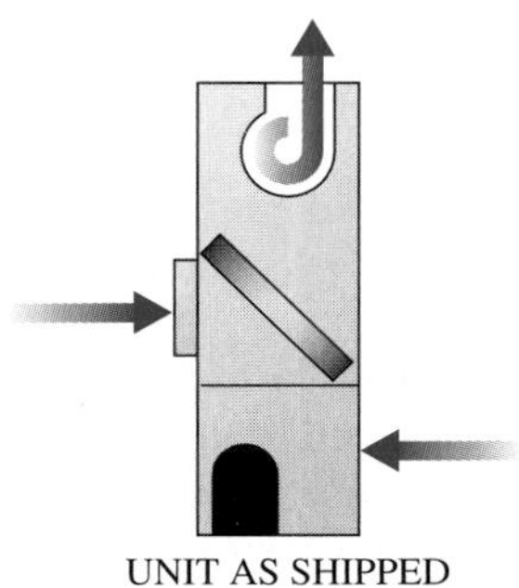

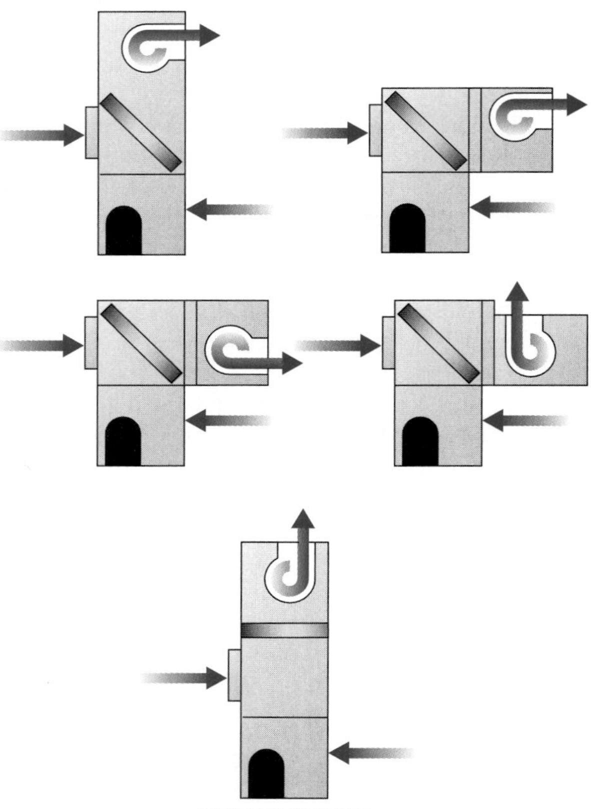

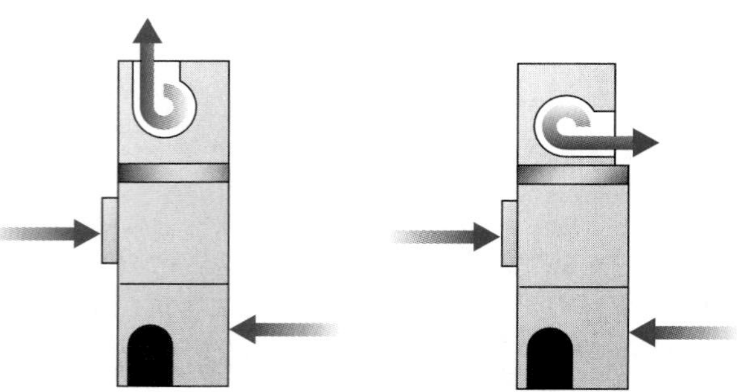

Figure A30-27 Duct arrangements for a vertical single-package conditioner. (The Trane Company)

A30-6.4 Performance

Typical capacities for these units are in the range of 3 to 15 tons under standard rating conditions. Air quantities range from 1,080 to 7,200 cfm. Units are designed to operate on 208–230-V, single-phase, 60-Hz power.

A30-7 SPLIT-SYSTEM CONDITIONERS

For many air-conditioning systems, it is not practical to place all components in a single package, particularly those that involve the use of air-cooled condensers. The air-cooled condenser must have access to outside air and can best be placed outside. For this reason split systems have been developed with the inside unit consisting of a fan-coil unit, with or without heating, an outside-mounted air-cooled condensing unit, and connecting refrigerant piping between the two.

Particularly for residential systems, using a split system offers the opportunity to add cooling to an existing heating unit where the necessary modifications are feasible.

A30-7.1 Residential Split Systems

Where the furnace exists (gas, oil, or electric) and the size of the duct work is adequate, a cooling coil may be added to the discharge outlet of the furnace. An air-cooled condensing unit is located outdoors on a suitable base. The two are connected by properly sized liquid and suction refrigerant lines. A typical installation is illustrated in Fig. A30-28.

Add-on Coils

Add-on coils are available in a number of configurations to fit various types of heating and fan-coil units, as shown in Fig. A30-29.

This illustration also shows the manner in which the position of the coils is related to the air-handling unit. Coils are supplied for upflow, horizontal, and downflow furnace applications.

Coil cabinets are insulated to prevent sweating and all have pans for collecting condensate runoff. Low-cost plastic hose may be used to pipe condensate drain water to the nearest drain. If no nearby drain is available, a small condensate pump may be installed to pump the water to a drain or to an outdoor disposal arrangement.

An important consideration in applying an add-on cooling coil to an existing furnace is the air resistance it adds to the furnace fan. During the heating cycle the coil is inactive and dry. In summer, when the unit is cooling

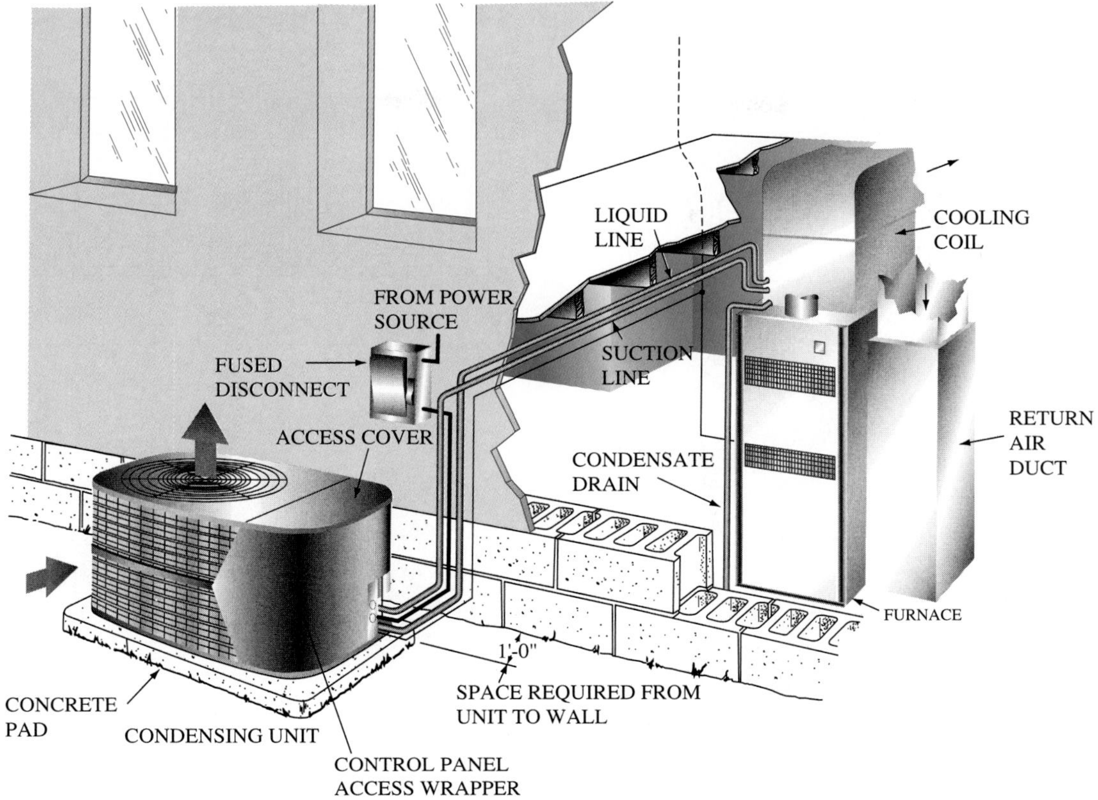

Figure A30-28 Residential add-on cooling to furnace installation.

and dehumidifying, the coil is wet. The wet coil will add an average of 0.20–0.30 in. WC static pressure loss. This added resistance will be sufficient to require a change in furnace fan speed. Also, an increase in motor horsepower will probably be needed. If the installation is a new system, a furnace fan should be selected which is capable of producing sufficient external static.

Air-cooled Condensing Unit

The outdoor air-cooled condensing unit of an add-on system (Fig. A30-30) consists of a compressor, condenser coil, a condenser fan, and the necessary electrical control box assembly. On residential condensing units, a fully hermetic compressor is used. It is sealed from dust and dirt and requires little ventilation.

Condenser Coil

The condenser coil is a finned-tube arrangement which varies in design by manufacturer. A large surface area is desirable, and many units offer almost a complete wrap-around coil to gain maximum area. The coil-tube depth is limited in order to reduce resistance to airflow.

Condenser Fan

The condenser fan also varies in design but is usually a propeller-type fan which can move large volumes of air while offering little resistance.

Airflow direction is a function of the cabinet and coil arrangement, and there is no one best arrangement. Most units, however, use the draw-through operation over the condenser coil. Outlet air, in its direction and velocity, can have an effect on surrounding plant life. Top discharge is the most common arrangement.

Fan motors are sealed or covered with rain shields. Fan blades are shielded for protection of hands and fingers.

Outdoor Installation

When installing a condensing unit at ground level, it is very important to provide a solid foundation (Fig. A30-31).

A concrete slab over a fill of gravel is recommended to minimize movement due to ground heaving. Where there is new construction, a foundation over unstable fill may settle sufficiently to place stress on or even break refrigerant lines.

Location of the outdoor condensing unit often is a compromise between several factors:

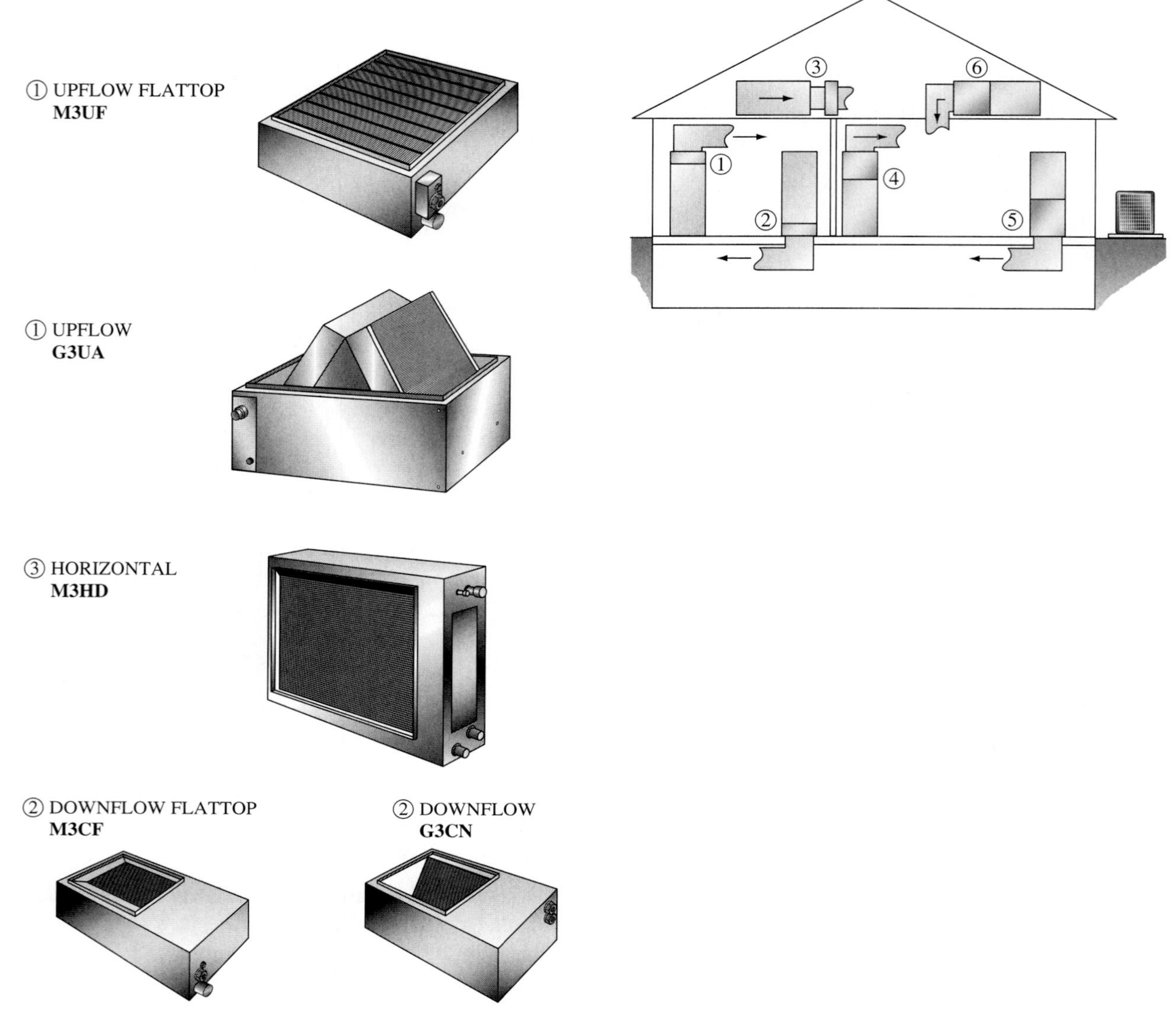

Figure A30-29 Types of evaporators for residential split systems. (Courtesy of York International Corp.)

- the actual available space
- the length of run for refrigerant lines
- the aesthetic effects on the home or landscaping
- noise/sounds effects on the home and neighbors

In general, avoid placement of the unit directly underneath a window or immediately adjacent to a patio. Locate it where plants provide a visual as well as sound-absorbing screen.

Noise

Noise is recognized as an environmental pollutant. Outdoor air-cooled condensing units are sound-producing mechanical devices, which some areas regulate. Early attempts to create local ordinances prompted action on the part of the ARI to establish in 1971 an industry method of equipment sound rating and application standards (which local communities could adopt).

ARI Standard 270 applies to the outdoor sections of factory-made air-conditioning and heat-pump equipment (unitary air conditioners). Under the program, all participating manufacturers are required to rate the sound power levels of their equipment in accordance with the technical specifications contained in this standard. Test results by the manufacturers are submitted to ARI for review and evaluation. Units are sound-rated with a single number, the *sound-rated number* (SRN). Typical ratings are between 14 and 24. ARI anticipates that the sound rating program will encourage manufacturers to produce quieter units.

Figure A30-30 Outdoor condensing unit for a residential split system. (Courtesy of York International Corp.)

Along with equipment rating, ARI Standard 270 contains recommended application procedures for using the sound rating to predict and control the sound level.

Refrigerant Piping

Refrigerant piping between the indoor coil and the outdoor condensing unit can take several forms. On residential type equipment, three methods of piping are:

1. Quick-connect couplings,
2. Compression fittings, and
3. Flared fittings.

These are non-soldered methods. Pump-down, evacuation, and refrigerant charging are not usually required.

The development of flexible, quick-connect, pre-charged refrigerant lines (Fig. A30-32) has played a major role in the growth and reliability of home air conditioning.

The liquid and suction lines are made of bendable-copper or coiled-steel tubing. Suction lines are factory covered with a foam-rubber insulation material. Each end of the tubing is fitted with a coupling half that mates with a coupling half on the equipment (Fig. A30-33). Both coupling halves have metal diaphragms that provide leak-proof seals. In addition, the coupling on the equipment contains a cutter blade and an intermediate synthetic rubber seal to prevent loss of refrigerant while the coupling is being connected. Tightening the union nut draws the coupling halves together (Fig. A30-34), piercing and folding both metal diaphragms back and opening the fluid passage. When fully coupled (Fig. A30-35) a parent metal seal forms a permanent leak-proof joint between the two coupling halves.

Note the service port for checking refrigerant pressure (Fig. A30-35). The port is equipped with a valve (not shown), similar to a tire valve (Schrader valve), that opens when depressed by gauge lines. When installing quick-connect pre-charged refrigerant lines, follow the manufacturer's installation directions on the radius of the bend for the size of tubing, lubrication of the coupling, and proper torque values when couplings are joined together. If excess tubing is present, form a loop or coil in a flat, horizontal manner. Do not form vertical loops, which will create oil traps.

Pre-charged lines are manufactured in various sizes and lengths from 10 to 50 feet. Plan the installation in order to have sufficient but not excessive tubing which adds cost and pressure drop to the system.

Another method of mechanically sealing refrigerant tubing is the compression-fitting technique (Fig. A30-36).

A coupling nut is slipped on the tubing, followed by a compression O-ring. When joined with the male adapter, the O-ring is compressed to form a tight and leakproof seal. The male adapter on the air-conditioning unit is also part of a service-valve assembly. Refrigerant tubing is shipped in coils that are clean and dehydrated. Removal of the seals and piping connection should be made as quickly as possible to minimize moisture condensation.

The third form of mechanical coupling is the flare connection which is generally limited to tubing sizes of ¾ in. OD. The tubing kit is factory cleaned and capped at both ends to minimize dirt and moisture penetration, but it is not pre-charged with refrigerant. The installer must be skilled in making the flare and quickly make the connection. If the system is open for more than 5 min., a complete procedure of purging, leak testing, full evacuation and recharging will be needed.

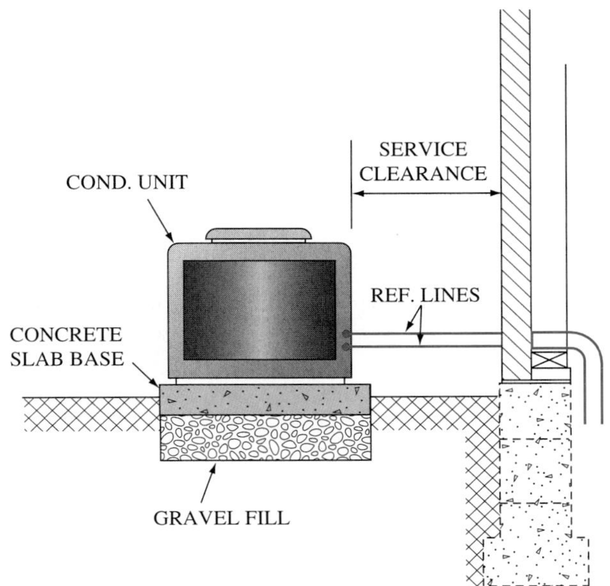

Figure A30-31 Outdoor condensing unit installed on a concrete slab.

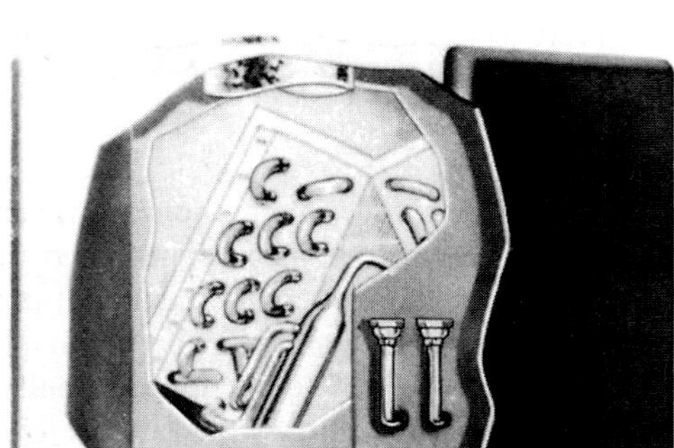

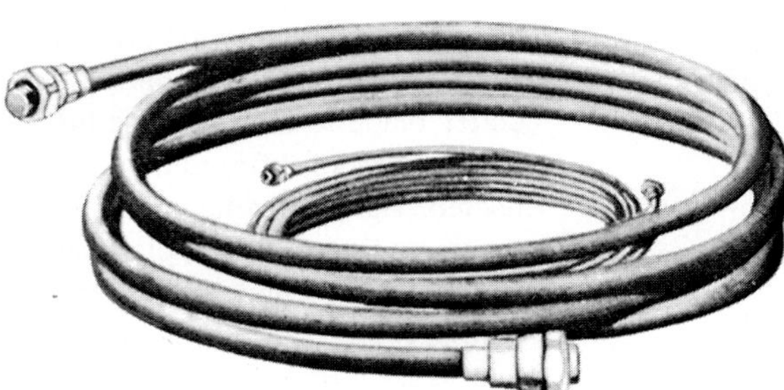

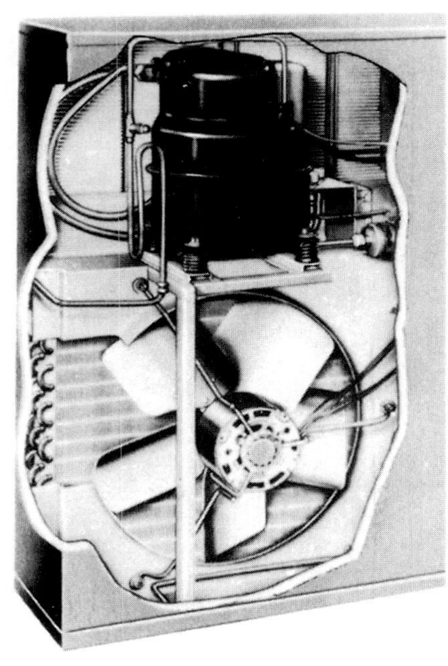

Figure A30-32 Pre-charged refrigerant lines. (Courtesy of Aeroquip Corporation)

Figure A30-33 Diaphragm-type piping coupling.

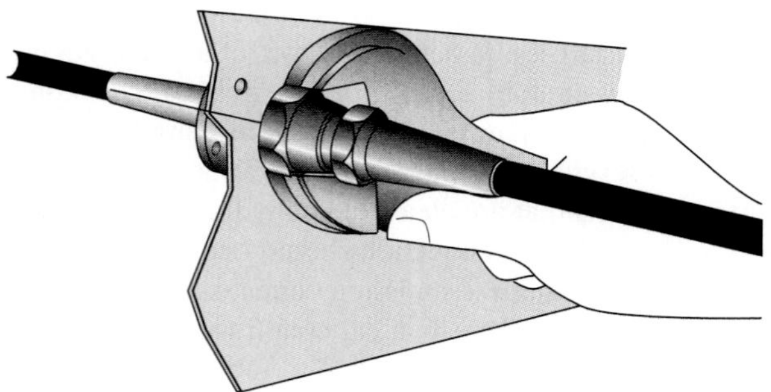

Figure A30-34 Tightening a piping coupling.

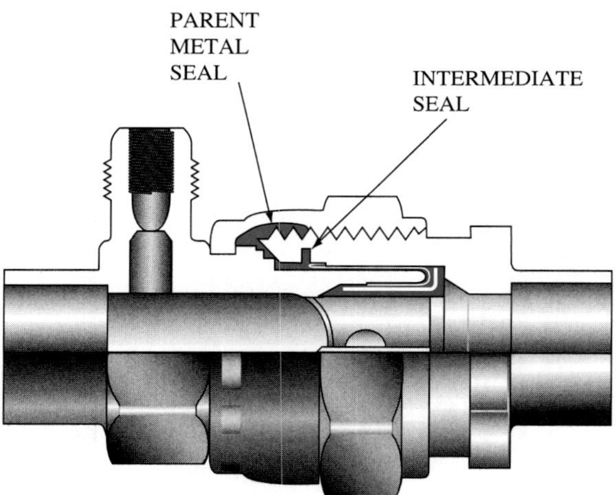

Figure A30-35 Cutaway view of a completed diaphragm coupling.

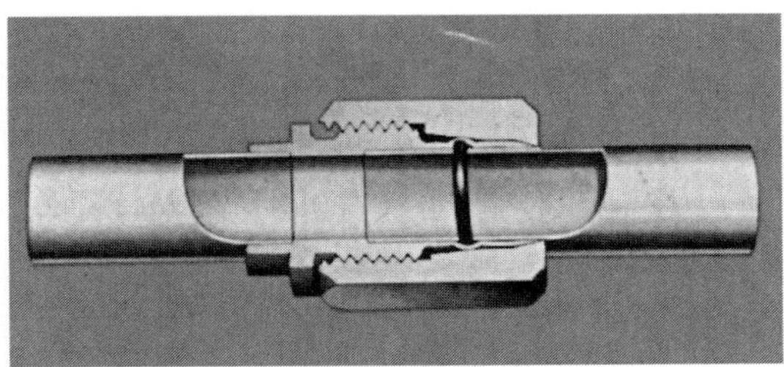

Figure A30-36 Compression fitting. (Courtesy of Carrier Corporation)

Condensing Unit and Evaporator

Single-compressor pre-charged systems are an economy of time and labor for the smaller systems of 5 to 7½ tons or less and with refrigerant lines up to 25 ft; however, these systems are not practical where large pipe sizes are required. The limiting size of soft-drawn copper tubing that can be formed or bent is about ⅞ in. OD. Unless dual circuits are employed, the larger systems must be field-fabricated with brazed connections.

In addition to the physical limitations of large pipe sizes, refrigerant characteristics limit both the length of line and height of refrigerant lift for split-system piping (Fig. A30-37).

When the evaporator is located above the condensing unit, there is a pressure loss in the liquid line due to the weight of the column of liquid and the friction of the tubing walls. The following table shows the results of these losses in pounds per square inch (psi) for R-22:

Vertical lift in ft	5	10	15	20	25	30
Static pressure loss (psi)	2½	5	7½	10	12½	15

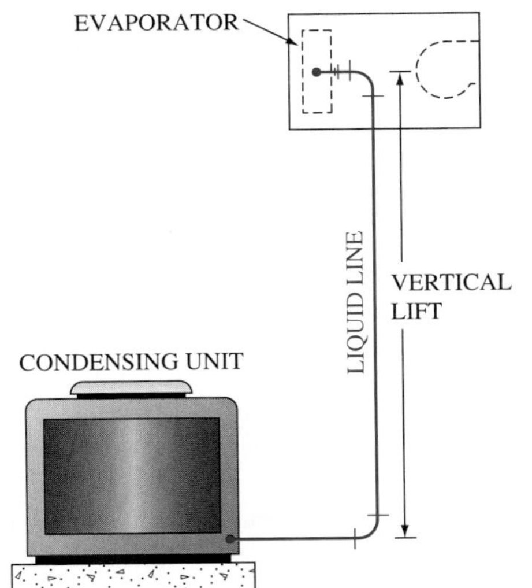

Figure A30-37 Piping from condensing unit to the evaporator.

Thirty feet of vertical lift is considered the maximum height. If the pressure loss is great enough, vapor will form in the liquid line prior to normal vaporization in the evaporator, adversely affecting operation and capacity.

Where the evaporator is above the condenser, the suction line should be pitched toward the condenser a minimum of 1 in. in 10 ft of run to ensure proper gravity oil return.

Wiring and piping a typical single-circuit add-on split system is shown in Fig. A30-38.

Electrical wiring consists of a line voltage of 208/240 V to the outdoor condensing unit through a fused outside-disconnect switch. The 115-V line voltage is supplied to the furnace, which furnishes power to the fan motor. For the lower control voltage, 115/24-V transformer is supplied. A 24-V combination heating- and cooling-room thermostat controls the on/off of cooling or heating through inter-locking relays.

It is considered good electrical practice to use permanent-split capacitor motors in small hermetic compressors for residential work. The auxiliary winding remains energized at all times during motor operation. The running capacitor is added to provide additional torque both during starting and running. Where low voltage fluctuation exists, it may be necessary to add a starting capacitor for extra starting power. Hard-start kits, which essentially consist of a starting capacitor, are available to overcome low-voltage starting problems and minimize light flicker.

The sizes of residential add-on split systems range from 1 ton to 7½ tons, the most popular being 2 to 3 tons in size. Models are available in multiples of 6000 Btu/hr (i.e., 12,000, 18,000, etc. Btu/hr). Most manufacturers rate and certify their equipment in accordance with ARI Standard 210, which is based on matching specific-size condensing units and cooling coils.

Split systems are also installed in new homes, apartments, and motels where the indoor equipment is a fan-coil unit rather than a furnace. Fig. A30-39 shows a ceiling evaporator-blower installed in a dropped ceiling, in the furred-down area above the closets, hallways, and bathroom. Room air is returned through a ceiling grille and conditioned air is discharged through high sidewall grilles located in each room.

Fan-coil units used in apartments and motels can also be supplied with electric resistance heaters for winter heating. Cooling sizes of ceiling evaporator-blowers range from 12,000 to 30,000 Btu/hr and are installed with quick-connect, pre-charged refrigerant lines. Condensate drain lines are carefully designed to prevent overflow that could result in damage to the ceiling.

A30-7.2 Commercial Split Systems

Split-system equipment up to 7½ tons in capacity may be classified as either commercial or residential. There is a wide range of applications in both markets using the same

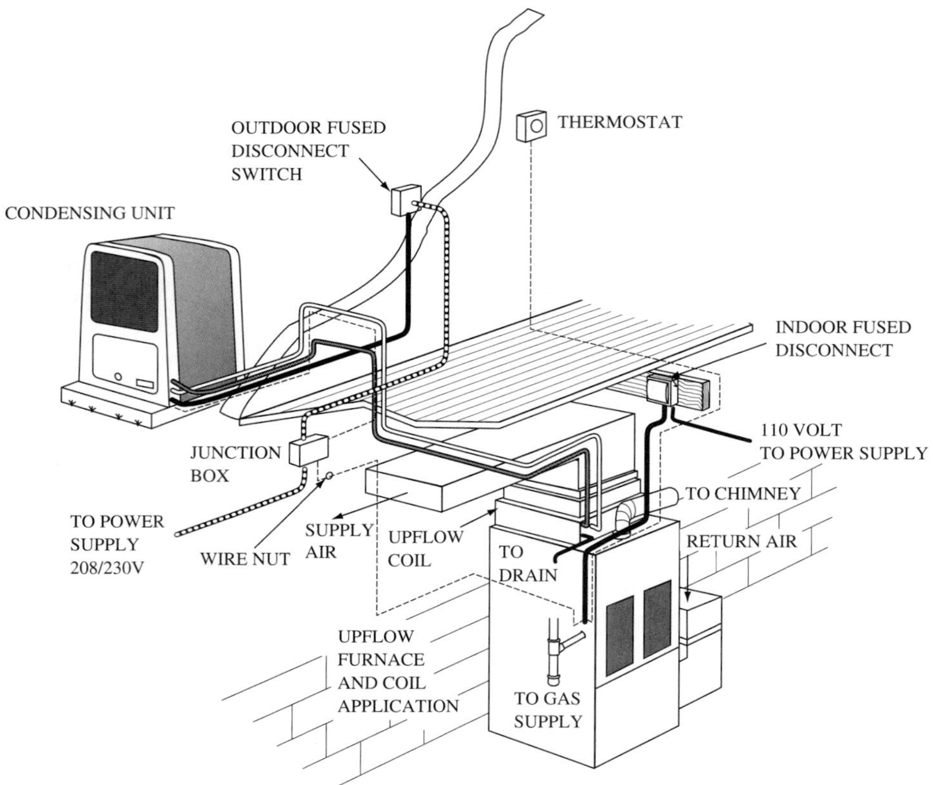

Figure A30-38 A single circuit add-on split-system installation.

product; however, above 7½ tons the application becomes distinctly commercial, and product designs use different components:

Soldered refrigerant lines and expansion valves are used rather than capillary-tube orifice-type metering devices;

Multiple compressors and condenser fans are used for capacity reduction and low-ambient operation;

More functional, heavier structural designs place less emphasis on appearance; and

Compressors all use three-phase electricity.

The air-cooled condensing unit, located outside the building, can be designed to discharge air either vertically or horizontally. Vertical discharge has several advantages:

1. These units expel large volumes of air which are best discharged in an upward direction.
2. The horizontal arrangement of fan blades is least affected by windmilling if the unit is off. This can be a problem if the fan tries to start with the blades rotating in the wrong direction.
3. The use of direct-drive multiple-propeller fans is almost universal in this type of equipment.
4. Ducting is not needed when equipment is located on the roof or at ground level in open space.
5. Fan noise directed upward is less objectionable.

Capacity control is necessary because commercial air-conditioning loads are rarely constant. Capacity reduction is accomplished in several ways. The most common are: (1) When multiple compressors are used for higher tonnages, staging or sequencing the operation is accomplished by cutting compressors on and off to match the load; (2) With single compressors, cylinder unloaders can be used to vary the pumping capacity down 25% or less. The machine starts almost unloaded and the cylinders are cut in as the heat load demands. Unloaded starting reduces power demand.

The centrifugal blower used for commercial systems must deliver a higher airflow and static pressure than for residential applications. A V-belt drive with an optional variable pitch motor pulley and motor horsepower is needed. The fan scroll can be rotated within the cabinet to obtain alternate discharge arrangements for greater application flexibility.

The evaporator coil(s) are circuited to match the number of compressors. Space is also provided for the inclusion of either non-freezable steam coils or hot-water coils for winter heating.

Ratings of matched condensing units with specific air handlers are published by the manufacturer and are certified under ARI Standard 280. A sample specification is shown in Fig. A30-40.

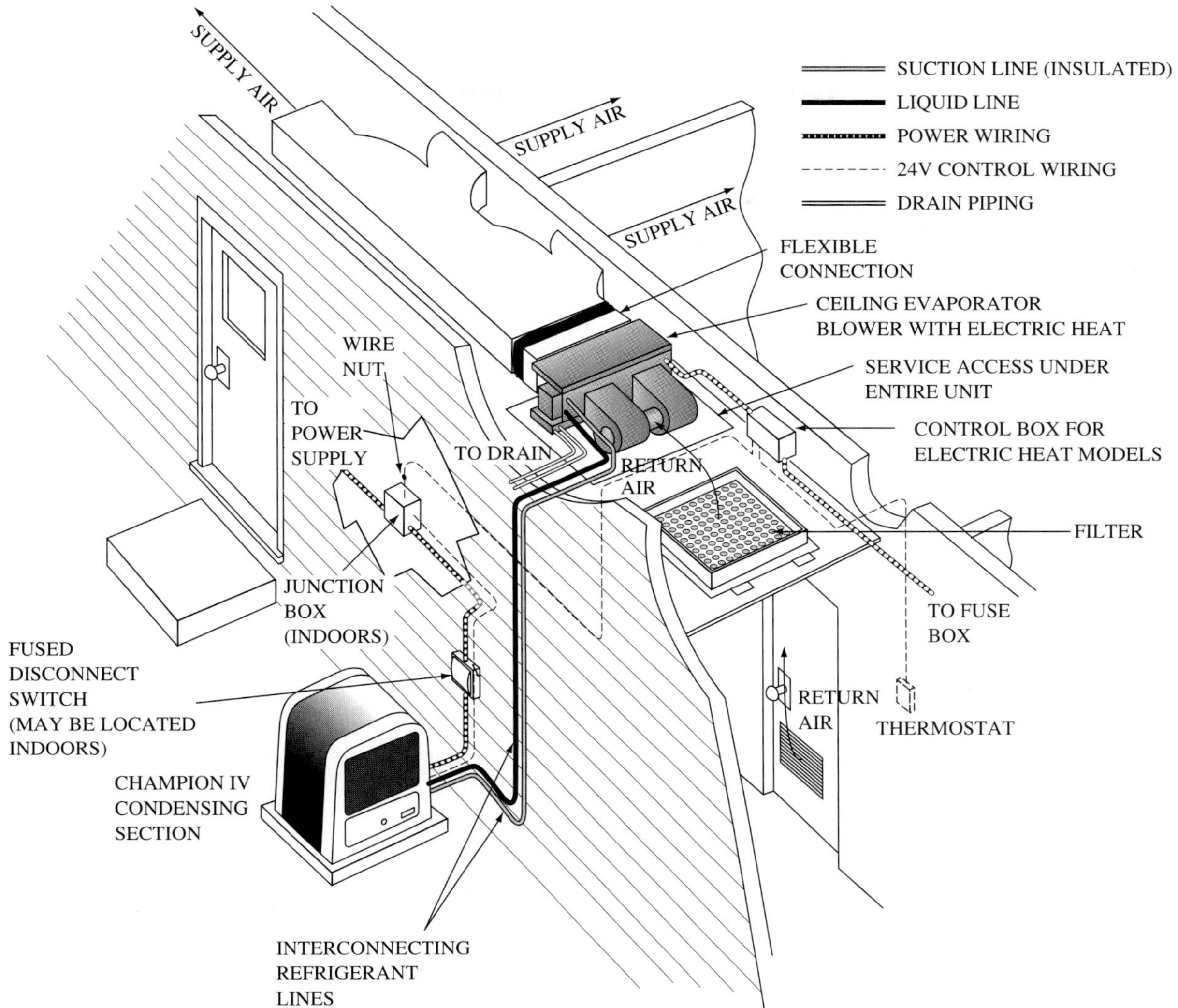

Figure A30-39 Ceiling evaporator-blower installation.

System		Air on Evaporator		Total Capacity (MBtu/hr)	WB Temp. off Evap. (°F)	Comp. and Cond. Fan Input (kW)	Sensible Capacity (MBtu/hr) Dry-Bulb Temp. on Evaporator (°F)				
		ft³/min	WB Temp. (°F)				70	75	80	85	90
CA 91	(EB92-B)	3300	72	101	65.9	10.4	—	40	56	69	89
	(EBV92-B)		67	93	60.6	9.9	43	57	73	88	91
	(C90UX[4])		62	86	55.4	9.4	57	73	84	86	86
	(92DX)		57	78	50.2	9.0	67	72	78	78	78
			72	—	—	—	—	—	—	—	—
	(EB122-B)	4400	67	98	60.6	10.2	43	59	77	93	98
	(EBV122-B)		62	91	55.4	9.7	61	79	87	91	91
	(122DX)		57	84	50.2	9.3	83	84	84	84	84

Cooling capacity—95°F DB air on condenser (MBtu/hr).

Figure A30-40 Performance specifications for a residential cooling system.

EXAMPLE

A system combination (condensing unit and air handler) has the following standards:

- Air volume based on 400 cfm/ton,
- 95°F outdoor air on the condenser,
- Air entering the evaporator at 80°F DBT, 67°F WBT.

Solution

The unit capacity can be read from the specifications as follows:

1. The total capacity as 93 MBtu/hr
2. Air leaving the evaporator at 60.6°F WBT
3. Total power input as 9.9 kW
4. The sensible heat capacity is 73 MBtu/hr

Performance tables of this type are supplied by the manufacturers to assist in the selection of standard products.

There are occasions, however, when non-standard combinations are desired. Manufacturers furnish capacity data on compressors and evaporator coils which can be plotted to arrive at a proper selection to produce a desired condition. These plots indicate the system balance in terms of the refrigerant operating temperature.

A knowledge of refrigerant piping for commercial split systems is important for the installation and servicing technician. The same applies to component parts of split systems. Detailed information on both of these systems is found elsewhere in the text.

A30-8 ROOFTOP CONDITIONERS

Rooftop conditioners are similar to single-package conditioners except that they are thoroughly weatherproofed and provide for duct access at the bottom of the unit. They are popular for air conditioning low-story commercial buildings because they offer a substantial savings of space within the building.

From the standpoint of the service technician, they are desirable because they offer plenty of access space around the unit for servicing. On many jobs, however, access to the roof is only by ladder and in bad weather the units may be difficult to reach and offer physical restraints in supplying needed tools and parts.

Rooftop self-contained air-conditioning units are commonly used on commercial installations. The sizes range from 3 tons to 130 tons of cooling capacity under standard rating conditions. Besides the difference in size of the components, individual units differ in the type of heating supplied with the package.

KNOWLEDGE EXTENSION

Larger packaged rooftop air conditioners now come with modern electronic direct digital controllers (DDCs) which will operate fan motors and compressors on and off automatically. For greater safety when servicing the equipment, lock out and tag out the equipment circuit breakers and/or disconnects. If these packaged units do not come with a terminal control panel to help diagnose and interface to the controls, you will probably require a laptop or personal computer. Most larger building-maintenance managers already have the ability from their office to monitor and control these units from their computers.

To learn more about these types of controls, contact the control vendors that originally installed the DDC controls and request service literature and materials.

For the units in the 3- to 25-ton range, gas-fired or electric heat can be supplied. Larger units can also be equipped with hot water or steam coils. When the units are located outside, adequate freeze protection needs to be provided where the unit contains water. Condensate drain pans must be free draining.

Since the most frequently used units are in the 3- to 25-ton range, this equipment will be described in the following sections.

Dual-compressor models are usually available starting at 7½ tons and higher. When dual compressors are furnished, dual refrigeration circuits are also supplied. This arrangement makes possible better performance ratings and increased energy savings at partial loads.

A30-8.1 Cabinet Construction

Cabinets are constructed of zinc-coated heavy-gauge steel and are weather tight. All services can be performed through access panels on one side. Supply and return ductwork connections can be made at the bottom or side of the unit. Roof curb frames are available for roof mounting.

A30-8.2 Air Filters

Most units in the 3- to 7½-ton range are provided with 1 in. throw-away filters. Larger sized units use 2-in. filters.

A30-8.3 Compressors

Reciprocating or scroll-type, direct-drive, hermetic compressors are used. Compressor motors are suction-gas cooled, protected with temperature- and current-sensitive overloads. Crankcase heaters are standard equipment.

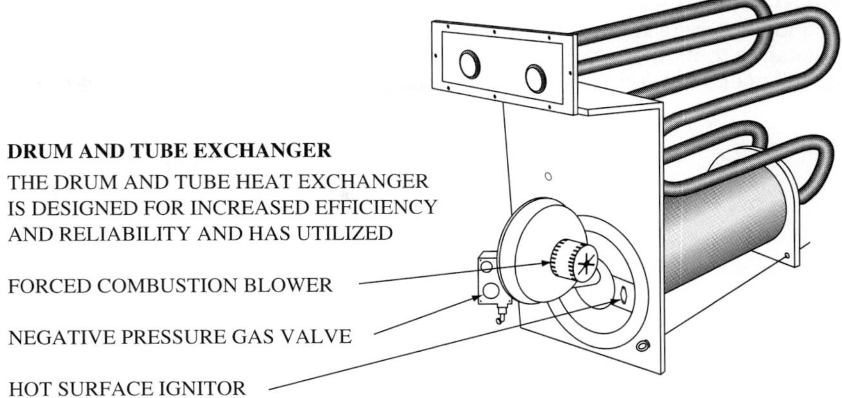

Figure A30-41 Drum-and-tube heat exchanger. (The Trane Company)

A30-8.4 Gas Heating

The heat exchanger is made of corrosion-resistant steel. One manufacturer uses a drum-and-tube design (Fig. A30-41). Units use a forced-combustion blower and hot-surface ignition. Gas will not ignite unless the blower is operating.

On an initial call for heat, the combustion air blower will purge the heat exchanger with fresh air. If the attempt to light the main flame is unsuccessful, the system will purge the heat exchanger again and start a second trial for ignition. If there are three unsuccessful trials for ignition, the entire heating system will lock out until it is manually reset.

A30-8.5 Condenser and Evaporator Fan

The condenser fan is a propeller type with a permanently lubricated, overload-protected motor. The evaporator fan is a centrifugal type, belt driven in most sizes, with adjustable sheaves. Units are capable of delivering nominal airflows at 1 in. external static pressure (ESP).

The unit can be equipped to handle a supply duct system using variable-air-volume (VAV) control as shown in Fig. A30-42.

A30-8.6 Controls

The newer units are provided with microprocessor controls for all 24 V functions. Control decisions are automatically made in response to the input from indoor- and outdoor-temperature sensors. Anti-short-cycle compressor controls are provided. The control system can be interfaced with a central direct digital control (DDC) system.

A30-8.7 Outside Air Dampers

Manually positioned outside-air dampers that can be adjusted to provide up to 25% outside air are supplied with a rain screen and hood for field installation.

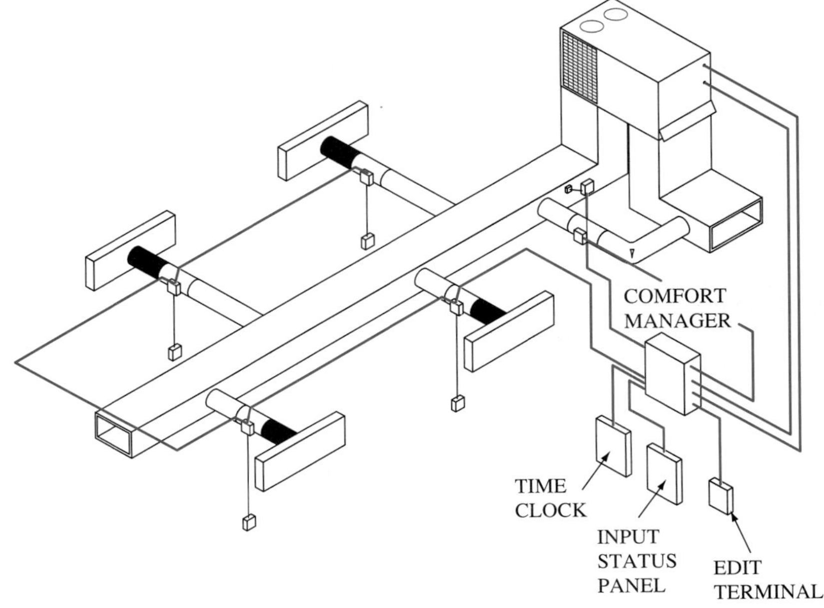

Figure A30-42 Variable-air-volume installation. (The Trane Company)

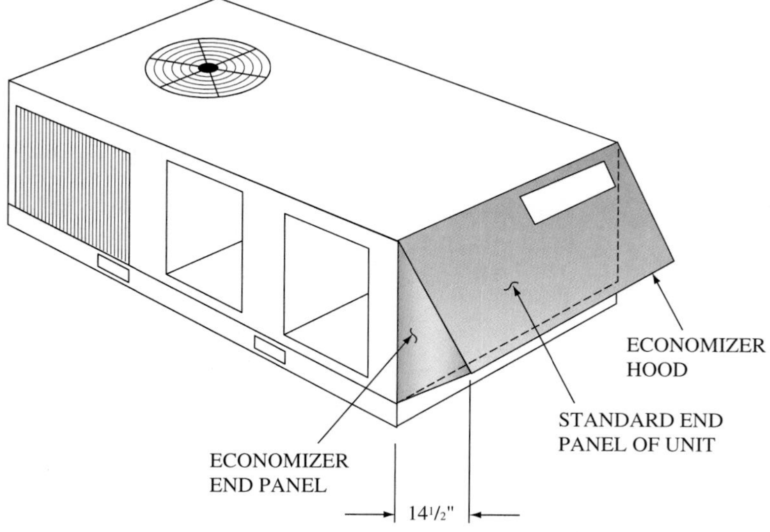

Figure A30-43 An economizer installation in a packaged rooftop conditioner. (The Trane Company)

A30-8.8 Economizer

An *economizer* is an optional arrangement for using outside air for cooling if conditions are feasible, as sensed by the control system. The alterations to the unit are shown in Fig. A30-43. The assembly includes a fully modulated 0–100% motor and dampers, barometric relief, minimum position setting, preset linkage, wiring harness and plug, and controls. Fig. A30-43 shows the outside-air opening for the economizer cycle. The barometric-relief damper provides automatic closing of the outside-air opening when the equipment is not operating.

A30-9 DESICCANT COOLING SYSTEMS

A *desiccant* is a material that absorbs moisture without causing a chemical change in the material. The material can thus be reactivated and reused by applying heat to the saturated product. Systems with desiccants are used for dehumidifying air. The term *desiccant cooling* is applied to these systems since dehumidification is a part of the air-conditioning (cooling) process. These systems can only be justified in relation to their ability to save air-conditioning energy.

An example of the use of a desiccant for an air-conditioning application is shown in Fig. A30-44. The figure illustrates a rotating wheel impregnated with a porous desiccant. The air passages through the wheel are divided into two parts, one for low-temperature outside air entering the building, and the other for hot, moist, exhaust air leaving the building.

When the hot moist air contacts the desiccant, the moisture is absorbed and at the same time produces heat (970 Btu/lb of moisture absorbed). As the wheel rotates into the outside airstream the heat held by the desiccant is given off, preheating the cool outside air before it enters the building. This is an efficient use of the desiccant in winter.

In summer the desiccant wheel can be used to dry the incoming air and eliminate the need to chill the air deeply to condense the moisture. This reduces the amount of energy required to produce cooling.

A30-9.1 Energy Recovery Ventilators

The unitary equipment that uses the desiccant wheel is the energy-recovery ventilator (ERV). An illustration of one

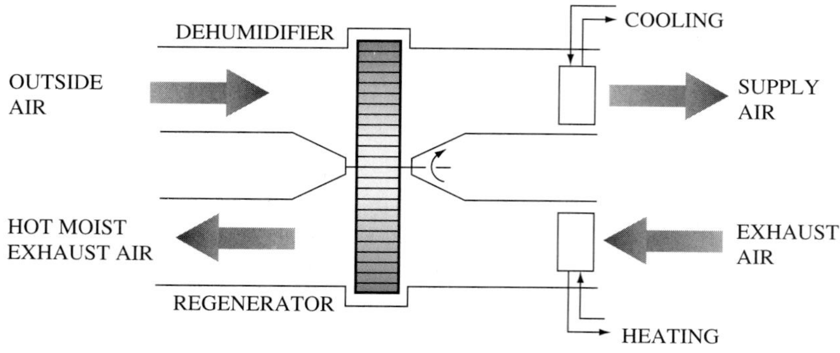

Figure A30-44 Rotating wheel with solid desiccant for transferring water vapor. (Courtesy of ENERGY ENGINEERING)

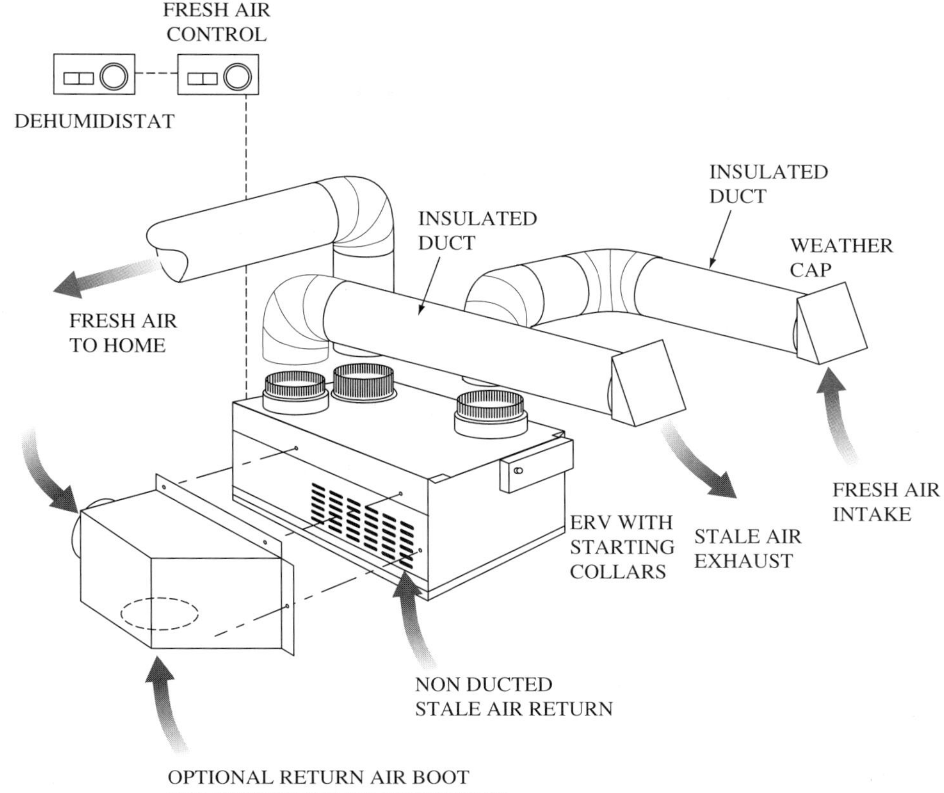

Figure A30-45 Energy-recovery ventilator (ERV).

of these units that has been developed for residential application is shown in Fig. A30-45. The desiccant wheel is designed to transfer water vapor from one airstream to another.

The wheel will also transfer heat. It will recover enough heat from exhaust air to warm incoming air to 60°F, when the outside air temperature is as low as 5°F. On sub-zero days, an electric frost control operates an electric preheater to warm incoming air a few more degrees. Different-sized heaters are available for different design-temperature ranges.

Another type of heat recovery ventilator unit is shown in Fig. A30-46. This unit is primarily used to transfer heat from one airstream to the other. In winter the transfer core uses the heat of indoor air to warm the incoming cool fresh air. In the warm months air-conditioned indoor air cools the incoming fresh air. The core recovers about 75% of the energy during this process and at the same time removes a substantial amount of humidity from the warm incoming air to reduce the air-conditioning requirement.

A typical installation schematic for an energy-recovery system is shown in Fig. A30-47.

Both of these recovery units are designed to provide a continuous supply of outside (ventilation) air to improve the indoor air quality (IAQ) of the building.

Figure A30-46 Fresh-air heat recovery ventilator. (Courtesy of Research Products Corporation)

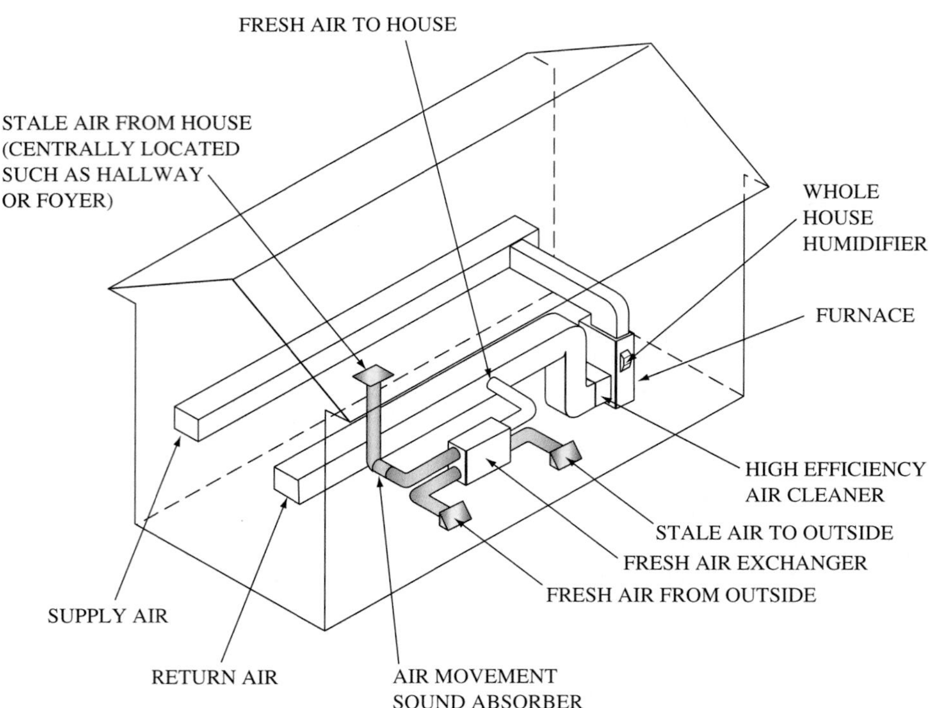

Figure A30-47 Installation diagram of heat recovery ventilator in a residence. (Courtesy of Research Products Corporation)

REVIEW

- Unitary air-conditioning equipment—Factory-built and tested systems, complete with controls, piping, wiring, and refrigerant; require only service connections and duct work to install. Assemblies include an evaporator, a compressor, and condenser, and may include a heating unit. A packaged air conditioner includes all components in one assembly and is self-contained.
- Types of unitary equipment:
  Room air conditioners—Provide semi-portable cooling, dehumidifying, and ventilation to an existing room.
  Console through-the-wall conditioner—Type of room cooler designed for permanent installation. It was developed to provide individual room cooling for hotels, motels, and offices. These units are known as packaged terminal air conditioners (PTACs), sometimes called incremental systems.
  Dehumidifier units—Small, portable, self-contained refrigeration systems designed to extract moisture from the air.
  Single-packaged conditioner—Horizontal or vertical packaged system designed for permanent installation, to condition larger spaces. 1½ to 15 tons
  Split system—Remote-connected air condenser, with evaporator mounted in air handler.
  - Commercial split systems (7½ horsepower and above):
    Use soldered refrigerant lines and TXV valves.
    Multiple compressors, unloaders, and condenser fans for capacity reduction control.
    Use three-phase electricity power supply.
    Rooftop air conditioner—Single-package conditioner, weatherproofed with duct access at the bottom of the unit, placed on flat roof with holes cut into the roof for duct access. May be single- or multi-zone, provided with conventional electrical or microprocessor controls.
    Desiccant cooling system—Uses a desiccant for dehumidifying and cooling the air.
- EER (Energy Efficiency Ratio)—Cooling output in Btu/hr divided by the power input in watts under standard rating conditions.
- Energy-recovery ventilator—Provides fresh air while filtering and providing dehumidification and heat transfer.
- Standard ARI rating conditions are based on 80°F DBT, 67°F WBT entering air and 95°F DBT outdoor ambient air.
- Standards are set up by:
  Canadian Standards Association (CSA)—performance standards.
  Underwriters Laboratory (UL)—electrical and safety standards.
  Air Conditioning and Refrigeration Institute (ARI)—performance standards.
  American Society of Testing and Materials Engineers (ASTME)—wind and rain infiltration standards.

Problems and Questions

1. Packaged terminal units are also called _____.
2. Rooftop combination heating and cooling units also divide into two descriptions based on air flow. What are they?
3. The use of outside air during mild weather for cooling is called _____ cycle.
4. What types of heating systems are used with window or through-the-wall console units?
5. What component differentiates a heat pump from an air conditioner, and provides the reverse cycle?
6. How can desiccant cooling units be used for heating?
7. What organization sets safety and electrical standards?
8. What organizations set the standards for performance of cooling units?
9. What are other names for packaged terminal air conditioners?
10. What does EER mean? What is SEER?
11. The size of unitary equipment ranges from small fractional-tonnage room coolers to large packaged rooftop units in the 100-ton range. True or false?
12. In numbers sold, rooftop air conditioners outsell all other types of unitary equipment. True or false?
13. PTAC is also known as a packaged terminal air conditioner. True or false?
14. Energy Efficiency Rating (EER) is equal to the cooling input in Btu/hr divided by the power input in watts under standard rating conditions. True or false?
15. If no condensate drain is available, a small condensate pump may be installed to pump the water to a drain or to an outdoor disposal arrangement. True or false?
16. The condenser fan on the split condensing unit is usually a:
    a. Propeller-type fan.
    b. Squirrel-type fan.
    c. Shaded-pole motor fan.
    d. None of the above.
17. Refrigerant piping on residential split systems includes:
    a. Quick-connect couplings.
    b. Compression fittings.
    c. Flared or soldered fittings.
    d. All of the above.
18. To adjust the load according to capacity, larger commercial split systems will use:
    a. Multiple compressors, unloaders, and condenser fans.
    b. Special refrigerant circuiting.
    c. Special economizer fan circuits.
    d. None of the above.
19. The controls on large rooftop packaged units can often be interfaced with:
    a. DDC system.
    b. Economizer controls.
    c. Outside air dampers.
    d. All of the above.
20. A desiccant is a material that:
    a. Absorbs the cold air and moves it to a location of less importance.
    b. Absorbs sensible heat from the refrigeration circuit.
    c. Absorbs moisture without causing a chemical change in the material.
    d. None of the above.

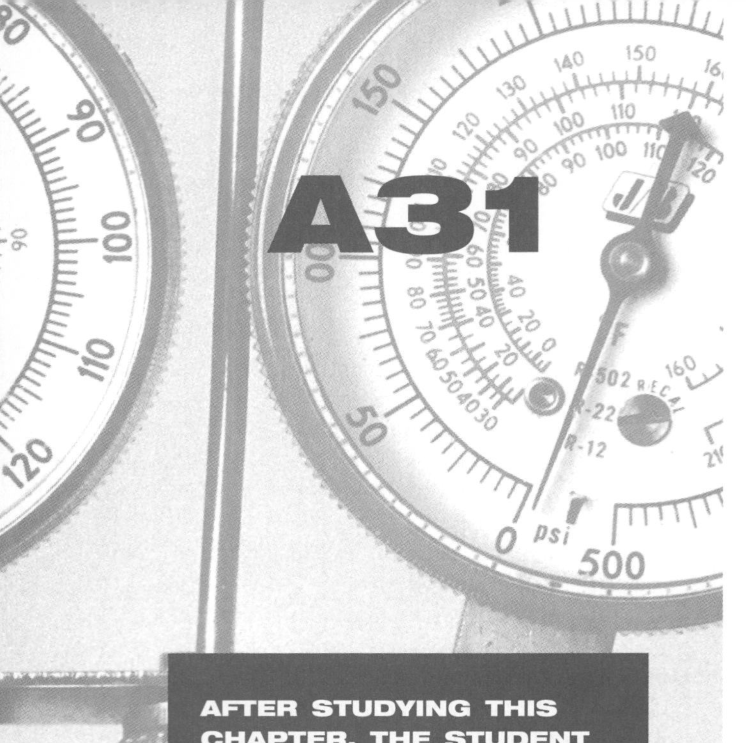

A31 Central Station Systems, Including Thermal Storage Systems

AFTER STUDYING THIS CHAPTER, THE STUDENT WILL BE ABLE TO:

- Describe and identify central-station air-conditioning systems, and how they differ from unitary systems.
- Explain water-chiller types and configurations.
- Describe boiler classifications for hot-water and steam applications.
- Identify the different types of air-distribution systems.
- Describe the purpose for circulating pumps and cooling towers.
- Explain control systems utilizing electric, electronic, pneumatic, or any combination.
- Describe direct digital-control systems.
- List types of ice-storage equipment and various configurations available.
- Explain the concept of thermal storage, its benefits, and limitations.
- Describe and explain the basic function of components of central-station systems: air distribution systems; expansion tanks; heat recovery systems; water chillers; water cooling towers.
- Explain the operation of a central station system.
- List the requirements for system control: electronic DDC; pneumatic; building systems.

A31-1 DESCRIPTION

A *central-station system* is also commonly called an applied system, applied equipment, applied machinery, and engineered system. Essentially, these are names adopted by the various companies that manufacture the equipment.

Central-station equipment is associated with installations where the cooling plant is located in the basement or in a penthouse on the roof of a multi-story building. The central station serves air-handling equipment and air-distribution systems throughout the building. Although size is not necessarily the crossover point between unitary and central station, it is usually acknowledged that central-station equipment starts at 25 to 50 tons and extends upward to multi-thousand-ton systems. Unitary equipment tapers off in the range of 50 to 75 tons, with a few packages of 100 tons or slightly more.

Another distinguishing difference is that central-station systems use the medium of chilled or heated liquid (usually water) to transfer heating and cooling to air-delivery systems, where the air is conditioned by heating and cooling coils. Unitary systems heat or cool air directly by contact with refrigerated or heated coils, without any intermediate heat transfer, and deliver it directly to the conditioned space.

Unitary equipment makes use of factory packaged, balanced and tested equipment components, requiring a minimum of on-site labor to be operational. Central-station systems are made up of separate equipment components, selected to operate together to provide the required performance. Typical system components include water chillers, hot-water generators, air handlers, pumps, cooling towers and control systems, field assembled into a complex air-conditioning system, requiring a variety of on-site trades and crafts. Central-station equipment is usually associated with *"plan-and-spec"* engineered systems, developed and put out by consulting engineering firms. Equipment is then selected and built to order to comply with these specifications. Delivery time for some of this equipment may run from a few months to as much as a year on complex units. The art of selecting and matching the components is done by skilled professional engineers, and the servicing technician does not have to be concerned with that function.

The service technician, however, is concerned with the system's installation, operation, and maintenance and therefore must learn the basics of the equipment being utilized and the system application. Let's take a look at the scope of a generic central station system.

A31-2 SYSTEM COMPONENTS

A typical system is shown in Fig. A31-1. The drawing shows the central-station air handler which heats or cools

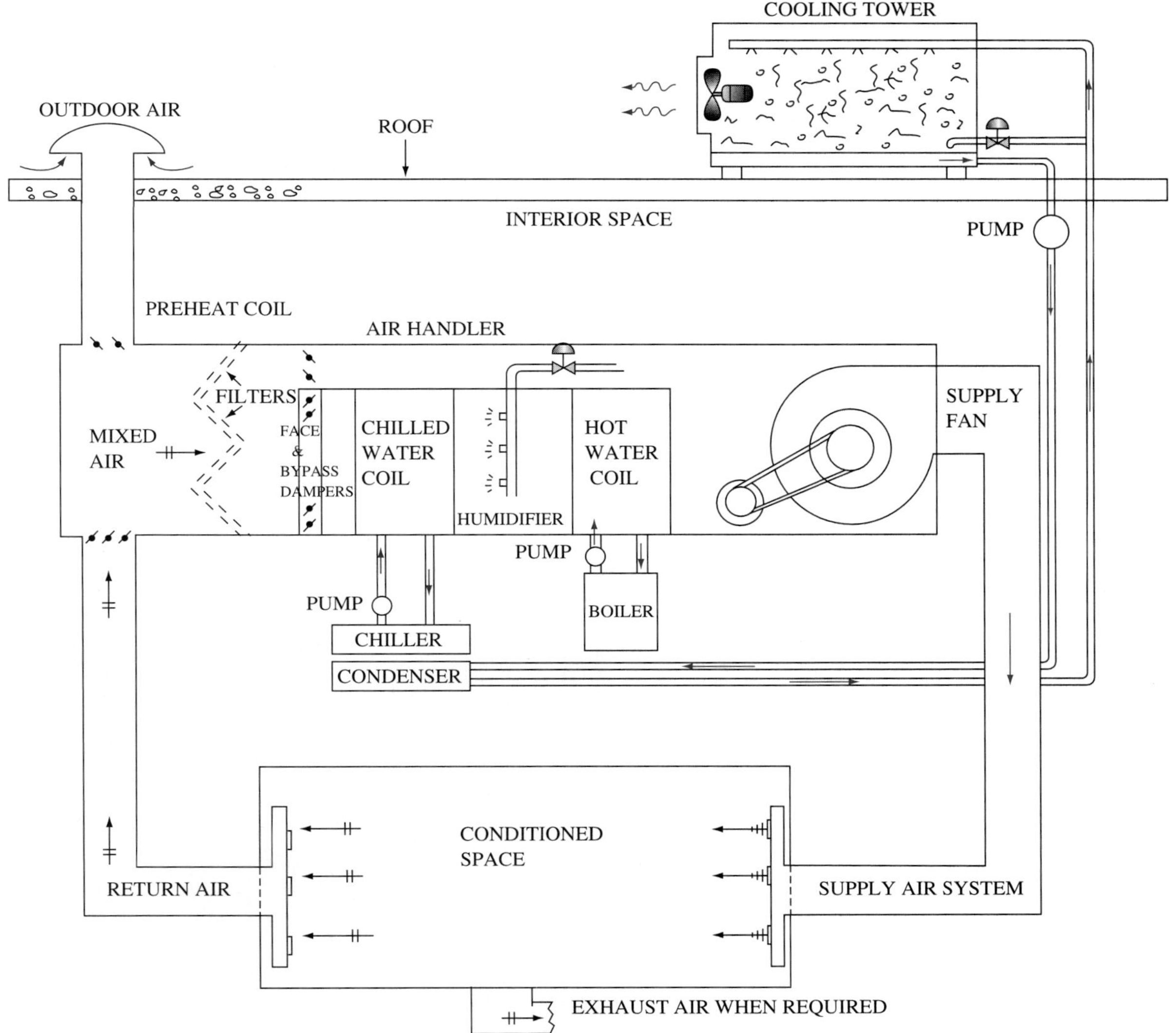

Figure A31-1 Conventional central-station system.

the air and includes the following components, as a minimum:

1. Water chillers
2. Boilers
3. Circulating water pumps
4. Cooling towers
5. Air-handling units
6. Air-distribution components
7. Control systems
8. Thermal storage

A31-3 WATER CHILLERS

The *water chiller* refrigerates water to approximately 43 to 45°F. The *chilled-water pump* circulates the chilled-water supply (CHWS) to the cooling coil in the air handler. The cooling coil normally has piping-isolation valves and a control valve. The heat absorbed by the cooling coil warms the water about 10°F at full load, with a chilled-water return (CHWR) temperature of 53 to 55°F.

There are various configurations of water chillers. Basically they differ in: (1) type of condenser, (2) type of compressor, and (3) size. Based on consideration for these parameters, we will be discussing the following types:

1. Water-cooled chillers
2. Air-cooled chillers
3. Centrifugal chillers
4. Screw-compressor chillers
5. Absorption chillers

A31-3.1 Water-cooled Chillers

The design and range of water-cooled chillers is related to the compressor and condensing medium (air or water). All chillers have compressors, liquid chillers, compressor starters, controls, and refrigerant- and oil-pressure gauges neatly assembled into a package.

Starting at the lower capacities, packaged water chillers use one or more *reciprocating compressors.* Fig. A31-2 shows an example of a small water-cooled chiller in the 20-ton range.

The larger packaged chillers, also water cooled, with reciprocating compressors, range upward from 40 to 200 tons (Fig. A31-3). These have multiple compressors that permit close control of capacity. They may also have standby compressors in the event of a malfunction. Where single compressors are used, these are equipped with cylinder unloading to allow capacity reduction and minimum starting-power requirements.

In addition to reciprocating compressors, the *scroll-type compressor* is being utilized increasingly in packaged chillers. These compressors use a pair of mating scroll-shaped surfaces to compress the refrigerant with pure rotary motion (Fig. A31-4).

The individual scroll compressors are very efficient and are used in sizes from 5 to 20 tons. Since they do not have any capacity control, they are usually used in multiples of 2, 3, or 4 compressors in a chiller unit for efficient operation.

Heat rejection is the function of a *shell-and-tube condenser* (Fig. A31-5). Water flows through the tubes, and refrigerant vapor fills the shell, condensing to a liquid. The liquid is collected in the bottom where it is subcooled 10 to 15°F for greater cooling capacity. In the water circuit, the coldest condenser water enters the lower part of the shell and circulates through the tubes. The water will make two or three passes through the shell before it is discharged. This is arranged by circuit baffles in the condenser heads. The higher the number of passes, the greater the pressure drop and pressure required to produce the re-

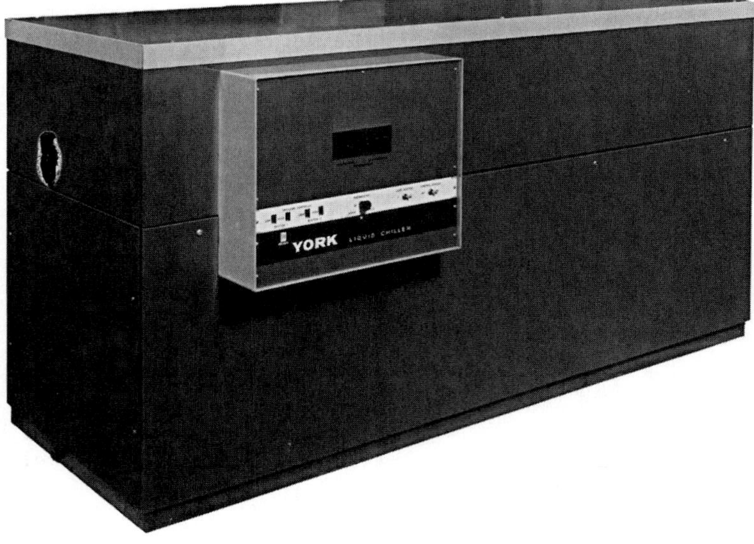

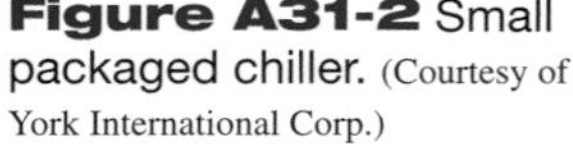
Figure A31-2 Small packaged chiller. (Courtesy of York International Corp.)

Figure A31-3 Large packaged chiller with reciprocating compressor.
(Courtesy of York International Corp.)

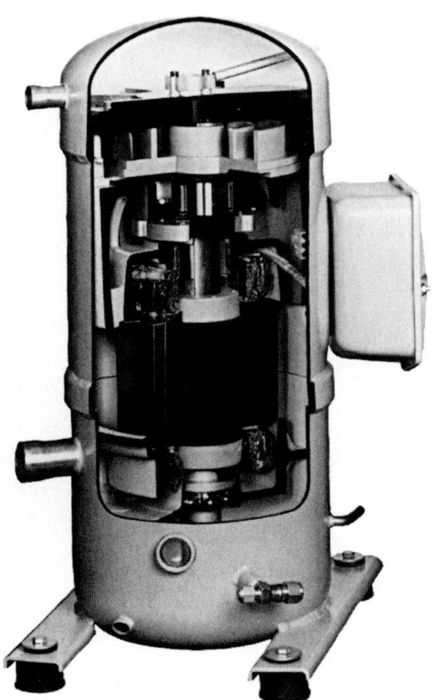

Figure A31-4 Cutaway view of a scroll compressor.
(Courtesy of The Trane Company)

quired flow rate. There are also cross baffles within the shell which serve to hold the tube bundles but also help to spread the refrigerant gas over the entire length of the shell. Condenser capacity is normally based on 85°F entering-water temperature with a 10°F rise.

A special sea water condenser is used for marine duty where salt water is the cooling medium. These tubes are made of cupronickel steel to withstand the salt corrosion effects.

The chiller (evaporator) shown in Fig. A31-6 is a direct-expansion type, associated with R-22 reciprocating compressors. Refrigerant flows through the tubes and will generally make two passes for standard operation, which gives a counter-flow arrangement versus the water-flow pattern.

The cooler shell and suction lines must be properly insulated to prevent sweating. This is done at the factory with a layer of closed-cell foam insulation prior to painting.

Both condenser and cooler shells must comply with ANSI B9.1 and the applicable ASME safety codes for pressure vessels.

Standard chiller ratings are based on ARI Standard 590: 44°F leaving-water temperature off the cooler at 105°F and 120°F condensing temperatures; and 95°F leaving water off the condenser with a 10°F rise. The 95°F condenser-water rating will produce a condensing temperature near 105°F. The rating point of 120°F condensing temperature is used for applications where remote air-cooled condensers (Fig. A31-7) are used instead of water-cooled types.

A31-3.2 Air-cooled Chillers

These chillers, as shown in Fig. A31-8, are complete packages which can be mounted on the roof or outside the building. They are similar to large air-cooled condensing

PRACTICING PROFESSIONAL SERVICES

Large compressors and mechanical rooms can be potentially hazardous to your hearing if they are operating at full speed. Whenever working in these areas, it is good practice to wear protective hearing devices to avoid permanent hearing damage. Also, be cautious of oil spills around moving parts which may cause you to slip or fall.

Becoming familiar with the compressor room will require the technician to identify all of the piping, moving parts, and associated controls. Always look for thermometers located in the chilled-water supply, chilled-water return, condenser-water supply, and condenser-water return. Expect to find a 10°F temperature difference during normal operation across the condenser and water chiller. Become familiar with normal suction, discharge, and oil pressures during normal conditions, partial-load conditions, and fully loaded conditions. Larger facilities will have a maintenance and operations log sheet which will provide standard operating conditions plus historical data and trends. Learn to utilize all available drawings, schematics, and manuals to better become acquainted with these systems.

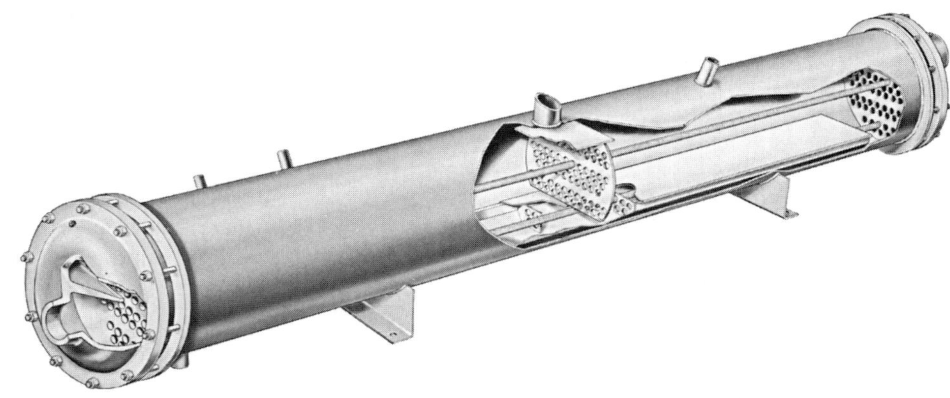

Figure A31-5 Shell-and-tube condenser. (Courtesy of York International Corp.)

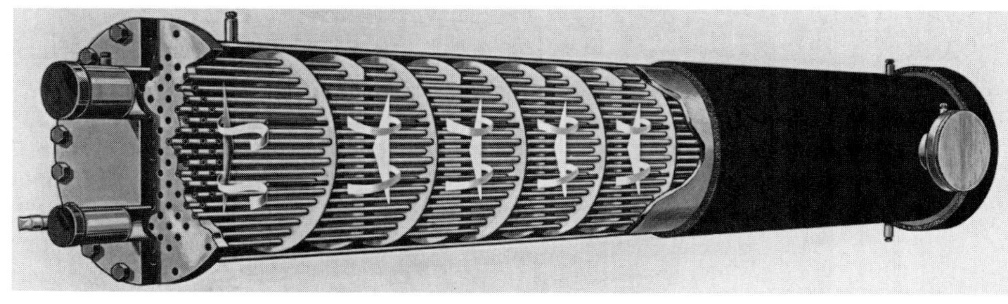

Figure A31-6 Shell-and-tube chiller. (Courtesy of York International Corp.)

units, except the cooler shell is suspended beneath the condenser coil and fan section. The chiller shell (evaporator) must be protected against freezing. Electric-heating elements are wrapped around the shell and then covered with a thick layer of insulation. Some manufacturers also add a final protective metal jacket that doubles as a good vapor barrier.

The sizes of reciprocating-compressor air-cooled packaged water chillers range from 10 to over 100 tons. These are also rated in accordance with ARI Standard 590 at 44°F leaving-chilled-water temperature and 95°F DB condenser entering-air temperature.

A31-3.3 Centrifugal Chillers

For very large installations the industry offers a range of hermetic centrifugal-compressor water chillers of up to 2000 tons in a single assembly. When used in multiples they handle applications of huge magnitude such as sports arenas, airports, and high-rise office buildings.

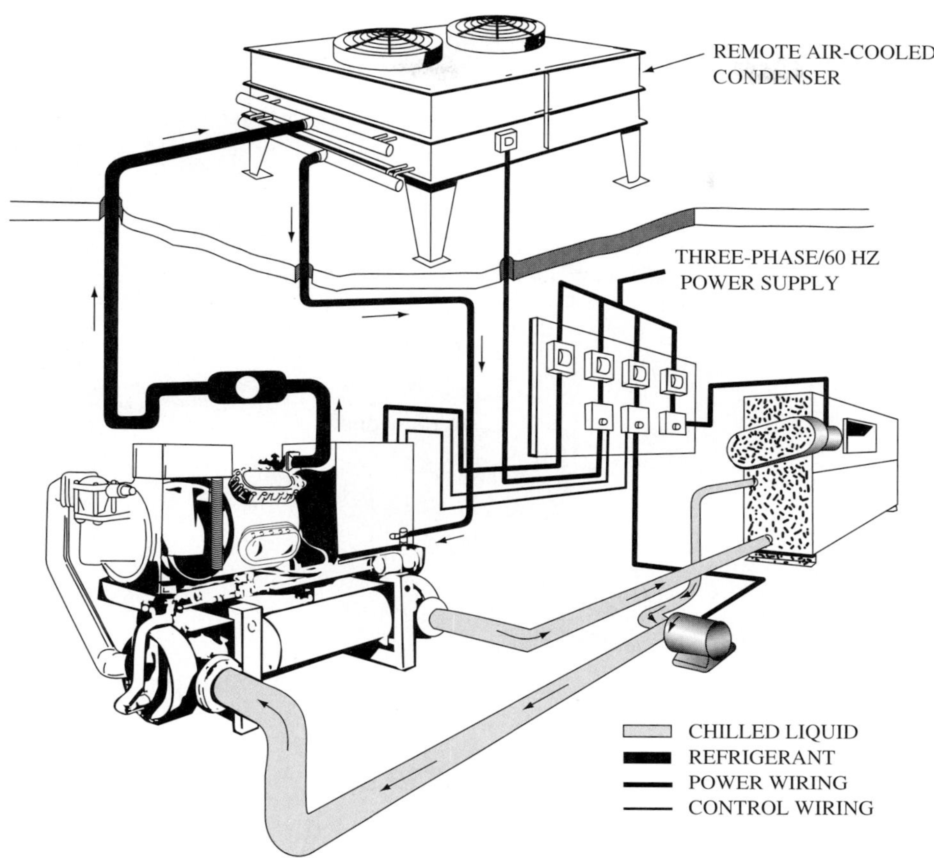

Figure A31-7 Air-cooled condenser installation with a packaged chiller.

Figure A31-8 Packed air-cooled water chiller.
(Courtesy of York International Corp.)

Hermetic centrifugal compressors (Fig. A31-9) vary in design and refrigerant use. Some are single-stage, others multi-stage. Some are direct-drive while others are gear-driven. The operating principle of all centrifugal compressors is the same: a rotating impeller is used to draw suction gas from the chiller (cooler) and compress it through a spiral discharge passage into a condenser. The speed of the impeller is a function of the design. Some gear-driven impellers reach speeds of 20,000 to 25,000 rpm. Capacity control is accomplished by a set of inlet vanes that throttle the suction gas to load or unload the impeller.

A complete water-cooled hermetic centrifugal-chiller assembly is illustrated in Fig. A31-10. This unit uses a combination chiller and condenser in one shell although they are separated internally. Chillers of this size and design do not use expansion valves. They use either a float control or a metering device to flood the chiller shell with liquid refrigerant.

Some centrifugal chillers have compressors with an open drive. Units of this type produce up to 5000 tons. The choice of drive can be gas, steam, diesel, or electric motor. These units are somewhat specialized and not as easily designed, selected and installed as are the complete packaged units.

A31-3.4 Screw Compressor Chillers

An important type of packaged water chiller uses the helical-rotary compressor, commonly known as a screw compressor because of the appearance of the rotors. The

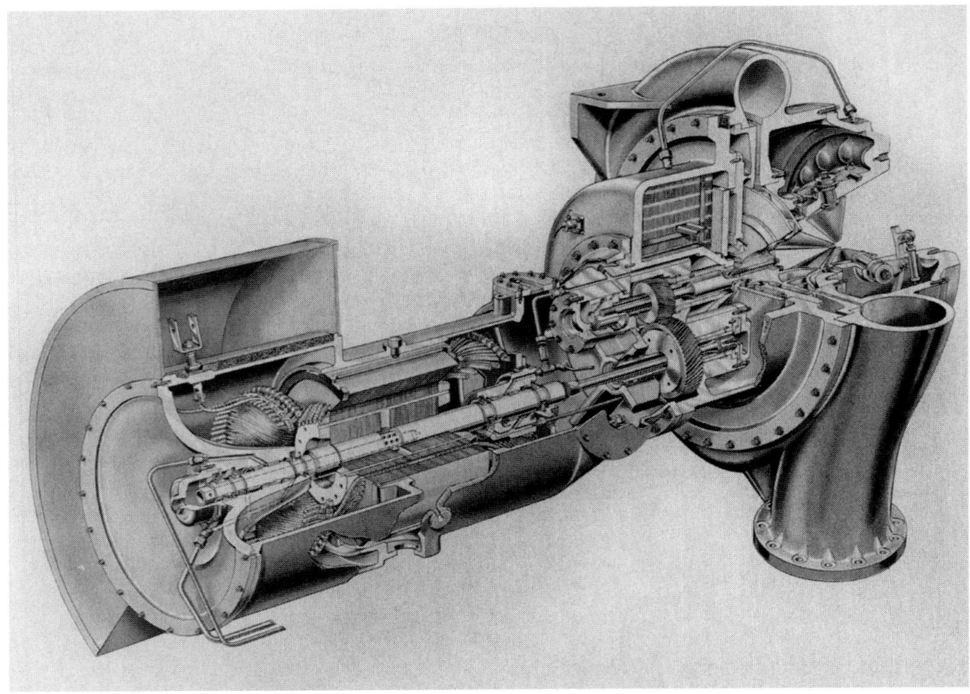

Figure A31-9 Hermetic centrifugal compressor and drive. (Courtesy of York International Corp.)

Figure A31-10 Hermetic centrifugal liquid chiller. (Courtesy of York International Corp.)

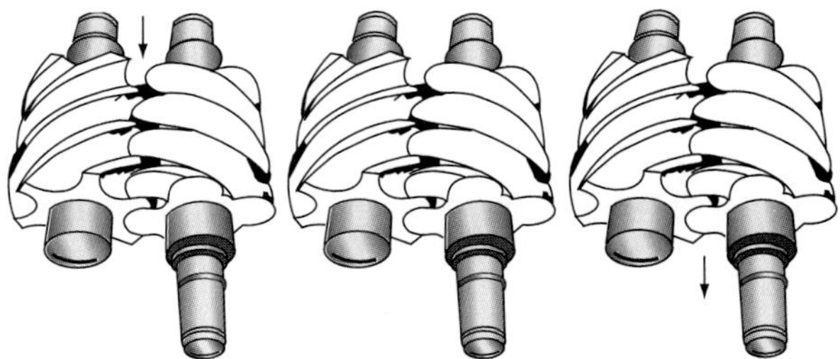

GAS DRAWN IN TO FILL THE INTERLOBE SPACE BETWEEN ADJACENT LOBES.

AS THE ROTORS ROTATE THE INTERLOBE SPACE MOVES PAST THE INLET PORT,WHICH SEALS THE INTER-LOBE SPACE. CONTINUED ROTA-TION PROGRES-SIVELY REDUCES THE SPACE OCCUPIED BY THE GAS CAUSING COMPRESSION.

WHEN THE INTER-LOBE SPACE BE-COMES EXPOSED TO THE OUTLET PORT THE GAS IS DISCHARGED.

Figure A31-11 Diagram of the rotors of a screw compressor.

twin-rotor screw compressor (Fig. A31-11) uses a mating pair of rotors with lobes which rotate much like a pair of gears.

During rotation the space or mesh between the lobes first expands to draw in the suction gas. At a point where the interlobe space is at maximum, the lobes seal off the inlet port. As the lobes rotate, the interlobe space becomes smaller as the gas is carried to the discharge end of the compressor. The refrigerant gas is internally compressed by this positive-displacement compressor until the rotors uncover the discharge port, where the compressed gas is discharged from the compressor.

Screw compressors are used because of their high capacity for a small unit, and their continuously variable (step-less) capacity control, typially modulating from 100% to 10% of full capacity. Being positive-displacement compressors, they can also be used with remote condensers, and have other piping flexibility.

Screw units are typically used with a high-pressure refrigerant, such as R-22. They are normally matched to shell-and-tube condensers and direct-expansion chillers, similar to those used on the reciprocating packaged chillers already reviewed.

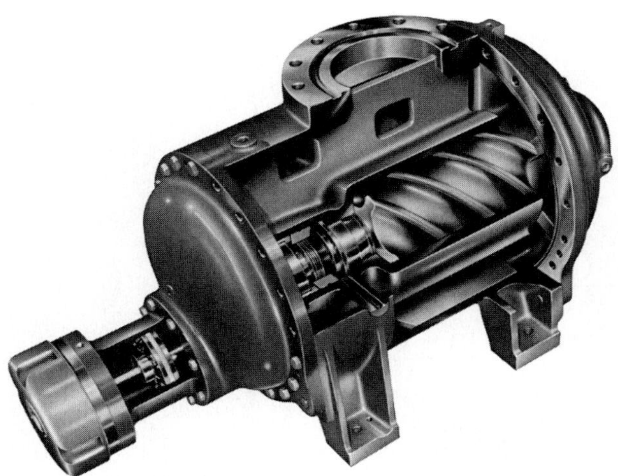

Figure A31-12 External-drive screw compressor.
(Courtesy of York International Corp.)

Screw compressors and their chiller packages were originally developed as effective units for the refrigeration needs of the food and chemical industries. They have been effectively adapted to the needs of the comfort air-conditioning market, with industrial-based technology.

Twin-rotor screw compressors come in a variety of configurations, depending on the manufacturer and application. The most common type is the horizontal open-drive unit, typically driven at 3500 rpm by an external motor (Fig. A31-12). Other units are of vertical-shaft construction, for minimum floor space usage. Many packaged chiller units use semi-hermetic construction on the compressor, eliminating the problems of shaft alignment and mechanical-shaft seal leakage.

Screw compressors also come in single-rotor designs, using gate rotors to seal the low-pressure from the high-pressure side of the compressor. A semi-hermetic single-rotor design is shown in Figure A31-13.

A31-3.5 Absorption Chillers

Unlike the conventional mechanical-compression refrigeration cycle used in all the other equipment discussed, an absorption chiller (Fig. A31-14) uses steam, hot water, or direct firing by natural gas as an energy source to produce a pressure differential in a generator section. Some absorption units are called double-effect, and generate both chilled water and hot water from the same unit. Typically, absorption systems are not as thermodynamically efficient as compression-cycle equipment. Since they can operate on waste steam or direct-fired natural gas, however, the actual costs of operation may be less than electrically driven equipment, depending on the relative energy costs. For more information, refer to the chapter on absorption.

In the review of chilling equipment, water is the fluid medium to be cooled and circulated, which is also true for most comfort-conditioning applications. These machines also cool other liquids such as brines which are circulated at low temperatures for refrigerating ice-skating rinks, freezing plants, and the chemical, drug, and petrochemical products.

Figure A31-13 Semi-hermetic screw compressor.
(Courtesy of McQuay International)

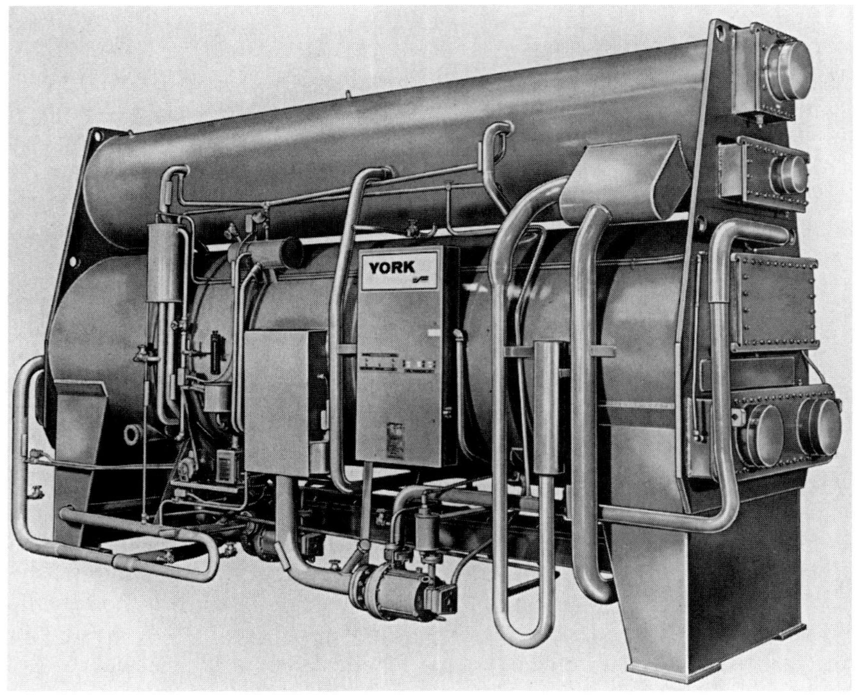

Figure A31-14 Large absorption chiller. (Courtesy of York International Corp.)

A31-4 BOILERS

For heating, a *hot-water generator,* commonly called a hot-water boiler, or just boiler, produces water at 180 to 200°F. The hot-water pump delivers this hot-water supply (HWS) to the heating coil in the air handler. As with the cooling coil, the heating coil has isolation and control valves. The hot water normally has a greater temperature change than the chilled water, giving up 20 to 40°F with a hot-water return (HWR) temperature as low as 140°F.

A **boiler** is a pressure vessel designed to transfer heat (produced by combustion or by electrical resistance) to a fluid, usually water. If the fluid being heated is air, the unit is called a furnace.

The heating surface of a boiler is the area of the fluid-backed surface exposed to the products of combustion, or fire-side surface. Boiler design provides for connections to a piping system which delivers heated fluid to the point of use and returns the fluid to the boiler.

A31-4.1 Boiler Classification

Types of boilers are classified in three ways: (1) working temperature/pressure, (2) fuel used, and (3) materials of construction.

Working Temperature/Pressure

Low-pressure boilers are constructed for a maximum working pressure of 15 psi steam and up to 160 psi hot water. Hot-water boilers are limited to 250°F operating temperature. If a package boiler is used, the temperature and pressure-limiting devices are supplied with the package; otherwise they must be applied during installation.

Medium- and high-pressure boilers operate above the levels of low-pressure boilers depending on the use to which they are applied.

Steam boilers are available for up to 50,000 lb/hr of steam. Many of them are used in central-station systems for heating medium and large commercial buildings which are beyond the range of furnaces. They are also used for industrial heating and industrial processing.

Water boilers are available from outputs of 50,000 Btu to 50,000,000 Btu. Many of these are low-pressure boilers and are used for space heating. Some are equipped with internal or external heat exchangers to supply domestic hot water.

Every steam and hot-water boiler is rated at a maximum working pressure determined by the ASME code under which it is constructed and tested.

Fuel Used

Boilers may be designed to burn coal, wood, various grades of fuel oil, various types of fuel gas, or operate as electric boilers. Each fuel has its own special firing arrangement depending on type of application.

Materials of Construction

Most boilers are made of cast iron or steel. Some small boilers are made of copper or copper-clad steel.

Cast-iron boilers are constructed of individually cast sections, assembled in groups or sections. Push or screw nipples or an external header join the section pressure-tight and provide passages for water, steam, and products of combustion.

Steel boilers are fabricated into one assembly of a given size, usually by welding. The heat exchangers for steel boilers can be fire-tube or water-tube design. The fire-tube construction, which is most common, has flue-gas passage space between the water holding sections. The water-tube construction uses water-filled tubes for the heat exchanger with the flue gases in contact with the external surface of the tubes.

Copper boilers are usually some variation of the water-tube type. Parallel-finned copper-tube coils with headers, and serpentine copper-tube units are the most common. Some are offered as residential wall-hung boilers.

Condensing boilers have recently been developed to improve the efficiency of boilers. Previously the flue gases were not allowed to condense in the boiler due to the corrosion it could cause. These new condensing boilers are constructed with heat exchangers made of materials to resist corrosion. Depending on the constituents of the fuel and the application, overall efficiencies as high as 97% have been achieved using this new construction.

Electric boilers are in a separate class. No combustion takes place and no flue passages are needed. Electric elements can be immersed in the boiler water.

The illustration Fig. A31-15 shows the configuration of a number of types of residential boilers. Figures (a) through (e) illustrate fire-tube designs, (f) and (g) are water-tube, and (h) is a condensing boiler.

Boiler Controls

Boiler controls regulate the fuel input in response to a control signal representing load change. Boiler controls include safety controls that shut off the fuel flow when unsafe conditions develop. Details on the control systems for various applications are given in the section of the text under Hydronic Heating.

A31-5 CIRCULATING WATER PUMPS

The major uses of pumps in an HVAC/R system are for: (1) pumping chilled or heated water, (2) pumping condenser water from the cooling tower, and (3) circulating water in a cooling tower or evaporative-condenser water circuit. These pumps are generally centrifugal types with configurations as shown in Fig. A31-16. It is common practice in central-station systems to provide identical pumps operating in parallel, one of which is a spare for critical uses.

Pumps are selected to perform a specific purpose, providing the proper flow at a pressure to overcome the resistance of the circuit in which they are placed. They are selected from *pump curves* provided by the manufacturer (Fig. A31-17). These curves show the flow in gpm for various head pressures measured in feet (1 foot = 2.31 psi). The curves also show the performance of the pump with different diameter impellers, as well as the required motor size, based on the desired performance. For example, using the pump curve in Fig. A31-17, if 320 gpm is required at a 42 ft head, a 7.75 in. impeller could be used with a 7.5 Hp motor.

It is important to understand the difference between an open- and closed-piping system in working with pumps. An **open system** has some part of the circuit open to the atmosphere, such as a cooling pump water circuit. In this type of circuit the height that the water must be lifted must be added to the friction loss of the piping to determine the pump head. The suction lift of the pump is also limited.

In a **closed system** all piping is in series with the pump, such as in a circulating hot water heating system. The fluid that is pumped out returns to the suction side of the pump. An expansion tank is required in the circuit to allow for the volume change of the fluid due to variations in temperature. The location of the expansion tank and pump are important, as shown in Fig. A31-18. In a closed system the system pressure can be regulated or limited by a pressure-relief valve and an automatic make-up valve.

Fig. A31-19 (p. 624) describes common pumping problems and presents solutions.

A31-6 COOLING TOWERS

The *cooling tower* delivers the condenser-water supply (CWS) at about 85°F, pumped by the condenser pump, to the water-cooled condenser on the water chiller. The condenser water is warmed by the rejected heat from the chiller to about 95°F, and returned through the condenser-water return (CWR) to the cooling tower, where it is cooled by evaporation in an airstream.

Cooling towers are an essential part of most central-station air-conditioning systems. For the smaller systems, even up to 100 tons, there has been a trend toward the use of remote air-cooled condensers. Their use is limited, however, to installations where the length of the refrigerant piping can be short enough to be practical. Most all large condensing units and large packaged chiller units use water-cooled condensers with cooling towers.

For a complete discussion on the application of cooling towers to central-system equipment, refer to the chapter on condensers. They are considered an accessory to the condenser since towers supply the means of disposing of the heat collected in the water cooled condensers.

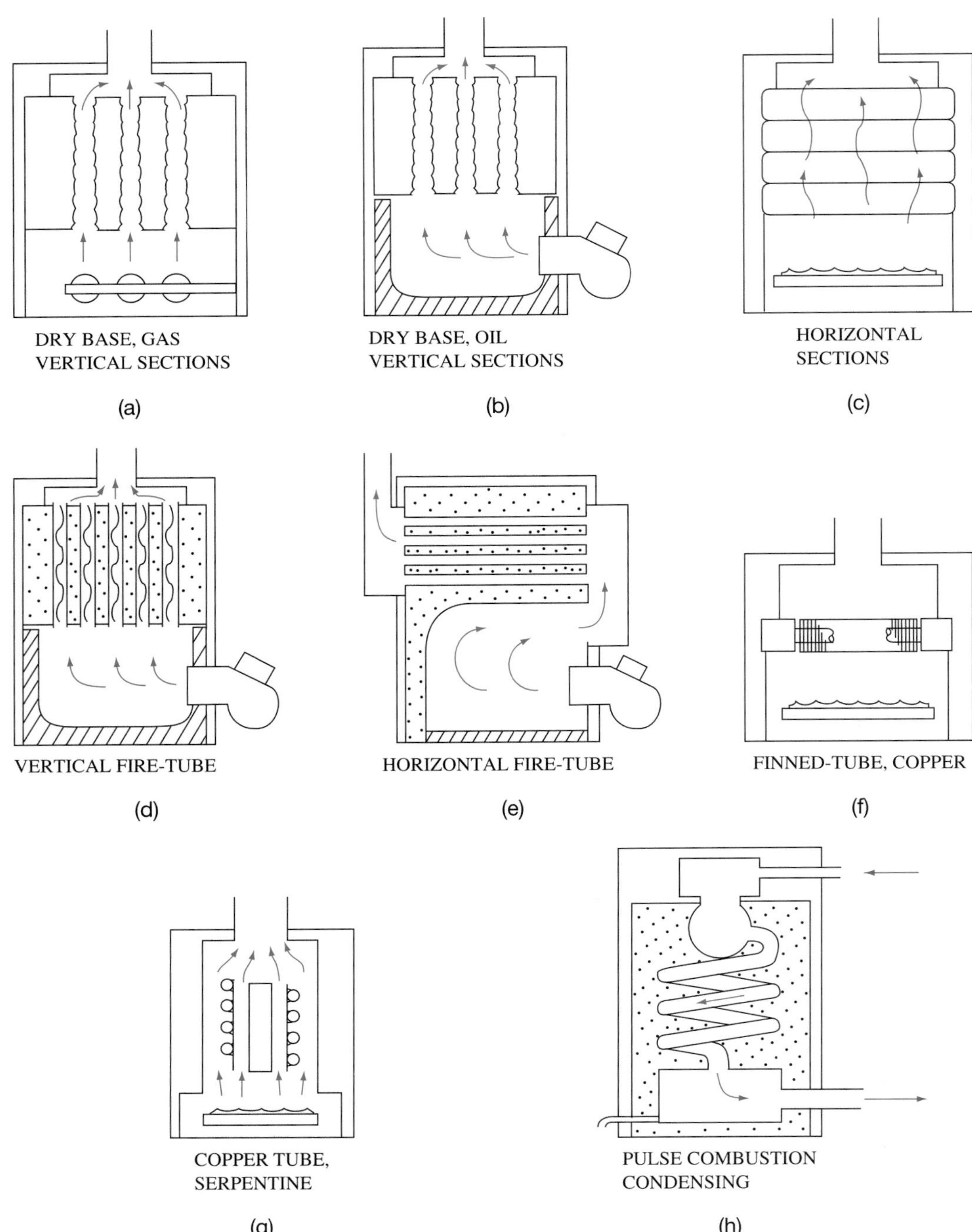

Figure A31-15 Types of residential boilers. (Copyright by the American Society of Heating, Refrigerating, and Air-Conditioning Engineers, Inc. Used by permission.)

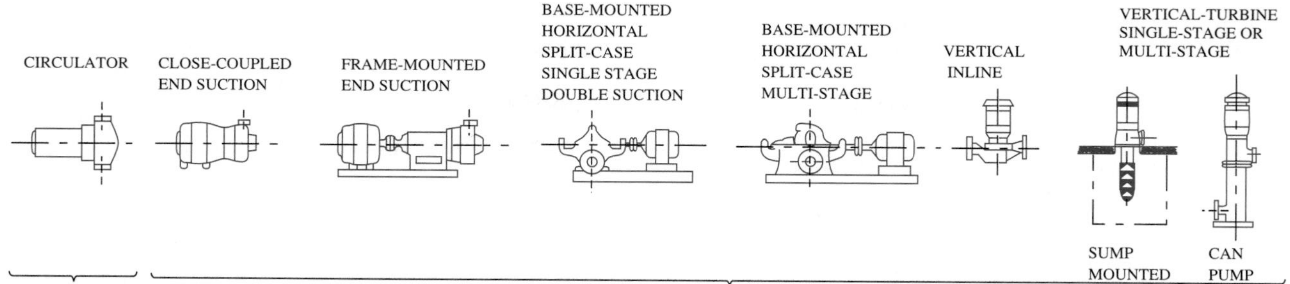

Figure A31-16 Typical centrifugal pumps. (Copyright by the American Society of Heating, Refrigerating, and Air-Conditioning Engineers, Inc. Used by permission.)

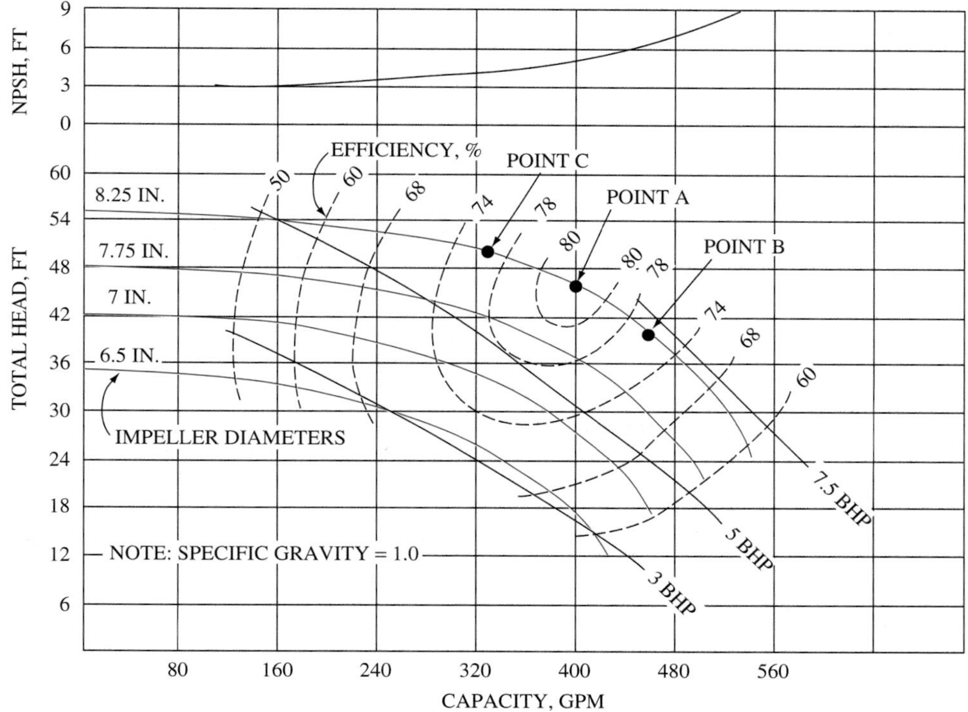

Figure A31-17 Typical pump performance curves provided by manufacturers. (Copyright by the American Society of Heating, Refrigerating, and Air-Conditioning Engineers, Inc. Used by permission.)

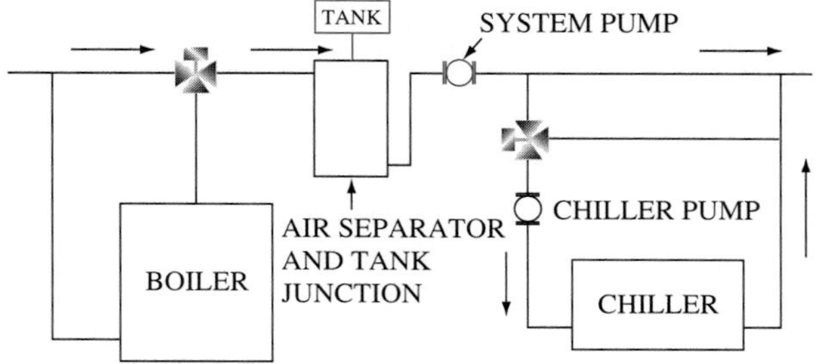

Figure A31-18 Correct expansion tank connection.

Complaint	Possible Cause	Recommended Action
Pump or system noise	Shaft misalignment	■ Check and realign.
	Worn coupling	■ Replace and realign.
	Worn pump/ motor bearings	■ Replace, check manufacturer's lubrication recommendations. ■ Check and realign shafts.
	Improper foundation or installation	■ Check foundation bolting or proper grouting. ■ Check possible shifting because of piping expansion/ contraction. ■ Realign shafts.
	Pipe vibration and/or strain caused by pipe expansion/ contraction	■ Inspect, alter, or add hangers and expansion provision to eliminate strain on pump(s).
	Water velocity	■ Check actual pump performance against specified, and reduce impeller diameter as required. ■ Check for excessive throttling by balance valves or control valves.
	Pump operating close to or beyond end point of performance curve	■ Check actual pump performance against specified, and reduce impeller diameter as required.
	Entrained air or low suction pressure	■ Check expansion tank connection to system relative to pump suction. ■ If pumping from cooling tower sump or reservoir, check line size. ■ Check actual ability of pump against installation requirements. ■ Check for vortex entraining air into suction line.
Inadequate or no circulation	Pump running backward (three-phase)	■ Reverse any two-motor leads.
	Broken pump coupling	■ Replace and realign.
	Improper motor speed	■ Check motor nameplate wiring and voltage.
	Pump (or impeller diameter) too small	■ Check pump selection (impeller diameter) against specified system requirements.
	Clogged strainer(s)	■ Inspect and clean screen.
	Clogged impeller	■ Inspect and clean.
	System not completely filled	■ Check setting of PRV fill valve. ■ Vent terminal units and piping high points.
	Balance valves or isolating valves improperly set	■ Check settings and adjust as required.
	Air-bound system	■ Vent piping and terminal units. ■ Check location of expansion tank connection line relative to pump suction. ■ Review provision for air elimination.
	Air entrainment	■ Check pump suction inlet conditions to determine if air is being entrained from suction tanks or sumps.
	Insufficient NPSHR	■ Check NPSHR of pump. ■ Inspect strainers and check pipe sizing and water temperature.

Figure A31-19 Pumping system trouble analysis guide. (Copyright by the American Society of Heating, Refrigerating, and Air-Conditioning Engineers, Inc. Used by permission.)

A31-7 AIR HANDLING UNITS

As shown in Fig. A31-20 the air-handling unit is the focal point of a number of subsystems that make up a typical central HVAC layout. Referring to the diagram, the subsystems are as follows.

A31-7.1 Heating

The heating circuit consists of the boiler, a pump, the heating coil, and the necessary piping.

A31-7.2 Cooling

The cooling circuit consists of a chiller, a pump, the cooling coil, and the necessary piping.

A31-7.3 Condenser Water

This circuit consists of a cooling tower, a pump, a water-cooled condenser (part of the chiller), and the necessary piping. This is actually a sub-circuit to the cooling circuit.

A31-7.4 Air Handling

This circuit has many parts. In tracing this circuit, we will start at the entrance to the air-handling unit. Air that enters the unit is made up of both return air from the conditioned space and outside ventilation air. In the air-handling unit, air passes through the filter, humidifier (not shown), various heat-transfer coils and into the fan. The fan delivers air through the distribution system to supply conditioned air to meet the space requirements.

Note that provision is made in the air system for the use of 100% outside air during an *economizer cycle*. The excess air is exhausted after it has served its purpose, through the air-relief opening.

Due to the importance of air-handling units and the many variations that occur to conform to different applications, refer to the next chapter, "Air Handling Units and Accessories," for further discussion.

A31-8 AIR DISTRIBUTION SYSTEMS

For simplicity, types of commercial air distribution may be classified as low-, medium-, or high-velocity systems.

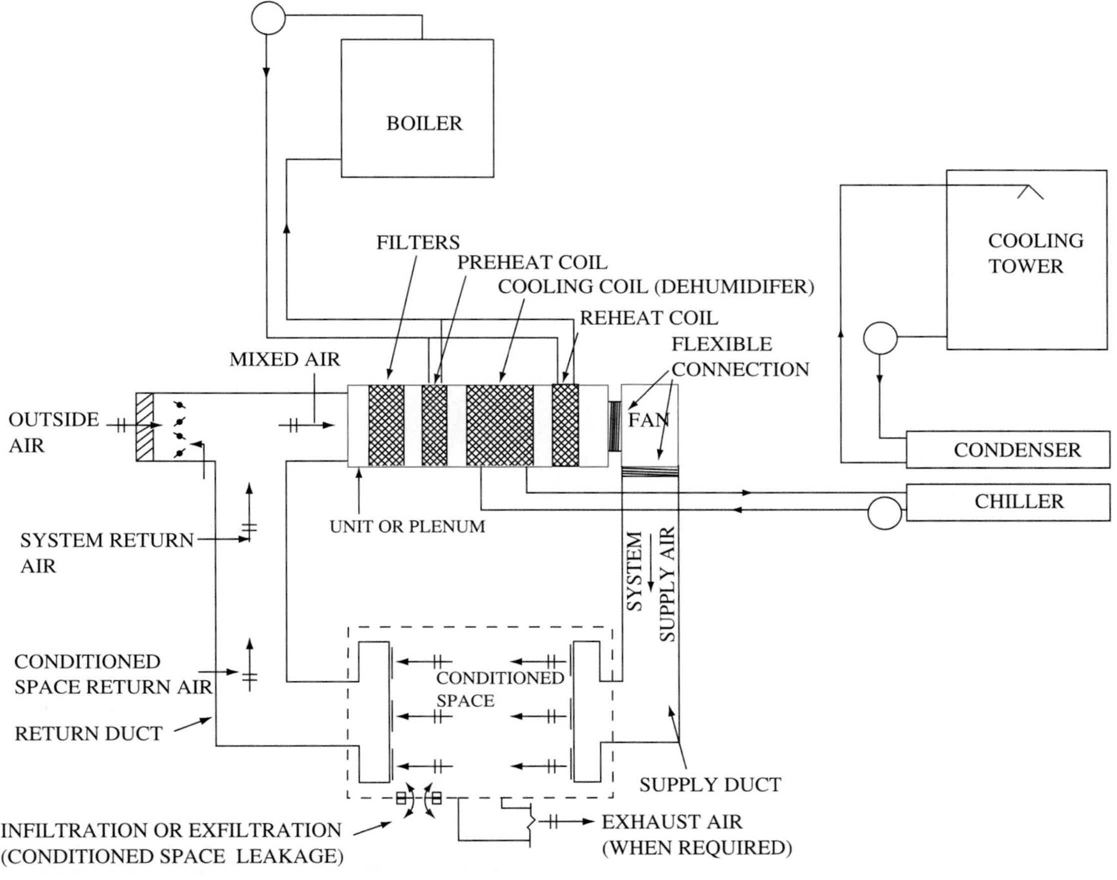

Figure A31-20 Typical central HVAC system showing air-handler unit.

Low-velocity systems are those associated with the application of smaller unitary-packaged and split-system units. The use of limited ductwork or none at all (free blow) is typical of the classification. Where ductwork is used, the external static pressure is held down to the range 0.25 to 0.50 in. WC. The use of concentric supply and return ducts, as shown in the illustration of the supermarket, is a common application. The type of duct design is the same as that of the equal friction method, which permits the prediction or control of total static.

A31-8.1 Air-distribution Applications

Air-distribution applications in central-station systems fall into three broad areas: interior or core areas, exterior or perimeter areas or zones, and entire-building applications.

Core areas have conditioning loads subject only to interior loads, such as lighting, people, and equipment. Consequently, they are basically cooling-only loads, except for top floors and/or warm-up cycles in extremely cold climates.

Perimeter zones are exposed to outer-building skin variables, such as wall and window loads, wind effects and exposure effects, as well as the items in the core areas. The perimeter zones have to handle a wide range of conditions, including such variances as high solar gain on a winter day requiring cooling, while shaded parts of the building require heating.

Entire-building applications are a combination of these systems, the selection of which depends on building size and economics. Small buildings frequently cannot justify one system for the core and another for the perimeter. In these instances, any one of the perimeter systems can be used for the entire building.

The systems are all of the type that use primary (supply) air from the air conditioner, and the maximum quantity needed is based on the maximum load conditions of the area.

A31-8.2 Core-area Applications

If the interior core of a building is a large open area, a *single-zone constant-volume system* (Fig. A31-21) can be used at a reasonable initial cost. A single heat-cool thermostat with automatic changeover provides year-round control of the coil temperatures. Top-floor and ground-floor systems will have heating for morning warm up. Intermediate floors usually only have provisions for cooling. The duct design follows conventional practice, with room-air distribution from the ceiling.

For core areas divided into smaller spaces, with variations in lighting and people loads, a *variable-air-volume* (VAV) *single-zone system* has typically been used. There are several approaches to this application.

Fig. A31-22 shows a constant-volume air handler connected to individual VAV terminals. The terminal may be an individual control box, a system-air-powered VAV slot diffuser, or a self-contained temperature-actuated VAV diffuser. In all these cases, the variable air volume is accomplished by throttling the airflow at the individual duct run, causing the constant-volume air handler to "back up" the fan curve to a new balance point of lower cfm at a higher static pressure. The controls for this type of system are self-contained within the air-distribution system, and do not require any more control than the thermostat noted above.

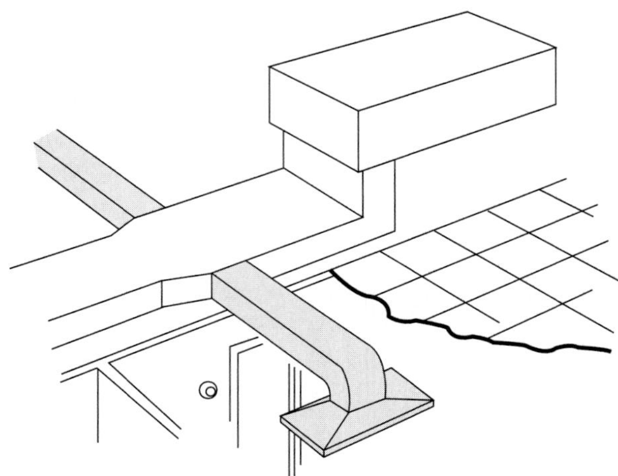

Figure A31-21 Single-zone system using constant volume.

Another approach for the core area is to use *VAV terminals,* with the capability of handling several duct runs from each terminal (Fig. A31-23). In this case, the air handler is also constant volume, but the VAV terminal throttles the air supply to the ducts, and dumps the remainder back into the ceiling-space return-air plenum. The space is controlled with variable-air volume, but the air handler is operating at constant volume. The control for this subsystem requires that the space thermostat controls an actuator

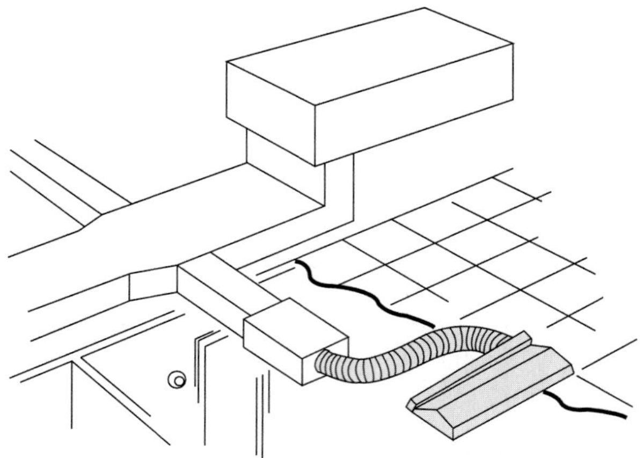

Figure A31-22 Single-zone system using variable air volume (VAV).

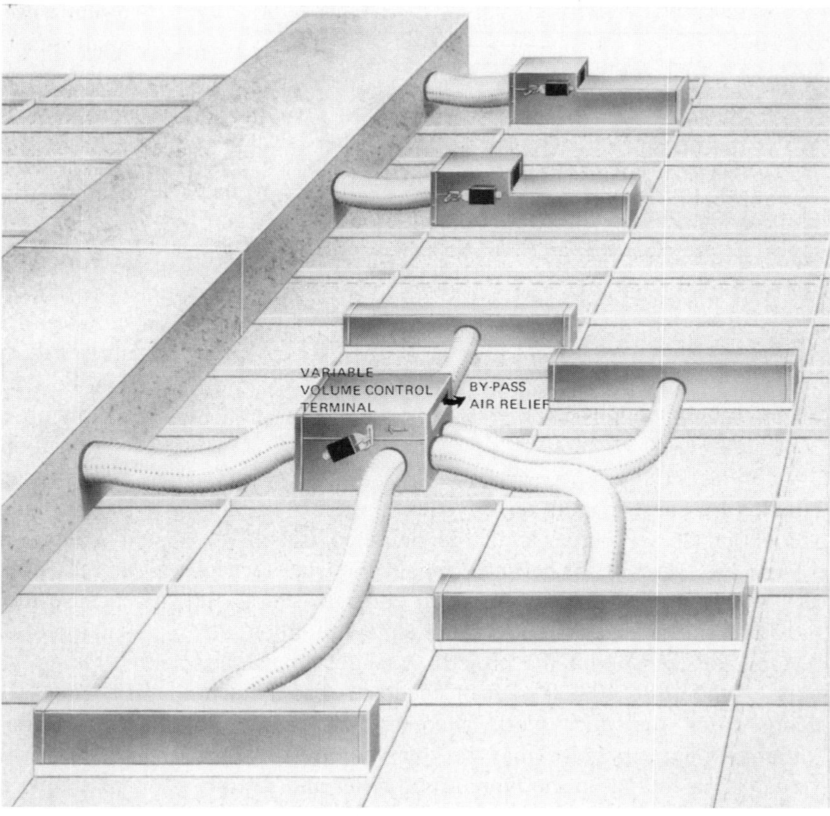

Figure A31-23 Variable-air-volume system using terminal units. (Courtesy of York International Corp.)

on the VAV terminal, as well as controls the cooling and heating demand.

In either of these cases, a single air handler is sufficient in the core zone and can provide proper comfort conditions. The air handler is typically constant-volume because the loads are relatively steady, regardless of the season or ambient air temperature.

An approach often seen on older systems, but prohibited by current energy codes, is commonly called *zone reheat.* As seen in Fig. A31-24, an air handler delivers *constant-volume, constant-temperature* air to the distribution system. Individual zones or branches have heaters in them, which reheat the cool air to a comfort level for each space. These heaters may be individually controlled electric heaters or hot-water coils. In many cases, these heaters are applied to a problem area in a system, to provide adequate heat where the original design was insufficient.

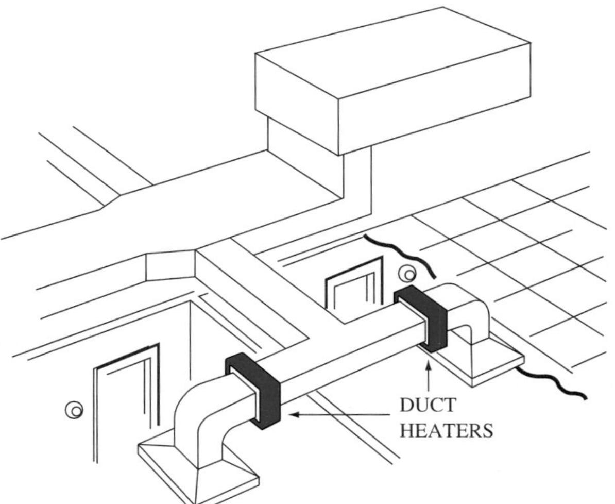

Figure A31-24 Constant-volume system using zone duct heaters.

A31-8.3 Perimeter-zone Applications

Perimeter zones require *terminal systems* which can handle the wide range of conditions, from the coldest morning warm up to the hottest solar-gain cooling load. In many systems, this range was handled by providing perimeter radiant or forced-air heat systems on the outer wall and under windows to furnish the necessary heat for the cold loads. A separate air system provided the necessary air movement, outside-air requirements, and the cooling-load responsibility. The air systems could be either constant or variable volume, but it has been found that only VAV really can provide the necessary control for comfort. In milder climates, the perimeter heating system is eliminated, without much loss of comfort. In some cases, the entire perimeter heating and cooling load is

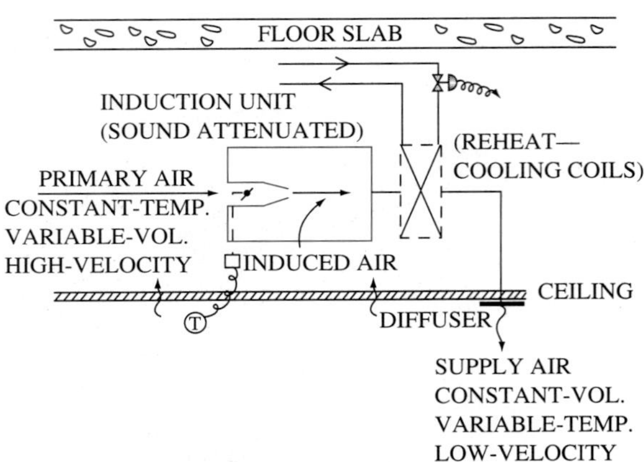

Figure A31-25 Induction terminal.

handled by console units on the outside wall, with individual outside-air sources and temperature controls.

The *VAV systems* for perimeter applications are generally provided with primary air from either a central air handler for the entire building, or an entire floor, depending on the size of the project. A single air handler may handle up to 40,000 sq ft of occupied space, but in many cases the duct work becomes unmanageable. Variable-volume systems may be low-velocity or high-velocity, depending on the type of controller and the terminal units used. The volume of primary air is automatically adjusted to the total cooling demand by duct sensors controlling fan volume controls.

The terminal units for VAV systems have gone through an evolution since the advent of VAV. Some earlier systems were *induction air-distribution systems.* Induction systems use a primary air system delivering a relatively small quantity of cooled high-velocity, high-pressure air to induction-room terminals. This saves space, since the ductwork is much smaller than conventional ducting. The terminal unit takes the high-velocity air through a nozzle arrangement (Fig. A31-25) which induces room air into the unit through a heating coil. The temperature control is both from the VAV of the primary air and the reheat and/or cooling coils in the unit.

Some of these systems were not VAV. Fig. A31-26 shows an induction room terminal which is constant volume, with temperature control achieved by reheat. Note that these terminal units do not rely on fans; the induced airflow provides the necessary secondary air.

Induction air systems have some trouble spots. First, high-pressure air requires very tight and sealed ductwork. High-pressure air leaks are very noisy. Second, these systems tend to use much energy, since the fan pressure requires considerable power; the VAV aspects are limited because the induction effect falls off quickly with a reduction of airflow.

Other VAV terminals for low-pressure systems are the system-powered boxes or diffusers (Fig. A31-23) which use system air pressure to control inflation of bladders which, in turn, control the air volume in the unit. These are simple in concept and effective at a reasonable cost. They provide VAV space conditioning, but usually without the benefit of variable-air-volume central-fan systems. Some of these diffuser units use self-contained, temperature-powered actuators to modulate the diffuser openings. These can provide fairly simple, reasonable-cost VAV to smaller systems.

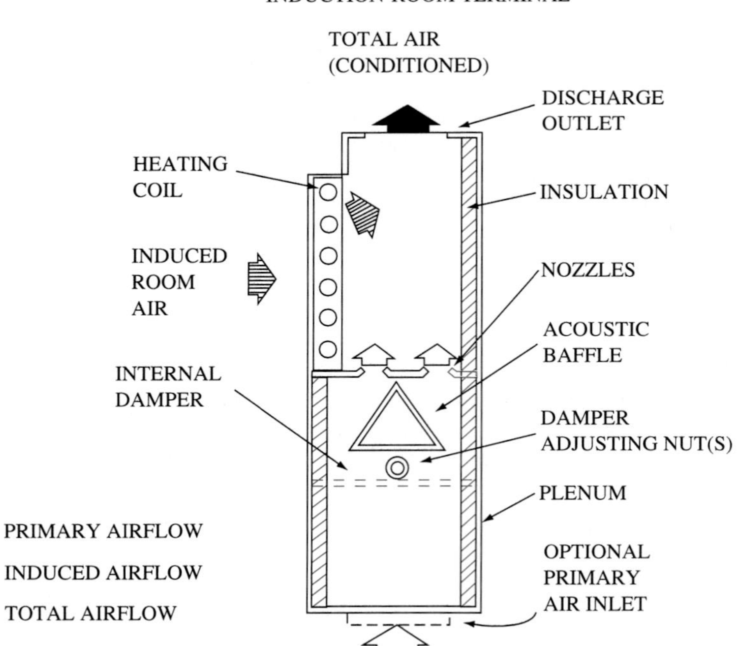

Figure A31-26 Sectional view of an induction-type room terminal unit.

Other systems provide cost-effective zone control with a combination of VAV duct dampers and a sophisticated temperature-control system. These allow the system to provide only heating or cooling at any given time. The demand of a number of zones are programmed to switch the system to that mode, for example, cooling. Any zones requiring heat are temporarily closed off. With a heating demand in place, the system controls will then switch the system back to heating, closing off the zones requiring cooling. The individual zone dampers are modulating, so there are no abrupt changes. A fan speed controller is part of the system, reducing airflow to a practical minimum, as allowable. These systems are generally designed for unitary equipment, but they can be used as part of a central system.

The later trend in VAV systems is to use a variable-air-supply primary system, with minimum airflows to meet ventilation requirements. The terminal units are fan-powered, providing constant volume in the conditioned zone, with primary air operating from 100% down to about 20% of total air requirements. The low-powered terminal fan provides terminal reheat at reduced air conditions for the zone. The primary air handler is variable-volume, constant-temperature, with the cooling-plant demand proportional to the airflow. The individual zone controls can be stand-alone systems, but usually are integrated through a DDC system controlling the overall system.

A31-8.4 Entire Building Applications

In the chapter on central-station air handlers, mention is made of multi-zone and double-duct applications. Before energy codes were enacted, these were the ultimate in zoned-comfort systems. By providing a blend of always-available heated and cooled air to any zone, with the addition of good pneumatic control systems, these systems supplied excellent building comfort. The major limitation of them was the cost of energy to provide simultaneous heating and cooling. Later, the heat rejected from the chiller was used to provide warm water for heating. Other heat recovery approaches were tried, but the fully energy-efficient approaches were not always the simplest or most economical to install. The primary differences in these two types of systems are noted below.

Multi-zone systems provide one trunk duct to each zone from the multi-zone air handler (Fig. A31-27). The air handler is a blow-through design with a hot and a cold deck. The airflow from each of these decks is fed through a set of dampers, 90 degrees opposed to each other on the same shaft. When the cooling damper is fully closed, the heating damper is fully open, and vice-versa. Most of the time, the dampers are modulating in response to a temperature signal from the zone. The overall system is low- to medium-pressure, constant-volume. The multi-zone's limitations, from a design standpoint, are that only a limited number of zones are available on

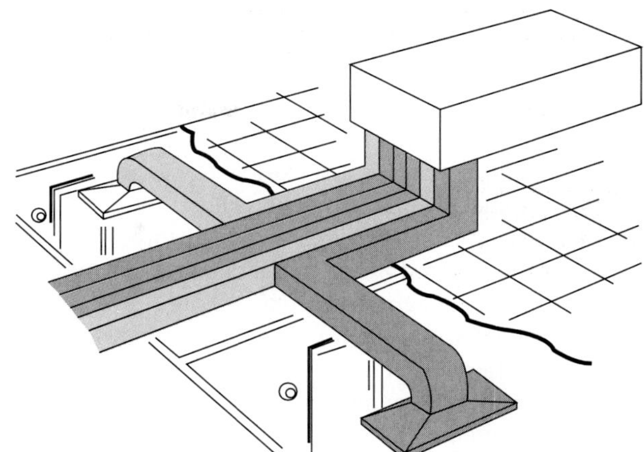

Figure A32-27 Constant-volume multi-zone unit.

the air handler. Consequently, adjustments are always needed to get an even balance. The smallest zone is typically 8–10% of the full load.

Double-duct systems were the ultimate in design flexibility, and very high in initial cost. Double-duct systems are high-velocity, high-pressure systems. Some are variable-volume, but most are constant volume. Two full-sized supply ducts are required for the system, one for cooling and one for heating (Fig. A31-28). The major benefit of these systems is that the two ducts each serve one terminal mixing box, which controls the air temperature for each zone. The mixing boxes are available in very small sizes, down to 200 cfm, allowing for zones as small as ½ ton. Zoning is limited by this small increment and the overall size of the system. The high-pressure air in the main supply ducts (4 to 5 in. WC) is reduced in the mixing boxes, so that the distribution ducting is normally low-pressure design, with low noise levels. Energy costs for these units run more than twice as much as current efficient VAV designs.

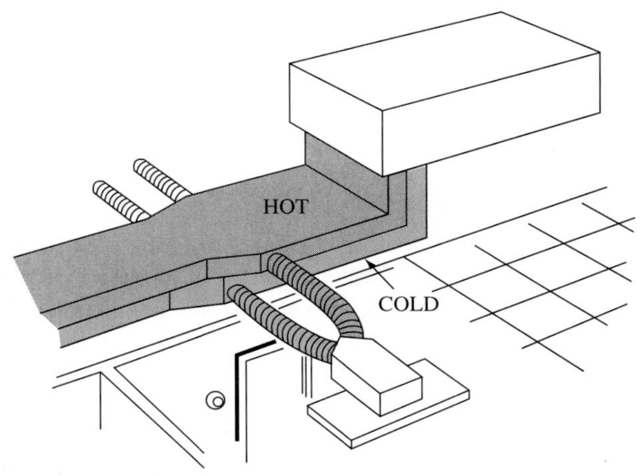

Figure A31-28 Double duct system.

A31-9 CONTROL SYSTEMS

The *control system* must direct the operation of all elements automatically. In large installations, the control system is usually a separate installation from the air-conditioning equipment. Controls may be electric, electronic, or pneumatic (air), or a combination of all three. The controls must be included in the initial construction stages to provide the total integration of the system.

In a basic control system for cooling, a thermostat operates the chiller to maintain a set leaving-water temperature. The condenser pump operates when the chiller is running. The cooling tower fan is controlled by the condenser-water temperature, as is the water-bypass control.

For the heating system, boiler firing is activated by the hot-water thermostat. Space thermostats and humidistats control the functions of the air handler.

A31-9.1 Electric Control Circuits

An example of a commercial and engineered system is a rooftop air-conditioning unit with gas-heating, low ambient control and an economizer cycle. It is common for the manufacturer to present the control information in three parts:

1. Basic-refrigeration circuit controls (Fig. A31-29)
2. Natural-gas-heating circuit controls (Fig. A31-30)
3. Legend, to identify the control components (Fig. A31-31)

The basic control circuit is line voltage, with wiring for the dual compressors and fan motors. The heating circuit is low voltage and includes the selector switch and thermostats. The legend defines the various control abbreviations.

A31-9.2 Refrigeration Circuit, High Voltage

Referring to Fig. A31-31, a power supply of 208/230 V, three-phase, 60 Hz has been selected for sizing fuses and the disconnect switch. From the manufacturer's table (Fig. A31-32), this model will require a 100-A disconnect and a maximum of 70-A dual element fuses. Assuming that the distance of the switch to the panel is 125 ft, the minimum wire size is 4 AWG 60°C wire.

In Fig. A31-29, compressor motors are three-phase, with external overloads (OL) in the motor-winding circuit. The evaporator motor is also three-phase, with inherent motor protection. All three motors are started by the contactors, 1M, 2M, and 3M, respectively.

The balance of the line voltage controls are single-phase to permit series circuit controls with ordinary control devices. The condenser fans in this illustration are single-phase, as well. Some units may use three-phase motors; they will require relays or contactors for control, as the compressors do. Note that the single-phase power is connected at terminals 1 and 3 on line 10. A transformer is shown, with a note showing its use in this unit for only 460-V power, to convert it to the control voltage of 240 V. Since the example is shown as 240-V main power, a transformer is not required. Other manufacturers use 120 V for control voltage, and will require a transformer (or a separate power source) for all primary power voltages.

Note that fuses 1FU and 2FU are provided to protect the control circuits whether they are from direct or transformed power. On 120-V control circuits there is no fuse on the secondary, or neutral side of the circuit. This maintains the protection of a grounded neutral for safety.

Dropping down to lines 11 and 12 of Fig. A31-29, note that the crankcase heaters (1HTR and 2HTR) come on when the crankcase thermostat closes in response to the oil temperature. The heaters can operate as long as the power is on, regardless of the system operation. It is therefore important that the power not be turned off in cold weather.

Both condenser fans have single-phase motors with external running capacitors (1RC and 2RC) and inherent motor protection. Since this unit is equipped with low ambient control, note that in line 13 the #2 condenser fan motor is energized by the fan relay (1R-1) to start and run. It can also be dropped out by the first low ambient thermostat (1TH).

If it were not for the economizer cycle, the #1 condenser fan motor would be wired the same as the #2 fan. But the economizer system works within the same temperature range as the first-stage low ambient control, and the two must be electrically interlocked as shown in lines 15 and 26.

1R-2 in line 26 is the #2 cooling relay and the key to energizing the circuit that feeds terminals 2 and 6 and the #1 condenser fan motor. When the low ambient accessory kit is used, however, the jumper between terminals 2 and 6 is removed, and the circuit shown in dashed lines is connected. Following line 26, the relay switch 1R-2 releases power to coil 6R and on to 2R (cooling contacts) and to contactor coil 2M, closing the #2 compressor contactor. 2R functions from the second stage of the cooling thermostat. The 6R coil is the key to furnishing power to line 25 through contact 6R. Line 25 has a low-pressure cutout (2LP), high-pressure cutout (2HP), electrical overload switches 4-0L and 5-0L, and thermal cutout 6-0L, all of which protect #2 compressor.

Note that line 18 is a similar protective circuit for compressor #1. The two lockout relays 6R and 7R are normally closed.

Without a complete explanation of the low ambient accessory, it can be noted that 2TH, line 20 (low ambient thermostat), is the key to controlling the #1 condenser fan motor through a system of time-delay relays 1TR, 2TR, and 3TR, eventually feeding to terminal 6 and the con-

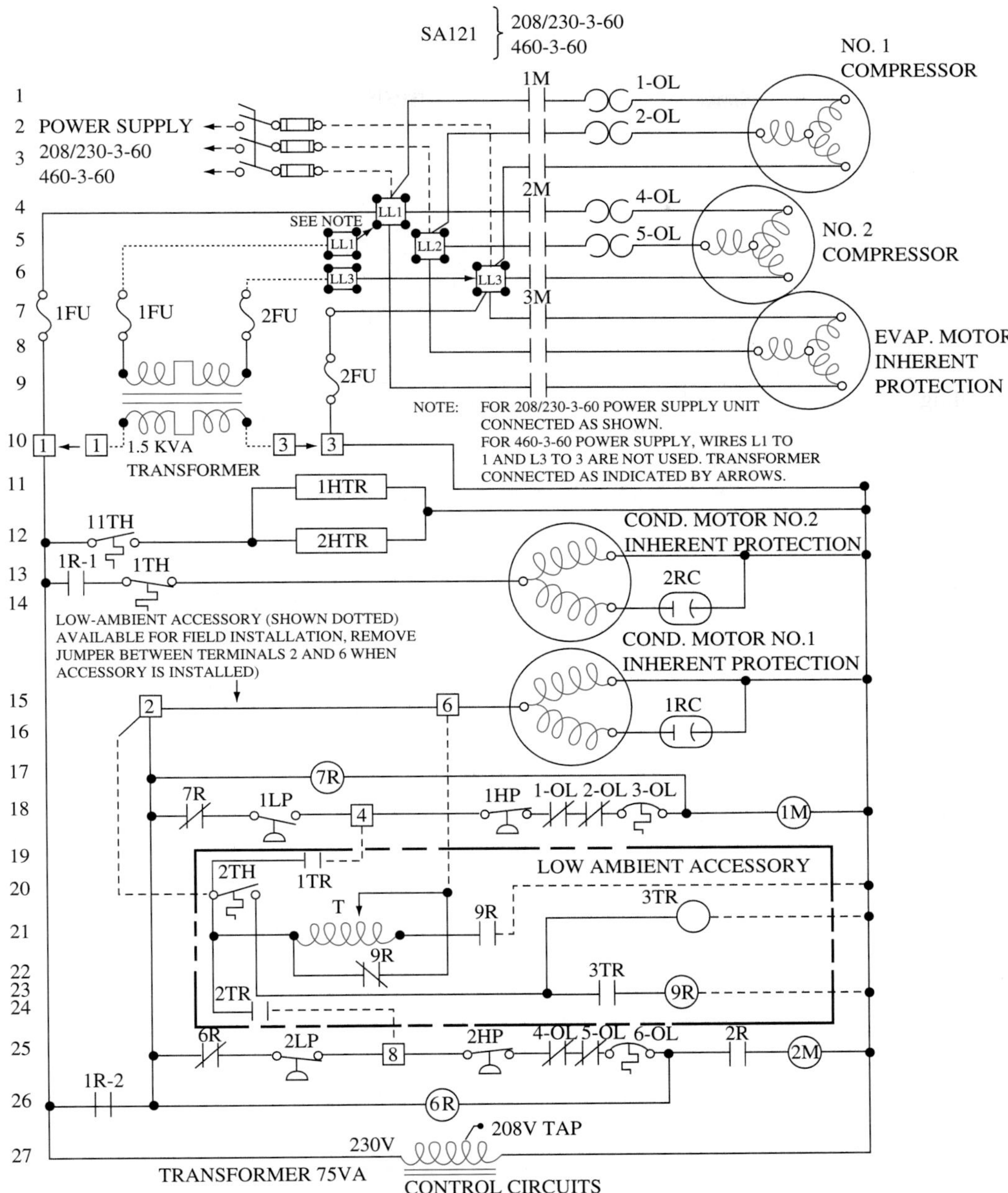

Figure A31-29 Line-voltage wiring diagram for a central-station air-conditioning system. (Courtesy of York International Corp.)

denser fan motor. T (line 20) is an auto-transformer fan-speed control that modulates the fan-motor rpm instead of providing a straight on-off. Note that it is put in the circuit when 9R switches reverse in response to the need for low ambient control. Otherwise, the fan speed is full rpm when the current goes around T, because 9R in line 21 is open. Line 27 is the primary line voltage side of the low-voltage control transformer. It is rated at 75 VA (volt-amps).

A31-9.3 Heating Circuit, Low Voltage

Referring to Fig. A31-30, line 1 is 24-V secondary, with fused protection based on amperage draw. The space thermostat has a COOL-OFF-HEAT-AUTO system selection. In automatic mode, the system will bring on heating, cooling, or the economizer system without manual selection. TC1 and TC2 are the first- and second-stage cooling stats. TH1 is the first-stage heating stat. The indoor fan selection is ON (continuous) or AUTO to cycle with the 3TH blower control.

The economizer control thermostat, 5TH in the mixing boxes, is wired to the C1 first-stage cooling circuit, so that #1 compressor cannot operate as long as it is open. This means that the temperature of the outside air is sufficient to provide cool air without refrigeration. But when the outside ambient is high enough to call for cooling, 5TH

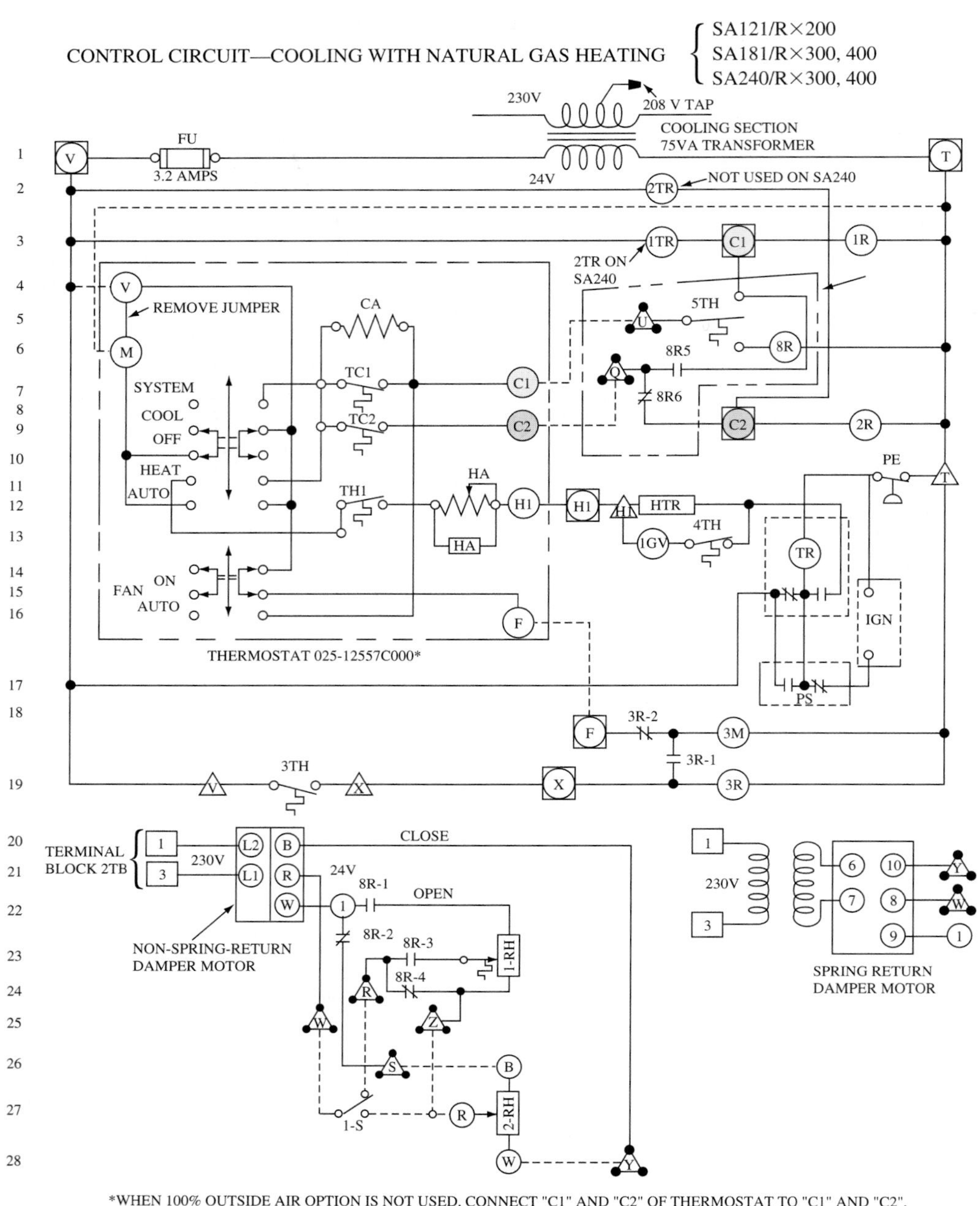

Figure A31-30 Low-voltage wiring diagram for a central-station heating and air-conditioning system. (Courtesy of York International Corp.)

closes, furnishing power to relay 8R and, through contacts 8R5 and 8R6, allowing both cooling and relays 1R and 2R to close on command of the cooling stats. 1R and 2R are time-delay relays in the cooling circuit that prevent both compressors from coming on at the same time.

In the heating mode, the economizer system is bypassed. Current from the TH1 stat (line 12) goes through the heat anticipator HA to terminal H1. From H1 it feeds to the gas valve 1GV, provided that the limit-control contact 4TH is closed. Another circuit on line 17 provides an alternate flow of current to the pilot safety switch, PS, and through its contacts feeds the IGN pilot-ignition transformer. Ignition of the pilot cannot take place unless there is sufficient gas pressure to close PE (pressure/electric) switch (line 11). At the same time the TR (time-delay relay) is also energized, the contacts below it are reversed, and the gas valve opens.

If the fan selector is in the ON position, current from terminal F goes through normally closed 3R-2 switch and energizes 3M (the contactor coil) for the indoor or evaporator fan. In the AUTO position the circuit is essentially bypassed. Current then (on line 19) flows through 3TH,

Common Legend for SA121, SA181, SA240 Elementary Wiring Diagrams			
CA	Anticipator, Cooling	1RC, 2RC / 3RC, 4RC	Running Capacitor
HA	Anticipator, Heating	S	Switch, Oil Pressure
T	Auto-Transformer, Speed Controller	1-S	Switch, Bypass
FU, 1FU, 2FU	Fuse	TC1	Thermostat, Cooling 1st Stage
1M, 2M	Contactor, Compressor	TC2	Thermostat, Cooling 2nd Stage
3M	Contactor, Blower Motor	TH1	Thermostat, Heating 1st Stage
1GV	Gas Valve	TH2	Thermostat, Heating 2nd Stage
2GV	Gas Valve, Second Stage	1TH, 2TH	Thermostat, Low Ambient Control
HTR	Heater in 3TH Fan Thermostat	3TH	Thermostat, Blower (Heat)
1HTR, 2HTR	Heater, Compressor Crankcase	4TH	Thermostat, Limit (Heat)
1HP, 2HP	High Pressure Cutout, Refrig.	5TH	Thermostat, Mixing Box
IGN	Ignition Trans. (For Pilot Relighter)	11TH	Thermostat, Crankcase Heater
PE	Low Gas Pressure Switch	TR	Time Delay Relay, Pilot Ignition
1LP, 2LP	Low Pressure Cutout, Refrig.	VFS	Venter Fan Sail Switch
1-OL, 2-OL / 3-OL, 4-OL / 5-OL, 6-OL	Overload Protectors, Compressor	10R, 11R	Venter Motor Relays
		1RH	Control Damper, Mixed Air
PS	Pilot Safety Switch	2RH	Control Damper Controller, Min. Position
R	Second Stage Gas Valve Relay	□	Terminal Block, 1TB
1R, 2R	Relay, Control Cooling	□	Terminal Block, 2TB
3R	Low Voltage Control Relay	▢	24-Volt Terminal Block, 3TB
4R, 5R	Relay, Control Electric Heat	△	Identified Connection in Heating Section
8R	Relay, Control, Mixing Box	△	0-100% Outside Air Terminal Block, 6TB (SA 121) 7TB (SA 181): 4TB (SA 240)
9R	Relay, Low Ambient Control		

Specific Legend for SA121, SA181, SA240 Elementary Wiring Diagram					
SA121		SA181		SA240	
6R	Relay, Lockout No. 2 System	6R	Relay, Lockout No. 2 System	6R	Relay, Lockout.
7R	Relay, Lockout No. 1 System	7R	Relay, Lockout No. 1 System	MP	Compr. Protection
1TR, 2TR	Time Delay Relay (Low Ambient Accessory)	1TR, 2TR	Time Delay Relay, Low Ambient Control	1TR	Time Delay Relay, Part Winding Start
3TR	Time Delay Relay, Cond. Fan			2TR	Time Delay Relay, Low Ambient Control
T	Auto-Transformer (Low Ambient Accessory)			3TR	Time Delay Relay, Oil Pressure Switch
				1-SOL	Solenoid, Compr. Unloader
				2-SOL	Solenoid, Evap. Unloader

Figure A31-31 Legend for the low-voltage and line. (Courtesy of York International Corp.)

the blower-control thermostat (if heat is present to close the switch), and to coil 3R, which then closes 3R-1 and opens 3R-2. Thus, 3M is again energized and the indoor blower comes on. It will remain on until 3TH opens to indicate heat in the plenum.

The power (240 V on lines 20 and 21) to operate the economizer outside-air damper is picked up at terminal blocks 1 and 3 from line 10 in Fig. A31-29. Two diagrams are shown. One is a damper motor with a non-spring return; the other has a spring return. With the spring return the motor will close completely in the event of power failure. If there is no spring return, the damper will close to a minimum position based on a minimum amount of outside air needed. Note that relay 8R, coil line 6, is the key to all switch actions: 8R-1, 8R-2, 8R-3, and 8R-4 position the control dampers 1-RH and 2-RH.

A31-9.4 Pneumatic Control Systems

Pneumatic control systems use compressed air to supply energy for the operation of valves, motors, relays and other pneumatic control equipment. Consequently the circuits consist of air piping, valves, orifices, and similar mechanical devices.

Pneumatic control systems offer some distinct advantages:

- They provide an excellent means of modulating control operation.

Model	Power Supply	Length Circuit One Way, Ft Up To	Min. Wire Size Copper AWG 60 C 2% Voltage Drop	Max. Fuse Size Dual Element	Disconnect Switch Size, Amps
SA91-25A	A	100 175 200 250	6 4 3 2	55	60
SA91-45A	A	175 250	10 8	30	30
208/230 V SA 121-258	A	175 200 250	125 3 2 1	4 70	100
SA 121-46B	A	200 250	8 6	40	60
SA181-25A	A	150 200 250	1 0 000	100	100
SA181-45A	A	250	4	60	60
SA240-25C	A	150 200 250	00 000 250MCM	175	200

Figure A31-32 Wire, fuse, and disconnect sizes for several sizes of central station air conditioners. (Courtesy of York International Corp.)

- They provide a wide variety of control sequences with relatively simple equipment.
- They are relatively free of operational problems.
- They cost less than electrical controls if the codes require electrical conduit.

Pneumatic controls are made up of the following elements:

1. A constant supply of clean, dry compressed air.
2. Air lines consisting of mains and branches, usually copper or plastic, to connect the control devices.
3. A series of controllers including thermostats, humidistats, humidity controllers, relays, and switches.
4. A series of controlled devices including motors and valves called operators or actuators.

The air source is usually an electrically driven compressor (Fig. A31-33), which is connected to a storage tank. The air pressure is maintained between fixed limits (usually between 20 and 35 psi for low-pressure systems). Air leaving the tank is filtered to remove the oil and dust. Many installations use a small refrigeration system to de-

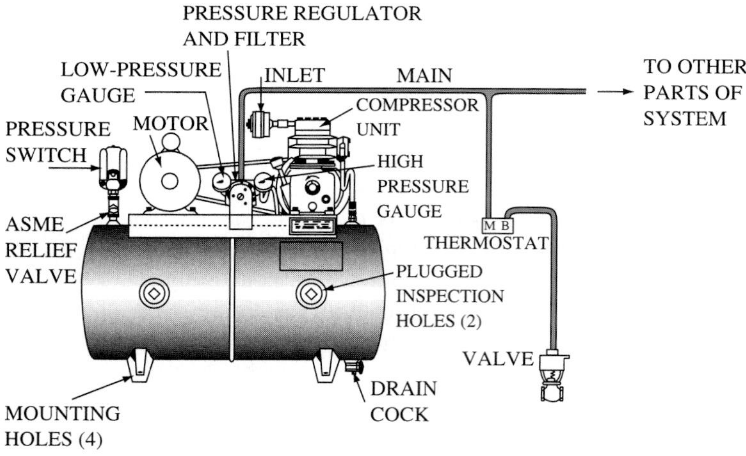

Figure A31-33 Air compressor for a pneumatic control system.

humidify the air. Pressure-reducing valves control the air pressure.

A31-9.5 How the Controls Operate

The controller function is to regulate the position of the controlled device. It does this by taking air from the supply main at a constant pressure and adjusting the delivered pressure according to the measured conditions.

One type of thermostat is the bleed type, shown in Fig. A31-34. The bimetal element reacts to the temperature and controls the branch-line bleed-off pressure. These thermostats do not have a wide range of control, therefore the branch line is often run to a relay that controls the action. Bleed controls cause a constant drain on the compressed air source.

Non-bleed controllers use air only when the branch line pressure is being increased. The air pressure is regulated by a system of valves (Fig. A31-35), which eliminates the constant bleeding characteristic of the bleed type unit. Valves C and D are controlled by the action of the bellows (A) resulting from the changes in room temperature. Although the exhaust is a bleeding action, it is relatively small and occurs only on a pressure increase.

Controlled devices, operators or actuators, are mostly pneumatic damper motors or valves. The principle of operation is the same for both. Fig. A31-36 is a diagram of a typical motor. The movement of the bellows as the branch line changes activates the lever arm or valve stem. The spring exerts an opposing force so that a balanced, controlled position can be stabilized. The motor arm L can be linked to a number of functions.

Fig. A31-37 shows a pictorial review of some of the functions in a pneumatic control system. There is always some crossover between the air devices and the electrical system. The device most widely used is the pneumatic/electric relay.

A31-9.6 Electronic Control Systems

Electronic control may also be used effectively for central-station equipment. Due to a number of advantages they are rapidly gaining in popularity:

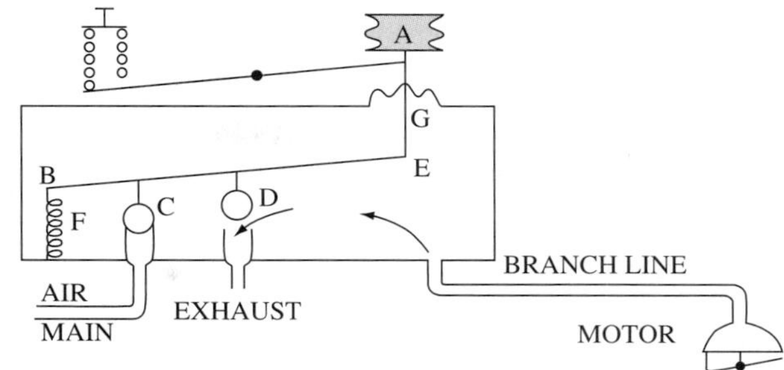

Figure A31-35 Diagram of a non-bleed pneumatic thermostat.

1. They have no moving parts.
2. The response is fast.
3. The regulatory element can easily be a reasonable distance from the sensing element, permitting:
   a. Adjustments made at a central location.
   b. Cleaner conditions at a central location than at the location of the sensing element.
   c. Easier temperature averaging.

In the Fundamentals of Electronic Devices and Circuits chapter, a basic overview of electronic circuits is presented. In this section, the emphasis is on the specific control functions of electronic controls. With a comprehension of these fundamentals, most electronic control circuits can be reduced to simple components. The interaction of these controls and the logic behind them is the heart of electronic control of engineered air-conditioning systems.

Only simple low-voltage connections are needed between the sensing element and the electric circuit. Flexibility is important since electronic circuits can be combined with both electric and pneumatic circuits to provide results that could not usually be achieved separately. Electronic circuits can coordinate temperature changes from several sources such as room, outdoor-air, and fan-discharge-air, and program action accordingly.

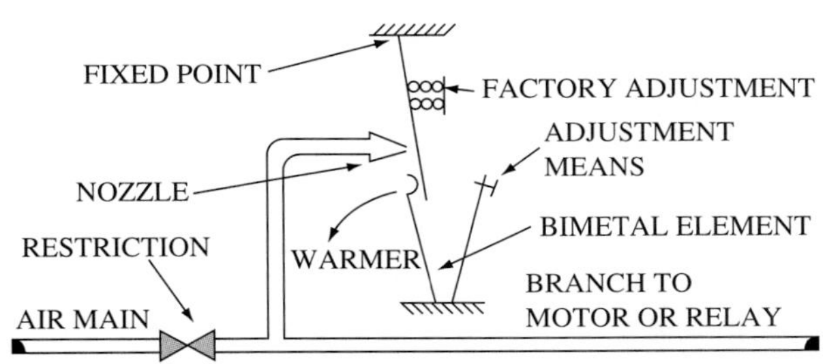

Figure A31-34 Diagram of a bleed-type pneumatic thermostat.

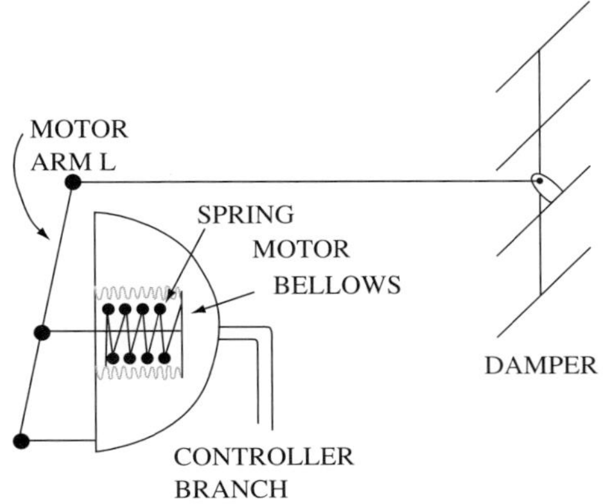

Figure A31-36 Pneumatic actuator with normally open damper.

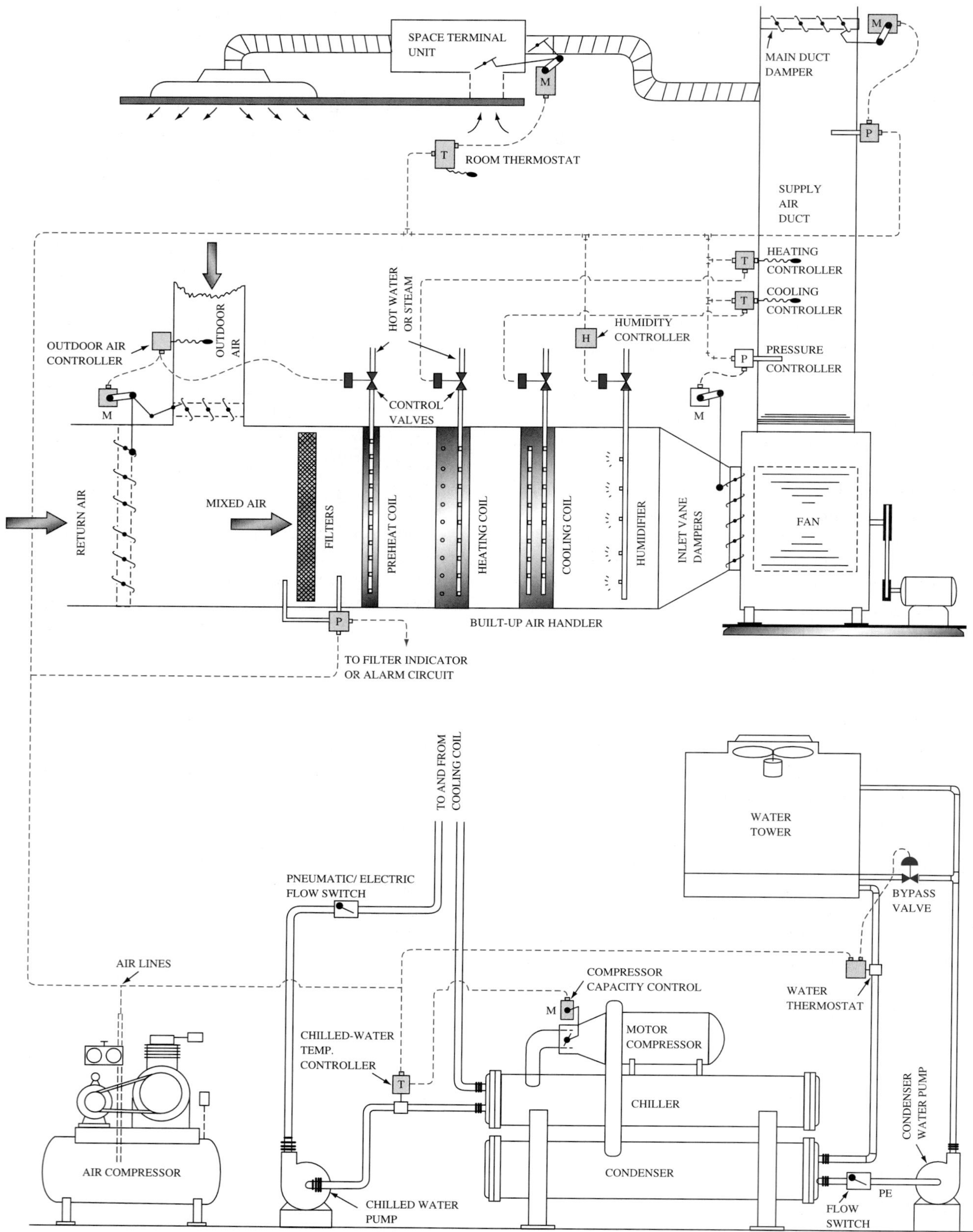

Figure A31-37 System diagram for a central station installation showing pneumatic controls.

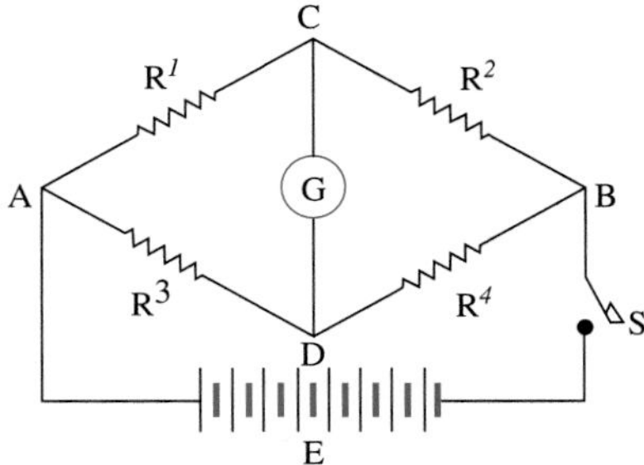

Figure A31-38 Wheatstone bridge circuit.

Electronic controls are based on the principle of the Wheatstone bridge (Fig. A31-38), which is composed of two series resistors (R_1 and R_2, R_3 and R_4), connected in parallel across a DC-voltage source. A galvanometer G (a sensitive indicator of electrical current) is connected across the parallel branches at junctions C and D between the series resistors. If switch S is closed, voltage E (DC battery) energizes both branches. If the potential at C equals the potential at D, the net potential difference is zero. When this condition exists, the bridge is in balance.

If the resistance of any one leg is changed, the galvanometer will register a flow of current. The bridge is now unbalanced.

If that resistance is changed as a result of a temperature reaction, we now have an electronic method of measuring current in relation to temperature change. With a few changes we can create an electronic main circuit such as in Fig. A31-39. A 15-V AC circuit replaces the DC battery. The galvanometer is replaced by a combination voltage-amplifier phase discriminator-switching-relay unit. Resistor R_2 is replaced by sensing element T of an electronic controller.

The purpose of the *voltage amplifier* is to take the small voltage from the bridge and increase its magnitude by stage amplification to do the work. *Phase discrimination* means determining the sensor action. In an electric bimetal thermostat, the mechanical movement is directly related to the temperature changes; however, the electronic sensing element is a non-moving part and the phase discriminator determines whether the signal will indicate a rise or fall in temperature. The *relay* then operates the final element action. Phase discrimination can be two-position or, with certain modifications, can be converted to a modular system.

The crossover point from the electronic to electric occurs at the output of the amplifier and relay signal. The motor that it operates is a conventional ON-OFF motor, or a proportional (modulating) electric motorized valve or damper actuator.

Electronic temperature-sensing elements are room thermostats, outdoor thermostats, insertion thermostats for ducts or insertion thermostats for liquids. The typical room thermostat is a coil of wire wound on a bobbin. The resistance of the wire varies directly with the temperature changes.

A *sensor* is any device that converts a non-electrical impulse such as sound, heat, light, or pressure into an electrical signal. Sensors have been developed to provide the necessary input for controlling the pressure in a duct or the relative humidity in a conditioned area. The sensor provides an input to a solid-state controller. The logic in the controller sends an output to some mechanical device to produce the required action.

For example, to control humidity, a pair of electronic sensors read wet- and dry-bulb temperatures. The logic in the controller converts these readings to a relative-humidity value. Based on the limits set up, the connected mechanical device is programmed to add or deduct moisture from the air to meet the requirements. The use of electronic equipment makes possible more accurate control of the space conditions than can be accomplished by pneumatic or electromechanical control equipment.

A31-9.7 Direct Digital Control Systems

In a direct digital-control (DDC) system, the computer acts as the primary control for all HVAC/R functions. Valves, dampers, fan speeds, etc., are all controlled by the computer without the use of conventional control devices such as thermostats, humidistats, timers, etc.

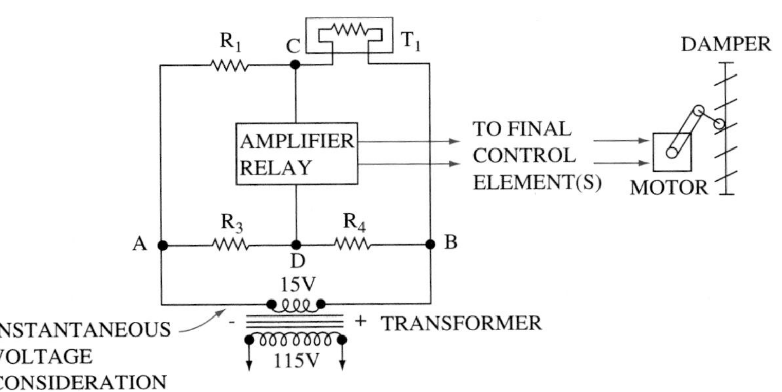

Figure A31-39 Electronic main bridge circuit.

The computer directly senses the building environmental conditions, and based on a user defined programmed set of instructions, initiates the proper control actions in the HVAC/R system. Direct digital control of HVAC/R components gives more accurate control and greater flexibility than other commonly used mechanical and electrical control devices. It also has the capability of coordinating inputs from a number of sensing devices and arriving at an output that takes into consideration numerous influencing factors. The following are some examples of the capabilities of the DDC control systems:

1. DDC system can control a VAV terminal box to discharge the proper air supply based on a variety of inputs such as dry-bulb temperature, relative humidity and mean radiant temperature. In this way, considering the total environmental conditions, a greater feeling of comfort is produced for the occupants in the space.
2. In a central-station system with a large supply fan, the microprocessor can regulate the speed of the fan to produce the required airflow using a minimum amount of power.
3. Computers are currently used to turn on chillers and boilers at an optimum time to recover from a period when the building is unconditioned.

A typical DDC system is shown in Fig. A31-40. A central *stand-alone controller* (SAC) is connected to a series of *remote control units* (RCUs). The SAC is wired to a computer terminal, a printer and modem (Fig. A31-41). The terminal is used for input communications by the operator. The printer is used to record any information concerning the operation of the system. The modem is an electronic device that permits the computer to communicate through the telephone lines.

The individual RCUs are located in the building near the equipment being controlled. These units have a series of input and output wiring connections that go to sensors and controls on the HVAC/R system that permit the SAC unit to control the operations of the system. Both the input and output functions are of two types: analog and digital. The *analog functions* supply or deliver modulated information. For example, an analog temperature sensor may be capable of reading temperatures between 0° and 100°F. This input could be converted by the computer to produce an analog output signal to control a damper to any position between fully open to fully closed.

The *digital function* is a binary or two-position function. For an input, the signal could monitor whether a switch is open or closed. For output, the binary signal could position the switch in either an ON or OFF position.

The SAC unit, RCU units, computer terminal, printer, and modem are all considered *hardware*. The control program, which is programmed for an individual system, is called *software*. The software is installed in the SAC unit at the time that the DDC takes over the operation of the system. The operator periodically monitors the operation of the system through the computer, notes the reports and alarms, and makes any changes in the program that are considered advantageous.

The DDC systems are often called *energy-management systems* since one of their main functions, and usually justification for their adoption, is saving energy. Special provision has been made in the selection of sensors to make possible continuous monitoring of the energy usage. The control system is set up to energize loads only when necessary and to use such features as free cooling (economizer operation) whenever possible.

The control center collects key operating data from the HVAC/R system and incorporates remote-control devices to supervise the system's operation. The elaborateness of the data center is related to the type and size of the system and to economic considerations. Some control centers have continuous scanners with alarm indicators to monitor refrigeration machines, oil and refrigerant pressures, chilled-water temperatures, air-filter conditions, low-water conditions in the boiler, and conventional-space temperature and humidity conditions in each zone.

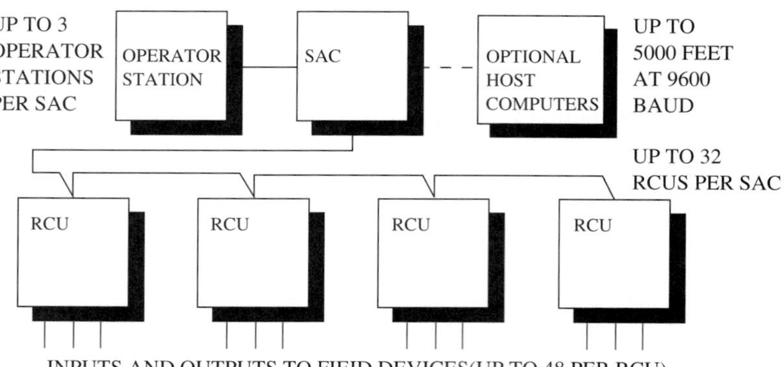

Figure A31-40 Data control center for a central station system.

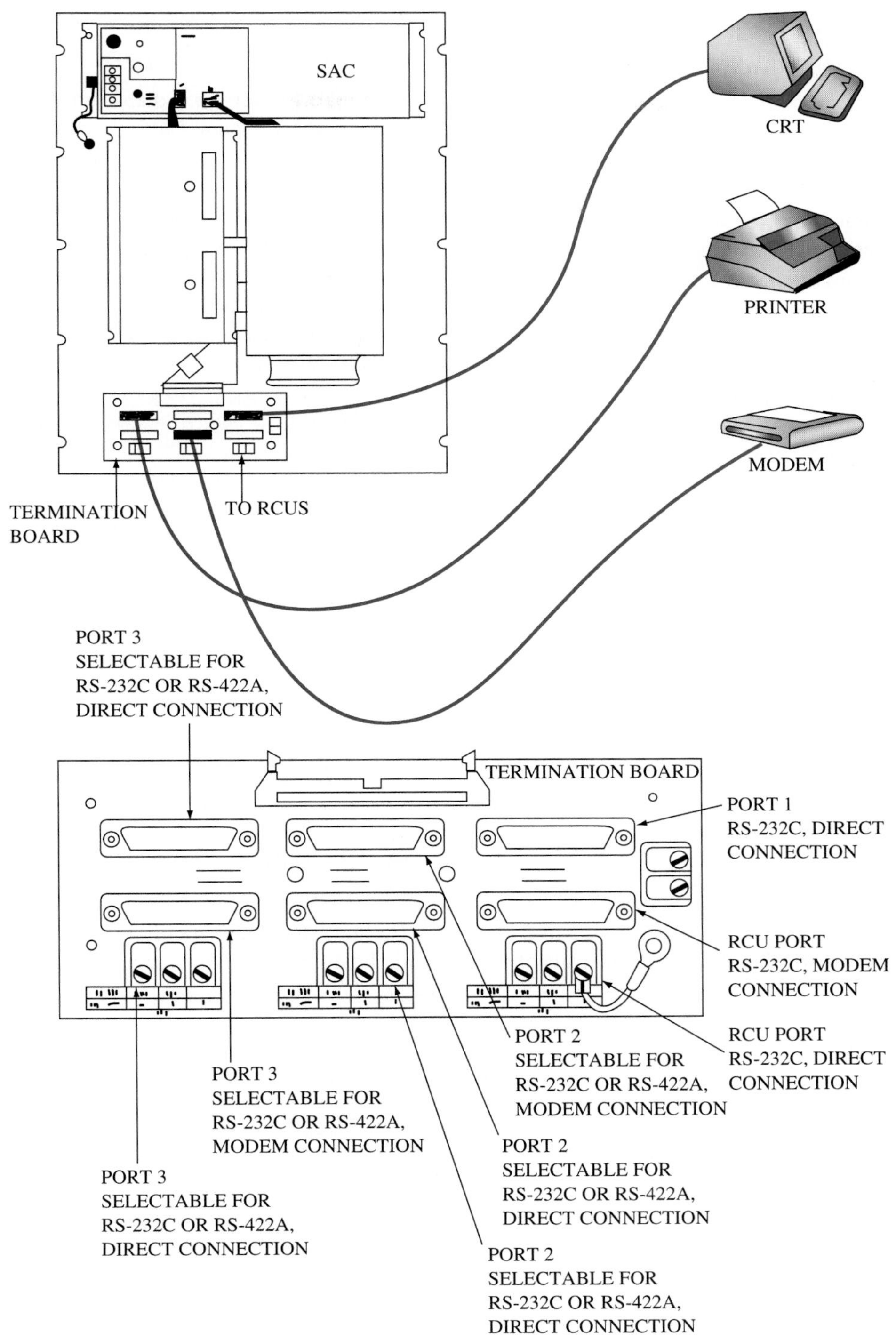

Figure A31-41 Stand-alone controller (SAC).

A31-10 THERMAL STORAGE SYSTEMS

Thermal storage is the temporary storage of high- or low-temperature energy for later use. Examples of thermal storage are the storage of solar energy for night heating, the storage of summer heat for winter use, the storage of winter ice for space cooling in the summer, and the storage of heat or coolness generated electrically during off-peak hours for use during subsequent peak rate hours. Most thermal-storage applications involve a 24-hour storage cycle, although weekly and seasonal cycles are also used. In the context of the energy-efficient systems applications, storage of cooling is of prime interest.

For HVAC/R purposes, water and phase-change materials (PCMs)—particularly ice—constitute the principal storage media. Water has the advantage of universal availability, low cost, and transportability through other system components. Ice has the advantage of approximately 80% less volume than that of water storage, for a temperature range of 18°F, which is the difference in temperature between a fully charged and a fully discharged storage vessel.

A31-10.1 Modes of Operation

There are five modes of operation of a cooling storage system which stores and releases thermal energy. These are illustrated in Fig. A31-42, and are listed below.

1. *Charging storage:* The refrigeration system is extracting heat from the storage vessel.
2. *Simultaneous recharging storage and live-load chilling:* Some of the refrigeration capacity is being used by the load at the same time the storage is being recharged.
3. *Live-load chilling:* Identical to normal water-chiller operation, providing cooling as needed.
4. *Discharging and live load chilling:* The refrigeration unit operates and the storage vessel is being discharged at the same time to satisfy the cooling-load requirement.
5. *Discharging:* Cooling needs are met only from the storage, with no refrigeration operating.

KNOWLEDGE EXTENSION

Ice thermal-storage systems have become more popular in recent years, but it may be difficult for the aspiring HVAC/R technician to learn more about these systems. A method to extend your knowledge of these systems would be to contact a mechanical engineering consulting firm in your area and ask if there are any buildings in your region which may be using this technology. If so, contact the building owner or maintenance manager and request a field trip to review the installation and operations. While in the building, study the blueprints and any control diagrams to help clarify the overall picture of the actual operations.

A31-10.2 Selecting the Method of Operation

Thermal storage, particularly ice storage, has its roots in applications involving short-duration loads, with rela-

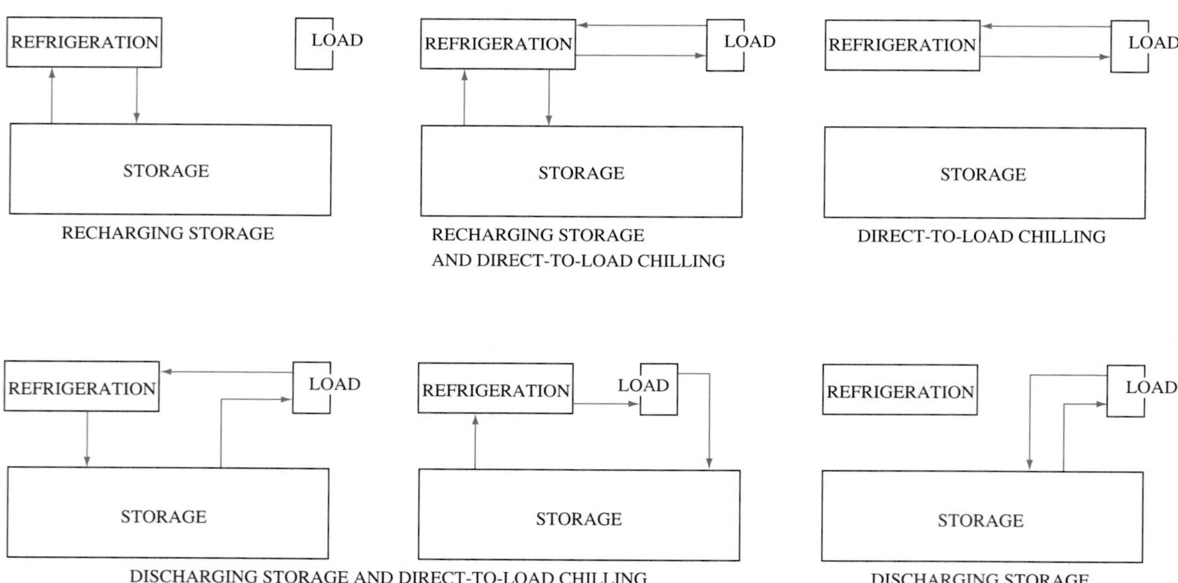

Figure A31-42 Operating strategies for ice-storage systems. (Copyright by the American Society of Heating, Refrigerating, and Air-Conditioning Engineers, Inc. Used by permission.)

tively long times between loads. The classic examples are churches, sports facilities, and older movie theaters. Current applications are driven by increasing electric utility rates, time-of-day rate schedules, and large demand charges for electric power.

The essential purpose of ice storage is to make ice while the power rates are lower. During periods when the rates are higher, either the power use is minimized by using ice and less refrigeration (load leveling, strategy 4 above), or using only the stored ice for cooling (full storage, strategy 5 above). The optimum combinations are derived from an analysis of all the elements involved, including:

1. Total cooling load.
2. Total daily ton-hours of load (the sum of each of the hourly loads in a day's use).
3. Available ice-recharging time.
4. Steady loads and temporary loads.
5. Energy cost, peak and off-peak, and hours of billing.
6. Space for storage.
7. Relative costs of type of systems.

Typical HVAC/R cooling-load profiles of a building are shown in Fig. A31-43.

Water-based thermal storage has certain specialized applications, and has been successfully applied many times. Problems are: (1) providing stratified storage temperatures, so that the recirculated, warmer water does not disturb the stored chilled water; (2) space and structural limitations; and (3) open systems requiring water treatment.

Ice-storage systems are practical on systems as low as 50 tons. Ice provides cold (33–34°F) water for cooling and has the following advantages:

1. Colder water allows a greater temperature rise, requiring less flow for the capacity.
2. Lower flows mean smaller pipes and less pumping cost.
3. Colder water allows lower leaving-air temperatures from cooling coils.
4. Colder supply air allows greater temperature rise, requiring less cfm.
5. Less cfm (airflow) allows smaller fans and duct work.

Applications utilizing ice-storage systems require modifications of conventional system designs.

A31-10.3 Types of Ice Storage Equipment

Ice-on-coil Systems

Ice-on-coil systems represent the oldest technology for ice storage currently in use. This type of unit is normally called an *ice builder,* since it forms ice on the outside of a

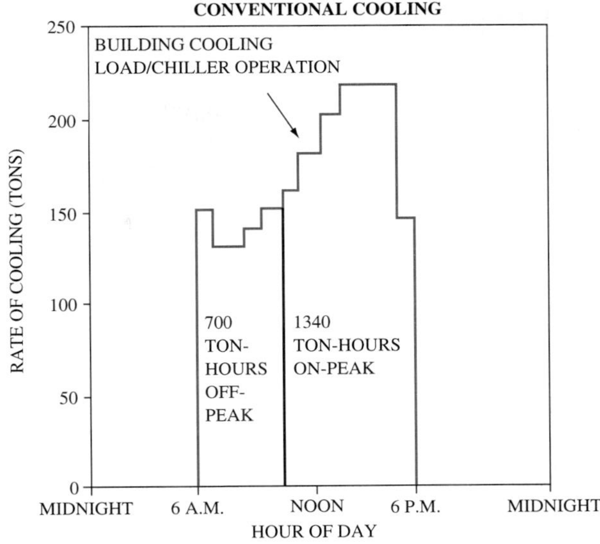

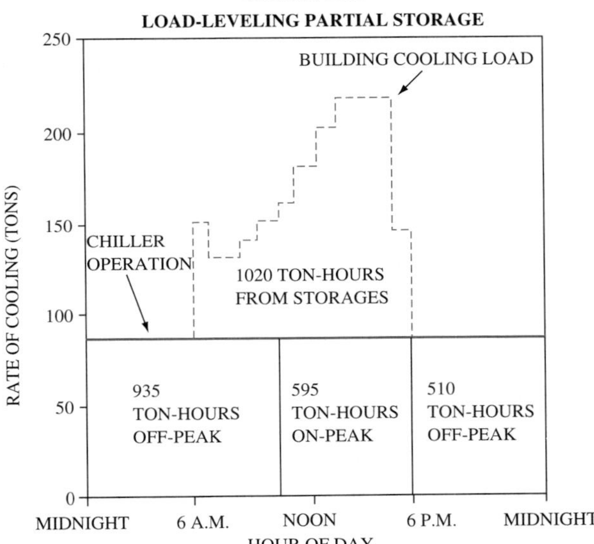

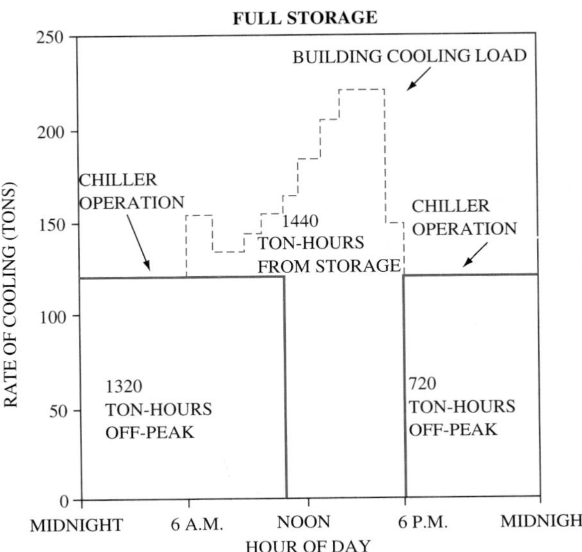

Figure A31-43 Cooling load profiles for ice-storage systems. (Copyright by the American Society of Heating, Refrigerating, and Air-Conditioning Engineers, Inc. Used by permission.)

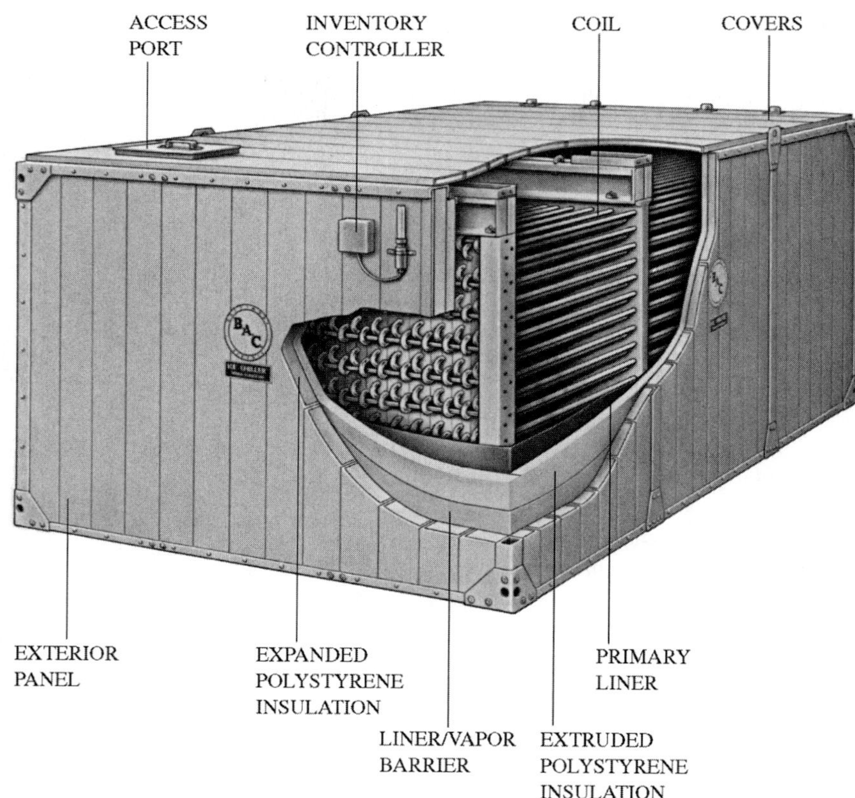

Figure A31-44 Ice builder construction. (Courtesy of Baltimore Aircoil Company)

pipe bundle, with refrigerant in the pipes (Fig. A31-44). Ice builders are sized for tons of refrigeration as a chiller, and for total capacity of ice (usually in tons of ice) formed on the pipes.

These are very efficient units, since the ice is formed on the outside of the pipes and is then melted from the outside of the ice surface back to the bare pipe. This keeps the melting ice in contact with the chilled water until the ice is gone. These units can be charged with a variety of direct refrigerants, including ammonia.

An ice-thickness control is required for these systems so that the ice builder does not bridge ice between pipes, or worse, freeze solid. Ice sensors are used to monitor ice thickness, and shut down the refrigeration when the total ice build is completed. The ice builder has water-flow baffling to create even water flow over the pipes to ensure uniform ice formation and melting.

Solid-ice Brine-coil Systems

Another method, using brine as a secondary refrigerant, is less efficient than the direct refrigeration system since a double heat exchange is required. The most common brine application uses a conventional packaged chiller to chill a glycol solution to about 25°F, to charge the system. It also requires a glycol-to-water system to interface with the chilled-water system. The offsetting factor is a very effective ice melt directly to a chilled-water system, providing simple operation, and some of the benefits of a cold chilled-water system noted above.

In this glycol system, plastic mats containing brine coils are tightly rolled and placed inside a cylindrical water tank (Figs. A31-45 and A31-46).

The mats occupy approximately 10% of the tank volume; another 10% of the volume is left empty to allow for the expansion of the water when it freezes, and the rest is filled with water. A brine solution, typically 25% ethylene glycol, is cooled by a packaged chiller and circulates through the coils and freezes water in the tank. Ice is built up to about ½ in. thickness on the coils. During discharge, the cool brine solution circulates through the system cooling loop and returns to the tank to be cooled again.

Control is very simple, since the water expands 9% upon freezing. The level in the vessels is monitored, and the refrigeration is turned off when the frozen level is reached.

The offsetting factors for these units are that they use conventional packaged chillers for their cooling, and the ice vessels are very economical to purchase, and very lightweight to install. The weight of the water/ice still needs to be considered when supporting the units, but the overall weight is significantly less than the ice builder units. These units are currently the most common units used for ice storage because of their overall cost of installation, simplicity of operation, and competitive energy savings.

Figure A31-45 Installation of a series of tanks using the solid-ice brine-coil system. (Courtesy of Calmac Manufacturing Corp.)

Ice in Containers

In this type of system, small water-filled containers are placed inside a tank, with the brine filling the tank, and circulating over the containers. Some containers are spherical, with dimples to allow for expansion on freezing (Fig. A31-47). Other units are rectangular, with flexible sides to allow for expansion. These containers are stacked in the storage tanks which can vary in shape and design. The piping in and out of the tank must create a flow path which is effective for both charging and discharging.

This container type of system operates much like the solid-ice brine-coil system above, in that the ice is formed from contact with a brine solution, and the ice is first melted away from the contact surface with the brine, requiring the ice to transfer its heat through water to the brine.

The economical factors for this type of system include storage-tank shape and configuration, and simple installation. The water-filled containers are heavier than the ice tanks of the solid-ice brine coils, but they are very flexible to use. The control monitors the liquid level of the tanks to determine the amount of ice formed. In the case of the ice in containers, the containers expand, raising the glycol level in the tank, which is monitored for control.

Ice-harvesting Systems

Ice-harvesting systems are another application of industrial ice makers/water chillers. These units build ice in sheets on the surface of vertical refrigerated plates. The ice is harvested by slightly warming the plates with refrigerant discharge gas, which causes the ice sheets to

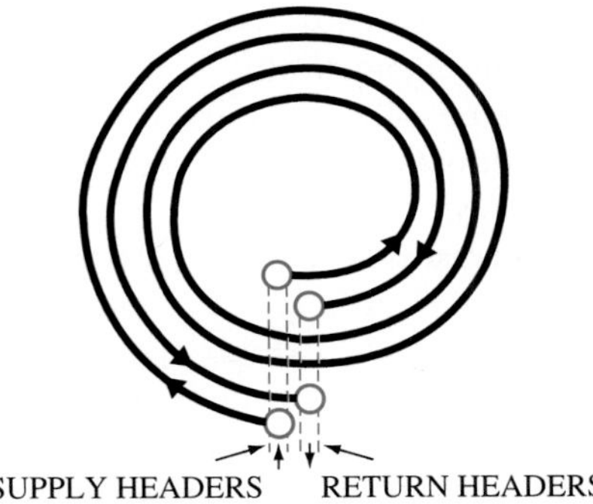

Figure A31-46 Internal construction of a solid-ice brine-coil tank.(Courtesy of Calmac Manufacturing Corp.)

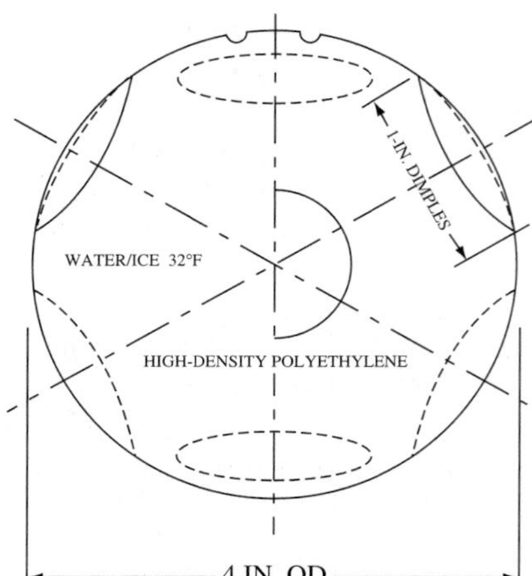

Figure A31-47 Spherical container for ice thermal storage.

break away from the plates and fall into a water-filled storage tank. The sheets of ice break up when they hit the water, providing many exposed ice surfaces for heat transfer to the water. The thin sheets of ice provide a very even discharge cycle, with consistently low water temperatures. Monitoring of the tank level determines refrigeration operation.

The storage tank must be constructed with weirs, to prevent the ice from getting into the circulation system. Since the ice floats and distributes itself in the tank, water-flow distribution is less critical than with some of the other systems.

The ice-harvest systems are designed to operate on direct-refrigerant applications only, which allows them highly efficient operation. A disadvantage is that the equipment is large and heavy, with limited configurations. They usually require field-piped refrigeration systems.

REVIEW

- Central-station system—Commonly called an applied system, applied equipment, applied machinery, and engineered system.
- Cooling and heating plant located in basement or penthouse of large building serves air-handling equipment and air-distribution systems throughout the building. Equipment size starts at 25 to 50 tons and extends upward to multi-thousand-ton systems. Central-station systems use chilled water or heated liquid (usually water) to transfer heating or cooling to air-delivery systems, where air is conditioned by heating and cooling coils.
- Typical systems include:
  Water chillers
  Hot-water generators (boilers)
  Air handlers
  Pumps
  Cooling towers
  Control systems
  Air-distribution components
  Thermal storage
- Water chiller regulates chilled water to approximately 43 to 45°F. Chillers may be either vapor-compression or absorption.
- Chiller types:
  Water-cooled chillers
  Air-cooled chillers
  Centrifugal chillers
  Screw-compressor chillers
  Absorption chillers
- Components of the compression-cycle chiller:
  Compressor
  Evaporator (chilled coil)
  Condenser (water or air)
  Metering device
- Compression chillers:
  High-pressure—R-22, R-12
  Low-pressure—R-11, R-123
- Condenser capacity is normally based on 85°F entering-water temperature within a 10°F rise.
- Unloading cylinders is used to control capacity of piston compressors.
- Centrifugal compressors use variable vanes to control capacity.
- Motor speed and gear-box speeds can also control capacity.
- Screw and scroll compressors are used with high-pressure systems.
- Air-cooled condensers are used when space does not allow cooling towers or where water usage may be a problem.
- Air-distribution systems:
  Low-velocity—0.25–0.50 in. WC
  Medium-velocity—2.0–3.0 in. WC
  High-velocity—>3.0 in. WC
  VAV—may be either medium- or high-velocity
  Single-zone—constant volume
  Multi-zone—VAV, constant volume
  Dual-duct—constant volume multizone, highest initial cost
- Control systems—Must direct the operation of all elements automatically.
  Electric—Basic system using electricity.
  Electronic—Advanced controls utilizing DDC technology.
  Pneumatic—Uses compressed air pressure to supply energy.
  Combination hybrid system—Mixes electric, electronic, and pneumatic controls.
- Thermal storage systems—Temporary storage of high- or low-temperature energy for later use.
  Ice-on-coil systems
  Solid-ice brine-coil systems

Ice in containers
Ice-harvesting systems

Problems and Questions

1. Central-station equipment is also called ____.
2. Central-station systems are ____ fabricated.
3. Heat transfer is by ____ to ____ to ____.
4. A conventional central system will generally include what five major components?
5. Chilled-water systems for comfort application produce cold water in the range ____ to ____°F.
6. In the winter heating season the boiler-water temperature is usually in the range ____ to ____°F.
7. Chillers up to a nominal 100 tons usually use ____ compressors.
8. What is meant by "number of passes" in a chiller or condenser?
9. Condensers used for seawater (marine) applications are usually made of ____.
10. Centrifugal compressors may be ____ driven or ____ driven.
11. Fan-coil unit piping arrangements are of three types. What are they?
12. Humidifiers used in large systems are of three types. What are they?
13. The air-distribution systems used with central-station equipment are classified as ____ systems.
14. On perimeter air-conditioning problems, the best system to use is usually a ____ heating and cooling system.
15. What is the biggest advantage of a double-duct system?
16. Central-station equipment is usually associated with "plan and spec" engineered systems. True or false?
17. When using chilled water, the cooling coil at the air handler warms the water about 20° Fahrenheit at full load. True or false?
18. Lower-capacity water chillers use reciprocating and scroll compressors. True or false?
19. The shell and tube condenser has refrigerant flowing through the tubes. True or false?
20. Special seawater condensers are made of stainless steel to withstand the salt corrosion effects. True or false?
21. Air-cooled chillers use air-cooled evaporators and water cooled condensers. True or false?
22. Centrifugal chillers can be single-stage or multi-stage compressors. True or false?
23. Screw compressor chillers:
    a. Use a mating pair of rotors with lobes which rotate much like a pair of gears.
    b. Are positive-displacement compressors.
    c. Have variable-capacity control from 100% to 10%.
    d. All of the above.
24. Absorption chillers:
    a. Use steam to generate a pressure differential in a generator section.
    b. Use hot water to generate a pressure differential in a generator section.
    c. Are direct-fired by natural gas to generate a pressure differential in a generator section.
    d. All of the above.
25. A hot-water boiler produces water at:
    a. 180 to 200°F.
    b. 180 to 212°F.
    c. 150 to 180°F.
    d. None of the above.
26. Multi-zone systems provide:
    a. A hot-air duct and cold-air duct to the multiple zones in the building.
    b. One trunk duct to each zone from the multi-zone air handler.
    c. A chiller-water and hot-water line to the multiple zones in the building.
    d. All of the above.
27. Water-based thermal storage problem(s) are:
    a. Providing stratified storage temperatures.
    b. Space and structural limitations.
    c. Open systems requiring water treatment.
    d. All of the above.

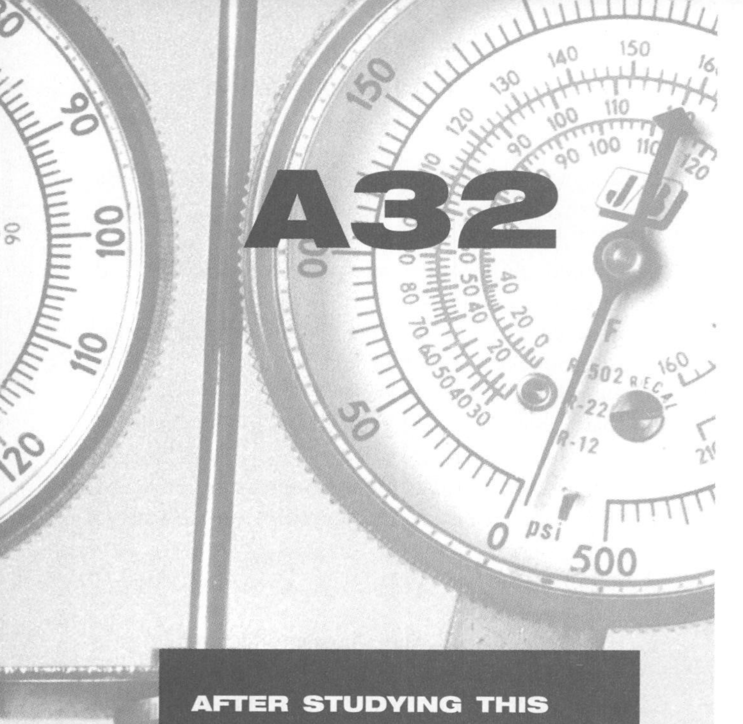

A32 Air-Handling Units and Accessories

AFTER STUDYING THIS CHAPTER, THE STUDENT WILL BE ABLE TO:

- Describe the functions of an air-handling unit.
- Describe the remote air-handling unit.
- Illustrate central-station air-handling units.
- Define the types of heating coils available.
- Explain the operation of the humidifier.
- Describe mechanical air filters.
- Illustrate the operation and benefits of the electronic air filters.
- Explain mixing dampers and exhaust dampers.
- Characterize the types of fans and motors used with air handlers.
- Apply the principles and applications of the fan equations.
- Identify different types of fans/blowers: centrifugal and axial.
- Determine the proper direction of rotation.
- Explain the difference between tubeaxial and vaneaxial.
- Identify the types of centrifugal fans/blowers: forward-curved and backward-curved.
- Identify types of mechanical filters: disposable; permanent foam, mesh, and fiber; high efficiency; HEPA; electrostatic.
- Describe operation of electronic air cleaners.

A32-1 TYPES OF AIR-HANDLING UNITS

The **air-handling unit** is that portion of the air-conditioning system that conveys the conditioned air to and from the building. Historically, these units distributed air by means of gravity action like space heaters, which are still used for certain applications. Most units, however, now use a fan (or blower) to force the air through the conditioning equipment and its distribution system.

Air-handling units for these forced-air systems are separated from the primary heating or cooling apparatus. These units form a part of a central-plant, built-up system which is flexible in makeup to fit the application.

There are two general types of air-handling units that are used for central plant systems: (1) ones with *remote air-handling units,* called fan-coil units, and (2) ones with *central-station air-handling units.*

Fan-coil unit systems consist of room units that are supplied with hot water or cold water directly from central plant boilers and chillers. Each one of these units has fans, water coils, filters, a fresh air supply, and a control system. They were primarily designed to supply conditioned air to individual rooms.

Central-station systems use one or more large air-handling units equipped with fans, heating and cooling coils and other accessories, and duct work to convey the conditioned air to remote areas.

A32-2 FAN-COIL UNITS

Fan-coil units for combination heating and cooling come in a variety of designs, the most familiar being the individual room conditioner (Fig. A32-1).

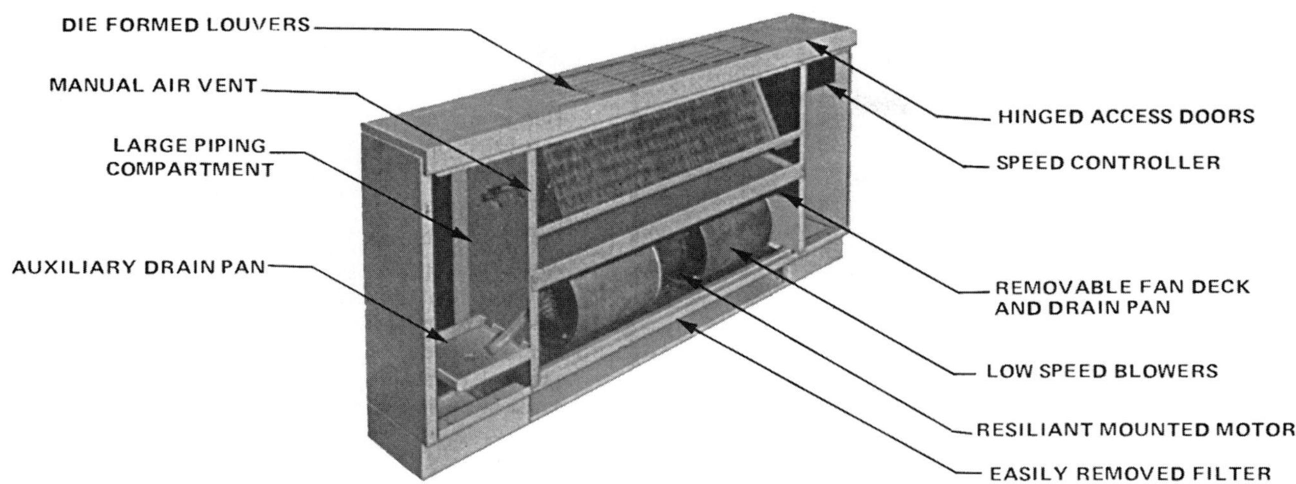

Figure A32-1 Typical fan coil unit, showing component parts. (Courtesy of Dunham-Bush, Inc.)

This type of unit consists of a filter, direct-driven centrifugal fan(s), and a coil suitable for handling chilled or hot water. The size of the unit is based on its cooling ability; it is usually more than adequate for heating. A variety of water-flow control packages are available for manual, semi-automatic, or fully automatic motorized or solenoid-valve operation.

Airflow is controlled by fan-speed adjustment—manual or automatic. Outside air is introduced through a dampered opening to the outside wall. The size of the cabinet is designated in cu ft of airflow and ranges from 200 to 1200 ft^3/min. These units may be installed on two-, three-, or four-pipe system designs as illustrated in Fig. A32-2.

With modifications to the cabinet, the same components are assembled in a horizontal ceiling-mounted version (Fig. A32-3). Where hot-water heating is not desirable for heating, electric-resistance heaters are installed in the cabinet on the leaving-air side of the cooling coil.

Another version of the room fan-coil unit is a vertical column design (Fig. A32-4) which may be installed exposed or concealed in the wall. These are placed in common walls between two apartments, motel rooms, etc. Water-piping risers are also included in the same wall cavity. They are not designed for ducting but can serve two rooms by adding air-supply grilles.

Small ducted fan-coil units (Fig. A32-5) with water coils may be installed in a drop ceiling or a closet, with ducts running to individual rooms in an apartment. These range in sizes from 800 to 2000 ft^3. The unit's cooling capacity is selected based on the desired ft^3/min.

Larger fan-coil units (Fig. A32-6, p. 650), used to condition offices, stores, etc., where common air distribution is feasible, resemble cabinets of self-contained store conditioners. They can be equipped with supply- and return-air grilles for in-spacing application, or they may be remotely located with ducting. Sizes range from 800 to 15,000 ft^3/min. In larger sizes, cabinet and fan-discharge arrangements permit flexible installations. These units approach the next category of equipment called *air handlers,* but, in general, they do not have the size and functions available in central-station air handlers.

Fan-coil units have a wide variety of applications. One advantage they have is the ability to provide individual room conditioning. This is valuable in applications for hospital rooms, motel rooms and individual private offices. An accessory fresh-air duct can be provided in the bottom rear of the unit to permit the entrance of outside

ON THE JOB

The following safety rules should be observed when servicing air-handling equipment.

- Never enter an air-handling unit or reach into it while the unit is running.
- Lock open and tag the electrical power supply while working on the motor or electric reheat coil.

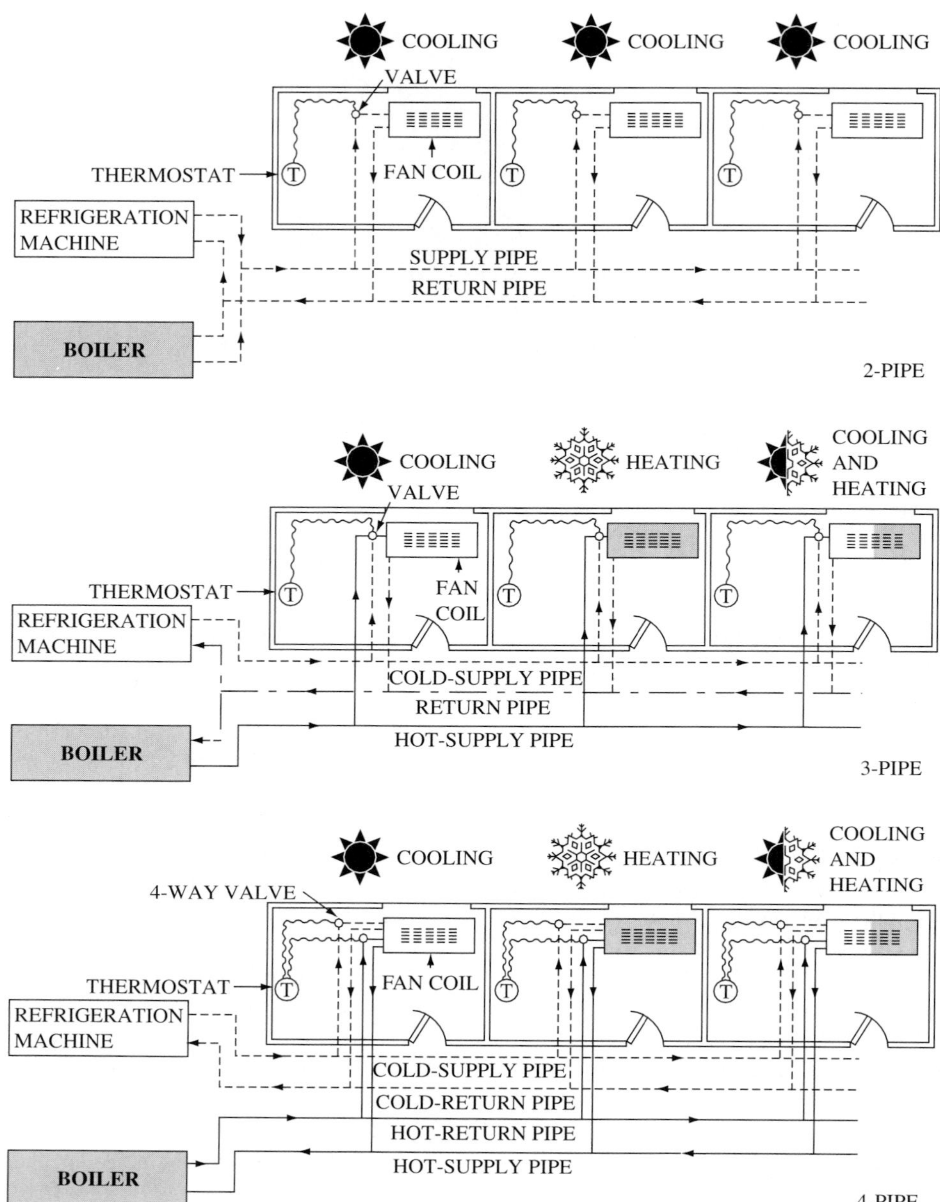

2-PIPE SYSTEM
EITHER HOT OR CHILLED WATER IS PIPED THROUGHOUT THE BUILDING TO A NUMBER OF FAN-COIL UNITS. ONE PIPE SUPPLIES WATER AND THE OTHER RETURNS IT. COOLING OPERATION IS ILLUSTRATED HERE.

3-PIPE SYSTEM
TWO SUPPLY PIPES, ONE CARRYING HOT AND THE OTHER CHILLED WATER, MAKE BOTH HEATING AND COOLING AVAILABLE AT ANY TIME NEEDED. ONE COMMON RETURN PIPE SERVES ALL FAN-COIL UNITS.

4-PIPE SYSTEM
TWO SEPARATE PIPING CIRCUITS – ONE FOR HOT AND ONE FOR CHILLED WATER. MODIFIED FAN COIL UNIT HAS A DOUBLE OR SPLIT COIL. PART OF THIS HEATS ONLY, PART COOLS ONLY.

Figure A32-2 Three types of piping arrangements for fan-coil units.

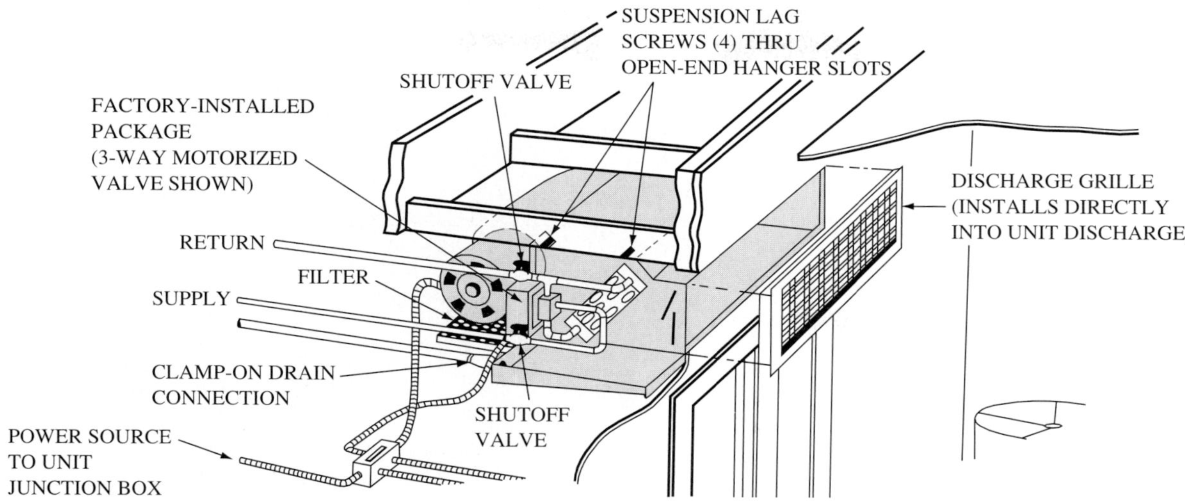

Figure A32-3 Ceiling-mounted fan-coil unit.

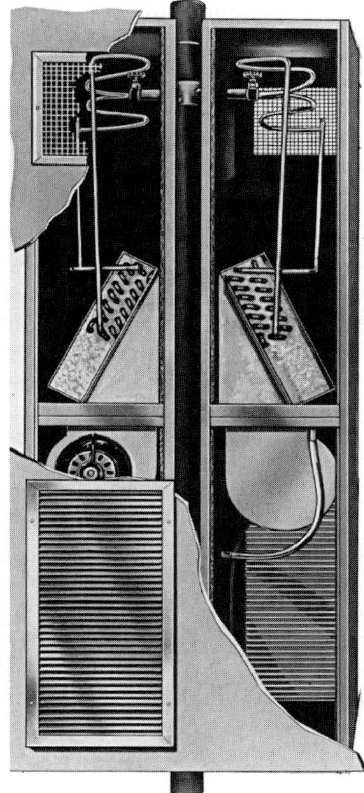

Figure A32-4 Vertical-column fan-coil unit. (Courtesy of York International Corp.)

Figure A32-5 Fan-coil unit arranged for duct work. (Courtesy of York International Corp.)

ventilation air. Each room unit can be individually controlled without mixing air from an adjacent room.

One very interesting application of fan-coil units is in small church air conditioning. Units are placed around the perimeter of the sanctuary, heating or cooling the exterior exposure of the building and gently supplying conditioned air to the occupants. The units are quiet, effective, and can be used the year around.

For the installation of fan-coil units that include cooling, it is important to provide a method of disposal for the dehumidification condensate from the cooling coil. This condensate falls into a condensate pan below the coil and is removed by gravity flow to an open drain.

Larger sized fan-coil units, sometimes called *unit ventilators,* are used for school-classroom conditioning. A single-unit ventilator is installed below window level in each classroom with extension supply-air ducts and grilles running along the sill the full length of the room. An opening in the back supplies outside air from a through-the-wall opening. The units include air filters to clean the air. Units are individually controlled and thus allow for differences in sun load, depending on the orientation of

Figure A32-6 Large-size air-handling unit. (Courtesy of Rheem Air Conditioning Division)

the room. They also meet the ventilation requirements of local codes for classrooms.

A32-3 CENTRAL-STATION AIR-HANDLING UNITS

An expanded view of a typical large air handler is shown in Fig. A32-7. Often these units are made in modules or sections that can be joined together on the job. This arrangement offers flexibility in the selection of components required for a specific installation.

A schematic diagram of a two-fan central-station air-handling system is shown in Fig. A32-8. On a large system, the reason for using the return-air fan is to be able to control the building pressurization. With a single fan, the suction pressure in the return system can create a negative pressure in many parts of the building. This can cause undesirable infiltration, affect the operation of doors, and put an extra resistance on the operation of the fan. It is desirable to maintain a slightly positive building pressure.

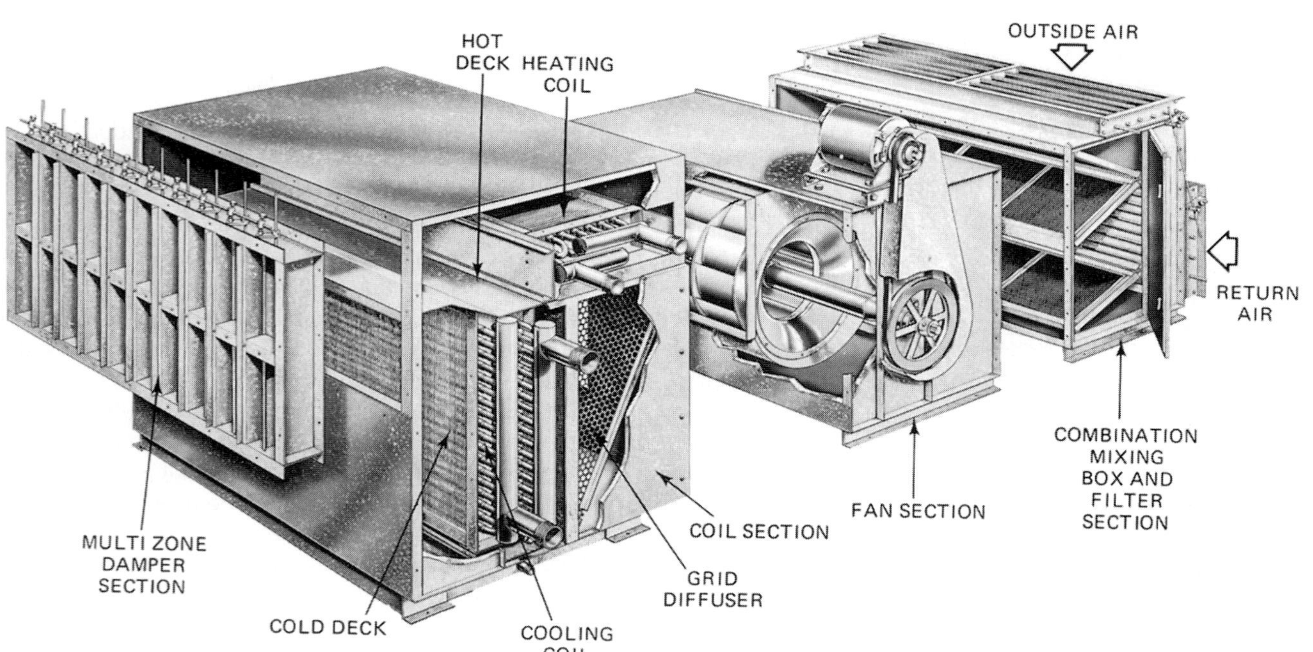

Figure A32-7 Central multizone air-handling unit, modular construction. (Courtesy of York International Corp.)

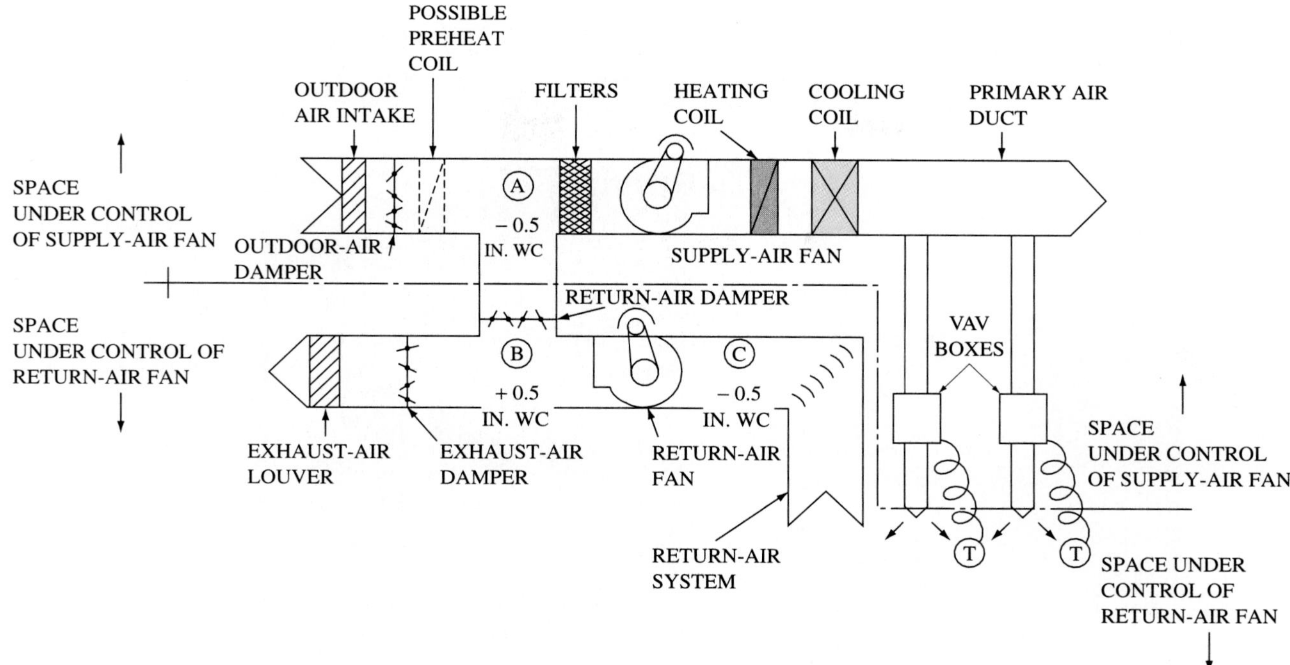

Figure A32-8 Schematic view of a central-station air-handling unit.

The return-air fan is installed in the duct work and is not part of the air-handling unit. Likewise the preheating coil is usually installed in the duct work near the entrance of the outside air into the building. The preheating coil is provided with some type of freeze protection. The air-handling unit itself consists of:

1. Heating and cooling coils
2. A humidifier
3. Mechanical air filters
4. Electronic air filters (optional)
5. Mixing dampers and (optional) exhaust dampers
6. Fans and motors

A32-3.1 Heating and Cooling Coils

With central air-handling units of this type it is not possible to operate both the heating and cooling coils at the same time, for two reasons: (1) the energy-conservation codes do not permit it, and (2) there would be no advantage since a single-zone unit can only provide heating or cooling at one time. The control system includes a provision for changing over from heating to cooling (and the reverse) when conditions warrant it.

The heating coil is usually selected for hot-water heating, although occasionally steam is used. The advantage of hot water is that it is easier to control. There can be parts of the building that never require heat, such as an interior zone. For those units, the heating coil may be optional depending on the application.

The chilled-water cooling coil is similar in construction to the hot-water heating coil. The cooling coil, however, will probably be deeper (more rows deep), since more surface is necessary for dehumidification.

A typical finned-tube water coil for heating and/or cooling is shown in Fig. A32-9. Water coils are available in depths of 1, 2, 3, 4, 5, 6, and 8 rows deep with various types of circuiting. They are free-draining and are provided with ¼ in. NPT fittings in the header for draining and venting. Tubes are staggered to improve thermal efficiency.

A32-3.2 Humidifiers

Humidity is the amount of water vapor within a given air space. *Absolute humidity* is the weight of the water vapor per pound of dry air. *Relative humidity* is the ratio of the actual amount of water vapor in the air to the amount that it would hold if 100% saturated (the dew point).

For example, referring to the psychrometric chart in Fig. A32-10, if the dry-bulb temperature is 72°F and the wet-bulb temperature is 56°F, the absolute humidity is 41 gr/lb, the relative humidity is 36%, and the dew point is 43°F.

There are four benefits in maintaining the proper humidity in the occupied space: (1) comfort, (2) preservation, (3) health, and (4) energy conservation.

Comfort is greatly affected by the relative humidity of air. At a temperature of 72°F, with 75% RH you would feel very uncomfortable. At the same temperature of 72°F

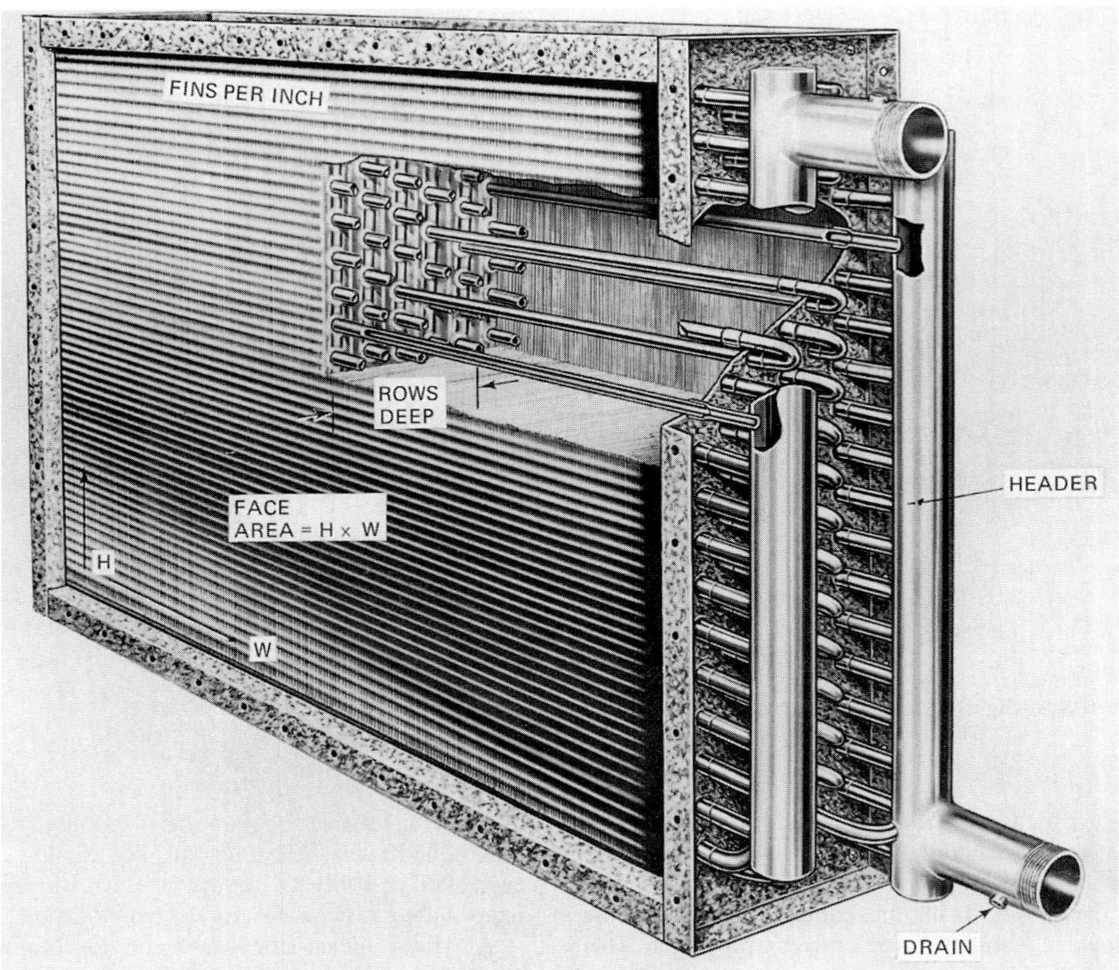

Figure A32-9 Typical water coil for a central-plant air-handling unit. (Courtesy of York International Corp.)

with 35% RH you would feel cozy. At 72°F and 10% RH you would probably feel cool. Too much or too little humidity is unpleasant. It is important to maintain the proper humidity.

For the *preservation* of hydroscopic materials such as wood, leather, paper, and cloth, a fixed amount of moisture is necessary to preserve their proper condition. These materials shrink when they dry out. If moisture loss is rapid, warping and cracking can take place. To retain their good quality, proper humidity must be maintained.

From a *health* standpoint, medical science reveals that the nasal mucus contains some 96% water. Doctors indicate that the drying out of the nasal tissues in winter helps to initiate the common cold. Maintaining proper humidity can greatly affect one's susceptibility to colds.

Low humidities can cause shrinkage of the framing around doors and windows, increasing the infiltration of outside air, causing increased energy usage. Also, proper humidity permits lower comfort temperatures, providing *energy conservation.*

One of the limiting factors in maintaining proper humidity is the *condensation* that occurs on windows in winter. The table shown in Fig. A32-11 indicates the outside temperature at which condensation will occur for various types of glass.

Not shown in the air handler illustrated (Fig. A32-7) are humidifiers that can be used to add moisture to the air. This would be done in the hot deck depending on the need and application. Several types are used (Figs. A32-12 through A32-14).

Water-spray types are used with hot-water heating and provide optimum performance in applications where the humidity level is fairly low and the most precise control is not required.

The steam-pan type is used when the introduction of steam directly into the airstream is undesirable. The vaporization of water from the pan provides moisture to the conditioned air.

The steam-grid type is highly recommended because it offers simplicity in construction and operation, and hu-

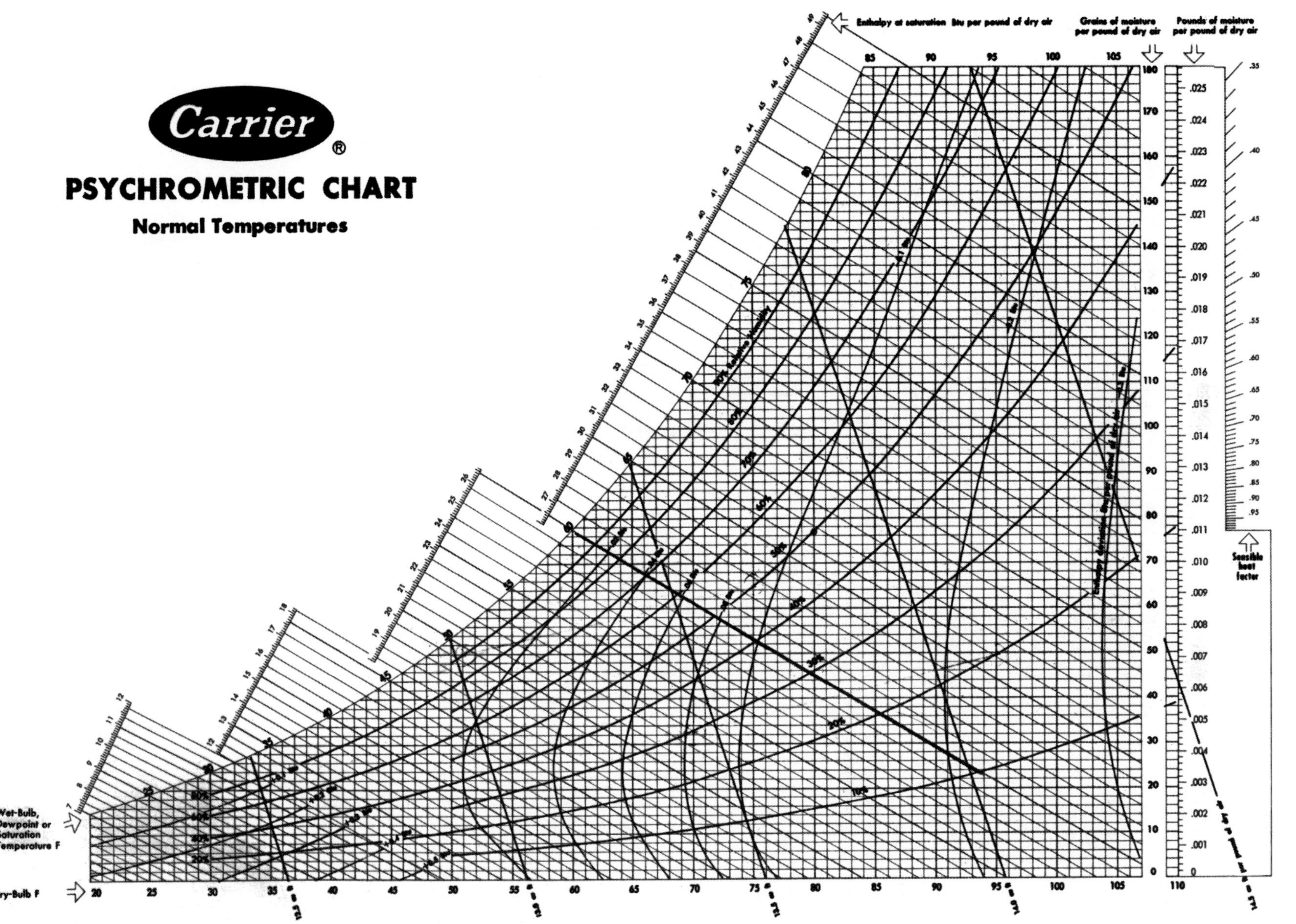

Figure A32-10 Psychrometric chart showing information that can be determined based on having dry- and wet-bulb temperatures. (Courtesy of Carrier Corporation)

Inside Temp. 70°F		Outside Temperature at which Condensation Will Probably Occur*	
INSIDE RH	SINGLE PANE	DOUBLE PANE	TRIPLE PANE
50%	43°	18°	10°
45%	38°	11°	0°
40%	34°	2°	–10°
35%	28°	–8°	–25°
30%	22°	–20°	–35°
25%	15°	–30°	
20%	8°		
15%	0°		
10%	–11°		

*With constant air circulation over the windows, a higher inside relative humidity can be maintained. Heavy drapes, closed blinds, etc., will have an adverse effect.

Figure A32-11 The effect of single-, double- and triple-pane windows on condensation temperatures. (Courtesy of Research Products Corporation)

midification can be closely controlled. A source of steam must be available.

A32-3.3 Mechanical Air Filters

Air contaminants are of many types, including dust, fumes, smokes, mists, fogs, vapors, and gases. On a particle-count basis, over 99% of the particles in a typical atmosphere are below 1 micron in size. The particle size for many of the airborne contaminants is shown in Fig. A32-15.

A micron is 1 millionth part of a meter or 1/25,400 part of an inch. A human hair is about 100 microns in thickness.

Referring to the chart in Fig. A32-15, in order to remove airborne bacteria, the filter must be capable of removing particles down to 0.3 microns in size. To block out tobacco smoke, the filter must remove particles down to 0.01 microns. The chart indicates that this removal can be accomplished with either a high-efficiency mechanical filter or an electronic filter. The common throw-away filters used in many furnaces are only capable of removing about 1-micron-size particles.

A filter becomes more efficient as it fills with particles, although the disadvantages far outweigh the advantages. Air filters must be cleaned or replaced when they become dirty.

Figure A32-12 Spray-type humidifier.

Figure A32-13 Pan-type humidifier.

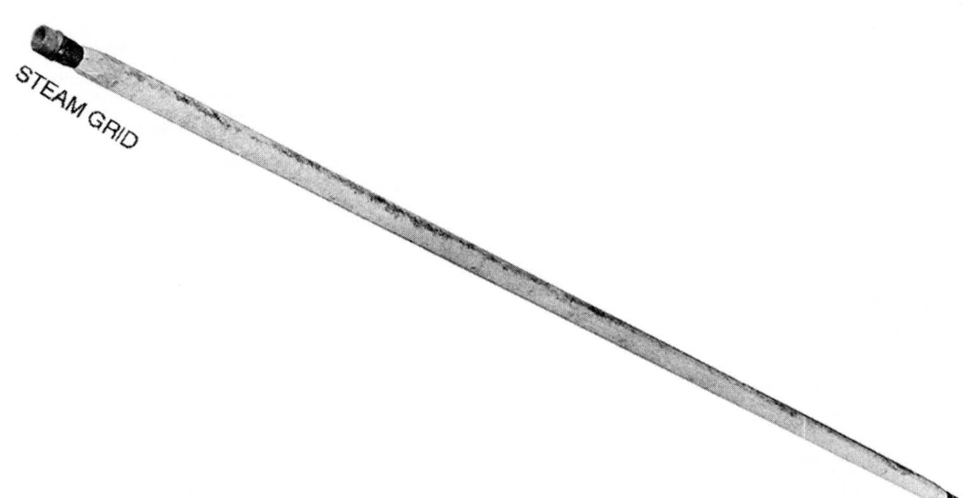

Figure A32-14 Steam-grid humidifier.

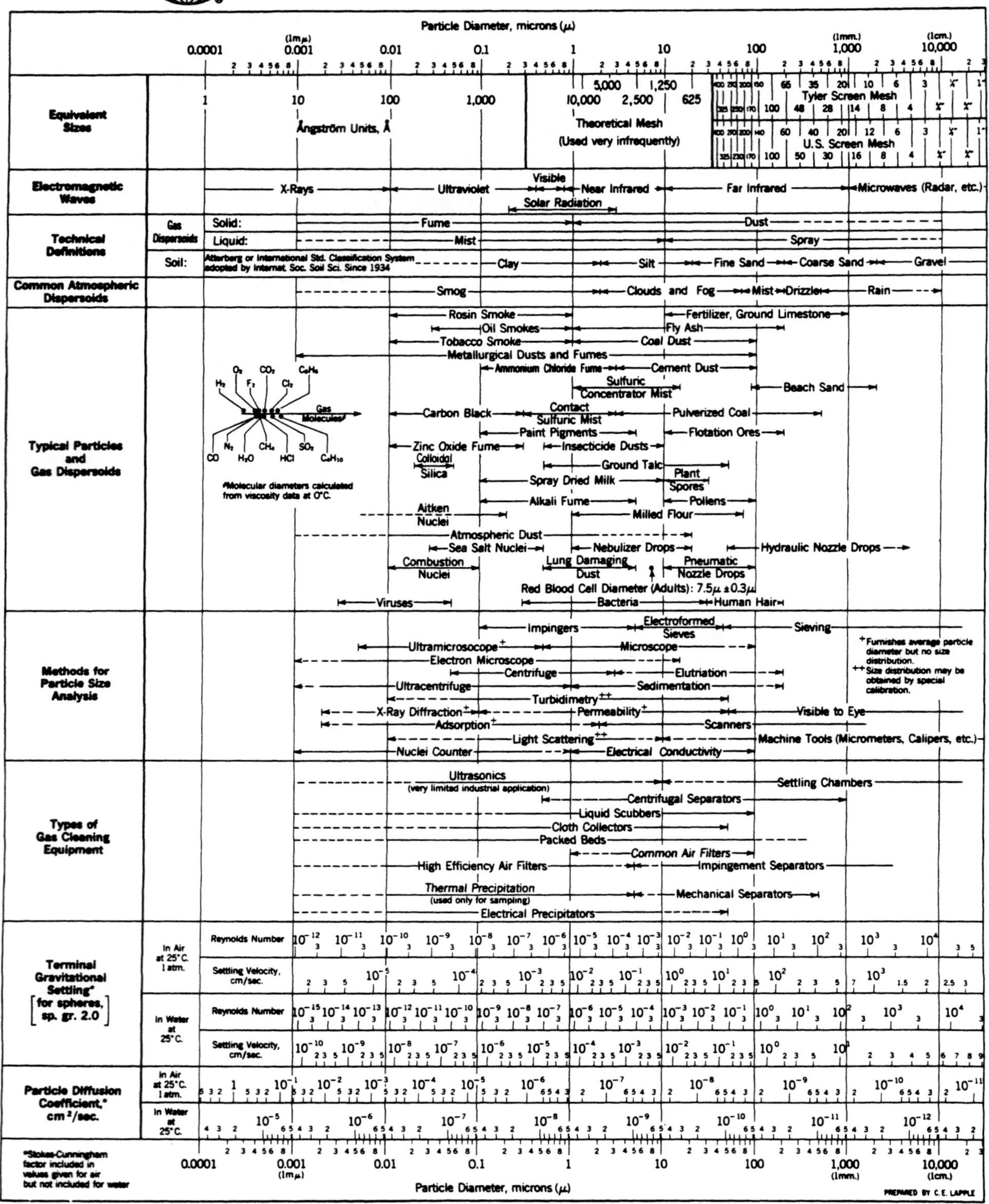

Figure A32-15 Characteristics of particles carried by the airstream. (C. E. Lapple, SRI JOURNAL, 5.94 (Third Quarter, 1961). Reprinted with permission of SRI International.)

Figure A32-16 V-type filter section. (Courtesy of Dunham-Bush, Inc.)

Figure A32-17 Flat-filter section. (Courtesy of Dunham-Bush, Inc.)

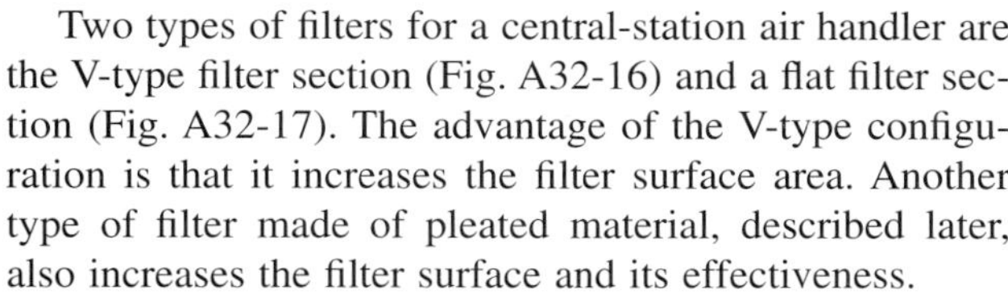

Two types of filters for a central-station air handler are the V-type filter section (Fig. A32-16) and a flat filter section (Fig. A32-17). The advantage of the V-type configuration is that it increases the filter surface area. Another type of filter made of pleated material, described later, also increases the filter surface and its effectiveness.

Rating Air Filters

There are three ways of rating air filters: by their (1) efficiency, (2) resistance, and (3) particulate-holding capacity.

The *efficiency* of a filter can be determined by using procedures set up by ASHRAE Test Standard 52-76 to compare input with output air. Tests are performed by the filter manufacturer.

The *resistance* can be measured with an ordinary manometer. It is common practice on central systems to include a differential-pressure-measuring device for the filter section. This instrument compares the air pressure on the upstream and downstream sides of the filter assembly. The gauge on the instrument indicates when the filters should be changed.

Tests are available for measuring the *particulate-holding capacity* of the filter. From this information the length of operating time before changing can be estimated for a given application. Filters made of pleated fabric, with depths of 6 to 12 in., have a large dust-holding capacity and are often used on large systems.

Panel-type Filters

Viscous impingement panel-type filters are made of coarse fibers with a high porosity. The filter media is coated with a viscous substance, such as oil, which acts as an adhesive to the airborne particles coming into contact with it.

A number of materials have been used as the filtering media, including coarse glass fibers, animal hair, vegetable fibers, synthetic fibers, metallic wool, expanded metals and foils, crimped screens, random-matted wire, and synthetic open-cell foams.

Servicing these filters depends on their construction. Disposable filters are made of inexpensive materials and are discarded after one usage. Permanent filters have metal frames to stand continued usage. Cleaning is done by washing with steam, water, or detergent.

Dry-type extended-surface filters are made of random fiber mats or blankets of varying thicknesses. Bonded glass fiber, wool felt, synthetics, and other materials are commonly used. Pleating of the media provides a larger filter surface area compared to the face area and a reasonable pressure drop. The efficiency of the dry filters is usually higher than the viscous coated type. In addition to their effectiveness, the dry filters have a larger dust-holding capacity and, therefore, longer periods of use.

The *high-efficiency particulate air* (HEPA) filters are the standard for clean-room design. They have an extended surface configuration with deep folds of submicron

ON THE JOB

A simple method used in commercial buildings to monitor the efficiency of the air filter-bank and coils is to install an inclined manometer with pressure-sensing probes on each side of the filter and/or coils. If pressure drops across the filters and coils exceed manufacturers recommendation, replace the filter bank. If replacing the filters does not bring the pressure drop down to original specifications, clean the coil.

glass-asbestos fiber paper. This fiber will operate at 250 fpm face velocity with a pressure drop of 0.50 to 1.00 in. of water or higher during their service life.

Renewable Media Filters

The *moving-curtain viscous impingement filter* is an automatic moving curtain with the random fiber medium, treated with a viscous material, and furnished in roll form. The material rolls down from the top of the unit. As the exposed area becomes saturated with dirt, a clean section is automatically rolled into place. The used portion is collected in a roll at the bottom and thrown away. A fresh roll is again placed at the top of the unit and the filtering continues.

The *moving-curtain dry-media filter* operates in a manner similar to the viscous impingement model. These filters are effectively used to remove the lint in the air in textile mills, dry-cleaning establishments, and lint and ink mist in press rooms.

A32-3.4 Electronic Air Filters

The electronic air filter uses the precipitation principle to collect airborne particles. There are three types of units used for commercial service: (1) ionizing-plate, (2) charged-media non-ionizing and (3) charged-media ionizing.

With the *ionized-plate unit,* the incoming air passes through a series of high-potential ionized wires that generate positive ions. These ions adhere to the dust particles carried by the airstream. The charged dust particles then pass through an electrical field, attracting the charged particles and removing them. The direct-current field is created by a 12,000 V_{DC} current that maintains 6,000 V_{DC} between the attracting plates. These plates offer little resistance to the airflow. For best results the airflow through the plates should be evenly distributed.

Conventional type pre-filters are used ahead of the electrostatic filters to screen out the larger particles in the air. These filters offer some resistance to the airflow in the range of 0.14 to 0.26 in. WC. In addition, 16 mesh screens are used to cover all outside air entrances to remove paper, leaves, insects and the like which would clog the air passages.

Collector plates are often used, coated with oil to act as an adhesive. The filters are cleaned by use of water sprayed on the plates in place. Suitable drains are provided in the bottom of the filter compartment. Due to the high voltage used by the electrostatic units, safety switches are provided which will turn off the power when the filter access door is opened.

The *charged-media non-ionizing filters* are very different in construction. These filters consist of a dielectric filtering medium, usually arranged in pleats, as in typical dry filters. The dielectric medium consists of glass-fiber mat, cellulose mat, or similar material, supported by a grid consisting of alternately grounded and charged members. The charged members are supplied with 12,000-V_{DC} power. Airborne particles that approach the field are polarized and drawn to the filaments of the fibers of the media.

This type of filter offers about 0.10 in. WC resistance to the flow of air when clean, with a face velocity of 250 fpm. This small resistance serves to equalize the flow of air over the exposed surface.

The *charged-media ionizing electronic filters* combine the effects of the other two designs. Dust is charged in a corona-discharge ionizer and collected on a charged-media filter mat. This construction increases the effectiveness of the filter but is more critical to operate successfully.

In using electronic air filters, two conditions of operation can cause operating problems: (1) space charge and (2) ozone.

The unit needs to be carefully built so that charged dirt particles do not escape into the filtered space *(space charge).* If they do, they can darken the walls faster than if no cleaning arrangement were used.

All high-voltage devices are capable of producing *ozone,* or O_3. When the unit is operating correctly the amount of ozone produced is well within the recommended limits. If the unit is continuously arcing, it may yield levels of ozone which are annoying and even poisonous. This is indicated by a strong ozone odor.

Filter Installation

Filters should be placed ahead of heating and cooling coils and other mechanical equipment to protect the system from dust. The published performance of filters is based on straight through unrestricted airflow. Failure of filters to give satisfactory service is usually the result of improper installation and maintenance.

A32-3.5 Mixing Dampers

Dampers installed on air-handling units control the flow of air through various parts of the system. The function of dampers depends on the design of the system. There are two general types of multiple-leaf dampers: *parallel-blade* and *opposing-blade.* Parallel-blade dampers tend to direct the air as they open, whereas opposing blades provide a more uniform airflow. Opposing blade dampers are usually constructed to more readily offer a positive close off. Most of the dampers now supplied are of the opposing-blade type.

Dampers can operate either in two positions, open or closed, or they can be modulated so that they can be positioned anywhere between fully open and fully closed, depending on the requirements of the control system.

Dampers can be linked together, so that when one opens the other closes. For example, in a mixing box such

PREVENTIVE MAINTENANCE

Modulating dampers should be inspected and maintained as part of a normal preventive maintenance program. This includes lubricating linkages, adjusting spring tension on damper actuators, and replacing damper gaskets as required to prevent leakage. Failure to maintain dampers will lead to dampers sticking in a failed position, not modulating, and not fully closing on shutdown periods. If dampers fail in the open position, allowing outside air to enter the building during heavy-load summer conditions, it will also cause the refrigeration equipment to work harder than normal and increase the total operating energy costs.

Fan blades on axial and centrifugal fans should also be inspected and cleaned periodically. If the blades accumulate any dirt or grease, fan efficiencies and performance deteriorate proportionately.

In dirty environments where filters are not adequately protecting the cooling and heating coils, the coils should be chemically cleaned annually using a commercial grade product.

as shown in Fig. A32-18, where the return air is mixed with outside air, the dampers can be linked and controlled, so that when the outside air is increased, the return air is decreased (and the reverse).

Most dampers on central-station air handlers are of the modulating type. This provides the control of the airflow needed for the wide variety of conditioning requirements. The following are some of the basic applications.

Mixed-air Control

To illustrate some of the many uses of dampers, refer to Fig. A32-19. Of primary importance is the mixing of return air and outside air. During normal operation only the return air and the minimum outside-air dampers are involved. The minimum outside-air damper is a two-position damper.

Figure A32-18 Return- and outside-air mixing box with dampers. (Courtesy of Dunham-Bush, Inc.)

Normally it is fully open to provide ventilation air to meet code requirements. The only time it closes is when the building is unoccupied or the system shut down.

Provision is made in the damper system and the control system to use additional outside air for free cooling (economizer cycle) when practical. A maximum outside-air damper is modulated along with the return-air and exhaust-air damper to maintain a slightly positive pressure in the building. The total air quantity through the air handler does not change but the proportions of outside air to return air do change to meet the control requirements.

Face and Bypass Control

One way to control the amount of cooling or heating is to use a face damper in series with the coil surface and a bypass damper in a duct connection to the return air, as shown in Fig. A32-19. With this arrangement the total air volume remains constant and the air entering the fan is a mixture of cooled or heated air and return air. The damper control modulates the air quantity from each source to match the load. The two sets of dampers are linked together either mechanically or electrically so that when one modulates toward the open position, the other modulates toward the closed position (and the reverse).

Variable-air-volume Control System

This system differs from the constant-volume/variable-temperature system since basically it maintains a nearly constant air temperature but matches the load by changing the air volume. This system became popular when attention was given to conserving energy without in any way decreasing the comfort level during partial load conditions.

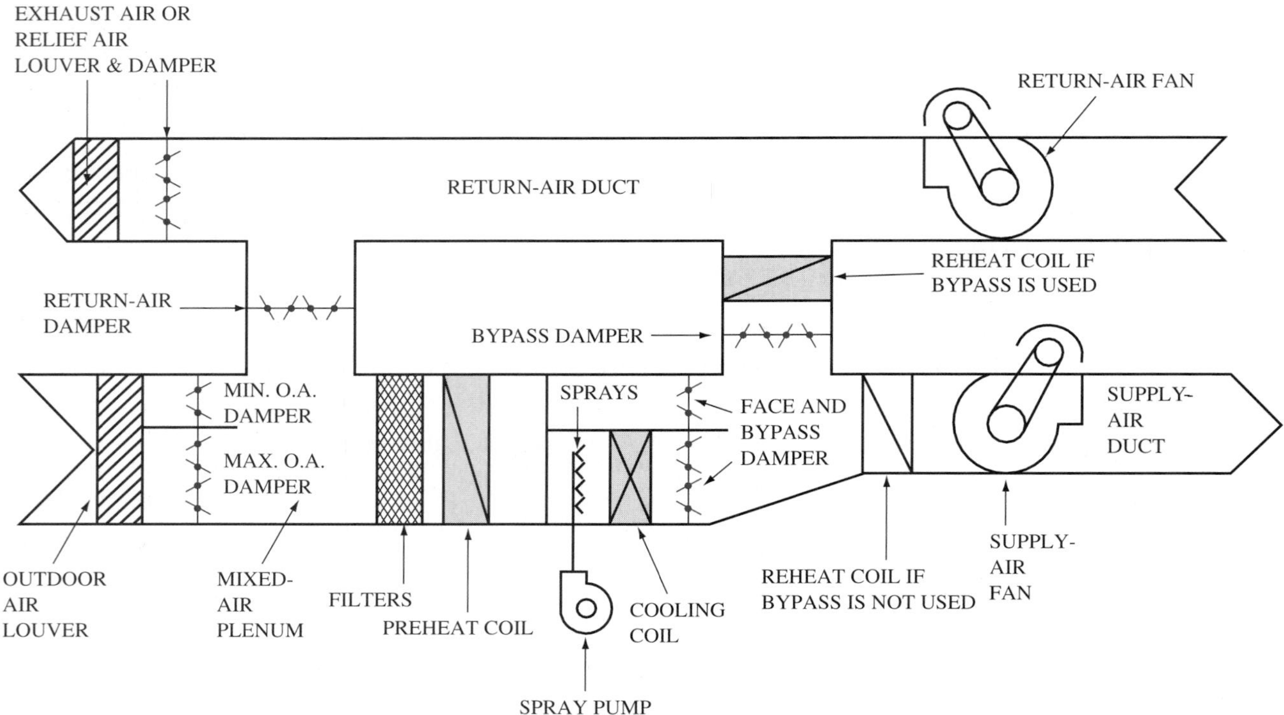

Figure A32-19 Schematic diagram of a typical air-handling unit showing the various types of damper arrangements.

One of the features of these systems is the use of a *variable-frequency fan-speed controller* that is applied to the air-handler fan. This electrical device, along with the necessary sensors, controls the speed of the central fan to match the total air volume required by the individual-zone terminal units. Since the power required to drive the fan is proportional to the cube of the fan speed, tremendous power savings can result from slowing down the fan when the extra air is not needed.

In addition to the arrangement for controlling the air volume at the central unit, the VAV systems use a *terminal unit* in each zone. A typical system of this type is shown in Fig. A32-20, along with a solid-state control system. Each terminal unit has an automatic air-volume-controlled damper to supply the needs of the zone (Fig. A32-21).

Multi-zone Units

A multi-zone unit is a specially designed air-handling unit that provides separate zone ducts for the distribution of conditioned air (Fig. A32-22). In addition, each zone is provided with mixing dampers to supply conditioned air to meet the requirements of the zone. These units operate on the basis of using dampers to mix the air from a heat source and a cold source to produce the required temperature.

A32-3.6 Fans and Motors

The proper operation of the air handling unit's fan and motor is essential to a good air-conditioning job. They not only need to be sized properly to begin with, but most jobs require in-the-field adjustment of these components to fit the actual conditions of the installation.

There are two basic type of fans used in central-station air handlers: (1) centrifugal (Fig. A32-23) and (2) axial (Fig. A32-24).

The *centrifugal fan* is much like a centrifugal pump, except that it handles air instead of water. It is used to "pump" small or large quantities of air through equipment that usually offers a relatively high resistance to the airflow. The fan is belt-driven with variable-sized pulley wheels on the fan shaft and motor shaft, so that the speed can be adjusted and the motor sized to meet the needs of the job.

There are two types of centrifugal fans: (a) *backward-curved blades* and (b) *forward-curved blades* (Fig. A32-25). The backward-curved blade fans usually run faster than the forward-curved type and therefore are noisier, but have the advantage of being non-overloading. The forward-curved blade fans are preferred because of their flexibility for adjustment and quieter operation.

The *axial fan* is used where large volumes of air are required at a relatively low static pressure. This fan also op-

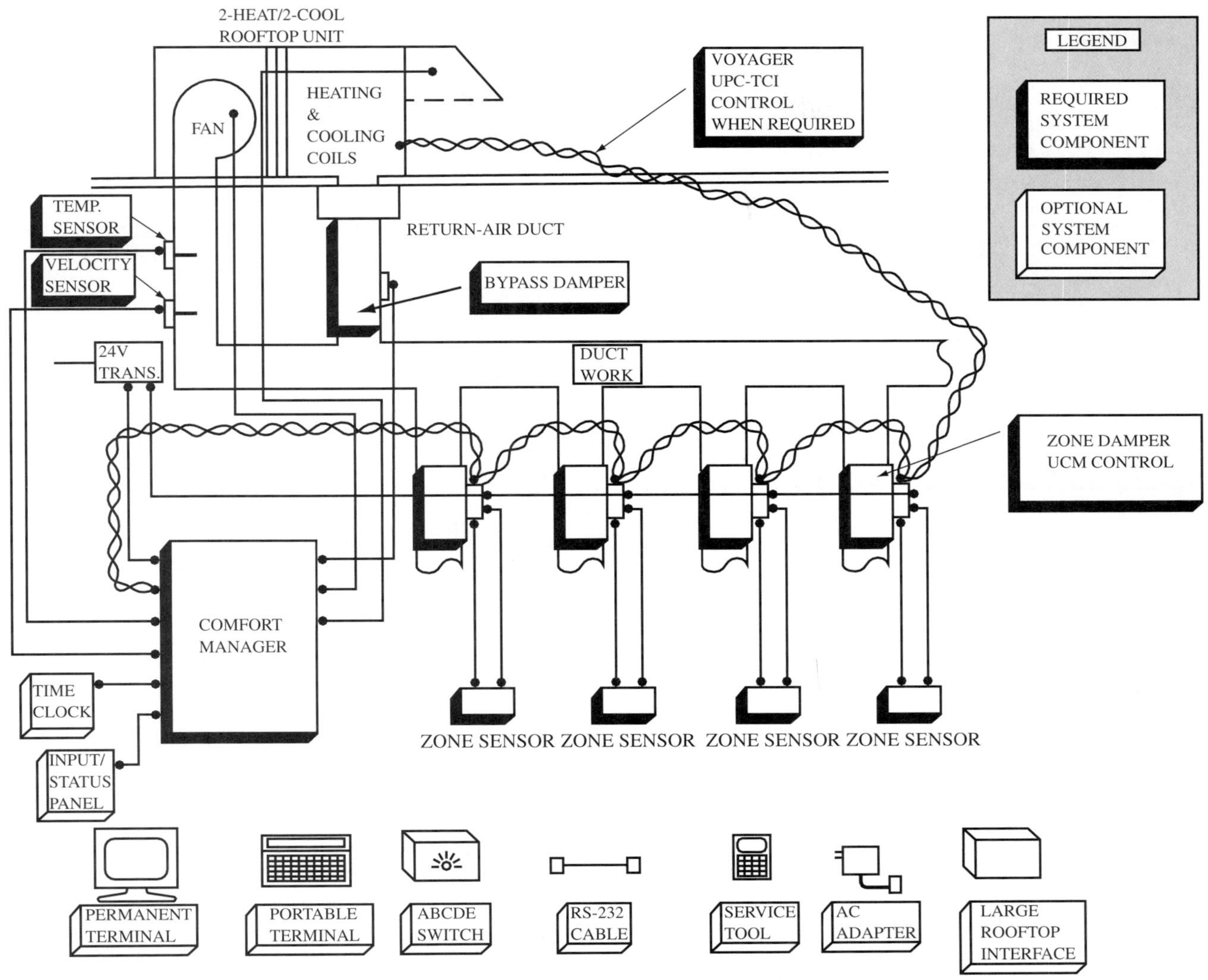

2-HEAT/2-COOL ROOFTOP UNIT
FAN
HEATING & COOLING COILS
VOYAGER UPC-TCI CONTROL WHEN REQUIRED
LEGEND
REQUIRED SYSTEM COMPONENT
OPTIONAL SYSTEM COMPONENT
TEMP. SENSOR
VELOCITY SENSOR
RETURN-AIR DUCT
BYPASS DAMPER
24V TRANS.
DUCT WORK
ZONE DAMPER UCM CONTROL
COMFORT MANAGER
TIME CLOCK
INPUT/ STATUS PANEL
ZONE SENSOR ZONE SENSOR ZONE SENSOR ZONE SENSOR
PERMANENT TERMINAL
PORTABLE TERMINAL
ABCDE SWITCH
RS-232 CABLE
SERVICE TOOL
AC ADAPTER
LARGE ROOFTOP INTERFACE

Figure A32-20 Schematic view of a typical variable-air-volume (VAV) system.

Figure A32-21 VAV terminal unit. (The Trane Company)

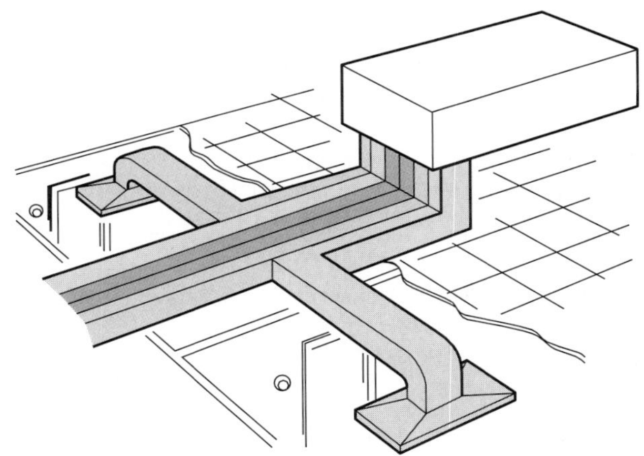

Figure A32-22 Constant-volume multi-zone heating and cooling unit.

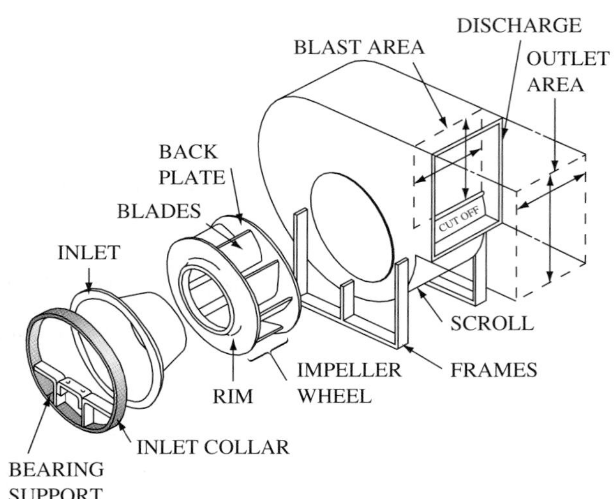

Figure A32-23 Exploded view of a centrifugal fan. (Copyright by the American Society of Heating, Refrigerating, and Air-Conditioning Engineers. Used by permission.)

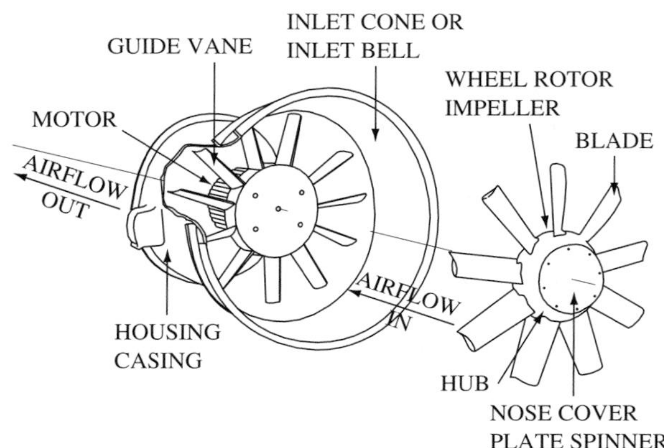

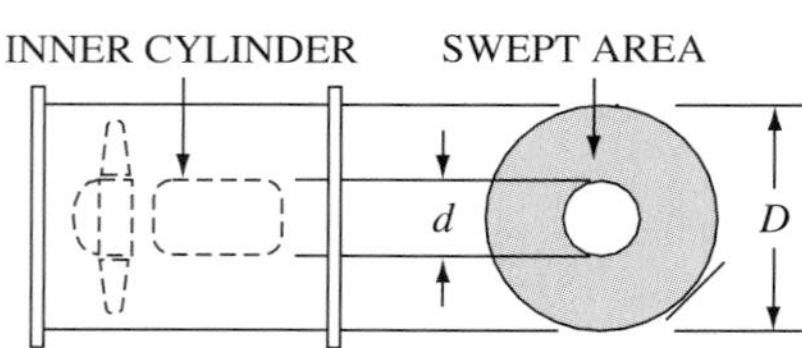

$$\text{SWEPT AREA RATIO} = 1 - \frac{d^2}{D^2} = 1 - \frac{\text{AREA OF INNER CYLINDER}}{\text{OUTLET AREA OF FAN}}$$

NOTE: THE SWEPT AREA RATIO IN AXIAL FANS IS EQUIVALENT TO THE BLAST AREA RATIO IN CENTRIFUGAL FANS.

Figure A32-24 Schematic drawings of axial fans. (Copyright by the American Society of Heating, Refrigerating, and Air-Conditioning Engineers. Used by permission.)

erates at high speeds and can be noisy. It therefore needs to be strategically located.

In adjusting the speed of centrifugal fans, the technician needs to be knowledgeable of the fan laws (Fig. A32-26) that govern the relation between speed, static pressure and horsepower requirements. A great deal of electrical energy can usually be saved by properly selecting the fan-motor size.

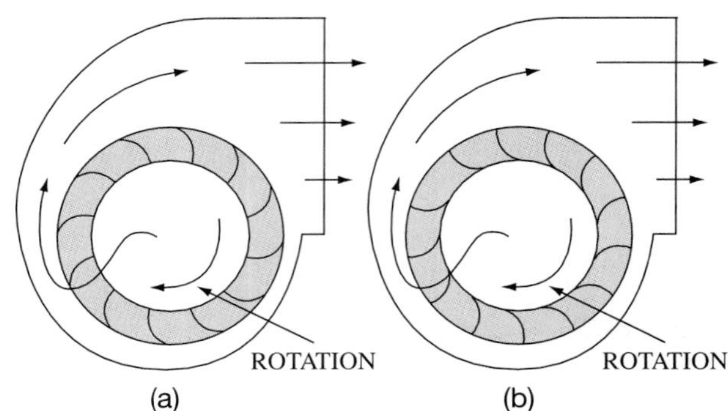

Figure A32-25 Schematic view of the construction of (a) backward-curved fan and (b) forward-curved fan.

Fan Equations

a) $\frac{cfm_2}{cfm_1} = \frac{rpm_2}{rpm_1}$ cfm = Cubic feet per minute

rpm = Revolutions per minute

b) $\frac{P_2}{P_1} = \left(\frac{rpm_2}{rpm_1}\right)^2$ P = Static or total pressure (in. WC)

c) $\frac{Bhp_2}{Bhp_1} = \left(\frac{rpm_2}{rpm_1}\right)^3$ Bhp = Brake horsepower

d) $\frac{\text{rpm (fan)}}{\text{rpm (motor)}} = \frac{\text{Pitch diam. motor pulley}}{\text{Pitch diam. fan pulley}}$

Figure A32-26 Fan laws showing the relationships between speed, air quantity, and motor requirements.

Squirrel-cage three-phase motors with laminated rotors are used for most central-station air-handling units. These motors have starting torques of 125 to 275% of full-load torque, depending on the design. The application voltage and ampere draw of the motor should always be measured on the job, under load conditions, to be certain the use is compatible with the nameplate rating.

REVIEW

- Air-handling unit—Conveys the conditioned air to and from the building.
- Types of air-handling units:
  Remote air-handling units (Fan-coil units)
  Central-station air-handling units
- Fan coil units—Room units supplied with hot water and/or chilled water directly from central plants. Each has fans, water coils, filters, fresh air supply, and control system. Designed to supply individual rooms or area.

- Central-station units—Use one or more large air-handling units equipped with fans, heating and cooling coils, filters, fresh-air supply and mixing, accessories, and duct work to convey the conditioned air to remote areas.
- Air-handling unit:
  Heating coil—Hot water or steam
  Cooling coil—Chilled water
  Humidifier—Water spray, steam pan, steam grid
  Mechanical air filters—Filter airborne contaminants
  - To remove airborne bacteria, filter must remove down to 0.3 microns
  - To remove tobacco smoke, filter must remove down to 0.01 microns.
  - Common throw-away filters only good to 1-micron-size particles.
  - Filters become more efficient as they fill with particles.

  Electronic air filters (optional)—Higher filtering capabilities
  - Ionizing-plate
  - Charged-media non-ionizing
  - Charged-media ionizing

  Mixing dampers and exhaust dampers (optional)—Ventilation, fresh air, economizing
  - Parallel-blade—tend to direct the air as they open
  - Opposed blade—more uniform flow, most common
  - 2-positioned or modulated
  - Applications for dampers include:
    Mixed air control
    Face and bypass control
    Variable air volume (VAV)
    Multi-zone units

  Fans and motors—Centrifugal, axial
  - Centrifugal—Forward-curved, backward-curved
  - Axial—Large volumes at relatively low static pressures
- Four benefits of humidity control:
  Comfort—Proper humidity control of total heat
  Preservation—Hydroscopic materials
  Health—Less susceptibility to colds, not drying out of nasal tissues
  Energy conservation—Lower comfort temperatures

Problems and Questions

1. Identify the types of mechanical filters.
2. What are the different types of fans/blowers?
3. What are the types of centrifugal fans/blowers?
4. Describe the operation of electronic filters.
5. What are the purposes of the air-handling system?
6. Name the types of humidifiers found in air-handling systems.
7. What is the principle behind variable-air-volume systems?
8. What are the components of a central-station air-handling unit?
9. Fan-coil units consist of room units that are supplied with hot water from central-plant boilers. True or false?
10. Central-station systems use one small air handler unit and duct work to convey the conditioned air to local areas. True or false?
11. Airflow on the fan coil unit is controlled by fan-speed adjustment, manual or automatic control. True or false?
12. Small-ducted fan-coil units with water coils may be installed in a drop ceiling or a closet, with ducts running to the individual rooms. True or false?
13. Unit ventilators are simply large-sized fan-coil units. True or false?
14. On a large central-station air-handling unit, the reason for using the return-air fan is to:
    a. Ventilate the building.
    b. Control building pressurization.
    c. Get the return air back to the fan unit.
    d. All of the above.
15. The advantage of hot-water heating coils versus steam is:
    a. Easier to clean and maintain.
    b. Easier to control.
    c. Cheaper to buy initially.
    d. None of the above.
16. Relative humidity is greatly affected by the:
    a. Temperature of the air.
    b. Cleanliness of the air.
    c. Wet-bulb temperature of the air.
    d. None of the above.
17. For a filter to remove tobacco smoke, it must filter down to:
    a. 0.01 microns.
    b. 0.3 microns.
    c. 1 micron.
    d. None of the above.
18. Most mixing dampers now supplied to the job site are:
    a. Two-positioned.
    b. Opposing-blade.
    c. Parallel-blade.
    d. None of the above.

A33 Absorption Refrigeration Systems

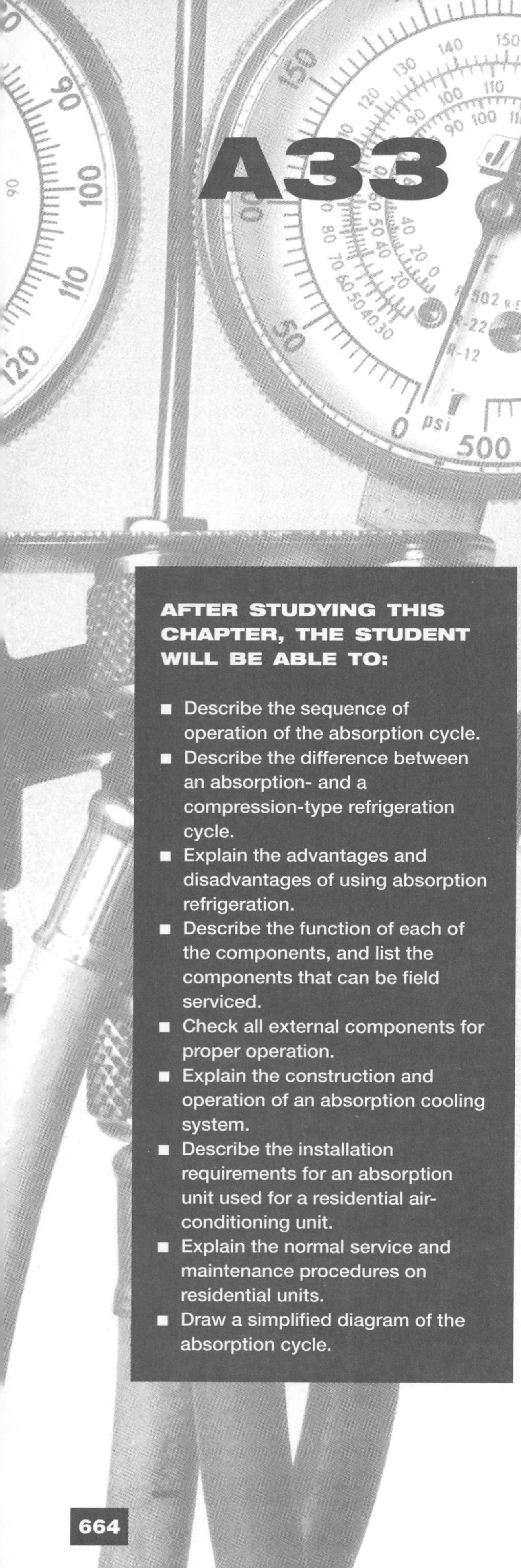

AFTER STUDYING THIS CHAPTER, THE STUDENT WILL BE ABLE TO:

- Describe the sequence of operation of the absorption cycle.
- Describe the difference between an absorption- and a compression-type refrigeration cycle.
- Explain the advantages and disadvantages of using absorption refrigeration.
- Describe the function of each of the components, and list the components that can be field serviced.
- Check all external components for proper operation.
- Explain the construction and operation of an absorption cooling system.
- Describe the installation requirements for an absorption unit used for a residential air-conditioning unit.
- Explain the normal service and maintenance procedures on residential units.
- Draw a simplified diagram of the absorption cycle.

A33-1 THE ABSORPTION PROCESS

The absorption process of producing cooling was first discovered by Michael Faraday in 1824. He observed that a chemical, silver chloride, absorbed ammonia vapor and at the same time the ammonia vessel became cold. He set up an experiment to demonstrate how this absorption process could produce cooling (Fig. A33-1). Heat was applied to the silver-chloride mixture and the vapor that was driven off by the heat was condensed in a water-cooled container. The condensed liquid was pure ammonia.

Then he removed the heater and the container of water (Fig. A33-2). The liquid ammonia began to boil violently as the vapor was again absorbed by the silver nitrate. The tube of boiling ammonia became very cold as the ammonia evaporated. The boiling action was produced by the lowering of the vapor pressure over the ammonia as vapor was constantly being absorbed. Heat for the reaction was being drawn from the surrounding air, cooling it and the container. The action stopped when the silver nitrate became saturated with the ammonia vapor.

The process just described was an intermittent method of producing refrigeration. Later when commercial systems were developed, using the absorption process, the equipment was designed for continuous operation. Other absorbants were found to be more practical for higher-capacity units. For example, systems of up to 25 tons of cooling now use water as the absorbant and ammonia as the refrigerant. Larger systems of up to 1500 tons use a lithium-bromide solution as an absorbent and water as a refrigerant.

Absorption units use heat rather than mechanical energy to produce cooling. Heat sources can be a number of types including solar, natural gas, LP gas, kerosene, steam, hot water, or electrical energy.

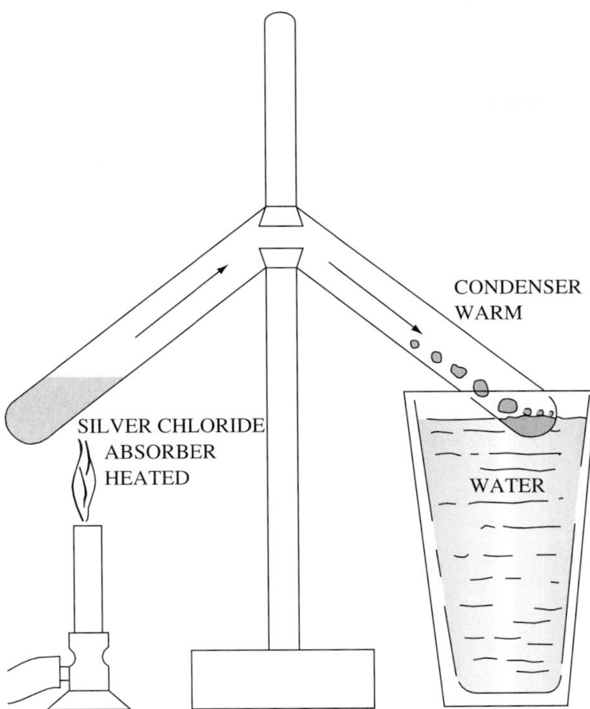

Figure A33-1 Faraday's experiment heating the solid absorbant and cooling the refrigerant vapor.

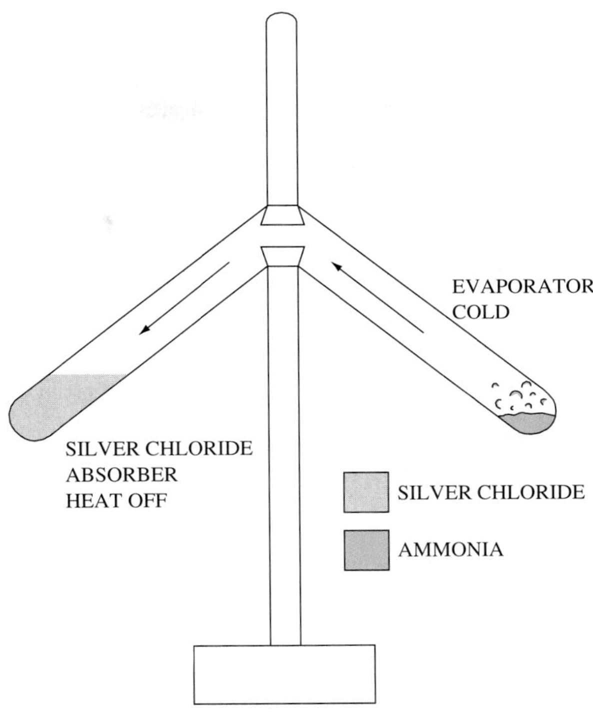

Figure A33-2 Faraday's experiment, absorbing the refrigerant and producing cooling.

A33-2 APPLICATIONS

Absorption units are used for small refrigerators, and residential and commercial air-conditioning units. Units range in size from the small refrigerator sizes to large-capacity units of 1500 tons used for commercial and industrial cooling. This type of system is used where the load is fairly continuous and uniform.

The advantages of using absorption units include the following:

1. *Environment.* Absorption units use refrigerants that have a zero ozone-depletion factor (ODF = 0).
2. *Power.* Units use heat rather than electrical energy, and therefore can be added without increasing the power load which could be too expensive or impractical.
3. *Service.* Absorption units have few moving parts. They require a minimum of service and maintenance. Their normal life is longer than that of electrical units.
4. *Operating cost.* In areas where the ratio of gas cost to electric cost is favorable, units are purchased based on a lower operating cost. In most areas, the larger sizes are more advantageous than the smaller sizes.

The characteristics that make absorption units less attractive are:

1. *Cost.* Absorption-unit equipment costs more than electric units having the same cooling capacity.
2. *Capacity reduction.* In the past, capacity reduction to match reduced load has been limited using absorption units. Newer systems, however, in the

PREVENTIVE MAINTENANCE

Smaller absorption units are normally factory-welded closed and require minimum maintenance. Larger commercial/industrial units will have access ports but often require full-time maintenance personnel performing routine checks and inspections.

Keeping condenser coils and evaporator coils clean will help extend the life of any absorption unit. Water inlet and outlet temperatures are good indications of normal operation. The heat source—gas, electric, or steam—should also be inspected periodically.

ON THE JOB

Since absorption units use gas or electricity to generate heat, it is important to inspect the supply side when troubleshooting the unit. Minimum gas pressures should be checked with a gas manometer upon startup. Electrical supply voltage should also be checked.

Whether the unit is in a recreational vehicle or a residence, the importance of leveling the base—front, back, and sideways—is extremely critical prior to start-up. Some older gas-fired units will maintain a constant pilot light using a standard thermocouple. If the thermocouple gets dirty or corroded, then the unit will not ignite. The thermocouple should always be checked for minimum voltage output to the main gas valve if suspected faulty.

Other possible problems to inspect when troubleshooting are the operating controls, specifically the thermostats and wiring connections. If the thermostat can be jumpered out to run the unit, then replace or repair it as needed.

10- to 25-ton range, now can be purchased with capacity reduction in 5-ton increments.

3. *Weight.* Absorption units are heavier than electric units of the same capacity.

The application of an ammonia system to residential and small commercial chillers, made in 3- to 5-ton sizes, will be discussed in the following sections.

A33-3 COMPONENTS

The operation of a modern absorption cycle is illustrated in Fig. A33-3. This simplified diagram shows the refrigeration cycle of an ammonia-water system. The principal component parts are labeled in the diagram as follows:

1. Generator
2. Condenser coil
3. Evaporator coil
4. Solution-cooled absorber (SCA)
5. Solution pump

The following is a brief description of how the system operates.

The solution in the sealed system is a mixture of ammonia and water: ammonia is the refrigerant and water is the absorbant. When the thermostat calls for cooling, a gas burner ignites, heating the generator (1), causing the solution in the generator to boil. During this boiling process, the ammonia is separated from the water. The ammonia vapor leaves the generator and enters the condenser coil (2) where it is condensed to a liquid. As the

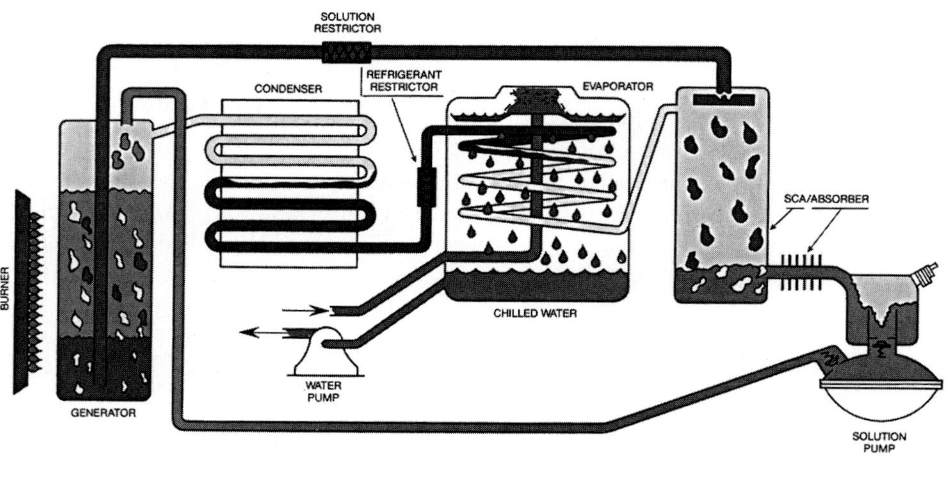

Figure A33-3 Simplified cycle of operation for absorption refrigeration. (Courtesy of The Robur Corporation)

liquid ammonia leaves the condenser, it passes through a restrictor, which lowers both the pressure and the temperature of the ammonia to approximately 34°F as it enters the evaporator coil (3).

The 34°F liquid ammonia removes heat from the chilled water dripping over the outer surface of the evaporator coil. As the chilled water is cooled, the liquid-ammonia refrigerant changes back to a gas due to the heat having been absorbed from the chilled water. This refrigerant gas now enters the absorber, or SCA (4).

Absorbent water, which was left in the generator as a weak solution, is heavier than the strong solution and sinks to the bottom. A drain is taken from the bottom of the generator, which returns the weak liquid back to the absorber. It passes through a pressure-reducing valve to maintain the difference in pressure between the two chambers. It is cooled and becomes saturated with ammonia again.

From the SCA the saturated solution enters a solution pump (5). The generator is at high pressure, so a mechanical pump must be used to pump the strong liquid to the generator against this pressure. The solution goes back to the generator and the cycle repeats itself.

A33-4 COMPARISON WITH A VAPOR-COMPRESSION SYSTEM

The absorption cycle has some similarities to the vapor-compression cycle and some differences. The condenser, refrigeration restrictor (metering device), and evaporator are similar on both systems. The compressor in the vapor-compression system has been replaced by the heater and generator in the absorption system. The absorption unit includes a number of additional accessories, such as the solution restrictor, the absorber, and solution pump, which are unique to this system.

In order to understand how these principles are applied to the construction of actual equipment, we will now review a unit of current production, an absorption chiller-heater.

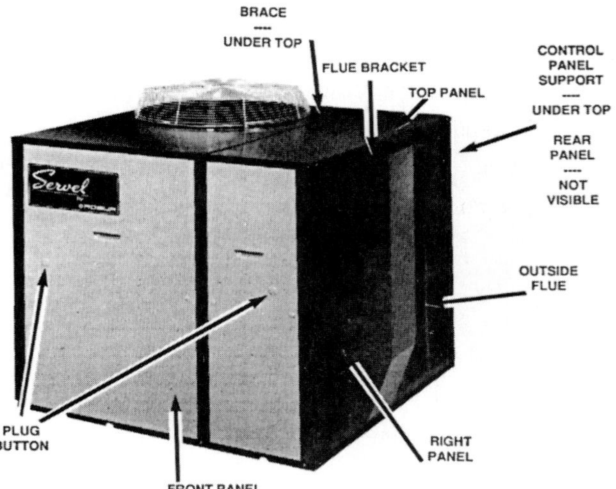

Figure A33-4 Packaged absorption chiller-heater. (Courtesy of The Robur Corporation)

A33-5 ABSORPTION CHILLER-HEATER

The unit is made either as a gas-fired self-contained chiller or a chiller-heater (Fig. A33-4). It is located outside the building with water piping (antifreeze added where necessary) and connected to a coil in a fan-coil unit (Fig. A33-5) or furnace (Fig. A33-6) inside the building. With

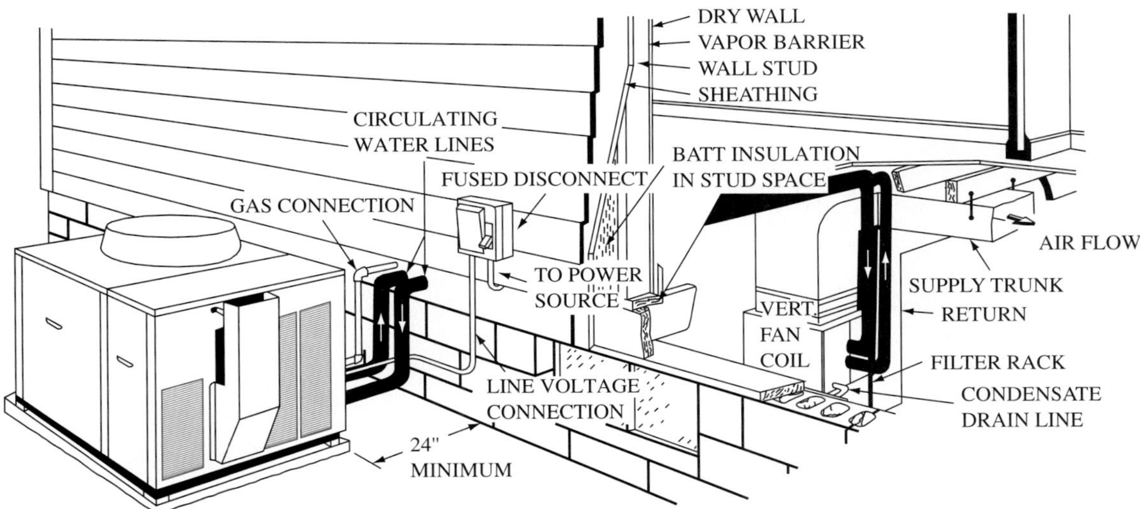

Figure A33-5 Installation diagram for an absorption unit connected to a fan-coil unit. (Courtesy of The Robur Corporation)

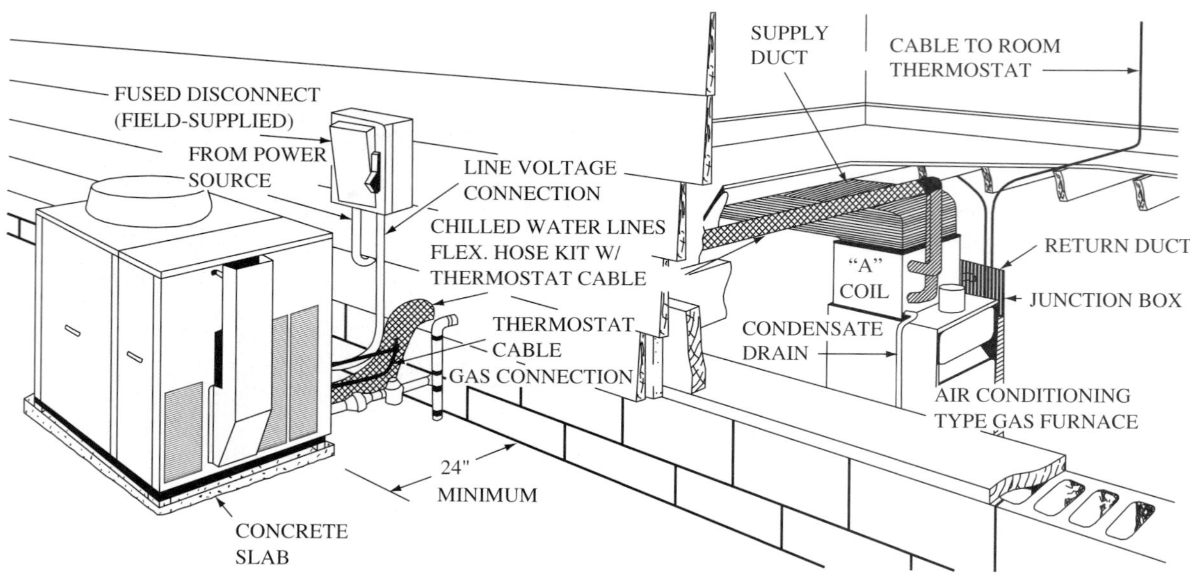

Figure A33-6 Installation diagram for an absorption unit connected to a coil in a furnace. (Courtesy of The Robur Corporation)

this arrangement, both cooling and heating are provided for air conditioning.

A view of the unit with the side panels removed is shown in Fig. A33-7. The chiller portion is on the left and the heater portion is on the right.

The refrigeration-circuit evaporator (Fig. A33-8) is used to *cool* water. The heater or boiler (Fig. A33-9) is used to *heat* water. Common piping is used for both the cooled and heated water. Two pumps are used on the chiller unit—a solution pump and a water pump, driven by a common electric-motor drive (Fig. A33-10). The chiller controls (Fig. A33-11) and the heater controls (Fig. A33-12) are mounted on the control panel.

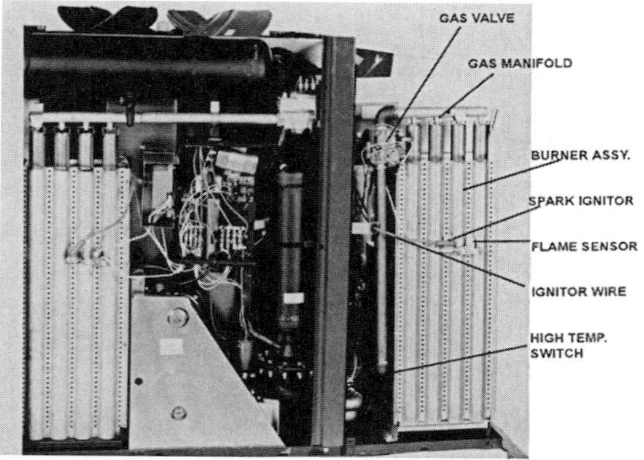

Figure A33-7 Absorption-type chiller-heater with side panels removed. (Courtesy of The Robur Corporation)

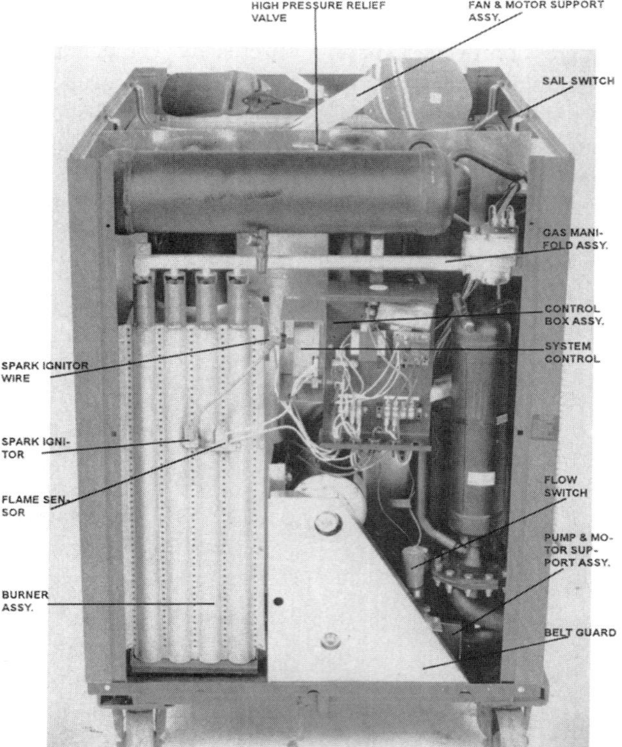

Figure A33-8 Cutaway view of the chiller portion of an absorption chiller-heater. (Courtesy of The Robur Corporation)

Figure A33-9 Cutaway view of the heater portion of an absorption chiller-heater. (Courtesy of The Robur Corporation)

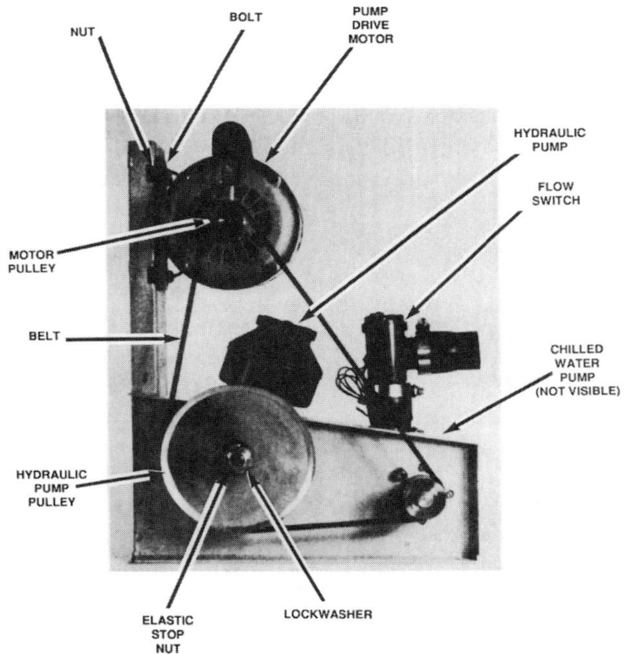

Figure A33-10 Chiller pump-drive system. (Courtesy of The Robur Corporation)

A33-5.1 Diagram of a Chiller

Fig. A33-3 shows a simplified diagram of the absorption cycle. From a practical standpoint, however, a number of other components are added for the production model, as shown in Fig. A33-13. These include:

1. A *levelizing chamber* and a *rectifier,* which are used to extract additional water from the refrigerant, not removed by the generator, to produce as pure a refrigerant vapor as possible.
2. A *two-circuit condenser coil,* where one circuit is used to condense the refrigerant, the other to cool the solution for use in the rectifier and to precool the refrigerant vapor on its way to the condenser.
3. A *condenser restrictor* lowers the pressure and temperature of the refrigerant liquid prior to its entrance into the liquid-vapor heat exchanger.
4. A *liquid-vapor heat exchanger* cools the refrigerant vapor leaving the evaporator.

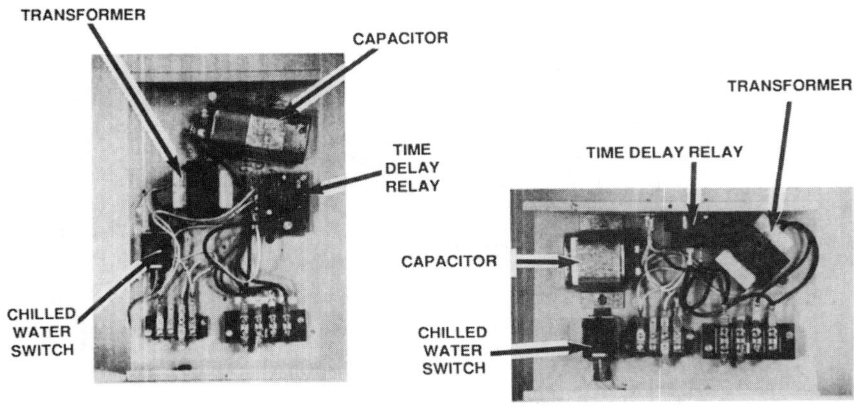

Figure A33-11 Chiller control module showing component parts. (Courtesy of The Robur Corporation)

Figure A33-12 Heater control module showing component parts. (Courtesy of The Robur Corporation)

A33-5.2 Installation

A number of installation requirements are important for an absorption unit. These are described below.

Freeze Protection

Antifreeze must be added to the water used in the entire water circuit and must remain there at all times. Part of this precaution is for protection during freezing weather and the other to protect against freezing temperatures in the evaporator during normal operation. Below is a table indicating the amount of antifreeze for various climatic conditions. Note the minimum amount is 10% by volume.

Lowest Expected Outdoor Temperature, °F	Permanent Antifreeze % by Volume
25	10
15	20
5	30
0	33
–5	35
–10	40
–20	45

Location

The chiller-heater should have a minimum of an 18-inch clearance on each side, for the necessary airflow for the condenser. It should also have a 24-inch minimum clear-

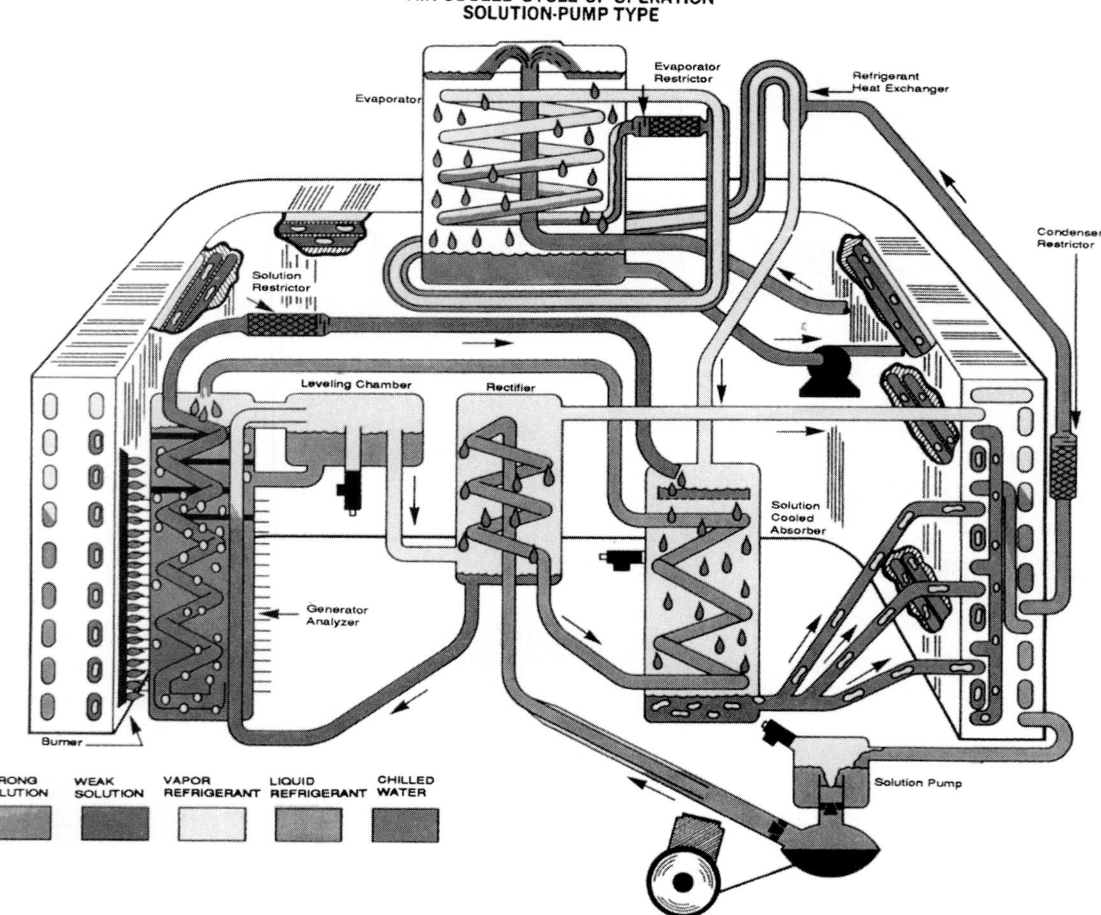

Figure A33-13 Detailed diagram of the operation of an absorption cycle. (Courtesy of the Robur Corporation)

ance at the front and back of the unit for maintenance and service. It should be mounted on a secure foundation, preferably a concrete pad.

Leveling

The unit should be installed level both front-to-back and side-to-side.

Water Piping

Details for the water piping for both down-flow and up-flow installations are shown in Fig. A33-14. Note also in this drawing the use of a thermometer well and a balancing valve in the coil piping.

Piping for chilled hot-water systems must be copper pipe or tubing.

Piping Insulation

All supply- and return-water lines should be insulated with ⅜-in. minimum Armaflex or similar material.

Utility Connections

A fused disconnect switch should be installed in the 220-V supply to the unit within sight of and not over 50 ft from the unit. Power wiring is connected to the terminal strip in the control box, with access from the side or back. The ground bar in the control box must have its own ground connection.

The gas inlet connection is through the rear panel.

A solution of water and antifreeze is used to fill the system at the time of startup.

A33-5.3 Operation and Controls

A typical control panel for the chiller is shown in Fig. A33-15 and for the heater in Fig. A33-16. A typical wiring diagram is shown in Fig. A33-17.

The sequence of operation of the control system is as follows:

1. *Cooling*. When the thermostat calls for operation:
    a. The time-delay relay (TDR) is energized.
    b. The TDR closes, starting pump and condenser-fan motors.
    c. With the generator high-temperature limit switch and chilled-water low-temperature limit switches closed, when the sail switch closes, the direct spark ignition (DSI) system will be energized.
    d. Through switches in the DSI control and flow switch, contact is made; the operator gas valve and the spark ignitor will be energized.

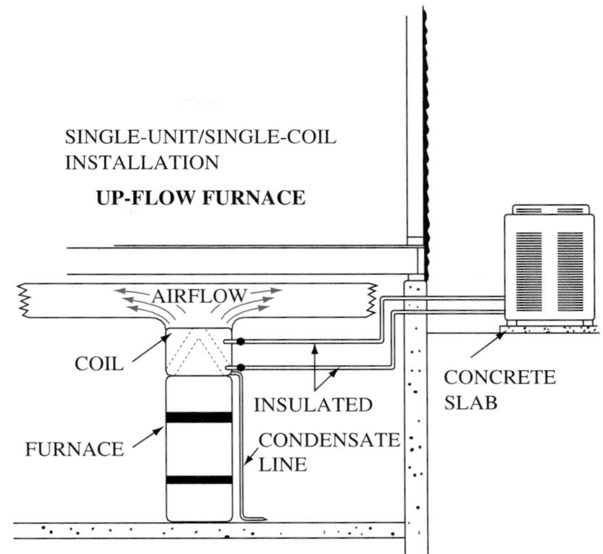

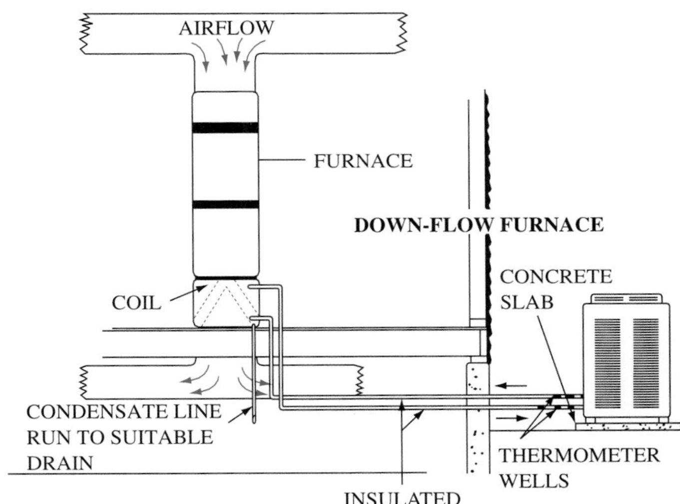

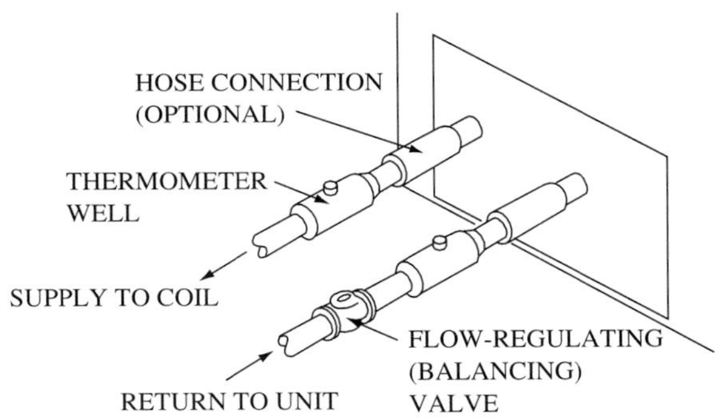

TYPICAL UNIT CONNECTIONS

Figure A33-14 Chilled-water piping installation for an absorption chiller. (Courtesy of The Robur Corporation)

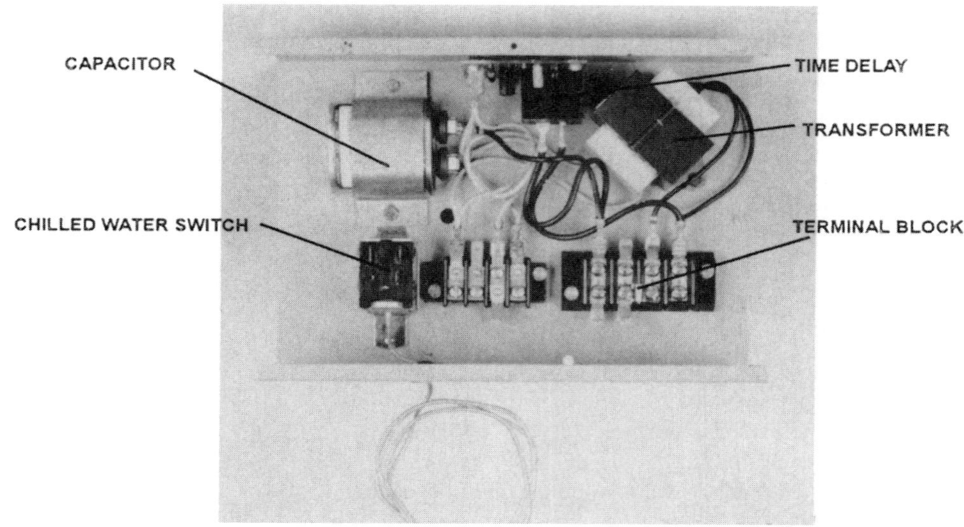

Figure A33-15
Physical location of chiller controls for the absorption chiller. (Courtesy of The Robur Corporation)

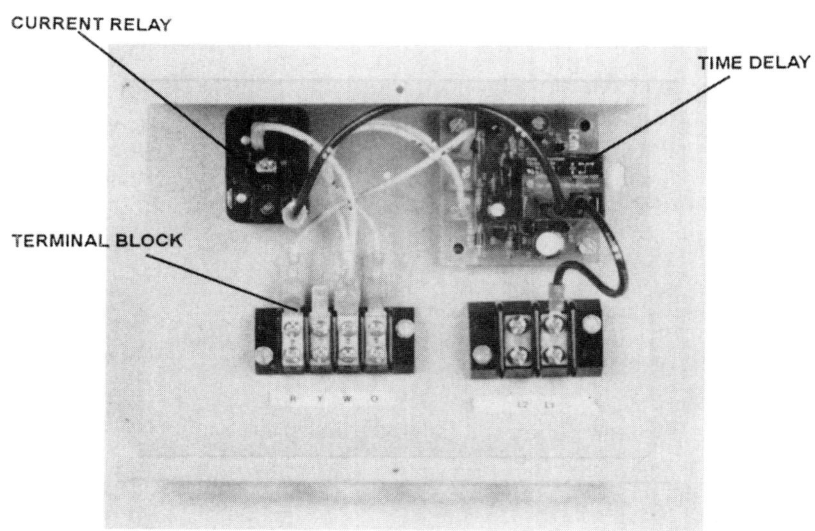

Figure A33-16
Physical location of heater controls for the absorption chiller. (Courtesy of The Robur Corporation)

   e. The gas should ignite within seconds. The DSI sensor will sense the flame and cut off the spark ignitor's operation.

   When the thermostat is satisfied, the DSI system is de-energized, but the pump and condenser fan motors will continue to operate until the TDR switch opens (3 to 4 minutes).

2. *Heating.* When the thermostat calls for operation:
   a. The TDR is energized.
   b. The TDR switch closes, starting the pump-blower motor.
   c. With the generator high-temperature limit switch and generator hot-water limit switch closed, when the current relay switch closes, the DSI system will be energized.
   d. Through switches in the DSI system control, the operator gas valve and the spark ignitor will be energized.
   e. The gas should ignite within seconds. The DSI sensor will sense the flame and cut off the spark ignitor's operation.

   When the thermostat is satisfied, the DSI system is de-energized, but the pump-blower motor will continue to operate until the TDR switch opens (approximately 2 minutes).

3. *Safety Operation:* Whenever the DSI control system is energized:

   The system has three trials for ignition. There will be a pre-purge period of approximately 15 seconds before sparking begins and the gas valve is energized. This pre-purge occurs on each trial for ignition. Sparking will occur after each purge cycle for 9 to 11 seconds. If the sensor does not sense a proper flame during the trials, the DSI system control will lock out to ignite the burner.

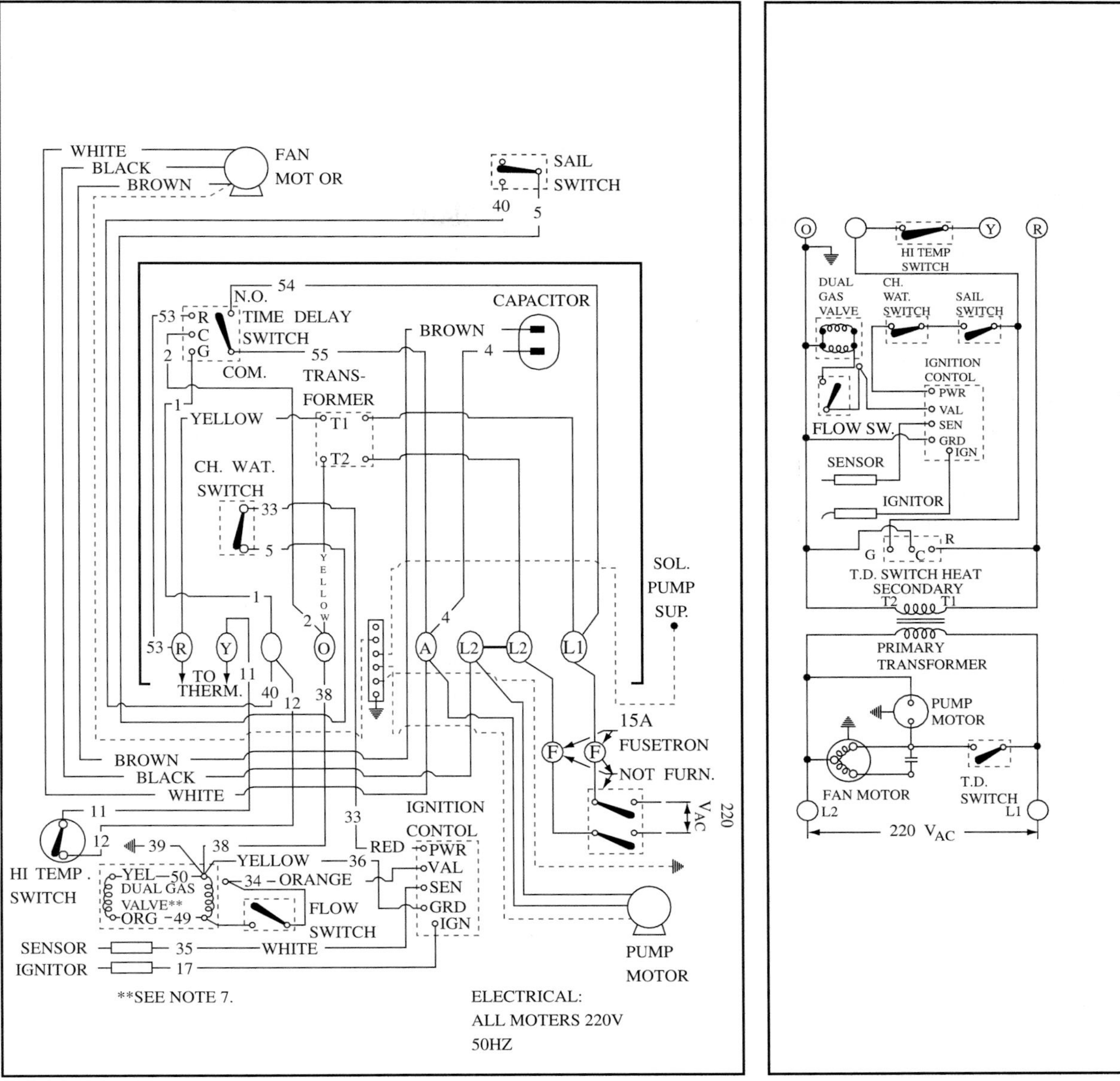

NOTES:

1. HEAVY SOLID AND BROKEN LINES INDICATE CONNECTIONS TO BE MADE AT THE TIME OF INSTALLATION.
2. POSITION OF ALL SWITCHES ARE SHOWN WITH CURRENT OFF.
3. SEE SERVICE HANDBOOK FOR EXPLANATION OF CONTROLS.
4. IF ANY OF THE ORIGINAL WIRE AS SUPPLIED WITH THE APPLIANCE MUST BE REPLACED, IT MUST BE REPLACED WITH THERMOPLASTIC 105° C WIRE, EXCEPT IGNITER WIRE 230° C, SENSOR WIRE AND FLAME CURRENT GROUND WIRE 200° C OR EQUIV.
5. ALL ELECTRICAL SWITCHES INTERRUPTING THE LINE VOLTAGE SUPPLY SHALL BE INSTALLED WITHIN SIGHT OF AND NOT OVER 50 FEET FROM THE UNIT.
6. UNIT MUST BE GROUNDED AS DESCRIBED IN INSTALLATION-ADJUSTMENT AND START-UP INSTRUCTIONS.
7. ROBERTSHAW 7000BDER VALVE SHOWN. OPTIONAL ROBERTSHAW 7200DER VALVE WILL NOT HAVE #49 AND #50 WIRES. #34 WIRE TO #1 TERMINAL AND #36, 38 AND 39 WIRES TO #2 TERMINAL ON OPTIONAL VALVE.

Figure A33-17 Wiring diagram for an absorption chiller-heater. (Courtesy of The Robur Corporation)

Items to Check	Found	Corrections Made	Corrections Made	Notes
Level (Front-to-back)				
Level (Side-to-side)				
Condenser-Absorber Condition				
Condenser Fan Height				
Chilled Water Level				
Chilled Water Flow Rate (gpm)				
Nat. Gas Btu Content				
Nat. Gas Specific Gravity				
Gas Press. (Water Column) Cooling				
Gas Press. (Water Column) Heating				
Hydraulic Fluid Level				
Belt Tension				
Solution Pump Pulsations (1 Min.)				
Ambient Temp. (Away from Unit)				
Condenser Air In—Right Side				
Condenser Air In—Left Side				
Condenser Air In—Back				
Air Volume at Air Handler				
°F Air In at Air Handler				
°F Air Out at Air Handler				
Chilled-Hot Water Temp.—In °F				
Chilled-Hot WaterTemp.—Out °F				
Sealed System Pressure—Purge				
Sealed System Pressure—Low Side				
Sealed System Pressure—High Side				

Figure A33-18 Service check list for an absorption chiller-heater. (Courtesy of The Robur Corporation)

GAS CHILLER & CHILLER-HEATER
SERVICE CHECKLIST, CONTINUED: CHILLER AND CHILLER-HEATER MODELS

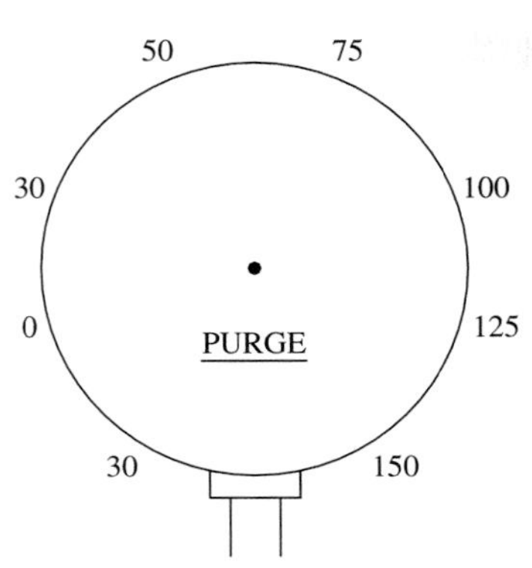

1. PURGE GAUGE PATTERN (BURNER OFF)

NOTE: THIS IS A PRELIMINARY CHECK ONLY

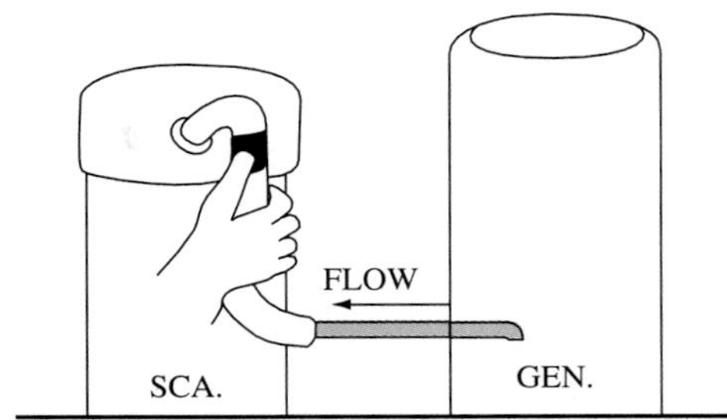

2. SOLUTION RESTRICTOR CHECK (BURNER ON) GRASP SOLUTION RESTRICTOR AND FEEL FOR SOLUTION FLOW THROUGH TUBE FROM GENERATOR END TO SCA.

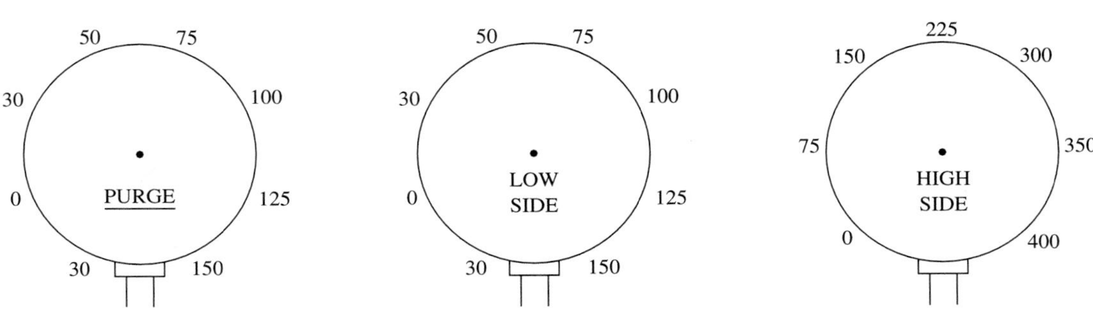

3. PRESSURE PATTERN OBSERVED. REFER TO SERVICE MANUAL FOR SYMPTOM PATTERNS.

DIAGNOSIS ____________ CORRECTION ____________ SOL. LEVEL ____________

Figure A33-18 (Continued)

A33-5.4 Service and Maintenance

Refrigeration technicians can perform electrical-control service and maintenance. Problems in the function of the absorption cycle need to be addressed by technicians with factory training. To analyze cycle problems, a check list (Fig. A33-18) should be filled out. With the help of this information, a specially trained technician can analyze the cause of the problem and proceed with a solution.

It is particularly important that proper safety procedures be followed in servicing the equipment just described, since the refrigerant is ammonia. Emergency and first-aid procedures should be posted and easily available. If the refrigerant and/or a refrigerant solution gets into the eyes, on the skin, is inhaled or ingested, emergency and first-aid procedures should be followed.

Due to the unique nature of this equipment, the technician should fully utilize the information provided by the manufacturers' service manuals in any situations requiring work to be done on the system. A recommended procedure for troubleshooting is to follow a flow chart such as the one shown in Fig. A33-19.

Preventive maintenance practice consists of a yearly inspection of the solution level and the tension on the pump drive belt, and making certain that all electrical connections are tight.

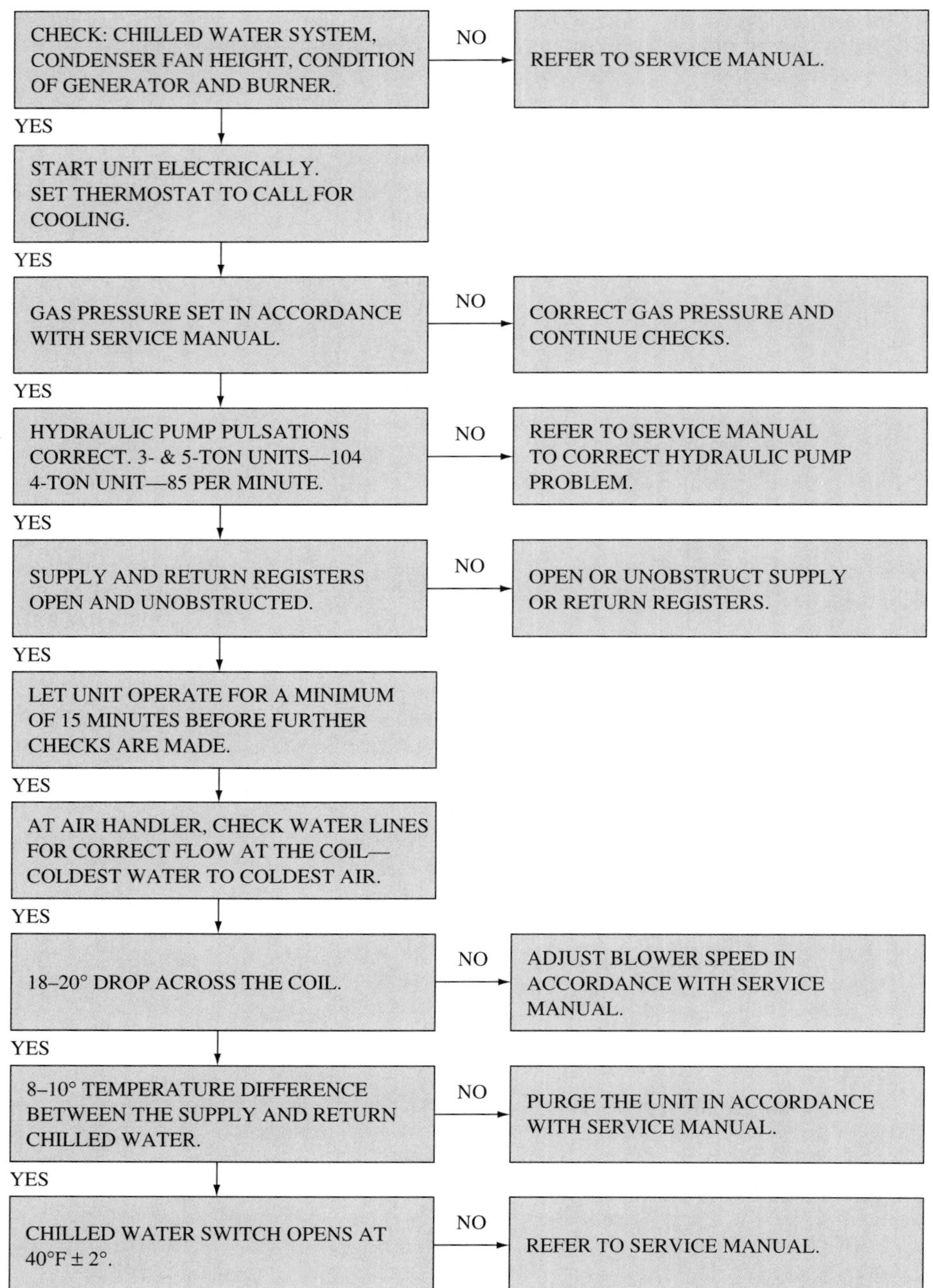

Figure A33-19 Service maintenance and adjustment flow chart for an absorption chiller. (Courtesy of The Robur Corporation)

REVIEW

- The absorption process was discovered by Michael Faraday in 1824.
- Advantages of absorption units:
  Environment—Refrigerants have a zero ozone depletion factor.
  Power—Use heat rather than electrical energy.
  Service—Absorption units have few moving parts and require a minimum of service and maintenance.
  Operating costs—In areas where the ratio of gas cost to electric cost is favorable, units have lower operating costs.
- Disadvantages of absorption units:
  Cost—Initial installation cost is higher.
  Capacity reduction—Capacity reduction to match load is limited.
  Weight—Absorption units are heavier.
- Principal components of an absorption unit:
  Generator—Heat is added to separate the lithium-bromide solution from the refrigerant vapor.

Condenser—Condenses the refrigerant vapor into a liquid.
Evaporator—Cools a liquid for use in process refrigeration or air conditioning.
Solution-cooled absorber (SCA)—Refrigerant vapor is absorbed, producing a dilute solution.
Solution pump—Pumps strong liquid solution back to generator.

- Installation requirements:
  Freeze protection—Prevents water from freezing.
  Location—Minimum 18-inch clearance on each side, with 24-inch clearance in front and back. Should be mounted on solid foundation, concrete preferred.
  Leveling—Front to back, side to side.
  Water piping—Thermometer wells and balancing valve.
  Piping insulation—Minimum ⅜ in.
  Utility connections—220-V supply with disconnect.
- Sequence of operations:
  1. Cooling
  2. Heating
  3. Safety operation
- Service and preventive maintenance:
  Annual inspections of solution level and pump-drive tension.
  Check electrical controls and connections.
- Auxiliary items:
  Heat exchanger–Heat exchange between the warm concentrated solution from the generator and dilute solution coming from the absorber.
  Fluid pump–Pumps fluid around the system.
  Purge unit–Removes non-condensible gases.
  Automatic decrystallization device–Prevents crystallization
  Solution control valve–Improves efficiency of the system at part load.
  Steam or hot water valve–Modulates to control flow to the generator tubes.
  Eductor–Provides for circulation of the lithium bromide over the absorber tubes.
  Control center–Controls unit functions and cycle.

Problems and Questions

1. Absorption refrigeration units use _____ as their energy source.
2. Are sound and vibration levels of absorption equipment higher or lower than in conventional systems?
3. What refrigerant is used in an absorption unit?
4. The most common absorbant is _____ .
5. Name the four main components of an absorption chiller.
6. Does the absorber contain the evaporator or the condenser?
7. The condenser is located within the _____ .
8. The function of the purge unit is to remove _____ gases.
9. Is the purge unit normally an automatic or a manual operation?
10. The control valve modulates the flow of steam or hot water to the generator. True or false?
11. The absorption process of cooling was discovered by Willis Carrier. True or false?
12. Commercial absorbers use water as the absorbant. True or false?
13. Absorbers use mechanical energy from a compressor to move the heat. True or false?
14. Absorption units are used strictly for commercial applications. True or false?
15. The typical absorption unit has a zero ozone depletion factor. True or false?
16. The reason absorption refrigeration use is sometimes less attractive is:
    a. Costs.
    b. Capacity reduction.
    c. Weight.
    d. All of the above.
17. During the boiling process the _____ is separated from the water.
    a. Ammonia.
    b. Refrigerant.
    c. Solution.
    d. None of the above.
18. The absorption unit does not use a vapor compressor, it uses a:
    a. Liquid compressor.
    b. Heater and generator.
    c. Solution restrictor.
    d. All of the above.
19. A condenser restrictor is used to:
    a. Raise the pressure and temperature of the liquid refrigerant.
    b. Lower the pressure and temperature of the vapor refrigerant.
    c. Lower the pressure and temperature of the liquid refrigerant.
    d. None of the above.
20. The gas-fired residential absorber utilizes a prepurge of about:
    a. 15 seconds.
    b. 30 seconds.
    c. 1 minute.
    d. None of the above.

SECTION 7

Heating Systems

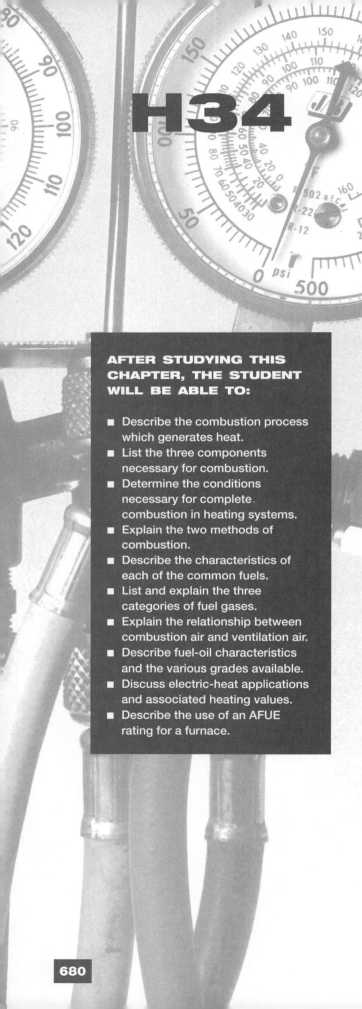

H34 Combustion and Fuels

AFTER STUDYING THIS CHAPTER, THE STUDENT WILL BE ABLE TO:

- Describe the combustion process which generates heat.
- List the three components necessary for combustion.
- Determine the conditions necessary for complete combustion in heating systems.
- Explain the two methods of combustion.
- Describe the characteristics of each of the common fuels.
- List and explain the three categories of fuel gases.
- Explain the relationship between combustion air and ventilation air.
- Describe fuel-oil characteristics and the various grades available.
- Discuss electric-heat applications and associated heating values.
- Describe the use of an AFUE rating for a furnace.

H34-1 COMBUSTION

Combustion is the chemical process by which oxygen is combined rapidly with a fuel to release energy in the form of heat. Other forms of energy are released at the same time, such as electromagnetic energy (light), which gives visibility to the flame.

Conventional hydrocarbon fuels contain primarily hydrogen and carbon in elemental form in various compounds. Their complete combustion produces mainly carbon dioxide and water; however, small quantities of carbon monoxide and partially reacted fuel constituents (gases and liquid or solid aerosols) may form. Most conventional fuels also contain small amounts of sulfur, oxidized to sulfur dioxide and trioxide during combustion, and non-combustible products such as ash and inert gases. Flue gas usually also contains excess air.

Combustion requires: (1) fuel, (2) heat, and (3) oxygen. The rate of combustion depends on:

1. The chemical reaction rate of the combustible fuel constituents with oxygen.
2. The rate at which oxygen is supplied to the fuel (mixing air and fuel).
3. The temperature in the combustion region.

The chemical reaction rate is fixed in the selection of the fuel. Increasing the mixing rate or the temperature increases the combustion rate.

With complete combustion, all hydrogen and carbon are oxidized into H_2O and CO_2. Generally for complete combustion, excess air must be supplied beyond the amount theoretically required to oxidize the fuel.

Incomplete combustion occurs when the fuel is not completely oxidized in the combustion process. Substances such as carbon monoxide and aldehydes are formed. This is caused by (1) insufficient fuel and air mixing, (2) insufficient air supply to the flame, (3) flame impingement on a cold surface, and (4) insufficient flame temperature. Incomplete combustion is not only inefficient, but hazardous because producing carbon monoxide is harmful to health.

The two most popular gaseous fuels are (1) natural gas, consisting primarily of methane and ethane, and (2) liquid petroleum which is usually propane or a mixture of propane and butane. The measurement of CO_2 in the flue gas is an indicator of the efficiency of the combustion process. The actual measured CO_2 needs to be compared to the ultimate (highest possible) CO_2 in order to be meaningful. The following table gives the ultimate CO_2 values that can be obtained for the chief constituents of these gaseous fuels:

Constituent	Chemical Formula	Ultimate CO_2 (%)
Methane	CH_4	11.73
Ethane	C_2H_6	13.18
Propane	C_3H_8	13.75
Butane	C_4H_{10}	14.05

Other interesting factors relating to the combustion of gaseous fuels are the flammability limits and the ignition temperatures. Fuels will burn in a continuous combustion process only when the volume percentages of fuel and air in a mixture are within their flammability limits and the mixture reaches ignition temperature. The following table gives these values for the chief constituents of gaseous fuels:

Constituent	Lower Flammability Limit (%)	Upper Flammability Limit (%)	Ignition Temperature (°F)
Methane	5.0	15.0	1301
Ethane	3.0	12.5	968 to 1166
Propane	2.1	10.1	871
Butane	1.86	8.41	761

There are two modes of combustion: (1) continuous and (2) pulse. *Continuous combustion* has been described above, and is most common.

Pulse combustion is used in some high-efficiency furnaces and burns fuel in small, discrete, fuel-air mixture volumes in a very rapid series of combustions. The introduction of the fuel-air mixture into the pulse combustor is controlled by mechanical valves. A typical combustor consists of one or more valves, a combustion chamber, exit pipe, and a control system (ignition means, fuel-metering devices, etc.). A flame trap at the combustion-chamber entrance may be used to prevent flashback. The combustion chamber and exit-pipe geometry determine the resonant frequency of the combustor. The new fuel-air charge is ignited by the residual combustion and/or heat.

The pulse combustors operate at 30 to 100 cycles per second and emit resonant sound. The pulses produce high convective heat-transfer rates.

The heating values of fuels can be of two types: (1) higher heat values, which include the heat produced in condensing the water vapor in the products of combustion, and (2) lower heat values, which do not include the latent heat produced by condensing the water vapor. The heating values for the common constituents of gaseous fuels are as follows:

Substance	Higher Heating Values (Btu/lb)	Lower Heating Values (Btu/lb)	Specific Volume (ft^3/lb)
Methane	23,875	21,495	23.6
Ethane	22,323	20,418	12.5
Propane	21,669	19,937	8.36
Butane	21,321	19,678	6.32

Heating requires the expenditure of some source of energy to raise the temperature of air or water, depending on the type of equipment used. One can classify energy sources for heating as: (1) gases, both natural gas and from liquid petroleum, (2) oil, (3) electricity, (4) other, including coal, wood, and solar. The first three are the principal sources used for domestic heating.

H34-2 GASES

Fuel gases are employed for various heating and air-conditioning (cooling) processes and fall into three broad categories: *natural, manufactured,* and *liquefied petroleum.* Stories about the discovery and use of gas date back as far as 2000 B.C. History notes that the Chinese piped gas from shallow wells through bamboo poles and boiled sea water to obtain salt.

The great advantage of *natural gas* over other fuels is its relative simplicity of production, transportation, and use. When a sufficiently large quantity is discovered, it is pumped from drill holes (wells) to processing plants, and on to refineries and industrial centers where it has a large number of diverse uses in addition to heating.

Natural gas comes from sedimentation of trillions of tiny organisms at the bottom of the sea, buried and initially chemically converted into dense organic material. Over millions of years, pressure and heat gradually cracked this material into lighter hydrocarbon compounds. Liquids were liberated first, then the gases. Compaction squeezed them from the source rock and each migrated into the more porous reservoir rocks until stopped by impermeable barriers. Pools of gas and oil developed which are the target of petroleum exploration. Most reserves of gas in the United States are not dissolved in or in contact with oil.

Natural gas, Fig. H34-1, is mostly methane, which consists of one carbon atom linked to four hydrogen atoms; and ethane, which consists of two carbon and six hydro-

```
     H                     H   H
     |                     |   |
   H-C-H                 H-C---C-H
     |                     |   |
     H                     H   H
```
METHANE CH_4 ETHANE C_2H_6

```
     H   H   H               H   H   H   H
     |   |   |               |   |   |   |
   H-C---C---C-H           H-C---C---C---C-H
     |   |   |               |   |   |   |
     H   H   H               H   H   H   H
```
PROPANE C_3H_8 BUTANE C_4H_{10}

Figure H34-1 Chemical components of common fuels.

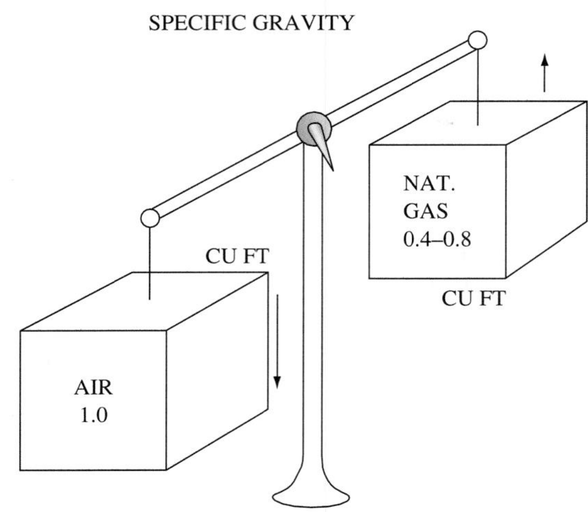

Figure H34-2 Weight of one cubic foot of air compared to one cubic foot of gas.

gen atoms. Liquid petroleum (LP) gases are propane and butane or a mixture of the two, and these fuel gases are obtained from natural gas or as a by-product of refining oil. From Fig. H34-1 you will note they contain more carbon and hydrogen atoms than natural gases, are thus heavier, and, as a result, have more heating value per cubic foot. Gas is a precious commodity and may someday not be available as a substitute for other fuels.

Manufactured gas, as the name states, is man-made as a by-product from other manufacturing operations. For example, in iron-making, large amounts of gases are produced that can be used as fuel gases. The use of manufactured fuel gases has declined greatly in the United States. Today over 99% of sales by gas distribution and transmission companies is natural gas; however, the man-made kind is still a popular fuel in Europe.

Mixed gas, as the name implies, is a man-made mixture of gases such as natural gas and manufactured gas.

It is important to know the density of a gas as expressed by its specific gravity. Compared to the standard, air (Fig. H34-2) which has a specific gravity of 1.0, natural gas ranges from 0.4 to 0.8 and thus is lighter than air. On the other hand, of the *liquid petroleum* gases, the specific gravity of propane is 1.5 and that of butane is 2.0, meaning they are heavier than air. The density of the gases is important because it affects the flow of the gas through orifices (small holes) and then to the burner. Should a leak develop in a gas pipe, natural gas will rise while LP gases will not and may drift to low spots and collect in pools, creating a hazard if open flames are present. Specific gravity also affects gas flow in supply pipes and the pressure needed to move the gas.

The heating value (or heat value) of a gas is the amount of heat released when 1 cu ft of the gas is completely burned (Fig. H34-3). Natural gas (which is largely methane) has a heating value of about 950 to 1150 Btu/ft^3. Propane has a heat value of approximately 2500 Btu/ft^3 and butane, about 3200 Btu/ft^3. The hourly rate of a heating value is thus the number of cubic feet burned in one hour times the heating value. For example, if a fur-

PRACTICING PROFESSIONAL SERVICES

Gas pressures for natural-gas and LP-gas heating appliances do not use the same burner pressures for normal operation. Typically, the natural-gas unit will use 3.5 in. WC and the LP unit will use 11 in. WC pressure. These pressures can be checked with a simple U-tube manometer. If the pressure at the burner is low, check the supply pressure from the main building regulator. If that pressure is also lower than normally expected, contact your gas company technical representative for a site inspection.

Whenever the flame appears yellow or has improper burning and abnormal color, perform the gas-pressure check. Improper flame patterns, however, are not always the result of low gas pressure. Be sure to check primary and secondary combustion air adjustments. Also, if the burner orifice(s) get dirty, they could affect the combustion process.

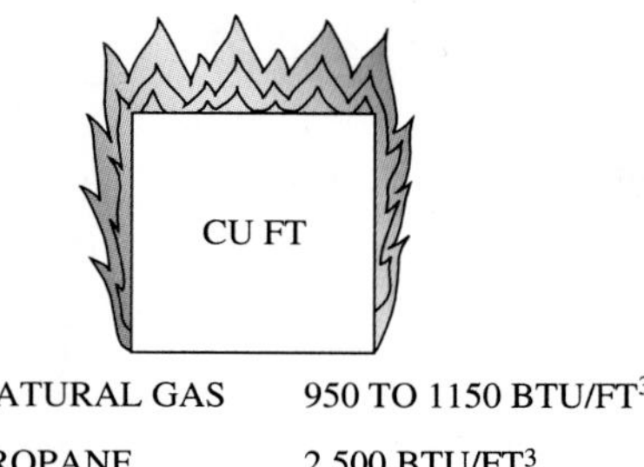

Figure H34-3 Heat delivered by gas fuels when burned.

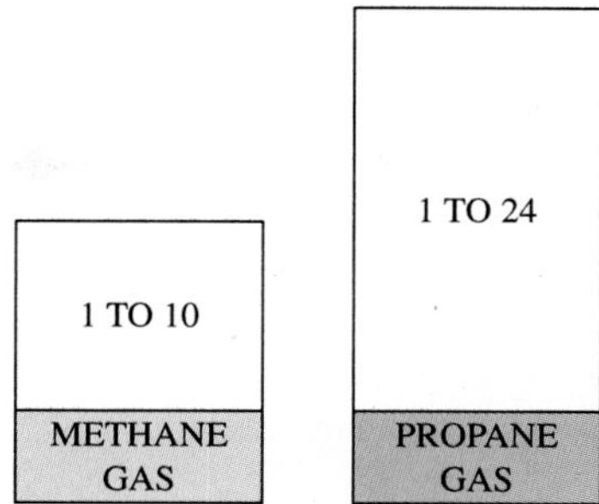

Figure H34-5 Amount of air required for combustion of methane compared to propane.

nace burns 75 ft^3 of natural gas in 1 hr and the heat value is 1000 Btu/ft^3, the total heating input is $75 \times 1000 =$ 75,000 Btu/hr. The actual output depends on the operating efficiency of the furnace. The exact heating value of gases in your local area can be obtained from the gas company or LP-gas distributor.

Propane is the LP gas mostly used for domestic heating. *Butane* has more agricultural and industrial applications.

A knowledge of combustion is closely related to understanding the importance of ventilation by the technician. Combustion (Fig. H34-4) takes place when fuel gases are burned in the presence of air. Methane gas combines with the oxygen and nitrogen present in the air and the resulting combustion reaction produces heat, with by-products of carbon dioxide, water vapor, and nitrogen. For each cu ft of methane gas, 10 cu ft of air is needed for complete combustion (Fig. H34-5). Although natural gas requires a 10–1 ratio of air, LP fuels require much more, due to the concentration (greater density) of carbon and hydrogen atoms. Liquid-petroleum combustion must have more than 24 cu ft of air per cu ft of gas to support proper combustion. When complete information on the fuel is not available, a frequently used value for estimating air requirements is that 0.9 cu ft is required for 100 Btu of fuel. Excess air is also required to ensure complete combustion.

The by-products, called flue products, are vented to the outside. In interior space, insufficient make-up air can produce hazardous conditions. If too little oxygen is supplied, part of the by-products will be dangerous carbon monoxide gas (CO) rather than harmless carbon dioxide gas (CO_2). Second, lack of make-up air causes poor flue action and spillage of combustion products into the room. These flue products in a living area cause serious problems. Specific recommendations on venting and make-up air will be covered later, but remember, combustion-heating ventilation is an essential element of the system's design, and central to indoor air quality.

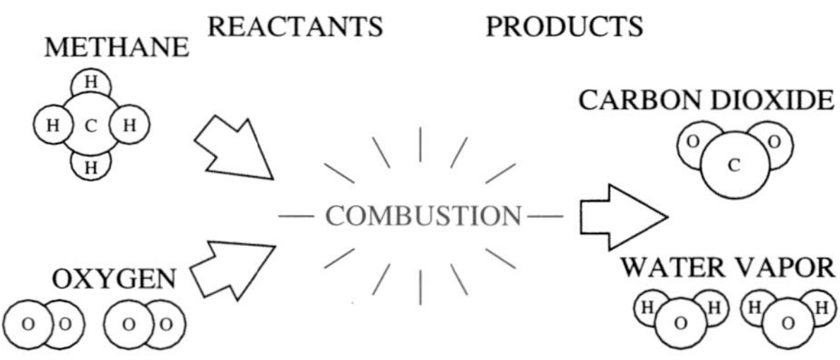

Figure H34-4 Combustion process when a fuel is burned.

H34-3 OILS

Fuel oil is a major source of heating energy for homes and commercial buildings. It is very popular in the northern and eastern sections of the country and is found in many rural areas where the availability or desirability of gas or electrical energy are limited. Fuel oils are mixtures of hydrocarbons derived from crude petroleum by various refining processes. They are classified by dividing them into grades according to their characteristics, mainly viscosity; however, other properties like flash point, pour point, water and sediment content, carbon residue, and ash, are important in the storage, handling, and types of burning equipment for oil. The *viscosity* determines whether the fuel oil can flow or be pumped through lines or if it can be atomized into small droplets.

For comfort-heating applications, we are primarily interested in two grades of fuel oil: No. 1 and No. 2 (Fig. H34-6), which contain 84 to 86% carbon, up to 1% sulfur, and mostly hydrogen in the remainder. The heavier grades, No. 4 and No. 5, have even higher carbon con-

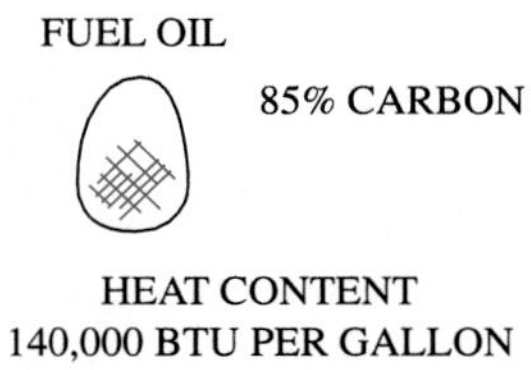

FUEL OIL

85% CARBON

HEAT CONTENT
140,000 BTU PER GALLON

Figure H34-6 Carbon content of fuel oil and quantity of heat produced.

PREVENTIVE MAINTENANCE

Oil furnaces as well as gas furnaces need routine inspections and maintenance performed annually prior to startup for the heating season. On oil furnaces, this includes changing oil filters, strainers, installing new nozzles, and checking ignition electrodes. On gas furnaces this includes checking ignition sequences, flame conditions, and condensate vent lines and drains. Both type of units should have their air filters replaced if needed, have plenum combination fan/limit switches and heat exchangers inspected for corrosion. A flue-gas analysis is also a good idea on both units to ensure proper combustion exists. This entails sampling the flue gas for CO_2, checking stack temperature, and checking the temperature rise across the heat exchanger.

tents, but also considerably more sulfur than permissible in domestic grades.

No. 1 grade fuel oil is considered premium quality and priced accordingly. It is used in room-type space heaters, which do not use high-pressure burners and depend on gravity flow, thus the need for the lower viscosity.

No. 2 grade is the standard heating oil sold by most oil-supply firms. It weighs between 6.870 and 7.296 lb/gal as compared to water at 8.34 lb/gal. No. 2 oil is used in equipment that has pressure-type atomizing, which includes most forced-air furnaces and boilers. The heating value is approximately 135,000 to 142,000 Btu/gal.

EXAMPLE

What is the heating effect of a furnace burning 5 gallons of No. 2 oil the first hour, at start-up?

Solution

5 gal/hr × 135,000 Btu/gal = 675,000 Btu/hr

H34-4 ELECTRICITY

The growth of electric heating came immediately after World War II and found its main use in areas of cheap public power such as the TVA (Tennessee Valley Authority) and in the Pacific Northwest. In both situations, the cost of power and the winter climate were favorable to the use of electric heating. In the late 1950s, when it became apparent that investor-owned private utilities were going to be faced with heavy summer-cooling loads, electric utilities and heating manufacturers began promoting electric heating to build up their winter load. Special rates and promotion programs like "Live Better Electrically" and "Total Electric Home" were introduced in the 1960s and the market developed rapidly. The situation in the mid-1970s changed somewhat due to higher and higher power-generating costs, which were passed on to the consumer. Also, the federal government directed that the use of special promotional programs be stopped. Even with these current obstacles, however, the use of electric heating continues to grow, due to availability, convenience, and other considerations. If nuclear generating stations increase in numbers, the cost of power may be stabilized and may even start to decline in the future.

The heating value of electric-resistance heat is easy to remember and calculate (Fig. H34-7). For each watt of power consumed, 3.4 Btu/hr of heat output will be generated. Resistance heating is 100% efficient; there are no losses such as those experienced with oil- and gas-combustion processes. Electric power is measured in kilowatts (1 kW is 1000 W). Therefore, consuming 1 kW for 1 hour is called a kilowatt-hour (kwh). And if the ratio is 3.4 Btuh/W, then a kilowatt of electric resistance heating will produce heat at the rate of 3,400 Btu/hr.

This discussion of electric-resistance heating does not include the all-electric heat pump. Its unique reverse-refrigerant cycle can produce heating with efficiency ratios of 1 input to 3 output under ideal conditions, and 1 to 2 or 2.5 under normal operation. That means for each unit of electricity put in, 2 to 2.5 units of heat are produced, compared to the 1:1 ratio of straight resistance heat. The operation of heat pumps will be explained more fully in subsequent chapters, but because of their ability to produce all-electric heating and cooling, they are a vital factor in future electric energy developments.

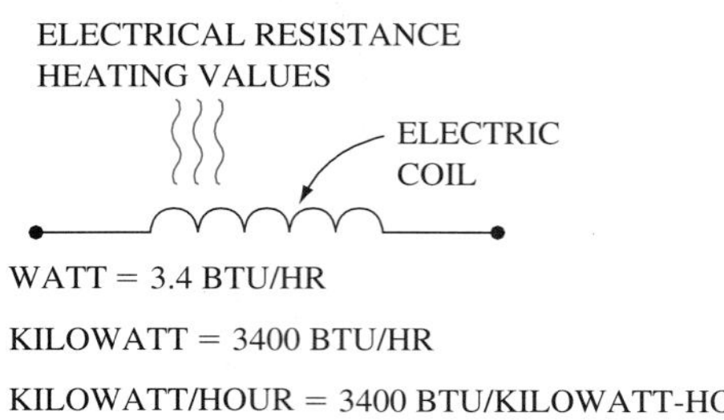

Figure H34-7 Amount of heat produced by an electric resistance heater.

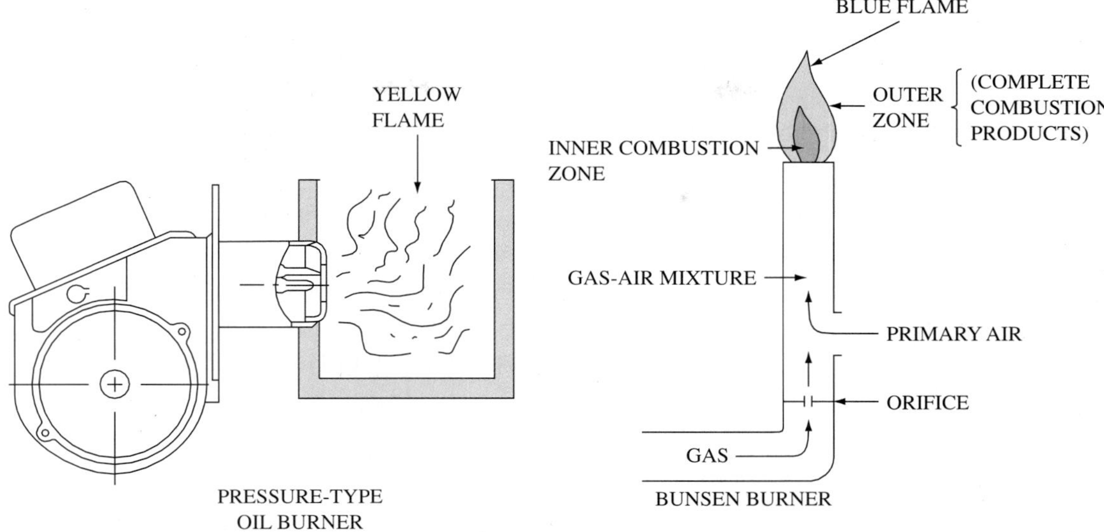

Figure H34-8 Types of flames produced by pressure-type oil burner and Bunsen burner.

H34-5 TYPES OF FLAMES

Basically there are two types of flames: (1) yellow and (2) blue (Fig. H34-8). The difference is mainly due to the manner in which the fuel is mixed with the air. A *yellow flame* is produced when gas is burned by igniting the gas gushing from an open end of a gas pipe, such as is common for ornamental lighting. Pressure-type oil burners burn with a yellow flame. Most modern gas burners burn with a blue flame. A *blue flame* is produced by a Bunsen burner such as used in a laboratory, where 50% of the air requirement is mixed with the gas prior to ignition. This part of the air supply is called primary air. The balance of the air (secondary air) is supplied during combustion to the exterior of the flame for complete combustion.

A good example of the use of primary and secondary air is shown in Fig. H34-9. This figure shows the primary air adjustment on a drilled port burner. The gas is supplied through a manifold and metered into the burner by a properly selected gas orifice. The air enters through adjustable openings around the gas orifice. The gas is under pressure of about 3½ in. WC in the manifold. The purpose of the venturi is to create a vacuum as the pressurized gas is forced through it, sucking in the primary air. The air is mixed with the gas as they pass through the venturi tube. The secondary air is supplied above the burner head. When the air supplies are properly adjusted for complete combustion, the burner operates with a blue flame.

H34-5.1 Annual Fuel Utilization Efficiency (AFUE)

Furnaces are rated for their annual fuel utilization efficiency. This rating is obtained by applying an equation developed by the National Institute of Standards and Technology (NIST) for 100% efficiency and deducting losses for exhausted latent and sensible heat, the effects of cycling, infiltration, and pilot-burner effect. The AFUE is determined for residential fan-type furnaces by using the ANSI/ASHRAE Standard 103-1988 method of testing.

The federal law, effective January 1, 1992, requires that all furnaces produced meet the minimum AFUE level of 78% efficiency.

H34-5.2 Practical Combustion Considerations

Air Pollution

The combustion processes constitute the largest single source of air pollution, which includes:

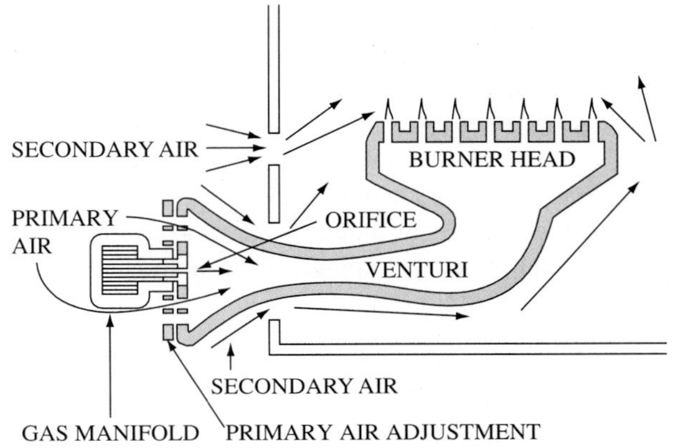

Figure H34-9 Primary and secondary air supply to a drilled-port gas burner. (Courtesy of Carrier Corporation)

1. Products of incomplete fuel combustion
   - Combustible aerosols, such as smoke and soot
   - Carbon monoxide
   - Gaseous hydrocarbons
2. Oxides of nitrogen, such as nitrous oxide and nitrogen dioxide
3. Emissions resulting from fuel contaminants
   - Primarily sulfur dioxide
   - Some sulfur trioxide
   - Ash
   - Trace metals
4. Emissions resulting from fuel additives

Some of the ways that the heating technician can help to reduce the health hazards due to air pollution are:

1. Properly test installed furnaces and adjust the fuel-burning device for highest efficiency.
2. Encourage the use of high-efficiency furnaces.
3. Encourage the conservation of energy by recommending proper insulation and reduction of infiltration.
4. Recommend the installation and operation of adequate outside air for ventilation. Check intake location to avoid re-entry of contaminants. Calculate the percentage of air required to determine if it meets guidelines:

$$\%\ \text{Outdoor air} = \frac{\text{No. of people} \times \text{cfm/person} \times 100}{\text{Total flow}}$$

5. Maintain clean plenum, air intakes, filters, ducts, and system components. Seal leaks.

Condensation and Corrosion

The heating technician must always be on the lookout for problems caused by condensation and corrosion produced by the flue gases. When the fuel-burning system cycles on and off to meet the demand, the flue passages cool down during the off cycle. When the system starts up again, condensate forms briefly on the surfaces until they are heated above the dew-point temperature. Low-temperature corrosion occurs in system components (heat exchangers, flues, vents, chimneys). The condensate includes such corrosive substances such as sulfides, chlorides and fluorides.

Corrosion increases as the condensate dwell time increases. One of the common evidences of this is the corrosion that takes place in the flue pipe between the furnace and the chimney. Even though a galvanized flue pipe is used, this piping needs to be replaced periodically due to the corrosion effect.

In the high-efficiency furnace where condensation is allowed to occur inside the furnace on a continuous basis, the flue passages must be constructed of corrosion-resistant materials such as stainless steel. The material used for condensate drains from high efficiency furnaces is usually corrosion-resistant plastic tubing. Some cities require neutralizing the condensate before it enters the public sewers.

Soot

Soot deposited on flue surfaces acts as an insulating layer, reducing the heat transfer and lowering the efficiency. Soot can also clog flues, reduce draft and available air, and prevent proper combustion. Proper burner adjustment can minimize soot accumulation.

Certification of Heating Equipment

Common to all types of heating apparatus sold domestically is the need to have these tested, certified, or listed by the proper regulatory agencies.

The American Gas Association (AGA) establishes the minimum construction safety and performance standards for gas heating equipment. The AGA maintains laboratories to examine and test furnaces, and also maintains a field-inspection service. Furnaces submitted and found to be in compliance are listed in the AGA Directory; they also bear the Blue Star Seal of Certification (Fig. H34-10).

Oil-heating equipment and electric-heating products are subject to examination, testing, and approval by UL (Underwriters' Laboratories, Inc.). Most people are familiar with the UL stamp (Fig. H34-11) appearing on everything from toasters to electric blankets, but UL also gets deeply involved in the approval and listing of heating and air-conditioning equipment, even dealing with large centrifugal machines 100 tons and over. Local city codes and inspectors are guided by UL standards, and failure to comply with them may be costly to the manufacturer and installer. Underwriters' Laboratories maintains testing laboratories for certain types of products; however, they often perform the necessary tests at the manufacturer's plant.

Manufacturers who actively sell their products in Canada seek CSA (Canadian Standards Association, Fig. H34-12) approvals as well. CSA is the Canadian counterpart to UL in the United States.

Figure H34-10 American Gas Association (AGA) seal. (Courtesy of American Gas Association)

Figure H35-11 Underwriters' Laboratories (UL) seal. (Underwriters' Laboratories, Inc. (UL))

Figure H35-12 Canadian Standards Assoc. (CSA) seal. (By permission of the Canadian Standards Association, 178 Rexdale Blvd., Etobicoke, Ontario M9W 1R3)

REVIEW

- Combustion is a chemical process by which oxygen is combined rapidly with a fuel to release energy in the form of heat and sometimes light also.
- Combustion requires fuel, heat, and oxygen.
- In complete combustion, all carbon and hydrogen are oxidized into H_2O and CO_2.
- The ultimate CO_2 is the highest percentage of CO_2 that can be obtained in the flue gas with complete combustion.
- There are two modes of combustion, continuous and pulse. Pulse combustion is used for certain high-efficiency furnaces.
- The four most common energy sources for heating are natural gas, LP gas, oil, and electricity.
- The heating value for the gases is given in Btu/ft^3, for oil in Btu/gal, and for electricity in Btu/W.
- In order to obtain the highest combustion efficiency, excess air needs to be supplied.
- There are two types of flames, yellow and blue. The difference is mainly due to the manner in which air is mixed with the fuel.
- Since the beginning of 1992, all new furnaces manufactured must produce an efficiency rating of 78% AFUE or higher.
- Combustion is a major cause of air pollution. The technician can be effective in maintaining healthy indoor air quality in a number of ways: regulating combustion devices, insuring adequate ventilation, and maintenance of system components for operational efficiencies.
- Condensation and corrosion—The technician should always be aware of the possibility of heat-exchanger corrosion. Common evidence of this is the flue pipe between the furnace and the chimney.
- Certification of heating equipment—The AGA establishes the minimum construction safety and performance standards for gas-heating equipment. Oil-heating and electric-heating products are subject to examination and approval by the UL. Manufacturers who sell their products in Canada seek CSA approval as well.

Problems and Questions

1. Define combustion.
2. What are the three requirements for combustion?
3. What controls the rate of combustion?
4. What chemicals in the flue indicate incomplete combustion?
5. What is measured in the flue to indicate the efficiency of the combustion process?
6. What are the two modes of combustion?
7. Why must by-products of combustion be vented?
8. What is the heating value of No. 2 fuel oil?
9. How much is the heating effect of a furnace burning No. 2 oil, burning 2 gallons of fuel in the first half hour of start-up?
10. Using electric heat, how many Btu/hr will be generated by a 25 kW heater? How much No. 2 fuel oil would generate the same heat?
11. Conventional heating fuels contain primarily hydrogen and carbon. True or false?
12. Combustion requires fuel, heat, and fire. True or false?
13. The measurement of CO_2 in the flue gas is an indicator of the efficiency of the combustion process. True or false?
14. Combustion can be achieved by continuous or intermittent pulsing. True or false?
15. Natural gas is mostly methane. True or false?
16. Incomplete combustion is caused by:
    a. Insufficient fuel and air mixing.
    b. Insufficient air supply to the flame.
    c. Flame impingement on a cold surface and insufficient flame temperature.
    d. All of the above.
17. Natural gas has a heating value of:
    a. 950 to 1,150 Btu/ft^3.
    b. 2,500 to 3,200 Btu/ft^3.
    c. 3,500 to 3,800 Btu/ft^3.
    d. None of the above.
18. Number _____ grade fuel oil is considered premium quality and priced accordingly.
    a. 1
    b. 2
    c. 3
    d. None of the above.

19. Electric resistance heat is:
    a. 95% efficient.
    b. 100% efficient.
    c. Based on 4.3 Btu/W of power.
    d. All of the above.

20. The AFUE rating:
    a. Certifies the annual fuel utilization efficiency.
    b. Was developed by the National Institute of Standards and Technology.
    c. Is determined by using the ANSI/ASHRAE Standard 103-1988.
    d. All of the above.

FIGURE **R14-16**

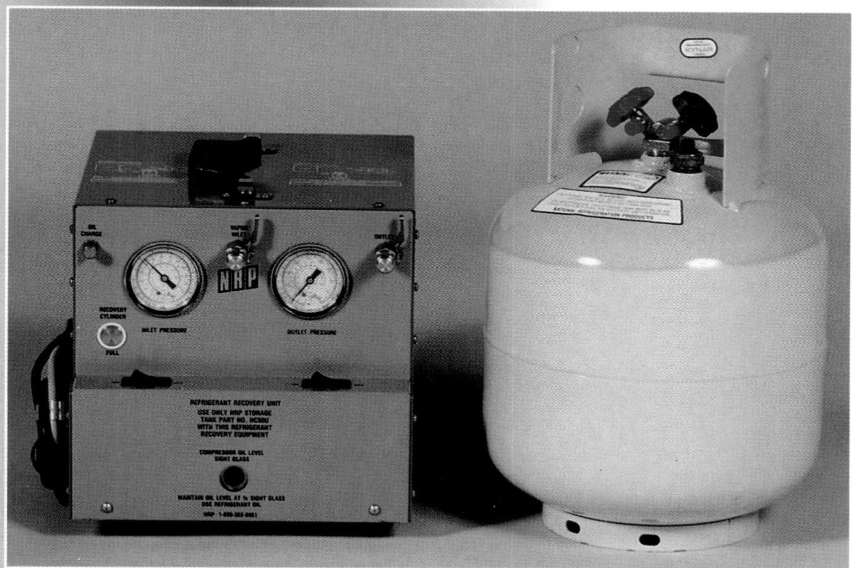

Typical recovery unit. (Courtesy of National Refrigeration Products, Bensalem, PA 19020. Model LV1CUL.)

FIGURE **A31-47**

Installation of a series of tanks using the solid-ice brine-coil system.

FIGURE **R6-50**

Digital micron gauge with 9V battery or AC.
(Courtesy of TIF Instruments, Inc.)

FIGURE **R6-69**

Insulation-resistance tester.
(Courtesy of TIF Instruments, Inc.)

FIGURE **R6-28**

Clamp-on digital wattmeter used for single phase, split phase or 3-phase.
(Courtesy of TIF Instruments, Inc.)

FIGURE **R6-21**

Digital multimeter with auto ranging.
(Courtesy of Amprobe)

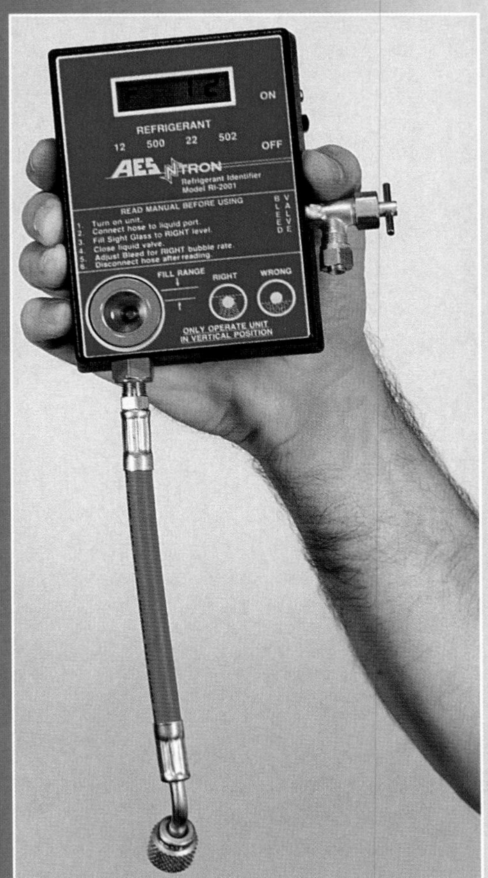

FIGURE **R6-66**

Refrigerant identifier for R-12, R-22, R-500, and R-502.
(Courtesy of AES-NTRON, Inc.)

FIGURE **R6-33**

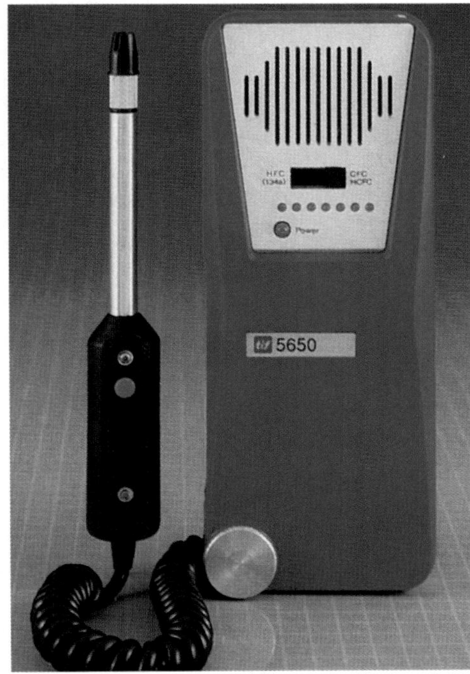

Pump style electronic leak detector.
(Courtesy of TIF Instruments, Inc.)

FIGURE **R6-56**

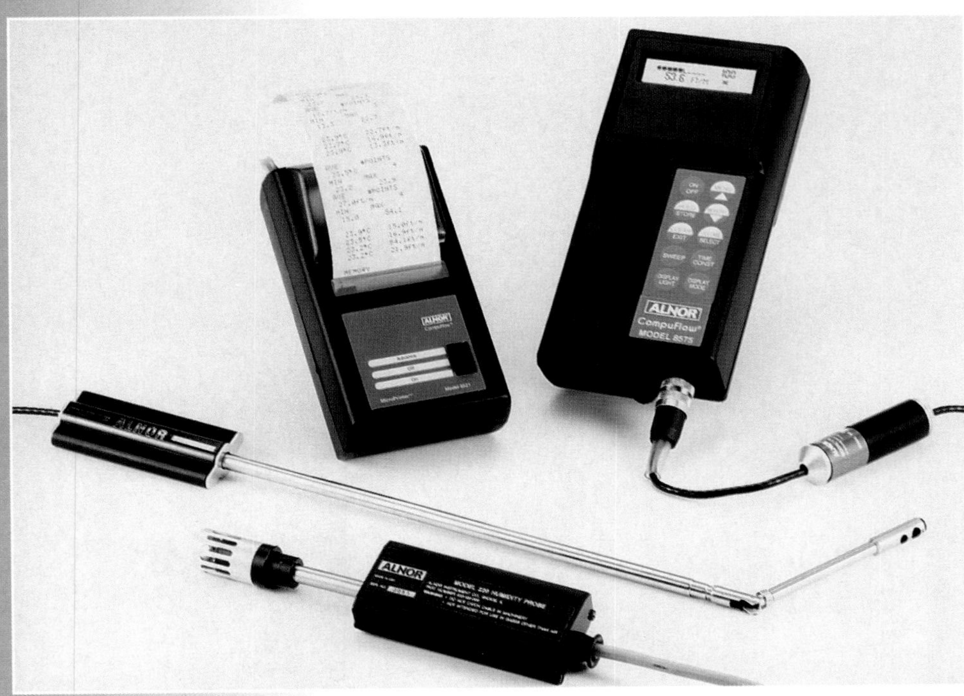

Hand held thermal anemometer, battery powered.
(Courtesy of Alnor Instruments Company)

FIGURE **A27-06**

Bolted semi-hermetic compressor.

Partial cutaway of a scroll compressor.
(Courtesy of Copeland Corporation)

FIGURE **R11-44**

FIGURE **R11-13**

Cutaway view of open type industrial duty compressor.
(Courtesy of Vilter Manufacturing Corp.)

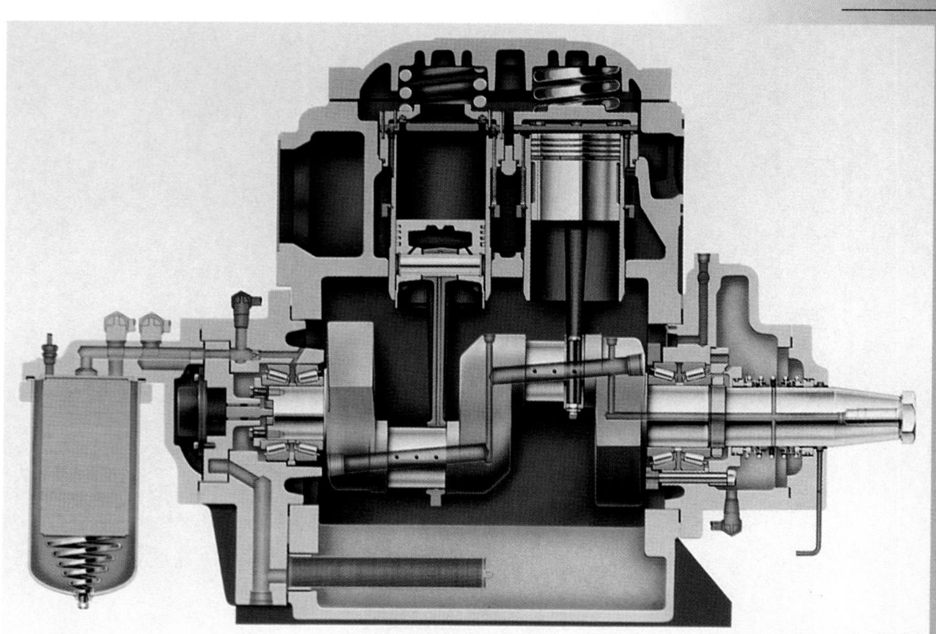

FIGURE **R11-50**

Tandem scroll compressor used for commercial air conditioning application. (Courtesy of Copeland Corporation)

FIGURE **H36-14**

Internal view of a York high-efficiency gas furnace.

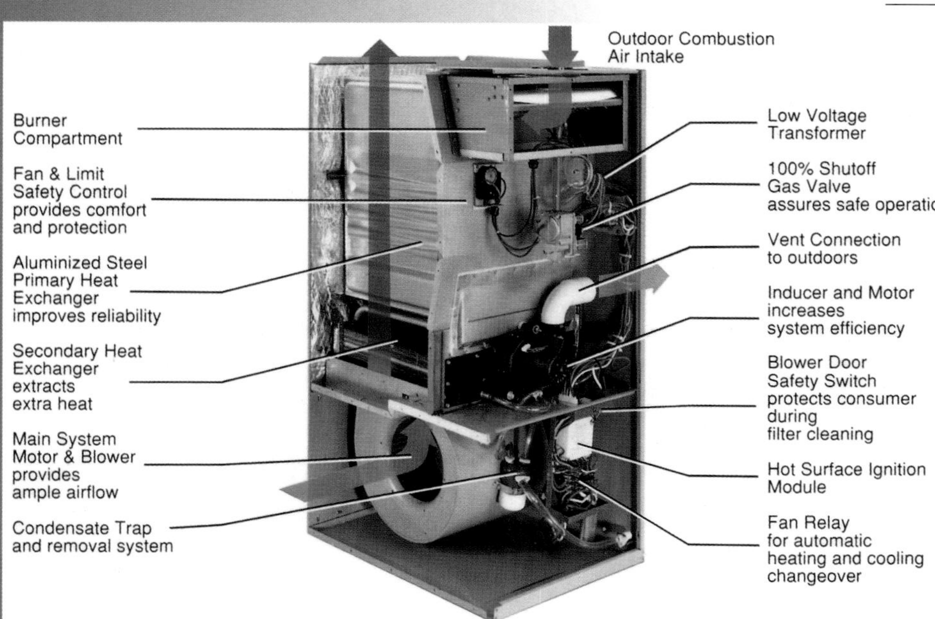

FIGURE **CR25-19**

Cutaway view of the evaporator of a flake-ice machine.

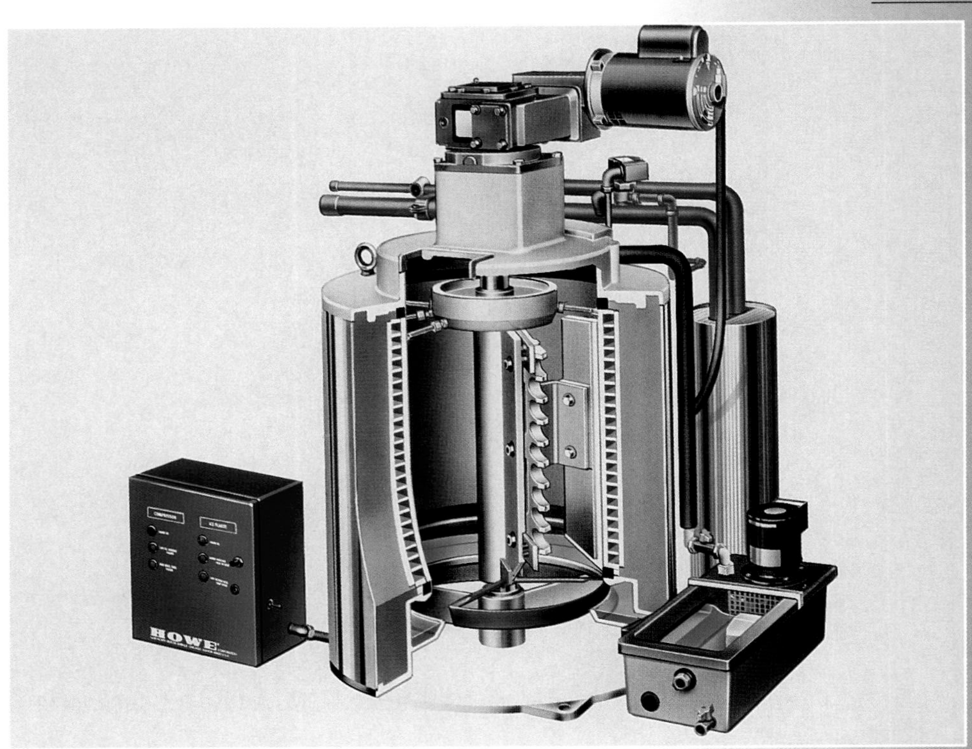

FIGURE **CR25-43**

Ammonia refrigeration system.

FIGURE **CR25-26**

Multi-deck dairy merchandiser.

FIGURE **CR25-24**

Meat and deli merchandiser cabinet.

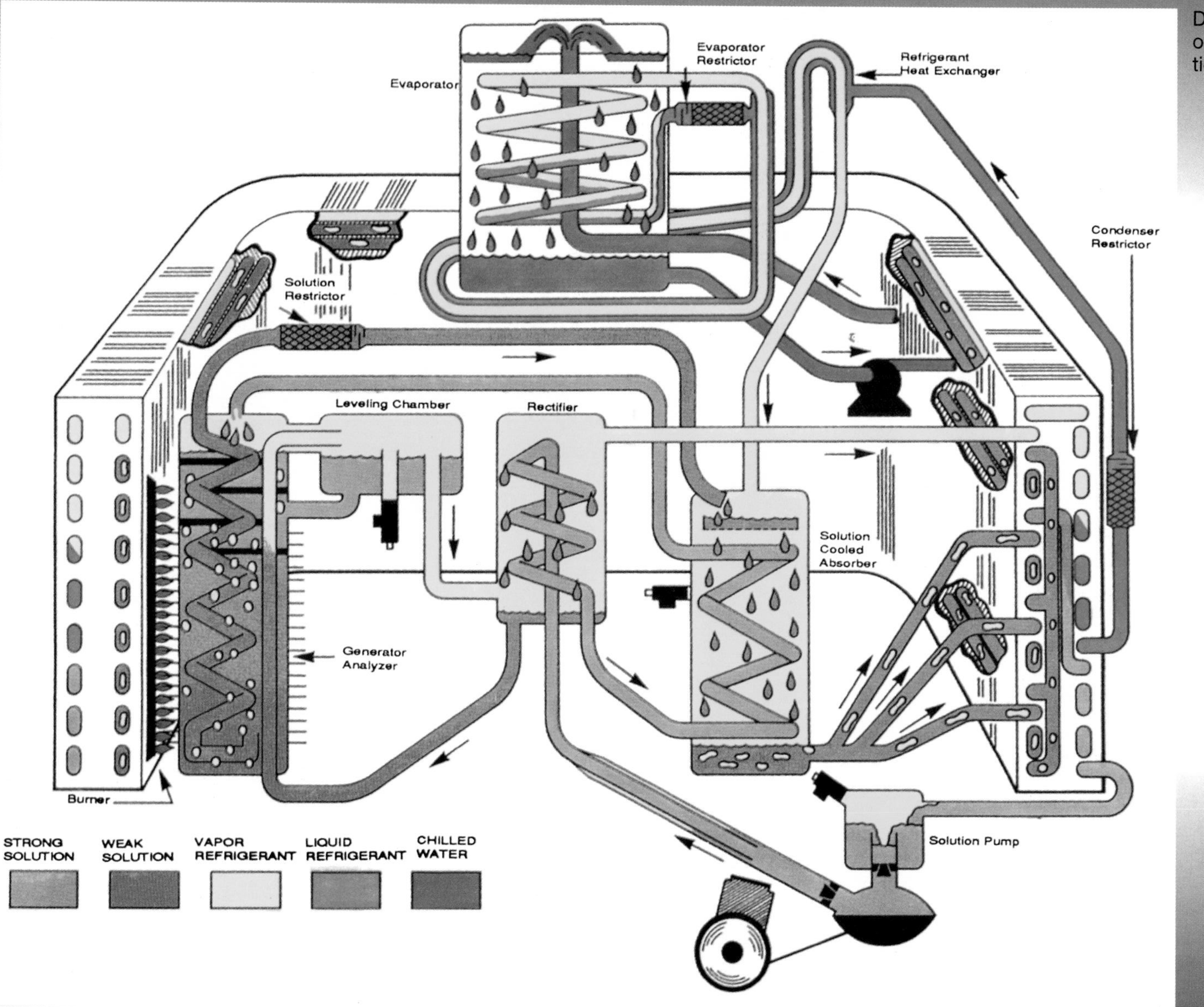

FIGURE **A33-13**

Detailed diagram of the operation of an absorption cycle.

Controls

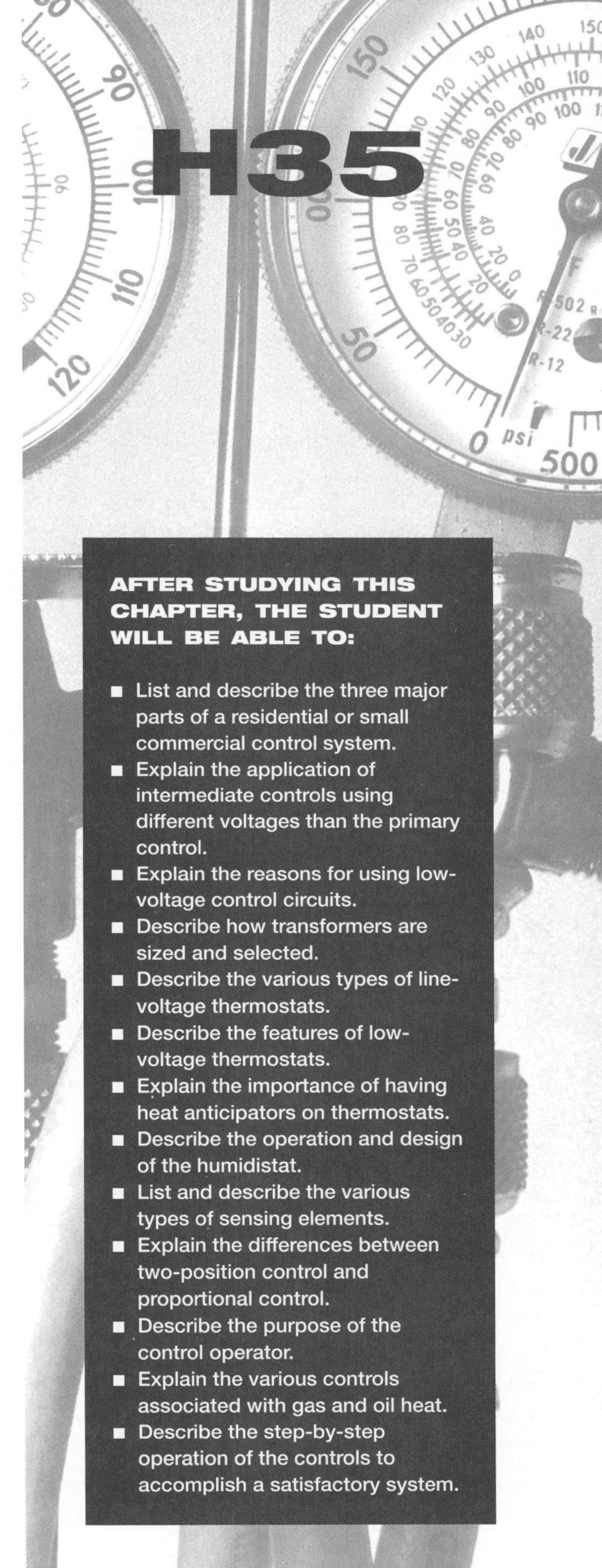

H35

AFTER STUDYING THIS CHAPTER, THE STUDENT WILL BE ABLE TO:

- List and describe the three major parts of a residential or small commercial control system.
- Explain the application of intermediate controls using different voltages than the primary control.
- Explain the reasons for using low-voltage control circuits.
- Describe how transformers are sized and selected.
- Describe the various types of line-voltage thermostats.
- Describe the features of low-voltage thermostats.
- Explain the importance of having heat anticipators on thermostats.
- Describe the operation and design of the humidistat.
- List and describe the various types of sensing elements.
- Explain the differences between two-position control and proportional control.
- Describe the purpose of the control operator.
- Explain the various controls associated with gas and oil heat.
- Describe the step-by-step operation of the controls to accomplish a satisfactory system.

H35-1 OVERVIEW

The development of *controls* and control systems has gone hand in hand with the development of overall heating and air-conditioning equipment. Particular controls have been developed to answer the need for one or more factors, including operation, safety, personal convenience, and economy of the equipment.

Very simply, a control system checks or regulates, within prescribed limits, the functions of an HVAC/R system. Such a system consists of three major parts:

1. A source of power to operate the control system,
2. A load or loads to utilize the power, to obtain the desired results, and
3. Controllers to obtain the desired levels of the end results.

A typical control system for a year-round air-conditioning system using gas for winter heating and refrigerated air for summer conditions is shown in Fig. H35-1. A power-type humidifier is included for winter and an electronic air cleaner for year-round air filtration. A multispeed blower motor in the heating (air-handling) unit is used to provide the best results in both the heating and cooling phases.

The operation of the system is under the master control of a room thermostat which controls a gas valve in the heating phase and the condensing unit's operation in the cooling phase. The blower in the gas furnace must operate intermittently in the heating phase, depending on the furnace supply-plenum temperature. In the cooling phase the blower operates continuously or with the operation of the condensing unit. A low-voltage room thermostat is used to control a high-voltage blower motor (120-V) and a higher voltage condensing unit (240-V).

When a device of one voltage is controlled by a device of another voltage, intermediate controls are required. These could be relays, contactors, or motor starters, depending on the load characteristics encountered.

In this control circuit, the power source is a transformer which is part of a plate-mounted relay/transformer assembly. The other component of the assembly is a relay containing two sets of single-pole, double-throw contacts. One set of contacts controls the speed of the blower motor

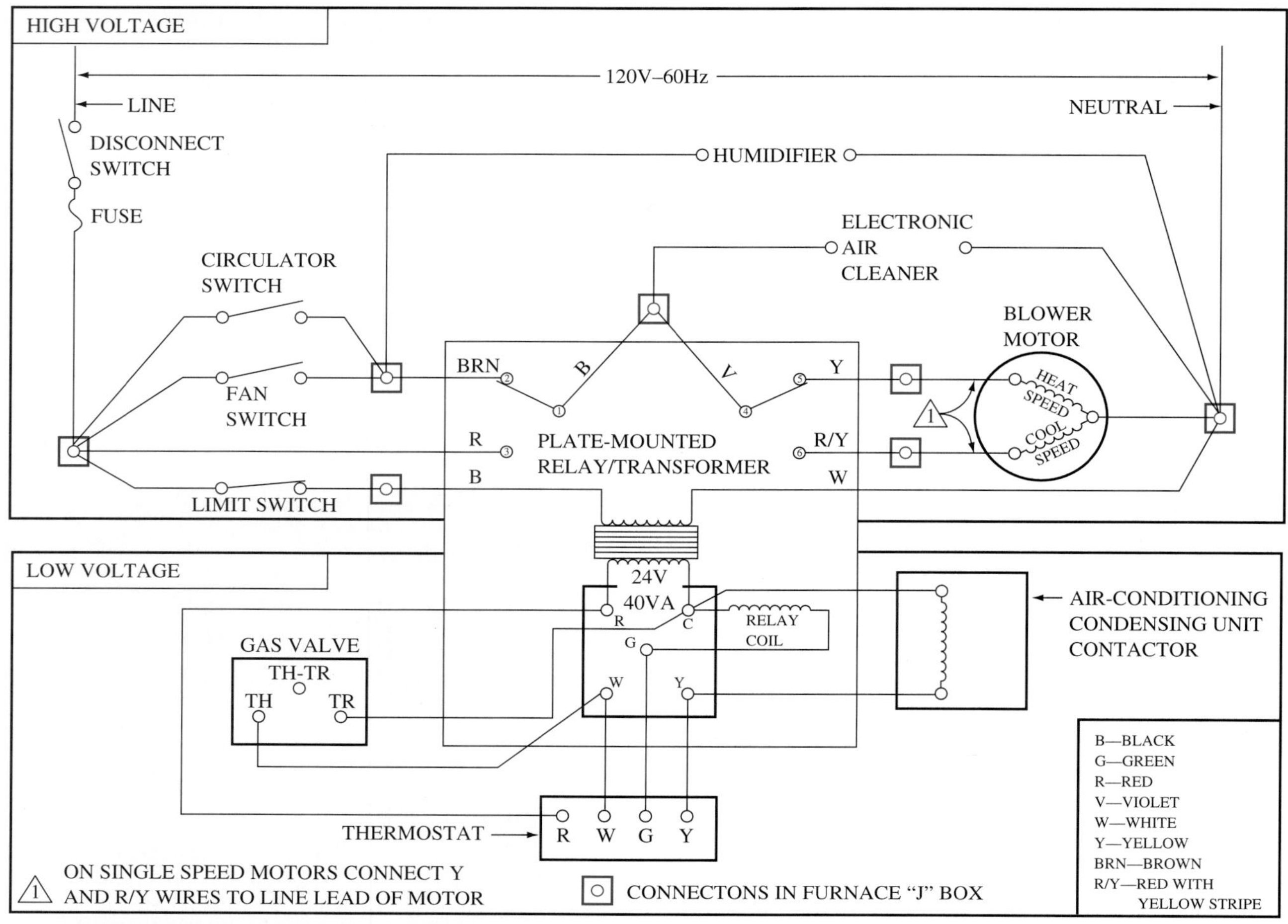

Figure H35-1 Typical control system for an air-conditioning system using gas heating and electric cooling.

PRACTICING PROFESSIONAL SERVICES

Many modern heat pumps and air conditioners are coming equipped from the factory with electronic boards to control specific functions. Whenever working on these units, it is necessary to get the service manual or technical sheet from the manufacturer's supply house. To help simplify the troubleshooting at the board level, identify four areas on the board: the input, output, board ground, and power supply. The input is probably an on/off signal from a thermostat or variable signal from a thermistor which indicates a change has occurred. The output is the signal coming out of the board to activate a relay or controlled device. Most solid-state circuit boards rely on direct current to operate the electronic components, so it must be properly grounded. The power supply required is most likely low voltage, either 24 V_{AC} or 5 V_{DC}. If all four of these areas on the board are being provided as specified from the factory, there is a good chance the board has failed. It is not normally repaired in the field and should be replaced. Prior to replacement, however, verify there are no electrical shorts or wiring problems which may have caused premature failure.

PREVENTIVE MAINTENANCE

Good preventive maintenance includes annual start-up and calibration on gas- and oil-heat equipment prior to the heating season. The units should be tested and verified during normal operation to determine the operating and safety controls are working as designed. This includes checking air filters, oiling motors, and testing the plenum switches. On gas equipment, specific inspections should include pilot ignition, startup, and flame condition. On oil equipment, inspections include replacing oil filters, strainers, checking electrodes, and cleaning the cad cell. It is also a good idea to open the cad-cell circuit or put black electrical tape over it to verify the flame failure control will shut down the burner during a no flame condition.

while the other set controls the operation of the blower motor—either under the control of the furnace fan switch or directly for cooling operation.

The master control is the room thermostat for both the heating and cooling operation. Intermediate controls consist of the fan switch and limit switch for heating operation and a circulator switch for continuous blower operation with humidification in the heating season.

The loads in this control circuit consist of the gas valve for field-input control to the heating unit, the condensing-unit contactor for cooling operation, the blower relay coil for control of the blower motor, and the blower motor for air circulation.

H35-2 CONTROL SYSTEM COMPONENTS

H35-2.1 Power Source

While the source of power to operate a control system is usually electricity, engineered systems also use electronic and pneumatic systems or a combination of all three. Here we will concentrate on electricity as the power source for residential and light commercial controls.

Heating and cooling control circuits can be designed to operate on either line voltage (115 to 120-V) or on low voltage, which is designated as a 24-V system. A low-voltage control circuit is superior to a line-voltage circuit because (1) the wiring is simplified and safer, and (2) low-voltage thermostats provide closer temperature control than do line-voltage thermostats.

A step-down or low-voltage transformer (Fig. H35-2) is used in heating and air-conditioning control systems to reduce line voltage to operate the control components. Inside a simple step-down transformer are two unconnected coils of insulated wire wound around a common iron core (Fig. H35-3). To go from 120-V (primary) to 24-V (secondary), there are five primary turns to one secondary turn. For a 240-V primary the ratio would be 10:1, etc. This, the *induction ratio,* is a direct proportion. Step-up transformers would be just the reverse.

In reducing the voltage, is any energy lost? No, because the value of the current on the secondary side would be five times higher than on the primary side, so that the power on both sides of the transformer remains the same—assuming a transformer that is 100% efficient. Thus, the ratio can be expressed by the formula:

volts × amperes = volts × amperes
(primary) (secondary)

or

$$V_p \times A_p = V_s \times A_s$$

EXAMPLE

A given transformer has a primary voltage of 120 V and a secondary voltage of 24 V. Assuming that the transformer is rated at 40 VA, what is the primary current and the secondary current?

Solution

Primary (amps) = 40 VA/120 V = 0.333A
Secondary (amps) = 40 VA/24 V = 1.67A

Therefore, using the formula:

$$120\text{ V} \times 0.0334\text{ A} = 24\text{ V} \times 1.67\text{ A}$$
$$40\text{ VA} = 40\text{ VA}$$

Transformers are available in a variety of voltages and capacities. The capacity refers to the amount of electrical current expressed in volt-amperes. A transformer for a control circuit must have a capacity rating sufficient to handle the current (amperage) requirements of the loads connected to the secondary. Twenty V-A transformers are usually found only in heating forced-air systems. Heavier V-A ratings are needed for air conditioning because electrical devices containing a coil and iron, such as solenoid valves and relays, have a power factor of approximately 50%. Thus, for secondary circuits with such controls, the capacity of a transformer must be equal to or greater than twice the total name-plate wattages of the connected loads.

Figure H35-2 AC voltage transformer for low-voltage controls. (Courtesy of Honeywell, Inc.)

The proper transformer rating for the electric control circuits will have been selected by the equipment manufacturer. If accessory equipment is added, however, the additional power draw must be considered and might possibly result in an increased rating. This situation is common, for example, when cooling is added to an existing furnace and where the original transformer is too small. When replacing a defective transformer, make sure that the rating is equal to or greater than the original equipment.

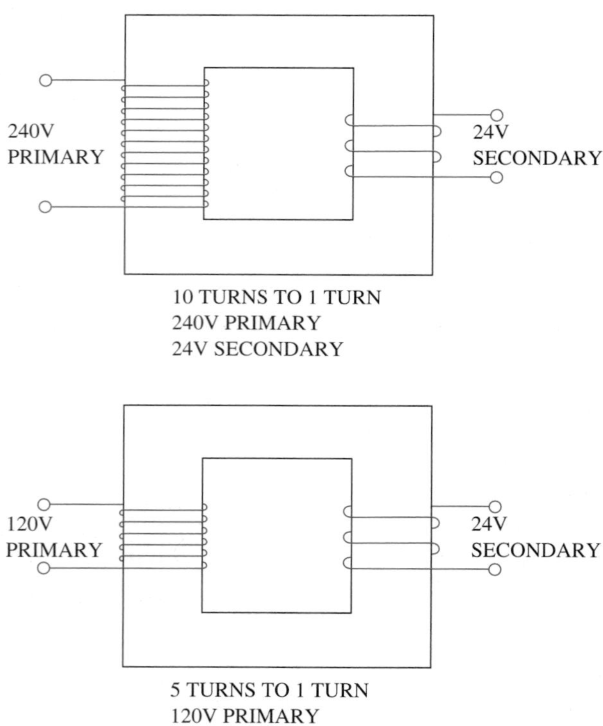

Figure H35-3 Construction of a step-down transformer.

H35-2.2 Thermostats

Line-voltage Stats

While the low-voltage thermostat is much more accurate than *line-voltage stats,* the line-voltage stat is used in direct electrical-resistance heating, for wall, baseboard, room air conditioners, etc.

Early typical line-voltage stats, sometimes called snap-action thermostats, consisted of bimetallic temperature-sensing elements (Fig. H35-4) made of two or more metallic alloys welded together, each having different coefficients of expansion when exposed to heat. One metal will expand more rapidly than the other, causing the bimetallic element to change its curvature with changes in temperature. The bimetallic element may be a straight form (cantilever), U-shaped, or spiral. Line voltage usually flows through the bimetallic element which has a moving contact that closes against a fixed contact point.

The distance *(h)* the moving contact travels is the differential range between off and on, and in a simple, low-cost line stat a change in room temperature of 35° or more may be required to make the moving contact travel the required distance. If a knob is attached to the fixed-contact screw, the temperature-sensing range can be adjusted with a fairly narrow range. The function of the permanent magnet is to snap the contacts closed when the moving contact comes within range, to help reduce the arcing of contacts.

This type of control can be used for heating or cooling, depending on whether the contact is made on a rise or a drop in room temperature. Dual-contact models can perform both functions.

Snap-action line voltage thermostats are satisfactory for limited control situations where close temperature swings are not economically justified. Also, line-voltage stats can suffer from dirty contacts or pitting of the contacts due to arcing when making or breaking contact. The control of a snap-action stat is also a delayed reaction. This means that by the time the bimetallic element has traveled its range and the heating or cooling equipment is shut down, the actual room temperature will overshoot the desired space temperature. The anticipation method (to be discussed later) is a solution to the above, but it is harder to build into high-voltage controls than into low-voltage designs.

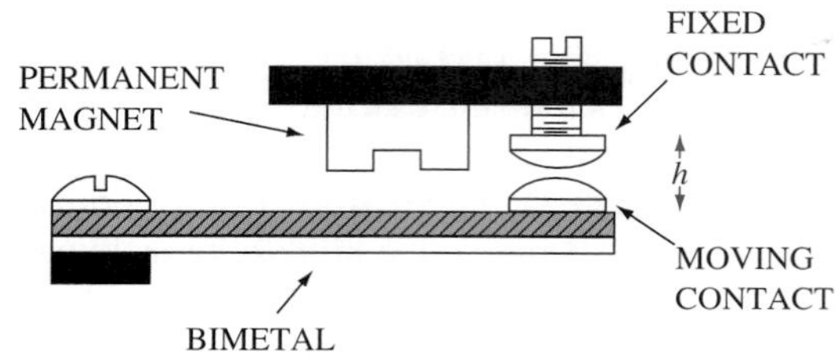

Figure H35-4 Snap-action thermostat.

Newer line-voltage thermostats for the direct control of electric heating make use of liquid-filled elements that respond to both ambient temperature and radiant heat. Also, the design includes slower-moving elements, which can handle direct resistance loads of up to 5000 W at 240 V_{AC}. The differential between the set point and room conditions is less than that of the snap action variety.

From an installation viewpoint, wall-mounted line voltage thermostats must be served by heavy-duty wiring. On commercial work, codes may require that they be run in conduit. Also, where many wires are involved, identification coding is difficult. Therefore, line-voltage stats are almost totally limited to space heaters and the room-and-store type of self-contained air-conditioning products.

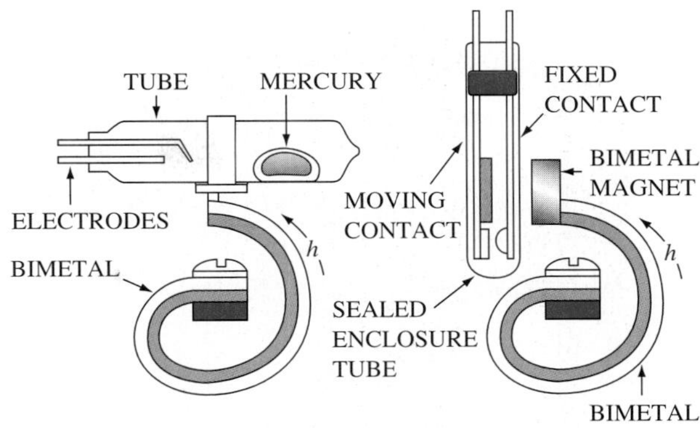

Figure H35-6 A single-action mercury-bulb thermometer on the left. A sealed tube with metal-to-metal contact, on the right.

Low-voltage Thermostats

This type of stat, as shown in Fig. H35-5, overcomes the limitations of line-voltage models and is almost universally used in modern central-system control circuits. First, the use of spiral-shaped, lightweight bimetallic elements increases the effective length and thus the sensitivity to temperature change. Second, the use of sealed contacts eliminates the problem of dirt and dust. Although there are minor variations among different manufacturers, the contacts are always sealed in a glass tube.

A single-action mercury bulb design is shown in Fig. H35-6. As the bimetallic strip expands and curves, the mercury fluid moves to the left, completing an electric circuit between the two electrodes which carry only 24 V. The differential gap between OFF and ON is very small: ¾ to 1°F from the set point. On the right is a sealed tube that has a metal-to-metal contact. The magnet provides the force that closes the contacts.

Combination heating and cooling thermostats may have two sets of contacts: two single-pole, single-throw mercury tubes, or one single-pole, double-throw mercury tube, to provide control of both the heating and cooling systems.

The subbase portion of the thermostat assembly (Fig. H35-7) not only provides a mounting base for the thermostat, but is also used to control the system operation through a series of electrical switches. The system (SYS) switch selects COOL, OFF, or HEAT. The blower operation is controlled by the fan switch, which is a simple two-position switch. When set in automatic, the fan will cycle with the furnace on heating. If in the ON position, the fan will run continuously.

More complex thermostats contain two dials that establish different control points for heating and cooling, with an automatic changeover from heating to cooling. Fig. H35-8 represents a two-stage heating and two-stage cooling schematic wiring diagram. Two-stage thermostats are common in heat pumps or rooftop equipment which use multiple compressors for cooling and two or more stages of heating. Note that seven electrical connections are required; however, M and V terminals are the same power source, so six wires are needed. With color-coded low-voltage wire, however, it is no problem to connect the thermostat to the mechanical equipment.

H35-2.3 Heat Anticipation

The sensitivity of room thermostats is affected by both system lag and operating differential. *System lag* is the

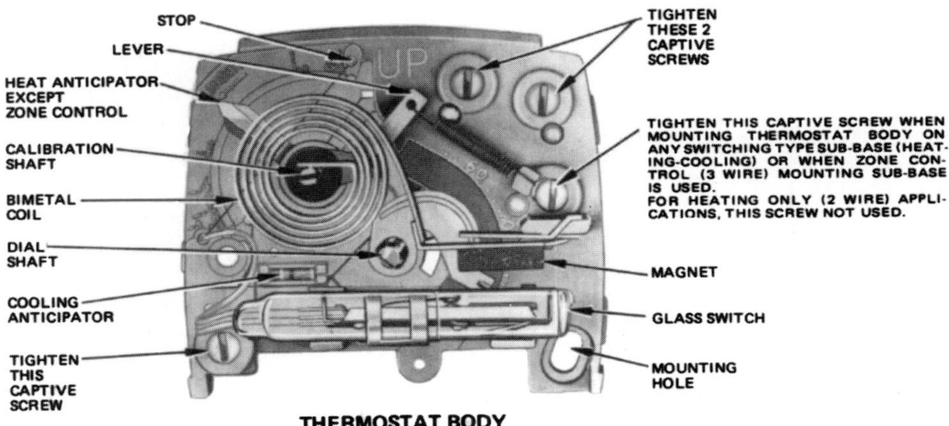

Figure H35-5 Low-voltage thermostat used in central-system control circuits. (Courtesy of Grayson Division of Robertshaw Controls)

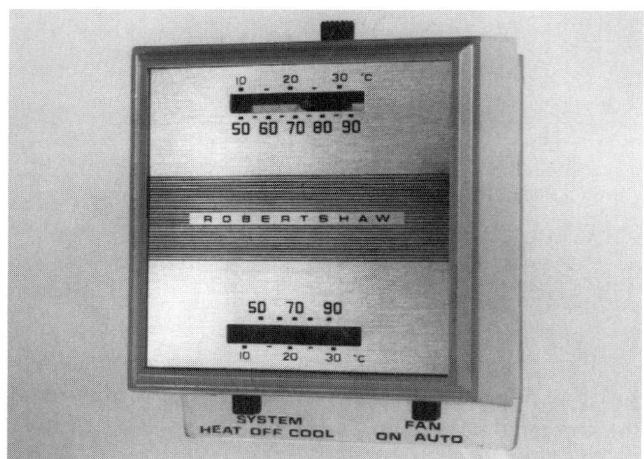

Figure H35-7 Subbase portion provides a mounting base for the thermostat and for system controls. (Courtesy of Grayson Division of Robertshaw Controls)

amount of time required for the heating or cooling system to produce a temperature change that is felt at the thermostat. The operating differential of a thermostat is the change in room-air temperature needed to open or close the thermostat contacts.

Heat anticipators are used in low-voltage thermostats to reduce the effects of the system-lag operating differential. They are simply small electrical resistors that generate a small amount of heat when current flows through. If the bimetallic strip in the thermostat is exposed to a small amount of "artificial" heat, the sensitivity of the thermostat will increase, because the bimetallic element reacts to changes in surrounding air temperature and will lead to actual changes in room temperature. If, for example, the room-air temperature increases a certain number of degrees in 5 minutes, the anticipated bimetallic element will feel this change in only 3 minutes, while the unanticipated bimetallic strip will feel the change of air temperature after 6 minutes.

The anticipated bimetallic strip will shut the furnace burner off before the actual room temperature reaches the thermostat set point. Anticipation results in closer room-

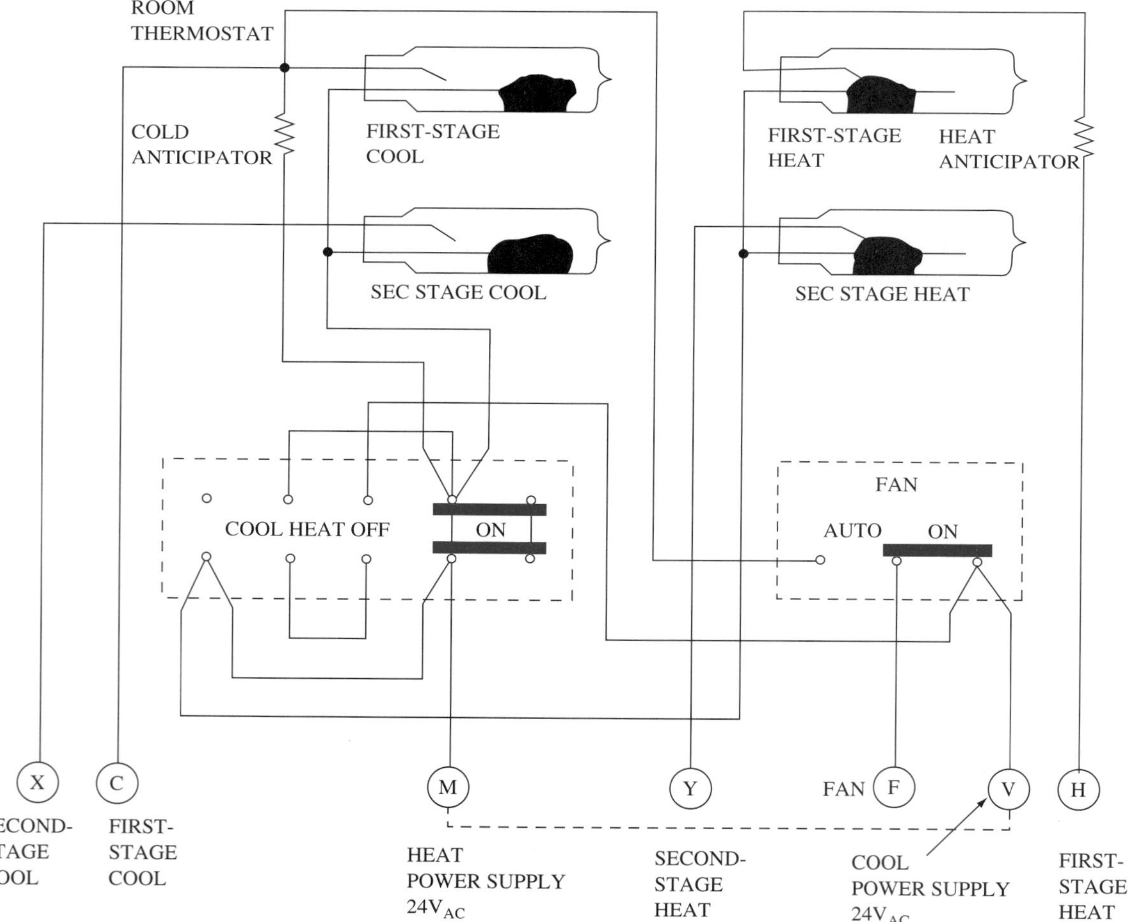

Figure H35-8 A two-stage heating and two-stage cooling schematic wiring diagram.

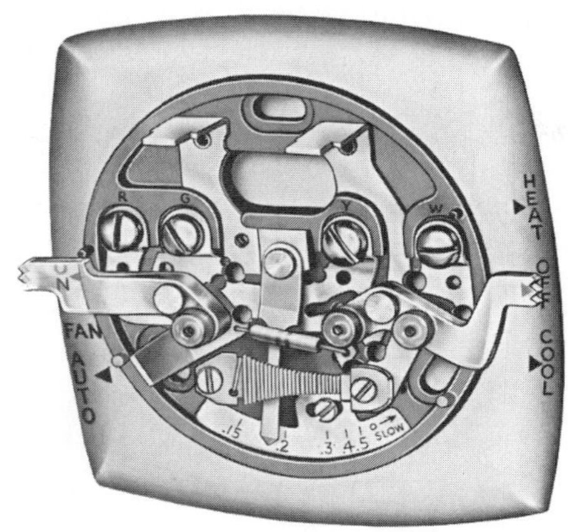

Figure H35-9 An adjustable heating anticipator.
(Photo courtesy of Johnson Controls/PENN)

temperature control and less overshooting of heating due to residual heat in the furnace heat exchanger that must be dissipated.

Some heating anticipators are fixed, while others are wire-wound variable resistors wired in series with the load. They are rated in amperes or fractions thereof and must be matched to the total load controlled by the thermostat. On the adjustable type, as illustrated in Fig. H35-9, the installer will position the sliding arm to the proper load rating. If on/off cycles are too long or too short, the system operation can be changed to give a faster or slower response.

H35-2.4 Cooling Anticipation

A similar technique is used for cold anticipation except that the cold anticipator is energized only when the cooling is off. The bimetallic element senses the anticipator heat and turns the system on again before the room temperature gets too warm; thus, it leads room air temperature.

H35-2.5 Humidistats

Another type of controller found in residential comfort applications is the wall-mounted *humidistat* (Fig. H35-10). It is very similar to the low-voltage thermostat, and it contains both a sensing element and a low-voltage electric switch.

The sensing element consists of either an exceptionally thin moisture-sensitive nylon ribbon or strands of human hair that react to changes in humidity. The movement of the sensing element is sufficient to make and break electrical contacts directly or when coupled with a mercury-type switch. Humidistats for mounting on duct work are also available.

A simple schematic wiring diagram for humidity control is shown in Fig. H35-11. In winter the elements contract and close the contact to the humidifier. In summer the humidistat expands in response to a rise in the relative humidity and closes the contacts to the dehumidifier (cooling unit) circuit. A summer/winter changeover switch selects the appropriate operation.

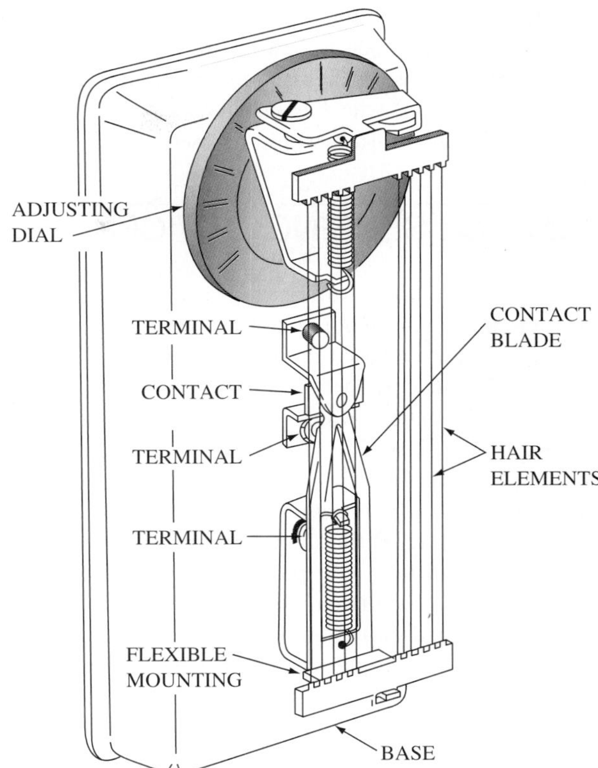

Figure H35-10 Construction of a wall-mounted humidistat.

The humidistat is not as precise a controller as a standard low-voltage room thermostat. Normally, a change of 5% relative humidity is required to actuate switching action. This is acceptable, however, for normal comfort applications since people cannot detect or react to changes

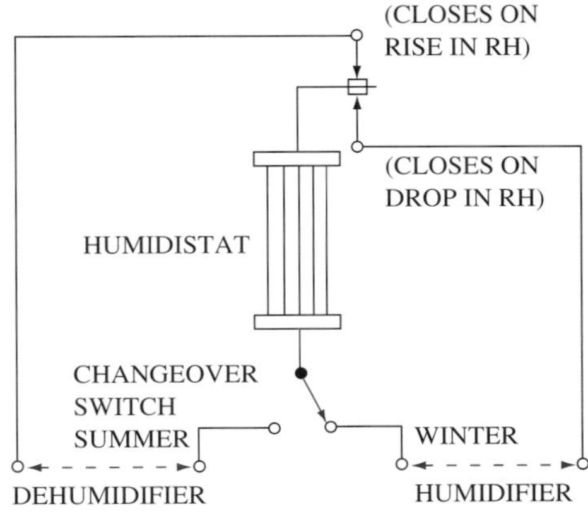

Figure H35-11 A simple schematic wiring diagram for year-round control of humidity.

in relative humidity with the same sensitivity as room temperature. More precise controls are available for specialized applications, such as computer rooms, libraries, and printing plants, where controllers such as wet-bulb and dew-point thermostats would be used.

H35-3 SENSING DEVICES

In addition to controls, various types of sensing devices react to temperature and/or pressure. Bimetallic temperature controls are available in various forms, as shown in Fig. H35-12. There are metal actuators that produce rotary movement, elongation, warping, bending, or snap action when exposed to heat. The mechanical movement is relatively large in response to small changes in temperature.

Bellows and *diaphragms* (Fig. H35-13) react to changes in pressure. Because of a relatively large internal volume, the bellows will produce greater mechanical movement, but the diaphragm is more accurate. The choice will depend on the amount of movement or sensitivity and the working pressures that must be contained.

The bellows and/or diaphragm can also be adapted to control temperature by the addition of a remote-bulb sensing element (Fig. H35-14) filled with a vapor, a vapor/liquid, or a solid/liquid. Heat will cause the fill to expand into the bellows or diaphragm cavity, thus creating mechanical movement.

Figure H35-12 Bimetal temperature controls or actuators.

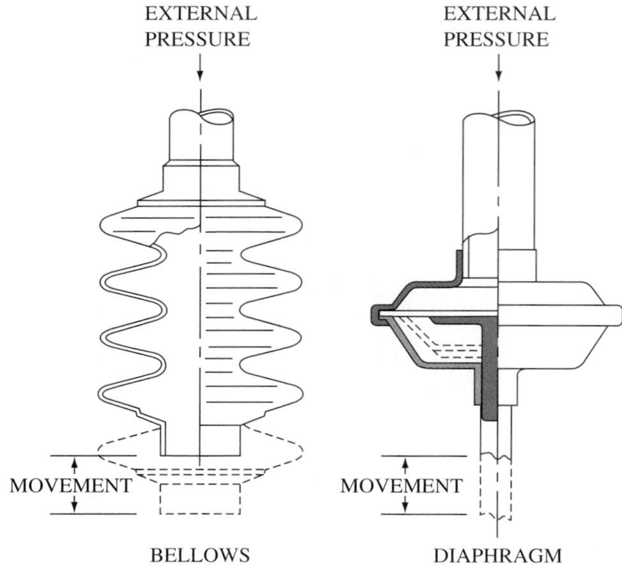

Figure H35-13 Bellows and diaphragm actuators.

The *Bourdon tube* (Fig. H35-15) is another sensing element that reacts to changes in pressure. It is an elliptical tube that is sealed and linked to a transducer. As pressure is applied to the pressure connection, the tube will tend to straighten and thus create a mechanical movement. Pressure can come from refrigerant, water, oil, steam, or any fluid that can be pressurized. The Bourdon tube may also be adapted to temperature sensing with a closed bulb filled with liquid or vapor. Bourdon tubes are used most extensively in gauges.

The *rod-and-tube sensing element* is constructed by placing an inert rod in an active tube (Fig. H35-16). Since the rod is anchored to one end of the tube, a temperature

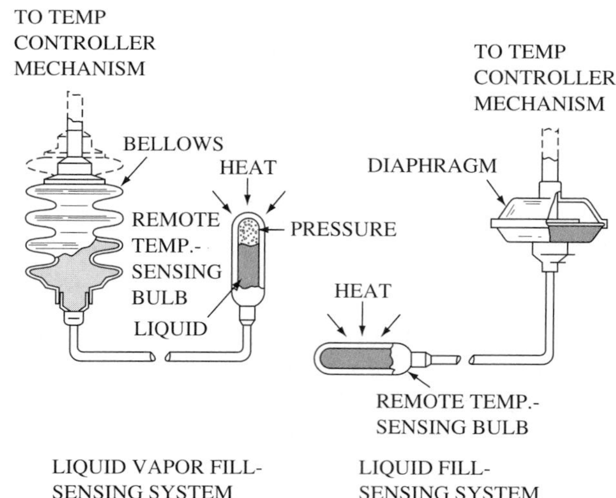

Figure H35-14 A remote-bulb sensing element for bellows and diaphragm.

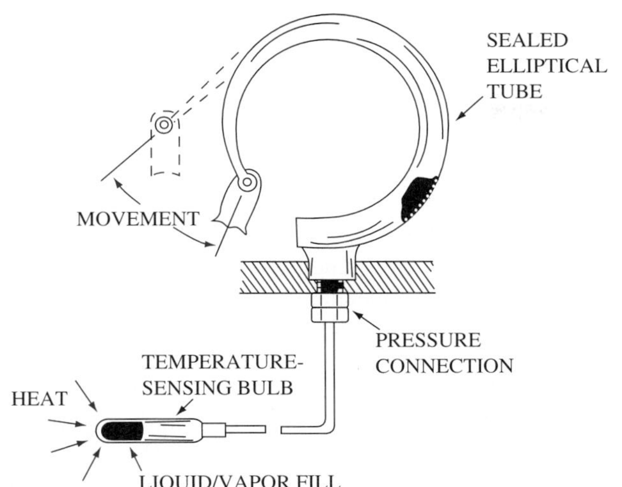

Figure H35-15 Bourdon-tube sensing element used in gauges.

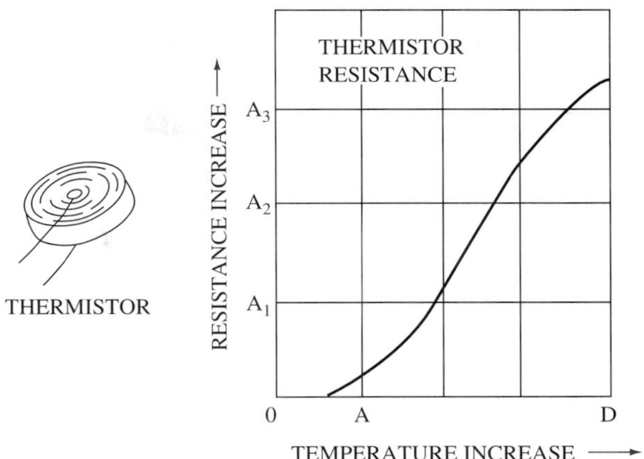

Figure H35-17 A thermistor increases its resistance as temperature increases.

change will cause a pushing or pulling movement that can be made to operate a switch.

Resistance elements are sensing elements designed to vary their electrical resistance when exposed to changes in temperature or light intensity. Thermistors, photoresistance cells, and photocells are typical. A *thermistor* is an electrical device that will increase its resistance to the current as its temperature increases (Fig. H35-17). Changes in current flow can be used to activate or deactivate a remote control for various uses.

The cadmium-sulfide cell (CdS or "cad" cell) or *photoresistance cell,* as shown in Fig. H35-18, is a resistance element that reacts to changes in light intensity and is highly resistive to the passage of electrical current when its sensing base is in complete darkness. When exposed to light, the electrical resistance declines proportionally to the intensity of the light.

Some of the more common controls that use the principle of these sensing elements are described in the section on residential and light commercial equipment.

In the area of safety, the high-limit thermal-cutout control is widely used (Fig. H35-19). It employs a snap-disk bimetallic strip to open an electrical circuit. These controls are common in electric space heaters, electric duct

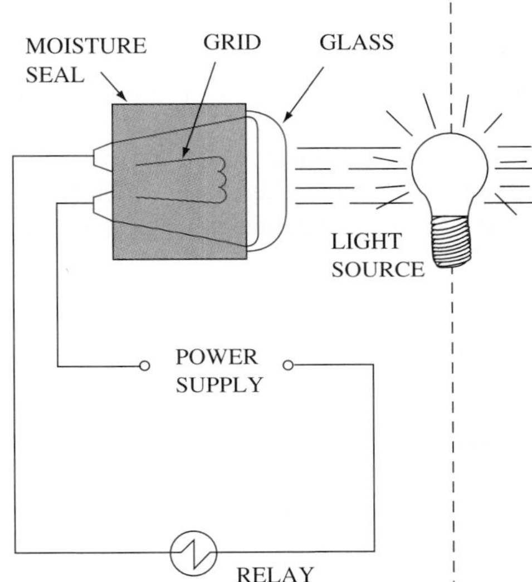

Figure H35-18 Cadmium-sulfide or photoresistance cell reacts to changes in light.

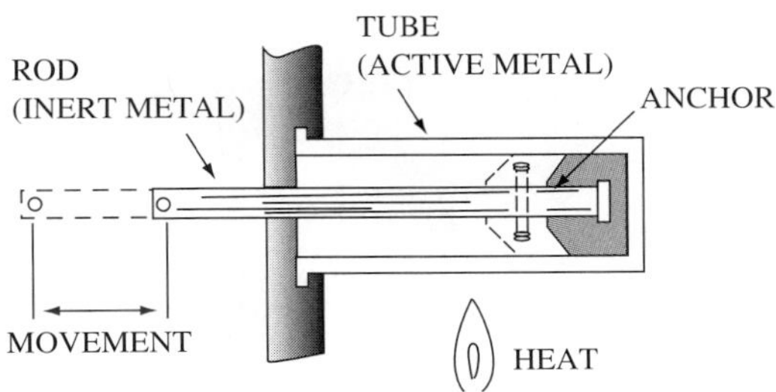

Figure H35-16 Rod-and-tube sensor construction.

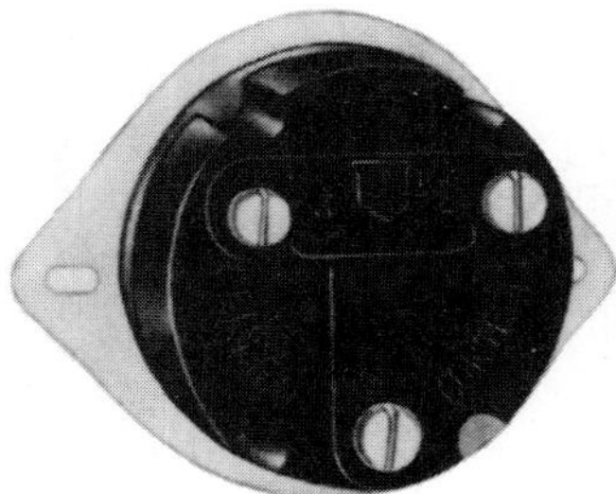

Figure H35-19 The high-limit thermal cutout control is used for safety in electrical appliances.

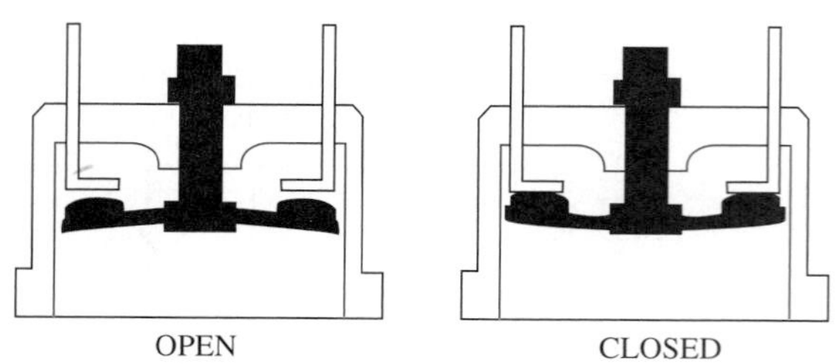

Figure H35-20 A current- and heat-sensing overload device used to protect hermetic compressors.

heaters, electric furnaces, heat pumps, and any application where the small, quarter-sized control can be mounted and exposed to heat sources.

The inherent protector used in hermetic compressors is an excellent example of bimetallic action. The overload (Fig. H35-20) is a current- and heat-sensing device and consists of a bimetallic contact in series with the motor-contactor coil-control circuit. Whenever the compressor motor is overheated, overloaded, or stalled, the heavy continuous current through the heater warps the bimetallic contact until it opens the motor-contactor low-voltage-relay-coil circuit and stops the flow of current to the compressor.

Another form of high-limit cutout, used with baseboard electric heaters, is the *lineal control,* which has a sealed vapor/liquid fill tube extending the entire length of the heater. If the heater is blocked by drapes, etc., the safety limit will open.

One type of *combination fan-and-limit control* for a typical gas or oil warm-air furnace (Fig. H35-21) uses the power created by the rotary movement of a helix bimetal. The rotating cam makes or breaks the separate fan-and-limit electrical contacts.

The combination fan-and-limit switch illustrated in Fig. H35-22 has a dual function. As a safety limit, some have fixed-limit temperature settings; others are adjustable (approximately 180 to 200°F is the usual range). This allows a 50 to 60°F rise above normal operation before it opens. The fan control switch is also a temperature-sensing device that is set to turn on the fan after the furnace has warmed up at least 10 to 15°F above room conditions, so that cold drafts are not experienced. It also stops the blower after the burner cuts off, so again there are no uncomfortable drafts. It is important to note that some systems employ constant fan operation and thus override this switch.

Other models may use spiral, flat bimetallic, or even liquid-filled elements. Some forms of duct-mounted limit controls use a rod and tube or a liquid-filled bulb to sense the air conditions.

A common control in hydronic heating is the *immersion controller,* which uses a liquid-filled bulb that operates a snap-acting switch (Fig. H35-23). Immersion controls are inserted directly into the boiler to detect water

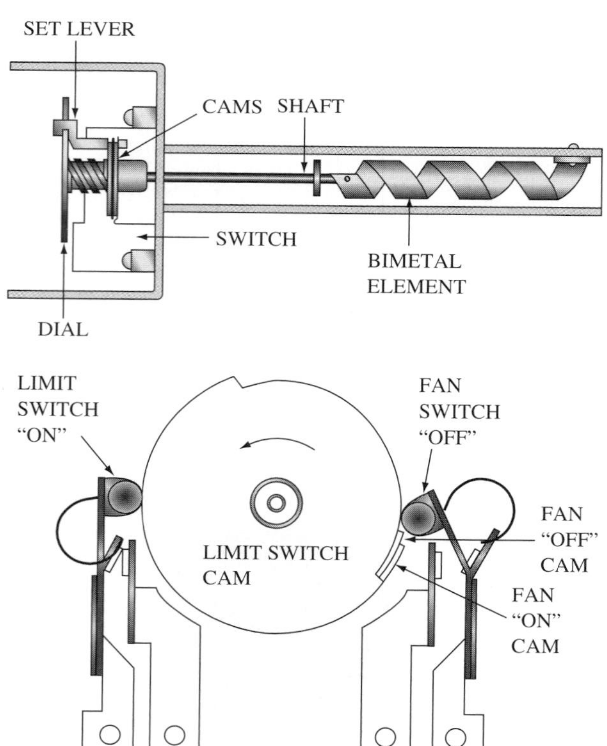

Figure H35-21 Construction of combination fan-and-limit control for gas or oil warm-air furnace. Shows how twisting motion of heated helix activates switches.

Figure H35-22 Combination fan-and-limit switch acts as a safety limit and temperature-sensing device. (Courtesy of Honeywell, Inc.)

Figure H35-23 Immersion controller uses a liquid-filled bulb to operate snap-action switch. (Courtesy of Honeywell, Inc.)

temperature and can be used as high limit cutouts and as low limit and circulator pump controls.

A prime example of the thermistor-sensing device is the *high-temperature cutout* on a hermetic compressor (Fig. H35-24). The thermistor, not much larger than an aspirin tablet, is embedded deep in the motor windings. It

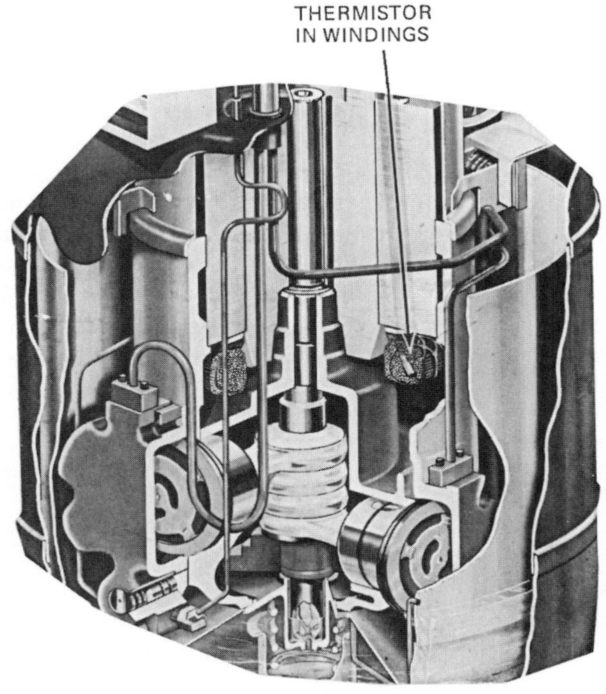

Figure H35-24 High-temperature cutout on a hermetic compressor is an example of the thermistor sensing device.

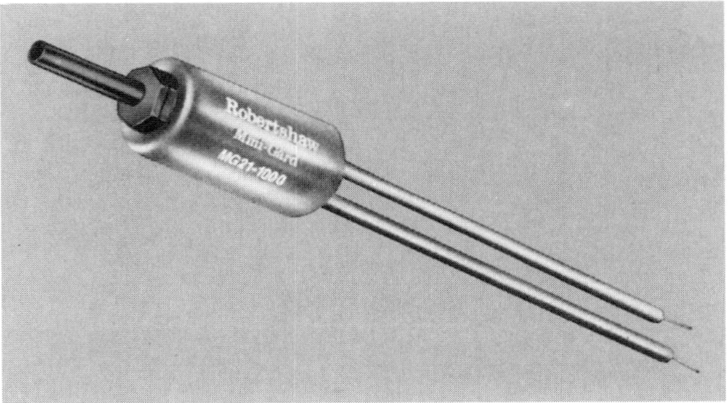

Figure H35-25 High-pressure cutout with fixed setting uses a diaphragm to sense refrigeration pressure. (Source: Robertshaw Controls)

senses overheating and will signal a transducer to shut down the compressor until normal temperatures are restored.

High-pressure and low-pressure controls for compressor and system operation (Fig. H35-25) utilize a diaphragm to sense the refrigeration pressure and react accordingly through a switch mechanism to shut down the operation if abnormal conditions exist. High-pressure cutout controls frequently must be manually reset in order for the system to operate. This prevents rapid short cycling which can eventually cause compressor overheating and possible damage.

Pressure controls can also be used on large compressors to detect the buildup of oil pressure. They are equipped with a time-delay mechanism (usually 1 minute or less), and if normal oil pressure is not achieved within the time period, the controller will stop the compressor before damage can occur. It can also be connected to an alarm indicator.

The *diaphragm-type of air-pressure switch* (Fig. H35-26) is used in heat pumps to start the defrost cycle. In that case they are used for measuring the differential pressure across the outdoor coil. These devices can also be used to measure positive pressure, as in a central-system air duct, and energize an alarm circuit when a change in air pressure is sensed.

Static-pressure regulators (diaphragm device) are also air controllers that are used in larger systems to measure static pressure, and then, through an electric circuit to a damper motor, to regulate the air damper and maintain a

Figure H35-26 Diaphragm-type of air-pressure switch.

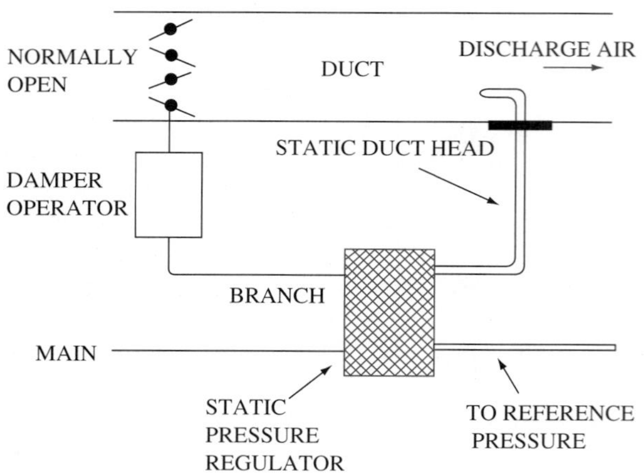

Figure H35-27 Schematic diagram of a pneumatic air-damper control, used in large systems to control static pressure.

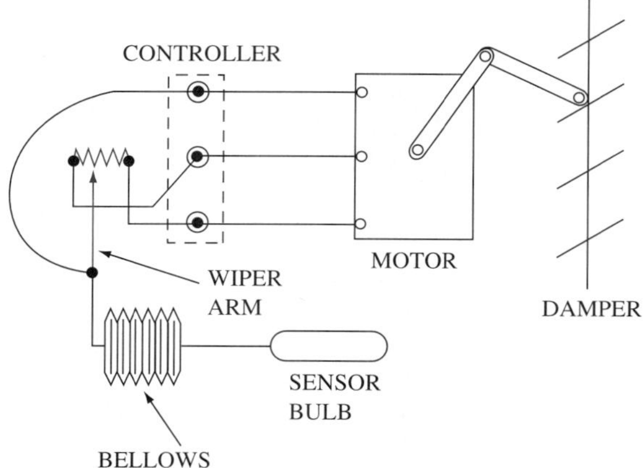

Figure H35-29 Diagram of potentiometer action showing movement of the damper in response to the sensed condition.

constant static pressure in the system as illustrated in Fig. H35-27. A static-pressure regulator can also be connected to a control that varies the central-fan inlet air vane to maintain constant fan performance.

A different form of sensing element not previously described but frequently used in residential and light commercial work is the *sail switch* (Fig. H35-28). As the name implies, it detects the movement of air by the use of a sail or paddle attached to an arm; the arm, in turn, is attached to an electric switch. The illustration is of a combination duct-mounted humidistat and sail switch. The humidistat mounts in the return-air duct of a forced-air system. The function of the sail switch is to shut down the humidifier when the supply air is cut off. Sail switches are also used as safety devices to detect airflow or its absence and cut off heating and/or cooling before abnormal system operation occurs.

Figure H35-28 Sail switch detects movement of air by use of a sail or paddle attached to arm of a switch. (Courtesy of Honeywell, Inc.)

Most of the controllers reviewed above are of the two-position variety: that is, they are OFF or ON, OPEN or CLOSED. But even within the scope of residential and light commercial work there is a need for modulating controls to vary an action somewhere in between fully open and closed, or OFF and ON. The *proportional controller* provides this. Proportional or modulating action is achieved by using a potentiometer to vary the position of the controlled device in proportion to any temperature (or pressure or flow) variation felt by the sensing element. A potentiometer (Fig. H35-29) is an adjustable three-wire variable resistor in conjunction with a wiper arm connected to the sensor. The wiper arm is the third wire contact. As the arm moves through the complete stroke (throttling range) the current flow changes in proportion to the resistance. This current flow is sent to a motor relay that actuates a damper motor or motorized valve to bring about the required temperature change in the controlled space.

H35-4 OPERATORS

Operators, both intermediate and final, are on the receiving end of the controller signal. An *intermediate operator* is one that controls another electrical circuit. A *final operator* will control or act on a product or final action.

The *electric relay* (Fig. H35-30) is the best example of an intermediate operator. Its function is to take a low-voltage signal from a controller, and through a magnetic

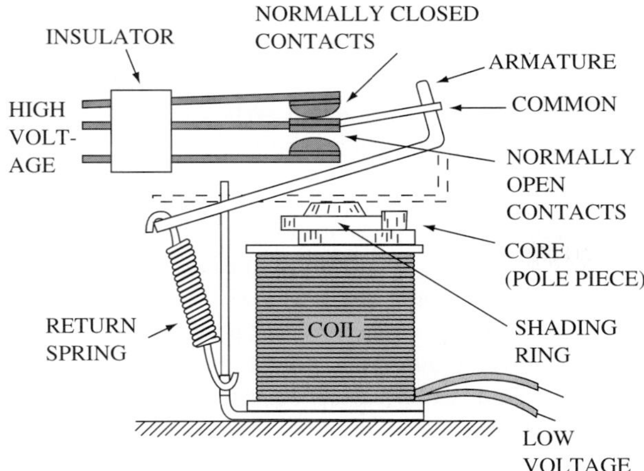

Figure H35-30 Electric relay showing how low-voltage signal from the controller through magnetic coil controls a high-voltage load such as a fan or compressor.

coil, close a set of higher-voltage contacts. These contacts in turn will start a fan or compressor or perhaps only supply high voltage to another control circuit. The coil which sets up the magnetic field is installed around a pole piece. The armature moves in response to coil energization to open or close the electrical contact. Return springs are used to pull the armature to the de-energized position.

Relays are used to control electrical loads that draw less than 20 amps, such as small motors, valves, ignition transformers, and small compressors. A contactor is a heavy-duty relay designed to handle over 20 amps, and in most cases is used on compressor and pump motors. The volt-amp draw of the commonly used relay and contactor coils is low; therefore, low-voltage transformers can supply enough power to operate these devices. A starter is a form of contactor or relay that contains overload protection.

Figure H35-31 Damper motor, responds to a command from a controller and operates a damper, valve or step controller. (Photo courtesy of Johnson Controls/PENN)

To complete the relay story, it is important to mention briefly a second type of electric relay: the *thermal relay,* which uses a resistance heater and a bimetallic element instead of a magnetic coil and switch. When the low-voltage circuit is energized, perhaps by thermostatic action, the low-voltage heater warms the bimetallic strip. This warps the strip and closes a switch on the high-voltage side. The chief advantage of this relay is its quiet action, which makes it well suited to low-capacity residential types of switching needs.

Sealed plug-in relays have been used extensively in commercial work and are becoming popular in residential use to simplify the addition of humidity control and electronic air cleaning. They look much like a square radio tube and plug into a control panel arrangement. Since the electrical contacts are not exposed to dust and dirt, reliability is greatly improved. Also, the plug-in arrangement permits quick replacement in the event of trouble.

Another form of operator is a motor (Fig. H35-31) that responds to a command of a controller and mechanically operates a damper, valve, or step controller that produces a final action. The crank arm is connected to the final apparatus by means of linkage rods. Motors are available for two-position response and may swing through a 160 to 180° arc from start to stop. Others respond to a modulating controller and operate in an intermediate position to bring the controlled temperature or pressure back to the set point. Motors vary in arc rotation from 15 seconds to 4 or 5 minutes depending on the application. They are also rated according to the torque force they can exert (in inch-pounds or the size in square feet of the damper it can handle).

Some models have a spring return to some normal setting in the event of power failure. Auxiliary switches can be included to perform a variety of other automatic switching combinations for secondary equipment.

The final operators in the example given in the beginning of the chapter are the air-conditioning condensing-unit contactor, the blower switching relay, and the fuel or energy control.

H35-5 FUEL CONTROLS

The fuel controls (gas or oil) and the energy controls (electrical sequences) are classified as *final operators.* Each is designed to control a specific energy source to provide safety as well as the desired results as demanded by the master control—the thermostat.

H35-6 GAS VALVES AND REGULATORS

At one time, the supplying of fuel gas to a heating unit was done by a combination of controls consisting of a

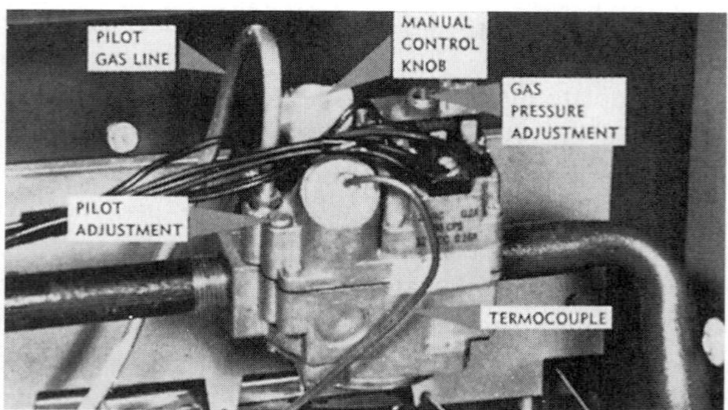

Figure H35-32 Component parts of a combination gas valve.

gas-pressure regulator and a solenoid valve. These controls were combined into a single valve assembly (Fig. H35-32) to meet the exacting standards of the American Gas Association (AGA) for proper ignition, input control, and quiet cutoff of gas-unit operation.

Valves have the following functions:

1. Manual control for ignition and normal operation;
2. Pilot supply, adjustment, and safety shutoff;
3. Pressure regulation of burner gas feed;
4. On/off electric solenoid valve controlled by the room thermostat.

Fig. H35-33 shows a basic sketch of these functions for a natural-gas valve. The schematic drawing of the valve shows the main diaphragm valve in the open condition that occurs during heat demand. When in this condition, the following assumptions can be made:

1. The schematic applies to a gas heating appliance with the pilot flame burning;
2. A thermocouple is connected to the automatic pilot magnet operator;
3. The lighting operation was previously performed to open the automatic pilot valve; and
4. The main gas cock has been turned to the ON position after the pilot is lit.

In the application, the 24-V operator and room thermostat are in series. Closure of the room thermostat switch on heat demand has energized the 24-V operator, which causes the armature to be attached to the pole face of the magnet, and results in a clockwise rotation of the armature as indicated by the arrows at the end of the armature. This rotation has overcome the valve spring and pulled the valve stem of the dual-operator valve downward, allowing the diaphragm above it to seat on the valve seat. The seating of this diaphragm shuts off the bypass porting.

Bleed gas can enter the actuator cavity only through the bleed orifice. Bleed gas is allowed to flow from the actuator cavity through the dual-operator valve and the reg-

UNITROL 7000ER SCHEMATIC

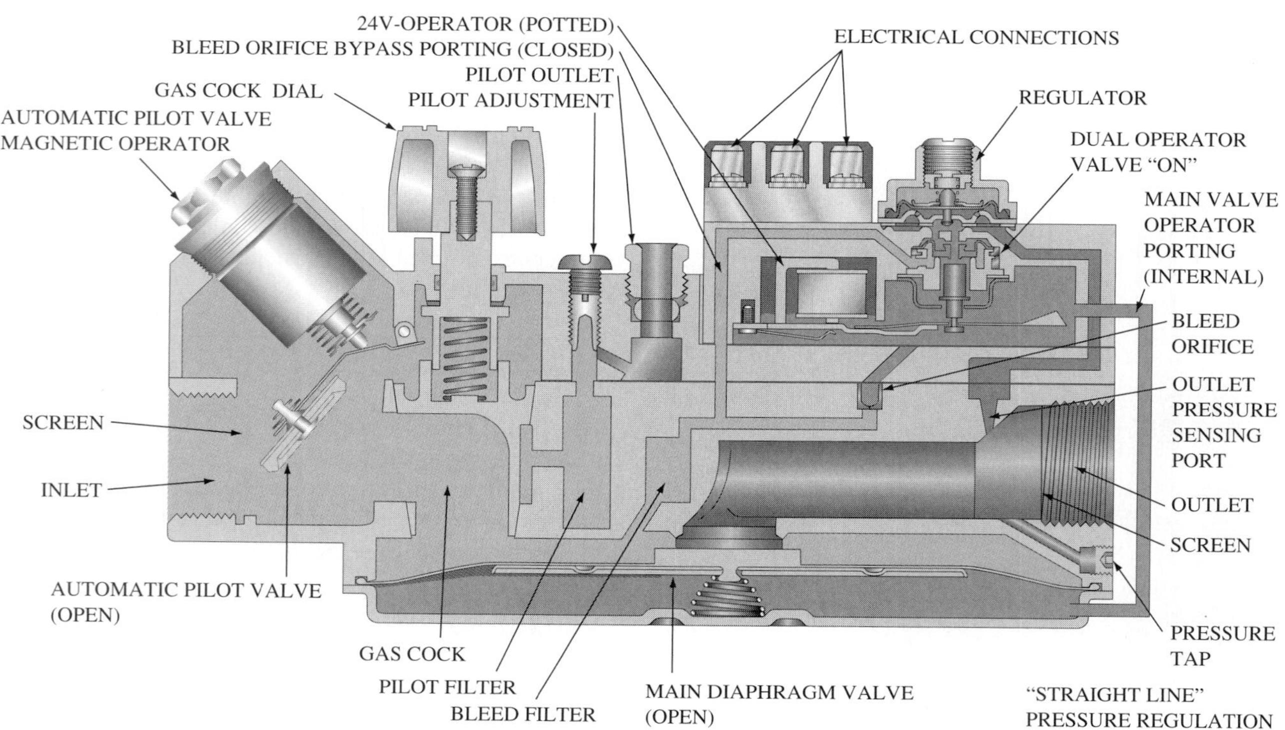

Figure H35-33 Cutaway of a combination gas valve in open position.

ulator to the outlet through the outlet pressure-sensing port. (The valve stem has a square cross section in a circular guide, allowing gas to pass between the stem and the guide.) The resultant drop in pressure within the actuator cavity and through the main diaphragm allows the inlet pressure above the main diaphragm to open the main diaphragm valve spring.

After opening (Fig. H35-33, main diaphragm valve open), straight-line pressure regulation is secured by the feedback of outlet pressure through the outlet-sensing port to the pressure regulator in the bleed line. A rise in pressure at the control outlet above the set pressure causes a proportional closure of the regulator valve in the bleed line. The proportional closure of the regulator valve causes a corresponding rise in pressure in the bleed line ahead of the regulator. This rise in pressure in the actuator cavity increases the pressure beneath the main diaphragm and causes a partial closure of the main diaphragm valve, lowering the control outlet pressure to the pressure setting. Upon a drop in control outlet pressure, a like decrease in bleed-line pressure transmitted to the underside of the main diaphragm through the action of the bleed line pressure regulator causes a proportional increase in the main valve opening to bring the delivered outlet pressure back up to the set pressure.

The schematic of the closed main diaphragm (Fig. H35-34) shows the action of the bypass valve in the dual-operator valve during a fast OFF response independent of the bleed orifice. Upon heating "satisfaction" the thermostat switch opens and the 24-V operator is de-energized. The return spring on the valve stem of the dual-operator valve then forces the stem upward, closing the center part of the small diaphragm above it and shutting off bleed gas to the bleed regulator and outlet-sensing port. The resultant counterclockwise rotation of the armature is indicated by the arrows at the ends of the armature. As the small diaphragm above the valve stem is forced upward, the bleed-orifice bypass porting is opened by the diaphragm, leaving the valve seat. The actuator cavity, main-valve operator port, and cavity beneath the diaphragm are rapidly exposed to full inlet pressure, which acts to close the main diaphragm valve. The bleed-orifice bypass port allows the pressure above and below the main diaphragm to be equalized rapidly, independently of the bleed orifice. This pressure is then equalized and the main valve is closed by the main valve spring.

The dual-operator valve that provides the bleed-orifice bypass porting control for the fast OFF response is an important feature. It helps to prevent flashback conditions. Flashback can cause pilot-outage problems as well as burner and appliance sooting. Relatively slow closing has been a problem on larger capacity applications. The dual-operator valve is an important factor in helping to eliminate this application problem.

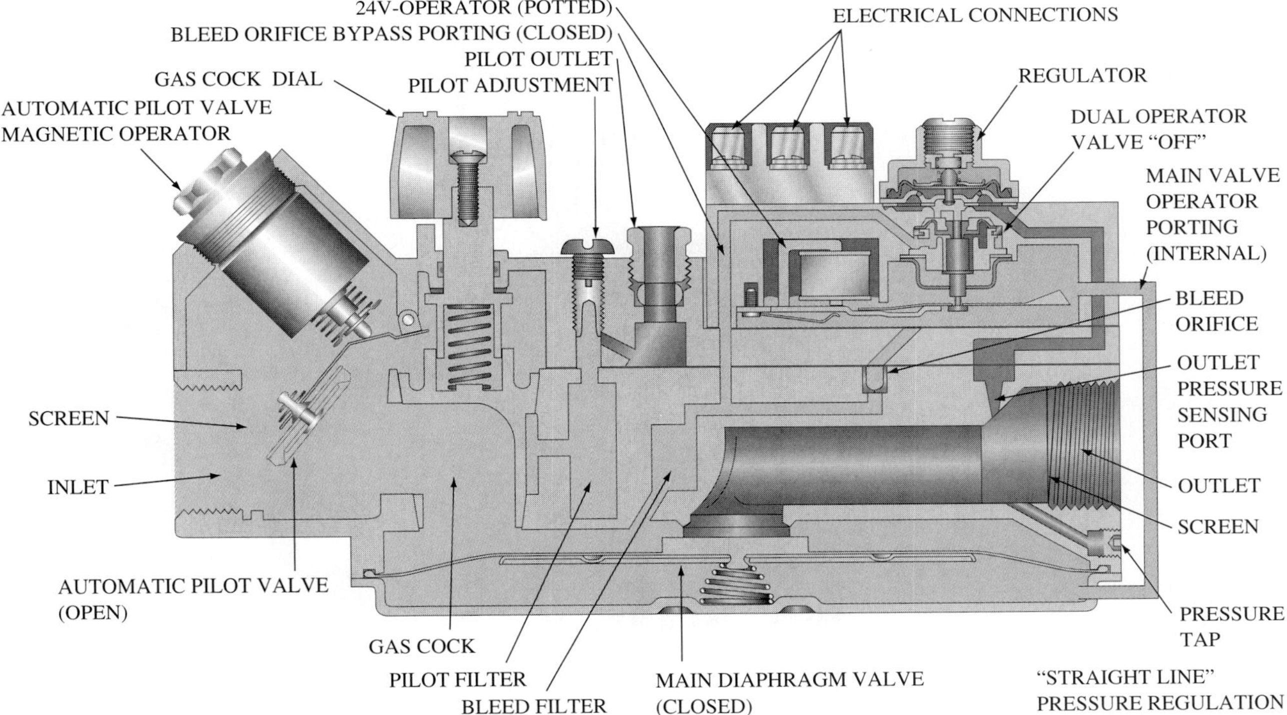

Figure H35-34 Cutaway of a combination gas valve in closed position.

Figure H35-35 Front view of a furnace showing the position of the electronic-ignition system. (Courtesy of Bard Manufacturing Company)

The valve has built-in protection against gas-line contaminants through the use of inlet and outlet screens for the main gas passages, a pilot filter for the pilot line, and a bleed filter for the bleed control line. These means protect the control from malfunction due to the entering of contaminants into control passages. In the highly unlikely case that the bleed-orifice restrictor becomes clogged with the main valve in the open or ON position in spite of the bleed filter protection, the dual-operator valve will enable the control to shut off the main gas valve when so signaled by the room thermostat or limit control.

On an LP furnace, the gas valves do not have the pressure-regulation function, but there is still 100% pilot shutoff in the event of pilot outage, which is most important. LP gas is heavier than air, and over a prolonged period enough pilot gas could collect to be a hazard.

The methods of pilot ignition and gas-valve control described above apply chiefly to the older residential furnaces, potentially requiring service. *Electronic-ignition systems,* developed for commercial systems, have been adapted to the newer residential furnaces and require an entirely different approach to service.

An electronic-ignition system is shown in Fig. H35-35. Pilot gas is fed to the assembly and burner orifice. The spark electrode is positioned to ignite the gas on a signal from the room thermostat. With the pilot ignited and burning, a sensing probe establishes an electric current sufficient to energize a relay that opens the main gas valve by

Figure H35-36 Component parts of an electric spark-ignition package. (Courtesy of White-Rodgers Division of Emerson Electric Company)

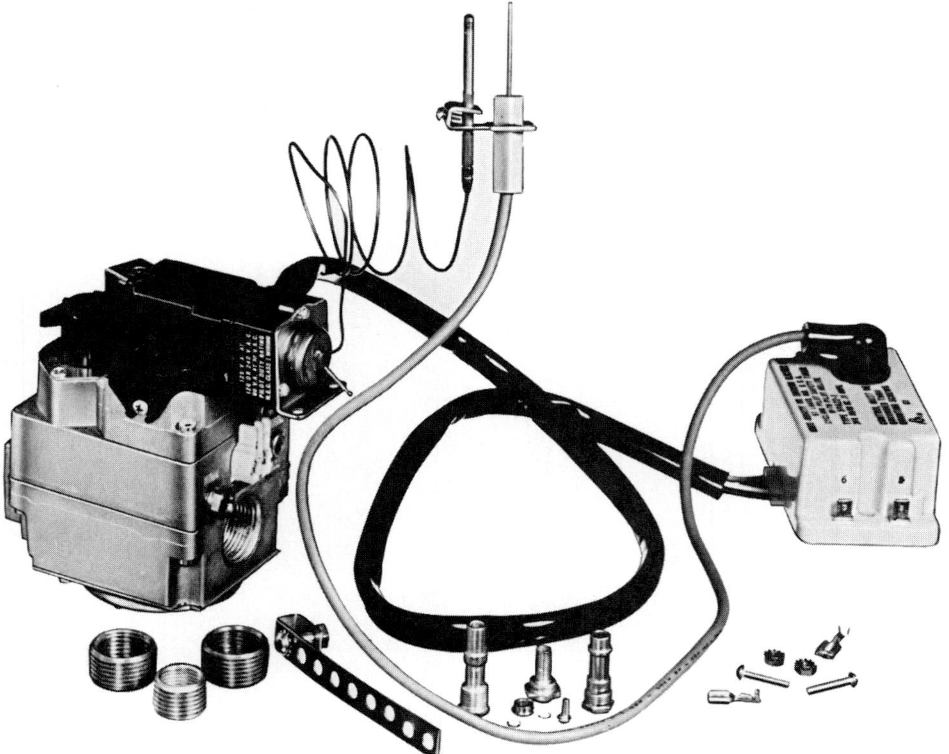

closing normally open electrical contacts. At the same time the relay's normally closed electrical contacts in the spark-ignition circuit open, terminating the spark. As long as the sensing probe recognizes the pilot flame, the relay coil will be energized, and normally closed contacts in the spark ignition system will remain open.

The changeover from the ignition phase to the operating (gas-valve-open) phase is done by a single-pole, double-throw switch using a filled actuator. Most use a liquid such as mercury to cause pressure rise in the switch with a rise in the temperature at the sensing bulb located in the pilot assembly. Fig. H35-36 shows the components used to accomplish this.

Another type of system, the *direct-spark ignition system,* uses an electricity-carrying flame-sensor rod mounted to have direct contact by the main burner flame. Because the gas flame will carry electrical energy by means of electrically charging the carbon atoms in the gas before combustion takes place, a current flow can be passed from the burner to a positive-charged flame rod. This current flow controls a circuit in the solid-state control module and keeps the gas valve energized. Fig. H35-37 shows the components used to accomplish this.

Upon a call from the room thermostat, both the main gas valve and the spark ignitor are activated. Allowing a predetermined time for main flame ignition, the ignition control module will shut down the lockout circuit and maintain burner operation if main flame ignition occurs in that period. The period may be from 4 to 21 seconds, depending on the model of control module used. Usually, the higher the input to the gas unit, the shorter the proving time.

If the main burner flame is not established in the set time, the control module automatically locks out. To reset the circuit, electrical power to the system must be cut off and then back on to start another cycle. Manual reset of such a system is used for maximum safety to the equipment and area sensed.

Because this system uses the main burner assembly as the ground terminal of the spark system, it is absolutely necessary that the gas burning unit be thoroughly grounded to the electrical-supply ground. It is wise and usually necessary to run a ground (green) wire from the power distribution panel to the unit to provide this positive ground. Because the white or neutral wire is a 120/240-V supply system, it is a current-carrying wire. It is not suitable for use as a unit ground.

Figure H35-37
Component parts of a direct-spark ignition kit. (Courtesy of Honeywell, Inc.)

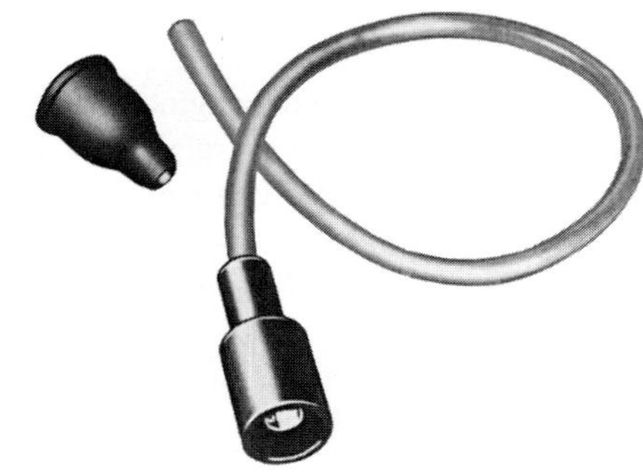

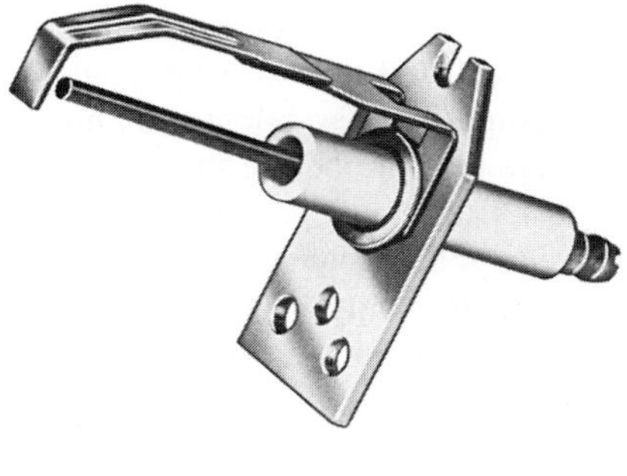

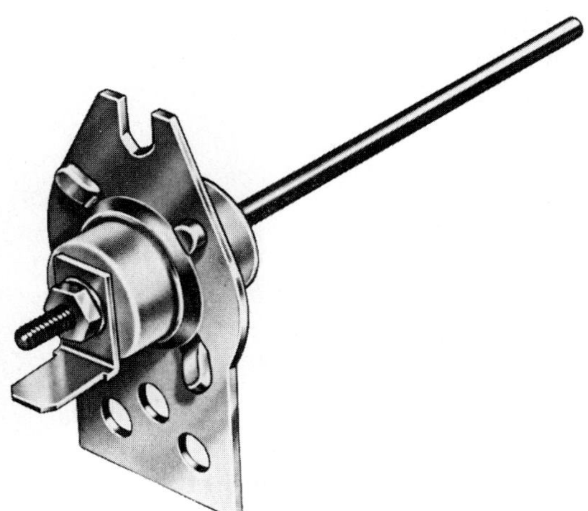

Figure H35-38 Hot-surface gas-fuel igniter. (Carrier Corporation—Residential Products)

H35-6.1 Hot-surface Ignition

A *hot surface ignition unit* (Fig. H35-38) is used for igniting the gas burners on many of the latest furnaces. This unit uses a material called *silicon carbide* which has a very high resistance to the current flow, and when energized reaches the ignition temperature of gas. It is very tough and will not burn up, something like a glow coil. The control system allows this material to reach the ignition temperature before the gas valve opens.

These units are powered by 120-V current and draw a considerable amount of current when energized; however, they are used only a few minutes per day and therefore do not materially increase the electric bill. If the burner fails to light after the gas is turned on for a few seconds, a safety lockout will occur to stop the flow of gas.

H35-6.2 Power Supply

If the heating and air-conditioning control circuit used an oil fired heating unit instead of a gas fired unit, a change in the thermostat connection arrangement would be required. The dotted-line jumper between terminals M and V (Fig. H35-8, p. 694) in the thermostat control setup would be removed and the thermostat would have two separate control circuits: one for cooling between V (cooling power) and C (first-stage cooling) and X (second-stage cooling); the second control circuit between M (heat power) and H (first-stage heat) and Y (second-stage heat).

This separation of the two control systems is necessary because the oil-burner control has its own power source. Therefore, the heating control system must be isolated from the air-conditioning control system with its own power source. Never connect two separate power sources together in the same control system. Overload and burnout will occur.

H35-7 PRIMARY OIL-BURNER CONTROLS

The original *primary oil-burner control,* known as a protector relay, was mounted in the smoke pipe. The bimetallic element in the sensing tube would react to a rise in flue-product temperature and cause the relay to keep the oil burner operating. Fig. H35-39 shows an example of

Figure H35-39 Oil-burner stack mounted primary control. (Courtesy of Honeywell, Inc.)

this control with the cover in place and the sensing (stack) element protruding from the rear.

Fig. H35-40 shows the same control with the front cover removed. The control has its own power-source transformer as well as temperature-actuation contacts and manual-reset safety switch. The operation of the control is by means of a slide clutch operating safety and holding contacts. The drive shaft at the top of the control provides the action for the slide clutch when moved by the expansion and contraction of the bimetallic element. Two relays are included, one to control the oil-burner ignition transformer (relay 1A) and the other to control the oil-burner motor (relay 2A). Low-voltage thermostat terminals on the lower left and high-voltage (120 V) on the lower right are separated by an insulated barrier. Fig. H35-41 shows the internal wiring arrangement of the control as well as the burner motor and ignition-transformer circuits, 120-V power supply, and 24-V thermostat circuit.

Following through the control circuit, the action of the cycle would be as follows: The thermostat, calling for heat, closes the circuit from the transformer through the safety switch, through the thermostat, through relay coil 1K, through the right-hand cold contact, the left-hand cold contact, and the safety switch heater to the other side of the transformer.

Relay 1K pulls in closing contact 1K1, powering the ignition transformer; 1K2 energizes relay coil 2K and closing contact 1K3 to the center or common of the cold contacts.

Figure H35-40 Oil-burner stack-mounted primary control, with cover removed. (Courtesy of Honeywell, Inc.)

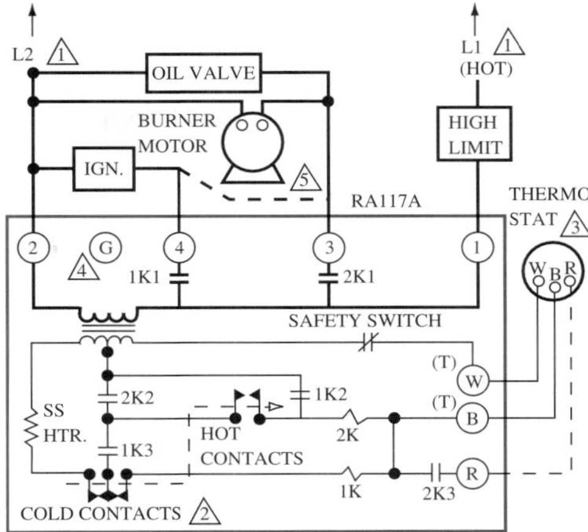

1. POWER SUPPLY. PROVIDE DISCONNECT MEANS AND OVERLOAD PROTECTION AS REQUIRED.
2. CONTACTS BREAK IN SEQUENCE ON RISE IN TEMP.
3. MAY BE CONTROLLED BY TWO-WIRE THERMOSTAT. CONNECT TO W AND B ONLY. TAPE LOOSE END OF RED WIRE, IF ANY.
4. CONTROL CASE MUST BE CONNECTED TO EARTH GROUND. USE GREEN GROUNDING SCREW PROVIDED.
5. TO REPLACE INTERMITTENT (FORMERLY CALLED CONSTANT) IGNITION DEVICE, WIRE IGNITION LEADWIRE TO TERMINAL 3 ON RA117A.

Figure H35-41 Wiring diagram of an oil burner primary control.

When relay coil 2K is energized, it pulls in, closing contact 2K1, energizing the oil burner motor and oil valve (if used). Contact 2K2 removes the relays from the series circuit through the safety switch heater, reducing the amount of heat produced in the heater. If this action does not occur, the safety switch will open in a very short period of time, possibly 3 to 5 seconds. By removing the relay current from the safety switch circuit, the switch delay time is increased to 60 to 90 seconds.

The start of the oil burner produces hot flue products from the heat exchanger over the bimetallic helix in the sensing tube. This rise in temperature forces the drive shaft forward, opening the left-hand cold contact and removing the safety switch heater from the circuit. A further rise in temperature (forward action of the drive shaft) closes the hot contact; this allows the oil-burner motor and oil valve (if used) to continue to operate by passing relay contact 1K2.

Finally, the drive shaft moves forward enough to open the right-hand cold contact, dropping out relay 1K. This stops the ignition and keeps the oil burner operating through the hot contacts. If a flame-out occurs or the burner shuts down from the action of the thermostat, the hot contact opens immediately upon a drop in flue gas temperature. Because contact 1K2 is open, the burner motor cannot operate until the cold contacts close, first the left side to energize the safety switch and then the right side to start the ignition/oil-burner motor cycle. By means

Figure H35-43 Oil-burner relay using cad cell sensing element, with cover removed. (Courtesy of Honeywell, Inc.)

of these lockout circuits, an explosive situation of spraying oil vapor into a white-hot refractory can be prevented.

The stack-mounted protector relay required extra wiring, as well as time and cost for manufacturing assembly or installation. To overcome this, a flame-detection relay was developed. Fig. H35-42 shows an oil-burner protector relay using a cadmium-cell flame detector instead of the bimetallic helix. It is mounted directly in the oil-burner blast tube directly behind the turbulator plate. The cell sees the light of the burner the instant the flame is established.

The interior arrangement of the relay using the light-sensitive cadmium cell is shown in Fig. H35-43. Two relays are used: 1K for control of the oil burner and ignition transformer, and 2K, the sensitive relay, controlled by the cadmium cell. Following the circuit action in the schematic in Fig. H35-44, when the thermostat calls for heat and closes the contact between T and T, current flows from the transformer through the thermostat through relay coil 1K, safety switch, timer contact 1, the safety switch heater, and contact 2K1 to the other side of the transformer. This puts a higher voltage through the coil and safety-switch heater. If the relay 1K fails to act, the safety switch will open in only a few seconds.

When relay 1K pulls in, however, contact 1K1 pulls in, energizing the oil-burner motor, oil valve (if used), and the ignition transformer. Also, contact 1K2 closes, bypassing the safety-switch heater, reducing its current draw, and increasing the safety-switch action time. If no flame is established, the heater will continue to receive voltage and heat until the safety-switch contact (SS1) opens and shuts down the system. The switch is manually reset.

When a flame is established, the light strikes the cadmium cell, and immediately the resistance of the cell is reduced. The amount of current through relay coil 2K increases and the sensitive relay pulls in. This opens contact 2K1, which de-energizes the safety switch heater and energizes the timer heater through contact 1K3, which closed when relay 1K pulled in. The timer heater now opens contact T1, shutting off the ignition transformer and the holding contact of the sensitive relay. This action will continue as long as burner operation is required.

If flame failure should occur, the cadmium-cell resistance will increase, current flow through the sensitive relay will decrease, and the relay will drop out. This opens contact 2K2, which opens the circuit to relay 1K, and the burner shuts down.

The burner cannot come on until the timer heater has cooled sufficiently to close the ignition contact T1 and the relay 1K circuit contact T2. This assures the unit sufficient time to vent the furnace heat exchanger of unburned vapors as well as assuring ignition at the next startup.

These are only two of the various types of controls manufactured and used on oil-fired equipment. The service person should collect and retain as much information as possible from all manufacturers and all types.

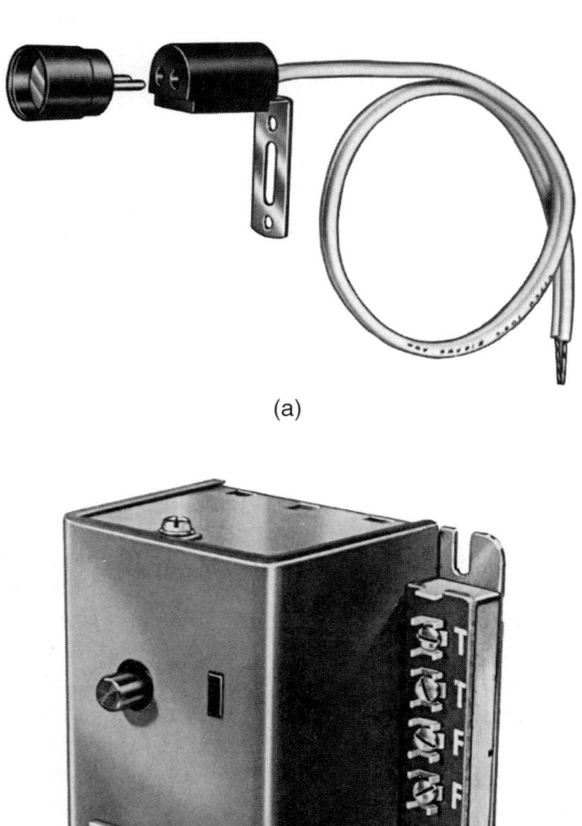

Figure H35-42 Oil-burner relay using cad cell sensing element. (Courtesy of Honeywell, Inc.)

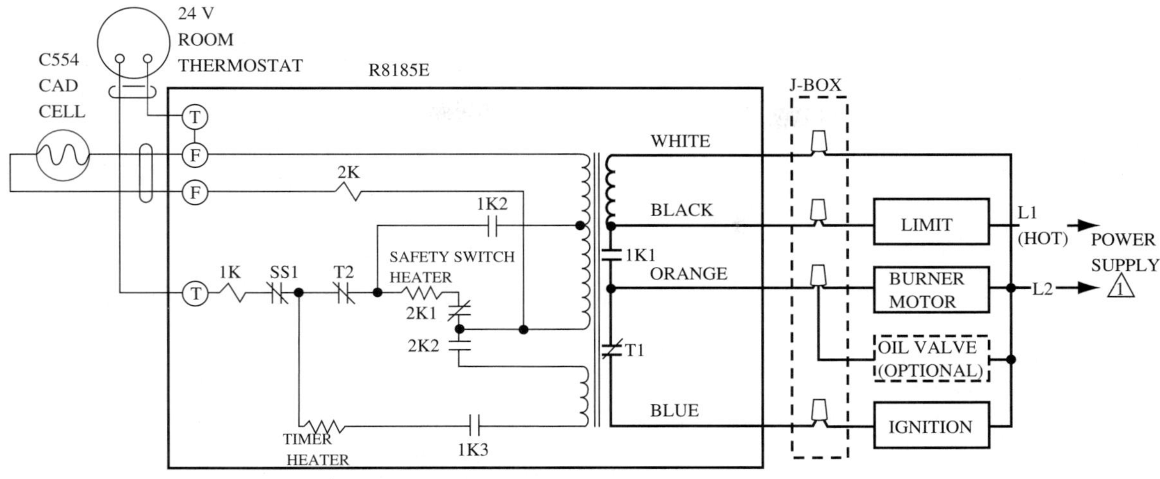

Figure H35-44 Oil-burning wiring diagram, using a cad cell relay.

H35-7.1 Oil Valves

The oil burner should reach operating speed and combustion air volume before oil is supplied and combustion is established. Also, at cutoff of the burner operation the best method of operation is to have instant cutoff of fuel and combustion.

Fig. H35-45 shows a delay-type *oil valve* that is installed between the oil-pump outlet and the burner fixing assembly. The valve, although wired to be energized at the same time as the oil burner motor (wired in parallel), has a delayed opening. This allows full operation of the burner blower and pump before oil flow is allowed. Using a thermistor in the valve-coil circuit, the thermistor limits the current to the coil on start. As the thermistor heats, the resistance drops and the current increases. After 8 to 10 seconds, the current increases sufficiently to cause the coil magnetic pull to open the valve. The valve acts the same as any other type of solenoid valve on cutoff. It closes immediately upon de-energizing the valve and motor. Thus, instant cutoff of combustion occurs.

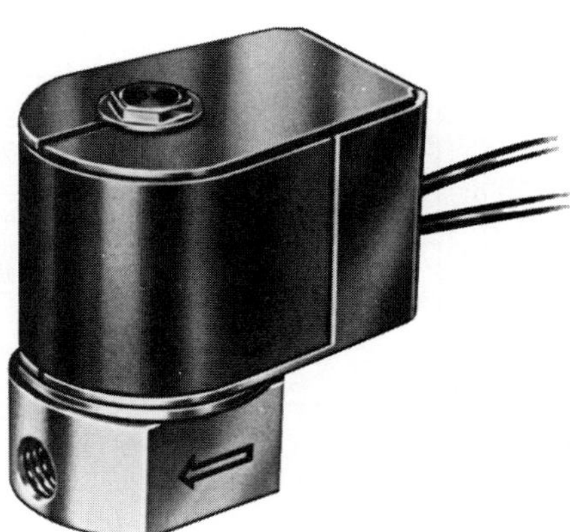

Figure H35-45 Delay-type oil valve which is placed in the oil line to the burner. Energized at the same time as the oil burner but has delayed opening. (Courtesy of Honeywell, Inc.)

H35-8 RESIDENTIAL AND SMALL COMMERCIAL CONTROL SYSTEMS

Some typical heating and cooling control systems will now be examined in order to become familiar with the function and operation of their component parts.

Different manufacturers will have variations in the physical wiring, but the function remains the same in all systems. With a firm understanding of the construction and use of schematic wiring diagrams, the student should be able to wire any manufacturer's product. The industry uses standardized electrical symbols, as shown in Fig. H35-46 which help the technician identify the common controls. Also, the legend designations shown in Fig. H35-47 are related alphabetically to the devices they represent. For example, R is a general relay while CR is a cooling relay, HR is a heating relay, and DR is a defrost relay. Legends may vary with the particular wiring diagram or the manufacturer's method of labeling.

Study the symbols and legend designations carefully, for they are important to the following discussion, and they will also serve later as a reference.

H35-8.1 Wire Size

Wire sizes and fusing for independent appliances will be determined by the manufacturers and specified on their wiring diagrams. Local codes should be examined to de-

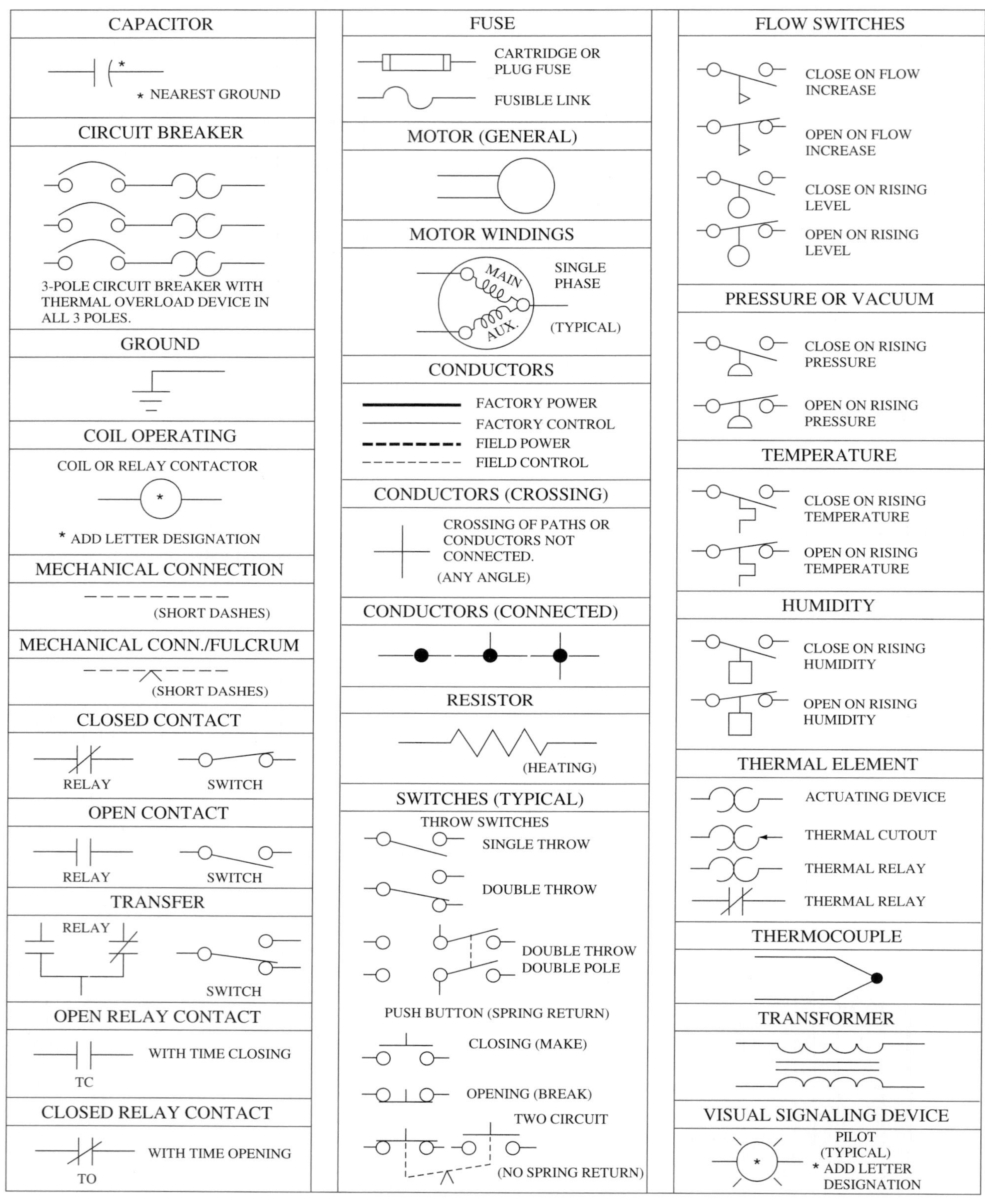

Figure H35-46 Electrical wiring symbols.

Relays	
R	Relay, General
CR	Cooling Relay
DR	Defrost Relay
FR	Fan Relay
IFR	Indoor Fan Relay
OFR	Outdoor Fan Relay
GR	Guardistor Relay
HR	Heating Relay
LR	Locking Relay (Lock-in or Lockout)
PR	Protection Relay (Relay in series with protective devices)
VR	Voltage Relay
TD	Time Delay Device
THR	Thermal Relay (type)
M	Contactor
MA	Auxiliary Contact

Solenoids	
S	Solenoid, General
CS	Capacity Solenoid
GS	Gas Solenoid
RS	Reversing Solenoid

Switches	
DI	Defrost Initiation
DT	Defrost Termination
DIT	Defrost Initiation-Defrost Termination (dual function device)
GP	Gas Pressure
HP	High Pressure
LP	Low Pressure
HLP	Combination High-Low Pressure
OP	Oil Pressure
RM	Reset, Manual
FS	Fan Switch
SS	System Switch
HS	Humidity Switch (Humidistat)
TA	Thermostat, Ambient
TC	Thermostat, Cooling
TH	Thermostat, Heating
TMA	Thermostat, Mixed Air
CT	Thermostat, Compressor Motor
HT	High Temperature
LT	Low Temperature
RT	Refrigerant Temperature
WT	Water Temperature

Miscellaneous	
C-HTR	Crankcase Heater
RES	Resistor
HTR	Heater
PC	Program Control
OL	Overload
L	Indicating Lamp
⊕	Manual Reset Device
+	Automatic Reset Device

Figure H35-47 Legend for the wiring diagrams.

termine whether or not conduit is required in bringing the wire from the main circuit breaker to the furnace or air conditioner. Some codes even require conduit for parts of the circuit on 120 V. Specification sheets will give the minimum wire and fuse sizes that will meet the requirements of the National Electric Code, and this wire size is based on 125% of the full-load current rating. The length of the wire run will also determine the wire size, and Fig. H35-48 shows the maximum length of two wire runs for various wire sizes. This table is based on holding the voltage drop to 3% at 240 V. For other voltages, the multipliers at the bottom should be used. If the required run is nearly equal to or somewhat more than the maximum run shown by the calculation of the table, the next larger wire size should be used as a safety factor. It is always possible to use larger wire than called for on the specifications, but a smaller wire size should never be used, as this will cause nuisance trips, affect the efficiency of operation, and create a safety hazard. All replacement wire should be the same size and type as the original wire and should be rated at 90 to 105°C.

H35-8.2 Heating Circuit

A simple schematic diagram for a typical gas-fired upflow air furnace is shown in Fig. H35-49. The wire conductors are represented by lines and the other components by symbols and letter designations. Note the very important legend which identifies these. Although other types of diagrams will be discussed later, the schematic is the type of diagram used by service personnel to analyze the system.

Only five major electrical devices are in the diagram:

1. the fan motor, which circulates the heated air;
2. the 24-V automatic-gas solenoid valve to control the flow of gas to the burner;
3. the combination blower-and-limit control which controls fan operation and governs the flow of current to the low-voltage control circuit;
4. the step-down transformer, which provides 24-V current to operate the automatic gas valve; and
5. the room thermostat, which directly controls the opening and closing of the automatic gas valve.

Maximum Length of Two-wire Run*

Wire size	Amperes												
	5	10	15	20	25	30	35	40	45	50	55	70	80
14	274	137	91										
12		218	145	109									
10			230	173	138	115							
8					220	182	156	138					
6								219	193	175	159		
4									309	278	253	199	
3										350	319	250	219
2											402	316	276
1												399	349
0												502	439
00													560

*To limit voltage drop to 3% at 240V. For other voltages use the following multipliers:

110V	0.458	220V	0.917
115V	0.479	230V	0.966
120V	0.50	250V	1.042
125V	0.521		

EXAMPLE: Find maximum run for #10 wire carrying 30 amp at 120V. . . . . . 115 × 0.5 = 57 ft.

NOTE: If the required length of run is nearly equal to, or even somewhat more than, the maximum run shown for a given wire size—select the next larger wire. This will provide a margin of safety. The recommended limit on voltage drop is 3%. Something less than the maximum is preferable.

Figure H35-48 Wire sizes based on length and current capacity.

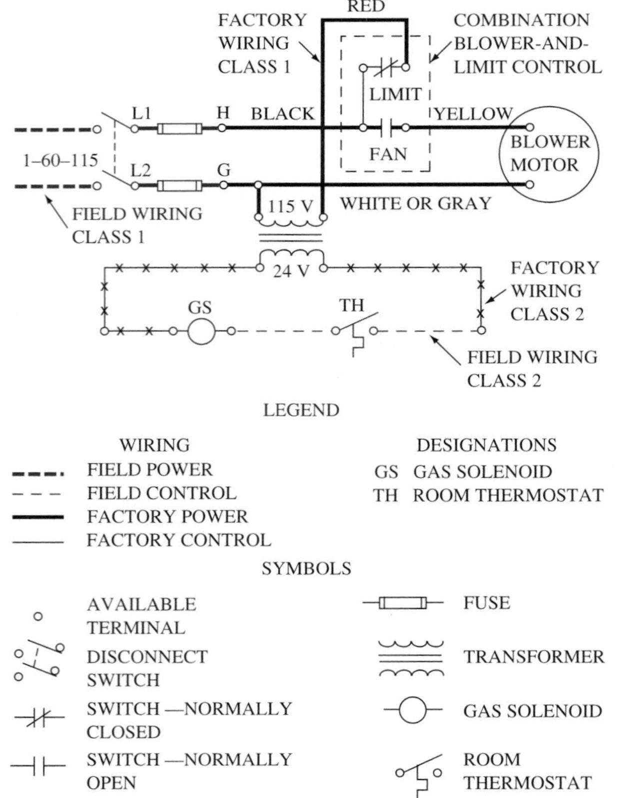

Figure H35-49 Heating circuit diagram for gas-fired up-flow furnace.

To build this same schematic diagram, begin with the power supply. The wires for the power supply are usually run by the installing technician. They must carry line voltage (single-phase, 60-Hz, 120-V) to operate the blower motor and are represented by heavy broken lines to indicate field wiring (Fig. H35-50).

As a protection to the circuit, the power supply must be run through a fused disconnect switch. This is the main switch to the entire system.

In a gas-fired heating system, the fan motor produces the heaviest load and is connected to the power supply (Fig. H35-51). Since the fan motor is clearly identified on the diagram, it is not necessary to add it to the legend. Internal windings are not shown.

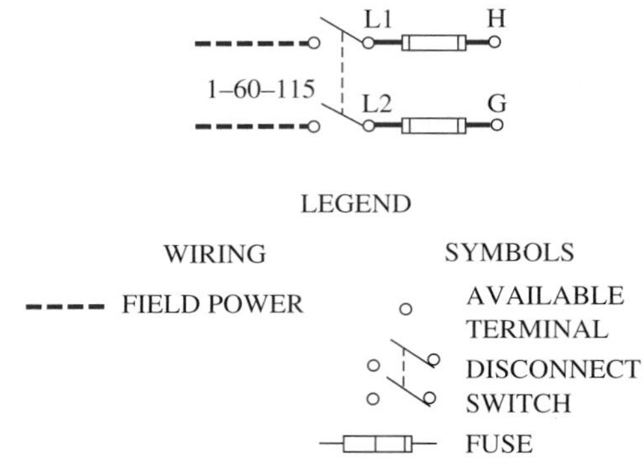

Figure H35-50 Diagram for a fused disconnect.

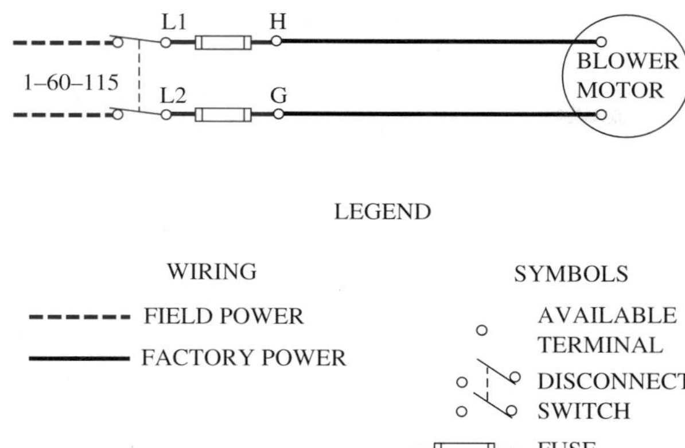

Figure H35-51 Diagram showing the fused disconnect wired to the blower motor.

To test the system at this point, close the disconnect switch. The fan motor runs, since a simple circuit is completed from L1 through the windings of the blower motor and back to neutral. Open the disconnect switch and continue the diagramming.

Since there must be automatic control of both the fan motor and the automatic-gas solenoid valve, connect the combination fan-and-limit control (Fig. H35-52). This control, as previously described, consists of a sensing element which feels the temperature of the heated air inside the furnace, and two adjustable switches: a fan switch and a limit switch.

The fan switch controls the starting and stopping of the blower motor. When the burner starts, the fan switch will not close to start the blower until the air temperature inside the furnace plenum chamber has warmed up to the FAN ON setting. When the burner stops, the fan switch remains closed, keeping the blower in operation until the air temperature inside the furnace cools down to the FAN OFF setting.

The limit switch is a safety control. It prevents the furnace from overheating. As long as the plenum-chamber temperature is below the setting of the limit switch, the switch remains closed. If the plenum-chamber temperature rises to the switch setting, it opens to deenergize the entire 24-V circuit and close the automatic-gas valve.

As illustrated, the combination fan-and-limit control are partially connected. The fan switch is wired into L1 in series with the blower-motor windings. If the disconnect switch is closed now, the fan motor will not run because the fan switch in the fan-motor circuit is open and will remain open until the temperature in the furnace warms up to the switch setting.

Small electrical devices, such as relays and solenoid valves, do very little work and require little current for operation. A step-down transformer to reduce line voltage to 24 V is required and properly installed in the wiring diagram (Fig. H35-53). One side is connected to neutral and the other side is connected to L1 through the limit switch. This places the limit switch in control of the entire 24-V circuit.

If the disconnect switch is closed, the blower motor still cannot run since the fan switch must remain open until the air temperature warms up. The transformer and

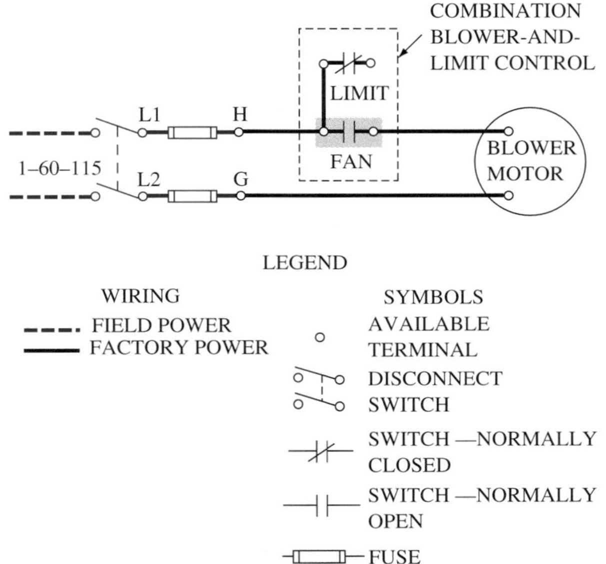

Figure H35-52 Wiring diagram showing the use of a fan limit control wired to a blower motor.

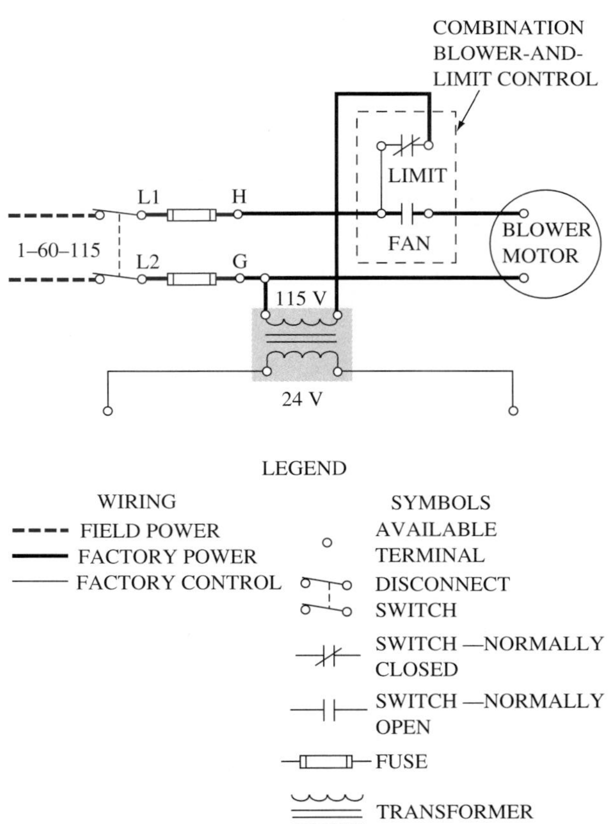

Figure H35-53 Schematic diagram showing the addition of a control-circuit transformer.

the 24-V power source, however, are energized immediately from L1 through the closed limit switch and back to L2. Although there is now a source of 24-V current, nothing can be accomplished until there is a circuit across the 24-V supply.

The automatic-gas solenoid valve and room thermostat are wired in series across the 24-V circuit (Fig. H35-54) since the opening and closing of the gas solenoid valve is controlled by the thermostat. The gas valve is factory-wired, indicated on the diagram as a light, unbroken line between terminals. Normally, the thermostat is located in a room away from the furnace and must be field-wired, as indicated by the light broken lines used to represent control circuit field wiring.

The circuit is now completely wired. Close the disconnect and limit switches and open the thermostat. The transformer and low-voltage circuit are immediately energized from L1, through the closed limit switch back to neutral. With the thermostat open, the automatic gas valve remains closed. The fan motor cannot operate because the fan switch is open, and it will remain open until the air temperature in the furnace warms up to the FAN ON setting.

When the thermostat calls for heat, the automatic-gas valve is energized (Fig. H35-55) and opens to admit gas to the burner. The fan switch would still be open so that the fan motor cannot run until the air temperature in the furnace plenum chamber warms up to the FAN ON setting.

Thermostat and gas-valve circuits are reopened when the room temperature setting is reached. Fan and blower contacts remain closed until the temperature of the heated air cools to the FAN OFF setting.

If, at any time during furnace operation, the plenum chamber overheats, the air temperature soon reaches the setting of the limit switch. This is an important safety switch which opens at the overheating setting to de-energize the entire 24-V circuit. When this happens, the automatic-gas valve closes, the burner is extinguished, and the thermostat is overridden. The fan switch remains closed and the blower continues to run as long as the heated air remains above the FAN OFF setting. The limit switch is manually reset by the service person after cor-

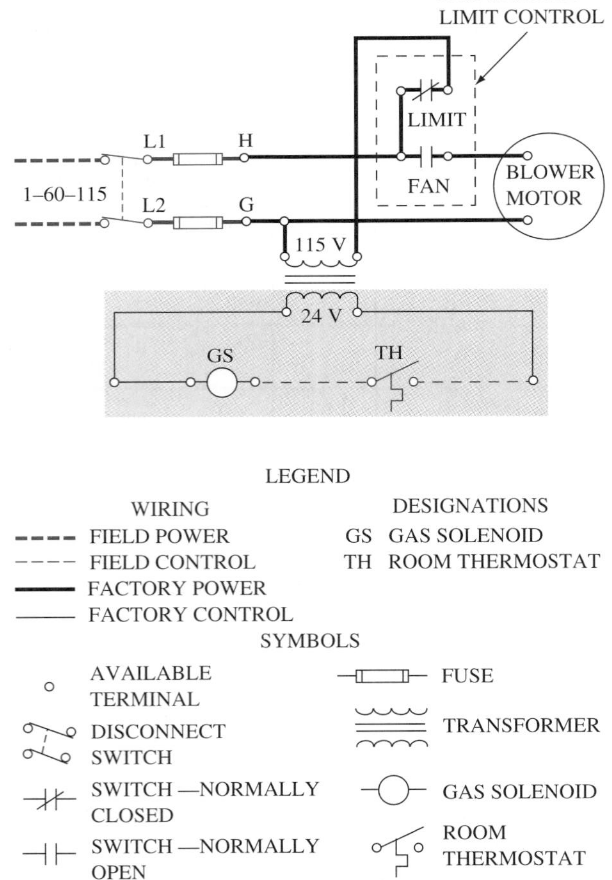

Figure H35-54 Schematic diagram of the low-voltage circuit showing the thermostat operating the gas valve.

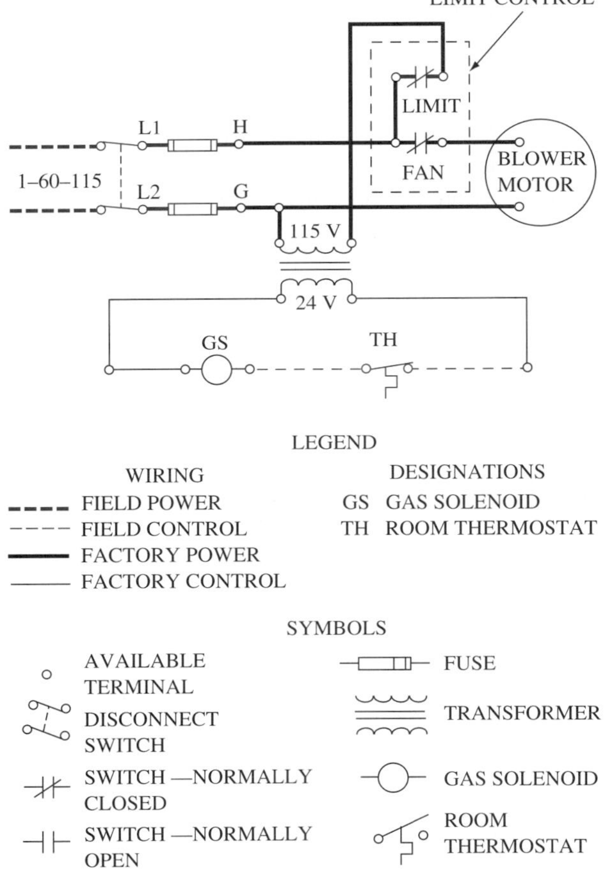

Figure H35-55 Schematic diagram showing the wiring of the limit control in series with the line-voltage side of the transformer.

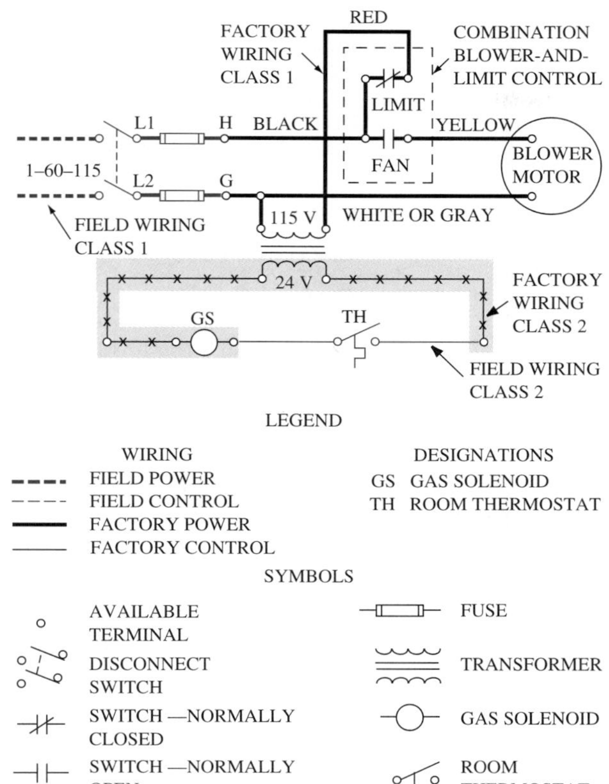

Figure H35-56 A diagram showing the gas-valve circuit separating factory wiring from field wiring.

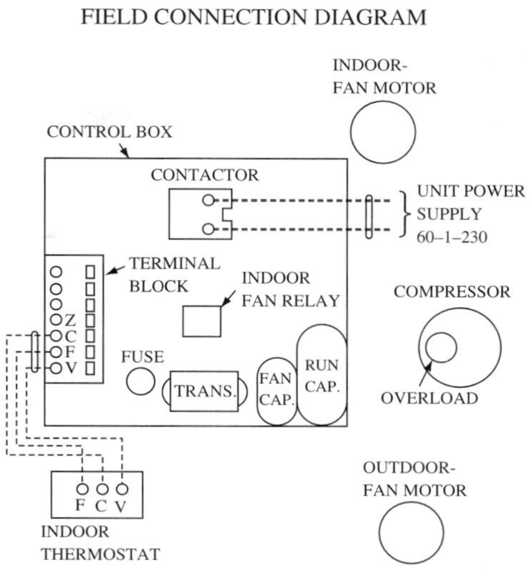

recting the problem. Causes of overheating are discussed in the trouble shooting section of this text.

The operating sequence just completed proves that the wiring diagram is correct. The disconnect switch controls the entire electrical system; the blower motor operates independently of the burner, under the control of the fan switch. The automatic-gas valve is cycled on and off by the room thermostat, under the control of the limit switch. The complete story of the electrical system, however, includes more than just the lines and symbols (Fig. H35-56). The legend clearly identifies all symbol and letter designations used in the diagram, and the notes call attention to important facts relating to the circuit.

H35-8.3 Cooling Circuit

Wiring diagrams are made for different purposes, and it is important to recognize each type and know the intended purpose of each. Two common types are the *field-connection diagram* and the *schematic diagram.*

The field-connection diagram (Fig. H35-57) identifies the various electrical controls in the control box and indicates the necessary field wiring by means of broken lines and shaded areas. The field-connection diagram is designed to instruct the installing electrician or technician

Figure H35-57 Field-connection diagram for an air-conditioning unit..

how to run the proper power supply to the unit and the correct wiring between sections if the unit consists of more than one section. Internal wiring is not shown since the principal use is for installation.

A typical schematic diagram for an air-cooled packaged residential air-conditioning unit is shown in Fig. H35-58. It is easy to see that this type of diagram contains more information than the field-connection diagram. It is used by service personnel to see how the system works and why. Most manufacturers include a copy of the diagram in the control cover or access panel of the unit. The load devices (motors) and the controls are represented by symbols, the circuits by lines. Note that some lines are heavy while others are light. The heavy lines represent wires in the power circuit which supplies higher-voltage current to the heavy electrical loads, such as motors. The light lines represent the control circuit which supplies low voltage through a step-down transformer to the light-load devices, like relay coils that are used only to actuate switches.

To build this cooling electrical circuit, component by component, start with the power supply to the unit (Fig. H35-59). Since these wires must be run by the installing electrician or technician in the field, they are represented by heavy broken lines. For protection, the power supply must pass through a properly sized, fused disconnect switch. Proper fuse sizes are found in the equipment installation manual. If located outdoors, the fused switch must be installed in a weatherproof enclosure. The power supply shown is single phase, 60-Hz, 230-V current, coming in through wires L1 and L2.

The heaviest load, in this case the compressor, is placed in the diagram next and connected to the power supply between L1 and L2. This is now a circuit and if the disconnect switch were closed manually, the compressor

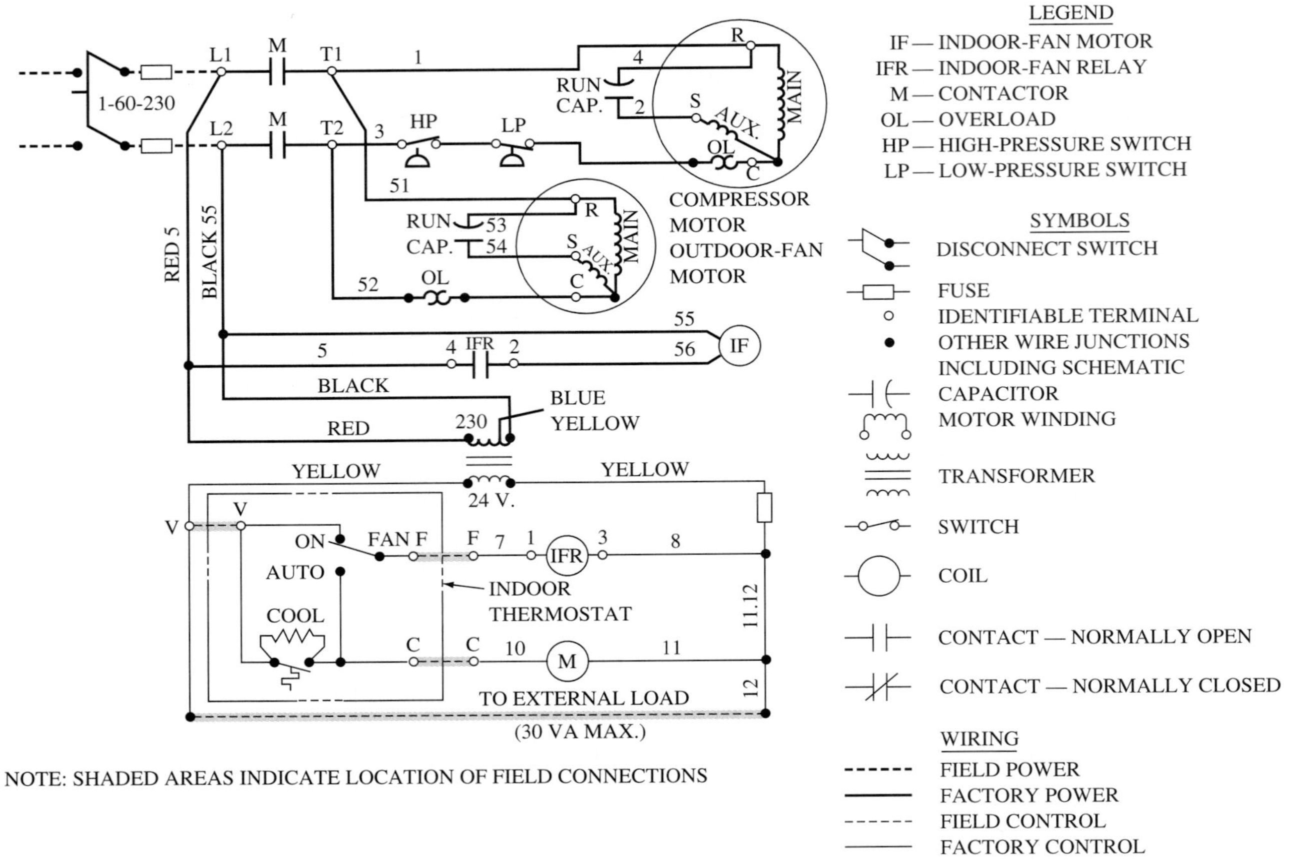

LEGEND
IF— INDOOR-FAN MOTOR
IFR— INDOOR-FAN RELAY
M— CONTACTOR
OL— OVERLOAD
HP— HIGH-PRESSURE SWITCH
LP— LOW-PRESSURE SWITCH

SYMBOLS
DISCONNECT SWITCH
FUSE
IDENTIFIABLE TERMINAL
OTHER WIRE JUNCTIONS INCLUDING SCHEMATIC
CAPACITOR
MOTOR WINDING
TRANSFORMER
SWITCH
COIL
CONTACT — NORMALLY OPEN
CONTACT — NORMALLY CLOSED

WIRING
FIELD POWER
FACTORY POWER
FIELD CONTROL
FACTORY CONTROL

NOTE: SHADED AREAS INDICATE LOCATION OF FIELD CONNECTIONS

Figure H35-58 A complete wiring diagram for the air-conditioning unit showing the connection of a FAN ON-AUTO switch and cooling thermostat.

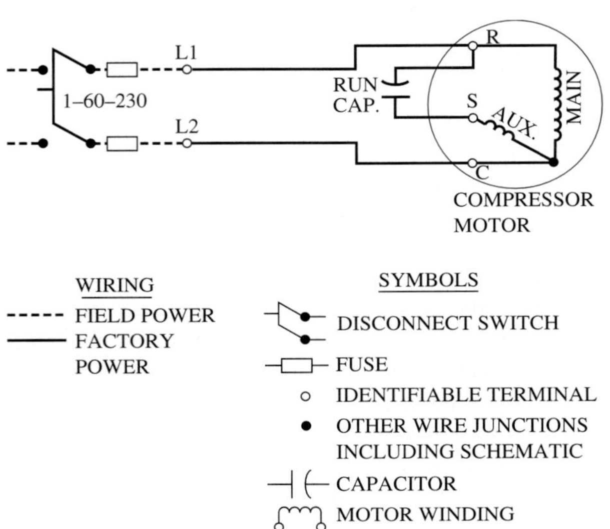

Figure H35-59 Diagram showing the wiring from the disconnect to the compressor motor.

would run; however, if the system must run automatically the circuit is incomplete.

Since this is an air-cooled system, an outdoor fan and an indoor fan are needed. These are put into the diagram next (Fig. H35-60). It is easy to see that both fans are connected across L1 and L2, and all three motors are in individual circuits of their own. Again, with manual operation, the wiring diagram is complete, and all components would operate with the disconnect switch closed. Electrical controls are needed for automatic operation.

Electrical controls use a low voltage, 24-V circuit, permitting the use of much smaller wire, making the circuit much safer for residential use, and most important, providing much closer control of system operation. A source of low-voltage current is a small step-down transformer, shown in the wiring diagram (Fig. H35-61). The sole function of the transformer is to reduce the 240-V current to 24 V, which is all the voltage needed to supply the necessary current to actuate the relays for automatic operation. A control installed across the 24-V source would complete the circuit.

Since an air-conditioning system is designed to maintain a comfortable temperature in the conditioned space, a low-voltage adjustable indoor cooling thermostat is used

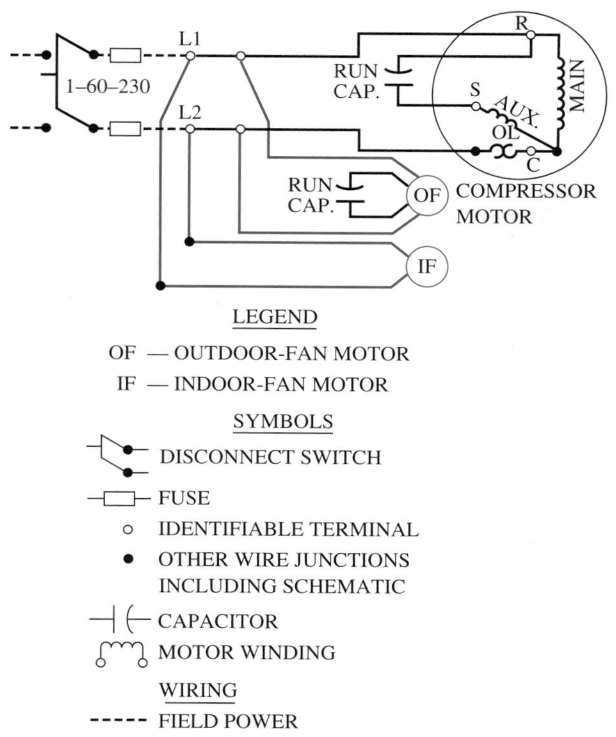

Figure H35-60 Wiring diagram showing the addition of the outside and inside fan motors.

as the primary system control in the wiring diagram (Fig. H35-62). Although there are many variations of indoor thermostats, the one illustrated has two internal switches: a fan switch with ON and AUTO positions, and a cooling control switch with positions marked OFF and COOL.

The fan switch controls the indoor fan only. In the ON position, the fan will operate continuously; in AUTO, fan operation is actually controlled by the position of the cooling control switch. The contacts of the cooling switch are sealed inside a tilting bulb partially filled with mercury. When the bulb is tilted so that mercury covers only one contact, the switch is open. When the bulb is positioned so that mercury covers both contacts, the switch is closed. The bulb is tilted by means of a bimetal element that warps as it feels a change in temperature, to either close or open the mercury switch. The symbol V is the common low-voltage terminal feeding through the fan switch to terminal F and through the mercury bulb to terminal C.

When installed in the electrical circuit, the common terminal V of the thermostat is connected to the L1 side of the low-voltage supply (Fig. H35-63) but nothing will happen until a circuit is completed to the other side of the 24-V supply. A few more controls are needed for automatic operation.

To control the compressor motor automatically, a contactor is required (Fig. H35-64). This can be described as

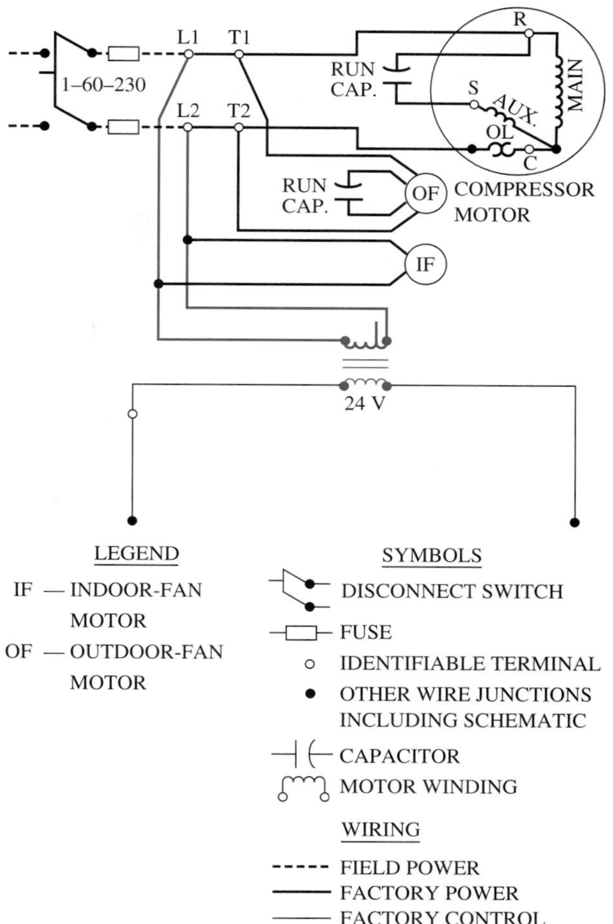

Figure H35-61 Wiring diagram showing the addition of a low-voltage control transformer.

an oversized relay with a switch large enough to carry the heavy current drawn by motors, and a magnetic coil strong enough to actuate the larger switches.

In order for the cooling switch in the thermostat to govern the action of the contactor in starting and stopping the compressor motor, the contactor coil M is wired in series with the cooling switch of the thermostat (Fig. H35-65, p. 720). The contactor switches must be connected in the power circuit so that they control both the compressor motor and outdoor-fan motor, but not the indoor fan; it must be able to operate independently. Note that the contactor coil and both contactor switches are marked M for identification. This simply means that coil M actuates the two switches marked M. Since both switches are normally open, they will close when the coil is energized and open when it is de-energized.

With the main disconnect switch closed now, current flows to the indoor-fan motor and the 24-V transformer immediately, but the compressor motor and outdoor fan motor cannot start. The thermostat cooling switch is open so no current can flow to contactor coil to close its switches to the compressor and outdoor-fan motors.

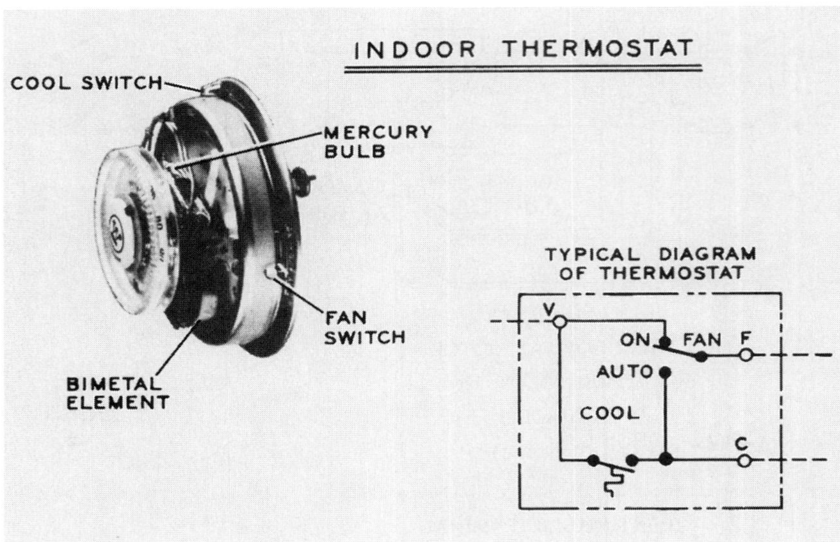

Figure H35-62 Room thermostat with cover removed. (Courtesy of Honeywell, Inc.)

When the thermostat cooling switch closes, calling for cooling (Fig. H35-66), current immediately flows through contactor coil M, and the two switches marked M close at the same time. This energizes and starts both the compressor motor and the outdoor-fan motor at the same time. When the thermostat is satisfied, the cooling switch again moves to the open position. The compressor and outdoor fan motors stop, but the indoor-fan motor continues to run. It will remain in operation as long as the disconnect is closed, regardless of the position of the fan switch.

Another control is needed for the indoor-fan motor, IF. A relay with a coil and a normally open switch, both marked IFR for indoor-fan relay, is used (Fig. H35-67, p. 721). Note that the coil is wired in series with the fan switch in the thermostat, and its switch is wired into the indoor-fan motor circuit.

With the fan switch in the ON position, low-voltage current now flows through the fan switch to energize relay coil IFR and close its switch in the indoor-fan power circuit. The indoor fan is now running, and it will continue to run as long as the fan switch remains in the ON position, regardless of the position of the cooling switch. The cooling system—the compressor and the outdoor fan—is free to cycle on and off as the cooling switch dictates, with no effect on the indoor fan.

The last change to the wiring diagram is to move the fan switch to AUTO (Fig. H35-68, p. 721). The thermostat cooling switch is open; it is not calling for cooling. When the disconnect switch is closed, current immediately flows through the transformer into the 24-V supply, but the fan switch is in AUTO and the cooling switch is open, so the 24-V circuit cannot be completed. Contactor coil M and indoor-fan relay coil IFR cannot be energized, so their switches all remain open, and none of the motors can operate.

Assume that the cooling switch closes and calls for cooling without changing the position of the fan switch or the disconnect (disconnect closed) and the fan is on AUTO (Fig. H35-69, p. 722). Current immediately flows through the cooling switch and through both the contactor coil M and the indoor-fan relay coil IFR at the same time. This closes both switches marked M in the power circuit to the compressor and outdoor-fan motors, as well as the switch marked IFR in the power circuit to the indoor-fan motor.

The compressor and outdoor-fan motors are energized and started through closed switches M. At the same time, the indoor-fan motor is also started through closed switch IFR. The entire system is now energized and all motors are running automatically.

When the thermostat is satisfied, the cooling switch opens the circuit automatically to both the contactor and indoor-fan relay coils, their respective switches immediately snap to their normally open positions, and all motors stop—compressor, outdoor fan, and indoor fan.

The wiring diagram is correct from the standpoint of operation, but safety controls must also be added to protect the system and improve its operating performance (Fig. H35-70, p. 722).

Overloads (OL) have been added to protect the compressor and outdoor-fan motor windings against excessive current. Overloads may be installed inside the motor housing as shown for the compressor motor or externally as shown for the outdoor-fan motor. Although the windings are not shown for the indoor-fan motor, all small single-phase motors are equipped with some type of protection.

As an added precaution against compressor operation under abnormal conditions such as excessive discharge pressure or dangerously low suction pressure, a high-pressure switch and a low-pressure switch have been wired in

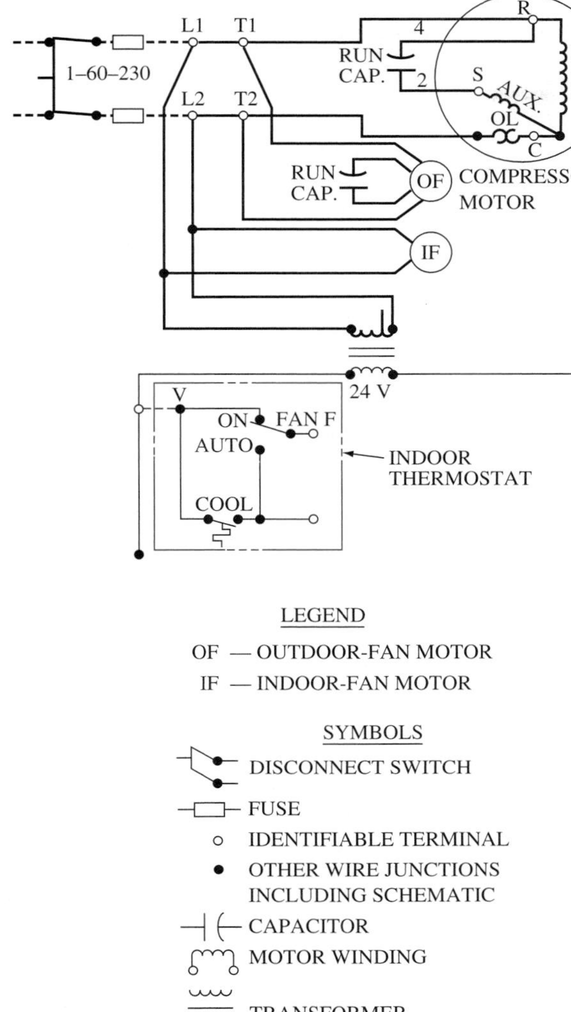

Figure H35-63 Wiring diagram showing the connection of the room thermostat to one side of the low-voltage circuit.

series with the compressor motor windings. The high-pressure switch (HP) is actuated by system discharge pressure to open and stop the compressor if the discharge pressure exceeds the switch setting. The low-pressure switch (LP) is actuated by system suction pressure to open and stop the compressor if the suction pressure falls below the switch setting. To protect the relay coils and wiring in the low-voltage control circuit, a fuse has been installed which is properly sized to open the 24-V circuit in case of excessive current.

Figure H35-64 Compressor contactor assembly. (Courtesy of Honeywell, Inc.)

For closer control of air temperature in the conditioned space, an anticipator has been wired across the terminals of the thermostat cooling switch. It is a fixed, non-adjustable type.

One relay coil may actuate many switches. Each switch may be located in a different section of the wiring diagram. As the diagram becomes larger and more relays are used, these switches become more difficult to locate. To make it easy to locate the switches of each relay, the ladder diagram is often used.

On the ladder diagram (Fig. H35-70, p. 722), each horizontal line is numbered in sequence from top to bottom on the left-hand side for line identification. On the right side of the diagram, opposite each relay coil, are shown the line numbers in which its switches are located. For example, the diagram shows that contactor coil M in line 16 has two switches located in lines 1 and 4, while relay coil IFR has switches located in line 9 and line 10.

The factory-installed wiring has also been color-coded or given a numerical designation by the manufacturer for ease of identification.

H35-8.4 Total-comfort Systems

For simplicity, this text will review a total-comfort system (TCS) using an up-flow gas furnace with single-speed fan operation, a split-system air-cooled condensing unit (with indoor coil), a combination heat/cool thermostat, a humidistat, and an electronic air cleaner.

First, the physical wiring and point-to-point connections are illustrated in Fig. H35-71 (p. 723). This is what the electrical installer will use as a guide during the installation. Note the need for two junction boxes to make the required interlocks. The low-voltage wiring is also detailed

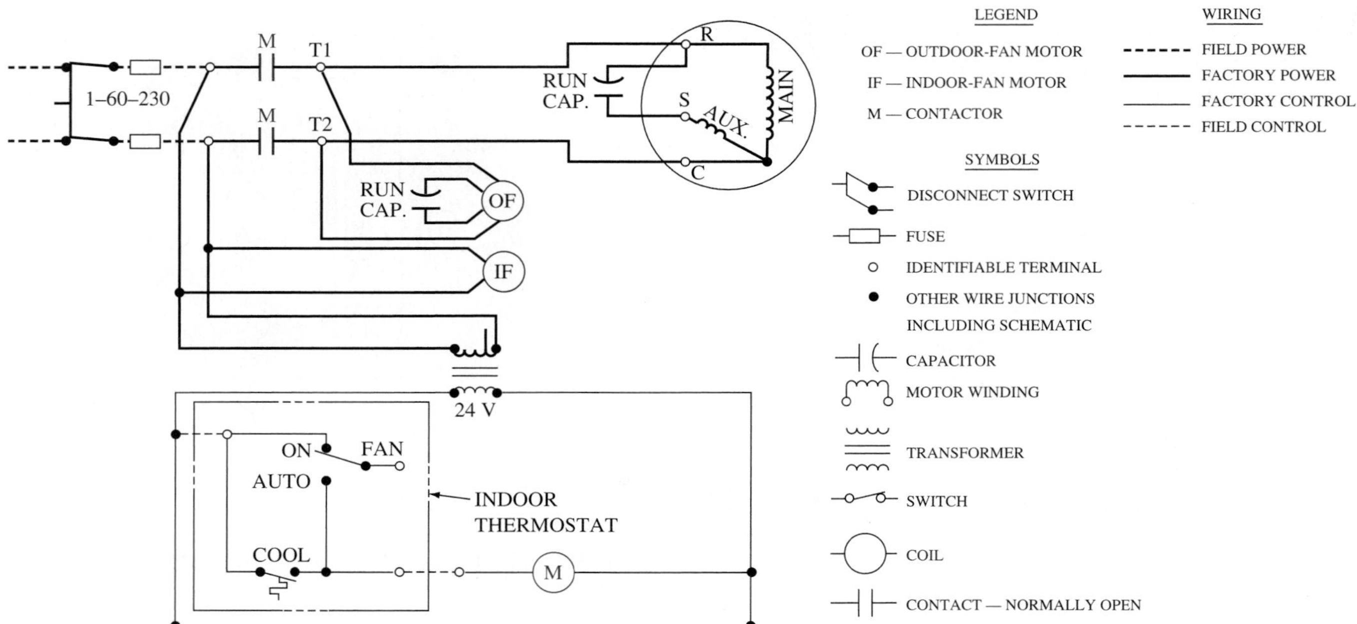

Figure H35-65 Wiring diagram showing the addition of the compressor contactor coil placed in the cooling circuit.

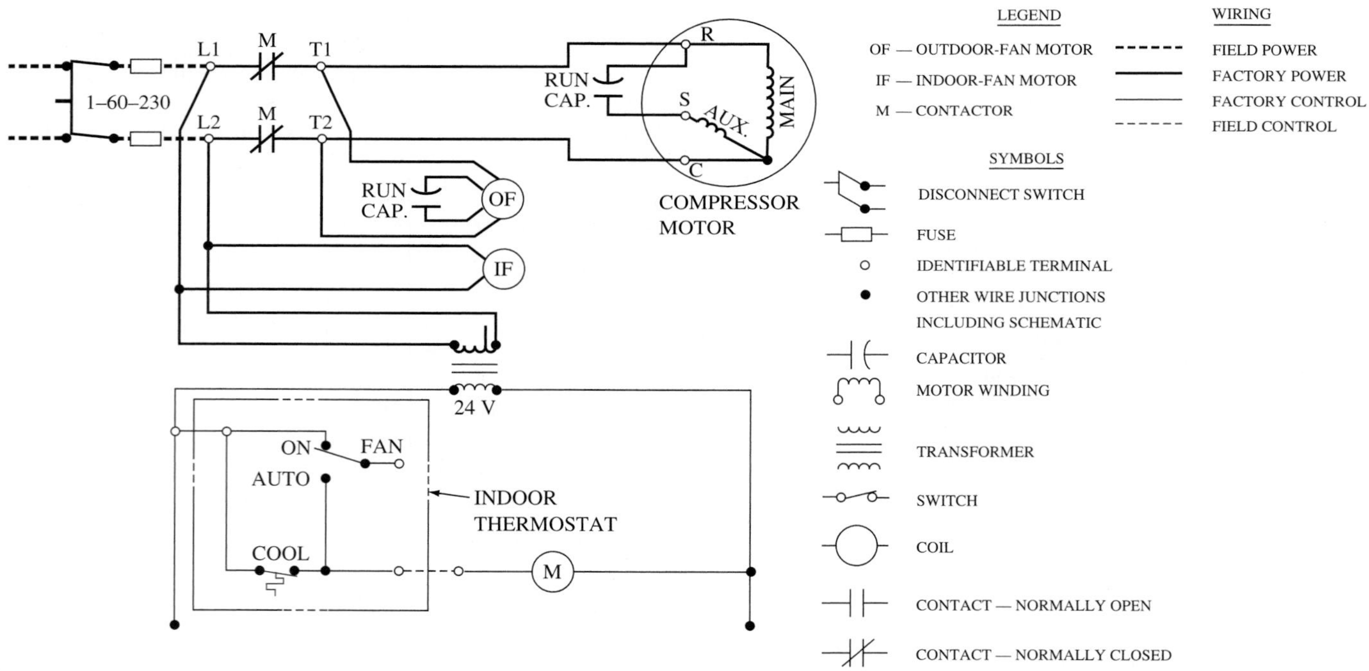

Figure H35-66 Wiring diagram with the cooling thermostat closed and the compressor contactor energized.

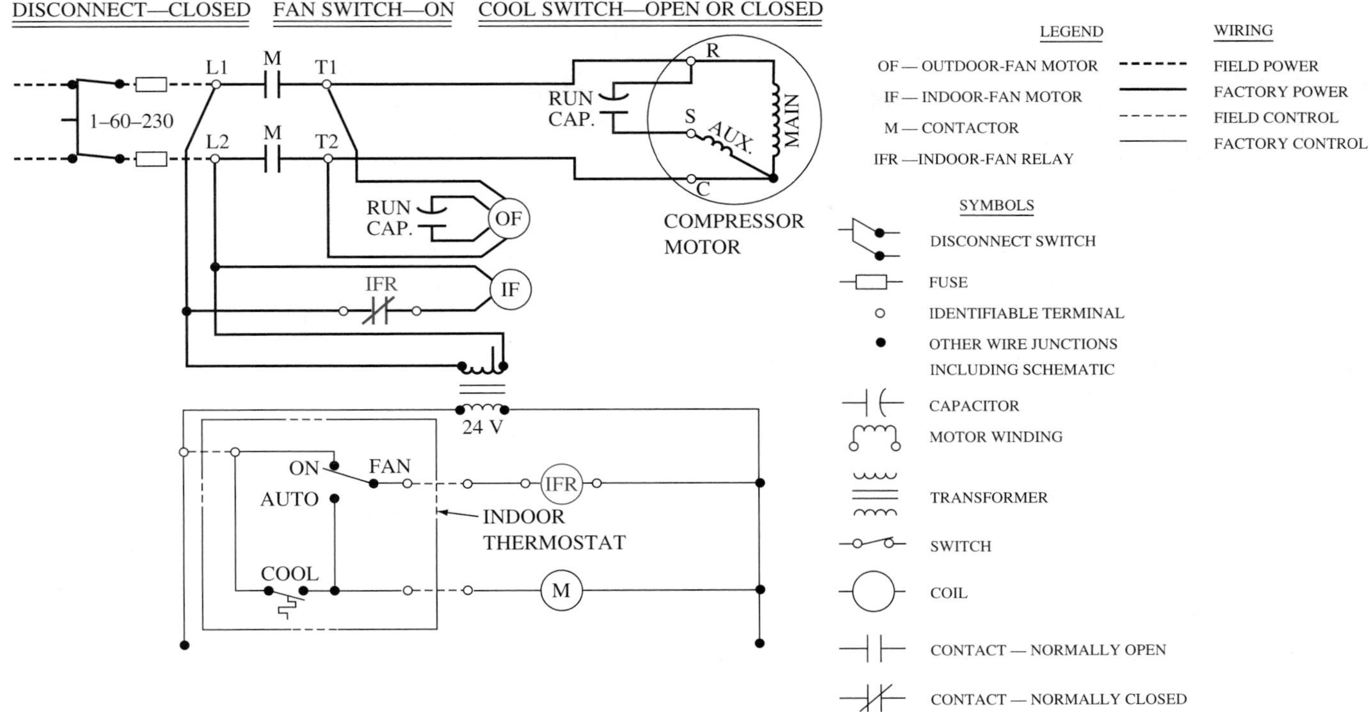

Figure H35-67 Wiring diagram showing the addition of the fan relay operated by the FAN ON switch.

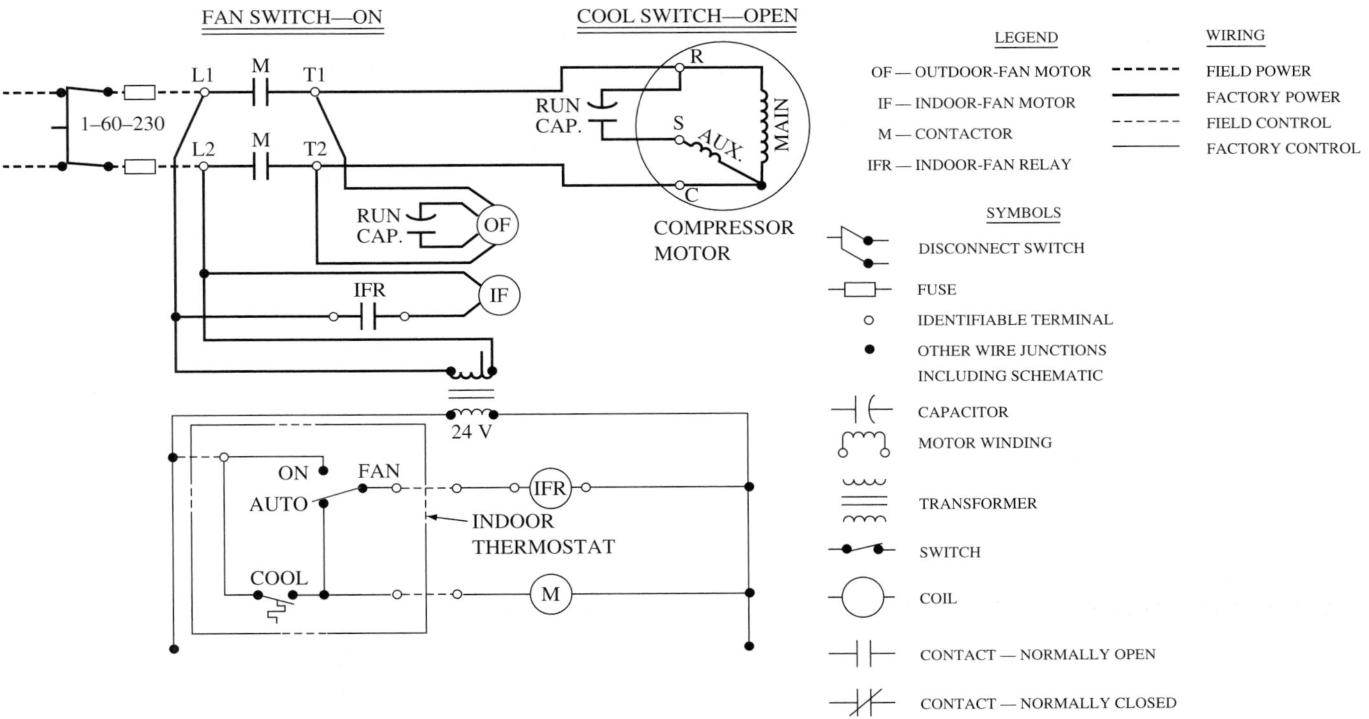

Figure H35-68 Fan switch in AUTO position.

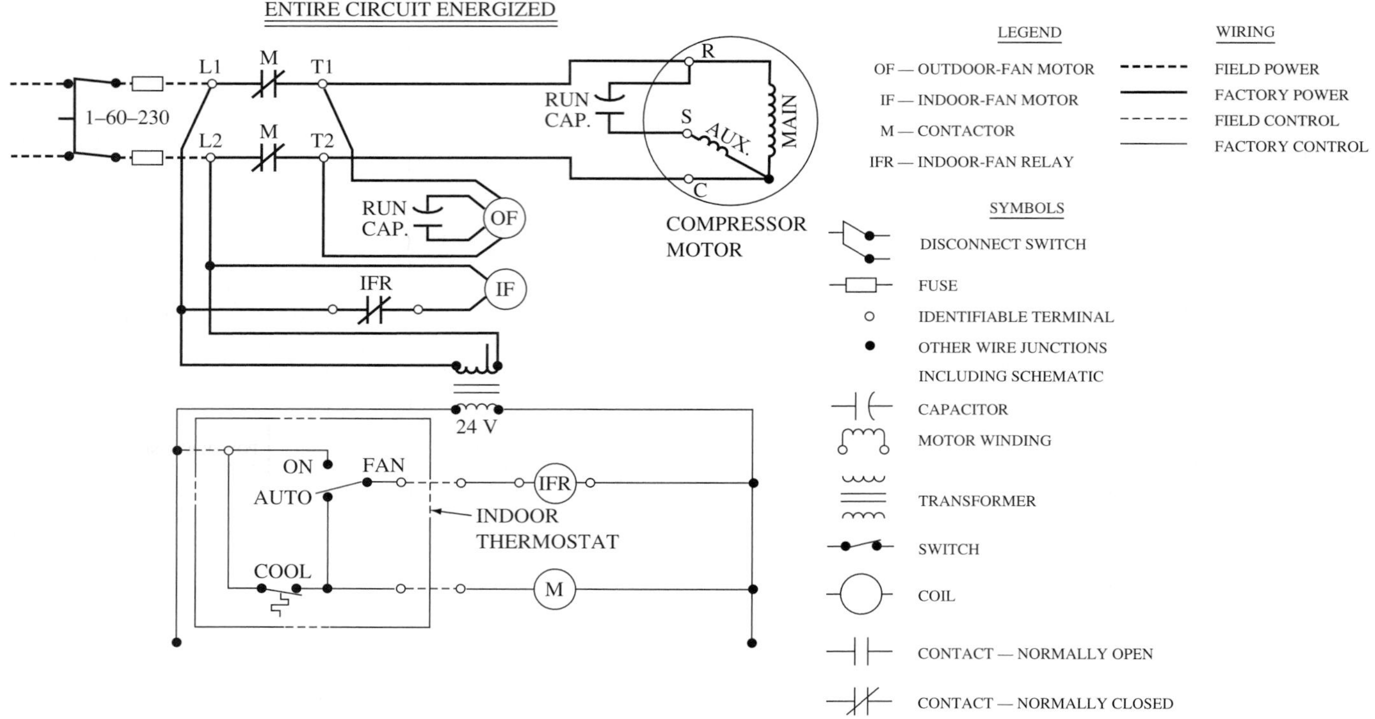

Figure H35-69 Wiring diagram showing fan in AUTO position and thermostat calling for cooling.

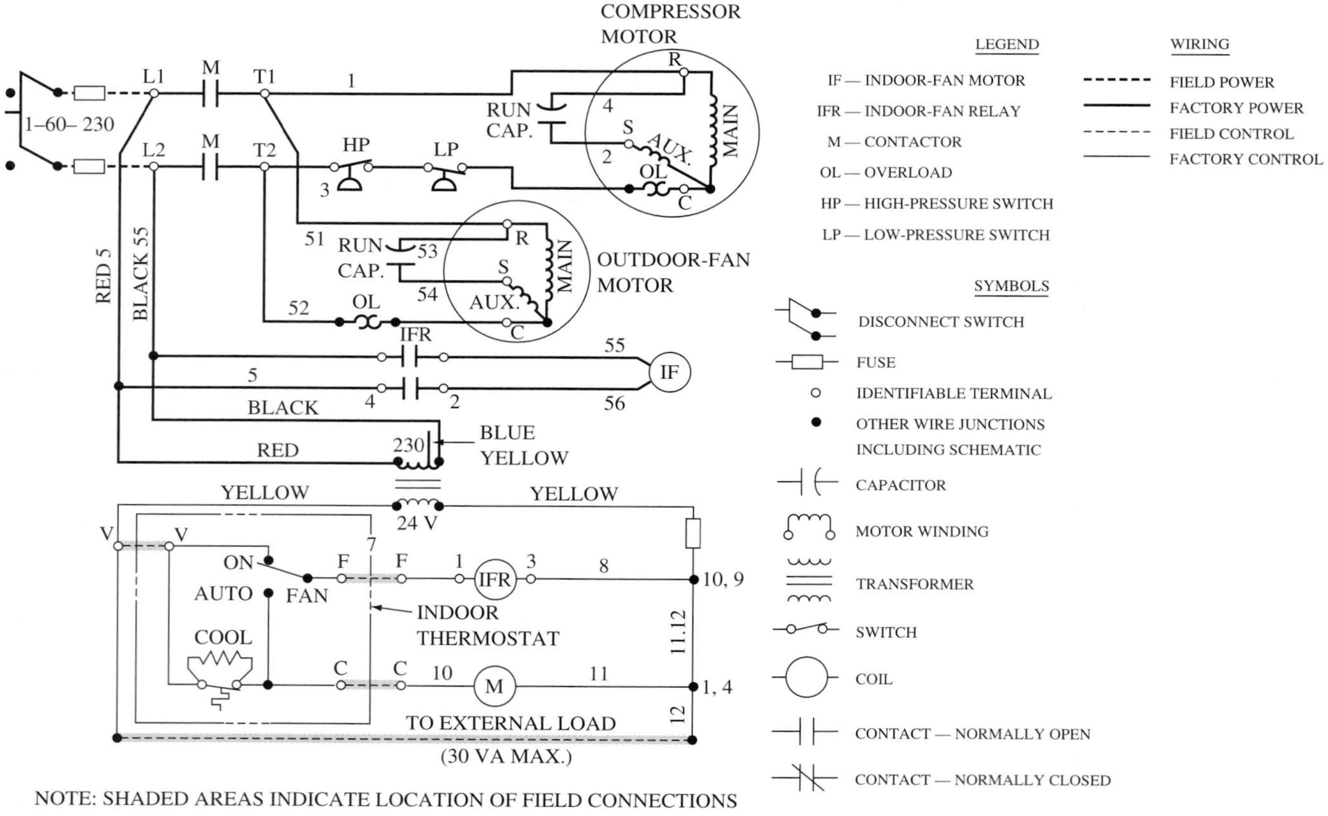

COMPRESSOR MOTOR
L1
M
T1
1
R
RUN CAP.
4
S
AUX.
MAIN
1–60– 230
L2
M
T2
HP
LP
2
OL
C
3
51
RUN CAP.
53
R
S
MAIN
OUTDOOR-FAN MOTOR
54
AUX.
RED 5
BLACK 55
52
OL
C
IFR
55
IF
5
BLACK
4
2
56
RED
230
BLUE
YELLOW
YELLOW
24 V
YELLOW
V
V
7
ON
F
F
1
IFR
3
8
10, 9
AUTO
FAN
INDOOR THERMOSTAT
11.12
COOL
C
C
10
M
11
1, 4
12
TO EXTERNAL LOAD (30 VA MAX.)
NOTE: SHADED AREAS INDICATE LOCATION OF FIELD CONNECTIONS

LEGEND
IF — INDOOR-FAN MOTOR
IFR — INDOOR-FAN RELAY
M — CONTACTOR
OL — OVERLOAD
HP — HIGH-PRESSURE SWITCH
LP — LOW-PRESSURE SWITCH

WIRING
FIELD POWER
FACTORY POWER
FIELD CONTROL
FACTORY CONTROL

SYMBOLS
DISCONNECT SWITCH
FUSE
IDENTIFIABLE TERMINAL
OTHER WIRE JUNCTIONS INCLUDING SCHEMATIC
CAPACITOR
MOTOR WINDING
TRANSFORMER
SWITCH
COIL
CONTACT — NORMALLY OPEN
CONTACT — NORMALLY CLOSED

Figure H35-70 Complete ladder diagram for the air-conditioning unit.

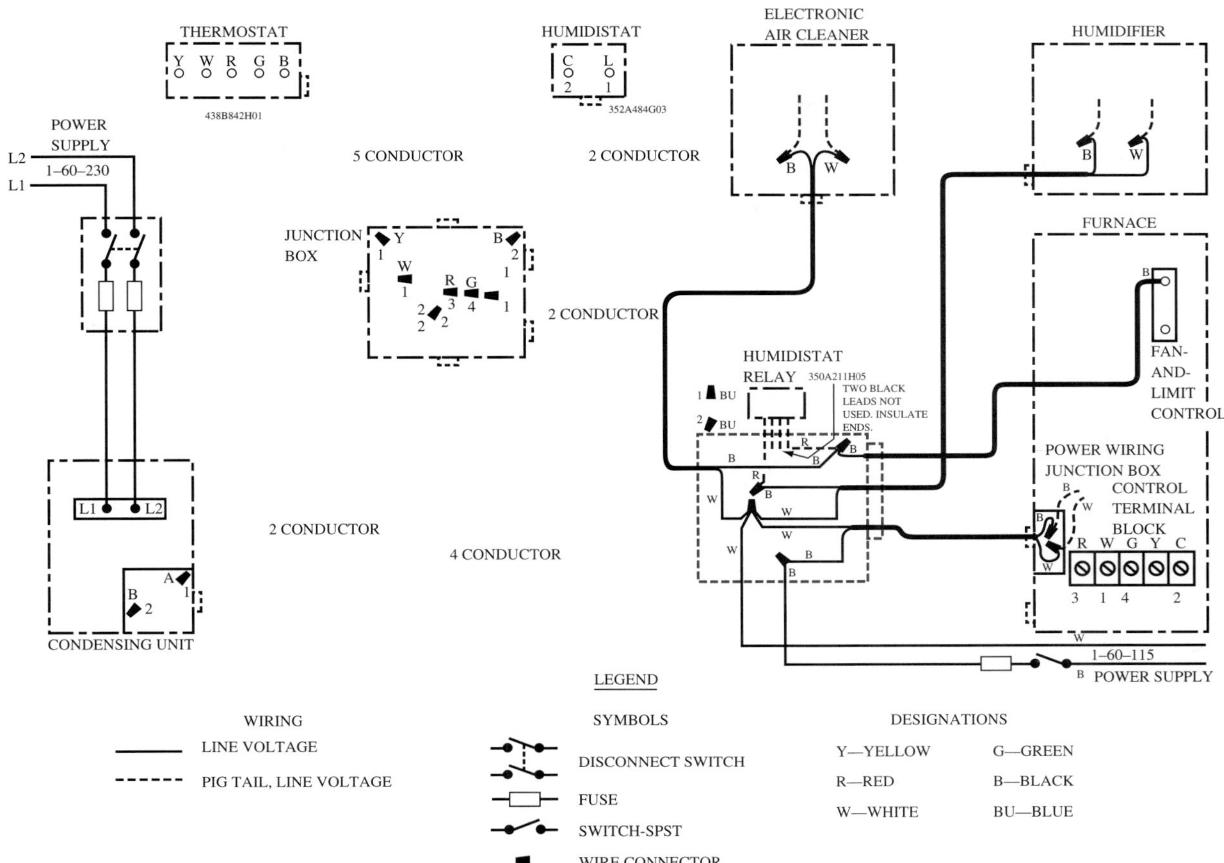

Figure H35-71 Connection diagram showing the wiring harnesses used to connect the terminals.

as to the number of conductors needed and the terminal or junction connections. Power wiring is black (B) and white (W), and ground conductors are sized according to the load and to meet local code requirements. The internal wiring of the furnace, condensing unit, and other components are not shown on the installation diagram.

The schematic ladder diagram for the same system is shown in Fig. H35-72. First locate and relate the major components to the previous drawing: the furnace blower motor, the humidifier, the electronic air cleaner, and the condensing unit. Note that the combination thermostat has a HEAT-OFF-COOL system selection and an ON-AUTO fan selection. Anticipators are not shown. The system diagram is shown on the heating mode.

Closing the disconnect switch (line 4) will produce a 120-V current to the transformer. Trace the black wire through the closed limit switch (LS) to the transformer primary coil and to the white power return (line 5). A low-voltage potential is created.

Low voltage can now be established through the fan switch ON position and energize the FR (fan relay) coil in line 9. This closes the FR switch in line 3, allowing line-voltage current to flow to the fan motor and to the humidifier relay (HU) in line 1. Thus the fan runs constantly. Note that the FS (furnace switch) in line 4 may or may not be closed, depending on whether there is sufficient heat to close its contacts. The gas valve (GS) operation in line 13 will depend on the activation of the mercury bulb by the room temperature. Note in line 15 that the HEAT position selector switch feeds low voltage to H (the humidistat), and if humidity is needed, it closes its contacts and energizes the humidistat relay coil (HU), also in line 15. The contacts of HU are located in line 1 and will activate the humidifier (fan motor or water solenoid valve). This is called *permissive humidification* by automatic demand from the humidistat on the heating cycle with the normal cycle of furnace fan. *Command humidification* would be a system where the humidistat can command fan-only operation, regardless of the heating/cooling mode. The electronic air cleaner simply parallels the fan operation and is energized any time the fan runs.

If the fan selection is AUTOMATIC, FR cannot be energized during heating, but the furnace fan cycles from its fan switch, line 4. On cooling, however, there is a contact to the automatic terminal and the fan cycles with the cooling thermostat. Also, on cooling there is a completed cir-

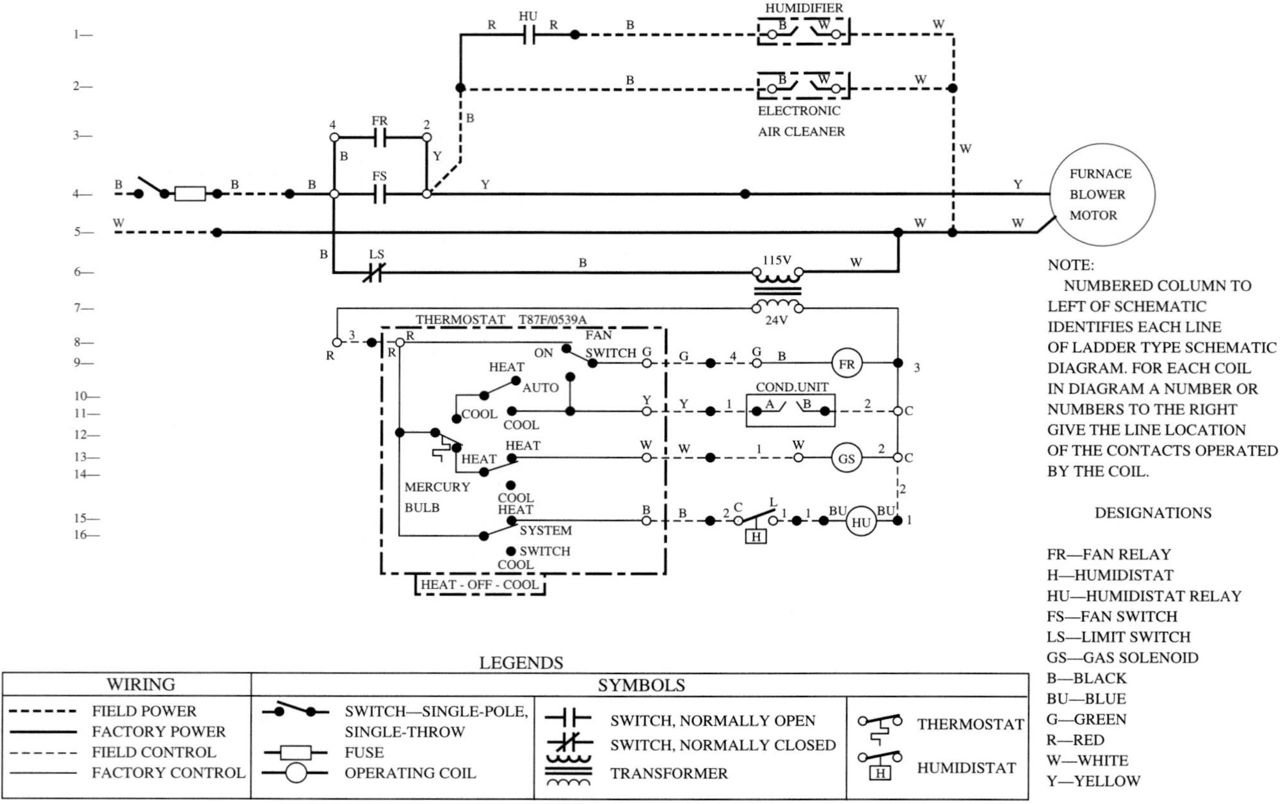

Figure H35-72 A complete schematic wiring diagram of the heating and air-conditioning unit.

cuit on line 11 to the condensing unit contactor terminals A and B. This pulls in the compressor and the outdoor condenser fan. They then cycle off and on with the room temperature mercury switch. Note that the middle system switch on cool drops out the gas valve and the lower switch drops out the humidistat. Note also that on cooling, the furnace fan can operate from the FR relay (ON or AUTO) and thus furnish power to the air cleaner whenever it runs.

This is how a simple but common TCS is installed and wired schematically. The introduction of furnace fan-speed control, two-stage cooling and heating thermostats, and command humidification, add sophistication and enlarge the diagram. These items do not necessarily add to its complexity if the schematic is analyzed step by step as to function and operation.

REVIEW

- Control system—Checks or regulates, within prescribed limits, the functions of an HVAC/R system
- Component parts of a control system:
  Power source—Electricity, electronics, pneumatic.
  Controllers—Obtain the desired levels of the end results.
  Operators (load or loads)—Use the power to obtain the desired results.
- Heating and cooling control circuits can be designed for either line- or low-voltage control.
- Step-down transformer—Reduces line voltage to low voltage. Transformers are rated in volt-amperes (va)
- Thermostats:
  Line voltage—direct resistant heating
  Low voltage—universal usage
- Components of thermostat:
  Base
  Subbase
  Cover
- Heat anticipation—Reduces the effect of system lag operating differential and overshoot of heating; variable resistance is adjustable.
- Cooling anticipation—Fixed resistance; reduces system drop.
- Humidistat—Controller that controls equipment by sensing humidity; may be wall or duct mounted.
- Types of temperature/pressure-sensing devices:
  Bimetallic—temperature
  Bellows and diaphragms—temperature or pressure

Bourdon tube—pressure
Rod and tube—temperature
Resistance elements—temperature
Static pressure regulator—pressure

- Controllers:
  Two-position—on/off switching
  Proportional—analog output
- Relays—Use a low-voltage signal to operate switches in line-voltage circuits.
- Operators:
  Motors
  Solenoids
  Valves
- Gas heat controls:
  Valves—Manual control, pilot light, pressure regulator, on/off control
  Ignition circuits—Electronic-direct spark ignition, hot surface
- Oil heat controls:
  Protector relays
  Cad cell
  Oil valves—instant on/off, delay type
- Schematic wiring diagrams—Include fan motor, automatic gas valve, control-voltage transformer, and other devices. If cooling is added a condensing unit contactor is required, to operate the compressor and condenser fan motor.

Problems and Questions

1. What are the three major elements in a control system?
2. Give an example of a common controller.
3. Give an example of a controlled load.
4. What is the most common power source in air-conditioning control circuits?
5. Define a power-type transformer.
6. When the voltage out is lower than the voltage in, the transformer is classified as a _____ transformer.
7. The change in voltage is determined by _____.
8. The change in amperage capacity is in a direct ratio with the change in voltage. True or false?
9. The capacity of a transformer is rated in _____, which is determined by _____.
10. What is the bimetallic principle?
11. Why are low-voltage thermostats used instead of high-voltage types?
12. Heating and cooling anticipators are actually small _____.
13. Is a heating anticipator connected in series with or parallel to the heating contact?
14. Is a cooling anticipator connected in series with or parallel to the cooling contact?
15. Are most heating thermostats fixed or adjustable?
16. By what is the setting of an adjustable heat anticipator determined?
17. What are other types of temperature control sensing elements?
18. A thermistor functions by changing its _____ with a change in temperature.
19. High-pressure controls are made in both _____ and _____ reset types.
20. What are the most common forms of bimetallic configuration used in controls?
21. What are the common forms of power elements used in pressure controls?
22. The control action of controls are of two types. What are they?
23. Define and give an example of an intermittent operator.
24. Explain the difference between a relay, a contactor, and a motor starter.
25. Name the five functions of a natural-gas combination gas valve.
26. What is the function of a combination gas valve that is not used with LP gas supply?
27. It is possible to pass an electrical current through a gas flame. True or false?
28. An oil burner is controlled by a device called a _____.
29. A stack-mounted oil-burner control is operated by a _____ , which is actuated by _____.
30. A burner-mounted oil-burner control is operated by a _____ , which is actuated by _____.
31. A delayed-action oil valve is used to ensure that _____.
32. Some heat anticipators are fixed, while others are wire-wound variable resistors wired in series with the load. True or false?
33. The proportional controller utilizes a modulating potentiometer which then goes to a two-position valve. True or false?
34. A combination gas valve provides
    a. pilot supply, adjustment, and safety shutoff.
    b. pressure regulation of burner gas feed.
    c. manual control for ignition and on/off electric solenoid valve.
    d. All of the above.
35. Hot-surface ignition units are used for ignition on modern gas furnaces
    a. which use high-voltage step-up transformers.
    b. which use 120-V power.
    c. which use 24-V power.
    d. None of the above.

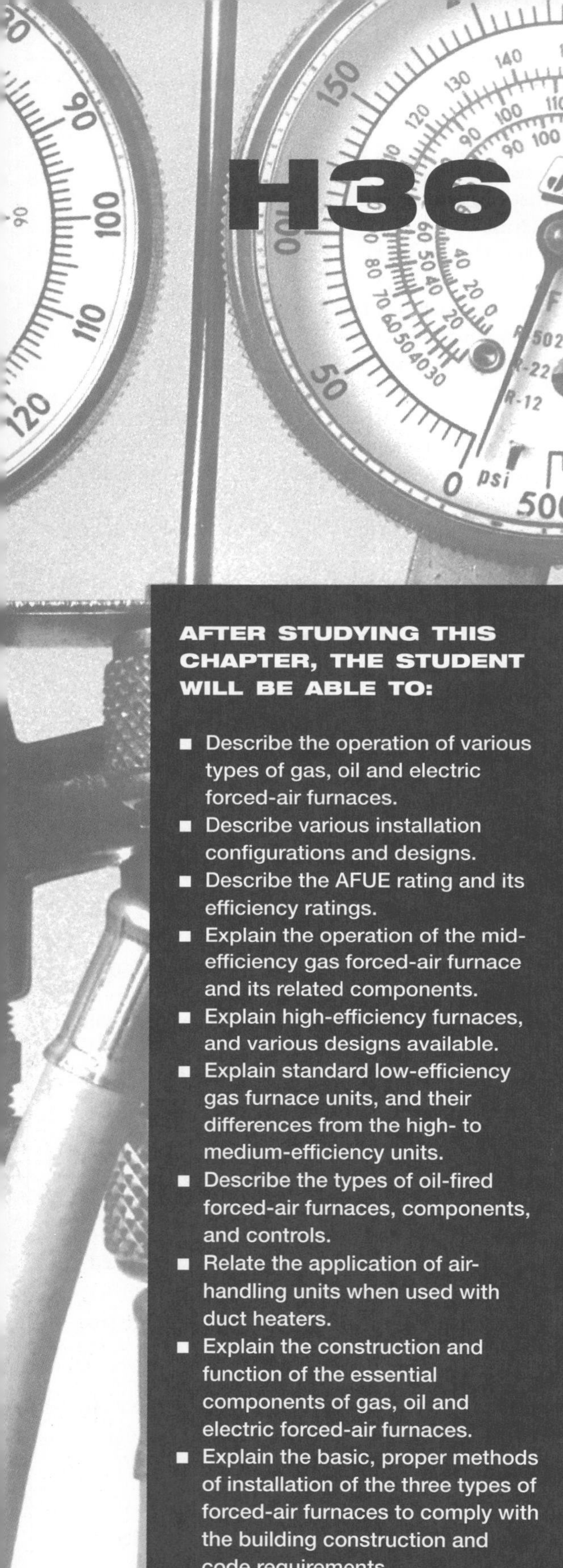

AFTER STUDYING THIS CHAPTER, THE STUDENT WILL BE ABLE TO:

- Describe the operation of various types of gas, oil and electric forced-air furnaces.
- Describe various installation configurations and designs.
- Describe the AFUE rating and its efficiency ratings.
- Explain the operation of the mid-efficiency gas forced-air furnace and its related components.
- Explain high-efficiency furnaces, and various designs available.
- Explain standard low-efficiency gas furnace units, and their differences from the high- to medium-efficiency units.
- Describe the types of oil-fired forced-air furnaces, components, and controls.
- Relate the application of air-handling units when used with duct heaters.
- Explain the construction and function of the essential components of gas, oil and electric forced-air furnaces.
- Explain the basic, proper methods of installation of the three types of forced-air furnaces to comply with the building construction and code requirements.

Warm-air Furnaces

H36-1 OVERVIEW

Residential furnaces are available in a variety of configurations to facilitate (1) installation locations and the (2) type of heat source. Installation requirements have also been responsible for a variety of mounting arrangements and airflow designs.

H36-1.1 Styles Based on Installation Location

Furnaces can be categorized into four styles: (1) up-flow, (2) lowboy, (3) down-flow or counter-flow, and (4) horizontal.

The *up-flow highboy* (Fig. H36-1) is most popular. Its narrow width and depth allow for location in first-floor closets and/or utility rooms. It can still be used in most basement applications for heating only or with cooling coils where head room space permits. Blowers are usually direct-drive multi-speed. Air intake can be either from the sides or from the bottom.

Lowboy furnaces (Fig. H36-2) are built low in height to accommodate areas where head room is minimal. These furnaces are approximately 4 ft high, providing for easy installation in a substandard basement. Supply and return ducts are mounted on top for easy attachment. Blowers are commonly belt driven. Many of these furnaces are sold for retrofitting older homes.

The *counter-flow* or *down-flow* furnace (Fig. H36-3) is similar in design and style to the highboy, except that the air intake and fan are at the top and the air discharge is at the bottom. These are widely used where duct systems are set in concrete or in a crawl space beneath the floor. A fireproof base is required when the furnace is installed on a combustible floor. An extra safety limit control is also used.

The fourth style, the *horizontal* furnace (Fig. H36-4), is adaptable to installation in low areas such as crawl spaces, attics, or partial basements. It requires no floor space.

Figure H36-1 Up-flow highboy furnace. (Courtesy of Bard Manufacturing Company)

Figure H36-2 Lowboy furnace. (Courtesy of Bard Manufacturing Company)

Intake air enters at one end and is discharged out the other end. Burners are usually field-changeable for left- or right-hand application.

Typical installations of the horizontal, counter-flow and up-flow furnaces are shown in Fig. H36-5.

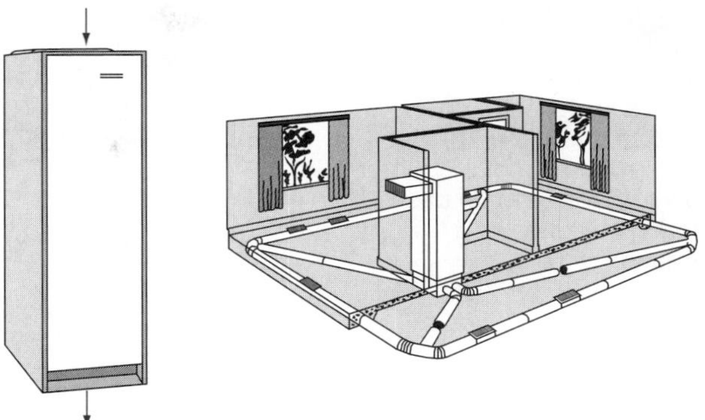

Figure H36-3 Counter-flow furnace.

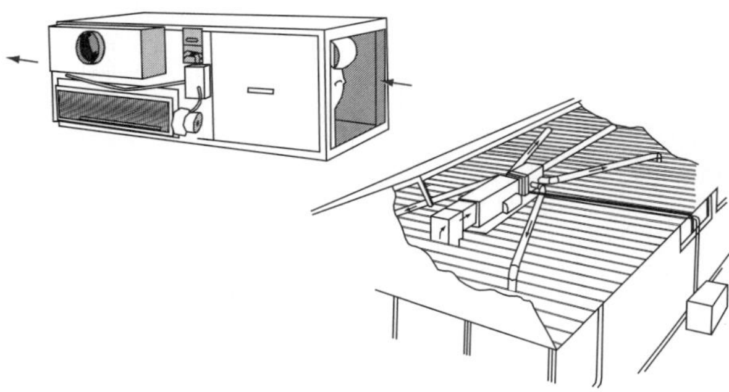

Figure H36-4 Horizontal furnace.

H36-1.2 Type of Heat Source

Furnaces may also be described in relation to the type of heat source: (1) fuel-burning furnaces and (2) electric furnaces.

In *fuel-burning furnaces,* fuel is burned in the combustion chamber. Circulating air passes over the outside surface of the heat exchanger. The products of combustion are vented to the atmosphere. Fuels used in these systems include (1) gas, both natural and LP, and (2) oil.

In an *electric furnace,* circulating air is directly heated by resistance-type heating elements or through a metal sheath that encloses the resistance element.

H36-2 GAS FORCED-AIR SYSTEMS

Due to the need to conserve energy, two distinctly new types of furnaces have been developed: (1) the mid-efficiency and (2) the high-efficiency condensing-type furnace.

The *mid-efficiency furnaces* (Fig. H36-6) meet the minimum requirement of 78% AFUE (Annual Fuel Utilization Efficiency) and have a flue gas temperature above the dew

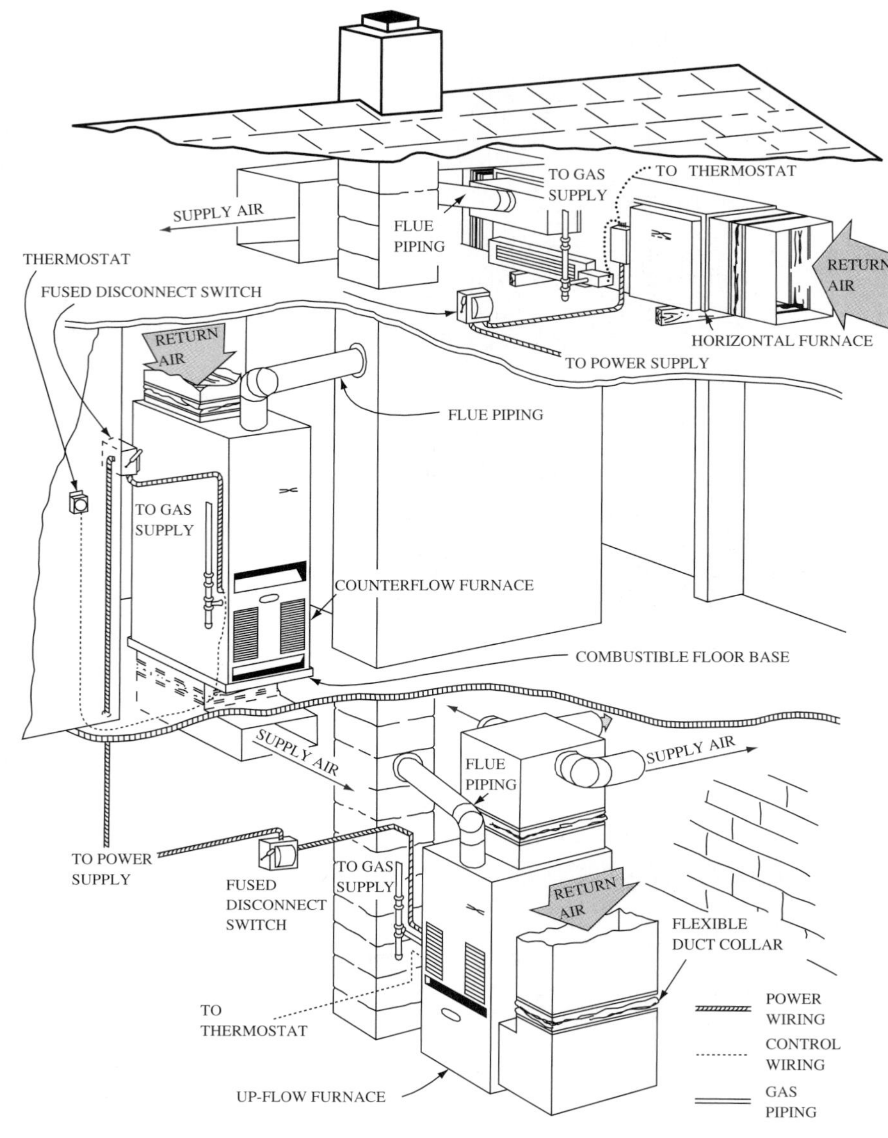

Figure H36-5
Composite diagram of three types of furnace installations: (1) Attic installation of a horizontal furnace; (2) first-floor installation of a counter-flow furnace; and (3) basement installation of a lowboy furnace.

PRACTICING PROFESSIONAL SERVICES

Safety should be emphasized when you work with warm-air furnaces which use combustible fuels such as oil or gas. Always inspect fittings for leaks by using soap bubbles or electronic leak detectors. If leaks are found, shut down the unit at the disconnect switch and repair the leak.

During ignition startup of the unit, do not look at the flame directly until it has ignited. Stand to the side of the unit and observe ignition. If there is flame roll-out, this will prevent burning your face or skin. Again, if there is a problem with ignition, shut down the unit at the disconnect and repair as necessary.

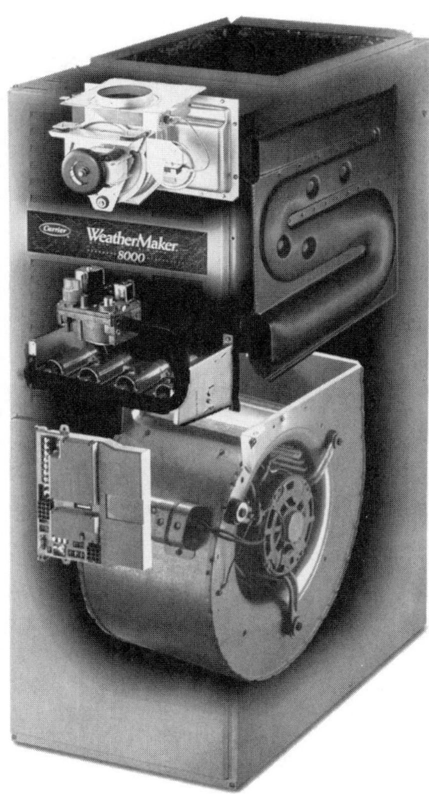

Figure H36-6 Internal view of the major components of a mid-efficiency gas furnace. (Courtesy of Carrier Corporation—Residential Products)

point. The *high-efficiency furnaces* permit condensation within the furnace to pick up the extra latent heat. Since certain components are different, they will be described separately.

Large numbers of the earlier "standard" type furnaces are still in use and require service. They will also be described in this text, although they are no longer being manufactured.

H36-2.1 Mid-efficiency Furnaces (80% AFUE)

A sketch of the internal components of a mid-efficiency furnace is shown in Fig. H36-7. The parts include:

1. The *inducer assembly,* with a forced-draft blower for propelling the flue gases through the heat exchanger and into the vent, as shown in Fig. H36-8. The inducer motor pulls the exact amount of air needed through the heat exchanger. Assembly is resilient-mounted to provide quiet operation.
2. The *pressure switch,* which provides a positive indication that the inducer fan is operating before permitting the fuel to be ignited.
3. The *gas-control valve,* which meters the gas fuel into the burner. The gas valve is the slow-opening type with provision for 100% shutoff to provide safe operation.
4. The *burner assembly,* which provides for proper mixing of fuel and air, hot-surface silicon carbide ignition, and combustion of the fuel. Aluminized steel burners are monoport, in-shot type.
5. The *blower-door safety switch,* which disconnects the power supply to the unit whenever the front access panel is removed.
6. The *control box,* which houses controls, including a microprocessor board that controls most operations and functions of the unit. It provides a blower delay in start-up and shutdown, while monitoring furnace performance.

   The technician can use this self-testing feature to identify a major component failure. The control board will check itself, then the inducer, silicon-carbide ignition, low- and high-speed blower operation, and humidifier connections. The control board includes a 3-amp fuse that protects the transformer and control board. The board also includes an LED status indicator light.
7. The *air-filter retainer,* which holds the air filter.
8. The *air filter,* which is either the repealable type or an optional electrostatic filter that can be applied at the return-air opening.
9. The *wrap-around casing,* which is a one-piece seamless construction. Finish is double protected: (1) a galvanized steel substrata provides resistance to rusting, and (2) the cabinet is prepainted with the same high-quality finish found on refrigerators.
10. The *heat exchanger,* which transfers the heat of combustion to the *air-distribution system.* This design uses four-pass aluminized steel construction and carries a 20-year limited warranty.
11. The *blower* and *blower motor,* which force the air over the external surface of the heat exchanger, to pick up the heat for delivery to the space being conditioned. The motor is a direct-drive multiple-speed type.

H36-2.2 Product Data (80% AFUE Furnace)

The following is typical information supplied by the manufacturer about the product. The technician needs to check the installation, either at the time of startup or whenever a service problem arises, to determine that the product has been properly applied. Some of this information appears on the nameplate, but not all of it. This data can be very useful.

An external wiring diagram for a mid-efficiency furnace is shown in Fig. H36-9. Although these units are primarily designed for heating, a number of accessories can be added (Fig. H36-10, p. 732), such as a (1) V-type cool-

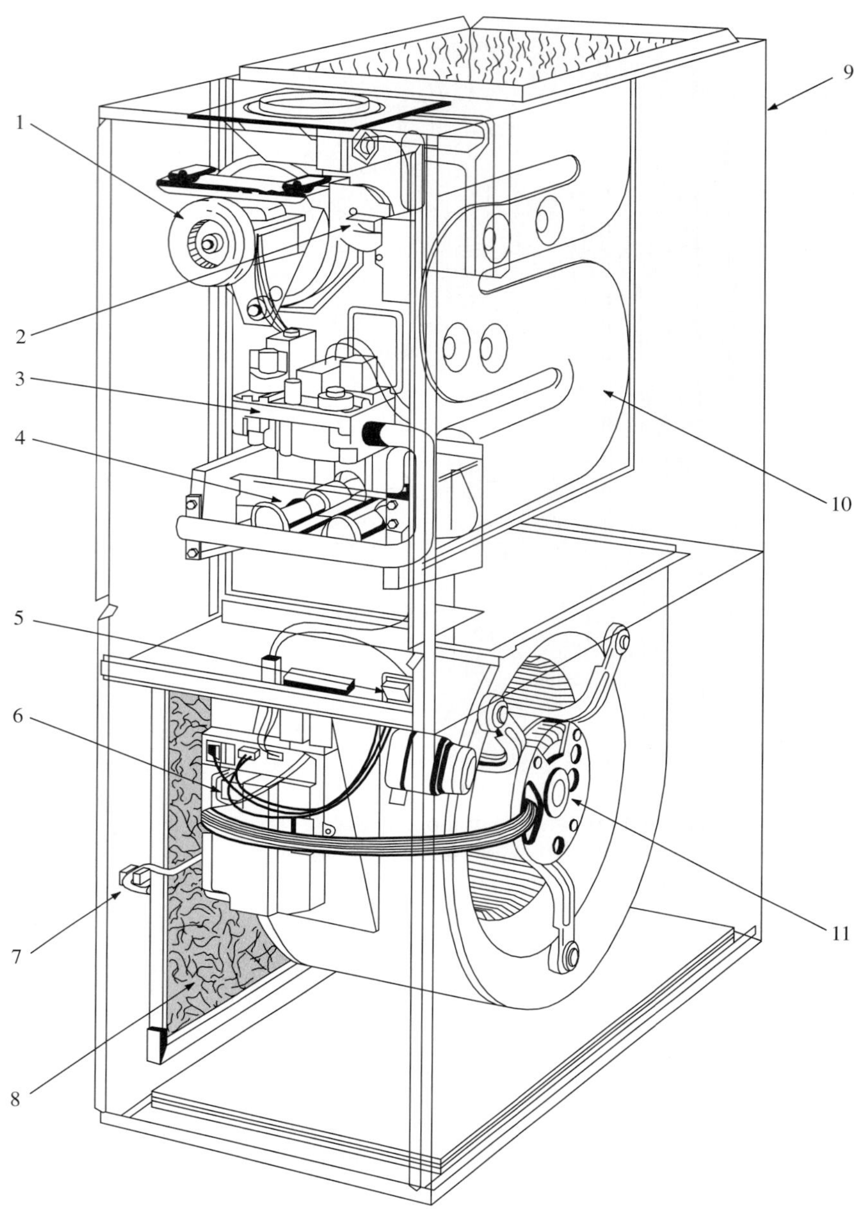

Figure H36-7 Internal view of a mid-efficiency gas furnace, showing the component parts. (Courtesy of Carrier Corporation)

1. INDUCER ASSEMBLY
2. PRESSURE SWITCH
3. GAS CONTROL VALVE
4. BURNER ASSEMBLY
5. BLOWER DOOR SAFETY SWITCH
6. CONTROL BOX
7. AIR FILTER RETAINER
8. AIR FILTER
9. WRAP-AROUND CASING
10. HEAT EXCHANGER
11. BLOWER AND BLOWER MOTOR

ON THE JOB

When working on a customer's furnace equipment, avoid tracking grease and/or dirt into her home after the work is complete. If you are in a basement and must exit the house across a clean floor or rug, be sure to inspect and clean the bottom of your shoes to ensure that you won't leave any tracks behind. If, in your previous job location, you were on a rooftop or in a dirty mechanical room wipe your shoes off prior to entering the customer's home. A little time spent ensuring your footwear is clean will help guarantee customer satisfaction.

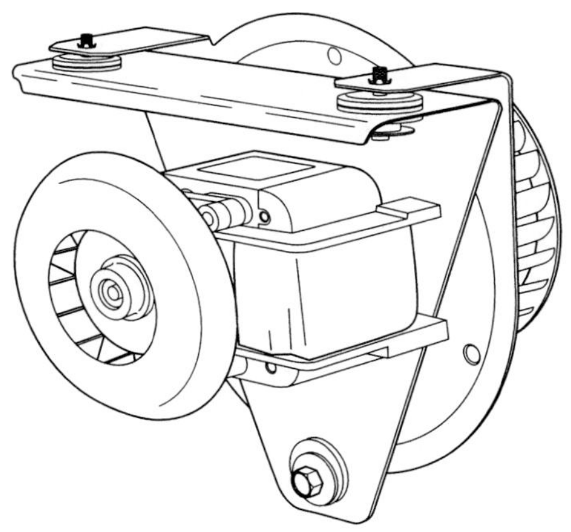

Figure H36-8 An inducer-blower used to draw the combustion gases through the furnace. (Courtesy of Carrier Corporation—Residential Products)

ing coil installed in the supply plenum, (2) a humidifier, installed on the supply duct, and (3) an electronic air cleaner, installed at the return-air entrance to the unit.

Due to the unique construction of these furnaces, there is a limit to the sizes available. The 80% AFUE units usually have output ranges between 35,000 and 124,000 Btu/hr. It has, therefore, been useful on larger jobs to install two of them linked together in a configuration called a *twin.* Twinning kits are available for field linking as shown in Fig. H36-11. A number of parts are required and the manufacturer's instructions must be carefully followed to coordinate the two systems.

H36-3 GAS HIGH-EFFICIENCY FURNACES (HEF)

High-efficiency or *condensing-type gas furnaces* are rated at 90% AFUE or better. They differ from the 80% AFUE mid-efficiency furnaces in that an extra heat exchanger

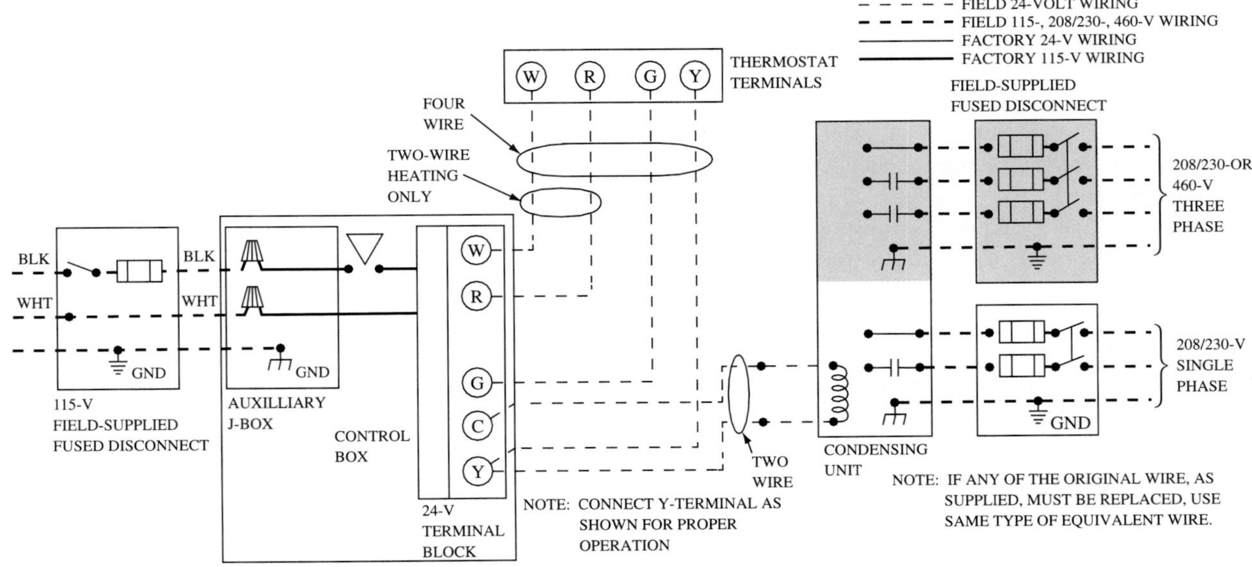

Figure H36-9 Typical schematic wiring diagram for a mid-efficiency furnace. (Courtesy of Carrier Corporation)

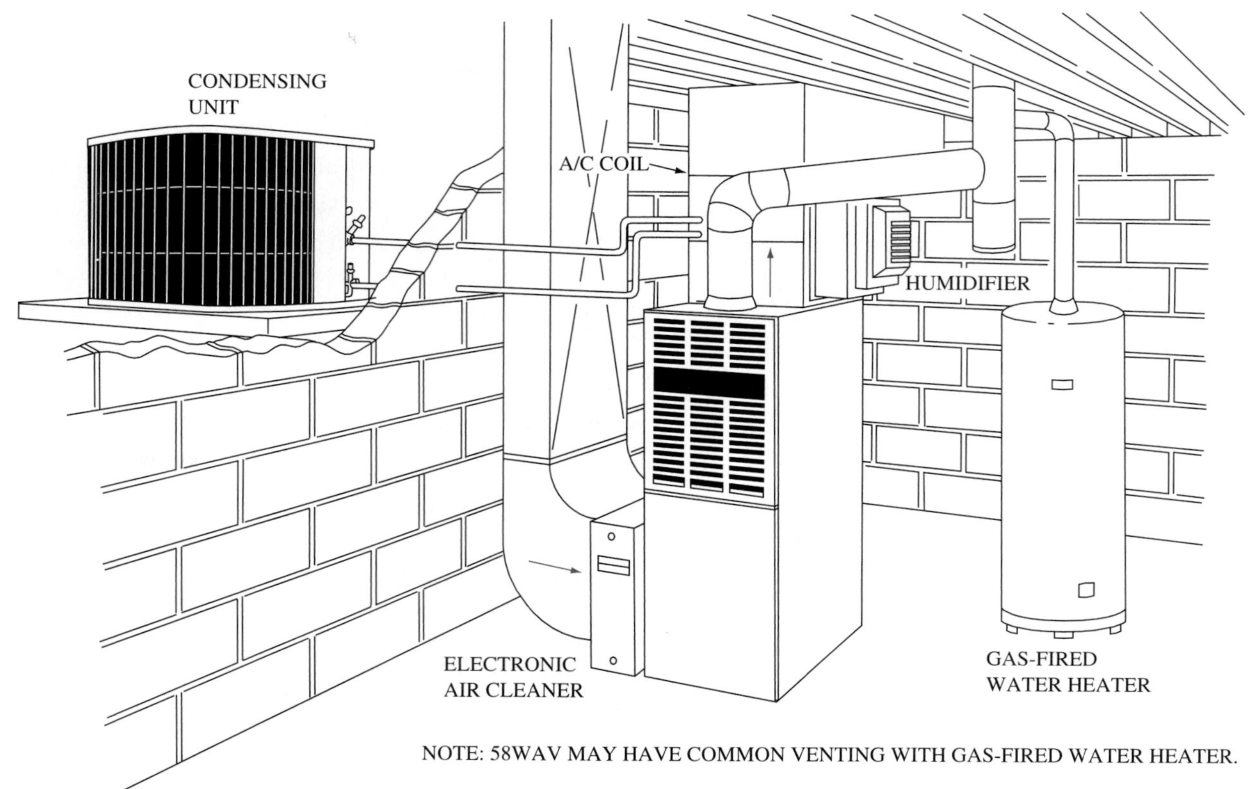

Figure H36-10 Typical installation of an up-flow furnace, with accessories. (Courtesy of Carrier Corporation)

(secondary heat exchanger) is added to extract more heat from the flue gases. This additional surface causes the moisture in the flue gas to condense. Additional heat is obtained by (1) further lowering the temperature of the flue gas and (2) picking up the heat rejected by condensing moisture.

This construction reduces the volume and temperature of the flue gas, making it possible to use smaller vent pipes and simplify venting arrangements. Since the temperature of the flue gas is lower, PVC vent piping can be used and vent outlets can be run to the side of a building, if more convenient.

A condensate line is required to dispose of the condensed water. This needs to be run to the public sewer lines. The condensate usually contains some contaminants. Local codes need to be examined to determine the possible need for a condensate neutralizer kit.

The following illustrations show the construction of a number of different types of high-efficiency furnaces:

1. Variable capacity, 96.6% AFUE furnace, Fig. H36-12;
2. Sealed combustion chamber, 92.6% AFUE furnace, Fig. H36-13; and
3. Tubular heat exchanger, 90% AFUE furnace, Fig. H36-14 and the pulse-type, Fig. H36-15 (p. 735).

The highest rating (96.6 AFUE) indicated above is obtained using a variable-capacity furnace with a two-stage gas valve, a variable-speed inducer fan, and a solid-state controlled blower motor. The speed of both fans is electronically controlled to match the air-volume requirements.

The pulse-type furnace is unique and operates entirely differently from the others. The essential parts of the pulse combustion unit and its sequence of operation is shown in Fig. H36-16 (p. 735).

Combustion takes place in a finned, cast-iron chamber. Note in the lower part of this chamber the gas intake, the air intake, the spark-plug ignition, and the flame sensor. At the time of initial combustion the spark plug ignites the gas-air mixture. The pressure in the chamber following combustion closes the gas and air intakes. The pressure buildup also forces the hot gases out of the combustion chamber, through the tailpipe, into the heat-exchanger exhaust decoupler, and into the heat-exchanger coil.

As the chamber empties after combustion, its pressure becomes negative, drawing in a new supply of gas and air for the next pulse of combustion. The flame remnants of the previous combustion ignite the new gas-air mixtures and the cycle continues. Once combustion is started, the purge fan and the spark ignitor are shut off. These pulse cycles occur about 60 to 70 times a second.

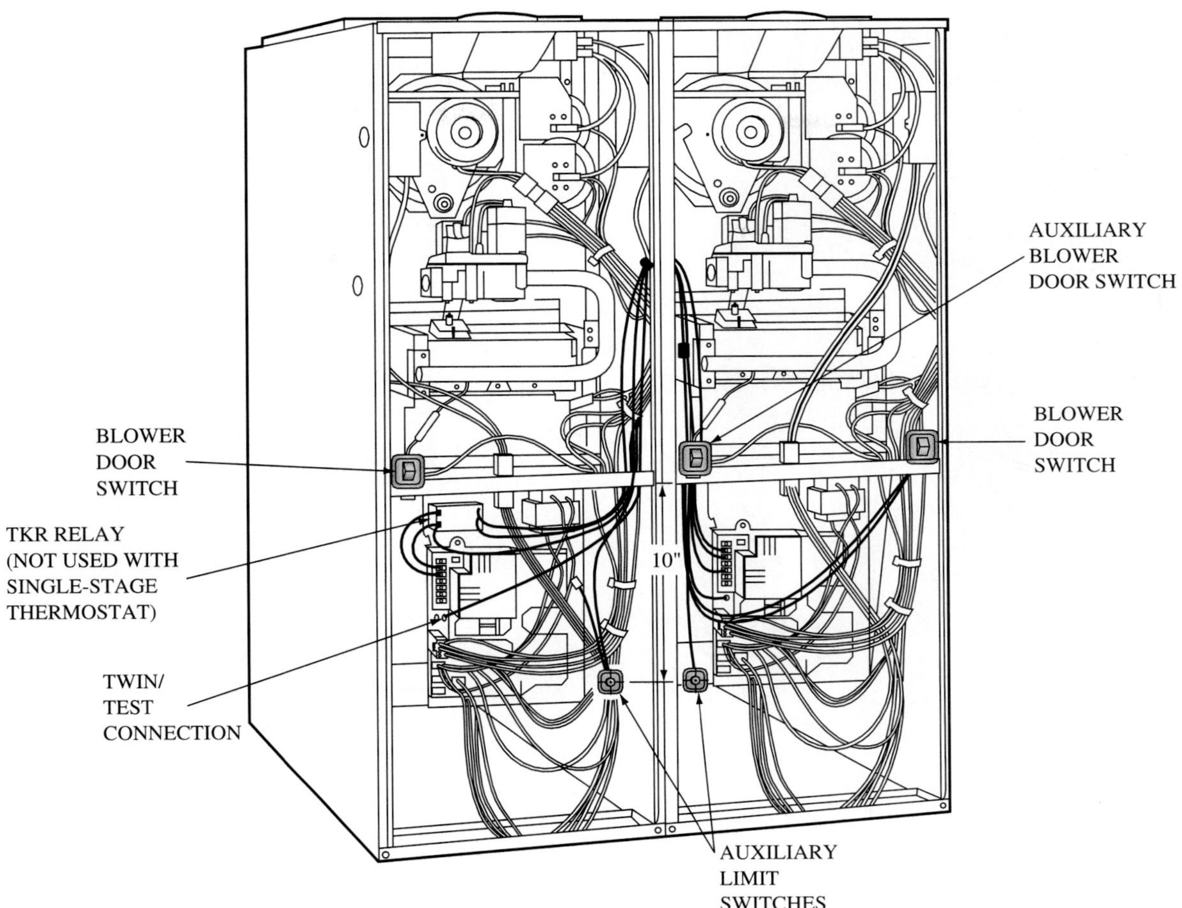

Figure H36-11 Twinning installation showing the method of installation. (Courtesy of Carrier Corporation)

The sequence of operation of the pulse furnace is as follows: When the room thermometer calls for heat, it initiates the operation of the purge fan, which runs for 34 seconds. This is followed by the turning on of the ignition and the opening of the gas valve. The flame sensor provides proof of ignition and de-energizes the purge fan and spark ignition.

When the thermostat is satisfied, the gas valve closes and the purge fan is turned on for 34 seconds. The furnace continues to operate until the temperature in the plenum reaches 90°F. In case the flame is lost before the thermostat is satisfied, the flame sensor will try to re-ignite the gas-air mixture three or five times before locking out. Should there be either a loss of intake gas or air, the furnace will shut down automatically.

Note that combustion air is piped with the same PVC pipe as used for the exhaust gases. The condensate liquids range from a pH of 4.0 to 6.0, which permits it to be drained into city sewers or septic tanks. The combustion mixture of gas and air are preset at the factory, and no field adjustments are necessary.

The 90–92% AFUE models have much in common. The following is a list of the components of a typical unit (Fig. H36-17, p. 736):

1. *Combustion-air intake connection.* This ensures contamination-free combustion air, accessible from either the right or left side.
2. *Burner sight glass.*
3. *Burner assembly* (inside). In-shot burners are energy-saving and operate with a safe hot-surface igniter.
4. *Redundant gas valve,* having one control for two internal shut off valves.
5. *Vent outlet.* PVC pipe is used to carry vent gases from either the right or left side of the furnace's combustion system.
6. *Inducer motor.* Hot flue gases are pulled through the heat exchangers. Negative pressure is maintained for additional safety.
7. *Blower-access-panel safety-interlock switch.*
8. *Air filter and retainer,* used for side return option.

Figure H36-12 Internal view of a Carrier high-efficiency AFUE 96% furnace. (Courtesy of Carrier Corporation—Residential Products)

Figure H36-14 Internal view of a Quatro high-efficiency gas furnace. (Courtesy of Quatro® by Consolidated Industries)

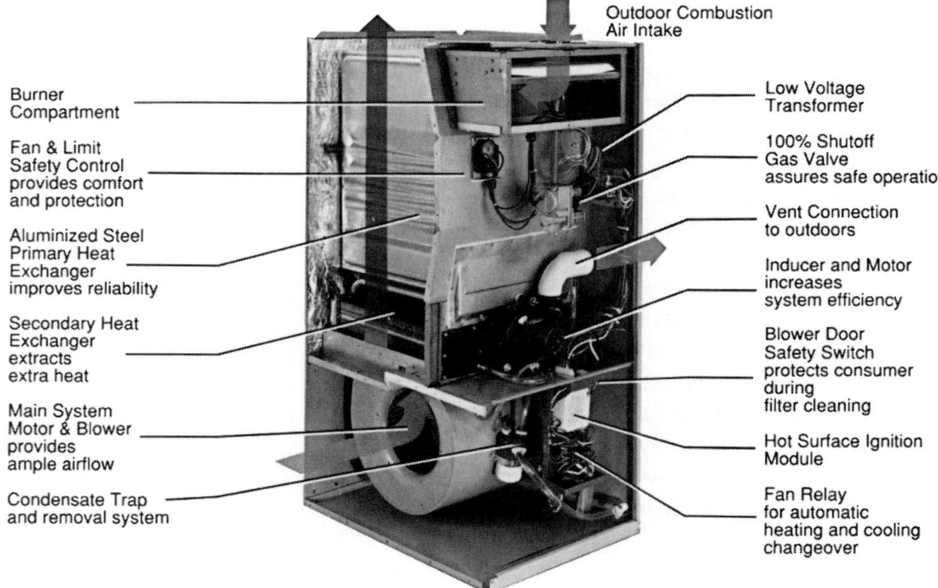

Figure H36-13 Internal view of a York high-efficiency gas furnace. (Courtesy of York International Corp.)

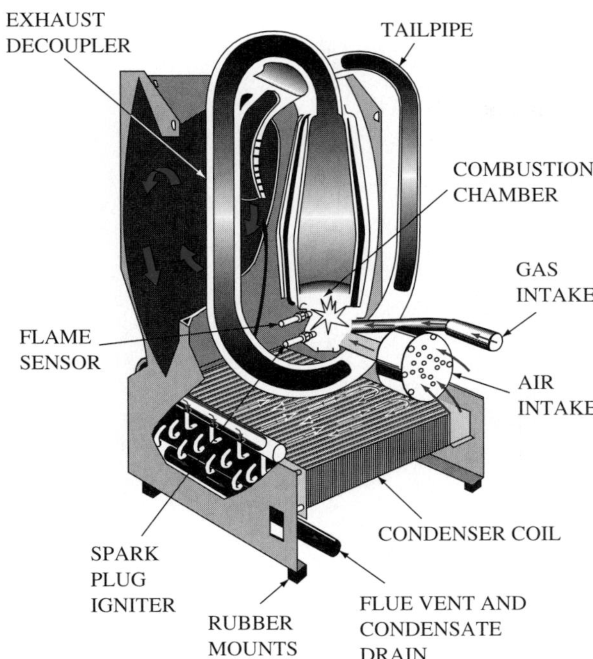

Figure H36-15 Internal view of a Lennox pulse-type high-efficiency gas furnace.

9. *Condensate drain connection.* This collects the condensed moisture from the combustion process.
10. *Heavy-duty blower.* Circulates air over heat exchangers for transfer to area to be heated.
11. *Secondary condensing heat exchanger* (inside). Removes additional heat through condensation. Made of durable polypropylene laminated steel.
12. *Primary serpentine heat exchanger* (inside). The S-shaped heat-flow design conserves fuel. It is constructed of corrosion-resistant aluminized steel.
13. *Control center.*
14. 3-amp *fuse* provides electrical and component protection.
15. *Light-emitting diode* (LED) on control center. Lights are coded for diagnosing furnace operation and service requirements.
16. *Pressure switch.* Ensures that flue products are adequately vented out of the furnace.
17. *Rollout switch* (manual reset) prevents overheating.
18. *Junction box* for 115-V electrical power system.
19. *Transformer* (24-V) behind the control center provides low voltage power for the control center and thermostat.

H36-3.1 Product Data (90%+ AFUE Furnaces)

The product data for the 90%+ AFUE furnaces covers categories similar to that of the 80% AFUE models, with a few exceptions. Since these are condensing-type furnaces, provision needs to be made for the combustion air and vent piping.

A typical table giving maximum allowable pipe length (ft) is as follows:

	Pipe Dia.	Number of 90° Elbows					
Unit Size	**(in.)**	**1**	**2**	**3**	**4**	**5**	**6**
040-08	1½	70	70	65	60	60	55
	2	70	70	70	70	70	70

These values are for altitudes of 0 to 2000 ft above sea level, using either two separate pipes or a 2-in. concentric vent.

The *concentric vent* (Fig. H36-18), is an interesting device. Since some manufacturers require 100% outside air for combustion on high-efficiency furnaces, this device al-

HOW A PULSE FURNACE WORKS

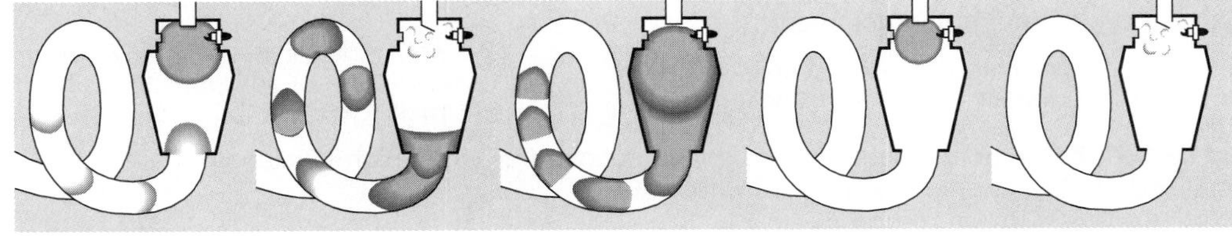

1 THE HEATING PROCESS BEGINS AS AIR AND GAS ENTER THE COMBUSTION CHAMBER AND MIX NEAR THE SPARK IGNITER.

2 A SPARK CREATES INITIAL COMBUSTION, SIMILAR TO THE WAY A SPARK PLUG WORKS IN YOUR CAR. THIS CAUSES A PRESSURE BUILD-UP INSIDE THE COMBUSTION CHAMBER.

3 THE INTERNAL PRESSURE RELIEVES ITSELF BY FORCING THE PRODUCTS OF COMBUSTION DOWN A TAILPIPE AND VENTING THEM OUTDOORS.

4 AS THE COMBUSTION CHAMBER EMPTIES, IT CREATES A VACUUM THAT PREPARES IT FOR THE NEXT IGNITION. AT THE SAME INSTANT, PRESSURE "PULSES" BACK FROM THE END OF THE TAILPIPE.

5 THE "PULSE" RETURNING FROM THE TAILPIPE RE-ENTERS THE COMBUSTION CHAMBER, CAUSING THE NEW GAS/AIR MIXTURE IN THE CHAMBER TO IGNITE. AS A RESULT, THE HEATING CYCLE IS REPEATED AT A RATE OF 60 TO 70 TIMES PER SECOND.

Figure H36-16 Diagram of the sequence of a pulse-type ignition cycle.

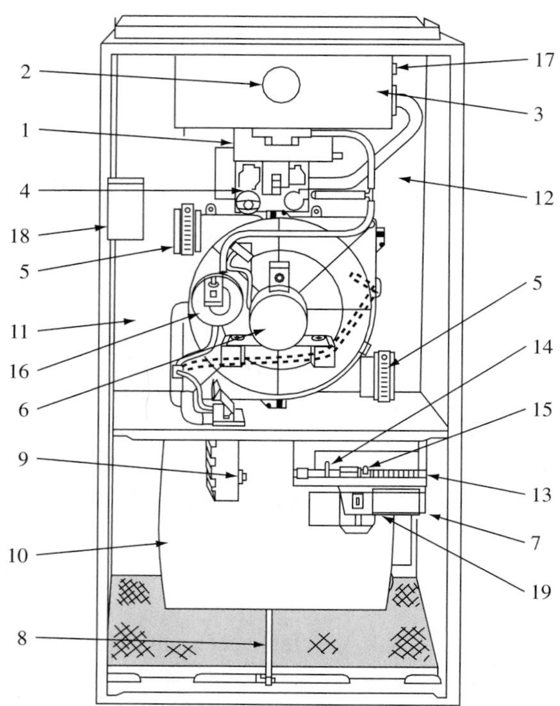

① COMBUSTION-AIR INTAKE CONNECTION TO ENSURE CONTAMINANT-FREE AIR (RIGHT OR LEFT SIDE).

② BURNER SIGHT GLASS FOR VIEWING BURNER FLAME.

③ BURNER ASSEMBLY (INSIDE). OPERATES WITH ENERGY-SAVING, INSHOT BURNERS AND HOT SURFACE IGNITOR FOR SAFE, DEPENDABLE HEATING.

④ REDUNDANT GAS VALVE. SAFE, EFFICIENT. FEATURES 1 GAS CONTROL WITH 2 INTERNAL SHUT OFF VALVES.

⑤ VENT OUTLET. USES PVC PIPE TO CARRY VENT GASES FROM THE FURNACE'S COMBUSTION SYSTEM (RIGHT OR LEFT SIDE).

⑥ INDUCER MOTOR. PULLS HOT FLUE GASES THROUGH THE HEAT EXCHANGERS. MAINTAINING NEGATIVE PRESSURE FOR ADDED SAFETY.

⑦ BLOWER ACCESS PANEL SAFETY INTERLOCK SWITCH.

⑧ AIR FILTER AND RETAINER. MAY BE USED FOR SIDE RETURN APPLICATION.

⑨ CONDENSATE DRAIN CONNECTION. COLLECTS MOISTURE CONDENSED DURING THE COMBUSTION PROCESS.

⑩ HEAVY-DUTY BLOWER. CIRCULATES AIR ACROSS THE HEAT EXCHANGERS TO TRANSFER HEAT INTO THE HOME.

⑪ SECONDARY CONDENSING HEAT EXCHANGER (INSIDE). WRINGS OUT MORE HEAT THROUGH CONDENSATION. CONSTRUCTED WITH POLYPROPYLENE-LAMINATED STEEL TO ENSURE DURABILITY.

⑫ PRIMARY SERPENTINE HEAT EXCHANGER (INSIDE). STRETCHES FUEL DOLLARS WITH THE S-SHAPED HEAT-FLOW DESIGN. SOLID CONSTRUCTION OF CORROSION-RESISTANT ALUMINIZED STEEL MEANS RELIABILITY.

⑬ CONTROL CENTER.

⑭ 3-AMP FUSE PROVIDES ELECTRICAL AND COMPONENT PROTECTION

⑮ LIGHT EMITTING DIODE (LED) ON CONTROL CENTER. CODE LIGHTS ARE FOR DIAGNOSING FURNACE OPERATION AND SERVICE REQUIREMENTS.

⑯ PRESSURE SWITCH ENSURES ADEQUATE FLOW OF FLUE PRODUCTS THROUGH FURNACE AND OUT VENT SYSTEM.

⑰ ROLLOUT SWITCH (MANUAL RESET) TO PREVENT OVERTEMPERATURE.

⑱ JUNCTION BOX FOR 115-V ELECTRICAL POWER SUPPLY.

⑲ TRANSFORMER (24-V) BEHIND CONTROL CENTER PROVIDES LOW-VOLTAGE POWER TO FURNACE CONTROL CENTER AND THERMOSTAT.

Figure H36-17 Component location and configuration of a Carrier high-efficiency gas furnace (AFUE 92%). (Courtesy of Carrier Corporation)

CONCENTRIC VENT
B IN. DIA PVC VENT/EXHAUST
C IN. DIA
$1^3/_{16}$
$1^1/_2$
D
A
B IN. DIA PVC INTAKE/COMBUSTION AIR

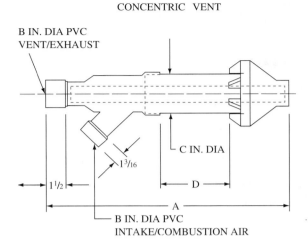

Figure H36-18 Diagram of a concentric vent. (Courtesy of Carrier Corporation)

lows both vent- and combustion-air pipes to terminate through a single exit in the roof or sidewall. One pipe runs inside the other, permitting venting through the inner pipe and combustion air to be drawn through the outer pipe.

There is also a limit on the length of an exposed insulated vent pipe (ft) in an unheated area, as follows:

Unit Size	Winter Temp. °F	Pipe Dia. (in.)	Insulation Thickness (in.) 0	⅜	½	¾	1
040-08	20	1½	31	56	63	70	70
	0	1½	16	34	39	47	54
	−20	1½	9	23	27	34	39

Typical installations of these units along with accessories and venting arrangements are shown in Figs. H36-19, H36-20, H36-21, and H36-22.

H36-3.2 Standard Gas Furnaces

Lower-efficiency gas furnaces refer to the older units now in the field, all of which were manufactured prior to 1992. Since so many are in use and will require service for some time, this section describes them and indicates differences from the newer designs.

A cross section of one of these units is shown in Fig. H36-23 (p. 740). One distinctive characteristic of the older design is the gravity flow of combustion gases from the burner, past the heat exchanger and into the vent. The newer units use an inducer fan. This gravity action of the flue gases required a different type of venting.

Another distinguishing characteristic of the older units is the use of a standing, continuous-flame gas pilot. (The

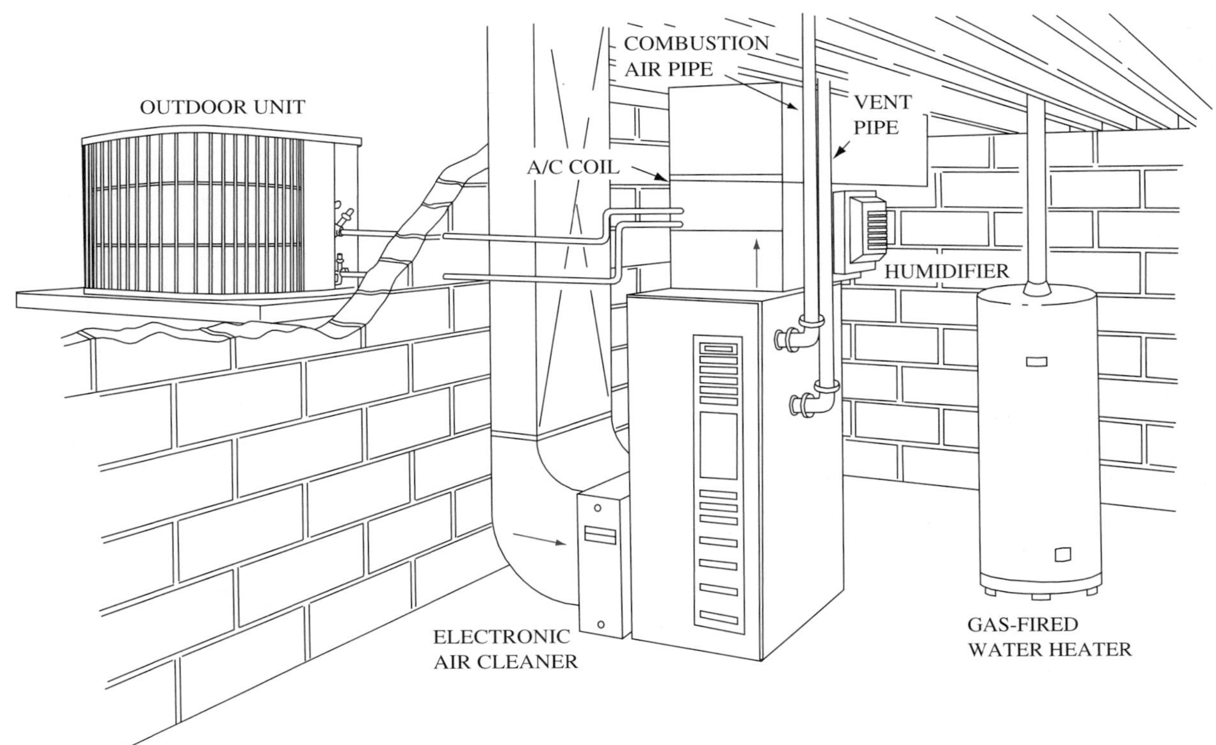

Figure H36-19 Typical application of an up-flow gas furnace. (Courtesy of Carrier Corporation)

PREVENTIVE MAINTENANCE

The following steps should be taken to properly maintain oil burners:

Replace the oil tank filter. Then shut off oil to the burner.

Remove the drawer assembly from the burner. Replace the nozzle. Clean electrodes, look at the porcelain for cracks or crazing, and replace if necessary. Inspect the front of the burner with a mirror for deterioration, and clean the slots on a flame-retention head. Adjust electrodes to factory-specified settings, then replace the drawer assembly.

Clean the blower wheel, making sure it is tight on the motor shaft. Pay attention to the air-intake slots on the air-adjustment shutter. While in the blower compartment check the pump coupling for alignment and tightness on the pump shaft.

Test ignition transformer for the correct voltage output to the electrodes. Assure there is good contact from the transformer to the buss bar and that the buss bars are in good condition.

Test the oil-pump outlet for pressure and inlet for proper vacuum. Clean the screen if the pump has one. Test the pump cutoff to verify a complete fuel shut off.

Check the oil burner motor and the amperage draw. The amp reading should be under the name plate rating.

Test the cad-cell primary control and cad-cell flame-detector eye.

Replace all the components. Turn the fuel on and start the burner.

Set up the burner for the most efficient operation, using a combustion analyzer.

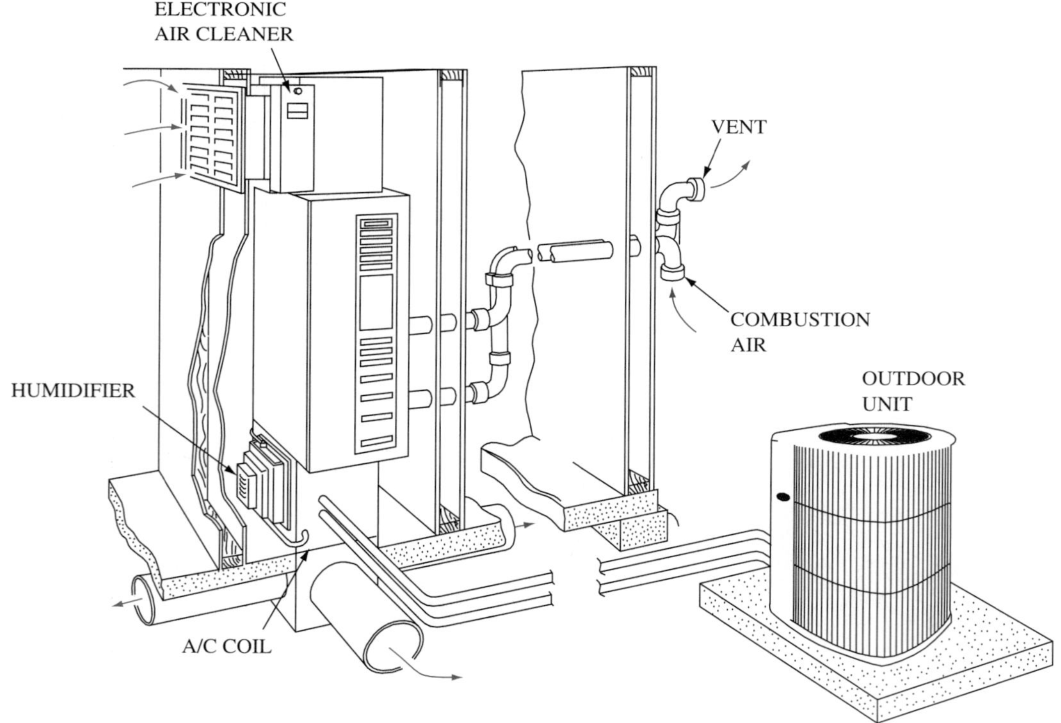

Figure H36-20 Typical application of a counter-flow gas furnace. (Courtesy of Carrier Corporation)

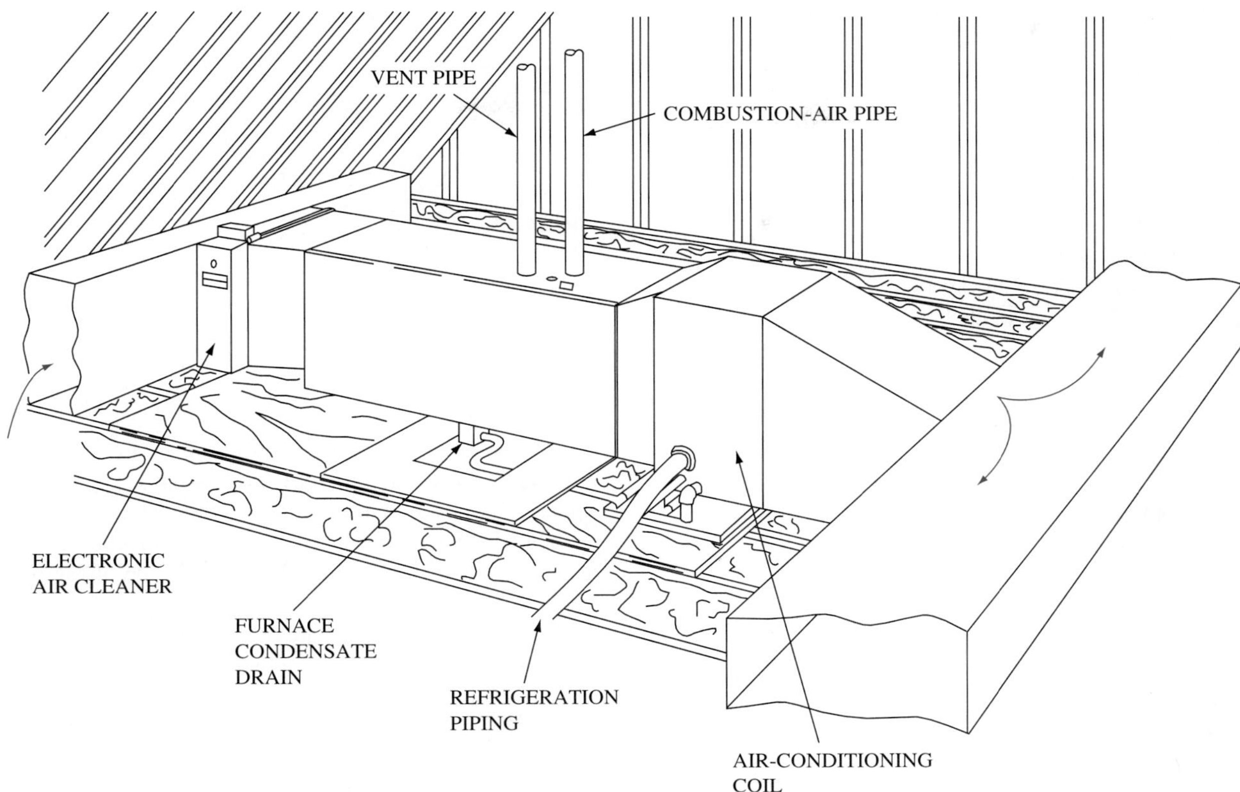

Figure H36-21 Typical attic application of a horizontal gas furnace. (Courtesy of Carrier Corporation)

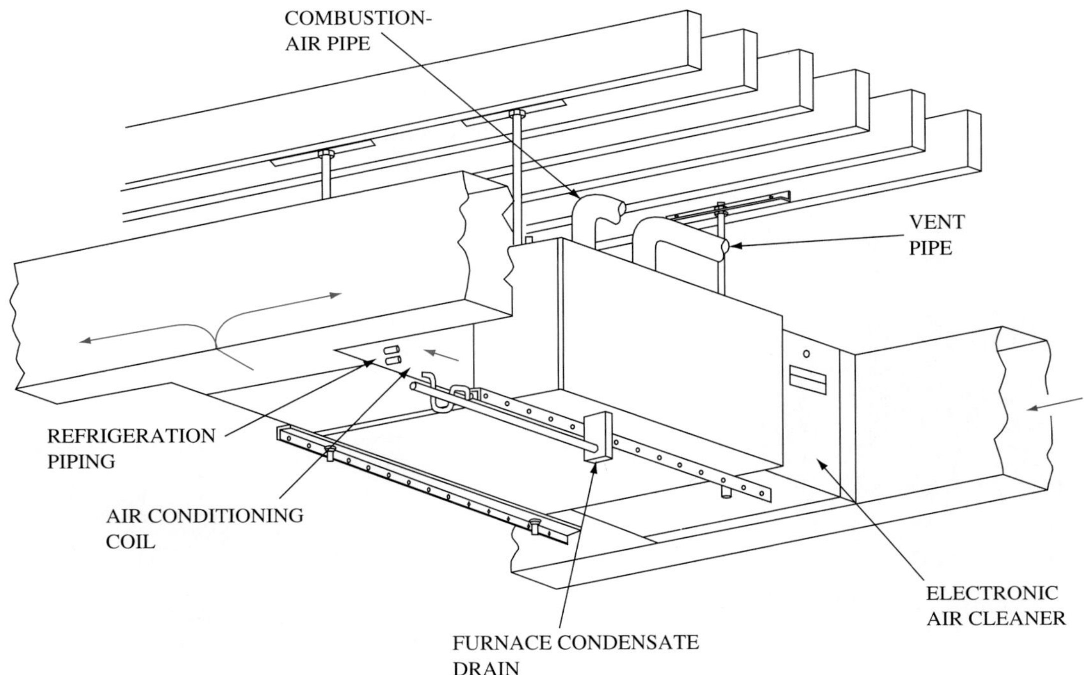

Figure H36-22 Typical crawlspace application of a horizontal gas furnace. (Courtesy of Carrier Corporation)

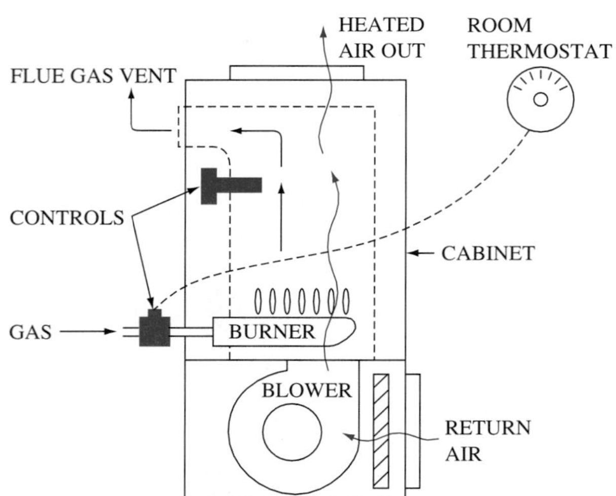

Figure H36-23 Cross section of a gas-up-flow warm-air furnace.

newer units use hot-surface igniters.) Along with the pilot is a runner attached to the main burner. The purpose of the runner is to carry the gas flame from the pilot to all of the burners. The pilot flame impinges on the thermocouple about ½ inch, which provides a safety arrangement for turning off the gas if the flame is ever extinguished.

The most common type of heat exchanger is made from two formed or stamped steel sheets welded together (Fig. H36-24). These sections are headered at the top and bottom and welded into a fixed position. The number of sections depends on the capacity of the furnace. The heat exchanger has a protective coating to prevent rusting. The contours of the heat-exchanger sections create turbulence to increase heat transfer.

Figure H36-24 Gas furnace heat exchanger. (Rheem Air Conditioning Division)

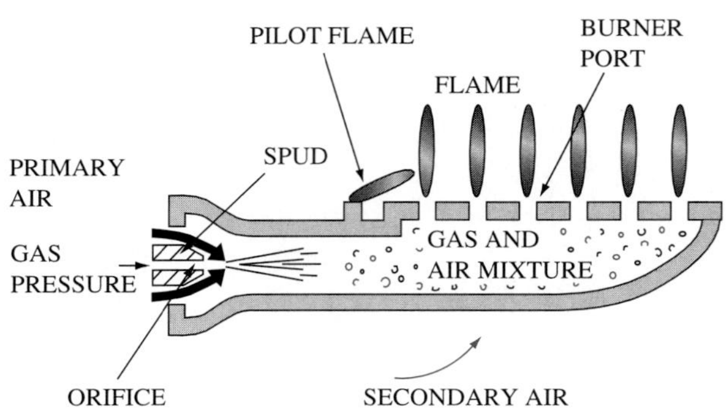

Figure H36-25 Cross section of a gas burner.

The burner fits into the opening of the heat exchanger. Fig. H36-25 shows a cross section of a typical atmospheric burner. Gas enters the burner cavity through the orifice, drilled precisely to meter the correct amount of gas. The gas is ejected into the venturi tube of the burner, sucking in the primary air. The air and gas are mixed and flow through the ports of the burner and are ignited by the pilot. The secondary air is added outside the ports and is controlled by the restrictor baffles designed into the heat exchanger. The primary air can be adjusted by positioning the air shutters as shown in Fig. H36-26 to produce a blue flame for most efficient burning.

The main gas valve for the gas supply or flow performs many functions:

1. Manual on and off control of the gas supply.
2. Pilot supply, adjustment, and safety shutoff.
3. Pressure regulation of the gas feed.
4. On/off electric solenoid valve controlled by the thermostat.

The blower (Fig. H36-27) delivers the heat from the furnace through the distribution system to the conditioned space. Most furnaces use a double-inlet centrifugal blower. The blower can be either direct driven or belt driven. Direct-drive blowers usually have multiple-speed motors. The air quantity delivered by a belt-driven blower can be regulated by adjusting the movable flange on the motor pulley (Fig. H36-28) or by changing the diameter

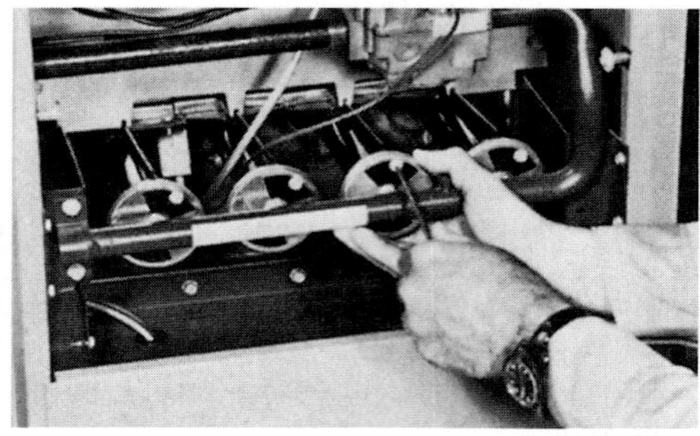

Figure H36-26 Adjusting the primary air on a gas furnace. (Courtesy of York International Corp.)

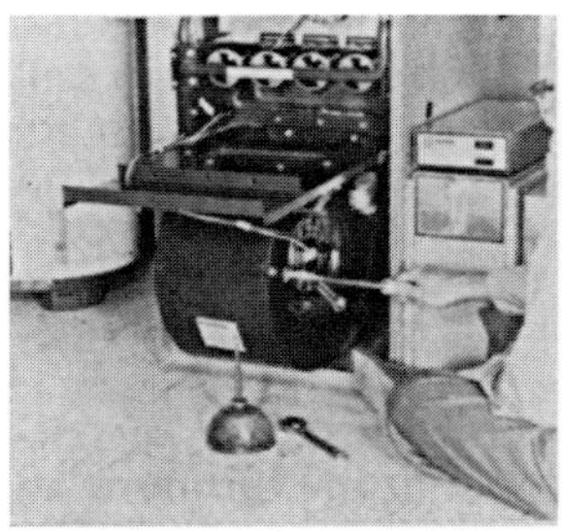

Figure H36-27 Blower assembly for a gas furnace. (Courtesy of York International Corp.)

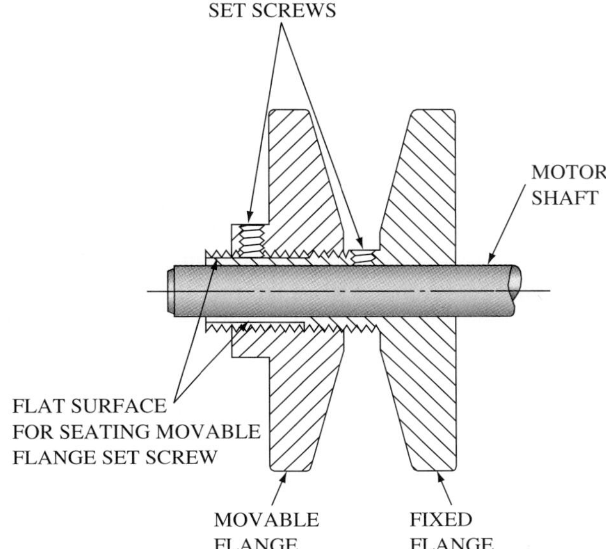

Figure H36-28 Adjustable-pitch motor pulley.

of the pulleys. The motor size is directly dependent on the required air quantity and the resistance offered by the distribution system.

The cabinet provides support for all the internal components, as well as access to the controls, filters, blower-drive components (if belts and pulleys are used), and includes the necessary openings for the combustion-air supply. The cabinet is insulated where exposed to the warm parts of the furnace. It encloses and provides relief openings for the draft diverter. The draft diverter is required by the safety code to divert back-draft flue gases from extinguishing the burner flame.

H36-4 OIL-FIRED FORCED-AIR SYSTEMS

Oil has been a popular fuel for heating in the northeast portion of the country for some time. It is widely used in rural areas where gas is not available and where the use of electricity is not practical or economical.

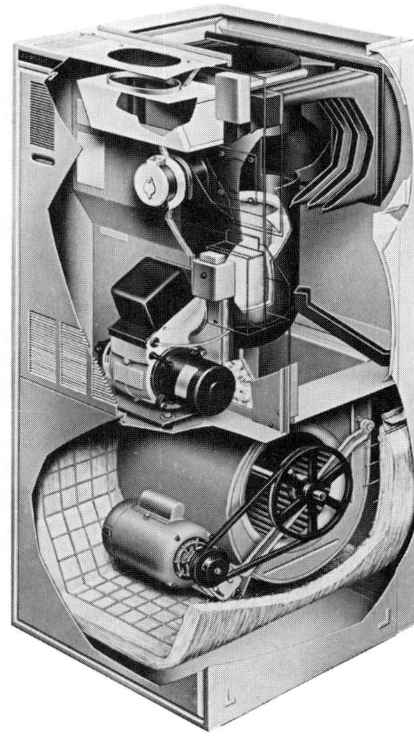

Figure H36-29 Internal construction of an up-flow oil furnace. (Courtesy of Lennox Industries, Inc.)

H36-4.1 Types of Furnaces

Like gas furnaces, oil forced-air furnaces are designed in four basic configurations: up-flow (Fig. H36-29), lowboy, down-flow (counter-flow), and horizontal models.

H36-4.2 Heat Exchanger

The typical oil-fired heat exchanger (Fig. H36-30) is a cylindrical shell of heavy-gauge steel in which combustion takes place. It offers additional surfaces for heat transfer from the products of combustion inside, to the air around the outside of the chamber. This type of heat exchanger is called a *drum and radiator.* The inside, containing the flame, is called the primary surface, and the outside is called the secondary surface.

Some manufacturers add baffles, flanges, fins, or ribs to the surfaces to provide faster heat transfer to the air passing over the surfaces.

The burner assembly is bolted to the heat exchanger. The firing assembly and blast tube extend into the primary surface in correct relationship to the combustion chamber or refractory. A flame-inspection port is provided just above the upper edge of the refractory. This is used to observe proper ignition and the flame, and to measure overfire draft for start-up and service operations.

Figure H36-30 Heat exchanger construction for an oil furnace. (Courtesy of Bard Manufacturing Company)

H36-4.3 Refractory

High temperatures are required in the flame area of the heat exchanger to produce maximum burning efficiency of the oil/air mixture. To obtain this high temperature, a reflective material, called the *refractory,* is installed around the combustion area. It is an insulation type of material designed to reach white-hot surface temperatures quickly, with minimum deterioration.

H36-4.4 Burner

The high-pressure atomizing gun burner (Fig. H36-31) is the type used on most residential and small commercial oil-fired heating systems. It has the following components:

1. Oil pump
2. Air blower
3. Electric motor
4. Ignition transformer
5. Blast tube, with nozzles and ignition system.

A cross section of the burner (Fig. H36-32) shows a detailed view of the operation. The pump and blower are driven from a common motor shaft. The oil pump is a gear arrangement. By changing a bypass plug in the housing, the pump can operate as a single-stage or a two-stage system. Single-stage operation is used where the oil tank is above the burner and gravity oil feed to the burner is permitted. Two-stage operation is needed where the oil tank is below the burner and the pump must lift the oil from the tank as well as furnish pressure to the nozzles. The pump supplies oil to the nozzle at 100 to 300 psi.

The burner blower includes a centrifugal wheel also mounted on the common motor shaft. It furnishes air through the blast tube and turbulator for proper combustion. The amount of air is controlled by a rotating shutter band on the blower housing section.

Oil is pumped to the nozzle which is mounted in an adapter. The orifice in the nozzle is factory-bored to produce the correct firing rate, and must not be altered or changed in any way. Firing rate is approximately 0.8

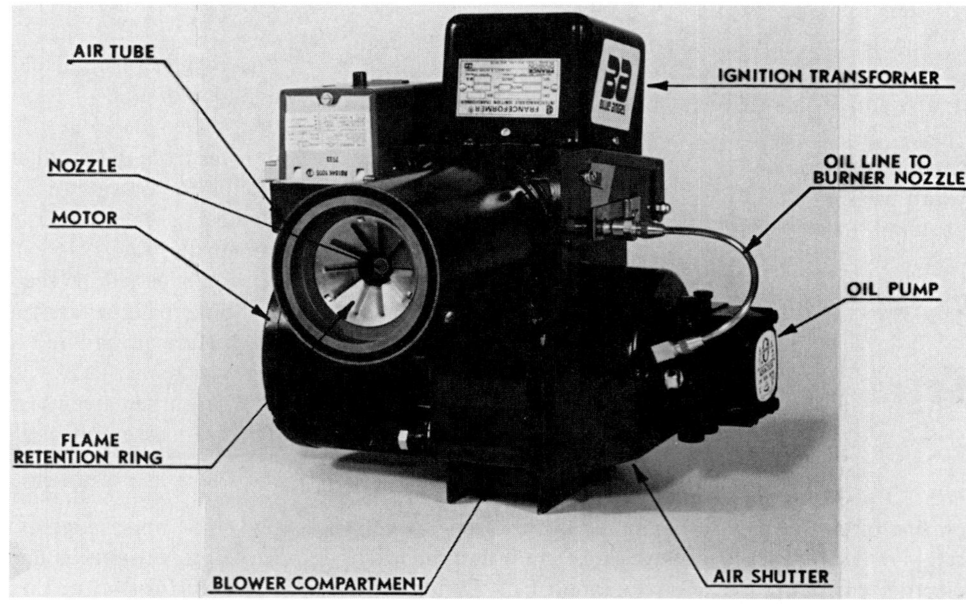

Figure H36-31 High-pressure atomizing gun-type oil burner. (Wayne Home Equipment Company)

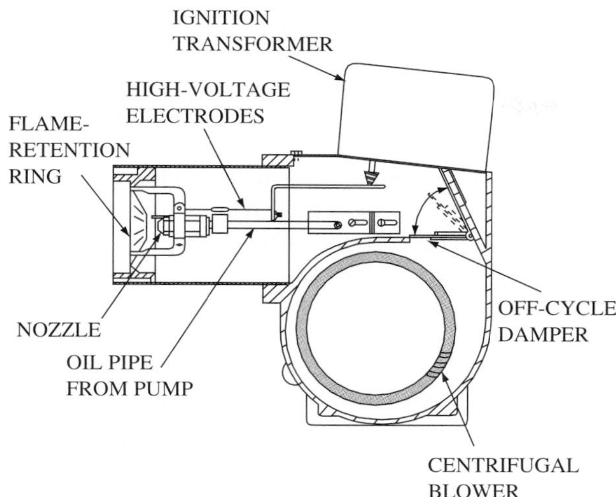

Figure H36-32 Cutaway view of the internal construction of an oil burner.

gal/hr for each 100,000 Btu/hr output of the furnace. Under high pressure, the oil is atomized into fine droplets and mixes with the primary air.

The air-deflector vane ring (turbulator) rotates or spins the atomized oil and air into the heat exchanger. Ignition by a high-voltage electric spark is continuous when the motor is on or interrupted on start-only combustion. The ignition transformer is located in the burner control compartment.

A safety device operates when there is an ignition failure. When there is no flame (light), a light-sensitive cad cell will stop the burner motor, thus preventing oil from flowing into the heat exchanger. Some furnaces have a lock-out system requiring the control to be reset manually before the burner can run again.

H36-4.5 Controls

A typical oil-fired forced-air furnace control system is shown in Fig. H36-33. The fan-and-limit control performs the same function as on a gas furnace. The safety limit, if

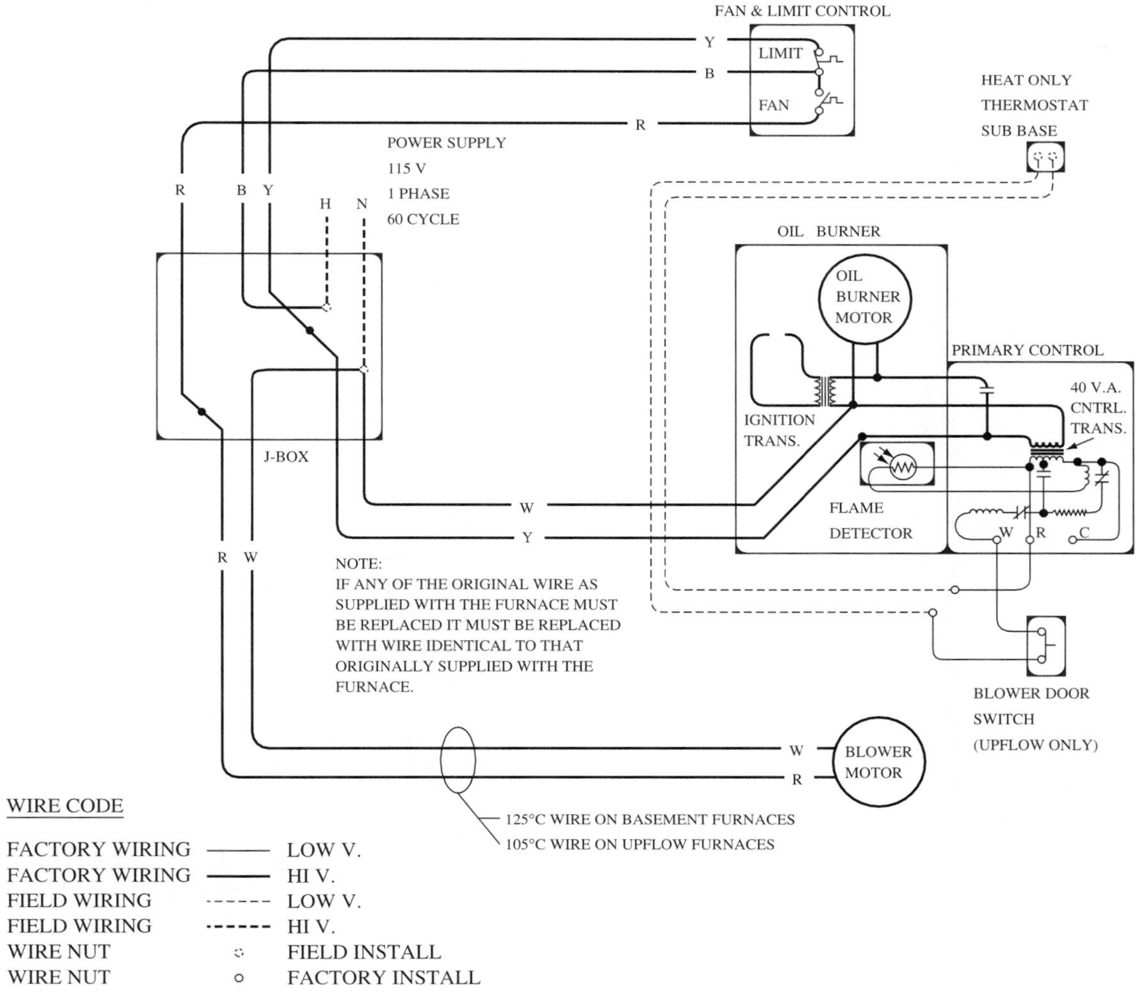

FAN & LIMIT CONTROL
Y
B
LIMIT
FAN
R
HEAT ONLY
THERMOSTAT
SUB BASE
POWER SUPPLY
115 V
1 PHASE
60 CYCLE
R B Y
H N
OIL BURNER
OIL
BURNER
MOTOR
PRIMARY CONTROL
40 V.A.
CNTRL.
TRANS.
IGNITION
TRANS.
J-BOX
W
Y
FLAME
DETECTOR
W R C
R W
NOTE:
IF ANY OF THE ORIGINAL WIRE AS SUPPLIED WITH THE FURNACE MUST BE REPLACED IT MUST BE REPLACED WITH WIRE IDENTICAL TO THAT ORIGINALLY SUPPLIED WITH THE FURNACE.
BLOWER DOOR
SWITCH
(UPFLOW ONLY)
W
R
BLOWER
MOTOR
125°C WIRE ON BASEMENT FURNACES
105°C WIRE ON UPFLOW FURNACES

WIRE CODE

FACTORY WIRING	———	LOW V.
FACTORY WIRING	———	HI V.
FIELD WIRING	- - - - -	LOW V.
FIELD WIRING	- - - - -	HI V.
WIRE NUT	∴	FIELD INSTALL
WIRE NUT	○	FACTORY INSTALL

Figure H36-33 Typical oil-burner control-system wiring diagram.

open, will stop the electrical current to the burner before overheating. The fan switch is set to cycle the furnace blower as desired. The space thermostat (24-V) feeds directly to the burner low-voltage secondary control circuit. This actuates the flame detector and relay, feeding the primary line voltage to the burner motor and ignition transformer.

The operation of the controls, as well as start-up, adjusting, and servicing the system are covered in subsequent chapters.

H36-4.6 Flue Venting

Oil-fired furnaces must have an ample supply of make-up air for combustion, and the methods of introducing make-up air for a gas furnace will also be adequate for oil equipment.

Masonry chimneys used for oil-fired furnaces should be constructed as specified in the National Building Code of the National Board of Fire Underwriters. Prefabricated lightweight metal chimneys (Fig. H36-34) are also available. These are rated for all fuels, Class A and Class B. Class A flues are used for solid and liquid fuels while Class B flues are made specifically for gas-fired equipment. The double-wall type is filled with insulation.

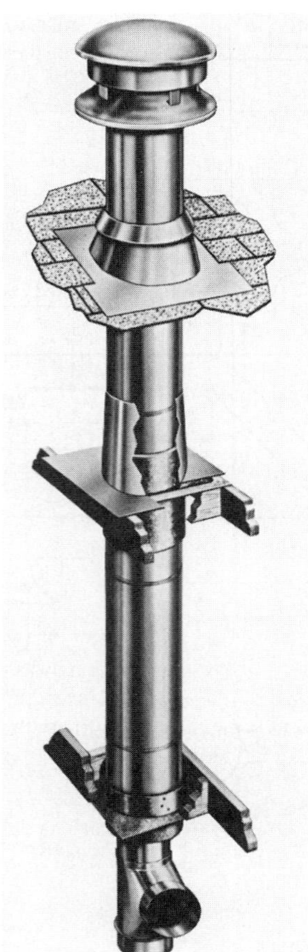

Figure H36-34 Factory-built, all-fuel chimney.
(Courtesy of Selkirk Metalbestos)

The above-roof dimension of the vent or chimney is the same for both gas and oil. Since oil-fired furnaces operate on positive pressure from the burner blower, it is important to have a chimney that will develop a minimum draft of 0.01 to 0.02 inches of WC as measured at the burner flame-inspection port. Consistency or stability of draft is also more critical, and the use of a *barometric damper* (Fig. H36-35) is required. The damper is usually installed in the horizontal vent pipe between the furnace and chimney. It may also be attached directly to the furnace flue outlet. The damper has a movable weight so that it can be set to counterbalance the suction and to maintain reasonably constant flue operation. It is adjusted while the furnace is in operation and the chimney is hot. The overfire draft at the burner should be in accordance with the equipment installation instructions.

H36-4.7 Clearances

Cabinet temperatures and flue-pipe temperatures run warmer for oil-burning equipment, so clearances from combustible material should be adjusted accordingly. One-in. clearances are common for the sides and rear of the cabinet, compared to zero clearances for many gas units. Flue-pipe clearances of 9 in. or more are needed, whereas only 6 in. of clearance are needed for gas.

Front clearance is principally space needed to remove the burner assembly. Floor bases over combustible surfaces are generally increased for oil as compared to gas. For example, in horizontal oil furnaces installed in attics, the installer must pay particular attention to recommendations related to adjoining surfaces.

Clearances are an important part of the compliance and inspection procedures required by Underwriters' Laboratories (UL) for securing their seal of approval.

H36-4.8 Oil Storage

The installation of the fuel-oil tank and connecting piping must conform to the standards of the National Fire Protection Association (NFPA) and to local code requirements. Regulations and space permitting, the oil tank can be located indoors as shown in Fig. H36-36. Note the use of shutoff valves and a filter to catch impurities.

If the oil tank is placed outdoors above ground, firm footings must be provided. The exposed tank and piping are subject to more condensation of water vapor and the possibility of freezing.

If the oil tank is installed below ground (Fig. H36-37), it is important to keep it filled with oil during periods when the ground is saturated with water. Otherwise, high groundwater may force it to float upward. A heavy concrete cover over the tank may be advisable. Copper piping may provide flexibility if ground movement should occur.

Figure H36-35 Barometric flue damper.
(Courtesy of The Field Controls Company)

Note the use of suction and return lines in the illustration. It is normal practice to use a two-stage fuel pump with a two-pipe system whenever the oil supply is below the level of the burner.

The oil-burner pump suction is measured in in. of mercury vacuum. A two-stage pump should never exceed a 15-in. vacuum. Generally, there is 1 in. of vacuum for each ft of vertical oil lift and 1 in. of vacuum for each 10 ft of horizontal run of supply piping.

H36-4.9 Ratings and Efficiencies

Oil-burning furnaces are limited in the number of sizes available due to the restrictions in available oil-burner

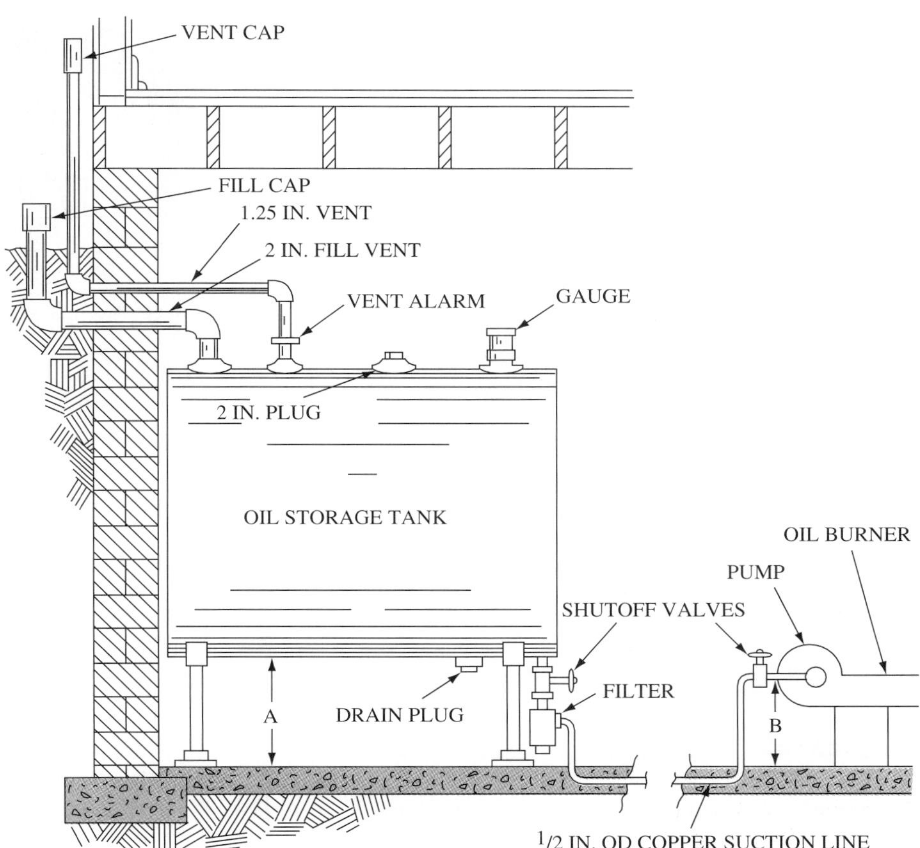

Figure H36-36 Indoor fuel-oil storage tank.

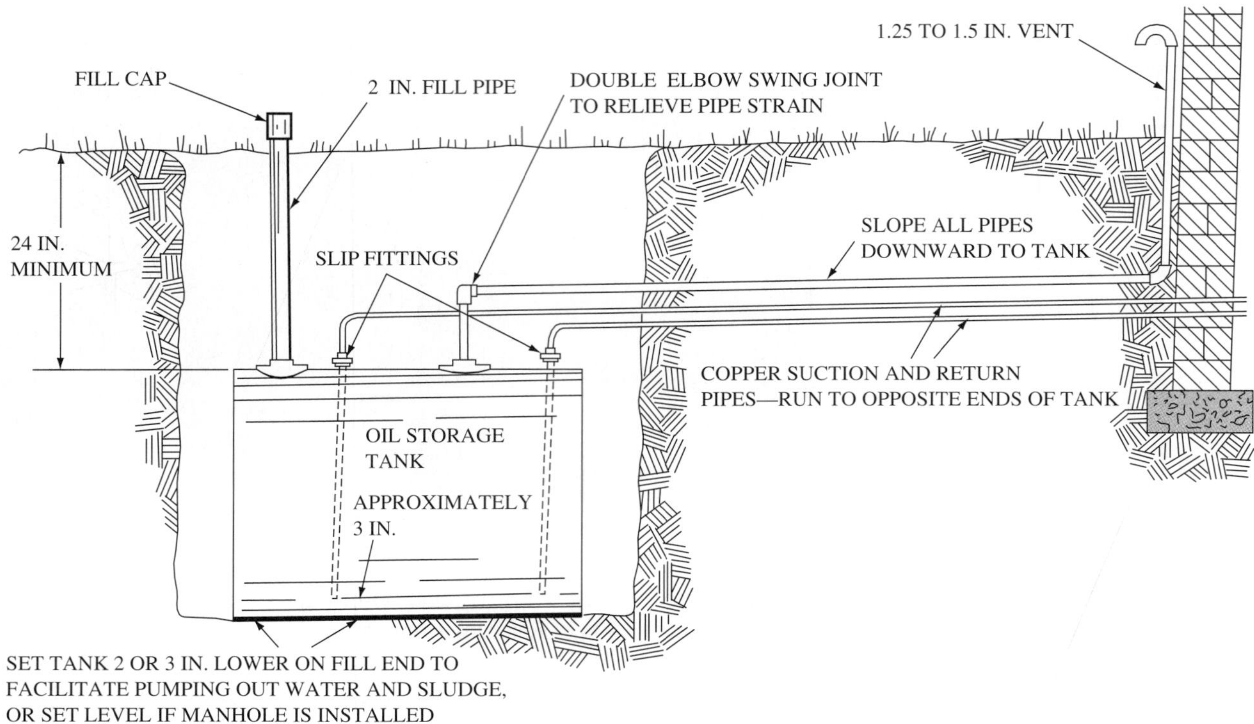

Figure H36-37 Underground fuel-oil storage tank installation.

nozzles. For example, the models available from a typical manufacturer are as follows:

Input (Btu/hr)	Output (Btu/hr)	AFUE% (ICS)	Application
70,000–154,000	56,000–125,000	80%	Up-flow
105,000–210,000	85,000–166,000	Up to 80.6%	Basement (Lowboy)
105,000–154,000	84,000–125,000	80%	Down-flow or Horizontal
175,000–210,000	134,000–165,000	Up to 74.3%	Horizontal

The efficiency ratings given in the above chart are for the mid-efficiency furnaces. These new units are very similar in construction to the standard furnaces previously manufactured with certain improvements to produce the new ratings. Some of the new features that have been added to increase the efficiency include:

1. A larger heat-exchanger drum.
2. Electronic fan control (Fig. H36-38) which turns blower on 60 to 90 seconds after thermostat demand.
3. Flame detector (cadmium sulfide).

H36-5 ELECTRIC FURNACES

The typical electrical furnace (Fig. H36-39) consists of a cabinet, blower compartment, filter, and resistance-heating section. Some manufacturers provide a cooling coil.

Figure H36-38 Electronic fan control center for an oil-fired furnace. (Courtesy of York International Corp.)

An electrical furnace has many advantages:

1. It is more compact than the equivalent gas or oil furnace.
2. Due to cooler surface temperature, it has "zero" clearance requirements on all sides. It may, therefore, be located in small spaces such as closets.
3. Since there is no combustion process, there are no requirements for venting pipes, chimney, or make-up air. This reduces building costs and simplifies installation.

Figure H36-39 External view of an electric furnace.

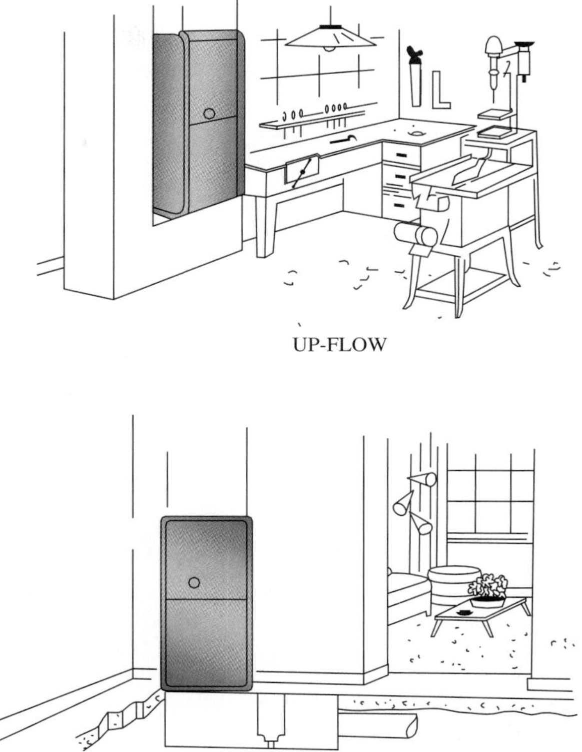

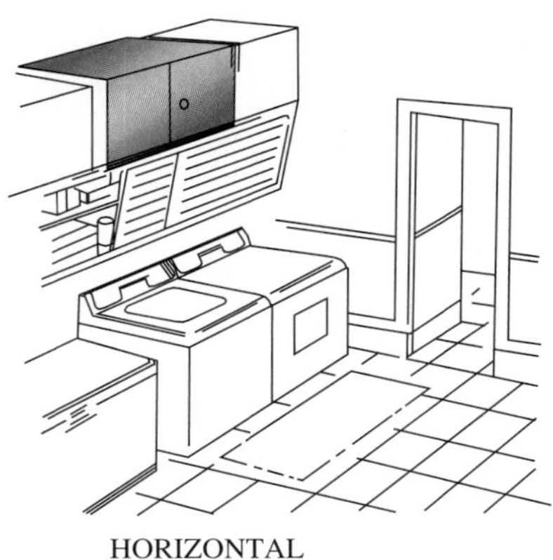

Figure H36-40 Various types of electric furnaces.

4. Units may be mounted for up, down, or horizontal airflow applications (Fig. H36-40).

The blower compartment (Fig. H36-41) usually houses a centrifugal multi-speed direct-drive fan. Larger units use a belt-driven fan. Airflow through an electric furnace has less resistance, and fan performance is more efficient. Mechanical air filters are usually used.

The heating section consists of banks of resistance heater coils of nickel-chrome wire held in place by ceramic spacers (Fig. H36-41). The heater resistance operates on 208- to 240-V power with output according to use.

The amount of heat per bank is a function of the amperage draw and/or staging. The amount of current that can be put on the line in one surge is limited by regulations, residences being limited to 200 amps. Total furnace output capacities range from 5 kW (17,000 Btu) to 35 kW (119,400 Btu/hr). At 35 kW the amperage draw on 240 V approaches 150 amps and, considering that 200 amps is the total service to a residence, this only leaves 50 amps for other electrical uses. Two electric furnaces would therefore be used, with sequencing and zoning.

The *sequence control system* is an important operation of the electric furnace. There are single-stage and two-stage thermostat systems, depending on the size of the unit. When there is a call for heat, the electric sequencer control closes its circuit and begins to heat up. The blower and first heater come on; there is a time delay before each additional heater stage is energized. This time delay (in seconds) is adequate to stagger power inrush.

Larger furnaces use two-stage thermostats. The first stage would bring on at least 50% of the total capacity. The second stage of the thermostat would respond only when full heating capacity is needed. With this added control, wide variations in indoor temperature are avoided.

While these design features are feasible for modifying the load, code assumes all installations are used continuously, for safety purposes.

Electrical heating elements are protected from any overheating that may be caused by fan failure or a blocked filter. High-limit switches sense air temperature and open the electrical circuit when overheating occurs. As a backup, some furnaces have fusible links wired in

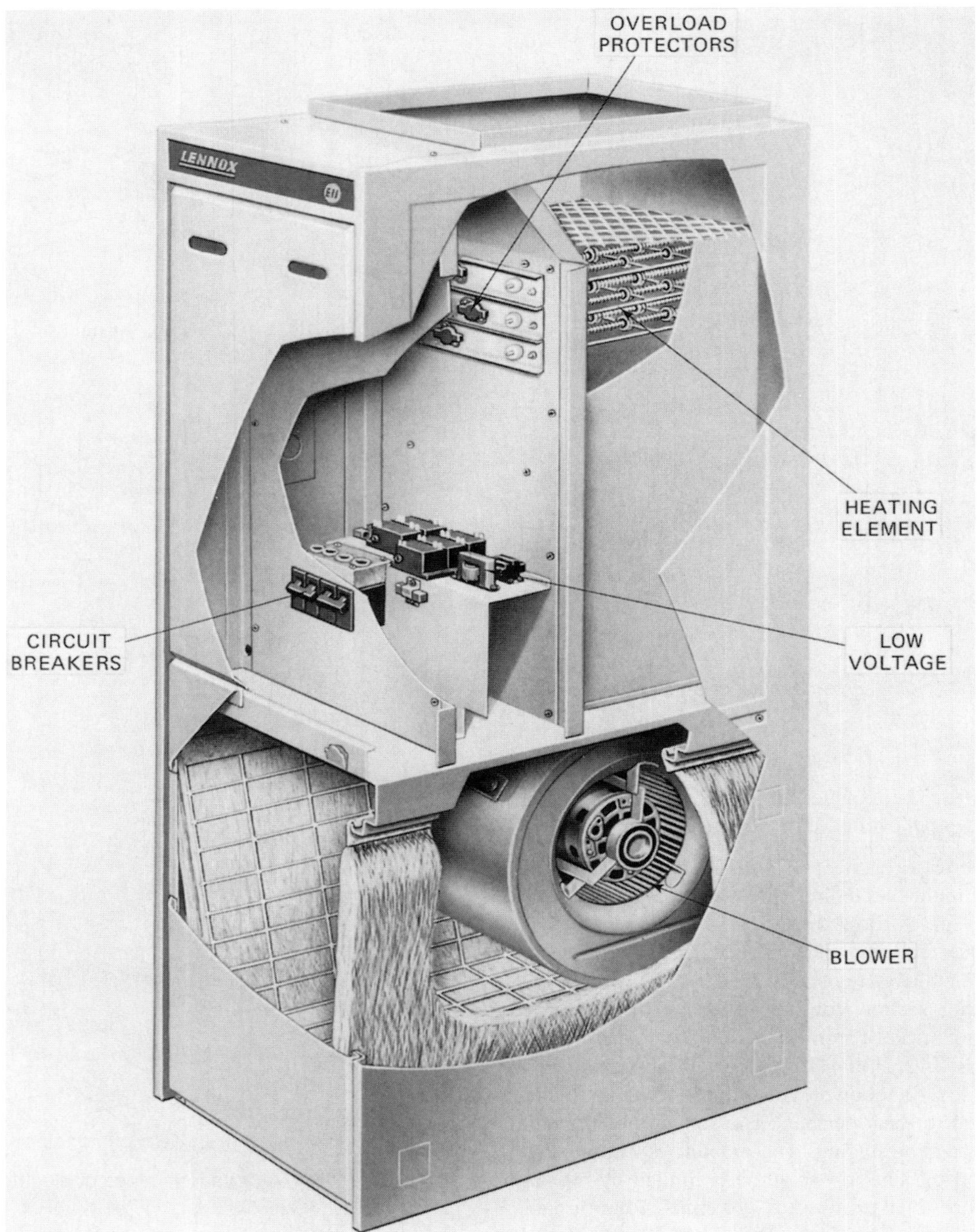

Figure H36-41 Internal diagram of an electric furnace. (Courtesy of Lennox Industries, Inc.)

series with the heater which melt at 300°F, opening the circuit.

Built-in internal fusing or circuit breakers are provided by some manufacturers in compliance with the National Electric Code K. This eliminates the need for external fuse boxes.

The air-distribution system for an electric furnace requires the following special considerations:

1. Air temperatures coming off the furnace are lower, as compared to gas and oil equipment. Temperatures of 120°F and below can create drafts if improperly introduced into conditioned space. Additional air diffusers are recommended.
2. Duct loss through unconditioned areas can be critical, so well-insulated ducts are a must to maintain comfort and reduce operating cost.

H36-5.1 Applications

One of the most useful applications of an electric furnace is its combination with a split-system heat pump. In many geographical locations where heat pumps are installed, electric furnaces supply the supplementary heat required, below the outside temperatures where heat pumps are ineffective.

For these installations the electric furnace acts as a fan-coil unit with the refrigerant coil installed in the plenum. The coil is connected by refrigerant tubing to the heat pump located on a slab outside the building. The thermostat located in the conditioned space controls the components of the complete system to produce the heating or cooling as required.

Air handling capabilities of electric furnaces are similar to other forced-air furnaces, using a multi-speed direct-drive blower, with capability for supplying the required air quantity at various external static pressures.

H36-6 AIR-HANDLING UNITS AND DUCT HEATERS

A variation of the electric furnace is an air-handling unit with duct-type electric heaters. The air handler consists of a blower housed in an insulated cabinet with openings for connections to supply and return ducts. Electric-resistance heaters are installed either in the primary supply trunk or in branch ducts leading from the main trunk to the rooms in the dwelling. There is comfort flexibility by zone control when heaters are installed in branch runs; i.e., room temperatures can be individually controlled. This system has not been as popular as the complete furnace package, primarily because it complicates installation requirements and adds costs.

Duct heaters are made to fit standard-size ducts and are equipped with overheating protection devices. Electric duct heaters can also be used to supplement other types of ducted heating systems, to add heat in remote duct runs or to beef up the system if the house has been enlarged. They may be connected to come on with the furnace blower, thus controlled by a room thermostat.

REVIEW

- Furnace configurations:
  - Up-flow—Installed in first-floor closets, utility rooms, or basements.
  - Lowboy—Low height to accommodate areas where head room is minimal.
  - Down-flow or counter-flow—Air intake and fan are at the top and discharge is on the bottom.
  - Horizontal—Adaptable to installation in low areas, crawl spaces, attics, or partial basements, overhead ceilings.
- Types of fuel:
  - Gas
    - Natural
    - LP
  - Oil
  - Electric
- Gas forced-air systems:
  - Mid-efficiency—80% AFUE
  - High-efficiency—90% AFUE or better
    - Fixed capacity
    - Sealed combustion
    - Tubular heat exchanger
    - Pulse
    - Variable capacity
  - Lower-efficiency
    - Standard (Applies to furnaces prior to 1992)
- Oil burner:
  - Oil pump
  - Air blower
  - Electric motor
  - Ignition transformer
  - Blast tube with nozzles, and ignition system
- Features added to increase oil burner efficiency:
  - Large heat exchanger
  - Electronic fan control
  - Flame detector
- Electric furnaces:
  - Advantages
    - More compact
    - Can be installed with zero clearance
    - No requirements for venting
    - May be mounted up, down, horizontal
  - Furnace outputs—5 kW (17,000 Btu/hr) to 35 kW (119,000 Btu/hr)
  - Larger furnace uses staging
    - 1st stage—50%
    - 2nd stage—Full capacity
  - Duct heaters—Zone control; may be used as reheat
- Air-handling units and duct heaters:
  - Heaters installed in supply or branch ducts.
  - Room temperatures individually controlled.
  - Supplements other heating sources.

Problems and Questions

1. The counterflow gas furnace is the most popular type. True or false?
2. The mid-efficiency gas-fired forced-air furnace has a 78% AFUE rating. True or false?
3. High-efficiency gas-fired forced-air furnaces burn at 92% or greater. True or false?

4. The highest rating gas furnace
   a. runs at 96.6% AFUE.
   b. uses a variable-capacity furnace with a two-stage gas valve, and variable-speed inducer motor.
   c. uses a solid-state controlled blower motor.
   d. All of the above.
5. The pulse furnace
   a. pulsations occur 60–70 times per second.
   b. utilizes pressure buildup to force the hot gases out of the combustion chamber.
   c. uses negative pressure to draw in a new supply of gas.
   d. All of the above.
6. One distinctive characteristic of the older standard gas furnace designs is the
   a. combustion blower fan method.
   b. gravity flow of combustion gases from the burner.
   c. method of electronic ignition.
   d. All of the above.
7. The installation of the fuel oil tank and connecting piping must conform to the
   a. NFPA.
   b. ARI Code.
   c. building code.
   d. None of the above.
8. At 35 kW, the amperage draw on 240 V is
   a. 50 amps.
   b. 100 amps.
   c. 150 amps.
   d. None of the above.
9. Name four styles of furnaces.
10. Describe the differences between fuel-burning furnaces and electric furnaces.

Space Heaters

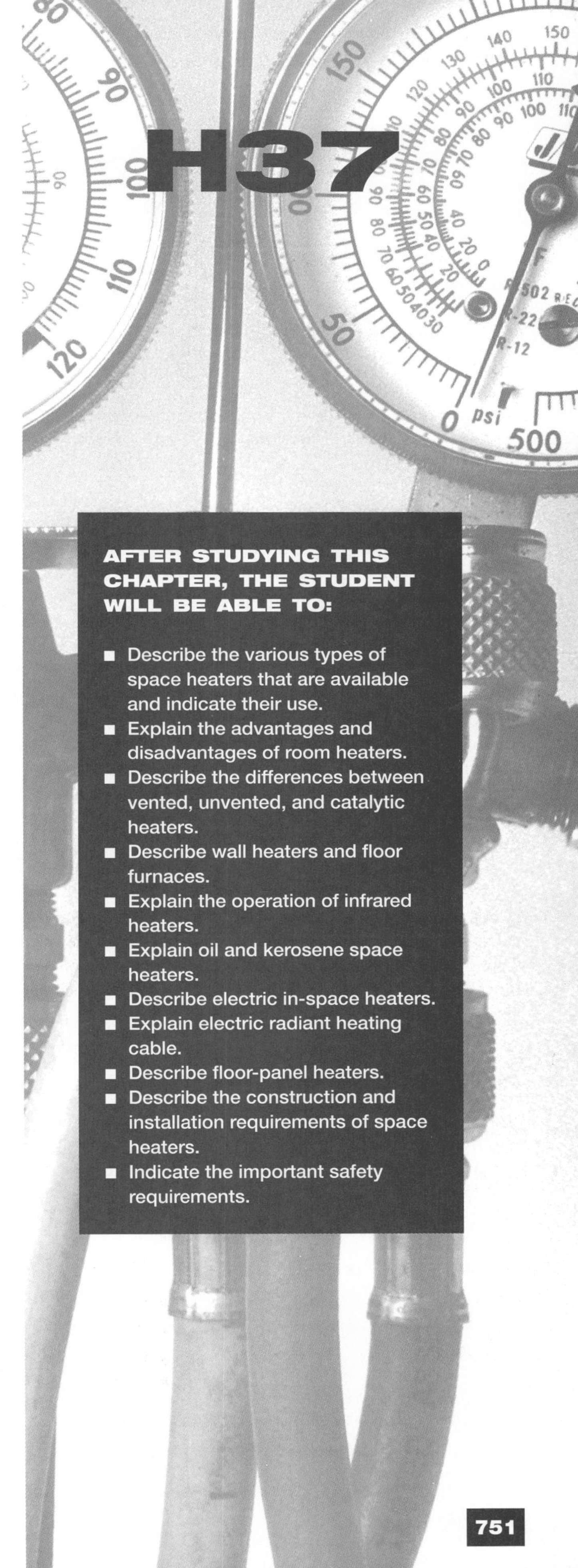

H37

AFTER STUDYING THIS CHAPTER, THE STUDENT WILL BE ABLE TO:

- Describe the various types of space heaters that are available and indicate their use.
- Explain the advantages and disadvantages of room heaters.
- Describe the differences between vented, unvented, and catalytic heaters.
- Describe wall heaters and floor furnaces.
- Explain the operation of infrared heaters.
- Explain oil and kerosene space heaters.
- Describe electric in-space heaters.
- Explain electric radiant heating cable.
- Describe floor-panel heaters.
- Describe the construction and installation requirements of space heaters.
- Indicate the important safety requirements.

OVERVIEW

Space heaters is a general term applied to "in-space" heating equipment where the fuel is converted to heat in the space to be heated. Such heaters may be permanently installed or portable and employ a combination of radiation, natural convection, and forced convection to transfer the heat produced. The energy source may be liquid, solid, gaseous, or electric.

H37-1 GAS SPACE HEATERS

Gas space heaters are available in a variety of styles—room heaters, wall furnaces and floor furnaces. Each has its own unique construction and application.

H37-1.1 Room Heaters

Room heaters are self-contained free-standing units that transfer the heat produced from fuel combustion to room air through a heat-exchanger system. The heat is transferred by radiation and convection without mixing the flue gas with the circulating room air. Since air for combustion is supplied from the room air, these units must be installed in a space that has sufficient infiltration to supply the necessary combustion air. Room heaters are made in a number of different constructions: vented, unvented, and catalytic.

Vented room heaters have an opening that is permanently connected to the chimney to convey the flue gas to the outdoors. The combustion gases pass through a heat exchanger which transfers the heat to the room air by radiation and convection. Cool room air enters the grille at the bottom of the unit and the heated air is distributed from the top (Fig. H37-1). These heaters often have a glass panel on the front to supply radiant heat. Room air is completely separated from the combustion gases.

Some room heaters are operated entirely by gravity. Others use a fan to circulate the room air through the unit.

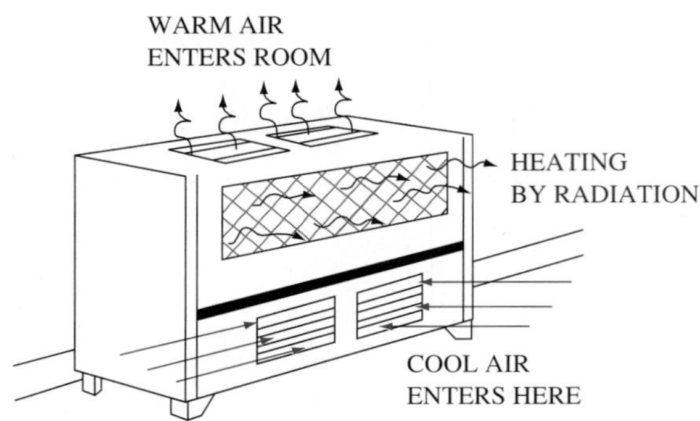

Figure H37-1 Room space heater. (Copyright by the American Society of Heating, Refrigerating, and Air-Conditioning Engineers, Inc. Used by permission.)

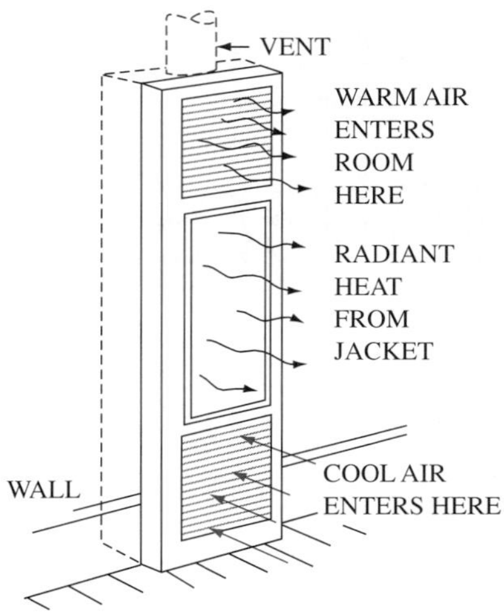

Figure H37-2 Wall furnace. (Copyright by the American Society of Heating, Refrigerating, and Air-Conditioning Engineers, Inc. Used by permission.)

The fan increases the efficiency of the heater and provides better distribution of the heated air.

Since the air for combustion is taken from the room, applications are limited to spaces where there is sufficient infiltrated air to supply the required combustion air. These units are available in sizes ranging from 10,000 to 75,000 Btu/hr.

Unvented room heaters discharge the products of combustion into the room. They are limited in application to commercial projects where the area is relatively open. They are often used during building construction to supply temporary heat. One type commonly used is called a "Salamander." It is portable and can be located where needed.

The *catalytic heater* transfers heat from a glowing heat exchanger. It has no flame. The heat exchanger is constructed of fibrous material impregnated with a catalytic substance that accelerates the oxidation of a gaseous fuel. Catalytic heaters transfer heat by radiation and convection. The surface temperature is below red heat, usually about 1200°F.

H37-1.2 Wall Furnaces

Wall furnaces are designed in a vertical configuration to fit into the stud space (Fig. H37-2). They heat a single room or have a rear boot to also supply heat to an adjacent room. The units are usually 6 or 8 in. deep, so that part of the cabinet protrudes into the room. The units are completely self-contained and operate from a room thermostat.

Some units have gravity air circulation, others use a small blower. Some units have conventional venting arrangements with a flue extending above the roof. Others, located on the outside wall, have direct vents.

These units pull the air for combustion from the outside and therefore can be placed in a tight room. Heating capacities are available from 6,000 to 65,000 Btu/hr.

H37-1.3 Floor Furnaces

Floor furnaces (Fig. H37-3) are constructed for suspension from a floor. The unit is constructed for supplying heated air through the center of the grille, with the return air entering through the outside corners, as shown in the illustration.

Combustion air is drawn from the outside of the building. The common application of these units is in a central room of a small house, where often a single unit is used to circulate heated air through the entire house.

ON THE JOB

Catalytic heaters burn and generate heat without a flame or vent piping. They still require oxygen to burn within the conditioned space. Whenever you work around these units, be aware that they do require oxygen and will take oxygen from the conditioned space. To prevent sickness and potential health problems, always supply adequate outside ventilation air to the space being heated. Never allow the customer to use these units within a sealed area, or suffocation will occur.

PREVENTIVE MAINTENANCE

Since many wall and/or floor units are installed in open areas, they are more susceptible to getting dirty and operating poorly. Good maintenance on these units should include oiling the motors (if provided) and checking the condition of the flame during normal mode. A good clean flame will have a strong blue color, but a dirty burner will have a yellow color. If cleaning the unit and burner ports does not remedy the condition of the flame, then install a U-tube manometer or gas-pressure measuring device to verify proper gas supply pressure to the flame. Normal pressure for natural gas is 3.5 in. of water column. For LP normal pressure is 11 in. of water column.

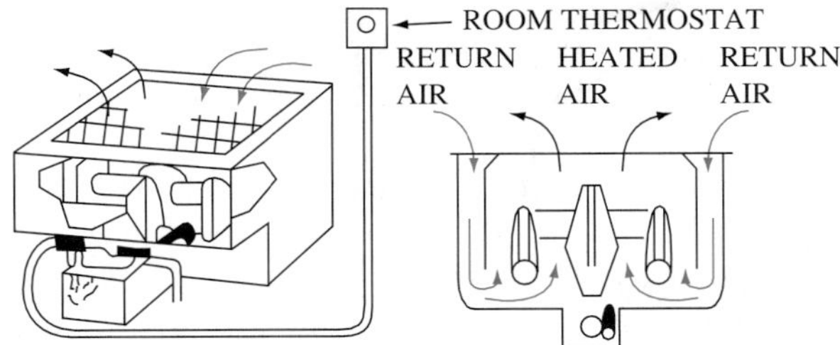

Figure H37-3 Floor furnace. (Copyright by the American Society of Heating, Refrigerating, and Air-Conditioning Engineers, Inc. Used by permission.)

H37-1.4 Minimum Efficiency Requirements

The National Appliance Energy Conservation Act of 1987 established minimum efficiency requirements for all gas-fired direct heating equipment, shown in Fig. H37-4. The AFUE values are obtained by the test methods set up by the Department of Energy.

Input, 1000 Btu/hr	Minimum AFUE, %	Input, 1000 Btu/hr	Minimum AFUE, %
Wall Furnace (with fan)		*Floor Furnace*	
<42	73	<37	56
>42	74	>37	57
Wall Furnace (gravity type)			
<10	59	*Room Heaters*	
>10–<12	60	<18	57
>12–<15	61	>18–<20	58
>15–<19	62	>20–<27	63
>19–<27	63	>27–<46	64
>27–<46	64	>46	65
>46	65		

Figure H37-4 Gas-fired direct-heating equipment efficiency requirements. (Copyright by the American Society of Heating, Refrigerating, and Air-Conditioning Engineers, Inc. Used by permission.)

H37-1.5 Controls

The thermostats used by space heaters are of two types:

1. *Wall thermostats.* These thermostats can be either 24-V or use millivolt power. They are selected to operate whatever gas valve is being used. A suitable source of power must be supplied.
2. *Built-in hydraulic thermostats.* These are made in two types: (1) a snap-action two-position thermostat with a liquid-filled capillary tube temperature-sensing element, and (2) a modulating-type thermostat, similar to the first type, except the temperature alters the position of the valve between off and fully open.

H37-2 INFRARED HEATERS

A *multi-mount electric infrared heater* (Fig. H37-5) has many applications, including total area heat, spot heat, and snow and ice control. They can be mounted horizontally, or at a 30°, 60°, or 90° angle. The double reflectors can be placed to provide a wide variety of patterns as shown in Fig. H37-6. Two elements are used per fixture. Three different lengths are available and units can be mounted end-to-end for additional lengths. Units are available in sizes 3200, 5000 and 7300 W.

Any surface or object should be 24 in. away from direct radiation from the unit. For surface mounting, UL re-

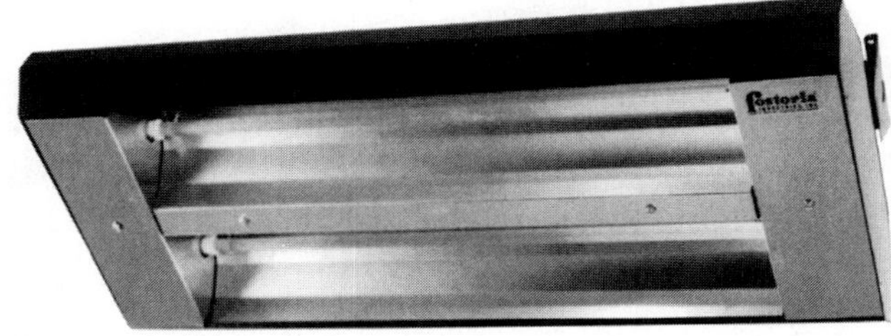

Figure H37-5 Multi-mount infrared heater. (Courtesy of Fostoria Industries, Inc.)

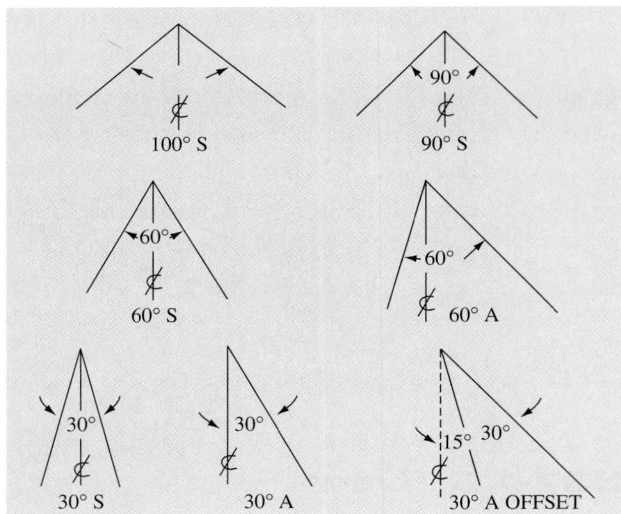

Figure H37-6 Reflector patterns for an infrared heater. (Courtesy of Fostoria Industries, Inc.)

quires fixtures to be at least 3 in. from the ceiling and 24 in. from a vertical surface. Rows of fixtures should be separated by a minimum of 36 in.

A typical *gas-fired infrared heater* installation for a warehouse or large open area is shown in Fig. H37-7. Heaters are located overhead, and radiate heat to the area below. By using radiation, heat is directed to the people and objects in the space that require warming, rather than wasting it to the air or other non-essential surfaces. This makes possible maintaining lower room temperatures, yet creating comfortable conditions and saving energy.

A more detailed view of these heaters is shown in Fig. H37-8. This system consists of individual burners ranging from 20,000 through 120,000 Btu/hr, fired in series, with branches connected to one vacuum pump. These systems are custom engineered to meet specific floor plans as well as heating requirements. They can be operated in a condensing mode for the ultimate in heating efficiency. Direct spark ignition is used. Combustion chambers can be either fabricated steel or optional cast iron. The tail pipe is porcelain-lined.

A factory-assembled control panel combines pre-purge and post-purge controls. Connections are provided for up to four low-voltage thermostats for zone temperature control. Units are fully vented to avoid release of combustion moisture inside the building.

An optional controller continuously monitors the demand for heat and adjusts firing cycles. Automatic setback can be programmed for nights, weekends, and holidays. Optional decorative grilles are available for drop ceilings. Various configurations can be supplied to match the requirements, as shown in Figs. H37-9 and H37-10.

High-intensity infrared heaters, using natural gas or LP, (Fig. H37-11), are available for direct radiant heat where needed.

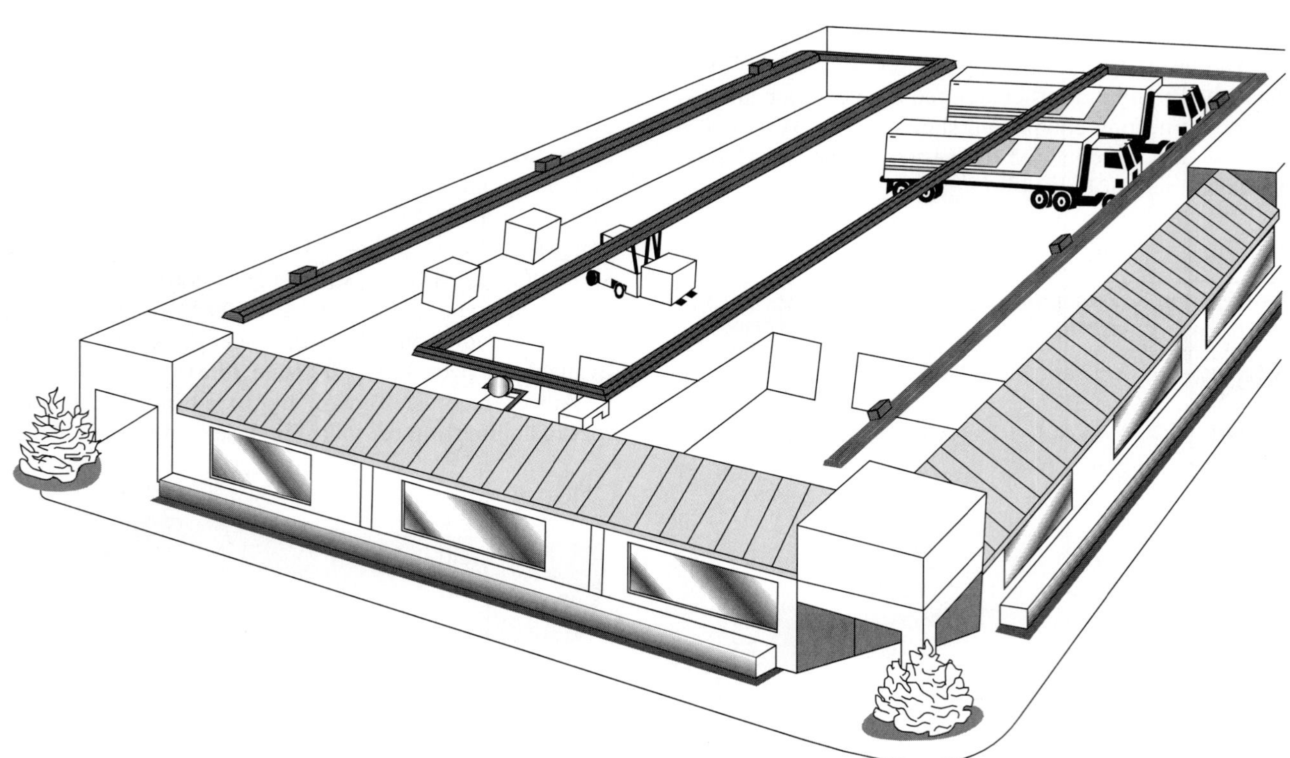

Figure H37-7 Infrared heating equipment installation. (Courtesy of Roberts Gordon, Inc.)

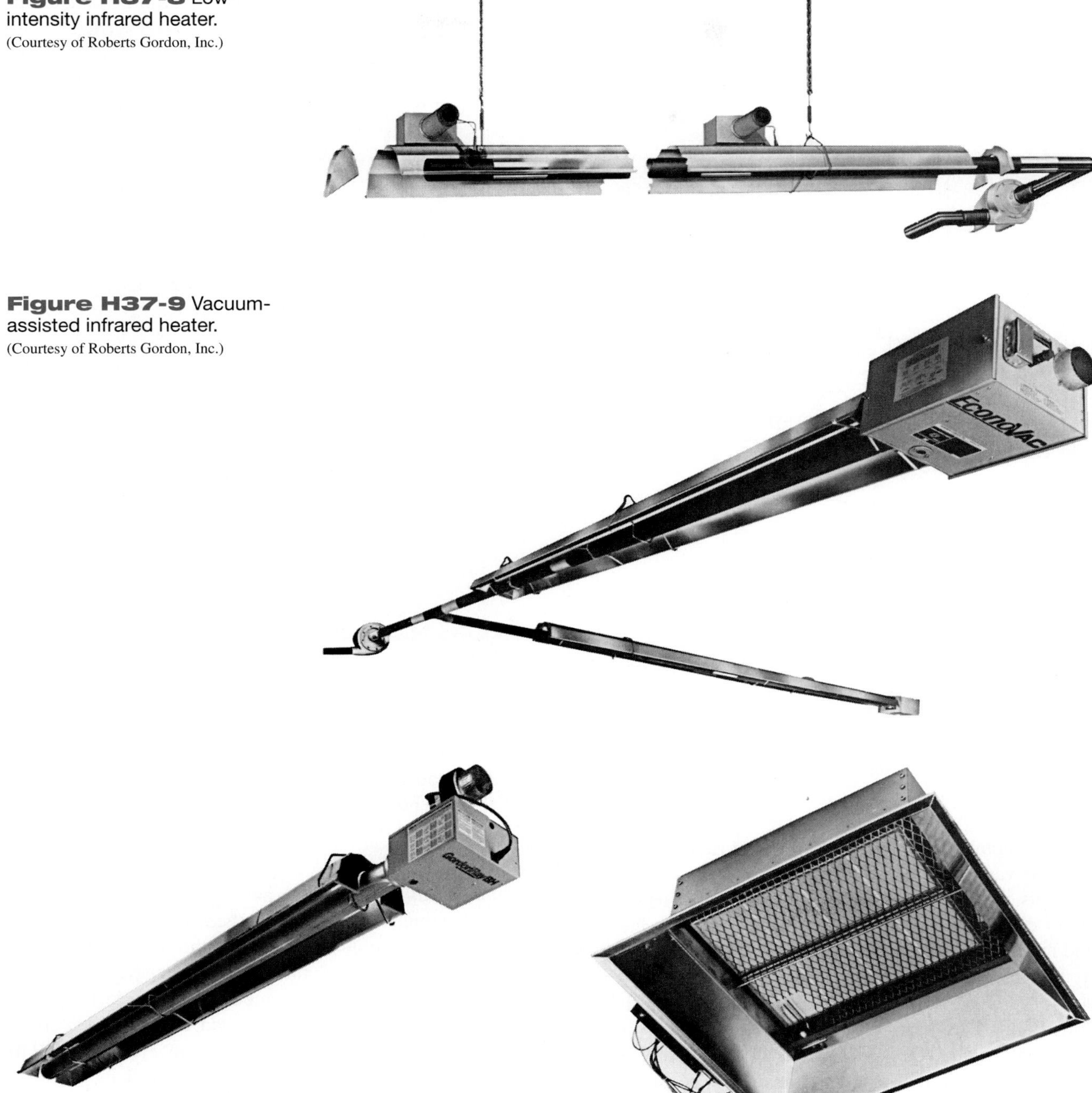

Figure H37-8 Low-intensity infrared heater. (Courtesy of Roberts Gordon, Inc.)

Figure H37-9 Vacuum-assisted infrared heater. (Courtesy of Roberts Gordon, Inc.)

Figure H37-10 Unitary infrared heater. (Courtesy of Roberts Gordon, Inc.)

Figure H37-11 High-intensity infrared heater. (Courtesy of Roberts Gordon, Inc.)

H37-3 VAPORIZING OIL POT HEATERS

Vaporizing-type oil-burning units use No. 1 grade fuel oil. A typical unit is shown in Fig. H37-12. The oil is burned in a bowl or pot (A). A constant-level metering device, with an adjustable needle, controls the flow of oil into the combustion chamber and pilot burner (B), through a burner oil supply pipe (C). The adjustable needle can be manually placed in an OFF, PILOT, or VARIABLE-FLOW setting. The oil flows by gravity from a 2- or 3-gallon tank attached to the unit or a larger tank located outside.

Oil is vaporized by the heat of combustion and by contact with the hot metal surfaces of the combustion cham-

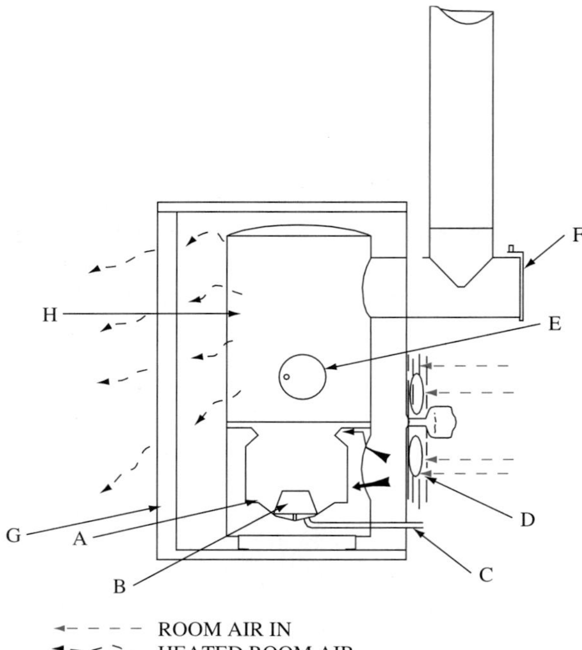

A — POT-TYPE BURNER COMBUSTION CHAMBER
B — PILOT BURNER
C — BURNER OIL SUPPLY PIPE
D — BURNER DRAFT BOOSTER — WARM AIR CIRCULATOR WITH SAFETY GUARD
E — LIGHTER AND CLEANOUT PORT
F — DRAFT CONTROLLER
G — PERFORATED METAL GRILLE
H — STEEL DRUM-TYPE HEAT EXCHANGER

Figure H37-12 Oil-fired space heater. (Copyright by the American Society of Heating, Refrigerating, and Air-Conditioning Engineers, Inc. Used by permission.)

Figure H37-13 Wall-mounted recessed space heater. (Source: Federal Pacific Electric)

ber. The combustion air enters from the room and mixes with the vaporized oil.

Units can be supplied for gravity air circulation or with an air circulator (D). A port opening for cleaning the unit and for access for lighting the pilot (E), is provided on the front of the drum-type heat exchanger. An automatic draft controller (F) is used on the flue pipe to maintain the required draft. Flue gases are vented to the outside. A perforated metal grille (G) is located on the lower front of the cabinet to admit return air for heating. The supply of heated air enters the room from the top of the unit. The steel heat exchanger (H) furnishes radiant heat to the room by heating the side walls of the cabinet. A thermostat can be used to operate the unit by turning on and off the unit at a selected firing rate.

H37-4 ELECTRIC IN-SPACE HEATERS

There are various types of electrical in-space heaters. Due to the fact that there are no products of combustion and no flue, their location is very flexible. These types are described:

1. Forced-air heaters
2. Suspended heaters
3. Cabinet-type heaters
4. Baseboard heaters
5. Ceiling radiant heating cable
6. Floor panels

Forced-air heaters add a measure of comfort to electric space heating. In homes and offices the recessed models (Fig. H37-13) combine style with the forced-air circulation, resulting in more uniform room temperature control. Most are wall-mounted but models are also available for recessing into the ceiling, and even under kitchen counters. Wall-mounted units provide up to 3000 W (10,240 Btu/hr) which can warm larger rooms or offices.

Suspended force-fan unit heaters (Fig. H37-14) are most functional, being used in homes (garages, workrooms, playrooms, etc.) where large capacity and positive air circulation are important as well as in commercial and industrial establishments of all kinds. They may be suspended in vertical or horizontal fashion with a number of control options such as wall switch only, thermostats, timers, and night setback operation. Capacities generally range from 3 to 12 kW and above. Discharge louvers may be set to regulate air-motion patterns. The heavy-duty construction of this heater category is geared for commercial application.

Cabinet-type space heaters (Fig. H37-15) are found in classrooms, corridors, foyers, and similar areas of com-

Figure H37-14 Suspended-type space heater.
(Source: Federal Pacific Electric)

mercial and institutional buildings where cooling is not a consideration. Heating elements go up to 24 kW, and the fans are usually the centrifugal type for quiet but high-volume airflow. Cabinets have provision for introducing outside air where use and/or local codes require minimum ventilation air. Outdoor air intake damper operation can be manual or automatic. Control systems can be simple one-unit operation or multi-zone control from a central control station.

The most frequently used electric space heater is the *convection baseboard model* (Fig. H37-16), installed in residential and commercial buildings. A cross section through the heater shows the convective airflow across the finned tube heating element. The contour of the casing provides warm air motion away from the walls, thus keeping them cooler and cleaner.

It is important that drapes do not block the air flow and cause overheating, or carpeting block inlet air. Most baseboard units have a built-linear type of thermal protection that prevents overheating. If blockage occurs, the safety limit stops the flow of electrical current. It cycles off and on until the blockage is removed. Baseboard heaters come in lengths of 2 to 10 ft. in the nominal 120-, 208-, 240-, and 277-V rating. Standard wattage per foot is 250 at the rated voltage. But lower heat output can be obtained by applying heaters on lower voltage.

For example, a standard 4-ft heater rated at 250 W/ft at 240 V will produce 1000 W of heat. Most manufacturers offer these reduced-output (low-density) models by changing the heating element to 187 W/ft. Low density is generally preferred by engineers and utilities and is recommended for greatest comfort. Controls may be incorporated in the baseboard or through the use of wall-mounted thermostats. Accessories such as corner boxes, electrical outlets, and plug-in outlets for room air conditioners are available.

Electric radiant heating cable (Fig. H37-17) was one of the most popular methods of heating early in the development of this technology. It is an invisible source of heat and does not interfere with the placement of furnishings. The source of warmth is spread evenly over the area of the room. It can be installed within either plaster or drywall ceilings. Staples hold the wire in place until the finish layer of plaster or drywall is applied. Wall-mounted thermostats control space temperature.

Floor panels (Fig. H37-18) can be constructed using perforated mats. These mats consist of PVC-insulated heating cable, woven in, or attached to metal or glass fiber mesh. Such assemblies are available in sizes from 2 to 100 ft^2, with various watt densities ranging from 15 to 25 W/ft^2.

Another effective method of slab heating uses mineral-insulated (MI) heating cable. MI cable is small-diameter,

Figure H37-15
Cabinet-type space heater.
(Source: Federal Pacific Electric)

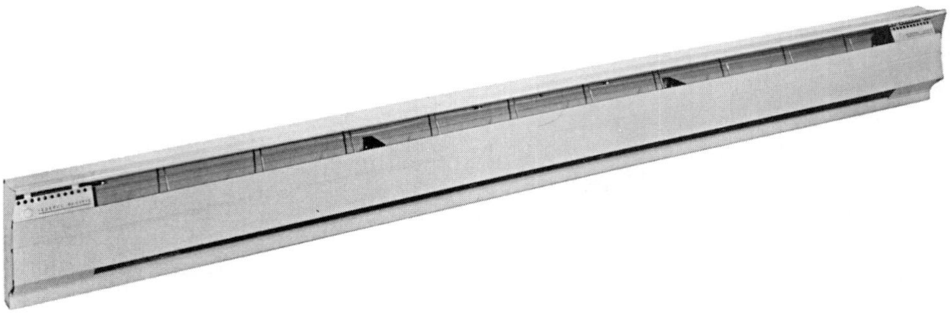

Figure H37-16 Baseboard-type space heater. (Source: Federal Pacific Electric)

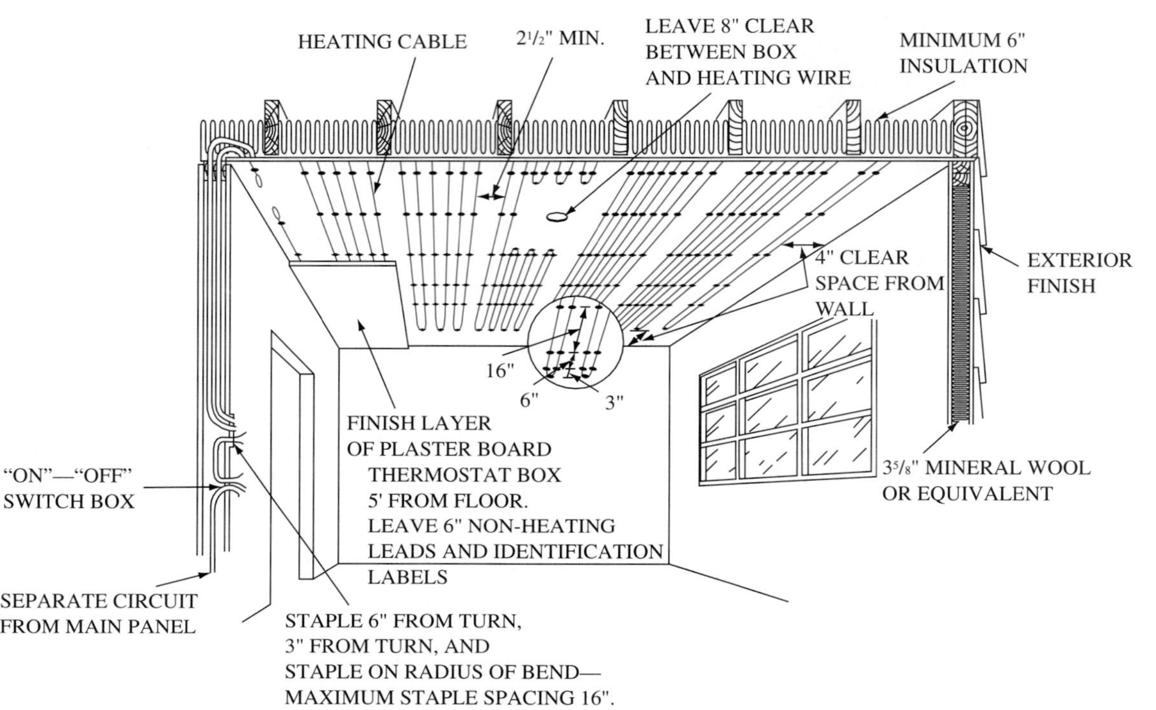

Figure H37-17 Diagram showing installation of ceiling cable for radiant heating.

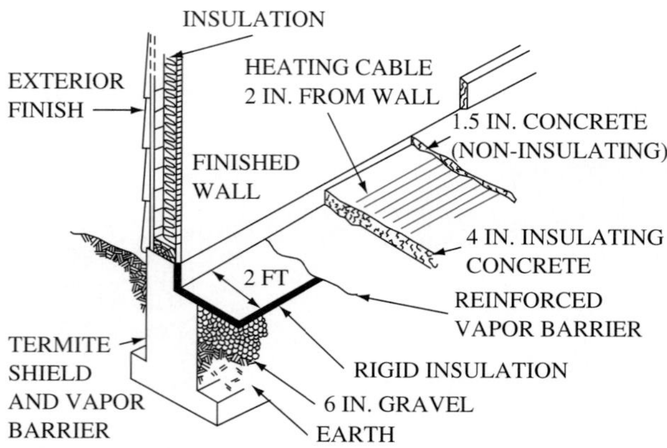

Figure H37-18 Diagram showing installation of electric-heating cable in the floor for radiant heating. (Copyright by the American Society of Heating, Refrigerating, and Air-Conditioning Engineers, Inc. Used by permission.)

highly durable, solid, electrical-resistance heating wire surrounded by compressed magnesium-oxide electrical insulation and enclosed in a metal sheath. Several MI cable constructions are available, such as single-conductor, double-conductor and double-cable, as well as custom-designed cables. Either of these constructions can be embedded in concrete.

REVIEW

- Space heater—"In-space" heating equipment where the fuel is converted to heat in the space to be heated. These heaters may be portable or installed, and employ a combination of radiation, natural convection, and forced convection to transfer the heat produced.

- Room heaters are made in a number of different constructions: vented, unvented, and catalytic.
- Vented room heaters have an opening that is permanently connected to the chimney to convey the flue gas to the outside.
- Unvented room heaters discharge the products of combustion into the room. They are limited in application to commercial projects where the area is relatively open. They may be used in construction areas to provide temporary heat.
- The catalytic heater transfers heat from a glowing heat exchanger. It has no flame. The heat exchanger is made of a fibrous material impregnated with a catalytic substance that accelerates the oxidation of a gaseous fuel. Heat is transferred by radiation and conduction.
- Wall furnaces have a vertical configuration to fit into the stud space of a wall. They heat a single room or two adjacent rooms. These units are self-contained and operate from a room thermostat.
- Floor furnaces are suspended from a floor, and supply the heated air through the center of the grille. Return air enters through the outside corners. Combustion air is drawn in from the outside of the house.
- Thermostats:
  Wall thermostats
    24 V
    Millivolt
  Built-in hydraulic
    Snap action—two-position
    Modulating—alters position between off and fully open.
- Gas-fired infrared heaters—Have many applications: total area heat, spot heat, and snow and ice control.
- Vaporizing oil pot heaters—Constant-level metering devices control flow; oil is burned in a bowl or pot.
- Electric in-space heaters:
  Forced-air heaters
  Suspended heaters
  Cabinet-type heaters
  Baseboard heaters
  Ceiling radiant heating
  Floor panels

Problems and Questions

1. Some room gas-heat units have a design which is unvented. True or false?
2. Catalytic heaters transfer heat from a glowing heat exchanger with no flame. True or false?
3. Floor units are constructed for suspension from a floor. True or false?
4. A two-stage valve has two positions, 100% open and 50% open. True or false?
5. The AFUE values are obtained by the test methods set up by ARI standards. True or false?
6. The thermostats used by space heaters
   a. use 24-V or millivolt power.
   b. are snap acting two-position.
   c. are the modulating type.
   d. All of the above.
7. Gas-fired infrared heaters can be mounted
   a. horizontally.
   b. at a 30 or 60° angle.
   c. at a 90° angle.
   d. All of the above.
8. Vaporizing oil pot heaters
   a. use No. 1 fuel oil.
   b. use No. 2 fuel oil.
   c. use No. 3 fuel oil.
   d. None of the above.
9. Force-fan electric unit heaters
   a. are installed to blow the air as an up-flow furnace.
   b. are installed to blow the air down as a down-flow furnace.
   c. may be installed in any direction.
   d. None of the above.

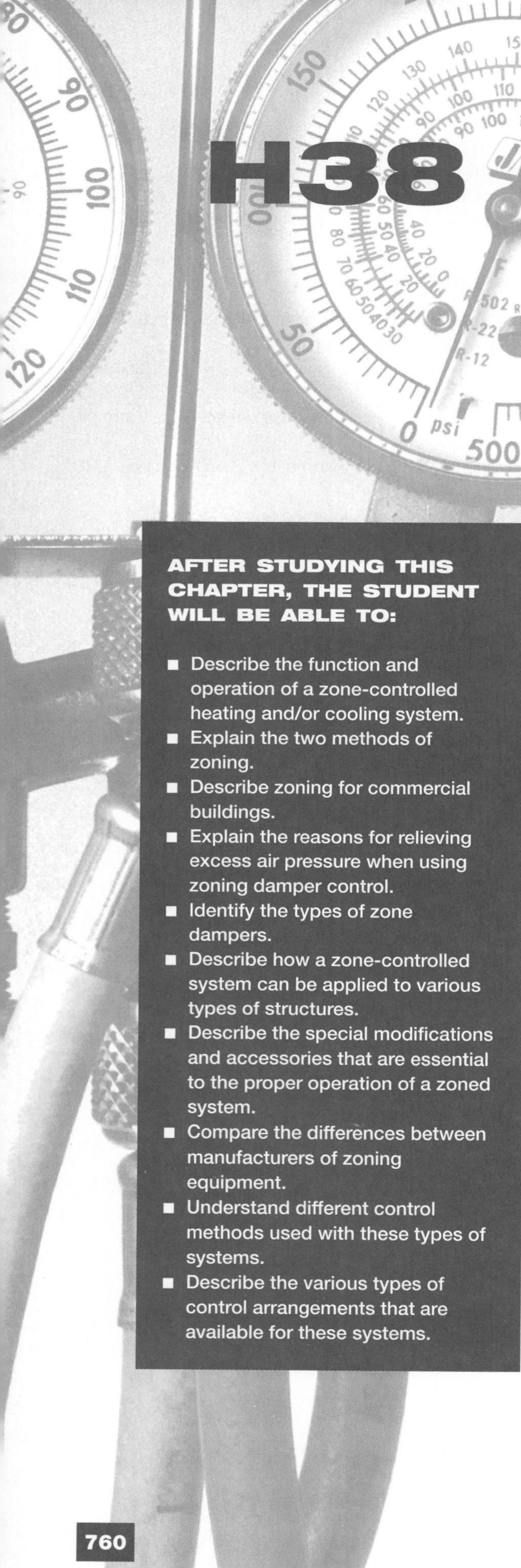

H38 Zone-Control Systems

AFTER STUDYING THIS CHAPTER, THE STUDENT WILL BE ABLE TO:

- Describe the function and operation of a zone-controlled heating and/or cooling system.
- Explain the two methods of zoning.
- Describe zoning for commercial buildings.
- Explain the reasons for relieving excess air pressure when using zoning damper control.
- Identify the types of zone dampers.
- Describe how a zone-controlled system can be applied to various types of structures.
- Describe the special modifications and accessories that are essential to the proper operation of a zoned system.
- Compare the differences between manufacturers of zoning equipment.
- Understand different control methods used with these types of systems.
- Describe the various types of control arrangements that are available for these systems.

H38-1 PRINCIPLES OF ZONING

Many heating and air-conditioning installations are installed in buildings where the floor plan is spread out, such as ranch-style and split-level houses. On these jobs it is not practical to control comfort throughout the entire area using one thermostat. These layouts require zoning, because the loads differ from one area to another. Zoning is a method of controlling the supply of conditioned air to each unique area to match the load requirements. This is done by separating the air supply to each unique area by the use of a zone damper in the duct work and controlling each zone damper with an individual thermostat.

Basically, there are two methods of zoning: (1) use zoning, and (2) building orientation zoning.

Use zoning is based on separating the areas depending on how they are used. For example, bedrooms can be on one zone and living areas on another. This is common for residential zoning.

Orientation zoning considers the directions the building faces. For example, the south side of a building would be on a separate zone from the north side. This type of zoning is common for commercial structures.

The following are some examples of zone controlled systems:

1. A 3-level split-level system is shown in Fig. H38-1. Each level of this residence has a separate trunk duct from the supply-air plenum with a zone damper installed. Each of these dampers is controlled from an individual thermostat located in that zone.
2. Zone control for a bi-level house is shown in Fig. H38-2. This system has three zones: living area, bedrooms and recreation room. The system uses a Honeywell Mastertrol, a 40-VA transformer, three zone dampers and thermostats.

ON THE JOB

When your customer complains about having trouble closing their house door on zone-control systems, this indicates that the pressure-relief device is either not working properly or is not calibrated. To test the static pressure on the unit, install a U-tube manometer in the supply duct work and operate the system. Go to all of the zone-control thermostats and set them to close down the dampers. Inspect the static pressure with the dampers open and closed. When they are closed, the pressure-relief dampers should open up to allow the pressure to escape to a secondary location. If this does not happen, check the manufacturer's specifications to verify proper operation and maintenance procedures. Never manually disconnect or force the relief dampers open because this will prevent the unit from delivering adequate air during heavy load periods.

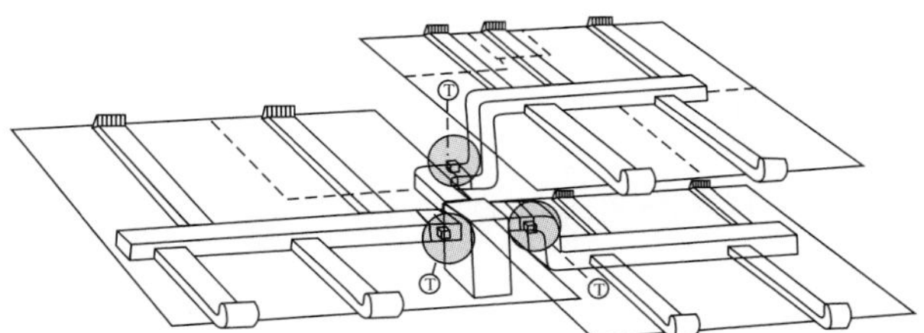

Figure H38-1 Three-zone split-level system.

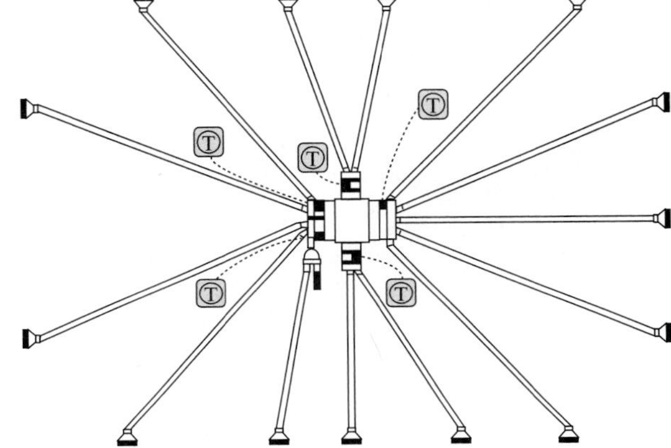

Figure H38-3 Five-zone radial system.

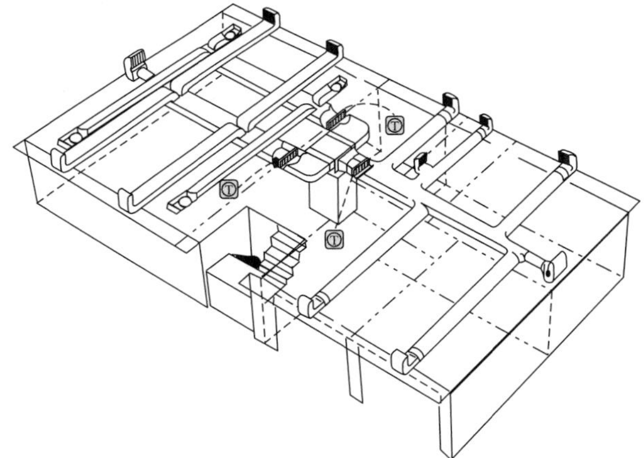

Figure H38-2 Zone control for a bi-level house.

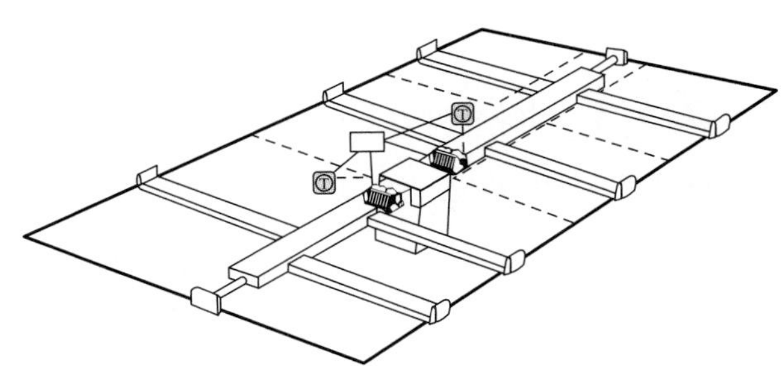

Figure H38-4 Two-zone ranch style house.

3. A room-by-room temperature control is shown in Fig. H38-3 for a five-zone radial system. These ducts can be located overhead or in the slab floor. If located in the floor, the system must be used for heating only, since cold air could cause floor condensation.
4. A two-zone ranch house is shown in Fig. H38-4. Each zone has a thermostat controlling a zone damper. This system can be installed using a Honeywell Mastertrol unit arranged for two zones.
5. A four-zone office building system is shown in Fig. H38-5. Using Honeywell equipment, this system requires a four-zone Mastertrol unit, 4 dampers and thermostats, and a 40-VA transformer.
6. A room-by-room comfort control layout is shown in Fig. H38-6. In this illustration each outlet has an automatic square-to-round transition damper.

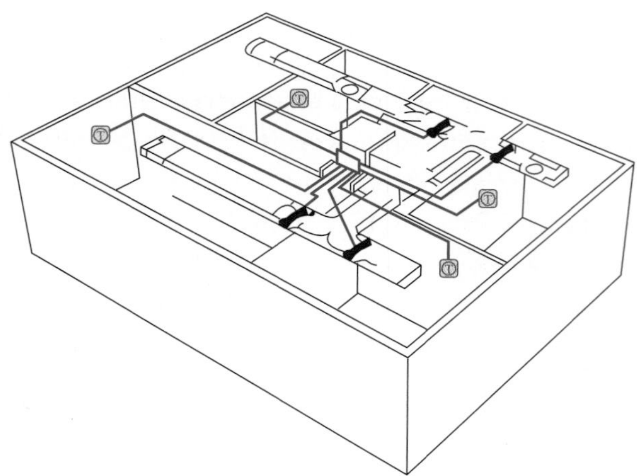

Figure H38-5 Four-zone professional office system.

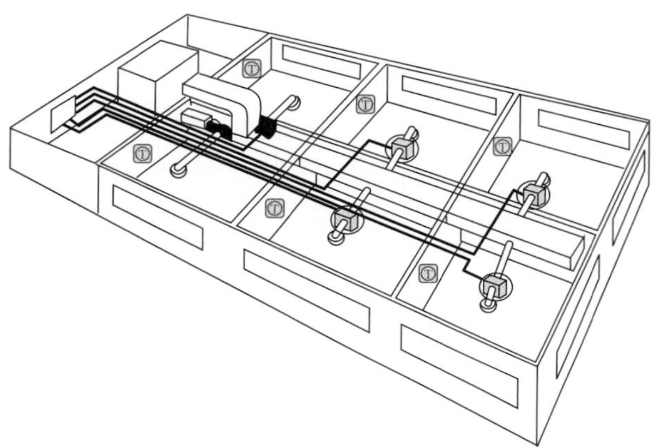

Figure H38-6 Individual room zone control system.

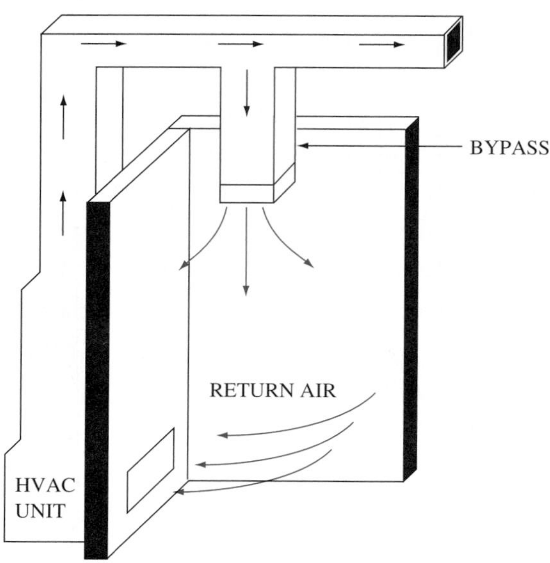

Figure H38-7 Bypassing air to a utility room.

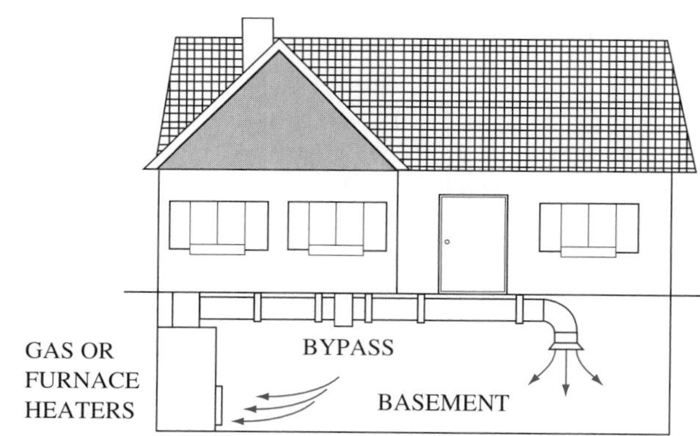

Figure H38-8 Relieving air pressure by dumping it into the basement.

H38-1.1 Relieving Excess Air Pressure

A problem that can occur on a system of thermostatically controlled zone dampers is the condition where some major zone dampers are closed off and the remaining zones are handling an excessive amount of air. This can cause air noise and poor control in the operating zones. The solution to this condition is to provide some means of relieving this excessive air pressure.

There are a number of ways to accomplish this:

1. Dump the excess air to another area such as the basement or utility room, as shown in Figs. H38-7 and H38-8. This is usually the preferred arrangement.
2. Excess air can be bypassed back into the return air duct as shown in Fig. H38-9. The disadvantage of this arrangement is that with prolonged use on heating the high-limit control may shut down the system, or on cooling the system may be shut down by the low-limit control. Thus, short-cycling can occur.
3. Undersize the zone dampers so that some leakage will occur through the damper even when it is closed. For example, a 22 in. × 8 in. damper could be used in a 24 in. × 8 in. duct. The disadvantage of this arrangement is that the performance is difficult to predict.
4. Oversize each zone duct to handle 60 to 70% of the total air. This arrangement can only be used on small jobs.

H38-1.2 Automatic Relief Dampers

In designing the control arrangement to relieve excess air pressure by dumping, described in (1) above, or bypassing, described in (2) above, automatic dampers can be used. There are two types:

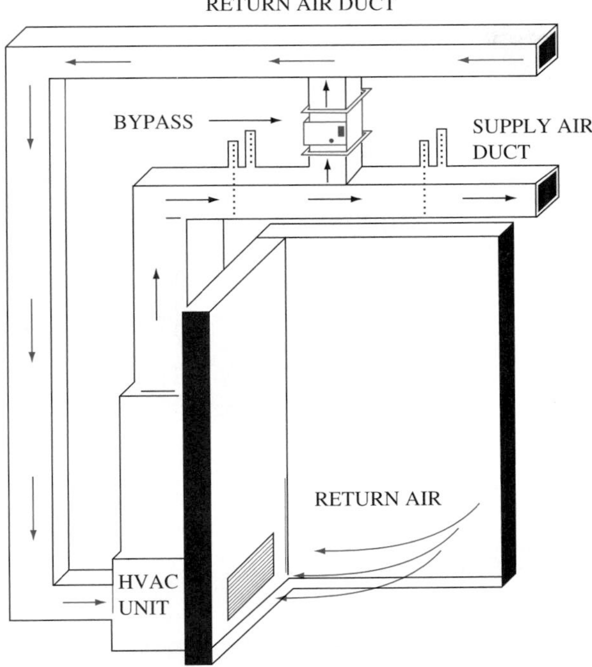

Figure H38-9 Relieving air pressure through bypassing air from the supply plenum back to the return.

1. Barometric static-pressure relief dampers require no electrical connections. They are used on low-pressure systems with static pressures usually below 0.5 in. WC. They are activated by the pressure within the duct.
2. Motorized static-pressure relief dampers are electrically or electronically operated in response to a signal from a remotely located duct pressurestat.

A single-blade static-pressure relief damper is shown in Fig. H38-10.

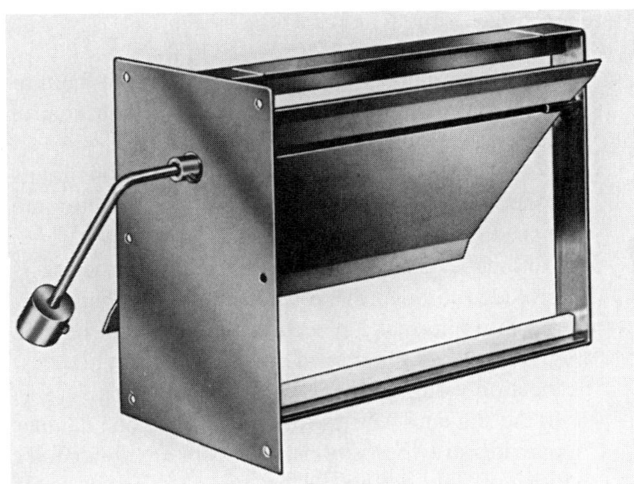

Figure H38-10 Static-pressure relief damper.
(Courtesy of Honeywell, Inc.)

During relief operations, it is important to maintain at least the minimum airflow through the air-handling unit; otherwise the coil performance will be unsatisfactory. The following formula can be used to determine the quantity of bypass air:

$$\begin{matrix}\text{air} & & \text{smallest} & & \text{leakage of all} & & \text{bypass} \\ \text{handler} & - & \text{zone peak} & - & \text{closed dampers} & = & \text{air flow} \\ \text{(cfm)} & & \text{(cfm)} & & \text{(cfm)} & & \text{(cfm)}\end{matrix}$$

H38-2 TYPES OF ZONE DAMPERS

There are many shapes, sizes, and styles of zone dampers designed to fit various requirements:

1. Opposed-blade damper for rectangular ducts (Fig. H38-11).
2. Round duct dampers (Fig. H38-12).
3. Multi-valve supply register dampers (Fig. H38-13).
4. Square ceiling-diffuser dampers (Fig. H38-14).
5. Round ceiling-diffuser dampers.
6. Floor-diffuser dampers.

Each of these has a specific application. Dampers can either be modulating or two-position type, depending on the type of operator and thermostat selected.

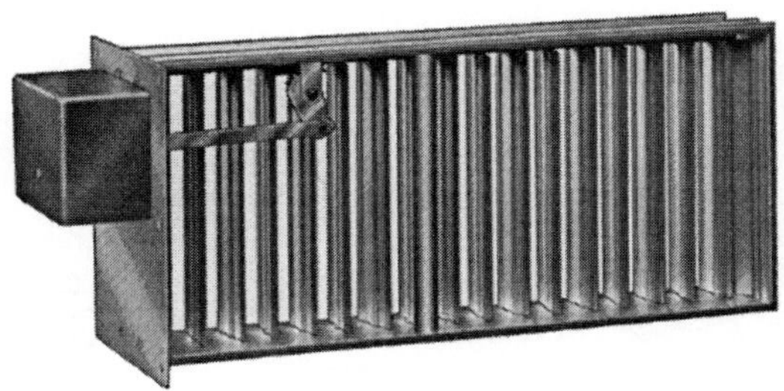

Figure H38-11 Zone damper for rectangular duct.
(Courtesy of Honeywell, Inc.)

Figure H38-12 Zone damper for a round duct.
(Courtesy of Honeywell, Inc.)

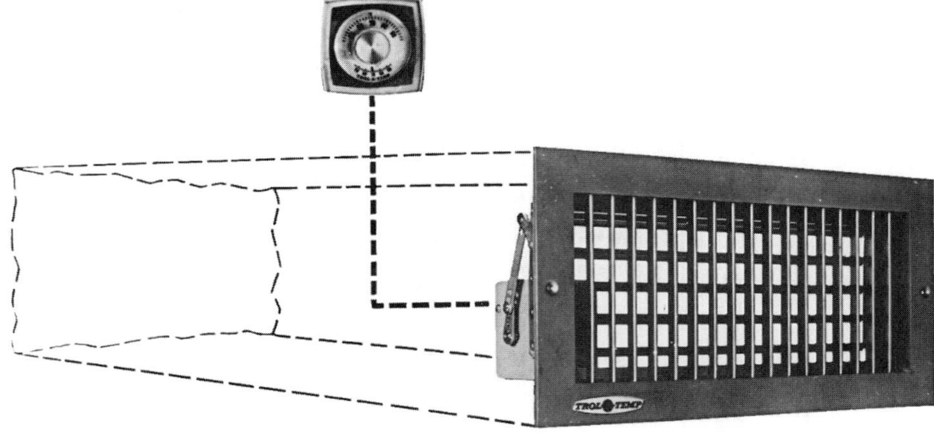

Figure H38-13 Register with a damper to regulate the air volume. (Courtesy of Trol-A-Temp Division of Trolex Corporation)

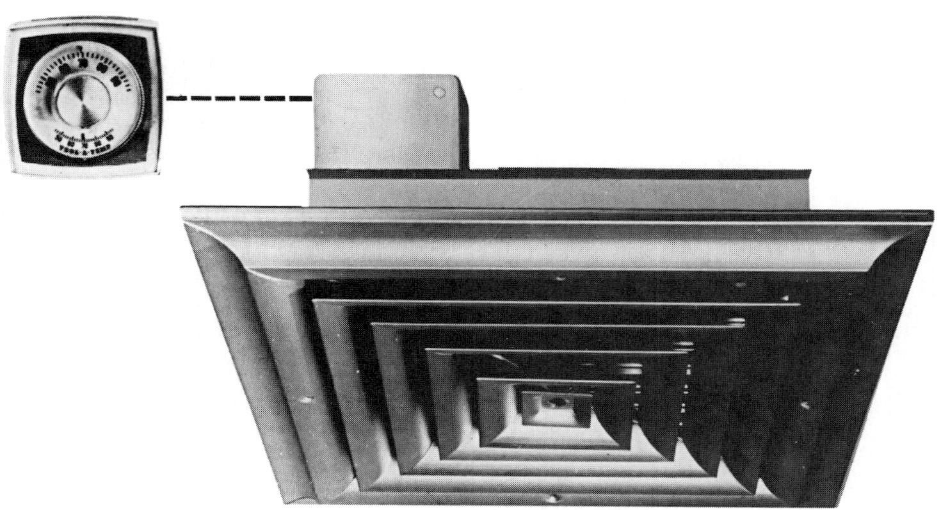

Figure H38-14 Square ceiling diffuser with damper. (Courtesy of Trol-A-Temp Division of Trolex Corporation)

H38-3 TYPES OF ZONING SYSTEMS

A number of manufacturers make control systems for installations requiring zoning. These systems differ in number of zones handled, types of applications, methods of control and types of adjustments provided. In this text, three of the commonly used systems will be described:

1. The Honeywell Trol-A-Temp zone-control system.
2. The Carrier Comfort Zone system.
3. The Lennox Harmony II system.

H38-3.1 Honeywell Trol-A-Temp Zone Control System

The Honeywell Trol-A-Temp zone-control system is an integrated set of devices for controlling a multi-zoned HVAC system. The control center is the Mastertrol automatic balancing system (MABS), shown in Fig. H38-15. The basic controls are capable of controlling a two- or three-zoned system. If a greater number of zones are required, Mastertrol add-a-zone (MAZ) panels are used. Here are some of the features of the system:

1. Each zone damper is controlled by a room thermostat which is electrically connected to the control center.
2. When any zone calls for heating (or cooling), its zone damper opens and zones not calling are closed.
3. Automatic changeover from heating to cooling is provided through the zone 1 thermostat. This thermostat is normally placed in the zone that has the greatest usage. In a residential installation, the best location would probably be the living room.
4. In the standard Mastertrol system, the zone damper operates in two positions. Dampers are either fully open or fully closed.
5. When all the zones are satisfied, all dampers go to a fully open position. This permits the fan to dissi-

Knowledge Extension

More residential buildings are being constructed with zone-temperature control systems than ever before. Some new home construction utilizes the "smart building" concept. The "smart building" integrates all normal building functions into one controller, including security, fire detection, lawn sprinklers, lights, and air conditioning. To learn more about these new "smart buildings" contact your local building supply house and ask for literature or any other promotional materials.

(a)

(b)

Figure H38-15 Honeywell Mastertrol Mark II zone-control system. (a, Courtesy of Honeywell Inc.; b, Courtesy of Trol-A-Temp Division of Trolex Corporation)

pate any residual heat that may be left in the furnace.

6. For the single-staged system, the temperature reversing thermostat (TRT) is used with a Mastertrol changeover subbase (MCRS) in zone 1. The subbase also causes changeover at all other zone thermostats.
7. For a multi-stage system, a Mastertrol two-stage (MCTS) thermostat is used in zone 1. For a heat-pump system, a Mastertrol heat pump (MCHP) thermostat is used in zone 1. All other zones use standard two-stage heating-cooling thermostats (HCT-2S).

Mode Switches

The MABS central control panel has three built-in mode switches: (1) Fan-on in-heat, (2) two-compressor, and (3) one-zone cooling.

The fan-on in-heat switch is used with heat-pump equipment. When any zone calls for heat, the switch can be turned on to permit bringing the fan and the heating on at the same time. The two-compressor switch changes the thermostat operation in zones 1 and 2 to operate the first-stage compressor and zone 3 thermostat to operate the second-stage compressor. The one-zone cooling switch permits individual control for cooling in the OFF position or only the zone 1 thermostat to control the cooling with all other zones locked open, in the ON position.

Accessory Equipment

The Trol-A-Temp system uses the following accessory equipment:

1. A multi-position fresh-air damper (MPFAD) and control can be provided, permitting the occupant to regulate the outside air intake.
2. A 7-day clock thermostat or a programmable thermostat with night set back is available for any zone.
3. If only two zones are required, a Mastertrol mini-zone (MM-2) can be used in place of a MABS central control panel, at a savings in cost.
4. A Mastertrol junior (MABS-JR) panel is available for heating-only or cooling-only systems.
5. A slave driver-control relay (SDCR) is available to control 2, 3, or 4 dampers simultaneously from a single thermostat.

Master Control Panel Wiring

A typical wiring diagram for a Mastertrol panel, with T87F thermostats and single-stage heating and cooling is shown in Fig. H38-16. The subbase on the zone 1 thermostat has provision for operating the changeover control.

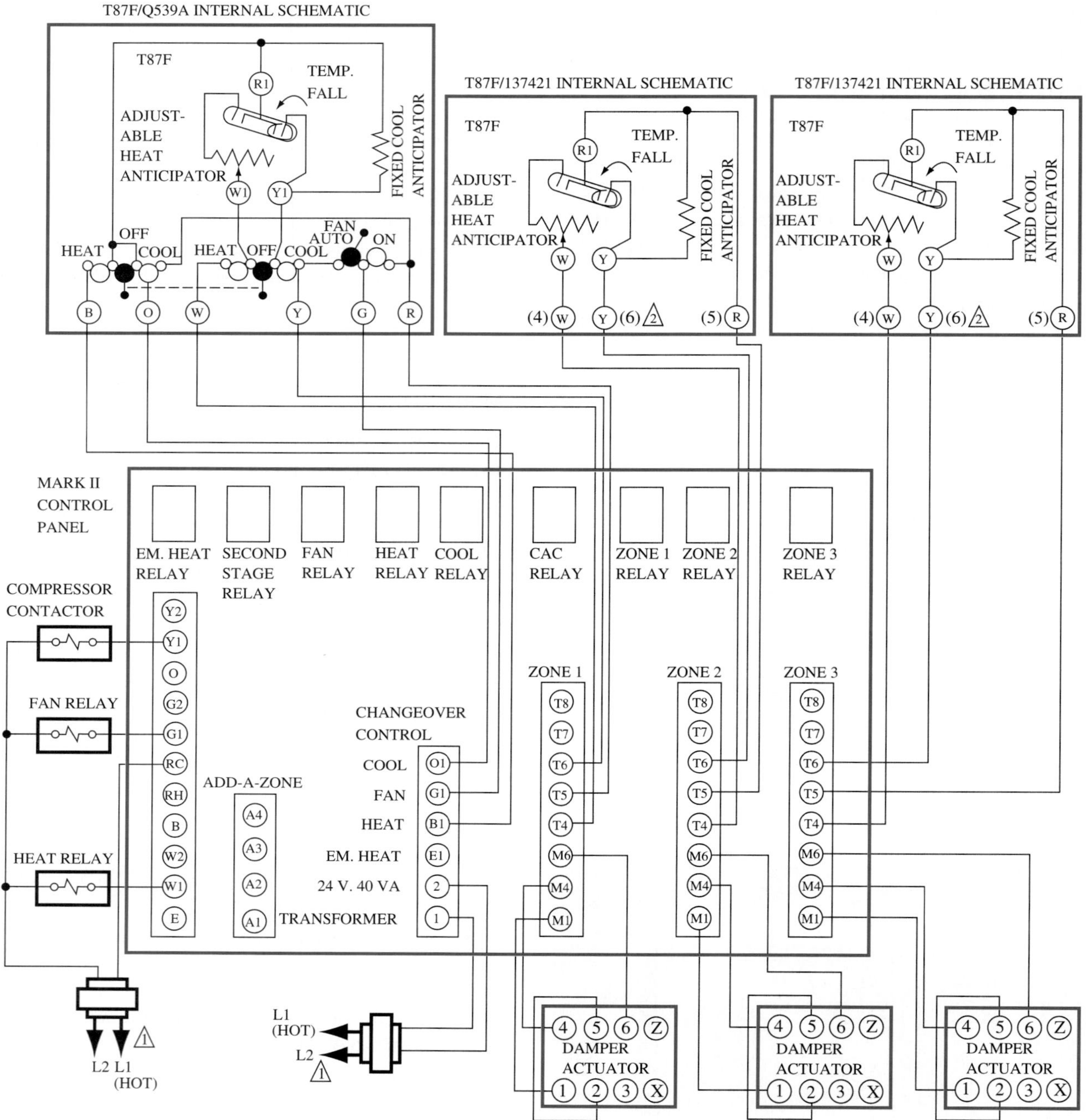

Figure H38-16 Wiring diagram for a Mastertrol panel. (Courtesy of Honeywell, Inc.)

H38-3.2 Carrier Comfort Zone System

The Carrier Comfort Zone system is primarily designed for residential applications, although it is often applied to small commercial installations. The basic system handles four zones of heating and/or cooling. It is recommended for systems requiring 5 tons of cooling (2000 cfm) or smaller with a maximum of 1 in. WC static pressure at the zone dampers. A typical four-zone layout is shown in Fig. H38-17.

Each damper is a modulating type controlled by a zone thermostat. All wiring is run to the comfort-zone center. All standard systems are supplied with an automatic barometric bypass damper to relieve excess air pressure when a reduced number of zones are calling for conditioned air.

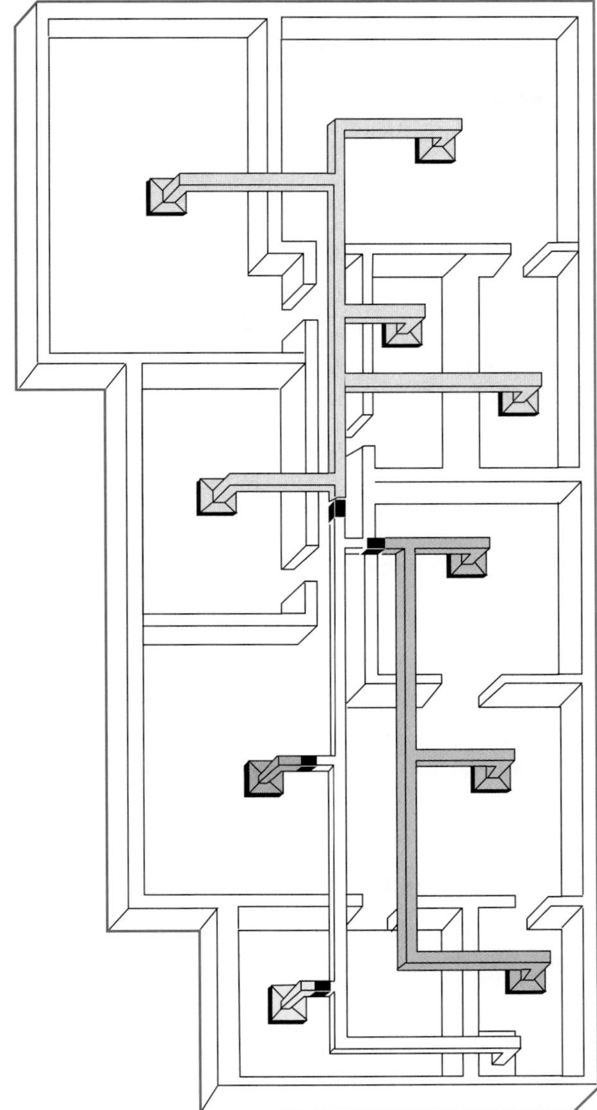

Figure H38-17 Four-zone Carrier Comfort Zone residential system. (Courtesy of Carrier Corporation)

The central controller panel is placed in zone 1. This is considered the most occupied zone of the building. In a residence it is the family room or the living room. It not only controls zone 1 but also provides programming for setting up the other zones. A duct-temperature sensor is placed between the bypass damper and the heating unit.

Operation of the System

The comfort-zone controller is used for programming the system. Through the use of this device the unique demands of each zone can be set. This includes room temperatures for day operation, night set back and other special conditions. It also provides access to service requirement information. A description of the principal parts of the controller and their function is as follows:

1. *Controller display,* shows the set points for the zone temperature, the zone temperature, and the programming information for adjusting the set points.
2. *Clock display,* indicates the current time and the day. During programming it shows the start times and the weekly periods.
3. *Programming adjustment buttons,* are used for programming weekly periods, start times for each zone and the setting of the clock.
4. *System switches,* provide the selection of heat, cool, fan, and emergency heat operation.
5. *Zone selector dial,* provides the selection of each zone and vacation mode. Optional area is for use in accessing information for installation and service.

Programming and Operation

The following is a more detailed description of the programming and operation required for this type of zoning system. Many features have been incorporated to increase the usefulness of the system, provided it is properly programmed and operated. A description of some of the essential programming is as follows:

Setting the System Switches. There are four system switches: heat, cool, fan, and emergency heat.

The heat switch can be placed in either the OFF or AUTO position. In the OFF position, the heat will not come on. In the AUTO position, heat will turn on when any zone is more than 1.5°F below the desired heat set point.

The cool switch can be placed in either the OFF or AUTO position. In the OFF position, the cooling will not come on. In the AUTO position, cooling will turn on when any zone is more than 1.5°F above the desired cool set point.

The fan switch can be placed in either the ON or AUTO position. In the ON position, the fan will stay on continuously (recommended). In the AUTO position, the fan will come on only when heating or cooling is required.

The emergency heat switch can be placed in either the OFF or ON position. In the OFF position, the electric strip heat will not come on unless there is a zone more than 2°F below the desired heat set point and the comfort trend demand has been satisfied. In the ON position, when any zone requires heat, the electric heat strip will turn on. The compressor will remain off.

Adjusting Set Points. When the controller display is set for normal operation, it displays the current zone-temperature set points and allows adjustment of these set points using the set point buttons.

By pressing both buttons simultaneously, the controller display will toggle between displaying set point values and the zone temperature. To view the setpoints for each zone, turn the zone selector dial to the zone number to be viewed.

Each zone has three weekly period settings (Weekdays, Saturday and Sunday) and up to 4 different possible combinations. Each zone can be programmed independently.

The vacation mode can be selected by turning the selector dial to Vacation. In this mode the entire building is maintained at the set points displayed on the controller.

Overriding Programming. To override the time programming, press any setpoint button. This adjusts the setpoint for the override period. It will remain in the new setting until the next time period. If the HOLD button is pressed, the setpoint will not change until the HOLD button is pressed the second time.

H38-3.3 Lennox Harmony II Zone Control System

The two principal components of the Lennox Harmony II zone-control system are the control panel (Fig. H38-18) and the control center (Fig. H38-19). The control panel is used for owner access to operate the system, along with the zone thermostats. The control center organizes the operation of the thermostats, dampers, and HVAC equipment to result in total comfort. The view shown includes the terminal connections and jumper blocks.

This control system manages the distribution of conditioned air to specific areas or zones in a house or small commercial building. The system is designed to operate with any of the following:

1. A Lennox high-efficiency pulse-type furnace used in combination with either a single- or two-speed condensing unit.
2. A Lennox variable-speed blower coil unit used in combination with a single- or two-speed heat pump.
3. A Lennox high-efficiency pulse-type furnace used in combination with a Lennox heat pump and Lennox heat-pump control.

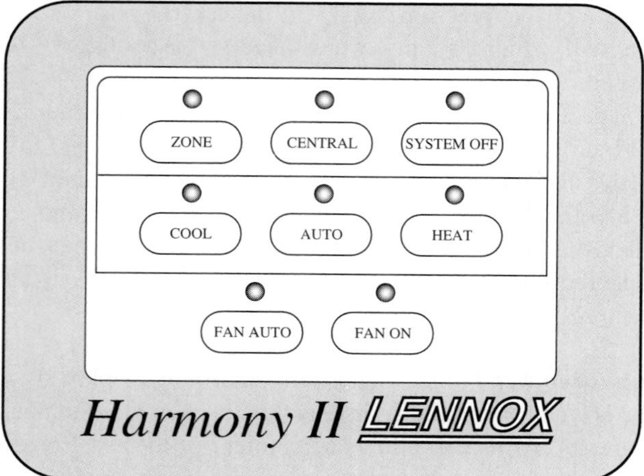

Figure H38-18 Control panel for the Lennox Harmony II zone-control system. (Courtesy of Lennox Industries, Inc.)

The system operates in either of two modes: *central control* or *zone control.* Central control allows the heating and cooling to condition all zones. In the zone-control mode, specific zones are conditioned only when a demand comes from that zone. LED lights on the panel indicate the operating mode.

The principal features of the system are:

1. The system controls the air volume of the blower, which eliminates the need for bypass dampers and discharge-temperature limit switches.
2. The system does not require programming.
3. The system does not use sensors.
4. There is no overriding or re-programming to condition a zone.
5. The standard components provide a balanced comfortable environment.

Some typical basement layouts showing damper locations for the reduced radial and extended plenum duct designs are shown in Figs. H38-20 and H38-21.

Accessory Components

Thermostats are required for each zone and may be programmable or conventional. The zone 1 thermostat is designated as the master thermostat.

Dampers and damper transformers are required for each zone. Dampers are motorized, using 24 V_{AC} to close and spring-return to open.

A discharge-air temperature probe monitors the supply air. It is used to gather temperature information across the airstream. It is not a limit switch.

A pressure switch is required for application with a heat pump. It acts as a safety switch in case of high head pressures during first- and second-stage heating.

System Operation—Zone/Central/ System Off Modes

In the Zone mode the system will respond to the demands of any zone. The only dampers that remain open are those supplying air to the demanding zones. The blower operates at the speed determined by the position of the CFM selection jumpers. In the Zone mode, it is recommended that the fan be placed in the Fan Auto position to minimize air mixing between zones.

In the Central mode, the system responds only to the master thermostat and all rooms receive conditioned air. When in the Central mode, the control panel should be set

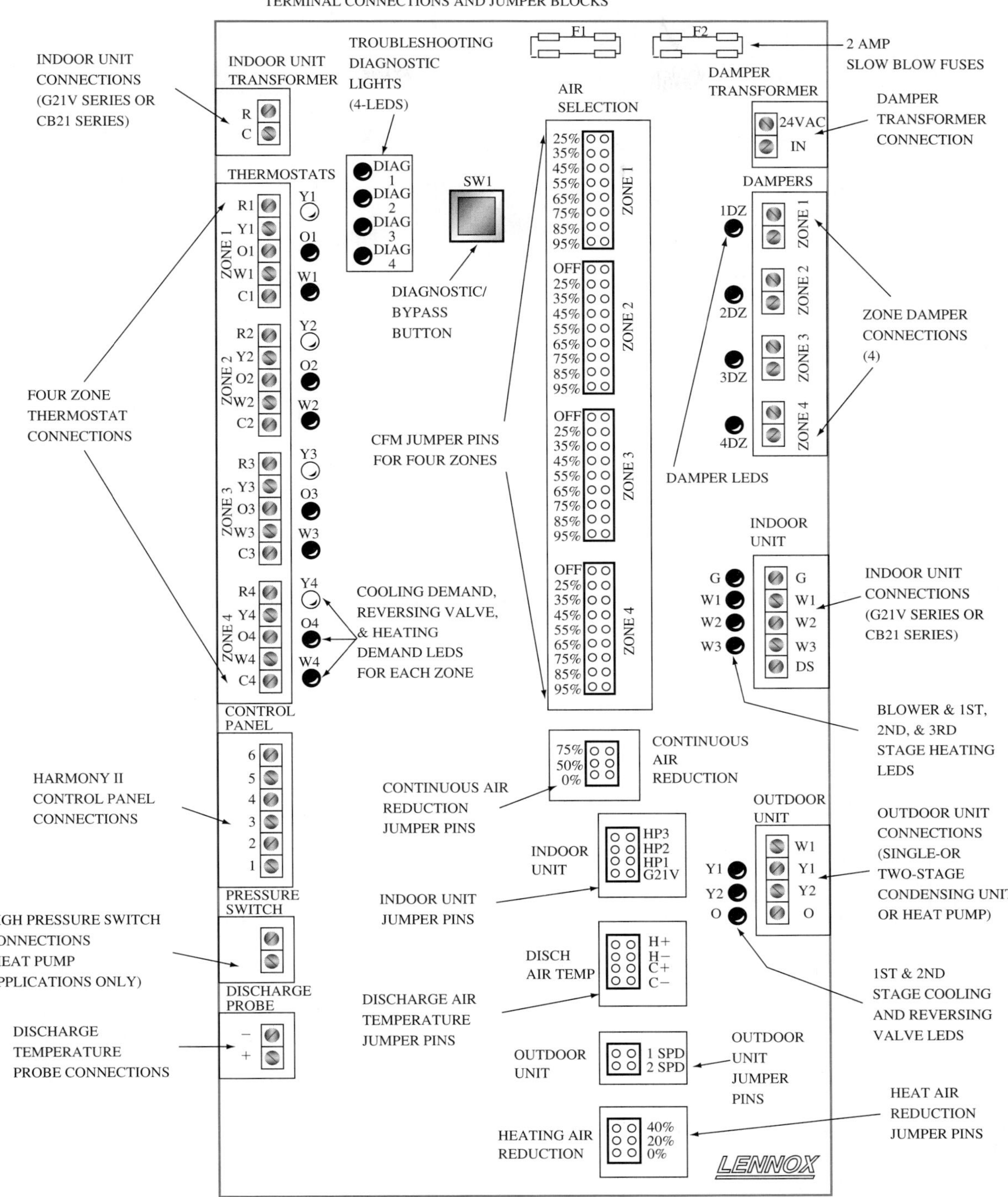

Figure H38-19 Terminal connections for the Lennox Harmony II control center. (Courtesy of Lennox Industries, Inc.)

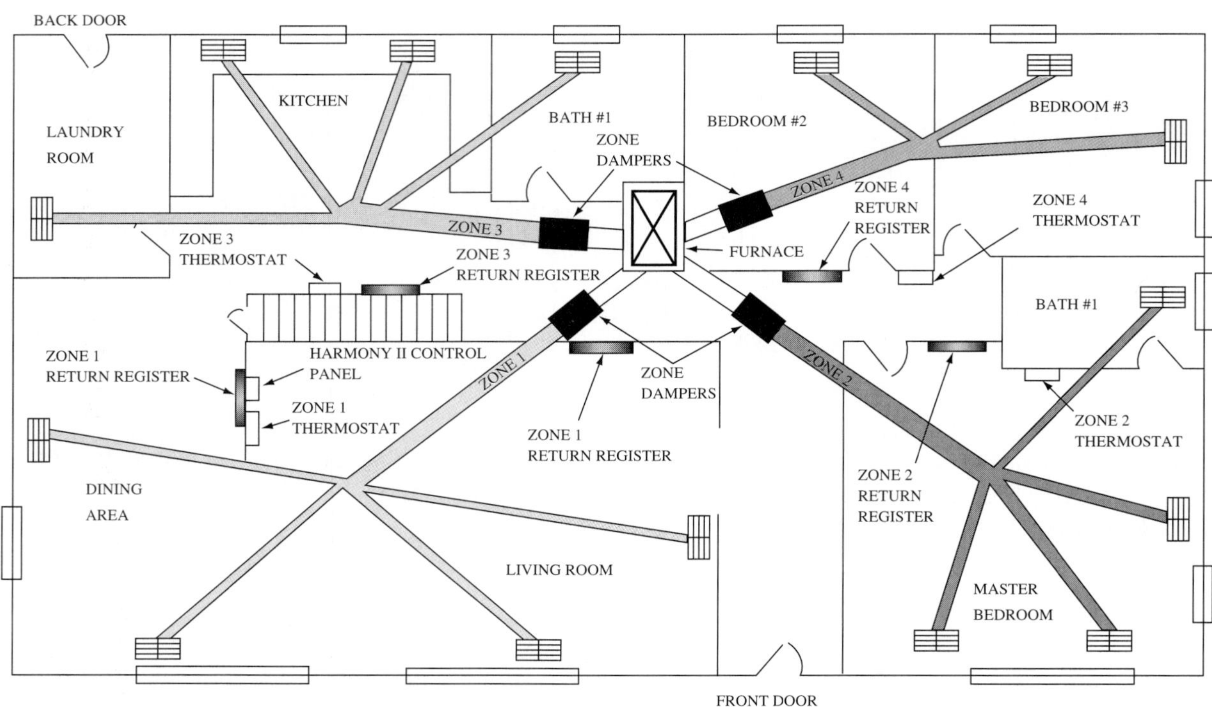

Figure H38-20 A typical radial zone duct design using the Harmony II control system. (Courtesy of Lennox Industries, Inc.)

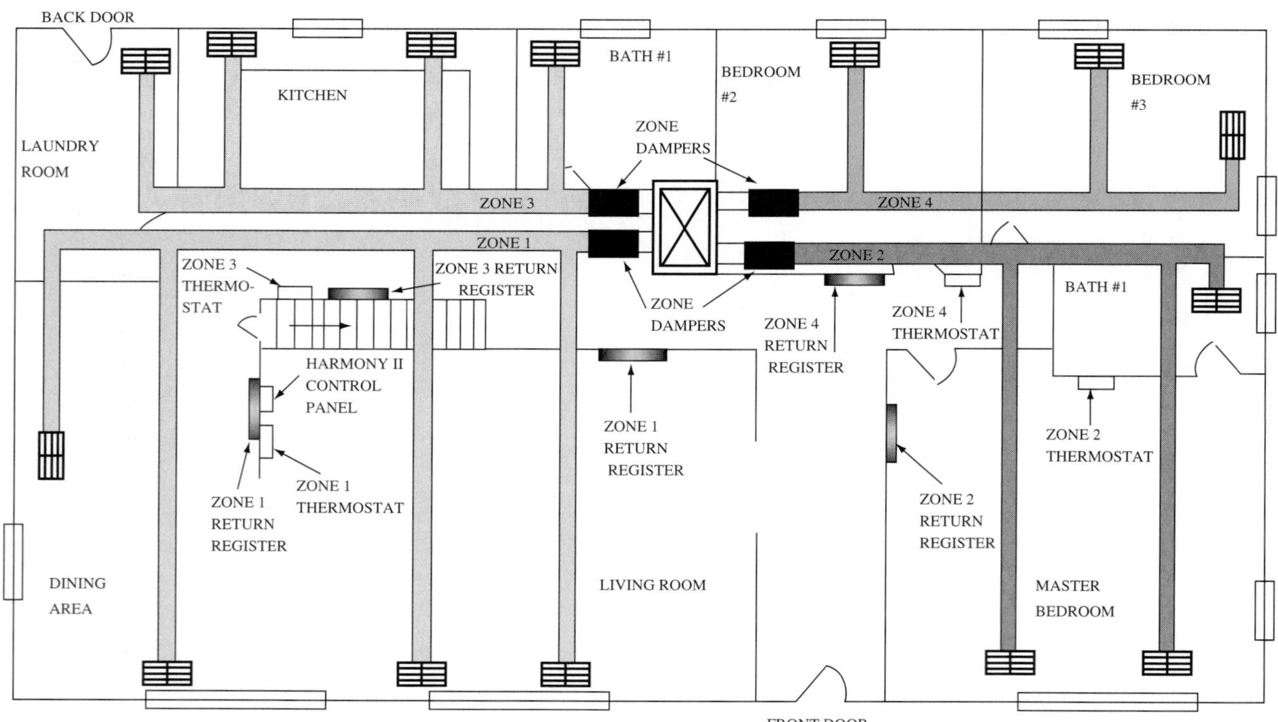

Figure H38-21 Typical extended plenum zone duct design using the Harmony II control system. (Courtesy of Lennox Industries, Inc.)

Delay	Time	Function
Blower Delay (gas heat only)	5½ min.	Gas furnace only. Dumps air into last zone called during cool down following heat demand.
Compressor Speed Change	4 + 1 min.	Between low speed and high speed in order to make sure high-speed demand is valid and to equalize refrigerant pressures. 1 minute due to TSC in outdoor unit. 4 minutes due to delay in Harmony control center. Compressor can cycle off anytime.
Compressor Off	5 min.	At end of demand. Equalizes pressure in refrigerant system and prevents short cycling.
Heat Staging (electric heat only)	2 min.	Between stages up or down.
Auto-changeover	20 min.	When opposing demands are present, Harmony II must work to satisfy current demand at least 20 min. If current demand is not satisfied after time has elapsed, system will changeover and satisfy opposing demand. On and Off delays above will also apply.

Figure H38-22 Table showing the details of the various time delay relays used in the Harmony II control system.

to Auto changeover so that the heat/cool selection can be made by the master thermostat.

In the System Off mode, the compressor and all heating equipment are turned off.

System Operation—Cool/Heat/Auto-Changeover Modes

When the Cool mode is selected, the system will respond to the cooling demands of the room thermostats. When the Heat mode is selected, the system will respond to the heating demands of the zone thermostats.

When the Auto mode is selected, the system will respond to both the heating and the cooling demands from the zone thermostats. For example, this will allow the system to heat in the morning and cool in the afternoon.

When in the Auto mode, there is a possibility that one zone may be calling for cooling at the same time that another zone is calling for heating. Opposing zones are satisfied on a first-come-first-serve basis. If opposing zones reach the control center at the same time, the heating will be supplied first. If opposing zones persist, the system will work to supply the current demand for a maximum of 20 minutes, then switch over and try to satisfy the opposing demand for a maximum of 20 minutes. When one or the other demand is satisfied, the system will work to satisfy the remaining demand.

System Operation—Fan Auto/Fan On Modes

In the Fan Auto mode, the blower will cycle on and off with the demand, delivering air to the calling zones. During gas or electric strip heating, the blower will continue after the demand until the heater is cooled sufficiently.

In the Fan On mode with demands satisfied, all dampers open and air is circulated to all zones. When a cooling or heating demand is present, the system responds by closing all dampers except the demanding zone.

Time Delays

Several different delay timers are used in the system as described in Fig. H38-22.

REVIEW

- Zone control is an energy-conservation method used on large buildings, ranch, and split-level houses, where it is not practical to heat or cool the entire structure at all times.
- Zoning is a method of controlling the supply of conditioned air to each unique area to match load requirements.
- Two methods of zoning are (1) use zoning, (2) building orientation zoning:
  Use zoning—Based on separating the areas depending on how they are used.
  Orientation zoning—Considers the direction the building faces. It is common in commercial structures.
- Relieving excess air pressure:
  Dump the excess air to another area.
  Bypass the excess air back into the return-air duct.

Undersize the zone dampers.
Oversize each zone duct.
Automatic relief dampers.
Barometric static pressure
Motorized static pressure

- Types of zone dampers:
  Opposed-blade damper
  Round-duct damper
  Multi-valve supply register damper
  Square ceiling-diffuser damper
  Round ceiling-diffuser damper
  Floor-diffuser damper
- Dampers may be modulating or two-position.
- Commonly used zone control systems:
  Honeywell Trol-A-Temp
  Carrier Comfort Zone
  Lennox Harmony II
- Honeywell Trol-A-Temp zone control system is an integrated set of devices for controlling a multi-zoned HVAC system.
  Control center—Mastertrol automatic-balancing system (MABS)
  Mastertrol add-a-zone (MAZ)—Used if a greater number of zones are required.
  Zone dampers
  Zone thermostat
  Single-stage—Temperature-reversing thermostat (TRT) with Mastertrol changeover subbase (MCRS) (installed in zone 1).
  Multi-stage—Mastertrol two-stage thermostat (MCTS), two-stage heating/cooling thermostat (H-CT-23).
  Heat pump—Mastertrol heat pump thermostat (MCHP)
  Automatic relief damper
  Accessory equipment
  Multi-position fresh-air damper (MPFAD)
  7-Day clock thermostat or programmable thermostat
  Mastertrol Minizone (MM-2) two-zone control panel (MABS)
  Mastertrol Junior (MABS-JR) heating only/cooling only
  Slave driver-control relay (SDCR)
- Carrier Comfort Zone—Primarily designed for residential applications/small commercial installations; handles up to 4 zones of heating/cooling. For systems requiring 5 tons of cooling or less with a maximum of 1 in. WC static pressure at the zone dampers.
  Zone dampers
  Zone thermostat
  Automatic barometric bypass damper
  Central controller panel (Placed in zone 1)
  Duct-temperature sensor
  Home access module (optional)
  Smart sensors (optional)
  Motorized bypass damper (optional)
- Lennox Harmony II zone-control system—Manages the distribution of conditioned air to specific zones of the structure.
  Control panel—Provides owner access to operate the system, zone thermostats.
  Control center—Organizes the operation of thermostats, dampers, and HVAC equipment.
  The Lennox Harmony II is designed to operate with:
  Lennox high-efficiency pulse-type furnace
  single or two-speed condenser unit
  Lennox variable-speed blower with single- or two-speed heat pump
  The Lennox Harmony II has two modes of control:
  Central control allows the heating/cooling to condition all zones.
  Zone control conditions specific zones when demand occurs.

Problems and Questions

1. Why would zone control be desirable?
2. What is meant by "zoning"?
3. What are the two methods of zoning?
4. What is the principle behind "use zoning"?
5. What is the principle of operation for "building orientation"?
6. If excess air pressure is not relieved, what kinds of problems may exist?
7. What are two common types of automatic relief dampers?
8. List the types of zone dampers.
9. What are the two operating modes for zone dampers?
10. Name three commonly used zone-control systems.
11. What are the two modes of control for the Lennox Harmony II zone-control system?
12. How do you size the quantity of bypass air?
13. How are zone dampers controlled?
14. Carrier Comfort Zone systems can handle up to _____ zones and are designed for systems requiring _____ tons or less of cooling.
15. Zone-control systems are _____, which allows operator parameters, and flexibility. They may also be interfaced with home computers through _____ access modules.
16. Orientation zoning considers the occupancy of the building. True or false?
17. A problem with zone dampers is that a method of relieving excess air pressure is needed. True or false?
18. Relieving air pressure can be accomplished by dumping the air to the basement or utility room. True or false?
19. Barometric static-pressure relief dampers require little electrical connections. True or false?
20. Zone dampers can only be modulating. True or false?

21. The Honeywell Trol-A-Temp control system automatically
    a. balances the system.
    b. controls two or three zones.
    c. integrates a set of devices.
    d. All of the above.
22. The Carrier Comfort Zone system is designed for
    a. residential and small commercial buildings.
    b. large residential and larger commercial buildings.
    c. primarily residential applications.
    d. All of the above.
23. The Carrier system requires how much static pressure at the dampers to operate?
    a. 0.5 in. WC
    b. 1 in. WC
    c. 1.25 in. WC
    d. None of the above.
24. The Lennox Harmony II zone-control system works well with
    a. Lennox pulse furnace.
    b. Lennox variable-speed blower coil.
    c. Lennox heat pump.
    d. All of the above.
25. Zone control is used
    a. to conserve energy.
    b. on larger buildings where it is not practical to heat or cool the entire building.
    c. on ranch and split-level houses.
    d. All of the above.

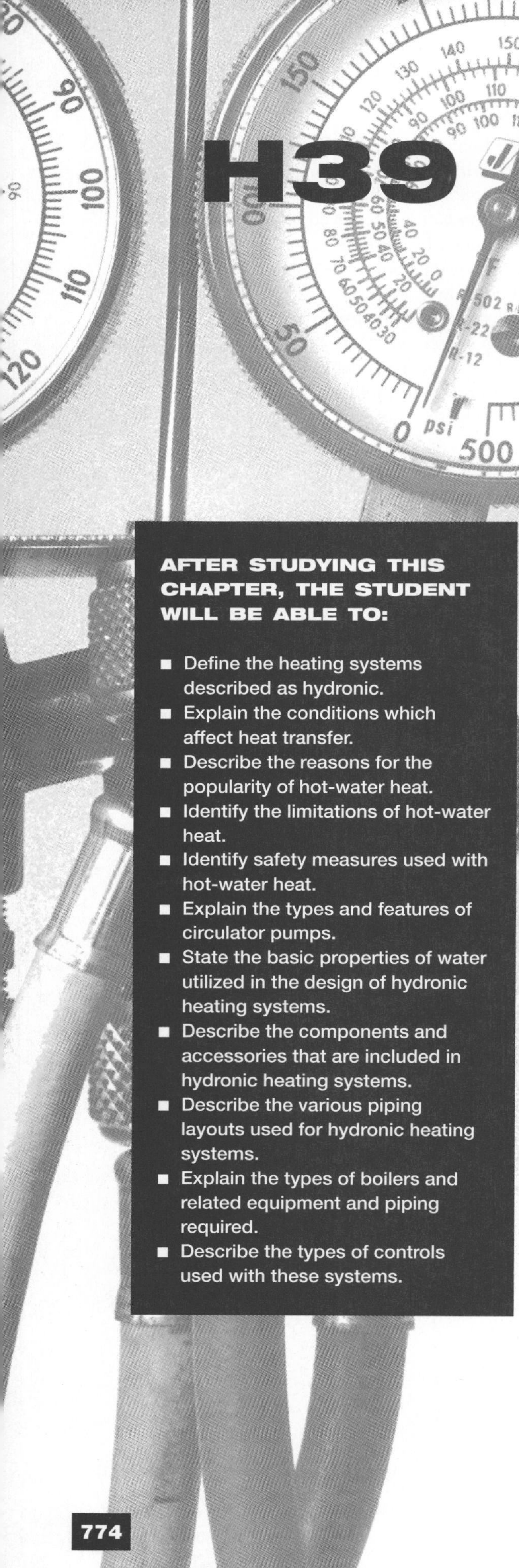

H39 Hydronic Heating

AFTER STUDYING THIS CHAPTER, THE STUDENT WILL BE ABLE TO:

- Define the heating systems described as hydronic.
- Explain the conditions which affect heat transfer.
- Describe the reasons for the popularity of hot-water heat.
- Identify the limitations of hot-water heat.
- Identify safety measures used with hot-water heat.
- Explain the types and features of circulator pumps.
- State the basic properties of water utilized in the design of hydronic heating systems.
- Describe the components and accessories that are included in hydronic heating systems.
- Describe the various piping layouts used for hydronic heating systems.
- Explain the types of boilers and related equipment and piping required.
- Describe the types of controls used with these systems.

H39-1 OVERVIEW

Hydronics can be defined as a science that utilizes water or steam to transfer heat from the source where it is produced, to an area where it can be used, through a closed system of piping. Hydronics is applied also to cooling systems; however, this chapter is confined to heating applications.

The amount of heat transfer that takes place is dependent on three conditions:

1. The temperature difference between the water and the surrounding medium.
2. The quantity of water flowing through the piping, expressed in gpm or gph.
3. The design characteristics of the heat exchanger.

Although water can be circulated by gravity, modern heating systems use forced circulation for two reasons:

1. Smaller pipe sizes can be used.
2. Systems are easier to control.

There are a number of reasons for the popularity of hot water heating systems, the greatest of which is comfort, based on the following features:

1. Heat is supplied at the base of the outside wall, warming the exterior exposure, by convection, to near room temperature. This prevents the uncomfortable feeling of radiating body heat to the cold outside walls.
2. The gravity circulation of warm air from the convectors supplies heat to the room by a gentle movement of air. This prevents uncomfortable drafts and air velocities that absorb moisture from the skin, creating a cooling effect.
3. The hot water heats and cools slowly, creating a "flywheel" effect. This prevents sudden changes in temperature and the "cold 70" condition experienced on a warm air installation.

From the above, the hot water system seems perfect. So why doesn't everyone have one? There are three reasons:

ON THE JOB

Problem: A customer calls and says there is always a trickling noise whenever the heat is on. You find it sounds like a bubbling brook. The temperature is 145°F and the pressure is 4 psig. Finding the highest radiator with an air vent, you open it to bleed the air out, but hardly any air comes out and then it stops completely. You know there's more air in the system. How are you going to get the air out?

Solution: The pressure is too low. The air cannot be bled from the system until the pressure is increased—usually to at least 12 psig in a residential system. There must be something wrong with the pressure-reducing valve or fill valve. Replacement of the fill valve may be necessary, although cleaning the internal screen and adjusting the pressure may solve the problem.

1. They cost more than an equivalent warm-air installation.
2. If humidification and air filtration are desired, a separate air system needs to be installed.
3. If cooling is desired as well as heating, dual systems are required.

H39-2 PROPERTIES OF WATER

In order to deal with hydronic heating systems, it is advantageous to review some of the properties of water. Fortunately, most tap water used in these closed systems is free from contamination and requires no special treatment. Certain properties, however, are critical and must be considered during application and installation, as follows:

1. Water expands or contracts as its temperature is changed. In a liquid state the change in volume with temperature is relatively small (but significant). When its temperature increases to the steam temperature, the change in volume is enormous. In closed hot-water heating systems, to allow for the expansion characteristics, certain safety measures are provided as follows:
   a. Both temperature- and pressure-limiting controls are used to maintain the water within safe conditions.
   b. A pressure-relief device is supplied in the piping to relieve excess pressure should it occur.
   c. An expansion tank is provided in the piping, partly filled with air, to permit normal expansion and contraction of the water.
2. Water will freeze into ice at temperatures around 32°F, depending on the pressure. When it does, it expands. Under these conditions, it can burst pipes or containers. Provisions need to be made to:
   a. Prevent the water from reaching the freezing point, or
   b. Use an antifreeze solution in the system in place of pure water, to lower the freezing point to a safe value.
3. Water has considerable weight (62.3 lb/ft^3), and can add weight to any container in which it is placed. Allowances often need to be made in the supporting structure for equipment containing water. Due to its weight, extra energy is required when it is raised from its original level.
4. Water creates friction as it flows through piping, fittings, and equipment. This frictional force must be considered when selecting the flow of a circulating pump.

EXAMPLE

Determine the total pump head required for selecting a cooling tower pump for the installation shown in Fig. H39-1, based on the following conditions:

Piping—75 ft of steel pipe
Fittings—10 standard elbows
4 gate valves
Static lift—5 ft
Water flow—15 gpm
Condenser pressure drop—30 ft

Solution
From Fig. H39-2, a 1¼-in. steel pipe is selected, with a head loss of 6.35 ft/100 ft of pipe.

EXAMPLE

From Fig. 39–3 (p. 778), the equivalent length of the piping and fittings is:

Item	Equivalent Length
75 ft of 1¼-in. steel pipe	75.0 ft
10 1¼-in. standard elbows @ 3.5	35.0 ft
4 1¼-in. open gate valve @ 0.74	2.96 ft
Total Equivalent Length	112.96 ft

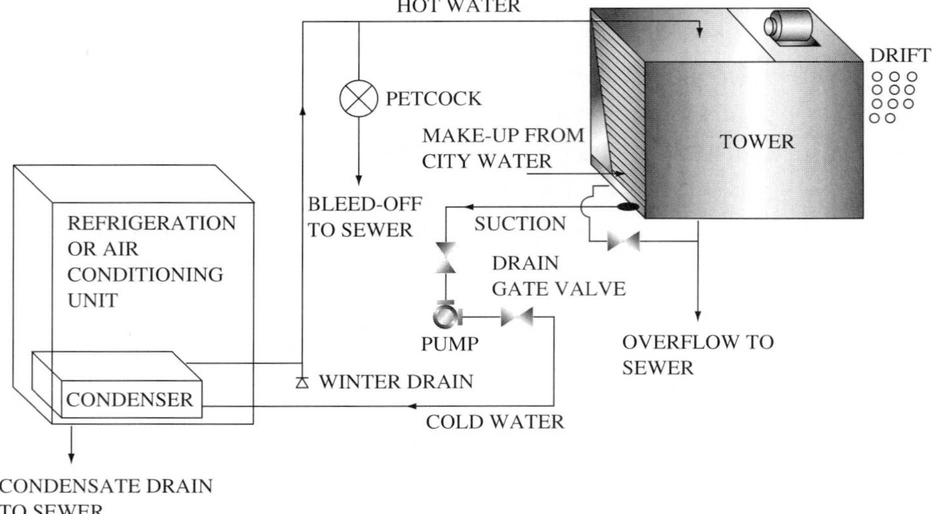

Figure H39-1 Diagram of a cooling tower piping arrangement.

Head loss in piping and fittings = (112.96/100) × 6.35
= 7.17 ft

Pressure Loss	Total Pump Head
Due to pipe and fittings	7.17 ft
Due to condenser (13 × 2.31)	30.00 ft
Due to static lift	5.00 ft
Total pump head	42.17 ft

H39-3 PIPING, PUMPS, AND ACCESSORIES

H39-3.1 Piping

The supply piping conveys heat from the source to the terminal units. A proportion of its heat is released and the return piping carries water back to the boiler for re-heating to its original discharge temperature.

A good piping system has the following qualities:

1. The piping is the correct size.
2. It is well supported, with hangers that permit adequate pipe expansion and contraction.
3. The piping is properly assembled with leak-tight joints.

Pipe sizing is usually performed by using a friction chart, as shown in Fig. H39-4. The water velocity should not exceed 5 ft/min. Fig. H39-4 gives the flow rates for water in copper tubing. This chart shows head loss in ft/100 ft on the vertical axis, flow rate in gpm on the horizontal axis. Pipe-velocity lines run upward to the left, and pipe-size lines run upward to the right. Having given the flow rate and the pipe velocity, the pipe size and the loss in head per 100 ft can be read from the chart.

EXAMPLE

Given a boiler with an output capacity of 100,000 Btu/hr, a supply water temperature of 210°F, a return water temperature of 190°F, and a piping velocity requirement of 4 fpm, what size copper pipe main should be used?

In order to convert the boiler output to gpm, the following formula is used:

gpm = boiler output in Btu/hr/(TD × 500)

where

TD = 210-190°F = 20°F

Substituting in the formula,

gpm = 100,000 Btu/hr (20°F × 500) = 10 gpm

Solution

Plot the intersection of the 10 gpm line and the 4 fpm line. At this point the chart indicates a 1-in. copper pipe.

Commonly used pipe supports consist of hanger rods which extend down from the ceiling, with loosely fitting circular openings for the pipe. This construction provides support, but permits the pipe to move during expansion or contraction.

Pipe anchors need to be provided at regular intervals to control and contain piping movements. These anchor supports are clamped tightly to the pipe. Where pipes are insulated, the hanger must surround the insulation without compressing it.

Piping materials for heating systems are normally steel or copper. Joints can be flanged, screwed or welded for steel pipes and brazed for copper. The joints must be tight and verified by approved tests.

Water Flow (gal/min)	Type of Pipe or Tubing	¾ in.		1 in.		1¼ in.		1½ in.		2 in.	
		Velocity (ft/sec)	Head Loss (ft/100 ft)	Velocity (ft/sec)	Head Loss (ft/100 ft)	Velocity (ft/sec)	Head Loss (ft/100 ft)	Velocity (ft/sec)	Head Loss (ft/100 ft)	Velocity (ft/sec)	Head Loss (ft/100 ft)
6	Std. steel	3.61	14.7	2.23	4.54						
	Copper type L	3.98	11.5	2.34	3.13						
9	Std. steel	5.42	31.1	3.34	9.72	1.93	2.75				
	Copper type L	5.96	24.2	3.50	6.63	2.30	2.38				
12	Std. steel			4.46	16.4	2.57	4.31	1.89	2.04		
	Copper type L			4.67	11.3	3.06	4.04	2.16	1.73		
15	Std. steel			5.57	24.9	3.22	6.35	2.36	3.22		
	Copper type L			5.84	17.1	3.83	6.12	2.70	2.62		
22	Std. steel					4.72	13.2	3.47	6.25	2.10	1.85
	Copper type L					5.21	12.5	3.96	5.57	2.28	1.40
30	Std. steel							4.73	11.1	2.87	3.29
	Copper type L							5.41	9.44	3.11	2.45
45	Std. steel									4.30	6.96
	Copper type L									4.66	5.20

Note: Data on friction losses based on information published in *Cameron Hydraulic Data* by Ingersoll Rand Company. Data based on clear water and reasonable corrosion and scaling.

Figure H39-2 Friction losses of standard steel and type L copper tubing in feet of head per 100 feet of pipe. (Data source: Ingersoll Rand Company)

Pipe Size (in.)	Gate Valve Fully Open	45 ° Elbow	Long Sweep Elbow or Run of Std. Tee	Std. Elbow or Run of Tee Reduced One-Half	Std. Tee through Side Outlet	Close Return Bend	Swing Check Valve Fully Open	Angle Valve Fully Open	Globe Valve Fully Open
¾	0.44	0.97	1.4	2.1	4.2	5.1	5.3	11.5	23.1
1	0.56	1.23	1.8	2.6	5.3	6.5	6.8	14.7	29.4
1¼	0.74	1.6	2.3	3.5	7.0	8.5	8.9	19.3	38.6
1½	0.86	1.9	2.7	4.1	8.1	9.9	10.4	22.6	45.2
2	1.10	2.4	3.5	5.2	10.4	12.8	13.4	29.0	58.0

Figure H39-3 Friction loss of fittings and valves in equivalent feet of pipe. (Data source: Crane Company)

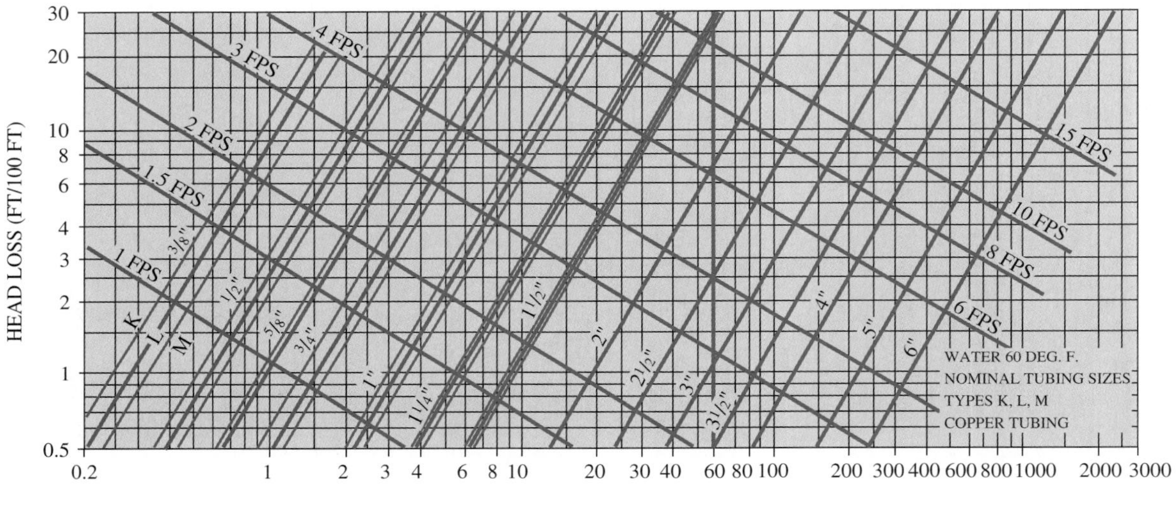

Figure H39-4 Friction losses of water flowing in copper tube showing pipe sizes and velocities. (Copyright by the American Society of Heating, Refrigerating, and Air-Conditioning Engineers, Inc. Used by permission.)

ON THE JOB

Problem: A customer complains that water is dripping on the floor. When you arrive you find the water is dripping from the relief valve. The temperature is 160°F and the pressure in the system is 18 psig. The pump is working okay. The relief valve is the correct size (30 psig) for the system. The expansion tank is not water bound. Why is the water dripping?

Solution: Relief valves play an important role in the operations of a boiler or hydronic heating system. They protect it from excessive pressure. With age, the seat in the relief valve hardens and develops a memory. Relief valves should be tested periodically. After completing the test, the valve may not return to the pretest seated position and water seepage can occur. When relief valves leak and are below the rated pressure, they need to be replaced. Never cap a relief valve to stop the seepage.

Valves

Valves are an important part of the system. They can be two-position or adjustable to regulate flow. Some of the types of valves and their uses are shown as follows:

Type	Function
Gate valve	Stops and starts fluid flow
Globe valve	Throttles or controls fluid flow
Ball valve	Same as globe valve
Butterfly valve	Same as globe valve
Check valve	Prevents backward flow in pipes
Safety or relief valve	Relieves the excessive pressure and/or temperature

Valves must be selected to operate satisfactorily with the pressure in the system in which they are located.

H39-3.2 Circulator Pumps

Pumps are used to force the flow of a certain quantity of liquid against the pressure drop of the system. The pumps used for hydronic systems are usually the centrifugal type.

One of the newer types of circulator pumps used for domestic and small commercial hot-water heating systems is shown in Fig. H39-5. The pump has a unique replaceable cartridge, which contains all the moving parts and allows the pump to be serviced instead of replacing the entire unit. It is self-lubricating and contains no mechanical seal.

The performance is indicated in Fig. H39-6. Six curves are shown for the various pumps that can be applied. For example, with a 0010 pump 18 gpm can be obtained with an 8-ft total head.

Figure H39-5 Centrifugal pump used for hydronic systems. (Courtesy of Taco, Inc.)

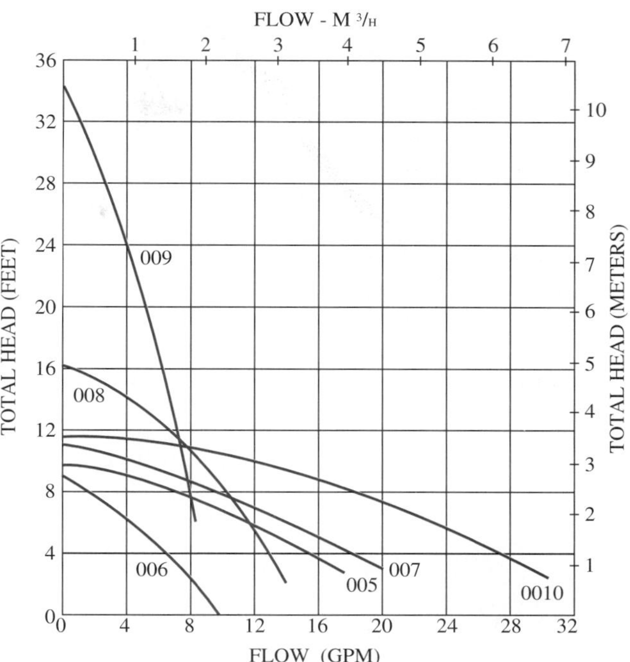

Figure H39-6 Typical performance of Taco "00" series centrifugal pumps. (Courtesy of Taco, Inc.)

Terminal Units

Terminal units transfer heat from hot water or steam to the various building areas. The heat supplied by these units is controlled to provide comfortable conditions.

Normally this equipment is rated in Btu/hr or MBtu/hr; however, these ratings can be stated in terms of equivalent direct radiation (EDR). For steam systems, 1 ft^2 EDR = 240 Btu, with 1 psig steam condensing in the unit. For hot water systems 1 ft^2 EDR = 150 Btu, based on an average water temperature of 170°F. Hot-water units are rated based on the water temperature drop through the unit which can vary between 10 and 30°F.

A number of types of terminal heating units are available, including: baseboard units, convectors, fan-coil units, and radiators.

Baseboard units are installed near the floor on the outside walls of each room. The heating element is finned-tube construction or cast iron. Heat is supplied by radiation and convection.

A typical finned-tube unit is shown in Fig. H39-7 along with the accessory pieces for corners and ends. Typical ratings are shown in Fig. H39-8. Dampers can be provided to regulate the airflow. The most common units have elements using copper tube and aluminum fins.

Convector units (Fig. H39-9, p. 783) can be free-standing or recessed. These units heat the room mainly by convection, with the air entering the bottom and leaving from the upper front. The free-standing model is supplied either with a flat or slanting top. An adjustable damper can be supplied, if desired.

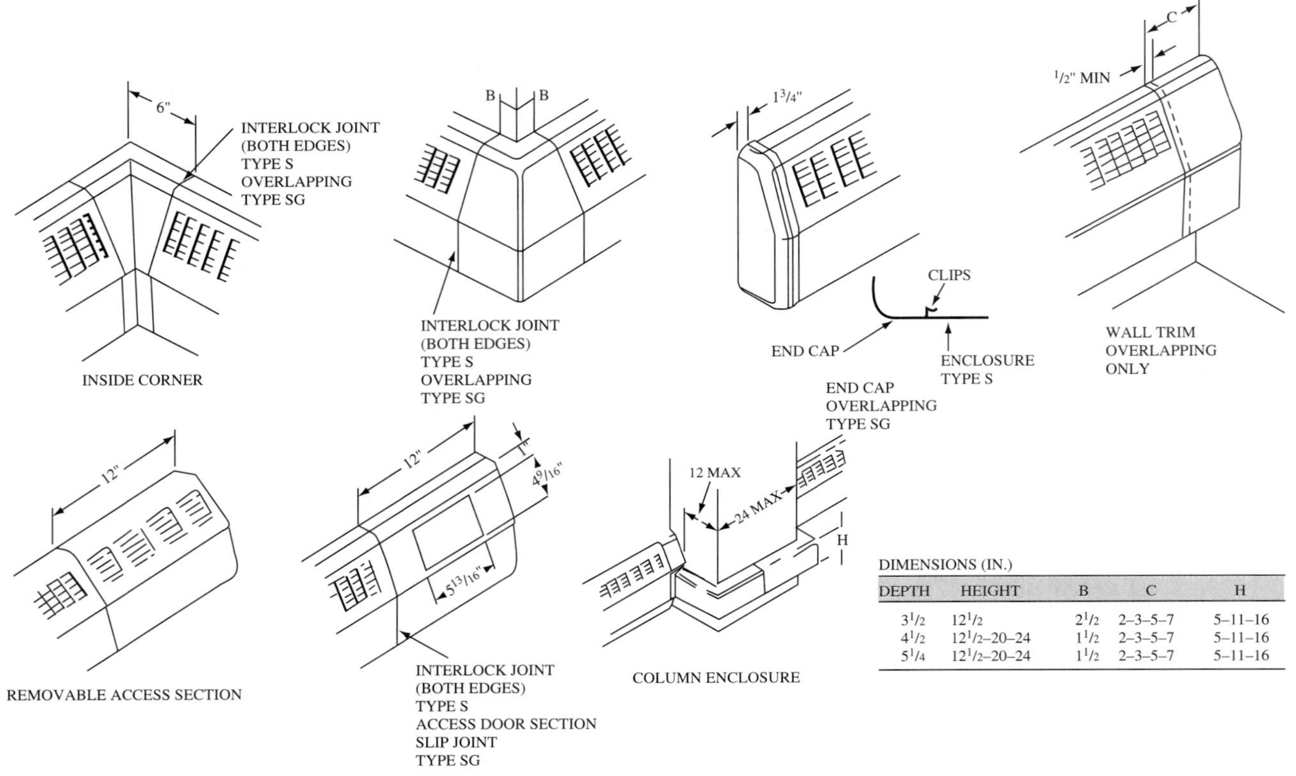

DEPTH	HEIGHT	B	C	H
3 1/2	12 1/2	2 1/2	2–3–5–7	5–11–16
4 1/2	12 1/2–20–24	1 1/2	2–3–5–7	5–11–16
5 1/4	12 1/2–20–24	1 1/2	2–3–5–7	5–11–16

Figure H39-7 Trim pieces for baseboard enclosures. (Courtesy of Dunham-Bush, Inc.)

A typical rating table for units with a 20°F hot-water temperature drop is shown in Fig. H39-10 (p. 784). A typical rating table for a steam unit using 215°F steam is shown in Fig. H39-11 (p. 785).

Fan-coil units (Fig. H39-12, p. 786), can serve the same function as convectors, but have additional features that increase their efficiency and usefulness. Some of these features are as follows:

1. The unit has a centrifugal fan, with manual speed control, for circulating the room air through the unit.
2. Air filters are standard equipment.
3. The unit can be used for both heating and cooling.
4. Various piping arrangements can be used for heating/cooling applications to facilitate operation.
5. Various automatic control arrangements can be used.
6. Outside air can be supplied through the back of the unit.

Units are available with air-volume ratings from 200 to 1200 cfm and hot water heating capacities from 7.6 to 112.0 MBtu/hr.

Radiators are primarily used for steam installations. They were used more frequently in past years. Fig. H39-13 (p. 786) gives the dimensions and ratings for small-tube cast-iron radiators. These have been replaced on new installations with more efficient heat-transfer units.

Radiant-heat panels, using hot water, have been used for heating from floors or ceilings. Typical installations use floor-panel temperatures of 80 to 85°F and ceiling-panel temperatures from 120 to 130°F. If water pipes are placed in concrete or plaster, the expansion coefficients of both must be the same. Pipes must be treated to prevent corrosion.

H39-4 TYPES OF PIPING SYSTEMS

Hot-water heating systems can be installed with a variety of piping systems. It is important that the technician be able to recognize each of the common systems, because only in so doing is he able to make adjustments where necessary and/or provide service.

Each of these systems controls the supply of hot water from the boiler to the terminal units (and the return) in such a way that comfortable conditions are produced.

The common systems of piping for hot water heating applications in residential and small commercial installations are as follows:

1. Series loop, single circuit
2. Series loop, double circuit
3. One-pipe venturi system, single circuit
4. One-pipe venturi system, double circuit

Tube Dia.	Fin Style	Fins Per Ft.	Tiers	Encl. Type	Average Water Temperature (°F)										Installed Height (inches)
					240	230	220	215	210	200	190	180	170	160	
1¼″	41S025	32	1	S512	1530	1390	1280	1220	1160	1050	950	840	740	650	16½
			1	S520	1650	1500	1390	1320	1250	1140	1030	910	800	700	24
			1	S524	1680	1530	1410	1340	1270	1150	1050	930	820	710	28
			2	S520	2460	2250	2070	1970	1870	1690	1540	1360	1200	1040	28
			2	S524	2590	2360	2170	2070	1970	1780	1610	1430	1260	1100	28
			3	S524	3180	2900	2670	2540	2410	2180	1980	1750	1550	1350	28
1¼″	41S025	40	1	S512	1690	1540	1420	1350	1280	1160	1050	930	820	720	16½
			1	S520	1840	1680	1540	1470	1400	1260	1150	1010	900	780	24
			1	S524	1875	1710	1580	1500	1420	1290	1170	1040	920	800	28
			2	S520	2710	2470	2280	2170	2060	1870	1690	1500	1320	1150	24
			2	S524	2790	2540	2340	2230	2120	1920	1740	1540	1360	1180	28
			3	S524	3210	2930	2700	2570	2440	2210	2000	1770	1570	1360	28
2″	42S025	24	1	S512	1290	1170	1080	1030	980	890	800	710	630	550	16½
			1	S520	1350	1230	1130	1080	1030	930	840	750	660	570	24
			1	S524	1450	1320	1220	1160	1100	1000	900	800	710	620	28
			2	S520	2060	1880	1730	1650	1570	1420	1290	1140	1010	870	24
			2	S524	2210	2020	1860	1770	1680	1520	1380	1220	1080	940	28
			3	S524	2810	2570	2360	2250	2140	1940	1760	1550	1370	1190	28
2″	42S025	32	1	S512	1510	1380	1270	1210	1130	1040	940	830	740	640	16½
			1	S520	1640	1490	1380	1310	1240	1130	1020	900	800	690	24
			1	S524	1750	1600	1470	1400	1330	1200	1090	970	850	740	28
			2	S520	2350	2140	1970	1880	1790	1620	1460	1300	1150	1000	24
			2	S524	2550	2910	2140	2040	1940	1750	1590	1410	1240	1080	28
			3	S524	2960	2700	2490	2370	2250	2040	1850	1640	1450	1260	28
1¼″	41S032	32	1	S512	1630	1480	1370	1300	1240	1120	1010	900	790	690	16½
			1	S520	1750	1600	1470	1400	1330	1200	1090	970	850	740	24
			1	S524	1780	1620	1490	1420	1350	1220	1110	980	870	750	28
			2	S520	2630	2400	2200	2100	2000	1800	1640	1450	1280	1110	24
			2	S524	2750	2510	2310	2200	2090	1890	1720	1520	1340	1170	28
			3	S524	3380	3080	2840	2700	2570	2320	2110	1860	1650	1430	28

Figure H39-8 Rating table for slope-top convectors with steel elements. (Courtesy of Dunham-Bush, Inc.)

Tube Dia.	Fin Style	Fins Per Ft.	Tiers	Encl. Type	Average Water Temperature (°F)										Installed Height (inches)
					240	230	220	215	210	200	190	180	170	160	
1¼″	41S032	40	1	S512	1790	1630	1500	1430	1360	1230	1120	990	870	760	16½
			1	S520	1950	1780	1640	1560	1480	1340	1220	1080	950	830	24
			1	S524	2000	1824	1680	1600	1520	1380	1250	1100	980	850	28
			2	S520	2890	2630	2430	2310	2190	1990	1800	1600	1410	1220	24
			2	S524	2960	2700	2490	2370	2250	2040	1850	1630	1450	1260	28
			3	S524	3410	3110	2870	2730	2590	2350	2130	1880	1670	1450	28
2″	42S032	24	1	S512	1360	1240	1140	1090	1040	940	850	750	670	580	16½
			1	S520	1440	1310	1210	1150	1090	990	900	790	700	610	24
			1	S524	1540	1400	1290	1230	1170	1060	960	850	750	650	28
			2	S520	2190	2000	1840	1750	1660	1510	1370	1210	1070	930	24
			2	S524	2350	2140	1970	1880	1790	1620	1470	1300	1150	1000	28
			3	S524	2990	2720	2510	2390	2270	2060	1860	1650	1460	1270	28
2″	42S032	32	1	S512	1610	1470	1350	1290	1230	1110	1010	890	790	680	16½
			1	S520	1740	1580	1460	1390	1320	1200	1080	960	850	740	24
			1	S524	1860	1700	1560	1490	1420	1280	1160	1030	910	790	28
			2	S520	2500	2280	2100	2000	1900	1720	1570	1380	1220	1060	24
			2	S524	2710	2470	2280	2170	2060	1860	1690	1500	1320	1150	28
			3	S524	3150	2860	2650	2520	2390	2170	1970	1740	1540	1340	28

All ratings are Btuh/linear ft and are based on 65°F entering air and water velocity of 3 ft/sec.
Elements are located with longest dimension in horizontal position.

Figure H39-8 Continued

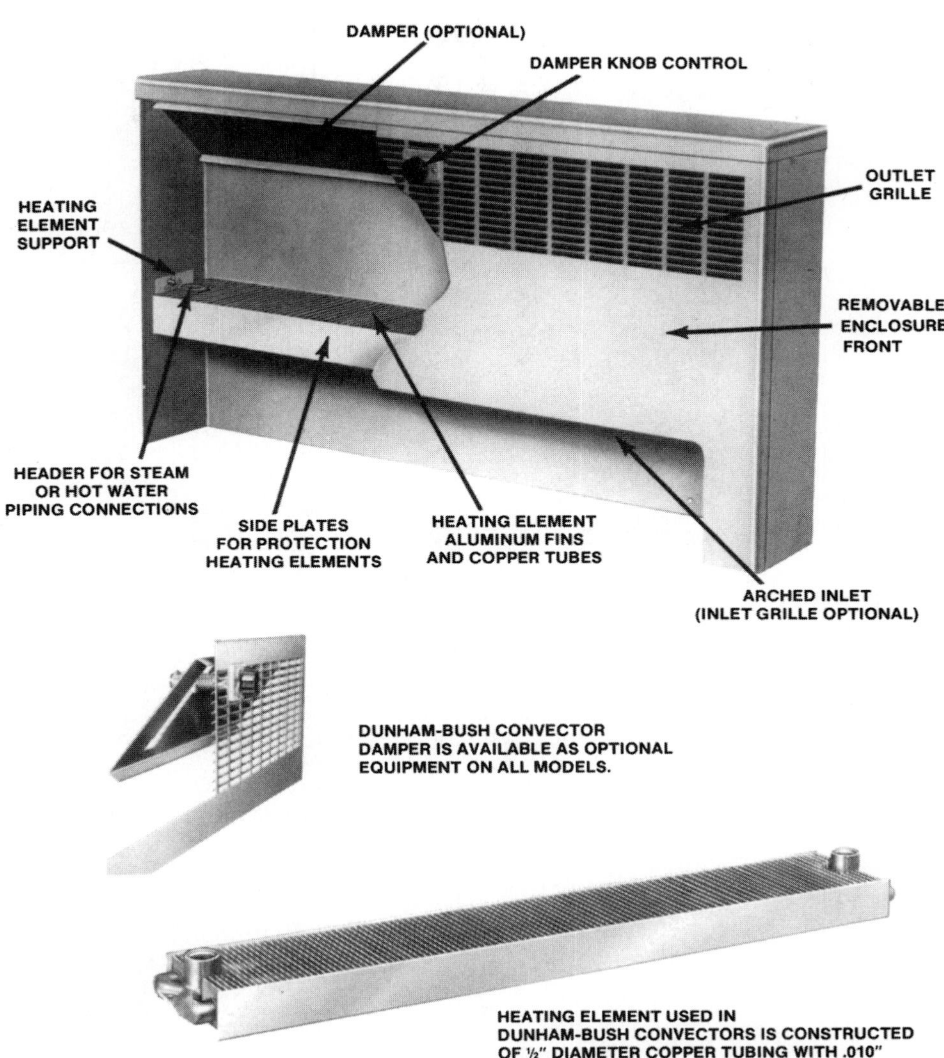

Figure H39-9 Typical convector.

5. Two-pipe system, direct return
6. Two-pipe system, reverse return
7. A zoned system, using pumps
8. A zoned system, using motorized valves.

Series Loop, Single Circuit

The *series-loop system* is probably the simplest of all the piping arrangements. A schematic diagram of the single circuit piping layout is shown in Fig. H39-14 (p. 787). In this piping system the main supply pipe from the boiler enters the first baseboard unit. The outlet from the first unit is connected directly to the inlet of the second unit. This arrangement is continued until all baseboard sections are connected in series. The outlet from the last section is connected to the circulator pump, which connects to the boiler. Thus, all the water flowing through the system passes through each unit.

In many cases the bathroom or kitchen does not have sufficient baseboard space and a convector is used in place of baseboard. The convector can be tied into the system using a venturi fitting as shown in the diagram.

When these systems are installed, the entire perimeter of the building has baseboard enclosures. The amount of finned-tube element placed in these enclosures is dependent on the heat loss requirements. Where the main pipe passes a doorway, an offset is made in the piping to run the main below the floor until it passes the door opening.

With this piping arrangement each downstream unit is supplied with cooler water than the preceding one. The actual hot water temperature drop between terminal units is about 2°F. To compensate for this, each successive unit is oversized about 2%.

The use of a series loop system reduces the cost of the installation to a minimum.

Series Loop, Double Circuit

On larger systems, better performance can be obtained by dividing the system into two (or more) approximately

Height (in.)	Depth (in.)	Length (in.)																
		16	20	24	28	32	36	40	44	48	56	64	72	80	88	96	104	112
Type FH 18	4	1.0	1.1	1.4	1.7	2.0	2.3	2.5	2.8	3.1	3.6	4.2	4.7	5.3	5.8	6.3	6.9	7.4
	6	1.4	1.7	2.1	2.5	2.9	3.3	3.7	4.2	4.6	5.4	6.3	6.9	7.7	8.5	9.3	10.1	10.9
	8	1.7	2.1	2.6	3.3	3.9	4.5	5.1	5.7	6.3	7.5	8.6	9.0	10.0	11.4	12.4	13.7	14.8
	10	2.1	2.4	2.9	3.5	4.1	4.8	5.5	6.3	7.0	8.5	10.0	10.4	11.6	12.9	14.1	16.5	17.1
Type FH 20 Type WH 14	4	1.1	1.2	1.6	2.0	2.3	2.6	3.0	3.3	3.7	4.3	5.0	5.6	6.3	6.9	7.6	8.3	9.0
	6	1.5	1.9	2.3	2.8	3.3	3.8	4.3	4.8	5.3	6.2	7.2	7.7	8.7	9.6	10.5	11.4	12.3
	8	1.9	2.3	3.0	3.6	4.2	4.8	5.4	6.1	6.7	7.9	9.2	10.0	11.2	12.5	13.7	14.8	16.0
	10	2.4	2.7	3.4	4.1	4.8	5.5	6.2	6.9	7.6	9.1	10.6	11.5	12.9	14.2	15.6	17.0	18.3
Type FH 24 Type WH 18	4	1.2	1.5	1.8	2.2	2.6	3.0	3.4	3.8	4.2	4.9	5.7	6.2	7.0	7.8	8.5	9.2	9.9
	6	1.8	2.2	2.8	3.4	3.9	4.7	5.0	5.6	6.2	7.3	8.4	9.4	10.4	11.4	12.4	13.4	14.5
	8	2.2	2.6	3.3	4.0	4.6	5.3	6.0	6.7	7.4	8.7	10.1	11.1	12.3	13.6	14.8	16.1	17.4
	10	2.7	3.0	3.7	4.5	5.3	6.0	6.8	7.6	8.3	9.9	11.4	12.6	14.1	15.5	17.0	18.5	20.0
Type FH 26 Type WH 20	4	1.3	1.5	1.9	2.3	2.6	3.1	3.4	3.9	4.3	5.0	5.7	6.6	7.3	8.1	8.9	9.6	10.4
	6	1.9	2.3	2.8	3.4	4.0	4.6	5.2	5.8	6.4	7.5	8.7	9.8	10.8	12.0	13.1	13.6	14.6
	8	2.4	2.6	3.3	3.7	4.8	5.4	6.2	6.9	7.6	9.0	10.4	11.7	13.0	14.4	15.6	17.1	18.5
	10	2.8	3.0	3.8	4.6	5.4	6.2	7.0	7.8	8.6	10.1	11.7	13.0	14.6	16.1	17.6	19.1	20.6
Type FH 32 Type WH 26	4	1.4	1.6	2.0	2.5	2.9	3.3	3.7	4.2	4.6	5.4	6.3	6.9	7.8	8.5	9.2	10.0	10.8
	6	2.1	2.5	3.1	3.8	4.4	5.1	5.7	6.4	7.0	8.3	9.6	10.5	11.6	12.7	14.0	14.9	16.0
	8	2.7	2.9	3.6	4.4	5.1	5.9	6.6	7.4	8.2	9.7	11.2	12.3	13.9	15.3	16.7	18.1	19.6
	10	3.1	3.3	4.1	5.0	5.9	6.7	7.6	8.4	9.3	11.0	12.7	14.5	15.5	17.1	19.5	20.4	21.4
Type FH 38 Type WH 32	4	1.5	1.7	2.1	2.3	3.0	3.5	3.9	4.4	4.8	5.8	6.7	7.4	8.3	9.1	9.9	10.8	11.6
	6	2.3	2.6	3.3	4.0	4.6	5.3	5.9	6.6	7.3	8.6	10.0	11.1	12.3	13.4	14.5	15.7	16.8
	8	2.8	3.0	3.8	4.6	5.4	6.2	6.9	7.7	8.5	10.1	11.7	12.9	14.5	15.9	17.4	18.9	20.3
	10	3.2	3.4	4.3	5.2	6.1	7.0	7.9	8.8	9.7	11.4	13.2	14.6	16.3	18.0	19.7	21.4	22.5
Type WH 38	4	1.6	1.7	2.2	2.7	3.2	3.6	4.1	4.6	5.0	6.1	7.0	8.0	9.0	9.7	10.7	11.5	12.2
	6	2.4	2.7	3.4	4.1	4.7	5.4	6.1	6.8	7.5	8.9	10.4	11.9	13.0	14.2	15.3	16.5	17.7
	8	3.0	3.1	3.9	4.7	5.5	6.4	7.2	8.0	8.8	10.4	12.0	13.6	15.2	16.6	18.1	19.6	21.1
	10	3.4	3.5	4.4	5.4	6.3	7.2	8.1	9.0	10.0	11.8	13.7	15.3	17.0	17.5	20.5	22.2	23.8

Figure H39-10 Typical hot water capacities for front outlet convector.

Height	Depth	Length (in.)																
(in.)	(in.)	16	20	24	28	32	36	40	44	48	56	64	72	80	88	96	104	112
18	4	8.4	9.3	11.6	13.8	16.1	18.3	20.6	22.8	25.0	29.6	34.2	38.5	42.7	47.3	51.4	55.7	60.3
	6	11.7	13.5	16.8	20.2	23.6	27.0	30.4	33.8	37.2	44.1	51.0	55.7	62.3	58.9	75.4	82.0	88.9
	8	13.5	16.8	21.5	26.6	31.7	36.6	41.4	46.3	51.1	62.1	70.2	73.4	81.4	92.3	101.1	111.7	120.4
	10	17.1	19.7	23.3	27.5	33.3	39.2	44.8	51.2	57.0	69.1	81.3	84.6	94.4	104.5	114.8	134.4	139.3
20	4	8.8	10.0	13.1	15.9	18.7	21.5	24.3	27.0	29.7	35.3	40.8	45.6	51.1	56.5	62.1	67.5	72.9
	6	12.4	15.1	19.0	23.0	27.0	30.8	34.9	38.8	42.7	50.6	58.3	63.0	70.5	78.0	85.5	93.0	100.3
	8	15.8	18.9	24.0	29.0	34.0	39.1	44.2	49.2	54.3	64.4	74.4	81.1	90.9	101.3	111.6	120.2	129.9
	10	19.2	22.0	27.7	33.4	39.2	45.0	50.8	56.5	62.1	73.6	86.1	93.8	104.7	115.8	126.7	137.9	148.8
24	4	9.6	11.8	15.0	18.1	21.3	24.4	27.5	30.7	33.8	40.0	46.3	50.7	57.1	63.5	68.9	74.8	80.4
	6	14.5	17.7	22.4	27.0	31.6	36.2	40.8	45.4	50.0	59.2	68.4	76.4	84.5	92.9	101.0	109.2	118.0
	8	18.2	21.0	26.5	32.0	37.5	43.1	48.7	54.2	59.8	70.9	82.0	90.9	100.1	110.2	120.4	131.0	141.3
	10	21.7	24.0	30.2	36.5	42.8	49.0	55.3	61.5	67.7	80.2	92.7	102.2	114.4	126.4	138.4	150.4	162.4
26	4	10.8	12.3	15.4	18.5	21.8	25.0	28.2	31.4	34.6	41.0	47.4	53.5	59.7	66.0	72.3	78.4	84.8
	6	15.8	18.3	23.1	27.9	32.6	37.4	42.2	47.0	56.9	61.2	70.8	79.3	88.3	97.2	106.1	110.7	118.8
	8	19.9	21.4	27.2	32.4	38.6	44.3	50.0	55.7	61.4	72.8	84.3	95.1	105.9	116.7	127.2	138.8	152.3
	10	23.1	24.6	31.0	37.5	43.9	50.4	56.8	63.2	69.6	82.4	95.3	105.7	118.3	130.6	142.9	155.3	167.6
32	4	11.4	13.0	16.5	20.0	23.4	26.9	30.3	33.8	37.2	44.1	51.0	56.3	63.5	68.9	75.0	81.2	87.8
	6	17.1	20.2	25.5	30.6	36.0	41.2	46.5	51.7	58.1	67.4	77.9	85.0	95.4	103.1	113.7	121.2	130.1
	8	21.6	23.2	29.4	35.5	41.6	47.8	54.0	60.2	66.4	78.8	91.0	100.3	112.9	124.2	135.5	147.4	159.6
	10	24.9	26.6	33.5	40.6	47.6	54.5	61.5	68.4	75.4	89.3	103.1	113.2	126.3	139.4	158.5	165.8	178.6
38	4	12.0	13.6	17.3	21.0	24.7	28.4	32.0	35.8	39.4	46.8	54.2	60.0	67.2	74.3	81.6	88.8	96.0
	6	18.6	21.0	26.5	32.0	37.4	42.8	48.3	53.8	59.2	70.1	81.2	89.9	110.7	111.3	122.3	132.9	143.6
	8	23.0	24.2	30.8	37.0	43.6	50.0	56.4	62.8	69.2	82.0	94.8	105.1	117.9	130.3	142.8	155.3	167.9
	10	26.2	27.7	34.9	42.3	49.6	56.8	64.0	71.2	78.6	92.9	107.4	118.8	132.6	146.5	160.4	174.3	183.2

Figure H39-11 Typical steam capacities for front-outlet convector.

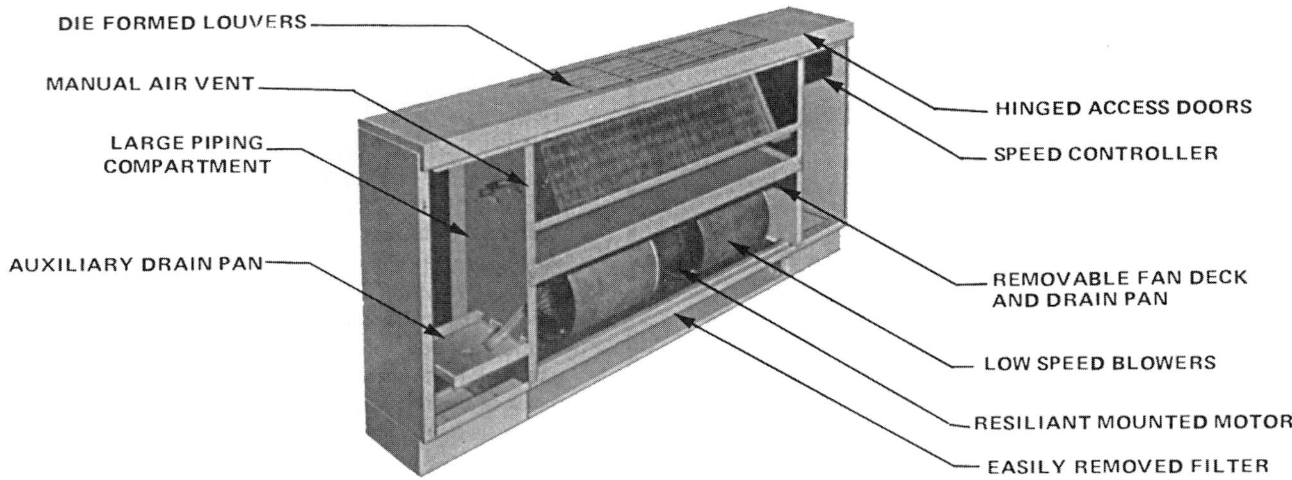

Figure H39-12 Typical fan-coil unit. (Courtesy of Dunham-Bush, Inc.)

equal circuits (Fig. H39-15). When this is done, balancing valves are installed in each branch to permit adjusting the flow.

One-pipe Venturi System, Single Circuit

The one-pipe venturi fitting system has a single main extending from the supply to the return connection of the boiler. The terminal units are fed by a supply and return branch connected to the main as shown in Fig. H39-16. The supply connection uses a standard tee and the return connection uses a special venturi fitting. The venturi creates a negative pressure at the main connection, drawing the necessary flow through the branch.

This system differs from the series loop in that only a portion of the total flow enters the terminal unit. There is still a temperature drop in the main as it reaches the location of successive units. The reduced flow, however, improves the performance of the terminal units.

One-pipe Venturi Unit, Double Circuit

On larger systems using this piping arrangement, better performance can be obtained by dividing the system into

Number of Tubes per Section	Catalog Rating per Section[a]		Section Dimensions A Height[c]	B Width Min.	B Width Max.	C Spacing[b]	D Leg Height[c]
	ft²	Btu/hr (W)	in. (mm)	in. (mm)	in. (mm)	in. (mm)	in. (mm)
3	1.6	384 (113)	25 (635)	3.25 (83)	3.50 (89)	1.75 (44)	2.50 (64)
	1.6	384 (113)	19 (483)	4.44 (113)	4.81 (122)	1.75 (44)	2.50 (64)
4	1.8	432 (127)	22 (559)	4.44 (113)	4.81 (122)	1.75 (44)	2.50 (64)
	2.0	480 (141)	25 (635)	4.44 (113)	4.81 (122)	1.75 (44)	2.50 (64)
5	2.1	504 (148)	22 (559)	5.63 (143)	6.31 (160)	1.75 (44)	2.50 (64)
	2.4	576 (169)	25 (635)	5.63 (143)	6.31 (160)	1.75 (44)	2.50 (64)
	2.3	552 (162)	19 (483)	6.81 (173)	8 (203)	1.75 (44)	2.50 (64)
6	3.0	720 (211)	25 (635)	6.81 (173)	8 (203)	1.75 (44)	2.50 (64)
	3.7	888 (260)	32 (813)	6.81 (173)	8 (203)	1.75 (44)	2.50 (64)

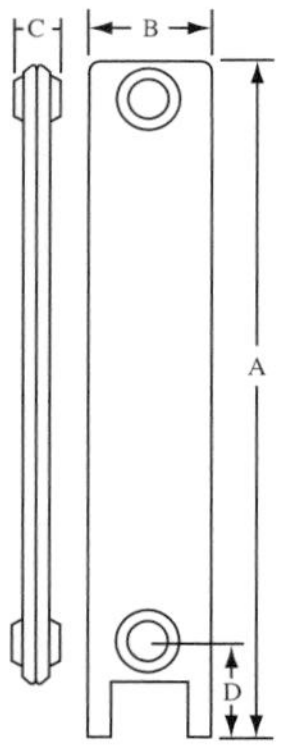

[a]These ratings are based on steam at 215 F (101.7°C) and air at 70 F (21.1°C). They apply only to installed radiators exposed in a normal manner, not to radiators installed behind enclosures, grills, or under shelves.
[b]Length equals number of sections multiplied by 1.75 in. (44 mm).
[c]Overall height and leg height, as produced by some manufacturers, are 1 in. (25 mm) greater than shown in Columns A and D. Radiators may be furnished without legs. Where greater than standard leg heights are required, this dimension shall be 4.5 in. (114 mm).

Figure H39-13 Ratings for small-tube cast-iron radiators. (Copyright by the American Society of Heating, Refrigerating, and Air-Conditioning Engineers, Inc. Used by permission.)

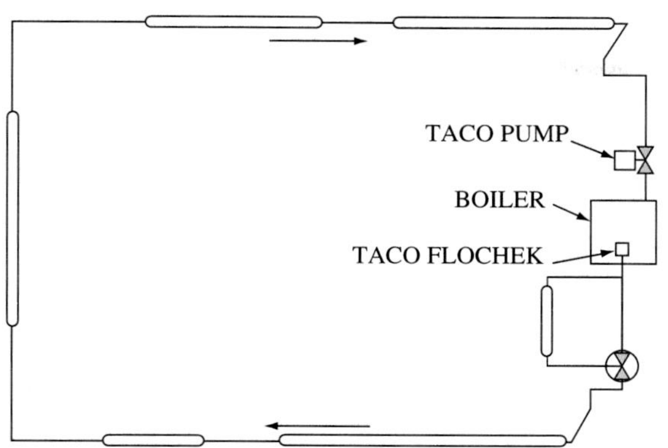

Figure H39-14 Series-loop piping layout, single circuit. (Courtesy of Taco, Inc.)

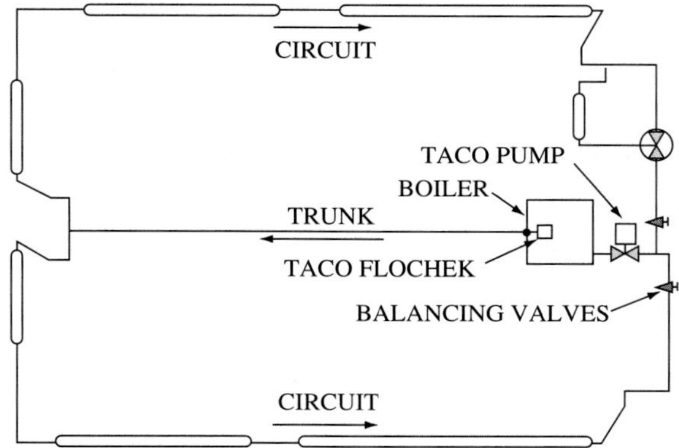

Figure H39-15 Series-loop piping layout, double circuit. (Courtesy of Taco, Inc.)

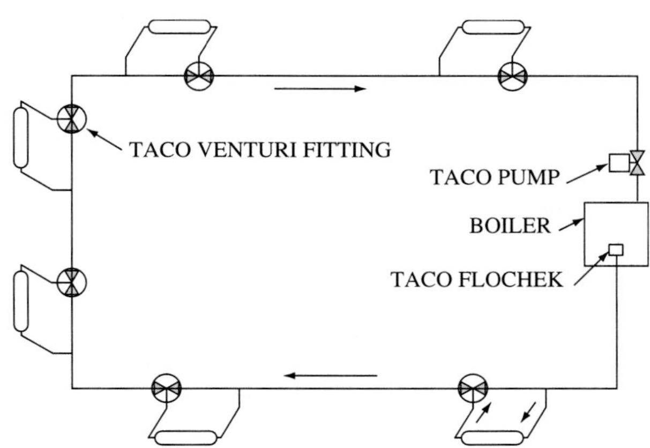

Figure H39-16 One-pipe venturi system, single circuit. (Courtesy of Taco, Inc.)

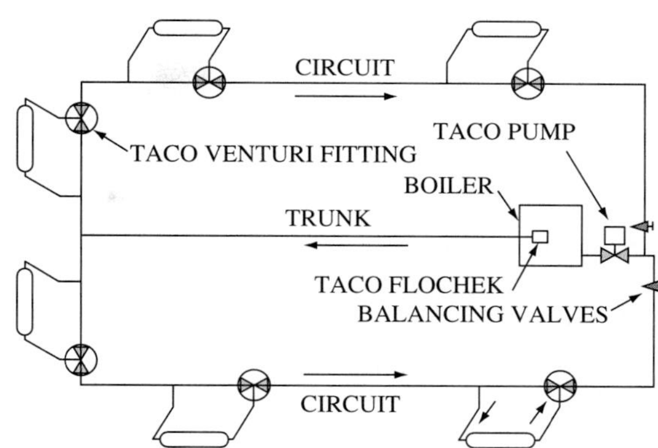

Figure H39-17 One-pipe venturi system, double circuit. (Courtesy of Taco, Inc.)

two (or more) approximately equal circuits (Fig. H39-17). When this is done balancing valves are installed in each branch to permit adjusting the flow.

Two-pipe System, Direct Return

In the two-pipe system, separate supply and return mains are used. The return water from the terminal units is collected and returned to the boiler. With this piping arrangement, each unit will receive the same supply water temperature.

With the direct return, the two mains are run side-by-side with a supply branch run to each unit from the supply main and a return branch run from each unit to the return main. The first unit taken off the supply main is the first unit on the return main before it reaches the boiler (Fig. H39-18). Obviously, the shortest piping is used on the unit nearest the boiler. The longest piping is used on the unit farthest from the boiler. This difference in piping pressure drop must be equalized by using a square head cock in either the supply or return branch to each unit.

The two-pipe direct-return system uses more pipe than the one-pipe system, however; it is considered to be more

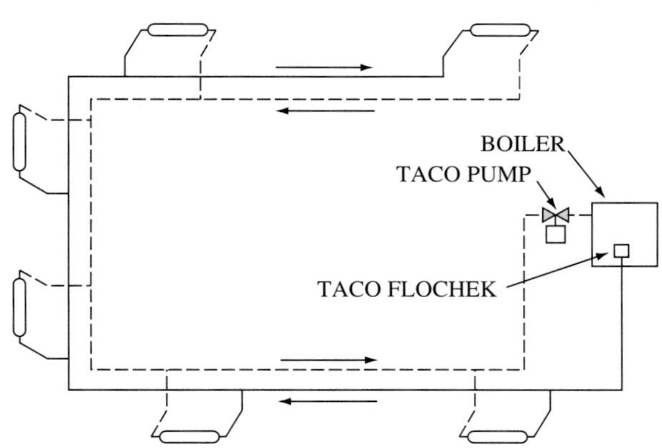

Figure H39-18 Two-pipe direct-return system. (Courtesy of Taco, Inc.)

efficient by providing better water distribution and better control. This system can also be designed using two or more circuits.

Two-pipe System, Reverse Return

The two-pipe reverse return system is similar to the direct-return system except the piping is arranged so that the same length of pipe (supply + return) is used for each terminal unit. This is accomplished, as shown in Fig. H39-19, by taking the unit nearest the boiler off the supply main first, connecting it to the return main last and proceeding on this basis to equalize the piping lengths for all units. This equalization of the piping loss to each unit provides a balanced condition so that each unit receives its proper share of water.

Zoned System, Using Pumps

Zoning is provided where it is desirable to control the supply of heat (hot water) to different areas of the building by separate thermostats. The areas selected can be determined by usage or building orientation. For example, to zone by usage, the living areas of a residence can be placed in one zone and the sleeping areas in another. To illustrate orientation zoning, a small commercial building can place all the rooms with southern exposure in one zone and the rooms with northern exposure in another. Whatever separation is made, allowance must be made in the piping and controls to provide this feature.

The piping needs to provide a separate hot-water supply circuit to the terminal units in each zone. The number of supply water circuits is the same as the number of zones. One way to control the flow of hot water in each zone is to provide a separate circulator for each zone as shown in Fig. H39-20. Any one of the systems described can be provided with zoning. The circulator in each zone responds to the requirements of the thermostat in that zone.

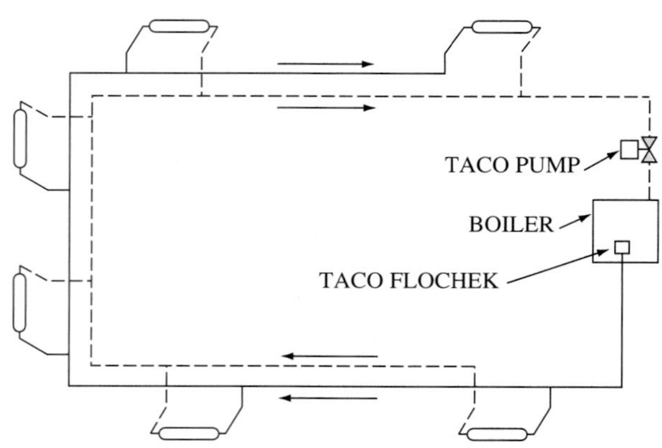

Figure H39-19 Two-pipe reverse-return system. (Courtesy of Taco, Inc.)

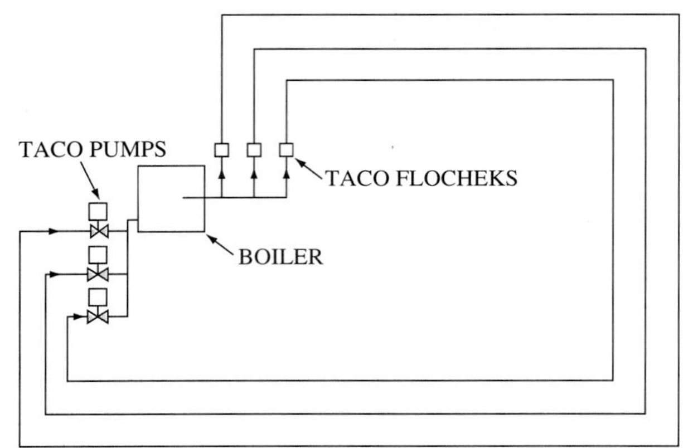

Figure H39-20 Zoned system using separate circulators for each zone. (Courtesy of Taco, Inc.)

On larger systems, the water in the main is circulated continuously, so that hot water is always available at each zone (Fig. H39-21). On smaller systems, the pump operates only on a call for heat from an individual zone thermostat.

Zoned System, Using Motorized Valves

The zoning arrangement using motorized valves is similar to the system with pump zoning, with the exception that the motorized valves control the flow in each zone (Fig. H39-22). Each supply circuit has a motorized valve controlled by the thermostat in that zone. A motorized valve is also used in a bypass around the circulator pump to relieve the pressure when only one zone valve is open.

H39-5 BOILERS AND RELATED EQUIPMENT

All boilers are used to heat water. Where hot water is used to transfer heat, the output of the boiler maintains a temperature below the steaming temperature. At sea-level atmospheric pressure, water boils at 212°F; at higher pressures the boiling point increases.

Where large amounts of heat need to be transferred, the boiler is used to produce steam. This is due to the fact that about 970 Btu/lb of heat can be added to water at 212°F and atmospheric pressure to change it into steam at the same temperature.

In general, for residential and small commercial installations, hot-water boilers are used rather than steam due to the size of the installations and accuracy of control that can be provided.

Hot-water boilers are constructed using either cast iron or steel. A typical gas-fired cast-iron boiler is shown in Fig. H39-23, and an oil-fired cast-iron boiler is shown in Fig. H39-24. Cast-iron boilers are made in sections held

PRACTICING PROFESSIONAL SERVICES

Old boilers and chilled water systems had insulation on their piping that contained asbestos. If the insulation is in good shape or has been encapsulated, it is relatively safe, but if you find it has deteriorated and is falling apart, take the appropriate caution. Follow all federal and state regulations for the procedure on removal, or hire only licensed people who can legally remove the asbestos. Make sure you have all the documentation for the removal, including how it was destroyed and proper disposal site licensing information. Only authorized personnel should remove asbestos.

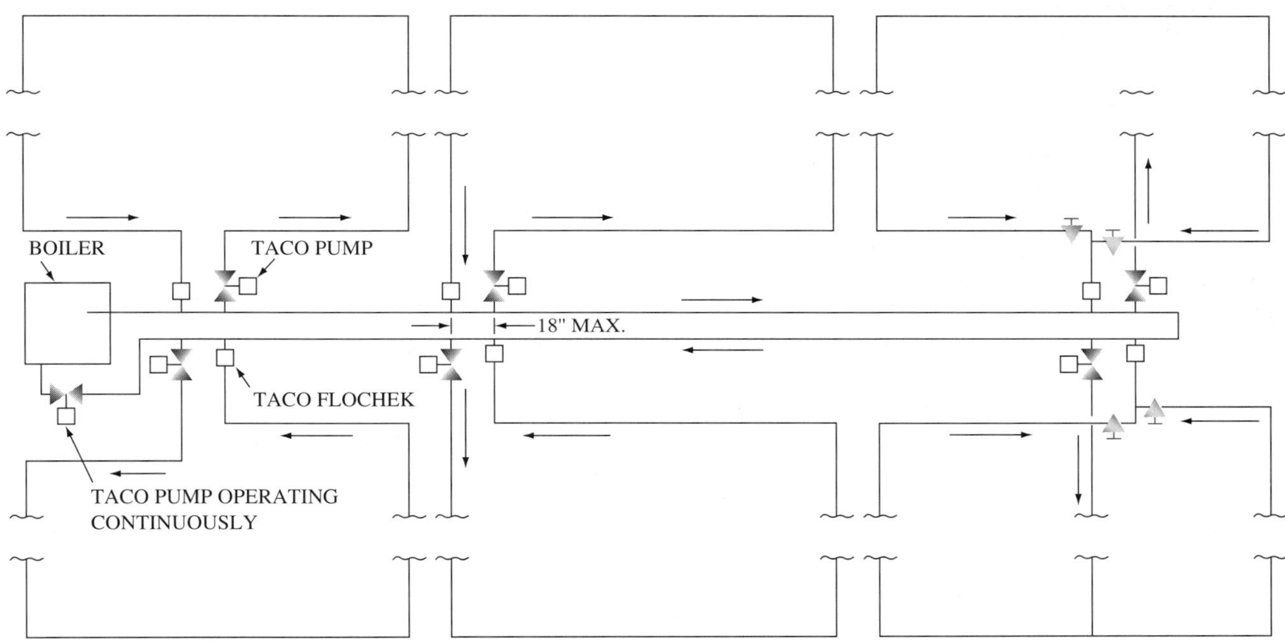

Figure H39-21 Large zone system with continuously operating main pump. (Courtesy of Taco, Inc.)

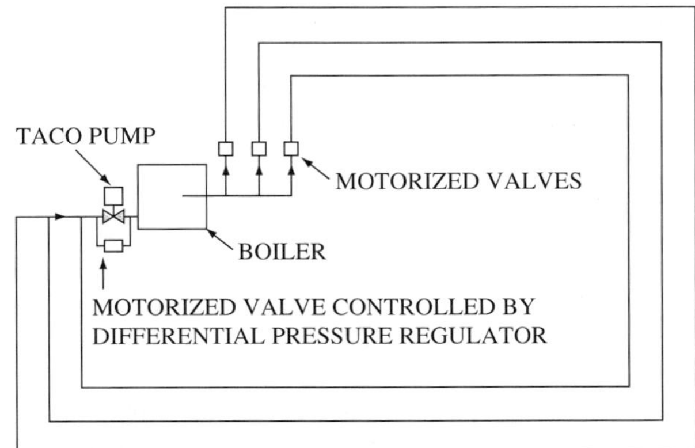

Figure H39-22 Zone system using motorized zone valves. (Courtesy of Taco, Inc.)

together by iron nipples. Additional sections increase the boiler capacity.

Both of these boilers are assembled and shipped as packages. The package usually includes the circulator pump and complete controls. The wall thermostat is shipped separately for field mounting and wiring. Some packages also include a diaphragm-type expansion tank and a check valve piped into the unit.

A steel boiler is shown in Fig. H39-25. This boiler is also shipped as a package with accessories and controls. The design illustrated is a high efficiency unit with a condensing-type heat exchanger producing efficiencies in the AFUE 90+% range.

A unique integrated home-comfort system, using a boiler, is shown in Fig. H39-26. It supplies many house-

Figure H39-23 Typical cast-iron gas boiler. (Courtesy of Mestek, Inc.)

Figure H39-24 Typical cast-iron oil boiler. (Courtesy of Mestek, Inc.)

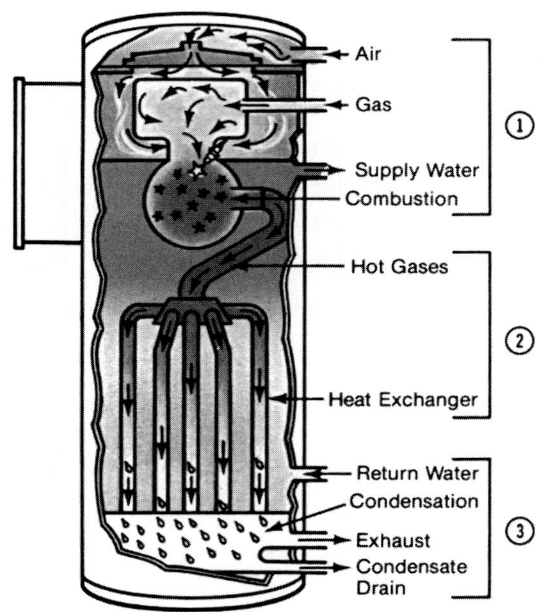

Figure H39-25 Steel boiler, gas-fired. (Courtesy of Mestek, Inc.)

hold heating functions as shown in Fig. H39-27. There are three connected units in the system:

1. A packaged gas-fired boiler.
2. A storage tank for domestic hot water.
3. An air-handling unit that can supply either heating or cooling. The cooling coil is arranged for connection to a remote condensing unit for operation as a split-system air conditioner or heat pump. When operated as a heat pump, supplementary heat can be supplied by the hot-water heating coil.

For larger hot-water installations, a high-efficiency pulse-type steel boiler is available, as shown in Fig. H39-28. This boiler produces efficiencies as high as AFUE 96%. Both combustion-air and gas-metering valves are standard equipment. A small fan is used to deliver combustion air on startup, but shuts off after ignition. The unit is primarily controlled by a microprocessor.

H39-6 PIPING DETAILS

A number of accessory devices are used in piping hot water systems, such as shutoff valves, air-purging fittings, automatic-fill valves, flow-control valves, expansion tanks, safety controls and gauges. Each of these devices has a special function in the proper operation of the system.

One of the main concerns in the selection and use of these accessories is to maintain a system of piping that contains only water and is free from air pockets. Air in the system can prevent proper operation of the pump and interfere with the transfer of heat in the terminal units.

Figure H39-26 Integrated home-comfort system. (Courtesy of GlowCore)

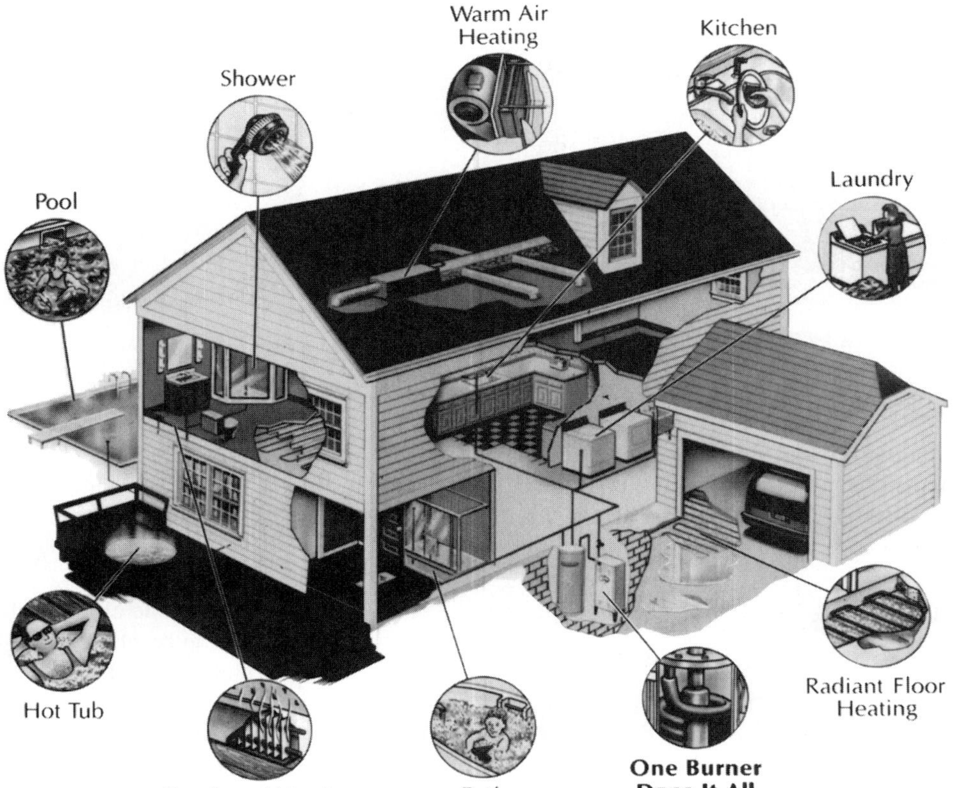

Figure H39-27 Household functions supported by integrated heating system. (Courtesy of GlowCore)

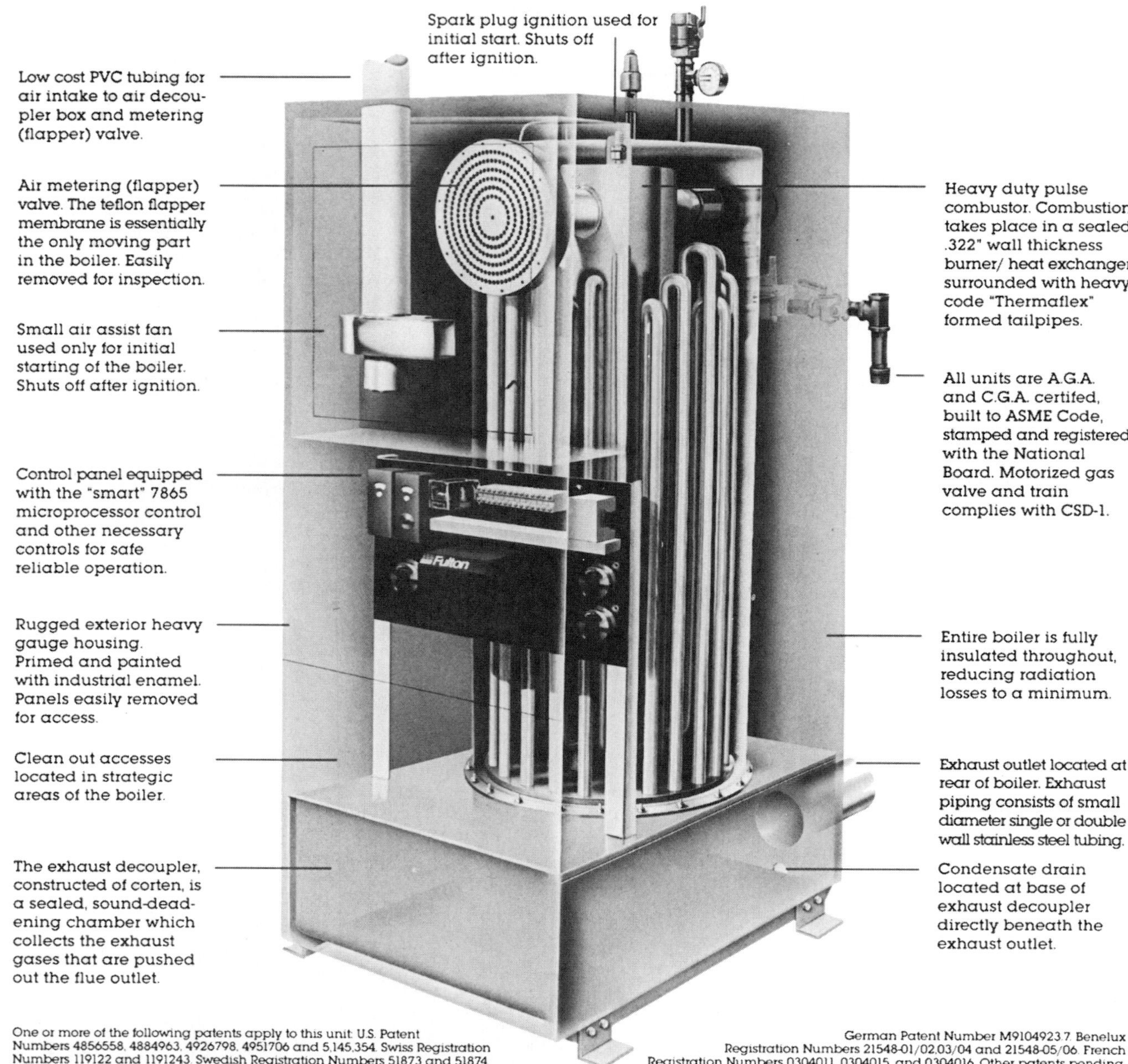

Spark plug ignition used for initial start. Shuts off after ignition.
Low cost PVC tubing for air intake to air decoupler box and metering (flapper) valve.
Air metering (flapper) valve. The teflon flapper membrane is essentially the only moving part in the boiler. Easily removed for inspection.
Small air assist fan used only for initial starting of the boiler. Shuts off after ignition.
Control panel equipped with the "smart" 7865 microprocessor control and other necessary controls for safe reliable operation.
Rugged exterior heavy gauge housing. Primed and painted with industrial enamel. Panels easily removed for access.
Clean out accesses located in strategic areas of the boiler.
The exhaust decoupler, constructed of corten, is a sealed, sound-deadening chamber which collects the exhaust gases that are pushed out the flue outlet.
Heavy duty pulse combustor. Combustion takes place in a sealed .322" wall thickness burner/ heat exchanger surrounded with heavy, code "Thermaflex" formed tailpipes.
All units are A.G.A. and C.G.A. certifed, built to ASME Code, stamped and registered with the National Board. Motorized gas valve and train complies with CSD-1.
Entire boiler is fully insulated throughout, reducing radiation losses to a minimum.
Exhaust outlet located at rear of boiler. Exhaust piping consists of small diameter single or double wall stainless steel tubing.
Condensate drain located at base of exhaust decoupler directly beneath the exhaust outlet.
Fulton
One or more of the following patents apply to this unit: U.S. Patent Numbers 4856558, 4884963, 4926798, 4951706 and 5,145,354. Swiss Registration Numbers 119122 and 1191243. Swedish Registration Numbers 51873 and 51874.
German Patent Number M9104923.7. Benelux Registration Numbers 21548-01/02,03/04 and 21548-05/06. French Registration Numbers 0304011, 0304015, and 0304016. Other patents pending.

Figure H39-28 Large high-efficiency steel boiler. (Courtesy of Fulton Boiler Works, Inc.)

Any air that forms in the system must be either directed to the expansion tank or vented. The system must be properly filled with water and the pump pressure maintained to produce the required flow.

H39-6.1 Location of the Expansion Tank

The expansion tank can be located on the suction side of the pump (Fig. H39-29) or the discharge side of the boiler (Fig. H39-30). Both the relief valve and the reducing valve should be connected to the boiler on the expansion tank side, as shown in the illustrations.

H39-6.2 Air Purging Arrangement

Air must be eliminated from the piping system. Air vents need to be placed at high points of the system and on terminal units. Even if the system is free from air when filled, air can enter by the heating of water during operation. Depending on the selection of accessories, air that collects during operation can be vented at the expansion

PREVENTIVE MAINTENANCE

When maintaining gas-fired boilers, start by checking the fire side of the boiler. Check the vent connect pipe, chimney, and chimney clean-out. Look for possible leaks at joints and corroded pipes in the venting system. The vent connect should extend into, but not beyond, the inside edge of the chimney or vent.

Inspect the burners and flue passages in the boiler. Remove the jacket panels, draft hood, and flue collector when brushing out the boiler. Brush the boiler flue passages horizontally (through the clean-out opening) and vertically (from the top of the boiler). Burners should be cleaned inside and out, and burner ports must be open and not rusted shut.

Visually inspect the pilot assembly for problems. Look at the thermocouple for deterioration at the tip and the sensor for cracks and crazing on the porcelain. Clean the pilot orifice and air port.

Inspect the boiler base and insulation.

Check wiring in and around the boiler. Look for loose terminals, frayed wires or bare wire, and replace if necessary. Follow the thermostat wire, looking for the same problems.

Check and adjust gas pressure and water pressure.

Test all limits and safeties according to the manufacturer's specifications.

Visually inspect water pipes for leaks. Oil the motor, check the pump coupling, and remove any air from the water side of the boiler.

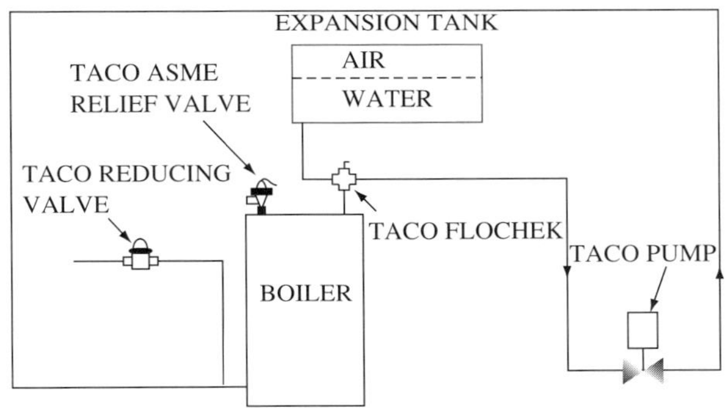

Figure H39-29 Recommended expansion tank hookup connected to suction side of pump. (Courtesy of Taco, Inc.)

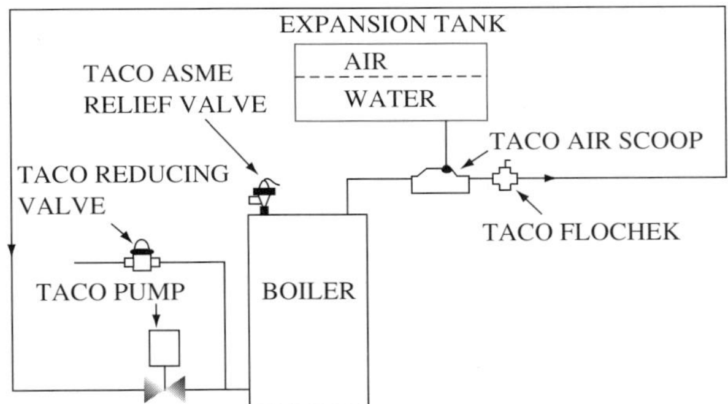

Figure H39-30 Recommended expansion tank hookup connected to discharge side of boiler. (Courtesy of Taco, Inc.)

tank. Some air remains in the expansion tank to act as a cushion for the expansion and contraction of water during temperature change.

H39-6.3 Water Volume Control

An expansion tank is placed in the piping to permit the expansion and contraction of the water volume. The tank holds both air and water. The air is compressed when the water volume expands and the air expands when the water volume decreases.

There are two types of expansion tanks: the diaphragm-type (Fig. H39-31) and the air-cushion compression-type (Fig. H39-32).

The diaphragm type has a flexible separator between the air space and the water space which moves during operation to allow for the changing volume of water. As shown in the illustration, the air removed from the system is released through a float vent.

The air-cushion compression-type feeds air into the tank through an airtrol tank fitting. This fitting permits the air bubbles to rise directly into the tank, but restricts the flow of water back into the tank. The air cushion provides adequate pressurization for all fluctuations of water volume.

H39-6.4 Filling the System

It is important in filling the system to supply a continuous flow of water to replace the air in the piping. A good method of doing this can be illustrated by referring to di-

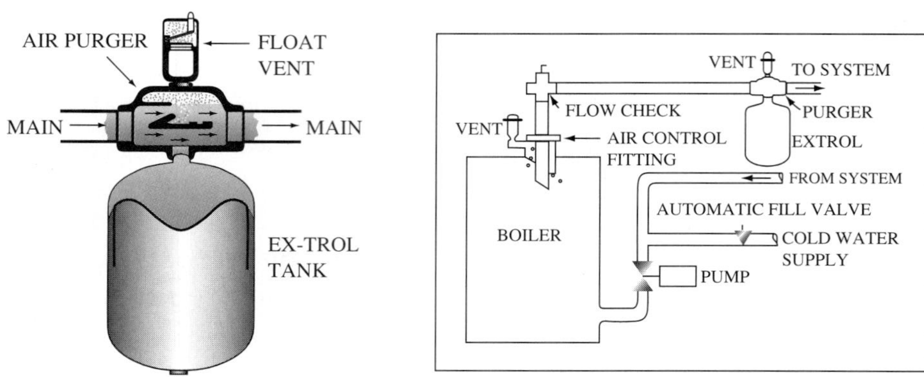

Figure H39-31 Diaphragm-type expansion tank.

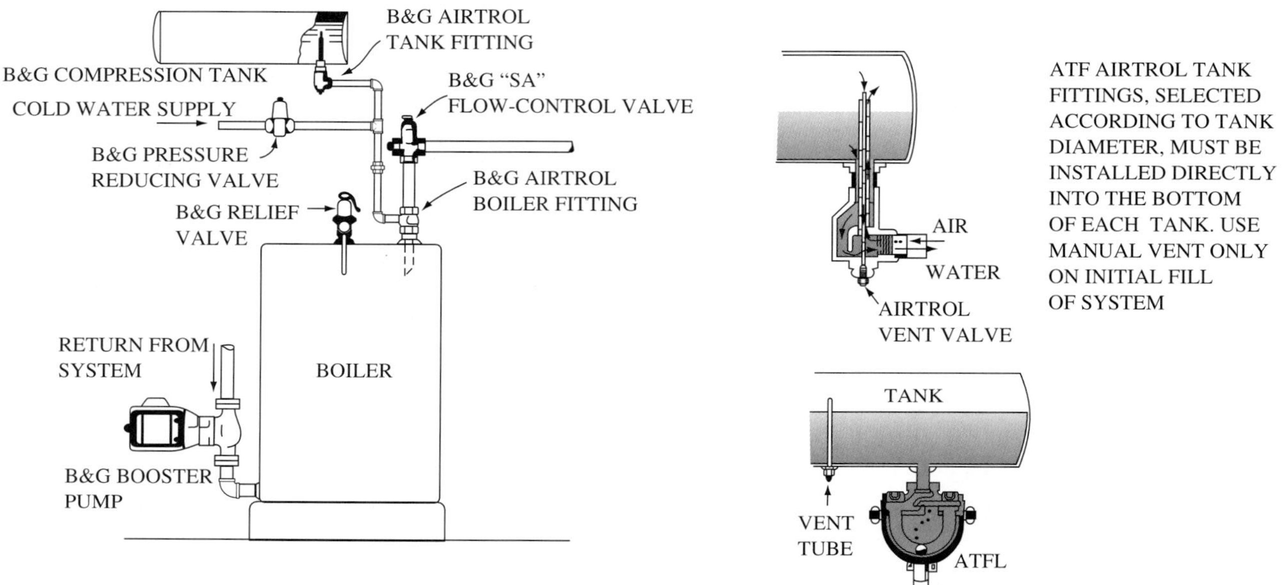

Figure H39-32 Compression-type expansion tank.

agram Fig. H39-33. The drain valve is opened, valve "B" is closed, and valve "A" is opened. This arrangement generally uses city water pressure to fill the piping. Care must be exercised to return the system to 12 psi when the automatic feed is placed in control.

H39-6.5 Safety Controls and Gauges

The location of the pressure-relief valve and the temperature-pressure gauge is shown in Fig. H39-33. The pressure-relief valve is usually set for 30 psig on a domestic or small commercial system and piping is arranged to discharge into an approved drain. The gauge measures water pressure (in psig or ft of head) and boiler water temperature.

H39-6.6 Controls

There are a number of ways residential or small commercial hot-water heating systems can be controlled:

1. An aquastat can be used to maintain boiler water temperature by controlling the fuel-burning device. A room thermostat starts and stops the circulator pump to supply the heat to the terminal units. This control arrangement is particularly suitable for systems where domestic hot water is heated by the boiler water.
2. The thermostat can be used to control both the fuel-burning device and the circulating pump. The advantage of this arrangement is that in mild weather the boiler water temperature is lower due

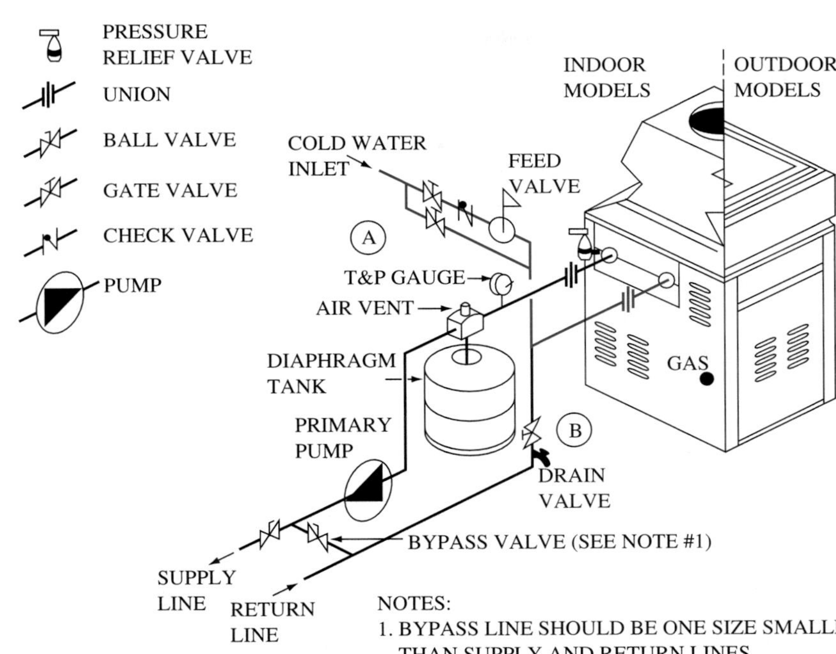

Figure H39-33 Diagram showing piping arrangement for filling the system.

to the shorter running time, producing more efficient operation.

3. For a system using zone valves, any zone calling for heat can start the circulator. Either of the two above arrangements can be used to control the boiler water temperature.

REVIEW

- Hydronics—The science that utilizes water or steam to transfer heat from a source where it is produced, to an area where it can be used, through a closed system of piping.
- The amount of heat transfer depends upon:
  The temperature difference between the water and the surrounding medium;
  The quantity of water flowing through the piping;
  The design characteristics of the heat exchanger.
- Modern heating systems use forced (pumped) circulation because:
  Smaller pipes can be used.
  Systems are easier to control.
- Hot-water heating systems are popular because of increased comfort and ease of control.
- Disadvantages to hydronic heating:
  Installation costs are higher;
  Separate systems are required for humidification and filtration;
  A dual system is required for cooling.
- Properties of water that are critical to hydronic heating:
  Water expands or contracts as temperature changes.
  Water will freeze at 32°F, depending on pressure.
  Water has considerable weight.
  Water creates friction.
- Good piping practices:
  The piping is the correct size.
  Piping is well supported with hangers.
  Piping is permitted adequate expansion and contraction.
  Piping is leak-tight.
  Piping is sized correctly.
  Pipe is properly insulated.
- Valves—Two-position or adjustable.
- Types of valves:
  Gate valve—Stops and starts flow.
  Globe valve—Throttles or controls flow.
  Ball valve—Throttles or controls flow.
  Butterfly valve—Throttles or controls flow.
  Check valve—Prevents backward flow.

Safety (relief) valve—Relieves excessive pressure and/or temperature.

- Circulator pumps—Used to force the flow of a certain quantity of liquid against the pressure drop of the system.
- Hydronic heating pumps are usually centrifugal type.
- Terminal units—Transfer heat from hot water or steam to the various building areas.
- Types of piping systems:
  Series loop, single circuit
  Series loop, double circuit
  One-pipe venturi system, single circuit
  One-pipe venturi system, double circuit
  Two-pipe system, direct return
  Two-pipe system, reverse return
  Zoned system, using pumps
  Zoned system, using motorized valves
- Boilers—Used to heat water:
  Hot water
  Steam
- Piping accessories:
  Valves—Shut off, automatic fill, flow control
  Air purge fittings
  Expansion tanks
  Safety controls and gauges
- Controls:
  Aquastat—Water sensing thermostats
  Thermostat—Controlling burner and circulator pumps
  Zone valves

Problems and Questions:

1. The term "hydronics" covers three types of heat-transfer systems. What are they?
2. The maximum boiler pressure and water temperature for a residential hot water boiler are _____ psig and _____ °F.
3. What are the four types of piping layout for low-temperature hot-water systems?
4. Describe a water tube boiler.
5. Define "hydronics".
6. Heat transfer in a hydronic system depends upon _____ .
7. Why do modern hydronic heating systems use forced (pumped) circulation?
8. What is the greatest reason for the popularity of hot-water heating systems?
9. What are some disadvantages to hydronic heating?
10. What properties of water are critical to hydronic heating?
11. List some good piping practices.
12. What type of pump is used in a hydronic heating system?
13. What type of valve prevents backflow?
14. What type of valve stops and starts flow?
15. What is the purpose of terminal units?
16. Hydronics can indicate hot water or steam heat. True or false?
17. The quantity of water in gpm affects heat transfer. True or false?
18. Hot water heat creates a "flywheel" effect. True or false?
19. Hydronic systems cost the same or less than heat pump systems. True or false?
20. Water creates friction as it flows through piping. True or false?
21. A good piping system
    a. is well supported with hangers for pipe expansion and contraction.
    b. is the correct size.
    c. is built with leak-tight joints.
    d. All of the above.
22. The water velocity through the pipes should not exceed
    a. 2.5 feet per minute.
    b. 5 feet per minute.
    c. 10 feet per minute.
    d. None of the above.
23. Circulator pumps on hydronic systems are usually the
    a. rotary vane type.
    b. centrifugal type.
    c. positive-displacement type.
    d. None of the above.
24. Which pipe system is the simplest of all the piping arrangements?
    a. Two-pipe system, direct return
    b. Two-pipe system, reverse return
    c. Zoned system, using pumps
    d. None of the above.
25. The expansion tank can be mounted
    a. on the suction or discharge side of the pump.
    b. at any location.
    c. separate from the mechanical room system.
    d. None of the above.

Testing and Balancing

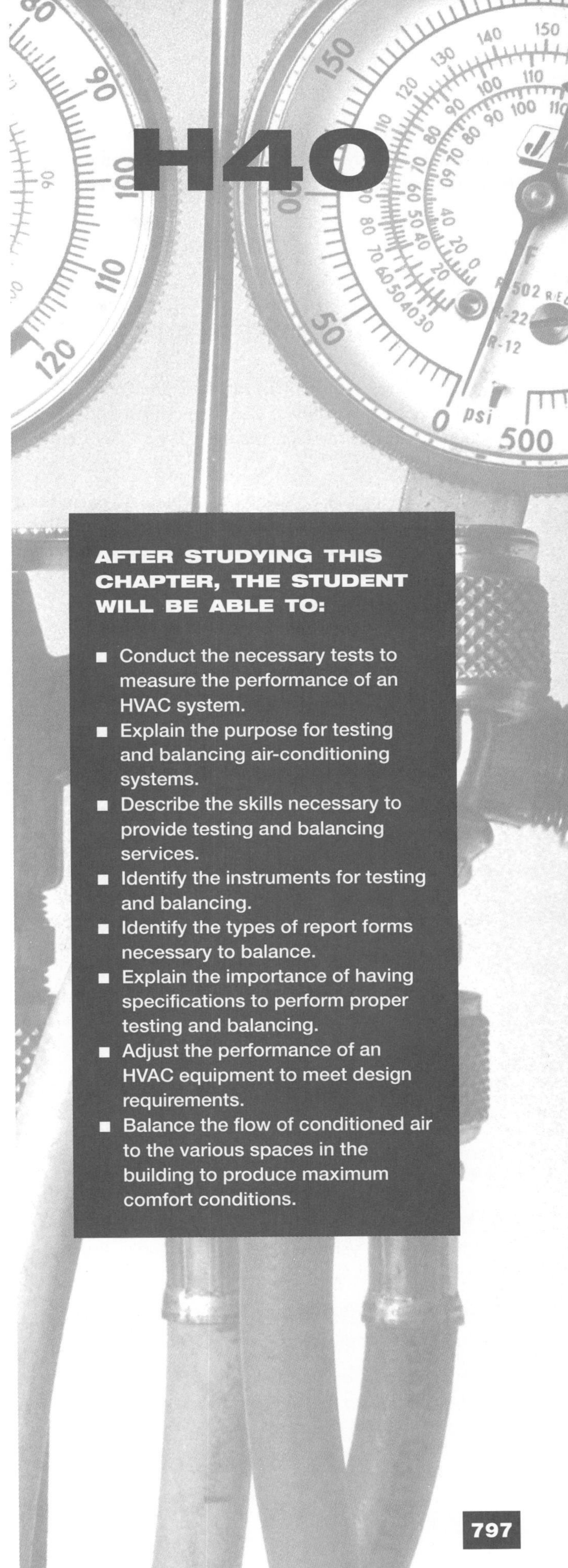

AFTER STUDYING THIS CHAPTER, THE STUDENT WILL BE ABLE TO:

- Conduct the necessary tests to measure the performance of an HVAC system.
- Explain the purpose for testing and balancing air-conditioning systems.
- Describe the skills necessary to provide testing and balancing services.
- Identify the instruments for testing and balancing.
- Identify the types of report forms necessary to balance.
- Explain the importance of having specifications to perform proper testing and balancing.
- Adjust the performance of an HVAC equipment to meet design requirements.
- Balance the flow of conditioned air to the various spaces in the building to produce maximum comfort conditions.

H40-1 OVERVIEW

After the system is installed, an important function of the technician is to test its performance and make whatever adjustments are necessary to produce design requirements. For an HVAC system this usually means providing comfortable conditions at the lowest operating cost.

The equipment must produce its rated capacity. The air quantity handled by the blower must be adequate to handle the load. The temperature drop or rise over the heat exchanger or coil must be in the proper range. The air must be delivered to individual rooms in proportion to the requirement.

It is seldom found that a system installed in the normal manner is completely satisfactory without performing a proper test-and-balance (TAB) procedure. This process provides a systematic checkout of the critical factors that influence the performance of the job.

In existing buildings, changes take place in the use of the system that require periodic TAB services. Many systems require rebalancing twice a year, when the system is changed over from heating to cooling and cooling to heating. Firms that specialize in this type of work also usually provide other maintenance services.

The technician that provides TAB services must be highly qualified. The person must be skillful and knowledgeable in the following:

1. Fundamentals of airflow
2. Fundamentals of hydronic flow
3. HVAC equipment
4. HVAC systems
5. Temperature-control systems
6. Refrigeration systems
7. Temperature measurement
8. Pressure measurement
9. Flow measurement
10. Troubleshooting

In larger organizations doing much TAB work, a team of people are used to speed up the process. Training programs are available for technicians who wish to specialize in this work. Many sizable new air-conditioning installations include a separate specification covering the TAB contract.

H40-2 GENERAL REQUIREMENTS

In addition to the requirement for trained technicians, certain other needs must be supplied to perform a successful program of testing and balancing. These other requirements include:

1. The testing and balancing technician must have tools to measure air flow, air pressure, air temperature, air velocity, humidity, motor amps, and water pressure.
2. Report forms
3. Specification for the work to be performed.

H40-2.1 Instruments

Instruments for Air and Hydronic Balancing

Function/ Measurement	Range	Minimum Accuracy
Rotation	0–5000 rpm	±2%
Temperature	-40°–120°F	Within ½ scale division
Temperature	0°–220°F.	
Electrical	0–600 V_{AC} 0–100 Amps	Within ½ scale division
	0–30 V_{DC} 0–Inf. Ohms	3% of full scale

Instruments for Air Balancing

Function/ Measurement	Range	Minimum Accuracy
Air Pressure	0–1.0 in. WC 0–10.0 in. WC	± .01 in. WC
Pitot Tube	18 in. 36 in.	N/A
Air Velocity	100–3000 fpm	± 10%
Humidity	10–90%	2% RH
Air Volume	0–1400 cfm	± 5%

Instruments for Hydronic Balancing

Function/ Measurement	Range	Minimum Accuracy
Hydraulic Pressure	0–30 psi 0–200 psi 30–to 60 psi	± 1% Full Scale
Hydronic Differential Pressure	0 to 36 in. WC	± 1% Full Scale

H40-2.2 Report Forms

The use of report forms offers a convenient means of recording data and provides a reminder of the items that should be measured. The number and details of the forms used is dependent on the type of job. As a minimum, the forms needed for an air-conditioning job include:

1. Apparatus description (Fig. H40-1)
2. Coil performance (Fig. H40-2)
3. Gas/oil heating apparatus (Fig. H40-3)
4. Fan performance (Fig. H40-4)

PRACTICING PROFESSIONAL SERVICES

Prior to testing and balancing the air-conditioning and heating ventilation system, always review the building owner's final "as-built" plans and written specifications to verify the mechanical engineer's original building intent. Many buildings which have been remodeled and redesigned will not satisfy the new heating or air-conditioning building loads. Remodeling can include adding computer rooms, lunch rooms, modifying office space, and/or moving duct work to satisfy the tenants. If an increased load was placed inside the building without modifying or increasing the size of the mechanical equipment, air balancing most likely will not fix the problem. When this situation occurs, the technician should discuss the problem with a mechanical consultant, a contractor, and the building owner to determine a cost-efficient remedy to the problem. If or when larger cooling and heating equipment is installed in the building, the technician should perform a thorough test and balance of the system.

AIR APPARATUS TEST REPORT

PROJECT ______________________ SYSTEM/UNIT ______________________

LOCATION ______________________ DATE ______________________

Unit Data	
Make/Model No.	
Type/Size	
Serial Number	
Arr./Class	
Discharge	
Make Sheave	
Sheave Diam/Bore	
No. Belts/ make/size	
No. Filters/type/size	

Motor Data	
Make/Frame	
Hp (W) rpm	
Volts/Phase/Hertz	
F.L. Amps/S.F.	
Make Sheave	
Make Sheave Diam/Bore	
Sheave ⊄ Distance	

Test Data	Design	Actual
Total cfm (L/s)		
Total S.P.		
Fan rpm		
Motor Volts		
Motor Amps $T_1/T_2/T_3$		
Outside Air cfm (L/s)		
Return Air cfm (L/s)		

Test Data	Design	Actual
Discharge S.P.		
Suction S.P.		
Reheat coil ΔS.P.		
Cooling Coil ΔS.P.		
Preheat Coil ΔS.P.		
Filters ΔS.P.		
Vortex Damp. Position		
Out. Air Damp. Position		
Ret. Air Damp. Position		

REMARKS:

TEST DATE ______________________ READINGS BY ______________________

Figure H40-1 Air-apparatus test report. (Courtesy of SMACNA)

APPARATUS COIL TEST REPORT

PROJECT ____________________

Coil Data	Coil No.		Coil No.		Coil No.		Coil No.	
System Number								
Location								
Coil Type								
No. Rows-Fins/In. (mm)								
Manufacturer								
Model Number								
Face Area, Ft.^{2} (m^2)								
Test Data	**Design**	**Actual**	**Design**	**Actual**	**Design**	**Actual**	**Design**	**Actual**
Air Qty., cfm (L/s)								
Air Vel., fpm (m/s)								
Press. Drop, In. (Pa)								
Out. Air DB/WB								
Ret. Air DB/WB								
Ent. Air DB/WB								
Lvg. Air DB/WB								
Air ΔT								
Water Flow, gpm (L/s)								
Press. Drop, psi (kPa)								
Ent. Water Temp.								
Lvg. Water Temp.								
Water ΔT								
Exp. Valve/Refrig.								
Refrig. Suction Pres.								
Refrig. Suction Temp.								
Inlet Steam Pres.								

REMARKS:

TEST DATE ____________________ READINGS BY ____________________

Figure H40-2 Coil-performance test report. (Courtesy of SMACNA)

GAS/OIL FIRED HEAT APPARATUS TEST REPORT

PROJECT ____________________

Unit Data	Unit No.		Unit No.		Unit No.		Unit No.	
System								
Location								
Make/Model								
Type/Size								
Serial Number								
Type Fuel/Input								
Output								
Ignition Type								
Burner Control								
Volts/Phase/Hertz								
Hp (W)/rpm								
F.L. Amps/S.F.								
Drive Data								
Test Data	**Design**	**Actual**	**Design**	**Actual**	**Design**	**Actual**	**Design**	**Actual**
CFM (L/s)								
Ent./Lvg. Air Temp								
Air Temp ΔT								
Ent. Lvg. Air Press.								
Air Press. ΔT								
Low Fire Input								
High Fire Input								
Manifold Press.								
High Limit Setting								
Operating Set Point								

REMARKS:

TEST DATE ____________________ READINGS BY ____________________

Figure H40-3 Gas/oil heating-performance report. (Courtesy of SMACNA)

FAN TEST REPORT

PROJECT ______________________________

Fan Data	Fan No.		Fan No.		Fan No.	
Location						
Service						
Manufacturer						
Model Number						
Serial Number						
Class						
Motor Make/Frame						
Motor Hp (W)/rpm						
Volts/Phase/Hertz						
F.L. Amps/S.F.						
Motor Sheave Make						
Motor Sheave Diam./Bore						
Fan Sheave Make						
Fan Sheave Diam./Bore						
No. Belts/Make/Size						
Sheave ℄ Distance						
Test Data	**Design**	**Actual**	**Design**	**Actual**	**Design**	**Actual**
CFM (L/s)						
Fan rpm						
Total S.P.						
Voltage						
Amperage $T_1/T_2/T_3$						

REMARKS:

TEST DATE ______________________ READINGS BY ______________________

Figure H40-4 Fan-performance test report. (Courtesy of SMACNA)

If a duct traverse is made, a separate form should be used.

5. Duct traverse (Fig. H40-5).

Other forms for the following may apply:

6. Air outlet test
7. Packaged chiller
8. Rooftop unit/heat pump
9. Compressor/condenser
10. Cooling tower
11. Pump
12. Boiler

H40-2.3 Specifications Assessment (Minimum Requirements)

Technicians should visit a job during construction to verify that all fittings, damper control devices, and valves are properly located and installed. They should examine the distribution system to see that it is free from obstructions, that moving parts are properly lubricated, and that valves and dampers are in open or in operating position. The technician should have access to the proper, recently calibrated test instruments and adjust volume dampers, variable speed drives, balancing valves and control devices to meet the requirements of the job specifications. All air and hydronic volumes in the distribution system must be adjusted within 10% of the specified quantities. The technician should document the tests and adjustments on suitable forms for review. A check will be made to assure that all national, state and local codes have been followed.

H40-3 INSTRUMENTS FOR TESTING

A complete set of well calibrated test instruments is essential for quality TAB work. The following are some of the most common uses and a description of the test instruments used:

1. Air (gas) flow measuring
2. Pressure measuring
3. Rotation measuring
4. Temperature measuring
5. Electrical measuring
6. Hydronic flow measuring

H40-3.1 Air (Gas) Flow Measuring

The *U-tube manometer* (Fig. H40-6) is used for measuring various types of gas flow. When not connected, both ends of the U-tube are open to the atmosphere. The tube is partly filled with liquid, so that in the idle (open) condition, both tubes register zero. When one tube is attached to a gas under pressure, such as air in a duct, both liquid levels in the tubes change position. The pressure measurement (which is proportional to flow), is indicated by the difference in height of the two columns.

These gauges are recommended for measuring pressures above 1.0 in. WC, across filters, coils, fans, terminal devices, and sections of ductwork.

The instruments that use mercury should be equipped with safety reservoirs to prevent accidental blowing of mercury into the airstream. Tubes must be clean to give accurate results.

The *inclined manometer* (Fig. H40-7) or the *inclined/vertical* manometer (Fig. H40-8), is useful in measuring static pressure, velocity pressure, and total pressure in a duct when connected to a Pitot tube (Fig. H40-9, p. 806). This combination of instruments is used in measuring air volume in a duct by making a traverse such as shown in Fig. H40-10 (p. 807).

The inclined and/or vertical manometer is usually constructed of molded transparent plastic so that it is easy to see the position of the measuring liquid. The inclined scale reads small pressures (below 1.0 in. WC) and the vertical scale reads large pressures (up to 10.0 in. WC). The inclined/vertical manometer is usually equipped with magnetic lugs so that it can be mounted conveniently on the side of the duct.

KNOWLEDGE EXTENSION

Testing and balancing of commercial and industrial buildings has become very important in recent years due to indoor air quality and ventilation issues and requirements. This has encouraged many mechanical contracting firms to start their own specialized air-testing and balancing service. Larger engineering firms and mechanical consultants have also hired staff to perform their own testing and balancing services. The increased growth in this area within the heating, air-conditioning, and ventilation industries has created many new employment opportunities for the student or aspiring apprentice. These jobs provide an indoor work environment, technical challenges, mobility, and good wages. Anyone seeking employment in this specialized industry should inquire within large mechanical contractors and consulting engineering firms to determine specific opportunities. It is also a good idea to become involved with organizations such as Air Conditioning Contractors of America (ACCA), Sheet Metal and Air Conditioning Contractors National Association (SMACNA), or National Environmental Balancing Bureau (NEEB) to learn more about the testing and air-balancing business.

Page ____ of ____

INSTRUCTIONS

METHOD NO. 1
(EQUAL AREA)

TO DETERMINE THE AVERAGE AIR VELOCITY IN SQUARE OR RECTANGULAR DUCTS, A PITOT TUBE TRAVERSE MUST BE MADE TO MEASURE THE VELOCITIES AT THE CENTER POINTS OF EQUAL AREAS OVER THE CROSS SECTION OF THE DUCT. THE NUMBER OF EQUAL AREAS SHOULD NOT BE LESS THAN 15, BUT NEED NOT BE MORE THAN 64. THE MAXIMUM DISTANCE BETWEEN CENTER POINTS, FOR LESS THAN 64 READINGS, SHOULD NOT BE MORE THAN 6 INCHES (150 MM). THE READINGS CLOSEST TO THE DUCTWALLS SHOULD BE TAKEN AT ONE-HALF OF THIS DISTANCE. FOR MAXIMUM ACCURACY, THE VELOCITY CORRESPONDING TO EACH VELOCITY PRESSURE MEASURED MUST BE DETERMINED AND THEN AVERAGED.

X
X/8 X/4 X/4 X/4 X/8
Y/8 Y/4 Y Y/4 Y/4 Y/8
PITOT TUBE NOZZLE LOCATION

METHOD NO. 2
(LOG)

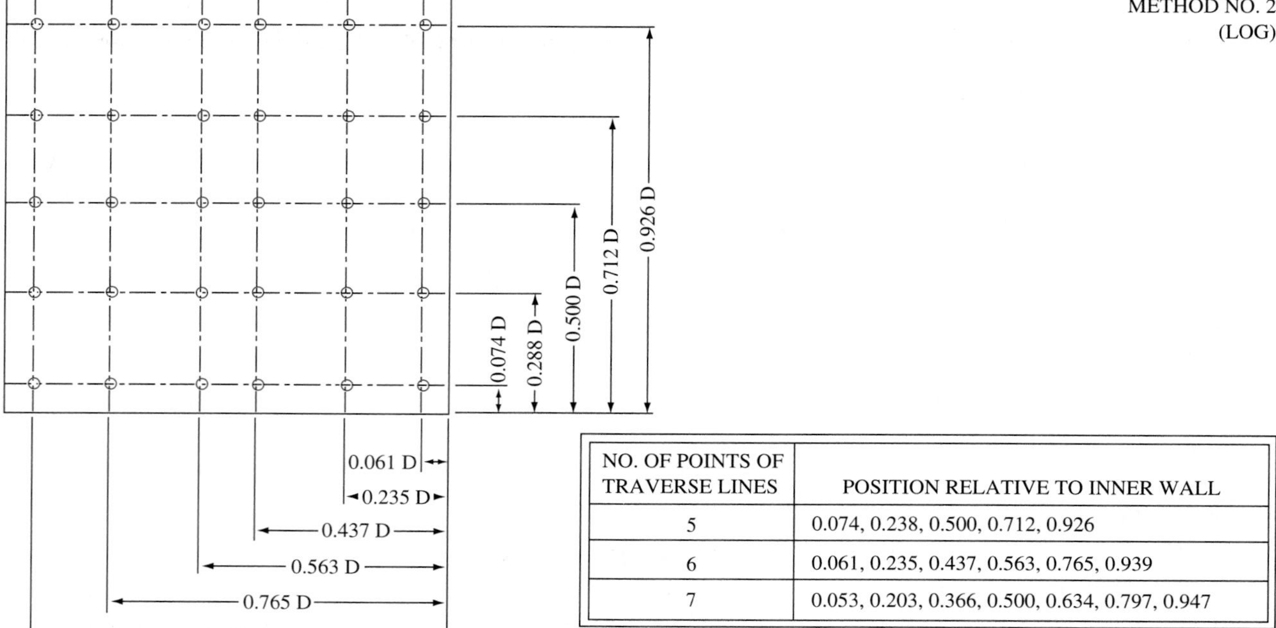

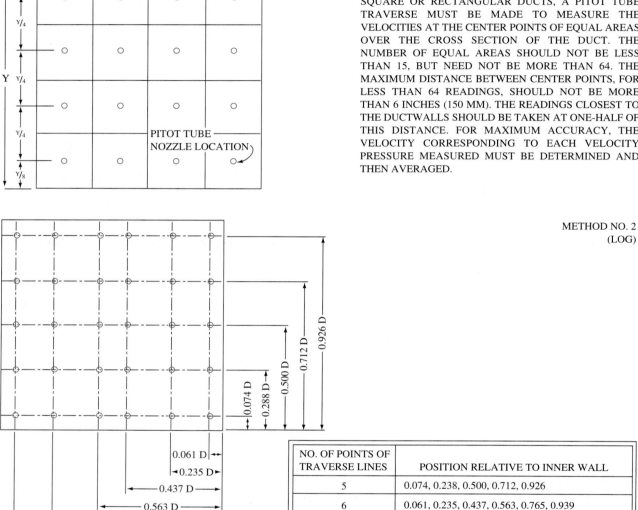

NO. OF POINTS OF TRAVERSE LINES	POSITION RELATIVE TO INNER WALL
5	0.074, 0.238, 0.500, 0.712, 0.926
6	0.061, 0.235, 0.437, 0.563, 0.765, 0.939
7	0.053, 0.203, 0.366, 0.500, 0.634, 0.797, 0.947

Figure H40-5 Rectangular duct traverse. (Courtesy of SMACNA)

The *Pitot tube* is constructed with a tube-in-a-tube design. When pointed against the air flow, the inner tube measures total air pressure, the outer tube measures static pressure and the difference between the two pressures, connected as shown in Fig. H40-7, is the velocity pressure. Convenient charts are available, shipped with the instrument, for converting velocity pressure to ft/min (fpm). Knowing the cross-sectional area of the duct in ft^2, the cfm can easily be calculated (cfm = A × fpm).

A traverse of the duct is important because a series of velocity readings can be taken and averaged to obtain a more accurate value.

It is usually necessary to use at least two manometers in measuring airflow, so that these readings can be taken

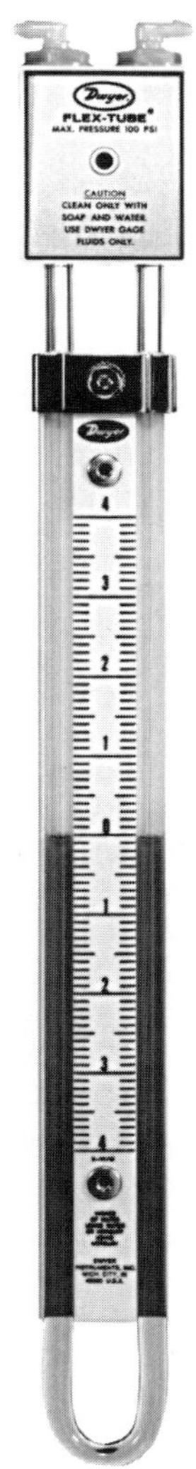

Figure H40-6 U-tube manometer equipped with over-pressure traps. (Courtesy of Dwyer Instruments, Inc.)

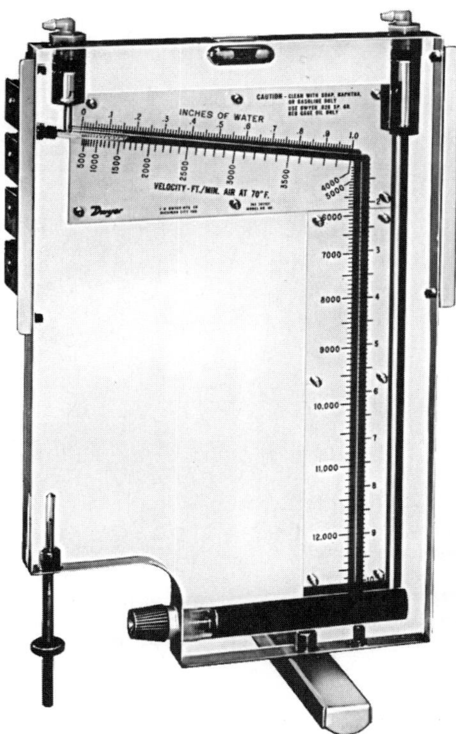

Figure H40-8 Inclined/vertical manometer. (Courtesy of Dwyer Instruments, Inc.)

Figure H40-7 Standard manometer connections to a Pitot tube. (Courtesy of SMACNA)

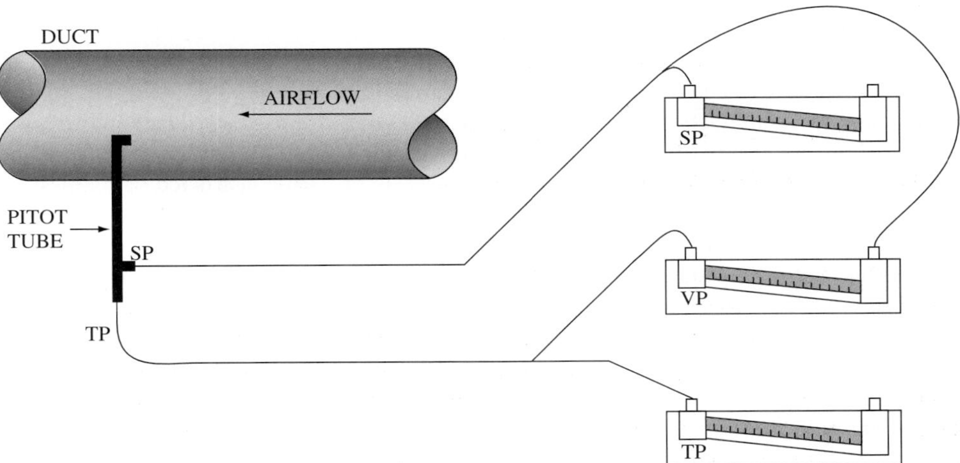

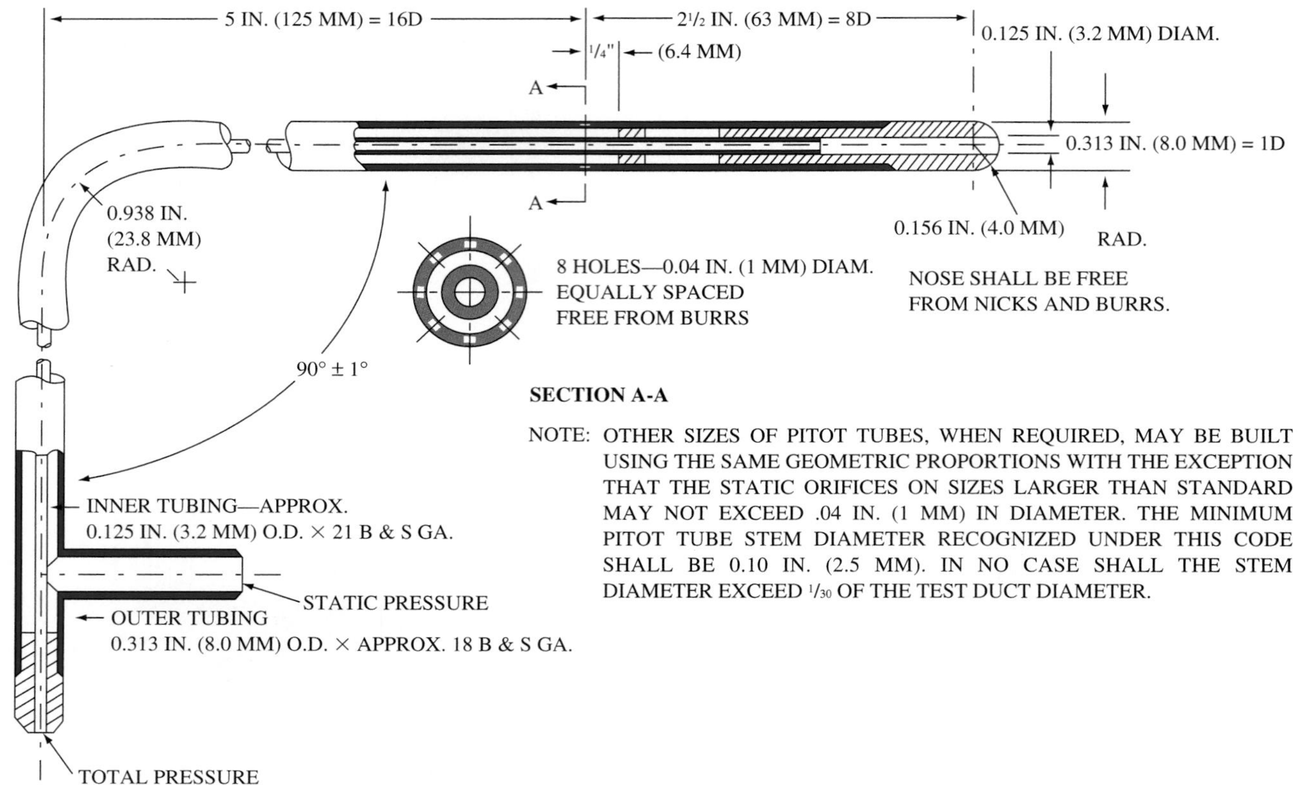

Figure H40-9 Detailed construction of a Pitot tube. (Courtesy of SMACNA)

simultaneously, since $VP = TP - SP$. Care should be taken to be certain that tube connections to the Pitot tube are not crimped and restricting the flow.

The *rotating-vane anemometer* (Fig. H40-11) consists of a lightweight air-propelled wheel, geared to dials that record the linear feet of airflow passing through the instrument. The instrument is placed in various portions of the airstream and readings taken for a measured amount of time. These readings are averaged and converted to fpm.

The *electronic rotating-vane anemometer* is much simpler to use. It is battery-operated and the readout can be either digital or analog. The digital instrument will automatically average the readings and display a velocity readout in fpm. The analog instrument is direct-reading with a choice of a number of different velocity scales.

The *Alnor Velometer* (Fig. H40-12) meets the requirements of most TAB work. The instrument set consists of the meter, measuring probes, range selectors and connecting hoses. Three velocity probes are provided: the low-flow, diffuser and Pitot tube. The low-flow probe is used to measure airflows below 300 fpm, such as encountered in open spaces. The diffuser probe automatically averages the flow over the face of the supply or return grille. The Pitot-tube probe is used to measure air velocities in ducts.

The *thermal anemometer* (Fig. H40-13, p. 809) operates on the principle that the resistance of a heated wire will change with its temperature. The probe is placed in the airstream and the velocity is indicated on the scale of the instrument. It is used for measuring very low velocities of air, such as found in a room, or it may be used in a duct to determine total flow.

The *flow hood* (Fig. H40-14, p. 809) is useful in accurately determining the total airflow in cfm from an outlet. It completely collects the air supply and directs it through a 1-ft^2 opening where a calibrated manometer provides a cfm readout. Hoods should be selected to as nearly as possible fit the diffuser being measured and have a tight seal around the opening.

H40-3.2 Pressure Measurement

The *calibrated pressure gauges* (Fig. H40-15, p. 809) use a Bourdon tube assembly for sensing pressure and are constructed of stainless steel, monel metal, or bronze. The dials are 3½- or 6-in. diameter and are available with pressure, vacuum, or compound gauges. They are used for checking pump pressures; coil, chiller and condenser pressure drops; and pressure drops across orifice plates, valves, and other flow-calibrated devices.

Most uses require determining the differential pressure drop. Two separate gauges can be used for this purpose or a single gauge with a rotating dial can be used as shown in Fig. H40-16 (p. 810). The gauge is installed with two valved connections. The high-pressure line is opened to

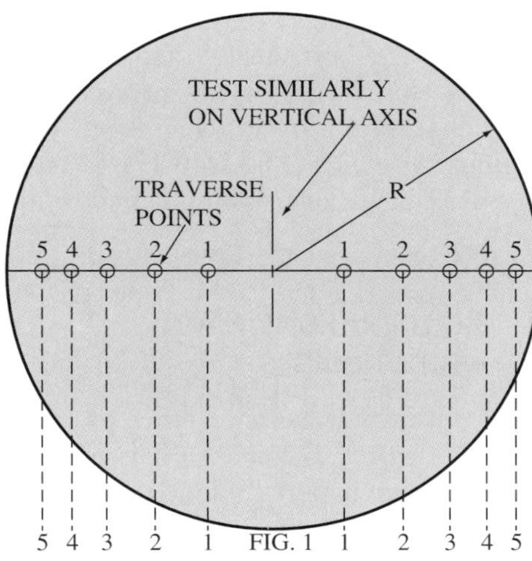

Fig.1: Locations for Pitot tube tip making a 10-point traverse across one circular pipe diameter. In making two traverses cross the pipe diameter, readings are taken at right angles to each other. The traverse points shown represent 5 annular zones of equal area.

Duct diam.		Readings in one diam.	Table1 Distances of Pitot Tube Tip from Pipe Center									
			Point 1		Point 2		Point 3		Point 4		Point 5	
inches	(mm)		inches	(mm)	inches	(mm)	inches	(mm)	inches	(mm)	inches	(mm)
3	75	6	.612	15.3	1.061	26.5	1.369	34.2				
4	100	6	.812	20.4	1.414	35.4	1.826	45.6				
5	125	6	1.021	25.5	1.768	44.2	2.285	57.1				
6	150	6	1.225	30.6	2.121	53.0	2.738	68.5				
7	175	6	1.429	35.7	2.475	61.9	3.195	80.0				
8	200	6	1.633	40.8	2.828	70.7	3.651	91.3				
9	225	6	1.837	45.9	3.182	79.5	4.108	102.7				
10	250	8	1.768	44.2	3.062	76.6	3.950	98.8	4.677	116.9		
12	300	8	2.122	53.0	3.674	91.1	4.740	118.6	5.612	140.3		
14	350	10	2.214	55.3	3.834	95.8	4.950	123.7	5.857	146.4	6.641	166.0
16	400	10	2.530	63.2	4.382	109.5	6.567	141.4	6.693	167.3	7.589	189.7
18	450	10	2.846	71.1	4.929	123.2	6.364	159.1	7.530	188.2	8.538	213.4
20	500	10	3.162	79.1	5.477	136.9	7.077	176.8	8.367	209.1	9.487	237.2
22	550	10	3.479	87.0	6.025	150.6	7.778	194.4	9.213	230.1	10.435	260.9
24	600	10	3.798	94.9	6.573	164.3	8.485	212.1	10.040	251.0	11.384	284.6
26	650	10	4.111	102.8	7.120	178.0	9.192	229.8	10.877	171.9	12.222	308.3
28	700	10	4.427	110.7	7.668	191.7	9.900	247.5	11.713	292.8	13.282	332.0
30	750	10	4.743	118.6	8.216	205.4	10.607	265.1	12.550	313.7	14.230	355.7
32	800	10	5.060	126.5	8.764	219.0	11.314	282.8	13.387	334.6	15.179	379.4
34	850	10	5.376	134.4	9.311	232.7	12.021	300.5	14.233	355.6	16.128	403.2

For distance of traverse points from pipe center for pipe diameters other than those given in Table 1, use constants in Table 2.

Readings in One Diameter	Table 2 Constants To Be Multiplied By Pipe Diameter For Distances of Pitot Tube Tip From Pipe Center				
	Point 1	Point 2	Point 3	Point 4	Point 5
6	.2041	.3535	.4564		
8	.1768	.3062	.3953	.4677	
10	.1581	.2738	.3535	.4183	.4743

Figure H40-10 Round duct traverse. (Courtesy of SMACNA)

Figure H40-11 Rotating-vane anemometer. (Courtesy of Alnor Instrument Company, Skokie, IL)

measure the high-side pressure, with the low-pressure line valve closed. The gauge dial is rotated to zero. Then the high-pressure valve is closed and the low-pressure line valve opened. The dial then reads the differential pressure.

A *differential pressure gauge* (Fig. H40-17) is available with two pressure connections and used as shown in the illustration.

H40-3.3 Rotation Measuring Instruments

A *tachometer* is an instrument for measuring the rotational speed of a shaft or wheel. The units of measurement are usually revolutions per minute (rpm). There are a number of types. Some have a digital readout. Others are mainly revolution counters and require a timing device.

The *chronometric tachometer* (Fig. H40-18) combines a revolution counter and a stop watch in one instrument. The spindle is placed in contact with the rotating shaft. This sets the meter hand at zero and starts the stop watch. After a fixed amount of time, usually 6 seconds, the counting mechanism is automatically uncoupled, and the instrument can be removed from the shaft and read.

The *optical tachometer* (Fig. H40-19, p. 811) uses a photocell to count the pulses as an object rotates. It includes a computerized circuit that produces a direct reading in rpm. It is completely portable since it is powered by batteries. It can be calibrated easily by directing it to a fluorescent light and comparing the reading against a 7200 rpm scale. A big advantage of this instrument is that it does not

Figure H40-12 Alnor velometer. (Courtesy of Alnor Instrument Company, Skokie, IL)

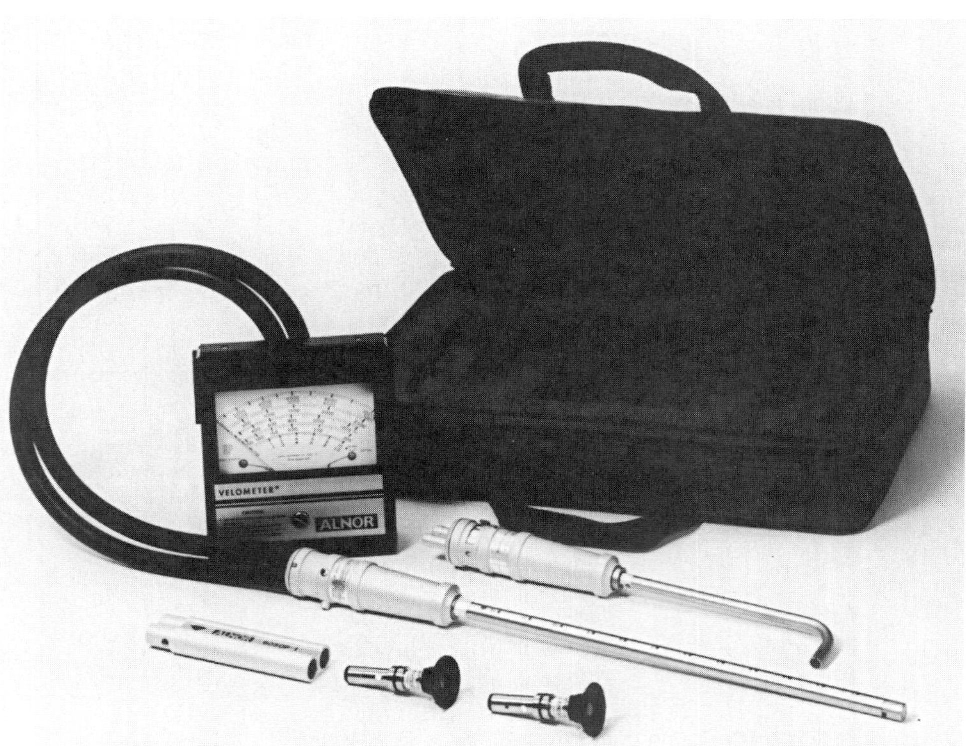

Figure H40-13 Thermal anemometer. (Courtesy of Alnor Instrument Company, Skokie, IL)

Figure H40-14 Flow-measuring hood. (Courtesy of Alnor Instrument Company, Skokie, IL)

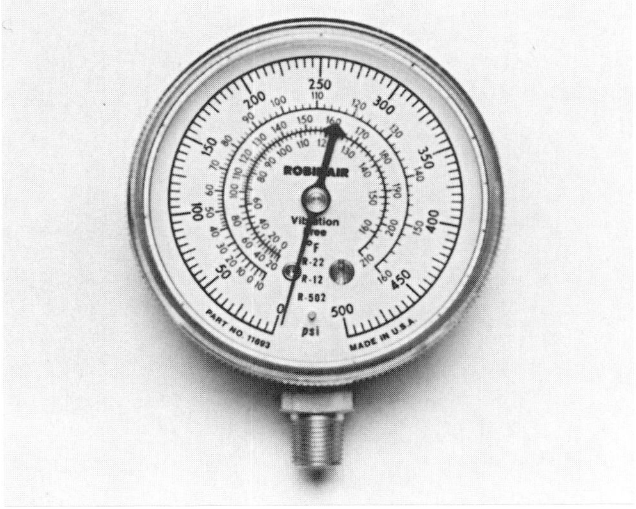

Figure H40-15 Calibrated pressure gauges. (Courtesy of Robinair Division, SPX Corporation)

need to make direct contact with the rotating device and can be used where the rotating shaft is not accessible.

The *stroboscope* (Fig. H40-20) is an electronic tachometer that uses an electronically flashing light. The frequency of the light flashes can be adjusted to equal the frequency of the rotating object. When the two are synchronized, the rotating object appears to be standing still. Care must be exercised in using it to prevent reading harmonics of the actual speed rather than the actual speed.

H40-3.4 Temperature Measurement

Glass-tube thermometers (Fig. H40-21) are available in a number of ranges, scale graduations, and lengths. They have a useful range from –40° to over 220°F.

Dial thermometers are available in two general types: stem (Fig. H40-22) and flexible capillary tube. Dial thermometers are usually bimetal operated, the bimetal being in spiral form within the insert tube. The flexible capillary tube model permits temperature measurements from a re-

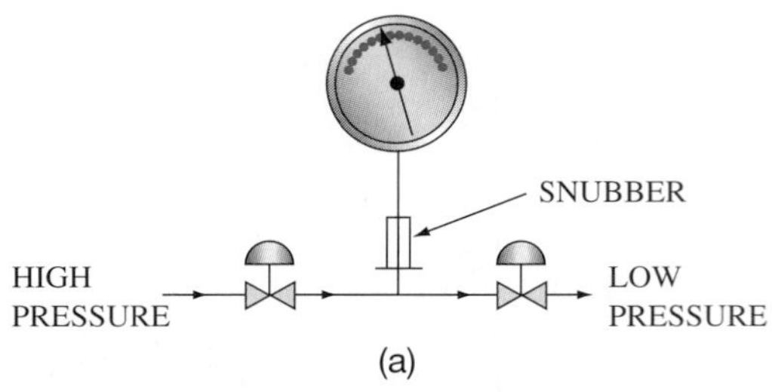

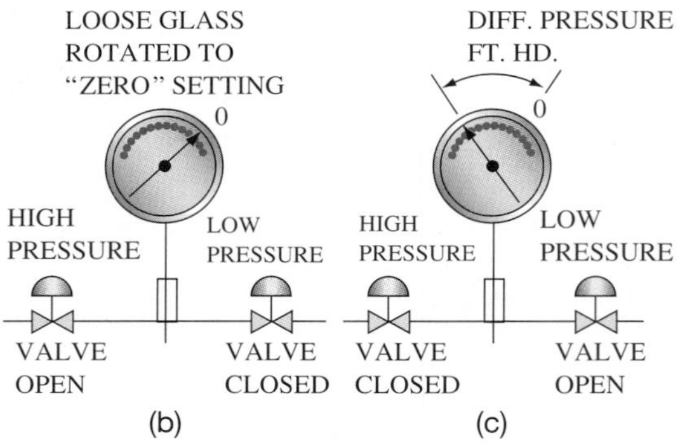

Figure H40-16 Single gauge for measuring differential pressures. (Courtesy of SMACNA)

Figure H40-18 Chronometric tachometer. (Reprinted with permission by AVO International)

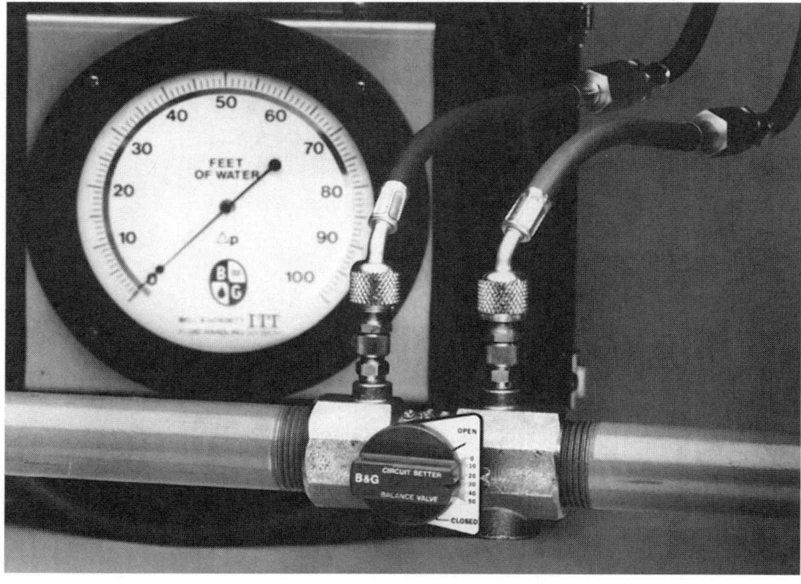

Figure H40-17
Differential pressure gauge with dual connections.
(Courtesy of ITT Bell & Gossett)

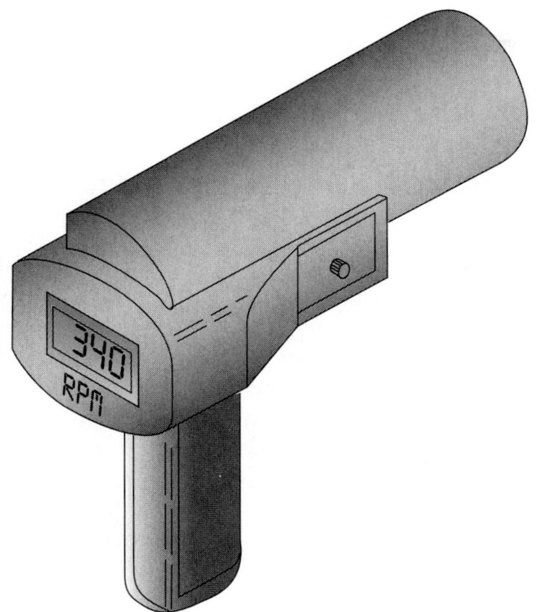

Figure H40-19 Digital optical tachometer.

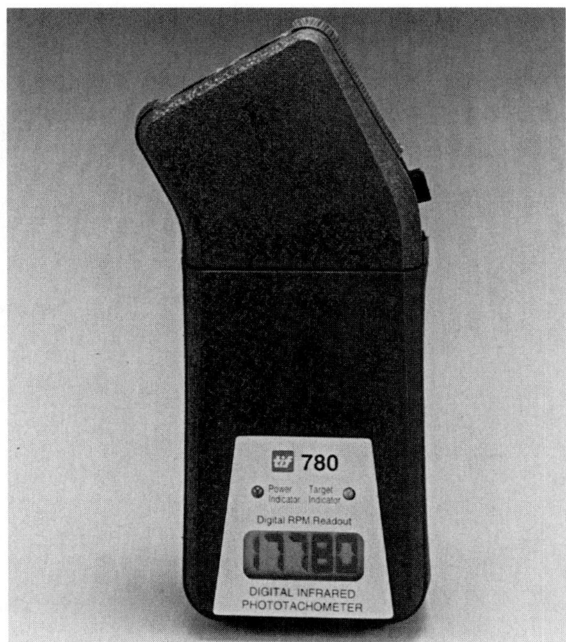

Figure H40-20 Stroboscope for measuring rotational speed. (Courtesy of TIF Instruments, Inc.)

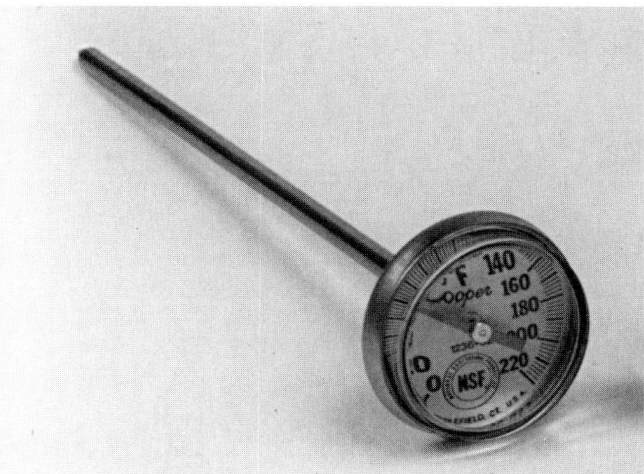

Figure H40-22 Dial thermometer. (Courtesy of Cooper Instrument Corp., Middlefield, CT)

mote location. The tubes are liquid- or gas-filled, which expands or contracts to operate a Bourdon-tube-held indicator. Dial thermometers usually have a longer reading time lag than glass tube thermometers.

Thermocouple thermometers (Fig. H40-23) use a thermocouple sensing device and a milli-voltmeter with a scale calibrated to read temperatures directly. They are useful for remote reading and can be obtained with multiple connections for convenience in reading temperatures at a number of locations.

Electronic thermometers (Fig. H40-24) have interchangeable probes to permit more accurate reading in selected ranges. There are a number of types: resistance temperature detectors (RTD), thermistors, thermocouple, and liquid crystal diode (LCD) or LED displays. The resistance type has a longer time lag than the thermocouple type.

Psychrometric Meters

The *sling psychrometer* consists of two mercury filled thermometers, one of which has a wick wetted by water surrounding the bulb. The frame that supports the two thermometers is hinged to permit revolving the wetted instrument in the air. The unit should be whirled at a rate of two revolutions per second for most accurate results. Readings are taken on both thermometers after the temperatures have stabilized. Readings indicate dry bulb and wet bulb temperatures, which can be plotted on a

Figure H40-21 Glass-tube thermometers. (Courtesy of Cooper Instrument Corp., Middlefield, CT)

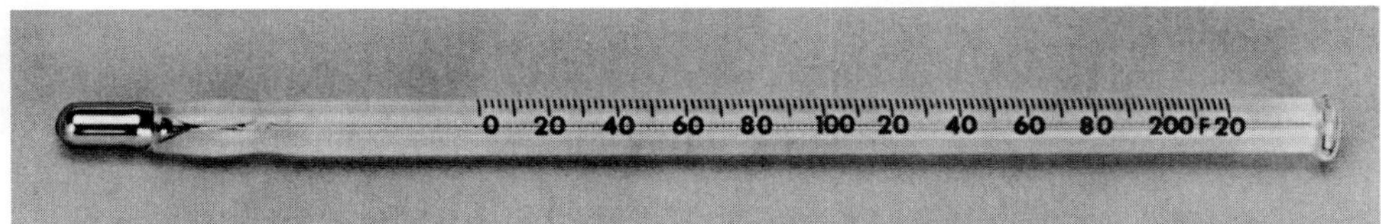

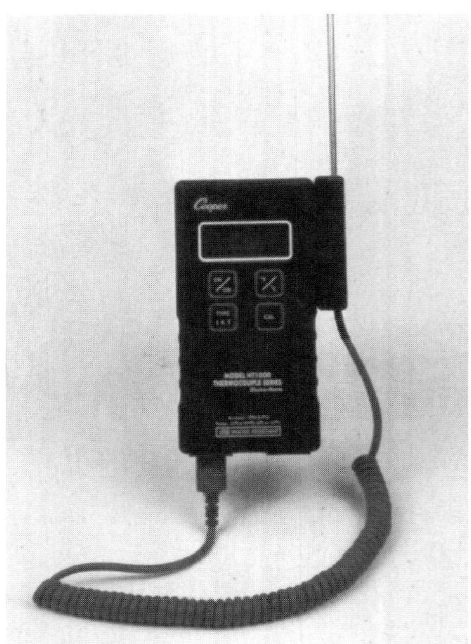

Figure H40-23 Thermocouple-type thermometer. (Courtesy of Cooper Instrument Corp., Middlefield, CT)

psychrometric chart to determine numerous properties of the air sampled.

The *electronic thermohygrometer* is usually constructed with a thin-film capacitance sensor. As the moisture content and temperature change, the resistance of the sensor changes proportionately. These instruments usually read directly in relative humidity. No wetted wick is necessary and the instrument can remain stationary when readings are taken.

H40-3.5 Electrical Measuring Instruments

The *clamp-on volt-ammeter* (Fig. H40-25) is useful for making electrical measurements in the field. Most meters have several scales in amperes, voltages, and ohms. The clamp-on transformer jaws permit reading current flow without disconnecting the circuit. Care must be exercised in reading amperes flow to only clamp onto one wire. Enclosing two wires may result in a zero reading. Separate leads, with DC battery in series, are required for resistance measurements.

H40-3.6 Hydronic Flow Measurements

The *orifice plate* and *venturi tube* (Figs. H40-26 and H40-27) each have a specific area reduction in the path of the flow. The pressure differential across the restriction is related to the velocity of the fluid. By accurately measuring the pressure drops at a range of flow rates, a graph is set up to provide data for future measurements in the same range.

Figure H40-24 Electronic thermometer. (Courtesy of Cooper Instrument Corp., Middlefield, CT)

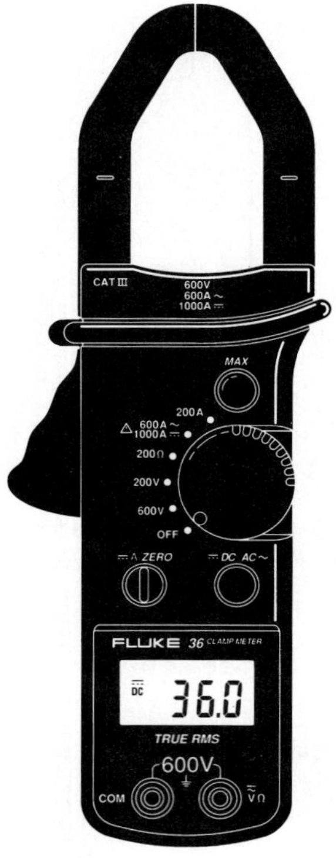

Figure H40-25 Clamp-on volt ammeter. (Reproduced with permission of Fluke Corporation)

ORIFICE PLATE

ORIFICE SIZE IDENTIFICATION

ORIFICE DIAMETER

AIR VENT HOLE; LOCATE AT TOP OF HORIZONTAL PIPE IF CARRYING WATER

DRAIN HOLE; LOCATE AT BOTTOM OF PIPE IF ORIFICE IS USED IN STEAM PIPE

(a)

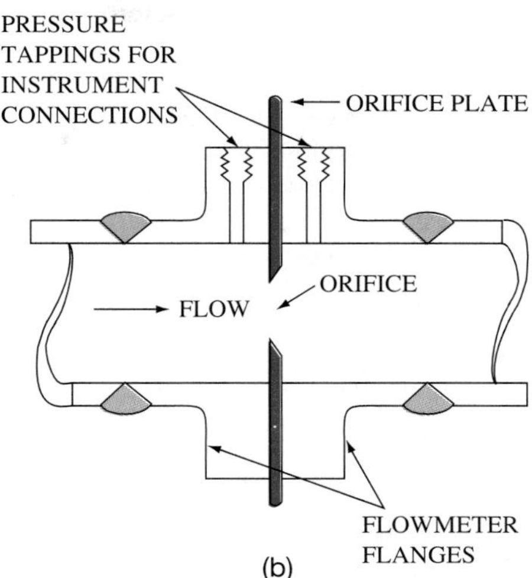

Figure H40-26 Orifice used as a measuring device. (Courtesy of SMACNA)

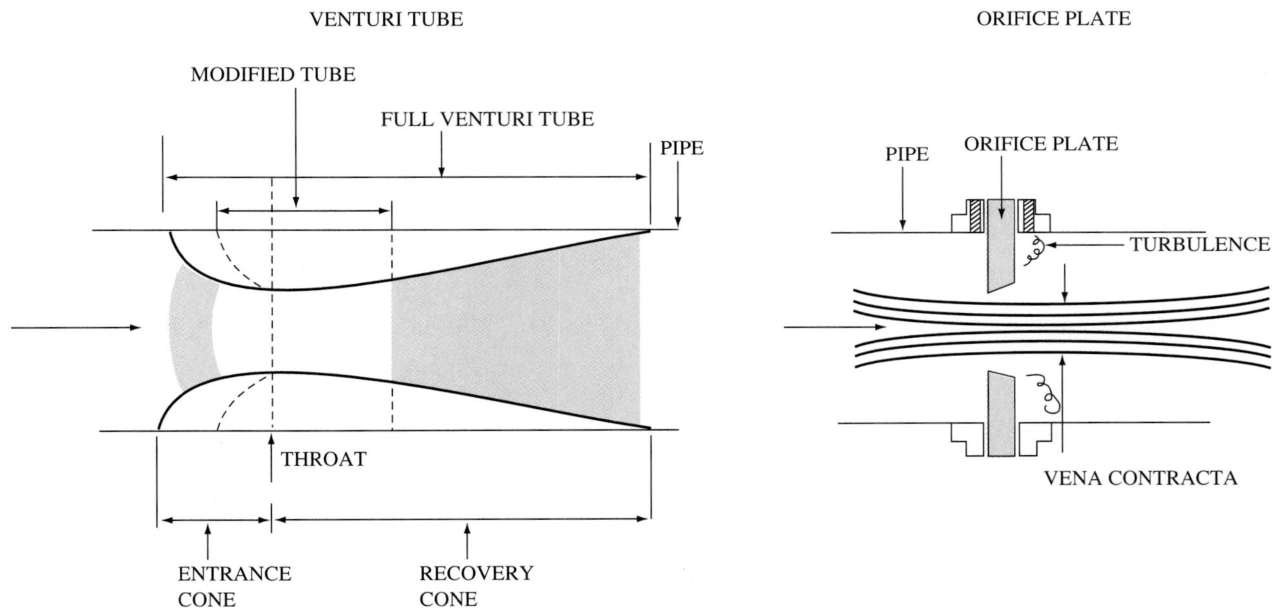

Figure H40-27 Venturi tube and orifice plate each create a flow restriction. (Courtesy of SMACNA)

H40-4 BALANCING THE SYSTEM

Balancing in its broadest meaning includes testing and adjusting an installed system to produce its design specifications. It is the final step in the HVAC project to prepare it for occupancy and use by the owner.

It is an important phase of the work, since many components, from many sources, have been assembled to perform a valuable service. Usually equipment adjustments and control settings are necessary to prepare for the desired operation.

Recommended procedures are described for balancing the following common systems:

1. Warm-air heating installation.
2. Heating/cooling split system.
3. Zoned-air system installation.
4. Hydronic-heating system.

H40-4.1 Balancing a Warm-air Heating System

It is easier to balance a heating system during cold weather and a cooling system during warm weather; however, this is not always possible. It is therefore recommended that the first balance be made when the job is completed or before occupancy, and if necessary, the final balance made during a period as near design temperature conditions as possible.

The following is a suggested procedure for balancing a warm-air heating system:

1. Collect complete information about the installation, including the specifications for the heating unit, details of the controls and electrical system, and a copy of the job layout, including the air quantities (cfm) required for each room.
2. Examine the entire installation. Look for problems that need to be corrected, such as leaky duct work, missing parts and incomplete wiring.
3. Open all dampers, turn the thermostat up so the unit will run continuously, and adjust the input to the furnace to meet the manufacturer's requirements.
4. Measure the temperature rise in the air passing through the furnace. Check the temperature rise with the manufacturer's specifications, to be certain it is within the required range. Based on the output rating of the furnace and the temperature rise, calculate the total cfm the unit is delivering, using the formula:

   $$\text{cfm} = \text{furnace output in Btu/hr}/(TD \times 1.08)$$

   where *TD* is the temperature rise in °F.
5. Adjust the airflow through each outlet to comply with the requirements, using either a flow-hood measuring device or an anemometer to measure the face velocity in ft/min (fpm). The cfm is calculated by the following formula,

   $$\text{cfm} = \text{free area of grille (ft}^2) \times \text{fpm}$$

6. Place thermometers in a suitable location in each room. Set the room thermostat to a reasonable testing temperature. Let the system operate for an hour or more until it balances out. Read the thermometer temperatures in each room.
7. Make any airflow adjustments necessary to produce even temperatures. Pay particular attention to outlets farthest from the furnace that may have a higher supply air temperature drop.
8. Reset the thermostat to the proper room temperature. Instruct the owner on operating the system. Record the information used in balancing the system, including the weather conditions on the day the job was balanced.

H40-4.2 Balancing a Heating/Cooling Split System

Balancing a split system consists of the following steps:

1. Collect complete information about the equipment, the controls, the electrical requirements, and the job layout.
2. Examine the entire installation. Look for problems that need to be corrected, such as leaky duct work, missing parts and incomplete wiring.
3. Measure the dry-bulb temperature drop across the cooling coil with the entering air temperature held as near 80°F as possible. The *TD* should be about 20°F. *TD* is the entering air temperature minus the leaving air temperature.
4. Determine the quantity of air used for cooling, based on the sensible cooling (H_s) requirement and the *TD* across the coil, in accordance with the following formula:

   $$\text{cfm} = H_s/(TD \times 1.08)$$

5. From the layout, record the cfm requirements for each outlet. If the total air required in the layout differs from the actual cfm determined in step 4 above, either adjust the unit fan or prorate the requirements, so that the two quantities match. Use the resulting cfm for balancing the job.
6. Adjust the airflow through each outlet to comply with the requirements, using either a flow-hood measuring device or an anemometer to measure the face velocity in ft/min (fpm). The cfm is calculated by the following formula:

   $$\text{cfm} = \text{Free area of grille (ft}^2) \times \text{fpm}$$

7. Place thermometers in a suitable location in each room. Set the room thermostat to a reasonable testing temperature. Let the system operate for an hour or more until it balances out. Read the thermometer temperatures in each room.
8. Make any airflow adjustments necessary to produce even temperatures. Pay particular attention to outlets farthest from the unit that may have a higher supply air temperature rise.
9. Reset the thermostat to the proper room temperature. Instruct the owner on operating the system. Record the information used in balancing the system, including the weather conditions on the day the job was balanced.

H40-4.3 Balancing a Zoned-air System Installation

The same procedure used for furnaces and split systems is followed for balancing a zoned-air system, except:

a. By adjusting zone dampers, the total cfm of the fan must be proportioned for the requirements of each zone.
b. The cfm for each outlet on the zone are adjusted to meet individual requirements when the zone cfm have been properly set.

H40-4.4 Balancing a Hydronic Heating System

This section applies to all types of piping layouts except the series loop. On the series-loop system the only adjustment that can be made is in the total flow using a balancing valve in the main return before it enters the boiler. For all other systems the following procedure is recommended:

1. Collect complete information about the installation, including the specifications for the boiler, the terminal units, the controls, the electrical system, and the job layout, including the water quantities (gpm) required for each room.
2. Examine the entire installation. Look for problems that need to be corrected, such as air traps, missing parts or incomplete wiring.
3. Examine the piping. Be certain that the system has been installed with the necessary devices for balancing. A minimum should include pressure taps on each side of the pump, a pump-flow volume-balancing valve and flow setters in the branch lines connecting each terminal unit.
4. Be sure all valves in the system are open. Start the system and measure the pressure drop across the pump. Refer to the manufacturer's pump data and determine the flow (gpm) under operating conditions.
5. Adjust the flow through each terminal unit to meet the requirements of the job. Recheck the pressure drop across the pump to verify the total flow requirements.
6. If the weather is suitable, check the temperatures in each room, with the thermostat set at design temperature. If a damper is provided, adjust the airflow in the terminal unit, or reset the branch water flow, to provide even room temperatures. Final adjustment may need to be made later during colder weather.
7. Instruct the owner in operating the system. Record the information used in balancing, including the weather conditions.

REVIEW

- The purpose of testing and balancing is to test performance of the system and make whatever adjustments are necessary to produce design requirements, producing comfortable conditions at the lowest operating costs.
- Equipment must produce its rated capacity.
- The air quantity must be able to handle load.
- The temperature drop or rise over the heat exchanger or coil must be in the proper range.
- The air delivered to individual rooms must be in proportion to load requirements.
- A test and balance procedure provides a systematic check out of the critical factors that influence performance.
- The technician that provides the test and balance must be highly qualified and be skillful and knowledgeable in:
  Fundamentals of airflow
  Fundamentals of hydronic flow
  HVAC equipment and systems
  Temperature-control systems
  Refrigeration systems
  Temperature measurement
  Pressure measurement
  Flow measurement
  Troubleshooting
- Test instruments:
  - Airflow measuring
    - U-tube manometer
    - Inclined manometer
    - Vertical manometer
    - Pitot-tube
    - Rotating-vane anemometer
    - Electronic rotating-vane anemometer
    - Deflecting-vane anemometer
    - Thermal anemometer
    - Flow hood
  - Pressure measurement
    - Calibrated pressure gauges
    - Differential pressure gauge
  - Rotation measurement
    - Tachometer
    - Chronometric tachometer
    - Optical tachometer
    - Stroboscope
  - Temperature measurement
    - Glass sterm thermometer
    - Dial thermometer
    - Thermocouple thermometer
    - Electronic thermometer
  - Psychrometric measurement
    - Sling psychrometer
    - Electronic thermohygrometer
  - Electrical measurements
    - Clamp-on volt-ammeter
  - Hydronic flow measurements
    - Venturi tube
    - Pressure/temp
- Balancing includes testing and adjusting an installed system to produce its design specifications. The final

step in an HVAC project is to prepare it for occupancy and use by owner.

- Balancing procedure (general):
    1. Collect complete information about the HVAC system.
    2. Examine the entire system (including piping, ducts accessories)
    3. Adjust input to the furnace under full load (max.) conditions.
    4. Take and record measurements.
    5. Determine requirements based on loads.
    6. Make adjustments.
    7. Survey zones for comfort conditions.
    8. Readjust as required.
    9. Reset controls to normal operation.
    10. Instruct the owner.

Problems and Questions

1. Why is measurement of the temperature drop of the air through the DX coil important?
2. The subcooling of the liquid refrigerant off the condenser must be measured to determine if the _____ is correct.
3. To measure the conditions of the return air to the coil, a _____ is used.
4. The reason for the incline of the pressure-indicating fluid and scale in an inclined manometer is to _____ .
5. The instruments that can be used to measure the air quantity from a register or grille are called an _____ and a _____ .
6. It is seldom found that a system installed in the normal manner is completely satisfactory without performing a proper test and balance procedure. True or false?
7. Many systems require rebalancing once a year when the air-conditioning system is turned over from cooling to heating. True or false?
8. If a duct traverse is made, a separate form should be used. True or false?
9. The U-tube manometer is used for measuring various types of air flow. True or false?
10. All air and hydronic volumes in the distribution system are adjusted within 5% of the specified quantities. True or false?
11. The Pitot tube is constructed with a tube-in-a-tube design. When pointed against the airflow, the
    a. outer tube measures total air pressure.
    b. inner tube measures static pressure.
    c. inner tube measures total air pressure.
    d. None of the above.
12. The flow hood is useful in accurately determining the
    a. total air flow in cfm from an outlet.
    b. total air flow in fpm from an outlet.
    c. total air flow in scfm from an outlet.
    d. None of the above.
13. The chronometric tachometer
    a. combines a revolution counter and a stop watch in one instrument.
    b. utilizes two different meters, a revolution counter, and a stop watch into two distinct separate instruments.
    c. is a special type of tachometer to measure sick buildings with poor indoor-air quality.
    d. All of the above.
14. The electronic thermohygrometer is used to
    a. provide wet-bulb and dry-bulb temperatures.
    b. provide relative humidity and dry-bulb temperatures.
    c. provide a direct reading in relative humidity.
    d. All of the above.
15. When balancing a warm-air heating system,
    a. collect specifications for the heating units.
    b. open all dampers and turn the thermostat up to make the unit run continuously.
    c. measure the temperature rise in the air passing through the furnace.
    d. All of the above

Filters and Humidifiers

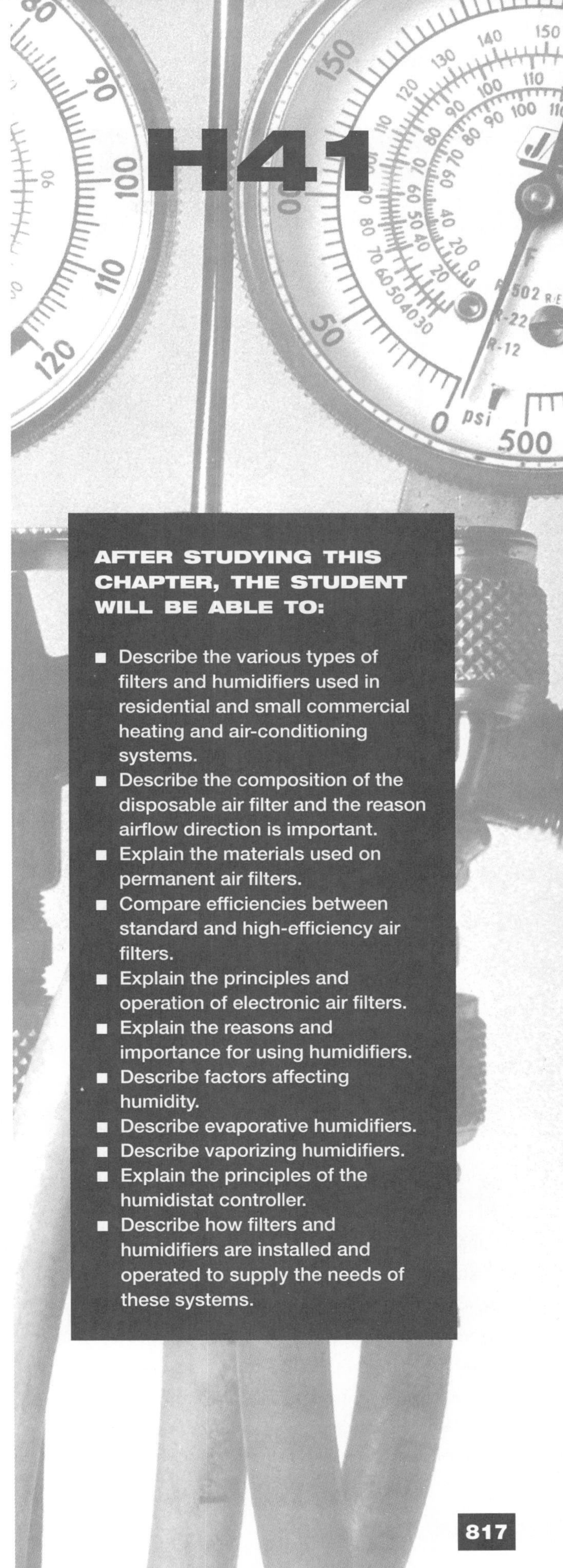

H41

AFTER STUDYING THIS CHAPTER, THE STUDENT WILL BE ABLE TO:

- Describe the various types of filters and humidifiers used in residential and small commercial heating and air-conditioning systems.
- Describe the composition of the disposable air filter and the reason airflow direction is important.
- Explain the materials used on permanent air filters.
- Compare efficiencies between standard and high-efficiency air filters.
- Explain the principles and operation of electronic air filters.
- Explain the reasons and importance for using humidifiers.
- Describe factors affecting humidity.
- Describe evaporative humidifiers.
- Describe vaporizing humidifiers.
- Explain the principles of the humidistat controller.
- Describe how filters and humidifiers are installed and operated to supply the needs of these systems.

H41-1 MECHANICAL FILTRATION

The term *mechanical filter* typically describes a fibrous material filter (Fig. H41-1) commonly called a throw-away or replacement type which comes as standard equipment on forced-air heating units. It consists of a cardboard frame with a metal grille to hold the filter material in place. The filter media material consists of continuous glass fibers packed and loosely woven that face the entering-air side and more dense packing and weave on the leaving air side. Air direction is clearly marked. Some makes also have the fiber media coated with an adhesive substance in order to attract and hold dust and dirt. Filter thickness is usually 1 in. for residential equipment.

The recommended maximum air velocity across the face should be approximately 500 ft/min, with a maximum allowable pressure drop of about 0.5 in. WC.

Many manufacturers and filter suppliers also offer a permanent or washable filter. It consists of a metal frame with a viscous impingement type of filter material supported by metal baffles with graduated openings or air passages. The filter material is coated with a thin spray of oil or adhesive to trap airborne particles. These filters may be removed, cleaned with detergent, dried, and recoated.

Another variation of dry filter is a pad or roll of material supported by a metal mesh frame (Figs. H41-2 and H41-3).

In commercial and industrial installations there are variations of wet and dry filters for special uses where larger volumes of air or filtrations of chemical substances are required. Applications involving paint in spray booths, commercial laundries, hospital operating rooms, etc., usually require the services of experts.

The *arrestance factor* (efficiency) of standard residential and light commercial mechanical filters can vary from 25% for a typical window air conditioner to 80–85% for the better heating and central air-conditioning equipment. This means the filters are effective in removing lint, hair,

Figure H41-1 Throw-away filter. (Courtesy of AAF International)

Figure H41-2 Replaceable filter insert. (Courtesy of AAF International)

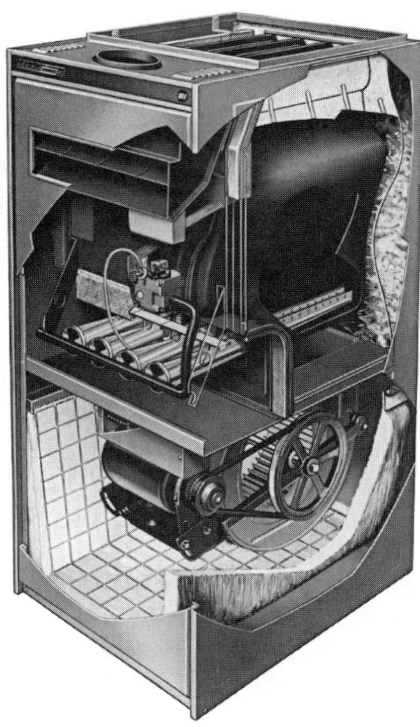

Figure H41-3 Wrap-around filter for a gas furnace. (Courtesy of Lennox Industries, Inc.)

large dust particles, and somewhat effective on common ragweed pollen. They are relatively ineffective on smoke and staining particles. For average use, clean filters do a creditable job.

Getting the homeowner to keep filters clean is not easy, and dirty filters are probably the main contributor to malfunctioning equipment. As the mechanical filter clogs with dirt, airflow is reduced to a point where the cooling coil will freeze, causing compressor failure. In a heating

PREVENTIVE MAINTENANCE

Humidifiers must have special preventive maintenance annually or bi-annually to clean water pans and associated equipment. This is especially true in areas with hard-water supply systems. This maintenance should include cleaning the humidifier housing, heating elements, supply-line float assemblies, and any other moving parts. Failure to routinely maintain this type of equipment will accelerate water-deposit accumulations which result in premature failure and customer dissatisfaction.

Air-filter maintenance is probably the most important component of a good preventive maintenance program, but often it is the most ignored by the homeowner or building manager. Dirty air-filtering media contributes to compressor failures faster than any other component within the air-conditioning system. Thus, air filters should be maintained or replaced at least bi-annually or more often, depending on ambient environmental conditions.

system, dirty filters can cause overheating and reduce the life of the heat exchanger or cause nuisance tripping of the limit switch. Operating costs increase with loss of efficiency.

H41-2 ELECTRONIC FILTRATION

Where greater filtering efficiency is desired (above 80%), the use of electronic air cleaners is highly recommended.

Electronic air cleaners use the principle of electrostatic precipitation in filtering air. These have been adapted to residential use from commercial and industrial applications, made possible by mass production and effective marketing. There are two types of electrostatic cleaners: the ionizing type and the charged media.

In *ionizing cleaners* (Fig. H41-4), positive ions are generated in the ionizing section by the high-potential ionizer wires. The ions flow across the airstream, striking and adhering to any dust particles carried by the airstream. These particles then pass into a collecting cell having charged and grounded plates. The charged particles are driven to the collecting plates by the force exerted by the electrical field on the charges they carry. The dust particles which reach the plates are thus removed from the airstream and held there until washed away. A pre-filter/screen is usually employed to catch large particles such as lint and hair.

In a typical electronic air cleaner design a DC current potential of 12,000 V or more is used to create the ionizing field, and a similar or slightly lower voltage is maintained between the plates of the collector cell. These voltages necessitate safety precautions which are designed into the equipment.

The power pack boosts the 110-V supply to 12,000 V by means of a high-voltage transformer, which uses solid-state rectifiers to convert AC to DC. Power consumption is 50 W or less.

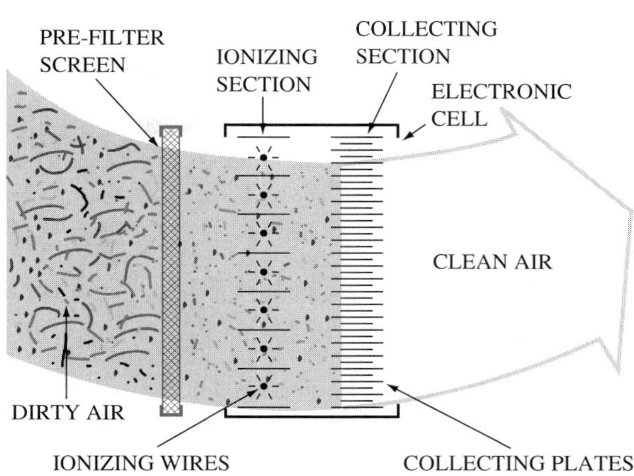

Figure H41-4 Ionizing air filter.

The collector cell (Fig. H41-5), made of aluminum plates, requires periodic cleaning. This can be done by soaking it in detergent or liquid soap, then hosing it down. Some cells are small enough to be cleaned in an automatic dishwasher.

The *charged-media* type of air cleaner has certain characteristics of both dry and electrostatic filters. It consists of a dielectric filtering medium, usually arranged in pleats or pads. Ionization is not used. The dielectric medium is

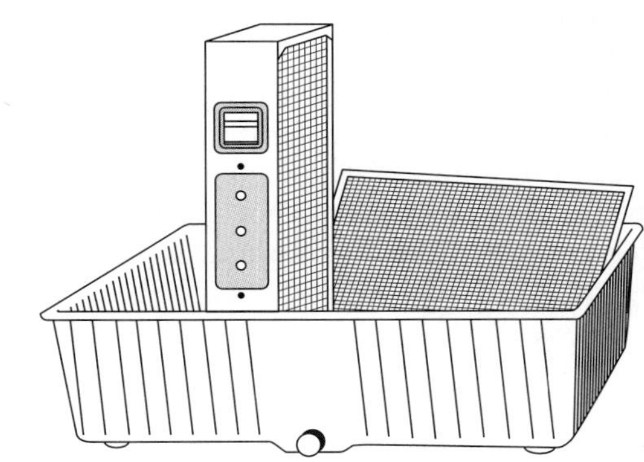

Figure H41-5 Cleaning an electronic filter.

ON THE JOB

Electronic air cleaners operate at extremely high voltages and should be treated with caution and safety at all times. Whenever servicing or working with this type of air cleaner, it is imperative to lock out and tag out all control panels, motor starters, and disconnects during maintenance. Failure to do so could result in electrical shock or harm to the service technician.

Since these units operate at voltages in excess of normal building supply voltages, the average voltmeter will not operate in these ranges. If servicing electronic air cleaners, consult the manufacturer's recommendations before using your standard VOM to test these units.

in contact with a gridwork consisting of alternately charged members held at very high voltage. An intense electrostatic field is created, and airborne particles approaching the field are polarized and drawn into the filaments or fibers of the media. This type of cleaner has much higher resistance to airflow, and the efficiency of the media is impaired by high humidity. At this time, charged-media air cleaners are no longer available.

How effective are electronic air cleaners? Most are capable of filtering airborne particles of dust, smoke, pollen, and bacteria ranging in size from diameters of 50 microns down to 0.03 micron. The micron is a unit of length equal to one-millionth of a meter or 1/25,400 of an inch. The dot made by a sharp pencil is about 200 microns in diameter. It takes special electronic microscopes to examine particles of this size.

The operating efficiency of electronic air cleaners is a function of airflow. With a fixed number of ionizer wires and collector plates, a particular model may be used over a range of airflow quantities (ft^3/min or cfm). It will be rated at a nominal cfm, and then for other efficiency ratings for the recommended span of operation. The higher the cfm of air and the resultant velocity, the lower the efficiency. Residential electrostatic air cleaners come in sizes of 800 to 2000 ft^3. Static pressure drop will run about 0.20 in. of WC at the nominal airflow rating point. Ratings are published in accordance with National Bureau of Standards Dust Spot Test and are certified under ARI standards.

The application of an air cleaner is most flexible (Fig. H41-6). When mounted at the furnace or air handler, it is installed in the return-air duct for upflow, downflow, or horizontal airflow. Position does not affect its operation or performance. When built-in washing is provided units must be installed vertically.

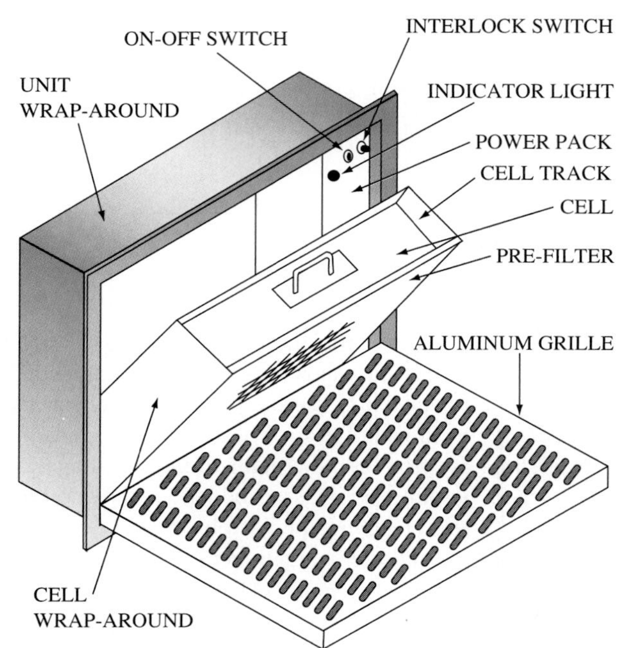

Figure H41-7 Wall filter located behind a grille.

Where furnaces are installed in very small closets or utility rooms and no duct space is available, some manufacturers offer a model (Fig. H41-7) which will recess into a wall and actually provide a decorative return-air louver that faces the living area. Service access is through the louver.

Some units have built-in washing without removing the cells. On an automatic control, washing action sprays water over the plates. The runoff is collected in the bottom of the unit, where it drains out.

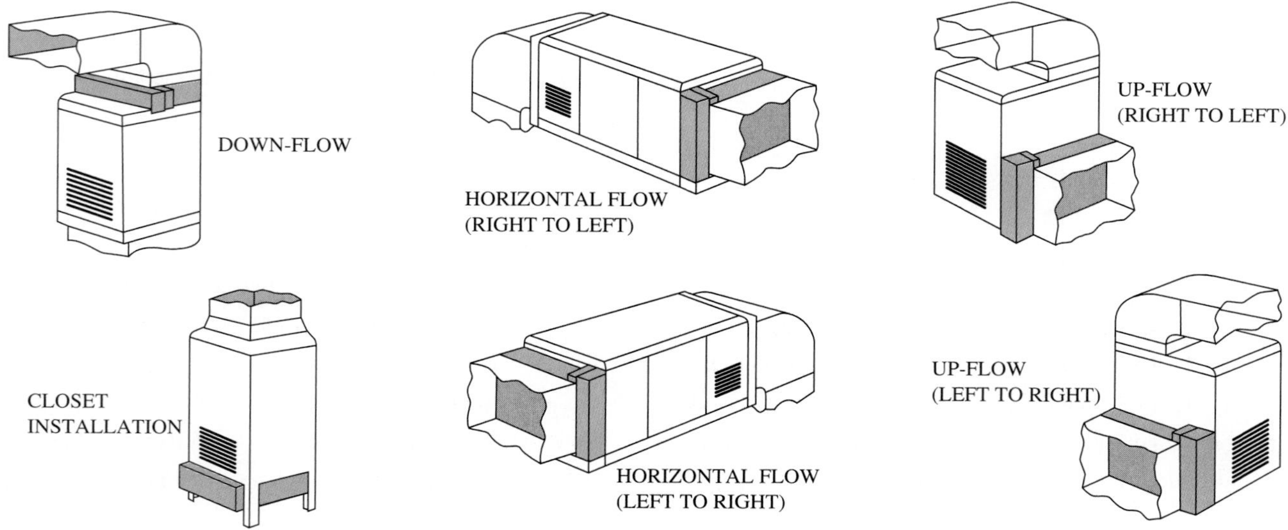

Figure H41-6 Electronic filter installations in various types of furnaces.

H41-3 TYPES OF HUMIDIFIERS

Humidifiers are used to add moisture to indoor air. The amount of moisture required depends on:

1. Outside temperatures
2. House construction
3. Amount of relative humidity that the interior of the house will withstand without condensation problems.

It is desirable to maintain 30 to 50% relative humidity. Too little humidity may cause furniture to crack. Too much humidity can cause condensation problems. From a comfort and health standpoint, a range as wide as 20 to 60% relative humidity is acceptable.

The colder the outside temperature, the greater the need for humidification. The amount of moisture the air will hold depends on its temperature. The cooler the outside air, the less moisture it contains. Air from the outside enters the house through infiltration. When outside air is warmed, its relative humidity is lowered unless moisture is added by humidification (Fig. H41-8).

For example, air at 20°F and 60% relative humidity contains 8 grains of moisture per lb of air. When air is heated to 72°F, it can hold 118 grains of moisture per lb. Thus, to maintain 60% relative humidity in a 72°F house, the grains of moisture per lb must be increased to 71 (118 × 0.6 = 71). One pound of air entering a house from the outside will require the addition of 63 grains of moisture (71 – 8 = 63). The amount of infiltration depends on the tightness of the windows and doors and other parts of the construction.

It is impractical in most buildings to maintain high relative humidity when the outside temperature is low. Condensation forms on the inside of the window when the surface temperature drops below the dew point of the air. (The dew point is the temperature at which water as vapor in the air has reached the saturation point, or 100% relative humidity.)

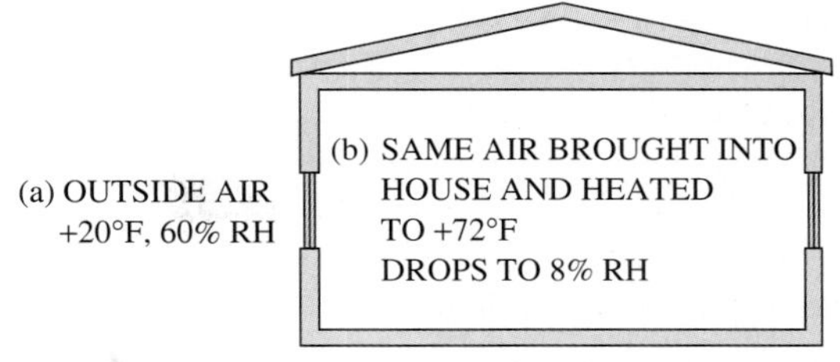

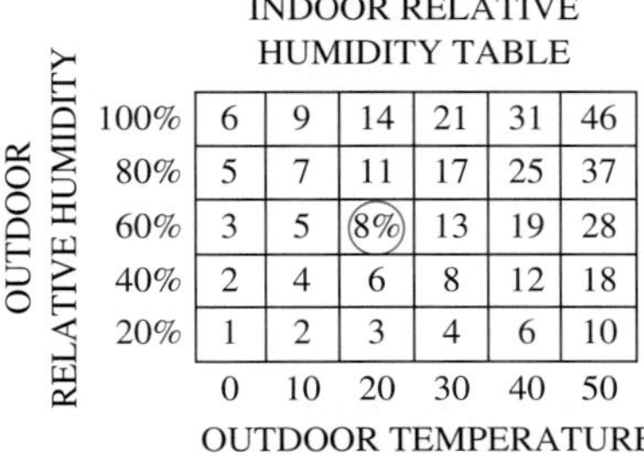

INDOOR RELATIVE HUMIDITY TABLE

OUTDOOR RELATIVE HUMIDITY						
100%	6	9	14	21	31	46
80%	5	7	11	17	25	37
60%	3	5	8%	13	19	28
40%	2	4	6	8	12	18
20%	1	2	3	4	6	10
OUTDOOR TEMPERATURE	0	10	20	30	40	50

Figure H41-8 Chart showing the effect of temperature on the relative humidity of air.

There are two types of humidifiers used with furnace installations:

1. Evaporative
2. Vaporizing

H41-3.1 Evaporative Humidifiers

Evaporative humidifiers provide a wetted surface which adds moisture to the heated air. There are three types of evaporative humidifiers:

1. Plate
2. Rotating-drum
3. Fan-powered

ON THE JOB

Problem: A customer complained of no humidity. You check the operation of the controls to the humidifier and find that nothing is functioning, and there is no control power. While troubleshooting, you find the transformer is bad. There were 120 V supplied to the humidifier's primary winding side of the low-voltage power transformer but *no* low voltage (24 V) at the secondary side. Is this a simple solution or not?

Solution: The low-voltage transformer must be replaced, but it is very common for them to be improperly wired. Normally, the transformer is powered by the same supply as the blower motor. Care must be taken when adding a low-voltage transformer on a multi-speed blower motor. Usually the low-speed tap of the motor is where the transformer is wired in parallel. When the high speed is energized, it can act as a generator and supply the transformer with the wrong voltage, and thus cause it to burn out. If this is the case, the transformer circuit must be interrupted by a switch from the fan relay and isolated.

ON THE JOB

Problem: A customer complained of low humidity. When you turn on the humidifier at the controls, something clicks but nothing happens. There is no water in it. What is the next step?

Solution: There are two main types of humidifiers: the drum type where there is a padded drum rotating in a water pan, and a panel type where water flows through a panel or spreader grid and air passes through the panel, picking up humidity.

In this case, the clue of no water leads to an investigation into the water supply. You find that the water supply line is plugged at the water solenoid valve and you must either clean or replace the valve.

Figure H41-9 Plate-type humidifier. (Source: Skuttle Indoor Air Quality Products)

The *plate evaporative humidifier* (Fig. H41-9) has a series of porous plates mounted in a rack. The lower section of the plates extends down into water contained in the pan. A float valve regulates the supply of water to maintain a constant level in the pan. The pan and plates are mounted in the warm-air plenum.

The *rotating-drum evaporative humidifier* (Fig. H41-10) has a slowly revolving drum covered with a polyurethane pad partially submerged in water. As the drum rotates, it absorbs water. The water level in the pan is maintained by a float valve. The humidifier is mounted on the side of the return-air plenum. Air from the supply plenum is ducted into the side of the humidifier. The air passes over the wetted surface, absorbs moisture, then goes into the return-air plenum.

Figure H41-10 Rotating-drum evaporative humidifier.

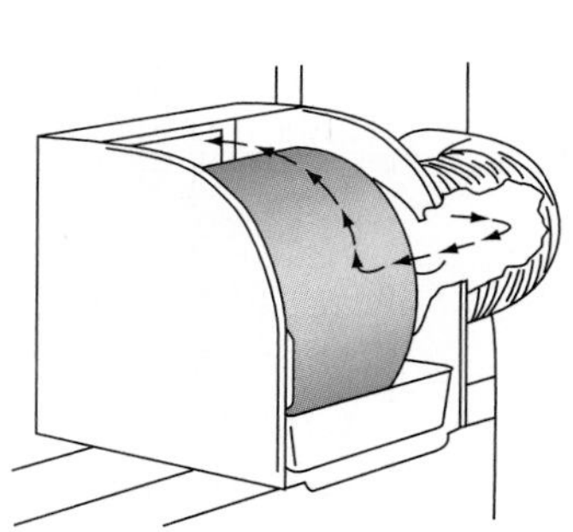

TYPICAL INSTALLATIONS

HORIZONTAL FURNACE

HIGHBOY FURNACE

LOWBOY FURNACE

COUNTER-FLOW FURNACE

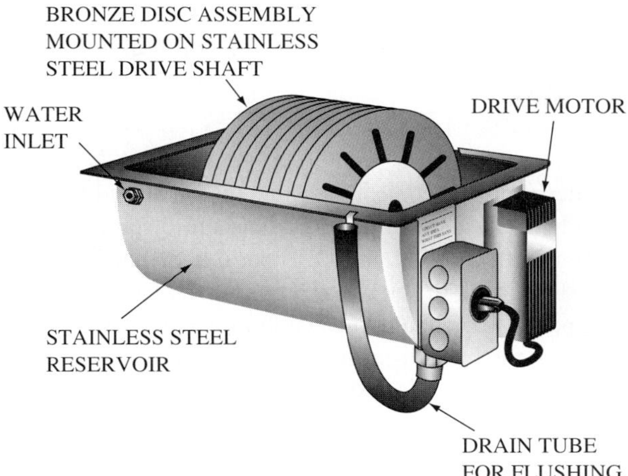

Figure H41-11 Rotating-plate evaporative humidifier.

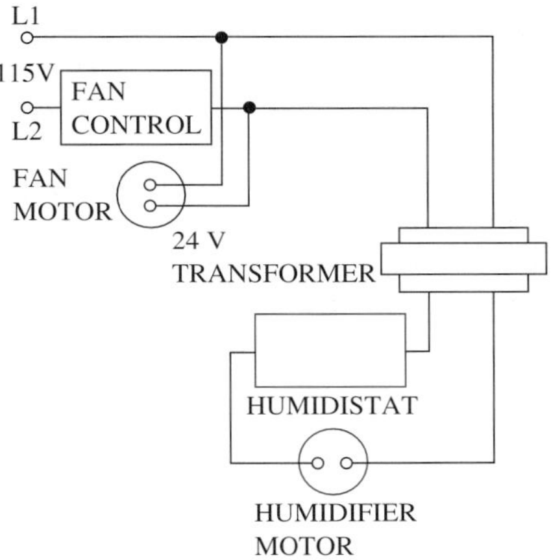

Figure H41-12 Wiring diagram for a humidifier with a single-speed motor.

The *rotating-plate evaporative humidifier* (Fig. H41-11) is similar to the drum type in that the water-absorbing material revolves; however, this type is normally mounted on the underside of the main warm-air supply duct.

The *fan-powered evaporative humidifier* is mounted on the supply-air plenum. Air is drawn in by the fan, forced over the wetted core, and delivered back into the supply-air plenum. The water flow over the core is controlled by a water valve. A humidistat is used to turn the humidifier on and off, controlling both the fan and the water valve. The control system is set up so that the humidifier can operate only when the furnace fan is running. Wiring diagrams are shown in Figs. H41-12 and H41-13.

H41-3.2 Vaporizing Humidifiers

The *vaporizing humidifier* (Fig. H41-14) uses an electrical heating element immersed in a water reservoir to evaporate moisture into the furnace supply-air plenum. A constant level of water is maintained in the reservoir. These humidifiers can operate even though the furnace is not supplying heat. The humidistat not only starts the water heater but also turns on the furnace fan if it is not running.

The *steam-powered humidifier* (Fig. H41-15), is a type of vaporizing humidifier. It is ideal for use with heat pumps, electric furnaces, high-efficiency furnaces and fur-

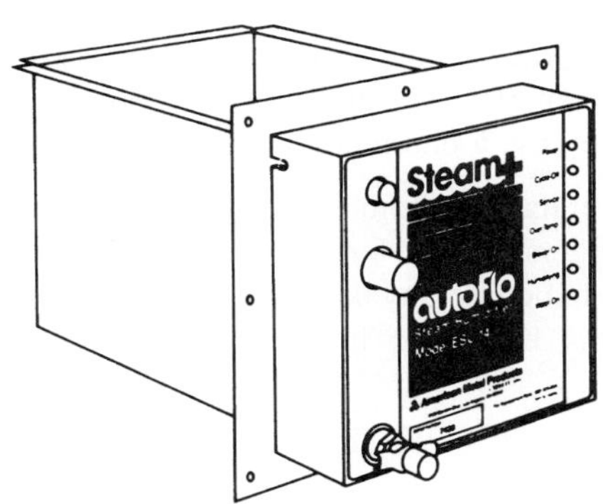

Figure H41-14 Vaporizing humidifier. (Courtesy of American Metal Products)

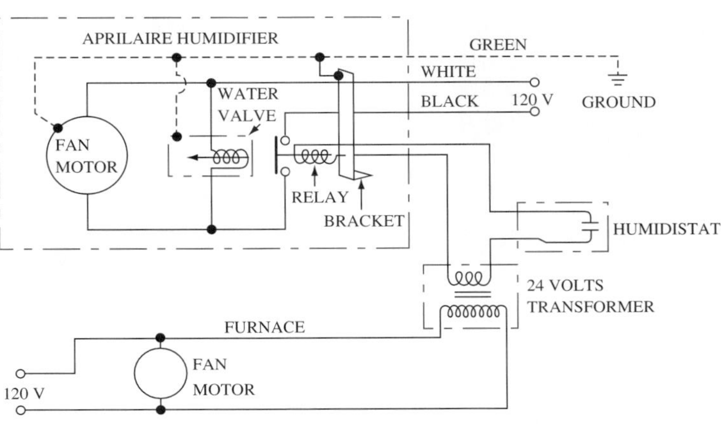

Figure H41-13 Wiring diagram for a fan-powered evaporative humidifier.

Figure H41-15 Steam humidifier. (Source: Skuttle Indoor Air Quality Products)

naces using night set-back thermostats. These units do not operate long enough or generate high-level temperatures necessary for evaporative type humidifiers. The internal water-temperature-sensing device operates the fan independently of the cooling or heating mode. Some of the features of these units are:

1. The thermal fan interlock control allows the unit to humidify the air without heat from the furnace.
2. Cleaning is made easy with a one-piece service drain petcock.
3. Low water safety cutoff switch is corrosion resistant and has a built-in overflow protection.
4. They use a minimum amount of water.

These units now feature a flushing timer and chlorine-removal filters. The water in the unit is flushed out every 12 hours, removing any solid materials that may have accumulated. The carbon filter removes chlorine from the water supply, thus eliminating the corrosive effects of chlorine on the unit, and reduces mineral build up.

H41-4 HUMIDISTATS

It is desirable to control the amount of humidity in the house by the use of a *humidistat.* A humidistat is a device that regulates the ON and OFF periods of humidification (Fig. H41-16). The setting of a humidistat can be changed to comply with changing outside air temperatures. If the setting on the humidistat is too high, it may cause "sweating" on interior walls and windows. As it gets colder outside, the humidistat setting may need to be lowered.

REVIEW

- Mechanical filter—A fibrous material; throw-away or replacement type.
- Permanent mechanical filter—Washable with metal frame. A viscous, impingement type of air filter supported by metal baffles with graduated openings or air passages.
- Recommended maximum air velocity across face of the filter is 500 ft/min with a maximum allowable pressure drop of about 0.5 in. WC.
- The arrestance (efficiency) of standard residential and light commercial mechanical filters can vary from 25% to 85%.
- Mechanical filters are effective in removing lint, hair, dust, and somewhat effective on small dust and common ragweed pollen.
- Home/building owners should be instructed on the importance of keeping filters clean.
- Electronic filters use the principle of electrostatic precipitation.

(a)

(b)

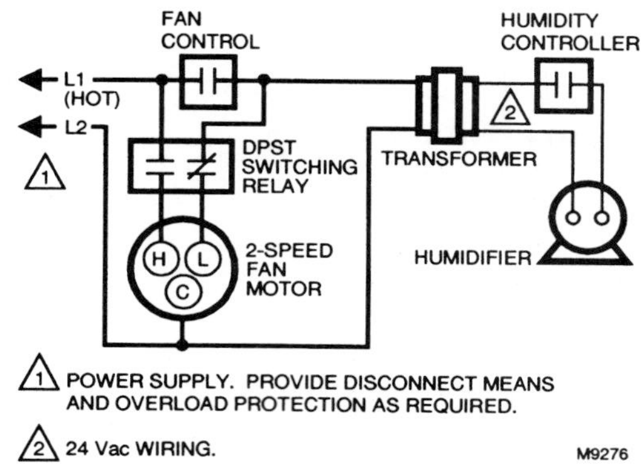

(c)

Figure H41-16 Two types of humidistats and a typical wiring diagram. (Courtesy of Honeywell, Inc.)

- Two types of electrostatic cleaners:
  Ionizing
  Charged-media
- Power pack—Boosts the 120-V supply to 12,000 V and converts the AC to DC.
- Ionizing section—Positive ions are generated by the high-potential ionized wires.
- Collector cell—Charged and grounded plates; requires periodic cleaning.
- Most electronic air cleaners are capable of filtering airborne particles in size from 50 microns down to 0.03 microns.
- Operating efficiency of electronic air cleaners is a function of airflow. The higher the cfm, the lower the efficiency.
- Ratings are published in accordance with National Bureau of Standards Dust Spot Test and certified under ARI standards.
- On an automatic control, washing action sprays water over the plates; the runoff is collected and drained in the bottom of the unit.
- Humidifiers are used to add moisture to indoor air. The amount depends upon:
  Outside temperatures—Colder outdoor temperatures create a greater need.
  House construction
  Amount of relative humidity that the interior will withstand.
- Desirable relative humidity: 30–60%
- Two types of furnace-installed humidifiers:
  Evaporative—Plate, rotating-drum, fan-powered
  Vaporizing—Electric, steam-powered
- Control for humidifier is called a humidistat.

Problems and Questions

1. List the types of mechanical filters.
2. Describe the operation of electronic air filters.
3. What is the arrestance of standard residential and light commercial mechanical filters?
4. What is the common name for non-cleanable mechanical filters?
5. Mechanical filters are generally effective up to _____%.
6. Filter maintenance is not important. True or false?
7. Electronic air cleaners are made in two types. What are they?
8. Operating voltage in an electronic filter can run as high as _____.
9. Electronic air cleaners are rated in _____.
10. What is a micron?
11. The power consumption of a typical electronic air cleaner is about _____ watts.
12. Position of the electronic air cleaner is not critical. True or false?
13. The collect cell is usually cleaned by _____.
14. What are the three types of commercial humidifiers?
15. What are the two types of furnace-installed humidifiers?
16. What type of device controls a humidifier?
17. What is the desirable humidity range?
18. What determines the amount of humidity needed?
19. Name the three types of evaporative humidifiers.
20. What does the term "relative humidity" mean?
21. Air direction is not critical with disposable air filters. True or false?
22. The recommended maximum air velocity across the face of the disposable air filter is 500 ft/min with about 0.5 in. of water column pressure drop. True or false?
23. Arrestance factor refers to the efficiency of the air filter. True or false?
24. In electronic cleaners, negative ions are generated in the ionizing section by the high-potential ionizer wires. True or false?
25. In a typical electronic air cleaner design a DC potential of 5000 volts or more is used to create the ionizing field. True or false?
26. The charged-media type of air cleaner
    a. consists of dielectric filtering media.
    b. does not use ionization.
    c. polarizes airborne particles and draws them into the filaments or fibers of the media.
    d. All of the above.
27. The higher the cfm and velocity of air across the electronic air cleaner,
    a. the lower the efficiency.
    b. the higher the efficiency.
    c. Will have no impact on the efficiency.
    d. All of the above.
28. The amount of moisture within the building depends on the
    a. outside temperatures.
    b. house construction.
    c. amount of relative humidity that the interior of the house will withstand without condensation problems.
    d. All of the above.
29. When outside air is warmed, its relative humidity is
    a. raised.
    b. lowered.
    c. not affected.
    d. All of the above.
30. The vaporizing humidifier
    a. can operate even though the furnace is not supplying heat.
    b. uses an electrical heating element.
    c. can be wired to turn on the furnace fan if the humidistat starts the humidifier.
    d. All of the above.

SECTION 8

Central Heat Pumps

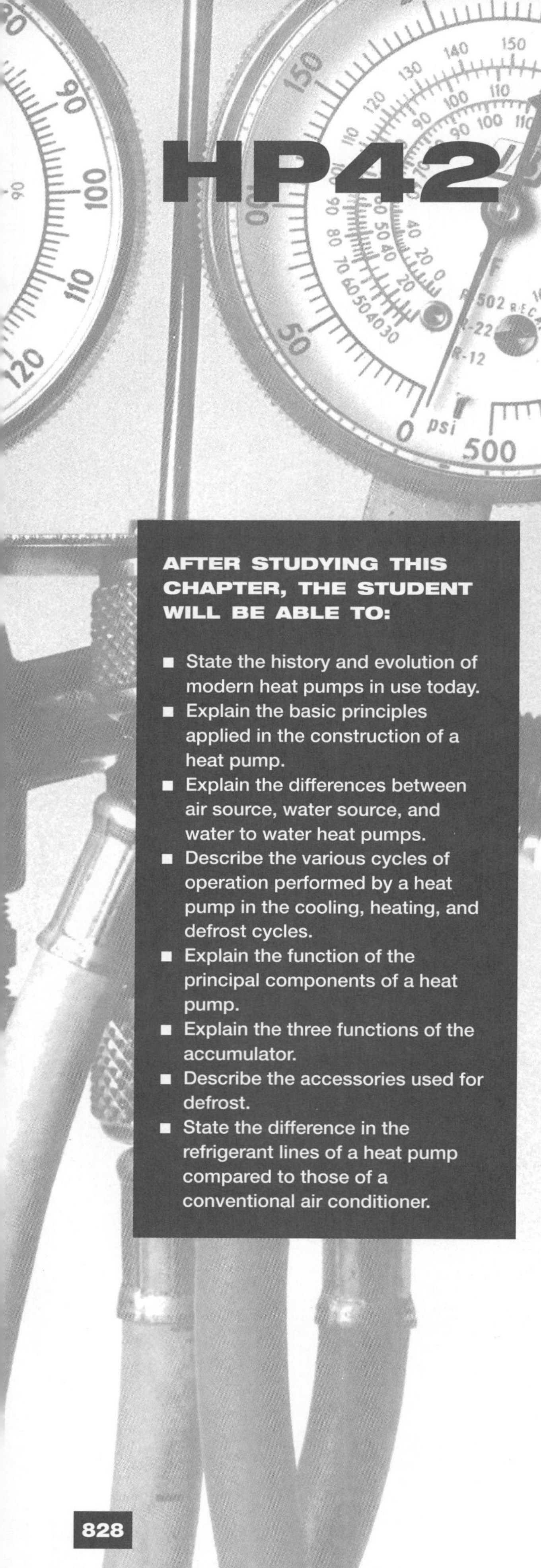

HP42 Basic Principles and Components

AFTER STUDYING THIS CHAPTER, THE STUDENT WILL BE ABLE TO:

- State the history and evolution of modern heat pumps in use today.
- Explain the basic principles applied in the construction of a heat pump.
- Explain the differences between air source, water source, and water to water heat pumps.
- Describe the various cycles of operation performed by a heat pump in the cooling, heating, and defrost cycles.
- Explain the function of the principal components of a heat pump.
- Explain the three functions of the accumulator.
- Describe the accessories used for defrost.
- State the difference in the refrigerant lines of a heat pump compared to those of a conventional air conditioner.

HP42-1 BRIEF HISTORY

Another source of heat for residential and small commercial applications became very popular with the increase in cost of electrical energy. This source, referred to as the *heat pump,* was actually developed by Lord Kelvin in 1852.

The first practically applied heat pump was installed in Scotland in 1927. Between 1927 and 1950, heat pumps were installed in hundreds of residential and small commercial applications throughout Europe and the southern part of the United States. Many of these installations were merely air-conditioning units converted to heat pumps by the addition of reversing valves and applicable controls.

Unfortunately, these units were also installed in colder climates not suited for this application. Heat pumps had a high failure rate and therefore acquired a bad reputation which, together with the relatively low cost of energy, almost destroyed this market.

With the rise in energy costs going into the 1970s, demand grew for an efficient means of heating with electrical energy, and the heat pump was reborn. This time, however, the units were designed for colder climates, with the necessary defrosting equipment as well as cold weather compressor protection.

HP42-2 BASIC PRINCIPLES

The heat pump is a refrigeration system that transfers heat from one place to another by the change in state of a liquid. In addition, the heat pump is able to reverse the action or direction of heat transfers. It can remove heat from an area for summer cooling and dispose of the heat into outside air, a water supply, or indirectly into earth or other material. By reversing the action, it will also remove heat from the outside air source, the water supply, or from the earth or other material and supply it to the occupied area.

Basically all refrigeration systems are heat pumps in that they transfer heat from a heat source at a low temperature to a heat sink or disposal means at a higher temperature. The title, heat pump, has only been given, however, to the system that actually provides a reverse-cycle refrigeration system. The principles that apply to this type of system will be discussed in this chapter.

HP42-3 HEAT-PUMP CYCLES

In the conventional refrigeration cooling cycle (Fig. H42-1) heat is absorbed in the indoor DX coil or evaporator and discharged by the outside air-cooled condenser. To reverse this process, the evaporator and condenser are not physically reversed. By means of a reversing valve, the refrigerant flow can be directed to make the process provide heating or cooling in the occupied area.

The heat-pump cycle is shown in Figs. H42-2 and H42-3. The coils are relabeled as indoor and outdoor as they are now dual-purpose. The outdoor coil is the condenser in the cooling cycle and the evaporator in the heating cycle. The indoor coil is the evaporator in the cooling cycle and the condenser in the heating cycle.

To accomplish the reversing of the refrigerant flow, a reversing valve is used in the suction and discharge lines between the compressor and the two coils. Check valves are also connected in parallel with the pressure-reducing devices to permit removing the devices from the circuit when they are not to be used.

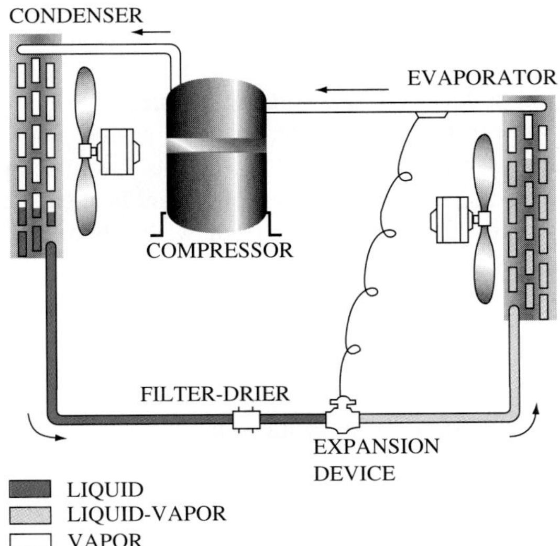

Figure HP42-1 A conventional refrigeration cycle.

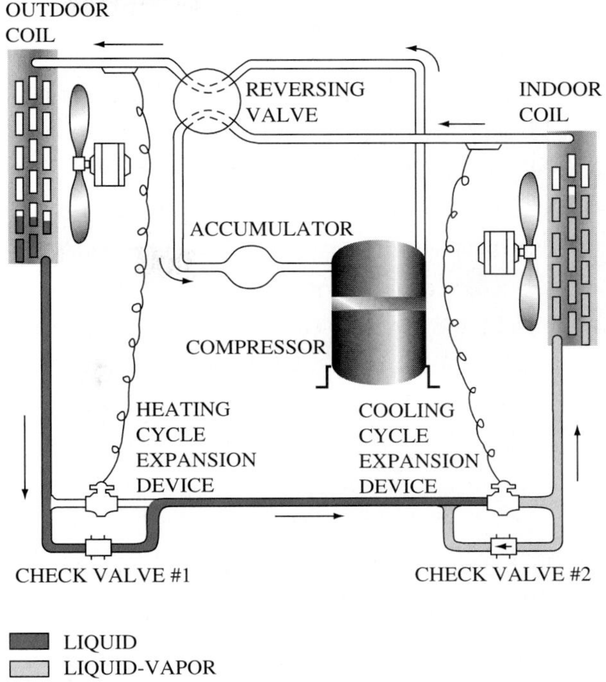

Figure HP42-2 Heat-pump cooling cycle.

Referring to the circuit in the cooling phase (Fig. HP42-2), the directional arrows show the high-pressure/high-temperature discharge gas being directed from the compressor to the outdoor coil (condenser), where it condenses

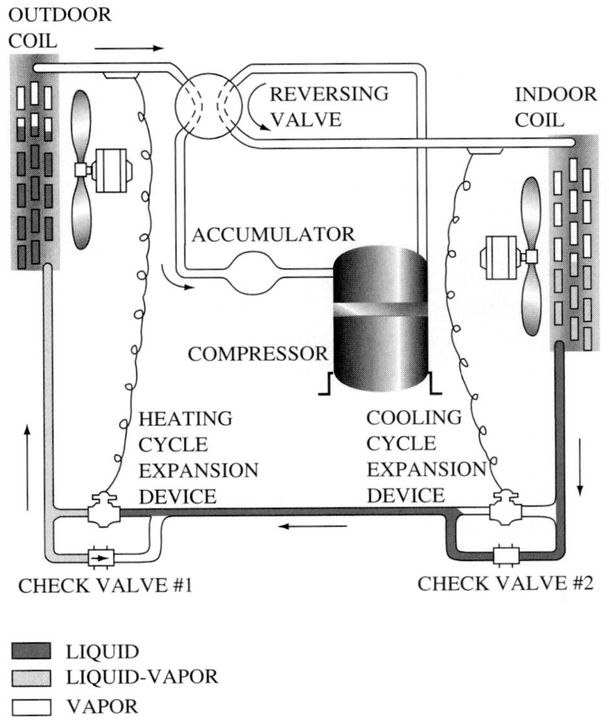

Figure HP42-3 Heat-pump heating cycle.

PREVENTIVE MAINTENANCE

Good preventive maintenance on heat pumps is critical to ensure proper airflow across the coils. Low airflow on indoor coils during winter months causes high head pressures and premature compressor damage due to oil breakdown. This problem is normally the result of a dirty air filter. The technician should change it and instruct the homeowner of the importance of keeping it clean. If the filter was dirty for a long time and problems persist after it is replaced, consider cleaning the coil. Proper airflow across the outdoor coil is equally important. Low airflow across the outdoor coil or evaporator in the winter causes liquid slugging and will also destroy the compressor. If the coil has an excessive amount of ice buildup, and the defrost controls are working properly as designed, adjust the defrost controller to a shorter duration to initiate defrost more frequently. Again, if the ice accumulation problem continues, consider cleaning the outdoor coil.

to the high-pressure subcooled liquid. To avoid the restriction of the pressure-reducing device connected to the outlet of the outdoor coil, a check valve is connected—allowing the liquid to bypass the pressure-reducing device.

The liquid refrigerant travels to the indoor-coil (evaporator) pressure-reducing device. Forced to flow through the device because the check valve (connected in parallel with the pressure-reducing device) has closed, the necessary pressure drop occurs to produce the heat absorption in the indoor coil. The refrigerant vapor produced in the indoor coil now travels through the reversing valve and accumulator to the compressor. The cycle is complete.

In the heat cycle (Fig. HP42-3) the reversing valve has changed position, changing the direction of gas flow. The high-pressure/high-temperature gas from the compressor flows through the reversing valve to the indoor coil. This coil now acts as a condenser, rejecting heat into the air from the conditioned area. The vapor is condensed, and the temperature is reduced to produce a high-pressure/medium-temperature subcooled liquid.

To remove the cooling-cycle pressure-reducing valve from the circuit, the check valve opens and allows the liquid refrigerant to flow around the pressure-reducing device. Continuing on through the liquid line, the liquid refrigerant is forced through the outdoor-coil pressure-reducing device by the closing of the check valve connected in parallel with the pressure-reducing device. Converted to a low-pressure/low-temperature liquid, the refrigerant flows into the outdoor coil, which now acts as an evaporator. Heat is picked up from the outdoor air by evaporation of the liquid refrigerant. This refrigerant vapor then flows through the vapor line, reversing valve, and accumulator to the compressor. The cycle is now complete.

Accumulators are recommended for refrigeration and air-conditioning systems; they are a necessity in heat pump systems. This will be covered more thoroughly under the subject of system defrost.

HP42-3.1 Air-Source Systems

The basic cycle is an *air-to-air heat pump* where heat is taken from and given up to air. In this system two air-type tube-and-fin heat exchangers are used. In the cooling cycle, the indoor coil collects heat from the indoor air supply and the outdoor coil gives up heat to the outdoor air. In the heating cycle, the indoor coil gives up heat to the indoor air-distribution system, and the outdoor coil collects heat.

A typical example of an air-to-air system is shown in Fig. HP42-4. The blower assembly in the fan-coil unit and the supply-and-return duct system provide the air through the inside coil. The propeller fan assembly in the outdoor section provides the air through the outdoor coil.

HP42-3.2 Water-Source Systems

Water-to-air heat pumps use a liquid heat exchanger in the high side or outdoor section of the heat-pump assembly. This section should be located in the heated area to prevent possible freeze-up if the outdoor ambient temperature should drop below 32°F. If a part of a packaged unit, it would be located in the conditioned area.

A typical water-to-air unit, with the casing panels removed, is shown in Fig. HP42-5. In its lower compartment, the double-tube water-to-refrigerant heat exchanger can be seen. Due to the variety of water conditions encountered, these heat exchangers are made of cupronickel or other corrosion-resistant material. The construction is efficient and compact, called tube-in-tube or coaxial. The rest of the refrigeration system (compressor, pressure-reducing devices, check valves, and reversing valve) is basically the same as in the air-to-air system.

HP42-3.3 Water-to-Water Systems

These types of heat-pump systems have been marketed in small unit sizes for water heating. They are mostly in the

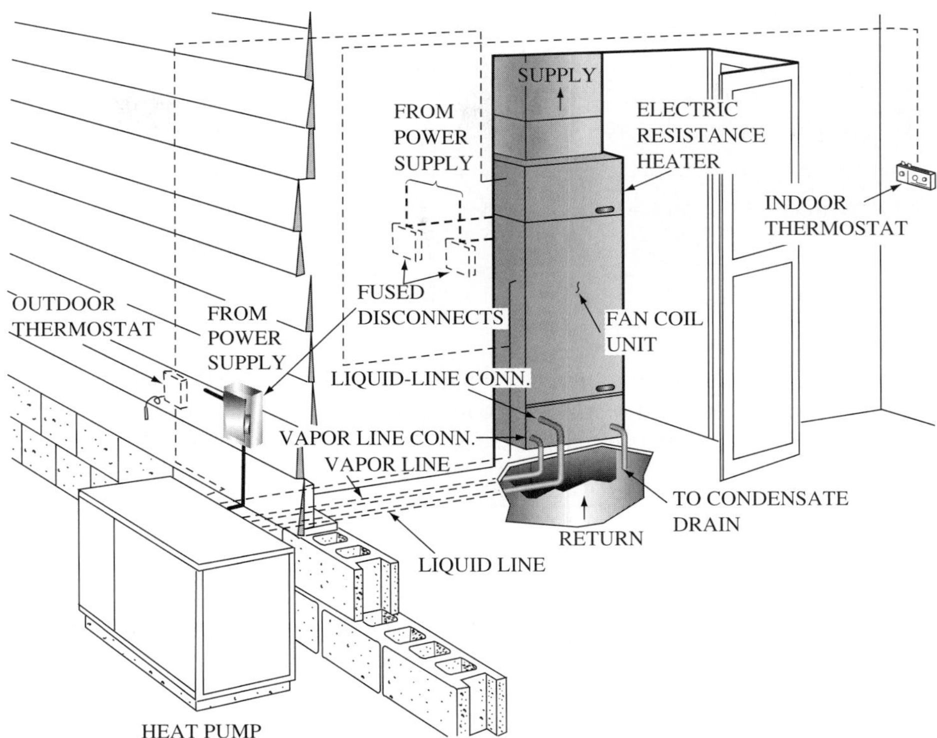

Figure HP42-4 Split-system heat pump.

capacity range of 6,000 to 12,000 Btu/hr and may be incorporated into a package unit along with a hot-water storage tank or as an add-on unit for existing storage tanks. Fig. HP42-6 shows a typical package unit that can be added to a hot-water heater. The unit becomes the primary heat source for domestic or small commercial hot-water use, with the heat source of the water heater as backup. The higher efficiency rating or COP (coefficient

Figure HP42-5 Water-source heat pump. (Courtesy of Bard Manufacturing Company)

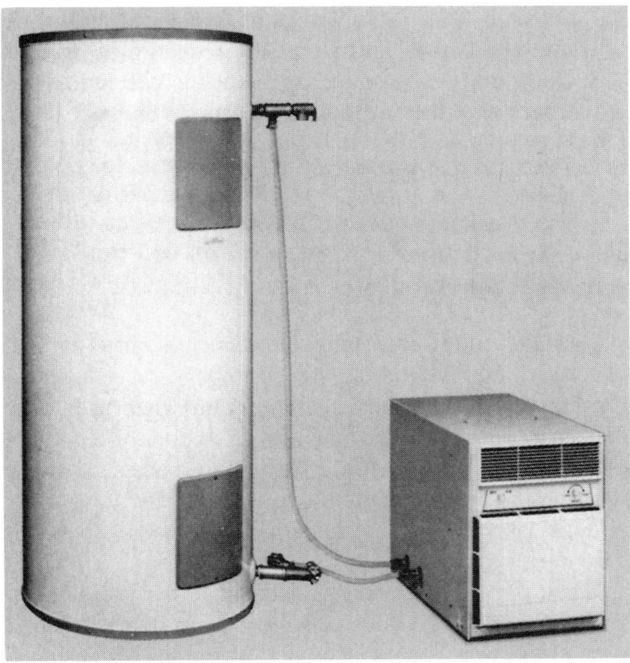

Figure HP42-6 Water-to-water heat pump. (Courtesy of York International Corp.)

ON THE JOB

If you go to a job site during the winter season and notice the outdoor coil excessively iced over, suspect defrost control problems. If the customer informs you that the outdoor unit runs even when it is below 10°F, check the unit's balance point. If the unit is operating below the balance point, not defrosting properly at low temperatures, and the controls are working properly, then install a low ambient lock-out control circuit to prevent compressor operation.

of performance) of the water-to-water heat pump will usually supply hot water at a lower energy cost. Installation and service are covered in later chapters.

HP42-4 COMPONENT PARTS

The basic heat-pump system is composed of two heat exchangers, a compressor for raising the refrigerant boiling point or condensing temperature, and pressure-reducing devices for lowering the refrigerant boiling point. Because the system operates at different temperature conditions in the heating and cooling modes, the pressure-reducing devices have to be matched to the coil that will act as the evaporator for the conditions to be encountered. For example, the pressure-reducing device on the inside coil must respond to air conditions entering the coil at 80°F DB and 50% RH as well as 95°F DB entering the outdoor coil (the condenser). The pressure-reducing device used in conjunction with the outdoor coil, when it is the evaporator on the heating cycle, must respond to 45°F air entering the coil with 70°F air entering the inside coil, which is now the condenser. Thus, pressure-reducing devices as well as the coils on heat exchangers are expected to operate differently than a straight air-conditioning system. Other components, having specific functions and not found in air-conditioning systems, are added to the heat pump.

HP42-4.1 Heat Exchangers

The inside coil is basically the same in air-to-air and water-to-water systems. Designed to act as a top feed of hot gas and a bottom outlet for liquid when used as a condenser, the coil then operates as a bottom-feed type of evaporator coil. Also, to keep flow resistance down when in the heating mode, these coils usually have more circuits carrying lighter loads than those carried by standard coils.

Their overall size and heat-exchanger surface area are larger than standard coils because their primary function is heat rejection as a condenser. Heat pumps are designed with heating as their prime function and cooling secondary. This has reduced field problems.

The standard heat pump operates like an air-conditioning unit with an oversized coil because of its larger size per 12,000 Btu/hr capacity. The total Btu/hr capacity may be higher, but the latent capacity is greatly reduced. As a result, higher humidity levels in the conditioned space will result from the heat-pump operation compared to standard air-conditioning systems.

Fig. HP42-7 shows the capillary tubes used for boiling-point reduction when the coil acts as the evaporator on the cooling cycle. Behind the liquid distributor feeding the capillary tubes is the check valve. On the heating cycles, when the coil acts as the condenser and the refrigerant flows in the reverse direction this valve opens to remove the capillary tube restriction from the circuit.

HP42-4.2 Air Source Outside Coils

The outside coil is also a top-feed/bottom-outlet coil operating as a condenser on the cooling cycle. Normally, condensers are circuited from top to bottom without regard to circuit loading because the operation is not affected by

Figure HP42-7 Heat pump style "A" inside coil. (Courtesy of Bard Manufacturing Company)

Figure HP42-8 Outdoor section of a heat pump. (Courtesy of Bard Manufacturing Company)

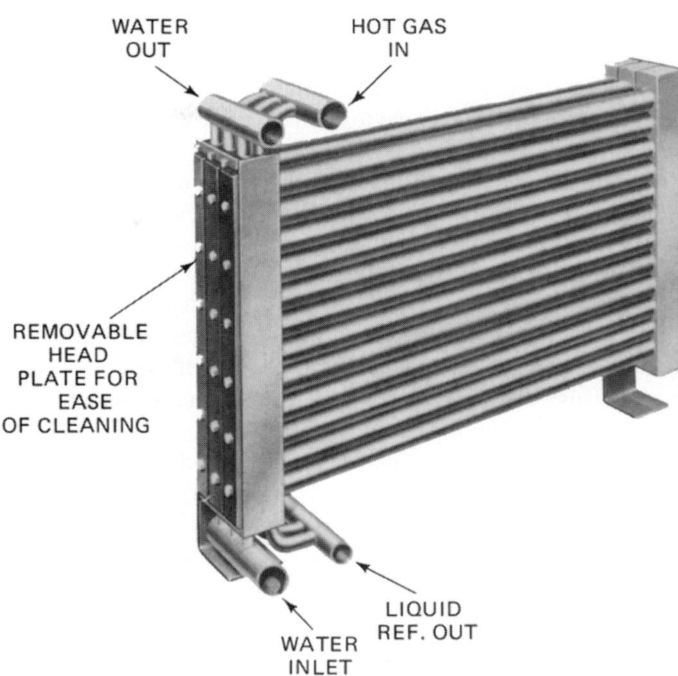

Figure HP42-9 Water-to-refrigerant coil. (Courtesy of Standard Refrigeration Company)

any circuit load unbalance. When used as the evaporator on the heating cycle, however, it must be circuited properly for highest capacity. Fig. HP42-8 shows an outside coil that has been circuited as a stacked double coil. Refrigerant flows from the pressure-reducing device into the bottom row of each section. Its section is then multiple circuited to provide the highest heat-absorbing capacity at the lowest possible pressure drop through the coil. In addition to the stacking of two sections, each section is also cross circuited to promote balanced loading through each section.

HP42-4.3 Water Source Outside Coils

The outside coil in water-source heat pumps are tube-in-tube heat exchangers that transfer heat from the compressor's high-pressure/high-temperature discharge vapor to the water in the cooling cycle. Heat is then transferred from the water to the low-pressure/low-temperature liquid refrigerant in the heating cycle.

These heat exchangers are usually either a stacked tube-in-tube (Fig. HP42-9) or a continuous tube-in-tube called a coaxial (Fig. HP42-10). In either case, the amount of surface involved is adequate to transfer the required amount of heat.

The stacked tube-in-tube condenser has removable head or end plates for cleaning the individual water tubes. When used in a water-source heat pump, capillary tubes are inserted through the liquid-out manifold to supply low-pressure/low-temperature liquid refrigerant into the bottom of each refrigerant circuit during the heating cycle.

Figure HP42-10 Water-source heat pump, with panels removed. (Courtesy of Bard Manufacturing Company)

The water flow through the device is reversed to provide the maximum capacity. Water enters the top and leaves through the bottom. Some sacrifice of capacity is necessary on the cooling cycle to produce peak capacity on the heating cycle.

Continuous tube-in-tube (coaxial) heat exchangers are chemically rather than mechanically cleanable. Fig. HP42-10 shows a coaxial heat exchanger in the right-hand side of the lower compartment. Connections with water-pressure gauge taps provide means of determining the limiting factor of the water tube. Another tube-in-tube heat exchanger is shown around the base of the compressor. This heat exchanger is used for heating domestic hot water on the cooling cycle.

In water-to-water systems, both heat exchangers would be of the coaxial type, with the necessary controls and water supply pumps to operate properly.

HP42-4.4 Reversing Valve

The selection of the heat-exchanger function to pick up heat (evaporator) or give up heat (condenser) is determined by the action of the reversing valve. A typical reversing valve is shown in Fig. HP42-11. Internally, it is composed of two pistons connected to a sliding block or cylinder with two openings. The diagrams (Figs. HP42-12 and HP42-13) show the position of the piston assembly in the heating mode and in the cooling mode.

The action of the piston is controlled by a solenoid valve that uses high-pressure compressor discharge vapor to move the piston left or right, depending on which mode is needed. With the compressor discharge connected directly to the center of the piston chamber, equal pressure is exerted on the internal surfaces of each of the piston ends. To create movement, a pressure difference is produced across the piston by bleeding the cylinder pressure into the suction side of the compressor.

The bleeding action is controlled by the action of the two-way solenoid valve. In Fig. HP42-12 the solenoid valve is de-energized and the control plunger is relaxed in the bottom position. The bottom port is closed and the top port is open to the middle common port. Thus, the top vent line is open to the equalizing line. The action that took place was the bleeding off of the pressure from the left cylinder chamber. When the pressure in this chamber was reduced sufficiently to cause the pressure difference across the piston to move the piston (usually 75 to 100

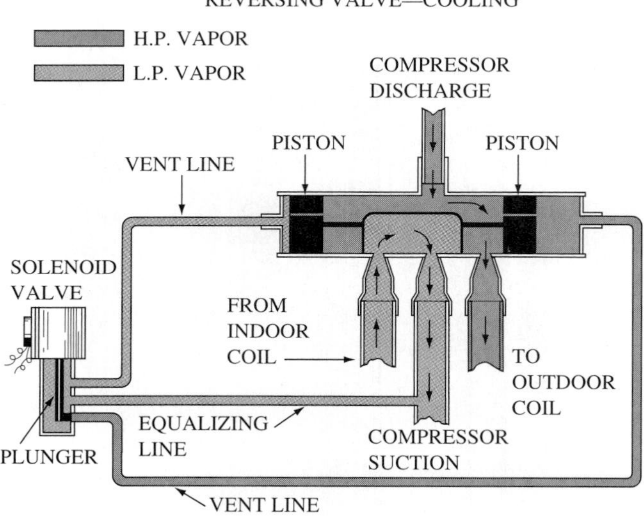

Figure HP42-12 Reversing valve in cooling position.

Figure HP42-11 Heat pump reversing valve.
(Courtesy of Ranco North America)

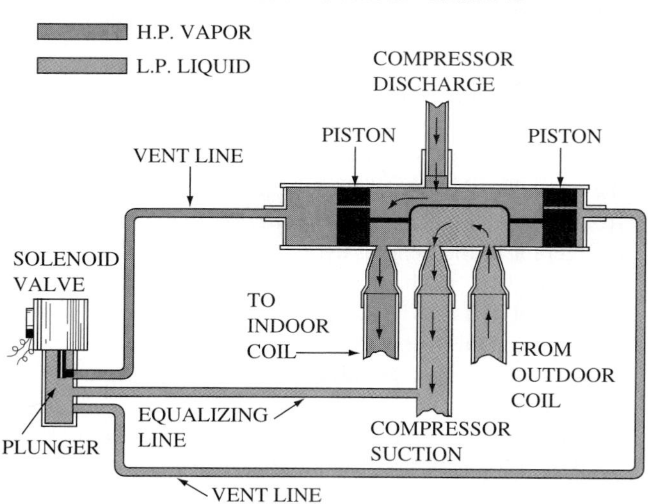

Figure HP42-13 Reversing valve in heating position.

psig), the piston traveled to the left. This caused the valve slider to open the left control port to the center port and the right control port to the piston pressure chamber. In this position the compressor discharge vapor travels to the indoor coil (condenser) and the vapor from the outdoor coil (evaporator) travels to the compressor suction.

In Fig. HP42-13 the solenoid valve coil is energized and the plunger has been lifted. This action has closed the vent line from the left end of the piston and opened the vent line from the right end of the piston. When the pressure in the right end of the piston has dropped sufficiently, the pressure difference across the piston will force the piston to the extreme right end.

This action moves the slide valve, opening the outdoor coil to the compressor discharge and the indoor coil to the compressor suction. The unit is now performing as a standard air-conditioning system. Not shown is a tip on the end of the piston which seals into a seat to prevent continuous bypass of the hot vapor from the cylinder through the piston bleed hole into the open vent line.

The piston bleed hole is a small opening through which the high-pressure gas can slowly find its way to regulate the speed of the piston travel. In the crossover action, too rapid a change in pressure could result in system shock and excessive noise.

HP42-4.5 Check Valve

The check valves in the heat pump circuit are vital to ensure that proper pressures and refrigerant boiling points are maintained to provide the heat absorption and transfer at the rated capacity. This is easily accomplished as these valves are either fully open or fully closed, depending on the direction of refrigerant flow. A swing-type check valve with a directional flow arrow is shown in Fig. HP42-14. When the refrigerant flows in the direction of the arrow, the valve opens and allows full flow with minimum resistance.

When the refrigerant is reversed, however, the check plate swings closed and stops the refrigerant flow. Thus, this action forces the refrigerant to flow through the pressure-reducing device and proper operation is obtained.

The first check valves used in residential and small commercial heat pumps were of the disk type. This valve used a flat disk over a circular flat set with a light pressure spring to aid in closing and still cause very little resistance to refrigerant flow. In normal situations, where moderate pressure forced the refrigerant through the check valve and little reverse pressure was exerted when the system was idle, the check valve performed adequately. In heat pump systems, where radical pressure differences occur rapidly, the disk would be forced to one side and hang up. This prevented the shut off of reverse refrigerant flow.

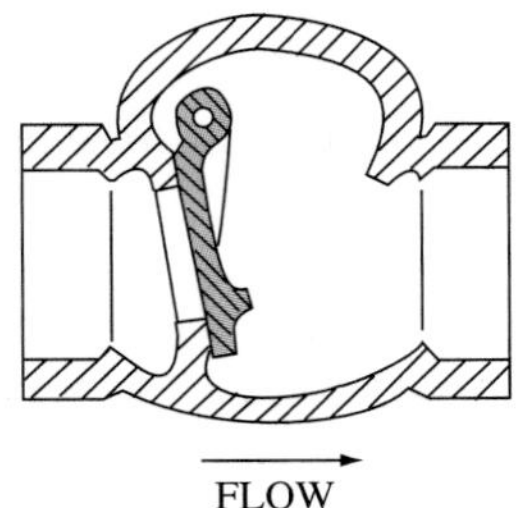

Figure HP42-14 Check valve.

As an improvement to the original version, a check valve using a steel ball instead of a flat disk was developed. This type of check valve can withstand the heavy reversing of refrigerant flow pressure, especially when the unit completes the defrost cycle.

Usually, two check valves are incorporated in each system, one connected in parallel with each pressure-reducing device on each of the inside and outside coils.

HP42-4.6 Pressure-Reducing Devices

As in standard air-conditioning systems, the two types of pressure-reducing devices used in heat pumps are thermostatic expansion valves and capillary tubes.

Capillary tubes in heat-pump systems are the same as those in air-conditioning systems when applied to the indoor coil. Those on the outdoor coil are sized for altogether different conditions of pressures and temperatures. On the heating cycle, the heat pump operates with 70°F air entering the condenser and –20 to 65°F air entering the evaporator. Manufacturer's specifications should be used for the size and length of the capillary tubing.

Thermal expansion (TXV) valves used in heat pumps are not interchangeable with those used in standard air-conditioning systems. The feeler bulb of the heat-pump TXV valve is clamped to the cool suction line in one operating mode and to the hot-gas line in the other mode. Thus, the valve must be able to operate when the coil is connected to the evaporator and yet stand the high temperatures of the hot-gas line when the coil is connected to the condenser. The power element of the valves has a special pressure-limiting charge to provide this range of operation.

HP42-4.7 Accumulators

Accumulators protect the compressor from liquid runout escaping the evaporator during light loads or air reduction problems. They are connected between the suction outlet of the reversing valve and the suction inlet of the compressor, shown in the top view of an outdoor section (Fig. HP42-15).

In the heat pump, the accumulator has three situations where compressor protection is needed:

1. Flood-back on the cooling cycle: if an air restriction causing light load and liquid runout occurs.

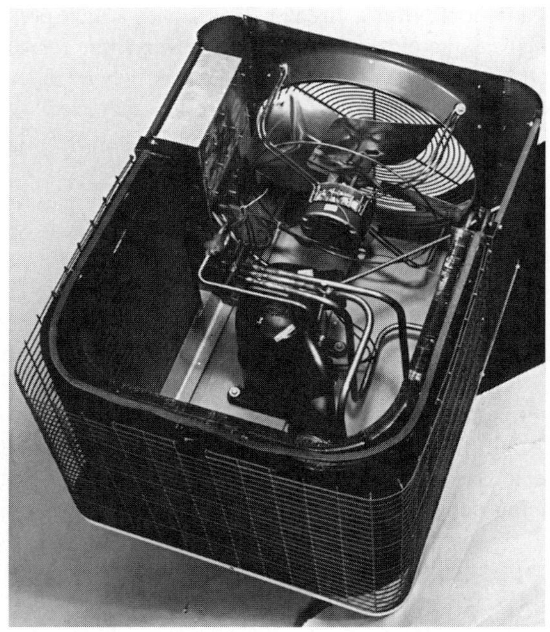

Figure HP42-15 Interior top view of the outdoor section of a heat pump. (Courtesy of Amana Refrigeration, Inc.)

2. Flood-back on the heating cycle: when excessive frost buildup on the outdoor coil occurs or if air troubles occur and cause liquid runout.
3. Termination of the defrost cycle: liquid flood-back will always occur when the defrost cycle is terminated.

To melt the frost and ice off the outdoor coil, the heat pump employs the hot-gas method. The system action is reversed, heat is picked up in the inside coil, is raised in pressure and temperature by the compressor, and is forced into the cold outside coil. With the outside blower or fan off, the coil rapidly heats up to melt the frost and/or ice on the outside surface of the coil. While doing this, the coil fills with condensed liquid refrigerant. Before the defrost operation is completed, and the liquid out of the bottom of the condenser reaches the proper termination temperature (55°F), pressure in the condenser can reach 350 psig and a temperature of 142°F.

When the defrost cycle terminates and the reversing valve switches over, this high pressure is relieved into the suction side of the compressor. Immediately, vapor will form deep in the condenser circuit and force liquid refrigerant out of the coil into the suction line. This action can be compared to removing the cap on the automobile radiator when the engine overheats.

Without the accumulator to catch and hold the liquid refrigerant, a liquid surge into the compressor can ruin valves. The protection of an accumulator in the heat-pump system has encouraged variations in the amount of refrigerant used, which is not be to tolerated. The refrigerant charge of a heat pump system is very critical.

HP42-4.8 Refrigerant Lines

The refrigerant lines in a heat-pump system are not the same as in an air-conditioning or refrigeration system, where the refrigerant always flows in the same direction: liquid refrigerant from condenser to pressure-reducing device; hot vapor from compressor to condenser; and cold vapor from the evaporator to the compressor.

In heat pumps the refrigerant flow reverses, depending on the operating mode. The liquid line is always the liquid line, regardless of the operating mode. It carries liquid refrigerant from condenser to evaporator in either mode because the heat exchangers change operating characteristics. The liquid pressure in this line is therefore always the high-side liquid operating pressure.

The large vapor line from the inside coil to the outside heat-pump section has a dual role. It is the cold-vapor line (suction line) in the cooling mode and the hot-vapor line (hot-gas line) in the heating mode. To check the operating pressures, the service technician must remember to connect a high-pressure gauge to this line to prevent damage to the gauge set.

The only sections of the refrigerant circulating systems that have single duty are: (1) the line between the reversing valve and the compressor inlet (suction line), and (2) the line from the compressor outlet to the reversing valve (hot-gas line). To be able to determine problems in the refrigerant system, four gauges are needed: (1) a compound gauge connected to the suction line, and standard gauges connected to the (2) hot gas line, (3) the liquid line, and the (4) vapor line to the outside coil.

In air-conditioning systems, the dual-purpose vapor line (suction line) is insulated to promote compressor life and reduce line sweating. In heat pumps, insulation of this dual purpose vapor line is also a necessity, to reduce heat loss between the compressor and the inside condenser. Lack of insulation can produce a loss in system capacity as high as 20%.

REVIEW

- Heat pump—A refrigeration system, that transfers heat from one place to another by the change of state of a liquid. Operates as a reverse-cycle air conditioning, heating a house in the winter, and cooling in the summer. Developed by Lord Kelvin in 1852.
- First applied heat pump was installed in Scotland in 1927.
- Between 1927 and 1950, heat pumps were installed in hundreds of residential and small commercial applications throughout Europe and the southern part of the United States.
- The rise of energy costs going into the 1970s, and the demand for an efficient means of heating with electrical energy, brought about a rebirth to the heat pump. The

units were redesigned for colder climates, and defrosting was added, as well as cold-weather compressor protection.

- Heat-pump cycles:
  Heating cycle
  Defrost cycle
  Cooling cycle
- Types of heat pumps:
  Air source
    Air-to-air
  Water source
    Water-to-air
    Water-to-water
- Heat pumps are designed with heating as their prime function and cooling secondary.
- Components of an air-to-air heat pump:
  Inside coil
  Outside coil
  Compressor
  Accumulator
  Check valves
  Reversing valve
  Pressure-reducing devices, with liquid distributor
- Air-source systems—Two tube-and-fin heat exchangers (coils) are used.
- Water source systems—Outside coil is a tube-in-tube heat exchanger, either stacked or continuous (coaxial).
- Water-to-water systems—Both heat exchangers (coils) are the coaxial type.
- Reversing valve—Solenoid-actuated, sliding block or cylinder, with two openings, energized for heat cycle, de-energized for cooling.
- Refrigerant lines—Refrigerant flow reverses in the liquid line, but the larger vapor line serves a dual purpose; it is a suction line in the cooling mode, and a hot-vapor line in the heating mode.

Problems and Questions

1. The heat pump was developed by _____ in _____.
2. Describe the difference in operation between an air-conditioning system and a heat pump.
3. The title "heat pump" has been given to a unit that is actually a _____.
4. The proper names for the two coils in the heat pump system are the _____ and _____ coils.
5. In the cooling mode, the evaporator is the _____ coil.
6. In the heating mode, the condenser is the _____ coil.
7. The change from heating mode to cooling mode, and vice versa, is accomplished by a device called a _____.
8. To protect the compressor from refrigerant surge, an _____ is used.
9. A tube-in-tube heat exchanger is called a _____ type.
10. Water-to-air heat pumps are always located in the heated area _____.
11. Generally, water-to-refrigerant heat exchangers used in water-to-air heat pumps are made of _____. Why?
12. The inside coil in heat pumps is always larger than the coil used in air conditioners because _____.
13. What effect does the larger coil have on the results in the cooling mode?
14. A reversing valve consists of two parts: the _____ and the _____.
15. The main valve in a reversing valve is moved by the pressure created by the _____.
16. The minimum pressure required to operate the main valve in a reversing valve is _____ to _____ psig.
17. The pilot valve operates the main valve by controlling the pressure applied to the valve. True or false?
18. The assembly consisting of the pressure-reducing device and the check valve is called a _____.
19. Two types of check valves have been used in heat pumps. What are they?
20. Which type of check valve has been the least trouble?
21. Name two types of pressure-reducing devices used in heat pumps.
22. A TXV valve used in heat pumps is not interchangeable with air-conditioning system valves because _____.
23. What are the three situations where accumulators are needed to protect the compressor?
24. Termination of the defrost cycle occurs when the liquid off the outside coil reaches _____ °F.
25. During the defrost cycle, condensing pressures and temperatures can reach _____ psig and _____ °F.
26. In the heat-pump system, the lines in which the refrigerant changes direction are the _____ and the _____.
27. The suction line is connected between the _____ and the _____.
28. The hot-gas line is always connected between the _____ and the _____.
29. In a heat pump, the two main purposes for insulation on the vapor line are _____ and _____.
30. The outdoor coil is the condenser in the heating cycle. True or false?
31. Water-to-air heat pumps use a liquid heat exchanger in the high side or outdoor section of the heat-pump assembly. True or false?
32. Water heat exchangers are made of cupronickel material. True or false?
33. Water-to-water systems are mostly in the
    a. 6,000 to 12,000 Btu/hr range.
    b. 12,000 to 18,000 Btu/hr range.
    c. 2 to 5 ton range.
    d. None of the above.

34. The reversing valve, to create movement, utilizes a
    a. floating piston.
    b. pressure difference across the piston by bleeding the cylinder pressure into the suction side of the compressor.
    c. two-way solenoid valve.
    d. All of the above.

35. The heat pump accumulator protects against:
    a. flood-back on the cooling cycle.
    b. flood-back on the heating cycle.
    c. termination on the defrost cycle; liquid flood-back will always occur when the defrost cycle is terminated.
    d. All of the above.

Air- and Water-Source Applications

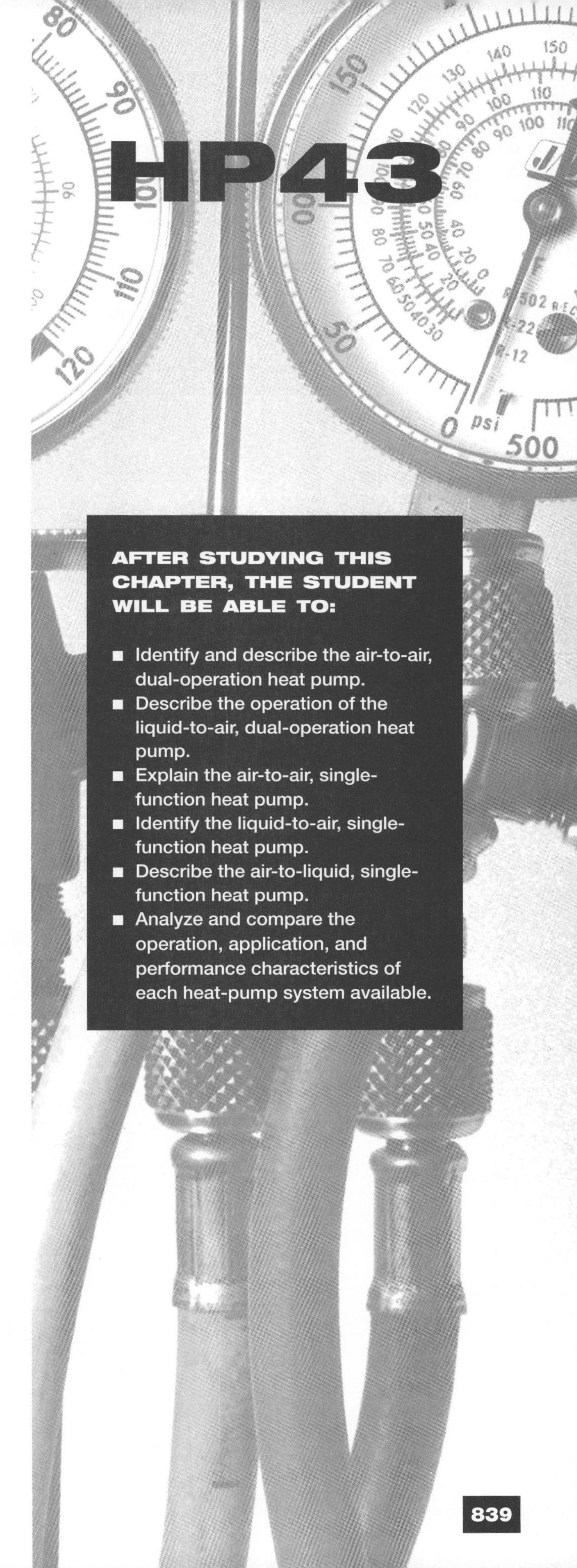

HP43

AFTER STUDYING THIS CHAPTER, THE STUDENT WILL BE ABLE TO:

- Identify and describe the air-to-air, dual-operation heat pump.
- Describe the operation of the liquid-to-air, dual-operation heat pump.
- Explain the air-to-air, single-function heat pump.
- Identify the liquid-to-air, single-function heat pump.
- Describe the air-to-liquid, single-function heat pump.
- Analyze and compare the operation, application, and performance characteristics of each heat-pump system available.

HP43-1 OVERVIEW

Heat-pump equipment is marketed in both *unitary* (complete package) and *split systems* (portion outdoors and portion indoors). The appearance of the equipment is exactly the same as that of its air-conditioning counterpart. The difference is in the design of the two coil assemblies involved, as well as the extra refrigerant flow controls needed for the reverse action.

The predominant product has been the package type, but the split or remote type has rapidly gained popularity due to its ease of installation and versatility. In both designs, the types of heat source are duplicated. The heat source used as well as reversing ability determines the general classification of the particular unit.

Heat pumps are classified by their operational systems as follows:

1. Air-to-air, dual-operation
2. Liquid-to-air, dual-operation
3. Air-to-air, single-operation
4. Liquid-to-air, single-operation
5. Air-to-liquid, single-operation

HP43-2 AIR-TO-AIR, DUAL-OPERATION

The *air-to-air dual-operation heat pump* is the predominant type of unit sold. Its capability to both heat and cool makes it a year-round application. Its popularity increases as energy costs increase. The air-to-air heat pump is marketed in both a package type, with all components in one package, and a split type, with an air handler in the area and the compressor/heat-exchanger assembly outdoors.

Fig. HP43-1 shows a typical self-contained or package heat pump, including a complete cabinet view as well as an interior view of the refrigerant circuit and electrical components. Included in the package view are the optional electric heat strips located in the blower discharge area. Some type of auxiliary heat is required except in semitropical climates such as that in southern Florida. The easiest type to incorporate into the heat-pump assembly are electric strip heaters.

A split system includes both the indoor air handler with an electric heat strip and the outdoor section. The electric heat strip is located in the upper section of the blower discharge area, behind the control assembly and enclosing panel.

Both types of equipment have their advantages and disadvantages. The package unit has no refrigerant-line installation cost but is limited in application situations. The split system can be used in a wide range of vertical, horizontal, or down-flow applications but installation cost is higher.

(a)

(b)

Figure HP43-1 Packaged heat pump. (a) External cabinet; (b) Internal components. (Courtesy of Bard Manufacturing Company)

KNOWLEDGE EXTENSION

Air- and ground-coupled heat pumps have gained immense customer popularity in recent years due to increased energy costs, and positive reports from consumers happy with their systems. The need for skilled technicians to install and service heat pumps has forced associations such as the Refrigeration Service Engineers Society (RSES) and the Air Conditioning Contractors of America (ACCA) to offer specialized heat-pump certification programs. The Air Conditioning and Refrigeration Institute (ARI) has also been providing education and guidance regarding heat pumps through the ARI Equipment Manufacturer Standards. Contact your local RSES or ACCA chapters and ARI to learn more about these certification programs.

HP43-3 LIQUID-TO-AIR, DUAL-OPERATION

A number of applications allow the use of a *liquid-to-air heat pump.* These uses are illustrated in Fig. HP43-2, as follows:

1. Water-loop heat-pump system
2. Groundwater heat-pump system
3. Surface-water heat-pump system
4. Closed-loop surface-water heat-pump system
5. Ground-coupled heat-pump system

The *water-loop system* uses water supplied by an evaporative cooler as a heat sink for heat-pump cooling and heating. Supplementary heat is furnished by a boiler when needed. Where freezing temperatures may be encountered, the water loop must contain an antifreeze solution. Some installations use a flat-plate heat exchanger between the outside and the inside water loop, to permit placing the antifreeze only in the outside loop.

The *groundwater system* uses well water as a heat sink for both heating and cooling. Most systems of this type return the water back to the ground through a re-injection well. Local codes must be examined relative to the permitted use and discharge of groundwater.

The *surface-water system* uses and returns water to a nearby lake, stream, or canal. Some of these systems use *closed-loop* arrangements where pipes or tubing are located on the surface of the water source to serve as a heat exchanger.

The *ground-coupled system* takes advantage of the massive thermal capacity of the earth which provides a temperature-stabilizing effect on the circulating water loop. The success of these installations requires a detailed knowledge of the climate, the site, the soil thermal characteristics, and the performance of the heat exchanger.

An upright self-contained unit is shown in Fig. HP43-3 and a horizontal type in Fig. HP43-4. This type of unit is located totally within the conditioned area. The only connection to the outside area would be to the water supply and disposal.

Fig. HP43-5 shows the interior parts arrangement of the vertical model shown in Fig. HP43-3. The upper section of the unit contains the inside coil with the necessary refrigerant controls as well as the blower and motor assembly for the inside air movement. The bottom compartment contains the motor/compressor assembly, the water-to-refrigerant heat exchanger, reversing valve, and the operating controls.

A liquid-to-air unit does not contain the defrost control system since it is not designed to operate with water-to-refrigerant heat-exchanger surfaces below 32°F. This unit employs a tube-in-tube or coaxial type of heat exchanger as well as a water-regulating valve to control the flow of water and a condensate pump for disposal of the inside coil condensate to drains elevated above the unit.

Fig. HP43-6 (p. 844) shows the interior parts arrangement of the horizontal unit shown in Fig. HP43-4.

HP43-4 AIR-TO-AIR, SINGLE-OPERATION

A single-operation heat pump provides a heating-only cycle. It therefore does not contain a reversing valve or the pressure-reducing device and check valve combination on the inside coil. It also does not have a check valve in conjunction with the pressure-reducing device on the outside coil. In areas where the cooling load is slight or nonexistent, the *air-to-air, single-operation unit* has a potentially strong market.

HP43-5 LIQUID-TO-AIR, SINGLE-OPERATION

The *liquid-to-air, single-operation unit* is designed for heating only in extreme northern locations, with water temperatures slightly above the freezing point. The water-to-refrigerant heat exchanger is larger, with more surface to provide the necessary heat transfer with very little temperature drop of the water through the heat exchange. Liquid-to-air units (dual operation) operate with an 8 to 10°F ΔT of the water through the heat exchanger. Liquid-to-air units (single operation) operate with only a 2 to 3°F

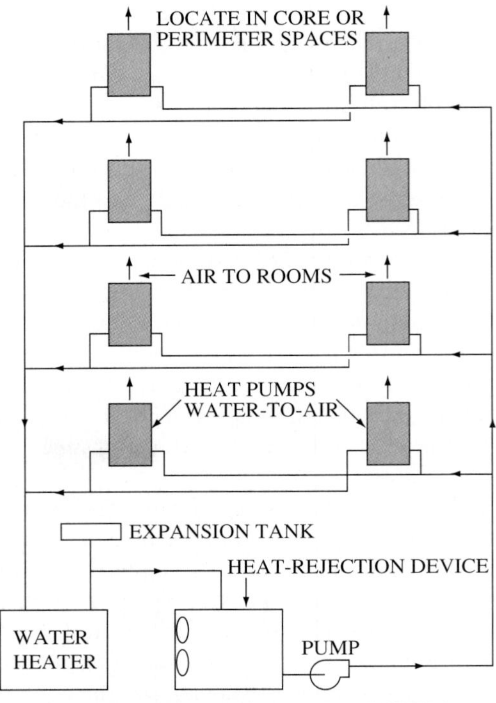

(1) WATER-LOOP HEAT-PUMP SYSTEM

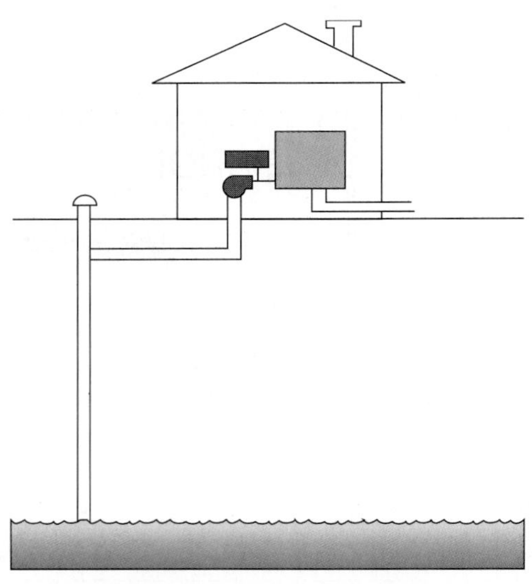
(2) GROUNDWATER HEAT-PUMP SYSTEM

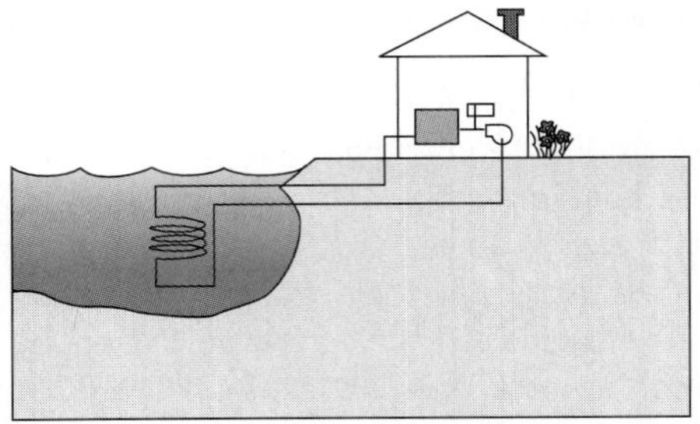
(3) SURFACE-WATER HEAT-PUMP SYSTEM

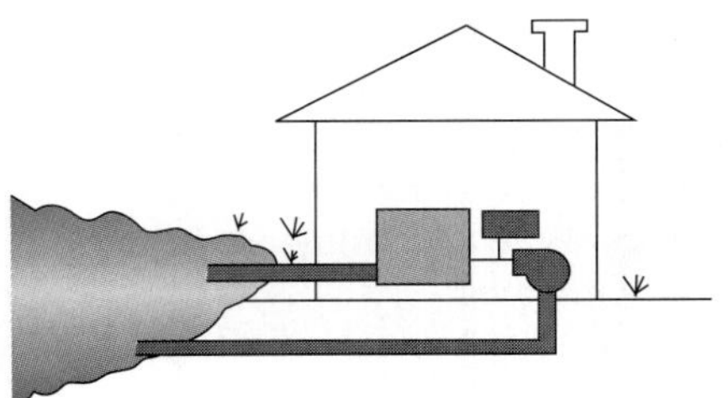
(4) CLOSED-LOOP SURFACE-WATER HEAT-PUMP SYSTEM

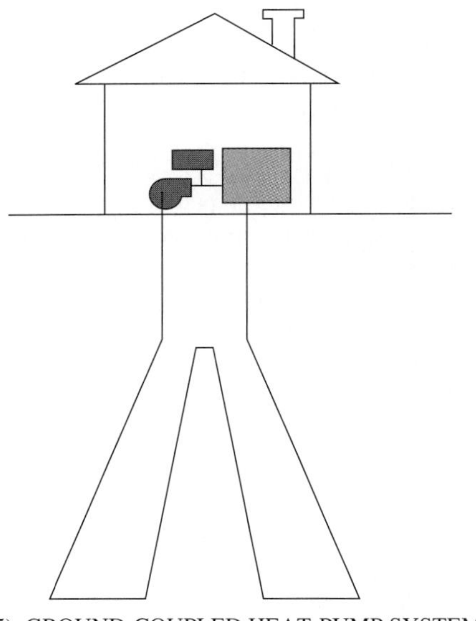
(5) GROUND-COUPLED HEAT-PUMP SYSTEM

Figure HP43-2 Various applications of the water-source heat pump. (Copyright by the American Society of Heating, Refrigerating, and Air-Conditioning Engineers. Used by permission.)

PRACTICING PROFESSIONAL SERVICES

Every weekday morning at 6 a.m. the furnace quit. It had a standard mercury thermostat. What was programming it?

The technician checked the gas pressure, micro-amp, did a complete check-test-start procedure—and found

The solution: The ground strap for the electrical panel was grounded to a cold water copper pipe. At 6 a.m. when the shower started, the copper sweated, lost the ground, lost the micro-amp signal, and went into lock-out.

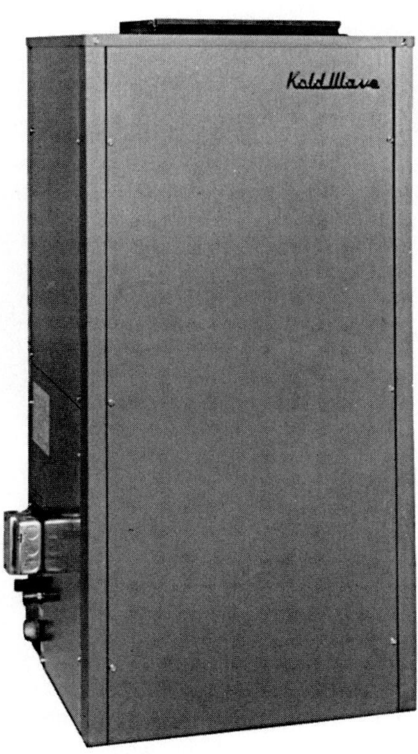

Figure HP43-3 Upright-style water-source heat pump. (Courtesy of Koldware Division of Mestek, Inc.)

Figure HP43-5 Upright-style water-source heat pump with panels removed. (Courtesy of Koldware Division of Mestek, Inc.)

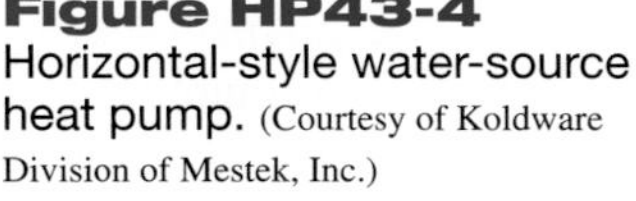

Figure HP43-4 Horizontal-style water-source heat pump. (Courtesy of Koldware Division of Mestek, Inc.)

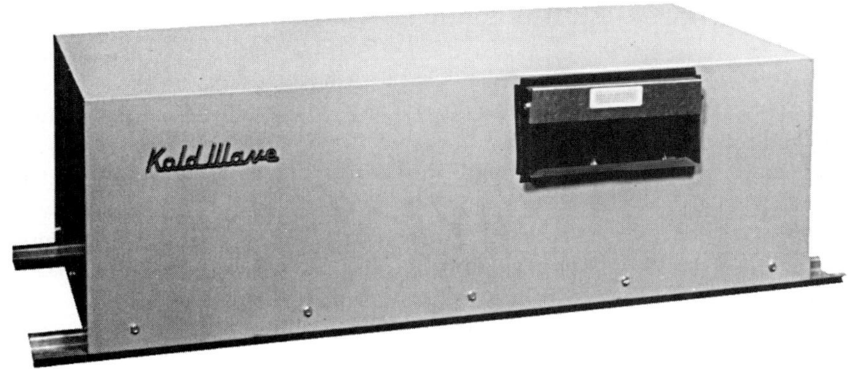

ON THE JOB

If the weather is below design temperature for an extended time but there are no calls from customers that their heating equipment cannot keep up with demand, the heating equipment has been oversized.

Efficient installations and lower costs result from:

1. The use of a recognized method of heat-loss/heat-gain calculations.
2. Heat-loss calculations performed by a qualified person.

Figure HP43-6 Horizontal-style water-source heat pump with panels removed. (Courtesy of Koldware Division of Mestek, Inc.)

ΔT in the water through the heat exchanger. Considerably more water is circulated, but the unit operates in the colder water zones.

HP43-6 AIR-TO-LIQUID, SINGLE-OPERATION

Most *air-to-liquid applications* have been for heating domestic hot water. The increase in the use of heated swimming pools has opened a market for large Btu/hr capacities in this type of heat pump. Fig. HP43-7 shows a smaller domestic water-heating unit. This unit is designed to sit on top of a standard electric water heater. This provides hot water by taking heat from the surrounding air and heating the water with an EER factor of 2.9—thus more hot water for less energy cost.

Some units are designed to be connected into the water-circulating system of the water heater. Shown in Fig. H43-8 is a remote type unit. In this application, plastic hose (copper or galvanized pipe where required by code) is used to connect the heat pump to the hot-water tank. The heat pump is set to maintain a higher water temperature than the settings of the thermostats of the gas or electric water heater. This means that the heat pump is the primary means of heating the water. The auxiliary electric or gas-heat source cuts in only if the heat pump cannot provide sufficient heating capacity for the amount of water drawn at that particular time.

Usually, the thermostat settings are 130°F for the heat pump and 120°F for the auxiliary heat source. The exact settings will vary with each type of equipment, and the manufacturer's instructions should be followed.

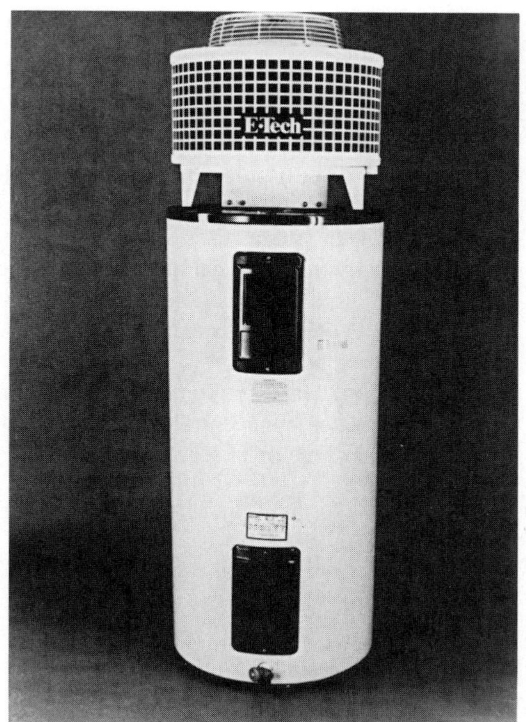

Figure HP43-7 Heat-pump water heater. (Courtesy of E-Tech, Inc.)

REVIEW

- Heat pumps are available in both unitary (complete package) and split systems.
- Heat pumps are classified by their operational systems:
  Air-to-air, dual-operation—Predominant type of unit sold, available in package or split system. Units come with auxiliary heat, normally electric heat.
  Liquid-to-air, dual—Water-loop heat pump, ground water, surface water, closed loop surface water, ground coupled. This unit takes advantage of massive thermal capacity of the earth, thus requiring large water or soil surface area.
  Air-to-air, single—Heating cycle only; no reversing valve. Applications include regions requiring slight or no cooling requirements.
  Liquid-to-air, single—Heating only; 2 to 3° delta temperature. Applications include extreme northern locations, with water temperatures slightly above the freezing point.
  Air-to-liquid, single—Heating only; hot-water heaters, swimming pools. Typically a domestic water-heater unit, it uses plastic hose, (copper or galvanized pipe where required by code) with auxiliary heat source to maintain thermostat settings when the unit cannot maintain sufficient heating capacity.

Figure HP43-8 Field piping for a heat-pump water heater used for supplementary hot water.

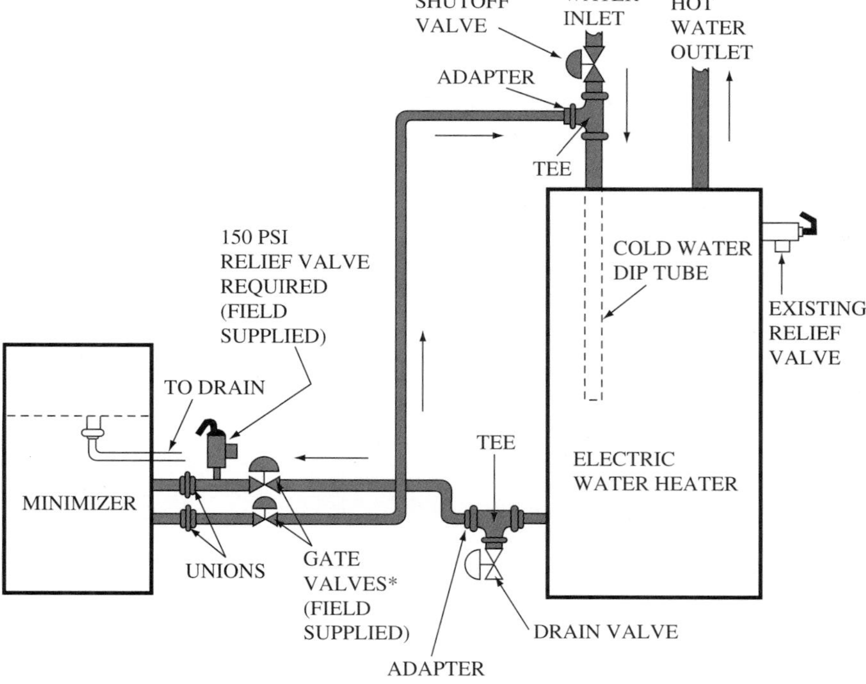

Problems and Questions

1. Heat pumps are produced in five classifications. What are they?
2. Air-to-air units are marketed in two types. What are they?
3. The most popular form of auxiliary heat is _____.
4. The advantage of a package-type unit is _____.
5. The advantage of the split-type system is _____.
6. Liquid-to-air units are manufactured in the same configurations as air-to-air with the same advantages. True or false?
7. Which type of unit is the easiest to use in a solar heating application?
8. Liquid-to-air units must be located in the conditioned area. Why?
9. On a liquid-to-air unit, two water-regulating valves are used. What is their purpose?
10. Liquid-to-air single-operation units operate with a 2 to 3°F ΔT of the water through the evaporator. Why is this?
11. In what market are air-to-liquid single-operation units primarily used?
12. The predominant heat-pump product is the air-to-air, dual-operation system. True or false?
13. The groundwater system uses well water as a heat sink for cooling-only applications. True or false?
14. The liquid-to-air, single-operation unit operates with an 8 to 10°F ΔT of the water through the heat exchanger. True or false?
15. The air-to-liquid heat-pump unit is applied primarily for heating water for central-heating distributed-piping systems. True or false?

Heat-Pump Controls

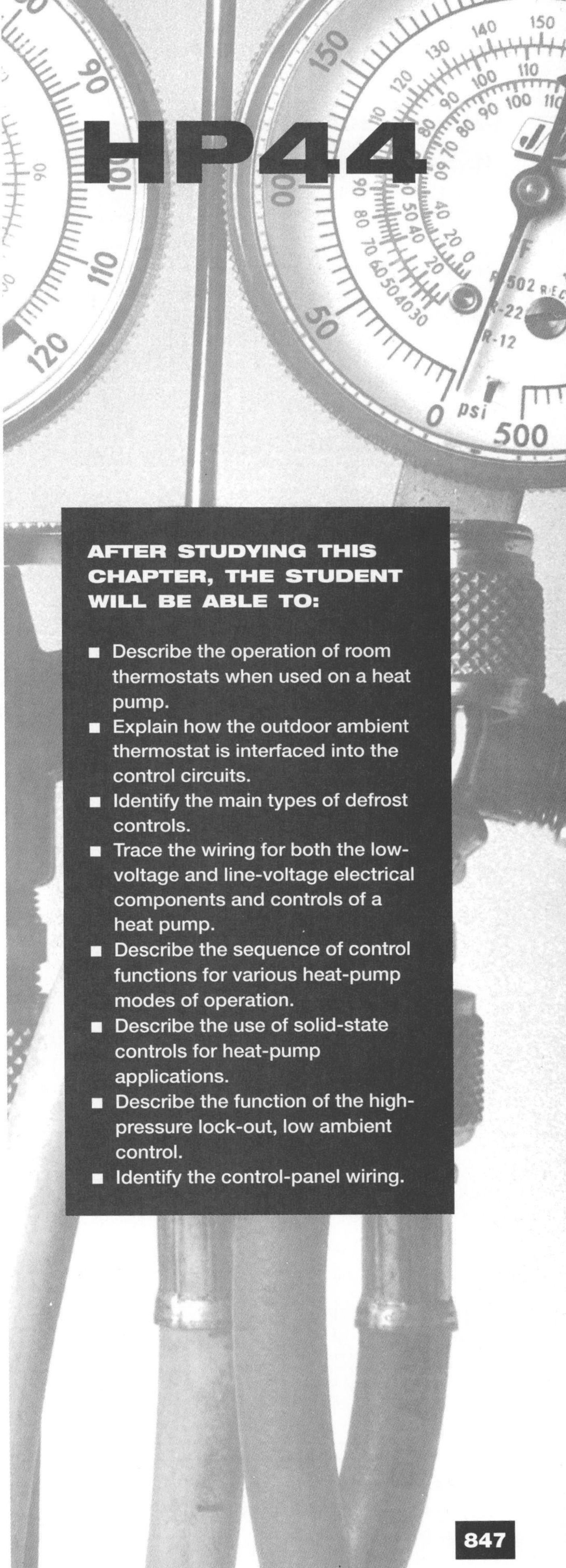

HP44

AFTER STUDYING THIS CHAPTER, THE STUDENT WILL BE ABLE TO:

- Describe the operation of room thermostats when used on a heat pump.
- Explain how the outdoor ambient thermostat is interfaced into the control circuits.
- Identify the main types of defrost controls.
- Trace the wiring for both the low-voltage and line-voltage electrical components and controls of a heat pump.
- Describe the sequence of control functions for various heat-pump modes of operation.
- Describe the use of solid-state controls for heat-pump applications.
- Describe the function of the high-pressure lock-out, low ambient control.
- Identify the control-panel wiring.

HP44-1 OVERVIEW

The purpose of this section is to describe and illustrate the operation of a typical control system for a heat pump. From previous sections we have learned that:

1. The heat pump is a unique type of conditioner which supplies both heating and cooling through its refrigeration cycle.
2. Improvements to components of the heat pump have extended its effective use to northern as well as southern applications. It is adaptable.
3. The heating output of the heat pump is directly related to the temperature: the greatest amount of heat is produced at the highest ambient temperatures. As the outside temperature drops, the heat output is reduced until supplementary heat is required to maintain design conditions.

To maximize the advantages of the heat pump, provision needs to be made in the control system to utilize the heat pump as much as possible during periods of high-efficiency heating and to use supplementary heat only when necessary.

The most common source of supplementary heat is electric heat, due to the ease in applying it and to its low equipment costs. Therefore, in the typical control system described in this section, the supplementary heat selected is electrical heat.

HP44-1.1 Controls

Controls will be discussed in the following categories:

1. Temperature controls using room thermostats
2. Defrost controls
   a. Pressure-time-temperature
   b. Defrost cycle operation
3. High-pressure lock-out
4. Low ambient control
5. Control-panel wiring

Knowledge Extension

Database For Chiller Performance/Service

Full information on how the machine is and was operating is essential to quality troubleshooting and servicing. Information on the following should be available from regular log entrees and taken at the time of servicing for objective analysis:

- Voltage
- Amperage
- Oil temperature in the sump
- Oil level
- Oil gauge pressure
- Evaporator gauge pressure
- Condenser gauge pressure
- Purge drum pressure (centrifugal)
- Condenser water-in temperature
- Condenser water-out temperature
- Refrigerant liquid temperature, condenser
- Chilled water pressure difference
- Chilled water-in temperature
- Chilled water-out temperature
- Refrigerant liquid temperature, evaporator
- Percent current

Two basic applications of thermostats exist in heat pump systems: the control of temperature, both heating and cooling, and the control of auxiliary heat as required.

HP44-1.2 Wiring Diagrams

Schematic wiring diagrams are used to present information on heat pump controls. The general layout of the diagrams is shown in Fig. HP44-1. There are three parts to the diagram: the left, middle and right parts.

1. The right part depicts line voltage and is used mainly for the refrigeration cycle to show the operation of the following loads: reversing valve, compressor motor, condenser-fan motor, defrost relay, timing motor, and auxiliary defrost relay.
2. The middle part depicts low voltage and is used mainly to show the operation of the thermostat circuits including the loads that the thermostat normally controls, as follows: compressor contactor, defrost control, reversing valve, fan relay, and supplemental heat.
3. The left part depicts line voltage and is used mainly for the supplementary electric heating and evaporator-fan motor. This part of the diagram also shows the low-voltage transformer.

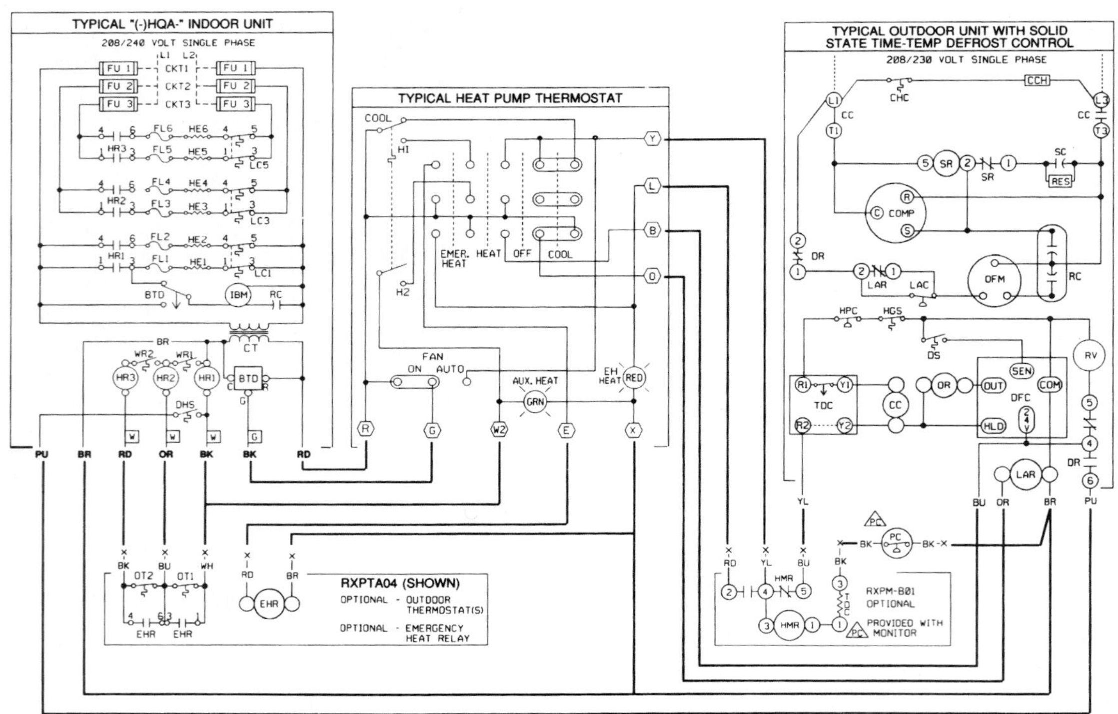

Figure HP44-1 Wiring diagram for a heat-pump system. (Courtesy of Rheem Air Conditioning Division)

ON THE JOB

On a retrofit job where you are installing a new heat pump to replace an older gas- or oil-fired furnace, it is important to instruct the homeowner on the normal operation of the new unit. Many homeowners will complain that the heat pump blows out cold air, but in reality, it is not cold air, just cooler air than their old furnace. If they continue to complain about drafts and cold air coming from the diffusers with their new heat pump, consider installing different diffuser covers to alter the air patterns into the room to offset the cool air feeling. Also instruct them not to switch the thermostat to emergency heat unless the heat pump does not warm the house in normal mode.

HP44-1.3 Control Functions

The schematic wiring diagrams have symbols, defined in the legends, to indicate the components of the control system and their function in the operation of the system. A number of diagrams are used so that various functions can be separated into easily understandable units that work together to create the automatic operation that the heat pump requires.

HP44-2 TEMPERATURE CONTROLS

HP44-2.1 Room Thermostats

The basic operation of a room thermostat was covered in a previous chapter. In heat-pump applications, thermostats perform the following functions:

1. Turn system on,
2. Turn system off,
3. Initiate cooling mode,
4. Initiate heating mode,
5. Initiate auxiliary heat stages, and
6. Allow for fan options.

A wiring diagram for a typical heat-pump thermostat that provides all these functions is shown in Fig. HP44-2. The system switch selects the system mode of operation: HEAT for heating-only operation; COOL for cooling-only operation. The fan switch in the AUTO position will cycle the indoor blower motor with the compressor during any mode of operation. In the ON position the inside blower motor will operate continuously regardless of the position of the system switch.

If the heat-pump system should fail, the occupant can substitute the backup heat or auxiliary heat for the system by moving the system switch to EMER. HEAT. This removes the compressor from the operation and uses backup heat as the primary heating source. At the same time an emergency heat light will energize and remains on as long as the switch is in the emergency position.

The diagrams that follow are highlighted to show specific functions of the heat pump. The first series of diagrams shows the functions initiated by various positions of the selector switch on the thermostat.

Position 1. Selector Switch OFF, Fan Switch AUTO

In position 1 (Fig. HP44-2), 240 V are supplied to the control transformer which will supply 24 V to the power (R) terminal of the thermostat. All switches in the thermostat control subbase are open and no function can result.

Position 2. Selector Switch OFF, Fan Switch ON

In position 2 (Fig. HP44-3), 24 V are supplied through the R circuit of the thermostat to the ON terminal of the fan

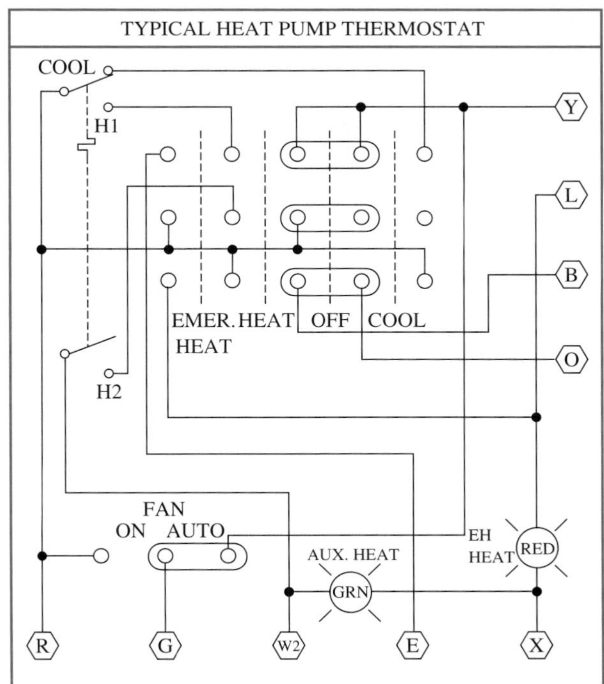

Figure HP44-2 Circuits activated, selector switch OFF, fan AUTO. (Courtesy of Rheem Air Conditioning Division)

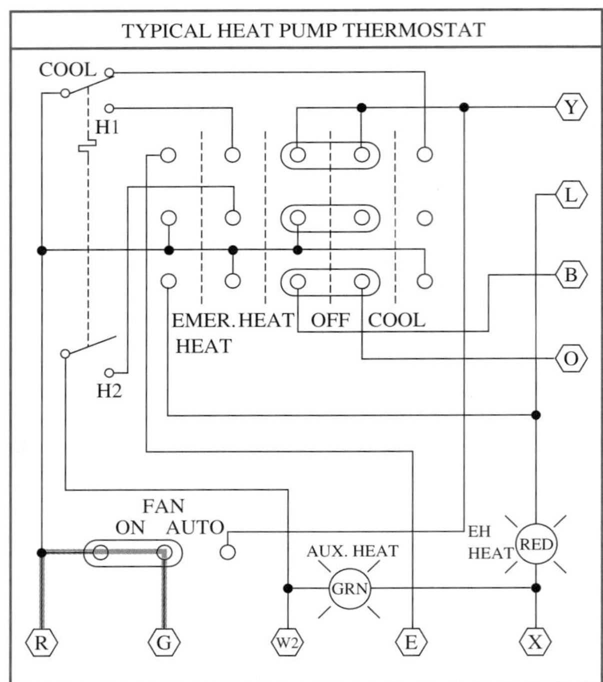

Figure HP44-3 Circuits activated, selector switch OFF, fan ON. (Courtesy of Rheem Air Conditioning Division)

switch, and the G terminal of the subbase to energize the fan relay. The indoor blower motor runs as long as the thermostat subbase fan switch is in the ON position.

Position 3. Selector Switch HEAT, Fan Switch AUTO

In position 3 (Fig. HP44-4), when the room temperature reaches 70°F (thermostat setting), the heating first-stage contact (H1) closes. This provides 24 V through the fan switch and the G circuit to the fan relay. In addition, power is supplied through the normal (HEAT) position of the system switch to the compressor contactor and the defrost and reversing valve circuit.

Second-Stage Heating Activated

If the area temperature should continue to drop because the heating load is greater than the capacity of the heat pump, the second-stage contact (H2) in the thermostat closes (Fig. HP44-5). This supplies 24 V of power through the W2 terminal of the subbase to the supplemental heat. The AUX HEAT light comes on when the supplemental heat is energized.

When the area temperature rises, the second stage of the thermostat opens to cut off the auxiliary heat before the heat pump is cut off. There is a 2°F difference between the first and second steps to prevent the reverse action of stopping the heat pump before stopping the auxiliary heat.

Position 4. Selector Switch COOLING, Fan Switch AUTO

In position 4 (Fig. HP44-6), when the area temperature rises above the temperature setting of the thermostat, the COOL temperature switch closes. This energizes the compressor contactor through the Y terminal.

Position 5. Emergency Heat Switch ON, Fan Switch AUTO

In position 5 (Fig. HP44-7), when the emergency heat switch is moved to the EMER. HEAT position the control circuit to the compressor contactor is opened and the supplementary heat becomes the only heat source. Placing a jumper wire from terminal E to terminal W2 on the subbase will allow the first stage of the thermostat to control the supplementary heat in the emergency heat mode.

HP44-3 DEFROST CONTROLS

When the heat-pump is in the heating mode, it is likely that moisture from the outside air will condense on the outdoor coil. If the coil temperature is below 32°F, the moisture turns to frost and sometimes ice. This must be removed, because the formation of frost restricts the airflow across the outdoor coil and reduces system capacity and performance greatly.

Of the different types of defrost systems tried, the most successful method has been by the hot-gas method or reversing the system to the cooling mode with the outdoor blower off. This means that the unit is operated as a cooling unit, taking heat from the indoor air (the occupied area during the defrost cycle). This, of course, causes a drop in area temperature during defrost. To overcome this, part of the auxiliary heat is turned on to prevent the air-temperature drop.

The defrost cycle on air-source heat pumps can be controlled in a number of ways, all related to initiating the defrost and terminating it. These methods are classified as follows:

1. Temperature-initiated/temperature-terminated
2. Pressure-initiated/temperature-terminated
3. Time-initiated/temperature-terminated
4. Pressure-and-time-initiated/temperature-terminated

HP44-3.1 Temperature-Initiated/Temperature-Terminated Cycle

Figure HP44-8 (p. 853) shows a wiring diagram for a time-temperature defrost control. The defrost cycle is initiated when the solid-state timing circuit is completed to the sensor. The sensor is in series with the coil of the defrost relay. When the sensed temperature falls to 28°F,

Figure HP44-4
Circuits activated, selector switch HEAT, fan AUTO.
(Courtesy of Rheem Air Conditioning Division)

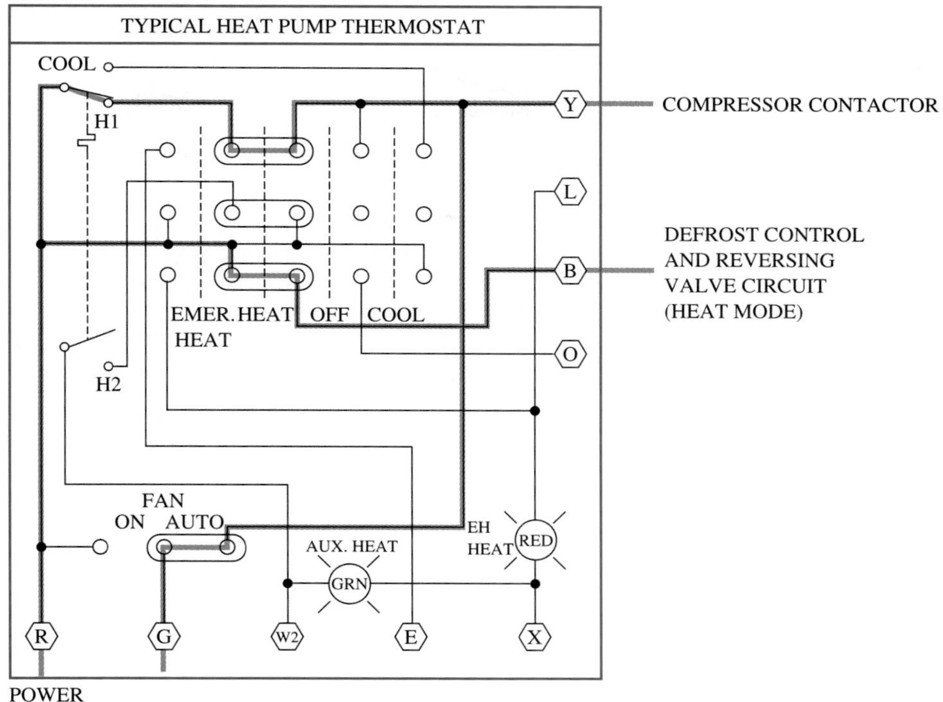

Figure HP44-5
Circuits activated, second stage heating. (Courtesy of Rheem Air Conditioning Division)

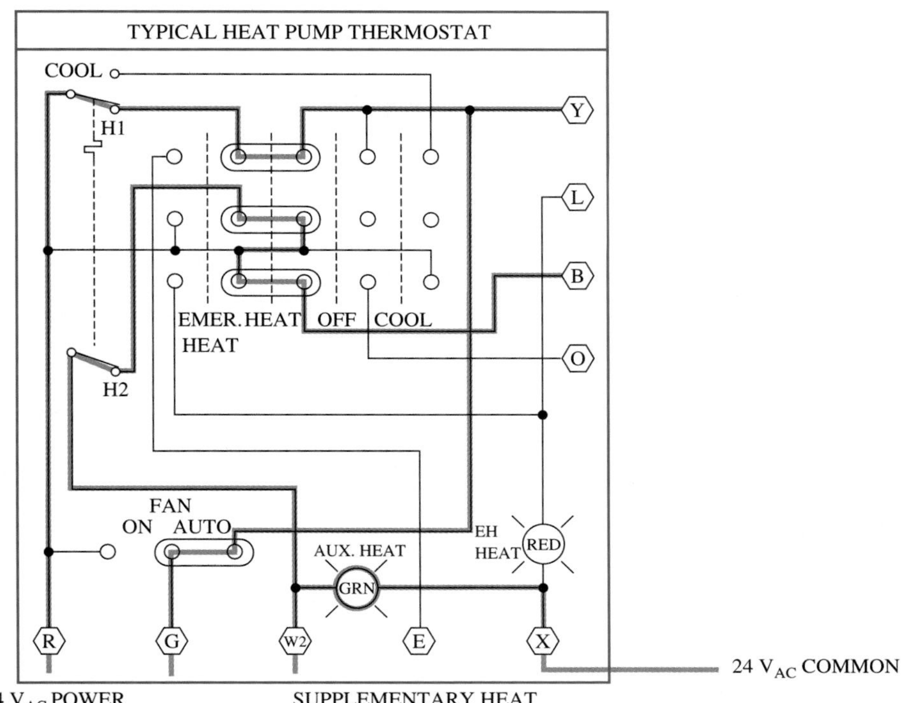

Figure HP44-6
Circuits activated, selector switch COOLING, fan AUTO. (Courtesy of Rheem Air Conditioning Division)

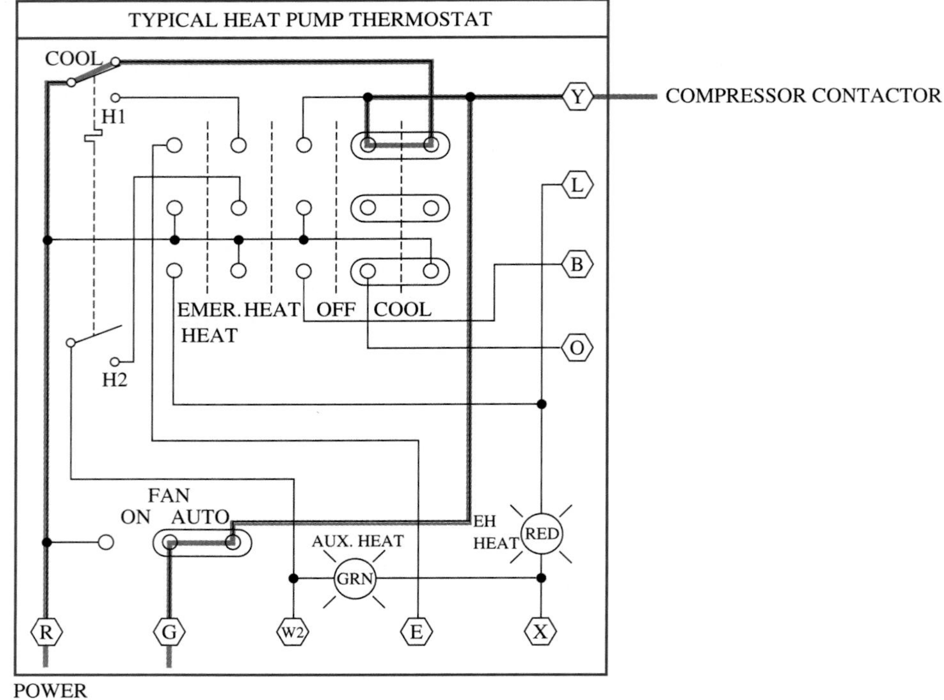

Figure HP44-7
Circuits activated, emergency heat ON, fan AUTO. (Courtesy of Rheem Air Conditioning Division)

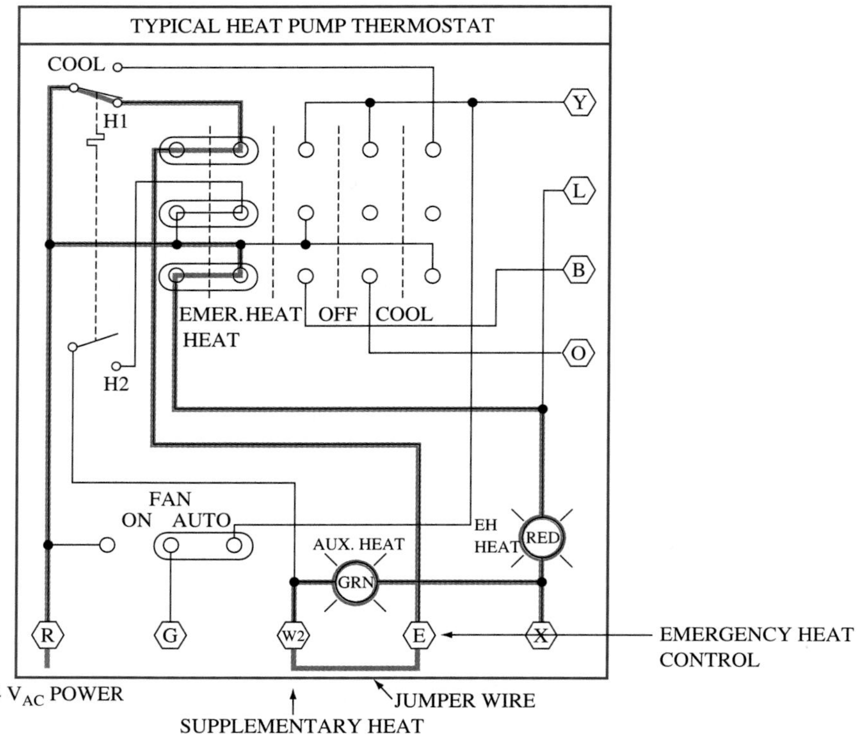

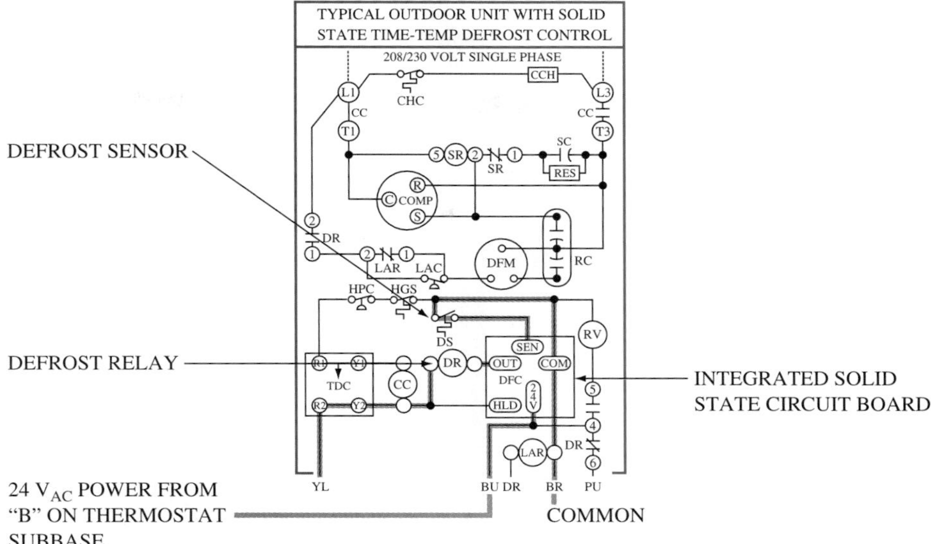

Figure HP44-8 Defrost control: temperature initiated; temperature terminated.
(Courtesy of Rheem Air Conditioning Division)

the sensor closes. This completes the 24 V circuit through terminal SEN on the defrost control, energizing the defrost relay coil through the OUT terminal.

HP44-3.2 Time-Initiated/Temperature-Terminated Defrost Cycle

In order to solve problems associated with the other defrost methods, the industry acted on the theory that the heat pump will require defrosting every 30 minutes of compressor operation. In dry climates, the time period may be 90 minutes. This time-initiated period requires that the terminating thermostat is closed, indicating that the coil temperature is below 26°F, cold enough to build up frost.

Fig. HP44-9 shows a typical timer installation in an electrical control panel. The defrost timer in the upper right hand corner is used in conjunction with the defrost relay and the terminating thermostat (Fig. HP44-10), fastened to the outside coil circuit. Two drive cams are used to determine the cycle time of the timer. Most of the factories ship the timer with the 30-minute cam installed and the 90-minute cam as a spare part item.

Fig. HP44-11 shows the 30-minute cam arrangement in the top picture and the double-cam 90-minute arrangement in the bottom section. In both arrangements, the action of the cam causes the contact point to close when the cam, turning clockwise, causes the bottom follower to drop into the notch. This closes the contact, energizing the relay. Approximately 30 seconds later the second follower drops, opening the circuit to the relay. The relay has a

PRACTICING PROFESSIONAL SERVICES

The hardest component on the heat pump to maintain and service is the control circuit, specifically the defrost controls. Before working on these units, thoroughly review the wiring schematics and determine the proper operating sequence. To test the operation of the defrost controls, it is necessary to force the unit into defrost in the heating mode. This will require simulating ice accumulation on the outdoor coil during low ambient conditions. If the unit is supplied with electronic defrost, refer to the manufacturer's instructions on how to advance the timing sequence. If the defrost control is mechanical-electrical, force the initiation switch to simulate defrost either by blocking airflow or inserting the temperature sensors into ice water. Remember to be patient when working with these controls and to not condemn a good component until it has been properly tested.

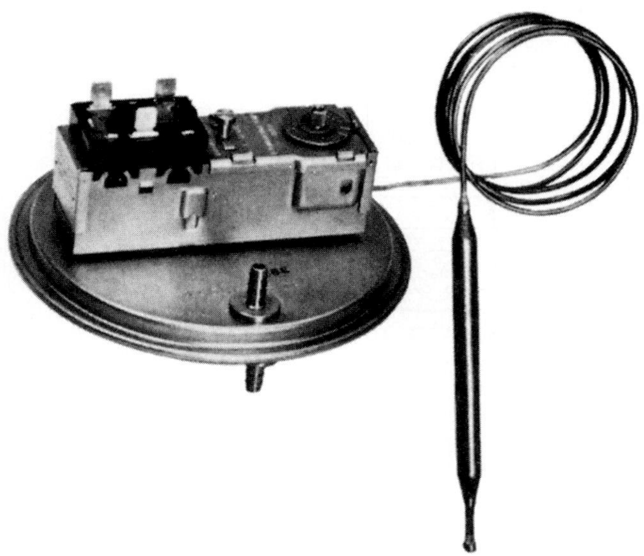

Figure HP44-9 Timer installation in a control panel. (Courtesy of Rheem Air Conditioning Division)

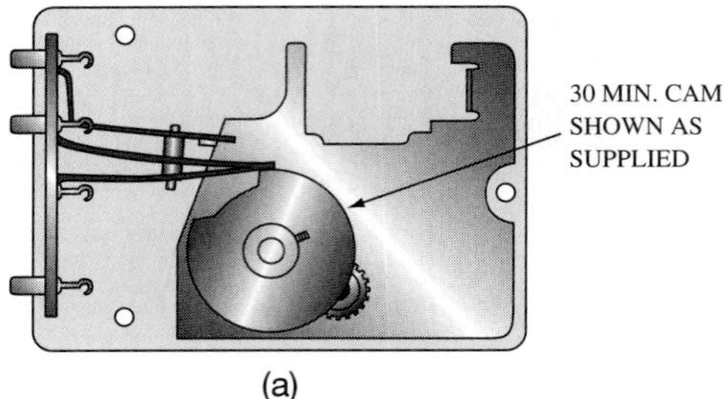
30 MIN. CAM
SHOWN AS
SUPPLIED

(a)

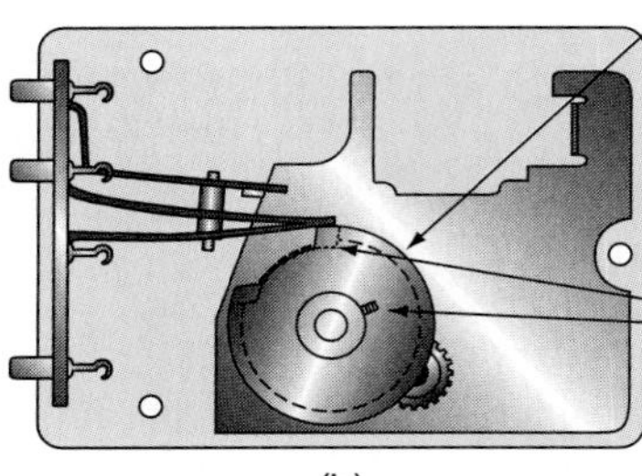
FOR 90-MIN. DEFROST CYCLE: REMOVE 30-MIN. CAM. INSTALL CAM SHIPPED LOOSE WITH UNIT ON BOTTOM AS SHOWN. REPLACE 30-MIN. CAM ON TOP.
IMPORTANT
NOTCH ON OUTER CAM MUST TRAIL NOTCH ON INNER CAM BY 1/8" AS SHOWN FOR PROPER CIRCUIT INTERRUPTION. TIGHTEN SET SCREW WITH 1/16" ALLEN WRENCH.

(b)

Figure HP44-11 Timer with cover removed, showing (a) 30-minute cam installed; (b) 90-minute cam installed.

holding contact that keeps it in the energized position when the timer contact opens.

Control-Panel Wiring

The defrost control circuit with the timer contacts and defrost-relay contacts are shown in Fig. HP44-12. The defrost relay controls the action of the outdoor-fan motor through the normally closed contact of terminal 4 and terminal 5. When the defrost relay is energized, these contacts open and the outdoor-fan motor stops.

The terminal combination of 1 and 3 is the holding circuit around the initiating contacts, terminals 3 and 4 of the defrost timer.

In addition to the relay stopping the outdoor-fan motor and the holding circuit, the relay also energizes the reversing valve through terminals 2 and 4 of the second defrost relay. This places the unit in the cooling mode to provide hot gas for defrost. This action continues until the defrost thermostat (Fig. HP44-11) opens at 55°F, signifying that the coil is free of frost or ice.

Figure HP44-10 Termination thermostat.

HP44-3.3 Pressure-and-Time Initiation/Temperature-Termination Cycle

This control arrangement holds the defrost to a minimum. A pressure switch used to measure the coil airflow resistance has been combined with a time clock and termination bulb to provide the following action:

1. No action until the coil temperature is below 26°F and the coil has reached 80% blockage from ice formation.
2. Adjusting the time between defrost cycles to allow a closer relationship between the length of the defrost cycle required and the frequency of the defrost cycle.
3. Terminating the defrost cycle in the case of high wind and cold outside-ambient temperatures.

HP44-3.4 Solid-State Controls

A solid-state control module is used by certain manufacturers to replace some of the standard electrical or mechanical controls. A typical control system of this type is shown in Fig. HP44-13.

This logic module responds to the demand signal of the thermostat, examines the input from four sensors (out-

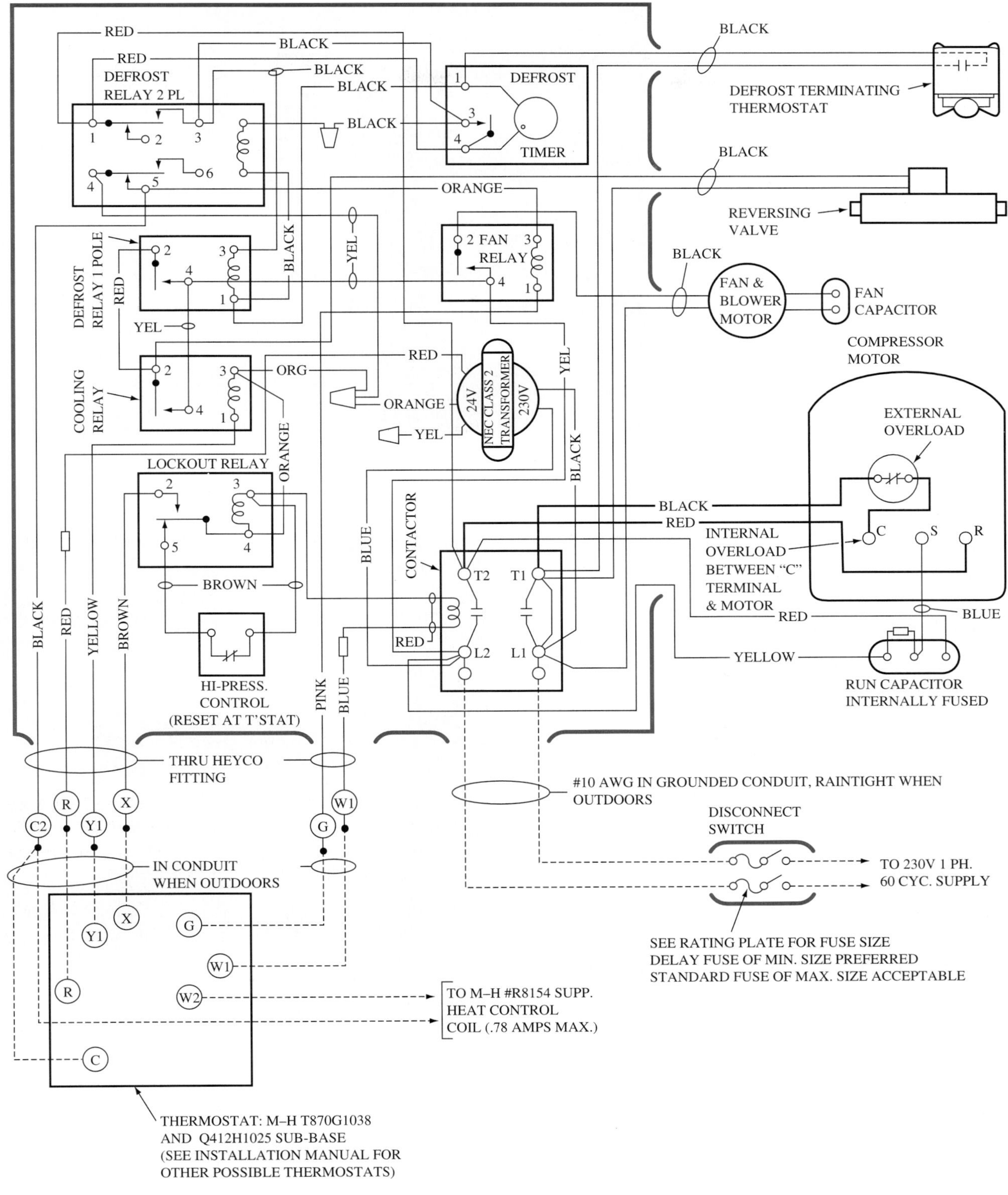

Figure HP44-12 Wiring diagram of a control panel. (Courtesy of Rheem Air Conditioning Division)

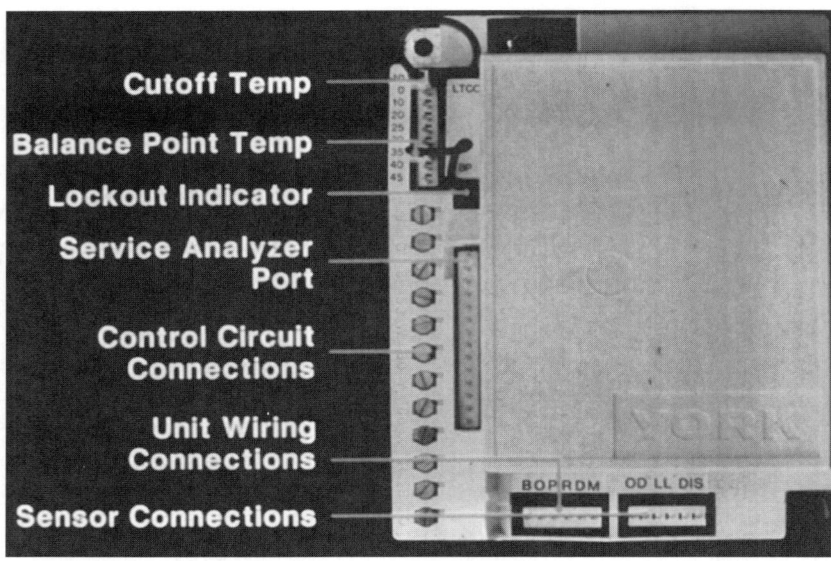

Figure HP44-13 Solid-state controls for heat pump. (Courtesy of York International Corporation)

door, discharge, defrost, and liquid) and determines when the heat pump or the supplementary heaters will operate.

Balance-Point Adjustment

The balance point is the lowest ambient temperature at which the heat pump can operate without the use of supplementary heat. The balance point is set by the local contractor based on:

1. Outdoor design temperature
2. Building heat loss
3. Unit capacity

Service Analyzer

The service analyzer, shown in Fig. HP44-14, is used by the service person to identify any malfunction in the operation of the heat pump. The analyzer indicates the service problem by a fault code.

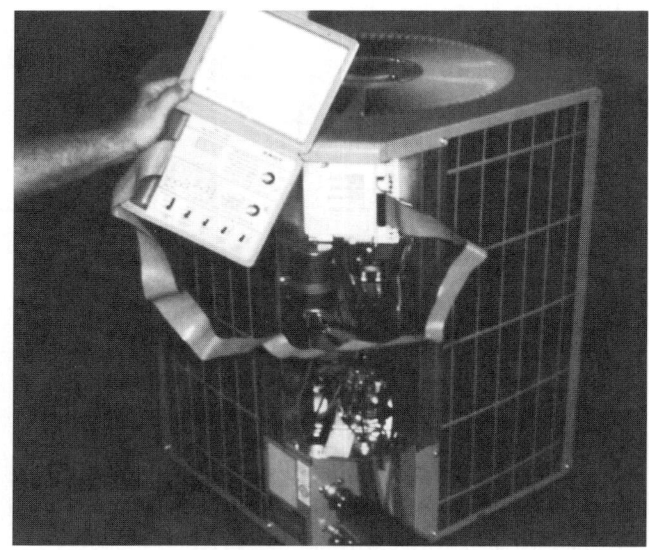

Figure HP44-14 Service analyzer for a solid-state control module for a heat pump. (Courtesy of York International Corporation)

HP44-4 LOW AMBIENT CONTROL

The low ambient control is a low-limit thermostat. It is placed in series with the control circuit, to cut out the heat pump when the ambient temperature is too low to produce useful heat. Below this temperature, all heating is performed by the secondary heat source. The setting at which this takes place is originally determined by the manufacturer, (usually about 0°F), but can be adjusted by the contractor to meet local conditions.

HP44-5 CONTROL-PANEL

The control-panel wiring for a heat pump is usually readily accessible, as shown in Fig. HP44-13. It is normally wired at the factory and furnished with the package. The technician must, however, enter the panel to make tests and perform service. The manufacturer will usually furnish only a schematic wiring diagram (not a connection diagram) to show how the panel is wired. Some manufacturers supply a layout of the panel showing the position of the various controls, which is helpful in tracing the circuits.

Fortunately the schematic diagram is sufficient to analyze or rewire the panel, if necessary. Where any major amount of work needs to be performed on the panel it is often desirable to draw a connection diagram. This can be done easily using the schematic as a guide and drawing the wiring connections on a copy of the panel layout.

REVIEW

- Temperature controls—Control temperature, both heating and cooling; also control auxiliary heat. Room thermostats
- Defrost controls—Initiate defrost cycle.
  Temperature-initiated/temperature-terminated—Compares temperature of outside coil with the temperature of air entering it. Increase in difference initiates defrost cycle.
  Time-temperature—Time initiates defrost, unit stays in defrost until coil is below 26°F.
  Pressure-initiated/temperature-terminated—Reacts to air pressure across coil, initiates defrost, then times through defrost cycle.
  Pressure-and-time-initiated/temperature-terminated—Reacts to differential pressure and time for initiation, and coil temperature for termination.
  Solid-state logic module—Uses demand signal from thermostat while examining input temperature signals from outdoor, discharge, defrost, and liquid line to determine proper control sequence.
- Low ambient control—Locks out system compressor circuit during low ambient temperature; many units shut off heat pump operation and use only supplemental heat when outside air is below 0°F.
- Control panel—Contains unit controls; normally factory-wired and furnished within package.
- Most successful method of defrosting—Hot-gas method, reversing the system to the cooling mode with outdoor blower off.
- Balance point of system is based on:
  Outdoor air temperature
  Building heat loss
  Unit capacity
- Balance point is the lowest ambient temperature at which the heat pump can operate without the use of supplementary heat.

Problems and Questions

1. Two types of temperature controls are used in heat pumps. What are they?
2. Four types of defrost-control systems have been used in heat pumps. What are they?
3. Normal heat-pump-system room thermostats have eight functions. What are they?
4. In two-stage mechanical thermostats, the difference in temperatures held by the two stages is usually _____ °F.
5. A minimum of 5°F is used between the heating and cooling levers of the standard mechanical room thermostat. Why?
6. This 5°F temperature period is called the _____.
7. The defrost relay has three separate functions. What are they?
8. What temperature is used by the pressure-initiated/temperature-terminated defrost system to terminate the defrost cycle?
9. In a time-initiation/temperature-termination defrost system, on what is the time the clock motor operates on based?
10. In a pressure-time-actuated/temperature-terminated defrost system, on what is the clock operating time based?
11. Manual reset of an automatic-reset high-pressure control is provided by a device called a _____.
12. When the lock-out relay is actuated in series with the contactor, why does the contactor drop out and the relay stay in?
13. The heating output of the heat pump is directly related to the temperature of the indoor air. True or false?
14. The most common source of supplementary heat is electric heat. True or false?
15. Solid-state control modules replace all of the electrical and mechanical controls. True or false?
16. The operation of the heat-pump thermostat is to
    a. turn the system on and off.
    b. initiate cooling and heating modes.
    c. initiate auxiliary heat stages and allow for fan options.
    d. All of the above.
17. The emergency heat switch on the thermostat
    a. removes the compressor from the control circuit.
    b. energizes the auxiliary heat, but lets the compressor run.
    c. energizes the second stage of auxiliary heat.
    d. All of the above.
18. The defrost-control circuit, temperature-initiated/temperature-terminated cycle
    a. compares temperature of the outside coil with the temperature of the air entering the coil.
    b. compares the temperature of the outside coil with the temperature of the liquid line.
    c. compares the temperature of the outside air with the temperature of the inside air.
    d. None of the above.
19. The solid-state logic control module
    a. monitors demand signal from thermostat.
    b. monitors outdoor and discharge temperatures.
    c. monitors defrost and liquid temperatures.
    d. All of the above.

SECTION 9

Installation and Start Up

145 Load Calculation

AFTER STUDYING THIS CHAPTER, THE STUDENT WILL BE ABLE TO:

- Define heat transmission and associated coefficients and conditions.
- Define insulation materials, types, and advantages of each.
- Explain the sun's effect on heat loads.
- Describe methods of establishing vapor seal.
- Interpret design temperatures, inside and outside, using ASHRAE charts.
- Explain infiltration sources and methods to compute the loads.
- Identify product-load considerations and factors.
- Identify supplementary loads and sources.
- Evaluate the selection of equipment for a refrigerated space.
- Calculate the refrigeration load for a space with a temperature above freezing.
- Calculate the refrigeration load for a space with a temperature below freezing.
- Select a suitable program to calculate the air-conditioning load for a residence or small commercial building.

145-1 REFRIGERATION LOAD

The total refrigeration load of the system as expressed in Btu/hr comes from many heat sources. For example, note the sources of heat in the refrigerated storage room of a supermarket, shown in Fig. I45-1. The sources may be categorized as follows:

1. Heat transmission.
   a. A temperature difference of 60°F exists between the 95°F outside air and the room temperature of 35°F which causes much heat conduction.
   b. The effect of the sun on the roof and walls results in radiation heat buildup.
2. Air infiltration.
   a. Air enters a room as a result of opening and closing the doors during normal working periods.
   b. Air enters the room through cracks in or around the structure and door seals.
   c. Air is purposely introduced for ventilation.
3. Product loads, or heat from the product(s) being stored:
   a. Dry or sensible heat, as from cooling a can of juice at room temperature to 35°F;
   b. A combination of dry and moist (latent) heat, as from produce;
   c. Heat from frozen products which have the latent heat of freezing;
   d. Some heat is also the result of chemical changes in the product, such as the ripening of fruit.
4. Supplementary loads are caused by heat-emitting objects such as electric lights, motors, and tools—and people.

Although refrigeration design engineers are primarily responsible for estimating the loads, refrigeration technicians should understand how these heat sources affect the operation of the system. They can then adjust the equip-

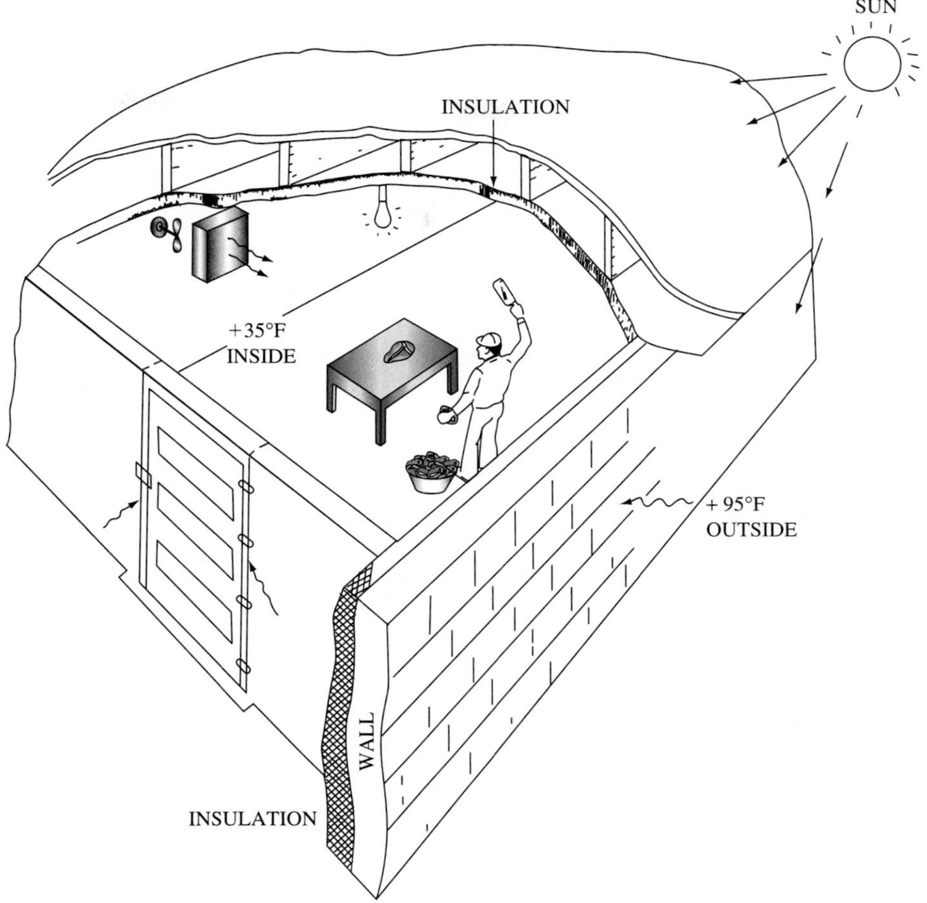

Figure I45-1
Refrigerated storeroom for supermarket.

ment to perform in a manner consistent with the design of the system.

The subjects of refrigeration load calculation and selection of equipment are presented under the following topics:

- Heat transmission
- Sun effect
- Design temperatures
- Air infiltration
- Product load
- Supplementary loads
- Total hourly load
- Selection of equipment
- Sample calculations

Typical data sheets for recording the information are shown in Fig. I45-2.

I45-1.1 Heat Transmission

The heat gain through walls, floors, and ceilings varies with:

1. Type of construction,
2. Area that is exposed to different temperatures,
3. Type and thickness of insulation, and
4. Temperature difference between the refrigerated space and the ambient air.

Thermal conductivity (k) varies directly with time, area, and temperature difference. It is expressed in Btu/hr, per sq ft of area, per in. of thickness, per °F of temperature difference.

In order to reduce heat transfer, the thermal conductivity factor (based on material composition) should be as small as possible and the insulating material as thick as economically feasible.

Heat transfer through any material is also affected by surface resistance to heat flow caused by the type of surface (rough or smooth), its position (vertical or horizontal), its reflective properties, and the rate of airflow over the surface.

Extensive testing has been done to determine accurate values for heat transfer through common building and structural materials. Certain materials have been found to have a high resistance to flow of heat and are good insulators while others are not.

Heat loss is measured by *resistance (R),* which is the resistance to heat flow of either 1 in. of material, or a specified thickness, of an air space, an air film, or an en-

- Load Survey & Estimating Data -

DESIGN AMBIENT: ______ °F DB, ______ °F WB, ______ %RH, ______ °F SUMMER GROUND TEMP.
(USE 55°F FOR INSULATED FREEZE FLOOR SLAB)

ROOM DESIGN: ______ °F DB, ______ °F WB, ______ %RH ______ °F WINTER DESIGN AMBIENT

ACCESS AREA: ______ °F DB, ______ °F WB, ______ %RH, (ANTE-RM/LOADING DOCK/OTHER)

ROOM DIM. OUTSIDE: ______ FT. W ______ FT. L ______ FT. H ______ TOTAL SQ. FT. (OUTSIDE SURFACE)

		Insulation (Table 6)			Wall Thick-ness	Adj. Area °F	Effective Wall TD	°F Sun Effect (Table 8)	Total TD	Overall Wall Heat Gain
	Type	Inches Thick	K Factor	U* Factor						BTU/24 hrs/sq. ft. (Table 6)
N. Wall										
S. Wall										
E. Wall										
W. Wall										
Ceiling										
Floor										

* 'U' – Factor = $\frac{K}{\text{Insul. Thickness (in.)}}$

REFRIG. DOOR(S): ______ VENT. FAN(S): ______

ROOM INT. VOL: ______ W × ______ L × ______ H = ______ CU. FT.
(INSIDE ROOM DIMENSION = OUTSIDE DIMENSION – WALL THICKNESSES)

FLOOR AREA ______ W × ______ L = ______ SQ. FT.

ELECTRICAL POWER ______ VOLTS, ______ PH. ______ HERTZ; CONTROL ______ VOLTS

TYPE CONTROL: ______

PRODUCT DATA AND CLASS OF PRODUCT: ______

	Amount of Product (Refer to Pg. 2 Item 3)			Product Temp. °F		Table 31				Table 42	
						Specific Heat					
Type Product	Amount Storage	Daily Turn-Over	Freezing or Cooling	Enter-ing	Final	Above Freeze	Below Freeze	Lat. Ht. Freeze Btu/lb.	Highest Product Freeze Temp.	Heat or Respir'n Btu/lb 24 Hr.	( ) Pull-Down ( ) Freezing Time Hrs.

EVAP. TD ______, TYPE DEFROST ☐ AIR, ☐ HOT GAS, ☐ ELECTRIC,

CLASS PRODUCT ______,

NO. OF DEFROSTS & TOTAL TIME PER 24 HRS. ______ NO., ______ HRS.

COMPRESSOR RUNNING TIME ______ HRS.

BOX USAGE ☐ AVERAGE, ☐ HEAVY, ☐ EXTRA HEAVY

PRODUCT LOAD AND ADDITIONAL INFORMATION: ______

PACKAGING ______ CONTAINERS ______ WGT ______ SP. HT. ______ (CONTAINER)

PALLETS: NO. ______ SIZE ______ WGT. EA. ______ SP. HT. ______

PRODUCT RACKS: NO. ______ MAT'L ______ WGT. EA. ______ SP. HT. ______

ESTIMATING PRODUCT LOADING CAPACITY OF ROOM

ESTIMATED PRODUCT LOADING = 0.40 × ______ CU. FT × ______ LBS./CU. FT. = ______ LBS:
(ROOM VOLUME) (LOADING DENSITY) (TABLE 45A OR B)

MISCELLANEOUS LOADS

PEOPLE NO. ______ HRS. ______

MOTORS (OTHER THAN EVAP. FAN)

USE: ______, ______ HP ______ HRS.
______, ______ HP. ______ HRS.

FORK LIFTS ______ NO., ______ HP, ______ HRS./DAY, OTHER ______
(REFER TO PG. 3, ITEM 4)

LIGHTS ______ WATTS/SQ. FT. (REFER TO PG. 3, ITEM 4)

Figure 145-2 Refrigeration load calculation sheet, blank form. (Courtesy of Dunham-Bush, Inc.)

Calculations				
I WALL LOSS (TRANSMISSION LOAD)				
SURFACE	TD	AREA OF SURFACE	ALL HEAT GAIN FACTOR TABLE 6	BTU/24 HRS
N. Wall		______ Ft. L × ______ Ft. H = ______ Sq. Ft. ×	______ =	
S. Wall		______ Ft. L × ______ Ft. H = ______ Sq. Ft. ×	______ =	
E. Wall		______ Ft. L × ______ Ft. H = ______ Sq. Ft. ×	______ =	
W. Wall		______ Ft. L × ______ Ft. H = ______ Sq. Ft. ×	______ =	
Ceiling		______ Ft. L × ______ Ft. W = ______ Sq. Ft. ×	______ =	
Floor		______ Ft. L × ______ Ft. W = ______ Sq. Ft. ×	______ =	
Box		(Table 14A, B, C) Total: Surface = ______ Sq. Ft. ×	______ =	
		I Total Wall Transmission Load BTU/24 HRS.	=	
II_{SF} (SHORT FORM) USAGE HEAT GAIN ( ) AVG. ( ) HVY. ( ) EX. HVY. COOLERS ONLY Refer to II_{LF} (Long Form) if application exceeds data shown in Table 15. ______ CU. FT. × ______ BTU/24 HR./CU. FT. (@ ______ TD) (INT. BOX VOLUME) (TABLE 14) (USAGE HEAT GAIN) (TABLE 15)			=	
NOTE: IF PRODUCT LOADS ARE UNUSUAL USE THE LONG FORM EX. HVY. = 1½ × HVY. USAGE			TOTAL I + II_{SF} =	
IF USAGE HEAT GAIN ABOVE IS USED DO NOT USE (II_{LF} III & IV)				
II_{LF} (LONG FORM) INFILTRATION (AIR CHANGE LOAD) ______ CU. FT. × ______ AIR CHANGES/24 HRS. × ______ SRVC. FACTOR × ______ BTU /CU. FT (TABLE 14) (TABLE 9 OR 10) (FROM NOTES TABLE 9 OR 10) (TABLE 11)			=	
		II INFILTRATION LOAD BTU/24 HRS.	=	
III PRODUCT LOAD				
PRODUCT TEMP. REDUCTION ABOVE FREEZING (SENSIBLE HEAT) ______ *LBS./DAY × ______ °F TEMP. REDUCTION × ______ SP. HT.			=	
PRODUCT FREEZING (LATENT HEAT LOAD) ______ *LBS./DAY × ______ BTU/LB. LATENT HEAT			=	
PRODUCT TEMP. REDUCTION BELOW FREEZING (SENSIBLE HEAT) ______ *LBS./DAY × ______ °F TEMP. REDUCTION × ______ SP. HT.			=	
HEAT OF RESPIRATION ______ LBS. PRODUCT (STORAGE) × ______ BTU/LB./24 HRS.			=	
MISCELLANEOUS PRODUCT LOADS (1) CONTAINERS (2) PALLETS (3) OTHER ______ LBS./DAY × ______ °F TEMP. REDUCTION × ______ SP. HT.			=	
______ LBS./DAY × ______ °F TEMP. REDUCTION × ______ SP. HT.			=	
		III TOTAL PRODUCT LOAD BTU/24 HRS.	=	
IV MISCELLANEOUS LOADS (a) LIGHTS ______ $Ft.^2$ Floor Area × ______ $Watts/Ft.^2$ × 3.41 Btu/Watt × ______ HRS/24 HRS = (1 TO 1½ WATTS/SQ. FT. IN STORAGE AREAS & 2 TO 3 FOR WORK AREAS)				
(b) OCCUPANCY ______ NO. OF PEOPLE × ______ BTU/HR. × ______ HRS. ______			=	
(c) MOTORS ______ BTU/HP/HR. × ______ HP × ______ HRS./24 HRS ______			=	
______ BTU/HP/HR. × ______ HP × ______ HRS./24 HRS ______			=	
(d) MATERIAL HANDLING ______ FORK LIFT(S) × ______ EQUIV. HP × 3100 BTU/HR./HP. × ______ HRS. OPERATION OTHER ______			= =	
*IF THE PRODUCT PULL-DOWN IS ACCOMPLISHED IN LESS THAN 24 HRS. THE DAILY PRODUCT WILL BE: LBS. PRODUCT × $\frac{\text{24 HRS.}}{\text{PULLDOWN HRS.}}$		IV TOTAL MISCEL. LOADS BTU/24 HRS.	=	
		TOTAL BTU LOAD I TO IV BTU/24 HRS.	=	
		ADD 10% SAFETY FACTOR	=	
		TOTAL BTU/24 HRS. WITH SAFETY FACTOR (NOT INCLUDING EVAP. FAN OR DEFROST HEAT LOADS) 24 HR. BASE REFRIGERATION LOAD	=	

Figure 145-2 Continued

Equipment Selection from Load Calculation Form

1. DETERMINE EVAP. TD REQUIRED FOR CLASS
   OF PRODUCT AND ROOM TEMP ________ °F (TD) (FROM LOAD SURVEY DATA)

2. DETERMINE COMPRESSOR RUNNING TIME BASED ON OPERATING TEMPERATURES
   AND DEFROST REQUIREMENTS ________ HRS. (FROM LOAD SURVEY DATA)

3. EVAPORATOR TEMP.°F = ________ – ________ = ________ °F
   (ROOM TEMP) (EVAP. TD**) **FROM LOAD SURVEY DATA

4. COMP. SUCT. TEMP.°F = ________ – ________ = ________ °F
   (EVAP. SUCT. TEMP) (SUCT. LINE LOSS)

BTU/24 HR. BASE REFRIGERATION LOAD WITH SAFETY FACTOR = ________
(NOT INCLUDING EVAPORATOR FAN OR DEFROST HEAT)

REFER TO PAGES 6 AND 7 "SELECTION OF REFRIGERATION EQUIPMENT TO DETERMINE PROCEDURE FOR EQUIPMENT SELECTION."

PRELIMINARY HOURLY LOAD = BTU/24 HR. (BASE LOAD) / HRS./DAY (COMP' RUNNING TIME) = ________ BTU/HR.

FAN HEAT LOAD ESTIMATE BTU/HR. =

______ QTY. (MOTORS) × ______ WATTS EA. (INPUT) × 3.41 BTU/WATT × ______ HRS. = ________ BTU/24 HRS.

OR ______ QTY. (MOTORS) × ______ HP EA. × ______ BTU/HP/HR. (TABLE 13) × ______ HRS. = ________ BTU/24 HRS.

DEFROST HEAT LOAD ESTIMATE BTU/HR =

______ QTY. EVAPS. × ______ WATTS EA. × ______ HRS. × 3.41 BTU/WATT × ______ DEFROST LOAD FACTOR*

= ________ BTU/24 HRS.

*USE 0.50 FOR ELECTRIC DEFROST 0.40 FOR HOT GAS DEFROST

BTU/24 HR. TOTAL LOAD = ________ (BASE LOAD) + ________ (FAN HEAT) + ________ (DEFROST HEAT) = ________ BTU/24 HRS.

OR (Refer to pg. 5) ________ (BASE LOAD) × ________ (BASE LOAD MULT.) = ________ BTU/24 HRS.

ACTUAL HOURLY LOAD = BTU/24 HRS. (TOTAL LOAD) / HRS/DAY (COMP. RUNNING TIME) = ________ BTU/HR.

Equipment Selection

	Compressor Units		Condensing Units		Evaporators		Condensers	
MODEL NO.								
QUANTITY								
BTU/HR. CAPACITY (EA.)								
CFM AIR VOLUME (EA.)								
	DESIGN	**ACTUAL**	**DESIGN**	**ACTUAL**	**DESIGN**	**ACTUAL**	**DESIGN**	**ACTUAL**
°F EVAPORATOR TEMP.								
°F EVAP TD								
°F SUCTION TEMP.								
°F CONDENSING TEMP.								
°F DESIGN AMBIENT TEMP.								
°F MIN. OPER. AMBIENT TEMP.								

Figure 145-2 Continued

tire assembly. Its value is expressed as °F temperature difference per Btu per hr per sq ft. A high R value indicates low heat-flow rates. The resistances of several components of a wall may be added together to obtain the total resistance:

$$R_t = R_1 + R_2 + R_3 + \ldots$$

Figure I45-3 lists some R values for common building materials in order to illustrate the differences in heat-flow characteristics. Extensive listings of R values are included in Appendix A.

The quantity of heat transmission *(Q)* through a substance or material is calculated by the formula:

$$Q = U \times A \times TD$$

where

Q = heat transfer, Btu/hr
U = overall heat transfer coefficient (Btu/hr/ft^2/°F)
A = area (ft^2)
TD = temperature difference between inside and outside design temperature and refrigerated-space design temperature

EXAMPLE

Calculate the heat flow through an 8-in. concrete block (cinder) wall (Fig. I45-4), 100 ft^2 in area, having a 60°F temperature difference between the inside and the outside.

Solution

The R value of an 8-in. concrete block wall (Fig. I45-3) is 1.72.

Therefore

$U = 1 \div R$ where R = total resistance of the individual components of the wall.
$= 1 \div 1.72$
$= 0.58$ Btu/hr/°F/ft^2
$Q = 0.58 \times 100$ ft$^2 \times 60$°F
$= 3480$ Btu/hr heat flow into the space

Add 6 in. of fiberglass insulation to the wall (Fig. I45-5) and recalculate the transmission load. Given that the R value for 1-in. insulation is 3.12:

$R = (6 \times 3.12) = 18.72.$
U factor $= 1 \div R$
$R_t = R_1$ (concrete block) $+ R_2$ (6 in. insulation)
$= 1.72 + 18.72$
$= 20.44$

Therefore,

$U = 1 \div 20.44 = 0.049$
$Q = 0.049 \times 100$ ft$^2 \times 60$°F $= 294$ Btu/hr

The example above demonstrates the cumulative effect of R values on determining the total wall resistance. It also shows the heat reduction that can be achieved through proper insulation, from 3480 Btu/hr to 294 Btu/hr. This would greatly reduce the size of the refriger-

Material	Density (lb/ft^3)	Mean Temp. (°F)	Conductivity, k	Conductance, C	Resistance, R Per inch	Resistance, R Overall
Insulating materials						
Mineral wool blanket	0.5	75	0.32		3.12	
Fiberglass blanket	0.5	75	0.32		3.12	
Corkboard	6.5–8.0	0	0.25		4.0	
Glass fiberboard	9.5–11.0	−16	0.21		4.76	
Expanded urethane, R-11		0	0.17		5.88	
Expanded polystyrene	1.0	0	0.24		4.17	
Mineral wool board	15.0	0	0.25		4.0	
Insulating roof deck, 2 in.		75		0.18		5.56
Mineral wool, loose fill	2.0–5.0	0	0.23		4.35	
Perlite, expanded	5.0–8.0	0	0.32		3.12	
Masonry materials						
Concrete, sand and gravel	140		12.0		0.08	
Brick, common	120	75	5.0		0.20	
Brick, face	130	75	9.0		0.11	
Hollow tile, two-cell, 6 in.		75		0.66		1.52
Concrete block, sand and gravel, 8 in.		75		0.90		1.11
Concrete block, cinder, 8 in.		75		0.58		1.72
Gypsum plaster, sand	105	75	5.6		0.18	

Figure I45-3 Typical heat transmission coefficients.

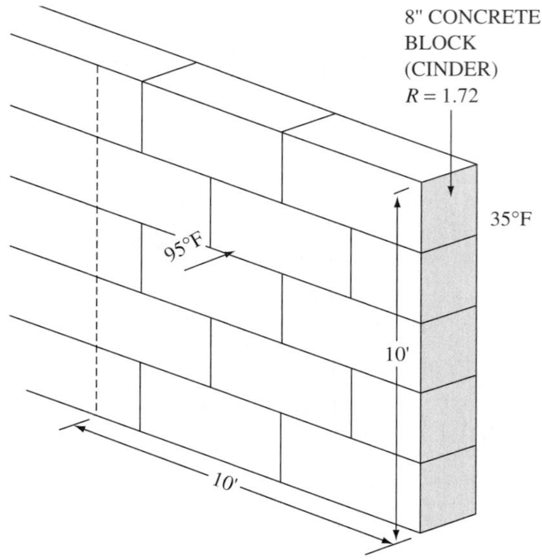

Figure I45-4 Heat flow through an 8-in. cinder block wall.

Figure I45-6 Loose-fill insulation.

ation equipment needed and the amount of the operating cost.

Insulation is the most effective method of reducing heat transmission. Types of insulation for various applications are: (1) loose fill, (2) flexible, (3) rigid or semi-rigid, (4) reflective, and (5) foamed-in-place.

Loose fill or *"blown" insulation* (Fig. I45-6) is used primarily in residential structures for reducing heat flow. *Flexible insulations* such as fiberglass, in bats or rolls

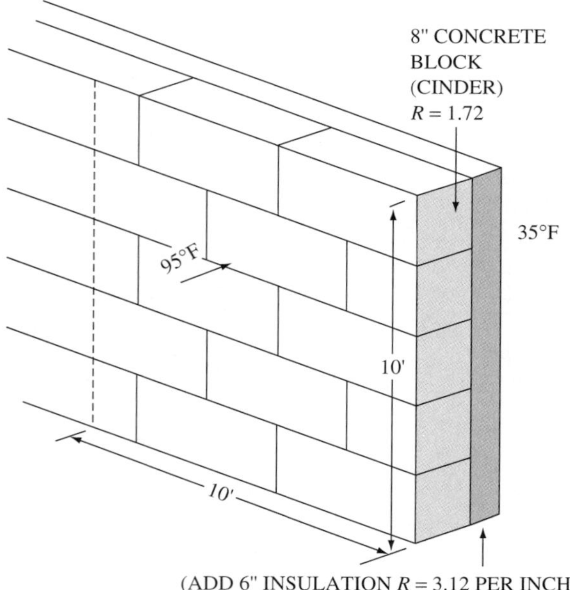

Figure I45-5 Heat flow through an 8-in. insulated cinder block wall.

(Fig. I45-7), are also common in new residential and commercial construction and come with a cover material such as craft paper which acts as a vapor barrier. Flexible insulations may be used to wrap ducts or to place between wall studs or ceiling joists.

Rigid and *semi-rigid insulations* (Fig. I45-8) are made of such materials as corkboard, polystyrene, foam glass, and polyurethane. These are available in boards and sheets, in various dimensions and forms. Some have a degree of structural strength.

The rigid and semi-rigid material is the type of insulation used in refrigeration equipment such as walk-in coolers, freezers, and display cases. Because of its density and structure, it has a built-in vapor barrier to moisture.

Foamed-in-place insulation (Fig. I45-9) is widely used for filling cavities that are hard to insulate. It is also used to cover drain pans and other places where temperature control and water seal are needed. Foamed insulation can be used in on-site, built-up refrigerated rooms in connection with rigid insulations.

For all types of insulation, moisture control is extremely important. Insulation should be dry when installed and be sealed perfectly so that it stays dry.

Fig. I45-10 shows the gradual temperature changes that take place within an insulating material from 90°F on the warm outside to 40°F on the cold inside.

In the example, the 90°F warm-air side has a dew-point temperature of 83°F. (At dew-point temperature air is saturated with moisture, and condensation from the vapor to a liquid occurs when the temperature is lowered). As illustrated, when there is no effective vapor barrier or seal on the warm side, water will start to condense inside the insulation as the temperature drops below the dew point. Water is a good conductor of heat, about 15 times as fast as fiberglass. Thus, if water gets into the insulation, the

(a)

(b)

Figure 145-7 Fiberglass insulation (a) rolls; (b) batts.

material's insulating value is greatly reduced, and physical problems also occur.

Vapor seals can be formed from various materials, including metal casings, metal foil, plastic film, and asphalt coverings. The selection depends on the application.

The ability of a material to resist water-vapor transmission is measured in *perms,* a term related to permeability. Vapor barriers of 1 perm or less have been found satisfactory for residential comfort heating and cooling work. In low-temperature commercial-refrigeration applications such as freezers, perm ratings of 0.10 and below may be accepted levels. As with insulation and heat flow, the resistance to vapor flow is a function of the composite of all the materials, as constructed, not just the rating of the vapor barrier itself.

One perm equals one grain (avoirdupois) of water vapor per hr flowing through one sq ft of a material induced by a vapor pressure difference of one inch of mercury across the two surfaces. One perm is the generally accepted maximum allowable permeance value for the residential building envelope in a 5000 heating degree-day climate.

The effectiveness of both insulation and vapor barriers are greatly reduced if any openings, however small, exist. Such openings may be caused by poor workmanship during construction and application, but they may

PRACTICING PROFESSIONAL SERVICES

Newer buildings with air-conditioning and heating equipment that will not handle the load adequately during extreme ambient conditions should be checked for proper equipment selection. Low-bid jobs sometimes will have contractors which use the minimum size equipment to seal the bid and get the contract. If the technician suspects the installed equipment is not sized properly, it should first be verified that the equipment is operating to its maximum capacity. This includes taking airflow readings, measuring temperatures, amps, and refrigeration pressures, and comparing the readings against the manufacturer's specifications. If the conclusive results determine undersized equipment, then the original installing contractor should be held responsible for the problem. If the job is out of warranty, the technician should consult with an applications engineer and redo the load calculation to verify the actual load requirements of the building. Next, the customer should be informed of the initial mistake and sold the proper unit.

Figure I45-8 Installing semi-rigid insulation.

Figure I45-9 Application of foam-type insulation.

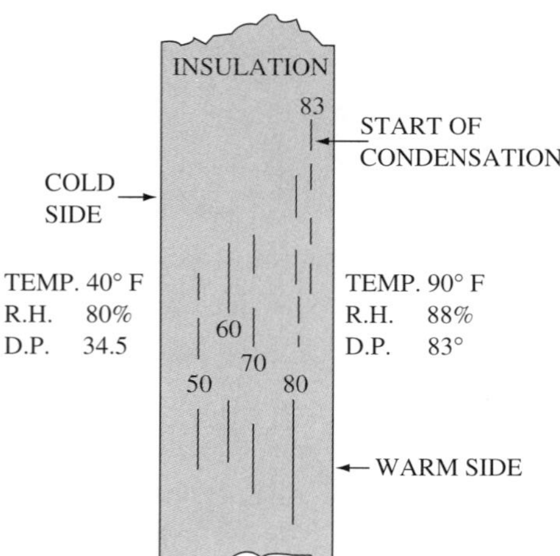

Figure I45-10 Gradual temperature changes in an insulated wall from outside to inside.

also result from negligence in sealing around openings for piping and wiring. This is the responsibility of the service technician.

I45-1.2 Sun Effect

The primary radiation factor in the refrigeration load is the heat gain from the sun's rays. If the walls of the refrigerated space are exposed to the sun, additional heat will be added to the heat load. A simplified method for estimating this effect is shown in Fig. I45-11.

The factors shown, in °F for various conditions and locations, are to be added to the normal temperature difference between indoor and outdoor design conditions.

I45-1.3 Design Temperatures

Two design temperatures must be determined to proceed with the load calculation: *outside* and *inside*.

Type of Structure	East Wall	South Wall	West Wall	Flat Roof
Dark-colored surfaces (Slate, tar, black asphalt shingles, and black paint)	8	5	8	20
Medium-colored surfaces (Brick, red tile, unpainted wood, dark cement)	6	4	6	15
Light-colored surfaces (White stone, light-colored cement, white asphalt shingles, white paint)	4	2	4	9

Figure I45-11 Temperature allowance for sun effect.

Recommended outside design conditions are the results of extensive studies by the National Weather Service. For air-conditioning and refrigeration applications the maximum load occurs during the hottest weather. It is neither economical nor practical, however, to design equipment for the hottest temperature, since it might last for only a few hours over a span of several years. The design temperature chosen is therefore less than the peak temperature.

A segment of the ASHRAE outdoor design data chart is shown in Fig. I45-12. Recommended design dry-bulb and wet-bulb temperatures are given for American states and major cities. The inside design conditions are the recommended storage conditions for the product being stored.

I45-1.4 Air Infiltration

Outside air entering the refrigerated space must be reduced to the storage temperature, thus increasing the refrigeration load. If the moisture content of the entering air is above that of the refrigerated space, the excess moisture will condense out of the air, also adding to the refrigeration load.

Traffic in and out of a refrigerator can vary with the size and volume of the refrigerator. Calculations for air infiltration from this source would therefore need to consider volume as well as the number of times refrigerator doors are opened.

Because of the many variables involved, it is difficult to calculate the total heat gain due to air infiltration. One method, the air-change method is based on the average number of air changes in a 24-hr period, compared to the refrigerator volume, as illustrated in Fig. I45-13 (p. 872).

This method is used for rooms where the temperature is above 32°F for average use. Values are increased by 2 where there is heavier use and higher temperatures. For storage at 0°F or below with decreased usage, the values are reduced.

Another means of computing infiltration is by the velocity of airflow through an open door. Charts are available that list average infiltration velocity based on door height and temperature difference. If the average time the door is opened each hour can be determined, the average hourly infiltration can be calculated.

Once the rate of infiltration in cu ft per hr has been determined by either method, the heat load can be calculated from the heat removed per cu ft, given in Fig. I45-14 (p. 873).

EXAMPLE

Assume that the volume of a refrigerated room is 1000 ft^3 and the storage temperature is 40°F, with an outside temperature of 95°F and 60% relative humidity. Calculate the heat load.

Solution

From Fig. I45-13 note that a 1000 ft^3 volume would average 17.5 air changes per 24 hrs. This would produce an infiltration of 17,500 ft^3 in 24 hrs (1000 × 17.5).

Referring to Fig. I45-14, for a room at 40°F with 95°F outside temperature and 60% RH, the Btu/ft^3 of heat is 2.62. Therefore, in 24 hrs the load would be 2.62 × 17,500, or 45,850 Btu. In 1 hr it is 45,850 divided by 24, or 1910 Btu/hr.

In systems where ventilation is provided by supply and/or exhaust fans, the ventilation load will replace the infiltration load if it is greater. The heat gain may be calculated on the basis of the ventilating-equipment air volume.

I45-1.5 Product Load

The *product load* is any heat gain from the product in the refrigerated space. The load may come from higher outside temperatures, from a chilling or freezing process, or from the *heat of respiration* of perishable products. It may also be the sum of various types of product loads.

To calculate the refrigeration product load for food products (solids and liquids), it is necessary to know their freezing points, specific heats, percent water, etc.

Sensible Heat Load Above Freezing

Most products are at a higher temperature than the storage temperature when placed in the refrigerator. Since many foods have a high water content, their reaction to a loss of heat is quite different above and below the freezing point due to the change in the state of water. *Specific heat* of a product is defined as the Btu required to raise the temperature of 1 lb of the substance 1°F.

The heat to be removed from a product to lower its temperature (above freezing) may be calculated as follows:

$$Q = W \times C \times (T_1 - T_2)$$

where

Q = Btu to be removed
W = weight of product in lbs
C = specific heat above freezing
T_1 = initial temperature, °F
T_2 = final temperature, °F (freezing or above)

For example, the heat to be removed in order to cool 1000 lb of veal (whose freezing point is 29°F) from 42°F to 29°F can be calculated as follows:

$$\begin{aligned} Q &= W \times C \times (T_1 - T_2) \\ &= 1000 \text{ lb} \times 0.71 \text{ specific heat (veal)} \times 13°\text{F} \\ &= 9230 \text{ Btu} \end{aligned}$$

Location	2½% Design Dry Bulb	Coincident Design Wet Bulb
ALABAMA		
Birmingham AP	94	75
Mobile AP	93	77
Montgomery AP	95	76
ALASKA		
Anchorage AP	68	58
Barrow (S)	53	50
Juneau AP	70	58
ARIZONA		
Flagstaff AP	82	55
Phoenix AP (S)	107	71
Tucson AP (S)	102	66
ARKANSAS		
El Dorado AP	96	76
Fayetteville AP	94	73
Little Rock AP (S)	96	77
CALIFORNIA		
Bakersfield AP	101	69
Eureka/Arcata AP	65	59
Los Angeles CO (S)	89	70
Sacramento AP	98	70
San Bernadino, Norton AFB	99	69
San Diego, AP	80	69
San Francisco CO	71	62
San Jose AP	81	65
COLORADO		
Denver AP	91	59
Durango	87	59
Fort Collins	91	59
CONNECTICUT		
Hartford, Brainard Field	88	73
New Haven AP	84	73
DELAWARE		
Wilmington AP	89	74
DISTRICT OF COLUMBIA		
Washington National AP	91	74
FLORIDA		
Gainesville AP (S)	93	77
Jacksonville AP	94	77
Key West AP	90	78
Miami AP (S)	90	77
Orlando AP	93	76
Tallahassee AP (S)	92	76
Tampa AP (S)	91	77
GEORGIA		
Atlanta AP (S)	92	74
Savannah-Travis AP	93	77
Valdosta-Moody AFB	94	77
HAWAII		
Hilo AP (S)	83	72
Honolulu AP	86	73
IDAHO		
Boise AP (S)	94	64
Coeur D'Alene AP	86	61
ILLINOIS		
Carbondale	93	77
Chicago CO	91	74
Springfield AP	92	74
INDIANA		
Evansville AP	93	75
Indianapolis AP (S)	90	74
South Bend AP	89	73
IOWA		
Burlington AP	91	75
Des Moines AP	91	74
Mason City AP	88	74
KANSAS		
Salina	100	74
Topeka AP	96	75
Wichita AP	98	73
KENTUCKY		
Bowling Green AP	92	75
Lexington AP (S)	91	73
Louisville AP	93	74
LOUISIANA		
Baton Rouge AP	93	77
New Orleans AP	92	78
Shreveport AP (S)	96	76
MAINE		
Bangor, Dow AFB	83	68
Caribou AP (S)	81	67
Portland (S)	84	71
MARYLAND		
Baltimore CO	89	76
Cumberland	89	74
Salisbury (S)	91	75
MASSACHUSETTS		
Boston AP (S)	88	71
New Bedford	82	71
Springfield, Westover AFB	87	71
Worcester AP	84	70
MICHIGAN		
Detroit	88	72
Grand Rapids AP	88	72
Marquette CO	81	69
MINNESOTA		
International Falls AP	83	68
Minneapolis/ St. Paul AP	89	73
Rochester AP	87	72
MISSISSIPPI		
Biloxi, Keesler AFB	92	79
Jackson AP	95	76
Tupelo	94	77
MISSOURI		
Kansas City AP	96	74
St. Louis CO	94	75
Springfield AP	93	74
MONTANA		
Billings AP	91	64
Butte AP	83	56
Great Falls AP (S)	88	60
NEBRASKA		
Lincoln CO (S)	95	74
Omaha AP	91	75
Scottsbluff AP	92	65
NEVADA		
Las Vegas AP (S)	106	65
Reno CO	93	60
NEW HAMPSHIRE		
Berlin	84	69
Concord AP	87	70
Manchester, Grenier AFB	88	71
NEW JERSEY		
Atlantic City CO	89	74
Newark AP	91	73
Trenton CO	88	74
NEW MEXICO		
Albuquerque AP (S)	94	61
Las Cruces	96	64
Santa Fe CO	88	61

Figure 145-12 Summer outside design conditions: United States, Canada. (Copyright by the American Society of Heating, Refrigerating, and Air-Conditioning Engineers, Inc. Used by permission.)

Location	2½% Design Dry Bulb	Coincident Design Wet Bulb
NEW YORK		
Albany CO	88	72
Buffalo AP	85	70
NYC-Central Park (S)	89	73
Rochester AP	88	71
Syracuse AP	87	71
NORTH CAROLINA		
Charlotte AP	93	74
Raleigh/Durham AP (S)	92	75
Wilmington AP	91	78
NORTH DAKOTA		
Bismark AP (S)	91	68
Fargo AP	89	71
Grands Forks AP	87	70
OHIO		
Cincinnati CO	90	72
Cleveland AP (S)	88	72
Columbus AP (S)	90	73
OKLAHOMA		
Muskogee AP	98	75
Oklahoma City AP	97	74
Tulsa AP	98	75
OREGON		
Eugene AP	89	66
Medford AP (S)	94	67
Portland AP	85	67
PENNSYLVANIA		
Allentown AP	88	72
Philadelphia AP	90	74
Pittsburgh CO	88	71
RHODE ISLAND		
Providence AP	86	72
SOUTH CAROLINA		
Charleston CO	92	78
Columbia AP	95	75
Greenville AP	91	74
SOUTH DAKOTA		
Aberdeen AP	91	72
Pierre AP	95	71
Sioux Falls AP	91	72
TENNESSEE		
Chattanooga AP	93	74
Memphis AP	95	76
Nashville AP (S)	94	74
TEXAS		
Corpus Christi AP	94	78
Dallas AP	100	75
Houston CO	95	77
Lubbock AP	96	69
San Antonio AP (S)	97	73
UTAH		
Cedar City AP	91	60
Provo	96	62
Salt Lake City AP (S)	95	62
VERMONT		
Barre	81	69
Burlington AP (S)	85	70
Rutland	84	70
VIRGINIA		
Norfolk AP	91	76
Richmond AP	92	76
Roanoke AP	91	72
WASHINGTON		
Bellingham AP	77	65
Seattle-Tacoma AP (S)	80	64
Spokane AP (S)	90	63
WEST VIRGINIA		
Charleston AP	90	73
Huntington CO	91	74
Wheeling	86	71
WISCONSIN		
Green Bay AP	85	72
Madison AP (S)	88	73
Milwaukee AP	87	73
WYOMING		
Casper AP	90	57
Cheyenne AP	86	58
Sheridan AP	91	62
ALBERTA		
Calgary AP	81	61
Edmonton AP	82	65
BRITISH COLUMBIA		
Vancouver AP (S)	77	66
Victoria CO	73	62
MANITOBA		
Flin Flon	81	66
Winnipeg AP (S)	86	71
NEW BRUNSWICK		
Fredericton AP (S)	85	69
Saint John AP	77	65
NEWFOUNDLAND		
Gander AP	79	65
St. John's AP (S)	75	65
NORTHWEST TERR.		
Fort Smith AP (S)	81	64
Yellowknife AP	77	61
NOVA SCOTIA		
Halifax AP (S)	76	65
Yarmouth AP	72	64
ONTARIO		
Sudbury AP	83	67
Thunder Bay AP	83	68
Toronto AP (S)	87	72
PRINCE EDWARD ISLAND		
Charlottetown AP (S)	78	68
QUEBEC		
Chicoutimi	83	68
Montreal AP (S)	85	72
Quebec AP	84	70
SASKATCHEWAN		
Regina AP	88	68
Saskatoon AP (S)	86	66
YUKON TERRITORY		
Whitehorse AP (S)	77	58

AP–Airport CO–City Office

Figure 145-12 Continued

Volume (cu ft)	Air Changes per 24 hr	Volume (cu ft)	Air Changes per 24 hr
200	44.0	6,000	6.5
300	34.5	8,000	5.5
400	29.5	10,000	4.9
500	26.0	15,000	3.9
600	23.0	20,000	3.5
800	20.0	25,000	3.0
1,000	17.5	30,000	2.7
1,500	14.0	40,000	2.3
2,000	12.0	50,000	2.0
3,000	9.5	75,000	1.6
4,000	8.2	100,000	1.4
5,000	7.2		

Note: For heavy usage multiply the above values by 2. For long storage multiply the above values by 0.6.

Figure I45-13 Average air changes per 24 hours, for storage rooms due to door openings and infiltration. (Above 32°F) (Copyright by the American Society of Heating, Refrigerating, and Air-Conditioning Engineers, Inc. Used by permission.)

Heat of Respiration

Products such as fresh fruits and vegetables, even when stored at above freezing temperatures, give off some heat due to respiration. *Respiration* is the oxidation process of ripening, and carbon dioxide and heat are by-products. This load varies with the type and temperature of the product. Tabulated values, given in Btu/lb/24 hr are shown in Fig. I45-15, and are applied to the total weight of the product stored, not just to the daily turn over. Use a value close to freezing (32°F).

Load Due to Latent Heat of Freezing

To calculate the heat removal required to freeze food products having a high percentage of water, only the water needs to be considered. The *latent heat of freezing* is found by multiplying the latent heat of water (144 Btu/lb) by the percentage of water in the food product.

EXAMPLE

Veal is 63% water and its latent heat is 91 Btu/lb (0.63 × 144 Btu/lb = 91 Btu). Calculate the latent heat of freezing 1000 lb of veal.

Solution

The heat to be removed from a product for the latent heat of freezing may be calculated as follows:

$$Q = W \times h_f$$

where

Q = Btu to be removed
W = weight of product, lb
h_f = latent heat of fusion, Btu/lb

The latent heat of freezing 1000 lb of veal at 29°F is:

$$\begin{aligned} Q &= W \times h_f \\ &= 1000 \text{ lb} \times 91 \text{ Btu/lb} \\ &= 91{,}000 \text{ Btu} \end{aligned}$$

Sensible Heat Load Below Freezing

Once the water content of a product has been frozen, sensible cooling can occur again in the same manner as freezing, with the exception that the ice in the product causes the specific heat to change. For example, note that the specific heat of veal above freezing is 0.71 while the specific heat of veal below freezing is 0.39.

The heat to be removed from a product to reduce its temperature below freezing is calculated as follows:

$$Q = W \times C_i \times (T_f - T_3)$$

where

Q = Btu to be removed
W = weight of product, lb
C_i = specific heat below freezing
T_f = freezing temperature
T_3 = final temperature

EXAMPLE

Calculate the heat to be removed in order to cool 1000 lb of veal from 29°F to 0°F.

Solution

$$\begin{aligned} Q &= W \times C_i \times (T_f - T_3) \\ &= 1000 \text{ lb} \times 0.39 \text{ specific heat} \times (29°\text{F} - 0°\text{F}) \\ &= 1000 \times 0.39 \times 29 \\ &= 11{,}310 \text{ Btu} \end{aligned}$$

Total product load is the sum of the individual calculations for the sensible heat above freezing, the latent heat of freezing, and the sensible heat below freezing. In the preceding example of 1000 lb of veal cooled and frozen from 42°F to 0°F, the total product load would be:

Sensible heat above freezing = 9,230 Btu
Latent heat of freezing = 91,000 Btu
Sensible heat below freezing = 11,310 Btu
Total product load = 111,540 Btu

If several different products are to be considered, separate calculations must be made for each item for an accurate estimate of the total product load. Note that in all the calculations above, the time factor is not considered, either for initial pull down or storage life.

Storage room temp °F	Temperature of Outside Air (°F)							
	85		90		95		100	
	Relative Humidity (Percent)							
	50	60	50	60	50	60	50	60
65	0.65	0.85	0.93	1.17	1.24	1.54	1.58	1.95
60	0.85	1.03	1.13	1.37	1.44	1.74	1.78	2.15
55	1.12	1.34	1.41	1.66	1.72	2.01	2.06	2.44
50	1.32	1.54	1.62	1.87	1.93	2.22	2.28	2.65
45	1.50	1.73	1.80	2.06	2.12	2.42	2.47	2.85
40	1.69	1.92	2.00	2.26	2.31	2.62	2.67	3.06
35	1.86	2.09	2.17	2.43	2.49	2.79	2.85	3.24
30	2.00	2.24	2.26	2.53	2.64	2.94	2.95	3.35

Storage room temp °F	Temperature of Outside Air (°F)							
	40		50		90		100	
	Relative Humidity (Percent)							
	70	80	70	80	50	60	50	60
30	0.24	0.29	0.58	0.66	2.26	2.53	2.95	3.35
25	0.41	0.45	0.75	0.83	2.44	2.71	3.14	3.54
20	0.56	0.61	0.91	0.99	2.62	2.90	3.33	3.73
15	0.71	0.75	1.06	1.14	2.80	3.07	3.51	3.92
10	0.85	0.89	1.19	1.27	2.93	3.20	3.64	4.04
5	0.98	1.03	1.34	1.42	3.12	3.40	3.84	4.27
0	1.12	1.17	1.48	1.56	3.28	3.56	4.01	4.43
–5	1.23	1.28	1.59	1.67	3.41	3.69	4.15	4.57
–10	1.35	1.41	1.73	1.81	3.56	3.85	4.31	4.74
–15	1.50	1.53	1.85	1.92	3.67	3.96	4.42	4.86
–20	1.63	1.68	2.01	2.09	3.88	4.18	4.66	5.10
–25	1.77	1.80	2.12	2.21	4.00	4.30	4.78	5.21
–30	1.90	1.95	2.29	2.38	4.21	4.51	4.90	5.44

Figure 145-14 Heat removed in cooling air to storage-room conditions (Btu per cu ft). (Copyright by the American Society of Heating, Refrigerating, and Air-Conditioning Engineers, Inc. Used by permission.)

145-1.6 Supplementary Loads

Heat gain from other sources must also be included in the total cooling load estimate. Some examples are:

1. Electric energy dissipated in the refrigerated space through lights and heaters (such as defrost) is converted into heat. One W of electric power equals 3.41 Btu.
2. For an electric motor in the refrigerated space, the following table gives the approximate Btu/hr

Motor HP	Btu/hr	Motor HP	Btu/hr
⅙	710	1	3,220
¼	1,000	1½	4,770
⅓	1,290	2	6,380
½	1,820	3	9,450
¾	2,680	5	15,600

3. For people working inside the refrigerated space, the following table gives the approximate heat generated. (Use the table for Heat Equivalent/Person—Btu/hr)

The total supplementary load is the sum of the individual factors contributing to it. For example, the total supplementary load in a refrigerated storeroom maintained at 0°F in which there are 300 W of electric lights, a 3 Hp motor driving a fan, and two people working continuously would be as follows:

$$300 \text{ W} \times 3.41 \text{ Btu/hr} = 1{,}023 \text{ Btu/hr}$$
$$3 \text{ Hp motor} = 9{,}450 \text{ Btu/hr}$$
$$2 \text{ people} \times 1300 \text{ Btu/hr} = 2{,}600 \text{ Btu/hr}$$
$$\text{Total supplementary load} = 13{,}073 \text{ Btu/hr}$$

Btu/lb/24 hrs.				
	Storage Temperature			
Product	32°F	40°F	60°F	°F Other
FRUITS				
Apples	.25–.450	.55–.80	1.5–3.4	
Apricots	0.55–.63	.70–1.0	2.33–3.74	
Avocados	–	–	6.6–15.35	
Bananas	–	–	2.3–2.75	@ 68° 4.2–4.6
Blackberries	1.70–2.52	5.91–5.0	7.71–15.97	
Blueberries	0.65–1.10	1.0–1.35	3.75–6.5	@ 70° 5.7–7.5
Cherries	0.65–0.90	1.4–1.45	5.5–6.6	
Cherries, Sour	0.63–1.44	1.41–1.45	3.0–5.49	
Cranberries	0.30–0.35	0.45–0.50	–	
Figs, Mission	–	1.18–1.45	2.37–3.52	
Gooseberries	0.74–0.96	1.33–1.48	2.37–3.52	
Grapefruit	0.20–0.50	0.35–0.65	1.1–2	
Grapes–American	0.30	0.60	1.75	
Grapes–European	0.15–0.20	0.35–0.65	1.10–1.30	
Lemons	0.25–0.45	0.30–0.95	1.15–2.50	
Limes	–	0.405	1.485	
Melons–Cantaloupes	0.55–0.63	0.96–1.11	3.70–4.22	
Melons–Honey Dew	–	0.45–0.55	1.2–1.65	
Oranges	0.20–0.50	0.65–0.8	1.85–2.6	
Peaches	0.45–0.70	0.70–1.0	3.65–4.65	
Pears	0.35–0.45	–	4.40–6.60	
Plums	0.20–0.35	0.45–0.75	1.20–1.40	
Raspberries	1.95–2.75	3.40–4.25	9.05–11.15	
Strawberries	1.35–1.90	1.80–3.40	7.80–10.15	
Tangerines	1.63	2.93	–	
VEGETABLES				
Artichokes (Globe)	2.48–4.93	3.48–6.56	8.49–15.90	
Asparagus	2.95–6.60	5.85–11.55	11.0–25.75	
Beans, Green or Snap	–	4.60–5.7	16.05–22.05	
Beans, Lima	1.15–1.6	2.15–3.05	11.0–13.7	
Beets, Topped	1.35	2.05	3.60	
Broccoli	3.75	5.50–8.80	16.9–25.0	
Brussels Sprouts	1.65–4.15	3.30–5.50	6.60–13.75	
Cabbage	0.60	0.85	2.05	
Carrots, Topped	1.05	1.75	4.05	
Cauliflower	1.80–2.10	2.10–2.40	4.70–5.40	
Celery	0.80	1.20	4.10	
Corn, Sweet	3.60–5.65	5.30–6.60	19.20	
Cucumbers	–	–	1.65–3.65	
Garlic	0.33–1.19	0.63–1.08	1.18–3.0	
Horseradish	0.89	1.19	3.59	
Kohlrabi	1.11	1.78	5.37	

Figure 145-15 Heat of respiration. (Courtesy of Dunham-Bush, Inc.)

Product	Btu/lb/24 hrs.			
	Storage Temperature			
	32°F	40°F	60°F	°F Other
VEGETABLES				
Leeks	1.04–1.78	2.15–3.19	9.08–12.82	
Lettuce, Head	1.15	1.35	3.95	
Lettuce, Leaf	2.25	3.20	7.20	
Mushrooms	3.10–4.80	7.80	–	@ 50° 11.0
Okra	–	6.05	15.8	
Olives	–	–	2.37–4.26	
Onions, Dry	0.35–0.55	0.90	1.20	
Onions, Green	1.15–2.45	1.90–7.50	7.25–10.70	
Peas, Green	4.10–4.20	6.60–8.0	19.65–22.25	
Peppers, Sweet	1.35	2.35	4.25	
Potatoes, Immature	–	1.30	1.45–3.4	
Potatoes, Mature	–	0.65–0.90	0.75–1.30	
Potatoes, Sweet	–	0.85	2.15–3.15	
Radishes with Tops	1.59–1.89	2.11–2.30	7.67–8.5	
Radishes, Topped	0.59–0.63	0.85–0.89	3.04–3.59	
Rhubarb, Topped	0.89–1.44	1.19–2.0	3.41–4.97	
Spinach	2.10–2.45	3.95–5.60	18.45–19.0	
Squash, Yellow	1.3–1.41	1.55–2.04	8.23–9.97	
Tomatoes, Mature Green	–	0.55	3.10	
Tomatoes, Ripe	0.50	0.65	2.8	
Turnips	0.95	1.10	2.65	
Vegetables, Mixed	2.0	–	–	
MISCELLANEOUS				
Caviar, Tub	–	–	1.91	
Cheese, American	–	–	2.34	
Camembert	–	–	2.46	
Limburger	–	–	2.46	
Roquefort	–	–	–	@ 45° 2.0
Swiss	–	–	2.33	
Flowers, Cut	0.24 Btu/24 Hrs/Sq Ft Floor Area			
Honey	–	0.71	–	
Hops	–	–	–	@ 35° 0.75
Malt	–	–	–	@ 50° 0.75
Maple Sugar	–	–	–	@ 45° 0.71
Maple Syrup	–	–	–	@ 45° 0.71
Nuts	0.074	0.185	0.37	
Nuts, Dried	–	–	–	@ 35° 0.50

Notes: All fruits and vegetables are living and give off heat in storage: If heat of respiration not given, an approximate value or average value should be used. For Btu/24 hrs./Ton/°F multiply by 2000.

Figure I45-15 Continued

145-1.7 Total Hourly Load

For refrigeration appliances produced in quantity, the load is usually specified by the manufacturer. The refrigeration equipment is pre-selected and may be already installed in the fixture.

If it must be estimated, the expected load should be calculated by determining the heat gain due to each of the factors contributing to the total load. The most accurate methods use forms and data available from the manufacturer for such purposes, and each factor is considered separately as well.

Refrigeration equipment is designed to function continuously, and normally the compressor operating time is determined by the requirements of the defrost system. The load is calculated on a 24-hr basis and the required hourly compressor capacity is determined by dividing the 24-hr load by the desired number of hours of compressor operation during the 24-hr period. A reasonable safety factor must be provided to enable the unit to recover rapidly after a temperature rise, and to allow for any load that might be larger than originally estimated.

In cases where the refrigerant evaporating temperature does not drop below 30°F, frost will not accumulate on the evaporator, and no defrost period is necessary. The compressor for such applications is chosen on the basis of an 18- to 20-hr period of operation.

For applications with storage temperatures of 35°F or higher and refrigerant temperatures low enough to cause frosting, it is common practice to defrost by stopping the compressor and allowing the return air to melt the ice from the coil. Compressors for such applications should be selected for 16- to 18-hr operation periods.

On low-temperature applications, some integrated means of defrosting must be provided. With normal defrost periods, 18-hr compressor operation is usually acceptable, although some systems are designed for continuous operation except during the defrost period.

An additional 5 to 10% safety factor is often added to load calculations as a conservative measure to be sure that the equipment will not be undersized. When data concerning the refrigeration load are uncertain, this practice may be desirable. Since, however, the compressor is sized on the basis of 16- or 18-hour operation, this provides a sizable safety factor. The load should be calculated on the basis of the peak demand at design conditions. Usually the design conditions are selected on the assumption that peak demand will occur no more than 1% of the hours during the summer months.

145-1.8 Selection of Equipment

The following information is important to know in the selection of equipment:

1. Daily cooling load
2. Running time for the compressors (hours per day)
3. Hourly cooling load in Btu/hr
4. Refrigerated room temperature
5. Refrigerated room relative humidity (%RH)
6. Coil operating temperature difference (*TD* = room temperature − evaporating temperature)
7. Evaporator temperature
8. Drop in suction line temperature due to pressure loss between coil and compressor
9. Condensing temperature of the compressor including discharge-line temperature drop to remote condenser
10. Available coil and compressor sizes

The following tables are useful in determining the required compressor and evaporator operating conditions:

a. Storage temperature and humidity conditions for various foods (Fig. I45-16). Note the moisture classification for the foods.
b. Coil temperature differences for four classes of foods (Fig. I45-17).
c. Compressor running time, shown in Fig. I45-18.

145-1.9 Sample Calculations

EXAMPLE

Calculate the load and select the equipment for a walk-in cooler having the following specifications:

Location:	Atlanta, GA
Room use:	Storage mixed vegetables
Outside dimensions:	12 ft × 14 ft × 9 ft
Room temperature:	35°F
Insulation:	3 in. molded polystyrene
Total room product load:	8000 lb
Daily delivery:	2000 lb precooled to 45°F
Containers:	Paper cartons (200 lbs)

Calculations are made as shown on the data sheet (Fig. I45-19). For condensing-unit performance see Fig. I45-20 and for evaporator performance see Fig. I45-21.

EXAMPLE

Calculate the load and select the equipment for a walk-in cooler having the following specifications:

Location:	Atlanta, GA
Room use:	Storage frozen beef
Outside dimensions:	10 ft × 12 ft × 9 ft
Room temperature:	0°F
Insulation:	5 in. molded polystyrene
Total room product load:	2000 lb
Daily delivery:	300 lb precooled to 40°F
Containers:	Paper cartons (200 lb)
Type of defrost:	Electric

Calculations are made as shown on the data sheet (Fig. I45-22). For condensing-unit performance, see Fig. I45-20; for evaporator performance see Fig. I45-23.

Product	Short-Time Storage		24–72 Hr. Storage	
	°F	Moisture	°F	Moisture
Vegetables	36–42	Class 2	32–36	Class 1
Fruits	36–42	Class 2	32–36	Class 2
Meats (cut)	34–38	Class 2	32–36	Class 2
Meats (carcass)	34–38	Class 3	32–36	Class 3
Poultry	32–36	Class 2	30–35	Class 1–2‡
Fish	35–40	Class 1	35–40	Class 1
Eggs	36–42	Class 2	31–35	Class 1
Butter, Cheese	38–45	Class 1*	35–40	Class 2–3†
Bottled beverages	35–45	Class 4	40–45	Class 4
Frozen Foods			0	Class 2

ASRE 1959 Data Book and ASHRAE 1962 Guide and Data book. Reprinted by permission.
*If not packaged. †If packaged ‡Freeze and hold at 0°F or below if held for more than 72 hours.

Figure 145-16 Storage temperature and humidity conditions for various food products. (Courtesy of Dunham-Bush, Inc.)

CLASSES OF FOOD

Class 1. Such products as eggs, unpackaged butter and cheese and most vegetables held for comparatively long periods of time. These products require very high relative humidity because it is necessary to effect a minimum of moisture evaporation during storage.

Class 2. Such foods as cut meats, fruits and similar products. These require high relative humidities but not as high as Class 1.

Class 3. Carcass meats and fruit such as melons which have tough skins. These products require only moderate relative humidities because they have surfaces whose rate of moisture evaporation is moderate.

Class 4. Canned goods, bottled goods, and other products which have a protective covering. These are products which need only low relative humidities or which are unaffected by humidity. Products from whose surfaces there is a very low rate of moisture evaporation or none at all, fall into this class.

Type Coils	Class 1	Class 2	Class 3	Class 4
Forced Air Coils	6–9°F	9–12°F	12–20°F	Above 20°F
Gravity Coils	14–18°F	18–22°F	21–28°F	27–37°F

Temperature difference is defined as average fixture temperature minus average refrigerant temperature.

Figure 145-17 Temperature differences between evaporator coil and room for various types of food products. (Courtesy of Dunham-Bush, Inc.)

Room Temp. °F	Evap. Temp. °F	Defrosting		Compressor Running Time Hrs.	Rel. Hum. %	Evap. °F TD
		Type	No./24 Hrs.			
Over 35	Over 30	None	–	18–20 & up	90 85 80 75	8–10 10–12 12–15 16–20
35 & up		Ambient	4	16		
35 to 25	Below 30	Elec. or Hot Gas	4	18–(20*)		
10 & less			6	(18*)–20	–	8–10

*Preferred compressor running time.

Figure 145-18 Compressor running times for various food-storage room temperatures. (Courtesy of Dunham-Bush, Inc.)

Location: Atlanta, Ga.
Room use: Storage of mixed vegetables
Outside dimensions: 12 × 14 × 9
Room temperature: 35°F

Insulation: 3 in. molded polystyrene
Total room load: 8000 lb
Daily delivery: 2000 lb precooled to 45°F
Containers: Paper cartons, total weight 200 lb

- Load Survey & Estimating Data -

DESIGN AMBIENT: 95 °F DB, ______ °F WB, 50 %RH, ______ °F SUMMER GROUND TEMP.
(USE 55°F FOR INSULATED FREEZE FLOOR SLAB)

ROOM DESIGN: 35 °F DB, ______ °F WB, 80 %RH ______ °F WINTER DESIGN AMBIENT

ACCESS AREA: 95 °F DB, ______ °F WB, 50 %RH, (ANTE-RM/LOADING DOCK/OTHER)

ROOM DIM. OUTSIDE: 12 FT. W 14 FT. L 9 FT. H 804 TOTAL FT². (OUTSIDE SURFACE)

	Insulation				Wall Thick-ness	Adj. Area °F	Effective Wall TD	Sun Effect (°F)	Total TD	Overall wall heat gain
	Type	Thick (inches)	K Factor	U* Factor						Btu/24 hrs/ft²
N. Wall	POLY.	3"	20	0.67	5"	95	60	■	60	96
S. Wall										
E. Wall										
W. Wall										
Ceiling										
Floor								■		

$$\text{*U – Factor} = \frac{K}{\text{Insul. Thickness (in.)}}$$

REFRIG. DOOR(S): (1) 7 × 4 VENT. FAN(S): ______

ROOM INT. VOL: 11 W × 13 L × 8 H = 1144 FT³.
(INSIDE ROOM DIMENSION = OUTSIDE DIMENSION – WALL THICKNESSES)

FLOOR AREA 11 W × 13 L = 143 FT².

ELECTRICAL POWER 240 VOLTS, 3 PH. 60 HZ: CONTROL: 120 V

TYPE CONTROL: ______

PRODUCT DATA AND CLASS OF PRODUCT:

Type Product	Amount of Product			Product Temp. °F		Specific Heat					
	Amount Storage	Daily Turn-Over	Freezing or Cooling	Enter-ing	Final	Above Freeze	Below Freeze	Lat. Ht. Freeze Btu/lb.	Highest Product Freeze Temp.	Heat or Respir'n Btu/lb 24 Hr.	( ) Pull-Down ( ) Freezing Time Hrs.
MIX. VEG.	8000	2000	COOLING	45	35	.9				2.0	

EVAP. TD 10 ; TYPE DEFROST: ☐ AIR, ☐ HOT GAS, ☐ ELECTRIC

CLASS PRODUCT 2 : ______;

NO. OF DEFROSTS—TOTAL TIME PER 24 HR; ______ HR.

COMPRESSOR RUNNING TIME: 16 HR.

BOX USAGE: ☐ AVERAGE, ☐ HEAVY, ☐ EXTRA HEAVY

PRODUCT LOAD AND ADDITIONAL INFORMATION 2000 lb. mixed vegetables per day, entering temperature 45°F

PACKAGING Paper cartons CONTAINERS ______ WGT. 200 lbs. SP. HT. .32 (CONTAINER)

PALLETS: NO. ______ SIZE ______ WGT. EA. ______ SP. HT. ______

PRODUCT RACKS: NO. ______ MAT'L ______ WGT. EA. ______ SP. HT. ______

ESTIMATING PRODUCT LOADING CAPACITY OF ROOM

ESTIMATED PRODUCT LOADING = 0.40 × ______ (ROOM VOLUME) FT³ × ______ (LOADING DENSITY) LBS./FT³ = ______ LB:

MISCELLANEOUS LOADS

PEOPLE: NO. ______ HR. ______

MOTORS (OTHER THAN EVAP. FAN)

USE: ______, ______ HP. ______ HR. ______
______, ______ HP. ______ HR. ______

FORK LIFTS: ______ NO., ______ HP ______ HR./DAY, OTHER ______

LIGHTS ______ W/ FT².

Figure 145-19 Load calculation data sheet for food storage rooms above freezing.

Calculations

I WALL LOSS (TRANSMISSION LOAD)				
SURFACE	TD	AREA OF SURFACE	WALL HEAT GAIN FACTOR	BTU/24 HR
N. Wall		____ Ft. L × ____ Ft. H = ____ Ft². ×	____ =	
S. Wall		____ Ft. L × ____ Ft. H = ____ Ft². ×	____ =	
E. Wall		____ Ft. L × ____ Ft. H = ____ Ft². ×	____ =	
W. Wall		____ Ft. L × ____ Ft. H = ____ Ft². ×	____ =	
Ceiling		____ Ft. L × ____ Ft. W = ____ Ft². ×	____ =	
Floor		____ Ft. L × ____ Ft. W = 804 Ft². ×	96 =	77,184
Box	60	Total: Surface = ____ Ft². ×	____ =	77,184
			I TOTAL WALL TRANSMISSION LOAD =	

	BTU/24 HR
II (LONG FORM) INFILTRATION (AIR CHANGE LOAD)	
1144 FT³. × 17.5 AIR CHANGES/24 HR × 1 SERVICE FACTOR × 2.49 BTU /FT³ =	49,850
II TOTAL INFILTRATION LOAD =	49,850
III PRODUCT LOAD	
PRODUCT TEMP. REDUCTION ABOVE FREEZING (SENSIBLE HEAT)	
2000 *LB/DAY × 10 °F TEMP. REDUCTION × .9 SP. HT. =	18,000
PRODUCT FREEZING (LATENT HEAT LOAD)	
____ *LB/DAY × ____ BTU/LB/LATENT HEAT =	
PRODUCT TEMP. REDUCTION BELOW FREEZING (SENSIBLE HEAT)	
____ *LB/DAY × ____ °F TEMP. REDUCTION × ____ SP. HT. =	
HEAT OF RESPIRATION	
8000 LB PRODUCT (STORAGE) × 2.0 BTU/LB/24 HR =	16,000
MISCELLANEOUS PRODUCT LOADS (1) CONTAINERS (2) PALLETS (3) OTHER	
200 LB/DAY × 10 °F TEMP. REDUCTION × .32 SP. HT. =	640
____ LB/DAY × ____ °F TEMP. REDUCTION × ____ SP. HT. =	
III TOTAL PRODUCT LOAD =	34,640
IV MISCELLANEOUS LOADS	
(a) LIGHTS ____ Ft.² Floor Area × ____ W/Ft.² × 3.41 Btu/W × ____ HR/24 HR (1 TO 1½ W/FT². IN STORAGE AREAS & 2 TO 3 FOR WORK AREAS) =	
(b) OCCUPANCY ____ NO. OF PEOPLE × ____ BTU/HR × ____ HR =	
(c) MOTORS ____ BTU/HP/HR × ____ HP × ____ HR/24 HR =	
____ BTU/HP/HR × ____ HP × ____ HR/24 HR =	
(d) MATERIAL HANDLING ____ FORK LIFT(S) × ____ EQUIV. HP × 3100 BTU/HR/HP × ____ HR OPERATION =	
OTHER ____ =	
IV TOTAL MISCELLANEOUS LOADS =	
TOTAL BTU LOAD I TO IV =	161,674
ADD 10% SAFETY FACTOR =	16,167
TOTAL BTU/24 HR WITH SAFETY FACTOR (NOT INCLUDING EVAP. FAN OR DEFROST HEAT LOADS) 24 HR. BASE REFRIGERATION LOAD =	177,841

*IF THE PRODUCT PULL-DOWN IS ACCOMPLISHED IN LESS THAN 24 HR THE DAILY PRODUCT WILL BE:

$$\text{POUNDS PRODUCT} \times \frac{\text{24 HRS.}}{\text{PULL-DOWN HOURS}}$$

Figure 145-19 Continued

Equipment Selection from Load Calculation Form

1. DETERMINE EVAP. TD REQUIRED FOR CLASS OF PRODUCT AND ROOM TEMP 10 °F (TD) (FROM LOAD SURVEY DATA)

2. DETERMINE COMPRESSOR RUNNING TIME BASED ON OPERATING TEMPERATURES AND DEFROST REQUIREMENTS 16 HRS. (FROM LOAD SURVEY DATA)

3. EVAPORATOR TEMP. °F = 35 (ROOM TEMP) − 10 [EVAP. TD] = 25 °F (FROM LOAD SURVEY DATA)]

4. COMP. SUCT. TEMP. °F = 25 (EVAP. SUCT. TEMP) − 2 (SUCT. LINE LOSS) = 23 °F

BTU/24 HR. BASE REFRIGERATION LOAD WITH SAFETY FACTOR = 177,641
(NOT INCLUDING EVAPORATOR FAN OR DEFROST HEAT)

PRELIMINARY HOURLY LOAD = BTU/24 HR (BASE LOAD) / HR/DAY (COMP. RUNNING TIME) = ______ BTU/HR

FAN HEAT LOAD ESTIMATE BTU/HR =

______ QTY. (MOTORS) × ______ WATTS EA. (INPUT) × 3.41 BTU/W × ______ HR = ______ BTU/24 HR

OR 1 QTY. (MOTORS) × 1/6 HP EA. × 4350 BTU/HP/HR × 24 HR = 17,400 BTU/24 HR

DEFROST HEAT LOAD ESTIMATE BTU/HR =

______ QTY. EVAPS. × ______ W EA. × ______ HR × 3.41 BTU/W × ______ DEFROST LOAD FACTOR*

= ______ BTU/HR

*USE 0.50 FOR ELECTRIC DEFROST, 0.40 FOR HOT GAS DEFROST

BTU/24 HR TOTAL LOAD = 177,641 (BASE LOAD) + 17,400 (FAN HEAT) + ______ (DEFROST HEAT) = 195,041 BTU/24 HR

OR ______ (BASE LOAD) × ______ (BASE LOAD MULT.) = ______ BTU/24 HR

ACTUAL HOURLY LOAD = 195041 BTU/24 HR (TOTAL LOAD) / 16 HRS/DAY (COMP. RUNNING TIME) = 12,190 BTU/HR

Equipment Selection

	Compressor Units		Condensing Units		Evaporators		Condensers	
MODEL NO.			154		WJ 120			
QUANTITY			1		1			
BTU/HR. CAPACITY (EA.)			12,600		12,000			
CFM AIR VOLUME (EA.)								
	DESIGN	**ACTUAL**	**DESIGN**	**ACTUAL**	**DESIGN**	**ACTUAL**	**DESIGN**	**ACTUAL**
°F EVAPORATOR TEMP.					25			
°F EVAP TD								
°F SUCTION TEMP.			25					
°F CONDENSING TEMP.								
°F DESIGN AMBIENT TEMP.			100					
°F MIN. OPER. AMBIENT TEMP.								

Figure 145-19 Continued

				Low Temp.				Commercial Temp.				High Temp.		
				Saturated Suction Temperature °F										
	60 HZ.	50 HZ.		-40°F	-30°F	-20°F	-10°F	0°F	10°F	20°F	25°F	30°F	40°F	50°F
				Suction Pressure — Psig or *Vacuum Inches of Mercury										
			Ambient Air	10.9*	5.5*	0.6	4.5	9.2	14.64	21.04	24.61	28.45	36.97	46.70
HP	AH/AWH	AH/AWH	Temp. °F	Btu/hr	Btu/hr	Btu/hr	Btu/hr	Btu/hr	Btu/hr	Btu/hr	Btu/hr	Btu/hr	Btu/hr	Btu/hr
1/2	Δ5HCL		90	–	1,370	1,900	2,600	3,350	4,150	5,000	5,400	6,000	6,900	7,800
			100	–	1,250	1,750	2,350	3,050	3,800	4,650	5,100	5,800	6,700	7,600
3/4	Δ8HCL		90	–	1,550	2,200	3,000	3,800	4,700	5,700	6,250	6,700	7,900	9,500
			100	–	1,400	2,000	2,700	3,500	4,400	5,400	5,950	6,500	7,700	9,000
	Δ9HCL		90	–	1,950	2,750	3,700	4,700	5,800	7,050	7,800	8,400	9,700	11,100
			100	–	1,750	2,500	3,400	4,350	5,400	6,500	7,200	7,700	8,900	10,200
1	Δ10HCL		90	–	2,250	3,200	4,350	5,550	6,900	8,300	9,150	9,800	11,400	13,000
			100	–	2,050	2,850	3,900	5,100	6,450	7,850	8,700	9,400	11,000	12,600
	Δ11HCL**		90	2,070	2,900	4,150	5,600	7,100	8,800	10,600	11,700	12,500	14,600	16,600
			100	1,850	2,650	3,800	5,100	6,600	8,400	10,000	11,100	12,000	14,100	16,100
11/2	15H	15H5	90	–	–	–	–	–	11,800	12,200	13,400	14,600	16,800	18,900
			100	–	–	–	–	–	9,700	11,600	12,600	13,800	15,900	18,000
	15C	15C5	90	–	–	–	7,100	9,300	11,500	14,500	16,000	–	–	–
			100	–	–	–	6,500	8,700	10,900	13,600	15,300	–	–	–
	15L	15L5	90	2,800	4,200	5,800	7,600	9,600	–	–	–	–	–	–
			100	2,500	3,700	5,400	7,000	8,900	–	–	–	–	–	–
2	20H	20H5	90	–	–	–	–	–	14,500	16,500	17,500	19,000	22,000	26,000
			100	–	–	–	–	–	12,000	14,500	15,500	17,000	20,000	23,000
	20C	20C5	90	–	–	–	8,600	11,300	15,400	17,500	19,400	–	–	–
			100	–	–	–	8,100	10,600	14,000	16,200	18,000	–	–	–
	20L	22L5	90	3,600	5,500	7,500	9,800	12,200	–	–	–	–	–	–
			100	3,400	5,000	7,000	9,100	11,400	–	–	–	–	–	–
3	30H	32H5	90	–	–	–	–	–	18,000	23,000	25,500	28,000	32,500	38,000
			100	–	–	–	–	–	17,000	21,500	24,000	26,000	30,500	35,400
	30C	30C5	90	–	–	–	–	18,600	23,000	28,700	32,000	–	–	–
			100	–	–	–	–	17,300	21,500	27,000	30,000	–	–	–
	32C		90	–	–	–	12,500	16,200	19,900	25,400	28,200	–	–	–
			100	–	–	–	11,500	15,000	18,500	24,000	26,400	–	–	–
	30L	30L5	90	6,000	8,500	12,000	16,000	20,500	–	–	–	–	–	–
			100	5,400	8,000	11,000	15,000	19,000	–	–	–	–	–	–
		41PL5	90	6,560	9,700	13,900	18,900	24,100	–	–	–	–	–	–
			100	6,070	9,000	12,400	17,500	22,500	–	–	–	–	–	–

Figure 145-20 Typical condensing unit performance data.

HP	60 HZ.	50 HZ.	Ambient Air Temp. °F	Low Temp.				Commercial Temp.				High Temp.		
				Saturated Suction Temperature °F										
				-40°F	-30°F	-20°F	-10°F	0°F	10°F	20°F	25°F	30°F	40°F	50°F
				Suction Pressure — Psig or *Vacuum Inches of Mercury										
				10.9*	5.5*	0.6	4.5	9.2	14.64	21.04	24.61	28.45	36.97	46.70
HP	AH/AWH	AH/AWH	Ambient Air Temp. °F	Btu/hr	Btu/hr	Btu/hr	Btu/hr	Btu/hr	Btu/hr	Btu/hr	Btu/hr	Btu/hr	Btu/hr	Btu/hr
5	50H	50H5	90	–	–	–	–	–	30,000	37,500	41,000	45,000	52,500	61,500
			100	–	–	–	–	–	29,300	36,000	39,000	42,700	50,000	58,500
	50C	51PC5	90	–	–	–	21,300	27,400	33,500	41,800	46,300	–	–	–
			100	–	–	–	18,600	24,600	30,600	38,300	43,000	–	–	–
	51PL		90	8,000	11,800	17,000	23,000	29,400	–	–	–	–	–	–
			100	7,400	11,000	15,100	21,300	27,400	–	–	–	–	–	–
		51PH5	90	–	–	–	–	–	37,200	46,500	50,800	55,800	65,100	76,300
			100	–	–	–	–	–	36,300	44,600	48,400	53,000	62,000	72,500
		62PC5	90	–	–	–	26,200	33,700	41,200	51,400	57,000	–	–	–
			100	–	–	–	22,900	30,200	37,600	47,100	52,900	–	–	–
		61PL5	90	9,800	14,500	20,900	28,300	36,200	–	–	–	–	–	–
			100	9,100	13,500	18,600	26,200	33,700	–	–	–	–	–	–
71/2	76PH/C		90	–	–	–	26,500	34,000	41,500	54,000	60,000	67,000	84,500	105,000
			100	–	–	–	23,900	32,400	38,000	50,500	56,000	62,500	78,500	96,000
	77PC	77PC5	90	–	–	–	32,000	41,000	54,000	67,000	75,000	–	–	–
			100	–	–	–	30,000	39,000	50,000	63,500	70,000	–	–	–
	76PL	75PL5	90	10,700	16,600	25,400	35,400	47,100	–	–	–	–	–	–
			100	9,700	15,500	23,300	32,600	43,500	–	–	–	–	–	–
		77PH5	90	–	–	–	–	–	50,400	66,400	73,800	82,400	104,000	129,000
			100	–	–	–	–	–	46,700	62,100	68,900	76,900	96,600	118,000
10	101PH	101PH5	90	–	–	–	–	–	68,900	90,600	100,700	112,300	141,700	176,100
			100	–	–	–	–	–	64,600	84,700	93,900	104,800	131,700	161,000
	101PC	101PC5	90	–	–	–	40,000	51,300	67,500	83,800	93,800	–	–	–
			100	–	–	–	37,500	48,800	62,500	79,400	87,500	–	–	–
	104PC		90	12,500	20,000	30,000	46,000	61,000	–	–	–	–	–	–
			100	11,750	18,800	28,200	43,300	57,400	–	–	–	–	–	–
7	D154PH		90	–	–	–	–	–	74,800	101,200	106,400	119,000	147,600	182,000
			100	–	–	–	–	–	70,600	92,700	99,300	111,000	137,000	166,200
	154PC†		90	–	–	–	42,700	56,800	71,200	95,900	–	–5	–	–
			100	–	–	–	39,400	52,500	67,200	89,600	–	–	–	–

† This unit will not operate at suction temperatures greater than 20°F. Δ AH units only. * Inches of mercury below one atmosphere.
** Not available for +30°F through +50°F SST 208v/1ø applications.

Figure 145–20 Continued

Wall Jet (WJ) Unit Coolers Series "C"

Wall Jet Unit Cooler Model	Unit Capacity BTU/HR @ 10°F TD	Unit Capacity BTU/HR @ 15°F TD	Fan CFM	Fan Size Inch	Motor HP 120/1/60†	Total Fan Mtr. Amps 120V	Optional Heat Exchanger Recommended	Connections** Coil Inlet	Connections** Suction O.D.	Connections** Drain O.D.	Approx. Shipping Weight Lbs.
WJ35*	3,500	5,250	625	10	1/20	.82	B 25XS	½ FL.	½	1⅛	40
WJ45*	4,500	6,750	725	10	1/20	.82	B 25 XS	½ FL.	½	1⅛	46
WJ65	6,500	9,750	1050	12	1/10	1.70	B 50XS	½ FL.	½	1⅛	52
WJ85	8,500	12,750	1300	12	1/10	1.70	B 75XS	½ FL.	¾	1⅛	60
WJ105	10,500	15,750	1600	14	⅙	1.70	B 75XS	½ FL.	¾	1⅛	66
WJ120	12,000	18,000	1800	14	⅙	1.70	B 75XS	½ FL.	¾	1⅛	72
WJ150	15,000	22,500	2375	16	⅙	1.70	B 120XS	½ FL.	⅞	1⅛	82
WJ180	18,000	27,000	2900	16	⅙	1.70	B 120XS	½ FL.	⅞	1⅛	88
WJ240	24,000	36,000	4000	16	½	3.00	B 200XS	½ FL.	1⅛	1⅛	108

Figure 145-21 Evaporator performance data above freezing.

145-2 RESIDENTIAL HEATING AND COOLING LOADS

Residential load calculations have certain unique characteristics that are best served by using a procedure especially designed for residential applications. Many procedures are available, including some available for computer operation. One method which has industry-wide acceptance is *Manual J: Load Calculation for Residential Winter and Summer Air Conditioning,* published by the Air Conditioning Contractors of America (ACCA). The two divisions of the manual, one for heat losses and the other for heat gains, provide practical procedures and data for making load calculations. The following material from Manual J illustrates this procedure:

145-2.1 Heat Loss

1. Outside and inside design conditions: The manual describes selecting the outside design conditions from ASHRAE weather data and recommends an inside design temperature of 70°F.
2. Building losses, which includes those associated with the building envelope such as:
   a. Heat loss through glass windows and doors by conduction.
   b. Heat loss through solid doors by conduction.
   c. Heat loss through walls exposed to outdoor temperatures or through walls below grade.
   d. Heat loss through partitions which separate spaces within the structure that are at different temperatures.
   e. Heat loss through ceilings to a colder room or to an attic.
   f. Heat loss through a roof-ceiling combination.
   g. Heat loss through floors to a colder basement, crawl space, or to the outside.
   h. Heat loss through on-grade slab floors or through basement floors.

PREVENTIVE MAINTENANCE

A building which will not handle the load during heavy load periods may not necessarily have undersized equipment. The equipment may only require cleaning and fine tuning or adjustments. Good preventive maintenance will increase efficiencies and bring back equipment to original design parameters. This includes adequate airflow across both evaporator and condenser coils, valves adjusted appropriately, and refrigerant pressures operating as normally intended. If airflow is low, replace filters and clean coils. If refrigerant pressures are low, find and fix the leaks and recharge as required. Upon completion of the preventive maintenance and service procedures, be sure to inform the homeowner of the progress and the findings.

Location: Atlanta, Ga.
Room use: Storage of frozen beef
Outside dimensions: 10 × 12 × 9
Room temperature: 0°F

Insulation: 5 in. molded polystyrene
Total room load: 2000 lb
Daily delivery: 300 lb precooled to 40°F
Containers: Paper cartons, total weight 200 lb
Type of defrost: electric

- Load Survey & Estimating Data -

DESIGN AMBIENT: 95 °F DB, ______ °F WB, 50 %RH, ______ °F SUMMER GROUND TEMP.
(USE 55°F FOR INSULATED FREEZE FLOOR SLAB)

ROOM DESIGN: ______ °F DB, 0 °F WB, ______ %RH ______ °F WINTER DESIGN AMBIENT

ACCESS AREA: 95 °F DB, 50 °F WB, ______ %RH, (ANTE-RM/LOADING DOCK/OTHER)

ROOM DIM. OUTSIDE: 10 FT. W 12 FT. L 9 FT. H 636 TOTAL FT². (OUTSIDE SURFACE)

	Insulation				Wall Thick-ness	Adj. Area °F	Effective Wall TD	Sun Effect (°F)	Total TD	Overall wall heat gain
	Type	Thick (inches)	K Factor	U* Factor						Btu/24 hrs/ft²
N. Wall	Polystyrene	5"	.20	.04	6	95	95		95	91
S. Wall										
E. Wall										
W. Wall										
Ceiling										
Floor										

$$*U - Factor = \frac{K}{\text{Insul. Thickness (in.)}}$$

REFRIG. DOOR(S): (1) - 7 × 4 VENT. FAN(S): ______

ROOM INT. VOL: 9 W × 11 L × 8 H = 792 FT³.
(INSIDE ROOM DIMENSION = OUTSIDE DIMENSION – WALL THICKNESSES)

FLOOR AREA 9 W × 11 L = 99 FT².

ELECTRICAL POWER 240 VOLTS, 3 PH. 60 HZ: CONTROL: 120 V

TYPE CONTROL: ______

PRODUCT DATA AND CLASS OF PRODUCT: ______

Type Product	Amount of Product			Product Temp. °F		Specific Heat		Lat. Ht. Freeze Btu/lb.	Highest Product Freeze Temp.	Heat or Respir'n Btu/lb 24 Hr.	() Pull-Down () Freezing Time Hrs.
	Amount Storage	Daily Turn-Over	Freezing or Cooling	Enter-ing	Final	Above Freeze	Below Freeze				
Beef	2000	300	Freezing	40	0	.80	.40	100	28		

EVAP. TD 10 ; TYPE DEFROST: ☐ AIR, ☐ HOT GAS, ☐ ELECTRIC

CLASS PRODUCT II : ______;

NO. OF DEFROSTS—TOTAL TIME PER 24 HR; 6 HR.

COMPRESSOR RUNNING TIME: 18 HR.

BOX USAGE: ☑ AVERAGE, ☐ HEAVY, ☐ EXTRA HEAVY

PRODUCT LOAD AND ADDITIONAL INFORMATION ______

PACKAGING ______ CONTAINERS ______ WGT. 200 lbs. ______ SP. HT. .32 ______ (CONTAINER)

PALLETS: NO. ______ SIZE ______ WGT. EA. ______ SP. HT. ______

PRODUCT RACKS: NO. ______ MAT'L ______ WGT. EA. ______ SP. HT. ______

ESTIMATING PRODUCT LOADING CAPACITY OF ROOM

ESTIMATED PRODUCT LOADING = 0.40 × ______ FT³ × ______ LBS./FT³ = ______ LB:
(ROOM VOLUME) (LOADING DENSITY)

MISCELLANEOUS LOADS

PEOPLE: NO. ______ HR. ______

MOTORS (OTHER THAN EVAP. FAN)

USE: ______, ______ HP. ______ HR. ______
______, ______ HP. ______ HR. ______

FORK LIFTS: ______ NO., ______ HP ______ HR./DAY, OTHER ______

LIGHTS ______ W/ FT².

Figure 145-22 Load calculation data sheet for food storage rooms below freezing.

Calculations

I WALL LOSS (TRANSMISSION LOAD				
		WALL HEAT GAIN		
SURFACE	TD	AREA OF SURFACE	FACTOR	BTU/24 HR
N. Wall		___ Ft. L × ___ Ft. H = ___ Ft^2. × ___	=	
S. Wall		___ Ft. L × ___ Ft. H = ___ Ft^2. × ___	=	
E. Wall		___ Ft. L × ___ Ft. H = ___ Ft^2. × ___	=	
W. Wall		___ Ft. L × ___ Ft. H = ___ Ft^2. × ___	=	
Ceiling		___ Ft. L × ___ Ft. W = ___ Ft^2. × ___	=	
Floor		___ Ft. L × ___ Ft. W = 636 Ft^2. × 91	=	57,876
Box	95	Total: Surface = ___ Ft^2. × ___	=	57,876
		I TOTAL WALL TRANSMISSION LOAD	=	

II (LONG FORM) INFILTRATION (AIR CHANGE LOAD)

792 FT^3. × 15.3 AIR CHANGES/24 HR × 1 SERVICE FACTOR × 3.28 BTU /FT^3 = 39,745

II TOTAL INFILTRATION LOAD = 39,745

III PRODUCT LOAD

PRODUCT TEMP. REDUCTION ABOVE FREEZING (SENSIBLE HEAT)

300 *LB/DAY × 12 °F TEMP. REDUCTION × .80 SP. HT. = 2880

PRODUCT FREEZING (LATENT HEAT LOAD)

300 *LB/DAY × 100 BTU/LB/LATENT HEAT = 30,000

PRODUCT TEMP. REDUCTION BELOW FREEZING (SENSIBLE HEAT)

300 *LB/DAY × 28 °F TEMP. REDUCTION × .40 SP. HT. = 3360

HEAT OF RESPIRATION

___ LB PRODUCT (STORAGE) × ___ BTU/LB/24 HR =

MISCELLANEOUS PRODUCT LOADS (1) CONTAINERS (2) PALLETS (3) OTHER

200 LB/DAY × 40 °F TEMP. REDUCTION × .32 SP. HT. = 2560

___ LB/DAY × ___ °F TEMP. REDUCTION × ___ SP. HT. =

III TOTAL PRODUCT LOAD = 38,800

IV MISCELLANEOUS LOADS

(a) LIGHTS ___ $Ft.^2$ Floor Area × ___ $W/Ft.^2$ × 3.41 Btu/W × ___ HR/24 HR =
(1 TO 1½ W/FT^2. IN STORAGE AREAS & 2 TO 3 FOR WORK AREAS)

(b) OCCUPANCY ___ NO. OF PEOPLE × ___ BTU/HR × ___ HR =

(c) MOTORS ___ BTU/HP/HR × ___ HP × ___ HR/24 HR =

___ BTU/HP/HR × ___ HP × ___ HR/24 HR =

(d) MATERIAL HANDLING
___ FORK LIFT(S) × ___ EQUIV. HP × 3100 BTU/HR/HP × ___ HR OPERATION =
OTHER ___ =

*IF THE PRODUCT PULL-DOWN IS ACCOMPLISHED IN LESS THAN 24 HR THE DAILY PRODUCT WILL BE:

$$\text{POUNDS PRODUCT} \times \frac{24\ \text{HR}}{\text{PULL-DOWN HOURS}}$$

IV TOTAL MISCELLANEOUS LOADS =

TOTAL BTU LOAD I TO IV = 136,421

ADD 10% SAFETY FACTOR = 13642

TOTAL BTU/24 HR WITH SAFETY FACTOR (NOT INCLUDING EVAP. FAN OR DEFROST HEAT LOADS) 24 HR. BASE REFRIGERATION LOAD = 150,063

Figure 145-22 Continued

Equipment Selection from Load Calculation Form

1. DETERMINE EVAP. TD REQUIRED FOR CLASS OF PRODUCT AND ROOM TEMP ___10___ °F (TD) (FROM LOAD SURVEY DATA)

2. DETERMINE COMPRESSOR RUNNING TIME BASED ON OPERATING TEMPERATURES AND DEFROST REQUIREMENTS ___18___ HRS. (FROM LOAD SURVEY DATA)

3. EVAPORATOR TEMP.°F = ___0___ (ROOM TEMP) – ___10___ [EVAP. TD) = ___–10___ °F (FROM LOAD SURVEY DATA)]

4. COMP. SUCT. TEMP. °F = ___–10___ (EVAP. SUCT. TEMP) – ___3___ (SUCT. LINE LOSS) = ___–13___ °F

BTU/24 HR. BASE REFRIGERATION LOAD WITH SAFETY FACTOR = ______
(NOT INCLUDING EVAPORATOR FAN OR DEFROST HEAT)

PRELIMINARY HOURLY LOAD = BTU/24 HR (BASE LOAD) / HR/DAY (COMP. RUNNING TIME) = ______ BTU/HR

FAN HEAT LOAD ESTIMATE BTU/HR =

___1___ QTY. (MOTORS) × ___210___ WATTS EA. (INPUT) × 3.41 BTU/W × ___24___ HR = ___17186___ BTU/24 HR

OR ______ QTY. (MOTORS) × ______ HP EA. × ______ BTU/HP/HR × ______ HR = ______ BTU/24 HR

DEFROST HEAT LOAD ESTIMATE BTU/HR =

___1___ QTY. EVAPS. × ___3380___ W EA. × ___1½___ HR × 3.41 BTU/W × ___.5___ DEFROST LOAD FACTOR*

= ___8644___ BTU/HR

*USE 0.50 FOR ELECTRIC DEFROST, 0.40 FOR HOT GAS DEFROST

BTU/24 HR TOTAL LOAD = ___150,063___ (BASE LOAD) + ___17,186___ (FAN HEAT) + ___8644___ (DEFROST HEAT) = ___175,893___ BTU/24 HR

OR ______ (BASE LOAD) × ______ (BASE LOAD MULT.) = ______ BTU/24 HRS.

ACTUAL HOURLY LOAD = 175,893 BTU/24 HRS. (TOTAL LOAD) / 18 HRS/DAY (COMP. RUNNING TIME) = ___9771___ BTU/HR

Equipment Selection

	Compressor Units		Condensing Units		Evaporators		Condensers	
MODEL NO.			32 c		NDE 105			
QUANTITY			1		1			
BTU/HR. CAPACITY (EA.)			11,500		10,500			
CFM AIR VOLUME (EA.)					1420			
	DESIGN	ACTUAL	DESIGN	ACTUAL	DESIGN	ACTUAL	DESIGN	ACTUAL
°F EVAPORATOR TEMP.					–10			
°F EVAP TD								
°F SUCTION TEMP.			–10					
°F CONDENSING TEMP.								
°F DESIGN AMBIENT TEMP.			100					
°F MIN. OPER. AMBIENT TEMP.								

Figure 145-22 Continued

Model	Capacity (Btu/hr)			Fan Data			Motor Data				Connections		
	@ 10°F TD	@ 12°F TD	@ 15°F TD	cfm	Qty.	Size	Watts*	Standard 208–240/1/60 Amps* @ 240V	Standard 208–240/1/60 Amps* @ 208V	Optional 115/1/60 Amps* @ 115V	Liquid Fl† (in.)	Suction O.D. (in.)	Drain O.D. (in.)
NDE 45	4,500	5,400	6,750	740	1	10″	105	.82	.76	1.5	½	½	1⅛
NDE 65	6,500	7,800	9,750	880	1	10″	105	.82	.76	1.5	½	½	1⅛
NDE 85	8,500	10,200	12,750	1,180	2	10″	210	1.64	1.52	3.0	½	⅝	1⅛
NDE 105	10,500	12,600	15,750	1,420	2	10″	210	1.64	1.52	3.0	½	⅝	1⅛
NDE 120	12,000	14,400	18,000	1,620	2	10″	210	1.64	1.52	3.0	½	⅝	1⅛
NDE 150	15,000	18,000	22,500	2,080	3	10″	315	2.46	2.28	4.5	½	⅝	1⅛
NDE 180	18,000	21,600	27,000	2,340	3	10″	315	2.46	2.28	4.5	⅝	⅞	1⅛
NDE 280	28,000	33,600	42,000	3,580	4	10″	420	3.24	3.04	6.0	⅝	1⅛	1⅛

†Externally equalized expansion valves must be used for all models. *Watts or Amps are total for all fan motors.

Model	Approx. Refrig. Charge lbs.	Optional Heat Exchanger Recommended	Approx. Shipping Weight lbs.	Defrost Electrical Data							
				Standard† 240/1/60		Optional 220/3/60		Optional 480/1/60		Optional 440/3/60	
				Amps	Watts	Amps	Watts	Amps	Watts	Amps	Watts
NDE 45	1.53	B 50 XS	80	6.6	1580	5.25	1450	3.30	1580	2.61	1450
NDE 65	2.24	B 75 XS	95	9.1	2180	7.15	2000	4.50	2180	3.41	2000
NDE 85	2.75	B 75 XS	110	11.6	2780	9.25	2540	5.80	2780	4.36	2540
NDE 105	3.56	B 120 XS	125	14.1	3380	11.30	3100	7.10	3380	5.30	3100
NDE 120	4.08	B 120 XS	140	16.6	3980	13.20	3640	8.30	3980	6.23	3640
NDE 150	5.10	B 200 XS	160	20.0	4780	15.90	4380	10.00	4780	7.50	4380
NDE 180	5.60	B 200 XS	175	21.5	5160	17.20	4720	10.80	5160	8.60	4720
NDE 280	8.60	B 500 XS	200	24.8	5960	19.70	5460	12.40	5960	9.85	5460

†Standard Stock Units

Electrical Defrost

Single-Phase Defrost			Three-Phase Defrost		
NDE Units	No. of Units	Kit No.	NDE Units	No. of Units	Kit No.
45 THRU 280	1	1D	45 THRU 280	1	4D
45 THRU 105	2*	1D	45 THRU 120	2*	4D
120 THRU 280	2*	22D	150 THRU 280	2*	23D

*Any combination of units may be used

Figure 145-23 Evaporator performance data below freezing, electric defrost.
(Courtesy of Dunham-Bush, Inc.)

   i. Heat loss due to infiltration through windows and doors or through cracks and penetrations in the building envelope.
3. System losses:
   a. Heat loss through ducts located in an unheated space.
   b. Ventilation air which must be heated before it is introduced into the space. (In older structures infiltration provided enough fresh air to the space making ventilation unnecessary. In newer structures, tighter construction may require ventilation.)
   c. Bathroom and kitchen exhaust systems tend to increase infiltration, but they are not a design factor because they are used intermittently.
   d. Normally, combustion air for gas- or oil-fired furnaces must be provided. In older homes, infiltration will meet the combustion air requirement. In newer, tighter homes it may be necessary to introduce combustion air to the burner or the furnace room. Generally, codes require one square inch of opening to a source of outside air for each 1,000 Btu/hr of fuel input. Local codes should be consulted.

For heating garages, the manual recommends a separate heater and control.

145-2.2 Heat-Transmission Multiplier

The *heat-transmission multiplier (HTM),* for various exposures, is the product of the heat transmission factor *(U)* multiplied by the temperature difference *(TD).* The infiltration *HTM* is calculated by dividing the winter infiltration Btu/hr by the total window and door area.

145-2.3 Heat Gain

The calculation of *heat gain* includes consideration of the following conditions:

1. The summer outside design temperature.
2. Radiation from the sun.
3. Heat and moisture given off by equipment and appliances.
4. Heat and moisture given off by people.
5. Heat and moisture gained by infiltration.

These conditions may or may not occur simultaneously.

145-2.4 Outside Design Conditions

The *outside design conditions* are selected from ASHRAE 97.5% weather data. These conditions are only exceeded 50 to 100 hrs from June through September. Using the 97.5% conditions prevents excessive oversizing of the equipment to handle a temporary high load.

145-2.5 Daily Range

The difference between the average high and low temperatures is the *daily range,* given for U.S. and Canada locations. The high usually occurs late in the afternoon and the low at about daybreak. The daily range affects the cooling load since a low night temperature can reduce the daytime load due to the storage factor. The daily range factors are given for three ranges: the "L" range below 15°F, the "M" range for temperatures between 15° and 25°F, and the "H" group which refers to a range above 25°F.

145-2.6 Storage

When the sun shines in a window, its radiant heat is released when it reaches the interior surfaces of the house. This heat is stored and later gradually heats room air. The same kind of action occurs on the exterior of structures. Radiant heat from the sun warms the surfaces and is stored. Gradually, at a later time, the heat reaches the interior where it warms the interior air. The net effect of *storage* is to delay and smooth out solar loads. The delayed effect of the sun's load is incorporated in the procedures used for calculating heat gain.

145-2.7 Building Gains

The *building gains* included in the calculations are as follows:

1. Heat gained by solar radiation through glass.
2. Heat transmitted through glass by conduction.
3. Heat transmitted through walls exposed to outside air.
4. Heat gained through partitions which separate conditioned and unconditioned spaces.
5. Heat gained through ceilings from the attic.
6. Heat gained through roofs and roof/ceiling combinations.
7. Heat gained through ceilings and floors that separate conditioned and unconditioned spaces.
8. Heat gained by infiltration through doors and windows or building envelope.
9. Heat produced by people.
10. Heat produced by lights.
11. Heat produced by appliances and equipment.

The *system gains* included in the calculations are as follows:

1. Heat gained through ducts located in unconditioned spaces.
2. Heat gains associated with ventilation air mechanically introduced through the system.
3. Bathroom and kitchen exhaust tend to increase infiltration. These devices only operate a small part of the time.

Walls below grade and slabs are not included in the heat gain calculations. Moisture loads due to the introduction of outside air or infiltration, people and appliances are included in the load.

I45-2.8 Data Sheets

Winter heating loads are summarized in Fig. I45-24. A sample data sheet filled in with typical calculations is shown in Fig. I45-25.

Summer cooling loads are summarized in Fig. I45-26. A sample data sheet filled in with typical calculations is shown in Fig. I45-27 (p. 892).

I45-3 COMMERCIAL HEATING AND COOLING LOAD

There are a number of similarities in load calculations for refrigerated rooms, residential buildings, and commercial establishments. Each requires the calculation of heat transmission through the enclosure, gain or loss due to infiltration or ventilation of outside air, and gains due to internal loads such as people or product.

The load calculations for each of these applications differs in respect to the major factor that comprises the load. For example, for refrigerated loads, a large factor in the total load is the *internal load.* To correctly calculate this load, the cooled product must be analyzed from a number of standpoints: What condition is the product in when it is brought into the room? Is it frozen in the room or just cooled? What is the nature of the product and what is the timing in bringing various amounts into the room?

In calculating residential loads, the condition of the structure is of major importance. Is the building low and spread out like a ranch-style house, or are the spaces compact as in an apartment house? Is the house well insulated and tightly constructed, or was the house built in the thirties with limited insulation and loosely fitted windows and doors?

For commercial load calculations, the most important consideration is the use of the space. Is it a restaurant with a high concentration of people consuming hot food or is it an office with many pieces of heat-producing equipment such as duplicating machines and computers? Is it a laboratory where closely controlling temperature and humidity are necessary or is it a "clean room" requiring highly efficient air filters?

In the discussion of commercial loads, one general application is selected that fits many small- and medium-sized commercial spaces, where unitary equipment can be applied. These units are used in restaurants, small offices, banks, retail stores, and many other commercial buildings. The load calculation process can use a simplified form. A sample of such a form, consisting of seven tables and data sheets for figuring both heating and cooling loads, is shown in Fig. I45-28 (p. 893). The step-by-step proce-

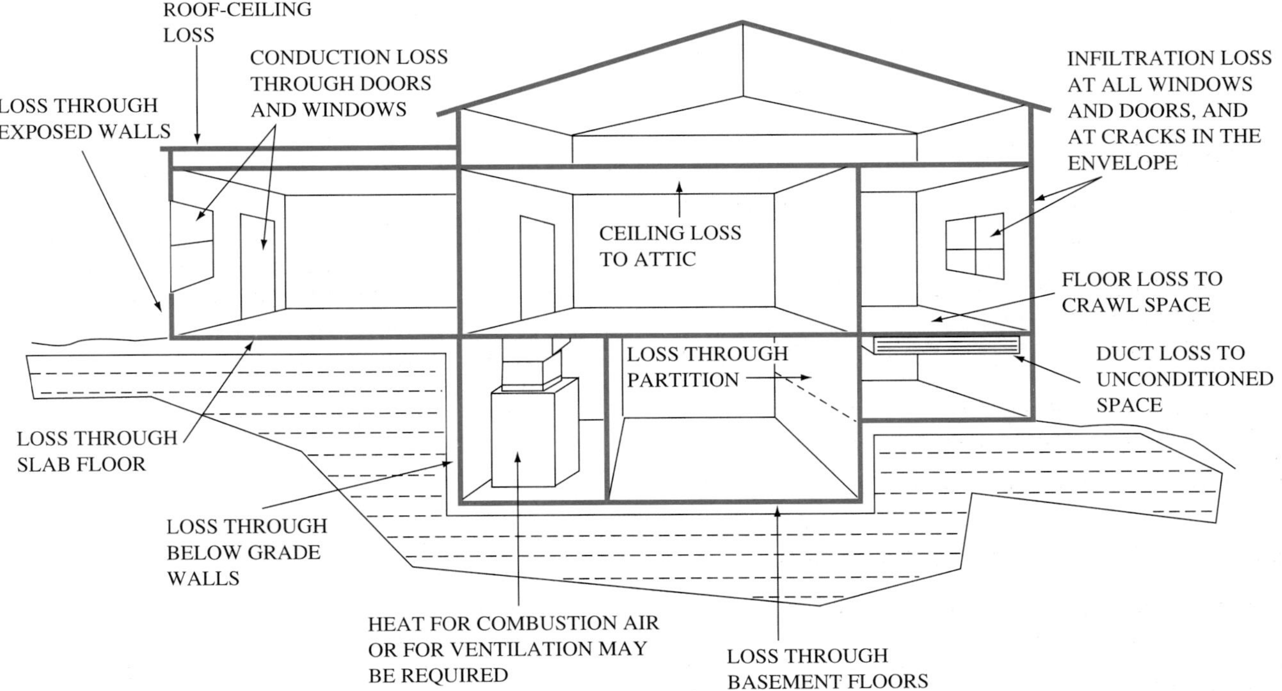

Figure I45-24 Winter heating loads. (Courtesy of ACCA)

#	Type of Exposure		Const No.	HTM Htg.	HTM Clg.	Entire House Area or Length	Entire House Btu/hr Htg.	Entire House Btu/hr Clg.	Living Area or Length	Living Btu/hr Htg.	Living Btu/hr Clg.	Dining Area or Length	Dining Btu/hr Htg.	Dining Btu/hr Clg.	Laundry Area or Length	Laundry Btu/hr Htg.	Laundry Btu/hr Clg.	Kitchen Area or Length	Kitchen Btu/hr Htg.	Kitchen Btu/hr Clg.	Bath-1 Area or Length	Bath-1 Btu/hr Htg.	Bath-1 Btu/hr Clg.
1	Name of Room					Entire House			1 Living			2 Dining			3 Laundry			4 Kitchen			5 Bath-1		
2	Running Ft. Exposed Wall					160			21			25			18			11			9		
3	Room Dimensions Ft.					51 × 29			21 × 14			7 × 18			7 × 11			11 × 11			9 × 11		
4	Ceiling Ht. Ft.	Directions Room Faces				8			8	West		8	North		8			8	East		8	East	
5	Gross	a	12-d			1280			168			200			144			88			72		
	Exposed	b	14-b			480																	
	Walls &	c	15-b			800																	
	Partitions	d																					
6	Windows	a	3-A	41.3		60	2478		40	1652		20	826										
	& Glass	b	2-C	48.8		20	976																
	Doors Htg.	c	2-A	35.6		105	3738											11	392		8	285	
		d																					
7	Windows	North																					
	& Glass	E&W																					
	Doors Clg.	South																					
		Basement																					
8	Other Doors		11-E	14.3		37	529								17	243							
9	Net	a	12-d	6.0		1078	6468		128	768		180	1080		127	762		77	462		64	384	
	Exposed	b	14-b	10.8		460	4968																
	Walls &	c	15-b	5.5		800	4400																
	Partitions	d																					
10	Ceilings	a	16-d	4.0		1479	5916		294	1176		126	504		77	308		121	484		99	396	
		b																					
11	Floors	a	21-a	1.8		1479	2662																
		b																					
12	Infiltration HTM			78.8		222	17490		40	3152		20	1576		17	1340		11	867		8	630	
13	Sub Total Btu/hr Loss = 6+8+9+10+11+12						49625			6748			3986			2653			2205			1695	
14	Duct Btu/hr Loss			0%		—			—			—			—			—					
15	Total Btu/hr Loss = 13 + 14						49625			6748			3986			2653			2205			1695	
16	People @ 300 & Appliances 1200																						
17	Sensible Btu/hr Gain = 7+8+9+10+11+12+16																						
18	Duct Btu/hr Gain				%																		
19	Total Sensible Gain = 17 + 18																						

Figure 145-25 Sample heat loss calculation sheet. (Courtesy of ACCA)

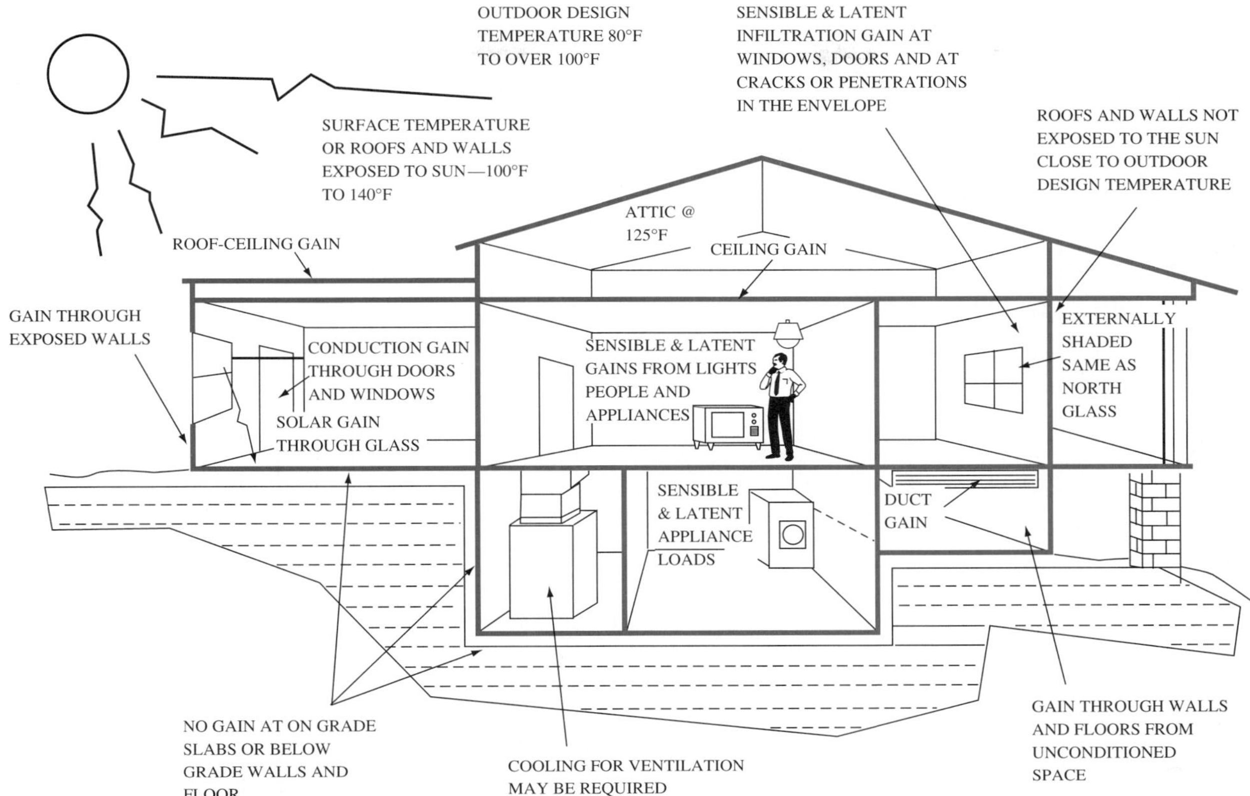

Figure I45-26 Summer cooling loads. (Courtesy of ACCA)

dure, using this simplified form, is described under the following topics: (Note: all table numbers, indicated below, refer to tables shown in Fig. I45-28.)

For calculating cooling loads:

A. Design conditions
B. Solar heat gain (Table 1)
C. Heat transmission (Table 2)
D. Heat gain from occupants (Table 3)
E. Heat gain from appliances (Table 4)
F. Infiltration/ventilation (Tables 5 and 6)
G. Duct heat gain (Table 7)

For calculating heating loads:

A. Heat transmission (Table 2)
B. Infiltration/ventilation (Tables 5 and 6)
C. Duct heat loss (Table 7)

I45-3.1 Procedure

The instructions for calculating cooling load, shown in Fig. I45-28, can be followed, and the cooling load data sheet filled out.

The instructions for calculating the heating load, shown in Fig. I45-28, can be followed and the heating load data sheet filled out.

In order to demonstrate how the simplified form is used, an example is shown using the following typical problem.

EXAMPLE

Calculate the cooling and heating load for the one-story office building shown in Fig. I45-29 (p. 898), using a free-standing Unitaire, based on the following conditions:

Cooling design: Outside, 95°F DB and 78°F WB
  Inside, 80°F DB and 67°F WB
Heating design: Outside, 0°F DB and 20 gr/lb
  Inside, 70°F DB and 32 gr/lb
Solar heat gain: Light inside shades
Heat transmission: Single glass
  Floors over unconditioned room
  Roof insulated with ceiling
Heat from occupants: 5 People, seated, light work
  320 W of fluorescent lights
Heat gain from appliances: 200 W from computers
Ventilation/infiltration: 20 cfm per person
Duct heat gain: None

(Text continues on p. 898.)

1	Name of Room					Entire House			1 Living			2 Dining			3 Laundry			4 Kitchen			5 Bath-1		
2	Running Ft. Exposed Wall					160			21			25			18			11			9		
3	Room Dimensions Ft.					51 × 29			21 × 14			7 × 18			7 × 11			11 × 11			9 × 11		
4	Ceiling Ht. Ft.	Directions Room Faces				8			8	West		8	North		8			8	East		8	East	
Type of Exposure			Const No.	HTM		Area or Length	Btu/hr		Area or Length	Btu/hr		Area or Length	Btu/hr		Area or Length	Btu/hr		Area or Length	Btu/hr		Area or Length	Btu/hr	
				Htg.	Clg.		Htg.	Clg.		Htg.	Clg.		Htg.	Clg.		Htg.	Clg.		Htg.	Clg.		Htg.	Clg.
5	Gross	a	12-d			1280			168			200			144			88			72		
	Exposed	b	14-b			480																	
	Walls &	c	15-b			800																	
	Partitions	d	13N			232																	
6	Windows	a																					
	& Glass	b																					
	Doors Htg.	c																					
		d																					
7	Windows	North			14	20		280				20		280									
	& Glass	E & W			44	115		5060	40		1760							11		484	8		352
	Doors Clg.	South			23	30		690															
		Basement			70/36	8/8		848															
8	Other Doors		10-e		3.5	37		130							17		60						
9	Net	a	12-d		1.5	1078		1617	128		192	180		270	127		191	77		116	64		96
	Exposed	b	14-b		1.6	233		373															
	Walls &	c	15-b		0																		
	Partitions	d	13-n		0																		
10	Ceilings	a	16-d		2.1	1479		3106	294		617	126		265	77		162	121		254	99		208
		b																					
11	Floors	a	21-a		0																		
		b	19-f		0																		
12	Infiltration HTM				9.0	218		1962	40		360	20		180	17		153	11		99	8		72
13	Sub Total Btu/hr Loss = 6+8+9+10+11+12																						
14	Duct Btu/hr Loss			%																			
15	Total Btu/hr Loss = 13 + 14																						
16	People @ 300 & Appliances 1200							3000	3		900	3		900			—			1200			—
17	Sensible Btu/hr Gain = 7+8+9+10+11+12+16							17066			3829			1895			566			2153			728
18	Duct Btu/hr Gain				%						—			—			—			—			—
19	Total Sensible Gain = 17 + 18							17066			3829			1895			566			2153			728

Figure 145-27 Sample heat gain calculation sheet. (Courtesy of ACCA)

Table 1 Solar Heat Gain (Btu/hr/sq ft sash area)

Direction Windows Face	N	NE & NW	E & W	SE & SW	S	Horiz.
Clear glass (single or double), no protection	40	130	200	160	100	
Shaded completely by awnings	12	48	56	45	29	265
Light inside shades or Venetian blinds	24	77	122	95	58	
Glass brick, no protection	16	52	80	64	40	

Table 2 Average Transmission Factors

Item	Factor [Btu per (hr) (sq ft) (F)]			
	Cooling Load		Heating Load	
Windows				
Single glass	1.06		1.13	
Double glass	0.61		0.65	
Doors	*		*	
Walls and interior partitions	0.32		0.32	
Ceiling under unconditioned room	0.24		0.29	
Floors:				
Over unconditioned room	0.29		0.24	
Over basement	0.20		0.23	
On ground	0.00		†	
Roofs:[‡]	No Ceiling	Ceiling	No Ceiling	Ceiling
Uninsulated, frame	[§]1.20	[§]1.07	0.48	0.33
Uninsulated, light masonry	[§]1.87	[§]1.60	0.70	0.41
Insulated	[§]0.80	[§]0.67	0.18	0.14

*For glass doors, use window factors; for all other doors, use wall factors.
†For slab floors on ground, use 50 Btuh per lineal foot of exposed building perimeter.
‡Factors based on flat-roof construction.
§These factors include an average allowance for increased temperature caused by sun load.
This table was extracted from ARI Standard 220-67
NOTE: Factors can be adjusted to actual U factors, if known.

Table 3 Heat Gain From Occupants*

Degree of Activity	Typical Application	Sensible Heat Btuh	Latent Heat Btuh
Seated at rest	Theater—Matinee Theater—Evening	225 245	105 105
Seated, very light office work	Offices, Hotels, Apartments	245	155
Moderately-active office work	Offices, Hotels, Apartments	250	200
Standing, light work; walking slowly	Dept. Store, Retail store	250	200
Walking; seated Standing; walking slowly	Drug Store Bank	250	250
Sedentary work	Restaurants[§]	275	275
Light bench work	Factory	275	475
Moderate dancing	Dance Hall	305	545
Walking, 3 mph Moderately-heavy work	Factory	375	625
Bowling[‡] Heavy Work	Bowling Alley Factory	580	870

§The adjusted total heat value for *sedentary work, restaurant* includes 60 Btu per hour for food per individual (30 Btuh sensible and 30 Btuh latent).
‡For *bowling,* figure one person per alley actually bowling and all others as sitting (400 Btuh) or standing (550 Btuh).
Note: The above values are based on 75 F room dry-bulb temperature. For 80 F room dry-bulb temperature, the total heat gain remains the same, but the sensible heat values should be decreased by approximately 20 per cent and the latent heat values increased accordingly.
This table was extracted with permission from the 1993 ASHRAE Handbook: Fundamentals.*

Figure 145-28 Load calculation for commercial buildings using unitary equipment. (The Trane Company)

Table 4 Recommended Rate of Heat Gain from Restaurant Equipment Located in Air-Conditioned Area

		Input Rating, Btu/h		Recommended Rate of Heat Gain, Btu/h			
				Without Hood			With Hood
Appliance	Size	Maximum	Standby	Sensible	Latent	Total	Sensible
Electric, No Hood Required							
Barbeque (pit), per pound of food capacity	80 to 300 lb	136	—	86	50	136	42
Barbeque (pressurized), per pound of food capacity	44 lb	327	—	109	54	163	50
Blender, per quart of capacity	1 to 4 qt	1550	—	1000	520	1520	480
Braising pan, per quart of capacity	108 to 140 qt	360	—	180	95	275	132
Cabinet (large hot holding)	16.2 to 17.3 ft^3	7100	—	610	340	960	290
Cabinet (large hot serving)	37.4 to 406 ft^3	6820	—	610	310	920	280
Cabinet (large proofing)	16 to 17 ft^3	693	—	610	310	920	280
Cabinet (small hot holding)	3.2 to 6.4 ft^3	3070	—	270	140	410	130
Cabinet (very hot holding)	17.3 ft^3	21000	—	1880	960	2830	850
Can opener		580	—	580	—	580	0
Coffee brewer	12 cup/2 brnrs	5660	—	3750	1910	5660	1810
Coffee heater, per boiling burner	1 to 2 brnrs	2290	—	1500	790	2290	720
Coffee heater, per warming burner	1 to 2 brnrs	340	—	230	110	340	110
Coffee/hot water boiling urn, per quart of capacity	11.6 qt	390	—	256	132	388	123
Coffee brewing urn (large), per quart of capacity	23 to 40 qt	2130	—	1420	710	2130	680
Coffee brewing urn (small), per quart of capacity	10.6 qt	1350	—	908	445	1353	416
Cutter (large)	18 in. bowl	2560	—	2560	—	2560	0
Cutter (small)	14 in. bowl	1260	—	1260	—	1260	0
Cutter and mixer (large)	30 to 48 qt	12730	—	12730	—	12730	0
Dishwasher (hood type, chemical sanitizing), per 100 dishes/h	950 to 2000 dishes/h	1300	—	170	370	540	170

This table was extracted with permission from the 1993 ASHRAE *Handbook—Fundamentals.*

Figure 145-28 Continued

Table 5 Infiltration

Type of Window	Description of Window	Wind Velocity (mph) Summer	Winter
		7½	15
Double-Hung	Total for average window, Nonweatherstripped, 1/16 in. crack and 3/64 in. clearance	14	39
Wood Sash	Ditto, Weatherstripped	8	24
(Unlocked)	Total for poorly fitted Window, Nonweather-stripped, 3/32 in. crack and 3/32 in. clearance	48	111
	Ditto, Weatherstripped	13	34
Double-Hung	Nonweatherstripped, locked.	33	70
Metal Sash	Nonweatherstripped, unlocked.	34	74
	Weatherstripped, unlocked.	13	32
Rolled	Industrial Pivoted, 1/16 in. crack.	80	176
Section	Architectural Projected, 3/64 in. crack	36	88
Steel	Residential casement, 1/32 in. crack.	23	52
Sash	Heavy casement section, Projected 1/32 in. crack.	16	38
	Description of Door		
Poorly fitted		96	222
Well fitted		48	111
Weatherstripped		24	56

More current information is available in the 1993 ASHRAE Handbook.

Table 6 Ventilation*

Application	Est. Max. Occupancy p/1000 ft²	Outdoor Air Requirements cfm/ person	cfm/ ft²
Food and beverage service			
Dining rooms	70	20	
Cafeteria, fast food	100	20	
Kitchens (cooking)	20	15	
Garages, service stations			
Enclosed parking garage			1.50
Auto repair rooms			1.50
Hotels, Motels, Resorts			*Room*
Bedrooms			30
Living rooms			30
Baths			35
Lobbies	30	15	
Conference rooms	50	20	
Assembly rooms	120	15	
Gambling casinos	120	30	
Offices			
Office space	7	20	
Reception area	60	15	
Data entry rooms	60	20	
Conference rooms	50	20	
Retail stores			*ft²*
Basement, street floors	30		.30
Upper floors	20		.20
Malls	20		.20
Elevators			1.00
Supermarkets	8	15	
Hardware, drugs, fabric	8	15	
Beauty	25	25	
Barber	25	15	
Sports			
Spectator areas	150	15	
Game rooms	70	25	
Ice arenas (playing areas)			.50
Swimming pools			.50
Gymnasium (playing floors)	30	20	
Bowling alleys (seating areas)	70	25	
Theaters			
Lobbies	150	20	
Auditorium	150	14	

This table was extracted with permission from Table 2, pp 8–9, from ASHRAE STANDARD 62-1989.

The ventilation air should be calculated as follows:

_________ cfm per person x _________ persons = _________ cfm

or

_________ cfm per sq ft x _________ sq ft = _________ cfm

Figure 145-28 Continued

Instructions: Following numbered paragraphs refer to items on load estimate section:

1. Insert design conditions and subtract to obtain differences. (Obtain outdoor design conditions from ASHRAE Handbook or ARI Standard 220-67 and specific humidity from Trane Psychrometic Chart.)
2. Insert window sash area (ft^2) and multiply by factor from Table 1. Use the largest solar gain for one side only.
3. Insert total area (ft^2) of window sash, net wall, partition, roof or ceiling and floor. Multiply by factors from Table 2, and dry bulb temperature differences.
4. Insert number of occupants and multiply by factors from Table 3. Insert wattage of lights in use and multiply by factor on form.
5. Insert Hp of motors in use and multiply by factor on form. Insert appliance heat gains from Table 4.
6. Calculate requirements for both infiltration (Table 5) and ventilation (Table 6), and use larger cfm. Use no less cfm than required by local codes and no less cfm than amount drawn from space by exhaust fans, if used.
7. Multiply duct heat gain by factor for insulation thickness (Table 7), multiply by duct length (ft) and divide by 100.
8. Add sensible and latent heat gains.
9. Add sensible and latent heat gains together to obtain Total Cooling Load.

Cooling Load

1. Design Conditions	Dry Bulb (°F)	Wet Bulb (°F)	Specific Humidity (gr/lb)
Outside			
Inside			
Difference		–	

ITEMS				COOLING LOAD. BTUH Sensible	COOLING LOAD. BTUH Latent
2. Sensible Heat Gain Through Glass (Table 1)					
Sq. Ft.	×	Factor			
			=		
			=		
			=		
			=		
Use largest single load from one side only. Transfer sum of all glass areas to window portion of transmission gain calculation (Step 3).					
3. Transmission Gain (Table 2)					
	Sq. Ft. ×	Factor ×	Dry Bulb Temp. Diff.		
Windows			=		
Walls			=		
Partitions			=		
Roof			=		
Ceiling			=		
Floor			=		
4. Internal Heat Gains—People, Lights (Table 3)					
	Number ×	Sensible Factor ×	Latent Factor		
People			– =		–
		–	=	–	
	Watts ×	Factor			
Lights		3.4	=		–
		*4.25	=		–
5. Internal Heat Gains—Other					
Motors	Horsepower ×	Factor			
		3393	=		–
Appliances (Use Table 4)			=		
			=		
			=		
6. Ventilation Or Infiltration (Table 5 & 6, Use Larger Quantity)					
CFM ×	Dry Bulb Temp. Diff. ×	Factor			
		1.08	=		–
CFM ×	Specific Humidity Diff. ×	Factor			
		0.67	=	–	
7. Duct Heat Gain (Table 7)					
Factor for Insulation Thickness ______ × Heat Gain ______ × Duct Length (ft.) ______ ÷ 100			=		–
8. Total Sensible And Latent Heat Gain			=		
9. Total Cooling Load			=		

*For use with fluorescent lamps with ballast within conditioned space.

Figure 145-28 Continued

Heat Load Calculations

Normally, heat gains from people, motors, solar radiation, etc., are not considered when calculating heat load, because they may not be available at peak load.

Instructions: Following numbered paragraphs refer to items in load estimate section, below.

1. Insert design conditions and subtract to obtain differences. (Obtain outdoor design conditions from ASHRAE Handbook, and specific humidity from Trane Psychrometric Chart.)
2. Insert area (ft.^{2}) of window sash, net wall, roof and floor and multiply by factors from Table 2, and inside-outside dry bulb differences.
3. Calculate requirements for both infiltration (Table 5) and ventilation (Table 6), and use larger cfm. Use no less cfm than required by local codes and no less cfm than amount drawn from space by exhaust fans, if used. Insert same cfm and multiply by specific humidity difference and factor on form for humidification load.
4. Multiply duct heat loss by factor for insulation thickness (Table 7), multiply by duct length (ft) and divide by 100.
5. Record heat load.

Table 7 Duct Heat Transfer vs Duct Air Flow at Various Temperature Differences for Ducts 100 Feet Long

TEMP. DIFF. (F)	Air Flow (cfm)							
	1000	2000	3000	4000	5000	6000	7000	8000
	Heat Transfer—Gain or Loss (Btu/hr)*							
20	900	1,200	1,480	1,725	1,975	2,190	2,300	2,400
40	1,800	2,450	3,000	3,500	3,900	4,300	4,625	4,900
60	2,575	3,500	4,375	5,125	5,800	6,375	6,900	7,325
80	3,525	4,775	5,900	6,900	7,800	8,525	9,225	9,800
100	4,375	5,925	7,325	8,575	9,650	10,625	11,500	12,300
120	5,375	7,275	8,950	10,400	11,675	12,800	13,775	14,675

*For 2 Inch Insulation, Use BTUH Values Shown.
For 1 Inch Insulation, Multiply by 1.8.
For No Insulation, Multiply by 7.6.

Heating Load

1. Design Conditions	Dry Bulb (F)	Specify Humidity gr./lb.
Outside		
Inside		
Difference		

2. Transmission Gain Items (Table 2)	Calculations			Heat Load (Btu/hr)
Windows	Sq. Ft. ×	Factor ×	Dry Bulb Temp. Diff.	
			=	
			=	
Walls			=	
Partitions			=	
Roof			=	
Floor			=	
Other			=	

3. Ventilation or Infiltration Items (Tables 5 and 6, use larger quantity)	Calculations			Heat Load (Btu/hr)
Sensible Load	CFM ×	Dry Bulb Temp. Diff. ×	Factor	
			1.08 =	
Humidification Load	CFM ×	Specific Humidity Diff. ×		
			0.67 =	

4. Duct Heat Loss Items (Table 7)	Calculations		Heat Load (Btu/hr)
Heat Loss ×	Factor for Insulation Thickness ×	Duct Length (ft.)	
		÷ 100 =	
5. Total Heating Load		=	

Figure 145-28 Continued

Figure I45-29 Sample building layout.

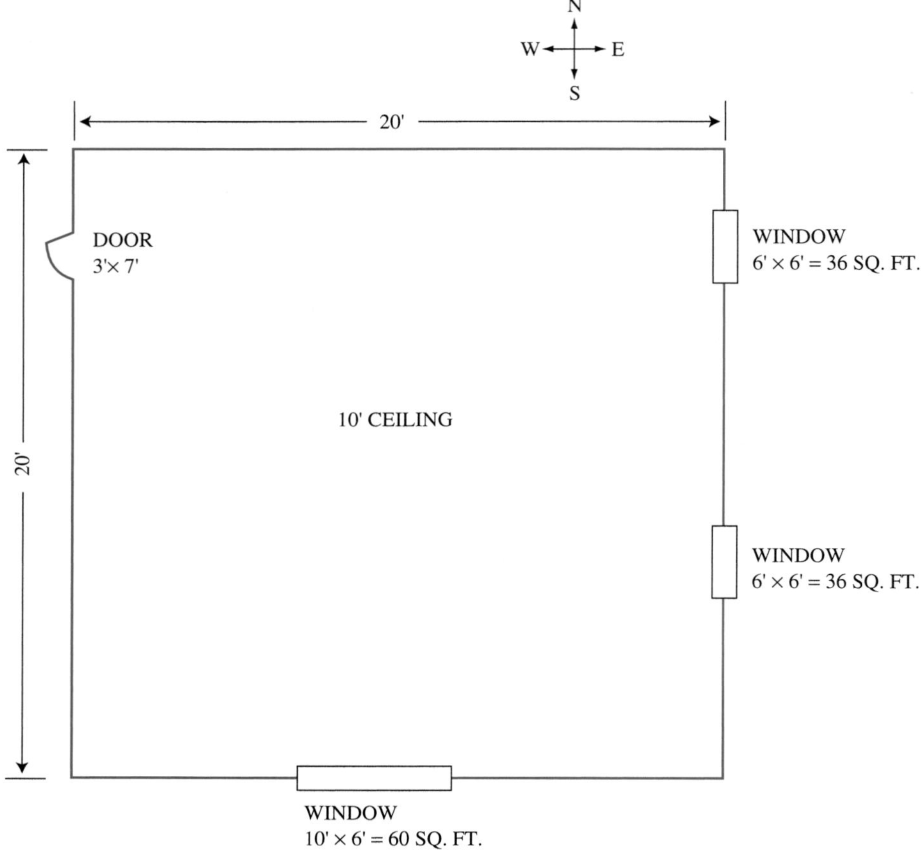

Solution

The cooling-load form is filled out using the above information and the tables that apply, shown in Fig. I45-30.

Note that loads are recorded in two separate columns, one for sensible heat gains and one for latent heat gains. The only loads that have latent heat gains are occupants and ventilation.

The following are a few explanations that will assist in filling out the form:

Design conditions: For specific humidity values in gr/lb, refer to the psychrometric chart Fig. I45-31.

Solar heat gain: Glass areas must be separated by the direction they are facing.

Heat transmission: The temperature difference comes from the determination in design conditions.

Heat gain from occupants: Type of activity influences the load factor.

Heat gain from appliances: The wattage used by the appliance is the basis for the load.

Infiltration/ventilation: Using 20 cfm per person makes ventilation a larger quantity than obtained by determining the air entering by infiltration and therefore is used in the calculations.

Duct heat gain: Not a factor on this job since there is no ductwork.

Now fill out the heating-load form (Fig. I45-28) using the above information and the tables that apply as shown in Fig. I45-32 (p. 901).

Cooling Load

1. Design Conditions	Dry Bulb (°F)	Wet Bulb (°F)	Specific Humidity (gr/lb)
Outside	95	76	125
Inside	80	67	78
Difference	15	–	47

ITEMS						COOLING LOAD. BTUH Sensible	Latent
2. Sensible Heat Gain Through Glass (Table 1)							
	Sq. Ft. ×	Factor					
E	72	122	=	8784		8784	
S	60	58	=	3480			
			=				
	132		=				
Use largest single load from one side only. Transfer sum of all glass areas to window portion of transmission gain calculation (Step 3).							
3. Transmission Gain (Table 2)							
	Sq. Ft. ×	Factor ×	Dry Bulb Temp. Diff.				
Windows	132	1.06	15	=	2099		
Walls	668	.32	15	=	3206		
Partitions				=			
Roof	400	.67	15	=	4020		
Ceiling				=			
Floor	400	.29	15	=	1740		
4. Internal Heat Gains—People, Lights (Table 3)							
People	Number ×	Sensible Factor ×	Latent Factor				
	5	245	–	=	1225	–	
	5	–	155	=	–	775	
Lights	Watts ×	Factor					
		3.4		=		–	
	320	*4.25		=	1360	–	
5. Internal Heat Gains—Other							
Motors	Horsepower ×	Factor					
		3393		=		–	
Appliances (Use Table 4)				=			
				=			
computer 200 × 3.4				=	680		
6. Ventilation or Infiltration (Table 5 & 6, Use Larger Quantity)							
CFM ×	Dry Bulb Temp. Diff. ×	Factor					
100	15	1.08		=	1620	–	
CFM ×	Specific Humidity Diff. ×	Factor					
100	47	0.67		=	–	3149	
7. Duct Heat Gain (Table 7)							
Factor for Insulation Thickness ______ × Heat Gain ______ × Duct Length (ft.) ______ ÷ 100				=		–	
8. Total Sensible and Latent Heat Gain				=	24,734	3924	
9. Total Cooling Load	=				28,658		

Figure I45-30 Cooling load form filled out. (The Trane Company)

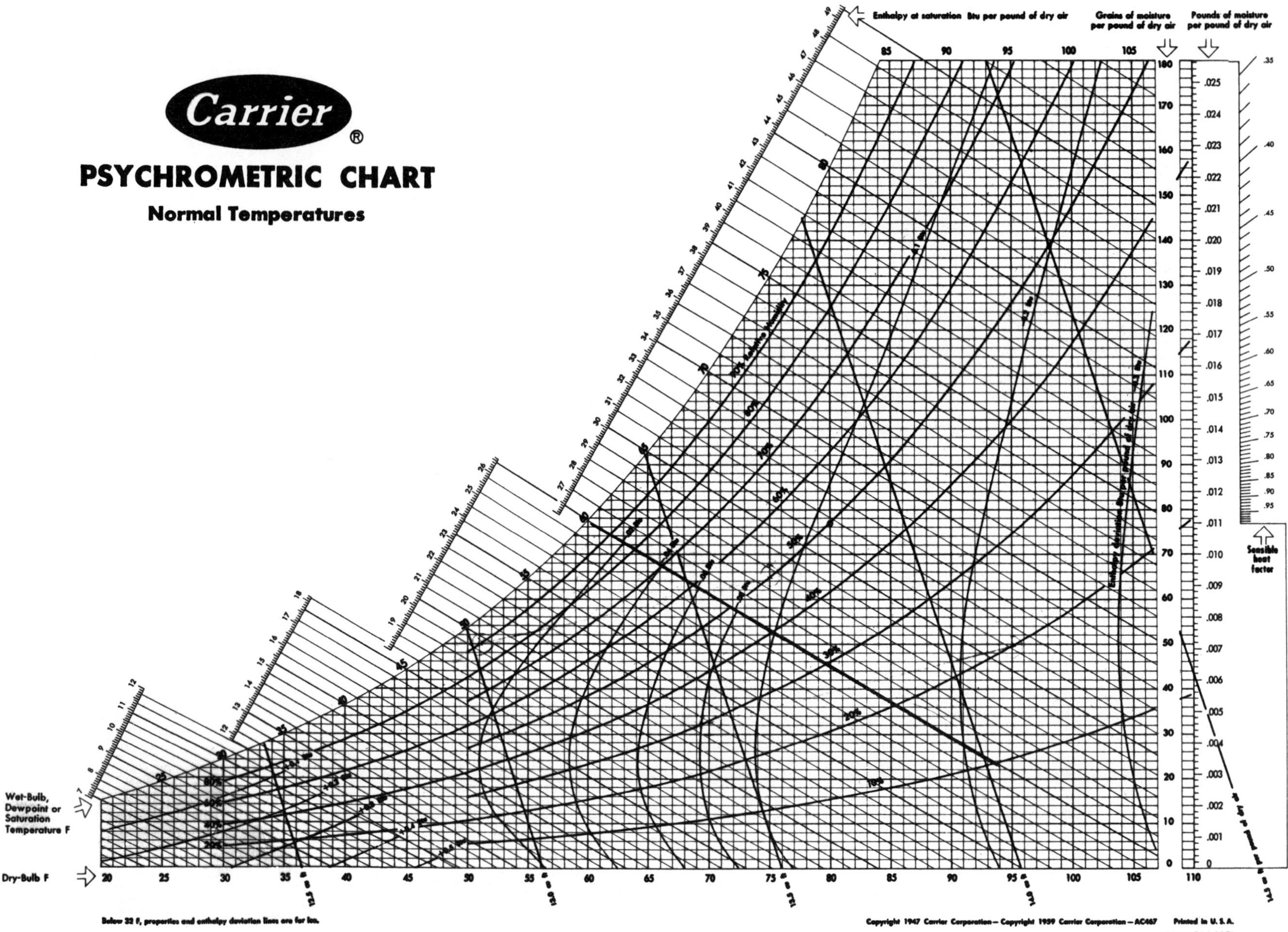

Figure 145-31 Psychrometric chart. (Courtesy of Carrier Corporation)

Heating Load

1. Design Conditions		Dry Bulb (F)		Specify Humidity gr./lb.
Outside		0°F		20
Inside		70°F		32
Difference	TD =	70°		12
2. Transmission Gain Items (Table 2)		**Calculations**		**Heat Load (Btu/hr)**
Windows	Sq. Ft. ×	Factor ×	Dry Bulb Temp. Diff.	
	132	1.13	70 =	101,441
			=	
Walls	668	.32	70 =	14,963
Partitions			=	
Roof	400	.14	70 =	3,920
Floor			=	
Other	400	.24	35 =	3,360
3. Ventilation or Infiltration Items (Tables 5 and 6, use larger quantity)		**Calculations**		**Heat Load (Btu/hr)**
Sensible Load	CFM ×	Dry Bulb Temp. Diff. ×	Factor	
	100	70	1.08 =	7560
Humidification Load	CFM ×	Specific Humidity Diff. ×		
	100	12	0.67 =	804
4. Duct Heat Loss Items (Table 7)		**Calculations**		**Heat Load (Btu/hr)**
Heat Loss ×	Factor for Insulation Thickness ×	Duct Length (ft.)		
			÷ 100 =	
5. Total Heating Load			=	41,048

Figure 145-32 Heating load form filled out. (The Trane Company)

REVIEW

- Refrigeration load:
  - Heat transmission—Based on type of construction, area exposed, insulation type/thickness, temperature difference.
  - Sun effect—Radiation factor; walls are exposed to the sun.
  - Design temperature—Two temperature considerations: outside and inside.
  - Air infiltration—Based on outside air leaking inside and traffic during use. Calculations determined by either air-change method or air velocity through open door.
  - Product load—Requires knowing product sensible load, specific heat above and below freezing, heat of respiration, and latent heat of freezing.
  - Supplementary loads—Lights, defrost heaters, electric motors, people.
  - Total hourly load—An accumulation of the individual loads based on 24-hr loads.
  - Equipment selection—Load is calculated on 24-hr basis and the required compressor capacity is determined by dividing the total load by desired compressor run-time operation.
  - Safety factor—Normally an additional 5 to 10% load increase.

- Residential heating and cooling loads:
  - Industry standard for residential calculation is the manual "J" by ACCA.
  - Load factors:
    Outside and inside design conditions.
    Building losses—Exposed glass, doors, walls, unheated partitions, ceilings, floors, and infiltration.
    System losses—Ducts, ventilation air, exhaust systems, furnace combustion air.
    Building gains—Solar radiation, exposed glass, walls, ceilings, doors, floors, infiltration, people, lights, and appliances/equipment.
- Commercial heating and cooling loads:
  - Similar to residential buildings except in the commercial building loads, interior space usage is large factor (i.e., restaurant, office, factory production, and humidity control)

Problems and Questions

1. The refrigeration load generally comes from four sources. What are they?
2. Product loads may be made up of either _____ or _____ heat or both.
3. Does insulation have a high or a low resistance value?
4. How much heat will be transmitted through 10 ft^2 of a 4-in.-thick common brick wall with no insulation if the temperature difference across the wall is 70°F?
5. Insulation is available in several different forms. Name them.
6. Define "design temperature."
7. Opening and closing the refrigerator door creates an _____ load.
8. The heat-removal rate of a product is determined by the _____ of the product.
9. Is the specific heat of a product higher above or below the freezing point of the product?
10. The latent heat load of a product is related to the percentage of its _____ content.
11. Thermal conductivity is the k factor. True or false?
12. $1/R$ is the same as the U factor. True or false?
13. The air change infiltration method is based on the average number of air changes in a one-hr period. True or false?
14. The sensible heat of a product is the same above or below freezing. True or false?
15. The industry standard for load calculations is the manual "D" by ACCA. True or false?
16. Heat is lost through a solid door by
    a. conduction.
    b. convection.
    c. radiation.
    d. None of the above.
17. System losses are through
    a. ducts located in unheated spaces.
    b. ventilation air.
    c. exhaust systems.
    d. All of the above.
18. The daily temperature range
    a. is normally high for most regions.
    b. is low, medium, or high.
    c. is 10 to 20°F.
    d. All of the above.
19. In the heat-gain calculations, walls below grade
    a. are not included.
    b. are heavy loads and must be considered.
    c. are light loads but should still be considered.
    d. None of the above.
20. For commercial load calculations, the most important consideration is the
    a. R value of the roof.
    b. use of the space.
    c. U value of the walls and glass.
    d. None of the above.
21. Using the data sheet, Fig. I45-2, calculate the refrigeration load for a walk-in cooler having the following characteristics:

Location:	Los Angeles, Calif.
Room use:	Storage ripe tomatoes
Outside dimensions:	10 ft × 10 ft × 9 ft
Room temperature:	45°F
Insulation:	3 in. molded polystyrene
Total room product load:	6000 lbs
Daily delivery:	1500 lbs precooled to 55°F.
Containers:	Paper cartons. Total weight 150 lbs.

Installation Techniques

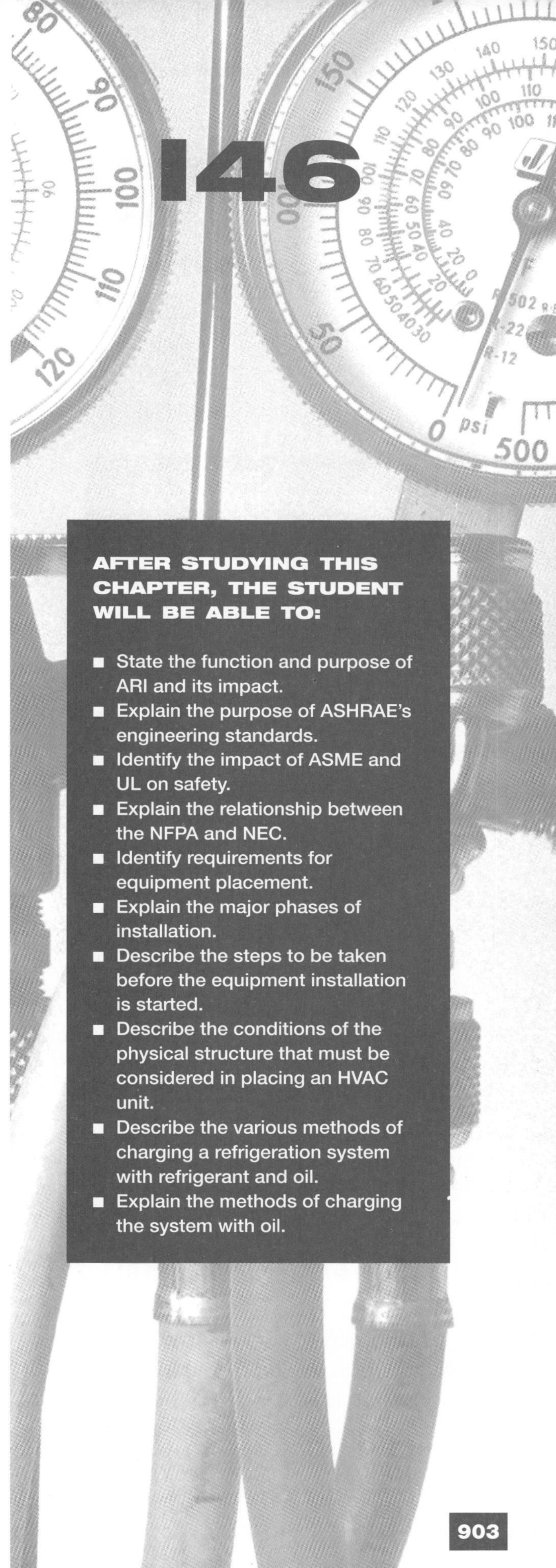

146

AFTER STUDYING THIS CHAPTER, THE STUDENT WILL BE ABLE TO:

- State the function and purpose of ARI and its impact.
- Explain the purpose of ASHRAE's engineering standards.
- Identify the impact of ASME and UL on safety.
- Explain the relationship between the NFPA and NEC.
- Identify requirements for equipment placement.
- Explain the major phases of installation.
- Describe the steps to be taken before the equipment installation is started.
- Describe the conditions of the physical structure that must be considered in placing an HVAC unit.
- Describe the various methods of charging a refrigeration system with refrigerant and oil.
- Explain the methods of charging the system with oil.

146-1 CODES, ORDINANCES AND STANDARDS

The statement is often made that, "All work should be performed in accordance with applicable national and local codes and standards." These regulations may apply to the design and performance of a product, its application, its installation, or safety considerations. Ordinances are set up locally to require conformity to national standards. The national standards, that apply to the HVAC industry, usually originate from ARI, ASHRAE, ASME, UL, or NFPA. The symbols that designate these organizations are shown in Fig. I46-1. The following descriptions indicate the scope of the regulations they produce.

146-1.1 ARI

The Air-Conditioning and Refrigeration Institute (ARI) has its headquarters in Arlington, Virginia. It is an association of manufacturers of refrigeration and air-conditioning equipment and allied products. Although it is a public relations and information center for industry data, one of the institute's most important functions is to establish product or application standards by which the associated members can design, rate, and apply their hardware. In some cases, the products are submitted for test and are subject to certification and listing in nationally published directories. The intent is to provide the user with equipment that meets a recognized standard.

146-1.2 ASHRAE

The American Society of Heating, Refrigeration, and Air-conditioning Engineers (ASHRAE) started in 1904 as the American Society of Refrigeration Engineers. Today its membership includes thousands of engineers and technicians from all areas of the industry. ASHRAE creates equipment standards, but its most important contribution probably has been the publication of a series of books that have become the reference bibles of the industry. These

Figure 146-1 Symbols of industry organizations that produce national standards for HVAC equipment.

include the Guide and Data Books for Equipment, Fundamentals, Applications, and Systems.

146-1.3 ASME

The American Society of Mechanical Engineers (ASME) is concerned primarily with codes and standards related to the safety aspects of pressure vessels.

146-1.4 UL

The Underwriters' Laboratories (UL) is a testing and code agency which specializes in the safety aspects of electrical products, while also including an overall review of some products. The UL seal on household appliances is a familiar one and it is also applied to the approval of refrigeration and air-conditioning equipment. Its scope of activity has expanded into large centrifugal-refrigeration machines.

UL approval for certain types of refrigeration and air-conditioning products may be mandatory by local electrical inspectors. Compliance with UL is the responsibility of the manufacturer, and the approved products are listed in a directory sent to all local agencies. Installation in accordance with the approved standards is the responsibility of the installer. Violation of these standards can cause a safety hazard as well as possibly voiding the user's insurance coverage should an accident or fire result. Installation procedures should therefore always conform to UL approval standards.

146-1.5 NFPA

Closely associated with the work of UL in terms of electrical safety is the National Electrical Code R sponsored by the National Fire Protection Association (NFPA). The original code was developed in 1897 as a united effort of various insurance, electrical, architectural and allied interests. Although it is called the National Electrical Code, its intent is to guide local parties in the proper application and installation of electrical devices. It is the backbone of most state or local electrical codes and ordinances. It is essential, therefore, that those involved with installation procedures be thoroughly familiar with the scope and content of the National Electrical Code Book and be able to find any information it contains.

Local codes are usually divided into: (1) electrical, (2) plumbing and refrigeration, and (3) other codes, such as sound control. Local inspectors enforce these codes based on the permits issued.

146-2 EQUIPMENT PLACEMENT

Despite what may seem to be a great many possibilities for positioning major cycle components during installations, three factors must be considered in the placement of equipment if satisfactory installation and proper operation are to be assured:

1. Ample space must be provided for air movement around air-cooled condensing equipment to and from the condenser.
2. All major components must be installed so that they may be serviced readily. When an assembly is

PRACTICING PROFESSIONAL SERVICES

When rigging equipment for installation, read the manufacturer's drawings and service installation for assembly and weight requirements. Check center of gravity, eyebolt holes, and special rigging instructions. Make sure rigging equipment has been inspected, is in good repair, and has OSHA approval. Look over the installation site for enough clearance when using crane, especially for overhead power lines.

ON THE JOB

Always use lifting lugs when provided and follow lifting and rigging instructions that come with the piece of equipment. Look out for your fellow workers, and know their location.

not easily accessible for service, the cost of service becomes excessive.

3. Vibration isolation must always be considered, not only in regard to the equipment itself, but also in relation to the interconnecting piping and sheet metal duct work. All manufacturers supply recommendations of the space required; these recommendations should be followed.

Fig. I46-2 shows a properly placed air-cooled condensing unit. Although the unit has been installed in an inside corner, sufficient room has been left for the passage of incoming air around and over the unit.

An example of a common error in positioning major system components is shown in Fig. I46-3. A shell-and-tube water cooled condenser has been placed in such a position that the entire condensing unit must be moved to replace a single condenser tube. Be sure to allow room for replacement of items such as compressors, fan motors, fans, and filters. High service costs are often attributable to the poor placement of system components.

Noise is also an important factor in the placement of air-cooled condensing equipment. Noise generated within the unit will be carried out by the discharge air. It is poor practice to "aim" the condenser discharge air in a direction where noise may be disturbing, such as neighboring office windows.

Vibrations set up by rotating assemblies such as compressors, fans, and fan motors can break refrigerant lines, cause structural building damage, and create noise. Vibration isolation is required on all refrigeration and air-conditioning equipment where noise or vibration may be disturbing. Almost all manufacturers use some form of vibration isolation in the production of their equipment. This is usually enough for the average installation; however, individual conditions must be investigated to determine whether more stringent measures need to be taken.

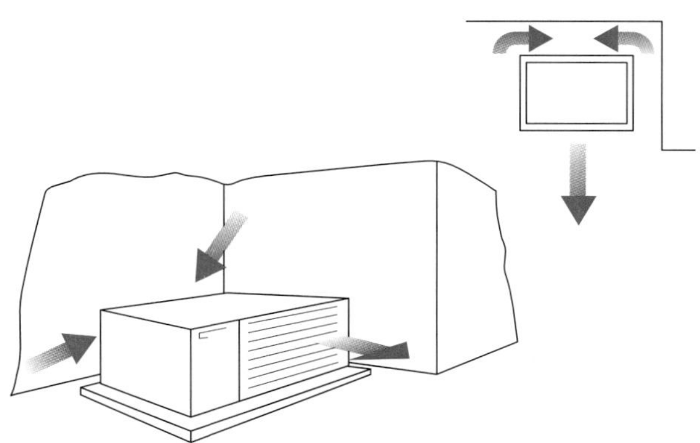

Figure I46-2 A properly placed air-cooled condensing unit.

Figure I46-3 Service accessibility is important in positioning major system components.

The compressor is considered the chief source of system vibration. Since it is good practice to isolate vibration at its source, compressor vibration isolation is essential during installation. One method of minimizing compressor vibration at the source is to bolt the compressor firmly to a solid isolated foundation. An example of this is the use of poured concrete pads. Stud bolts are positioned in the concrete when it is poured and the compressor or condensing unit is bolted firmly to the pad, as shown in Fig. I46-4.

When the compressor or condensing unit is to be installed on the roof or in the upper stories of a multi-story building, *vibration isolators* as shown at the upper right in Fig. I46-4 may be used. This type of isolator is usually available from the manufacturer of the unit, and in many cases, it is standard equipment.

An isolation pad designed specifically for vibration dampening is shown at the lower left in Fig. I46-4. This type of material is designed to dampen the vibration from a given amount of weight per square inch of area. As the right amount and type of this material can be properly selected only when both weight and vibration cycle are taken into consideration, a competent engineer should be consulted.

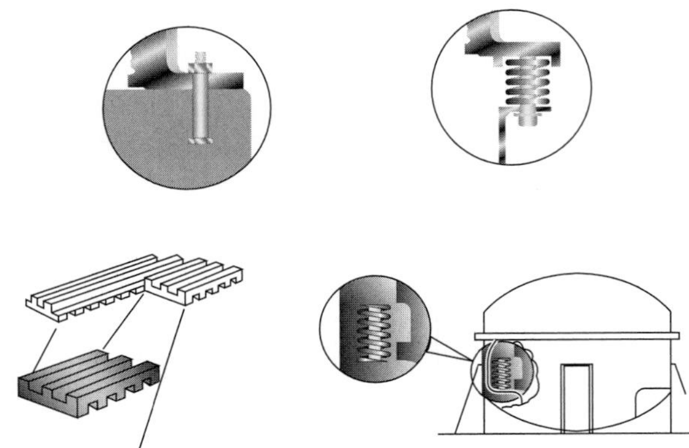

Figure I46-4 Types of vibration isolators used for compressors.

A final method of vibration isolation, used with small hermetic compressors, is shown at the lower right in Fig. I46-4. In this case the compressor is spring-mounted within the hermetic shell.

Fig. I46-5 illustrates a vibration eliminator inserted in the discharge line of the compressor. This eliminator is designed to absorb compressor discharge-line pulsation before it creates noise or breaks refrigerant lines. The eliminator consists of a flexible corrugated metal hose core with an overall metal braid. The braid will allow some linear movement of the flexible material, but no expansion or contraction. This type of eliminator is normally placed in the compressor discharge line as close to the compressor as practical. It is particularly effective on installations where the compressor and condenser are on different bases, yet quite close together.

The *vibration isolator* should be placed in the line so that the movement it absorbs is in a plane at right angles to the device. Do not place this unit in a position that will put tension on it, as its useful life will be shortened.

Some system components, such as the evaporator, may be suspended from the ceiling. As the air-handling unit containing the evaporator will also normally contain a fan and fan motor, it too is a possible source of vibration. Most manufacturers isolate the fan and fan motor inside the unit with rubber mounts. If this supplies sufficient vibration isolation, the unit may be directly connected to the ceiling. When more isolation is required, the same method used with compressors is effective. Coil springs support the air-handling unit for additional isolation.

I46-3 OTHER MAJOR PHASES OF INSTALLATION

In addition to the placement of major system components, most refrigeration system installations have three other major phases:

1. Erection of piping (both refrigerant and water)
2. Making electrical connections
3. Erection of duct work

The erection of refrigerant piping is a primary responsibility of the refrigeration installer, along with the placement of major system components. Although not always the responsibility of the refrigeration installer, electrical and duct work are important parts of most installations. The refrigeration installer should be familiar with good electrical and air-duct installation techniques, since quite often, particularly with small equipment, the refrigeration installer is called upon to do the electrical wiring and make connections to duct work.

I46-3.1 Piping

The art of making flared and soldered connections in copper tubing has been discussed, as well as the procedures for sizing lines, installing traps, etc., but several important points should be remembered while actually erecting the system piping.

When hard copper tubing is selected for refrigerant piping, it is recommended that low-temperature silver alloy brazing materials be used. These alloys have flow points ranging from 1,100 to 1,400°F.

To attain these temperatures, oxyacetylene welding equipment is required. Fig. I46-6 shows the equipment necessary for this type of brazing. Both oxygen and acetylene tanks with gauges and reducing valves are needed. Also shown is the torch used with the tanks. At the right is a bottle of dry nitrogen also equipped with gauges and reducing valves. The use of nitrogen is recommended, since it serves to keep the interior of the pipe clean during the brazing process. During low-temperature brazing, the surface of the copper will reach a temperature at which the metal will react with oxygen in the air to form a copper-oxide scale. If this scale forms on the interior surface of the pipe, it might be washed off by the refrigerant and circulated in the system. The scale can clog strainers or capillary tubes and will damage orifices.

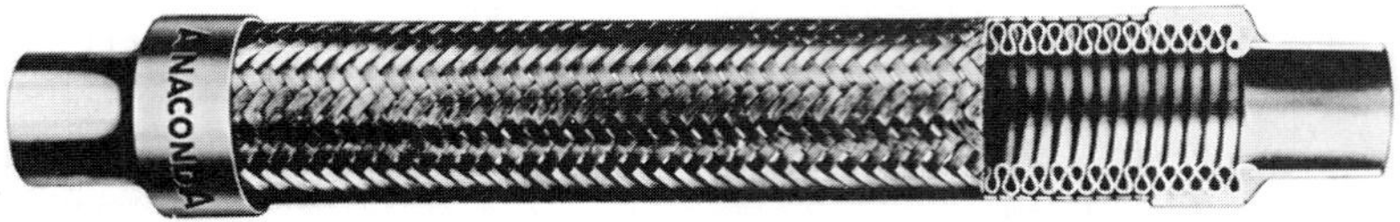

Figure I46-5 A vibration eliminator. (Courtesy of Anamet Industrial, Inc.)

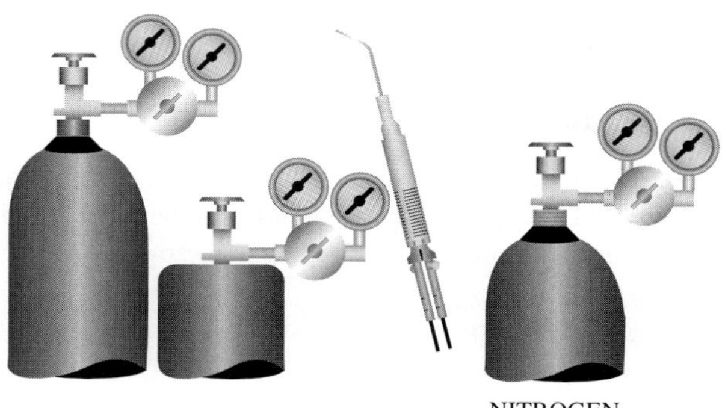

Figure I46-6 Equipment necessary for oxyacetylene welding or brazing.

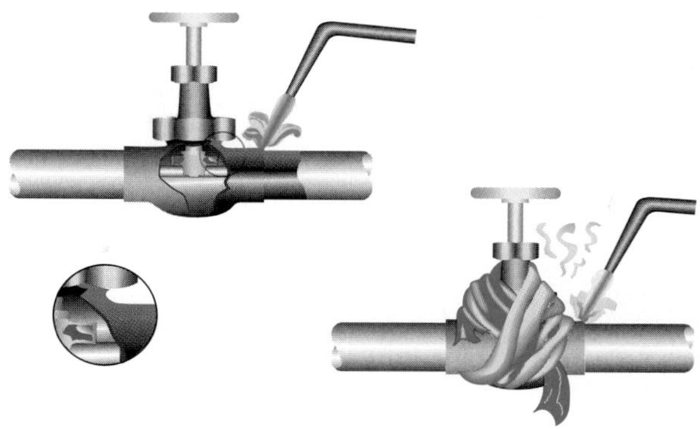

Figure I46-8 Protecting valves from overheating during brazing operation.

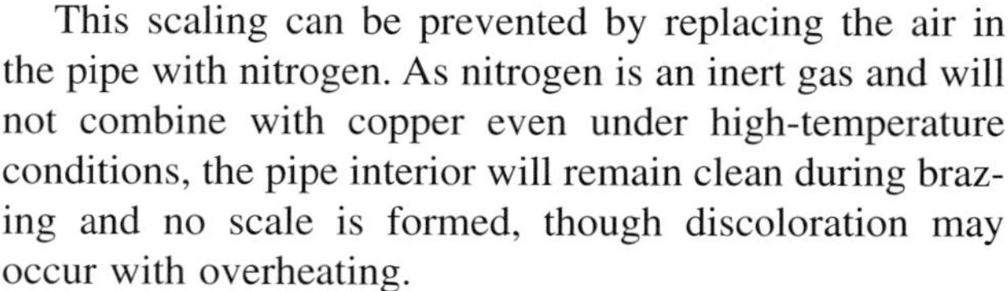

This scaling can be prevented by replacing the air in the pipe with nitrogen. As nitrogen is an inert gas and will not combine with copper even under high-temperature conditions, the pipe interior will remain clean during brazing and no scale is formed, though discoloration may occur with overheating.

Fig. I46-7 shows a method by which nitrogen may be introduced into the refrigerant tubing during the brazing operation. Instead of connecting the liquid line to the "king" valve or liquid-line stop valve, connect the liquid line directly to the nitrogen bottle as shown. By use of the reducing valve on the nitrogen bottle, a slight pressure is admitted to the tubing. This is just enough pressure to assure that air will be forced from the pipe. As shown in the inset, if the nitrogen flow can be felt on the palm of the hand, the flow is sufficient. This nitrogen pressure is kept in the tubing throughout the entire brazing operation, thus assuring an oxygen-free pipe.

Clean pipe is essential in refrigeration installation. Therefore, nitrogen is an extremely important part of the brazing operation as it assures scale-free interior pipe walls.

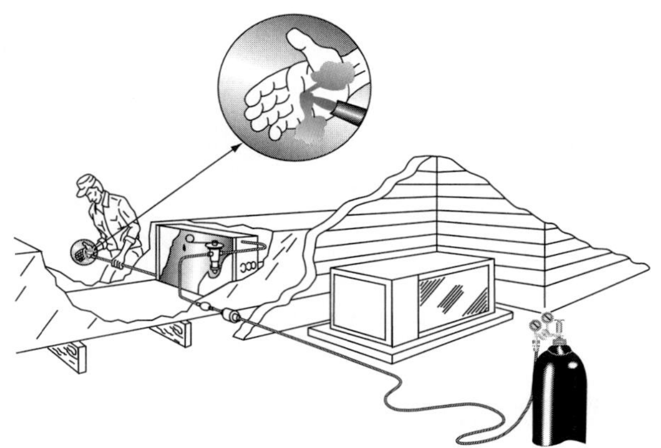

Figure I46-7 Brazing operation using nitrogen in the tube to eliminate scaling.

The temperatures during brazing operations can warp metals and burn or distort plastic valve seats. It is important that brazing heat does not reach metal or plastic components that might be damaged. Fig. I46-8 shows the results of brazing heat damage and also one method of assuring this will not happen. The valve at the upper left was not protected from brazing heat; the plastic seat has been damaged. Also, the seat holder has been warped, as shown in the inset. The valve obviously will not operate properly and would require immediate replacement. The valve at the bottom right, however, has a wet rag wrapped around the body. The water absorbs the heat that flows to the valve body during the brazing operation. By keeping the rag wet, the valve and its component parts are protected from heat damage.

In many cases, the suction line of the refrigerant cycle installation will run through a non-refrigerated or a non-air-conditioned area. The outside surface temperature of this pipe is frequently below the dew point of the surrounding air. In this case the moisture in the air will condense on the exterior of the pipe. This will create problems where this continuous moisture drips and can be annoying. When this condensation is expected or does occur, the pipe should be insulated; thus the condensation is prevented.

The insulation must be of good quality so that the temperature of its exterior surface will never drop below the dew point of the surrounding area. It must also be well sealed, so that air and the moisture it contains cannot flow through the insulation to the pipe, thus causing condensation underneath the insulation.

At the right in Fig. I46-9 is an example of a typical pipe hanger that might be found on a small commercial-refrigeration installation. This hanger serves as a vibration isolator as well as a pipe supporter. This example is being used on an insulated line. A short length of light-gauge rustproof metal has been wrapped around the insulated pipe. A metal strap has been attached to this length of metal. In some cases it is merely wrapped around the

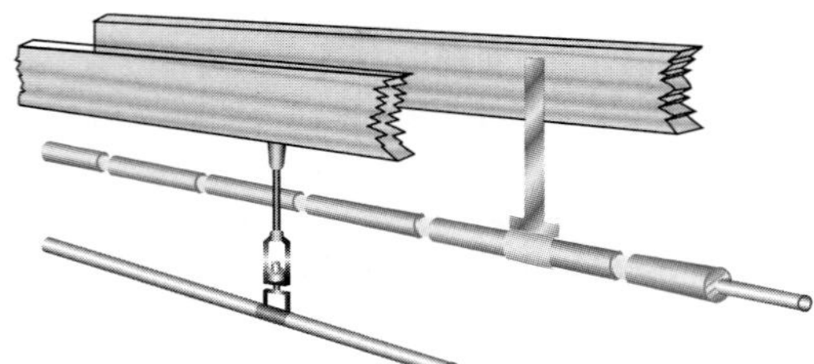

Figure I46-9 Pipe hangers used to support insulated pipes.

length of sheet metal. The free end of the strap is then fastened to the joist or ceiling.

The insulation in such a hanger will act as the vibration isolator. The purpose of the short length of metal is to prevent the thin strap from cutting the insulation.

The hanger shown on the left (Fig. I46-9) has a height-adjustment feature for leveling or pitching the pipe if required for oil flow. This type of hanger can be used with or without insulation.

I46-3.2 Electrical Connections

The service technician who installs the refrigeration cycle is sometimes responsible for the final wiring connections between the installed unit and the fused disconnect switch shown in Fig. I46-10. All electrical power to the refrigeration unit must pass through this switch. When this switch is pulled, or opened, all electrical power to the unit must be disconnected. This same disconnect switch also contains fuses, which will interrupt the flow of current whenever a severe electrical overload occurs. This mechanism is a protection against fires and explosions and also against electrical shocks to people.

Electrical codes, both national and local, are made to protect property and life, and should always be followed. All refrigeration-cycle installations having electrical connections are governed by national or local electrical codes. For example, electrical codes require that the fused disconnect switch in Fig. I46-10 always be placed within sight of the unit that receives the power passing through the switch.

When electrical circuits are to be connected by the refrigeration installer, wiring should be made to assure good electrical contacts. Lug-type connectors are recommended.

When braided wire is used, a single wire may separate and create a potential electrical hazard. Loose wires might contact other wires or "ground," causing electrical problems. Braided wire should be looped into the size required, and soft solder should be allowed to flow over and coat the braided loop. This will assure good contact and eliminate the hazard of wire separation.

I46-3.3 Duct Work

The refrigeration system installer is frequently required to make the final connection between the evaporator or air-handling unit and the duct work (Fig. I46-11). This final connection is made with a flexible connector that will eliminate vibration transmission from the air handler to the duct work. It must be installed correctly. If, as shown in the upper left in the illustration, it is too loose, it will drop into the airstream and obstruct normal airflow. It may also flap and create noise. If too tight it will stretch, harden, and disintegrate over a period of time, causing leaks. If the canvas is damp and installed too tightly, the duct work might be pulled out of alignment. The center of Fig. I46-11 shows the proper application: the canvas is loose enough to absorb vibration and not so loose as to interfere with airflow.

I46-3.4 Final Checking

When the refrigeration system has been completely assembled and all electrical and duct connections com-

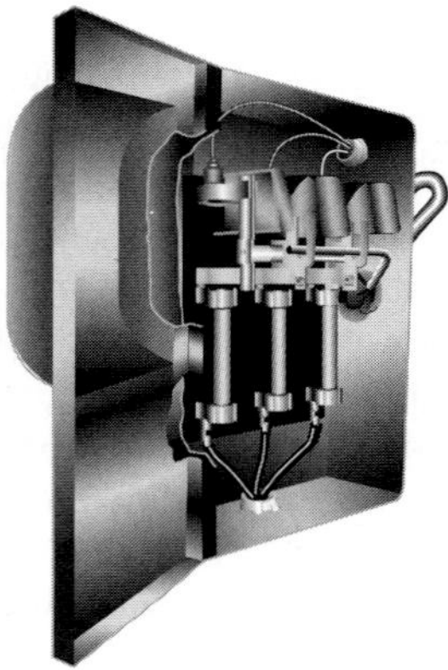

Figure I46-10 Fused disconnect switch for the power supply.

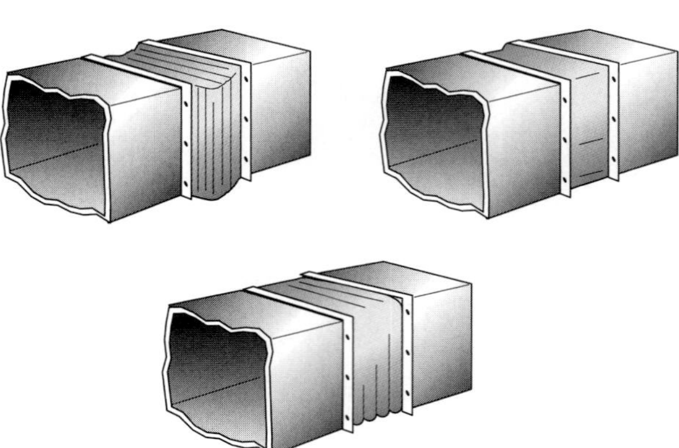

Figure I46-11 Flexible duct connector to eliminate vibration.

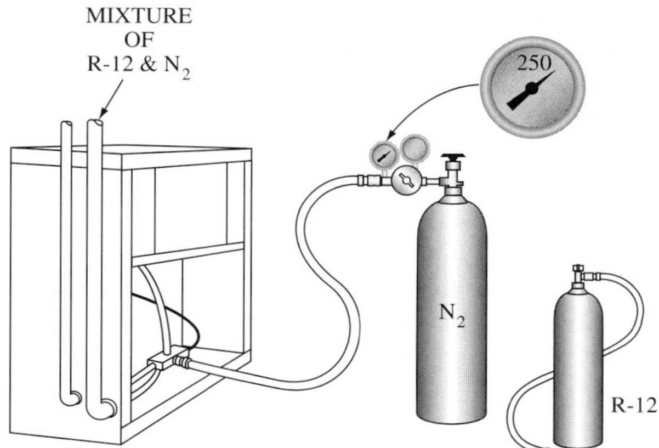

Figure I46-12 Leak-testing using nitrogen to pressurize the piping.

pleted, a number of important steps must still be taken before the equipment is started:

1. The unit must be leak-tested and charged.
2. All belts must be checked for tension and alignment.
3. There must be an electrical motor check of the direction of rotation.
4. The bearings must be oiled.
5. The power source must be checked to be absolutely sure that the correct power will be applied to the unit.
6. Field leak detection of the refrigerant is generally done using an electronic leak detector.

After all interconnecting tubing has been assembled, some refrigerant is introduced into the system as a gas. Although a leak test could be made at this time, some refrigerants do not exert sufficient pressure at room temperature to assure trustworthy results. By using nitrogen, the pressure in the system may be built up to approximately 250 psig, at which pressure a true leak test can be made (Fig. I46-12). The mixture of refrigerant and nitrogen inside the unit will cause a reaction on the detector if a leak is present. Since nitrogen is an inert gas, the system should be evacuated after it has been determined to be leak-free.

Following the leak test, the system is ready for charging or dehydration.

I46-4 CHARGING THE SYSTEM WITH REFRIGERANT

The system may be charged with refrigerant as a liquid or gas. Although a number of factors may affect the method of charging, the most important is the quantity of refrigerant involved.

Relatively speaking, most refrigerant containers, regardless of size, have a small outlet. In small units, sufficient gas vapor may be passed through this outlet to complete the full charging of the unit in a reasonable length of time. On large units, however, the time required for the proper amount of vapor to pass through this small outlet may take so long that gas charging is impractical. When this is the case, liquid charging is used. The small orifice will pass a much greater weight of liquid refrigerant in any given amount of time.

The point at which the refrigerant will actually enter the system is determined by one basic consideration: whether the unit is to be charged with refrigerant in a gas or liquid form.

Under normal charging conditions the refrigerant cylinder will be at ambient temperature and corresponding pressure. As most field refrigerant charging is done with the unit operating, the refrigerant cylinder pressure will usually be below the head pressure of the condenser and above the back pressure of the evaporator.

The metering-device orifice is not large enough to make gas charging on the high side of the device practical because of the time consumed. Therefore, gas is normally charged into the system after the metering device. This charging can be done at any convenient point between the metering device and the compressor.

If the refrigerant is to be charged as a liquid, the charging should take place ahead of the metering device to protect the compressor from damage due to liquid flooding.

With open-type and serviceable hermetic systems that are small enough to make gas charging practical, or where only a small amount of refrigerant is required, charging is usually done through the suction service valve on the compressor. Since welded hermetic systems are normally small enough for gas charging, pinch-off tubes are made available at the compressor for this purpose. These pinch-off tubes are connected into the suction side of the compressor and are designed to allow the compressor to pump gas directly from a refrigerant drum.

In units of this type, the original charge is generally done at the factory, but occasionally the charge must be replaced in the field. When this is necessary, the pinch-off tube may be cut and a connection placed on the tube. The unit is then charged through the tube, which is repinched and brazed. Occasionally, valves are placed permanently on this pinch-off line. This is practical where the pinch-off tube is either very short or inaccessible.

There are also devices that are designed to puncture a refrigerant line for charging purposes. The device then remains on the line and serves as a valve for either future charging or as a gauge connection.

Another device used for charging the hermetic circuit is a type of valve called a *Schrader fitting.* This valve contains a plunger which, when depressed, opens the circuit. A special adapter on the charging hose will depress the plunger when the hose is firmly attached to the valve.

Refrigerant will then flow through the valve into the system. When the charging hose is removed, the plunger returns to its original position and reseals the refrigerant circuit. The valve cap should always be replaced after servicing.

There are four methods for determining whether the proper amount of refrigerant is introduced into the system:

1. Charging by sight glass
2. Weighing the charge
3. Charging by head pressures
4. Using a frost line.

All four methods are used extensively in commercial refrigeration work, and the choice of method depends on the size and type of system involved. "Critical-charge" methods, common to small residential air-conditioning systems are not included; they are critical to ounces of refrigerant charge.

146-4.1 Charging by Sight Glass

In properly charged systems there should always be a solid flow of liquid refrigerant (no bubbles) to the metering device. A sight glass will indicate the presence of gas by bubbles in the glass.

To charge properly by the sight-glass method, refrigerant is added in the usual manner. Following the addition of a reasonably large percentage of the estimated charge of gas or liquid, the refrigerant cylinder is shut off and the system allowed to "settle down" for a period of time. When the operating pressures have stabilized, the sight glass is again observed. If bubbles are still present, additional charge is introduced slowly until the bubbles disappear. If, after stabilizing the pressures, no bubbles are present when the unit is operating under maximum load conditions, the unit is properly charged.

146-4.2 Weighing the Charge

A common method of determining whether the proper amount of charge is being introduced into the system is by "weighing" the charge. This is the most accurate method of adding a full charge when the required charge is known. The method is quite simple, shown in Fig. I46-13.

With the system evacuated and ready to receive the refrigerant, the refrigerant supply bottle is weighed. On the left the scales show a total weight of 190 lb. If the system requires a charge of 40 lb, this amount of gas or liquid is released from the bottle into the system. The refrigerant supply bottle is again weighed. The total weight of refrig-

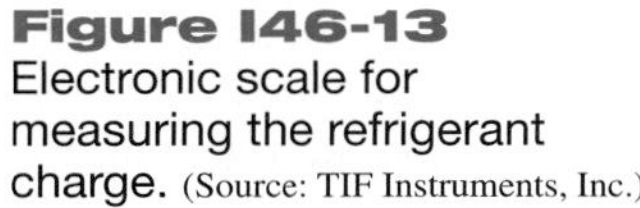

Figure I46-13 Electronic scale for measuring the refrigerant charge. (Source: TIF Instruments, Inc.)

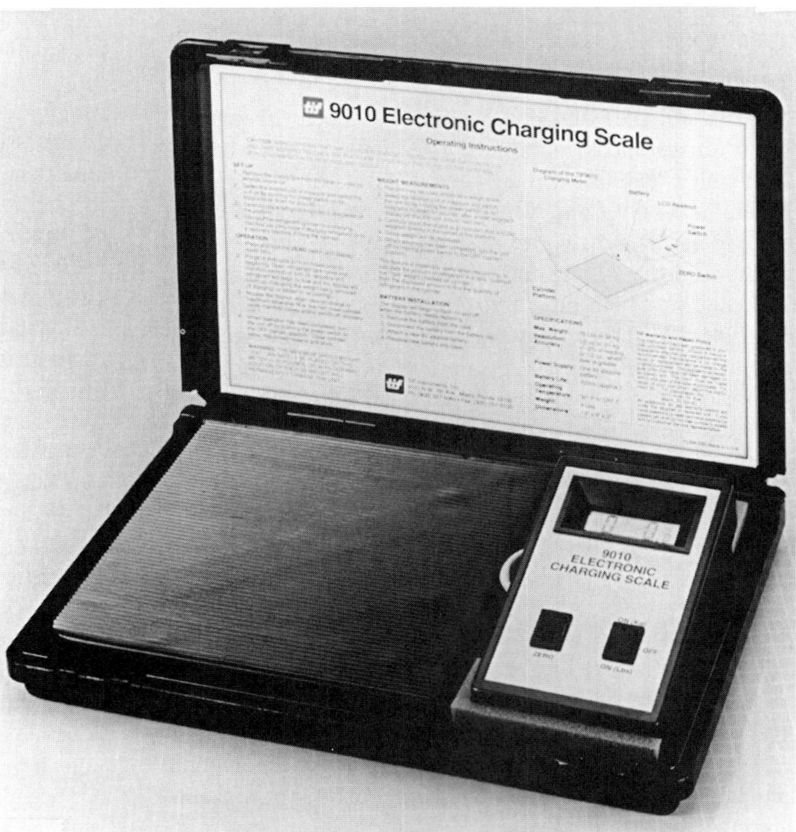

erant and bottle should now be 190 − 40, or 150 lb, as shown at the right in Fig. I46-13.

This method is used mostly on packaged equipment and then only when it requires a complete charge. The method is of little use when only a portion of the charge is required. Rarely is the exact charge known under such conditions.

I46-4.3 Charging by Head Pressures

A third way of charging, which may be used with factory-designed and balanced packages, is the head-pressure method. From test data, the factory determines the proper head pressures under various evaporator loads or temperatures. This information is furnished to the installer either as a graph or a table, which consists of a list of possible low-side pressures along with the proper head pressures supplied by the test data of the unit.

After charging with an estimated amount as directed by the factory, the unit is run for enough time to allow the pressures to settle and stabilize. Pressure readings are then compared to the manufacturer's chart.

If the actual head pressure is below the pressure that the chart indicates is correct, refrigerant is added. If the actual head pressure is above that indicated by the chart, refrigerant should be removed.

Fig. I46-14 shows how the head pressure method is used. An original charge as suggested by the factory has been introduced into the unit. The unit is run long enough for pressure stabilization, and the back pressure is found to be 5 psig. This point is on the horizontal axis at point A. The chart is then read upward vertically until the intersection with the outdoor temperature line at B. From B, read horizontally to the left until point C is reached. This is the head pressure under which the unit should be operating when the back pressure is 5 psig. The chart shows this figure to be 244. If the discharge-pressure-gauge reading had been below this figure, refrigerant would have to be added until the gauge pressure is brought up to 244 psig. If the discharge-pressure-gauge reading had been above this figure, refrigerant would have to be removed until the pressure was brought down to 244 psig.

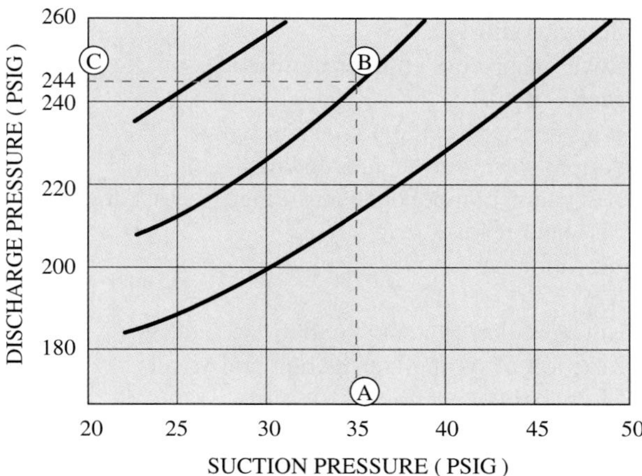

Figure I46-14 Determining the refrigerant charge from a chart showing suction and discharge pressures.

I46-4.4 Frost-line Method of Charging

This method can only be used in small hermetic systems that use capillary tubes. Such systems are not too common in commercial refrigeration work. When a system of this type is operated without evaporator load, the back pressures will normally drop below the freezing temperature and frost will form on the coil.

In Fig. I46-15 the evaporator load has been removed by placing a piece of cardboard over the coil face, thus shutting off the airflow. Since the load has been removed from the evaporator, the refrigerant will not evaporate as rapidly, and some will pass through the evaporator and evaporate in the suction line. Tests have shown that under these conditions a properly charged unit will normally frost to within a few inches of the compressor. By factory testing, this final frost point can be determined and given to the installer.

By recreating the frost line on the suction line, the installer can determine the proper charge. If the frost line does not reach the point designated by the factory tests, more refrigerant should be added. If the frost goes beyond the point designated by the factory, refrigerant should be removed.

In most systems, some overcharging can be tolerated, but an undercharge is rarely acceptable. Overcharging will create high head pressure and high temperature, with all the resultant problems such as motor overloading, sludge formation, and compressor-valve failure. High head pressures can also result in poor load control, with liquid refrigerant flooding to the compressor.

Although the biggest problem with undercharging is that of capacity, it may also create frost conditions on the evaporator in higher-temperature refrigeration equipment, and may also cause high evaporator superheats. As some hermetic compressor motors depend on suction gas for cooling, they can be damaged by overheating due to high suction-gas temperatures.

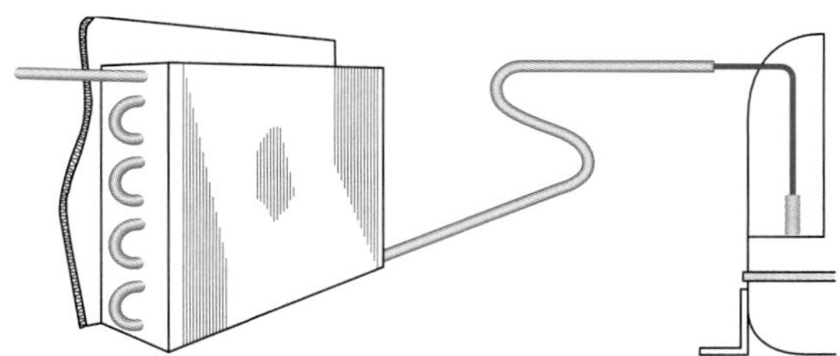

Figure I46-15 Determining the refrigerant charge from the position of the frost line.

Both overcharging and undercharging should be avoided, since either condition can do serious harm or destroy system components.

I46-5 CHARGING THE SYSTEM WITH OIL

The proper amount of oil can be measured into the system in several ways:

1. In a new system, it can be measured or weighed in. Unit installation instructions include the compressor oil requirements in either weight or liquid measurements. This method is also applicable following a compressor overhaul, when all of the oil has been removed from the compressor; however, it should be used only when the system has no oil in it.
2. The dip-stick method is used primarily with small, vertical-shafted, hermetic compressors. Some larger, open types of compressors may also have openings designed for the use of a dip stick. The manufacturer's recommendations of the correct level should always be followed.
3. The compressor crankcase sight glass is used after the system has operated for a period of time under normal conditions to determine the proper oil level. This procedure will assure proper oil return to the crankcase. It will also allow the oil lines and reservoirs to fill and give the refrigerant an opportunity to absorb its normal operating oil content, if applicable.

When a compressor is replaced, the new unit should be charged with the same amount of oil as the old unit.

Oil is normally introduced into a refrigerant system by one of two methods (Fig. I46-16). It may be poured in as shown on the left, providing the compressor crankcase is at atmospheric pressure. This method is normally used prior to dehydration since it will expose the compressor crankcase interior to air and the moisture the air contains.

On the right in Fig. I46-16 the method normally used with an operating unit is shown. In this case the crankcase is pumped down below atmospheric pressure and the oil is drawn in. When this method is used, the tube in the container subjected to air pressure should never be allowed to get close enough to the surface of the oil to draw air. As shown, the tube is well below the level of the oil in the container.

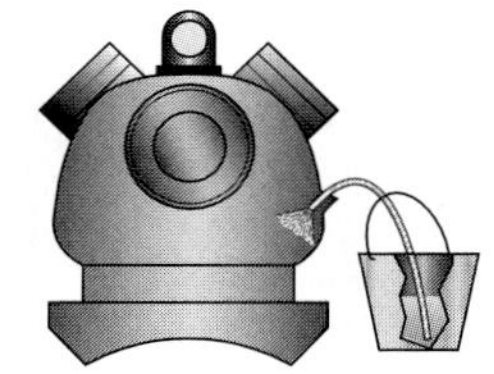

Figure I46-16 Two methods of charging oil into the compressor.

Oil charging of replacement welded hermetic compressors should be done according to the manufacturer's recommendations, and depends on whether the replacement compressor has been shipped with or without an oil charge.

There are three precautions to take in charging or removing oil:

1. Use clean, dry oil. Hermetically sealed oil containers are available and should be used.
2. Pressure must be controlled when the crankcase is opened to the atmosphere. Too much pressure can force oil out through the opening rapidly and create quite a mess.
3. System overcharging should be avoided. Not only will this create the possibility of oil slugs damaging the compressor, but it also can hinder the performance of the refrigerant in the evaporator. Oil overcharging will also cause liquid refrigerant to return to the compressor from the evaporator.

REVIEW

- National Standards originate from:
  ARI—Product or application.
  ASHRAE—Equipment standards, references.
  ASME—Codes and standards related to safety of pressure vessels.
  UL—Testing and code agency; safety aspects of electrical products.
  NFPA—National Electrical Code R sponsored by National Fire Protection Association.
- Local codes usually divided into:
  Electrical
  Plumbing and refrigeration
  Sound control
- Local inspectors enforce these codes on the permits issued.
- Equipment placement considerations:
  Ample space for air movement
  All major components positioned so that they can be serviced readily
  Vibration
  Noise
- Refrigeration system installations:
  Erection of piping (refrigerant and water)
  Electrical connections
  Erection of duct work
- Dry nitrogen is used during brazing operations thus assuring an oxygen-free piping and scale-free interior pipe walls.

- Electrical codes, both national and local, are made to protect property and life.
- Before start-up:
  The unit must be leak-tested and charged.
  All belts must be checked for tension and alignment.
  Electrical motor check (direction of rotation) must be done.
  Bearings must be oiled.
  Power source must be checked to be absolutely sure that the correct power will be applied to the unit.
- Refrigerant charging:
  Vapor charge—Low side.
  Liquid charge—Liquid line ahead of the metering device.
- Four methods of determining proper refrigerant charge:
  Sight glass
  Weighing the charge
  Superheat and subcooling/head pressure
  Frost line (high and medium temperature system)
- Charging oil:
  New system—Measured or weighed in
  Dip-stick method
  Crankcase sight glass
- Three precautions in charging or removing oil:
  Use clean dry oil.
  Pressure must be controlled.
  System overcharging should be avoided.

Problems and Questions

1. The National Electrical Code Book is published by _____.
2. UL approvals are not generally required on refrigeration or air-conditioning equipment. True or false?
3. What is the difference between codes and standards?
4. When installing equipment, placement is important to allow for _____.
5. Vibration of equipment has been eliminated by the manufacturer and is no problem in the field. True or false?
6. Dry nitrogen is used in the brazing process to reduce _____.
7. Suction lines running outside the refrigerated area will generally require insulation to eliminate _____.
8. To prevent vibration transmission in metal ducts, a _____ connection is used.
9. Belt-driven equipment which is too tight will draw high amps. True or false?
10. The final step in any installation is to instruct the customer in the operation of the equipment. True or false?
11. ARI is a public relations and information center for industry data. True or false?
12. The NFPA is responsible for the NEC. True or false?
13. When hard copper tubing is selected for refrigerant piping, it is recommended that high-temperature silver brazing be used. True or false?
14. Scaling inside piping is prevented by flushing with dry nitrogen. True or false?
15. Electrical codes are made to protect property first and life second. True or false?
16. Equipment installation should:
    a. Include ample space for air movement.
    b. Be installed so that they are serviced readily.
    c. Have vibration isolation.
    d. All of the above.
17. After all interconnecting refrigeration tubing has been assembled, some refrigerant is:
    a. Introduced into the system as a liquid.
    b. Introduced into the system as a gas.
    c. Not to be used due to new EPA regulations.
    d. None of the above.
18. Dehydration is accomplished through:
    a. Installing good liquid-line filter-driers.
    b. Installing good liquid-line and suction-line filters.
    c. Installing good filter-driers and deep evacuation.
    d. All of the above.
19. To determine the proper amount of refrigerant, the technician can:
    a. Charge by sight glass.
    b. Weigh the refrigerant.
    c. Charge by head pressures or frost line.
    d. All of the above.
20. When removing oil from a refrigeration system and adding new oil:
    a. Use clean, dry oil.
    b. Control crankcase pressure.
    c. Don't overfill or overcharge.
    d. All of the above.

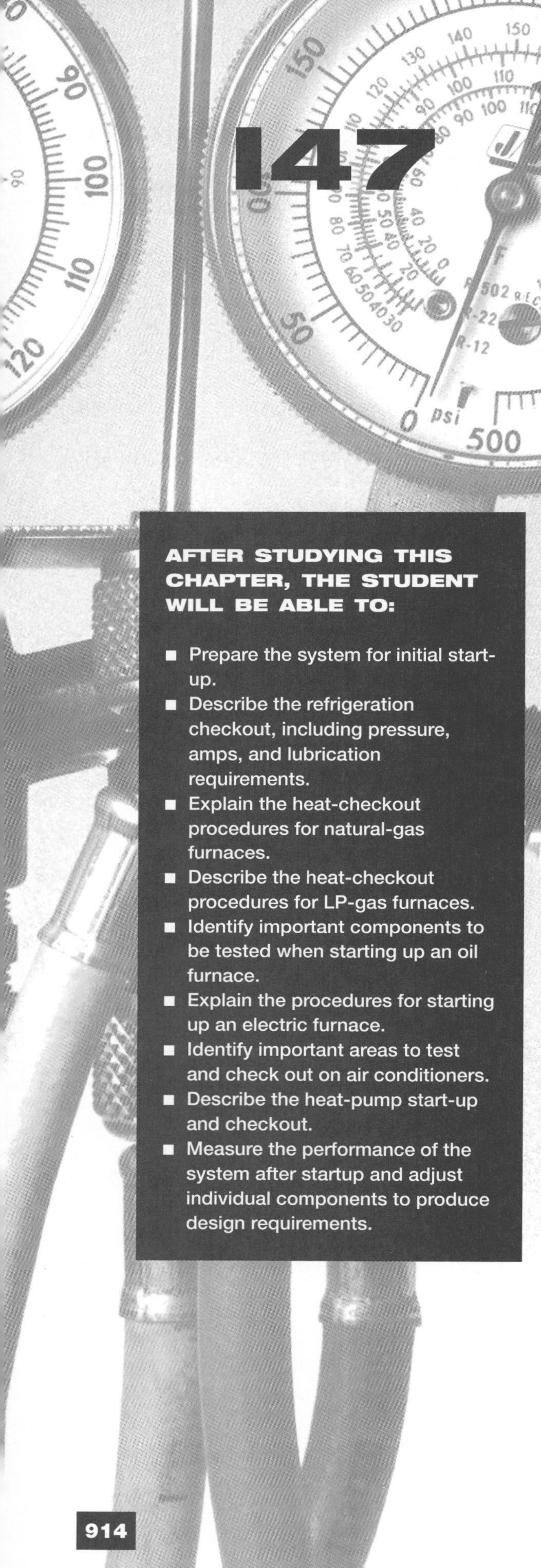

147 Installation Start-up, Checkout, and Operation

AFTER STUDYING THIS CHAPTER, THE STUDENT WILL BE ABLE TO:

- Prepare the system for initial start-up.
- Describe the refrigeration checkout, including pressure, amps, and lubrication requirements.
- Explain the heat-checkout procedures for natural-gas furnaces.
- Describe the heat-checkout procedures for LP-gas furnaces.
- Identify important components to be tested when starting up an oil furnace.
- Explain the procedures for starting up an electric furnace.
- Identify important areas to test and check out on air conditioners.
- Describe the heat-pump start-up and checkout.
- Measure the performance of the system after startup and adjust individual components to produce design requirements.

OVERVIEW

The sole purpose of proper start-up, checkout, and operation is to have the system produce the design conditions in the most efficient manner. This means that each unit must produce its rated capacity, deliver the correct amount of air, produce the correct temperature rise or drop over the heat exchanger, distribute air to various parts of the system in keeping with load conditions, and include a properly functioning control system.

I47-1 THE REFRIGERATION SYSTEM

Care must be taken to prevent damage to the refrigeration system during initial start-up (Fig. I47-1). Valve positions must be checked to make certain that only proper and safe pressures will occur at the compressor. The compressor-discharge valve must always be opened prior to start. Low-side valves must be adjusted in such a way as to assure neither excessively high nor low pressures in the compressor suction. Pressures should be carefully observed and regulated until the unit is operating normally. When starting a new system, try to keep the operating pressures as close to normal as practical.

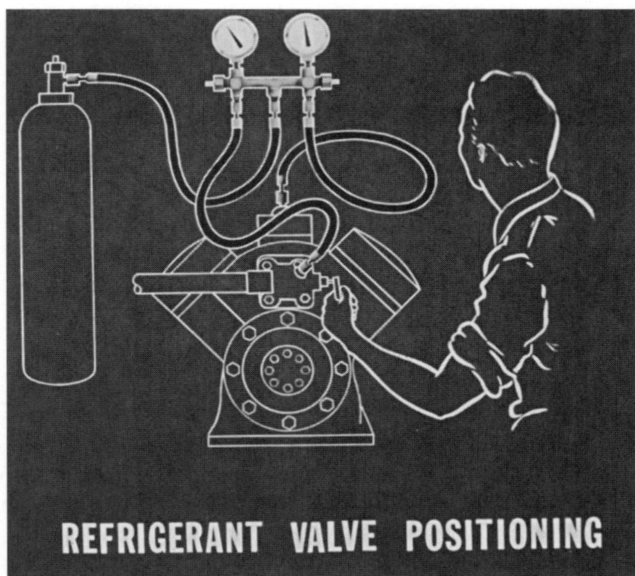

Figure 147-1 Throttling the suction valve on start-up.

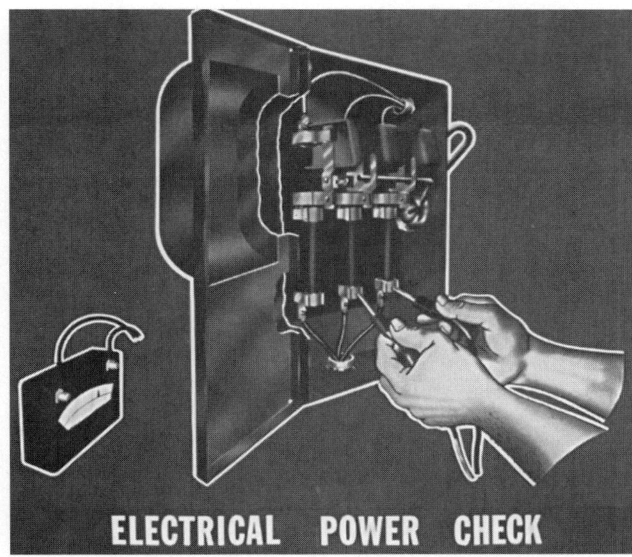

Figure 147-2 Checking the electric power.

When the installed equipment is belt-driven, both the alignment and tension of the belt must be checked prior to start-up. Improperly aligned belts show excessive wear and have an extremely short life. Loose belts wear rapidly, slap and frequently slip. Belts that are too tight may cause extreme motor-bearing and fan-bearing wear. Belt alignment is simple to check. A straightedge laid along the side of the pulley and flywheel will show any misalignment immediately.

The belt tension should also be checked. The belt stretches very little, so the belt should be set at the right tension during initial start-up. The correct tension allows about one inch of deflection on each side of the belt when squeezed by the fingers.

It is good installation practice to check electric-motor nameplates against the available voltage prior to initial start-up. Fig. 147-2 shows the voltage at the disconnect switch being checked for comparison with the electric motor nameplate. It is also good practice to determine if a phase unbalance exists in polyphase units prior to the final wiring connection.

When the system has been thoroughly checked for leaks, properly charged, valves positioned for start-up, belts aligned, motors oiled, and wiring checked, the unit can be started.

Following start-up the unit must be checked to make certain it is doing the job it was designed to do efficiently. This includes determining electrical load characteristics. No installation should ever be considered complete until it has been proven it can do the job it was designed to do.

Following the completion of tests to determine that the unit is performing satisfactorily, the final cleanup should be started. All excess tubing, wire, scrap metal, and other parts should be removed and the unit left in a condition that will indicate the installer's pride of workmanship. The equipment should be an asset to the area in which it is located.

The final step in any installation is to instruct the customer in the operation of the equipment. There are probably as many service complaints because of a lack of customer understanding of the new equipment as from any other reason. Customers should be clearly and concisely shown how their equipment can be made to operate properly and efficiently. Maintenance requirements should be explained, including such things as filter changes and motor coiling. Customers who are thoroughly familiar with the way their equipment should operate and what they can do to keep it that way are going to be satisfied customers.

147-2 THE HEATING SYSTEM

Heating start-up, checkout, and operation will be discussed in separate sections relating to three leading domestic energy sources: (1) natural gas, (2) LP gas, and (3) oil.

147-2.1 Natural Gas

To check, test, and adjust a gas-burning unit for highest operating efficiency, the unit must have the proper amount of input, the proper adjustment of the burners, the correct amount of combustion air, proper venting, and the correct amount of air for heat distribution.

Knowledge Extension

Often overlooked, automatic temperature-control valves may be the reason for improper heating and cooling attributed to the system design or control system. Critical sizing is essential to the proper transfer of heat-exchange fluids in a building. Improper sizing is poorly understood by maintenance people and usually the last thing they would find at fault. Blame is usually placed on the wall thermostat or the duct stat for poor temperature control, then on the design system.

Basic information from the manufacturer on the types and how they work helps avoid problems. Kits to repack and/or rebuild leaking valves are available when problems arise. They are relatively inexpensive compared to the cost of a new valve. Surprisingly, few service people use these kits. Why?

1. Service and maintenance people may be unaware that these kits exist.
2. It may be simpler and quicker to replace a valve; however, weight may be a factor: a 4-in. valve can weigh 205 lb, requiring two or three service people to handle the valve. A better solution would be to rebuild it.
3. Distributors may not stock them because of low demand and lengthy time required to get the kits.
4. Servicing may be for maintenance or because of a problem.

For maintenance packing may be required and this may be a routine procedure. If the valve is leaking, the service person would need a rebuild kit based on changing the disc, stem, and packing—almost the entire assembly as an alternative to replacing the valve.

It is almost impossible to tell when a valve will fail because the service each valve performs varies from one installation to another. Service companies and maintenance operations can establish regular schedules for repacking valves. A well-kept log of the system's performance can provide prediction charts when valve actuators are likely to fail.

To arrive at the correct gas input, two important factors must be known about the gas:

1. Heat content, in Btu/ft^3
2. Specific gravity

Since natural gas is a mixture of methane and ethane, various sources have different heat contents ranging between 945 and 1121 Btu/ft^3. The specific gravity also varies between 0.56 and 0.72. These characteristics relate to sizing the piping and determining the amount of gas to be supplied to each burner.

Heating units are constructed to use gas-pressure regulators set at an output of 3½ in. WC. Burner designs will allow operation between 3 and 4 in. WC manifold pressure, but these pressures should not be exceeded.

Checking the Gas Input

Fig. I47-3 shows the index or dial of a typical domestic gas meter. Included are two test dials, one for ½ ft^3 per revolution and the other for 2 ft^3 per revolution. To determine if the correct amount of gas is being fed to the heating unit, it is only necessary to find the feed rate through the meter. For accuracy, all other appliances must be turned off. If the pilot lights of the other appliances are turned off, be sure to relight before leaving. Usually, the requirements of pilot lights are so small that they are ignored.

Gas flow through the meter is determined by the time it takes the test dials to turn one revolution. To determine this time, the following formula can be used:

$$\text{time (sec/ft}^3\text{)} = \frac{\text{seconds per hour}}{\text{ft}^3\text{/hr of gas}}$$

In the example shown above, the formula would be set up as follows:

$$\text{sec/ft}^3 = \frac{60 \times 60}{142.8} = \frac{3600}{142.8} = 25.2 \text{ sec}$$

It will require 25.2 sec for 1 ft^3 of gas to go through the meter. Thus, the ½ ft^3 dial would require 12.6 seconds per revolution, and the 2 ft^3 dial, 50.4 seconds per revolution. A good stopwatch is recommended or a digital watch timing function.

Rather than use the formula, Fig. I47-4 shows a table of revolution timing for various test dial sizes for various inputs. Using this table, the 142.8 ft^3/hr of gas would cause the ½ ft^3 dial to turn between 12 and 13 sec, the 1 ft^3 dial between 25 and 26 sec.

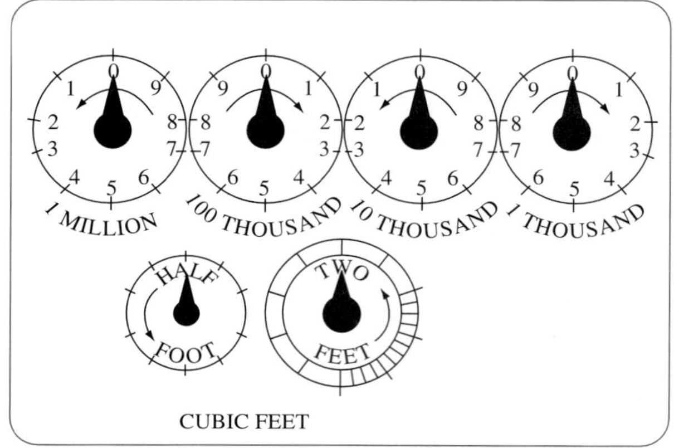

Figure I47-3 Typical domestic gas meter dials in ft^3.

Seconds for One Revolution	Cubic Feet Per Hour				Seconds for One Revolution	Cubic Feet Per Hour			
	Size of Test Meter Dial					Size of Test Meter Dial			
	One-Half Cu. Ft.	One Cu Ft.	Two Cu Ft.	Five Cu Ft.		One-Half Cu. Ft.	One Cu Ft.	Two Cu Ft.	Five Cu Ft.
10	180	360	720	1,800	52	35	69	138	346
11	164	327	655	1,636	53	34	68	136	340
12	150	300	600	1,500	54	33	67	133	333
13	138	277	555	1,385	55	33	65	131	327
14	129	257	514	1,286	56	32	64	129	321
15	120	240	480	1,200	57	32	63	126	316
16	112	225	450	1,125	58	31	62	124	310
17	106	212	424	1,059	59	30	61	122	305
18	100	200	400	1,000	60	30	60	120	300
19	95	189	379	947	62	29	58	116	290
20	90	180	360	900	64	29	56	112	281
21	86	171	343	857	66	29	54	109	273
22	82	164	327	818	68	28	53	106	265
23	78	157	313	783	70	26	51	103	257
24	75	150	300	750	72	25	50	100	250
25	72	144	288	720	74	24	48	97	243
26	69	138	277	692	76	24	47	95	237
27	67	133	267	667	78	23	46	92	231
28	64	129	257	643	80	22	45	90	225
29	62	124	248	621	82	22	44	88	220
30	60	120	240	600	84	21	43	86	214
31	58	116	232	581	86	21	42	84	209
32	56	113	225	563	88	20	41	82	205
33	55	109	218	545	90	20	40	80	200
34	53	106	212	529	94	19	38	76	192
35	51	103	206	514	98	18	37	74	184
36	50	100	200	500	100	18	36	72	180
37	49	97	195	486	104	17	35	69	173
38	47	95	189	474	108	17	33	67	167
39	46	92	185	462	112	16	32	64	161
40	45	90	180	450	116	15	31	62	155
41	44	88	176	440	120	15	30	60	150
42	43	86	172	430	130	14	28	55	138
43	42	84	167	420	140	13	26	51	129
44	41	82	164	410	150	12	24	48	120
45	40	80	160	400	160	11	22	45	112
46	39	78	157	391	170	11	21	42	106
47	38	77	153	383	180	10	20	40	100
48	37	75	150	375	190	9.5	19	38	95
49	37	73	147	367	200	9	18	36	90
50	36	72	144	360	210	8.5	17	34	86
51	35	71	141	353	220	8	16	33	82

Note: To convert to Btu per hour multiply by the Btu heating value of gas used.

Figure 147-4 Gas input to the meter (ft^3/hr).

If adjustments in the gas flow need to be made to produce the required input, it can be done by adjusting the gas pressure regulator, as long as the input pressure is within the 3- to 4-in. range. If the adjustment cannot be made within this range, the gas orifice size for the burners needs to be changed. The manufacturer's installation data supplies information for changing the orifice size.

I47-2.2 Liquid Petroleum (LP) Gas

Propane and butane are obtained from crude-oil wells and do occur in the pure state. Although mixed, they are separable by condensing them at their respective boiling points. As a result, propane gives a fairly constant 2522 Btu/ft^3 and butane yields 3261 Btu/ft^3.

The LP gas industry has established a set manifold pressure for all LP gas-burning appliances—11 inches of WC. Therefore, it is only necessary to determine that the LP supply system is large enough to maintain 11 in. of WC at the units when the total connected load is operating. The LP supplier will provide the tank installation, pressure-regulating devices, and piping to the furnace, but the service person should be familiar with the hook up procedure.

There are two basic systems, as illustrated in Fig. I47-5. The single system on the right, which is used most frequently in residential work, uses only one pressure regulator located at the tank. The two-stage system shown on the left is used mainly in commercial work, where the number of appliances and volume of gas must be greater. Therefore, the line pressure at the outlet of the tank regulator on the two-stage system is 10 to 15 psig, whereas it is 11 in. of WC on the single-stage system. The main supply line will carry more ft^3/hr at the higher pressure, and this is smaller. Sometimes it is necessary to connect from a single-stage to a two-stage system if the supply lines are installed undersized.

With the single-stage systems, the pressure on the inlet of the regulator will be direct tank pressure and this will vary with fuel and temperature.

To ensure that the gas will flow from the tank into the supply system to the heating unit, the tank pressure must at all times be higher than the line pressure required to supply the system. To maintain a pressure of 11 in. of WC at the heating unit, allowing a pressure loss in the line of ½ in. of WC, the minimum tank pressure would be 2 psig. This means that the minimum outside temperature must be considered when selecting the mixture of LP fuel.

Fig. I47-6 shows the tank pressure that will occur at temperatures of −30 to +110°F for mixtures of propane and butane. From this table it can be seen that butane is not usable below +40°F and even propane will not develop sufficient pressures below −30°F. In extremely cold climates, tank heaters are used to ensure adequate fuel supply.

Combustion Air

Many burners have an adjustment for the primary air (the air that is mixed with the fuel in the burner before combustion). It should be set to produce a soft blue flame with yellow tips. Too much primary air will produce a hard, blue flame that will waste fuel. Insufficient primary air will produce a long, yellow flame that will waste fuel and possibly form carbon monoxide and soot (Fig. I47-7).

Most high-efficiency furnaces use an in-shot-type gas burner (Fig. I47-8) and an induced draft fan. On these burners normally the only adjustment that is made is on the gas-pressure regulator which should be set between 3.2 and 3.8 in. WC. These burners do not have a primary air adjustment. Regulation of the gas flow is by the gas pressure and the size of the orifice. The flame characteristic is a clear blue.

PREVENTIVE MAINTENANCE

From a service standpoint, two major problem areas in hydronic systems are (a) water-logged steel expansion tanks without a bladder or diaphragm, and (b) air traps within the boiler system.

Clean tanks to remove sediment and to provide an air cushion for water expansion. To clean the tank:

1. Run a drain hose from the drain valve to a floor drain.
2. Shut off the supply valve to the inlet of the tank to create a vacuum in the tank as it is drained.
3. When drained, close the drain valve and open the air vent on top of tank to provide air cushion in tank.
4. Close the air vent and refill the tank to the half-full line.

Air traps are noisy, interfere with circulation, and cause flow switches to drop out. Install automatic air vents in rises of the piping and make use of air scoops.

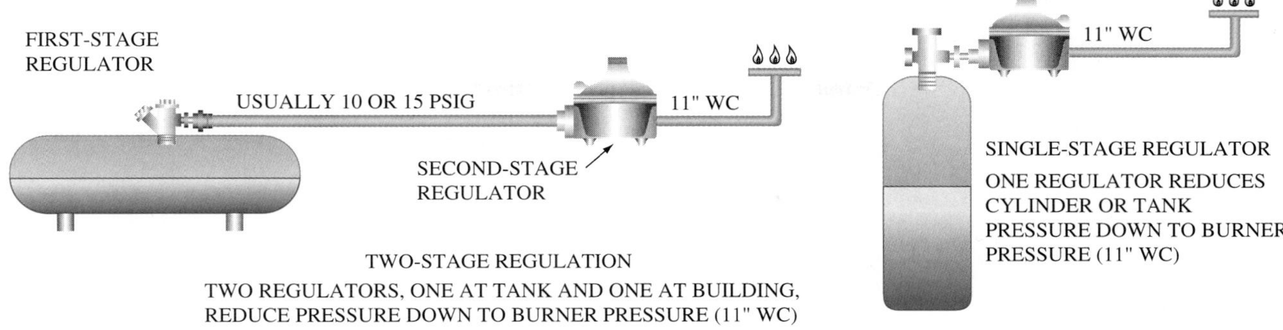

Figure I47-5 LP-gas supply systems.

	Vapor Pressure (psig)														
	Outside Temperature (°Fahrenheit)														
	−30	−20	−10	0	10	20	30	40	50	60	70	80	90	100	110
100% Propane	6.8	11.5	17.5	24.5	34	42	53	65	78	93	110	128	150	177	204
70% Propane 30% Butane	—	4.7	9	15	20.5	28	36.5	46	56	68	82	96	114	134	158
50% Propane 50% Butane	—	—	3.5	7.6	12.3	17.8	24.5	32.4	41	50	61	74	88	104	122
70% Butane 30% Propane	—	—	—	2.3	5.9	10.2	15.4	21.5	28.5	36.5	45	54	66	79	93
100% Butane	—	—	—	—	—	—	—	3.1	6.9	11.5	17	23	30	38	47

Figure I47-6 LP-gas tank pressures.

After checking and setting the Btu/hr input and adjusting the primary air shutters, the vent-draft condition should be checked.

Air Temperature Rise

The temperature rise of the air through the heating unit should be 80°F, or whatever temperature rise is specified by the manufacturer. Fig. I47-9 shows the insertion of dial-type thermometers into the supply and return plenums of the heating unit. Use of a sheet-metal scratch awl will allow the formation of a hole large enough to take the ⅛-in.-diameter stem of the dial thermometer and allow closure of the holes with sheet-metal screws after the test is completed.

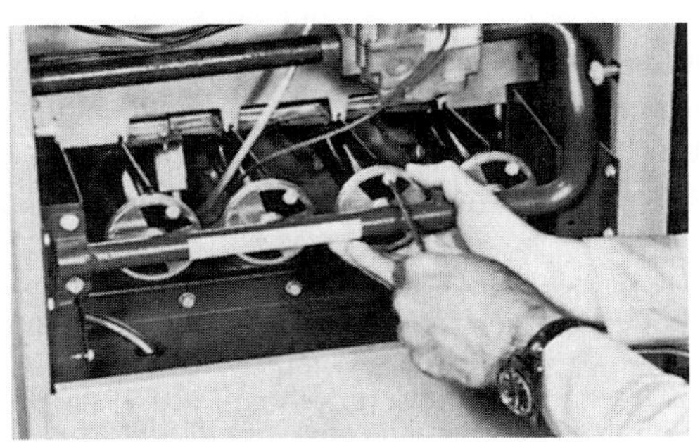

Figure I47-7 Adjusting primary air on a gas burner.
(Courtesy of York International Corp.)

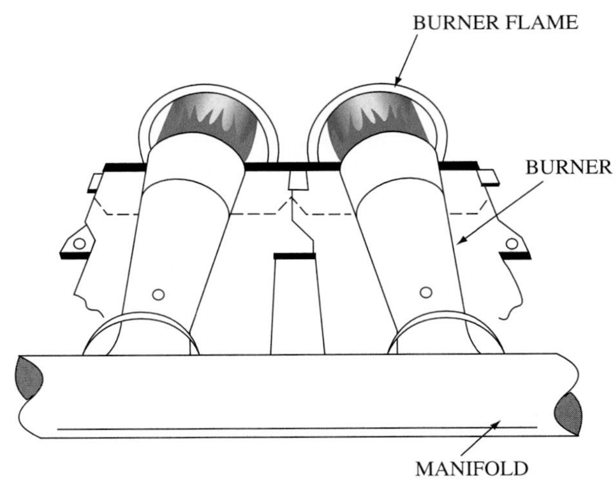

Figure I47-8 In-shot gas burner.

Figure I47-9 Measuring the air-temperature rise through a furnace. (Courtesy of York International Corp.)

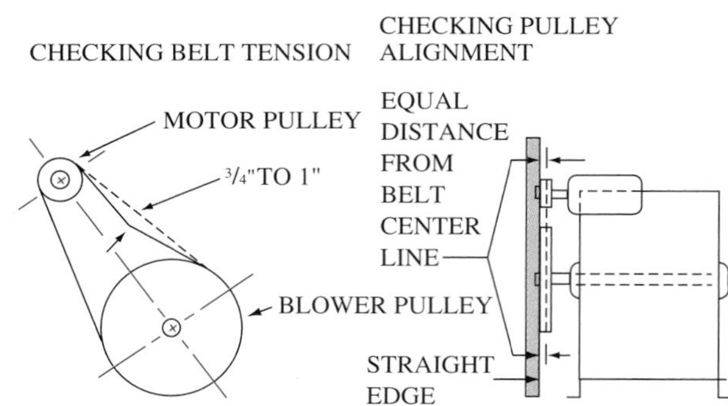

Figure I47-10 Aligning pulleys and tightening belts.

The supply-air thermometer should be located far enough away from the surface of the heat exchanger to reduce the effect on the thermometer of radiant heat from the heat exchanger. This is usually a minimum distance of 12 in. If the thermometer can be located in the main duct off the supply-air pressure, the radiant effect will be practically eliminated.

After the unit has operated long enough for the supply-air thermometer to hold a steady reading (stabilized), the return-air temperature should be subtracted from the supply-air temperature and the temperature rise recorded. If the temperature rise is below 80°F, the heating unit blower is moving too much air, and the speed should be reduced. If the temperature rise is above 80°F, the blower speed should be increased.

The procedure for this would depend on the type of drive used on the blower. Belt-driven blowers are adjusted by changing the size of the motor or driving pulley, direct-drive blowers by changing the electrical connections to the motor. Belt-driven blowers use a combination of adjustable motor pulley (the driving pulley), blower pulley (the driven pulley), and drive belt. The speed of the blower is adjusted by changing the spread of the flanges of the driving pulley. Opening the pulley by spreading the flanges allows the belt to ride lower in the pulley, thus reducing the effective diameter of the pulley. This in turn reduces the pulley diameter rotation between the two flanges and reduces the blower speed.

Closing the pulley spread increases the drive pulley diameter, resulting in an increase in blower speed. The driven pulley usually has two set-screw flats to allow adjustment of the pulley in half-turn increments. Use these flats—do not drive the set screw into the pulley adjustment threads. This malpractice ruins the chance for future adjustment.

After the blower speed has been set, the belt tension and alignment should be checked (see Fig. I47-10). The belt tension should be tight enough to avoid slippage but not too tight to cause excessive wear. Approximately ¾ to 1 in. of play should be allowed for each 12-in. distance between the motor and blower shafts. Check the pulley alignment by using a straightedge (yardstick).

On direct-drive blowers, the choice of blower speeds is limited to the number of speeds built into the blower motor. On some heating-only units, the blower has only one speed and no choice of temperature rise is available. If air conditioning is added to the unit, a change in blower motor or blower assembly is required to accommodate the additional pressure loss through the coil.

If the heating unit is equipped with a multi-speed blower motor, observe which fan speed lead is connected: high, medium, or low. For the initial start-up it is recommended that the fan be set for medium speed, subject to verification when the temperature rise is obtained.

In either case, increasing the blower speed will increase the cfm of air through the unit and lower the temperature rise. Conversely, lowering the blower speed will decrease the cfm through the unit and increase the temperature rise.

When increasing the blower speed, which increases the load in the motor, the amperage draw of the motor will also increase. A clamp-type ammeter should be used to check the motor operating amperage. If the required cfm causes the motor to draw more than its amperage rating, a motor or blower assembly of larger capacity will have to be substituted.

Efficiency Testing

With the heating-unit input set, the burners properly adjusted, and the unit operating at as close to desirable temperature rise as possible, the unit can be tested for operating efficiency. The objective of efficiency testing is to obtain as high an efficiency rating of the heating unit as possible, taking into account the operating cost, the equipment operating life, and the comfort obtained in the conditioned area.

It is necessary to reach adjustments that achieve a balance between operating cost (efficiency) and comfort. Standards that are used for efficiency testing are:

1. *Input.* The unit must be supplied with the correct amount of fuel within 90 to 100% of its rated capacity.

PREVENTIVE MAINTENANCE

COMMON MISTAKES IN CHILLER MAINTENANCE

1. Location: Chiller placed too close to heat, such as a boiler. Can cause rupture disk to blow when heat raises pressure of idle chiller.
2. Low temperatures of cooling tower water: If too cold, can cause loss of oil, stacking of refrigerant in the condenser, and starvation of the evaporator. May be mistaken for low refrigerant supply.
3. Thermal shock: Running too hot water through a unit injures the tubes possibly distorting them and inhibiting flow. Caused by using valving systems to divert boiler water to the chiller.
4. Running system at high gpm: Appears more efficient when run at higher than design specifications gpm, but this can erode tubes, requiring replacement, and increase cost of pump operation. Results: wear and tear without savings.
5. Ignoring design specifications: Intuitive operation rather than following the manufacturer's optimum operating parameters can seriously impair chiller performance.
6. Underestimating purges (centrifugal chillers): Frequent purges indicates refrigerant leaks. This wastes refrigerant, causes low refrigerant levels, allows contaminants and moisture to enter the chiller—all of which shortens the life of the machine. The tighter the machine, the more efficient no matter what type of purge is used.
7. Adding too much oil: This can result in carryover in the machine, inducing impeller damage to the compressor. The amount of oil is directly related to the size of the load: at full load oil is added if the oil level falls below midpoint in the oil sump sight glass. Other times, a lower level is justified.
8. Failure to clean filters and strainers: Both in the oil and water sides of the chiller plugged filters reduce the capacity of the chiller.
9. Neglect of cooling towers: Can cause condenser tube fouling. Tubes can be cleaned by scrubbing or chemically. If not flushed after chemical cleaning, pipes become perforated.
10. Poor record-keeping procedures: Without accurate, adequate, and regular logging throughout the life of the chiller, potential problems cannot be identified early or abnormal performance or a long-term trend interpreted with confidence.

2. *Burner primary-air adjustment.* Soft blue fire without yellow color or flame lift.
3. *Air-temperature rise.* As close to 80°F temperature rise of the air through the unit as the blower design will allow.
4. *Fan-control settings.* With fan on, at 125 to 130°F; with fan off, at 100 to 105°F. With units having duct runs over 50 ft long, a fan or setting of 150°F is recommended to overcome drafts upon start-up.
5. *CO_2.* The percentage of CO_2 in the flue gas is an indicator of the overall combustion efficiency. It may be measured with CO_2 analyzer shown in Fig. I47-11. Readings above 10% indicate incomplete combustion and possible production of carbon monoxide. Readings below 8.5% indicate too much combustion air.
6. *Stack temperature.* Stack temperature should be 475 to 500°F for low efficiency furnaces. It may be measured as shown in Fig. I47-12. Excessive stack

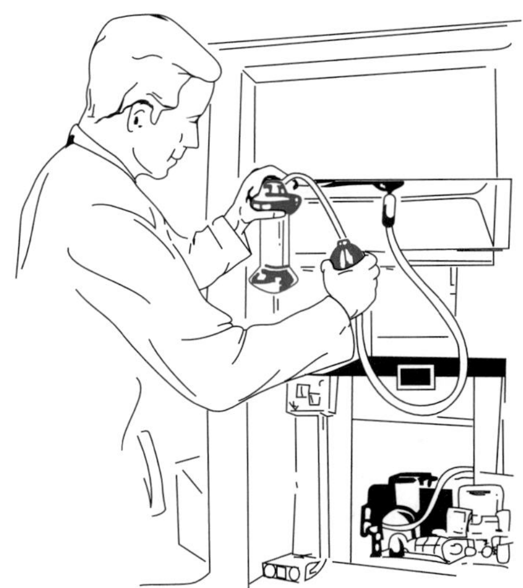

Figure I47-11 Testing the percent of CO_2 in the flue gas.

Figure I47-12 Checking the flue gas temperature.

temperature indicates reduced heat-exchanger efficiency. Stack temperature below 475°F may indicate too much combustion air.

A means of recording such information is necessary so that it will become a permanent part of the unit operating and service history. This efficiency check sheet should include the following information:

Input

1. Type of gas: Nat. _____ Mixed _____ Mfg. _____ Prop. _____ Bu. _____
2. Heat content (Btu/ft^3) ____________________
3. Specific gravity of the gas: ____________________
4. Main burner orifice drill size: Found _____ Left _____
5. Manifold pressure (in. WC): Found _____ Left _____
6. Meter test dial size: _____ ft^3 per rev.
7. Seconds required per rev. of test dial: Found _____ Left _____

Primary-Air Adjustment

1. Flame before adjustment: Sharp blue _____ Soft blue _____ Yellow tips _____
2. Flame after adjustment: Soft blue _____

Air-Temperature Rise

	First test	Second test	Left
Supply-Air Temperature	_____	_____	_____
Return-Air Temperature	_____	_____	_____
Air temperature rise	_____	_____	_____

CO_2

First test _____ %
Second test _____ %
Left _____ %

Stack-Temperature Rise

	First test	Second test	Left
Stack temperature	_____	_____	_____
Combustion air temperature	_____	_____	_____
Stack temperature rise	_____	_____	_____

Combustion Efficiency

% Efficiency	_____	_____	_____

Gas-burning equipment of standard design should always be capable of 75 to 80% efficiency. Unless the unit is of a higher-efficiency design (when the manufacturer's instructions and settings must be followed), an efficiency above 80% could adversely affect the draft of the unit as well as cause condensation of moisture in the chimney or flue pipe and on the surfaces of the heat exchanger. If the efficiency results are less than this range, the test should be repeated, checking and setting the proper input.

I47-2.3 Oil

Flue Venting

Oil-fired furnaces operate on positive pressure from the burner blower, and it is most important to have a chimney that will develop a minimum draft of 0.01 to 0.02 in. of WC as measured at the burner flame inspection port. Consistency or stability of draft is also more critical, and the use of a barometric damper (Fig. I47-13) is required. The damper is usually installed in the horizontal vent pipe between the furnace and chimney. Some manufacturers attach them directly to the furnace flue outlet. The damper has a movable weight so that it can be set to counterbalance the suction and to maintain reasonably constant flue operation. It is adjusted while the furnace is in operation and the chimney is hot. The overfire draft at the burner should be in accordance with the equipment installation instructions.

Clearances

Cabinet and flue-pipe temperatures run warmer for oil-burning equipment, and the clearances from combustible material should be adjusted accordingly. One-inch clearances are common for the sides and rear of the cabinet, as opposed to zero clearances for many gas units. Flue-pipe clearances of 9 in. or more are needed, whereas only 6 in. of clearance is needed for gas.

Front clearance is generally determined by the space needed to remove the burner assembly. Non-combustible floor bases for oil are generally increased as compared to gas. Horizontal oil furnaces are particularly important in respect to attic installations and the installer must check the recommendations carefully.

Figure 147-13
Barometric damper. (Courtesy of Field Controls Company)

ON THE JOB

Talk to the people who insure boilers and you get a pretty good perspective on what goes wrong with boilers, and why. These provide good direction for maintenance and troubleshooting. The following are some of the common problems encountered:

1. Equipment gets out of calibration and is not checked. Controls are disconnected because the alarms are a nuisance.
2. Sight glass is dirty. Deposits left on the glass fog it and even show a water level stain which is mistaken for the true (inadequate) water level.
3. Tube leaks. These cause system failure or forced outages. On thick-walled components like steam drum, headers, and piping connections there is a serious safety factor. Tube life is difficult to predict and deterioration is difficult to see from the outside. An internal examination method is required.
4. Burners: Cheap oil leaves deposits which clog the nozzles. Stubborn deposits are "punched" out, enlarging the nozzle, endangering operations and operator. Also, burners are difficult to light so operators open up holes on burner tips. When the boiler shuts down on safety trip, the operator has jumpered out the flame-safety system. After repeated tries to fire, the accumulated fuel explodes.
5. Deaerator tanks: Often located in a messy, corrosive place, they deteriorate. Leaks let air into the boiler, causing pitting. It is connected to the steam-supply system which finds leaks and becomes dangerous. Visual inspection is not good enough to see cracks and leaks through the accumulated scale. The National Association of Corrosion Engineers has issued a paper that details the places to look, techniques to use, and frequency.
6. Low water cut-offs: They look rugged, don't have many moving parts, and can appear OK when not used on a regular basis. Water chemicals, sludge, and other debris lodged in them may cause freeze-up. Then, without warning, something may explode.

The moral is: Good maintenance requires a competent service technician or a boiler manufacturer's manual to identify what should be done. Then, a checklist should be developed and followed. Problems should be added to it as they arise.

Oil furnaces are rated and listed by UL, and clearances are a vital part of this inspection and compliance procedure along with many other safety considerations.

Start-up

Fig. I47-14 shows a typical wiring diagram for an oil-fired heating-only installation using a thermostat controlling a cad cell type of primary control with a blower door switch in the low-voltage circuit as an air-supply safety switch. The primary control is controlling the oil-burner motor and ignition transformer through the limit control to prevent overheating the unit.

The final function is the operation of the system blower motor by the fan portion of the fan limit control. Properly set, this control will close at 125 to 130°F to start the blower motor and open at 95 to 100°F to stop the blower motor.

Operating Sequence

The step-by-step operation of the unit would be as follows:

1. With the blower compartment door closed and the blower-door switch closed, the room thermostat, upon a call for heat in the conditioned area, will close. This causes current flow from the 40-VA transformer through the control circuits and the burner-motor relay. The relay pulls the contact closed.
2. The closed contacts cause a current flow from the H or hot side of the supply line through the limit switch to the burner motor and ignition transformer. In addition, a current flow in the 24-V circuit exists through the dark cad cell and the safety switch heater.
3. Immediately, the ignition transformer establishes a spark across the electrodes located above the noz-

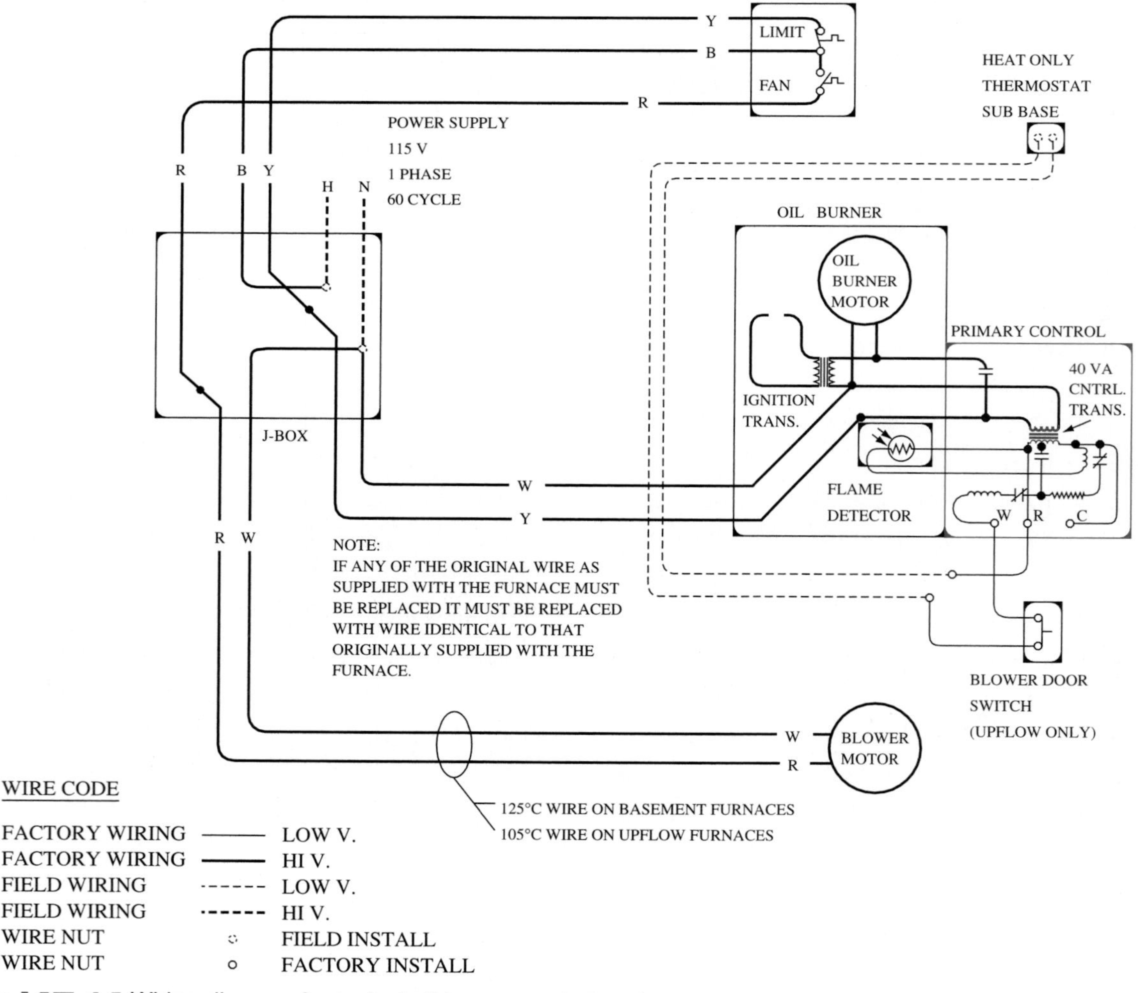

Figure I47-14 Wiring diagram for typical oil-burner control system.

zle. The spark at this time is only ⅛-in. long, the distance between the electrode tips.

4. The blower motor is energized at the same time as the ignition transformer. It reaches full-load speed within 1 sec. When the full amount of air is delivered by the blower, the ignition spark is blown forward into the oil spray and the oil is ignited. This occurs directly in front of the nozzle and quickly expands into a full burst of fire in the combustion chamber.
5. As soon as the light from the fire reaches the cad cell located in the fire tube, the electrical resistance of the cell increases. This increase in resistance reduces the current flow through the safety-switch heater and prevents the safety switch from opening.
6. The flame efficiency increases as the refractory temperature rises. At a white-hot surface temperature of the refractory, the burner is operating at peak efficiency. At this point efficiency tests can be taken. To reach these conditions, the burner should be allowed to operate at least 5 min.

The initial system start-up will require more than just turning on the burner. The first requirement will be to purge the fuel unit of air. This procedure would be different for the single-pipe system where the fuel supply is above the burner, and the two-pipe system where the fuel supply is below the burner.

On the single-pipe system with gravity feed (the supply is above the burner) it is only necessary to purge the air from the oil filter, fuel line, and fuel unit. The purge valve is located under the valve and the unit is off. It is only necessary to open this valve until the air in the system starts to flow out. Usually, this is one-half to one turn of the valve. When the air is out and oil starts to flow out, close the valve to prevent oil leakage.

In a two-pipe system, the pump will force the air down the oil-return line to the tank and the system is self-purging. It may take longer than the cycle time of the safety switch to purge the system and the burner may cut off before flame is established. It is then necessary to allow the safety switch to cool and be reset, and the burner started again. This should be repeated until the oil lines are clean and fire is established.

Efficiency Testing

When fire is established and the refractory obtains white heat, efficiency testing can be done on the unit. As with the gas unit, certain information is necessary. Again, a test sheet is desirable to record the conditions to which the burner was adjusted, for future reference. The following information should be included on the test sheet:

1. Make of oil burner ____________________
2. Model No. ____________ Serial No. ____________
3. Nozzle: Size (gal/hr) ______ Type ____________
   Angle ________
4. Refractory: Shape ____________________

Operation

1. Overfire draft ____________________
2. Stack draft ____________________
3. Heat exchanger flow resistance ____________
4. CO_2% ______________
5. Net stack temperature ______________
6. Efficiency ______________
7. Smoke number ____________________

PRACTICING PROFESSIONAL SERVICES

Compressor failure may be prevented by:

1. Opening up and diagnosing the problem(s) of every compressor that fails rather than replacing it.
2. Locating the system problem, repairing it or informing the owner of the situation and action required.
3. Establishing procedures that support quality work, customer satisfaction and confidence:
   a. Evacuation procedures and levels conform to current regulations.
   b. Complete toolkit to do any job required.
   c. Findings and actions logged to show measures taken, for any future work required.
   d. Follow up of repairs in 24 hours to test operation of compressor system.
   e. Restoration and cleanup of jobsite upon completion of work.
   f. Monitors and recorders available to log potential problem units, to compare with operations of other similar compressor systems analyzed.
   g. Documentation of training and hands-on experience with compressors in support of diagnosis made.

With quality service the service person builds quality relationships with the boss and the customers.

8. Air temperature (supply plenum) ____________
9. Air temperature (return plenum) ____________
10. Air temperature (rise) ____________

Although it seems unnecessary, it is wise to record on the test sheet the make, model number, and serial number of the burner for future situations. Recording the nozzle size, spray angle, and type forces the checking of the unit for proper input as well as spray angle for the shape of the refractory. Round or square refractories will take the 80 or 90° spray angle; a rectangular refractory, the 45 to 60° angle, depending on the length of the refractory. With the unit operating and refractory up to white heat, operating tests can be made.

Overfire Draft

The first test should be *over-fire draft.* Using a draft gauge such as those pictured in Fig. I47-15 and Fig. I47-16 and a ¼-in. hole in the pressure-relief door over the burner (sometimes called the observation door), insert the test pipe for the gauge through the hole until the end of the pipe is beyond the inner edge of the combustion chamber. To accomplish this, it is sometimes necessary to substitute a longer piece of ¼-in. copper tubing for the gauge-probe pipe. This test should result in a minimum pressure of 0.02 to –0.04 in. of WC to ensure the proper draft for best operation. If the manufacturer of the unit does not specify the over-fire draft, it is best to start at a draft control setting of –0.04 in. of WC.

Some manufacturers specify stack draft for a setting. This means that the flow resistance of the heat exchanger must be taken into account for burner operation. By measuring both the over-fire draft and stack draft on the new unit, the design draft resistance can be determined. To produce a negative pressure of a given amount over the fire, the stack draft must be a greater negative amount. The difference between the over-fire draft and stack draft will be the heat-exchanger-draft flow resistance. For example, if it is necessary to have a stack draft of –0.06 in. of WC to produce an over-fire draft of –0.04 in. of WC, the heat-exchanger resistance would be 0.02 in. of WC. When checking the unit performance after a period of operation, if the heat exchanger flow resistance has doubled, it is necessary to mechanically clean the soot from the heat-exchanger flue passages.

Figure I47-15 Draft gauge for measuring draft over the oil-burner flame. (Courtesy of Bacharach, Inc.)

Figure I47-16 Stack draft gauge. (Courtesy of Bacharach, Inc.)

A negative pressure must be maintained on the heat exchanger to prevent the products of combustion from being forced into the occupied area. Not only do they carry free carbon or soot, but they also contain a high percentage of CO. This is especially true at startup. Therefore, a minimum over-fire draft of –0.02 in. of WC is usually required for proper operation, with –0.04 in. of WC to be on the safe side.

CO_2

As in the gas furnace, the CO_2 sample is taken far enough ahead of the barometric draft control to obtain a good sample of flue gas without outside air mixture.

Again, this may mean a longer sampling tube on the analyzer. Use the CO_2 analyzer (Fig. I47-17), following the manufacturer's instructions. The following results should be obtained:

1. Old-style gun burners: Burners with no special air-handling parts other than an end cone and a stabilizer. A CO_2 reading of 7 to 9% should be obtained unless a CO_2 reading in this range results in more than a No. 2 smoke. If so, the CO_2 reading should

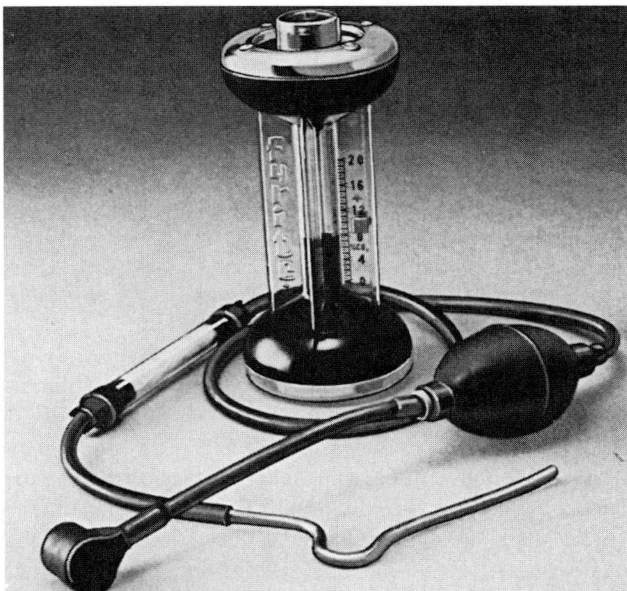

Figure I47-17 Carbon dioxide (CO_2) analyzer.
(Courtesy of Bacharach, Inc.)

be reduced until the smoke test results in below a No. 2 smoke.

2. Newer-style gun burners: Burners with special air-handling parts should be set in the range 9% to 11%. This should result in less than a No. 2 smoke (closer to a No. 1 smoke).
3. Flame-retention gun burners: These burners should be set in the range of 10 to 12% with the same smoke test results.
4. Rotary burners: These burners should be set the same as old-style gun burners.
5. Rotary wall flame burners: Higher CO_2 settings are available, up to 13.5%, with these burners, but the maximum of No. 2 on the smoke test is required.

Smoke Testing

A *smoke tester,* such as the one shown in Fig. I47-18, uses a pump piston to draw flue products through a filter inserted in the head of the pump between the sample tube connection and the piston body. This particular instrument requires ten slow strokes of the piston to draw the required amount of flue products through the filter. The filtered sample is compared to numbered rings on the test card and should never be higher than the No. 2 ring. If higher, the burner is receiving insufficient air and is not burning the carbon sufficiently. This will quickly produce carbon deposits and plug heat exchanger passes. If less than the No. 1 ring, too much air is being allowed in the burner. The CO_2 content is too low and too much heat is being forced out of the heat exchanger. This results in considerably lower efficiency and much higher operating cost.

Figure I47-18 Smoke tester. (Courtesy of Bacharach, Inc.)

Stack Temperature

A high-temperature thermometer, such as that shown in Fig. I47-19, is inserted in the same opening in which the stack draft was taken. The temperature of the flue products will also determine the efficiency of the heating unit. The cleaner the heat exchanger, the more able it is to remove heat from the flue products as they pass through. If the CO_2 content is correct but the stack temperature is too high, this is usually a sign of excessive carbon deposit on the heat exchanger surfaces. To determine the next stack temperature, subtract the temperature of the air entering the burner from the stack temperature.

Calculation of Efficiency

Use an efficiency calculator as shown in Fig. I47-20. The horizontal slider is set so that the net stack temperature to the nearest 50°F is shown in the net stack temperature window. The vertical slider is then adjusted so that the tip of the arrow is at the CO_2 determined by the analyzer test. The efficiency of the unit is then shown in the arrow in the vertical slider.

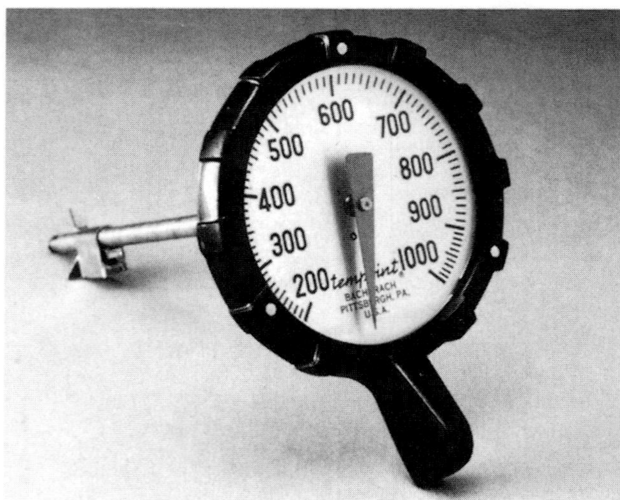

Figure I47-19 High temperature stack thermometer. (Courtesy of Bacharach, Inc.)

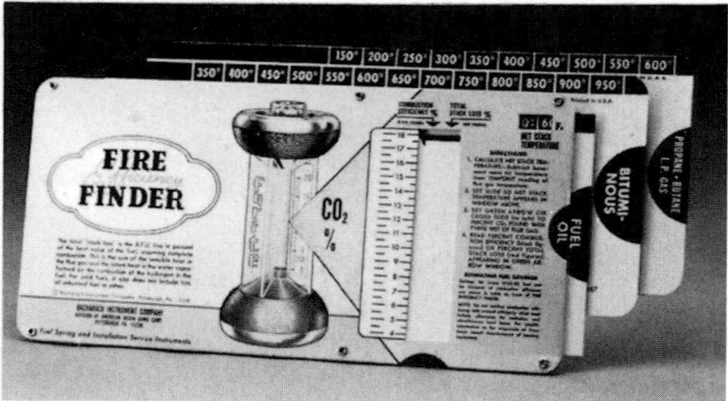

Figure I47-20 Oil-burner efficiency calculator. (Courtesy of Bacharach, Inc.)

The efficiency results should be within the smoke test ratings as described above or the manufacturer's specifications. If not, the unit should be examined for the cause of the difference. Usually, a cleaning of the heat exchanger is required.

I47-3 AIR CONDITIONING

To obtain the highest performance from the cooling unit it is necessary for the technician to give special attention to the following and make adjustments where necessary:

1. Airflow and temperature drop over the coil
2. Refrigerant charge.
3. Capacity of the system.
4. Adaptability to local conditions.

It is assumed that the loads have been carefully calculated and that the equipment has been selected to match the loads.

I47-3.1 Adjusting the Airflow

Residential and packaged commercial air conditioning units are commonly supplied with multi-speed motors. The wiring diagram will suggest the proper speed selection. For most applications, the factory setting will produce a supply air temperature of between 52 and 58°F. If the supply air temperature is below 52°F, the airflow may be too low. On a direct expansion system, this can produce a suction pressure below 32°F, and can cause frosting on the evaporator coil. Check for obstructions in the duct work, such as a piece of interior duct insulation that has come loose or a crushed flexible duct. If the duct work is clear, but just has a higher-than-normal pressure drop, the fan speed may need to be increased.

On field assembled systems using air handlers, the plans will specify the required cfm. This should have been measured and verified prior to start-up (see Chapter H41, "Testing and Balancing").

I47-3.2 Determining the Refrigerant Charge

Most manufacturers supply operating data on their equipment (Fig. I47-21) indicating the total, sensible, and latent heat removal rating, at various outdoor dry-bulb and indoor wet-bulb temperatures. For these same conditions, they supply the operating suction and discharge pressures of the equipment. This information is given so that the technician can match the actual conditions on the job with the performance conditions shown on the manufacturer's chart.

Methods for determining refrigerant charge were covered in a previous chapter. The following is a summary of the various methods:

1. Observing the condition of the sight glass located on the inlet side of the evaporator.
2. Measuring the superheat in the refrigerant leaving the evaporator (for capillary tube and fixed orifice systems only).
3. Measuring the subcooling, in the refrigerant leaving the condenser (for thermal expansion systems only).
4. Measuring the dry bulb temperature drop across the evaporator.
5. Adding the required charges of each of the components.

WCC018F—B at 600 cfm
Capacities are net (indoor fan heat deducted) Btu/Hr/1000

O.D. D.B.	I.D. W.B.	Total Cap.	Sens. Cap. at Entering D.B. Temp. 72	74	76	78	80	Compr. kW	App. Dew Pt.
85	59	15.6	12.0	13.1	14.1	15.1	15.9*	1.23	44.2
	63	17.0	10.1	11.1	12.1	13.2	14.2	1.27	48.0
	67	18.4	7.9	8.9	9.9	11.0	12.0	1.30	52.1
	71	19.9	5.6	6.7	7.7	8.7	9.8	1.33	56.3
90	59	15.5	12.0	13.0	14.0	15.1	15.8*	1.29	44.3
	63	16.8	10.0	11.0	12.1	13.1	14.1	1.32	48.1
	67	18.2	7.8	8.8	9.9	10.9	11.9	1.35	52.2
	71	19.6	5.6	6.6	7.6	8.7	9.7	1.39	56.5
95	59	15.4	11.9	13.0	14.0	15.0	15.7*	1.35	44.4
	63	16.7	9.9	11.0	12.0	13.0	14.1	1.38	48.2
	67	18.0	7.7	8.8	9.8	10.8	11.9	1.41	52.4
	71	19.4	5.5	6.5	7.5	8.6	9.6	1.44	56.6
100	59	15.1	11.8	12.8	13.9	14.9	15.5*	1.41	44.6
	63	16.4	9.8	10.8	11.9	12.9	13.9	1.44	48.5
	67	17.6	7.6	8.6	9.7	10.7	11.7	1.47	52.7
	71	19.0	5.4	6.4	7.4	8.5	9.5	1.50	56.9
105	59	14.8	11.7	12.7	13.8	14.8	15.2*	1.48	44.9
	63	16.0	9.7	10.7	11.8	12.8	13.8	1.50	48.7
	67	17.3	7.5	8.5	9.5	10.6	11.6	1.53	52.9
	71	18.6	5.2	6.3	7.3	8.3	9.4	1.55	57.2
115	59	14.3	11.5	12.5	13.6	14.4*	14.8*	1.61	45.3
	63	15.4	9.5	10.5	11.5	12.6	13.6	1.63	49.2
	67	16.6	7.2	8.3	9.3	10.3	11.4	1.65	53.4
	71	17.8	5.0	6.0	7.0	8.1	9.1	1.67	57.7

Correction Factors—Other Airflows
(multiply or add as indicated)

Airflow	525	675
Total Cap.	×0.98	×1.02
Sens. Cap.	×0.94	×1.06
Compr. kW	×0.98	×1.01
A.D.P.	−1.3	+1.1

Values at ARI Rating Conditions

Total Net Capacity = 18000 Btu/Hr
Airflow = 600 cfm
App. Dew Pt. = 52.4°F
Compressor Power = 1410 W
I.D. Fan Power = 210 W
O.D. Fan Power = 260 W
S.E.E.R. = 10.00 Btuh/W
*Dry coil condition (total capacity = Sensible capacity)
Total Capacity, Comp. kW and App. Dew Pt. are Valid Only for
All Temperatures in Degrees F.

Figure 147-21 Typical cooling-unit performance data. (The Trane Company)

147-3.3 Determining the System Capacity

The actual cooling being done through the evaporator may be determined by measuring the entering and leaving air conditions, and the air flow across the coil. They are then used in the following formula previously discussed in Chapter A29:

$$Q_t = 4.45 \times \text{cfm} \times \Delta h$$

where

Q_t = the total (sensible and latent) cooling being done
cfm = airflow across the evaporator coil
Δh = the change of enthalpy of the air across the coil.

147-3.4 Other Useful Measurements

Other useful measurements can be made that indicate losses in performance on specific jobs. The technician must constantly be on the lookout for conditions that waste energy and reduce efficiency. Some of these additional tests that can be made are:

1. Measure the difference in temperature between the return air at the grille compared to return-air temperature as it enters the unit. This difference should not exceed 2°F. If it does, the return duct needs to be insulated or there may be openings in the duct that need to be closed.
2. Measuring the refrigerant superheat in the evaporator may indicate that the expansion valve needs adjusting. The superheat should not exceed 10°F.
3. The compressor-discharge pressure can be compared to the liquid-line pressure. This pressure drop should not exceed 3 psig. An excessive drop indicates discharge piping that is too small or too high a pressure drop in the condenser.
4. Measure the degrees of subcooling in the liquid line. A safe value is 10°F. Any value under this may permit flash gas to form and reduce the capacity of the expansion valve.

147-4 HEAT PUMPS

Since there are a number of different types of heat pumps, the start-up, checkout, and operation is presented for each basic design. Where the unit is used for dual operation, separate tests are needed for heating and cooling. Also, a dual unit has operational limits. It cannot be operated and checked out in both phases of the operation unless the outside temperature is in the narrow overlapping range of temperatures for each phase of operation.

In the cooling cycle, operation is possible down to 65°F. In the heating cycle, operation is possible up to 70°F. Therefore the initial checkout of the unit will be in either the heating or cooling cycle.

147-4.1 Air-to-Air, Dual Operation

To calculate the net capacity on the cooling cycle, the same method is used for the heat pump as used for an air-conditioning unit. This is possible since the reversing valve is in a position that permits the refrigeration cycle to operate in the normal way. Test methods can also be used for the heat pump, similar to those described for the air-conditioning unit.

To calculate the heating capacity, a number of quantities need to be added:

1. The heat picked up by the outside coil (Fig. I47-22).
2. The heat from the electrical energy of the compressor and blower motor (Watts × 3.415).
3. Any supplementary heat that is added, where necessary, that the heat pump cannot supply (Watts × 3.415).

By testing, however, the gross heating capacity can be determined by knowing the cfm handled by the blower and the dry bulb temperature rise through the unit, using the standard formula:

$$H_s = \text{cfm} \times TD \times 1.08$$

147-4.2 Liquid-to-Air, Dual Operation

On the cooling cycle, the liquid-to-refrigerant coil is the liquid-cooled condenser. The gross capacity on the cooling cycle can be calculated from the flow (gpm) through the water circuit and the temperature rise, using the following formula:

$$HT = \text{gpm} \times TD \times 500$$

To obtain the net cooling capacity, the motor heat must be deducted. The motor heat is calculated by

$$HM = \text{watts} \times 3.415 \times PF$$

To calculate the net cooling capacity,

$$HC = HT - HM$$

where,

HT = Gross capacity
HM = Motor heat
HC = Net cooling capacity
PF = Power factor

WCC018F—B at 600 cfm

O.D. Temp. F.	Heating Capacity (Btuh/1000) at Indicated Indoor Dry Bulb Temp.				Total Power in Kilowatts at Indicated Indoor Dry Bulb Temp.			
	60	70	75	80	60	70	75	80
–3	7.14	6.91	6.80	6.68	1.31	1.37	1.39	1.42
2	8.35	8.08	7.94	7.80	1.39	1.45	1.48	1.51
7	9.55	9.24	9.08	8.93	1.47	1.53	1.56	1.60
12	10.8	10.4	10.2	10.1	1.55	1.62	1.65	1.68
17	12.0	11.6	11.4	11.2	1.63	1.70	1.74	1.77
22	13.5	13.1	12.8	12.6	1.68	1.76	1.79	1.83
27	15.1	14.6	14.3	14.1	1.73	1.81	1.85	1.89
32	16.6	16.1	15.8	15.5	1.79	1.87	1.91	1.95
37	18.2	17.6	17.3	17.0	1.84	1.92	1.96	2.01
42	19.7	19.1	18.7	18.4	1.89	1.98	2.02	2.06
47	18.6	18.0	17.7	17.4	1.68	1.76	1.80	1.84
52	19.7	19.1	18.7	18.4	1.69	1.77	1.81	1.85
57	20.8	20.1	19.8	19.5	1.70	1.78	1.82	1.86
62	21.9	21.2	20.9	20.5	1.71	1.79	1.83	1.87
67	23.1	22.3	21.9	21.5	1.72	1.80	1.84	1.88
72	24.2	23.4	23.0	22.6	1.73	1.81	1.85	1.89

Correction Factors—Other Airflows
(value at 600 cfm times corr. factor = value at new airflow)

Airflow	525	675
Heating Cap.	×0.98	×1.01
Compr. kW	×1.02	×0.99

Values at ARI Rating Conditions of:
70&47/43 (High Temp. Point)
70&17/15 (Low Temp. Point)
Airflow = 600 cfm
Heating Cap. (High Temp.) = 18000 Btuh
Heating Cap. (Low Temp.) = 11568 Btuh
Compr. Power (High Temp.) = 1290 Watts
Compr. Power (Low Temp.) = 1230 Watts
Hspf (min DHR) = 6.80
Coeff. of Perf. (High Temp.) = 3.00
Coeff. of Perf. (Low Temp.) = 2.00
Outdoor Fan Power = 260 Watts
Indoor Fan Power = 210 Watts

Figure 147-22 Typical heating performance of a heat pump. (The Trane Company)

EXAMPLE

A liquid-to-air heat pump has a condenser using 4 gpa with a temperature rise of 16°F. What is its gross cooling capacity?

The compressor draws 2784 watts with a power factor of 0.9. What is the net cooling capacity?

Solution

Using the above formulas:

$$HT = 4 \times 16 \times 500 = 32{,}000 \text{ Btu/hr}$$
$$HM = 2784 \times 3.415 \times 0.9 = 8556 \text{ Btu/hr}$$
$$HC = 32000 - 8556 = 23{,}444 \text{ Btu/hr}$$

To obtain the heating capacity, the motor heat is added to the heat picked up by the outside coil.

EXAMPLE

A liquid-to-air heat pump has an outside coil that uses 5 gpm with a 9°F *TD* and a compressor that uses 2552 W at a power factor of 0.9. What is the heating capacity?

Solution

Using the standard formulas,

$$HO = 5 \times 9 \times 500 = 22{,}500 \text{ Btu/hr}$$
$$HM = 2552 \times 3.415 \times 0.9 = 7844 \text{ Btu/hr}$$
$$HM = 22{,}500 + 7844 = 30{,}344 \text{ Btu/hr}$$

where,

HO = Heat absorbed from outside coil

147-4.3 Air-to-Air, Single Operation

Several manufacturers market an air-to-air single-operation unit with a heating-only cycle. These units do not have reversing valves but do have defrost control systems. The start-up, checkout and operation of this type of system is the same as the dual operation on the heating cycle. The only difference is that the vapor line to the inside coil is permanently the hot-gas line and the pressure drop through the reversing valve is eliminated.

147-4.4 Liquid-to-Air, Single Operation

Like the air-to-air unit, the liquid-to-air unit is manufactured in a heating-only system. No reversing valve is included and defrost controls are not needed. The start-up, checkout and operation are covered in the dual operation section.

147-4.5 Air-to-Liquid, Single Operation

The air-to-liquid, single-operation heat pump units are used for water heating. An exterior view of one of these add-on units is shown in Fig. I47-23. It is designed to be directly connected to the domestic hot-water system as shown in Fig. I47-24. The capacity of one of these units can be determined by measuring the time it takes to raise the temperature of a given quantity of water.

For example, if the unit can heat a 30 gal. tank of water from 55°F to 75°F in one hour, the capacity is calculated using the formula

$$Q = 8.33 \times \text{gal} \times TD$$

where,

Q = the heating capacity of the unit, Btu/hr
8.33 = conversion factor from gal to lb
TD = the temperature rise of the water in one hour, °F

Using the data above,

$$Q = 8.33 \times 30 \times (75\text{-}55)$$
$$= 4998 \text{ Btu/hr}$$

Note that you don't need to waste a whole hour to determine the temperature rise. For example, you can measure the temperature rise for 15 minutes, and then multiply that number by four to use in the above formula.

Figure 147-23 Water-heating heat pump. (Courtesy of York International Corp.)

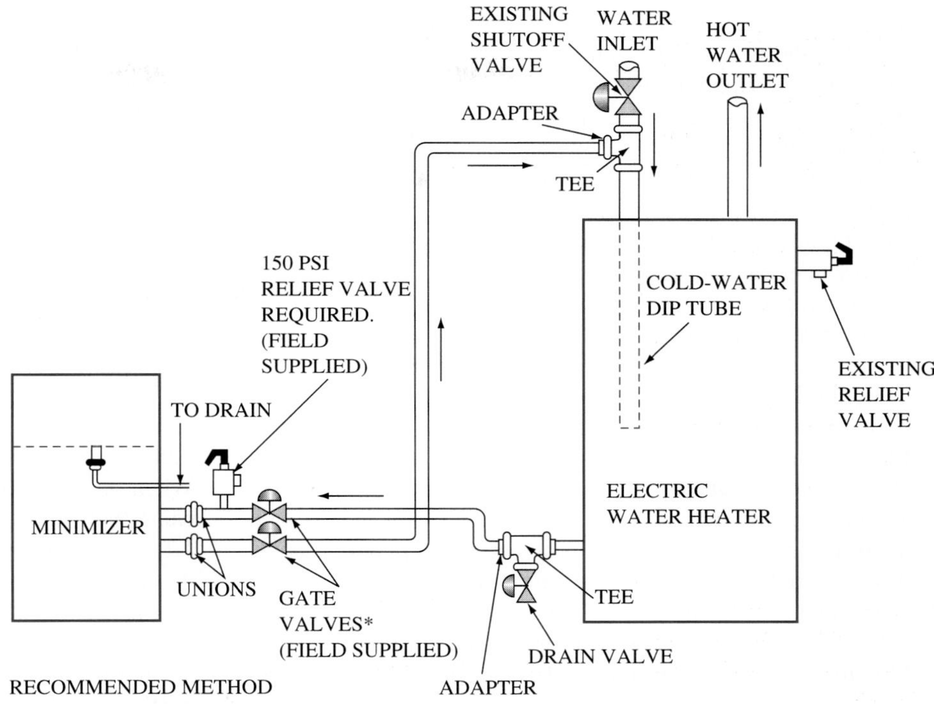

Figure 147-24 Field piping for a water-heating heat pump. (Courtesy of York International Corp.)

REVIEW

- The purpose of proper start-up, checkout, and operations is to have the system produce the design conditions in the most efficient manner. Key items include:
  Rated capacity
  Correct airflow
  Correct temperature rise/drop
  Correct distribution
  Control system functioning properly
- Safety note—Care must be taken during initial start-ups to prevent damage to equipment or cause a safety hazard to personnel.
- Refrigeration system start-up:
  Valve positions must be checked.
  Pressures should be carefully observed and regulated.
  Keep operating pressures as close to normal as possible.
  Check belt tensions and alignment.
  Lubricate.
  Perform maintenance outlined in installation instructions.
  Check motor nameplates against available voltage.
  On three-phase systems, check phase relationship and check for unbalance.
  Test control circuits.
  Following start-up, check operation parameters, compare to design specifications.
  Final clean-up.
  Instruct customer.
- Heating system start-up:
  Determine correct pressure input, gas or oil
  Conduct visual inspections.
  Make adjustment to gas flow.
  Adjust primary air.
  Check and adjust vent draft.
  Test heat rise.
  Adjust and check blower speed.
  Test fan and limit switches.
  Check air-supply temperature.
  Change filters.
  On belt drive units, check belt tension and alignment.
  Lubricate.
  Perform maintenance as outlined in installation instructions.
  Check electrical voltages and amperes.
  Conduct efficiency test.
  Check clearances.
  Check sequencers and heaters on electrical units.
  Clean up.
  Instruct customer.
- Air-conditioning system start-up:
  The technician must check and adjust as necessary the following:
    Airflow
    Temperature drop across the coil

Refrigerant charge
Capacity of the system
Adaptability to local conditions

- Methods to test and adjust airflow:
  Air temperature change
  Air duct velocity
  Total cubic feet per min. measurement
  Portable flow-hood measuring instrument
- Determining the refrigerant charge:
  Follow manufacturer's operating data for:
  - suction pressure
  - liquid-line pressure
  - suction temperature (superheat)
  - liquid-line temperature (subcooling)
  - sensible- and latent-heat removal

  Check sight glass.
  Measure superheat.
  Measure subcooling.
  Measure dry bulb temperature drop across evaporator.
  Add required charge per manufacturer's specifications.
- Determine system capacity:
  Measure net cooling capacity at the evaporator.
  Measure net cooling capacity at the condenser.

Problems and Questions

1. Why must valve positions be checked first during system start-up?
2. What is the correct belt tension?
3. What damage could occur from belts that are too tight?
4. What does misalignment of belts do?
5. Where can one find out what maintenance is required during start-up?
6. When installing equipment, placement is important to allow for _____.
7. The frost-line method of checking the refrigerant charge applies to all systems. True or false?
8. What is the most accurate method of refrigerant charging the system?
9. To check and test a gas heating unit, the first thing that must be established is _____.
10. The amount of gas to be burned in a particular unit will depend on _____ and _____.
11. The Btu/ft^3 of gas in your area is _____.
12. The recommended manifold pressure of gas heating units is _____ in. of WC with an allowance range from _____ to _____ in. of WC.
13. How many cubic feet of gas would be burned in a 125,000-Btu input unit using 1050-Btu/ft^3 gas?
14. When checking the input of a 150,000-Btu/hr unit, using 950-Btu/ft^3 gas, by timing the meter using the 1-ft^3 dial, how many seconds would it take for the dial to make 1 revolution?
15. The operating manifold pressure of an LP heating unit burning propane or butane is _____ in. of WC.
16. In an atmospheric-type burner, the air for the combustion that mixes with the gas inside the burner is called _____.
17. The best operating flame on an atmospheric-type burner is usually one that is _____.
18. Define the "operating efficiency" of a gas-fired heating unit.
19. What are the advantages and disadvantages of increasing the amount of air over the heat exchanger?
20. Name the oil spray patterns of the nozzles used in residential oil burners.

SECTION 10

Trouble-shooting

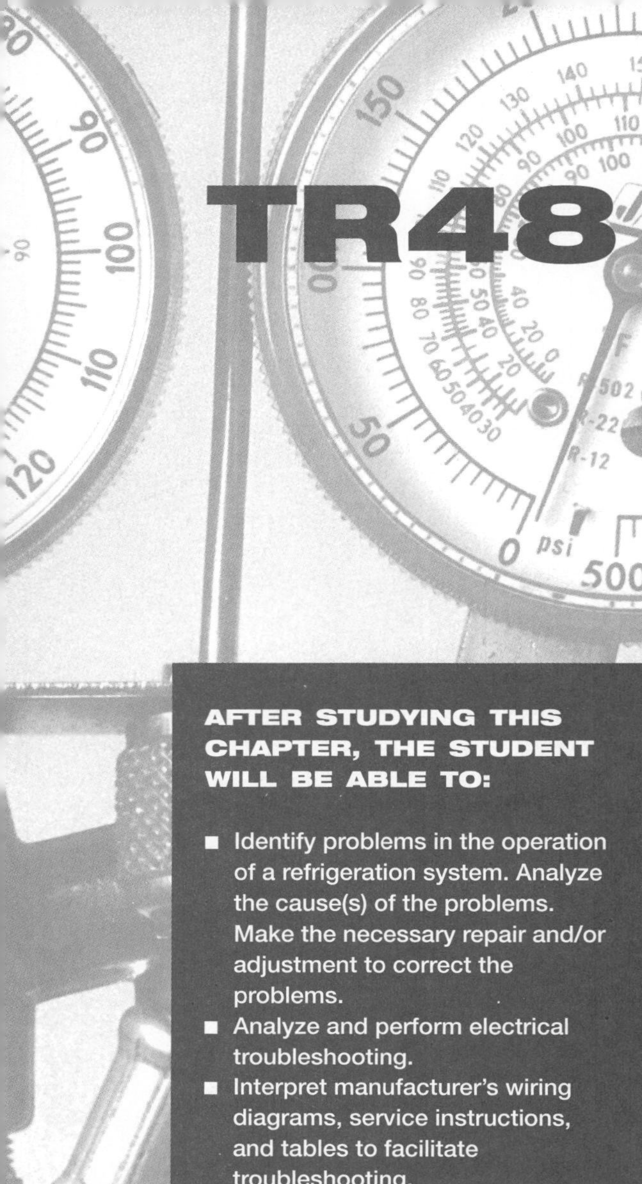

TR48 Trouble-shooting Refrigeration Systems

AFTER STUDYING THIS CHAPTER, THE STUDENT WILL BE ABLE TO:

- Identify problems in the operation of a refrigeration system. Analyze the cause(s) of the problems. Make the necessary repair and/or adjustment to correct the problems.
- Analyze and perform electrical troubleshooting.
- Interpret manufacturer's wiring diagrams, service instructions, and tables to facilitate troubleshooting.
- Implement fault-isolation diagrams and diagnostic tests to isolate component failures.
- Demonstrate efficient means of troubleshooting mechanical problems.
- State preliminary power-on and power-off inspections.
- Identify factors causing shortened compressor life.
- Analyze the causes of a failed compressor.
- Demonstrate the troubleshooting procedures to test a dead compressor.
- Identify problems and the troubleshooting process for evaporators and condensers.
- State the problems associated with metering devices.
- Troubleshoot operating problems that specifically apply to hermetically sealed systems.

TR48-1 OVERVIEW

Any operating mechanical/electrical equipment will at some time require service. The repair work necessary to place the equipment back on-line is one of the important functions of the HVAC/R technician. The basic term that describes this type of work is *troubleshooting*. Troubleshooting is the process of determining the cause of an equipment malfunction and performing corrective measures. Depending on the problem, this may require a high degree of knowledge, experience and skill.

Basically there are two types of problems: (1) electrical, and (2) mechanical, although there is much overlapping. Whatever the nature of the problem, it is good practice to follow a logical, structured, systematic approach. In this manner the correct solution is usually found in the shortest possible time.

Since the greatest number of malfunction problems are electrical, it is common practice to perform electrical troubleshooting (including controls) first. If the problem is mechanical, the electrical analysis will usually point the technician in that direction.

TR48-2 ELECTRICAL TROUBLESHOOTING

In preparing to analyze electrical problems, four items are important to know or recognize:

1. The operating sequence of the unit.
2. The functions of the equipment that are working and those that are not working.

3. The electrical test instruments that are needed to analyze the problem.
4. The power circuit driving the system.

The operating sequence is usually supplied by the manufacturer in the service instructions or it can be determined by the technician by studying the schematic wiring diagram. The functions of the operating and non-operating equipment are determined by examination and testing. Necessary test instruments include: the voltmeter, the clamp-on ammeter, the ohmmeter, and the capacitor tester. The technician must be proficient in the use of these instruments.

The power circuit is the first to be examined, because power must be available to operate the loads. For example, on a refrigeration system with an air-cooled condenser, the two principal loads that must be energized are the compressor motor and two fan motors. Before proceeding with anything else, the technician must be certain that the proper power can be supplied to loads.

Each electrical circuit has one or more switches that start or stop the operation of a load. This switching operation is called the control function. In troubleshooting, when a load is not working, the technician must determine whether the problem is in the load itself or in the switches that control the load.

To assist in analyzing an operational problem of the unit, the manufacturer furnishes one or more of the following:

1. Wiring diagrams
2. Installation and service instructions
3. Troubleshooting tables
4. Fault-isolation diagrams
5. Diagnostic tests

The *wiring diagrams* usually consist of connection diagrams and schematic diagrams. The *connection diagram* shows the wires to the various electrical component terminals in their approximate location on the unit. This is the diagram that the technician must use to locate the test points. The *schematic diagram* separates each circuit to clearly indicate the function of switches that control each load. This is the diagram that the technician uses to determine the sequence of operation for the system.

The installation and service instructions supply a wide variety of information that the manufacturer believes is necessary to properly install and service the unit. This bulletin includes the wiring diagram, the sequence of operation, and any notes or cautions that need to be observed in using them.

The troubleshooting tables (Fig. TR48-1) are helpful as a guide to corrective action. By a process of elimination, these tables offer a quick way to solve a service problem. The process of elimination permits the technician to examine each suggested remedy and disregard ones that do not apply or are impractical, leaving only the solution(s) that fits the problem.

SYMPTOM	PROBABLE CAUSE	REMEDY
1. IFM NOT RUNNING	1. LINE FUSE(S) BAD; IFC BAD; IFR NOT PICKING UP	1. REPLACE FUSE(S); CHECK IFC; CHECK -A/C SWITCH -THERMOSTAT
2. ETC.	2. ETC.	2. ETC.

Figure TR48-1 Typical troubleshooting table.

A *fault-isolation diagram* (Fig. TR48-2) starts with a failure symptom and goes through a logical decision-action process to isolate the failure.

Diagnostic tests (Fig. TR48-3) can be conducted on electronic circuit boards, at points indicated by the manufacturer, to check voltages or other essential information critical to the operation of the unit.

Some electronically controlled systems have automatic testing features which indicate by code number a mal-

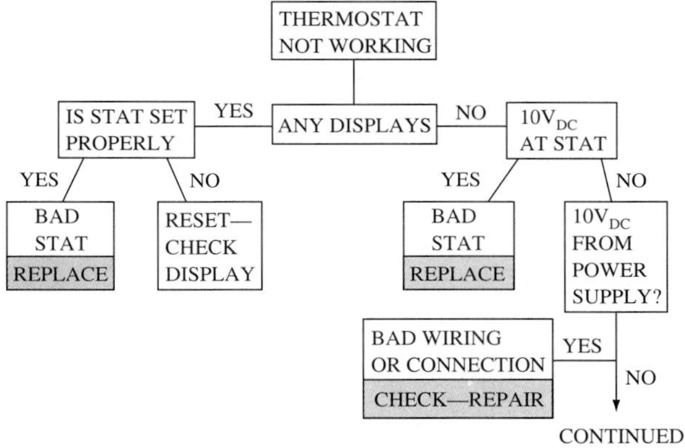

Figure TR48-2 Typical fault-isolation diagram.

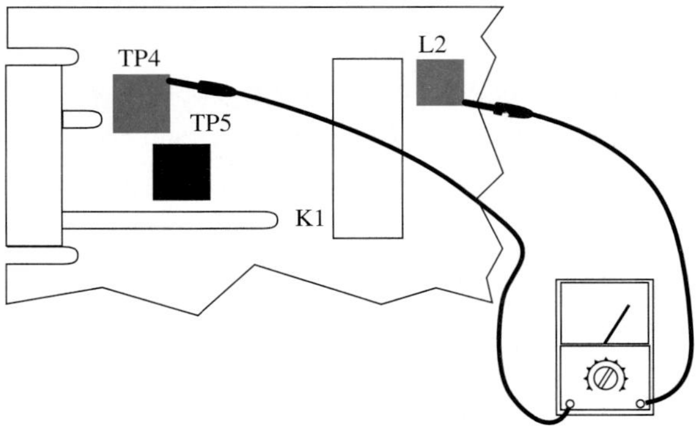

Figure TR48-3 Typical diagnostic test.

Figure TR48-4 Fault message.

function in the operation of the equipment (Fig. TR48-4). Further tests are usually required to determine the action that is required.

The procedure for electrical troubleshooting is to:

1. First select the proper test instrument. If the unit will not operate, test with an ohmmeter. If parts of the unit operate, test with a voltmeter.
2. Select the circuits that contain electrical devices that are not functioning.
3. Test the switches and loads in that circuit until the problem device is found.
4. Repair or replace the defective equipment.

TR48-3 MECHANICAL REFRIGERATION TROUBLESHOOTING

The most efficient means of troubleshooting mechanical problems in the operation of refrigeration systems is a systematic approach. Shortcuts are possible, depending on the problem, the type of system, and the experience of the technician, but it is usually helpful to follow a step-by-step procedure. Here are the steps:

1. Collect information about the problem.
   a. A description of the problem when the service call was received.
   b. Direct information about the problem by discussion with the customer.
   c. Conduct a preliminary power-off visual system inspection.
   d. Conduct a preliminary power-on system inspection.
2. Read and calculate the system's vital signs.
   a. Read and record vital signs, including suction and discharge pressures for type of refrigerant being used.
   b. Calculate the refrigerant liquid subcooling at the metering device.
   c. Calculate the refrigerant gas superheat at the compressor.
3. Compare typical versus actual values.
   a. Determine typical values for the conditions and system.
   b. Compare typical with actual conditions
4. Consult troubleshooting aids.
   a. Perform basic system analysis. Using a basic analysis guide (Fig. TR48-5), select possible system problems, based on a comparison of the five actual to typical vital values shown in the guide.
   b. Using the manufacturer's troubleshooting information (Fig. TR48-6), perform a detailed analysis. Eliminate possible causes of the problem by test or observation, and select the cause that fits the condition.

PRACTICING PROFESSIONAL SERVICES

Learning to troubleshoot refrigeration equipment requires a certain level of mechanical competency, but equally important, it requires patience and perseverance to locate the correct problem. Many times the technician will condemn the first problem without locating the cause of a problem. For example, a compressor has an electrical winding burn out to ground and blows the fuses. The technician may assume an electrical problem caused the electrical failure, but there could be other system problems which may cause an electrical failure. Possible problems include: 1) a flood-back due to poor airflow, 2) a faulty metering device, and 3) high head pressures due to condenser airflow problems causing acidic oil. A refrigerant flood-back can cause the oil to be washed out of the crankcase, destroying the compressor bearings. If the bearings do not support the rotor, it will drop onto the stator winding and cause electrical failure.

It is very important to troubleshoot the problem and the cause of the problem. This will improve with time and experience for the young technician.

System Problem	Discharge Pressure	Suction Pressure	Superheat	Subcooling	Amps
Overcharge	↑	↑	↓	↑	↑
Undercharge	↓	↓	↑	↓	↓
Liquid Restriction (drier)	↓	↓	↑	↑	↓
Low Evaporator Airflow	↓	↓	↓	↑	↓
Dirty Condenser	↑	↑	↓	↓	↑
Low Outside Ambient Temperature	↓	↓	↑	↑	↓
Inefficient Compressor	↓	↑	↑	↑	↓
TXV Feeler Bulb Charge Lost	↓	↓	↑	↑	↓
TXV Feeler Bulb Loose Mounted	↑	↑	↓	↓	↑
Poorly Insulated Feeler Bulb	↑	↑	↓	↓	↑

Figure TR48-5 Basic system analysis.

General Problem Category	Symptoms	Possible Causes
Plugged Filter-Drier (Liquid Line)	■ Starved Evaporator Symptoms (See Evaporator Sheet) ■ Compressor Cycles on Low Pressure Switch	■ Dirty Refrigeration System ■ Improper Evacuation/Dehydration ■ Metal Chips, Scale, etc., in System from Installation
Wet Filter-Drier	■ Moisture-Indicating Sight Glass Shows Wet ■ Valves Stick Intermittently and System Cycles Off from Internal Ice Blockage ■ Sealed-Tube Test of Refrigerant Shows Wet	■ System Refrigerant Leak ■ Improper Evacuation/Dehydration ■ Leaking Water-Cooled Condenser Tubes ■ Filter-Drier Exposed to Air Before Installation
Undersized Filter-Drier	■ Low System Capacity ■ Low Compressor Power Draw (kW) ■ Low Saturated Suction Temperature ■ Low Saturated Condensing Temperature ■ High Discharge Gas Superheat ■ Flash Gas in Liquid Line Sight Glass ■ High Liquid Refrigerant Subcooling	■ Bad Design on Field-Piped Systems
Crankcase Heater Inoperative	■ Flooded Start: ▪ High Compressor Power Draw (kW) ▪ Noisy Operation ▪ Excessive Compressor Vibration ▪ Overheating of Compressor ■ Violent Oil Foaming (Visible in Compressor Sight Glass)	■ Never Switched On ■ Heater Element Broken ■ Control Circuit Problem ■ Electrical Power or Control Connection Loose or Corroded
Oil Separator Trapping Oil	■ Oil Level Low on Compressor Sight Glass ■ High Compressor Power Draw (kW) ■ Compressor Overheating ■ Compressor Noisy	■ Sludge Blocking Oil Separator Float Valve Orifice ■ Oil Separator Float Assembly Faulty
Oil Separator Float Valve Stuck Open	■ High Saturated Suction Temperature ■ High Saturated Condensing Temperature ■ High Compressor Power Draw (kW) ■ Flooded Start	■ Debris at Oil Separator Orifice Keeps Float Valve from Seating Properly ■ Faulty Float Assembly ■ Liquid Refrigerant Migrates Through Separator to Compressor Oil Sump at Shutdown
4-Way (Reversing) Valve Damaged or Stuck—Heating	■ Low or No Heating Capacity ■ Compressor Cycles Intermittently on Internal Overload ■ Compressor Runs but Low on Heating Capacity	■ Sludge in Refrigeration System ■ Acids in Refrigeration System ■ Improper Evacuation/Dehydration ■ 4-Way Valve Leaks Internally
4-Way (Reversing) Valve Damaged or Stuck—Cooling	■ Low or No Cooling Capacity ■ Compressor Cycles Intermittently on Internal Overload ■ Compressor Runs but Low on Cooling Capacity ■ High Saturated Suction Temperature ■ Low Saturated Condensing Temperature	■ Sludge in Refrigeration System ■ Acids in Refrigeration System ■ Improper Evacuation/Dehydration ■ 4-Way Valve Leaks Internally

Figure TR48-6 Troubleshooting chart for refrigeration cycle accessories.

KNOWLEDGE EXTENSION

Many compressor manufacturers such as Copeland, Tecumseh, and Bristol provide operation and maintenance manuals for a minimal fee to the service technician. These manuals have troubleshooting guides, charts, wiring diagrams, and specific operating conditions. They are invaluable to any technician servicing and troubleshooting refrigeration equipment and are available at your local refrigeration supply house or by contacting the manufacturer's representative. Several of these companies also offer specialized educational seminars about their product line for service personnel to learn more about proper operation and maintenance of their product. Contact the manufacturer's training department to find out about a seminar in your area.

In using these steps to diagnose the cause of the problem, the answer may be found in the first two steps, eliminating the need to go further. A difficult problem may require completing all four steps. Proceed through the steps only so far as necessary to find the cause of the problem.

For example, in step 1c, preliminary power-off inspection, one of the following causes for low capacity may be found:

1. Dirty or missing filters
2. Dirty or loose fan
3. Loose belts
4. Dirty or corroded coil or fins
5. Loose or uninsulated TXV bulb
6. Damaged interconnecting piping

In step 1d, preliminary power-on inspection, one of the following conditions may indicate the source of the problem:

1. Incorrect fan-rotation direction
2. Insufficient air circulation
3. Noise from a loose pulley wheel
4. Odor from an overheated transformer
5. Hot spots on a bearing

In order to illustrate the use of the step-by-step procedure to locate the problem, follow the example below:

EXAMPLE

A customer using a split-system air conditioner reports that the compressor is running but that the cooling is inadequate. What should you do?

Solution

Proceed with step-by-step analysis and use the basic symptom analysis (step 4a) to solve the problem. Tests and calculations indicate the following:

1. Discharge pressure is high.
2. Suction pressure is high.
3. Superheat is low.
4. Subcooling is high.
5. Amps are high

Referring to the chart, this condition could be caused by only one condition: an overcharge of refrigerant. Now that the problem is identified, the manufacturer's charging charts can be used to adjust the charge for the present indoor and outdoor conditions.

TR48-4 TROUBLESHOOTING THE COMPRESSOR

Compressors built today are expected to provide many years of constant, trouble-free, quiet operation. In many applications the compressor is required to run 24 hours per day, 365 days a year. Such continuous operation, however, is often not as hard on a compressor as is a cycling operation, where temperatures constantly change and oil is not maintained at a constant viscosity.

The compressor must not only be designed to withstand normal operating conditions, but also occasional abnormal conditions such as liquid slugging and excessive discharge pressure. Compressors have been designed to take extra punishment and yet function properly. Most compressor failures are caused by system faults and not from operating fatigue. The degree of skill technicians use to install, operate, and maintain the equipment will ultimately determine the actual life expectancy of the system, particularly the compressor's. It is therefore helpful to review some of the factors that shorten the life of a compressor.

TR48-4.1 Loss of Efficiency

The loss of efficiency of a compressor is usually an indication that the compressor is being subjected to system problems that are wearing some of the component parts. For a reciprocating machine this can result from a number of conditions:

1. If liquid enters the compressor, the efficiency and resulting capacity will be seriously affected. The

PREVENTIVE MAINTENANCE

A contractor, concentrating on service work, has many customers who do not buy traditional service contracts or maintenance agreements. They join the company's "Energy Club." The membership certificate is impressive enough to be framed.

Benefits include two service calls (spring and fall), customary discounts, and priority during temperature extremes. Along with service is education on preventive maintenance, such as the importance of changing filters.

Rising costs of replacing modules, repairing complicated electronic systems and cleaning required by environmental contaminants are persuasive reasons for joining the "Club" as an insurance policy. Members are made to feel prestigious by prompt and efficient service people.

physical damage reduces the effectiveness of the internal parts.

2. Leaking discharge valves reduce the pumping efficiency and cause the crankcase pressure to rise, increasing the load on the machine.
3. Leaking suction valves seriously affect the compressor efficiency (and capacity) especially at lower temperature applications.
4. Loose pistons cause excessive blow-by and lack of compression.
5. Worn bearings, especially loose connecting rods and wrist pins, prevent the pistons from rising as far as they should on the compression stroke. This has the effect of reducing the clearance volume and results in excessive re-expansion.
6. Belt slippage on belt-driven units.

TR48-4.2 Motor Overloading

When the compressor is not performing satisfactorily, the motor load sometimes provides a clue to the trouble. Either an exceptionally high or exceptionally low motor load is an indication of improper operation. Here are some of the causes of motor overloading:

1. Mechanical problems such as loose pistons, improper suction valve operation, or excessive clearance volume usually lead to reduction in motor load.
2. Another common problem is a restricted suction chamber or inlet screen (caused by system contaminants). The result is much lower actual pressure in the cylinders at the end of the suction stroke than the pressure in the suction line as registered on the suction gauge. If so, an abnormally low motor load will result.
3. Improper discharge valve operation, partially restricting ports in the valve plate (which do not show up on the discharge pressure gauge), and tight pistons will usually be accompanied by high motor load.
4. Abnormally high suction temperatures created by an excess load will cause a high motor load.
5. Abnormally high condensing temperatures, created by problems associated with the condenser, will also lead to high motor load.
6. Low voltage at the compressor, whether the source is the power supply or excessive line loss, will contribute to high motor loading.

TR48-4.3 Noisy Operation

Noisy operation usually indicates that something is wrong. There may be some noisy condition outside the compressor or something defective or badly worn in the compressor itself. Before changing the compressor, a check should be made to determine the cause of the noise. Here are some possible causes outside the compressor:

1. Liquid slugging. Make sure that only superheated vapor enters the compressor.
2. Oil slugging. Possibly oil is being trapped in the evaporator or suction line and is intermittently coming back in slugs to the compressor.
3. Loose flywheel (on belt-driven units).
4. Improperly adjusted compressor mountings. In externally mounted hermetic type compressors, the feet of the compressor may be bumping the studs. The hold-down nuts may not be backed off sufficiently, or springs may be too weak, thus allowing the compressor to bump against the base.

TR48-4.4 Compressor Noises

Noises coming from the inside of the compressor may be one of the following:

1. Insufficient lubrication. The oil level may be too low for adequate lubrication of all bearings. If an oil pump is incorporated, it may not be operating properly, or it may have failed entirely. Oil ports may be plugged by foreign matter or oil sludged from moisture and acid in the system.
2. Excessive oil level. The oil level may be high enough to cause excessive oil pumping or slugging.
3. Tight piston or bearing. A tight piston or bearing can cause another bearing to knock—even though it has proper clearance. Sometimes in a new compressor such a condition will "wear in" after a few hours of running. In a compressor that has been in operation for some time, a tight piston or bearing may be due to copper plating, resulting from moisture in the system.
4. Defective internal mounting. In an internally spring-mounted compressor, the mountings may be bent, causing the compressor body to bump against the shell.
5. Loose bearings. A loose connecting rod, wrist pin, or main bearing will naturally create excessive noise. Misalignment of main bearings, shaft to crankpins or eccentrics, main bearings to cylinder walls, etc., can also cause noise and rapid wear.
6. Broken valves. A broken suction or discharge valve may lodge in the top of a piston and hit the valve plate at the end of each compressor stroke. Chips, scale, or any foreign material lying on a piston head can cause the same result.
7. Loose rotor or eccentric. In hermetic compressors a loose rotor on the shaft can cause play between the key and the keyway, resulting in noisy operation. If the shaft and eccentric are not integral, a loose locking device can be the cause of knocking.
8. Vibrating discharge valves. Some compressors, under certain conditions, especially at low suction pressure, have inherent noise which is due to vibration of the discharge reed or disc on the compression stroke. No damage will result, but if the noise is objectionable, some modification of the discharge valve may be available from the compressor manufacturer.
9. Gas pulsation. Under certain conditions noise may be emitted from the evaporator, a condenser, or suction line. It might appear that a knock and/or a whistling noise is being transmitted and amplified through the suction line or discharge tube. Actually, there may be no mechanical knock, but merely a pulsation caused by the intermittent suction and compression stroke, coupled with certain phenomena associated with the size and length of refrigerant lines, the number of bends, and other factors.

TR48-5 ANALYSIS OF A DEAD COMPRESSOR

This exercise assumes that the power system has been checked and that the thermostat is calling for cooling, so that even though the proper power is available at the compressor, it will not start. The compressor may not even "hum" when power is applied to it or it may "hum" and cycle on overload. In any event, it is the duty of the technician to analyze the problem, locate the cause and provide a remedy. Here are some of the causes of dead compressors:

1. Open control contacts
2. Overload contacts tripped
3. Improper wiring
4. Overload cutout destroyed
5. Shortage of refrigerant
6. Low voltage
7. Start capacitor defective or wrong one
8. Run capacitor defective or wrong one
9. Start relay defective or wrong one
10. High head pressure
11. Compressor winding burnt out
12. Piping restriction

To locate the cause, the electrical system and, if necessary, the refrigeration system must be thoroughly tested. The best procedure usually is to check the electrical system first. The procedure is outlined below.

TR48-5.1 Prepare for Testing

Disconnect the power supply and remove the refrigerant from the system. All electrical devices such as relays, capacitors, and external overloads must be disconnected.

TR48-5.2 Locate the Compressor Motor Terminals

Find the two terminals that provide the highest resistance reading. That reading represents the combined resistance of two windings, from R to S. The remaining terminal is the common (C) terminal. Put one ohmmeter lead on the C terminal and find which of the remaining terminals gives the highest resistance reading. That will be the start winding and therefore the S terminal. The remaining terminal is the R terminal (Fig. TR48-7).

TR48-5.3 Check Motor Windings with an Ohmmeter

Read the resistance across each pair of terminals. For example, a typical set of readings for a good single-phase compressor could be as follows:

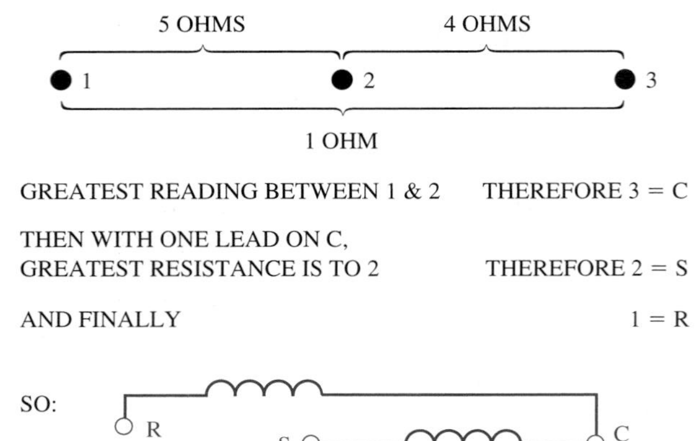

Figure TR48-7 Measuring resistances of compressor windings.

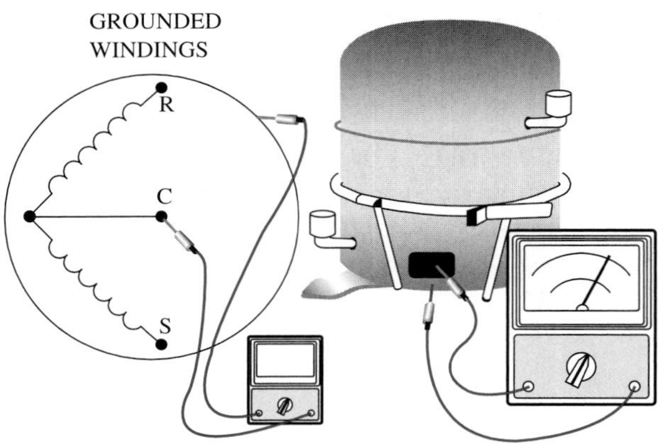

Figure TR48-9 Testing for grounded motor windings

C-to-S = 4 ohms

C-to-R = 1 ohm

S-to-R = 5 ohms

Note that the S-to-R reading is the sum of the other two and that the C-to-S resistance is always higher than C-to-R. A good rule of thumb is that the resistance of the start winding is three to five times that of the run winding.

If "zero" is read for any one of these pairs, there is a shorted winding and a defective compressor (Fig. TR48-8). If "infinity" is read for any one of these pairs, there is an open winding and a defective compressor.

To test for a grounded winding, the ohmmeter must be capable of measuring a very high resistance. It needs to have a scale that can be set to R × 100,000. One lead of the ohmmeter is placed on a compressor terminal and the other on the housing (Fig. TR48-9). The resistance between the winding and the housing ranges from 1 to 3 million ohms for an ungrounded winding. When placing the meter lead on the housing be sure that the meter is making good contact. A coat of paint, a layer of dirt, or corrosion can hide a grounded winding.

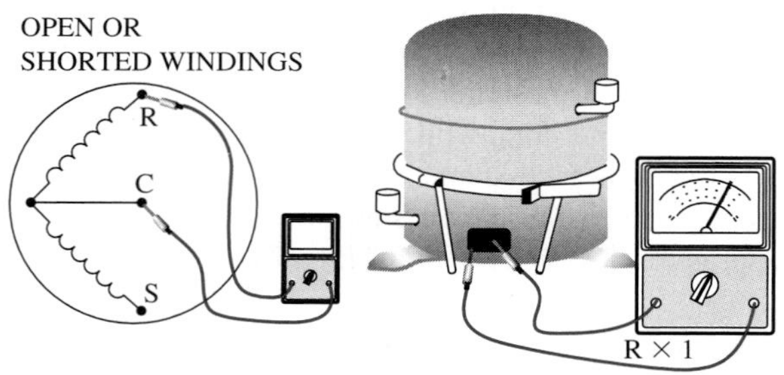

- ZERO RESISTANCE = SHORTED WINDING
- LOW RESISTANCE = GOOD WINDING
- INFINITY = OPEN WINDING

Figure TR48-8 Testing for shorted or open windings.

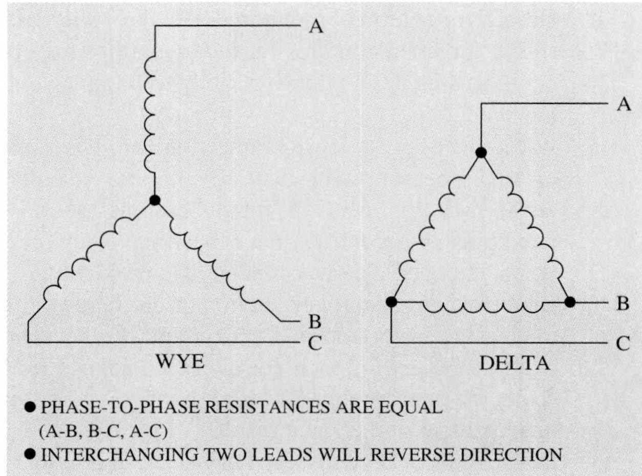

Figure TR48-10 Three-phase motors.

In testing three-phase compressor motors, a good motor will show equal resistances between each of the terminals (Fig. TR48-10). Testing the windings is done in the same way as for single-phase motors. One caution: after the testing is done, be sure to reconnect the windings exactly as they were. Interchanging any two windings of a three-phase motor will reverse the direction of rotation. Also, when checking windings of a compressor that is under pressure, make the meter connections "upstream" from the terminals.

TR48-5.4 Test the Capacitors and Relays

Capacitors

If the windings of the compressor motor are satisfactory, the problem may be in one of the capacitors or the relay. There are two types of capacitors, the start capacitor and the run capacitor (Fig. TR48-11). The procedure for testing the capacitors is as follows:

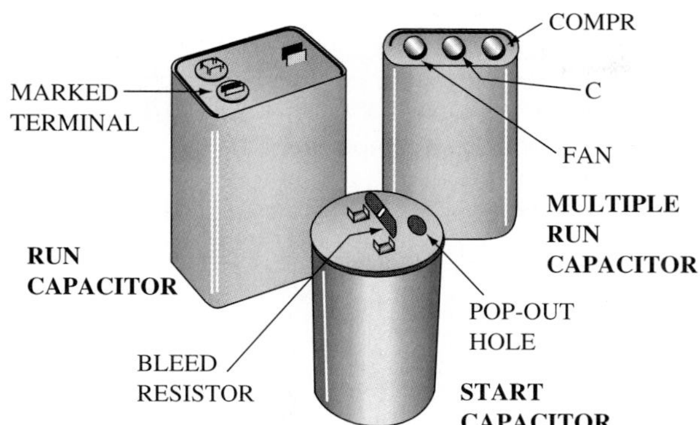

Figure TR48-11 Various types of capacitors.

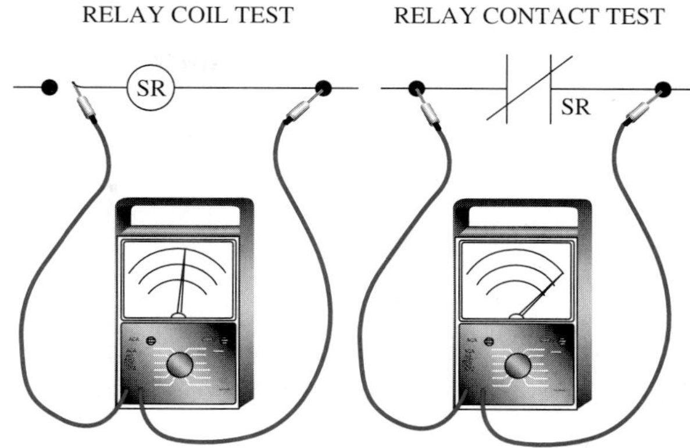

Figure TR48-12 Testing relays.

1. Discharge the capacitor. Never place your fingers across the terminals of a capacitor. Do not place a direct short across the terminals of a capacitor—this can damage the capacitor. Place it in a protective case and connect a 20,000-ohm, 2-W resistor across the terminals and gradually bleed off the charge.
2. Disconnect the circuit wiring and connect the leads of the ohmmeter across the terminals and observe the meter needle. If the capacitor is OK, the needle will make a rapid swing toward zero and slowly return to infinity. If the capacitor has an internal short, the needle will fall back part of the way and stop.
3. A capacitor analyzer is used to read the microfarad value of the capacitor and detect insulation breakdown under load conditions. In this case, the meter indicates a "dead short."

Almost all run capacitors have a marked terminal (Fig. TR48-11). This terminal should be connected to the motor run terminal and one leg of the power circuit. With this arrangement, an internal short circuit to the capacitor case will blow the system fuses without passing the destructive short circuit current through the motor start winding.

Some run capacitors have three terminals and are used in the circuits of two motors. They are actually two capacitors in the same container. They are tested in the same way as single capacitors.

Many start capacitors are built with a bleed resistor connected between the terminals. This resistor discharges the capacitor after it is switched out of the circuit.

A "pop-out" hole on the start capacitor allows insulation expansion if the capacitor is overheated. If the hole is ruptured, the capacitor should be replaced.

Testing Start Relays

The start relay removes the start capacitor from the circuit when the motor reaches a certain rpm. If it is necessary to replace the relay, an identical replacement must be used, because these relays are built with unique characteristics to match the motor with which they are used. When replaced they must be located in the same position, and wired exactly the same as the original relay.

The coil and the switch are tested separately, as shown in Fig. TR48-12. Disconnect the relay from the circuit. In testing the coil it should register a substantial resistance. Use the R × 100 scale on the ohmmeter. In testing the contacts, the ohmmeter should read zero, indicating that the switch is closed. Relay switches usually fail in the open position. When they fail by sticking closed, the start capacitor or the start winding of the compressor motor will usually be damaged.

TR48-6 TROUBLESHOOTING EVAPORATORS AND CONDENSERS

TR48-6.1 Evaporators

The following is a list of some of the refrigeration system problems associated with evaporators:

1. Low airflow
2. Excessive airflow
3. Uneven airflow over coils
4. Low refrigerant supply
5. Uneven refrigerant distribution to coil circuits
6. Low water flow in cooler (water cooling evaporator for chillers)
7. Uneven water flow through cooler
8. Low refrigerant supply to cooler

The symptoms and possible causes for these conditions are shown in Fig. TR48-13. The technician, by process of elimination, uses this information to arrive at the cause of a particular problem. The following information will be helpful in analyzing some of the common problems.

A dirty filter is probably the number one cause of low airflow, particularly if the owner has not provided periodic

General Problem Category	Symptoms	Possible Causes
Low Airflow	■ Low Saturated Suction Temperature ■ Low Suction-Gas Superheat ■ Low Saturated Condensing Temperature ■ Low Compressor Power Draw (kW) ■ Low Supply-Air Temperature ■ Low System Capacity ■ High Space-Air Temperature ■ Iced or Frosted Evaporator ■ Compressor Liquid Floodback ■ Compressor Slugging	■ Dirty Evaporator Coil (No Filters?) ■ Badly Bent Evaporator Fins ■ Dirty Filters
Excessive Airflow	■ High Supply-Air Temperature ■ High Saturated Suction Temperature ■ High Compressor Power Draw (kW) ■ Noisy Air System ■ Noisy Air Handler ■ Water Dripping from Fan and Supply Ductwork Near Air Handler	■ Fan-Motor Speed Set Too High ■ Wrong Fan Drive Package and/or Setting ■ Undersized Coil (Applied Air Handlers) ■ Condensate Carries Over Air Handler Drain Pan into Fan and Supply Ductwork
Uneven Airflow Over Coil	■ Low System Capacity ■ Low Saturated Suction Temperature ■ Uneven Condensate Coverage Over Coil Surface ■ Uneven Coil Surface Temperature ■ Refrigerant Floodback to Compressor ■ Compressor Slugging	■ Bad Duct Design Near Evaporator Coil ■ Coil Placement Improper ■ Air Turbulence at Coil ■ Lack of Necessary Air Baffling Near Coil ■ Obstruction within Air Handler ■ Obstruction in Duct Work Near Air Handler ■ Mismatched Coil and Air Handler
Low Refrigerant Supply	■ Low System Capacity ■ Low Saturated Suction Temperature ■ High Suction-Gas Superheat ■ Low Compressor Power Draw (kW) ■ Low Saturated Condensing Temperature ■ Measurable Temperature Drop in Liquid Line ■ Visible Bubbles in Liquid-Line Sight Glass ■ High Supply-Air Temperature ■ Frosted or Iced Evaporator ■ High Discharge-Gas Superheat	■ System Undercharged ■ Liquid Line Kinked or Crushed ■ Evaporator Tube Crushed (Especially Return Bends) ■ System Refrigerant Leak ■ Malfunctioning Metering Device ■ TXV Power Element Low on Charge ■ Undersized Metering Device ■ Undersized Distributor Nozzle ■ Head Pressure Control Faulty at Low Outdoor-Ambient Temperatures ■ TXV Plugged or Stuck Closed ■ Plugged Distributor Oil Nozzle ■ Free Water in System Forms Ice ■ Filter-Drier Plugged
Uneven Refrigerant Distribution to Coil Circuits or to Cooler Circuits	■ Low System Capacity ■ Low Saturated Suction Temperature ■ Little or No Suction-Gas Superheat ■ TXV Hunts ■ Compressor Floodback ■ Compressor Slugging ■ Uneven Coil Surface Temperature ■ Uneven Condensate Formation on Evaporator ■ Frost on Some Areas of Evaporator but not on Others	■ Plugged Evaporator Feeder Tube(s) ■ Kinked or Crushed Feeder Tube(s) ■ Partially Blocked Distributor ■ Oversized Distributor ■ Oversized Distributor Nozzle (Applied Air Handlers) ■ Improperly Installed Distributor (Applied Air Handlers) ■ Crushed Evaporator Tube (Especially Return Bends) ■ Plugged Evaporator (or Cooler) Circuit

Figure TR48-13 Troubleshooting chart for evaporators.

General Problem Category	Symptoms	Possible Causes
Low Water Flow in Cooler (Water-Cooling Evaporator for Chillers)	■ Low Saturated Suction Temperature ■ Low Suction-Gas Superheat ■ Low Saturated Condensing Temperature ■ Low Compressor Power Draw (kW) ■ Low Leaving Chilled Water Temperature ■ Low System Capacity ■ High Space Temperature ■ High Temperature Drop Between Entering and Leaving Chilled Water ■ Chiller Shuts Down Intermittently (Even Though Thermostat Calls for Cooling) on Low Leaving Water Safety Thermostat	■ Chilled-Water Pump Undersized ■ Faulty Pump Motor ■ Damaged or Blocked Pump Impeller ■ Blocked Chilled-Water Line or Valve ■ Water Baffle(s) in D-X Cooler Misplaced Blocking Flow ■ Excessive Water Scaling (Flooded Cooler) ■ Reverse Chilled-Water Pump Rotation ■ Blockage in Chilled Water Piping ■ Water-Flow Control Valve Restricting Flow
Uneven Water Flow Through Cooler	■ Low System Capacity ■ Low Saturated Suction Temperature ■ Compressor Floodback ■ Compressor Slugging ■ High Leaving Chilled-Water Temperature ■ Low Temperature Drop Between Entering and Leaving Chilled Water	■ (D-X Cooler): ▪ Misplaced or Broken Baffle(s) ▪ Excess Air in Water System ▪ Debris Inside Shell of Cooler ■ (Flooded Cooler): ▪ Badly Scaled Water Tube(s) ▪ Kinked or Crushed Water Tube(s) ▪ Plugged Water Tube(s) or Water Box
Low Refrigerant Supply to Cooler	■ Low System Capacity ■ High Leaving Chilled-Water Temperature ■ Low Saturated Suction Temperature ■ High Suction Gas Superheat ■ Low Compressor Power Draw (kW) ■ Low Saturated Condensing Temperature ■ Space Temperature Too Warm ■ Chiller Compressor Cycles Off Intermittently on Low-Pressure Switch	■ System Undercharged ■ Head-Pressure Control Not Working at Low Outdoor Ambient Temperatures ■ Refrigerant System Leak ■ Flooded Cooler: ▪ Refrigerant Flow from Condenser Blocked ▪ Cooler Refrigerant Supply Valve Stuck ■ D-X Cooler: ▪ Liquid Line or Accessories Plugged ▪ Liquid Line Kinked or Crushed ▪ TXV Power Element Low on Charge ▪ TXV Plugged or Stuck ▪ Refrigerant Distributor or Nozzle Plugged ▪ Electronic Expansion Valve Faulty or Microprocessor Problem ▪ Frost-Pinched Cooler Tubes

Figure TR48-13 Continued

maintenance of air filters. On larger systems a differential-pressure-sensing meter can be used to indicate the pressure drop through the filters when filter changing is necessary. A responsible person is required to perform the cleaning or replacement necessary periodically. Filters provide a valuable function but do require maintenance.

Excessive airflow can lower the efficiency, interfere with the comfort level and/or produce noisy operation. Excessive airflow is indicated by a low air-temperature drop over the evaporator coil. For proper airflow the temperature drop should normally be between 18 and 22°F. Adjusting the airflow should be a remedial measure in the checkout procedure. Since the power (watts) supplied to the blower motor varies as the cube of the blower rpm, a large savings can be effected on most jobs by supplying the proper air quantity. Too high an airflow reduces the dehumidifying function of an air-conditioning system and therefore interferes with the comfort level. The humidity can be tested by measuring the leaving air wet-bulb temperature. The noise factor should be greatly improved by reducing the fan speed.

A low refrigerant supply can often be traced to a plugged filter-drier. This is a comparatively easy item to check. Due to the restriction in the filter-drier, some vaporization of the refrigerant can occur, lowering the temperature at the outlet. By sensing a temperature drop through the drier, the restriction is located.

A low water flow through a cooler will be evidenced by a low water temperature leaving the chiller. This is usually caused by a restriction in the water line, a water pump with the wrong impeller rotation, or a defective pump. Many times a shutoff valve is found that has been improperly opened and is creating the problem. As soon as the problem is found, a relatively simple solution can often solve the problem.

TR48-6.2 Condensers

The following is a list of some of the refrigeration problems associated with condensers:

1. High head pressure.
2. Refrigerant charge incorrect
3. Low head pressure

The symptoms and possible causes for these conditions are shown in Fig. TR48-14. The technician, by process of elimination, uses this information to arrive at the cause of

General Problem Category	Symptoms	Possible Causes
High Head Pressure (Saturated Condensing Temperature or Saturated Discharge Temperature)	■ Compressor Cycles Off Intermittently on High-Pressure Switch while System Calls for Cooling ■ Compressor Cycles Off Intermittently on Compressor-Motor Protection Switch ■ High Saturated Condensing Temperature ■ High Discharge-Gas Superheat ■ Compressor Overheats ■ Compressor Seizure ■ Compressor Motor Burnout ■ High Compressor Power Draw (kW) ■ Low System Capacity ■ Saturated Suction Temperature Normal to High ■ Excessive Condenser Water-Flow Rate	■ Faulty Head Pressure Control Device ■ Dirty Condenser Coil ■ Faulty Condenser Fan Motor ■ Extensive Fin Damage ■ Condenser Air Recirculation ■ Dirty Condenser Fan ■ Condenser Airflow Blocked ■ Prevailing Winds Prohibit Proper Airflow Across Coil ■ Backward Condenser Fan Rotation ■ Slipping Condenser Fan Belt ■ Bent or Broken Condenser Fan Blade(s) ■ Scaled Water-Cooled Condenser Tubes ■ Faulty Condenser Water Pump ■ Damaged Water Pump Impeller ■ Plugged Condenser Water Lines or Screens ■ Condenser Water Valve Stuck Closed ■ Cooling-Tower Problems ■ Condenser Vapor Locked by Undersized or Poorly Laid Out Refrigerant Condensate Line which Prevents Refrigerant from Freely Draining to Receiver ■ System Overcharged ■ Noncondensible Gases Present
Refrigerant Charge Incorrect	■ High Head Pressure ■ High Liquid Subcooling ■ Low System Capacity ■ High Saturated Suction Temperature ■ High Compressor Power Draw (kW)	■ System Overcharged
	■ Low Head Pressure ■ Low Saturated Suction Temperature ■ Low System Capacity ■ Low or Nonexistent Liquid Subcooling ■ Flash Gas at Metering-Device Inlet	■ System Undercharged
Low Head Pressure	■ Low Saturated Condensing Temperature ■ Low System Capacity ■ Low Saturated Suction Temperature ■ Low Compressor Power Draw (kW)	■ Faulty Head-Pressure Control Device ■ Refrigerant System Leak ■ Undercharged System ■ Condenser-Water Valve Stuck Open

Figure TR49-14 Troubleshooting chart for condensers.

a particular problem. The following information will be helpful in analyzing some of the common problems.

A high head pressure can often be traced to noncondensible gases present in the refrigerant. The saturated condensing refrigerant has an equivalent condensing pressure, as shown in the charts. Reading a higher condensing pressure on the equipment is an indication that noncondensibles are present. These noncondensibles, usually air in the system, need to be removed for the system to operate efficiently.

A high head pressure can also be due to a dirty condenser surface. Occasionally an operator will spray an air-cooled condenser coil with water during extremely hot weather to increase its capacity. As a result a deposit is left on the coil that fills the fin space and reduces the capacity of the condenser. This deposit is hard to remove, usually removed with acid and a stiff brush. It can also be so destructive that the coil must be replaced.

If the system has excess refrigerant, depending on the system, too much of the condenser can be filled with liquid not leaving enough room for condensation. This can cause high head pressure. The overcharge needs to be removed to have the condenser operate normally.

Low head pressure is often an indication of lack of refrigerant due to a leak. If so, the leak needs to be located with a suitable leak detector (usually electronic), the leak repaired, and the system recharged with the correct amount of refrigerant.

TR48-7 TROUBLESHOOTING METERING DEVICES

The following is a list of some of the refrigeration system problems associated with metering devices:

1. Evaporator overfeeding ("flooding")
2. Evaporator underfeeding ("starving")
3. Thermal expansion valve (TXV) hunting
4. Distributor nozzles unevenly feeding

The symptoms and possible causes for these conditions are shown in Fig. TR48-15. The technician, by process of elimination, uses this information to arrive at the cause of a particular problem. The following information will be helpful in analyzing some of the common problems.

Evaporator flooding can be caused by a low superheat setting or loose sensing bulb. When this occurs, the application and installation of the valve needs to be carefully checked. Most valves are set at the factory for 10°F superheat. The coil may have a high pressure drop and require an external equalizer to operate properly. The sensing bulb must be tightly held to the suction line and the connection insulated if it is picking up stray heat from other sources.

A starved evaporator may be caused by an undercharged system. If this is the condition look for a refrigerant leak, particularly if the system has been operating satisfactorily for some time.

A hunting TXV is usually a sign that the expansion valve is too large for the application. Most valves operate best for loads down to 50% of their capacity. A valve that is too large will hunt too much of the time. Hunting is the condition where the valve continually opens and closes rather than reaches a stabilized condition.

When the distributor nozzle is furnished with the TXV it must be sized properly to fit the load and to comply with the installation requirements. The tubes of the coil can be inspected to see that all circuits are being fed equally. If not, a properly sized nozzle needs to be applied.

ON THE JOB

Here are some classic quotes from the field that provide insight into professional success.

"Fix the problem, not the blame."

"One employee submitted eight problems that were client-specific—not relating to the work but to things noticed at the client's site. We forwarded these to the client and he was very grateful."

"We rely on a three-point plan: hire the right people, provide training aimed at their advancement, and recognize accomplishment."

"Create your own success by caring about the success of your customers."

"Sometimes the most valuable tool in the truck is your ears: Find out what the customer is really trying to tell you."

"The customer is always right—trite, but everlastingly true."

"Never forget a customer—never let a customer forget you."

"Courtesy is no substitute for efficiency, but adds to it enormously."

General Problem Category	Symptoms	Possible Causes
Evaporator Overfeed ("Flooding")	■ High Saturated Suction Temperature ■ Low Suction-Gas Superheat ■ Liquid Floodback ■ Compressor Slugging ■ Compressor Overheats ■ High Compressor Power Draw (kW) ■ Compressor Failure ■ Compressor Pumps Improperly ■ TXV Hunts	■ System Overcharge (Fixed Metering Device) ■ Oversized Metering Device ■ TXV Stuck Open ■ TXV Superheat Setting Too Low ■ TXV Type Wrong for Refrigerant in System ■ Uninsulated TXV Sensing Bulb in Warm Area ■ Loose TXV Sensing Bulb ■ Incorrectly Located TXV Sensing Bulb ■ Partial Load Too Low for Metering Device ■ Excess Oil Circulating in System ■ High Head Pressure (Fixed Metering Device)
Evaporator Underfeed ("Starvation")	■ Low System Capacity ■ Low Saturated Suction Temperature ■ High Suction-Gas Superheat ■ Low Compressor Power Draw (kW) ■ Low Saturated Condensing Temperature ■ High Discharge-Gas Superheat ■ High Supply-Air Temperature ■ Iced or Frosted Evaporator	■ System Undercharged ■ Undersized Metering Device ■ Plugged Metering Device ■ Plugged Distributor or Nozzle ■ Undersized Distributor or Nozzle (TXV Jobs) ■ Kinked or Crushed Capillary Tube ■ TXV Stuck in Closed Position ■ TXV Power Element Low on Charge ■ Wrong TXV for Refrigerant in System ■ Plugged or Crushed TXV External Equalizer Line ■ TXV Superheat Setting Too High ■ Incorrect TXV Sensing Bulb Location ■ Free Water in System Forms Ice and Blocks Refrigerant Flow ■ Low Head Pressure (Fixed Metering Device) ■ Faulty Head Pressure Control Device
TXV Hunting	■ Saturated Suction Temperature Oscillates High then Low, in a Cyclical Fashion ■ Suction Gas Superheat Oscillates High and Low in a Cyclical Fashion ■ Compressor Power Draw (kW) Oscillates High and Low, in a Cyclical Fashion ■ Intermittent Floodback and Compressor Slugging ■ Unstable Supply Air Temperature ■ Unstable Evaporator Surface Temperature	■ Oversized TXV ■ Improper Part-Load Control Operation Loads TXV Too Lightly ■ Very Light Cooling Load ■ Rapid Cooling Load Changes ■ Rapidly Changing High-Side Pressure ■ Intermittent Flashing in Liquid Line ■ Incorrect Evaporator Circuiting Selected (Applied Air Handlers)
Distributor Nozzles (TXV Applications)	■ Evaporator Underfeed (See Symptoms Above)	■ Undersized Distributor Nozzle (Quite Unlikely on Comfort Work)
	■ Evaporator Unevenly Fed by Refrigerant (See Symptoms on Evaporator Sheet)	■ Oversized Nozzle ■ Nozzle Not Sized for Low Load Stability ■ Faulty Part-Load Control Sequence for Evaporator Sections

Figure TR48-15 Troubleshooting chart for metering devices.

TR48-7.1 Other Areas for Troubleshooting

In the above descriptions, troubleshooting certain major components of the system have been suggested. If additional areas need to be examined to find the problem, here are some further areas to troubleshoot:

1. Refrigeration cycle accessories
2. Refrigeration piping
3. The quality of the installation process
4. The quality of the evacuation/dehydration process

TR48-8 SEALED SYSTEM TROUBLESHOOTING

A hermetically sealed refrigeration system is one that is completely enclosed by welding or soldering that prevents the escape of refrigerant or the entrance of air. Further, these systems use a hermetic compressor and a capillary-tube metering device, requiring no external adjustments. The units are assembled under controlled conditions, evacuated, charged and tested at the factory. Everything possible is done by the manufacturer to provide units with long, trouble-free service. With this assurance many manufacturers offer a 5-year limited warranty on the sealed system. A warranty of this type applies only if the equipment is given normal and proper use. The unit must be correctly installed and certain items of maintenance need to be performed, such as keeping the condenser cleaned.

Like all mechanical equipment, with continuous use there will come a time when the system must be opened for service. The following topics pertain to information for troubleshooting these systems when they fail:

1. Refrigerant leaks
2. Restrictions in the refrigerant system
3. Installation problems

TR48-8.1 Refrigerant Leaks

Refrigerant leaks on a refrigeration system can be either inward or outward. If the system operates below atmospheric pressure, an inward leak will cause air and moisture to enter the cycle. This will cause an increase in discharge pressure and temperature, which increases corrosion and can stop the operation quite quickly. An inward leak can be more serious than an outward leak.

Several methods of outward leak detection can be used with any halocarbon refrigerant. The best method is usually the *electronic leak detector* (Fig. TR48-16). It is usually a time saver over some of the former methods.

The *halide torch* (Fig. TR48-17) is a fast and reliable method for use with halocarbon refrigerants. The fuel that operates the torch is either methyl alcohol or a hydrocarbon such as butane or propane. The flame heats a copper element. Air to support combustion is drawn through the tube. The free end of the tube is passed over any suspected leak areas. When a halocarbon vapor passes over the hot copper element, the flame changes from normal color to bright green or purple.

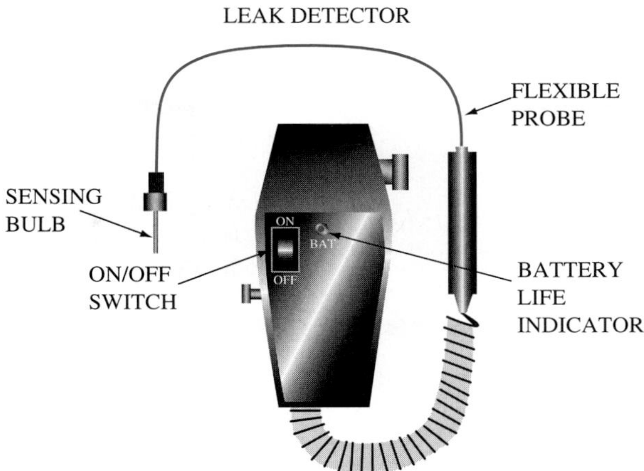

Figure TR48-16 Electronic refrigerant leak detector.

A halide torch should only be used in well ventilated areas. When working in direct sunlight, it is sometimes hard to see the color change.

Another reliable method of leak detection is *soap-bubble testing* (Fig. TR48-18). One method is to paint the suspected leak area with the soap solution and then, using a bright light, watch for the formation of bubbles.

Another method of testing is to pressurize the system with dry air and submerge the suspected leak area in water, and watch for bubbles. The system pressure should be at least 50 psig.

To repair a leak in a hermetic system, some type of access valve(s) needs to be installed. *Piercing valves* (Fig. TR48-19) are handy for penetrating refrigerant lines that have no service valve. The valve is bolted or brazed to the

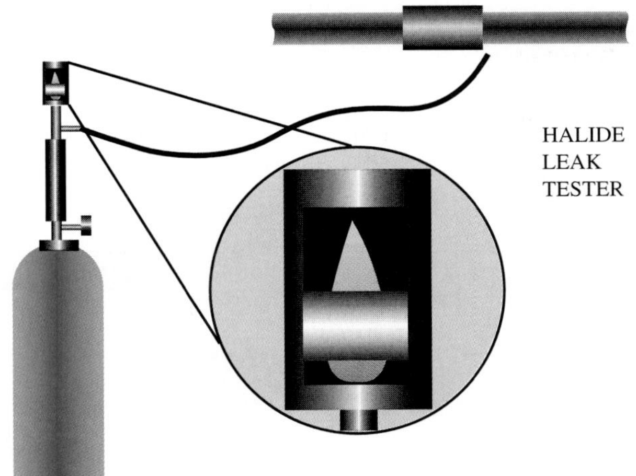

Figure TR48-17 Halide refrigerant leak detector.

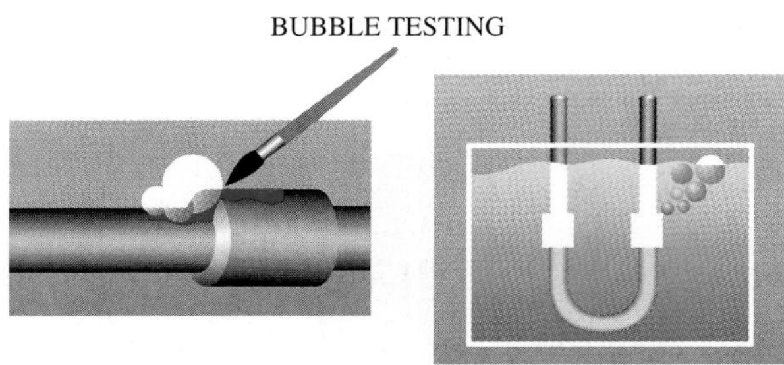

Figure TR48-18 The use of soap bubbles as a leak detector.

line, depending on the type of valve. A needle in the valve penetrates the tubing when the stem is turned down. This permits installing a gauge manifold and proceeding with recovery of the remaining refrigerant, and repairing, evacuating, and recharging the system.

TR48-8.2 Restrictions in the System

In a hermetically sealed system a restriction in the refrigerant circuit sometimes develops. This can be caused by kinked or blocked tubing, a moisture restriction, or a blocked filter-drier. Occasionally a capillary tube is bent and kinked, or contaminants may be in the system that block the opening of the tube, preventing refrigerant flow.

If moisture is in the system and the evaporator is operating below freezing temperatures, ice can form at the metering device, stopping the flow of refrigerant. If the filter-drier is full of waste materials collected from the system, it can cause a restriction in the flow of refrigerant.

A restriction is usually easy to diagnose, since it prevents the normal flow of refrigerant. The discharge pressure, the suction pressure, and the amperage all drop. The superheat and the subcooling both increase.

Normally, restrictions take place in the liquid line. When they do there is a decided temperature drop across the restriction.

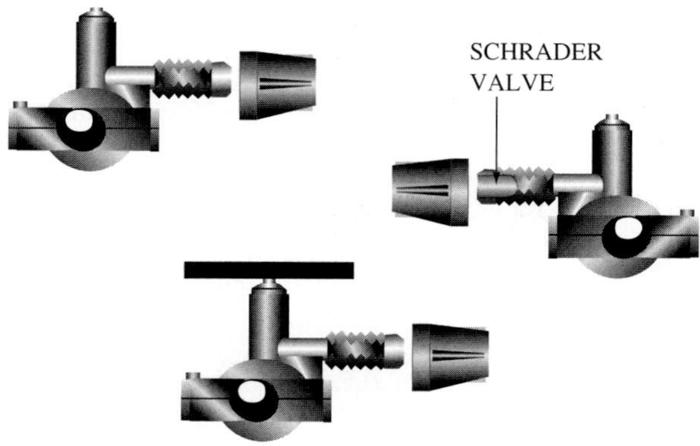

Figure TR48-19 The use of piercing valves for access to a hermetic system.

Whenever a restriction does occur, its location and cause must be found. Remedial action depends on the nature of the restriction. Kits are available for cleaning out a plugged capillary tube; alternatively, the exact replacement can be installed. A plugged filter-drier can be replaced.

TR48-8.3 Installation Problems

Problems in the installation of hermetically sealed systems can be minimized by following specific installation instructions supplied by the manufacturer. Since there are many types and uses of hermetic systems, only general information can be given here. For example, there is a considerable difference between the proper installation of a room cooler and a domestic refrigerator, although they both contain hermetic systems.

Generally, installation instructions for units with hermetic refrigeration systems include the following topics:

1. Location of equipment
2. Leveling the equipment
3. Electrical and plumbing connections
4. Adding accessories
5. Necessary maintenance

The equipment location must provide the proper clearances to supply ventilation air for an air-cooled condenser, accessibility for service, and suitable support for the unit. High head pressure would be an indication that the outside air was not properly circulating through the air-cooled condenser. This could be caused by improper clearances or a dirty condenser. If necessary, the unit should be relocated. Occasionally the condenser is located outside the building where shrubbery has grown to restrict the air into the condenser. Vegetation should be trimmed away from the unit and this noted for future maintenance.

Units need to be installed in a level position, so that condensate will flow to the drain opening for proper disposal. If the drain pan becomes dirty or the opening for condensate removal is blocked, these areas must be cleaned to permit proper flow after levelling the unit.

The unit needs to be located so that electrical and plumbing connections can be properly applied. Usually a trapped drain is required and one must be provided. Otherwise, under certain conditions the drainage will flow in the reverse direction, causing the unit to flood. Electrical connections must be kept dry and accessible.

Sometimes a unit can be improved by adding certain accessories. In some cases the manufacturer has available an air deflector to direct the moving air over the condenser. Without the accessory, the prevailing wind may pass through the condenser in the wrong direction, not cooling the unit and causing high head pressure.

Very often, the installation instructions will indicate certain maintenance that must be provided for the equipment to work properly. For example, an air-cooled condenser coil must be cleaned periodically. Maintenance records should be examined to see whether this has been done, what history the problem may have, and what previous actions have been taken.

REVIEW

- Troubleshooting—The ability of a service technician to determine the cause of an equipment malfunction and to perform corrective measures.
- Troubleshooting requires a high degree of knowledge, experience, and skill.
- Two types of system problems—Electrical and mechanical.
- Key to being a good "troubleshooter" is using a logical, structural, and systematic approach.
- The greatest number of malfunctions are electrical.
- To analyze electrical problems you must know:
  The operating sequence of the unit
  Functions of the equipment
  What is working and what is not
  Electrical test instruments
  What power circuit is driving the system
- Knowledge of operating sequence:
  Usually supplied by the manufacturer in the service and installation instructions.
  Can be determined by the technician by studying the schematic.
- Functions of the equipment:
  Determined by examination and testing
- Required test instruments:
  Voltmeter
  Clamp-on ammeter
  Ohmmeter
- Power circuit:
  First to be examined, because power must be available to operate the loads
- Troubleshooting references:
  Wiring diagrams
  Installation and service instructions
  Troubleshooting tables
  Fault-isolation diagrams
  Diagnostic tests
- Wiring diagrams:
  Connection diagrams—Used to hook up unit or find a component, wires, or terminals.
  Schematic diagrams (ladder)—Used to troubleshoot and determine the sequence of operation.
- Some electronically controlled systems or PLCs have automatic diagnostics, that indicate by code, a malfunction in the operation of the equipment. Further tests are required to confirm and determine the action required.
- Electrical troubleshooting procedure:
  Collect information about the problem.
  Determine if the problem is electrical or mechanical.
  Select the proper test instrument.
  Eliminate good circuits.
  Select the circuits for testing.
  Test the switches and loads.
  Identify faulty component.
  Rerun test on equipment.
- Mechanical troubleshooting procedure:
  Collect information about the problem.
  Determine if the problem is electrical or mechanical.
  Read and calculate the system's "vital" signs.
  Compare typical versus actual values.
  Consult troubleshooting aids.
  Perform system analysis.
  Eliminate possible causes of the problem.
  Select the cause that fits the condition.
  Confirm diagnosis.
  Repair or replace fault.
  Rerun test on equipment.
- Compressor malfunctions:
  Loss of efficiency
  Motor overloading
  Noisy operation
  Stuck compressor
  Motor burnout
- Analysis of "dead" compressor:
  Open control contacts
  Overload tripped
  Improper wiring
  Overload cutout destroyed
  Shortage of refrigerant
  Low voltage
  Start capacitor defective or wrong one
  Run capacitor defective or wrong one
  High head pressure
  Compressor winding burnout
  Piping restriction
- Problems associated with evaporators:
  Low airflow
  Excessive airflow
  Uneven airflow
  Low refrigerant supply
  Uneven refrigerant distribution
  Low water in cooler (chiller)
  Uneven water flow (chiller)
- Problems associated with condensers:
  High head pressure
  Refrigerant charge incorrect
  Low head pressure
- Problems associated with metering devices:
  Overfeeding
  Underfeeding
  TXV hunting
  Distributor feeding unevenly
- Sealed system troubleshooting includes:
  Refrigerant leaks
  Restrictions in the refrigerant system
  Installation problems

Problems and Questions

1. Basically, there are two types of problems when troubleshooting. They are _____ and _____, although there is much overlapping.
2. Does reduced evaporator airflow cause high or low suction pressure?
3. On an evaporator using a TXV valve, will reducing the airflow cause the coil superheat to increase, decrease, or stay the same?
4. On an evaporator using a capillary tube, will reducing the airflow cause the coil superheat to increase, decrease, or stay the same?
5. If an expansion valve fails by losing the feeder bulb charge, will the suction pressure rise or fall?
6. Location of the expansion valve feeder bulb has no effect on valve operation. True or false?
7. What is the most common cause of high head pressure?
8. What is the most common cause of low suction pressure?
9. What is the easiest way to check for a clogged capillary tube on multitube coils?
10. With the unit off and system at ambient temperature, the pressure in the system should equal the _____.
11. The greatest number of malfunction problems are mechanical. True or false?
12. The operating sequence is usually supplied by the manufacturer in the service instructions or it can be determined by studying the schematic wiring diagram. True or false?
13. A fault-isolation diagram starts with a failure symptom and goes through a logical decision-action process to isolate the failure. True or false?
14. An inoperative crankcase heater will cause high overheating of the compressor on start-up due to flooded start. True or false?
15. On a residential air conditioner, when discharge pressure is high, suction pressure is low, and amps are high, this is caused by an overcharge of refrigerant. True or false?
16. Leaking discharge valves:
    a. Reduce the pumping efficiency.
    b. Cause crankcase pressure to rise.
    c. Increase the load on the machine.
    d. All of the above.
17. In preparing to analyze electrical problems, it is important to know or recognize:
    a. The operating sequence of the unit.
    b. The functions of the equipment that are working and those that are not working.
    c. Electrical test instruments needed and the power circuit driving the system.
    d. All of the above.
18. When using an ohmmeter to check motor windings, the greatest resistance will be:
    a. Between C and S
    b. Between R and S
    c. Between C and R
    d. None of the above.
19. Some run capacitors have three terminals. They are:
    a. Using the third terminal as a ground.
    b. Actually two capacitors in the same container.
    c. Used on three-phase compressors to assist start and run efficiencies.
    d. None of the above.
20. A starved evaporator may be caused by:
    a. An undercharged system.
    b. A restricted filter-drier.
    c. A refrigerant leak.
    d. None of the above

Trouble-shooting Air-Conditioning Systems

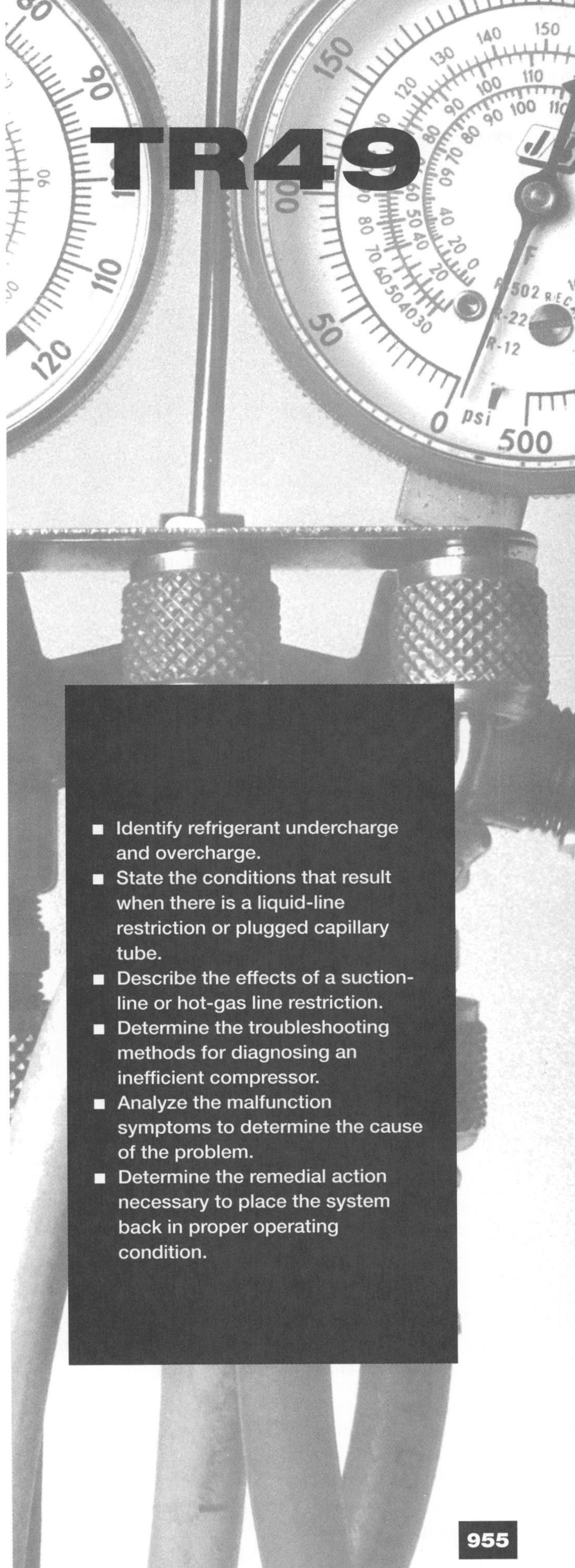

AFTER STUDYING THIS CHAPTER, THE STUDENT WILL BE ABLE TO:

- Identify common operating problems associated with residential and small commercial air-conditioning systems.
- Identify useful instruments necessary to troubleshoot air-conditioning systems.
- State air-system problems and remedies.
- Explain refrigeration-system problems pertaining to refrigerant quantity and refrigerant flow rate.
- Identify expected temperatures for the DX coil, condensing temperature, and refrigerant subcooling temperatures.
- Explain the causes and remedies for insufficient, unbalanced, or excessive loads.
- Describe the effects of low and high ambient conditions.
- Identify refrigerant undercharge and overcharge.
- State the conditions that result when there is a liquid-line restriction or plugged capillary tube.
- Describe the effects of a suction-line or hot-gas line restriction.
- Determine the troubleshooting methods for diagnosing an inefficient compressor.
- Analyze the malfunction symptoms to determine the cause of the problem.
- Determine the remedial action necessary to place the system back in proper operating condition.

TR49-1 SAFETY

It is important in troubleshooting that the technician pay utmost attention to safety measures. The following general measures should be strictly followed on every job to provide for personal safety:

1. Wear safety glasses and gloves when handling refrigerants or when brazing.
2. Recover or recycle refrigerant using an approved device.
3. Shut off all power when working on electrical equipment.
4. If the work must be done while the electrical equipment is energized, remove all watches and rings to reduce the risk of shock.
5. Always read the specific safety recommendations in the manufacturer's installation and service literature.

TR49-2 GENERAL

Technicians servicing air-conditioning equipment find that system performance can be greatly improved not only by good installation practices but also by good maintenance. Careful reference to the manufacturer's installation instructions is important when the system is installed. Good maintenance keeps the equipment running in its original efficient manner.

In troubleshooting air-conditioning systems, some of the most useful instruments are:

1. A gauge manifold for measuring suction and discharge pressures at the compressor.
2. A minimum of five thermometers or an electronic thermometer with provision for connecting to at least five remote temperature sensors (see below).
3. A sling psychrometer for measuring wet- and dry-bulb temperatures to identify the conditions of the supply and return air.
4. A clamp-on ammeter with capability of reading amperes, voltages, and ohms.

The thermometers are used for measuring the following temperatures:

1. Supply air
2. Return air
3. Condenser entering air
4. Suction line at the coil outlet or at the compressor
5. Liquid line at the condenser outlet

By measuring the suction-line temperature, the superheat can be determined. Since the suction pressure has been measured, the evaporating temperature can be read from a pressure-temperature chart. The superheat is the difference between the suction-line temperature and the evaporating temperature.

By measuring the liquid-line temperature, the subcooling can be determined. Since the discharge pressure has been measured, the condensing temperature can be read from the pressure-temperature chart. The subcooling is the difference between the liquid-line temperature and the condensing temperature.

TR49-3 PROBLEM ANALYSIS

In this chapter the discussion is specifically directed to problems and solutions that apply to air-conditioning systems. For those solutions that apply to the refrigeration system, please refer to chapter TR48.

A typical troubleshooting job begins with a unit that was originally checked out and put into proper operating condition; has been running satisfactorily for a period of time; and has developed a problem. Problems in an air conditioning system are classified in two categories: *air-system problems* and *refrigeration-system problems.*

TR49-4 AIR-SYSTEM PROBLEMS

The primary problem that can occur in the air category is a reduction in quantity. Air-handling systems do not suddenly increase in capacity, that is, increase the amount of air across the coil. On the other hand, the refrigeration system does not suddenly increase in heat-transfer ability. The first check is therefore the temperature drop of the air through the DX coil. After measuring the return- and supply-air temperatures and subtracting to get the temperature drop, is it higher or lower than it should be?

This means that "what it should be" has to be determined first. This is done by using the sling psychrometer to measure and determine the return-air wet-bulb temperature and relative humidity. From this the proper temperature drop across the coil can be determined from the chart in Fig. TR49-1.

Using the required temperature drop as compared to the actual temperature drop, the problem can be classified as either an air problem or a refrigerant problem. If the actual temperature drop is greater than the required temperature drop, the air quantity has been reduced; look for problems in the air-handling system. These could be:

1. Air filters
2. Blower motor and drive
3. Unusual restrictions in the duct system

TR49-4.1 Air Filters

Air filters of the throwaway type should be replaced at least twice each year, at the beginning of both the cooling and heating seasons. In some areas where dust is high, they may have to be replaced as often as every 30 days. In

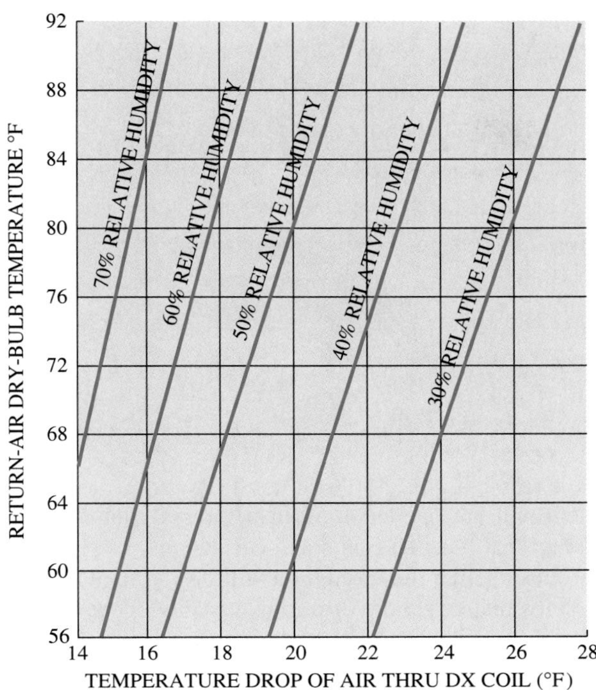

Figure TR49-1 Air-temperature drop for various return-air conditions.

commercial and industrial applications, a regular schedule of maintenance must be worked out for best performance and longest equipment life. Because this is the most common problem of air failure, check the filtering system first.

TR49-4.2 Blower Motor and Drive

Check the blower motor and drive in the case of belt-driven blowers to make sure that:

1. The blower motor is properly lubricated and operating freely.
2. The blower wheel is clean. The blades could be loaded with dust and dirt or other debris. If the wheel is dirty, it must be removed and cleaned. Do not try brushing only, because a poor cleaning job will cause an imbalance to occur in the wheel. Extreme vibration in the wheel and noise will result. This could cause deterioration of the wheel.
3. On belt-driven blowers, the blower bearing must be lubricated and operating freely.
4. The blower drive belt must be in good condition and properly adjusted. Cracked or heavily glazed belts must be replaced. Heavy glazing can be caused by too much tension on the belt, driving the

ON THE JOB

Problem: Not enough cooling, and conditions are hot and humid. You arrive at the job and find the condenser fan and compressor running. The furnace fan is running and the filter is clean enough. You take a temperature reading across the evaporator coil and it is 14°F. You checked the pressures and found the low side is 78 psig and the high side is 310. What do you do now?

Solution: The pressures are high. Check the condenser coil. The surface of the coil is slightly dirty but a closer look between the fins exposes a build up of lint inside of the coil. A good cleaning with a strong spray from a hose and some coil cleaner caused a drop in discharge pressure from 310 to 240. (It is still slightly high due to high ambient temperatures and the load inside.) The cleaning also increased the temperature drop across the coil from 14 to 17°F. Once the humidity inside the house drops the temperature drop will probably increase to 18 or 19°F.

.

ON THE JOB

Problem: Not enough cooling. The service person arrived and checked the operation. Temperatures were 85°F outside and 80°F inside. The suction pressure was 50 psig and the discharge 220 psig. Everything was running. The temperature drop across the evaporator was 26°F. The filter was just changed and the customer said it was dirty. Very little cool air was coming out of the registers. What do you do now?

Solution: After carefully exposing the evaporator he found there was substantial dirt covering the bottom of the evaporator coil. Sometimes the customers get embarrassed about the dirty fllters or no filters and put a clean one in just before you get there. If allowed to run this way the evaporator coil would eventually form a layer of ice on its surface and block the alr completely.

ON THE JOB

Problem: No cooling. You arrive on the job. The first thing you notice is ice on the suction line. There is ice on the compressor. You check the temperature on both sides of the evaporator coil. The temperature drop is only 3°F. You check the pressures and find 45 psig on the low side and 175 psig on the high side. You look in through a register on the plenum and you can see there is ice all over the evaporator coil. What do you do?

Solution: Don't jump to any conclusions such as adding freon. Check for causes related to airflow, such as dirty filters, loose belts, closed registers, or carpet over registers. Don't rule out anything unusual like paper in the blower wheel. Turn off the air conditioner and run the blower only to melt the ice on the coil. If there is no airflow problem, check for leaks in the system and then make repairs.

belt down into the pulleys. Proper adjustment requires the ability to depress the belt midway between the pulleys approximately 1 in. for each 12 in. between the pulley shaft centers.

TR49-4.3 Unusual Restrictions in Duct Systems

Placing furniture or carpeting over return-air grilles reduces the air available for the blower to handle. Shutting off the air to unused areas will reduce the air over the coil. Covering a return-air grille to reduce the noise from the centrally located furnace or air handler may reduce the objectionable noise, but it also drastically affects the operation of the system by reducing the air quantity.

Collapse of the return-air duct system will affect the entire duct system performance. Air leaks in the return duct will raise the return-air temperature and reduce the temperature drop across the coil.

TR49-5 REFRIGERATION-SYSTEM PROBLEMS

When the temperature drop across the coil is less than required, this means that the heat-handling capacity of the system has been reduced.

These problems can be simply divided into two categories: (1) *refrigerant quantity,* and (2) *refrigerant flow rate.* If the system has the correct amount of refrigerant charge and refrigerant is flowing at the desired rate, the system should work properly and deliver rated capacity. Any problems in either category will affect the temperatures and pressures that will occur in the unit when the correct amount of air is supplied over the DX coil for the capacity of the unit. Obviously, if the system is empty of refrigerant, a leak has occurred, and it must be found and repaired. The system must be evacuated thoroughly and recharged with the correct amount of refrigerant. If the system will not operate at all, it is probably an electrical problem that must be found and corrected.

In this chapter the discussion will be confined to those problems that affect the operating capacity of the system. The system will start and run but will not produce satisfactory results. This means that the amount of heat picked up in the coil plus the amount of motor heat added and the total rejected from the condenser is not the total heat quantity the unit is designed to handle. To determine the problem, all the information listed in Section TR49-2 must be measured. These results compared to normal operating results will generally identify the problem. The use of the word "normal" does not imply a fixed set of pressures and temperatures. These will vary with each make and model of the system. A few temperatures are fairly consistent throughout the industry and can be used for comparison:

1. DX-coil operating temperature
2. Condensing-unit condensing temperature
3. Refrigerant subcooling

These items must also be modified according to the energy efficiency ratio (EER) of the unit. The reason for this is that the amount of evaporation and condensing surface designed into the unit are the main factors in efficiency rating. A larger condensing surface results in a lower condensing temperature and a higher energy efficiency ratio (EER). A larger evaporating surface results in a higher suction pressure and a higher EER. The letters EER stand for Energy Efficiency Ratio, which is a measure of efficiency. It is calculated by dividing the net capacity of the unit in Btu/hr by the watts input.

TR49-5.1 DX-Coil Operating Temperature

Normal coil operating temperatures can be found by subtracting the design coil split from the average air temperature going through the coil. The coil split will vary with the system design.

Systems in the EER range of 7.0 to 8.0 will have design splits in the range 25 to 30°F. Systems in the EER range of 8.0 to 9.0 will have design splits in the range 20 to 25°. Systems with 9.0+ EER ratings will have design splits in the range 15 to 20°F. The formula used for determining coil operating temperatures is:

$$COT = \left(\frac{EAT + LAT}{2}\right) - \text{split}$$

where

COT = coil operating temperature
EAT = temperature of air entering the coil
LAT = temperature of air leaving the coil

The latter two temperatures added together and divided by 2 will give the average air temperature. This is also referred to as the *mean temperature difference* (MTD).

"Split" is the design split according to the EER rating. For example, a unit having an entering air condition of 80° DB and a 20°F temperature drop across the evaporator coil, will have an operating coil temperature determined as follows:

EXAMPLE

For an EER rating of 7.0 to 8.0:

$$COT = \frac{80 + 60}{2} - 25 \text{ to } 30° = 40 \text{ to } 45°\text{F}$$

For an EER rating of 8.0 to 9.0:

$$COT = \frac{80 + 60}{2} - 20 \text{ to } 25° = 45 \text{ to } 50°\text{F}$$

For an EER rating of 9.0+:

$$COT = \frac{80 + 60}{2} - 15 \text{ to } 20° = 50 \text{ to } 55°\text{F}$$

This demonstrates that the operating coil temperature changes with the EER rating of the unit.

TR49-5.2 Condensing Unit Condensing Temperature

The amount of surface in the condenser affects the condensing temperature the unit must develop to operate at rated capacity. The variation in the size of the condenser also affects the production cost and price of the unit. The smaller the condenser, the lower the price, but also the lower the efficiency (EER) rating. In the same EER ratings used for the DX coil, at 95°F outside ambient, the 7.0 to 8.0 EER category will operate in the 25 to 30° condenser split range, the 8.0 to 9.0 EER category in the 20 to 25° condenser split range, and the 9.0+ EER category in the 15 to 20° condenser split range.

This means that when the air entering the condenser is at 95°F, the formula for finding the condensing temperature would be:

$$RCT = EAT + \text{split}$$

where

RCT = refrigerant condensing temperature
EAT = temperature of the air entering the condenser
split = design temperature difference between the entering air temperature and the condensing temperatures of the hot high-pressure vapor from the compressor.

EXAMPLE

Using the formula with 95°F *EAT,* the split for the various EER systems would be:

For an EER rating of 7.0 to 8.0

$$RCT = 95° + 25 \text{ to } 30° = 120 \text{ to } 125°\text{F}$$

For an EER rating of 8.0 to 9.0

$$RCT = 95° + 20 \text{ to } 25° = 115 \text{ to } 120°\text{F}$$

For an EER rating of 9.0+

$$RCT = 95° + 15 \text{ to } 20° = 110 \text{ to } 115°\text{F}$$

This demonstrates that operating head pressures vary not only from changes in outdoor temperatures but with the different EER ratings.

TR49-5.3 Refrigerant Subcooling

The amount of subcooling produced in the condenser is determined primarily by the quantity of refrigerant in the system. The temperature of the air entering the condenser and the load in the DX coil will have only a small effect on the amount of subcooling produced. The amount of refrigerant in the system has the predominant effect. Therefore, regardless of EER ratings, the unit should have, if properly charged, a liquid subcooled to 15 to 20°F. High outdoor temperatures will produce the lower subcooled liquid because of the reduced quantity of refrigerant in the liquid state in the system. More refrigerant will stay in the vapor state to produce the higher pressure and condensing temperatures needed to eject the required amount of heat.

TR49-6 ANALYZING PROBLEMS

Using the information obtained from the two pressure gauges, a minimum of five thermometers, the sling psychrometer, and a clamp-type ammeter, we can analyze the system problems by using the chart in Fig. TR49-2.

Probable Cause	Lowside (Suction) Pressure (psig)	D.X. Coil Superheat (°F)	Highside (Hotgas) Pressure (psig)	Condenser Liquid Subcooling (°F)	Cond. Unit Amperage Draw (Amps)
1 Insufficient or unbalanced load	Low	Low	Low	Normal	Low
2 Excessive load	High	High	High	Normal	High
3 Low ambient (cond. entering air °F)	Low	High	Low	Normal	Low
4 High ambient (cond. entering air °F)	High	High	High	Normal	High
5 Refrigerant undercharge	Low	High	Low	Low	Low
6 Refrigerant overcharge	High	Low	High	High	High
7 Liquid line restriction	Low	High	High	High	Low
8 Plugged capillary tube	Low	High	High	High	Low
9 Suction line restriction	Low	High	Low	Normal	Low
10 Hot gas line restriction	High	High	High	Normal	High
11 Inefficient compressor	High	High	Low	Low	Low

Figure TR49-2 Troubleshooting chart for refrigeration and air-conditioning systems, showing symptoms and probable causes.

The figure shows that there are 11 probable causes of trouble in an air-conditioning system. After each probable cause is the reaction that the cause would have on the refrigeration system low-side or suction pressure, the DX-coil superheat, the high-side or discharge pressure, the amount of subcooling of the liquid leaving the condenser, and the amperage draw of the condensing unit.

TR49-6.1 Insufficient or Unbalanced Load

Insufficient air over the DX coil would be indicated by a greater-than-desired temperature drop through the coil. An unbalanced load on the DX coil would also give the opposite indication; some of the circuits of the DX coil would be overloaded, while others would be lightly loaded. This would result in a mixture of air off the coil that would cause a reduced temperature drop of the air mixture. The lightly loaded sections of the DX coil would allow liquid refrigerant to leave the coil and enter the suction manifold and suction line.

In TXV valve systems, the liquid refrigerant passing the feeler bulb of the TXV valve would cause the valve to close down. This would reduce the operating temperature and capacity of the DX coil as well as lower the suction pressure. This reduction would be very pronounced. The DX-coil operating superheat would be very low, probably zero, because of the liquid leaving some of the sections of the DX coil.

High-side or discharge pressure would be low due to the reduced load on the compressor, reduced amount of refrigerant vapor pumped, and reduced heat load on the condenser. Condenser liquid subcooling would be on the high side of the normal range because of the reduction in refrigerant demand by the TXV valve. Condensing-unit amperage draw would be down due to the reduced load.

In systems using capillary tubes, the unbalanced load would produce a lower temperature drop of the air through the DX coil because the amount of refrigerant supplied by the capillary tubes would not be reduced, therefore, the system pressure (boiling point) would be approximately the same.

The DX-coil superheat would drop to zero with flood-out of the refrigerant into the suction line. Under extreme cases of unbalance, liquid return to the compressor could cause compressor damage. The reduction in heat gathered in the DX coil and the lowering of the refrigerant vapor to the compressor will lower the load on the compressor. The compressor discharge pressure (hot-gas pressure) will be reduced.

The flow rate of the refrigerant will be only slightly reduced because of the lower head pressure. The sub-cooling of the refrigerant will be in the normal range. The amperage draw of the condensing unit will be slightly lower because of the reduced load on the compressor and reduction in head pressure.

TR49-6.2 Excessive Load

In this case the opposite effect exists. The temperature drop of the air through the coil will be low, so the unit cannot cool the air as much as it should. Air is moving through the coil at too high a velocity. There is the possibility that the temperature of the air entering the coil is higher than the return air from the conditioned area. This could be from leaks in the return-air system drawing air from unconditioned areas.

The excessive load raises the suction pressure. The refrigerant is evaporating at a rate faster than the pumping rate of the compressor. The superheat developed in the coil will be as follows:

ON THE JOB

Problem: Not enough cooling. When you arrive you find everything working, including the compressor, the condenser fan, and the furnace fan. The filters are clean. You take a temperature reading on each side of the evaporator coil and hook up your gauges to the service ports. The temperature drop is 9°F and the pressures on your gauges read 82 psig on the suction side and 160 psig on the discharge side. What should you do next?

Solution: The indicators point to a faulty compressor. The valves inside the compressor are leaking back through the compressor and it cannot build up enough pressure to do the job. If allowed to run like this, eventually the compressor will overheat or burn out, which would make repairs even more costly. If the compressor is hermetic or sealed, the only choice would be replacement. If it is a semi-hermetic compressor, the valves can be replaced.

1. If the system uses a TXV valve, the superheat will be normal to slightly high. The valve will operate at a higher flow rate to attempt to maintain superheat settings.
2. If the system uses capillary tubes, the superheat will be high. The capillary tubes cannot feed enough increase in refrigerant quantity to keep the DX coil fully active.

The high-side or discharge pressure will be high. The compressor will pump more vapor because of the increase in suction pressure. The condenser must handle more heat and will develop a higher condensing temperature to eject the additional heat. A higher condensing temperature means higher high-side pressure. The quantity of liquid in the system has not changed, nor is the refrigerant flow restricted. The liquid subcooling will be in the normal range. The amperage draw of the unit will be high because of the additional load on the compressor.

TR49-6.3 Low Ambient (Condenser Entering Air) Temperature

In this case, the condenser heat-transfer rate is excessive, producing an excessively low discharge pressure. As a result, the suction pressure will be low because the amount of refrigerant through the pressure-reducing device will be reduced. This reduction will reduce the amount of liquid refrigerant supplied to the DX coil. The coil will produce less vapor and the suction pressure drops.

The decrease in the flow rate into the coil reduces the amount of active coil, and a higher superheat results. In addition, the reduced system capacity will decrease the amount of heat removed from the air. There will be higher temperature and relative humidity in the conditioned area and the high-side pressure will be low. This starts a reduction in system capacity. The amount of subcooling of the liquid will be in the normal range. The quantity of liquid in the condenser will be higher, but the heat transfer rate of the lower temperatures is less. This will result in a subcooling in the normal range. The amperage draw of the condensing unit will be less. The compressor is doing less work.

The amount of drop in the condenser ambient air temperature that the air-conditioning system will tolerate depends on the type of pressure-reducing device in the system. Systems using capillary tubes will have a gradual reduction in capacity as the outside ambient drops from 95°F. This gradual reduction occurs down to 65°F. Below this temperature the capacity loss is drastic, and some means of maintaining head pressure must be employed. The most reliable means is control of air through the condenser via dampers in the airstream or a variable-speed condenser fan.

Systems that use TXV valves will maintain higher capacity down to an ambient temperature of 35°F. Below this temperature, controls must be used. The control of cfm through the condenser using dampers or the condenser-fan speed control can also be used. In larger TXV-valve systems, liquid quantity in the condenser is used to control head pressure.

TR49-6.4 High Ambient (Condenser Entering Air) Temperature

The higher the temperature of the air entering the condenser, the higher the condensing temperature of the refrigerant vapor to eject the heat in the vapor. The higher the condensing temperature, the higher the head pressure. The suction pressure will be high for two reasons: 1) the pumping efficiency of the compressor will be less; and 2) the higher temperature of the liquid will increase the amount of flash gas in the coil, further reducing the system efficiency.

The amount of superheat produced in the coil will be different in a TXV-valve system and a capillary-tube system. In the TXV-valve system the valve will maintain superheat close to the limits of its adjustment range even though the actual temperatures involved will be higher. In a capillary-tube system, the amount of superheat produced in the coil is the reverse of the temperature of the air through the condenser. The flow rate through the capillary tubes is directly affected by the head pressure. The higher the air temperature, the higher the head pressure and the higher the flow rate. As a result of the higher flow rate, the subcooling is lower.

Fig. TR49-3 shows the superheat that will be developed in a properly charged air-conditioning system using capillary tubes. Do not attempt to charge a capillary system below 65°F, as system operating characteristics become very erratic.

The head pressure will be high at the higher ambient temperatures because of the higher condensing temperatures required. The condenser liquid subcooling will be in the lower portion of the normal range. The amount of liquid refrigerant in the condenser will be reduced slightly because more will stay in the vapor state to produce the higher pressure and condensing temperature. The amperage draw of the condensing unit will be high.

TR49-6.5 Refrigerant Undercharge

A shortage of refrigerant in the system means less liquid refrigerant in the DX coil to pick up heat, and lower suction pressure. The smaller quantity of liquid supplied the DX coil means less active surface in the coil for vaporizing the liquid refrigerant, and more surface to raise vapor temperature. The superheat will be high. There will be less vapor for the compressor to handle and less head for the condenser to reject, lower high-side pressure, and lower condensing temperature.

The amount of subcooling will be below normal to none, depending on the amount of undercharge. The system operation is usually not affected very seriously until the subcooling is zero and hot gas starts to leave the condenser, together with the liquid refrigerant. The amperage draw of the condensing unit will be slightly less than normal.

Outdoor Air Temperature Entering Condenser Coil (°F)	Superheat (°F)
65	30
75	25
80	20
85	18
90	15
95	10
105 & above	5

Figure TR49-3 The effects of outdoor (ambient) temperature on superheat.

TR49-6.6 Refrigerant Overcharge

An overcharge of refrigerant will affect the system in different ways, depending on the pressure-reducing device used in the system and the amount of overcharge.

TXV-Valve Systems

In systems using a TXV valve, the valve will attempt to control the refrigerant flow into the coil to maintain the superheat setting of the valve. However, the extra refrigerant will back up into the condenser, occupying some of the heat transfer area that would otherwise be available for condensing. As a result, the discharge pressure will be slightly higher than normal, the liquid subcooling will be high, and the unit amperage draw will be high. The suction pressure and DX coil superheat will be normal. Excessive overcharging will cause even higher head pressure, and hunting of the TXV valve.

For TXV-valve systems with excessive overcharge:

1. The suction pressure will be high. Not only does the reduction in compressor capacity (due to higher head pressure) raise the suction pressure, but the higher pressure will cause the TXV valve to overfeed on its opening stroke. This will cause a wider range of "hunt" of the valve.
2. The DX coil superheat will be very erratic from the low normal range to liquid out of the coil.
3. The high-side or discharge pressure will be extremely high.
4. Subcooling of the liquid will also be high because of the excessive liquid in the condenser.
5. The condensing unit amperage draw will be higher because of the extreme load on the compressor motor.

Capillary-Tube Systems

The amount of refrigerant in the capillary tube system has a direct effect on system performance. An overcharge has a greater effect than an undercharge, but both affect system performance, efficiency (EER) and operating cost.

Figs. TR49-4 through TR49-6 show how the performance of an air-conditioning system is affected by an incorrect amount of refrigerant charge.

Shown in Fig. TR49-4, at 100% of correct charge (55 oz), the unit developed a net capacity of 26,200 Btu/hr. When the amount of charge was varied 5% in either direction, the capacity dropped as the charge varied. Removing 5% (3 oz) of refrigerant reduced the net capacity to 25,000 Btu/hr. Another 5% (2.5 oz) reduced the capacity to 22,000 Btu/hr. From there on the reduction in capacity became very drastic: 85% (8 oz), 18,000 Btu/hr; 80% (11 oz), 13,000 Btu/hr; and 75% (14 oz), 8000 Btu/hr.

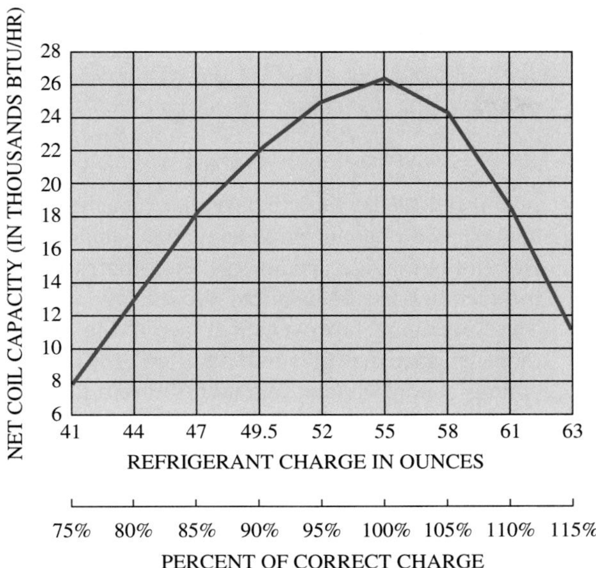

Figure TR49-4 The effect of the refrigerant charge on the capacity of the unit.

Addition of overcharge had the same effect but at a greater reduction rate. The addition of 3 oz of refrigerant (5%) reduced the next capacity to 24,600 Btu/hr; 6 oz added (10%) reduced the capacity to 19,000 Btu/hr; and 8 oz added (15%) dropped the capacity to 11,000 Btu/hr. This shows that overcharging of a unit has a greater effect per ounce of refrigerant than does undercharging.

Fig. TR49-5 is a chart showing the amount of electrical energy the unit will demand because of pressure created

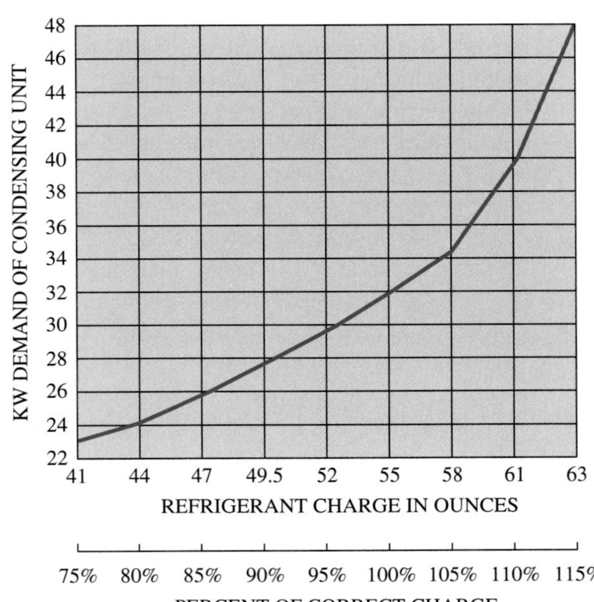

Figure TR49-5 The effect of the refrigerant charge on the kW demand of the condensing unit.

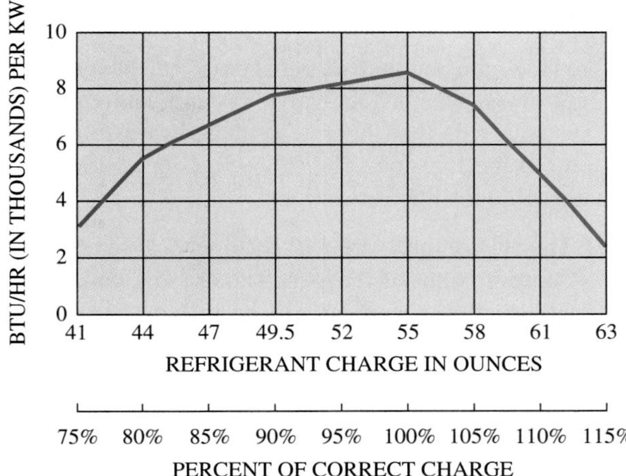

Figure TR49-6 The effect of the refrigerant charge on the Btu/hr/kW ratio.

by the amount of refrigerant in the system, with the only variable being the refrigerant charge. At 100% of charge (55 oz) the unit required 32 kW. As the charge was reduced, the wattage demand also dropped, 29.6 kW at 95% (3 oz), 27.6 kW at 90% (6.5 oz), 25.7 kW at 85% (8 oz), 25 kW at 80% (11 oz), and 22.4 kW at 75% (14 oz short of correct charge). When the unit was overcharged, the wattage required went up. At 3 oz (5% overcharge) the wattage required was 34.2 kW; at 6 oz (10% overcharge), 39.5 kW; and at 8 oz (15% overcharge), 48 kW.

Fig. TR49-6 shows the efficiency of the unit (EER rating) based on the Btu/hr capacity of the system versus the wattage demand of the condensing unit. At correct charge (55 oz) the efficiency (EER rating) of the unit was 8.49. As the refrigerant was reduced, the EER rating dropped to 8.22 at 95% of charge, 7.97 at 90%, 7.03 at 85%, 5.2 at 80%, and 3.57 at 75% of full refrigerant charge. When refrigerant was added, adding 5% (3 oz) the EER rating dropped to 7.19. At 10% (6 oz) the EER was 4.8, and at 15% overcharge (8 oz) the EER was 2.29. From these charts the only conclusion is that the capillary-tube systems must be charged to the correct charge with only a –5% tolerance.

The effect of overcharge produces a high suction pressure because the refrigerant flow to the DX coil increases. Suction superheat will decrease because of the additional quantity to the DX coil. At approximately 8 to 10% of overcharge, the suction superheat becomes zero and liquid refrigerant will leave the DX coil. This will cause flooding of the compressor and greatly increases the chance of compressor failure. The high-side or discharge pressure will be high because of the extra refrigerant in the condenser. Liquid subcooling will also be high for the same reason. The wattage draw will increase due to the greater amount of vapor pumped as well as the higher compressor discharge pressure.

TR49-6.7 Liquid-Line Restriction

This reduces the amount of refrigerant to the pressure-reducing device. Both TXV-valve systems and capillary-tube systems will then operate with reduced refrigerant flow rate to the DX coil. The following observations can be made of liquid-line restrictions:

1. The suction pressure will be low because of the reduced amount of refrigerant to the DX coil.
2. The suction superheat will be high because of the reduced active portion of the coil, allowing more coil surface for increasing the vapor temperature as well as reducing the refrigerant boiling point.
3. The high-side or discharge pressure will be low because of the reduced load on the compressor.
4. Liquid subcooling will be high. The liquid refrigerant will accumulate in the condenser. It cannot flow out at the proper rate because of the restriction. As a result, the liquid will cool more than desired.
5. The amperage draw of the condensing unit will be low.

TR49-6.8 Plugged Capillary Tube or Feeder Tube

Either a plugged capillary tube or plugged feeder tube between the TXV-valve distributor and the coil will cause part of the coil to be inactive. The system will then be operating with an undersized coil, resulting in the following:

1. The suction pressure will be low because the coil capacity has been reduced.
2. The suction superheat will be high in the capillary tube systems. The reduced amount of vapor produced in the coil and resultant reduction in suction pressure will reduce compressor capacity, head pressure, and the flow rate of the remaining active capillary tubes.
3. The high-side or discharge pressure will be low.
4. Liquid subcooling will be high; the liquid refrigerant will accumulate in the condenser.
5. The unit amperage draw will be low.

In TXV-valve systems, the following will result:

1. A plugged feeder tube reduces the capacity of the coil. The coil cannot provide enough vapor to satisfy the pumping capacity of the compressor and the suction pressure balances out at a low pressure.
2. The superheat, however, will be in the normal range because the valve will adjust to the lower operating conditions and maintain the setting superheat range.
3. The high-side or discharge pressure will be low because of the reduced load on the compressor and condenser.
4. Liquid subcooling will be high because of the liquid refrigerant accumulating in the condenser.
5. The amperage draw of the condensing unit will be low.

TR49-6.9 Suction-Line Restriction

This could be caused by a plugged suction line strainer, a kink in the suction line, or a solder joint fitted with solder. It results in a high pressure drop between the DX coil and the compressor.

1. The suction pressure, if measured at the condensing unit end of the suction line, will be low.
2. The superheat, as measured by suction-line temperature at the DX coil and suction pressure (boiling point) at the condensing unit, will be extremely high.
3. The high-side or discharge pressure will be low because of reduced load on the compressor.
4. The low suction and discharge pressure usually indicate a refrigerant shortage. Warning: The liquid

ON THE JOB

Problem: Not enough cooling (insufficient cooling does not always lead to the air conditioner). You arrive on the job and find everything running. The condenser fan and compressor are running. The temperature drop across the evaporator coil is 24°F. The pressures are 60 psig on the suction side and 220 psig on the discharge side. The suction temperature is 39°F. There is a humidifier hooked to the plenum with a bypass to the return air. The humidifier is turned off. It appears to be an airflow problem. Where do you go from here?

Solution: First check for obvious problems: air filter, loose belt, or lack of return air. All that is left is the humidifier. You see there is no damper in the humidifier bypass line to close it off during the air conditioner cycle. Install a damper to keep cool, low-temperature, supply air in the plenum from short circuiting into the return air.

subcooling is normal to slightly above normal. This indicates a surplus of refrigerant in the condenser. Most of the refrigerant is in the coil, where the evaporation rate is low due to the higher operating pressure in the coil.

5. The amperage draw of the condensing unit would be low because of the light load on the compressor.

TR49-6.10 Hot Gas-Line Restriction

The high-side or compressor discharge pressure will be high if measured at the compressor outlet or low if measured at the condenser outlet or liquid line. In either case the compressor amperage draw will be high.

Therefore:

1. The suction pressure is high due to reduced pumping capacity of the compressor.
2. The DX-coil superheat is high because the suction pressure is high.
3. The high-side pressure is high when measured at the compressor discharge or low when measured at the liquid line.
4. Liquid subcooling is in the high end of normal range.
5. Even with all of this, the compressor amperage draw is above normal. All symptoms point to an extreme restriction in the hot-gas line. This problem is easily found when the discharge pressure is measured at the compressor discharge.

Where the measuring point is the liquid line at the condenser outlet, the facts are easily misinterpreted. High suction pressure and low discharge pressure will usually be interpreted as an inefficient compressor. The amperage draw of the compressor must be measured. The high amperage draw indicates that the compressor is operating against a high discharge pressure. A restriction apparently exists between the outlet of the compressor and the pressure measuring point.

TR49-6.11 Inefficient Compressor

This problem is last on the list because it is the least likely to be a problem. When the compressor will not pump the required amount of refrigerant vapor:

1. The suction pressure will balance out higher than normal.
2. The DX-coil superheat will be high.
3. The high-side or discharge pressure will be extremely low.
4. Liquid subcooling will be low because not much heat will be in the condenser. The condensing temperature will therefore be close to the entering air temperature.
5. The amperage draw of the condensing unit will be extremely low, indicating that the compressor is doing very little work.

REVIEW

- When troubleshooting, the technician should follow the following personal safety measures:
  Wear safety glasses and gloves.
  Recover, recycle, or reclaim refrigerants using approved devices.
  Shut off all power when working on electrical equipment.
  If the work must be done with power applied, remove all watches and rings.
  Always read manufacturer's safety recommendations.
- In troubleshooing air-conditioning equipment, the minimum set of test instruments should include:
  A gauge manifold
  At least five thermometers
  A sling psychrometer
  A clamp-on ammeter
- For analyzing the problem, the following temperature readings should be taken:
  Supply air
  Return air
  Ambient air
  Suction line
  Liquid line
- From these temperature readings and the pressure data obtained using the gauge manifold, the actual performance of the equipment can be compared to normal operation.
- In testing an operating system to determine the cause of the problem, the following actual readings should be compared with typical (normal) readings:
  Suction pressure
  Evaporator superheat
  Discharge pressure
  Condenser subcooling
  Amperage draw
- Air system problems:
  Air filters
  Blower motor drive
  Unusual restrictions in the duct system
- Refrigeration temperature checks:
  DX-coil operating temperatures
  Condensing-unit condensing temperature
  Refrigerant subcooling
- Symptoms of refrigeration problems
  Insufficient or unbalanced load
  Excessive load
  Low ambient temperature
  High ambient temperature
  Refrigerant undercharge

Refrigerant overcharge
Liquid-line restriction
Plugged capillary tube
Suction-line restriction
Hot-gas line restriction
Inefficient compressor
  Higher than normal suction pressure
  Low discharge pressure
  High DX-coil superheat
  Low liquid subcooling
  Low compressor amps

From this comparison of operational data, reference to a troubleshooting chart, such as shown in Fig. TR49-2, will assist in locating the cause of the problem.

As soon as the cause is verified, in most cases the remedy is self evident. For example, if a wiring connection is loose, it needs to be tightened. When a motor is burned out, it needs to be replaced. Where replacement or service of specialized parts is required, usually the manufacturer provides detailed instructions for performing the work. When maintenance is required, such as cleaning a dirty coil or replacing worn out belts, helpful instructions will be found in the section of this manual on maintenance.

Problems and Questions

1. What instruments are needed to properly diagnose problems in an air-conditioning system?
2. At what locations are air temperature measurements required?
3. If the air temperature drop through the coil is higher than when the unit was previously left, is the problem in the refrigeration or the air part of the system?
4. Problems in the refrigeration portion of the system can be divided into two categories. What are they?
5. Define "EER."
6. EER is found by dividing the _____ by the _____.
7. How many probable causes of trouble are there in an air-conditioning system?
8. If the Btu/hr load is increased on the evaporator, will the superheat of the evaporator increase or decrease on a TXV valve coil and a capillary tube coil?
9. What effect does low outside ambient temperature have on the capacity of the air-conditioning unit? Why?
10. What is the minimum outside operating temperature of a capillary-tube system and a TXV-valve system?
11. The minimum outside temperature at which a capillary tube system can be properly charged is _____ .
12. What is the easiest way to determine if a unit does not have the proper amount of refrigerant charge?
13. Which has the greatest adverse effect on the capacity of a system, an overcharge or an undercharge of refrigerant?
14. The charge quantity tolerance of a capillary tube system is _____ .
15. To properly diagnose trouble in the refrigeration system, five operating characteristics must be known. What are they?
16. A sling psychrometer is used during troubleshooting for measuring the dry-bulb temperature and relative humidity of the supply and return air. True or false?
17. By measuring the suction-line temperature, the superheat can be determined. True or false?
18. Problems in an air-conditioning system are classified in two categories: air-circuit problems and electrical problems. True or false?
19. Dirty air filters create reduced air flow, which creates increased suction pressure. True or false?
20. Heavy glazing on belts can be caused by too much tension on the belt. True or false?
21. The formula used for determining DX-coil operating temperatures is:
    a. $COT = EAT + RT -$ split
    b. $COT = EAT + LAT -$ split
    c. EER $= RAT + AT -$ split
    d. None of the above.
22. The amount of surface in the condenser affects:
    a. The condensing temperature the unit must develop to operate at rated capacity.
    b. The price.
    c. The EER.
    d. All of the above.
23. The unit should have, regardless of EER ratings, a subcooling temperature of:
    a. 5 to 10 °F.
    b. 15 to 20 °F.
    c. 20 to 30 °F.
    d. None of the above.
24. During excessive loads, when using the capillary tube, the superheat will be:
    a. Higher than normal.
    b. Normal operating temperatures.
    c. Lower than normal.
    d. None of the above.
25. The inefficient compressor will have:
    a. A higher than normal suction pressure.
    b. High DX coil superheat and low compressor amps.
    c. A lower than normal discharge pressure.
    d. All of the above.

Trouble-shooting Heating Systems

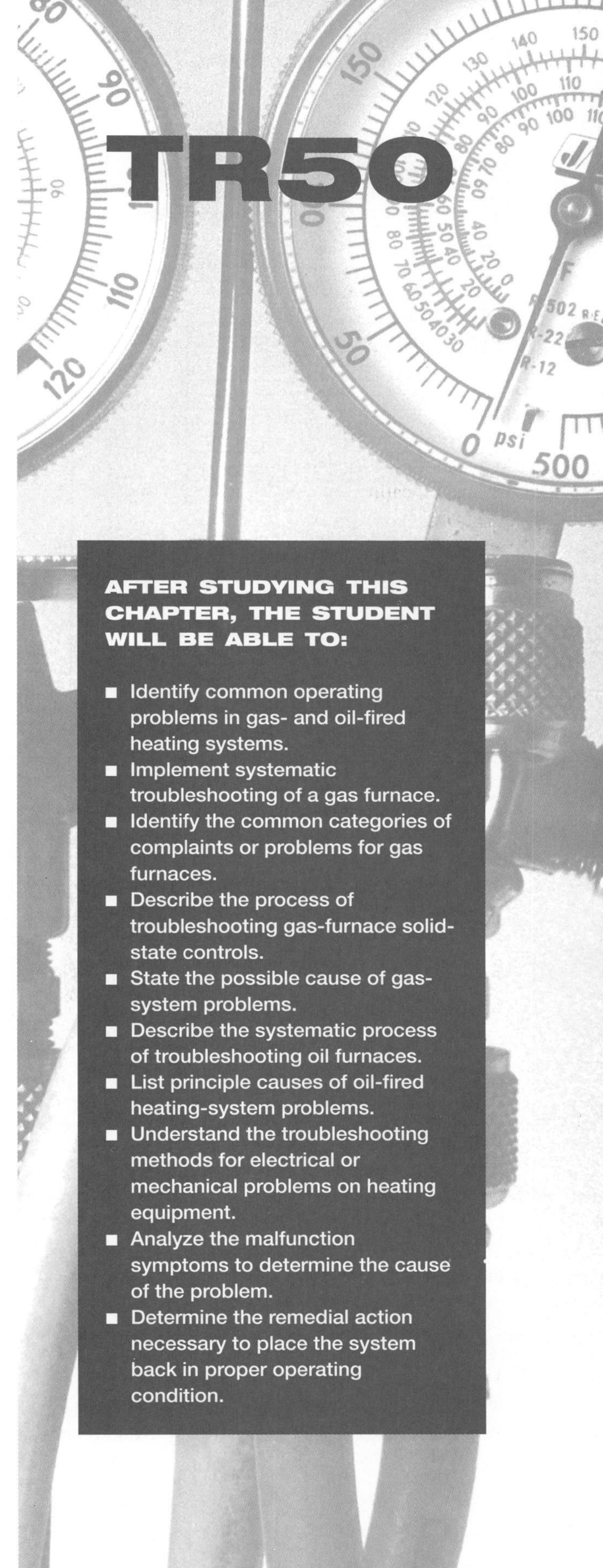

AFTER STUDYING THIS CHAPTER, THE STUDENT WILL BE ABLE TO:

- Identify common operating problems in gas- and oil-fired heating systems.
- Implement systematic troubleshooting of a gas furnace.
- Identify the common categories of complaints or problems for gas furnaces.
- Describe the process of troubleshooting gas-furnace solid-state controls.
- State the possible cause of gas-system problems.
- Describe the systematic process of troubleshooting oil furnaces.
- List principle causes of oil-fired heating-system problems.
- Understand the troubleshooting methods for electrical or mechanical problems on heating equipment.
- Analyze the malfunction symptoms to determine the cause of the problem.
- Determine the remedial action necessary to place the system back in proper operating condition.

TR50-1 OVERVIEW

Any heating service organization is aware of the accumulation of service calls during the fall of the year. When homeowners first try to start their systems, the problems reveal themselves. The most serious of these is "no heat." Whenever a no-heat call is received, it receives top priority. It means that for some reason the furnace will not run. The technician must solve the problem as quickly as possible.

Preliminary information is essential. What type of furnace is being used? What type of fuel is being supplied? Has the unit operated properly in the past or is it a new installation? What has the customer done to get the unit started? Does any part of the unit run or is it completely "dead"?

TR50-2 A SYSTEMATIC ANALYSIS

TR50-2.1 78% AFUE Gas Furnaces

The first thing to check is always the power supply. Is proper power being supplied to the unit? A voltmeter is a handy tool to use in checking the power supply. The second item to check is the thermostat. Is the thermostat calling for heat? If both of these conditions are satisfactory, the wiring diagram should be referenced.

Fig. TR50-1 shows a typical wiring diagram for a down-flow gas furnace that meets the 78% AFUE requirement. The top portion of the diagram shows the

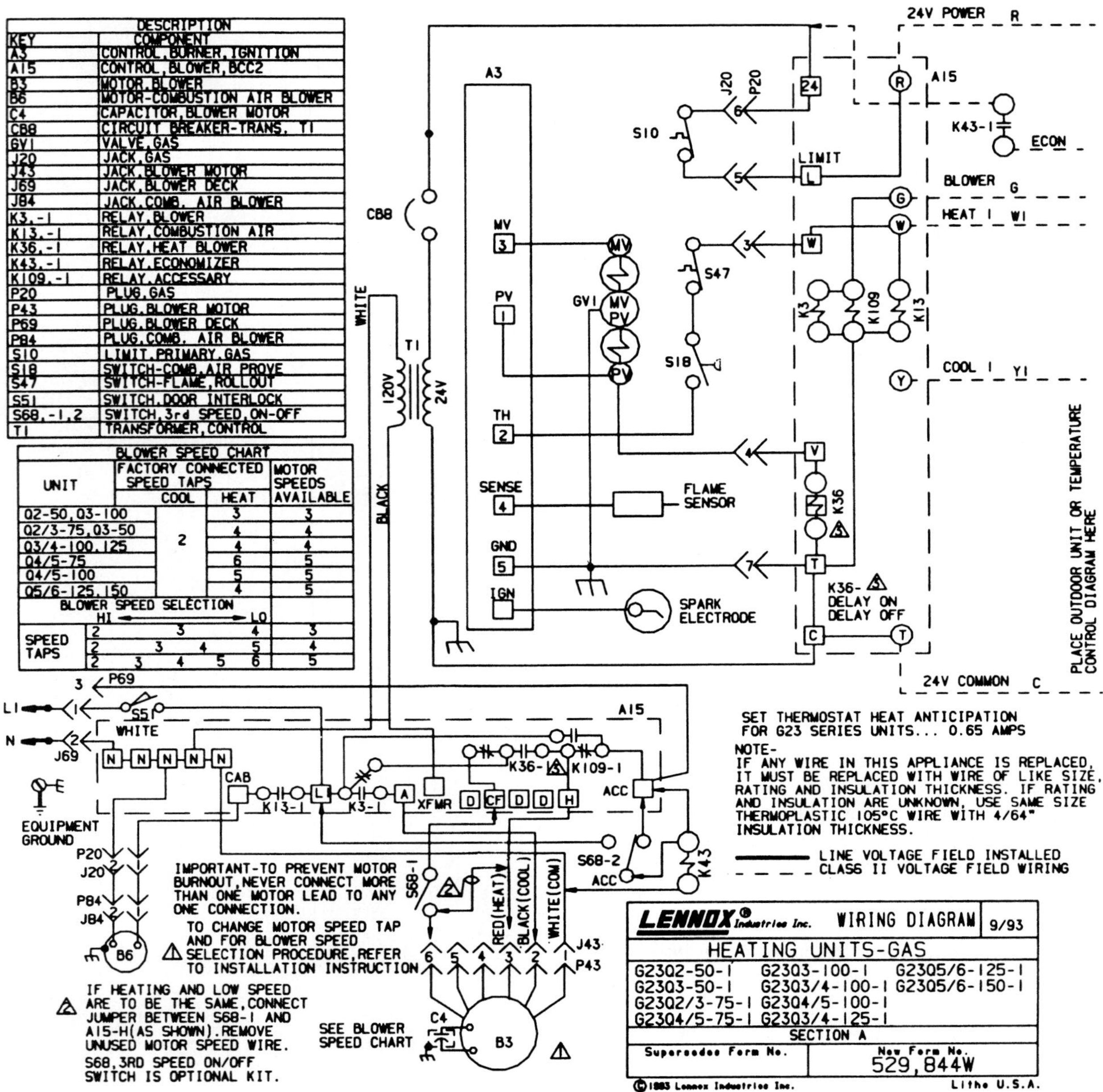

Figure TR50-1 Wiring diagram for a down-flow gas furnace. (Courtesy of Lennox Industries, Inc.)

ON THE JOB

Problem: You have arrived on the job to find that the high-efficiency condensing furnace was not heating. Investigation reveals that the pressure switch has opened the control circuit. Why? Some standing water was found in the vestibule and in the blower compartment. There was no water coming from the drain line. The drain was hooked up with clear plastic soft hose. What do you do next?

Solution: Remove the drain line from the furnace, and make sure it is clear. Water may pour out and all over the floor. Be prepared to have a bucket ready to catch water and remember that this is an acidic solution. Use safety precautions to protect yourself and the customer's property. Remove any traps and make sure that everything in the drain system is clean and free to drain.

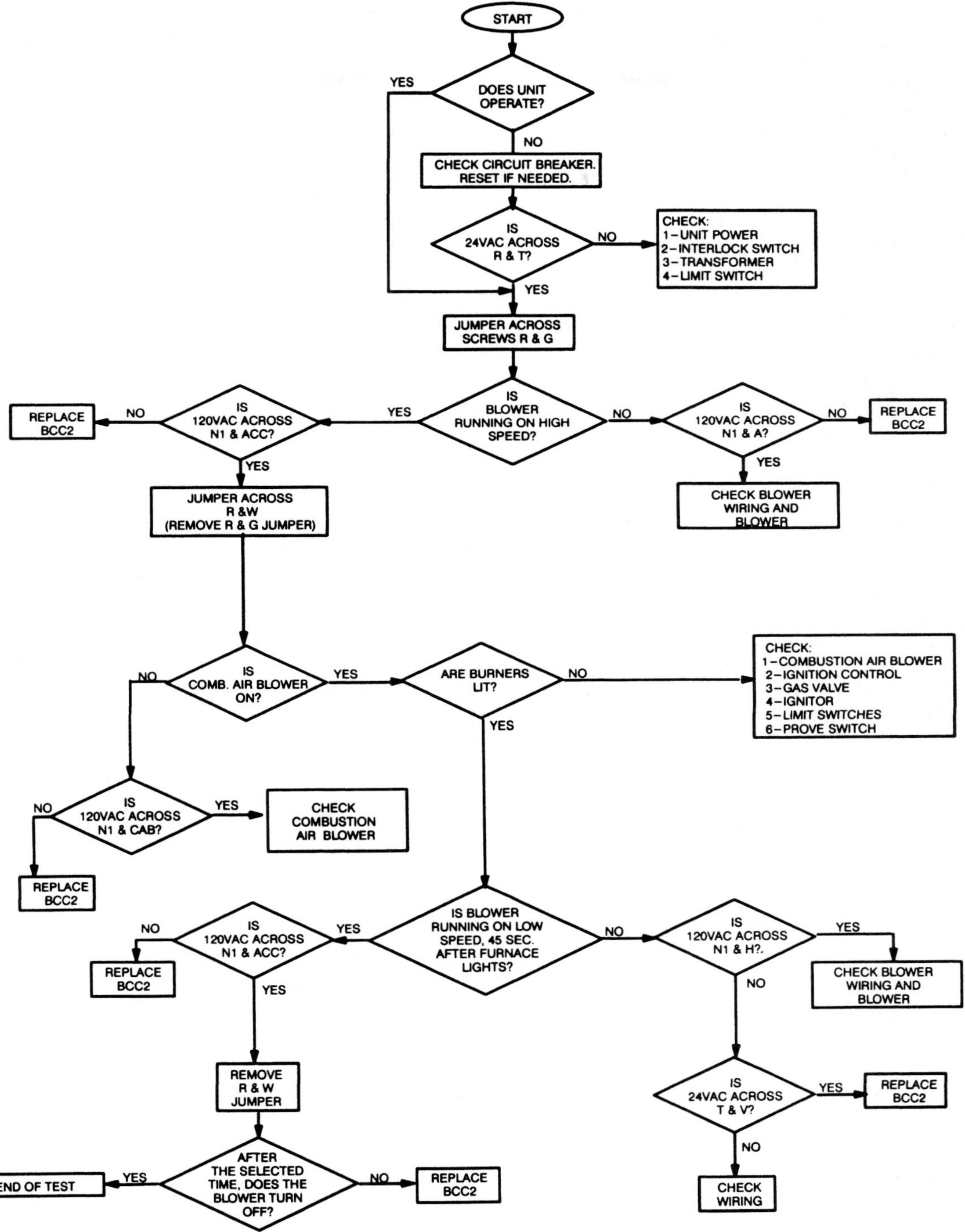

Figure TR50-2 Troubleshooting flowchart. (Courtesy of Lennox Industries, Inc.)

low-voltage wiring. The thermostat terminal connections to the furnace are shown on the top right. Toward the left in the diagram are the blower controls. Further to the left are the burner and ignition controls. At the extreme left is the control-voltage transformer and the legend. In the lower part of the diagram are the line-voltage connections to the circulating air blower and the combustion air blower.

Many technicians find it helpful to use a troubleshooting flowchart such as the one shown in Fig. TR50-2. The advantage of this procedure is the systematic process that it provides to help locate the problem.

TR50-2.2 Gas Furnaces with Solid-State Controls

Many of the newer type furnaces, particularly the high-efficiency furnaces, use microprocessor controls and circuit boards (Fig. TR50-3). The diagram on the left shows the component connections to the various parts of the printed circuit board. The diagram on the right shows the line-voltage wiring at the top and the low-voltage wiring, including the central processing unit (CPU), at the bottom.

One helpful feature of the solid-state control system is its ability to diagnose its own service problems. For example, in the system just described, a component test is available which allows all components, except the gas valve, to run for a short period of time to reveal any service problems or indicate a component failure.

As a preparation for troubleshooting, it is important for the technician to be completely familiar with the operating sequence of the furnace. This makes it possible to see what parts are operating, those that should be, and those not operating. The following is the sequence of operations for the heating mode of the furnace described in the wiring diagram (Fig. TR50-3):

1. When the thermostat calls for heat, the R-W circuit closes. The furnace control system performs a self-check, verifies that the pressure switch contacts are open, and starts the inducer motor.
2. Pre-purge period. As the inducer motor comes up to speed, the pressure-switch contacts close to begin a 15-sec prepurge period.
3. Ignitor warm-up. At the end of the pre-purge period, the ignitor is energized for a 17-sec ignitor warm-up period.
4. Ignition sequence. When the ignitor warm-up period is completed, the gas valve opens, permitting gas flow to the burners where it is ignited. After 5 sec, the ignitor is de-energized and a 2-sec flame-sensing period begins.
5. The HUM (humidifier) terminal is energized with the gas valve.
6. Flame sensing. When the burner flame is sensed, the control begins the blower-on delay period and continues to hold the gas valve open. If the burner flame is not sensed, the control closes the gas valve and repeats the ignition cycle.

   Note: Ignition process will repeat three additional times before a lock-out occurs. Lock-out automatically resets after 3 hrs, or can be manually reset by turning the 115-V power off for 3 sec minimum, then on again.
7. Blower-on delay. Approximately 40 sec after the burner flame is proven, the blower motor is energized on the heating speed. Simultaneously the humidifier and electronic air-filter terminals are energized. The jumper is on pin 6 and pin 8; the blower-on delay period is 60 sec.
8. Blower-off delay. When the thermostat is satisfied, the circuit between R and W is broken, de-energizing the gas valve and stopping gas flow to the burners. The blower motor, and EAC remain energized 90, 135, 180, or 225 sec, depending on the blower-off time selection. The furnace is factory set for 135 sec blower-off delay period.
9. Post purge. The inducer motor remains energized 5 sec after the burners are extinguished. If the jumper is on pins 6 and 9, the post purge period is 15 sec.

TR50-2.3 Standard Furnaces

Since many thousands of standard furnaces are in residential use, manufactured before the 78% AFUE minimum requirements, the service technician must be prepared to troubleshoot these units. Some important differences from the newer designs are natural draft venting and standing gas pilots. The troubleshooting information on standard furnaces is presented under the following topics:

1. General procedures
2. Gas-system problems
3. Oil-system problems

ON THE JOB

Problem: Moisture buildup on windows. You checked the furnace heat exchanger for cracks. Because of your experience you know there is moisture in flue gases and cracks in the heat exchanger emit moisture-laden flue gas into the conditioned space. There were no cracks visible. What else could you check? There was a humidifier. It was not running.

Solution: In some situations, excessive plants in a small house can cause moisture build up. Remodeling and weather sealing the house with a vapor barrier and caulking joints can sometimes make a house too airtight, trapping moisture. Another option would be a fault in the humidifier. It is possible that the control is out of calibration or water is leaking through the valve, causing moisture to be added every time the furnace runs.

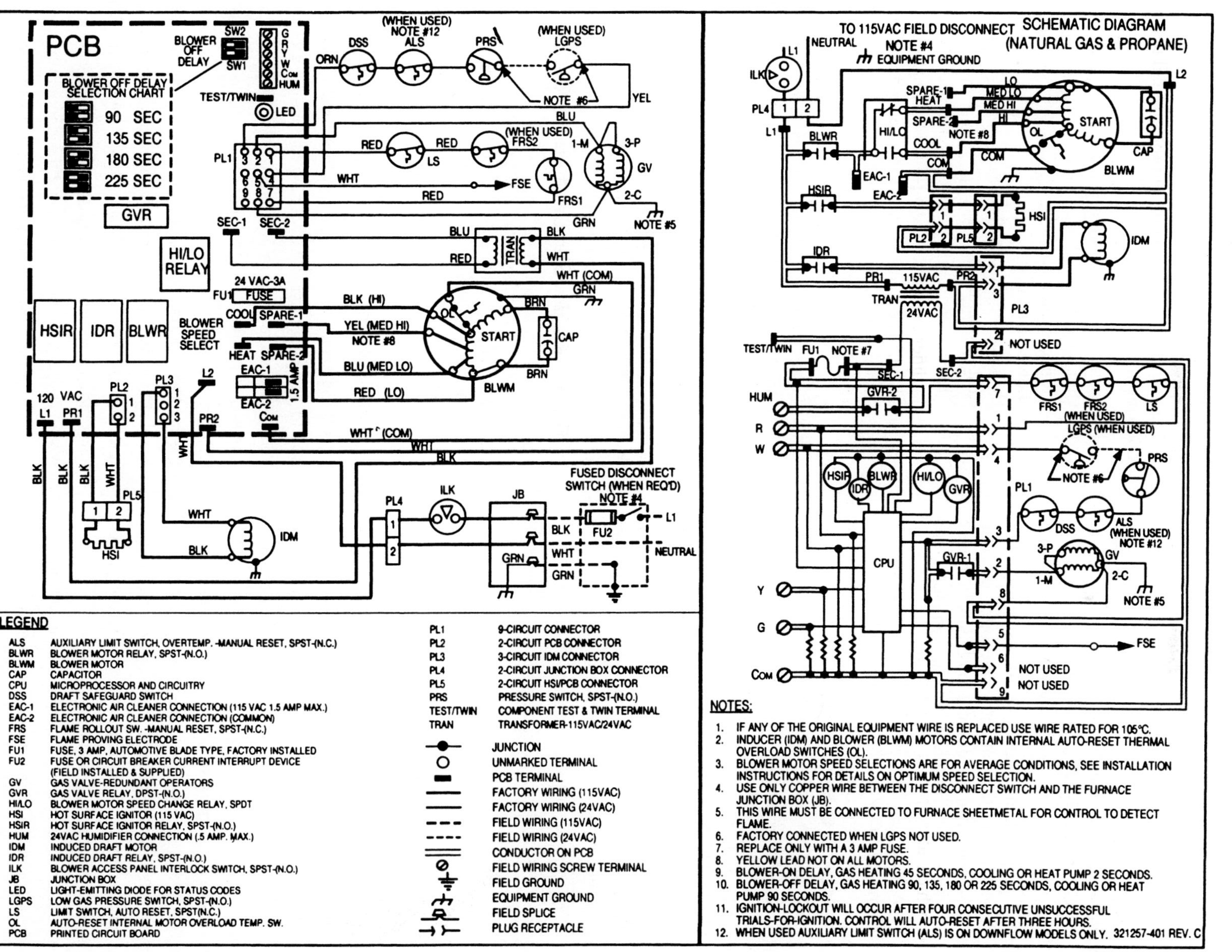

Figure TR50-3 Wiring diagram of a gas furnace using microprocessor controls.

Many of the procedures covered in this section also apply to the newer units. The technician will find that the procedures described here can also be used for troubleshooting all types of oil-fired equipment.

TR50-3 GENERAL PROCEDURES

A suggested progression of analysis for troubleshooting starts with identifying the common categories of complaints or problems, then the possible problems in system categories (gas or oil), followed by the symptoms and causes of specific problems. Causes can be either electrical or mechanical and related to the use of gas or oil equipment.

The categories of complaints or problems fall under the following headings:

1. Entire system operation
2. Unit operation
3. Burner operation
4. Blower operation
5. Heat-exchanger complaints
6. Cost of operation
7. Noise

ON THE JOB

Problem: No heat (oil). When you arrive and do a visual inspection, you find that the furnace primary control is locked out on the safety, but everything else looks good. There is a call for heat at the thermostat. Next you press the reset and the burner starts but there is no flame. The next step would be to determine whether or not you have oil. Check the tank. What does the gauge say? Where is the tank? The tank is outside and the oil filter is also outside, the temperature is 12°F. What is the next step?

Solution: The technician loosened the flare fitting on the oil filter and found no oil coming out even though there was oil in the tank. On occasion, moisture builds up in the oil tank, and on very cold days, it can freeze in the oil filter. This blocks the flow of oil. The tech replaces the oil filter where the blockage is and checks to make sure there is flow through it. He tries again and the unit fires up adequately.

ON THE JOB

Problem: No heat (oil). When you arrive you find the unit not running and in safety lock-out at the flame relay. You do a visual check for clues before you reset the flame relay. There are some black marks in the chamber. There are also some black streaks around the registers that look like dirt. When you reset the unit, the flame fires up, but you notice when the blower comes on, the flame changes to a very smokey yellow. What should you be looking for?

Solution: Whenever the blower interferes with the flame or the draft, causing poor flame, this usually means a crack in the heat exchanger. This problem usually requires replacement of the furnace or the heat exchanger when available.

It is important to find the crack, because it is expensive for the customer to buy a new furnace on a hunch. You may have to remove a register in the plenum or cut a hole in it to verify that the heat exchanger is cracked. Using a mirror and flashlight will help when inspecting the heat exchanger. In this case the crack was on the back side, where it is difficult to see without a mirror. Another clue to look for is discoloration by the wall registers. Check the indoor air quality for carbon monoxide. Inform the customer of the dangers of carbon monoxide. The customer should not stay there and the furnace should be shut off.

TR50-4 GAS-SYSTEM PROBLEMS

The possible causes for gas-system problems follow.

Problems	Possible Causes (See the following number references for further information)
Will not start	Season switch open
	Room thermostat set too low
	Disconnect switch open
	Blown fuse
	Limit control open
	Control transformer burned out (1)
	Open circuit in thermostat
	Pilot outage (2)
	Gas valve stuck open/closed (3)
	Safety pilot burned out
	No fuel
Runs, but short cycles	Improper heat anticipator setting (4)
	Cycles on limit control (5)
	Gas input too low
Room temperature high	Thermostat setting too high
	Improper heat anticipator setting
Runs continuously	Short in thermostat circuit
	Gas valve stuck open
Blower cycles after thermostat is satisfied	Blower cfm adjustment
	Fan control setting
Blows cool air at start	Blower cfm adjustment
	Fan control setting (6)
Start-up/cool-down noise	Expansion noise in heat exchanger (11)
	Duct expansion
	Oil-can effect (12)
Noise from vibration	Blower wheel unbalanced or out of line
	Pulleys unbalanced (10)
	Defective blower belt
	Bearings burned out
Odor	Burning rust on start
	Improper venting
	Cracked heat exchanger
High fuel cost	Improper input
	Improper burner adjustment (7)
	Improper unit sizing
High electrical cost	Improper motor load
	Defective motor
	Incorrect motor speed
No fuel	No line pressure
	Defective regulator
	Supply valve closed
Gas valve won't open	Transformer burned out
	Open circuit in thermostat
	Gas valve stuck open/closed
	Defective safety pilot
Gas valve short cycles	Improper heat anticipator setting
	Improper limit control setting
	Dirty air filters
	Restriction in air supply
Delayed ignition	Improper burner adjustment
	Improper input
	Delayed valve opening
	Low line pressure
Pilot outage	Pilot orifice burned out
	Thermocouple burned out
	Safety pilot burned out
	Low line pressure
	Drafts
Extinction pop	Improper air adjustment
	Improper burner adjustment
	Improper orifice alignment
	Poor valve cutoff
Burns inside burner	Improper air adjustment
	Improper input
	Extinction pop
	Leaking gas valve
Flame lift	Improper input
	Improper air adjustment
Flame roll out	Restriction in heat exchanger
	Improper venting
	Improper combustion air supply (8)
Yellow fire/carbon deposit	Improper air adjustment
	Improper input
Flashback	Improper input
	Improper air adjustment
	Improper venting (9)
	Restriction in heat exchanger
Intermittent blower operation	Improper cfm adjustment
	Improper fan control setting
Heat exchanger burn out	Improper input
	Chemical atmosphere
	Improper position of burners
Resonance (Pipe organ effect)	Resonance unit design

Many of the solutions to these problems are evident when the cause is determined. Below, however, is some further information regarding some of the remedies that can be used. Numbers refer to references above.

(1) Control transformer burned out. An ohmmeter should be used to determine whether the primary or the sec-

ondary of the transformer is burned out. The cause is probably an overload, which needs to be corrected before the transformer is replaced. If the overload cannot be reduced, a larger transformer must be used for the replacement.
(2) Pilot outage. One of the most common causes of pilot failure is improper impingement of the flame on the thermocouple. See Fig. TR50-4. This figure illustrates the proper size and location of the flame. Sometimes the pilot is extinguished by the gas burner during lighting. It is actually blown out by the burner flame. To correct this it is often necessary to reposition the pilot to a more favorable location.
(3) Gas valve stuck open or closed. A malfunctioning gas valve should be replaced.
(4) Improper heat-anticipator setting. The anticipator should be set at the amount of current traveling in the control circuit when the unit is operating. The current can be measured using a multiplier coil and clamp-on ammeter.
(5) Cycling on the limit control. Occasionally a limit control will weaken and lower the operating range of the control. The normal range is to cut off between 140°F for a counter-flow unit to 160 to 220°F on upright and horizontal units. If the control is cycling at a lower range, replace it.
(6) Fan control settings. Almost all gas-fired units operate best with fan control settings of 125 to 130°F fan-on and 100 to 105°F fan-off. If the unit blows cold air on startup, the fan-on temperature can be changed to 145 to 150°F.
(7) Improper burner adjustment. A properly adjusted burner will have approximately 40% of the combustion air mixing with 100% of the gas in the burner and 60% of the air mixing with the flame above the burner to complete the combustion process. Fig. TR50-5 shows a typical burner arrangement. As the gas is emitted from the orifice it expands and hits the proper place in the throat of the burner venturi. This produces maximum pull of primary air into the burner. Setting the burner for the correct flame condition will mean a minimum opening in the primary air control.

Fig. TR50-6 shows the size of the opening in an ordinary butterfly-type air control. When the burner and orifice are aligned properly and the burner is working correctly, a small opening in the primary air control will produce the soft blue fire with slightly yellow tips that gives best overall unit performance.

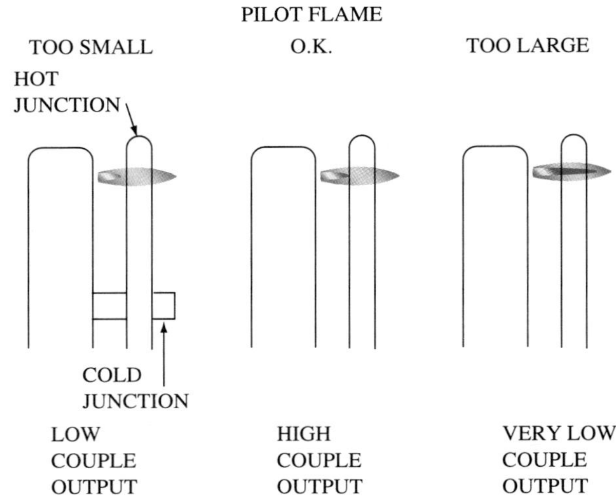

Figure TR50-4 Proper size for a pilot flame.

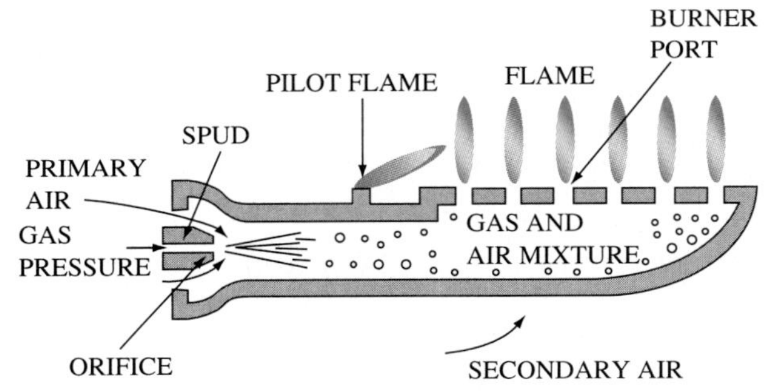

Figure TR50-5 Primary and secondary air supply to the gas burner.

ON THE JOB

Problem: No heat. You arrive and the customer complains about the smell. They shut off the power. When you check the furnace out you find that it starts satisfactorily. The burners light satisfactorily. You wait, and the burners later shut off. The furnace is hot but nothing is coming through the registers. There is also the smell of something hot or burning. What will you do?

Solution: The filter could be plugged, but it would have to be very plugged to cause an odor. A further check proved the blower motor to be seized up. The motor had not been oiled for a long time. There was enough damage to the motor to warrant replacement. If the windings of the motor had not been scorched by repeated overheating, lubrication at this time may save it, and the motor's life may be extended.

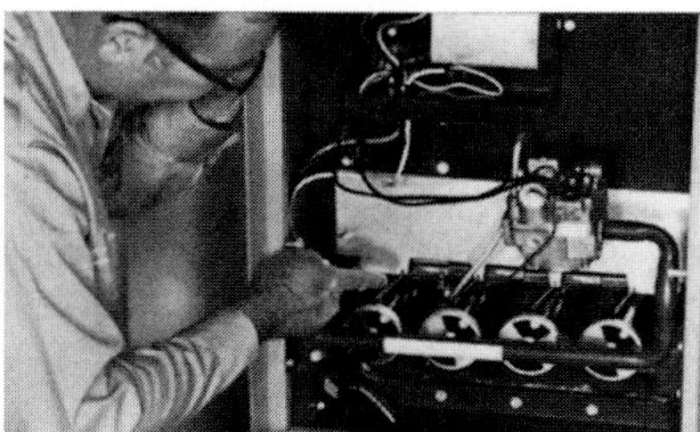

Figure TR50-6 Primary air-control opening size.

An all-blue sharp fire is receiving too much primary air. This means that there is less radiant heat to heat the lower portion of the heat exchanger. Also, the excess air drives the flame products from the heat exchanger before good transfer of heat occurs from flue product to heat exchanger. Flue product temperatures rise and unit efficiency drops. If the primary air is reduced too much, heavy yellow tips of improperly burning carbon are produced.

These are much lower temperatures and do not produce the heat. Unit efficiency therefore drops. The carbon can be released from the flame and collected in the heat exchanger to cause sooting and plugging of the flue passages. The proper setting of the primary air quantity is the beginning step in producing high unit operating efficiency.

Improper setting of the primary air shutter can also contribute to pilot outage by producing extinction flash back, called *extinction pop.*

When the burner is operating, the gas/air mixture is blowing upward through the burner port at a given speed or velocity (determined by the burner design and type of gas). There is also a downward force or burning velocity which is equalized by the gas/air outward velocity when burning is taking place. If, however, the burning velocity were to increase due to shutoff of the gas supply, the flame could approach the burner. Either the burner would absorb the heat below the combustion point and extinguish the fire or the fire would burn down through the burner port and ignite the mixture in the burner. This ignition produces the extinction pop.

Fig. TR50-7 shows what happens to the gas flame after the gas valve shuts off. At the moment of shutoff the gas/air mixture inside the burner is at a negative (below atmospheric) pressure. At full fire there is a full cone and a full tail of flame. Immediately after the gas valve closes, the burner pressure partially collapses. The full collapse of the fire down to the burner followed by extinction occurs when the burner absorbs the heat from the fire.

If the speed of gas burning is too high due to too much primary air, the flame does not collapse as rapidly as it should. The negative pressure is insufficient and there is an explosion or extinction pop within the burner when carry through occurs. This could cause a pressure wave over the pilot that blows the pilot out. A properly adjusted burner greatly reduces the chances of pilot outage.

(8) Improper combustion air supply. Excess air is needed for proper combustion, even though changes in gas pressure, heat content of the gas, and barometric pressure may occur. Draft conditions also change with barometric pressure and wind conditions.

When a unit encounters insufficient combustion air, the flame tends to become hazy and erratic and may even roll over the edge of the burner and out the burner pouch opening. The flame will seek air. Fig. TR50-8 shows the effect of insufficient secondary air causing a floating flame.

(9) Improper venting. The mixture of gas and air produces a mixture of water, carbon dioxide, nitrogen, and excess air. All of this has to be removed from the heat exchanger. This removal process is called *venting.*

There are two types of venting: active (power) venting and atmospheric or passive (gravity) venting. *Active venting* uses a mechanical device such as motor-driven blow-

ON THE JOB

Problem: Call-back, no heat, pilot out again. The customer said that you just replaced the thermocouple a month ago. The thermocouple looks okay, the connection to the gas valve is tight. The millivolt meter shows a reading at 14 mV on a closed-circuit test. The pilot flame is good. This is odd. When you start the furnace it starts up okay. Has anything changed? The customer tells you that about a year ago they had a problem and the service technician at that time replaced the gas-control valve. The valve appears to be working, but the numbers on the valve do not cross reference to this furnace. Is it the same as the old one, and will it work?

Solution: When you have occasional pilot-outage problems, and there doesn't seem to be any cause, check the gas valve. Gas valves have different valve opening speeds. When the valve is replaced with a type that is the wrong opening speed, it takes a few more seconds to stabilize the flame. Also check the pilot dropout; a good valve will only drop out the pilot flame when the millivolt reading falls below six mV.

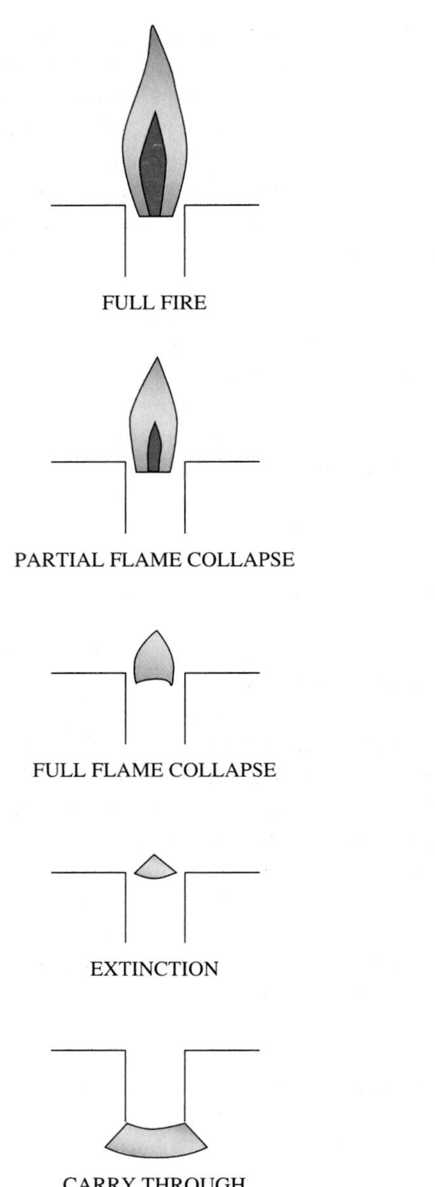

Figure TR50-7 Types of flame action at the gas burner.

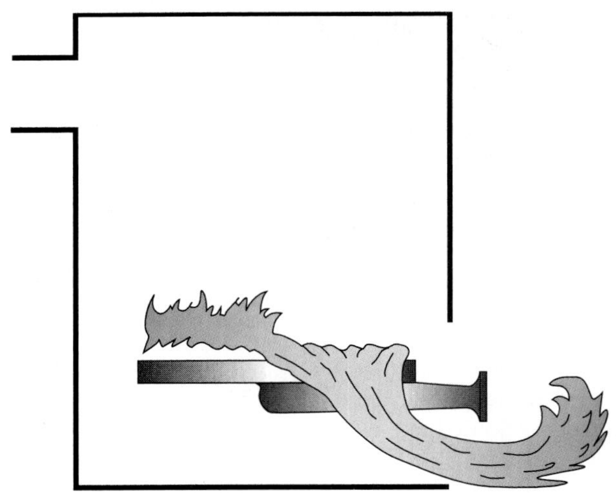

Figure TR50-8 Rollout, due to insufficient secondary air.

ers to either draw flue products from the heat exchanger or to force combustion air into the heat exchanger. The most popular type is the draw type, where the blower is mounted on the flue outlet and creates the negative pressure in the outlet of the heat exchanger to get the desired combustion efficiency. Because the pressure difference is caused by mechanical power, wind and/or atmospheric conditions have little effect on the venting performance.

In *atmospheric* or *passive venting,* hot flue gases pass from the heat exchanger into a flue pipe, chimney, or vent stack. The driving force for a passive vent is obtained from the hot gases rising in the surrounding cooler air. The amount of force depends on the temperature of the hot gases and the height of the gravity vent. The hotter the gases and/or the higher the vent, the greater the amount of driving force or pull that is produced. Also, the greater the pull, the more secondary air is drawn through the heat exchanger.

Enough air must be drawn through to provide complete combustion as well as complete venting of the flue products. If too much air is drawn out of the heat exchanger before the correct amount of heat extraction is done, this results in higher flue temperatures and reduced unit efficiency.

If the passive vent pipe were connected directly to the flue outlet, the amount of air drawn through the heat exchanger would vary with factors like the pull of the vent stack, the wind effect on the vent stack, and outside temperature. Control of the venting rate on the heat exchanger would be impossible. Further, under some atmospheric conditions it may be possible to have a higher pressure at the outlet of the vent than the combustion process can overcome. This can produce poor combustion with the production of CO as well as the CO_2 and H_2O.

To overcome the effect of atmospheric conditions, all units use an opening in the venting system called a *draft diverter.* Fig. TR50-9 shows four typical heating unit draft diverters. These all consist of an opening from the flue outlet of the heating unit, an opening into the vent pipe, and a relief opening to the surrounding atmosphere.

Fig. TR50-10 shows the operation of a typical draft diverter under no-wind conditions and with updraft and downdraft conditions. The amount of flue products (1), the dilution air entering the relief opening (2), and the amount of vent gases (3) are indicated by the length of the arrows. With normal venting some air is pulled into the draft diverter by the pull of the passive vent. The mixture of flue products and surrounding air (called dilution air) that blow up the vent is called *vent gas.* The action of the draft diverter is to break the effect of the vent by introducing

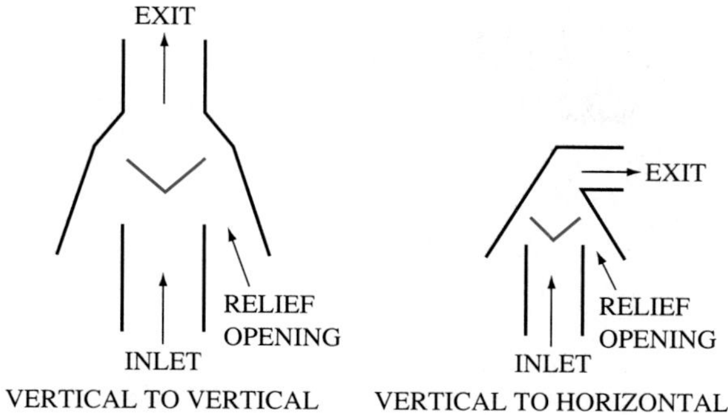

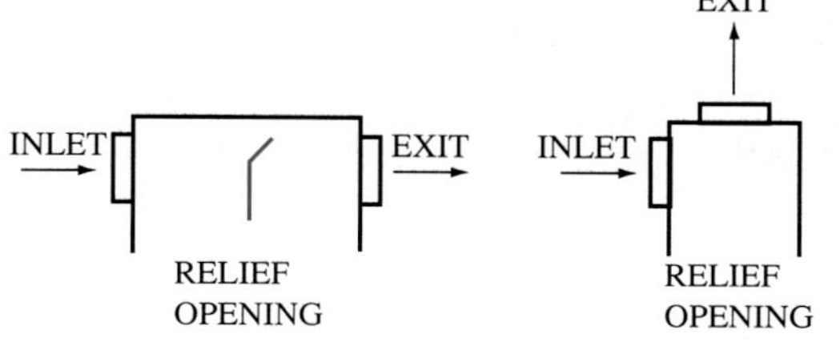

Figure TR50-9 Typical gas appliance draft diverters.

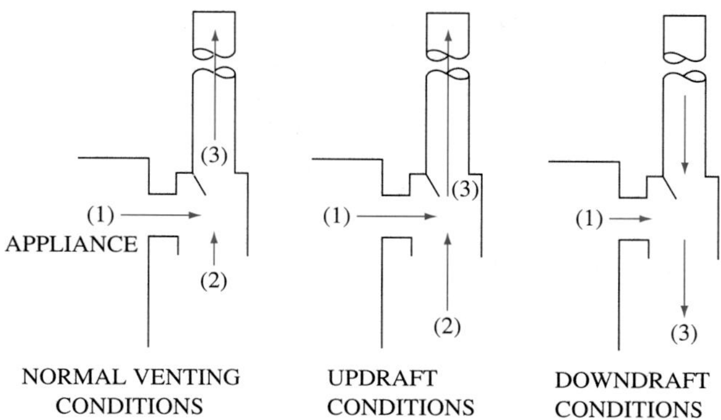

Figure TR50-10 Operation of a draft diverter under various wind conditions.

surrounding air and neutralizing the pull at the flue outlet. The heat exchanger then operates at approximately equal pressure from burner pouch opening to flue outlet. The amount of air for combustion is then controlled by the flue restrictors.

If the conditions surrounding the vent stack increase the stack pull, additional air is drawn into the draft diverter to compensate for the increased pull. There is little effect on the heat-exchanger performance.

Under conditions where the vent stack pull is reduced or even reversed, creating a downdraft, all combustion flue products are forced into the surrounding area. In addition, the increased pressure in the flue outlet will reduce the flow through the heat exchanger. This can cause incomplete combustion and produce odors carried by the gases and moisture produced by the combustion process. Even though no odors may result, the large amount of moisture produced in the combustion process can accumulate in the occupied area and create adverse living conditions or possibly structural damage.

To check for proper operation of the vent system, use a candle placed below the bottom edge of the diverter opening. With the unit operating and up to temperature, the candle flame should bend in the direction of the opening in the diverter. If the flame is neutral, the draft is on the weak side. Possibly the vent stack is not high enough or large enough. If the candle flame bends outward, a draft problem definitely exists that must be corrected. If the vent stack cannot be lengthened or enlarged, a forced draft unit must be installed to overcome the problem.

(10) Blower drive. Belt-driven blowers have a higher probability of vibration problems than direct-drive blowers due to the additional parts involved. The most common problem is due to belt tension. It is commonly believed that the tighter the belt, the better the performance, but the opposite is true. The tighter the belt, the harder the motor has to work to get the belt in and out of the pulleys. The belt should therefore be as loose as possible without slipping on startup.

Fig. TR50-11 gives the test for proper belt tension. It should be possible to easily depress the belt midway be-

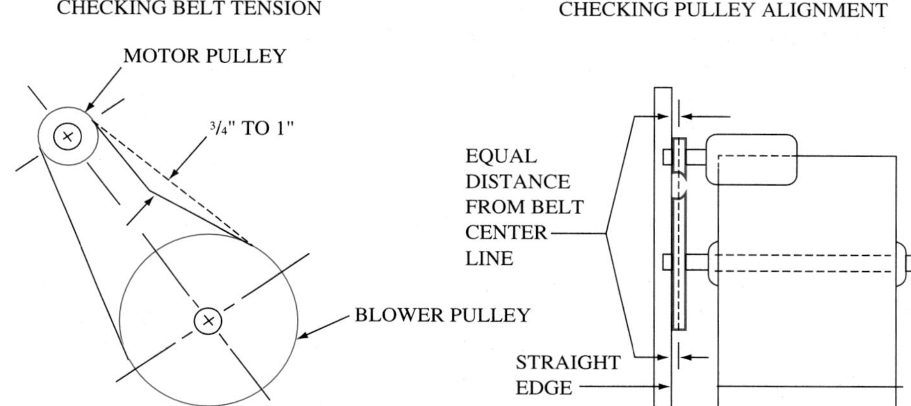

Figure TR50-11 Aligning pulleys and checking belt tension.

Figure TR50-12 Clamshell-type gas heat exchanger. (Rheem Air Conditioning Division)

Figure TR50-13 Drum-and-radiator heat exchanger.

tween the motor shaft and blower shafts ¾ to 1 in. for each 12 in. of distance between the shafts. Alignment of the motor and blower pulleys is important to keep vibration to a minimum as well as to reduce wear on the sides of the belt.

Finally, each pulley, both motor and blower, should be checked for running true. Any warpage that creates wobble in the pulley requires replacement of the pulley.

(11) Expansion noise in heat exchanger. Fig. TR50-12 shows a three-section unit, each section composed of a right-hand and a left-hand drawn-steel "clamshell" welded together. The sections are then welded into an assembly by fastening to the front mounting plate and rear retainer strap. Sometimes in the welding process, stresses will be set up if the two metals are at different temperatures when the bond is made. This results in expansion noises, ticking, and popping as the heat exchanger heats and cools. Most of the time these noises are muffled by the unit casing and duct system to a level where they are not objectionable.

In extreme cases, it is possible to reduce the noises by operating the unit with the blower disconnected, allowing the limit control to turn the unit off and on. This should be done through several cycles of the limit control. Cycling on this extreme heat will cause metal stretch beyond the normal operation range and eliminate the sound. If this does not produce satisfactory results, the only cure is to change the heat exchanger.

(12) Oil-can effect. This effect is caused by the sudden movement of a flat metal surface where a forming stress has been left in the surface. This stress causes the metal to have a slightly concave or convex position rather than a flat plane surface.

Temperature change will cause a stress increase in the material until the metal rapidly changes position to the opposite of its original position. This change will produce a loud "bang". Duct work is very prone to this action and must be cross-broken over any large panel areas. Heat exchangers such as the one shown in Fig. TR50-13 will usually have a large flat surface. Unless this surface is cross-broken, it is subject to the oil-can effect. The best correction is removal of the panel and cross-break it to relieve temperature stresses.

TR50-5 OIL-SYSTEM PROBLEMS

Principal causes of oil-fired-system problems follow.

Problems	Possible Causes (see the following number references for further information)
Will not start	Season switch open Room thermostat improperly set Safety switch open on protector relay Fuse blown Limit control open Protector relay transformer defective (1) Protector relay defective (2)
Starts, but will not continue to run	A. No fire established No fuel (6)

Problem	Possible Cause
	No ignition
	Protector relay not functioning
	Nozzle defective
	Fuel pump defective
	B. Fire established
	Defective oil-burner component
	Defective protector relay
Runs, but short cycles	Improper heat-anticipator setting
	Cycling on limit control
Runs, but room temperature too high	High setting of thermostat
	Improper heat-anticipator setting
Runs continuously	Short in thermostat circuit
	Stuck contacts in protector relay
Blower cycles after thermostat is satisfied	Incorrect blower cfm
	Incorrect fan control setting
	Heat exchange heavy with soot
Blows cold air on start	Incorrect blower cfm
	Incorrect fan control setting
Start-up/cool-down noise	Expansion noise in heat exchanger
	Duct expansion
Noise/vibration	Burner pulsation
	Blower wheel unbalanced
	Blower drive problems
Odor	Oil odor on unit start-up
	Improper venting
	Cracked heat exchanger
High fuel cost	Improper input
	Improper burner adjustment
High electric cost	Improper load on blower motor
Smoke or odor from observation door	Improper draft or draft-control setting
	Improper venting
	Improper input
	Delayed ignition
Burner pulses	Improper draft
	Improper venting
Burner cuts off on safety switch	A. No fire established
	No fuel
	No ignition
	Protector relay defective
	Components of burner defective
	B. Fire established
	Components of burner defective
Delayed ignition	Carbon deposit on firing head
	Improper electrode adjustment (3)
	Cracked electrode insulator (4)
	Ignition leads burned out
	Ignition transformer burned out (5)
Burns inside burner after cutoff	Fuel unit defective
Carbon deposits on refractory	Defective nozzle
Noisy flame	Improper air adjustment
	Improper nozzle
Fuel pump sings	Fuel unit defective (7)
Heat-exchanger burnout	Input too high
	Defective refractory
	Defective nozzle
Blower short cycles	Input too low
	Blower cfm too high
	Fan control defective

Many of the solutions to these problems are evident when the cause is determined. Below, however, is further information regarding some of the remedies that can be used. Numbers refer to references above.

(1) Protector relay transformer burned out. It is usually not possible to open the protector relay to reach the transformer to test it. It is not necessary since corrections to the transformer are done through the T-T terminals on the outside terminal board.

Fig. TR50-14 shows the inside wiring diagram of a typical protector-relay circuit. With the control wires removed from terminals T-T, a circuit from the top terminal

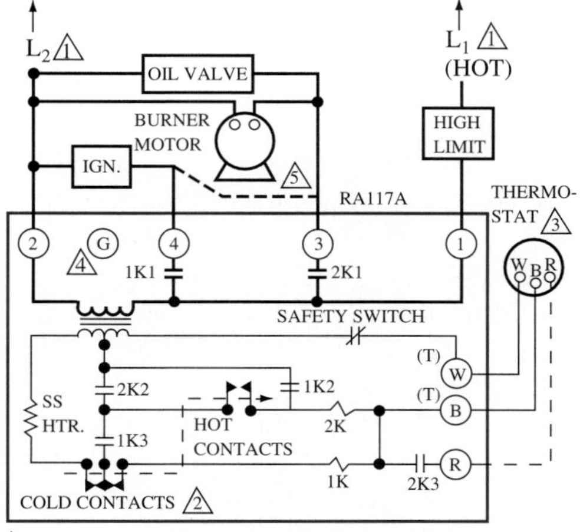

Figure TR50-14 Wiring diagram for an oil-fired furnace.

T, through the safety switch (SS), the transformer, normally closed contact (2K1), the safety-switch heater, and relay coil (1K) should put 24 V across terminals T-T. With 120 V across the black and white leads to the relay, if no voltage is at T-T, replace the protector relay. If there is voltage at T-T and the thermostat will not operate the relay, check the thermostat and subbase on the control cable.

(2) Protector relay defective. There are two types of protector relays: (a) a bimetallic control in the hot flue gases of a smoke pipe mount, and (b) a cad cell which operates the relay from the light of the fire.

(a) Bimetallic controls. This type of control is clutch-operated to move the hot and cold controls through their proper sequence. A sharp blow to the control can release this clutch and throw the control out of sequence. The clutch also wears and can loosen the hold on the contact fingers.

If the control refuses to operate, pull the drive shaft lever (Fig. TR50-15) forward until the stop is reached. Slowly release the drive shaft lever to the cold position. This should close the contacts and the unit should operate. If not, the bimetallic element could be jammed or the contacts defective. To check the bimetallic element, remove the protector relay from the vent stack and check it. Usually carbon (soot) buildup through the bimetallic helix will be the cause of jamming the drive-shaft lever. When cleaning the helix, be careful not to bend or break it.

With a clean helix, if the hot and cold contacts still do not hold, the clutch and contact leaves are worn and the entire control should be replaced.

Figure TR50-15 Oil burner control type RA117A, cover removed. (Courtesy of Honeywell, Inc.)

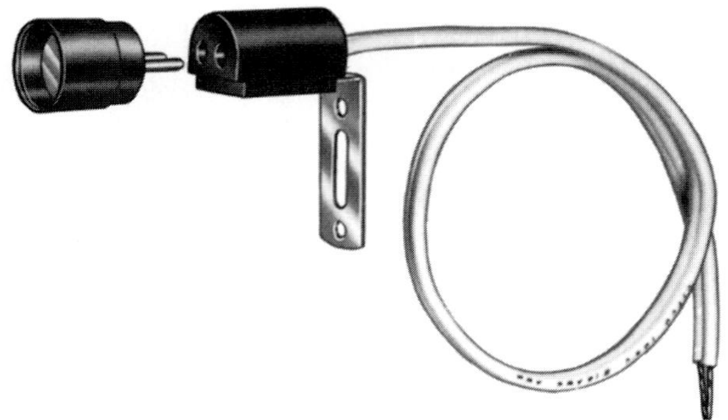

Figure TR50-16 Light-sensitive cadmium sulfide cell. (Courtesy of Honeywell, Inc.)

(b) Cad cell. This type of protector relay uses a light-sensitive cadmium sulfide flame detector mounted to the firing tube of the oil burner (Fig. TR50-16). When the cell is exposed to light, its resistance is very low, which allows current to flow through the cell. This current is sufficient to pull in the sensitive relay in the protector relay (Fig. TR50-17).

When the relay pulls in, it opens the circuit to the safety switch heater and prevents cutout of the burner. If the cell is not exposed to the light of the fire or if the cell becomes so dirty or covered with soot, it will not allow the current to pull in the sensitive relay; the safety-switch heater remains in the circuit until the safety switch opens. This breaks the thermostat circuit and the burner stops (Fig. TR50-18).

The safety switch is manually reset. When checking the protector relay for repeated burner cutoff even though the flame is established, make sure that the cad cell is clean. The cad cell can also be checked using an ohm-

Figure TR50-17 Internal view of cad cell, oil burner protector relay. (Courtesy of Honeywell, Inc.)

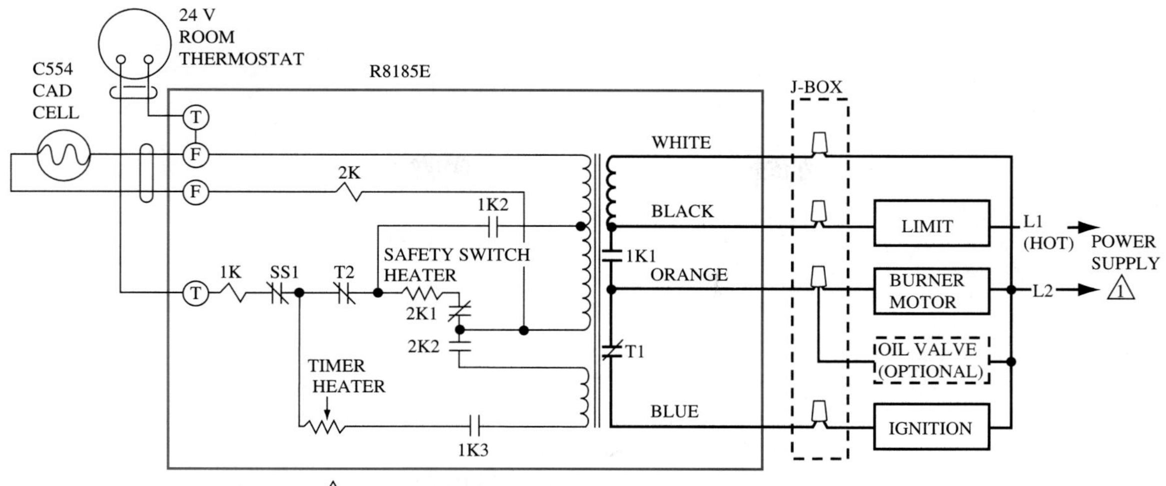

Figure TR50-18 Oil burner wiring schematic diagram using cad-type protector relay.

meter. Connect the ohmmeter leads across the leads of the cad cell. If the cad cell is exposed to light, the resistance will be less than 10 ohms. Placing a finger over the cell, cutting off the light will raise the resistance to 10,000 ohms or higher.

If the cad cell checks out, check the protector relay by placing a jumper wire across the F-F terminals of the protector relay. The burner should start and continue to operate. If it cuts off, the timer contacts and heater circuit are defective and the relay should be replaced.

Before starting the burner, check for liquid oil in the bottom of the refractory. The customer knows that the burner should start if the reset button is pressed. If the reset button has been pressed a number of times, it is possible to have a considerable quantity of oil sprayed into the combustion chamber before the owner calls for help.

If any oil is pooling in the bottom of the refractory, it must be removed by soaking it up with sponges or rags. When reaching into the refractory, make sure that the power to the burner is off and your arm is covered. A fire extinguisher must be within reach and the observation door secured open. When the flame is established, considerable fire will develop until all the oil is burned out of the bottom and where it has soaked into the refractory. If it is a large unit of 2.5 gal/hr input or larger, it is advisable to call the fire department for standby before lighting the burner.

(3) Improper electrode adjustment. To establish the spark necessary to ignite the oil, the electrodes have to be close enough together to present a minimum gap resistance to the 12,000 V supplied by the ignition transformer. To keep the electrodes out of the oil spray and still have the arc flame blown into the oil spray to ignite the oil spray, the electrodes must be high enough above the hole in the nozzle as well as the proper distance ahead of the nozzle end. These dimensions will vary with each manufacturer's unit. Fig. TR50-19 shows the settings of three different burners that have been used on a particular manufacturer's oil-fired unit.

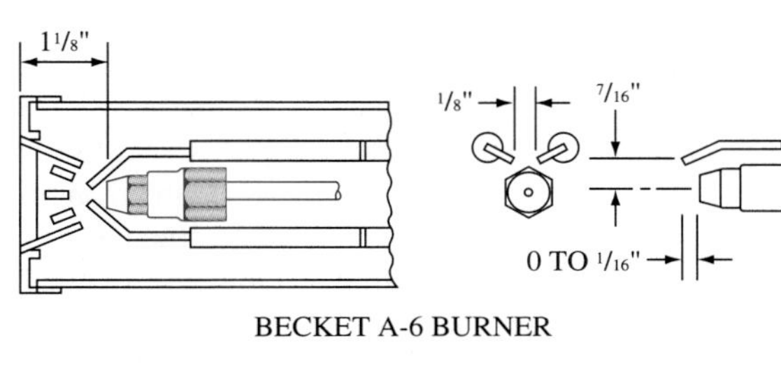

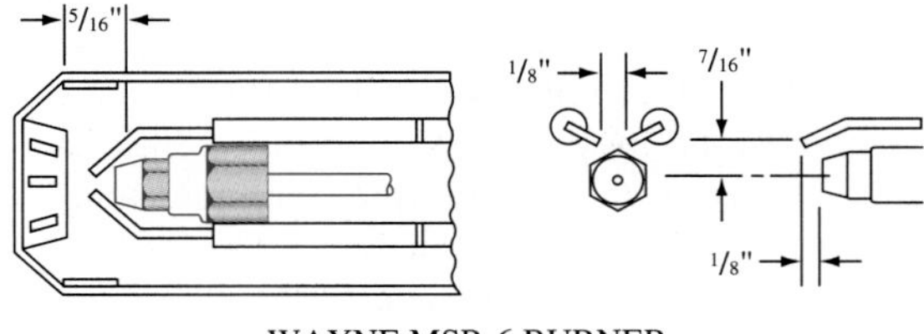

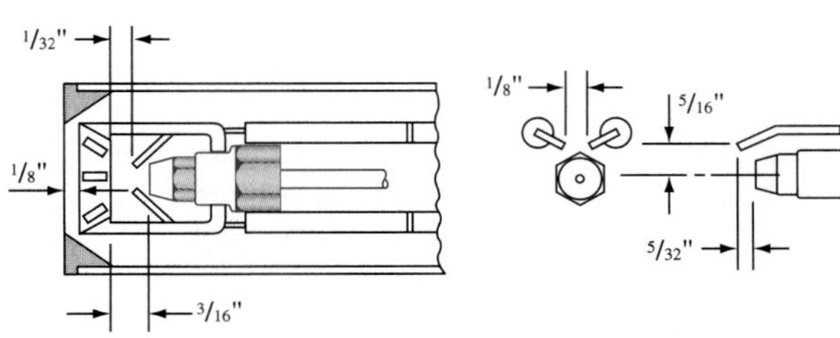

Figure TR50-19 Oil burner electrode spacing dimensions. (Courtesy of Rheem Air Conditioning Division).

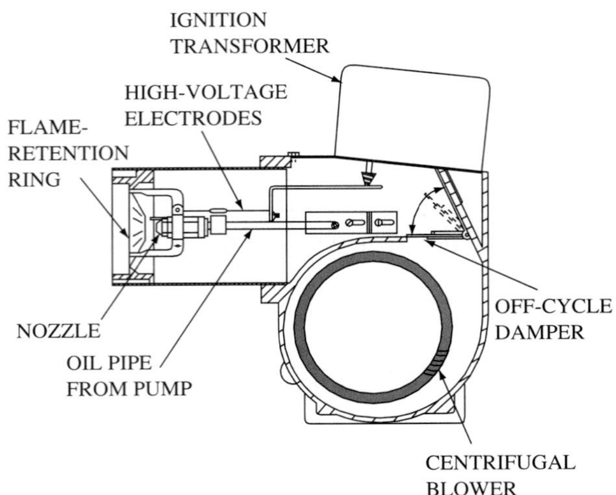

Figure TR50-20 Oil burner cutaway view.

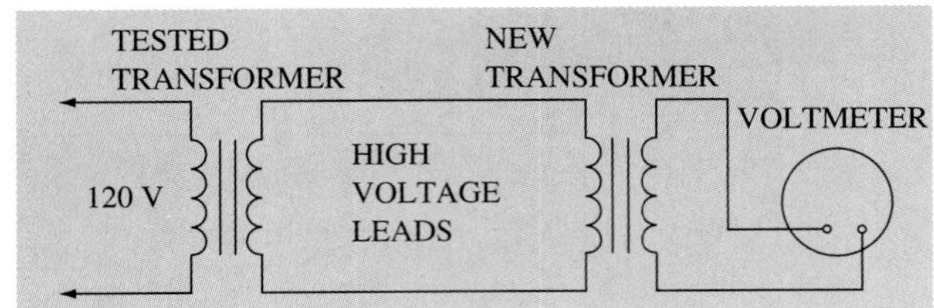

Figure TR50-21 Transformer test wiring diagram.

Another dimension is the distance of the nozzle end from the end of the firing-tube air turbulator. This dimension is important to keep oil spray from impinging on the turbulator and causing carbon buildup and still have the nozzle back far enough to keep the effect of the heat of the fire to a minimum.

The three different burners have only one common dimension: the size of the gap between the electrodes, which is ⅛ in. The electrode height and forward distance vary. When setting electrode positions, the manufacturer's specifications must be followed.

(4) Cracked electrode insulators. Fig. TR50-20 shows the position of the high-voltage electrodes in the firing assembly. The electrodes are held in a clamp device to ensure stability in the proper position. Clamped around the ceramic insulator, they hold the electrode and yet insulate the spark voltage from the grounded assembly. The electrode must insulate against 6000 V (one-half the spark voltage) and still stand up against the heat of the burner and combustion chamber when the burner shuts off.

The ceramic insulators are hard and brittle and crack easily. If any twisting or bending pressure is applied to change the position of the electrodes, loosen the electrode clamps. Do not attempt to bend the electrode wires; they are harder than the ceramic. Also, when tightening the clamps, they should not be over-tightened.

If any fire cracks or crazing are noted in the surface of the insulators, replace them. Do not take a chance on the old ones.

(5) Ignition transformer defective. Ignition transformers are 120 to 12,000 V with a grounded center tap on the high-voltage side. With 12,000 V between the terminals and 6,000 V from either terminal to ground, attempting to check the transformer by producing a spark with a wire, screwdriver, or other shorting means can be dangerous.

In time, the heat of operation dries the transformer, and cracks develop. Moisture enters the assembly and is absorbed into the windings. Producing shorts between the windings, the output voltage is lowered to the point of failure of the spark across the ignition gap, and faulty ignition results.

To check a transformer, the best way is with a high-voltage meter in the range of 10,000 to 15,000 V. If this meter is not available, two 120-V voltmeters and a new ignition transformer can be substituted. Fig. TR50-21 shows the wiring diagram for this test.

The secondary of each of the transformers (high-voltage terminals) is connected together to make the new transformer a step-down load of the test transformer. With 120 V applied to the test transformer (measured with one of the voltmeters), the output of the new or testing transformer should be within 10% of the applied voltage. In addition, the output voltage should hold steady. If there is more than a 10% difference or if the output voltage varies, the original transformer has internal shorts and should be replaced.

(6) No fuel. Before taking the burner apart, first check the quantity of oil in the supply tank (the oil gauge on the inside tank or the dip rod for the outside buried tank). Second, make sure that the tank outlet valve is open. Third, close the tank outlet valve, open the filter cartridge case, and put in a new filter cartridge. One is needed 99% of the time. Bleed the air from the filter cartridge after assembly.

If this procedure still does not supply oil to the burner and produce fire, remove the oil-burner nozzle and check the nozzle filter. If this is plugged, replace the entire nozzle assemble with one of like capacity, spray angle, and cone type.

If these steps do not produce oil flow when the unit runs, check the inlet screen of the fuel unit. Fig. TR50-22 shows a single-stage fuel unit cutaway. Part no. 3 indicates the fine mesh screen that filters the oil supply before it enters the pump gear assembly. This screen can be removed, cleaned in fresh fuel oil, and replaced. Do not leave this screen out of the pump.

After any of the portions of the fuel supply system have been opened, the system must be purged of air. A two-pipe system will automatically purge itself of air. The single-pipe gravity-feed system does not have an automatic purge feature. Purging must therefore be done.

Figure TR50-22 Fuel unit, cutaway view. (Courtesy of Suntec Industries, Inc., Rockford, IL)

With a short piece of plastic hose from the valve to a suitable container and the valve open, operate the unit until a clean stream of fuel oil is emitted from the hose. It may be necessary to reset the protector relay safety switch several times before the supply line is completely purged.

If it is not possible to obtain a flow of clear fuel oil, it is possible that the unit is receiving air through a line or fitting leak. This must be corrected to provide proper burner operation.

(7) Fuel unit defective. The most common problem in fuel units is poor cutoff of the oil supply when the unit cycles off. Correct cutoff will provide instantaneous cutoff of the oil to the fuel pipe and nozzle when the pump pressure drops to 80% of the operating pressure. At 100 psig operating pressure the cutoff pressure is 80 psig. This is accomplished as the spring forces the control piston in the fuel unit against the fuel outlet seat and cuts off the flow of oil. The nozzle pressure drops immediately (Fig. TR50-23).

If a particle buildup occurs on the face of the neoprene seat on the end of the piston, this prevents good full-circle contact of the neoprene disk against the seat. Instead of

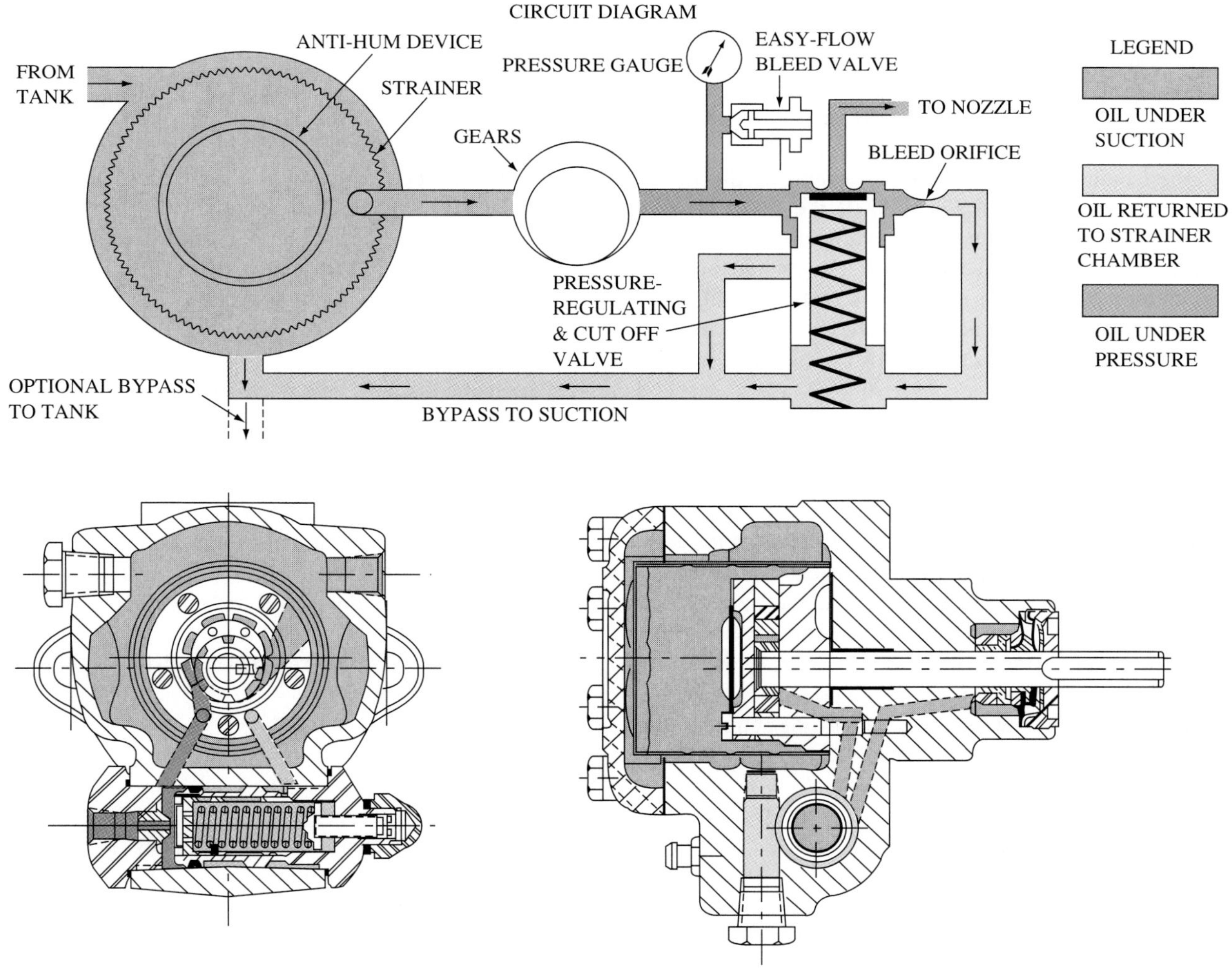

Figure TR50-23 Fuel unit, circuit diagram. (Courtesy of Suntec Industries, Inc., Rockford, IL)

Figure TR50-24 Delayed action oil valve used with fuel unit. (Courtesy of Suntec Industries, Inc., Rockford, IL)

positive cutoff, leakage occurs and the pressure gradually decreases in the nozzle. This gradual pressure reduction causes oil flow from the nozzle after the unit stops and the air supply disappears. The oil now burns with very little combustion air, producing a very smoky flame.

Carbon builds up on the turbulator end of the firing tube as well as the firing assembly. If the burner is not slanted at least 2° downward toward the combustion chamber, burning oil can flow back toward the blower. This can burn or smoke up the cad cell, with resulting cutoff of the burner on the safety switch. The correction for this is to clean the piston chamber of the fuel unit as well as the intake screen.

A persistent case of this problem due to the quality of fuel-oil supply can be reduced by the use of double filtering of the oil before it reaches the unit and the use of a *delayed-action oil valve* (Fig. TR50-24).

Located in the fuel line between the outlet of the fuel unit and the firing head, this valve provides a time delay between the time the blower starts and the start of oil spray. This ensures that air for combustion as well as airflow through the heat exchanger is established before combustion starts. The valve also provides instant cutoff of oil pressure to the nozzle, regardless of the action of the fuel unit. This is a highly recommended accessory for any oil-fired unit.

REVIEW

- Three types of gas furnaces in troubleshooting:
  1. The 78% AFUE or better units:
     Forced draft
     Higher efficiency
     Hot surface ignition
     Monoport in-shot burners
  2. The solid-state controlled units:
     Use microprocessor control
     Use printed circuit boards
     Self diagnostic mechanisms
  3. The standard units:
     No longer manufactured
     Manufactured before the 78% AFUE minimum requirements.
     Natural draft venting
     Standing pilots
     Multiple port burners
- Operating problems can be electrical, mechanical, or combination—mechanical and electrical
- Common categories of complaints or problems for gas heat:
  Entire system operation
  Unit operation
  Burner operation
  Blower operation
  Heat-exchanger complaints
  Costs of operation
  Noise
- Possible causes for gas-heating problems:
  Control transformer burned out
  Pilot outage
  Gas valve stuck open or closed
  Improper heat-anticipator setting
  Cycling on the limit-control or fan-control settings
  Improper burner adjustment
  Improper combustion-air supply
  Improper venting
  Blower drive
  Expansion noise in heat exchanger
  Oil-can effect
- Common categories of complaints or problems for oil heat:
  Will not start
  Starts but will not continue to run
  Runs but short cycles
  Runs but room temperatures are too high
  Runs continuously
  Blower cycles after thermostat is satisfied
  Blows cold air on start
  Start-up/cool-down noise
  Noise/vibration
  Odor
  High fuel costs
- Possible causes for oil-system problems:
  Protector relay transformer burned out
  Protector relay defective
  Bad bimetallic control
  Broken cad cell
  Improper electrode adjustment

Cracked electrode insulators
Ignition transformer defective
No fuel
Fuel unit defective

- Use troubleshooting charts when possible

Problems and Questions

1. What is the first item to check if the heating unit will not operate?
2. What is the first item to check if there is no heat because the gas valve will not open?
3. What would be the easiest way to check for an open circuit in the thermostat or subbase?
4. There is voltage at the gas valve terminals but the valve will not open. What should the next test be?
5. The room temperature is much higher or lower than the thermostat setting. What should be checked?
6. The LP-gas branch line supply to an outside heating unit should always be taken off the top of the main-line. Why?
7. The average thermocouple used in gas-fired heating units should develop between _____ and _____ mV.
8. To reduce the possibility of water, sediment, rust, etc., from the supply system causing damage, a device called a _____ must be installed in the piping ahead of the unit.
9. Fire in the main burner or control/compartment is usually caused by _____.
10. An all-blue sharp fire is caused by _____.
11. The effect of an all-blue sharp fire is _____.
12. What is the chief cause of burner extinction pop?
13. The input tolerance for a gas-fired heating unit is + _____ % to − _____%.
14. A major cause of fan short cycling is _____.
15. Two causes of the gas valve short cycling even though the fan runs steady are _____ and _____.
16. The proper fan control settings are _____ to _____ °F on and _____ to _____ °F off.
17. The first indication of dirty air filters is _____.
18. What is the easiest way to check for proper draft at the diverter?
19. The main cause of the rusting out of heat exchangers is chlorine or fluorine in the combustion air. What two common products are a source of these?
20. What three methods can be used to cure resonance or pipe organ effect?
21. What kind of control device does a stack-mounted protector relay use to operate the burner sequence?
22. What kind of control device does a burner-mounted protector relay use to operate the burner sequence?
23. When called on to fix a burner that fails to keep operating, what is the first item to check?
24. With 120-V power to the oil burner and the thermostat turned as high as it will go, if an oil burner will not run, what is the easiest way to check the thermostat circuit?
25. What is the most common cause of ignition failure?
26. The easiest way to adjust the electrodes is to bend the wires. True or false?
27. What is the most common cause of lack of fuel supply to the oil burner?
28. The input tolerance of an oil-fired heating unit is _____ to _____% of rated input.
29. When at ambient temperature, a flue or chimney must be able to produce a draft of _____ or better.
30. Oil burner nozzles are cleanable with a wire brush. True or false?
31. What causes carbon buildup in the turbulator end of a firing tube?
32. Chemical cleaners may be used to clean a heat exchanger heavily filled with soot. True or false?
33. If oil fumes are encountered in the occupied area after the unit starts, what is the first thing that should be checked?
34. What is the most common cause of pulsation when an oil-fired unit starts?
35. On gas furnaces with solid-state controls, the flame sensing controls do not sense flame on ignition; they will repeat the process twice before shutting down on lock-out. True or false?
36. Flame roll-out on gas furnaces is caused by a restriction in the heat exchanger, improper venting, or improper combustion air supply. True or false?
37. Premature burn out of heat exchangers is probably caused by faulty workmanship in the factory. True or false?
38. The oil-can effect happens when:
    a. The oil can runs dry when oiling the blower bearings.
    b. The heated duct work has not been cross-broken.
    c. The oil furnace gets too hot for too long.
    d. None of the above.
39. When the cad cell is exposed to light its resistance is:
    a. Very high
    b. Very low
    c. Unaffected
    d. None of the above.
40. To check a high-voltage transformer, the best way is to:
    a. Produce a spark using a wire or screwdriver.
    b. Use a high-voltage meter in the range of 10,000 to 15,000 V.
    c. Use a high-voltage meter in the range of 7,500 to 10,000 V.
    d. None of the above.

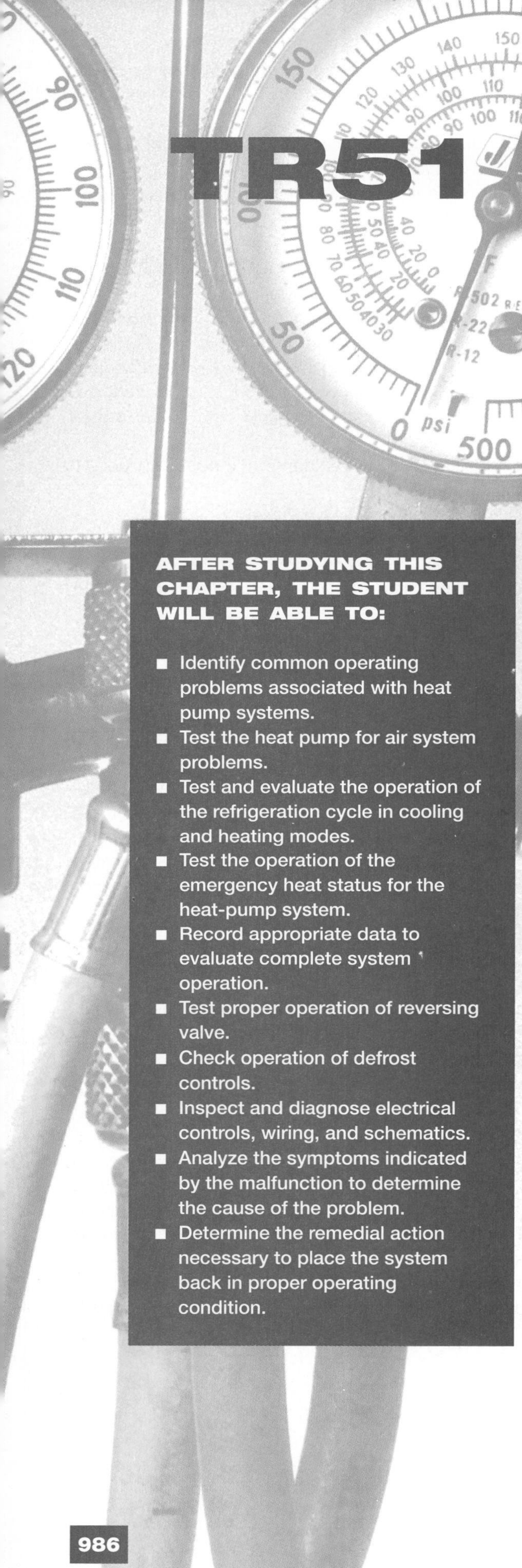

TR51

Trouble-shooting Heat Pump Systems

AFTER STUDYING THIS CHAPTER, THE STUDENT WILL BE ABLE TO:

- Identify common operating problems associated with heat pump systems.
- Test the heat pump for air system problems.
- Test and evaluate the operation of the refrigeration cycle in cooling and heating modes.
- Test the operation of the emergency heat status for the heat-pump system.
- Record appropriate data to evaluate complete system operation.
- Test proper operation of reversing valve.
- Check operation of defrost controls.
- Inspect and diagnose electrical controls, wiring, and schematics.
- Analyze the symptoms indicated by the malfunction to determine the cause of the problem.
- Determine the remedial action necessary to place the system back in proper operating condition.

TR51-1 GENERAL

The heat pump is unique in design compared to other air-conditioning apparatus in that it operates year-round, both in summer and winter, cooling in one season and heating in the other. The refrigeration cycle operates for a longer period of the year and can develop service problems more frequently.

Since heat pumps are a form of air conditioning, the technician's basic understanding of the fundamentals of air conditioning apply. Air is the principal medium used for transferring heat. The correct quantity must be delivered at the correct temperature over the variations of winter and summer to provide comfortable conditions.

In troubleshooting heat pumps, the technician must become thoroughly familiar with the controls and operating sequence of the particular design being serviced.

Each design potentially has its own selection of components and method of operation. To assist in finding a problem, many manufacturers describe helpful procedures in their installation and service bulletins.

It is important in troubleshooting heat pumps to find out as much as possible about the problem before starting a test procedure. Probably the most important question to ask first is: "Does the unit run or not run?" If the unit does not run, the first step is to check the power supply. If the unit does run, find out what components run and what components do not run. If all the components operate, then investigate whether the problem occurs on the heating cycle or the cooling cycle.

Basically, the service problems with heat pumps fall into two main categories: air circulation problems and refrigeration problems, each of which is discussed in the following sections.

TR51-2 AIR-SYSTEM PROBLEMS

When in the cooling cycle, the inside coil must be able to produce the correct amount of sensible and latent capacity to produce the desired room conditions. Assuming that the system has previously been set to the required temperature drop, the first check is the temperature drop through the evaporator. If the temperature drop has increased, the amount of air through the coil has decreased. The capacity of the refrigeration system will not increase; therefore, a reduction in the cfm has occurred.

If the unit is in the heating cycle, the temperature rise of the air through the inside coil, now the condenser, indicates a reduction of air through the coil. The reduction can be severe enough to cause the unit to cut out on the high head pressure lock-out relay system. Repetitive cutout bringing on the lock-out light on the thermostat should be investigated by a competent service technician before damage to the compressor results.

The heat pump operates on a year-round basis, and the amount of air through the inside coil is more critical than in a heating or air conditioning system. Throwaway air filters should be replaced every 30 days. It is advisable to use electronic air cleaners with heat pumps due to their low static resistance even when dirty.

Inspection of the blower and drive should be done at least once a year. Both blower motor and blower bearings should be lubricated with no more than ten drops of No. 20 electric motor oil. This oil is a detergent-free oil. Use of automobile oil is discouraged because it has detergent (soap) which coats the outer surface of the sintered bronze bearing and prevents oil passage through the bearing.

Unusual restrictions in the duct system, such as closing off unused rooms and placement of furniture or carpeting over supply and/or return grilles, will cause coil frosting in the cooling cycle and unit cutoff in the heating cycle. This practice has a greater effect in the heating cycle than in the cooling cycle.

Failure of the duct system also has a greater effect in the heating cycle than in the cooling cycle. With 70°F in the occupied area, the supply-air temperature will only be in the range of 100 to 105°F when the outdoor temperature is 60 to 65°F, and down to the 85°F range when it is −10°F outside. Any leakage in the duct system will therefore seriously reduce the capacity of the unit to handle the heating load. As a result, the operating cost will be higher than normal.

TR51-3 REFRIGERATION-SYSTEM PROBLEMS

When the temperature drop through the inside coil is less than it should be on either the cooling cycle or the heating cycle, the refrigeration system should be suspected. As in an air-conditioning system, it can be classified into (1) refrigerant quantity or (2) refrigerant flow rate. If the system has the proper amount of refrigerant and refrigerant is flowing at the desired rate, the system has to work properly and deliver rated capacity. Any problem in either category will affect the temperatures and pressures that will occur in the unit when the correct amount of air is supplied over the inside coil for the capacity of the unit.

If the system is low on refrigerant, the problem is a leak. It must be found and repaired. The system must then be evacuated thoroughly and recharged with the correct amount of refrigerant. If the system will simply not operate, the problem is probably electrical which must be located and corrected.

To compare the air-conditioning system with the heat-pump system, various systems are shown in Figs. TR51-1, TR51-2, and TR51-3. The only difference between the heat pump and the air-conditioning system is the addition

PRACTICING PROFESSIONAL SERVICES

Problem: The customer calls your service manager complaining that their new heat pump is putting off steam during mornings in the early spring season. What is the problem and the remedy?

Solution: You arrive at the job site the following morning and inspect the unit. You evaluate the operation of the heat pump and wait for a defrost cycle. All pressures and temperatures are normal. You get your sling psychrometer and check the dry-bulb and wet-bulb temperatures to determine the relative humidity. You notice that during the early morning when the outside air humidity levels are high in the spring, the condensate during defrost turns to steam. This is a normal condition and is not a problem. Inform the customer of the situation and ensure the operation of the heat pump is working as designed.

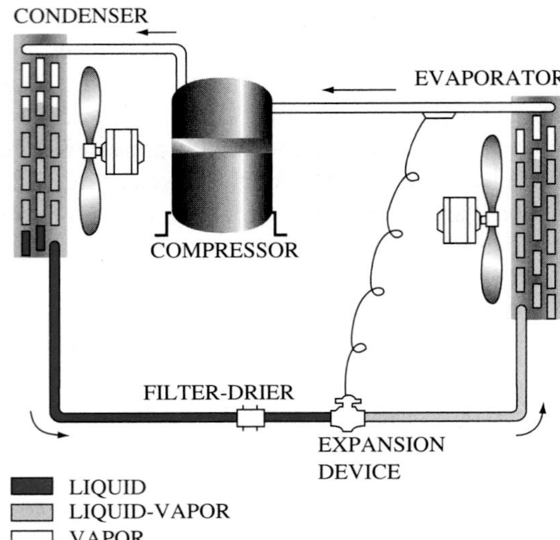

Figure TR51-1 Schematic diagram of a conventional air-conditioning/refrigeration system.

of a reversing valve, two check valves, and a second pressure-reducing device.

In Fig. TR51-1 a conventional air-conditioning system is shown. This system shows the refrigerant flow from the discharge of the compressor to the condenser, the outside coil. The refrigerant condenses and flows from the outside coil through the liquid line, the filter-drier, and pressure-reducing device to the evaporator, the inside coil. The refrigerant expands, picking up heat in the evaporator, and then flows as a vapor to the compressor.

In Fig. TR51-2 the same action takes place except that the following devices have been added.

A *reversing valve* has been added to the suction and discharge lines to enable the system to reverse the flow of refrigerant in the system. The position of the reversing valve still directs the hot gas from the compressor to the outside coil and the vapor from the inside coil to the compressor.

In addition, a *check valve* has been installed around the pressure-reducing device. It feeds refrigerant to the inside coil.

The check valve is connected to prevent flow around the pressure-reducing device during cooling operations. It opens during the heating operation to eliminate the restriction of the pressure-reducing device.

A *pressure-reducing device* has also been added to the outside coil for this coil to operate as an evaporator during the heating cycle. A check valve is also connected around this pressure-reducing device to remove its pressure drop during the cooling cycle. The refrigerant flow is the same as for the air-conditioning unit, and this configuration is used for cooling.

In Fig. TR51-3, the reversing valve has changed position. The hot refrigerant vapor flows to the inside coil, the condenser. Giving up heat to the air supplied to the occu-

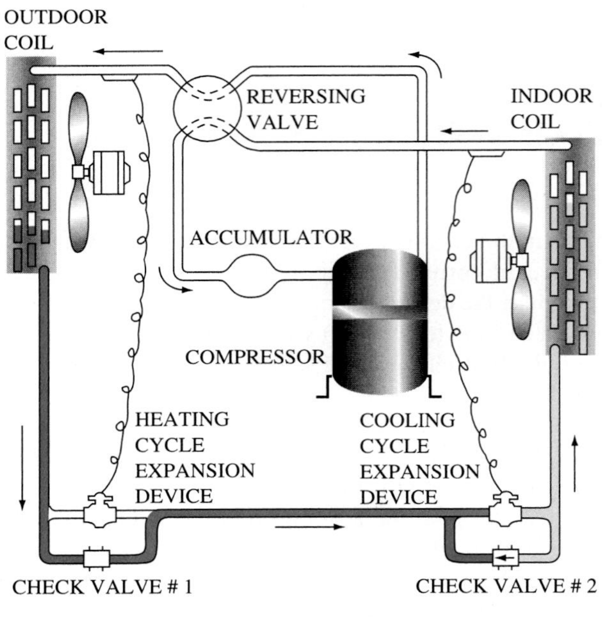

Figure TR51-2 Schematic diagram of a heat-pump system used for cooling, showing the position of the reversing valve.

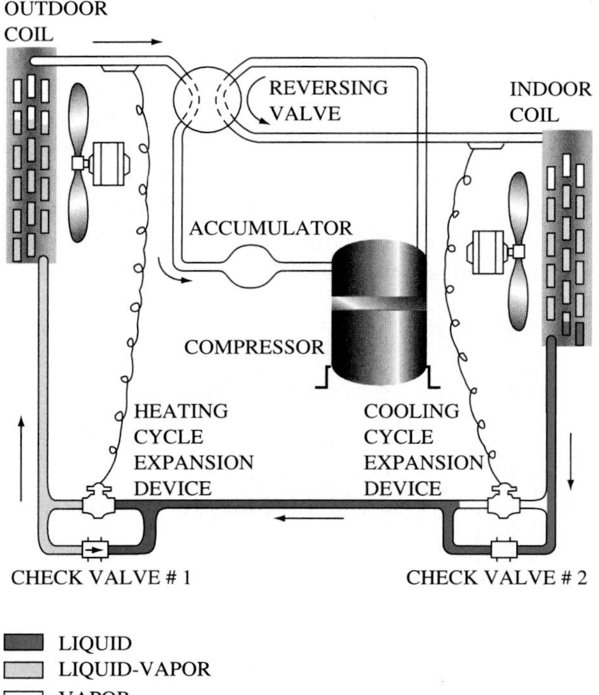

Figure TR51-3 Schematic diagram of a heat-pump system used for heating, showing the position of the reversing valve.

pied area, the hot refrigerant vapor cools, condenses, and flows out of the bottom of the condenser. With the flow reversed, the check valve opens and allows the liquid refrigerant to flow around the pressure-reducing device on the inside coil to eliminate any pressure loss. The liquid refrigerant flows in the reverse direction to the check valve on the outside coil. The refrigerant is forced to flow through the pressure-reducing device on the outside coil (now the evaporator). Here the liquid refrigerant, at a lower pressure and boiling point, vaporizes and picks up heat from the outside air. The vaporized refrigerant flows through the reversing valve and the accumulator to the compressor.

Note that the only reversing of the refrigerant flow is from the reversing valve through the coils, pressure-reducing devices, and check valves. The refrigerant vapor always flows from the reversing valve, accumulator, and compressor back to the reversing valve.

TR51-3.1 Refrigerant Flow Problems

Check Valves

Check valves have the capability of sticking in either the open or closed position. The system will therefore work correctly in any of the cycles, depending on the valve's location in the system. If the valve is doing what it is supposed to, the system will operate properly. If not, the problem will show up. Common problems include the following:

1. The check valve on the inside coil sticks in the open position. The system will operate properly in the heating cycle. The valve is supposed to be open in this cycle. In the cooling cycle, however, no pressure-reducing device is in the circuit, so the refrigerant will flood through the evaporator, suction pressure will be high, discharge pressure will be low, and the accumulator will be filled with liquid refrigerant. It may be flooding back to the compressor if the liquid line is short or in a packaged heat pump.
2. The check valve on the inside coil sticks in a closed position. The system will operate properly in the cooling cycle. The valve is supposed to be closed. In the heating cycle, the suction pressure will be much lower than normal. In units using capillary tubes, the suction pressure will be the result of two capillary tubes in series. In systems using TXV valves, the valve will close and the compressor could pull the suction pressure into a vacuum. At the same time, the discharge pressure will be low due to little vapor for the compressor to pump. Before adding gas to the system, check the system by switching to the opposite cycle.
3. The check valve on the outside coil sticks in a closed position. The system will operate properly on heating. The valve is supposed to be closed. On cooling, low suction pressure and discharge pressure will result.
4. The check valve on the outside coil sticks open. The system will operate on cooling but not on heating. The outside check valve is supposed to be open on the cooling cycle.

Usually, a check valve can be released from the stuck-open position by means of a magnet placed against the outlet end of the valve. Moving the magnet toward the center will force the ball or flapper to move to the seat. If in the stuck-closed position, place the magnet at the middle of the valve and move it to the outlet end. If this doesn't work, replace the valve. To reduce the possibility of future problems, use a ball-type check. Before installing the new valve, however, shake it. If it rattles, install it. If it does not rattle, it is already stuck and will not function properly. Take it back to the supplier.

Reversing Valves

Fig. TR51-4 shows the exterior view of a reversing valve. The tube connection is always connected to the compressor discharge. The bottom middle connection is always connected to the compressor suction connection. When an accumulator is in the circuit, this connection is to the accumulator inlet. The accumulator outlet is then connected to the compressor suction. This puts the accumulator upstream from the compressor, and provides surge protection in either the heating or cooling mode.

The right and left connections are to the outlets of the inside and outside coils. Which connection goes where depends on whether the operating coil is energized on heating or cooling. Most heat pumps are designed to operate in colder climates, where the heating operating hours are more than the cooling operating hours. To reduce the

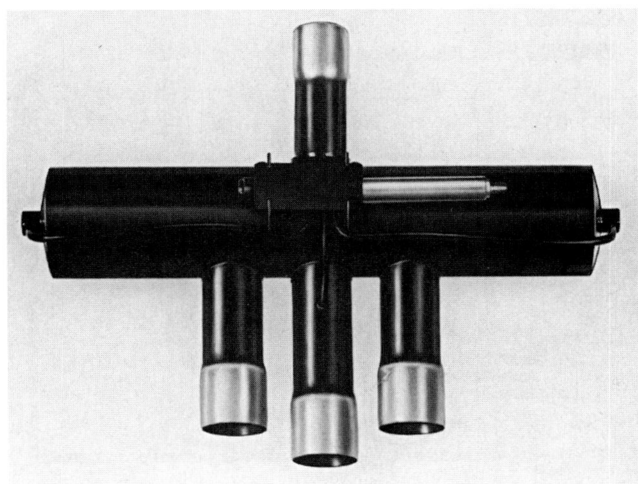

Figure TR51-4 Reversing valve. (Courtesy of Ranco North America)

operating coil on-time, the coil is energized during the cooling season. In this case, the flow through the coil would be from the left connection to the middle and from the compressor discharge through the right-hand connection. The left-hand connection would be to the inside coil (the condenser) and the right-hand connection to the outside coil (the evaporator).

With the coil energized, the valve slide assembly would be in the opposite position. The hot gas from the compressor would flow through the left-hand connection to the outside coil, now the condenser. The cold gas from the inside coil (now the evaporator) flows through the right-hand connection to the middle connection to the accumulator and on to the compressor.

If the cycling of the valve is reversed, the coil is energized during the heating cycle and the outside connections would be reversed. Before checking the position of the valve operation, the electrical control system should be checked to determine the operating requirements of the valve.

Reviewing Fig. TR51-5, the solenoid valve is actually a single-port, double-throw valve. The center connection is to the suction line and relieves pressure out of the valve. The bottom port has been closed. This port is normally closed. Because this port is closed, compressor discharge pressure builds up behind the main valve piston and in this line.

The top port is open and the pressure in this line has drained into the suction line. Because discharge pressure is on the right side of the left piston in the main valve, and suction pressure on the left side, the piston has been forced to the left side of the main cylinder. This moves the bypass valve to cover the middle and left hand outlets. Not shown is the V point on the end of the piston that seats in the outlet, sent to shut off gas flow through the piston port into the vent line.

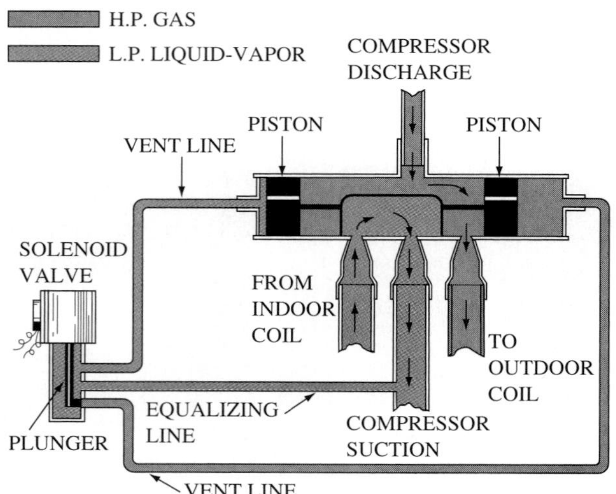

Figure TR51-5 Reversing valve in the cooling position. (Courtesy of Ranco North America)

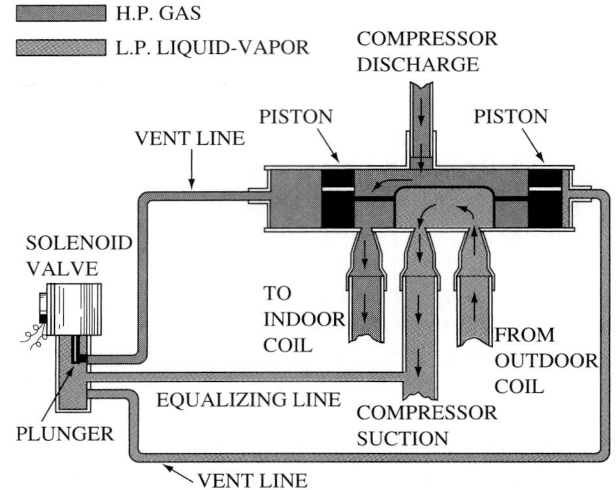

Figure TR51-6 Reversing valve in the heating position. (Courtesy of Ranco North America)

When the pilot valve (solenoid valve) is energized (Fig. TR51-6), the ports in the pilot valve reverse; the lower valve opens and the upper valve closes. This lowers the pressure in the right end of the main cylinder to suction pressure. The pressure difference that develops across the right-hand piston forces the piston to the right end of the main cylinder. The bypass valve is moved to cover the center and right-hand outlets. When the piston stroke is completed, the gas flow is shut off by the V point entering the outlet valve seat.

This shows that the main valve piston is controlled by the action of the pilot valve draining the pressure off the ends of the main valve piston to the suction side of the compressor. The valve works on the compressor differential pressure. The minimum pressure required to operate the valve is 75 to 100 psig. This means that the refrigeration system must be fully charged with refrigerant and operating long enough to develop a 75-psig difference between suction and discharge pressure.

Problems in reversing valves are either electrical or mechanical. Electrical problems are confined to the solenoid coil on the pilot valve. When the solenoid coil is supposedly energized and nothing happens, test the coil as follows:

1. Make sure that voltage is applied to the coil. Some units have the coil in the 240-V portion of the system, while others use it in the 24-V portion. Check the wiring diagram for coil voltage before applying the leads of the voltmeter. Also, start the test with the voltmeter set to the higher range. This helps reduce meter burnout.
2. With voltage applied to the coil, remove the coil-holding nut and attempt to pull the coil off the pilot valve plunger casing. If you feel a resistance to removing the coil, the coil is active. If no pull is felt,

the coil is dead. Shut off the power, remove the coil heads, and check for continuity with the ohmmeter. If open, replace the coil. If a circuit exists, check the leads and the connections for continuity.

If the coil is active, when removing the coil a "click" should be heard when the pilot plunger returns to the normally closed position. When replacing the coil, a click should also be heard when the pilot plunger is lifted to the open position. If no clicks are heard, the pilot valve is stuck. The only repair is to replace the reversing valve.

3. When the pilot valve checks out satisfactorily and the main valve does not shift, make sure that suction and discharge show more than a 100-psig difference. If the valve is in the cooling position, block off the air to the condenser with plastic on the inlet face of the coil. Allow the unit to operate until the condensing temperature rises to 130°F. With the unit operating, cycle the valve on and off several times. This will usually free the main valve to operate again. If no results are achieved, change the valve.

When changing a reversing valve, after completely removing the refrigerant from the system, be sure to read the installation instructions supplied with the valve. Replace the valve with one of a comparable size. Always position the valve so that the main piston is in a horizontal position and the pilot valve is higher than the main valve. This is to keep oil from gathering in the pilot valve and affecting its operation.

The valve body must always be protected from heat by wrapping the body with some type of thermoplastic material. The maximum temperature the valve body will tolerate is 250°F.

TR51-4 PROBLEM ANALYSIS

Various refrigeration, electrical, and air distribution problems relating to air-conditioning systems have been discussed in previous chapters and should be referenced whenever applicable. In this chapter we will deal with only those problems uniquely related to heat pumps.

TR51-4.1 Refrigeration System

The primary refrigeration problem occurs when the unit will not change cycle, heating to cooling, or vice versa, regardless of thermostat switch settings. The changeover of the system is controlled by the action of the reversing valve. Malfunction of this control can be either electrical or mechanical.

Electrical

Either a high-voltage or low-voltage valve must have applied voltage with ±10% of the design voltage range in order for the control to operate properly. The operating coil should also be able to produce the proper magnetic pull to operate the pilot valve.

Mechanical

The main valve piston may be stuck in one end of the cylinder because it is not getting the proper pressure changes from the pilot valve. This can be determined by touching the coil connection tubes with the coil energized and de-energized. Any sticking of the pilot valve plugs in the small tubes between the pilot valve and the ends of the main valve, or seizure of the main valve, requires replace-

ON THE JOB

Problem: You have arrived at the job site during the hottest day of summer to find that the unit is not cooling adequately. Upon inspecting the equipment, you notice that the suction line back to the compressor is covered with thick ice. You also notice that the inside coil is completely iced over, restricting airflow into the building. What is the possible problem and how do you resolve it?

Solution: There is a lack of airflow across the indoor coil, the evaporator. Finding the actual restriction is not possible until you melt the ice off the coil. Set the thermostat to fan only and cooling off to shut down the compressor. After the unit is defrosted, inspect for possible airflow restrictions. The most likely problem is a neglected and dirty air filter. Replace the air filter and start up the unit. If ice begins to accumulate again, shut down the unit and clean the indoor coil. It probably also accumulated dirt when the air filter was dirty. After cleaning the coil, restart the unit and check the operation.

ment of the valve. Repair of the valve is not justified because of the time cost and high percentage of failure repeat.

If the unit works fine on heating but poorly on cooling, the problem is usually found in the action of the check valves. If the suction pressure is higher than reasonable for the temperature of the air leaving the coil, the check valve on the inside coil may be stuck open. This removes the pressure-reducing device from the circuit and refrigerant is flooding through the coil. Superheat will be low or nonexistent, and the temperature of the accumulator, reversing valve tubes, and possibly the compressor shell, will be very low.

If the suction pressure is lower than normal, the check valve on the outside coil is stuck closed. This keeps the pressure-reducing device in the circuit. If it is a TXV valve, the valve will close and shut off the refrigerant flow to the liquid line. The suction pressure will drop extremely low, possibly into a vacuum.

If the pressure-reducing devices are capillary tubes, refrigerant flow will take place but much less than normal. The unit will have two capillary tube sets in series with the higher-pressure reduction of both sets. On some units, the head pressure is measured on the liquid line between the check valve and the pressure-reducing valve sets. With TXV valves, the head pressure gauge will indicate a loss of refrigerant. With capillary tubes, the head pressure will be low, along with a low suction pressure. In either case, do not add refrigerant until the subcooling has been checked. The subcooling will be found to be extremely high because the refrigerant has accumulated in the condenser.

TR51-4.2 Defrost System

To diagnose problems in the defrost system, the type of defrost system must be determined. The four most predominant systems are covered here. On those systems used by individual manufacturers, the manufacturer should be contacted for service information.

Temperature-Differential Defrost System

1. The unit will not go into defrost. This is the most common complaint on this control. The control is operated by the temperature difference between the entering-air temperature of the outside coil (the evaporator) and the coil operating temperature.

   The predominant reason that the unit will not go into defrost is that the coil will not reach a low enough temperature to provide an increase in the control differential to activate the defrost cycle. The major reason for this is that the head pressure is too high, which forces the suction pressure and coil boiling point up. Only a 2°F rise in boiling point is necessary to require complete coverage of the coil with frost and ice before initiating the defrost cycle. Rather than correct the air problem—the duct system may be inadequate or the occupants dislike the low-temperature discharge air—the service technician will often replace the control. Worse yet, the service person may try to adjust the control, which is not possible without closely controlled water baths.

   The correct solution to this is to restore the correct flow of air over the indoor coil to design conditions. This will lower the head pressure and the suction pressure.
2. The unit goes into defrost with very little frost on the bottom of the coil. The location of the coil temperature sensing bulb is critical. If the bulb is located too far to the bottom of the coil and too close to the entering low-pressure liquid refrigerant, the lower temperature of this portion of the coil will promote premature initiation of the defrost cycle. The cure is to move the coil temperature bulb up one return bend at a time until the correct location is obtained. Once the defrost cycle is initiated, the coil has to reach a preset temperature to terminate the defrost cycle. Usually, this temperature is 50 to 60°F.
3. The unit does not defrost the entire coil; an ice ring builds up around the bottom of the coil. The ice buildup indicates that the coil temperature bulb is located too high on the coil. The bulb is reaching termination temperature before the coil is completely defrosted. The bulb should be lowered one return bend at a time until complete defrosting takes place.

Knowledge Extension

The hardest area to learn about troubleshooting a heat pump is the defrost control circuit. The newer heat pumps are shipped from the factory with microprocessor-based defrost-control circuit boards and associated sensors. If these units do not come with sequence of operation and troubleshooting charts, consult the manufacturer for service literature. It is nearly impossible to diagnose these advanced control circuits without the assistance of these factory manuals. To learn more about specific heat pumps, consider attending a factory-sponsored or trade association training seminar on heat pump operation and maintenance.

Pressure-Temperature Defrost System

In this system, the pressure differential across the coil initiates the defrost cycle and the temperature of the liquid leaving the coil terminates it. The liquid-line thermostat also controls the possibility of a defrost cycle.

A thermostat fastened to the liquid line at the bottom outlet of the coil is exposed to the expanding liquid refrigerant when the unit is operating in the heating cycle. If the evaporator coil is operating with an entering liquid of 26°F or higher, there is little chance that frost will form on the coil. If this temperature is below 26°F, the thermostat closes and completes the circuit through the defrost relay to the differential-pressure switch.

When the frost buildup causes enough resistance to close the differential pressure switch, the defrost cycle is initiated. When the liquid leaving the coil reaches 55°F, the circuit is broken, the defrost relay circuit opens, and the relay drops out. This opens the holding circuit of the relay and completes the switchover to the heating cycle.

1. The unit will not go into defrost. Two possible causes can prevent the unit from going into defrost. The most probable cause is that the termination thermostat has come loose from the coil outlet pipe. The thermostat cannot reach 29°F or below to close and allow the defrost circuit to initiate. The other probable cause is plugged tubes to the differential-pressure switch. Insects sometimes plug these pipes.
2. The unit goes into defrost with very little frost on the bottom of the coil. Because the pressure drop of the air through the coil is the determining factor for initiating the defrost cycle, both the cleanliness of the coil and the frost buildup make up this pressure drop. Therefore, the more dirt buildup on the coil, the less frost has to form to reach the required pressure drop to initiate the defrost cycle. When the unit starts defrosting with little frost buildup, clean the coil.
3. The unit does not defrost the entire coil; an ice ring forms at the bottom of the coil. The fact that the ice ring forms indicates that the defrost cycle is interrupted before the defrost cycle is completed. The most common cause of this is too much reheat when the unit goes into defrost, which brings on the reheat as part of the defrost cycle. The amount of reheat should not exceed the sensible capacity of the heat-pump cooling cycle.

   When the coil is defrosted, the termination thermostat breaks the power to the defrost relay. The holding contact in the relay that keeps the relay energized is broken. The system must now develop the necessary frost buildup to close the pressure switch.

   If the power supply to the defrost circuit is interrupted, the same action occurs. Therefore, if the reheat is great enough to warm the occupied area and cause the room thermostat to open before the defrost cycle is completed, the incomplete defrost cycle will leave an ice buildup at the bottom of the coil. The alternate thawing and freezing of the water held on the coil by the ice coating will cause collapse of the coil tubes and loss of refrigerant.

Time-Temperature Defrost System

1. The unit will not go into defrost. This system uses a timer to control the defrost cycle under the control of the termination thermostat. The power to the timer is supplied from the compressor circuit. The timer therefore operates even when the compressor is operating and the termination thermostat is below 26°F.

   The most common cause of this problem is a loose termination thermostat. A poor contact between the thermostat and the coil tube prevents the thermostat temperature from dropping below 26°F. If contact is made, the thermostat should be checked to determine if it closes at a temperature below 26°F. Immersion in an ice and saltwater bath using an immersion thermometer will accomplish this. The timer motor should also be checked for proper motor operation.
2. The unit goes into defrost with very little frost on the bottom of the coil. The timer may be operating on too short a cycle time. Most timers have 30-min. and 90-min. cycle cams. Convert the timer from the 30-min. cam to the 90-min. cam.
3. The unit does not defrost the entire coil, an ice ring builds up around the bottom of the outside coil. The correction to this problem is to reduce the amount of reheat used in the defrost cycle to an amount not more than the sensible capacity of the unit on the cooling cycle.

Pressure-Time-Temperature Defrost System

This system uses a differential-pressure switch to measure the amount of frost on the outside coil. When the pressure drop through the coil reaches a preset amount, the switch closes the timer circuit.

If the termination thermostat is below 26°F, the timer will operate. This timer usually requires 5 min. of pressure switch closure time to initiate the defrost cycle by energizing the defrost relay. When the termination thermostat reaches 55°F, the power to the defrost relay is interrupted and the unit changes over to the heating cycle.

1. The unit will not go into defrost. The most common cause of this is a loose or poorly connected termination bulb of the defrost control. It is very important that the bulb be securely fastened with heat-transfer compound on the joint, insulated, and sealed from moisture. Ice can build up at this location and loosen the bulb fastening. The second

cause can be failure of the pressure switch due to dirt and/or insects in the pressure switch tube connections. There is always the possibility of failure of the timer control, but this is remote. Check the previous two items before replacing any parts.

2. The unit goes into defrost with very little frost on the bottom of the coil. The major cause of this action is a dirty outside coil. Its frost-free resistance is so high that it takes very little increase in pressure drop to start the defrost cycle. A thorough coil cleaning is in order.
3. The unit does not defrost the entire coil. This system has an automatic defrost cycle termination anywhere from 5 to 12 min., depending on the frost accumulation rate of the outside weather conditions.

REVIEW

- Categories of heat-pump problems:
  Air-system circulation
  Refrigeration
  Electrical or defrost controls
- Heat-pump temperature drop across indoor coil can help identify problems:
  During cooling season, excessive temperature drop can be caused by:
    Dirty air filters
    Dirty coil surface
    Blower drive problems
  During cooling season, insufficient temperature drop can be caused by:
    Refrigerant quantity
    Refrigerant flow rate
  During heating season, excessive temperature rise can be caused by:
    Reduction of air through coil
  During heating season, insufficient temperature rise can be caused by:
    Refrigerant quantity
    Refrigerant flow rate
- Refrigeration problems:
  Low on charge
  Reversing valve stuck or not switching
  Check valve failure—stuck open or closed
  Flow problem—possible restriction
  Compressor worn out/inefficient
- Electrical or defrost controls:
  Check supply and secondary voltages
  If problem with defrost on the outside coil:
    Determine type of defrost involved:
      Temperature-differential
      Pressure-temperature
      Time-temperature
      Pressure-time-temperature
  Check electrical and mechanical connections.
  Check controls initiating and terminating defrost.
  Check time setting and adjust accordingly.
  Check all other controls—ambient thermostats, safeties, indoor thermostats.

Problems and Questions

1. Fastening a thermometer to the center tube from the reversing valve to the compressor and from the evaporator coil to the reversing valve would be a possible check for what type of problem?
2. The only lines in a heat pump in which the refrigerant flows in one direction regardless of the operating mode are the _____ and the _____.
3. Before condemning a part or parts of a heat-pump system, what must be done to the system?
4. When operating the dual-operation unit in the cooling mode, which check valve will be open and which closed?
5. The check valve on the inside coil sticks open. In which mode will the problem show up?
6. The check valve on the outside coil sticks closed. In which mode will the problem show up?
7. In the heating mode, the head pressure reading taken on the liquid line is low, the suction pressure is low, but the compressor amperage draw is high. Performance in the cooling mode is good. What is the trouble in the system?
8. Placing your hand on the center tube of the reversing valve, you find the temperature higher than that on the line from the evaporator. What is the problem?
9. How much refrigerant must be in the system to check the operation of the check valve?
10. What is the minimum pressure difference across a reversing valve for it to operate?
11. When changing a reversing valve, what is the maximum temperature the valve body will take?
12. Upon hooking up the low-voltage control system and turning the unit on, the thermostat will bring on the first-stage electric heat instead of the compressor. What is the problem?
13. The control that controls the operation of the second stage of the auxiliary heat is called the _____.
14. The outdoor temperature at which the heat loss of the building equals the capacity of the heat pump is called the _____.
15. The air-to-air heat pump operates with an ice ring at the bottom of the outside coil. What is the most likely problem?
16. Most termination thermostats open at what temperature?
17. On temperature-termination defrost controls, what is the most likely cause of failure to go into the defrost cycle?

18. On a pressure-initiation defrost system, the unit goes into defrost with very little frost on the coil. What is the problem?
19. What is the maximum amount of reheat that should be used?
20. The correct quantity of air is not as critical as the correct refrigerant charge on the heat pump. True or false?
21. Basically, the service problems with heat pumps fall into two main categories: air-circulation problems and refrigeration problems. True or false?
22. When in the cooling cycle, if the temperature drop through the evaporator has dropped, the amount of air through the coil has increased. True or false?
23. Unusual restrictions in the duct system, such as closing off unused rooms, will cause coil frosting in the cooling cycle. True or false?
24. The check valve is connected to prevent flow around the pressure-reducing device. True or false?
25. When the check valve on the inside coil sticks in the open position, the system:
    a. Will operate properly in the heating cycle.
    b. Will operate properly in the cooling cycle.
    c. Will operate as normal except during defrost.
    d. None of the above.
26. The bottom middle tube on the four-way reversing valve is always connected to the:
    a. Hot-gas defrost line.
    b. Compressor discharge line.
    c. Compressor suction line.
    d. None of the above.
27. The maximum temperature the four-way reversing valve will tolerate during repairs is:
    a. 200°F.
    b. 250°F.
    c. 300°F.
    d. None of the above.
28. In a pressure-temperature defrost system:
    a. The pressure differential across the coil initiates defrost.
    b. The discharge pressure on the coil initiates defrost.
    c. The saturated discharge pressure temperature initiates defrost.
    d. None of the above.
29. The most common cause of failure to go into the defrost cycle when using a pressure-time-temperature defrost system is:
    a. A broken timer control.
    b. A loose or poorly connected termination bulb.
    c. A restricted differential-pressure switch at the coil.
    d. None of the above.

SECTION 11

Preventive Maintenance

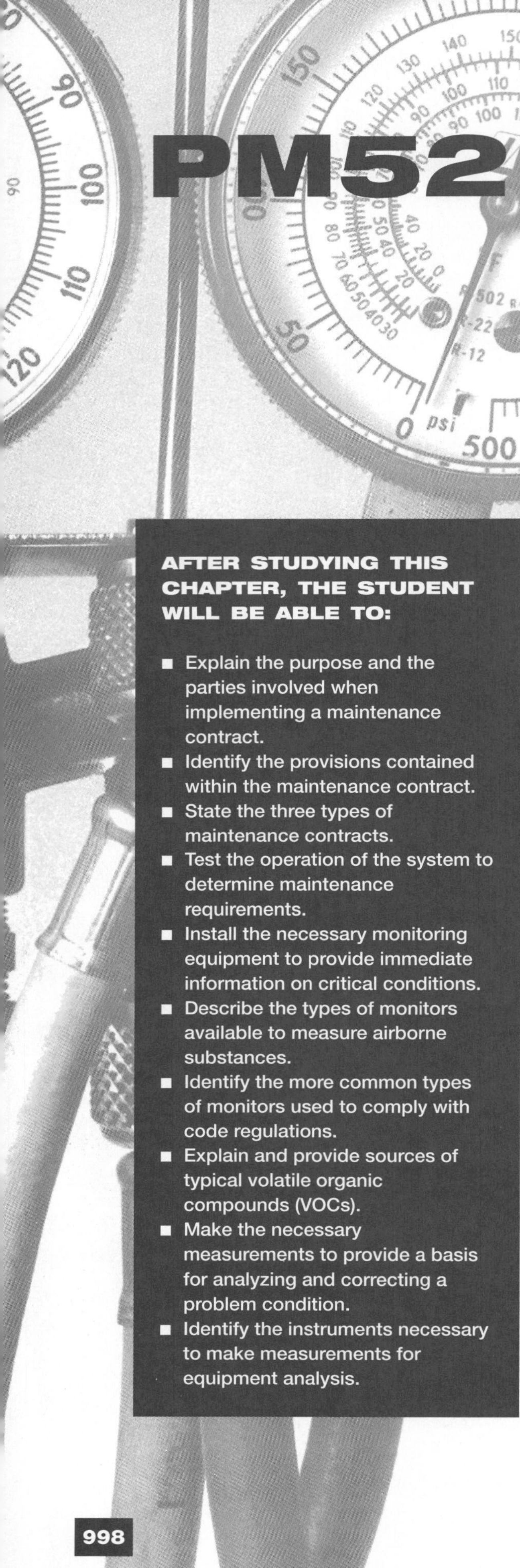

AFTER STUDYING THIS CHAPTER, THE STUDENT WILL BE ABLE TO:

- Explain the purpose and the parties involved when implementing a maintenance contract.
- Identify the provisions contained within the maintenance contract.
- State the three types of maintenance contracts.
- Test the operation of the system to determine maintenance requirements.
- Install the necessary monitoring equipment to provide immediate information on critical conditions.
- Describe the types of monitors available to measure airborne substances.
- Identify the more common types of monitors used to comply with code regulations.
- Explain and provide sources of typical volatile organic compounds (VOCs).
- Make the necessary measurements to provide a basis for analyzing and correcting a problem condition.
- Identify the instruments necessary to make measurements for equipment analysis.

Verifying Operation

PM52-1 MAINTENANCE CONTRACTS

Many contractors offer maintenance contracts for providing periodic customer service to keep the equipment in good working order and to prevent major breakdowns. These contracts usually contain provisions for some of the following items:

1. Period of time covered by the contract
2. Equipment covered by the agreement
3. Work to be performed by the contractor, such as:
   a. Preparing the system for seasonal operation
   b. Performing routine maintenance
4. Materials to be provided by the contractor included in the contract price, and those materials for which the customer will be charged extra.
5. Emergency service
6. Labor to be provided by the contractor included in the contract price, and labor for which the customer will be charged extra.
7. Items that the customer agrees to do, such as:
   a. Operate the equipment in accordance with instructions provided.
   b. Notify contractor of any unusual operating conditions.
   c. Permit only contractor's personnel to work on the equipment.
8. Exclusions of items not covered by the agreement, such as:
   a. Plumbing services beyond the equipment proper.
   b. Electrical service beyond the disconnect switch.
   c. Repairs due to flood, freezing, fires, vandalism, strikes, and other causes beyond the contractor's control.
   d. Expenses caused by obsolescence or insurance requirements.
9. Warranties on parts and labor.
10. Prices for work covered by the agreement.

Basically there are three types of maintenance contracts. Each carries a different price range, as follows:

1. Inspection and lubrication. This is a minimum-cost contract. It includes specified materials, such as filters, but normally materials and labor needed for repair are extra.
2. Full labor. This is similar to the above, except that labor is included and materials are extra. Sometimes labor above a certain limit is extra. This is not a common agreement.
3. Full maintenance. This contract includes inspections as well as labor and materials for repair. To limit the liability to the contractor, limits are often placed on the amount or types of repair. For example, compressors and old equipment due for replacement are excluded.

PM52-2 SYSTEM OPERATIONAL CHECKS

An important procedure in providing maintenance on an existing heating and cooling system is to determine whether or not the system is operating properly. This can be done by checking the sequence of operations to see that it conforms with the design requirements.

For an exercise, use an up-flow induced-combustion gas furnace which can be installed with a split-system condensing unit or heat pump. It also can be equipped with a humidifier (HUM) and/or an electrostatic air cleaner (EAC). The wiring diagram and legend for this unit is shown in Fig. PM52-1 and the control board in Fig. PM52-2. The sequence of operation for the different modes are discussed below.

PM52-2.1 Heating Mode

When the thermostat calls for heat, the R-W circuit closes. The furnace control system performs a self check, verifies that the pressure switch contacts are open, and starts the induced-fan motor. The following process then occurs:

1. The inducer fan runs, and the pressure switch contacts close to initiate a 15-sec pre-purge period.
2. At the end of the pre-purge period, the ignitor is energized for a 17-sec ignitor warm-up period.
3. When the ignitor warm-up period is completed, the gas valve opens, and the burners are ignited. After 5 sec, the ignitor is de-energized and a 2-sec flame-sensing period begins.
4. The HUM terminal is energized with the gas valve.
5. When the burner flame is sensed, the control begins the blower-on delay period and continues holding the gas valve open. If the flame is not sensed, the control closes the gas valve and repeats the ignition cycle.
6. Approximately 40 sec after the blower flame is proven, the blower motor is energized on heating speed.
7. When the thermostat is satisfied, the circuit between R-W is broken, de-energizing the gas valve. The gas valve closes. The blower continues to run for 135 sec for the blower-off delay.
8. The inducer blower remains energized 5 sec after the burners are extinguished. This is the post-purge cycle.

PM52-2.2 Cooling Mode

When the thermostat calls for cooling, the R-G and R-Y circuits close. The R-Y circuit starts the outdoor condensing unit and the combined R-G and R-Y circuits start the furnace blower motor on cooling speed.

When the thermostat is satisfied, R-G and R-Y circuits are opened. The furnace blower continues to operate on the cooling speed for an additional 90 sec.

PM52-2.3 Heat-Pump Mode

When installed with a heat pump, the furnace control automatically changes the timing sequence to avoid long

PRACTICING PROFESSIONAL SERVICES

The best person to sell maintenance contracts is very often the service technician performing routine and emergency repairs during weekends and late evenings. He is the first contact person with the customer, and if accepted by the customer, the best salesperson working for the company. The technician should, upon completing the repair of the equipment, explain the advantages of service contracts and preventive maintenance. If the customer is interested in learning more about specific options and maintenance programs from the company, and their questions are beyond the scope of the service technician, a maintenance contract salesperson should call on the customer to provide the extended details.

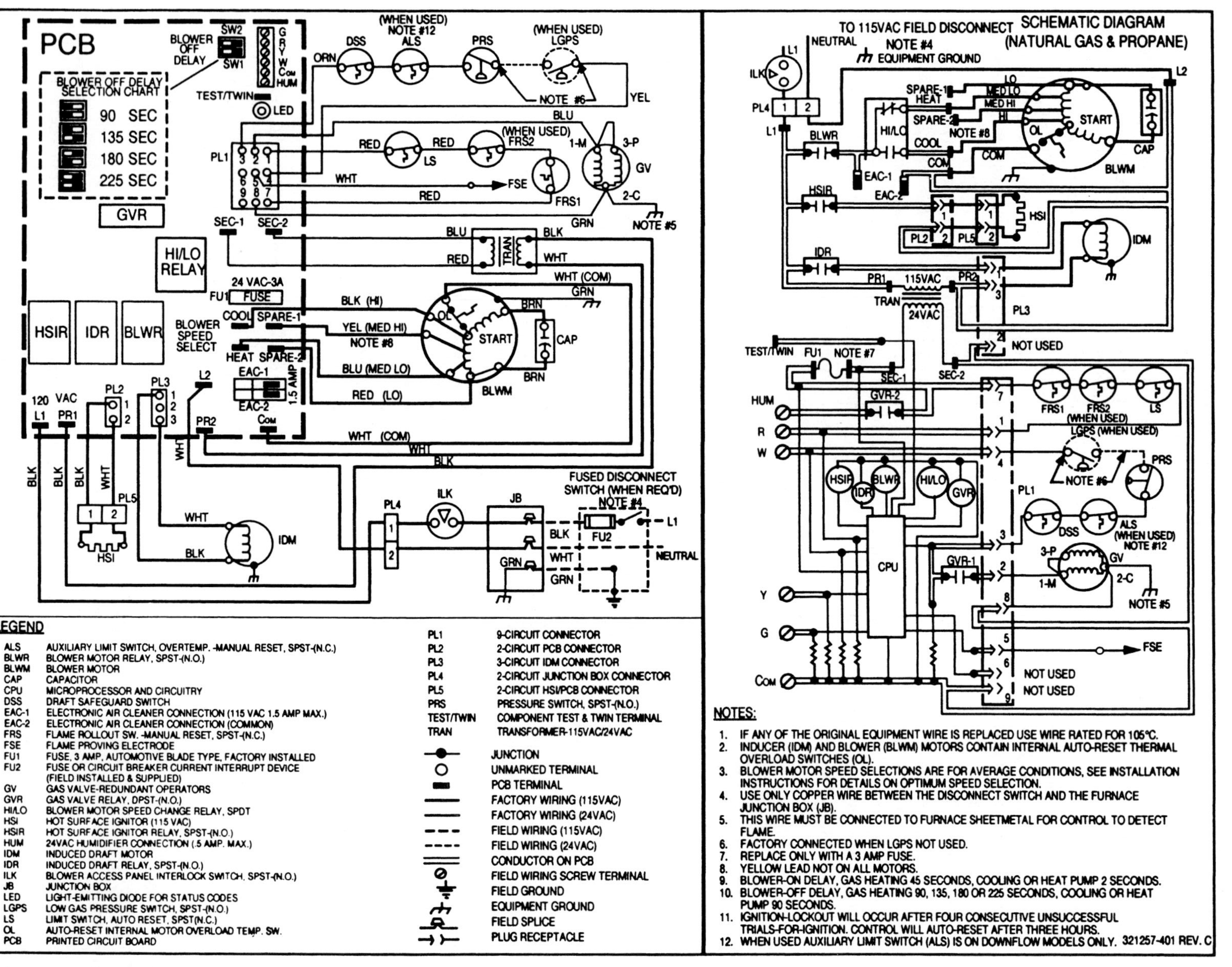

Figure PM52-1 Wiring diagram for a gas furnace designed for use with a condensing unit or heat pump.

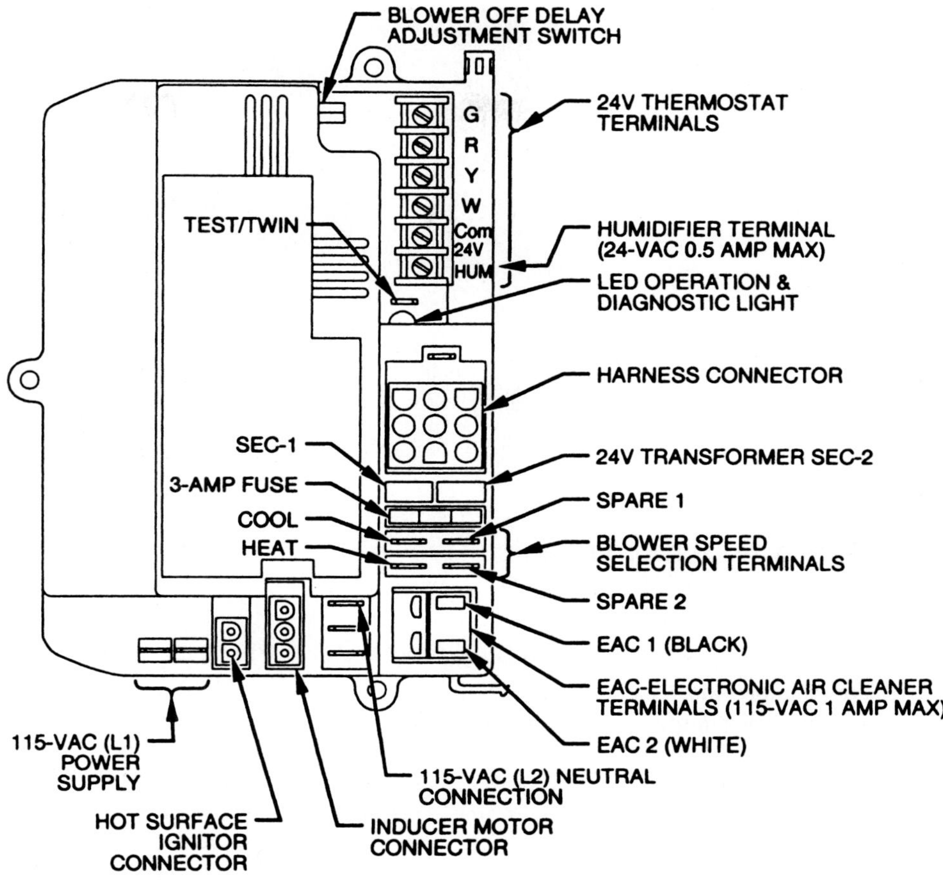

Figure PM52-2
Control board for heating/cooling system.

blower-off time during demand defrost cycles. When the W-Y-G thermostat input is received the control changes the blower to heating speed or starts the blower, if it is off, to begin the heating cycle. The blower remains on until the end of the pre-purge period, and then shuts off until the end of the ignition warm-up. The blower then comes back on at heating speed.

When the W input signal disappears, the control begins the normal inducer post-purge period and the blower changes to cooling speed after a 1-sec delay. If the W-Y-G signal disappears at the same time, the blower remains on for the selected heating blower-off delay period and the inducer goes through its normal post-purge period. If W-Y inputs disappear, leaving the G signal input, the control goes into continuous blower and the inducer remains on for the normal post-purge period.

Any time the control senses false flame, the control locks out of the heating mode. This occurs because the control cannot sense the W input due to the false flame signal, and as a result sees only the Y input and goes into the cooling mode, blower-off delay. All other control functions remain in standard format.

If any problems are found during the system check out, they can either be corrected at this stage or held for more information. Testing and measuring are required to proceed with further analysis.

PM52-3 MONITORING EQUIPMENT

One useful device now being applied to operating systems is the monitor. This equipment senses certain problems at an early stage, notifying operating personnel, making possible immediate repair. For example, a refrigerant monitor can sense a leak as small as 1–10 ppm, and initiate three levels of alarms. This serves to protect the operation of the system as well as prevent costly loss of refrigerant.

Monitors are available for a wide variety of airborne substances, including:

Space temperature	Oxygen
Airflow velocity	Humidity
Sulfur oxides	Pressure
Hydrogen sulfide	Ozone
Carbon dioxide	Formaldehyde
Nitrous oxides	Combustibles
Noise levels	Refrigerants
Carbon monoxide	Radon
Lead and mercury	Vapors

Figure PM52-3
Typical refrigerant monitor.
(Courtesy of General Analysis Corp.)

PM52-3.1 Airborne Particulates/Volatile Organic Compounds

Some of these monitors have been available for many years. For example, most supermarkets continually monitor the storage temperature in critical areas. Various levels of alarms are used. If a system improperly shuts down when the store is closed, the alarm system will activate phone calls to a listing of technicians until it reaches someone.

Recently, there has been an increased use of monitors due to code regulations and improvements in the technology. Some of the newer, more common solid-state monitoring systems are:

1. Refrigerant monitors
2. Indoor air-quality monitors
3. Carbon dioxide monitors

Refrigerant Monitors

A typical *refrigerant monitor* is shown in Fig. PM52-3 and installed in an equipment room as shown in Fig. PM52-4. This model can sample four different locations at the same time. For example, if an equipment room with four chillers is being monitored for refrigerant leaks, a sensor can be located beneath each unit. Since the refrigerant vapor is heavier than air, this is the quickest place to sense the leak. A typical cycle time for sampling all four points is less than 5 min.

Three levels of programmable alarms are common to all sampling locations. Alarms have adjustable set points up to 1000 ppm. Each alarm set point has a corresponding relay contact. An optional expansion board provides additional contacts, so each sample point has its own dedicated alarm relays. In case of sensor malfunction a separate alarm will be energized, indicating that repair or replacement of the monitor is required.

Figure PM52-4
Refrigerant monitor installed in the equipment room.
(Courtesy of General Analysis Corp.)

ON THE JOB

Problem: A customer at a large office building complains that the building air-conditioning system must not be working properly because the occupants are always complaining that they have headaches and are leaving work early each day. What are the possible problems and remedies?

Solution: Upon arriving at the building the service technician notices that the indoor air-ventilation system was recently modified with the intake over the service loading ramp. Further investigations reveal that each morning at 7 a.m. when the cafeteria supply truck arrives, the outside air-ventilation dampers are fully open to help cool down the building, but close down to minimum position about 8:30 a.m. Thus, the problem with the building was that the exhaust fumes were getting into the building early in the morning through the ventilation system. The only viable and least costly remedy was to move the ventilation duct work to a different location to provide clean fresh air into the building. To further ensure the problem was rectified, contract with the building owner to install as part of a preventive maintenance program an indoor air-quality monitoring system inside the building. This will verify minimum ventilation standards for occupied buildings are achieved.

The monitor shown offers both selective and nonselective refrigerant monitoring. It can easily be set to monitor a specific refrigerant or configured to monitor more than one refrigerant. The infrared-based optical system gives it exceptionally high selectivity.

The following installation guidelines should be followed for monitors:

1. The monitor unit should be mounted indoors in an area with minimal variations of humidity and temperature. The display should be readily visible for servicing and calibration.
2. The nylon or copper sampling tube is effective within 500 ft of the monitor and should be located 12 to 18 in. above the floor in an area where refrigerant vapors are likely to concentrate.
3. Airflow dynamics are critical in choosing an optimum location. Consider airflow patterns and activity levels.
4. Alarm contacts can be hard wired to initiate audible or visual alarms, activate equipment-room ventilation systems, and signal building management systems.

Indoor Air-Quality Monitors

The indoor air-quality (IAQ) monitors sense common volatile organic compounds (VOCs) that contaminate indoor air and cause "sick building syndrome" and building-related illness. A VOCplus™ portable air quality monitor of this type is shown in Fig. PM52-5. The entire package consists of a sensor and a controller. When the sensor indicates a pollution problem, the controller activates some mechanical device, such as a fan, to relieve the problem condition.

Utilizing a solid-state oxidation/reduction-type sensing element, the sensor can detect over 100 VOCs. When the total pollution concentration exceeds a preset level, a signal is sent to the controller to operate a pollution-control device. The IAQ sensor and controller regulates the air quality according to acceptable pollution levels as simply as a thermostat regulates space temperature by operating a furnace.

Typical VOCs include:

Ammonia	Acetic acid	Acetone
Acetylene	Amyl alcohol	Benzene
Butane	Butyl alcohol	Butyl formulate
Butylamine	Butylene	Chloroform
Chloro benzene	Carbon monoxide	Carbon tetrachloride

Common sources of VOCs include photocopying materials, paints, gasoline, building materials, refrigerants, and cleaning compounds. The critical levels are established by federal regulatory agencies and enforced by the Environmental Protection Agency (EPA). While they are in the range of 3 mg/m^3, current levels should be identified. Increased ventilation is frequently used to control concentrations.

Applications for IAQ monitors include:

Manufacturing plants	Printing facilities
Conference rooms	Hospitals
Laboratories	Schools
Office buildings	Department stores

Types of apparatus that can be activated by the controller include:

Air cleaners	Outside air dampers
Space ventilators	HVAC/R systems

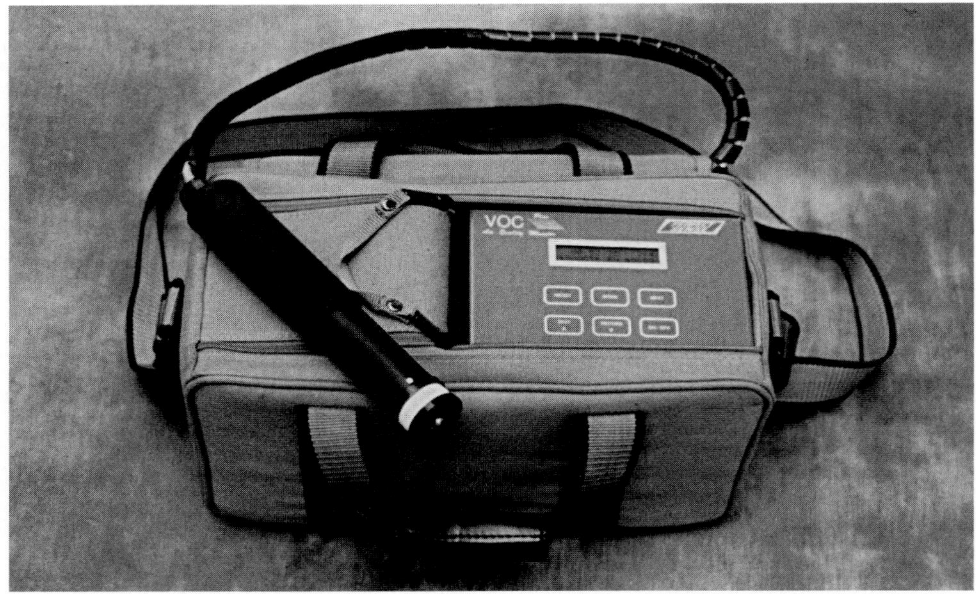

Figure PM52-5 Indoor air-quality monitor. (Courtesy of Perfect Sense, Inc.)

Odor control equipment	Energy recovery ventilators
Alarm indicators	Building control systems

Carbon Dioxide Monitors

The measurement of carbon dioxide within a space has been acknowledged by ASHRAE as a good indicator of the ventilation rate in relation to the occupancy. When people are confined within a space, adequate ventilation is essential to health. Ventilation monitors continually measure the CO_2 in the space and can be used to activate ventilation equipment and maintain the required amount of outside air.

Studies indicate that ventilation control based on CO_2 levels from occupancy requirements saves energy costs. This is compared to supplying a fixed supply of outside air based on a percentage of the total air circulated. Further, by introducing the proper amount of outside air and exhausting the "stale" air, the detrimental effects of floating toxins and organic compounds are reduced. Adequate ventilation can reduce the spread of common viruses and bacteria that lead to temporary sickness, downtime, and loss of productivity.

The CO_2 monitoring device is a non-dispersing infrared carbon dioxide monitor/controller. The diffusion gas sampling chamber is surrounded by a semi-permeable membrane and has an electrically pulsed infrared source and no moving parts.

The device monitors the CO_2 level in the occupied space, comparing it to any user-selected set point between 0 and 5,000 ppm. If the level reaches the set point, the ventilation equipment is automatically turned on. It provides an electrical output to the business management computer of 0–5 VDC or 4-20 mA. This produces a record of the CO_2 level. The sensor is located: (a) in the occupied space to be conditioned, or (b) in a return-air duct from the space. It is recommended that at least one sensor be used for each 25,000 sq ft of habitable space. A calibration kit is available, or factory re-calibration can be provided.

PM52-4 MEASUREMENT AND ANALYSIS

An important part of performing preventive maintenance is to make sufficient measurements to determine the operating condition of the equipment. An accurate comparison needs to be made between the proper performance and the actual performance after a definite period of operation.

PM52-4.1 Test Instruments

To make these measurements the technician must be equipped to use the necessary test instruments. Some of the essential instruments are described below.

VOM (Volt, Ohm, Milliamp) Meter

These instruments have typical ranges as follows:

DC volts:	0.25 to 1000
AC volts:	10 to 1000
Ohms:	R × 1, R × 10, R × 100
DC milliamps:	0.05 to 250
AC amps (with transducer):	0 to 100

These instruments can be supplied in either digital or analog type. The analog type is useful for observing changes and variations in readings. The digital type is accurate and rugged: most digital models will withstand a 5-ft drop on a concrete floor.

Clamp-on AC Ammeters

These meters permit the technician to read line currents to fan motors, compressors, and electric heaters without contacting or disconnecting the wiring. Typical scales available include:

Amps:	0.5 to 1000
Volts:	0.1 to 1000
Ohms:	0.1 to 800

Terminal leads are used in measuring volts and ohms.

Insulation Resistance Tester (Megger)

These instruments are used to test motor windings using a potential of 500 to 1000 V. They are much more reliable for checking hermetic compressor windings for grounds than an ohmmeter powered by only a few volts of DC. They may be battery-powered, AC-powered, or operated from an attached hand-crank generator. They will read resistances of 100 megohms and higher. Normally, a compressor-winding resistance to ground of under 1 megohm is a suspect for a defective compressor.

Leak Detectors

The use of the electronic leak detector is recommended. These detectors are sensitive and reliable and can be secured for servicing all halocarbon refrigerants (HFC, HCFC, and CFCs).

Gauge Manifold Set

This is an essential instrument including both high-pressure and compound gauges. Analog and digital gauges are available. The four-valve models have increased versatility compared to the two-valve models.

Manometers

Manometers are useful in diagnosing problems in the airflow on air-conditioning units. Plastic tube manometers with magnetic clips are available for reading static pressures up to 10 in. WC and below 0.1 in. WC. When used with a Pitot tube they can reliably read air velocities in duct work. Used alone they can read static pressure differentials across filters, coils, and fans. They can also read air-discharge and return static pressures. These tests can indicate problems such as plugged coils, dirty filters, fans running backwards, and closed fire dampers.

For a dry-type differential-pressure gauge, the diaphragm-type magnetic-helix coupled (Magnehelic) models are recommended. These gauges have a simple frictionless movement and resist shock, vibration, and overpressure.

PM52-4.2 Design Maintenance to Fit the Job

A good preventive maintenance program must be designed to fit the individual job equipment. Each type of system has certain critical functions that must be maintained in good working order. For most systems, a program should be arranged for at least two inspections per year.

For example, assume the equipment is an automatic flake-ice machine. A flaker uses a continuous ice-making process. The ice is formed on the inside of the evaporator, cracked and broken away, moved upward, and extruded by a slowly turning auger. This exerts much pressure on the auger and gear motor assembly. The bearing system is critical to the life of the flaker. It is the component that most often fails. These failures can be largely prevented by the application of proper preventive maintenance. The semiannual cleaning and bearing inspection is the best insurance the customer can receive against expensive repairs and downtime.

Twelve Items to Check at Each Inspection

With reference to HVAC/R equipment in general, twelve maintenance checks should be covered during each inspection. They include, but are not limited to the following:

1. Record suction pressures.
2. Record discharge pressures.
3. Record amperage draw of the compressor.
4. Observe condition of the air-cooled condenser, if applicable. Clean it if necessary.
5. Record room temperatures or box temperatures if commercial refrigeration is involved.
6. Observe condition of the evaporator. Clean if necessary.
7. Check all electrical connections. Tighten if necessary.
8. Record ambient temperatures.
9. Oil any motors where applicable.
10. Record voltage at the compressor.
11. Check overall condition of the equipment such as oil leaks, water leaks and refrigerant leaks, where applicable. Schedule repairs where necessary.
12. Check condition of any special components that are unique to that equipment for faulty operation. Schedule repairs or adjustments where necessary.

All recorded information should be entered on an inspection sheet. A copy should be given to the customer, and a copy should be kept in the contractor's file for future reference. These reports are useful in detecting any potential problems by comparing the last inspections with the current inspections.

An important part of any preventive maintenance program is to ensure that the system operates at peak efficiency. By following the above check list, the overall system performance can be determined. For example, a high discharge pressure and a low suction pressure will affect the operating efficiencies of the system.

Another good indicator of the system's performance is the amperage draw of the compressor. Normally, the higher the amperage draw, the higher the load on the compressor. An abnormal amperage draw could be an indication of an internal problem in the compressor.

Early detection of refrigerant leaks is important in preventing downtime and saving resources. By checking the suction and discharge pressures along with the amperage draw of the compressor, a low refrigerant charge can be detected. Whenever and wherever possible install a monitor.

Be sure to check the condition of an air-cooled condenser. A dirty condenser can raise the head pressure, and with increased blockage, can cause the unit to cutout.

An experienced technician always takes an overall look at the equipment. An oil spot on the piping may be an indication of a refrigerant leak. The wiring connections need to be checked for tightness. The entire area should be kept clean. Good preventive maintenance is a benefit to the customer and the contractor.

REVIEW

- Maintenance contract—An agreement between the service organization and the customer to provide periodic service to keep the equipment in good working order and to prevent major breakdowns.
- Maintenance contracts usually contain provisions for:
  Period of time covered by contract
  Equipment covered by the agreement
  Work to be performed by the contractor
  Materials included/excluded
  Emergency services
  Labor provided/labor charged
  Items customer agrees to do
  Exclusions not covered by the agreement
  Warranties
  Prices for work covered by the agreement
- Three types of maintenance contracts:
  Inspection and lubrication
  Full labor
  Full maintenance
- System operational check:
  Determine whether or not the system is operating properly.
    Observe sequence of operation
    Measure system parameters
    Compare readings with design requirements
    Check each mode of operation
    Measure electrical values for reference
    Test limit and safety controls
- Use of monitors and alarms has increased due to codes, regulations, and improvements in technology.
- Some common types of monitoring/alarm systems:
  Refrigerant monitors
  Indoor air-quality monitor
  Carbon dioxide monitor
  Space temperature
  Humidity
  Noise levels
  Pressure
  Oxygen
  Radon
  Fuel gas
- Many of the monitoring devices can be linked to a building-management system. When a sensor indicates a problem, the controller actuates some mechanical device, such as an exhaust fan, to relieve the problem.
- The measurement of CO_2 in a building space can be used as an indicator of adequate ventilation. It can be used to activate ventilation equipment to maintain the required amount of outside air.
- Adequate ventilation as defined by ASHRAE's Standard 15-1994 (1000 ppm, TVOC high level concentration 800 ppm), reduces the detrimental effects of floating toxins and organic compounds. Adequate ventilation can also reduce the spread of common viruses and bacteria.
- Maintenance items for HVAC/R equipment:
  Suction pressure
  Discharge pressure
  Amperage draw of the compressor
  Condition of the condenser
  Room temperatures or box temperatures
  Condition of the evaporator
  All electrical connections
  Ambient temperature
  Voltage at compressor
  Lubrication of motors where applicable
  Overall condition of the equipment
  Condition of any special components

Problems and Questions

1. What is the purpose of maintenance contracts?
2. What are the incentives to the contractor to provide maintenance contracts?
3. What is the purpose of "system operational checks"?
4. What are the steps in "system operational checks"?
5. List the three types of maintenance contracts.
6. What is the primary reason for increased use of monitoring equipment in HVAC/R?

7. What is measured or monitored in building space as an indicator of adequate ventilation?
8. What does ASHRAE Standard 15-1994 (*Safety Code for Mechanical Refrigeration*) set as a limit for building CO_2?
9. What is the limit for toxic volatile organic compounds (TVOC)?
10. Refrigerant leak detectors (monitors) are required in equipment rooms and are linked to automatically activate what device?
11. Maintenance contracts normally have some provision for emergency service. True or false?
12. A maintenance contract will not typically cover costs and expenses associated with obsolescence or insurance requirements. True or false?
13. The full maintenance contract provides the customer with maximum equipment protection while maintaining minimum costs. True or false?
14. A refrigerant monitor can sense a refrigerant leak as small as 1-5 ppm. True or false?
15. When sensing refrigerant leaks, the quickest place to sense the leak is directly above the compressor piping. True or false?
16. A modern solid-state monitoring system:
    a. Includes refrigerant monitors.
    b. Includes indoor air-quality monitors.
    c. Includes carbon dioxide monitors.
    d. All of the above.
17. Volatile organic compounds contaminate the indoor air and cause:
    a. Mildew to become present around windows and high moisture areas.
    b. Reduced outside supply ventilation air.
    c. Sick building syndrome.
    d. None of the above.
18. Studies indicate that ventilation control based on _____ levels from occupancy requirements saves energy costs.
    a. Oxygen
    b. Carbon dioxide
    c. Carbon monoxide
    d. None of the above.
19. The insulation resistance tester device is called the:
    a. Ohmmeter
    b. Motor insulation tester
    c. Megger
    d. None of the above.
20. Manometers when used with the _____ tube can reliably read air velocities in ductwork.
    a. Pitot
    b. Static
    c. Velocity
    d. None of the above.

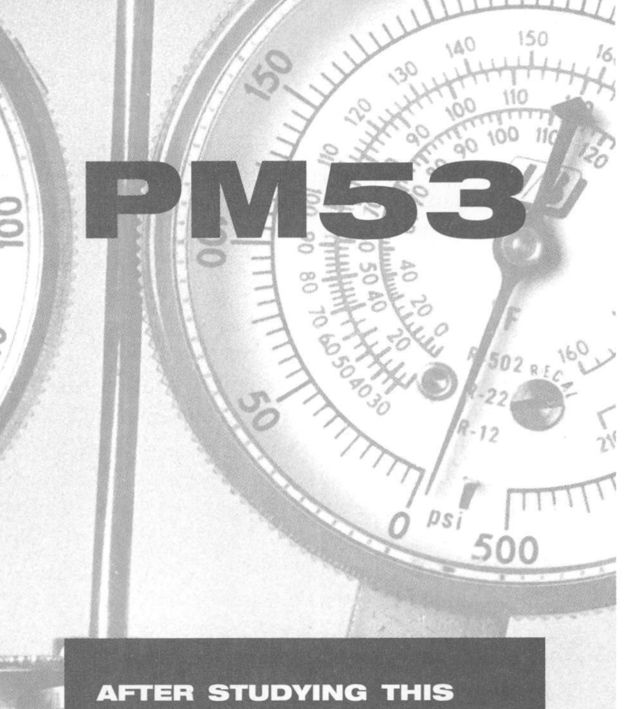

PM53 Preventive Maintenance

AFTER STUDYING THIS CHAPTER, THE STUDENT WILL BE ABLE TO:

- Identify special tools and equipment for preventive maintenance.
- Describe routine maintenance operations performed at each inspection for refrigeration/air-conditioning.
- Describe preventive maintenance requirements for electric, oil, and gas heating systems.
- Identify routine maintenance of the air and water components for hydronic heating.
- Explain the maintenance required on heat pumps for optimum year-round efficiency.
- Describe the methods to check and clean coils and electronic air filters.
- Identify the importance of replacing and/or cleaning filters.
- Check and inspect belt-drive assemblies.
- Perform sound-level monitoring.
- Identify the importance of proper lubrication methods.
- Perform visual inspections and calibrations as required.
- Understand the importance of scheduling parts to be changed prior to failure.
- Perform safety checks on the equipment as necessary.

PM53-1 OVERVIEW

In this section we have reached the point where the components of the system have been identified, the system has been operated, and vital measurements have been made, and a list of preventive maintenance tasks have been set up that need to be performed.

Some of the special tools and servicing equipment that the technician will find useful beyond the usual assortment of combination open/box wrenches, flare-nut wrenches, adjustable wrenches, screwdrivers, and pliers, are the following:

1. Spray units for cleaning coils, either hand operated or pressure pump types. Spray units handle cleaning detergents which dissolve accumulations on the surface and between the fins of evaporator coils, heating coils, and condenser coils.
2. Refrigerant recovery and reclaim units, which are essential in removing refrigerant from the system to permit the repair and replacement of components.
3. Refrigerant measuring units, such as weight-charging meters or dial-a-charge units, to accurately charge package units and monitor the refrigerant used for replacement.
4. Nitrogen pressure regulator, for use with a tank of nitrogen, to purge tubing being brazed or for use in pressurizing a system for testing.
5. Torque wrench, for working on open and semi-hermetic compressors. These wrenches ensure tightening bolts and nuts to the manufacturer's requirements.
6. One quarter-inch socket-wrench set and driver bits, for adjusting valves.
7. Nut drivers, to save time in opening enclosures.
8. Pinch-off tools, to close compressor pinch-off tubes used in servicing hermetic systems.

PM53-2 PREVENTIVE MAINTENANCE ROUTINES

PM53-2.1 Refrigeration/Air-Conditioning Maintenance

Every maintenance contractor should perform a series of routine maintenance operations at the time of each inspection. The following list of maintenance items used by one contractor is an example:

1. Clean and flush condensate drain pans and drain lines. Clean, lubricate, and test condensate pumps.
2. Check fan belts. Replace worn or cracked belts with new ones. Adjust belt tension.
3. Lubricate bearings. For oil-type bearings, use non-detergent oil (not automotive engine oil) of weight (viscosity) specified by bearing or motor manufacturers. Technicians find that plastic containers with extended spouts are useful for reaching out-of-the-way oil cups.

   For grease-type bearings, use grade and type of grease specified. Do not over-grease. Forcing excessive grease into bearings can damage the grease seals and lead to eventual failure. Where a drain or vent plug is provided, follow instructions and allow excess lubricant to escape.
4. Clean up the equipment. Wipe up any oil or grease spills. Clean equipment makes it easy to spot oil leaks, which sometimes indicate refrigerant leaks and bearing problems. Be sure to tighten packing glands on refrigerant valves and replace valve caps.
5. Check crankcase oil heaters. Follow the instructions for turning on the heater for 12 to 24 hrs before attempting to start the compressor.
6. Check and tighten all power connections. Loose connections will overheat, damaging wire insulation and could cause motor failures.
7. Examine starter contacts for burning or excessive pitting. Check for free operation of the armature without binding. Rebuild or replace defective units.
8. Examine the air filters. Replace dirty disposable filters or replace with washable filters. Ordinary disposable air filters only have an efficiency rating of 5 to 10%. Often it is possible to install a better filter with 30% cleaning efficiency (same size), at only a modest increase in cost. Most owners appreciate an improvement in indoor air quality.
9. Examine the water pumps. Mechanical seals do not require service. Packed stuffing boxes usually need repacking periodically. Do not tighten the packing gland so tight that no dripping occurs. This can score the shaft or burn the packing rings. Some dripping is desired and normal.
10. Check the refrigeration circuit. Attach a gauge manifold and record operating pressures. Convert refrigerant pressures to their corresponding refrigerant temperatures. Check the refrigerant sight glass (if provided) to see that the charge is adequate. Check compressor oil sight glass (if

ON THE JOB

The following is a partial list of items, forms, and records needed for a professional service technician:

- Customer's Service Information
  - Name
  - Address
  - Phone number
  - Service complaint
  - Dispatch time
  - Arrival time
  - Completed time
- Forms
  - Refrigerant use log
  - Accidental refrigerant venting report
  - Refrigerant removal form
  - Material safety data sheets
  - Truck parts inventory
  - Truck tool inventory
  - Maintenance agreement forms
  - Time sheets
  - Service tickets
  - Sales lead forms
  - Reference manuals

ON THE JOB

Complete all forms, service tickets and warranty claims at the service call or job site. This is the best time to recall what was done, parts used and time spent on the job. Remember this is part of the time that should be charged to the job. Doing your paper work at home, in the shop or giving it to the dispatcher or service manager should not be done. Incomplete forms, or forms that are lost, filled out incorrectly or in a messy manner cost you and your service organization money.

provided) to verify that the oil charge is adequate.

11. Check the cooling towers or evaporative condensers. Clean the sumps before filling. Scrape and paint rusty spots. Make sure all spray nozzles or openings are clear and functioning properly.
12. Check the operating temperatures of the system. Many experienced technicians can make a preliminary check by feeling various parts of the system to determine any abnormal temperatures.
13. Check the condition of the evaporator and air-cooled condenser coils. They must be clean for the system to operate properly.
14. Always shut off the power before servicing equipment. Technicians tend to become complacent and forget the hazard of electric shock. Play safe, always shut off the power before servicing.
15. Clean-up. A dirty littered workplace can be hazardous and certainly makes a poor impression on the owner or manager. Old belts, dirty filters, broken controls, empty refrigerant drums and oil cans should be removed from the premises or placed in the dumpster. A good clean-up lets people know the technician is professional.

PM53-2.2 Heating Systems Maintenance

Heating-system preventive maintenance is much like air conditioning insofar as common elements such as filters, fans, motors, and belt drives are concerned. In addition to these common air-side components, the other necessary maintenance is associated with the type of fuel used or the manner in which the heat is supplied. Airside maintenance is discussed under air-conditioning maintenance (above). The unique features related to the fuels that require maintenance are as follows:

Electric Heating

Electric heating can be panel-type, baseboard-type or forced air. For the system to maintain its original efficiency, the heat exchanger must be kept clean. Periodic cleaning is particularly important on finned-tube baseboards. Due to the gentle airflow over the heating elements, no air filter can be used. Dust and lint can accumulate in the fins and restrict heat transfer.

Special attention also is necessary wherever electrical contacts are exposed. This condition exists on certain

PREVENTIVE MAINTENANCE

When maintaining thermostats, make sure the thermostat is level. For some thermostats this is critical; for others it is just an aesthetic point for the customer. Controls are subject to daily use, and vibrations can cause the wires to come loose. Tighten the lugs with every inspection. Contacts should be kept clean. To clean the contact (except sealed-glass types) use a business card or match book cover between the points. Never use abrasive, or files on the contacts. Check calibration, follow manufacturer's recommendations. The thermostat should be installed away from drafts, heat sources such as lights, outer walls, and entry doors. Batteries should be replaced annually or more often if needed. Check anticipator or cycling control for proper cycling, and adjust if necessary. Boiler, electric heat, gas and oil furnaces, and air conditioners all require different cycling rates for human comfort. After all the adjustments have been made, observe the thermostat operation, letting the control go through a complete cycle. This will ensure that the control is operating correctly before you leave.

types of thermostats. They need to be inspected and cleaned when necessary. Dirty contacts can prevent the thermostat and other controls from working properly.

The fuses on individual heating elements sometimes fail, keeping the circuit open. This problem may appear to be a burned-out element, when simply replacing the fuse will put the element back in service.

Gas Heating

The gas should burn with a clear blue flame. There should not be any yellow tips on the flame. If there are, and they cannot be cleared up by adjustment, they are probably due to rust, dust, or scale on the burners or manifold. Occasionally this material can be removed by a brush and a vacuum cleaner. If the material adheres tightly, however, it may be necessary to remove the burner and use a pressure air hose to blow down through the burner ports. Goggles should be worn any time high-pressure air is used for cleaning.

Standing gas pilots can be a problem. The pilot provides several functions: (1) A flame to ignite the main burner, (2) A support to hold the thermocouple in proper position, and (3) A supply of heat for the thermocouple to produce the millivoltage necessary to hold in the safety cutoff. A steady flame should impinge on the thermocouple about ½ inch. Occasionally these units need to be cleaned. Compressed air usually removes unwanted material that interferes with the flame.

If the furnace uses a natural draft vent, the draft action can be checked by holding a lighted match near the vent opening. The draft should draw the flame toward the opening. Any obstructions need to be removed.

On high-efficiency furnaces, the induced-draft blower may need periodic lubrication. The condensing portion of the secondary heat exchanger may be a finned surface that should be inspected for possible obstructions. The condensate drain must be kept free from any blockage.

It is usually desirable to observe the operational sequence of the furnace from the call for heat by the thermostat to the actual supply of heat; the same holds true for the reverse action, when the thermostat is satisfied. The burner should light properly on start-up, with delayed starting of the fan. On shut down, the burner flame is extinguished and the fan continues to run until the residual heat is removed. On high-efficiency furnaces there is a purge cycle operated by the induced draft fan which precedes the lighting of the burner. The limit control setting can be checked by operating the burner with the fan disconnected. This is a good safety procedure.

Oil-fired Heating

This equipment requires regular maintenance in order to function properly. Burning oil is a more involved process than burning gas. The liquid oil must be heated, vaporized, and mixed with the correct amount of air to burn.

Special equipment has been designed for burning oil. Oil is delivered to a spray nozzle through tubing under a pressure of 100 psi. A forced-draft blower supplies combustion air to the oil spray. The oil is ignited by a high-voltage spark positioned close to the spray. The flame retention ring controls the shape of the flame to fit the insulated combustion chamber. The draft over the fire is held constant at a negative pressure of –0.04 in. WC using a counterbalanced draft regulator. Safety controls are provided to shut off the flow of oil in case the flame is extinguished. The burner must operate efficiently or soot can accumulate and interfere with proper operation.

The service technician must constantly be on the lookout for leaks in the oil lines. Oil can leak out through any kind of an opening in a pressurized oil line.

If the system uses a tank submerged in the ground with an oil line operating in a vacuum, air or water can leak in, mix with the oil and create a fuel burning problem. It may be desirable to eliminate the outside tank and replace it with an inside tank.

Water can accumulate in an oil tank due to condensation, particularly when the tank is nearly empty during the

PREVENTIVE MAINTENANCE

For blowers, always clean the blower compartment once each season. Vacuum the compartment and blower wheel, and if necessary scrape each fin of the wheel to remove dirt buildup. Follow the motor manufacturer's recommendation for lubricating the motor. Check wiring, tighten up each wire lug, look for cracked or frayed wires, and replace if necessary. Check pulleys and sheaves, and tighten set screws. Pulleys should be aligned, with proper belt tension. Belts should be free of wear and cracks. Lubricate blower bearings; if bearings are permanently sealed, they require no further lubrication. Replace the air filter and inspect blower door to insure it is sealing tight.

summer months. Each year the technician should check to see if any accumulation of water is in a submerged tank, requiring corrective action. For this reason, it is considered good practice to keep the tank filled with oil when the system is not in use.

It is also good practice to provide preventive maintenance in the fall before the heating system is put into service. Routine service requires checking the entire combustion process and making adjustments or replacing parts where necessary. Annual replacement of the oil filter is standard procedure.

The annual maintenance for an oil burner should include the following:

1. Burner assembly
   a. Clean fan blades, fan housing and screen.
   b. Oil motor with a few drops of SAE No. 10 oil.
   c. Clean pump strainer.
   d. Adjust oil pressure to 100 psi.
   e. Check oil pressure cutoff.
   f. Conduct combustion test and adjust air to burner for best efficiency.
2. Nozzle assembly
   a. Replace nozzle.
   b. Clean nozzle assembly.
   c. Check ceramic insulators for hairline cracks and replace if necessary.
   d. Check location of electrodes and adjust if necessary.
   e. Replace cartridges in oil-line strainers.
3. Ignition system and controls
   a. Test transformer spark.
   b. Clean thermostat contacts.
   c. Clean control elements that may be contaminated with soot, especially those that protrude into the furnace or flue pipe.
   d. Check system electrically.
4. Furnace
   a. Clean combustion chamber and flue passages.
   b. Clean furnace fan blades.
   c. Oil fan motor.
   d. Replace air filter.

After this work is completed, run the furnace through a complete cycle and check all safety controls. Clean up the exterior of the furnace and the area around it.

Caution: If the unit runs out of oil or has a leak in the oil lines, the fuel pump can become air-bound and not pump oil. To correct this, the air must be bled from the pump and replaced with oil. On a one-pipe system, this is done by loosening or removing the plug on the port side of the intake. Start the furnace and run until oil flows out the opening, then turn it off and replace the plug. The system can then be put back into operation. A two-pipe system is considered to be self-priming; however, if it fails to prime, follow the procedure just described for a one-pipe system.

Hydronic Heating

The preventive maintenance for hydronic systems involves both air and water components. Terminal units, where the heat is transferred to the space, must be kept clean. Fan motors must be properly lubricated.

One of the problems encountered in hydronic systems is air collecting in the top of a room convector, interfering with the flow of water in the system. Normally it is standard practice to install manual or automatic air vents at the highest points of the system. These vents can be used to exhaust the air from the system.

Occasionally, depending on the system and the accessories, the expansion tank will become waterlogged and lose its cushion of air. Air needs to be pumped back into the tank or the tank needs to be replaced with a design, such as the diaphragm type, where the air cushion is maintained.

If the system overheats, it may be due to maintaining too high a boiler water temperature. Either the operating control needs to be reset or a control arrangement selected that will vary the boiler water temperature with the changes in outside temperature.

Circulator pumps may require maintenance. Periodic lubrication is needed on some types. Others have couplings connecting the motor to the pump shaft, where the alignment needs to be checked.

If the water being circulated in the system is contaminated, the system shou 1 be drained and refilled with treated fresh water. Water treatment must comply with local requirements.

Heat Pumps

Airflow maintenance on a heat pump is similar to that of an air-conditioning unit. Reference can therefore be made to the information supplied above.

Where the heat pump involves a split system, the technician should give special attention to the condition of the outside coil. Any blockage can interfere with airflow and cause a reduction in capacity. The outside coil must be kept clean.

At least once a year the compressor contactor should be examined for pitting or loose connection. This device is used more frequently on a heat pump than on an air-conditioning system and shows wear much quicker. Damaged contactors or wiring with poor insulation should be replaced.

Another critical condition that should be examined is the frost build up on the outdoor coil during the heating cycle. Unless the frost is removed, the capacity of the unit will be reduced and the safety controls can shut the unit down completely. The defrost cycle can be simulated on most units by disconnecting the fan and operating the unit in the heating cycle. The defrost cycle should begin when the coil temperature reaches 26°F and should terminate when the coil temperature reaches 50°F.

PM53-3 COIL MAINTENANCE

Although the filters remove much of the dust and dirt carried by an air conditioning system, foreign materials do collect on finned-tube coils and need to be removed. Heat and moisture on specific coils can increase the holding capacity for these materials that gradually block the flow of air and reduce the heat-transmission rate.

One of the best ways to clean a finned coil on the job is to use a pressure spray of cleaning material to loosen dirt, lint, oil, and other grime and remove it from the area. The cleaning material is sprayed on under a pressure of 100 psi or more. It effectively removes the unwanted material and uses a relatively small amount of cleaning fluid. A spray gun with a fairly large orifice (at least 1/16 in. diameter) should be used. The following points need to be considered when using this procedure:

1. Some portable means of pressure needs to be provided. This can be an air compressor or a small CO_2 cylinder with a pressure-reducing valve.
2. The cleaning material must be nonflammable and nontoxic.
3. The cleaning material must not have an objectionable odor and the room must be well ventilated.
4. The operator of the hose or spray must be protected to prevent being scalded with hot water or steam, having debris blown into the hair or eyes, getting harsh chemicals on the skin, or being injured by a whipping hose. The use of rubber gloves, goggles, boots, and special clothing may be advisable under certain conditions where such risks exist.
5. If electrical machinery is to be cleaned, the cleaning material must not contain water. Both sides of the electrical line must be disconnected before the cleaning takes place.
6. Where dust and lint are loose, the use of a vacuum cleaner is recommended.

PM53-4 MAINTENANCE OF ELECTRONIC AIR FILTERS

Electronic-air-filter screens and cells must be cleaned periodically with detergent and hot water. Some designs include an automatic washing arrangement with the filter screens and cells in place. Other designs require the removal of the screens and cells for cleaning. The frequency of cleaning depends upon the application. After initial start-up, the filters should be inspected frequently. Once the rate of build-up has been observed, the appropriate frequency for regular cleaning is easily determined.

The filter-washing procedure for units without an automatic built-in washer is as follows:

1. Turn the electronic filter, blower, and furnace power off.
2. Remove lint screens and ionizing collecting cells.
3. Clean using hot soapy water, and rinse thoroughly.
4. Replace lint screens and ionizing collecting cells after they have dried thoroughly.

A properly operating unit will be indicated by black water when the cell is washed. If a cheese cloth placed over the air-outlet grille becomes discolored, the electronic air filter is not working properly.

PM53-5 FILTER REPLACEMENT

One of the most neglected items of maintenance on air-conditioning systems is the periodic cleaning or replacement of air filters. As the mechanical filter clogs with dirt, it can reduce airflow to the point where the cooling coil will freeze and possibly lead to a compressor failure. In a heating system, dirty filters can cause overheating and reduce the life of the heat exchanger, or it can cause nuisance tripping of the limit switch. Dirty filters increase operating cost as the efficiency goes down.

The filter(s) furnished on new HVAC units are often the throwaway type. They are easily accessible by removing a panel on the unit. They will then slide out and can be replaced by a similar filter. The filter material consists of continuous glass fibers packed and loosely woven that face the entering-air side, and more dense packing and weave on the leaving-air side. Air direction is clearly marked. Filters are usually 1 in. thick for residential equipment. Some filters have fiber material coated with an adhesive substance in order to attract and hold dust and dirt.

Many technicians replace these with a permanent or washable filter which has longer life. It consists of a metal frame with a washable viscous-impingement type of material supported by metal baffles and graduated openings or air passages. When these filters are removed and cleaned with detergent, they are dried and re-coated before reinstallation.

PM53-6 BELT-DRIVE ASSEMBLY CHECKING

When the installed equipment is belt-driven, both the alignment and the tension of the belt must be checked. Improperly aligned belts show excessive wear and have an extremely short life. Loose belts wear rapidly, slap, and frequently slip. Belts that are too tight may cause excessive motor-bearing and fan-bearing wear. Belt alignment is simple to check; a straightedge laid along side of the pulley and flywheel will show any misalignment immediately.

Belt tension should be checked. The belt stretches a little, so the belt should be set at the right tension. The correct tension allows 1 in. deflection on each side of the belt.

PM53-7 SOUND-LEVEL MONITORING

Unacceptable levels of noise and vibrations from HVAC/R systems can be avoided once the technician becomes familiar with some fundamental principles about sound and vibration. This includes knowing the characteristics of sound, the pathways sounds take, the effect of sound on people, and the factors to apply in achieving satisfactory installation and operation in any given location.

Sound is generally considered a longitudinal wave travelling in a fluid medium such as air or water. It is identified by its speed, frequency, and wave length, which are interrelated. Sound is generated by a vibrating body or a turbulent air or water stream. HVAC/R systems produce both. The number of cycles per second, expressed in Hz, is the frequency, and defines the audible levels of sound as ranging between 20 Hz to 20 kHz. Within any given range, the frequency is also characterized by its sound power output, expressed as octave or ⅓ octave bands (ANSI standards for rating the acoustics, for example, of an operating room).

Noise is any unwanted sound or a broadband of sound having no distinguishable frequency characteristics. Broadbands are often used to mask other sounds.

The intensity of sound is measured in a number of ways. In acoustics, where the quality of sounds in a room is being analyzed, the strength of sound sources, levels and attenuation are measured by the *decibel* (dB) unit. Since the human ear is pressure-sensitive, a decibel scale for sound pressure is based on pressure squared being proportional to intensity. The unit used for this measurement is the micropascal (μPa) and the threshold of hearing is about 20 μPa. The human ear can tolerate a broad range of sound pressures before reaching the level of pain. Fig. PM53-1 shows some typical sound sources, their sound pressure levels and the human reactions to them.

The quality of sound, as well as the loudness, is also important. Diffusers are often used to fill in the frequencies of equipment sounds which have unpleasant low or high tones. This becomes complex when a number of such sources are present.

Measuring sound levels is accomplished with a battery-operated, hand-held sound level meter with an attached filter set. The meter has a microphone, internal electronic circuits, and a readout display to measure the pressure at any given location.

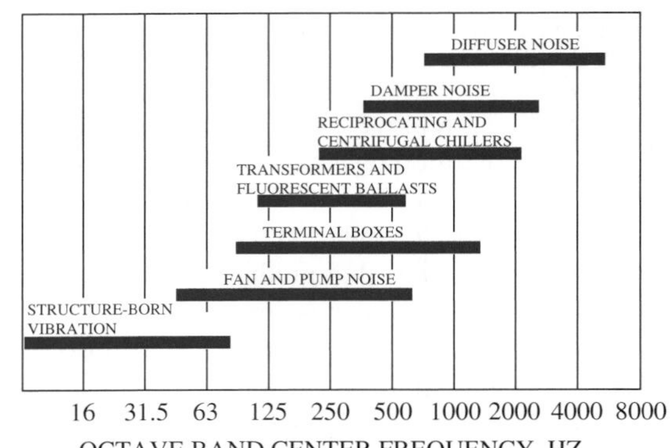

Figure PM53-1 Sound frequencies of various types of mechanical and electrical equipment. (Copyright by the American Society of Heating, Refrigerating, and Air-Conditioning Engineers, Inc. Used by permission.)

In measuring sound, a type "A" filter is placed in the microphone circuit to reduce the intensity of the low frequencies. A typical instrument for measuring sound is shown in Fig. PM53-2. This is known as a *Noise Dosimeter.* It measures the sound level in dBA units (decibels, using an "A" filter). It is supplied with an integrated data logger. The stored data can be viewed on a computer monitor, or a hard copy of the data can be produced on a printer attached to the unit. It records up to 31 hrs at 1-min. intervals of average and maximum readings for each period.

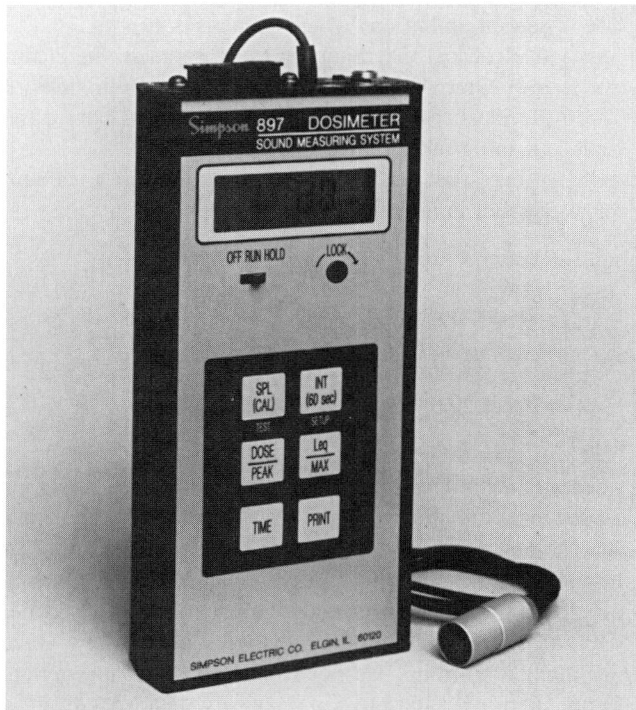

Figure PM53-2 Noise dosimeter with data logger. (Courtesy of Simpson Electric Company)

The technician is concerned with reducing noise and vibration from HVAC equipment wherever possible. Noise can be disturbing to the customer. Vibration can create noise, but in many cases the most serious effect is that it can be destructive.

PM53-8 LUBRICATION

Electric motors that require periodic lubrication should be checked and oiled with care. Motors can be damaged by over-oiling.

When a motor is over-oiled, oil runs out of the bearing into the housing and is sprayed off the rotor throughout the interior of the motor. This coats the windings, terminals, and starting switch, causing gumming, carbonizing the moving parts, and often causing motor failure. It is better to oil the motor every two years than to do it more often than required.

The type of oil is also very important. The oil should be heavy enough to provide proper lubrication. It should not contain soaps, detergents, or any other additives. Three-in-one, Finol, or any other rust-preventive oil should never be used in blower or fan motors. Also, ordinary automotive engine oil should never be used, regardless of the make or grade. Use electric motor oil of a weight or grade specified on the motor. Generally, this is No. 10 weight.

PM53-9 VISUAL INSPECTION

Most service and maintenance technicians make a visual inspection of the entire system before applying instruments or starting their work. Many problems are evident through visual indicators, and considerable time can be saved by this procedure.

Here are some of the system conditions that point to problems or the need for maintenance:

1. Water leaks. A condensate pan may have a clogged drain and water will overflow on the floor.
2. A dirty or obstructed condenser coil. Outside condensing units are subject to air blockage due to the growth of shrubbery. There also may have been construction work in the area and airborne dirt has collected on the coil.
3. A motor or transformer may be discolored due to excessive heat. This can be an indication of an overload.
4. An excessive operating noise. This could be a broken valve on the compressor or a loose pulley wheel, requiring further investigation.
5. Reduced airflow. Dirty filters reduce the airflow and the capacity of the system.

Here are some of the refrigeration conditions that point to problems or the need for maintenance:

1. Short cycling on the low-pressure control. This may indicate a restriction in the liquid line that needs further investigation.
2. Oil on a refrigerant line. This may indicate a refrigerant leak.
3. Bubbles in the refrigerant sight glass. This is usually a sign that the system is short of refrigerant.
4. Low oil level in the compressor sight glass. Oil probably needs to be added to the compressor crankcase.
5. A kink in a refrigerant line or capillary tube. This usually creates a restriction and can indicate a vibration problem that may later cause a break in the tubing.

PM53-10 CALIBRATION

It is important in using test instruments to be certain that they are indicating accurate readings. Some means needs to be provided for periodically checking most instruments.

Instruments do vary widely in their need and method of calibration. Some instruments, such as a liquid manometer, do not require re-calibration unless fluid is lost. Under normal usage they retain their original accuracy.

Some instruments, such as an ohmmeter, require re-calibration every time they are used. Each ohmmeter has a calibrating wheel that is turned as required to "zero in" the instrument before it is used.

Some instruments, such as a pressure gauge, have a calibration adjustment on the face of the instrument which is used occasionally, particularly after the gauge has been subjected to pressures beyond the scale reading.

Instruments, such as a refrigerant monitor, can be calibrated in the field using a special calibration kit furnished by the manufacturer.

Other instruments, such as an electronic thermometer, need to be sent into the factory for calibration.

Any time an instrument is used or purchased, the technician should determine the type of calibration required for that particular instrument. By comparison with another instrument of the same kind, it may be feasible to determine whether or not the instrument is reading correctly. The instructions for re-calibration usually are supplied with printed material that comes with the instrument or can be obtained from the manufacturer.

PM53-11 SCHEDULING PARTS CHANGES

During a preventive maintenance call, the technician may find certain parts that need changing. Some of these are normal-period parts changes, such as replacing the oil filter and oil-burner nozzle in the fall before the seasonal

ON THE JOB

Here are some tips for safe driving:

Keep your truck in good repair, have a well organized storage area, and wash it frequently.

Check tires, oil, etc. Make a check list to be used as a daily, weekly, and monthly check.

Follow the speed limit. Slow down on curves. If in doubt about weather conditions and headlights, turn on your lights. Allow a greater braking distance between you and the car ahead in wet, icy, or snowy road conditions.

Have a first aid kit in the truck.

Don't drink and drive. Alcohol accounts for half of all traffic deaths each year.

Carry a fire extinguisher and traffic warning sign kit.

During winter months in cold climates carry extra winter clothes and a winter survival kit, containing blankets, flashlight, and snacks just in case you get stranded.

Drive defensively. When available take a defensive driving class.

Have an accident kit in the truck. Know what information your company, and insurance carrier will permit to be given out. Have a list of the police department, fire department and other emergency numbers.

Watch out for changing weather conditions.

start-up. Other parts changes may be needed due to wear or to prevent certain problems from recurring.

For example, the technician may find on a semi-hermetic compressor that the valves are leaking and should be replaced. This may be caused by the liquid refrigerant slugging the compressor. As a result the service person may conclude that a suction-line accumulator needs to be installed. The problem is not solved by just replacing the valve plate, but the cause must be corrected to prevent recurrence.

Properly scheduling the work requires securing the approval of the owner for the extra expense, securing the necessary parts and tools, and arranging for a return date.

One condition to be avoided is making a parts change without proper analysis (trial-and-error method). For example, if a unit is stopped by the high limit control, the problem is not usually solved by changing the limit control. The cause may be an excessive load. When the load is reduced, the unit is found to run in a satisfactory manner. A well trained technician is sure that the part is needed before replacing it.

The preventive maintenance technician should carry with him all the parts and tools required for normal maintenance. He or she should have a record of the size required for such items as replacement air filters. If a return call is necessary to complete the work, the technician should be well equipped with materials and tools to efficiently complete the job.

PM53-12 SAFETY CHECKS

Too often accidents occur because safety controls fail to function. An important part of the technician's job is to be certain that everything possible is done to provide for safety to the occupants, the workers, and the equipment.

Each system has a series of devices and work practices designed to provide safety. These range from overload protection to devices that stop the system if a dangerous condition exists.

Here is an example of a protective device that needs to be checked by the preventive maintenance technician during routine maintenance. Every boiler installation has at least one pressure-relief device that relieves the excess pressure if it builds up to limiting levels. Since these devices are seldom (if ever) used, corrosion can form to increase their relief-pressure limits. The maintenance technician can provide an important service by manually opening these valves and flushing them out. If any sticking is found in their operation, they should be re-piped and replaced. They must function if ever needed.

Another example of an important safety device that must be checked during routine maintenance is the "no-flame" control on a furnace. If the burner fails to light when fuel and ignition are supplied, the fuel must shut off. This can be tested both on an oil burner and a gas burner. If the safety equipment is defective, it should be replaced.

The monitoring of environmental conditions involving HVAC systems is increasing to prevent the occurrence of health problems. Detecting operational problems at an early stage is a key to preventing health problems.

REVIEW

- Special tools and servicing equipment:
  Spray units for cleaning coils
  Refrigerant-recovery/recycle units
  Refrigerant-measuring units
  Nitrogen pressure regulator and nitrogen tank
  Torque wrench

Specialty wrenches and nut drivers
Pinch-off tools

- Refrigeration/air-conditioning maintenance routines:
  Clean and flush condensate drains.
  Check fan belts.
  Lubricate bearings.
  Clean up equipment.
  Check crankcase oil heaters.
  Check and tighten all power connections.
  Examine starter contacts.
  Examine air filters.
  Examine water pumps.
  Check the refrigeration circuit.
  Check cooling towers or evaporative condensers.
  Check the operating temperatures of the system.
  Check the condition of evaporator and condenser coils.
  Always shut off power before servicing equipment.
  Clean up work place when done.
- Heating system maintenance routines:
  Check electric heat— elements, airflow, controls.
  Check gas heat—ignition, draft, operation.
  Check oil heat—oil pressure, draft, burners, filters, tanks.
  Check hydronic heat—air and water components, pumps, water level.
  Check heat pumps—Same inspections as air conditioner; check defrost operation.
- Check and clean coils.
- Check and clean electronic air filters.
- Replace/clean filters.
- Check belt/drive assembly.
- Monitor sound levels.
- Lubricate bearing only as necessary—Do not over-oil.
- Visually inspect all equipment.
- Calibrate instruments and controls.
- Schedule parts changes.
- Perform safety checks on all controls and devices.

Problems and Questions

1. What are some of the special tools and servicing equipment that the technician will find useful to perform preventive maintenance?
2. During preventive maintenance, why is it important to "clean and flush condensate drain pans and drain lines"?
3. Describe maintenance requirements for fan belts.
4. Why is it important that electrical connections be tight?
5. In examining a refrigeration system, you find oil on the compressor near a fitting or valve. What could this mean?
6. Each year before the cooling season, what must be done to cooling towers or evaporative condensers?
7. What is a good technique for cleaning air-cooled condenser coils?
8. It is good practice to provide preventive maintenance in the fall before the heating system is put into service. True or false?
9. Why is it important that the technician do a visual inspection and test heat exchangers on heating equipment?
10. What could happen if damaged contactor contacts or wiring with poor insulation are not replaced?
11. The purpose of using nitrogen when brazing is to keep the system pressurized preventing contaminates from getting inside the system. True or false?
12. When lubricating bearings which are oil type, use detergent oil. True or false?
13. Turn on crankcase heaters 12 to 24 hrs before attempting to start the compressor. True or false?
14. Ordinary disposable air filters have an efficiency rating of 5 to 10%. True or false?
15. Gas-heating equipment should have a clear blue flame with a small yellow tip on the flame. True or false?
16. A steady pilot flame on a gas furnace should impinge on the thermocouple about:
    a. ¼ in.
    b. ½ in.
    c. ¾ in.
    d. None of the above.
17. The oil pressure on the oil furnace pump should be:
    a. 3.5 in. WC.
    b. 80 psi.
    c. 100 psi.
    d. None of the above.
18. Occasionally on hydronic systems, the expansion tank will become:
    a. Waterlogged.
    b. Airlogged.
    c. Overheated.
    d. None of the above.
19. One of the best ways to clean a finned coil on the job is to use:
    a. A wire brush and hot soapy water.
    b. Pressure spray of cleaning material.
    c. Portable air compressor and high-pressure air.
    d. None of the above.
20. Sound-level monitoring, when concerned with the human ear, is measured using the units:
    a. dB.
    b. Micropascals.
    c. Octave bands.
    d. None of the above.

Appendix A

Tables A-1 through A-12 courtesy of DuPont Fluoroproducts.

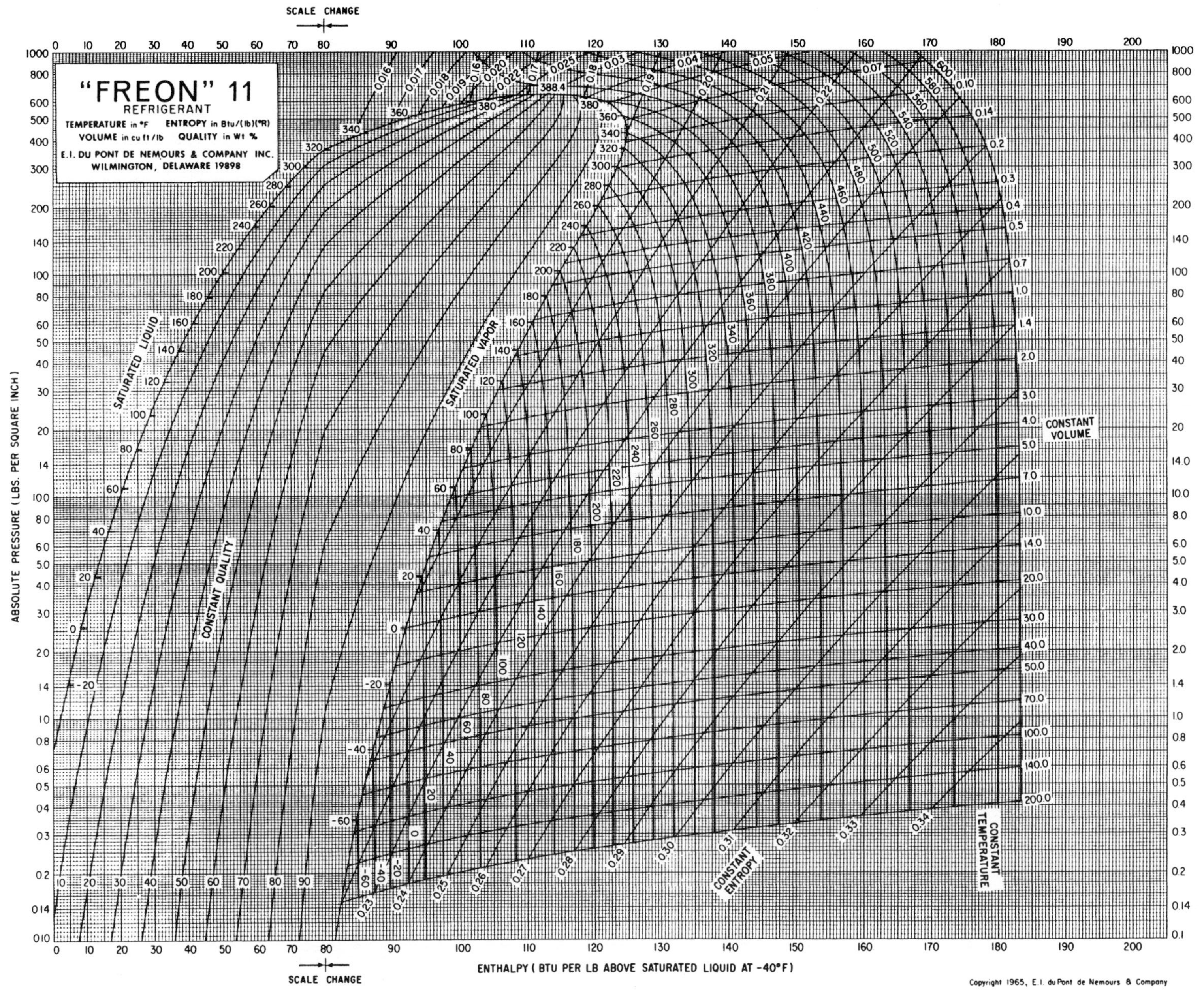

Table A-1 I-P Pressure-Heat Diagram for R-11

TEMP.	PRESSURE		VOLUME cu ft/lb		DENSITY lb/cu ft		ENTHALPY Btu/lb			ENTROPY Btu/(lb)(°R)		TEMP.
°F	PSIA	PSIG	LIQUID v_f	VAPOR v_g	LIQUID $1/v_f$	VAPOR $1/v_g$	LIQUID h_f	LATENT h_{fg}	VAPOR h_g	LIQUID s_f	VAPOR s_g	°F
−58	0.38715	29.13297*	0.0097455	80.831	102.61	0.012371	−3.634	88.714	85.080	−0.008848	0.21201	−58
−57	0.40215	29.10241*	0.0097524	78.002	102.54	0.012820	−3.432	88.628	85.197	−0.008346	0.21176	−57
−56	0.41765	29.07086*	0.0097594	75.288	102.46	0.013282	−3.230	88.543	85.313	−0.007845	0.21150	−56
−55	0.43365	29.03829*	0.0097664	72.684	102.39	0.013758	−3.028	88.458	85.430	−0.007346	0.21125	−55
−54	0.45016	29.00466*	0.0097735	70.185	102.32	0.014248	−2.826	88.373	85.547	−0.006848	0.21100	−54
−53	0.46720	28.96997*	0.0097805	67.785	102.24	0.014752	−2.624	88.288	85.664	−0.006351	0.21075	−53
−52	0.48479	28.93417*	0.0097875	65.482	102.17	0.015271	−2.423	88.203	85.781	−0.005855	0.21050	−52
−51	0.50292	28.89724*	0.0097946	63.269	102.10	0.015806	−2.221	88.119	85.898	−0.005361	0.21026	−51
−50	0.52163	28.85916*	0.0098017	61.144	102.02	0.016355	−2.019	88.034	86.015	−0.004868	0.21002	−50
−49	0.54092	28.81988*	0.0098088	59.102	101.95	0.016920	−1.817	87.950	86.133	−0.004376	0.20979	−49
−48	0.56080	28.77940*	0.0098159	57.139	101.88	0.017501	−1.615	87.865	86.250	−0.003885	0.20955	−48
−47	0.58130	28.73766*	0.0098230	55.253	101.80	0.018099	−1.413	87.781	86.368	−0.003395	0.20932	−47
−46	0.60243	28.69466*	0.0098301	53.439	101.73	0.018713	−1.211	87.697	86.485	−0.002907	0.20909	−46
−45	0.62419	28.65034*	0.0098373	51.695	101.65	0.019344	−1.010	87.612	86.603	−0.002419	0.20886	−45
−44	0.64661	28.60469*	0.0098444	50.017	101.58	0.019993	−0.808	87.528	86.721	−0.001933	0.20864	−44
−43	0.66970	28.55768*	0.0098516	48.404	101.51	0.020660	−0.606	87.444	86.839	−0.001448	0.20842	−43
−42	0.69348	28.50926*	0.0098588	46.851	101.43	0.021344	−0.404	87.360	86.957	−0.000964	0.20820	−42
−41	0.71797	28.45941*	0.0098660	45.356	101.36	0.022048	−0.202	87.276	87.075	−0.000482	0.20798	−41
−40	0.74317	28.40809*	0.0098733	43.917	101.28	0.022770	0.000	87.193	87.193	0.000000	0.20776	−40
−39	0.76911	28.35527*	0.0098805	42.532	101.21	0.023511	0.202	87.109	87.311	0.000481	0.20755	−39
−38	0.79581	28.30092*	0.0098878	41.198	101.14	0.024273	0.404	87.025	87.429	0.000960	0.20734	−38
−37	0.82327	28.24500*	0.0098951	39.914	101.06	0.025054	0.606	86.942	87.548	0.001438	0.20713	−37
−36	0.85153	28.18748*	0.0099023	38.676	100.99	0.025856	0.808	86.858	87.666	0.001915	0.20693	−36
−35	0.88059	28.12832*	0.0099097	37.483	100.91	0.026679	1.010	86.775	87.784	0.002392	0.20673	−35
−34	0.91047	28.06748*	0.0099170	36.333	100.84	0.027523	1.212	86.691	87.903	0.002867	0.20652	−34
−33	0.94119	28.00493*	0.0099243	35.225	100.76	0.028389	1.414	86.608	88.022	0.003341	0.20633	−33
−32	0.97277	27.94063*	0.0099317	34.157	100.69	0.029277	1.616	86.524	88.140	0.003814	0.20613	−32
−31	1.0052	27.8745*	0.0099391	33.127	100.61	0.030187	1.818	86.441	88.259	0.004285	0.20593	−31
−30	1.0386	27.8066*	0.0099464	32.133	100.54	0.031120	2.020	86.358	88.378	0.004756	0.20574	−30
−29	1.0729	27.7369*	0.0099539	31.175	100.46	0.032077	2.222	86.275	88.497	0.005226	0.20555	−29
−28	1.1081	27.6652*	0.0099613	30.250	100.39	0.033058	2.425	86.192	88.616	0.005695	0.20536	−28
−27	1.1442	27.5915*	0.0099687	29.357	100.31	0.034063	2.627	86.108	88.735	0.006162	0.20518	−27
−26	1.1814	27.5159*	0.0099762	28.496	100.24	0.035093	2.829	86.025	88.854	0.006629	0.20499	−26
−25	1.2195	27.4383*	0.0099837	27.664	100.16	0.036148	3.031	85.942	88.974	0.007095	0.20481	−25
−24	1.2586	27.3586*	0.0099911	26.861	100.09	0.037228	3.233	85.859	89.093	0.007559	0.20463	−24
−23	1.2988	27.2768*	0.0099987	26.086	100.01	0.038335	3.436	85.776	89.212	0.008023	0.20446	−23
−22	1.3401	27.1928*	0.010006	25.337	99.938	0.039468	3.638	85.694	89.331	0.008485	0.20428	−22
−21	1.3824	27.1067*	0.010014	24.613	99.863	0.040628	3.840	85.611	89.451	0.008947	0.20411	−21
−20	1.4258	27.0183*	0.010021	23.914	99.788	0.041816	4.043	85.528	89.570	0.009408	0.20393	−20
−19	1.4703	26.9276*	0.010029	23.239	99.712	0.043032	4.245	85.445	89.690	0.009867	0.20376	−19
−18	1.5160	26.8346*	0.010036	22.586	99.637	0.044276	4.448	85.362	89.810	0.010326	0.20360	−18
−17	1.5629	26.7392*	0.010044	21.954	99.561	0.045549	4.650	85.279	89.929	0.010783	0.20343	−17
−16	1.6109	26.6414*	0.010052	21.344	99.486	0.046852	4.852	85.197	90.049	0.011240	0.20327	−16
−15	1.6602	26.5411*	0.010059	20.754	99.410	0.048184	5.055	85.114	90.169	0.011696	0.20310	−15
−14	1.7106	26.4383*	0.010067	20.183	99.334	0.049547	5.258	85.031	90.289	0.012151	0.20294	−14
−13	1.7624	26.3330*	0.010075	19.630	99.259	0.050941	5.460	84.948	90.408	0.012605	0.20279	−13
−12	1.8154	26.2250*	0.010082	19.096	99.183	0.052367	5.663	84.866	90.528	0.013058	0.20263	−12
−11	1.8697	26.1144*	0.010090	18.579	99.107	0.053824	5.865	84.783	90.648	0.013510	0.20247	−11
−10	1.9254	26.0011*	0.010098	18.079	99.031	0.055314	6.068	84.700	90.768	0.013961	0.20232	−10
−9	1.9824	25.8850*	0.010106	17.594	98.955	0.056837	6.271	84.617	90.888	0.014411	0.20217	−9
−8	2.0408	25.7661*	0.010113	17.125	98.879	0.058393	6.474	84.535	91.008	0.014860	0.20202	−8
−7	2.1006	25.6443*	0.010121	16.671	98.803	0.059984	6.677	84.452	91.129	0.015308	0.20187	−7
−6	2.1618	25.5197*	0.010129	16.231	98.726	0.061609	6.879	84.369	91.249	0.015756	0.20173	−6
−5	2.2245	25.3920*	0.010137	15.805	98.650	0.063270	7.082	84.287	91.369	0.016202	0.20158	−5
−4	2.2887	25.2614*	0.010145	15.393	98.574	0.064966	7.285	84.204	91.489	0.016648	0.20144	−4

*Inches of mercury below one atmosphere

Table A-2 I-P Properties of Saturated Liquid and Vapor for R-11

TEMP.	PRESSURE		VOLUME cu ft/lb		DENSITY lb/cu ft		ENTHALPY Btu/lb			ENTROPY Btu/(lb)(°R)		TEMP.
°F	PSIA	PSIG	LIQUID v_f	VAPOR v_g	LIQUID $1/v_f$	VAPOR $1/v_g$	LIQUID h_f	LATENT h_{fg}	VAPOR h_g	LIQUID s_f	VAPOR s_g	°F
−3	2.3544	25.1277*	0.010153	14.993	98.497	0.066699	7.488	84.121	91.609	0.017093	0.20130	−3
−2	2.4216	24.9908*	0.010160	14.605	98.421	0.068468	7.691	84.038	91.730	0.017536	0.20116	−2
−1	2.4904	24.8508*	0.010168	14.230	98.345	0.070275	7.894	83.956	91.850	0.017979	0.20102	−1
0	2.5607	24.7076*	0.010176	13.866	98.268	0.072119	8.098	83.873	91.970	0.018421	0.20088	0
1	2.6327	24.5610*	0.010184	13.513	98.191	0.074003	8.301	83.790	92.091	0.018863	0.20075	1
2	2.7063	24.4112*	0.010192	13.171	98.115	0.075925	8.504	83.707	92.211	0.019303	0.20062	2
3	2.7816	24.2578*	0.010200	12.839	98.038	0.077887	8.707	83.624	92.332	0.019743	0.20049	3
4	2.8586	24.1011*	0.010208	12.517	97.961	0.079890	8.910	83.542	92.452	0.020181	0.20036	4
5	2.9373	23.9408*	0.010216	12.205	97.884	0.081933	9.114	83.459	92.572	0.020619	0.20023	5
6	3.0178	23.7770*	0.010224	11.902	97.807	0.084018	9.317	83.376	92.693	0.021056	0.20010	6
7	3.1000	23.6095*	0.010232	11.608	97.730	0.086145	9.521	83.293	92.814	0.021492	0.19998	7
8	3.1841	23.4383*	0.010240	11.323	97.653	0.088315	9.724	83.210	92.934	0.021927	0.19985	8
9	3.2700	23.2634*	0.010248	11.046	97.576	0.090528	9.928	83.127	93.055	0.022362	0.19973	9
10	3.3578	23.0847*	0.010257	10.778	97.498	0.092785	10.131	83.044	93.175	0.022795	0.19961	10
11	3.4475	22.9020*	0.010265	10.517	97.421	0.095087	10.335	82.961	93.296	0.023228	0.19949	11
12	3.5391	22.7155*	0.010273	10.263	97.344	0.097433	10.539	82.878	93.416	0.023660	0.19937	12
13	3.6327	22.5249*	0.010281	10.017	97.266	0.099826	10.742	82.795	93.537	0.024091	0.19925	13
14	3.7283	22.3303*	0.010289	9.7785	97.189	0.10226	10.946	82.712	93.658	0.024521	0.19914	14
15	3.8259	22.1316*	0.010297	9.5465	97.111	0.10475	11.150	82.628	93.778	0.024951	0.19903	15
16	3.9256	21.9286*	0.010306	9.3210	97.033	0.10728	11.354	82.545	93.899	0.025379	0.19891	16
17	4.0274	21.7214*	0.010314	9.1019	96.956	0.10987	11.558	82.462	94.020	0.025807	0.19880	17
18	4.1313	21.5099*	0.010322	8.8890	96.878	0.11250	11.762	82.379	94.140	0.026234	0.19869	18
19	4.2373	21.2940*	0.010331	8.6821	96.800	0.11518	11.966	82.295	94.261	0.026661	0.19859	19
20	4.3456	21.0736*	0.010339	8.4810	96.722	0.11791	12.170	82.212	94.382	0.027086	0.19848	20
21	4.4560	20.8487*	0.010347	8.2855	96.644	0.12069	12.374	82.128	94.502	0.027511	0.19837	21
22	4.5687	20.6192*	0.010356	8.0954	96.566	0.12353	12.578	82.045	94.623	0.027935	0.19827	22
23	4.6837	20.3851*	0.010364	7.9106	96.488	0.12641	12.783	81.961	94.744	0.028358	0.19817	23
24	4.8010	20.1462*	0.010372	7.7308	96.409	0.12935	12.987	81.878	94.864	0.028781	0.19806	24
25	4.9207	19.9026*	0.010381	7.5560	96.331	0.13235	13.191	81.794	94.985	0.029202	0.19796	25
26	5.0428	19.6541*	0.010389	7.3859	96.253	0.13539	13.396	81.710	95.106	0.029623	0.19787	26
27	5.1672	19.4006*	0.010398	7.2205	96.174	0.13850	13.600	81.626	95.227	0.030043	0.19777	27
28	5.2942	19.1422*	0.010406	7.0595	96.096	0.14165	13.805	81.542	95.347	0.030463	0.19767	28
29	5.4236	18.8786*	0.010415	6.9028	96.017	0.14487	14.009	81.458	95.468	0.030881	0.19758	29
30	5.5556	18.6100*	0.010423	6.7503	95.938	0.14814	14.214	81.374	95.588	0.031299	0.19748	30
31	5.6901	18.3361*	0.010432	6.6019	95.860	0.15147	14.419	81.290	95.709	0.031716	0.19739	31
32	5.8273	18.0569*	0.010441	6.4574	95.781	0.15486	14.624	81.206	95.830	0.032133	0.19730	32
33	5.9670	17.7723*	0.010449	6.3168	95.702	0.15831	14.828	81.122	95.950	0.032548	0.19721	33
34	6.1094	17.4823*	0.010458	6.1798	95.623	0.16182	15.033	81.038	96.071	0.032963	0.19712	34
35	6.2546	17.1868*	0.010466	6.0465	95.544	0.16539	15.238	80.953	96.192	0.033377	0.19703	35
36	6.4025	16.8857*	0.010475	5.9166	95.464	0.16902	15.443	80.869	96.312	0.033791	0.19694	36
37	6.5532	16.5789*	0.010484	5.7901	95.385	0.17271	15.648	80.784	96.433	0.034204	0.19686	37
38	6.7067	16.2664*	0.010493	5.6668	95.306	0.17647	15.853	80.700	96.553	0.034616	0.19677	38
39	6.8630	15.9480*	0.010501	5.5468	95.226	0.18029	16.059	80.615	96.674	0.035027	0.19669	39
40	7.0223	15.6238*	0.010510	5.4298	95.147	0.18417	16.264	80.530	96.794	0.035437	0.19660	40
41	7.1844	15.2936*	0.010519	5.3158	95.067	0.18812	16.469	80.445	96.915	0.035847	0.19652	41
42	7.3496	14.9574*	0.010528	5.2047	94.988	0.19214	16.675	80.361	97.035	0.036256	0.19644	42
43	7.5178	14.6150*	0.010537	5.0964	94.908	0.19622	16.880	80.276	97.156	0.036665	0.19636	43
44	7.6890	14.2664*	0.010545	4.9908	94.828	0.20037	17.086	80.190	97.276	0.037073	0.19628	44
45	7.8633	13.9115*	0.010554	4.8879	94.748	0.20459	17.291	80.105	97.396	0.037480	0.19621	45
46	8.0407	13.5502*	0.010563	4.7876	94.668	0.20887	17.497	80.020	97.517	0.037886	0.19613	46
47	8.2213	13.1825*	0.010572	4.6897	94.588	0.21323	17.702	79.935	97.637	0.038292	0.19606	47
48	8.4051	12.8083*	0.010581	4.5943	94.508	0.21766	17.908	79.849	97.757	0.038697	0.19598	48
49	8.5922	12.4274*	0.010590	4.5012	94.428	0.22216	18.114	79.764	97.877	0.039101	0.19591	49
50	8.7825	12.0399*	0.010599	4.4105	94.347	0.22673	18.320	79.678	97.998	0.039505	0.19584	50
51	8.9762	11.6456*	0.010608	4.3219	94.267	0.23138	18.526	79.592	98.118	0.039907	0.19577	51

*Inches of mercury below one atmosphere

Table A-2 Continued

TEMP.	PRESSURE		VOLUME cu ft/lb		DENSITY lb/cu ft		ENTHALPY Btu/lb			ENTROPY Btu/(lb)(°R)		TEMP.
°F	PSIA	PSIG	LIQUID v_f	VAPOR v_g	LIQUID $1/v_f$	VAPOR $1/v_g$	LIQUID h_f	LATENT h_{fg}	VAPOR h_g	LIQUID s_f	VAPOR s_g	°F
52	9.1733	11.2444*	0.010617	4.2355	94.186	0.23610	18.732	79.506	98.238	0.040310	0.19570	52
53	9.3737	10.8362*	0.010626	4.1512	94.106	0.24089	18.938	79.420	98.358	0.040711	0.19563	53
54	9.5776	10.4211*	0.010635	4.0690	94.025	0.24576	19.144	79.334	98.478	0.041112	0.19556	54
55	9.7850	9.9988*	0.010645	3.9887	93.944	0.25071	19.350	79.248	98.598	0.041512	0.19549	55
56	9.9960	9.5693*	0.010654	3.9103	93.864	0.25573	19.556	79.162	98.718	0.041912	0.19542	56
57	10.210	9.133*	0.010663	3.8338	93.783	0.26083	19.763	79.075	98.838	0.042311	0.19536	57
58	10.429	8.688*	0.010672	3.7592	93.702	0.26602	19.969	78.989	98.958	0.042709	0.19529	58
59	10.650	8.237*	0.010681	3.6863	93.620	0.27128	20.176	78.902	99.078	0.043107	0.19523	59
60	10.876	7.778*	0.010691	3.6151	93.539	0.27662	20.382	78.816	99.198	0.043504	0.19517	60
61	11.105	7.311*	0.010700	3.5456	93.458	0.28204	20.589	78.729	99.317	0.043900	0.19511	61
62	11.338	6.836*	0.010709	3.4777	93.377	0.28754	20.795	78.642	99.437	0.044295	0.19505	62
63	11.575	6.354*	0.010719	3.4114	93.295	0.29313	21.002	78.555	99.557	0.044690	0.19499	63
64	11.816	5.864*	0.010728	3.3467	93.213	0.29880	21.209	78.468	99.676	0.045085	0.19493	64
65	12.061	5.366*	0.010737	3.2834	93.132	0.30456	21.416	78.380	99.796	0.045478	0.19487	65
66	12.309	4.859*	0.010747	3.2216	93.050	0.31040	21.623	78.293	99.916	0.045872	0.19481	66
67	12.562	4.345*	0.010756	3.1612	92.968	0.31633	21.830	78.205	100.035	0.046264	0.19475	67
68	12.819	3.822*	0.010766	3.1022	92.886	0.32235	22.037	78.118	100.154	0.046656	0.19470	68
69	13.080	3.291*	0.010775	3.0445	92.804	0.32846	22.244	78.030	100.274	0.047047	0.19464	69
70	13.345	2.752*	0.010785	2.9882	92.722	0.33465	22.451	77.942	100.393	0.047437	0.19459	70
71	13.614	2.204*	0.010795	2.9331	92.640	0.34094	22.658	77.854	100.512	0.047827	0.19454	71
72	13.887	1.647*	0.010804	2.8792	92.557	0.34731	22.866	77.766	100.631	0.048217	0.19448	72
73	14.165	1.081*	0.010814	2.8266	92.475	0.35378	23.073	77.678	100.751	0.048605	0.19443	73
74	14.447	0.507*	0.010823	2.7751	92.392	0.36034	23.280	77.589	100.870	0.048993	0.19438	74
75	14.733	0.037	0.010833	2.7248	92.310	0.36700	23.488	77.501	100.989	0.049381	0.19433	75
76	15.024	0.328	0.010843	2.6756	92.227	0.37375	23.696	77.412	101.108	0.049768	0.19428	76
77	15.319	0.623	0.010853	2.6275	92.144	0.38059	23.903	77.323	101.226	0.050154	0.19423	77
78	15.619	0.923	0.010862	2.5804	92.061	0.38754	24.111	77.234	101.345	0.050540	0.19419	78
79	15.924	1.228	0.010872	2.5344	91.978	0.39458	24.319	77.145	101.464	0.050925	0.19414	79
80	16.233	1.537	0.010882	2.4893	91.895	0.40172	24.527	77.056	101.583	0.051309	0.19409	80
81	16.546	1.850	0.010892	2.4453	91.812	0.40895	24.735	76.967	101.701	0.051693	0.19405	81
82	16.865	2.169	0.010902	2.4021	91.728	0.41629	24.943	76.877	101.820	0.052076	0.19400	82
83	17.188	2.492	0.010912	2.3600	91.645	0.42374	25.151	76.787	101.938	0.052459	0.19396	83
84	17.516	2.820	0.010922	2.3187	91.561	0.43128	25.359	76.698	102.057	0.052841	0.19391	84
85	17.848	3.152	0.010932	2.2783	91.477	0.43893	25.567	76.608	102.175	0.053222	0.19387	85
86	18.186	3.490	0.010942	2.2388	91.394	0.44668	25.776	76.518	102.293	0.053603	0.19383	86
87	18.529	3.833	0.010952	2.2001	91.310	0.45453	25.984	76.427	102.411	0.053983	0.19379	87
88	18.876	4.180	0.010962	2.1622	91.226	0.46250	26.192	76.337	102.529	0.054363	0.19375	88
89	19.229	4.533	0.010972	2.1251	91.142	0.47057	26.401	76.246	102.647	0.054742	0.19371	89
90	19.587	4.891	0.010982	2.0888	91.057	0.47874	26.610	76.156	102.765	0.055121	0.19367	90
91	19.950	5.254	0.010992	2.0532	90.973	0.48703	26.818	76.065	102.883	0.055499	0.19363	91
92	20.318	5.622	0.011002	2.0184	90.889	0.49543	27.027	75.974	103.001	0.055876	0.19359	92
93	20.691	5.995	0.011013	1.9844	90.804	0.50394	27.236	75.883	103.119	0.056253	0.19355	93
94	21.070	6.374	0.011023	1.9510	90.719	0.51256	27.445	75.791	103.236	0.056629	0.19352	94
95	21.454	6.758	0.011033	1.9183	90.635	0.52130	27.654	75.700	103.354	0.057005	0.19348	95
96	21.843	7.147	0.011044	1.8863	90.550	0.53015	27.863	75.608	103.471	0.057380	0.19345	96
97	22.238	7.542	0.011054	1.8549	90.465	0.53912	28.072	75.516	103.588	0.057755	0.19341	97
98	22.638	7.942	0.011064	1.8241	90.379	0.54820	28.281	75.424	103.706	0.058129	0.19338	98
99	23.044	8.348	0.011075	1.7940	90.294	0.55740	28.491	75.332	103.823	0.058502	0.19334	99
100	23.456	8.760	0.011085	1.7645	90.209	0.56672	28.700	75.240	103.940	0.058875	0.19331	100
101	23.873	9.177	0.011096	1.7356	90.123	0.57616	28.909	75.147	104.057	0.059248	0.19328	101
102	24.296	9.600	0.011106	1.7073	90.038	0.58572	29.119	75.055	104.174	0.059620	0.19325	102
103	24.724	10.028	0.011117	1.6795	89.952	0.59540	29.329	74.962	104.290	0.059991	0.19322	103
104	25.159	10.463	0.011128	1.6523	89.866	0.60521	29.538	74.869	104.407	0.060362	0.19319	104
105	25.599	10.903	0.011138	1.6256	89.780	0.61514	29.748	74.776	104.524	0.060732	0.19316	105
106	26.045	11.349	0.011149	1.5995	89.694	0.62519	29.958	74.682	104.640	0.061102	0.19313	106

*Inches of mercury below one atmosphere

Table A-2 Continued

TEMP.	PRESSURE		VOLUME cu ft/lb		DENSITY lb/cu ft		ENTHALPY Btu/lb			ENTROPY Btu/(lb)(°R)		TEMP.
°F	PSIA	PSIG	LIQUID v_f	VAPOR v_g	LIQUID $1/v_f$	VAPOR $1/v_g$	LIQUID h_f	LATENT h_{fg}	VAPOR h_g	LIQUID s_f	VAPOR s_g	°F
107	26.497	11.801	0.011160	1.5739	89.608	0.63538	30.168	74.589	104.757	0.061471	0.19310	107
108	26.956	12.260	0.011170	1.5487	89.522	0.64569	30.378	74.495	104.873	0.061840	0.19307	108
109	27.420	12.724	0.011181	1.5241	89.435	0.65612	30.588	74.401	104.989	0.062208	0.19304	109
110	27.890	13.194	0.011192	1.4999	89.349	0.66669	30.798	74.307	105.105	0.062575	0.19301	110
111	28.367	13.671	0.011203	1.4763	89.262	0.67739	31.009	74.213	105.221	0.062942	0.19299	111
112	28.850	14.154	0.011214	1.4530	89.175	0.68822	31.219	74.118	105.337	0.063309	0.19296	112
113	29.339	14.643	0.011225	1.4302	89.088	0.69919	31.429	74.023	105.453	0.063675	0.19294	113
114	29.834	15.138	0.011236	1.4079	89.001	0.71029	31.640	73.929	105.569	0.064041	0.19291	114
115	30.336	15.640	0.011247	1.3860	88.914	0.72152	31.851	73.833	105.684	0.064406	0.19289	115
116	30.844	16.148	0.011258	1.3645	88.826	0.73289	32.061	73.738	105.800	0.064770	0.19286	116
117	31.359	16.663	0.011269	1.3434	88.739	0.74440	32.272	73.643	105.915	0.065134	0.19284	117
118	31.880	17.184	0.011280	1.3227	88.651	0.75605	32.483	73.547	106.030	0.065498	0.19281	118
119	32.408	17.712	0.011291	1.3024	88.564	0.76784	32.694	73.451	106.145	0.065861	0.19279	119
120	32.943	18.247	0.011303	1.2824	88.476	0.77977	32.905	73.355	106.260	0.066223	0.19277	120
121	33.484	18.788	0.011314	1.2629	88.388	0.79184	33.116	73.259	106.375	0.066585	0.19275	121
122	34.032	19.336	0.011325	1.2437	88.300	0.80406	33.328	73.162	106.490	0.066947	0.19273	122
123	34.587	19.891	0.011336	1.2249	88.211	0.81642	33.539	73.066	106.605	0.067308	0.19271	123
124	35.149	20.453	0.011348	1.2064	88.123	0.82893	33.751	72.969	106.719	0.067669	0.19269	124
125	35.718	21.022	0.011359	1.1882	88.034	0.84159	33.962	72.872	106.834	0.068029	0.19267	125
126	36.294	21.598	0.011371	1.1704	87.946	0.85440	34.174	72.774	106.948	0.068388	0.19265	126
127	36.877	22.181	0.011382	1.1529	87.857	0.86735	34.385	72.677	107.062	0.068747	0.19263	127
128	37.467	22.771	0.011394	1.1358	87.768	0.88046	34.597	72.579	107.176	0.069106	0.19261	128
129	38.064	23.368	0.011405	1.1189	87.679	0.89372	34.809	72.481	107.290	0.069464	0.19259	129
130	38.668	23.972	0.011417	1.1024	87.589	0.90714	35.021	72.383	107.404	0.069822	0.19257	130
131	39.280	24.584	0.011429	1.0861	87.500	0.92071	35.233	72.284	107.518	0.070179	0.19256	131
132	39.899	25.203	0.011440	1.0702	87.410	0.93443	35.446	72.186	107.631	0.070536	0.19254	132
133	40.525	25.829	0.011452	1.0545	87.321	0.94832	35.658	72.087	107.745	0.070892	0.19252	133
134	41.159	26.463	0.011464	1.0391	87.231	0.96237	35.870	71.988	107.858	0.071248	0.19251	134
135	41.801	27.105	0.011476	1.0240	87.141	0.97657	36.083	71.888	107.971	0.071604	0.19249	135
136	42.450	27.754	0.011488	1.0091	87.051	0.99094	36.296	71.789	108.084	0.071959	0.19248	136
137	43.106	28.410	0.011499	0.99456	86.960	1.0055	36.508	71.689	108.197	0.072313	0.19246	137
138	43.771	29.075	0.011511	0.98023	86.870	1.0202	36.721	71.589	108.310	0.072667	0.19245	138
139	44.443	29.747	0.011523	0.96615	86.779	1.0350	36.934	71.489	108.423	0.073021	0.19243	139
140	45.123	30.427	0.011536	0.95232	86.689	1.0501	37.147	71.388	108.535	0.073374	0.19242	140
141	45.810	31.114	0.011548	0.93873	86.598	1.0653	37.360	71.287	108.648	0.073727	0.19241	141
142	46.506	31.810	0.011560	0.92538	86.507	1.0806	37.574	71.186	108.760	0.074079	0.19239	142
143	47.210	32.514	0.011572	0.91226	86.415	1.0962	37.787	71.085	108.872	0.074431	0.19238	143
144	47.921	33.225	0.011584	0.89937	86.324	1.1119	38.001	70.983	108.984	0.074783	0.19237	144
145	48.641	33.945	0.011597	0.88670	86.232	1.1278	38.214	70.881	109.096	0.075134	0.19236	145
146	49.369	34.673	0.011609	0.87424	86.141	1.1439	38.428	70.779	109.207	0.075484	0.19235	146
147	50.105	35.409	0.011621	0.86200	86.049	1.1601	38.642	70.677	109.319	0.075835	0.19233	147
148	50.850	36.154	0.011634	0.84996	85.957	1.1765	38.856	70.574	109.430	0.076184	0.19232	148
149	51.603	36.907	0.011646	0.83813	85.864	1.1931	39.070	70.472	109.542	0.076534	0.19231	149
150	52.364	37.668	0.011659	0.82650	85.772	1.2099	39.284	70.368	109.653	0.076883	0.19230	150
151	53.133	38.437	0.011671	0.81507	85.680	1.2269	39.499	70.265	109.764	0.077231	0.19229	151
152	53.912	39.216	0.011684	0.80383	85.587	1.2440	39.713	70.161	109.874	0.077579	0.19228	152
153	54.698	40.002	0.011697	0.79278	85.494	1.2614	39.928	70.057	109.985	0.077927	0.19227	153
154	55.494	40.798	0.011709	0.78191	85.401	1.2789	40.142	69.953	110.096	0.078275	0.19227	154
155	56.298	41.602	0.011722	0.77122	85.308	1.2966	40.357	69.849	110.206	0.078622	0.19226	155
156	57.111	42.415	0.011735	0.76071	85.214	1.3146	40.572	69.744	110.316	0.078968	0.19225	156
157	57.933	43.237	0.011748	0.75037	85.121	1.3327	40.787	69.639	110.426	0.079314	0.19224	157
158	58.763	44.067	0.011761	0.74021	85.027	1.3510	41.003	69.533	110.536	0.079660	0.19223	158
159	59.603	44.907	0.011774	0.73021	84.933	1.3695	41.218	69.428	110.646	0.080006	0.19223	159
160	60.451	45.755	0.011787	0.72037	84.839	1.3882	41.433	69.322	110.755	0.080351	0.19222	160
161	61.309	46.613	0.011800	0.71070	84.745	1.4071	41.649	69.215	110.865	0.080695	0.19221	161

Table A-2 Continued

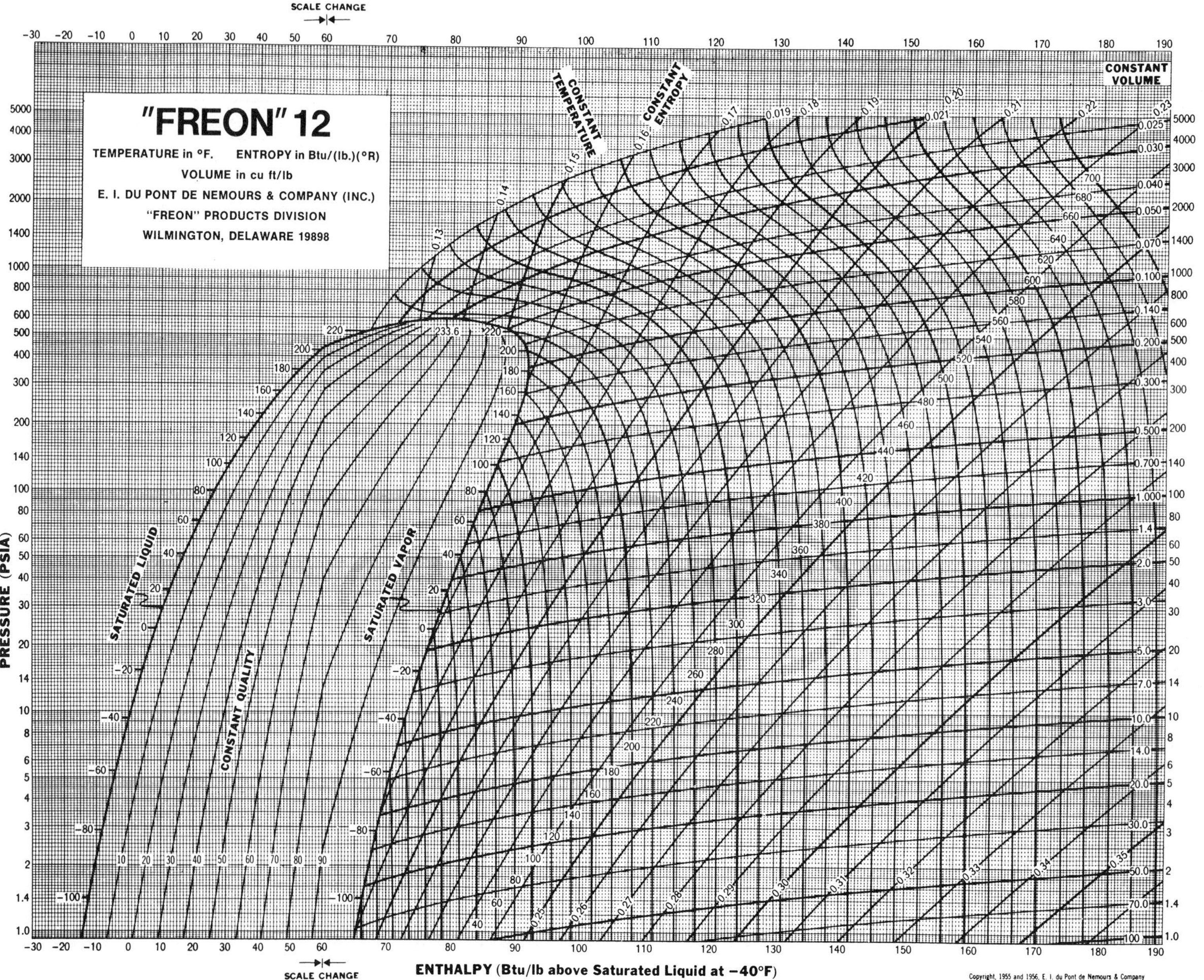

Table A-3 I-P Pressure-Heat Diagram for R-12

TEMP.	PRESSURE		VOLUME cu ft/lb		DENSITY lb/cu ft		ENTHALPY Btu/lb			ENTROPY Btu/(lb)(° R)		TEMP.
°F	PSIA	PSIG	LIQUID v_f	VAPOR v_g	LIQUID $1/v_f$	VAPOR $1/v_g$	LIQUID h_f	LATENT h_{fg}	VAPOR h_g	LIQUID s_f	VAPOR s_g	°F
−40	9.3076	10.9709*	0.010564	3.8750	94.661	0.25806	0	72.913	72.913	0	0.17373	−40
−39	9.5530	10.4712*	0.010575	3.7823	94.565	0.26439	0.2107	72.812	73.023	0.000500	0.17357	−39
−38	9.8035	9.9611*	0.010586	3.6922	94.469	0.27084	0.4215	72.712	73.134	0.001000	0.17343	−38
−37	10.059	9.441*	0.010596	3.6047	94.372	0.27741	0.6324	72.611	73.243	0.001498	0.17328	−37
−36	10.320	8.909*	0.010607	3.5198	94.275	0.28411	0.8434	72.511	73.354	0.001995	0.17313	−36
−35	10.586	8.367*	0.010618	3.4373	94.178	0.29093	1.0546	72.409	73.464	0.002492	0.17299	−35
−34	10.858	7.814*	0.010629	3.3571	94.081	0.29788	1.2659	72.309	73.575	0.002988	0.17285	−34
−33	11.135	7.250*	0.010640	3.2792	93.983	0.30495	1.4772	72.208	73.685	0.003482	0.17271	−33
−32	11.417	6.675*	0.010651	3.2035	93.886	0.31216	1.6887	72.106	73.795	0.003976	0.17257	−32
−31	11.706	6.088*	0.010662	3.1300	93.788	0.31949	1.9003	72.004	73.904	0.004469	0.17243	−31
−30	11.999	5.490*	0.010674	3.0585	93.690	0.32696	2.1120	71.903	74.015	0.004961	0.17229	−30
−29	12.299	4.880*	0.010685	2.9890	93.592	0.33457	2.3239	71.801	74.125	0.005452	0.17216	−29
−28	12.604	4.259*	0.010696	2.9214	93.493	0.34231	2.5358	71.698	74.234	0.005942	0.17203	−28
−27	12.916	3.625*	0.010707	2.8556	93.395	0.35018	2.7479	71.596	74.344	0.006431	0.17189	−27
−26	13.233	2.979*	0.010719	2.7917	93.296	0.35820	2.9601	71.494	74.454	0.006919	0.17177	−26
−25	13.556	2.320*	0.010730	2.7295	93.197	0.36636	3.1724	71.391	74.563	0.007407	0.17164	−25
−24	13.886	1.649*	0.010741	2.6691	93.098	0.37466	3.3848	71.288	74.673	0.007894	0.17151	−24
−23	14.222	0.966*	0.010753	2.6102	92.999	0.38311	3.5973	71.185	74.782	0.008379	0.17139	−23
−22	14.564	0.270*	0.010764	2.5529	92.899	0.39171	3.8100	71.081	74.891	0.008864	0.17126	−22
−21	14.912	0.216	0.010776	2.4972	92.799	0.40045	4.0228	70.978	75.001	0.009348	0.17114	−21
−20	15.267	0.571	0.010788	2.4429	92.699	0.40934	4.2357	70.874	75.110	0.009831	0.17102	−20
−19	15.628	0.932	0.010799	2.3901	92.599	0.41839	4.4487	70.770	75.219	0.010314	0.17090	−19
−18	15.996	1.300	0.010811	2.3387	92.499	0.42758	4.6618	70.666	75.328	0.010795	0.17078	−18
−17	16.371	1.675	0.010823	2.2886	92.399	0.43694	4.8751	70.561	75.436	0.011276	0.17066	−17
−16	16.753	2.057	0.010834	2.2399	92.298	0.44645	5.0885	70.456	75.545	0.011755	0.17055	−16
−15	17.141	2.445	0.010846	2.1924	92.197	0.45612	5.3020	70.352	75.654	0.012234	0.17043	−15
−14	17.536	2.840	0.010858	2.1461	92.096	0.46595	5.5157	70.246	75.762	0.012712	0.17032	−14
−13	17.939	3.243	0.010870	2.1011	91.995	0.47595	5.7295	70.141	75.871	0.013190	0.17021	−13
−12	18.348	3.652	0.010882	2.0572	91.893	0.48611	5.9434	70.036	75.979	0.013666	0.17010	−12
−11	18.765	4.069	0.010894	2.0144	91.791	0.49643	6.1574	69.930	76.087	0.014142	0.16999	−11
−10	19.189	4.493	0.010906	1.9727	91.689	0.50693	6.3716	69.824	76.196	0.014617	0.16989	−10
−9	19.621	4.925	0.010919	1.9320	91.587	0.51759	6.5859	69.718	76.304	0.015091	0.16978	−9
−8	20.059	5.363	0.010931	1.8924	91.485	0.52843	6.8003	69.611	76.411	0.015564	0.16967	−8
−7	20.506	5.810	0.010943	1.8538	91.382	0.53944	7.0149	69.505	76.520	0.016037	0.16957	−7
−6	20.960	6.264	0.010955	1.8161	91.280	0.55063	7.2296	69.397	76.627	0.016508	0.16947	−6
−5	21.422	6.726	0.010968	1.7794	91.177	0.56199	7.4444	69.291	76.735	0.016979	0.16937	−5
−4	21.891	7.195	0.010980	1.7436	91.074	0.57354	7.6594	69.183	76.842	0.017449	0.16927	−4
−3	22.369	7.673	0.010993	1.7086	90.970	0.58526	7.8745	69.075	76.950	0.017919	0.16917	−3
−2	22.854	8.158	0.011005	1.6745	90.867	0.59718	8.0898	68.967	77.057	0.018388	0.16907	−2
−1	23.348	8.652	0.011018	1.6413	90.763	0.60927	8.3052	68.859	77.164	0.018855	0.16897	−1
0	23.849	9.153	0.011030	1.6089	90.659	0.62156	8.5207	68.750	77.271	0.019323	0.16888	0
1	24.359	9.663	0.011043	1.5772	90.554	0.63404	8.7364	68.642	77.378	0.019789	0.16878	1
2	24.878	10.182	0.011056	1.5463	90.450	0.64670	8.9522	68.533	77.485	0.020255	0.16869	2
3	25.404	10.708	0.011069	1.5161	90.345	0.65957	9.1682	68.424	77.592	0.020719	0.16860	3
4	25.939	11.243	0.011082	1.4867	90.240	0.67263	9.3843	68.314	77.698	0.021184	0.16851	4
5	26.483	11.787	0.011094	1.4580	90.135	0.68588	9.6005	68.204	77.805	0.021647	0.16842	5
6	27.036	12.340	0.011107	1.4299	90.030	0.69934	9.8169	68.094	77.911	0.022110	0.16833	6
7	27.597	12.901	0.011121	1.4025	89.924	0.71300	10.033	67.984	78.017	0.022572	0.16824	7
8	28.167	13.471	0.011134	1.3758	89.818	0.72687	10.250	67.873	78.123	0.023033	0.16815	8
9	28.747	14.051	0.011147	1.3496	89.712	0.74094	10.467	67.762	78.229	0.023494	0.16807	9
10	29.335	14.639	0.011160	1.3241	89.606	0.75523	10.684	67.651	78.335	0.023954	0.16798	10
11	29.932	15.236	0.011173	1.2992	89.499	0.76972	10.901	67.539	78.440	0.024413	0.16790	11
12	30.539	15.843	0.011187	1.2748	89.392	0.78443	11.118	67.428	78.546	0.024871	0.16782	12
13	31.155	16.459	0.011200	1.2510	89.285	0.79935	11.336	67.315	78.651	0.025329	0.16774	13
14	31.780	17.084	0.011214	1.2278	89.178	0.81449	11.554	67.203	78.757	0.025786	0.16765	14
15	32.415	17.719	0.011227	1.2050	89.070	0.82986	11.771	67.090	78.861	0.026243	0.16758	15

* Inches of mercury below one atmosphere

Table A-4 I-P Properties of Saturated Liquid and Vapor for R-12

TEMP.	PRESSURE		VOLUME cu ft/lb		DENSITY lb/cu ft		ENTHALPY Btu/lb			ENTROPY Btu/(lb)(° R)		TEMP.
°F	PSIA	PSIG	LIQUID v_f	VAPOR v_g	LIQUID $1/v_f$	VAPOR $1/v_g$	LIQUID h_f	LATENT h_{fg}	VAPOR h_g	LIQUID s_f	VAPOR s_g	°F
15	32.415	17.719	0.011227	1.2050	89.070	0.82986	11.771	67.090	78.861	0.026243	0.16758	15
16	33.060	18.364	0.011241	1.1828	88.962	0.84544	11.989	66.977	78.966	0.026699	0.16750	16
17	33.714	19.018	0.011254	1.1611	88.854	0.86125	12.207	66.864	79.071	0.027154	0.16742	17
18	34.378	19.682	0.011268	1.1399	88.746	0.87729	12.426	66.750	79.176	0.027608	0.16734	18
19	35.052	20.356	0.011282	1.1191	88.637	0.89356	12.644	66.636	79.280	0.028062	0.16727	19
20	35.736	21.040	0.011296	1.0988	88.529	0.91006	12.863	66.522	79.385	0.028515	0.16719	20
21	36.430	21.734	0.011310	1.0790	88.419	0.92679	13.081	66.407	79.488	0.028968	0.16712	21
22	37.135	22.439	0.011324	1.0596	88.310	0.94377	13.300	66.293	79.593	0.029420	0.16704	22
23	37.849	23.153	0.011338	1.0406	88.201	0.96098	13.520	66.177	79.697	0.029871	0.16697	23
24	38.574	23.878	0.011352	1.0220	88.091	0.97843	13.739	66.061	79.800	0.030322	0.16690	24
25	39.310	24.614	0.011366	1.0039	87.981	0.99613	13.958	65.946	79.904	0.030772	0.16683	25
26	40.056	25.360	0.011380	0.98612	87.870	1.0141	14.178	65.829	80.007	0.031221	0.16676	26
27	40.813	26.117	0.011395	0.96874	87.760	1.0323	14.398	65.713	80.111	0.031670	0.16669	27
28	41.580	26.884	0.011409	0.95173	87.649	1.0507	14.618	65.596	80.214	0.032118	0.16662	28
29	42.359	27.663	0.011424	0.93509	87.537	1.0694	14.838	65.478	80.316	0.032566	0.16655	29
30	43.148	28.452	0.011438	0.91880	87.426	1.0884	15.058	65.361	80.419	0.033013	0.16648	30
31	43.948	29.252	0.011453	0.90286	87.314	1.1076	15.279	65.243	80.522	0.033460	0.16642	31
32	44.760	30.064	0.011468	0.88725	87.202	1.1271	15.500	65.124	80.624	0.033905	0.16635	32
33	45.583	30.887	0.011482	0.87197	87.090	1.1468	15.720	65.006	80.726	0.034351	0.16629	33
34	46.417	31.721	0.011497	0.85702	86.977	1.1668	15.942	64.886	80.828	0.034796	0.16622	34
35	47.263	32.567	0.011512	0.84237	86.865	1.1871	16.163	64.767	80.930	0.035240	0.16616	35
36	48.120	33.424	0.011527	0.82803	86.751	1.2077	16.384	64.647	81.031	0.035683	0.16610	36
37	48.989	34.293	0.011542	0.81399	86.638	1.2285	16.606	64.527	81.133	0.036126	0.16604	37
38	49.870	35.174	0.011557	0.80023	86.524	1.2496	16.828	64.406	81.234	0.036569	0.16598	38
39	50.763	36.067	0.011573	0.78676	86.410	1.2710	17.050	64.285	81.335	0.037011	0.16592	39
40	51.667	36.971	0.011588	0.77357	86.296	1.2927	17.273	64.163	81.436	0.037453	0.16586	40
41	52.584	37.888	0.011603	0.76064	86.181	1.3147	17.495	64.042	81.537	0.037893	0.16580	41
42	53.513	38.817	0.011619	0.74798	86.066	1.3369	17.718	63.919	81.637	0.038334	0.16574	42
43	54.454	39.758	0.011635	0.73557	85.951	1.3595	17.941	63.796	81.737	0.038774	0.16568	43
44	55.407	40.711	0.011650	0.72341	85.836	1.3823	18.164	63.673	81.837	0.039213	0.16562	44
45	56.373	41.677	0.011666	0.71149	85.720	1.4055	18.387	63.550	81.937	0.039652	0.16557	45
46	57.352	42.656	0.011682	0.69982	85.604	1.4289	18.611	63.426	82.037	0.040091	0.16551	46
47	58.343	43.647	0.011698	0.68837	85.487	1.4527	18.835	63.301	82.136	0.040529	0.16546	47
48	59.347	44.651	0.011714	0.67715	85.371	1.4768	19.059	63.177	82.236	0.040966	0.16540	48
49	60.364	45.668	0.011730	0.66616	85.254	1.5012	19.283	63.051	82.334	0.041403	0.16535	49
50	61.394	46.698	0.011746	0.65537	85.136	1.5258	19.507	62.926	82.433	0.041839	0.16530	50
51	62.437	47.741	0.011762	0.64480	85.018	1.5509	19.732	62.800	82.532	0.042276	0.16524	51
52	63.494	48.798	0.011779	0.63444	84.900	1.5762	19.957	62.673	82.630	0.042711	0.16519	52
53	64.563	49.867	0.011795	0.62428	84.782	1.6019	20.182	62.546	82.728	0.043146	0.16514	53
54	65.646	50.950	0.011811	0.61431	84.663	1.6278	20.408	62.418	82.826	0.043581	0.16509	54
55	66.743	52.047	0.011828	0.60453	84.544	1.6542	20.634	62.290	82.924	0.044015	0.16504	55
56	67.853	53.157	0.011845	0.59495	84.425	1.6808	20.859	62.162	83.021	0.044449	0.16499	56
57	68.977	54.281	0.011862	0.58554	84.305	1.7078	21.086	62.033	83.119	0.044883	0.16494	57
58	70.115	55.419	0.011879	0.57632	84.185	1.7352	21.312	61.903	83.215	0.045316	0.16489	58
59	71.267	56.571	0.011896	0.56727	84.065	1.7628	21.539	61.773	83.312	0.045748	0.16484	59
60	72.433	57.737	0.011913	0.55839	83.944	1.7909	21.766	61.643	83.409	0.046180	0.16479	60
61	73.613	58.917	0.011930	0.54967	83.823	1.8193	21.993	61.512	83.505	0.046612	0.16474	61
62	74.807	60.111	0.011947	0.54112	83.701	1.8480	22.221	61.380	83.601	0.047044	0.16470	62
63	76.016	61.320	0.011965	0.53273	83.580	1.8771	22.448	61.248	83.696	0.047475	0.16465	63
64	77.239	62.543	0.011982	0.52450	83.457	1.9066	22.676	61.116	83.792	0.047905	0.16460	64
65	78.477	63.781	0.012000	0.51642	83.335	1.9364	22.905	60.982	83.887	0.048336	0.16456	65
66	79.729	65.033	0.012017	0.50848	83.212	1.9666	23.133	60.849	83.982	0.048765	0.16451	66
67	80.996	66.300	0.012035	0.50070	83.089	1.9972	23.362	60.715	84.077	0.049195	0.16447	67
68	82.279	67.583	0.012053	0.49305	82.965	2.0282	23.591	60.580	84.171	0.049624	0.16442	68
69	83.576	68.880	0.012071	0.48555	82.841	2.0595	23.821	60.445	84.266	0.050053	0.16438	69
70	84.888	70.192	0.012089	0.47818	82.717	2.0913	24.050	60.309	84.359	0.050482	0.16434	70

Table A-4 Continued

TEMP.	PRESSURE		VOLUME cu ft/lb		DENSITY lb/cu ft		ENTHALPY Btu/lb			ENTROPY Btu/(lb)(°R)		TEMP.
°F	PSIA	PSIG	LIQUID v_f	VAPOR v_g	LIQUID $1/v_f$	VAPOR $1/v_g$	LIQUID h_f	LATENT h_{fg}	VAPOR h_g	LIQUID s_f	VAPOR s_g	°F
70	84.888	70.192	0.012089	0.47818	82.717	2.0913	24.050	60.309	84.359	0.050482	0.16434	70
71	86.216	71.520	0.012108	0.47094	82.592	2.1234	24.281	60.172	84.453	0.050910	0.16429	71
72	87.559	72.863	0.012126	0.46383	82.467	2.1559	24.511	60.035	84.546	0.051338	0.16425	72
73	88.918	74.222	0.012145	0.45686	82.341	2.1889	24.741	59.898	84.639	0.051766	0.16421	73
74	90.292	75.596	0.012163	0.45000	82.215	2.2222	24.973	59.759	84.732	0.052193	0.16417	74
75	91.682	76.986	0.012182	0.44327	82.089	2.2560	25.204	59.621	84.825	0.052620	0.16412	75
76	93.087	78.391	0.012201	0.43666	81.962	2.2901	25.435	59.481	84.916	0.053047	0.16408	76
77	94.509	79.813	0.012220	0.43016	81.835	2.3247	25.667	59.341	85.008	0.053473	0.16404	77
78	95.946	81.250	0.012239	0.42378	81.707	2.3597	25.899	59.201	85.100	0.053900	0.16400	78
79	97.400	82.704	0.012258	0.41751	81.579	2.3951	26.132	59.059	85.191	0.054326	0.16396	79
80	98.870	84.174	0.012277	0.41135	81.450	2.4310	26.365	58.917	85.282	0.054751	0.16392	80
81	100.36	85.66	0.012297	0.40530	81.322	2.4673	26.598	58.775	85.373	0.055177	0.16388	81
82	101.86	87.16	0.012316	0.39935	81.192	2.5041	26.832	58.631	85.463	0.055602	0.16384	82
83	103.38	88.68	0.012336	0.39351	81.063	2.5413	27.065	58.488	85.553	0.056027	0.16380	83
84	104.92	90.22	0.012356	0.38776	80.932	2.5789	27.300	58.343	85.643	0.056452	0.16376	84
85	106.47	91.77	0.012376	0.38212	80.802	2.6170	27.534	58.198	85.732	0.056877	0.16372	85
86	108.04	93.34	0.012396	0.37657	80.671	2.6556	27.769	58.052	85.821	0.057301	0.16368	**86**
87	109.63	94.93	0.012416	0.37111	80.539	2.6946	28.005	57.905	85.910	0.057725	0.16364	87
88	111.23	96.53	0.012437	0.36575	80.407	2.7341	28.241	57.757	85.998	0.058149	0.16360	88
89	112.85	98.15	0.012457	0.36047	80.275	2.7741	28.477	57.609	86.086	0.058573	0.16357	89
90	114.49	99.79	0.012478	0.35529	80.142	2.8146	28.713	57.461	86.174	0.058997	0.16353	90
91	116.15	101.45	0.012499	0.35019	80.008	2.8556	28.950	57.311	86.261	0.059420	0.16349	91
92	117.82	103.12	0.012520	0.34518	79.874	2.8970	29.187	57.161	86.348	0.059844	0.16345	92
93	119.51	104.81	0.012541	0.34025	79.740	2.9390	29.425	57.009	86.434	0.060267	0.16341	93
94	121.22	106.52	0.012562	0.33540	79.605	2.9815	29.663	56.858	86.521	0.060690	0.16338	94
95	122.95	108.25	0.012583	0.33063	79.470	3.0245	29.901	56.705	86.606	0.061113	0.16334	95
96	124.70	110.00	0.012605	0.32594	79.334	3.0680	30.140	56.551	86.691	0.061536	0.16330	96
97	126.46	111.76	0.012627	0.32133	79.198	3.1120	30.380	56.397	86.777	0.061959	0.16326	97
98	128.24	113.54	0.012649	0.31679	79.061	3.1566	30.619	56.242	86.861	0.062381	0.16323	98
99	130.04	115.34	0.012671	0.31233	78.923	3.2017	30.859	56.086	86.945	0.062804	0.16319	99
100	131.86	117.16	0.012693	0.30794	78.785	3.2474	31.100	55.929	87.029	0.063227	0.16315	100
101	133.70	119.00	0.012715	0.30362	78.647	3.2936	31.341	55.772	87.113	0.063649	0.16312	101
102	135.56	120.86	0.012738	0.29937	78.508	3.3404	31.583	55.613	87.196	0.064072	0.16308	102
103	137.44	122.74	0.012760	0.29518	78.368	3.3877	31.824	55.454	87.278	0.064494	0.16304	103
104	139.33	124.63	0.012783	0.29106	78.228	3.4357	32.067	55.293	87.360	0.064916	0.16301	104
105	141.25	126.55	0.012806	0.28701	78.088	3.4842	32.310	55.132	87.442	0.065339	0.16297	105
106	143.18	128.48	0.012829	0.28303	77.946	3.5333	32.553	54.970	87.523	0.065761	0.16293	106
107	145.13	130.43	0.012853	0.27910	77.804	3.5829	32.797	54.807	87.604	0.066184	0.16290	107
108	147.11	132.41	0.012876	0.27524	77.662	3.6332	33.041	54.643	87.684	0.066606	0.16286	108
109	149.10	134.40	0.012900	0.27143	77.519	3.6841	33.286	54.478	87.764	0.067028	0.16282	109
110	151.11	136.41	0.012924	0.26769	77.376	3.7357	33.531	54.313	87.844	0.067451	0.16279	110
111	153.14	138.44	0.012948	0.26400	77.231	3.7878	33.777	54.146	87.923	0.067873	0.16275	111
112	155.19	140.49	0.012972	0.26037	77.087	3.8406	34.023	53.978	88.001	0.068296	0.16271	112
113	157.27	142.57	0.012997	0.25680	76.941	3.8941	34.270	53.809	88.079	0.068719	0.16268	113
114	159.36	144.66	0.013022	0.25328	76.795	3.9482	34.517	53.639	88.156	0.069141	0.16264	114
115	161.47	146.77	0.013047	0.24982	76.649	4.0029	34.765	53.468	88.233	0.069564	0.16260	115
116	163.61	148.91	0.013072	0.24641	76.501	4.0584	35.014	53.296	88.310	0.069987	0.16256	116
117	165.76	151.06	0.013097	0.24304	76.353	4.1145	35.263	53.123	88.386	0.070410	0.16253	117
118	167.94	153.24	0.013123	0.23974	76.205	4.1713	35.512	52.949	88.461	0.070833	0.16249	118
119	170.13	155.43	0.013148	0.23647	76.056	4.2288	35.762	52.774	88.536	0.071257	0.16245	119
120	172.35	157.65	0.013174	0.23326	75.906	4.2870	36.013	52.597	88.610	0.071680	0.16241	120
121	174.59	159.89	0.013200	0.23010	75.755	4.3459	36.264	52.420	88.684	0.072104	0.16237	121
122	176.85	162.15	0.013227	0.22698	75.604	4.4056	36.516	52.241	88.757	0.072528	0.16234	122
123	179.13	164.43	0.013254	0.22391	75.452	4.4660	36.768	52.062	88.830	0.072952	0.16230	123
124	181.43	166.73	0.013280	0.22089	75.299	4.5272	37.021	51.881	88.902	0.073376	0.16226	124
125	183.76	169.06	0.013308	0.21791	75.145	4.5891	37.275	51.698	88.973	0.073800	0.16222	125

Table A-4 Continued

TEMP.	PRESSURE		VOLUME cu ft/lb		DENSITY lb/cu ft		ENTHALPY Btu/lb			ENTROPY Btu/(lb)(°R)		TEMP.
°F	PSIA	PSIG	LIQUID v_f	VAPOR v_g	LIQUID $1/v_f$	VAPOR $1/v_g$	LIQUID h_f	LATENT h_{fg}	VAPOR h_g	LIQUID s_f	VAPOR s_g	°F
125	183.76	169.06	0.013308	0.21791	75.145	4.5891	37.275	51.698	88.973	0.073800	0.16222	125
126	186.10	171.40	0.013335	0.21497	74.991	4.6518	37.529	51.515	89.044	0.074225	0.16218	126
127	188.47	173.77	0.013363	0.21207	74.836	4.7153	37.785	51.330	89.115	0.074650	0.16214	127
128	190.86	176.16	0.013390	0.20922	74.680	4.7796	38.040	51.144	89.184	0.075075	0.16210	128
129	193.27	178.57	0.013419	0.20641	74.524	4.8448	38.296	50.957	89.253	0.075501	0.16206	129
130	195.71	181.01	0.013447	0.20364	74.367	4.9107	38.553	50.768	89.321	0.075927	0.16202	130
131	198.16	183.46	0.013476	0.20091	74.209	4.9775	38.811	50.578	89.389	0.076353	0.16198	131
132	200.64	185.94	0.013504	0.19821	74.050	5.0451	39.069	50.387	89.456	0.076779	0.16194	132
133	203.15	188.45	0.013534	0.19556	73.890	5.1136	39.328	50.194	89.522	0.077206	0.16189	133
134	205.67	190.97	0.013563	0.19294	73.729	5.1829	39.588	50.000	89.588	0.077633	0.16185	134
135	208.22	193.52	0.013593	0.19036	73.568	5.2532	39.848	49.805	89.653	0.078061	0.16181	135
136	210.79	196.09	0.013623	0.18782	73.406	5.3244	40.110	49.608	89.718	0.078489	0.16177	136
137	213.39	198.69	0.013653	0.18531	73.243	5.3965	40.372	49.409	89.781	0.078917	0.16172	137
138	216.01	201.31	0.013684	0.18283	73.079	5.4695	40.634	49.210	89.844	0.079346	0.16168	138
139	218.65	203.95	0.013715	0.18039	72.914	5.5435	40.898	49.008	89.906	0.079775	0.16163	139
140	221.32	206.62	0.013746	0.17799	72.748	5.6184	41.162	48.805	89.967	0.080205	0.16159	140
141	224.00	209.30	0.013778	0.17561	72.581	5.6944	41.427	48.601	90.028	0.080635	0.16154	141
142	226.72	212.02	0.013810	0.17327	72.413	5.7713	41.693	48.394	90.087	0.081065	0.16150	142
143	229.46	214.76	0.013842	0.17096	72.244	5.8493	41.959	48.187	90.146	0.081497	0.16145	143
144	232.22	217.52	0.013874	0.16868	72.075	5.9283	42.227	47.977	90.204	0.081928	0.16140	144
145	235.00	220.30	0.013907	0.16644	71.904	6.0083	42.495	47.766	90.261	0.082361	0.16135	145
146	237.82	223.12	0.013941	0.16422	71.732	6.0895	42.765	47.553	90.318	0.082794	0.16130	146
147	240.65	225.95	0.013974	0.16203	71.559	6.1717	43.035	47.338	90.373	0.083227	0.16125	147
148	243.51	228.81	0.014008	0.15987	71.386	6.2551	43.306	47.122	90.428	0.083661	0.16120	148
149	246.40	231.70	0.014043	0.15774	71.211	6.3395	43.578	46.904	90.482	0.084096	0.16115	149
150	249.31	234.61	0.014078	0.15564	71.035	6.4252	43.850	46.684	90.534	0.084531	0.16110	150
151	252.24	237.54	0.014113	0.15356	70.857	6.5120	44.124	46.462	90.586	0.084967	0.16105	151
152	255.20	240.50	0.014148	0.15151	70.679	6.6001	44.399	46.238	90.637	0.085404	0.16099	152
153	258.19	243.49	0.014184	0.14949	70.500	6.6893	44.675	46.012	90.687	0.085842	0.16094	153
154	261.20	246.50	0.014221	0.14750	70.319	6.7799	44.951	45.784	90.735	0.086280	0.16088	154
155	264.24	249.54	0.014258	0.14552	70.137	6.8717	45.229	45.554	90.783	0.086719	0.16083	155
156	267.30	252.60	0.014295	0.14358	69.954	6.9648	45.508	45.322	90.830	0.087159	0.16077	156
157	270.39	255.69	0.014333	0.14166	69.770	7.0592	45.787	45.088	90.875	0.087600	0.16071	157
158	273.51	258.81	0.014371	0.13976	69.584	7.1551	46.068	44.852	90.920	0.088041	0.16065	158
159	276.65	261.95	0.014410	0.13789	69.397	7.2523	46.350	44.614	90.964	0.088484	0.16059	159
160	279.82	265.12	0.014449	0.13604	69.209	7.3509	46.633	44.373	91.006	0.088927	0.16053	160
161	283.02	268.32	0.014489	0.13421	69.019	7.4510	46.917	44.130	91.047	0.089371	0.16047	161
162	286.24	271.54	0.014529	0.13241	68.828	7.5525	47.202	43.885	91.087	0.089817	0.16040	162
163	289.49	274.79	0.014570	0.13062	68.635	7.6556	47.489	43.637	91.126	0.090263	0.16034	163
164	292.77	278.07	0.014611	0.12886	68.441	7.7602	47.777	43.386	91.163	0.090710	0.16027	164
165	296.07	281.37	0.014653	0.12712	68.245	7.8665	48.065	43.134	91.199	0.091159	0.16021	165
166	299.40	284.70	0.014695	0.12540	68.048	7.9743	48.355	42.879	91.234	0.091608	0.16014	166
167	302.76	288.06	0.014738	0.12370	67.850	8.0838	48.647	42.620	91.267	0.092059	0.16007	167
168	306.15	291.45	0.014782	0.12202	67.649	8.1950	48.939	42.360	91.299	0.092511	0.16000	168
169	309.56	294.86	0.014826	0.12037	67.447	8.3080	49.233	42.097	91.330	0.092964	0.15992	169
170	313.00	298.30	0.014871	0.11873	67.244	8.4228	49.529	41.830	91.359	0.093418	0.15985	170
171	316.47	301.77	0.014917	0.11710	67.038	8.5394	49.825	41.562	91.387	0.093874	0.15977	171
172	319.97	305.27	0.014963	0.11550	66.831	8.6579	50.123	41.290	91.413	0.094330	0.15969	172
173	323.50	308.80	0.015010	0.11392	66.622	8.7783	50.423	41.015	91.438	0.094789	0.15961	173
174	327.06	312.36	0.015058	0.11235	66.411	8.9007	50.724	40.736	91.460	0.095248	0.15953	174
175	330.64	315.94	0.015106	0.11080	66.198	9.0252	51.026	40.455	91.481	0.095709	0.15945	175
176	334.25	319.55	0.015155	0.10927	65.983	9.1518	51.330	40.171	91.501	0.096172	0.15936	176
177	337.90	323.20	0.015205	0.10775	65.766	9.2805	51.636	39.883	91.519	0.096636	0.15928	177
178	341.57	326.87	0.015256	0.10625	65.547	9.4114	51.943	39.592	91.535	0.097102	0.15919	178
179	345.27	330 57	0.015308	0.10477	65.326	9.5446	52.252	39.297	91.549	0.097569	0.15910	179
180	349.00	334.30	0.015360	0.10330	65.102	9.6802	52.562	38.999	91.561	0.098039	0.15900	180

Table A-4 Continued

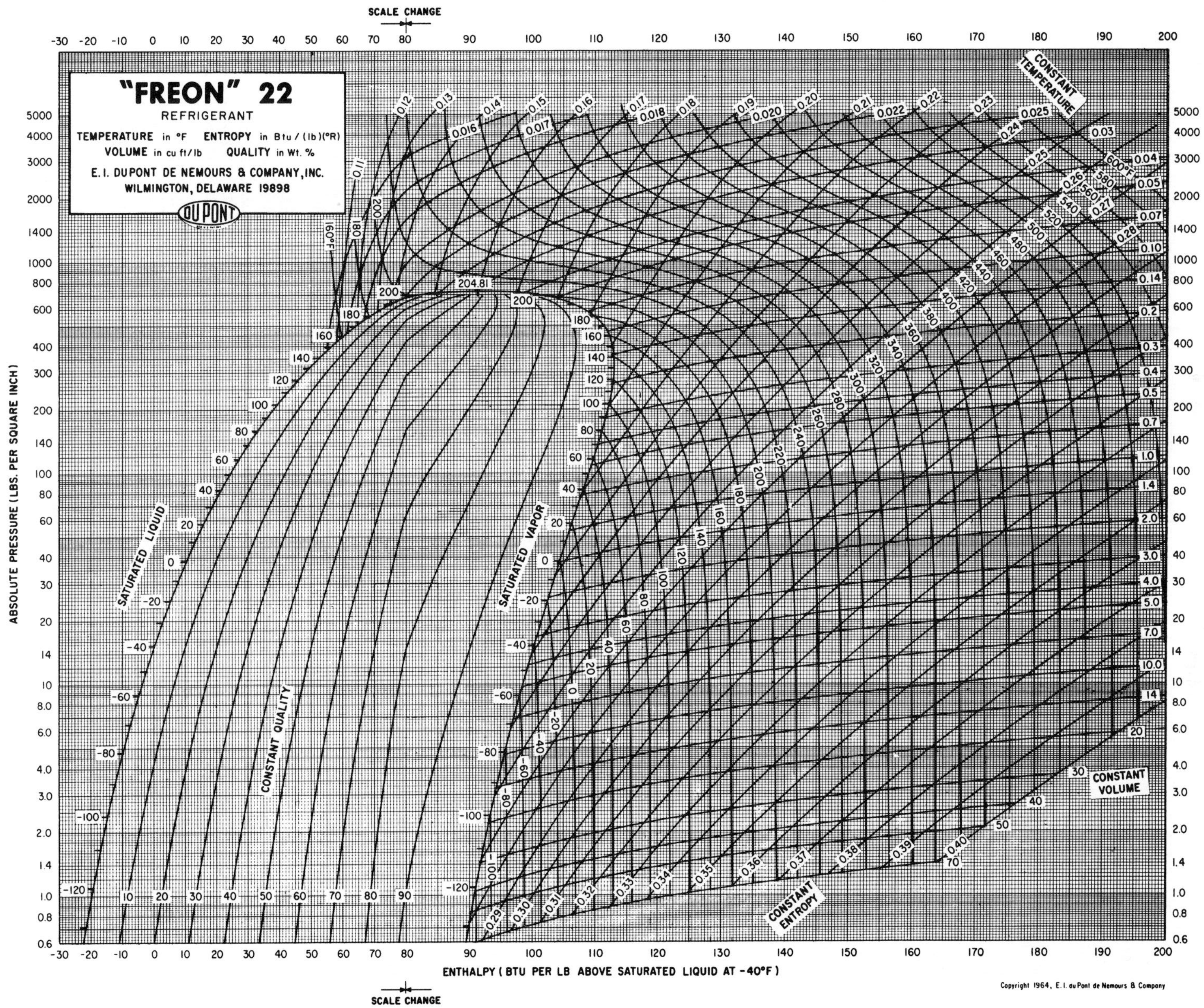

Table A-5 I-P Pressure-Heat Diagram for R-22

TEMP.	PRESSURE		VOLUME cu ft/lb		DENSITY lb/cu ft		ENTHALPY Btu/lb			ENTROPY Btu/(lb)(°R)		TEMP.
°F	PSIA	PSIG	LIQUID v_f	VAPOR v_g	LIQUID $1/v_f$	VAPOR $1/v_g$	LIQUID h_f	LATENT h_{fg}	VAPOR h_g	LIQUID s_f	VAPOR s_g	°F
−45	13.354	2.732*	0.011298	3.7243	88.507	0.26851	−1.260	100.963	99.703	−0.00301	0.24046	−45
−44	13.712	2.002*	0.011311	3.6334	88.407	0.27523	−1.009	100.823	99.814	−0.00241	0.24014	−44
−43	14.078	1.258*	0.011324	3.5452	88.307	0.28207	−0.757	100.683	99.925	−0.00181	0.23982	−43
−42	14.451	0.498*	0.011337	3.4596	88.207	0.28905	−0.505	100.541	100.036	−0.00120	0.23951	−42
−41	14.833	0.137	0.011350	3.3764	88.107	0.29617	−0.253	100.399	100.147	−0.00060	0.23919	−41
−40	15.222	0.526	0.011363	3.2957	88.006	0.30342	0.000	100.257	100.257	0.00000	0.23888	−40
−39	15.619	0.923	0.011376	3.2173	87.905	0.31082	0.253	100.114	100.367	0.00060	0.23858	−39
−38	16.024	1.328	0.011389	3.1412	87.805	0.31835	0.506	99.971	100.477	0.00120	0.23827	−38
−37	16.437	1.741	0.011402	3.0673	87.703	0.32602	0.760	99.826	100.587	0.00180	0.23797	−37
−36	16.859	2.163	0.011415	2.9954	87.602	0.33384	1.014	99.682	100.696	0.00240	0.23767	−36
−35	17.290	2.594	0.011428	2.9256	87.501	0.34181	1.269	99.536	100.805	0.00300	0.23737	−35
−34	17.728	3.032	0.011442	2.8578	87.399	0.34992	1.524	99.391	100.914	0.00359	0.23707	−34
−33	18.176	3.480	0.011455	2.7919	87.297	0.35818	1.779	99.244	101.023	0.00419	0.23678	−33
−32	18.633	3.937	0.011469	2.7278	87.195	0.36660	2.035	99.097	101.132	0.00479	0.23649	−32
−31	19.098	4.402	0.011482	2.6655	87.093	0.37517	2.291	98.949	101.240	0.00538	0.23620	−31
−30	19.573	4.877	0.011495	2.6049	86.991	0.38389	2.547	98.801	101.348	0.00598	0.23591	−30
−29	20.056	5.360	0.011509	2.5460	86.888	0.39278	2.804	98.652	101.456	0.00657	0.23563	−29
−28	20.549	5.853	0.011523	2.4887	86.785	0.40182	3.061	98.503	101.564	0.00716	0.23534	−28
−27	21.052	6.536	0.011536	2.4329	86.682	0.41103	3.318	98.353	101.671	0.00776	0.23506	−27
−26	21.564	6.868	0.011550	2.3787	86.579	0.42040	3.576	98.202	101.778	0.00835	0.23478	−26
−25	22.086	7.390	0.011564	2.3260	86.476	0.42993	3.834	98.051	101.885	0.00894	0.23451	−25
−24	22.617	7.921	0.011578	2.2746	86.372	0.43964	4.093	97.899	101.992	0.00953	0.23423	−24
−23	23.159	8.463	0.011592	2.2246	86.269	0.44951	4.352	97.746	102.098	0.01013	0.23396	−23
−22	23.711	9.015	0.011606	2.1760	86.165	0.45956	4.611	97.593	102.204	0.01072	0.23369	−22
−21	24.272	9.576	0.011620	2.1287	86.061	0.46978	4.871	97.439	102.310	0.01131	0.23342	−21
−20	24.845	10.149	0.011634	2.0826	85.956	0.48018	5.131	97.285	102.415	0.01189	0.23315	−20
−19	25.427	10.731	0.011648	2.0377	85.852	0.49075	5.391	97.129	102.521	0.01248	0.23289	−19
−18	26.020	11.324	0.011662	1.9940	85.747	0.50151	5.652	96.974	102.626	0.01307	0.23262	−18
−17	26.624	11.928	0.011677	1.9514	85.642	0.51245	5.913	96.817	102.730	0.01366	0.23236	−17
−16	27.239	12.543	0.011691	1.9099	85.537	0.52358	6.175	96.660	102.835	0.01425	0.23210	−16
−15	27.865	13.169	0.011705	1.8695	85.431	0.53489	6.436	96.502	102.939	0.01483	0.23184	−15
−14	28.501	13.805	0.011720	1.8302	85.326	0.54640	6.699	96.344	103.043	0.01542	0.23159	−14
−13	29.149	14.453	0.011734	1.7918	85.220	0.55810	6.961	96.185	103.146	0.01600	0.23133	−13
−12	29.809	15.113	0.011749	1.7544	85.114	0.56999	7.224	96.025	103.250	0.01659	0.23108	−12
−11	30.480	15.784	0.011764	1.7180	85.008	0.58207	7.488	95.865	103.353	0.01717	0.23083	−11
−10	31.162	16.466	0.011778	1.6825	84.901	0.59436	7.751	95.704	103.455	0.01776	0.23058	−10
− 9	31.856	17.160	0.011793	1.6479	84.795	0.60685	8.015	95.542	103.558	0.01834	0.23033	− 9
− 8	32.563	17.867	0.011808	1.6141	84.688	0.61954	8.280	95.380	103.660	0.01892	0.23008	− 8
− 7	33.281	18.585	0.011823	1.5812	84.581	0.63244	8.545	95.217	103.762	0.01950	0.22984	− 7
− 6	34.011	19.315	0.011838	1.5491	84.473	0.64555	8.810	95.053	103.863	0.02009	0.22960	− 6
− 5	34.754	20.058	0.011853	1.5177	84.366	0.65887	9.075	94.889	103.964	0.02067	0.22936	− 5
− 4	35.509	20.813	0.011868	1.4872	84.258	0.67240	9.341	94.724	104.065	0.02125	0.22912	− 4
− 3	36.277	21.581	0.011884	1.4574	84.150	0.68615	9.608	94.558	104.166	0.02183	0.22888	− 3
− 2	37.057	22.361	0.011899	1.4283	84.042	0.70012	9.874	94.391	104.266	0.02241	0.22864	− 2
− 1	37.850	23.154	0.011914	1.4000	83.933	0.71431	10.142	94.224	104.366	0.02299	0.22841	− 1
0	38.657	23.961	0.011930	1.3723	83.825	0.72872	10.409	94.056	104.465	0.02357	0.22817	0
1	39.476	24.780	0.011945	1.3453	83.716	0.74336	10.677	93.888	104.565	0.02414	0.22794	1
2	40.309	25.613	0.011961	1.3189	83.606	0.75822	10.945	93.718	104.663	0.02472	0.22771	2
3	41.155	26.459	0.011976	1.2931	83.497	0.77332	11.214	93.548	104.762	0.02530	0.22748	3
4	42.014	27.318	0.011992	1.2680	83.387	0.78865	11.483	93.378	104.860	0.02587	0.22725	4
5	42.888	28.192	0.012008	1.2434	83.277	0.80422	11.752	93.206	104.958	0.02645	0.22703	5
6	43.775	29.079	0.012024	1.2195	83.167	0.82003	12.022	93.034	105.056	0.02703	0.22680	6
7	44.676	29.980	0.012040	1.1961	83.057	0.83608	12.292	92.861	105.153	0.02760	0.22658	7
8	45.591	30.895	0.012056	1.1732	82.946	0.85237	12.562	92.688	105.250	0.02818	0.22636	8
9	46.521	31.825	0.012072	1.1509	82.835	0.86892	12.833	92.513	105.346	0.02875	0.22614	9

*Inches of mercury below one atmosphere

Table A-6 I-P Properties of Saturated Liquid and Vapor for R-22

TEMP.	PRESSURE		VOLUME cu ft/lb		DENSITY lb/cu ft		ENTHALPY Btu/lb			ENTROPY Btu/(lb)(°R)		TEMP.
°F	PSIA	PSIG	LIQUID v_f	VAPOR v_g	LIQUID $1/v_f$	VAPOR $1/v_g$	LIQUID h_f	LATENT h_{fg}	VAPOR h_g	LIQUID s_f	VAPOR s_g	°F
10	47.464	32.768	0.012088	1.1290	82.724	0.88571	13.104	92.338	105.442	0.02932	0.22592	10
11	48.423	33.727	0.012105	1.1077	82.612	0.90275	13.376	92.162	105.538	0.02990	0.22570	11
12	49.396	34.700	0.012121	1.0869	82.501	0.92005	13.648	91.986	105.633	0.03047	0.22548	12
13	50.384	35.688	0.012138	1.0665	82.389	0.93761	13.920	91.808	105.728	0.03104	0.22527	13
14	51.387	36.691	0.012154	1.0466	82.276	0.95544	14.193	91.630	105.823	0.03161	0.22505	14
15	52.405	37.709	0.012171	1.0272	82.164	0.97352	14.466	91.451	105.917	0.03218	0.22484	15
16	53.438	38.742	0.012188	1.0082	82.051	0.99188	14.739	91.272	106.011	0.03275	0.22463	16
17	54.487	39.791	0.012204	0.98961	81.938	1.0105	15.013	91.091	106.105	0.03332	0.22442	17
18	55.551	40.855	0.012221	0.97144	81.825	1.0294	15.288	90.910	106.198	0.03389	0.22421	18
19	56.631	41.935	0.012238	0.95368	81.711	1.0486	15.562	90.728	106.290	0.03446	0.22400	19
20	57.727	43.031	0.012255	0.93631	81.597	1.0680	15.837	90.545	106.383	0.03503	0.22379	20
21	58.839	44.143	0.012273	0.91932	81.483	1.0878	16.113	90.362	106.475	0.03560	0.22358	21
22	59.967	45.271	0.012290	0.90270	81.368	1.1078	16.389	90.178	106.566	0.03617	0.22338	22
23	61.111	46.415	0.012307	0.88645	81.253	1.1281	16.665	89.993	106.657	0.03674	0.22318	23
24	62.272	47.576	0.012325	0.87055	81.138	1.1487	16.942	89.807	106.748	0.03730	0.22297	24
25	63.450	48.754	0.012342	0.85500	81.023	1.1696	17.219	89.620	106.839	0.03787	0.22277	25
26	64.644	49.948	0.012360	0.83978	80.907	1.1908	17.496	89.433	106.928	0.03844	0.22257	26
27	65.855	51.159	0.012378	0.82488	80.791	1.2123	17.774	89.244	107.018	0.03900	0.22237	27
28	67.083	52.387	0.012395	0.81031	80.675	1.2341	18.052	89.055	107.107	0.03958	0.22217	28
29	68.328	53.632	0.012413	0.79604	80.558	1.2562	18.330	88.865	107.196	0.04013	0.22198	29
30	69.591	54.895	0.012431	0.78208	80.441	1.2786	18.609	88.674	107.284	0.04070	0.22178	30
31	70.871	56.175	0.012450	0.76842	80.324	1.3014	18.889	88.483	107.372	0.04126	0.22158	31
32	72.169	57.473	0.012468	0.75503	80.207	1.3244	19.169	88.290	107.459	0.04182	0.22139	32
33	73.485	58.789	0.012486	0.74194	80.089	1.3478	19.449	88.097	107.546	0.04239	0.22119	33
34	74.818	60.122	0.012505	0.72911	79.971	1.3715	19.729	87.903	107.632	0.04295	0.22100	34
35	76.170	61.474	0.012523	0.71655	79.852	1.3956	20.010	87.708	107.719	0.04351	0.22081	35
36	77.540	62.844	0.012542	0.70425	79.733	1.4199	20.292	87.512	107.804	0.04407	0.22062	36
37	78.929	64.233	0.012561	0.69221	79.614	1.4447	20.574	87.316	107.889	0.04464	0.22043	37
38	80.336	65.640	0.012579	0.68041	79.495	1.4697	20.856	87.118	107.974	0.04520	0.22024	38
39	81.761	67.065	0.012598	0.66885	79.375	1.4951	21.138	86.920	108.058	0.04576	0.22005	39
40	83.206	68.510	0.012618	0.65753	79.255	1.5208	21.422	86.720	108.142	0.04632	0.21986	40
41	84.670	69.974	0.012637	0.64643	79.134	1.5469	21.705	86.520	108.225	0.04688	0.21968	41
42	86.153	71.457	0.012656	0.63557	79.013	1.5734	21.989	86.319	108.308	0.04744	0.21949	42
43	87.655	72.959	0.012676	0.62492	78.892	1.6002	22.273	86.117	108.390	0.04800	0.21931	43
44	89.177	74.481	0.012695	0.61448	78.770	1.6274	22.558	85.914	108.472	0.04855	0.21912	44
45	90.719	76.023	0.012715	0.60425	78.648	1.6549	22.843	85.710	108.553	0.04911	0.21894	45
46	92.280	77.584	0.012735	0.59422	78.526	1.6829	23.129	85.506	108.634	0.04967	0.21876	46
47	93.861	79.165	0.012755	0.58440	78.403	1.7112	23.415	85.300	108.715	0.05023	0.21858	47
48	95.463	80.767	0.012775	0.57476	78.280	1.7398	23.701	85.094	108.795	0.05079	0.21839	48
49	97.085	82.389	0.012795	0.56532	78.157	1.7689	23.988	84.886	108.874	0.05134	0.21821	49
50	98.727	84.031	0.012815	0.55606	78.033	1.7984	24.275	84.678	108.953	0.05190	0.21803	50
51	100.39	85.69	0.012836	0.54698	77.909	1.8282	24.563	84.468	109.031	0.05245	0.21785	51
52	102.07	87.38	0.012856	0.53808	77.784	1.8585	24.851	84.258	109.109	0.05301	0.21768	52
53	103.78	89.08	0.012877	0.52934	77.659	1.8891	25.139	84.047	109.186	0.05357	0.21750	53
54	105.50	90.81	0.012898	0.52078	77.534	1.9202	25.429	83.834	109.263	0.05412	0.21732	54
55	107.25	92.56	0.012919	0.51238	77.408	1.9517	25.718	83.621	109.339	0.05468	0.21714	55
56	109.02	94.32	0.012940	0.50414	77.282	1.9836	26.008	83.407	109.415	0.05523	0.21697	56
57	110.81	96.11	0.012961	0.49606	77.155	2.0159	26.298	83.191	109.490	0.05579	0.21679	57
58	112.62	97.93	0.012982	0.48813	77.028	2.0486	26.589	82.975	109.564	0.05634	0.21662	58
59	114.46	99.76	0.013004	0.48035	76.900	2.0818	26.880	82.758	109.638	0.05689	0.21644	59
60	116.31	101.62	0.013025	0.47272	76.773	2.1154	27.172	82.540	109.712	0.05745	0.21627	60
61	118.19	103.49	0.013047	0.46523	76.644	2.1495	27.464	82.320	109.785	0.05800	0.21610	61
62	120.09	105.39	0.013069	0.45788	76.515	2.1840	27.757	82.100	109.857	0.05855	0.21592	62
63	122.01	107.32	0.013091	0.45066	76.386	2.2190	28.050	81.878	109.929	0.05910	0.21575	63
64	123.96	109.26	0.013114	0.44358	76.257	2.2544	28.344	81.656	110.000	0.05966	0.21558	64

Table A-6 Continued

TEMP.	PRESSURE		VOLUME cu ft/lb		DENSITY lb/cu ft		ENTHALPY Btu/lb			ENTROPY Btu/(lb)(°R)		TEMP.
°F	PSIA	PSIG	LIQUID v_f	VAPOR v_g	LIQUID $1/v_f$	VAPOR $1/v_g$	LIQUID h_f	LATENT h_{fg}	VAPOR h_g	LIQUID s_f	VAPOR s_g	°F
65	125.93	111.23	0.013136	0.43663	76.126	2.2903	28.638	81.432	110.070	0.06021	0.21541	65
66	127.92	113.22	0.013159	0.42981	75.996	2.3266	28.932	81.208	110.140	0.06076	0.21524	66
67	129.94	115.24	0.013181	0.42311	75.865	2.3635	29.228	80.982	110.209	0.06131	0.21507	67
68	131.97	117.28	0.013204	0.41653	75.733	2.4008	29.523	80.755	110.278	0.06186	0.21490	68
69	134.04	119.34	0.013227	0.41007	75.601	2.4386	29.819	80.527	110.346	0.06241	0.21473	69
70	136.12	121.43	0.013251	0.40373	75.469	2.4769	30.116	80.298	110.414	0.06296	0.21456	70
71	138.23	123.54	0.013274	0.39751	75.336	2.5157	30.413	80.068	110.480	0.06351	0.21439	71
72	140.37	125.67	0.013297	0.39139	75.202	2.5550	30.710	79.836	110.547	0.06406	0.21422	72
73	142.52	127.83	0.013321	0.38539	75.068	2.5948	31.008	79.604	110.612	0.06461	0.21405	73
74	144.71	130.01	0.013345	0.37949	74.934	2.6351	31.307	79.370	110.677	0.06516	0.21388	74
75	146.91	132.22	0.013369	0.37369	74.799	2.6760	31.606	79.135	110.741	0.06571	0.21372	75
76	149.15	134.45	0.013393	0.36800	74.664	2.7174	31.906	78.899	110.805	0.06626	0.21355	76
77	151.40	136.71	0.013418	0.36241	74.528	2.7593	32.206	78.662	110.868	0.06681	0.21338	77
78	153.69	138.99	0.013442	0.35691	74.391	2.8018	32.506	78.423	110.930	0.06736	0.21321	78
79	155.99	141.30	0.013467	0.35151	74.254	2.8449	32.808	78.184	110.991	0.06791	0.21305	79
80	158.33	143.63	0.013492	0.34621	74.116	2.8885	33.109	77.943	111.052	0.06846	0.21288	80
81	160.68	145.99	0.013518	0.34099	73.978	2.9326	33.412	77.701	111.112	0.06901	0.21271	81
82	163.07	148.37	0.013543	0.33587	73.839	2.9774	33.714	77.457	111.171	0.06956	0.21255	82
83	165.48	150.78	0.013569	0.33083	73.700	3.0227	34.018	77.212	111.230	0.07011	0.21238	83
84	167.92	153.22	0.013594	0.32588	73.560	3.0686	34.322	76.966	111.288	0.07065	0.21222	84
85	170.38	155.68	0.013620	0.32101	73.420	3.1151	34.626	76.719	111.345	0.07120	0.21205	85
86	172.87	158.17	0.013647	0.31623	73.278	3.1622	34.931	76.470	111.401	0.07175	0.21188	86
87	175.38	160.69	0.013673	0.31153	73.137	3.2100	35.237	76.220	111.457	0.07230	0.21172	87
88	177.93	163.23	0.013700	0.30690	72.994	3.2583	35.543	75.968	111.512	0.07285	0.21155	88
89	180.50	165.80	0.013727	0.30236	72.851	3.3073	35.850	75.716	111.566	0.07339	0.21139	89
90	183.09	168.40	0.013754	0.29789	72.708	3.3570	36.158	75.461	111.619	0.07394	0.21122	90
91	185.72	171.02	0.013781	0.29349	72.564	3.4073	36.466	75.206	111.671	0.07449	0.21106	91
92	188.37	173.67	0.013809	0.28917	72.419	3.4582	36.774	74.949	111.723	0.07504	0.21089	92
93	191.05	176.35	0.013836	0.28491	72.273	3.5098	37.084	74.690	111.774	0.07559	0.21072	93
94	193.76	179.06	0.013864	0.28073	72.127	3.5621	37.394	74.430	111.824	0.07613	0.21056	94
95	196.50	181.80	0.013893	0.27662	71.980	3.6151	37.704	74.168	111.873	0.07668	0.21039	95
96	199.26	184.56	0.013921	0.27257	71.833	3.6688	38.016	73.905	111.921	0.07723	0.21023	96
97	202.05	187.36	0.013950	0.26859	71.685	3.7232	38.328	73.641	111.968	0.07778	0.21006	97
98	204.87	190.18	0.013979	0.26467	71.536	3.7783	38.640	73.375	112.015	0.07832	0.20989	98
99	207.72	193.03	0.014008	0.26081	71.386	3.8341	38.953	73.107	112.060	0.07887	0.20973	99
100	210.60	195.91	0.014038	0.25702	71.236	3.8907	39.267	72.838	112.105	0.07942	0.20956	100
101	213.51	198.82	0.014068	0.25329	71.084	3.9481	39.582	72.567	112.149	0.07997	0.20939	101
102	216.45	201.76	0.014098	0.24962	70.933	4.0062	39.897	72.294	112.192	0.08052	0.20923	102
103	219.42	204.72	0.014128	0.24600	70.780	4.0651	40.213	72.020	112.233	0.08107	0.20906	103
104	222.42	207.72	0.014159	0.24244	70.626	4.1247	40.530	71.744	112.274	0.08161	0.20889	104
105	225.45	210.75	0.014190	0.23894	70.472	4.1852	40.847	71.467	112.314	0.08216	0.20872	105
106	228.50	213.81	0.014221	0.23549	70.317	4.2465	41.166	71.187	112.353	0.08271	0.20855	106
107	231.59	216.90	0.014253	0.23209	70.161	4.3086	41.485	70.906	112.391	0.08326	0.20838	107
108	234.71	220.02	0.014285	0.22875	70.005	4.3715	41.804	70.623	112.427	0.08381	0.20821	108
109	237.86	223.17	0.014317	0.22546	69.847	4.4354	42.125	70.338	112.463	0.08436	0.20804	109
110	241.04	226.35	0.014350	0.22222	69.689	4.5000	42.446	70.052	112.498	0.08491	0.20787	110
111	244.25	229.56	0.014382	0.21903	69.529	4.5656	42.768	69.763	112.531	0.08546	0.20770	111
112	247.50	232.80	0.014416	0.21589	69.369	4.6321	43.091	69.473	112.564	0.08601	0.20753	112
113	250.77	236.08	0.014449	0.21279	69.208	4.6994	43.415	69.180	112.595	0.08656	0.20736	113
114	254.08	239.38	0.014483	0.20974	69.046	4.7677	43.739	68.886	112.626	0.08711	0.20718	114
115	257.42	242.72	0.014517	0.20674	68.883	4.8370	44.065	68.590	112.655	0.08766	0.20701	115
116	260.79	246.10	0.014552	0.20378	68.719	4.9072	44.391	68.291	112.682	0.08821	0.20684	116
117	264.20	249.50	0.014587	0.20087	68.554	4.9784	44.718	67.991	112.709	0.08876	0.20666	117
118	267.63	252.94	0.014622	0.19800	68.388	5.0506	45.046	67.688	112.735	0.08932	0.20649	118
119	271.10	256.41	0.014658	0.19517	68.221	5.1238	45.375	67.384	112.759	0.08987	0.20631	119

Table A-6 Continued

TEMP. °F	PRESSURE PSIA	PRESSURE PSIG	VOLUME cu ft/lb LIQUID v_f	VOLUME cu ft/lb VAPOR v_g	DENSITY lb/cu ft LIQUID $1/v_f$	DENSITY lb/cu ft VAPOR $1/v_g$	ENTHALPY Btu/lb LIQUID h_f	ENTHALPY Btu/lb LATENT h_{fg}	ENTHALPY Btu/lb VAPOR h_g	ENTROPY Btu/(lb)(°R) LIQUID s_f	ENTROPY Btu/(lb)(°R) VAPOR s_g	TEMP. °F
120	274.60	259.91	0.014694	0.19238	68.054	5.1981	45.705	67.077	112.782	0.09042	0.20613	120
121	278.14	263.44	0.014731	0.18963	67.885	5.2734	46.036	66.767	112.803	0.09098	0.20595	121
122	281.71	267.01	0.014768	0.18692	67.714	5.3498	46.368	66.456	112.824	0.09153	0.20578	122
123	285.31	270.62	0.014805	0.18426	67.543	5.4272	46.701	66.142	112.843	0.09208	0.20560	123
124	288.95	274.25	0.014843	0.18163	67.371	5.5058	47.034	65.826	112.860	0.09264	0.20542	124
125	292.62	277.92	0.014882	0.17903	67.197	5.5856	47.369	65.507	112.877	0.09320	0.20523	125
126	296.33	281.63	0.014920	0.17648	67.023	5.6665	47.705	65.186	112.891	0.09375	0.20505	126
127	300.07	285.37	0.014960	0.17396	66.847	5.7486	48.042	64.863	112.905	0.09431	0.20487	127
128	303.84	289.14	0.014999	0.17147	66.670	5.8319	48.380	64.537	112.917	0.09487	0.20468	128
129	307.65	292.95	0.015039	0.16902	66.492	5.9164	48.719	64.208	112.927	0.09543	0.20449	129
130	311.50	296.80	0.015080	0.16661	66.312	6.0022	49.059	63.877	112.936	0.09598	0.20431	130
131	315.38	300.68	0.015121	0.16422	66.131	6.0893	49.400	63.543	112.943	0.09654	0.20412	131
132	319.29	304.60	0.015163	0.16187	65.949	6.1777	49.743	63.206	112.949	0.09711	0.20393	132
133	323.25	308.55	0.015206	0.15956	65.766	6.2674	50.087	62.866	112.953	0.09767	0.20374	133
134	327.23	312.54	0.015248	0.15727	65.581	6.3585	50.432	62.523	112.955	0.09823	0.20354	134
135	331.26	316.56	0.015292	0.15501	65.394	6.4510	50.778	62.178	112.956	0.09879	0.20335	135
136	335.32	320.63	0.015336	0.15279	65.207	6.5450	51.125	61.829	112.954	0.09936	0.20315	136
137	339.42	324.73	0.015381	0.15059	65.017	6.6405	51.474	61.477	112.951	0.09992	0.20295	137
138	343.56	328.86	0.015426	0.14843	64.826	6.7374	51.824	61.123	112.947	0.10049	0.20275	138
139	347.73	333.04	0.015472	0.14629	64.634	6.8359	52.175	60.764	112.940	0.10106	0.20255	139
140	351.94	337.25	0.015518	0.14418	64.440	6.9360	52.528	60.403	112.931	0.10163	0.20235	140
141	356.19	341.50	0.015566	0.14209	64.244	7.0377	52.883	60.038	112.921	0.10220	0.20214	141
142	360.48	345.79	0.015613	0.14004	64.047	7.1410	53.238	59.670	112.908	0.10277	0.20194	142
143	364.81	350.11	0.015662	0.13801	63.848	7.2461	53.596	59.298	112.893	0.10334	0.20173	143
144	369.17	354.48	0.015712	0.13600	63.647	7.3529	53.955	58.922	112.877	0.10391	0.20152	144
145	373.58	358.88	0.015762	0.13402	63.445	7.4615	54.315	58.543	112.858	0.10449	0.20130	145
146	378.02	363.32	0.015813	0.13207	63.240	7.5719	54.677	58.159	112.836	0.10507	0.20109	146
147	382.50	367.81	0.015865	0.13014	63.034	7.6842	55.041	57.772	112.813	0.10564	0.20087	147
148	387.03	372.33	0.015917	0.12823	62.825	7.7985	55.406	57.380	112.787	0.10622	0.20065	148
149	391.59	376.89	0.015971	0.12635	62.615	7.9148	55.774	56.985	112.758	0.10681	0.20042	149
150	396.19	381.50	0.016025	0.12448	62.402	8.0331	56.143	56.585	112.728	0.10739	0.20020	150
151	400.84	386.14	0.016080	0.12265	62.187	8.1536	56.514	56.180	112.694	0.10797	0.19997	151
152	405.52	390.83	0.016137	0.12083	61.970	8.2763	56.887	55.771	112.658	0.10856	0.19974	152
153	410.25	395.56	0.016194	0.11903	61.751	8.4011	57.261	55.358	112.619	0.10915	0.19950	153
154	415.02	400.32	0.016252	0.11726	61.529	8.5284	57.638	54.939	112.577	0.10974	0.19926	154
155	419.83	405.13	0.016312	0.11550	61.305	8.6580	58.017	54.515	112.533	0.11034	0.19902	155
156	424.68	409.99	0.016372	0.11376	61.079	8.7901	58.399	54.087	112.485	0.11093	0.19878	156
157	429.58	414.88	0.016434	0.11205	60.849	8.9247	58.782	53.652	112.435	0.11153	0.19853	157
158	434.52	419.82	0.016497	0.11035	60.617	9.0620	59.168	53.213	112.381	0.11213	0.19828	158
159	439.50	424.80	0.016561	0.10867	60.383	9.2020	59.557	52.767	112.324	0.11273	0.19802	159
160	444.53	429.83	0.016627	0.10701	60.145	9.3449	59.948	52.316	112.263	0.11334	0.19776	160
161	449.59	434.90	0.016693	0.10537	59.904	9.4907	60.341	51.858	112.199	0.11395	0.19750	161
162	454.71	440.01	0.016762	0.10374	59.660	9.6395	60.737	51.394	112.131	0.11456	0.19723	162
163	459.87	445.17	0.016831	0.10213	59.413	9.7915	61.136	50.923	112.060	0.11518	0.19696	163
164	465.07	450.37	0.016902	0.10054	59.163	9.9467	61.538	50.446	111.984	0.11580	0.19668	164
165	470.32	455.62	0.016975	0.098956	58.909	10.106	61.943	49.961	111.904	0.11642	0.19640	165
166	475.61	460.92	0.017050	0.097393	58.651	10.268	62.351	49.469	111.820	0.11705	0.19611	166
167	480.95	466.26	0.017126	0.095844	58.390	10.434	62.763	48.969	111.732	0.11768	0.19581	167
168	486.34	471.65	0.017204	0.094309	58.125	10.603	63.178	48.461	111.639	0.11831	0.19552	168
169	491.78	477.08	0.017285	0.092787	57.855	10.777	63.596	47.945	111.541	0.11895	0.19521	169
170	497.26	482.56	0.017367	0.091279	57.581	10.955	64.019	47.419	111.438	0.11959	0.19490	170
171	502.79	488.09	0.017451	0.089783	57.303	11.138	64.445	46.885	111.330	0.12024	0.19458	171
172	508.37	493.67	0.017538	0.088299	57.019	11.325	64.875	46.340	111.216	0.12089	0.19425	172
173	513.99	499.30	0.017627	0.086827	56.731	11.517	65.310	45.786	111.096	0.12155	0.19392	173
174	519.67	504.97	0.017719	0.085365	56.438	11.714	65.750	45.221	110.970	0.12222	0.19358	174

Table A-6 Continued

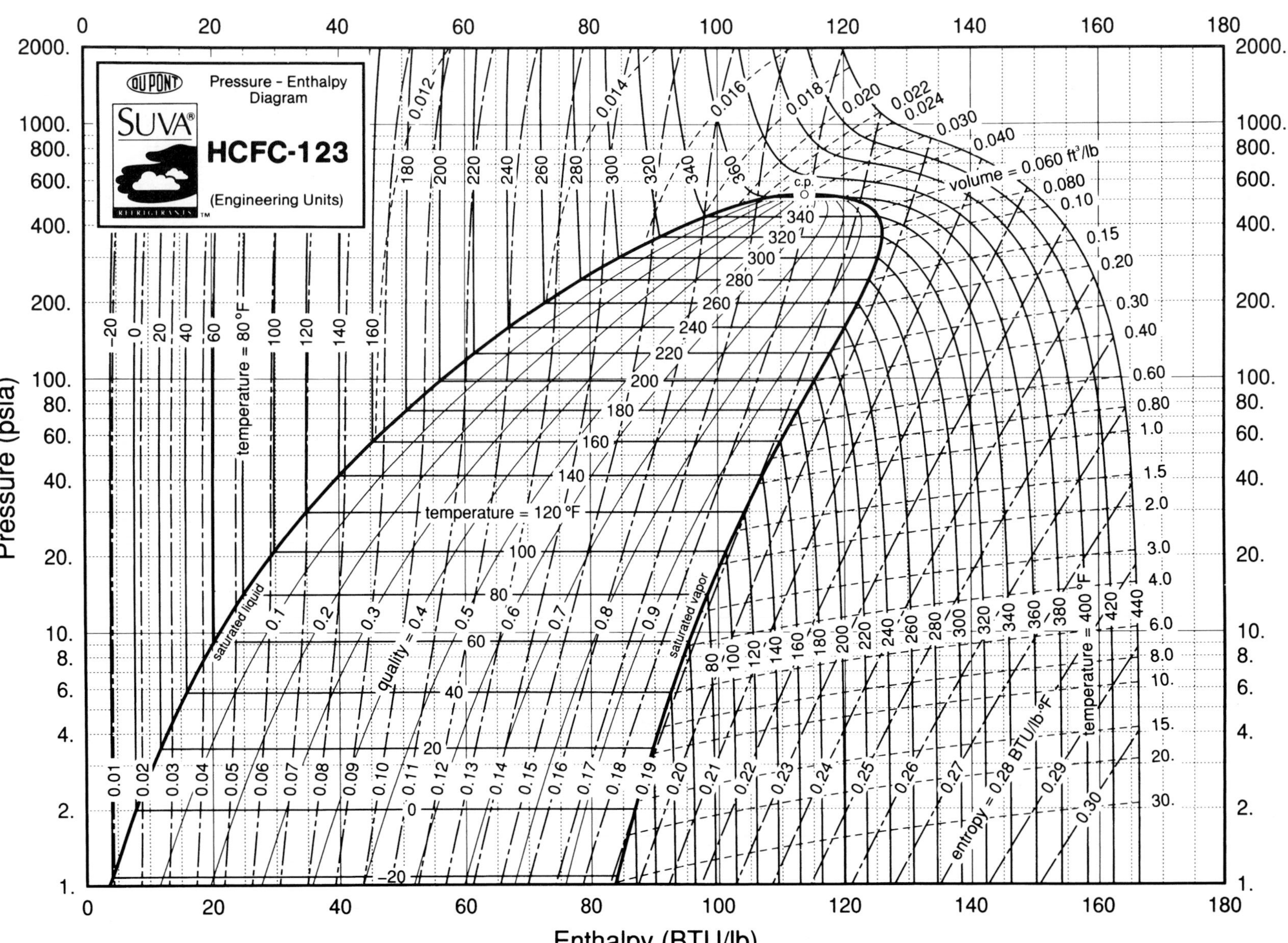

Table A-7 I-P Pressure-Heat Diagram for R-123

TEMP. °F	PRESSURE psia	VOLUME ft³/lb LIQUID v_f	VAPOR v_g	DENSITY lb/ft³ LIQUID $1/v_f$	VAPOR $1/v_g$	ENTHALPY Btu/lb LIQUID h_f	LATENT h_{fg}	VAPOR h_g	ENTROPY Btu/(lb)(°R) LIQUID s_f	VAPOR s_g	TEMP. °F
−30	0.776	0.0100	38.6100	100.0	0.026	2.0	80.7	82.7	0.0047	0.1925	−30
−29	0.803	0.0100	37.4532	99.9	0.027	2.2	80.6	82.8	0.0052	0.1924	−29
−28	0.830	0.0100	36.2319	99.8	0.028	2.4	80.6	83.0	0.0056	0.1923	−28
−27	0.859	0.0100	35.0877	99.7	0.029	2.6	80.5	83.1	0.0061	0.1922	−27
−26	0.888	0.0100	34.0136	99.7	0.029	2.8	80.5	83.3	0.0065	0.1921	−26
−25	0.918	0.0100	33.0033	99.6	0.030	3.0	80.4	83.4	0.0070	0.1920	−25
−24	0.948	0.0100	32.0513	99.5	0.031	3.2	80.4	83.5	0.0074	0.1919	−24
−23	0.980	0.0101	31.0559	99.4	0.032	3.4	80.3	83.7	0.0078	0.1918	−23
−22	1.012	0.0101	30.1205	99.4	0.033	3.5	80.3	83.8	0.0083	0.1917	−22
−21	1.046	0.0101	29.2398	99.3	0.034	3.7	80.2	83.9	0.0087	0.1916	−21
−20	1.080	0.0101	28.3286	99.2	0.035	3.9	80.2	84.1	0.0091	0.1915	−20
−19	1.115	0.0101	27.4725	99.1	0.036	4.1	80.1	84.2	0.0096	0.1914	−19
−18	1.151	0.0101	26.6667	99.1	0.038	4.3	80.1	84.4	0.0100	0.1913	−18
−17	1.188	0.0101	25.9067	99.0	0.039	4.5	80.0	84.5	0.0104	0.1912	−17
−16	1.227	0.0101	25.1889	98.9	0.040	4.7	80.0	84.6	0.0108	0.1911	−16
−15	1.266	0.0101	24.4499	98.8	0.041	4.9	79.9	84.8	0.0113	0.1910	−15
−14	1.306	0.0101	23.7530	98.8	0.042	5.1	79.9	84.9	0.0117	0.1909	−14
−13	1.347	0.0101	23.0415	98.7	0.043	5.2	79.8	85.1	0.0121	0.1908	−13
−12	1.390	0.0101	22.3714	98.6	0.045	5.4	79.8	85.2	0.0125	0.1907	−12
−11	1.433	0.0102	21.7391	98.5	0.046	5.6	79.7	85.3	0.0130	0.1906	−11
−10	1.478	0.0102	21.1416	98.4	0.047	5.8	79.7	85.5	0.0134	0.1905	−10
−9	1.524	0.0102	20.5339	98.4	0.049	6.0	79.6	85.6	0.0138	0.1905	−9
−8	1.571	0.0102	19.9601	98.3	0.050	6.2	79.6	85.8	0.0142	0.1904	−8
−7	1.619	0.0102	19.4175	98.2	0.052	6.4	79.5	85.9	0.0146	0.1903	−7
−6	1.669	0.0102	18.8679	98.1	0.053	6.6	79.5	86.0	0.0151	0.1902	−6
−5	1.720	0.0102	18.3486	98.1	0.055	6.8	79.4	86.2	0.0155	0.1901	−5
−4	1.772	0.0102	17.8571	98.0	0.056	6.9	79.4	86.3	0.0159	0.1901	−4
−3	1.825	0.0102	17.3611	97.9	0.058	7.1	79.3	86.5	0.0163	0.1900	−3
−2	1.880	0.0102	16.8919	97.8	0.059	7.3	79.3	86.6	0.0167	0.1899	−2
−1	1.936	0.0102	16.4474	97.8	0.061	7.5	79.2	86.7	0.0171	0.1898	−1
0	1.993	0.0102	16.0000	97.7	0.063	7.7	79.2	86.9	0.0176	0.1898	0
1	2.052	0.0102	15.5521	97.6	0.064	7.9	79.1	87.0	0.0180	0.1897	1
2	2.113	0.0103	15.1515	97.5	0.066	8.1	79.1	87.2	0.0184	0.1896	2
3	2.174	0.0103	14.7493	97.4	0.068	8.3	79.0	87.3	0.0188	0.1896	3
4	2.238	0.0103	14.3472	97.4	0.070	8.5	79.0	87.4	0.0192	0.1895	4
5	2.303	0.0103	13.9860	97.3	0.072	8.7	78.9	87.6	0.0196	0.1894	5
6	2.369	0.0103	13.6054	97.2	0.074	8.9	78.9	87.7	0.0200	0.1894	6
7	2.437	0.0103	13.2626	97.1	0.075	9.1	78.8	87.9	0.0205	0.1893	7
8	2.507	0.0103	12.9199	97.0	0.077	9.3	78.7	88.0	0.0209	0.1893	8
9	2.578	0.0103	12.5786	97.0	0.080	9.4	78.7	88.1	0.0213	0.1892	9
10	2.651	0.0103	12.2549	96.9	0.082	9.6	78.6	88.3	0.0217	0.1892	10
11	2.726	0.0103	11.9474	96.8	0.084	9.8	78.6	88.4	0.0221	0.1891	11
12	2.802	0.0103	11.6414	96.7	0.086	10.0	78.5	88.6	0.0225	0.1890	12
13	2.880	0.0103	11.3507	96.7	0.088	10.2	78.5	88.7	0.0230	0.1890	13
14	2.960	0.0104	11.0619	96.6	0.090	10.4	78.4	88.9	0.0234	0.1889	14
15	3.042	0.0104	10.7759	96.5	0.093	10.6	78.4	89.0	0.0238	0.1889	15
16	3.126	0.0104	10.5152	96.4	0.095	10.8	78.3	89.1	0.0242	0.1888	16
17	3.212	0.0104	10.2459	96.3	0.098	11.0	78.3	89.3	0.0246	0.1888	17
18	3.299	0.0104	10.0000	96.3	0.100	11.2	78.2	89.4	0.0251	0.1888	18
19	3.389	0.0104	9.7466	96.2	0.103	11.4	78.1	89.6	0.0255	0.1887	19
20	3.480	0.0104	9.5147	96.1	0.105	11.6	78.1	89.7	0.0259	0.1887	20
21	3.574	0.0104	9.2764	96.0	0.108	11.8	78.0	89.8	0.0263	0.1886	21
22	3.669	0.0104	9.0498	95.9	0.111	12.0	78.0	90.0	0.0267	0.1886	22
23	3.767	0.0104	8.8339	95.9	0.113	12.2	77.9	90.1	0.0272	0.1885	23
24	3.867	0.0104	8.6207	95.8	0.116	12.4	77.8	90.3	0.0276	0.1885	24
25	3.969	0.0105	8.4104	95.7	0.119	12.6	77.8	90.4	0.0280	0.1885	25
26	4.073	0.0105	8.2102	95.6	0.122	12.9	77.7	90.6	0.0284	0.1884	26
27	4.180	0.0105	8.0192	95.5	0.125	13.1	77.6	90.7	0.0288	0.1884	27
28	4.288	0.0105	7.8309	95.5	0.128	13.3	77.6	90.8	0.0293	0.1884	28
29	4.400	0.0105	7.6453	95.4	0.131	13.5	77.5	91.0	0.0297	0.1883	29

Table A-8 I-P Properties of Saturated Liquid and Vapor for R-123

TEMP.	PRESSURE	VOLUME ft³/lb		DENSITY lb/ft³		ENTHALPY Btu/lb			ENTROPY Btu/(lb)(°R)		TEMP.
°F	psia	LIQUID v_f	VAPOR v_g	LIQUID $1/v_f$	VAPOR $1/v_g$	LIQUID h_f	LATENT h_{fg}	VAPOR h_g	LIQUID s_f	VAPOR s_g	°F
30	4.513	0.0105	7.4627	95.3	0.134	13.7	77.5	91.1	0.0301	0.1883	30
31	4.629	0.0105	7.2886	95.2	0.137	13.9	77.4	91.3	0.0305	0.1883	31
32	4.747	0.0105	7.1225	95.1	0.140	14.1	77.3	91.4	0.0310	0.1882	32
33	4.868	0.0105	6.9541	95.0	0.144	14.3	77.3	91.6	0.0314	0.1882	33
34	4.991	0.0105	6.7935	95.0	0.147	14.5	77.2	91.7	0.0318	0.1882	34
35	5.117	0.0105	6.6401	94.9	0.151	14.7	77.1	91.9	0.0322	0.1882	35
36	5.245	0.0105	6.4893	94.8	0.154	14.9	77.1	92.0	0.0327	0.1881	36
37	5.376	0.0106	6.3412	94.7	0.158	15.1	77.0	92.1	0.0331	0.1881	37
38	5.509	0.0106	6.1958	94.6	0.161	15.4	76.9	92.3	0.0335	0.1881	38
39	5.646	0.0106	6.0569	94.6	0.165	15.6	76.9	92.4	0.0339	0.1881	39
40	5.785	0.0106	5.9207	94.5	0.169	15.8	76.8	92.6	0.0344	0.1881	40
41	5.927	0.0106	5.7904	94.4	0.173	16.0	76.7	92.7	0.0348	0.1880	41
42	6.071	0.0106	5.6593	94.3	0.177	16.2	76.6	92.9	0.0352	0.1880	42
43	6.219	0.0106	5.5340	94.2	0.181	16.4	76.6	93.0	0.0357	0.1880	43
44	6.369	0.0106	5.4142	94.1	0.185	16.7	76.5	93.2	0.0361	0.1880	44
45	6.522	0.0106	5.2938	94.1	0.189	16.9	76.4	93.3	0.0365	0.1880	45
46	6.679	0.0106	5.1787	94.0	0.193	17.1	76.4	93.4	0.0370	0.1880	46
47	6.838	0.0107	5.0659	93.9	0.197	17.3	76.3	93.6	0.0374	0.1879	47
48	7.000	0.0107	4.9554	93.8	0.202	17.5	76.2	93.7	0.0378	0.1879	48
49	7.166	0.0107	4.8497	93.7	0.206	17.7	76.1	93.9	0.0383	0.1879	49
50	7.334	0.0107	4.7461	93.6	0.211	18.0	76.1	94.0	0.0387	0.1879	50
51	7.506	0.0107	4.6425	93.6	0.215	18.2	76.0	94.2	0.0391	0.1879	51
52	7.681	0.0107	4.5455	93.5	0.220	18.4	75.9	94.3	0.0396	0.1879	52
53	7.860	0.0107	4.4484	93.4	0.225	18.6	75.8	94.5	0.0400	0.1879	53
54	8.041	0.0107	4.3535	93.3	0.230	18.9	75.7	94.6	0.0404	0.1879	54
55	8.226	0.0107	4.2626	93.2	0.235	19.1	75.7	94.7	0.0409	0.1879	55
56	8.415	0.0107	4.1736	93.1	0.240	19.3	75.6	94.9	0.0413	0.1879	56
57	8.607	0.0107	4.0866	93.1	0.245	19.5	75.5	95.0	0.0417	0.1879	57
58	8.802	0.0108	4.0016	93.0	0.250	19.8	75.4	95.2	0.0422	0.1879	58
59	9.001	0.0108	3.9185	92.9	0.255	20.0	75.3	95.3	0.0426	0.1879	59
60	9.203	0.0108	3.8388	92.8	0.261	20.2	75.3	95.5	0.0430	0.1879	60
61	9.410	0.0108	3.7594	92.7	0.266	20.4	75.2	95.6	0.0435	0.1879	61
62	9.619	0.0108	3.6832	92.6	0.272	20.7	75.1	95.8	0.0439	0.1879	62
63	9.833	0.0108	3.6075	92.6	0.277	20.9	75.0	95.9	0.0443	0.1879	63
64	10.050	0.0108	3.5348	92.5	0.283	21.1	74.9	96.1	0.0448	0.1879	64
65	10.272	0.0108	3.4638	92.4	0.289	21.4	74.8	96.2	0.0452	0.1879	65
66	10.497	0.0108	3.3944	92.3	0.295	21.6	74.8	96.3	0.0457	0.1879	66
67	10.726	0.0108	3.3267	92.2	0.301	21.8	74.7	96.5	0.0461	0.1879	67
68	10.958	0.0109	3.2605	92.1	0.307	22.0	74.6	96.6	0.0465	0.1879	68
69	11.195	0.0109	3.1959	92.0	0.313	22.3	74.5	96.8	0.0470	0.1879	69
70	11.436	0.0109	3.1328	92.0	0.319	22.5	74.4	96.9	0.0474	0.1879	70
71	11.682	0.0109	3.0713	91.9	0.326	22.7	74.3	97.1	0.0479	0.1879	71
72	11.931	0.0109	3.0111	91.8	0.332	23.0	74.2	97.2	0.0483	0.1879	72
73	12.184	0.0109	2.9525	91.7	0.339	23.2	74.2	97.4	0.0487	0.1879	73
74	12.442	0.0109	2.8952	91.6	0.345	23.4	74.1	97.5	0.0492	0.1880	74
75	12.704	0.0109	2.8393	91.5	0.352	23.7	74.0	97.7	0.0496	0.1880	75
76	12.970	0.0109	2.7840	91.4	0.359	23.9	73.9	97.8	0.0501	0.1880	76
77	13.241	0.0109	2.7307	91.3	0.366	24.2	73.8	98.0	0.0505	0.1880	77
78	13.517	0.0110	2.6788	91.3	0.373	24.4	73.7	98.1	0.0509	0.1880	78
79	13.796	0.0110	2.6274	91.2	0.381	24.6	73.6	98.2	0.0514	0.1880	79
80	14.081	0.0110	2.5780	91.1	0.388	24.9	73.5	98.4	0.0518	0.1880	80
81	14.369	0.0110	2.5291	91.0	0.395	25.1	73.4	98.5	0.0523	0.1881	81
82	14.663	0.0110	2.4814	90.9	0.403	25.3	73.3	98.7	0.0527	0.1881	82
83	14.961	0.0110	2.4355	90.8	0.411	25.6	73.2	98.8	0.0531	0.1881	83
84	15.264	0.0110	2.3901	90.7	0.418	25.8	73.1	99.0	0.0536	0.1881	84
85	15.572	0.0110	2.3452	90.6	0.426	26.1	73.1	99.1	0.0540	0.1881	85
86	15.885	0.0110	2.3020	90.6	0.434	26.3	73.0	99.3	0.0545	0.1882	86
87	16.203	0.0111	2.2594	90.5	0.443	26.6	72.9	99.4	0.0549	0.1882	87
88	16.525	0.0111	2.2183	90.4	0.451	26.8	72.8	99.6	0.0553	0.1882	88
89	16.853	0.0111	2.1777	90.3	0.459	27.0	72.7	99.7	0.0558	0.1882	89

Table A-8 Continued

TEMP.	PRESSURE	VOLUME ft³/lb		DENSITY lb/ft³		ENTHALPY Btu/lb			ENTROPY Btu/(lb)(°R)		TEMP.
°F	psia	LIQUID v_f	VAPOR v_g	LIQUID $1/v_f$	VAPOR $1/v_g$	LIQUID h_f	LATENT h_{fg}	VAPOR h_g	LIQUID s_f	VAPOR s_g	°F
90	17.186	0.0111	2.1381	90.2	0.468	27.3	72.6	99.9	0.0562	0.1883	90
91	17.523	0.0111	2.0991	90.1	0.476	27.5	72.5	100.0	0.0567	0.1883	91
92	17.866	0.0111	2.0610	90.0	0.485	27.8	72.4	100.1	0.0571	0.1883	92
93	18.215	0.0111	2.0239	89.9	0.494	28.0	72.3	100.3	0.0576	0.1883	93
94	18.568	0.0111	1.9877	89.8	0.503	28.3	72.2	100.4	0.0580	0.1884	94
95	18.927	0.0111	1.9524	89.8	0.512	28.5	72.1	100.6	0.0584	0.1884	95
96	19.291	0.0112	1.9175	89.7	0.522	28.7	72.0	100.7	0.0589	0.1884	96
97	19.661	0.0112	1.8836	89.6	0.531	29.0	71.9	100.9	0.0593	0.1884	97
98	20.036	0.0112	1.8501	89.5	0.541	29.2	71.8	101.0	0.0598	0.1885	98
99	20.417	0.0112	1.8179	89.4	0.550	29.5	71.7	101.2	0.0602	0.1885	99
100	20.803	0.0112	1.7857	89.3	0.560	29.7	71.6	101.3	0.0606	0.1885	100
101	21.195	0.0112	1.7547	89.2	0.570	30.0	71.5	101.5	0.0611	0.1886	101
102	21.593	0.0112	1.7241	89.1	0.580	30.2	71.4	101.6	0.0615	0.1886	102
103	21.997	0.0112	1.6943	89.0	0.590	30.5	71.3	101.8	0.0620	0.1886	103
104	22.406	0.0112	1.6650	88.9	0.601	30.7	71.2	101.9	0.0624	0.1887	104
105	22.821	0.0113	1.6364	88.8	0.611	31.0	71.1	102.0	0.0628	0.1887	105
106	23.242	0.0113	1.6082	88.8	0.622	31.2	71.0	102.2	0.0633	0.1887	106
107	23.670	0.0113	1.5808	88.7	0.633	31.5	70.9	102.3	0.0637	0.1888	107
108	24.103	0.0113	1.5540	88.6	0.644	31.7	70.8	102.5	0.0642	0.1888	108
109	24.542	0.0113	1.5277	88.5	0.655	32.0	70.6	102.6	0.0646	0.1888	109
110	24.988	0.0113	1.5017	88.4	0.666	32.2	70.5	102.8	0.0650	0.1889	110
111	25.440	0.0113	1.4765	88.3	0.677	32.5	70.4	102.9	0.0655	0.1889	111
112	25.898	0.0113	1.4518	88.2	0.689	32.7	70.3	103.1	0.0659	0.1890	112
113	26.362	0.0113	1.4276	88.1	0.701	33.0	70.2	103.2	0.0664	0.1890	113
114	26.833	0.0114	1.4037	88.0	0.712	33.2	70.1	103.4	0.0668	0.1890	114
115	27.310	0.0114	1.3805	87.9	0.724	33.5	70.0	103.5	0.0672	0.1891	115
116	27.794	0.0114	1.3576	87.8	0.737	33.7	69.9	103.6	0.0677	0.1891	116
117	28.284	0.0114	1.3353	87.7	0.749	34.0	69.8	103.8	0.0681	0.1891	117
118	28.781	0.0114	1.3134	87.6	0.761	34.3	69.7	103.9	0.0686	0.1892	118
119	29.285	0.0114	1.2918	87.6	0.774	34.5	69.6	104.1	0.0690	0.1892	119
120	29.796	0.0114	1.2708	87.5	0.787	34.8	69.5	104.2	0.0694	0.1893	120
121	30.313	0.0114	1.2502	87.4	0.800	35.0	69.4	104.4	0.0699	0.1893	121
122	30.837	0.0115	1.2300	87.3	0.813	35.3	69.2	104.5	0.0703	0.1894	122
123	31.368	0.0115	1.2102	87.2	0.826	35.5	69.1	104.7	0.0707	0.1894	123
124	31.906	0.0115	1.1908	87.1	0.840	35.8	69.0	104.8	0.0712	0.1894	124
125	32.451	0.0115	1.1716	87.0	0.854	36.0	68.9	105.0	0.0716	0.1895	125
126	33.004	0.0115	1.1529	86.9	0.867	36.3	68.8	105.1	0.0721	0.1895	126
127	33.563	0.0115	1.1346	86.8	0.881	36.6	68.7	105.2	0.0725	0.1896	127
128	34.130	0.0115	1.1166	86.7	0.896	36.8	68.6	105.4	0.0729	0.1896	128
129	34.703	0.0115	1.0989	86.6	0.910	37.1	68.5	105.5	0.0734	0.1897	129
130	35.285	0.0116	1.0817	86.5	0.925	37.3	68.3	105.7	0.0738	0.1897	130
131	35.873	0.0116	1.0646	86.4	0.939	37.6	68.2	105.8	0.0742	0.1898	131
132	36.470	0.0116	1.0480	86.3	0.954	37.8	68.1	106.0	0.0747	0.1898	132
133	37.073	0.0116	1.0317	86.2	0.969	38.1	68.0	106.1	0.0751	0.1899	133
134	37.684	0.0116	1.0156	86.1	0.985	38.4	67.9	106.3	0.0755	0.1899	134
135	38.303	0.0116	0.9999	86.0	1.000	38.6	67.8	106.4	0.0760	0.1899	135
136	38.930	0.0116	0.9844	85.9	1.016	38.9	67.7	106.5	0.0764	0.1900	136
137	39.565	0.0117	0.9693	85.8	1.032	39.1	67.5	106.7	0.0768	0.1900	137
138	40.207	0.0117	0.9545	85.7	1.048	39.4	67.4	106.8	0.0773	0.1901	138
139	40.857	0.0117	0.9398	85.6	1.064	39.7	67.3	107.0	0.0777	0.1901	139
140	41.515	0.0117	0.9255	85.5	1.081	39.9	67.2	107.1	0.0781	0.1902	140
141	42.181	0.0117	0.9115	85.4	1.097	40.2	67.1	107.3	0.0786	0.1902	141
142	42.856	0.0117	0.8977	85.3	1.114	40.4	67.0	107.4	0.0790	0.1903	142
143	43.538	0.0117	0.8841	85.2	1.131	40.7	66.8	107.5	0.0794	0.1903	143
144	44.229	0.0117	0.8708	85.2	1.148	41.0	66.7	107.7	0.0799	0.1904	144
145	44.928	0.0118	0.8578	85.1	1.166	41.2	66.6	107.8	0.0803	0.1904	145
146	45.635	0.0118	0.8449	85.0	1.184	41.5	66.5	108.0	0.0807	0.1905	146
147	46.351	0.0118	0.8323	84.9	1.202	41.7	66.4	108.1	0.0812	0.1905	147
148	47.075	0.0118	0.8199	84.8	1.220	42.0	66.2	108.3	0.0816	0.1906	148
149	47.808	0.0118	0.8078	84.7	1.238	42.3	66.1	108.4	0.0820	0.1906	149

Table A-8 Continued

TEMP. °F	PRESSURE psia	VOLUME ft³/lb LIQUID v_f	VOLUME ft³/lb VAPOR v_g	DENSITY lb/ft³ LIQUID $1/v_f$	DENSITY lb/ft³ VAPOR $1/v_g$	ENTHALPY Btu/lb LIQUID h_f	ENTHALPY Btu/lb LATENT h_{fg}	ENTHALPY Btu/lb VAPOR h_g	ENTROPY Btu/(lb)(°R) LIQUID s_f	ENTROPY Btu/(lb)(°R) VAPOR s_g	TEMP. °F
150	48.549	0.0118	0.7959	84.6	1.257	42.5	66.0	108.5	0.0824	0.1907	150
151	49.299	0.0118	0.7841	84.5	1.275	42.8	65.9	108.7	0.0829	0.1907	151
152	50.058	0.0119	0.7726	84.4	1.294	43.1	65.8	108.8	0.0833	0.1908	152
153	50.825	0.0119	0.7613	84.3	1.314	43.3	65.6	109.0	0.0837	0.1909	153
154	51.602	0.0119	0.7502	84.2	1.333	43.6	65.5	109.1	0.0842	0.1909	154
155	52.387	0.0119	0.7393	84.1	1.353	43.9	65.4	109.2	0.0846	0.1910	155
156	53.181	0.0119	0.7285	84.0	1.373	44.1	65.3	109.4	0.0850	0.1910	156
157	53.985	0.0119	0.7180	83.8	1.393	44.4	65.1	109.5	0.0854	0.1911	157
158	54.797	0.0119	0.7077	83.7	1.413	44.6	65.0	109.7	0.0859	0.1911	158
159	55.619	0.0120	0.6974	83.6	1.434	44.9	64.9	109.8	0.0863	0.1912	159
160	56.450	0.0120	0.6875	83.5	1.455	45.2	64.8	109.9	0.0867	0.1912	160
161	57.290	0.0120	0.6776	83.4	1.476	45.4	64.6	110.1	0.0871	0.1913	161
162	58.140	0.0120	0.6680	83.3	1.497	45.7	64.5	110.2	0.0876	0.1913	162
163	58.999	0.0120	0.6585	83.2	1.519	46.0	64.4	110.4	0.0880	0.1914	163
164	59.868	0.0120	0.6491	83.1	1.541	46.2	64.3	110.5	0.0884	0.1914	164
165	60.746	0.0120	0.6399	83.0	1.563	46.5	64.1	110.6	0.0888	0.1915	165
166	61.634	0.0121	0.6309	82.9	1.585	46.8	64.0	110.8	0.0892	0.1916	166
167	62.532	0.0121	0.6220	82.8	1.608	47.0	63.9	110.9	0.0897	0.1916	167
168	63.439	0.0121	0.6133	82.7	1.631	47.3	63.8	111.1	0.0901	0.1917	168
169	64.357	0.0121	0.6047	82.6	1.654	47.6	63.6	111.2	0.0905	0.1917	169
170	65.284	0.0121	0.5963	82.5	1.677	47.8	63.5	111.3	0.0909	0.1918	170
171	66.221	0.0121	0.5880	82.4	1.701	48.1	63.4	111.5	0.0914	0.1918	171
172	67.169	0.0122	0.5798	82.3	1.725	48.4	63.2	111.6	0.0918	0.1919	172
173	68.126	0.0122	0.5718	82.2	1.749	48.6	63.1	111.7	0.0922	0.1919	173
174	69.094	0.0122	0.5639	82.1	1.773	48.9	63.0	111.9	0.0926	0.1920	174
175	70.072	0.0122	0.5561	82.0	1.798	49.2	62.8	112.0	0.0930	0.1921	175
176	71.060	0.0122	0.5485	81.9	1.823	49.4	62.7	112.2	0.0934	0.1921	176
177	72.059	0.0122	0.5410	81.8	1.849	49.7	62.6	112.3	0.0939	0.1922	177
178	73.068	0.0122	0.5336	81.7	1.874	50.0	62.5	112.4	0.0943	0.1922	178
179	74.088	0.0123	0.5263	81.6	1.900	50.2	62.3	112.6	0.0947	0.1923	179
180	75.118	0.0123	0.5192	81.4	1.926	50.5	62.2	112.7	0.0951	0.1923	180
181	76.159	0.0123	0.5121	81.3	1.953	50.8	62.1	112.8	0.0955	0.1924	181
182	77.211	0.0123	0.5052	81.2	1.980	51.0	61.9	113.0	0.0959	0.1924	182
183	78.274	0.0123	0.4984	81.1	2.007	51.3	61.8	113.1	0.0964	0.1925	183
184	79.348	0.0123	0.4916	81.0	2.034	51.6	61.6	113.2	0.0968	0.1926	184
185	80.432	0.0124	0.4850	80.9	2.062	51.9	61.5	113.4	0.0972	0.1926	185
186	81.528	0.0124	0.4785	80.8	2.090	52.1	61.4	113.5	0.0976	0.1927	186
187	82.635	0.0124	0.4721	80.7	2.118	52.4	61.2	113.6	0.0980	0.1927	187
188	83.753	0.0124	0.4659	80.6	2.147	52.7	61.1	113.8	0.0984	0.1928	188
189	84.882	0.0124	0.4596	80.5	2.176	52.9	61.0	113.9	0.0988	0.1928	189
190	86.023	0.0124	0.4536	80.4	2.205	53.2	60.8	114.0	0.0993	0.1929	190
191	87.175	0.0125	0.4475	80.2	2.234	53.5	60.7	114.2	0.0997	0.1929	191
192	88.339	0.0125	0.4416	80.1	2.264	53.7	60.6	114.3	0.1001	0.1930	192
193	89.514	0.0125	0.4358	80.0	2.295	54.0	60.4	114.4	0.1005	0.1931	193
194	90.700	0.0125	0.4301	79.9	2.325	54.3	60.3	114.6	0.1009	0.1931	194
195	91.899	0.0125	0.4244	79.8	2.356	54.6	60.1	114.7	0.1013	0.1932	195
196	93.109	0.0126	0.4189	79.7	2.387	54.8	60.0	114.8	0.1017	0.1932	196
197	94.331	0.0126	0.4134	79.6	2.419	55.1	59.9	115.0	0.1021	0.1933	197
198	95.565	0.0126	0.4080	79.5	2.451	55.4	59.7	115.1	0.1025	0.1933	198
199	96.811	0.0126	0.4027	79.3	2.483	55.6	59.6	115.2	0.1029	0.1934	199
200	98.069	0.0126	0.3975	79.2	2.516	55.9	59.4	115.3	0.1034	0.1934	200
201	99.340	0.0126	0.3923	79.1	2.549	56.2	59.3	115.5	0.1038	0.1935	201
202	100.622	0.0127	0.3872	79.0	2.583	56.5	59.1	115.6	0.1042	0.1935	202
203	101.917	0.0127	0.3822	78.9	2.616	56.7	59.0	115.7	0.1046	0.1936	203
204	103.224	0.0127	0.3773	78.8	2.650	57.0	58.8	115.9	0.1050	0.1937	204
205	104.544	0.0127	0.3725	78.7	2.685	57.3	58.7	116.0	0.1054	0.1937	205
206	105.876	0.0127	0.3677	78.5	2.720	57.6	58.6	116.1	0.1058	0.1938	206
207	107.221	0.0128	0.3630	78.4	2.755	57.8	58.4	116.2	0.1062	0.1938	207
208	108.578	0.0128	0.3583	78.3	2.791	58.1	58.3	116.4	0.1066	0.1939	208
209	109.948	0.0128	0.3538	78.2	2.827	58.4	58.1	116.5	0.1070	0.1939	209

Table A-8 Continued

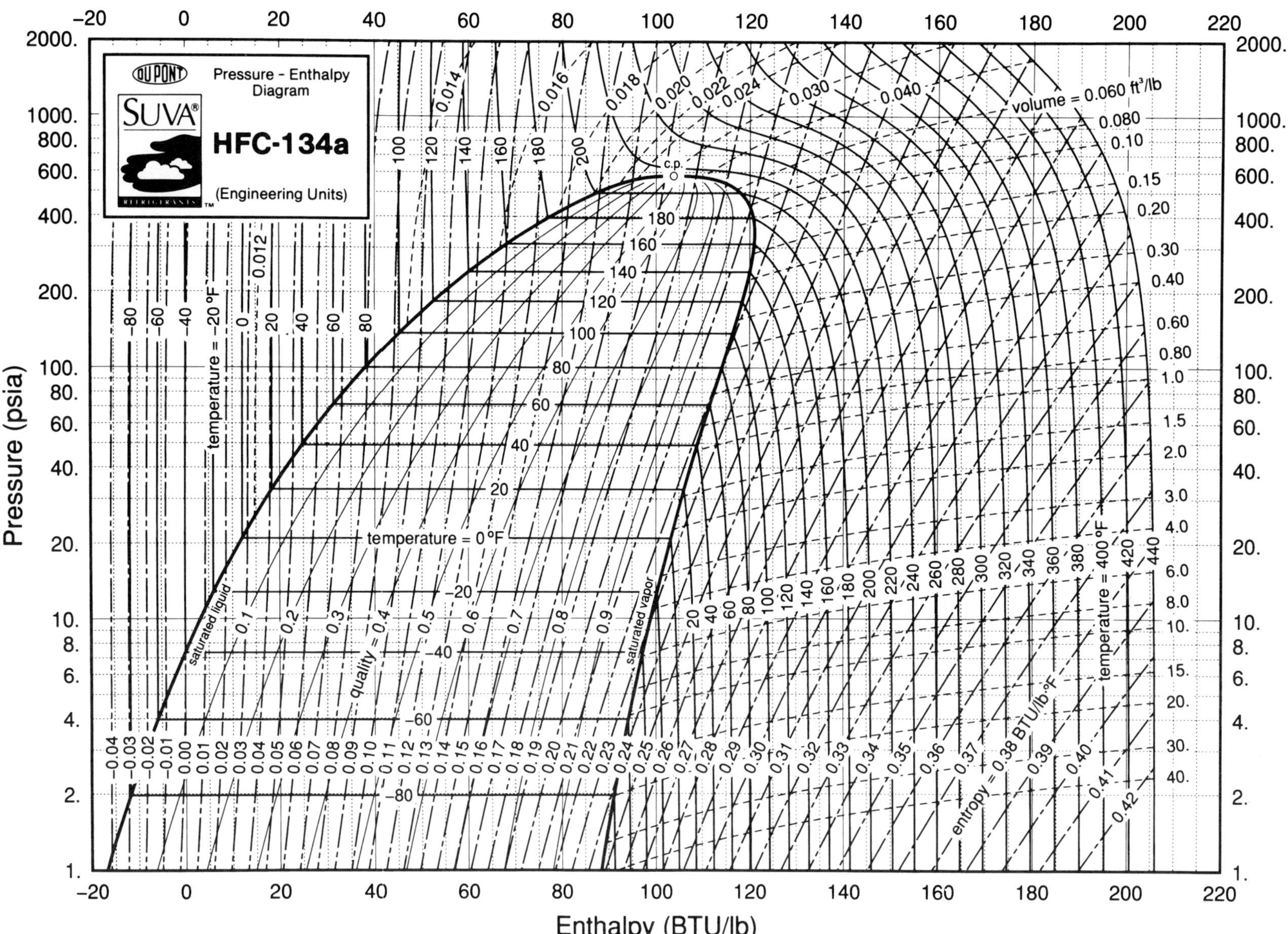

Table A-9 I-P Pressure-Heat Diagram for R-134a

TEMP.	PRESSURE	VOLUME ft³/lb		DENSITY lb/ft³		ENTHALPY Btu/lb			ENTROPY Btu/(lb)(°R)		TEMP.
°F	psia	LIQUID v_f	VAPOR v_g	LIQUID $1/v_f$	VAPOR $1/v_g$	LIQUID h_f	LATENT h_{fg}	VAPOR h_g	LIQUID s_f	VAPOR s_g	°F
−30	9.851	0.0115	4.4366	87.31	0.2254	3.0	95.7	98.7	0.0070	0.2297	−30
−29	10.126	0.0115	4.3234	87.21	0.2313	3.3	95.5	98.8	0.0077	0.2296	−29
−28	10.407	0.0115	4.2141	87.11	0.2373	3.6	95.4	99.0	0.0084	0.2294	−28
−27	10.694	0.0115	4.1068	87.01	0.2435	3.9	95.2	99.1	0.0091	0.2292	−27
−26	10.987	0.0115	4.0032	86.91	0.2498	4.2	95.1	99.3	0.0098	0.2291	−26
−25	11.287	0.0115	3.9032	86.81	0.2562	4.5	94.9	99.4	0.0105	0.2289	−25
−24	11.594	0.0115	3.8066	86.71	0.2627	4.8	94.8	99.6	0.0112	0.2288	−24
−23	11.906	0.0115	3.7120	86.61	0.2694	5.1	94.6	99.7	0.0119	0.2286	−23
−22	12.226	0.0116	3.6206	86.51	0.2762	5.4	94.5	99.9	0.0126	0.2284	−22
−21	12.553	0.0116	3.5323	86.41	0.2831	5.7	94.3	100.0	0.0132	0.2283	−21
−20	12.885	0.0116	3.4471	86.30	0.2901	6.0	94.2	100.2	0.0139	0.2281	−20
−19	13.225	0.0116	3.3625	86.20	0.2974	6.3	94.0	100.3	0.0146	0.2280	−19
−18	13.572	0.0116	3.2819	86.10	0.3047	6.6	93.9	100.5	0.0153	0.2278	−18
−17	13.927	0.0116	3.2031	86.00	0.3122	6.9	93.7	100.6	0.0160	0.2277	−17
−16	14.289	0.0116	3.1270	85.89	0.3198	7.2	93.6	100.8	0.0167	0.2276	−16
−15	14.659	0.0117	3.0525	85.79	0.3276	7.5	93.4	100.9	0.0174	0.2274	−15
−14	15.035	0.0117	2.9806	85.69	0.3355	7.8	93.3	101.1	0.0180	0.2273	−14
−13	15.420	0.0117	2.9104	85.59	0.3436	8.1	93.1	101.2	0.0187	0.2271	−13
−12	15.812	0.0117	2.8417	85.48	0.3519	8.4	92.9	101.4	0.0194	0.2270	−12
−11	16.212	0.0117	2.7762	85.38	0.3602	8.7	92.8	101.5	0.0201	0.2269	−11
−10	16.620	0.0117	2.7115	85.28	0.3688	9.0	92.6	101.7	0.0208	0.2267	−10
−9	17.037	0.0117	2.6490	85.17	0.3775	9.3	92.5	101.8	0.0214	0.2266	−9
−8	17.461	0.0118	2.5880	85.07	0.3864	9.7	92.3	102.0	0.0221	0.2265	−8
−7	17.893	0.0118	2.5291	84.97	0.3954	10.0	92.1	102.1	0.0228	0.2264	−7
−6	18.334	0.0118	2.4716	84.86	0.4046	10.3	92.0	102.3	0.0235	0.2262	−6
−5	18.784	0.0118	2.4155	84.76	0.4140	10.6	91.8	102.4	0.0241	0.2261	−5
−4	19.242	0.0118	2.3613	84.65	0.4235	10.9	91.7	102.5	0.0248	0.2260	−4
−3	19.709	0.0118	2.3084	84.55	0.4332	11.2	91.5	102.7	0.0255	0.2259	−3
−2	20.184	0.0118	2.2568	84.44	0.4431	11.5	91.3	102.8	0.0262	0.2258	−2
−1	20.669	0.0119	2.2065	84.34	0.4532	11.8	91.2	103.0	0.0268	0.2256	−1
0	21.163	0.0119	2.1580	84.23	0.4634	12.1	91.0	103.1	0.0275	0.2255	0
1	21.666	0.0119	2.1106	84.13	0.4738	12.4	90.9	103.3	0.0282	0.2254	1
2	22.178	0.0119	2.0644	84.02	0.4844	12.7	90.7	103.4	0.0288	0.2253	2
3	22.700	0.0119	2.0194	83.91	0.4952	13.0	90.5	103.6	0.0295	0.2252	3
4	23.231	0.0119	1.9755	83.81	0.5062	13.4	90.4	103.7	0.0302	0.2251	4
5	23.772	0.0119	1.9327	83.70	0.5174	13.7	90.2	103.9	0.0308	0.2250	5
6	24.322	0.0120	1.8911	83.60	0.5288	14.0	90.0	104.0	0.0315	0.2249	6
7	24.883	0.0120	1.8508	83.49	0.5403	14.3	89.9	104.2	0.0322	0.2248	7
8	25.454	0.0120	1.8113	83.38	0.5521	14.6	89.7	104.3	0.0328	0.2247	8
9	26.034	0.0120	1.7727	83.27	0.5641	14.9	89.5	104.4	0.0335	0.2245	9
10	26.625	0.0120	1.7355	83.17	0.5762	15.2	89.4	104.6	0.0342	0.2244	10
11	27.227	0.0120	1.6989	83.06	0.5886	15.5	89.2	104.7	0.0348	0.2244	11
12	27.839	0.0121	1.6633	82.95	0.6012	15.9	89.0	104.9	0.0355	0.2243	12
13	28.462	0.0121	1.6287	82.84	0.6140	16.2	88.9	105.0	0.0362	0.2242	13
14	29.095	0.0121	1.5949	82.73	0.6270	16.5	88.7	105.2	0.0368	0.2241	14
15	29.739	0.0121	1.5620	82.63	0.6402	16.8	88.5	105.3	0.0375	0.2240	15
16	30.395	0.0121	1.5298	82.52	0.6537	17.1	88.3	105.5	0.0381	0.2239	16
17	31.061	0.0121	1.4984	82.41	0.6674	17.4	88.2	105.6	0.0388	0.2238	17
18	31.739	0.0122	1.4678	82.30	0.6813	17.7	88.0	105.7	0.0395	0.2237	18
19	32.428	0.0122	1.4380	82.19	0.6954	18.1	87.8	105.9	0.0401	0.2236	19
20	33.129	0.0122	1.4090	82.08	0.7097	18.4	87.7	106.0	0.0408	0.2235	20
21	33.841	0.0122	1.3806	81.97	0.7243	18.7	87.5	106.2	0.0414	0.2234	21
22	34.566	0.0122	1.3528	81.86	0.7392	19.0	87.3	106.3	0.0421	0.2233	22
23	35.302	0.0122	1.3259	81.75	0.7542	19.3	87.1	106.5	0.0427	0.2233	23
24	36.050	0.0122	1.2995	81.64	0.7695	19.6	87.0	106.6	0.0434	0.2232	24
25	36.810	0.0123	1.2737	81.52	0.7851	20.0	86.8	106.7	0.0440	0.2231	25
26	37.583	0.0123	1.2486	81.41	0.8009	20.3	86.6	106.9	0.0447	0.2230	26
27	38.368	0.0123	1.2240	81.30	0.8170	20.6	86.4	107.0	0.0453	0.2229	27
28	39.166	0.0123	1.2000	81.19	0.8333	20.9	86.2	107.2	0.0460	0.2229	28
29	39.977	0.0123	1.1766	81.08	0.8499	21.2	86.1	107.3	0.0467	0.2228	29

Table A-10 I-P Properties of Saturated Liquid and Vapor for R-134a

TEMP. °F	PRESSURE psia	VOLUME ft³/lb LIQUID v_f	VOLUME ft³/lb VAPOR v_g	DENSITY lb/ft³ LIQUID $1/v_f$	DENSITY lb/ft³ VAPOR $1/v_g$	ENTHALPY Btu/lb LIQUID h_f	ENTHALPY Btu/lb LATENT h_{fg}	ENTHALPY Btu/lb VAPOR h_g	ENTROPY Btu/(lb)(°R) LIQUID s_f	ENTROPY Btu/(lb)(°R) VAPOR s_g	TEMP. °F
30	40.800	0.0124	1.1538	80.96	0.8667	21.6	85.9	107.4	0.0473	0.2227	30
31	41.636	0.0124	1.1315	80.85	0.8838	21.9	85.7	107.6	0.0480	0.2226	31
32	42.486	0.0124	1.1098	80.74	0.9011	22.2	85.5	107.7	0.0486	0.2226	32
33	43.349	0.0124	1.0884	80.62	0.9188	22.5	85.3	107.9	0.0492	0.2225	33
34	44.225	0.0124	1.0676	80.51	0.9367	22.8	85.2	108.0	0.0499	0.2224	34
35	45.115	0.0124	1.0472	80.40	0.9549	23.2	85.0	108.1	0.0505	0.2223	35
36	46.018	0.0125	1.0274	80.28	0.9733	23.5	84.8	108.3	0.0512	0.2223	36
37	46.935	0.0125	1.0080	80.17	0.9921	23.8	84.6	108.4	0.0518	0.2222	37
38	47.866	0.0125	0.9890	80.05	1.0111	24.1	84.4	108.6	0.0525	0.2221	38
39	48.812	0.0125	0.9705	79.94	1.0304	24.5	84.2	108.7	0.0531	0.2221	39
40	49.771	0.0125	0.9523	79.82	1.0501	24.8	84.1	108.8	0.0538	0.2220	40
41	50.745	0.0125	0.9346	79.70	1.0700	25.1	83.9	109.0	0.0544	0.2219	41
42	51.733	0.0126	0.9173	79.59	1.0902	25.4	83.7	109.1	0.0551	0.2219	42
43	52.736	0.0126	0.9003	79.47	1.1107	25.8	83.5	109.2	0.0557	0.2218	43
44	53.754	0.0126	0.8837	79.35	1.1316	26.1	83.3	109.4	0.0564	0.2217	44
45	54.787	0.0126	0.8675	79.24	1.1527	26.4	83.1	109.5	0.0570	0.2217	45
46	55.835	0.0126	0.8516	79.12	1.1742	26.7	82.9	109.7	0.0576	0.2216	46
47	56.898	0.0127	0.8361	79.00	1.1960	27.1	82.7	109.8	0.0583	0.2216	47
48	57.976	0.0127	0.8210	78.88	1.2181	27.4	82.5	109.9	0.0589	0.2215	48
49	59.070	0.0127	0.8061	78.76	1.2405	27.7	82.3	110.1	0.0596	0.2214	49
50	60.180	0.0127	0.7916	78.64	1.2633	28.0	82.1	110.2	0.0602	0.2214	50
51	61.305	0.0127	0.7774	78.53	1.2864	28.4	81.9	110.3	0.0608	0.2213	51
52	62.447	0.0128	0.7634	78.41	1.3099	28.7	81.8	110.5	0.0615	0.2213	52
53	63.604	0.0128	0.7498	78.29	1.3337	29.0	81.6	110.6	0.0621	0.2212	53
54	64.778	0.0128	0.7365	78.16	1.3578	29.4	81.4	110.7	0.0628	0.2211	54
55	65.963	0.0128	0.7234	78.04	1.3823	29.7	81.2	110.9	0.0634	0.2211	55
56	67.170	0.0128	0.7106	77.92	1.4072	30.0	81.0	111.0	0.0640	0.2210	56
57	68.394	0.0129	0.6981	77.80	1.4324	30.4	80.8	111.1	0.0647	0.2210	57
58	69.635	0.0129	0.6859	77.68	1.4579	30.7	80.6	111.3	0.0653	0.2209	58
59	70.892	0.0129	0.6739	77.56	1.4839	31.0	80.4	111.4	0.0659	0.2209	59
60	72.167	0.0129	0.6622	77.43	1.5102	31.4	80.2	111.5	0.0666	0.2208	60
61	73.459	0.0129	0.6507	77.31	1.5369	31.7	80.0	111.6	0.0672	0.2208	61
62	74.769	0.0130	0.6394	77.19	1.5640	32.0	79.7	111.8	0.0678	0.2207	62
63	76.096	0.0130	0.6283	77.06	1.5915	32.4	79.5	111.9	0.0685	0.2207	63
64	77.440	0.0130	0.6175	76.94	1.6194	32.7	79.3	112.0	0.0691	0.2206	64
65	78.803	0.0130	0.6069	76.81	1.6477	33.0	79.1	112.2	0.0698	0.2206	65
66	80.184	0.0130	0.5965	76.69	1.6764	33.4	78.9	112.3	0.0704	0.2205	66
67	81.582	0.0131	0.5863	76.56	1.7055	33.7	78.7	112.4	0.0710	0.2205	67
68	83.000	0.0131	0.5764	76.44	1.7350	34.0	78.5	112.5	0.0717	0.2204	68
69	84.435	0.0131	0.5666	76.31	1.7649	34.4	78.3	112.7	0.0723	0.2204	69
70	85.890	0.0131	0.5570	76.18	1.7952	34.7	78.1	112.8	0.0729	0.2203	70
71	87.363	0.0131	0.5476	76.05	1.8260	35.1	77.9	112.9	0.0735	0.2203	71
72	88.855	0.0132	0.5384	75.93	1.8573	35.4	77.6	113.0	0.0742	0.2202	72
73	90.366	0.0132	0.5294	75.80	1.8889	35.7	77.4	113.2	0.0748	0.2202	73
74	91.897	0.0132	0.5206	75.67	1.9210	36.1	77.2	113.3	0.0754	0.2201	74
75	93.447	0.0132	0.5119	75.54	1.9536	36.4	77.0	113.4	0.0761	0.2201	75
76	95.016	0.0133	0.5034	75.41	1.9866	36.8	76.8	113.5	0.0767	0.2200	76
77	96.606	0.0133	0.4950	75.28	2.0201	37.1	76.6	113.7	0.0773	0.2200	77
78	98.215	0.0133	0.4868	75.15	2.0541	37.4	76.3	113.8	0.0780	0.2200	78
79	99.844	0.0133	0.4788	75.02	2.0885	37.8	76.1	113.9	0.0786	0.2199	79
80	101.494	0.0134	0.4709	74.89	2.1234	38.1	75.9	114.0	0.0792	0.2199	80
81	103.164	0.0134	0.4632	74.75	2.1589	38.5	75.7	114.1	0.0799	0.2198	81
82	104.855	0.0134	0.4556	74.62	2.1948	38.8	75.4	114.3	0.0805	0.2198	82
83	106.566	0.0134	0.4482	74.49	2.2312	39.2	75.2	114.4	0.0811	0.2197	83
84	108.290	0.0134	0.4409	74.35	2.2681	39.5	75.0	114.5	0.0817	0.2197	84
85	110.050	0.0135	0.4337	74.22	2.3056	39.9	74.8	114.6	0.0824	0.2196	85
86	111.828	0.0135	0.4267	74.08	2.3436	40.2	74.5	114.7	0.0830	0.2196	86
87	113.626	0.0135	0.4198	73.95	2.3821	40.5	74.3	114.9	0.0836	0.2196	87
88	115.444	0.0135	0.4130	73.81	2.4211	40.9	74.1	115.0	0.0843	0.2195	88
89	117.281	0.0136	0.4064	73.67	2.4607	41.2	73.8	115.1	0.0849	0.2195	89

Table A-10 Continued

TEMP. °F	PRESSURE psia	VOLUME ft³/lb LIQUID v_f	VOLUME ft³/lb VAPOR v_g	DENSITY lb/ft³ LIQUID $1/v_f$	DENSITY lb/ft³ VAPOR $1/v_g$	ENTHALPY Btu/lb LIQUID h_f	ENTHALPY Btu/lb LATENT h_{fg}	ENTHALPY Btu/lb VAPOR h_g	ENTROPY Btu/(lb)(°R) LIQUID s_f	ENTROPY Btu/(lb)(°R) VAPOR s_g	TEMP. °F
90	119.138	0.0136	0.3999	73.54	2.5009	41.6	73.6	115.2	0.0855	0.2194	90
91	121.024	0.0136	0.3935	73.40	2.5416	41.9	73.4	115.3	0.0861	0.2194	91
92	122.930	0.0137	0.3872	73.26	2.5829	42.3	73.1	115.4	0.0868	0.2193	92
93	124.858	0.0137	0.3810	73.12	2.6247	42.6	72.9	115.5	0.0874	0.2193	93
94	126.809	0.0137	0.3749	72.98	2.6672	43.0	72.7	115.7	0.0880	0.2193	94
95	128.782	0.0137	0.3690	72.84	2.7102	43.4	72.4	115.8	0.0886	0.2192	95
96	130.778	0.0138	0.3631	72.70	2.7539	43.7	72.2	115.9	0.0893	0.2192	96
97	132.798	0.0138	0.3574	72.56	2.7981	44.1	71.9	116.0	0.0899	0.2191	97
98	134.840	0.0138	0.3517	72.42	2.8430	44.4	71.7	116.1	0.0905	0.2191	98
99	136.906	0.0138	0.3462	72.27	2.8885	44.8	71.4	116.2	0.0912	0.2190	99
100	138.996	0.0139	0.3408	72.13	2.9347	45.1	71.2	116.3	0.0918	0.2190	100
101	141.109	0.0139	0.3354	71.99	2.9815	45.5	70.9	116.4	0.0924	0.2190	101
102	143.247	0.0139	0.3302	71.84	3.0289	45.8	70.7	116.5	0.0930	0.2189	102
103	145.408	0.0139	0.3250	71.70	3.0771	46.2	70.4	116.6	0.0937	0.2189	103
104	147.594	0.0140	0.3199	71.55	3.1259	46.6	70.2	116.7	0.0943	0.2188	104
105	149.804	0.0140	0.3149	71.40	3.1754	46.9	69.9	116.9	0.0949	0.2188	105
106	152.039	0.0140	0.3100	71.25	3.2256	47.3	69.7	117.0	0.0955	0.2187	106
107	154.298	0.0141	0.3052	71.11	3.2765	47.6	69.4	117.1	0.0962	0.2187	107
108	156.583	0.0141	0.3005	70.96	3.3282	48.0	69.2	117.2	0.0968	0.2186	108
109	158.893	0.0141	0.2958	70.81	3.3806	48.4	68.9	117.3	0.0974	0.2186	109
110	161.227	0.0142	0.2912	70.66	3.4337	48.7	68.6	117.4	0.0981	0.2185	110
111	163.588	0.0142	0.2867	70.51	3.4876	49.1	68.4	117.5	0.0987	0.2185	111
112	165.974	0.0142	0.2823	70.35	3.5423	49.5	68.1	117.6	0.0993	0.2185	112
113	168.393	0.0142	0.2780	70.20	3.5977	49.8	67.8	117.7	0.0999	0.2184	113
114	170.833	0.0143	0.2737	70.05	3.6539	50.2	67.6	117.8	0.1006	0.2184	114
115	173.298	0.0143	0.2695	69.89	3.7110	50.5	67.3	117.9	0.1012	0.2183	115
116	175.790	0.0143	0.2653	69.74	3.7689	50.9	67.0	117.9	0.1018	0.2183	116
117	178.297	0.0144	0.2613	69.58	3.8276	51.3	66.8	118.0	0.1024	0.2182	117
118	180.846	0.0144	0.2573	69.42	3.8872	51.7	66.5	118.1	0.1031	0.2182	118
119	183.421	0.0144	0.2533	69.26	3.9476	52.0	66.2	118.2	0.1037	0.2181	119
120	186.023	0.0145	0.2494	69.10	4.0089	52.4	65.9	118.3	0.1043	0.2181	120
121	188.652	0.0145	0.2456	68.94	4.0712	52.8	65.6	118.4	0.1050	0.2180	121
122	191.308	0.0145	0.2419	68.78	4.1343	53.1	65.4	118.5	0.1056	0.2180	122
123	193.992	0.0146	0.2382	68.62	4.1984	53.5	65.1	118.6	0.1062	0.2179	123
124	196.703	0.0146	0.2346	68.46	4.2634	53.9	64.8	118.7	0.1068	0.2178	124
125	199.443	0.0146	0.2310	68.29	4.3294	54.3	64.5	118.8	0.1075	0.2178	125
126	202.211	0.0147	0.2275	68.13	4.3964	54.6	64.2	118.8	0.1081	0.2177	126
127	205.008	0.0147	0.2240	67.96	4.4644	55.0	63.9	118.9	0.1087	0.2177	127
128	207.834	0.0147	0.2206	67.80	4.5334	55.4	63.6	119.0	0.1094	0.2176	128
129	210.688	0.0148	0.2172	67.63	4.6034	55.8	63.3	119.1	0.1100	0.2176	129
130	213.572	0.0148	0.2139	67.46	4.6745	56.2	63.0	119.2	0.1106	0.2175	130
131	216.485	0.0149	0.2107	67.29	4.7467	56.5	62.7	119.2	0.1113	0.2174	131
132	219.429	0.0149	0.2075	67.12	4.8200	56.9	62.4	119.3	0.1119	0.2174	132
133	222.402	0.0149	0.2043	66.95	4.8945	57.3	62.1	119.4	0.1125	0.2173	133
134	225.405	0.0150	0.2012	66.77	4.9700	57.7	61.8	119.5	0.1132	0.2173	134
135	228.438	0.0150	0.1981	66.60	5.0468	58.1	61.5	119.6	0.1138	0.2172	135
136	231.502	0.0151	0.1951	66.42	5.1248	58.5	61.2	119.6	0.1144	0.2171	136
137	234.597	0.0151	0.1922	66.24	5.2040	58.8	60.8	119.7	0.1151	0.2171	137
138	237.723	0.0151	0.1892	66.06	5.2844	59.2	60.5	119.8	0.1157	0.2170	138
139	240.880	0.0152	0.1864	65.88	5.3661	59.6	60.2	119.8	0.1163	0.2169	139
140	244.068	0.0152	0.1835	65.70	5.4491	60.0	59.9	119.9	0.1170	0.2168	140
141	247.288	0.0153	0.1807	65.52	5.5335	60.4	59.6	120.0	0.1176	0.2168	141
142	250.540	0.0153	0.1780	65.34	5.6192	60.8	59.2	120.0	0.1183	0.2167	142
143	253.824	0.0153	0.1752	65.15	5.7064	61.2	58.9	120.1	0.1189	0.2166	143
144	257.140	0.0154	0.1726	64.96	5.7949	61.6	58.6	120.1	0.1195	0.2165	144
145	260.489	0.0154	0.1699	64.78	5.8849	62.0	58.2	120.2	0.1202	0.2165	145
146	263.871	0.0155	0.1673	64.59	5.9765	62.4	57.9	120.3	0.1208	0.2164	146
147	267.270	0.0155	0.1648	64.39	6.0695	62.8	57.5	120.3	0.1215	0.2163	147
148	270.721	0.0156	0.1622	64.20	6.1642	63.2	57.2	120.4	0.1221	0.2162	148
149	274.204	0.0156	0.1597	64.01	6.2604	63.6	56.8	120.4	0.1228	0.2161	149

Table A-10 Continued

TEMP.	PRESSURE	VOLUME ft³/lb		DENSITY lb/ft³		ENTHALPY Btu/lb			ENTROPY Btu/(lb)(°R)		TEMP.
°F	psia	LIQUID v_f	VAPOR v_g	LIQUID $1/v_f$	VAPOR $1/v_g$	LIQUID h_f	LATENT h_{fg}	VAPOR h_g	LIQUID s_f	VAPOR s_g	°F
150	277.721	0.0157	0.1573	63.81	6.3584	64.0	56.5	120.5	0.1234	0.2160	150
151	281.272	0.0157	0.1548	63.61	6.4580	64.4	56.1	120.5	0.1240	0.2159	151
152	284.857	0.0158	0.1525	63.41	6.5593	64.8	55.7	120.6	0.1247	0.2158	152
153	288.477	0.0158	0.1501	63.21	6.6625	65.2	55.4	120.6	0.1253	0.2157	153
154	292.131	0.0159	0.1478	63.01	6.7675	65.6	55.0	120.6	0.1260	0.2156	154
155	295.820	0.0159	0.1455	62.80	6.8743	66.0	54.6	120.7	0.1266	0.2155	155
156	299.544	0.0160	0.1432	62.59	6.9831	66.4	54.3	120.7	0.1273	0.2154	156
157	303.304	0.0160	0.1410	62.38	7.0940	66.9	53.9	120.7	0.1279	0.2153	157
158	307.100	0.0161	0.1388	62.17	7.2068	67.3	53.5	120.8	0.1286	0.2152	158
159	310.931	0.0161	0.1366	61.96	7.3218	67.7	53.1	120.8	0.1293	0.2151	159
160	314.800	0.0162	0.1344	61.74	7.4390	68.1	52.7	120.8	0.1299	0.2150	160
161	318.704	0.0163	0.1323	61.52	7.5584	68.5	52.3	120.9	0.1306	0.2149	161
162	322.646	0.0163	0.1302	61.30	7.6801	69.0	51.9	120.9	0.1312	0.2148	162
163	326.625	0.0164	0.1281	61.08	7.8042	69.4	51.5	120.9	0.1319	0.2146	163
164	330.641	0.0164	0.1261	60.86	7.9308	69.8	51.1	120.9	0.1326	0.2145	164
165	334.696	0.0165	0.1241	60.63	8.0600	70.2	50.7	120.9	0.1332	0.2144	165
166	338.788	0.0166	0.1221	60.40	8.1917	70.7	50.3	120.9	0.1339	0.2142	166
167	342.919	0.0166	0.1201	60.16	8.3262	71.1	49.8	120.9	0.1346	0.2141	167
168	347.089	0.0167	0.1182	59.93	8.4635	71.5	49.4	120.9	0.1352	0.2140	168
169	351.298	0.0168	0.1162	59.69	8.6037	72.0	49.0	120.9	0.1359	0.2138	169
170	355.547	0.0168	0.1143	59.45	8.7470	72.4	48.5	120.9	0.1366	0.2137	170
171	359.835	0.0169	0.1124	59.20	8.8934	72.8	48.1	120.9	0.1373	0.2135	171
172	364.164	0.0170	0.1106	58.95	9.0431	73.3	47.6	120.9	0.1380	0.2133	172
173	368.533	0.0170	0.1087	58.70	9.1961	73.7	47.2	120.9	0.1386	0.2132	173
174	372.942	0.0171	0.1069	58.45	9.3527	74.2	46.7	120.9	0.1393	0.2130	174
175	377.393	0.0172	0.1051	58.19	9.5129	74.6	46.2	120.8	0.1400	0.2128	175
176	381.886	0.0173	0.1033	57.92	9.6770	75.1	45.7	120.8	0.1407	0.2126	176
177	386.421	0.0173	0.1016	57.66	9.8451	75.5	45.2	120.8	0.1414	0.2125	177
178	390.998	0.0174	0.0998	57.39	10.0173	76.0	44.7	120.7	0.1421	0.2123	178
179	395.617	0.0175	0.0981	57.11	10.1939	76.5	44.2	120.7	0.1428	0.2121	179
180	400.280	0.0176	0.0964	56.83	10.3750	76.9	43.7	120.7	0.1435	0.2119	180
181	404.987	0.0177	0.0947	56.55	10.5609	77.4	43.2	120.6	0.1442	0.2116	181
182	409.738	0.0178	0.0930	56.26	10.7518	77.9	42.7	120.5	0.1449	0.2114	182
183	414.533	0.0179	0.0913	55.96	10.9481	78.4	42.1	120.5	0.1456	0.2112	183
184	419.373	0.0180	0.0897	55.66	11.1498	78.8	41.6	120.4	0.1464	0.2109	184
185	424.258	0.0181	0.0880	55.35	11.3575	79.3	41.0	120.3	0.1471	0.2107	185
186	429.189	0.0182	0.0864	55.04	11.5713	79.8	40.4	120.2	0.1478	0.2104	186
187	434.167	0.0183	0.0848	54.72	11.7916	80.3	39.8	120.1	0.1486	0.2102	187
188	439.192	0.0184	0.0832	54.40	12.0189	80.8	39.2	120.0	0.1493	0.2099	188
189	444.264	0.0185	0.0816	54.07	12.2536	81.3	38.6	119.9	0.1501	0.2096	189
190	449.384	0.0186	0.0800	53.73	12.4962	81.8	38.0	119.8	0.1508	0.2093	190
191	454.552	0.0187	0.0784	53.38	12.7472	82.3	37.4	119.7	0.1516	0.2090	191
192	459.757	0.0189	0.0769	53.02	13.0072	82.8	36.7	119.5	0.1523	0.2087	192
193	465.026	0.0190	0.0753	52.65	13.2769	83.4	36.0	119.4	0.1531	0.2083	193
194	470.346	0.0191	0.0738	52.27	13.5570	83.9	35.3	119.2	0.1539	0.2080	194
195	475.717	0.0193	0.0722	51.88	13.8484	84.4	34.6	119.1	0.1547	0.2076	195
196	481.139	0.0194	0.0707	51.48	14.1522	85.0	33.9	118.9	0.1555	0.2072	196
197	486.614	0.0196	0.0691	51.07	14.4693	85.5	33.1	118.7	0.1563	0.2068	197
198	492.142	0.0197	0.0676	50.64	14.8012	86.1	32.3	118.4	0.1572	0.2063	198
199	497.724	0.0199	0.0660	50.20	15.1493	86.7	31.5	118.2	0.1580	0.2059	199
200	503.361	0.0201	0.0645	49.73	15.5155	87.3	30.7	118.0	0.1589	0.2054	200
201	509.054	0.0203	0.0629	49.25	15.9020	87.9	29.8	117.7	0.1597	0.2049	201
202	514.805	0.0205	0.0613	48.75	16.3113	88.5	28.9	117.4	0.1606	0.2043	202
203	520.613	0.0207	0.0597	48.22	16.7466	89.1	27.9	117.0	0.1616	0.2037	203
204	526.481	0.0210	0.0581	47.66	17.2121	89.8	26.9	116.7	0.1625	0.2031	204
205	532.410	0.0212	0.0565	47.06	17.7129	90.5	25.8	116.3	0.1635	0.2024	205
206	538.402	0.0215	0.0548	46.42	18.2558	91.2	24.7	115.8	0.1645	0.2016	206
207	544.458	0.0219	0.0531	45.73	18.8499	91.9	23.4	115.3	0.1656	0.2008	207
208	550.581	0.0222	0.0513	44.98	19.5084	92.7	22.1	114.8	0.1667	0.1998	208
209	556.773	0.0227	0.0494	44.14	20.2504	93.5	20.6	114.1	0.1680	0.1988	209

Table A-10 Continued

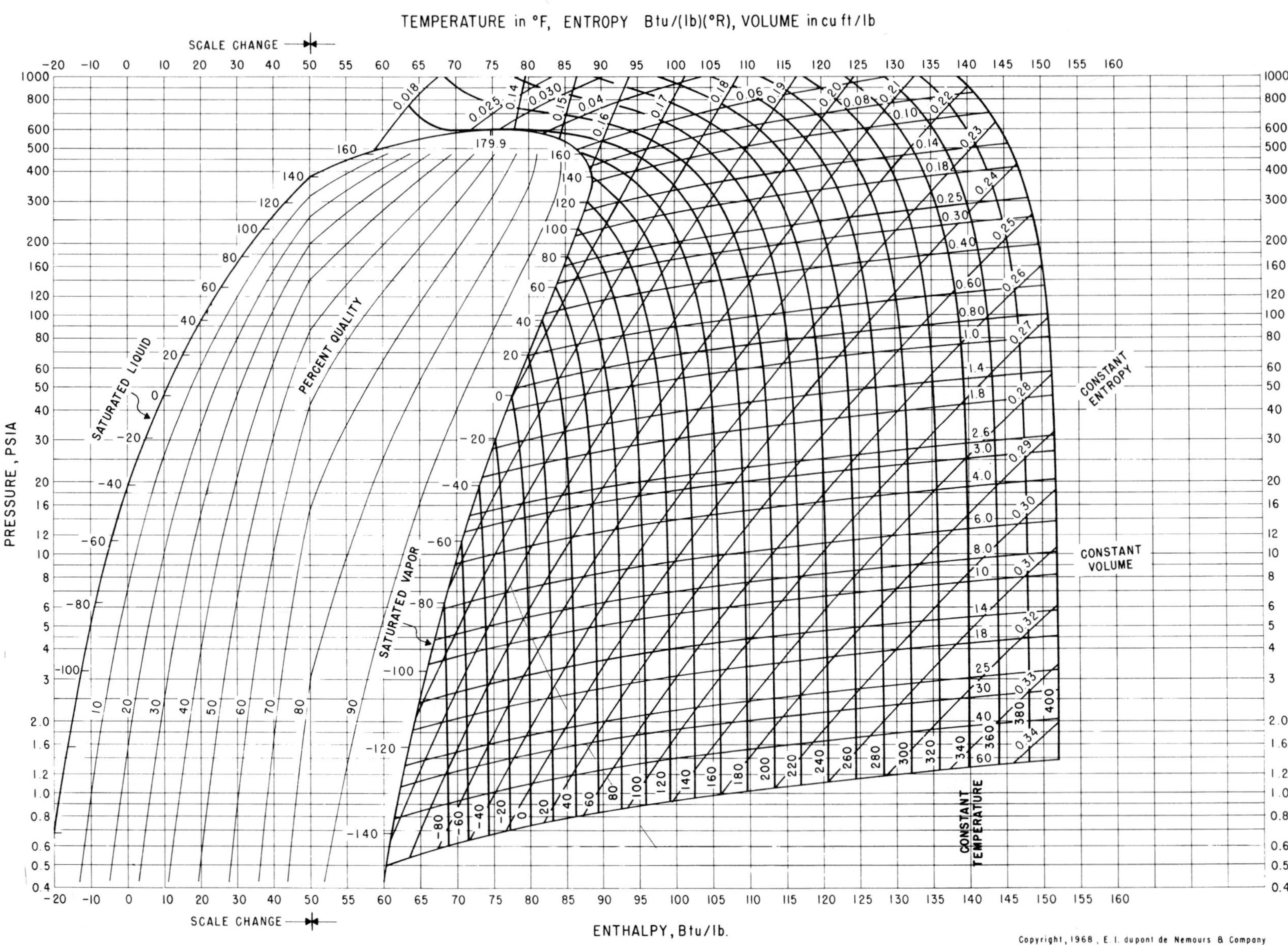

Table A-11 I-P Pressure-Heat Diagram for R-502

TEMP. °F	PRESSURE		VOLUME cu ft/lb		DENSITY lb/cu ft		ENTHALPY Btu/lb			ENTROPY Btu/(lb)(°R)		TEMP. °F
	PSIA	PSIG	LIQUID	VAPOR	LIQUID	VAPOR	LIQUID	LATENT	VAPOR	LIQUID	VAPOR	
−40	18.802	4.106	0.010941	2.0453	91.393	0.48891	0.000	73.114	73.114	0.00000	0.17421	−40
−39	19.269	4.573	0.010955	1.9987	91.280	0.50030	0.228	73.003	73.232	0.00054	0.17408	−39
−38	19.746	5.050	0.010968	1.9535	91.167	0.51189	0.457	72.892	73.350	0.00108	0.17395	−38
−37	20.231	5.535	0.010982	1.9094	91.053	0.52370	0.687	72.780	73.468	0.00163	0.17381	−37
−36	20.726	6.030	0.010996	1.8666	90.939	0.53571	0.918	72.667	73.585	0.00217	0.17368	−36
−35	21.230	6.534	0.011010	1.8249	90.825	0.54795	1.149	72.553	73.702	0.00271	0.17355	−35
−34	21.744	7.048	0.011023	1.7844	90.711	0.56040	1.381	72.438	73.819	0.00325	0.17342	−34
−33	22.267	7.571	0.011037	1.7449	90.597	0.57307	1.613	72.323	73.936	0.00380	0.17330	−33
−32	22.800	8.104	0.011051	1.7065	90.482	0.58597	1.846	72.206	74.053	0.00434	0.17317	−32
−31	23.343	8.647	0.011065	1.6691	90.367	0.59909	2.080	72.089	74.170	0.00488	0.17305	−31
−30	23.897	9.201	0.011080	1.6328	90.252	0.61244	2.315	71.971	74.286	0.00543	0.17293	−30
−29	24.460	9.764	0.011094	1.5973	90.136	0.62602	2.550	71.853	74.403	0.00597	0.17281	−29
−28	25.033	10.337	0.011108	1.5628	90.021	0.63984	2.785	71.733	74.519	0.00651	0.17269	−28
−27	25.617	10.921	0.011122	1.5292	89.905	0.65390	3.021	71.613	74.635	0.00706	0.17257	−27
−26	26.212	11.516	0.011137	1.4965	89.789	0.66819	3.258	71.492	74.751	0.00760	0.17245	−26
−25	26.817	12.121	0.011151	1.4646	89.673	0.68273	3.496	71.370	74.866	0.00815	0.17234	−25
−24	27.433	12.737	0.011166	1.4336	89.556	0.69752	3.734	71.247	74.982	0.00869	0.17222	−24
−23	28.059	13.363	0.011180	1.4033	89.439	0.71256	3.973	71.124	75.097	0.00923	0.17211	−23
−22	28.697	14.001	0.011195	1.3739	89.322	0.72785	4.212	71.000	75.212	0.00978	0.17200	−22
−21	29.346	14.650	0.011210	1.3451	89.205	0.74340	4.452	70.875	75.327	0.01032	0.17189	−21
−20	30.006	15.310	0.011224	1.3171	89.088	0.75920	4.693	70.749	75.442	0.01087	0.17178	−20
−19	30.678	15.982	0.011239	1.2898	88.970	0.77527	4.934	70.622	75.556	0.01141	0.17167	−19
−18	31.361	16.665	0.011254	1.2632	88.852	0.79160	5.176	70.494	75.671	0.01196	0.17156	−18
−17	32.056	17.360	0.011269	1.2373	88.734	0.80820	5.418	70.366	75.785	0.01250	0.17146	−17
−16	32.762	18.066	0.011284	1.2120	88.615	0.82507	5.661	70.237	75.899	0.01305	0.17135	−16
−15	33.480	18.784	0.011299	1.1873	88.496	0.84222	5.905	70.107	76.012	0.01359	0.17125	−15
−14	34.211	19.515	0.011315	1.1632	88.377	0.85964	6.149	69.976	76.126	0.01414	0.17115	−14
−13	34.954	20.258	0.011330	1.1397	88.258	0.87734	6.394	69.844	76.239	0.01469	0.17105	−13
−12	35.709	21.013	0.011345	1.1168	88.138	0.89533	6.640	69.712	76.352	0.01523	0.17095	−12
−11	36.476	21.780	0.011361	1.0945	88.018	0.91361	6.886	69.578	76.465	0.01578	0.17085	−11
−10	37.256	22.560	0.011376	1.0727	87.898	0.93218	7.133	69.444	76.577	0.01632	0.17075	−10
− 9	38.049	23.353	0.011392	1.0514	87.778	0.95104	7.380	69.309	76.690	0.01687	0.17066	− 9
− 8	38.854	24.158	0.011408	1.0307	87.657	0.97020	7.628	69.173	76.802	0.01741	0.17056	− 8
− 7	39.673	24.977	0.011423	1.0104	87.536	0.98966	7.877	69.037	76.914	0.01796	0.17047	− 7
− 6	40.504	25.808	0.011439	0.99065	87.415	1.0094	8.126	68.899	77.025	0.01851	0.17037	− 6
− 5	41.349	26.653	0.011455	0.97134	87.293	1.0295	8.376	68.761	77.137	0.01905	0.17028	− 5
− 4	42.207	27.511	0.011471	0.95247	87.172	1.0498	8.626	68.622	77.248	0.01960	0.17019	− 4
− 3	43.078	28.382	0.011487	0.93406	87.050	1.0705	8.877	68.482	77.359	0.02014	0.17010	− 3
− 2	43.964	29.268	0.011503	0.91607	86.927	1.0916	9.129	68.341	77.470	0.02069	0.17001	− 2
− 1	44.863	30.167	0.011520	0.89851	86.804	1.1129	9.381	68.199	77.580	0.02124	0.16992	− 1
0	45.775	31.079	0.011536	0.88135	86.681	1.1346	9.633	68.056	77.690	0.02178	0.16983	0
1	46.702	32.006	0.011552	0.86458	86.558	1.1566	9.887	67.913	77.800	0.02233	0.16975	1
2	47.643	32.947	0.011569	0.84821	86.434	1.1789	10.141	67.769	77.910	0.02287	0.16966	2
3	48.599	33.903	0.011586	0.83220	86.310	1.2016	10.395	67.624	78.019	0.02342	0.16958	3
4	49.568	34.872	0.011602	0.81657	86.186	1.2246	10.650	67.478	78.128	0.02397	0.16949	4
5	50.553	35.857	0.011619	0.80129	86.062	1.2479	10.906	67.331	78.237	0.02451	0.16941	5
6	51.552	36.856	0.011636	0.78635	85.937	1.2716	11.162	67.183	78.346	0.02506	0.16933	6
7	52.565	37.869	0.011653	0.77175	85.811	1.2957	11.419	67.035	78.454	0.02561	0.16925	7
8	53.594	38.898	0.011670	0.75748	85.686	1.3201	11.676	66.885	78.562	0.02615	0.16917	8
9	54.638	39.942	0.011687	0.74353	85.560	1.3449	11.934	66.735	78.670	0.02670	0.16909	9
10	55.697	41.001	0.011704	0.72988	85.434	1.3700	12.193	66.584	78.777	0.02724	0.16901	10
11	56.771	42.075	0.011722	0.71654	85.307	1.3955	12.452	66.432	78.885	0.02779	0.16893	11
12	57.861	43.165	0.011739	0.70349	85.180	1.4214	12.711	66.280	78.991	0.02834	0.16885	12
13	58.967	44.271	0.011757	0.69073	85.053	1.4477	12.972	66.126	79.098	0.02888	0.16878	13
14	60.088	45.392	0.011774	0.67825	84.925	1.4743	13.232	65.971	79.204	0.02943	0.16870	14

Table A-12 I-P Properties of Saturated Liquid and Vapor for R-502

TEMP. °F	PRESSURE		VOLUME cu ft/lb		DENSITY lb/cu ft		ENTHALPY Btu/lb			ENTROPY Btu/(lb)(°R)		TEMP. °F
	PSIA	PSIG	LIQUID	VAPOR	LIQUID	VAPOR	LIQUID	LATENT	VAPOR	LIQUID	VAPOR	
15	61.225	46.529	0.011792	0.66604	84.797	1.5013	13.494	65.816	79.310	0.02997	0.16863	15
16	62.379	47.683	0.011810	0.65410	84.669	1.5288	13.756	65.660	79.416	0.03052	0.16855	16
17	63.548	48.852	0.011828	0.64241	84.540	1.5566	14.018	65.503	79.521	0.03107	0.16848	17
18	64.734	50.038	0.011846	0.63098	84.411	1.5848	14.281	65.345	79.626	0.03161	0.16841	18
19	65.936	51.240	0.011864	0.61979	84.282	1.6134	14.545	65.186	79.731	0.03216	0.16833	19
20	67.155	52.459	0.011883	0.60884	84.152	1.6424	14.809	65.026	79.836	0.03270	0.16826	20
21	68.391	53.695	0.011901	0.59812	84.022	1.6719	15.073	64.866	79.940	0.03325	0.16819	21
22	69.643	54.947	0.011920	0.58762	83.891	1.7017	15.339	64.704	80.043	0.03379	0.16812	22
23	70.913	56.217	0.011938	0.57735	83.760	1.7320	15.604	64.542	80.147	0.03434	0.16805	23
24	72.199	57.503	0.011957	0.56730	83.629	1.7627	15.871	64.379	80.250	0.03488	0.16799	24
25	73.503	58.807	0.011976	0.55745	83.497	1.7938	16.138	64.215	80.353	0.03543	0.16792	25
26	74.824	60.128	0.011995	0.54781	83.365	1.8254	16.405	64.050	80.455	0.03597	0.16785	26
27	76.163	61.467	0.012014	0.53837	83.232	1.8574	16.673	63.884	80.557	0.03652	0.16778	27
28	77.520	62.824	0.012033	0.52912	83.099	1.8898	16.941	63.717	80.659	0.03706	0.16772	28
29	78.894	64.198	0.012053	0.52007	82.966	1.9228	17.210	63.550	80.760	0.03761	0.16765	29
30	80.286	65.590	0.012072	0.51120	82.832	1.9561	17.480	63.381	80.861	0.03815	0.16759	30
31	81.697	67.001	0.012092	0.50251	82.698	1.9900	17.750	63.212	80.962	0.03870	0.16752	31
32	83.126	68.430	0.012111	0.49399	82.563	2.0242	18.020	63.042	81.062	0.03924	0.16746	32
33	84.573	69.877	0.012131	0.48565	82.428	2.0590	18.291	62.871	81.162	0.03979	0.16739	33
34	86.038	71.342	0.012151	0.47748	82.292	2.0943	18.563	62.698	81.262	0.04033	0.16733	34
35	87.522	72.826	0.012171	0.45947	82.156	2.1300	18.835	62.526	81.361	0.04087	0.16727	35
36	89.026	74.330	0.012192	0.46162	82.020	2.1662	19.107	62.352	81.460	0.04142	0.16721	36
37	90.548	75.852	0.012212	0.45393	81.883	2.2029	19.381	62.177	81.558	0.04196	0.16715	37
38	92.089	77.393	0.012233	0.44638	81.746	2.2401	19.654	62.001	81.656	0.04250	0.16708	38
39	93.649	78.953	0.012253	0.43899	81.608	2.2779	19.928	61.825	81.754	0.04305	0.16702	39
40	95.229	80.533	0.012274	0.43175	81.469	2.3161	20.203	61.647	81.851	0.04359	0.16696	40
41	96.828	82.132	0.012295	0.42464	81.330	2.3549	20.478	61.469	81.948	0.04413	0.16690	41
42	98.447	83.751	0.012316	0.41767	81.191	2.3941	20.754	61.290	82.044	0.04468	0.16684	42
43	100.08	85.38	0.012337	0.41084	81.051	2.4339	21.030	61.110	82.140	0.04522	0.16678	43
44	101.74	87.04	0.012359	0.40414	80.911	2.4743	21.307	60.928	82.235	0.04576	0.16673	44
45	103.42	88.72	0.012380	0.39757	80.770	2.5152	21.584	60.746	82.331	0.04630	0.16667	45
46	105.12	90.42	0.012402	0.39113	80.629	2.5566	21.861	60.563	82.425	0.04685	0.16661	46
47	106.84	92.14	0.012424	0.38481	80.487	2.5986	22.140	60.380	82.520	0.04739	0.16655	47
48	108.58	93.88	0.012446	0.37861	80.345	2.6412	22.418	60.195	82.613	0.04793	0.16649	48
49	110.34	95.64	0.012468	0.37252	80.202	2.6843	22.697	60.009	82.707	0.04847	0.16644	49
50	112.12	97.42	0.012490	0.36655	80.058	2.7280	22.977	59.822	82.800	0.04901	0.16638	50
51	113.92	99.22	0.012513	0.36070	79.914	2.7723	23.257	59.635	82.892	0.04955	0.16632	51
52	115.74	101.05	0.012536	0.35495	79.769	2.8172	23.538	59.446	82.984	0.05009	0.16627	52
53	117.59	102.89	0.012558	0.34931	79.624	2.8627	23.819	59.257	83.076	0.05063	0.16621	53
54	119.45	104.75	0.012581	0.34377	79.479	2.9088	24.100	59.066	83.167	0.05117	0.16616	54
55	121.34	106.64	0.012605	0.33834	79.332	2.9555	24.382	58.875	83.257	0.05171	0.16610	55
56	123.25	108.55	0.012628	0.33301	79.185	3.0028	24.665	58.682	83.347	0.05225	0.16605	56
57	125.18	110.48	0.012652	0.32777	79.038	3.0508	24.948	58.489	83.437	0.05279	0.16599	57
58	127.13	112.43	0.012675	0.32263	78.890	3.0994	25.231	58.294	83.526	0.05333	0.16594	58
59	129.10	114.41	0.012699	0.31759	78.741	3.1486	25.515	58.099	83.615	0.05387	0.16588	59
60	131.10	116.40	0.012723	0.31263	78.592	3.1985	25.799	57.903	83.703	0.05441	0.16583	60
61	133.12	118.42	0.012748	0.30777	78.441	3.2491	26.084	57.705	83.790	0.05495	0.16577	61
62	135.16	120.46	0.012772	0.30299	78.291	3.3004	26.370	57.507	83.877	0.05549	0.16572	62
63	137.22	122.53	0.012797	0.29830	78.140	3.3523	26.656	57.308	83.964	0.05602	0.16566	63
64	139.31	124.61	0.012822	0.29369	77.988	3.4049	26.942	57.107	84.050	0.05656	0.16561	64
65	141.42	126.72	0.012847	0.28916	77.835	3.4582	27.229	56.906	84.135	0.05710	0.16556	65
66	143.55	128.86	0.012872	0.28471	77.682	3.5122	27.516	56.704	84.220	0.05764	0.16550	66
67	145.71	131.01	0.012898	0.28034	77.528	3.5669	27.804	56.500	84.304	0.05817	0.16545	67
68	147.89	133.19	0.012924	0.27605	77.373	3.6224	28.092	56.296	84.388	0.05871	0.16539	68
69	150.09	135.39	0.012950	0.27183	77.217	3.6786	28.380	56.091	84.471	0.05925	0.16534	69

Table A-12 Continued

TEMP. °F	PRESSURE		VOLUME cu ft/lb		DENSITY lb/cu ft		ENTHALPY Btu/lb			ENTROPY Btu/(lb)(°R)		TEMP. °F
	PSIA	PSIG	LIQUID	VAPOR	LIQUID	VAPOR	LIQUID	LATENT	VAPOR	LIQUID	VAPOR	
70	152.32	137.62	0.012976	0.26769	77.061	3.7356	28.669	55.884	84.554	0.05978	0.16529	70
71	154.57	139.87	0.013003	0.26362	76.904	3.7933	28.959	55.676	84.636	0.06032	0.16523	71
72	156.84	142.15	0.013029	0.25961	76.747	3.8517	29.249	55.468	84.717	0.06085	0.16518	72
73	159.14	144.44	0.013056	0.25568	76.588	3.9110	29.539	55.258	84.798	0.06139	0.16512	73
74	161.46	146.77	0.013083	0.25181	76.429	3.9711	29.830	55.047	84.878	0.06193	0.16507	74
75	163.81	149.11	0.013111	0.24801	76.269	4.0319	30.122	54.835	84.958	0.06246	0.16502	75
76	166.18	151.49	0.013139	0.24428	76.108	4.0936	30.414	54.622	85.037	0.06299	0.16496	76
77	168.58	153.88	0.013167	0.24060	75.947	4.1561	30.706	54.408	85.115	0.06353	0.16491	77
78	171.00	156.30	0.013195	0.23699	75.784	4.2194	30.999	54.193	85.193	0.06406	0.16485	78
79	173.45	158.75	0.013223	0.23344	75.621	4.2836	31.292	53.977	85.270	0.06460	0.16480	79
80	175.92	161.22	0.013252	0.22995	75.457	4.3487	31.586	53.759	85.346	0.06513	0.16474	80
81	178.41	163.72	0.013281	0.22651	75.292	4.4146	31.880	53.541	85.421	0.06566	0.16469	81
82	180.94	166.24	0.013310	0.22313	75.126	4.4815	32.175	53.321	85.496	0.06620	0.16463	82
83	183.48	168.79	0.013340	0.21981	74.959	4.5492	32.470	53.100	85.570	0.06673	0.16458	83
84	186.06	171.36	0.013370	0.21654	74.791	4.6179	32.766	52.878	85.644	0.06726	0.16452	84
85	188.66	173.96	0.013400	0.21333	74.622	4.6874	33.062	52.654	85.717	0.06780	0.16446	85
86	191.28	176.59	0.013431	0.21017	74.453	4.7580	33.359	52.430	85.789	0.06833	0.16441	86
87	193.94	179.24	0.013462	0.20705	74.282	4.8295	33.656	52.204	85.860	0.06886	0.16435	87
88	196.62	181.92	0.013493	0.20399	74.111	4.9020	33.953	51.977	85.930	0.06939	0.16429	88
89	199.32	184.62	0.013524	0.20098	73.938	4.9755	34.251	51.748	86.000	0.06992	0.16423	89
90	202.05	187.36	0.013556	0.19801	73.764	5.0500	34.550	51.519	86.069	0.07045	0.16418	90
91	204.81	190.12	0.013588	0.19509	73.590	5.1255	34.849	51.288	86.137	0.07098	0.16412	91
92	207.60	192.90	0.013621	0.19222	73.414	5.2021	35.148	51.055	86.204	0.07151	0.16406	92
93	210.41	195.72	0.013654	0.18939	73.237	5.2798	35.448	50.822	86.271	0.07205	0.16400	93
94	213.25	198.56	0.013687	0.18661	73.059	5.3585	35.749	50.587	86.336	0.07258	0.16394	94
95	216.12	201.43	0.013721	0.18387	72.880	5.4384	36.050	50.350	86.401	0.07311	0.16388	95
96	219.02	204.32	0.013755	0.18117	72.700	5.5194	36.352	50.113	86.465	0.07364	0.16382	96
97	221.94	207.25	0.013789	0.17852	72.519	5.6015	36.654	49.874	86.528	0.07417	0.16375	97
98	224.90	210.20	0.013824	0.17590	72.336	5.6848	36.956	49.633	86.590	0.07470	0.16369	98
99	227.88	213.18	0.013859	0.17333	72.152	5.7693	37.259	49.391	86.651	0.07522	0.16363	99
100	230.89	216.19	0.013895	0.17079	71.967	5.8550	37.563	49.147	86.711	0.07575	0.16356	100
101	233.93	219.23	0.013931	0.16829	71.781	5.9419	37.867	48.902	86.770	0.07628	0.16350	101
102	237.00	222.30	0.013967	0.16583	71.593	6.0301	38.172	48.656	86.828	0.07681	0.16343	102
103	240.09	225.40	0.014004	0.16340	71.405	6.1196	38.477	48.407	86.885	0.07734	0.16337	103
104	243.22	228.52	0.014042	0.16101	71.214	6.2104	38.783	48.158	86.941	0.07787	0.16330	104
105	246.38	231.68	0.014079	0.15866	71.023	6.3026	39.090	47.906	86.997	0.07840	0.16323	105
106	249.56	234.87	0.014118	0.15634	70.829	6.3961	39.398	47.653	87.051	0.07893	0.16316	106
107	252.78	238.08	0.014157	0.15405	70.635	6.4910	39.705	47.398	87.103	0.07946	0.16309	107
108	256.02	241.33	0.014196	0.15180	70.439	6.5873	40.013	47.142	87.155	0.07998	0.16302	108
109	259.30	244.60	0.014236	0.14958	70.241	6.6851	40.322	46.883	87.206	0.08051	0.16295	109
110	262.61	247.91	0.014277	0.14739	70.042	6.7843	40.631	46.623	87.255	0.08104	0.16288	110
111	265.94	251.25	0.014318	0.14523	69.841	6.8851	40.942	46.361	87.303	0.08157	0.16281	111
112	269.31	254.62	0.014359	0.14311	69.639	6.9875	41.252	46.098	87.350	0.08210	0.16273	112
113	272.71	258.02	0.014401	0.14101	69.435	7.0914	41.564	45.832	87.396	0.08263	0.16265	113
114	276.15	261.45	0.014444	0.13894	69.229	7.1970	41.876	45.564	87.441	0.08316	0.16258	114
115	279.61	264.91	0.014488	0.13690	69.022	7.3042	42.189	45.294	87.484	0.08368	0.16250	115
116	283.10	268.41	0.014532	0.13489	68.812	7.4132	42.503	45.022	87.525	0.08421	0.16242	116
117	286.63	271.94	0.014576	0.13290	68.601	7.5239	42.817	44.748	87.566	0.08474	0.16234	117
118	290.19	275.50	0.014622	0.13095	68.388	7.6364	43.132	44.472	87.605	0.08527	0.16225	118
119	293.78	279.09	0.014668	0.12902	68.173	7.7507	43.448	44.194	87.642	0.08580	0.16217	119
120	297.41	282.71	0.014715	0.12711	67.956	7.8669	43.765	43.913	87.678	0.08633	0.16208	120
121	301.07	286.37	0.014762	0.12523	67.737	7.9850	44.082	43.630	87.713	0.08686	0.16199	121
122	304.76	290.06	0.014811	0.12337	67.515	8.1052	44.401	43.344	87.746	0.08739	0.16190	122
123	308.49	293.79	0.014860	0.12154	67.292	8.2274	44.720	43.056	87.777	0.08792	0.16181	123
124	312.25	297.55	0.014910	0.11973	67.066	8.3516	45.040	42.766	87.807	0.08845	0.16172	124

Table A-12 Continued

TEMP. °F	PRESSURE		VOLUME cu ft/lb		DENSITY lb/cu ft		ENTHALPY Btu/lb			ENTROPY Btu/(lb)(°R)		TEMP. °F
	PSIA	PSIG	LIQUID	VAPOR	LIQUID	VAPOR	LIQUID	LATENT	VAPOR	LIQUID	VAPOR	
125	316.04	301.35	0.014961	0.11795	66.838	8.4781	45.361	42.472	87.834	0.08899	0.16163	125
126	319.87	305.17	0.015013	0.11618	66.608	8.6067	45.684	42.176	87.860	0.08952	0.16153	126
127	323.73	309.04	0.015065	0.11444	66.375	8.7377	46.007	41.878	87.885	0.09005	0.16143	127
128	327.63	312.94	0.015119	0.11272	66.140	8.8710	46.331	41.576	87.907	0.09058	0.16133	128
129	331.57	316.87	0.015174	0.11102	65.901	9.0067	46.656	41.271	87.928	0.09112	0.16122	129
130	335.54	320.84	0.015229	0.10934	65.661	9.1449	46.983	40.963	87.946	0.09165	0.16112	130
131	339.55	324.85	0.015286	0.10769	65.417	9.2858	47.310	40.652	87.962	0.09219	0.16101	131
132	343.59	328.89	0.015344	0.10605	65.171	9.4293	47.639	40.337	87.977	0.09273	0.16090	132
133	347.67	332.97	0.015403	0.10443	64.921	9.5755	47.969	40.019	87.989	0.09326	0.16078	133
134	351.79	337.09	0.015463	0.10283	64.669	9.7246	48.300	39.697	87.998	0.09380	0.16067	134
135	355.94	341.24	0.015524	0.10124	64.413	9.8767	48.633	39.372	88.006	0.09434	0.16055	135
136	360.13	345.44	0.015587	0.099682	64.154	10.031	48.967	39.043	88.011	0.09488	0.16042	136
137	364.36	349.67	0.015651	0.098133	63.892	10.190	49.303	38.709	88.013	0.09543	0.16030	137
138	368.63	353.94	0.015716	0.096601	63.626	10.351	49.641	38.372	88.013	0.09597	0.16017	138
139	372.94	358.25	0.015783	0.095086	63.356	10.516	49.980	38.030	88.010	0.09652	0.16004	139
140	377.29	362.60	0.015852	0.093586	63.082	10.685	50.320	37.683	88.004	0.09706	0.15990	140
141	381.68	366.98	0.015922	0.092101	62.805	10.857	50.663	37.331	87.995	0.09761	0.15976	141
142	386.11	371.41	0.015993	0.090630	62.523	11.033	51.008	36.975	87.983	0.09817	0.15962	142
143	390.58	375.88	0.016067	0.089174	62.237	11.213	51.354	36.613	87.968	0.09877	0.15947	143
144	395.09	380.39	0.016142	0.087732	61.946	11.398	51.703	36.245	87.949	0.09928	0.15931	144
145	399.64	384.95	0.016220	0.086303	61.650	11.587	52.054	35.872	87.927	0.09983	0.15916	145
146	404.24	389.54	0.016299	0.084886	61.350	11.780	52.408	35.493	87.901	0.10039	0.15899	146
147	408.88	394.18	0.016381	0.083481	61.044	11.978	52.764	35.107	87.871	0.10096	0.15882	147
148	413.56	398.86	0.016465	0.082088	60.732	12.181	53.123	34.714	87.838	0.10152	0.15865	148
149	418.29	403.59	0.016551	0.080706	60.415	12.390	53.485	34.314	87.800	0.10209	0.15847	149
150	423.06	408.35	0.016641	0.079335	60.092	12.604	53.850	33.907	87.757	0.10267	0.15828	150
151	427.87	413.18	0.016732	0.077973	59.762	12.824	54.218	33.491	87.710	0.10325	0.15809	151
152	432.74	418.04	0.016827	0.076620	59.425	13.051	54.590	33.067	87.657	0.10383	0.15789	152
153	437.65	422.95	0.016925	0.075276	59.082	13.284	54.966	32.633	87.599	0.10442	0.15769	153
154	442.60	427.91	0.017026	0.073939	58.730	13.524	55.346	32.190	87.536	0.10501	0.15747	154
155	447.61	432.91	0.017131	0.072610	58.371	13.772	55.730	31.736	87.467	0.10561	0.15725	155
156	452.66	437.97	0.017240	0.071286	58.002	14.027	56.119	31.271	87.391	0.10622	0.15701	156
157	457.77	443.07	0.017353	0.069968	57.625	14.292	56.513	30.794	87.308	0.10683	0.15677	157
158	462.92	448.22	0.017470	0.068655	57.237	14.565	56.913	30.304	87.218	0.10746	0.15652	158
159	468.13	453.43	0.017593	0.067345	56.839	14.848	57.320	29.800	87.120	0.10809	0.15625	159
160	473.38	458.69	0.017721	0.066037	56.429	15.142	57.732	29.281	87.013	0.10872	0.15598	160
161	478.70	464.00	0.017854	0.064730	56.007	15.448	58.153	28.744	86.898	0.10937	0.15569	161
162	484.06	469.37	0.017994	0.063423	55.571	15.767	58.581	28.190	86.772	0.11004	0.15538	162
163	489.48	474.79	0.018141	0.062114	55.121	16.099	59.019	27.616	86.635	0.11071	0.15506	163
164	494.96	480.27	0.018296	0.060802	54.654	16.446	59.466	27.019	86.486	0.11140	0.15472	164
165	500.50	485.80	0.018460	0.059484	54.169	16.811	59.925	26.398	86.324	0.11210	0.15436	165
166	506.10	491.40	0.018634	0.058158	53.665	17.194	60.397	25.749	86.146	0.11283	0.15398	166
167	511.75	497.06	0.018818	0.056822	53.138	17.598	60.883	25.069	85.953	0.11357	0.15358	167
168	517.47	502.78	0.019016	0.055472	52.586	18.026	61.385	24.355	85.740	0.11434	0.15314	168
169	523.26	508.56	0.019228	0.054104	52.005	18.482	61.906	23.599	85.506	0.11514	0.15268	169
170	529.11	514.41	0.019458	0.052714	51.392	18.970	62.449	22.798	85.248	0.11597	0.15217	170
171	535.03	520.33	0.019708	0.051295	50.739	19.495	63.019	21.941	84.960	0.11684	0.15163	171
172	541.01	526.32	0.019983	0.049839	50.041	20.064	63.619	21.018	84.638	0.11775	0.15103	172
173	547.07	532.38	0.020288	0.048335	49.288	20.688	64.258	20.016	84.274	0.11873	0.15037	173
174	553.20	538.51	0.020633	0.046769	48.465	21.381	64.944	18.913	83.858	0.11978	0.14962	174
175	559.41	544.72	0.021029	0.045118	47.552	22.163	65.693	17.681	83.374	0.12092	0.14878	175
176	565.70	551.01	0.021496	0.043348	46.518	23.069	66.525	16.273	82.798	0.12219	0.14779	176
177	572.08	557.38	0.022071	0.041396	45.307	24.156	67.480	14.607	82.087	0.12365	0.14659	177
178	578.54	563.84	0.022829	0.039137	43.803	25.551	68.635	12.515	81.151	0.12542	0.14505	178
179	585.09	570.39	0.023993	0.036203	41.677	27.621	70.210	9.518	79.729	0.12784	0.14275	179
179.889	591.00	576.30	0.028571	0.028571	35.000	35.000	74.654	0.000	74.654	0.13476	0.13476	179.889

Table A-12 Continued

Glossary of Technical Terms

Absolute Humidity: The amount of moisture actually in a given volume of air.

Absolute Pressure: Gauge pressure plus atmospheric pressure.

Absolute Temperature: Temperature measured from absolute zero.

Absolute Zero: Temperature at which all molecular motion ceases (–460°F and –273°C).

Absorbent: Substance which has the ability to take up or absorb another substance.

Absorber: A device containing liquid for absorbing refrigerant vapor or other vapors.

Accumulator: Storage tank which receives liquid refrigerant from evaporator and prevents it from flowing into suction line.

Activated Carbon: Specially processed carbon commonly used to clean air.

Adiabatic Compression: Compressing refrigerant gas without removing or adding heat.

Adsorbent: Substance which has property to hold molecules of fluids without causing a chemical or physical change.

Air Binding: A condition in which a bubble or other pocket of air is present in a pipeline that prevents the desired flow in the pipeline.

Air Changes: The amount of air leakage through a building in terms of the number of building volumes exchanged.

Air Conditioner: Device used to control temperature, humidity, cleanliness, and movement of air in conditioned space.

Air Cushion Tank: A closed tank that allows for water expansion without creating excessive pressure.

Air Shutter: An adjustable shutter on the primary air openings of a burner, which is used to control the amount of combustion air.

Air Vent: A valve installed at the high points in a hot water system to eliminate air from the system.

Aldehyde: A class of compounds, which can be produced during incomplete combustion of a fuel gas.

Allen-Type Screw: Screw with recessed head designed to be turned with hex shaped wrench.

Alternating Current: Abbreviated ac. Current that reverses polarity or direction periodically. It rises from zero to maximum strength and returns to zero in one direction then goes to similar variation in the opposite direction. This is a cycle which is repeated at a fixed frequency. It can be single phase, two phase, three phase, and poly phase. Its advantage over direct or undirectional current is that its voltage can be stepped up by transformers to the high values which reduce transmission costs.

Alternator: A machine which converts mechanical energy into alternating current.

Ambient Temperature: Temperature of fluid (usually air) which surrounds an object.

American Wire Gauge: Abbreviated AWG. A system of numbers which designate cross-sectional area of wire. As the diameter gets smaller, the number gets larger, e.g., AWG #14 = 0.0641 in., AWG #12 = 0.0808 in.

Ammeter: An electric meter used to measure current.

Ampere: Unit of electric current equivalent to flow of one coulomb per second.

Ampere-Turn: Abbreviated AT or NI. Unit of magnetizing force produced by a current flow of one ampere through one turn of wire in a coil.

Amplitude: The maximum instantaneous value of alternating current or voltage. It can be in either a positive or negative direction.

Anemometer: Instrument for measuring the rate of flow of air.

Angle of Lag or Lead: Phase angle difference between two sinusoidal wave forms having the same frequency.

Annealing: Process of heat treating metal to obtain desired properties of softness and ductility.

Anticipator: A heater used to adjust thermostat operation to produce a closer temperature differential than the mechanical capability of the control.

Armature: The moving or rotating component of a motor, generator, relay or other electromagnetic device.

ASME: American Society of Mechanical Engineers.

Aspect Ratio: Ratio of length to width of rectangular duct.

Aspirating Psychrometer: A device which draws sample of air through it for humidity measurement purposes.

Atom: Smallest particle of element.

Atomic Weight: The number of protons in an atom of a material is classified as its *atomic weight.*

Atomize: Process of changing a liquid to a fine spray.

Attenuate: Decrease or lessen in intensity.

Automatic Defrost: System of removing ice and frost from evaporators automatically.

Automatic Expansion Valve (AEV): Pressure controlled valve used as a metering device.

Autotransformer: A transformer in which both primary and secondary coils have turns in common. Step-up or stepdown of voltage is accomplished by taps on common winding.

B

Back Pressure: Pressure in low side of refrigerating system; also called suction pressure or low side pressure.
Back Seat: The position of a service valve that is all the way counter-clockwise.
Ball Check Valve: Valve assembly call which permits flow of fluid in one section only.
Ball Valve: A valve in which modulation or shut-off is accomplished by a one-quarter turn of a ball that has an opening through it.
Barometer: Instrument for measuring atmospheric pressure.
Baseboard: A terminal unit resembling the base trim of a house.
Battery: Two or more primary or secondary electrically interconnected cells.
Baudelot Cooler: Heat exchanger in which water flows by gravity over the outside of the tubes or plates.
Bellows: Corrugated cylindrical container which moves as pressures change.
Bending Spring: Coil spring which is used to keep tube from collapsing while bending it.
Bimetal Strip: Two dissimilar metals with unequal expansion rates, welded together.
Bleed off: The continuous or intermittent wasting of a small-fraction of the water to a cooling tower to prevent the buildup of scale-forming chemicals.
Bleed-Valve: Valve which permits a minimum fluid flow when valve is closed.
Blower: A centrifugal fan.
Boiler Horsepower: The equivalent evaporation of 34.5 lb of water per hr from and at 212°F. This is equal to a heat output of 970.3 × 34.5 = 33,475 Btu/h.
Boiling Temperature: Temperature at which a fluid changes from a liquid to a gas.
Bore: Inside diameter of a cylindrical hole.
Bourdon Tube: Thin walled tube of circular shape, which tends to straighten as pressure inside is increased.
Brazing: Soldering with a filler material whose melting temperature is higher than 800°F.
Break: Electrical discontinuity in the circuit generally resulting from the operation of a switch or circuit breaker.
Breaker Strip: Strip of plastic used to cover joint between outside case and inside liner of refrigerator.
Brine: Water saturated with chemical such as salt.
British Thermal Unit (Btu): Quantity of heat required to change temperature of one pound of water one degree F.
Bull Head: The installation of a pipe tee in such a way that water enters (or leaves) the tee at both ends of the run (the straight through section of the tee) and leaves (or enters) through the side connection only.
Bunsen-Type Burner: A gas burner in which combustion air is premixed with the gas supply within the burner body before the gas burns on the burner port.
Burner: A device for the final conveyance of gas, or a mixture of gas and air, to the combustion zone.
Burnout: Accidental passage of high voltage through an electrical circuit or device which causes damage.

C

Cadmium Cell: A device whose resistance changes according to the amount of light sensed.
Calibrate: To adjust an indicator so that it correctly indicates the variable sensed.
Calorie: Heat required to raise temperature of one gram of water one degree celsius.
Capacitor: Type of electrical storage device used in starting and/or running circuits on many electric motors.
Capillary Tube: A type of refrigerant control consisting of several feet of tubing having small inside diameter.
Carbon Dioxide (CO_2): A non-toxic product of combustion.
Carbon Monoxide: A chemical resulting from incomplete combustion (CO). It is odorless, colorless, and toxic.
Cascade System: One having two or more refrigerant circuits, each with a compressor, condenser, and evaporator, where the evaporator of one circuit cools the condenser of the other (lower-temperature) circuit.
Celsius: The metric system temperature scale.
Centimeter: Metric unit of linear measurement which equals 0.3937 in.
Centrifugal Compressor: Compressor which compresses gaseous refrigerants by centrifugal force.
Cfm: Cubic feet per minute.
Charge: The amount of refrigerant in a system.
Charging Cylinder: A device for charging a predetermined weight of refrigerant into a system.
Check Valve: A one-direction valve.
Chimney Effect: The tendency of air or gas in a duct or other vertical passage to rise when heated.
Circuit: A tubing, piping or electrical wire installation which permits flow from the energy source back to energy source.
Circuit Breaker: A device that senses current flow, and opens when its rated current flow is exceeded.
Circulator: A pump.
Clearance: Space in cylinder not occupied by piston at end of compression stroke.
Closed Loop: Any cycle in which the primary medium is always enclosed and repeats the same sequence of events.
Coefficience of Performance (COP): Usable energy output divided by the energy input.
Coil: A wound conductor, which creates a strong magnetic field with current passage.
Cold: Cold is the absence of heat.
Cold Junction: That part of the thermoelectric system which absorbs heat.
Combustion: The rapid oxidation of fuel gases accompanied by the production of heat.
Combustion Air: Air supplied for the combustion of a fuel.
Combustion Products: Constituents resulting from the combustion of a fuel gas with the oxygen in air, including the inerts, but excluding excess air.
Comfort Chart: Chart used in air-conditioning to show the dry bulb temperature and humidity for human comfort.
Commercial Buildings: Such buildings as stores, shops, restaurants, motels, and large apartment buildings.
Commutator: A ring of copper segments insulated from each other and connecting the armature and brushes of a motor or generator. It passes power into or from the brushes.

Compound Gauge: Pressure gauge that has scales both above and below atmospheric pressure.

Compound Refrigerating System: System which has several compressors in series.

Compression Ratio: The absolute discharge pressure divided by the absolute suction pressure for a compressor.

Compression Tank: (See Air Cushion Tank)

Compressor: The pump of a refrigerating mechanism which draws a vacuum or low pressure on cooling side of refrigerant cycle and squeezes or compresses the gas into the high pressure or condensing side of the cycle.

Compressor Displacement: The volume discharged by a compressor in one rotation of the crankshaft.

Compressor Seal: Leakproof seal between crankshaft and compressor body.

Condensible: A gas which can be easily converted to liquid form.

Condensate Pump: Device used to remove fluid condensate.

Condensation: Liquid which forms when a gas is cooled below its dew point.

Condenser: The part of refrigeration mechanism which receives hot, high pressure refrigerant gas from compressor and cools it until it returns to liquid state.

Condenser Water Pump: Forced water moving device used to move water through condenser.

Condensing Pressure: The refrigerant pressure inside the condenser coil.

Condensing Temperature: The temperature at which refrigerant condensation is taking place inside the condenser.

Condensing Unit: Refrigeration unit consisting of a compressor, condenser, and controls.

Conductance (Thermal): "C" factor—The time rate of heat flow per unit area through a material.

Conduction (Thermal): Particle to particle transmission of heat.

Conductivity: The ability of a substance to allow the flow of heat or electricity.

Conductor: Material or substance which readily passes electricity.

Connected Load: The sum of the capacities or continuous ratings of the load-consuming apparatus connected to a supplying system.

Connecting Rod: That part of compressor which connects piston to crankshaft.

Constrictor: Tube or orifice used to restrict flow.

Contact: The part of a switch or relay that carries current.

Contactor: A device for making or breaking load-carrying contacts by a pilot circuit through a magnetic coil.

Contaminant: A substance (dirt, moisture, etc.) foreign to refrigerant or refrigerant oil in system.

Controller: Measures the difference between sensed output and desired output and initiates a response to correct the difference.

Controls: Devices designed to regulate the gas, air, water or electricity.

Convection: Transfer of heat by means of movement of a fluid.

Cooling Anticipator: A resistor in a room thermostat that causes the cooling cycle to begin prematurely.

Cooling Tower: Device which cools water by water evaporation in air. Water is cooled to wet bulb temperature of air.

Coulomb: An electrical unit of charge, one coulomb per second equals one ampere or 6.25×10^{18} electrons past a given point in one second.

Counter EMF: Counter electromotive force; the EMF induced in a coil, which opposes applied voltage.

Counterflow: Flow in opposite directions.

"Cracking" a Valve: Opening valve a small amount.

Crankshaft Seal: Leakproof joint between crankshaft and compressor body.

Current: The flow of electrons through a conductor.

Cut-In: The temperature or pressure at which an automatic control switch closes.

Cut-Out: The temperature or pressure at which an automatic control switch opens.

Cylinder Head: Part which encloses compression end of compressor cylinder.

Damper: Valve for controlling airflow.

Dead Band: A range of temperature in a heating/cooling system in which no heating or cooling is supplied.

Decibel: Unit used for measuring relative loudness of sounds.

Defrost Control: A control system on a heat pump used to detect buildup of frost on the outside coil during the heating cycle and cause system reversal for hot gas defrost of the coil.

Defrost Cycle: Refrigerating cycle in which evaporator frost and ice accumulation is melted.

Defrost Timer: Device connected into electrical circuit which shuts unit off long enough to permit ice and frost accumulation on evaporator to melt.

Degree-Day: Unit that represents one degree of difference from given point in average outdoor temperature of one day.

Dehydrate: To remove water.

Delta Connection: The connection in a three-phase system in which terminal connections are triangular similar to the Greek letter delta.

Demand: The size of any load generally averaged over a specified interval of time.

Demand (Billing): The demand upon which billing to a customer is based.

Demand Meter: An instrument used to measure peak kilowatt-hour consumption.

Density: Weight per unit volume.

Desiccant: Substance used to collect and hold moisture.

Design Load: The amount of heating or cooling required to maintain inside conditions when the outdoor conditions are at design temperature.

Dew Point: Temperature at which vapor (at 100 percent humidity) begins to condense.

Diaphragm: Flexible membrane.

Dielectric: An insulator-nonconductor.

Dies (Thread): Tool used to cut external threads.

Differential: As applied to refrigeration and heating: difference between "cut-in" and "cut-out" temperature or pressure.

Diffuser: Air distribution outlet designed to direct airflow into a room.

Dilution Air: Air which enters a draft hood and mixes with the flue gases.

Diode: A device that will carry current in one direction but not the reverse direction.

Direct Current: Abbreviated dc. Electric current that flows only in one direction.

Direct Expansion Evaporator: An evaporator containing liquid and vapor refrigerant.

Direct Return: A two pipe system in which the first terminal unit taken off the supply main is the first unit connected to the return main.

Disconnecting Switch: A knife switch that opens a circuit.

Domestic Hot Water: The heated water used for cooking, washing, etc.

Double-Pole Switch: Simultaneously opens and closes two wires of a circuit.

Double Thickness Flare: Tubing end which has been formed into two-wall thickness.

Dowel Pin: Pin pressed through two assembled parts to ensure accurate alignment.

Downdraft: Downward flow of flue gas.

Draft Gauge: Instrument used to measure air pressure.

Drier: A substance or device used to remove moisture from a refrigeration system.

Drift: Entrained water carried from a cooling tower by wind.

Drilled Port Burner: A burner in which the ports have been formed by drilled holes.

Drip Pan: Pan-shaped panel or trough used to collect condensate from evaporator coil.

Dry Bulb Temperature: Air temperature as indicated by ordinary thermometer.

Dry Ice: Solid carbon dioxide.

Dynamometer: Device for measuring power.

E

E: Symbol for volts.

Eccentric: A circle or disk mounted off center on a shaft.

Effective Area: Actual net flow area.

Efficiency (electrical): A percentage value denoting the ratio of power output to power input.

Electric Field: A magnetic region in space.

Electric Heating: Heating system in which heat is produced from electrical resistance units.

Electrolyte: A solution of a substance (liquid or paste) that is capable of conducting electricity.

Electromagnet: A magnet created by the flow of electricity through a coil of wire.

Electromotive Force: Abbreviated EMF. The difference in potential electrical energy between two points.

Electronics: Field of science dealing with electron devises and their uses.

Electrostatic Filter: Type of filter which gives particles of dust electric charge.

End Bell: End structure of electric motor which usually holds motor bearings.

End Play: Slight movement of shaft along center line.

Energy Efficiency Ratio (EER): Btu/hr of cooling produced per watt of electrical input.

Enthalpy: Total amount of heat in one pound of a substance.

Epoxy (Resins): A synthetic plastic adhesive.

EPR (Evaporation Pressure Regulator): An automatic pressure regulating valve to maintain a predetermined pressure in the evaporator.

Equalizer Tube: Device used to maintain equal pressure or equal liquid levels between two containers.

Evaporation: A term applied to the changing of a liquid to a gas.

Evaporative Condenser: A device which uses a water spray to cool a condenser..

Evaporator: Part of a refrigerating mechanism in which the refrigerant evaporizes and absorbs heat.

Evaporator, Dry Type: An evaporator into which refrigerant is fed from a pressure reducing device.

Evaporator Fan: Fan which moves air through the evaporator.

Evaporator, Flooded: An evaporator containing liquid refrigerant at all times.

Excess Air: Air which is in excess of that which is required for complete combustion of the gas.

Exfiltration: Air flow outward through a room.

Exhaust Opening: Any opening through which air is removed from a space.

Expansion Joint: A fitting or piping arrangement designed to relieve stress caused by expansion of piping.

Expansion Tank: (See Air Cushion Tank)

Expansion Valve: A device in refrigerating system which maintains a pressure difference between the high side and low side.

Expendable Refrigerant System: System which discards the refrigerant after it has evaporated.

Extended Surface: Heat transfer surface, one side of which is increased in area by the use of fins.

F

Fahrenheit: The common scale of temperature measurement in the English system of units.

Fan-Coil: A terminal unit consisting of a finned-tube coil and a fan in a single enclosure.

Farad: A large unit of measurement for capacitance.

Feedback: The transfer of energy output back to input.

Female Thread: A thread on the inside of a pipe or fitting.

Ferrous: Objects made of iron or steel.

Field: The space involving the magnetic lines of force.

Filter: Device for removing small particles from a fluid.

Fin Comb: Comb-like device used to straighten the metal fins on coils.

Fire Tube Boiler: A steel boiler in which the hot gases from combustion are circulated through tubes which are surrounded by water.

Flare: An angle formed at the end of a tube.

Flare Nut: Fitting used to clamp tubing flare against another fitting.

Flashback: The movement of the gas flame down through the burner port upon shutdown of the gas supply.

Flash Gas: Gas formed by a sudden reduction in pressure.

Flash Point: Temperature at which an oil will give off sufficient vapor to support a flash flame.

Float Valve: Type of valve which is operated by sphere which floats on liquid surface and controls level of liquid.

Floc Point: The temperature at which the wax in oil will start to separate from the oil.

Flooded System: Type of refrigerating system in which liquid refrigerant fills evaporator.

Flow Switch: An automatic switch that senses fluid flow.

Flue: A passage in the chimney to carry flue gas.

Flue Gases: Products of combustion and excess air.

Fluid: Substance in a liquid or gaseous state.

Flux-Brazing, Soldering: Substance applied to surfaces to be joined to free them from oxides.

Flux (Electrical): The electric or magnetic lines of force in a region.

Flux, Magnetic: Lines of force of a magnet.

Foaming: Formation of a foam in an oil-refrigerant mixture due to rapid evaporation of refrigerant dissolved in the oil.

Foot of Water: A measure of pressure.

Foot Pound: A unit of work.

Forced Convection: Transfer of heat resulting from forced movement of liquid or gas by means of fan or pump.

Forced Draft Burner: A burner in which combustion air is supplied by a blower.

Free Area: The total minimum area of the openings in a grille.

Freeze-Up: 1-The Formation of ice in the refrigerant control device which may stop the flow of refrigerant into the evaporator. 2-Frost formation on a coil may stop the airflow through the coil.

Freezing: Change of state from liquid to solid.

Freezing Point: The temperature at which a liquid will solidify upon removal of heat.

Freon: Trade name for a family of synthetic chemical refrigerants manufactured by DuPont, Inc.

Frequency: The number of complete cycles in a unit of time.

Friction Head: In a hydronic system, the loss in pressure resulting from the flow of water in the piping system.

Frost Back: Condition in which liquid refrigerant flows from evaporator into suction line.

Frost Free Refrigerator: A refrigerated cabinet which operates with an automatic defrost.

Full-load Amperage (FLA): The amount of amperage an inductive load (motor) will draw at its full design load.

Furnace: That part of a warm air heating system in which combustion takes place.

Fuse: Electrical safety device consisting of strip of fusible metal in circuit which melts when the circuit is overloaded.

Fusible Plug: A plug or a mode with a metal of low melting temperature to release pressures in case of fire.

G

Galvanic: Current generated chemically, which results when two dissimilar conductors are immersed in an electrolyte.

Galvanic Action: Corrosion action between two metals of different electronic activity.

Gang: Mechanical connection of two or more switches.

Gasket: A resilient material used between mating surfaces to provide a leakproof seal.

Gas—Noncondensible: A gas which will not form into a liquid under pressure-temperature conditions.

Generator: A machine that converts mechanical energy into electrical energy. (Also see Alternator)

Gravity System: A heating system in which the distribution of the warm air or water relies upon the buoyancy of that air or water. There is no fan or pump.

Grille: An ornamental or louvered opening placed at the end of an air passageway.

Gross Output: A rating applied to boilers. It is the total quantity of heat which the boiler will deliver and at the same time meet all limitations of applicable testing and rating codes.

Ground: Connection between an electrical circuit and earth.

Ground Wire: An electrical wire which will safely conduct electricity from a structure into the ground.

H

Halide Refrigerants: Family of refrigerants containing halogen chemicals.

Halide Torch: Type of torch used to detect halogen refrigerant leaks.

Head: As used in this course, head refers to a pressure difference. (See Pressure Head, Pump Head, Available Head)

Head (Total): In flowing fluid, the sum of the static and velocity pressures at the point of measurement.

Header: A piping arrangement for inter-connecting two or more supply or return tappings of a boiler.

Head Pressure: Pressure which exists in condensing side of refrigerating system.

Head-Pressure Control: A pressure-activated control that prevents the head pressure from falling too low.

Head, Static: Pressure of fluid expressed in terms of height of column of the fluid, such as water or mercury.

Head, Velocity: In flowing fluid, height of fluid equivalent to its velocity pressure.

Heat: Form of energy the addition of which causes substances to rise in temperature; energy associated with random motion of molecules.

Heat Anticipator: A resistor in a room thermostat that shuts the heating cycle off prematurely.

Heat Exchanger: Device used to transfer heat from a warm or hot surface to a cold or cooler surface.

Heat Lag: When a substance is heated on one side, it takes time for the heat to travel through the substance.

Heat Load: The amount of heat that must be removed from the air conditioned or refrigerated space in order to maintain the design temperature.

Heat Loss: The rate of heat transfer from a heated building to the outdoors.

Heat of Compression: The heat added to the refrigerant by the work being done by the compressor.

Heat of Fusion: The heat released in changing a substance from a liquid state to a solid state.

Heat of Respiration: When carbon dioxide and water are given off by foods in storage.

Heat Pump: Air-conditioning system that is reversible so as to be able to remove heat from or add heat to a given space.

Heat Sink: The material in which the refrigeration system discards unwanted heat.

Heat Transfer: Movement of heat from one substance to another.

Heating Value: The number of Btu produced by the combustion of one cubic foot of gas.

Henry: Abbreviated H. The unit of inductance.

Hermetic: Completely sealed.

Hermetic Motor: Compressor drive motor sealed within same casing which contains compressor.

High Side: Parts of a refrigerating system which are under condensing or high side pressure.

High Velocity System: Usually large commercial or industrial air distribution systems designed to operate with static pressures of 6 to 9 inches water gauge.

Horsepower: A unit of power equal to 33,000 foot pounds of work per minute.

Hot Gas Defrost: A defrosting system in which hot refrigerant gas from the high side is directed through evaporator.

Hot Gas Line: The line that carries the hot discharge gas from the compressor to the condenser.

Hot Junction: That part of thermoelectric circuit which releases heat.

Humidifiers: Device used to add humidity.

Humidistat: An electrical control which is operated by changing humidity.

Humidity: Moisture in air.

Hydrocarbon: Any of a number of compounds composed of carbon and hydrogen.

Hydrometer: Floating instrument used to measure specific gravity of a liquid.

Hydronics: Pertaining to heating or cooling with water or vapor.

Hygrometer: An instrument used to measure degree of moisture in atmosphere.

Hygroscopic: Ability of a substance to absorb and retain moisture.

I

Idler: A pulley used on some belt drives to provide the proper belt tension.

Ignition Temperature: The minimum temperature at which combustion can be started.

Ignition Transformer: A transformer designed to provide a high voltage current. Used in many heating systems to ignite fuel.

Impedence: A type of electrical resistance that is only present in alternating current circuits.

Impeller: Rotating part of a centrifugal pump.

Induced Draft Burner: A burner which depends on draft induced by a fan or blower at the flue outlet to draw in combustion air and vent flue gases.

Inductance: The characteristic of an alternating current circuit to oppose a change in current flow.

Induction: The act that produces induced voltage in an object by exposure to a magnetic field.

Induction Motor: An AC motor which operates on principle of rotating magnetic field. Rotor has no electrical connection, but receives electrical energy by transformer action from field windings.

Inductive Load: A device that uses electrical energy to produce motion.

Inductive Reactance: Opposition, measured in ohms, to an alternating or pulsating current.

Inerts: Non-combustible substances in a fuel, or in flue gases, such as nitrogen or carbon dioxide.

Infiltration: The air that leaks into the refrigerated or air conditioned space.

Infrared Lamp: An electrical device which emits infrared rays; invisible rays just beyond red in the visible spectrum.

In Phase: The condition existing when two waves of the same frequency have their maximum and minimum values of like polarity at the same instant.

Insulation, Thermal: Substance used to retard or slow flow of heat through wall or partition.

Interlock: A safety device that allows power to a circuit only after a predetermined function has taken place.

IR Drop: Voltage drop resulting from current flow through a resistor.

Isothermal: At constant temperature.

J

Joule: A measure of heat in the metric system.

Journal, Crankshaft: Part of shaft which contacts the bearing.

Junction Box: Group of electrical terminals housed in protective box or container.

K

Kelvin Scale (K): Thermometer scale on which unit of measurement equals the centigrade degree and according to which absolute zero is 0°, the equivalent of –273.16° C.

Kilopascal: A measure of pressure in the metric system.

Kilowatt: Unit of electrical power, equal to 1000 watts.

Kilowatt-Hour: A unit for measuring electrical energy equal to 3,413 Btu.

King-Valve: The valve located at the outlet of the receiver.

L

Latent Heat: Heat characterized by a change of state of the substance concerned.

Limit Control: Control used to open or close electrical circuits as temperature or pressure limits are reached.

Line Voltage: Voltage existing at wall outlets or terminals of a power line system above 30 V.

Liquefied Petroleum Gases: The terms "Liquefied Petroleum Gases," "LPG," and "LP Gas" mean and include any fuel gas which is composed predominantly of any of the following hydrocarbons, or mixtures of them: propane, propylene, normal butane or isobutane and butylenes.

Liquid Line: The tube which carries liquid refrigerant from the condenser or liquid receiver to the pressure reducing device.

Liquid Nitrogen: Nitrogen in liquid form.

Liquid Receiver: Cylinder connected to condenser outlet for storage of liquid refrigerant in a system.

Liter: Metric unit of volume.

LNG: Liquefied natural gas. Natural gas which has been cooled until it becomes a liquid.

Load: A device that converts electrical energy to another form of usable energy.

Locked Rotor Amperage (LRA): The amount of energy a motor will draw under stalled conditions.

Lock-Out Relay: A control scheme that prevents the restarting of a compressor after a safety control opens, even after the safety control resets itself.

Louvers: Sloping, overlapping boards or metal plates intended to permit ventilation and shed falling water.

Low Pressure Cut-Out: Device used to keep low side evaporating pressure from dropping below certain pressure.

Low Side: That portion of a refrigerating system which is under the lowest evaporating pressure.

Low Voltage Control: Controls designed to operate at voltages of 20 to 30 V.

M

Magnetic Field: The space in which a magnetic force exists.

Make-up Air: The air which is supplied to a building to replace air that has been removed by an exhaust system.

Make-up Water: The water required to replace the water lost from a cooling tower by evaporation, drift, and bleedoff.

Male Thread: A thread on the outside of a pipe or fitting.

Manifold Gauge: A device constructed to hold compound and high pressure gauges and valved to control flow of fluids through it.

Manometer: Instrument for measuring pressure of gases and vapors.

Mass: A quantity of matter cohering together to make one body which is usually of indefinite shape.

Mean Temperature Difference: The average temperature between the temperature before process begins and the temperature after process is completed.

Mechanical Cycle: Cycle which is a repetitive series of mechanical events.

Mega: Prefix for one million; e.g., megohm—one million ohms.

Megohm: 1,000,000 ohms of electrical resistance.

Melting Point: Temperature at atmospheric pressure, at which a substance will melt.

Mercaptan: An odorant that gives natural gas its characteristic odor.

Mercoid Bulb: An electrical circuit switch which uses a small quantity of mercury in a sealed glass tube to make or break electrical contact with terminals within the tube.

Meter: Metric unit of linear measurement.

Methane: A hydrocarbon gas with the formula CH_4, the principal component of natural gases.

MFD: Abbreviation for microfarad.

Micro: A combining form denoting one millionth.

Microfarad: The unit of measurement for capacitance, equal to a millionth of a farad.

Micron: Unit of length in metric system; a thousandth part of one millimeter.

Micron Gauge: Instrument for measuring vacuums very close to a perfect vacuum.

Milli: A combining form denoting one thousandth; example, millivolt, one thousandth of a volt.

Modulating: A type of device or control which tends to adjust by increments (minute changes) rather than by either full on or full off operation.

Moisture Indicator: Instrument used to measure moisture content of a refrigerant.

Molecule: The smallest portion of an element or compound which retains the identity and characteristics of the element or compound.

Mollier Diagram: Graph of refrigerant pressure, heat, and temperature properties.

Motor Burnout: Condition in which the insulation of electric motor has deteriorated by overheating.

Motor Control: Device to start and/or stop a motor at certain temperature or pressure conditions.

Mullion: Stationary part of a structure between two doors.

Multimeter: A volt-ohm meter.

N

National Electrical Code: Abbreviated NEC. A code of electrical rules based on fire underwriters' requirements for interior electric wiring.

Natural Convection: Circulation of a gas or liquid due to difference in density resulting from temperature differences.

Natural Gas: Any gas found in the earth, as opposed to gases which are manufactured.

Needle Valve: A valve used to accurately control very low flow rates of fluids.

Neoprene: A synthetic rubber which is resistant to hydrocarbon oil and gas.

Nominal Size Tubing: Tubing measurement which has an inside diameter the same as iron pipe of the same stated size.

Noncondensible Gas: Gas which does not change into a liquid at operating temperatures and pressures.

Nonferrous: Group of metals and metal alloys which contain no iron; also metals other than iron or steel. In heating systems the principal non-ferrous metals are copper and aluminum.

Normally Closed: Switch contacts closed with the circuit deenergized.

Normally Open: Switch contacts open with the circuit deenergized.

Nozzle: The device on the end of the oil fuel pipe used to form the oil into fine droplets by forcing the oil through a small hole to cause the oil to breakup.

Odorant: A substance added to an otherwise odorless, colorless and tasteless gas to give warning of gas leakage and to aid in leak detection.

Off Cycle: That part of a refrigeration cycle when the system is not operating.

Ohm: A unit of resistance.

Oil Binding: Physical condition when an oil layer on top of refrigerant liquid hinders it from evaporating at its normal pressure-temperature condition.

Oil Rings: Expanding rings mounted in grooves and piston; designed to prevent oil from moving into compression chamber.

Oil Separator: A device used to separate refrigerant oil from refrigerant gas and return the oil to crankcase of compressor.

One-Pipe Fitting: A specially designed tee for use in a one-pipe system to connect the supply or return branch into a circuit.

One-Pipe System: A forced hot-water system using one continuous pipe or main from the boiler supply to the boiler return.

Open Compressor: One with a separate motor or drive.

Orifice: Accurate size opening for controlling fluid flow.

Orifice Spud: A removable plug containing an orifice which determines the quantity of gas that will flow.

Oscilloscope: A fluorescent coated tube which visually shows an electrical wave.

Outside Air: Air exterior to refrigerated or conditioned space.

Overload: Load greater than load for which system or mechanism was intended.

Overload Protector: A device which will stop operation of unit if dangerous conditions arise.
Ozone: A gaseous form of oxygen (O_3).

Packaged Boiler: A boiler having all components assembled as a unit.
Packaged Unit: A complete refrigeration or air conditioning system in which all the components are factory assembled into a single unit.
Parallel: Circuit connected so current has two or more paths to follow.
Parallel Circuit: An electrical circuit in which there is more than one different path across the power supply.
Partial Pressures: Condition where two or more gases occupy a space and each one creates part of the total pressure.
Pascal's Law: A pressure imposed upon a fluid is transmitted equally in all directions.
Perm: The unit of permeance.
Permeance: The ratio of water vapor flow to the vapor pressure difference.
pH: A term based on the hydrogen ion concentration in water, which denotes whether the water is acid, alkaline, or neutral.
Photoelectricity: A physical action wherein an electrical flow is generated by light waves.
Pilot: A small flame which is used to ignite the gas at the main burner.
Pilot Generator: A device used to generate voltage in a millivolt heating system.
Pilot Safety Valve: A valve that will shut off main gas if the pilot flame is not proved.
Pilot Switch: A control used in conjunction with gas burners. Its function is to prevent operation of the burner in the event of pilot failure.
Pinch-Off Tool: Device used to press walls of a tubing together until fluid flow ceases.
Piston: Close fitting part which moves up and down in a cylinder.
Piston Displacement: Volume obtained by multiplying area of cylinder bore by length of piston stroke.
Pitch: Pipe slope.
Pitot Tube: A device that senses static and total pressure.
Polarity: The condition denoting direction of current flow.
Polystyrene: Plastic used as an insulation in some refrigerator cabinet structures.
Ponded Roof: Flat roof designed to hold quantity of water.
Potential: The amount of voltage between points of a circuit.
Potentiometer: A variable resistor.
Pour Point (Oil): Lowest temperature at which oil will pour.
Power: Time rate at which work is done or energy emitted.
Power Element: Sensitive element of a temperature operated control.
Power Factor: The rate of actual power as measured by a wattmeter in an alternating circuit to the apparent power determined by multiplying amperes by volts.
Pressure: Force per unit area.
Pressure Limiter: Device which remains closed until a certain pressure is reached and then opens and releases fluid to another part of system.
Pressure Motor Control: A device which opens and closes on electrical circuit as pressures change to desired pressures.
Pressure-Operated Altitude (POA) Valve: Device which maintains a constant low side pressure independent of altitude of operation.
Pressure-Reducing Device: The device used to produce a reduction in pressure and corresponding boiling point before the refrigerant is introduced into the evaporator.
Pressure Reducing Valve: A diaphragm operated valve installed in the make-up water line of a hot water heating system.
Pressure Regulator: A device for controlling a uniform outlet gas pressure.
Pressure Relief Valve: A device for protecting a tank from excessive pressure by opening at a predetermined pressure.
Primary Air: The combustion air introduced into a burner which mixes with the gas before it reaches the port.
Primary Control: One type of operating controller for an oil burner.
Primary Voltage: The voltage of the circuit supplying power to a transformer.
Process Tube: Length of tubing fastened to hermetic unit dome, used for servicing unit.
Propane: A hydrocarbon fuel.
psi: Pressure measured in pounds per square inch.
psia: Pressure measured in pounds per square inch absolute.
psig: A symbol or initials used to indicate pressure in pounds per square inch gauge.
Psychrometer or Wet Bulb Hygrometer: An instrument for measuring dry-bulb and wet-bulb temperature.
Psychrometric Chart: A chart that shows relationship between the temperature and moisture content of the air.
PTC: Positive temperature coefficient; a thermistat used as a start relay.
Pump: A motor driven device used to mechanically circulate water in the system.
Pump Down: A service procedure where the refrigerant is pumped into the receiver.
Pump Head: The difference in pressure on the supply and intake sides of the pump created by the operation of the pump.
Purging: Releasing compressed gas to atmosphere through some part or parts for the purpose of removing contaminants from that part or parts.
Pyrometer: Instrument for measuring temperatures.

Quenching: Submerging hot solid object in cooling fluid.
Quick Connect Coupling: A device which permits fast connecting of two fluid lines.

Radiant Heating: A heating system in which only the heat radiated from panels is effective.
Radiation: Transfer of heat by heat rays.
Radiator: A heating unit exposed to view within the room or space to be heated.
Radiator Valve: A valve installed on a terminal unit to manually control the flow of water through the unit.

Range: Pressure or temperature settings of a control.

Rankine Scale: Absolute Fahrenheit scale.

Reactance: Opposition to alternating current by either inductance or capacitance or both.

Reciprocating: Action in which the motion is back and forth in a straight line.

Recirculated Air: Return air passed through the conditioner before being again supplied to the conditioned space.

Rectifier, Electric: An electrical device for converting ac into dc.

Reducing Fitting: A pipe fitting designed to change from one pipe size to another.

Reed Valve: Thin flat tempered steel plate fastened at one end.

Refractory: A material that can withstand very high temperature.

Refrigerant: Substance used in refrigerating mechanism to absorb heat in evaporator coil and to release its heat in a condenser.

Refrigerant Charge: Quantity of refrigerant in a system.

Refrigerating Effect: The amount of heat in Btu/h the system is capable of transferring.

Refrigeration: The process of transferring heat from one place to another.

Refrigeration Oil: Specially prepared oil used in refrigerator mechanism. It circulates with referigerant.

Register: Combination grille and damper assembly covering on an air opening or end of an air duct.

Relative Humidity: A ratio of the weight of moisture that air actually contains at a certain temperature as compared to the amount that it could contain if it were saturated.

Relay: Electrical mechanism which uses small current in control circuit to operate a valve switch in operating circuit.

Relief Opening: The opening in a draft hood to permit ready escape to the atmosphere of flue products.

Relief Valve: Safety device designed to open before dangerous pressure is reached.

Repulsion-Start Induction Motor: Type of motor which has an electrical winding on the rotor for starting-purposes.

Resistance: The opposition to current flow by a physical conductor.

Resonance: The pipe organ effect produced by a gas furnace when the frequency of the burner flame combustion and the pressure wave distance in the burner pouch are in exact synchronization.

Return Air: Air returned from conditioned or refrigerated space.

Return Piping: That portion of the piping system that carries water from the terminal units back to the boiler.

Reverse Acting Control: A switch controlled by temperature and designed to open on temperature drop and close on temperature rise.

Reverse Cycle Defrost: Method of heating evaporator for defrosting purposes by using valves to move hot gas from compressor into evaporator.

Reverse Return: A two-pipe system in which the return connections from the terminal units into the return main are made in the reverse order from that in which the supply connections are made.

Reversing Valve: The control used to regulate the flow of refrigerant in a heat pump.

Rheostat: An adjustable or variable resistor.

Rich Mixture: A mixture of gas and air containing too much fuel or too little air for complete combustion of the gas.

Rollout: A condition where flame rolls out of a combustion chamber when the burner is turned on.

Rotary Compressor: Mechanism which pumps fluid by using rotating motion.

Rotor: Rotating part of a mechanism.

Running Winding: Electrical winding of motor which has current flowing through it during normal operation of motor.

Run-Out: This term generally applies to the horizontal portion of branch circuits.

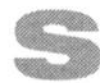

Saddle Valve: Valve body shaped so it may be silver brazed to refrigerant tubing surface.

Safety Control: Device which will stop the refrigerating unit if unsafe pressures and/or temperatures are reached.

Safety Plug: Device which will release the contents of a container above normal pressure conditions and before rupture pressures are reached.

Saturated Vapor: A vapor condition which will result in condensation into droplets of liquid as vapor temperature is reduced.

Saturation: A condition existing when a substance contains maximum of another substance for that temperature and pressure.

Schrader Valve: Spring loaded device which permits fluid flow when a center pin is depressed.

Scotch Yoke: Mechanism used to change reciprocating motion into rotary motion or vice-versa.

SCR: A solid state device (silicon controlled rectifier) used to modulate the capacity of an electric heating element.

Secondary Air: Combustion air externally supplied to a burner flame at the point of combustion.

Secondary Voltage: The output, or load-supply voltage, of a transformer.

Second Law of Thermodynamics: Heat will flow only from material at certain temperature to material at lower temperature.

Semihermetic Compressor: A serviceable hermetic compressor.

Sensible Heat: A term used in heating and cooling to indicate any portion of heat which changes only the temperature of the substances involved; also heat which changes the temperature of a substance without changing its form.

Sequencer: A controller used in electric heat systems to stagger the operation of the heating elements.

Series: A circuit with one continuous path for current flow.

Serviceable Hermetic: Hermetic unit housing containing motor and compressor assembled by use of bolts or threads.

Service Valve: Device used by service technicians to check pressures and change refrigerating units.

Shaded Pole Motor: A small ac motor used for light start loads.

Shaft Seal: A device used to prevent leakage between shaft and housing.

Short Circuit: A low-resistance connection (usually accidental and undesirable) between two parts of an electrical circuit.

Short Cycling: Refrigerating system that starts and stops more frequently than it should.

Sight Glass: Glass tube or glass window in refrigerating mechanism which shows amount of refrigerant, or oil in system.

Silica Gel: Chemical compound used as a drier, which has ability to absorb moisture.

Silver Brazing: Brazing process in which brazing alloy contains some silver as part of joining alloy.

Sling Psychrometer: Humidity measuring device with wet and dry bulb thermometers.

Slugging: A condition where liquid refrigerant is entering an operating compressor.

Smoke Test: Test made to determine completeness of combustion.

Soft Flame: A flame partially deprived of primary air such that the combustion zone is extended and inner cone is ill-defined.

Solar Heat: Heat from visible and invisible energy waves from the sun.

Soldering: Joining two metals by adhesion of a low melting temperature metal (less than 800°F).

Solenoid: A movable plunger activated by an electromagnetic coil.

Solenoid Valve: Valve actuated by magnetic action by means of an electrically energized coil.

Soot: A black substance, mostly consisting of small particles of carbon, which can result from incomplete combustion.

Specific Gravity: For a liquid or solid, the ratio of its density compared to water. For a vapor, the ratio of its density to air.

Specific Heat: The heat absorbed (or given up) by a unit mass of a substance when its temperature is increased (or decreased) by 1-degree common units: Btu per (pound) (Fahrenheit degree), calories per (gram) (centigrade degree). For gases, both specific heat at constant pressure (c_p) and specific heat at constant volume (c_v) are frequently used. In air-conditioning, c_p is usually used.

Specific Volume: The volume of a substance per unit mass: the reciprocal of density units: cubic feet per pound, cubic centimeters per gram, etc.

Split-Phase Motor: Motor with two stator windings.

Split System: Refrigeration or air-conditioning installation which places condensing unit remote from evaporator.

Squirrel Cage: Fan which has blades parallel to fan axis and moves air at right angles to fan axis.

Standard Conditions: Temperature of 68°F, pressure of 29.92 in. of Hg and relative humidity of 30 percent.

Starting Relay: An electrical device which connects and/or disconnects starting winding of electric motor.

Starting Winding: Winding in electric motor used only during brief period when motor is starting.

Static Pressure: The pressure exerted against the inside of a duct or pipe in all directions.

Stator: Stationary part of electric motor.

Steam Jet Refrigeration: Refrigerating system which uses a steam venturi to create high vacuum (low pressure) on a water container causing water to evaporate at low temperature.

Steam Trap: A device that will prevent the flow of steam, but will allow the flow of condensate.

Strainer: A screen used to retain solid particles while liquid passes through.

Stratification: Condition in which air lies in temperature layers.

Subcooling: Cooling of liquid refrigerant below its condensing temperature.

Sublimation: Condition where a substance changes from a solid to a gas without becoming a liquid.

Suction Line: Tube or pipe used to carry refrigerant gas from evaporator to compressor.

Suction Service Valve: A two-way manual-operated valve located at the inlet to compressor.

Superheat: Temperature of vapor above boiling temperature of its liquid at that pressure.

Surge Tank: Container connected to a refrigerating system which increases gas volume and reduces rate of pressure change.

Swaging: Enlarging one tube end so end of other tube of same size will fit within.

Swash Plate-Wobble Plate: Device used to change rotary motion to reciprocating motion.

Sweating: Condensation of moisture from air on cold surface.

Sweet Water: Tap water.

T

Tankless Water Heater: An indirect water heater designed to operate without a hot water storage tank.

Tap (Screw Thread): Tool used to cut internal threads.

Temperature: Degree of hotness or coldness as measured by a thermometer.

Terminal Units: Radiators, convectors, baseboard, unit heaters, finned tube, etc.

Therm: A unit of heat having a value of 100,000 Btu.

Thermal Conductivity: The ability of a material to transmit heat.

Thermal Resistance: The resistance a material offers to the transmission of heat.

Thermistor: A semiconductor which has electrical resistance that varies with temperature.

Thermocouple: Device which generates electricity, using principle that if two dissimilar metals are welded together and junction is heated, a voltage will develop across open ends.

Thermoelectric Refrigeration: A refrigerator mechanism which depends on Peletier effect.

Thermometer: Device for measuring temperatures.

Thermopile: A pilot generator.

Thermostat: Switch responsive to temperature.

Thermostatic Expansion Valve: A valve operated by temperature and pressure within evaporator coil.

Thermostatic Valve: Valve controlled by thermostatic elements.

Throttling: Expansion of gas through orifice.

Ton: Refrigerating effect equal to the melting of one ton of ice in 24 hours.

Torque: Turning or twisting force.

Torque Wrenches: Wrenches which may be used to measure torque applied.

Total Pressure: Also called impact pressure. The sum of the velocity pressure and the static pressure.

Transformer: A device designed to change voltage.

Turbulent Flow: The movement of a liquid or vapor in a pipe in a constantly churning and mixing fashion.

Turndown: The ratio of maximum to minimum input rates.

Two-Pipe System: A hot-water heating system using one pipe from the boiler to supply heated water to the terminal units, and a second pipe to return the water from the terminal units back to the boiler.

U

"U" Factor: Thermal conductivity.

Ultraviolet: Invisible radiation waves with frequencies shorter than wave lengths of visible light and longer than X-Ray.

Underwriters Laboratories: Abbreviated UL. Underwriters Laboratories, Inc. maintain and operate laboratories for the examination and testing of devices.

Unit Heater: A fan and motor, a heating element, and an enclosure hung from a ceiling or wall.

Unit Ventilator: A terminal unit in which a fan is used to mechanically circulate air over the heating coil.

Urethane Foam: Type of insulation which is foamed in between inner and outer walls of display case.

Vacuum: Pressure below atmospheric pressure.

Vacuum Pump: Special high efficiency compressor used for creating high vacuums.

Valve: Device used for controlling fluid flow.

Valve Plate: Part of compressor located between top of compressor body and head which contains compressor valves.

Vapor Barrier: Thin plastic or metal foil sheet used to prevent water vapor from penetrating insulating material.

Variable Pitch Pulley: Pulley which can be adjusted to provide different pulley ratios.

V-Belt: Type of drive belt.

Velocimeter: Instrument used to measure air velocities.

Velocity Pressure: The pressure exerted in direction of flow.

Vent Gases: Products of combustion.

Ventilation: The introduction of outdoor air into a building by mechanical means.

Venturi: A section in a pipe or a burner body that narrows down and then flares out again.

Viscosity: Measure of a fluid's ability to flow.

Volt: The unit of electrical potential or pressure.

Voltage Relay: One that functions at a predetermined voltage value.

Volumetric Efficiency: Ratio of the actual performance of a compressor and calculated performance.

VOM: A meter that measures voltage and resistance (ohms).

W

Walk-In Cooler: Large commercial refrigerated room.

Water Column: Abbreviated as WC. A unit used for expressing pressure.

Water Tube Boiler: A hot-water boiler in which the water is circulated through the tubes and the hot gases from combustion of the fuel are circulated around the tubes.

Watt: A unit of electrical power.

Z

Zone: That portion of a building whose temperature is controlled by a single thermostat.

Index

A

B

I

J

K

L

M

Q

R

S

T

U

V